P9-DHF-001

INTRODUCTORY CIRCUIT ANALYSIS

EIGHTH EDITION

INTRODUCTORY CIRCUIT ANALYSIS

Robert L. Boylestad

Prentice Hall
Upper Saddle River, New Jersey ||| Columbus, Ohio

Library of Congress Cataloging-in-Publication Data

Boylestad, Robert L.
Introductory circuit analysis / Robert L. Boylestad.—8th ed.
p. cm.
Includes index.
ISBN 0-13-235904-9 (hardcover)
1. Electric circuits. 2. Electric circuit analysis—Data processing 3. PSpice. I. Title
TK454.B68 1997
621.319′2—dc20 96-2199
CIP

Editor: Linda Ludewig
Developmental Editor: Carol Hinklin Robison
Production Editor: Rex Davidson
Text Designer: Rebecca M. Bobb
Design Coordinator: Jill E. Bonar
Cover Designer: Brian Deep
Production Manager: Pamela D. Bennett
Marketing Manager: Debbie Yarnell

This book was set in Times Roman by The Clarinda Company and was printed and bound by Von Hoffmann Press, Inc. The cover was printed by Von Hoffmann Press, Inc.

Printed in the United States of America

10 9 8 7 6 5 4 3 2 1

ISBN: 0-13-235904-9

Prentice-Hall International (UK) Limited, *London*
Prentice-Hall of Australia Pty. Limited, *Sydney*
Prentice-Hall Canada Inc., *Toronto*
Prentice-Hall Hispanoamericana, S. A., *Mexico*
Prentice-Hall of India Private Limited, *New Delhi*
Prentice-Hall of Japan, Inc., *Tokyo*
Simon & Schuster Asia Pte. Ltd., *Singapore*
Editora Prentice-Hall do Brasil, Ltda., *Rio de Janeiro*

To Else Marie
Alison, Mark, Kelcy, and Morgan
Stacey and Douglas
and Eric

Preface

It has been thirty years—three decades—since the original contract for *Introductory Circuit Analysis* was signed with the Charles E. Merrill Publishing Company—the only publishing house of the number I approached that was willing to take the gamble. I recall one reviewer of the original manuscript from another publishing company commenting that the text had so many examples and helpful hints that he was concerned about what he was going to do in class—it left little work for the instructor. The result was a negative review—a review I wish I had saved.

Through the years the production cycle continues to generate experiences that are difficult to plan for, but in general the responses made to these unexpected surprises and the results have always been very satisfying primarily because of the wonderful people at the Charles E. Merrill Publishing Company (now Prentice Hall). Their contributions to the quality, layout, and design of the text have always been an important factor in the success of the publication.

Students continue to be my best and most informative critics. The cover of my file folder for each class is covered with their comments about areas that need revision and about possible new ways to present material or interesting new problems. Letters I've received from users are always appreciated, and some have been quite helpful in defining content and finding corrections. The reviewers of each edition have been excellent, with some overextending themselves to help contribute to the content. In the early years, when each edition was sent to press, I felt that the text now had all the elements I had hoped for in the first edition. However, as the years have passed, it has become more apparent that improvements can always be made, and new material will surface that must be included. It's now becoming more and more difficult to make the 5% to 10% content change because the length of the text leaves me with difficult decisions about what may have to be cut. As I now complete the 8th edition, I already have changes ready for the 9th edition. For instance, I feel sure that PSpice DOS will give way to the Windows version over the next year or two. When I first encountered the Windows version, I was a skeptic and immediately pointed out to my peers all the functions it could not perform as well as the DOS version. However, with the 6.2 version and many hours of exposure, I found that some of my original assumptions were wrong, and now I use only the Windows version. The DOS version appears in this edition to provide a bridge for current users. The content of this edition with regard to PSpice was developed to provide all the details students need to carry them through the dc and ac sequence. In other words, details are sufficient to remove the need for a supplement. I believe that holding a short session with the students at the computer facilities at the beginning of the term and establishing a place for them to go with questions is sufficient to carry them through most of the programs in the text. I am always pleased to find that students who begin our program with deficiencies in math or English can pick up on the computer fairly quickly. For many, I believe, the computer provides an avenue that develops self-confidence and immediately makes them feel like part of the mainstream of the class. C++ was added to this edition in reaction to numerous requests from users and the fact that it is now the choice for a rapidly growing number of academic programs. The C++ content, which is limited but sufficient, reveals some of the language's characteristics and at the very least makes the student aware of its existence. In addition, MathCAD was added for many of the same reasons as C++. The growing number of users in all the technologies suggested that its power be demonstrated at appropriate points throughout the text. Finally, the TI-85 calculator was included in depth in the ac section to demonstrate its power with complex numbers. For someone (like me) who started out with the slide rule and tables to work with complex numbers, this calculator made the operation downright friendly. The result is that there are numerous examples in the use of the calculator in the ac section with the same color highlighting that is used for the software packages.

For this edition all the chapter headings remain the same, with only nine new sections. However, the above should not suggest that this is simply the 7th edition with a new cover. Changes occur in every chapter and material in each chapter has been rewritten, updated, or expanded. Some of the new topics include significant figures; accuracy and rounding off; safety considerations; GFCI breakers; grounding; supermesh and supernodes; more experimental procedures; recognition factors for capacitors and inductors; computer hard disks; Hall effect sensors; initial values (L and C); summary tables; and crossover networks. Some of the areas expanded include potentiometers; the voltage and current divider rules; internal resistance; measurement techniques; open and short circuits; series-parallel networks; nodal analysis; Thévenin's theorem; dependent sources; power; parallel resonance; dB instrumentation; compensation probes; and problems. As I look back on my marked-up master copy of my text, it surprises me that the area that always seems to need the most attention is the introductory material. Obviously, it is critical that students clearly understand the fundamental concepts to establish a strong foundation for the material to follow. One important change that occurs throughout the text is that every figure now has a caption to define its content and purpose. In the past only about 20% had captions. I believe that these captions will be an aid to students as they proceed through the descriptions and examples.

Ancillaries

The laboratory manual was carefully reviewed and modified to ensure that the data obtained were appropriate and resulted in a clear verification of the theory—in total about a 10% modification. A number of the ac experiments were simplified by using peak-to-peak values throughout to avoid the repetitive process of converting from rms to peak-to-peak values and back again. In addition, procedures, questions, and tables were modified to clarify the procedure and ensure that the goal of each operation was met and understood.

Other ancillaries that complement this text include an Instructor's Solutions Manual, Problem Supplement, Test Item File, PH Custom Test (DOS), Transparency Masters, Transparencies, PSpice/Electronics Workbench Data Disk (circuit simulator), and a selection of Bergwall Instructional Videos.

As with every edition, there is an extensive list of individuals who contributed to the content, quality, and accuracy of the text. At my own institution, the subject matter and advice offered by Jerry Sitbon were invaluable in establishing a more practical orientation to the text. His wife, Mrs. Catherine Sitbon, born in Paris, France, was instrumental in obtaining the information from Ecole Polytechnique on Leon Thévenin for this edition. Professor James L. Antonakos of Broome Community College was particularly helpful with the C++ content, bringing all his years of experience with the program and the publishing world to help develop the coverage that now appears in the 8th edition. Professor William Boettcher of Albuquerque V-TI, a longtime friend of both myself and the book, was very helpful with the MathCAD content that now plays an important role in the text. Let me also extend my deepest appreciation to the long list of contributors from both the educational and industrial communities that appears below. Their input, criticism, support, and assistance were instrumental in establishing the priorities for this edition of the text.

Acknowledgments

Professor Derek Abbot—University of Adelaide, Australia
Professor G. Thomas Bellarmine—University of West Florida
Bill Boettcher—Albuquerque Technical-Vocational Institute
Professor Lester W. Cory—Southeastern Massachusetts University
Professor Gerald L. Doutt—DeVry Institute of Technology–Kansas City
John Dunbar—DeVry Institute of Technology–Atlanta
Kenneth Frament—DeVry Institute of Technology–Phoenix
Professor Bernard Guss—Pennsylvania State University
Robert Herrick—Purdue University
Ernest Joerg
Karen Karger—Tektronix, Inc.
Ms. Mary Kuykendall—Hall of History Foundation
Professor M. David Luneau, Jr.—University of Arkansas at Little Rock
Leei Mao—Greenville Technical College
Professor Edward F. McBrien—Arizona State University
Ms. Carol Parcels—Hewlett-Packard Corp.
Vic Quiros—DeVry Institute of Technology–Phoenix
Madame Martine Roudeix—Ecole Polytechnique, Palaiseau, France
Monsieur Raymond Josue Seckel—National Library of France
Professor Paul T. Svatik—Owens Community College
Mrs. Barbara Sweeney—AT&T archives
Ms. Marie-Christine Thooris—Ecole Polytechnique, Palaiseau, France
Ms. Kathy Truesdell—Texas Instruments
Dr. Domingo L. Uy—Hampton University

I am very fortunate to be working with a very talented team at the Ohio office of Prentice-Hall. I refuse to think about going through the publishing cycle without people like Dave Garza to oversee the project, Rex Davidson to keep me sane and control the flow, and Carol Robison to coordinate and keep track of the infinite number of details. With three publications to complete in the last 18 months, it would have been impossible without their support and assistance. The copyeditor, Ms. Marianne L'Abbate, was a wonderful addition to the team and managed to make the process as smooth as possible, as did the excellent work of proofreaders Maggie Shaffer and Lois Porter. In addition, I want to specifically thank Linda Ludewig, Electronics Editor; Debbie Yarnell, Marketing Manager; Ruta K. Fiorino and Maureen Henry, Advertising; and Jayne Demsky, Publishing Representative, for their efforts in making the text a success.

My best wishes for a successful and productive school year and the best of health.

Contents

1

Introduction

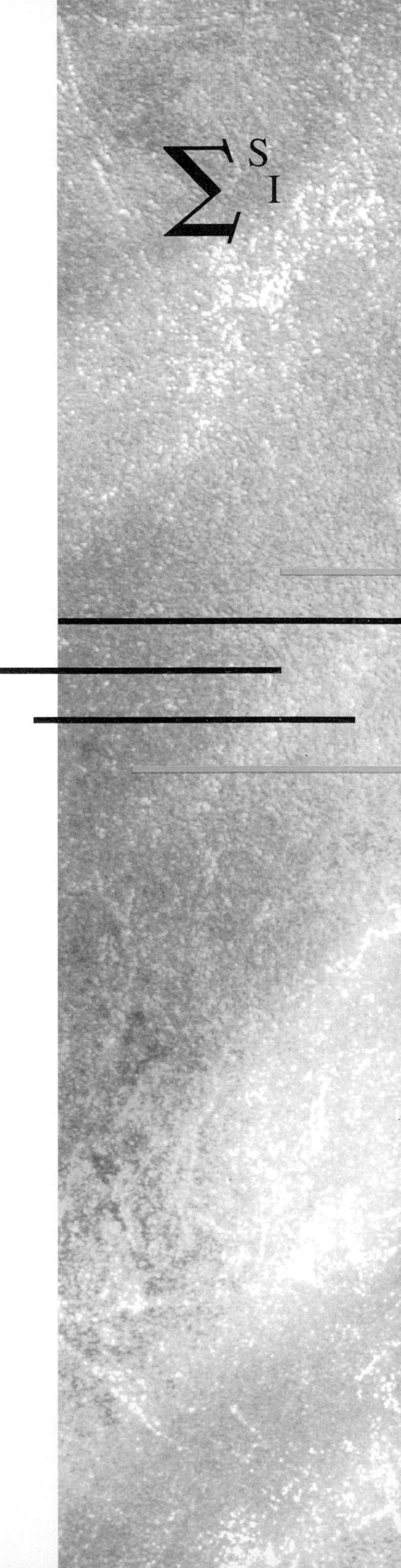

1.1 THE ELECTRICAL/ELECTRONICS INDUSTRY

The recent surge in the technologies on Wall Street is clear evidence that the electrical/electronics industry is one that will have a sweeping impact on future development in a wide range of areas that affect our life style, general health, and capabilities. Can you think of a field today, even those in which people try to minimize technical ties, that does not, at the very least, seek to broaden its horizons through the use of some technical innovation such as recording, duplication, computing, or data-handling instrumentation?

Every facet of our lives seems touched by developments that appear to surface at an ever-increasing rate. For the layperson, the most obvious improvement of recent years has been the reduced size of electrical/electronics systems. Televisions are now small enough to be hand-held and have a battery capability that allows them to be more portable. Computers with significant memory capacity are now smaller than this textbook. The size of radios is limited simply by the ability to read the numbers on the face of the dial. Hearing aids are no longer visible, and pacemakers are significantly smaller and more reliable. All the reduction in size is due primarily to a marvelous development of the last few decades—the integrated circuit (IC). First developed in the late 1950s, the IC has now reached a point where it can cut $\frac{1}{2}$-micrometer lines. Consider that some 50,000 of these lines would fit within 1 in. Try to visualize breaking down an inch into 100 divisions and then consider 1000 or 25,000 divisions—an incredible achievement. The integrated circuit of Fig. 1.1 has over 1.2 million components, yet is only about $\frac{1}{2}$ in. on each side.

It is natural to wonder what the limits to growth may be when we consider the changes over the last few decades. Rather than following a steady growth curve that would be somewhat predictable, the industry is subject to surges that revolve around significant developments in the field. Present indications are that the level of miniaturization will continue, but at a more moderate pace. Interest has turned toward increasing the quality and yield levels (percentage of good integrated circuits in the production process).

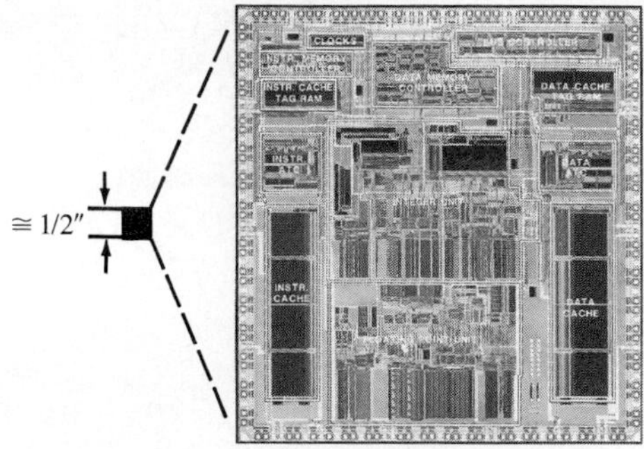

FIG. 1.1
Integrated circuit. (Courtesy of Motorola Semiconductor Products.)

History reveals that there have been peaks and valleys in industry growth but that revenues continue to rise at a steady rate and funds set aside for research and development continue to command an increasing share of the budget. The field changes at a rate that requires constant retraining of employees from the entry to the director level. Many companies have instituted their own training programs and have encouraged local universities to develop programs to ensure that the latest concepts and procedures are brought to the attention of their employees. A period of relaxation could be disastrous to a company dealing in competitive products.

No matter what the pressures on an individual in this field may be to keep up with the latest technology, there is one saving grace that becomes immediately obvious: Once a concept or procedure is clearly and correctly understood, it will bear fruit throughout the career of the individual at any level of the industry. For example, once a fundamental equation such as Ohm's law (Chapter 4) is understood, it will not be *replaced* by another equation as more advanced theory is considered. It is a relationship of fundamental quantities that can have application in the most advanced setting. In addition, once a procedure or method of analysis is understood, it usually can be applied to a wide (if not infinite) variety of problems, making it unnecessary to learn a different technique for each slight variation in the system. The content of this text is such that every morsel of information will have application in more advanced courses. It will not be replaced by a different set of equations and procedures unless required by the specific area of application. Even then, the new procedures will usually be an expanded application of concepts already presented in the text.

It is paramount therefore that the material presented in this introductory course be clearly and precisely understood. It is the foundation for the material to follow and will be applied throughout your working days in this growing and exciting field.

1.2 A BRIEF HISTORY

In the sciences, once a hypothesis is proven and accepted, it becomes one of the building blocks of that area of study, permitting additional investigation and development. Naturally, the more pieces of a puzzle available, the more obvious the avenue toward a possible solution. In fact, history demonstrates that a single development may provide the key that will result in a mushroom effect that brings the science to a new plateau of understanding and impact.

If the opportunity presents itself, read one of the many publications reviewing the history of this field. Space requirements are such that only a brief review can be provided here. There are many more contributors than could be listed, and their efforts have often provided important keys to the solution of some very important concepts.

As noted earlier, there were periods characterized by what appeared to be an explosion of interest and development in particular areas. As you will see from the discussion of the late 1700s and the early 1800s, inventions, discoveries, and theories came fast and furiously. Each new concept broadened the possible areas of application until it becomes almost impossible to trace developments without picking a particular area of interest and following it through. In the review, as you read about the development of the radio, television, and computer, keep in mind that similar progressive steps were occurring in the areas of the telegraph, the telephone, power generation, the phonograph, appliances, and so on.

There is a tendency when reading about the great scientists, inventors, and innovators to believe that their contribution was a totally individual effort. In many instances, this was not the case. In fact, many of the great contributors were friends or associates and provided support and encouragement in their efforts to investigate various theories. At the very least, they were aware of one another's efforts to the degree possible in the days when a letter was often the best form of communication. In particular, note the closeness of the dates during periods of rapid development. One contributor seemed to spur on the efforts of the others or possibly provided the key needed to continue with the area of interest.

In the early stages, the contributors were not electrical, electronic, or computer engineers as we know them today. In most cases, they were physicists, chemists, mathematicians, or even philosophers. In addition, they were not from one or two communities of the Old World. The home country of many of the major contributors listed below is provided to show that almost every established community had some impact on the development of the fundamental laws of electrical circuits.

As you proceed through the remaining chapters of the text, you will find that a number of the units of measurement bear the name of major contributors in those areas—*volt* after Count Alessandro Volta, *ampere* after André Ampère, *ohm* after Georg Ohm, and so forth—fitting recognition for their important contributions to the birth of a major field of study.

Time charts indicating a limited number of major developments are provided in Fig. 1.2, primarily to identify specific periods of rapid development and to reveal how far we have come in the last few decades. In essence, the current state of the art is a result of efforts that

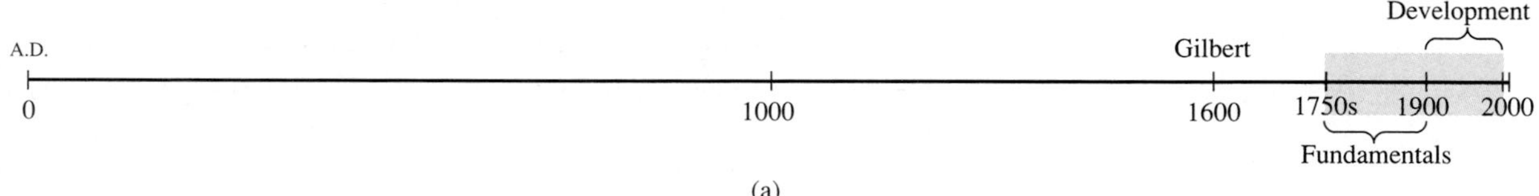

(a)

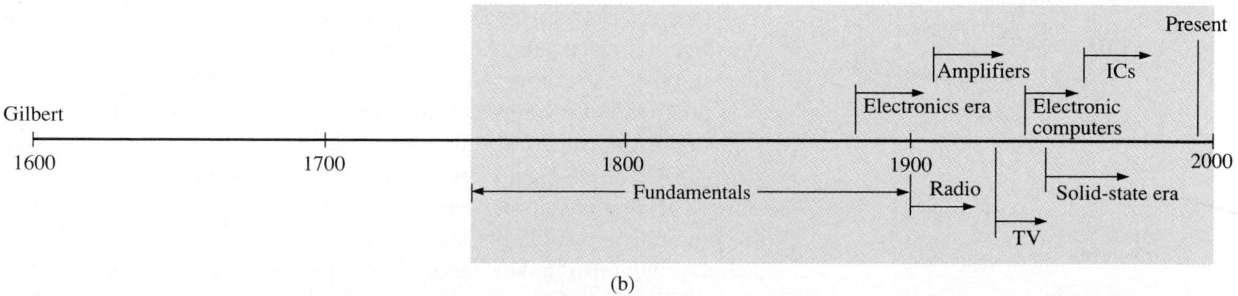

(b)

FIG. 1.2
Time charts: (a) long-range; (b) expanded.

began in earnest some 250 years ago, with progress in the last 100 years almost exponential.

As you read through the following brief review, try to sense the growing interest in the field and the enthusiasm and excitement that must have accompanied each new revelation. Although you may find some of the terms used in the review new and essentially meaningless, the remaining chapters will explain them thoroughly.

The Beginning

The phenomenon of static electricity has been toyed with since antiquity. The Greeks called the fossil resin substance so often used to demonstrate the effects of static electricity *elektron,* but no extensive study was made of the subject until William Gilbert researched the event in 1600. In the years to follow, there was a continuing investigation of electrostatic charge by many individuals such as Otto von Guericke, who developed the first machine to generate large amounts of charge, and Stephen Gray, who was able to transmit electrical charge over long distances on silk threads. Charles DuFay demonstrated that charges either attract or repel each other, leading him to believe that there were two types of charge—a theory we subscribe to today with our defined positive and negative charges.

There are many who believe that the true beginnings of the electrical era lie with the efforts of Pieter van Musschenbroek and Benjamin Franklin. In 1745, van Musschenbroek introduced the *Leyden jar* for the storage of electrical charge (the first capacitor) and demonstrated electrical shock (and therefore the power of this new form of energy). Franklin used the Leyden jar some seven years later to establish that lightning is simply an electrical discharge, and expanded on a number of other important theories including the definition of the two types of charge as *positive* and *negative.* From this point on, new discoveries and theories seemed to occur at an increasing rate as the number of individuals performing research in the area grew.

In 1784, Charles Coulomb demonstrated in Paris that the force between charges is inversely related to the square of the distance between the charges. In 1791, Luigi Galvani, professor of anatomy at the University of Bologna, Italy, performed experiments revealing that electricity is present in every animal. The first *voltaic cell,* with its ability to produce electricity through the chemical action of a metal dissolving in an acid, was developed by another Italian, Alessandro Volta, in 1799.

The fever pitch continued into the early 1800s with Hans Christian Oersted, a Swedish professor of physics, announcing in 1820 a relationship between magnetism and electricity that serves as the foundation for the theory of *electromagnetism* as we know it today. In the same year, a French physicist, André Ampère, demonstrated that there are magnetic effects around every current-carrying conductor and that current-carrying conductors can attract and repel each other just like magnets. In the period 1826 to 1827, a German physicist, Georg Ohm, introduced an important relationship between potential, current, and resistance which we now refer to as Ohm's law. In 1831, an English physicist, Michael Faraday, demonstrated his theory of *electromagnetic induction,* whereby a changing current in one coil can induce a changing current in another coil, even though the two coils are not directly connected. Professor Faraday also did extensive work on a storage device he called the condenser, which we refer to today as a capacitor. He introduced the idea of adding a dielectric between the plates of a capacitor to increase the storage capacity (Chapter 10). James Clerk Maxwell, a Scottish professor of natural philosophy, performed extensive mathematical analyses to develop what are currently called *Maxwell's equations,* which support the efforts of Faraday linking electric and magnetic effects. Maxwell also developed the *electromagnetic theory of light* in 1862, which, among other things, revealed that electromagnetic waves travel through air at the velocity of light (186,000 miles per second or 3×10^8 meters per second). In 1888, a German physicist, Heinrich Rudolph Hertz, through experimentation with lower-frequency electromagnetic waves (microwaves), substantiated Maxwell's predictions and equations. In the mid 1800s, Professor Gustav Robert Kirchhoff introduced a series of laws of voltages and currents that find application at every level and area of this field (Chapters 5 and 6). In 1895, another German physicist, Wilhelm Röntgen, discovered electromagnetic waves of high frequency, commonly called X rays today.

By the end of the 1800s, a significant number of the fundamental equations, laws, and relationships had been established and various fields of study, including electronics, power generation, and calculating equipment, started to develop in earnest.

The Age of Electronics

Radio The true beginning of the electronics era is open to debate and is sometimes attributed to efforts by early scientists in applying potentials across evacuated glass envelopes. However, many trace the beginning to Thomas Edison, who added a metallic electrode to the vacuum of the tube and discovered that a current was established between the metal electrode and the filament when a positive voltage was applied to the metal electrode. The phenomenon, demonstrated in 1883, was referred to as the *Edison effect*. In the period to follow, the transmission of radio waves and the development of the radio received widespread attention. In 1887, Heinrich Hertz, in his efforts to verify Maxwell's equations, transmitted radio waves for the first time in his laboratory. In 1896, an Italian scientist, Guglielmo Marconi (often called the father of the radio), demonstrated that telegraph signals could be sent through the air over long distances (2.5 kilometers) using a grounded antenna. In the same year, Aleksandr Popov sent what might have been the first radio message some 300 yards. The message was the name *"Heinrich Hertz"* in respect for Hertz's earlier contributions. In 1901, Marconi established radio communication across the Atlantic.

In 1904, John Ambrose Fleming expanded on the efforts of Edison to develop the first diode, commonly called *Fleming's valve*—actually the first of the *electronic devices*. The device had a profound impact on the design of detectors in the receiving section of radios. In 1906, Lee De Forest added a third element to the vacuum structure and created the first amplifier, the triode. Shortly thereafter, in 1912, Edwin Armstrong built the first regenerative circuit to improve receiver capabilities and then used the same contribution to develop the first nonmechanical oscillator. By 1915 radio signals were being transmitted across the United States, and in 1918 Armstrong applied for a patent for the superheterodyne circuit employed in virtually every television and radio to permit amplification at one frequency rather than at the full range of incoming signals. The major components of the modern-day radio were now in place and sales in radios grew from a few million dollars in the early 1920s to over $1 billion by the 1930s. The 1930s were truly the golden years of radio, with a wide range of productions for the listening audience.

Television The 1930s were also the true beginnings of the television era, although development on the picture tube began in earlier years with Paul Nipkow and his *electrical telescope* in 1884 and John Baird and his long list of successes, including the transmission of television pictures over telephone lines in 1927 and over radio waves in 1928, and simultaneous transmission of pictures and sound in 1930. In 1932, NBC installed the first commercial television antenna on top of the Empire State Building in New York City, and RCA began regular broadcasting in 1939. The war slowed development and sales, but in the mid 1940s the number of sets grew from a few thousand to a few million. Color television became popular in the early 1960s.

Computers The earliest computer system can be traced back to Blaise Pascal in 1642 with his mechanical machine for adding and subtracting numbers. In 1673 Gottfried Wilhelm von Leibniz used the *Leibniz wheel* to add multiplication and division to the range of operations, and in 1823 Charles Babbage developed the *Difference Engine* to add the mathematical operations of sine, cosine, logs, and several others. In the years to follow, improvements were made, but the system remained primarily mechanical until the 1930s when electromechanical systems using components such as relays were introduced. It was not until the 1940s that totally electronic systems became the new wave. It is interesting to note that, even though IBM was formed in 1924, it did not enter the computer industry until 1937. An entirely electronic system known as Eniac was dedicated at the University of Pennsylvania in 1946. It contained 18,000 tubes and weighed 30 tons but was several times faster than most electromechanical systems. Although other vacuum tube systems were built, it was not until the birth of the solid-state era that computer systems experienced a major change in size, speed, and capability.

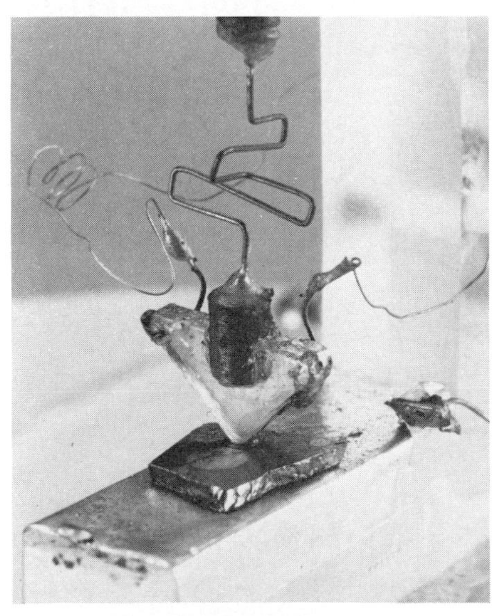

FIG. 1.3
The first transistor. (Courtesy of AT&T, Bell Laboratories.)

The Solid-State Era

In 1947, physicists William Shockley, John Bardeen, and Walter H. Brattain of Bell Telephone Laboratories demonstrated the point-contact *transistor* (Fig. 1.3), an amplifier constructed entirely of solid-state materials with no requirement for a vacuum, glass envelope, or heater voltage for the filament. Although reluctant at first due to the vast amount of material available on the design, analysis, and synthesis of tube networks, the industry eventually accepted this new technology as the wave of the future. In 1958 the first *integrated circuit* (IC) was developed at Texas Instruments, and in 1961 the first commercial integrated circuit was manufactured by the Fairchild Corporation.

It is impossible to review properly the entire history of the electrical/electronics field in a few pages. The effort here, both through the discussion and the time graphs of Fig. 1.2, was to reveal the amazing progress of this field in the last 50 years. The growth appears to be truly exponential since the early 1900s, raising the interesting question, Where do we go from here? The time chart suggests that the next few

decades will probably contain many important innovative contributions that may cause an even faster growth curve than we are now experiencing.

1.3 UNITS OF MEASUREMENT

It is vital that the importance of units of measurement be understood and appreciated early in the development of a technically oriented background. Too frequently their effect on the most basic substitution is ignored. Consider, for example, the following very fundamental physics equation:

$$v = \frac{d}{t} \qquad \begin{aligned} v &= \text{velocity} \\ d &= \text{distance} \\ t &= \text{time} \end{aligned} \tag{1.1}$$

Assume, for the moment, that the following data are obtained for a moving object:

$$d = 4000 \text{ ft}$$
$$t = 1 \text{ min}$$

and v is desired in miles per hour. Often, without a second thought or consideration, the numerical values are simply substituted into the equation, with the result here that

$$v = \frac{d}{t} = \frac{4000 \text{ ft}}{1 \text{ min}} = \cancel{4000 \text{ mi/h}}$$

As indicated above, the solution is totally incorrect. If the result is desired in *miles per hour,* the unit of measurement for distance must be *miles* and that for time, *hours*. In a moment, when the problem is analyzed properly, the extent of the error will demonstrate the importance of ensuring that

the numerical value substituted into an equation must have the unit of measurement specified by the equation.

The next question is normally, How do I convert the distance and time to the proper unit of measurement? A method will be presented in a later section of this chapter, but for now it is given that

$$1 \text{ mi} = 5280 \text{ ft}$$
$$4000 \text{ ft} = 0.7576 \text{ mi}$$
$$1 \text{ min} = \tfrac{1}{60} \text{ h} = 0.0167 \text{ h}$$

Substituting into Eq. (1.1), we have

$$v = \frac{d}{t} = \frac{0.7576 \text{ mi}}{0.0167 \text{ h}} = 45.37 \text{ mi/h}$$

which is significantly different from the result obtained before.

To complicate the matter further, suppose the distance is given in kilometers, as is now the case on many road signs. First, we must realize that the prefix *kilo* stands for a multiplier of 1000 (to be introduced in Section 1.5), and then we must find the conversion factor between kilometers and miles. If this conversion factor is not readily available, we must be able to make the conversion between units using the conversion factors between meters and feet or inches, as described in Section 1.6.

Before substituting numerical values into an equation, try to mentally establish a reasonable range of solutions for comparison purposes. For instance, if a car travels 4000 ft in 1 min, does it seem reasonable that the speed would be 4000 mi/h? Obviously not! This self-checking procedure is particularly important in this day of the hand-held calculator, when ridiculous results may be accepted simply because they appear on the digital display of the instrument.

Finally,

if a unit of measurement is applicable to a result or piece of data, then it must be applied to the numerical value.

To state that $v = 45.37$ without including the unit of measurement *mi/h* is meaningless.

Equation (1.1) is not a difficult one. A simple algebraic manipulation will result in the solution for any one of the three variables. However, in light of the number of questions arising from this equation, the reader may wonder if the difficulty associated with an equation will increase at the same rate as the number of terms in the equation. In the broad sense, this will not be the case. There is, of course, more room for a mathematical error with a more complex equation, but once the proper system of units is chosen and each term properly found in that system, there should be very little added difficulty associated with an equation requiring an increased number of mathematical calculations.

In review, before substituting numerical values into an equation, be absolutely sure of the following:

1. ***Each quantity has the proper unit of measurement as defined by the equation.***
2. ***The proper magnitude of each quantity as determined by the defining equation is substituted.***
3. ***Each quantity is in the same system of units (or as defined by the equation).***
4. ***The magnitude of the result is of a reasonable nature when compared to the level of the substituted quantities.***
5. ***The proper unit of measurement is applied to the result.***

1.4 SYSTEMS OF UNITS

In the past, the *systems of units* most commonly used were the English and metric, as outlined in Table 1.1. Note that while the English system is based on a single standard, the metric is subdivided into two interrelated standards: the MKS and CGS. Fundamental quantities of these systems are compared in Table 1.1 along with their abbreviations. The MKS and CGS systems draw their names from the units of measurement used with each system; the MKS system uses *M*eters, *K*ilograms, and *S*econds, while the CGS system uses *C*entimeters, *G*rams, and *S*econds.

Understandably, the use of more than one system of units in a world that finds itself continually shrinking in size, due to advanced technical developments in communications and transportation, would introduce unnecessary complications to the basic understanding of any technical data. The need for a standard set of units to be adopted by all nations has become increasingly obvious. The International Bureau of Weights and Measures located at Sèvres, France, has been the host for the General Conference of Weights and Measures, attended by representatives

TABLE 1.1
Comparison of the English and metric systems of units.

English	Metric		SI
	MKS	CGS	
Length:			
Yard (yd) (0.914 m)	Meter (m) (39.37 in.) (100 cm)	Centimeter (cm) (2.54 cm = 1 in.)	Meter (**m**)
Mass:			
Slug (14.6 kg)	Kilogram (kg) (1000 g)	Gram (g)	Kilogram (**kg**)
Force:			
Pound (lb) (4.45 N)	Newton (N) (100,000 dynes)	Dyne	Newton (**N**)
Temperature:			
Fahrenheit (°F) $\left(= \frac{9}{5}°C + 32\right)$	Celsius or centigrade (°C) $\left(= \frac{5}{9}(°F - 32)\right)$	Centigrade (°C)	Kelvin (**K**) K = 273.15 + °C
Energy:			
Foot-pound (ft-lb) (1.356 joules)	Newton-meter (N-m) or joule (J) (0.7376 ft-lb)	Dyne-centimeter or erg (1 joule = 10^7 ergs)	Joule (**J**)
Time:			
Second (s)	Second (s)	Second (s)	Second (**s**)

from all nations of the world. In 1960, the General Conference adopted a system called Le Système International d'Unités (International System of Units), which has the international abbreviation SI. Since then, it has been adopted by the Institute of Electrical and Electronic Engineers, Inc. (IEEE) in 1965 and by the United States of America Standards Institute in 1967 as a standard for all scientific and engineering literature.

For comparison, the SI units of measurement and their abbreviations appear in Table 1.1. These abbreviations are those usually applied to each unit of measurement, and they were carefully chosen to be the most effective. Therefore, it is important that they be used whenever applicable to ensure universal understanding. Note the similarities of the SI system to the MKS system. This text will employ, whenever possible and practical, all of the major units and abbreviations of the SI system in an effort to support the need for a universal system. Those readers requiring additional information on the SI system should contact the information office of the American Society for Engineering Education (ASEE).*

Figure 1.4 should help the reader develop some feeling for the relative magnitudes of the units of measurement of each system of units. Note in the figure the relatively small magnitude of the units of measurement for the CGS system.

A standard exists for each unit of measurement of each system. The standards of some units are quite interesting.

*American Society for Engineering Education (ASEE), One Dupont Circle, Suite 400, Washington, D.C. 20036.

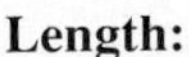

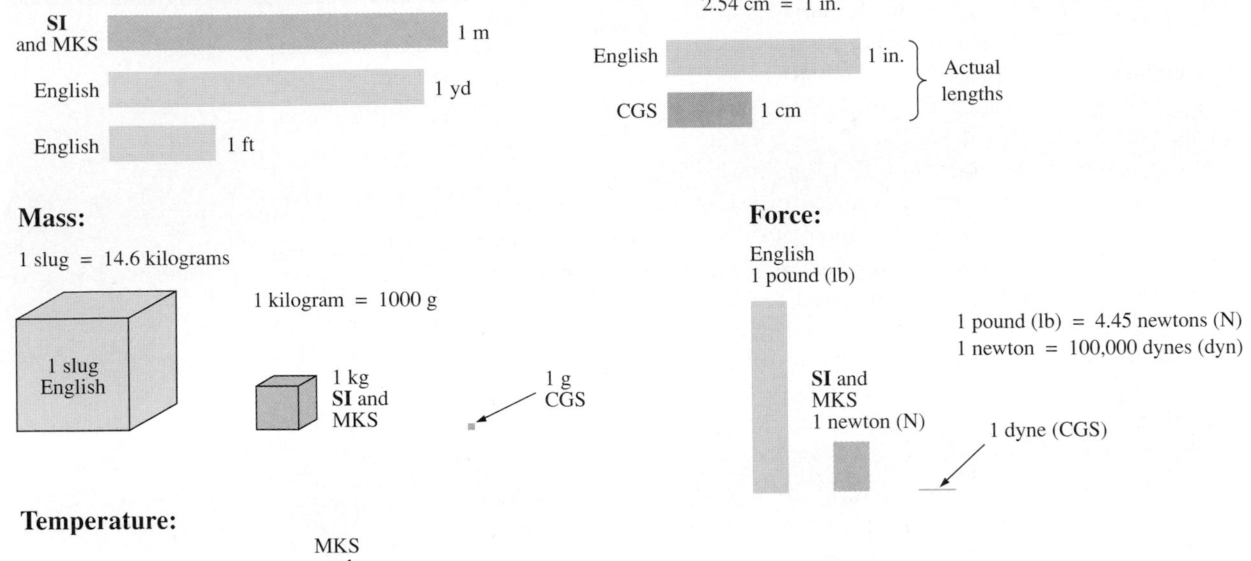

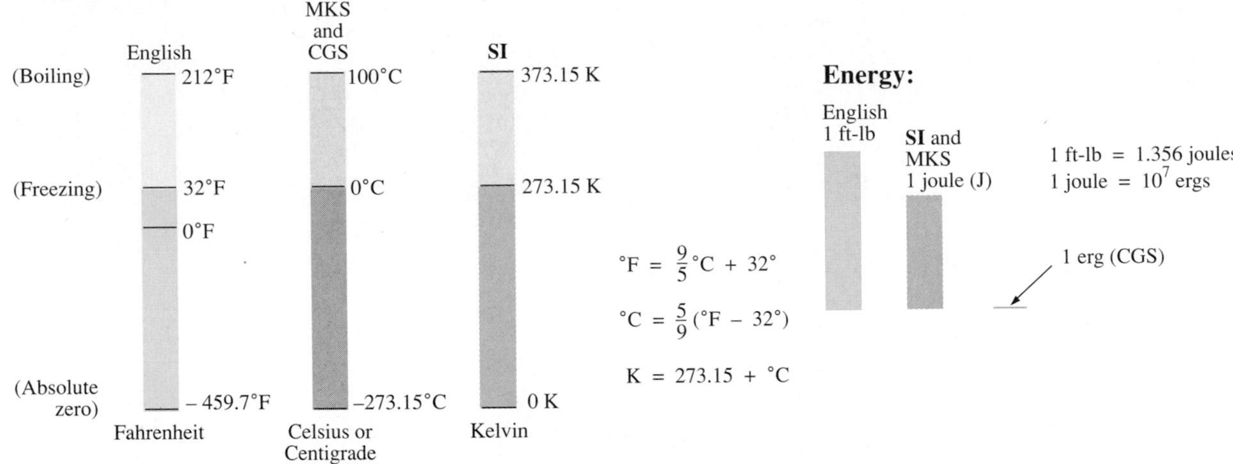

FIG. 1.4
Comparison of units of the various systems of units.

The meter was originally defined in 1790 to be 1/10,000,000 the distance between the equator and either pole at sea level, a length preserved on a platinum-iridium bar at the International Bureau of Weights and Measures at Sèvres, France.

The meter is now defined with reference to the speed of light in a vacuum, which is 299,792,458 m/s.

The kilogram is defined as a mass equal to 1000 times the mass of one cubic centimeter of pure water at 4°C.

This standard is preserved in the form of a platinum-iridium cylinder in Sèvres.

The *second* was originally defined as 1/86,400 of the mean solar day. However, since Earth's rotation is slowing down by almost 1 second every 10 years,

the second was redefined in 1967 as 9,192,631,770 periods of the electromagnetic radiation emitted by a particular transition of cesium atom.

1.5 SIGNIFICANT FIGURES, ACCURACY, AND ROUNDING OFF

This section will emphasize the importance of being aware of the source of a piece of data, how a number appears, and how it should be treated. Too often we write numbers in various forms with little concern for the format used, the number of digits that should be included, and the resulting accuracy of the calculations performed.

For instance, measurements of 22.1″ and 22.10″ imply different levels of accuracy. The first suggests that the measurement was made by an instrument accurate only to the tenths place; the latter was obtained with instrumentation capable of reading to the hundredths place. The use of zeros in a number, therefore, must be treated with care and the implications must be understood.

In general, there are two types of numbers, *exact* and *approximate*. Exact numbers are precise to the exact number of digits presented, just as we know that there are 12 apples in a dozen and not 12.1. Throughout the text the numbers that appear in the descriptions, diagrams, and examples are considered *exact*, so that a battery of 100 V can be written as 100.0 V, 100.00 V, and so on, since it is 100 V at any level of accuracy. The additional zeros were not included for the purposes of clarity. However, in the laboratory environment, where measurements are continually being taken and the level of accuracy can vary from one instrument to another, it is important to understand how to work with the results. Any reading obtained in the laboratory should be considered *approximate*. The analog scales with their pointers may be difficult to read, and even though the digital meter provides only specific digits on its display, it is limited to the number of digits it can provide, leaving us to wonder about the less significant digits not appearing on the display.

The accuracy of a reading can be determined by the number of *significant figures (digits)* present. Significant digits are those integers (0 to 9) that can be assumed to be accurate for the measurement being made. The result is that all nonzero numbers are considered significant, with zeros being significant in only some cases. For instance, the zeros in 1005 are considered significant because they define the size of the number and are surrounded by nonzero digits. However, for a number such as 0.064, the two zeros are not considered significant because they are used only to define the location of the decimal point and not the accuracy of the reading. For the number 0.4020, the zero to the left of the decimal point is not significant but the other two are because they define the magnitude of the number and the fourth-place accuracy of the reading.

When adding approximate numbers, it is important to be sure the accuracy of the readings is consistent throughout. To add a quantity accurate only to the tenths place to a number accurate to the thousandths place will result in a total having accuracy only to the tenths place. One cannot expect the reading with the higher level accuracy to improve the reading with only tenths-place accuracy.

In the addition or subtraction of approximate numbers, the entry with the lowest level of accuracy determines the format of the solution.

For the multiplication and division of approximate numbers, the result has the same number of significant figures as the number with the least number of significant figures.

For approximate numbers (and exact, for that matter) there is often a need to *round off* the result; that is, you must decide on the appropriate level of accuracy and alter the result accordingly. The accepted procedure is simply to note the digit following the last to appear in the rounded-off form, and add a 1 to the last digit if it is greater than or equal to 5, and leave it alone if it is less than 5. For example, $3.186 \cong 3.19 \cong 3.2$, depending on the level of accuracy desired. The symbol $\cong$ appearing means *approximately equal to.*

EXAMPLE 1.1 Perform the indicated operations with the following approximate numbers and round off to the appropriate level of accuracy.

a. $532.6 + 4.02 + 0.036 = 536.656 \cong$ **536.7** (as determined by 532.6)
b. $0.04 + 0.003 + 0.0064 = 0.0494 \cong$ **0.05** (as determined by 0.04)
c. $4.632 \times 2.4 = 11.1168 \cong$ **11** (as determined by the two significant digits of 2.4)
d. $3.051 \times 802 = 2446.902 \cong$ **2450** (as determined by the three significant digits of 802)
e. $1402/6.4 = 219.0625 \cong$ **220** (as determined by the two significant digits of 6.4)
f. $0.0046/0.05 = 0.0920 \cong$ **0.09** (as determined by the one significant digit of 0.05)

1.6 POWERS OF TEN

It should be apparent from the relative magnitude of the various units of measurement that very large and very small numbers will frequently be encountered in the sciences. To ease the difficulty of mathematical operations with numbers of such varying size, *powers of ten* are usually employed. This notation takes full advantage of the mathematical properties of powers of ten. The notation used to represent numbers that are integer powers of ten is as follows:

$$1 = 10^0 \qquad 1/10 = 0.1 = 10^{-1}$$
$$10 = 10^1 \qquad 1/100 = 0.01 = 10^{-2}$$
$$100 = 10^2 \qquad 1/1000 = 0.001 = 10^{-3}$$
$$1000 = 10^3 \qquad 1/10{,}000 = 0.0001 = 10^{-4}$$

In particular, note that $10^0 = 1$, and, in fact, any quantity to the zero power is 1 ($x^0 = 1$, $1000^0 = 1$, and so on). Also, note that the numbers in the list that are greater than 1 are associated with positive powers of ten, and numbers in the list that are less than 1 are associated with negative powers of ten.

A quick method of determining the proper power of 10 is to place a caret mark to the right of the numeral 1 wherever it may occur; then count from this point to the number of places to the right or left before arriving at the decimal point. Moving to the right indicates a positive power of ten, whereas moving to the left indicates a negative power. For example,

$$10{,}000.0 = 1\underbrace{0{,}000}_{1\ 2\ 3\ 4}. = 10^{+4}$$

$$0.00001 = 0.\underbrace{00001}_{5\ 4\ 3\ 2\ 1} = 10^{-5}$$

Some important mathematical equations and relationships pertaining to powers of ten are listed below, along with a few examples. In each case, n and m can be any positive or negative real number.

$$\frac{1}{10^n} = 10^{-n} \qquad \frac{1}{10^{-n}} = 10^n \tag{1.2}$$

Equation (1.2) clearly reveals that shifting a power of ten from the denominator to the numerator, or the reverse, requires simply changing the sign of the power.

EXAMPLES

$$\frac{1}{1000} = \frac{1}{10^{+3}} = 10^{-3}$$

$$\frac{1}{0.00001} = \frac{1}{10^{-5}} = 10^{+5}$$

The product of powers of ten:

$$(10^n)(10^m) = 10^{(n+m)} \tag{1.3}$$

EXAMPLES

$$(1000)(10{,}000) = (10^3)(10^4) = 10^{(3+4)} = 10^7$$

$$(0.00001)(100) = (10^{-5})(10^2) = 10^{(-5+2)} = 10^{-3}$$

The division of powers of ten:

$$\frac{10^n}{10^m} = 10^{(n-m)} \tag{1.4}$$

EXAMPLES

$$\frac{100{,}000}{100} = \frac{10^5}{10^2} = 10^{(5-2)} = 10^3$$

$$\frac{1000}{0.0001} = \frac{10^3}{10^{-4}} = 10^{(3-(-4))} = 10^{(3+4)} = 10^7$$

Note the use of parentheses in the second example to ensure that the proper sign is established between operators.

The power of powers of ten:

$$(10^n)^m = 10^{(nm)} \tag{1.5}$$

EXAMPLES

$$(100)^4 = (10^2)^4 = 10^{(2)(4)} = 10^8$$
$$(1000)^{-2} = (10^3)^{-2} = 10^{(3)(-2)} = 10^{-6}$$
$$(0.01)^{-3} = (10^{-2})^{-3} = 10^{(-2)(-3)} = 10^6$$

Basic Arithmetic Operations

Let us now examine the use of powers of ten to perform some basic arithmetic operations using numbers that are not just powers of ten. The number 5000 can be written as $5 \times 1000 = 5 \times 10^3$, and the number 0.0004 can be written as $4 \times 0.0001 = 4 \times 10^{-4}$. Of course, 10^5 can also be written as 1×10^5 if it clarifies the operation to be performed.

Addition and Subtraction To perform addition or subtraction using powers of ten, the power of ten *must be the same for each term;* that is,

$$A \times 10^n \pm B \times 10^n = (A \pm B) \times 10^n \tag{1.6}$$

EXAMPLES

$$\begin{aligned} 6300 + 75{,}000 &= (6.3)(1000) + (75)(1000) \\ &= 6.3 \times 10^3 + 75 \times 10^3 \\ &= (6.3 + 75) \times 10^3 \\ &= \mathbf{81.3 \times 10^3} \end{aligned}$$

$$\begin{aligned} 0.00096 - 0.000086 &= (96)(0.00001) - (8.6)(0.00001) \\ &= 96 \times 10^{-5} - 8.6 \times 10^{-5} \\ &= (96 - 8.6) \times 10^{-5} \\ &= \mathbf{87.4 \times 10^{-5}} \end{aligned}$$

Multiplication In general,

$$(A \times 10^n)(B \times 10^m) = (A)(B) \times 10^{n+m} \tag{1.7}$$

revealing that the *operations with the powers of ten can be separated from the operation with the multipliers.*

EXAMPLES

$$\begin{aligned} (0.0002)(0.000007) &= [(2)(0.0001)][(7)(0.000001)] \\ &= (2 \times 10^{-4})(7 \times 10^{-6}) \\ &= (2)(7) \times (10^{-4})(10^{-6}) \\ &= \mathbf{14 \times 10^{-10}} \end{aligned}$$

$$\begin{aligned} (340{,}000)(0.00061) &= (3.4 \times 10^5)(61 \times 10^{-5}) \\ &= (3.4)(61) \times (10^5)(10^{-5}) \\ &= 207.4 \times 10^0 \\ &= \mathbf{207.4} \end{aligned}$$

Division In general,

$$\frac{A \times 10^n}{B \times 10^m} = \frac{A}{B} \times 10^{n-m} \tag{1.8}$$

revealing again that the *operations with the powers of ten can be separated from the same operation with the multipliers.*

EXAMPLES

$$\frac{0.00047}{0.002} = \frac{47 \times 10^{-5}}{2 \times 10^{-3}} = \left(\frac{47}{2}\right) \times \left(\frac{10^{-5}}{10^{-3}}\right)$$
$$= \mathbf{23.5 \times 10^{-2}}$$

$$\frac{690{,}000}{0.00000013} = \frac{69 \times 10^4}{13 \times 10^{-8}} = \left(\frac{69}{13}\right) \times \left(\frac{10^4}{10^{-8}}\right)$$
$$= \mathbf{5.31 \times 10^{12}}$$

Powers In general,

$$(A \times 10^n)^m = A^m \times 10^{nm} \tag{1.9}$$

which again permits the separation of the *operation with the powers of ten from the multipliers.*

EXAMPLES

$$(0.00003)^3 = (3 \times 10^{-5})^3 = (3)^3 \times (10^{-5})^3$$
$$= \mathbf{27 \times 10^{-15}}$$

$$(90{,}800{,}000)^2 = (9.08 \times 10^7)^2 = (9.08)^2 \times (10^7)^2$$
$$= \mathbf{82.4464 \times 10^{14}}$$

Fixed-Point, Floating-Point, Scientific, and Engineering Notation

There are, in general, four ways in which numbers appear when using a computer or calculator. If powers of ten are not employed, they are written in the *fixed* or *floating point notation.* The fixed-point format requires that the decimal point appear in the same place each time. In the floating-point format, the decimal point will appear in a location defined by the number to be displayed. Most computers and calculators permit a choice of fixed- or floating-point notation. In the fixed format, the user can choose the level of accuracy for the output as tenths place, hundredths place, thousandths place, and so on. Every output will then fix the decimal point to one location, such as the following examples using thousandths place accuracy:

$$\frac{1}{3} = \mathbf{0.333} \qquad \frac{1}{16} = \mathbf{0.063} \qquad \frac{2300}{2} = \mathbf{1150.000}$$

If left in the floating-point format, the results will appear as follows for the above operations:

$$\frac{1}{3} = \mathbf{0.333333333333} \qquad \frac{1}{16} = \mathbf{0.0625} \qquad \frac{2300}{2} = \mathbf{1150}$$

Powers of ten will creep into the fixed- or floating-point notation if the number is too small or too large to be displayed properly.

Scientific (also called *standard*) *notation* and *engineering notation* make use of powers of 10 with restrictions on the mantissa (multiplier) or scale factor (power of the power of ten). Scientific notation requires that the decimal point appear directly after the first digit greater than or equal to 1 but less than 10. A power of ten will then appear with the number (usually following the power notation E), even if it has to be to the zero power. A few examples:

$$\frac{1}{3} = \mathbf{3.33333333333E{-}1} \qquad \frac{1}{16} = \mathbf{6.25E{-}2} \qquad \frac{2300}{2} = \mathbf{1.15E3}$$

Within the scientific notation, the fixed- or floating-point format can be chosen. In the above examples, floating was employed. If fixed is chosen and set at the thousandths point accuracy, the following will result for the above operations:

$$\frac{1}{3} = \mathbf{3.333E{-}1} \qquad \frac{1}{16} = \mathbf{6.250E{-}2} \qquad \frac{2300}{2} = \mathbf{1.150E3}$$

The last format to be introduced is *engineering notation,* which specifies that all powers of ten must be multiples of 3, and the mantissa must be greater than or equal to 1 but less than 1000. This restriction on the powers of ten is due to the fact that specific powers of ten have been assigned prefixes that will be introduced in the next few paragraphs. Using scientific notation in the floating-point mode will result in the following for the above operations:

$$\frac{1}{3} = \mathbf{333.333333333E{-}3} \qquad \frac{1}{16} = \mathbf{62.5E{-}3} \qquad \frac{2300}{2} = \mathbf{1.15E3}$$

Using engineering notation with three-place accuracy will result in the following:

$$\frac{1}{3} = \mathbf{333.333E{-}3} \qquad \frac{1}{16} = \mathbf{62.500E{-}3} \qquad \frac{2300}{2} = \mathbf{1.150E3}$$

Prefixes

Specific powers of ten in engineering notation have been assigned prefixes and symbols, as appearing in Table 1.2. They permit easy recognition of the power of ten and an improved channel of communication between technologists.

TABLE 1.2

Multiplication Factors	SI Prefix	SI Symbol
1 000 000 000 000 = 10^{12}	tera	T
1 000 000 000 = 10^{9}	giga	G
1 000 000 = 10^{6}	mega	M
1 000 = 10^{3}	kilo	k
0.001 = 10^{-3}	milli	m
0.000 001 = 10^{-6}	micro	μ
0.000 000 001 = 10^{-9}	nano	n
0.000 000 000 001 = 10^{-12}	pico	p

EXAMPLES

$$\begin{aligned}1{,}000{,}000 \text{ ohms} &= 1 \times 10^6 \text{ ohms}\\ &= 1 \text{ megohm } (\text{M}\Omega)\\ 100{,}000 \text{ meters} &= 100 \times 10^3 \text{ meters}\\ &= 100 \text{ kilometers (km)}\\ 0.0001 \text{ second} &= 0.1 \times 10^{-3} \text{ second}\\ &= 0.1 \text{ millisecond (ms)}\\ 0.000001 \text{ farad} &= 1 \times 10^{-6} \text{ farad}\\ &= 1 \text{ microfarad } (\mu\text{F})\end{aligned}$$

Here are a few examples with numbers that are not strictly powers of ten.

EXAMPLES

a. 41,200 m is equivalent to 41.2×10^3 m = 41.2 kilometers = **41.2 km.**
b. 0.00956 J is equivalent to 9.56×10^{-3} J = 9.56 millijoules = **9.56 mJ.**
c. 0.000768 s is equivalent to 768×10^{-6} s = 768 microseconds = **768 μs.**
d. $\dfrac{8400 \text{ m}}{0.06} = \dfrac{8.4 \times 10^3 \text{ m}}{6 \times 10^{-2}} = \left(\dfrac{8.4}{6}\right) \times \left(\dfrac{10^3}{10^{-2}}\right)\text{m}$
$= 1.4 \times 10^5 \text{ m} = 140 \times 10^3 \text{ m} = 140$ kilometers = **140 km**
e. $(0.0003)^4 \text{ s} = (3 \times 10^{-4})^4 \text{ s} = 81 \times 10^{-16} \text{ s}$
$= 0.0081 \times 10^{-12} \text{ s} = 0.008$ picoseconds = **0.0081 ps**

1.7 CONVERSION BETWEEN LEVELS OF POWERS OF TEN

It is often necessary to convert from one power of ten to another. For instance, if a meter measures kilohertz (kHz), it may be necessary to find the corresponding level in megahertz (MHz), or if time is measured in milliseconds (ms), it may be necessary to find the corresponding time in microseconds (μs) for a graphical plot. The process is not a difficult one if we simply keep in mind that an increase or decrease in the power of ten must be associated with the opposite effect on the multiplying factor. The procedure is best described by a few examples.

EXAMPLES

a. Convert 20 kHz to megahertz.

Solution: In the power-of-ten format:

$$20 \text{ kHz} = 20 \times 10^3 \text{ Hz}$$

The conversion requires that we find the multiplying factor to appear in the space below:

Increase by 3

$$20 \times 10^3 \text{ Hz} \Rightarrow \_\_\_ \times 10^6 \text{ Hz}$$

Decrease by 3

Since the power of ten will be *increased* by a factor of *three,* the multiplying factor must be *decreased* by moving the decimal point *three* places to the left, as shown below:

$$\underbrace{020.}_{3} = 0.02$$

and $$20 \times 10^3\ \text{Hz} = 0.02 \times 10^6\ \text{Hz} = \mathbf{0.02\ MHz}$$

b. Convert 0.01 ms to microseconds.

Solution: In the power-of-10 format:

$$0.01\ \text{ms} = 0.01 \times 10^{-3}\ \text{s}$$

Reduce by 3

and $$0.01 \times 10^{-3}\ \text{s} = \underline{\quad} \times 10^{-6}\ \text{s}$$

Increase by 3

Since the power of ten will be *reduced* by a factor of three, the multiplying factor must be *increased* by moving the decimal point three places to the right, as follows:

$$0.\underbrace{010}_{3} = 10$$

and $$0.01 \times 10^{-3}\ \text{s} = 10 \times 10^{-6}\ \text{s} = \mathbf{10\ \mu s}$$

There is a tendency when comparing -3 to -6 to think the power of ten has increased, but keep in mind when making your judgment about increasing or decreasing the magnitude of the multiplier that 10^{-6} is a great deal smaller than 10^{-3}.

c. Convert 0.002 km to millimeters.

Solution:

Reduce by 6

$$0.002 \times 10^3\ \text{m} \Rightarrow \underline{\quad} \times 10^{-3}\ \text{m}$$

Increase by 6

In this example we have to be very careful because the difference between $+3$ and -3 is a factor of 6, requiring that the multiplying factor be modified as follows:

$$0.\underbrace{002000}_{6} = 2000$$

and $$0.002 \times 10^3\ \text{m} = 2000 \times 10^{-3}\ \text{m} = \mathbf{2000\ mm}$$

1.8 CONVERSION WITHIN AND BETWEEN SYSTEMS OF UNITS

The conversion within and between systems of units is a process that cannot be avoided in the study of any technical field. It is an operation, however, that is performed incorrectly so often that this section was included to provide one approach that, if applied properly, will lead to the correct result.

There is more than one method of performing the conversion process. In fact, some people prefer to determine mentally whether the

conversion factor is multiplied or divided. This approach is acceptable for some elementary conversions, but it is risky with more complex operations.

The procedure to be described here is best introduced by examining a relatively simple problem such as converting inches to meters. Specifically, let us convert 48 in. (4 ft) to meters.

If we multiply the 48 in. by a factor of 1, the magnitude of the quantity remains the same:

$$48 \text{ in.} = 48 \text{ in.}(1) \qquad \textbf{(1.10)}$$

Let us now look at the conversion factor, which is the following for this example:

$$1 \text{ m} = 39.37 \text{ in.}$$

Dividing both sides of the conversion factor by 39.37 in. will result in the following format:

$$\frac{1 \text{ m}}{39.37 \text{ in.}} = (1)$$

Note that the end result is that the ratio 1 m/39.37 in. equals 1, as it should since they are equal quantities. If we now substitute this factor (1) into Eq. (1.10), we obtain

$$48 \text{ in.}(1) = 48 \cancel{\text{in.}}\left(\frac{1 \text{ m}}{39.37 \cancel{\text{in.}}}\right)$$

which results in the cancellation of inches as a unit of measure and leaves meters as the unit of measure. In addition, since the 39.37 is in the denominator, it must be divided into the 48 to complete the operation:

$$\frac{48}{39.37} \text{ m} = \mathbf{1.219 \text{ m}}$$

Let us now review the method, which has the following steps:

1. ***Set up the conversion factor to form a numerical value of (1) with the unit of measurement to be removed from the original quantity in the denominator.***
2. ***Perform the required mathematics to obtain the proper magnitude for the remaining unit of measurement.***

EXAMPLES

a. Convert 6.8 min to seconds.

Solution: The conversion factor is

$$1 \text{ min} = 60 \text{ s}$$

Since the minute is to be removed as the unit of measurement, it must appear in the denominator of the (1) factor, as follows:

Step 1:
$$\left(\frac{60 \text{ s}}{1 \text{ min}}\right) = (1)$$

Step 2:
$$6.8 \text{ min}(1) = 6.8 \cancel{\text{min}}\left(\frac{60 \text{ s}}{1 \cancel{\text{min}}}\right) = (6.8)(60) \text{ s}$$
$$= \mathbf{408 \text{ s}}$$

b. Convert 0.24 m to centimeters.

Solution: The conversion factor is

$$1\text{ m} = 100\text{ cm}$$

Since the meter is to be removed as the unit of measurement, it must appear in the denominator of the (1) factor as follows:

Step 1: $$\left(\frac{100\text{ cm}}{1\text{ m}}\right) = 1$$

Step 2: $$0.24\text{ m}(1) = 0.24\,\cancel{\text{m}}\left(\frac{100\text{ cm}}{1\,\cancel{\text{m}}}\right) = (0.24)(100)\text{ cm}$$

$$= \mathbf{24\ cm}$$

The products (1)(1) and (1)(1)(1) are still 1. Using this fact, we can perform a series of conversions in the same operation.

EXAMPLES

a. Determine the number of minutes in half a day.

Solution: Working our way through from days to hours to minutes, always ensuring that the unit of measurement to be removed is in the denominator, will result in the following sequence:

$$0.5\,\cancel{\text{day}}\left(\frac{24\,\cancel{\text{h}}}{1\,\cancel{\text{day}}}\right)\left(\frac{60\text{ min}}{1\,\cancel{\text{h}}}\right) = (0.5)(24)(60)\text{ min}$$

$$= \mathbf{720\ min}$$

b. Convert 2.2 yards to meters.

Solution: Working our way through from yards to feet to inches to meters will result in the following:

$$2.2\,\cancel{\text{yards}}\left(\frac{3\,\cancel{\text{ft}}}{1\,\cancel{\text{yard}}}\right)\left(\frac{12\,\cancel{\text{in.}}}{1\,\cancel{\text{ft}}}\right)\left(\frac{1\text{ m}}{39.37\,\cancel{\text{in.}}}\right) = \frac{(2.2)(3)(12)}{39.37}\text{ m}$$

$$= \mathbf{2.012\ m}$$

The following examples are variations of the above in practical situations.

EXAMPLES In Europe and Canada, and many other locations throughout the world, the speed limit is posted in kilometers per hour. How fast in miles per hour is 100 km/h?

$$\left(\frac{100\text{ km}}{\text{h}}\right)(1)(1)(1)(1)$$

$$= \left(\frac{100\,\cancel{\text{km}}}{\text{h}}\right)\left(\frac{1000\,\cancel{\text{m}}}{1\,\cancel{\text{km}}}\right)\left(\frac{39.37\,\cancel{\text{in.}}}{1\,\cancel{\text{m}}}\right)\left(\frac{1\,\cancel{\text{ft}}}{12\,\cancel{\text{in.}}}\right)\left(\frac{1\text{ mi}}{5280\,\cancel{\text{ft}}}\right)$$

$$= \frac{(100)(1000)(39.37)}{(12)(5280)}\ \frac{\text{mi}}{\text{h}}$$

$$= \mathbf{62.14\ mi/h}$$

Many travelers use 0.6 as a conversion factor to simplify the math involved; that is,

$$(100\text{ km/h})(0.6) \cong 60\text{ mi/h}$$

and $$(60\text{ km/h})(0.6) \cong 36\text{ mi/h}$$

Determine the speed in miles per hour of a competitor who can run a 4-min mile.

Inverting the factor 4 min/1 mi to 1 mi/4 min, we can proceed as follows:

$$\left(\frac{1\text{ mi}}{4\not{\text{min}}}\right)\left(\frac{60\not{\text{min}}}{\text{h}}\right) = \frac{60}{4}\text{ mi/h} = \mathbf{15\text{ mi/h}}$$

1.9 SYMBOLS

Throughout the text, various symbols will be employed that the reader may not have had occasion to use. Some are defined in Table 1.3, and others will be defined in the text as the need arises.

TABLE 1.3

Symbol	Meaning
$\neq$	Not equal to $6.12 \neq 6.13$
$>$	Greater than $4.78 > 4.20$
$\gg$	Much greater than $840 \gg 16$
$<$	Less than $430 < 540$
$\ll$	Much less than $0.002 \ll 46$
$\geq$	Greater than or equal to $x \geq y$ is satisfied for $y = 3$ and $x > 3$ or $x = 3$
$\leq$	Less than or equal to $x \leq y$ is satisfied for $y = 3$ and $x < 3$ or $x = 3$
$\cong$	Approximately equal to $3.14159 \cong 3.14$
Σ	Sum of $\Sigma\,(4 + 6 + 8) = 18$
$\lvert\ \rvert$	Absolute magnitude of $\lvert a \rvert = 4$, where $a = -4$ or $+4$
$\therefore$	Therefore $x = \sqrt{4} \quad \therefore x = \pm 2$
$\equiv$	By definition Establishes a relationship between two or more quantities

1.10 CONVERSION TABLES

Conversion tables such as those appearing in Appendix B can be very useful when time does not permit the application of methods described in this chapter. However, even though such tables appear easy to use, frequent errors occur because the operations appearing at the head of the table are not performed properly. In any case, when using such tables, try to establish mentally some order of magnitude for the quantity to be determined as compared to the magnitude of the quantity in its original set of units. This simple operation should prevent several impossible results that may occur if the conversion operation is improperly applied.

For example, consider the following from such a conversion table:

To convert from	To	Multiply by
Miles	Meters	1.609×10^3

A conversion of 2.5 mi to meters would require that we multiply 2.5 by the conversion factor; that is,

$$2.5\text{ mi}(1.609 \times 10^3) = 4.0225 \times 10^3\text{ m}$$

A conversion from 4000 m to miles would require a division process:

$$\frac{4000\text{ m}}{1.609 \times 10^3} = 2486.02 \times 10^{-3} = 2.48602\text{ mi}$$

In each of the above, there should have been little difficulty realizing that 2.5 mi would convert to a few thousand meters and 4000 m would be only a few miles. As indicated above, this kind of anticipatory thinking will eliminate the possibility of ridiculous conversion results.

1.11 COMPUTER ANALYSIS

The use of computers in the educational process has been growing exponentially in the past few years. There are very few texts at this introductory level that fail to include some discussion of current popular computer techniques. In fact, the very accreditation of a technology program may be a function of the depth to which computer methods are incorporated in the program.

There is no question that a basic knowledge of computer methods is something that the graduating student should carry away from a 2-year or 4-year program. Industry is now expecting students to have a basic knowledge of computer jargon and some hands-on experience.

For some students the thought of having to learn how to use a computer will result in an insecure, uncomfortable feeling normally associated with outright fear. Be assured, however, that through the proper learning experience and exposure, the computer can become a very "friendly," useful, and supportive tool in the development and application of your technical skills in a professional environment.

For the new student of computers, there are two general directions that can be taken to develop the necessary computer skills: the study of languages or the use of software packages.

Languages

There are several languages that provide a direct line of communication with the computer and the operations it can perform. A language is a set of symbols, letters, words, or statements that the user can enter into the computer. The computer system will "understand" these entries and perform them in the order established by a series of commands called a *program.* The program tells the computer what to do on a sequential, line-by-line basis in the same order a student would perform the calculations in longhand. The computer can respond only to the commands entered by the user. This requires that the programmer understand fully the sequence of operations and calculations required to obtain a particular solution. In other words, the computer can only respond to the user's input—it does not have some mysterious way of providing solutions unless told how to obtain those solutions. A lengthy analysis can result in a program having hundreds or thousands of lines. Once written, the program has to be checked carefully to be sure the results have meaning and are valid for an expected range of input variables. Writing a program can, therefore, be a long, tedious process, but keep in mind that once the program is tested and true, it can be stored in memory for future use. The user can be assured that any future results obtained have a high degree of accuracy but require a minimum expenditure of energy and time. Some of the popular languages applied in the electrical/electronics field today include C++, BASIC, PASCAL and FORTRAN. Each has its own set of commands and statements to communicate with the computer, but each can be used to perform the same type of analysis.

This text includes C++ and BASIC (Beginning All-purpose Symbolic Instruction Code) in its development because of the growing popularity of C++ and the friendly format of BASIC. The C language was first developed at Bell Laboratories to establish an efficient communication link between the user and the machine language of the central processing unit (CPU) of a computer. The language has grown in popularity throughout industry and education because it has the characteristics of a high-level language (easily understood by the user) with an efficient link to the computer's operating system. The C++ language was introduced as an extension of the C language to assist in the writing of complex programs using an enhanced, modular top-down approach. Both C++ and BASIC use English words, but clearly the BASIC language is easier to follow at first exposure. The general appearance of a BASIC program is also quite different, with its numbered lines versus the unnumbered, general flow of a C++ program. Of course, the price you pay with the increased initial friendliness of

BASIC is the less efficient link with the CPU, which requires additional operating time and a more sophisticated link to the computer hardware.

In any event, it is not assumed that the coverage of C++ or BASIC in this text is sufficient to permit the writing of additional programs. Their inclusion is meant as an introduction only: to reveal the appearance and characteristics of each, and to follow the development of some simple programs. A proper exposure to C++ or BASIC would require a course in itself, or at least a comprehensive supplemental program to fill in the many gaps of this text's presentation.

Software Packages

The second approach to computer analysis avoids the need to know a particular language; in fact, the user may not be aware of which language was used to write the programs within the package. All that is required is a knowledge of how to input the network parameters, define the operations to be performed, and extract the results; the package will do the rest. The individual steps toward a solution are beyond the needs of the user—all the user needs is an idea of how to get the network parameters into the computer and how to extract the results. Herein lie two of the concerns of the author with packaged programs—obtaining a solution without the faintest idea of how the solution was obtained, or whether the results are valid or way off base. It is imperative that the student realize that the computer should be used as a tool to assist the user—it must not be allowed to control the scope and potential of the user! Therefore, as we progress through the chapters of the text, be sure concepts are clearly understood before turning to the computer for support and efficiency.

Each software package has a *menu,* which defines the range of application of the package. Once entered into the computer, the system is preprogrammed to perform all the functions appearing in the menu. Be aware, however, that if a particular type of analysis is requested that is not on the menu, the software package cannot provide the desired results. The package is limited solely to those maneuvers developed by the team of programmers who developed the software package. In such situations the user must turn to another software package or write a program using one of the languages listed above.

In broad terms, if a software package is available to perform a particular analysis, then it should be used rather than developing routines. Most popular software packages are the result of many hours of effort by teams of programmers with years of experience. However, if the results are not in the desired format or the software package does not provide all the desired results, then the user's innovative talents should be put to use to develop a software package. As noted above, any program the user writes that passes the tests of range and accuracy can be considered a software package of his or her authorship for future use.

The software package to be employed in this text is PSpice (DOS and Windows versions), which is an educational version of a larger commercial version referred to simply as SPICE (Simulation Program with Integrated Circuit Emphasis). In the previous edition of this text, the discussion was limited solely to the DOS version. However, in the last year or so, the Windows version has grown so popular that it was necessary to dedicate equal attention to this new approach. A photograph of the new 6.2 version of PSpice in CD-ROM (as received from

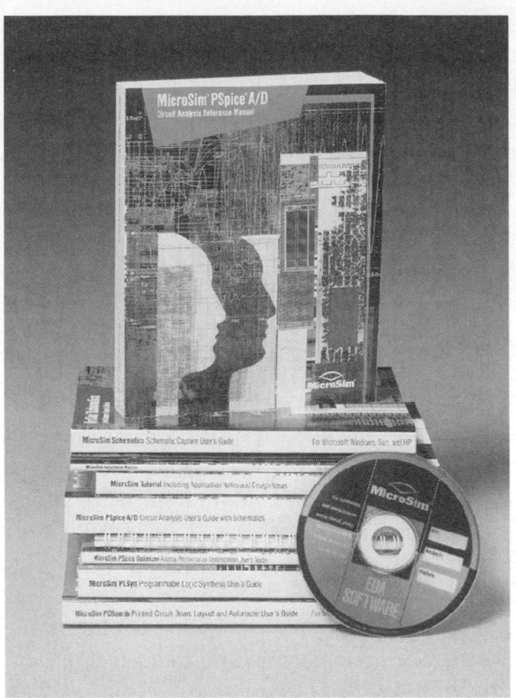

FIG. 1.5
PSpice design center. (Courtesy of MicroSim Corp.)

MicroSim Corporation) appears in Fig. 1.5. It is also available in the 3.5″ disc format. Appendix A provides a condensed list of information regarding the PSpice package.

The scope of the coverage was limited to the content of this text to permit as much detail as possible for the range of coverage. The PSpice menu is an extensive one and can perform most of the procedures described in this text. In those instances where the capability was not built in, the author has turned to the C++ or BASIC language and a detailed program. In fact, in numerous instances the same analysis is performed using a programming language and PSpice to demonstrate the salient differences between the application of a language and a software package.

In many ways the coverage of PSpice in this text is sufficient to act as the reference for the software package when applied to dc and ac networks. With a proper introduction to the equipment and support available when necessary, most areas of the text can be successfully examined and the problems completed using PSpice in either the DOS or Windows format. The manuals provided with the PSpice package have improved with the latest versions and can expand on the coverage of specific commands and operations. Other software packages available include BREADBOARD and ELECTRONICS WORKBENCH, both of which employ the Windows format.

Computer simulation and methods are an important, integral part of the text and should not be treated as superfluous material of the lowest priority. Once a basic concept is understood, take the time to investigate computer methods and start to develop a familiarity with the terminology and basic format of a program or use of a software package. It will be time well spent in preparation for the instruction you will eventually receive on computer systems.

PROBLEMS

Note: More difficult problems are denoted by an asterisk (*) throughout the text.

SECTION 1.2 A Brief History

1. Visit your local library (at school or home) and describe the extent to which it provides literature and computer support for the technologies—in particular, electricity, electronics, electromagnetics, and computers.
2. Choose an area of particular interest in this field and write a very brief report on the history of the subject.
3. Choose an individual of particular importance in this field and write a very brief review of his or her life and important contributions.

SECTION 1.3 Units of Measurement

4. Determine the distance in feet traveled by a car moving at 50 mi/h for 1 min.
5. How many hours would it take a person to walk 12 mi if the average pace is 15 min/mile?

SECTION 1.4 Systems of Units

6. Are there any relative advantages associated with the metric system as compared to the English system with respect to length, mass, force, and temperature? If so, explain.
7. Which of the four systems of units appearing in Table 1.1 has the smallest units for length, mass, and force? When would this system be used most effectively?

*8. Which system of Table 1.1 is closest in definition to the SI system? How are the two systems different? Why do you think the units of measurement for the SI system were chosen as listed in Table 1.1? Give the best reasons you can without referencing additional literature.

9. What is room temperature (68°F) in the MKS, CGS, and SI systems?
10. How many foot-pounds of energy are associated with 1000 J?
11. How many centimeters are there in $\frac{1}{2}$ yd?

SECTION 1.6 Powers of Ten

12. Express the following numbers as powers of 10:
    a. 10,000 **b.** 0.0001
    c. 1000 **d.** 1,000,000
    e. 0.0000001 **f.** 0.00001
13. Using only those powers of 10 listed in Table 1.2, express the following numbers in what seems to you the most logical form for future calculations:
    a. 15,000 **b.** 0.03000
    c. 7,400,000 **d.** 0.0000068
    e. 0.00040200 **f.** 0.0000000002
14. Perform the following operations and express your answer as a power of ten:
    a. $4200 + 6{,}800{,}000$
    b. $9 \times 10^4 + 3.6 \times 10^3$
    c. $0.5 \times 10^{-3} - 6 \times 10^{-5}$
    d. $1.2 \times 10^3 + 50{,}000 \times 10^{-3} - 0.006 \times 10^5$
15. Perform the following operations and express your answer as a power of ten:
    a. $(100)(100)$ **b.** $(0.01)(1000)$
    c. $(10^3)(10^6)$ **d.** $(1000)(0.00001)$
    e. $(10^{-6})(10{,}000{,}000)$ **f.** $(10{,}000)(10^{-8})(10^{35})$
16. Perform the following operations and express your answer as a power of ten:
    a. $(50{,}000)(0.0003)$
    b. 2200×0.08
    c. $(0.000082)(0.00007)$
    d. $(30 \times 10^{-4})(0.0002)(7 \times 10^8)$
17. Perform the following operations and express your answer as a power of ten:
    a. $\dfrac{100}{1000}$ **b.** $\dfrac{0.01}{100}$
    c. $\dfrac{10{,}000}{0.00001}$ **d.** $\dfrac{0.0000001}{100}$
    e. $\dfrac{10^{38}}{0.000100}$ **f.** $\dfrac{(100)^{1/2}}{0.01}$
18. Perform the following operations and express your answer as a power of ten:
    a. $\dfrac{2000}{0.00008}$ **b.** $\dfrac{0.00408}{60{,}000}$
    c. $\dfrac{0.000215}{0.00005}$ **d.** $\dfrac{78 \times 10^9}{4 \times 10^{-6}}$
19. Perform the following operations and express your answer as a power of ten:
    a. $(100)^3$ **b.** $(0.0001)^{1/2}$
    c. $(10{,}000)^8$ **d.** $(0.00000010)^9$
20. Perform the following operations and express your answer as a power of ten:
    a. $(2.2 \times 10^3)^3$
    b. $(0.0006 \times 10^2)^4$
    c. $(0.004)(6 \times 10^2)^2$
    d. $((2 \times 10^{-3})(0.8 \times 10^4)(0.003 \times 10^5))^3$
21. Perform the following operations and express your answer in scientific notation:
    a. $(-0.001)^2$ **b.** $\dfrac{(100)(10^{-4})}{10}$
    c. $\dfrac{(0.001)^2(100)}{10{,}000}$ **d.** $\dfrac{(10^2)(10{,}000)}{0.001}$
    e. $\dfrac{(0.0001)^3(100)}{1{,}000{,}000}$ ***f.** $\dfrac{[(100)(0.01)]^{-3}}{[(100)^2][0.001]}$

*22. Perform the following operations and express your answer in engineering notation:

a. $\dfrac{(300)^2(100)}{10^4}$ b. $[(40{,}000)^2][(20)^{-3}]$

c. $\dfrac{(60{,}000)^2}{(0.02)^2}$ d. $\dfrac{(0.000027)^{1/3}}{210{,}000}$

e. $\dfrac{[(4000)^2][300]}{0.02}$

f. $[(0.000016)^{1/2}][(100{,}000)^5][0.02]$

g. $\dfrac{[(0.003)^3][(0.00007)^2][(800)^2]}{[(100)(0.0009)]^{1/2}}$ (a challenge)

SECTION 1.7 Conversion Between Levels of Powers of Ten

23. Fill in the blanks of the following conversions:

a. $6 \times 10^3 =$ ____ $\times 10^6$

b. $4 \times 10^{-4} =$ ____ $\times 10^{-6}$

c. $50 \times 10^5 =$ ____ $\times 10^3 =$ ____ $\times 10^6$ $=$ ____ $\times 10^9$

d. $30 \times 10^{-8} =$ ____ $\times 10^{-3} =$ ____ $\times 10^{-6}$ $=$ ____ $\times 10^{-9}$

24. Perform the following conversions:

a. 2000 μs to milliseconds
b. 0.04 ms to microseconds
c. 0.06 μF to nanofarads
d. 8400 ps to microseconds
e. 0.006 km to millimeters
f. 260×10^3 mm to kilometers

SECTION 1.8 Conversion Within and Between Systems of Units

For Problems 25 to 27, convert the following:

25. a. 1.5 min to seconds
b. 0.04 h to seconds
c. 0.05 s to microseconds
d. 0.16 m to millimeters
e. 0.00000012 s to nanoseconds
f. 3,620,000 s to days
g. 1020 mm to meters

26. a. 0.1 μF (microfarad) to picofarads
b. 0.467 km to meters
c. 63.9 mm to centimeters
d. 69 cm to kilometers
e. 3.2 h to milliseconds
f. 0.016 mm to μm
g. 60 sq cm (cm^2) to square meters (m^2)

*27. a. 100 in. to meters
b. 4 ft to meters
c. 6 lb to newtons
d. 60,000 dyn to pounds
e. 150,000 cm to feet
f. 0.002 mi to meters (5280 ft = 1 mi)
g. 7800 m to yards

28. What is a mile in feet, yards, meters, and kilometers?

29. Calculate the speed of light in miles per hour using the defined speed of Section 1.4.

30. Find the velocity in miles per hour of a mass that travels 50 ft in 20 s.

31. How long in seconds will it take a car traveling at 100 mi/h to travel the length of a football field (100 yd)?

32. Convert 6 mi/h to meters per second.

33. If an athlete can row at a rate of 50 m/min, how many days would it take to cross the Atlantic ($\cong$3000 mi)?

34. How long would it take a runner to complete a 10-km race if a pace of 6.5 min/mi were maintained?

35. Quarters are about 1 in. in diameter. How many would be required to stretch from one end of a football field to the other (100 yd)?

36. Compare the total time in hours to cross the United States ($\cong$3000 mi) at an average speed of 55 mi/h versus an average speed of 65 mi/h. What is your reaction to the total time required versus the safety factor?

*37. Find the distance in meters that a mass traveling at 600 cm/s will cover in 0.016 h.

*38. Each spring there is a race up 86 floors of the 102-story Empire State Building in New York City. If you were able to climb 2 steps/second, how long would it take you to reach the 86th floor if each floor is 14 ft. high and each step is about 9 in.?

*39. The record for the race in Problem 38 is 10 minutes, 47 seconds. What was the racer's speed in min/mi for the race?

*40. If the race of Problem 38 were a horizontal distance, how long would it take a runner who can run 5-minute miles to cover the distance? Compare this with the record speed of Problem 39. Gravity is certainly a factor to be reckoned with!

SECTION 1.10 Conversion Tables

41. Using Appendix B, determine the number of

a. Btu in 5 J of energy.
b. cubic meters in 24 oz of a liquid.
c. seconds in 1.4 days.
d. pints in 1 m^3 of a liquid.

SECTION 1.11 Computer Analysis

42. Investigate the availability of computer courses and computer time in your curriculum. Which languages are commonly used and which software packages are popular?

43. Develop a list of five popular computer languages with a few characteristics of each. Why do you think some languages are better for the analysis of electric circuits than others?

GLOSSARY

BASIC A language that employs familiar English phrases to direct the operation of a computer.

C++ A computer language having an efficient communication link between the user and the machine language of the central processing unit (CPU) of a computer.

CGS system The system of units employing the *C*entimeter, *G*ram, and *S*econd as its fundamental units of measure.

Difference engine One of the first mechanical calculators.

Edison effect Establishing a flow of charge between two elements in an evacuated tube.

Electromagnetism The relationship between magnetic and electrical effects.

Eniac The first totally electronic computer.

Fleming's value The first of the electronic devices, the diode.

Integrated circuit (IC) A subminiature structure containing a vast number of electronic devices designed to perform a particular set of functions.

Joule (J) A unit of measurement for energy in the SI or MKS system. Equal to 0.7378 foot-pound in the English system and 10^7 ergs in the CGS system.

Kelvin (K) A unit of measurement for temperature in the SI system. Equal to 273.15 + °C in the MKS and CGS systems.

Kilogram (kg) A unit of measure for mass in the SI and MKS systems. Equal to 1000 grams in the CGS system.

Language A communication link between user and computer to define the operations to be performed and the results to be displayed or printed.

Leyden jar One of the first charge storage devices.

Menu A computer-generated list of choices for the user to determine the next operation to be performed.

Meter (m) A unit of measure for length in the SI and MKS systems. Equal to 1.094 yards in the English system and 100 centimeters in the CGS system.

MKS system The system of units employing the *M*eter, *K*ilogram, and *S*econd as its fundamental units of measure.

Newton (N) A unit of measurement for force in the SI and MKS systems. Equal to 100,000 dynes in the CGS system.

Pound (lb) A unit of measurement for force in the English system. Equal to 4.45 newtons in the SI or MKS system.

Program A sequential list of commands, instructions, etc., to perform a specified task using a computer.

PSpice A software package designed to analyze various dc, ac, and transient electrical and electronic systems.

Scientific notation A method for describing very large and very small numbers through the use of powers of 10, which requires that the multiplier be a number between 1 and 10.

Second (s) A unit of measurement for time in the SI, MKS, English, and CGS systems.

SI system The system of units adopted by the IEEE in 1965 and the USASI in 1967 as the International System of Units (*S*ystème *I*nternational d'Unités).

Slug A unit of measure for mass in the English system. Equal to 14.6 kilograms in the SI or MKS system.

Software package A computer program designed to perform specific analysis and design operations or generate results in a particular format.

Static electricity Stationary charge in a state of equilibrium.

Transistor The first semiconductor amplifier.

Voltaic cell A storage device that converts chemical to electrical energy.

2

Current and Voltage

2.1 ATOMS AND THEIR STRUCTURE

A basic understanding of the fundamental concepts of current and voltage requires a degree of familiarity with the atom and its structure. The simplest of all atoms is the hydrogen atom, made up of two basic particles, the *proton* and the *electron,* in the relative positions shown in Fig. 2.1(a). The *nucleus* of the hydrogen atom is the proton, a positively charged particle. *The orbiting electron carries a negative charge that is equal in magnitude to the positive charge of the proton.* In all other ele-

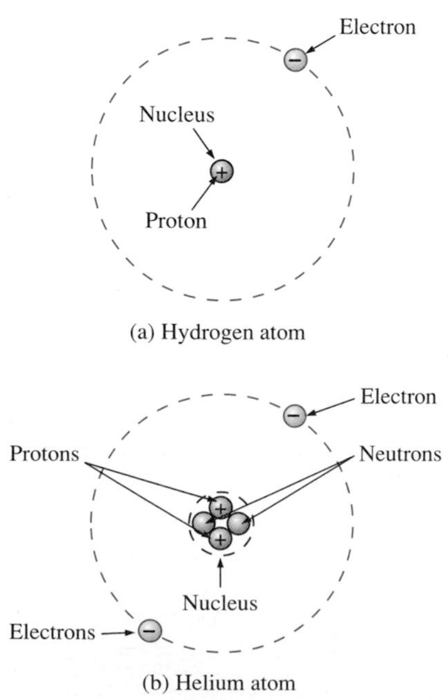

FIG. 2.1
The hydrogen and helium atoms.

ments, the nucleus also contains *neutrons,* which are slightly heavier than protons and have no electrical charge. The helium atom, for example, has two neutrons in addition to two electrons and two protons, as shown in Fig. 2.1(b). *In all neutral atoms the number of electrons is equal to the number of protons.* The mass of the electron is 9.11×10^{-28} g, and that of the proton and neutron is 1.672×10^{-24} g. The mass of the proton (or neutron) is therefore approximately 1836 times that of the electron. The radii of the proton, neutron, and electron are all of the order of magnitude of 2×10^{-15} m.

For the hydrogen atom, the radius of the smallest orbit followed by the electron is about 5×10^{-11} m. The radius of this orbit is approximately 25,000 times that of the basic constituents of the atom. This is approximately equivalent to a sphere the size of a dime revolving about another sphere of the same size more than a quarter of a mile away.

Different atoms will have various numbers of electrons in the concentric shells about the nucleus. The first shell, which is closest to the nucleus, can contain only two electrons. If an atom should have three electrons, the third must go to the next shell. The second shell can contain a maximum of eight electrons; the third, 18; and the fourth, 32; as determined by the equation $2n^2$, where n is the shell number. These shells are usually denoted by a number ($n = 1, 2, 3, \ldots$) or letter ($n = k, l, m, \ldots$).

Each shell is then broken down into subshells, where the first subshell can contain a maximum of two electrons; the second subshell, six electrons; the third, 10 electrons; and the fourth, 14; as shown in Fig. 2.2. The subshells are usually denoted by the letters *s, p, d,* and *f,* in that order, outward from the nucleus.

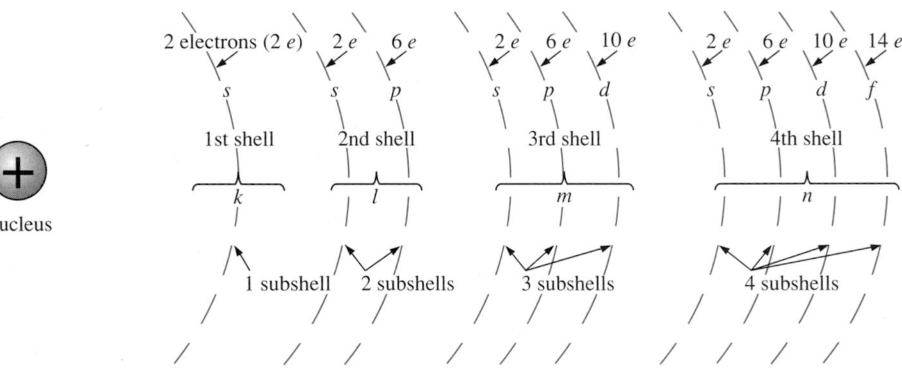

FIG. 2.2
Shells and subshells of the atomic structure.

It has been determined by experimentation that *unlike charges attract, and like charges repel.* The force of attraction or repulsion between two charged bodies Q_1 and Q_2 can be determined by Coulomb's law:

$$F \text{ (attraction or repulsion)} = \frac{kQ_1Q_2}{r^2} \quad \text{(newtons, N)} \qquad \textbf{(2.1)}$$

where F is in newtons, k = a constant = $9.0 \times 10^9\ \text{N}\cdot\text{m}^2/\text{C}^2$, Q_1 and Q_2 are the charges in coulombs (to be introduced in Section 2.2), and r is the distance in meters between the two charges. In particular, note the squared r term in the denominator, resulting in rapidly decreasing levels of F for increasing values of r. (See Fig. 2.3.)

In the atom, therefore, electrons will repel each other, and protons and electrons will attract each other. Since the nucleus consists of many positive charges (protons), a strong attractive force exists for the electrons in orbits close to the nucleus [note the effects of a large charge Q and a small distance r in Eq. (2.1)]. As the distance between the nucleus and the orbital electrons increases, the binding force diminishes until it reaches its lowest level at the outermost subshell (largest r). Due to the weaker binding forces, less energy must be expended to remove an electron from an outer subshell than from an inner subshell. Also, it is generally true that electrons are more readily removed from atoms having outer subshells that are incomplete *and,* in addition, possess few electrons. These properties of the atom that permit the removal of electrons under certain conditions are essential if motion of charge is to be created. Without this motion, this text could venture no further—our basic quantities rely on it.

Copper is the most commonly used metal in the electrical/electronics industry. An examination of its atomic structure will help identify why it has such widespread applications. The copper atom (Fig. 2.4) has one more electron than needed to complete the first three shells. This incomplete outermost subshell, possessing only one electron, and the distance between this electron and the nucleus reveal that the twenty-ninth electron is loosely bound to the copper atom. If this twenty-ninth electron gains sufficient energy from the surrounding medium to leave its parent atom, it is called a *free electron.* In one cubic inch of copper at room temperature, there are approximately $1.4 \times 10^{+24}$ free electrons. Other metals that exhibit the same properties as copper, but to a different degree, are silver, gold, aluminum, and tungsten. Additional discussion of conductors and their characteristics can be found in Section 3.2.

French (Angoulème, Paris)
(1736–1806)
Scientist and Inventor
Military Engineer in the West Indies

Courtesy of the Smithsonian Institution Photo No. 52,597

Attended the engineering school at Mezieres, the first such school of its kind. Formulated *Coulomb's law,* which defines the force between two electrical charges and is, in fact, one of the principal forces in atomic reactions. Performed extensive research on the friction encountered in machinery and windmills and the elasticity of metal and silk fibers.

FIG. 2.3
Charles Augustin de Coulomb.

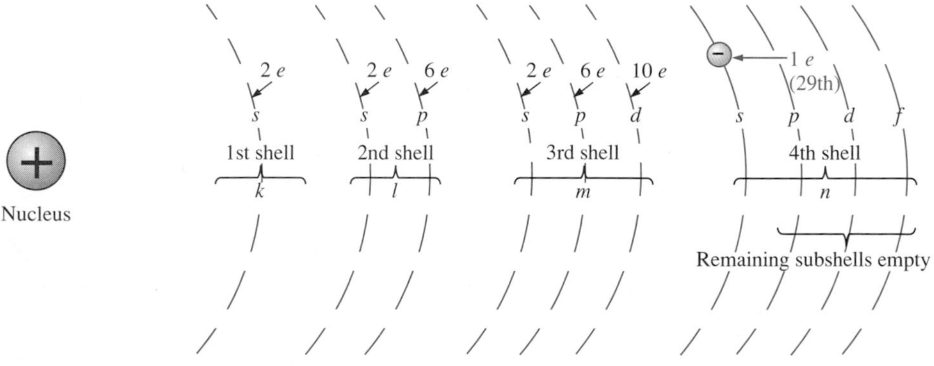

FIG. 2.4
The copper atom.

2.2 CURRENT

Consider a short length of copper wire cut with an imaginary perpendicular plane, producing the circular cross-section shown in Fig. 2.5. At room temperature with no external forces applied, there exists within the copper wire the random motion of free electrons created by

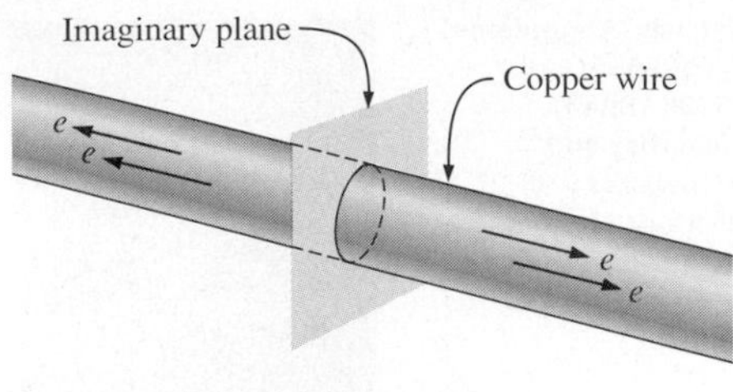

FIG. 2.5
Random motion of electrons in a copper wire with no external "pressure" (voltage) applied.

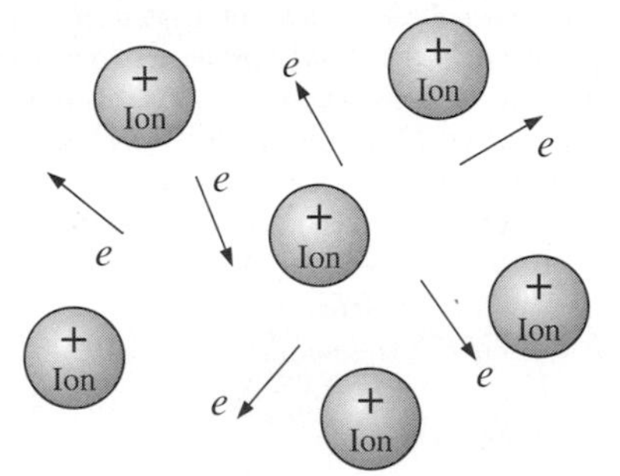

FIG. 2.6
Random motion of free electrons in an atomic structure.

the thermal energy that the electrons gain from the surrounding medium. When an atom loses its free electron, it acquires a net positive charge and is referred to as a *positive ion*. The free electron is able to move within these positive ions and leave the general area of the parent atom, while the positive ions only oscillate in a mean fixed position. For this reason,

the free electron is the charge carrier in a copper wire or any other solid conductor of electricity.

An array of positive ions and free electrons is depicted in Fig. 2.6. Within this array, the free electrons find themselves continually gaining or losing energy by virtue of their changing direction and velocity. Some of the factors responsible for this random motion include (1) the collisions with positive ions and other electrons, (2) the attractive forces for the positive ions, and (3) the force of repulsion that exists between electrons. This random motion of free electrons is such that over a period of time, the number of electrons moving to the right across the circular cross-section of Fig. 2.5 is exactly equal to the number passing over to the left.

With no external forces applied, the net flow of charge in a conductor in any one direction is zero.

Let us now connect copper wire between two battery terminals and a light bulb, as shown in Fig. 2.7, to create the simplest of electric circuits. The battery, at the expense of chemical energy, places a net positive charge at one terminal and a net negative charge on the other. The instant the final connection is made, the free electrons (of negative charge) will drift toward the positive terminal, while the positive ions left behind in the copper wire will simply oscillate in a mean fixed position. The negative terminal is a "supply" of electrons to be drawn from when the electrons of the copper wire drift toward the positive terminal.

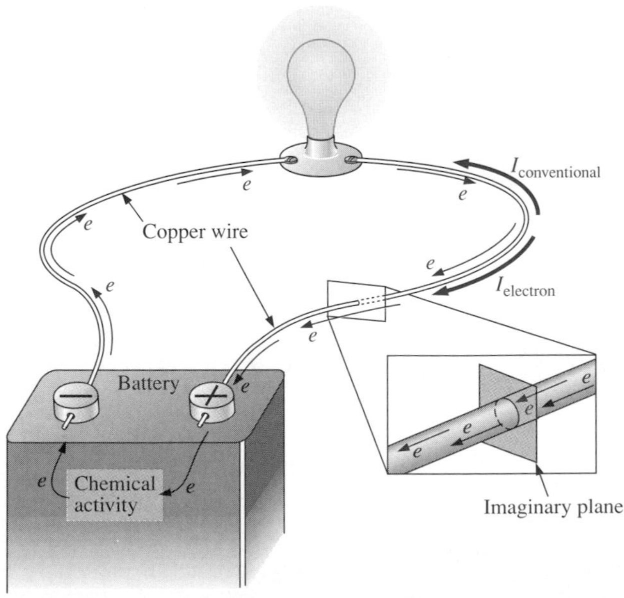

FIG. 2.7
Basic electric circuit.

The chemical activity of the battery will absorb the electrons at the positive terminal and maintain a steady supply of electrons at the negative terminal. The flow of charge (electrons) through the bulb will heat up the filament of the bulb to the point that will glow red hot and emit the desired light.

If 6.242×10^{18} *electrons* drift at uniform velocity through the imaginary circular cross-section of Fig. 2.7 in 1 *second*, the flow of charge, or *current*, is said to be 1 *ampere* (A) in honor of André-Marie Ampère (Fig. 2.8). The discussion of Chapter 1 clearly reveals that this is an enormous number of electrons passing through the surface in 1 second. The current associated with only a few electrons per second would be inconsequential and of little practical value. To establish numerical values that permit immediate comparisons between levels, a coulomb (C) of charge was defined as the total charge associated with 6.242×10^{18} electrons. The charge associated with one electron can then be determined from

$$\text{Charge/electron} = Q_e = \frac{1\text{ C}}{6.242 \times 10^{18}} = 1.6 \times 10^{-19}\text{ C}$$

The current in amperes can now be calculated using the following equation:

$$I = \frac{Q}{t} \qquad \begin{aligned} I &= \text{amperes (A)} \\ Q &= \text{coulombs (C)} \\ t &= \text{seconds (s)} \end{aligned} \qquad (2.2)$$

The capital letter *I* was chosen from the French word for current: *intensité*. The SI abbreviation for each quantity in Eq. (2.2) is provided to the right of the equation. The equation clearly reveals that for equal time intervals, the more charge that flows through the wire, the heavier the current.

Through algebraic manipulations, the other two quantities can be determined as follows:

$$Q = It \qquad \text{(coulombs, C)} \qquad (2.3)$$

and

$$t = \frac{Q}{I} \qquad \text{(seconds, s)} \qquad (2.4)$$

EXAMPLE 2.1 The charge flowing through the imaginary surface of Fig. 2.7 is 0.16 C every 64 ms. Determine the current in amperes.

Solution: Eq. (2.2):

$$I = \frac{Q}{t} = \frac{0.16\text{ C}}{64 \times 10^{-3}\text{ s}} = \frac{160 \times 10^{-3}\text{ C}}{64 \times 10^{-3}\text{ s}} = \mathbf{2.50\text{ A}}$$

EXAMPLE 2.2 Determine the time required for 4×10^{16} electrons to pass through the imaginary surface of Fig. 2.7 if the current is 5 mA.

Solution: Determine Q:

$$4 \times 10^{16}\text{ electrons}\left(\frac{1\text{ C}}{6.242 \times 10^{18}\text{ electrons}}\right) = 0.641 \times 10^{-2}\text{ C}$$
$$= 0.00641\text{ C} = 6.41\text{ mC}$$

French (Lyon, Paris)
(1775–1836)
Mathematician and Physicist
Professor of Mathematics, École Polytechnique in Paris

Courtesy of the Smithsonian Institution Photo No. 76,524

On September 18, 1820 introduced a new field of study, *electrodynamics,* devoted to the effect of electricity in motion, including the interaction between currents in adjoining conductors and the interplay of the surrounding magnetic fields. Constructed the first *solenoid* and demonstrated how it could behave like a magnet (the first *electromagnet*). Suggested the name *galvanometer* for an instrument designed to measure current levels.

FIG. 2.8
André Marie Ampère.

Calculate t [Eq. (2.4)]:

$$t = \frac{Q}{I} = \frac{6.41 \times 10^{-3}\text{ C}}{5 \times 10^{-3}\text{ A}} = \mathbf{1.282\ s}$$

A second glance at Fig. 2.7 will reveal that two directions of charge flow have been indicated. One is called *conventional flow,* and the other is called *electron flow.* This text will deal only with conventional flow for a variety of reasons, including the fact that it is the most widely used at educational institutions and in industry, it is employed in the design of all electronic device symbols, and it is the popular choice for all major computer software packages. The flow controversy is a result of an assumption made at the time electricity was discovered that the positive charge was the moving particle in metallic conductors. Be assured that the choice of conventional flow will not create great difficulty and confusion in the chapters to follow. Once the direction of I is established, the issue is dropped and the analysis can continue without confusion.

Safety Considerations

It is important to realize that even small levels of current through the human body can cause serious, dangerous side effects. Experimental results reveal that the human body begins to react to currents of only a few milliamperes. Although most individuals can withstand currents up to perhaps 10 mA for very short periods of time without serious side effects, any current over 10 mA should be considered dangerous. In fact, currents of 50 mA can cause severe shock, and currents of over 100 mA can be fatal. In most cases the skin resistance of the body when dry is sufficiently high to limit the current through the body to relatively safe levels for voltage levels typically found in the home. However, be aware that when the skin is wet due to perspiration, bathing, etc., or the skin barrier is broken due to an injury, the skin resistance drops dramatically and current levels could rise to dangerous levels for the same voltage shock. In general, therefore, simply remember that *water and electricity don't mix.* Granted, there are safety devices in the home today [such as the ground-fault-current-interrupt (GFCI) breaker to be introduced in Chapter 4] that are designed specifically for use in wet areas such as the bathroom and kitchen, but accidents happen. Treat electricity with respect—not fear.

2.3 VOLTAGE

The flow of charge described in the previous section is established by an external "pressure" derived from the energy that a mass has by virtue of its position: *potential energy.*

Energy, by definition, is the *capacity to do work.* If a mass (m) is raised to some height (h) above a reference plane, it has a measure of potential energy expressed in *joules* (J) that is determined by

$$W \text{ (potential energy)} = mgh \qquad \text{(joules, J)} \qquad \textbf{(2.5)}$$

where g is the gravitational acceleration (9.754 m/s^2). This mass now has the "potential" to do work such as crush an object placed on the ref-

erence plane. If the weight is raised further, it has an increased measure of potential energy and can do additional work. There is an obvious *difference in potential* between the two heights above the reference plane.

In the battery of Fig. 2.7, the internal chemical action will establish (through an expenditure of energy) an accumulation of negative charges (electrons) on one terminal (the negative terminal) and positive charges (positive ions) on the other (the positive terminal). A "positioning" of the charges has been established that will result in a *potential difference* between the terminals. If a conductor is connected between the terminals of the battery, the electrons at the negative terminal have sufficient potential energy to overcome collisions with other particles in the conductor and the repulsion from similar charges to reach the positive terminal to which they are attracted.

Charge can be raised to a higher potential level through the expenditure of energy from an external source, or it can lose potential energy as it travels through an electrical system. In any case, by definition:

A potential difference of 1 volt (V) exists between two points if 1 joule (J) of energy is exchanged in moving 1 coulomb (C) of charge between the two points.

The unit of measurement volt was chosen to honor Alessandro Volta (Fig. 2.9).

Pictorially, if one joule of energy (1 J) is required to move the one coulomb (1 C) of charge of Fig. 2.10 from position x to position y, the potential difference or voltage between the two points is one volt (1 V). If the energy required to move the 1 C of charge increases to 12 J due to additional opposing forces, then the potential difference will increase to 12 V. Voltage is therefore an indication of how much energy is involved in moving a charge between two points in an electrical system. Conversely, the higher the voltage rating of an energy source such as a battery, the more energy will be available to move charge through the system. Note in the above discussion that two points are always involved when talking about voltage or potential difference. In the future, therefore, it is very important to keep in mind that

a potential difference or voltage is always measured between two points in the system. Changing either point may change the potential difference between the two points under investigation.

In general, the potential difference between two points is determined by

$$V = \frac{W}{Q} \quad \text{(volts)} \tag{2.6}$$

Through algebraic manipulations, we have

$$W = QV \quad \text{(joules)} \tag{2.7}$$

and

$$Q = \frac{W}{V} \quad \text{(coulombs)} \tag{2.8}$$

Italian (Como, Pavia)
(1745–1827)
Physicist
Professor of Physics,
Pavia, Italy

Courtesy of the Smithsonian Institution Photo No. 55,393

Began electrical experiments at the age of 18 working with other European investigators. Major contribution was the development of an electrical energy source from chemical action in 1800. For the first time electrical energy was available on a continuous basis and could be used for practical purposes. Developed the first *condenser* known today as the *capacitor*. Was invited to Paris to demonstrate the *voltaic cell* to Napoleon. The International Electrical Congress meeting in Paris in 1881 honored his efforts by choosing the *volt* as the unit of measure for electromotive force.

FIG. 2.9
Count Alessandro Volta.

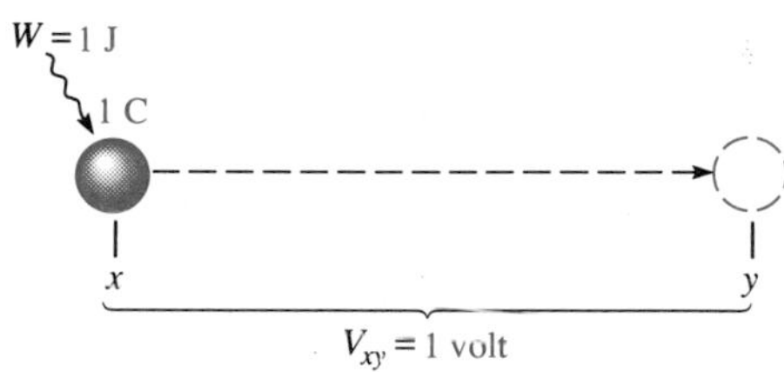

FIG. 2.10
Defining the unit of measurement for voltage.

EXAMPLE 2.3 Find the potential difference between two points in an electrical system if 60 J of energy are expended by a charge of 20 C between these two points.

Solution: Eq. (2.6):

$$V = \frac{W}{Q} = \frac{60\text{ J}}{20\text{ C}} = \mathbf{3\ V}$$

EXAMPLE 2.4 Determine the energy expended moving a charge of 50 μC through a potential difference of 6 V.

Solution: Eq. (2.7):

$$W = QV = (50 \times 10^{-6}\text{ C})(6\text{ V}) = 300 \times 10^{-6}\text{ J} = \mathbf{300\ \mu J}$$

Notation plays a very important role in the analysis of electrical and electronic systems. To distinguish between sources of voltage (batteries and the like) and losses in potential across dissipative elements, the following notation will be used:

E for voltage sources (volts)
V for voltage drops (volts)

An occasional source of confusion is the terminology applied to this subject matter. Terms commonly encountered include *potential, potential difference, voltage, voltage difference* (*drop* or *rise*), and *electromotive force.* As noted in the description above, some are used interchangeably. The following definitions are provided as an aid in understanding the meaning of each term:

Potential: The voltage at a point with respect to another point in the electrical system. Typically the reference point is ground, which is at zero potential.

Potential difference: The algebraic difference in potential (or voltage) between two points of a network.

Voltage: When isolated, like potential, the voltage at a point with respect to some reference such as ground (0 V).

Voltage difference: The algebraic difference in voltage (or potential) between two points of the system. A voltage drop or rise is as the terminology would suggest.

Electromotive force (emf): The force that establishes the flow of charge (or current) in a system due to the application of a difference in potential. This term is not applied that often in today's literature but is primarily associated with sources of energy.

In summary, the applied potential difference (in volts) of a voltage source in an electric circuit is the "pressure" to set the system in motion and "cause" the flow of charge or current through the electrical system. A mechanical analogy of the applied voltage is the pressure applied to the water in a main. The resulting flow of water through the system is likened to the flow of charge through an electric circuit. Without the applied pressure from the spigot, the water will simply sit in the hose, just as the electrons of a copper wire do not have a general direction without an applied voltage.

2.4 FIXED (dc) SUPPLIES

The terminology *dc* employed in the heading of this section is an abbreviation for *direct current,* which encompasses the various electrical systems in which there is a *unidirectional* ("one direction") flow of charge. A great deal more will be said about this terminology in the chapters to follow. For now, we will consider only those supplies that provide a fixed voltage or current.

dc Voltage Sources

Since the dc voltage source is the more familiar of the two types of supplies, it will be examined first. The symbol used for all dc voltage supplies in this text appears in Fig. 2.11. The relative lengths of the bars indicate the terminals they represent.

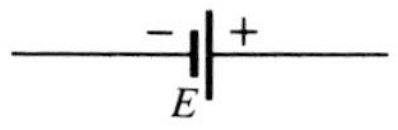

Dc voltage sources can be divided into three broad categories: (1) batteries (chemical action), (2) generators (electromechanical), and (3) power supplies (rectification).

– | + E

FIG. 2.11
Symbol for a dc voltage source.

Batteries

General Information For the layperson, the battery is the most common of the dc sources. By definition, a battery (derived from the expression "battery of cells") consists of a combination of two or more similar *cells,* a cell being the fundamental source of electrical energy developed through the conversion of chemical or solar energy. All cells can be divided into the *primary* or *secondary* types. The secondary is rechargeable, whereas the primary is not. That is, the chemical reaction of the secondary cell can be reversed to restore its capacity. The two most common rechargeable batteries are the lead-acid unit (used primarily in automobiles) and the nickel-cadmium battery (used in calculators, tools, photoflash units, shavers, and so on). The obvious advantage of the rechargeable unit is the reduced costs associated with not having to continually replace discharged primary cells.

All the cells appearing in this chapter except the *solar cell,* which absorbs energy from incident light in the form of photons, establish a potential difference at the expense of chemical energy. In addition, each has a positive and a negative *electrode* and an *electrolyte* to complete the circuit between electrodes within the battery. The electrolyte is the contact element and the source of ions for conduction between the terminals.

Alkaline and Lithium–Iodine Primary Cells The popular alkaline primary battery employs a powdered zinc anode (+); a potassium (alkali metal) hydroxide electrolyte; and a manganese dioxide, carbon cathode (−) as shown in Fig. 2.12(a). In particular, note in Fig. 2.12(b) that the larger the cylindrical unit, the higher the current capacity. The lantern is designed primarily for long-term use. Figure 2.13 shows two lithium–iodine primary units with an area of application and a rating to be introduced later in this section.

Lead–Acid Secondary Cell For the secondary lead-acid unit appearing in Fig. 2.14, the electrolyte is sulfuric acid and the electrodes are spongy lead (Pb) and lead peroxide (PbO_2). When a load is applied to the battery terminals, there is a transfer of electrons from the spongy lead electrode to the lead peroxide electrode through the load. This

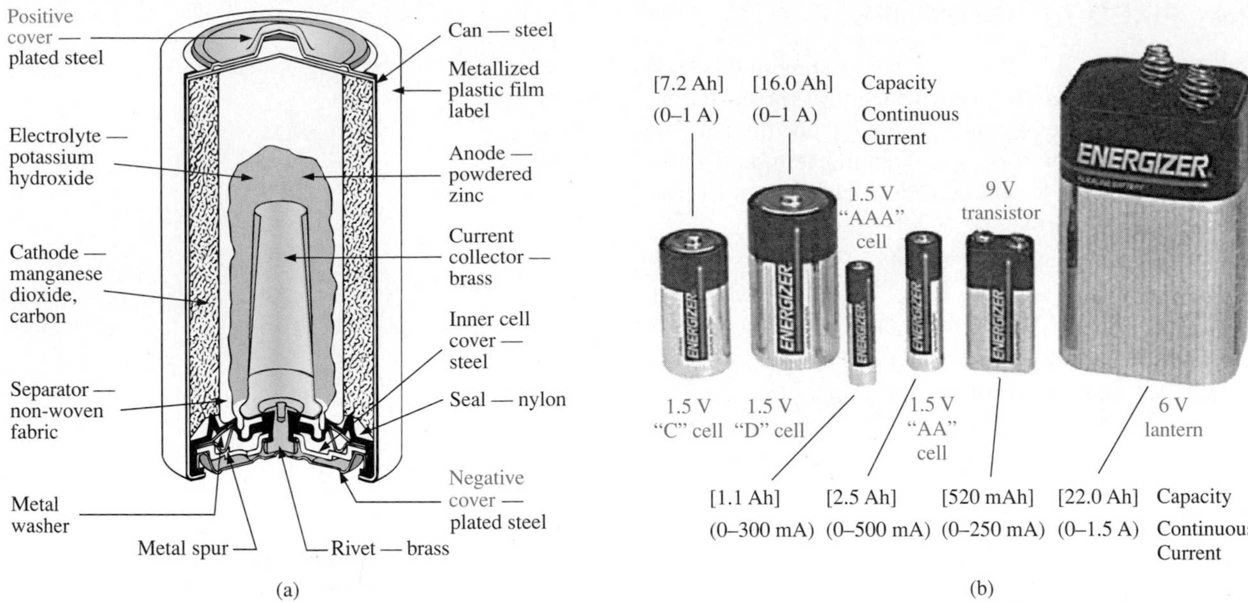

FIG. 2.12
(a) Cutaway of cylindrical Energizer alkaline cell; (b) Eveready Energizer primary cells. (Courtesy of Eveready Battery Company, Inc.)

(a) Lithiode™ lithium-iodine cell
2.8 V, 870 mAh
Long-life power sources with printed circuit board mounting capability

(b) Lithium-iodine pacemaker cell
2.8 V, 2.0 Ah

FIG. 2.13
Lithium–iodine primary cells. (Courtesy of Catalyst Research Corp.)

transfer of electrons will continue until the battery is completely discharged. The discharge time is determined by how diluted the acid has become and how heavy the coating of lead sulfate is on each plate. The state of discharge of a lead storage cell can be determined by measuring the specific gravity of the electrolyte with a hydrometer. The specific gravity of a substance is defined to be the ratio of the weight of a given volume of the substance to the weight of an equal volume of water at 4°C. For fully charged batteries, the specific gravity should be somewhere between 1.28 and 1.30. When the specific gravity drops to about 1.1, the battery should be recharged.

Since the lead storage cell is a secondary cell, it can be recharged at any point during the discharge phase simply by applying an external dc source across the cell that will pass current through the cell in a direction opposite to that in which the cell supplied current to the load. This

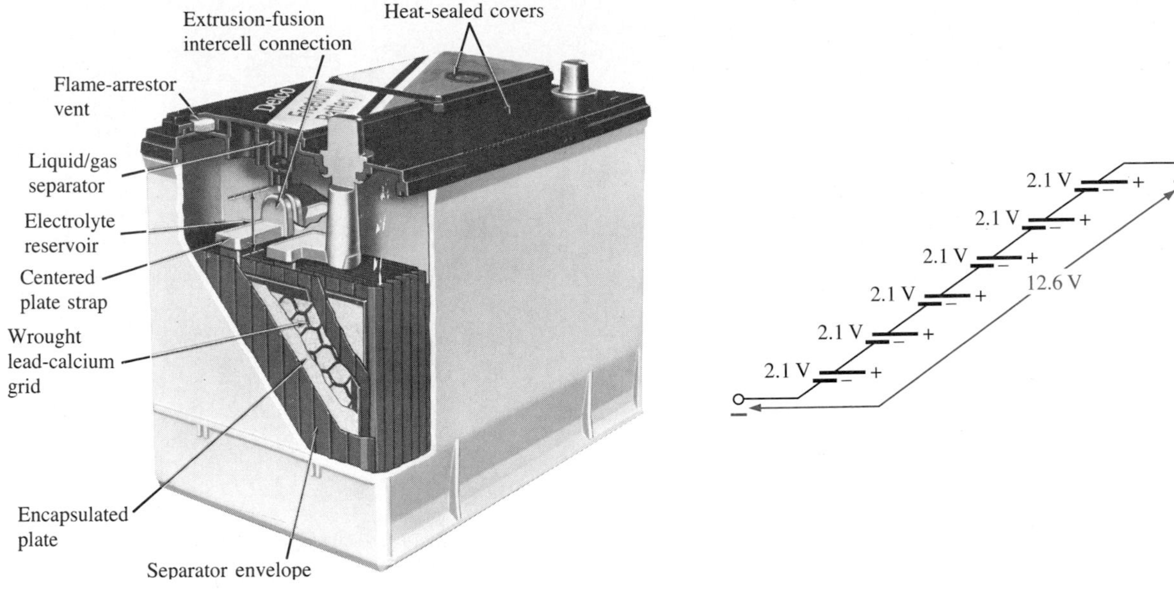

FIG. 2.14

Maintenance-free 12-V (actually 12.6-V) lead–acid battery. (Courtesy of Delco-Remy, a division of General Motors Corp.)

will remove the lead sulfate from the plates and restore the concentration of sulfuric acid.

The output of a lead storage cell over most of the discharge phase is about 2.1 V. In the commercial lead storage batteries used in the automobile, 12.6 V can be produced by six cells in series, as shown in Fig. 2.14. In general, lead–acid storage batteries are used in situations where a high current is required for relatively short periods of time. At one time all lead-acid batteries were vented. Gases created during the discharge cycle could escape, and the vent plugs provided access to replace the water or electrolyte and to check the acid level with a hydrometer. The use of a grid made from a wrought lead–calcium alloy strip rather than the lead–antimony cast grid commonly used has resulted in maintenance–free batteries such as that appearing in Fig. 2.14. The lead–antimony structure was susceptible to corrosion, overcharge, gasing, water usage, and self-discharge. Improved design with the lead–calcium grid has either eliminated or substantially reduced most of these problems.

It would seem with all the years of technology surrounding batteries that smaller, more powerful units would now be available. However, when it comes to the electric car, which is slowly gaining interest and popularity throughout the world, the lead–acid battery is still the primary source of power. A "station car," manufactured in Norway and used on a test basis in San Francisco for typical commuter runs, has a total weight of 1650 pounds, with 550 pounds (a third of its weight) for the lead-acid rechargeable batteries. Although the "station car" will travel at speeds of 55 mph, its range is limited to 65 miles on a charge. It would appear that long-distance travel with significantly reduced weight factors for the batteries will depend on an innovative new approach to battery design.

Nickel–Cadmium Secondary Cell The nickel–cadmium battery is a rechargeable battery that has been receiving enormous interest and

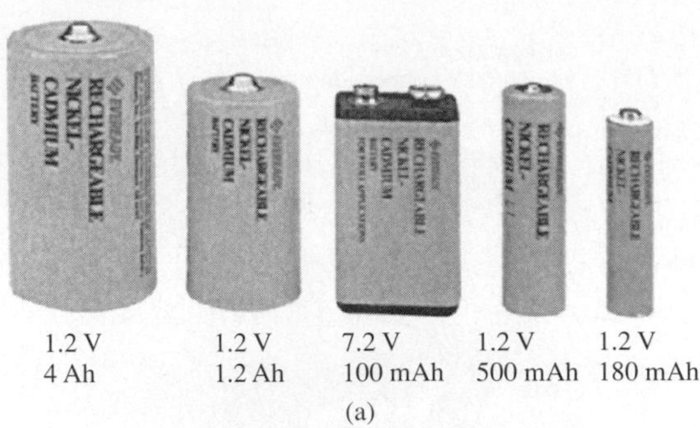

(a)

Eveready® BH 500 cell
1.2 V, 500 mAh
App: Where vertical height is severe limitation

(b)

FIG. 2.15
Rechargeable nickel–cadmium batteries. (Courtesy of Eveready Batteries.)

development in recent years. For applications such as flashlights, shavers, portable televisions, power drills, and so on, the nickel–cadmium (Ni-Cad) battery of Fig. 2.15 is the secondary battery of choice because the current levels are lower and the period of continuous drain is usually longer. A typical nickel–cadmium battery can survive over 1000 charge/discharge cycles over a period of time that can last for years.

It is important to recognize that when an appliance or system calls for a Ni-Cad battery, a primary cell should not be used. The appliance or system may have an internal charging network that would be dysfunctional with a primary cell. In addition, be aware that all Ni-Cad batteries are about 1.2 V per cell, while the most common primary cells are typically 1.5 V per cell. There is some ambiguity about how often a secondary cell should be recharged. For the vast majority of situations the battery can be used until there is some indication that the energy level is low, such as a dimming light from a flashlight, less power from a drill, or a blinking light if one is provided with the equipment. Keep in mind that secondary cells do have some "memory." If they are recharged continuously after being used for a short period of time, they may begin to believe they are short-term units and actually fail to hold the charge for the rated period of time. In any event, always try to avoid a "hard" discharge, which results when every bit of energy is drained from a cell. Too many hard discharge cycles will reduce the cycle life of the battery. Finally, be aware that the charging mechanism for nickel–cadmium cells is quite different from that for lead–acid batteries. The nickel–cadmium battery is charged by a constant current source, with the terminal voltage staying pretty steady through the entire charging cycle. The lead–acid battery is charged by a constant voltage source, permitting the current to vary as determined by the state of the battery. The capacity of the Ni-Cad battery increases almost linearly throughout most of the charging cycle. One may find that Ni-Cad batteries are relatively warm when charging. The lower the capacity level of the battery when charging, the higher the temperature of the cell. As the battery approaches rated capacity, the temperature of the cell approaches room temperature.

Nickel–Hydrogen and Nickel–Metal Hydride Secondary Cells Two other types of secondary cell include the nickel–hydrogen and nickel–metal hydride cells. The nickel–hydrogen cell is currently limited primarily to space vehicle applications where high-energy-density batteries are required that are rugged and reliable and can withstand a high number of charge/discharge cycles over a relatively long period of time. The nickel–metal hydride cell is actually a hybrid of the nickel–cadmium and nickel–hydrogen cells, combining the positive characteristics of each to create a product with a high power level in a small package that has a long cycle life. Although relatively expensive, this hybrid is a valid option for applications such as portable computers, as shown in Fig. 2.16.

10.8 V, 2.9 Ah,
600 mA (monochrome display),
900 mA (color display)

FIG. 2.16
Nickel–metal hydride (Ni-MH) battery for the IBM lap-top computer.

Solar Cell A high-density, 40-W solar cell appears in Fig. 2.17 with some of its associated data and areas of application. Since the maximum available wattage in an average bright sunlit day is 100 mW/cm^2 and conversion efficiencies are currently between 10% and 14%, the maximum available power per square centimeter from most commercial units is between 10 mW and 14 mW. For a square meter, however, the return would be 100 W to 140 W. A more detailed description of the solar cell will appear in your electronics courses. For now it is important to realize that a fixed illumination of the solar cell will provide a fairly steady dc voltage for driving various loads, from watches to automobiles.

40-W, high-density solar module
100-mm × 100-mm (4″ × 4″) square cells are used to provide maximum power in a minimum of space. The 33 series cell module provides a strong 12-V battery charging current for a wide range of temperatures (−40°C to 60°C)

FIG. 2.17
Solar module. (Courtesy of Motorola Semiconductor Products.)

Ampere-Hour Rating Batteries have a capacity rating given in ampere-hours (Ah) or milliampere-hours (mAh). Some of these ratings are included in the above figures. A battery with an ampere-hour rating of 100 will theoretically provide a steady current of 1 A for 100 h, 2 A for 50 h, 10 A for 10 h, and so on, as determined by the following equation:

$$\text{Life (hours)} = \frac{\text{ampere-hour rating (Ah)}}{\text{amperes drawn (A)}} \tag{2.9}$$

Two factors that affect this rating, however, are the temperature and the rate of discharge. The disc-type EVEREADY® BH 500 cell appearing in Fig. 2.15 has the terminal characteristics appearing in Fig. 2.18. Figure 2.18 reveals that

the capacity of a dc battery decreases with an increase in the current demand

and

the capacity of a dc battery decreases at relatively (compared to room temperature) low and high temperatures.

For the 1-V unit of Fig. 2.18(a), the rating is just above 500 mAh at a discharge current of 100 mA, but drops to about 300 mAh at about 1 A. For a unit that is less than $1\frac{1}{2}$ in. in diameter and less than $\frac{1}{2}$ in. in thickness, however, these are excellent terminal characteristics. Figure

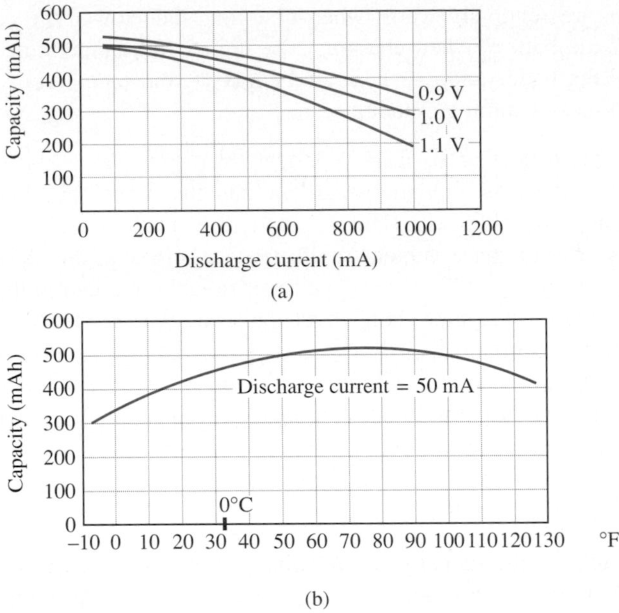

FIG. 2.18
EVEREADY® BH 500 cell characteristics: (a) capacity versus discharge current; (b) capacity versus temperature. (Courtesy of Eveready Batteries.)

2.18(b) reveals that the maximum mAh rating (at a current drain of 50 mA) occurs at about 75°F (≅ 24°C), or just above average room temperature. Note how the curve drops to the right and left of this maximum value. We are all aware of the reduced "strength" of a battery at low temperatures. Note that it has dropped to almost 300 mAh at about −8°F.

Another curve of interest appears in Fig. 2.19. It provides the expected cell voltage at a particular drain over a period of hours of use. It is noteworthy that the loss in hours between 50 mA and 100 mA is much greater than between 100 mA and 150 mA, even though the increase in current is the same between levels. In general,

the terminal voltage of a dc battery decreases with the length of the discharge time at a particular drain current.

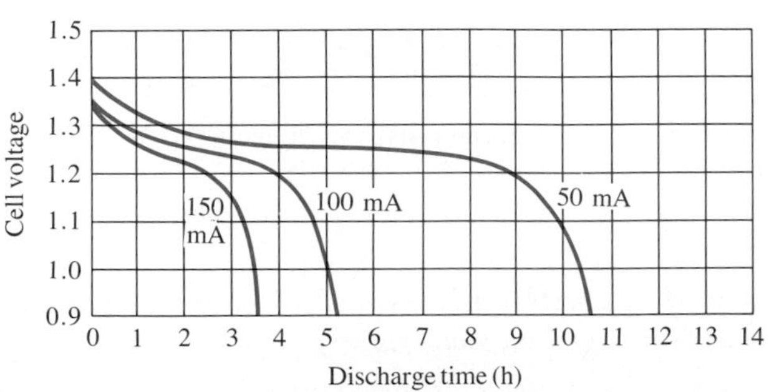

FIG. 2.19
EVEREADY® BH 500 cell discharge curves. (Courtesy of Eveready Batteries.)

EXAMPLE 2.5

a. Determine the capacity in milliampere-hours and life in minutes for the 0.9-V BH 500 cell of Fig. 2.18(a) if the discharge current is 600 mA.

b. At what temperature will the mAh rating of the cell of Fig. 2.18(b) be 90% of its maximum value if the discharge current is 50 mA?

Solutions:

a. From Fig. 2.18(a), the capacity at 600 mA is about 450 mAh. Thus, from Eq. (2.9),

$$\text{Life} = \frac{450\text{ mAh}}{600\text{ mA}} = 0.75\text{ h} = \mathbf{45\text{ min}}$$

b. From Fig. 2.18(b), the maximum is approximately 520 mAh. The 90% level is therefore 468 mAh, which occurs just above freezing, or **1°C,** and at the higher temperature of **45°C.**

Generators The dc generator is quite different, both in construction (Fig. 2.20) and in mode of operation, from the battery. When the shaft of the generator is rotating at the nameplate speed due to the applied torque of some external source of mechanical power, a voltage of rated value will appear across the external terminals. The terminal voltage and power-handling capabilities of the dc generator are typically higher than those of most batteries, and its lifetime is determined only by its construction. Commercially used dc generators are typically of the 120-V or 240-V variety. As pointed out earlier in this section, for the purposes of this text, no distinction will be made between the symbols for a battery and a generator.

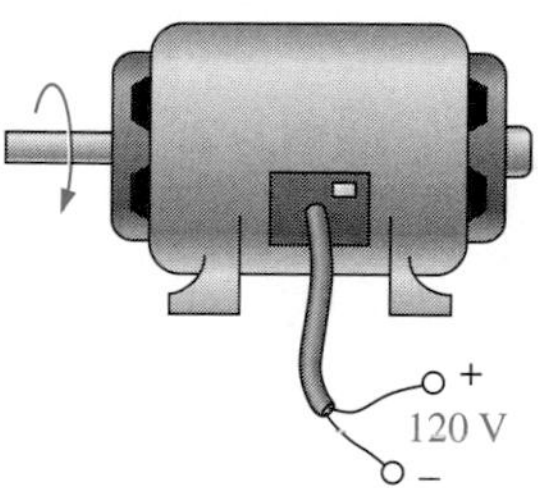

FIG. 2.20
dc generator.

Power Supplies The dc supply encountered most frequently in the laboratory employs the *rectification* and *filtering* processes as its means toward obtaining a steady dc voltage. Both processes will be covered in detail in your basic electronics courses. In total, a time-varying voltage (such as ac voltage available from a home outlet) is converted to one of a fixed magnitude. A dc laboratory supply of this type appears in Fig. 2.21.

Most dc laboratory supplies have a regulated, adjustable voltage output with three available terminals, as indicated in Figs. 2.21 and 2.22(a). The symbol for ground or zero potential (the reference) is also shown in Fig. 2.22(a). If 10 V above ground potential are required, then the connections are made as shown in Fig. 2.22(b). If 15 V below ground potential are required, then the connections are made as shown in Fig. 2.22(c). If connections are as shown in Fig. 2.22(d), we say we have a "floating" voltage of 5 V since the reference level is not in-cluded. Seldom is the configuration of Fig. 2.22(d) employed since it fails to protect the operator by providing a direct low-resistance path to ground and to establish a common ground for the system. In any case, the positive and negative terminals must be part of any circuit configuration.

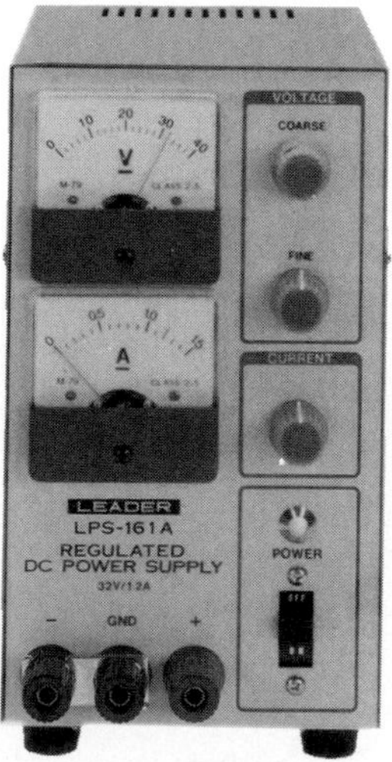

FIG. 2.21
dc laboratory supply. (Courtesy of Leader Instruments Corporation.)

dc Current Sources

The wide variety of types of, and applications for the dc voltage source has resulted in its becoming a rather familiar device, the characteristics of which are understood, at least basically, by the layperson. For example, it is common knowledge that a 12-V car battery has a terminal volt-

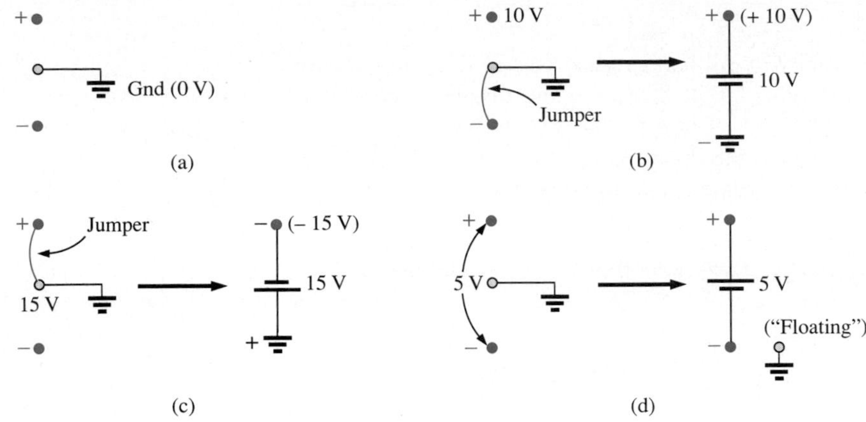

FIG. 2.22
dc laboratory supply: (a) available terminals; (b) positive voltage with respect to (w.r.t.) ground; (c) negative voltage w.r.t. ground; (d) floating supply.

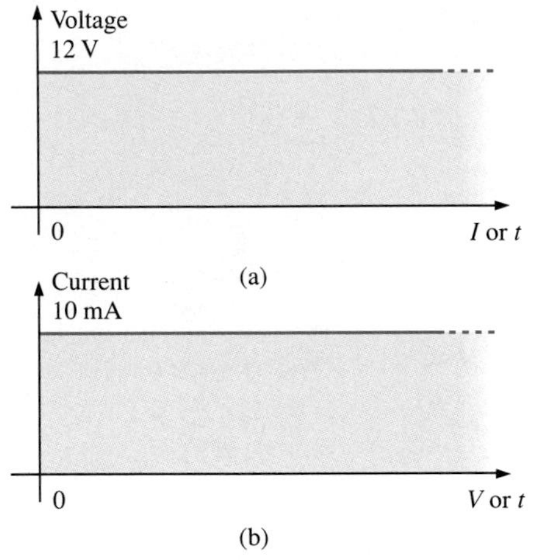

FIG. 2.23
Terminal characteristics: (a) ideal voltage source; (b) ideal current source.

age (at least approximately) of 12 V, even though the current drain by the automobile may vary under different operating conditions. In other words, *a dc voltage source will provide ideally a fixed terminal voltage, even though the current demand from the electrical/electronic system may vary,* as depicted in Fig. 2.23(a). A dc current source is the dual of the voltage source; that is,

the current source will supply, ideally, a fixed current to an electrical/electronic system, even though there may be variations in the terminal voltage as determined by the system,

as depicted in Fig. 2.23(b). Do not become alarmed if the concept of a current source is strange and somewhat confusing at this point. It will be covered in great detail in later chapters. Also, additional exposure will be provided in basic electronics courses.

2.5 CONDUCTORS AND INSULATORS

Different wires placed across the same two battery terminals will allow different amounts of charge to flow between the terminals. Many factors, such as the density, mobility, and stability characteristics of a material, account for these variations in charge flow. In general, however,

conductors are those materials that permit a generous flow of electrons with very little external force (voltage) applied.

In addition,

good conductors typically have only one electron in the valence (most distant from the nucleus) ring.

Since copper is used most frequently, it serves as the standard of comparison for the relative conductivity in Table 2.1. Note that aluminum, which has seen some commercial use, has only 61% of the conductivity level of copper, but keep in mind that this must be weighed against the cost and weight factors.

Insulators are those materials that have very few free electrons and require a large applied potential (voltage) to establish a measurable current level.

TABLE 2.1
Relative conductivity of various materials.

Metal	Relative Conductivity (%)
Silver	105
Copper	100
Gold	70.5
Aluminum	61
Tungsten	31.2
Nickel	22.1
Iron	14
Constantan	3.52
Nichrome	1.73
Calorite	1.44

A common use of insulating material is for covering current-carrying wire, which, if uninsulated, could cause dangerous side effects. Power-line repair people wear rubber gloves and stand on rubber mats as safety measures when working on high-voltage transmission lines. A number of different types of insulators and their applications appear in Fig. 2.24.

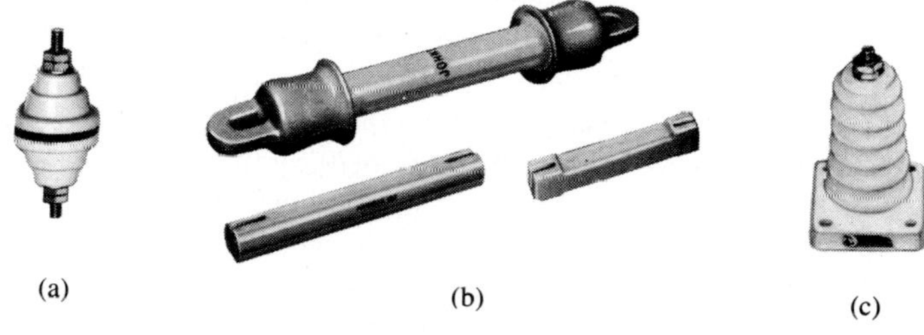

(a) (b) (c)

FIG. 2.24
Insulators: (a) insulated thru-panel bushings; (b) antenna strain insulators; (c) porcelain stand-off insulators. (Courtesy of Herman H. Smith, Inc.)

It must be pointed out, however, that even the best insulator will break down (permit charge to flow through it) if a sufficiently large potential is applied across it. The breakdown strengths of some common insulators are listed in Table 2.2. According to this table, for insu-

TABLE 2.2
Breakdown strength of some common insulators.

Material	Average Breakdown Strength (kV/cm)
Air	30
Porcelain	70
Oils	140
Bakelite	150
Rubber	270
Paper (paraffin-coated)	500
Teflon	600
Glass	900
Mica	2000

lators with the same geometric shape, it would require 270/30 = 9 times as much potential to pass current through rubber as compared to air and approximately 67 times as much voltage to pass current through mica as through air.

2.6 SEMICONDUCTORS

Semiconductors are a specific group of elements that exhibit characteristics between those of insulators and conductors.

The term *semi,* included in the terminology, has the dictionary definition of *half, partial,* or *between,* as defined by its use. The entire electronics industry is dependent on this class of materials since the electronic devices and integrated circuits (ICs) are constructed of semiconductor materials. Although *silicon* (Si) is the most extensively employed material, *germanium* (Ge) and *gallium arsenide* (GaAs) are also used in many important devices.

Semiconductor materials typically have four electrons in the outermost valence ring.

Semiconductors are further characterized as being photoconductive and having a negative temperature coefficient. Photoconductivity is a phenomenon where the photons (small packages of energy) from incident light can increase the carrier density in the material and thereby the charge flow level. A negative temperature coefficient reveals that the resistance (a characteristic to be described in detail in the next chapter) will decrease with an increase in temperature (opposite to that of most conductors). A great deal more will be said about semiconductors in the chapters to follow and in your basic electronics courses.

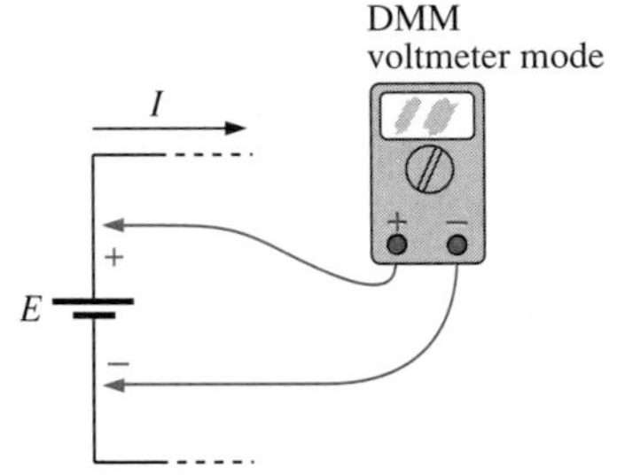

FIG. 2.25
Voltmeter connection for an up-scale (+) reading.

2.7 AMMETERS AND VOLTMETERS

It is important to be able to measure the current and voltage levels of an operating electrical system to check its operation, isolate malfunctions, and investigate effects impossible to predict on paper. As the names imply, *ammeters* are used to measure current levels, and *voltmeters,* the potential difference between two points. If the current levels are usually of the order of milliamperes, the instrument will typically be referred to as a milliammeter, and if the current levels are in the microampere range, as a microammeter. Similar statements can be made for voltage levels. Throughout the industry, voltage levels are measured more frequently than current levels primarily because the former does not require that the network connections be disturbed.

The potential difference between two points can be measured by simply connecting the leads of the meter *across the two points,* as indicated in Fig. 2.25. An up-scale reading is obtained by placing the positive lead of the meter to the point of higher potential of the network and the common or negative lead to the point of lower potential. The reverse connection will result in a negative reading or a below-zero indication.

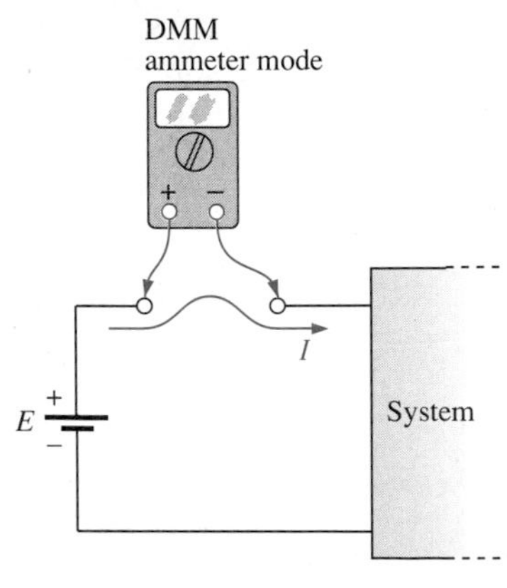

FIG. 2.26
Ammeter connection for an up-scale (+) reading.

Ammeters are connected as shown in Fig. 2.26. Since ammeters measure the rate of flow of charge, the meter must be placed in the network such that the charge will flow through the meter. The only way this can

be accomplished is to open the path in which the current is to be measured and place the meter between the two resulting terminals. For the configuration of Fig. 2.26, the voltage source lead (+) must be disconnected from the system and the ammeter inserted as shown. An up-scale reading will be obtained if the polarities on the terminals of the ammeter are such that the current of the system enters the positive terminal.

The introduction of any meter into an electrical/electronic system raises a concern about whether the meter will affect the behavior of the system. This question and others will be examined in Chapters 5 and 6 after additional terms and concepts have been introduced. For the moment, let it be said that since voltmeters and ammeters do not have internal sources, they will affect the network when introduced for measurement purposes. The design of each, however, is such that the impact is minimized.

There are instruments designed to measure just current or just voltage levels. However, the most common laboratory meters include the *volt-ohm-milliammeter* (VOM) and the *digital multimeter* (DMM) of Figs. 2.27 and 2.28, respectively. Both instruments will measure voltage and current and a third quantity, resistance, to be introduced in the next chapter. The VOM uses an analog scale, which requires interpreting the position of a pointer on a continuous scale, while the DMM provides a display of numbers with decimal point accuracy determined by the chosen scale. Comments on the characteristics and use of various meters will be made throughout the text. However, the major study of meters will be left for the laboratory sessions.

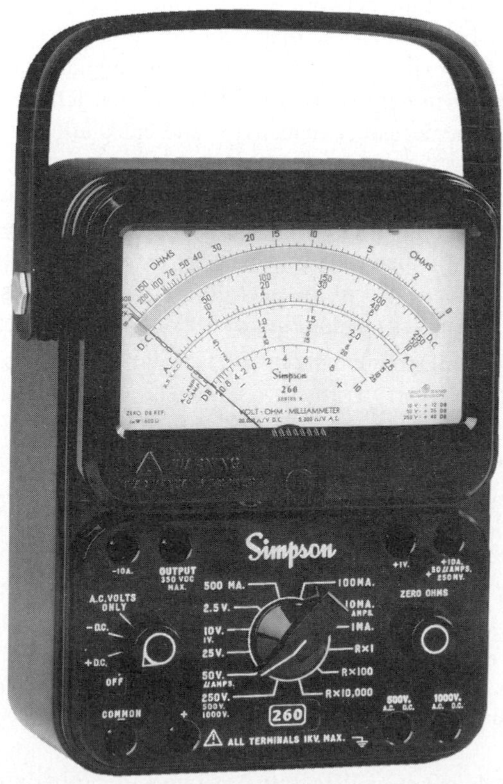

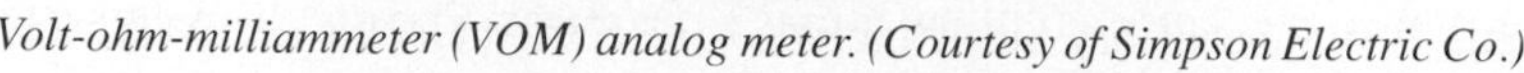
FIG. 2.27
Volt-ohm-milliammeter (VOM) analog meter. (Courtesy of Simpson Electric Co.)

FIG. 2.28
Digital multimeter (DMM). (Courtesy of John Fluke Mfg. Co. Inc.)

2.8 COMPUTER ANALYSIS

In an effort to develop familiarity with the software packages, the early chapters of the text will review how some of the basic elements of a network are entered into the computer. The entry of the basic elements using a programming language will be left for the example programs. In this chapter, since the independent voltage source was introduced, we review the procedure for entering the information.

The key word in entering any data into a computer is *format.* The data *must* be entered in a specified manner, or it will be rejected or, worse yet, misinterpreted. There is also no room for sloppiness when entering data. One misplaced comma or a wrong letter can invalidate entered data or completely change their magnitude.

PSpice does not distinguish between upper- and lowercase letters, but since computer programs are traditionally written in uppercase, we use uppercase throughout this text. Some of the following comments may be a repetition of material appearing in Appendix A, but they are repeated for completeness. In addition, keep in mind throughout the text that the computer content is not meant to make you an expert at using PSpice or a programming language but simply to provide a surface treatment for familiarity and comparison purposes. For both approaches, literature on each should be consulted using those avenues available to you through your educational institution.

PSpice (DOS)

In PSpice (DOS) all the network elements are entered in a specific format, followed by a statement that indicates what output information you require. You will find that the first few applications of PSpice (DOS) will result in very short lists of input data, letting the software package do all the necessary maneuvers and calculations.

In PSpice (DOS) every element is defined between terminals called *nodes.* At this stage in your development, suffice it to say that a node is simply a connection point between one element and another. Obviously, therefore, since every dc voltage source has two terminals to be connected in the network, it has two nodes to be specified. The basic format for entering a dc voltage source into PSpice is the following:

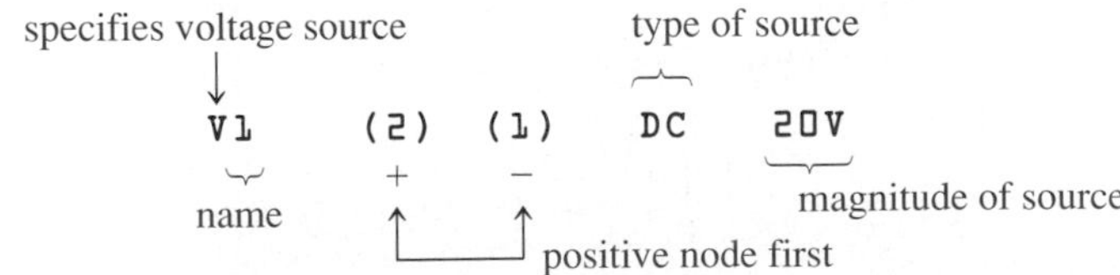

The input must begin with the uppercase letter V, followed by whatever *name* you may want to give the source. Typically the name is limited to 8 characters, which can include numbers and letters. The next number is the node corresponding to the positive side of the battery, as shown in Fig. 2.29(a). Node numbers can have any value between (and including) 0 and 9999, but they do not have to be sequential in their use as long as the network is properly defined. The DC reveals that it is a dc source, and the 20 V is its magnitude. The unit V is unnecessary but is included for clarity. In PSpice (DOS) any

letters that follow a number are ignored, but they are often included for the user and future reference. The spacing between items is not important, but an even spacing provides a pleasing format. The basic format is certainly not that hard to follow and should present little difficulty in entering or reading. For instance, the following defines the voltage source of Fig. 2.29(b).

```
VDCELL   48   24   DC   1.5V
```

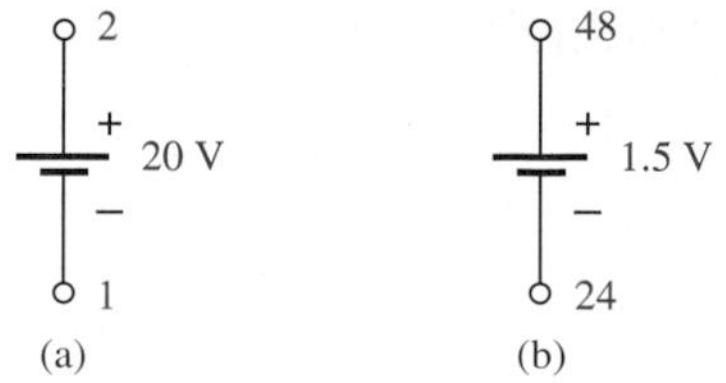

FIG. 2.29
Terminal voltage notation for PSpice (DOS).

PSpice (Windows)

In PSpice (Windows) elements such as the voltage source are retrieved from files within the software package and placed on a schematic with their labels and controlling parameters. As the network is "drawn," the computer becomes aware of the structure and the important parameters of each element for the analysis to follow. In other words, there is no need to assign nodes and be concerned about resulting polarities or current directions. It will take some practice to draw the networks quickly and accurately, but be assured that in time the process will move quite rapidly. The obvious advantage of the Windows approach is that a schematic of the network will be available immediately when the total network has been entered rather than simply a listing of nodal references. There will be some repetition of the procedure in the first few chapters to provide a complete picture with each chapter. In later chapters, however, this will not be the case because of the amount of new material that has to be presented.

PSpice (Windows) is initiated by calling up the **Program Manager**, choosing the **Design Center Eval** window, and then double-clicking the **Schematics** icon. Once the schematics window appears, take note of the main menu at the top of the schematic and proceed with the following sequence to obtain a dc voltage source: **Draw-Get New Part-Browse-source.slb-VDC-OK.** The dc voltage source will appear on the schematic, and it can then be moved to the desired location. A left click will set the element. If no additional sources are to be set, a right click will complete the operation. If other sources are present, they can be set one after another with the left click. Now the label and value have to be set. Double-clicking on the current label **V1** will bring up the **Edit Reference Designator** dialog box, in which the label E can be typed. Click **OK** and the new label will appear on the schematic. An additional left click will remove the defining rectangular structures. To change the value, double-click on the current value of **0 V** and obtain the **Set Attribute Value** dialog box. Enter the desired value of 20 V, followed by **OK,** and the new value will appear on the schematic. Another left click will again remove the defining rectangles. If the label and value must be moved, simply left click on the label or value to create a small rectangle around the parameter. Then click on the small rectangle and drag (holding the clicker down) the parameter to the desired location. Then release the clicker and the new placement is set. An additional left click will remove the surrounding rectangular area. The result is the source of Fig. 2.30. Rotation of the source can be accomplished by the sequence **Ctrl-R** or **Edit-Rotate.** Negative sources can be created by either rotating the source or entering its value as a negative number.

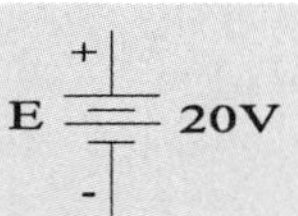

FIG. 2.30
A dc voltage source using PSpice (Windows).

PROBLEMS

SECTION 2.1 Atoms and Their Structure

1. The number of orbiting electrons in aluminum and silver is 13 and 47, respectively. Draw the electronic configuration, including all the shells and subshells, and discuss briefly why each is a good conductor.
2. Find the force of attraction between a proton and an electron separated by a distance equal to the radius of the smallest orbit followed by an electron (5×10^{-11} m) in a hydrogen atom.
3. Find the force of attraction in newtons between the charges Q_1 and Q_2 in Fig. 2.31 when
   a. $r = 1$ m
   b. $r = 3$ m
   c. $r = 10$ m
   (Note how quickly the force drops with an increase in r.)

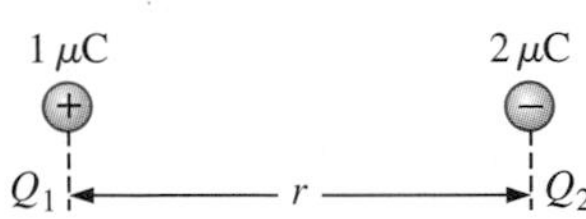

FIG. 2.31
Problem 3.

*4. Find the force of repulsion in newtons between Q_1 and Q_2 in Fig. 2.32 when
   a. $r = 1$ mi
   b. $r = 0.01$ m
   c. $r = 1/16$ in.

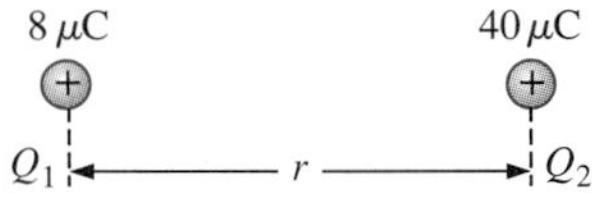

FIG. 2.32
Problem 4.

*5. Plot the force of attraction (in newtons) versus separation (in meters) for two charges of 2 mC and -4 μC. Set r to 0.5 m and 1 m, followed by 1-m intervals to 10 m. Comment on the shape of the curve. Is it linear or nonlinear? What does it tell you about the force of attraction between charges as they are separated? What does it tell you about any function plotted against a squared term in the denominator?

6. Determine the distance between two charges of 20 μC if the force between the two charges is 3.6×10^4 N.

*7. Two charged bodies, Q_1 and Q_2, when separated by a distance of 2 m, experience a force of repulsion equal to 1.8 N.
   a. What will the force of repulsion be when they are 10 m apart?
   b. If the ratio $Q_1/Q_2 = 1/2$, find Q_1 and Q_2 ($r = 10$ m).

SECTION 2.2 Current

8. Find the current in amperes if 650 C of charge pass through a wire in 50 s.
9. If 465 C of charge pass through a wire in 2.5 min, find the current in amperes.
10. If a current of 40 A exists for 1 min, how many coulombs of charge have passed through the wire?
11. How many coulombs of charge pass through a lamp in 2 min if the current is constant at 750 mA?
12. If the current in a conductor is constant at 2 mA, how much time is required for 4600×10^{-6} C to pass through the conductor?
13. If $21.847 \times 10^{+18}$ electrons pass through a wire in 7 s, find the current.
14. How many electrons pass through a conductor in 1 min if the current is 1 A?
15. Will a fuse rated at 1 A "blow" if 86 C pass through it in 1.2 min?

*16. If $0.784 \times 10^{+18}$ electrons pass through a wire in 643 ms, find the current.

*17. Which would you prefer?
   a. A penny for every electron that passes through a wire in 0.01 μs at a current of 2mA, or
   b. A dollar for every electron that passes through a wire in 1.5 ns if the current is 100 μA.

SECTION 2.3 Voltage

18. What is the voltage between two points if 96 mJ of energy are required to move 50×10^{18} electrons between the two points?
19. If the potential difference between two points is 42 V, how much work is required to bring 6 C from one point to the other?
20. Find the charge Q that requires 96 J of energy to be moved through a potential difference of 16 V.
21. How much charge passes through a battery of 22.5 V if the energy expended is 90 J?
22. If a conductor with a current of 200 mA passing through it converts 40 J of electrical energy into heat in 30 s, what is the potential drop across the conductor?

*23. Charge is flowing through a conductor at the rate of 420 C/min. If 742 J of electrical energy are converted to heat in 30 s, what is the potential drop across the conductor?

*24. The potential difference between two points in an electric circuit is 24 V. If 0.4 J of energy were dissipated in a period of 5 ms, what would the current be between the two points?

SECTION 2.4 Fixed (dc) Supplies

25. What current will a battery with an Ah rating of 200 theoretically provide for 40 h?
26. What is the Ah rating of a battery that can provide 0.8 A for 76 h?
27. For how many hours will a battery with an Ah rating of 32 theoretically provide a current of 1.28 A?

28. Find the mAh rating of the EVEREADY® BH 500 battery at 100°F and 0°C at a discharge current of 50 mA using Fig. 2.18(b).

29. Find the mAh rating of the 1.0 V EVEREADY® BH 500 battery if the current drain is 550 mA using Fig. 2.18(a). How long will it supply this current?

30. For how long can 50 mA be drawn from the battery of Fig. 2.19 before its terminal voltage drops below 1 V? Determine the number of hours at a drain current of 150 mA, and compare the ratio of drain current to the resulting ratio of hours of availability.

31. A standard 12-V car battery has an ampere-hour rating of 40 Ah, whereas a heavy-duty battery has a rating of 60 Ah. How would you compare the energy levels of each and the available current for starting purposes?

***32.** Using the relevant equations of the past few sections, determine the available energy (in joules) from the Eveready battery of Fig. 2.15(b).

***33.** A portable television using a 12-V, 3-Ah rechargeable battery can operate for a period of about 5.5 h. What is the average current drawn during this period? What is the energy expended by the battery in joules?

34. Discuss briefly the difference among the three types of dc voltage supplies (batteries, rectification, and generators).

35. Compare the characteristics of a dc current source with those of a dc voltage source. How are they similar and how are they different?

SECTION 2.5 Conductors and Insulators

36. Discuss two properties of the atomic structure of copper that make it a good conductor.

37. Name two materials not listed in Table 2.1 that are good conductors of electricity.

38. Explain the terms *insulator* and *breakdown strength.*

39. List three uses of insulators not mentioned in Section 2.5.

SECTION 2.6 Semiconductors

40. What is a semiconductor? How does it compare with a conductor and an insulator?

41. Consult a semiconductor electronics text and note the extensive use of germanium and silicon semiconductor materials. Review the characteristics of each material.

SECTION 2.7 Ammeters and Voltmeters

42. What are the significant differences in the way ammeters and voltmeters are connected?

43. If an ammeter reads 2.5 A for a period of 4 min, determine the charge that has passed through the meter.

44. Between two points in an electric circuit, a voltmeter reads 12.5 V for a period of 20 s. If the current measured by an ammeter is 10 mA, determine the energy expended and the charge that flowed between the two points.

GLOSSARY

Ammeter An instrument designed to read the current through elements in series with the meter.

Ampere (A) The SI unit of measurement applied to the flow of charge through a conductor.

Ampere-hour rating The rating applied to a source of energy that will reveal how long a particular level of current can be drawn from that source.

Cell A fundamental source of electrical energy developed through the conversion of chemical or solar energy.

Conductors Materials that permit a generous flow of electrons with very little voltage applied.

Copper A material possessing physical properties that make it particularly useful as a conductor of electricity.

Coulomb (C) The fundamental SI unit of measure for charge. It is equal to the charge carried by 6.242×10^{18} electrons.

Coulomb's law An equation defining the force of attraction or repulsion between two charges.

dc current source A source that will provide a fixed current level even though the load to which it is applied may cause its terminal voltage to change.

dc generator A source of dc voltage available through the turning of the shaft of the device by some external means.

Direct current Current having a single direction (unidirectional) and a fixed magnitude over time.

Ductility The property of a material that allows it to be drawn into long thin wires.

Electrolytes The contact element and the source of ions between the electrodes of the battery.

Electron The particle with negative polarity that orbits the nucleus of an atom.

Free electron An electron unassociated with any particular atom, relatively free to move through a crystal lattice structure under the influence of external forces.

Insulators Materials in which a very high voltage must be applied to produce any measurable current flow.

Malleability The property of a material that allows it to be worked into many different shapes.

Neutron The particle having no electrical charge, found in the nucleus of the atom.

Node A terminal point between elements of a network.

Nucleus The structural center of an atom that contains both protons and neutrons.

Positive ion An atom having a net positive charge due to the loss of one of its negatively charged electrons.

Potential difference The algebraic difference in potential (or voltage) between two points in an electrical system.

Potential energy The energy that a mass possesses by virtue of its position.

Primary cell Sources of voltage that cannot be recharged.

Proton The particle of positive polarity found in the nucleus of the atom.

Rectification The process by which an ac signal is converted to one that has an average dc level.

Secondary cell Sources of voltage that can be recharged.

Semiconductor A material having a conductance value between that of an insulator and that of a conductor. Of significant importance in the manufacture of semiconductor electronic devices.

Solar cell Sources of voltage available through the conversion of light energy (photons) into electrical energy.

Specific gravity The ratio of the weight of a given volume of a substance to the weight of an equal volume of water at 4°C.

Volt (V) The unit of measurement applied to the difference in potential between two points. If one joule of energy is required to move one coulomb of charge between two points, the difference in potential is said to be one volt.

Voltmeter An instrument designed to read the voltage across an element or between any two points in a network.

3

Resistance

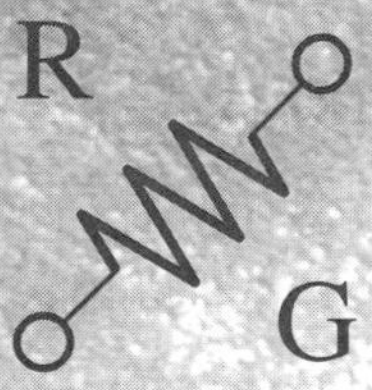

3.1 INTRODUCTION

The flow of charge through any material encounters an opposing force similar in many respects to mechanical friction. This opposition, due to the collisions between electrons and between electrons and other atoms in the material, *which converts electrical energy into heat,* is called the *resistance* of the material. The unit of measurement of resistance is the *ohm,* for which the symbol is Ω, the capital Greek letter omega. The circuit symbol for resistance appears in Fig. 3.1 with the graphic abbreviation for resistance (R).

R

FIG. 3.1
Resistance symbol and notation.

The resistance of any material with a uniform cross-sectional area is determined by the following four factors:

1. ***Material***
2. ***Length***
3. ***Cross-sectional area***
4. ***Temperature***

The chosen material, with its unique molecular structure, will react differentially to pressures to establish current through its core. Conductors that permit a generous flow of charge with little external pressure will have low resistance levels, while insulators will have high resistance characteristics.

As one might expect, the longer the path the charge must pass through, the higher the resistance level, whereas the larger the area (and therefore available room), the lower the resistance. Resistance is thus directly proportional to length and inversely proportional to area.

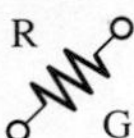

As the temperature of most conductors increases, the increased motion of the particles within the molecular structure makes it increasingly difficult for the "free" carriers to pass through, and the resistance level increases.

At a fixed temperature of 20°C (room temperature), the resistance is related to the other three factors by

$$R = \rho \frac{l}{A} \quad \text{(ohms, } \Omega\text{)} \qquad \textbf{(3.1)}$$

where ρ (Greek letter rho) is a characteristic of the material called the *resistivity*, l is the length of the sample, and A is the cross-sectional area of the sample.

The units of measurement substituted into Eq. (3.1) are related to the application. For circular wires, units of measurement are usually defined as in Section 3.2. For most other applications involving important areas such as integrated circuits, the units are as defined in Section 3.4.

3.2 RESISTANCE: CIRCULAR WIRES

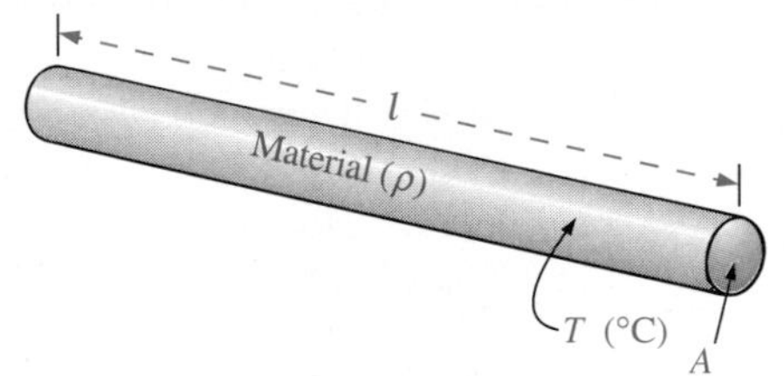

FIG. 3.2
Factors affecting the resistance of a conductor.

For a circular wire, the quantities appearing in Eq. (3.1) are defined by Fig. 3.2. For two wires of the same physical size at the same temperature, as shown in Fig. 3.3(a),

the higher the resistivity, the more the resistance.

As indicated in Fig. 3.3(b),

the longer the length of a conductor, the more the resistance.

Figure 3.3(c) reveals for remaining similar determining variables that

the smaller the area of a conductor, the more the resistance.

Finally, Fig. 3.3(d) states that for metallic wires of identical construction and material

the higher the temperature of a conductor, the more the resistance.

For circular wires, the quantities of Eq. (3.1) have the following units:

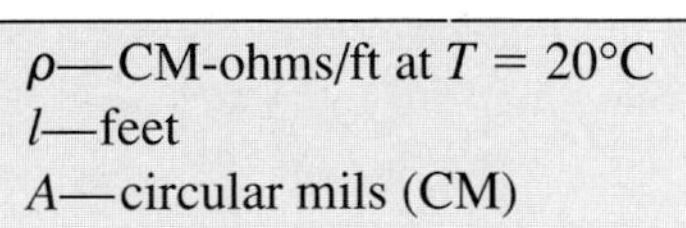

ρ—CM-ohms/ft at $T = 20°\text{C}$
l—feet
A—circular mils (CM)

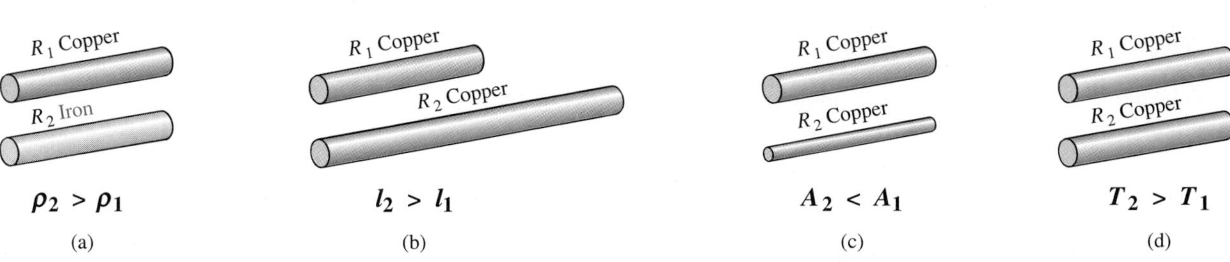

FIG. 3.3
Cases in which $R_2 > R_1$. For each case, all remaining parameters that control the resistance level are the same.

Note that the area of the conductor is measured in *circular mils* and *not* in *square meters, inches,* and so on, as determined by the equation

$$\text{Area (circle)} = \pi r^2 = \frac{\pi d^2}{4} \qquad \begin{matrix} r = \text{radius} \\ d = \text{diameter} \end{matrix} \qquad (3.2)$$

The *mil* is a unit of measurement for length and is related to the inch by

$$\mathbf{1\ mil = \frac{1}{1000}\ in.}$$

or

$$\mathbf{1000\ mils = 1\ in.}$$

By definition,

a wire with a diameter of 1 mil has an area of 1 circular mil (CM), as shown in Fig. 3.4.

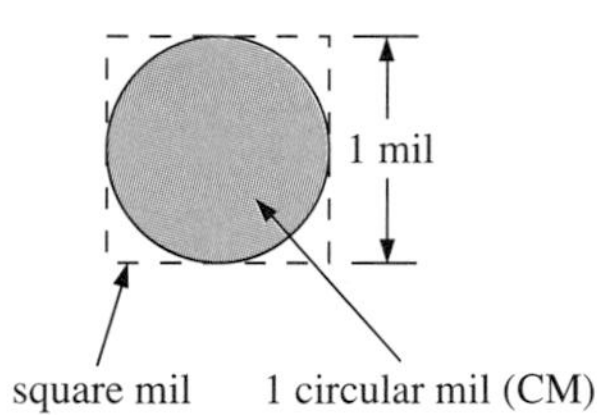

FIG. 3.4
Defining the circular mil (CM).

One square mil was superimposed on the 1-CM area of Fig. 3.4 to clearly show that the square mil has a larger surface area than the circular mil.

Applying the above definition to a wire having a diameter of 1 mil, and applying Eq. (3.2), we have:

$$A = \frac{\pi d^2}{4} = \frac{\pi}{4}(1\ \text{mil})^2 = \frac{\pi}{4}\ \text{sq mils} \overset{\text{by definition}}{\equiv} 1\ \text{CM}$$

Therefore,

$$\mathbf{1\ CM = \frac{\pi}{4}\ sq\ mils} \qquad (3.3a)$$

or

$$\mathbf{1\ sq\ mil = \frac{4}{\pi}\ CM} \qquad (3.3b)$$

Dividing Eq. (3.3b) through will result in

$$1\ \text{sq mil} = \frac{4}{\pi}\ \text{CM} = 1.273\ \text{CM}$$

which certainly agrees with the pictorial representation of Fig. 3.4. For a wire with a diameter of N mils (where N can be any positive number),

$$A = \frac{\pi d^2}{4} = \frac{\pi N^2}{4}\ \text{sq mils}$$

Substituting the fact that $4/\pi$ CM $= 1$ sq mil, we have

$$A = \frac{\pi N^2}{4}\ (\text{sq mils}) = \left(\frac{\cancel{\pi} N^2}{\cancel{4}}\right)\left(\frac{\cancel{4}}{\cancel{\pi}}\ \text{CM}\right) = N^2\ \text{CM}$$

Since $d = N$, the area in circular mils is simply equal to the diameter in mils square; that is,

$$A_{\text{CM}} = (d_{\text{mils}})^2 \qquad (3.4)$$

Verification that an area can simply be the diameter squared is provided in part by Fig. 3.5 for diameters of 2 and 3 mils. Although some areas are not circular, they have the same area as 1 circular mil.

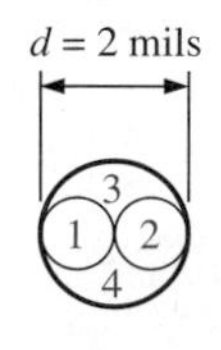

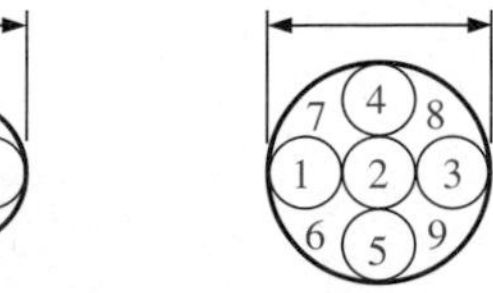

$A = (2\ \text{mils})^2 = 4$ CM $A = (3\ \text{mils})^2 = 9$ CM

FIG. 3.5
Verification of Eq. (3.4): $A_{CM} = (d_{mils})^2$.

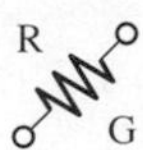

In the future, therefore, to find the area in circular mils, the diameter must first be converted to mils. Since 1 mil = 0.001 in., if the diameter is given in inches, simply move the decimal point three places to the right. For example,

$$0.02 \text{ in.} = 0.020 \text{ mils} = 20 \text{ mils}$$

If in fractional form, first convert to decimal form and then proceed as above. For example,

$$\frac{1}{8} \text{ in.} = 0.125 \text{ in.} = 125 \text{ mils}$$

The constant ρ (resistivity) is different for every material. Its value is the resistance of a length of wire 1 ft by 1 mil in diameter, measured at 20°C (Fig. 3.6). The unit of measurement for ρ can be determined from Eq. (3.1) by first solving for ρ and then substituting the units of the other quantities. That is,

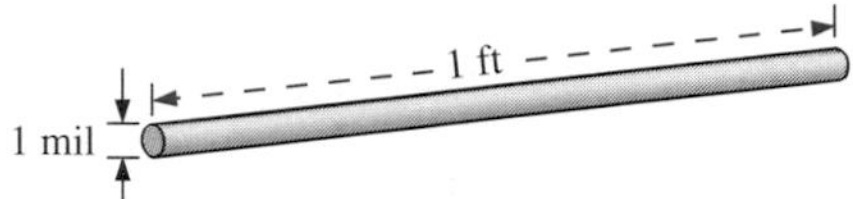

FIG. 3.6
Defining the constant ρ (resistivity).

$$\rho = \frac{AR}{l}$$

and

$$\text{Units of } \rho = \frac{\mathbf{CM \cdot \Omega}}{\mathbf{ft}}$$

The resistivity ρ is also measured in ohms per mil-foot, as determined by Fig. 3.6, or *ohm-meters* in the SI system of units. Some typical values of ρ are provided in Table 3.1.

TABLE 3.1
The resistivity of various materials.

Material	ρ @ 20°C
Silver	9.9
Copper	**10.37**
Gold	14.7
Aluminum	17.0
Tungsten	33.0
Nickel	47.0
Iron	74.0
Constantan	295.0
Nichrome	600.0
Calorite	720.0
Carbon	21,000.0

EXAMPLE 3.1 What is the resistance of a 100-ft length of copper wire with a diameter of 0.020 in. at 20°C?

Solution:

$$\rho = 10.37 \frac{\text{CM} \cdot \Omega}{\text{ft}} \qquad 0.020 \text{ in.} = 20 \text{ mils}$$

$$A_{\text{CM}} = (d_{\text{mils}})^2 = (20 \text{ mils})^2 = 400 \text{ CM}$$

$$R = \rho\frac{l}{A} = \frac{(10.37 \text{ CM} \cdot \Omega/\text{ft})(100 \text{ ft})}{400 \text{ CM}}$$

$$R = \mathbf{2.59\ \Omega}$$

EXAMPLE 3.2 An undetermined number of feet of wire have been used from the carton of Fig. 3.7. Find the length of the remaining copper wire if it has a diameter of 1/16 in. and a resistance of 0.5 Ω.

Solution:

$$\rho = 10.37 \text{ CM} \cdot \Omega/\text{ft} \qquad \frac{1}{16} \text{ in.} = 0.0625 \text{ in.} = 62.5 \text{ mils}$$

$$A_{\text{CM}} = (d_{\text{mils}})^2 = (62.5 \text{ mils})^2 = 3906.25 \text{ CM}$$

$$R = \rho\frac{l}{A} \Rightarrow l = \frac{RA}{\rho} = \frac{(0.5\ \Omega)(3906.25 \text{ CM})}{10.37 \dfrac{\text{CM} \cdot \Omega}{\text{ft}}} = \frac{1953.125}{10.37}$$

$$l = \mathbf{188.34\ ft}$$

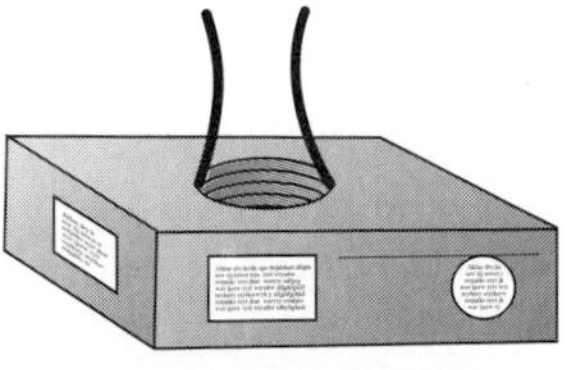

FIG. 3.7
Example 3.2.

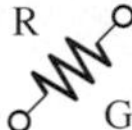

EXAMPLE 3.3 What is the resistance of a copper bus-bar, as used in the power distribution panel of a high-rise office building, with the dimensions indicated in Fig. 3.8?

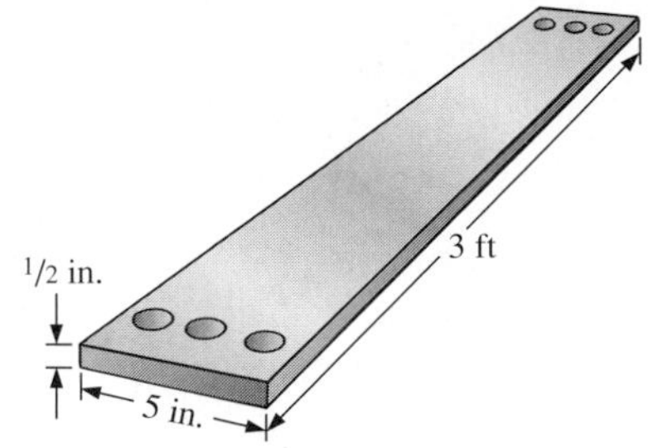

FIG. 3.8
Example 3.3.

Solution:

$$A_{CM}\begin{cases} 5.0 \text{ in.} = 5000 \text{ mils} \\ \dfrac{1}{2} \text{ in.} = 500 \text{ mils} \\ A = (5000 \text{ mils})(500 \text{ mils}) = 2.5 \times 10^6 \text{ sq mils} \\ \quad = 2.5 \times 10^6 \cancel{\text{sq mils}} \left(\dfrac{4/\pi \text{ CM}}{1 \cancel{\text{sq mil}}}\right) \\ A = 3.185 \times 10^6 \text{ CM} \end{cases}$$

$$R = \rho\frac{l}{A} = \frac{(10.37 \text{ CM}\cdot\Omega/\text{ft})(3 \text{ ft})}{3.185 \times 10^6 \text{ CM}} = \frac{31.110}{3.185 \times 10^6}$$

$$R = \mathbf{9.768 \times 10^{-6}\ \Omega}$$

(quite small, 0.000009768 Ω)

We will find in the chapters to follow that the less the resistance of a conductor, the lower the losses in conduction from the source to the load. Similarly, since resistivity is a major factor in determining the resistance of a conductor, the lower the resistivity, the lower the resistance for the same size conductor. Table 3.1 would suggest therefore that silver, copper, gold, and aluminum would be the best conductors and the most common. In general, there are other factors, however, such as malleability (ability of a material to be shaped), ductility (ability of a material to be drawn into long, thin wires), temperature sensitivity, resistance to abuse, and, of course, cost, that must all be weighed when choosing a conductor for a particular application.

In general, copper is the most widely used material because it is quite malleable, ductile, and available; has good thermal characteristics; and is less expensive than silver or gold. It is certainly not cheap, however. Wiring is removed quickly from buildings to be torn down, for example, to extract the copper. At one time aluminum was introduced for general wiring because it is cheaper than copper, but its thermal characteristics created some difficulties. It was found that the heating due to current flow and the cooling that occurred when the circuit was turned off resulted in expansion and contraction of the aluminum wire to the point where connections could eventually work themselves loose and dangerous side effects could result. Aluminum is still used today, however, in areas such as integrated circuit manufacturing and in situations where the connections can be made secure. Silver and gold are, of course, much more expensive than copper or aluminum, but there are places where the cost is justified. Silver has excellent plating characteristics for surface preparations, and gold is used quite extensively in integrated circuits. Tungsten has a resistivity three times that of copper, but there are occasions when its physical characteristics (durability, hardness) are the overriding considerations.

3.3 WIRE TABLES

The wire table was designed primarily to standardize the size of wire produced by manufacturers throughout the United States. As a result,

TABLE 3.2
American Wire Gage (AWG) sizes.

	AWG #	Area (CM)	Ω/1000 ft at 20°C	Maximum Allowable Current for RHW Insulation (A)*
(4/0)	**0000**	211,600	0.0490	**230**
(3/0)	**000**	167,810	0.0618	**200**
(2/0)	**00**	133,080	0.0780	**175**
(1/0)	**0**	105,530	0.0983	**150**
	1	83,694	0.1240	**130**
	2	66,373	0.1563	**115**
	3	52,634	0.1970	**100**
	4	41,742	0.2485	**85**
	5	33,102	0.3133	—
	6	26,250	0.3951	**65**
	7	20,816	0.4982	—
	8	16,509	0.6282	**50**
	9	13,094	0.7921	—
	10	10,381	0.9989	**30**
	11	8,234.0	1.260	—
	12	6,529.0	1.588	**20**
	13	5,178.4	2.003	—
	14	4,106.8	2.525	**15**
	15	3,256.7	3.184	
	16	2,582.9	4.016	
	17	2,048.2	5.064	
	18	1,624.3	6.385	
	19	1,288.1	8.051	
	20	1,021.5	10.15	
	21	810.10	12.80	
	22	642.40	16.14	
	23	509.45	20.36	
	24	404.01	25.67	
	25	320.40	32.37	
	26	254.10	40.81	
	27	201.50	51.47	
	28	159.79	64.90	
	29	126.72	81.83	
	30	100.50	103.2	
	31	79.70	130.1	
	32	63.21	164.1	
	33	50.13	206.9	
	34	39.75	260.9	
	35	31.52	329.0	
	36	25.00	414.8	
	37	19.83	523.1	
	38	15.72	659.6	
	39	12.47	831.8	
	40 ↓	9.89 ↓	1049.0	

*Not more than three conductors in raceway, cable, or direct burial.

Source: Reprinted by permission from NFPA No. SPP-6C, National Electrical Code®, copyright © 1996, National Fire Protection Association, Quincy, MA 02269. This reprinted material is not the complete and official position of the NFPA on the referenced subject which is represented only by the standard in its entirety. *National Electrical Code* is a registered trademark of the National Fire Protection Association, Inc., Quincy, MA for a triennial electrical publication. The term *National Electrical Code,* as used herein, means the triennial publication constituting the National Electrical Code and is used with permission of the National Fire Protection Association.

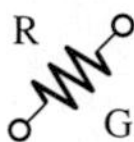

the manufacturer has a larger market and the consumer knows that standard wire sizes will always be available. The table was designed to assist the user in every way possible; it usually includes data such as the cross-sectional area in circular mils, diameter in mils, ohms per 1000 feet at 20°C, and weight per 1000 feet.

The American Wire Gage (AWG) sizes are given in Table 3.2 for solid round copper wire. A column indicating the maximum allowable current in amperes, as determined by the National Fire Protection Association, has also been included.

The chosen sizes have an interesting relationship: For every drop in 3 gage numbers, the area is doubled; and for every drop in 10 gage numbers, the area increases by a factor of 10.

Examining Eq. (3.1), we note also that *doubling the area cuts the resistance in half, and increasing the area by a factor of 10 decreases the resistance of 1/10 the original,* everything else kept constant.

The actual sizes of the gage wires listed in Table 3.2 are shown in Fig. 3.9 with a few of their areas of application. A few examples using Table 3.2 follow.

EXAMPLE 3.4 Find the resistance of 650 ft of #8 copper wire ($T = 20°C$).

Solution: For #8 copper wire (solid), $\Omega/1000$ ft at 20°C = 0.6282 Ω, and

$$650\ \text{ft}\left(\frac{0.6282\ \Omega}{1000\ \text{ft}}\right) = \mathbf{0.408\ \Omega}$$

EXAMPLE 3.5 What is the diameter, in inches, of a #12 copper wire?

Solution: For #12 copper wire (solid), $A = 6529.9$ CM, and

$$d_{\text{mils}} = \sqrt{A_{\text{CM}}} = \sqrt{6529.9\ \text{CM}} \cong 80.81\ \text{mils}$$
$$d = \mathbf{0.0808\ in.}\ \text{(or close to 1/12 in.)}$$

EXAMPLE 3.6 For the system of Fig. 3.10, the total resistance of *each* power line cannot exceed 0.025 Ω, and the maximum current to be drawn by the load is 95 A. What gage wire should be used?

Solution:

$$R = \rho\frac{l}{A} \Rightarrow A = \rho\frac{l}{R} = \frac{(10.37\ \text{CM}\cdot\Omega/\text{ft})(100\ \text{ft})}{0.025\ \Omega} = 41{,}480\ \text{CM}$$

Using the wire table, we choose the wire with the next largest area, which is #4, to satisfy the resistance requirement. We note, however, that 95 A must flow through the line. This specification requires that #3 wire be used since the #4 wire can carry a maximum current of only 85 A.

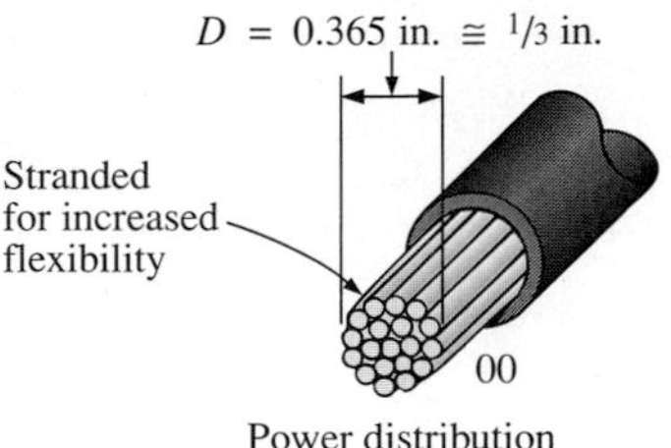

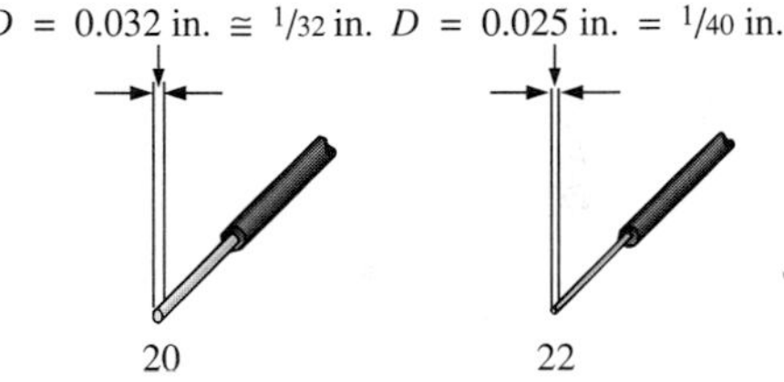

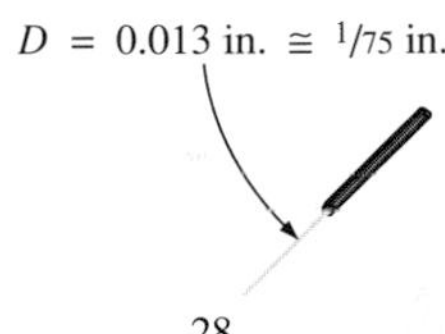

FIG. 3.9
Popular wire sizes and some of their areas of application.

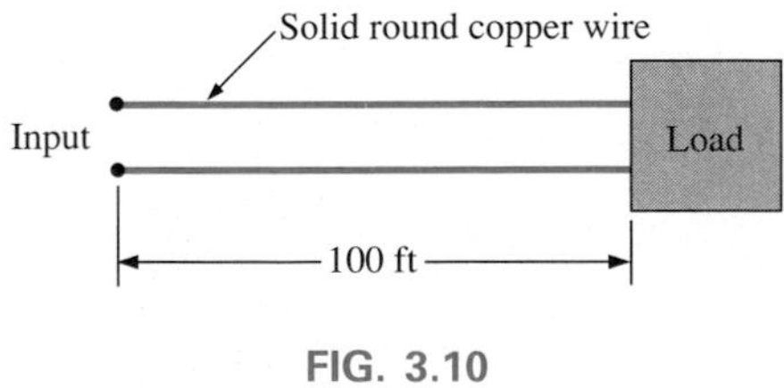

FIG. 3.10
Example 3.6.

3.4 RESISTANCE: METRIC UNITS

The design of resistive elements for various areas of application, including thin-film resistors and integrated circuits, uses metric units for the quantities of Eq. (3.1). In SI units, the resistivity would be measured in ohm-meters, the area in square meters, and the length in

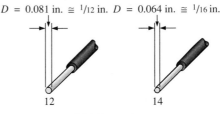

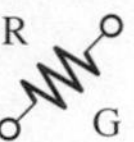

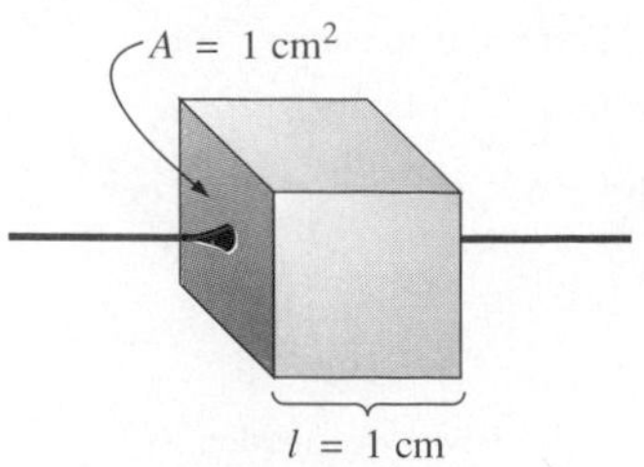

FIG. 3.11
Defining ρ in ohm-centimeters.

TABLE 3.3
Resistivity (ρ) of various materials in ohm-centimeters.

Silver	1.645×10^{-6}
Copper	$\mathbf{1.723 \times 10^{-6}}$
Gold	2.443×10^{-6}
Aluminum	2.825×10^{-6}
Tungsten	5.485×10^{-6}
Nickel	7.811×10^{-6}
Iron	12.299×10^{-6}
Tantalum	15.54×10^{-6}
Nichrome	99.72×10^{-6}
Tin oxide	250×10^{-6}
Carbon	3500×10^{-6}

meters. However, the meter is generally too large a unit of measure for most applications, and so the centimeter is usually employed. The resulting dimensions for Eq. (3.1) are therefore

ρ—ohm-centimeters l—centimeters A—square centimeters

The units for ρ can be derived from

$$\rho = \frac{RA}{l} = \frac{\Omega \cdot \text{cm}^2}{\text{cm}} = \Omega \cdot \text{cm}$$

The resistivity of a material is actually the resistance of a sample such as that appearing in Fig. 3.11. Table 3.3 provides a list of values of ρ in ohm-centimeters. Note that the area now is expressed in square centimeters, which can be determined using the basic equation $A = \pi d^2/4$, eliminating the need to work with circular mils, the special unit of measure associated with circular wires.

EXAMPLE 3.7 Determine the resistance of 100 ft of #28 copper telephone wire if the diameter is 0.0126 in.

Solution: Unit conversions:

$$l = 100 \cancel{\text{ft}}\left(\frac{12\ \cancel{\text{in.}}}{1\ \cancel{\text{ft}}}\right)\left(\frac{2.54\text{ cm}}{1\ \cancel{\text{in.}}}\right) = 3048\text{ cm}$$

$$d = 0.0126\text{ in.}\left(\frac{2.54\text{ cm}}{1\text{ in.}}\right) = 0.032\text{ cm}$$

Therefore,

$$A = \frac{\pi d^2}{4} = \frac{(3.1416)(0.032\text{ cm})^2}{4} = 8.04 \times 10^{-4}\text{ cm}^2$$

$$R = \rho\frac{l}{A} = \frac{(1.723 \times 10^{-6}\ \Omega \cdot \text{cm})(3048\text{ cm})}{8.04 \times 10^{-4}\text{ cm}^2} \cong \mathbf{6.5\ \Omega}$$

Using the units for circular wires and Table 3.2 for the area of a #28 wire, we find

$$R = \rho\frac{l}{A} = \frac{(10.37\text{ CM} \cdot \Omega/\text{ft})(100\text{ ft})}{159.79\text{ CM}} \cong \mathbf{6.5\ \Omega}$$

EXAMPLE 3.8 Determine the resistance of the thin-film resistor of Fig. 3.12 if the *sheet resistance* R_s (defined by $R_s = \rho/d$) is 100 Ω.

Solution: For deposited materials of the same thickness, the sheet resistance factor is usually employed in the design of thin-film resistors.

Equation (3.1) can be written

$$R = \rho\frac{l}{A} = \rho\frac{l}{dw} = \left(\frac{\rho}{d}\right)\left(\frac{l}{w}\right) = R_s\frac{l}{w}$$

where l is the length of the sample and w is the width. Substituting into the above equation yields

$$R = R_s\frac{l}{w} = \frac{(100\ \Omega)(0.6\text{ cm})}{0.3\text{ cm}} = \mathbf{200\ \Omega}$$

as one might expect since $l = 2w$.

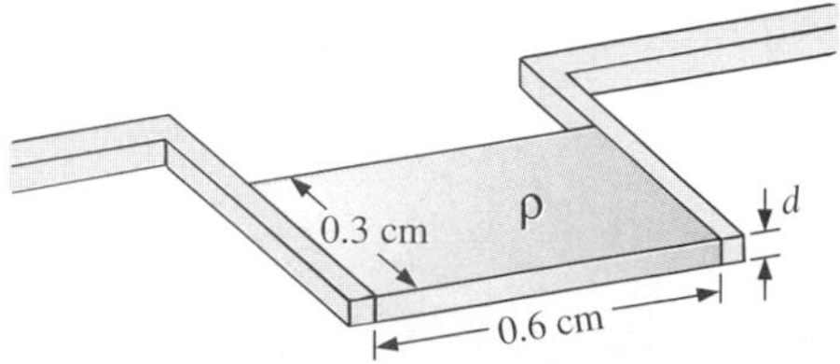

FIG. 3.12
Thin-film resistor (note Fig. 3.24).

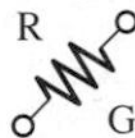

The conversion factor between resistivity in circular mil-ohms per foot and ohm-centimeters is the following:

$$\rho\ (\Omega\cdot\text{cm}) = (1.662 \times 10^{-7}) \times (\text{value in CM}\cdot\Omega/\text{ft})$$

For example, for copper $\rho = 10.37\ \text{CM}\cdot\Omega/\text{ft}$:

$$\begin{aligned}\rho\ (\Omega\cdot\text{cm}) &= 1.662 \times 10^{-7}(10.37\ \text{CM}\cdot\Omega/\text{ft}) \\ &= 1.723 \times 10^{-6}\ \Omega\cdot\text{cm}\end{aligned}$$

as indicated in Table 3.3.

The resistivity in IC design is typically in ohm-centimeter units, although tables often provide ρ in ohm-meters or microhm-centimeters. Using the conversion technique of Chapter 1, we find that the conversion factor between ohm-centimeters and ohm-meters is the following:

$$1.723 \times 10^{-6}\ \Omega\cdot\not{\text{cm}}\left[\frac{1\ \text{m}}{100\ \not{\text{cm}}}\right] = \frac{1}{100}[1.723 \times 10^{-6}]\ \Omega\cdot\text{m}$$

or the value in ohm-meters is 1/100 the value in ohm-centimeters, and

$$\rho\ (\Omega\cdot\text{m}) = \left(\frac{1}{100}\right) \times (\text{value in } \Omega\cdot\text{cm})$$

Similarly:

$$\rho\ (\mu\Omega\cdot\text{cm}) = (10^6) \times (\text{value in } \Omega\cdot\text{cm})$$

For comparison purposes, typical values of ρ in ohm-centimeters for conductors, semiconductors, and insulators are provided in Table 3.4.

TABLE 3.4
Comparing levels of ρ in $\Omega\cdot$cm.

Conductor	Semiconductor	Insulator
Copper 1.723×10^{-6}	Ge 50	In general: 10^{15}
	Si 200×10^3	
	GaAs 70×10^6	

In particular, note the power of 10 difference between conductors and insulators (10^{21})—a difference of huge proportions. There is a significant difference in levels of ρ for the list of semiconductors, but the power of 10 difference between the conductor and insulator levels is at least 10^6 for each of the semiconductors listed.

3.5 TEMPERATURE EFFECTS

Temperature has a significant effect on the resistance of conductors, semiconductors, and insulators.

Conductors

In conductors there is a generous number of free electrons, and any introduction of thermal energy will have little impact on the total num-

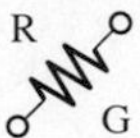

ber of free carriers. In fact the thermal energy will only increase the intensity of the random motion of the particles within the material and make it increasingly difficult for a general drift of electrons in any one direction to be established. The result is that

for good conductors, an increase in temperature will result in an increase in the resistance level. Consequently, conductors have a positive temperature coefficient.

The plot of Fig. 3.13(a) has a positive temperature coefficient.

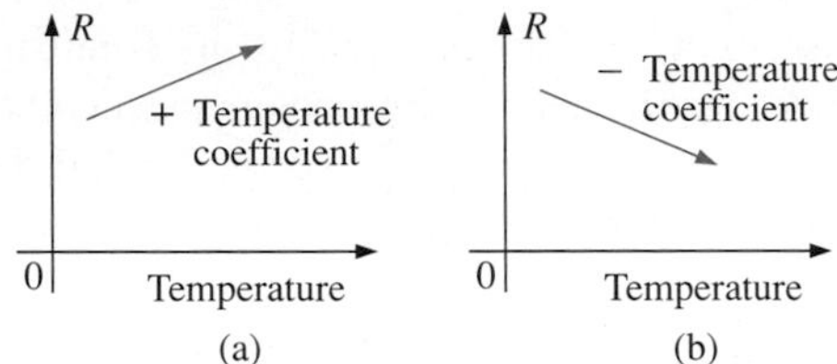

FIG. 3.13
(a) Positive temperature coefficient—conductors; (b) negative temperature coefficient—semiconductors.

Semiconductors

In semiconductors an increase in temperature will impart a measure of thermal energy to the system that will result in an increase in the number of free carriers in the material for conduction. The result is that

for semiconductor materials, an increase in temperature will result in a decrease in the resistance level. Consequently, semiconductors have negative temperature coefficients.

The thermistor and photoconductive cell of Sections 3.10 and 3.11 of this chapter are excellent examples of semiconductor devices with negative temperature coefficients. The plot of Fig. 3.13(b) has a negative temperature coefficient.

Insulators

As with semiconductors, an increase in temperature will result in a decrease in the resistance of an insulator. The result is a negative temperature coefficient.

Inferred Absolute Temperature

Figure 3.14 reveals that for copper (and most other metallic conductors), the resistance increases almost linearly (in a straight-line relationship) with an increase in temperature. Since temperature can have such a pronounced effect on the resistance of a conductor, it is important that we have some method of determining the resistance at any temperature within operating limits. An equation for this purpose can be obtained by approximating the curve of Fig. 3.14 by the straight dashed line that intersects the temperature scale at $-234.5°C$. Although the actual curve extends to *absolute zero* ($-273.15°C$, or 0 K), the straight-line approximation is quite accurate for the normal operating temperature range. At two different temperatures, t_1 and t_2, the resistance of copper is R_1 and R_2, as indicated on the curve. Using a property of similar triangles, we may develop a mathematical relationship between these values of resis-

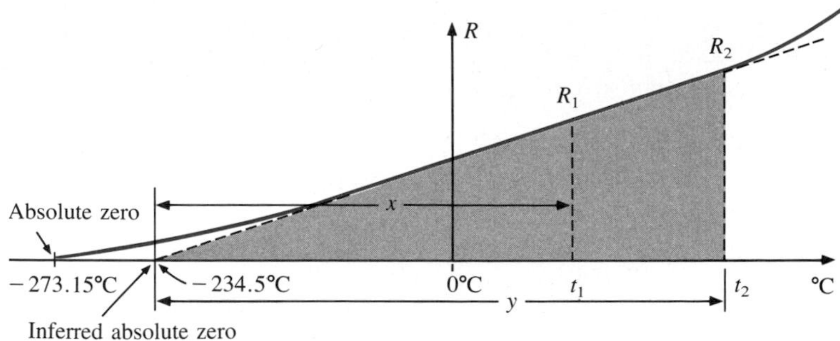

FIG. 3.14
Effect of temperature on the resistance of copper.

tances at different temperatures. Let x equal the distance from $-234.5°\text{C}$ to t_1 and y the distance from $-234.5°\text{C}$ to t_2, as shown in Fig. 3.14. From similar triangles,

$$\frac{x}{R_1} = \frac{y}{R_2}$$

or

$$\frac{234.5 + t_1}{R_1} = \frac{234.5 + t_2}{R_2} \tag{3.5}$$

The temperature of $-234.5°\text{C}$ is called the *inferred absolute temperature* of copper. For different conducting materials, the intersection of the straight-line approximation will occur at different temperatures. A few typical values are listed in Table 3.5.

The minus sign does not appear with the inferred absolute temperature on either side of Eq. (3.5) because x and y are the *distances* from $-234.5°\text{C}$ to t_1 and t_2, respectively, and therefore are simply magnitudes. For t_1 and t_2 less than zero, x and y are less than $-234.5°\text{C}$ and the distances are the differences between the inferred absolute temperature and the temperature of interest.

Equation (3.5) can easily be adapted to any material by inserting the proper inferred absolute temperature. It may therefore be written as follows:

$$\frac{|T| + t_1}{R_1} = \frac{|T| + t_2}{R_2} \tag{3.6}$$

where $|T|$ indicates that the inferred absolute temperature of the material involved is inserted as a positive value in the equation. In general, therefore, associate the sign only with t_1 and t_2.

TABLE 3.5
Inferred absolute temperatures.

Material	°C
Silver	−243
Copper	**−234.5**
Gold	−274
Aluminum	−236
Tungsten	−204
Nickel	−147
Iron	−162
Nichrome	−2,250
Constantan	−125,000

EXAMPLE 3.9 If the resistance of a copper wire is 50 Ω at 20°C, what is its resistance at 100°C (boiling point of water)?

Solution: Eq. (3.5):

$$\frac{234.5°\text{C} + 20°\text{C}}{50\ \Omega} = \frac{234.5°\text{C} + 100°\text{C}}{R_2}$$

$$R_2 = \frac{(50\ \Omega)(334.5°\text{C})}{254.5°\text{C}} = \mathbf{65.72\ \Omega}$$

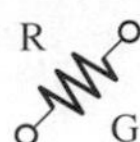

EXAMPLE 3.10 If the resistance of a copper wire at freezing (0°C) is 30 Ω, what is its resistance at −40°C?

Solution: Eq. (3.5):

$$\frac{234.5°\text{C} + 0}{30\ \Omega} = \frac{234.5°\text{C} - 40°\text{C}}{R_2}$$

$$R_2 = \frac{(30\ \Omega)(194.5°\text{C})}{234.5°\text{C}} = \mathbf{24.88\ \Omega}$$

EXAMPLE 3.11 If the resistance of an aluminum wire at room temperature (20°C) is 100 mΩ (measured by a milliohmmeter), at what temperature will its resistance increase to 120 mΩ?

Solution: Eq. (3.5):

$$\frac{236°\text{C} + 20°\text{C}}{100\ \text{m}\Omega} = \frac{236°\text{C} + t_2}{120\ \text{m}\Omega}$$

and

$$t_2 = 120\ \text{m}\Omega\left(\frac{256°\text{C}}{100\ \text{m}\Omega}\right) - 236°\text{C}$$

$$t_2 = \mathbf{71.2°C}$$

Temperature Coefficient of Resistance

There is a second popular equation for calculating the resistance of a conductor at different temperatures. Defining

$$\alpha_{20} = \frac{1}{|T| + 20°\text{C}} \qquad (\Omega/°\text{C}/\Omega) \qquad \textbf{(3.7)}$$

as the *temperature coefficient of resistance* at a temperature of 20°C, and R_{20} as the resistance of the sample at 20°C, the resistance R at a temperature t is determined by

$$R = R_{20}[1 + \alpha_{20}(t - 20°\text{C})] \qquad \textbf{(3.8)}$$

The values of α_{20} for different materials have been evaluated, and a few are listed in Table 3.6.

Equation (3.8) can be written in the following form:

$$\alpha_{20} = \frac{\left(\dfrac{R - R_{20}}{t - 20°\text{C}}\right)}{R_{20}} = \frac{\dfrac{\Delta R}{\Delta T}}{R_{20}}$$

from which the units of Ω/°C/Ω for α_{20} are defined.

Since $\Delta R/\Delta T$ is the slope of the curve of Fig. 3.14, we can conclude that

the higher the temperature coefficient of resistance for a material, the more sensitive the resistance level to changes in temperature.

Referring to Table 3.5, we find that copper is more sensitive to temperature variations than is silver, gold, or aluminum, although the differences are quite small. The slope defined by α_{20} for constantan is so small the curve is almost horizontal.

TABLE 3.6
Temperature coefficient of resistance for various conductors at 20°C.

Material	Temperature Coefficient (α_{20})
Silver	0.0038
Copper	**0.00393**
Gold	0.0034
Aluminum	0.00391
Tungsten	0.005
Nickel	0.006
Iron	0.0055
Constantan	0.000008
Nichrome	0.00044

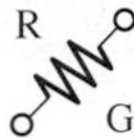

Since R_{20} of Eq. (3.8) is the resistance of the conductor at 20°C and $t - 20°C$ is the change in temperature from 20°C, Eq. (3.8) can be written in the following form:

$$R = \rho\frac{l}{A}[1 + \alpha_{20}\,\Delta T] \tag{3.9}$$

providing an equation for resistance in terms of all the controlling parameters.

PPM/°C

For resistors, as for conductors, resistance changes with a change in temperature. The specification is normally provided in parts per million per degree Celsius (PPM/°C), providing an immediate indication of the sensitivity level of the resistor to temperature. For resistors, a 5000-PPM level is considered high, whereas 20 PPM is quite low. A 1000-PPM/°C characteristic reveals that a 1° change in temperature will result in a change in resistance equal to 1000 PPM, or 1000/1,000,000 = 1/1000 of its nameplate value—not a significant change for most applications. However, a 10° change would result in a change equal to 1/100 (1%) of its nameplate value, which is becoming significant. The concern, therefore, lies not only with the PPM level but with the range of expected temperature variation.

In equation form, the change in resistance is given by

$$\Delta R = \frac{R_{\text{nominal}}}{10^6}(\text{PPM})(\Delta T) \tag{3.10}$$

where R_{nominal} is the nameplate value of the resistor at room temperature and ΔT is the change in temperature from the reference level of 20°C.

EXAMPLE 3.12 For a 1-kΩ carbon composition resistor with a PPM of 2500, determine the resistance at 60°C.

Solution:

$$\Delta R = \frac{1000\ \Omega}{10^6}(2500)(60°\text{C} - 20°\text{C})$$
$$= 100\ \Omega$$

and

$$R = R_{\text{nominal}} + \Delta R = 1000\ \Omega + 100\ \Omega$$
$$= \mathbf{1100\ \Omega}$$

3.6 SUPERCONDUCTORS

Introduction

There is no question that the field of electricity/electronics has to be one of the most exciting of the 20th century. Even though new developments appear almost weekly from extensive research and development activities, every once in a while there is some very special step forward

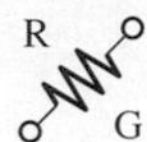

that has the whole field at the edge of its seat waiting to see what might develop in the near future. Such a level of excitement and interest surrounds the research drive to develop a room-temperature *superconductor*—an advance that will rival the introduction of semiconductor devices such as the transistor (to replace tubes), wireless communication, or the electric light. The implications of such a development are so far-reaching that it is difficult to forecast the vast impact it will have on the entire field.

The intensity of the research effort throughout the world today to develop a room-temperature superconductor is described by some researchers as "unbelievable, contagious, exciting, and demanding" but an adventure in which they treasure the opportunity to be involved. Progress in the field since 1986 suggests that the use of superconductivity in commercial applications will grow quite rapidly in the next few decades. It is also hinted that room-temperature superconductors may be a reality by the year 2000. It is indeed an exciting era full of growing anticipation! Why this interest in superconductors? What are they all about? In a nutshell,

superconductors are conductors of electric charge that, for all practical purposes, have zero resistance.

Cooper Effect

In a conventional conductor, electrons travel at average speeds in the neighborhood of 1000 mi/s (they can cross the United States in about 3 seconds), even though Einstein's theory of relativity suggests that the maximum speed of information transmission is the speed of light, or 186,000 mi/s. The relatively slow speed of conventional conduction is due to collisions with other atoms in the material, repulsive forces between electrons (like charges repel), thermal agitation that results in indirect paths due to the increased motion of the neighboring atoms, impurities in the conductor, and so on. In the superconductive state, there is a pairing of electrons, denoted by the *Cooper effect,* in which electrons travel in pairs and help each other maintain a significantly higher velocity through the medium. In some ways this is like "drafting" by competitive cyclists or runners. There is an oscillation of energy between partners or even "new" partners (as the need arises) to ensure passage through the conductor at the highest possible velocity with the least total expenditure of energy.

Ceramics

Even though the concept of superconductivity first surfaced in 1911, it was not until 1986 that the possibility of superconductivity at room temperature became a renewed goal of the research community. For some 74 years superconductivity could be established only at temperatures colder than 23 K. (Kelvin temperature is universally accepted as the unit of measurement for temperature for superconductive effects. Recall that $K = 273.15^\circ + {}^\circ C$, so a temperature of 23 K is $-250^\circ C$, or $-418^\circ F$.) In 1986, however, physicists Alex Muller and George Bednorz of the IBM Zurich Research Center found a ceramic material, lanthanum barium copper oxide, that exhibited superconductivity at 30 K. Although it would not appear to be a significant step forward, it introduced a new direction to the research effort and spurred others to

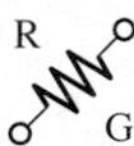

improve on the new standard. In October 1987 both scientists received the Nobel prize for their contribution to an important area of development.

In just a few short months, Professors Paul Chu of the University of Houston and Man Kven Wu of the University of Alabama raised the temperature to 95 K using a superconductor of yttrium barium copper oxide. The result was a level of excitement in the scientific community that brought research in the area to a new level of effort and investment. The major impact of such a discovery was that liquid nitrogen (boiling point of 77 K) could now be used to bring the material down to the required temperature rather than liquid helium, which boils at 4 K. The result is a tremendous saving in the cooling expense since liquid helium is at least ten times more expensive than liquid nitrogen. Pursuing the same direction, some success has been achieved at 125 K and 162 K using a thallium compound (unfortunately, however, thallium is a very poisonous substance). Figure 3.15 clearly reveals the tremendous change in the success curve since 1911 and also suggests that room-temperature success in the not-too-distant future is a strong possibility. However, the compound will probably not be one that has already surfaced and may in fact be of a totally different nature. Although the curve levels off at the 162 K level, the research effort continues with vigor to identify other superconducting compounds that may permit higher temperatures and to expand the industrial application of the materials already available.

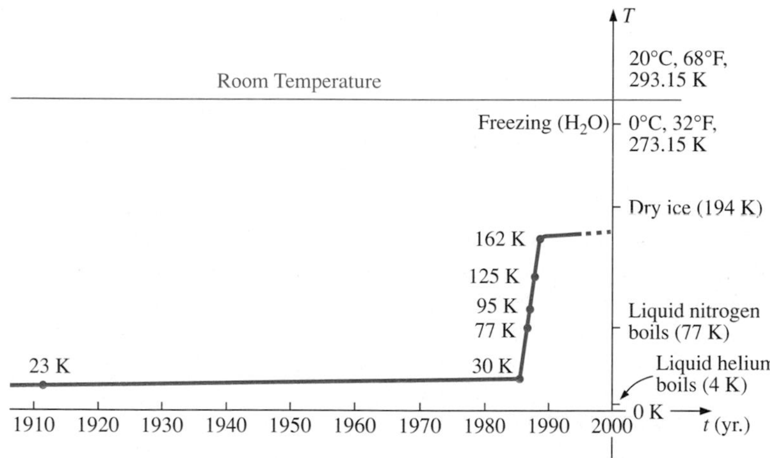

FIG. 3.15
Rising temperatures of superconductors.

The fact that ceramics have provided the recent breakthrough in superconductivity is probably a surprise when you consider that they are also an important class of insulators. However, the ceramics that exhibit the characteristics of superconductivity are compounds that include copper, oxygen, and rare earth elements such as yttrium, lanthanum, and thallium. There are also indicators that the current compounds may be limited to a maximum temperature of 200 K (about 100 K short of room temperature), leaving the door wide open to innovative approaches to compound selection. The temperature at which a superconductor reverts back to the characteristics of a conventional

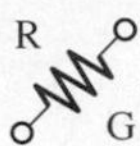

conductor is called the *critical temperature,* denoted by T_c. Note in Fig. 3.16 how the resistivity level changes abruptly at T_c. The sharpness of the transition region is a function of the purity of the sample. Long listings of critical temperatures for a variety of tested compounds can be found in reference materials providing tables of a wide variety to support research in physics, chemistry, geology, and related fields. Two such publications include the CRC (The Chemical Rubber Co.) *Handbook of Tables for Applied Engineering Science* and the CRC *Handbook of Chemistry and Physics.*

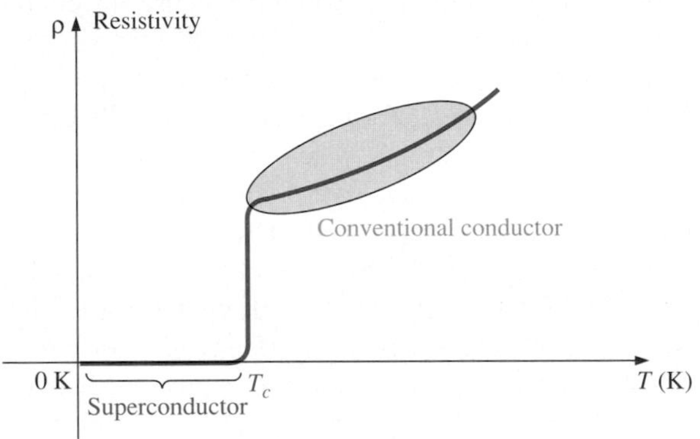

FIG. 3.16
Defining the critical temperature T_c.

Even though ceramic compounds have established higher transition temperatures, there is concern about their brittleness and current density limitations. In the area of integrated circuit manufacturing, current density levels must equal or exceed 1 MA/cm^2, or 1 million amperes through a cross-sectional area about one-half the size of a dime. Recently IBM attained a level of 4 MA/cm^2 at 77 K, permitting the use of superconductors in the design of some new-generation, high-speed computers.

TIB Relationship

There are, in fact, three factors linked together in the development of a superconductor for extensive practical applications at room temperature—*temperature, current density,* and *magnetic field strength* (Chapter 11). As one factor (such as temperature) is pushed to its limit, the other two (in this case, current density and magnetic flux density) will eventually drop off rather sharply, as shown in Fig. 3.17. In particular, note the temperature T_1 in Fig. 3.17, which defines current density and magnetic field strength less than maximum values. In addition, both continue to drop rather rapidly as the temperature is increased toward its critical value. Fortunately, the magnetic field strength currently available at workable superconductor temperatures is high enough (in excess of 250T at low temperatures and 100T at higher temperatures) for the majority of applications. The major concern for current materials lies in ensuring sufficient current density at the temperature of interest.

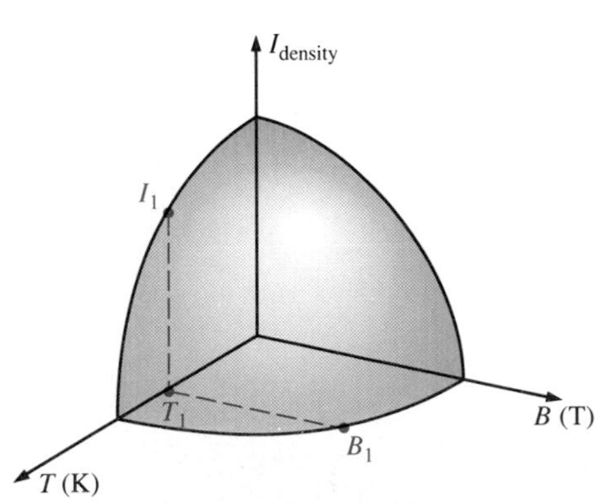

FIG. 3.17
The TIB relationship.

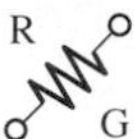

Meissner Effect

A true physical indicator of whether superconduction has been established is beautifully demonstrated by the Meissner effect (Fig. 3.18(a)). At temperatures above the critical temperature, magnetic lines of force can pass through a conductor, as shown in Fig. 3.18(a). When the temperature of the conductor has been brought down to a level where superconductivity can be established, external magnetic lines of force cannot pass through the superconductor, and the magnet will levitate (float) above the superconductor, as shown in Fig. 3.18(b), with a photograph of the actual phenomenon shown in Fig. 3.18(c). The inability of external magnetic flux lines to pass through a material in the superconducting state has extensive applications in sensors, which will be described in the Applications section.

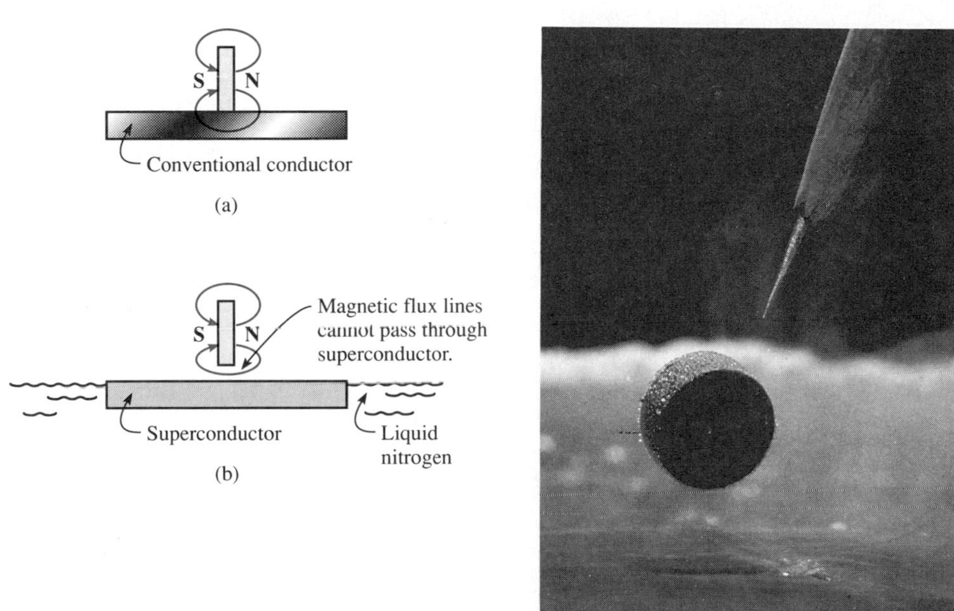

FIG. 3.18
Demonstrating the Meissner effect. (Courtesy of IBM.)

Applications

Although room-temperature success has not been attained, there are numerous applications for some of the superconductors developed thus far. It is simply a matter of balancing the additional cost against the results obtained or deciding whether any results at all can be obtained without the use of this zero-resistance state. Some research efforts require high-energy accelerators or strong magnets attainable only with superconductive materials. Superconductivity is currently applied in the design of 300-mi/h Meglev trains (trains that ride on a cushion of air established by opposite magnetic poles), in powerful motors and generators, in nuclear magnetic resonance imaging systems to obtain cross-sectional images of the brain (and other parts of the body), in the design of computers with operating speeds four times that of conventional systems, and in improved power distribution systems.

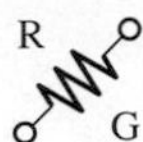

Through the use of the Josephson effect (left to the student as a research activity), there are magnetic field detectors known as SQUIDs (*s*uperconducting *qu*antum *i*nterference *d*evices) that can measure magnetic fields thousands of times smaller than those measured with conventional methods. Applications of such devices range from medicine to geology. Since the human skull distorts electric currents and not magnetic fields, SQUIDs can be used to detect extremely small magnetic fields that can provide important diagnostic information about the patient. In geology they can be used to detect magnetic fields that reveal the presence of specific minerals or even oil and water.

The range of future uses for superconductors is a function of the success physicists have in raising the operating temperature and how well they can utilize the successes obtained thus far. However, it would appear that it is only a matter of time (the eternal optimist) before magnetically levitated trains increase in number, improved medical diagnostic equipment is available, computers operate at much higher speeds, high-efficiency power and storage systems are available, and transmission systems operate at very high efficiency levels due to this area of developing interest. Only time will reveal the impact that this new direction will have on the quality of life.

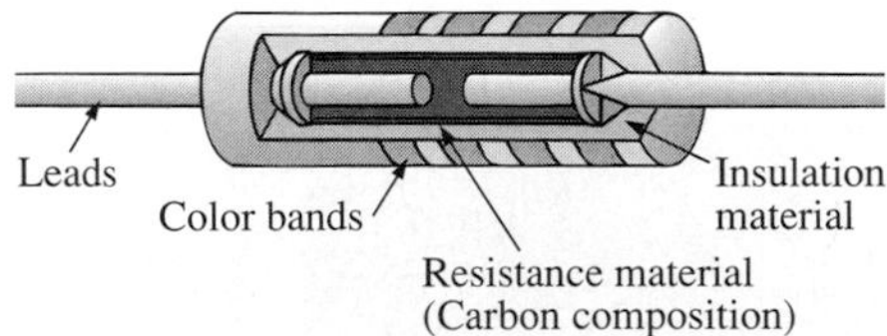

FIG. 3.19
Fixed composition resistor.

3.7 TYPES OF RESISTORS

Fixed Resistors

Resistors are made in many forms, but all belong in either of two groups: fixed or variable. The most common of the low-wattage, fixed-type resistors is the molded carbon composition resistor. The basic construction is shown in Fig. 3.19.

The relative sizes of all fixed and variable resistors change with the wattage (power) rating, increasing in size for increased wattage ratings in order to withstand the higher currents and dissipation losses. The relative sizes of the molded composition resistors for different wattage ratings are shown in Fig. 3.20. Resistors of this type are readily available in values ranging from 2.7 Ω to 22 MΩ.

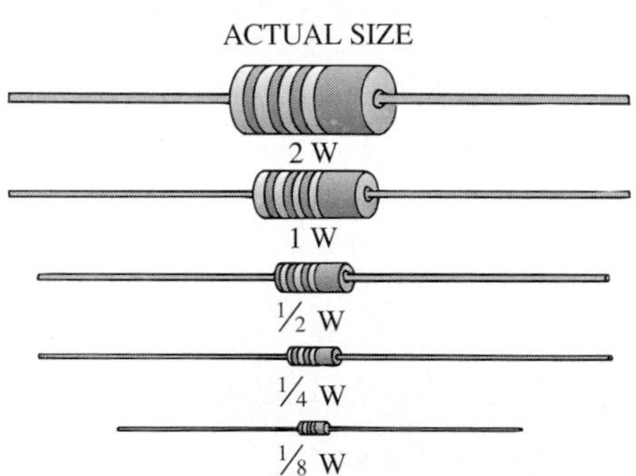

FIG. 3.20
Fixed composition resistors of different wattage ratings.

The temperature-versus-resistance curve for a 10,000-Ω and 0.5-MΩ composition-type resistor is shown in Fig. 3.21. Note the small percent resistance change in the normal temperature operating range. Several

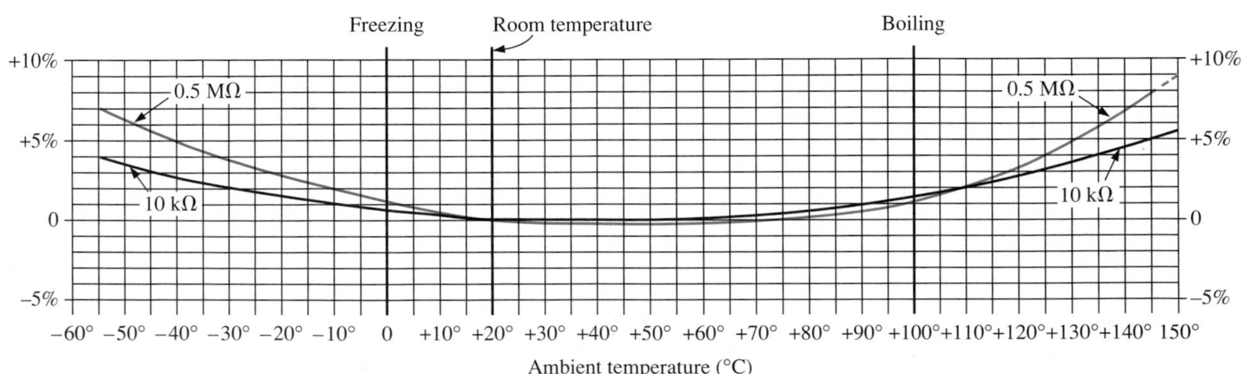

FIG. 3.21
Curves showing percentage temporary resistance changes from +20°C values. (Courtesy of Allen-Bradley Co.)

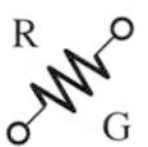

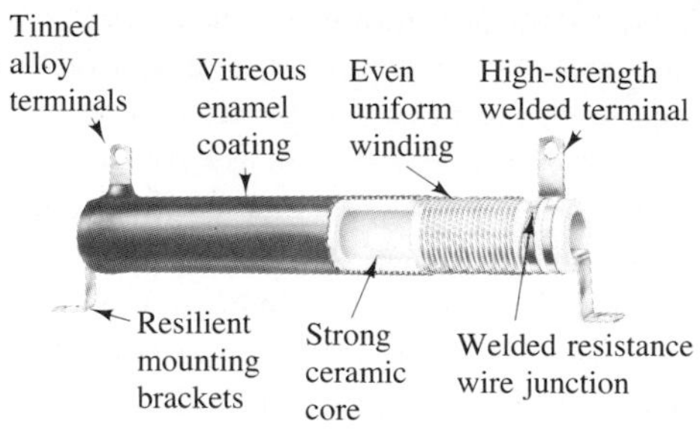

(a) Vitreous-enameled wire-wound resistor
App: All types of equipment

(b) High-voltage cermet film resistors (on a high grade ceramic body).
App: For high-voltage applications up to 10 kV requiring high levels of stability.

(c) Metal-film precision resistors
App: Where high stability, low temperature coefficient, and low noise level desired

FIG. 3.22
Fixed resistors. [Parts (a) and (c) courtesy of Ohmite Manufacturing Co. Part (b) courtesy of Philips Components Inc.]

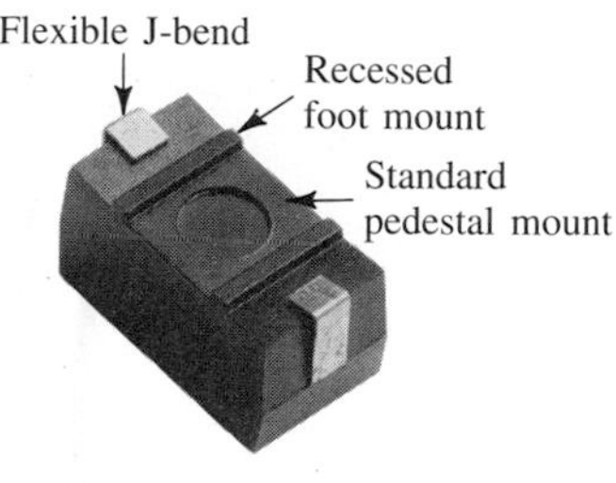

(a) Surface mount power resistor ideal for printed circuit boards. Patented J-bends eliminate need for solder connections. (0.8 W to 3 W in wire-wound, film, or power film construction)

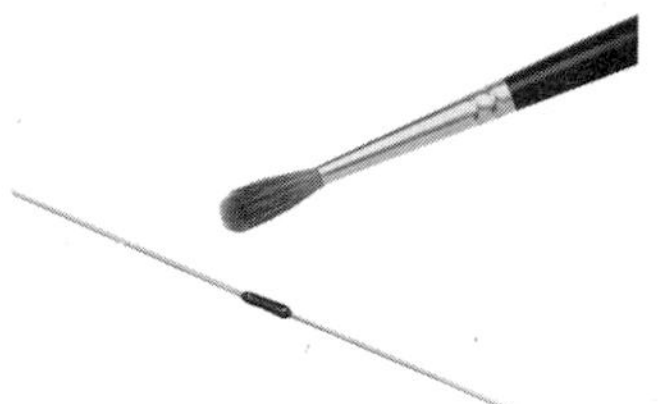

(b) Precision power wire-wound resistors with ratings as high as 2 W and tolerances as low as 0.05%. Temperature coefficients as low as 20 ppm/°C are also available.

(c) Thick-film chip resistors for design flexibility with hybrid circuitry. Pre-tinned, gold or silver electrodes available. Operating temperature range −55°C to +150°C.

FIG. 3.23
Miniature fixed resistors. [Part (a) courtesy of Ohmite Manufacturing Co. Parts (b) and (c) courtesy of Dale Electronics, Inc.]

other types of fixed resistors using high resistance wire or metal films are shown in Fig. 3.22.

The miniaturization of parts—used quite extensively in computers—requires that resistances of different values be placed in very small packages. Some examples appear in Fig. 3.23.

For use with printed circuit boards, fixed resistor networks in a variety of configurations are available in miniature packages, such as those shown in Fig. 3.24. The figure includes a photograph of three different casings and the internal resistor configuration for the single in-line structure to the right.

FIG. 3.24
Thick-film resistor networks. (Courtesy of Dale Electronics, Inc.)

Variable Resistors

Variable resistors, as the name implies, have a terminal resistance that can be varied by turning a dial, knob, screw, or whatever seems appropriate for the application. They can have two or three terminals, but most have three terminals. If the two- or three-terminal device is used as a variable resistor, it is usually referred to as a *rheostat.* If the three-terminal device is used for controlling potential levels, it is then commonly called a *potentiometer.* Even though a three-terminal device can

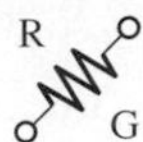

be used as a rheostat or potentiometer (depending on how it is connected), it is typically called a *potentiometer* when listed in trade magazines or requested for a particular application.

The symbol for a three-terminal potentiometer appears in Fig. 3.25(a). When used as a variable resistor (or rheostat), it can be hooked up in one of two ways, as shown in Fig. 3.25(b) and (c). In Fig. 3.25(b), points *a* and *b* are hooked up to the circuit, and the remaining terminal is left hanging. The resistance introduced is determined by that portion of the resistive element between points *a* and *b*. In Fig. 3.25(c), the resistance is again between points *a* and *b*, but now the remaining resistance is "shorted-out" (effect removed) by the connection from *b* to *c*. The universally accepted symbol for a rheostat appears in Fig. 3.25(d).

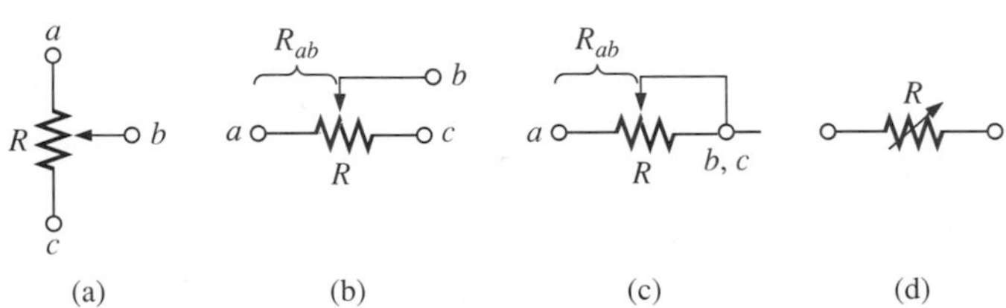

FIG. 3.25
Potentiometer: (a) symbol; (b) and (c) rheostat connections; (d) rheostat symbol.

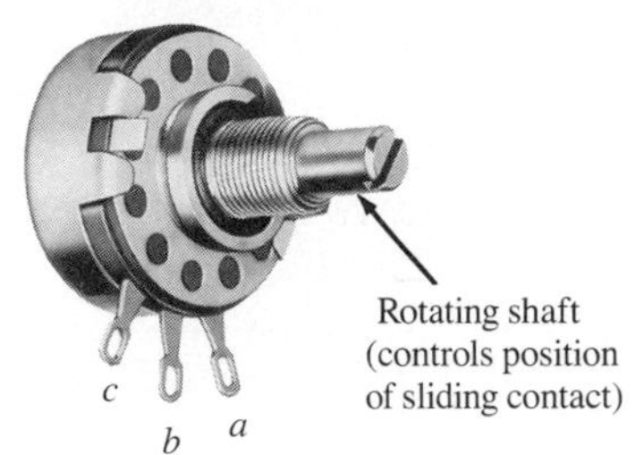

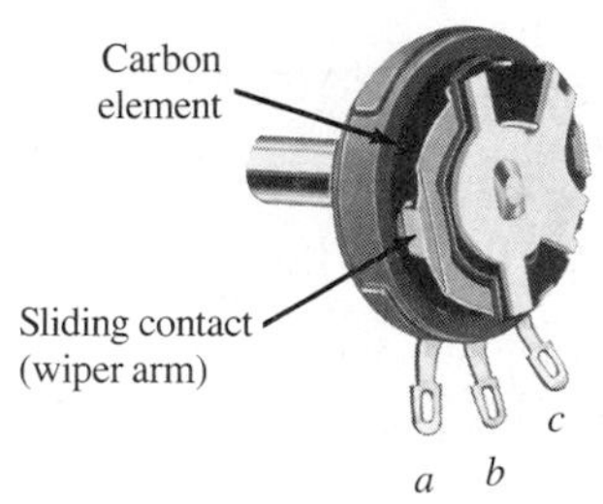

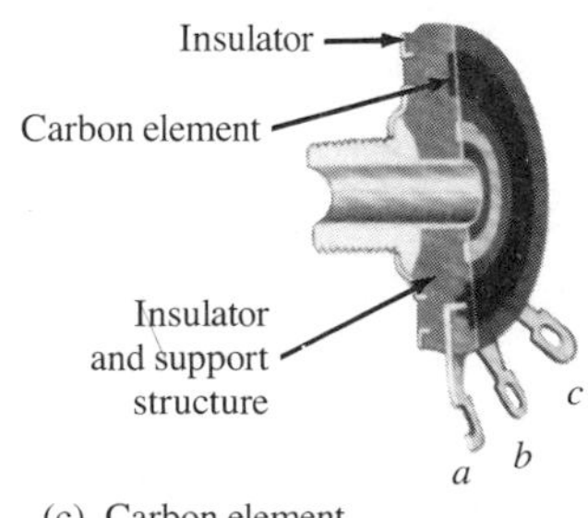

FIG. 3.26
Molded-composition-type potentiometer. (Courtesy of Allen-Bradley Co.)

Most potentiometers have three terminals in the relative positions shown in Fig. 3.26. The knob, dial, or screw in the center of the housing controls the motion of a contact that can move along the resistive element connected between the outer two terminals. The contact is connected to the center terminal, establishing a resistance from movable contact to each outer terminal.

The resistance between the outside terminals a and c of Fig. 3.27(a) (and Fig. 3.26) is always fixed at the full rated value of the potentiometer, regardless of the position of the wiper arm b.

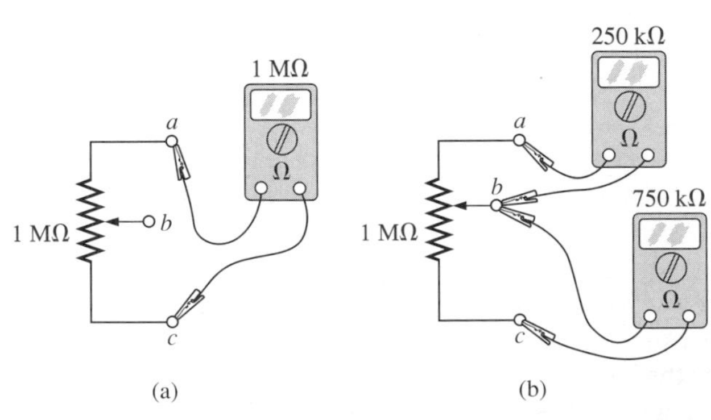

FIG. 3.27
Terminal resistance of a potentiometer: (a) between outside terminals; (b) among all three terminals.

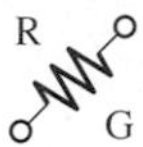

In other words, the resistance between terminals a and c of Fig. 3.27(a) for a 1-MΩ potentiometer will always be 1 MΩ, no matter how we turn the control element and move the contact. In Fig. 3.27(a) the center contact is not part of the network configuration.

The resistance between the wiper arm and either outside terminal can be varied from a minimum of 0 Ω to a maximum value equal to the full rated value of the potentiometer.

In Fig. 3.27(b) the wiper arm has been placed 1/4 of the way down from point a to point c. The resulting resistance between points a and b will therefore be 1/4 of the total, or 250 kΩ (for a 1-MΩ potentiometer), and the resistance between b and c will be 3/4 of the total, or 750 kΩ.

The sum of the resistances between the wiper arm and each outside terminal will equal the full rated resistance of the potentiometer.

This was demonstrated by Fig. 3.27(b), where 250 kΩ + 750 kΩ = 1 MΩ. Specifically:

$$R_{ac} = R_{ab} + R_{bc} \qquad (3.11)$$

Therefore, as the resistance from the wiper arm to one outside contact increases, the resistance between the wiper arm and the other outside terminal must decrease accordingly. For example, if R_{ab} of a 1-kΩ potentiometer is 200 Ω, then the resistance R_{bc} must be 800 Ω. If R_{ab} is further decreased to 50 Ω, then R_{bc} must increase to 950 Ω, and so on.

The molded carbon composition potentiometer is typically applied in networks with smaller power demands, and it ranges in size from 20 Ω to 22 MΩ (maximum values). Other commercially available potentiometers appear in Fig. 3.28.

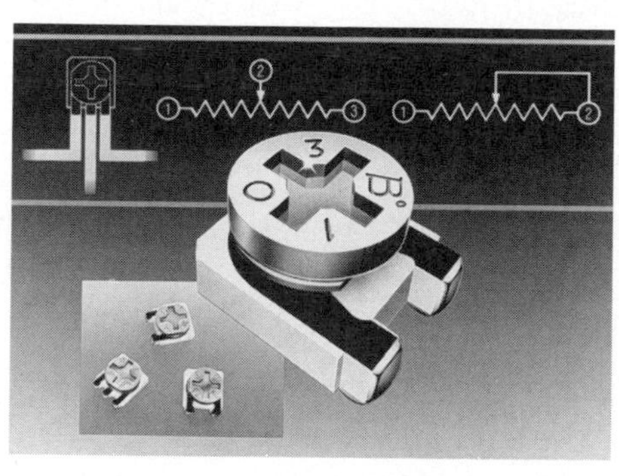

(a)

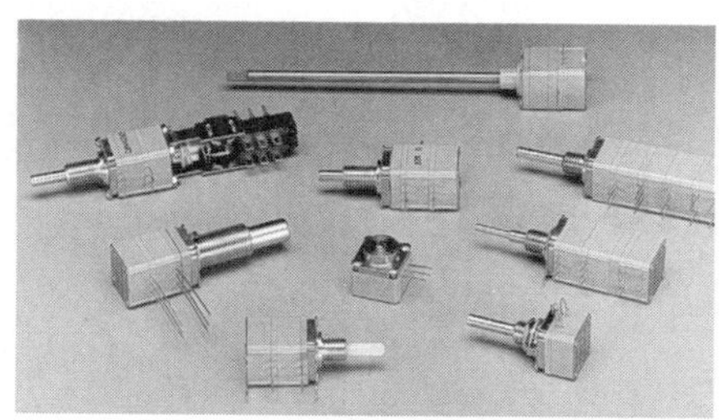

(b)

FIG. 3.28
Potentiometers: (a) 4-mm (≈5/32″) trimmer. (Courtesy of Bourns, Inc.); (b) conductive plastic and cermet element. (Courtesy of Clarostat Mfg. Co.)

When the device is used as a potentiometer, the connections are as shown in Fig. 3.29. It can be used to control the level of V_{ab}, V_{bc}, or both, depending on the application. Additional discussion of the potentiometer in a loaded situation can be found in the chapters that follow.

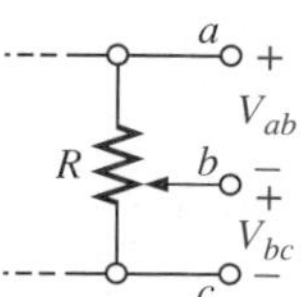

FIG. 3.29
Potentiometer control of voltage levels.

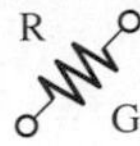

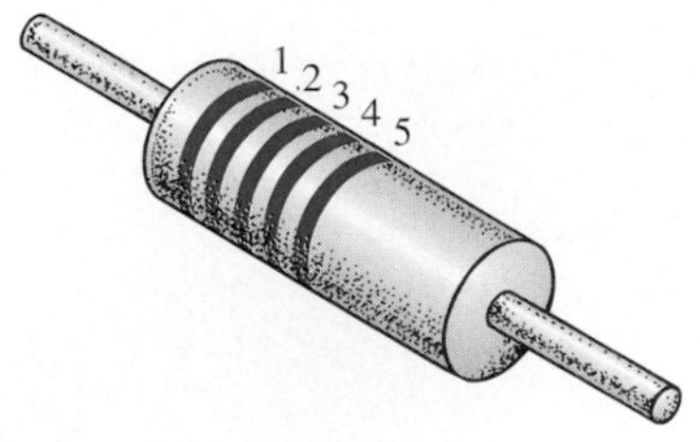

FIG. 3.30
Color coding—fixed molded composition resistor.

3.8 COLOR CODING AND STANDARD RESISTOR VALUES

A wide variety of resistors, fixed or variable, are large enough to have their resistance in ohms printed on the casing. There are some, however, that are too small to have numbers printed on them, so a system of color coding is used. For the fixed molded composition resistor, four or five color bands are printed on one end of the outer casing, as shown in Fig. 3.30. Each color has the numerical value indicated in Table 3.7. The color bands are always read from the end that has the band closest to it, as shown in Fig. 3.30. The first and second bands represent the first and second digits, respectively. The third band determines the power-of-10 multiplier for the first two digits (actually the number of zeros that follow the second digit) or a multiplying factor if gold or silver. The fourth band is the manufacturer's tolerance, which is an indication of the precision by which the resistor was made. If the fourth band is omitted, the tolerance is assumed to be ±20%. The fifth band is a reliability factor, which gives the percentage of failure per 1000 hours of use. For instance, a 1% failure rate would reveal that one out of every 100 (or 10 out of every 1000) will fail to fall within the tolerance range after 1000 hours of use.

TABLE 3.7
Resistor color coding.

Bands 1–3	Band 3	Band 4	Band 5
0 Black	0.1 Gold } multiplying factors	5% Gold	1% Brown
1 Brown	0.01 Silver } multiplying factors	10% Silver	0.1% Red
2 Red		20% No band	0.01% Orange
3 Orange			0.001% Yellow
4 Yellow			
5 Green			
6 Blue			
7 Violet			
8 Gray			
9 White			

EXAMPLE 3.13 Find the range in which a resistor having the following color bands must exist to satisfy the manufacturer's tolerance:

a.

1st band	2nd band	3rd band	4th band	5th band
Gray	Red	Black	Gold	Brown
8	2	0	±5%	1%

82 Ω ± 5% (1% reliability)

Since 5% of 82 = 4.10, the resistor should be within the range 82 Ω ± 4.10 Ω, or *between 77.90 and 86.10 Ω.*

b.

1st band	2nd band	3rd band	4th band	5th band
Orange	White	Gold	Silver	No color
3	9	0.1	±10%	

3.9 Ω ± 10% = 3.9 ± 0.39 Ω

The resistor should lie somewhere *between 3.51 and 4.29 Ω.*

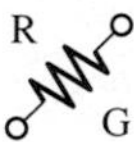

One might expect that resistors would be available for a full range of values such as 10 Ω, 20 Ω, 30 Ω, 40 Ω, 50 Ω, and so on. However, this is not the case with some typical commercial values, such as 27 Ω, 56 Ω, and 68 Ω. This may seem somewhat strange and out of place. There is a reason for the chosen values, which is best demonstrated by examining the list of standard values of commercially available resistors in Table 3.8. The values in boldface blue are available with 5%, 10%, and 20% tolerances, making them the most common of the commercial variety. The values in boldface black are typically available with 5% and 10% tolerances, and those in normal print are available only in the 5% variety. If we separate the values available into tolerance levels, we have Table 3.9, which clearly reveals how few are available up to 100 Ω with 20% tolerances.

An examination of the impact of the tolerance level will now help explain the choice of numbers for the commercial values. Take the sequence 47 Ω–68 Ω–100 Ω, which are all available with 20% tolerances. In Fig. 3.31(a), the tolerance band for each has been determined and plotted on a single axis. Take note that, with this tolerance (which is all the manufacturer will guarantee), the full range of resistor values is available from 37.6 Ω to 120 Ω. In other words, the manufacturer is guaranteeing the full range, using the tolerances to fill in the gaps. Dropping to the 10% level introduces the 56-Ω and 82-Ω resistors to fill in the gaps, as shown in Fig. 3.31(b). Dropping to the 5% level would require additional resistor values to fill in the gaps. In total, therefore, the resistor values were chosen to ensure that the full range was covered, as determined by the tolerances employed. Of course, if a

TABLE 3.8
Standard values of commercially available resistors.

Ohms (Ω)					Kilohms (kΩ)		Megohms (MΩ)	
0.10	**1.0**	**10**	**100**	**1000**	**10**	**100**	**1.0**	**10.0**
0.11	1.1	11	110	1100	11	110	1.1	11.0
0.12	**1.2**	**12**	**120**	**1200**	**12**	**120**	**1.2**	**12.0**
0.13	1.3	13	130	1300	13	130	1.3	13.0
0.15	**1.5**	**15**	**150**	**1500**	**15**	**150**	**1.5**	**15.0**
0.16	1.6	16	160	1600	16	160	1.6	16.0
0.18	**1.8**	**18**	**180**	**1800**	**18**	**180**	**1.8**	**18.0**
0.20	2.0	20	200	2000	20	200	2.0	20.0
0.22	**2.2**	**22**	**220**	**2200**	**22**	**220**	**2.2**	**22.0**
0.24	2.4	24	240	2400	24	240	2.4	
0.27	**2.7**	**27**	**270**	**2700**	**27**	**270**	**2.7**	
0.30	3.0	30	300	3000	30	300	3.0	
0.33	**3.3**	**33**	**330**	**3300**	**33**	**330**	**3.3**	
0.36	3.6	36	360	3600	36	360	3.6	
0.39	**3.9**	**39**	**390**	**3900**	**39**	**390**	**3.9**	
0.43	4.3	43	430	4300	43	430	4.3	
0.47	**4.7**	**47**	**470**	**4700**	**47**	**470**	**4.7**	
0.51	5.1	51	510	5100	51	510	5.1	
0.56	**5.6**	**56**	**560**	**5600**	**56**	**560**	**5.6**	
0.62	6.2	62	620	6200	62	620	6.2	
0.68	**6.8**	**68**	**680**	**6800**	**68**	**680**	**6.8**	
0.75	7.5	75	750	7500	75	750	7.5	
0.82	**8.2**	**82**	**820**	**8200**	**82**	**820**	**8.2**	
0.91	9.1	91	910	9100	91	910	9.1	

TABLE 3.9
Standard values and their tolerances.

±5%	±10%	±20%
10	**10**	**10**
11		
12	**12**	
13		
15	**15**	**15**
16		
18	**18**	
20		
22	**22**	**22**
24		
27	**27**	
30		
33	**33**	**33**
36		
39	**39**	
43		
47	**47**	**47**
51		
56	**56**	
62		
68	**68**	**68**
75		
82	**82**	
91		

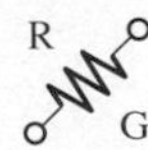

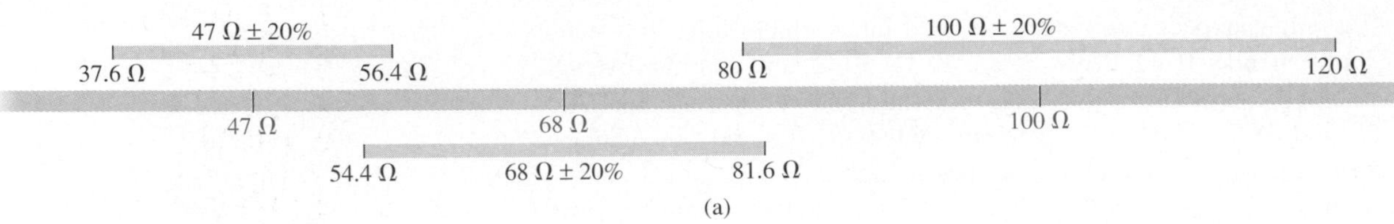

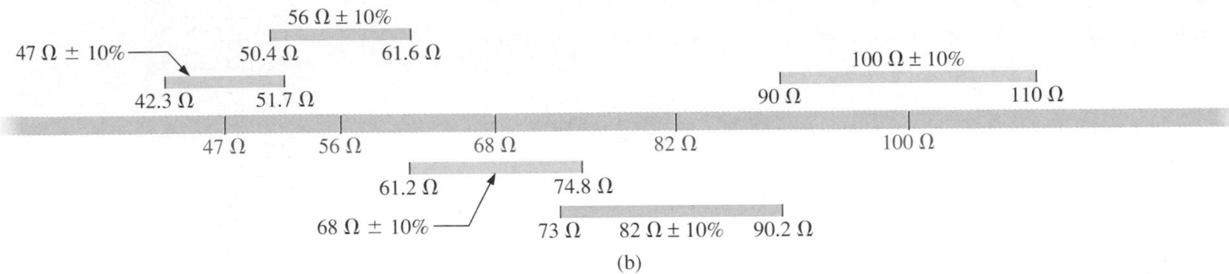

FIG. 3.31
Guaranteeing the full range of resistor values for the given tolerance: (a) 20%; (b) 10%.

specific value is desired but is not one of the standard values, combinations of standard values will often result in a total resistance very close to the desired level. If this approach is still not satisfactory, a potentiometer can be set to the exact value and then inserted in the network.

Throughout the text you will find that many of the resistor values are not standard values. This was done to reduce the mathematical complexity, which might deter from or cloud the procedure or analysis technique being introduced. In the problem sections, however, standard values are frequently employed to ensure that the reader starts to become familiar with the commercial values available.

3.9 CONDUCTANCE

By finding the reciprocal of the resistance of a material, we have a measure of how well the material will conduct electricity. The quantity is called *conductance*, has the symbol G, and is measured in *siemens* (S) (note Fig. 3.32). In equation form, conductance is

$$G = \frac{1}{R} \qquad \text{(siemens, S)} \qquad \textbf{(3.12)}$$

A resistance of 1 MΩ is equivalent to a conductance of 10^{-6} S, and a resistance of 10 Ω is equivalent to a conductance of 10^{-1} S. The larger the conductance, therefore, the less the resistance and the greater the conductivity.

In equation form, the conductance is determined by

$$G = \frac{A}{\rho l} \qquad \text{(S)} \qquad \textbf{(3.13)}$$

indicating that increasing the area or decreasing either the length or the resistivity will increase the conductance.

German (Lenthe, Berlin)
(1816–1892)
Electrical Engineer
Telegraph Manufacturer: Siemens & Halske AG

Bettman Archives
Photo Number 336.19

Developed an *electroplating process* during a brief stay in prison for acting as a second in a dual between fellow officers of the Prussian army. Inspired by the electronic telegraph invented by Sir Charles Wheatstone in 1817, he improved on the design and proceeded to lay cable with the help of his brother Carl across the Mediterranean and from Europe to India. His inventions included the first *self-excited generator,* which depended on the *residual* magnetism of its electromagnet rather than an inefficient permanent magnet. In 1888 he was raised to the rank of nobility with the addition of *von* to his name. The current firm of Siemens AG has manufacturing outlets in some 35 countries with sales offices in some 125 countries.

FIG. 3.32
Werner von Siemens.

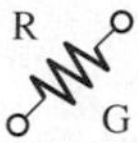

EXAMPLE 3.14 What is the relative increase or decrease in conductivity of a conductor if the area is reduced by 30% and the length is increased by 40%? The resistivity is fixed.

Solution: Eq. (3.11):

$$G_i = \frac{A_i}{\rho_i l_i}$$

with the subscript i for the initial value. Using the subscript n for new value:

$$G_n = \frac{A_n}{\rho_n l_n} = \frac{0.70A_i}{\rho_i(1.4l_i)} = \frac{0.70}{1.4}\frac{A_i}{\rho_i l_i} = \frac{0.70}{1.4\,G_i}$$

and $\boldsymbol{G_n = 0.5G_i}$

3.10 OHMMETERS

The *ohmmeter* is an instrument used to perform the following tasks and several other useful functions:

1. Measure the resistance of individual or combined elements
2. Detect open-circuit (high-resistance) and short-circuit (low-resistance) situations
3. Check continuity of network connections and identify wires of a multilead cable
4. Test some semiconductor (electronic) devices

For most applications, the ohmmeters used most frequently are the ohmmeter section of a VOM or DMM. The details of the internal circuitry and the method of using the meter will be left primarily for a laboratory exercise. In general, however, the resistance of a resistor can be measured by simply connecting the two leads of the meter across the resistor, as shown in Fig. 3.33. There is no need to be concerned about which lead goes on which end; the result will be the same in either case since resistors offer the same resistance to the flow of charge (current) in either direction. If the VOM is employed, a switch must be set to the proper resistance range, and a nonlinear scale (usually the top scale of the meter) must be properly read to obtain the resistance value. The DMM also requires choosing the best scale setting for the resistance to be measured, but the result appears as a numerical display, with the proper placement of the decimal point as determined by the chosen scale. When measuring the resistance of a single resistor, it is usually best to remove the resistor from the network before making the measurement. If this is difficult or impossible, at least one end of the resistor must not be connected to the network, or the reading may include the effects of the other elements of the system.

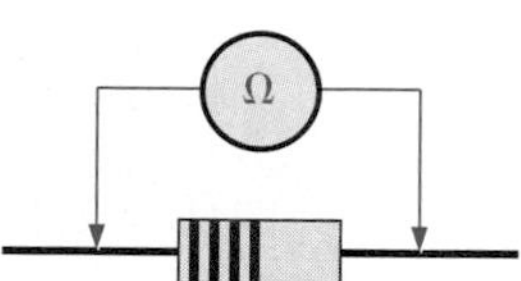

FIG. 3.33
Measuring the resistance of a single element.

If the two leads of the meter are touching in the ohmmeter mode, the resulting resistance is zero. A connection can be checked as shown in Fig. 3.34 by simply hooking up the meter to either side of the connection. If the resistance is zero, the connection is secure. If other than zero, it could be a weak connection, and, if infinite, there is no connection at all.

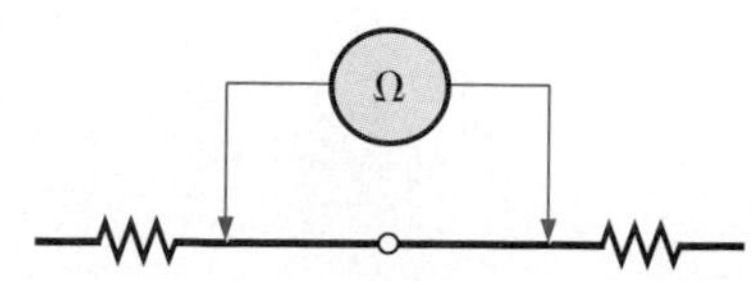

FIG. 3.34
Checking the continuity of a connection.

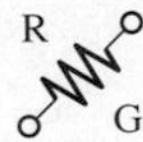

If one wire of a harness is known, a second can be found as shown in Fig. 3.35. Simply connect the end of the known lead to the end of any other lead. When the ohmmeter indicates zero ohms (or very low resistance), the second lead has been identified. The above procedure can also be used to determine the first known lead by simply connecting the meter to any wire at one end and then touching all the leads at the other end until a zero-ohm indication is obtained.

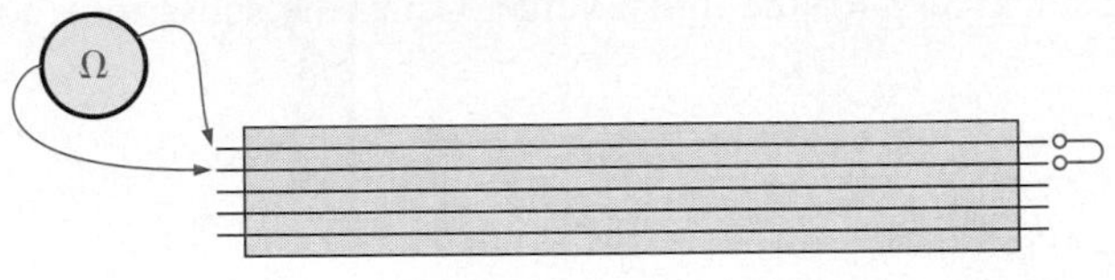

FIG. 3.35
Identifying the leads of a multilead cable.

Preliminary measurements of the condition of some electronic devices such as the diode and transistor can be made using the ohmmeter. The meter can also be used to identify the terminals of such devices.

One important note about the use of any ohmmeter:

Never hook up an ohmmeter to a live circuit!

The reading will be meaningless and you may damage the instrument. The ohmmeter section of any meter is designed to pass a small sensing current through the resistance to be measured. A large external current could damage the movement and would certainly throw off the calibration of the instrument. In addition,

never store an ohmmeter in the resistance mode.

The two leads of the meter could touch and the small sensing current could drain the internal battery. VOMs should be stored with the selector switch on the highest voltage range, and the selector switch of DMMs should be in the off position.

3.11 THERMISTORS

The *thermistor* is a two-terminal semiconductor device whose resistance, as the name suggests, is temperature sensitive. A representative characteristic appears in Fig. 3.36 with the graphic symbol for the device. Note the nonlinearity of the curve and the drop in resistance from about 5000 Ω to 100 Ω for an increase in temperature from 20°C to 100°C. The decrease in resistance with an increase in temperature indicates a negative temperature coefficient.

The temperature of the device can be changed internally or externally. An increase in current through the device will raise its temperature, causing a drop in its terminal resistance. Any externally applied heat source will result in an increase in its body temperature and a drop in resistance. This type of action (internal or external) lends itself well to control mechanisms. Many different types of thermistors are shown in Fig. 3.37. Materials employed in the manufacture of thermistors include oxides of cobalt, nickel, strontium, and manganese.

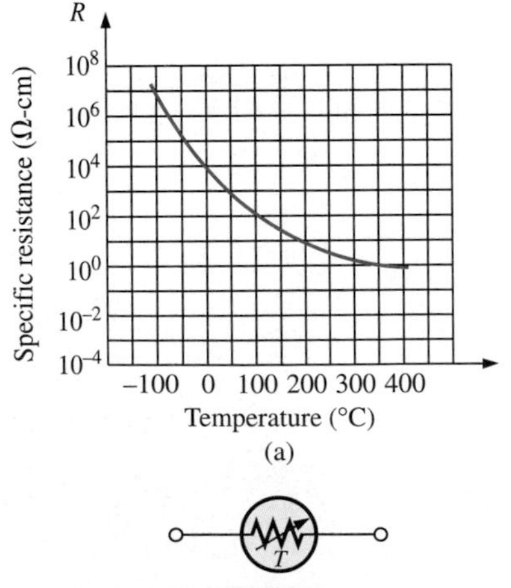

FIG. 3.36
Thermistor: (a) characteristics; (b) symbol.

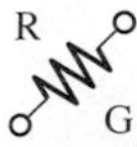

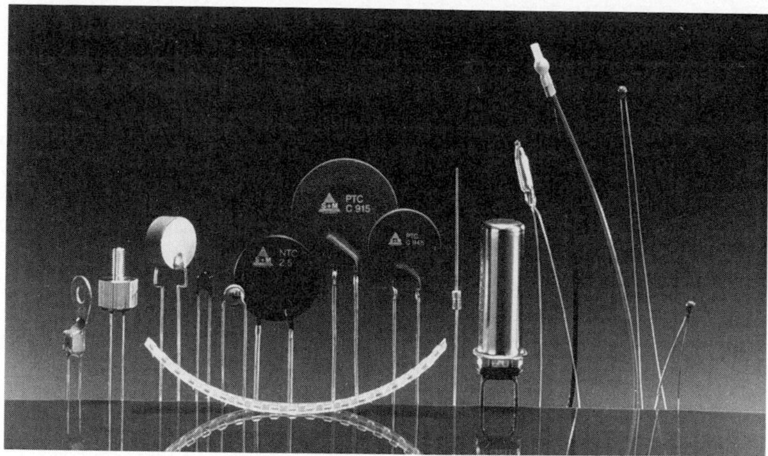

FIG. 3.37
NTC (negative temperature coefficient) and PTC (positive temperature coefficient) thermistors. (Courtesy of Siemens Components, Inc.)

Note the use of a log scale (to be discussed in Chapter 21) in Fig. 3.36 for the vertical axis. The log scale permits the display of a wider range of specific resistance levels than a linear scale such as the horizontal axis. Note that it extends from 0.0001 Ω·cm to 100,000,000 Ω·cm over a very short interval. The log scale is used for both the vertical and the horizontal axis of Fig. 3.38, which appears in the next section.

3.12 PHOTOCONDUCTIVE CELL

The *photoconductive cell* is a two-terminal semiconductor device whose terminal resistance is determined by the intensity of the incident light on its exposed surface. As the applied illumination increases in intensity, the energy state of the surface electrons and atoms increases, with a resultant increase in the number of "free carriers" and a corresponding drop in resistance. A typical set of characteristics and the photoconductive cell's graphic symbol appear in Fig. 3.38. Note the negative illumination coefficient. Several cadmium sulfide photoconductive cells appear in Fig. 3.39.

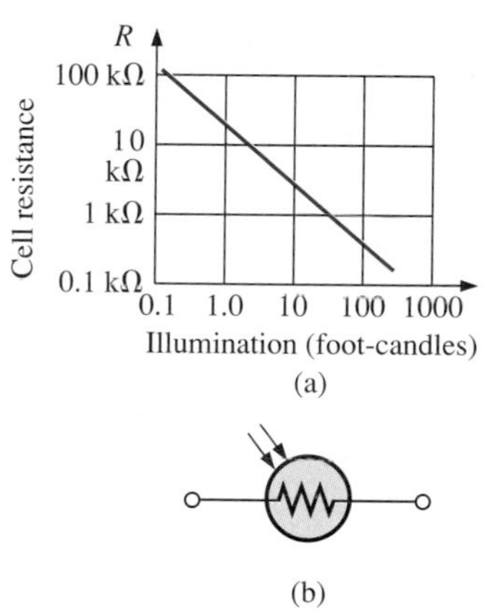

FIG. 3.38
Photoconductive cell: (a) characteristics; (b) symbol.

3.13 VARISTORS

Varistors are voltage-dependent, nonlinear resistors used to suppress high-voltage transients; that is, their characteristics are such as to limit the voltage that can appear across the terminals of a sensitive device or system. A typical set of characteristics appears in Fig. 3.40(a), along with a linear resistance characteristic for comparison purposes. Note that at a particular "firing voltage," the current rises rapidly but the voltage is limited to a level just above this firing potential. In other words, the magnitude of the voltage that can appear across this device cannot exceed that level defined by its characteristics. Through proper design techniques this device can therefore limit the voltage appearing across sensitive regions of a network. The current is simply limited by the network to which it is connected. A photograph of a number of commercial units appears in Fig. 3.40(b).

FIG. 3.39
Photoconductive cells. (Courtesy of EG&G VACTEC, Inc.)

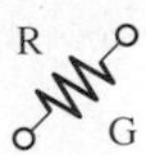

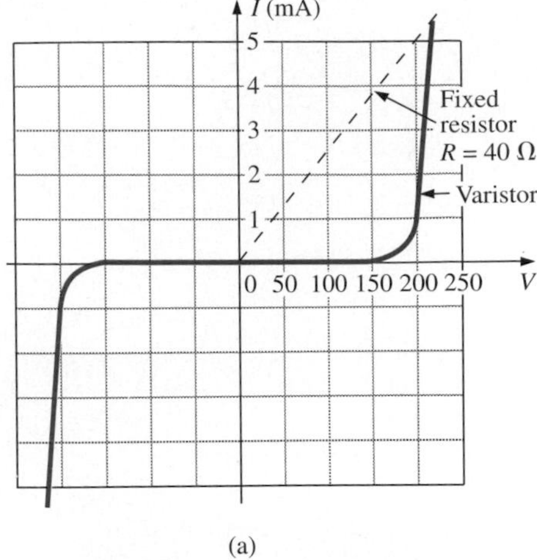

(a)

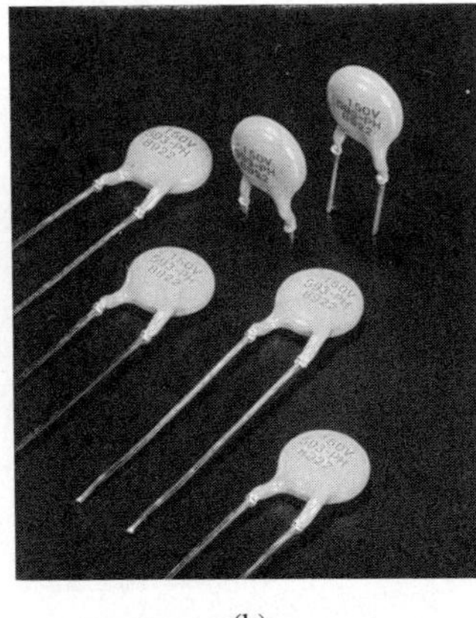

(b)

FIG. 3.40
Varistors available with maximum dc voltage ratings between 18 V and 615 V. (Courtesy of Philips Components, Inc.)

3.14 COMPUTER ANALYSIS

The input of resistive elements will give us a chance to review the power-of-10 notation. Virtually all computer systems use power-of-10 notation, although some specify scientific notation (decimal point to the right of the first whole number between 1 and 10 (such as 5.67×10^5), and others ask for floating-point notation (decimal point at any location, such as 0.567×10^6 or 567.0×10^3). Both PSpice and BASIC use the floating-point notation; C++ and PASCAL have it as an option. Thus, you could see any one of the following forms for the number 1000 (using the uppercase letter E to designate power-of-10 format):

1000 1000.0 1E3 .001E6 0.1E4 10E2

Negative numbers simply have a negative sign in front of the number.

PSpice (DOS)

In PSpice specific suffixes (following the number) are given to frequently used powers of 10, as listed below. Since uppercase letters are specified for a particular purpose, you must be very careful in their use for other applications.

$$
\begin{aligned}
F &= 10^{-15} \\
P &= 10^{-12} \\
N &= 10^{-9} \\
U &= 10^{-6} \\
M &= 10^{-3} \\
K &= 10^{+3} \\
MEG &= 10^{+6} \\
G &= 10^{+9} \\
T &= 10^{+12}
\end{aligned}
$$

Using the above notation, the following are equivalent:

2E6 2MEG 2E3K .002G 2000K

Note, in particular, that milli and mega are differentiated by using an uppercase letter M for milli and uppercase letters MEG for mega.

The format for the input of a 2-kΩ resistor is the following:

```
              Assumed polarities for VR4
Required        (+) (-)
   ↓
  R4             4   3    2K
 Name                   Magnitude
```

The entry is very similar to that of a voltage source except the first letter must now be capital *R*. The name follows immediately using numbers or letters and then the node assignment. Even though resistors do not have polarity, as does a dc voltage source, the voltage across the resistor will have a particular polarity. In many cases the actual polarity will not be known when the circuit is entered, but an assumption is made that the output will confirm or reverse. For the above entry it is assumed that if the voltage across R4 is determined, node 4 will be at a higher potential than node 3. If correct, the output magnitude will have no sign at all (signifying +), but if the actual polarity is the reverse, the output will include a negative sign. In any case, do not be overly concerned about the polarity when the resistor values are entered. Make a

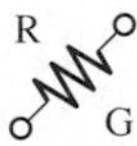

reasonable assumption about the polarity and then let the computer package determine the actual result.

PSpice (Windows)

When using PSpice (Windows) the resistor is placed directly on the schematic, and its parameters are changed by clicking on the label or value. The resistor is under the analog.slb library and is obtained using the sequence **Draw-Get New Part-Browse-analog.slb-R-OK.** The resistor will then appear on the screen and can be moved with the mouse to any location. A left click of the mouse will leave the resistor in the chosen location, but another resistor will appear that can again be moved with the mouse. Additional resistors can be placed in the same manner, and when all are in place, a right click of the mouse will end the process. The label for each can then be changed by double-clicking on the current label and entering the new label in the **Edit Reference Designator** dialog box. Enter R followed by **OK,** and the new label is applied on the schematic. An additional click will clear away the defining rectangles on the schematic. The value can be changed by double-clicking on the current value (1 kΩ) and entering the new value in the **Set Attribute Value** dialog box. Entering 2.2K-**OK** will then result in the display of Fig. 3.41. If the resistor has to be moved at any time, simply click on the symbol once and then click again, but this time hold the clicker down until the resistor is in the new location. Then simply release the clicker and the resistor remains in its new location.

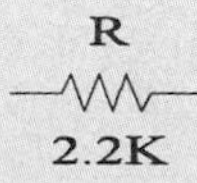

FIG. 3.41
A resistive element in PSpice (Windows).

Programming Language (C++, BASIC, PASCAL, etc.)

Since each language will have its own form for the resistive element, the format will be presented within the context of a program.

PROBLEMS

SECTION 3.2 Resistance: Circular Wires

1. Convert the following to mils:
   a. 0.5 in. **b.** 0.01 in.
   c. 0.004 in. **d.** 1 in.
   e. 0.02 ft **f.** 0.01 cm
2. Calculate the area in circular mils (CM) of wires having the following diameters:
   a. 0.050 in. **b.** 0.016 in.
   c. 0.30 in. **d.** 0.1 cm
   e. 0.003 ft **f.** 0.0042 m
3. The area in circular mils is
   a. 1600 CM **b.** 900 CM
   c. 40,000 CM **d.** 625 CM
   e. 7.75 CM **f.** 81 CM
   What is the diameter of each wire in inches?
4. What is the resistance of a copper wire 200 ft long and 0.01 in. in diameter ($T = 20°C$)?
5. Find the resistance of a silver wire 50 yd long and 0.0045 in. in diameter ($T = 20°C$).
6. **a.** What is the area in circular mils of an aluminum conductor that is 80 ft long with a resistance of 2.5 Ω?
   b. What is its diameter in inches?
7. A 2.2-Ω resistor is to be made of nichrome wire. If the available wire is 1/32 in. in diameter, how much wire is required?
8. **a.** What is the area in circular mils of a copper wire that has a resistance of 2.5 Ω and is 300 ft long ($T = 20°C$)?
   b. Without working out the numerical solution, determine whether the area of an aluminum wire will be smaller or larger than that of the copper wire. Explain.
   c. Repeat (b) for a silver wire.
9. In Fig. 3.42, three conductors of different materials are presented.
   a. Without working out the numerical solution, which section would appear to have the most resistance? Explain.
   b. Find the resistance of each section and compare with the result of (a) ($T = 20°C$).

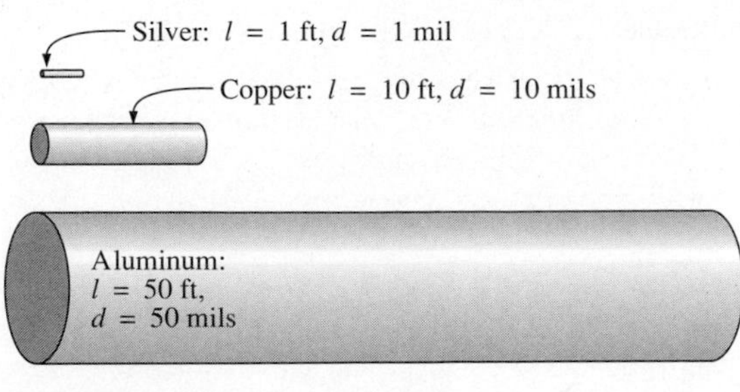

FIG. 3.42
Problem 9.

10. A wire 1000 ft long has a resistance of 0.5 kΩ and an area of 94 CM. Of what material is the wire made ($T = 20°C$)?

*11. a. What is the resistance of a copper bus-bar with the dimensions shown ($T = 20°C$) in Fig. 3.43?
 b. Repeat (a) for aluminum and compare the results.
 c. Without working out the numerical solution, determine whether the resistance of the bar (aluminum or copper) will increase or decrease with an increase in length. Explain your answer.
 d. Repeat (c) for an increase in cross-sectional area.

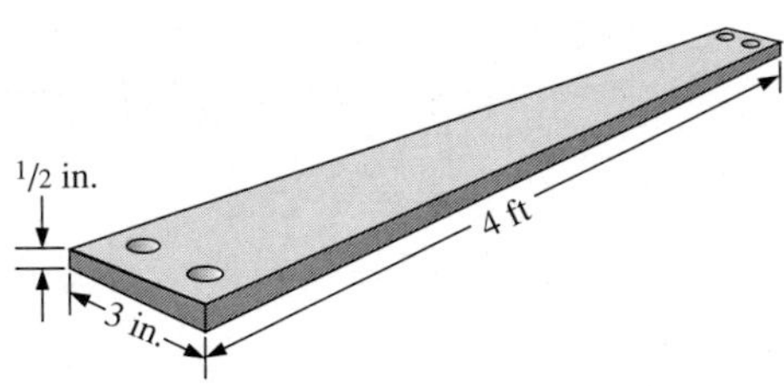

FIG. 3.43
Problem 11.

12. Determine the increase in resistance of a copper conductor if the area is reduced by a factor of 4 and the length is doubled. The original resistance was 0.2 Ω. The temperature remains fixed.

*13. What is the new resistance level of a copper wire if the length is changed from 200 ft to 100 yd, the area is changed from 40,000 CM to 0.04 in.^{2}, and the original resistance was 518.5 mΩ?

SECTION 3.3 Wire Tables

14. a. Using Table 3.2, find the resistance of 450 ft of #11 and #14 AWG wires.
 b. Compare the resistances of the two wires.
 c. Compare the areas of the two wires.

15. a. Using Table 3.2, find the resistance of 1800 ft of #8 and #18 AWG wires.
 b. Compare the resistances of the two wires.
 c. Compare the areas of the two wires.

16. a. For the system of Fig. 3.44, the resistance of each line cannot exceed 0.006 Ω, and the maximum current drawn by the load is 110 A. What gage wire should be used?
 b. Repeat (a) for a maximum resistance of 0.003 Ω, d = 30 ft, and a maximum current of 110 A.

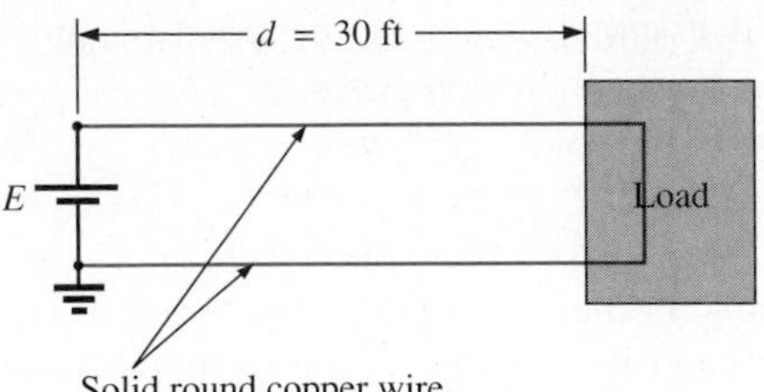

FIG. 3.44
Problem 16.

*17. a. From Table 3.2, determine the maximum permissible current density (A/CM) for an AWG #0000 wire.
 b. Convert the result of (a) to A/in.^{2}.
 c. Using the result of (b), determine the cross-sectional area required to carry a current of 5000 A.

SECTION 3.4 Resistance: Metric Units

18. Using metric units, determine the length of a copper wire that has a resistance of 0.2 Ω and a diameter of 1/10 in.

19. Repeat Problem 11 using metric units; that is, convert the given dimensions to metric units before determining the resistance.

20. If the sheet resistance of a tin oxide sample is 100 Ω, what is the thickness of the oxide layer?

21. Determine the width of a carbon resistor having a sheet resistance of 150 Ω if the length is 1/2 in. and the resistance is 500 Ω.

*22. Derive the conversion factor between ρ (CM·Ω/ft) and ρ (Ω·cm) by
 a. Solving for ρ for the wire of Fig. 3.45 in CM·Ω/ft.
 b. Solving for ρ for the same wire of Fig. 3.45 in Ω·cm by making the necessary conversions.
 c. Use the equation $\rho_2 = k\rho_1$ to determine the conversion factor k if ρ_1 is the solution of part (a) and ρ_2 the solution of part (b).

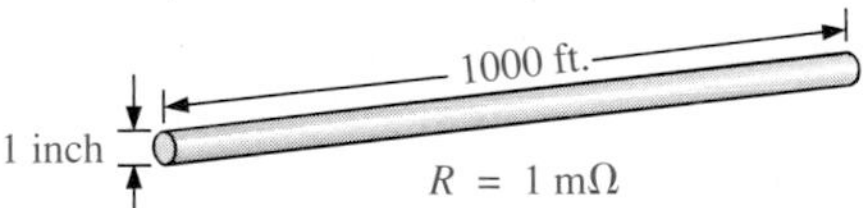

FIG. 3.45
Problem 22.

SECTION 3.5 Temperature Effects

23. The resistance of a copper wire is 2 Ω at 10°C. What is its resistance at 60°C?

24. The resistance of an aluminum bus-bar is 0.02 Ω at 0°C. What is its resistance at 100°C?

25. The resistance of a copper wire is 4 Ω at 70°F. What is its resistance at 32°F?

26. The resistance of a copper wire is 0.76 Ω at 30°C. What is its resistance at −40°C?

27. If the resistance of a silver wire is 0.04 Ω at −30°C, what is its resistance at 0°C?

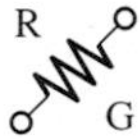

*28. a. The resistance of a copper wire is 0.002 Ω at room temperature (68°F). What is its resistance at 32°F (freezing) and 212°F (boiling)?
b. For (a), determine the change in resistance for each 10° change in temperature between room temperature and 212°F.

29. a. The resistance of a copper wire is 0.92 Ω at 4°C. At what temperature (°C) will it be 1.06 Ω?
b. At what temperature will it be 0.15 Ω?

*30. a. If the resistance of a 1000-ft length of copper wire is 10 Ω at room temperature (20°C), what will its resistance be at 50 K (Kelvin units) using Eq. (3.6)?
b. Repeat part (a) for a temperature of 38.65 K. Comment on the results obtained by reviewing the curve of Fig. 3.14.
c. What is the temperature of absolute zero in Fahrenheit units?

31. a. Verify the value of α_{20} for copper in Table 3.6 by substituting the inferred absolute temperature into Eq. (3.7).
b. Using Eq. (3.8) find the temperature at which the resistance of a copper conductor will increase to 1 Ω from a level of 0.8 Ω at 20°C.

32. Using Eq. (3.8), find the resistance of a copper wire at 16°C if its resistance at 20°C is 0.4 Ω.

*33. Determine the resistance of a 1000-ft coil of #12 copper wire sitting in the desert at a temperature of 115°F.

34. A 22-Ω wire-wound resistor is rated at +200 PPM for a temperature range of −10°C to +75°C. Determine its resistance at 65°C.

35. Determine the PPM rating of the 10-kΩ carbon composition resistor of Fig. 3.21 using the resistance level determined at 90°C.

SECTION 3.6 Superconductors

36. Visit your local library and find a table listing the critical temperatures for a variety of materials. List at least five materials with the critical temperatures that are not mentioned in this text. Choose a few materials that have relatively high critical temperatures.

37. Find at least one article on the application of superconductivity in the commercial sector and write a short summary, including all interesting facts and figures.

*38. Using the required 1-MA/cm^2 density level for IC manufacturing, what would the resulting current be through a #12 house wire? Compare the result obtained with the allowable limit of Table 3.2.

*39. Research the SQUID magnetic field detector and review its basic mode of operation and an application or two.

SECTION 3.7 Types of Resistors

40. a. What is the approximate increase in size from a 1-W to a 2-W carbon resistor?
b. What is the approximate increase in size from a 1/2-W to a 2-W carbon resistor?

41. If the 10-kΩ resistor of Fig. 3.21 is exactly 10 kΩ at room temperature, what is its approximate resistance at −30°C and 100°C (boiling)?

42. Repeat Problem 41 at a temperature of 120°F.

43. If the resistance between the outside terminals of a linear potentiometer is 10 kΩ, what is its resistance between the wiper (movable) arm and an outside terminal if the resistance between the wiper arm and the other outside terminal is 3.5 kΩ?

44. If the wiper arm of a linear potentiometer is one-quarter the way around the contact surface, what is the resistance between the wiper arm and each terminal if the total resistance is 25 kΩ?

*45. Show the connections required to establish 4 kΩ between the wiper arm and one outside terminal of a 10-kΩ potentiometer while having only zero ohms between the other outside terminal and the wiper arm.

SECTION 3.8 Color Coding and Standard Resistor Values

46. Find the range in which a resistor having the following color bands must exist to satisfy the manufacturer's tolerance:

	1st band	2nd band	3rd band	4th band
a.	green	blue	orange	gold
b.	red	red	brown	silver
c.	brown	black	black	—

47. Find the color code for the following 10% resistors:
a. 220 Ω
b. 4700 Ω
c. 68 kΩ
d. 9.1 MΩ

48. Is there an overlap in coverage between 20% resistors? That is, determine the tolerance range for a 10-Ω 20% resistor and a 15-Ω 20% resistor, and note whether their tolerance ranges overlap.

49. Repeat Problem 48 for 10% resistors of the same value.

SECTION 3.9 Conductance

50. Find the conductance of each of the following resistances:
a. 0.086 Ω
b. 4 kΩ
c. 2.2 MΩ
Compare the three results.

51. Find the conductance of 1000 ft of #18 AWG wire made of
a. copper
b. aluminum
c. iron

*52. The conductance of a wire is 100 S. If the area of the wire is increased by 2/3 and the length is reduced by the same amount, find the new conductance of the wire if the temperature remains fixed.

SECTION 3.10 Ohmmeters

53. How would you check the status of a fuse with an ohmmeter?

54. How would you determine the on and off states of a switch using an ohmmeter?

55. How would you use an ohmmeter to check the status of a light bulb?

SECTION 3.11 Thermistors

***56.** **a.** Find the resistance of the thermistor having the characteristics of Fig. 3.36 at −50°C, 50°C, and 200°C. Note that it is a log scale. If necessary, consult a reference with an expanded log scale.
b. Does the thermistor have a positive or negative temperature coefficient?
c. Is the coefficient a fixed value for the range −100°C to 400°C? Why?
d. What is the approximate rate of change of ρ with temperature at 100°C?

SECTION 3.12 Photoconductive Cell

***57.** **a.** Using the characteristics of Fig. 3.38, determine the resistance of the photoconductive cell at 10 and 100 foot-candle illumination. As in Problem 56, note that it is a log scale.
b. Does the cell have a positive or negative illumination coefficient?
c. Is the coefficient a fixed value for the range 0.1 to 1000 foot-candles? Why?
d. What is the approximate rate of change of R with illumination at 10 foot-candles?

SECTION 3.13 Varistors

58. **a.** Referring to Fig. 3.40(a), find the terminal voltage of the device at 0.5, 1, 3, and 5 mA.
b. What is the total change in voltage for the indicated range of current levels?
c. Compare the ratio of maximum to minimum current levels above to the corresponding ratio of voltage levels.

GLOSSARY

Absolute zero The temperature at which all molecular motion ceases; −273.15°C.

Circular mil (CM) The cross-sectional area of a wire having a diameter of one mil.

Color coding A technique employing bands of color to indicate the resistance levels and tolerance of resistors.

Conductance (G) An indication of the relative ease with which current can be established in a material. It is measured in siemens (S).

Cooper effect The "pairing" of electrons as they travel through a medium.

Inferred absolute temperature The temperature through which a straight-line approximation 84for the actual resistance-versus-temperature curve will intersect the temperature axis.

Meissner effect The levitation of a magnet above a superconductor.

Negative temperature coefficient of resistance The value revealing that the resistance of a material will decrease with an increase in temperature.

Ohm (Ω) The unit of measurement applied to resistance.

Ohmmeter An instrument for measuring resistance levels.

Photoconductive cell A two-terminal semiconductor device whose terminal resistance is determined by the intensity of the incident light on its exposed surface.

Positive temperature coefficient of resistance The value revealing that the resistance of a material will increase with an increase in temperature.

Potentiometer A three-terminal device through which potential levels can be varied in a linear or nonlinear manner.

PPM/°C Temperature sensitivity of a resistor in parts per million per degree Celsius.

Resistance A measure of the opposition to the flow of charge through a material.

Resistivity (ρ) A constant of proportionality between the resistance of a material and its physical dimensions.

Rheostat An element whose terminal resistance can be varied in a linear or nonlinear manner.

Sheet resistance Defined by ρ/d for thin-film and integrated circuit design.

SQUID Superconducting quantum interference device.

Superconductor Conductors of electric charge that have for all practical purposes zero ohms.

Thermistor A two-terminal semiconductor device whose resistance is temperature sensitive.

Varistor A voltage-dependent, nonlinear resistor used to suppress high-voltage transients.

4

Ohm's Law, Power, and Energy

4.1 OHM'S LAW

Consider the following relationship:

$$\boxed{\text{Effect} = \frac{\text{cause}}{\text{opposition}}} \tag{4.1}$$

Every conversion of energy from one form to another can be related to this equation. In electric circuits, the *effect* we are trying to establish is the flow of charge, or *current*. The *potential difference*, or voltage, between two points is the *cause* ("pressure"), and the opposition is the *resistance* encountered.

Substituting these terms into Eq. (4.1) results in

$$\text{Current} = \frac{\text{potential difference}}{\text{resistance}}$$

and

$$\boxed{I = \frac{E}{R}} \qquad \text{(amperes, A)} \tag{4.2}$$

Equation (4.2) is known as *Ohm's law* in honor of Georg Simon Ohm (Fig. 4.1). The law clearly reveals that for a fixed resistance, the greater the voltage (or pressure) across a resistor, the more the current, and the more the resistance for the same voltage, the less the current. In other words, the current is proportional to the applied voltage and inversely proportional to the resistance.

By simple mathematical manipulations, the voltage and resistance can be found in terms of the other two quantities:

$$\boxed{E = IR} \qquad \text{(volts, V)} \tag{4.3}$$

German (Erlangen, Cologne)
(1789–1854)
Physicist and Mathematician
Professor of Physics, University of Cologne

Courtesy of the Smithsonian Institution Photo No. 51,145

In 1827, developed one of the most important laws of electric circuits: *Ohm's law.* When first introduced, the supporting documentation was considered lacking and foolish, causing him to lose his teaching position and search for a living doing odd jobs and some tutoring. It took some 22 years for his work to be recognized as a major contribution to the field. He was then awarded a chair at the University of Munich and received the Copley Medal of the Royal Society in 1841. His research also extended into the areas of molecular physics, acoustics, and telegraphic communication.

FIG. 4.1
Georg Simon Ohm.

$$R = \frac{E}{I} \quad \text{(ohms, } \Omega\text{)} \tag{4.4}$$

The three quantities of Eqs. (4.2) through (4.4) are defined by the simple circuit of Fig. 4.2. The current I of Eq. (4.2) results from applying a dc supply of E volts across a network having a resistance R. Equation (4.3) determines the voltage E required to establish a current I through a network with a total resistance R, and Eq. (4.4) provides the resistance of a network that results in a current I due to an impressed voltage E.

Note in Fig. 4.2 that the voltage source "pressures" current in a direction that passes from the negative to the positive terminal of the battery. This will always be the case for single-source circuits. The effect of more than one source in the network will be examined in the chapters to follow. The symbol for the voltage of the battery (a source of electrical energy) is the uppercase letter E, whereas the loss in potential energy across the resistor is given the symbol V. The polarity of the voltage drop across the resistor is as defined by the applied source because the two terminals of the battery are connected directly across the resistive element.

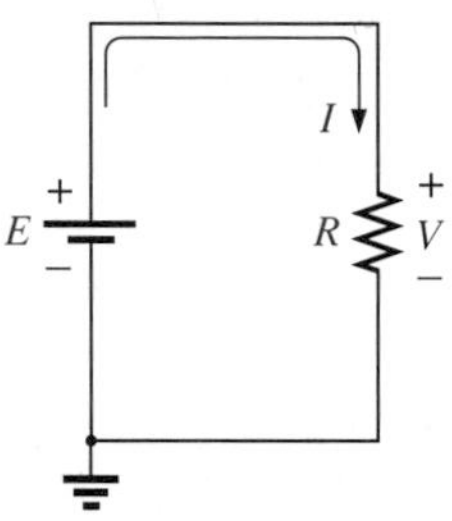

FIG. 4.2
Basic circuit.

EXAMPLE 4.1 Determine the current resulting from the application of a 9-V battery across a network with a resistance of 2.2 Ω.

Solution: Eq. (4.2):

$$I = \frac{E}{R} = \frac{9\text{ V}}{2.2\ \Omega} = \mathbf{4.09\ A}$$

EXAMPLE 4.2 Calculate the resistance of a 60-W bulb if a current of 500 mA results from an applied voltage of 120 V.

Solution: Eq. (4.4):

$$R = \frac{E}{I} = \frac{120\text{ V}}{500 \times 10^{-3}\text{ A}} = \mathbf{240\ \Omega}$$

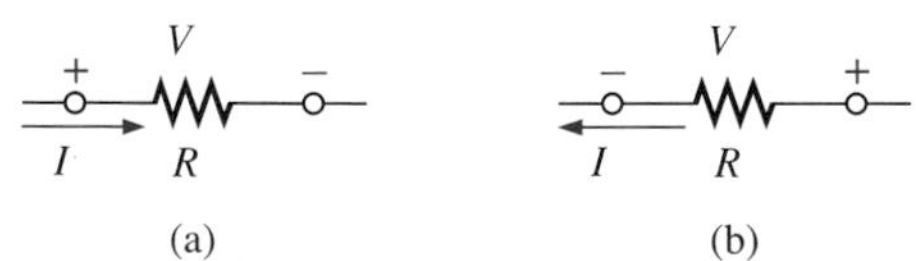

FIG. 4.3
Defining polarities.

For an isolated resistive element, the polarity of the voltage drop is as shown in Fig. 4.3(a) for the indicated current direction. A reversal in current will reverse the polarity, as shown in Fig. 4.3(b). In general, the flow of charge is from a high (+) to a low (−) potential. Polarities as established by current direction will become increasingly important in the analysis to follow.

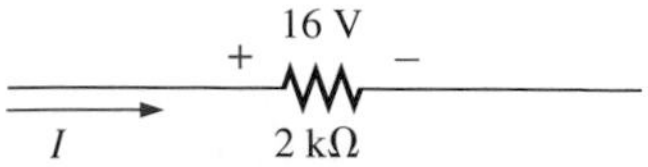

16 V
\+ –
I
2 kΩ

FIG. 4.4
Example 4.3.

EXAMPLE 4.3 Calculate the current through the 2-kΩ resistor of Fig. 4.4 if the voltage drop across it is 16 V.

Solution:

$$I = \frac{V}{R} = \frac{16\text{ V}}{2 \times 10^{3}\ \Omega} = \mathbf{8\ mA}$$

EXAMPLE 4.4 Calculate the voltage that must be applied across the soldering iron of Fig. 4.5 to establish a current of 1.5 A through the iron if its internal resistance is 80 Ω.

Solution:

$$E = IR = (1.5 \text{ A})(80 \ \Omega) = \mathbf{120 \ V}$$

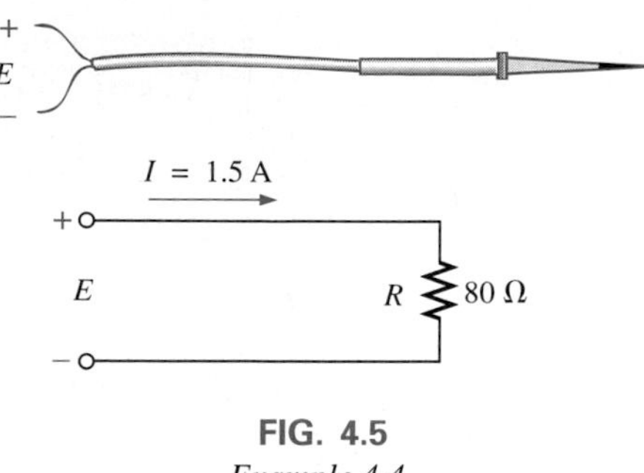

FIG. 4.5
Example 4.4.

4.2 PLOTTING OHM'S LAW

Graphs, characteristics, plots, and the like, play an important role in every technical field as a mode through which the broad picture of the behavior or response of a system can be conveniently displayed. It is therefore critical to develop the skills necessary both to read data and to plot them in such a manner that they can be interpreted easily.

For most sets of characteristics of electronic devices, the current is represented by the vertical axis (ordinate), and the voltage by the horizontal axis (abscissa), as shown in Fig. 4.6. First note that the vertical axis is in

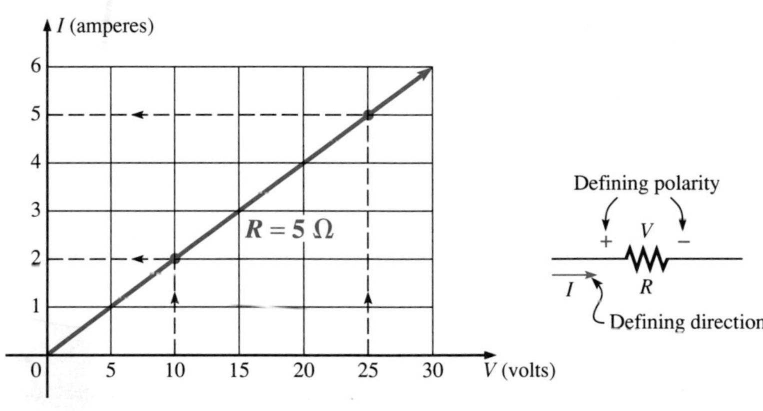

FIG. 4.6
Plotting Ohm's law.

amperes and the horizontal axis is in volts. For some plots, I may be in milliamperes (mA), microamperes (μA), or whatever is appropriate for the range of interest. The same is true for the levels of voltage on the horizontal axis. Note also that the chosen parameters require that the spacing between numerical values of the vertical axis be different from that of the horizontal axis. The linear (straight-line) graph reveals that the resistance is not changing with current or voltage level; rather, it is a fixed quantity throughout. The current direction and the voltage polarity appearing to the right of Fig. 4.6 are the defined direction and polarity for the provided plot. If the current direction is opposite to the defined direction, the region below the horizontal axis is the region of interest for the current I. If the voltage polarity is opposite to that defined, the region to the left of the current axis is the region of interest. For the standard fixed resistor, the first quadrant, or region, of Fig. 4.6 is the only region of interest. There are many devices, however, that you will encounter in your electronics courses that will use the other quadrants of a graph.

Once a graph such as Fig. 4.6 is developed, the current or voltage at any level can be found from the other quantity by simply using the resulting plot. For instance, at $V = 25$ V, if a vertical line is drawn on Fig. 4.6 to the

curve as shown, the resulting current can be found by drawing a horizontal line over to the current axis, where a result of 5 A is obtained. Similarly, at $V = 10$ V, a vertical line to the plot and a horizontal line to the current axis will result in a current of 2 A, as determined by Ohm's law.

If the resistance of a plot is unknown, it can be determined at any point on the plot since a straight line indicates a fixed resistance. At any point on the plot, find the resulting current and voltage and simply substitute into the following equation:

$$R_{dc} = \frac{V}{I} \tag{4.5}$$

To test Eq. (4.5) consider a point on the plot where $V = 20$ V and $I = 4$ A. The resulting resistance is $R_{dc} = V/I = 20\text{ V}/4\text{ A} = 5\ \Omega$. For comparison purposes, a 1-Ω and 10-Ω resistor were plotted on the graph of Fig. 4.7. Note that the less the resistance, the steeper the slope (closer to the vertical axis) of the curve.

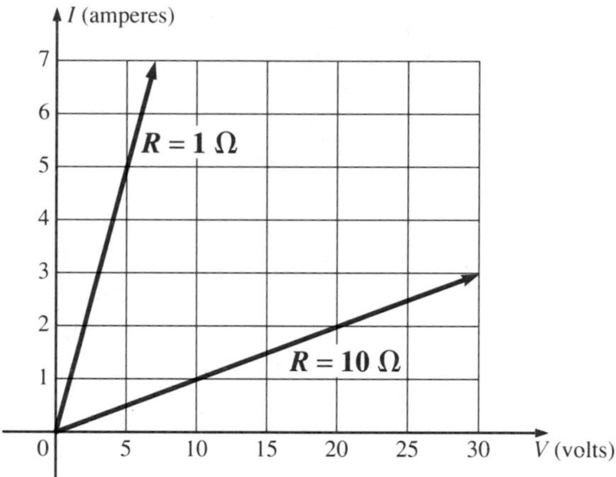

FIG. 4.7
Demonstrating on an I-V plot that the less the resistance, the steeper is the slope.

If we write Ohm's law in the following manner and relate it to the basic straight-line equation

$$\begin{array}{c} I = \dfrac{1}{R} \cdot E + 0 \\ \downarrow \quad \downarrow \quad \downarrow \quad \downarrow \\ y = m \cdot x + b \end{array}$$

we find that the slope is equal to 1 divided by the resistance value, as indicated by the following:

$$m = \text{slope} = \frac{\Delta y}{\Delta x} = \frac{\Delta I}{\Delta V} = \frac{1}{R} \tag{4.6}$$

where Δ signifies a small, finite change in the variable.

Equation (4.6) clearly reveals that the greater the resistance, the less the slope. If written in the following form, Eq. (4.6) can be used to determine the resistance from the linear curve:

$$R = \frac{\Delta V}{\Delta I} \quad \text{(ohms)} \tag{4.7}$$

The equation states that by choosing a particular ΔV (or ΔI), the corresponding ΔI (or ΔV, respectively) can be obtained from the graph, as shown in Fig. 4.8, and the resistance can be determined. It the plot is a straight line, Eq. (4.7) will provide the same result no matter where the equation is applied. However, if the plot curves at all, the resistance will change.

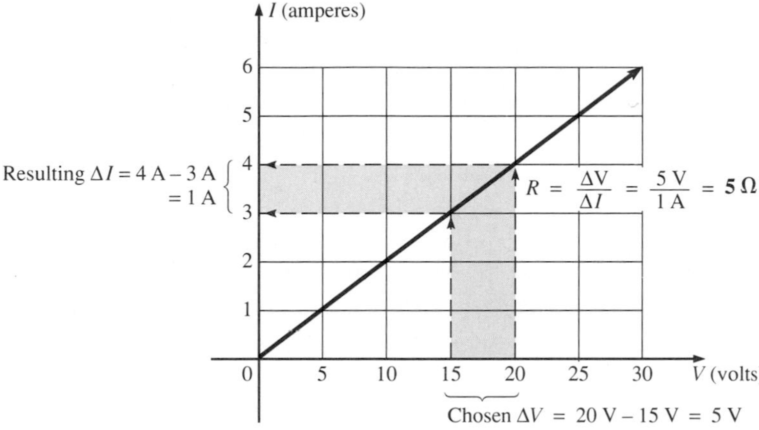

FIG. 4.8
Applying Eq. (4.6).

EXAMPLE 4.5 Determine the resistance associated with the curve of Fig. 4.9 using Eqs. (4.5) and (4.7), and compare results.

Solution: At $V = 6$ V, $I = 3$ mA, and

$$R_{dc} = \frac{V}{I} = \frac{6\text{ V}}{3\text{ mA}} = \mathbf{2\ k\Omega}$$

For the interval between 6 V and 8 V,

$$R = \frac{\Delta V}{\Delta I} = \frac{2\text{ V}}{1\text{ mA}} = \mathbf{2\ k\Omega}$$

The results are equivalent.

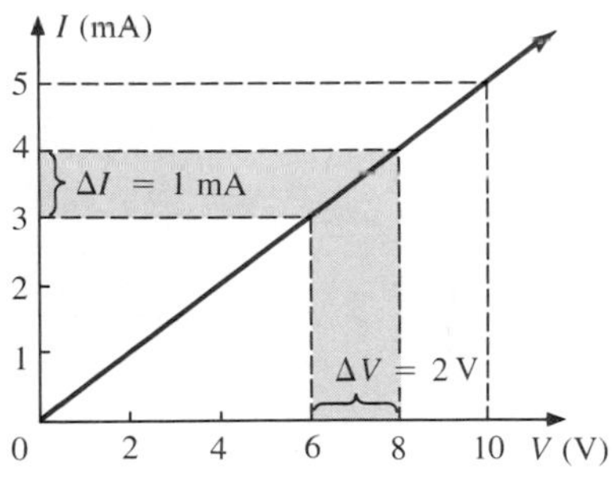

FIG. 4.9
Example 4.5.

Before leaving the subject, let us first investigate the characteristics of a very important semiconductor device called the *diode,* which will be examined in detail in basic electronics courses. This device will ideally act like a low resistance path to current in one direction and a high resistance path to current in the reverse direction, much like a switch that will pass current in only one direction. A typical set of characteristics appears in Fig. 4.10. Without any mathematical calculations, the closeness of the characteristic to the voltage axis for negative values of applied voltage indicates that this is the low conductance (high resistance, switch opened) region. Note that this region extends to approximately 0.7 V positive. However, for values of applied voltage greater than 0.7 V, the vertical rise in the characteristics indicates a high conductivity (low resistance, switch closed) region. Application of Ohm's law will now verify the above conclusions.

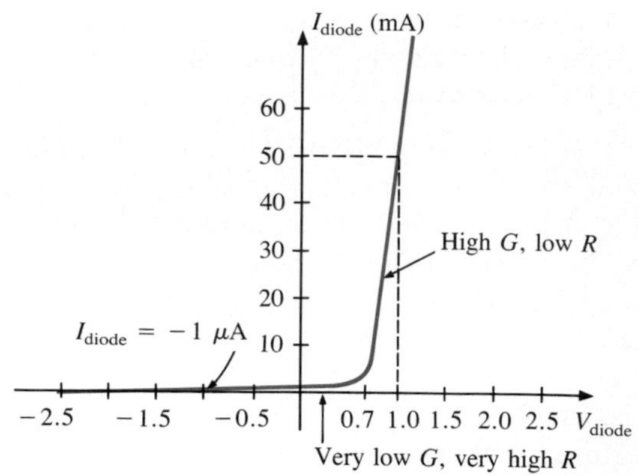

FIG. 4.10
Semiconductor diode characteristic.

At $V = +1$ V,

$$R_{diode} = \frac{V}{I} = \frac{1\text{ V}}{50\text{ mA}} = \frac{1\text{ V}}{50 \times 10^{-3}\text{ A}}$$

$$= 20\ \Omega$$

(a relatively low value for most applications)

At $V = -1$ V,

$$R_{diode} = \frac{V}{I} = \frac{1\text{ V}}{1\ \mu\text{A}}$$

$$= 1\ \text{M}\Omega$$

(which is often represented by an open-circuit equivalent)

Scottish (Greenock, Birmingham)
(1736–1819)
Instrument maker and inventor
Elected fellow of the Royal Society of London in 1785

Courtesy of the Smithsonian Institution Photo No. 30, 391

In 1757, at the age of 21, used his innovative talents to design mathematical instruments such as the *quadrant, compass,* and various *scales.* In 1765, introduced the use of a separate *condenser* to increase the efficiency of steam engines. In the years to follow he received a number of important patents on improved engine design, including a rotary motion for the steam engine (versus the reciprocating action) and a double-action engine, in which the piston pulled as well as pushed in its cyclic motion. Introduced the term *horsepower* as the average power of a strong dray (small cart) horse over a full working day.

FIG. 4.11
James Watt.

4.3 POWER

Power is an indication of how much work (the conversion of energy from one form to another) can be accomplished in a specified amount of time, that is, a *rate* of doing work. For instance, a large motor has more power than a small motor because it can convert more electrical energy into mechanical energy in the same period of time. Since converted energy is measured in *joules* (J) and time in seconds (s), power is measured in joules/second (J/s). The electrical unit of measurement for power is the watt (W), defined by

$$1 \text{ watt (W)} = 1 \text{ joule/second (J/s)} \tag{4.8}$$

In equation form, power is determined by

$$P = \frac{W}{t} \quad \text{(watts, W, or joules/second, J/s)} \tag{4.9}$$

with the energy W measured in joules and the time t in seconds.

Throughout the text, the abbreviation for energy (W) can be distinguished from that for the watt (W) by the fact that one is in italics while the other is in roman. In fact, all variables in the dc section appear in italics while the units appear in roman.

The unit of measurement, the watt, is derived from the surname of James Watt (Fig. 4.11), who was instrumental in establishing the standards for power measurements. He introduced the *horsepower* (hp) as a measure of the average power of a strong dray horse over a full working day. It is approximately 50% more than can be expected from the average horse. The horsepower and watt are related in the following manner:

$$1 \text{ horsepower} \cong 746 \text{ watts}$$

The power delivered to, or absorbed by, an electrical device or system can be found in terms of the current and voltage by first substituting Eq. (2.7) into Eq. (4.9):

$$P = \frac{W}{t} = \frac{QV}{t} = V\frac{Q}{t}$$

But

$$I = \frac{Q}{t}$$

so that

$$P = VI \quad \text{(watts)} \tag{4.10}$$

By direct substitution of Ohm's law, the equation for power can be obtained in two other forms:

$$P = VI = V\left(\frac{V}{R}\right)$$

and

$$P = \frac{V^2}{R} \quad \text{(watts)} \tag{4.11}$$

or

$$P = VI = (IR)I$$

and

$$P = I^2R \quad \text{(watts)} \tag{4.12}$$

The result is that the power absorbed by the resistor of Fig. 4.12 can be found directly depending on the information available. In other words, if the current and resistance are known, it pays to use Eq. (4.12) directly, and if V and I are known, Eq. (4.10) is appropriate. It saves having to apply Ohm's law before determining the power.

Power can be delivered or absorbed as defined by the polarity of the voltage and the direction of the current. For all dc voltage sources, power is being *delivered* by the source if the current has the direction appearing in Fig. 4.13(a). Note that the current has the same direction as established by the source in a single-source network. If the current direction and polarity are as shown in Fig. 4.13(b) due to a multisource network, the battery is absorbing power much like when a battery is being charged.

For resistive elements, all the power delivered is dissipated in the form of heat because the voltage polarity is defined by the current direction (and vice versa), and current will always enter the terminal of higher potential corresponding with the absorbing state of Fig. 4.13(b). A reversal of the current direction in Fig. 4.12 will also reverse the polarity of the voltage across the resistor and match the conditions of Fig. 4.13(b).

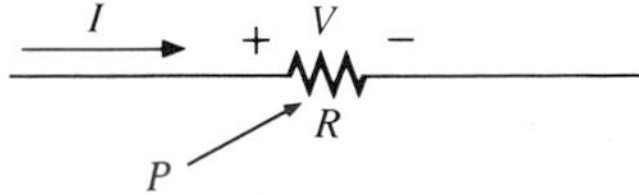

FIG. 4.12
Defining the power to a resistive element.

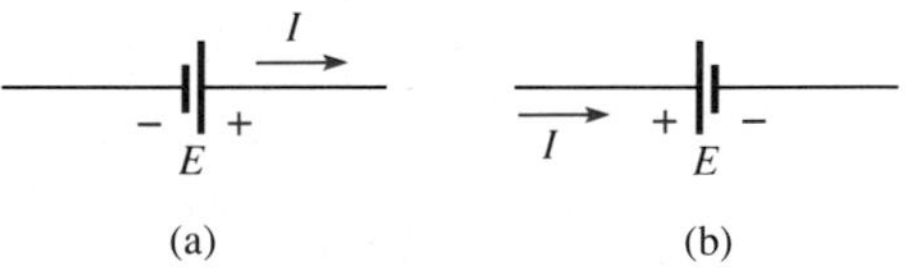

FIG. 4.13
Battery power: (a) supplied; (b) absorbed.

The magnitude of the power delivered or absorbed by a battery is given by

$$P = EI \quad \text{(watts)} \tag{4.13}$$

with E the battery terminal voltage and I the current through the source.

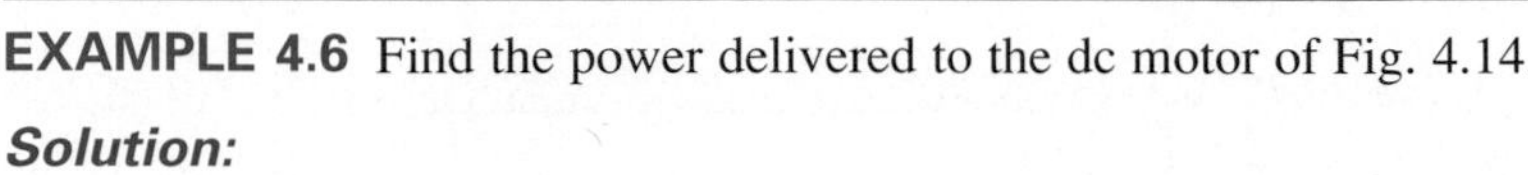

EXAMPLE 4.6 Find the power delivered to the dc motor of Fig. 4.14.

Solution:

$$P = VI = (120\ \text{V})(5\ \text{A}) = 600\ \text{W} = \mathbf{0.6\ kW}$$

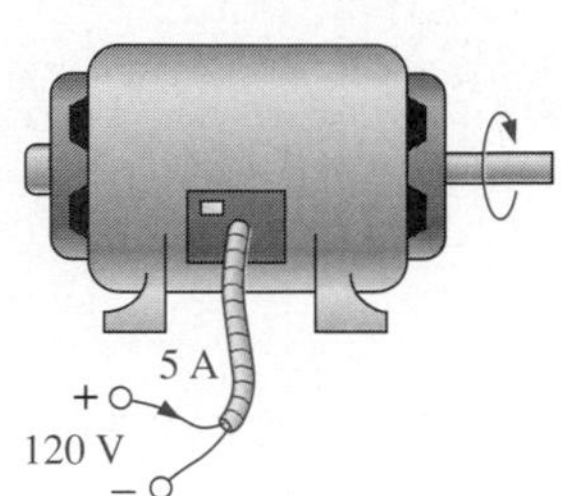

FIG. 4.14
Example 4.6.

EXAMPLE 4.7 What is the power dissipated by a 5-Ω resistor if the current is 4 A?

Solution:

$$P = I^2R = (4\ \text{A})^2(5\ \Omega) = \mathbf{80\ W}$$

EXAMPLE 4.8 The I-V characteristics of a light bulb are provided in Fig. 4.15. Note the nonlinearity of the curve, indicating a wide range in resistance of the bulb with applied voltage as defined by the discussion of Section 4.2. If the rated voltage is 120 V, find the wattage rating of the bulb. Also calculate the resistance of the bulb under rated conditions.

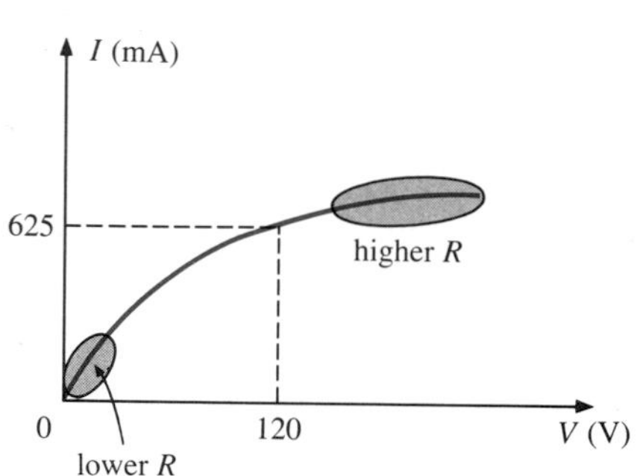

FIG. 4.15
The nonlinear I-V characteristics of a 75-W light bulb.

Solution: At 120 V,

$$I = 0.625\ \text{A}$$

and

$$P = VI = (120\ \text{V})(0.625\ \text{A}) = \mathbf{75\ W}$$

At 120 V,

$$R = \frac{V}{I} = \frac{120\ \text{V}}{0.625\ \text{A}} = \mathbf{192\ \Omega}$$

Sometimes the power is given and the current or voltage must be determined. Through algebraic manipulations, an equation for each variable is derived as follows:

$$P = I^2R \Rightarrow I^2 = \frac{P}{R}$$

and

$$I = \sqrt{\frac{P}{R}} \quad \text{(amperes)} \tag{4.14}$$

$$P = \frac{V^2}{R} \Rightarrow V^2 = PR$$

and

$$V = \sqrt{PR} \quad \text{(volts)} \tag{4.15}$$

EXAMPLE 4.9 Determine the current through a 5-kΩ resistor when the power dissipated by the element is 20 mW.

Solution: Eq. (4.14):

$$I = \sqrt{\frac{P}{R}} = \sqrt{\frac{20 \times 10^{-3}\text{ W}}{5 \times 10^{3}\ \Omega}} = \sqrt{4 \times 10^{-6}} = 2 \times 10^{-3}\text{ A}$$
$$= \mathbf{2\ mA}$$

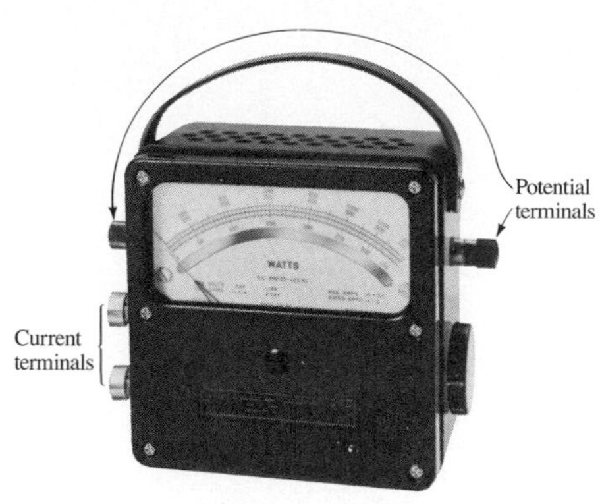

FIG. 4.16
Wattmeter. (Courtesy of Electrical Instrument Service, Inc.)

4.4 WATTMETERS

As one might expect, there are instruments that can measure the power delivered by a source and to a dissipative element. One such instrument appears in Fig. 4.16. Since power is a function of both the current and the voltage levels, four terminals must be connected as shown in Fig. 4.17 to measure the power to the resistor R.

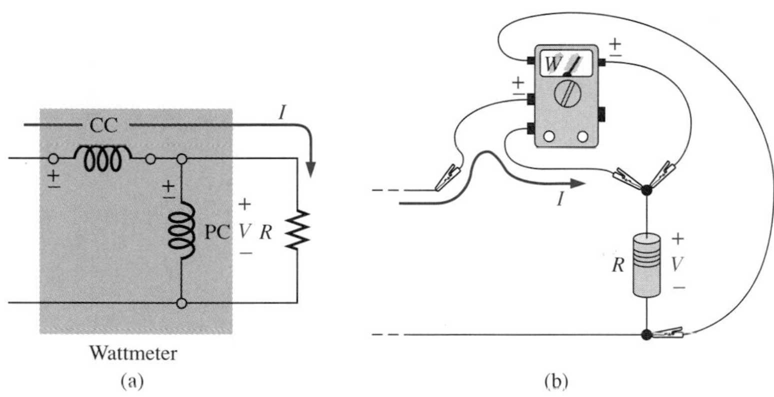

FIG. 4.17
Wattmeter connections.

If the current coils (CC) and potential coils (PC) of the wattmeter are connected as shown in Fig. 4.17, there will be an up-scale reading on the wattmeter. A reversal of either coil will result in a below-zero indication. Three voltage terminals may be available on the voltage side to permit a choice of voltage levels. On most wattmeters, the current terminals are physically larger than the voltage terminals for safety reasons and to ensure a solid connection.

4.5 EFFICIENCY

A flowchart for the energy levels associated with any system that converts energy from one form to another is provided in Fig. 4.18. Take particular note of the fact that the output energy level must always be less than the applied energy due to losses and storage within the system. The best one can hope for is that W_o and W_i are relatively close in magnitude.

Conservation of energy requires that

$$\text{Energy input} = \text{energy output} + \text{energy lost or stored in the system}$$

Dividing both sides of the relationship by t gives

$$\frac{W_{\text{in}}}{t} = \frac{W_{\text{out}}}{t} + \frac{W_{\text{lost or stored by the system}}}{t}$$

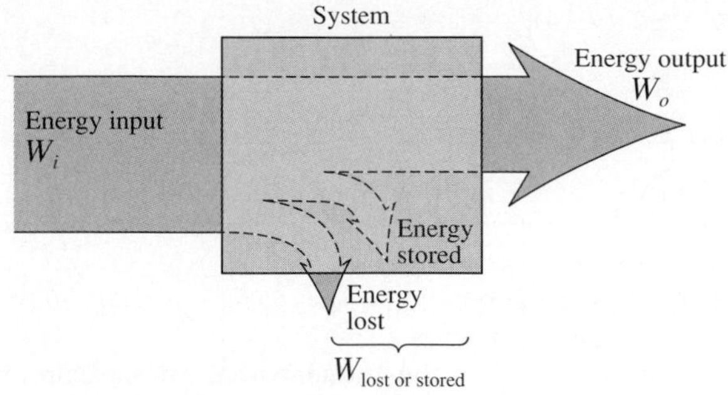

FIG. 4.18
Energy flow through a system.

Since $P = W/t$, we have the following:

$$P_i = P_o + P_{\text{lost or stored}} \quad \text{(W)} \tag{4.16}$$

The efficiency (η) of the system is then determined by the following equation:

$$\text{Efficiency} = \frac{\text{power output}}{\text{power input}}$$

and

$$\eta = \frac{P_o}{P_i} \quad \text{(decimal number)} \tag{4.17}$$

where η (lowercase Greek letter eta) is a decimal number. Expressed as a percentage,

$$\eta\% = \frac{P_o}{P_i} \times 100\% \quad \text{(percent)} \tag{4.18}$$

In terms of the input and output energy, the efficiency in percent is given by

$$\eta\% = \frac{W_o}{W_i} \times 100\% \quad \text{(percent)} \tag{4.19}$$

The maximum possible efficiency is 100%, which occurs when $P_o = P_i$, or when the power lost or stored in the system is zero. Obviously, the greater the internal losses of the system in generating the necessary output power or energy, the lower the net efficiency.

EXAMPLE 4.10 A 2-hp motor operates at an efficiency of 75%. What is the power input in watts? If the applied voltage is 220 V, what is the input current?

Solution:

$$\eta\% = \frac{P_o}{P_i} \times 100\%$$

$$0.75 = \frac{(2 \text{ hp})(746 \text{ W/hp})}{P_i}$$

and $$P_i = \frac{1492\text{ W}}{0.75} = \mathbf{1989.33\text{ W}}$$

$$P_i = EI \text{ or } I = \frac{P_i}{E} = \frac{1989.33\text{ W}}{220\text{ V}} = \mathbf{9.04\text{ A}}$$

EXAMPLE 4.11 What is the output in horsepower of a motor with an efficiency of 80% and an input current of 8 A at 120 V?

Solution:

$$\eta\% = \frac{P_o}{P_i} \times 100\%$$

$$0.80 = \frac{P_o}{(120\text{ V})(8\text{ A})}$$

and $$P_o = (0.80)(120\text{ V})(8\text{ A}) = 768\text{ W}$$

with $$768\text{ W}\left(\frac{1\text{ hp}}{746\text{ W}}\right) = \mathbf{1.029\text{ hp}}$$

EXAMPLE 4.12 If $\eta = 0.85$, determine the output energy level if the applied energy is 50 J.

Solution:

$$\eta = \frac{W_o}{W_i} \Rightarrow W_o = \eta W_i$$

$$= (0.85)(50\text{ J})$$
$$= \mathbf{42.5\text{ J}}$$

The very basic components of a generating (voltage) system are depicted in Fig. 4.19. The source of mechanical power is a structure such as a paddlewheel that is turned by water rushing over the dam. The gear train will then ensure that the rotating member of the generator is turning at rated speed. The output voltage must then be fed through a transmission system to the load. For each component of the system, an

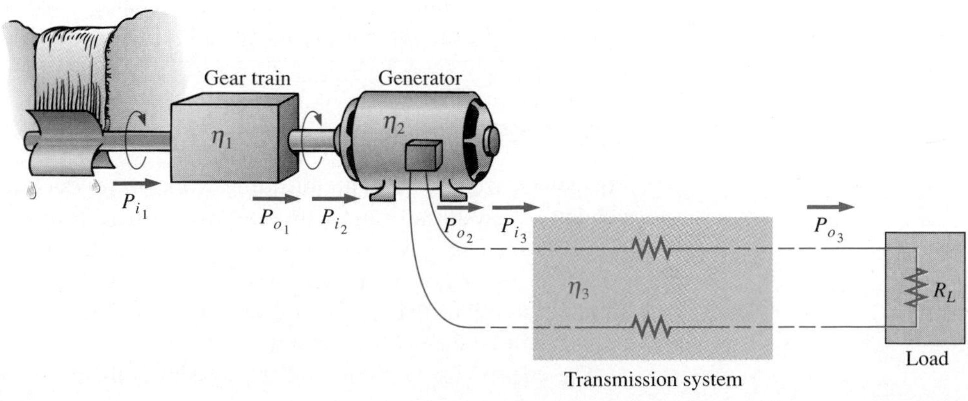

FIG. 4.19
Basic components of a generating system.

input and output power have been indicated. The efficiency of each system is given by

$$\eta_1 = \frac{P_{o_1}}{P_{i_1}} \qquad \eta_2 = \frac{P_{o_2}}{P_{i_2}} \qquad \eta_3 = \frac{P_{o_3}}{P_{i_3}}$$

If we form the product of these three efficiencies,

$$\eta_1 \cdot \eta_2 \cdot \eta_3 = \frac{P_{o_1}}{P_{i_1}} \cdot \frac{P_{o_2}}{P_{i_2}} \cdot \frac{P_{o_3}}{P_{i_3}}$$

and substitute the fact that $P_{i_2} = P_{o_1}$ and $P_{i_3} = P_{o_2}$, we find that the quantities indicated above will cancel, resulting in P_{o_3}/P_{i_1}, which is a measure of the efficiency of the entire system. In general, for the representative cascaded system of Fig. 4.20,

$$\eta_{\text{total}} = \eta_1 \cdot \eta_2 \cdot \eta_3 \cdots \eta_n \tag{4.20}$$

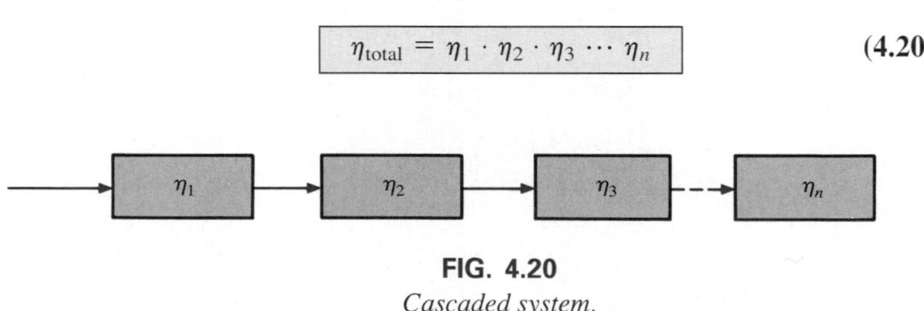

FIG. 4.20
Cascaded system.

EXAMPLE 4.13 Find the overall efficiency of the system of Fig. 4.19 if $\eta_1 = 90\%$, $\eta_2 = 85\%$, and $\eta_3 = 95\%$.

Solution:

$$\eta_T = \eta_1 \cdot \eta_2 \cdot \eta_3 = (0.90)(0.85)(0.95) = 0.727, \text{ or } \mathbf{72.7\%}$$

EXAMPLE 4.14 If the efficiency η_1 drops to 40%, find the new overall efficiency and compare the result with that obtained in Example 4.13.

Solution:

$$\eta_T = \eta_1 \cdot \eta_2 \cdot \eta_3 = (0.40)(0.85)(0.95) = 0.323, \text{ or } \mathbf{32.3\%}$$

Certainly 32.3% is noticeably less than 72.7%. The total efficiency of a cascaded system is therefore determined primarily by the lowest efficiency (weakest link) and is less than (or equal to if the remaining efficiencies are 100%) the least efficient link of the system.

4.6 ENERGY

For power, which is the rate of doing work, to produce an energy conversion of any form, it must be *used over a period of time.* For example, a motor may have the horsepower to run a heavy load, but unless the motor is *used* over a period of time, there will be no energy conversion. In addition, the longer the motor is used to drive the load, the greater will be the energy expended.

The energy lost or gained by any system is therefore determined by

$$W = Pt \qquad \text{(wattseconds, Ws, or joules)} \tag{4.21}$$

Since power is measured in watts (or joules per second) and time in seconds, the unit of energy is the *wattsecond* or *joule* (note Fig. 4.21) as indicated above. The wattsecond, however, is too small a quantity for most practical purposes, so the *watthour* (Wh) and *kilowatthour* (kWh) were defined, as follows:

$$\text{Energy (Wh)} = \text{power (W)} \times \text{time (h)} \tag{4.22}$$

$$\text{Energy (kWh)} = \frac{\text{power (W)} \times \text{time (h)}}{1000} \tag{4.23}$$

Note that the energy in kilowatthours is simply the energy in watthours divided by 1000. To develop some sense for the kilowatthour energy level, consider that 1 kWh is the energy dissipated by a 100-W bulb in 10 h.

The *kilowatthour meter* is an instrument for measuring the energy supplied to the residential or commercial user of electricity. It is normally connected directly to the lines at a point just prior to entering the power distribution panel of the building. A typical set of dials is shown in Fig. 4.22(a) with a photograph of an analog kilowatthour meter. As indicated, each power of 10 below a dial is in kilowatthours. The more rapidly the aluminum disc rotates, the greater the energy demand. The dials are connected through a set of gears to the rotation of this disc. A solid-state digital meter with an extended range of capabilities appears in Fig. 4.22(b).

British (Salford, Manchester)
(1818–1889)
Physicist
Honorary doctorates from the Universities of Dublin and Oxford

Bettmann Archive Photo Number 076800P

Contributed to the important fundamental *law of conservation of energy* by establishing that various forms of energy, whether they be electrical, mechanical, or heat, are in the same family and can be exchanged from one form to another. In 1841 introduced *Joule's law,* which stated that the heat developed by electric current in a wire is proportional to the product of the current squared and the resistance of the wire (I^2R). He further determined that the heat emitted was equivalent to the power absorbed and therefore heat is a form of energy.

FIG. 4.21
James Prescott Joule.

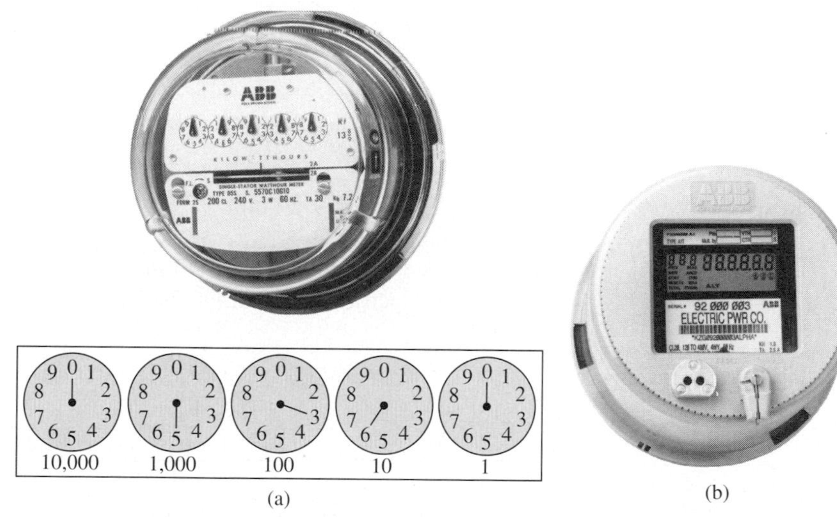

FIG. 4.22
Kilowatthour meters: (a) analog; (b) digital (Courtesy of ABB Electric Metering Systems.)

EXAMPLE 4.15 For the dial positions of Fig. 4.22(a), calculate the electricity bill if the previous reading was 4650 kWh and the average cost is 9¢ per kilowatthour.

Solution:

$$5360 \text{ kWh} - 4650 \text{ kWh} = 710 \text{ kWh used}$$

$$710 \cancel{\text{kWh}}\left(\frac{9¢}{\cancel{\text{kWh}}}\right) = \mathbf{\$63.90}$$

EXAMPLE 4.16 How much energy (in kilowatthours) is required to light a 60-W bulb continuously for 1 year (365 days)?

Solution:

$$W = \frac{Pt}{1000} = \frac{(60 \text{ W})(24 \text{ h/day})(365 \text{ days})}{1000} = \frac{525{,}600 \text{ Wh}}{1000}$$
$$= \mathbf{525.60\ kWh}$$

EXAMPLE 4.17 How long can a 205-W television set be on before using more than 4 kWh of energy?

Solution:

$$W = \frac{Pt}{1000} = \frac{(W)(1000)}{P} \Rightarrow t \text{ (hours)} = \frac{(W)(1000)}{P}$$
$$= \frac{(4 \text{ kWh})(1000)}{205 \text{ W}} = \mathbf{19.51\ h}$$

EXAMPLE 4.18 What is the cost of using a 5-hp motor for 2 h if the rate is 9¢ per kilowatthour?

Solution:

$$W \text{ (kilowatthours)} = \frac{Pt}{1000} = \frac{(5 \text{ hp} \times 746 \text{ W/hp})(2 \text{ h})}{1000} = 7.46 \text{ kWh}$$
$$\text{Cost} = (7.46 \text{ kWh})(9¢/\text{kWh}) = \mathbf{67.14¢}$$

EXAMPLE 4.19 What is the total cost of using all of the following at 9¢ per kilowatthour?

A 1200-W toaster for 30 min
Six 50-W bulbs for 4 h
A 400-W washing machine for 45 min
A 4800-W electric clothes dryer for 20 min

Solution:

$$W = \frac{(1200 \text{ W})(\frac{1}{2} \text{ h}) + (6)(50 \text{ W})(4 \text{ h}) + (400 \text{ W})(\frac{3}{4} \text{ h}) + (4800 \text{ W})(\frac{1}{3} \text{ h})}{1000}$$
$$= \frac{600 \text{ Wh} + 1200 \text{ Wh} + 300 \text{ Wh} + 1600 \text{ Wh}}{1000} = \frac{3700 \text{ Wh}}{1000}$$
$$W = 3.7 \text{ kWh}$$
$$\text{Cost} = (3.7 \text{ kWh})(9¢/\text{kWh}) = \mathbf{33.3¢}$$

The chart in Fig. 4.23 shows the average cost per kilowatthour as compared to the kilowatthours used per customer. Note that the cost today is above the level of 1926 and the average customer uses more than 20 times as much electrical energy in a year. Keep in mind that the chart of Fig. 4.23 is the average cost across the nation. Some states have average rates close to 5¢ per kilowatthour, whereas others approach 12¢ per kilowatthour.

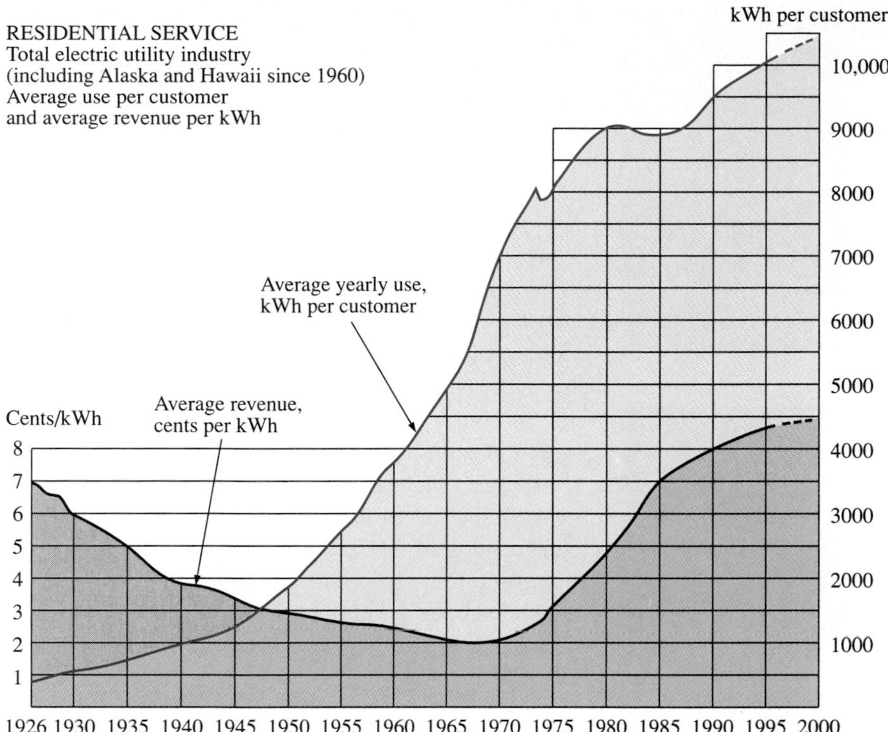

FIG. 4.23

Cost per kWh and average kWh per customer versus time. (Courtesy of Edison Electric Institute.)

Table 4.1 lists some common household items with their typical wattage ratings. It might prove interesting for the reader to calculate the cost of operating some of these appliances over a period of time using the chart in Fig. 4.23 to find the cost per kilowatthour.

TABLE 4.1

Typical wattage ratings of some common household appliances.

Appliance	Wattage Rating	Appliance	Wattage Rating
Air conditioner	860	Microwave oven	800
Blow dryer	1,300	Phonograph	75
Cassette player/		Projector	1,200
recorder	5	Radio	70
Clock	2	Range (self-cleaning)	12,200
Clothes dryer (electric)	4,800	Refrigerator (automatic	
Coffee maker	900	defrost)	1,800
Dishwasher	1,200	Shaver	15
Fan:		Stereo equipment	110
Portable	90	Sun lamp	280
Window	200	Toaster	1,200
Heater	1,322	Trash compactor	400
Heating equipment:		TV (color)	250
Furnace fan	320	Videocassette recorder	110
Oil-burner motor	230	Washing machine	400
Iron, dry or steam	1,100	Water heater	2,500

Courtesy of General Electric Co.

4.7 CIRCUIT BREAKERS, GFCIs, AND FUSES

The incoming power to any large industrial plant, heavy equipment, simple circuit in the home, or meters used in the laboratory must be limited to ensure that the current through the lines is not above the rated value. Otherwise, the conductors or the electrical or electronic equipment may be damaged or dangerous side effects such as fire or smoke may result. To limit the current level, fuses or circuit breakers are installed where the power enters the installation, such as in the panel in the basement of most homes at the point where the outside feeder lines enter the dwelling. The fuses of Fig. 4.24 have an internal metallic conductor through which the current will pass; it will begin to melt if the current through the system exceeds the rated value printed on the casing. Of course, if it melts through, the current path is broken and the load in its path is protected.

(a)

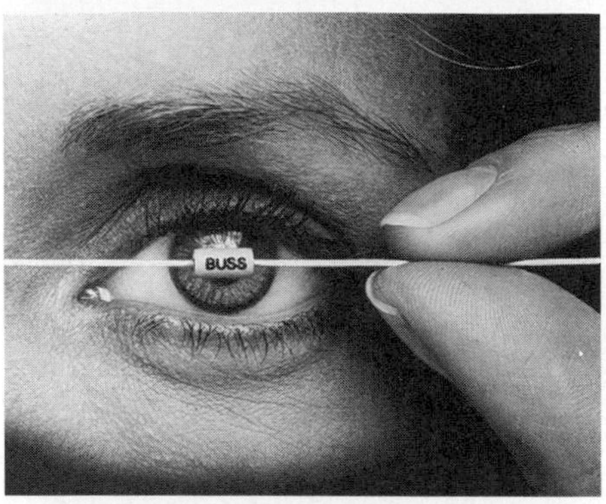

(b)

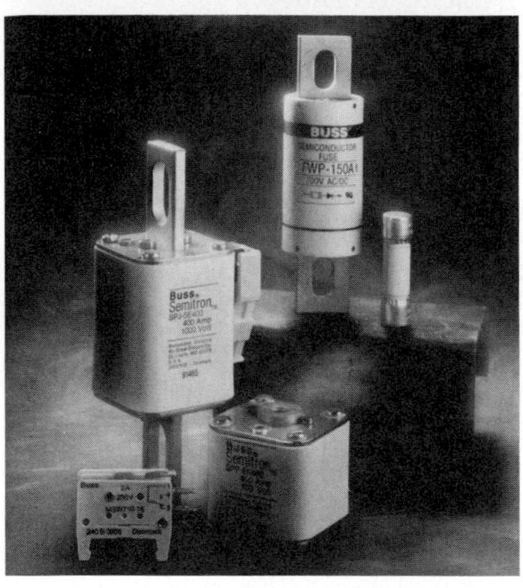

(c)

FIG. 4.24
Fuses: (a) CC-TRON® (0–10 A); (b) subminiature solid matrix; (c) Semitron (0–600 A). (Courtesy of Bussman Manufacturing Co.)

In homes built in recent years, fuses have been replaced by circuit breakers such as those appearing in Fig. 4.25. When the current exceeds rated conditions, an electromagnet in the device will have sufficient strength to draw the connecting metallic link in the breaker out of the circuit and open the current path. When conditions have been corrected, the breaker can be reset and used again.

FIG. 4.25
Circuit breakers. (Courtesy of Potter and Brumfield Division, AMF, Inc.)

The current National Electrical Code requires that outlets in the bathroom and other sensitive areas be of the ground fault current interrupt (GFCI) variety. They are designed to trip more quickly than the standard circuit breaker. The units in Fig. 4.26 trip in 500 ms ($\frac{1}{2}$ s). It has been determined that 6 mA is the maximum level that most individuals should be exposed to for a short period of time and not be seriously injured. A current higher than 11 mA can cause involuntary muscle contractions that could prevent a person from letting go of the conductor and possibly cause him or her to enter a state of shock. Higher currents lasting more than a second can cause the heart to go into fibrillation and possibly cause death in a few minutes. The GFCI is able to react as quickly as it does by sensing the difference between the input and output currents to the outlet. They should be the same if everything is working properly. An errant path such as through an individual establishes a difference in the two current levels and causes the breaker to trip and disconnect the power source.

4.8 COMPUTER ANALYSIS

This section will begin to reveal the differences between a language and a software package by examining the use of Ohm's law in its three forms. Using a programming language, it is possible to write a program that will

ask the user which quantity he or she would like to calculate, ask for the other two known quantities, and solve for the desired unknown—in fact, the program to be described shortly will perform all of the above to a high degree of accuracy and reliability. In PSpice, however, the package is designed primarily to solve for specific voltages and currents. There is no built-in facility to question the user about his or her interests or to input specific equations into the analysis routines—they are cast in stone within the package. In total, therefore, the software package is designed to perform specific tasks and provide output data in a variety of forms, whereas a language can be used to establish an interactive mode with the user and provide data in any format desired by the user.

The PSpice analysis to follow will be limited to finding the current through a resistor using Ohm's law, whereas a BASIC program (to follow) will determine *I*, *V*, or *R* depending on the input data.

FIG. 4.26
Ground fault current interrupter (GFCI) 125 V ac, 60 Hz, 15-A outlet. (Courtesy of Leviton, Inc.)

PSpice (DOS)

The earlier sections on computers were limited to the input format for dc voltage sources and resistors. We now look at the basic format of the *input file* (the name given to the lines of information fed into the computer) and how the data will appear in the *output file* (the manner in which the package will display the desired results).

The input file *must* contain three components as appearing in Fig. 4.27—the *title line*, the *network description*, and the *.END statement*. As indicated in Fig. 4.27, the first line *must* be the title line. It is constructed of characters (letters, numbers, and so on) in any format that normally catalogues the file for future reference and reveals the analysis to be performed. The next necessary component of the input file is the network description. Elements can be entered in any order as long as the nodes are properly defined. It is always advisable to label all the nodes of the network before beginning the input process. It is vitally important, however, to remember that

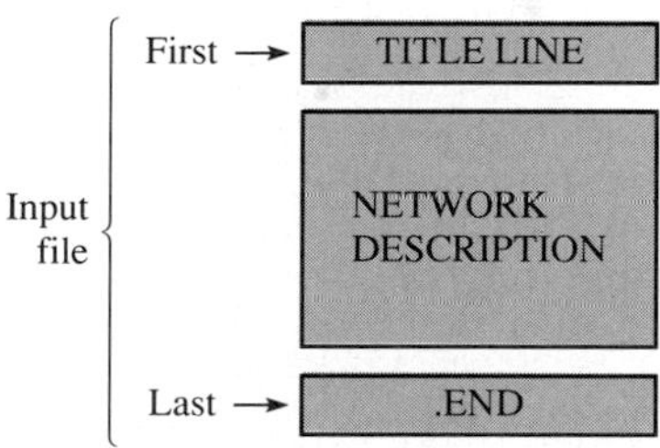

FIG. 4.27
PSpice input file.

in every network one node must be designated as the reference node and assigned the number 0. It is usually the node connected to ground potential (zero volts), with all other nodal voltages taken with reference to this defined reference level.

For example, the voltage output by the computer for node 1 is the voltage from node 1 to ground. A voltage V(8) is the voltage from node 8 to node 0 (or ground potential). Unless specified, PSpice will automatically generate the nodal voltages as defined above.

The last line of any input file must be the .END command, to let the software package know that the network is fully described and all the information the package requires has been provided. *Don't forget* the period before the .END statement. Its absence will invalidate the entry.

Let us now look at the analysis of the simple network of Fig. 4.28, with one voltage source and a resistor. There are two nodes (or connection points) between the two elements. As required, one is specified as the reference level 0 (in this case, the ground or zero-volt level), and the other, node 1. Note the relative simplicity of the input file, with a title line, two element entries, and an .END statement.

```
Chapter 4-Ohm's Law
VE 1 0 12V
R 1 0 4K
.END
```

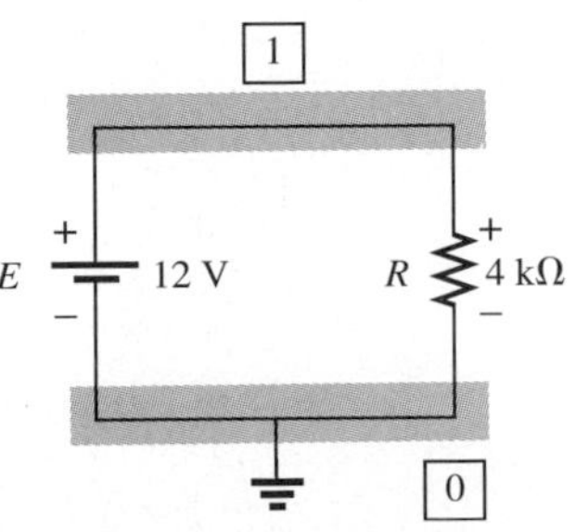

FIG. 4.28
Network to be analyzed using PSpice (DOS).

Once the input file is entered and saved, PSpice can be run, and the output file of Fig. 4.29 will result. Note how the title line and network description have been placed within the framework of a heading controlled by PSpice. The title CIRCUIT DESCRIPTION simply heads the network description to follow, and the SMALL-SIGNAL BIAS SOLUTION simply reveals that a solution is to follow. The heading SMALL SIGNAL will have more meaning when you begin your electronics courses, but for now simply relate the phase BIAS SOLUTION to mean dc solution.

```
Chapter 4 - Ohm's Law

 ****     CIRCUIT DESCRIPTION

******************************************************************************

VE 1 0 12V
R  1 0 4K
.END
```

```
Chapter 4 - Ohm's Law

 ****     SMALL SIGNAL BIAS SOLUTION       TEMPERATURE =   27.000 DEG C

******************************************************************************

NODE   VOLTAGE     NODE   VOLTAGE     NODE   VOLTAGE     NODE   VOLTAGE

(    1)   12.0000


    VOLTAGE SOURCE CURRENTS
    NAME            CURRENT

    VE             -3.000E-03

    TOTAL POWER DISSIPATION   3.60E-02  WATTS
```

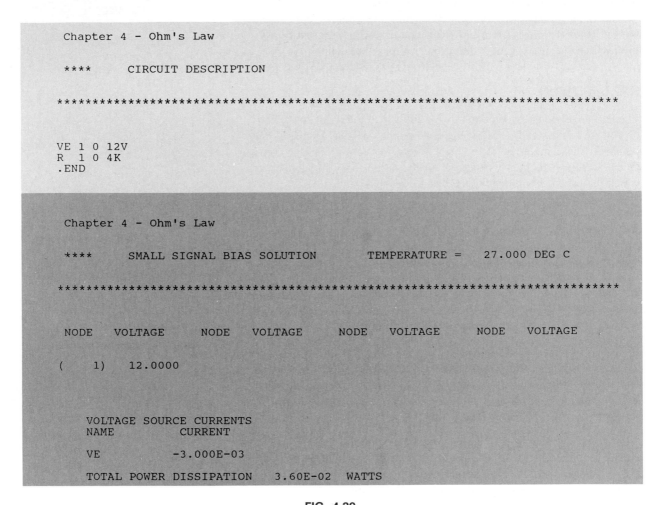

FIG. 4.29
Input and output files for the analysis of the network of Fig. 4.28.

In the solution, note that there is only one nodal voltage provided since there is only one in the original network description. The current provided is the voltage source current, but since the current through the battery of Fig. 4.28 is the same as that through the resistor, we have the desired current through the resistor. For all elements of a network, PSpice defines the direction of the current as from the + node to the − node. Since the current through the source is opposite to this defined direction, a minus sign appears in the printout. Using powers of 10, the resulting current is 3 mA. In this network, since the power supplied by the battery is equal to the power absorbed by the resistor, the power

level listed can be applied to either element. Its magnitude is 36 mW. The final line of the output file provides the time in seconds required to perform the necessary analysis.

A savings in paper and space can be accomplished by adding a control line .OPTIONS NOPAGE to the input file. It will eliminate repetitious items from the output file.

PSpice (Windows)

Since this is the first circuit to be analyzed using schematics, the details of obtaining the required elements will be repeated in detail. The newest version of PSpice (Windows) has icons across the header of the display for performing a variety of common operations. However, because older versions are still used, the approach throughout the text will be to use the standard paths for obtaining and placing elements and controlling the output display.

The chosen circuit will be the same as the one appearing in the PSpice (DOS) analysis to permit a comparison between the two approaches. The voltage source of Fig. 4.30 is obtained using the following sequence: **Draw-Get New Part-Browse-source.slb-VDC,** which can then be clicked (left click) in place. Since only one source is present, the operation can be ended with a right click of the mouse. By double-clicking the label **V1** on the screen, the **Edit Reference Designator** box, in which the chosen label E can be entered, will appear followed by **OK.** One more left click of the mouse and the defining box structures disappear and the new label is set. The source value can then be set with a double-click of the **0 V** on the screen and entering 12 V in the **Set Attribute Value** dialog box. Click **OK** and the 12 V will replace the **0 V** on the display. One more left click will clear the screen of the defining rectangular areas. If the new label or value has to be moved, simply left click once on the E or 12 V to obtain the small enclosure around the quantity. Left click on the box, move to the desired location, and release the clicker. Left click once more to clear the screen of the control boxes.

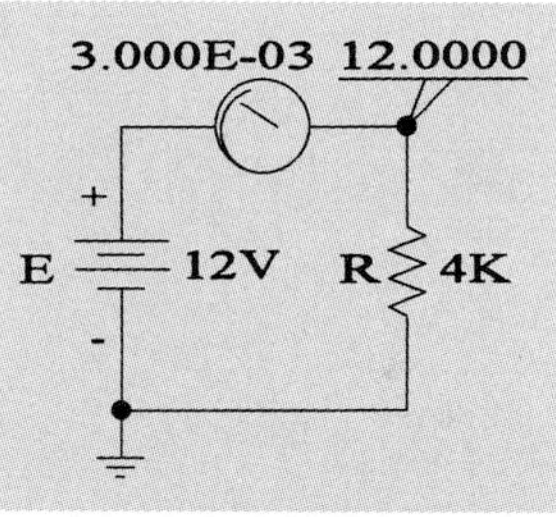

FIG. 4.30
PSpice (Windows) application of Ohm's law.

The resistive element is obtained by **Draw-Get New Part-Browse-analog.slb-R-OK** and placed as described for the voltage source. Its value and label can then be changed as described for the voltage source. Turning the resistor is accomplished by holding down the **Ctrl** key and pressing the **R** key. Each click of the **R** key will rotate the resistor 90°. This combination of **Ctrl-R** can be applied to any element.

The elements can then be connected using **Draw-Wire,** which results in a pencil on the display that will be used to "draw" the required connections. With the pencil at the starting point, left click once to initiate the wire. Move the pencil across the desired path and end the placement with a right click. The pencil will still be present for other connections that need to be made. An additional right click will end the process.

A ground connection must be added to every network to provide a reference for all the voltages that will appear on the display. The standard earth ground is obtained through **Draw-Get New Part-Browse-port.slb-EGND-OK** and placed as above.

Finally, **Draw-Get New Part-Browse-special.slb-VIEWPOINT** (or **IPROBE**) will result in a marker that can be placed as shown in Fig. 4.30 to display the voltage at that point when an analysis is initiated. **IPROBE** will result in the meter symbol shown in Fig. 4.30 to

display the current in the branch in which it is located. It is important to note the manner in which the **IPROBE** symbol is entered on the schematic. For an upscale (positive) reading, the symbol should be inserted to ensure that current enters the symbol on the circular scale side and leaves the symbol on the pointer side. Otherwise, a negative sign will result for the measured current.

Before analyzing the network, the schematic must be saved. **Analysis-Setup** results in the **Analysis Setup** dialog box, in which **Bias Point Detail** should be enabled. **Close-Analysis-Simulate** executes the analysis. Shortly thereafter, a PSpice dialog box appears, reviewing the activities under way. Once complete, the output response can be obtained through **File-Examine Output,** which will include a long list of information about the structure of the circuit, including the **SMALL SIGNAL BIAS SOLUTION** appearing in Fig. 4.31. Note that since the voltage across the source and resistor is the same, the 12-V level appears twice. The voltage source current appears below with the current through the resistor. Note again that they are both the same, but one is negative and the other positive. This is a result of the fact that current is defined as flowing from the positive to the negative terminal of an element. For the voltage source, therefore, the defined direction is opposite to the actual direction and minus sign results. Finally the output file also provides the total power dissipated by the network, which in this case also equals the power supplied by the battery.

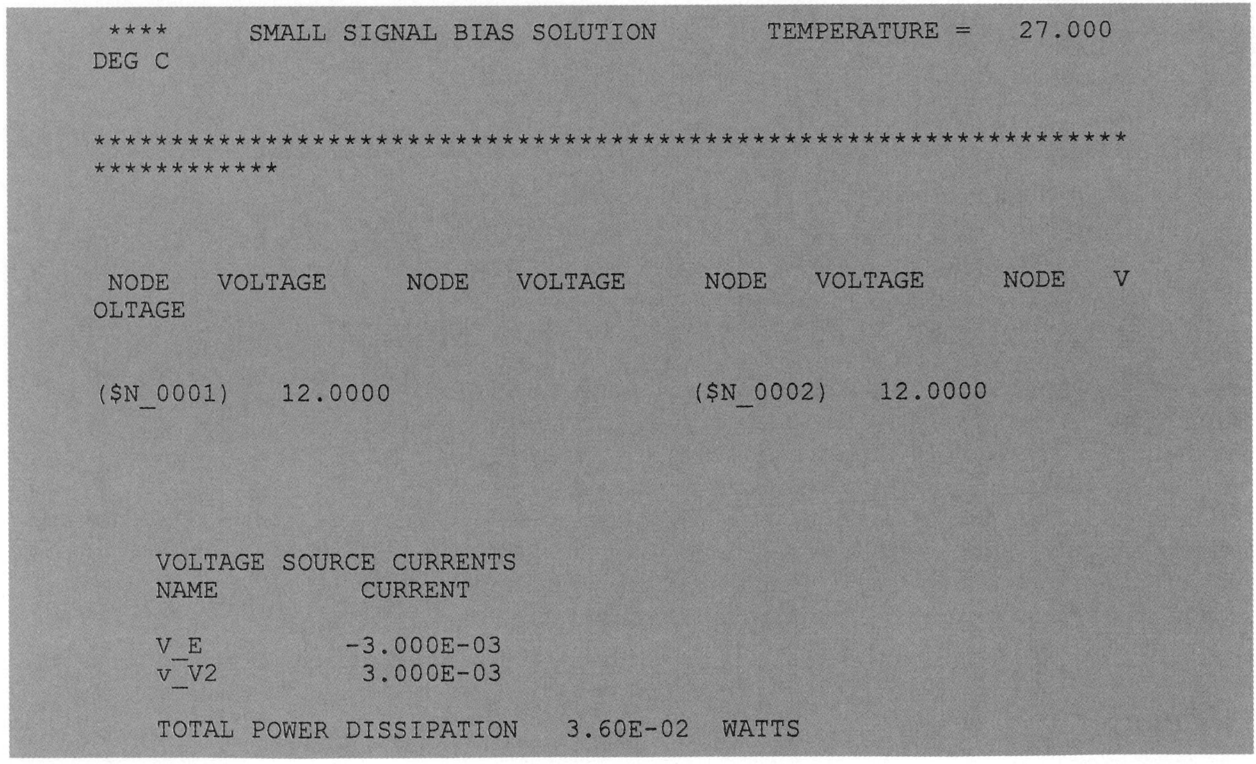

```
 ****     SMALL SIGNAL BIAS SOLUTION        TEMPERATURE =   27.000
DEG C

***********************************************************************
************



 NODE   VOLTAGE     NODE   VOLTAGE     NODE   VOLTAGE     NODE   V
OLTAGE


($N_0001)   12.0000                   ($N_0002)   12.0000




    VOLTAGE SOURCE CURRENTS
    NAME          CURRENT

    V_E          -3.000E-03
    v_V2          3.000E-03

    TOTAL POWER DISSIPATION   3.60E-02  WATTS
```

FIG. 4.31
Output file for the PSpice (Windows) analysis of the network of Fig. 4.30.

Removing the PSpice dialog box will result in the display of Fig. 4.30, with the voltage indicated at the **VIEWPOINT** and the current level of 3 mA indicated by **IPROBE.** In most cases, a dc analysis will

employ **VIEWPOINT**s and **IPROBE**s to display the desired results, with the output file reserved for a careful review of defined nodes, etc., as appearing in the DOS approach.

Programming Language (C++, BASIC, PASCAL, etc.)

The first complete program in BASIC is provided in Fig. 4.32. Everything you see is either entered into the computer or printed out by the computer, except the brackets at the left of the line numbers with the printed explanation of the function performed by those sections. First note that the commands, statements, and so on, found on each line use English words and phrases to indicate the operation to be performed. The REM statement (from the word *REMark*) simply indicates that a descriptive statement is being made about the program. The PRINT statement tells the system to print the characters between the quotes on the screen. The INPUT commands request the value of the variable appearing on the same line. The equations on lines 440, 640, and 840 carry out the operation to be performed in each module defined by the brackets at the left of the program. Lines 400 through 460 will request *I* and *R* and calculate the voltage. Line 450 will print the solution, while

```
10 REM ***** PROGRAM 4-1 *****
20 REM ******************************************
30 REM Program demonstrates selecting various forms
40 REM of equations
50 REM ******************************************
60 REM
```

Equation Selection

```
100 PRINT : PRINT "Select which form of Ohm's law equation"
110 PRINT "you wish to use."
120 PRINT
130 PRINT TAB(10); "(1) V=I*R"
140 PRINT TAB(10); "(2) I=V/R"
150 PRINT TAB(10); "(3) R=V/I"
160 PRINT TAB(20);
170 INPUT "choice="; C
180 IF C < 1 OR C > 3 THEN GOTO 100
190 ON C GOSUB 400, 600, 800
200 PRINT : PRINT
```

Continue?

```
210 INPUT "More (YES or NO)"; A$
220 IF A$ = "YES" THEN 100
230 PRINT "Have a good day"
240 END
```

$V = IR$

```
400 REM Accept input of I,R and output V
410 PRINT : PRINT "Enter the following data:"
420 INPUT "I="; I
430 INPUT "R="; R
440 V = I * R
450 PRINT "Voltage is "; V; "volts"
460 RETURN
```

$I = \frac{V}{R}$

```
600 REM Accept input of V,R and output I
610 PRINT "Enter the following data:"
620 INPUT "V="; V
630 INPUT "R="; R
640 I = V / R
650 PRINT "Current is "; I; "amperes"
660 RETURN
```

$R = \frac{V}{I}$

```
800 REM Accept input of V,I and output R
810 PRINT "Enter the following data:"
820 INPUT "V="; V
830 INPUT "I="; I
840 R = V / I
850 PRINT "Resistance is "; R; "ohms"
860 RETURN
```

FIG. 4.32
Program 4.1.

```
Select which form of Ohm's law equation
you wish to use.

         (1) V=I*R
         (2) I=V/R
         (3) R=V/I
                   choice=? 2
Enter the following data:
V=? 12
R=? 4E3
Current is  .003 amperes


More (YES or NO)? YES

Select which form of Ohm's law equation
you wish to use.

         (1) V=I*R
         (2) I=V/R
         (3) R=V/I
                   choice=? 1

Enter the following data:
I=? 2E-3
R=? 5.6E3
Voltage is  11.2 volts


More (YES or NO)? YES

Select which form of Ohm's law equation
you wish to use.

         (1) V=I*R
         (2) I=V/R
         (3) R=V/I
                   choice=? 3
Enter the following data:
V=? 48
I=? 0.025
Resistance is  1920 ohms


More (YES or NO)? NO
Have a good day
```

FIG. 4.32
(continued)

line 460 will "return" the program to line 210 to determine if a second calculation is to be performed. The module from line 600 to line 660 will calculate the current *I*, and the module from line 800 to line 860 will determine the resistance from the input voltage and current. Lines 100 through 200 permit a selection of the form of Ohm's law to be applied. Three runs of the program are provided in the figure to reveal the ormat of the request for data and the output response.

If you start with line (location) 10 and perform the operations indicated for each succeeding line, you will find that the program proceeds in much the same manner as if done by longhand. For instance, line 170 asks for your choice of which form of Ohm's law you would like to apply from lines 130 through 160. If the choice is C = 2 (I = V/R), line 190 will define line 600 as the next line to be performed. At 600 a sequence of lines follows that calculates I and prints out the result. The RETURN statement brings us back to line 200 for a two-line space (PRINT:PRINT), followed by the question of whether we want to continue. If so, we return to line 100 for a second choice of equations.

This longer section on computers begins to clarify the difference between the use of a language and a software package. Using a language requires that the entire program be written line by line, although a program does provide options on the type of analysis that can be performed and an interactive mode between user and machine. Using software provides the benefits of the many hours of writing and testing performed by a team of experienced programmers but limits the type of analysis we can perform and specifies the form of the input and output information.

PROBLEMS

SECTION 4.1 Ohm's Law

1. What is the potential drop across a 6-Ω resistor if the current through it is 2.5 A?
2. What is the current through a 72-Ω resistor if the voltage drop across it is 12 V?
3. How much resistance is required to limit the current to 1.5 mA if the potential drop across the resistor is 6 V?
4. At starting, what is the current drain on a 12-V car battery if the resistance of the starting motor is 0.056 Ω?
5. If the current through a 0.02-MΩ resistor is 3.6 μA, what is the voltage drop across the resistor?
6. If a voltmeter has an internal resistance of 15 kΩ, find the current through the meter when it reads 62 V.
7. If a refrigerator draws 2.2 A at 120 V, what is its resistance?
8. If a clock has an internal resistance of 7.5 kΩ, find the current through the clock if it is plugged into a 120-V outlet.
9. A washing machine is rated at 4.2 A at 120 V. What is its internal resistance?
10. If a soldering iron draws 0.76 A at 120 V, what is its resistance?
11. The input current to a transistor is 20 μA. If the applied (input) voltage is 24 mV, determine the input resistance of the transistor.
12. The internal resistance of a dc generator is 0.5 Ω. Determine the loss in terminal voltage across this internal resistance if the current is 15 A.

*13. a. If an electric heater draws 9.5 A when connected to a 120-V supply, what is the internal resistance of the heater?
    b. Using the basic relationships of Chapter 2, how much energy is converted in 1 h?

SECTION 4.2 Plotting Ohm's Law

14. Plot the linear curves of a 100-Ω and 0.5-Ω resistor on the graph of Fig. 4.6. If necessary, reproduce the graph.
15. Sketch the characteristics of a device that has an internal resistance of 20 Ω from 0 V to 10 V and an internal resistance of 2 Ω for higher voltages. Use the axes of Fig. 4.6. If necessary, reproduce the graph.
16. Plot the linear curves of a 2-kΩ and 50-kΩ resistor on the graph of Fig. 4.6. Use a horizontal scale that extends from 0 V to 20 V and a vertical axis scaled off in milliamperes. If necessary, reproduce the graph.
17. What is the change in voltage across a 2-kΩ resistor established by a change in current of 400 mA through the resistor?

*18. a. Using the axis of Fig. 4.10, sketch the characteristics of a device that has an internal resistance of 500 Ω from 0 V to 1 V and 50 Ω between 1 and 2 V. Its resistance then changes to −20 Ω for higher voltages. The result is a set of characteristics very similar to those of an electronic device called a tunnel diode.
    b. Using the above characteristics, determine the resulting current at voltages of 0.7 V, 1.5 V, and 2.5 V.

SECTION 4.3 Power

19. If 420 J of energy are absorbed by a resistor in 7 min, what is the power to the resistor?
20. The power to a device is 40 joules per second (J/s). How long will it take to deliver 640 J?
21. a. How many joules of energy does a 2-W nightlight dissipate in 8 h?
    b. How many kilowatthours does it dissipate?
22. A resistor of 10 Ω has charge flowing through it at the rate of 300 coulombs per minute (C/min). How much power is dissipated?
23. How long must a steady current of 2 A exist in a resistor that has 3 V across it to dissipate 12 J of energy?
24. What is the power delivered by a 6-V battery if the charge flows at the rate of 48 C/min?
25. The current through a 4-Ω resistor is 7 mA. What is the power delivered to the resistor?
26. The voltage drop across a 3-Ω resistor is 9 mV. What is the power input to the resistor?
27. If the power input to a 4-Ω resistor is 64 W, what is the current through the resistor?
28. A 1/2-W resistor has a resistance of 1000 Ω. What is the maximum current that it can safely handle?
29. A 2.2-kΩ resistor in a stereo system dissipates 42 mW of power. What is the voltage across the resistor?
30. A dc battery can deliver 45 mA at 9 V. What is the power rating?

31. What are the "hot" resistance level and current rating of a 120-V, 100-W bulb?

32. What are the internal resistance and voltage rating of a 450-W automatic washer that draws 3.75 A?

33. A calculator with an internal 3-V battery draw 0.4 mW when fully functional.
   a. What is the current demand from the supply?
   b. If the calculator is rated to operate 500 h on the same battery, what is the ampere-hour rating of the battery?

34. A 20-kΩ resistor has a rating of 100 W. What are the maximum current and the maximum voltage that can be applied to the resistor?

*35. a. Plot power versus current for a 100-Ω resistor. Use a power scale from 0 to 1 W and a current scale from 0 to 100 mA with divisions of 0.1 W and 10 mA, respectively.
   b. Is the curve linear or nonlinear?
   c. Using the resulting plot, determine the current at a power level of 500 mW.

*36. A small portable black-and-white television draws 0.455 A at 9 V.
   a. What is the power rating of the television?
   b. What is the internal resistance of the television?
   c. What is the energy converted in 6 h of typical battery life?

*37. a. If a home is supplied with a 120-V, 100-A service, find the maximum power capability.
   b. Can the homeowner safely operate the following loads at the same time?
   A 5-hp motor
   A 3000-W clothes dryer
   A 2400-W electric range
   A 1000-W steam iron

SECTION 4.5 Efficiency

38. What is the efficiency of a motor that has an output of 0.5 hp with an input of 450 W?

39. The motor of a power saw is rated 68.5% efficient. If 1.8 hp are required to cut a particular piece of lumber, what is the current drawn from a 120-V supply?

40. What is the efficiency of a dryer motor that delivers 1 hp when the input current and voltage are 4 A and 220 V, respectively?

41. A stereo system draws 2.4 A at 120 V. If the audio output power is 50 W,
   a. How much power is lost in the form of heat in the system?
   b. What is the efficiency of the system?

42. If an electric motor having an efficiency of 87% and operating off a 220-V line delivers 3.6 hp, what input current does the motor draw?

43. A motor is rated to deliver 2 hp.
   a. If it runs on 110 V and is 90% efficient, how many watts does it draw from the power line?
   b. What is the input current?
   c. What is the input current if the motor is only 70% efficient?

44. An electric motor used in an elevator system has an efficiency of 90%. If the input voltage is 220 V, what is the input current when the motor is delivering 15 hp?

45. A 2-hp motor drives a sanding belt. If the efficiency of the motor is 87% and that of the sanding belt 75% due to slippage, what is the overall efficiency of the system?

46. If two systems in cascade each have an efficiency of 80% and the input energy is 60 J, what is the output energy?

47. The overall efficiency of two systems in cascade is 72%. If the efficiency of one is 0.9, what is the efficiency in percent of the other?

*48. If the total input and output power of two systems in cascade are 400 W and 128 W, respectively, what is the efficiency of each system if one has twice the efficiency of the other?

49. a. What is the total efficiency of three systems in cascade with efficiencies of 98%, 87%, and 21%?
   b. If the system with the least efficiency (21%) were removed and replaced by one with an efficiency of 90%, what would be the percentage increase in total efficiency?

50. a. Perform the following conversions:
   1 Wh to joules
   1 kWh to joules
   b. Based on the results of part (a), discuss when it is more appropriate to use one unit versus the other.

SECTION 4.6 Energy

51. A 10-Ω resistor is connected across a 15-V battery.
   a. How many joules of energy will it dissipate in 1 min?
   b. If the resistor is left connected for 2 min instead of 1 min, will the energy used increase? Will the power dissipation level increase?

52. How much energy in kilowatthours is required to keep a 230-W oil-burner motor running 12 h a week for 5 months?

53. How long can a 1500-W heater be on before using more than 10 kWh of energy?

54. How much does it cost to use a 30-W radio for 3 h at 8¢ per kilowatthour?

55. a. In 10 h an electrical system converts 500 kWh of electrical energy into heat. What is the power level of the system?
   b. If the applied voltage is 208 V, what is the current drawn from the supply?
   c. If the efficiency of the system is 82%, how much energy is lost or stored in 10 h?

56. a. At 9¢ per kilowatthour, how long can one play a 250-W color television for $1?
   b. For $1, how long can one use a 4.8-kW dryer?
   c. Compare the results of parts (a) and (b) and comment on the effect of the wattage level on the relative cost of using an appliance.

57. What is the total cost of using the following at 9¢ per kilowatthour?

860-W air conditioner for 24 h
4800-W clothes dryer for 30 min
400-W washing machine for 1 h
1200-W dishwasher for 45 min

*58. What is the total cost of using the following at 9¢ per kilowatthour?
110-W stereo set for 4 h
1200-W projector for 20 min
60-W tape recorder for 1.5 h
150-W color television set for 3 h 45 min

SECTION 4.8 Computer Analysis

PSpice (DOS)

59. Write an input file for a network like the one shown in Fig. 4.26 with $E = 400$ mV and $R = 0.04$ MΩ. Use the appropriate control line to limit the amount of paper required for the output file.

60. Write an input file for a network like the one shown in Fig. 4.26, except reverse the polarity of the battery. Use $E = 0.02$ V and $R = 240\ \Omega$. Keep the paper used for the output file to a minimum.

PSpice (Windows)

61. Repeat Problem 59 using PSpice (Windows).
62. Repeat Problem 60 using PSpice (Windows).

Programming Language (C++, BASIC, PASCAL, etc.)

63. Write a program to calculate the cost of using five different appliances for varying lengths of time if the cost is 9¢ per kilowatthour.
64. Request I, R, and t and determine V, P, and W. Print out the results with the proper units.

GLOSSARY

Circuit breaker A two-terminal device designed to ensure that current levels do not exceed safe levels. If "tripped," it can be reset with a switch or a reset button.

Diode A semiconductor device whose behavior is much like that of a simple switch; that is, it will pass current ideally in only one direction when operating within specified limits.

Efficiency (η) A ratio of output to input power that provides immediate information about the energy-converting characteristics of a system.

Energy *(W)* A quantity whose change in state is determined by the product of the rate of conversion (P) and the period involved (t). It is measured in joules (J) or wattseconds (Ws).

Fuse A two-terminal device whose sole purpose is to ensure that current levels in a circuit do not exceed safe levels.

Horsepower (hp) Equivalent to 746 watts in the electrical system.

Input file The lines of information fed into the computer to define the system to be analyzed and the operations to be performed.

Kilowatthour meter An instrument for measuring kilowatthours of energy supplied to a residential or commercial user of electricity.

Ohm's law An equation that establishes a relationship among the current, voltage, and resistance of an electrical system.

Output file The manner in which the results of a computer run are displayed.

Power An indication of how much work can be done in a specified amount of time; a *rate* of doing work. It is measured in joules/second (J/s) or watts (W).

Wattmeter An instrument capable of measuring the power delivered to an element by sensing both the voltage across the element and the current through the element.

5

Series Circuits

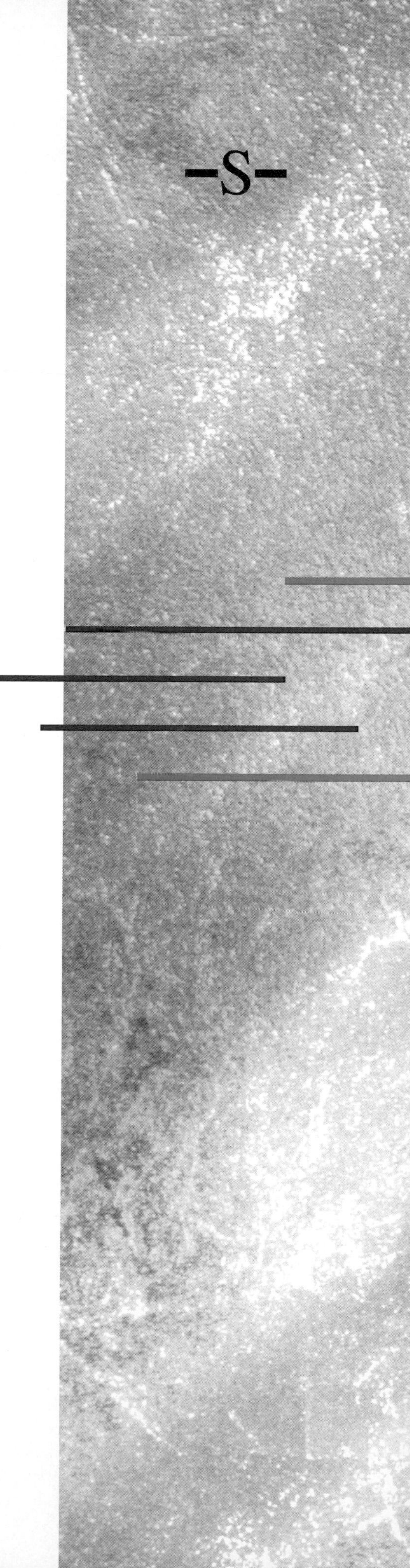

5.1 INTRODUCTION

Two types of current are readily available to the consumer today. One is *direct current* (dc), in which ideally the flow of charge (current) does not change in magnitude (or direction) with time. The other is *sinusoidal alternating current* (ac), in which the flow of charge is continually changing in magnitude (and direction) with time. The next few chapters are an introduction to circuit analysis purely from a dc approach. The methods and concepts will be discussed in detail for direct current; when possible, a short discussion will suffice to cover any variations we might encounter when we consider ac in the later chapters.

The battery of Fig. 5.1, by virtue of the potential difference between its terminals, has the ability to cause (or "pressure") charge to flow through the simple circuit. The positive terminal attracts the electrons through the wire at the same rate at which electrons are supplied by the negative terminal. As long as the battery is connected in the circuit and maintains its terminal characteristics, the current (dc) through the circuit will not change in magnitude or direction.

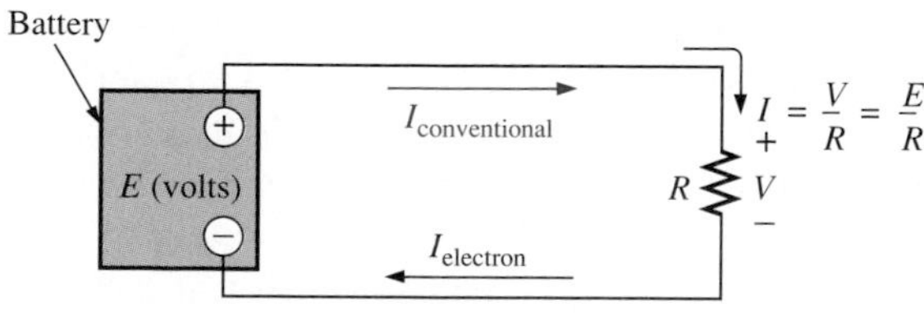

FIG. 5.1
Introducing the basic components of an electric circuit.

If we consider the wire to be an ideal conductor (that is, having no opposition to flow), the potential difference V across the resistor will equal the applied voltage of the battery: V (volts) = E (volts).

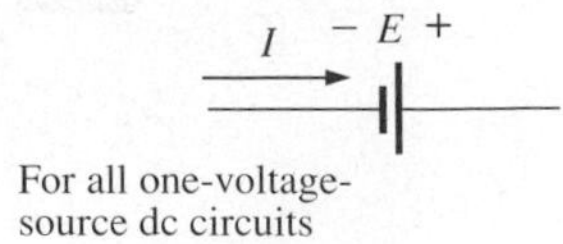

FIG. 5.2
Defining the direction of conventional flow for single-source dc circuits.

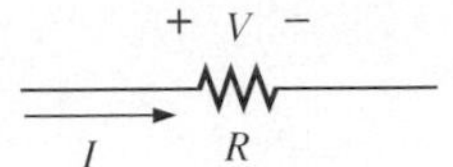

FIG. 5.3
Defining the polarity resulting from a conventional current I through a resistive element.

The current is limited only by the resistor R. The higher the resistance, the less the current, and conversely, as determined by Ohm's law.

By convention (as discussed in Chapter 2), the direction of $I_{conventional}$ as shown in Fig. 5.1 is opposite to that of electron flow ($I_{electron}$). Also, the uniform flow of charge dictates that the direct current I be the same everywhere in the circuit. By following the direction of conventional flow, we notice that there is a rise in potential across the battery (− to +), and a drop in potential across the resistor (+ to −). For single-voltage-source dc circuits, conventional flow always passes from a low potential to a high potential when passing through a voltage source, as shown in Fig. 5.2. However, conventional flow always passes from a high to a low potential when passing through a resistor for any number of voltage sources in the same circuit, as shown in Fig. 5.3.

The circuit of Fig. 5.1 is the simplest possible configuration. This chapter and the chapters to follow will add elements to the system in a very specific manner to introduce a range of concepts that will form a major part of the foundation required to analyze the most complex system. Be aware that the laws, rules, and so on, introduced in Chapters 5 and 6 will be used throughout your studies of electrical, electronic, or computer systems. They will not be dropped for a more advanced set as you progress to more sophisticated material. It is therefore critical that the concepts be understood thoroughly and that the various procedures and methods be applied with confidence.

5.2 SERIES CIRCUITS

A *circuit* consists of any number of elements joined at terminal points, providing at least one closed path through which charge can flow. The circuit of Fig. 5.4(a) has three elements joined at three terminal points (a, b, and c) to provide a closed path for the current I.

Two elements are in series if

1. ***They have only one terminal in common (i.e., one lead of one is connected to only one lead of the other).***
2. ***The common point between the two elements is not connected to another current-carrying element.***

In Fig. 5.4(a), the resistors R_1 and R_2 are in series because they have *only* point b in common. The other ends of the resistors are connected elsewhere in the circuit. For the same reason, the battery E and resistor R_1 are in series (terminal a in common) and the resistor R_2 and the battery E are in series (terminal c in common). Since all the elements are in series, the network is called a *series circuit*. Two common examples of series connections include the tying of small pieces of rope together to form a longer rope and the connecting of pipes to get water from one point to another.

If the circuit of Fig. 5.4(a) is modified such that a current-carrying resistor R_3 is introduced, as shown in Fig. 5.4(b), the resistors R_1 and R_2 are no longer in series due to a violation of part (b) of the above definition of series elements.

The current is the same through series elements.

For the circuit of Fig. 5.4(a), therefore, the current I through each resistor is the same as that through the battery. The fact that the current is

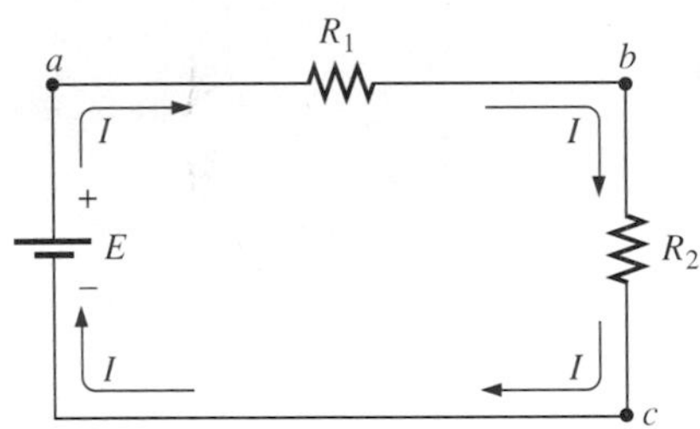

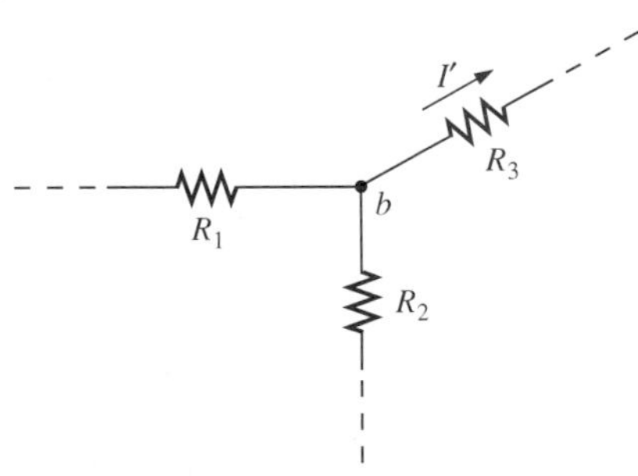

FIG. 5.4
(a) Series circuit, (b) situation in which R_1 and R_2 are not in series.

the same through series elements is often used as a path to determine whether two elements are in series or to confirm a conclusion.

A *branch* of a circuit is any portion of the circuit that has one or more elements in series. In Fig. 5.4(a), the resistor R_1 forms one branch of the circuit, the resistor R_2 another, and the battery E a third.

The total resistance of a series circuit is the sum of the resistance levels.

In Fig. 5.4(a), for example, the total resistance (R_T) is equal to $R_1 + R_2$. Note that the total resistance is actually the resistance "seen" by the battery as it "looks" into the series combination of elements as shown in Fig. 5.5.

In general, to find the total resistance of N resistors in series, the following equation is applied:

$$R_T = R_1 + R_2 + R_3 + \cdots + R_N \qquad \text{(ohms, } \Omega) \qquad \textbf{(5.1)}$$

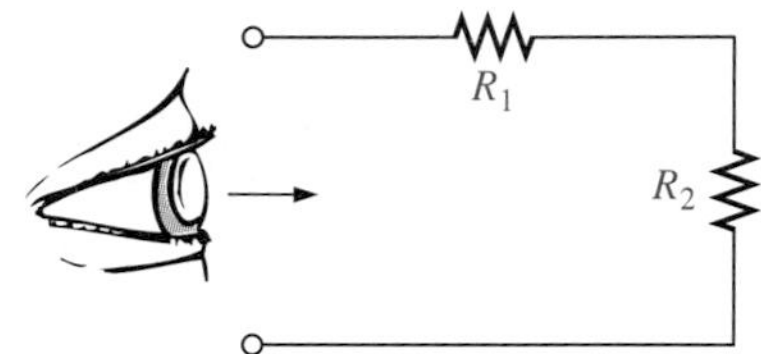

FIG. 5.5
Resistance "seen" by source.

Once the total resistance is known the circuit of Fig. 5.4(a) can be redrawn as shown in Fig. 5.6, clearly revealing that the only resistance the source "sees" is the total resistance. It is totally unaware of how the elements are connected to establish R_T. Once R_T is known, the current drawn from the source can be determined using Ohm's law, as follows:

$$I_s = \frac{E}{R_T} \qquad \text{(amperes, A)} \qquad \textbf{(5.2)}$$

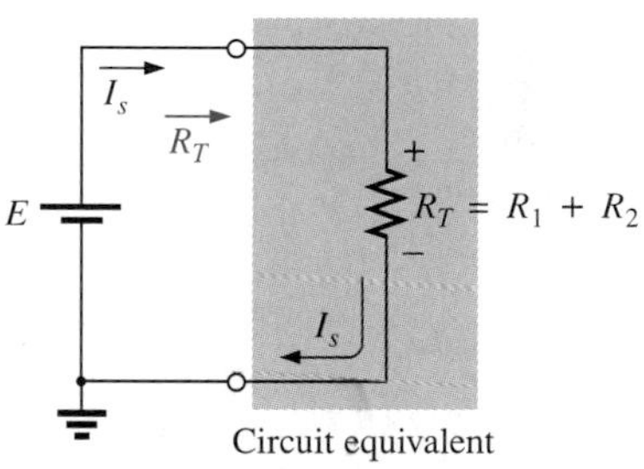

FIG. 5.6
Replacing the series resistors R_1 and R_2 of Fig. 5.5 with the total resistance.

Since E is fixed, the magnitude of the source current will be totally dependent on the magnitude of R_T. A larger R_T will result in a relatively small value of I_s, while lesser values of R_T will result in increased current levels.

The fact that the current is the same through each element of Fig. 5.4(a) permits a direct calculation of the voltage across each resistor using Ohm's law; that is,

$$V_1 = IR_1, V_2 = IR_2, V_3 = IR_3, \cdots, V_N = IR_N \qquad \text{(volts, V)} \qquad \textbf{(5.3)}$$

The power delivered to each resistor can then be determined using any one of three equations as listed below for R_1.

$$P_1 = V_1 I_1 = I_1^2 R_1 = \frac{V_1^2}{R_1} \qquad \text{(watts, W)} \qquad \textbf{(5.4)}$$

The power delivered by the source is

$$P_{\text{del}} = EI \qquad \text{(watts, W)} \qquad \textbf{(5.5)}$$

The total power delivered to a resistive network is equal to the total power dissipated by the resistive elements.

That is,

$$P_{\text{del}} = P_1 + P_2 + P_3 + \cdots + P_N \qquad \textbf{(5.6)}$$

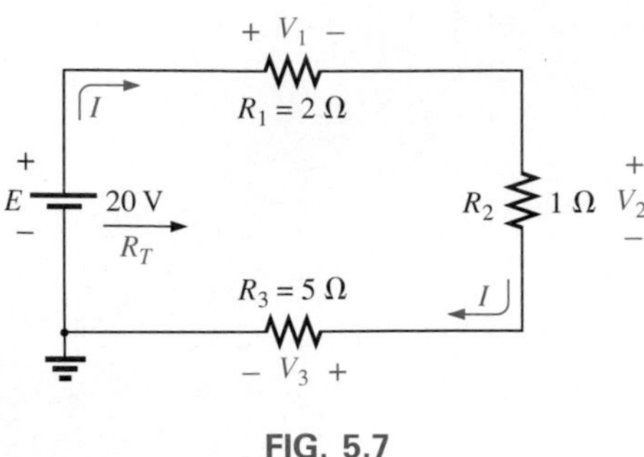

FIG. 5.7
Example 5.1.

EXAMPLE 5.1

a. Find the total resistance for the series circuit of Fig. 5.7.
b. Calculate the source current I_s.
c. Determine the voltages V_1, V_2, and V_3.
d. Calculate the power dissipated by R_1, R_2, and R_3.
e. Determine the power delivered by the source and compare it to the sum of the power levels of part (d).

Solutions:

a. $R_T = R_1 + R_2 + R_3 = 2\ \Omega + 1\ \Omega + 5\ \Omega = \mathbf{8\ \Omega}$

b. $I_s = \dfrac{E}{R_T} = \dfrac{20\ \text{V}}{8\ \Omega} = \mathbf{2.5\ A}$

c. $V_1 = IR_1 = (2.5\ \text{A})(2\ \Omega) = \mathbf{5\ V}$
$V_2 = IR_2 = (2.5\ \text{A})(1\ \Omega) = \mathbf{2.5\ V}$
$V_3 = IR_3 = (2.5\ \text{A})(5\ \Omega) = \mathbf{12.5\ V}$

d. $P_1 = V_1I_1 = (5\ \text{V})(2.5\ \text{A}) = \mathbf{12.5\ W}$
$P_2 = I_2^2R_2 = (2.5\ \text{A})^2(1\ \Omega) = \mathbf{6.25\ W}$
$P_3 = V_3^2/R_3 = (12.5\ \text{V})^2/5\ \Omega = \mathbf{31.25\ W}$

e. $P_{\text{del}} = EI = (20\ \text{V})(2.5\ \text{A}) = \mathbf{50\ W}$
$P_{\text{del}} = P_1 + P_2 + P_3$
$50\ \text{W} = 12.5\ \text{W} + 6.25\ \text{W} + 31.25\ \text{W}$
$50\ \text{W} = 50\ \text{W}$ (checks)

To find the total resistance of N resistors of the same value in series, simply multiply the value of *one* of the resistors by the number in series; that is,

$$R_T = NR \tag{5.7}$$

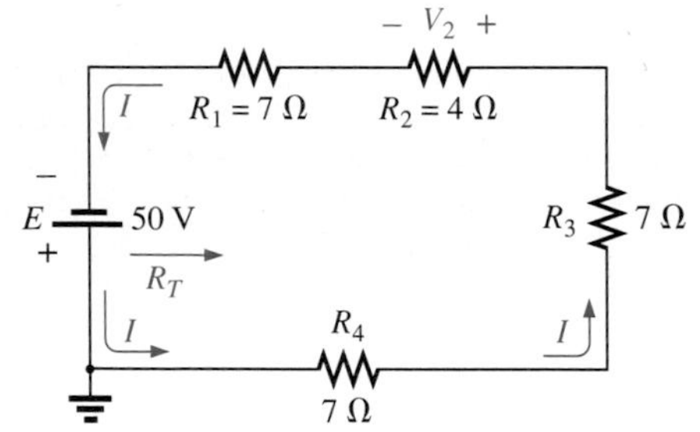

FIG. 5.8
Example 5.2.

EXAMPLE 5.2 Determine R_T, I, and V_2 for the circuit of Fig. 5.8.

Solution: Note the current direction as established by the battery and the polarity of the voltage drops across R_2 as determined by the current direction. Since $R_1 = R_3 = R_4$,

$$R_T = NR_1 + R_2 = (3)(7\ \Omega) + 4\ \Omega = 21\ \Omega + 4\ \Omega = \mathbf{25\ \Omega}$$

$$I = \frac{E}{R_T} = \frac{50\ \text{V}}{25\ \Omega} = \mathbf{2\ A}$$

$$V_2 = IR_2 = (2\ \text{A})(4\ \Omega) = \mathbf{8\ V}$$

Examples 5.1 and 5.2 are straightforward substitution-type problems that are relatively easy to solve with some practice. Example 5.3, however, is evidence of another type of problem that requires a firm grasp of the fundamental equations and an ability to identify which equation to use first. The best preparation for this type of exercise is simply to work through as many problems of this kind as possible.

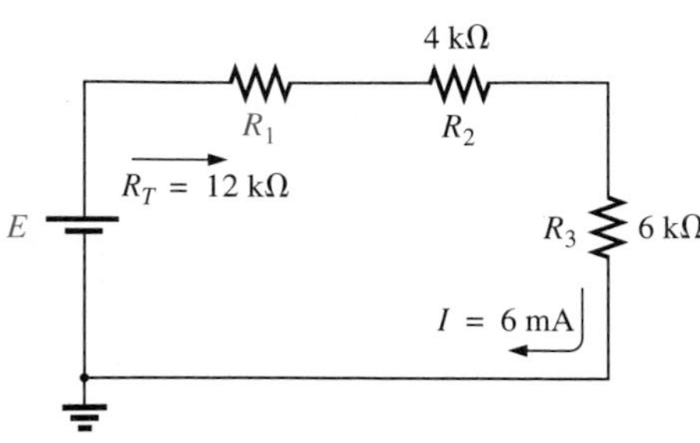

FIG. 5.9
Example 5.3.

EXAMPLE 5.3 Given R_T and I, calculate R_1 and E for the circuit of Fig. 5.9.

Solution:

$$R_T = R_1 + R_2 + R_3$$
$$12\text{ k}\Omega = R_1 + 4\text{ k}\Omega + 6\text{ k}\Omega$$
$$R_1 = 12\text{ k}\Omega - 10\text{ k}\Omega = \mathbf{2\text{ k}\Omega}$$
$$E = IR_T = (6 \times 10^{-3}\text{ A})(12 \times 10^3\ \Omega) = \mathbf{72\text{ V}}$$

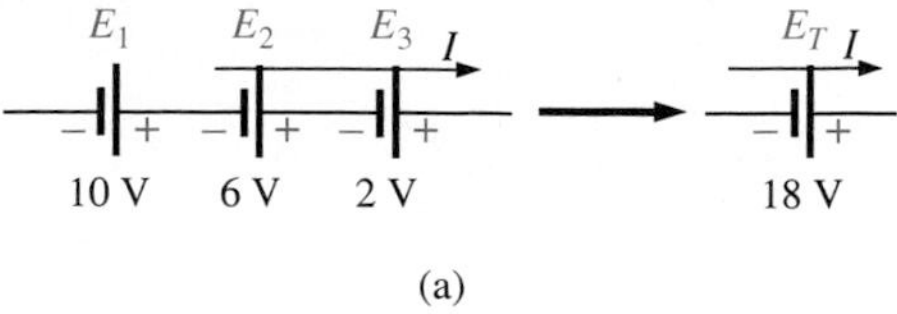

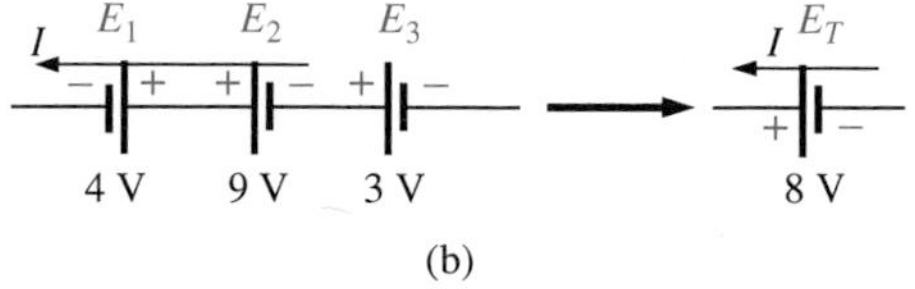

FIG. 5.10
Reducing series dc voltage sources to a single source.

5.3 VOLTAGE SOURCES IN SERIES

Voltage sources can be connected in series, as shown in Fig. 5.10, to increase or decrease the total voltage applied to a system. The net voltage is determined simply by summing the sources with the same polarity and subtracting the total of the sources with the opposite "pressure." The net polarity is the polarity of the larger sum.

In Fig. 5.10(a), for example, the sources are all "pressuring" current to the right, so the net voltage is

$$E_T = E_1 + E_2 + E_3 = 10\text{ V} + 6\text{ V} + 2\text{ V} = 18\text{ V}$$

as shown in the figure. In Fig. 5.8(b), however, the greater "pressure" is to the left, with a net voltage of

$$E_T = E_2 + E_3 - E_1 = 9\text{ V} + 3\text{ V} - 4\text{ V} = 8\text{ V}$$

and the polarity shown in the figure.

German (Königsberg, Berlin)
(1824–1887)
Physicist
Professor of Physics, University of Heidelberg

Courtesy of the Smithsonian Institution
Photo No. 58,283

Although a contributor to a number of areas in the physics domain, he is best known for his work in the electrical area with his definition of the relationships between the currents and voltages of a network in 1847. Did extensive research with German chemist Robert Bunsen (developed the *Bunsen burner*), resulting in the discovery of the important elements of *cesium* and *rubidium.*

FIG. 5.11
Gustav Robert Kirchhoff.

5.4 KIRCHHOFF'S VOLTAGE LAW

Note Fig. 5.11.

Kirchhoff's voltage law (KVL) states that the algebraic sum of the potential rises and drops around a closed loop (or path) is zero.

A *closed loop* is any continuous path that leaves a point in one direction and returns to that same point from another direction without leaving the circuit. In Fig. 5.12, by following the current, we can trace a continuous path that leaves point a through R_1 and returns through E without leaving the circuit. Therefore, $abcda$ is a closed loop. For us to be able to apply Kirchhoff's voltage law, the summation of potential rises and drops must be made in one direction around the closed loop.

For uniformity, the clockwise (CW) direction will be used throughout the text for all applications of Kirchhoff's voltage law. Be aware, however, that the same result will be obtained if the counterclockwise (CCW) direction is chosen and the law applied correctly.

A plus sign is assigned to a potential rise (− to +) and a minus sign to a potential drop (+ to −). If we follow the current in Fig. 5.12 from point a, we first encounter a potential drop V_1 (+ to −) across R_1 and then another potential drop V_2 across R_2. Continuing through the voltage source, we have a potential rise E (− to +) before returning to point a. In symbolic form, where Σ represents summation, $\circlearrowright$ the closed loop, and V the potential drops and rises, we have

$$\Sigma_{\circlearrowright} V = 0 \quad \text{(Kirchhoff's voltage law in symbolic form)} \qquad \textbf{(5.8)}$$

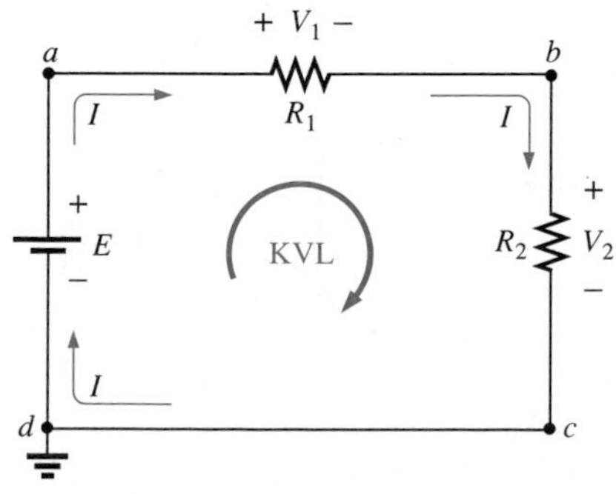

FIG. 5.12
Applying Kirchhoff's voltage law to a series dc circuit.

which for the circuit of Fig. 5.12 yields (clockwise direction, following the current I and starting at point d):

$$+E - V_1 - V_2 = 0$$

or
$$E = V_1 + V_2$$

revealing that

the applied voltage of a series circuit equals the sum of the voltage drops across the series elements.

Kirchhoff's voltage law can also be stated in the following form:

$$\boxed{\Sigma_{\circlearrowright} V_{\text{rises}} = \Sigma_{\circlearrowright} V_{\text{drops}}} \qquad \textbf{(5.9)}$$

which in words states that the sum of the rises around a closed loop must equal the sum of the drops in potential. The text will emphasize the use of Eq. (5.8), however.

If the loop were taken in the counterclockwise direction starting at point a, the following would result:

$$\Sigma_{\circlearrowright} V = 0$$
$$-E + V_2 + V_1 = 0$$

or, as before,
$$E = V_1 + V_2$$

The application of Kirchhoff's voltage law need not follow a path that includes current-carrying elements.

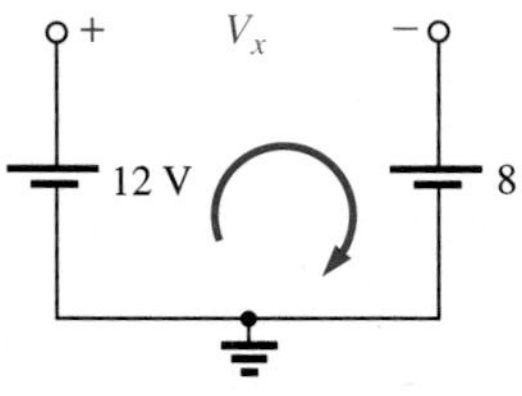

For example, in Fig. 5.13 there is a difference in potential between points a and b, even though the two points are not connected by a current-carrying element. Application of Kirchhoff's voltage law around the closed loop will result in a difference in potential of 4 V between the two points. That is, using the clockwise direction:

$$+12\text{ V} - V_x - 8\text{ V} = 0$$

and
$$V_x = \mathbf{4\ V}$$

FIG. 5.13
Demonstration that a voltage can exist between two points not connected by a current-carrying conductor.

EXAMPLE 5.4 Determine the unknown voltages for the networks of Fig. 5.14.

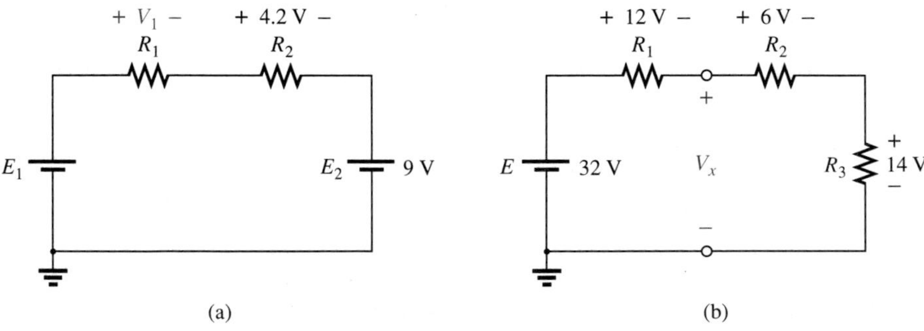

FIG. 5.14
Example 5.4.

Solutions: When applying Kirchhoff's voltage law, be sure to concentrate on the polarities of the voltage rise or drop rather than the type of element. In other words, do not treat a voltage drop across a resistive element differently from a voltage drop across a source. If the polarity dictates that a drop has occurred, that is the important fact when applying the law. In Fig. 5.14(a), for instance, if we choose the clockwise direction, we will find that there is a drop across the resistors R_1 and R_2 and a drop across the source E_2. All will therefore have a minus sign when Kirchhoff's voltage law is applied.

Application of Kirchhoff's voltage law to the circuit of Fig. 5.14(a) in the clockwise direction will result in

$$+E_1 - V_1 - V_2 - E_2 = 0$$

and
$$V_1 = E_1 - V_2 - E_2 = 16\text{ V} - 4.2\text{ V} - 9\text{ V}$$
$$= \mathbf{2.8\ V}$$

The result clearly indicates that there was no need to know the values of the resistors or the current to determine the unknown voltage. There was sufficient information carried by the other voltage levels to permit a determination of the unknown.

In Fig. 5.14(b) the unknown voltage is not across a current-carrying element. However, as indicated in the paragraphs above, Kirchhoff's voltage law is not limited to current-carrying elements. In this case there are two possible paths for finding the unknown. Using the clockwise path, including the voltage source E, will result in

$$+E - V_1 - V_x = 0$$

and
$$V_x = E - V_1 = 32\text{ V} - 12\text{ V}$$
$$= \mathbf{20\ V}$$

Using the clockwise direction for the other loop involving R_2 and R_3 will result in

$$+V_x - V_2 - V_3 = 0$$

and
$$V_x = V_2 + V_3 = 6\text{ V} + 14\text{ V}$$
$$= \mathbf{20\ V}$$

matching the result above.

EXAMPLE 5.5 Find V_1 and V_2 for the network of Fig. 5.15.

Solution: For path 1, starting at point a in a clockwise direction:

$$+25\text{ V} - V_1 + 15\text{ V} = 0$$

and
$$V_1 = \mathbf{40\ V}$$

For path 2, starting at point a in a clockwise direction:

$$-V_2 - 20\text{ V} = 0$$

and
$$V_2 = \mathbf{-20\ V}$$

The minus sign simply indicates that the actual polarities of the potential difference are opposite the assumed polarity indicated in Fig. 5.15.

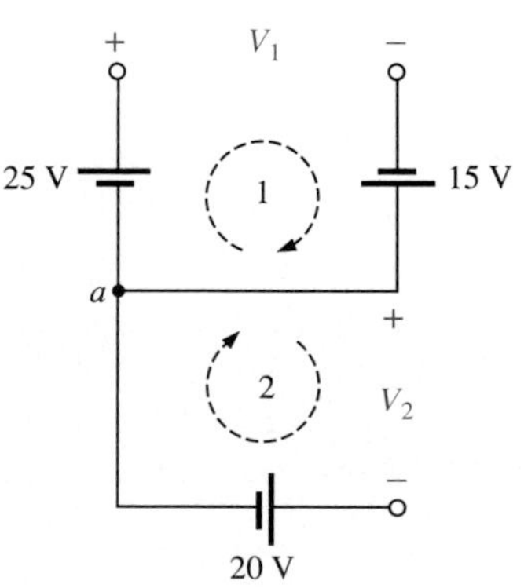

FIG. 5.15
Example 5.5.

The next example will emphasize the fact that it is the polarities of the voltage rise or drop that are the important parameters when applying Kirchhoff's voltage law and not the type of element involved.

EXAMPLE 5.6 Using Kirchhoff's voltage law, determine the unknown voltages for the network of Fig. 5.16.

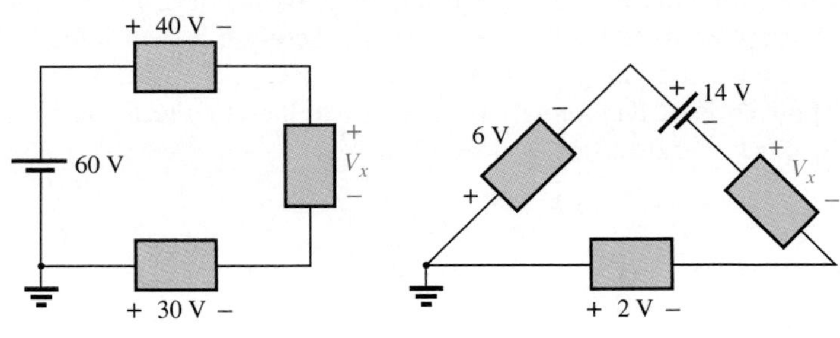

FIG. 5.16
Example 5.6.

Solutions: Note in each circuit that there are various polarities across the unknown elements since they can contain any mixture of components. Applying Kirchhoff's voltage law to the network of Fig. 5.16(a) in the clockwise direction will result in

$$60\text{ V} - 40\text{ V} - V_x + 30\text{ V} = 0$$

and
$$V_x = 60\text{ V} + 30\text{ V} - 40\text{ V} = 90\text{ V} - 40\text{ V}$$
$$= \mathbf{50\text{ V}}$$

In Fig. 5.16(b) the polarity of the unknown voltage is not provided. In such cases, make an assumption about the polarity and apply Kirchhoff's voltage law as before. If the result has a plus sign, the assumed polarity was correct. If it has a minus sign, the magnitude is correct but the assumed polarity has to be reversed. In this case if we assume *a* to be positive and *b* to be negative, an application of Kirchhoff's voltage law in the clockwise direction will result in

$$-6\text{ V} - 14\text{ V} - V_x + 2\text{ V} = 0$$

and
$$V_x = -20\text{ V} + 2\text{ V}$$
$$= \mathbf{-18\text{ V}}$$

Since the result is negative, we know that *a* should be negative and *b* should be positive, but the magnitude of 5 V is correct.

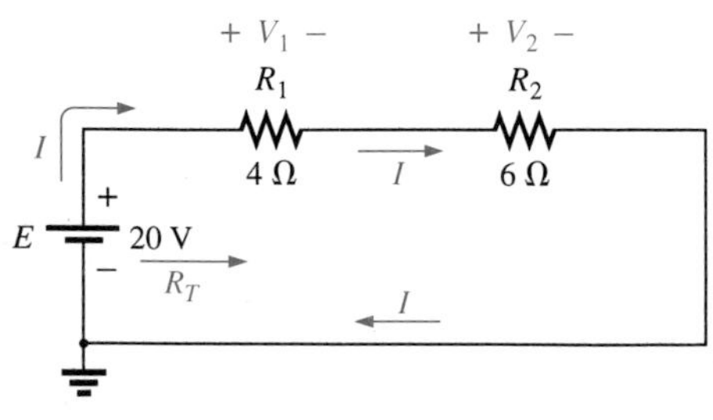

FIG. 5.17
Example 5.7.

EXAMPLE 5.7 For the circuit of Fig. 5.17:

a. Find R_T.
b. Find I.
c. Find V_1 and V_2.
d. Find the power to the 4-Ω and 6-Ω resistors.
e. Find the power delivered by the battery, and compare it to that dissipated by the 4-Ω and 6-Ω resistors combined.
f. Verify Kirchhoff's voltage law (clockwise direction).

Solutions:

a. $R_T = R_1 + R_2 = 4\ \Omega + 6\ \Omega = \mathbf{10\ \Omega}$

b. $I = \dfrac{E}{R_T} = \dfrac{20\text{ V}}{10\ \Omega} = \mathbf{2\text{ A}}$

c. $V_1 = IR_1 = (2\text{ A})(4\ \Omega) = \mathbf{8\ V}$
$V_2 = IR_2 = (2\text{ A})(6\ \Omega) = \mathbf{12\ V}$

d. $P_{4\Omega} = \dfrac{V_1^2}{R_1} = \dfrac{(8\text{ V})^2}{4} = \dfrac{64}{4} = \mathbf{16\ W}$

$P_{6\Omega} = I^2R_2 = (2\text{ A})^2(6\ \Omega) = (4)(6) = \mathbf{24\ W}$

e. $P_E = EI = (20\text{ V})(2\text{ A}) = \mathbf{40\ W}$
$P_E = P_{4\Omega} + P_{6\Omega}$
$40\text{ W} = 16\text{ W} + 24\text{ W}$
$40\text{ W} = 40\text{ W}$ (checks)

f. $\Sigma_{\circlearrowright} V = +E - V_1 - V_2 = 0$
$E = V_1 + V_2$
$20\text{ V} = 8\text{ V} + 12\text{ V}$
$20\text{ V} = 20\text{ V}$ (checks)

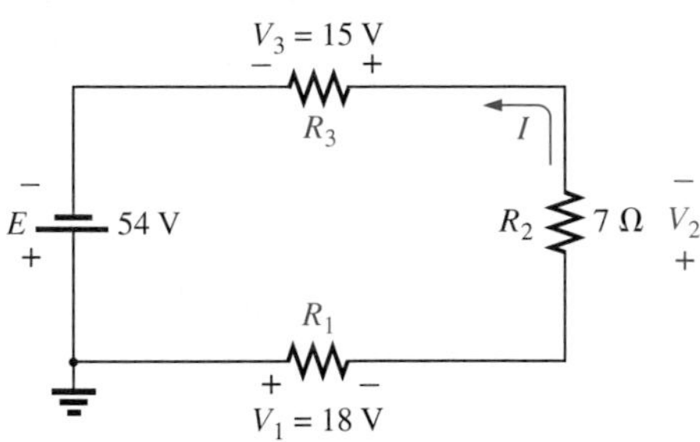

FIG. 5.18
Example 5.8.

EXAMPLE 5.8 For the circuit of Fig. 5.18:
a. Determine V_2 using Kirchhoff's voltage law.
b. Determine I.
c. Find R_1 and R_3.

Solutions:
a. Kirchhoff's voltage law (clockwise direction):

$$-E + V_3 + V_2 + V_1 = 0$$

or

$$E = V_1 + V_2 + V_3$$

and $V_2 = E - V_1 - V_3 = 54\text{ V} - 18\text{ V} - 15\text{ V} = \mathbf{21\ V}$

b. $I = \dfrac{V_2}{R_2} = \dfrac{21\text{ V}}{7\ \Omega} = \mathbf{3\ A}$

c. $R_1 = \dfrac{V_1}{I} = \dfrac{18\text{ V}}{3\text{ A}} = \mathbf{6\ \Omega}$

$R_3 = \dfrac{V_3}{I} = \dfrac{15\text{ V}}{3\text{ A}} = \mathbf{5\ \Omega}$

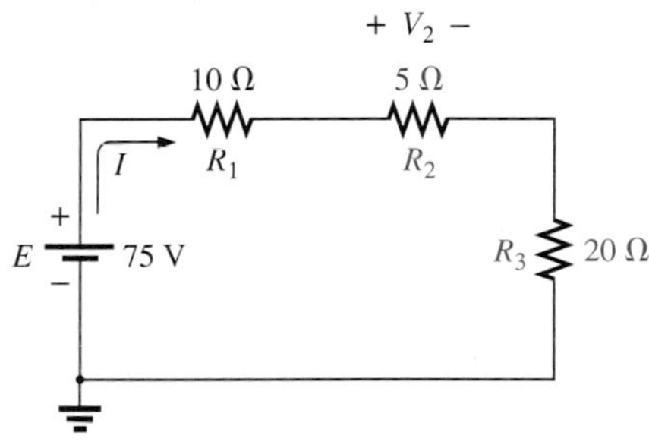

FIG. 5.19
Series dc circuit with elements to be interchanged.

5.5 INTERCHANGING SERIES ELEMENTS

The elements of a series circuit can be interchanged without affecting the total resistance, current, or power to each element. For instance, the network of Fig. 5.19 can be redrawn as shown in Fig. 5.20 without affecting I or V_2. The total resistance R_T is 35 Ω in both cases, and I = 70 V/35 Ω = 2 A. The voltage $V_2 = IR_2 = (2\text{ A})(5\ \Omega) = 10\text{ V}$ for both configurations.

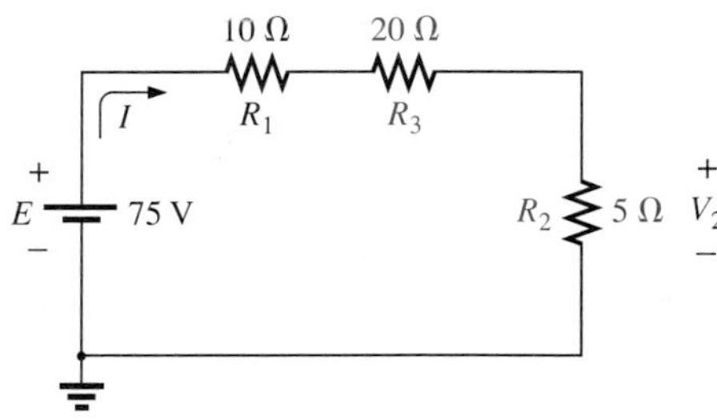

FIG. 5.20
Circuit of Fig. 5.19 with R_2 and R_3 interchanged.

EXAMPLE 5.9 Determine I and the voltage across the 7-Ω resistor for the network of Fig. 5.21.

Solution: The network is redrawn in Fig. 5.22.

$$R_T = (2)(4\ \Omega) + 7\ \Omega = 15\ \Omega$$

$$I = \frac{E}{R_T} = \frac{37.5\text{ V}}{15\ \Omega} = \mathbf{2.5\ A}$$

$$V_{7\Omega} = IR = (2.5\text{ A})(7\ \Omega) = \mathbf{17.5\ V}$$

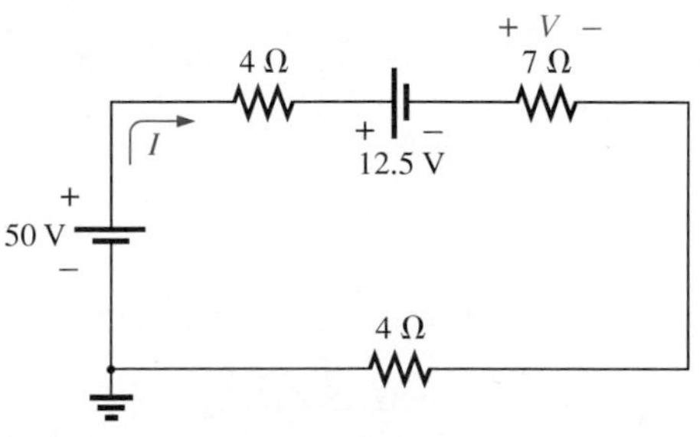

FIG. 5.21
Example 5.9.

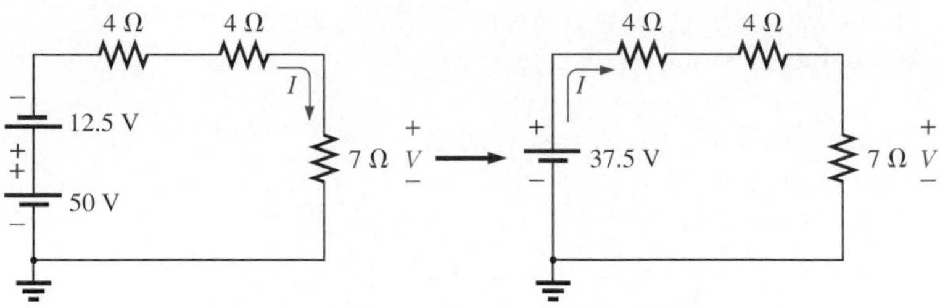

FIG. 5.22
Redrawing the circuit of Fig. 5.21.

5.6 VOLTAGE DIVIDER RULE

In a series circuit,

the voltage across the resistive elements will divide as the magnitude of the resistance levels.

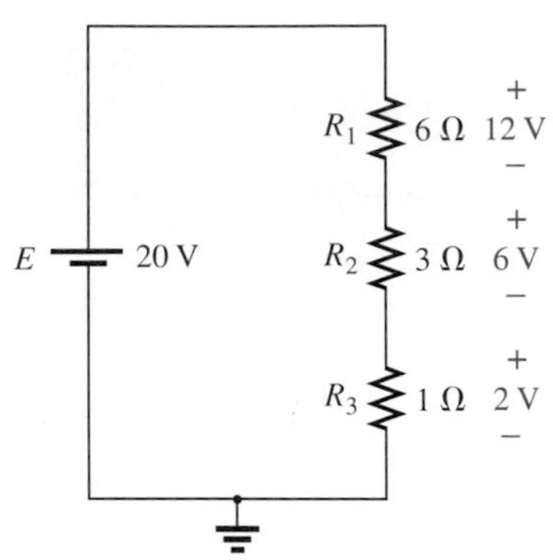

FIG. 5.23
Revealing how the voltage will divide across series resistive elements.

For example, the voltages across the resistive elements of Fig. 5.23 are provided. The largest resistor of 6 Ω captures the bulk of the applied voltage, while the smallest resistor R_3 has the least. Note in addition that, since the resistance level of R_1 is 6 times that of R_3, the voltage across R_1 is 6 times that of R_3. The fact that the resistance level of R_2 is 3 times that of R_1 results in three times the voltage across R_2. Finally, since R_1 is twice R_2, the voltage across R_1 is twice that of R_2. In general, therefore, the voltage across series resistors will have the same ratio as their resistance levels.

It is particularly interesting to note that, if the resistance levels of all the resistors of Fig. 5.23 are increased by the same amount, as shown in Fig. 5.24, the voltage levels will all remain the same. In other words, even though the resistance levels were increased by a factor of 1 million, the voltage ratios remain the same. Clearly, therefore, it is the ratio of resistor values that counts when it comes to voltage division and not the relative magnitude of all the resistors. The current level of the network will be severely affected by the change in resistance level from Fig. 5.23 to Fig. 5.24, but the voltage levels will remain the same.

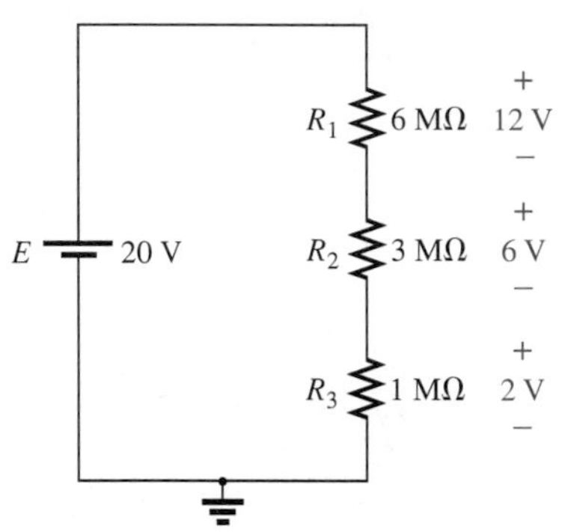

FIG. 5.24
The ratio of the resistive values determines the voltage division of a series dc circuit.

Based on the above, a first glance at the series network of Fig. 5.25 should suggest that the major part of the applied voltage will appear across the 1-MΩ resistor and very little across the 100-Ω resistor. In fact 1 MΩ = (1000)1 kΩ = (10,000)100 Ω, revealing that $V_1 = 1000V_2 = 10{,}000V_3$.

Solving for the current and then the three voltage levels will result in

$$I = \frac{E}{R_T} = \frac{100\text{ V}}{1{,}001{,}100\ \Omega} \cong 99.89\ \mu\text{A}$$

and

$$V_1 = IR_1 = (99.89\ \mu\text{A})(1\ \text{M}\Omega) = \mathbf{99.89\ V}$$
$$V_2 = IR_2 = (99.89\ \mu\text{A})(1\ \text{k}\Omega) = \mathbf{99.89\ mV} = 0.09989\text{ V}$$
$$V_3 = IR_3 = (99.89\ \mu\text{A})(100\ \Omega) = \mathbf{9.989\ mV} = 0.009989\text{ V}$$

clearly substantiating the above conclusions. For the future, therefore, use this approach to estimate the share of the input voltage across series elements to act as a check against the actual calculations or to simply obtain an estimate with a minimum of effort.

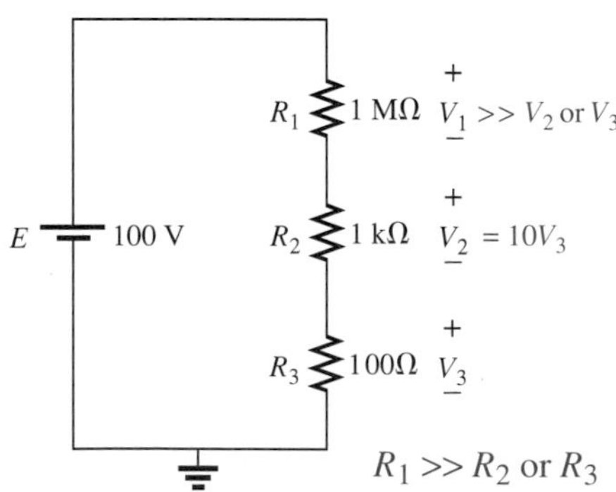

FIG. 5.25
The largest of the series resistive elements will capture the major share of the applied voltage.

In the above discussion the current was determined before the voltages of the network were determined. There is, however, a method referred to as the *voltage divider rule* that permits determining the voltage levels without first finding the current. The rule can be derived by analyzing the network of Fig. 5.26.

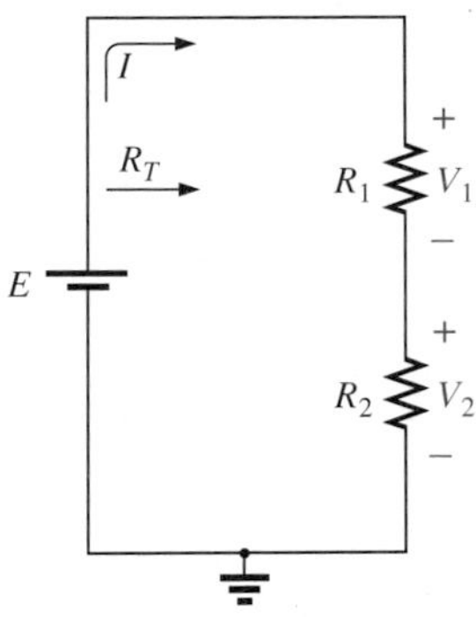

FIG. 5.26
Developing the voltage divider rule.

$$R_T = R_1 + R_2$$

and
$$I = \frac{E}{R_T}$$

Applying Ohm's law:

$$V_1 = IR_1 = \left(\frac{E}{R_T}\right)R_1 = \frac{R_1E}{R_T}$$

with
$$V_2 = IR_2 = \left(\frac{E}{R_T}\right)R_2 = \frac{R_2E}{R_T}$$

Note that the format for V_1 and V_2 is

$$\boxed{V_x = \frac{R_xE}{R_T}} \qquad \text{(voltage divider rule)} \qquad \textbf{(5.10)}$$

where V_x is the voltage across R_x, E is the impressed voltage across the series elements, and R_T is the total resistance of the series circuit.

In words, the *voltage divider rule* states that

the voltage across a resistor in a series circuit is equal to the value of that resistor times the total impressed voltage across the series elements divided by the total resistance of the series elements.

EXAMPLE 5.10 Determine the voltage V_1 for the network of Fig. 5.27.

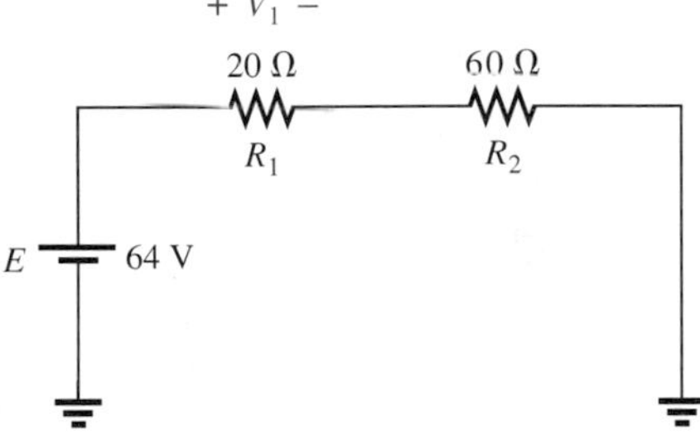

FIG. 5.27
Example 5.10.

Solution: Eq. (5.10):

$$V_1 = \frac{R_1E}{R_T} = \frac{R_1E}{R_1 + R_2} = \frac{(20\ \Omega)(64\ \text{V})}{20\ \Omega + 60\ \Omega} = \frac{1280\ \text{V}}{80} = \mathbf{16\ V}$$

EXAMPLE 5.11 Using the voltage divider rule, determine the voltages V_1 and V_3 for the series circuit of Fig. 5.28.

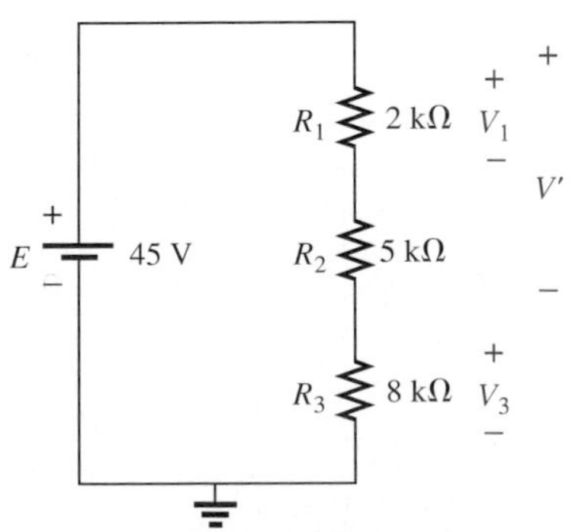

FIG. 5.28
Example 5.11.

Solution:

$$V_1 = \frac{R_1E}{R_T} = \frac{(2\ \text{k}\Omega)(45\ \text{V})}{2\ \text{k}\Omega + 5\ \text{k}\Omega + 8\ \text{k}\Omega} = \frac{(2\ \text{k}\Omega)(45\ \text{V})}{15\ \text{k}\Omega}$$

$$= \frac{(2 \times 10^3\ \Omega)(45\ \text{V})}{15 \times 10^3\ \Omega} = \frac{90\ \text{V}}{15} = \mathbf{6\ V}$$

$$V_3 = \frac{R_3E}{R_T} = \frac{(8\ \text{k}\Omega)(45\ \text{V})}{15\ \text{k}\Omega} = \frac{(8 \times 10^3\ \Omega)(45\ \text{V})}{15 \times 10^3\ \Omega}$$

$$= \frac{360\ \text{V}}{15} = \mathbf{24\ V}$$

The rule can be extended to the voltage across two or more series elements if the resistance in the numerator of Eq. (5.10) is expanded to include the total resistance of the series elements that the voltage is to be found across (R'); that is,

$$V' = \frac{R'E}{R_T} \qquad \text{(volts)} \qquad \textbf{(5.11)}$$

EXAMPLE 5.12 Determine the voltage V' in Fig. 5.28 across resistors R_1 and R_2.

Solution:

$$V' = \frac{R'E}{R_T} = \frac{(2\text{ k}\Omega + 5\text{ k}\Omega)(45\text{ V})}{15\text{ k}\Omega} = \frac{(7\text{ k}\Omega)(45\text{ V})}{15\text{ k}\Omega} = \mathbf{21\text{ V}}$$

There is also no need for the voltage E in the equation to be the source voltage of the network. For example, if V is the total voltage across a number of series elements such as those shown in Fig. 5.29, then

$$V_{2\Omega} = \frac{(2\ \Omega)(27\text{ V})}{4\ \Omega + 2\ \Omega + 3\ \Omega} = \frac{54\text{ V}}{9} = \mathbf{6\text{ V}}$$

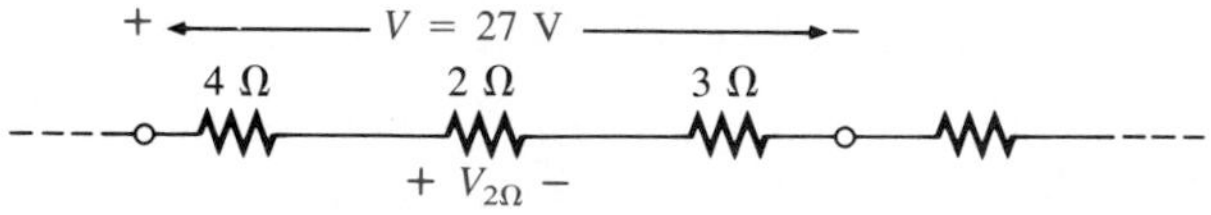

+ ← $V = 27$ V → –

4 Ω 2 Ω 3 Ω

+ $V_{2\Omega}$ –

FIG. 5.29
The total voltage across series elements need not be an independent voltage source.

EXAMPLE 5.13 Design the voltage divider of Fig. 5.30 such that $V_{R_1} = 4V_{R_2}$.

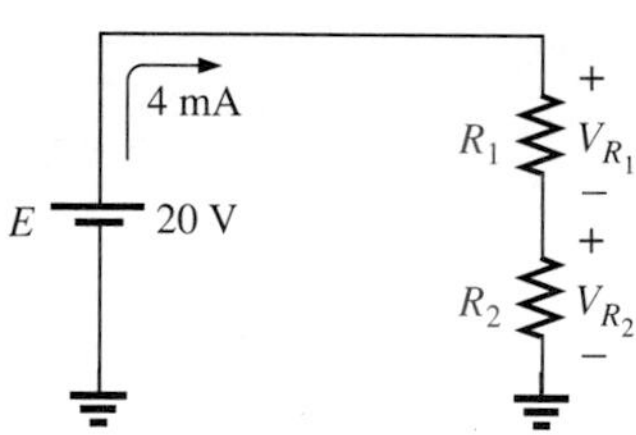

FIG. 5.30
Example 5.13.

Solution: The total resistance is defined by

$$R_T = \frac{E}{I} = \frac{20\text{ V}}{4\text{ mA}} = 5\text{ k}\Omega$$

Since $V_{R_1} = 4V_{R_2}$,

$$R_1 = 4R_2$$

Thus $$R_T = R_1 + R_2 = 4R_2 + R_2 = 5R_2$$

and $$5R_2 = 5\text{ k}\Omega$$
$$R_2 = \mathbf{1\text{ k}\Omega}$$

and $$R_1 = 4R_2 = \mathbf{4\text{ k}\Omega}$$

5.7 NOTATION

Notation will play an increasingly important role in the analysis to follow. It is important, therefore, that we begin to examine the notation used throughout the industry.

Voltage Sources and Ground

Except for a few special cases, electrical and electronic systems are grounded for reference and safety purposes. The symbol for the ground connection appears in Fig. 5.31 with its defined potential level—zero volts. None of the circuits discussed thus far have contained the ground connection. If Fig. 5.4(a) were redrawn with a grounded supply, it might appear as shown in Fig. 5.32(a) or (b) or (c). In either case, it is understood that the negative terminal of the battery and the bottom of the resistor R_2 are at ground potential. Although Fig. 5.32(c) shows no connection between the two grounds, it is recognized that such a connection exists for the continuous flow of charge. If $E = 12$ V, then point a is 12 V positive with respect to ground potential, and 12 V exist across the series combination of resistors R_1 and R_2. If a voltmeter placed from point b to ground reads 4 V, then the voltage across R_2 is 4 V, with the higher potential at point b.

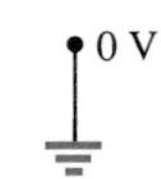

FIG. 5.31
Ground potential.

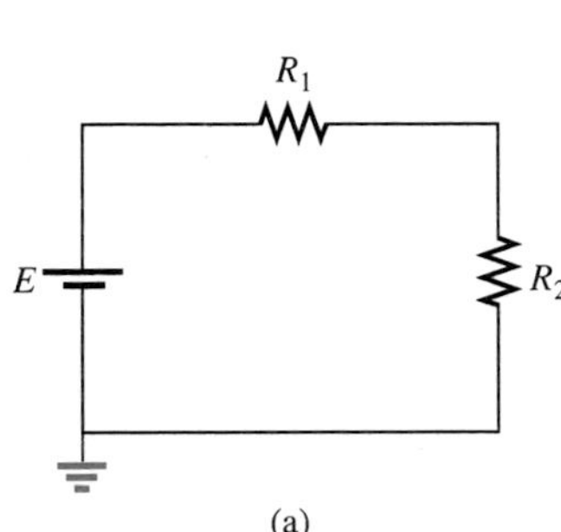

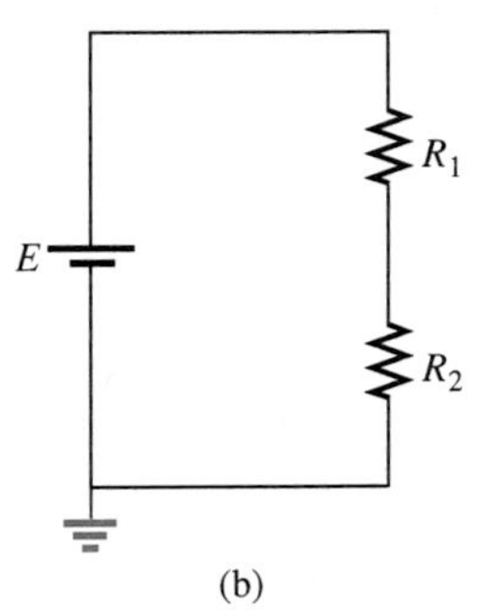

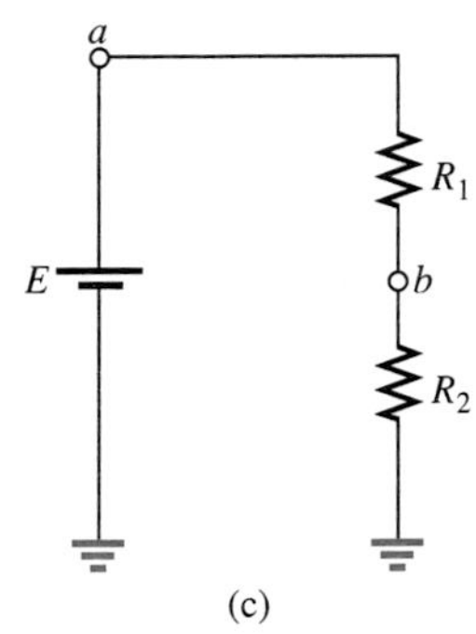

FIG. 5.32
Three ways to sketch the same series dc circuit.

On large schematics where space is at a premium and clarity is important, voltage sources may be indicated as shown in Figs. 5.33(a) and 5.34(a) rather than as illustrated in Figs. 5.33(b) and 5.34(b). In addition, potential levels may be indicated as in Fig. 5.35, to permit a rapid check of the potential levels at various points in a network with respect to ground to ensure that the system is operating properly.

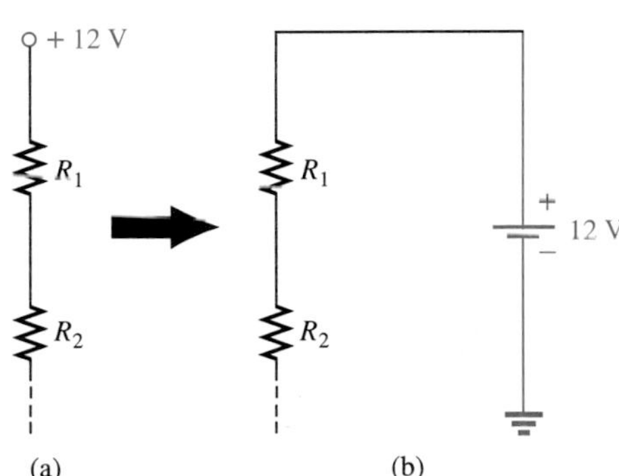

FIG. 5.33
Replacing the special notation for a dc voltage source with the standard symbol.

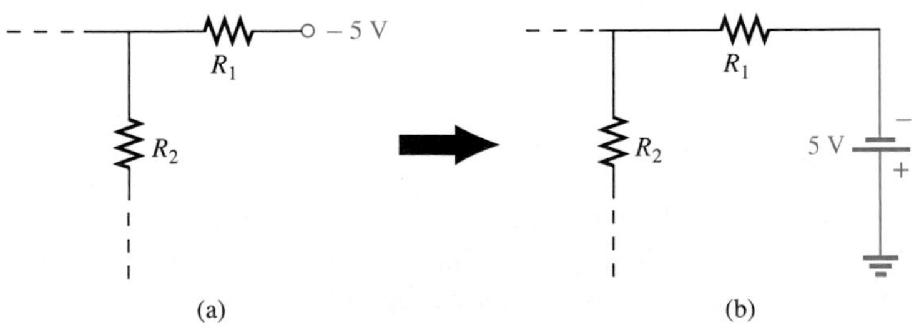

FIG. 5.34
Replacing the notation for a negative dc supply with the standard notation.

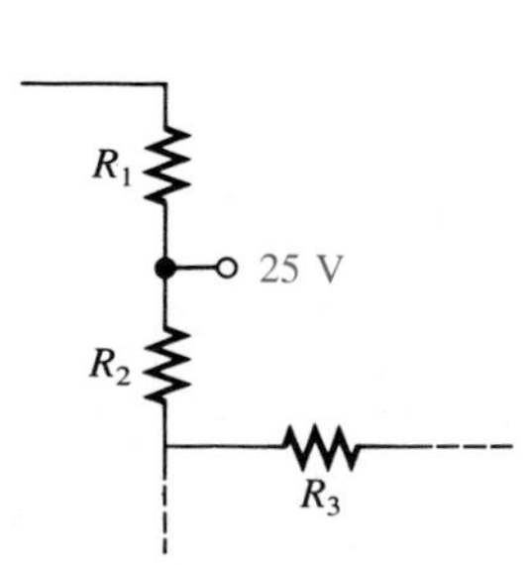

FIG. 5.35
The expected voltage level at a particular point in a network of the system is functioning properly.

Double-Subscript Notation

The fact that voltage is an *across* variable and exists between two points has resulted in a double-subscript notation that defines the first subscript as the higher potential. In Fig. 5.36(a), the two points that define the voltage across the resistor R are denoted by a and b. Since a is the first subscript for V_{ab}, point a must have a higher potential than point b if V_{ab} is to have a positive value. If, in fact, point b is at a higher potential than point a, V_{ab} will have a negative value, as indicated in Fig. 5.35(b).

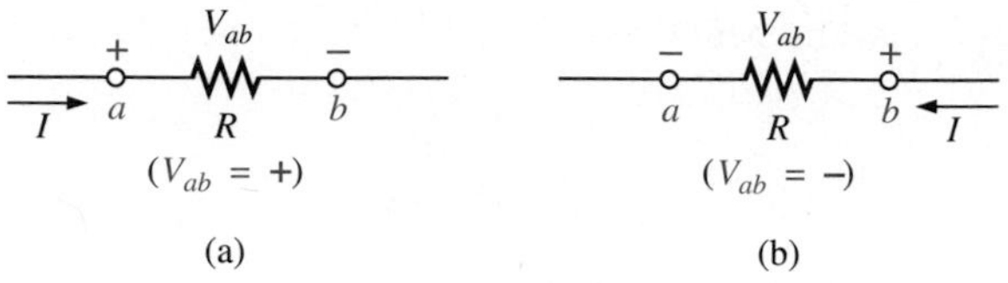

V_{ab} + – a b R I (V_{ab} = +) (a)

V_{ab} – + a b R I (V_{ab} = –) (b)

FIG. 5.36
Defining the sign for double-subscript notation.

In summary:

The double-subscript notation V_{ab} specifies point* a *as the higher potential. If this is not the case, a negative sign must be associated with the magnitude of V_{ab}.

In other words,

***the voltage V_{ab} is the voltage at point* a *with respect to (w.r.t.) point* b.**

Single-Subscript Notation

If point b of the notation V_{ab} is specified as ground potential (zero volts), then a single-subscript notation can be employed that provides the voltage at a point with respect to ground.

In Fig. 5.37, V_a is the voltage from point a to ground. In this case it is obviously 10 V since it is right across the source voltage E. The voltage V_b is the voltage from point b to ground. Because it is directly across the 4-Ω resistor, $V_b = 4$ V.

In summary:

The single-subscript notation V_a specifies the voltage at point* a *with respect to ground (zero volts). If the voltage is less than zero volts, a negative sign must be associated with the magnitude of V_a.

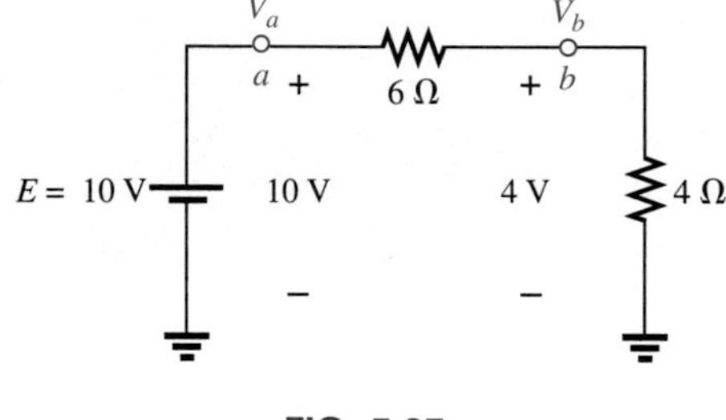

FIG. 5.37
Defining the use of single-subscript notation for voltage levels.

General Comments

A particularly useful relationship can now be established that will have extensive applications in the analysis of electronic circuits. For the above notational standards, the following relationship exists:

$$V_{ab} = V_a - V_b \tag{5.12}$$

In other words, if the voltage at points a and b is known with respect to ground, then the voltage V_{ab} can be determined using the above equation. In Fig. 5.37, for example,

$$V_{ab} = V_a - V_b = 10\text{ V} - 4\text{ V} = 6\text{ V}$$

EXAMPLE 5.14 Find the voltage V_{ab} for the conditions of Fig. 5.38.

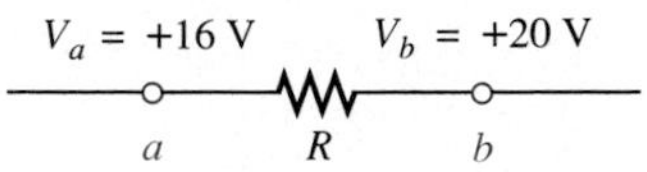

Solution: Applying Eq. (5.12):

$$V_{ab} = V_a - V_b = 16\text{ V} - 20\text{ V}$$
$$= \mathbf{-4\text{ V}}$$

$V_a = +16$ V $V_b = +20$ V

a R b

FIG. 5.38
Example 5.14.

Note the negative sign to reflect the fact that point b is at a higher potential than point a.

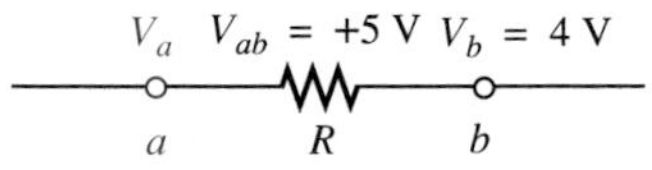

V_a $V_{ab} = +5$ V $V_b = 4$ V

a R b

FIG. 5.39
Example 5.15.

EXAMPLE 5.15 Find the voltage V_a for the configuration of Fig. 5.39.

Solution: Applying Eq. (5.12):

$$V_{ab} = V_a - V_b$$

and

$$V_a = V_{ab} + V_b = 5\text{ V} + 4\text{ V}$$
$$= \mathbf{9\text{ V}}$$

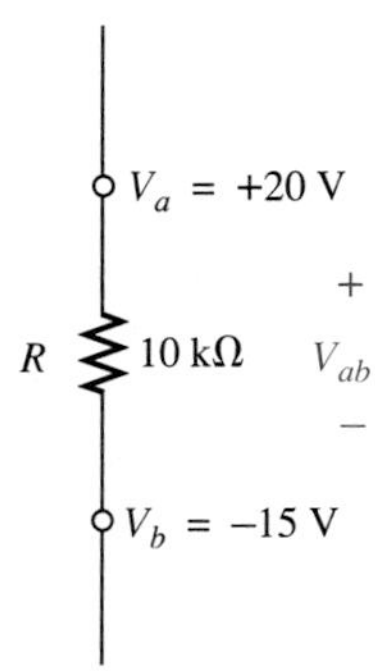

FIG. 5.40
Example 5.16.

EXAMPLE 5.16 Find the voltage V_{ab} for the configuration of Fig. 5.40.

Solution: Applying Eq. (5.12):

$$V_{ab} = V_a - V_b = 20\text{ V} - (-15\text{ V}) = 20\text{ V} + 15\text{ V}$$
$$= \mathbf{35\text{ V}}$$

Note in Example 5.16 the care that must be taken with the signs when applying the equation. The voltage is dropping from a high level of +20 V to a negative voltage of −15 V. As shown in Fig. 5.41, this represents a drop in voltage of 35 V. In some ways it's like going from a positive checking balance of $20 to owing $15; the total expenditure is $35.

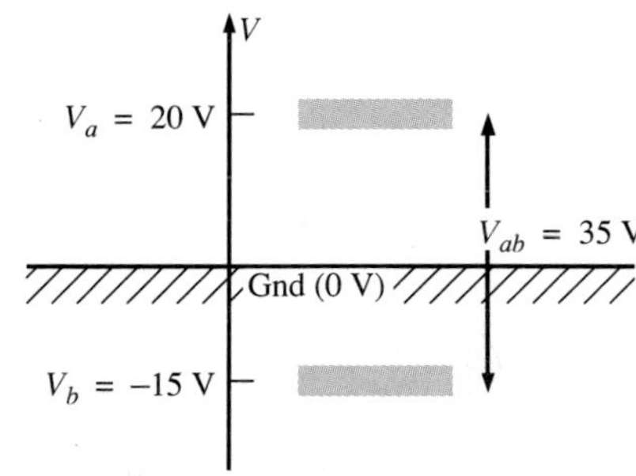

FIG. 5.41
The impact of positive and negative voltages on the total voltage drop.

EXAMPLE 5.17 Find the voltages V_b, V_c, and V_{ac} for the network of Fig. 5.42.

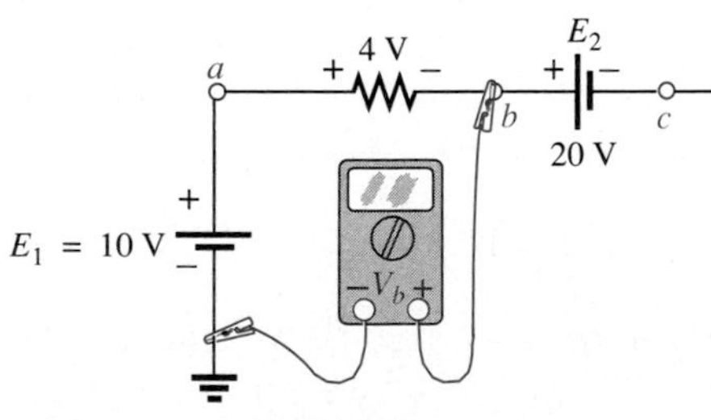

FIG. 5.42
Example 5.17.

Solution: Starting at ground potential (zero volts), we proceed through a rise of 10 V to reach point a and then pass through a drop in potential of 4 V to point b. The result is that the meter will read

$$V_b = +10\text{ V} - 4\text{ V} = \mathbf{6\text{ V}}$$

as clearly demonstrated by Fig. 5.43.

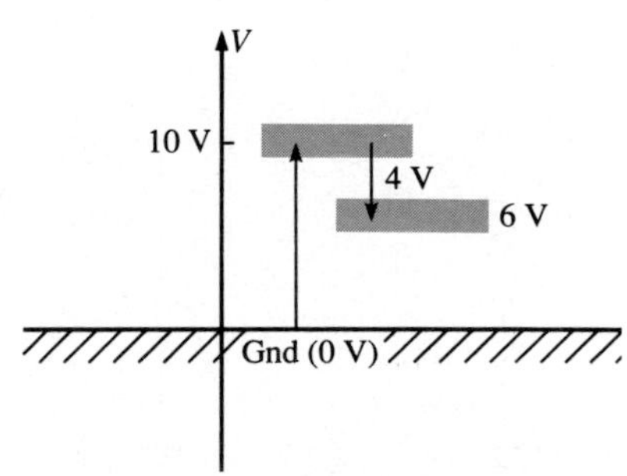

FIG. 5.43
Determining V_b using the defined voltage levels.

If we then proceed to point c, there is an additional drop of 20 V, resulting in

$$V_c = V_b - 20\text{ V} = 6\text{ V} - 20\text{ V} = \mathbf{-14\text{ V}}$$

as shown in Fig. 5.44.

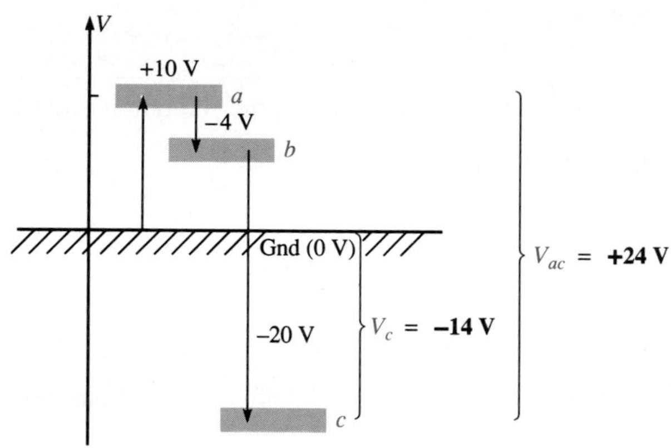

FIG. 5.44
Review of the potential levels for the circuit of Fig. 5.42.

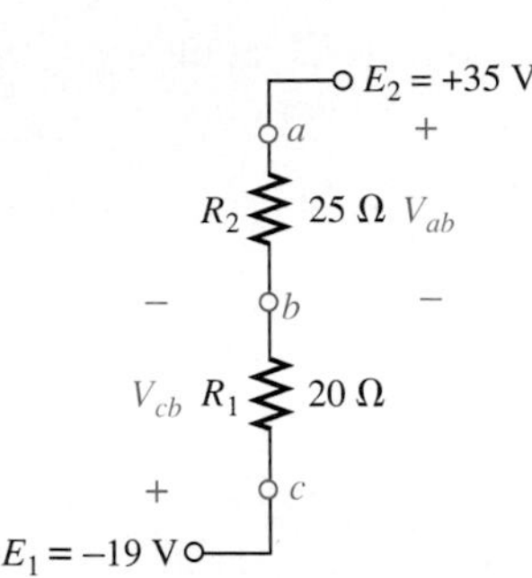

FIG. 5.45
Example 5.18.

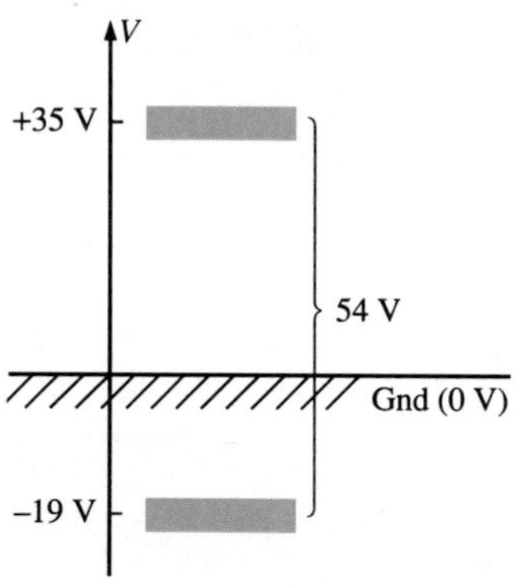

FIG. 5.46
Determining the total voltage drop across the resistive elements of Fig. 5.45.

The voltage V_{ac} can be obtained using Eq. (5.12) or by simply referring to Fig. 5.44:

$$V_{ac} = V_a - V_c = 10\text{ V} - (-14\text{ V})$$
$$= \mathbf{24\text{ V}}$$

EXAMPLE 5.18 Determine V_{ab}, V_{cb}, and V_c for the network of Fig. 5.45.

Solution: There are two ways to approach this problem. The first is to sketch the diagram of Fig. 5.46 and note that there is a 54-V drop across the series resistors R_1 and R_2. The current can then be determined using Ohm's law and the voltage levels as follows:

$$I = \frac{54\text{ V}}{45\ \Omega} = 1.2\text{ A}$$

$$V_{ab} = IR_2 = (1.2\text{ A})(25\ \Omega) = \mathbf{30\text{ V}}$$
$$V_{cb} = -IR_1 = -(1.2\text{ A})(20\ \Omega) = \mathbf{-24\text{ V}}$$
$$V_c = E_1 = \mathbf{-19\text{ V}}$$

The other approach is to redraw the network as shown in Fig. 5.47 to clearly establish the aiding effect of E_1 and E_2 and solve the resulting series circuit.

$$I = \frac{E_1 + E_2}{R_T} = \frac{19\text{ V} + 35\text{ V}}{45\ \Omega} = \frac{54\text{ V}}{45\ \Omega} = 1.2\text{ A}$$

and $\quad V_{ab} = \mathbf{30\text{ V}}, \quad V_{cb} = \mathbf{-24\text{ V}}, \quad V_c = \mathbf{-19\text{ V}}$

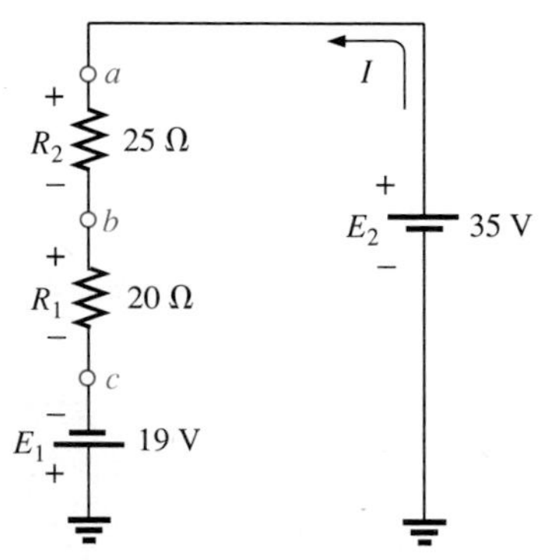

FIG. 5.47
Redrawing the circuit of Fig. 5.45 using standard dc voltage supply symbols.

EXAMPLE 5.19 Using the voltage divider rule, determine the voltages V_1 and V_2 of Fig. 5.48.

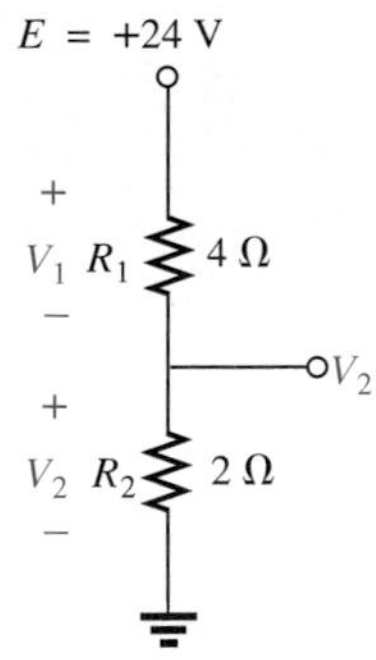

FIG. 5.48
Example 5.19.

Solution: Redrawing the network with the standard battery symbol will result in the network of Fig. 5.49. Applying the voltage divider rule,

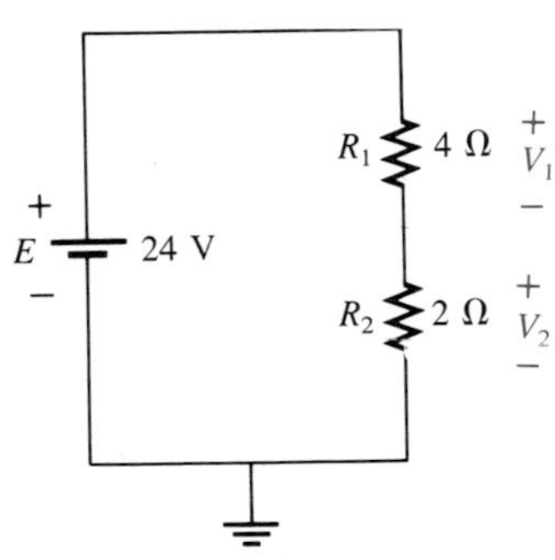

FIG. 5.49
Circuit of Fig. 5.48 redrawn.

$$V_1 = \frac{R_1E}{R_1 + R_2} = \frac{(4\ \Omega)(24\ \text{V})}{4\ \Omega + 2\ \Omega} = \mathbf{16\ V}$$

$$V_2 = \frac{R_2E}{R_1 + R_2} = \frac{(2\ \Omega)(24\ \text{V})}{4\ \Omega + 2\ \Omega} = \mathbf{8\ V}$$

EXAMPLE 5.20 For the network of Fig. 5.50:

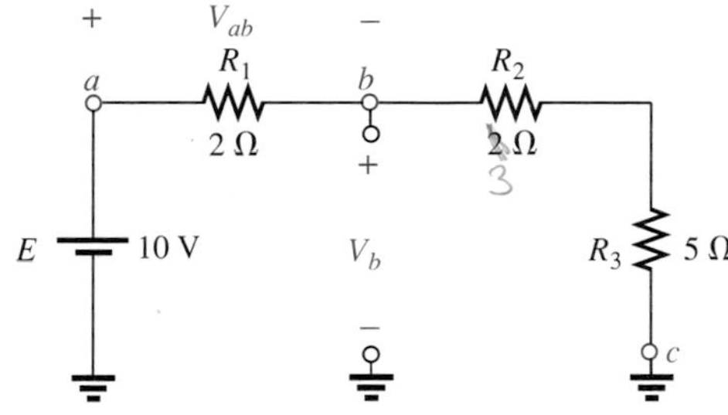

FIG. 5.50
Example 5.20.

a. Calculate V_{ab}.
b. Determine V_b.
c. Calculate V_c.

Solutions:

a. Voltage divider rule:

$$V_{ab} = \frac{R_1E}{R_T} = \frac{(2\ \Omega)(10\ \text{V})}{2\ \Omega + 3\ \Omega + 5\ \Omega} = \mathbf{+2\ V}$$

b. Voltage divider rule:

$$V_b = V_{R_2} + V_{R_3} = \frac{(R_2 + R_3)E}{R_T} = \frac{(3\ \Omega + 5\ \Omega)(10\ \text{V})}{10\ \Omega} = \mathbf{8\ V}$$

or $$V_b = V_a - V_{ab} = E - V_{ab} = 10\ \text{V} - 2\ \text{V} = \mathbf{8\ V}$$

c. V_c = ground potential = **0 V**

5.8 INTERNAL RESISTANCE OF VOLTAGE SOURCES

Every source of voltage, whether a generator, battery, or laboratory supply as shown in Fig. 5.51(a), will have some internal resistance. The equivalent circuit of any source of voltage will therefore appear as shown in Fig. 5.51(b). In this section, we will examine the effect of the internal resistance on the output voltage so that any unexpected changes in terminal characteristics can be explained.

In all the circuit analyses to this point, the ideal voltage source (no internal resistance) was used [see Fig. 5.52(a)]. The ideal voltage source has no internal resistance and an output voltage of E volts with no load or full load. In the practical case [Fig. 5.52(b)], where we con-

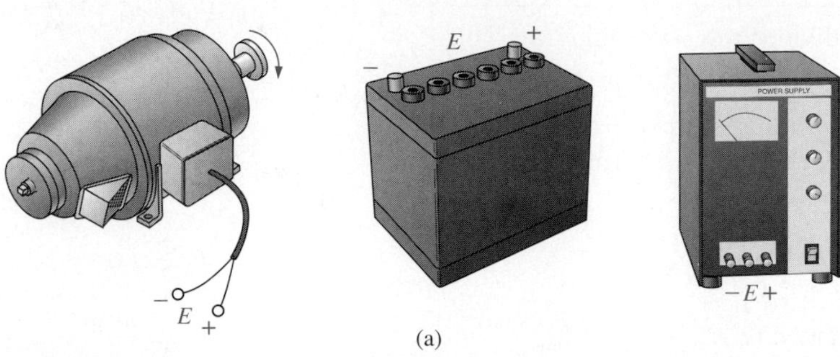

FIG. 5.51
(a) Sources of dc voltage; (b) equivalent circuit.

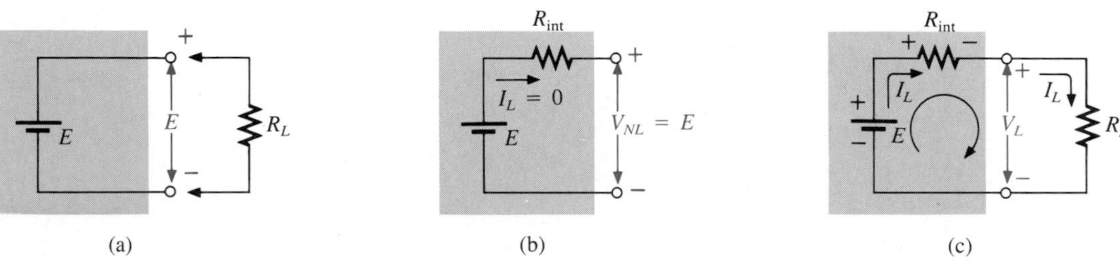

FIG. 5.52
Voltage source: (a) ideal, $R_{int} = 0\ \Omega$; (b) determining V_{NL}; (c) determining R_{int}.

sider the effects of the internal resistance, the output voltage will be E volts only when no-load ($I_L = 0$) conditions exist. When a load is connected [Fig. 5.52(c)], the output voltage of the voltage source will decrease due to the voltage drop across the internal resistance.

By applying Kirchhoff's voltage law around the indicated loop of Fig. 5.52(c), we obtain

$$E - I_L R_{int} - V_L = 0$$

or, since

$$E = V_{NL}$$

we have

$$V_{NL} - I_L R_{int} - V_L = 0$$

and

$$\boxed{V_L = V_{NL} - I_L R_{int}} \qquad \textbf{(5.13)}$$

If the value of R_{int} is not available, it can be found by first solving for R_{int} in the equation just derived for V_L; that is,

$$R_{int} = \frac{V_{NL} - V_L}{I_L} = \frac{V_{NL}}{I_L} - \frac{I_L R_L}{I_L}$$

and

$$\boxed{R_{int} = \frac{V_{NL}}{I_L} - R_L} \qquad \textbf{(5.14)}$$

A plot of the output voltage versus current appears in Fig. 5.53 for the dc generator having the circuit representation of Fig. 5.51(b). Note that any increase in load demand, starting at any level, causes an additional drop in terminal voltage due to the increasing loss in potential across the internal resistance. At maximum current, denoted by I_{FL}, the

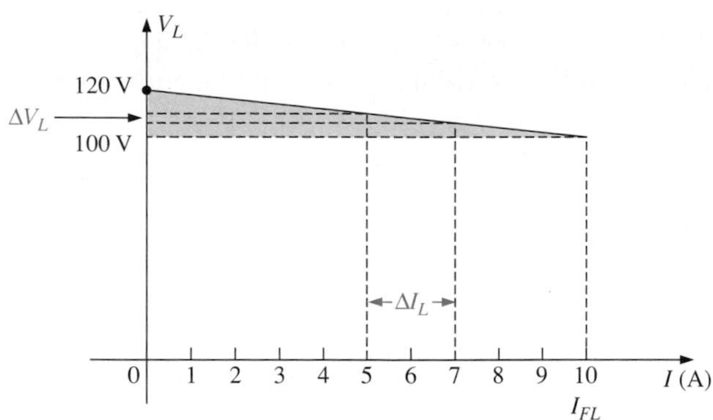

FIG. 5.53
V_L versus I_L for a dc generator with $R_{int} = 2\ \Omega$.

voltage across the internal resistance is $V_{\text{int}} = I_{FL}R_{\text{int}} = (10\ \text{A})(2\ \Omega) = 20\ \text{V}$, and the terminal voltage has dropped to 100 V—a significant difference when you can ideally expect a 120-V generator to provide the full 120 V if you stay below the listed full-load current. Eventually, if the load current were permitted to increase without limit, the voltage across the internal resistance would equal the supply voltage, and the terminal voltage would be zero. The larger the internal resistance, the steeper is the slope of the characteristics of Fig. 5.53. In fact, for any chosen interval of voltage or current, the magnitude of the internal resistance is given by

$$R_{\text{int}} = \frac{\Delta V_L}{\Delta I_L} \tag{5.15}$$

For the chosen interval of 5–7 A ($\Delta I_L = 2$ A) on Fig. 5.53, ΔV_L is 4 V, and $R_{\text{int}} = \Delta V_L/\Delta I_L = 4\ \text{V}/2\ \text{A} = 2\ \Omega$.

A direct consequence of the loss in output voltage is a loss in power delivered to the load. Multiplying both sides of Eq. (5.13) by the current I_L in the circuit, we obtain

$$\underset{\text{Power to load}}{I_L V_L} = \underset{\text{Power output by battery}}{I_L V_{NL}} - \underset{\text{Power loss in the form of heat}}{I_L^2 R_{\text{int}}} \tag{5.16}$$

EXAMPLE 5.21 Before a load is applied, the terminal voltage of the power supply of Fig. 5.54(a) is set to 40 V. When a load of 500 Ω is attached, as shown in Fig. 5.54(b), the terminal voltage drops to 38.5 V. What happened to the remainder of the no-load voltage, and what is the internal resistance of the source?

Solution: The difference of 40 V − 38.5 V = 1.5 V now appears across the internal resistance of the source. The load current is 38.5 V/0.5 kΩ = 77 mA. Applying Eq. (5.14),

$$R_{\text{int}} = \frac{V_{NL}}{I_L} - R_L = \frac{40\ \text{V}}{77\ \text{mA}} - 0.5\ \text{k}\Omega$$
$$= 519.48\ \Omega - 500\ \Omega = \mathbf{19.48\ \Omega}$$

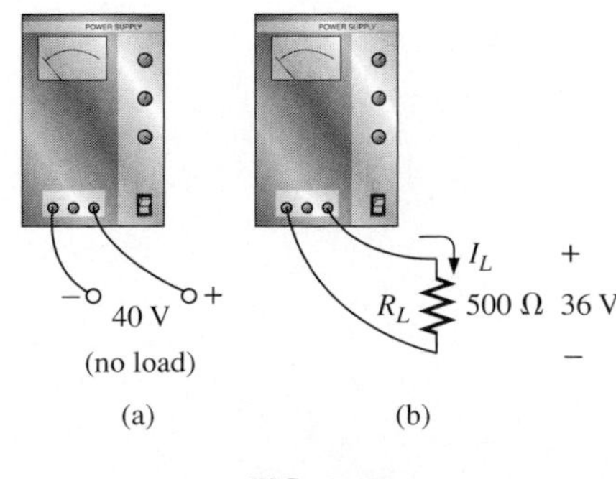

FIG. 5.54
Example 5.21.

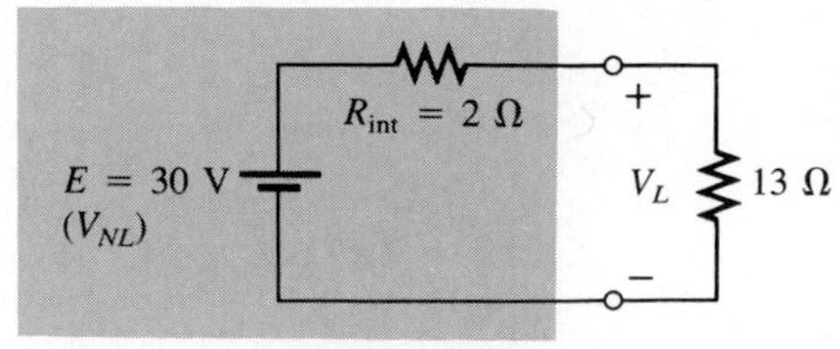

FIG. 5.55
Example 5.22.

EXAMPLE 5.22 The battery of Fig. 5.55 has an internal resistance of 2 Ω. Find the voltage V_L and the power lost to the internal resistance if the applied load is a 13-Ω resistor.

Solution:

$$I_L = \frac{30\text{ V}}{2\ \Omega + 13\ \Omega} = \frac{30\text{ V}}{15\ \Omega} = 2\text{ A}$$

$$V_L = V_{NL} - I_L R_{\text{int}} = 30\text{ V} - (2\text{ A})(2\ \Omega) = \mathbf{26\ V}$$

$$P_{\text{lost}} = I_L^2 R_{\text{int}} = (2\text{ A})^2(2\ \Omega) = (4)(2) = \mathbf{8\ W}$$

Procedures for measuring R_{int} will be described in Section 5.10.

5.9 VOLTAGE REGULATION

For any supply, ideal conditions dictate that for the range of load demand (I_L), the terminal voltage remain fixed in magnitude. In other words, if a supply is set for 12 V, it is desirable that it maintain this terminal voltage, even though the current demand on the supply may vary. A measure of how close a supply will come to ideal conditions is given by the voltage regulation characteristic. By definition, the voltage regulation of a supply between the limits of full-load and no-load conditions (Fig. 5.56) is given by the following:

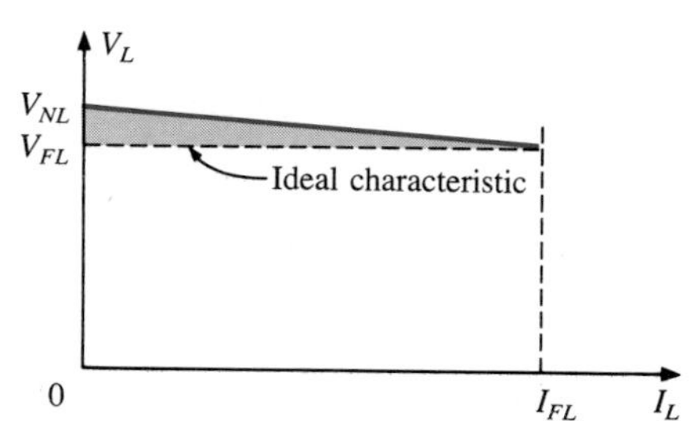

FIG. 5.56
Defining voltage regulation.

$$\text{Voltage regulation } (VR)\% = \frac{V_{NL} - V_{FL}}{V_{FL}} \times 100\% \tag{5.17}$$

For ideal conditions, $V_{FL} = V_{NL}$ and $VR\% = 0$. Therefore, *the smaller the voltage regulation, the less the variation in terminal voltage with change in load.*

It can be shown with a short derivation that the voltage regulation is also given by

$$VR\% = \frac{R_{\text{int}}}{R_L} \times 100\% \tag{5.18}$$

In other words, the smaller the internal resistance for the same load, the smaller the regulation and the more ideal the output.

EXAMPLE 5.23 Calculate the voltage regulation of a supply having the characteristics of Fig. 5.53.

Solution:

$$VR\% = \frac{V_{NL} - V_{FL}}{V_{FL}} \times 100\% = \frac{120\text{ V} - 100\text{ V}}{100\text{ V}} \times 100\%$$

$$= \frac{20}{100} \times 100\% = \mathbf{20\%}$$

EXAMPLE 5.24 Determine the voltage regulation of the supply of Fig. 5.54.

Solution:

$$VR\% = \frac{R_{\text{int}}}{R_L} \times 100\% = \frac{19.48\ \Omega}{500\ \Omega} \times 100\% \cong \mathbf{3.9\%}$$

5.10 MEASUREMENT TECHNIQUES

In Chapter 2, it was noted that ammeters are inserted in the branch in which the current is to be measured. We now realize that such a condition specifies that

ammeters are placed in series with the branch in which the current is to be measured

as shown in Fig. 5.57.

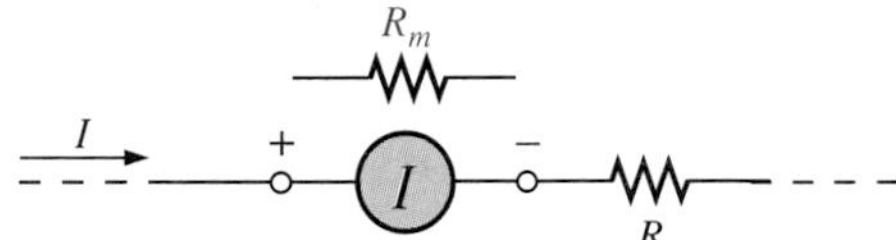

FIG. 5.57
Series connection of an ammeter.

If the ammeter is to have minimal impact on the behavior of the network, its resistance should be very small (ideally zero ohms) compared to the other series elements of the branch such as the resistor R of Fig. 5.57. If the meter resistance approaches or exceeds 10% of R, it would naturally have a significant impact on the current level it is measuring. It is also noteworthy that the resistances of the separate current scales of the same meter are usually not the same. In fact, the meter resistance normally increases with decreasing current levels. However, for the majority of situations one can simply assume that the internal ammeter resistance is small enough compared to the other circuit elements that it can be ignored.

For an up-scale (analog meter) or positive (digital meter) reading, an ammeter must be connected with current entering the positive terminal of the meter and leaving the negative terminal, as shown in Fig. 5.58. Since most meters employ a red lead for the positive terminal and a black lead for the negative, simply ensure that current enters the red lead and leaves the black one.

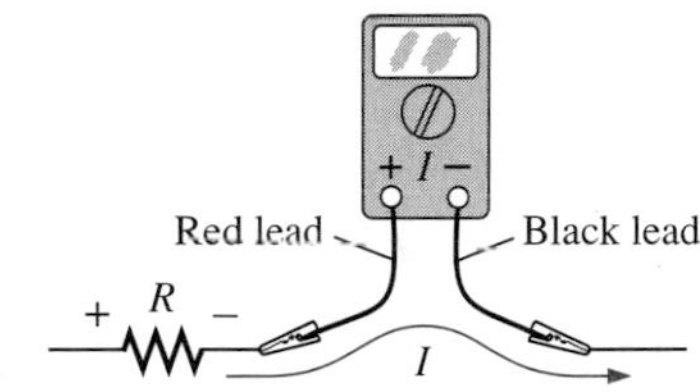

FIG. 5.58
Connecting an ammeter for an up-scale (positive) reading.

Voltmeters are always hooked up across the element for which the voltage is to be determined.

An up-scale or positive reading on a voltmeter is obtained by being sure that the positive terminal (red lead) is connected to the point of higher potential and the negative terminal (black lead) is connected to the lower potential, as shown in Fig. 5.59.

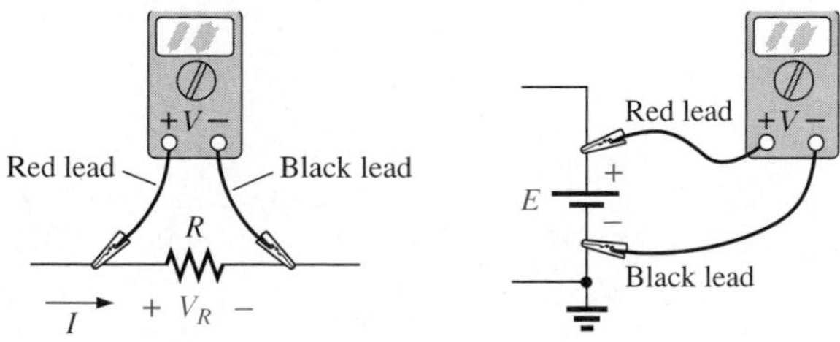

FIG. 5.59
Hooking up a voltmeter to obtain an up-scale (positive) reading.

For the double-subscript notation, always hook up the red lead to the first subscript and the black lead to the second; that is, to measure the voltage V_{ab} in Fig. 5.60, the red lead is connected to point a and the black lead is connected to point b. For single-subscript notation, hook up the red lead to the point of interest and the black lead to ground, as shown in Fig. 5.60 for V_a and V_b.

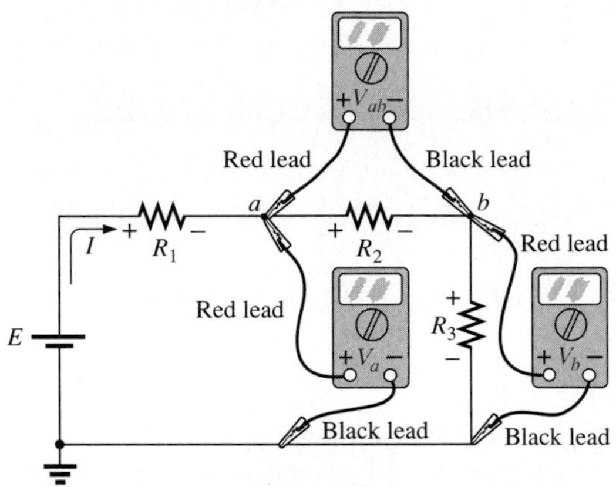

FIG. 5.60
Measuring voltages with double- and single-subscript notation.

The internal resistance of a supply cannot be measured with an ohmmeter due to the voltage present. However, the no-load voltage can be measured by simply hooking up the voltmeter as shown in Fig. 5.61(a). Do not be concerned about the apparent path for current that the meter seems to provide by completing the circuit. The internal resistance of the meter is usually sufficiently high to ensure that the resulting current is so small that it can be ignored. (Voltmeter loading effects will be discussed in detail in Section 6.9.) An ammeter could then be placed directly across the supply, as shown in Fig. 5.61(b), to measure the short-circuit current I_{SC} and R_{int} as determined by Ohm's law: $R_{int} = E_{NL}/I_{sc}$. However, since the internal resistance of the supply may be very low, performing the measurement could result in high current levels that could damage the meter and supply, and possibly cause dangerous side effects. The setup of Fig. 5.61(b) is therefore *not* suggested. A better approach would be to apply a resistive load that will result in a supply current of about half the maximum rated value and measure the terminal voltage. Then use Eq. (5.14).

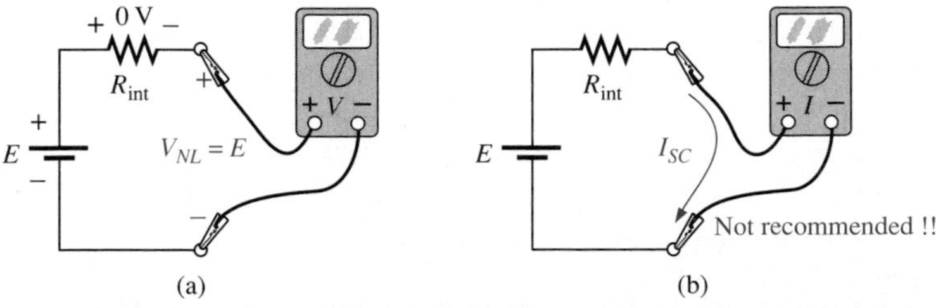

FIG. 5.61
(a) Measuring the no-load voltage E; (b) measuring the short-circuit current.

-S-

5.11 COMPUTER ANALYSIS

PSpice (DOS)

Let us now build on the foundation established in the last chapter to analyze series dc circuits. A new command or two may be introduced, but in general, the input file for PSpice will remain fairly straightforward. If you want to limit the output file of an analysis using PSpice to specific values or control the format of the data, the following command statement must be included somewhere between the title line and the .END statement:

.DC Source Name Starting Value Final Value Increment

First note that after the required .DC notation (to specify dc analysis), the remainder of the line is concerned with the applied dc source. The .DC command permits the analysis of a dc network for a *range* of source voltages that extend from the starting value to the final value in increments that are under the user's control.

If the analysis is for a fixed source, the following format can be employed where the starting and final values are the same. The increment is required in the format but is ignored by the package. Any number can be used for the increment, but we will use the number 1 throughout this text.

.DC Source Name Starting Value Starting Value 1

It is important to realize that once entered, the .DC command overrides the line of the network description for the dc voltage source. However, the voltage source must still appear in the network description.

Once the .DC command is employed, output data can be obtained only by specifying the quantities desired and the format. The following command will provide an output file for a dc analysis:

.PRINT DC Desired Quantities

Note the continued use of the period at the beginning of important control lines of the input file.

Let us now use the above to analyze the series network of Fig. 5.62. First the nodes are labeled with the ground connection defined by the 0 label. Note that PSpice requires that the nodes be defined numerically rather than using the letters *a, b, c,* etc., as employed in the text material.

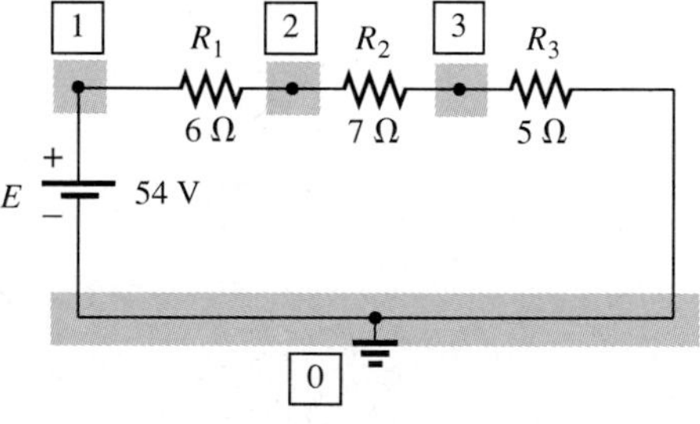

FIG. 5.62
Applying PSpice (DOS) to a series dc circuit.

The title line and circuit description follow the formats defined in the previous chapters. Now, however, we want to limit our output to the voltages V_2 and V_{R_2} and the series current I. This control of the output file requires the use of the .DC analysis line, which specifies the name of the source as VE, the beginning and ending values at 54 V (a fixed value), and the 1 for completeness. The .PRINT statement specifies the desired output values V(2), V(R2), and I(R1). Obviously, since it is a series circuit, I(R1) = I(R2) = I(R3). In addition V(R2) = V(2,3). The .OPTIONS NOPAGE command continues to minimize paper use. The output file appears in Fig. 5.63 with the requested values in the order appearing in the .PRINT command. The voltage V_2 is 36 V, V_{R_2} = 21 V, and $I = E/R_T$ = 54 V/18 Ω = 3 A, which will all be verified by the programs to follow.

```
Chapter 5 - Three Series Resistors

 ****     CIRCUIT DESCRIPTION

******************************************************************************


VE 1 0 54V
R1 1 2 6
R2 2 3 7
R3 3 0 5
.DC VE 54 54 1
.PRINT DC V(2) V(R2) I(R1)
.OPTIONS NOPAGE
.END

 ****     DC TRANSFER CURVES                    TEMPERATURE =   27.000 DEG C


  VE           V(2)         V(R2)        I(R1)

  5.400E+01    3.600E+01    2.100E+01    3.000E+00
```

FIG. 5.63
Input and output files for the circuit of Fig. 5.62.

PSpice (Windows)

The same series network analyzed using PSpice (DOS) will now be investigated using PSpice (Windows) to permit a comparison between the two approaches. First, the supply will be set with **Draw-Get New Part-Browse-source.slb-VDC-OK,** followed by a left click to place the source and a right click to end the process. Double-clicking the label **V1** will result in the **Edit Reference Designator,** which permits changing the label to E, as shown in Fig. 5.64. After typing in E and clicking **OK,** the new label will appear on the schematic in a small box. By clicking on the box and holding the clicker down, the new label can be moved to an appropriate position. Releasing the clicker will leave it in the chosen place. One more click anywhere on the schematic, and the rectangular defining areas will disappear. The magnitude of the source can now be changed from the given 0 V to 54 V by double-clicking the **0 V** to obtain the **Set Attribute Value** dialog box and entering 54. An **OK** followed by the placement procedure described above, and the voltage source is set.

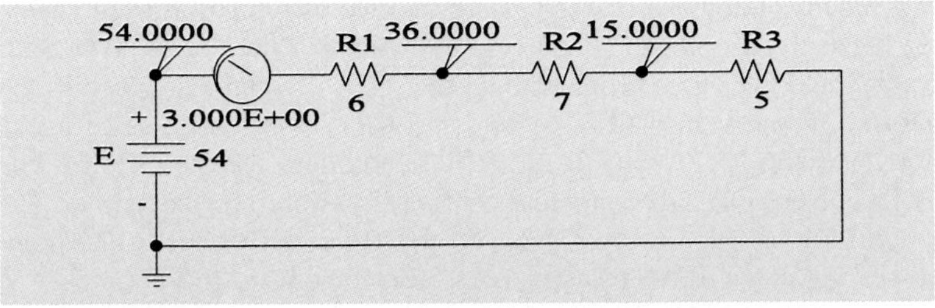

FIG. 5.64
Applying schematics to the series network of Fig. 5.62.

Next the sequence **Draw-Get New Part-Browse-analog.slb-R-OK** will result in a resistor appearing on the schematic. A single left click and the resistor is set. However, three resistors are needed. Rather than return to the library for another resistor, simply move the new resistor that automatically appears and click it in place. Repeat the process once more and then end the placement of resistive elements with a right-side click of the mouse. If you need to move a resistor once all of them are placed, simply click the resistor symbol once on the screen and then click it again but hold the clicker down. Move to the new position and release the clicker—it is set in the new position. Since we want to know the current, **Draw-Get New Part-Browse-special.slb-IPROBE-OK** will result in an ammeter symbol that can be placed in the circuit. Once it appears on the screen, use **Ctrl-R** to turn it once so the circular scale of the symbol is to the left and the pointer to the right. **IPROBE** applies a positive sign to the current entering the scale side of the symbol and leaving the pointer. Now move all the elements so they are nicely spaced, like Fig. 5.63, and connect the elements with **Draw-Wire.** Move the pencil to the beginning of the desired line and, after a left click, draw the desired line. At the end of the line, a single click will end the line and the connection will be in place. Finally, a ground connection must be introduced (for all networks) from the **port.slb** library using **EGND.** Once it appears, move it to the appropriate position, click in place, and then end the process with a right click. The resistor values can now be set by double-clicking the magnitude of a resistor and entering the desired value in the dialog box that appears.

The analysis is set up with **Analysis-Setup-Bias Point Detail-Close** and it is initiated by **Analysis-Simulate.** Once the analysis is complete, the results will appear on the schematic as shown in Fig. 5.63. Since the voltages are with respect to ground, the voltage across an element is the difference between the levels at each end of the element. The voltage across R_1 is therefore 54 V − 36 V = 18 V with 36 V − 15 V = 21 V across R_2 and 15 V across R_3. As indicated by **IPROBE,** the current is 3 A.

If the voltage across a resistor or the power to an element is desired, start with the following sequence: **Analysis-Setup-DC Sweep.** Choose **Voltage Source-Linear-Name:E** followed by **Start Value: 54, End Value: 54, Increment: 1.** Then be sure the small box next to **DC Sweep** is enabled. Following **Close,** choose **Probe Setup** and select **Automatically Run Probe After Simulation-OK.** Initiate the analysis with **Analysis-Simulate** and wait for the **Probe** heading. Then choose **Trace-Add-Add Traces-Alias Names** (to be sure all possibilities are available), followed by a click in the **Trace Command** box. By typing in **V(R1:1)-V(R1:2)-OK,** the voltage across the resistor R_1 will be displayed on the screen as shown in Fig. 5.65. Since it is difficult to read with the scale provided, use **Tools-Cursor-Display**, and an intersection will appear on the screen at 54 V with the result in the **Cursor** dialog box at the bottom of the display. Note in the small narrow box the numbers 54 and 18, which reveal that when $E = 54$ V, the voltage V_{R_1} is 18 V. If the power to R_1 were desired, the following entry would be made in the **Trace Command** box: **(V(R1:1)-V(R1:2))*I(R1),** with the cursor revealing a result of 54 W.

C++

We will now turn to the C++ language and review a program designed to perform the same analysis just performed using PSpice. As noted in

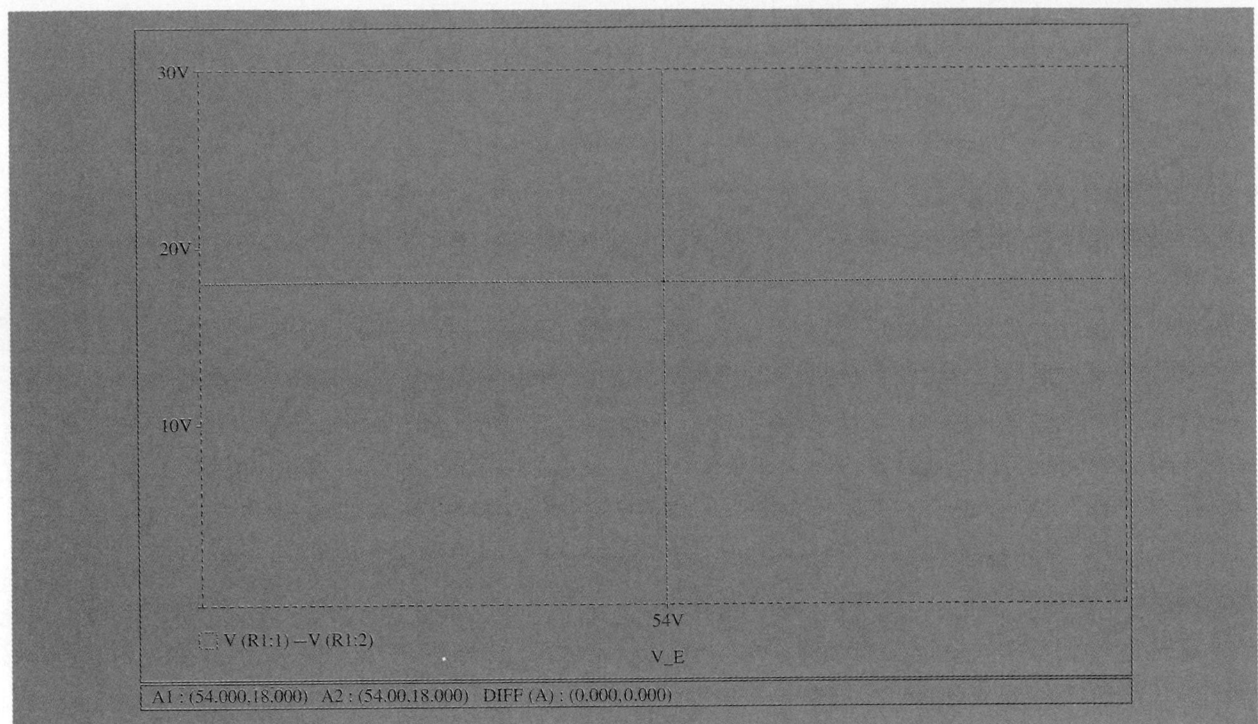

FIG. 5.65
Using Probe to display the voltage across the resistor R_1 of Fig. 5.64.

earlier chapters, do not expect to understand all the details of how the program was written and why specific paths were taken. The purpose here is simply to expose the reader to the general characteristics of a program using this increasingly popular programming language.

First take note of all the double forward slashes // in the program of Fig. 5.66. They are used to identify comments in the program that will not be recognized by the compiler when the program is run. They can also be used to remind the programmer about specific objectives to be met at a particular point in the program or the reason for specific entries. In this example, however, the primary purpose was to enlighten the reader about the purpose of a particular entry or operation.

The *#include* tells the computer to include the file to follow in the C++ program. The *<iostream.h>* is a header file that sets up the input–output path between the program and the disk operating system. The *class* format defines the data type (in this case all floating points, which means that a decimal point is included), and the *public* within the { } reveals that the variables *value, voltage,* and *power* are accessible for operations outside the data structure.

Note that the *main* ( ) part of the program extends all the way down to the bottom, as identified by the braces { }. Within this region all the parameters of the network will be given values, all the calculations will be made, and finally all the results will be provided. Next, three resistor objects are established. *Rtotal* is defined as a floating variable, and a voltage source object is introduced. The values of the resistors are then entered, and the total resistance is calculated. Next, through *cout,* the total resistance is printed out using the *Rtotal* just calculated. The \n" at the end of the *cout* line calls for a carriage return to prepare for the next *cout* statement.

Heading · Preprocessor directive · Define variables and data type · Establish objects and R_T · Assign values · Calculate R_T · Display R_T · Define E · $I = E/R_T$ · Display I · Calculate V_R · Display V_R · Calculate P_R · Display P_R · Calculate P_E · Display P_E · Body of program

```
//C++ Series Circuit Analysis

#include <iostream.h>                    //needed for input/output

class resistor {                         //define resistor class
public:                                  //allow access to variables in class
        float value;                     //resistance in ohms
        float voltage;                   //voltage across resistor
        float power;                     //power used by resistor
};

class voltage_source {                   //define voltage source class
public:
        float voltage;                   //source voltage
        float current;                   //source current
        float power;                     //power supplied by source
};

main()                                   //execution begins here
{
        resistor R1, R2, R3;            //create three resistor objects
        float Rtotal;                   //total resistance variable
        voltage_source V1;              //create voltage source object

        R1.value = 6;                   //assign resistance values
        R2.value = 7;
        R3.value = 5;
        Rtotal = R1.value + R2.value + R3.value;  //find atotal resistance
        cout << "The total resistance is " << Rtotal << " Ohms.\n";

        V1.voltage = 54;                                 //assign source voltage
        V1.current = V1.voltage / Rtotal;                //find circuit current
        cout << "The circuit current is " << V1.current << " Amperes.\n";

        R1.voltage = V1.current * R1.value;              //find resistor voltages
        R2.voltage = V1.current * R2.value;
        R3.voltage = V1.current * R3.value;
        cout << "The voltage across R1 is " << R1.voltage << " Volts.\n";
        cout << "The voltage across R2 is " << R2.voltage << " Volts.\n";
        cout << "The voltage across R3 is " << R3.voltage << " Volts.\n";

        R1.power = V1.current * R1.voltage;              //find resistor powers
        R2.power = V1.current * R2.voltage;
        R3.power = V1.current * R3.voltage;
        cout << "The power to R1 is " << R1.power << " Watts.\n";
        cout << "The power to R2 is " << R2.power << " Watts.\n";
        cout << "The power to R3 is " << R3.power << " Watts.\n";

        V1.power = V1.voltage * V1.current;              //find total power
        cout << "The total power is " << V1.power << " Watts.\n";
}
```

FIG. 5.66

C++ program designed to perform a complete analysis of the network of Fig. 5.62 or Fig. 5.64.

On the next line, the magnitude of the voltage source is introduced, followed by the calculation of the source current, which is then printed out on the next line. The voltage across each resistor is then calculated and printed out by the succeeding lines. Finally, the various powers are calculated and printed out.

When run, the output will appear as shown in Fig. 5.67 with the same results obtained using PSpice. As noted above, do not be perplexed by all the details of why certain lines appear as they do. Like everything, with proper instruction and experience, it will all become fairly obvious. Do note, however, that the first few lines set up the analysis to be performed by telling the computer the type of operations that need to be handled and the format of the data to be entered and expected. There is then a main part of the program where all the entries, calculations, and outputs are performed. When run, the flow of this program is top–down; that is, one step follows the other without loops back to certain points (an option to be described in a later program). There was no need to number the lines or to include detailed instructions. If all the comments were removed, the actual program would be quite compact and straightforward, with most of the body of the program being *cout* statements.

```
The total resistance is 18 Ohms.
The circuit current is 3 Amperes.
The voltage across R1 is 18 Volts.
The voltage across R2 is 21 Volts.
The voltage across R3 is 15 Volts.
The power to R1 is 54 Watts.
The power to R2 is 63 Watts.
The power to R3 is 45 Watts.
The total power is 162 Watts.
```

FIG. 5.67
Output results for the C++ program of Fig. 5.66.

BASIC

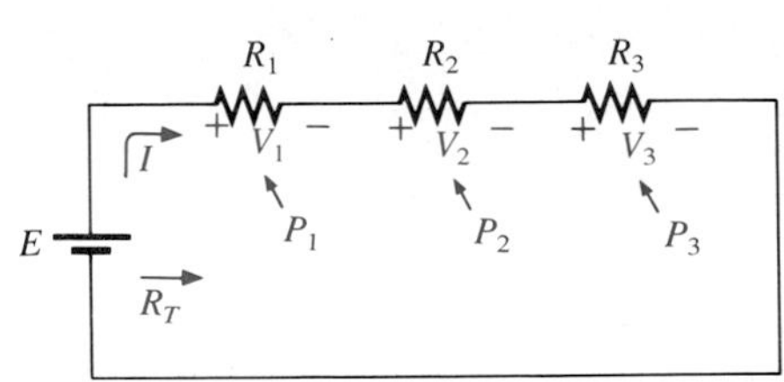

FIG. 5.68
Defining the parameters for a BASIC program.

The network of Fig. 5.62 (or Fig. 5.64) will be analyzed once more using BASIC to complete the comparison between four fairly popular methods of circuit analysis. The parameters to be defined in the BASIC program of Fig. 5.69 appear in the defining network of Fig. 5.68. When the program is run, it will request the network parameters and provide the output of the run in Fig. 5.69. Since the parameters are entered, the program is applicable for any resistor and source values. The total resistance R_T is calculated on line 150; I, on line 180; the voltage across each resistor, on line 230; and the power to each resistor, on line 250. In addition, Kirchhoff's voltage law is applied on line 300 to show that the applied voltage equals the sum of the voltage drops. The total power supplied or dissipated is then calculated on line 310.

A review of the BASIC program reveals that it is a fairly neat package, the command statements use English words that are easily understood, the sequence of operations is quite logical, and in general the program seems friendlier at first inspection than the C++ program. However, one must understand that, from an implementation standpoint, C++ provides a lower-level access to the computer system's hardware. In other words, the communication link between a C++ program and the computer hardware that will perform the program operations is more direct than that with a BASIC program. The result is a savings in overall system design and an increase in operating speeds. Both languages have their place in technology, with the user's overall goal being the deciding factor.

```
    10 REM ***** PROGRAM 5-1 *****
    20 REM ***********************************************
    30 REM Analysis of series resistor network
    40 REM ***********************************************
    50 REM
    100 PRINT : PRINT "Enter resistor values for up to 3 resistors"
    110 PRINT "in series (enter 0 if no resistor):"
    120 INPUT "R1="; R1
    130 INPUT "R2="; R2
    140 INPUT "R3="; R3
RT  150 RT = R1 + R2 + R3
    160 PRINT : PRINT "The total resistance is RT="; RT; "ohms"
    170 PRINT : INPUT "Enter value of supply voltage, E="; E
I   180 I = E / RT
    190 PRINT
    200 PRINT "Supply current is, I="; I; "amperes"
    210 PRINT
    220 PRINT "The voltage drop across each resistor is:"
Vx  230 V1 = I * R1: V2 = I * R2: V3 = I * R3
    240 PRINT "V1="; V1; "volts   V2="; V2; "volts   V3="; V3; "volts"
Px  250 P1 = I ^ 2 * R1: P2 = I ^ 2 * R2: P3 = I ^ 2 * R3
    260 PRINT
    270 PRINT "The power dissipated by each resistor is:"
    280 PRINT "P1="; P1; "watts", "P2="; P2; "watts", "P3="; P3; "watts"
    290 PRINT
    300 PRINT "Total voltage around loop is, V1+V2+V3="; V1 + V2 + V3; "volts"
    310 PRINT "and total power dissipated, P1+P2+P3="; P1 + P2 + P3; "watts"
    320 END
```

```
Enter resistor values for up to 3 resistors
in series (enter 0 if no resistor):
R1=? 6
R2=? 7
R3=? 5

The total resistance is RT= 18 ohms

Enter value of supply voltage, E=? 54

Supply current is, I= 3 amperes

The voltage drop across each resistor is:
V1= 18 volts   V2= 21 volts   V3= 15 volts

The power dissipated by each resistor is:
P1= 54 watts   P2= 63 watts   P3= 45 watts

Total voltage around loop is, V1+V2+V3= 54 volts
and total power dissipated, P1+P2+P3= 162 watts
Ok
```

FIG. 5.69

Input and output files for a complete analysis of the circuit of Fig. 5.68 using BASIC.

PROBLEMS

SECTION 5.2 Series Circuits

1. Find the total resistance and current I for each circuit of Fig. 5.70.

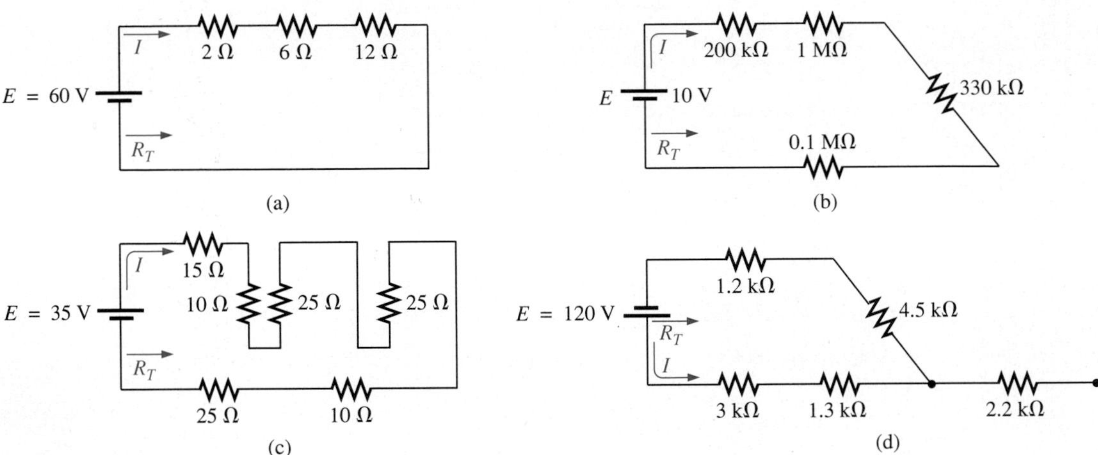

FIG. 5.70
Problem 1.

2. For the circuits of Fig. 5.71, the total resistance is specified. Find the unknown resistances and the current I for each circuit.

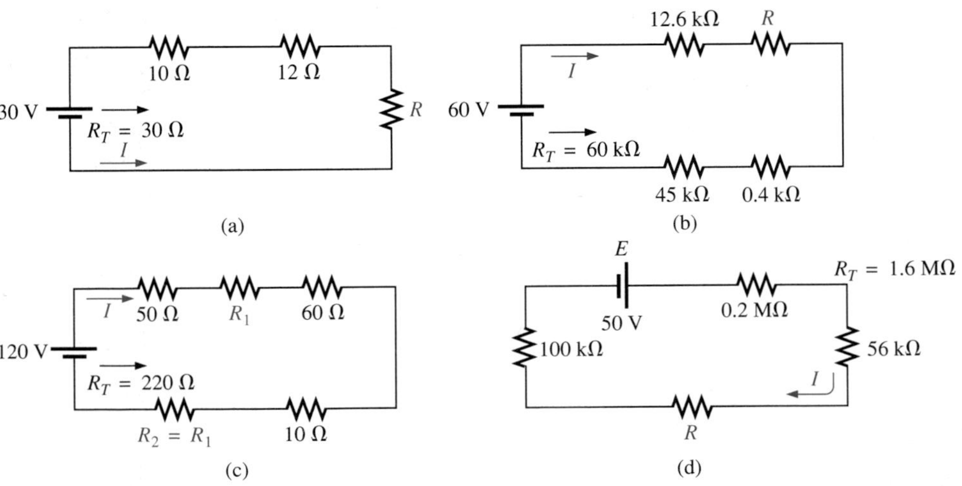

FIG. 5.71
Problem 2.

3. Find the applied voltage E necessary to develop the current specified in each network of Fig. 5.72.

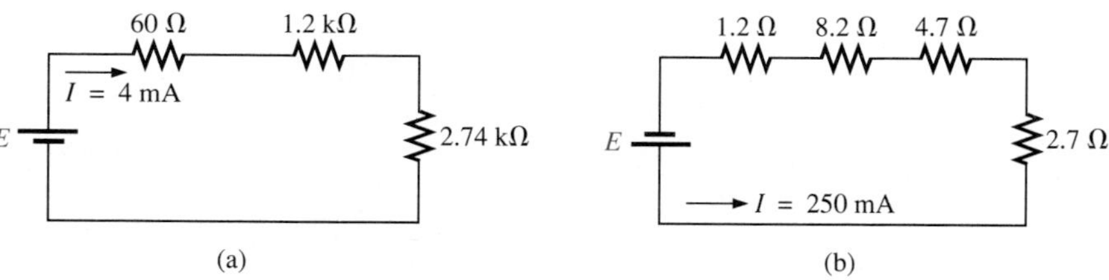

FIG. 5.72
Problem 3.

*4. For each network of Fig. 5.73, determine the current I, the source voltage E, the unknown resistance, and the voltage across each element.

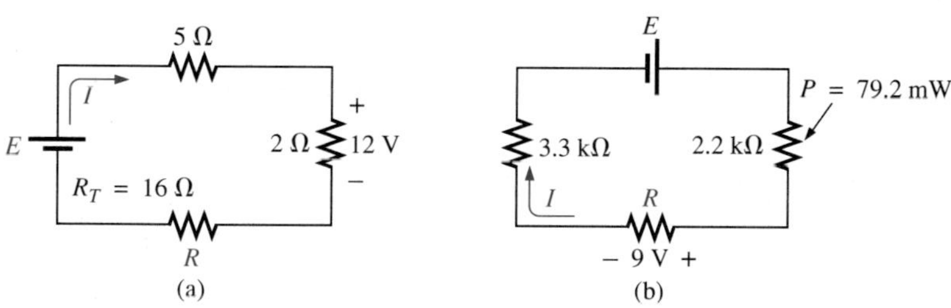

FIG. 5.73
Problem 4.

SECTION 5.3 Voltage Sources in Series

5. Determine the current I and its direction for each network of Fig. 5.74. Before solving for I, redraw each network with a single voltage source.

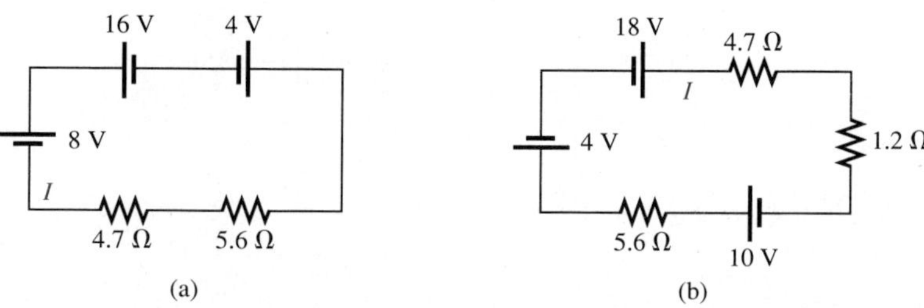

FIG. 5.74
Problem 5.

*6. Find the unknown voltage source and resistor for the networks of Fig. 5.75. Also indicate the direction of the resulting current.

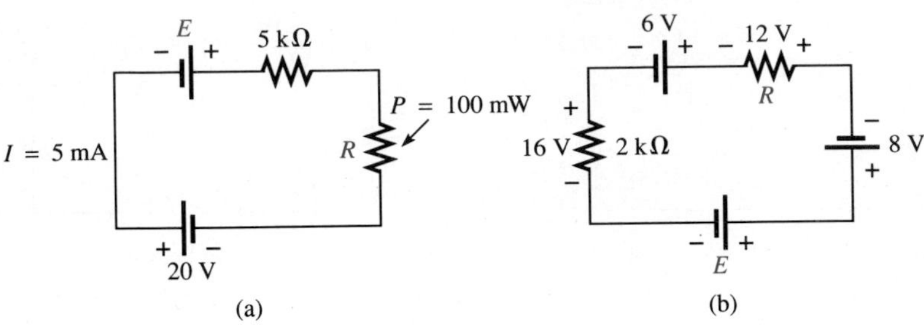

FIG. 5.75
Problem 6.

SECTION 5.4 Kirchhoff's Voltage Law

7. Find V_{ab} with polarity for the circuits of Fig. 5.76. Each box can contain a load or a power supply, or a combination of both.

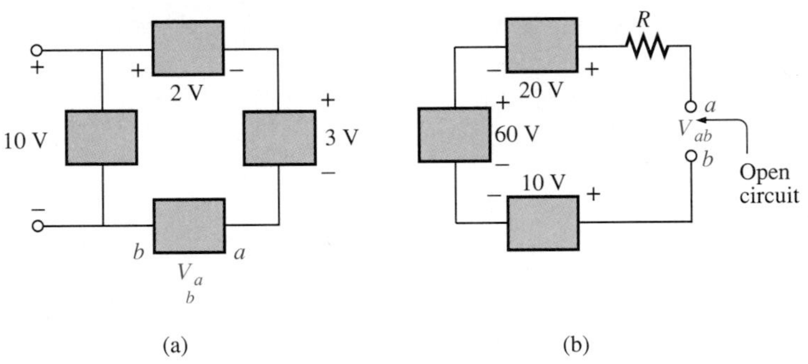

FIG. 5.76
Problem 7.

8. Although the networks of Fig. 5.77 are not simply series circuits, determine the unknown voltages using Kirchhoff's voltage law.

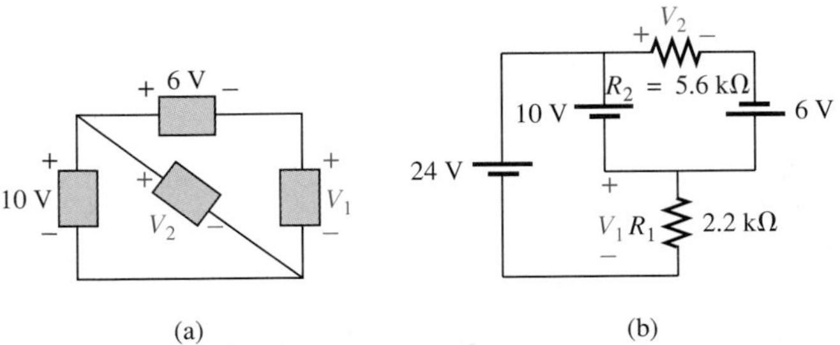

FIG. 5.77
Problem 8.

9. Determine the current I and the voltage V_1 for the network of Fig. 5.78.

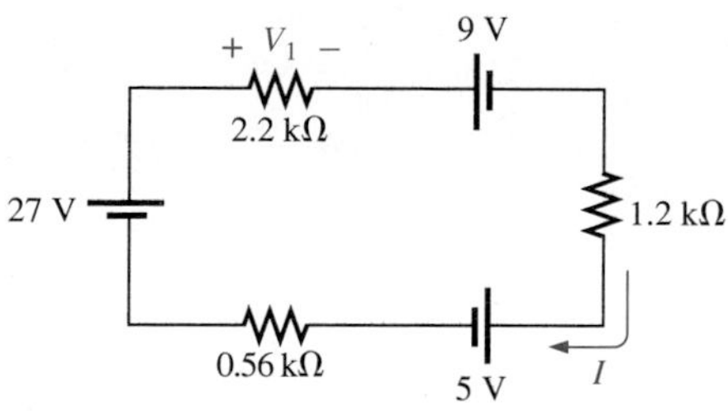

FIG. 5.78
Problem 9.

10. For the circuit of Fig. 5.79:

a. Find the total resistance, current, and unknown voltage drops.

b. Verify Kirchhoff's voltage law around the closed loop.

c. Find the power dissipated by each resistor, and note whether the power delivered is equal to the power dissipated.

d. If the resistors are available with wattage ratings of 1/2, 1, and 2 W, what minimum wattage rating can be used for each resistor in this circuit?

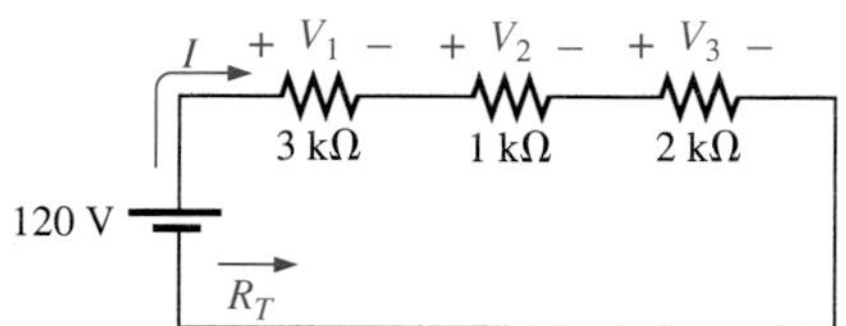

FIG. 5.79
Problem 10.

11. Repeat Problem 10 for the circuit of Fig. 5.80.

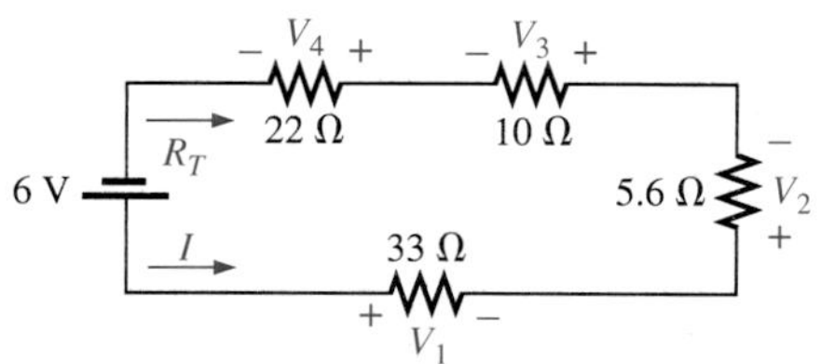

FIG. 5.80
Problem 11.

***12.** Find the unknown quantities in the circuits of Fig. 5.81 using the information provided.

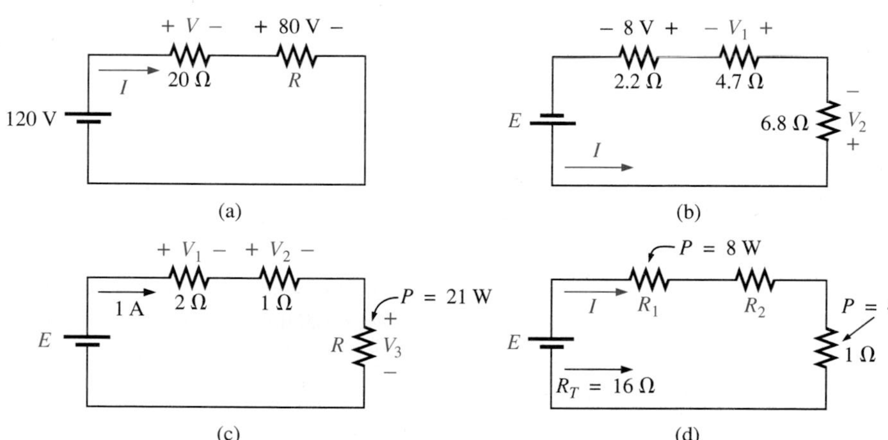

FIG. 5.81
Problem 12.

13. Eight holiday lights are connected in series as shown in Fig. 5.82.

a. If the set is connected to a 120-V source, what is the current through the bulbs if each bulb has an internal resistance of $28\frac{1}{8}\ \Omega$?

b. Determine the power delivered to each bulb.

c. Calculate the voltage drop across each bulb.

d. If one bulb burns out (that is, the filament opens), what is the effect on the remaining bulbs?

FIG. 5.82
Problem 13.

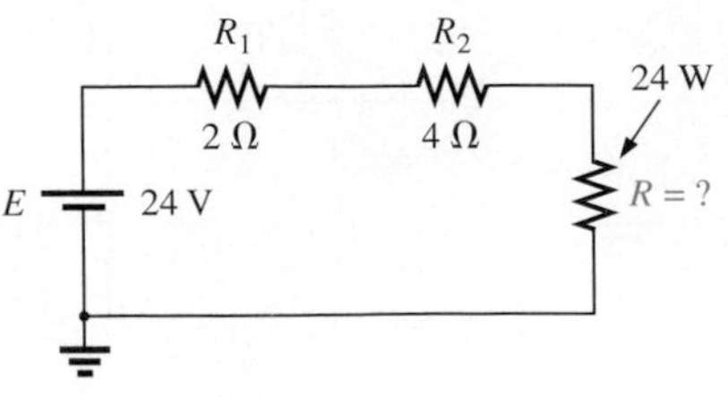

FIG. 5.83
Problem 14.

*14. For the conditions specified in Fig. 5.83, determine the unknown resistance.

SECTION 5.6 Voltage Divider Rule

15. Using the voltage divider rule, find V_{ab} (with polarity) for the circuits of Fig. 5.84.

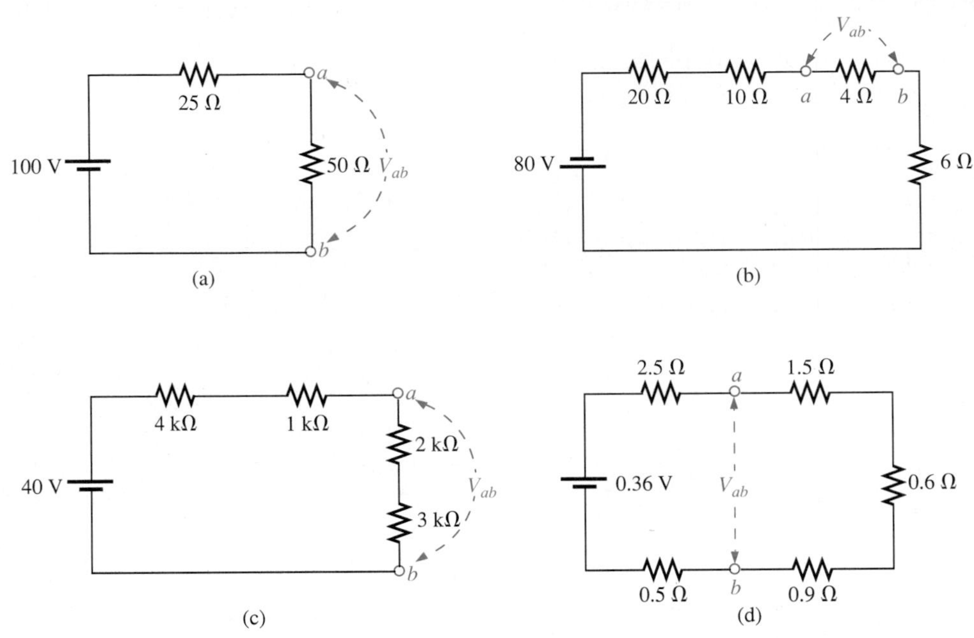

FIG. 5.84
Problem 15.

16. Find the unknown resistance using the voltage divider rule and the information provided for the circuits of Fig. 5.85.

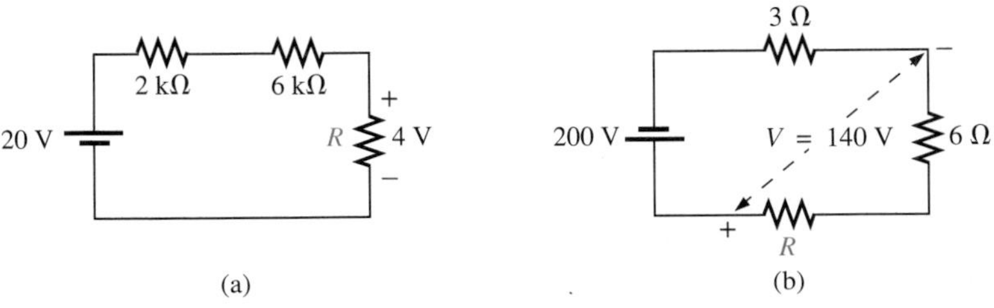

FIG. 5.85
Problem 16.

17. Referring to Fig. 5.86:
    a. Determine V_2 by simply noting that $R_2 = 3R_1$.
    b. Calculate V_3.
    c. Noting the magnitude of V_3 as compared to V_2 or V_1, determine R_3 by inspection.
    d. Calculate the source current I.
    e. Calculate the resistance R_3 using Ohm's law and compare it to the result of part (c).

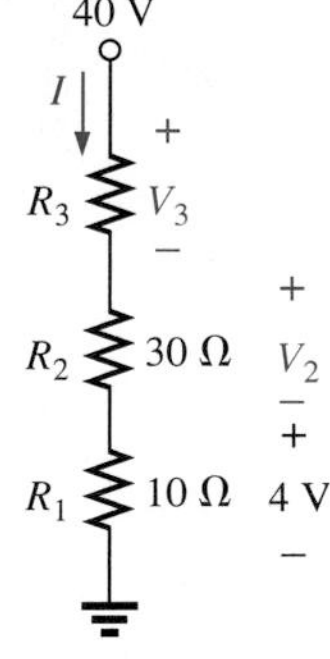

FIG. 5.86
Problem 17.

18. Given the information appearing in Fig. 5.87, find the level of resistance for R_1 and R_3.

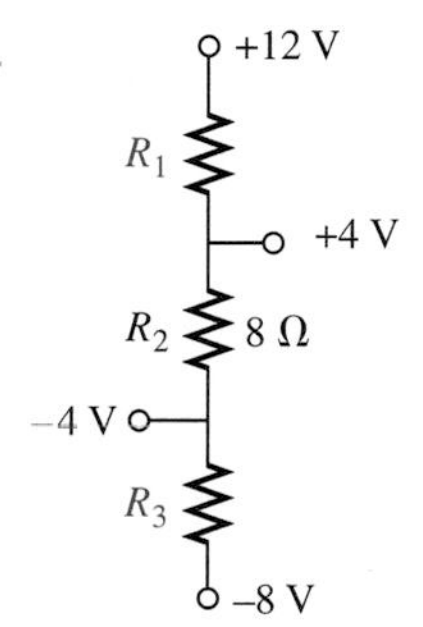

FIG. 5.87
Problem 18.

19. a. Design a voltage divider circuit that will permit the use of an 8-V, 50-mA bulb in an automobile with a 12-V electrical system.
    b. What is the minimum wattage rating of the chosen resistor if (1/4)-W, (1/2)-W and 1-W resistors are available?

20. Determine the values of R_1, R_2, R_3, and R_4 for the voltage divider of Fig. 5.88 if the source current is 16 mA.

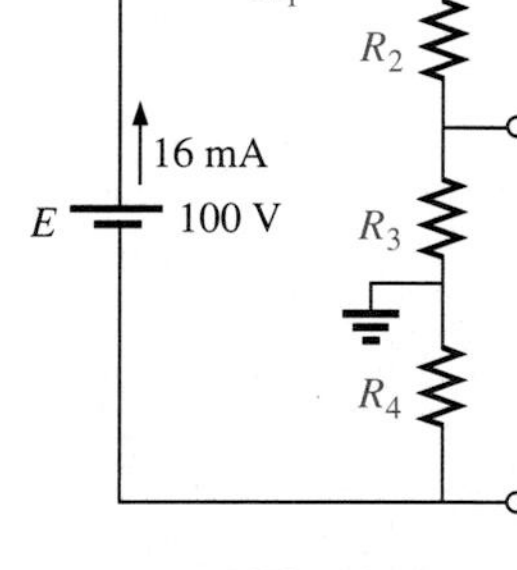

FIG. 5.88
Problem 20.

21. Design the voltage divider of Fig. 5.89 such that $V_{R_1} = (1/5)V_{R_2}$ if $I = 4$ mA.

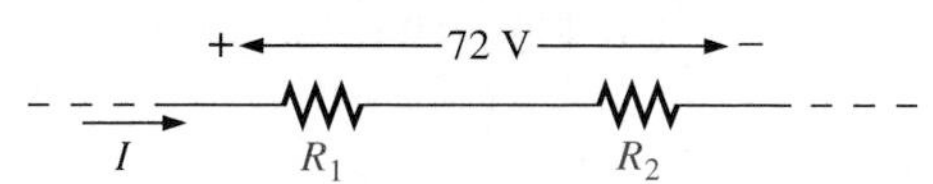

FIG. 5.89
Problem 21.

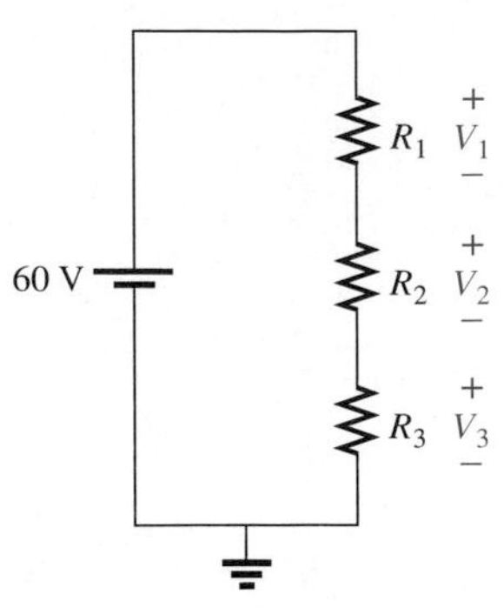

FIG. 5.90
Problem 22.

22. Find the voltage across each resistor of Fig. 5.90 if $R_1 = 2R_3$ and $R_2 = 7R_3$.

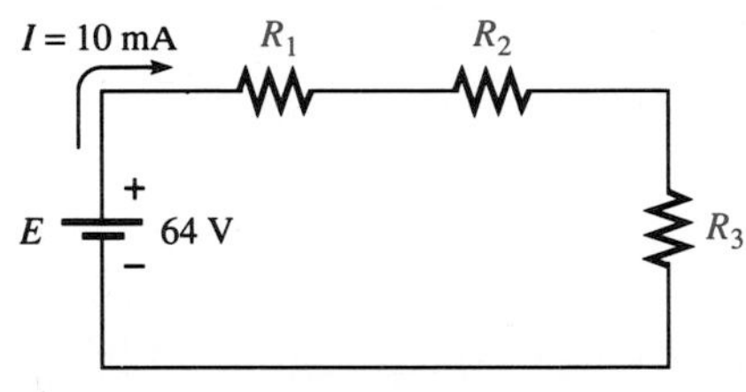

FIG. 5.91
Problem 23.

23. **a.** Design the circuit of Fig. 5.91 such that $V_{R_2} = 3V_{R_1}$ and $V_{R_3} = 4V_{R_2}$.
b. If the current I is reduced to 10 μA, what are the new values of R_1, R_2, and R_3? How do they compare to the results of part (a)?

SECTION 5.7 Notation

24. Determine the voltages V_a, V_b, and V_{ab} for the networks of Fig. 5.92.

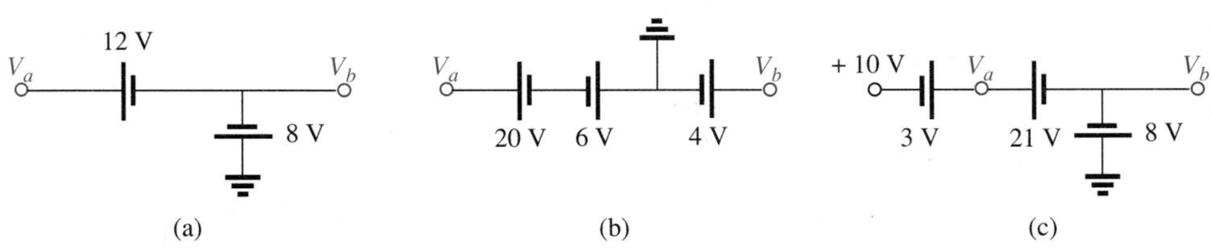

FIG. 5.92
Problem 24.

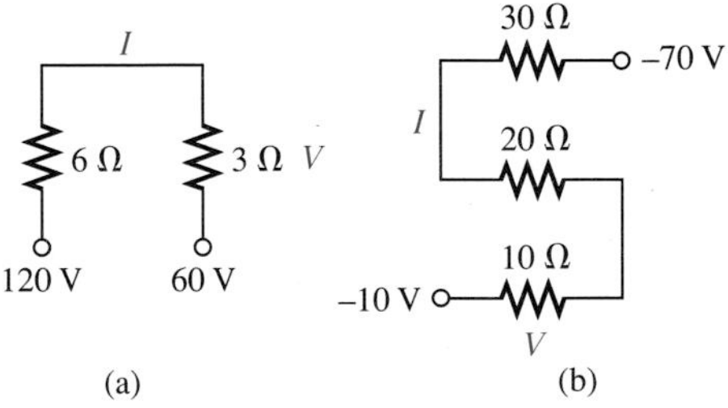

FIG. 5.93
Problem 25.

25. Determine the current I (with direction) and the voltage V (with polarity) for the networks of Fig. 5.93.

26. Determine the voltages V_a and V_1 for the networks of Fig. 5.94.

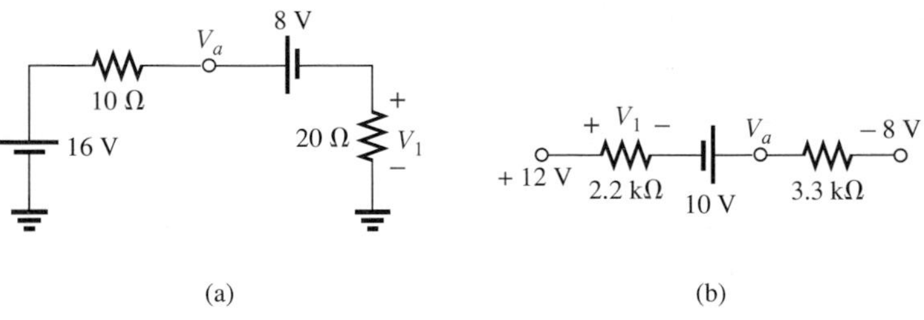

FIG. 5.94
Problem 26.

***27.** For the network of Fig. 5.95, determine the voltages:
a. V_a, V_b, V_c, V_d, V_e
b. V_{ab}, V_{dc}, V_{cb}
c. V_{ac}, V_{db}

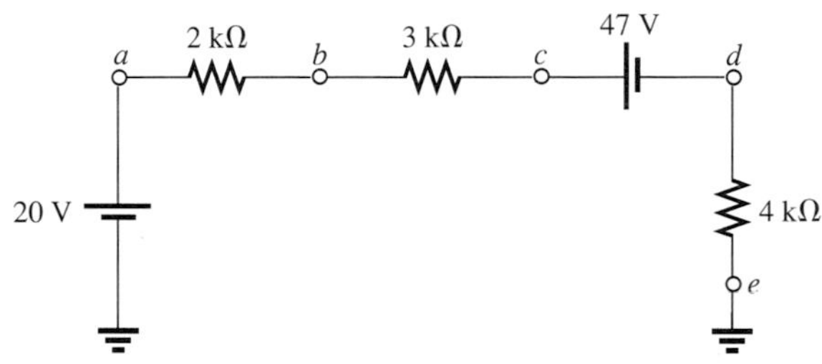

FIG. 5.95
Problem 27.

***28.** For the network of Fig. 5.96, determine the voltages:
a. V_a, V_b, V_c, V_d
b. V_{ab}, V_{cb}, V_{cd}
c. V_{ad}, V_{ca}

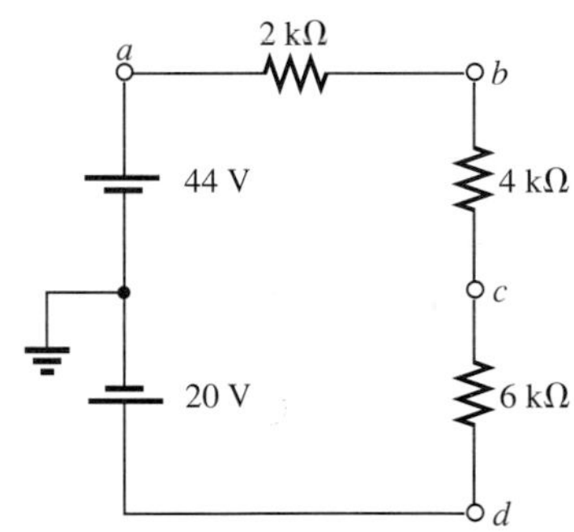

FIG. 5.96
Problem 28.

***29.** For the integrated circuit of Fig. 5.97, determine V_0, V_4, V_7, V_{10}, V_{23}, V_{30}, V_{67}, V_{56}, and I (magnitude and direction).

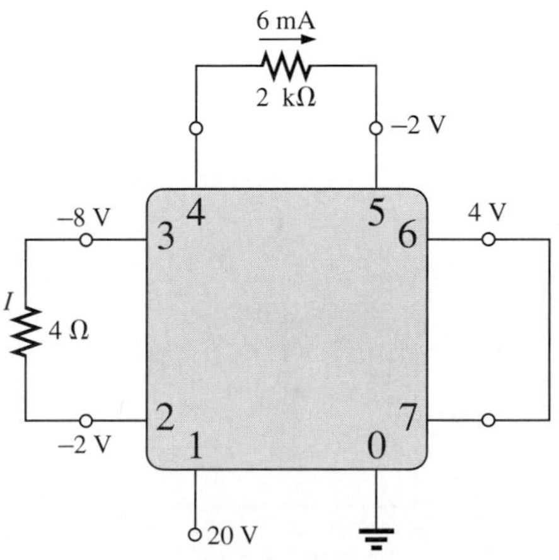

FIG. 5.97
Problem 29.

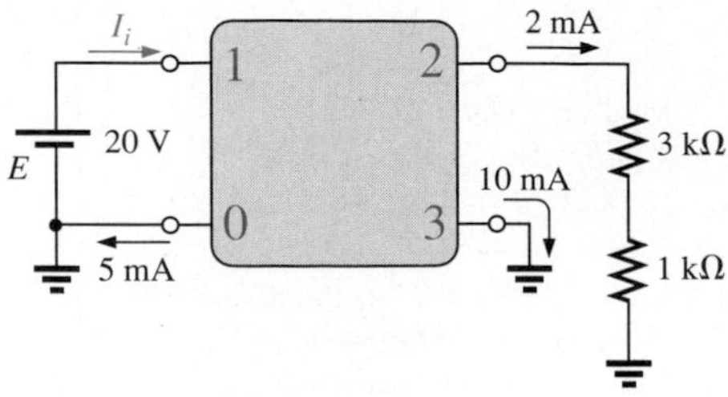

FIG. 5.98
Problem 30.

***30.** For the integrated circuit of Fig. 5.98, determine V_0, V_{03}, V_2, V_{23}, V_{12}, and I_i.

SECTION 5.8 Internal Resistance of Voltage Sources

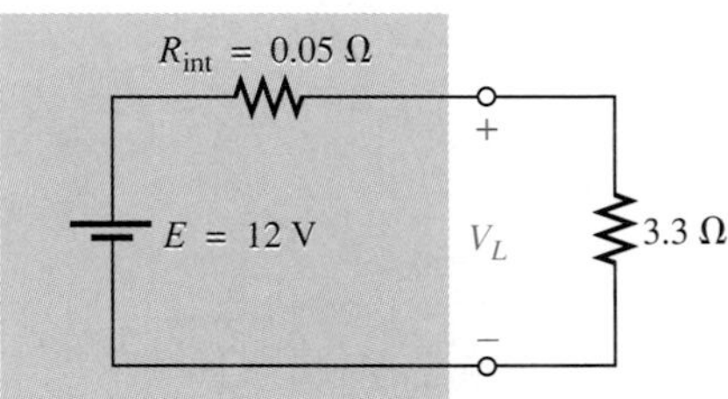

FIG. 5.99
Problem 32.

31. Find the internal resistance of a battery that has a no-load output voltage of 60 V and supplies a current of 2 A to a load of 28 Ω.

32. Find the voltage V_L and the power loss in the internal resistance for the configuration of Fig. 5.99.

33. Find the internal resistance of a battery that has a no-load output voltage of 6 V and supplies a current of 10 mA to a load of 1/2 kΩ.

SECTION 5.9 Voltage Regulation

34. Determine the voltage regulation for the battery of Problem 31.

35. Calculate the voltage regulation for the supply of Fig. 5.99.

SECTION 5.11 Computer Analysis

PSpice (DOS)

36. Write an input file to provide the voltages and current for the network of Fig. 5.79.

37. Write an input file to solve for the voltages across the resistors of Fig. 5.96.

PSpice (Windows)

38. Using schematics, determine the current I and the voltage across each resistor for the network of Fig. 5.70(a).

39. Using schematics, determine the voltage V_{ab} for the network of Fig. 5.84(d).

Programming Language (C++, BASIC, PASCAL, etc.)

40. Write a program to determine the total resistance of any number of resistors in series.

41. Write a program that will apply the voltage divider rule to either resistor of a series circuit with a single source and two series resistors.

42. Write a program to tabulate the current and power to the resistor R_L of the network of Fig. 5.100 for a range of values for R_L from 1 Ω to 20 Ω. Print out the value of R_L that results in maximum power to R_L.

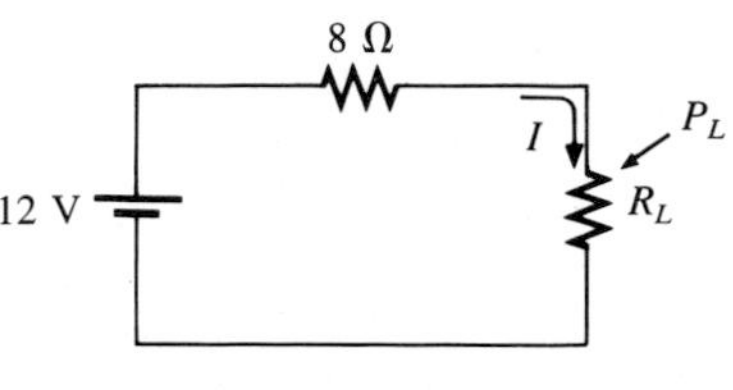

FIG. 5.100
Problem 42.

GLOSSARY

Branch The portion of a circuit consisting of one or more elements in series.

Circuit A combination of a number of elements joined at terminal points providing at least one closed path through which charge can flow.

Closed loop Any continuous connection of branches that allows tracing of a path that leaves a point in one direction and returns to that same point from another direction without leaving the circuit.

Conventional current flow A defined direction for the flow of charge in an electrical system that is opposite to that of the motion of electrons.

Electron flow The flow of charge in an electrical system having the same direction as the motion of electrons.

Internal resistance The inherent resistance found internal to any source of energy.

Kirchhoff's voltage law The algebraic sum of the potential rises and drops around a closed loop (or path) is zero.

Series circuit A circuit configuration in which the elements have only one point in common and each terminal is not connected to a third, current-carrying element.

Voltage divider rule A method by which a voltage in a series circuit can be determined without first calculating the current in the circuit.

Voltage regulation (VR) A value, given as a percent, that provides an indication of the change in terminal voltage of a supply with a change in load demand.

6

Parallel Circuits

6.1 INTRODUCTION

There are two network configurations that form the framework for some of the most complex network structures. A clear understanding of each will pay enormous dividends as more complex methods and networks are examined. The series connection was discussed in detail in the last chapter. We will now examine the *parallel* connection and all the methods and laws associated with this important configuration.

6.2 PARALLEL ELEMENTS

Two elements, branches, or networks are in parallel if they have two points in common.

In Fig. 6.1, for example, elements 1 and 2 have terminals *a* and *b* in common; they are therefore in parallel.

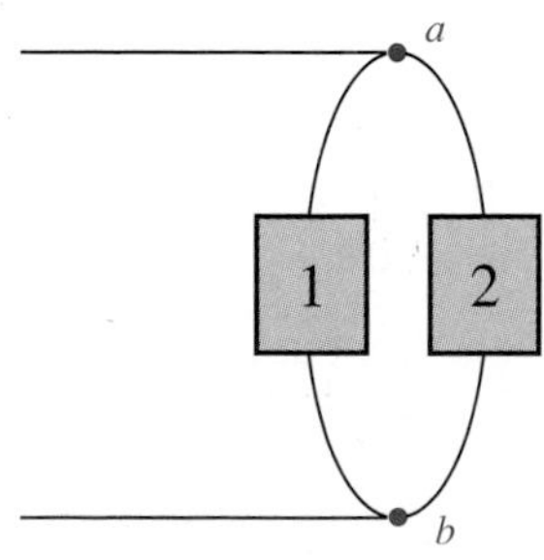

FIG. 6.1
Parallel elements.

In Fig. 6.2, all the elements are in parallel because they satisfy the above criterion. Three configurations are provided to demonstrate how the parallel networks can be drawn. Do not let the squaring of the con-

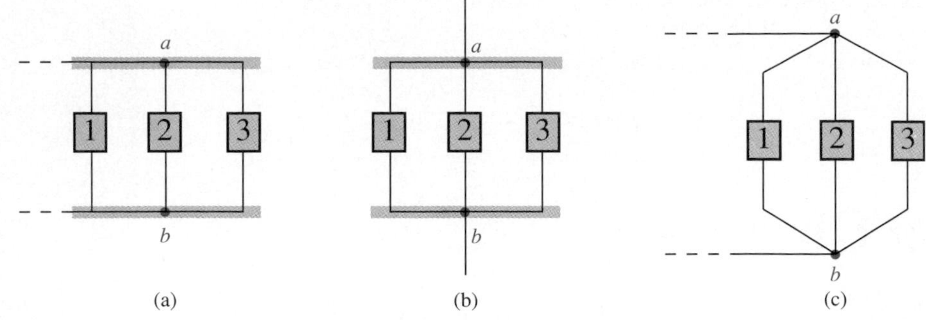

FIG. 6.2
Different ways in which three parallel elements may appear.

nection at the top and bottom of Fig. 6.2(a) and (b) cloud the fact that all the elements are connected to one terminal point at the top and bottom, as shown in Fig. 6.2(c).

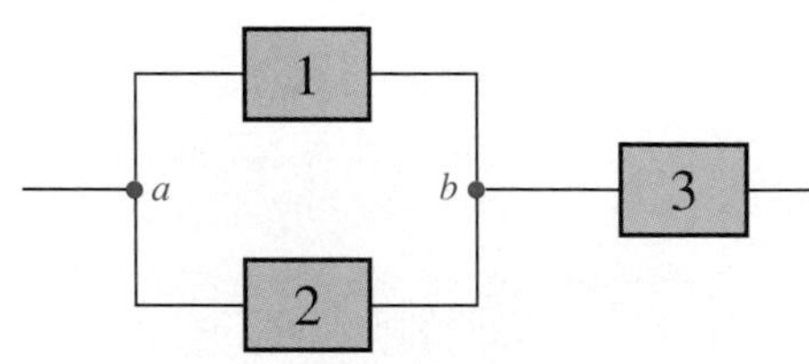

FIG. 6.3
Network in which 1 and 2 are in parallel and 3 is in series with the parallel combination of 1 and 2.

In Fig. 6.3, elements 1 and 2 are in parallel because they have terminals *a* and *b* in common. The parallel combination of 1 and 2 is then in series with element 3 due to the common terminal point *b*.

In Fig. 6.4, elements 1 and 2 are in series due to the common point *a*, but the series combination of 1 and 2 is in parallel with element 3 as defined by the common terminal connections at *b* and *c*.

In Figs. 6.1 through 6.4, the numbered boxes were used as a general symbol representing either single resistive elements, batteries, or complex network configurations.

Common examples of parallel elements include the rungs of a ladder, the tying of more than one rope between two points to increase the strength of the connection, and the use of pipes between two points to split the water between the two points at a ratio determined by the area of the pipes.

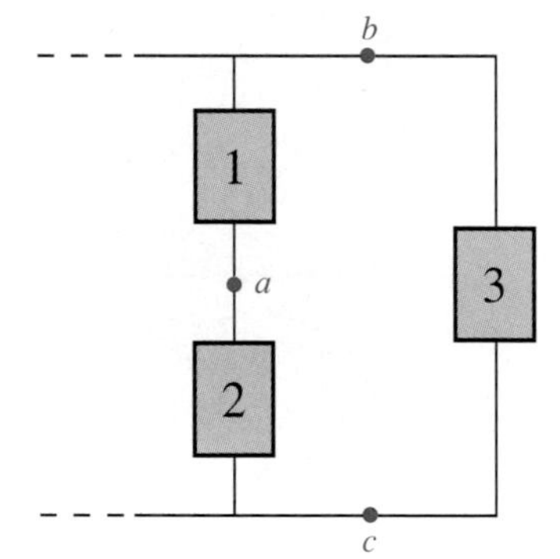

FIG. 6.4
Network in which 1 and 2 are in series and 3 is in parallel with the series combination of 1 and 2.

6.3 TOTAL CONDUCTANCE AND RESISTANCE

Recall that for series resistors, the total resistance is the sum of the resistor values.

For parallel elements, the total conductance is the sum of the individual conductances.

That is, for the parallel network of Fig. 6.5, we write

$$G_T = G_1 + G_2 + G_3 + \cdots + G_N \tag{6.1}$$

Since increasing levels of conductance will establish higher current levels, the more terms appearing in Eq. (6.1), the higher the input cur-

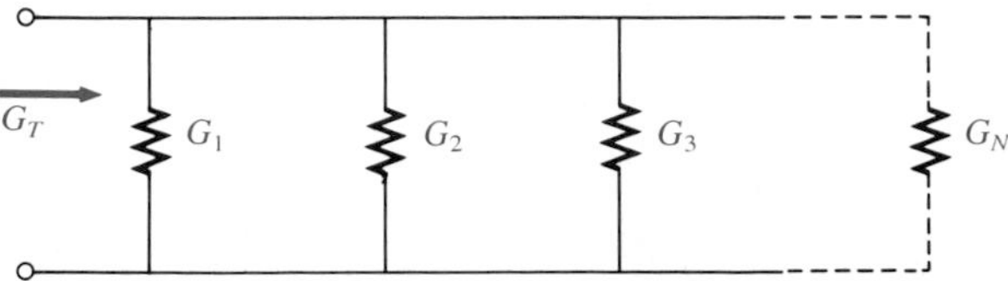

FIG. 6.5
Determining the total conductance of parallel conductances.

rent level. In other words, as the number of resistors in parallel increases, the input current level will increase for the same applied voltage—the opposite effect of increasing the number of resistors in series.

Substituting resistor values for the network of Fig. 6.5 will result in the network of Fig. 6.6. Since $G = 1/R$, the total resistance for the network can be determined by direct substitution into Eq. (6.1):

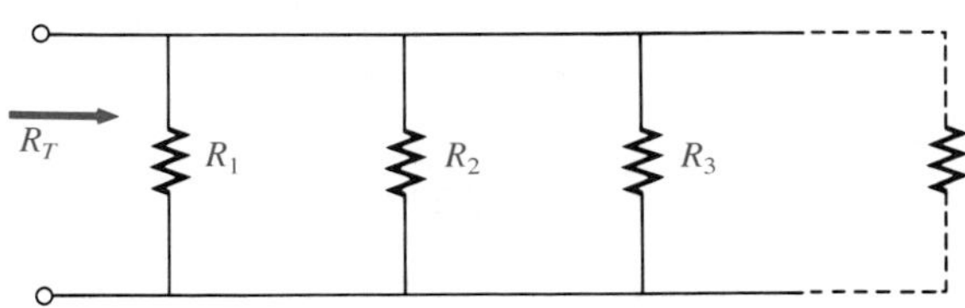

FIG. 6.6
Determining the total resistance of parallel resistors.

$$\frac{1}{R_T} = \frac{1}{R_1} + \frac{1}{R_2} + \frac{1}{R_3} + \cdots + \frac{1}{R_N} \qquad \textbf{(6.2)}$$

Note that the equation is for 1 divided by the total resistance rather than the total resistance. Once the sum of the terms to the right of the equals sign has been determined, it will than be necessary to divide the result into 1 to determine the total resistance. The following examples will demonstrate the additional calculations introduced by the inverse relationship.

EXAMPLE 6.1 Determine the total conductance and resistance for the parallel network of Fig. 6.7.

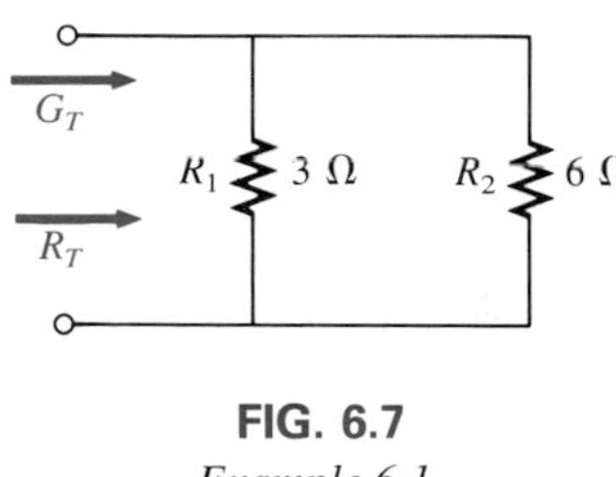

FIG. 6.7
Example 6.1.

Solution:

$$G_T = G_1 + G_2 = \frac{1}{3\ \Omega} + \frac{1}{6\ \Omega} = 0.333\ \text{S} + 0.167\ \text{S} = \mathbf{0.5\ S}$$

and
$$R_T = \frac{1}{G_T} = \frac{1}{0.5\ \text{S}} = \mathbf{2\ \Omega}$$

EXAMPLE 6.2 Determine the effect on the total conductance and resistance of the network of Fig. 6.7 if another resistor of 10 Ω were added in parallel with the other elements.

Solution:

$$G_T = 0.5\ \text{S} + \frac{1}{10\ \Omega} = 0.5\ \text{S} + 0.1\ \text{S} = \mathbf{0.6\ S}$$

$$R_T = \frac{1}{G_T} = \frac{1}{0.6\ \text{S}} \cong \mathbf{1.667\ \Omega}$$

Note, as mentioned above, that adding additional terms increases the conductance level and decreases the resistance level.

EXAMPLE 6.3 Determine the total resistance for the network of Fig. 6.8.

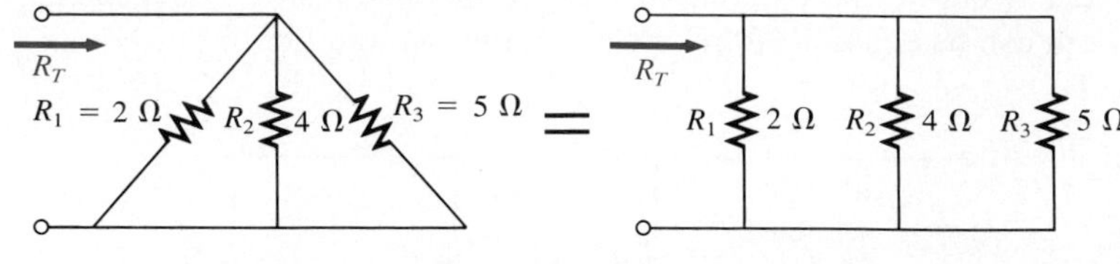

FIG. 6.8
Example 6.3.

Solution:

$$\begin{aligned}\frac{1}{R_T} &= \frac{1}{R_1} + \frac{1}{R_2} + \frac{1}{R_3} \\ &= \frac{1}{2\ \Omega} + \frac{1}{4\ \Omega} + \frac{1}{5\ \Omega} = 0.5\ \text{S} + 0.25\ \text{S} + 0.2\ \text{S} \\ &= 0.95\ \text{S}\end{aligned}$$

and
$$R_T = \frac{1}{0.95\ \text{S}} = \mathbf{1.053\ \Omega}$$

The above examples demonstrate an interesting and useful (for checking purposes) characteristic of parallel resistors:

The total resistance of parallel resistors is always less than the value of the smallest resistor.

In addition, the wider the spread in numerical value between two parallel resistors, the closer the total resistance will be to the smaller resistor. For instance, the total resistance of 3 Ω in parallel with 6 Ω is 2 Ω, as demonstrated in Example 6.1. However, the total resistance of 3 Ω in parallel with 60 Ω is 2.85 Ω, which is much closer to the value of the smaller resistor.

For *equal* resistors in parallel, the equation becomes significantly easier to apply. For N equal resistors in parallel, Eq. (6.2) becomes

$$\begin{aligned}\frac{1}{R_T} &= \underbrace{\frac{1}{R} + \frac{1}{R} + \frac{1}{R} + \cdots + \frac{1}{R}}_{N} \\ &= N\left(\frac{1}{R}\right)\end{aligned}$$

and
$$\boxed{R_T = \frac{R}{N}} \qquad \textbf{(6.3)}$$

In other words, the total resistance of N parallel resistors of equal value is the resistance of *one* resistor divided by the number (N) of parallel elements.

For conductance levels, we have

$$\boxed{G_T = NG} \qquad \textbf{(6.4)}$$

EXAMPLE 6.4

a. Find the total resistance of the network of Fig. 6.9.
b. Calculate the total resistance for the network of Fig. 6.10.

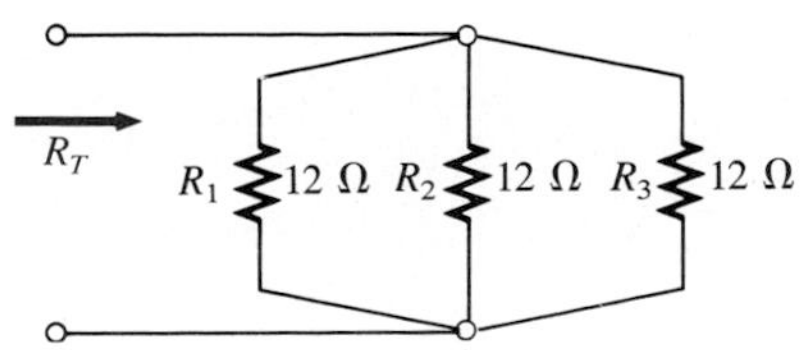

FIG. 6.9
Example 6.4: three parallel resistors of equal value.

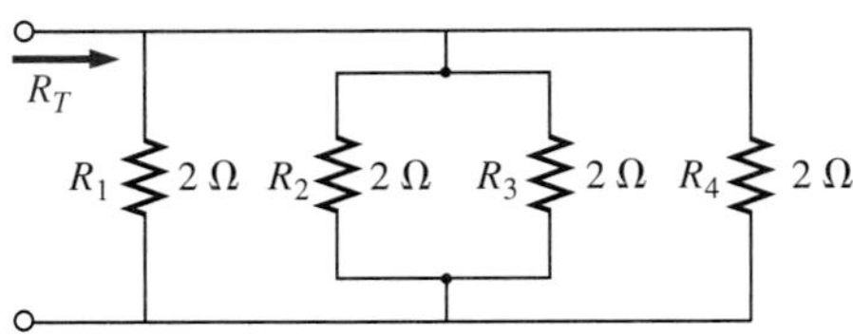

FIG. 6.10
Example 6.4: four parallel resistors of equal value.

Solutions:

a. Fig. 6.9 is redrawn in Fig. 6.11:

$$R_T = \frac{R}{N} = \frac{12\ \Omega}{3} = \mathbf{4\ \Omega}$$

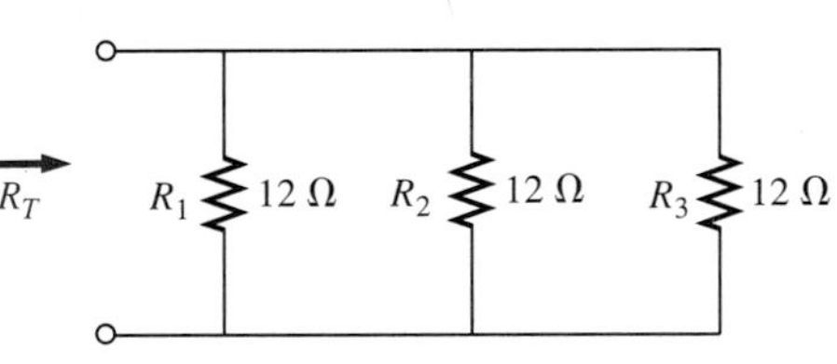

FIG. 6.11
Redrawing the network of Fig. 6.9.

b. Fig. 6.10 is redrawn in Fig. 6.12:

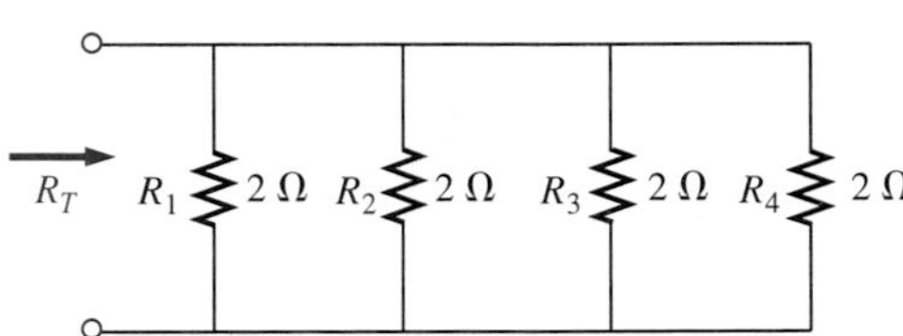

FIG. 6.12
Redrawing the network of Fig. 6.10.

$$R_T = \frac{R}{N} = \frac{2\ \Omega}{4} = \mathbf{0.5\ \Omega}$$

In the vast majority of situations, only two or three parallel resistive elements need to be combined. With this in mind, the following equations were developed to reduce the negative effects of the inverse relationship when determining R_T.

For two parallel resistors, we write

$$\frac{1}{R_T} = \frac{1}{R_1} + \frac{1}{R_2}$$

Multiplying the top and bottom of each term of the right side of the equation by the other resistor will result in

$$\frac{1}{R_T} = \left(\frac{R_2}{R_2}\right)\frac{1}{R_1} + \left(\frac{R_1}{R_1}\right)\frac{1}{R_2} = \frac{R_2}{R_1R_2} + \frac{R_1}{R_1R_2}$$

$$= \frac{R_2 + R_1}{R_1R_2}$$

and

$$R_T = \frac{R_1R_2}{R_1 + R_2} \tag{6.5}$$

In words,

the total resistance of two parallel resistors is the product of the two divided by their sum.

For three parallel resistors, the equation becomes

$$R_T = \frac{R_1R_2R_3}{R_1R_2 + R_1R_3 + R_2R_3} \tag{6.6}$$

with the denominator showing all the possible product combinations of the resistors taken two at a time. An alternative to Eq. (6.6) is to simply apply Eq. (6.5) twice, as will be demonstrated in Example 6.6.

EXAMPLE 6.5 Repeat Example 6.1 using Eq. (6.5).

Solution:

$$R_T = \frac{R_1R_2}{R_1 + R_2} = \frac{(3\ \Omega)(6\ \Omega)}{3\ \Omega + 6\ \Omega} = \frac{18\ \Omega}{9} = \mathbf{2\ \Omega}$$

EXAMPLE 6.6 Repeat Example 6.3 using Eq. (6.6).

Solution:

$$\begin{aligned} R_T &= \frac{R_1R_2R_3}{R_1R_2 + R_1R_3 + R_2R_3} \\ &= \frac{(2\ \Omega)(4\ \Omega)(5\ \Omega)}{(2\ \Omega)(4\ \Omega) + (2\ \Omega)(5\ \Omega) + (4\ \Omega)(5\ \Omega)} \\ &= \frac{40\ \Omega}{8 + 10 + 20} = \frac{40\ \Omega}{38} = \mathbf{1.053\ \Omega} \end{aligned}$$

Applying Eq. (6.5) twice yields:

$$R'_T = 2\ \Omega \parallel 4\ \Omega = \frac{(2\ \Omega)(4\ \Omega)}{2\ \Omega + 4\ \Omega} = \frac{4}{3}\ \Omega$$

$$R_T = R'_T \parallel 5\ \Omega = \frac{\left(\frac{4}{3}\ \Omega\right)(5\ \Omega)}{\frac{4}{3}\ \Omega + 5\ \Omega} = \mathbf{1.053\ \Omega}$$

Recall that series elements can be interchanged without affecting the magnitude of the total resistance or current. In parallel networks,

parallel elements can be interchanged without changing the total resistance or input current.

Note in the next example how redrawing the network can often clarify which operations and equations should be applied.

EXAMPLE 6.7 Calculate the total resistance of the parallel network of Fig. 6.13.

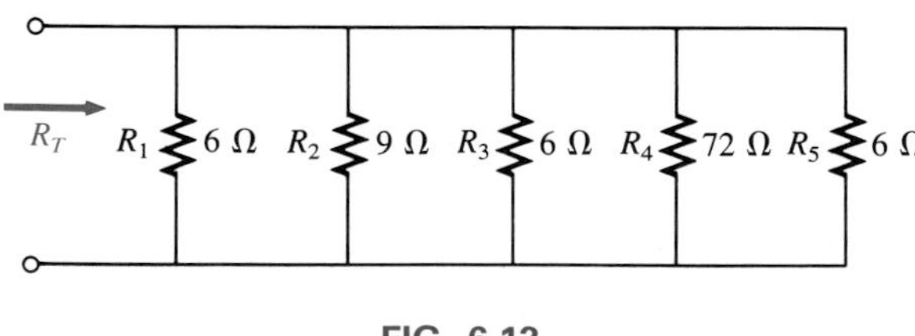

FIG. 6.13
Example 6.7.

Solution: The network is redrawn in Fig. 6.14:

$$R'_T = \frac{R}{N} = \frac{6\ \Omega}{3} = 2\ \Omega$$

$$R''_T = \frac{R_2R_4}{R_2 + R_4} = \frac{(9\ \Omega)(72\ \Omega)}{9\ \Omega + 72\ \Omega} = \frac{648\ \Omega}{81} = 8\ \Omega$$

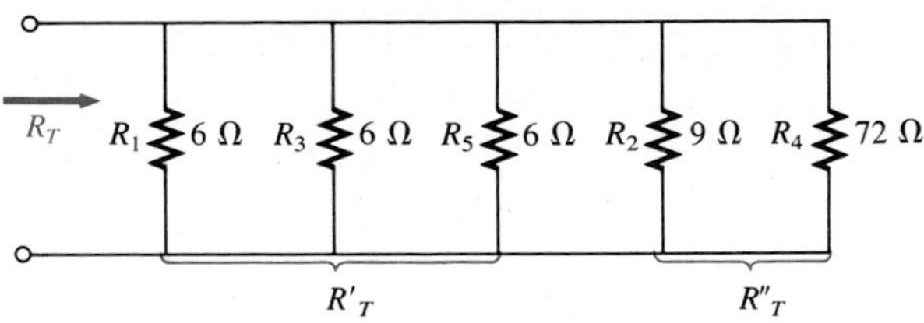

FIG. 6.14
Network of Fig. 6.13 redrawn.

and
$$R_T = R'_T \parallel R''_T$$
↑ In parallel with

$$R_T = \frac{R'_T R''_T}{R'_T + R''_T} = \frac{(2\ \Omega)(8\ \Omega)}{2\ \Omega + 8\ \Omega} = \frac{16\ \Omega}{10} = \mathbf{1.6\ \Omega}$$

The preceding examples show direct substitution, where once the proper equation is defined, it is only a matter of plugging in the numbers and performing the required algebraic maneuvers. The next two examples have a design orientation, where specific network parameters are defined and the circuit elements have to be determined.

EXAMPLE 6.8 Determine the value of R_2 in Fig. 6.15 to establish a total resistance of 9 kΩ.

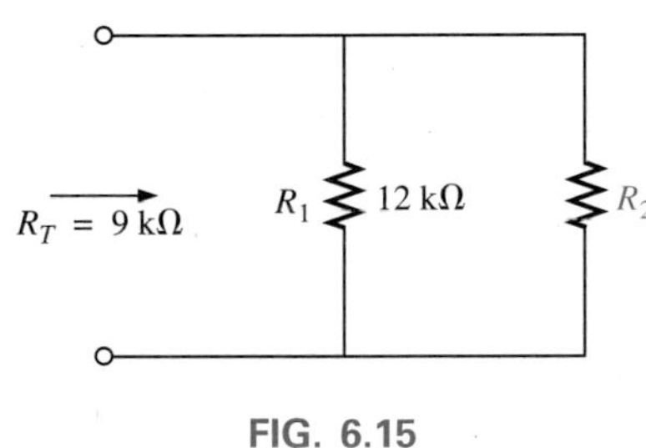

FIG. 6.15
Example 6.8.

Solution:

$$R_T = \frac{R_1 R_2}{R_1 + R_2}$$
$$R_T(R_1 + R_2) = R_1 R_2$$
$$R_T R_1 + R_T R_2 = R_1 R_2$$
$$R_T R_1 = R_1 R_2 - R_T R_2$$
$$R_T R_1 = (R_1 - R_T) R_2$$

and
$$\boxed{R_2 = \frac{R_T R_1}{R_1 - R_T}} \qquad \textbf{(6.7)}$$

Substituting values:

$$R_2 = \frac{(9\ \text{k}\Omega)(12\ \text{k}\Omega)}{12\ \text{k}\Omega - 9\ \text{k}\Omega}$$
$$= \frac{108\ \text{k}\Omega}{3} = \mathbf{36\ k\Omega}$$

EXAMPLE 6.9 Determine the values of R_1, R_2, and R_3 in Fig. 6.16 if $R_2 = 2R_1$ and $R_3 = 2R_2$ and the total resistance is 16 kΩ.

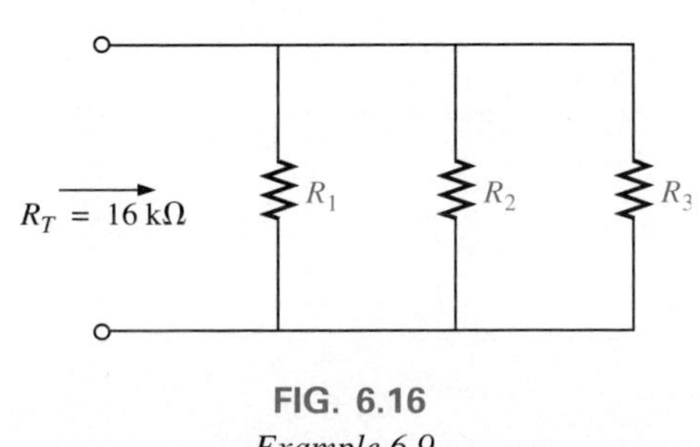

FIG. 6.16
Example 6.9.

Solution:

$$\frac{1}{R_T} = \frac{1}{R_1} + \frac{1}{R_2} + \frac{1}{R_3}$$
$$\frac{1}{16\ \text{k}\Omega} = \frac{1}{R_1} + \frac{1}{2R_1} + \frac{1}{4R_1}$$

since $R_3 = 2R_2 = 2(2R_1) = 4R_1$

and $$\frac{1}{16\text{ k}\Omega} = \frac{1}{R_1} + \frac{1}{2}\left(\frac{1}{R_1}\right) + \frac{1}{4}\left(\frac{1}{R_1}\right)$$

$$\frac{1}{16\text{ k}\Omega} = 1.75\left(\frac{1}{R_1}\right)$$

with $R_1 = 1.75(16\text{ k}\Omega) = \mathbf{28\text{ k}\Omega}$

Recall for series circuits that the total resistance will always increase as additional elements are added in series.

For parallel resistors, the total resistance will always decrease as additional elements are added in parallel.

The next example demonstrates this unique characteristic of parallel resistors.

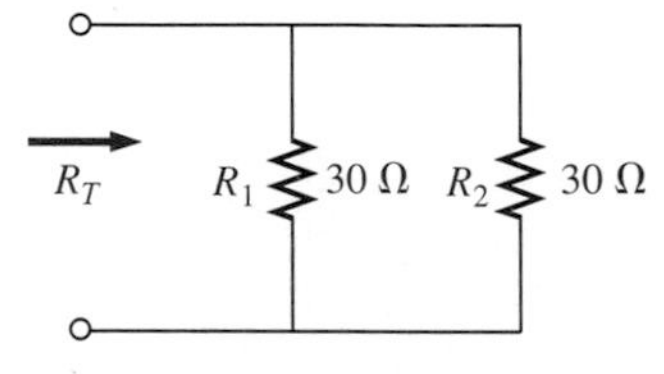

FIG. 6.17
Example 6.10: Two equal, parallel resistors.

EXAMPLE 6.10

a. Determine the total resistance of the network of Fig. 6.17.
b. What is the effect on the total resistance of the network of Fig. 6.17 if an additional resistor of the same value is added, as shown in Fig. 6.18?
c. What is the effect on the total resistance of the network of Fig. 6.17 if a very large resistance is added in parallel, as shown in Fig. 6.19?
d. What is the effect on the total resistance of the network of Fig. 6.17 if a very small resistance is added in parallel, as shown in Fig. 6.20?

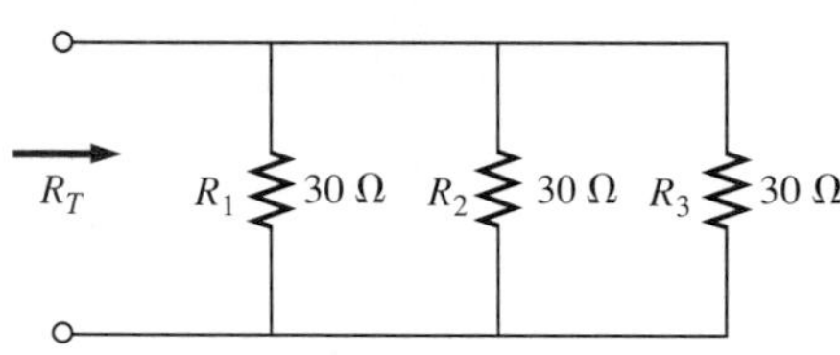

FIG. 6.18
Adding a third parallel resistor of equal value to the network of Fig. 6.17.

Solutions:

a. $R_T = 30\ \Omega \parallel 30\ \Omega = \dfrac{30\ \Omega}{2} = \mathbf{15\ \Omega}$

b. $R_T = 30\ \Omega \parallel 30\ \Omega \parallel 30\ \Omega = \dfrac{30\ \Omega}{3} = \mathbf{10\ \Omega} < 15\ \Omega$

R_T decreased

c. $R_T = 30\ \Omega \parallel 30\ \Omega \parallel 1\text{ k}\Omega = 15\ \Omega \parallel 1\text{ k}\Omega$

$= \dfrac{(15\ \Omega)(1000\ \Omega)}{15\ \Omega + 1000\ \Omega} = \mathbf{14.778\ \Omega} < 15\ \Omega$

Small decrease in R_T

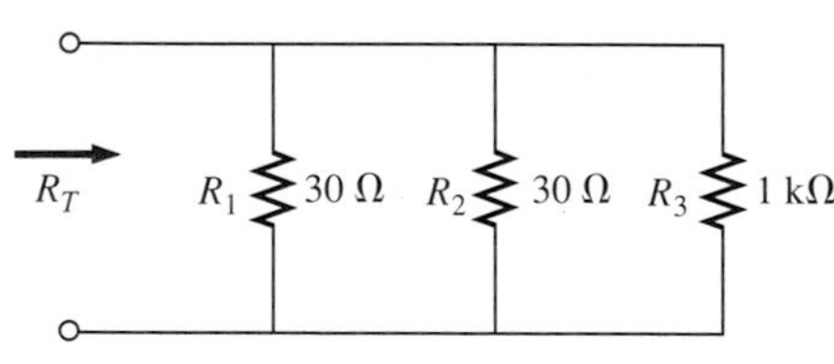

FIG. 6.19
Adding a much larger parallel resistor to the network of Fig. 6.17.

d. $R_T = 30\ \Omega \parallel 30\ \Omega \parallel 0.1\ \Omega = 15\ \Omega \parallel 0.1\ \Omega$

$= \dfrac{(15\ \Omega)(0.1\ \Omega)}{15\ \Omega + 0.1\ \Omega} = \mathbf{0.099\ \Omega} < 15\ \Omega$

Significant decrease in R_T

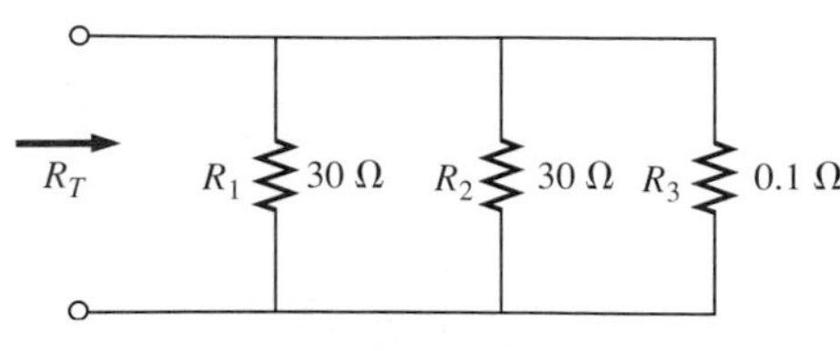

FIG. 6.20
Adding a much smaller parallel resistor to the network of Fig. 6.17.

In each case the total resistance of the network decreased with the increase of an additional parallel resistive element, no matter how large or small. Note also that the total resistance is also smaller than the smallest parallel element.

6.4 PARALLEL NETWORKS

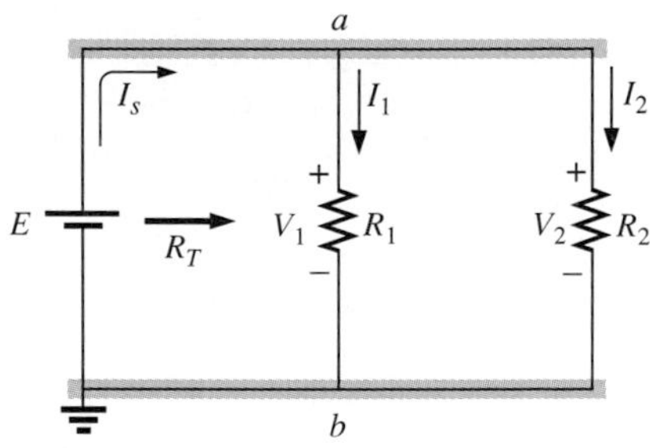

FIG. 6.21
Parallel network.

The network of Fig. 6.21 is the simplest of parallel networks. All the elements have terminals a and b in common. The total resistance is determined by $R_T = R_1R_2/(R_1 + R_2)$, and the source current by $I_s = E/R_T$. Throughout the text, the subscript s will be used to denote a property of the source. Since the terminals of the battery are connected directly across the resistors R_1 and R_2, the following should be obvious:

The voltage across parallel elements is the same.

Using this fact will result in

$$V_1 = V_2 = E$$

and

$$I_1 = \frac{V_1}{R_1} = \frac{E}{R_1}$$

with

$$I_2 = \frac{V_2}{R_2} = \frac{E}{R_2}$$

If we take the equation for the total resistance and multiply both sides by the applied voltage, we obtain

$$E\left(\frac{1}{R_T}\right) = E\left(\frac{1}{R_1} + \frac{1}{R_2}\right)$$

and

$$\frac{E}{R_T} = \frac{E}{R_1} + \frac{E}{R_2}$$

Substituting the Ohm's law relationships appearing above, we find that the source current

$$I_s = I_1 + I_2$$

permitting the following conclusion:

For single-source parallel networks, the source current (I_s) is equal to the sum of the individual branch currents.

The power dissipated by the resistors and delivered by the source can be determined from

$$P_1 = V_1I_1 = I_1^2R_1 = \frac{V_1^2}{R_1}$$

$$P_2 = V_2I_2 = I_2^2R_2 = \frac{V_2^2}{R_2}$$

$$P_s = EI_s = I_s^2R_T = \frac{E^2}{R_T}$$

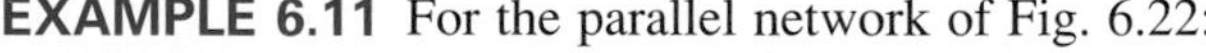

EXAMPLE 6.11 For the parallel network of Fig. 6.22:

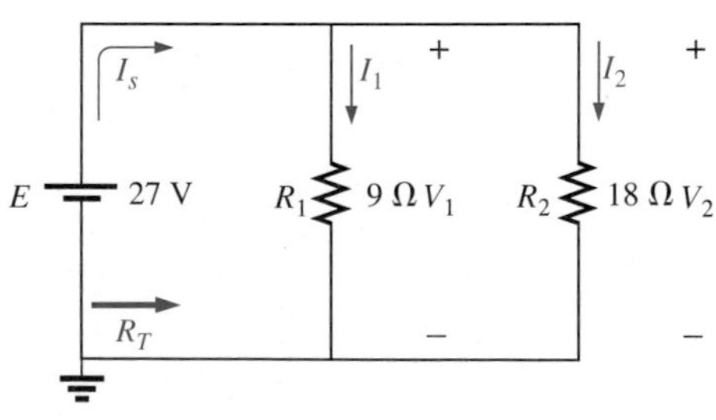

FIG. 6.22
Example 6.11.

a. Calculate R_T.
b. Determine I_s.
c. Calculate I_1 and I_2 and demonstrate that $I_s = I_1 + I_2$.
d. Determine the power to each resistive load.
e. Determine the power delivered by the source and compare it to the total power dissipated by the resistive elements.

Solutions:

a. $R_T = \dfrac{R_1R_2}{R_1 + R_2} = \dfrac{(9\ \Omega)(18\ \Omega)}{9\ \Omega + 18\ \Omega} = \dfrac{162\ \Omega}{27} = \mathbf{6\ \Omega}$

b. $I_s = \dfrac{E}{R_T} = \dfrac{27\ \text{V}}{6\ \Omega} = \mathbf{4.5\ A}$

c. $$I_1 = \frac{V_1}{R_1} = \frac{E}{R_1} = \frac{27\text{ V}}{9\ \Omega} = \mathbf{3\ A}$$

$$I_2 = \frac{V_2}{R_2} = \frac{E}{R_2} = \frac{27\text{ V}}{18\ \Omega} = \mathbf{1.5\ A}$$

$$I_s = I_1 + I_2$$

$$4.5\text{ A} = 3\text{ A} + 1.5\text{ A}$$

$$\mathbf{4.5\ A = 4.5\ A} \quad \text{(checks)}$$

d. $P_1 = V_1I_1 = EI_1 = (27\text{ V})(3\text{ A}) = \mathbf{81\ W}$
$P_2 = V_2I_2 = EI_2 = (27\text{ V})(1.5\text{ A}) = \mathbf{40.5\ W}$

e. $P_s = EI_s = (27\text{ V})(4.5\text{ A}) = \mathbf{121.5\ W}$
$P_s = P_1 + P_2 = 81\text{ W} + 40.5\text{ W} = \mathbf{121.5\ W}$

EXAMPLE 6.12 Given the information provided in Fig. 6.23:

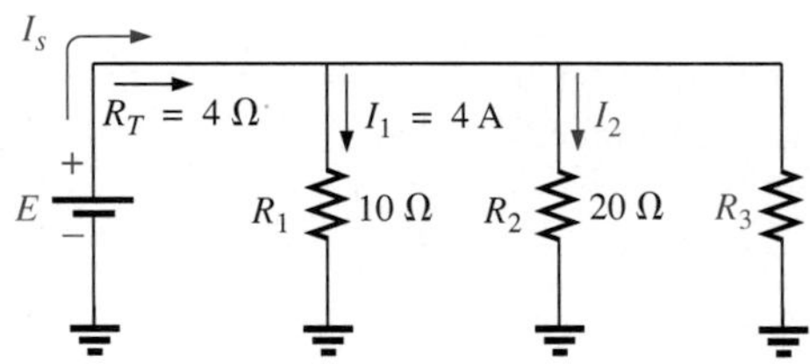

FIG. 6.23
Example 6.12.

a. Determine R_3.
b. Calculate E.
c. Find I_s.
d. Find I_2.
e. Determine P_2.

Solutions:

a. $$\frac{1}{R_T} = \frac{1}{R_1} + \frac{1}{R_2} + \frac{1}{R_3}$$

$$\frac{1}{4\ \Omega} = \frac{1}{10\ \Omega} + \frac{1}{20\ \Omega} + \frac{1}{R_3}$$

$$0.25\text{ S} = 0.1\text{ S} + 0.05\text{ S} + \frac{1}{R_3}$$

$$0.25\text{ S} = 0.15\text{ S} + \frac{1}{R_3}$$

$$\frac{1}{R_3} = 0.1\text{ S}$$

$$R_3 = \frac{1}{0.1\text{ S}} = \mathbf{10\ \Omega}$$

b. $E = V_1 = I_1R_1 = (4\text{ A})(10\ \Omega) = \mathbf{40\ V}$

c. $$I_s = \frac{E}{R_T} = \frac{40\text{ V}}{4\ \Omega} = \mathbf{10\ A}$$

d. $$I_2 = \frac{V_2}{R_2} = \frac{E}{R_2} = \frac{40\text{ V}}{20\ \Omega} = \mathbf{2\ A}$$

e. $P_2 = I_2^2R_2 = (2\text{ A})^2(20\ \Omega) = \mathbf{80\ W}$

6.5 KIRCHHOFF'S CURRENT LAW

Kirchhoff's voltage law provides an important relationship among voltage levels around any closed loop of a network. We now consider Kirchhoff's current law, which provides an equally important relationship among current levels at any junction.

Kirchhoff's current law (KCL) states that the algebraic sum of the currents entering and leaving an area, system, or junction is zero.

In other words,

the sum of the currents entering an area, system, or junction must equal the sum of the currents leaving the area, system, or junction.

In equation form:

$$\Sigma I_{\text{entering}} = \Sigma I_{\text{leaving}} \tag{6.8}$$

In Fig. 6.24, for instance, the shaded area can enclose an entire system, a complex network, or simply a junction of two or more paths. In each case the current entering must equal that leaving, as witnessed by the fact that

$$\begin{aligned} I_1 + I_4 &= I_2 + I_3 \\ 4\text{ A} + 8\text{ A} &= 2\text{ A} + 10\text{ A} \\ 12\text{ A} &= 12\text{ A} \end{aligned}$$

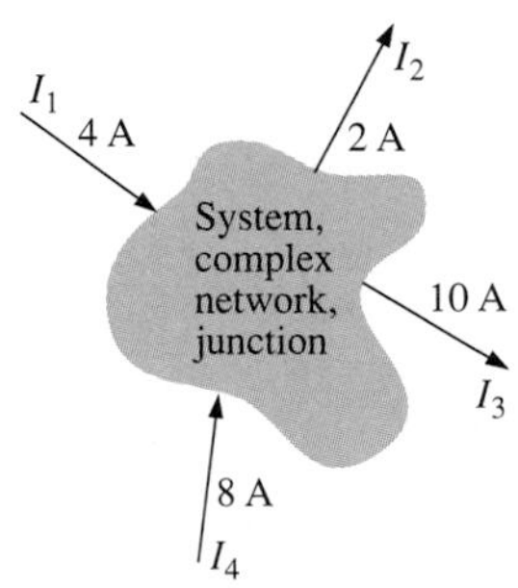

FIG. 6.24
Introducing Kirchhoff's current law.

The most common application of the law will be at the junction of two or more paths of current flow, as shown in Fig. 6.25. For some students it is difficult initially to determine whether a current is entering or leaving a junction. One approach that may help is to picture yourself as standing on the junction and treating the path currents as arrows. If the arrow appears to be heading toward you, as is the case for I_1 in Fig. 6.25, then it is entering the junction. If you see the tail of the arrow (from the junction) as it travels down its path away from you, it is leaving the junction, as is the case for I_2 and I_3 in Fig. 6.25.

Applying Kirchhoff's current law to the junction of Fig. 6.25:

$$\begin{aligned} \Sigma I_{\text{entering}} &= \Sigma I_{\text{leaving}} \\ 6\text{ A} &= 2\text{ A} + 4\text{ A} \\ 6\text{ A} &= 6\text{ A} \quad \text{(checks)} \end{aligned}$$

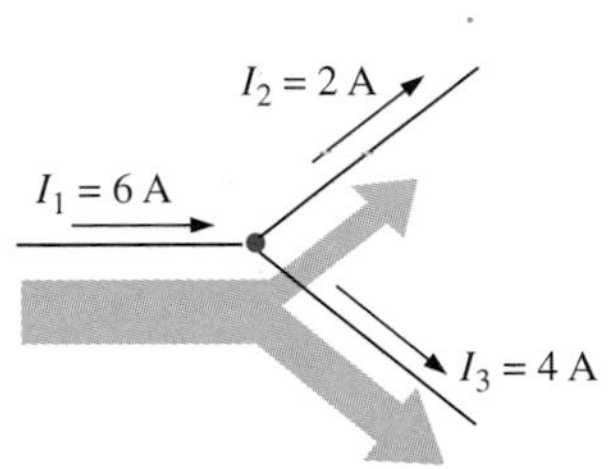

FIG. 6.25
Demonstrating Kirchhoff's current law.

In the next two examples, unknown currents can be determined by applying Kirchhoff's current law. Simply remember to place all current levels entering a junction to the left of the equals sign and the sum of all currents leaving a junction to the right of the equals sign. The water-in-the-pipe analogy is an excellent one for supporting and clarifying the preceding law. Quite obviously, the sum total of the water entering a junction must equal the total of the water leaving the exit pipes.

EXAMPLE 6.13 Determine the currents I_3 and I_4 of Fig. 6.26 using Kirchhoff's current law.

Solution: We must first work with junction *a* since the only unknown is I_3. At junction *b* there are two unknowns, and both cannot be determined from one application of the law.

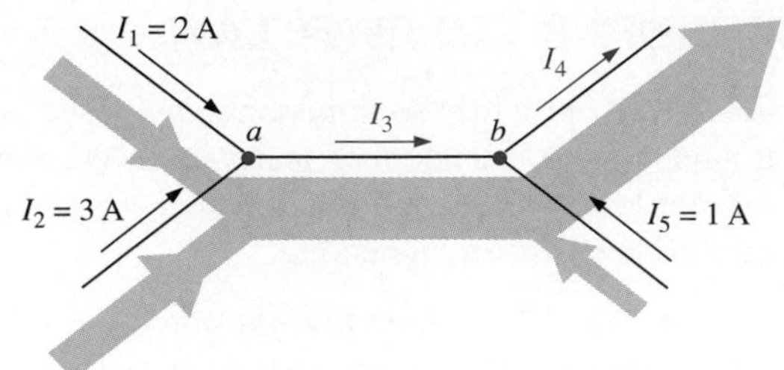

FIG. 6.26
Example 6.13.

At *a:*

$$\Sigma I_{\text{entering}} = \Sigma I_{\text{leaving}}$$
$$I_1 + I_2 = I_3$$
$$2\text{ A} + 3\text{ A} = I_3$$
$$I_3 = \mathbf{5\text{ A}}$$

At *b:*

$$\Sigma I_{\text{entering}} = \Sigma I_{\text{leaving}}$$
$$I_3 + I_5 = I_4$$
$$5\text{ A} + 1\text{ A} = I_4$$
$$I_4 = \mathbf{6\text{ A}}$$

EXAMPLE 6.14 Determine I_1, I_3, I_4, and I_5 for the network of Fig. 6.27.

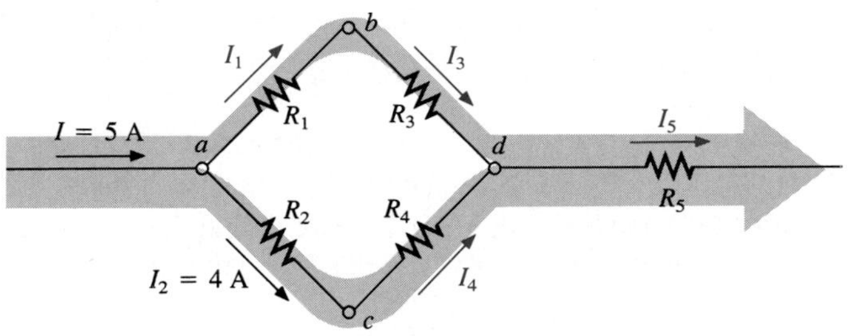

FIG. 6.27
Example 6.14.

Solution: At *a:*

$$\Sigma I_{\text{entering}} = \Sigma I_{\text{leaving}}$$
$$I = I_1 + I_2$$
$$5\text{ A} = I_1 + 4\text{ A}$$

Subtracting 4 A from both sides gives

$$5\text{ A} - 4\text{ A} = I_1 + 4\not{\text{A}} - 4\not{\text{A}}$$
$$I_1 = 5\text{ A} - 4\text{ A} = \mathbf{1\text{ A}}$$

At *b:*

$$\Sigma I_{\text{entering}} = \Sigma I_{\text{leaving}}$$
$$I_1 = I_3 = \mathbf{1\text{ A}}$$

as it should, since R_1 and R_3 are in series and the current is the same in series elements.

At c:

$$I_2 = I_4 = \mathbf{4\ A}$$

for the same reasons given for junction b.

At d:

$$\Sigma I_{\text{entering}} = \Sigma I_{\text{leaving}}$$
$$I_3 + I_4 = I_5$$
$$1\text{ A} + 4\text{ A} = I_5$$
$$I_5 = \mathbf{5\ A}$$

If we enclose the entire network, we find that the current entering is $I = 5$ A; the net current leaving from the far right is $I_5 = 5$ A. The two must be equal since the net current entering any system must equal that leaving.

EXAMPLE 6.15 Determine the currents I_3 and I_5 of Fig. 6.28 through applications of Kirchhoff's current law.

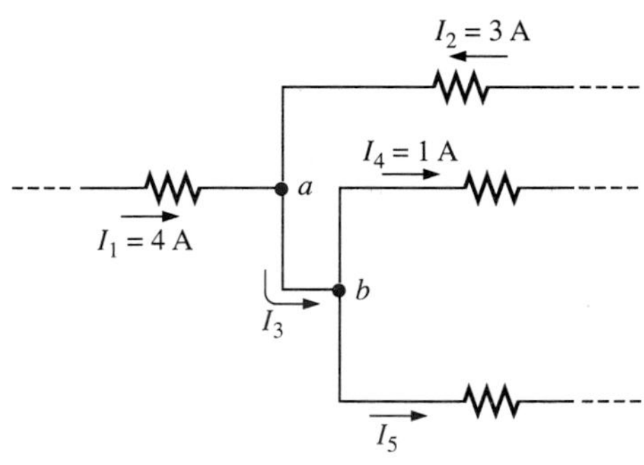

FIG. 6.28
Example 6.15.

Solution: Note that since node b has two unknown quantities and node a has only one, we must first apply Kirchhoff's current law to node a. The result can then be applied to node b. For node a,

$$I_1 + I_2 = I_3$$
$$4\text{ A} + 3\text{ A} = I_3$$

and

$$I_3 = \mathbf{7\ A}$$

For node b,

$$I_3 = I_4 + I_5$$
$$7\text{ A} = 1\text{ A} + I_5$$

and

$$I_5 = 7\text{ A} - 1\text{ A} = \mathbf{6\ A}$$

EXAMPLE 6.16 Find the magnitude and direction of the currents I_3, I_4, I_6, and I_7 for the network of Fig. 6.29. Even though the elements are not in series or parallel, Kirchhoff's current law can be applied to determine all the unknown currents.

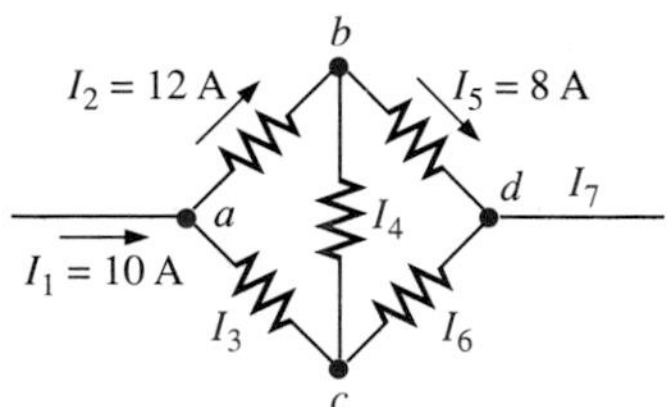

FIG. 6.29
Example 6.16.

Solution: Considering the overall system, we know that the current entering must equal that leaving. Therefore,

$$I_7 = I_1 = \mathbf{10\ A}$$

Since 10 A are entering node a and 12 A are leaving, I_3 must be supplying current to the node. Applying Kirchhoff's current law at node a,

$$I_1 + I_3 = I_2$$
$$10\text{ A} + I_3 = 12\text{ A}$$

and

$$I_3 = 12\text{ A} - 10\text{ A} = \mathbf{2\ A}$$

At node b, since 12 A are entering and 8 A are leaving, I_4 must be leaving. Therefore,

$$I_2 = I_4 + I_5$$
$$12\text{ A} = I_4 + 8\text{ A}$$

and

$$I_4 = 12\text{ A} - 8\text{ A} = \mathbf{4\ A}$$

At node c, I_3 is leaving at 2 A and I_4 is entering at 4 A, requiring that I_6 be leaving. Applying Kirchhoff's current law at node c,

$$I_4 = I_3 + I_6$$
$$4\text{ A} = 2\text{ A} + I_6$$

and $$I_6 = 4\text{ A} - 2\text{ A} = \mathbf{2\text{ A}}$$

As a check at node *d*,

$$I_5 + I_6 = I_7$$
$$8\text{ A} + 2\text{ A} = 10\text{ A}$$
$$10\text{ A} = 10\text{ A} \quad \text{(checks)}$$

Looking back at Example 6.11, we find that the current entering the top node is 4.5 A and the current leaving the node is $I_1 + I_2 = 3\text{ A} + 1.5\text{ A} = 4.5\text{ A}$. For Example 6.12, we have

$$I_s = I_1 + I_2 + I_3$$
$$10\text{ A} = 4\text{ A} + 2\text{ A} + I_3$$

and $$I_3 = 10\text{ A} - 6\text{ A} = 4\text{ A}$$

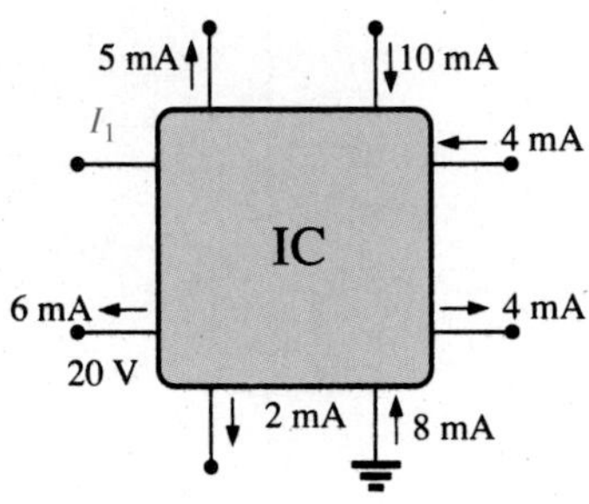

FIG. 6.30
Integrated circuit.

The application of Kirchhoff's current law is not limited to networks where all the internal connections are known or visible. For instance, all the currents of the integrated circuit of Fig. 6.30 are known except I_1. By treating the system as a single node, we can apply Kirchhoff's current law using the following table to ensure an accurate listing of all known quantities:

I_i	I_o
10 mA	5 mA
4 mA	4 mA
8 mA	2 mA
22 mA	6 mA
	17 mA

Noting the total input current versus that leaving clearly reveals that I_1 is a current of 22 mA − 17 mA = 5 mA leaving the system.

6.6 CURRENT DIVIDER RULE

As the name suggests, the *current divider rule* (CDR) will determine how the current entering a set of parallel branches will split between the elements.

For two parallel elements of equal value, the current will divide equally.

For parallel elements with different values, the smaller the resistance, the greater the share of input current.

For parallel elements of different values, the current will split with a ratio equal to the inverse of their resistor values.

For example, if one of two parallel resistors is twice the other, then the current through the larger resistor will be half the other.

In Fig. 6.31, since I_1 is 1 mA and R_1 is six times that of R_3, the current through R_3 must be 6 mA (without making any other calculations including the total current or the actual resistance levels). For R_2 the current must be 2 mA since R_1 is twice R_2. The total current must then be the sum of I_1, I_2, and I_3, or 9 mA. In total, therefore, knowing only the current through R_1, we were able to find all the other currents of the configuration without knowing anything more about the network.

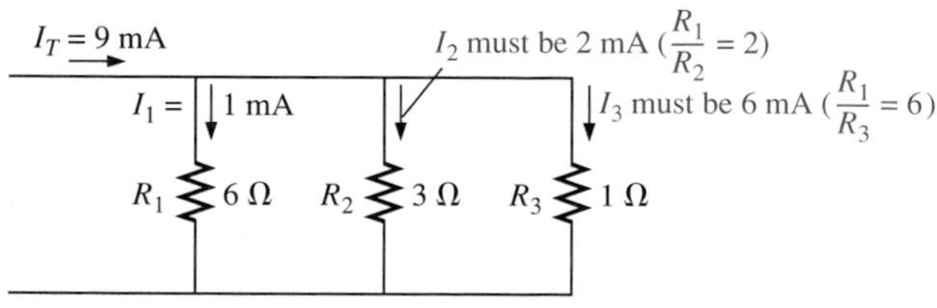

FIG. 6.31
Demonstrating how current will divide between unequal resistors.

For networks in which only the resistor values are given along with the input current, the *current divider rule* should be applied to determine the various branch currents. It can be derived using the network of Fig. 6.32.

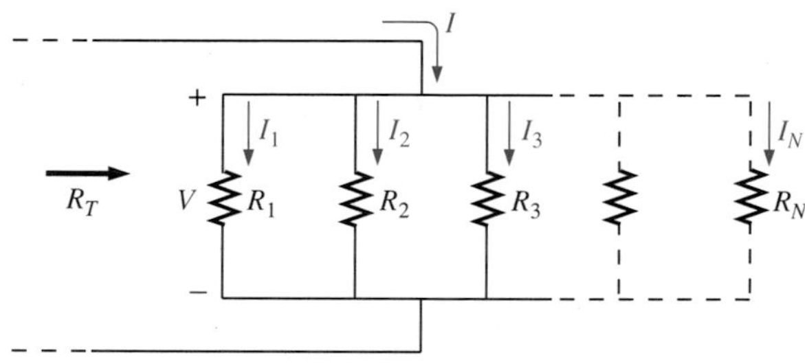

FIG. 6.32
Deriving the current divider rule.

The input current I equals V/R_T, where R_T is the total resistance of the parallel branches. Substituting $V = I_xR_x$ into the above equation, where I_x refers to the current through a parallel branch of resistance R_x, we have

$$I = \frac{V}{R_T} = \frac{I_xR_x}{R_T}$$

and

$$I_x = \frac{R_T}{R_x}I \tag{6.9}$$

which is the general form for the current divider rule. In words, the current through any parallel branch is equal to the product of the *total* resistance of the parallel branches and the input current divided by the resistance of the branch through which the current is to be determined.

For the current I_1,

$$I_1 = \frac{R_T}{R_1}I$$

and for I_2,

$$I_2 = \frac{R_T}{R_2}I$$

and so on.

For the particular case of *two parallel resistors,* as shown in Fig. 6.33,

$$R_T = \frac{R_1R_2}{R_1 + R_2}$$

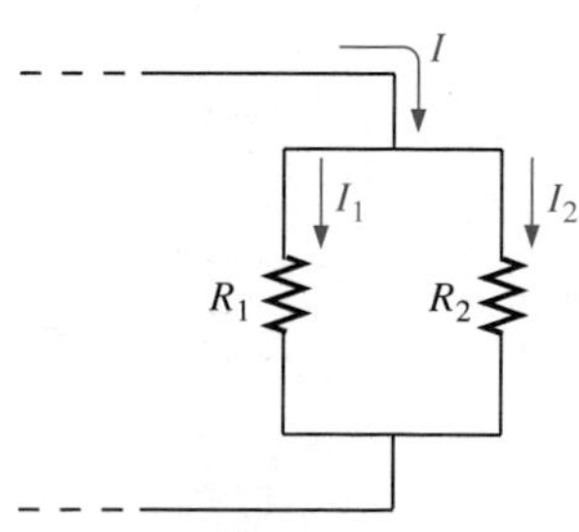

FIG. 6.33
Developing an equation for current division between two parallel resistors.

and $$I_1 = \frac{R_T}{R_1} I = \frac{\dfrac{R_1 R_2}{R_1 + R_2}}{R_1} I$$

Note difference in subscripts.

and $$I_1 = \frac{R_2 I}{R_1 + R_2} \tag{6.10}$$

Similarly for I_2,

$$I_2 = \frac{R_1 I}{R_1 + R_2} \tag{6.11}$$

In words, for two parallel branches, the current through either branch is equal to the product of the *other* parallel resistor and the input current divided by the *sum* (not total parallel resistance) of the two parallel resistances.

EXAMPLE 6.17 Determine the current I_2 for the network of Fig. 6.34 using the current divider rule.

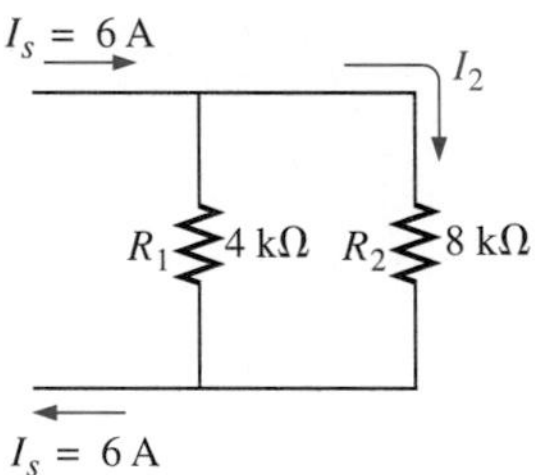

FIG. 6.34
Example 6.17.

Solution:

$$I_2 = \frac{R_1 I_s}{R_1 + R_2} = \frac{(4\text{ k}\Omega)(6\text{ A})}{4\text{ k}\Omega + 8\text{ k}\Omega} = \frac{4}{12}(6\text{ A}) = \frac{1}{3}(6\text{ A})$$
$$= \mathbf{2\text{ A}}$$

EXAMPLE 6.18 Find the current I_1 for the network of Fig. 6.35.

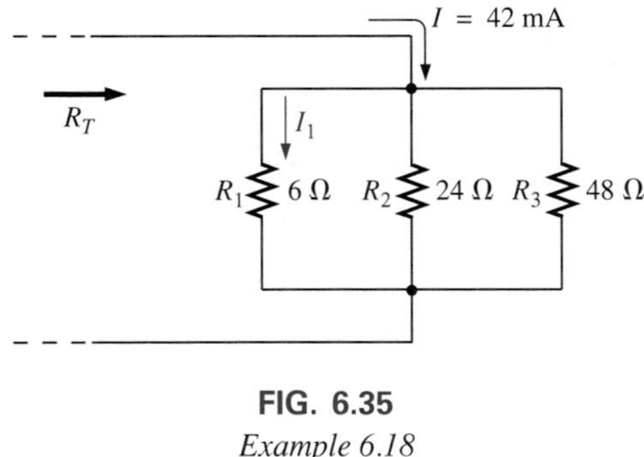

FIG. 6.35
Example 6.18

Solution: There are two options for solving this problem. The first is to use Eq. (6.9) as follows:

$$\frac{1}{R_T} = \frac{1}{6\ \Omega} + \frac{1}{24\ \Omega} + \frac{1}{48\ \Omega} = 0.1667\text{ S} + 0.0417\text{ S} + 0.0208\text{ S}$$
$$= 0.2292\text{ S}$$

and $$R_T = \frac{1}{0.2292\text{ S}} = 4.363\ \Omega$$

with $$I_1 = \frac{R_T}{R_1} I = \frac{4.363\ \Omega}{6\ \Omega}(42\text{ mA}) = \mathbf{30.54\text{ mA}}$$

or apply (Eq. (6.10) once after combining R_2 and R_3 as follows:

$$24\ \Omega \parallel 48\ \Omega = \frac{(24\ \Omega)(48\ \Omega)}{24\ \Omega + 48\ \Omega} = 16\ \Omega$$

and
$$I_1 = \frac{16\ \Omega(42\ \text{mA})}{16\ \Omega + 6\ \Omega} = \mathbf{30.54\ mA}$$

Both generated the same answer, leaving you with a choice for future calculations involving more than two parallel resistors.

EXAMPLE 6.19 Determine the magnitude of the currents I_1, I_2, and I_3 for the network of Fig. 6.36.

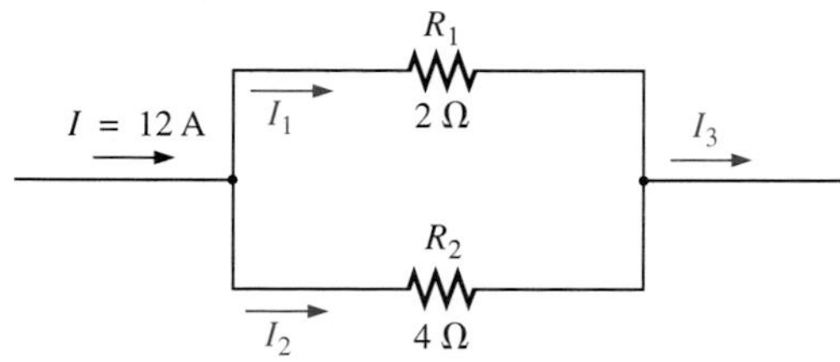

FIG. 6.36
Example 6.19.

Solution: By Eq. (6.10), the current divider rule,

$$I_1 = \frac{R_2 I}{R_1 + R_2} = \frac{(4\ \Omega)(12\ \text{A})}{2\ \Omega + 4\ \Omega} = \mathbf{8\ A}$$

Applying Kirchhoff's current law,

$$I = I_1 + I_2$$

and
$$I_2 = I - I_1 = 12\ \text{A} - 8\ \text{A} = \mathbf{4\ A}$$

or, using the current divider rule again,

$$I_2 = \frac{R_1 I}{R_1 + R_2} = \frac{(2\ \Omega)(12\ \text{A})}{2\ \Omega + 4\ \Omega} = \mathbf{4\ A}$$

The total current entering the parallel branches must equal that leaving. Therefore,

$$I_3 = I = \mathbf{12\ A}$$

or
$$I_3 = I_1 + I_2 = 8\ \text{A} + 4\ \text{A} = \mathbf{12\ A}$$

EXAMPLE 6.20 Determine the resistance R_1 to effect the division of current in Fig. 6.37.

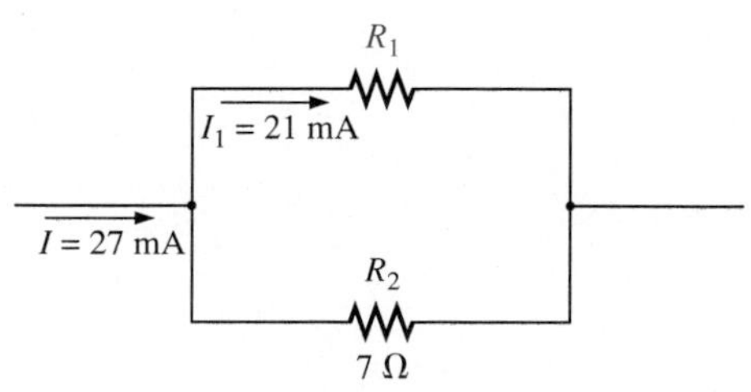

FIG. 6.37
Example 6.20.

Solution: Applying the current divider rule,

$$I_1 = \frac{R_2 I}{R_1 + R_2}$$

and
$$(R_1 + R_2)I_1 = R_2 I$$
$$R_1 I_1 + R_2 I_1 = R_2 I$$
$$R_1 I_1 = R_2 I - R_2 I_1$$
$$R_1 = \frac{R_2(I - I_1)}{I_1}$$

Substituting values:

$$R_1 = \frac{7\ \Omega(27\text{ mA} - 21\text{ mA})}{21\text{ mA}}$$

$$= 7\ \Omega\left(\frac{6}{21}\right) = \frac{42\ \Omega}{21} = \mathbf{2\ \Omega}$$

An alternative approach is

$$I_2 = I - I_1 \qquad \text{(Kirchhoff's current law)}$$
$$= 27\text{ mA} - 21\text{ mA} = 6\text{ mA}$$
$$V_2 = I_2R_2 = (6\text{ mA})(7\ \Omega) = 42\text{ mV}$$
$$V_1 = I_1R_1 = V_2 = 42\text{ mV}$$

and

$$R_1 = \frac{V_1}{I_1} = \frac{42\text{ mV}}{21\text{ mA}} = \mathbf{2\ \Omega}$$

From the examples just described, note the following:

Current seeks the path of least resistance.

That is,

1. More current passes through the smaller of two parallel resistors.
2. The current entering any number of parallel resistors divides into these resistors as the inverse ratio of their ohmic values. This relationship is depicted in Fig. 6.38.

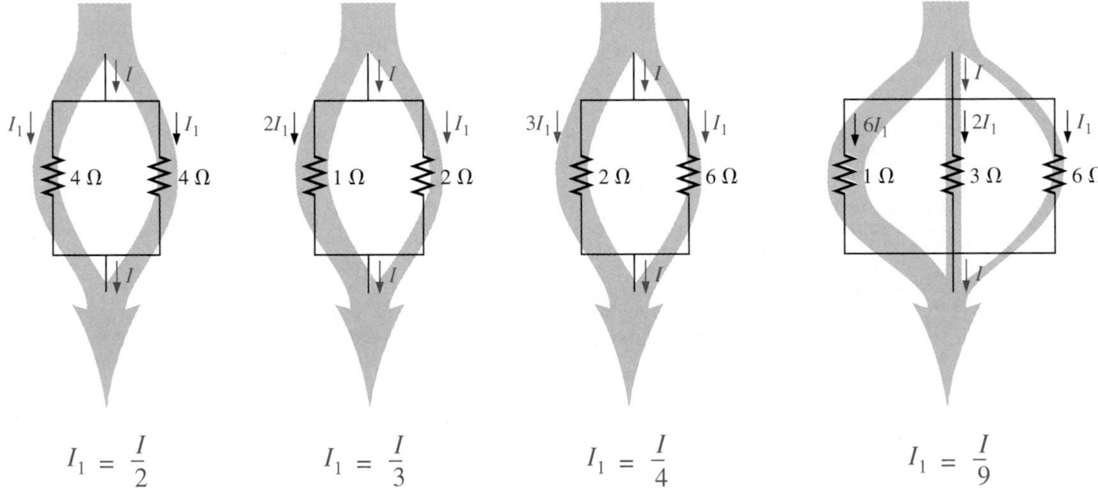

FIG. 6.38
Current division through parallel branches.

6.7 VOLTAGE SOURCES IN PARALLEL

Voltage sources are placed in parallel as shown in Fig. 6.39 only if they have the same voltage rating. The primary reason for placing two or more batteries in parallel of the same terminal voltage would be to increase the current rating (and, therefore, the power rating) of the source. As shown in Fig. 6.39, the current rating of the combination is determined by $I_s = I_1 + I_2$ at the same terminal voltage. The resulting power rating is twice that available with one supply.

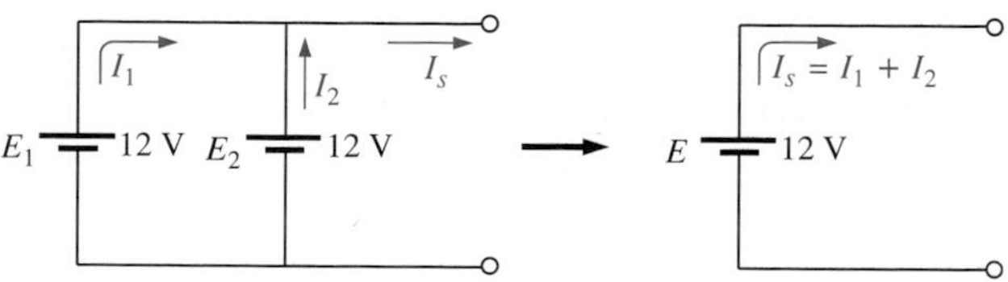

FIG. 6.39
Parallel voltage sources.

If two batteries of different terminal voltages are placed in parallel, both would be left ineffective or damaged because the terminal voltage of the larger battery would try to drop rapidly to that of the lower supply. Consider two lead-acid car batteries of different terminal voltage placed in parallel, as shown in Fig. 6.40.

The relatively small internal resistances of the batteries are the only current-limiting elements of the resulting series circuit. The current

$$I = \frac{E_1 - E_2}{R_{int_1} + R_{int_2}} = \frac{12\text{ V} - 6\text{ V}}{0.03\ \Omega + 0.02\ \Omega} = \frac{6\text{ V}}{0.05\ \Omega} = \mathbf{120\ A}$$

far exceeds the continuous drain rating of the larger supply, resulting in a rapid discharge of E_1 and a destructive impact on the smaller supply.

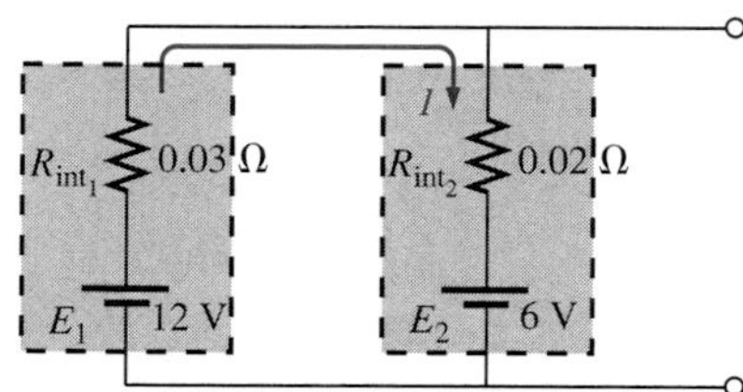

FIG. 6.40
Parallel batteries of different terminal voltages.

6.8 OPEN AND SHORT CIRCUITS

Open circuits and short circuits can often cause more confusion and difficulty in the analysis of a system than standard series or parallel configurations. This will become more obvious in the chapters to follow when we apply some of the methods and theorems.

An *open circuit* is simply two isolated terminals not connected by an element of any kind, as shown in Fig. 6.41(a). Since a path for conduction does not exist, the current associated with an open circuit must always be zero. The voltage across the open circuit, however, can be any value, as determined by the system it is connected to. In summary, therefore,

an open circuit can have a potential difference (voltage) across its terminals, but the current is always zero amperes.

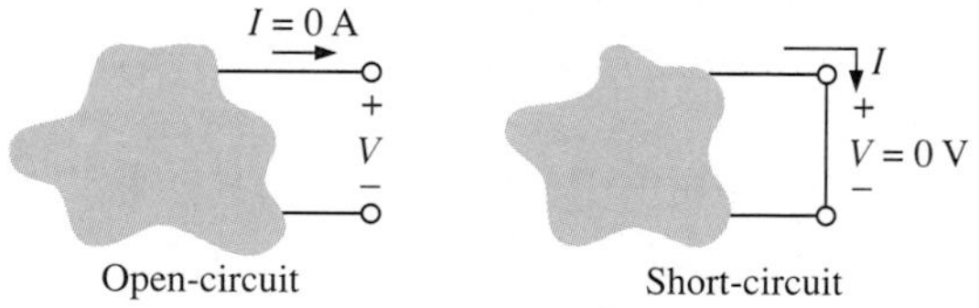

FIG. 6.41
Two special network configurations.

In Fig. 6.42, an open circuit exists between terminals *a* and *b*. As shown in the figure, the voltage across the open-circuit terminals is the supply voltage, but the current is zero due to the absence of a complete circuit.

A *short circuit* is a very low resistance, direct connection between two terminals of a network, as shown in Fig. 6.41(b). The current through the short circuit can be any value, as determined by the system it is connected to, but the voltage across the short circuit will always be zero volts because the resistance of the short circuit is assumed to be essentially zero ohms and $V = IR = I(0\ \Omega) = 0$ V.

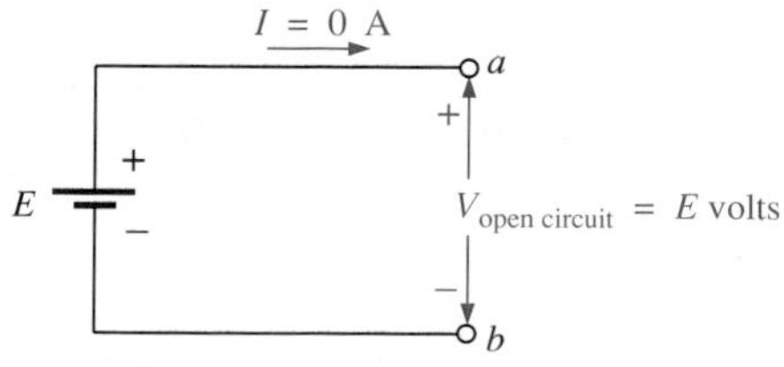

FIG. 6.42
Demonstrating the characteristics of an open circuit.

In summary, therefore,

a short circuit can carry a current of a level determined by the external circuit, but the potential difference (voltage) across its terminals is always zero volts.

In Fig. 6.43(a), the current through the 2-Ω resistor is 5 A. If a short circuit should develop across the 2-Ω resistor, the total resistance of the parallel combination of the 2-Ω resistor and the short (of essentially zero ohms) is $2\ \Omega \parallel 0\ \Omega = \dfrac{(2\ \Omega)(0\ \Omega)}{2\ \Omega + 0\ \Omega} = 0\ \Omega$, and the current will rise to very high levels, as determined by Ohm's law:

$$I = \frac{E}{R} = \frac{10\text{ V}}{0\ \Omega} \rightarrow \infty\text{ A}$$

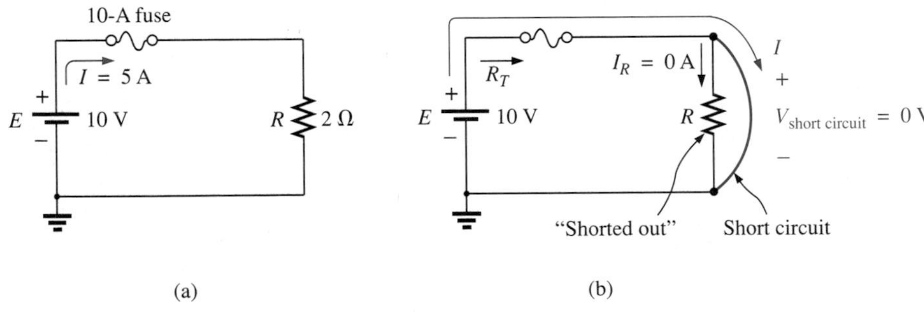

FIG. 6.43
Demonstrating the effect of a short circuit on current levels.

The effect of the 2-Ω resistor has effectively been "shorted out" by the low-resistance connection. The maximum current is now limited only by the circuit breaker or fuse in series with the source.

For the layperson, the terminology *short circuit* or *open circuit* is usually associated with dire situations such as power loss, smoke, or fire. However, in network analysis both can play an integral role in determining specific parameters about a system. Most often, however, if a short-circuit condition is to be established, it is accomplished with a *jumper*—a lead of negligible resistance to be connected between the points of interest. Establishing an open circuit simply requires making sure the terminals of interest are isolated from each other.

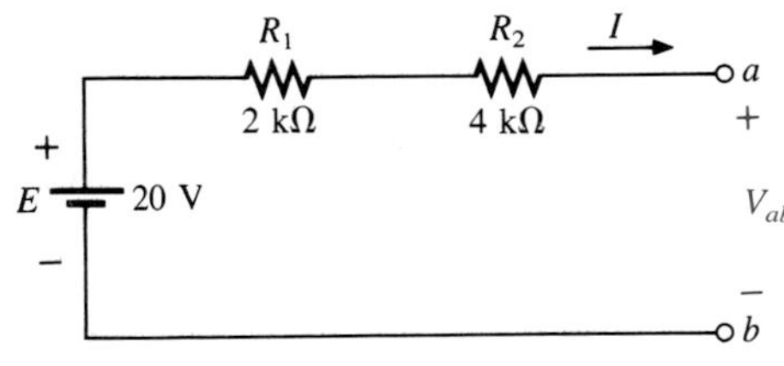

FIG. 6.44
Example 6.21.

EXAMPLE 6.21 Determine the voltage V_{ab} for the network of Fig. 6.44.

Solution: The open circuit requires that I be zero amperes. The voltage drop across both resistors is therefore zero volts since $V = IR = (0)R = 0$ V. Applying Kirchhoff's voltage law around the closed loop,

$$V_{ab} = E = \mathbf{20\ V}$$

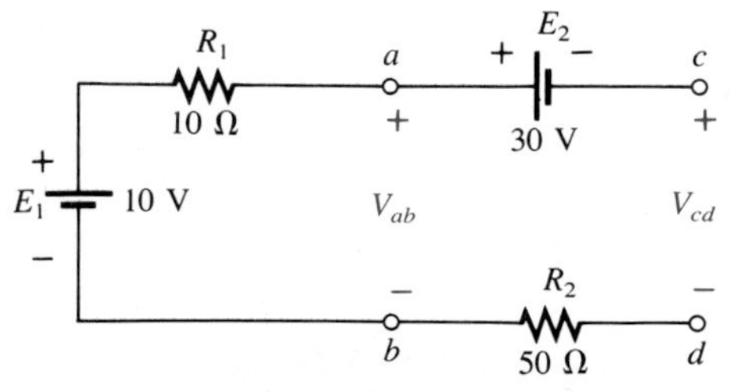

FIG. 6.45
Example 6.22.

EXAMPLE 6.22 Determine the voltages V_{ab} and V_{cd} for the network of Fig. 6.45.

Solution: The current through the system is zero amperes due to the open circuit, resulting in a 0-V drop across each resistor. Both resistors can therefore be replaced by short circuits, as shown in Fig. 6.46. The voltage V_{ab} is then directly across the 10-V battery, and

$$V_{ab} = E_1 = \mathbf{10\ V}$$

The voltage V_{cd} requires an application of Kirchhoff's voltage law:

$$+E_1 - E_2 - V_{cd} = 0$$

or
$$V_{cd} = E_1 - E_2 = 10\ \text{V} - 30\ \text{V} = \mathbf{-20\ V}$$

The negative sign in the solution simply indicates that the actual voltage V_{cd} has the opposite polarity of that appearing in Fig. 6.45.

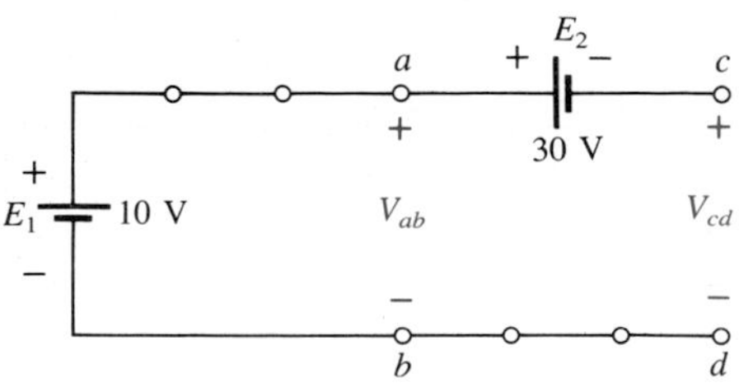

FIG. 6.46
Circuit of Fig. 6.45 redrawn.

EXAMPLE 6.23 Determine the unknown voltage and current for each network of Fig. 6.47.

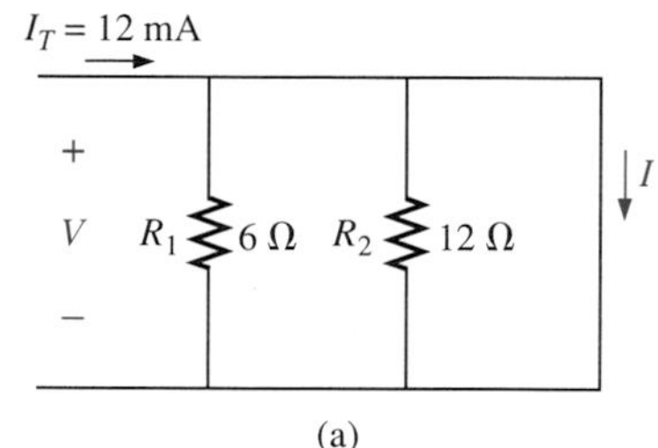

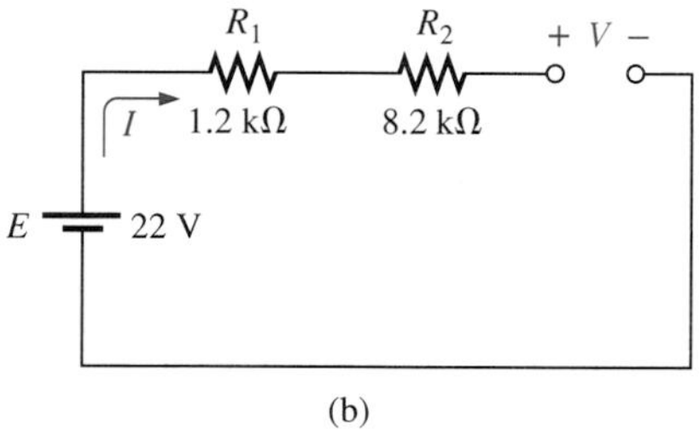

FIG. 6.47
Example 6.23.

Solutions: For the network of Fig. 6.47(a), the current I_T will take the path of least resistance, and, since the short-circuit condition at the end of the network is the least resistance path, all the current will pass through the short circuit. This conclusion can be verified using Eq. (6.9). The voltage across the network is the same as that across the shortcircuit and will be zero volts, as shown in Fig. 6.48(a).

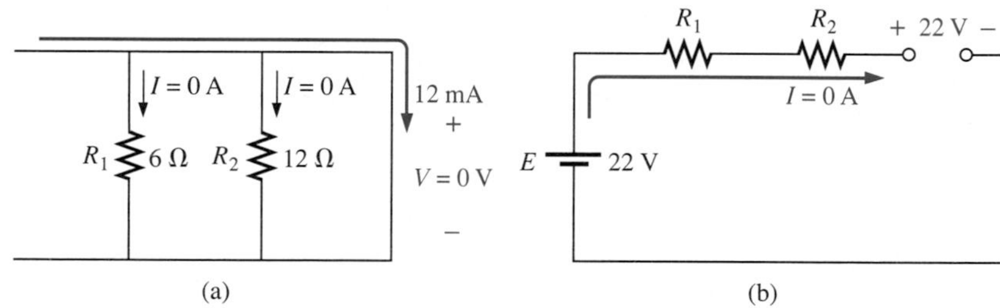

FIG. 6.48
Solutions to Example 6.23.

For the network of Fig. 6.47(b), the open-circuit condition requires that the current be zero amperes. The voltage drops across the resistors must therefore be zero volts, as determined by Ohm's law ($V_R = IR = (0)R = 0$ V), with the resistors simply acting as a connection from the supply to the open circuit. The result is that the open-circuit voltage will be $E = 22$ V, as shown in Fig. 6.48(b).

EXAMPLE 6.24 Calculate the current I and the voltage V for the network of Fig. 6.49.

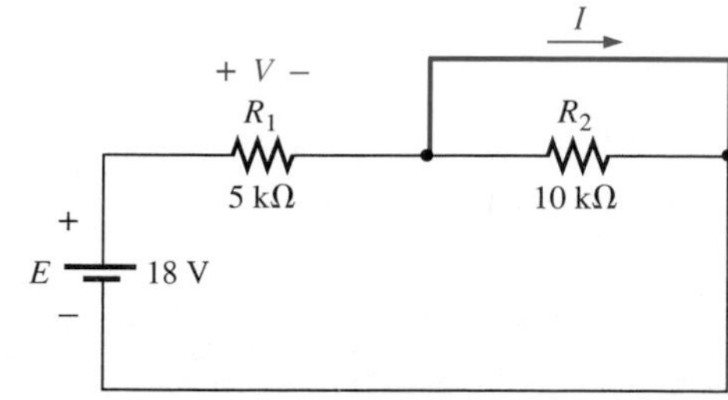

FIG. 6.49
Example 6.24.

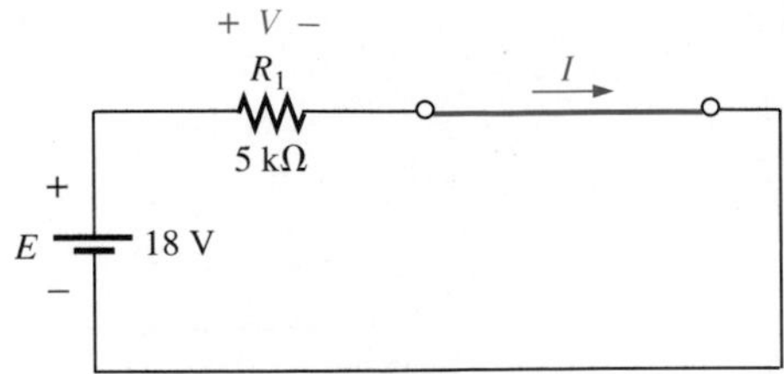

FIG. 6.50
Network of Fig. 6.49 redrawn.

Solution: The 10-kΩ resistor has been effectively shorted out by the jumper, resulting in the equivalent network of Fig. 6.50. Using Ohm's law,

$$I = \frac{E}{R_1} = \frac{18\text{ V}}{5\text{ k}\Omega} = \mathbf{3.6\ mA}$$

and

$$V = E = \mathbf{18\ V}$$

EXAMPLE 6.25 Determine V and I for the network of Fig. 6.51 if the resistor R_2 is shorted out.

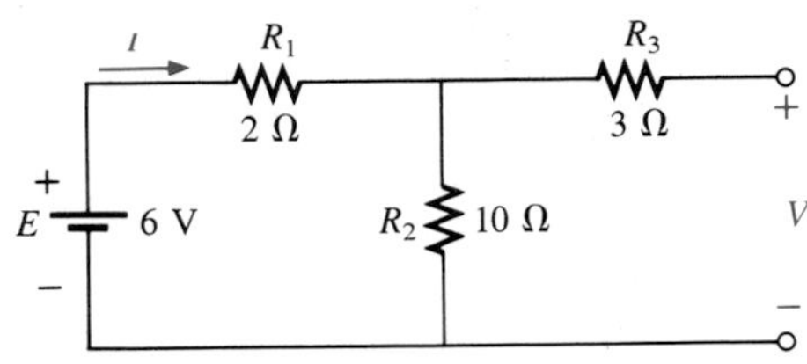

FIG. 6.51
Example 6.25.

Solution: The redrawn network appears in Fig. 6.52. The current through the 3-Ω resistor is zero due to the open circuit, causing all the current I to pass through the jumper. Since $V_{3\Omega} = IR = (0)R = 0$ V, the voltage V is directly across the short, and

$$V = \mathbf{0\ V}$$

with

$$I = \frac{E}{R_1} = \frac{6\text{ V}}{2\ \Omega} = \mathbf{3\ A}$$

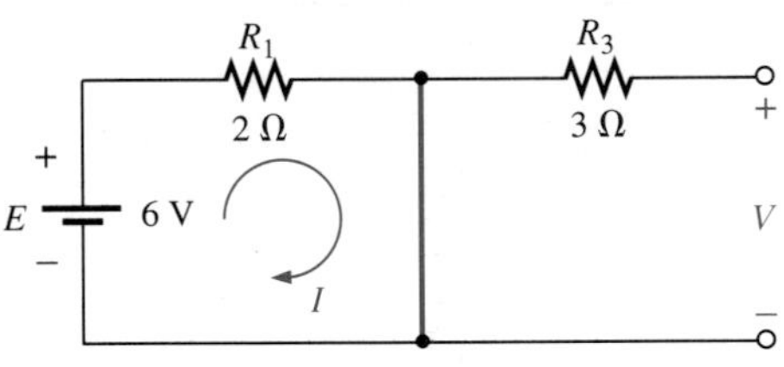

FIG. 6.52
Network of Fig. 6.51 with R_2 replaced by a jumper.

6.9 VOLTMETERS: LOADING EFFECT

In Chapters 2 and 5, it was noted that voltmeters are always placed across an element to measure the potential difference. We now realize that this connection is synonymous with placing the voltmeter in parallel with the element. The insertion of a meter in parallel with a resistor results in a combination of parallel resistors as shown in Fig. 6.53. Since the resistance of two parallel branches is always less than the smaller parallel resistance, the resistance of the voltmeter should be as large as possible (ideally infinite). In Fig. 6.53, a DMM with an internal resistance of 11 MΩ is measuring the voltage across a 10-kΩ resistor. The total resistance of the combination is

$$R_T = 10\text{ k}\Omega \parallel 11\text{ M}\Omega = \frac{(10^4\ \Omega)(11 \times 10^6\ \Omega)}{10^4\ \Omega + (11 \times 10^6\ \Omega)} = 9.99\text{ k}\Omega$$

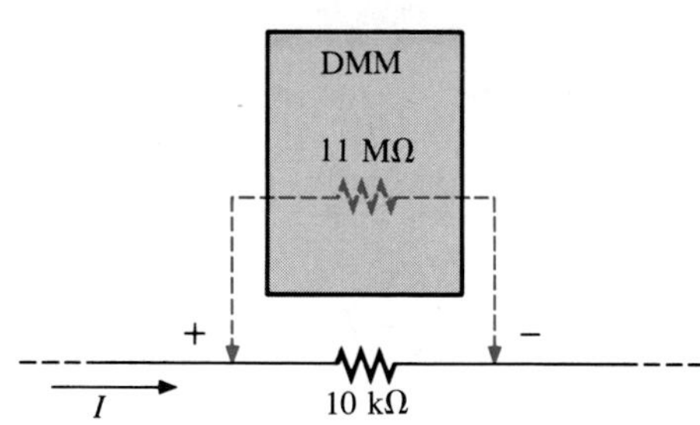

FIG. 6.53
Voltmeter loading.

and we find that the network is essentially undisturbed. However, if we use a VOM with an internal resistance of 50 kΩ on the 2.5-V scale, the parallel resistance is

$$R_T = 10\text{ k}\Omega \parallel 50\text{ k}\Omega = \frac{(10^4\ \Omega)(50 \times 10^3\ \Omega)}{10^4\ \Omega + (50 \times 10^3\ \Omega)} = 8.33\text{ k}\Omega$$

and the behavior of the network will be altered somewhat since the 10-kΩ resistor will now appear to be 8.33 kΩ to the rest of the network.

The loading of a network by the insertion of meters is not to be taken lightly, especially in research efforts where accuracy is a primary consideration. It is good practice always to check the meter resistance level against the resistive elements of the network before making measurements. A factor of 10 between resistance levels will usually provide fairly accurate meter readings for a wide range of applications.

Most DMMs have internal resistance levels in excess of 10 MΩ on all voltage scales, while the internal resistance of VOMs is sensitive to the chosen scale. To determine the resistance of each scale setting of a VOM in the voltmeter mode, simply multiply the maximum voltage of the scale setting by the ohm/volt (Ω/V) rating of the meter, normally found at the bottom of the face of the meter.

For a typical ohm/volt rating of 20,000, the 2.5-V scale would have an internal resistance of

$$(2.5 \text{ V})(20{,}000 \ \Omega/\text{V}) = 50 \text{ k}\Omega$$

whereas for the 100-V scale, it would be

$$(100 \text{ V})(20{,}000 \ \Omega/\text{V}) = 2 \text{ M}\Omega$$

and for the 250-V scale,

$$(250 \text{ V})(20{,}000 \ \Omega/\text{V}) = 5 \text{ M}\Omega$$

EXAMPLE 6.26 For the relatively simple network of Fig. 6.54:

a. What is the open-circuit voltage V_{ab}?
b. What will a DMM indicate if it has an internal resistance of 11 MΩ? Compare your answer to the results of part (a).
c. Repeat part (b) for a VOM with an Ω/V rating of 20,000 on the 100-V scale.

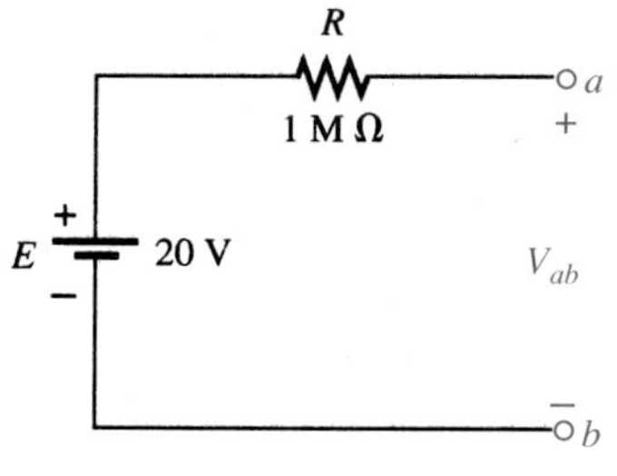

FIG. 6.54
Example 6.26.

Solutions:

a. $V_{ab} =$ **20 V.**
b. The meter will complete the circuit as shown in Fig. 6.55; using the voltage divider rule,

$$V_{ab} = \frac{11 \text{ M}\Omega(20 \text{ V})}{11 \text{ M}\Omega + 1 \text{ M}\Omega} = \mathbf{18.33 \text{ V}}$$

c. For the VOM, the internal resistance of the meter is

$$R_m = 100 \text{ V}(20{,}000 \ \Omega/\text{V}) = 2 \text{ M}\Omega$$

and
$$V_{ab} = \frac{2 \text{ M}\Omega(20 \text{ V})}{2 \text{ M}\Omega + 1 \text{ M}\Omega} = \mathbf{13.33 \text{ V}}$$

revealing the need to consider carefully the internal resistance of the meter in some instances.

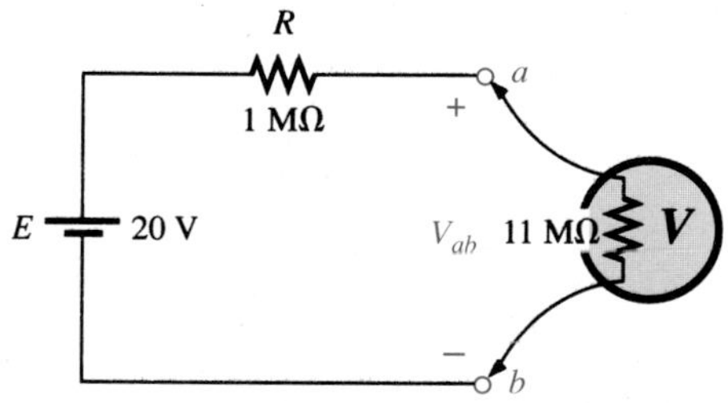

FIG. 6.55
Applying a DMM to the circuit of Fig. 6.54.

6.10 TROUBLESHOOTING TECHNIQUES

The art of *troubleshooting* is not limited solely to electrical or electronic systems. In the broad sense,

troubleshooting is a process by which acquired knowledge and experience are employed to localize a problem and offer or implement a solution.

There are many reasons why the simplest electrical circuit does not operate correctly. A connection may be open, the measuring instruments may need calibration, the power supply may not be on or may have been connected incorrectly to the circuit, an element may not be performing correctly due to earlier damage or poor manufacturing, a fuse may have blown, and so on. Unfortunately, a defined sequence of steps does not exist for identifying the wide range of problems that can surface in an electrical system. It is only through experience and a clear understanding of the basic laws of electric circuits that one can expect to become proficient at quickly locating the cause of an erroneous output.

It should be fairly obvious, however, that the first step in checking a network or identifying a problem area is to have some idea of the expected voltage and current levels. For instance, the circuit of Fig. 6.56 should have a current in the low milliampere range, with the majority of the supply voltage across the 8-kΩ resistor. However, as

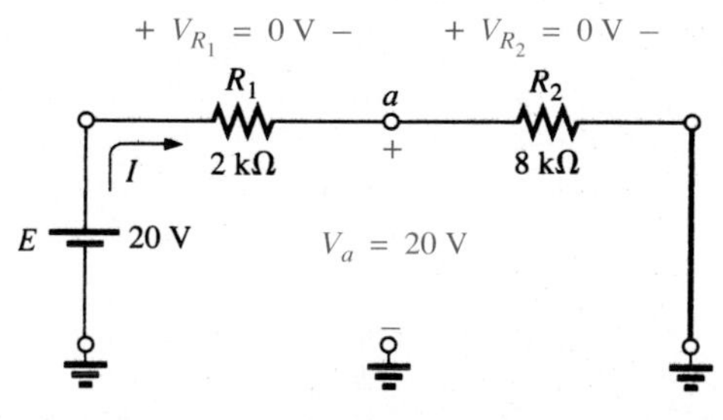

FIG. 6.56
A malfunctioning network.

indicated in Fig. 6.56, $V_{R_1} = V_{R_2} = 0$ V and $V_a = 20$ V. Since $V = IR$, the results immediately suggest that $I = 0$ A and an open circuit exists in the circuit. The fact that $V_a = 20$ V immediately tells us that the connections are true from the ground of the supply to point a. The open circuit must therefore exist between R_1 and R_2 or at the ground connection of R_2. An open circuit at either point will result in $I = 0$ A and the readings obtained previously. Keep in mind that, even though $I = 0$ A, R_1 does form a connection between the supply and point a. That is, if $I = 0$ A, $V_{R_1} = IR_2 = (0)R_2 = 0$ V, as obtained for a short circuit.

In Fig. 6.56, if $V_{R_1} \cong 20$ V and V_{R_2} is quite small ($\cong 0.08$ V), it first suggests that the circuit is complete, a current does exist, and a problem surrounds the resistor R_2. R_2 is not shorted out since such a condition would result in $V_{R_2} = 0$ V. A careful check of the inserted resistor reveals that an 8-Ω resistor was employed rather than the 8-kΩ resistor called for—an incorrect reading of the color code. Perhaps in the future an ohmmeter should be used to check a resistor to validate the color-code reading or to ensure that its value is still in the prescribed range set by the color code.

There will be occasions when frustration may develop. You've checked all the elements, and all the connections appear tight. The supply is on and set at the proper level; the meters appear to be functioning correctly. In situations such as this, experience becomes a key factor. Perhaps you can recall when a recent check of a resistor revealed that the internal connection (not externally visible) was a "make or break" situation or that the resistor was damaged earlier by excessive current levels, so its actual resistance was much lower than called for by the color code. Recheck the supply! Perhaps the terminal voltage was set correctly, but the current control knob was left in the zero or minimum position. Is the ground connection stable? The questions that arise may seem endless. However, take heart in the fact that with experience comes an ability to localize problems more rapidly. Of course, the more complicated the system, the longer the list of possibilities, but it is often possible to identify a particular area of the system that is behaving improperly before checking individual elements.

6.11 COMPUTER ANALYSIS

The computer analysis of parallel dc networks is very similar to that applied to series circuits in the previous chapter. However, in this case the voltage will be the same across all elements, and the current will be determined by the resistance of the parallel branch. In addition to the standard analysis, the effect of varying the applied voltage will be demonstrated for both PSpice (DOS) and PSpice (Windows).

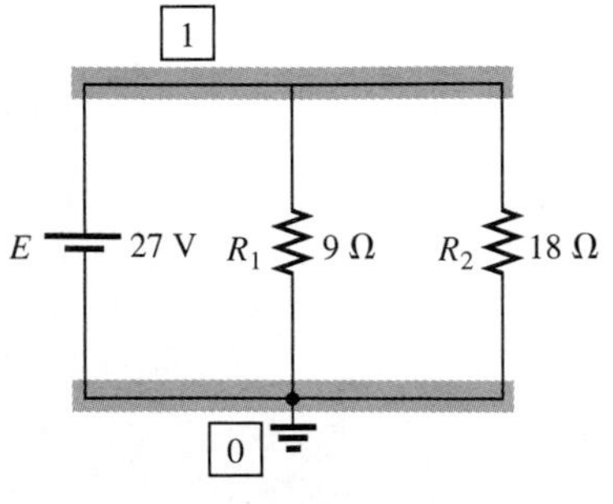

FIG. 6.57
Network to be examined using PSpice (DOS).

PSpice (DOS)

The input file for the parallel network of Fig. 6.57 is provided in Fig. 6.58. Note again the use of the .DC control line to specify the output file desired. The .PRINT line specifies only I_1 and I_2 for the output file, and the .OPTIONS NOPAGE line continues to minimize wasted space. The results obtained for the parameters entered support the results obtained in Example 6.11.

The second input file (Fig. 6.59) is different from the above only because the dc battery voltage is incremented from 20 V to 30 V in

```
Chapter 6 - Two Parallel Resistors

 ****     CIRCUIT DESCRIPTION

******************************************************************************


VE 1 0 27V
R1 1 0 9
R2 1 0 18
.DC VE 27 27 1
.PRINT DC I(R1) I(R2)
.OPTIONS NOPAGE
.END

 ****     DC TRANSFER CURVES                    TEMPERATURE =    27.000 DEG C


 VE            I(R1)         I(R2)

  2.700E+01    3.000E+00    1.500E+00
```

FIG. 6.58
Input and output files for the network of Fig. 6.57.

```
Chapter 6 - Two Parallel Resistors

 ****     CIRCUIT DESCRIPTION

******************************************************************************


VE 1 0 27V
R1 1 0 9
R2 1 0 18
.DC VE 20 30 1
.PRINT DC I(R1) I(R2)
.OPTIONS NOPAGE
.END

 ****     DC TRANSFER CURVES                    TEMPERATURE =    27.000 DEG C


 VE            I(R1)         I(R2)

  2.000E+01    2.222E+00    1.111E+00
  2.100E+01    2.333E+00    1.167E+00
  2.200E+01    2.444E+00    1.222E+00
  2.300E+01    2.556E+00    1.278E+00
  2.400E+01    2.667E+00    1.333E+00
  2.500E+01    2.778E+00    1.389E+00
  2.600E+01    2.889E+00    1.444E+00
  2.700E+01    3.000E+00    1.500E+00
  2.800E+01    3.111E+00    1.556E+00
  2.900E+01    3.222E+00    1.611E+00
  3.000E+01    3.333E+00    1.667E+00
```

FIG. 6.59
Input and output files for the network of Fig. 6.57 with E varying from 20 V to 30 V.

increments of 1 V. Note how the currents increase in magnitude with an increase in voltage and how the results obtained at E = 27 V = 2.700E+01 match those obtained above.

PSpice (Windows)

A parallel network with a wide range of resistor values will now be investigated using schematics. The calling up and placement of all the elements including **IPROBE** were described in Chapter 5. The resistor values, as appearing in the schematic of Fig. 6.60, extend from 1.2 MΩ down to 22 Ω. Note in particular the manner in which **IPROBE** was placed for each branch to be sure a positive value was obtained for all the currents of the network. Following the "drawing" of the network, the analysis is initiated with **Analysis-Simulate,** and the results appear as shown on Fig. 6.60. The placement of the results may not be as neat as that shown in Fig. 6.60, but they will be very close to the **IPROBE** symbol and accessible simply by clicking on the value. The neat, clean appearance of Fig. 6.60 is the result of moving things around to make everything fit once the results appeared. Note how nicely the floating point solutions can express numbers that range from 20 μA all the way up to 2.182 A. In addition, note how quickly the current increases with a decrease in resistor value. The range in resistor value suggests, by inspection, that the total resistance will be very close to the smallest resistor of 22 Ω. Using Ohm's law and the source current of 2.204 A results in a total resistance of $R_T = E/I_s$ = 48 V/2.204 A = 21.78 Ω, confirming the above conclusion.

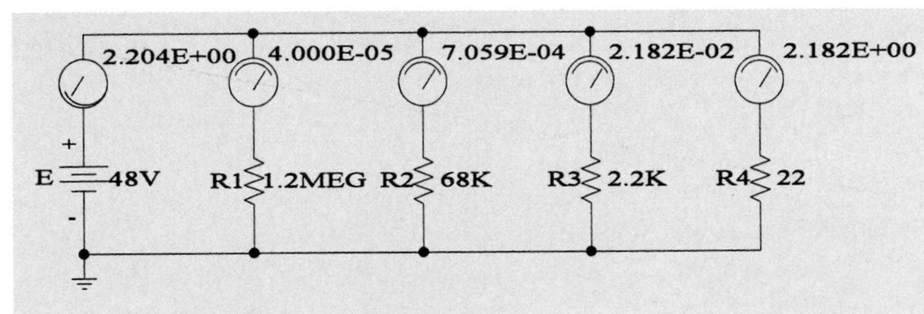

FIG. 6.60
Applying PSpice (Windows) to a parallel network with a wide range of parallel resistor values.

PROBLEMS

SECTION 6.2 Parallel Elements

1. For each configuration of Fig. 6.61, determine which elements are in series or parallel.

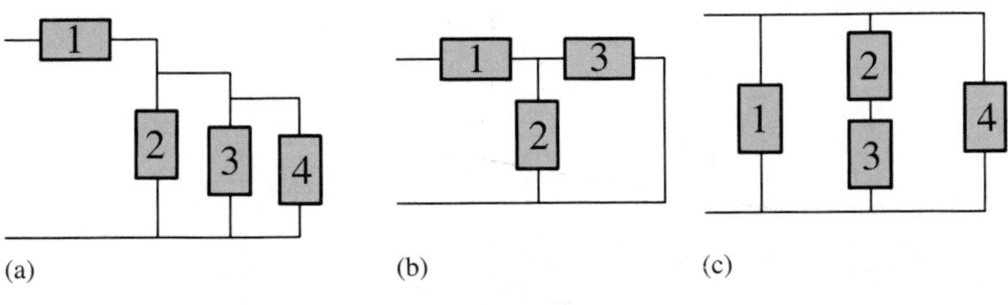

FIG. 6.61
Problem 1.

2. For the network of Fig. 6.62:
   a. Which elements are in parallel?
   b. Which elements are in series?
   c. Which branches are in parallel?

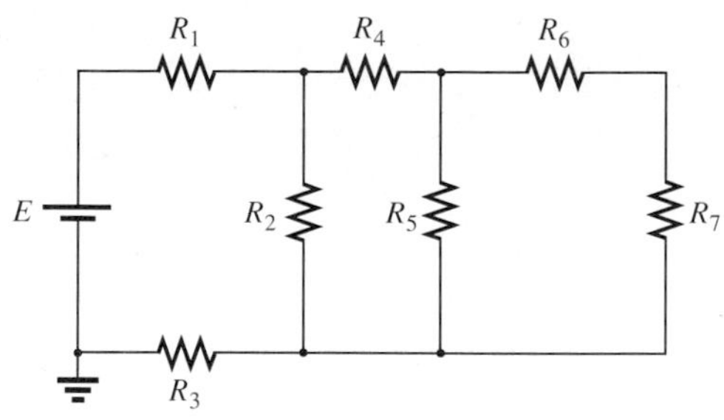

FIG. 6.62
Problem 2.

SECTION 6.3 Total Conductance and Resistance

3. Find the total conductance and resistance for the networks of Fig. 6.63.

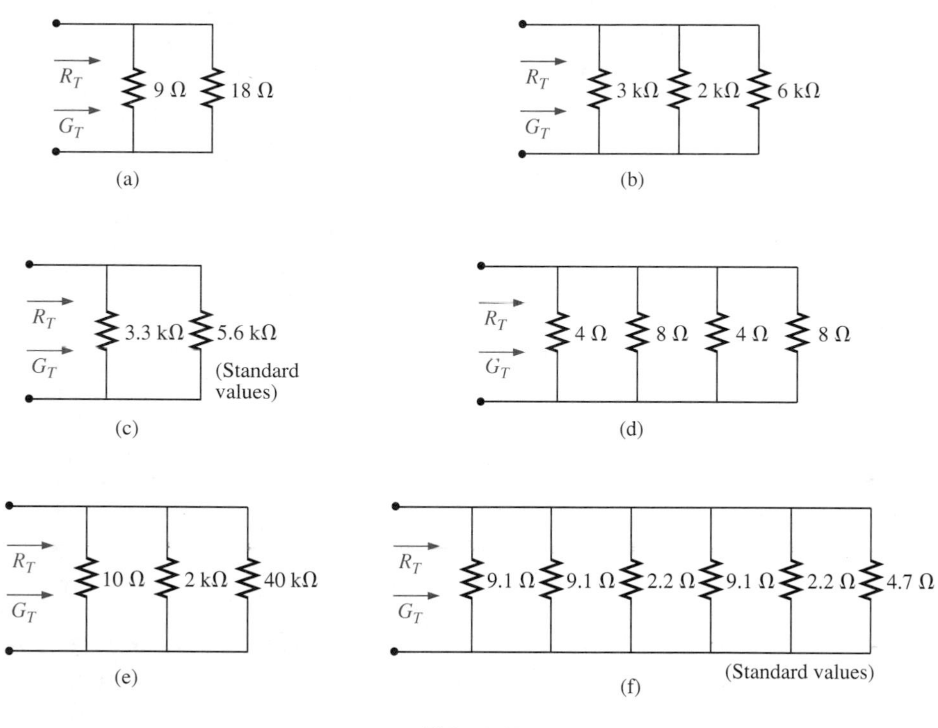

FIG. 6.63
Problem 3.

4. The total conductance of the networks of Fig. 6.64 is specified. Find the value in ohms of the unknown resistances.

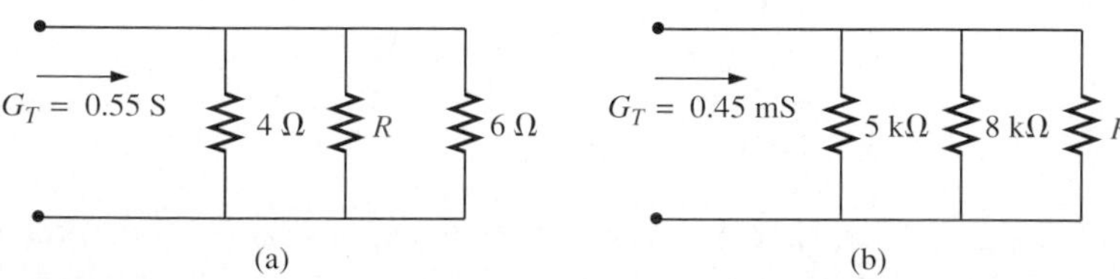

FIG. 6.64
Problem 4.

5. The total resistance of the circuits of Fig. 6.65 is specified. Find the value in ohms of the unknown resistances.

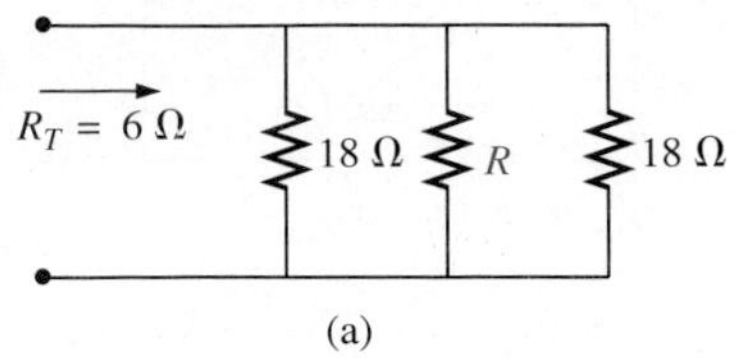

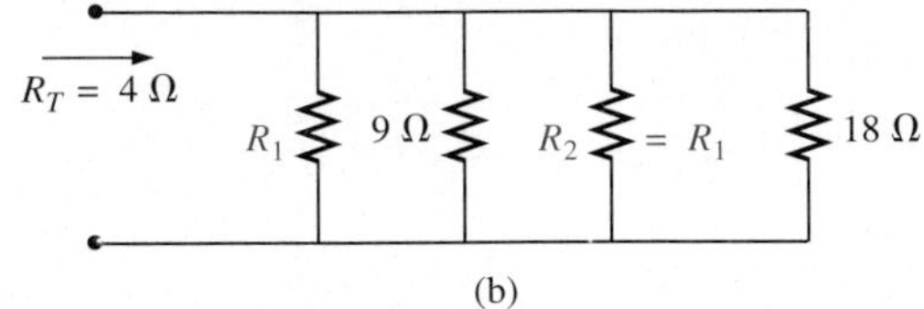

FIG. 6.65
Problem 5.

***6.** Determine the unknown resistors of Fig. 6.66 given the fact that $R_2 = 5R_1$ and $R_3 = (1/2)R_1$.

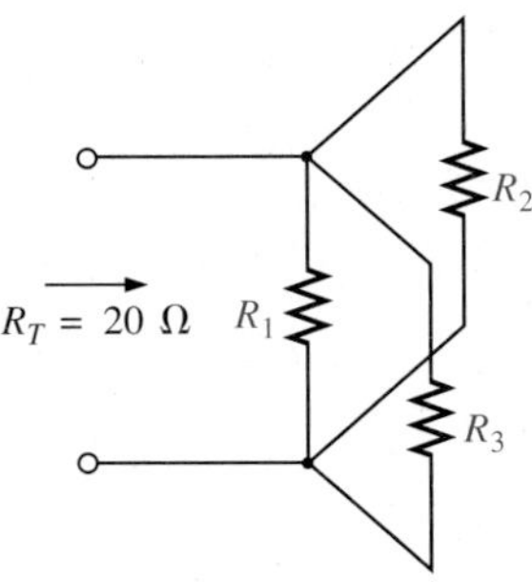

FIG. 6.66
Problem 6.

***7.** Determine R_1 for the network of Fig. 6.67.

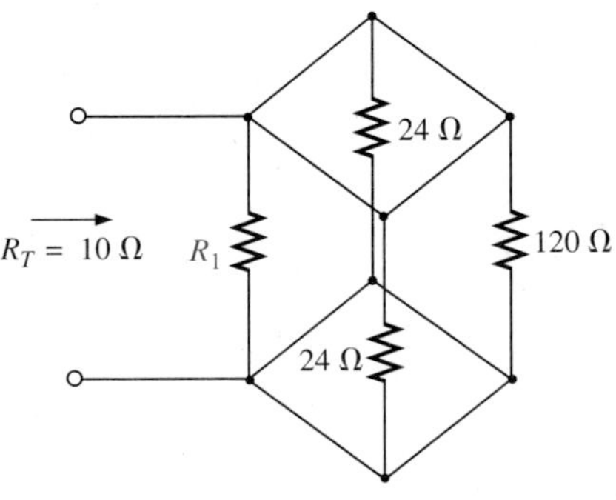

FIG. 6.67
Problem 7.

SECTION 6.4 Parallel Networks

8. For the network of Fig. 6.68:
   a. Find the total conductance and resistance.
   b. Determine I_s and the current through each parallel branch.
   c. Verify that the source current equals the sum of the parallel branch currents.
   d. Find the power dissipated by each resistor, and note whether the power delivered is equal to the power dissipated.
   e. If the resistors are available with wattage ratings of 1/2, 1, 2, and 50 W, what is the minimum wattage rating for each resistor?

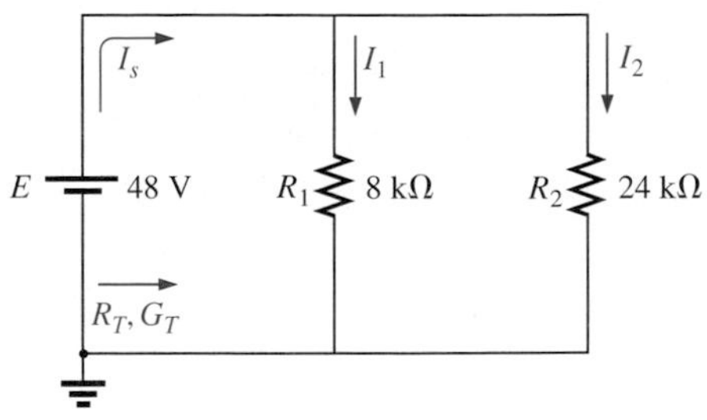

FIG. 6.68
Problem 8.

9. Repeat Problem 8 for the network of Fig. 6.69.

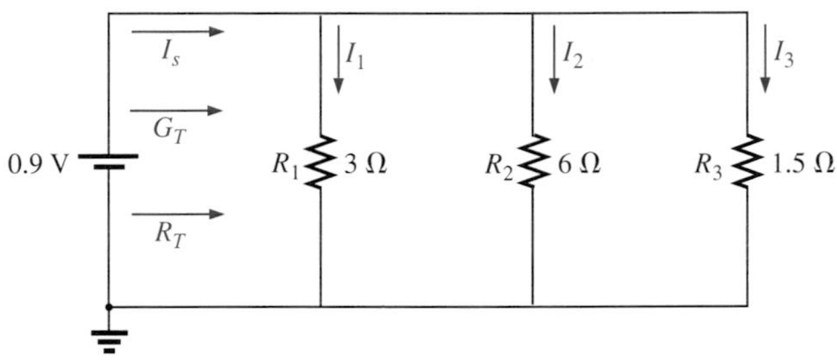

FIG. 6.69
Problems 9 and 37.

10. Repeat Problem 8 for the network of Fig. 6.70 constructed of standard resistor values.

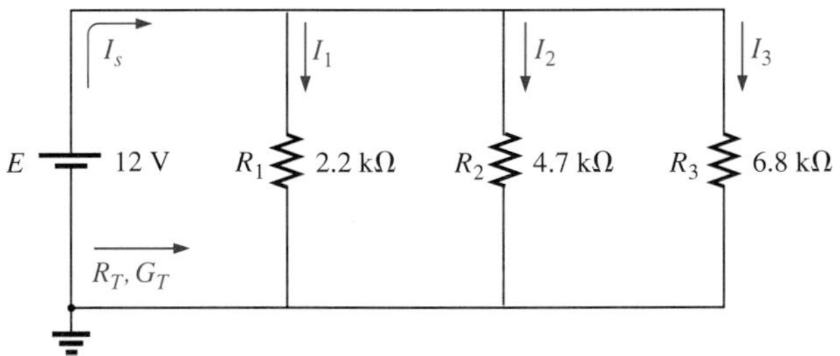

FIG. 6.70
Problem 10.

11. There are eight holiday lights connected in parallel as shown in Fig. 6.71.
   a. If the set is connected to a 120-V source, what is the current through each bulb if each bulb has an internal resistance of 1.8 kΩ?
   b. Determine the total resistance of the network.
   c. Find the power delivered to each bulb.
   d. If one bulb burns out (that is, the filament opens), what is the effect on the remaining bulbs?
   e. Compare the parallel arrangement of Fig. 6.71 to the series arrangement of Fig. 5.82. What are the relative advantages and disadvantages of the parallel system as compared to the series arrangement?

FIG. 6.71
Problem 11.

12. A portion of a residential service to a home is depicted in Fig. 6.72.
   a. Determine the current through each parallel branch of the network.
   b. Calculate the current drawn from the 120-V source. Will the 20-A circuit breaker trip?

c. What is the total resistance of the network?

d. Determine the power supplied by the 120-V source. How does it compare to the total power of the load?

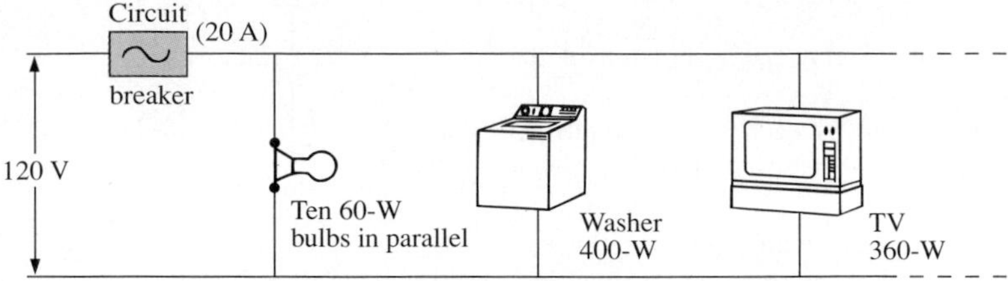

FIG. 6.72
Problems 12 and 39.

13. Determine the currents I_1 and I_s for the networks of Fig. 6.73.

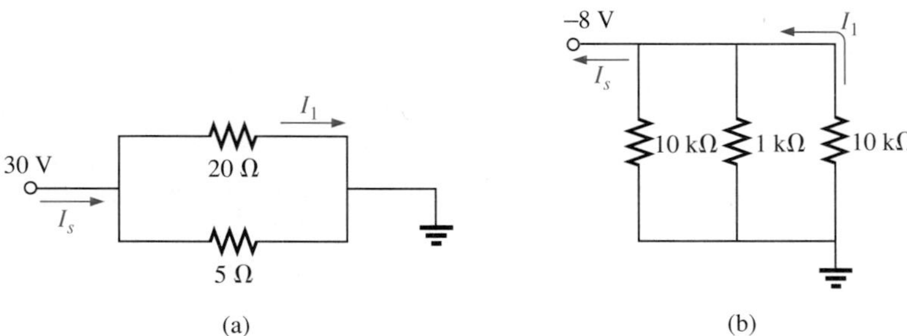

FIG. 6.73
Problem 13.

14. Using the information provided, determine the resistance R_1 for the network of Fig. 6.74.

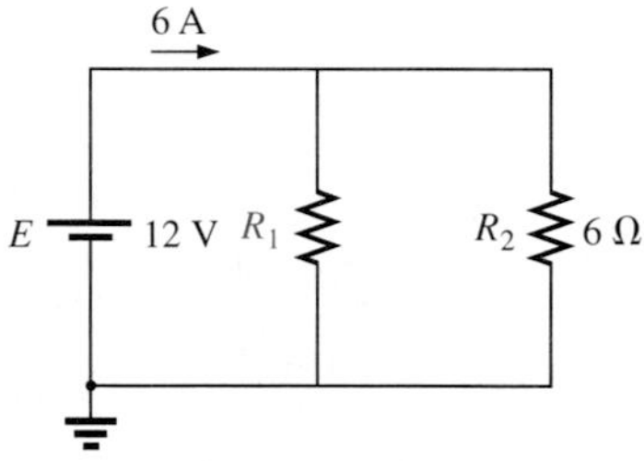

FIG. 6.74
Problem 14.

*15. Determine the power delivered by the dc battery in Fig. 6.75.

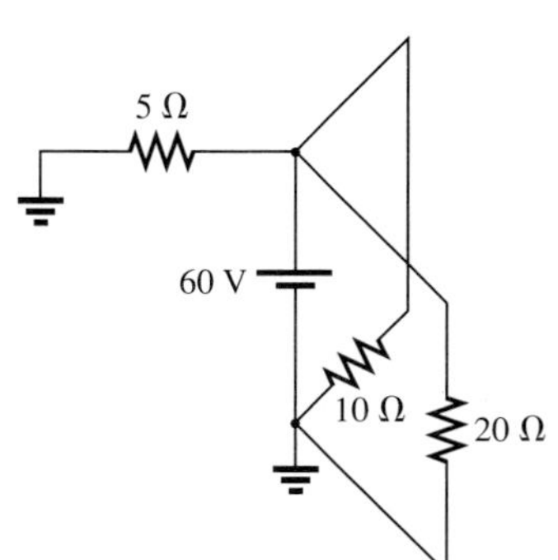

FIG. 6.75
Problems 15 and 38.

***16.** For the network of Fig. 6.76:
   a. Find the current I_1.
   b. Calculate the power dissipated by the 4-Ω resistor.
   c. Find the current I_2.

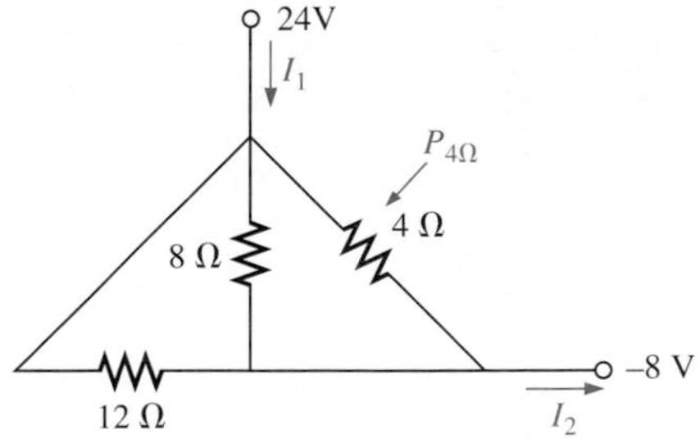

FIG. 6.76
Problem 16.

***17.** For the network of Fig. 6.77:
   a. Find the current I.
   b. Determine the voltage V.
   c. Calculate the source current I_s.

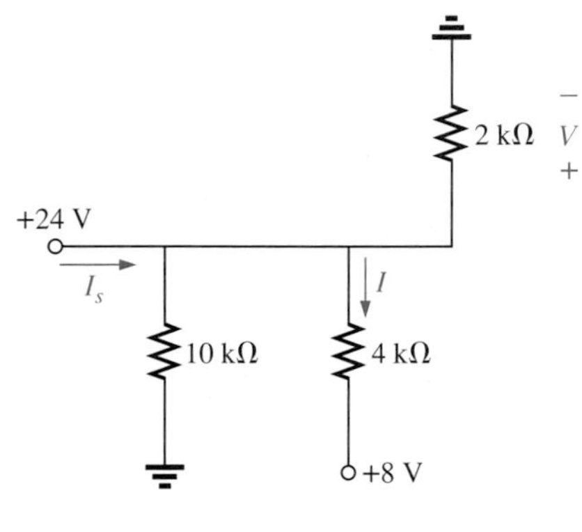

FIG. 6.77
Problem 17.

SECTION 6.5 Kirchhoff's Current Law

18. Find all unknown currents and their directions in the circuits of Fig. 6.78.

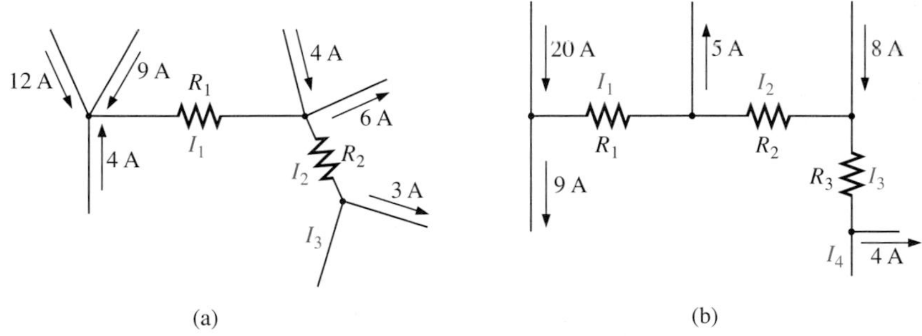

FIG. 6.78
Problems 18 and 40.

*19. Using Kirchhoff's current law, determine the unknown currents for the networks of Fig. 6.79.

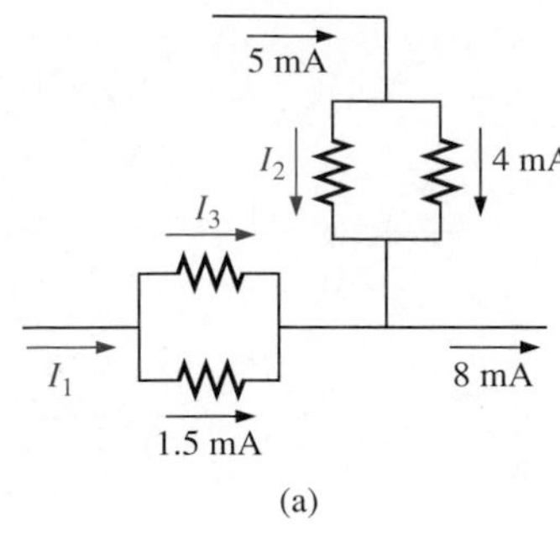

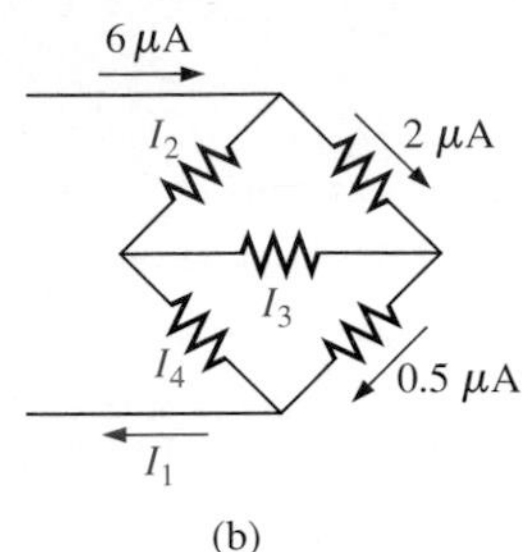

FIG. 6.79
Problem 19.

20. Using the information provided in Fig. 6.80, find the branch resistors R_1 and R_3, the total resistance R_T, and the voltage source E.

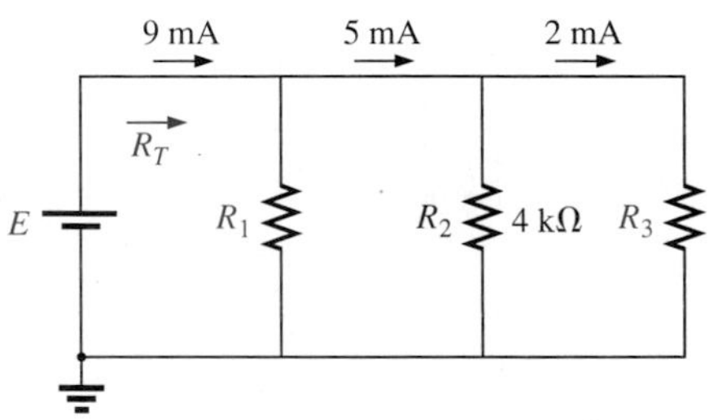

FIG. 6.80
Problem 20.

*21. Find the unknown quantities for the circuits of Fig. 6.81 using the information provided.

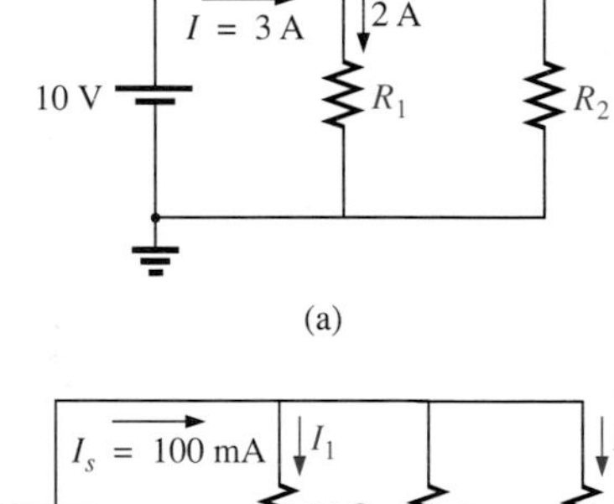

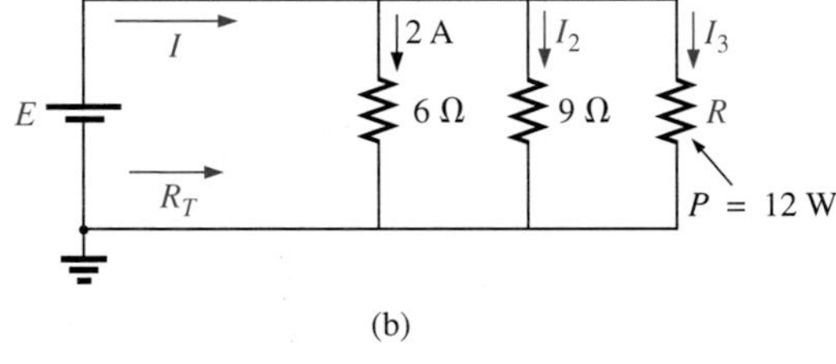

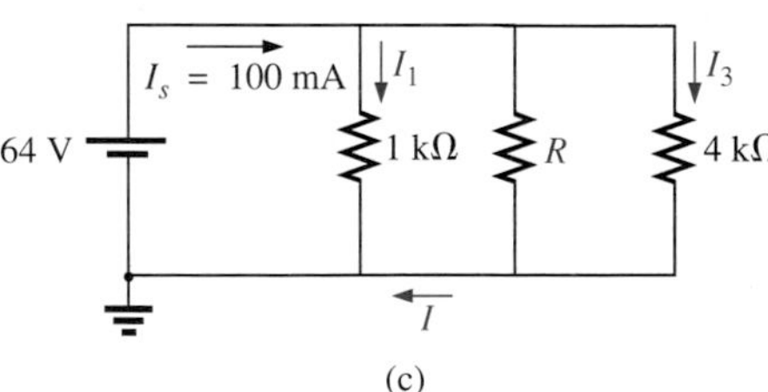

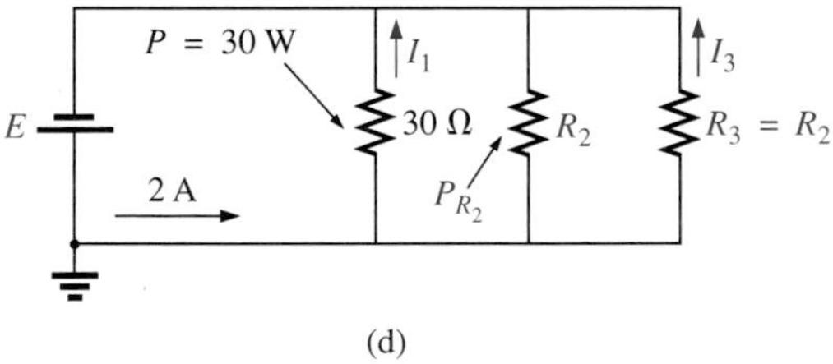

FIG. 6.81
Problem 21.

SECTION 6.6 Current Divider Rule

22. Using the information provided in Fig. 6.82, determine the current through each branch using simply the ratio of parallel resistor values. Then determine the total current I_T.

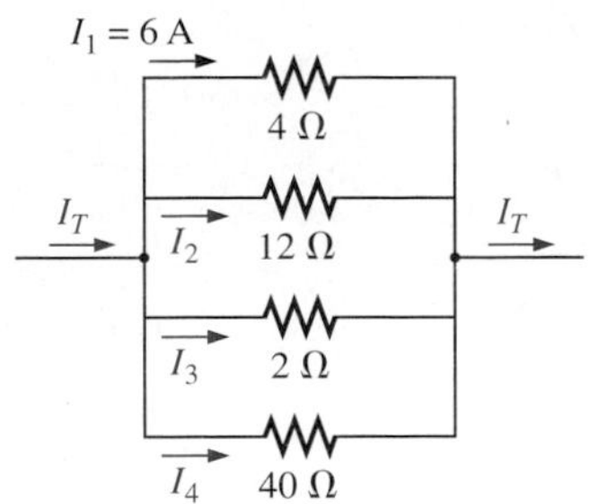

FIG. 6.82
Problem 22.

23. Using the current divider rule, find the unknown currents for the networks of Fig. 6.83.

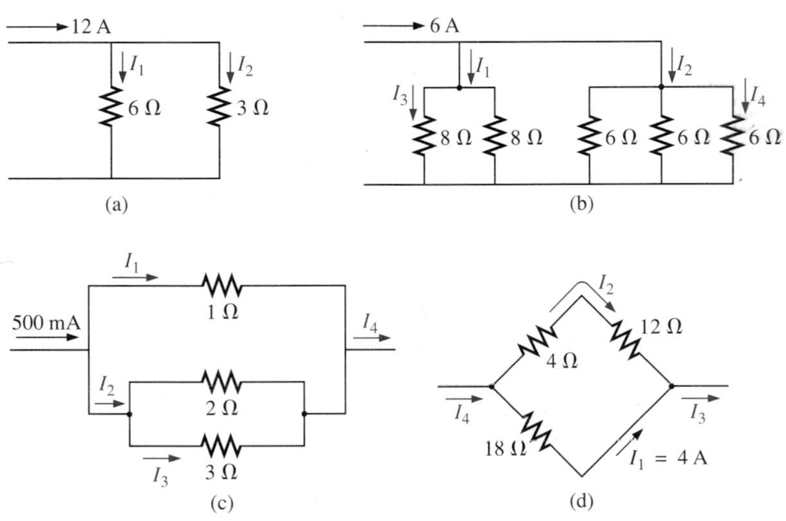

FIG. 6.83
Problem 23.

***24.** Parts (a), (b), and (c) of this problem should be done by inspection—that is, mentally. The intent is to obtain a solution without a lengthy series of calculations. For the network of Fig. 6.84:

a. What is the approximate value of I_1 considering the magnitude of the parallel elements?

b. What are the ratios I_1/I_2 and I_3/I_4?

c. What are the ratios I_2/I_3 and I_1/I_4?

d. Calculate the current I_1 and compare it to the result of part (a).

e. Determine the current I_4 and calculate the ratio I_1/I_4. How does the ratio compare to the result of part (c)?

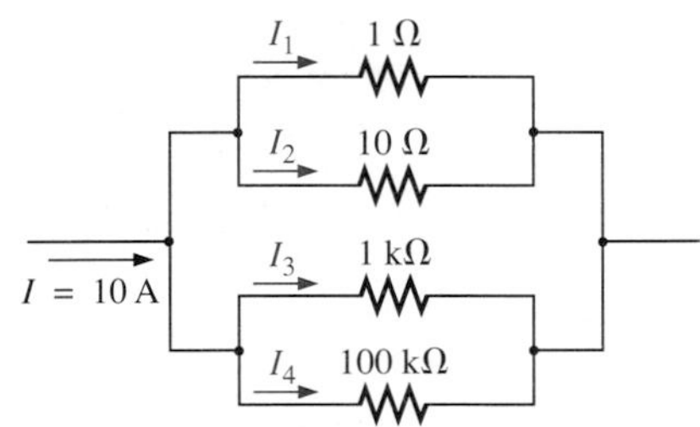

FIG. 6.84
Problem 24.

25. Find the unknown quantities using the information provided for the networks of Fig. 6.85.

(a)

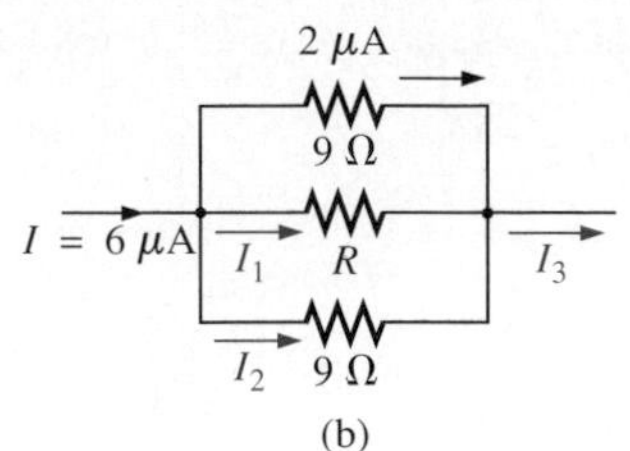

(b)

FIG. 6.85
Problem 25.

*26. Calculate the resistor R for the network of Fig. 6.86 that will ensure the current $I_1 = 3I_2$.

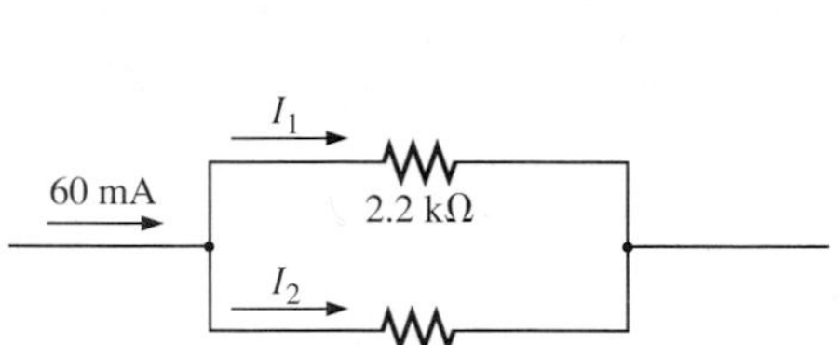

FIG. 6.86
Problem 26.

*27. Design the network of Fig. 6.87 such that $I_2 = 4I_1$ and $I_3 = 3I_2$.

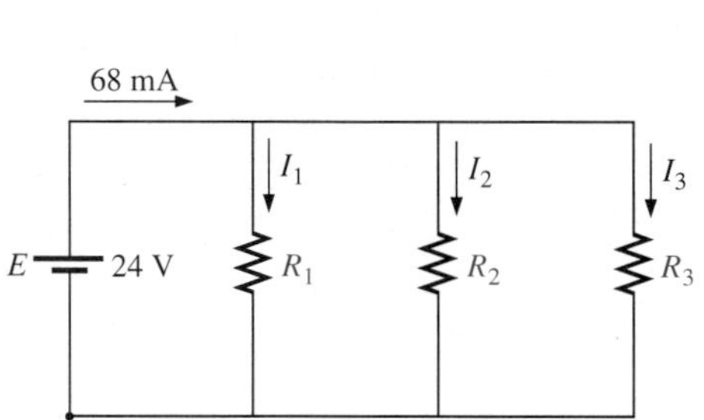

FIG. 6.87
Problem 27.

SECTION 6.7 Voltage Sources in Parallel

28. Assuming identical supplies, determine the currents I_1 and I_2 for the network of Fig. 6.88.

29. Assuming identical supplies, determine the current I and resistance R for the parallel network of Fig. 6.89.

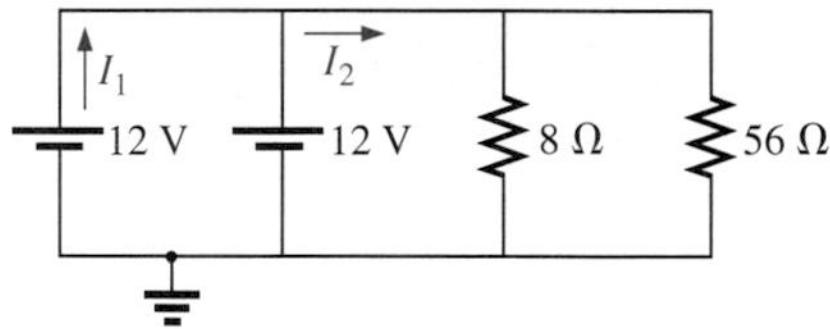

FIG. 6.88
Problem 28.

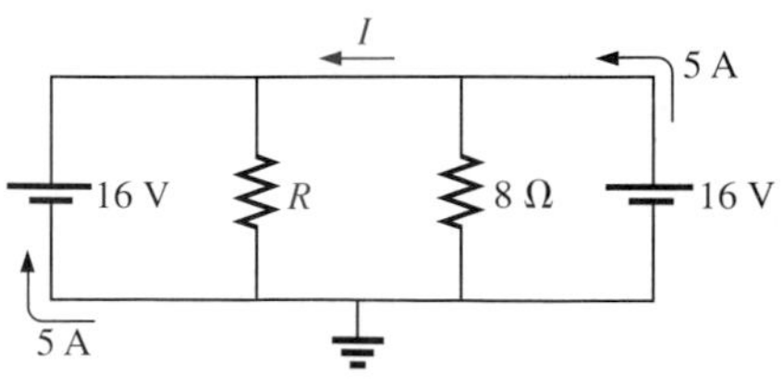

FIG. 6.89
Problem 29.

SECTION 6.8 Open and Short Circuits

30. For the network of Fig. 6.90:

a. Determine I_s and V_L.

b. Determine I_s if R_L is shorted out.

c. Determine V_L if R_L is replaced by an open circuit.

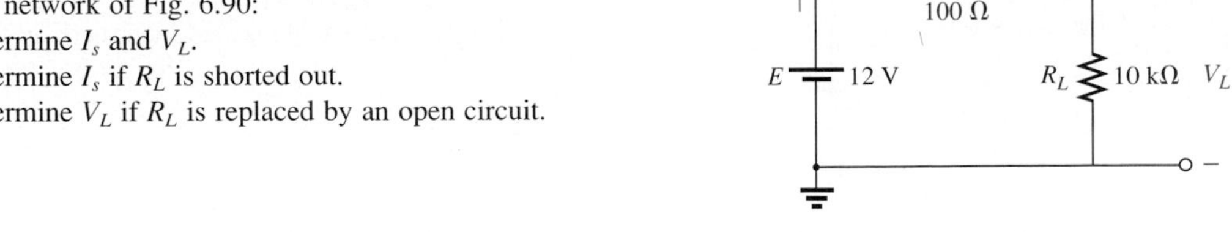

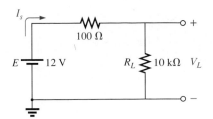

FIG. 6.90
Problem 30.

31. For the network of Fig. 6.91:

a. Determine the open-circuit voltage V_L.

b. If the 2.2-kΩ resistor is short circuited, what is the new value of V_L?

c. Determine V_L if the 4.7-kΩ resistor is replaced by an open circuit.

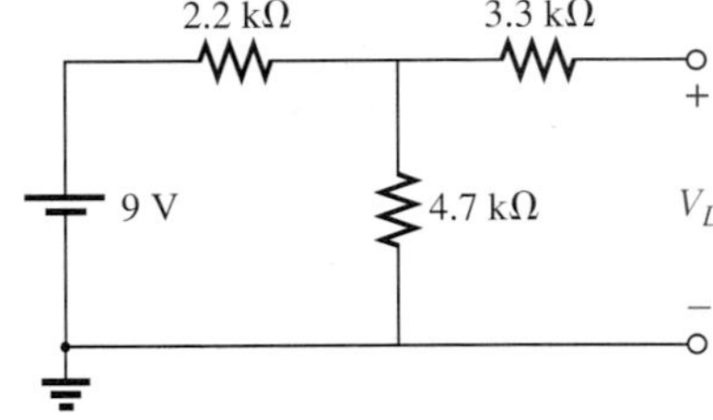

FIG. 6.91
Problem 31.

***32.** For the network of Fig. 6.92, determine

a. The short-circuit currents I_1 and I_2.

b. The voltages V_1 and V_2.

c. The source current I_s.

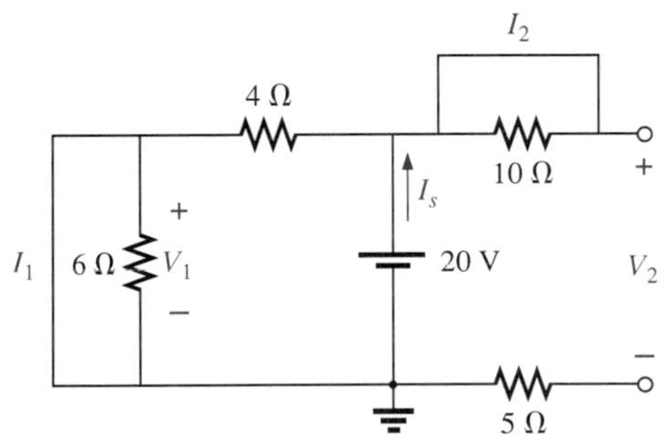

FIG. 6.92
Problem 32.

SECTION 6.9 Voltmeters: Loading Effect

33. For the network of Fig. 6.93:

a. Determine the voltage V_2.

b. Determine the reading of a DMM having an internal resistance of 11 MΩ when used to measure V_2.

c. Repeat part (b) with a VOM having an ohm/volt rating of 20,000 using the 10-V scale. Compare the results of parts (b) and (c). Explain any difference.

d. Repeat part (c) with $R_1 = 100$ kΩ and $R_2 = 200$ kΩ.

e. Based on the above, can you make any general conclusions about the use of a voltmeter?

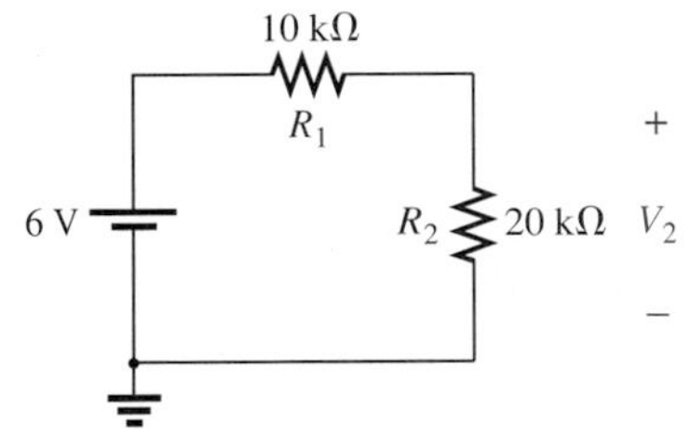

FIG. 6.93
Problems 33 and 42.

SECTION 6.10 Troubleshooting Techniques

34. Based on the measurements of Fig. 6.94, determine whether the network is operating correctly. If not, try to determine why.

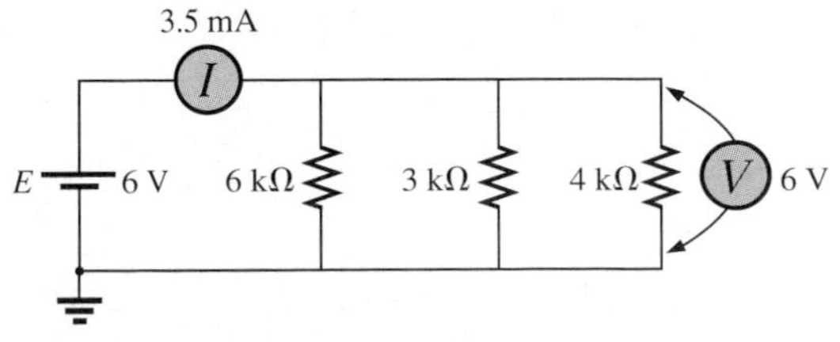

FIG. 6.94
Problem 34.

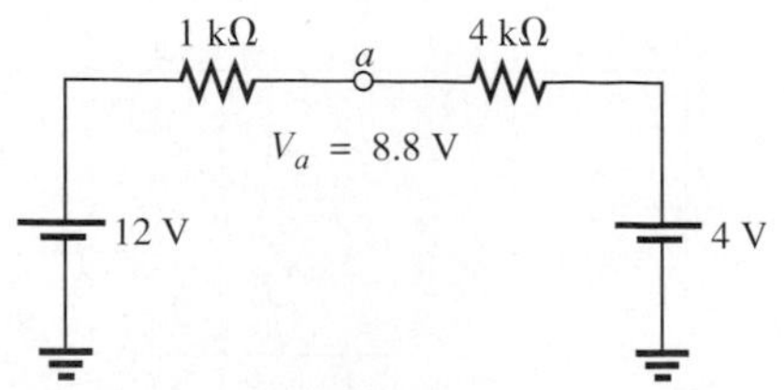

FIG. 6.95
Problem 35.

35. Referring to the network of Fig. 6.95, is $V_a = 8.8$ V the correct reading for the given configuration? If not, which element has been connected incorrectly in the network?

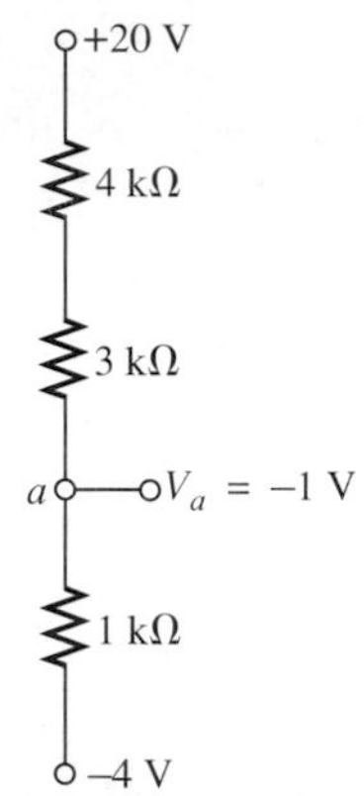

FIG. 6.96
Problem 36.

36. **a.** The voltage V_a for the network of Fig. 6.96 is -1 **V**. If it suddenly jumped to 20 V, what could have happened to the circuit structure? Localize the problem area.
 b. If the voltage V_a is 6 V rather than -1 V, try to explain what is wrong about the network construction.

SECTION 6.11 Computer Analysis

PSpice (DOS)

37. Write an input file to solve for the current through each resistor of Fig. 6.69.

38. Write an input file to find the source current and the current through each resistor of Fig. 6.75.

PSpice (Windows)

39. Using schematics, determine all the currents for the network of Fig. 6.72.

40. Using schematics, determine the unknown quantities for the network of Fig. 6.78.

Programming Language (C++, BASIC, PASCAL, etc.)

41. Write a program to determine the total resistance and conductance of any number of elements in parallel.

42. Write a program that will tabulate the voltage V_2 of Fig. 6.93 measured by a VOM with an internal resistance of 200 kΩ as R_2 varies from 10 kΩ to 200 kΩ in increments of 10 kΩ.

GLOSSARY

Current divider rule A method by which the current through parallel elements can be determined without first finding the voltage across those parallel elements.

Kirchhoff's current law The algebraic sum of the currents entering and leaving a node is zero.

Node A junction of two or more branches.

Ohm/volt rating A rating used to determine both the current sensitivity of the movement and the internal resistance of the meter.

Open circuit The absence of a direct connection between two points in a network.

Parallel circuit A circuit configuration in which the elements have two points in common.

Short circuit A direct connection of low resistive value that can significantly alter the behavior of an element or system.

7

Series-Parallel Networks

7.1 SERIES-PARALLEL NETWORKS

A firm understanding of the basic principles associated with series and parallel circuits is a sufficient background to approach most series-parallel networks (a network being a combination of any number of series and parallel elements) with *one* source of voltage. Multisource networks are considered in detail in Chapters 8 and 9. In general,

series-parallel networks are networks that contain both series and parallel circuit configurations.

One can become proficient in the analysis of series-parallel networks only through exposure, practice, and experience. In time the path to the desired unknown becomes more obvious as one recalls similar configurations and the frustration resulting from choosing the wrong approach. There are a few steps that can be helpful in getting started on the first few exercises, although the value of each will become apparent only with experience.

General Approach

1. Take a moment to study the problem "in total" and make a brief mental sketch of the overall approach you plan to use. The result may be time- and energy-saving shortcuts.
2. Next examine each region of the network independently before tying them together in series-parallel combinations. This will usually simplify the network and possibly reveal a direct approach toward obtaining one or more desired unknowns. It also eliminates many of the errors that might result due to the lack of a systematic approach.
3. Redraw the network as often as possible with the reduced branches and undisturbed unknown quantities to maintain clarity and provide the reduced networks for the trip back to unknown quantities from the source.

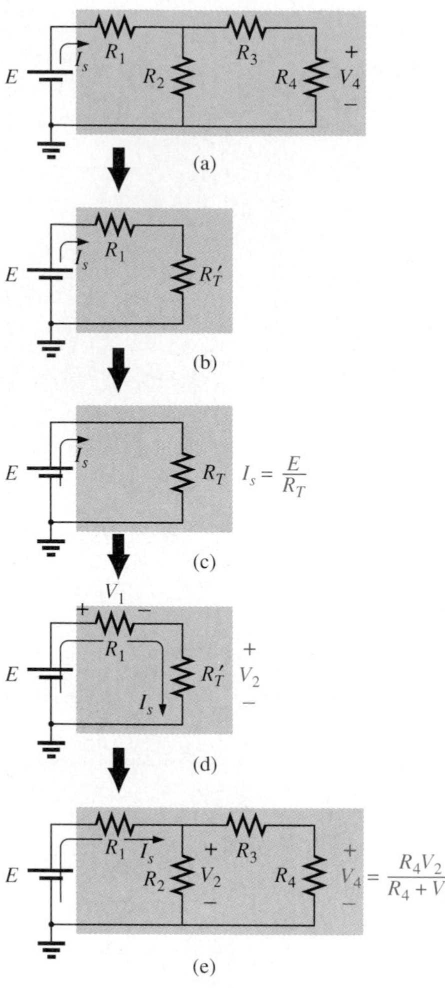

FIG. 7.1
Introducing the reduce and return approach.

4. When you have a solution, check that it is reasonable by considering the magnitudes of the energy source and the elements in the network. If it does not seem reasonable, either solve the circuit using another approach or check over your work very carefully.

Reduce and Return Approach

For many single-source, series-parallel networks, the analysis is one that works back to the source, determines the source current, and then finds its way to the desired unknown. In Fig. 7.1(a), for instance, the voltage V_4 is desired. The absence of a single series or parallel path to V_4 from the source immediately reveals that the methods introduced in the last two chapters cannot be applied here. First, series and parallel elements must be combined to establish the reduced circuit of Fig. 7.1(b). Then series elements are combined to form the simplest of configurations in Fig. 7.1(c). The source current can now be determined using Ohm's law, and we can proceed back through the network as shown in Fig. 7.1(d). The voltage V_2 can be determined and then the original network can be redrawn, as shown in Fig. 7.1(e). Since V_2 is now known, the voltage divider rule can be used to find the desired voltage V_4. Because of the similarities between the networks of Figs. 7.1(a) and 7.1(e), and between 7.1(b) and 7.1(d), the networks drawn during the reduction phase are often used for the return path. Although all the details of the analysis were not described above, the general procedure for a number of series-parallel network problems employs the procedure described above—work back for I_s and then follow the return path for the specific unknown. Not every problem will follow this path; some will have simpler, more direct solutions, but the reduce and return approach will handle one type of problem that does surface over and over again.

Block Diagram Approach

The block diagram approach will be employed throughout to emphasize the fact that combinations of elements, not simply single resistive elements, can be in series or parallel. The approach will also reveal the number of seemingly different networks that have the same basic structure and therefore can involve similar analysis techniques.

Initially, there will be some concern about identifying series and parallel elements and branches and choosing the best procedure to follow toward a solution. However, as you progress through the examples and try a few problems, a common path toward most solutions will surface that can actually make the analysis of such systems an interesting, enjoyable experience.

In Fig. 7.2, blocks B and C are in parallel (points b and c in common), and the voltage source E is in series with block A (point a in common). The parallel combination of B and C is also in series with A and the voltage source E due to the common points b and c, respectively.

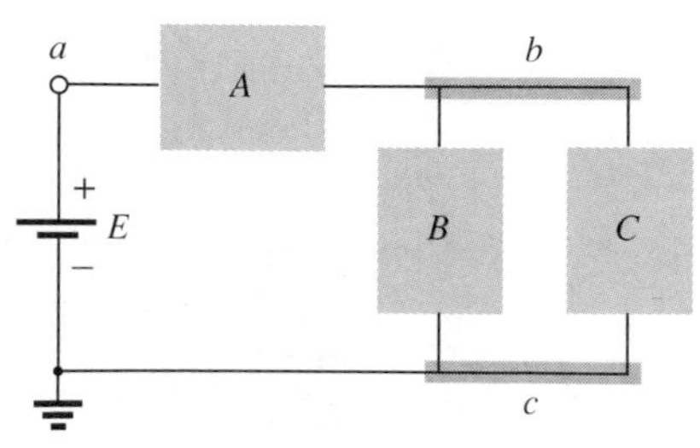

FIG. 7.2
Introducing the block diagram approach.

To ensure that the analysis to follow is as clear and uncluttered as possible, the following notation will be used for series and parallel combinations of elements. For series resistors R_1 and R_2, a comma will be inserted between their subscript notations, as shown here:

$$R_{1,2} = R_1 + R_2$$

For parallel resistors R_1 and R_2, the parallel symbol will be inserted between their subscript notations, as follows:

$$R_{1\|2} = R_1 \| R_2 = \frac{R_1 R_2}{R_1 + R_2}$$

EXAMPLE 7.1 If each block of Fig. 7.2 were a single resistive element, the network of Fig. 7.3 might result.

The parallel combination of R_B and R_C results in

$$R_{B\|C} = R_B \| R_C = \frac{(12\text{ k}\Omega)(6\text{ k}\Omega)}{12\text{ k}\Omega + 6\text{ k}\Omega} = 4\text{ k}\Omega$$

The equivalent resistance $R_{B\|C}$ is then in series with R_A and the total resistance "seen" by the source is

$$R_T = R_A + R_{B\|C}$$
$$= 2\text{ k}\Omega + 4\text{ k}\Omega = \mathbf{6\text{ k}\Omega}$$

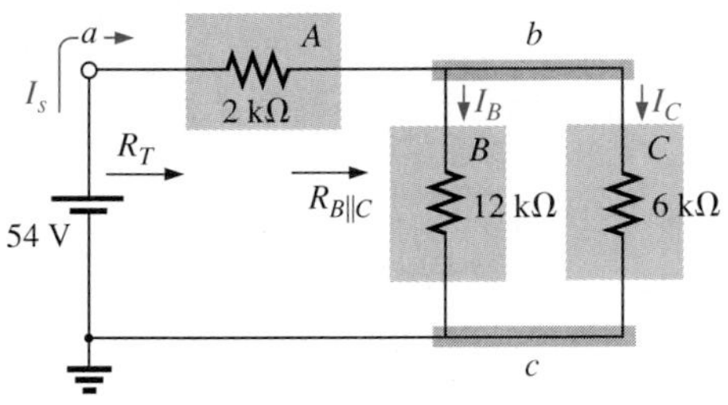

FIG. 7.3
Example 7.1.

The result is an equivalent network, as shown in Fig. 7.4, permitting the determination of the source current I_s.

$$I_s = \frac{E}{R_T} = \frac{54\text{ V}}{6\text{ k}\Omega} = \mathbf{9\text{ mA}}$$

and, since the source and R_A are in series,

$$I_A = I_s = 9\text{ mA}$$

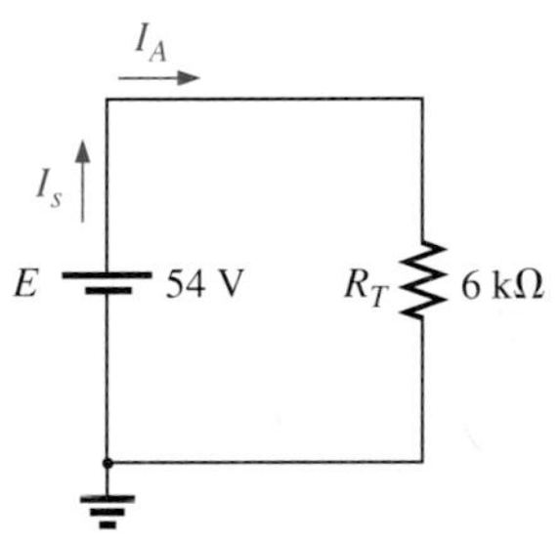

FIG. 7.4
Reduced equivalent of Fig. 7.3.

We can then use the equivalent network of Fig. 7.5 to determine I_B and I_C using the current divider rule:

$$I_B = \frac{6\text{ k}\Omega(I_s)}{6\text{ k}\Omega + 12\text{ k}\Omega} = \frac{6}{18}I_s = \frac{1}{3}(9\text{ mA}) = \mathbf{3\text{ mA}}$$

$$I_C = \frac{12\text{ k}\Omega(I_s)}{12\text{ k}\Omega + 6\text{ k}\Omega} = \frac{12}{18}I_s = \frac{2}{3}(9\text{ mA}) = \mathbf{6\text{ mA}}$$

or, applying Kirchhoff's current law,

$$I_C = I_s - I_B = 9\text{ mA} - 3\text{ mA} = \mathbf{6\text{ mA}}$$

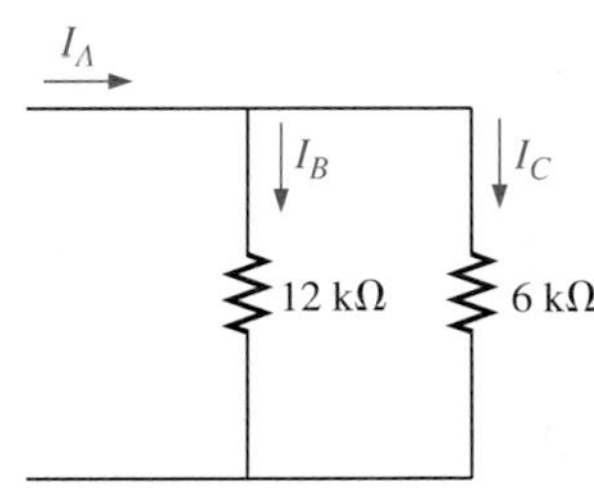

FIG. 7.5
Determining I_B and I_C for the network of Fig. 7.3.

Note in the preceding solution that we worked back to the source to obtain the source current or total current supplied by the source. The remaining unknowns were then determined by working back through the network to find the other unknowns.

EXAMPLE 7.2 It is also possible that the blocks *A*, *B*, and *C* of Fig. 7.2 contain the elements and configurations of Fig. 7.6. Working with each region:

A: $R_A = 4\ \Omega$

B: $R_B = R_2 \| R_3 = R_{2\|3} = \frac{R}{N} = \frac{4\ \Omega}{2} = 2\ \Omega$

C: $R_C = R_4 + R_5 = R_{4,5} = 0.5\ \Omega + 1.5\ \Omega = 2\ \Omega$

Blocks *B* and *C* are still in parallel and

$$R_{B\|C} = \frac{R}{N} = \frac{2\ \Omega}{2} = 1\ \Omega$$

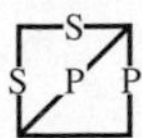

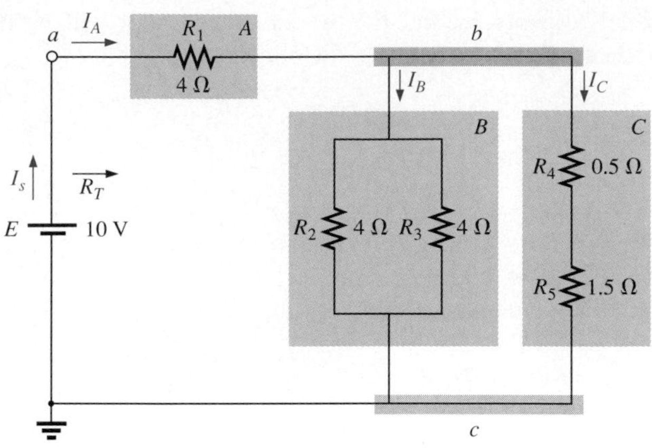

FIG. 7.6
Example 7.2.

with

$$R_T = R_A + R_{B\|C} = 4\ \Omega + 1\ \Omega = \mathbf{5\ \Omega}$$

(Note the similarity between this equation and that obtained for Example 7.1.)

and
$$I_s = \frac{E}{R_T} = \frac{10\ \text{V}}{5\ \Omega} = \mathbf{2\ A}$$

We can find the currents I_A, I_B, and I_C using the reduction of the network of Fig. 7.6 (recall Step 3) as found in Fig. 7.7. Note that I_A, I_B, and I_C are the same in Figs. 7.6 and 7.7 and therefore also appear in Fig. 7.7. In other words, the currents I_A, I_B, and I_C of Fig. 7.7 will have the same magnitude as the same currents of Fig. 7.6.

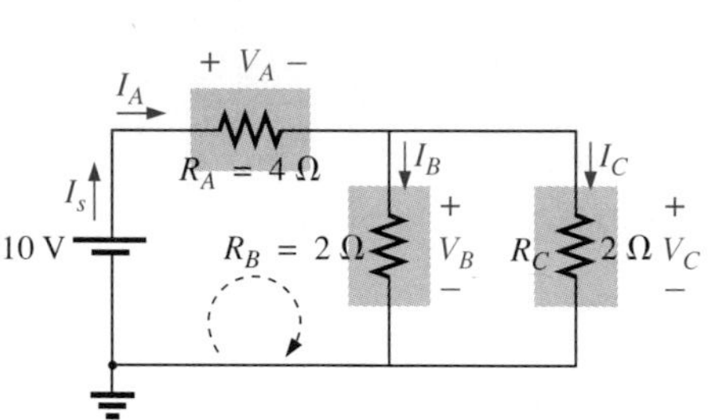

FIG. 7.7
Reduced equivalent of Fig. 7.6.

$$I_A = I_s = \mathbf{2\ A}$$

and
$$I_B = I_C = \frac{I_A}{2} = \frac{I_s}{2} = \frac{2\ \text{A}}{2} = \mathbf{1\ A}$$

Returning to the network of Fig. 7.6, we have

$$I_{R_2} = I_{R_3} = \frac{I_B}{2} = \mathbf{0.5\ A}$$

The voltages V_A, V_B, and V_C from either figure are

$$V_A = I_A R_A = (2\ \text{A})(4\ \Omega) = \mathbf{8\ V}$$
$$V_B = I_B R_B = (1\ \text{A})(2\ \Omega) = \mathbf{2\ V}$$
$$V_C = V_B = \mathbf{2\ V}$$

Applying Kirchhoff's voltage law for the loop indicated in Fig. 7.7, we obtain

$$\Sigma_{\circlearrowright} V = E - V_A - V_B = 0$$
$$E = V_A + V_B = 8\ \text{V} + 2\ \text{V}$$

or
$$10\ \text{V} = 10\ \text{V} \quad \text{(checks)}$$

EXAMPLE 7.3 Another possible variation of Fig. 7.2 appears in Fig. 7.8.

$$R_A = R_{1\|2} = \frac{(9\ \Omega)(6\ \Omega)}{9\ \Omega + 6\ \Omega} = \frac{54\ \Omega}{15} = 3.6\ \Omega$$

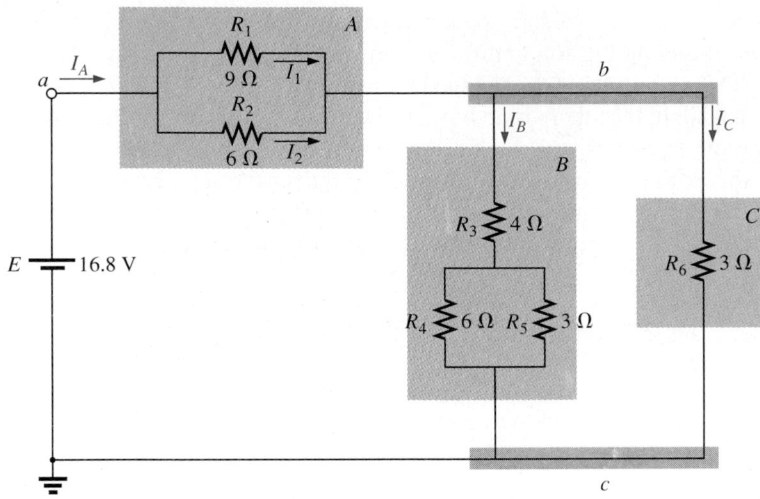

FIG. 7.8
Example 7.3.

$$R_B = R_3 + R_{4\|5} = 4\ \Omega + \frac{(6\ \Omega)(3\ \Omega)}{6\ \Omega + 3\ \Omega} = 4\ \Omega + 2\ \Omega = 6\ \Omega$$

$$R_C = 3\ \Omega$$

The network of Fig. 7.8 can then be redrawn in reduced form, as shown in Fig. 7.9. Note the similarities between this circuit and those of Fig. 7.3 and 7.7.

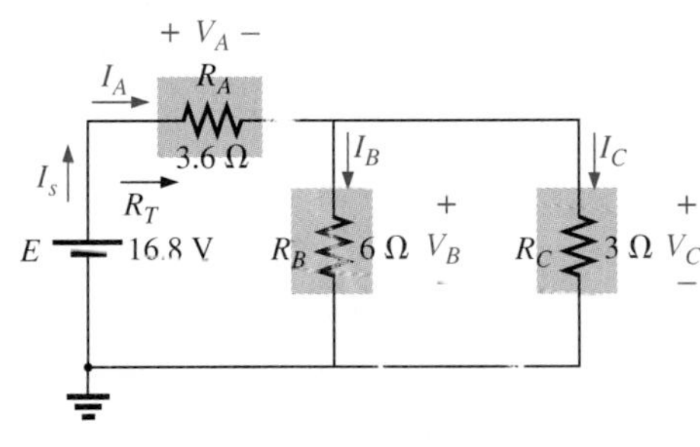

FIG. 7.9
Reduced equivalent of Fig. 7.8.

$$R_T = R_A + R_{B\|C} = 3.6\ \Omega + \frac{(6\ \Omega)(3\ \Omega)}{6\ \Omega + 3\ \Omega}$$

$$= 3.6\ \Omega + 2\ \Omega = \mathbf{5.6\ \Omega}$$

$$I_s = \frac{E}{R_T} = \frac{16.8\ \text{V}}{5.6\ \Omega} = \mathbf{3\ A}$$

$$I_A = I_s = \mathbf{3\ A}$$

Applying the current divider rule yields

$$I_B = \frac{R_C I_A}{R_C + R_B} = \frac{(3\ \Omega)(3\ \text{A})}{3\ \Omega + 6\ \Omega} = \frac{9\ \text{A}}{9} = \mathbf{1\ A}$$

By Kirchhoff's current law,

$$I_C = I_A - I_B = 3\ \text{A} - 1\ \text{A} = \mathbf{2\ A}$$

By Ohm's law,

$$V_A = I_A R_A = (3\ \text{A})(3.6\ \Omega) = \mathbf{10.8\ V}$$

$$V_B = I_B R_B = V_C = I_C R_C = (2\ \text{A})(3\ \Omega) = \mathbf{6\ V}$$

Returning to the original network (Fig. 7.8) and applying the current divider rule,

$$I_1 = \frac{R_2 I_A}{R_2 + R_1} = \frac{(6\ \Omega)(3\ \text{A})}{6\ \Omega + 9\ \Omega} = \frac{18\ \text{A}}{15} = \mathbf{1.2\ A}$$

By Kirchhoff's current law,

$$I_2 = I_A - I_1 = 3\ \text{A} - 1.2\ \text{A} = \mathbf{1.8\ A}$$

Figures 7.3, 7.6, and 7.8 are only a few of the infinite variety of configurations that the network can assume starting with the basic arrangement of Fig. 7.2. They were included in our discussion to emphasize the

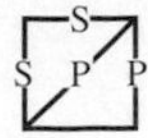

importance of considering each region of the network independently before finding the solution for the network as a whole.

There are a variety of ways in which the blocks of Fig. 7.2 can be arranged. In fact, there is no limit on the number of series-parallel configurations that can appear within a given network. In reverse, the block diagram approach can be used effectively to reduce the apparent complexity of a system by identifying the major series and parallel components of the network. This approach will be demonstrated in the next few examples.

7.2 DESCRIPTIVE EXAMPLES

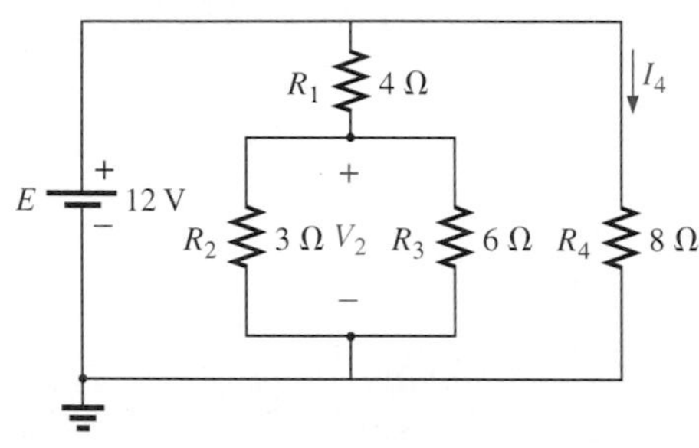

FIG. 7.10
Example 7.4.

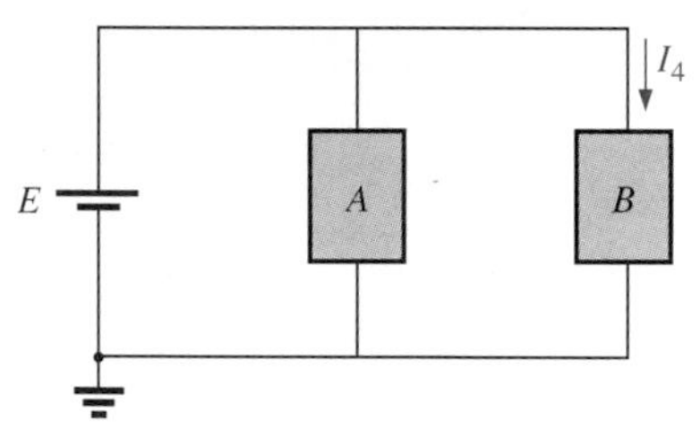

FIG. 7.11
Block diagram of Fig. 7.10.

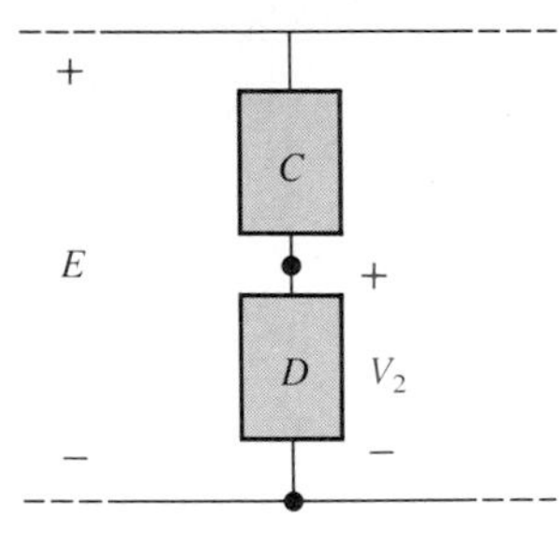

FIG. 7.12
Alternative block diagram for the first parallel branch of Fig. 7.10.

EXAMPLE 7.4 Find the current I_4 and the voltage V_2 for the network of Fig. 7.10.

Solution: In this case, particular unknowns are requested instead of a complete solution. It would, therefore, be a waste of time to find all the currents and voltages of the network. The method employed should concentrate on obtaining only the unknowns requested. With the block diagram approach, the network has the basic structure of Fig. 7.11, clearly indicating that the three branches are in parallel and the voltage across *A* and *B* is the supply voltage. The current I_4 is now immediately obvious as simply the supply voltage divided by the resultant resistance for *B*. If desired, block *A* could be broken down further, as shown in Fig. 7.12, to identify *C* and *D* as series elements, with the voltage V_2 capable of being determined using the voltage divider rule once the resistance of *C* and *D* is reduced to a single value. This is an example of how a mental sketch of the approach might be made before applying laws, rules, and so on, to avoid dead ends and growing frustration.

Applying Ohm's law,

$$I_4 = \frac{E}{R_B} = \frac{E}{R_4} = \frac{12\text{ V}}{8\ \Omega} = \mathbf{1.5\ A}$$

Combining the resistors R_2 and R_3 of Fig. 7.10 will result in

$$R_D = R_2 \parallel R_3 = 3\ \Omega \parallel 6\ \Omega = \frac{(3\ \Omega)(6\ \Omega)}{3\ \Omega + 6\ \Omega} = \frac{18\ \Omega}{9} = 2\ \Omega$$

and, applying the voltage divider rule,

$$V_2 = \frac{R_D E}{R_D + R_C} = \frac{(2\ \Omega)(12\text{ V})}{2\ \Omega + 4\ \Omega} = \frac{24\text{ V}}{6} = \mathbf{4\ V}$$

EXAMPLE 7.5 Find the indicated currents and voltages for the network of Fig. 7.13.

Solution: Again, only specific unknowns are requested. When the network is redrawn, it will be particularly important to note which unknowns are preserved and which will have to be determined using the original configuration. The block diagram of the network may appear as shown in Fig. 7.14, clearly revealing that *A* and *B* are in series. Note in this form the number of unknowns that have been preserved. The voltage V_1 will be the same across the three parallel branches of Fig. 7.13, and V_5 will be the same across R_4 and R_5. The unknown currents I_2 and I_4 are lost since they represent the currents through only one of the parallel branches. However, once V_1 and V_5 are known, the required currents can be found using Ohm's law.

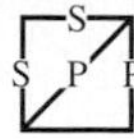

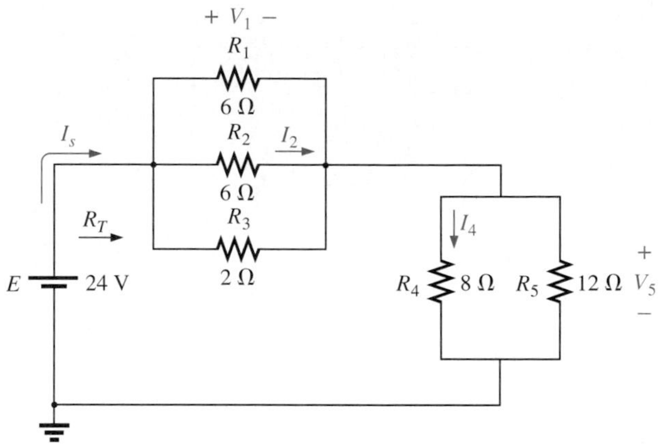

FIG. 7.13
Example 7.5.

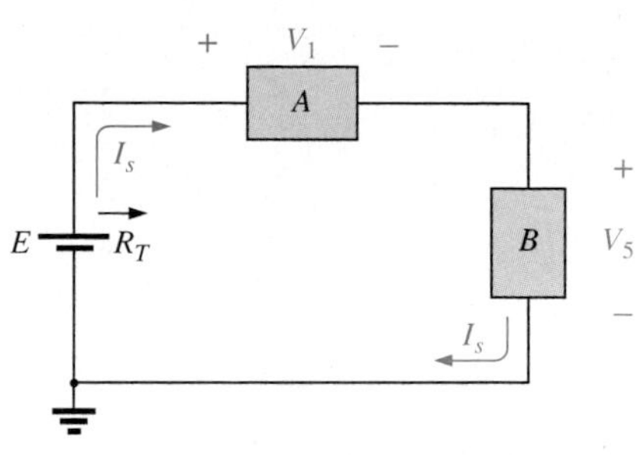

FIG. 7.14
Block diagram for Fig. 7.13.

$$R_{1\|2} = \frac{R}{N} = \frac{6\ \Omega}{2} = 3\ \Omega$$

$$R_A = R_{1\|2\|3} = \frac{(3\ \Omega)(2\ \Omega)}{3\ \Omega + 2\ \Omega} = \frac{6\ \Omega}{5} = 1.2\ \Omega$$

$$R_B = R_{4\|5} = \frac{(8\ \Omega)(12\ \Omega)}{8\ \Omega + 12\ \Omega} = \frac{96\ \Omega}{20} = 4.8\ \Omega$$

The reduced form of Fig. 7.13 will then appear as shown in Fig. 7.15, and

$$R_T = R_{1\|2\|3} + R_{4\|5} = 1.2\ \Omega + 4.8\ \Omega = \mathbf{6\ \Omega}$$

$$I_s = \frac{E}{R_T} = \frac{24\ \text{V}}{6\ \Omega} = \mathbf{4\ A}$$

with

$$V_1 = I_s R_{1\|2\|3} = (4\ \text{A})(1.2\ \Omega) = \mathbf{4.8\ V}$$

$$V_5 = I_s R_{4\|5} = (4\ \text{A})(4.8\ \Omega) = \mathbf{19.2\ V}$$

Applying Ohm's law,

$$I_4 = \frac{V_5}{R_4} = \frac{19.2\ \text{V}}{8\ \Omega} = \mathbf{2.4\ A}$$

$$I_2 = \frac{V_2}{R_2} = \frac{V_1}{R_2} = \frac{4.8\ \text{V}}{6\ \Omega} = \mathbf{0.8\ A}$$

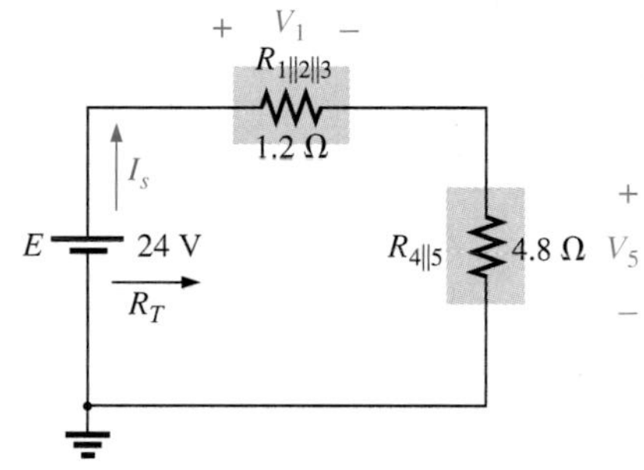

FIG. 7.15
Reduced form of Fig. 7.13.

The next example demonstrates that unknown voltages do not have to be across elements but can exist between any two points in a network. In addition, the importance of redrawing the network in a more familiar form is clearly revealed by the analysis to follow.

EXAMPLE 7.6

a. Find the voltages V_1, V_3, and V_{ab} for the network of Fig. 7.16.
b. Calculate the source current I_s.

Solutions: This is one of those situations where it might be best to redraw the network before beginning the analysis. Since combining both sources will not affect the unknowns, the network is redrawn as shown in Fig. 7.17, establishing a parallel network with the total source voltage across each parallel branch. The net source voltage is the difference between the two with the polarity of the larger.

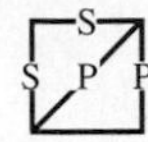

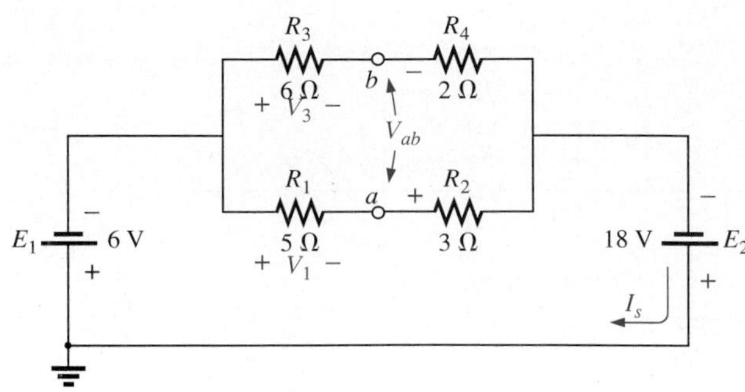

FIG. 7.16
Example 7.6.

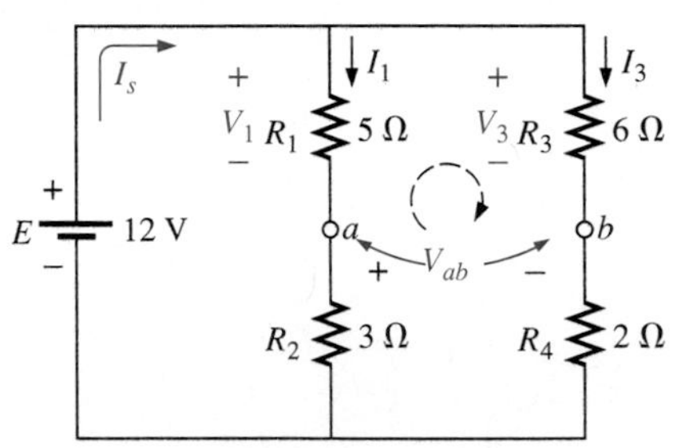

FIG. 7.17
Network of Fig. 7.16 redrawn.

a. Note the similarities with Fig. 7.12, permitting the use of the voltage divider rule to determine V_1 and V_3:

$$V_1 = \frac{R_1E}{R_1 + R_2} = \frac{(5\ \Omega)(12\ \text{V})}{5\ \Omega + 3\ \Omega} = \frac{60\ \text{V}}{8} = \mathbf{7.5\ V}$$

$$V_3 = \frac{R_3E}{R_3 + R_4} = \frac{(6\ \Omega)(12\ \text{V})}{6\ \Omega + 2\ \Omega} = \frac{72\ \text{V}}{8} = \mathbf{9\ V}$$

The open-circuit voltage V_{ab} is determined by applying Kirchhoff's voltage law around the indicated loop of Fig. 7.17 in the clockwise direction starting at terminal a.

$$+V_1 - V_3 + V_{ab} = 0$$

and $$V_{ab} = V_3 - V_1 = 9\ \text{V} - 7.5\ \text{V} = \mathbf{1.5\ V}$$

b. By Ohm's law,

$$I_1 = \frac{V_1}{R_1} = \frac{7.5\ \text{V}}{5\ \Omega} = 1.5\ \text{A}$$

$$I_3 = \frac{V_3}{R_3} = \frac{9\ \text{V}}{6\ \Omega} = 1.5\ \text{A}$$

Applying Kirchhoff's current law,

$$I_s = I_1 + I_3 = 1.5\ \text{A} + 1.5\ \text{A} = \mathbf{3\ A}$$

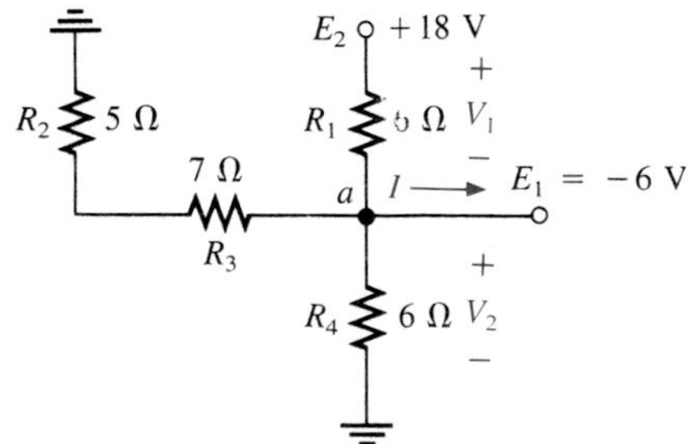

FIG. 7.18
Example 7.7.

EXAMPLE 7.7 For the network of Fig. 7.18, determine the voltages V_1 and V_2 and the current I.

Solution: It would indeed be difficult to analyze the network in the form of Fig. 7.18 with the symbolic notation for the sources and the reference or ground connection in the upper left-hand corner of the diagram. However, when the network is redrawn as shown in Fig. 7.19, the unknowns and the relationship between branches become significantly clearer. Note the common connection of the grounds and the replacing of the terminal notation by actual supplies.

It is now obvious that

$$V_2 = -E_1 = \mathbf{-6\ V}$$

The minus sign simply indicates that the chosen polarity for V_2 in Fig. 7.18 is opposite to the actual voltage. Applying Kirchhoff's voltage law to the loop indicated, we obtain

$$-E_1 + V_1 - E_2 = 0$$

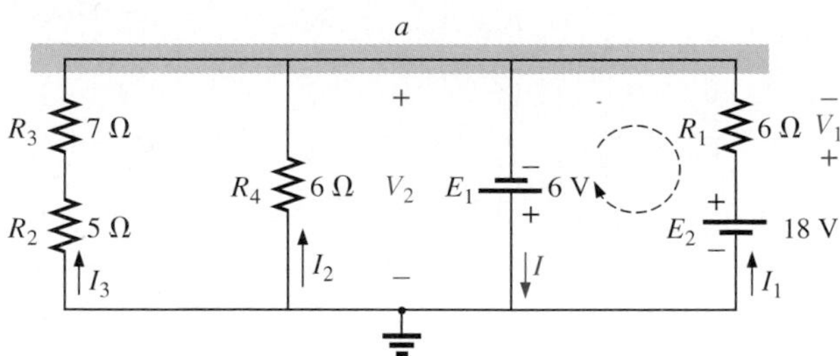

FIG. 7.19
Network of Fig. 7.18 redrawn.

and $\qquad V_1 = E_2 + E_1 = 18\text{ V} + 6\text{ V} = \mathbf{24\ V}$

Applying Kirchhoff's current law to node a yields

$$\begin{aligned} I &= I_1 + I_2 + I_3 \\ &= \frac{V_1}{R_1} + \frac{E_1}{R_4} + \frac{E_1}{R_2 + R_3} \\ &= \frac{24\text{ V}}{6\ \Omega} + \frac{6\text{ V}}{6\ \Omega} + \frac{6\text{ V}}{12\ \Omega} \\ &= 4\text{ A} + 1\text{ A} + 0.5\text{ A} \\ I &= \mathbf{5.5\ A} \end{aligned}$$

The next example is clear evidence of the fact that techniques learned in the current chapters will have far-reaching applications and will not be dropped for improved methods. Even though the transistor has not been introduced in this text, the dc levels of a transistor network can be examined using the basic rules and laws introduced in the early chapters of this text.

EXAMPLE 7.8 For the transistor configuration of Fig. 7.20, in which V_B and V_{BE} have been provided:

a. Determine the voltage V_E and the current I_E.
b. Calculate V_1.
c. Determine V_{BC} using the fact that the approximation $I_C = I_E$ is often applied to transistor networks.
d. Calculate V_{CE} using the information obtained in parts (a) through (c).

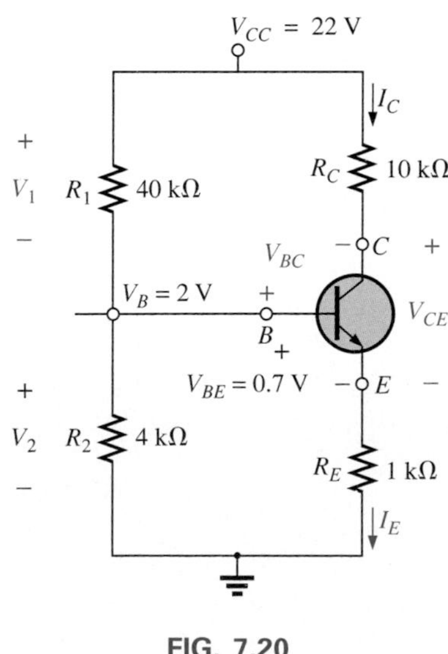

FIG. 7.20
Example 7.8.

Solutions:

a. From Fig. 7.20, we find

$$V_2 = V_B = 2\text{ V}$$

Writing Kirchhoff's voltage law around the lower loop yields

$$V_2 - V_{BE} - V_E = 0$$

or $\qquad V_E = V_2 - V_{BE} = 2\text{ V} - 0.7\text{ V} = \mathbf{1.3\ V}$

and $\qquad I_E = \dfrac{V_E}{R_E} = \dfrac{1.3\text{ V}}{1000\ \Omega} = \mathbf{1.3\ mA}$

b. Applying Kirchhoff's voltage law to the input side (left-hand region of the network) will result in:

$$V_2 + V_1 - V_{CC} = 0$$

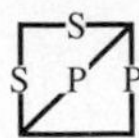

and $\qquad V_1 = V_{CC} - V_2$

but $\qquad V_2 = V_B$

and $\qquad V_1 = V_{CC} - V_2 = 22\text{ V} - 2\text{ V} = \mathbf{20\text{ V}}$

c. Redrawing the section of the network of immediate interest will result in Fig. 7.21, where Kirchhoff's voltage law yields:

$$V_C + V_{R_C} - V_{CC} = 0$$

and $\qquad V_C = V_{CC} - V_{R_C} = V_{CC} - I_C R_C$

but $\qquad I_C = I_E$

and $\qquad V_C = V_{CC} - I_E R_C = 22\text{ V} - (1.3\text{ mA})(10\text{ k}\Omega)$
$\qquad\qquad = 9\text{ V}$

Then $\qquad V_{BC} = V_B - V_C$
$\qquad\qquad = 2\text{ V} - 9\text{ V}$
$\qquad\qquad = \mathbf{-7\text{ V}}$

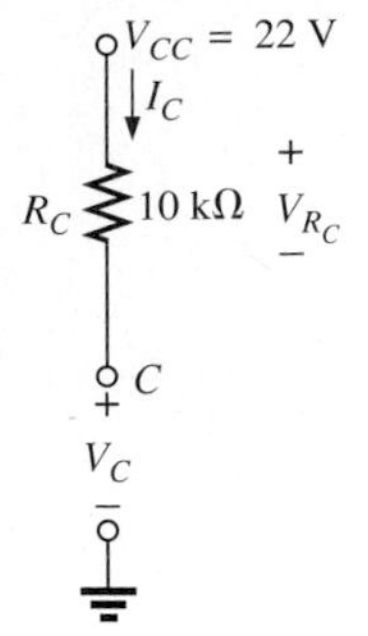

FIG. 7.21
Determining V_C for the network of Fig. 7.20.

d. $\qquad V_{CE} = V_C - V_E$
$\qquad\qquad = 9\text{ V} - 1.3\text{ V}$
$\qquad\qquad = \mathbf{7.7\text{ V}}$

EXAMPLE 7.9 Calculate the indicated currents and voltage of Fig. 7.22.

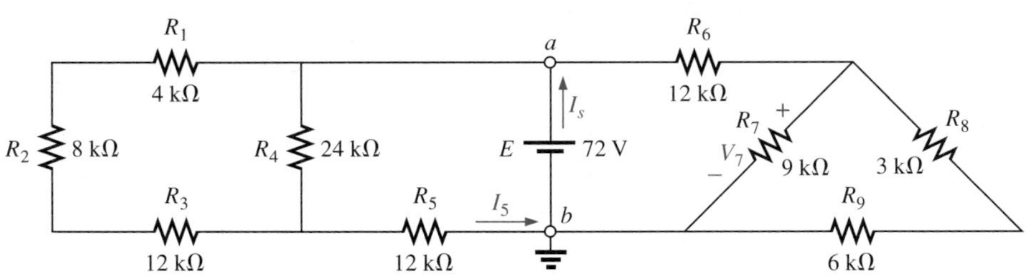

FIG. 7.22
Example 7.9.

Solution: Redrawing the network after combining series elements yields Fig. 7.23, and

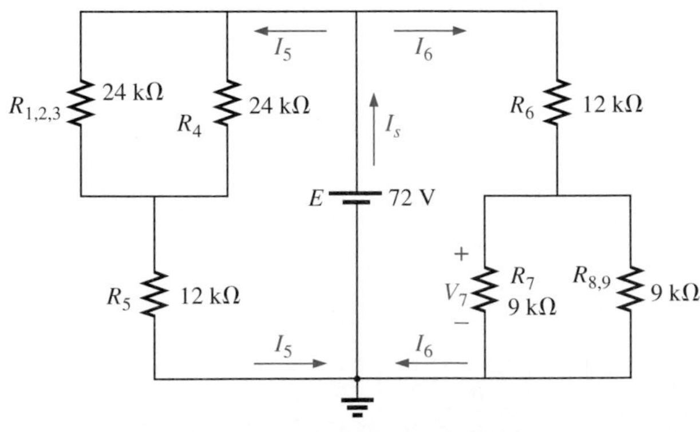

FIG. 7.23
Network of Fig. 7.22 redrawn.

$$I_5 = \frac{E}{R_{(1,2,3)\|4} + R_5} = \frac{72 \text{ V}}{12 \text{ k}\Omega + 12 \text{ k}\Omega} = \frac{72 \text{ V}}{24 \text{ k}\Omega} = \mathbf{3 \text{ mA}}$$

with

$$V_7 = \frac{R_{7\|(8,9)}E}{R_{7\|(8,9)} + R_6} = \frac{(4.5 \text{ k}\Omega)(72 \text{ V})}{4.5 \text{ k}\Omega + 12 \text{ k}\Omega} = \frac{324 \text{ V}}{16.5} = \mathbf{19.6 \text{ V}}$$

$$I_6 = \frac{V_7}{R_{7\|(8,9)}} = \frac{19.6 \text{ V}}{4.5 \text{ k}\Omega} = \mathbf{4.35 \text{ mA}}$$

and $$I_s = I_5 + I_6 = 3 \text{ mA} + 4.35 \text{ mA} = \mathbf{7.35 \text{ mA}}$$

Since the potential difference between points *a* and *b* of Fig. 7.22 is fixed at *E* volts, the circuit to the right or left is unaffected if the network is reconstructed as shown in Fig. 7.24.

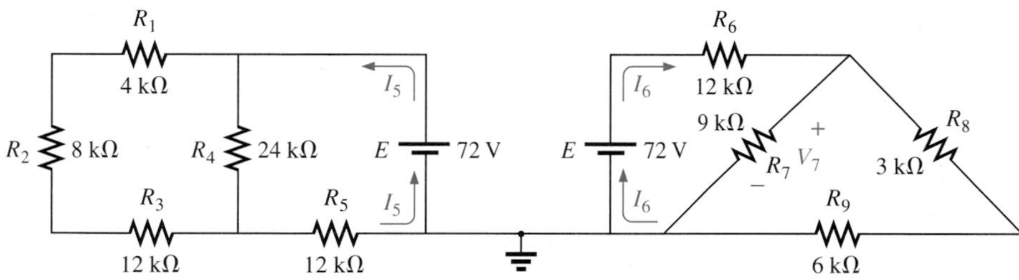

FIG. 7.24
An alternative approach to Example 7.9.

We can find each quantity required, except I_s, by analyzing each circuit independently. To find I_s, we must find the source current for each circuit and add it as in the above solution; that is, $I_s = I_5 + I_6$.

EXAMPLE 7.10 This example demonstrates the power of Kirchhoff's voltage law by determining the voltages V_1, V_2, and V_3 for the network of Fig. 7.25. For path 1 of Fig. 7.26,

$$E_1 - V_1 - E_3 = 0$$

and $$V_1 = E_1 - E_3 = 20 \text{ V} - 8 \text{ V} = \mathbf{12 \text{ V}}$$

For path 2,

$$E_2 - V_1 - V_2 = 0$$

and $$V_2 = E_2 - V_1 = 5 \text{ V} - 12 \text{ V} = \mathbf{-7 \text{ V}}$$

indicating that V_2 has a magnitude of 7 V but a polarity opposite to that appearing in Fig. 7.25. For path 3,

$$V_3 + V_2 - E_3 = 0$$

and $$V_3 = E_3 - V_2 = 8 \text{ V} - (-7 \text{ V}) = 8 \text{ V} + 7 \text{ V} = \mathbf{15 \text{ V}}$$

Note that the polarity of V_2 was maintained as originally assumed, requiring that -7 V be substituted for V_2.

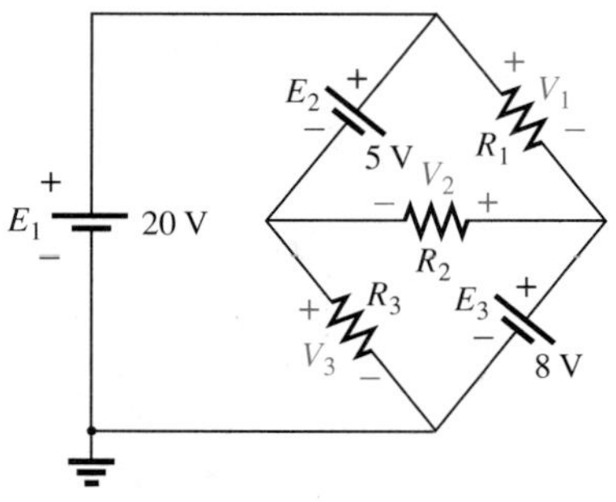

FIG. 7.25
Example 7.10.

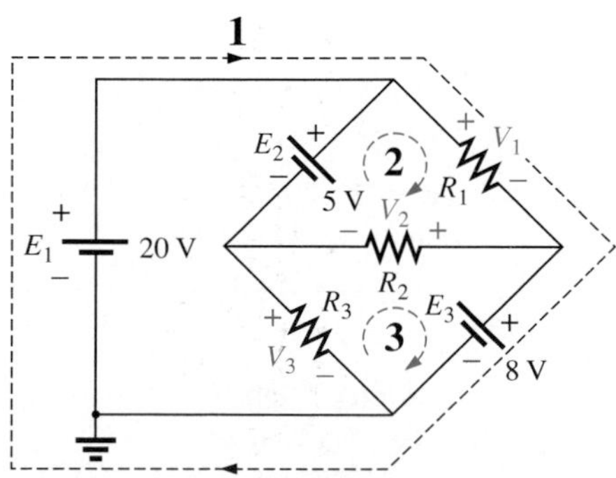

FIG. 7.26
Defining the paths for Kirchhoff's voltage law.

7.3 LADDER NETWORKS

A three-section *ladder* network appears in Fig. 7.27. The reason for the terminology is quite obvious for the repetitive structure. There are basically two approaches used to solve networks of this type.

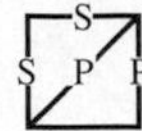

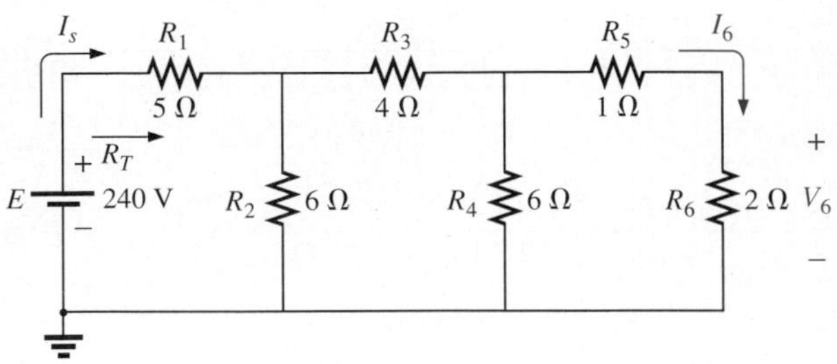

FIG. 7.27
Ladder network.

Method 1

Calculate the total resistance and resulting source current, and then work back through the ladder until the desired current or voltage is obtained. This method is now employed to determine V_6 in Fig. 7.27.

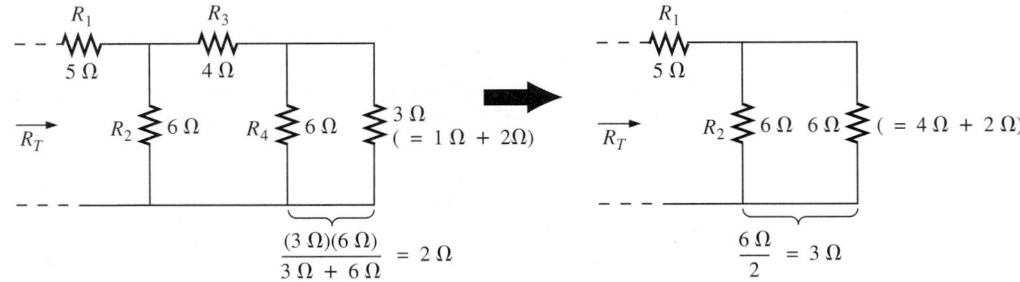

FIG. 7.28
Working back to the source to determine R_T for the network of Fig. 7.27.

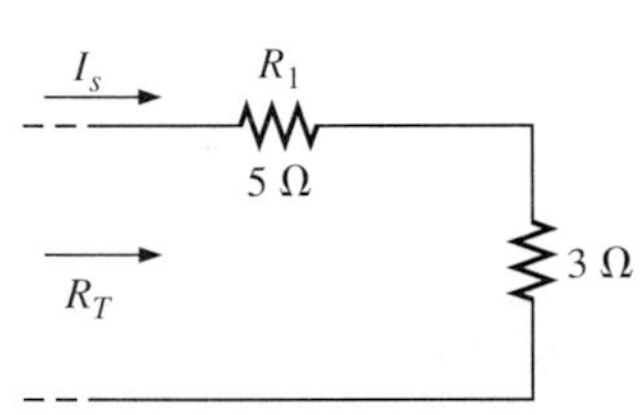

FIG. 7.29
Calculating R_T and I_s.

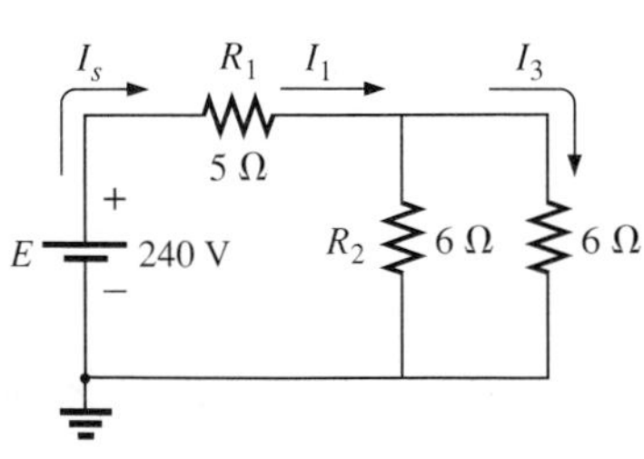

FIG. 7.30
Working back toward I_6.

Combining parallel and series elements as shown in Fig. 7.28 will result in the reduced network of Fig. 7.29, and

$$R_T = 5\ \Omega + 3\ \Omega = 8\ \Omega$$

$$I_s = \frac{E}{R_T} = \frac{240\ \text{V}}{8\ \Omega} = 30\ \text{A}$$

Working our way back to I_6 (Fig. 7.30), we find that

$$I_1 = I_s$$

and
$$I_3 = \frac{I_s}{2} = \frac{30\ \text{A}}{2} = 15\ \text{A}$$

and, finally (Fig. 7.31),

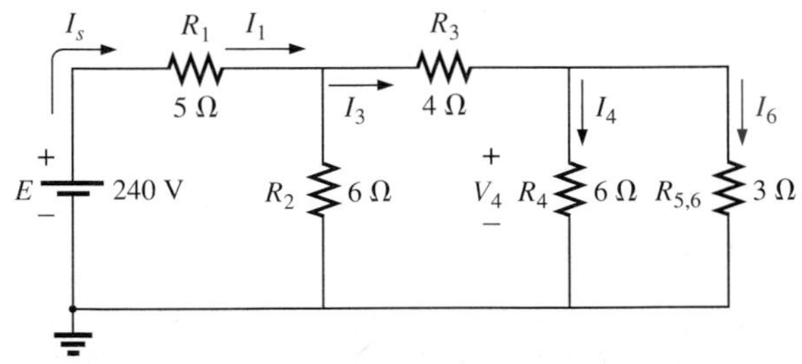

FIG. 7.31
Calculating I_6.

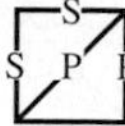

$$I_6 = \frac{(6\ \Omega)I_3}{6\ \Omega + 3\ \Omega} = \frac{6}{9}(15\ \text{A}) = 10\ \text{A}$$

and $$V_6 = I_6R_6 = (10\ \text{A})(2\ \Omega) = \mathbf{20\ V}$$

Method 2

Assign a letter symbol to the last branch current and work back through the network to the source, maintaining this assigned current or other current of interest. The desired current can then be found directly. This method can best be described through the analysis of the same network considered above in Fig. 7.27, redrawn in Fig. 7.32.

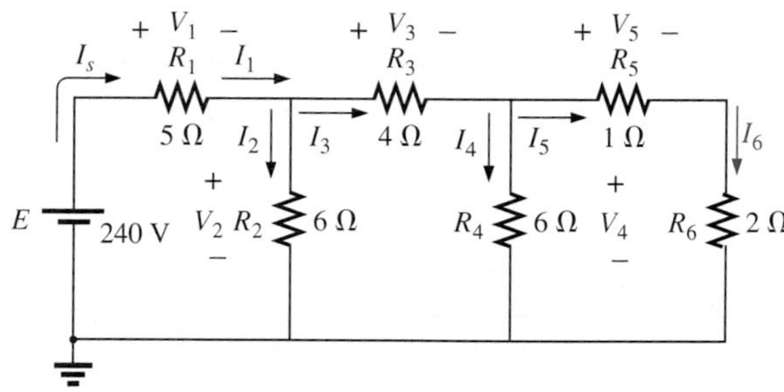

FIG. 7.32
An alternative approach for ladder networks.

The assigned notation for the current through the final branch is I_6:

$$I_6 = \frac{V_4}{R_5 + R_6} = \frac{V_4}{1\ \Omega + 2\ \Omega} = \frac{V_4}{3\ \Omega}$$

or $$V_4 = (3\ \Omega)I_6$$

so that $$I_4 = \frac{V_4}{R_4} = \frac{(3\ \Omega)I_6}{6\ \Omega} = 0.5I_6$$

and $$I_3 = I_4 + I_6 = 0.5I_6 + I_6 = 1.5I_6$$

$$V_3 = I_3R_3 = (1.5I_6)(4\ \Omega) = (6\ \Omega)I_6$$

Also, $$V_2 = V_3 + V_4 = (6\ \Omega)I_6 + (3\ \Omega)I_6 = (9\ \Omega)I_6$$

so that $$I_2 = \frac{V_2}{R_2} = \frac{(9\ \Omega)I_6}{6\ \Omega} = 1.5I_6$$

and $$I_s = I_2 + I_3 = 1.5I_6 + 1.5I_6 = 3I_6$$

with $$V_1 = I_1R_1 = I_sR_1 = (5\ \Omega)I_s$$

so that $$E = V_1 + V_2 = (5\ \Omega)I_s + (9\ \Omega)I_6$$
$$= (5\ \Omega)(3I_6) + (9\ \Omega)I_6 = (24\ \Omega)I_6$$

and $$I_6 = \frac{E}{24\ \Omega} = \frac{240\ \text{V}}{24\ \Omega} = 10\ \text{A}$$

with $$V_6 = I_6R_6 = (10\ \text{A})(2\ \Omega) = \mathbf{20\ V}$$

as was obtained using method 1.

7.4 VOLTAGE DIVIDER SUPPLY (UNLOADED AND LOADED)

The term *loaded* appearing in the title of this section refers to the application of an element, network, or system to a supply that will draw current from the supply. As pointed out in Section 5.8, the application of a load can affect the terminal voltage of the supply.

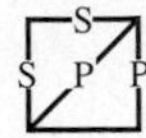

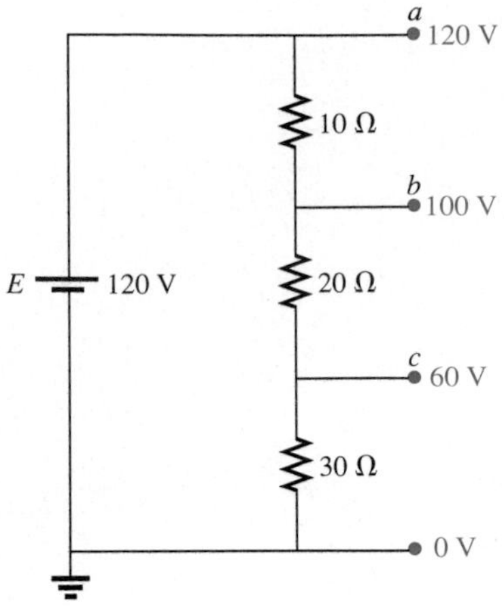

FIG. 7.33
Voltage divider supply.

Through a voltage divider network such as the one in Fig. 7.33, a number of terminal voltages can be made available from a single supply. The voltage levels shown (with respect to ground) are determined by a direct application of the voltage divider rule. Figure 7.33 reflects a no-load situation due to the absence of any current-drawing elements connected between terminals *a*, *b*, or *c* and ground.

The larger the resistance level of the applied loads compared to the resistance level of the voltage divider network, the closer the resulting terminal voltage to the no-load levels. In other words, the lower the current demand from a supply, the closer the terminal characteristics are to the no-load levels.

To demonstrate the validity of the above statement, let us consider the network of Fig. 7.33 with resistive loads that are the average value of the resistive elements of the voltage divider network, as shown in Fig. 7.34.

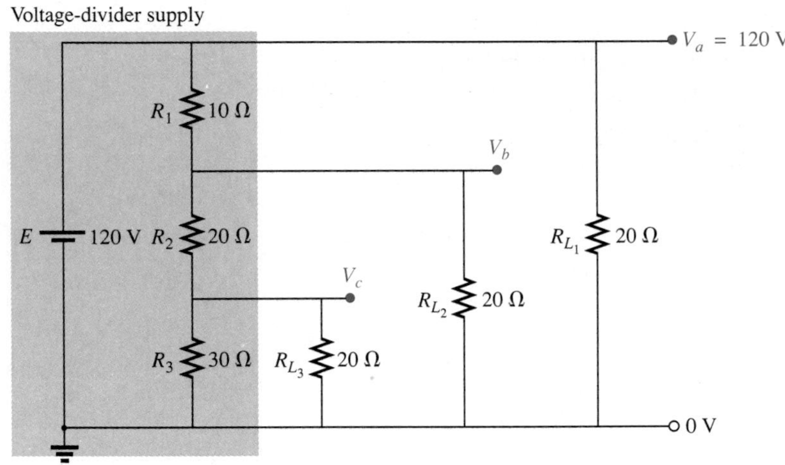

FIG. 7.34
Voltage divider supply with loads equal to the average value of the resistive elements that make up the supply.

The voltage V_a is unaffected by the load R_{L_1} since the load is in parallel with the supply voltage E. The result is $V_a = 120$ V, which is the same as the no-load level. To determine V_b, we must first note that R_3 and R_{L_3} are in parallel and $R'_3 = R_3 \parallel R_{L_3} = 30\ \Omega \parallel 20\ \Omega = 12\ \Omega$. The parallel combination $R'_2 = (R_2 + R'_3) \parallel R_{L_2} = (20\ \Omega + 12\ \Omega) \parallel 20\ \Omega = 32\ \Omega \parallel 20\ \Omega = 12.31\ \Omega$. Applying the voltage divider rule gives

$$V_b = \frac{(12.31\ \Omega)(120\ \text{V})}{12.31\ \Omega + 10\ \Omega} = 66.21\ \text{V}$$

versus 100 V under no-load conditions.

The voltage V_c is

$$V_c = \frac{(12\ \Omega)(66.21\ \text{V})}{12\ \Omega + 20\ \Omega} = 24.83\ \text{V}$$

versus 60 V under no-load conditions.

The effect of load resistors close in value to the resistor employed in the voltage divider network is, therefore, to decrease significantly some of the terminal voltages.

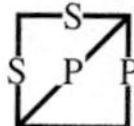

If the load resistors are changed to the 1-kΩ level, the terminal voltages will all be relatively close to the no-load values. The analysis is similar to the above, with the following results:

$$V_a = 120\text{ V}, \qquad V_b = 98.88\text{ V}, \qquad V_c = 58.63\text{ V}$$

If we compare current drains established by the applied loads, we find for the network of Fig. 7.34 that

$$I_{L_2} = \frac{V_{L_2}}{R_{L_2}} = \frac{66.21\text{ V}}{20\ \Omega} = 3.31\text{ A}$$

and for the 1-kΩ level,

$$I_{L_2} = \frac{98.88\text{ V}}{1\text{ k}\Omega} = 98.88\text{ mA} < 0.1\text{ A}$$

As noted above in the highlighted statement, the more the current drain, the greater the change in terminal voltage with the application of the load. This is certainly verified by the fact that I_{L_2} is about 33.5 times larger with the 20-Ω loads.

The next example is a design exercise. The voltage and current rating of each load is provided, along with the terminal ratings of the supply. The required voltage divider resistors must be found.

EXAMPLE 7.11 Determine R_1, R_2, and R_3 for the voltage divider supply of Fig. 7.35. Can 2-W resistors be used in the design?

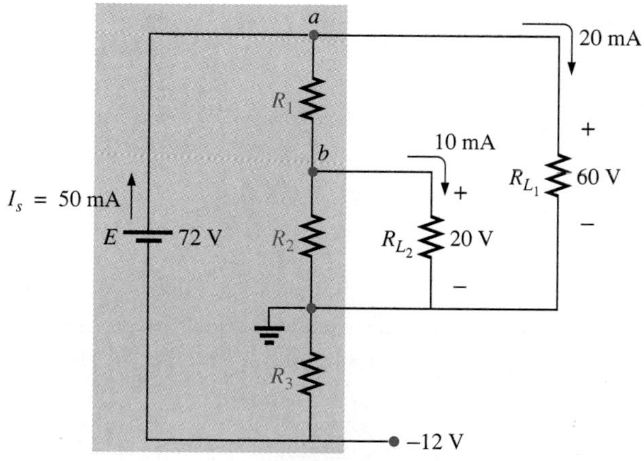

FIG. 7.35
Example 7.11.

Solution: R_3:

$$R_3 = \frac{V_{R_3}}{I_{R_3}} = \frac{V_{R_3}}{I_s} = \frac{12\text{ V}}{50\text{ mA}} = \mathbf{240\ \Omega}$$

$$P_{R_3} = (I_{R_3})^2 R_3 = (50\text{ mA})^2\, 240\ \Omega = 0.6\text{ W} < 2\text{ W}$$

R_1: Applying Kirchhoff's current law to node a:

$$I_s - I_{R_1} - I_{L_1} = 0$$

and $$I_{R_1} = I_s - I_{L_1} = 50\text{ mA} - 20\text{ mA}$$
$$= 30\text{ mA}$$

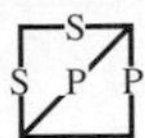

$$R_1 = \frac{V_{R_1}}{I_{R_1}} = \frac{V_{L_1} - V_{L_2}}{I_{R_1}} = \frac{60 \text{ V} - 20 \text{ V}}{30 \text{ mA}} = \frac{40 \text{ V}}{30 \text{ mA}}$$
$$= \mathbf{1.33\ k\Omega}$$
$$P_{R_1} = (I_{R_1})^2 R_1 = (30 \text{ mA})^2\, 1.33 \text{ k}\Omega = 1.197 \text{ W} < 2 \text{ W}$$

R_2: Applying Kirchhoff's current law at node *b:*

$$I_{R_1} - I_{R_2} - I_{L_2} = 0$$

and
$$I_{R_2} = I_{R_1} - I_{L_2} = 30 \text{ mA} - 10 \text{ mA}$$
$$= 20 \text{ mA}$$
$$R_2 = \frac{V_{R_2}}{I_{R_2}} = \frac{20 \text{ V}}{20 \text{ mA}} = \mathbf{1\ k\Omega}$$
$$P_{R_2} = (I_{R_2})^2 R_2 = (20 \text{ mA})^2\, 1 \text{ k}\Omega = 0.4 \text{ W} < 2 \text{ W}$$

Since P_{R_1}, P_{R_2}, and P_{R_3} are less than 2 W, 2-W resistors can be used for the design.

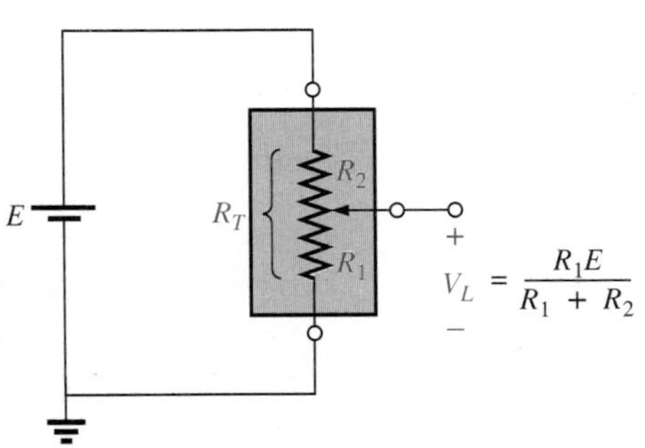

FIG. 7.36
Unloaded potentiometer.

7.5 POTENTIOMETER LOADING

For the unloaded potentiometer of Fig. 7.36, the output voltage is determined by the voltage divider rule, with R_T in the figure representing the total resistance of the potentiometer. Too often it is assumed that the voltage across a load connected to the wiper arm is determined solely by the potentiometer and the effect of the load can be ignored. This is definitely not the case, as is demonstrated in the next few paragraphs.

When a load is applied as shown in Fig. 7.37, the output voltage V_L is now a function of the magnitude of the load applied since R_1 is not as shown in Fig. 7.36 but is instead the parallel combination of R_1 and R_L.

The output voltage is now:

$$V_L = \frac{R'E}{R' + R_2} \qquad \text{with } R' = R_1 \parallel R_L \qquad \textbf{(7.1)}$$

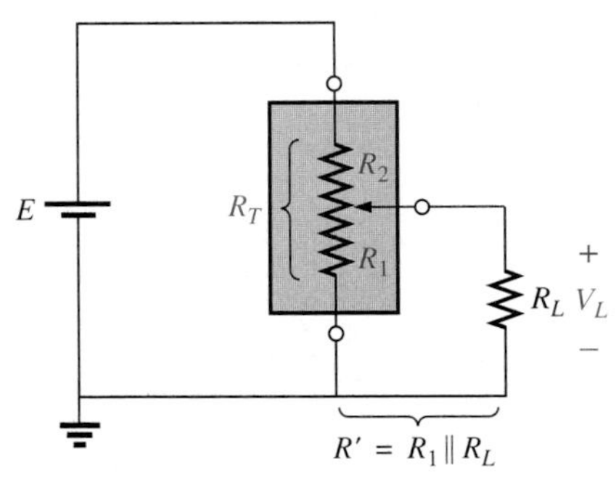

FIG. 7.37
Loaded potentiometer.

If it is desired to have good control of the output voltage V_L through the controlling dial, knob, screw, or whatever, it is advisable to choose a load or potentiometer that satisfies the following relationship:

$$\boxed{R_L \geq R_T} \qquad \textbf{(7.2)}$$

For example, if we disregard Eq. (7.2) and choose a 1-MΩ potentiometer with a 100-Ω load and set the wiper arm to 1/10 the total resistance, as shown in Fig. 7.38, then

$$R' = 100 \text{ k}\Omega \parallel 100\ \Omega = 99.9\ \Omega$$

and
$$V_L = \frac{99.9\ \Omega(10 \text{ V})}{99.9\ \Omega + 900 \text{ k}\Omega} \cong 0.001 \text{ V} = 1 \text{ mV}$$

which is extremely small compared to the expected level of 1 V.

In fact, if we move the wiper arm to the midpoint,

$$R' = 500 \text{ k}\Omega \parallel 100\ \Omega = 99.98\ \Omega$$

and
$$V_L = \frac{(99.98\ \Omega)(10 \text{ V})}{99.98\ \Omega + 500 \text{ k}\Omega} \cong 0.002 \text{ V} = 2 \text{ mV}$$

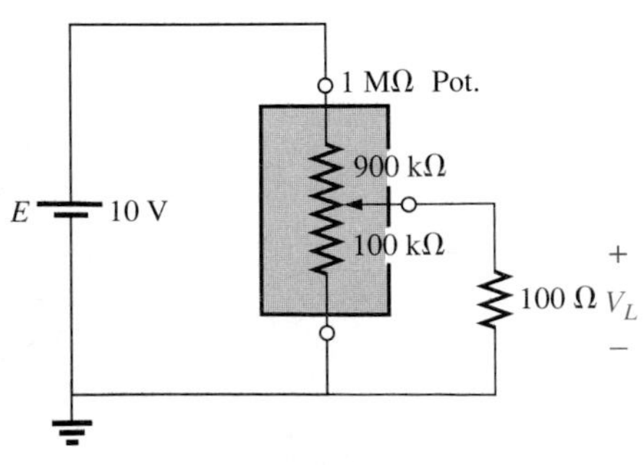

FIG. 7.38
$R_T > R_L$.

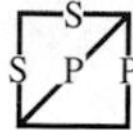

which is negligible compared to the expected level of 5 V. Even at $R_1 = 900\ \text{k}\Omega$, V_L is simply 0.01 V, or 1/1000 of the available voltage.

Using the reverse situation of $R_T = 100\ \Omega$ and $R_L = 1\ \text{M}\Omega$ and the wiper arm at the 1/10 position, as in Fig. 7.39, we find

$$R' = 10\ \Omega \parallel 1\ \text{M}\Omega \cong 10\ \Omega$$

and
$$V_L = \frac{10\ \Omega(10\ \text{V})}{10\ \Omega + 90\ \Omega} = 1\ \text{V}$$

as desired.

For the lower limit (worst-case design) of $R_L = R_T = 100\ \Omega$, as defined by Eq. (7.2) and the halfway position of Fig. 7.37,

$$R' = 50\ \Omega \parallel 100\ \Omega = 33.33\ \Omega$$

and
$$V_L = \frac{33.33\ \Omega(10\ \text{V})}{33.33\ \Omega + 50\ \Omega} \cong 4\ \text{V}$$

It may not be the ideal level of 5 V, but at least 40% of the voltage E has been achieved at the halfway position rather than the 0.02% obtained with $R_L = 100\ \Omega$ and $R_T = 1\ \text{M}\Omega$.

In general, therefore, try to establish a situation for potentiometer control in which Eq. (7.2) is satisfied to the highest degree possible.

Someone might suggest that we make R_T as small as possible to bring the percent result as close to the ideal as possible. Keep in mind, however, that the potentiometer has a power rating, and for networks such as Fig. 7.39, $P_{\text{max}} \cong E^2/R_T = (10\ \text{V})^2/100\ \Omega = 1\ \text{W}$. If R_T is reduced to $10\ \Omega$, $P_{\text{max}} = (10\ \text{V})^2/10\ \Omega = 10\ \text{W}$, which would require a *much larger* unit.

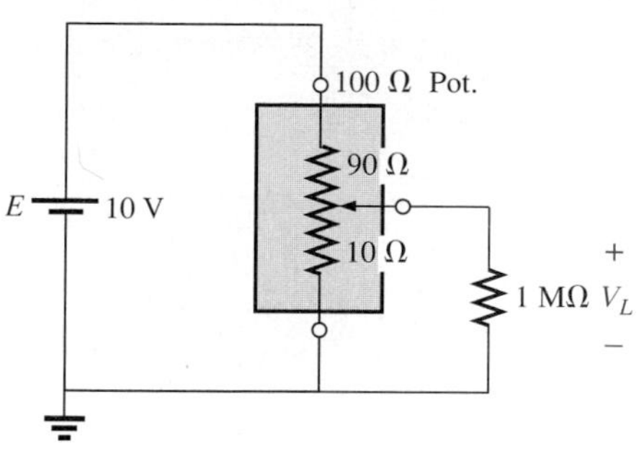

FIG. 7.39
$R_L > R_T$.

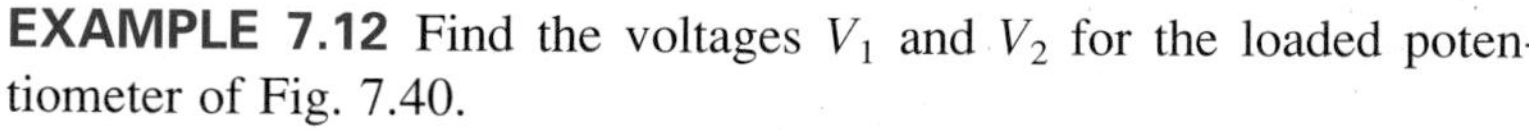

EXAMPLE 7.12 Find the voltages V_1 and V_2 for the loaded potentiometer of Fig. 7.40.

Solution:

Ideal (no load): $$V_1 = \frac{4\ \text{k}\Omega(120\ \text{V})}{10\ \text{k}\Omega} = 48\ \text{V}$$

$$V_2 = \frac{6\ \text{k}\Omega(120\ \text{V})}{10\ \text{k}\Omega} = 72\ \text{V}$$

Loaded: $$R' = 4\ \text{k}\Omega \parallel 12\ \text{k}\Omega = 3\ \text{k}\Omega$$

$$R'' = 6\ \text{k}\Omega \parallel 30\ \text{k}\Omega = 5\ \text{k}\Omega$$

$$V_1 = \frac{3\ \text{k}\Omega(120\ \text{V})}{8\ \text{k}\Omega} = \mathbf{45\ V}$$

$$V_2 = \frac{5\ \text{k}\Omega(120\ \text{V})}{8\ \text{k}\Omega} = \mathbf{75\ V}$$

The ideal and loaded voltage levels are so close that the design can be considered a good one for the applied loads. A slight variation in the position of the wiper arm will establish the ideal voltage levels across the two loads.

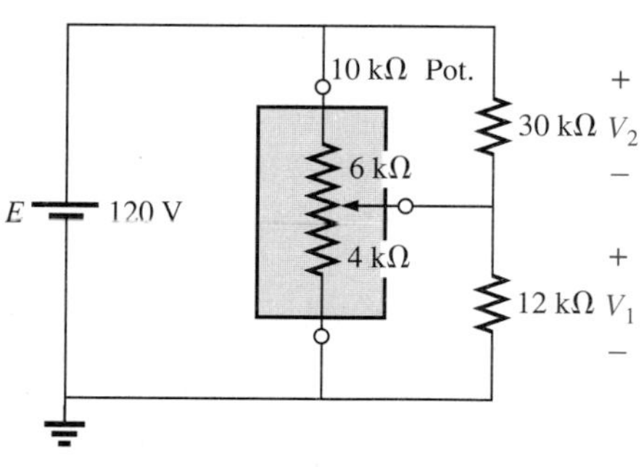

FIG. 7.40
Example 7.12.

7.6 AMMETER, VOLTMETER, AND OHMMETER DESIGN

Now that the fundamentals of series, parallel, and series-parallel networks have been introduced, we are prepared to investigate the funda-

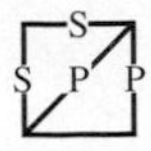

mental design of an ammeter, voltmeter, and ohmmeter. Our design of each will employ the d'Arsonval analog movement of Fig. 7.41. The movement consists basically of an iron-core coil mounted on bearings between a permanent magnet. The helical springs limit the turning motion of the coil and provide a path for the current to reach the coil. When a current is passed through the movable coil, the fluxes of the coil and permanent magnet will interact to develop a torque on the coil that will cause it to rotate on its bearings. The movement is adjusted to indicate zero deflection on a meter scale when the current through the coil is zero. The direction of current through the coil will then determine whether the pointer will display an up-scale or below-zero indication. For this reason, ammeters and voltmeters have an assigned polarity on their terminals to ensure an up-scale reading.

FIG. 7.41
d'Arsonval analog movement. (Courtesy of Weston Instruments, Inc.)

D'Arsonval movements are usually rated by current and resistance. The specifications of a typical movement might be 1 mA, 50 Ω. The 1 mA is the *current sensitivity* (*CS*) of the movement, which is the current required for a full-scale deflection. It will be denoted by the symbol I_{CS}. The 50 Ω represents the internal resistance (R_m) of the movement. A common notation for the movement and its specifications is provided in Fig. 7.42.

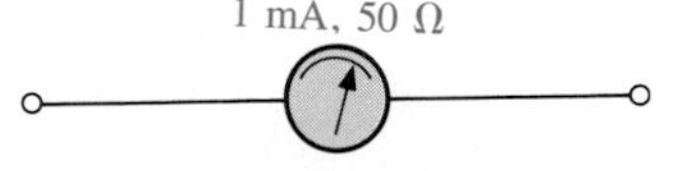

FIG. 7.42
Movement notation.

The Ammeter

The maximum current that the d'Arsonval movement can read independently is equal to the current sensitivity of the movement. However, higher currents can be measured if additional circuitry is introduced. This additional circuitry, as shown in Fig. 7.43, results in the basic construction of an ammeter.

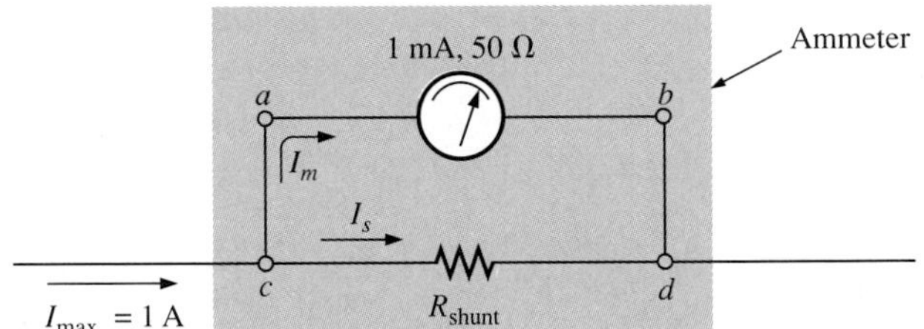

FIG. 7.43
Basic ammeter.

The resistance R_{shunt} is chosen for the ammeter of Fig. 7.43 to allow 1 mA to flow through the movement when a maximum current of 1 A enters the ammeter. If less than 1 A should flow through the ammeter, the movement will have less than 1 mA flowing through it and will indicate less than full-scale deflection.

Since the voltage across parallel elements must be the same, the potential drop across *a-b* in Fig. 7.43 must equal that across *c-d*; that is,

$$(1\text{ mA})(50\ \Omega) = R_{shunt}I_s$$

Also, I_s must equal 1 A − 1 mA = 999 mA if the current is to be limited to 1 mA through the movement (Kirchhoff's current law). Therefore,

$$(1\text{ mA})(50\ \Omega) = R_{shunt}(999\text{ mA})$$

$$R_{shunt} = \frac{(1\text{ mA})(50\ \Omega)}{999\text{ mA}}$$

$$\cong 0.05\ \Omega$$

In general,

$$R_{\text{shunt}} = \frac{R_m I_{CS}}{I_{\text{max}} - I_{CS}} \tag{7.3}$$

One method of constructing a multirange ammeter is shown in Fig. 7.44, where the rotary switch determines the R_{shunt} to be used for the maximum current indicated on the face of the meter. Most meters employ the same scale for various values of maximum current. If you read 375 on the 0–5 mA scale with the switch on the 5 setting, the current is 3.75 mA; on the 50 setting, the current is 37.5 mA; and so on.

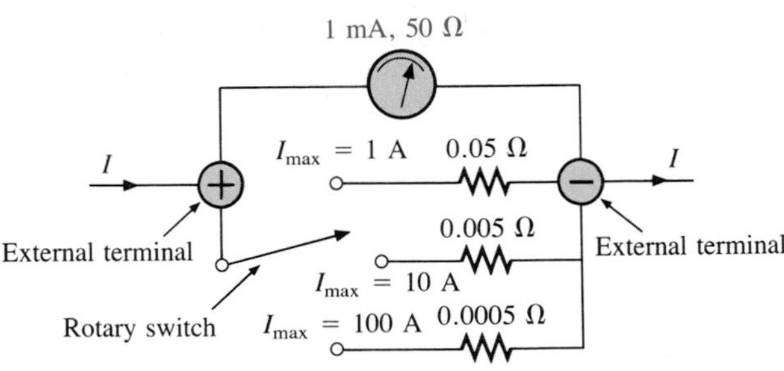

FIG. 7.44
Multirange ammeter.

The Voltmeter

A variation in the additional circuitry will permit the use of the d'Arsonval movement in the design of a voltmeter. The 1-mA, 50-Ω movement can also be rated as a 50-mV (1 mA × 50 Ω), 50-Ω movement, indicating that the maximum voltage that the movement can measure independently is 50 mV. The millivolt rating is sometimes referred to as the *voltage sensitivity* (*VS*). The basic construction of the voltmeter is shown in Fig. 7.45.

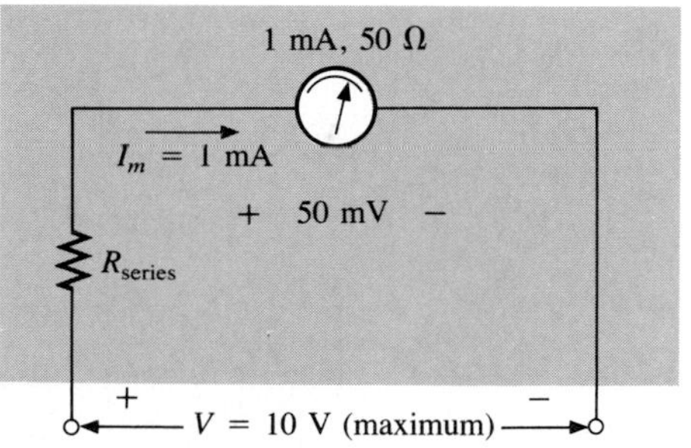

FIG. 7.45
Basic voltmeter.

The R_{series} is adjusted to limit the current through the movement to 1 mA when the maximum voltage is applied across the voltmeter. A lesser voltage would simply reduce the current in the circuit and thereby the deflection of the movement.

Applying Kirchhoff's voltage law around the closed loop of Fig. 7.45, we obtain

$$[10\text{ V} - (1\text{ mA})(R_{\text{series}})] - 50\text{ mV} = 0$$

or

$$R_{\text{series}} = \frac{10\text{ V} - (50\text{ mV})}{1\text{ mA}} = 9950\ \Omega$$

In general,

$$R_{\text{series}} = \frac{V_{\text{max}} - V_{VS}}{I_{CS}} \tag{7.4}$$

One method of constructing a multirange voltmeter is shown in Fig. 7.46. If the rotary switch is at 10 V, $R_{\text{series}} = 9.950\text{ k}\Omega$; at 50 V, $R_{\text{series}} = 40\text{ k}\Omega + 9.950\text{ k}\Omega = 49.950\text{ k}\Omega$; and at 100 V, $R_{\text{series}} = 50\text{ k}\Omega + 40\text{ k}\Omega + 9.950\text{ k}\Omega = 99.950\text{ k}\Omega$.

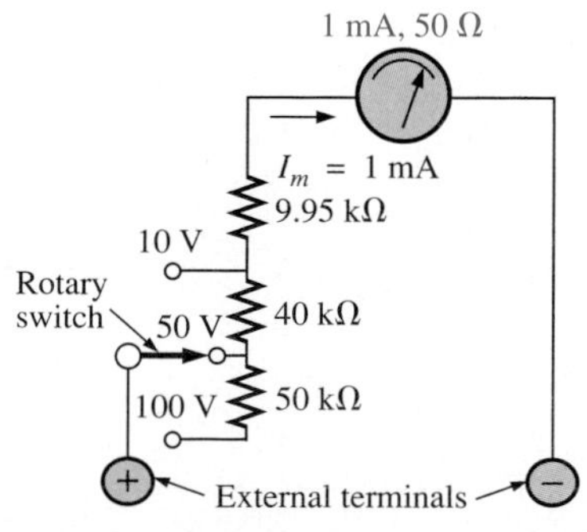

FIG. 7.46
Multirange voltmeter.

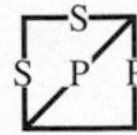

The Ohmmeter

In general, ohmmeters are designed to measure resistance in the low, mid-, or high range. The most common is the *series ohmmeter,* designed to read resistance levels in the midrange. It employs the series configuration of Fig. 7.47. The design is quite different from that of the ammeter or voltmeter because it will show a full-scale deflection for zero ohms and no deflection for infinite resistance.

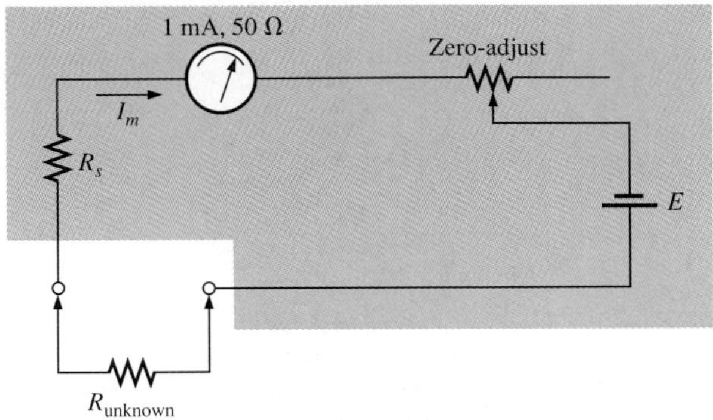

FIG. 7.47
Series ohmmeter.

To determine the series resistance R_s, the external terminals are shorted (a direct connection of zero ohms between the two) to simulate zero ohms, and the zero-adjust is set to half its maximum value. The resistance R_s is then adjusted to allow a current equal to the current sensitivity of the movement (1 mA) to flow in the circuit. The zero-adjust is set to half its value so that any variation in the components of the meter that may produce a current more or less than the current sensitivity can be compensated for. The current I_m is

$$I_m\,(\text{full scale}) = I_{CS} = \frac{E}{R_s + R_m + \dfrac{\text{zero-adjust}}{2}} \tag{7.5}$$

and

$$R_s = \frac{E}{I_{CS}} - R_m - \frac{\text{zero-adjust}}{2} \tag{7.6}$$

If an unknown resistance is then placed between the external terminals, the current will be reduced, causing a deflection less than full scale. If the terminals are left open, simulating infinite resistance, the pointer will not deflect since the current through the circuit is zero.

An instrument designed to read very low values of resistance appears in Fig. 7.48. Because of its low-range capability, the network design must be a great deal more sophisticated than described above. It employs electronic components that eliminate the inaccuracies introduced by lead and contact resistances. It is similar to the above system in the sense that it is completely portable and does require a dc battery to establish measurement conditions. Special leads are employed to limit any introduced resistance levels. The maximum scale setting can be set as low as 0.00352 (3.52 mΩ).

FIG. 7.48
Milliohmmeter. (Courtesy of Keithley Instruments, Inc.)

The megohmmeter is an instrument for measuring very high resistance values. Its primary function is to test the insulation found in power transmission systems, electrical machinery, transformers, and so on. To measure the high-resistance values, a high dc voltage is established by a hand-driven generator. If the shaft is rotated above some set value, the output of the generator will be fixed at one selectable voltage, typically 250, 500, or 1000 V. A photograph of the commercially available tester is shown in Fig. 7.49. For this instrument, the range is zero to 5000 MΩ.

FIG. 7.49
Megohmmeter. (Courtesy of AEMC Corp.)

7.7 GROUNDING

Although usually treated too lightly in most introductory electrical or electronics texts, the impact of the ground connection and how it can provide a measure of safety to a design are very important topics. Ground potential is 0 V at every point in a network that has a ground symbol. Since they are all at the same potential, they can all be connected together, but for purposes of clarity most are left isolated on a large schematic. On a schematic the voltage levels provided are always with respect to ground. A system can therefore be checked quite rapidly by simply connecting the black lead of the voltmeter to

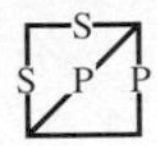

the ground connection and placing the red lead at the various points where the typical operating voltage is provided. A close match normally implies that that portion of the system is operating properly. Even though a major part of the discussion to follow includes ac systems that will not be introduced until Chapter 13, the content is such that the background established thus far will be sufficient to understand the material to be presented. The concept of grounding is one that should be introduced at the earliest opportunity for safety and theoretical reasons.

There are various types of grounds depending on the application. An *earth ground* is one that is connected directly to the earth by a low-impedance connection. The entire surface of the earth is defined to have a potential of 0 V. It is the same at every point because there are sufficient conductive agents in the soil such as water and electrolytes to ensure that any difference in voltage on the surface is equalized by a flow of charge between the two points. Every home has an earth ground, usually established by a long conductive rod driven into the ground and connected to the power panel. The electrical code requires a direct connection from earth ground to the cold-water pipes of a home for safety reasons. A "hot" wire touching a cold-water pipe draws sufficient current because of the low-impedance ground connection to throw the breaker. Otherwise, someone in the bathroom could pick up the voltage when they touch the cold-water faucet and risk bodily harm. Because water is a conductive agent, any area of the home with water such as the bathroom or kitchen is of particular concern. Most electrical systems are connected to earth ground primarily for safety reasons. All the power lines in a laboratory, at industrial locations, or in the home are connected to earth ground.

A second type is referred to as a *chassis ground,* which may be *floating* or connected directly to an earth ground. A chassis ground simply stipulates that the chassis has a reference potential for all points of the network. If the chassis is not connected to earth potential (0 V), it is said to be floating and can have any other reference voltage for the other voltages to be compared to. For instance, if the chassis is sitting at 120 V, all measured voltages of the network will be referenced to this level. A reading of 32 V between a point in the network and the chassis ground will therefore actually be at 152 V with respect to earth potential. Most high-voltage systems are not left floating, however, because of loss of the safety factor. For instance, if someone should touch the chassis and be standing on a suitable ground, the full 120 V would fall across that individual.

Grounding can be particularly important when working with numerous pieces of measuring equipment in the laboratory. For instance, the supply and oscilloscope in Fig. 7.50(a) are each connected directly to an earth ground through the negative terminal of each. If the oscilloscope is connected as shown in Fig. 7.50(a) to measure the voltage V_{R_1}, a dangerous situation will develop. The grounds of each piece of equipment are connected together through the earth ground and they effectively short out the resistor. Since the resistor is the primary current-controlling element in the network, the current will rise to a very high level and possibly damage the instruments or cause dangerous side effects. In this case the supply or scope should be used in the floating mode, or the resistors interchanged as shown in Fig. 7.50(b). In Fig. 7.50(b) the grounds have a common point and do not affect the structure of the network.

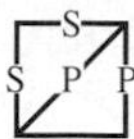

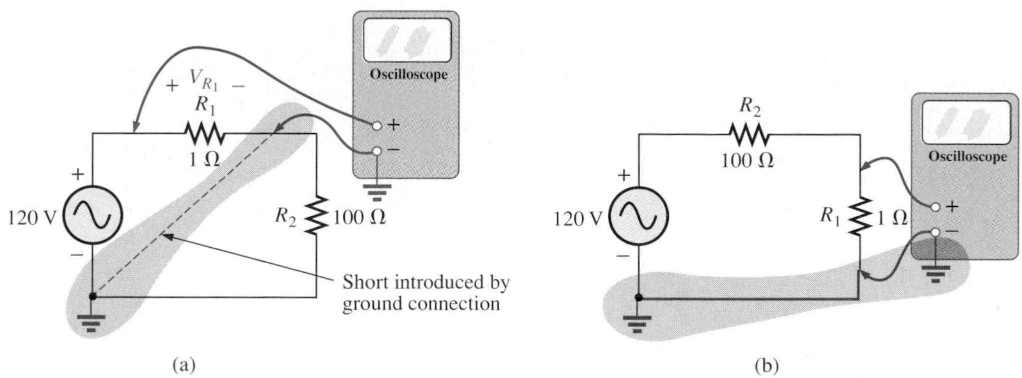

FIG. 7.50
Demonstrating the effect of the oscilloscope ground on the measurement of the voltage across the resistor R_1.

The National Electrical Code requires that the "hot" (or feeder line) that carries current to a load be *black,* and the line (called the neutral) that carries the current back to the supply be *white*. Three wire conductors have a ground wire that must be *green* or *bare,* which will ensure a common ground but is not designed to carry current. The components of a three-prong extension cord and wall outlet are shown in Fig. 7.51. Note that on both fixtures the connection to the hot lead is smaller than the return leg and that the ground connection is partially circular.

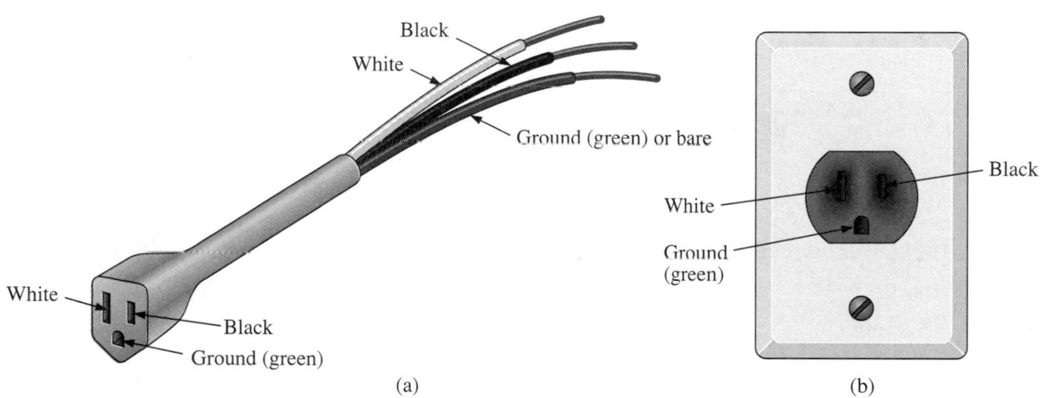

FIG. 7.51
Three-wire conductors: (a) extension cord: (b) home outlet.

The complete wiring diagram for a household outlet is shown in Fig. 7.52. Note that the current through the ground wire is zero and that both the return wire and ground wire are connected to an earth ground. The full current to the loads flows through the feeder and return lines.

The importance of the ground wire in a three-wire system can be demonstrated by the toaster in Fig. 7.53 rated 1200 W at 120 V. From the power equation $P = EI$, the current drawn under normal operating conditions is $I = P/E = 1200\text{ W}/120\text{ V} = 10\text{ A}$. If a two-wire line were employed as shown in Fig. 7.53(a), the 20-A breaker would be quite comfortable with the 10-A current and the system would perform normally. However, if abuse to the feeder (or return line) caused it to become frayed and to touch the metal housing of the toaster, the situation depicted in Fig. 7.53(b) would result. The housing would become "hot," yet the breaker would not "pop" because the current would still be the

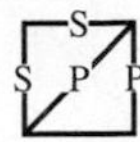

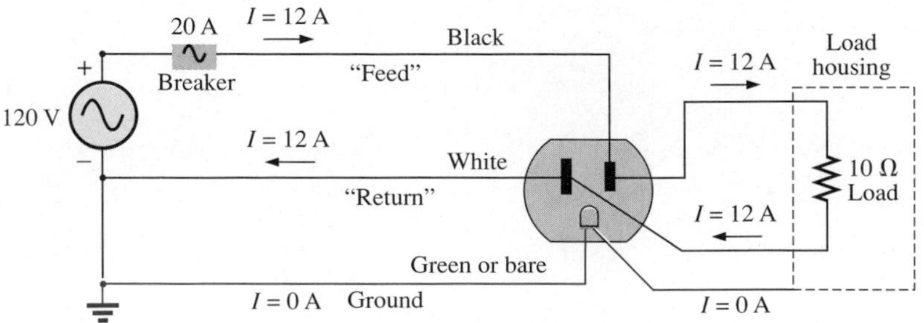

FIG. 7.52
Complete wiring diagram for a household outlet with a 10-Ω load.

rated 10 A. However, a dangerous condition would not exist because anyone touching the toaster would feel the full 120 V to ground. If the ground wire were attached to the chassis as shown in Fig. 7.53(c), a low-resistance path would be created between the short-circuit point and ground, and the current would jump to very high levels. The breaker would "pop," and the user would be warned that a problem exists.

Although the above does not cover all possible areas of concern with proper grounding or introduce all the nuances associated with the effect of grounds on a system's performance, it should provide an awareness of the importance of understanding its impact. Additional comment on the effects of grounding will be made in the chapters to follow as the need arises.

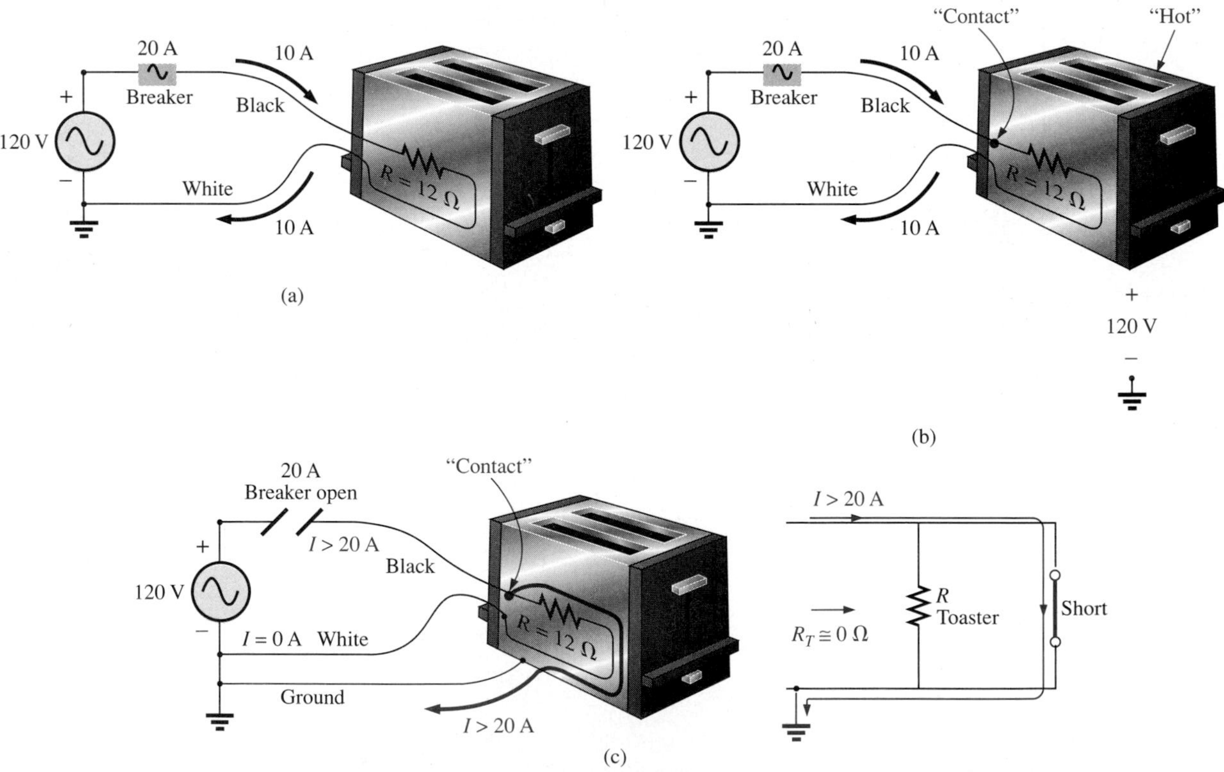

FIG. 7.53
Demonstrating the importance of a properly grounded appliance: (a) ungrounded; (b) ungrounded and undesirable contact; (c) grounded appliance with undesirable contact.

7.8 COMPUTER ANALYSIS

PSpice (DOS)

The series-parallel network to be analyzed using PSpice is the same network examined in Example 7.2 and redrawn in Fig. 7.54. A program in a language such as C++ or BASIC would require a solution of the network that would sequentially calculate R_T, followed by the source current and then the best path back to the desired unknowns. The resulting program would require a number of lines and would obviously take some time to write and test. The beauty of the PSpice package now becomes more obvious since the input file (Fig. 7.55) is not that difficult or lengthy, and all the desired results would be available in a few seconds of computer time. As an instructor, however, my concern remains that you can use the package to obtain the solution and yet

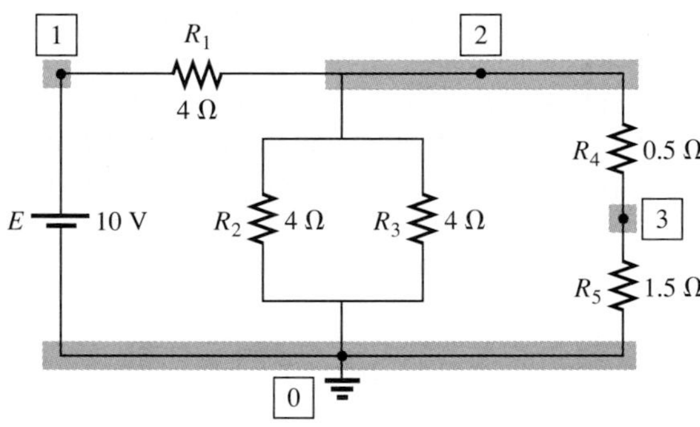

FIG. 7.54
Applying PSpice (DOS) to the network of Example 7.2.

```
Chapter 7 - Multi R Network

 ****     CIRCUIT DESCRIPTION

****************************************************************************


VE 1 0 10V
R1 1 2 4
R2 2 0 4
R3 2 0 4
R4 2 3 0.5
R5 3 0 1.5
.DC VE 10 10 1
.PRINT DC I(R1) I(R2) V(3,0)
.OPTIONS NOPAGE
.END

 ****     DC TRANSFER CURVES                    TEMPERATURE =   27.000 DEG C


  VE          I(R1)         I(R2)         V(3,0)

  1.000E+01   2.000E+00     5.000E-01     1.500E+00
```

FIG. 7.55
Input and output files for the network of Fig. 7.54.

have no idea (or a very weak understanding of) how to solve the network longhand. Do not let computer techniques place the "cart before the horse." Always try to have some understanding of the basic maneuvers a computer is going to go through before simply accepting the computer output as gospel. You do not have to be an expert in the subject, but at least be aware of the general process and the implication of the results obtained.

The output file reveals that

$$I_{R_1} = I_E = 2\text{ A}, \qquad I_{R_2} = I_{R_3} = 0.5\text{ A}, \quad \text{and} \quad V_{R_5} = 1.5\text{ V}$$

as obtained in Example 7.2.

PSpice (Windows)

Schematics will now be used to verify the results of Example 7.11. The calculated resistor values will be substituted and the voltage and current levels checked to see if they match the specified values for the network. The elements of the network are introduced and their values set as described in Chapter 5. In particular, note in the schematic of Fig. 7.56 that R_1 is set at 1.3333 kΩ rather than the example solution of 1.33 kΩ. When running the program, it was found that the **VIEWPOINT** and **IPROBE** levels would not be a perfect match unless we extended R_1 to four places. With $R_1 = 1.3333$ kΩ, the voltage across R_{L_1} was 59.9990 V, across R_{L_2} it was 20.0030 V, and across R_3 it was −12.0010 V. The analysis clearly verifies the results of Example 7.11.

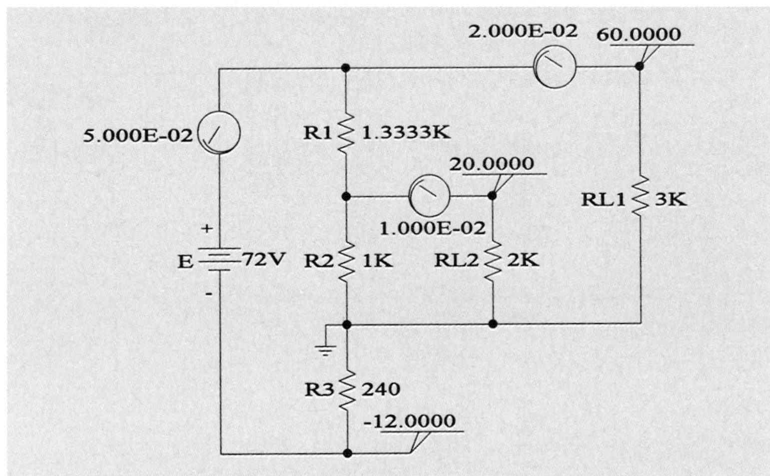

FIG. 7.56
Applying PSpice (Windows) to the network of Example 7.11.

C++

The C++ program to be introduced will perform a detailed analysis of the network of Fig. 7.57 (appearing as Fig. 7.27 in the text). Once all the parameters are introduced, the program will print out R_T, I_s, I_3, and I_6. The order of the program is exactly the same as that of a longhand solution. In Fig. 7.57, $R_a = R_5 + R_6$ is first determined, followed by $R_b = R_4 \parallel R_a$ and $R_c = R_3 + R_b$, with $R_d = R_2 \parallel R_c$ and $R_T = R_1 + R_d$. Then $I_s = E/R_T$ with I_3 and I_6 is determined by the current divider rule.

The program begins with a heading and preprocessor directive. The *<iostream.h>* header file sets up the input–output path between the pro-

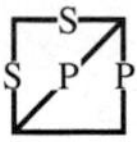

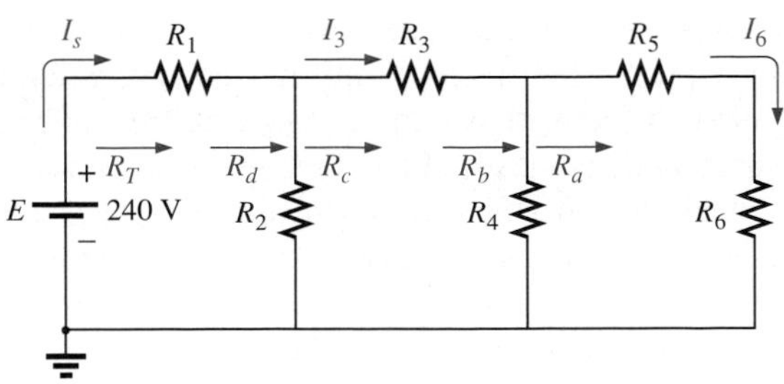

FIG. 7.57
Ladder network to be analyzed using C++.

```
//C++ Series-Parallel Circuit Analysis

#include <iostream.h>               //needed for input/output

main()                              //execution begins here
{
        float R1, R2, R3, R4, R5, R6;  //declare circuit resistors
        float Ra, Rb, Rc, Rd, Rt;      //declare equivalent resistances
        float E;                       //declare voltage source
        float Is, I3, I6;              //declare circuit currents

        cout << "Enter R1: ";          //get all circuit values
        cin >> R1;
        cout << "Enter R2: ";
        cin >> R2;
        cout << "Enter R3: ";
        cin >> R3;
        cout << "Enter R4: ";
        cin >> R4;
        cout << "Enter R5: ";
        cin >> R5;
        cout << "Enter R6: ";
        cin >> R6;
        cout << "Enter E: ";
        cin >> E;

        Ra = R5 + R6;                  //calculate the total resistance
        Rb = R4 * Ra / (R4 + Ra);
        Rc = R3 + Rb;
        Rd = R2 * Rc / (R2 + Rc);
        Rt = R1 + Rd;
        cout << "\n";
        cout << "The total resistance is " << Rt << " ohms.\n";

        Is = E / Rt;                   //calculate circuit currents
        I3 = Is * R2 / (R2 + Rc);
        I6 = I3 * R4 / (R4 + Ra);
        cout << "The source current is " << Is << " Amperes.\n";
        cout << "I3 equals " << I3 << " Amperes.\n";
        cout << "I6 equals " << I6 << " Amperes.\n";
}
```

Margin annotations: Heading; Preprocessor directive; Define variables and data type; Request and obtain network parameters; Find R_T; Display R_T; Calculate I_3 and I_6; Display I_3 and I_6; Body of program

FIG. 7.58
C++ program to analyze the ladder network of Fig. 7.57.

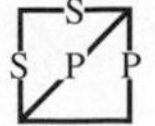

gram and the disk-operating system. The *main* ( ) part of the program, defined by the braces { }, includes all the remaining commands and statements. First, the network parameters and quantities to be determined are defined as floating point variables. Next, the *cout* and *cin* commands are used to obtain the resistor values and source voltage from the user. The total resistance is then determined in the order described above, followed by a carriage return "*n*" and a printout of the value. Then the currents are determined and printed out by the last three lines.

The program (Fig. 7.58) is quite straightforward and with experience not difficult to write. In addition, consider the benefits of having a program on file that can solve any ladder network having the configuration of Fig. 7.57. For the parameter values of Fig. 7.27, the printout will appear as shown in Fig. 7.59, confirming the results of Section 7.3. If an element is missing, simply insert a short-circuit or open-circuit equivalent, whichever is appropriate. For instance, if R_5 and R_6 are absent, leaving a two-loop network, simply plug in very large values for R_5 and R_6 compared to the other elements of the network, and they will appear as open-circuit equivalents in the analysis. This is demonstrated in the run of Fig. 7.60 with a negative supply of 60 V. The results have negative signs for the currents because the defined direction in the program has the opposite direction. The current I_6 is essentially zero amperes, as it should be if R_5 and R_6 do not exist. If R_1, R_3, R_5, or R_6 were the only resistive element to be missing, a short-circuit equivalent would be inserted. If R_2 or R_4 were the only missing element, the open-circuit equivalent would be substituted.

```
Enter R1: 5
Enter R2: 6
Enter R3: 4
Enter R4: 6
Enter R5: 1
Enter R6: 2
Enter E: 240

The total resistance is 8 ohms.
The source current is 30 Amperes.
I3 equals 15 Amperes.
I6 equals 10 Amperes.
```

FIG. 7.59

C++ response to an analysis of the ladder network of Fig. 7.57 with the parameter values of Fig. 7.27.

```
Enter R1: 10
Enter R2: 220
Enter R3: 12
Enter R4: 100
Enter R5: 1e30
Enter R6: 1e30
Enter E: -60

The total resistance is 84.216866 ohms.
The source current is -0.712446 Amperes.
I3 equals -0.472103 Amperes.
I6 equals -2.360515e-29 Amperes.
```

FIG. 7.60

C++ response to an analysis of the ladder network of Fig. 7.57 without the elements R_5 and R_6.

PROBLEMS

SECTION 7.2 Descriptive Examples

1. Which elements of the following networks (Fig. 7.61) are in series or parallel? In other words, which elements of the following networks have the same current (series) or voltage (parallel)? Restrict your decision to single elements and do not include combined elements.

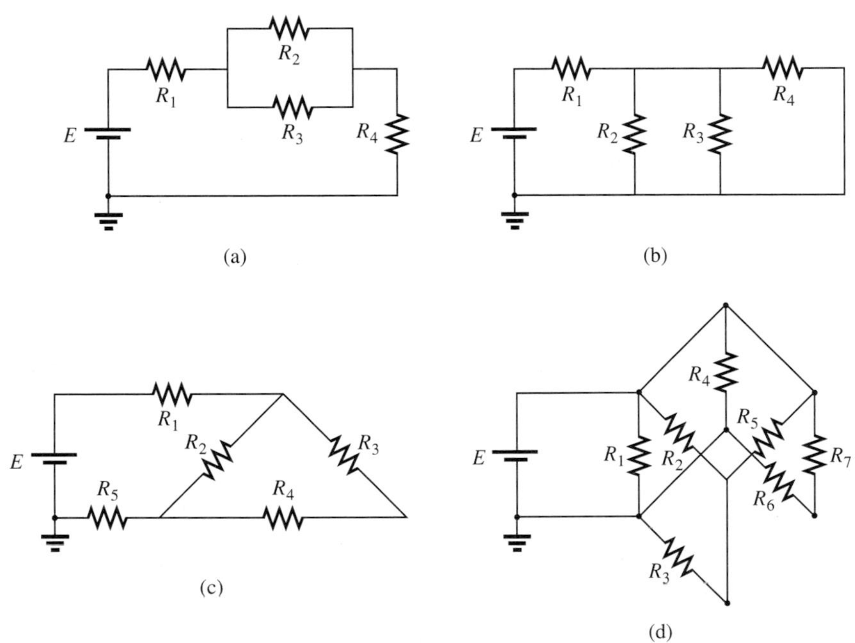

FIG. 7.61
Problem 1.

2. Determine R_T for the networks of Fig. 7.62.

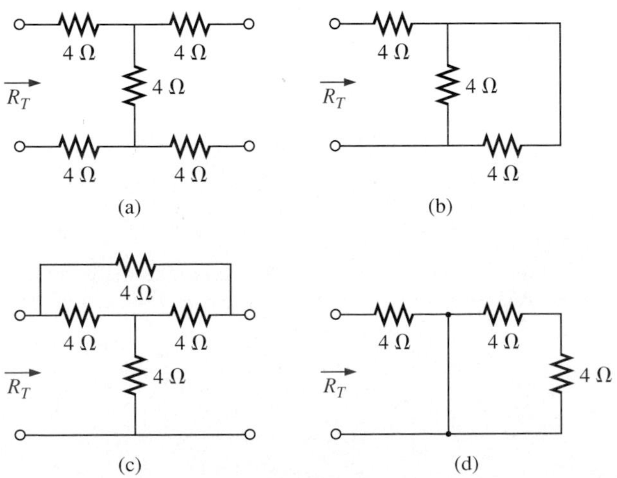

FIG. 7.62
Problem 2.

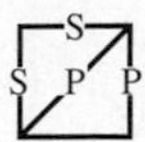

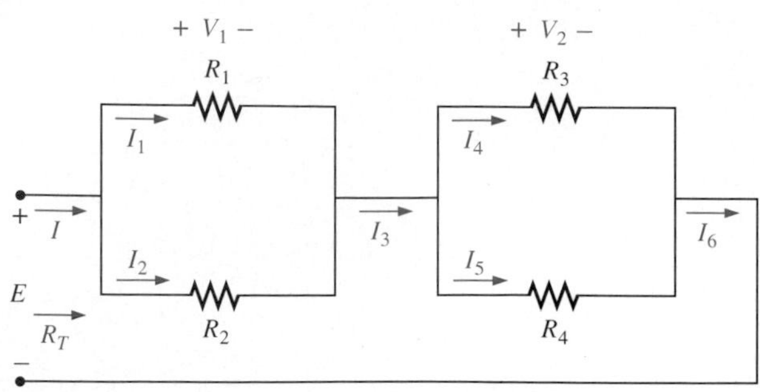

FIG. 7.63
Problem 3.

3. For the network of Fig. 7.63:
   a. Does $I = I_3 = I_6$? Explain.
   b. If $I = 5$ A and $I_1 = 2$ A, find I_2.
   c. Does $I_1 + I_2 = I_4 + I_5$? Explain.
   d. If $V_1 = 6$ V and $E = 10$ V, find V_2.
   e. If $R_1 = 3\ \Omega$, $R_2 = 2\ \Omega$, $R_3 = 4\ \Omega$, and $R_4 = 1\ \Omega$, what is R_T?
   f. If the resistors have the values given in part (e) and $E = 10$ V, what is the value of I in amperes?
   g. Using values given in parts (e) and (f), find the power delivered by the battery E and dissipated by the resistors R_1 and R_2.

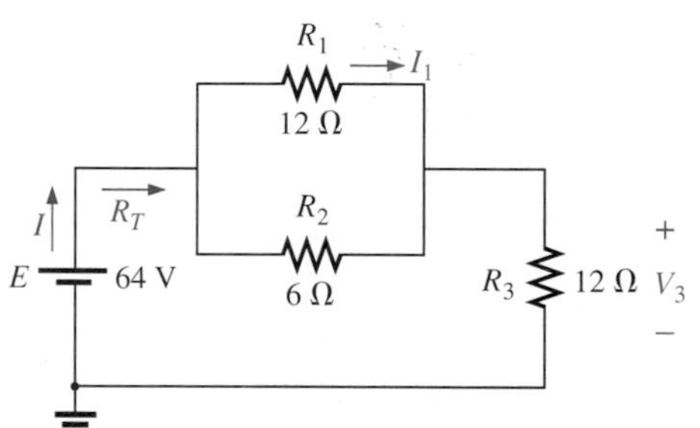

FIG. 7.64
Problem 4.

4. For the network of Fig. 7.64:
   a. Calculate R_T.
   b. Determine I and I_1.
   c. Find V_3.

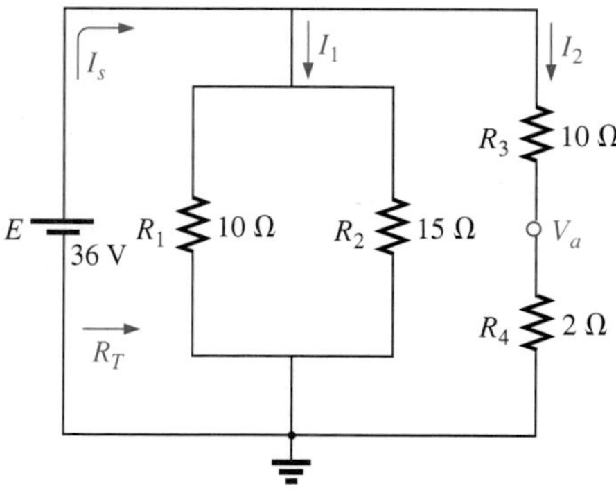

FIG. 7.65
Problem 5.

5. For the network of Fig. 7.65:
   a. Determine R_T.
   b. Find I_s, I_1, and I_2.
   c. Calculate V_a.

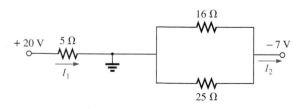

FIG. 7.66
Problem 6.

6. Determine the currents I_1 and I_2 for the network of Fig. 7.66.

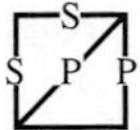

7. **a.** Find the magnitude and direction of the currents I, I_1, I_2, and I_3 for the network of Fig. 7.67.
   b. Indicate their direction on Fig. 7.67.

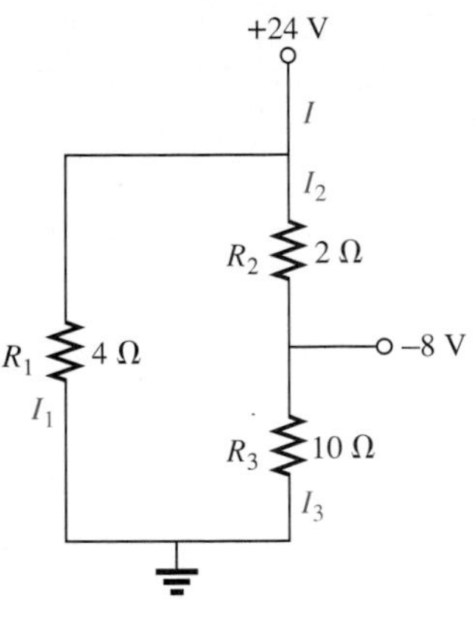

FIG. 7.67
Problem 7.

*8. For the network of Fig. 7.68:
   a. Determine the currents I_s, I_1, I_3, and I_4.
   b. Calculate V_a and V_{bc}.

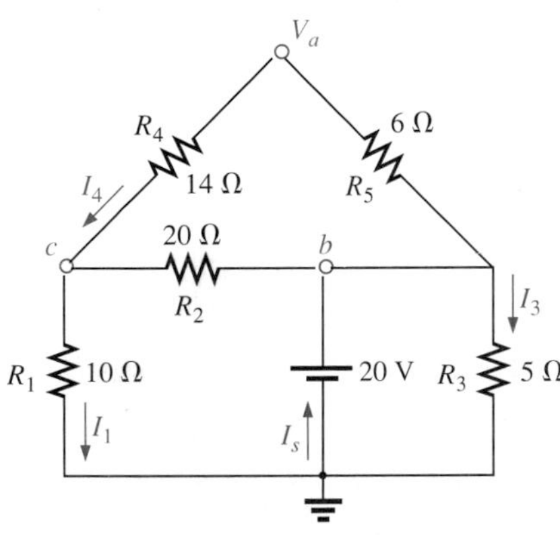

FIG. 7.68
Problem 8.

9. For the network of Fig. 7.69:
   a. Determine the current I_1.
   b. Calculate the currents I_2 and I_3.
   c. Determine the voltage levels V_a and V_b.

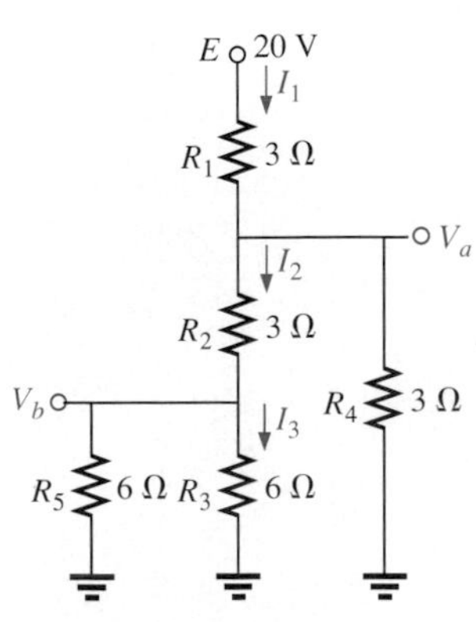

FIG. 7.69
Problem 9.

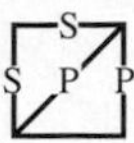

10. For the network of Fig. 7.70:

a. Find the currents I and I_6.

b. Find the voltages V_1 and V_5.

c. Find the power delivered to the 6-kΩ resistor.

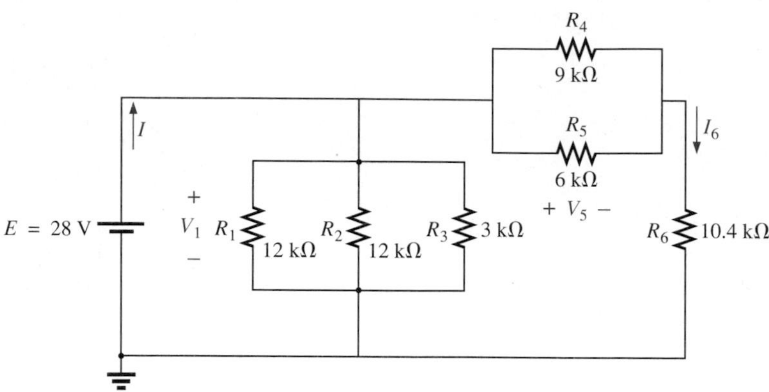

FIG. 7.70
Problem 10.

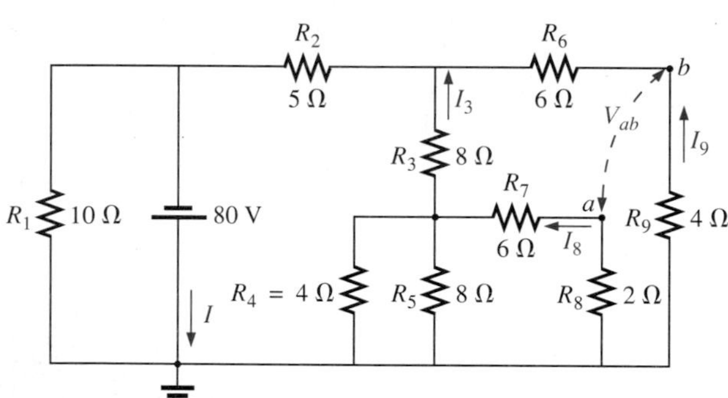

FIG. 7.71
Problem 11.

***11.** For the series-parallel network of Fig. 7.71:

a. Find the current I.

b. Find the currents I_3 and I_9.

c. Find the current I_8.

d. Find the voltage V_{ab}.

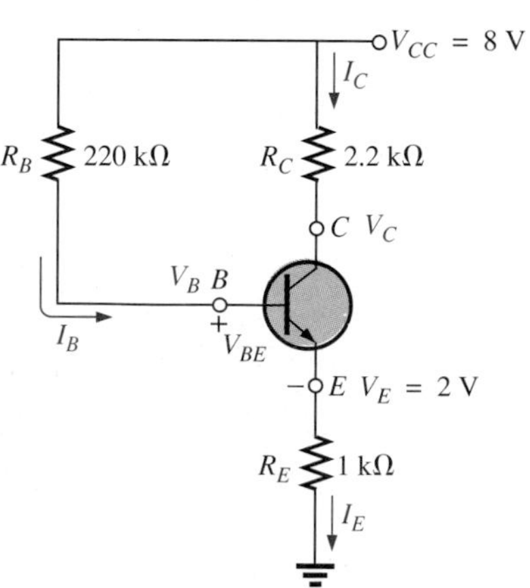

FIG. 7.72
Problem 12.

***12.** Determine the dc levels for the transistor network of Fig. 7.72 using the fact that $V_{BE} = 0.7$ V, $V_E = 2$ V, and $I_C = I_E$. That is,

a. Determine I_E and I_C.

b. Calculate I_B.

c. Determine V_B and V_C.

d. Find V_{CE} and V_{BC}.

*13. The network of Fig. 7.73 is the basic biasing arrangement for the *field-effect transistor* (FET), a device of increasing importance in electronic design. (*Biasing* simply means the application of dc levels to establish a particular set of operating conditions.) Even though you may be unfamiliar with the FET, you can perform the following analysis using only the basic laws introduced in this chapter and the information provided on the diagram.
   a. Determine the voltages V_G and V_S.
   b. Find the currents I_1, I_2, I_D, and I_S.
   c. Determine V_{DS}.
   d. Calculate V_{DG}.

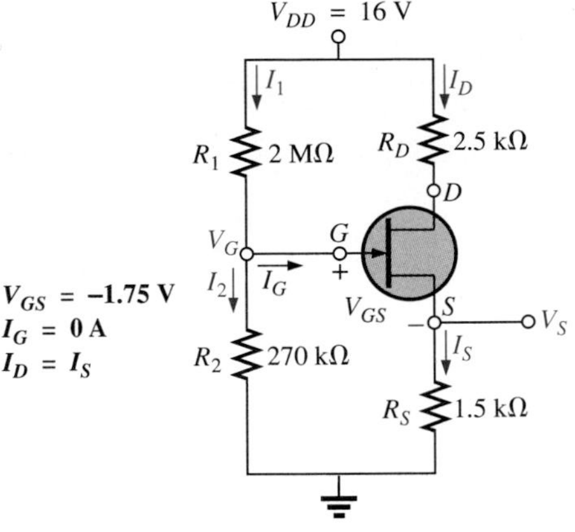

FIG. 7.73
Problem 13.

*14. For the network of Fig. 7.74:
   a. Determine R_T.
   b. Calculate V_a.
   c. Find V_1.
   d. Calculate V_2.
   e. Determine I (with direction).

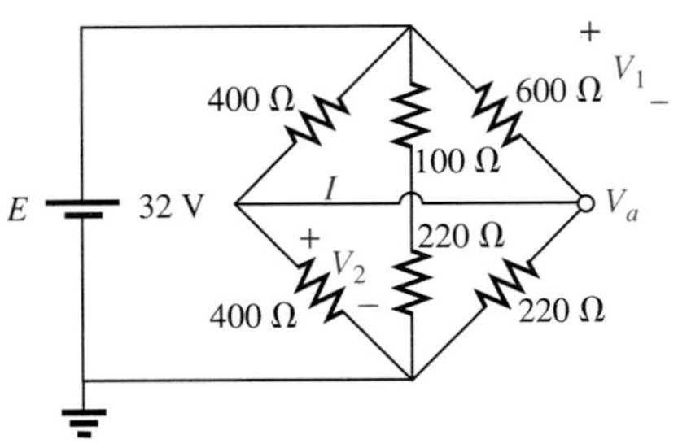

FIG. 7.74
Problem 14.

15. For the network of Fig. 7.75:
   a. Determine the current I.
   b. Find V.

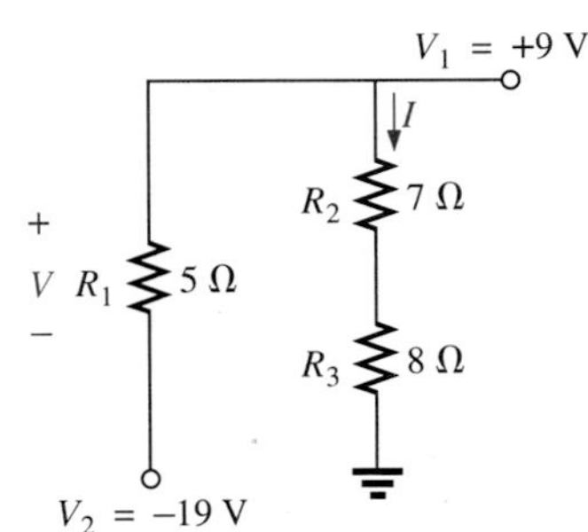

FIG. 7.75
Problem 15.

*16. Determine the current I and the voltages V_a, V_b, and V_{ab} for the network of Fig. 7.76.

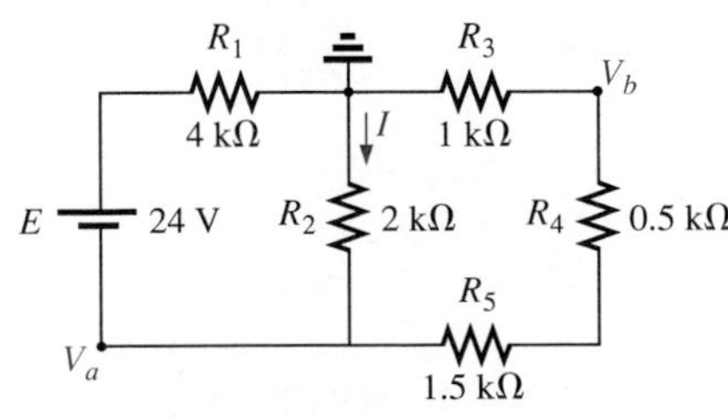

FIG. 7.76
Problem 16.

17. For the configuration of Fig. 7.77:
    a. Find the currents I_2, I_6, and I_8.
    b. Find the voltages V_4 and V_8.

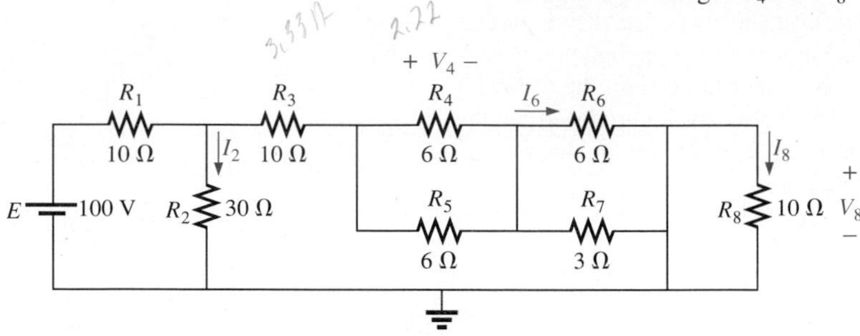

FIG. 7.77
Problem 17.

18. Determine the voltage V and the current I for the network of Fig. 7.78.

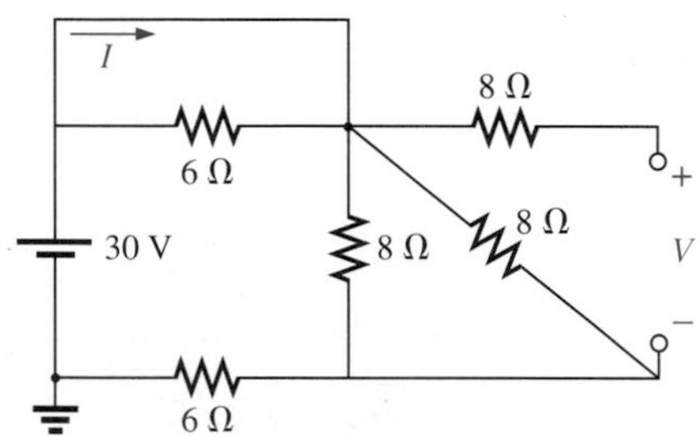

FIG. 7.78
Problem 18.

*19. For the network of Fig. 7.79:
    a. Determine R_T by combining resistive elements.
    b. Find V_1 and V_4.
    c. Calculate I_3 (with direction).
    d. Determine I_s by finding the current through each element and then applying Kirchhoff's current law. Then calculate R_T from $R_T = E/I_s$ and compare with the solution of part (a).

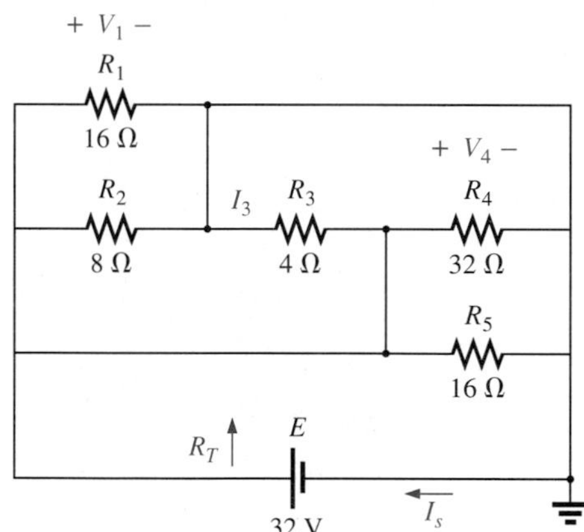

FIG. 7.79
Problem 19.

20. For the network of Fig. 7.80:
    a. Determine the voltage V_{ab}. (*Hint:* Just use Kirchhoff's voltage law.)
    b. Calculate the current I.

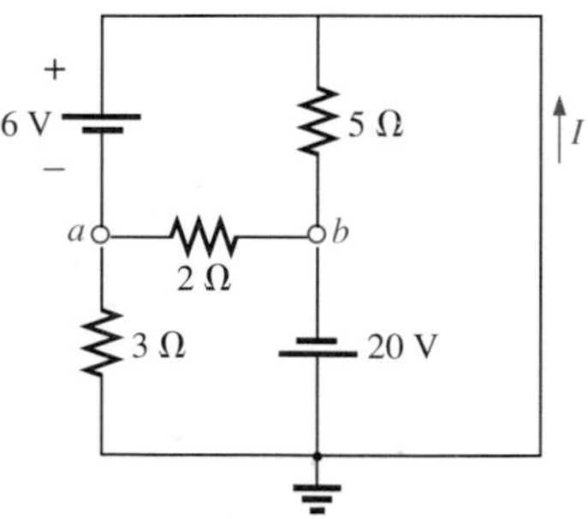

FIG. 7.80
Problem 20.

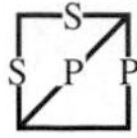

*21. For the network of Fig. 7.81:
  a. Determine the current I.
  b. Calculate the open-circuit voltage V.

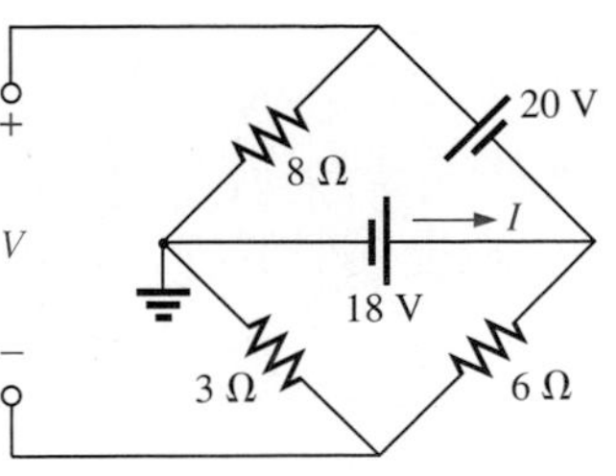

FIG. 7.81
Problem 21.

*22. For the network of Fig. 7.82, find the resistance R_3 if the current through it is 2 A.

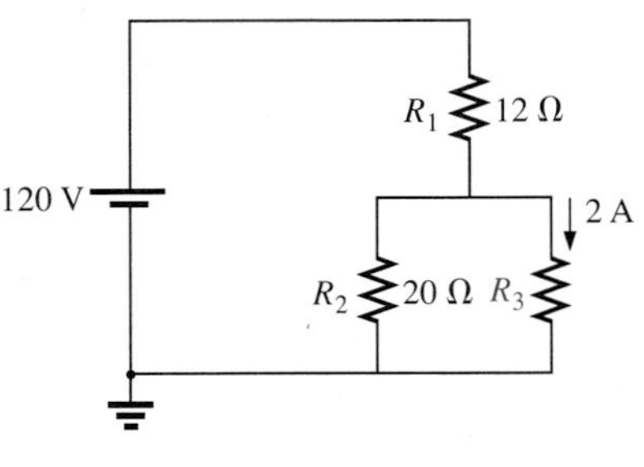

FIG. 7.82
Problem 22.

*23. If all the resistors of the cube (Fig. 7.83) are 10 Ω, what is the total resistance? (*Hint:* Make some basic assumptions about current division through the cube.)

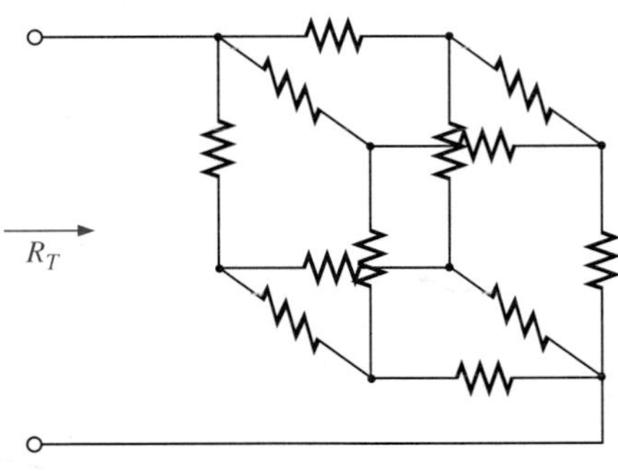

FIG. 7.83
Problem 23.

*24. Given the voltmeter reading $V = 27$ V in Fig. 7.84:
  a. Is the network operating properly?
  b. If not, what could be the cause of the incorrect reading?

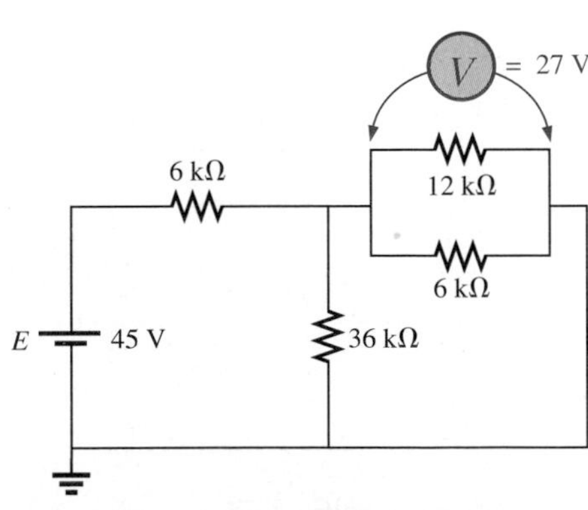

FIG. 7.84
Problem 24.

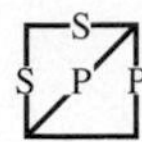

SECTION 7.3 Ladder Networks

25. For the ladder network of Fig. 7.85:
    a. Find the current I.
    b. Find the current I_7.
    c. Determine the voltages V_3, V_5, and V_7.
    d. Calculate the power delivered to R_7 and compare it to the power delivered by the 240-V supply.

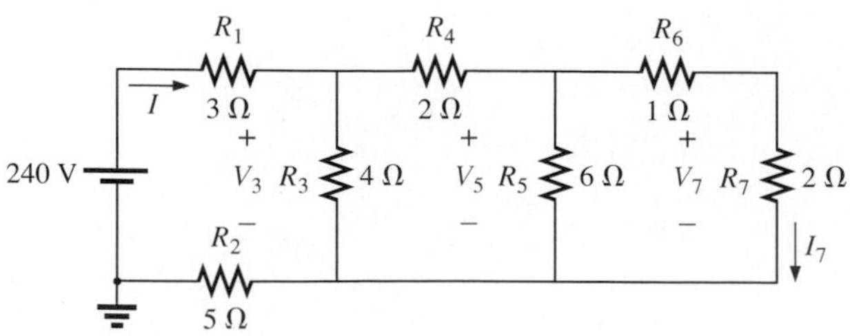

FIG. 7.85
Problem 25.

26. For the ladder network of Fig. 7.86:
    a. Determine R_T.
    b. Calculate I.
    c. Find I_8.

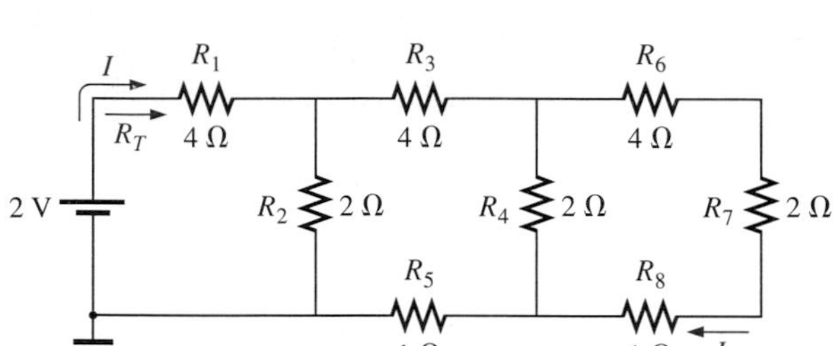

FIG. 7.86
Problem 26.

*27. Determine the power delivered to the 10-Ω load of Fig. 7.87.

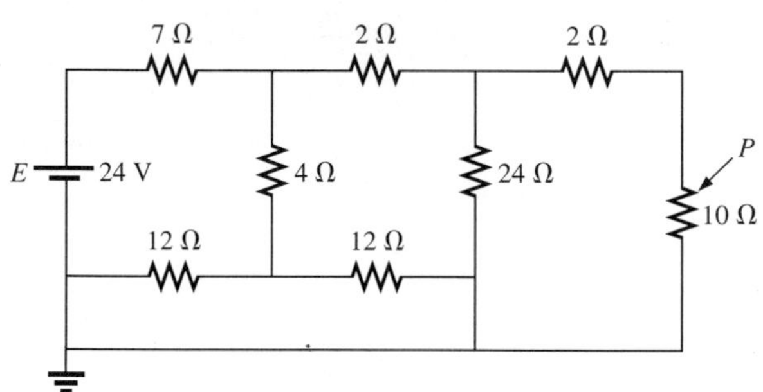

FIG. 7.87
Problem 27.

*28. For the multiple ladder configuration of Fig. 7.88:
    a. Determine I.
    b. Calculate I_4.
    c. Find I_6.
    d. Find I_{10}.

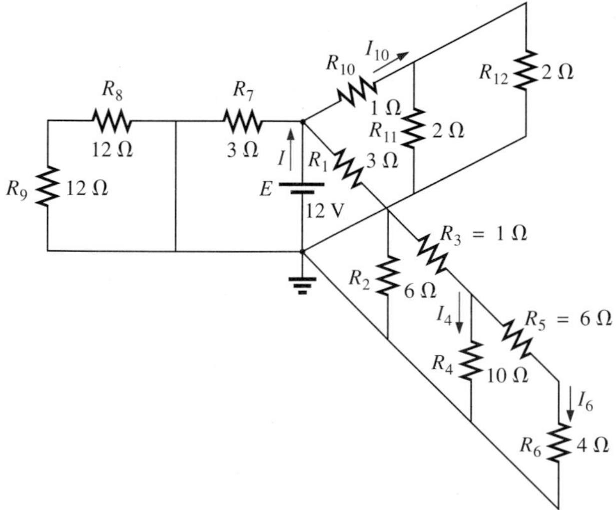

FIG. 7.88
Problem 28.

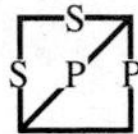

SECTION 7.4 Voltage Divider Supply (Unloaded and Loaded)

29. Given the voltage divider supply of Fig. 7.89:

   a. Determine the supply voltage E.

   b. Find the load resistors R_{L_2} ad R_{L_3}.

   c. Determine the voltage divider resistors R_1, R_2, and R_3.

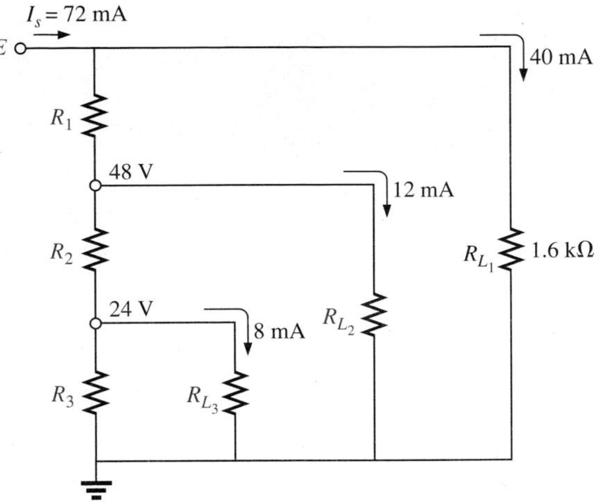

FIG. 7.89
Problem 29.

***30.** Determine the voltage divider supply resistors for the configuration of Fig. 7.90. Also determine the required wattage rating for each resistor and compare their levels.

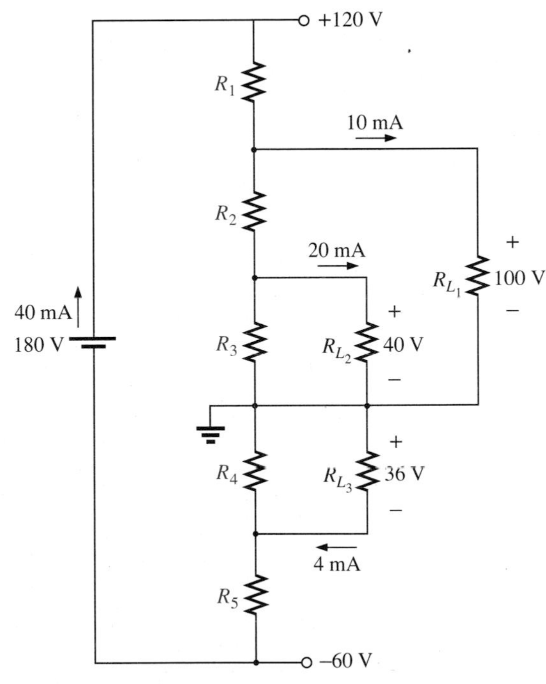

FIG. 7.90
Problem 30.

SECTION 7.5 Potentiometer Loading

***31.** For the system of Fig. 7.91:

   a. At first exposure, does the design appear to be a good one?

   b. In the absence of the 10-kΩ load, what are the values of R_1 and R_2 to establish 3 V across R_2?

   c. Determine the values of R_1 and R_2 when the load is applied and compare to the results of part (b).

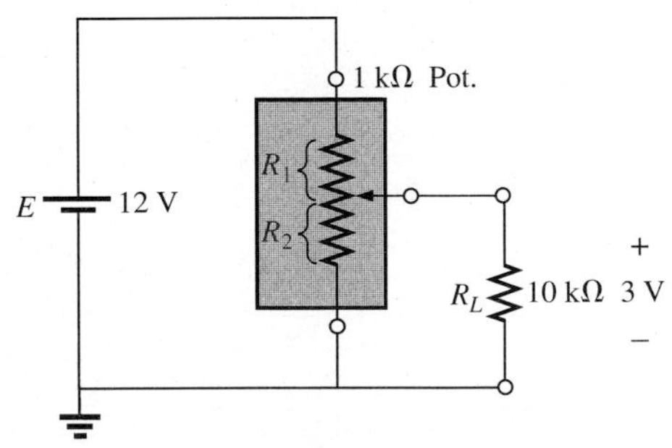

FIG. 7.91
Problem 31.

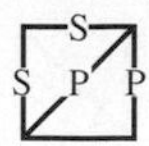

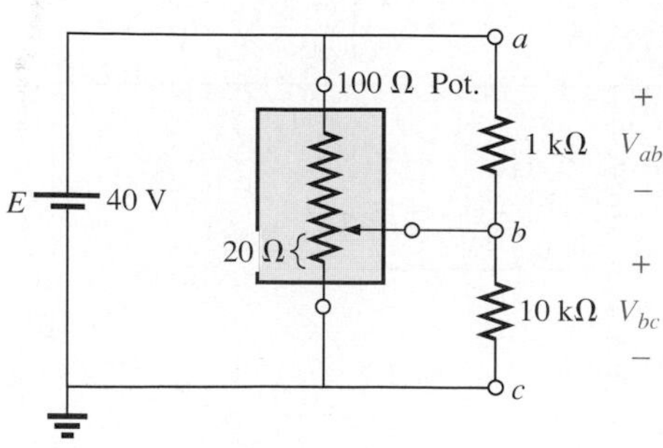

FIG. 7.92
Problem 32.

***32.** For the potentiometer of Fig. 7.92:
  a. What are the voltages V_{ab} and V_{bc} with no load applied?
  b. What are the voltages V_{ab} and V_{bc} with the indicated loads applied?
  c. What is the power dissipated by the potentiometer under the loaded conditions of Fig. 7.92?
  d. What is the power dissipated by the potentiometer with no loads applied? Compare to the results of part (c).

SECTION 7.6 Ammeter, Voltmeter, and Ohmmeter Design

33. A d'Arsonval movement is rated 1 mA, 100 Ω.
  a. What is the current sensitivity?
  b. Design a 20-A ammeter using the above movement. Show the circuit and component values.

34. Using a 50-μA, 1000-Ω d'Arsonval movement, design a multirange milliammeter having scales of 25, 50, and 100 mA. Show the circuit and component values.

35. A d'Arsonval movement is rated 50 μA, 1000 Ω.
  a. Design a 15-V dc voltmeter. Show the circuit and component values.
  b. What is the ohm/volt rating of the voltmeter?

36. Using a 1-mA, 100-Ω d'Arsonval movement, design a multirange voltmeter having scales of 5, 50, and 500 V. Show the circuit and component values.

37. A digital meter has an internal resistance of 10 MΩ on its 0.5-V range. If you had to build a voltmeter with a d'Arsonval movement, what current sensitivity would you need if the meter were to have the same internal resistance on the same voltage scale?

***38.** **a.** Design a series ohmmeter using a 100-μA, 1000-Ω movement; a zero-adjust with a maximum value of 2 kΩ; a battery of 3 V; and a series resistor whose value is to be determined.
  b. Find the resistance required for full-scale, 3/4-scale, 1/2-scale, and 1/4-scale deflection.
  c. Using the results of part (b), draw the scale to be used with the ohmmeter.

39. Describe the basic construction and operation of the megohmmeter.

***40.** Determine the reading of the ohmmeter for the configuration of Fig. 7.93.

SECTION 7.8 Computer Analysis

PSpice (DOS)

41. Write the input file for the network of Fig. 7.8 and request I_1, I_3, and I_6.

42. Write the input file for the network of Fig. 7.18 and print out V_1, V_2, and I.

PSpice (Windows)

43. Using schematics, determine V_1, V_3, V_{ab}, and I_s for the network of Fig. 7.16.

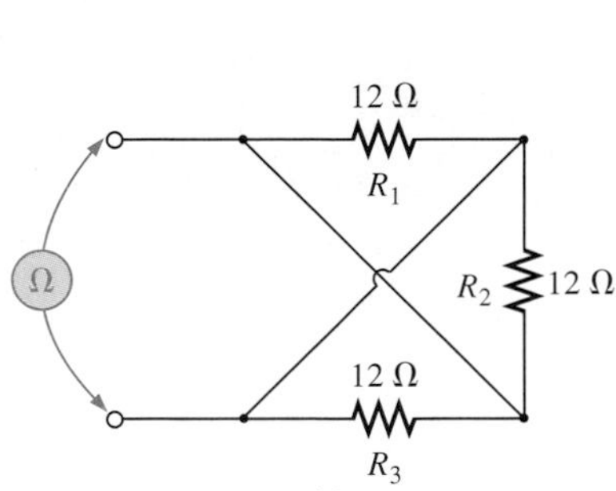

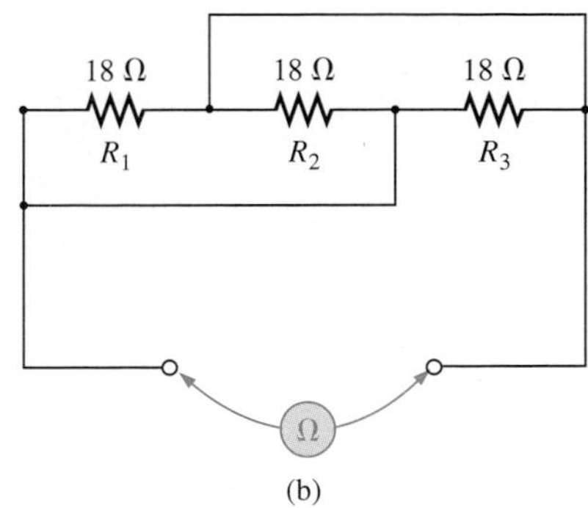

FIG. 7.93
Problem 40.

44. Using schematics, determine I_s, I_5, and V_7 for the network of Fig. 7.22.

Programming Language (C++, BASIC, PASCAL, etc.)

45. Write a program that will find the complete solution for the network of Fig. 7.6. That is, given all the parameters of the network, calculate the current, voltage, and power to each element.

46. Write a program to find all the quantities of Example 7.8 given the network parameters.

GLOSSARY

d'Arsonval movement An iron-core coil mounted on bearings between a permanent magnet. A pointer connected to the movable core indicates the strength of the current passing through the coil.

Ladder network A network that consists of a cascaded set of series-parallel combinations and has the appearance of a ladder.

Megohmmeter An instrument for measuring very high resistance levels, such as in the megohm range.

Series ohmmeter A resistance-measuring instrument in which the movement is placed in series with the unknown resistance.

Series-parallel network A network consisting of a combination of both series and parallel branches.

Transistor A three-terminal semiconductor electronic device that can be used for amplification and switching purposes.

8

Methods of Analysis and Selected Topics (dc)

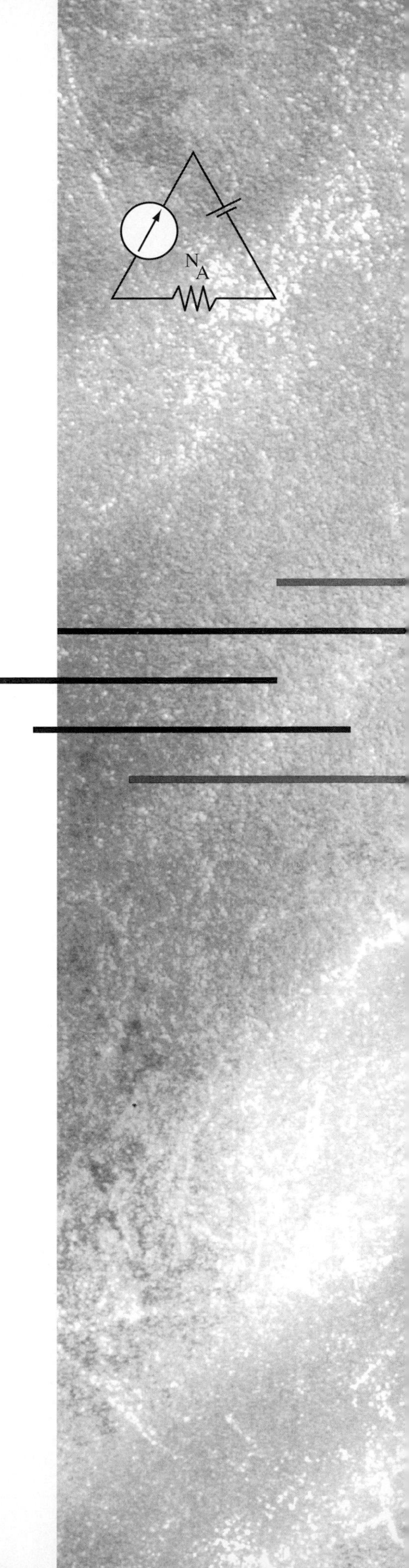

8.1 INTRODUCTION

The circuits described in the previous chapters had only one source or two or more sources in series or parallel present. The step-by-step procedure outlined in those chapters cannot be applied if the sources are not in series or parallel. There will be an interaction of sources that will not permit the reduction technique used in Chapter 7 to find quantities such as the total resistance and source current.

Methods of analysis have been developed that allow us to approach, in a systematic manner, a network with any number of sources in any arrangement. Fortunately, these methods can also be applied to networks with only one source. The methods to be discussed in detail in this chapter include *branch-current analysis, mesh analysis,* and *nodal analysis.* Each can be applied to the same network. The "best" method cannot be defined by a set of rules but can be determined only by acquiring a firm understanding of the relative advantages of each. All the methods can be applied to *linear bilateral* networks. The term *linear* indicates that the characteristics of the network elements (such as the resistors) are independent of the voltage across or current through them. The second term, *bilateral,* refers to the fact that there is no change in the behavior or characteristics of an element if the current through or voltage across the element is reversed. Of the three methods listed above, the branch-current method is the only one not restricted to bilateral devices. Before discussing the methods in detail, we shall consider the current source and conversions between voltage and current sources. At the end of the chapter we shall consider bridge networks and Δ-Y and Y-Δ conversions. Chapter 9 will present the important theorems of network analysis that can also be employed to solve networks with more than one source.

8.2 CURRENT SOURCES

The concept of the current source was introduced in Section 2.4 with the photograph of a commercially available unit. We must now investi-

gate its characteristics in greater detail so that we can properly determine its effect on the networks to be examined in this chapter.

The current source is often referred to as the *dual* of the voltage source. A battery supplies a *fixed* voltage and the source current can vary, but the current source supplies a *fixed* current to the branch in which it is located, while its terminal voltage may vary as determined by the network to which it is applied. Note from the above that *duality* simply implies an interchange of current and voltage to distinguish the characteristics of one source from the other.

The interest in the current source is due primarily to semiconductor devices such as the transistor. In the basic electronics courses, you will find that the transistor is a current-controlled device. In the physical model (equivalent circuit) of a transistor used in the analysis of transistor networks, there appears a current source as indicated in Fig. 8.1. The symbol for a current source appears in Fig. 8.1. The direction of the arrow within the circle indicates the direction in which current is being supplied.

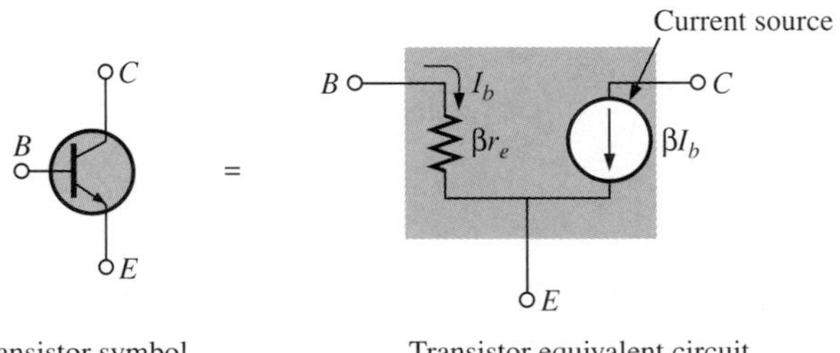

FIG. 8.1
Current source within the transistor equivalent circuit.

For further comparison, the terminal characteristics of an *ideal dc* voltage and current source are presented in Fig. 8.2, *ideal* implying perfect sources, or no internal losses sensitive to the demand from the applied load. Note that for the voltage source, the terminal voltage is fixed at E volts independent of the direction of the current I. The direction and magnitude of I will be determined by the network to which the supply is connected.

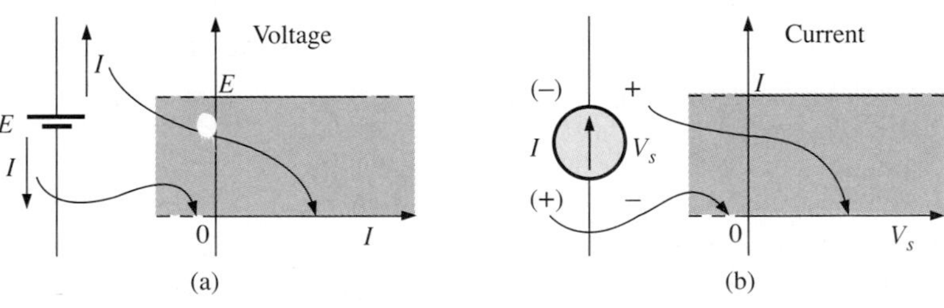

FIG. 8.2
Comparing the characteristics of an ideal voltage and current source.

The characteristics of the ideal source, shown in Fig. 8.2(b), reveal that the magnitude of the supply current is independent of the polarity of the voltage across the source. The polarity and magnitude of the source voltage V_s will be determined by the network to which the source is connected.

For all one-voltage-source networks the current will have the direction indicated to the right of the battery in Fig. 8.2(a). For all single-

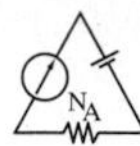

current-source networks, it will have the polarity indicated to the right of the current source in Fig. 8.2(b).

In review:

A current source determines the current in the branch in which it is located

and

the magnitude and polarity of the voltage across a current source are a function of the network to which it is applied.

EXAMPLE 8.1 Find the source voltage V_s and the current I_1 for the circuit of Fig. 8.3.

Solution:

$$I_1 = I = \mathbf{10\ mA}$$
$$V_s = V_1 = I_1R_1 = (10\ \text{mA})(20\ \text{k}\Omega) = \mathbf{200\ V}$$

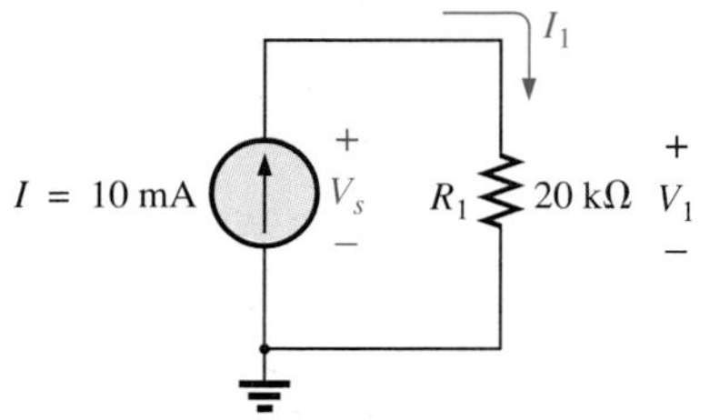

FIG. 8.3
Example 8.1.

EXAMPLE 8.2 Find the voltage V_s and the currents I_1 and I_2 for the network of Fig. 8.4.

Solution:

$$V_s = E = \mathbf{12\ V}$$
$$I_2 = \frac{V_R}{R} = \frac{E}{R} = \frac{12\ \text{V}}{4\ \Omega} = \mathbf{3\ A}$$

Applying Kirchhoff's current law:

$$I = I_1 + I_2$$

and
$$I_1 = I - I_2 = 7\ \text{A} - 3\ \text{A} = \mathbf{4\ A}$$

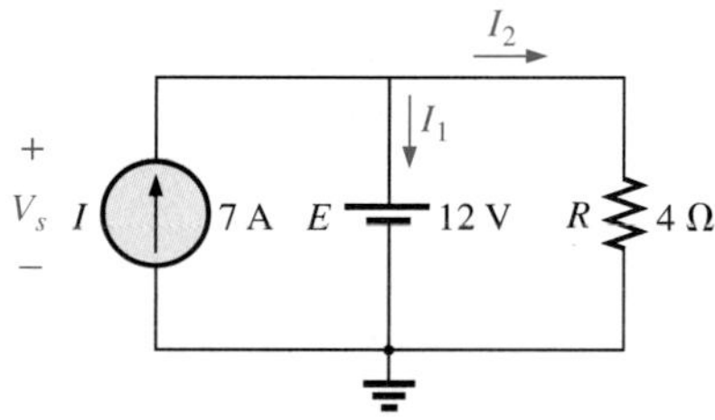

FIG. 8.4
Example 8.2.

EXAMPLE 8.3 Determine the current I_1 and the voltage V_s for the network of Fig. 8.5.

Solution: Using the current divider rule:

$$I_1 = \frac{R_2I}{R_2 + R_1} = \frac{(1\ \Omega)(6\ \text{A})}{1\ \Omega + 2\ \Omega} = \mathbf{2\ A}$$

The voltage V_1 is

$$V_1 = I_1R_1 = (2\ \text{A})(2\ \Omega) = 4\ \text{V}$$

and, applying Kirchhoff's voltage law,

$$+V_s - V_1 - 20\ \text{V} = 0$$

and
$$V_s = V_1 + 20\ \text{V} = 4\ \text{V} + 20\ \text{V} = \mathbf{24\ V}$$

Note the polarity of V_s as determined by the multisource network.

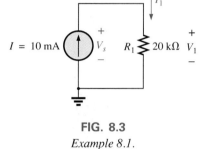

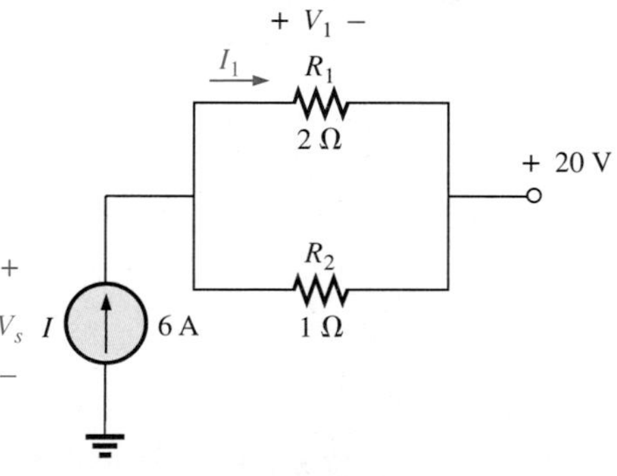

FIG. 8.5
Example 8.3.

8.3 SOURCE CONVERSIONS

The current source described in the previous section is called an *ideal source* due to the absence of any internal resistance. In reality, all

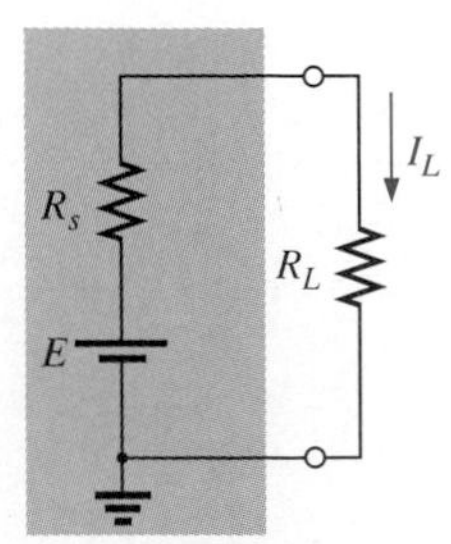

FIG. 8.6
Practical voltage source.

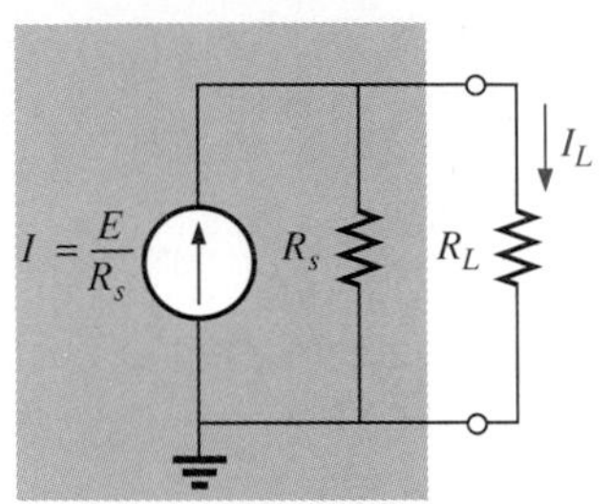

FIG. 8.7
Practical current source.

sources—whether they be voltage or current—have some internal resistance in the relative positions shown in Figs. 8.6 and 8.7. For the voltage source, if $R_s = 0\ \Omega$ or is so small compared to any series resistor that it can be ignored, then we have an "ideal" voltage source. For the current source, if $R_s = \infty\ \Omega$ or is large enough compared to other parallel elements that it can be ignored, then we have an "ideal" current source.

If the internal resistance is included with either source, then that source can be converted to the other type using the procedure to be described in this section. Since it is often advantageous to make such a maneuver, this entire section is devoted to being sure the steps are understood. It is important to realize, however, as we proceed through this section that

source conversions are equivalent only at their external terminals.

The internal characteristics of each are quite different.

We want the equivalence to ensure that the applied load of Figs. 8.6 and 8.7 will receive the same current, voltage, and power from each source and in effect not know, or care, which source is present.

In Fig. 8.6 if we solve for the load current I_L, we obtain

$$I_L = \frac{E}{R_s + R_L} \tag{8.1}$$

If we multiply this by a factor of 1, which we can choose to be R_s/R_s, we obtain

$$I_L = \frac{(1)E}{R_s + R_L} = \frac{(R_s/R_s)E}{R_s + R_L} = \frac{R_s(E/R_s)}{R_s + R_L} = \frac{R_sI}{R_s + R_L} \tag{8.2}$$

If we define $I = E/R_s$, Eq. (8.2) is the same as that obtained by applying the current divider rule to the network of Fig. 8.7. The result is an equivalence between the networks of Fig. 8.6 and Fig. 8.7 that simply requires that $I = E/R_s$ and the series resistor R_s of Fig. 8.6 be placed in parallel, as in Fig. 8.7. The validity of this is demonstrated in the first example of this section.

For clarity, the equivalent sources, *as far as terminals* a *and* b *are concerned,* are repeated in Fig. 8.8 with the equations for converting in either direction. Note, as just indicated, that the resistor R_s is the same in each source; only its position changes. The current of the current source or the voltage of the voltage source is determined using Ohm's law and the parameters of the other configuration. It was pointed out in some detail in Chapter 6 that every source of voltage has some internal series resistance. *For the current source, some internal parallel resistance will always exist in the practical world.* However, in many cases, it is an

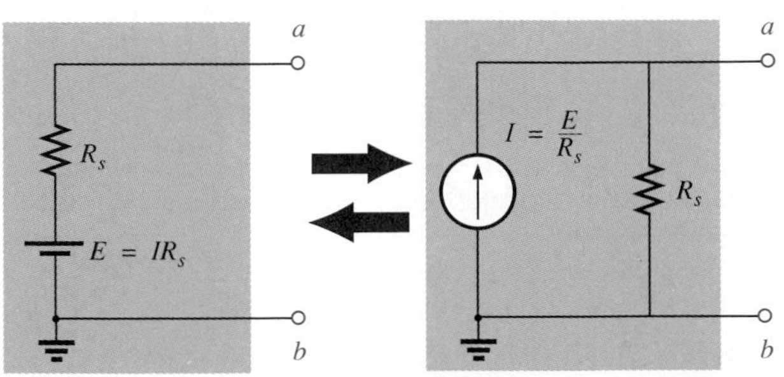

FIG. 8.8
Source conversion.

excellent approximation to drop the internal resistance of a source due to the magnitude of the elements of the network to which it is applied. For this reason, in the analyses to follow, voltage sources may appear without a series resistor, and current sources may appear without a parallel resistance. Realize, however, that to perform a conversion from one type of source to another, a voltage source must have a resistor in series with it, and a current source must have a resistor in parallel.

EXAMPLE 8.4

a. Convert the voltage source of Fig. 8.9(a) to a current source and calculate the current through the 4-Ω load for each source.
b. Replace the 4-Ω load with a 1-kΩ load and calculate the current I_L for the voltage source.
c. Repeat the calculation of part (b) assuming the voltage source is ideal ($R_s = 0\ \Omega$) because R_L is so much larger than R_s. Is this one of those situations where assuming the source is ideal is an appropriate approximation?

Solutions:

a. See Fig. 8.9.

$$\text{Fig. 8.9(a):}\quad I_L = \frac{E}{R_s + R_L} = \frac{6\text{ V}}{2\ \Omega + 4\ \Omega} = \mathbf{1\ A}$$

$$\text{Fig. 8.9(b):}\quad I_L = \frac{R_s I}{R_s + R_L} = \frac{(2\ \Omega)(3\text{ A})}{2\ \Omega + 4\ \Omega} = \mathbf{1\ A}$$

b. $$I_L = \frac{E}{R_s + R_L} = \frac{6\text{ V}}{2\ \Omega + 1\text{ k}\Omega} \cong \mathbf{5.99\ mA}$$

c. $$I_L = \frac{E}{R_L} = \frac{6\text{ V}}{1\text{ k}\Omega} = \mathbf{6\ mA} \cong 5.99\text{ mA}$$

Yes, $R_L \gg R_s$ (voltage source).

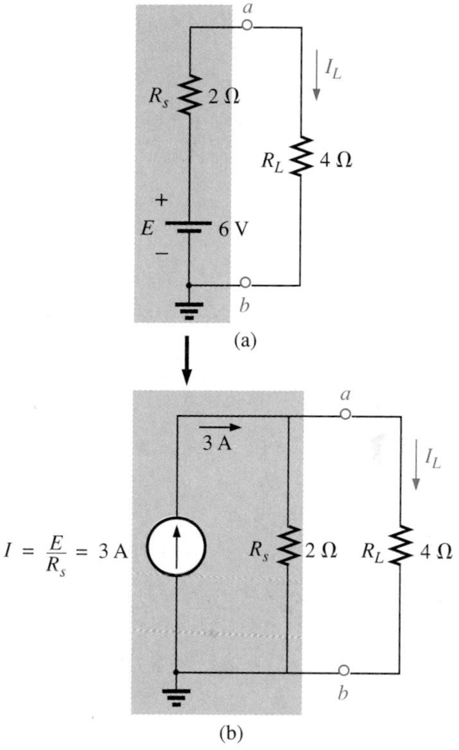

FIG. 8.9
Example 8.4.

EXAMPLE 8.5

a. Convert the current source of Fig. 8.10(a) to a voltage source and find the load current for each source.
b. Replace the 6-kΩ load with a 10-Ω load and calculate the current I_L for the current source.
c. Repeat the calculation of part (b) assuming the current source is ideal ($R_s = \infty\ \Omega$) because R_L is so much smaller than R_s. Is this one of those situations where assuming the source is ideal is an appropriate approximation?

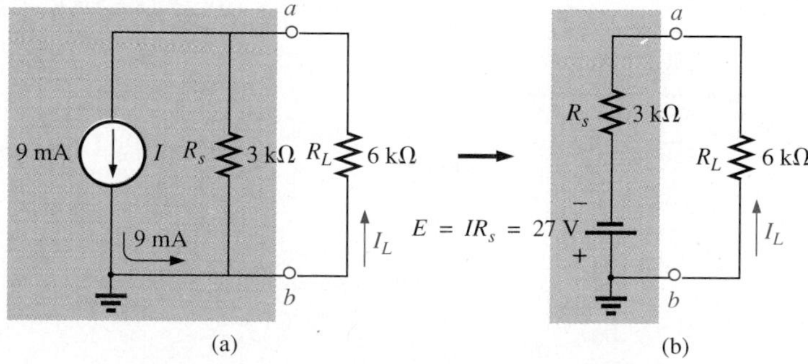

FIG. 8.10
Example 8.5.

Solutions:

a. See Fig. 8.10.

$$\text{Fig. 8.10(a):}\quad I_L = \frac{R_sI}{R_s + R_L} = \frac{(3\text{ k}\Omega)(9\text{ mA})}{3\text{ k}\Omega + 6\text{ k}\Omega} = \mathbf{3\text{ mA}}$$

$$\text{Fig. 8.10(b):}\quad I_L = \frac{E}{R_s + R_L} = \frac{27\text{ V}}{3\text{ k}\Omega + 6\text{ k}\Omega} = \frac{27\text{ V}}{9\text{ k}\Omega} = \mathbf{3\text{ mA}}$$

b. $I_L = \dfrac{R_sI}{R_s + R_L} = \dfrac{(3\text{ k}\Omega)(9\text{ mA})}{3\text{ k}\Omega + 10\ \Omega} = \mathbf{8.97\text{ mA}}$

c. $I_L = I = \mathbf{9\text{ mA}} \cong 8.97\text{ mA}$

Yes, $R_s \gg R_L$ (current source).

8.4 CURRENT SOURCES IN PARALLEL

If two or more current sources are in parallel, they may all be replaced by one current source having the magnitude and direction of the resultant, which can be found by summing the currents in one direction and subtracting the sum of the currents in the opposite direction. The new parallel resistance is determined by methods described in the discussion of parallel resistors in Chapter 5. Consider the following examples.

EXAMPLE 8.6 Reduce the parallel current sources of Figs. 8.11 and 8.12 to a single current source.

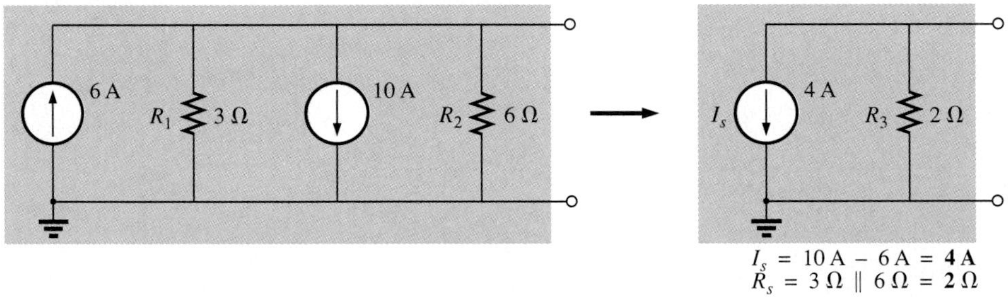

FIG. 8.11
Example 8.6.

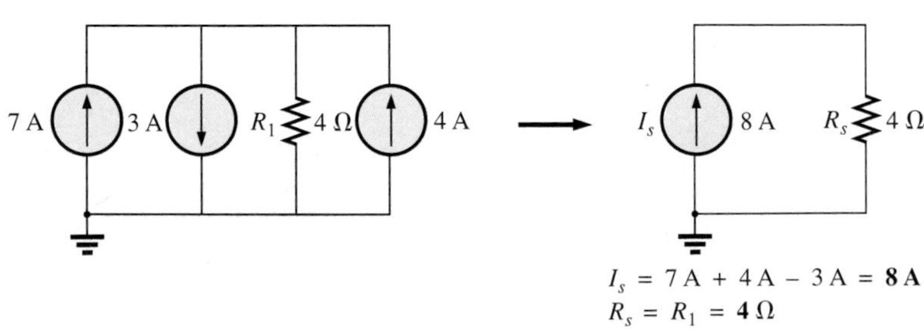

FIG. 8.12
Example 8.6.

Solution: Note the solution in each figure.

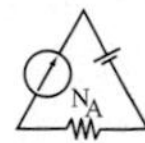

EXAMPLE 8.7 Reduce the network of Fig. 8.13 to a single current source and calculate the current through R_L.

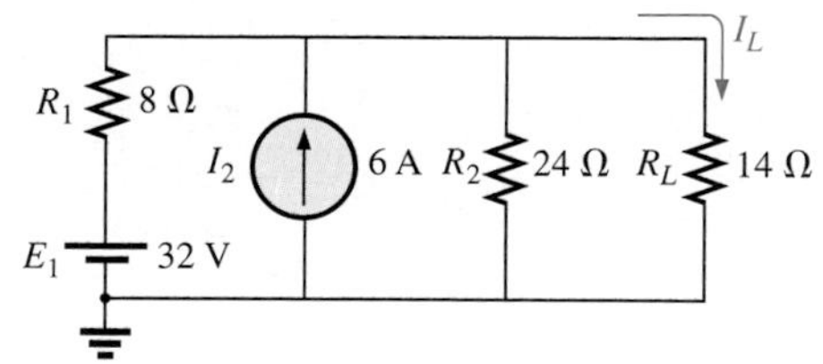

FIG. 8.13
Example 8.7.

Solution: In this example, the voltage source will first be converted to a current source as shown in Fig. 8.14. Combining current sources,

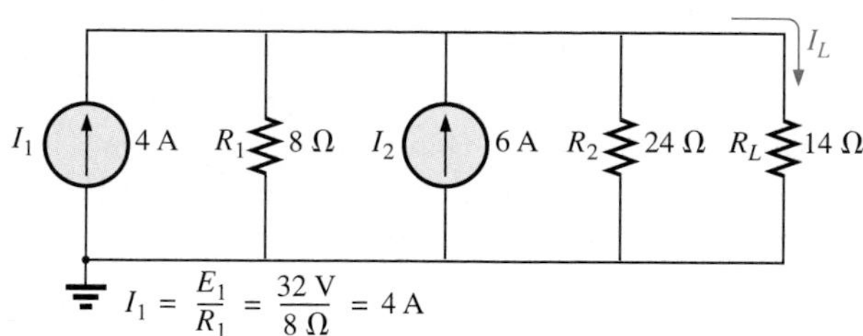

FIG. 8.14
Network of Fig. 8.13 following the conversion of the voltage source to a current source.

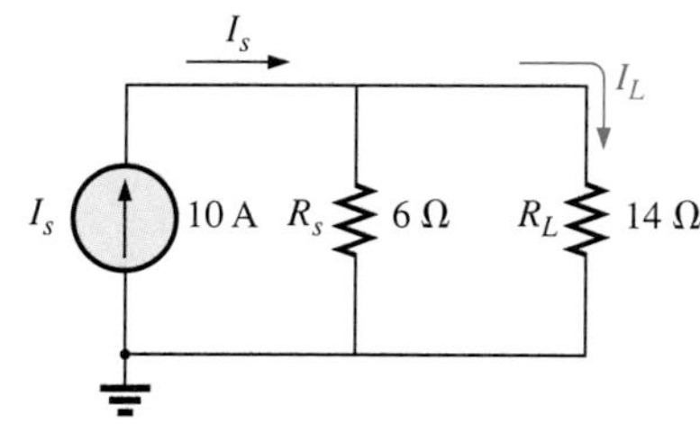

FIG. 8.15
Network of Fig. 8.14 reduced to its simplest form.

$$I_s = I_1 + I_2 = 4\text{ A} + 6\text{ A} = \mathbf{10\ A}$$

and
$$R_s = R_1 \parallel R_2 = 8\ \Omega \parallel 24\ \Omega = \mathbf{6\ \Omega}$$

Applying the current divider rule to the resulting network of Fig. 8.15,

$$I_L = \frac{R_s I_s}{R_s + R_L} = \frac{(6\ \Omega)(10\text{ A})}{6\ \Omega + 14\ \Omega} = \frac{60\text{ A}}{20} = \mathbf{3\ A}$$

EXAMPLE 8.8 Determine the current I_2 in the network of Fig. 8.16.

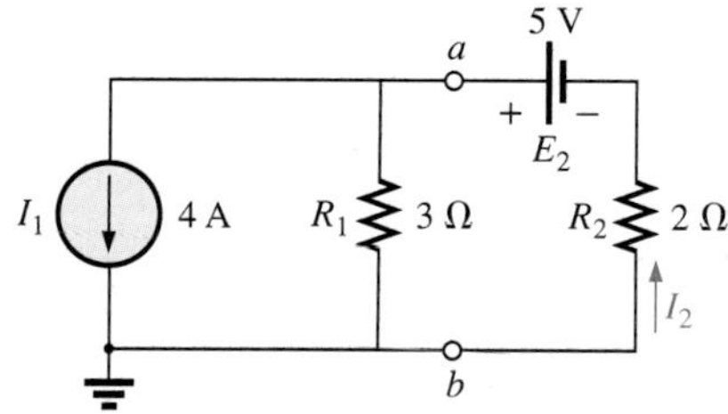

FIG. 8.16
Example 8.8.

Solution: Although it might appear that the network cannot be solved using methods introduced thus far, one source conversion as shown in Fig. 8.17 will result in a simple series circuit:

$$E_s = I_1 R_1 = (4\text{ A})(3\ \Omega) = 12\text{ V}$$

and
$$R_s = R_1 = 3\ \Omega$$

and
$$I_2 = \frac{E_s + E_2}{R_s + R_2} = \frac{12\text{ V} + 5\text{ V}}{3\ \Omega + 2\ \Omega} = \frac{17\text{ V}}{5\ \Omega} = \mathbf{3.4\ A}$$

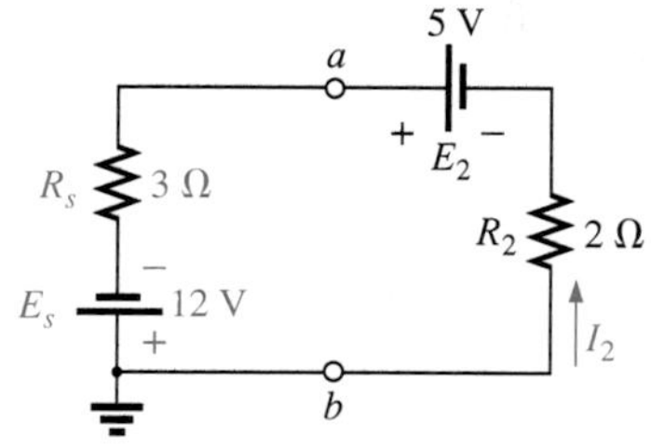

FIG. 8.17
Network of Fig. 8.16 following the conversion of the current source to a voltage source.

8.5 CURRENT SOURCES IN SERIES

The current through any branch of a network can be only single-valued. For the situation indicated at point *a* in Fig. 8.18, we find by application of Kirchhoff's current law that the current leaving that point is greater than that entering—an impossible situation. Therefore,

current sources of different current ratings are not connected in series,

just as voltage sources of different voltage ratings are not connected in parallel.

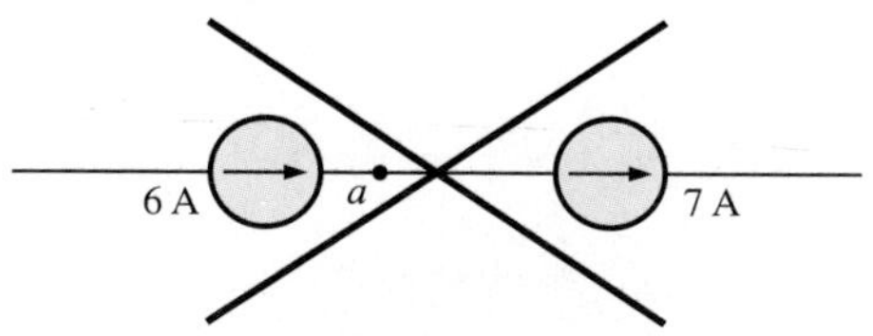

FIG. 8.18
Invalid situation.

8.6 BRANCH-CURRENT ANALYSIS

We will now consider the first in a series of methods for solving networks with two or more sources. Once this method is mastered, there is

no linear dc network for which a solution cannot be found. Keep in mind that networks with two isolated voltage sources cannot be solved using the approach of Chapter 7. For additional evidence of this fact, try solving for the unknown elements of Example 8.9 using the methods introduced in Chapter 7. The network of Fig. 8.21 can be solved using the source conversions described in the last section, but the method to be described in this section has applications far beyond the configuration of this network. The most direct introduction to a method of this type is to list the series of steps required for its application. There are four steps, as indicated below. Before continuing, understand that this method will produce the current through each branch of the network, the *branch current*. Once this is known, all other quantities, such as voltage or power, can be determined.

1. ***Assign a distinct current of arbitrary direction to each branch of the network.***
2. ***Indicate the polarities for each resistor as determined by the assumed current direction.***
3. ***Apply Kirchhoff's voltage law around each closed, independent loop of the network.***

The best way to determine how many times Kirchhoff's voltage law will have to be applied is to determine the number of "windows" in the network. The network of Example 8.9 has a definite similarity to the two-window configuration of Fig. 8.19(a). The result is a need to apply Kirchhoff's voltage law twice. For networks with three windows, as shown in Fig. 8.19(b), three applications of Kirchhoff's voltage law are required, and so on.

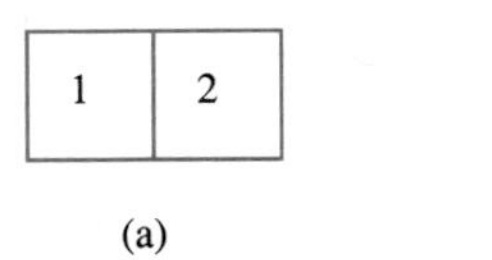

(a)

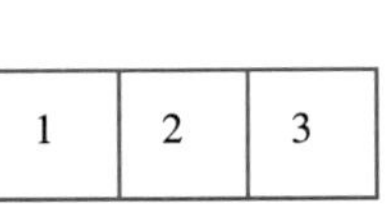

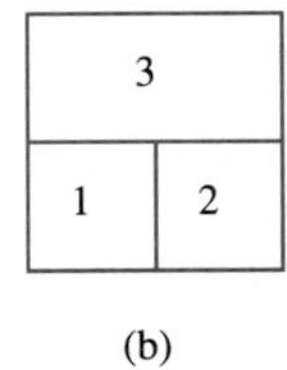

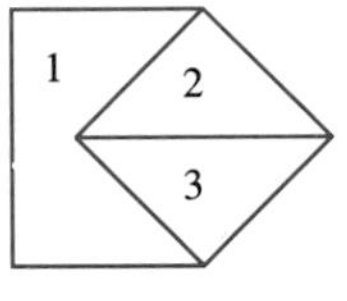

(b)

FIG. 8.19
Determining the number of independent closed loops.

4. ***Apply Kirchhoff's current law at the minimum number of nodes that will include all the branch currents of the network.***

The minimum number is one less than the number of independent nodes of the network. For the purposes of this analysis, a node is a junction of two or more branches, where a branch is any combination

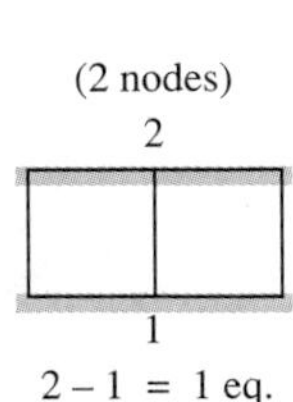

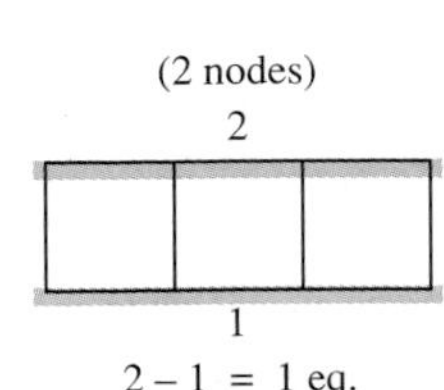

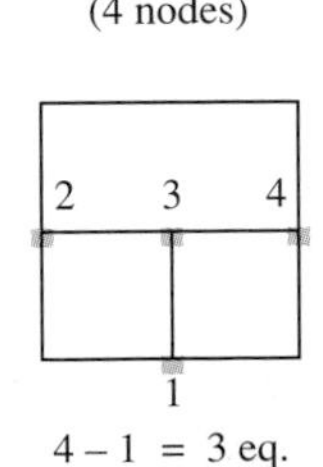

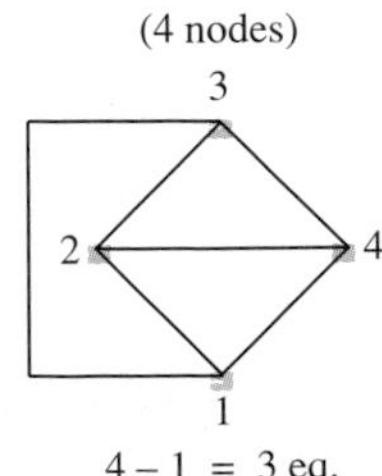

FIG. 8.20
Determining the number of applications of Kirchhoff's current law required.

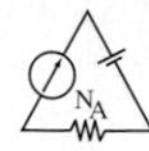

of series elements. Figure 8.20 defines the number of applications of Kirchhoff's current law for each configuration of Fig. 8.19.

5. ***Solve the resulting simultaneous linear equations for assumed branch currents.***

It is assumed that the use of determinants to solve for the currents I_1, I_2, and I_3 is understood and is a part of the student's mathematical background. If not, a detailed explanation of the procedure is provided in Appendix C. Calculators and computer software packages such as MathCad can find the solutions quickly and accurately.

EXAMPLE 8.9 Apply the branch-current method to the network of Fig. 8.21.

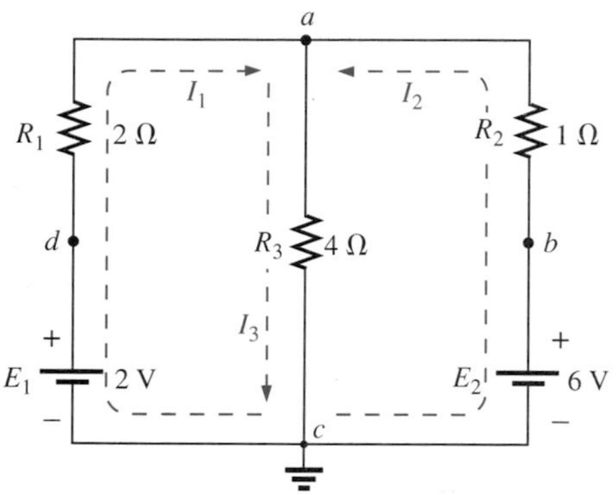

FIG. 8.21
Example 8.9.

Solution:

Step 1: Since there are three distinct branches (*cda, cba, ca*), three currents of arbitrary directions (I_1, I_2, I_3) are chosen, as indicated in Fig. 8.21. The current directions for I_1 and I_2 were chosen to match the "pressure" applied by sources E_1 and E_2, respectively. Since both I_1 and I_2 enter node a, I_3 is leaving.

Step 2: Polarities for each resistor are drawn to agree with assumed current directions, as indicated in Fig. 8.22.

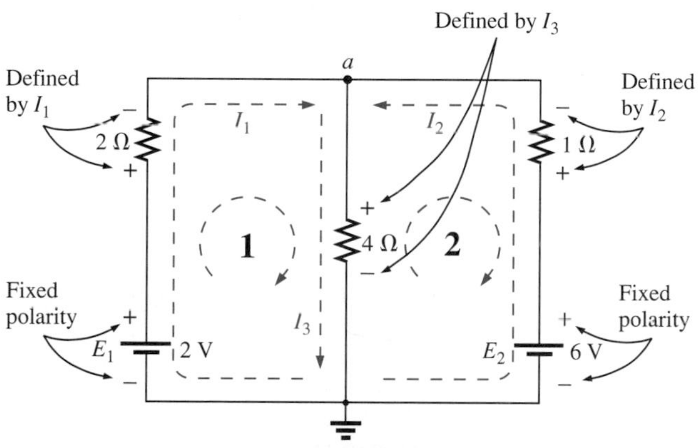

FIG. 8.22
Inserting the polarities across the resistive elements as defined by the chosen branch currents.

Step 3: Kirchhoff's voltage law is applied around each closed loop (1 and 2) in the clockwise direction:

$$\text{loop 1: } \Sigma_{\circlearrowright} V = \underbrace{+E_1}_{\text{Rise in potential}} - \underbrace{V_{R_1} - V_{R_3}}_{\text{Drop in potential}} = 0$$

$$\text{loop 2: } \Sigma_{\circlearrowright} V = \underbrace{+V_{R_3} + V_{R_2}}_{\text{Rise in potential}} \underbrace{- E_2}_{\text{Drop in potential}} = 0$$

and

$$\text{loop 1: } \Sigma_{\circlearrowright} V = \underbrace{+2\text{ V}}_{\substack{\text{Battery}\\ \text{potential}}} - \underbrace{(2\ \Omega)I_1}_{\substack{\text{Voltage drop}\\ \text{across 2-}\Omega\\ \text{resistor}}} - \underbrace{(4\ \Omega)I_3}_{\substack{\text{Voltage drop}\\ \text{across 4-}\Omega\\ \text{resistor}}} = 0$$

$$\text{loop 2: } \Sigma_{\circlearrowright} V = (4\ \Omega)I_3 + (1\ \Omega)I_2 - 6\text{ V} = 0$$

Step 4: Applying Kirchhoff's current law at node a (in a two-node network, the law is applied at only one node),

$$I_1 + I_2 = I_3$$

Step 5: There are three equations and three unknowns (units removed for clarity):

$$\begin{aligned} 2 - 2I_1 - 4I_3 &= 0 \\ 4I_3 + 1I_2 - 6 &= 0 \\ I_1 + I_2 &= I_3 \end{aligned} \qquad \text{Rewritten:} \quad \begin{aligned} 2I_1 + 0 + 4I_3 &= 2 \\ 0 + I_2 + 4I_3 &= 6 \\ I_1 + I_2 - I_3 &= 0 \end{aligned}$$

Using third-order determinants (Appendix C), we have

$$I_1 = \frac{\begin{vmatrix} 2 & 0 & 4 \\ 6 & 1 & 4 \\ 0 & 1 & -1 \end{vmatrix}}{D = \begin{vmatrix} 2 & 0 & 4 \\ 0 & 1 & 4 \\ 1 & 1 & -1 \end{vmatrix}} = -1\text{ A}$$

A negative sign in front of a branch current indicates only that the actual current is in the direction opposite to that assumed.

$$I_2 = \frac{\begin{vmatrix} 2 & 2 & 4 \\ 0 & 6 & 4 \\ 1 & 0 & -1 \end{vmatrix}}{D} = 2\text{ A}$$

$$I_3 = \frac{\begin{vmatrix} 2 & 0 & 2 \\ 0 & 1 & 6 \\ 1 & 1 & 0 \end{vmatrix}}{D} = 1\text{ A}$$

Using the student edition of the computer software package MathCad 2.5, the solution for I_1 would appear as follows:

$$A := \begin{bmatrix} 2 & 0 & 4 \\ 6 & 1 & 4 \\ 0 & 1 & -1 \end{bmatrix} \qquad B := \begin{bmatrix} 2 & 0 & 4 \\ 0 & 1 & 4 \\ 1 & 1 & -1 \end{bmatrix} \qquad I1 := \frac{|A|}{|B|} \qquad I1 = -1$$

MATHCAD 8.1

The format obviously compares directly with the standard determinant, except that each determinant must be separately defined and the desired unknown calculated as indicated above. The space involved and the time to enter the determinants are obviously quite minimal when you consider how quickly the results will appear and how much accuracy is obtained.

Instead of using third-order determinants, we could reduce the three equations to two by substituting the third equation in the first and second equations:

$$2 - 2I_1 - 4\overbrace{(I_1 + I_2)}^{I_3} = 0 \qquad 2 - 2I_1 - 4I_1 - 4I_2 = 0$$

$$4\overbrace{(I_1 + I_2)}^{I_3} + I_2 - 6 = 0 \qquad 4I_1 + 4I_2 + I_2 - 6 = 0$$

or

$$\begin{aligned} -6I_1 - 4I_2 &= -2 \\ +4I_1 + 5I_2 &= +6 \end{aligned}$$

Multiplying through by −1 in the top equation yields

$$\begin{aligned} 6I_1 + 4I_2 &= +2 \\ 4I_1 + 5I_2 &= +6 \end{aligned}$$

and using determinants,

$$I_1 = \frac{\begin{vmatrix} 2 & 4 \\ 6 & 5 \end{vmatrix}}{\begin{vmatrix} 6 & 4 \\ 4 & 5 \end{vmatrix}} = \frac{10 - 24}{30 - 16} = \frac{-14}{14} = \mathbf{-1\,A}$$

Using the TI-85 calculator:

det[[2,4][6,5]]/det[[6,4][4,5]] (ENTER)	−1

CALC. 8.1

Note the det (determinant) obtained from a Math listing under a MATRX menu and the fact that each determinant must be determined individually. The first set of brackets within the overall determinant brackets of the first determinant defines the first row of the determinant, while the second set of brackets within the same determinant defines the second row. A comma separates the entries for each row. Obviously, the time to learn how to enter the parameters is minimal when you consider the savings in time and the accuracy obtained.

$$I_2 = \frac{\begin{vmatrix} 6 & 2 \\ 4 & 6 \end{vmatrix}}{14} = \frac{36 - 8}{14} = \frac{28}{14} = \mathbf{2\,A}$$

$$I_3 = I_1 + I_2 = -1 + 2 - \mathbf{1\,A}$$

It is now important that the impact of the results obtained be understood. The currents I_1, I_2, and I_3 are the actual currents in the branches in which they were defined. A negative sign in the solution simply reveals that the actual current has the opposite direction than initially defined—the magnitude is correct. Once the actual current directions and their magnitudes are inserted in the original network, the various voltages and power levels can be determined. For Example 8.9, the actual current directions and their magnitudes have been entered on the original network in Fig. 8.23. Note that the current through the series

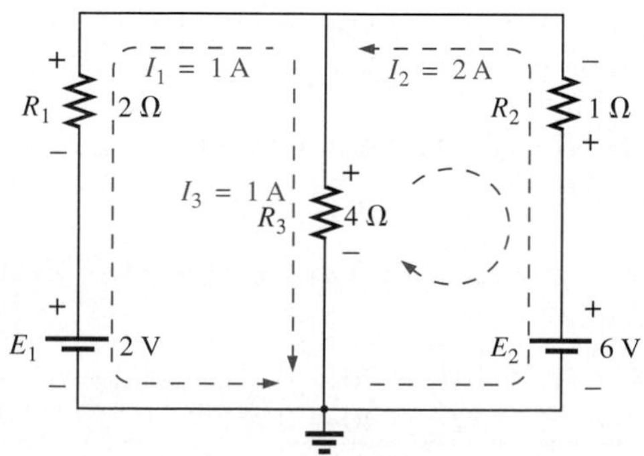

FIG. 8.23
Reviewing the results of the analysis of the network of Fig. 8.21.

elements R_1 and E_1 is 1 A; the current through R_3, 1 A; and the current through the series elements R_2 and E_2, 2 A. Due to the minus sign in the solution, the direction of I_1 is opposite to that shown in Fig. 8.21. The voltage across any resistor can now be found using Ohm's law, and the power delivered by either source or to any one of the three resistors can be found using the appropriate power equation.

Applying Kirchhoff's voltage law around the loop indicated in Fig. 8.23,

$$\Sigma_{\circlearrowright} V = +(4\ \Omega)I_3 + (1\ \Omega)I_2 - 6\text{ V} = 0$$

or

$$(4\ \Omega)I_3 + (1\ \Omega)I_2 = 6\text{ V}$$

and

$$(4\ \Omega)(1\text{ A}) + (1\ \Omega)(2\text{ A}) = 6\text{ V}$$
$$4\text{ V} + 2\text{ V} = 6\text{ V}$$
$$6\text{ V} = 6\text{ V} \qquad \text{(checks)}$$

EXAMPLE 8.10 Apply branch-current analysis to the network of Fig. 8.24.

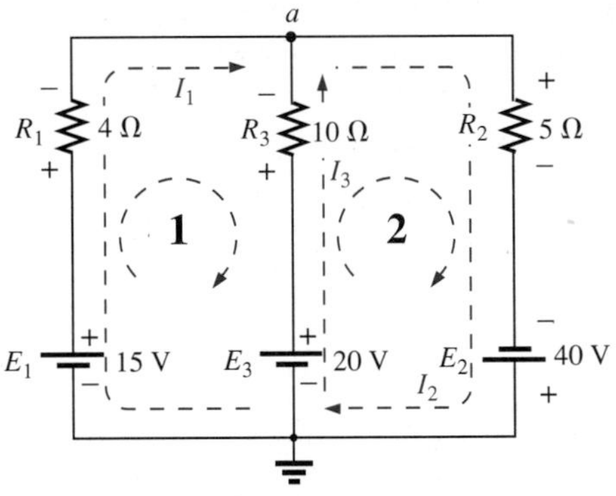

FIG. 8.24
Example 8.10.

Solution: Again, the current directions were chosen to match the "pressure" of each battery. The polarities are then added and Kirchhoff's voltage law is applied around each closed loop in the clockwise direction. The result is as follows:

$$\text{loop 1:}\quad +15\text{ V} - (4\ \Omega)I_1 + (10\ \Omega)I_3 - 20\text{ V} = 0$$
$$\text{loop 2:}\quad +20\text{ V} - (10\ \Omega)I_3 - (5\ \Omega)I_2 + 40\text{ V} = 0$$

Applying Kirchhoff's current law at node a,

$$I_1 + I_3 = I_2$$

Substituting the third equation into the other two yields (with units removed for clarity):

$$\left.\begin{aligned} 15 - 4I_1 + 10I_3 - 20 &= 0 \\ 20 - 10I_3 - 5(I_1 + I_3) + 40 &= 0 \end{aligned}\right\} \text{Substituting for } I_2 \text{ (since it occurs only once in the two equations)}$$

or

$$-4I_1 + 10I_3 = 5$$
$$-5I_1 - 15I_3 = -60$$

Multiplying the lower equation by -1, we have

$$\begin{aligned} -4I_1 + 10I_3 &= 5 \\ 5I_1 + 15I_3 &= 60 \end{aligned}$$

$$I_1 = \frac{\begin{vmatrix} 5 & 10 \\ 60 & 15 \end{vmatrix}}{\begin{vmatrix} -4 & 10 \\ 5 & 15 \end{vmatrix}} = \frac{75 - 600}{-60 - 50} = \frac{-525}{-110} = \mathbf{4.773\ A}$$

$$I_3 = \frac{\begin{vmatrix} -4 & 5 \\ 5 & 60 \end{vmatrix}}{-110} = \frac{-240 - 25}{-110} = \frac{-265}{-110} = \mathbf{2.409\ A}$$

$$I_2 = I_1 + I_3 = 4.773 + 2.409 = \mathbf{7.182\ A}$$

revealing that the assumed directions were the actual directions, with I_2 equal to the sum of I_1 and I_3.

8.7 MESH ANALYSIS (GENERAL APPROACH)

The second method of analysis to be described is called *mesh analysis.* The term *mesh* is derived from the similarities in appearance between the closed loops of a network and a wire mesh fence. Although this approach is on a more sophisticated plane than the branch-current method, it incorporates many of the ideas just developed. Of the two methods, mesh analysis is the one more frequently applied today. Branch-current analysis is introduced as a stepping stone to mesh analysis because branch currents are initially more "real" to the student than the loop currents employed in mesh analysis. Essentially, the mesh-analysis approach simply eliminates the need to substitute the results of Kirchhoff's current law into the equations derived from Kirchhoff's voltage law. It is now accomplished in the initial writing of the equations. The systematic approach outlined below should be followed when applying this method.

1. ***Assign a distinct current in the clockwise direction to each independent, closed loop of the network. It is not absolutely necessary to choose the clockwise direction for each loop current. In fact, any direction can be chosen for each loop current with no loss in accuracy, as long as the remaining steps are followed properly. However, by choosing the clockwise direction as a standard, we can develop a shorthand method (Section 8.8) for writing the required equations that will save time and possibly prevent some common errors.***

This first step is accomplished most effectively by placing a loop current *within* each "window" of the network, as demonstrated in the previous section, to ensure that they are all independent. There are a variety of other loop currents that can be assigned. In each case, however, be sure that the information carried by any one loop equation is not included in a combination of the other network equations. This is the crux of the terminology: *independent.* No matter how you choose your loop currents, the number of loop currents required is always equal to the number of windows of a planar (no-crossovers) network. On occasion a network may appear to be nonplanar. However, a

redrawing of the network may reveal that it is, in fact, planar. Such may be the case in one or two problems at the end of the chapter.

Before continuing to the next step, let us ensure that the concept of a loop current is clear. For the network of Fig. 8.25, the loop current I_1 is the branch current of the branch containing the 2-Ω resistor and 2-V battery. The current through the 4-Ω resistor is not I_1, however, since there is also a loop current I_2 through it. Since they have opposite directions, $I_{4\Omega}$ equals the difference between the two, $I_1 - I_2$ or $I_2 - I_1$, depending on which you choose to be the defining direction. In other words, *a loop current is a branch current only when it is the only loop current assigned to that branch.*

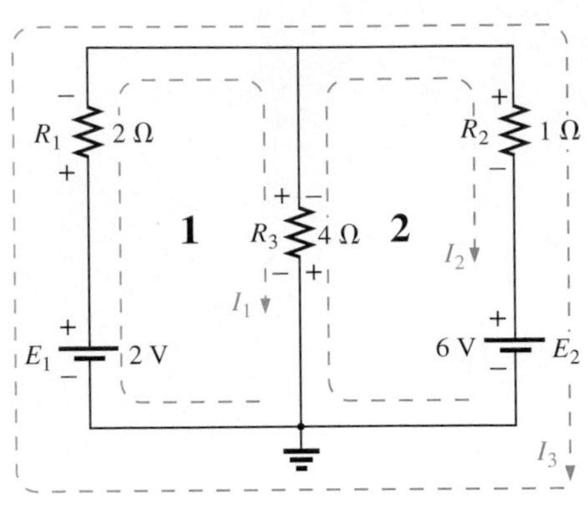

FIG. 8.25
Defining the mesh currents for a "two-window" network.

2. ***Indicate the polarities within each loop for each resistor as determined by the assumed direction of loop current for that loop. Note the requirement that the polarities be placed within each loop. This requires, as shown in Fig. 8.25, that the 4-Ω resistor have two sets of polarities across it.***
3. ***Apply Kirchhoff's voltage law around each closed loop in the clockwise direction. Again, the clockwise direction was chosen to establish uniformity and prepare us for the method to be introduced in the next section.***
   a. ***If a resistor has two or more assumed currents through it, the total current through the resistor is the assumed current of the loop in which Kirchhoff's voltage law is being applied, plus the assumed currents of the other loops passing through in the same direction, minus the assumed currents through in the opposite direction.***
   b. ***The polarity of a voltage source is unaffected by the direction of the assigned loop currents.***
4. ***Solve the resulting simultaneous linear equations for the assumed loop currents.***

EXAMPLE 8.11 Consider the same basic network as in Example 8.9 of the preceding section, now appearing in Fig. 8.25.

Solution:

Step 1: Two loop currents (I_1 and I_2) are assigned in the clockwise direction in the windows of the network. A third loop (I_3) could have been included around the entire network, but the information carried by this loop is already included in the other two.

Step 2: Polarities are drawn within each window to agree with assumed current directions. Note that for this case, the polarities across the 4-Ω resistor are the opposite for each loop current.

Step 3: Kirchhoff's voltage law is applied around each loop in the clockwise direction. Keep in mind as this step is performed that the law is concerned only with the magnitude and polarity of the voltages around the closed loop and not with whether a voltage rise or drop is due to a battery or resistive element. The voltage across each resistor is determined by $V = IR$, and for a resistor with more than one current through it, the current is the loop current of the loop being examined plus or minus the other loop currents as determined by their directions. If clockwise applications of Kirchhoff's voltage law are always chosen, the other loop currents will always be subtracted from the loop current of the loop being analyzed.

loop 1: $+E_1 - V_1 - V_3 = 0$ (clockwise starting at point a)

$$+2\text{ V} - (2\ \Omega)\,I_1 - \overbrace{(4\ \Omega)(\underbrace{I_1 - I_2}_{\substack{\text{Total current}\\ \text{through}\\ 4\text{-}\Omega\text{ resistor}}})}^{\text{Voltage drop across } 4\text{-}\Omega \text{ resistor}} = 0$$

Subtracted since I_2 is opposite in direction to I_1.

loop 2: $-V_3 - V_2 - E_2 = 0$ (clockwise starting at point b)

$$-(4\ \Omega)(I_2 - I_1) - (1\ \Omega)I_2 - 6\text{ V} = 0$$

Step 4: The equations are then rewritten as follows (without units for clarity):

$$\begin{aligned} \text{loop 1:}\quad & +2 - 2I_1 - 4I_1 + 4I_2 = 0 \\ \text{loop 2:}\quad & -4I_2 + 4I_1 - 1I_2 - 6 = 0 \end{aligned}$$

and

$$\begin{aligned} \text{loop 1:}\quad & +2 - 6I_1 + 4I_2 = 0 \\ \text{loop 2:}\quad & -5I_2 + 4I_1 - 6 = 0 \end{aligned}$$

or

$$\begin{aligned} \text{loop 1:}\quad & -6I_1 + 4I_2 = -2 \\ \text{loop 2:}\quad & +4I_1 - 5I_2 = +6 \end{aligned}$$

Applying determinants will result in

$$I_1 = \mathbf{-1\ A} \quad \text{and} \quad I_2 = \mathbf{-2\ A}$$

The minus signs indicate that the currents have a direction opposite to that indicated by the assumed loop current.

The actual current through the 2-V source and 2-Ω resistor is therefore 1 A in the other direction, and the current through the 6-V source and 1-Ω resistor is 2 A in the opposite direction indicated on the circuit. The current through the 4-Ω resistor is determined by the following equation from the original network:

$$\begin{aligned} \text{loop 1:}\quad I_{4\Omega} &= I_1 - I_2 = -1\text{ A} - (-2\text{ A}) = -1\text{ A} + 2\text{ A} \\ &= \mathbf{1\ A} \quad (\text{in the direction of } I_1) \end{aligned}$$

The outer loop (I_3) and *one* inner loop (either I_1 or I_2) would also have produced the correct results. This approach, however, will often lead to errors since the loop equations may be more difficult to write. The best method of picking these loop currents is to use the window approach.

EXAMPLE 8.12 Find the current through each branch of the network of Fig. 8.26.

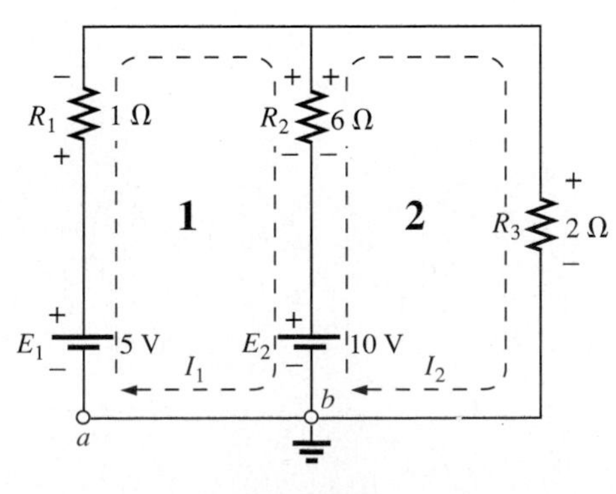

FIG. 8.26
Example 8.12.

Solution:

Steps 1 and 2 are as indicated in the circuit. Note that the polarities of the 6-Ω resistor are different for each loop current.

Step 3: Kirchhoff's voltage law is applied around each closed loop in the clockwise direction:

loop 1: $+E_1 - V_1 - V_2 - E_2 = 0$ (clockwise starting at point a)

$$+5\text{ V} - (1\ \Omega)I_1 - (6\ \Omega)(I_1 - I_2) - 10\text{ V} = 0$$

I_2 flows through the 6-Ω resistor in the direction opposite to I_1.

loop 2: $E_2 - V_2 - V_3 = 0$ (clockwise starting at point b)
$$+10\text{ V} - (6\ \Omega)(I_2 - I_1) - (2\ \Omega)I_2 = 0$$

The equations are rewritten as

$$\left.\begin{aligned} 5 - I_1 - 6I_1 + 6I_2 - 10 &= 0 \\ 10 - 6I_2 + 6I_1 - 2I_2 &= 0 \end{aligned}\right\} \begin{aligned} -7I_1 + 6I_2 &= 5 \\ +6I_1 - 8I_2 &= -10 \end{aligned}$$

$$I_1 = \frac{\begin{vmatrix} 5 & 6 \\ -10 & -8 \end{vmatrix}}{\begin{vmatrix} -7 & 6 \\ 6 & -8 \end{vmatrix}} = \frac{-40 + 60}{56 - 36} = \frac{20}{20} = \mathbf{1\ A}$$

$$I_2 = \frac{\begin{vmatrix} -7 & 5 \\ 6 & -10 \end{vmatrix}}{20} = \frac{70 - 30}{20} = \frac{40}{20} = \mathbf{2\ A}$$

Since I_1 and I_2 are positive and flow in opposite directions through the 6-Ω resistor and 10-V source, the total current in this branch is equal to the difference of the two currents in the direction of the larger:

$$I_2 > I_1 \qquad (2\text{ A} > 1\text{ A})$$

Therefore, $I_{R_2} = I_2 - I_1 = 2\text{ A} - 1\text{ A} = \mathbf{1\ A}$ in the direction of I_2.

It is sometimes impractical to draw all the branches of a circuit at right angles to one another. The next example demonstrates how a portion of a network may appear due to various constraints. The method of analysis does not change with this change in configuration.

EXAMPLE 8.13 Find the branch currents of the network of Fig. 8.27.

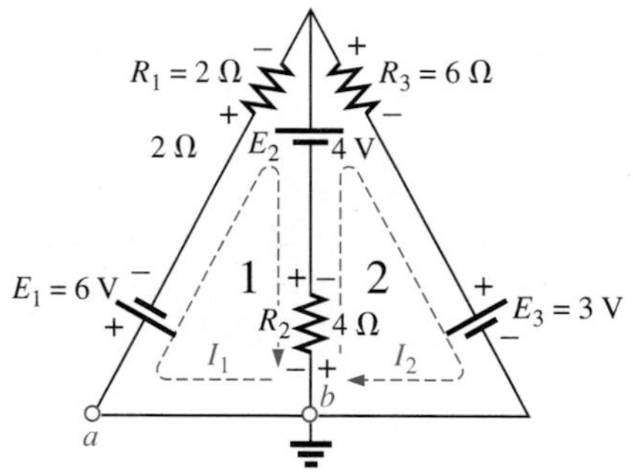

Solution:

Steps 1 and 2 are as indicated in the circuit.

Step 3: Kirchhoff's voltage law is applied around each closed loop:

loop 1: $-E_1 - I_1R_1 - E_2 - V_2 = 0$ (clockwise from point a)
$$-6\text{ V} - (2\ \Omega)I_1 - 4\text{ V} - (4\ \Omega)(I_1 - I_2) = 0$$

loop 2: $-V_2 + E_2 - V_3 - E_3 = 0$ (clockwise from point b)
$$-(4\ \Omega)(I_2 - I_1) + 4\text{ V} - (6\ \Omega)(I_2) - 3\text{ V} = 0$$

FIG. 8.27
Example 8.13.

which are rewritten as

$$\left.\begin{aligned} -10 - 4I_1 - 2I_1 + 4I_2 &= 0 \\ +1 + 4I_1 - 4I_2 - 6I_2 &= 0 \end{aligned}\right\} \begin{aligned} -6I_1 + 4I_2 &= +10 \\ +4I_1 - 10I_2 &= -1 \end{aligned}$$

or, by multiplying the top equation by -1, we obtain

$$\begin{aligned} 6I_1 - 4I_2 &= -10 \\ 4I_1 - 10I_2 &= -1 \end{aligned}$$

and
$$I_1 = \frac{\begin{vmatrix} -10 & -4 \\ -1 & -10 \end{vmatrix}}{\begin{vmatrix} 6 & -4 \\ 4 & -10 \end{vmatrix}} = \frac{100 - 4}{-60 + 16} = \frac{96}{-44} = \mathbf{-2.182\ A}$$

Using the TI-85 calculator:

det[[−10,−4][−1,−10]]/det[[6,−4][4,−10]] (ENTER)	−2.182

CALC. 8.2

$$I_2 = \frac{\begin{vmatrix} 6 & -10 \\ 4 & -1 \end{vmatrix}}{-44} = \frac{-6 + 40}{-44} = \frac{34}{-44} = \mathbf{-0.773\ A}$$

The current in the 4-Ω resistor and 4-V source for loop 1 is

$$\begin{aligned} I_1 - I_2 &= -2.182\text{ A} - (-0.773\text{ A}) \\ &= -2.182\text{ A} + 0.773\text{ A} \\ &= \mathbf{-1.409\ A} \end{aligned}$$

revealing that it is 1.409 A in a direction opposite (due to the minus sign) to I_1 in loop 1.

Supermesh Currents

On occasion there will be current sources in the network to which mesh analysis is to be applied. In such cases one can convert the current source to a voltage source (if a parallel resistor is present) and proceed as before or utilize a *supermesh* current and proceed as follows.

Start as before and assign a mesh current to each independent loop, including the current sources, as if they were resistors or voltage sources. Then mentally remove the current sources (replace with open-circuit equivalents) and apply Kirchhoff's voltage law to all the remaining independent paths of the network using the mesh currents just defined. Any resulting open window, including two or more mesh currents, is said to be the path of a *supermesh* current. Then relate the chosen mesh currents of the network to the independent current sources of the network and solve for the mesh currents. The next example will clarify the definition of a *supermesh* current and the procedure.

EXAMPLE 8.14 Using mesh analysis, determine the currents of the network of Fig. 8.28.

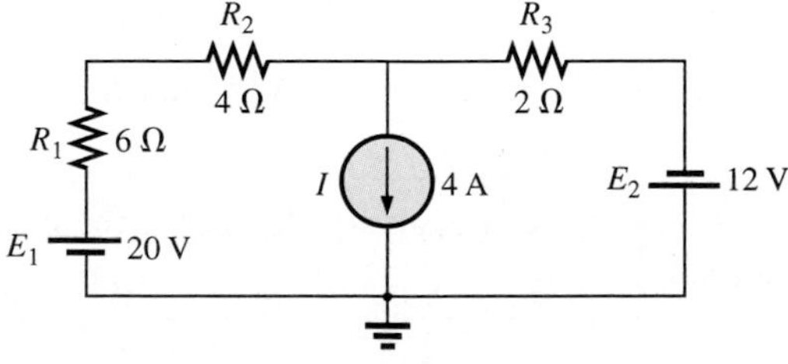

FIG. 8.28
Example 8.14.

Solution: First, the mesh currents for the network are defined, as shown in Fig. 8.29. Then the current source is mentally removed, as shown in Fig. 8.30, and Kirchhoff's voltage law is applied to the resulting network. The single path now including the effects of two mesh currents is referred to as the path of a *supermesh* current.

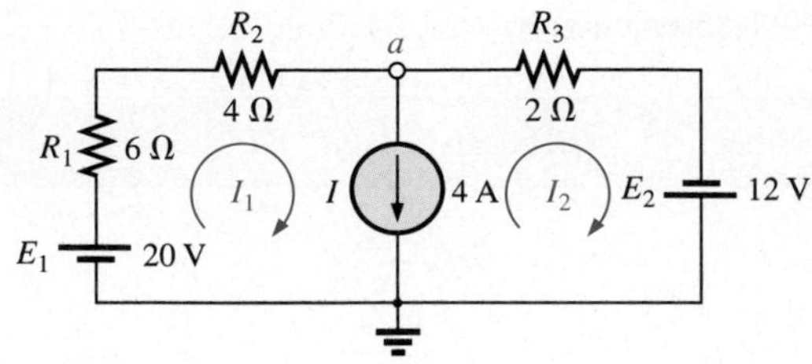

FIG. 8.29
Defining the mesh currents for the network of Fig. 8.28.

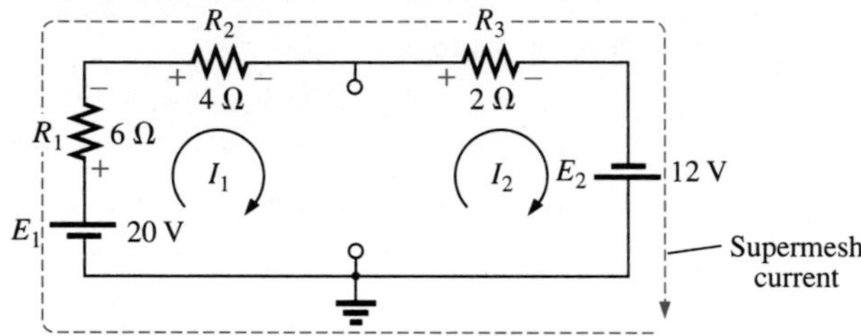

FIG. 8.30
Defining the supermesh current.

Applying Kirchhoff's law:

$$20\text{ V} - I_1(6\ \Omega) - I_1(4\ \Omega) - I_2(2\ \Omega) + 12\text{ V} = 0$$

or
$$10I_1 + 2I_2 = 32$$

Node a is then used to relate the mesh currents and the current source using Kirchhoff's current law:

$$I_1 = I + I_2$$

The result is two equations and two unknowns:

$$\begin{aligned} 10I_1 + 2I_2 &= 32 \\ I_1 - I_2 &= 4 \end{aligned}$$

Applying determinants:

$$I_1 = \frac{\begin{vmatrix} 32 & 2 \\ 4 & -1 \end{vmatrix}}{\begin{vmatrix} 10 & 2 \\ 1 & -1 \end{vmatrix}} = \frac{(32)(-1) - (2)(4)}{(10)(-1) - (2)(1)} = \frac{40}{12} = \mathbf{3.33\ A}$$

and
$$I_2 = I_1 - I = 3.33\text{ A} - 4\text{ A} = \mathbf{-0.67\ A}$$

EXAMPLE 8.15 Using mesh analysis, determine the currents for the network of Fig. 8.31.

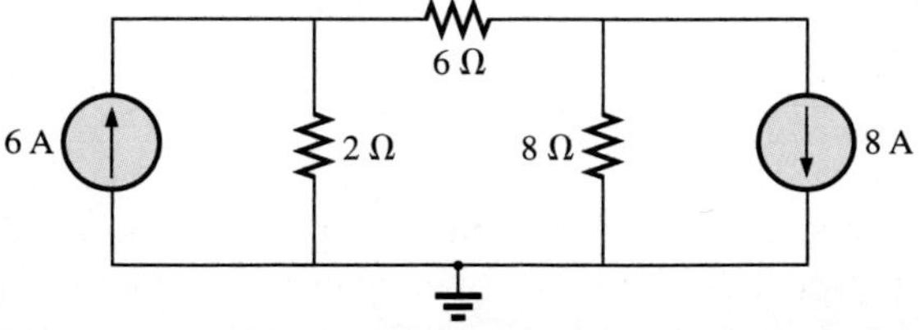

FIG. 8.31
Example 8.15.

Solution: The mesh currents are defined in Fig. 8.32. The current sources are removed, and the single supermesh path is defined in Fig. 8.33.

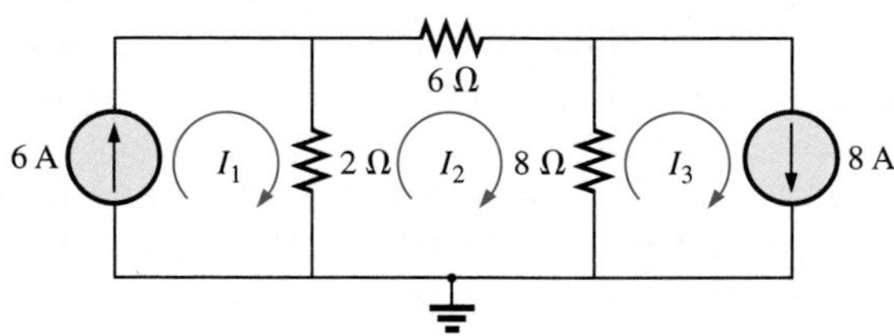

FIG. 8.32
Defining the mesh currents for the network of Fig. 8.31.

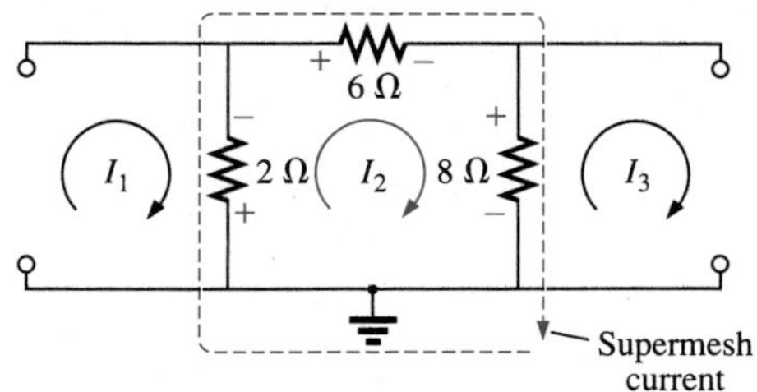

FIG. 8.33
Defining the supermesh current for the network of Fig. 8.31.

Applying Kirchhoff's voltage law around the supermesh path:

$$-V_{2\Omega} - V_{6\Omega} - V_{8\Omega} = 0$$
$$-(I_2 - I_1)2\ \Omega - I_2(6\ \Omega) - (I_2 - I_3)8\ \Omega = 0$$
$$-2I_2 + 2I_1 - 6I_2 - 8I_2 + 8I_3 = 0$$
$$2I_1 - 16I_2 + 8I_3 = 0$$

Introducing the relationship between the mesh currents and the current sources:

$$I_1 = 6\text{ A}$$
$$I_3 = 8\text{ A}$$

results in the following solutions:

$$2I_1 - 16I_2 + 8I_3 = 0$$
$$2(6\text{ A}) - 16I_2 + 8(8\text{ A}) = 0$$

and $$I_2 = \frac{76\text{ A}}{16} = \mathbf{4.75\ A}$$

Then $$I_{2\Omega}\downarrow = I_1 - I_2 = 6\text{ A} - 4.75\text{ A} = \mathbf{1.25\ A}$$

and $$I_{8\Omega}\uparrow = I_3 - I_2 = 8\text{ A} - 4.75\text{ A} = \mathbf{3.25\ A}$$

8.8 MESH ANALYSIS (FORMAT APPROACH)

Now that the basis for the mesh-analysis approach has been established, we will now examine a technique for writing the mesh equations more rapidly and usually with fewer errors. As an aid in introducing the procedure, the network of Example 8.12 (Fig. 8.26) has been redrawn in Fig. 8.34 with the assigned loop currents. (Note that each loop current has a clockwise direction.)

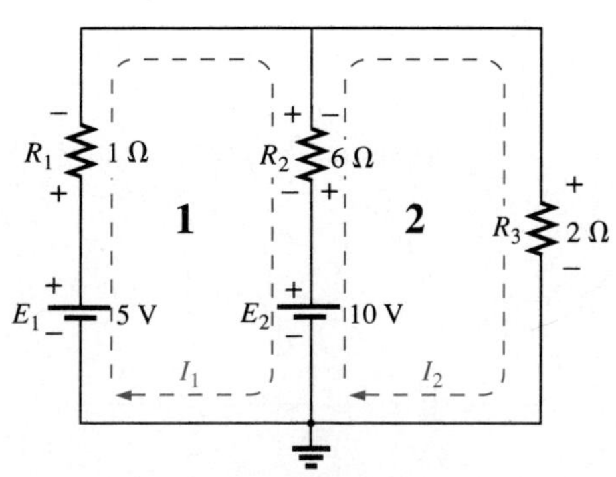

FIG. 8.34
Network of Fig. 8.26 redrawn with assigned loop currents.

The equations obtained are

$$\begin{aligned}-7I_1 + 6I_2 &= 5\\ 6I_1 - 8I_2 &= -10\end{aligned}$$

which can also be written as

$$\begin{aligned}7I_1 - 6I_2 &= -5\\ 8I_2 - 6I_1 &= 10\end{aligned}$$

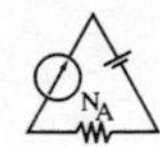

and expanded as

Col. 1	Col. 2	Col. 3
$(1 + 6)I_1 -$	$6I_2$	$= (5 - 10)$
$(2 + 6)I_2 -$	$6I_1$	$= 10$

Note in the above equations that column 1 is composed of a loop current times the sum of the resistors through which that loop current passes. Column 2 is the product of the resistors common to another loop current times that other loop current. Note that in each equation, this column is subtracted from column 1. Column 3 is the *algebraic* sum of the voltage sources through which the loop current of interest passes. A source is assigned a positive sign if the loop current passes from the negative to the positive terminal, and a negative value is assigned if the polarities are reversed. The comments above are correct only for a standard direction of loop current in each window, the one chosen being the clockwise direction.

The above statements can be extended to develop the following *format approach* to mesh analysis:

1. ***Assign a loop current to each independent, closed loop (as in the previous section) in a clockwise direction.***
2. ***The number of required equations is equal to the number of chosen independent, closed loops. Column 1 of each equation is formed by summing the resistance values of those resistors through which the loop current of interest passes and multiplying the result by that loop current.***
3. ***We must now consider the mutual terms, which, as noted in the examples above, are always subtracted from the first column. A mutual term is simply any resistive element having an additional loop current passing through it. It is possible to have more than one mutual term if the loop current of interest has an element in common with more than one other loop current. This will be demonstrated in an example to follow. Each term is the product of the mutual resistor and the other loop current passing through the same element.***
4. ***The column to the right of the equality sign is the algebraic sum of the voltage sources through which the loop current of interest passes. Positive signs are assigned to those sources of voltage having a polarity such that the loop current passes from the negative to the positive terminal. A negative sign is assigned to those potentials for which the reverse is true.***
5. ***Solve the resulting simultaneous equations for the desired loop currents.***

Before considering a few examples, be aware that since the column to the right of the equal sign is the algebraic sum of the voltages sources in that loop, *the format approach can be applied only to networks in which all current sources have been converted to their equivalent voltage source.*

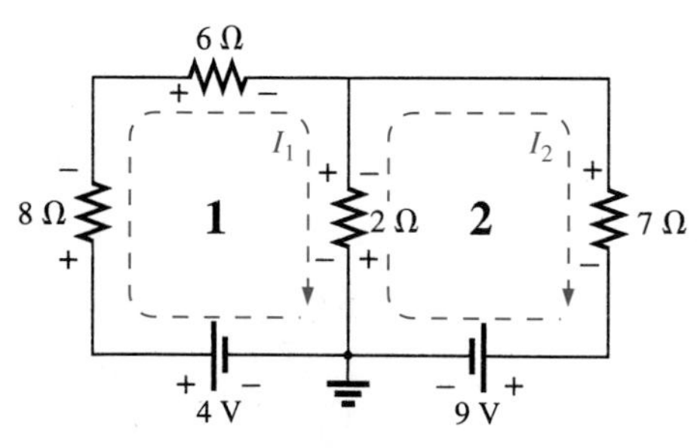

FIG. 8.35
Example 8.16.

EXAMPLE 8.16 Write the mesh equations for the network of Fig. 8.35 and find the current through the 7-Ω resistor.

Solution:

Step 1: As indicated in Fig. 8.35, each assigned loop current has a clockwise direction.

Steps 2 to 4:

$$I_1: \quad (8\,\Omega + 6\,\Omega + 2\,\Omega)I_1 - (2\,\Omega)I_2 = 4\text{ V}$$
$$I_2: \quad (7\,\Omega + 2\,\Omega)I_2 - (2\,\Omega)I_1 = -9\text{ V}$$

and

$$16I_1 - 2I_2 = 4$$
$$9I_2 - 2I_1 = -9$$

which, for determinants, are

$$16I_1 - 2I_2 = 4$$
$$-2I_1 + 9I_2 = -9$$

and

$$I_2 = I_{7\Omega} = \frac{\begin{vmatrix} 16 & 4 \\ -2 & -9 \end{vmatrix}}{\begin{vmatrix} 16 & -2 \\ -2 & 9 \end{vmatrix}} = \frac{-144 + 8}{144 - 4} = \frac{-136}{140}$$
$$= \mathbf{-0.971\text{ A}}$$

EXAMPLE 8.17 Write the mesh equations for the network of Fig. 8.36.

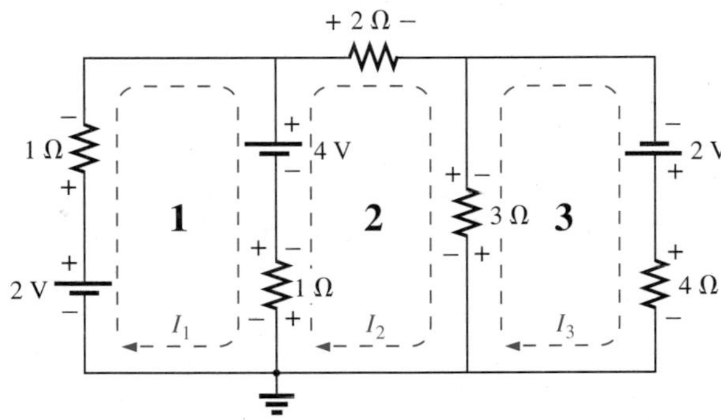

FIG. 8.36
Example 8.17.

Solution:

Step 1: Each window is assigned a loop current in the clockwise direction:

I_1 does not pass through an element mutual with I_3.

$$I_1: \quad (1\,\Omega + 1\,\Omega)I_1 - (1\,\Omega)I_2 + 0 = 2\text{ V} - 4\text{ V}$$
$$I_2: \quad (1\,\Omega + 2\,\Omega + 3\,\Omega)I_2 - (1\,\Omega)I_1 - (3\,\Omega)I_3 = 4\text{ V}$$
$$I_3: \quad (3\,\Omega + 4\,\Omega)I_3 - (3\,\Omega)I_2 + 0 = 2\text{ V}$$

I_3 does not pass through an element mutual with I_1.

Summing terms yields

$$2I_1 - I_2 + 0 = -2$$
$$6I_2 - I_1 - 3I_3 = 4$$
$$7I_3 - 3I_2 + 0 = 2$$

which are rewritten for determinants as

$$
\begin{aligned}
2I_1 - I_2 + 0 &= -2 \\
-I_1 + 6I_2 - 3I_3 &= 4 \\
0 - 3I_2 + 7I_3 &= 2
\end{aligned}
$$

(diagonals labeled c, b, a)

Note that the coefficients of the a and b diagonals are equal. This *symmetry* about the c axis will always be true for equations written using the format approach. It is a check on whether the equations were obtained correctly.

We will now consider a network with only one source of voltage to point out that mesh analysis can be used to advantage in other than multisource networks.

EXAMPLE 8.18 Find the current through the 10-Ω resistor of the network of Fig. 8.37.

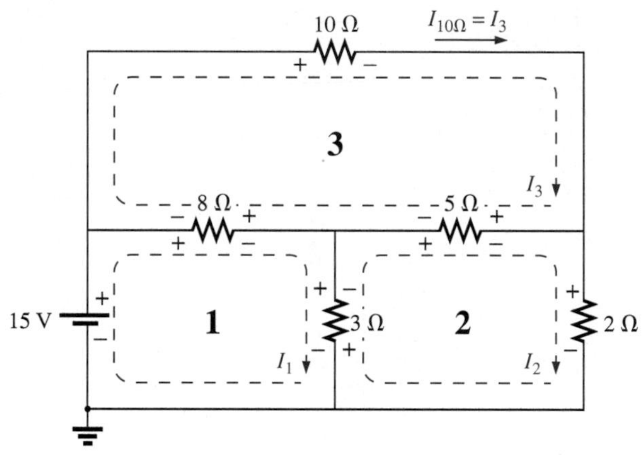

FIG. 8.37
Example 8.18.

Solution:

$$
\begin{aligned}
I_1&: \quad (8\,\Omega + 3\,\Omega)I_1 - (8\,\Omega)I_3 - (3\,\Omega)I_2 = 15\text{ V} \\
I_2&: \quad (3\,\Omega + 5\,\Omega + 2\,\Omega)I_2 - (3\,\Omega)I_1 - (5\,\Omega)I_3 = 0 \\
I_3&: \quad (8\,\Omega + 10\,\Omega + 5\,\Omega)I_3 - (8\,\Omega)I_1 - (5\,\Omega)I_2 = 0
\end{aligned}
$$

$$
\begin{aligned}
11I_1 - 8I_3 - 3I_2 &= 15 \\
10I_2 - 3I_1 - 5I_3 &= 0 \\
23I_3 - 8I_1 - 5I_2 &= 0
\end{aligned}
$$

or

$$
\begin{aligned}
11I_1 - 3I_2 - 8I_3 &= 15 \\
-3I_1 + 10I_2 - 5I_3 &= 0 \\
-8I_1 - 5I_2 + 23I_3 &= 0
\end{aligned}
$$

and

$$
I_3 = I_{10\Omega} = \frac{\begin{vmatrix} 11 & -3 & 15 \\ -3 & 10 & 0 \\ -8 & -5 & 0 \end{vmatrix}}{\begin{vmatrix} 11 & -3 & -8 \\ -3 & 10 & -5 \\ -8 & -5 & 23 \end{vmatrix}} = \mathbf{1.220\ A}
$$

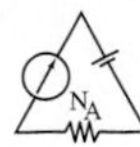

Using MathCad 2.5:

```
I1 := 1        I2 := 1        I3 := 1

All equation variables need to be defined using solution "Guess Values".

Given

      (8 + 3)·I1 - 8·I3 - 3·I2 ≈ 15          Enter "approx" symbol
                                             using <alt=> as required
    (3 + 5 + 2)·I2 - 3·I1 - 5·I3 ≈ 0         when using this solution
                                             format.
    (8 + 10 + 5)·I3 - 8·I1 - 5·I2 ≈ 0

                     [2.63271]
Find(I1,I2,I3)  =    [1.39983]      Amps
                     [1.22003]
```

MATHCAD 8.2

Note in the above MathCad solution that it was unnecessary to establish the determinant format. The solutions were obtained directly from the three simultaneous equations. Imagine the amount of time saved using MathCad compared to the longhand solution. Be absolutely sure, however, that all the entries are exact. One lost sign, symbol, or number will invalidate the solution completely.

Using the TI-85 calculator:

det[[11,−3,15][−3,10,0][−8,−5,0]]/det[[11,−3,−8][−3,10,−5][−8,−5,23]] (ENTER)	1.220

CALC. 8.3

This display certainly requires some care in entering the correct sequence of brackets in the required format, but it is still a rather neat, compact format.

8.9 NODAL ANALYSIS (GENERAL APPROACH)

Recall from the development of loop analysis that the general network equations were obtained by applying Kirchhoff's voltage law around each closed loop. We will now employ Kirchhoff's current law to develop a method referred to as *nodal analysis.*

A *node* is defined as a junction of two or more branches. If we now define one node of any network as a reference (that is, a point of zero potential or ground), the remaining nodes of the network will all have a fixed potential relative to this reference. For a network of N nodes, therefore, there will exist $(N-1)$ nodes with a fixed potential relative to the assigned reference node. Equations relating these nodal voltages can be written by applying Kirchhoff's current law at each of the $(N-1)$ nodes. To obtain the complete solution of a network, these nodal voltages are then evaluated in the same manner in which loop currents were found in loop analysis.

The nodal analysis method is applied as follows:

1. ***Determine the number of nodes within the network.***
2. ***Pick a reference node and label each remaining node with a subscripted value of voltage: V_1, V_2, and so on.***
3. ***Apply Kirchhoff's current law at each node except the reference. Assume all unknown currents leave the node for each application of Kirchhoff's current law. In other words, for each node, don't be influenced by the direction an unknown current for another node may have had. Each node is to be treated as a separate entity, independent of the application of Kirchhoff's current law to the other nodes.***
4. ***Solve the resulting equations for the nodal voltages.***

A few examples will clarify the procedure defined by step 3. It will initially take some practice writing the equations for Kirchhoff's current law correctly, but in time the advantage of assuming all the currents leave a node rather than identifying a specific direction for each branch will become obvious. (The same type of advantage is associated with assuming that all the mesh currents are clockwise when applying mesh analysis.)

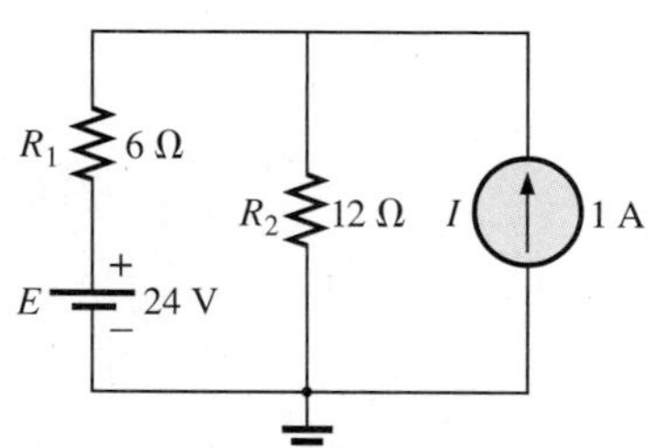

FIG. 8.38
Example 8.19.

EXAMPLE 8.19 Apply nodal analysis to the network of Fig. 8.38.

Solution:

Steps 1 and 2: The network has two nodes, as shown in Fig. 8.39. The lower node is defined as the reference node at ground potential (zero volts) and the other V_1, the voltage from node 1 to ground.

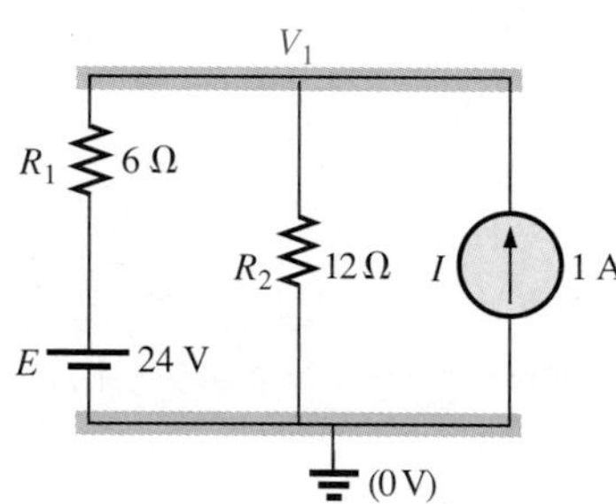

FIG. 8.39
Network of Fig. 8.38 with assigned nodes.

Step 3: I_1 and I_2 are defined as leaving the node in Fig. 8.40, and Kirchhoff's current law is applied as follows:

$$I = I_1 + I_2$$

The current I_2 is related to the nodal voltage V_1 by Ohm's law:

$$I_2 = \frac{V_{R_2}}{R_2} = \frac{V_1}{R_2}$$

The current I_1 is also determined by Ohm's law as follows:

$$I_1 = \frac{V_{R_1}}{R_1}$$

with $$V_{R_1} = V_1 - E$$

Substituting into the Kirchhoff's current law equation:

$$I = \frac{V_1 - E}{R_1} + \frac{V_1}{R_2}$$

and rearranging, we have:

$$I = \frac{V_1}{R_1} - \frac{E}{R_1} + \frac{V_1}{R_2} = V_1\left(\frac{1}{R_1} + \frac{1}{R_2}\right) - \frac{E}{R_1}$$

or $$V_1\left(\frac{1}{R_1} + \frac{1}{R_2}\right) = \frac{E}{R_1} + I$$

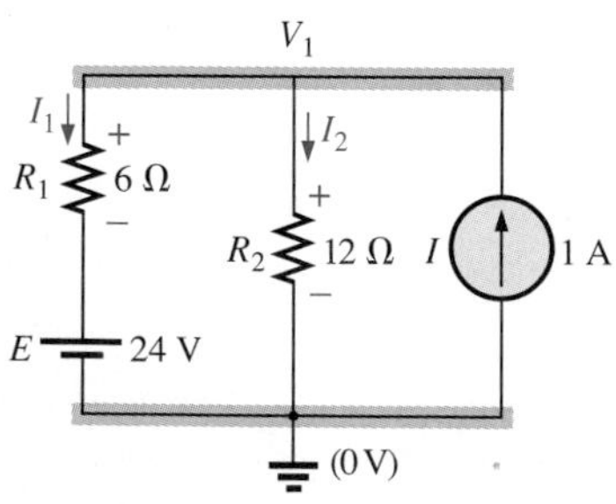

FIG. 8.40
Applying Kirchhoff's current law to the node V_1.

Substituting numerical values,

$$V_1\left(\frac{1}{6\ \Omega} + \frac{1}{12\ \Omega}\right) = \frac{24\ \text{V}}{6\ \Omega} + 1\ \text{A} = 4\ \text{A} + 1\ \text{A}$$

$$V_1\left(\frac{1}{4\ \Omega}\right) = 5\ \text{A}$$

$$V_1 = \mathbf{20\ V}$$

The currents I_1 and I_2 can then be determined using the preceding equations:

$$I_1 = \frac{V_1 - E}{R_1} = \frac{20\ \text{V} - 24\ \text{V}}{6\ \Omega} = \frac{-4\ \text{V}}{6\ \Omega}$$
$$= \mathbf{-0.667\ A}$$

The minus sign indicates simply that the current I_1 has a direction opposite to that appearing in Fig. 8.40.

$$I_2 = \frac{V_1}{R_2} = \frac{20\ \text{V}}{12\ \Omega} = \mathbf{1.667\ A}$$

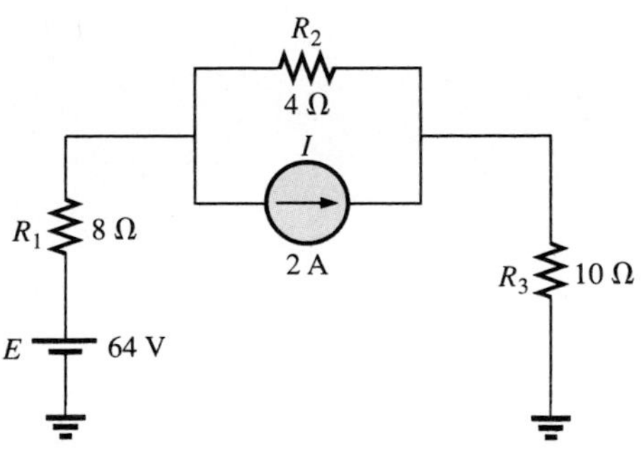

FIG. 8.41
Example 8.20.

EXAMPLE 8.20 Apply nodal analysis to the network of Fig. 8.41.

Solution:

Steps 1 and 2: The network has three nodes, as defined in Fig. 8.42, with the bottom node again defined as the reference node (at ground potential, or zero volts) and the other nodes V_1 and V_2.

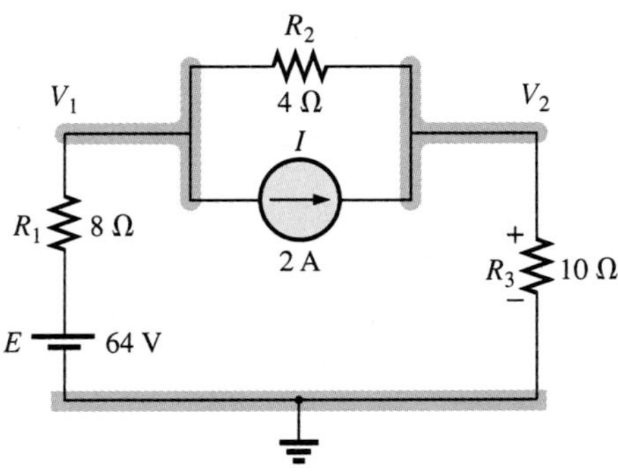

FIG. 8.42
Defining the nodes for the network of Fig. 8.41.

Step 3: For node V_1 the currents are defined as shown in Fig. 8.43, and Kirchhoff's current law is applied:

$$0 = I_1 + I_2 + I$$

with
$$I_1 = \frac{V_1 - E}{R_1}$$

and
$$I_2 = \frac{V_{R_2}}{R_2} = \frac{V_1 - V_2}{R_2}$$

so that
$$\frac{V_1 - E}{R_1} + \frac{V_1 - V_2}{R_2} + I = 0$$

or
$$\frac{V_1}{R_1} - \frac{E}{R_1} + \frac{V_1}{R_2} - \frac{V_2}{R_2} + I = 0$$

and
$$V_1\left(\frac{1}{R_1} + \frac{1}{R_2}\right) - V_2\left(\frac{1}{R_2}\right) = -I + \frac{E}{R_1}$$

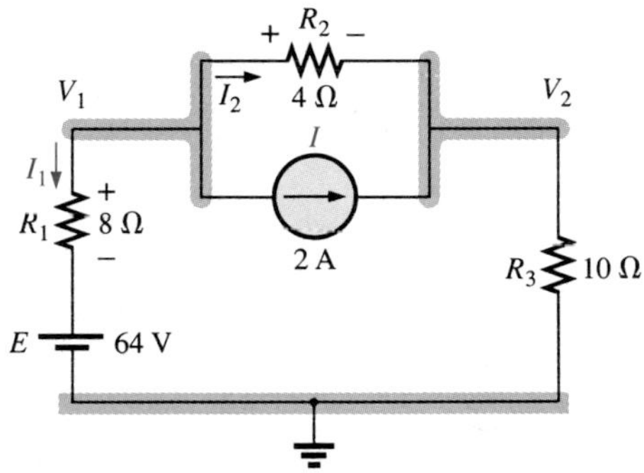

FIG. 8.43
Applying Kirchhoff's current law to node V_1.

Substituting values:

$$V_1\left(\frac{1}{8\ \Omega} + \frac{1}{4\ \Omega}\right) - V_2\left(\frac{1}{4\ \Omega}\right) = -2\ \text{A} + \frac{64\ \text{V}}{8\ \Omega} = 6\ \text{A}$$

For node V_2 the currents are defined as shown in Fig. 8.44, and Kirchhoff's current law is applied:

$$I = I_2 + I_3$$

with
$$I = \frac{V_2 - V_1}{R_2} + \frac{V_2}{R_3}$$

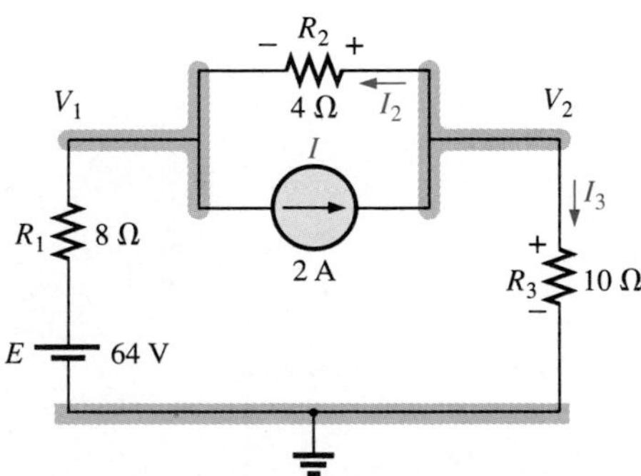

FIG. 8.44
Applying Kirchhoff's current law to node V_2.

or
$$I = \frac{V_2}{R_2} - \frac{V_1}{R_2} + \frac{V_2}{R_3}$$

and
$$V_2\left(\frac{1}{R_2} + \frac{1}{R_3}\right) - V_1\left(\frac{1}{R_2}\right) = I$$

Substituting values:

$$V_2\left(\frac{1}{4\ \Omega} + \frac{1}{10\ \Omega}\right) - V_1\left(\frac{1}{4\ \Omega}\right) = 2\text{ A}$$

Step 4: The result is two equations and two unknowns:

$$V_1\left(\frac{1}{8\ \Omega} + \frac{1}{4\ \Omega}\right) - V_2\left(\frac{1}{4\ \Omega}\right) = 6\text{ A}$$
$$-V_1\left(\frac{1}{4\ \Omega}\right) + V_2\left(\frac{1}{4\ \Omega} + \frac{1}{10\ \Omega}\right) = 2\text{ A}$$

which become

$$0.375V_1 - 0.25V_2 = 6$$
$$-0.25V_1 + 0.35V_2 = 2$$

Using determinants,

$$V_1 = \mathbf{37.818\ V}$$
$$V_2 = \mathbf{32.727\ V}$$

Since E is greater than V_1, the current I_1 flows from ground to V_1 and is equal to

$$I_{R_1} = \frac{E - V_1}{R_1} = \frac{64\text{ V} - 37.818\text{ V}}{8\ \Omega} = \mathbf{3.273\ A}$$

The positive value for V_2 results in a current I_{R_3} from node V_2 to ground equal to

$$I_{R_3} = \frac{V_{R_3}}{R_3} = \frac{V_2}{R_3} = \frac{32.727\text{ V}}{10\ \Omega} = \mathbf{3.273\ A}$$

Since V_1 is greater than V_2, the current I_{R_2} flows from V_1 to V_2 and is equal to

$$I_{R_2} = \frac{V_1 - V_2}{R_2} = \frac{37.818\text{ V} - 32.727\text{ V}}{4\ \Omega} = \mathbf{1.273\ A}$$

Using MathCad, the nodal voltages would be determined as follows:

```
V1 := 1          V2 := 1

Given

V1 - 64   V1 - V2              V2    V2 - V1
------- + -------  ≈ -2        --  + -------  ≈ 2
   8         4                 10       4

                 [37.818]   Volts
Find(V1,V2)  =   [32.727]   Volts
```

MATHCAD 8.3

EXAMPLE 8.21 Determine the nodal voltages for the network of Fig. 8.45.

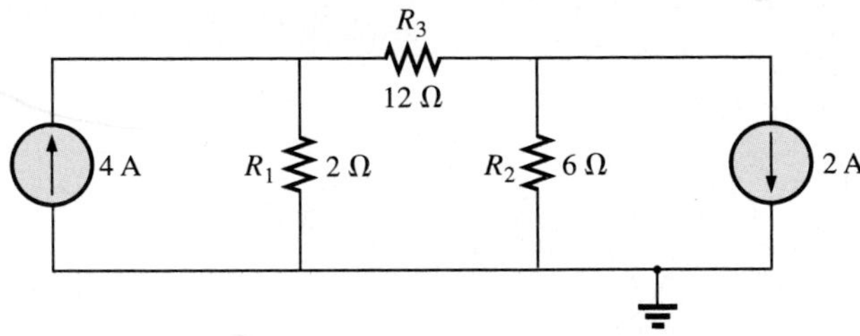

FIG. 8.45
Example 8.21.

Solution:

Steps 1 and 2: As indicated in Fig. 8.46.

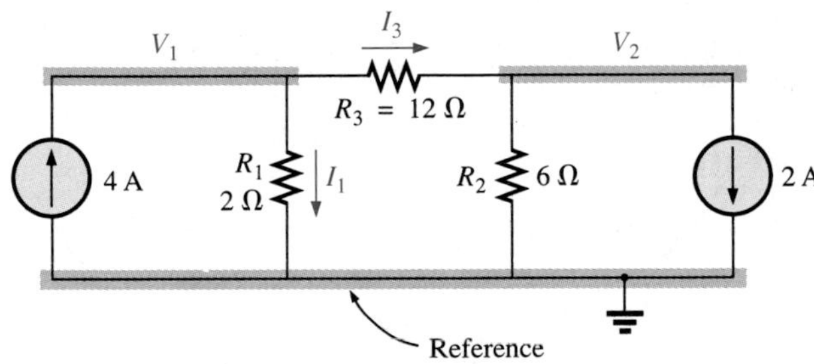

FIG. 8.46
Defining the nodes and applying Kirchhoff's current law to the node V_1.

Step 3: Included in Fig. 8.46 for the node V_1. Applying Kirchhoff's current law:

$$4\text{ A} = I_1 + I_3$$

and

$$4\text{ A} = \frac{V_1}{R_1} + \frac{V_1 - V_2}{R_3} = \frac{V_1}{2\ \Omega} + \frac{V_1 - V_2}{12\ \Omega}$$

Expanding and rearranging:

$$V_1\left(\frac{1}{2\ \Omega} + \frac{1}{12\ \Omega}\right) - V_2\left(\frac{1}{12\ \Omega}\right) = 4\text{ A}$$

For node V_2 the currents are defined as in Fig. 8.47.

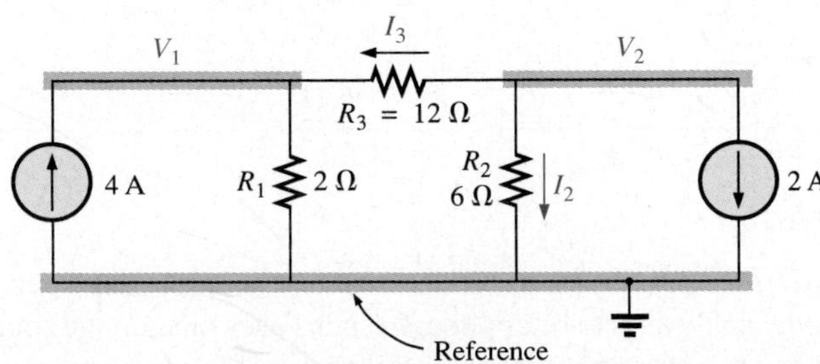

FIG. 8.47
Applying Kirchhoff's current law to the node V_2.

Applying Kirchhoff's current law:

$$0 = I_3 + I_2 + 2\text{ A}$$

and $$\frac{V_2 - V_1}{R_3} + \frac{V_2}{R_2} + 2\text{ A} = 0 \longrightarrow \frac{V_2 - V_1}{12\ \Omega} + \frac{V_2}{6\ \Omega} + 2\text{ A} = 0$$

Expanding and rearranging:

$$V_2\left(\frac{1}{12\ \Omega} + \frac{1}{6\ \Omega}\right) - V_1\left(\frac{1}{12\ \Omega}\right) = -2\text{ A}$$

resulting in two equations and two unknowns (numbered for later reference):

$$\begin{aligned} V_1\left(\frac{1}{2\ \Omega} + \frac{1}{12\ \Omega}\right) - V_2\left(\frac{1}{12\ \Omega}\right) &= +4\text{ A} \\ V_2\left(\frac{1}{12\ \Omega} + \frac{1}{6\ \Omega}\right) - V_1\left(\frac{1}{12\ \Omega}\right) &= -2\text{ A} \end{aligned} \quad \textbf{(8.3)}$$

producing

$$\left.\begin{aligned} \frac{7}{12}V_1 - \frac{1}{12}V_2 &= +4 \\ -\frac{1}{12}V_1 + \frac{3}{12}V_2 &= -2 \end{aligned}\right\} \quad \begin{aligned} 7V_1 - V_2 &= 48 \\ -1V_1 + 3V_2 &= -24 \end{aligned}$$

and $$V_1 = \frac{\begin{vmatrix} 48 & -1 \\ -24 & 3 \end{vmatrix}}{\begin{vmatrix} 7 & -1 \\ -1 & 3 \end{vmatrix}} = \frac{120}{20} = \mathbf{+6\ V}$$

$$V_2 = \frac{\begin{vmatrix} 7 & 48 \\ -1 & -24 \end{vmatrix}}{20} = \frac{-120}{20} = \mathbf{-6\ V}$$

Since V_1 is greater than V_2, the current through R_3 passes from V_1 to V_2. Its value is:

$$I_{R_3} = \frac{V_1 - V_2}{R_3} = \frac{6\text{ V} - (-6\text{ V})}{12\ \Omega} = \frac{12\text{ V}}{12\ \Omega} = \mathbf{1\ A}$$

The fact that V_1 is positive results in a current I_{R_1} from V_1 to ground equal to:

$$I_{R_1} = \frac{V_{R_1}}{R_1} = \frac{V_1}{R_1} = \frac{6\text{ V}}{2\ \Omega} = \mathbf{3\ A}$$

Finally, since V_2 is negative, the current I_{R_2} flows from ground to V_2 and is equal to:

$$I_{R_2} = \frac{V_{R_2}}{R_2} = \frac{V_2}{R_2} = \frac{6\text{ V}}{6\ \Omega} = \mathbf{1\ A}$$

Supernode

On occasion there will be independent voltage sources in the network to which nodal analysis is to be applied. In such cases the voltage source can be converted to a current source (if a series resistor is present) and proceed as before, or introduce the concept of a *supernode* and proceed as follows.

Start as before and assign a nodal voltage to each independent node of the network, including each independent voltage source as if it were a resistor or current source. Then mentally replace the independent voltage sources with short-circuit equivalents and apply Kirchhoff's current law to the defined nodes of the network. Any node including the effect of elements tied only to *other* nodes is referred to as a *supernode* (since it has an additional number of terms). Finally, relate the defined nodes to the independent voltage sources of the network and solve for the nodal voltages. The next example will clarify the definition of *supernode*.

EXAMPLE 8.22 Determine the nodal voltages V_1 and V_2 of Fig. 8.48 using the concept of a *supernode*.

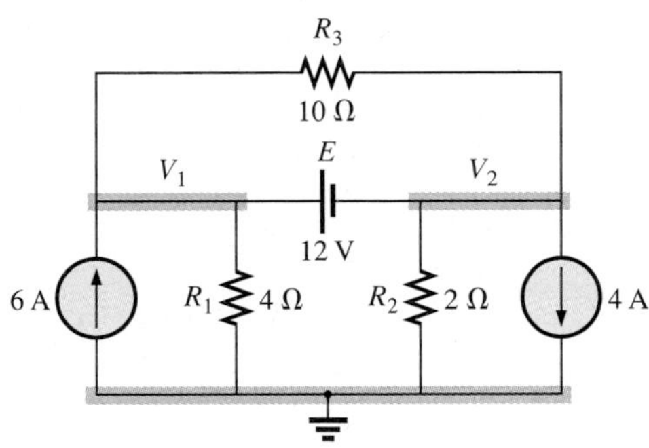

FIG. 8.48
Example 8.22.

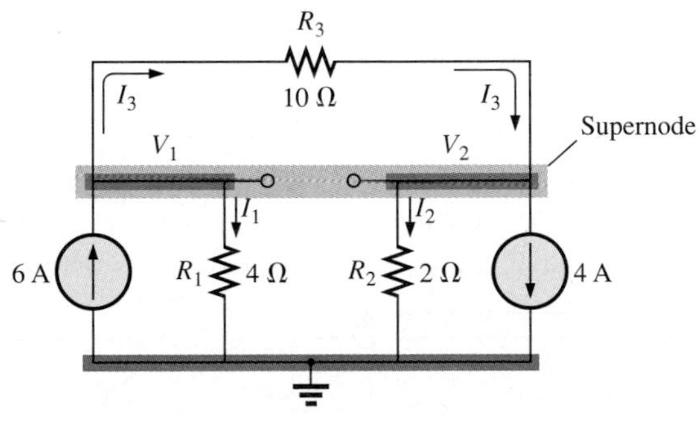

FIG. 8.49
Defining the supernode for the network of Fig. 8.48.

Solution: Replacing the independent voltage source of 12 V with a short-circuit equivalent will result in the network of Fig. 8.49. Even though the mental application of a short-circuit equivalent is discussed above, it would be wise in the early stage of development to redraw the network as shown in Fig. 8.49. The result is a single supernode for which Kirchhoff's current law must be applied. Be sure to leave the other defined nodes in place and use them to define the currents from that region of the network. In particular, note that the current I_3 will leave the supernode at V_1 and then enter the same supernode at V_2. It must therefore appear twice when applying Kirchhoff's current law, as shown below:

$$\Sigma I_i = \Sigma I_o$$

$$6\text{ A} + I_3 = I_1 + I_2 + 4\text{ A} + I_3$$

or $$I_1 + I_2 = 6\text{ A} - 4\text{ A} = 2\text{ A}$$

Then, $$\frac{V_1}{R_1} + \frac{V_2}{R_2} = 2\text{ A}$$

and $$\frac{V_1}{4\ \Omega} + \frac{V_2}{2\ \Omega} = 2\text{ A}$$

Relating the defined nodal voltages to the independent voltage source:

$$V_1 - V_2 = E = 12\text{ V}$$

which results in two equations and two unknowns:

$$\begin{aligned} 0.25V_1 + 0.5V_2 &= 2 \\ V_1 - \ \ V_2 &= 12 \end{aligned}$$

Substituting:

$$V_1 = V_2 + 12$$

$$0.25(V_2 + 12) + 0.5V_2 = 2$$

and $$0.75V_2 = 2 - 3 = -1$$

so that $$V_2 = \frac{-1}{0.75} = \mathbf{-1.333\ V}$$

and $$V_1 = V_2 + 12\text{ V} = -1.333\text{ V} + 12\text{ V} = \mathbf{+10.667\ V}$$

The current of the network can then be determined as follows:

$$I_1\downarrow = \frac{V}{R_1} = \frac{10.667\text{ V}}{4\ \Omega} = \mathbf{2.667\ A}$$

$$I_2\uparrow = \frac{V_2}{R_2} = \frac{1.333\text{ V}}{2\ \Omega} = \mathbf{0.667\ A}$$

$$\underset{\rightarrow}{I_3} = \frac{V_1 - V_2}{10\ \Omega} = \frac{10.667\text{ V} - (-1.333\text{ V})}{10\ \Omega} = \frac{12\text{ V}}{10\ \Omega} = \mathbf{1.2\ A}$$

A careful examination of the network at the beginning of the analysis would have revealed that the voltage across the resistor R_3 must be 12 V and I_3 must be equal to 1.2 A.

8.10 NODAL ANALYSIS (FORMAT APPROACH)

A close examination of Eq. (8.3) appearing in Example 8.21 reveals that the subscripted voltage at the node in which Kirchhoff's current law is applied is multiplied by the sum of the conductances attached to that node. Note also that the other nodal voltages within the same equation are multiplied by the negative of the conductance between the two nodes. The current sources are represented to the right of the equals sign with a positive sign if they supply current to the node and with a negative sign if they draw current from the node.

These conclusions can be expanded to include networks with any number of nodes. This will allow us to write nodal equations rapidly and in a form that is convenient for the use of determinants. A major requirement, however, is that *all voltage sources must first be converted to current sources before the procedure is applied.* Note the parallelism between the following four steps of application and those required for mesh analysis in Section 8.8:

1. ***Choose a reference node and assign a subscripted voltage label to the (N − 1) remaining nodes of the network.***
2. ***The number of equations required for a complete solution is equal to the number of subscripted voltages (N − 1). Column 1 of each equation is formed by summing the conductances tied to the node of interest and multiplying the result by that subscripted nodal voltage.***
3. ***We must now consider the mutual terms that, as noted in the preceding example, are always subtracted from the first column. It is possible to have more than one mutual term if the nodal voltage of current interest has an element in common with more than one other nodal voltage. This will be demonstrated in an example to follow. Each mutual term is the product of the mutual conductance and the other nodal voltage tied to that conductance.***
4. ***The column to the right of the equality sign is the algebraic sum of the current sources tied to the node of interest. A current source is assigned a positive sign if it supplies current to a node and a negative sign if it draws current from the node.***
5. ***Solve the resulting simultaneous equations for the desired voltages.***

Let us now consider a few examples.

EXAMPLE 8.23 Write the nodal equations for the network of Fig. 8.50.

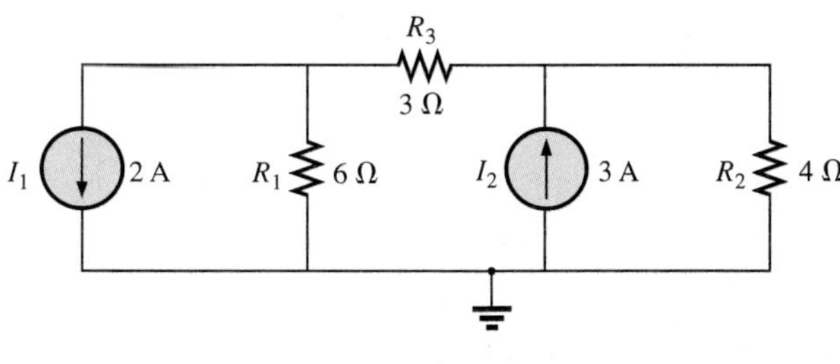

FIG. 8.50
Example 8.23.

Solution:

Step 1: The figure is redrawn with assigned subscripted voltages in Fig. 8.51.

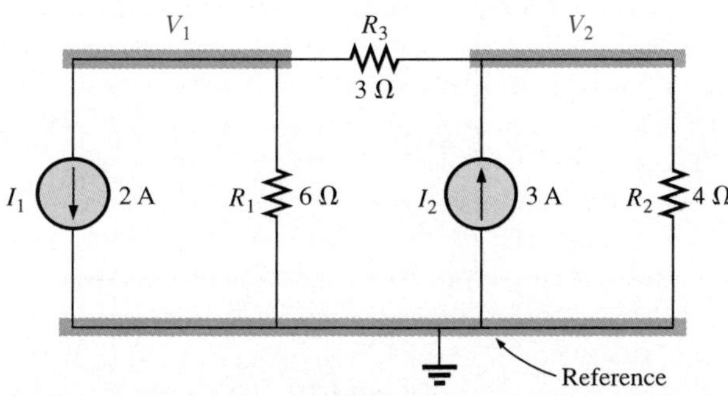

FIG. 8.51
Defining the nodes for the network of Fig. 8.50.

Steps 2 to 4:

Drawing current from node 1

$$V_1: \underbrace{\left(\frac{1}{6\,\Omega} + \frac{1}{3\,\Omega}\right)}_{\substack{\text{Sum of}\\ \text{conductances}\\ \text{connected}\\ \text{to node 1}}} V_1 - \underbrace{\left(\frac{1}{3\,\Omega}\right)}_{\substack{\text{Mutual}\\ \text{conductance}}} V_2 = -2\text{ A}$$

Supplying current to node 2

$$V_2: \underbrace{\left(\frac{1}{4\,\Omega} + \frac{1}{3\,\Omega}\right)}_{\substack{\text{Sum of}\\ \text{conductances}\\ \text{connected}\\ \text{to node 2}}} V_2 - \underbrace{\left(\frac{1}{3\,\Omega}\right)}_{\substack{\text{Mutual}\\ \text{conductance}}} V_1 = +3\text{ A}$$

and

$$\frac{1}{2}V_1 - \frac{1}{3}V_2 = -2$$

$$-\frac{1}{3}V_1 + \frac{7}{12}V_2 = 3$$

EXAMPLE 8.24 Find the voltage across the 3-Ω resistor of Fig. 8.52. by nodal analysis.

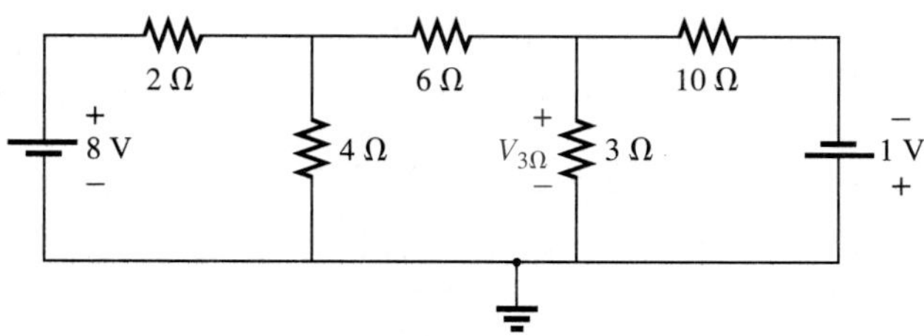

FIG. 8.52
Example 8.24.

Solution: Converting sources and choosing nodes (Fig. 8.53), we have

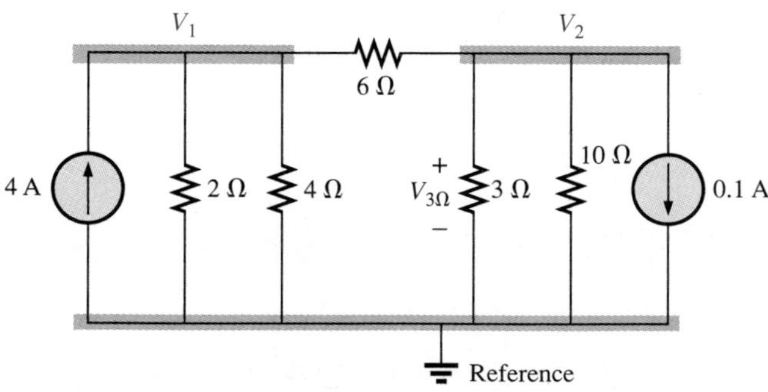

FIG. 8.53
Defining the nodes for the network of Fig. 8.52.

$$\left.\begin{aligned}\left(\frac{1}{2\,\Omega}+\frac{1}{4\,\Omega}+\frac{1}{6\,\Omega}\right)V_1-\left(\frac{1}{6\,\Omega}\right)V_2&=+4\text{ A}\\ \left(\frac{1}{10\,\Omega}+\frac{1}{3\,\Omega}+\frac{1}{6\,\Omega}\right)V_2-\left(\frac{1}{6\,\Omega}\right)V_1&=-0.1\text{ A}\end{aligned}\right\}$$

$$\frac{11}{12}V_1-\frac{1}{6}V_2=4$$

$$-\frac{1}{6}V_1+\frac{3}{5}V_2=-0.1$$

resulting in

$$\begin{aligned}11V_1-2V_2&=+48\\ -5V_1+18V_2&=-3\end{aligned}$$

and

$$V_2=V_{3\Omega}=\frac{\begin{vmatrix}11 & 48\\ -5 & -3\end{vmatrix}}{\begin{vmatrix}11 & -2\\ -5 & 18\end{vmatrix}}=\frac{-33+240}{198-10}=\frac{207}{188}=\mathbf{1.101\ V}$$

As demonstrated for mesh analysis, nodal analysis can also be a very useful technique for solving networks with only one source.

EXAMPLE 8.25 Using nodal analysis, determine the potential across the 4-Ω resistor in Fig. 8.54.

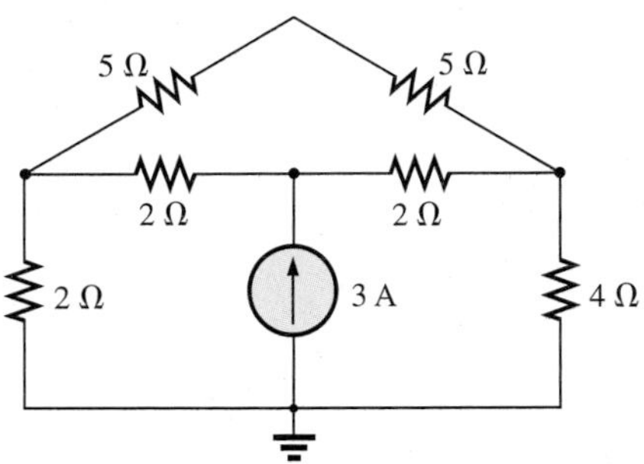

FIG. 8.54
Example 8.25.

Solution: The reference and four subscripted voltage levels were chosen as shown in Fig. 8.55. A moment of reflection should reveal that for any difference in potential between V_1 and V_3, the current through and the potential drop across each 5-Ω resistor will be the same. Therefore, V_4 is simply a midvoltage level between V_1 and V_3 and is known if V_1 and V_3 are available. We will therefore not include it in a nodal voltage and will redraw the network as shown in Fig. 8.56. Understand, however, that V_4 can be included if desired, although four nodal voltages will result rather than the three to be obtained in the solution of this problem.

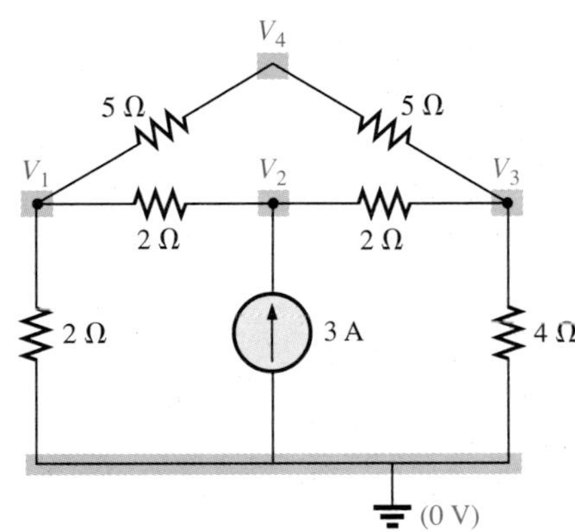

FIG. 8.55
Defining the nodes for the network of Fig. 8.54.

$$V_1:\quad\left(\frac{1}{2\,\Omega}+\frac{1}{2\,\Omega}+\frac{1}{10\,\Omega}\right)V_1-\left(\frac{1}{2\,\Omega}\right)V_2-\left(\frac{1}{10\,\Omega}\right)V_3=0$$

$$V_2:\quad\left(\frac{1}{2\,\Omega}+\frac{1}{2\,\Omega}\right)V_2-\left(\frac{1}{2\,\Omega}\right)V_1-\left(\frac{1}{2\,\Omega}\right)V_3=3\text{ A}$$

$$V_3:\quad\left(\frac{1}{10\,\Omega}+\frac{1}{2\,\Omega}+\frac{1}{4\,\Omega}\right)V_3-\left(\frac{1}{2\,\Omega}\right)V_2-\left(\frac{1}{10\,\Omega}\right)V_1=0$$

which are rewritten as

$$\begin{aligned}1.1V_1-0.5V_2-0.1V_3&=0\\ V_2-0.5V_1-0.5V_3&=3\\ 0.85V_3-0.5V_2-0.1V_1&=0\end{aligned}$$

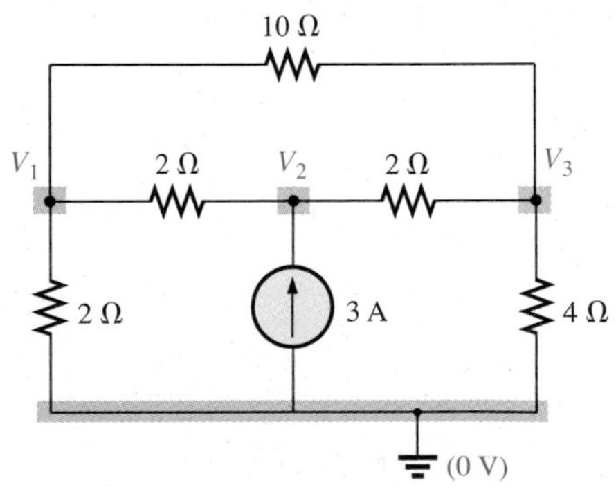

FIG. 8.56
Reducing the number of nodes for the network of Fig. 8.54 by combining the two 5-Ω resistors.

For determinants,

$$\begin{aligned} 1.1V_1 - 0.5V_2 - 0.1V_3 &= 0 \\ -0.5V_1 + 1V_2 - 0.5V_3 &= 3 \\ -0.1V_1 - 0.5V_2 + 0.85V_3 &= 0 \end{aligned}$$

Before continuing, note the symmetry about the major diagonal in the equation above. Recall a similar result for mesh analysis. Examples 8.23 and 8.24 also exhibit this property in the resulting equations. Keep this thought in mind as a check on future applications of nodal analysis.

$$V_3 = V_{4\Omega} = \frac{\begin{vmatrix} 1.1 & -0.5 & 0 \\ -0.5 & +1 & 3 \\ -0.1 & -0.5 & 0 \end{vmatrix}}{\begin{vmatrix} 1.1 & -0.5 & -0.1 \\ -0.5 & +1 & -0.5 \\ -0.1 & -0.5 & +0.85 \end{vmatrix}} = \mathbf{4.645\ V}$$

Using MathCad 2.5:

```
V1 := 1        V2 := 1        V3 := 1

Given

(1/2 + 1/2 + 1/10)·V1 - (1/2)·V2 - (1/10)·V3 ≈ 0

(1/2 + 1/2)·V2 - (1/2)·V1 - (1/2)·V3 ≈ 3

(1/10 + 1/2 + 1/4)·V3 - (1/2)·V2 - (1/10)·V1 ≈ 0

                    [3.677]
Find(V1,V2,V3)  =   [7.161]     Volts
                    [4.645]
```

MATHCAD 8.4

The next example has only one source applied to a ladder network.

EXAMPLE 8.26 Write the nodal equations and find the voltage across the 2-Ω resistor for the network of Fig. 8.57.

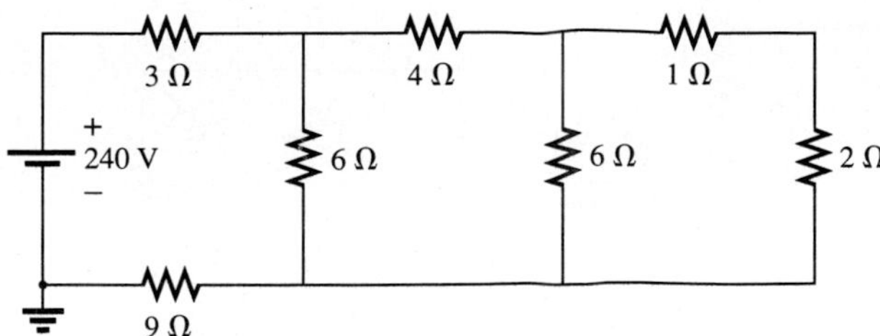

FIG. 8.57
Example 8.26.

Solution: The nodal voltages are chosen as shown in Fig. 8.58.

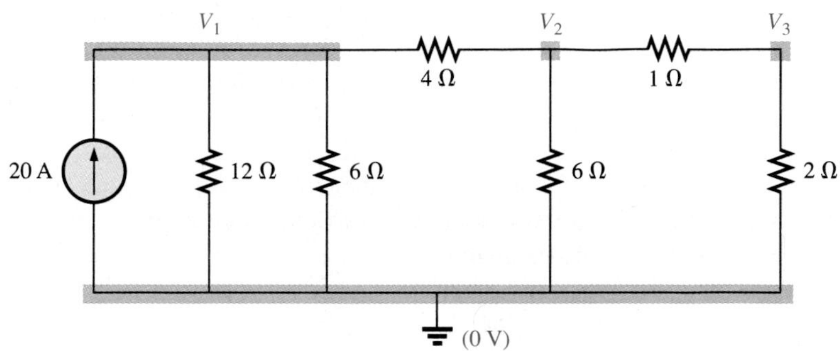

FIG. 8.58
Converting the voltage source to a current source and defining the nodes for the network of Fig. 8.57.

$$\begin{array}{ll}
V_1: & \left(\frac{1}{12\,\Omega}+\frac{1}{6\,\Omega}+\frac{1}{4\,\Omega}\right)V_1-\left(\frac{1}{4\,\Omega}\right)V_2+\quad 0 \quad = 20\text{ V}\\
V_2: & \left(\frac{1}{4\,\Omega}+\frac{1}{6\,\Omega}+\frac{1}{1\,\Omega}\right)V_2-\left(\frac{1}{4\,\Omega}\right)V_1-\left(\frac{1}{1\,\Omega}\right)V_3 = 0\\
V_3: & \left(\frac{1}{1\,\Omega}+\frac{1}{2\,\Omega}\right)V_3-\left(\frac{1}{1\,\Omega}\right)V_2+\quad 0 \quad = 0
\end{array}$$

and

$$\begin{array}{rrr}
0.5V_1 & -0.25V_2 & +0 = 20\\
-0.25V_1 & +\frac{17}{12}V_2 & -1V_3 = 0\\
0 & -1V_2 & +1.5V_3 = 0
\end{array}$$

Note the symmetry present about the major axis. Application of determinants reveals that

$$V_3 = V_{2\Omega} = \mathbf{10.667\text{ V}}$$

8.11 BRIDGE NETWORKS

This section introduces the bridge network, a configuration that has a multitude of applications. In the chapters to follow, it will be employed in both dc and ac meters. In the electronics courses it will be encountered early in the discussion of rectifying circuits employed in converting a varying signal to one of a steady nature (such as dc). There are a

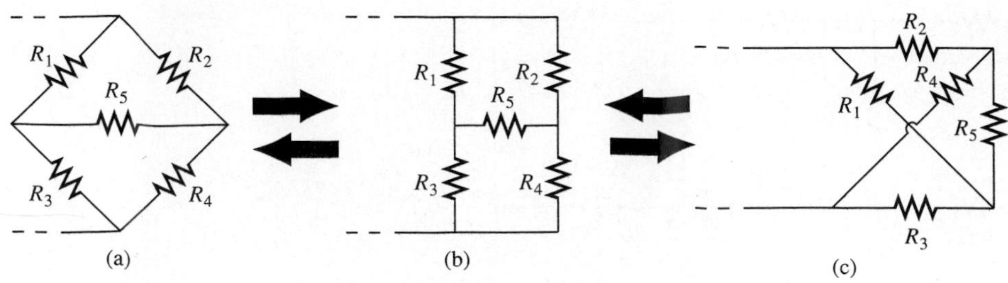

FIG. 8.59
Various formats for a bridge network.

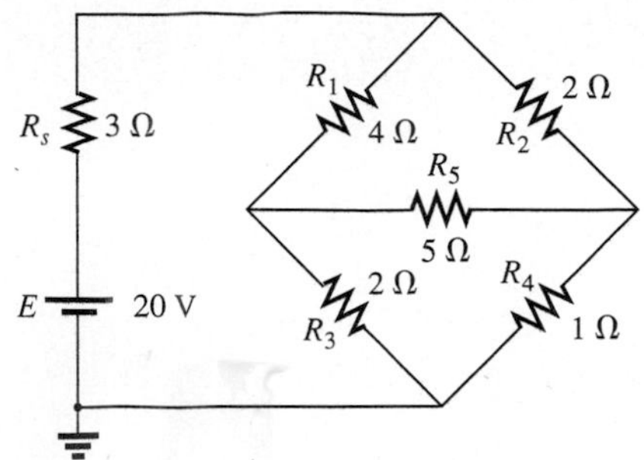

FIG. 8.60
Standard bridge configuration.

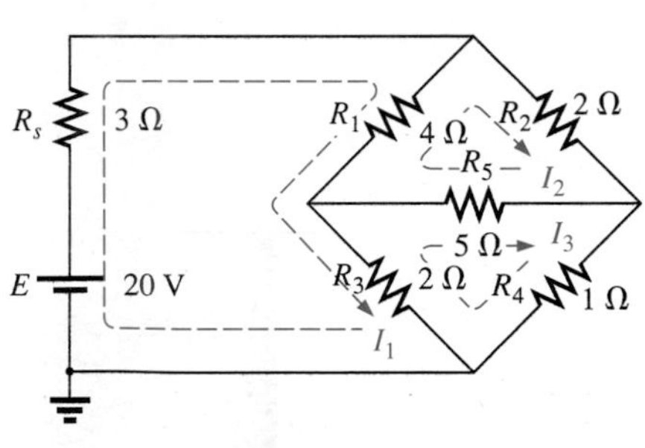

FIG. 8.61
Assigning the mesh currents to the network of Fig. 8.60.

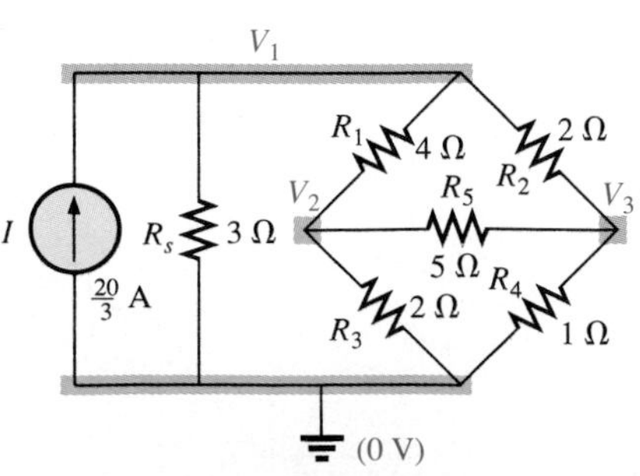

FIG. 8.62
Defining the nodal voltages for the network of Fig. 8.60.

number of other areas of application that require some knowledge of ac networks, which will be discussed later.

The bridge network may appear in one of the three forms as indicated in Fig. 8.59. The network of Fig. 8.59(c) is also called a symmetrical lattice network if $R_2 = R_3$ and $R_1 = R_4$. Figure 8.59(c) is an excellent example of how a planar network can be made to appear nonplanar. For the purposes of investigation, let us examine the network of Fig. 8.60 using mesh and nodal analysis.

Mesh analysis (Fig. 8.61) yields

$$\begin{aligned}(3\,\Omega + 4\,\Omega + 2\,\Omega)I_1 - (4\,\Omega)I_2 - (2\,\Omega)I_3 &= 20\text{ V}\\(4\,\Omega + 5\,\Omega + 2\,\Omega)I_2 - (4\,\Omega)I_1 - (5\,\Omega)I_3 &= 0\\(2\,\Omega + 5\,\Omega + 1\,\Omega)I_3 - (2\,\Omega)I_1 - (5\,\Omega)I_2 &= 0\end{aligned}$$

and

$$\begin{aligned}9I_1 - 4I_2 - 2I_3 &= 20\\-4I_1 + 11I_2 - 5I_3 &= 0\\-2I_1 - 5I_2 + 8I_3 &= 0\end{aligned}$$

with the result that

$$\begin{aligned}I_1 &= \mathbf{4\text{ A}}\\I_2 &= \mathbf{2.667\text{ A}}\\I_3 &= \mathbf{2.667\text{ A}}\end{aligned}$$

The net current through the 5-Ω resistor is

$$I_{5\Omega} = I_2 - I_3 = \mathbf{2.667\text{ A}} - \mathbf{2.667\text{ A}} = \mathbf{0\text{ A}}$$

Nodal analysis (Fig. 8.62) yields

$$\begin{aligned}\left(\frac{1}{3\,\Omega} + \frac{1}{4\,\Omega} + \frac{1}{2\,\Omega}\right)V_1 - \left(\frac{1}{4\,\Omega}\right)V_2 - \left(\frac{1}{2\,\Omega}\right)V_3 &= \frac{20}{3}\text{ A}\\\left(\frac{1}{4\,\Omega} + \frac{1}{2\,\Omega} + \frac{1}{5\,\Omega}\right)V_2 - \left(\frac{1}{4\,\Omega}\right)V_1 - \left(\frac{1}{5\,\Omega}\right)V_3 &= 0\\\left(\frac{1}{5\,\Omega} + \frac{1}{2\,\Omega} + \frac{1}{1\,\Omega}\right)V_3 - \left(\frac{1}{2\,\Omega}\right)V_1 - \left(\frac{1}{5\,\Omega}\right)V_2 &= 0\end{aligned}$$

and

$$\begin{aligned}\left(\frac{1}{3\,\Omega} + \frac{1}{4\,\Omega} + \frac{1}{2\,\Omega}\right)V_1 - \left(\frac{1}{4\,\Omega}\right)V_2 - \left(\frac{1}{2\,\Omega}\right)V_3 &= \frac{20}{3}\text{ A}\\-\left(\frac{1}{4\,\Omega}\right)V_1 + \left(\frac{1}{4\,\Omega} + \frac{1}{2\,\Omega} + \frac{1}{5\,\Omega}\right)V_2 - \left(\frac{1}{5\,\Omega}\right)V_3 &= 0\\-\left(\frac{1}{2\,\Omega}\right)V_1 - \left(\frac{1}{5\,\Omega}\right)V_2 + \left(\frac{1}{5\,\Omega} + \frac{1}{2\,\Omega} + \frac{1}{1\,\Omega}\right)V_3 &= 0\end{aligned}$$

Note the symmetry of the solution.

With the TI-85 calculator, the top part of the determinant is determined by the following (take note of the calculations within parentheses):

det[[20/3,−1/4,−1/2][0,(1/4+1/2+1/5),−1/5][0,−1/5,(1/5+1/2+1/1)]] (ENTER)	10.5

CALC. 8.4

with the bottom of the determinant determined by:

det[[(1/3+1/4+1/2),−1/4,−1/2][−1/4,(1/4+1/2+1/5),−1/5][−1/2,−1/5,(1/5+1/2+1/1)]] (ENTER)	1.312

CALC. 8.5

Finally

10.5/1.312 (ENTER)	8

CALC. 8.6

and $V_1 = \mathbf{8\ V}$

Similarly, $V_2 = \mathbf{2.667\ V}$ and $V_3 = \mathbf{2.667\ V}$

and the voltage across the 5-Ω resistor is

$$V_{5\Omega} = V_2 - V_3 = \mathbf{2.667\ V} - \mathbf{2.667\ V} = \mathbf{0\ V}$$

Since $V_{5\Omega} = 0$ V, we can insert a short in place of the bridge arm without affecting the network behavior. (Certainly $V = IR = I \cdot (0) = 0$ V.) In Fig. 8.63, a short circuit has replaced the resistor R_5, and the voltage across R_4 is to be determined. The network is redrawn in Fig. 8.64, and

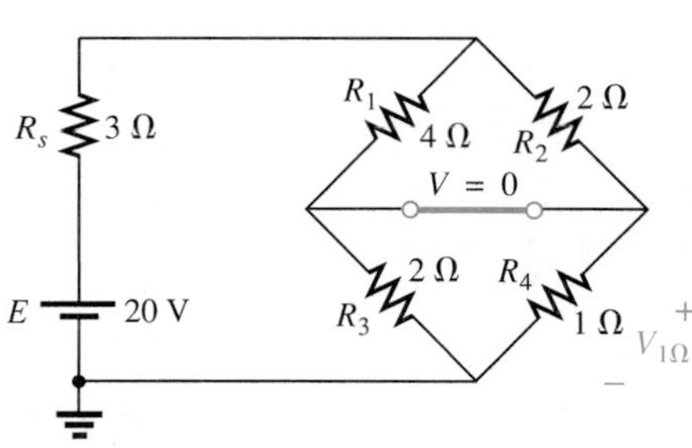

FIG. 8.63
Substituting the short-circuit equivalent for the balance arm of a balanced bridge.

$$V_{1\Omega} = \frac{(2\ \Omega \parallel 1\ \Omega)20\ \text{V}}{(2\ \Omega \parallel 1\ \Omega) + (4\ \Omega \parallel 2\ \Omega) + 3\ \Omega} \quad \text{(voltage divider rule)}$$

$$= \frac{\frac{2}{3}(20\ \text{V})}{\frac{2}{3} + \frac{8}{6} + 3} = \frac{\frac{2}{3}(20\ \text{V})}{\frac{2}{3} + \frac{4}{3} + \frac{9}{3}}$$

$$= \frac{2(20\ \text{V})}{2 + 4 + 9} = \frac{40\ \text{V}}{15} = \mathbf{2.667\ V}$$

as obtained earlier.

We found through mesh analysis that $I_{5\Omega} = 0$ A, which has as its equivalent an open circuit as shown in Fig. 8.65(a). (Certainly $I = V/R = 0/(\infty\ \Omega) = 0$ A.) The voltage across the resistor R_4 will again be determined and compared with the result above.

The network is redrawn after combining series elements, as shown in Fig. 8.65(b), and

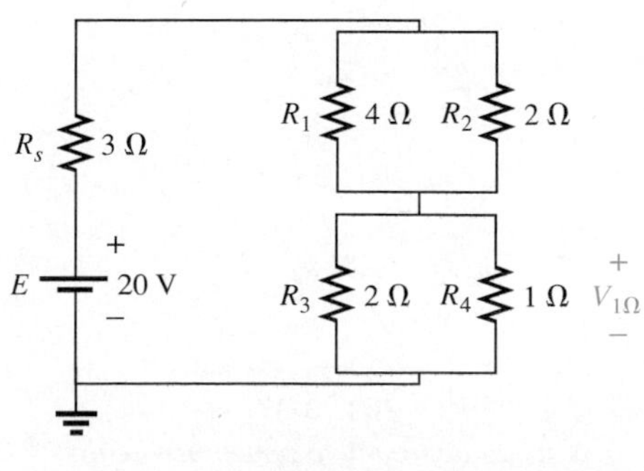

FIG. 8.64
Redrawing the network of Fig. 8.63.

$$V_{3\Omega} = \frac{(6\ \Omega \parallel 3\ \Omega)(20\ \text{V})}{6\ \Omega \parallel 3\ \Omega + 3\ \Omega} = \frac{2\ \Omega(20\ \text{V})}{2\ \Omega + 3\ \Omega} = 8\ \text{V}$$

and
$$V_{1\Omega} = \frac{1\ \Omega(8\ \text{V})}{1\ \Omega + 2\ \Omega} = \frac{8\ \text{V}}{3} = \mathbf{2.667\ V}$$

as above.

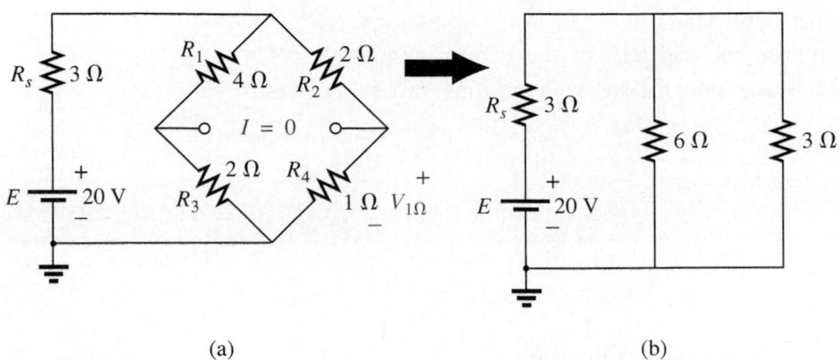

FIG. 8.65
Substituting the open-circuit equivalent for the balance arm of a balanced bridge.

The condition $V_{5\Omega} = 0$ V or $I_{5\Omega} = 0$ A exists only for a particular relationship between the resistors of the network. Let us now derive this relationship using the network of Fig. 8.66, in which it is indicated that $I = 0$ A and $V = 0$ V. Note that resistor R_s of the network of Fig. 8.65 will not appear in the following analysis.

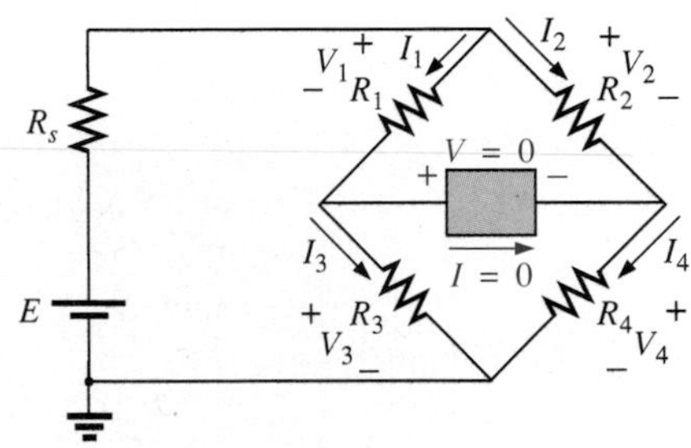

FIG. 8.66
Establishing the balance criteria for a bridge network.

The bridge network is said to be *balanced* when the condition of $I = 0$ A or $V = 0$ V exists.

If $V = 0$ V (short circuit between *a* and *b*), then

$$V_1 = V_2$$

and
$$I_1R_1 = I_2R_2$$

or
$$I_1 = \frac{I_2R_2}{R_1}$$

In addition, when $V = 0$ V,

$$V_3 = V_4$$

and
$$I_3R_3 = I_4R_4$$

If we set $I = 0$ A, then $I_3 = I_1$ and $I_4 = I_2$, with the result that the above equation becomes

$$I_1R_3 = I_2R_4$$

Substituting for I_1 from above yields

$$\left(\frac{I_2R_2}{R_1}\right)R_3 = I_2R_4$$

or, rearranging, we have

$$\boxed{\frac{R_1}{R_3} = \frac{R_2}{R_4}} \qquad \textbf{(8.4)}$$

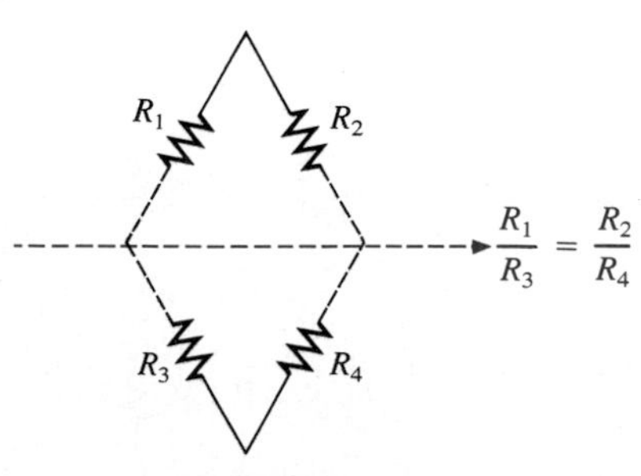

FIG. 8.67
A visual approach to remembering the balance condition.

This conclusion states that if the ratio of R_1 to R_3 is equal to that of R_2 to R_4, the bridge will be balanced, and $I = 0$ A or $V = 0$ V. A method of memorizing this form is indicated in Fig. 8.67.

For the example above, $R_1 = 4\ \Omega$, $R_2 = 2\ \Omega$, $R_3 = 2\ \Omega$, $R_4 = 1\ \Omega$, and

$$\frac{R_1}{R_3} = \frac{R_2}{R_4} \rightarrow \frac{4\ \Omega}{2\ \Omega} = \frac{2\ \Omega}{1\ \Omega} = 2$$

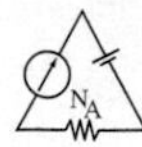

The emphasis in this section has been on the balanced situation. Understand that if the ratio is not satisfied, there will be a potential drop across the balance arm and a current through it. The methods just described (mesh and nodal analysis) will yield any and all potentials or currents desired, just as they did for the balanced situation.

8.12 Y-Δ (T-π) AND Δ-Y (π-T) CONVERSIONS

Circuit configurations are often encountered in which the resistors do not appear to be in series or parallel. Under these conditions, it may be necessary to convert the circuit from one form to another to solve for any unknown quantities if mesh or nodal analysis is not applied. Two circuit configurations that often account for these difficulties are the wye (Y) and delta (Δ), depicted in Fig. 8.68(a). They are also referred to as the tee (T) and pi (π), respectively, as indicated in Fig. 8.68(b). Note that the pi is actually an inverted delta.

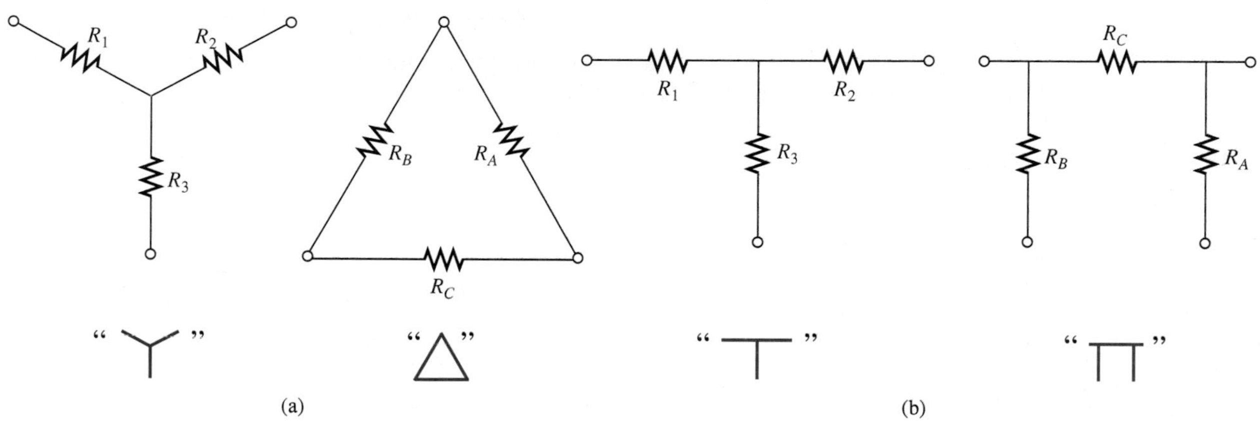

FIG. 8.68
The Y (T) and Δ (π) configurations.

The purpose of this section is to develop the equations for converting from Δ to Y, or vice versa. This type of conversion will normally lead to a network that can be solved using techniques such as those described in Chapter 7. In other words, in Fig. 8.69, with terminals *a*, *b*, and *c* held fast, if the wye (Y) configuration were desired *instead of* the inverted delta (Δ) configuration, all that would be necessary is a direct application of the equations to be derived. The phrase *instead of* is emphasized to ensure that it is understood that only one of these configurations is to appear at one time between the indicated terminals.

It is our purpose (referring to Fig. 8.69) to find some expression for R_1, R_2, and R_3 in terms of R_A, R_B, and R_C, and vice versa, that will ensure that the resistance between any two terminals of the Y configuration will be the same with the Δ configuration inserted in place of the Y configuration (and vice versa). If the two circuits are to be equivalent, the total resistance between any two terminals must be the same. Consider terminals *a*-*c* in the Δ-Y configurations of Fig. 8.70.

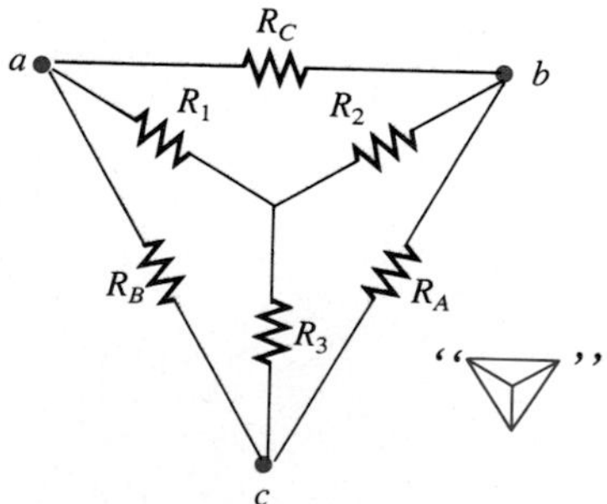

FIG. 8.69
Introducing the concept of Δ-Y or Y-Δ conversions.

Let us first assume that we want to convert the Δ (R_A, R_B, R_C) to the Y (R_1, R_2, R_3). This requires that we have a relationship for R_1, R_2, and R_3 in terms of R_A, R_B, and R_C. If the resistance is to be the same between terminals *a*-*c* for both the Δ and the Y, the following must be true:

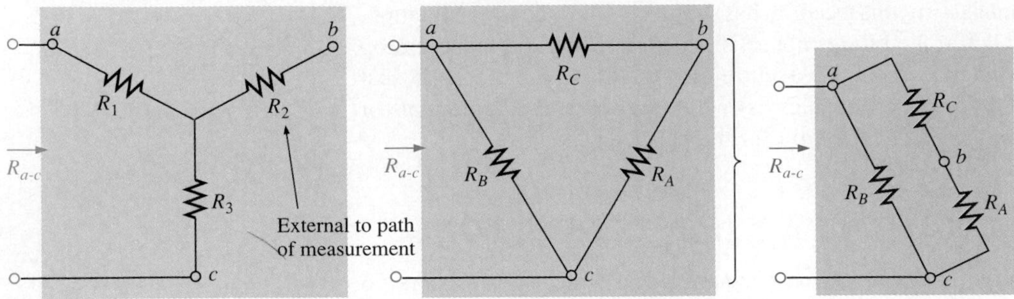

FIG. 8.70
Finding the resistance $R_{a\text{-}c}$ for the Y and Δ configurations.

$$R_{a\text{-}c}\ (\text{Y}) = R_{a\text{-}c}\ (\Delta)$$

so that

$$R_{a\text{-}c} = R_1 + R_3 = \frac{R_B(R_A + R_C)}{R_B + (R_A + R_C)} \tag{8.5a}$$

Using the same approach for *a-b* and *b-c*, we obtain the following relationships:

$$R_{a\text{-}b} = R_1 + R_2 = \frac{R_C(R_A + R_B)}{R_C + (R_A + R_B)} \tag{8.5b}$$

and

$$R_{b\text{-}c} = R_2 + R_3 = \frac{R_A(R_B + R_C)}{R_A + (R_B + R_C)} \tag{8.5c}$$

Subtracting Eq. (8.5a) from Eq. (8.5b), we have

$$(R_1 + R_2) - (R_1 + R_3) = \left(\frac{R_C R_B + R_C R_A}{R_A + R_B + R_C}\right) - \left(\frac{R_B R_A + R_B R_C}{R_A + R_B + R_C}\right)$$

so that

$$R_2 - R_3 = \frac{R_A R_C - R_B R_A}{R_A + R_B + R_C} \tag{8.5d}$$

Subtracting Eq. (8.5d) from Eq. (8.5c) yields

$$(R_2 + R_3) - (R_2 - R_3) = \left(\frac{R_A R_B + R_A R_C}{R_A + R_B + R_C}\right) - \left(\frac{R_A R_C - R_B R_A}{R_A + R_B + R_C}\right)$$

so that

$$2R_3 = \frac{2R_B R_A}{R_A + R_B + R_C}$$

resulting in the following expression for R_3 in terms of R_A, R_B, and R_C:

$$R_3 = \frac{R_A R_B}{R_A + R_B + R_C} \tag{8.6a}$$

Following the same procedure for R_1 and R_2, we have

$$R_1 = \frac{R_B R_C}{R_A + R_B + R_C} \tag{8.6b}$$

and $$R_2 = \frac{R_A R_C}{R_A + R_B + R_C} \quad \textbf{(8.6c)}$$

Note that each resistor of the Y is equal to the product of the resistors in the two closest branches of the Δ divided by the sum of the resistors in the Δ.

To obtain the relationships necessary to convert from a Y to a Δ, first divide Eq. (8.6a) by Eq. (8.6b):

$$\frac{R_3}{R_1} = \frac{(R_A R_B)/(R_A + R_B + R_C)}{(R_B R_C)/(R_A + R_B + R_C)} = \frac{R_A}{R_C}$$

or $$R_A = \frac{R_C R_3}{R_1}$$

Then divide Eq. (8.6a) by Eq. (8.6c):

$$\frac{R_3}{R_2} = \frac{(R_A R_B)/(R_A + R_B + R_C)}{(R_A R_C)/(R_A + R_B + R_C)} = \frac{R_B}{R_C}$$

or $$R_B = \frac{R_3 R_C}{R_2}$$

Substituting for R_A and R_B in Eq. (8.6c) yields

$$R_2 = \frac{(R_C R_3/R_1)R_C}{(R_3 R_C/R_2) + (R_C R_3/R_1) + R_C}$$
$$= \frac{(R_3/R_1)R_C}{(R_3/R_2) + (R_3/R_1) + 1}$$

Placing these over a common denominator, we obtain

$$R_2 = \frac{(R_3 R_C/R_1)}{(R_1R_2 + R_1R_3 + R_2R_3)/(R_1R_2)}$$
$$= \frac{R_2 R_3 R_C}{R_1R_2 + R_1R_3 + R_2R_3}$$

and $$R_C = \frac{R_1R_2 + R_1R_3 + R_2R_3}{R_3} \quad \textbf{(8.7a)}$$

We follow the same procedure for R_B and R_A:

$$R_A = \frac{R_1R_2 + R_1R_3 + R_2R_3}{R_1} \quad \textbf{(8.7b)}$$

and $$R_B = \frac{R_1R_2 + R_1R_3 + R_2R_3}{R_2} \quad \textbf{(8.7c)}$$

Note that the value of each resistor of the Δ is equal to the sum of the possible product combinations of the resistances of the Y divided by the resistance of the Y farthest from the resistor to be determined.

Let us consider what would occur if all the values of a Δ or Y were the same. If $R_A = R_B = R_C$, Eq. (8.6a) would become (using R_A only) the following:

$$R_3 = \frac{R_A R_B}{R_A + R_B + R_C} = \frac{R_A R_A}{R_A + R_A + R_A} = \frac{R_A^2}{3R_A} = \frac{R_A}{3}$$

and, following the same procedure,

$$R_1 = \frac{R_A}{3} \qquad R_2 = \frac{R_A}{3}$$

In general, therefore,

$$R_Y = \frac{R_\Delta}{3} \tag{8.8a}$$

or

$$R_\Delta = 3R_Y \tag{8.8b}$$

which indicates that *for a Y of three equal resistors, the value of each resistor of the Δ is equal to three times the value of any resistor of the Y*. If only two elements of a Y or a Δ are the same, the corresponding Δ or Y of each will also have two equal elements. The converting of equations will be left as an exercise for the reader.

The Y and the Δ will often appear as shown in Fig. 8.71. They are then referred to as a *tee* (T) and *pi* (π) network. The equations used to convert from one form to the other are exactly the same as those developed for the Y and Δ transformation.

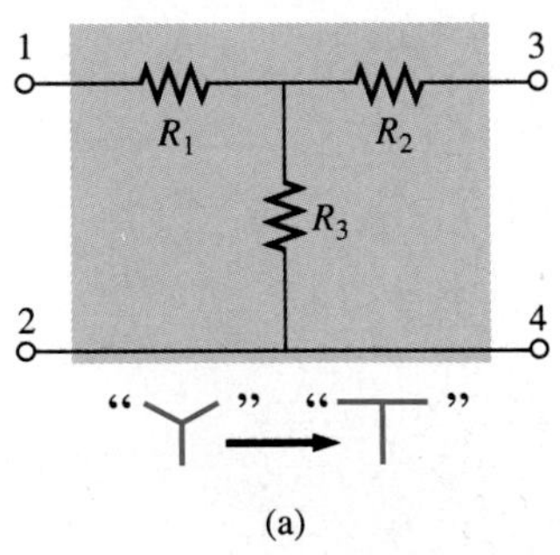

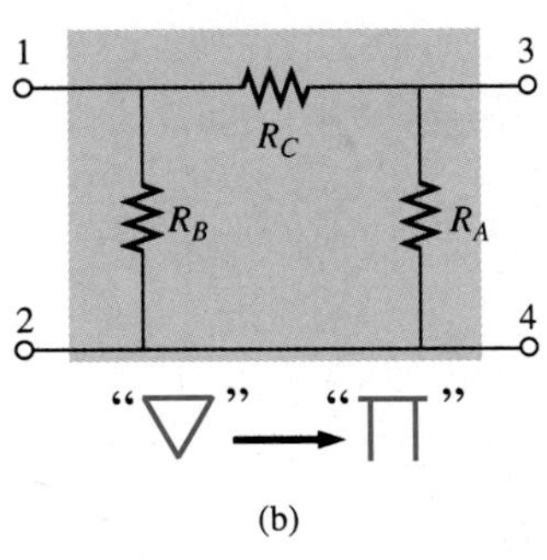

FIG. 8.71
The relationship between the Y and T configurations and the Δ and π configurations.

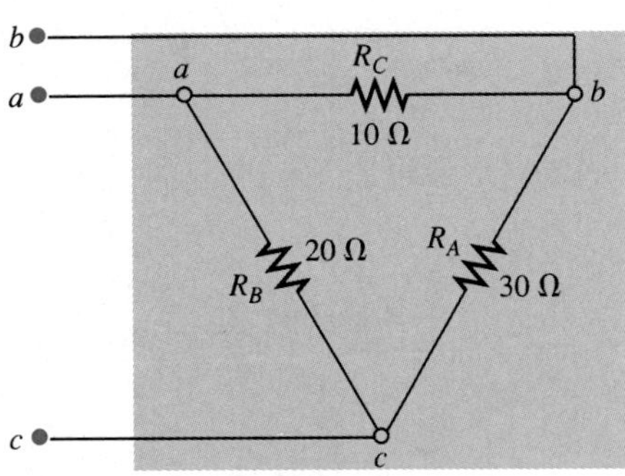

FIG. 8.72
Example 8.27.

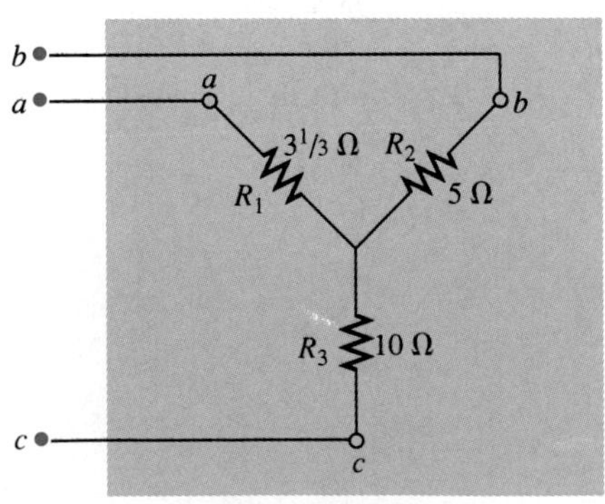

FIG. 8.73
The Y equivalent for the Δ of Fig. 8.72.

EXAMPLE 8.27 Convert the Δ of Fig. 8.72 to a Y.

Solution:

$$R_1 = \frac{R_B R_C}{R_A + R_B + R_C} = \frac{(20\ \Omega)(10\ \Omega)}{30\ \Omega + 20\ \Omega + 10\ \Omega} = \frac{200\ \Omega}{60} = \mathbf{3\tfrac{1}{3}\ \Omega}$$

$$R_2 = \frac{R_A R_C}{R_A + R_B + R_C} = \frac{(30\ \Omega)(10\ \Omega)}{60\ \Omega} = \frac{300\ \Omega}{60} = \mathbf{5\ \Omega}$$

$$R_3 = \frac{R_A R_B}{R_A + R_B + R_C} = \frac{(20\ \Omega)(30\ \Omega)}{60\ \Omega} = \frac{600\ \Omega}{60} = \mathbf{10\ \Omega}$$

The equivalent network is shown in Fig. 8.73.

EXAMPLE 8.28 Convert the Y of Fig. 8.74 to a Δ.

Solution:

$$R_A = \frac{R_1R_2 + R_1R_3 + R_2R_3}{R_1}$$
$$= \frac{(60\ \Omega)(60\ \Omega) + (60\ \Omega)(60\ \Omega) + (60\ \Omega)(60\ \Omega)}{60\ \Omega}$$
$$= \frac{3600\ \Omega + 3600\ \Omega + 3600\ \Omega}{60} = \frac{10{,}800\ \Omega}{60}$$
$$R_A = \mathbf{180\ \Omega}$$

However, the three resistors for the Y are equal, permitting the use of Eq. (8.8) and yielding

$$R_\Delta = 3R_Y = 3(60\ \Omega) = 180\ \Omega$$

and $$R_B = R_C = \mathbf{180\ \Omega}$$

The equivalent network is shown in Fig. 8.75.

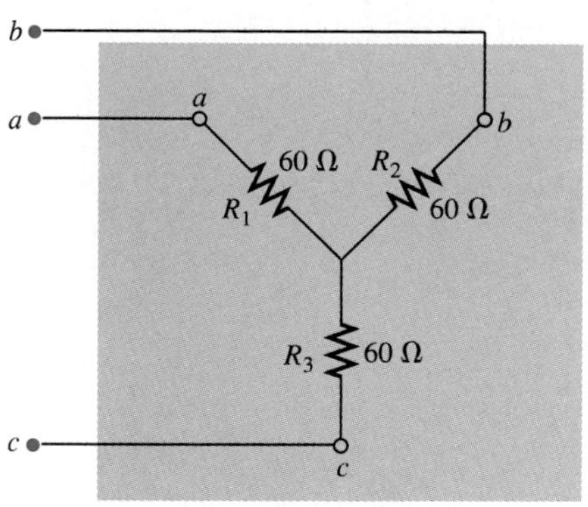

FIG. 8.74
Example 8.28.

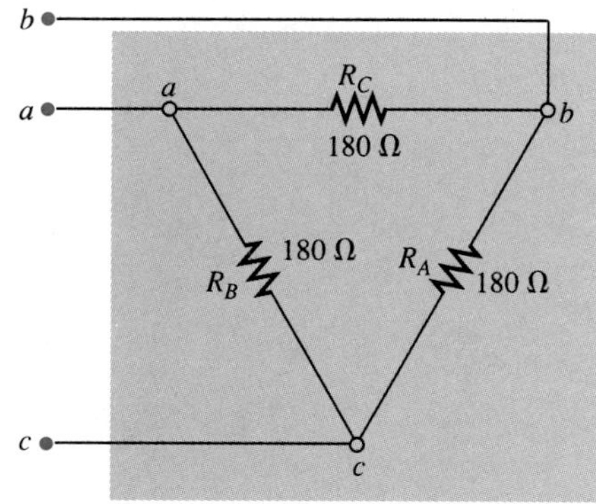

FIG. 8.75
The Δ equivalent for the Y of Fig. 8.74.

EXAMPLE 8.29 Find the total resistance of the network of Fig. 8.76, where $R_A = 3\ \Omega$, $R_B = 3\ \Omega$, and $R_C = 6\ \Omega$.

Solution:

Two resistors of the Δ were equal; therefore, two resistors of the Y will be equal.

$$R_1 = \frac{R_BR_C}{R_A + R_B + R_C} = \frac{(3\ \Omega)(6\ \Omega)}{3\ \Omega + 3\ \Omega + 6\ \Omega} = \frac{18\ \Omega}{12} = \mathbf{1.5\ \Omega}$$
$$R_2 = \frac{R_AR_C}{R_A + R_B + R_C} = \frac{(3\ \Omega)(6\ \Omega)}{12\ \Omega} = \frac{18\ \Omega}{12} = \mathbf{1.5\ \Omega}$$
$$R_3 = \frac{R_AR_B}{R_A + R_B + R_C} = \frac{(3\ \Omega)(3\ \Omega)}{12\ \Omega} = \frac{9\ \Omega}{12} = \mathbf{0.75\ \Omega}$$

Replacing the Δ by the Y, as shown in Fig. 8.77, yields

$$R_T = 0.75\ \Omega + \frac{(4\ \Omega + 1.5\ \Omega)(2\ \Omega + 1.5\ \Omega)}{(4\ \Omega + 1.5\ \Omega) + (2\ \Omega + 1.5\ \Omega)}$$
$$= 0.75\ \Omega + \frac{(5.5\ \Omega)(3.5\ \Omega)}{5.5\ \Omega + 3.5\ \Omega}$$
$$= 0.75\ \Omega + 2.139\ \Omega$$
$$R_T = \mathbf{2.889\ \Omega}$$

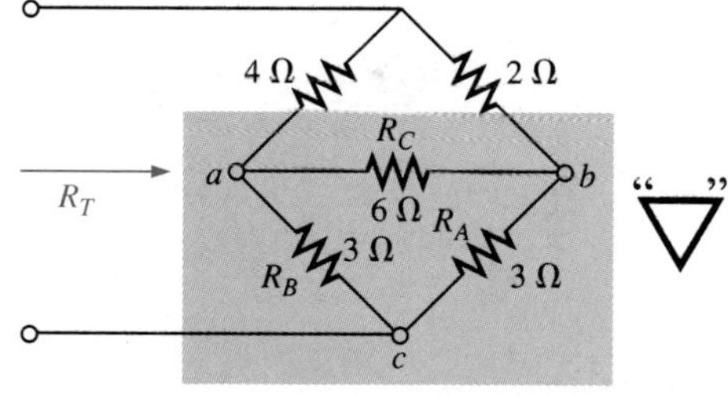

FIG. 8.76
Example 8.29.

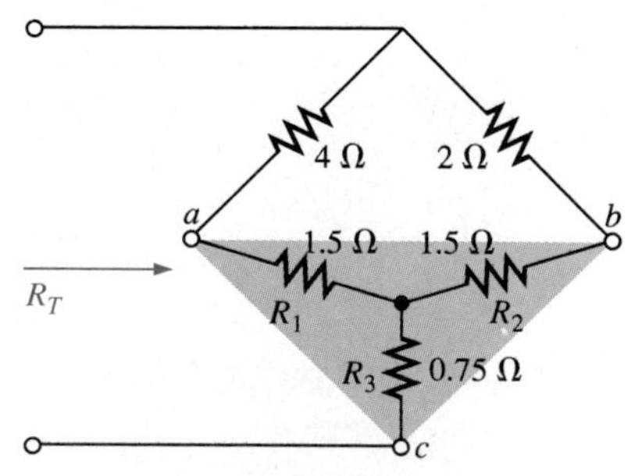

FIG. 8.77
Substituting the Y equivalent for the bottom Δ of Fig. 8.76.

EXAMPLE 8.30 Find the total resistance of the network of Fig. 8.78.

Solution: Since all the resistors of the Δ or Y are the same, Eqs. (8.8a) and (8.8b) can be used to convert either form to the other.

a. *Converting the Δ to a Y.* Note: When this is done, the resulting *d′* of the new Y will be the same as the point *d* shown in the original figure, only because both systems are "balanced." That is, the resistance in each branch of each system has the same value:

$$R_Y = \frac{R_\Delta}{3} = \frac{6\ \Omega}{3} = 2\ \Omega \qquad \text{(Fig. 8.79)}$$

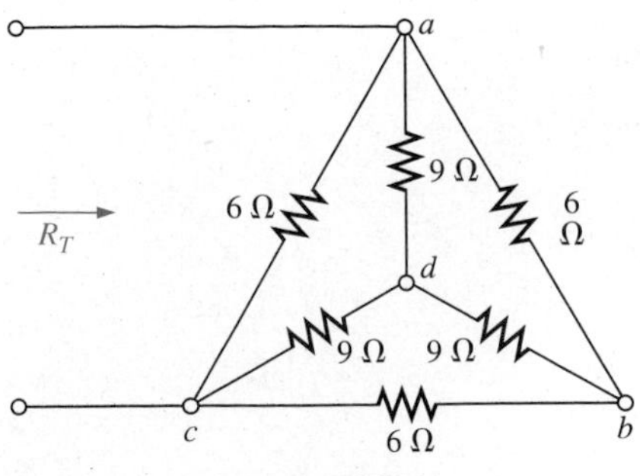

FIG. 8.78
Example 8.30.

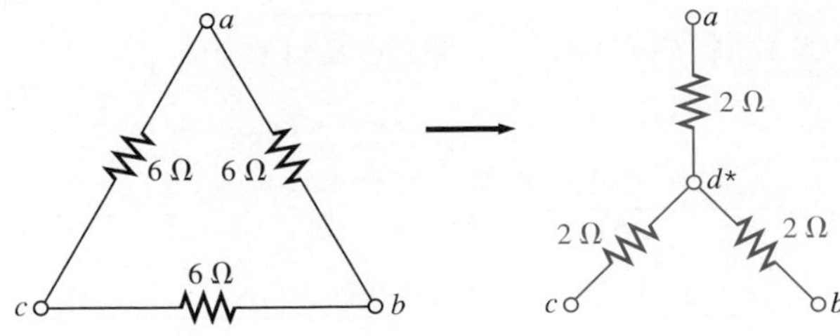

FIG. 8.79
Converting the Δ configuration of Fig. 8.78 to a Y configuration.

The network then appears as shown in Fig. 8.80.

$$R_T = 2\left[\frac{(2\ \Omega)(9\ \Omega)}{2\ \Omega + 9\ \Omega}\right] = \mathbf{3.2727\ \Omega}$$

b. *Converting the Y to a Δ.*

$$R_\Delta = 3R_Y = (3)(9\ \Omega) = 27\ \Omega \qquad \text{(Fig. 8.81)}$$

$$R'_T = \frac{(6\ \Omega)(27\ \Omega)}{6\ \Omega + 27\ \Omega} = \frac{162\ \Omega}{33} = 4.9091\ \Omega$$

$$R_T = \frac{R'_T\,(R'_T + R'_T)}{R'_T + (R'_T + R'_T)} = \frac{R'_T 2R'_T}{3R'_T} = \frac{2R'_T}{3}$$

$$= \frac{2(4.9091\ \Omega)}{3} = \mathbf{3.2727\ \Omega}$$

which checks with the previous solution.

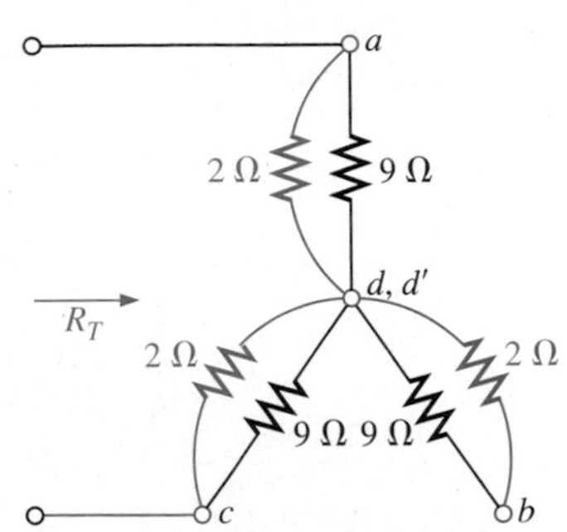

FIG. 8.80
Substituting the Y configuration for the converted Δ into the network of Fig. 8.78.

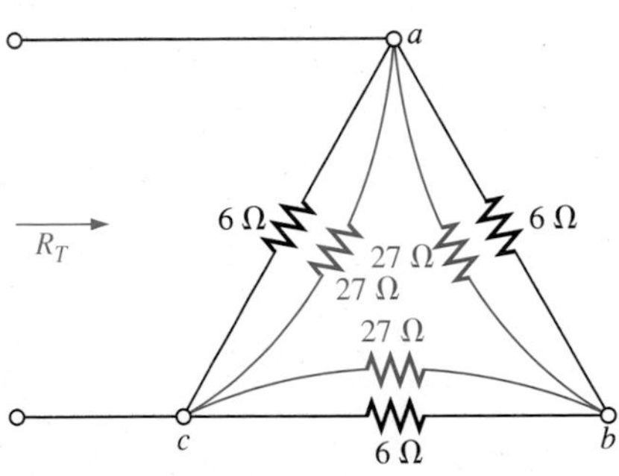

FIG. 8.81
Substituting the converted Y configuration into the network of Fig. 8.78.

8.13 COMPUTER ANALYSIS

PSpice (DOS)

The network appearing in Example 8.9 (Fig. 8.21) is redrawn in Fig. 8.82 with the nodes defined for a PSpice input file. The input file of Fig. 8.83 is requesting the nodal voltage V_2 and the branch currents I_1, I_2, and I_3. Since the current through R_1 is the same as the mesh current of that window, I_1 is both the branch and mesh current. The same is true for I_2 in the other window. The results reveal that $V_2 = V_{R_3} = +4$ V, that the branch and mesh current through R_1 is 1 A from node 1 to node 2, and that the branch and mesh current through R_2 is 2 A from node 3 to 2. The branch current through R_3 is 1 A from node 2 to our reference node (0). Keep in mind that a positive result for a current in PSpice indicates that the flow direction is from the first to second nodes of the input file for the resistor between the two nodes. A negative sign indicates the reverse. The results, in total, as obtained using PSpice are the same as those obtained in Example 8.9.

The input file of Fig. 8.85 is for the network of Fig. 8.84, which appeared as Fig. 8.53. It is the same network as that just examined but with the voltage sources converted to current sources. There are only two nodal voltages to be determined for Fig. 8.84, but note that the result for the voltage across the 4-Ω resistor is the same for either approach and corresponds exactly with the result of Example 8.23.

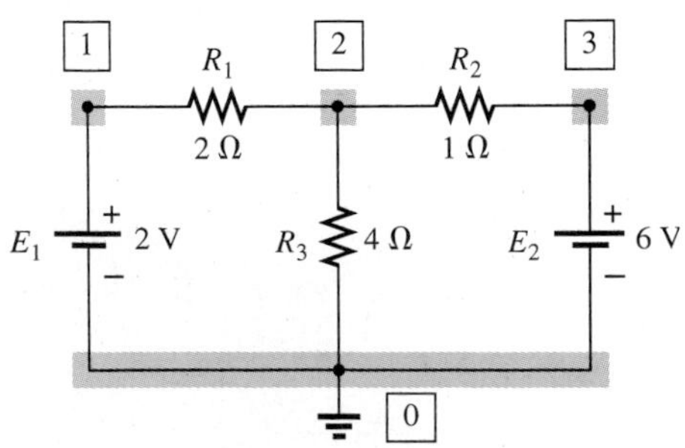

FIG. 8.82
Defining the nodes for a PSpice (DOS) analysis.

```
Chapter 8 - Two loop circuit

 ****      CIRCUIT DESCRIPTION

******************************************************************************

VE1 1 0 2V
VE2 3 0 6V
R1  2 1 2
R2  2 3 1
R3  2 0 4
.DC VE1 2 2 1
.PRINT DC V(2) I(R1) I(R2) I(R3)
.OPTIONS NOPAGE
.END

 ****      DC TRANSFER CURVES                    TEMPERATURE =   27.000 DEG C

  VE1          V(2)          I(R1)         I(R2)         I(R3)

   2.000E+00    4.000E+00    1.000E+00   -2.000E+00    1.000E+00
```

FIG. 8.83
The input and output files for the PSpice (DOS) analysis of Fig. 8.82.

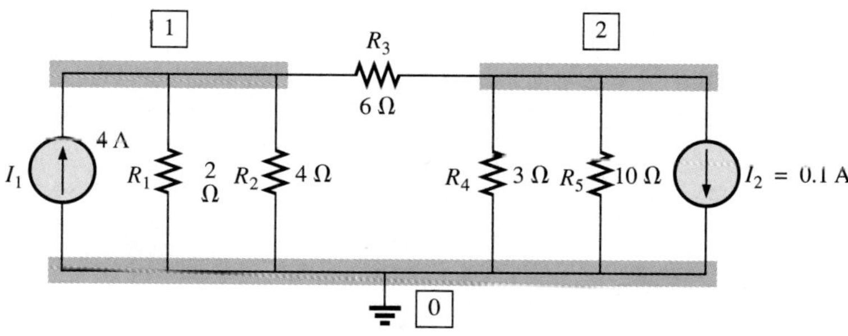

FIG. 8.84
Defining the nodes for a PSpice (DOS) analysis.

```
Chapter 8 - Three loop circuit

 ****      CIRCUIT DESCRIPTION

******************************************************************************

IS1 0 1 4
IS2 2 0 .1
R1  1 0 2
R2  1 0 4
R3  1 2 6
R4  2 0 3
R5  2 0 10
.DC IS2 .1 .1 1
.PRINT DC V(1) V(2)
.OPTIONS NOPAGE
.END
```

FIG. 8.85
The input file for the PSpice (DOS) analysis of Fig. 8.84.

```
****      DC TRANSFER CURVES                    TEMPERATURE =    27.000 DEG C

 IS2          V(1)          V(2)

  1.000E-01    4.564E+00     1.101E+00
```

FIG. 8.85 (continued)
The output file for the PSpice (DOS) analysis of Fig. 8.84.

PSpice (Windows)

The first application of schematics will be a verification of the results of Example 8.18. All the elements of creating the schematic of Fig. 8.86 have been presented in earlier chapters. Note that the desired mesh current is displayed using **IPROBE** obtained from the sequence: **Draw-Get New Part-Browse-special.slb-IPROBE.** Following **Analysis-Simulate,** the magnitude of the current is displayed as 1.22 A, which matches the solution of Example 8.18. Note the manner in which the **IPROBE** was installed to ensure a positive answer. Recall that it can be turned with the same **Ctrl-R** used for resistors and other elements. The nodal voltages were also obtained using **VIEWPOINTS,** which were obtained with the same sequence employed for **IPROBE.** Note that two of the nodal voltages were set to the left and right of the schematic to avoid having the solutions cross the network. This was accomplished by simply adding lines using **Draw-Wire.**

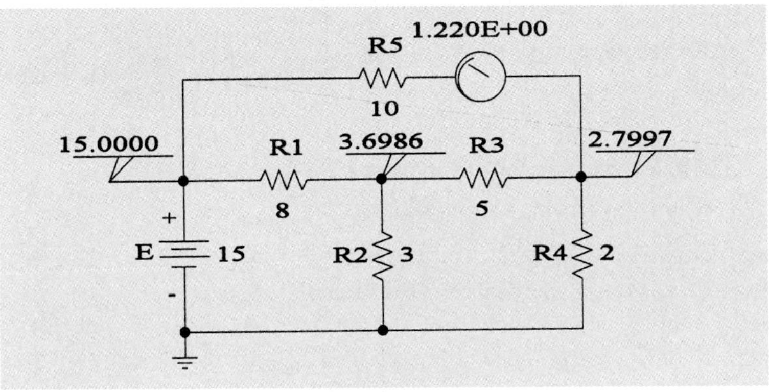

FIG. 8.86
Determining the mesh current $I_{10\Omega}$ and the nodal voltages for the network of Fig. 8.37 (Example 8.18).

The second network to be analyzed includes a current source, which can be obtained through the sequence **Draw-Get New Part-Browse-source.slb-IDC-OK.** It appears in the schematic of Fig. 8.87 representing the bridge network of Fig. 8.62. The **IDC** label was removed by clicking the label and then using the sequence **Edit-Delete.** It can also be done with the delete key on the keyboard after being identified. The 6.667 A was set by first clicking the 0-A default value and responding to **Set Attribute Value.** The designator for the source was changed by double-clicking the default label of **I1** and setting the **Reference Designator** to *I*. The source can be turned by the usual **Ctrl-R.** The remaining components and **VIEWPOINTS** were set as described in previous

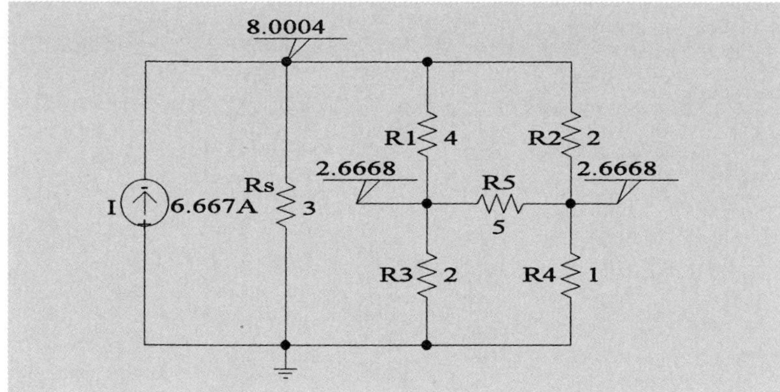

FIG. 8.87
Determining the nodal voltage for the bridge network of Fig. 8.62.

examples. The results obtained by **Analysis-Simulate** clearly indicate that the voltage across the resistor R5 is 0 V and the bridge is balanced. The nodal voltage of 8 V reveals that the current source and parallel resistor could be replaced by an ideal voltage source of 8 V without affecting the behavior of the network.

BASIC

The natural sequence of steps in either the mesh or nodal format approach simplifies the writing of the BASIC program required to solve single- or multisource networks.

A BASIC program to analyze the basic two-loop network of Fig. 8.88 is provided in Fig. 8.89. The resulting network equations are

$$I_1(R_1 + R_3) - I_2R_3 = E_1$$
$$-I_1R_3 + I_2(R_2 + R_3) = -E_2$$

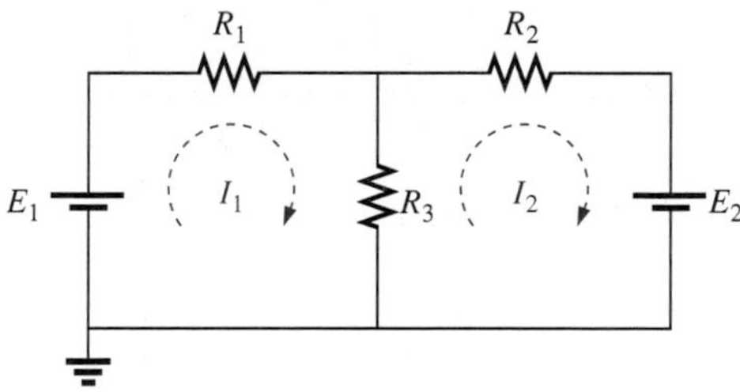

FIG. 8.88
Defining the mesh currents for a "two-window" network.

with the mesh currents calculated using determinants on lines 200 through 220. Note that the program requests the network parameters on lines 130 through 170 and prints out the result, as directed by lines 230 through 250. The values of the network are not specified in Fig. 8.88 or in the program, so once the program is run and the parameters are requested, any values can be substituted and a correct solution will be obtained. In PSpice, however, a new input file must be made up for each change of input parameters. The question is then whether it takes

```
10 REM ***** PROGRAM 8-1 *****
20 REM ******************************************
30 REM Program to evaluate the loop currents for a
40 REM 2-loop network.
50 REM ******************************************
60 REM
100 PRINT "For a 2-loop network"
110 PRINT "enter the following data:"
120 PRINT
130 INPUT "R1="; R1
140 INPUT "R2="; R2
150 INPUT "R3="; R3
160 INPUT "Voltage, E1="; E1
170 INPUT "Voltage, E2="; E2
180 PRINT
190 REM Calculate I1 and I2
200 D = R1 * R2 + R1 * R3 + R2 * R3
210 I1 = (E1 * (R2 + R3) - E2 * R3) / D
220 I2 = (-E2 * (R1 + R3) + E1 * R3) / D
230 PRINT "The loop currents are:"
240 PRINT "I1="; I1; "A"
250 PRINT "I2="; I2; "A"
260 END
```

```
For a 2-loop network
enter the following data:

R1=? 2
R2=? 1
R3=? 4
Voltage, E1=? 2
Voltage, E2=? 6

The loop currents are:
I1=-1 A
I2=-2 A
Ok


RUN
For a 2-loop network
enter the following data:

R1=? 1E3
R2=? 2.2E3
R3=? 3.3E3
Voltage, E1=? -5.4
Voltage, E2=? 8.6

The loop currents are:
I1=-4.551724E-03 A
I2=-4.294671E-03 A
```

FIG. 8.89
A BASIC program designed to calculate the mesh currents for the network of Fig. 8.88.

longer to input parameters using BASIC or to input a whole new file in PSpice. In many cases it will probably be a toss-up, but remember that it takes a great deal more time to write the BASIC program initially than to obtain a solution with PSpice. Obviously, we can debate this issue from many viewpoints. The important thing, however, is that you realize the trade-offs with each and have a working knowledge of both software packages and an appropriate language.

The first run employed the same values as Example 8.11, whereas the second run includes resistors in the kilohm range and a reversed source.

PROBLEMS

SECTION 8.2 Current Sources

1. Find the voltage V_{ab} (with polarity) across the ideal current source of Fig. 8.90.

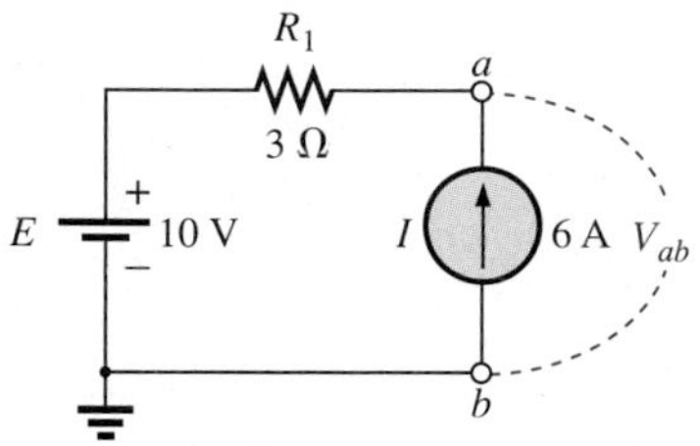

FIG. 8.90
Problem 1.

2. **a.** Determine V for the current source of Fig. 8.91(a) with an internal resistance of 10 kΩ.
   b. The source of part (a) is approximated by an ideal current source in Fig. 8.91(b) since the source resistance is much larger than the applied load. Determine the resulting voltage V for Fig. 8.91(b) and compare to that obtained in part (a). Is the use of the ideal current source a good approximation?

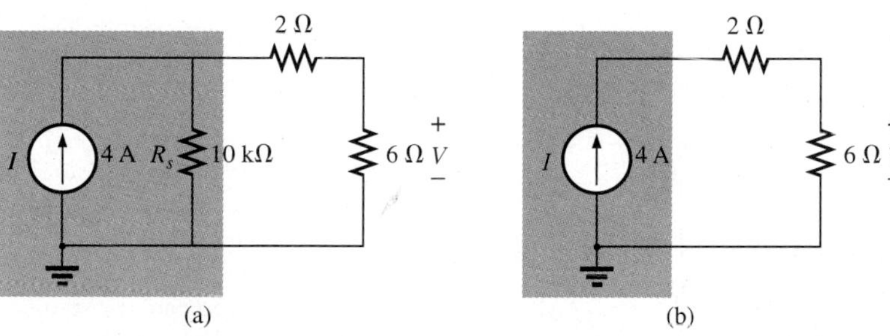

FIG. 8.91
Problem 2.

3. For the network of Fig. 8.92:
   a. Find the currents I_1 and I_s.
   b. Find the voltages V_s and V_3.
4. Find the voltage V_3 and the current I_2 for the network of Fig. 8.93.

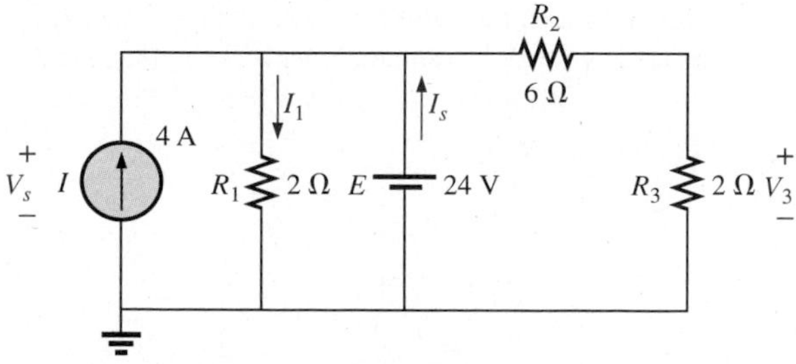

FIG. 8.92
Problem 3.

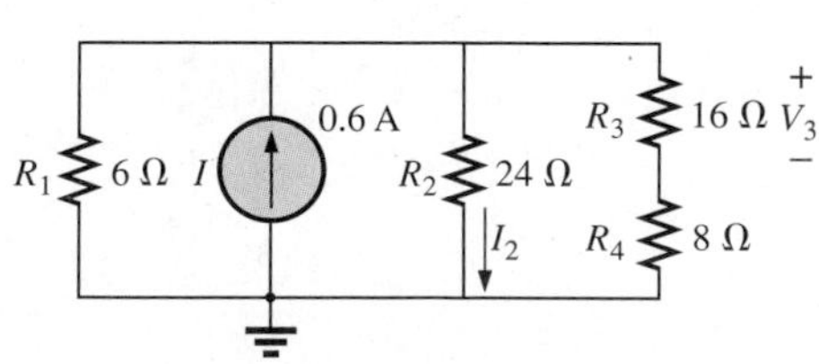

FIG. 8.93
Problem 4.

SECTION 8.3 Source Conversions

5. Convert the voltage sources of Fig. 8.94 to current sources.

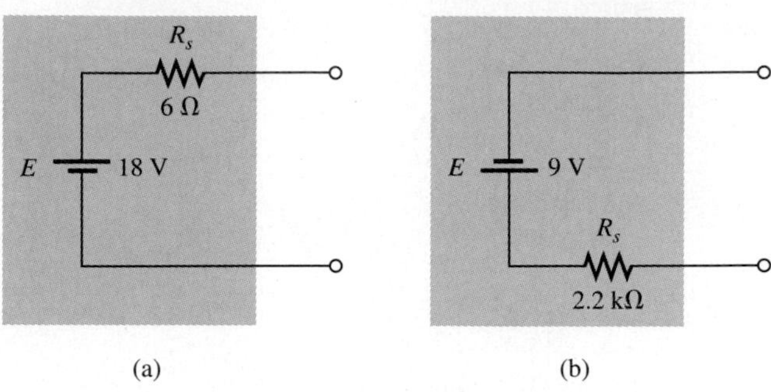

FIG. 8.94
Problem 5.

6. Convert the current sources of Fig. 8.95 to voltage sources.

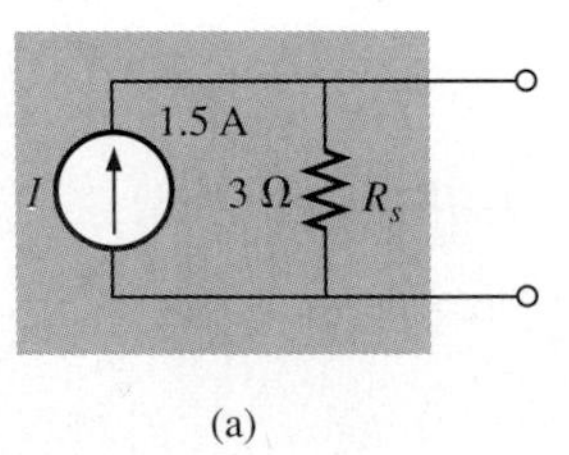

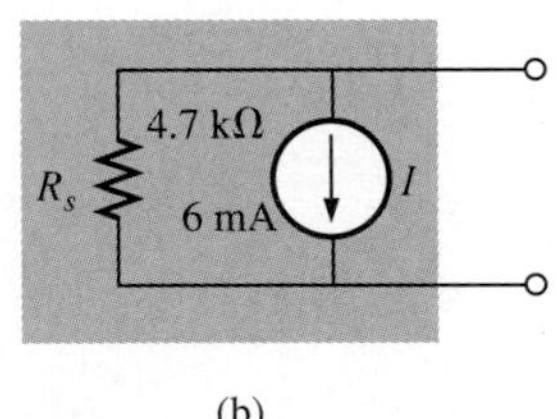

FIG. 8.95
Problem 6.

7. For the network of Fig. 8.96:
   a. Find the current through the 2-Ω resistor.
   b. Convert the current source and 4-Ω resistor to a voltage source, and again solve for the current in the 2-Ω resistor. Compare the results.

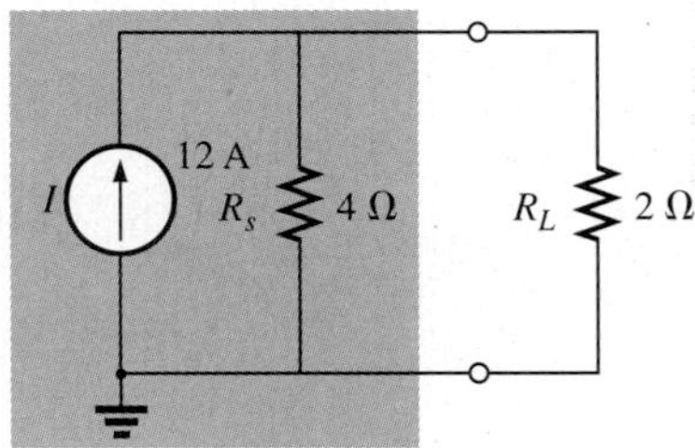

FIG. 8.96
Problem 7.

8. For the configuration of Fig. 8.97:
   a. Convert the current source and 6.8-Ω resistor to a voltage source.
   b. Find the magnitude and direction of the current I_1.
   c. Find the voltage V_{ab} and the polarity of points a and b.

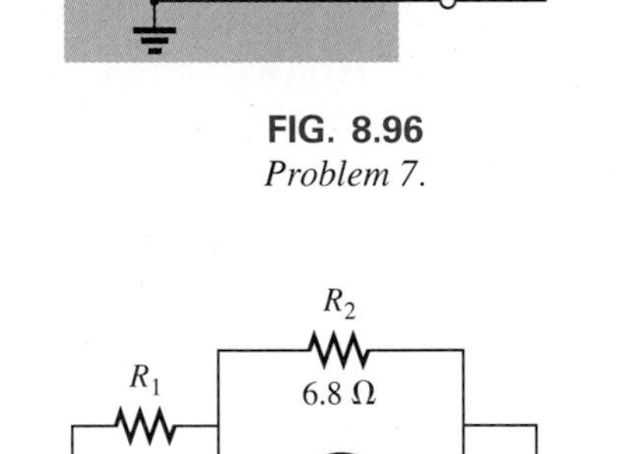

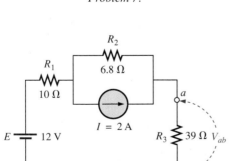

FIG. 8.97
Problem 8.

SECTION 8.4 Current Sources in Parallel

9. Find the voltage V_2 and the current I_1 for the network of Fig. 8.98.

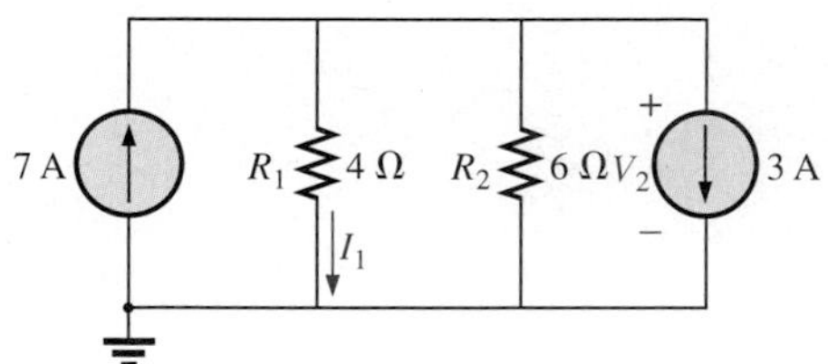

FIG. 8.98
Problem 9.

10. **a.** Convert the voltage sources of Fig. 8.99 to current sources.

b. Find the voltage V_{ab} and the polarity of points a and b.

c. Find the magnitude and direction of the current I.

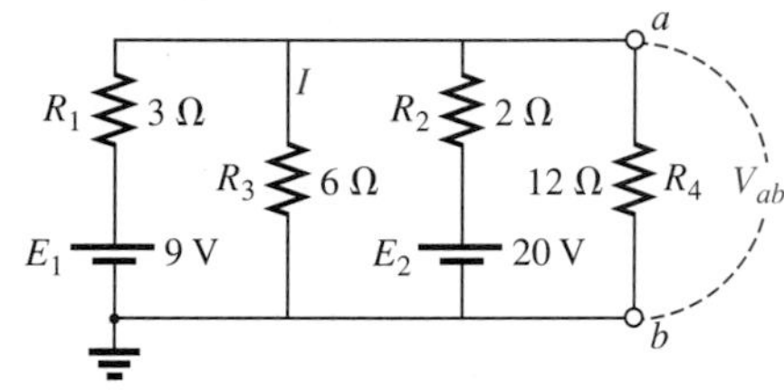

FIG. 8.99
Problem 10.

11. For the network of Fig. 8.100:

a. Convert the voltage source to a current source.

b. Reduce the network to a single current source and determine the voltage V_1.

c. Using the results of part (b), determine V_2.

d. Calculate the current I_2.

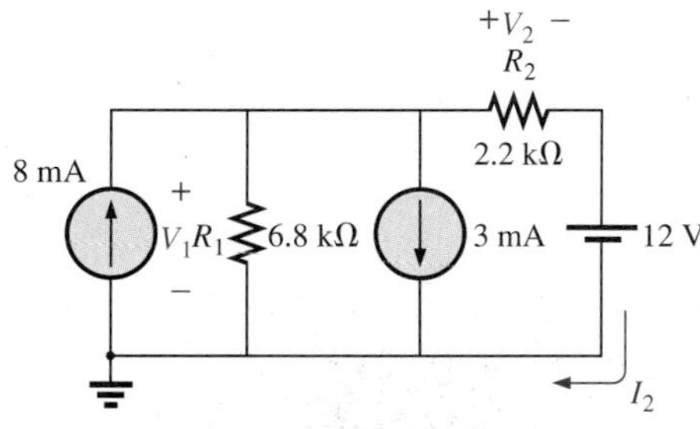

FIG. 8.100
Problem 11.

SECTION 8.6 Branch-Current Analysis

12. Using branch-current analysis, find the magnitude and direction of the current through each resistor for the networks of Fig. 8.101.

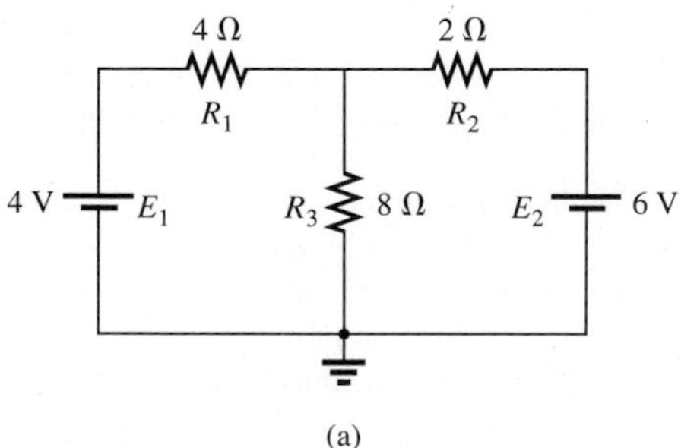

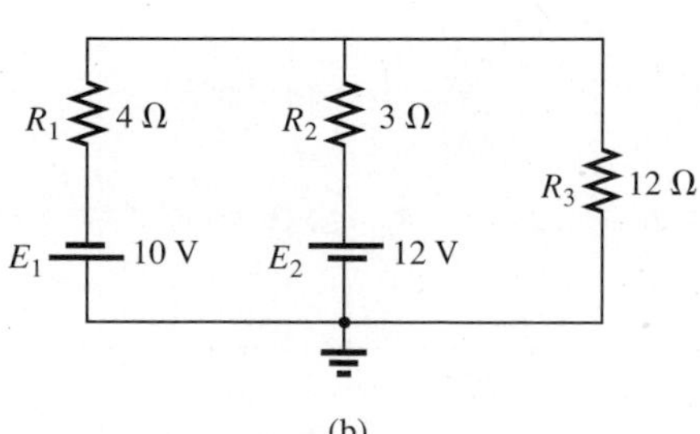

FIG. 8.101
Problems 12, 17, 25, 54, and 57.

***13.** Using branch-current analysis, find the current through each resistor for the networks of Fig. 8.102. The resistors are all standard values.

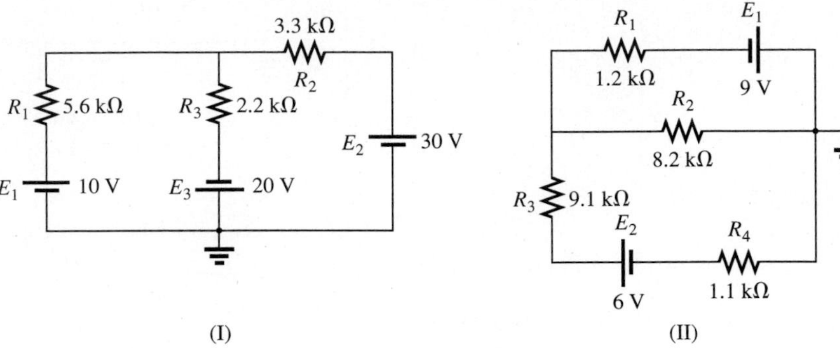

FIG. 8.102
Problems 13, 18, and 26.

***14.** For the networks of Fig. 8.103, determine the current I_2 using branch-current analysis, and then find the voltage V_{ab}.

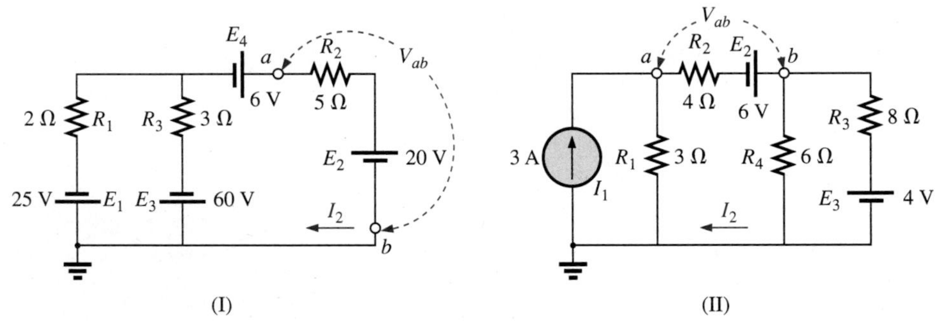

FIG. 8.103
Problems 14, 19, and 27.

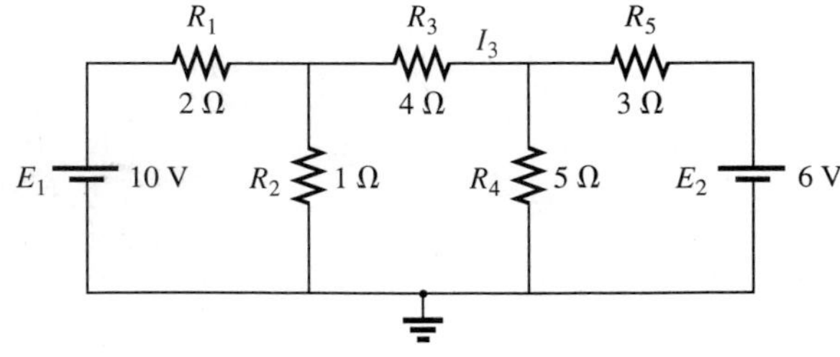

FIG. 8.104
Problems 15 and 28.

***15.** For the network of Fig. 8.104:

a. Write the equations necessary to solve for the branch currents.

b. By substitution of Kirchhoff's current law, reduce the set to three equations.

c. Rewrite the equations in a format that can be solved using third-order determinants.

d. Solve for the branch current through the resistor R_3.

***16.** For the transistor configuration of Fig. 8.105:

a. Solve for the currents I_B, I_C, and I_E using the fact that $V_{BE} = 0.7$ V and $V_{CE} = 8$ V.

b. Find the voltages V_B, V_C, and V_E with respect to ground.

c. What is the ratio of output current I_C to input current I_B? (*Note:* In transistor analysis this ratio is referred to as the dc beta of the transistor (β_{dc}).)

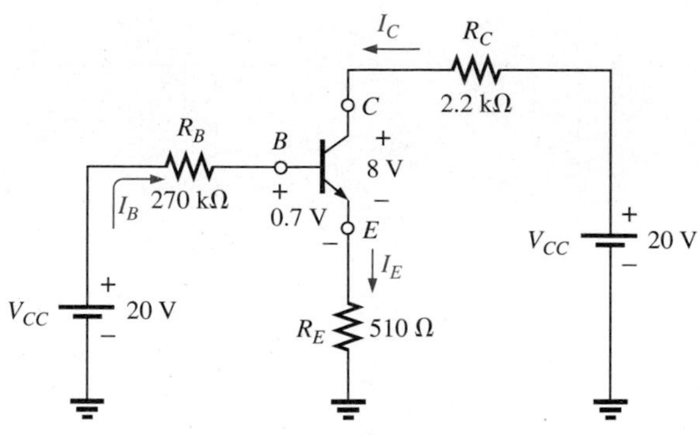

FIG. 8.105
Problem 16.

SECTION 8.7 Mesh Analysis (General Approach)

17. Find the current through each resistor for the networks of Fig. 8.101.

18. Find the current through each resistor for the networks of Fig. 8.102.

19. Find the mesh currents and the voltage V_{ab} for each network of Fig. 8.103. Use clockwise mesh currents.

20. a. Find the current I_3 for the network of Fig. 8.104 using mesh analysis.

b. Based on the results of part (a), how would you compare the application of mesh analysis to the branch-current method?

***21.** Using mesh analysis, determine the current through the 5-Ω resistor for each network of Fig. 8.106. Then determine the voltage V_a.

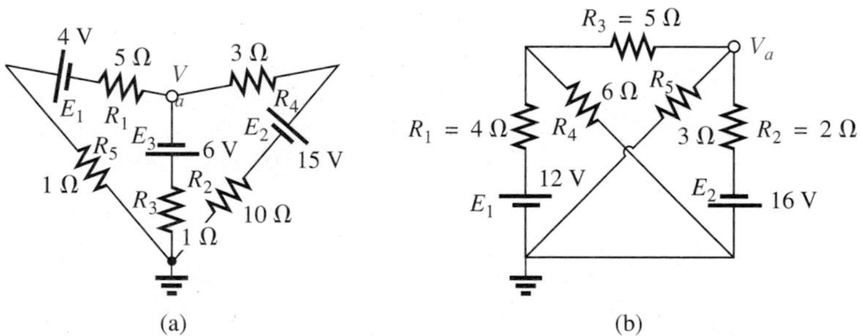

FIG. 8.106
Problems 21 and 29.

***22.** Write the mesh equations for each of the networks of Fig. 8.107 and, using determinants, solve for the loop currents in each network. Use clockwise mesh currents.

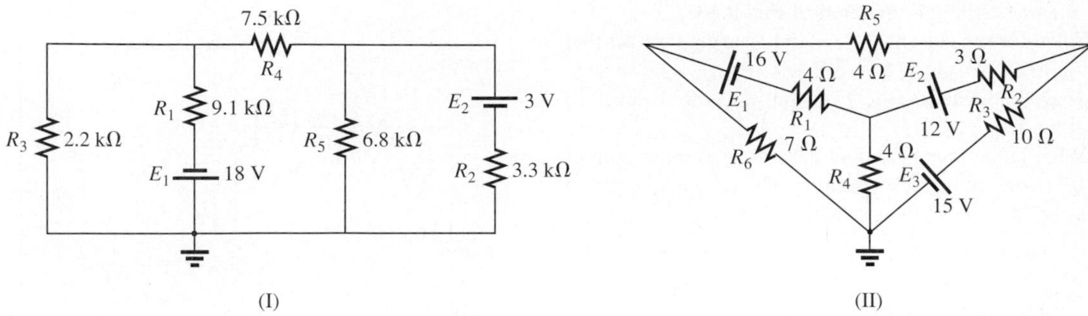

FIG. 8.107
Problems 22, 30, and 55.

***23.** Write the mesh equations for each of the networks of Fig. 8.108 and, using determinants, solve for the loop currents in each network.

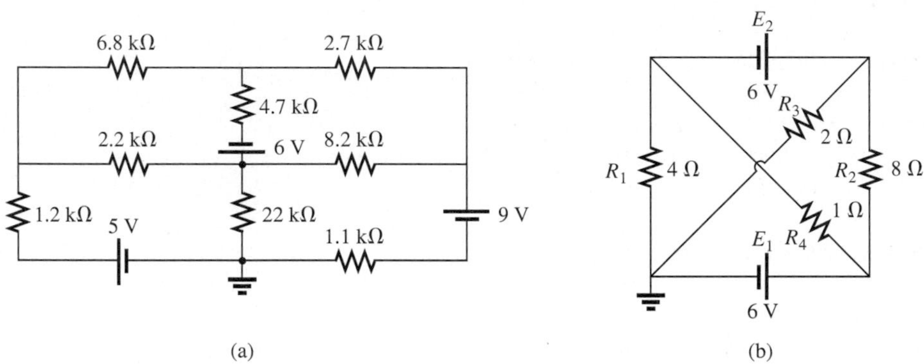

FIG. 8.108
Problems 23, 31, and 58.

***24.** Using the supermesh approach, find the current through each element of the networks of Fig. 8.109.

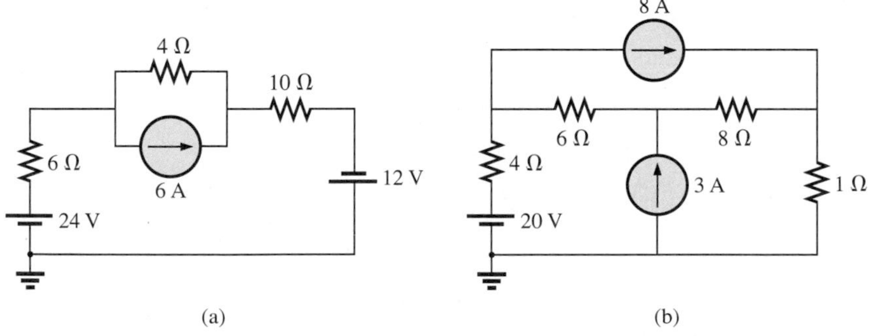

FIG. 8.109
Problem 24.

SECTION 8.8 Mesh Analysis (Format Approach)

25. Using the format approach, write the mesh equations for the networks of Fig. 8.101. Is symmetry present? Using determinants, solve for the mesh currents.

26. a. Using the format approach, write the mesh equations for the networks of Fig. 8.102.
b. Using determinants, solve for the mesh currents.
c. Determine the magnitude and direction of the current through each resistor.

27. a. Using the format approach, write the mesh equations for the networks of Fig. 8.103.
b. Using determinants, solve for the mesh currents.
c. Determine the magnitude and direction of the current through each resistor.

28. Determine the current I_3 for the network of Fig. 8.104 using mesh analysis and compare to the solution of Problem 15.

29. Using mesh analysis, determine $I_{5\Omega}$ and V_a for the network of Fig. 8.106(b).

30. Using mesh analysis, determine the mesh currents for the networks of Fig. 8.107.

31. Using mesh analysis, determine the mesh currents for the networks of Fig. 8.108.

SECTION 8.9 Nodal Analysis (General Approach)

32. Write the nodal equations for the networks of Fig. 8.110 and, using determinants, solve for the nodal voltages. Is symmetry present?

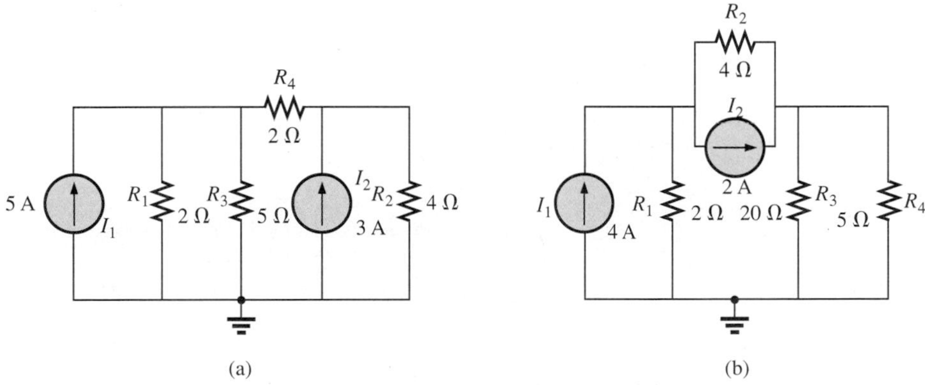

FIG. 8.110
Problems 32 and 38.

33. a. Write the nodal equations for the networks of Fig. 8.111.
b. Using determinants, solve for the nodal voltages.
c. Determine the magnitude and polarity of the voltage across each resistor.

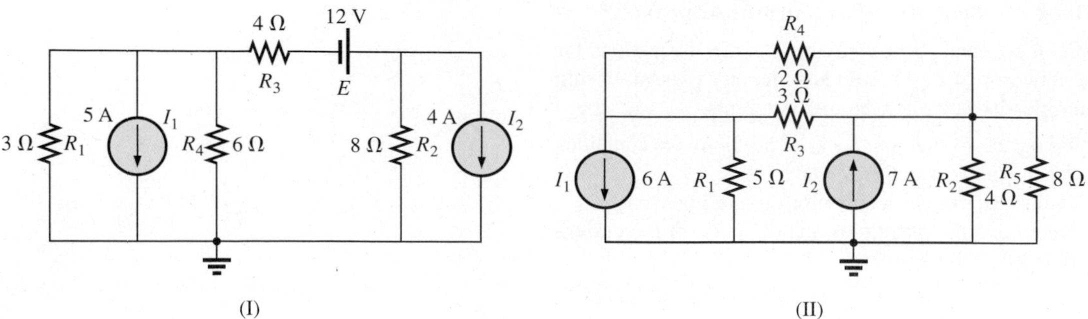

FIG. 8.111
Problems 33, 39, and 56.

34. **a.** Write the nodal equations for the networks of Fig. 8.107.
    b. Using determinants, solve for the nodal voltages.
    c. Determine the magnitude and polarity of the voltage across each resistor.

*35. For the networks of Fig. 8.112, write the nodal equations and solve for the nodal voltages.

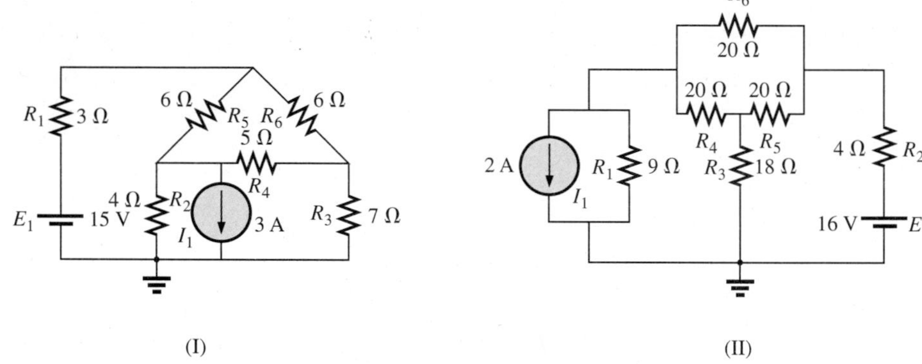

FIG. 8.112
Problems 35 and 40.

36. **a.** Determine the nodal voltages for the networks of Fig. 8.113.
    b. Find the voltage across each current source.

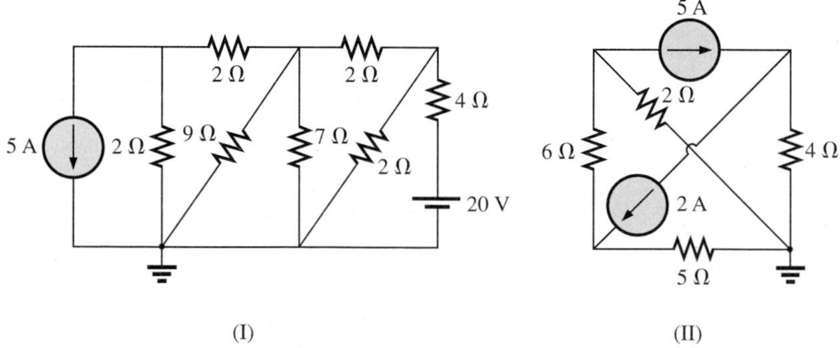

FIG. 8.113
Problems 36 and 41.

*37. Using the supernode approach, determine the nodal voltages for the networks of Fig. 8.114.

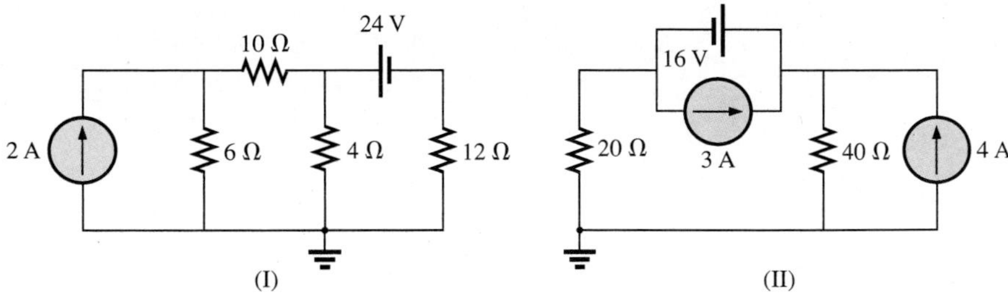

FIG. 8.114
Problems 37 and 59.

SECTION 8.10 Nodal Analysis (Format Approach)

38. Using the format approach, write the nodal equations for the networks of Fig. 8.110. Is symmetry present? Using determinants, solve for the nodal voltages.

39. a. Write the nodal equations for the networks of Fig. 8.111.
b. Solve for the nodal voltages.
c. Find the magnitude and polarity of the voltage across each resistor.

40. a. Write the nodal equations for the networks of Fig. 8.112.
b. Solve for the nodal voltages.
c. Find the magnitude and polarity of the voltage across each resistor.

41. Determine the nodal voltages for the networks of Fig. 8.113. Then determine the voltage across each current source.

SECTION 8.11 Bridge Networks

42. For the bridge network of Fig. 8.115:
a. Write the mesh equations using the format approach.
b. Determine the current through R_5.
c. Is the bridge balanced?
d. Is Eq. (8.4) satisfied?

43. For the network of Fig. 8.115:
a. Write the nodal equations using the format approach.
b. Determine the voltage across R_5.
c. Is the bridge balanced?
d. Is Eq. (8.4) satisfied?

44. For the bridge of Fig. 8.116:
a. Write the mesh equations using the format approach.
b. Determine the current through R_5.
c. Is the bridge balanced?
d. Is Eq. (8.4) satisfied?

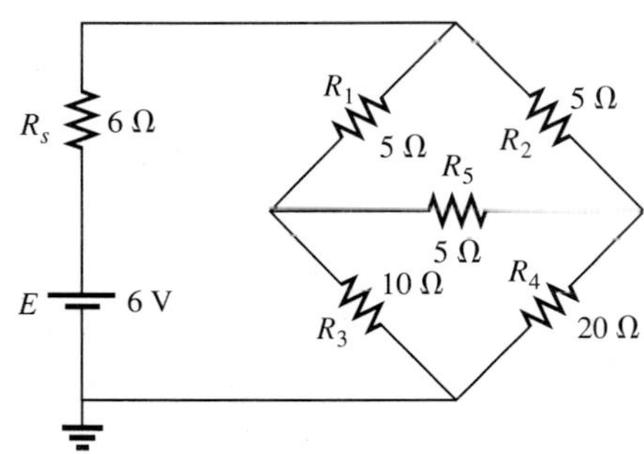

FIG. 8.115
Problems 42 and 43.

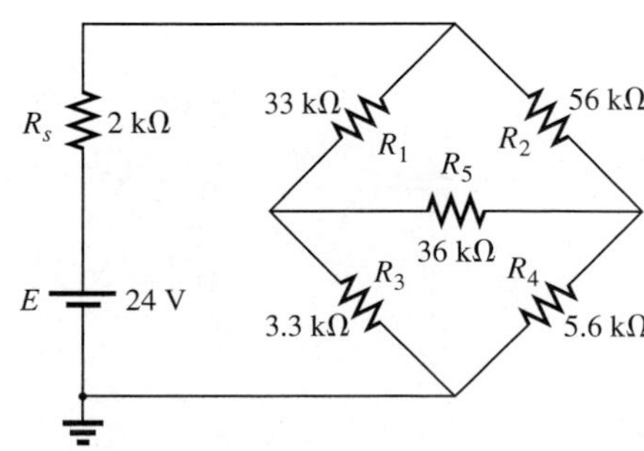

FIG. 8.116
Problems 44 and 45.

45. For the bridge network of Fig. 8.116:
    a. Write the nodal equations using the format approach.
    b. Determine the current across R_5.
    c. Is the bridge balanced?
    d. Is Eq. (8.4) satisfied?
46. Write the nodal equations for the bridge configuration of Fig. 8.117. Use the format approach.

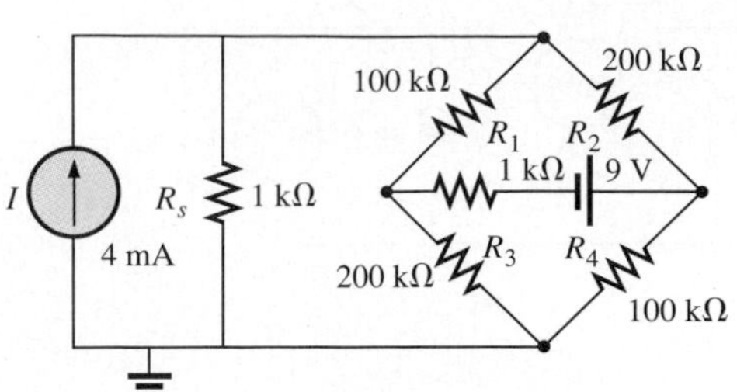

FIG. 8.117
Problem 46.

*47. Determine the current through the source resistor R_s of each network of Fig. 8.118 using either mesh or nodal analysis. Discuss why you chose one method over the other.

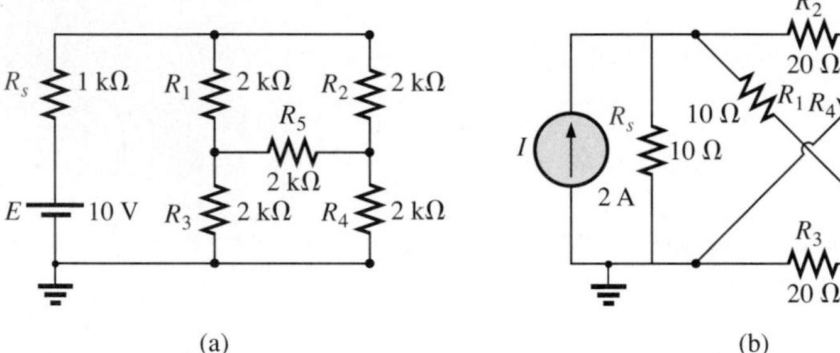

FIG. 8.118
Problem 47.

SECTION 8.12 Y-Δ (T-π) and Δ-Y (π-T) Conversions

48. Using a Δ-Y or Y-Δ conversion, find the current I in each of the networks of Fig. 8.119.

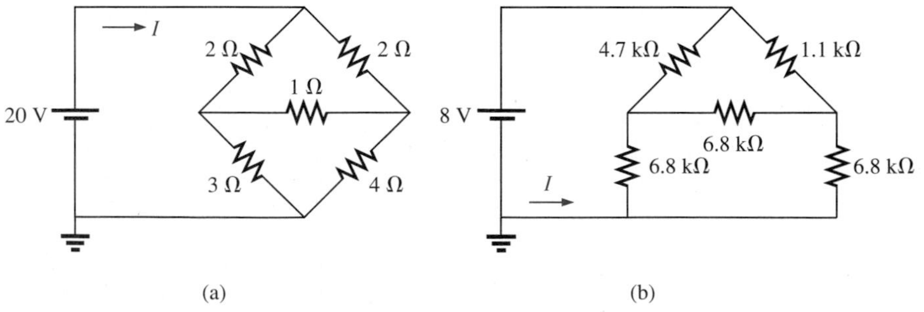

FIG. 8.119
Problem 48.

*49. Repeat Problem 48 for the networks of Fig. 8.120.

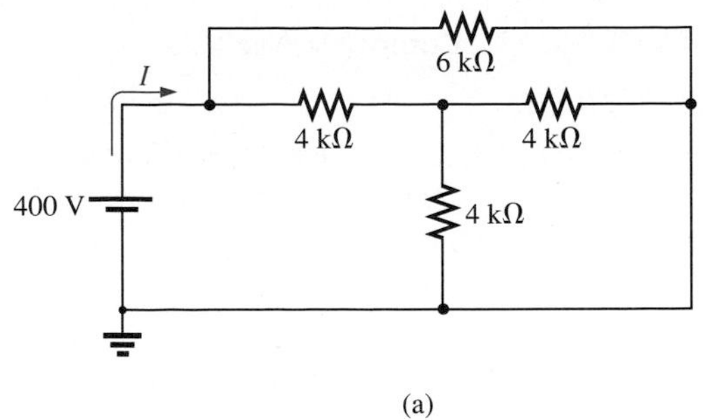

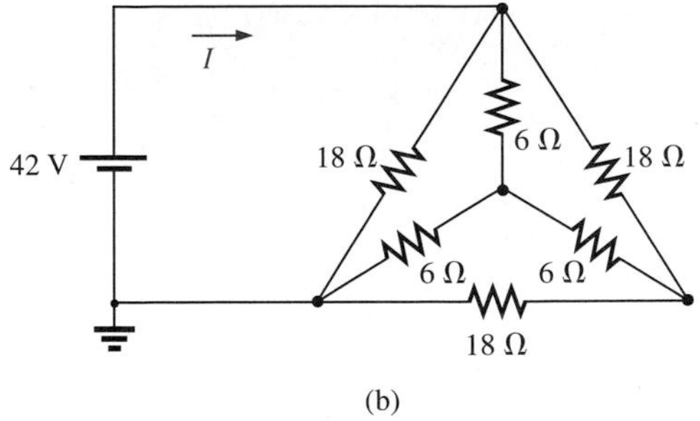

FIG. 8.120
Problem 49.

*50. Determine the current I for the network of Fig. 8.121.

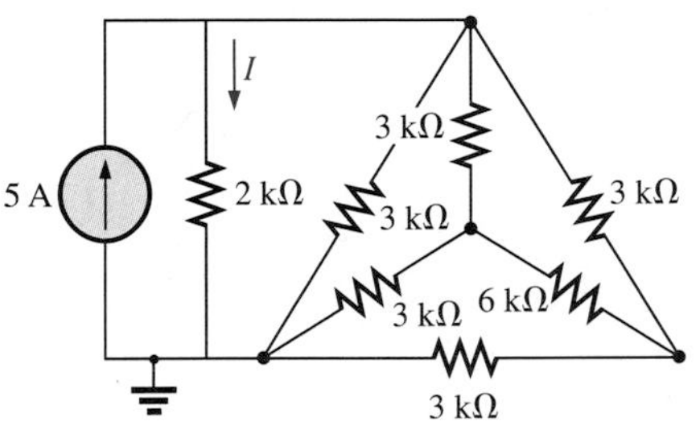

FIG. 8.121
Problem 50.

*51. **a.** Replace the T configuration of Fig. 8.122 (composed of 6-kΩ resistors) with a π configuration.
b. Solve for the source current I_{s_1}.

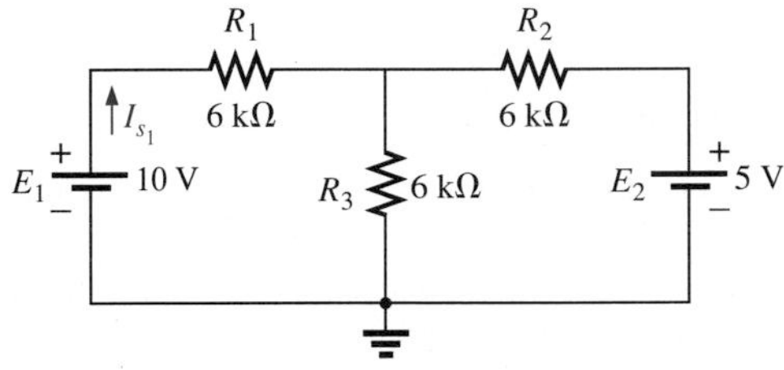

FIG. 8.122
Problem 51.

*52. **a.** Replace the π configuration of Fig. 8.123 (composed of 3-kΩ resistors) with a T configuration.
b. Solve for the source current I_s.

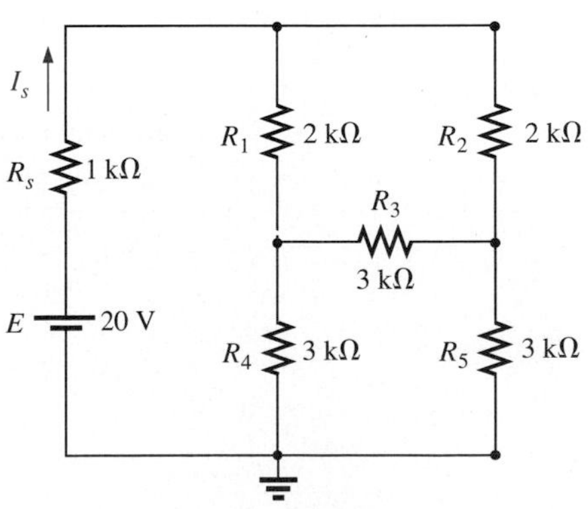

FIG. 8.123
Problem 52.

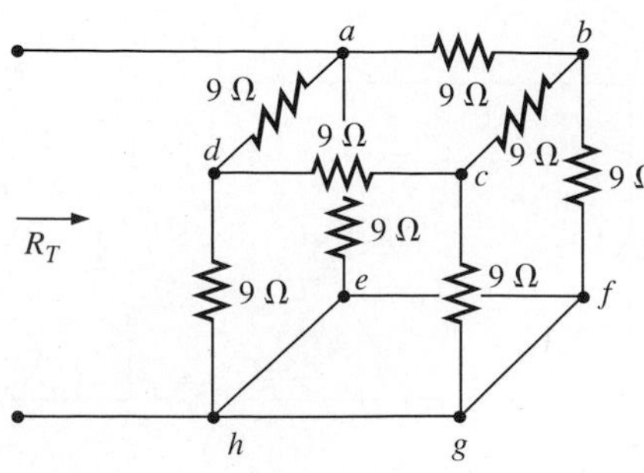

FIG. 8.124
Problem 53.

***53.** Using Y-Δ or Δ-Y conversions, determine the total resistance of the network of Fig. 8.124.

SECTION 8.13 Computer Analysis

PSpice (DOS)

54. Write the PSpice input file for the network of Fig. 101(b) that will print out the three branch currents.

55. Write the PSpice input file for the network of Fig. 8.107(I) that will print out the voltage V_{R_4} and current I_{R_2}.

56. Write the PSpice input file for the network of Fig. 8.111(II) that will print out V_{R_4}.

PSpice (Windows)

57. Using schematics, find the current through each element of Fig. 8.101.

***58.** Using schematics, find the mesh currents for the network of Fig. 8.108(a).

***59.** Using schematics, determine the nodal voltages for the network of Fig. 8.114(II).

Programming Language (C++, BASIC, PASCAL, etc.)

60. Given two simultaneous equations, write a program to solve for the unknown variables.

***61.** Write a program to solve for both mesh currents of the network of Fig. 8.25 (for any component values) using mesh analysis and determinants.

***62.** Write a program to solve for the nodal voltages of the network of Fig. 8.42 (for any component values) using nodal analysis and determinants.

GLOSSARY

Branch-current method A technique for determining the branch currents of a multiloop network.

Bridge network A network configuration typically having a diamond appearance in which no two elements are in series or parallel.

Current sources Sources that supply a fixed current to a network and have a terminal voltage dependent on the network to which they are applied.

Delta (Δ), pi (π) configuration A network structure that consists of three branches and has the appearance of the Greek letter delta (Δ) or pi (π).

Determinants method A mathematical technique for finding the unknown variables of two or more simultaneous linear equations.

Mesh analysis A technique for determining the mesh (loop) currents of a network that results in a reduced set of equations compared to the branch-current method.

Mesh (loop) current A labeled current assigned to each distinct closed loop of a network that can, individually or in combination with other mesh currents, define all of the branch currents of a network.

Nodal analysis A technique for determining the node voltages of a network.

Node A junction of two or more branches in a network.

Wye (Y), tee (T) configuration A network structure that consists of three branches and has the appearance of the capital letter Y or T.

9

Network Theorems

9.1 INTRODUCTION

This chapter will introduce the important fundamental theorems of network analysis. Included are the *superposition, Thévenin's, Norton's, maximum power transfer, substitution, Millman's,* and *reciprocity* theorems. We will consider a number of areas of application for each. A thorough understanding of each theorem is important because a number will be applied repeatedly in the material to follow.

9.2 SUPERPOSITION THEOREM

The superposition theorem, like the methods of the last chapter, can be used to find the solution to networks with two or more sources that are not in series or parallel. The most obvious advantage of this method is that it does not require the use of a mathematical technique such as determinants to find the required voltages or currents. Instead, each source is treated independently, and the algebraic sum is found to determine a particular unknown quantity of the network.

The superposition theorem states the following:

The current through, or voltage across, an element in a linear bilateral network is equal to the algebraic sum of the currents or voltages produced independently by each source.

When applying the theorem it is possible to consider the effects of two sources at the same time and reduce the number of networks that have to be analyzed, but, in general,

$$\begin{array}{c}\text{Number of networks}\\\text{to be analyzed}\end{array} = \begin{array}{c}\text{Number of}\\\text{independent sources}\end{array} \qquad \textbf{(9.1)}$$

To consider the effects of each source independently requires that sources be removed and replaced without affecting the final result. To remove a voltage source when applying this theorem, the difference in potential between the terminals of the voltage source must be set to zero (short circuited); removing a current source requires that its terminals be opened (open circuit). Any internal resistance or conductance associated with the displaced sources is not eliminated but must still be considered.

Figure 9.1 reviews the various substitutions required when removing an ideal source, and Fig. 9.2 reviews the substitutions with practical sources that have an internal resistance.

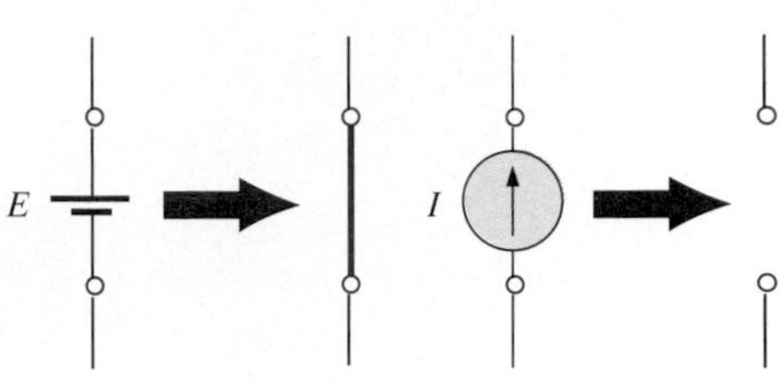

FIG. 9.1
Removing the effects of ideal sources.

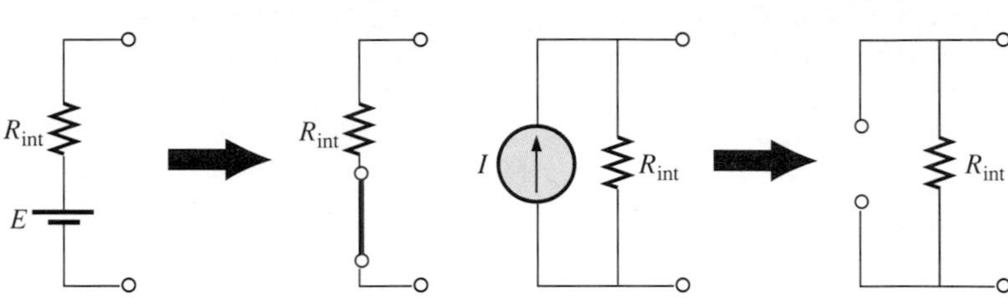

FIG. 9.2
Removing the effects of practical sources.

The total current through any portion of the network is equal to the algebraic sum of the currents produced independently by each source. That is, for a two-source network, if the current produced by one source is in one direction, while that produced by the other is in the opposite direction through the same resistor, *the resulting current is the difference of the two and has the direction of the larger.* If the individual currents are in the same direction, *the resulting current is the sum of two in the direction of either current.* This rule holds true for the voltage across a portion of a network as determined by polarities, and it can be extended to networks with any number of sources.

The superposition principle is not applicable to power effects since the power loss in a resistor varies as the square (nonlinear) of the current or voltage. For instance, the current through the resistor R of Fig. 9.3(a) is I_1 due to one source of a two-source network. The current through the same resistor due to the other source is I_2 as shown in Fig. 9.3(b). Applying the superposition theorem, the total current through the resistor due to both sources is I_T, as shown in Fig. 9.3(c) with

$$I_T = I_1 + I_2$$

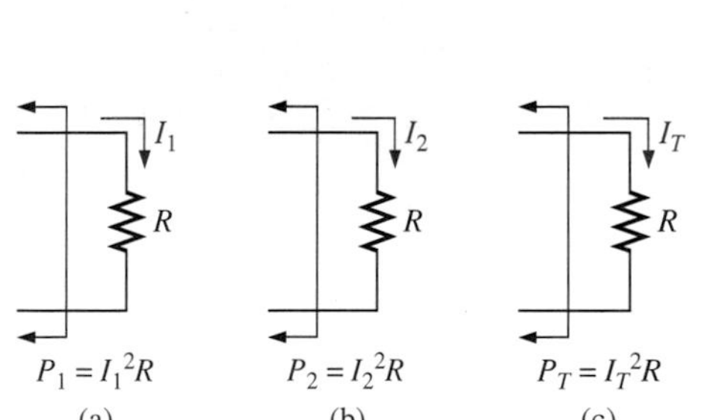

FIG. 9.3
Demonstration of the fact that superposition is not applicable to power effects.

The power delivered to the resistor in Fig. 9.3(a) is

$$P_1 = I_1^2R$$

while the power delivered to the same resistor in Fig. 9.3(b) is

$$P_2 = I_2^2R$$

If we assume that the total power delivered in Fig. 9.3(c) can be obtained by simply adding the power delivered due to each source, we find that

$$P_T = P_1 + P_2 = I_1^2R + I_2^2R = I_T^2R$$

or

$$I_T^2 = I_1^2 + I_2^2$$

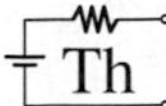

This final relationship between current levels is incorrect, however, as can be demonstrated by taking the total current determined by the superposition theorem and squaring it as follows:

$$I_T^2 = (I_1 + I_2)^2 = I_1^2 + I_2^2 + 2I_1I_2$$

which is certainly different from the expression obtained from the addition of power levels.

In general, therefore,

the total power delivered to a resistive element must be determined using the total current through or the total voltage across the element and cannot be determined by a simple sum of the power levels established by each source.

EXAMPLE 9.1 Determine I_1 for the network of Fig. 9.4.

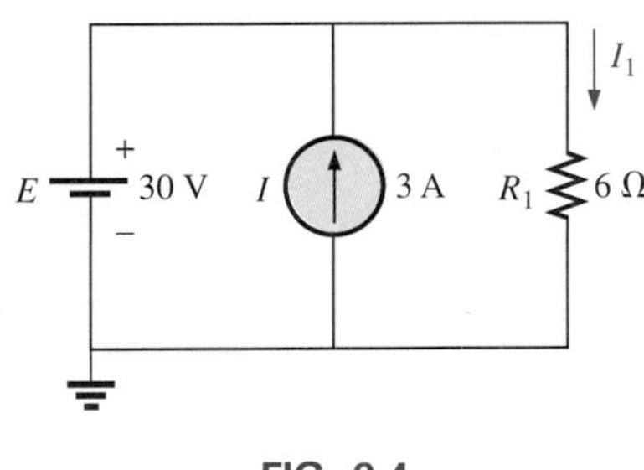

FIG. 9.4
Example 9.1.

Solution: Setting $E = 0$ V for the network of Fig. 9.4 results in the network of Fig. 9.5(a), where a short-circuit equivalent has replaced the 30-V source.

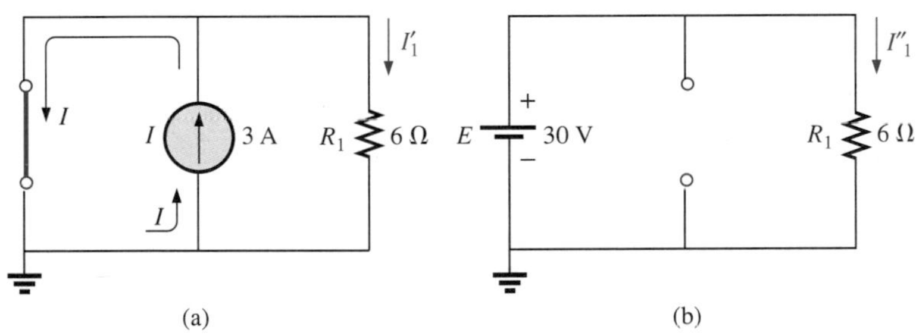

FIG. 9.5
(a) The contribution of I to I_1. (b) The contribution of E to I_1.

As shown in Fig. 9.5(a), the source current will choose the short-circuit path, and $I'_1 = 0$ A. If we applied the current divider rule,

$$I'_1 = \frac{R_{sc}I}{R_{sc} + R_1} = \frac{(0\ \Omega)I}{0\ \Omega + 6\ \Omega} = 0\text{ A}$$

Setting I to zero amperes will result in the network of Fig. 9.5(b), with the current source replaced by an open circuit. Applying Ohm's law,

$$I''_1 = \frac{E}{R_1} = \frac{30\text{ V}}{6\ \Omega} = 5\text{ A}$$

Since I'_1 and I''_1 have the same defined direction in Figs. 9.5(a) and (b), the current I_1 is the sum of the two, and

$$I_1 = I'_1 + I''_1 = 0\text{ A} + 5\text{ A} = \mathbf{5\text{ A}}$$

Note in this case that the current source has no effect on the current through the 6-Ω resistor. The voltage across the resistor must be fixed at 30 V because they are parallel elements.

EXAMPLE 9.2 Using superposition, determine the current through the 4-Ω resistor of Fig. 9.6. Note that this is a two-source network of the type considered in Chapter 8.

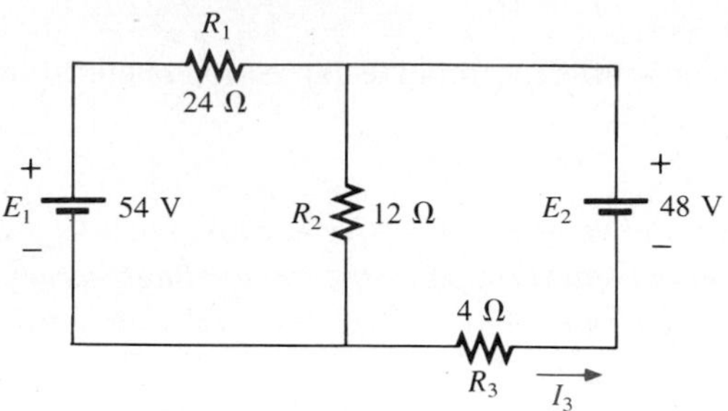

FIG. 9.6
Example 9.2.

Solution: *Considering the effects of a 54-V source (Fig. 9.7):*

$$R_T = R_1 + R_2 \| R_3 = 24\ \Omega + 12\ \Omega \| 4\ \Omega = 24\ \Omega + 3\ \Omega = 27\ \Omega$$

$$I = \frac{E_1}{R_T} = \frac{54\ \text{V}}{27\ \Omega} = 2\ \text{A}$$

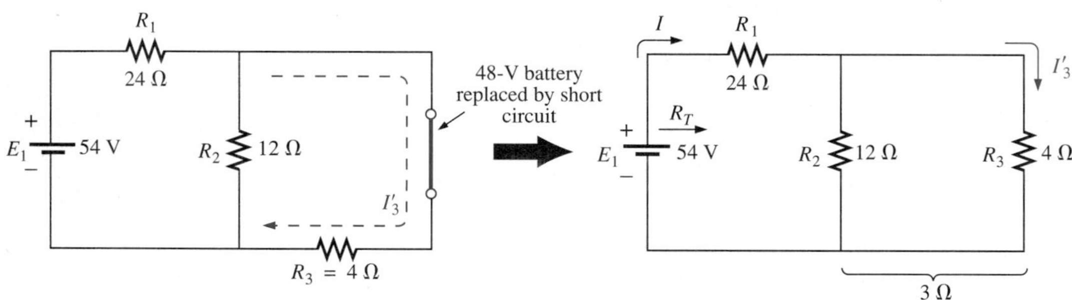

FIG. 9.7
The effect of E_1 on the current I_3.

Using the current divider rule,

$$I'_3 = \frac{R_2 I}{R_2 + R_3} = \frac{(12\ \Omega)(2\ \text{A})}{12\ \Omega + 4\ \Omega} = \frac{24\ \text{A}}{16} = 1.5\ \text{A}$$

Considering the effects of the 48-V source (Fig. 9.8):

$$R_T = R_3 + R_1 \| R_2 = 4\ \Omega + 24\ \Omega \| 12\ \Omega = 4\ \Omega + 8\ \Omega = 12\ \Omega$$

$$I''_3 = \frac{E_2}{R_T} = \frac{48\ \text{V}}{12\ \Omega} = 4\ \text{A}$$

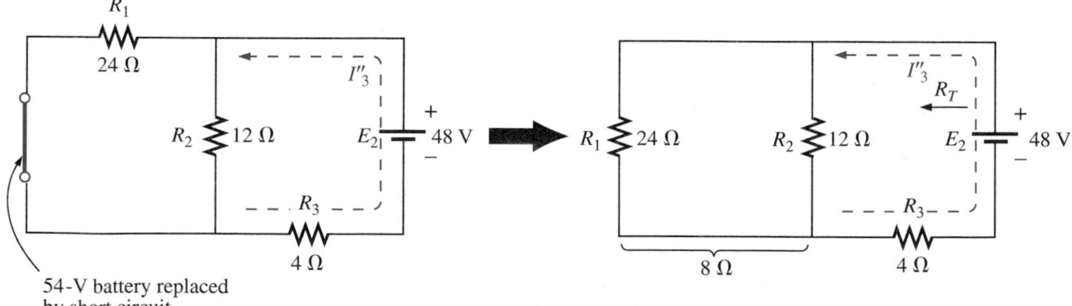

FIG. 9.8
The effect of E_2 on the current I_3.

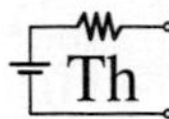

The total current through the 4-Ω resistor (Fig. 9.9) is

$$I_3 = I''_3 - I'_3 = 4\text{ A} - 1.5\text{ A} = \mathbf{2.5\text{ A}} \qquad (\text{direction of } I''_3)$$

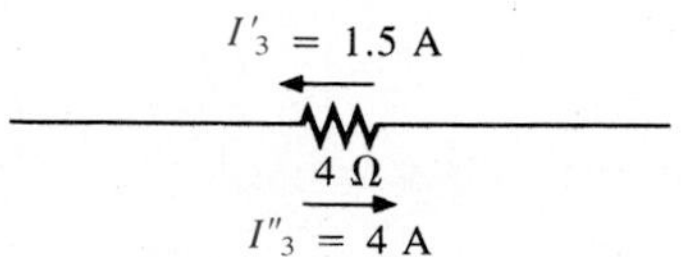

FIG. 9.9
The resultant current for I_3.

EXAMPLE 9.3

a. Using superposition, find the current through the 6-Ω resistor of the network of Fig. 9.10.

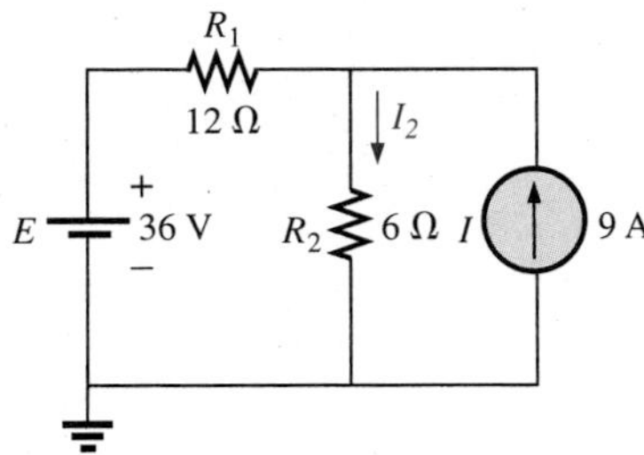

FIG. 9.10
Example 9.3.

b. Demonstrate that superposition is not applicable to power levels.

Solutions: a. *Considering the effect of the* 36-V *source (Fig.* 9.11):

$$I'_2 = \frac{E}{R_T} = \frac{E}{R_1 + R_2} = \frac{36\text{ V}}{12\,\Omega + 6\,\Omega} = 2\text{ A}$$

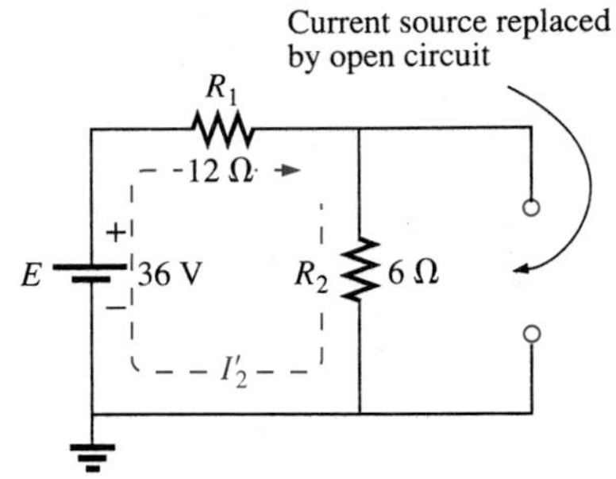

FIG. 9.11
The contribution of E to I_2.

Considering the effect of the 9-A *source (Fig.* 9.12):
Applying the current divider rule,

$$I''_2 = \frac{R_1 I}{R_1 + R_2} = \frac{(12\,\Omega)(9\text{ A})}{12\,\Omega + 6\,\Omega} = \frac{108\text{ A}}{18} = 6\text{ A}$$

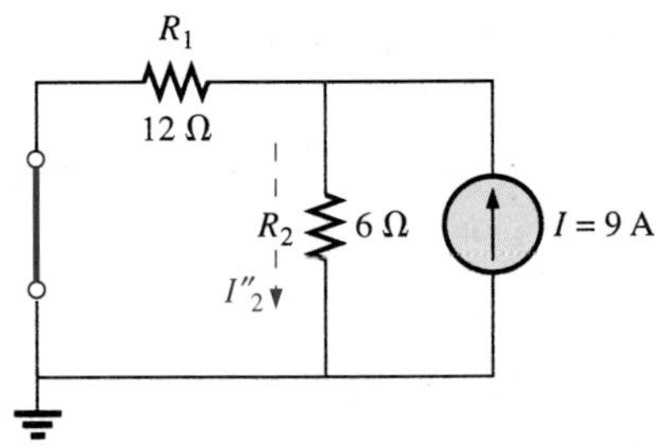

FIG. 9.12
The contribution of I to I_2.

The total current through the 6-Ω resistor (Fig. 9.13) is

$$I_2 = I'_2 + I''_2 = 2\text{ A} + 6\text{ A} = \mathbf{8\text{ A}}$$

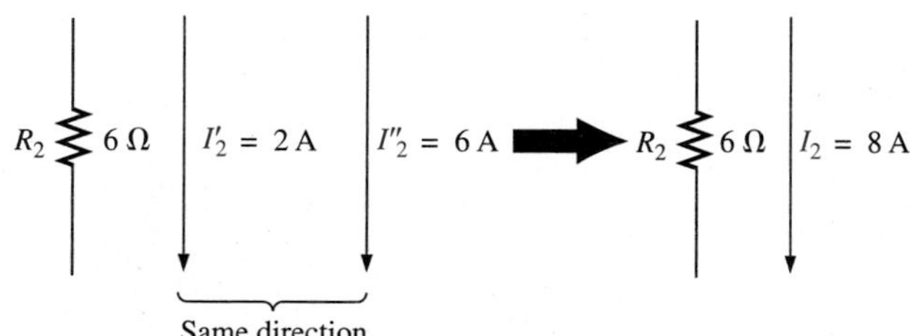

FIG. 9.13
The resultant current for I_2.

b. The power to the 6-Ω resistor is

$$\text{Power} = I^2R = (8\text{ A})^2(6\,\Omega) = \mathbf{384\text{ W}}$$

The calculated power to the 6-Ω resistor due to each source, *misusing* the principle of superposition, is

$$P_1 = (I'_2)^2R = (2\text{ A})^2(6\,\Omega) = 24\text{ W}$$
$$P_2 = (I''_2)^2R = (6\text{ A})^2(6\,\Omega) = 216\text{ W}$$
$$P_1 + P_2 = 240\text{ W} \neq 384\text{ W}$$

This results because 2 A + 6 A = 8 A, but

$$(2\text{ A})^2 + (6\text{ A})^2 \neq (8\text{ A})^2$$

As mentioned previously, the superposition principle is not applicable to power effects since power is proportional to the square of the current or voltage (I^2R or V^2/R).

Figure 9.14 is a plot of the power delivered to the 6-Ω resistor versus current.

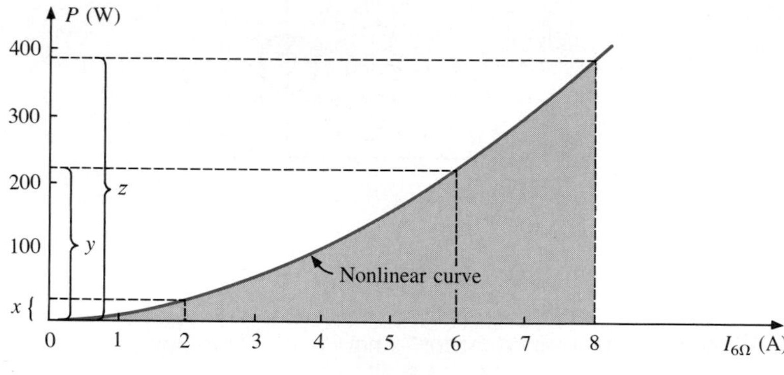

FIG. 9.14
Plotting the power delivered to the 6-Ω resistor versus current through the resistor.

Obviously, $x + y \neq z$, or 24 W + 216 W ≠ 384 W, and superposition does not hold. However, for a linear relationship, such as that between the voltage and current of the fixed-type 6-Ω resistor, superposition can be applied, as demonstrated by the graph of Fig. 9.15, where $a + b = c$, or 2 A + 6 A = 8 A.

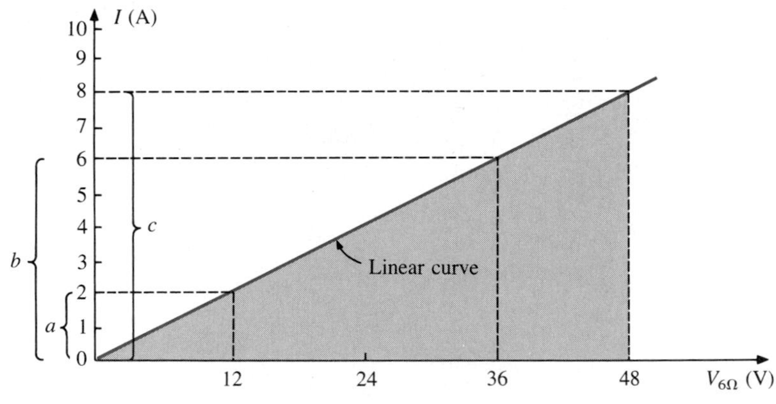

FIG. 9.15
Plotting I versus V for the 6-Ω resistor.

EXAMPLE 9.4 Using the principle of superposition, find the current I_2 through the 12-kΩ resistor of Fig. 9.16.

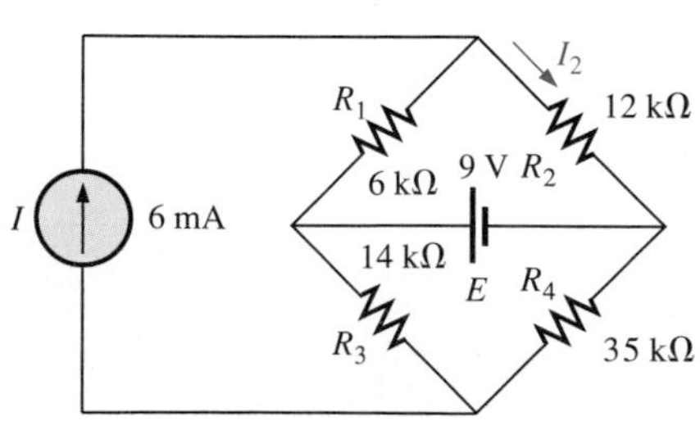

FIG. 9.16
Example 9.4.

Solution: *Considering the effect of the* 6-mA *current source (Fig.* 9.17):

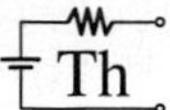

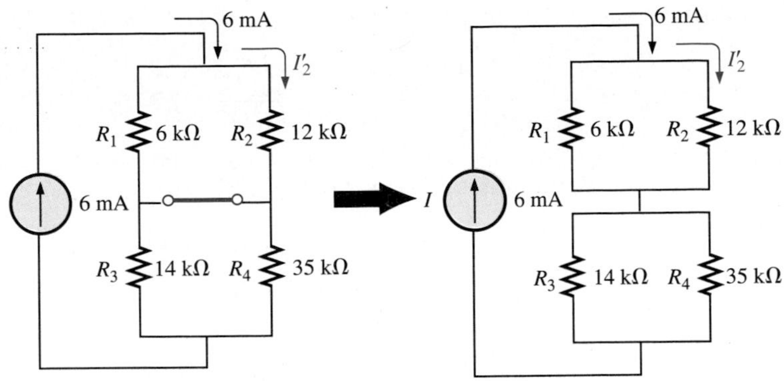

FIG. 9.17
The effect of the current source I on the current I_2.

Current divider rule:

$$I'_2 = \frac{R_1 I}{R_1 + R_2} = \frac{(6\text{ k}\Omega)(6\text{ mA})}{6\text{ k}\Omega + 12\text{ k}\Omega} = 2\text{ mA}$$

Considering the effect of the 9-V *voltage source (Fig.* 9.18*):*

$$I''_2 = \frac{E}{R_1 + R_2} = \frac{9\text{ V}}{6\text{ k}\Omega + 12\text{ k}\Omega} = 0.5\text{ mA}$$

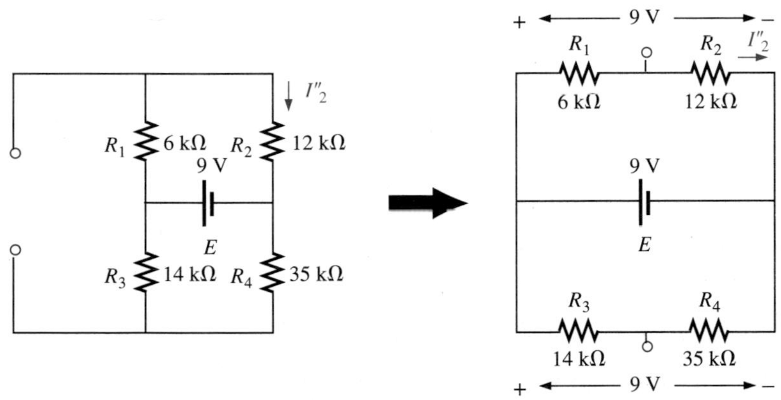

FIG. 9.18
The effect of the voltage source E on the current I_2.

Since I'_2 and I''_2 have the same direction through R_2, the desired current is the sum of the two:

$$\begin{aligned} I_2 &= I'_2 + I''_2 \\ &= 2\text{ mA} + 0.5\text{ mA} \\ &= \mathbf{2.5\text{ mA}} \end{aligned}$$

EXAMPLE 9.5 Find the current through the 2-Ω resistor of the network of Fig. 9.19. The presence of three sources will result in three different networks to be analyzed.

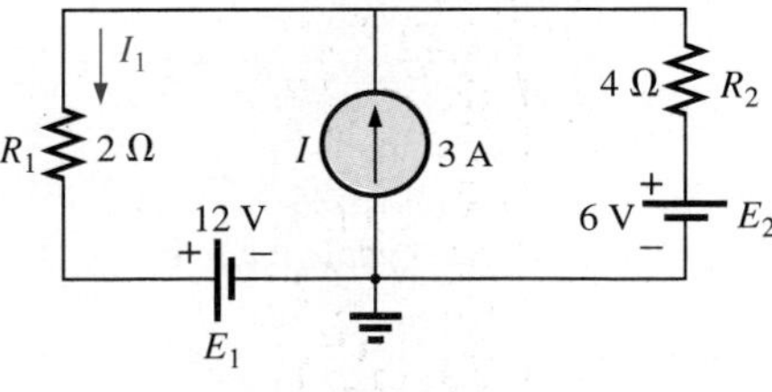

FIG. 9.19
Example 9.5.

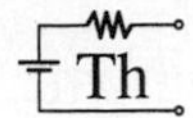

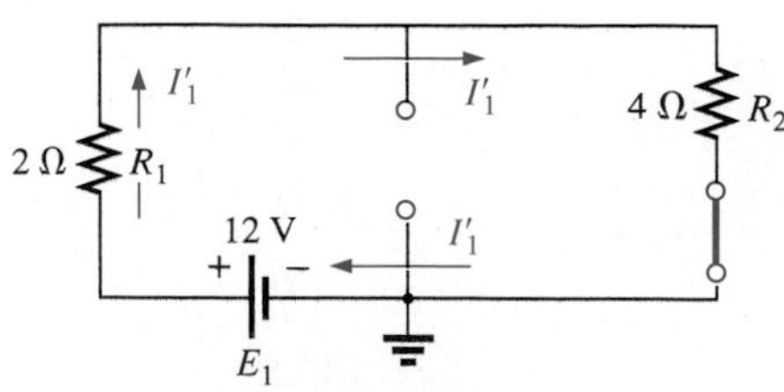

FIG. 9.20
The effect of E_1 on the current I.

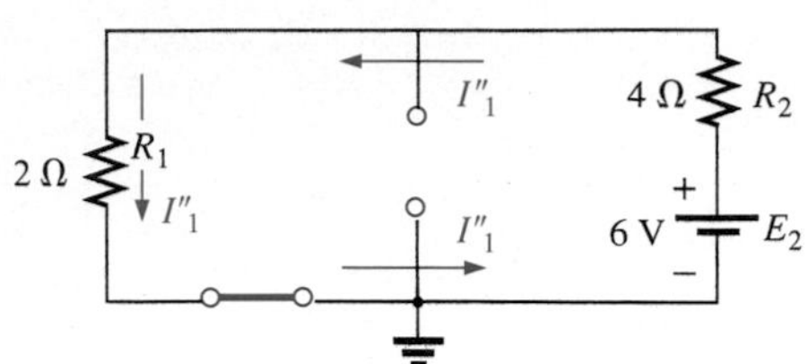

FIG. 9.21
The effect of E_2 on the current I_1.

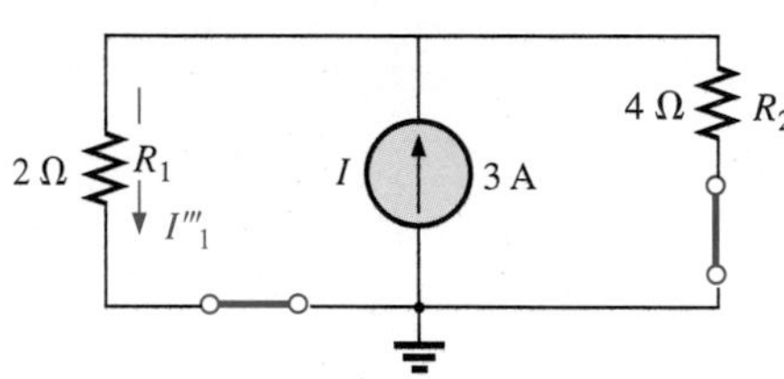

FIG. 9.22
The effect of I on the current I_1.

Solution: *Considering the effect of the* 12-V *source (Fig.* 9.20):

$$I'_1 = \frac{E_1}{R_1 + R_2} = \frac{12\text{ V}}{2\ \Omega + 4\ \Omega} = \frac{12\text{ V}}{6\ \Omega} = 2\text{ A}$$

Considering the effect of the 6-V *source (Fig.* 9.21):

$$I''_1 = \frac{E_2}{R_1 + R_2} = \frac{6\text{ V}}{2\ \Omega + 4\ \Omega} = \frac{6\text{ V}}{6\ \Omega} = 1\text{ A}$$

Considering the effect of the 3-A *source (Fig.* 9.22):
Applying the current divider rule,

$$I'''_1 = \frac{R_2 I}{R_1 + R_2} = \frac{(4\ \Omega)(3\text{ A})}{2\ \Omega + 4\ \Omega} = \frac{12\text{ A}}{6} = 2\text{ A}$$

The total current through the 2-Ω resistor appears in Fig. 9.23, and

Same direction as I_1 in Fig. 9.19 — Opposite direction to I_1 in Fig. 9.19

$$\begin{aligned} I_1 &= \overbrace{I''_1 + I'''_1} - I'_1 \\ &= 1\text{ A} + 2\text{ A} - 2\text{ A} = \mathbf{1\text{ A}} \end{aligned}$$

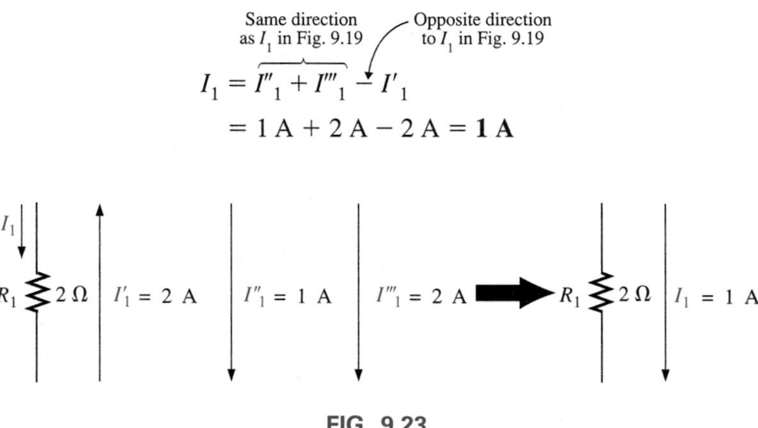

FIG. 9.23
The resultant current I_1.

9.3 THÉVENIN'S THEOREM

Thévenin's theorem states the following:

Any two-terminal linear bilateral dc network can be replaced by an equivalent circuit consisting of a voltage source and a series resistor, as shown in Fig. 9.24.

In Fig. 9.25(a), for example, the network within the container has only two terminals available to the outside world, labeled *a* and *b*. It is possible using Thévenin's theorem to replace everything in the container with one source and one resistor, as shown in Fig. 9.25(b), and maintain the same terminal characteristics at terminals *a* and *b*. That is, any load connected to terminals *a* and *b* will not know whether it is hooked up to the network of Fig. 9.25(a) or Fig. 9.25(b). The load will receive the same current, voltage, and power from either configuration of Fig. 9.25. Throughout the discussion to follow, however, always keep in mind that

the Thévenin equivalent circuit provides an equivalence at the terminals only—the internal construction and characteristics of the original network and the Thévenin equivalent are usually quite different.

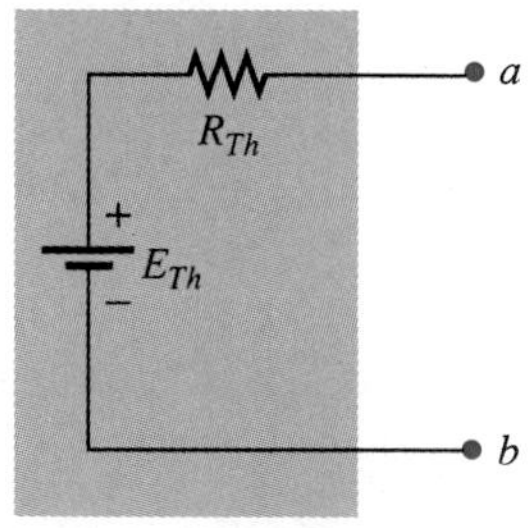

FIG. 9.24
Thévenin equivalent circuit.

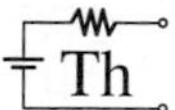

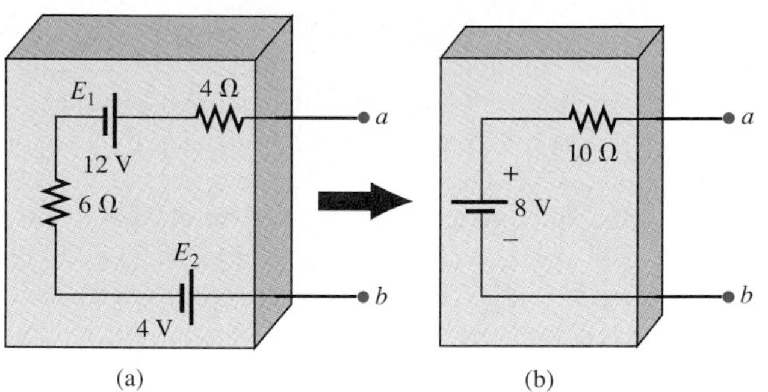

FIG. 9.25
The effect of applying Thévenin's theorem.

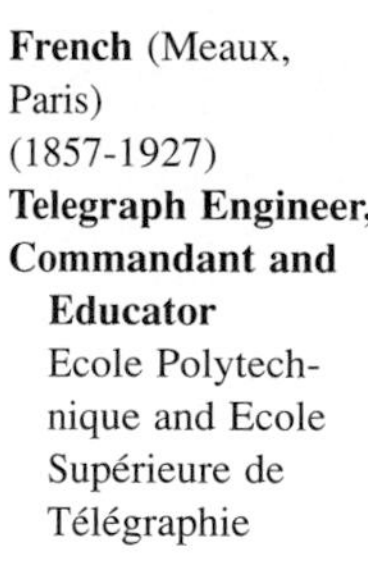

French (Meaux, Paris)
(1857-1927)
Telegraph Engineer, Commandant and Educator
Ecole Polytechnique and Ecole Supérieure de Télégraphie

Courtesy of the Bibliothèque Ecole Polytechnique, Paris, France

Although active in the study and design of telegraphic systems (including underground transmission), cylindrical condensers (capacitors), and electromagnetism, he is best known for a theorem first presented in the French Journal of Physics-Theory and Applications in 1883. It appeared under the heading of "Sur un nouveau théorème d'électricité dynamique" (On a new theorem of dynamic electricity") and was originally referred to as the equivalent generator theorem. There is some evidence that a similar theorem was introduced by Hermann von Helmholtz in 1853. However, Professor Helmholtz applied the theorem to animal physiology and not to communication or generator systems, and therefore has not received the credit in this field that he might deserve. In the early 1920s AT&T did some pioneering work using the equivalent circuit and may have initiated the reference to the theorem as simply Thévenin's theorem. In fact, Edward L. Norton, an engineer at AT&T at the time introduced a current source equivalent of the Thévenin equivalent currently referred to as the Norton equivalent circuit. As an aside, Commandant Thévenin was an avid skier and in fact was commissioner of an international ski competition in Chamonix, France in 1912.

LEON-CHARLES THÉVENIN

For the network of Fig. 9.25(a), the Thévenin equivalent circuit can be found quite directly by simply combining the series batteries and resistors. Note the exact similarity of the network of Fig. 9.25(b) with the Thévenin configuration of Fig. 9.24. The method described below will allow us to extend the procedure just applied to more complex configurations and still end up with the relatively simple network of Fig. 9.24.

In most cases, there will be other elements connected to the right of terminals a and b in Fig. 9.25. To apply the theorem, however, the network to be reduced to the Thévenin equivalent form must be isolated as shown in Fig. 9.25 and the two "holding" terminals identified. Once the proper Thévenin equivalent circuit has been determined, the voltage, current, or resistance readings between the two "holding" terminals will be the same whether the original or Thévenin equivalent circuit is connected to the left of terminals a and b in Fig. 9.25. Any load connected to the right of terminals a and b of Fig. 9.25 will receive the same voltage or current with either network.

This theorem achieves two important objectives. First, as was true for all the methods previously described, it allows us to find any particular voltage or current in a linear network with one, two, or any other number of sources. Second, we can concentrate on a specific portion of a network by replacing the remaining network with an equivalent circuit. In Fig. 9.26, for example, by finding the Thévenin equivalent circuit for the network in the shaded area, we can quickly calculate the change in current through or voltage across the variable resistor R_L for the various values that it may assume. This is demonstrated in Example 9.6.

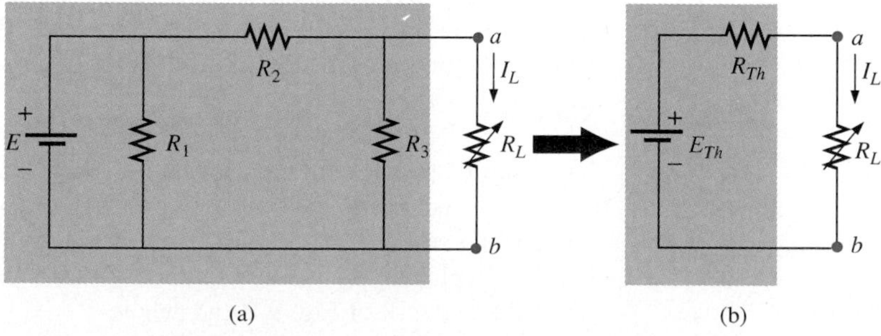

FIG. 9.26
Substituting the Thévenin equivalent circuit for a complex network.

Before we examine the steps involved in applying this theorem, it is important that an additional word be included here to ensure that the implications of the Thévenin equivalent circuit are clear. In Fig. 9.26, the entire network, except R_L, is to be replaced by a single series resistor and battery as shown in Fig. 9.24. The values of these two elements of the Thévenin equivalent circuit must be chosen to ensure that the resistor R_L will react to the network of Fig. 9.26(a) in the same manner as to the network of Fig. 9.26(b). In other words, the current through or voltage across R_L must be the same for either network for any value of R_L.

The following sequence of steps will lead to the proper value of R_{Th} and E_{Th}.

Preliminary:

1. ***Remove that portion of the network across which the Thévenin equivalent circuit is to be found. In Fig. 9.26(a), this requires that the load resistor R_L be temporarily removed from the network.***
2. ***Mark the terminals of the remaining two-terminal network. (The importance of this step will become obvious as we progress through some complex networks.)***

R_{Th}:

3. ***Calculate R_{Th} by first setting all sources to zero (voltage sources are replaced by short circuits and current sources by open circuits) and then finding the resultant resistance between the two marked terminals. (If the internal resistance of the voltage and/or current sources is included in the original network, it must remain when the sources are set to zero.)***

E_{Th}:

4. ***Calculate E_{Th} by first returning all sources to their original position and finding the open-circuit voltage between the marked terminals. (This step is invariably the one that will lead to the most confusion and errors. In all cases, keep in mind that it is the open-circuit potential between the two terminals marked in step 2.)***

Conclusion:

5. ***Draw the Thévenin equivalent circuit with the portion of the circuit previously removed replaced between the terminals of the equivalent circuit. This step is indicated by the placement of the resistor R_L between the terminals of the Thévenin equivalent circuit as shown in Fig. 9.26(b).***

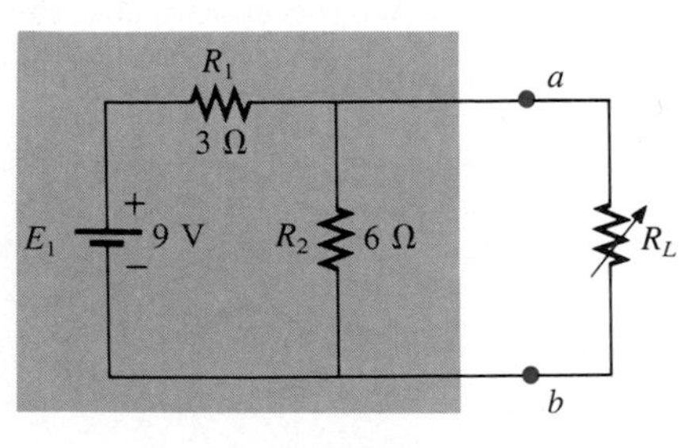

FIG. 9.27
Example 9.6.

EXAMPLE 9.6 Find the Thévenin equivalent circuit for the network in the shaded area of the network of Fig. 9.27. Then find the current through R_L for values of 2 Ω, 10 Ω, and 100 Ω.

Solution:

Steps 1 and 2 produce the network of Fig. 9.28. Note that the load resistor R_L has been removed and the two "holding" terminals have been defined as *a* and *b*.

Step 3: Replacing the voltage source E_1 with a short-circuit equivalent yields the network of Fig. 9.29(a), where

$$R_{Th} = R_1 \parallel R_2 = \frac{(3\ \Omega)(6\ \Omega)}{3\ \Omega + 6\ \Omega} = \mathbf{2\ \Omega}$$

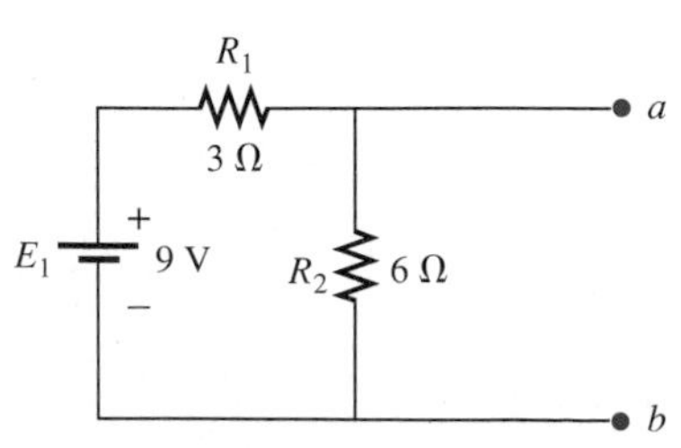

FIG. 9.28
Identifying the terminals of particular importance when applying Thévenin's theorem.

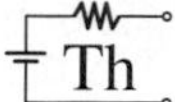

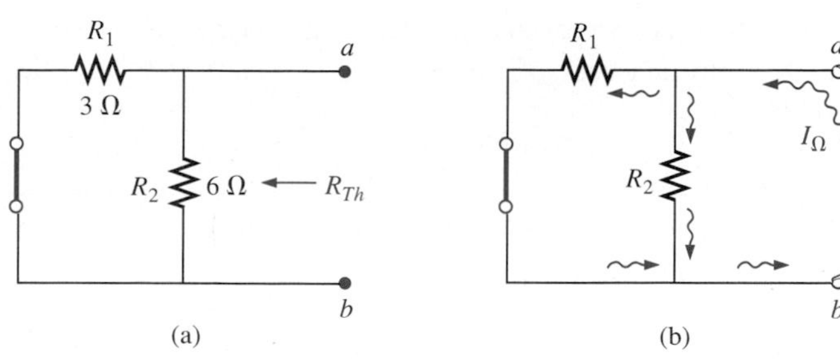

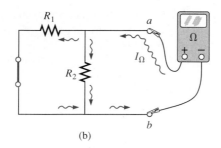

FIG. 9.29
Determining R_{Th} for the network of Fig. 9.28.

The importance of the two marked terminals now begins to surface. They are the two terminals across which the Thévenin resistance is measured. It is no longer the total resistance as seen by the source, as determined in the majority of problems of Chapter 7. If some difficulty develops when determining R_{Th} with regard to whether the resistive elements are in series or parallel, consider recalling that the ohmmeter sends out a trickle current into a resistive combination and senses the level of the resulting voltage to establish the measured resistance level. In Fig. 9.29(b), the trickle current of the ohmmeter approaches the network through terminal *a*, and when it reaches the junction of R_1 and R_2, it splits as shown. The fact that the trickle current splits and then recombines at the lower node reveals that the resistors are in parallel as far as the ohmmeter reading is concerned. In essence, the path of the sensing current of the ohmmeter has revealed the manner in which the resistors are connected to the two terminals of interest and how the Thévenin resistance should be determined. Keep the above in mind as you work through the various examples of this section.

Step 4: Replace the voltage source (Fig. 9.30). For this case, the open-circuit voltage E_{Th} is the same as the voltage drop across the 6-Ω resistor. Applying the voltage divider rule,

$$E_{Th} = \frac{R_2E_1}{R_2 + R_1} = \frac{(6\ \Omega)(9\ \text{V})}{6\ \Omega + 3\ \Omega} = \frac{54\ \text{V}}{9} = \mathbf{6\ V}$$

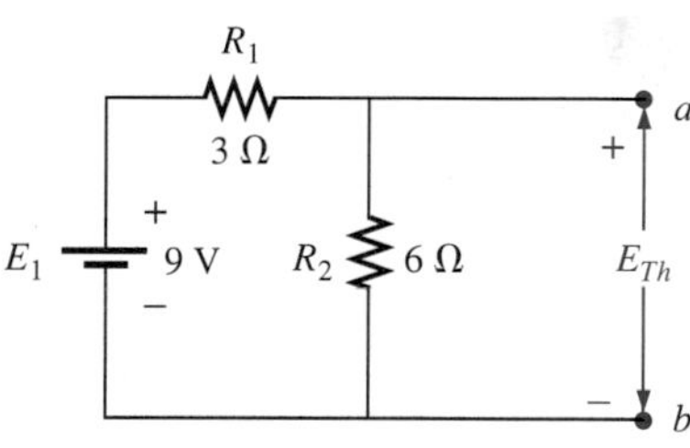

FIG. 9.30
Determining E_{Th} for the network of Fig. 9.28.

It is particularly important to recognize that E_{Th} is the open-circuit potential between points *a* and *b*. Remember that an open circuit can have any voltage across it, but the current must be zero. In fact, the current through any element in series with the open circuit must be zero also. The use of a voltmeter to measure E_{Th} appears in Fig. 9.31. Note that it is placed directly across the resistor R_2 since E_{Th} and V_{R_2} are in parallel.

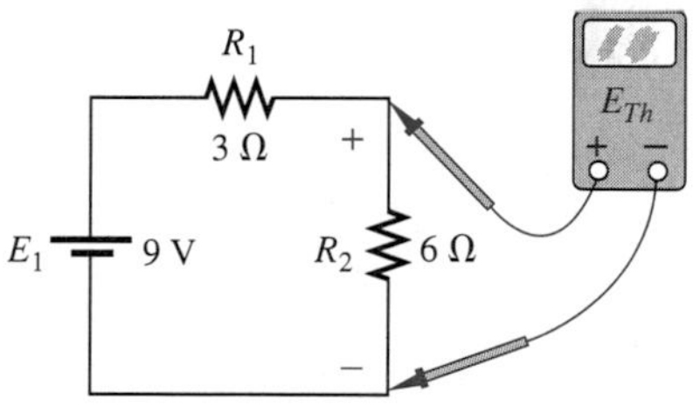

FIG. 9.31
Measuring E_{Th} for the network of Fig. 9.28.

Step 5 (Fig. 9.32):

$$I_L = \frac{E_{Th}}{R_{Th} + R_L}$$

$$R_L = 2\ \Omega: \quad I_L = \frac{6\ \text{V}}{2\ \Omega + 2\ \Omega} = \mathbf{1.5\ A}$$

$$R_L = 10\ \Omega: \quad I_L = \frac{6\ \text{V}}{2\ \Omega + 10\ \Omega} = \mathbf{0.5\ A}$$

$$R_L = 100\ \Omega: \quad I_L = \frac{6\ \text{V}}{2\ \Omega + 100\ \Omega} = \mathbf{0.059\ A}$$

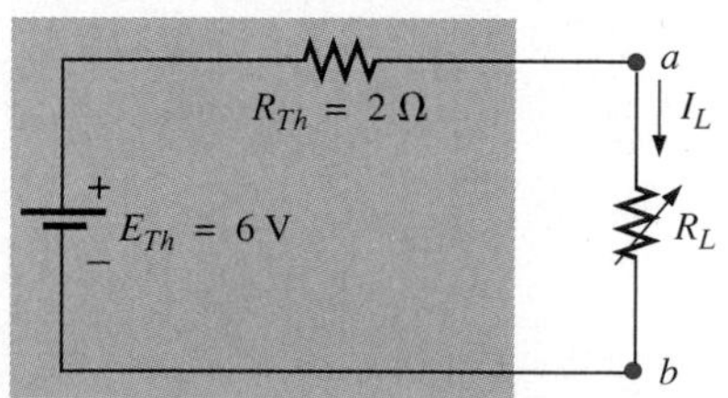

FIG. 9.32
Substituting the Thévenin equivalent circuit for the network external to R_L in Fig. 9.27.

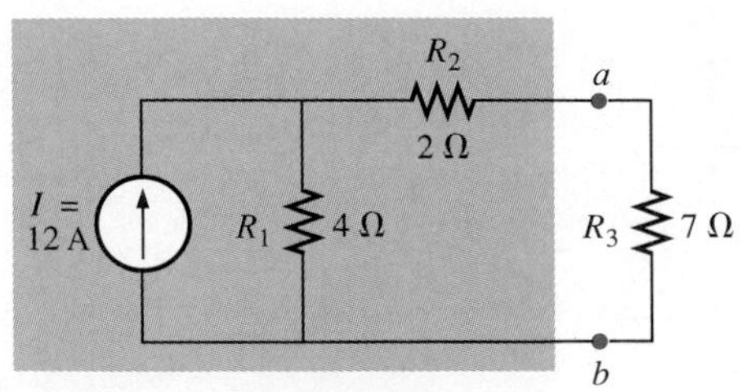

FIG. 9.33
Example 9.7.

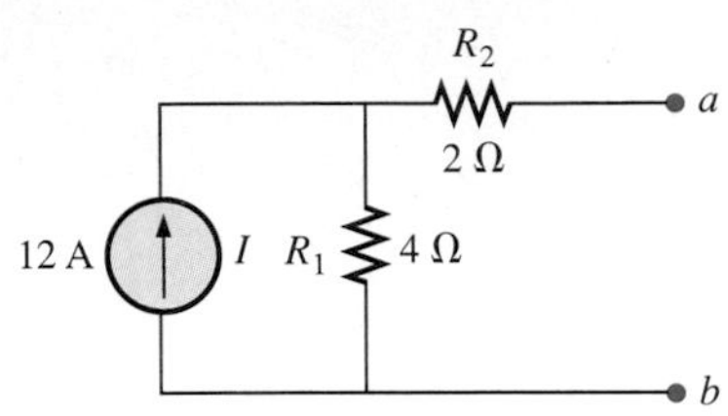

FIG. 9.34
Establishing the terminals of particular interest for the network of Fig. 9.33.

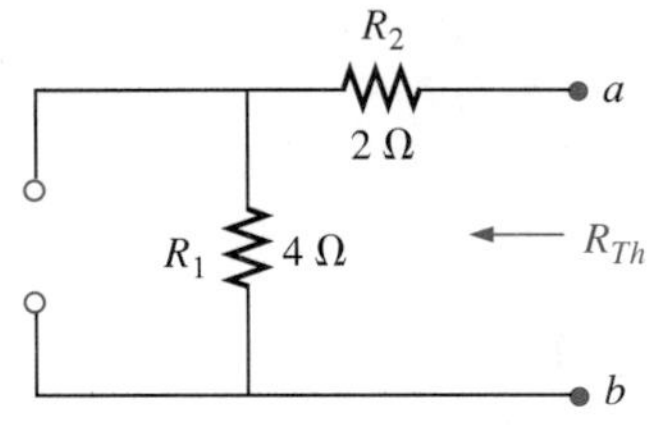

FIG. 9.35
Determining R_{Th} for the network of Fig. 9.34.

If Thévenin's theorem were unavailable, each change in R_L would require that the entire network of Fig. 9.27 be reexamined to find the new value of R_L.

EXAMPLE 9.7 Find the Thévenin equivalent circuit for the network in the shaded area of the network of Fig. 9.33.

Solution:

Steps 1 and 2 are shown in Fig. 9.34.

Step 3 is shown in Fig. 9.35. The current source has been replaced with an open-circuit equivalent and the resistance determined between terminals *a* and *b*.

In this case an ohmmeter connected between terminals *a* and *b* would send out a sensing current that would flow directly through R_1 and R_2 (at the same level). The result is that R_1 and R_2 are in series and the Thévenin resistance is the sum of the two.

$$R_{Th} = R_1 + R_2 = 4\ \Omega + 2\ \Omega = \mathbf{6\ \Omega}$$

Step 4 (Fig. 9.36): In this case, since an open circuit exists between the two marked terminals, the current is zero between these terminals and through the 2-Ω resistor. The voltage drop across R_2 is, therefore,

$$V_2 = I_2R_2 = (0)R_2 = 0\ \text{V}$$

and $$E_{Th} = V_1 = I_1R_1 = IR_1 = (12\ \text{A})(4\ \Omega) = \mathbf{48\ V}$$

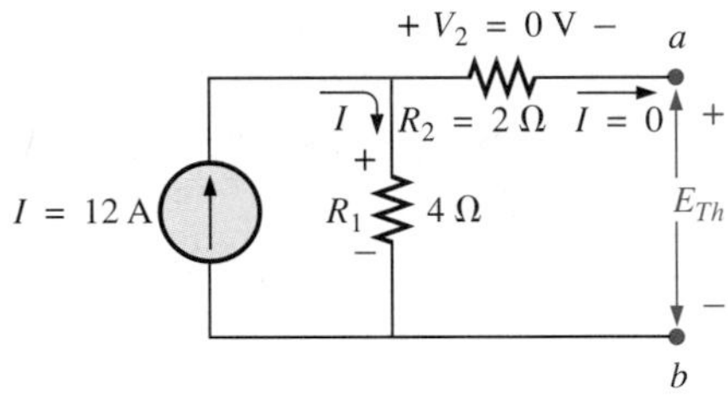

FIG. 9.36
Determining E_{Th} for the network of Fig. 9.34.

Step 5 is shown in Fig. 9.37.

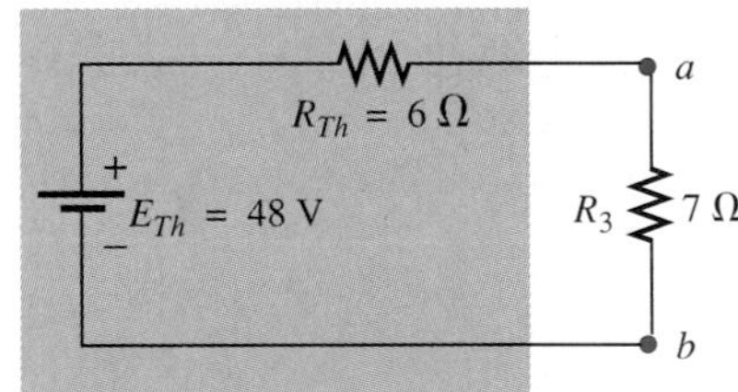

FIG. 9.37
Substituting the Thévenin equivalent circuit in the network external to the resistor R_3 of Fig. 9.33.

EXAMPLE 9.8 Find the Thévenin equivalent circuit for the network in the shaded area of the network of Fig. 9.38. Note in this example that

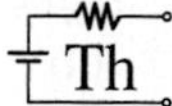

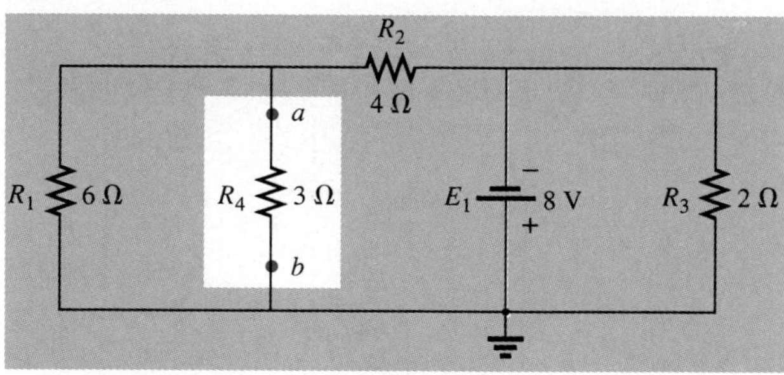

FIG. 9.38
Example 9.8.

there is no need for the section of the network to be preserved to be at the "end" of the configuration.

Solution:

Steps 1 and 2: See Fig. 9.39.

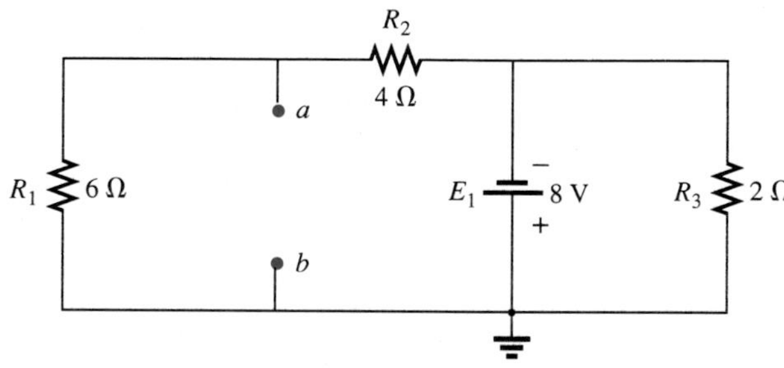

FIG. 9.39
Identifying the terminals of particular interest for the network of Fig. 9.38.

Step 3: See Fig. 9.40. Steps 1 and 2 are relatively easy to apply, but now we must be careful to "hold" onto the terminals a and b as the Thévenin resistance and voltage are determined. In Fig. 9.40, all the remaining elements turn out to be in parallel, and the network can be redrawn as shown.

$$R_{Th} = R_1 \parallel R_2 = \frac{(6\ \Omega)(4\ \Omega)}{6\ \Omega + 4\ \Omega} = \frac{24\ \Omega}{10} = \mathbf{2.4\ \Omega}$$

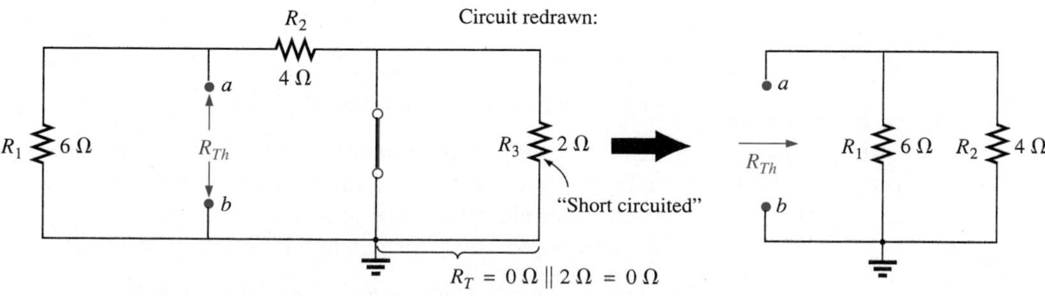

FIG. 9.40
Determining R_{Th} for the network of Fig. 9.39.

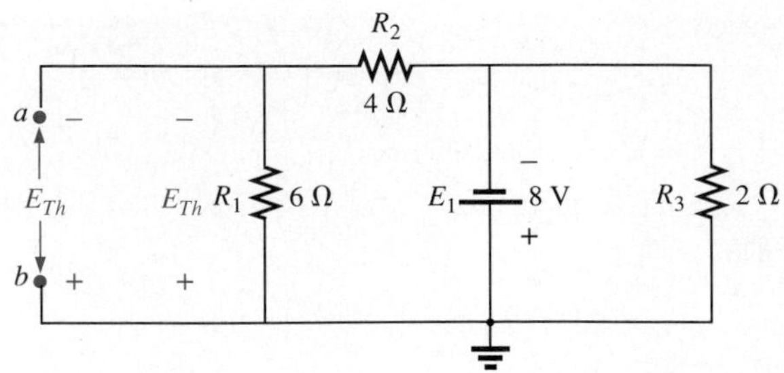

FIG. 9.41
Determining E_{Th} for the network of Fig. 9.39.

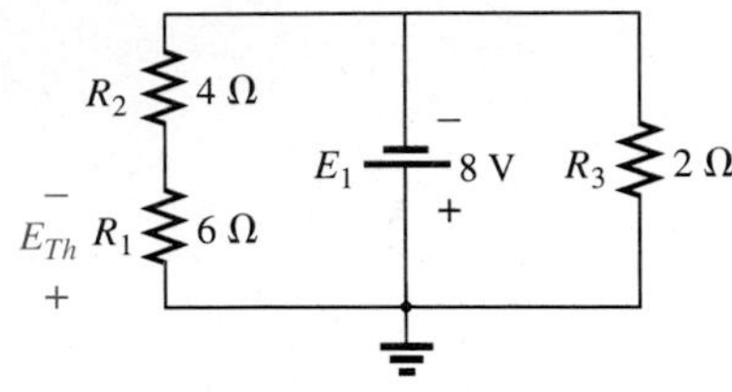

FIG. 9.42
Network of Fig. 9.41 redrawn.

Step 4: See Fig. 9.41. In this case, the network can be redrawn as shown in Fig. 9.42, and since the voltage is the same across parallel elements, the voltage across the series resistors R_1 and R_2 is E_1, or 8 V. Applying the voltage divider rule,

$$E_{Th} = \frac{R_1 E_1}{R_1 + R_2} = \frac{(6\ \Omega)(8\ \text{V})}{6\ \Omega + 4\ \Omega} = \frac{48\ \text{V}}{10} = \mathbf{4.8\ V}$$

Step 5: See Fig. 9.43.

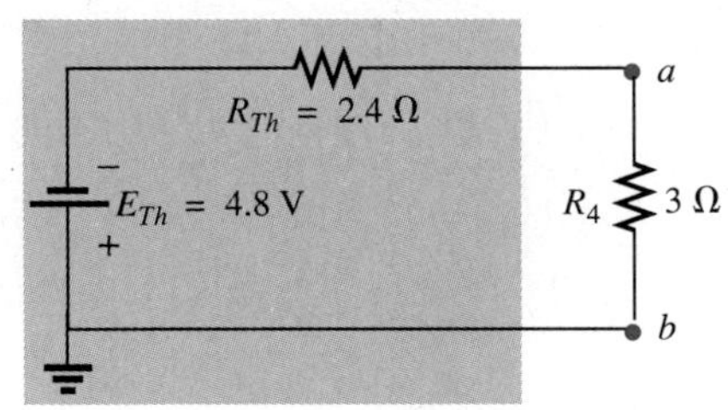

FIG. 9.43
Substituting the Thévenin equivalent circuit for the network external to the resistor R_4 of Fig. 9.38.

The importance of marking the terminals should be obvious from Example 9.8. Note that there is no requirement that the Thévenin voltage have the same polarity as the equivalent circuit originally introduced.

EXAMPLE 9.9 Find the Thévenin equivalent circuit for the network in the shaded area of the bridge network of Fig. 9.44.

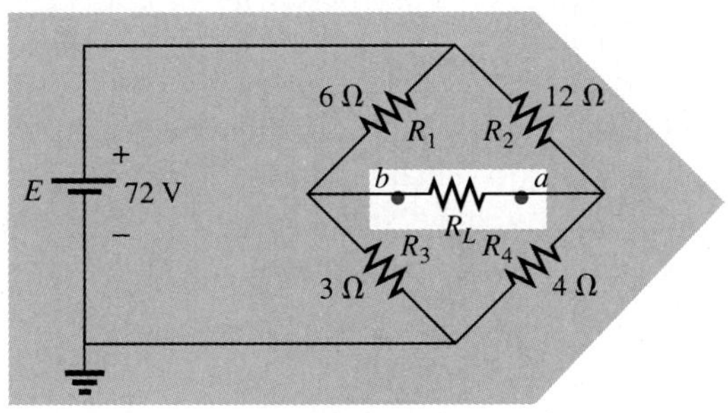

FIG. 9.44
Example 9.9.

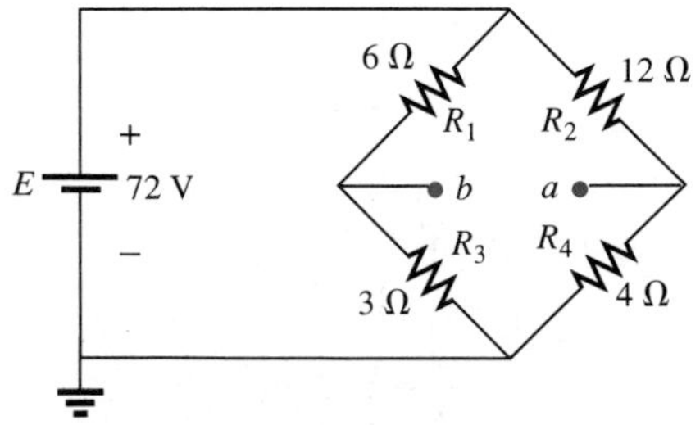

FIG. 9.45
Identifying the terminals of particular interest for the network of Fig. 9.44.

Solution:

Steps 1 and 2 are shown in Fig. 9.45.

Step 3: See Fig. 9.46. In this case, the short-circuit replacement of the voltage source *E* provides a direct connection between *c* and *c′* of Fig. 9.46(a), permitting a "folding" of the network around the horizontal line of *a-b* to produce the configuration of Fig. 9.46(b).

$$\begin{aligned} R_{Th} = R_{a\text{-}b} &= R_1 \parallel R_3 + R_2 \parallel R_4 \\ &= 6\ \Omega \parallel 3\ \Omega + 4\ \Omega \parallel 12\ \Omega \\ &= 2\ \Omega + 3\ \Omega = \mathbf{5\ \Omega} \end{aligned}$$

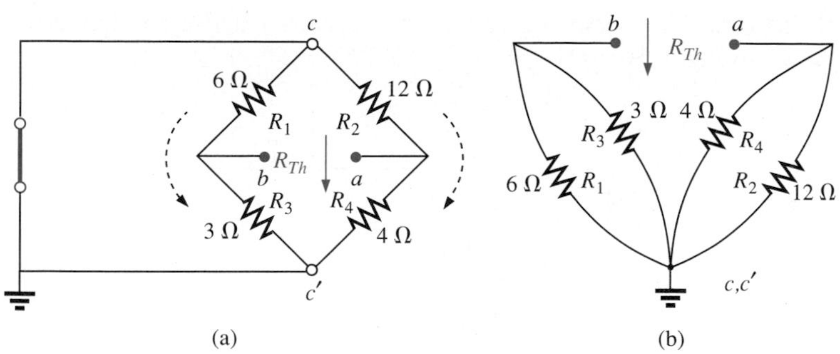

FIG. 9.46
Solving for R_{Th} for the network of Fig. 9.45.

Step 4: The circuit is redrawn in Fig. 9.47. The absence of a direct connection between a and b results in a network with three parallel branches. The voltages V_1 and V_2 can therefore be determined using the voltage divider rule:

$$V_1 = \frac{R_1 E}{R_1 + R_3} = \frac{(6\ \Omega)(72\ \text{V})}{6\ \Omega + 3\ \Omega} = \frac{432\ \text{V}}{9} = 48\ \text{V}$$

$$V_2 = \frac{R_2 E}{R_2 + R_4} = \frac{(12\ \Omega)(72\ \text{V})}{12\ \Omega + 4\ \Omega} = \frac{864\ \text{V}}{16} = 54\ \text{V}$$

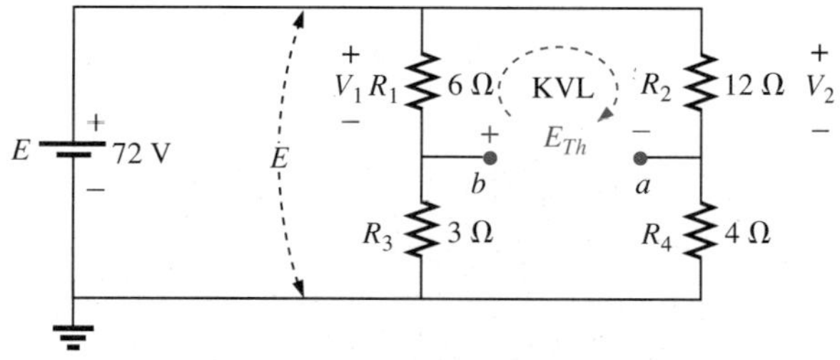

FIG. 9.47
Determining E_{Th} for the network of Fig. 9.45.

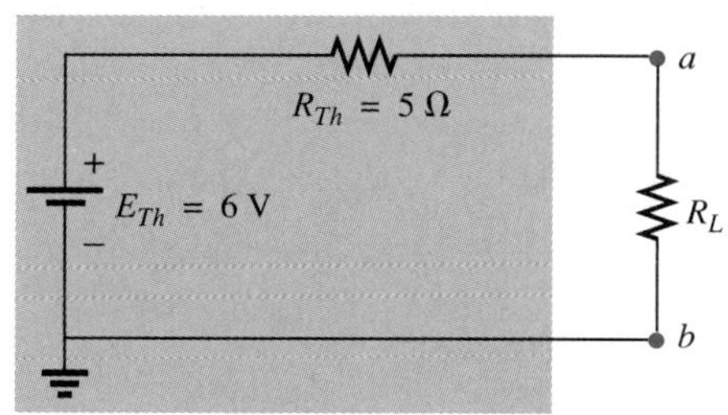

FIG. 9.48
Substituting the Thévenin equivalent circuit for the network external to the resistor R_L of Fig. 9.44.

Assuming the polarity shown for E_{Th} and applying Kirchhoff's voltage law to the top loop in the clockwise direction will result in

$$\Sigma_{\circlearrowright} V = +E_{Th} + V_1 - V_2 = 0$$

and $$E_{Th} = V_2 - V_1 = 54\ \text{V} - 48\ \text{V} = \mathbf{6\ V}$$

Step 5 is shown in Fig. 9.48.

Thévenin's theorem is not restricted to a single passive element, as shown in the preceding examples, but can be applied across sources, whole branches, portions of networks, or any circuit configuration, as shown in the following example. It is also possible that one of the methods previously described, such as mesh analysis or superposition, may have to be used to find the Thévenin equivalent circuit.

EXAMPLE 9.10 (Two sources) Find the Thévenin circuit for the network within the shaded area of Fig. 9.49.

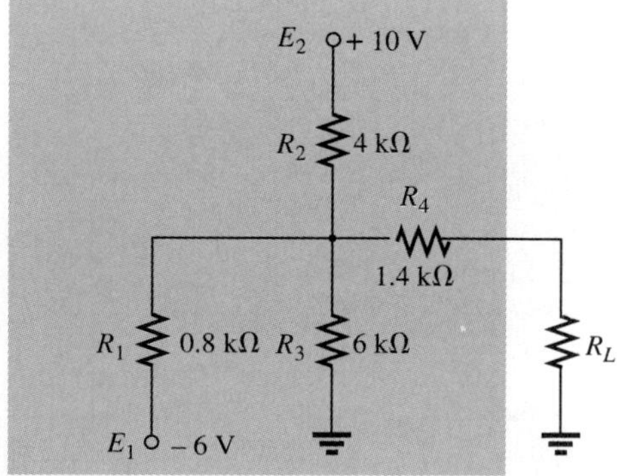

FIG. 9.49
Example 9.10.

Solution: The network is redrawn and steps 1 and 2 are applied as shown in Fig. 9.50.

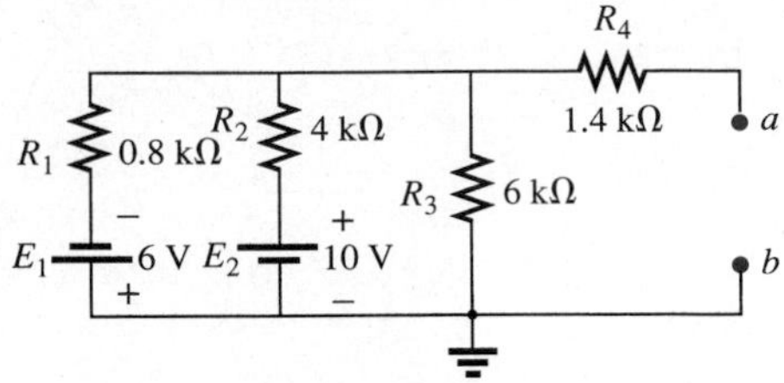

FIG. 9.50
Identifying the terminals of particular interest for the network of Fig. 9.49.

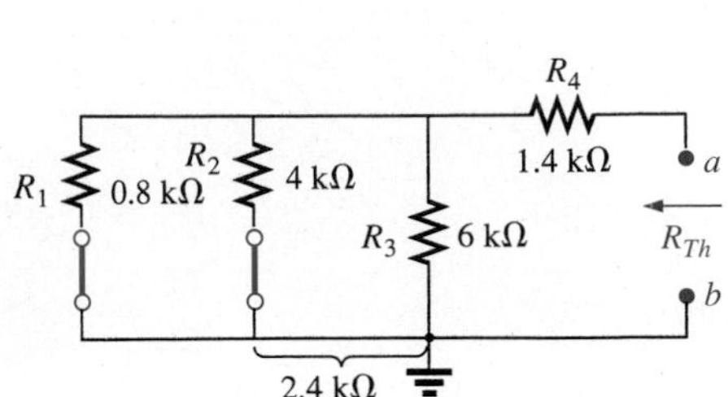

FIG. 9.51
Determining R_{Th} for the network of Fig. 9.50.

Step 3: See Fig. 9.51.

$$
\begin{aligned}
R_{Th} &= R_4 + R_1 \parallel R_2 \parallel R_3 \\
&= 1.4\text{ k}\Omega + 0.8\text{ k}\Omega \parallel 4\text{ k}\Omega \parallel 6\text{ k}\Omega \\
&= 1.4\text{ k}\Omega + 0.8\text{ k}\Omega \parallel 2.4\text{ k}\Omega \\
&= 1.4\text{ k}\Omega + 0.6\text{ k}\Omega \\
&= \mathbf{2\ k\Omega}
\end{aligned}
$$

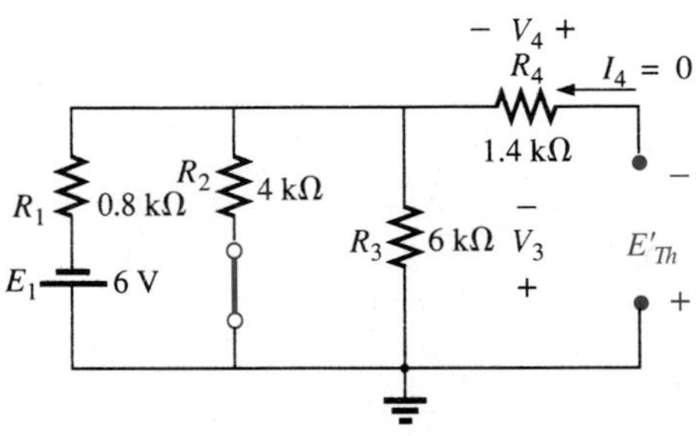

FIG. 9.52
Determining the contribution to E_{Th} from the source E_1 for the network of Fig. 9.50.

Step 4: Applying superposition, we will consider the effects of the voltage source E_1 first. Note Fig. 9.52. The open circuit requires that $V_4 = I_4R_4 = (0)R_4 = 0$ V, and

$$E'_{Th} = V_3$$

$$R'_T = R_2 \parallel R_3 = 4\text{ k}\Omega \parallel 6\text{ k}\Omega = 2.4\text{ k}\Omega$$

Applying the voltage divider rule,

$$V_3 = \frac{R'_T E_1}{R'_T + R_1} = \frac{(2.4\text{ k}\Omega)(6\text{ V})}{2.4\text{ k}\Omega + 0.8\text{ k}\Omega} = \frac{14.4\text{ V}}{3.2} = 4.5\text{ V}$$

and $$E'_{Th} = V_3 = 4.5\text{ V}$$

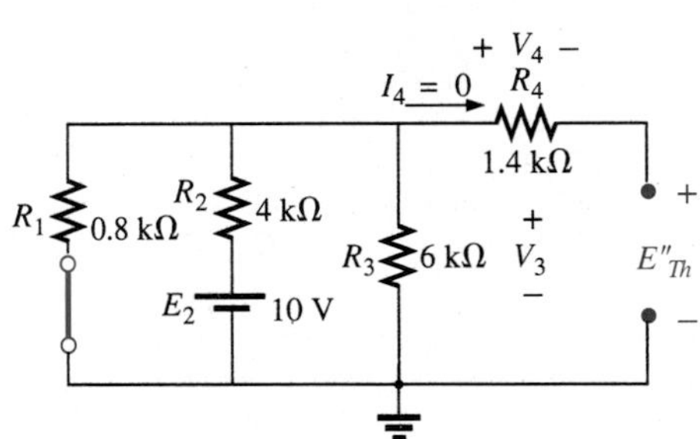

FIG. 9.53
Determining the contribution to E_{Th} from the source E_2 for the network of Fig. 9.50.

For the source E_2, the network of Fig. 9.53 will result. Again, $V_4 = I_4R_4 = (0)R_4 = 0$ V, and

$$E''_{Th} = V_3$$

$$R'_T = R_1 \parallel R_3 = 0.8\text{ k}\Omega \parallel 6\text{ k}\Omega = 0.706\text{ k}\Omega$$

and $$V_3 = \frac{R'_T E_2}{R'_T + R_2} = \frac{(0.706\text{ k}\Omega)(10\text{ V})}{0.706\text{ k}\Omega + 4\text{ k}\Omega} = \frac{7.06\text{ V}}{4.706} = 1.5\text{ V}$$

and $$E''_{Th} = V_3 = 1.5\text{ V}$$

Since E'_{Th} and E''_{Th} have opposite polarities,

$$
\begin{aligned}
E_{Th} &= E'_{Th} - E''_{Th} \\
&= 4.5\text{ V} - 1.5\text{ V} \\
&= \mathbf{3\ V} \qquad \text{(polarity of } E'_{Th}\text{)}
\end{aligned}
$$

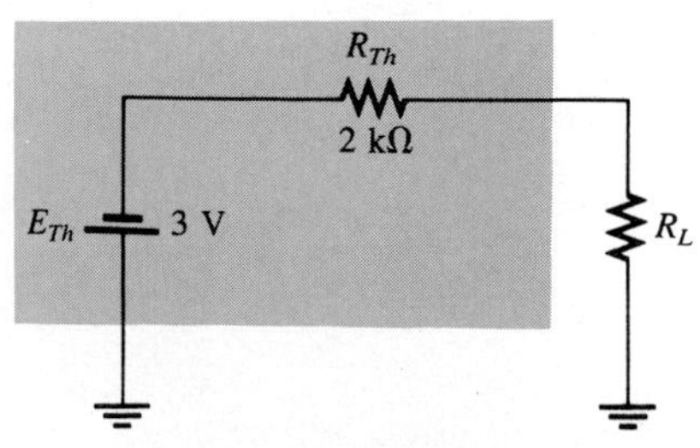

FIG. 9.54
Substituting the Thévenin equivalent circuit for the network external to the resistor R_L of Fig. 9.49.

Step 5: See Fig. 9.54.

Experimental Procedures

There are two popular experimental procedures for determining the parameters of a Thévenin equivalent network. The procedure for measuring the Thévenin voltage is the same for each, but the approach for determining the Thévenin resistance is quite different for each.

Direct Measurement of E_{Th} and R_{Th} For any physical network, the value of E_{Th} can be determined experimentally by measuring the open-circuit voltage across the load terminals, as shown in Fig. 9.55; $E_{Th} = V_{oc} = V_{ab}$. The value of R_{Th} can then be determined by completing the network with a variable R_L such as the potentiometer of Fig. 9.56(b). R_L can then be varied until the voltage appearing across the load is one-half the open-circuit value, or $V_L = E_{Th}/2$. For the series circuit of Fig. 9.56(a), when the load voltage is reduced to one-half the open-circuit level, the voltage across R_{Th} and R_L must be the same. If we read the value of R_L [as shown in Fig. 9.56(c)] that resulted in the preceding calculations, we will also have the value of R_{Th}, since $R_L = R_{Th}$ if V_L equals the voltage across R_{Th}.

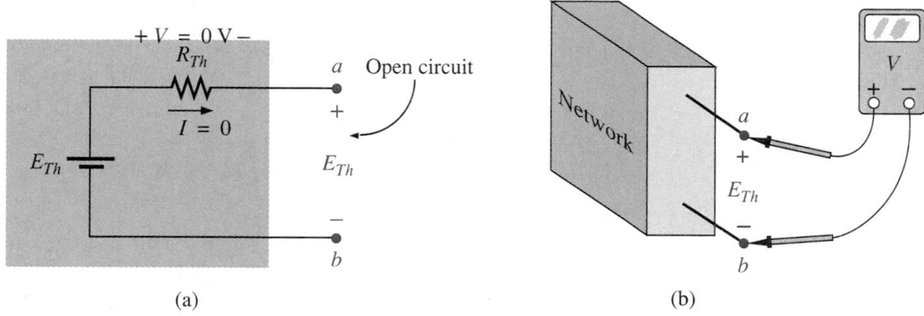

FIG. 9.55
Determining E_{Th} experimentally.

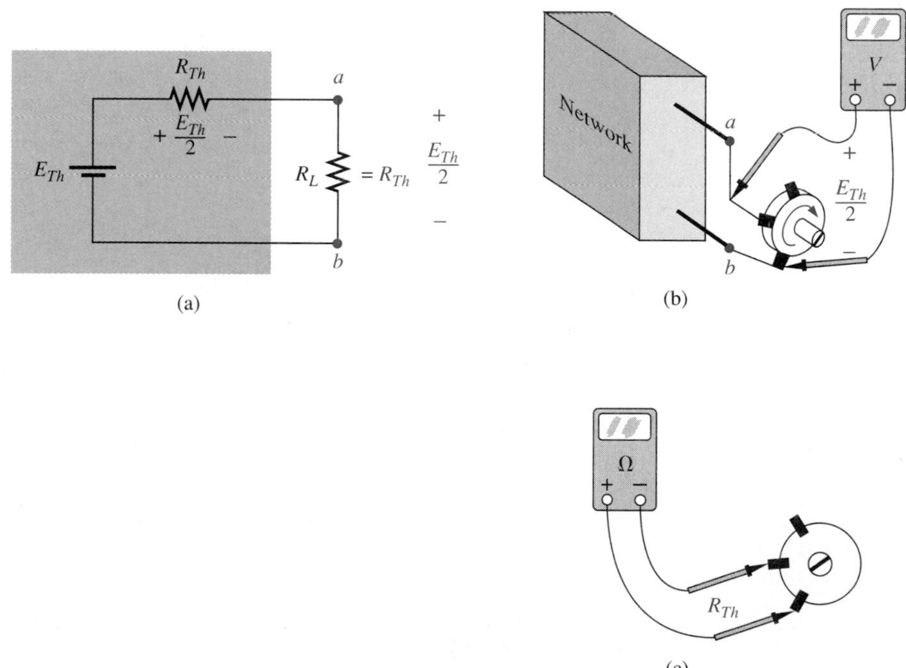

FIG. 9.56
Determining R_{Th} experimentally.

Measuring V_{oc} and I_{sc} The Thévenin voltage is again determined by measuring the open-circuit voltage across the terminals of interest;

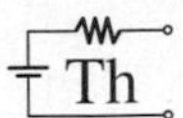

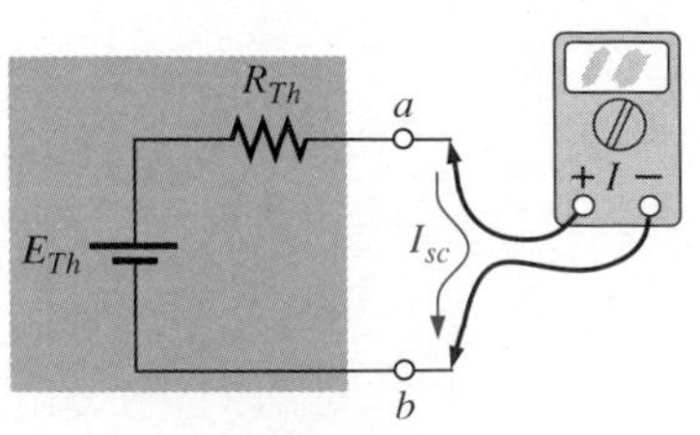

FIG. 9.57
Measuring I_{sc}.

that is, $E_{Th} = V_{oc}$. To determine R_{Th}, a short-circuit condition is established across the terminals of interest, as shown in Fig. 9.57, and the current through the short circuit is measured with an ammeter. Using Ohm's law, we find that the short-circuit current is determined by

$$I_{sc} = \frac{E_{Th}}{R_{Th}}$$

and the Thévenin resistance by

$$R_{Th} = \frac{E_{Th}}{I_{sc}}$$

However, $E_{Th} = V_{oc}$ resulting in the following equation for R_{Th}:

$$R_{Th} = \frac{V_{oc}}{I_{sc}} \tag{9.2}$$

American (Rockland, Maine; Summit, New Jersey)
(1898–1983)
Electrical Engineer, Scientist, Inventor
Department Head: Bell Laboratories
Fellow: Acoustical Society and Institute of Radio Engineers

Courtesy of AT&T Archives

Although primarily interested in communications circuit theory and the transmission of data at high speeds over telephone lines, Edward L. Norton is best remembered for development of the dual of Thévenin's equivalent circuit, currently referred to as Norton's equivalent circuit. In fact, Norton and his associates at AT&T in the early 1920s are recognized as some of the first to perform pioneering work applying Thévenin's equivalent circuit and who referred to this concept simply as Thévenin's theorem. In 1926 he proposed the equivalent circuit using a current source and parallel resistor to assist in the design of recording instrumentation that was primarily current driven. He began his telephone career in 1922 with the Western Electric Company's Engineering Department, which later became Bell Laboratories. His areas of active research included network theory, acoustical systems, electromagnetic apparatus, and data transmission. A graduate of MIT and Columbia University, he held nineteen patents on his work.

Edward L. Norton

9.4 NORTON'S THEOREM

It was demonstrated in Section 8.3 that every voltage source with a series internal resistance has a current source equivalent. The current source equivalent of the Thévenin network (which, you will note, satisfies the above conditions), as shown in Fig. 9.58, can be determined by Norton's theorem. It can also be found through the conversions of Section 8.3.

The theorem states the following:

Any two-terminal linear bilateral dc network can be replaced by an equivalent circuit consisting of a current source and a parallel resistor, as shown in Fig. 9.58.

The discussion of Thévenin's theorem with respect to the equivalent circuit can also be applied to the Norton equivalent circuit. The steps leading to the proper values of I_N and R_N are now listed.

Preliminary:

1. ***Remove that portion of the network across which the Norton equivalent circuit is found.***
2. ***Mark the terminals of the remaining two-terminal network.***

R_N:

3. ***Calculate R_N by first setting all sources to zero (voltage sources are replaced with short circuits and current sources with open circuits) and then finding the resultant resistance between the two marked terminals. (If the internal resistance of the voltage and/or current sources is included in the original network, it must remain when the sources are set to zero.) Since $R_N = R_{Th}$, the procedure and value obtained using the approach described for Thévenin's theorem will determine the proper value of R_N.***

I_N:

4. ***Calculate I_N by first returning all sources to their original position and then finding the short-circuit current between the marked terminals. It is the same current that would be measured by an ammeter placed between the marked terminals.***

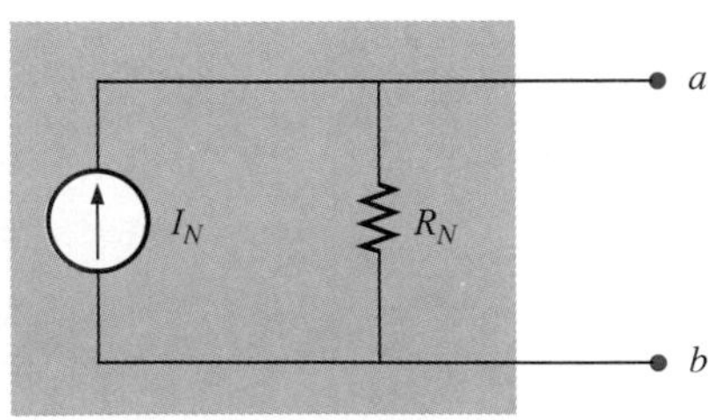

FIG. 9.58
Norton equivalent circuit.

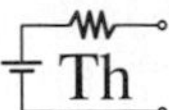

Conclusion:

5. ***Draw the Norton equivalent circuit with the portion of the circuit previously removed replaced between the terminals of the equivalent circuit.***

The Norton and Thévenin equivalent circuits can also be found from each other by using the source transformation discussed earlier in this chapter and reproduced in Fig. 9.59.

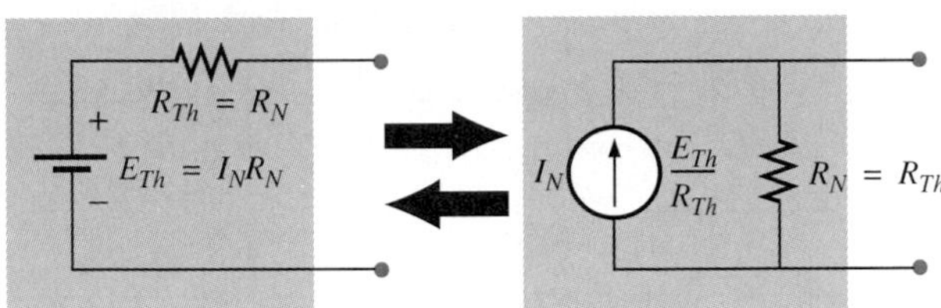

FIG. 9.59
Converting between Thévenin and Norton equivalent circuits.

EXAMPLE 9.11 Find the Norton equivalent circuit for the network in the shaded area of Fig. 9.60.

Solution:

Steps 1 and 2 are shown in Fig. 9.61.

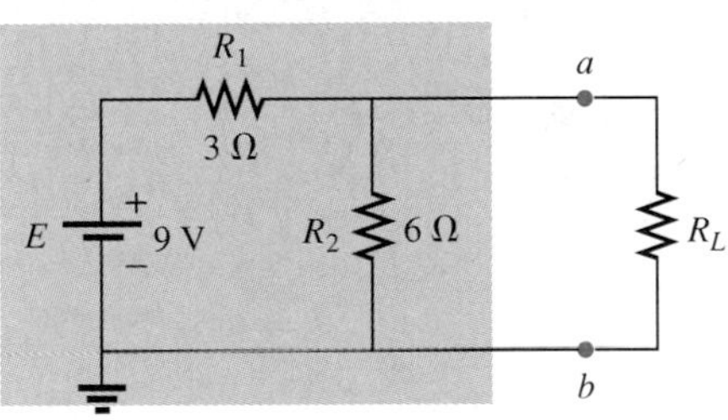

FIG. 9.60
Example 9.11.

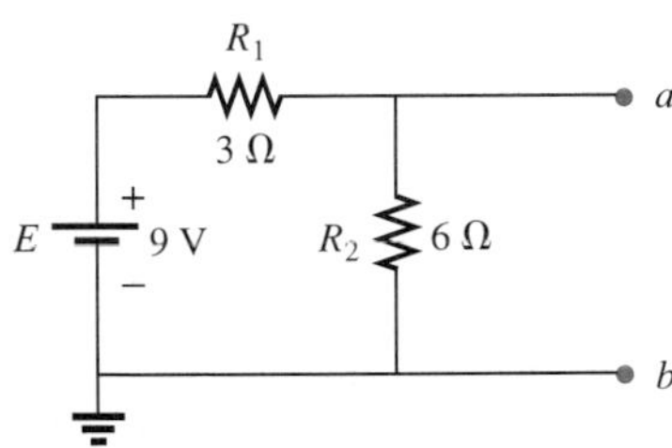

FIG. 9.61
Identifying the terminals of particular interest for the network of Fig. 9.60.

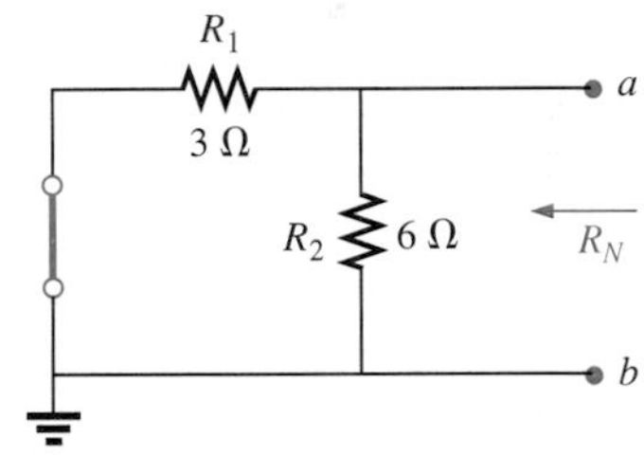

FIG. 9.62
Determining R_N for the network of Fig. 9.61.

Step 3 is shown in Fig. 9.62, and

$$R_N = R_1 \| R_2 = 3\ \Omega \| 6\ \Omega = \frac{(3\ \Omega)(6\ \Omega)}{3\ \Omega + 6\ \Omega} = \frac{18\ \Omega}{9} = \mathbf{2\ \Omega}$$

Step 4 is shown in Fig. 9.63, clearly indicating that the short-circuit connection between terminals *a* and *b* is in parallel with R_2 and eliminates its effect. I_N is therefore the same as through R_1, and the full battery voltage appears across R_1 since

$$V_2 = I_2 R_2 = (0)6\ \Omega = 0\ \text{V}$$

Therefore,

$$I_N = \frac{E}{R_1} = \frac{9\ \text{V}}{3\ \Omega} = \mathbf{3\ A}$$

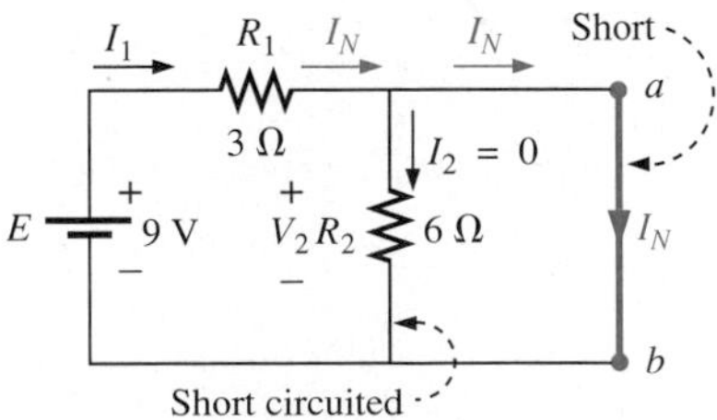

FIG. 9.63
Determining I_N for the network of Fig. 9.61.

Step 5: See Fig. 9.64. This circuit is the same as the first one considered in the development of Thévenin's theorem. A simple conversion indicates that the Thévenin circuits are, in fact, the same (Fig. 9.65).

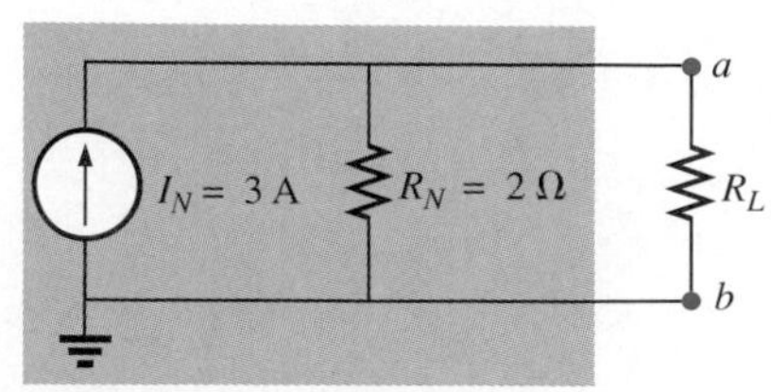

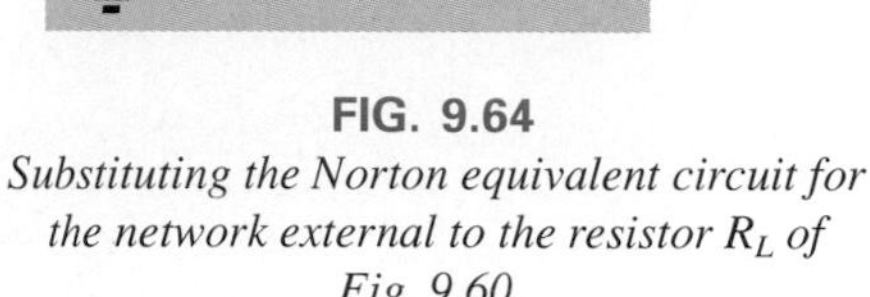

FIG. 9.64
Substituting the Norton equivalent circuit for the network external to the resistor R_L of Fig. 9.60.

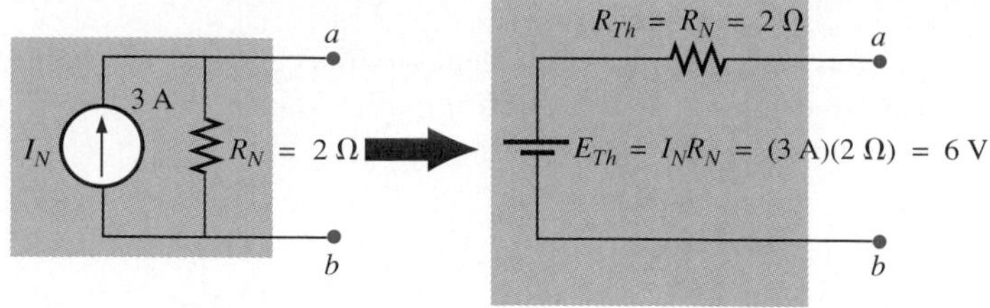

FIG. 9.65
Converting the Norton equivalent circuit of Fig. 9.64 to a Thévenin equivalent circuit.

EXAMPLE 9.12 Find the Norton equivalent circuit for the network external to the 9-Ω resistor in Fig. 9.66.

Solution:

Steps 1 and 2: See Fig. 9.67.

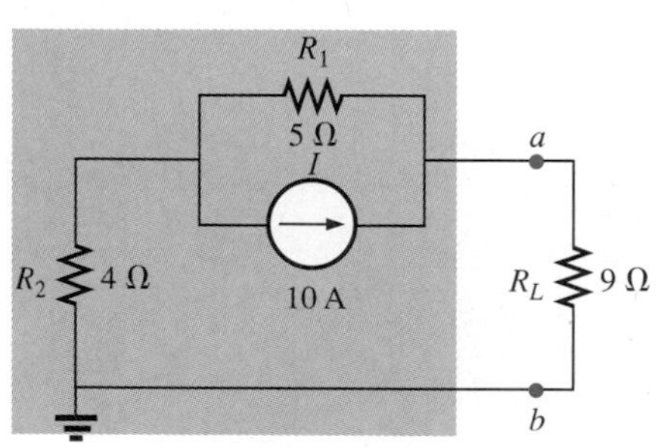

FIG. 9.66
Example 9.12.

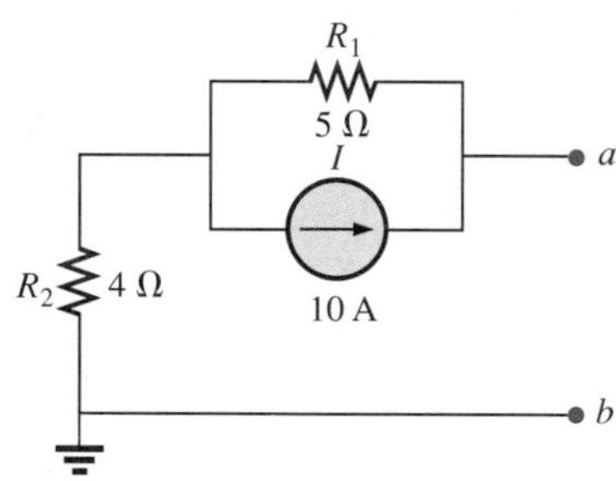

FIG. 9.67
Identifying the terminals of particular interest for the network of Fig. 9.66.

Step 3: See Fig. 9.68, and

$$R_N = R_1 + R_2 = 5\ \Omega + 4\ \Omega = \mathbf{9\ \Omega}$$

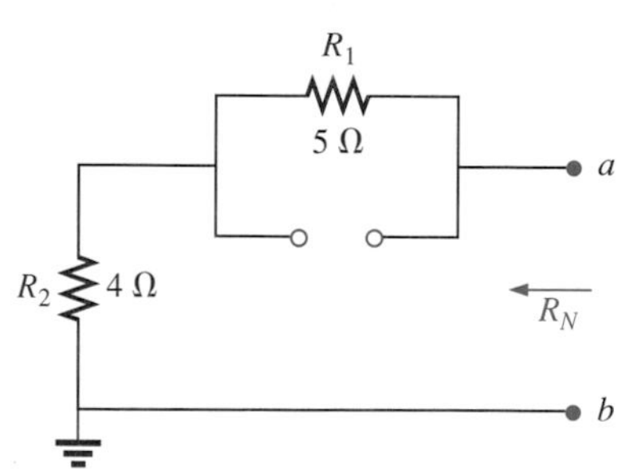

FIG. 9.68
Determining R_N for the network of Fig. 9.67.

Step 4: As shown in Fig. 9.69, the Norton current is the same as the current through the 4-Ω resistor. Applying the current divider rule,

$$I_N = \frac{R_1 I}{R_1 + R_2} = \frac{(5\ \Omega)(10\ \text{A})}{5\ \Omega + 4\ \Omega} = \frac{50\ \text{A}}{9} = \mathbf{5.556\ A}$$

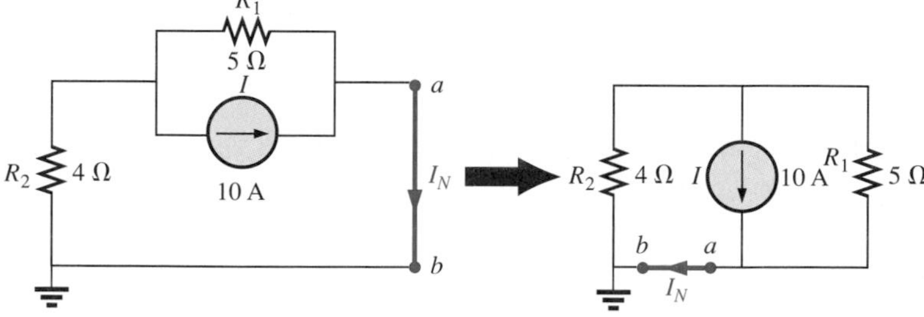

FIG. 9.69
Determining I_N for the network of Fig. 9.67.

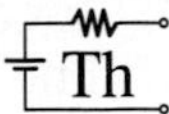

Step 5: See Fig. 9.70.

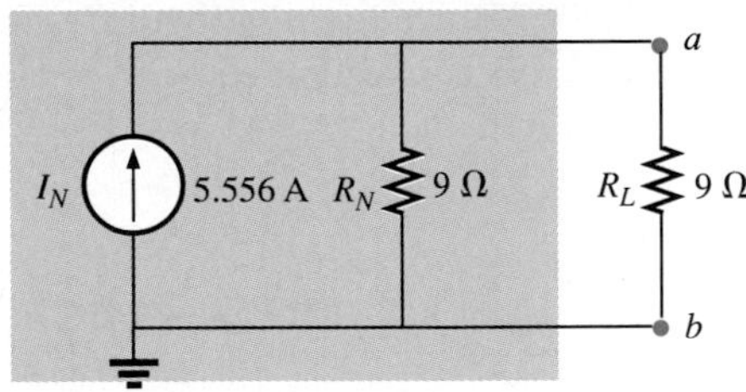

FIG. 9.70
Substituting the Norton equivalent circuit for the network external to the resistor R_L of Fig. 9.66

EXAMPLE 9.13 (Two sources) Find the Norton equivalent circuit for the portion of the network to the left of *a-b* in Fig. 9.71.

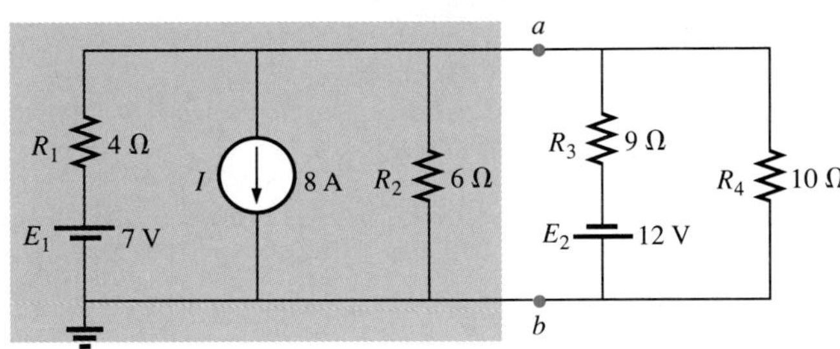

FIG. 9.71
Example 9.13.

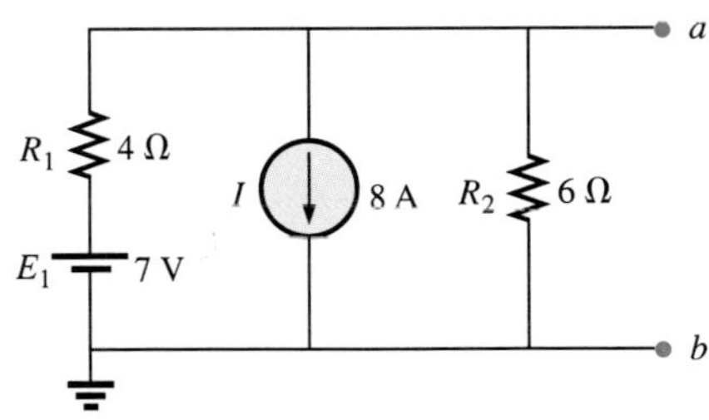

FIG. 9.72
Identifying the terminals of particular interest for the network of Fig. 9.71.

Solution:

Steps 1 and 2: See Fig. 9.72.

Step 3 is shown in Fig. 9.73, and

$$R_N = R_1 \parallel R_2 = 4\ \Omega \parallel 6\ \Omega = \frac{(4\ \Omega)(6\ \Omega)}{4\ \Omega + 6\ \Omega} = \frac{24\ \Omega}{10} = \mathbf{2.4\ \Omega}$$

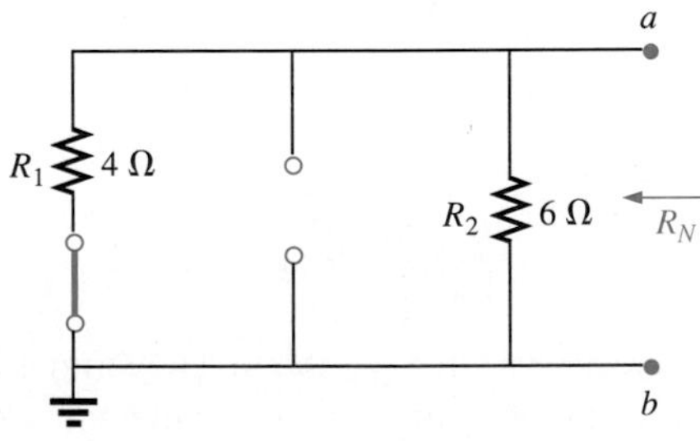

FIG. 9.73
Determining R_N for the network of Fig. 9.72.

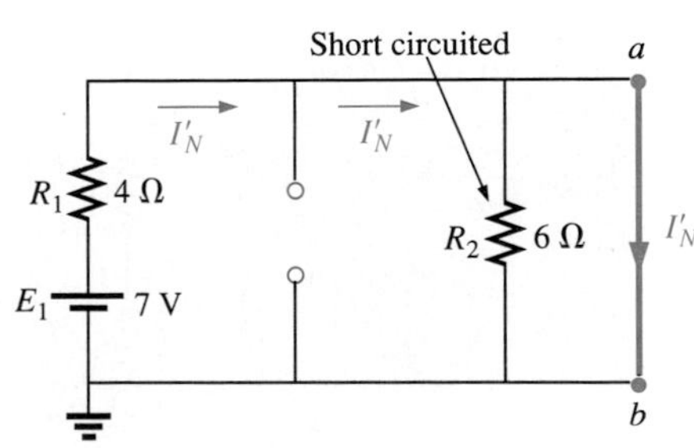

FIG. 9.74
Determining the contribution to I_N from the voltage source E_1.

Step 4: (Using superposition) For the 7-V battery (Fig. 9.74),

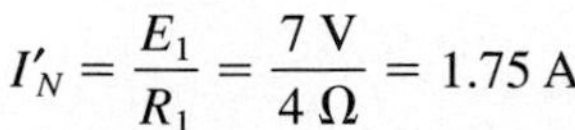

$$I'_N = \frac{E_1}{R_1} = \frac{7\text{ V}}{4\ \Omega} = 1.75\text{ A}$$

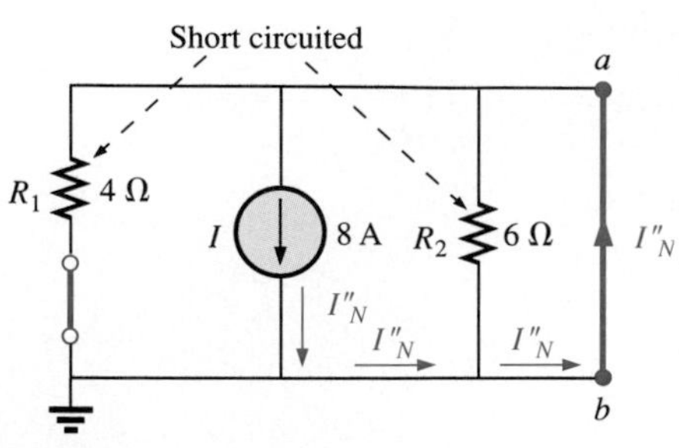

FIG. 9.75
Determining the contribution to I_N from the current source I.

For the 8-A source (Fig. 9.75), we find that both R_1 and R_2 have been "short circuited" by the direct connection between *a* and *b*, and

$$I''_N = I = 8\text{ A}$$

The result is

$$I_N = I''_N - I'_N = 8\text{ A} - 1.75\text{ A} = \mathbf{6.25\ A}$$

Step 5: See Fig. 9.76.

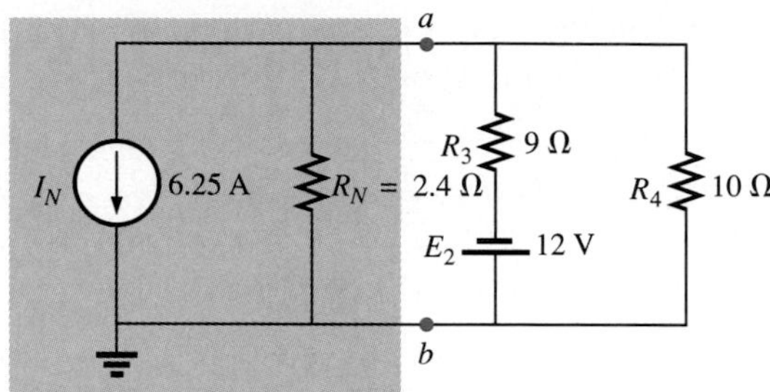

FIG. 9.76
Substituting the Norton equivalent circuit for the network to the left of terminals a-b in Fig. 9.71.

Experimental Procedure

The Norton current is measured in the same way as described for the short-circuit current for the Thévenin network. Since the Norton and Thévenin resistances are the same, the same procedures can be employed as described for the Thévenin network.

9.5 MAXIMUM POWER TRANSFER THEOREM

The maximum power transfer theorem states the following:

A load will receive maximum power from a linear bilateral dc network when its total resistive value is exactly equal to the Thévenin resistance of the network as "seen" by the load.

For the network of Fig. 9.77, maximum power will be delivered to the load when

$$R_L = R_{Th} \tag{9.3}$$

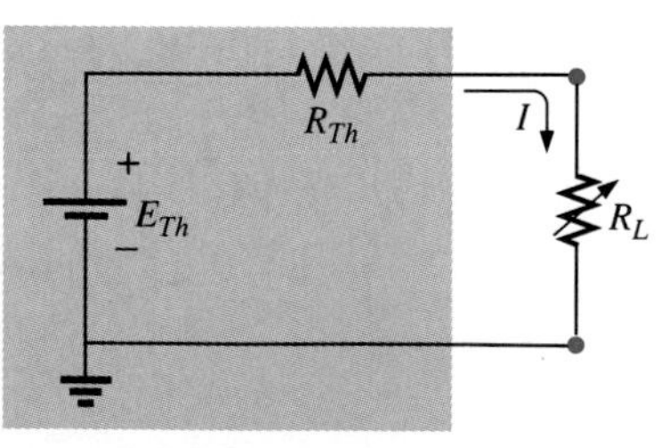

FIG. 9.77
Defining the conditions for maximum power to a load using the Thévenin equivalent circuit.

From past discussions, we realize that a Thévenin equivalent circuit can be found across any element or group of elements in a linear bilateral dc network. Therefore, if we consider the case of the Thévenin equivalent circuit with respect to the maximum power transfer theorem, we are, in essence, considering the *total* effects of any network across a resistor R_L, such as in Fig. 9.77.

For the Norton equivalent circuit of Fig. 9.78, maximum power will be delivered to the load when

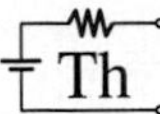

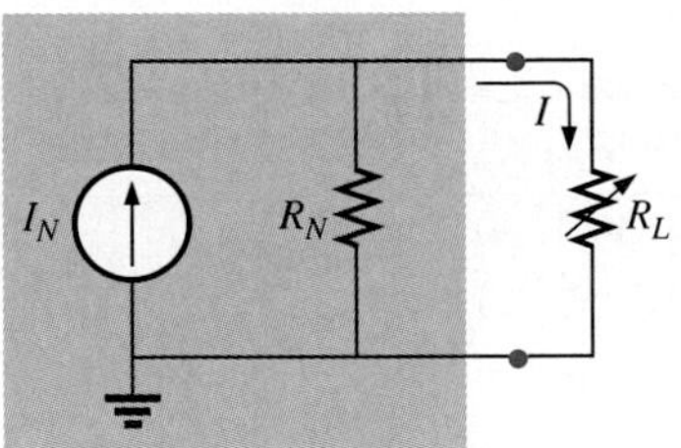

FIG. 9.78
Defining the conditions for maximum power to a load using the Norton equivalent circuit.

$$\boxed{R_L = R_N} \tag{9.4}$$

This result [Eq. (9.4)] will be used to its fullest advantage in the analysis of transistor networks, where the most frequently applied transistor circuit model employs a current source rather than a voltage source.

For the network of Fig. 9.77,

$$I = \frac{E_{Th}}{R_{Th} + R_L}$$

and
$$P_L = I^2 R_L = \left(\frac{E_{Th}}{R_{Th} + R_L}\right)^2 R_L$$

so that
$$P_L = \frac{E_{Th}^2 R_L}{(R_{Th} + R_L)^2}$$

Let us now consider an example where $E_{Th} = 60\text{ V}$ and $R_{Th} = 9\ \Omega$, as shown in Fig. 9.79.

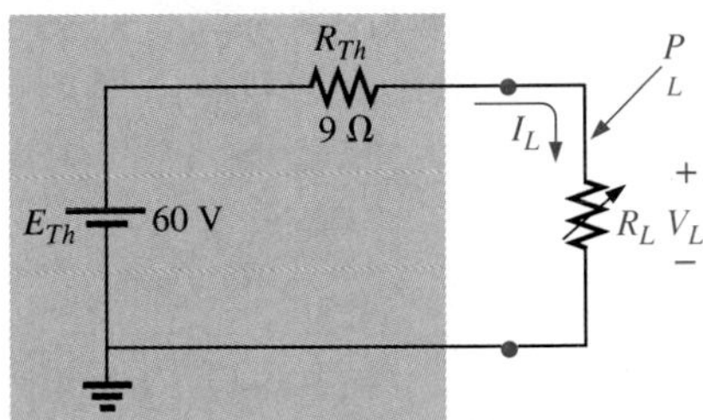

FIG. 9.79
Thévenin equivalent network to be used to validate the maximum power transfer theorem.

The power to the load is determined by:

$$P_L = \frac{E_{Th}^2 R_L}{(R_{Th} + R_L)^2} = \frac{3600 R_L}{(9\ \Omega + R_L)^2}$$

with
$$I_L = \frac{E_{Th}}{R_{Th} + R_L} = \frac{60\text{ V}}{9\ \Omega + R_L}$$

and
$$V_L = \frac{R_L(60\text{ V})}{R_{Th} + R_L} = \frac{R_L(60\text{ V})}{9\ \Omega + R_L}$$

A tabulation of P_L for a range of values of R_L yields Table 9.1. A plot of P_L versus R_L using the data of Table 9.1 will result in the plot of Fig. 9.80 for the range $R_L = 0.1\ \Omega$ to $30\ \Omega$.

TABLE 9.1

R_L (Ω)	P_L (W)		I_L (A)		V_L (V)	
0.1	4.35		6.59		0.66	
0.2	8.51		6.52		1.30	
0.5	19.94		6.32		3.16	
1	36.00		6.00		6.00	
2	59.50		5.46		10.91	
3	75.00		5.00		15.00	
4	85.21		4.62		18.46	
5	91.84	Increase	4.29	Decrease	21.43	Increase
6	96.00		4.00		24.00	
7	98.44		3.75		26.25	
8	99.65 ↓		3.53 ↓		28.23 ↓	
9(R_{Th})	100.00 (Maximum)		3.33 ($I_{max}/2$)		30.00 ($E_{Th}/2$)	
10	99.72		3.16		31.58	
11	99.00		3.00		33.00	
12	97.96		2.86		34.29	
13	96.69		2.73		35.46	
14	95.27		2.61		36.52	
15	93.75		2.50		37.50	
16	92.16		2.40		38.40	
17	90.53	Decrease	2.31	Decrease	39.23	Increase
18	88.89		2.22		40.00	
19	87.24		2.14		40.71	
20	85.61		2.07		41.38	
25	77.86		1.77		44.12	
30	71.00		1.54		46.15	
40	59.98		1.22		48.98	
100	30.30		0.55		55.05	
500	6.95		0.12		58.94	
1000	3.54 ↓		0.06 ↓		59.47 ↓	

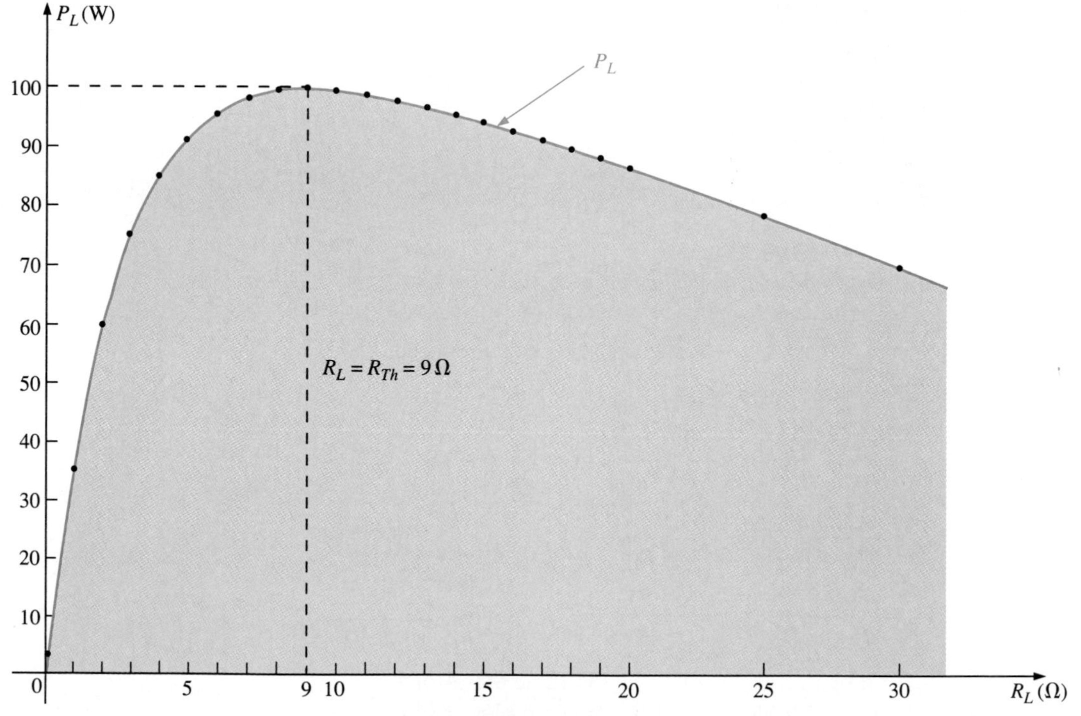

FIG. 9.80
P_L versus R_L for the network of Fig. 9.79.

Note, in particular, that P_L is, in fact, a maximum when $R_L = R_{Th} = 9\ \Omega$. The power curve increases more rapidly toward its maximum value than it decreases after the maximum point, clearly revealing that a small change in load resistance for levels of R_L below R_{Th} will have a more dramatic effect on the power delivered than similar changes in R_L above the R_{Th} level.

If we plot V_L and I_L versus the same resistance scale (Fig. 9.81), we find that both change nonlinearly, with the terminal voltage increasing with an increase in load resistance as the current decreases. Note again that the most dramatic changes in V_L and I_L occur for levels of R_L less than R_{Th}. As pointed out on the plot, when $R_L = R_{Th}$, $V_L = E_{Th}/2$ and $I_L = I_{max}/2$, with $I_{max} = E_{Th}/R_{Th}$.

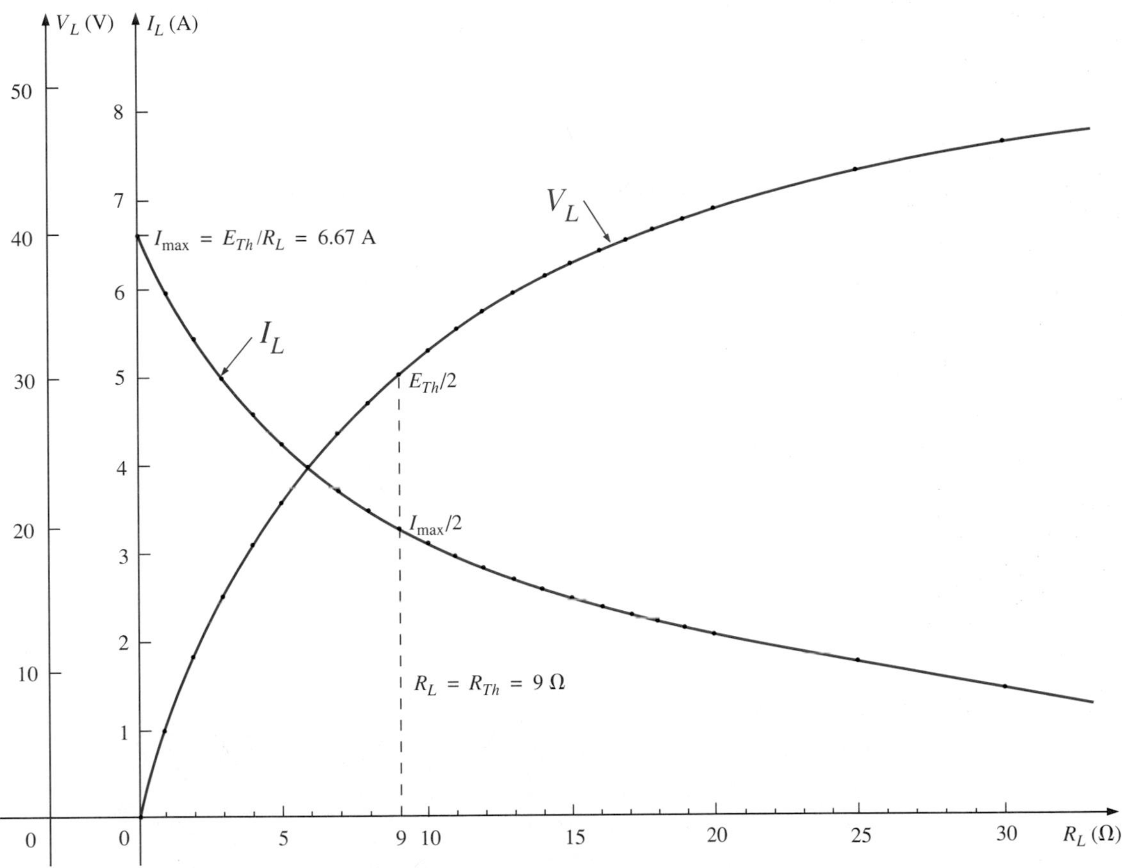

FIG. 9.81
V_L and I_L versus R_L for the network of Fig. 9.79.

The dc operating efficiency of a system is defined by the ratio of the power delivered to the load to the power supplied by the source; that is,

$$\eta\% = \frac{P_L}{P_s} \times 100\% \tag{9.5}$$

For the situation defined by Fig. 9.77,

$$\eta\% = \frac{P_L}{P_s} \times 100\% = \frac{I_L^2 R_L}{I_L^2 R_T} \times 100\%$$

and
$$\eta\% = \frac{R_L}{R_{Th} + R_L} \times 100\%$$

For R_L that is small compared to R_{Th}, $R_{Th} \gg R_L$ and $R_{Th} + R_L \cong R_{Th}$, with

$$\eta\% \cong \frac{R_L}{R_{Th}} \times 100\% = \underbrace{\left(\frac{1}{R_{Th}}\right)}_{\text{Constant}} R_L \times 100\% = kR_L \times 100\%$$

The resulting percentage efficiency, therefore, will be relatively low (since k is small) and will increase almost linearly as R_L increases.

For situations where the load resistance R_L is much larger than R_{Th}, $R_L \gg R_{Th}$ and $R_{Th} + R_L \cong R_L$.

$$\eta\% = \frac{R_L}{R_L} \times 100\% = 100\%$$

The efficiency therefore increases linearly and dramatically for small levels of R_L and then begins to level off as it approaches the 100% level for very large values of R_L, as shown in Fig. 9.82. Keep in mind, however, that the efficiency criterion is sensitive only to the ratio of P_L to P_s and not to their actual levels. At efficiency levels approaching 100%, the power delivered to the load may be so small as to have little practical value. Note the low level of power to the load in Table 9.1 when $R_L = 1000\ \Omega$ even though the efficiency level will be

$$\eta\% = \frac{R_L}{R_{Th} + R_L} \times 100\% = \frac{1000}{1009} \times 100\% = 99.11\%$$

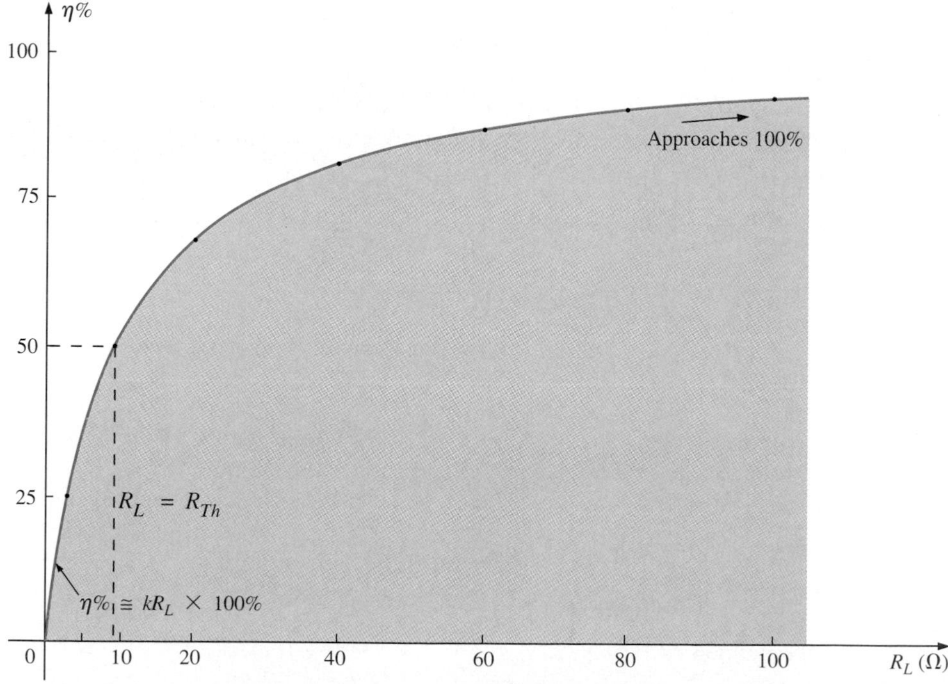

FIG. 9.82
Efficiency of operation versus increasing values of R_L.

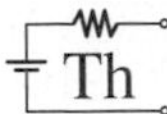

When $R_L = R_{Th}$,

$$\eta\% = \frac{R_L}{R_{Th} + R_L} \times 100\% = \frac{R_L}{2R_L} \times 100\% = \mathbf{50\%}$$

Under maximum power transfer conditions, therefore, P_L is a maximum, but the dc efficiency is only 50%; that is, only half the power delivered by the source is getting to the load.

A relatively low efficiency of 50% can be tolerated in situations where power levels are relatively low, such as in a wide variety of electronic systems. However, when large power levels are involved, such as at generating stations, efficiencies of 50% would not be acceptable. In fact, a great deal of expense and research is dedicated to raising power-generating and transmission efficiencies a few percentage points. Raising an efficiency level of a 10-mega-kW power plant from 94% to 95% (a 1% increase) can save 0.1 mega-kW, or 100 million watts, of power—an enormous saving!

Consider a change in load levels from 9 Ω to 20 Ω. In Fig. 9.80, the power level has dropped from 100 W to 85.61 W (a 14.4% drop), but the efficiency has increased substantially to 69% (a 38% increase), as shown in Fig. 9.82. For each application, therefore, a balance point must be identified where the efficiency is sufficiently high without reducing the power to the load to insignificant levels.

Figure 9.83 is a semilog plot of P_L and the power delivered by the source $P_s = E_{Th}I_L$ versus R_L for $E_{Th} = 60$ V and $R_{Th} = 9\ \Omega$. A semilog graph employs one log scale and one linear scale, as implied by the prefix *semi,* meaning *half.* Log scales are discussed in detail in Chapter 21. For the moment, note the wide range of R_L permitted using the log scale as compared to Figs. 9.80 through 9.82.

It is now quite clear that the P_L curve has only one maximum (at $R_L = R_{Th}$), whereas P_s decreases for every increase in R_L. In particular, note that for low levels of R_L, only a small portion of the power delivered by the source makes it to the load. In fact, even when $R_L = R_{Th}$, the source is generating twice the power absorbed by the load. For values of R_L greater than R_{Th}, the two curves approach each other until eventually they are essentially the same at high levels of R_L. For the range $R_L = R_{Th} = 9\ \Omega$ to $R_L = 100\ \Omega$, P_L and P_s are relatively close in magnitude, suggesting that this would be an appropriate range of operation, since a majority of the power delivered by the source is getting to the load and the power levels are still significant.

The power delivered to R_L under maximum power conditions ($R_L = R_{Th}$) is

$$I = \frac{E_{Th}}{R_{Th} + R_L} = \frac{E_{Th}}{2R_{Th}}$$

$$P_L = I^2R_L = \left(\frac{E_{Th}}{2R_{Th}}\right)^2 R_{Th} = \frac{E_{Th}^2 R_{Th}}{4R_{Th}^2}$$

and

$$P_{L_{\max}} = \frac{E_{Th}^2}{4R_{Th}} \qquad \text{(watts, W)} \qquad \mathbf{(9.6)}$$

For the Norton circuit of Fig. 9.78,

$$P_{L_{\max}} = \frac{I_N^2 R_N}{4} \qquad \text{(W)} \qquad \mathbf{(9.7)}$$

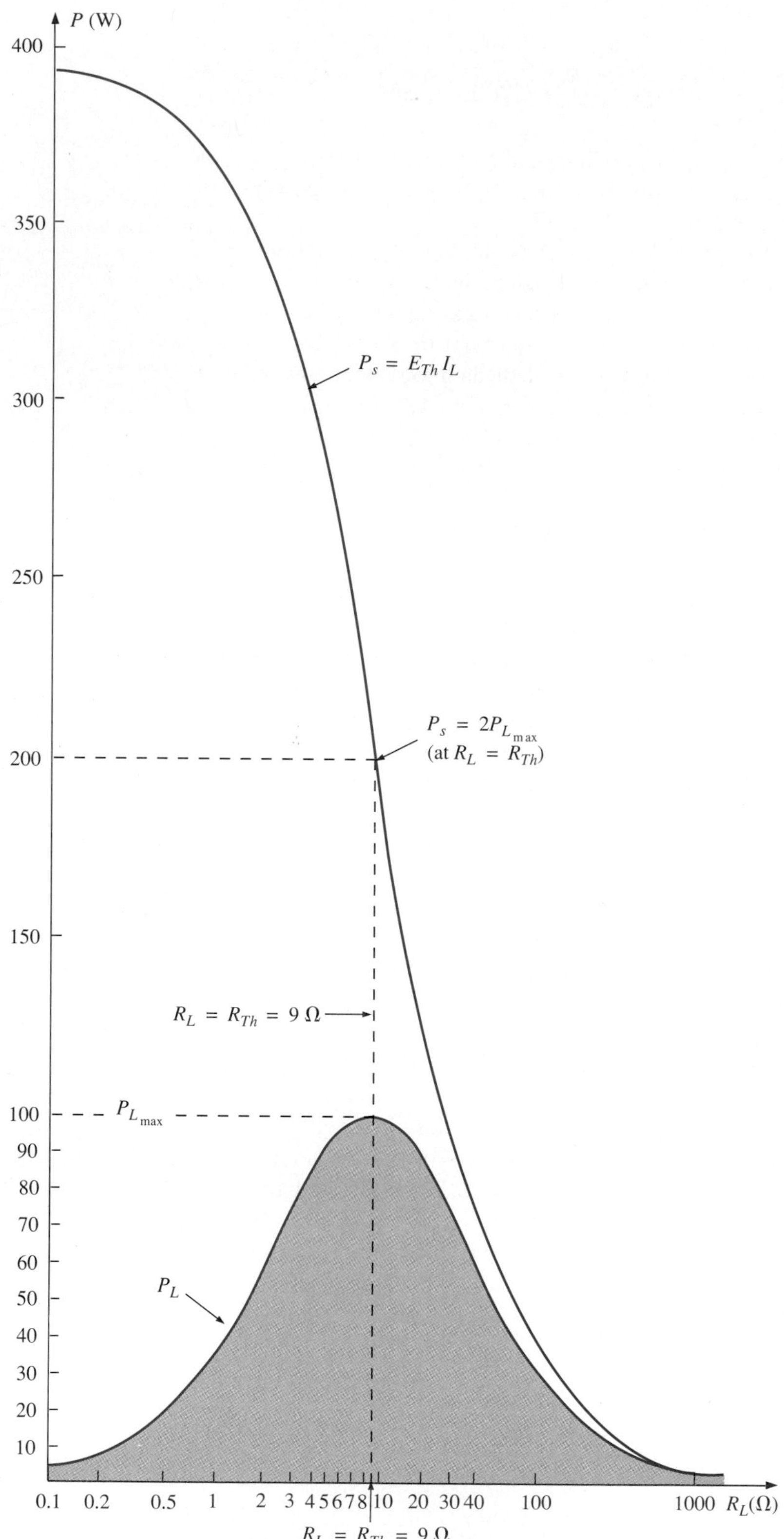

FIG. 9.83

P_s and P_L versus R_L for the network of Fig. 9.79.

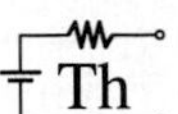

EXAMPLE 9.14 A dc generator, battery, and laboratory supply are connected to a resistive load R_L in Fig. 9.84(a), (b), and (c), respectively.

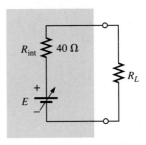

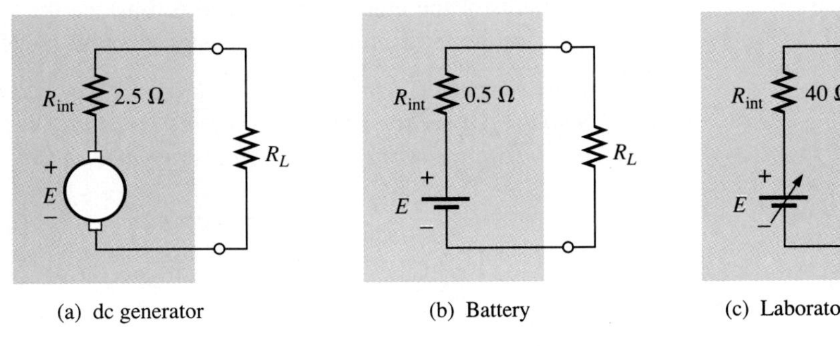

FIG. 9.84
Example 9.14.

a. For each, determine the value of R_L for maximum power transfer to R_L.
b. Determine R_L for 75% efficiency.

Solution:

a. For the dc generator,

$$R_L = R_{Th} = R_{int} = \mathbf{2.5\ \Omega}$$

For the battery,

$$R_L = R_{Th} = R_{int} = \mathbf{0.5\ \Omega}$$

For the laboratory supply,

$$R_L = R_{Th} = R_{int} = \mathbf{40\ \Omega}$$

b. For the dc generator,

$$\eta = \frac{P_o}{P_s} \ (\eta \text{ in decimal form})$$

$$\eta = \frac{R_L}{R_{Th} + R_L}$$

$$\eta(R_{Th} + R_L) = R_L$$

$$\eta R_{Th} + \eta R_L = R_L$$

$$R_L(1 - \eta) = \eta R_{Th}$$

and

$$\boxed{R_L = \frac{\eta R_{Th}}{1 - \eta}} \tag{9.8}$$

$$R_L = \frac{0.75(2.5\ \Omega)}{1 - 0.75} = \mathbf{7.5\ \Omega}$$

For the battery,

$$R_L = \frac{0.75(0.5\ \Omega)}{1 - 0.75} = \mathbf{1.5\ \Omega}$$

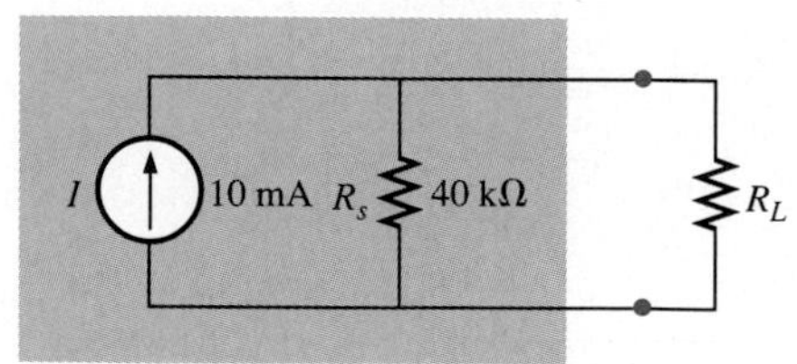

FIG. 9.85
Example 9.15.

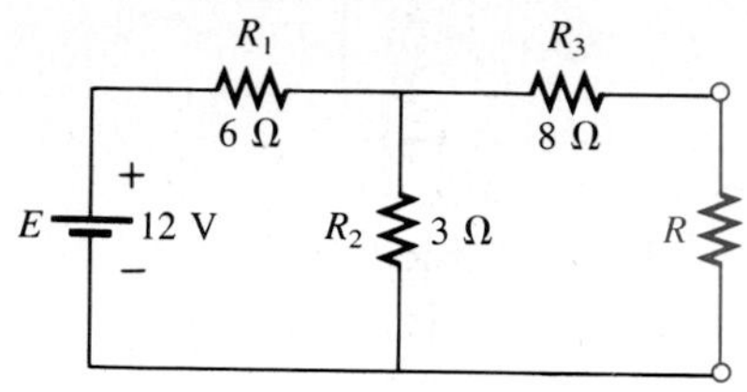

FIG. 9.86
Example 9.16.

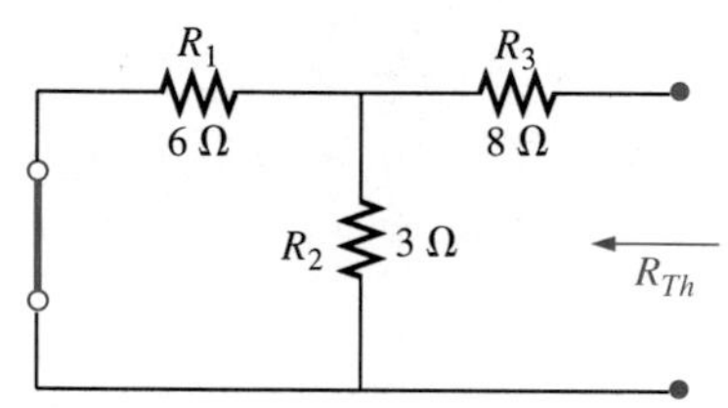

FIG. 9.87
Determining R_{Th} for the network external to the resistor R of Fig. 9.86.

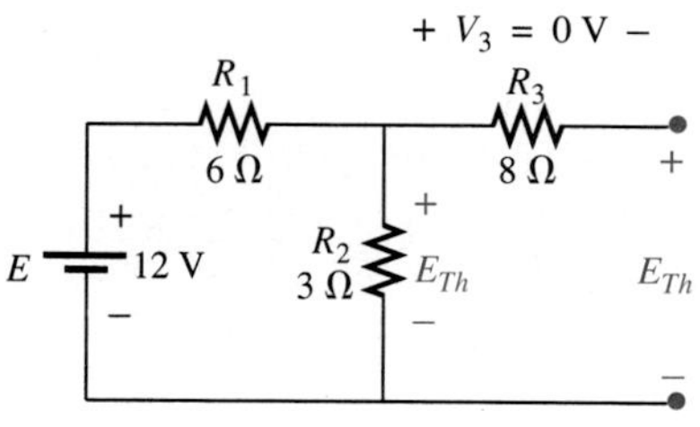

FIG. 9.88
Determining E_{Th} for the network external to the resistor R of Fig. 9.86.

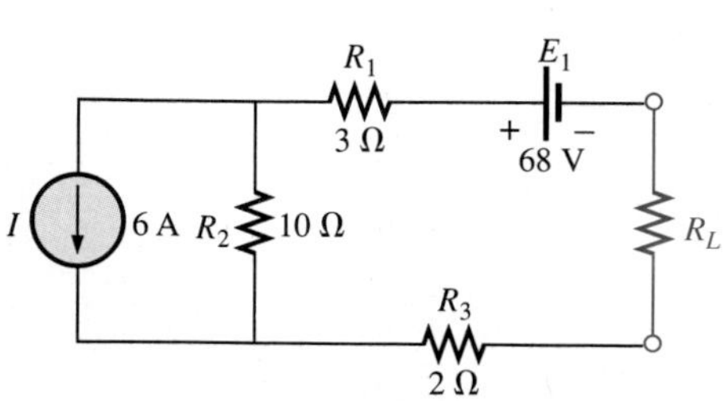

FIG. 9.89
Example 9.17.

For the laboratory supply,

$$R_L = \frac{0.75(40\ \Omega)}{1 - 0.75} = \mathbf{120\ \Omega}$$

The results of the preceding example reveal that the following modified form of the maximum power transfer theorem is valid:

For loads connected directly to a dc voltage supply, maximum power will be delivered to the load when the load resistance is equal to the internal resistance of the source; that is, when

$$R_L = R_{\text{int}} \tag{9.9}$$

EXAMPLE 9.15 Analysis of a transistor network resulted in the reduced configuration of Fig. 9.85. Determine the R_L necessary to transfer maximum power to R_L, and calculate the power of R_L under these conditions.

Solution: Eq. (9.4):

$$R_L = R_s = \mathbf{40\ k\Omega}$$

Eq. (9.7):

$$P_{L_{\max}} = \frac{I_N^2 R_N}{4} = \frac{(10\ \text{mA})^2(40\ \text{k}\Omega)}{4} = \mathbf{1\ W}$$

EXAMPLE 9.16 For the network of Fig. 9.86, determine the value of R for maximum power to R, and calculate the power delivered under these conditions.

Solution: See Fig. 9.87;

$$R_{Th} = R_3 + R_1 \parallel R_2 = 8\ \Omega + \frac{(6\ \Omega)(3\ \Omega)}{6\ \Omega + 3\ \Omega} = 8\ \Omega + 2\ \Omega$$

and

$$R = R_{Th} = \mathbf{10\ \Omega}$$

See Fig. 9.88;

$$E_{Th} = \frac{R_2 E}{R_2 + R_1} = \frac{(3\ \Omega)(12\ \text{V})}{3\ \Omega + 6\ \Omega} = \frac{36\ \text{V}}{9} = \mathbf{4\ V}$$

and, by Eq. (9.6),

$$P_{L_{\max}} = \frac{E_{Th}^2}{4R_{Th}} = \frac{(4\ \text{V})^2}{4(10\ \Omega)} = \mathbf{0.4\ W}$$

EXAMPLE 9.17 Find the value of R_L in Fig. 9.89 for maximum power to R_L, and determine the maximum power.

Solution: See Fig. 9.90;

$$R_{Th} = R_1 + R_2 + R_3 = 3\ \Omega + 10\ \Omega + 2\ \Omega = 15\ \Omega$$

and

$$R_L = R_{Th} = \mathbf{15\ \Omega}$$

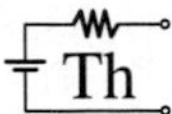

Note Fig. 9.91, where

$$V_1 = V_3 = 0\text{ V}$$

and
$$V_2 = I_2R_2 = IR_2 = (6\text{ A})(10\ \Omega) = 60\text{ V}$$

Applying Kirchhoff's voltage law,

$$\Sigma_{\circlearrowright} V = -V_2 - E_1 + E_{Th} = 0$$

and
$$E_{Th} = V_2 + E_1 = 60\text{ V} + 68\text{ V} = 128\text{ V}$$

Thus,
$$P_{L_{max}} = \frac{E_{Th}^2}{4R_{Th}} = \frac{(128\text{ V})^2}{4(15\ \Omega)} = \mathbf{273.07\ W}$$

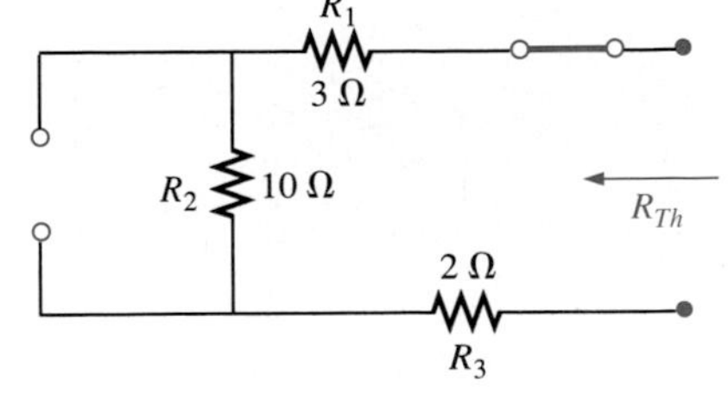

FIG. 9.90
Determining R_{Th} for the network external to the resistor R_L of Fig. 9.89.

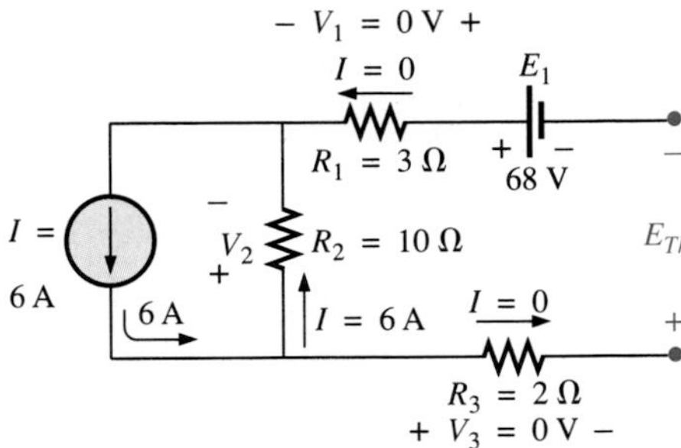

FIG. 9.91
Determining E_{Th} for the network external to the resistor R_L of Fig. 9.89.

9.6 MILLMAN'S THEOREM

Through the application of Millman's theorem, any number of parallel voltage sources can be reduced to one. In Fig. 9.92, for example, the three voltage sources can be reduced to one. This would permit finding the current through or voltage across R_L without having to apply a method such as mesh analysis, nodal analysis, superposition, and so on. The theorem can best be described by applying it to the network of Fig. 9.92. There are basically three steps included in its application.

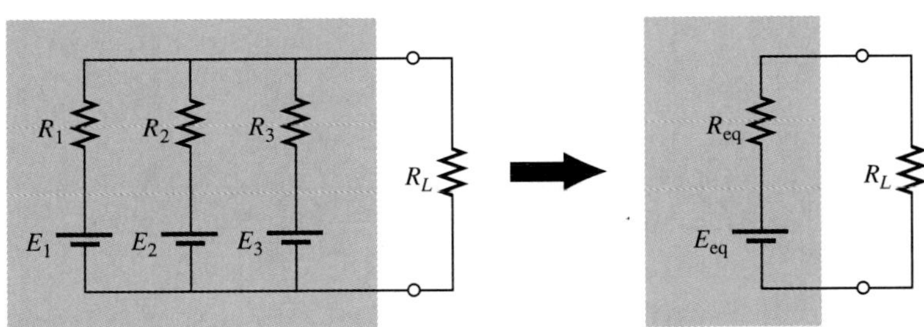

FIG. 9.92
Demonstrating the effect of applying Millman's theorem.

Step 1: Convert all voltage sources to current sources as outlined in Section 8.3. This is performed in Fig. 9.93 for the network of Fig. 9.92.

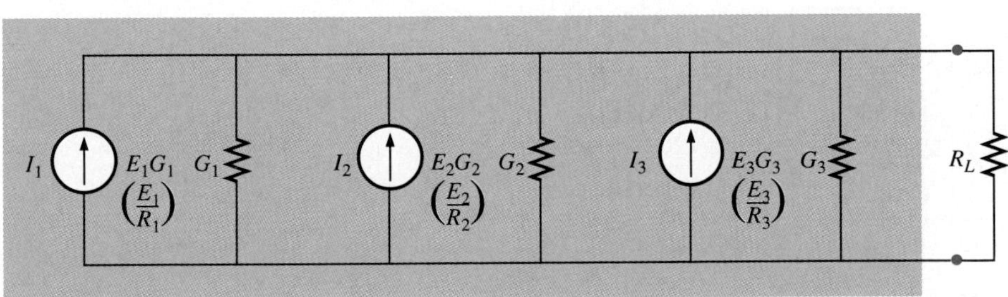

FIG. 9.93
Converting all the sources of Fig. 9.92 to current sources.

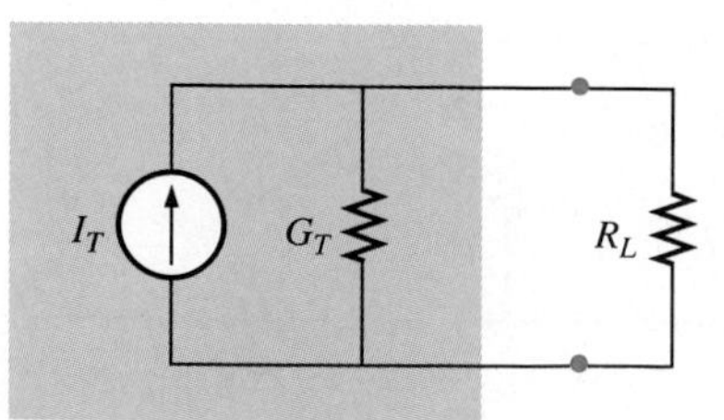

FIG. 9.94
Reducing all the current sources of Fig. 9.93 to a single current source.

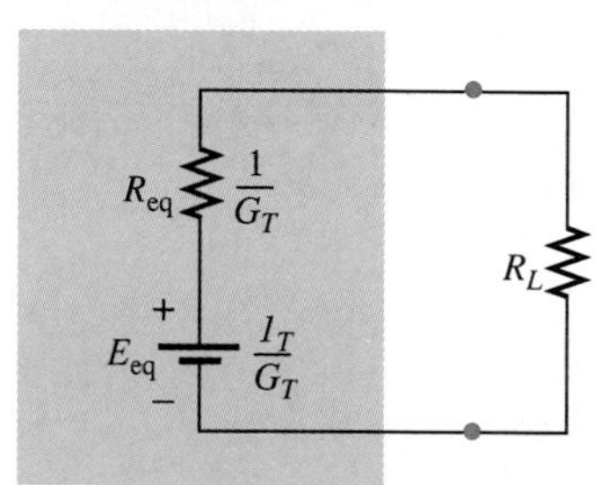

FIG. 9.95
Converting the current source of Fig. 9.94 to a current source.

Step 2: Combine parallel current sources as described in Section 8.4. The resulting network is shown in Fig. 9.94, where

$$I_T = I_1 + I_2 + I_3 \qquad G_T = G_1 + G_2 + G_3$$

Step 3: Convert the resulting current source to a voltage source, and the desired single-source network is obtained, as shown in Fig. 9.95.

In general, Millman's theorem states that for any number of parallel voltage sources,

$$E_{eq} = \frac{I_T}{G_T} = \frac{\pm I_1 \pm I_2 \pm I_3 \pm \cdots \pm I_N}{G_1 + G_2 + G_3 + \cdots + G_N}$$

or

$$E_{eq} = \frac{\pm E_1G_1 \pm E_2G_2 \pm E_3G_3 \pm \cdots \pm E_NG_N}{G_1 + G_2 + G_3 + \cdots + G_N} \qquad \textbf{(9.10)}$$

The plus and minus signs appear in Eq. (9.10) to include those cases where the sources may not be supplying energy in the same direction. (Note Example 9.18.)

The equivalent resistance is

$$R_{eq} = \frac{1}{G_T} = \frac{1}{G_1 + G_2 + G_3 + \cdots + G_N} \qquad \textbf{(9.11)}$$

In terms of the resistance values,

$$E_{eq} = \frac{\pm \dfrac{E_1}{R_1} \pm \dfrac{E_2}{R_2} \pm \dfrac{E_3}{R_3} \pm \cdots \pm \dfrac{E_N}{R_N}}{\dfrac{1}{R_1} + \dfrac{1}{R_2} + \dfrac{1}{R_3} + \cdots + \dfrac{1}{R_N}} \qquad \textbf{(9.12)}$$

and

$$R_{eq} = \frac{1}{\dfrac{1}{R_1} + \dfrac{1}{R_2} + \dfrac{1}{R_3} + \cdots + \dfrac{1}{R_N}} \qquad \textbf{(9.13)}$$

The relatively few direct steps required may result in the student's applying each step rather than memorizing and employing Eqs. (9.10) through (9.13).

EXAMPLE 9.18 Using Millman's theorem, find the current through and voltage across the resistor R_L of Fig. 9.96.

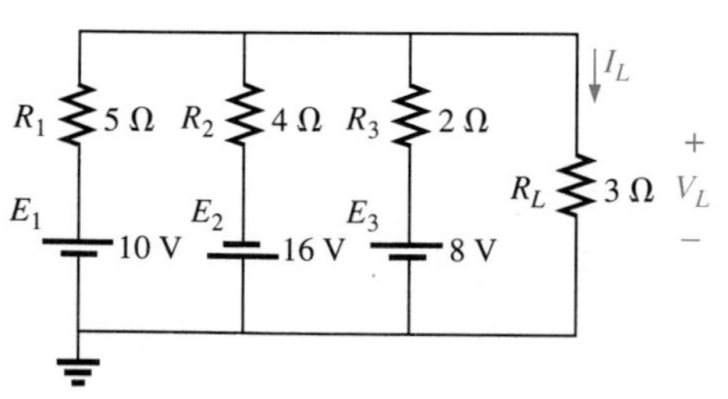

FIG. 9.96
Example 9.18.

Solution: By Eq. (9.12),

$$E_{eq} = \frac{+\dfrac{E_1}{R_1} - \dfrac{E_2}{R_2} + \dfrac{E_3}{R_3}}{\dfrac{1}{R_1} + \dfrac{1}{R_2} + \dfrac{1}{R_3}}$$

The minus sign is used for E_2/R_2 because that supply has the opposite polarity of the other two. The chosen reference direction is therefore

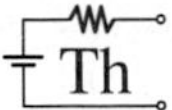

that of E_1 and E_3. The total conductance is unaffected by the direction, and

$$E_{eq} = \frac{+\dfrac{10\text{ V}}{5\ \Omega} - \dfrac{16\text{ V}}{4\ \Omega} + \dfrac{8\text{ V}}{2\ \Omega}}{\dfrac{1}{5\ \Omega} + \dfrac{1}{4\ \Omega} + \dfrac{1}{2\ \Omega}} = \frac{2\text{ A} - 4\text{ A} + 4\text{ A}}{0.2\text{ S} + 0.25\text{ S} + 0.5\text{ S}}$$

$$= \frac{2\text{ A}}{0.95\text{ S}} = \mathbf{2.105\ V}$$

with $$R_{eq} = \frac{1}{\dfrac{1}{5\ \Omega} + \dfrac{1}{4\ \Omega} + \dfrac{1}{2\ \Omega}} = \frac{1}{0.95\text{ S}} = \mathbf{1.053\ \Omega}$$

The resultant source is shown in Fig. 9.97, and

$$I_L = \frac{2.105\text{ V}}{1.053\ \Omega + 3\ \Omega} = \frac{2.105\text{ V}}{4.053\ \Omega} = \mathbf{0.519\ A}$$

with $$V_L = I_L R_L = (0.519\text{ A})(3\ \Omega) = \mathbf{1.557\ V}$$

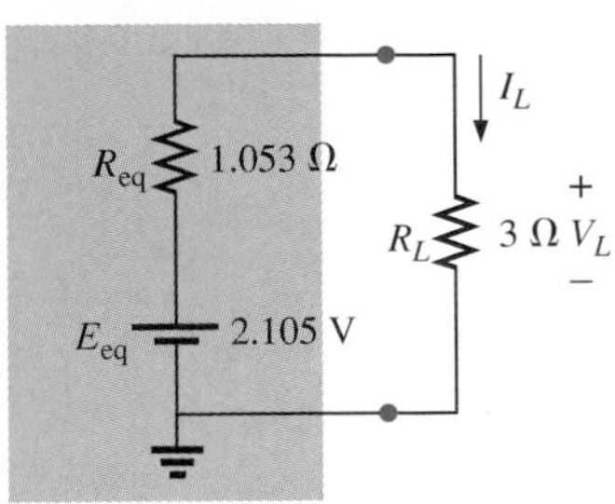

FIG. 9.97
The result of applying Millman's theorem to the network of Fig. 9.96.

EXAMPLE 9.19 Let us now consider the type of problem encountered in the introduction to mesh and nodal analysis in Chapter 8. Mesh analysis was applied to the network of Fig. 9.98 (Example 8.12). Let us now use Millman's theorem to find the current through the 2-Ω resistor and compare the results.

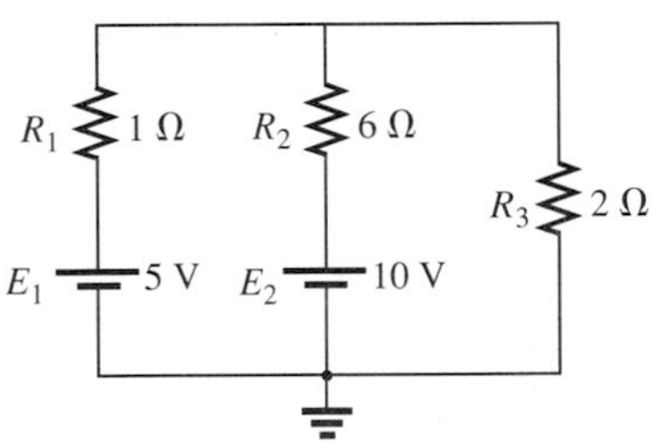

FIG. 9.98
Example 9.19.

Solution:

a. Let us first apply each step and, in the (b) solution, Eq. (9.12). Converting sources yields Fig. 9.99. Combining sources and parallel conductance branches (Fig. 9.100) yields

$$I_T = I_1 + I_2 = 5\text{ A} + \frac{5}{3}\text{ A} = \frac{15}{3}\text{ A} + \frac{5}{3}\text{ A} = \frac{20}{3}\text{ A}$$

$$G_T = G_1 + G_2 = 1\text{ S} + \frac{1}{6}\text{ S} = \frac{6}{6}\text{ S} + \frac{1}{6}\text{ S} = \frac{7}{6}\text{ S}$$

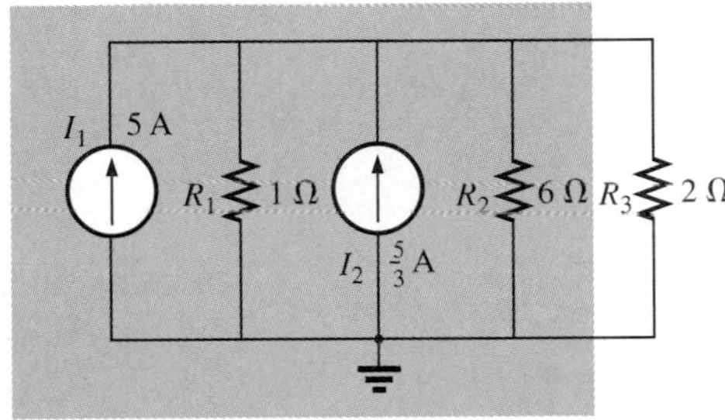

FIG. 9.99
Converting the sources of Fig. 9.98 to current sources.

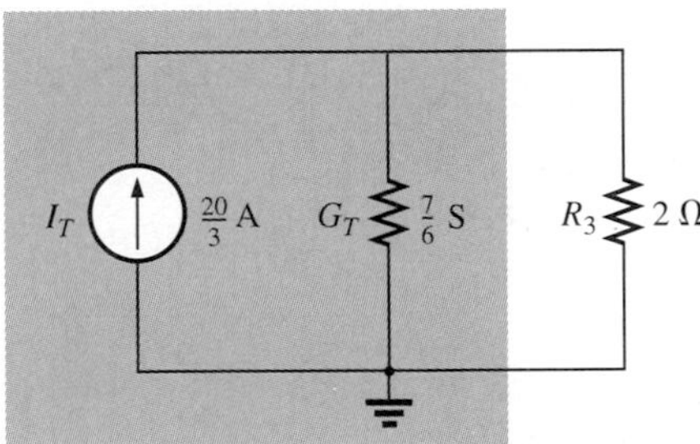

FIG. 9.100
Reducing the current sources of Fig. 9.99 to a single source.

Converting the current source to a voltage source (Fig. 9.101), we obtain

$$E_{eq} = \frac{I_T}{G_T} = \frac{\dfrac{20}{3}\text{ A}}{\dfrac{7}{6}\text{ S}} = \frac{(6)(20)}{(3)(7)}\text{ V} = \mathbf{\frac{40}{7}\ V}$$

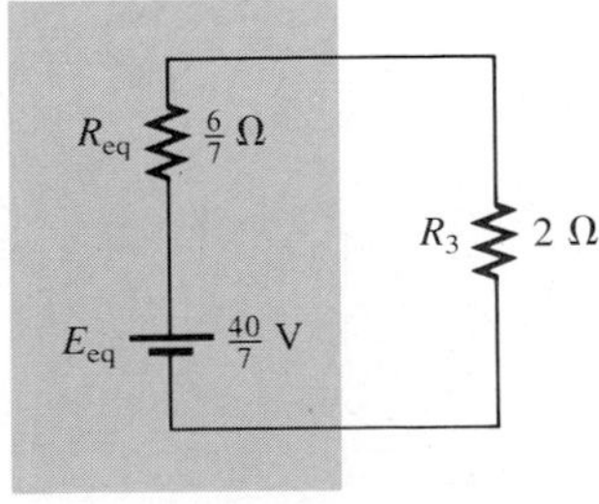

FIG. 9.101
Converting the current source of Fig. 9.100 to a voltage source.

and $$R_{eq} = \frac{1}{G_T} = \frac{1}{\frac{7}{6}\,\text{S}} = \mathbf{\frac{6}{7}\,\Omega}$$

so that

$$I_{2\Omega} = \frac{E_{eq}}{R_{eq} + R_3} = \frac{\frac{40}{7}\,\text{V}}{\frac{6}{7}\,\Omega + 2\,\Omega} = \frac{\frac{40}{7}\,\text{V}}{\frac{6}{7}\,\Omega + \frac{14}{7}\,\Omega} = \frac{40\,\text{V}}{20\,\Omega} = \mathbf{2\,A}$$

which agrees with the result obtained in Example 8.18.

b. Let us now simply apply the proper equation, Eq. (9.12):

$$E_{eq} = \frac{+\frac{5\,\text{V}}{1\,\Omega} + \frac{10\,\text{V}}{6\,\Omega}}{\frac{1}{1\,\Omega} + \frac{1}{6\,\Omega}} = \frac{\frac{30\,\text{V}}{6\,\Omega} + \frac{10\,\text{V}}{6\,\Omega}}{\frac{6}{6\,\Omega} + \frac{1}{6\,\Omega}} = \mathbf{\frac{40}{7}\,V}$$

and

$$R_{eq} = \frac{1}{\frac{1}{1\,\Omega} + \frac{1}{6\,\Omega}} = \frac{1}{\frac{6}{6\,\Omega} + \frac{1}{6\,\Omega}} = \frac{1}{\frac{7}{6}\,\text{S}} = \mathbf{\frac{6}{7}\,\Omega}$$

which are the same values obtained above.

The dual of Millman's theorem (Fig. 9.92) appears in Fig. 9.102. It can be shown that I_{eq} and R_{eq}, as in Fig. 9.102, are given by

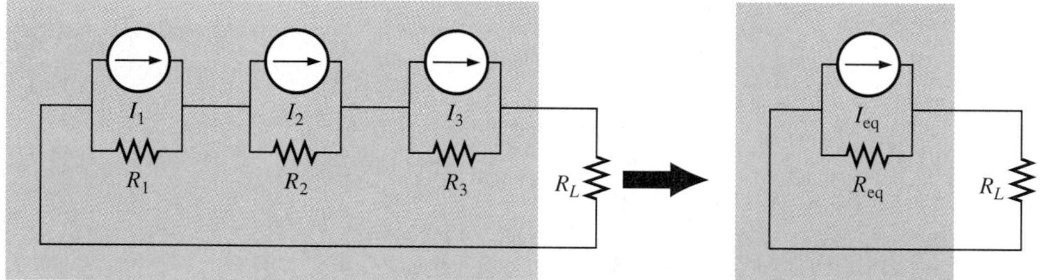

FIG. 9.102
The dual effect of Millman's theorem.

$$I_{eq} = \frac{\pm I_1R_1 \pm I_2R_2 \pm I_3R_3}{R_1 + R_2 + R_3} \qquad \textbf{(9.14)}$$

and $$R_{eq} = R_1 + R_2 + R_3 \qquad \textbf{(9.15)}$$

The derivation will appear as a problem at the end of the chapter.

9.7 SUBSTITUTION THEOREM

The substitution theorem states the following:

If the voltage across and current through any branch of a dc bilateral network are known, this branch can be replaced by any combination

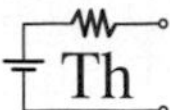

of elements that will maintain the same voltage across and current through the chosen branch.

More simply, the theorem states that for branch equivalence, the terminal voltage and current must be the same. Consider the circuit of Fig. 9.103, in which the voltage across and current through the branch *a-b* are determined. Through the use of the substitution theorem, a number of equivalent *a-a′* branches are shown in Fig. 9.104.

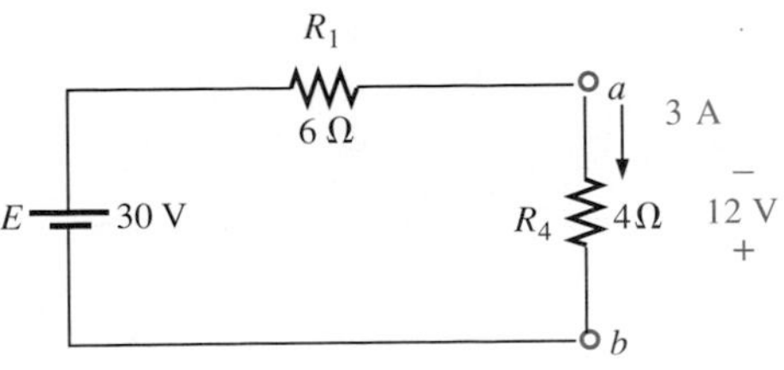

FIG. 9.103
Demonstrating the effect of the substitution theorem.

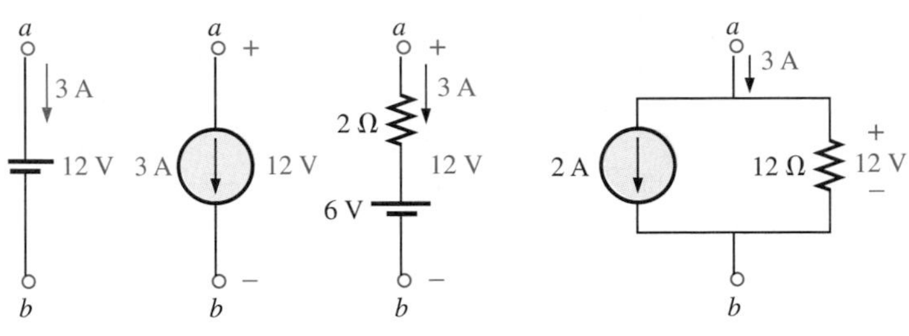

FIG. 9.104
Equivalent branches for the branch a-b of Fig. 9.103.

Note that for each equivalent, the terminal voltage and current are the same. Also consider that the response of the remainder of the circuit of Fig. 9.103 is unchanged by substituting any one of the equivalent branches. As demonstrated by the single-source equivalents of Fig. 9.104, *a known potential difference and current in a network can be replaced by an ideal voltage source and current source, respectively.*

Understand that this theorem cannot be used to *solve* networks with two or more sources that are not in series or parallel. For it to be applied, a potential difference or current value must be known or found using one of the techniques discussed earlier. One application of the theorem is shown in Fig. 9.105. Note that in the figure the known potential difference V was replaced by a voltage source, permitting the isolation of the portion of the network including R_3, R_4, and R_5. Recall that this was basically the approach employed in the analysis of the ladder network as we worked our way back toward the terminal resistance R_5.

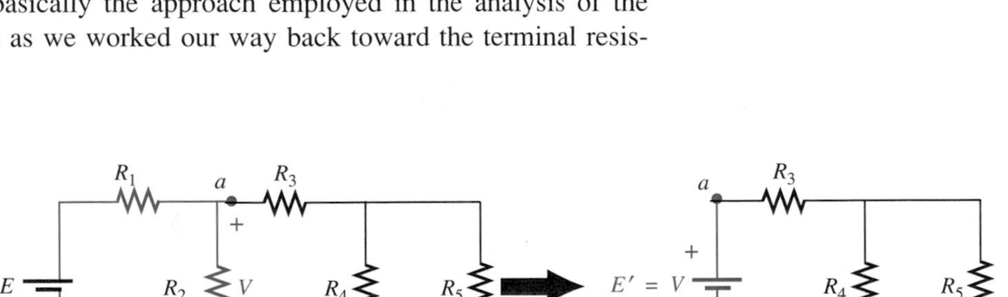

FIG. 9.105
Demonstrating the effect of knowing a voltage at some point in a complex network.

The current source equivalence of the above is shown in Fig. 9.106, where a known current is replaced by an ideal current source, permitting the isolation of R_4 and R_5.

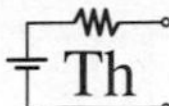

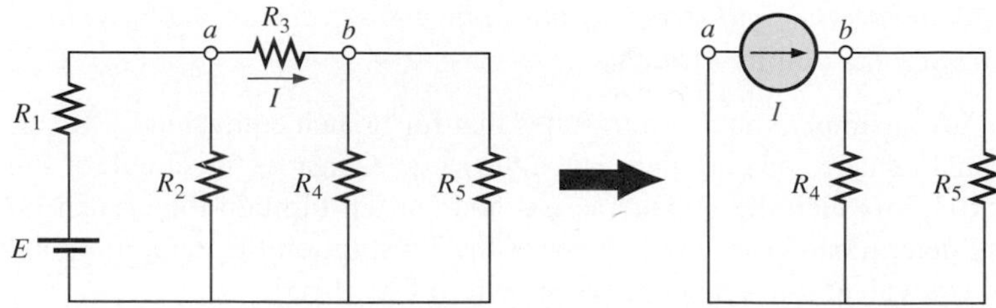

FIG. 9.106
Demonstrating the effect of knowing a current at some point in a complex network.

You will also recall from the discussion of bridge networks that $V = 0$ and $I = 0$ were replaced by a short circuit and an open circuit, respectively. This substitution is a very specific application of the substitution theorem.

9.8 RECIPROCITY THEOREM

The reciprocity theorem is applicable only to single-source networks. It is, therefore, not a theorem employed in the analysis of multisource networks described thus far. The theorem states the following:

The current I in any branch of a network, due to a single voltage source E anywhere else in the network, will equal the current through the branch in which the source was originally located if the source is placed in the branch in which the current I was originally measured.

In other words, the location of the voltage source and the resulting current may be interchanged without a change in current. The theorem requires that the polarity of the voltage source have the same correspondence with the direction of the branch current in each position.

In the representative network of Fig. 9.107(a), the current I due to the voltage source E was determined. If the position of each is interchanged as shown in Fig. 9.107(b), the current I will be the same value as indicated. To demonstrate the validity of this statement and the theorem, consider the network of Fig. 9.108, in which values for the elements of Fig. 9.107(a) have been assigned.

The total resistance is

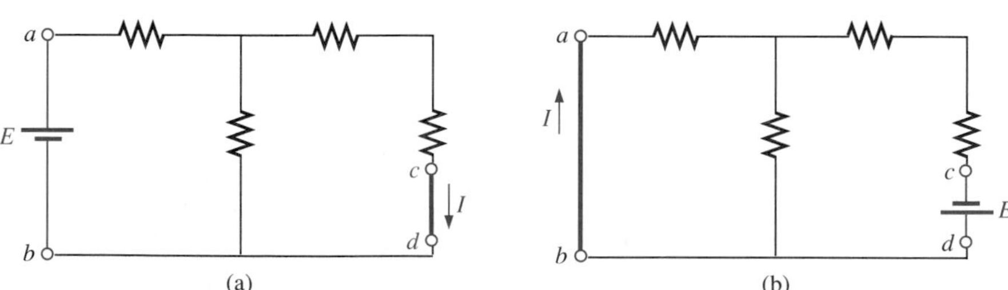

FIG. 9.107
Demonstrating the impact of the reciprocity theorem.

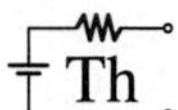

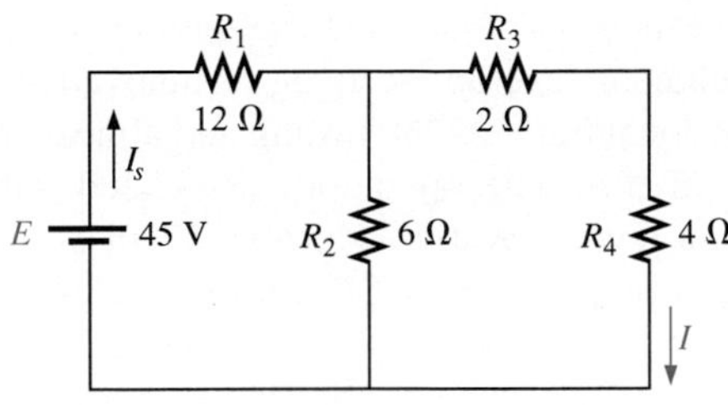

FIG. 9.108
Finding the current I due to a source E.

$$R_T = R_1 + R_2 \| (R_3 + R_4) = 12\,\Omega + 6\,\Omega \| (2\,\Omega + 4\,\Omega)$$
$$= 12\,\Omega + 6\,\Omega \| 6\,\Omega = 12\,\Omega + 3\,\Omega = 15\,\Omega$$

and
$$I_s = \frac{E}{R_T} = \frac{45\text{ V}}{15\,\Omega} = 3\text{ A}$$

with
$$I = \frac{3\text{ A}}{2} = \mathbf{1.5\text{ A}}$$

For the network of Fig. 9.109, which corresponds to that of Fig. 9.107(b), we find

$$R_T = R_4 + R_3 + R_1 \| R_2$$
$$= 4\,\Omega + 2\,\Omega + 12\,\Omega \| 6\,\Omega = 10\,\Omega$$

and
$$I_s = \frac{E}{R_T} = \frac{45\text{ V}}{10\,\Omega} = 4.5\text{ A}$$

so that
$$I = \frac{(6\,\Omega)(4.5\text{ A})}{12\,\Omega + 6\,\Omega} = \frac{4.5\text{ A}}{3} = \mathbf{1.5\text{ A}}$$

which agrees with the above.

The uniqueness and power of such a theorem can best be demonstrated by considering a complex, single-source network such as the one shown in Fig. 9.110.

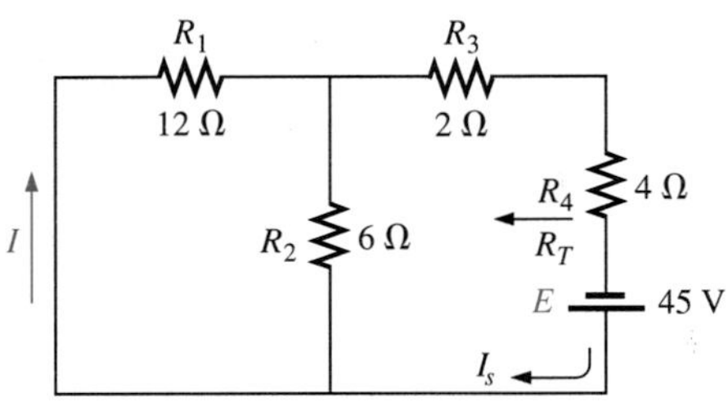

FIG. 9.109
Interchanging the location of E and I of Fig. 9.108 to demonstrate the validity of the reciprocity theorem.

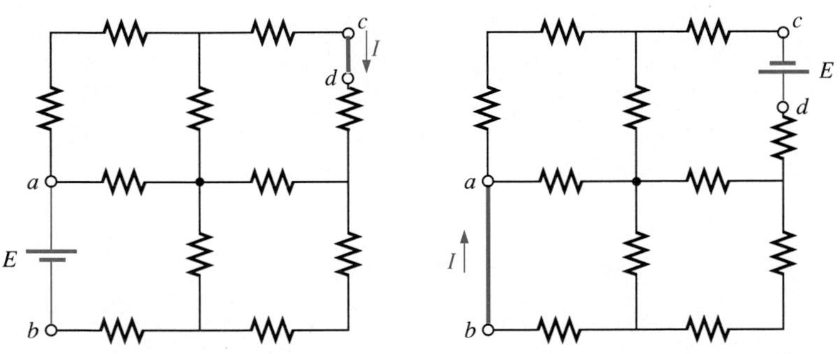

FIG. 9.110
Demonstrating the power and uniqueness of the reciprocity theorem.

9.9 COMPUTER ANALYSIS

Once the mechanics of applying a software package or language are understood, the opportunity to be creative and innovative presents

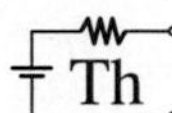

itself. Through years of exposure and trial and error experiences, professional programmers develop a catalog of innovative techniques that are not only functional but very interesting and almost artistic in nature. Now that some of the basic operations associated with PSpice have been introduced, a few innovative maneuvers will be made in the programs to follow. A program in BASIC will be included, but the analysis will follow a format similar to that encountered in earlier chapters, with the only major variation being the tabulation of a number of quantities for a changing variable.

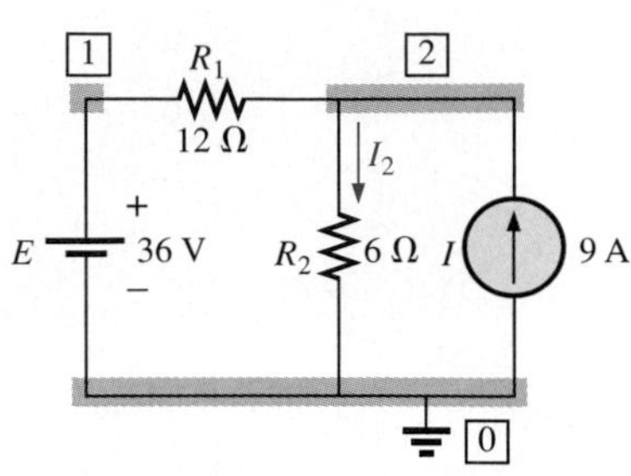

FIG. 9.111
Assigning nodes to the network of Example 9.3.

PSpice (DOS)

Superposition Let us now apply superposition to the network of Fig. 9.111, which appeared earlier as Example 9.3, to permit a comparison of the resulting solutions. The input file for the entire network is provided in Fig. 9.112 with the results for the current I_{R_2}.

To find the component of I_{R_2} due solely to the voltage source, the only required modification of the input file of Fig. 9.112 is to set the

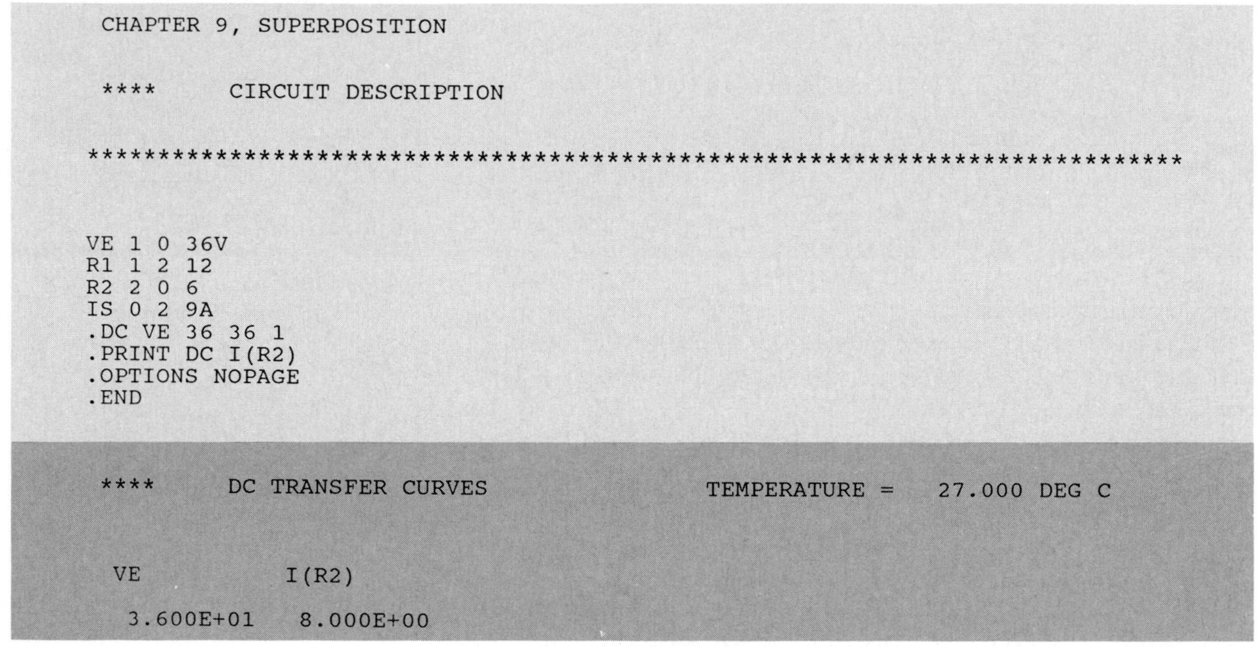

```
CHAPTER 9, SUPERPOSITION

 ****      CIRCUIT DESCRIPTION

******************************************************************************

VE 1 0 36V
R1 1 2 12
R2 2 0 6
IS 0 2 9A
.DC VE 36 36 1
.PRINT DC I(R2)
.OPTIONS NOPAGE
.END

 ****      DC TRANSFER CURVES                    TEMPERATURE =   27.000 DEG C

  VE             I(R2)

   3.600E+01    8.000E+00
```

FIG. 9.112
The input and output files for the network of Fig. 9.111 with both sources present.

current source to zero amperes, as shown in the input file of Fig. 9.113. The result is that $I'_2 = 2$ A, as obtained in Example 9.3.

For the effects of the current source, the input voltage is set to zero, as shown in the input file of Fig. 9.114. Note, however, that the .DC statement must be rewritten for the source that will have a specific magnitude. The result for I''_2 is 6 A, further confirming the results of Example 9.3. The sum of I'_2 and I''_2 results in a total current of 8 A, which matches the result obtained in Fig. 9.112.

```
CHAPTER 9, SUPERPOSITION

****     CIRCUIT DESCRIPTION

******************************************************************************

VE 1 0 36V
R1 1 2 12
R2 2 0 6
IS 0 2 0A
.DC VE 36 36 1
.PRINT DC I(R2)
.OPTIONS NOPAGE
.END

****     DC TRANSFER CURVES                        TEMPERATURE =    27.000 DEG C

 VE           I(R2)

  3.600E+01   2.000E+00
```

FIG. 9.113

The input and output files for the network of Fig. 9.111 with just the voltage source present.

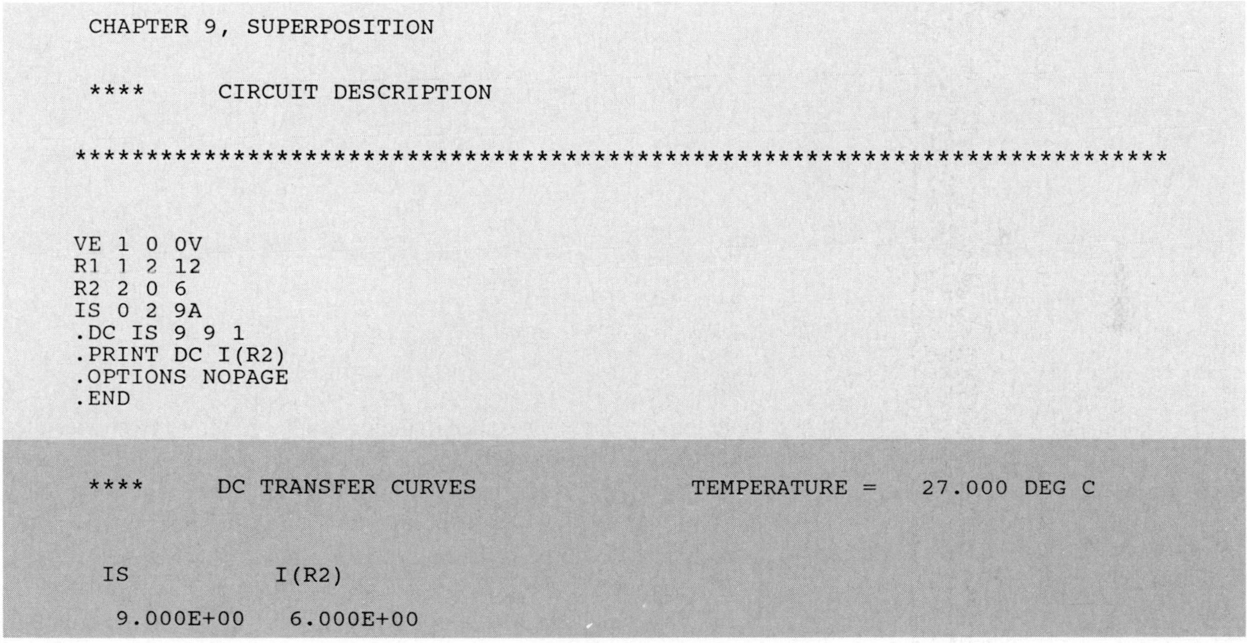

```
CHAPTER 9, SUPERPOSITION

****     CIRCUIT DESCRIPTION

******************************************************************************

VE 1 0 0V
R1 1 2 12
R2 2 0 6
IS 0 2 9A
.DC IS 9 9 1
.PRINT DC I(R2)
.OPTIONS NOPAGE
.END

****     DC TRANSFER CURVES                        TEMPERATURE =    27.000 DEG C

 IS           I(R2)

  9.000E+00   6.000E+00
```

FIG. 9.114

The input and output files for the network of Fig. 9.111 with just the current source present.

Thévenin's Theorem The application of Thévenin's theorem requires an interesting maneuver to determine the Thévenin resistance. It is a maneuver, however, that has application beyond Thévenin's theorem whenever a resistance level is required. The network to be analyzed appears in Fig. 9.115 and is the same analyzed in Example 9.10.

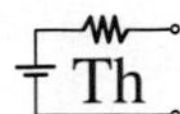

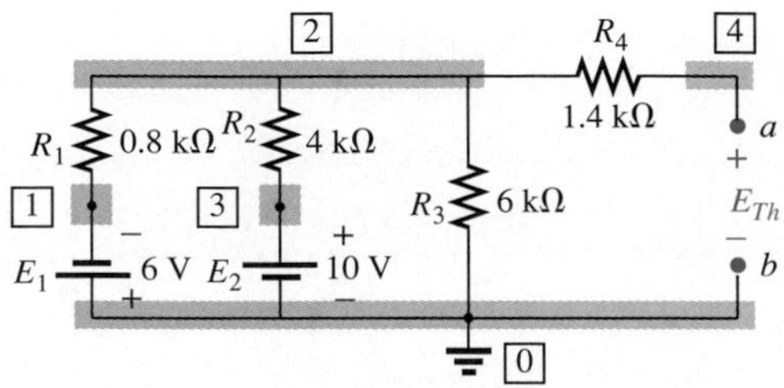

FIG. 9.115
Using PSpice to determine E_{Th} for the network of Example 9.10.

```
Chapter 9 - Thevenin's Theorem E(th)

****      CIRCUIT DESCRIPTION

******************************************************************************

V1 0 1 DC 6V
V2 3 0 DC 10V
R1 1 2 .8K
R2 2 3 4K
R3 2 0 6K
R4 2 4 1.4K
RL 4 0 1E30
.DC V1 6 6 1
.PRINT DC V(4,0)
.OPTIONS NOPAGE
.END

****      DC TRANSFER CURVES                    TEMPERATURE =   27.000 DEG C

  V1           V(4,0)

  6.000E+00   -3.000E+00
```

FIG. 9.116
The input and output files for the network of Fig. 9.115.

Note the very large resistance in the input file of Fig. 9.116 between terminals 4 and 0 to simulate an open circuit. PSpice does not recognize "dangling nodes" requiring at least one dc path to the reference level (in our case, ground). It also requires that each node appear twice in the input file (to accomplish the preceding for resistive circuits).

The output file of Fig. 9.116 reveals that the open-circuit voltage E_{Th} is -3 V to match the solution of Example 9.10. The minus sign simply reveals that node 4 is at a lower potential than the reference node 0.

To determine R_{Th}, a current source of 1 A is applied to the network of Fig. 9.115, as shown in Fig. 9.117(a) after setting the voltage sources to zero volts. In Fig. 9.117(b) the resistance R can be determined using Ohm's law and the resulting voltage V_R, as follows:

$$R = \frac{V_R}{I} = \frac{V_R}{1\text{ A}} \tag{9.16}$$

In Eq. (9.16), since $I = 1$ A the magnitude of R in ohms is the same as the magnitude of the voltage V_R. Therefore, when V(4, 0) is requested

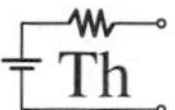

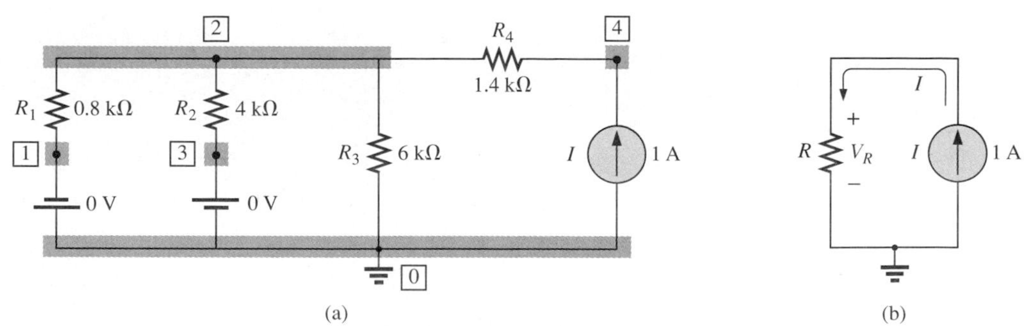

FIG. 9.117
Setting up the network of Fig. 9.115 to determine R_{Th}.

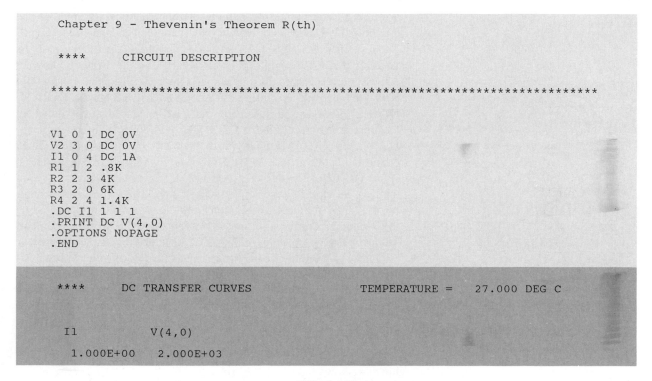

```
Chapter 9 - Thevenin's Theorem R(th)

****      CIRCUIT DESCRIPTION

**************************************************************************

V1 0 1 DC 0V
V2 3 0 DC 0V
I1 0 4 DC 1A
R1 1 2 .8K
R2 2 3 4K
R3 2 0 6K
R4 2 4 1.4K
.DC I1 1 1 1
.PRINT DC V(4,0)
.OPTIONS NOPAGE
.END

****      DC TRANSFER CURVES                    TEMPERATURE =   27.000 DEG C

 I1             V(4,0)

 1.000E+00    2.000E+03
```

FIG. 9.118
The input and output files for the network of Fig. 9.117.

in the input file of Fig. 9.118, you are in actuality also obtaining the magnitude of R_{Th}. The output file of Fig. 9.118 printed out a level of 2 kV for V_R, which we will read as $R_{Th} = 2\ \text{k}\Omega$, as obtained in Example 9.10.

PSpice (Windows)

All of the analysis performed in this chapter using PSpice (DOS) can be performed in PSpice (Windows) using a very similar approach. This session of PSpice (Windows) will introduce the procedure for sweeping a parameter through a series of values and plotting a quantity using **Probe** against the changing variable. The network of Fig. 9.79, for instance, has a variable resistor with values ranging from 1 to 30 Ω. For each value the power to the load is determined to set up the data for the plot of Fig. 9.80. The schematics version of Fig. 9.79 appears in Fig.

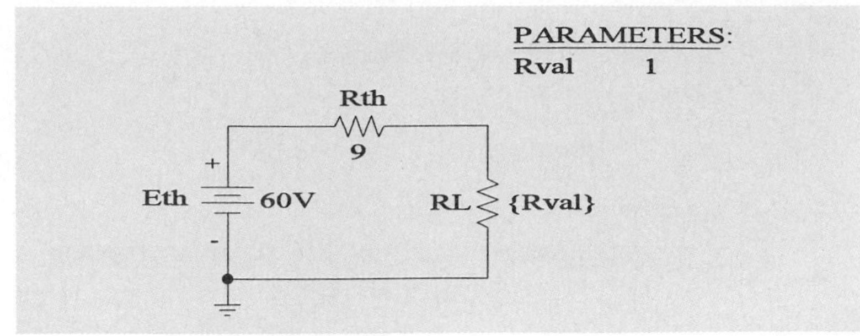

FIG. 9.119
Schematics version of Fig. 9.79.

9.119, with one noticeable difference from previous schematics—the load resistor **RL** does not have a fixed value but is labeled as **{Rval}.** The resistor was assigned this format by simply double-clicking on the value of the resistor and typing in **{Rval}.** The label possibilities are endless, but the braces must appear as shown. The sequence **Draw-Get New Part-Browse-special.slb-PARAM-OK** will generate a **PARAMETER** symbol on the schematic that can be placed in a convenient location. Once placed, the letters **PARAMETERS:** will appear. Double-clicking **PARAMETERS:** will generate a dialog box on which **Name 1** must be chosen as **Rval** followed by **Save Attr.** Then the first value **VALUE 1** should be chosen as 1, to represent the first value for the load resistor, followed by **Save Attr-OK.** The information regarding the range of values to be assigned to the load resistor is entered by the sequence **Analysis-Setup-DC Sweep-Global Parameter-Linear.** The **DC Sweep** specifies a series of dc calculations, while **Linear** establishes a fixed interval between values for **RL.** If **Start Value** is set at 1 Ω, the resulting plot will also start at 1 rather than 0. Since 0 is an invalid entry, enter .001 for the **Start Value.** A plot point will then appear at about 0 Ω. The **Final Value** must then be set at 30.001, or the last calculation will be made at 29 Ω rather than 30 Ω to match Fig. 9.80. The **Increment** will be set at 1 to obtain a plot point at 31 points. Click a final **OK** and the analysis can commence with **Analysis-Simulate.**

Once the **Probe** schematic appears, **Trace-Add-Trace Command-V(RL:1)*I(RL)-OK** will generate a plot of the power to the load versus the resistor value, as shown in Fig. 9.120. The notation **V(RL:1)** and **I(RL)** is established by schematics and appears in the listing of the dialog box. The plot confirms the results of Fig. 9.80 with a maximum value of 100 W. The vertical line at the maximum value was obtained by **Tools-Cursor-Display,** followed by a left click of the mouse on the schematic. A return to **Tools-Cursor-Max** will set the intersection of the cursor at the maximum value of the curve. The **Cursor** dialog box at the bottom of the plot reveals that the maximum value occurs at 9 Ω and has a value of 100 W. The **Probe** response will set the horizontal axis to a maximum of 35 Ω, causing a gap in the plot at the high end. This can be changed to a 0–30 range by **Plot-X-Axis Setting-User Defined-0 to 30.**

A plot of V_L versus load resistance can be obtained by **Trace-Add-V(RL:1)-OK,** as shown in Fig. 9.121. The **Tools-Cursor-Display,** followed by a left-side click of the mouse, will generate a **Cursor** that can be set at 9 Ω to reveal the magnitude of the voltage at this point. As noted, it is 30 V, which is one-half the supply voltage (as it should be

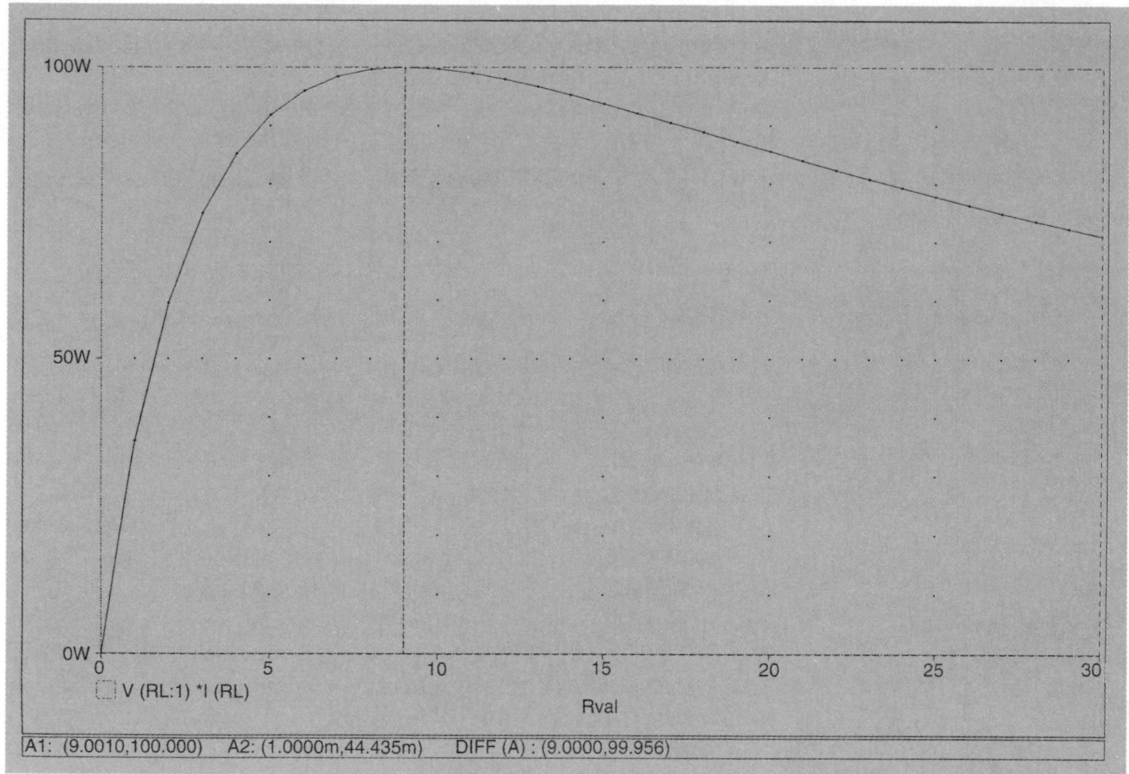

FIG. 9.120
Power to R_L versus R_L for the network of Fig. 9.119.

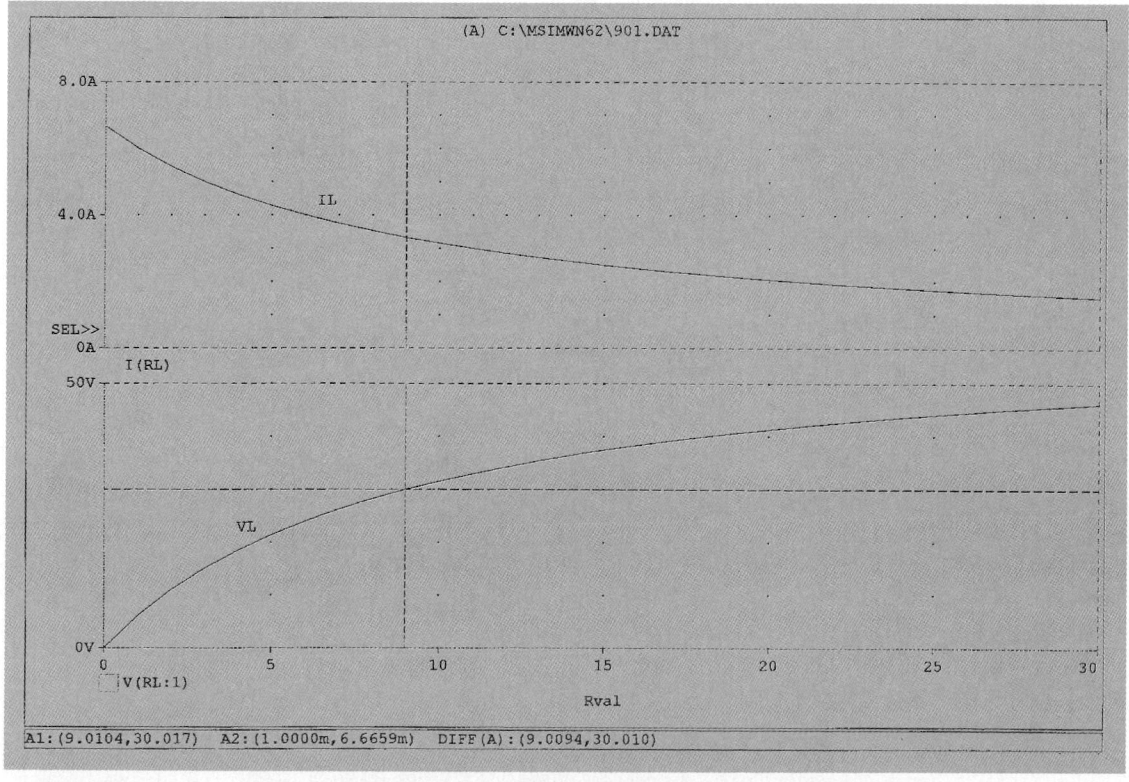

FIG. 9.121
V_L and I_L versus R_L for the network of Fig. 9.119.

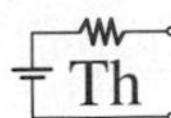

for maximum power transfer). The label **VL** is added by **Tools-Label-Text-VL-OK.** The plot of I_L versus resistance can be added by **Plot-Add Plot-Trace-Add-I(RL)-OK,** with the label added as above. The horizontal scale was also reduced to a maximum of 30 Ω. When working with a schematic with more than one plot, be sure to note the location of the **SEL≫** notation. It specifies which plot is to be altered or is under investigation. A simple click on the plot of interest will move the **SEL≫** accordingly.

BASIC

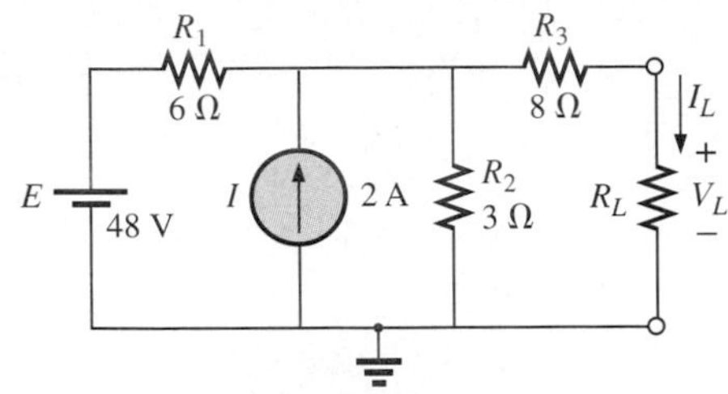

FIG. 9.122
Network for which E_{Th} and R_{Th} are to be determined using BASIC.

The BASIC program of Fig. 9.123 will also analyze the network of Fig. 9.122, but it will also provide the values of E_{Th} and R_{Th} and tabulate a list of output variables that extends from $R_{Th}/5$ to $2R_{Th}$ in increments of $R_{Th}/5$ (line 330). Since R_{Th} is 10 Ω, the range of R_L is from 2 Ω to 20 Ω in increments of 2 Ω. Lines 130 through 180 request the network parameters, whereas lines 200 and 210 calculate R_{Th}. E_{Th} is determined by lines 220 through 260 using superposition, and both E_{Th} and R_{Th} are printed out by lines 270 through 290. Lines 310 and 320 provide a heading for the printout, with the TAB command simply specifying the spaces from the left edge of the paper and between columns. The range of R_L is specified by line 330, and all the required calculations are per-

```
10 REM ***** PROGRAM 9-1 *****
20 REM *******************************************
30 REM Program to tabulate changes in load levels for
40 REM a range of load values using Thevenin's theorem
50 REM *******************************************
60 REM
100 PRINT "For the network of Fig. 9.122"
110 PRINT "enter the following data:"
120 PRINT
130 INPUT "R1="; R1: REM Enter 0 if resistor non-existent
140 INPUT "R2="; R2: REM Enter 1E30 if resistor non-existent
150 INPUT "R3="; R3: REM Enter 0 if resistor non-existent
160 INPUT "RL="; RL
170 INPUT "Supply voltage, E="; E
180 INPUT "and supply current, I="; I
190 PRINT
200 REM Determine Rth
210 RT = R3 + R1 * R2 / (R1 + R2)
220 REM Use superposition to determine Eth
230 E1 = R2 * E / (R1 + R2)
240 I2 = R2 * I / (R1 + R2)
250 E2 = R1 * R2 * I / (R1 + R2)
260 ET = E1 + E2
270 PRINT "Using Thevenin's Theorem:"
280 PRINT "Rth="; RT; "ohms"
290 PRINT "and Eth="; ET; "volts"
300 PRINT
310 PRINT TAB(7); "RL"; TAB(15); "IL"; TAB(25); "VL";
320 PRINT TAB(35); "PL"; TAB(45); "PD"; TAB(55); "n%"
330 FOR RL = RT / 5 TO 2 * RT STEP RT / 5
340 IL = ET / (RT + RL)
350 VL = IL * RL
360 PL = IL ^ 2 * RL
370 PD = ET * IL
380 N = 100 * PL / PD
390 IF RL = RT THEN PRINT "Rth=";
400 PRINT TAB(5); RL; TAB(13); IL; TAB(23); VL;
410 PRINT TAB(33); PL; TAB(43); PD; TAB(53); N
420 NEXT RL
430 END
```

FIG. 9.123
Input and output files for the network of Fig. 9.122 using BASIC.

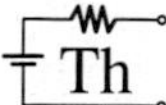


```
For the network of Fig. 9.122
enter the following data:

R1=? 6
R2=? 3
R3=? 8
RL=? 10
Supply voltage, E=? 48
and supply current, I=? 2

Using Thevenin's Theorem:
Rth= 10 ohms
and Eth= 20 volts

         RL       IL         VL         PL         PD         n%
          2     1.666667   3.333333   5.555556   33.33333   16.66667
          4     1.428572   5.714286   8.163265   28.57143   28.57143
          6     1.25       7.5        9.375      25         37.5
          8     1.111111   8.888889   9.876544   22.22222   44.44445
    Rth= 10     1          10         10         20         50
         12     .9090909   10.90909   9.917356   18.18182   54.54546
         14     .8333333   11.66667   9.722221   16.66667   58.33333
         16     .7692308   12.30769   9.467456   15.38462   61.53847
         18     .7142858   12.85714   9.183674   14.28572   64.28571
         20     .6666667   13.33333   8.888889   13.33333   66.66666
```

FIG. 9.123. (continued)

formed by lines 340 through 360. Line 390 specifies that if the value of R_L for a particular loop is equal to R_{Th}, then the comment "R_{Th}=" should be added to the printout as shown. The magnitude of all the quantities is printed out by lines 400 and 410, with line 420 sending the program back to line 330 to repeat the calculations for the next value of R_L.

Note that the maximum power is obtained at $R_L = R_{Th} = 10\ \Omega$ and the efficiency is 50% under maximum power transfer conditions. PD is the power delivered by the source, whereas PL is the power to the load.

PROBLEMS

SECTION 9.2 Superposition Theorem

1. a. Using superposition, find the current through each resistor of the network of Fig. 9.124.
   b. Find the power delivered to R_1 for each source.
   c. Find the power delivered to R_1 using the total current through R_1.
   d. Does superposition apply to power effects? Explain.

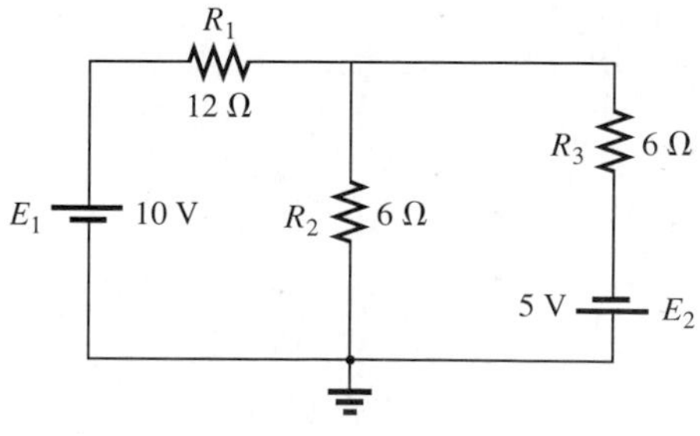

FIG. 9.124
Problem 1.

2. Using superposition, find the current I through the 10-Ω resistor for each of the networks of Fig. 9.125.

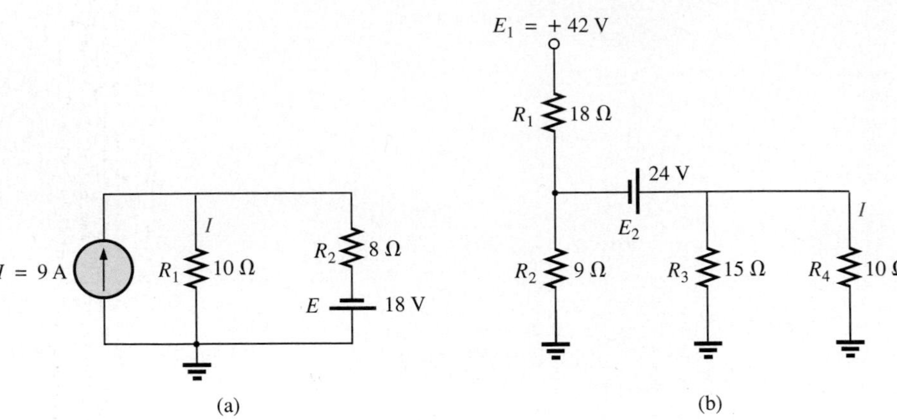

FIG. 9.125
Problems 2, 37, and 44.

*3. Using superposition, find the current through R_1 for each network of Fig. 9.126.

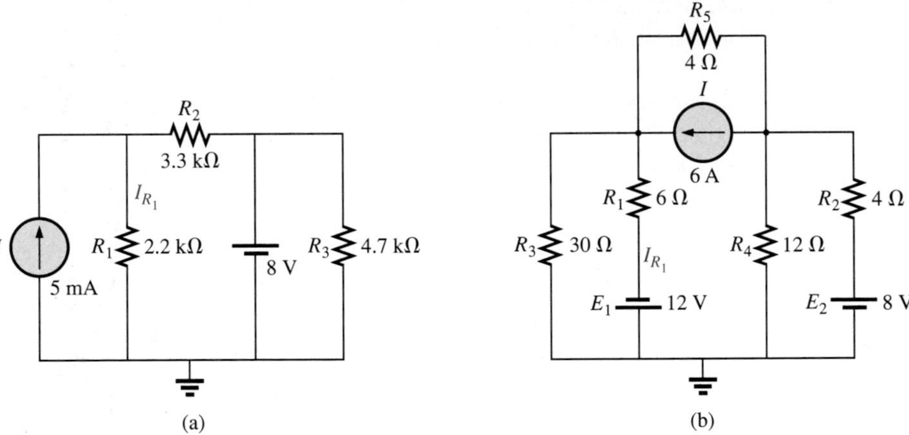

FIG. 9.126
Problem 3.

4. Using superposition, find the voltage V_2 for the network of Fig. 9.127.

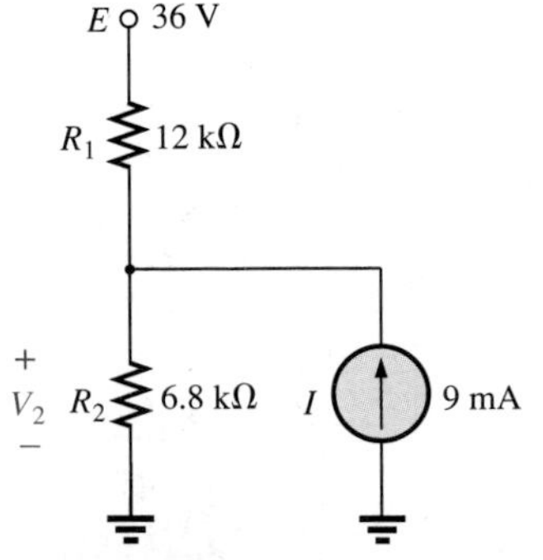

FIG. 9.127
Problems 4 and 40.

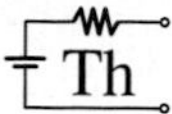

SECTION 9.3 Thévenin's Theorem

5. **a.** Find the Thévenin equivalent circuit for the network external to the resistor R of Fig. 9.128.
   b. Find the current through R when R is 2, 30, and 100 Ω.

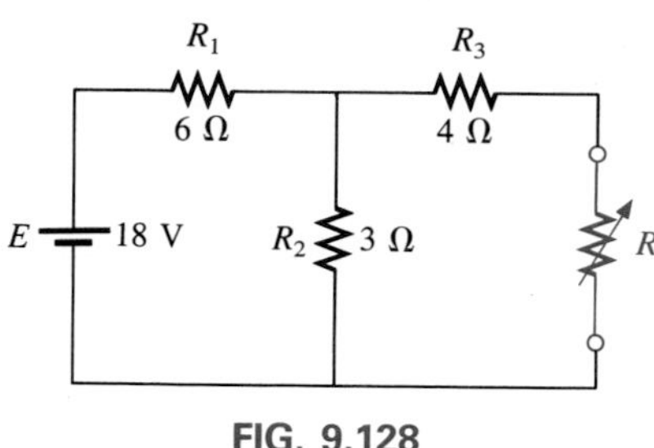

FIG. 9.128
Problem 5.

6. **a.** Find the Thévenin equivalent circuit for the network external to the resistor R in each of the networks of Fig. 9.129.
   b. Find the power delivered to R when R is 2 Ω and 100 Ω.

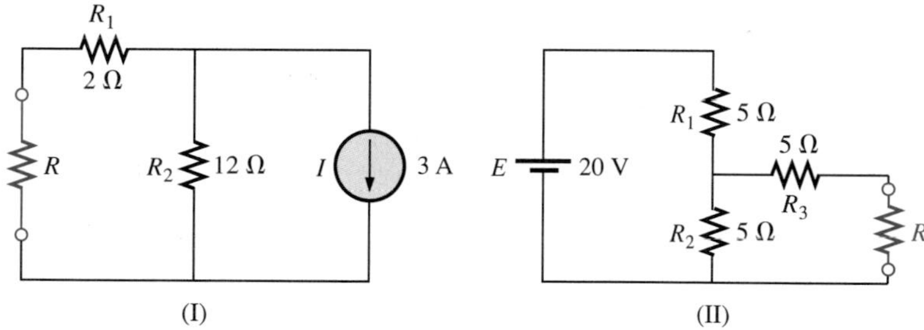

FIG. 9.129
Problems 6, 13, 19, and 38.

7. Find the Thévenin equivalent circuit for the network external to the resistor R in each of the networks of Fig. 9.130.

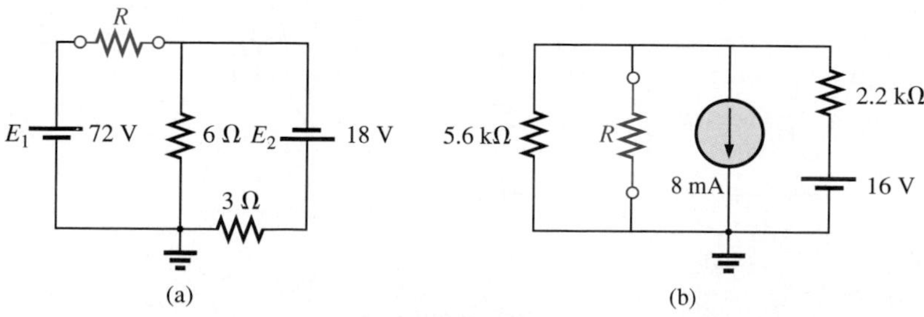

FIG. 9.130
Problems 7, 14, and 20.

***8.** Find the Thévenin equivalent circuit for the network external to the resistor R in each of the networks of Fig. 9.131.

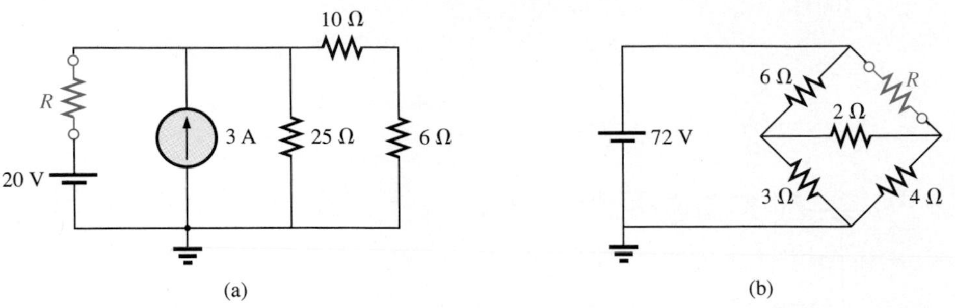

FIG. 9.131
Problems 8, 15, 21, 39, 41, 42, and 45.

***9.** Find the Thévenin equivalent circuit for the portions of the networks of Fig. 9.132 external to points a and b.

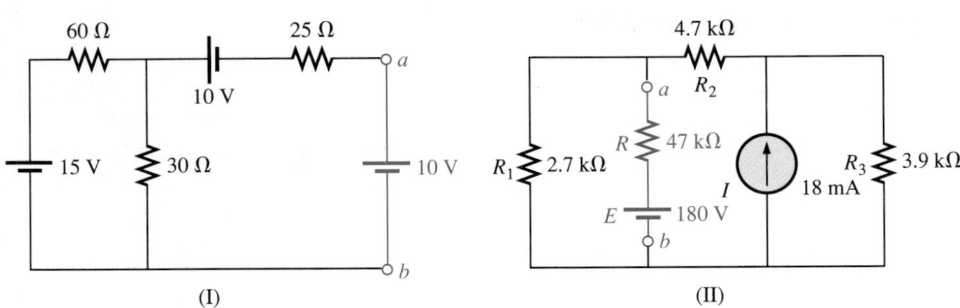

FIG. 9.132
Problems 9 and 16.

***10.** Determine the Thévenin equivalent circuit for the network external to the resistor R in both networks of Fig. 9.133.

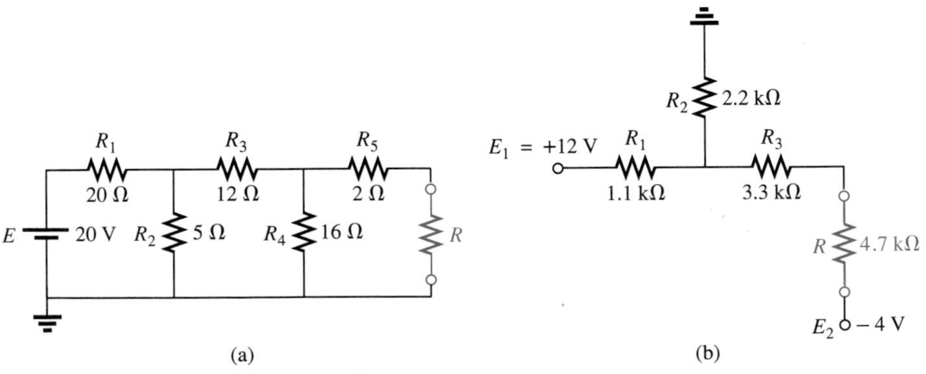

FIG. 9.133
Problems 10 and 17.

*11. For the network of Fig. 9.134, find the Thévenin equivalent circuit for the network external to the load resistor R_L.

*12. For the transistor network of Fig. 9.135,
  a. Find the Thévenin equivalent circuit for that portion of the network to the left of the base (B) terminal.
  b. Using the fact that $I_C = I_E$ and $V_{CE} = 8$ V, determine the magnitude of I_E.
  c. Using the results of parts (a) and (b), calculate the base current I_B if $V_{BE} = 0.7$ V.
  d. What is the voltage V_C?

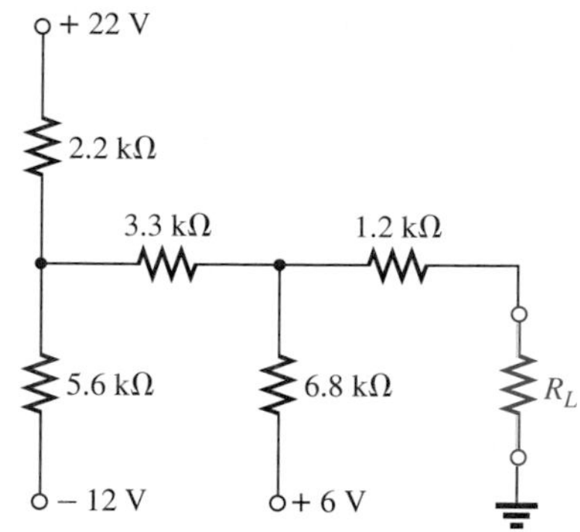

FIG. 9.134
Problem 11.

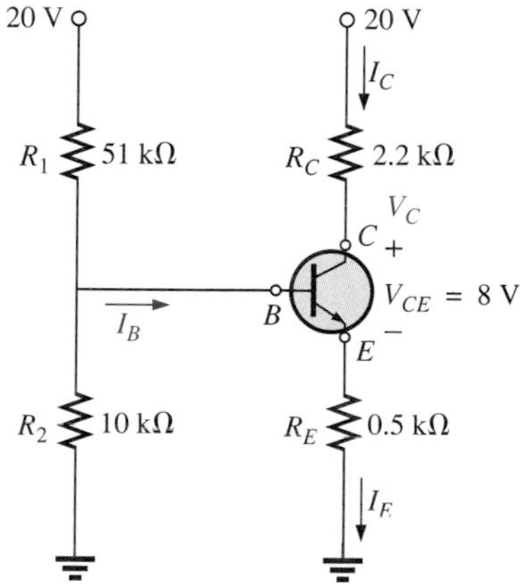

FIG. 9.135
Problem 12.

SECTION 9.4 Norton's Theorem

13. Find the Norton equivalent circuit for the network external to the resistor R in each network of Fig. 9.129.

14. a. Find the Norton equivalent circuit for the network external to the resistor R for each network of Fig. 9.130.
  b. Convert to the Thévenin equivalent circuit and compare your solution for E_{Th} and R_{Th} to that appearing in the solutions for Problem 7.

15. Find the Norton equivalent circuit for the network external to the resistor R for each network of Fig. 9.131.

16. a. Find the Norton equivalent circuit for the network external to the resistor R for each network of Fig. 9.132.
  b. Convert to the Thévenin equivalent circuit and compare your solution for E_{Th} and R_{Th} to that appearing in the solutions for Problem 9.

17. Find the Norton equivalent circuit for the network external to the resistor R for each network of Fig. 9.133.

18. Find the Norton equivalent circuit for the portions of the networks of Fig. 9.136 external to branch a-b.

SECTION 9.5 Maximum Power Transfer Theorem

19. a. For each network of Fig. 9.129, find the value of R for maximum power to R.
  b. Determine the maximum power to R for each network.

20. a. For each network of Fig. 9.130, find the value of R for maximum power to R.
  b. Determine the maximum power to R for each network.

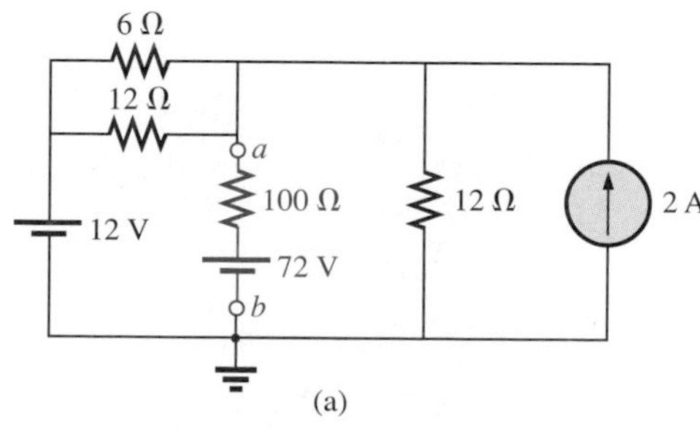

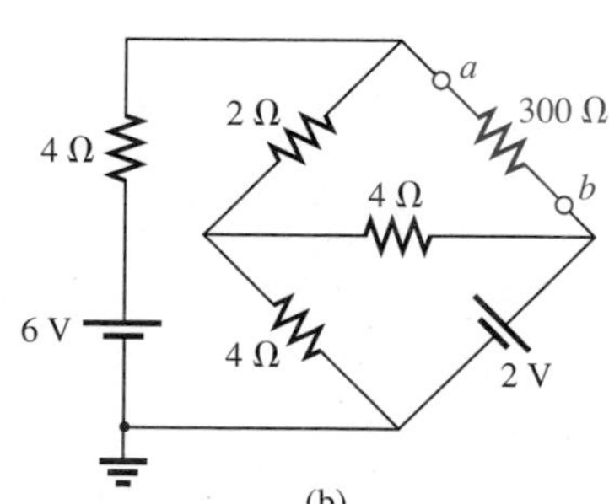

FIG. 9.136
Problems 18 and 43.

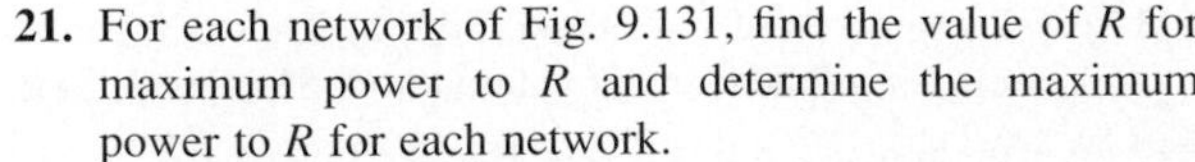

21. For each network of Fig. 9.131, find the value of R for maximum power to R and determine the maximum power to R for each network.

22. **a.** For the network of Fig. 9.137, determine the value of R for maximum power to R.
    b. Determine the maximum power to R.
    c. Plot a curve of power to R versus R for R equal to $\frac{1}{4}$, $\frac{1}{2}$, $\frac{3}{4}$, 1, $1\frac{1}{4}$, $1\frac{1}{2}$, $1\frac{3}{4}$, and 2 times the value obtained in part (a).

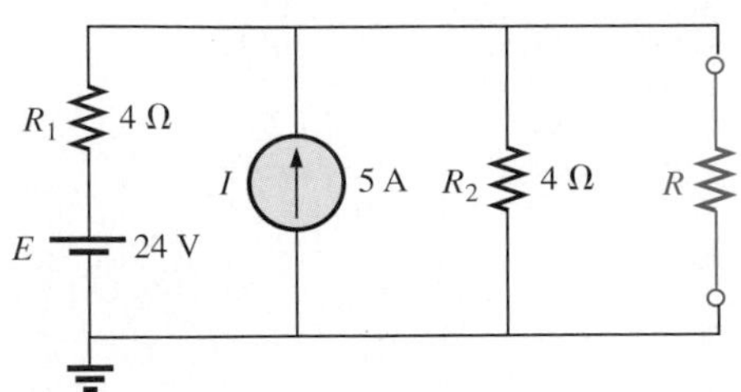

FIG. 9.137
Problems 22 and 46.

*23. Find the resistance R_1 of Fig. 9.138 such that the resistor R_4 will receive maximum power. Think!

*24. **a.** For the network of Fig. 9.139, determine the value R_2 for maximum power to R_4.
    b. Is there a general statement that can be made about situations such as those presented here and in Problem 23?

*25. For the network of Fig. 9.140, determine the level of R that will ensure maximum power to the 100-Ω resistor.

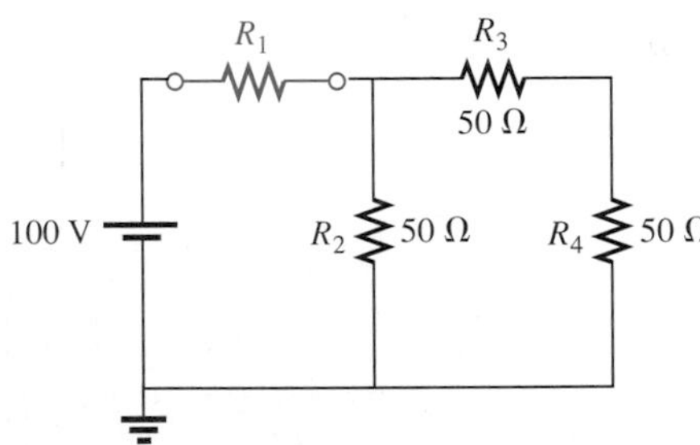

FIG. 9.138
Problem 23.

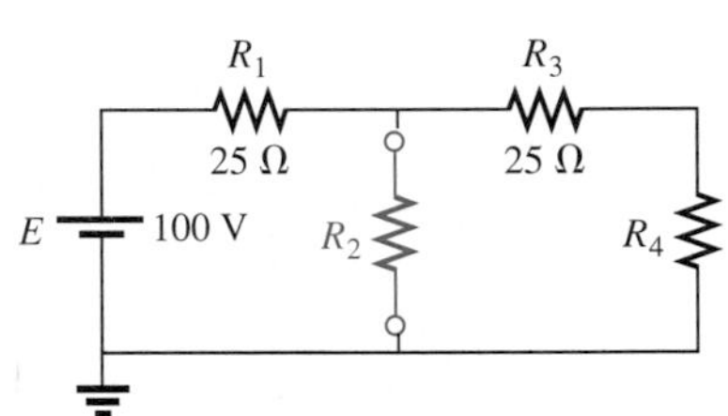

FIG. 9.139
Problem 24.

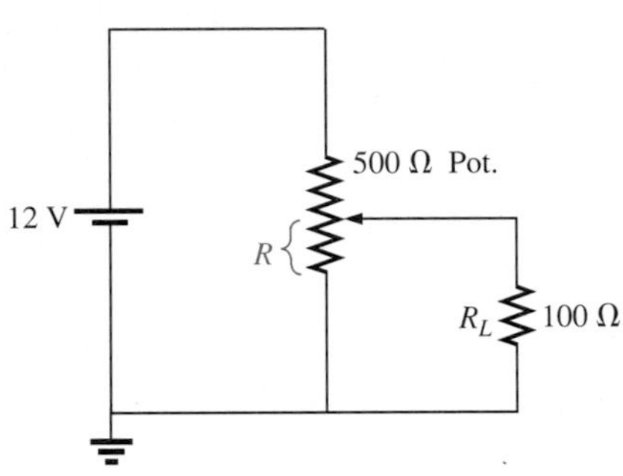

FIG. 9.140
Problem 25.

SECTION 9.6 Millman's Theorem

26. Using Millman's theorem, find the current through and voltage across the resistor R_L of Fig. 9.141.

27. Repeat Problem 26 for the network of Fig. 9.142.

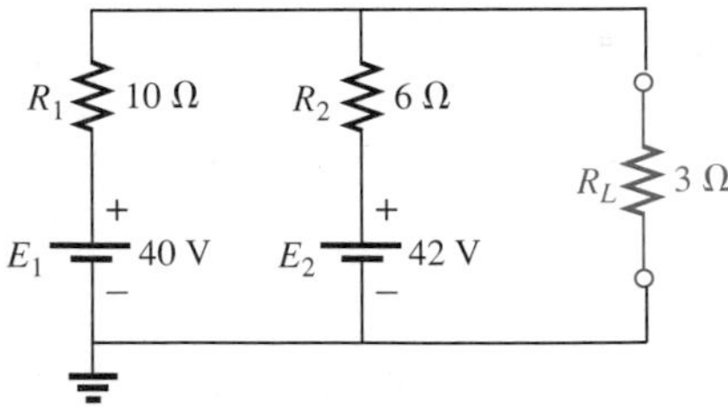

FIG. 9.141
Problem 26.

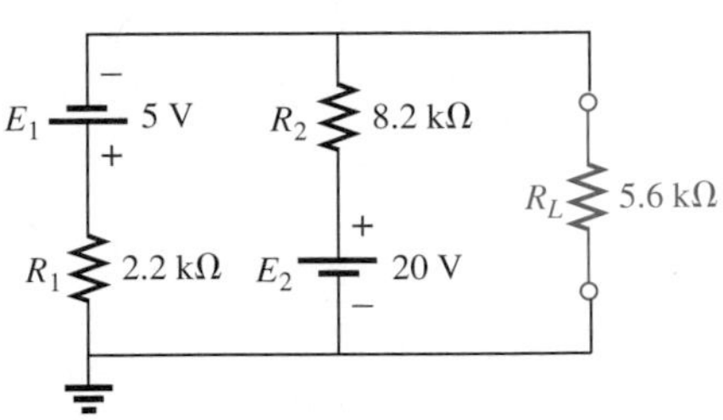

FIG. 9.142
Problem 27.

28. Repeat Problem 26 for the network of Fig. 9.143.

29. Using the dual of Millman's theorem, find the current through and voltage across the resistor R_L of Fig. 9.144.

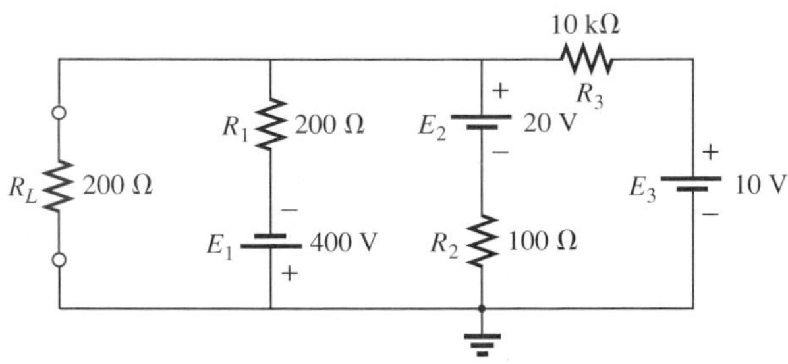

FIG. 9.143
Problem 28.

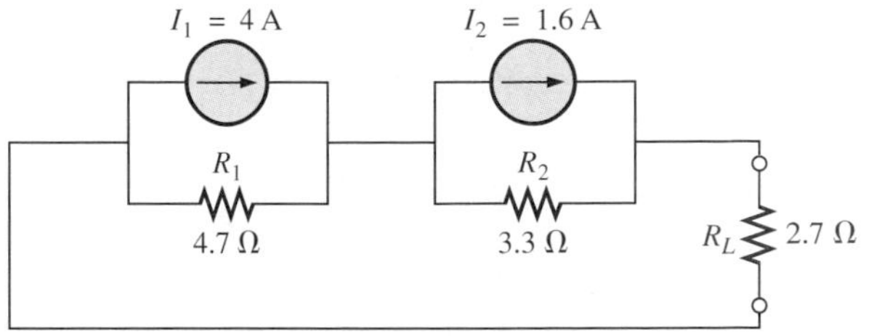

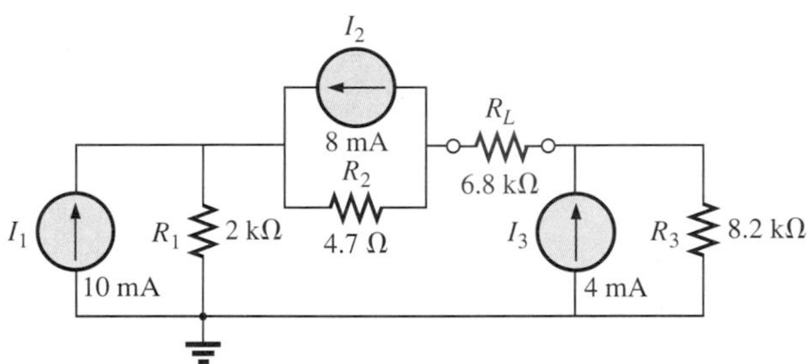

FIG. 9.144
Problem 29.

*30. Repeat Problem 29 for the network of Fig. 9.145.

FIG. 9.145
Problem 30.

SECTION 9.7 Substitution Theorem

31. Using the substitution theorem, draw three equivalent branches for the branch *a-b* of the network of Fig. 9.146.

32. Repeat Problem 31 for the network of Fig. 9.147.

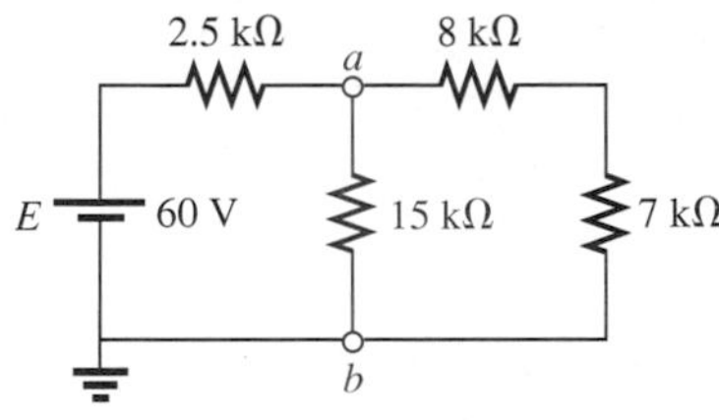

FIG. 9.146
Problem 31.

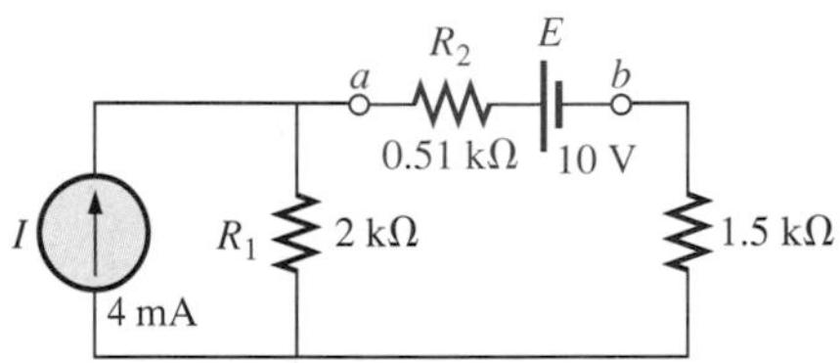

FIG. 9.147
Problem 32.

*33. Repeat Problem 31 for the network of Fig. 9.148. Be careful!

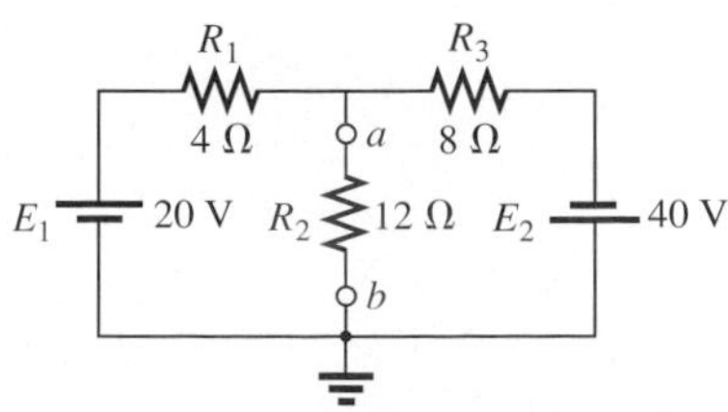

FIG. 9.148
Problem 33.

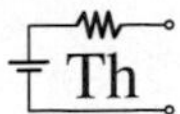

SECTION 9.8 Reciprocity Theorem

34. **a.** For the network of Fig. 9.149(a), determine the current I.
 b. Repeat part (a) for the network of Fig. 9.149(b).
 c. Is the reciprocity theorem satisfied?

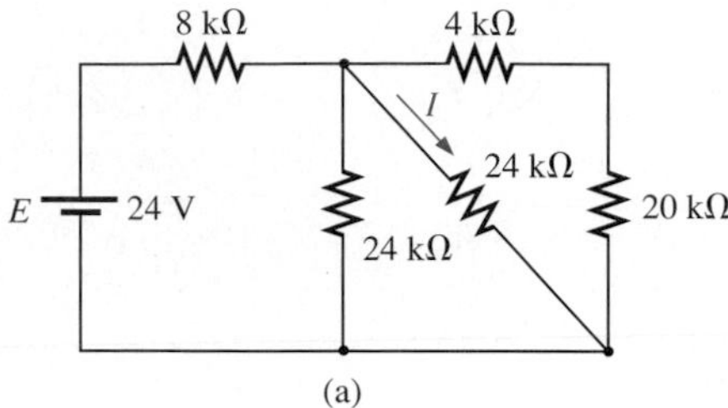

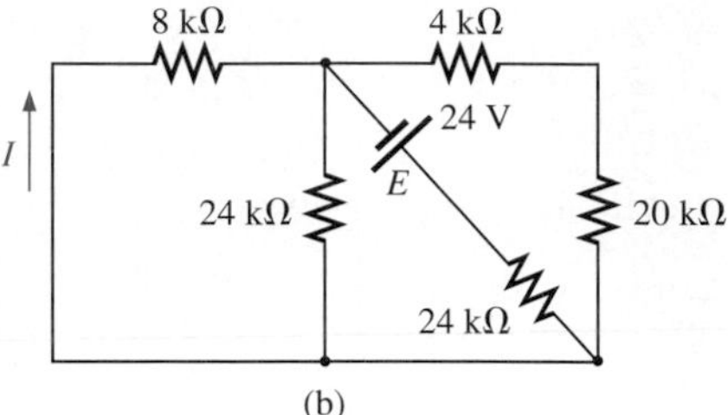

FIG. 9.149
Problem 34.

35. Repeat Problem 34 for the networks of Fig. 9.150.

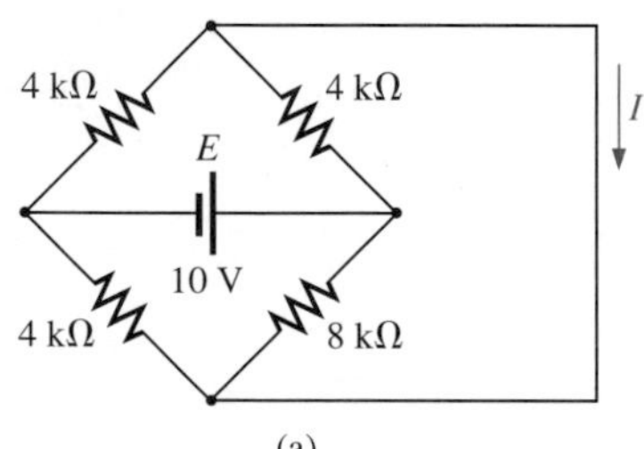

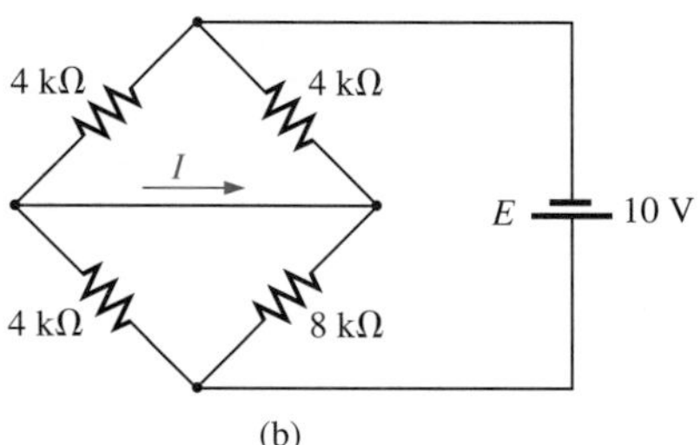

FIG. 9.150
Problem 35.

36. **a.** Determine the voltage V for the network of Fig. 9.151(a).
 b. Repeat part (a) for the network of Fig. 9.151(b).
 c. Is the dual of the reciprocity theorem satisfied?

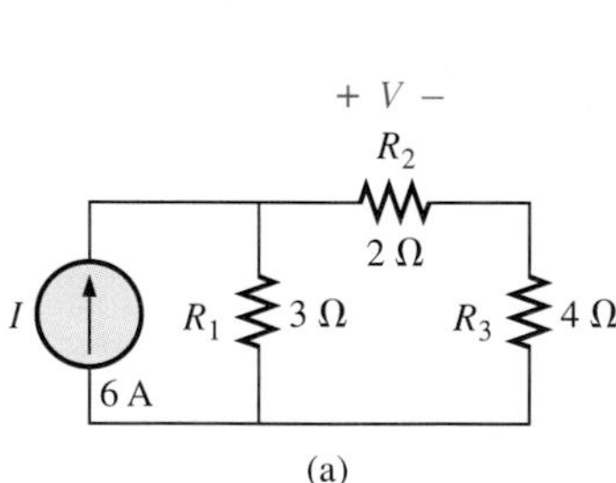

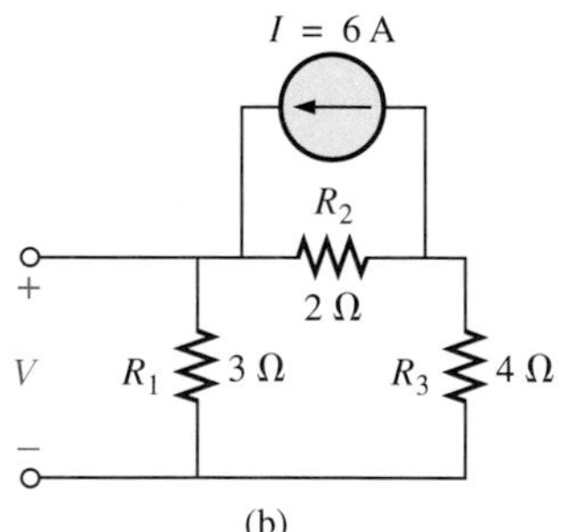

FIG. 9.151
Problem 36.

SECTION 9.9 Computer Analysis

PSpice (DOS)

37. Write the input files required to determine the current I and its components for the network of Fig. 9.125(b) using superposition.

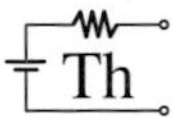

38. Write the input file to determine the Thévenin equivalent circuit for the network of Fig. 9.129(II) external to the resistor R.

39. Write the input files necessary to determine the Norton equivalent circuit for the network external to the resistor R in Fig. 9.131(a).

PSpice (Windows)

40. Using schematics, determine the voltage V_2 and its components for the network of Fig. 9.127.

41. Using schematics, determine the Thévenin equivalent circuit for the network of Fig. 9.131(b).

*42. a. Using schematics, plot the power delivered to the resistor R of Fig. 9.131(a) for R having values from 1 to 50 Ω.
   b. From the plot, determine the value of R resulting in maximum power to R and the maximum power to R.
   c. Compare the results of part (a) to the numerical solution.
   d. Plot V_R and I_R versus R and find the value of each under maximum power conditions.

*43. Change the 300-Ω resistor of Fig. 9.136(b) to a variable resistor and plot the power delivered to the resistor versus values of the resistor. Determine the range of resistance by trial and error rather than first performing a longhand calculation. Determine the Norton equivalent circuit from the results. The Norton current can be determined from the maximum power level.

Programming Language (C++, BASIC, PASCAL, etc.)

44. Write a program to determine the current through the 10-Ω resistor of Fig. 9.125(a) (for any component values) using superposition.

45. Write a program to perform the analysis required for Problem 8, network (b) for any component values.

*46. Write a program to perform the analysis of Problem 22 and tabulate the power to R for the values listed in part (c).

GLOSSARY

Maximum power transfer theorem A theorem used to determine the load resistance necessary to ensure maximum power transfer to the load.

Millman's theorem A method employing source conversions that will permit the determination of unknown variables in a multiloop network.

Norton's theorem A theorem that permits the reduction of any two-terminal linear dc network to having a single current source and parallel resistor.

.PROBE A PSpice command for obtaining output files that contain plots and data values not otherwise available.

Reciprocity theorem A theorem that states that for single-source networks, the current in any branch of a network, due to a single voltage source in the network, will equal the current through the branch in which the source was originally located if the source is placed in the branch in which the current was originally measured.

Substitution theorem A theorem that states that if the voltage across and current through any branch of a dc bilateral network are known, the branch can be replaced by any combination of elements that will maintain the same voltage across and current through the chosen branch.

Superposition theorem A network theorem that permits considering the effects of each source independently. The resulting current and/or voltage is the algebraic sum of the currents and/or voltages developed by each source independently.

Thévenin's theorem A theorem that permits the reduction of any two-terminal linear dc network to one having a single voltage source and series resistor.

10

Capacitors

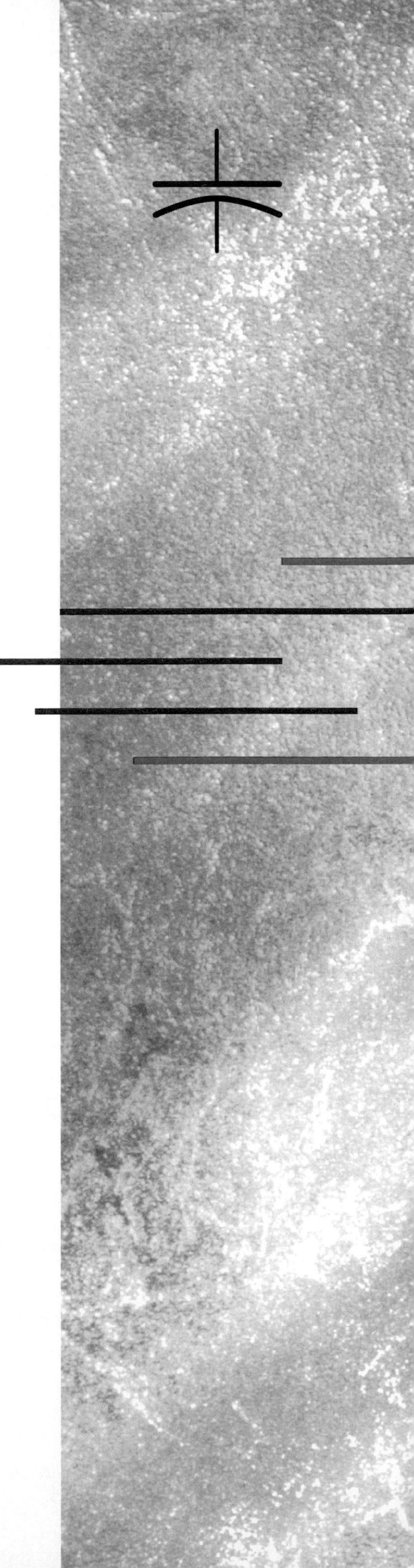

10.1 INTRODUCTION

Thus far, the only passive device appearing in the text has been the resistor. We will now consider two additional passive devices called the *capacitor* and the *inductor,* which are quite different from the resistor in purpose, operation, and construction.

Unlike the resistor, both elements display their total characteristics only when a change in voltage or current is made in the circuit in which they exist. In addition, if we consider the *ideal* situation, they do not dissipate energy like the resistor but store it in a form that can be returned to the circuit whenever required by the circuit design.

Proper treatment of each requires that we devote this entire chapter to the capacitor and Chapter 12 to the inductor. Since electromagnetic effects are a major consideration in the design of inductors, this topic will be covered in Chapter 11.

10.2 THE ELECTRIC FIELD

Recall from Chapter 2 that a force of attraction or repulsion exists between two charged bodies. We shall now examine this phenomenon in greater detail by considering the electric field that exists in the region around any charged body. This electric field is represented by electric flux lines, which are drawn to indicate the strength of the electric field at any point around the charged body; that is, the denser the lines of flux, the stronger the electric field. In Fig. 10.1, the electric field strength is stronger at position a than at position b because the flux lines are denser at a than at b. The symbol for electric flux is the Greek letter ψ (psi). The flux per unit area (flux density) is represented by the capital letter D and is determined by

$$D = \frac{\psi}{A} \quad \text{(flux/unit area)} \tag{10.1}$$

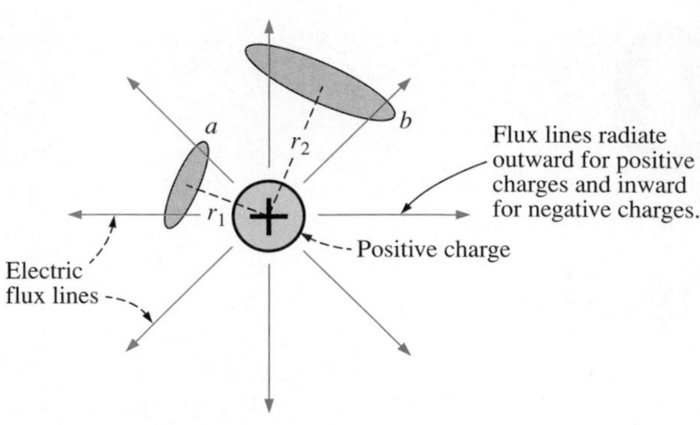

FIG. 10.1
Flux distribution from an isolated positive charge.

The larger the charge Q in coulombs, the greater the number of flux lines extending or terminating per unit area, independent of the surrounding medium. Twice the charge will produce twice the flux per unit area. The two can therefore be equated:

$$\psi \equiv Q \qquad \text{(coulombs, C)} \qquad \textbf{(10.2)}$$

By definition, the *electric field strength* at a point is the force acting on a unit positive charge at that point; that is,

$$\mathscr{E} = \frac{F}{Q} \qquad \text{(newtons/coulomb, N/C)} \qquad \textbf{(10.3)}$$

The force exerted on a unit positive charge ($Q_2 = 1$ C), by a charge Q_1, r meters away, as determined by Coulomb's law is

$$F = \frac{kQ_1Q_2}{r^2} = \frac{kQ_1(1)}{r^2} = \frac{kQ_1}{r^2} \qquad (k = 9 \times 10^9 \text{ N·m}^2/\text{C}^2)$$

Substituting this force F into Eq. (10.3) yields

$$\mathscr{E} = \frac{F}{Q_2} = \frac{kQ_1/r^2}{1}$$

$$\mathscr{E} = \frac{kQ_1}{r^2} \qquad \text{(N/C)} \qquad \textbf{(10.4)}$$

We can therefore conclude that the electric field strength at any point distance r from a point charge of Q coulombs is directly proportional to the magnitude of the charge and inversely proportional to the distance squared from the charge. The squared term in the denominator will result in a rapid decrease in the strength of the electric field with distance from the point charge. In Fig. 10.1, substituting distances r_1 and r_2 into Eq. (10.4) will verify our previous conclusion that the electric field strength is greater at a than at b.

Electric flux lines always extend from a positively charged to a negatively charged body, always extend or terminate perpendicular to the charged surfaces, and never intersect.

For two charges of similar and opposite polarities, the flux distribution would appear as shown in Fig. 10.2.

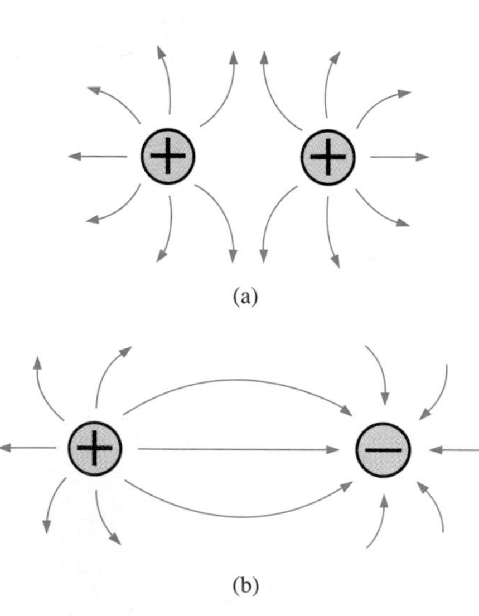

FIG. 10.2
Electric flux distribution: (a) like charges; (b) opposite charges.

The attraction and repulsion between charges can now be explained in terms of the electric field and its flux lines. In Fig. 10.2(a), the flux lines are not interlocked but tend to act as a buffer, preventing attraction and causing repulsion. Since the electric field strength is stronger (flux lines denser) for each charge the closer we are to the charge, the more we try to bring the two charges together, the stronger will be the force of repulsion between them. In Fig. 10.2(b), the flux lines extending from the positive charge are terminated at the negative charge. A basic law of physics states that electric flux lines always tend to be as short as possible. The two charges will therefore be drawn to each other. Again, the closer the two charges, the stronger the attraction between the two charges due to the increased field strengths.

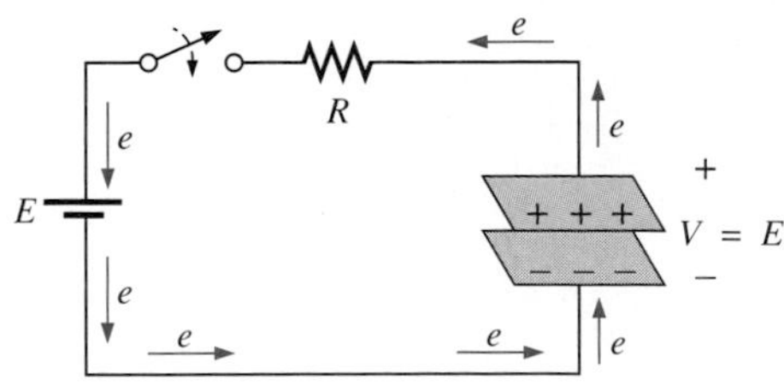

FIG. 10.3
Fundamental charging network.

10.3 CAPACITANCE

Up to this point we have considered only isolated positive and negative spherical charges, but the analysis can be extended to charged surfaces of any shape and size. In Fig. 10.3, for example, two parallel plates of a conducting material separated by an air gap have been connected through a switch and a resistor to a battery. If the parallel plates are initially uncharged and the switch is left open, no net positive or negative charge will exist on either plate. The instant the switch is closed, however, electrons are drawn from the upper plate through the resistor to the positive terminal of the battery. There will be a surge of current at first, limited in magnitude by the resistance present. The level of flow will then decline, as will be demonstrated in the sections to follow. This action creates a net positive charge on the top plate. Electrons are being repelled by the negative terminal through the lower conductor to the bottom plate at the same rate they are being drawn to the positive terminal. This transfer of electrons continues until the potential difference across the parallel plates is exactly equal to the battery voltage. The final result is a net positive charge on the top plate and a negative charge on the bottom plate, very similar in many respects to the two isolated charges of Fig. 10.2(b).

This element, constructed simply of two parallel conducting plates separated by an insulating material (in this case, air), is called a *capacitor. Capacitance* is a measure of a capacitor's ability to store charge on its plates—in other words, its storage capacity.

A capacitor has a capacitance of 1 farad if 1 coulomb of charge is deposited on the plates by a potential difference of 1 volt across the plates.

The farad is named after Michael Faraday (Fig. 10.4), a nineteenth-century English chemist and physicist. The farad, however, is generally too large a measure of capacitance for most practical applications, so the microfarad (10^{-6}) or picofarad (10^{-12}) is more commonly used. Expressed as an equation, the capacitance is determined by

$$C = \frac{Q}{V} \qquad \begin{aligned} C &= \text{farads (F)} \\ Q &= \text{coulombs (C)} \\ V &= \text{volts (V)} \end{aligned} \qquad \textbf{(10.5)}$$

Different capacitors for the same voltage across their plates will acquire greater or lesser amounts of charge on their plates. Hence the capacitors have a greater or lesser capacitance, respectively.

A cross-sectional view of the parallel plates is shown with the distribution of electric flux lines in Fig. 10.5(a). The number of flux lines per

English (London)
(1791–1867)
Chemist and Electrical Experimenter
Honorary Doctorate from Oxford in 1832

Courtesy of the
Smithsonian Institution
Photo No. 51,147

An experimenter with no formal education, he began his research career at the Royal Institute in London as a laboratory assistant. Intrigued by the interaction between electrical and magnetic effects, he discovered *electromagnetic induction,* demonstrating that electrical effects can be generated from a magnetic field (the birth of the generator as we know it today). He also discovered *self-induced currents* and introduced the concept of *lines and fields of magnetic force.* Having received over a hundred academic and scientific honors, he became a Fellow of the Royal Society in 1824 at the young age of 32.

FIG. 10.5 *Michael Faraday.*

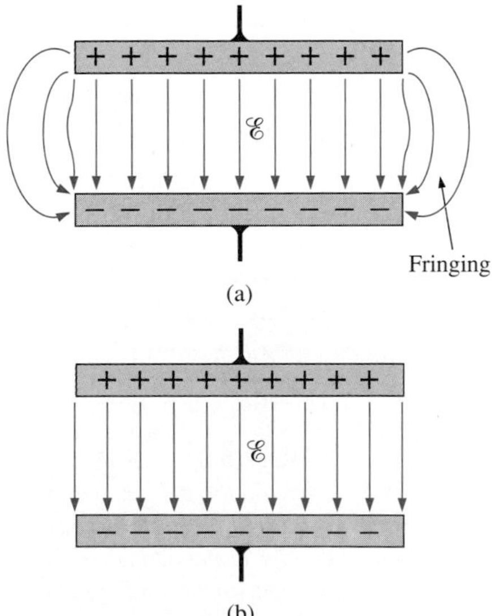

FIG. 10.5
Electric flux distribution between the plates of a capacitor: (a) including fringing; (b) ideal.

unit area (D) between the two plates is quite uniform. At the edges, the flux lines extend outside the common surface area of the plates, producing an effect known as *fringing*. This effect, which reduces the capacitance somewhat, can be neglected for most practical applications. For the analysis to follow, we will assume that all the flux lines leaving the positive plate will pass directly to the negative plate within the common surface area of the plates [Fig. 10.5(b)].

If a potential difference of V volts is applied across the two plates separated by a distance of d, the electric field strength between the plates is determined by

$$\mathscr{E} = \frac{V}{d} \quad \text{(volts/meter, V/m)} \qquad \textbf{(10.6)}$$

The uniformity of the flux distribution in Fig. 10.5(b) also indicates that the electric field strength is the same at any point between the two plates.

Many values of capacitance can be obtained for the same set of parallel plates by the addition of certain insulating materials between the plates. In Fig. 10.6(a), an insulating material has been placed between a set of parallel plates having a potential difference of V volts across them.

Since the material is an insulator, the electrons within the insulator are unable to leave the parent atom and travel to the positive plate. The positive components (protons) and negative components (electrons) of each atom do shift, however [as shown in Fig. 10.6(a)], to form *dipoles*.

When the dipoles align themselves as shown in Fig. 10.6(a), the material is *polarized*. A close examination within this polarized material will indicate that the positive and negative components of adjoining dipoles are neutralizing the effects of each other [note the dashed area in Fig. 10.6(a)]. The layer of positive charge on one surface and the negative charge on the other are not neutralized, however, resulting in the establishment of an electric field within the insulator [$\mathscr{E}_{\text{dielectric}}$, Fig. 10.6(b)]. The net electric field between the plates ($\mathscr{E}_{\text{resultant}} = \mathscr{E}_{\text{air}} - \mathscr{E}_{\text{dielectric}}$) would therefore be reduced due to the insertion of the dielectric.

The purpose of the dielectric, therefore, is to create an electric field to oppose the electric field set up by free charges on the parallel plates. For this reason, the insulating material is referred to as a *dielectric, di* for *opposing* and *electric* for *electric field.*

In either case—with or without the dielectric—if the potential across the plates is kept constant and the distance between the plates is fixed, the net electric field within the plates must remain the same, as determined by the equation $\mathscr{E} = V/d$. We just ascertained, however, that the net electric field between the plates would decrease with insertion of the dielectric for a fixed amount of free charge on the plates. To compensate and keep the net electric field equal to the value determined by V and d, more charge must be deposited on the plates. [Look ahead to Eq. (10.11).] This additional charge for the same potential across the plates increases the capacitance, as determined by the following equation:

$$C\uparrow = \frac{Q\uparrow}{V}$$

For different dielectric materials between the same two parallel plates, different amounts of charge will be deposited on the plates. But

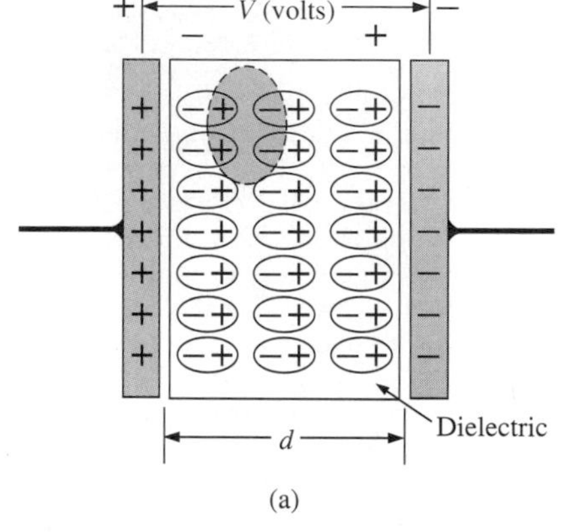

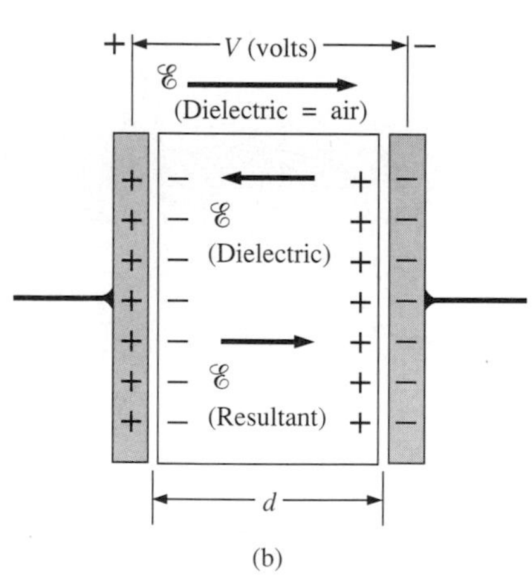

FIG. 10.6
Effect of a dielectric on the field distribution between the plates of a capacitor: (a) alignment of dipoles in the dielectric; (b) electric field components between the plates of a capacitor with a dielectric present.

$\psi \equiv Q$, so the dielectric is also determining the number of flux lines between the two plates and consequently the flux density ($D = \psi/A$) since A is fixed.

The ratio of the flux density to the electric field intensity in the dielectric is called the *permittivity* of the dielectric:

$$\epsilon = \frac{D}{\mathscr{E}} \quad \text{(farads/meter, F/m)} \tag{10.7}$$

It is a measure of how easily the dielectric will "permit" the establishment of flux lines within the dielectric. The greater its value, the greater the amount of charge deposited on the plates and, consequently, the greater the flux density for a fixed area.

For a vacuum, the value of ϵ (denoted by ϵ_o) is 8.85×10^{-12} F/m. The ratio of the permittivity of any dielectric to that of a vacuum is called the *relative permittivity*, ϵ_r. It simply compares the permittivity of the dielectric to that of air. In equation form,

$$\epsilon_r = \frac{\epsilon}{\epsilon_o} \tag{10.8}$$

The value of ϵ for any material, therefore, is

$$\epsilon = \epsilon_r \epsilon_o$$

Note that ϵ_r is a dimensionless quantity. The relative permittivity, or *dielectric constant*, as it is often called, is provided in Table 10.1 for various dielectric materials.

Substituting for D and $\mathscr{E}$ in Eq. (10.7), we have

$$\epsilon = \frac{D}{\mathscr{E}} = \frac{\psi/A}{V/d} = \frac{Q/A}{V/d} = \frac{Qd}{VA}$$

But

$$C = \frac{Q}{V}$$

and, therefore,

$$\epsilon = \frac{Cd}{A}$$

TABLE 10.1

Relative permittivity (dielectric constant) of various dielectrics.

Dielectric	ϵ_r (Average Values)
Vacuum	1.0
Air	1.0006
Teflon	2.0
Paper, paraffined	2.5
Rubber	3.0
Transformer oil	4.0
Mica	5.0
Porcelain	6.0
Bakelite	7.0
Glass	7.5
Distilled water	80.0
Barium-strontium titanite (ceramic)	7500.0

and

$$C = \epsilon \frac{A}{d} \qquad \text{(F)} \tag{10.9}$$

or

$$C = \epsilon_o \epsilon_r \frac{A}{d} = 8.85 \times 10^{-12} \epsilon_r \frac{A}{d} \qquad \text{(F)} \tag{10.10}$$

where A is the area in square meters of the plates, d is the distance in meters between the plates, and ϵ_r is the relative permittivity. The capacitance, therefore, will be greater if the area of the plates is increased, or the distance between the plates is decreased, or the dielectric is changed so that ϵ_r is increased.

Solving for the distance d in Eq. (10.9), we have

$$D = \frac{\epsilon A}{C}$$

and substituting into Eq. (10.6) yields

$$\mathscr{E} = \frac{V}{d} = \frac{V}{\epsilon A / C} = \frac{CV}{\epsilon A}$$

But $Q = CV$, and therefore

$$\mathscr{E} = \frac{Q}{\epsilon A} \qquad \text{(V/m)} \tag{10.11}$$

which gives the electric field intensity between the plates in terms of the permittivity ϵ, the charge Q, and the surface area A of the plates. The ratio

$$\frac{C = \epsilon A/d}{C_o = \epsilon_o A/d} = \frac{\epsilon}{\epsilon_o} = \epsilon_r$$

or

$$C = \epsilon_r C_o \tag{10.12}$$

which, in words, states that for the same set of parallel plates, the capacitance using a dielectric (of relative permittivity ϵ_r) is ϵ_r times that obtained for a vacuum (or air, approximately) between the plates. This relationship between ϵ_r and the capacitances provides an excellent experimental method for finding the value of ϵ_r for various dielectrics.

EXAMPLE 10.1 Determine the capacitance of each capacitor on the right side of Fig. 10.7.

Solutions:

a. $C = 3(5\ \mu\text{F}) = \mathbf{15\ \mu F}$

b. $C = \frac{1}{2}(0.1\ \mu\text{F}) = \mathbf{0.05\ \mu F}$

c. $C = 2.5(20\ \mu\text{F}) = \mathbf{50\ \mu F}$

d. $C = (5)\frac{4}{(1/8)}(1000\ \text{pF}) = (160)(1000\ \text{pF}) = \mathbf{0.16\ \mu F}$

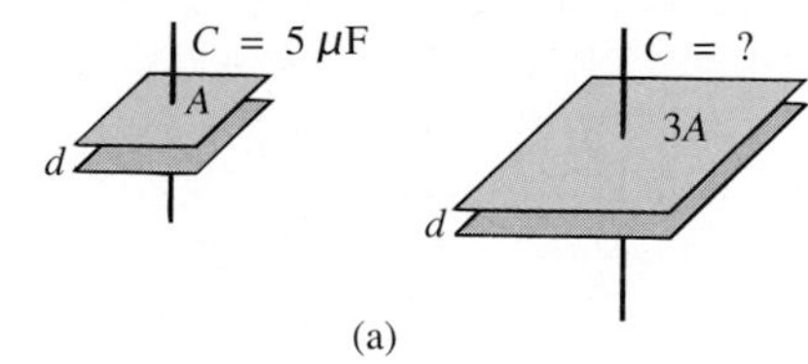

(a)

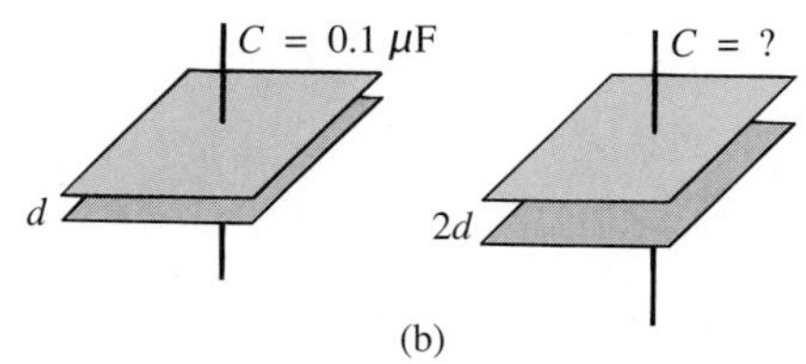

(b)

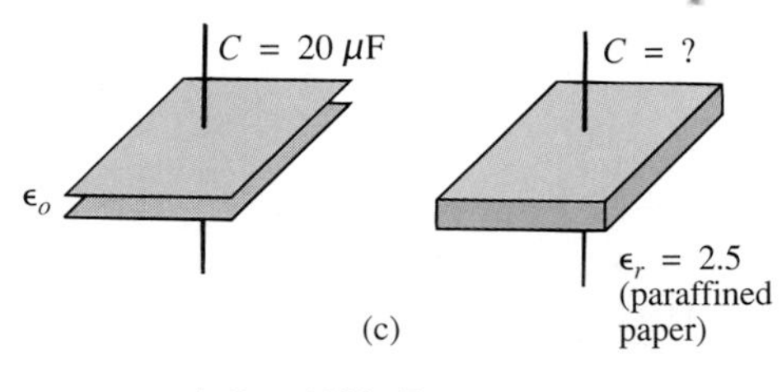

(c)

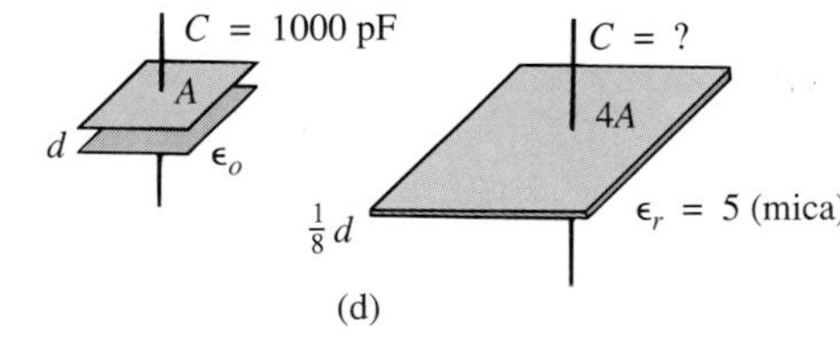

(d)

FIG. 10.7
Example 10.1.

EXAMPLE 10.2 For the capacitor of Fig. 10.8:

a. Determine the capacitance.
b. Determine the electric field strength between the plates if 450 V are applied across the plates.
c. Find the resulting charge on each plate.

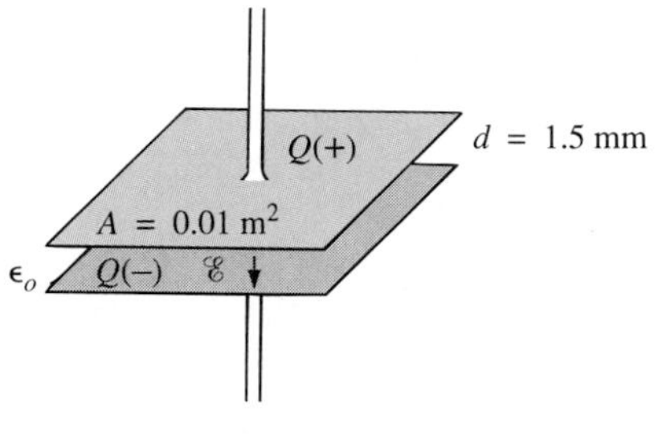

FIG. 10.8
Example 10.2.

Solutions:

a. $C_o = \dfrac{\epsilon_o A}{d} = \dfrac{(8.85 \times 10^{-12}\text{ F/m})(0.01\text{ m}^2)}{1.5 \times 10^{-3}\text{ m}} = 59.0 \times 10^{-12}\text{ F}$

$= \mathbf{59\ pF}$

b. $\mathscr{E} = \dfrac{V}{d} = \dfrac{450\text{ V}}{1.5 \times 10^{-3}\text{ m}}$

$\cong \mathbf{300 \times 10^3\ V/m}$

c. $C = \dfrac{Q}{V}$

or $\quad Q = CV = (59.0 \times 10^{-12}\text{ F})(450\text{ V})$

$= 26.550 \times 10^{-9}\text{ C}$

$= \mathbf{26.55\ nC}$

EXAMPLE 10.3 A sheet of mica 1.5 mm thick having the same area as the plates is inserted between the plates of Example 10.2.

a. Find the electric field strength between the plates.
b. Find the charge on each plate.
c. Find the capacitance.

Solutions:

a. $\mathscr{E}$ is fixed by

$$\mathscr{E} = \frac{V}{d} = \frac{450 \text{ V}}{1.5 \times 10^3 \text{ m}}$$
$$\cong \mathbf{300 \times 10^3 \text{ V/m}}$$

b. $\mathscr{E} = \dfrac{Q}{\epsilon A}$ or

$$Q = \epsilon\mathscr{E}A = \epsilon_r\epsilon_o\mathscr{E}A$$
$$= (5)(8.85 \times 10^{-12} \text{ F/m})(300 \times 10^3 \text{ V/m})(0.01 \text{ m}^2)$$
$$= 132.75 \times 10^{-9} \text{ C} = \mathbf{132.75 \text{ nC}}$$

(five times the amount for air between the plates)

c. $C = \epsilon_r C_o$

$$= (5)(59 \times 10^{-12} \text{ F}) = \mathbf{295 \text{ pF}}$$

10.4 DIELECTRIC STRENGTH

For every dielectric there is a potential that, if applied across the dielectric, will break the bonds within the dielectric and cause current to flow. The voltage required per unit length (electric field intensity) to establish conduction in a dielectric is an indication of its *dielectric strength* and is called the *breakdown voltage*. When breakdown occurs, the capacitor has characteristics very similar to those of a conductor. A typical example of breakdown is lightning, which occurs when the potential between the clouds and the earth is so high that charge can pass from one to the other through the atmosphere, which acts as the dielectric.

The average dielectric strengths for various dielectrics are tabulated in volts/mil in Table 10.2 (1 mil = 0.001 in.). The relative permittivity appears in parentheses to emphasize the importance of considering both factors in the design of capacitors. Take particular note of barium-strontium titanite and mica.

TABLE 10.2
Dielectric strength of some dielectric materials.

Dielectric	Dielectric Strength (Average Value), in Volts/Mil	(ϵ_r)
Air	75	(1.006)
Barium-strontium titanite (ceramic)	75	(7500)
Porcelain	200	(6.0)
Transformer oil	400	(4.0)
Bakelite	400	(7.0)
Rubber	700	(3.0)
Paper, paraffined	1300	(2.5)
Teflon	1500	(2.0)
Glass	3000	(7.5)
Mica	5000	(5.0)

EXAMPLE 10.4 Find the maximum voltage that can be applied across a 0.2-μF capacitor having a plate area of 0.3 m^2. The dielectric is porcelain. Assume a linear relationship between the dielectric strength and the thickness of the dielectric.

Solution:

$$C = 8.85 \times 10^{-12}\epsilon_r \frac{A}{d}$$

or
$$d = \frac{8.85\epsilon_r A}{10^{12}C} = \frac{(8.85)(6)(0.3\text{ m}^2)}{(10^{12})(0.2 \times 10^{-6}\text{ F})} = 7.965 \times 10^{-5}\text{ m}$$

$$\cong \mathbf{79.65\ \mu m}$$

Converting millimeters to mils, we have

$$79.65\ \mu\text{m}\left(\frac{10^{-6}\text{ m}}{\mu\text{m}}\right)\left(\frac{39.371\text{ in.}}{\text{m}}\right)\left(\frac{1000\text{ mils}}{1\text{ in.}}\right) = 3.136\text{ mils}$$

$$\text{Dielectric strength} = 200\text{ V/mil}$$

Therefore,
$$\left(\frac{200\text{ V}}{\text{mil}}\right)(3.136\text{ mils}) = \mathbf{627.20\ V}$$

10.5 LEAKAGE CURRENT

Up to this point, we have assumed that the flow of electrons will occur in a dielectric only when the breakdown voltage is reached. This is the ideal case. In actuality, there are free electrons in every dielectric due in part to impurities in the dielectric and forces within the material itself.

When a voltage is applied across the plates of a capacitor, a leakage current due to the free electrons flows from one plate to the other. The current is usually so small, however, that it can be neglected for most practical applications. This effect is represented by a resistor in parallel with the capacitor, as shown in Fig. 10.9(a), whose value is typically more than 100 megohms (MΩ). There are some capacitors, however, such as the electrolytic type, that have high leakage currents. When charged and then disconnected from the charging circuit, these capacitors lose their charge in a matter of seconds because of the flow of charge (leakage current) from one plate to the other [Fig. 10.9(b)].

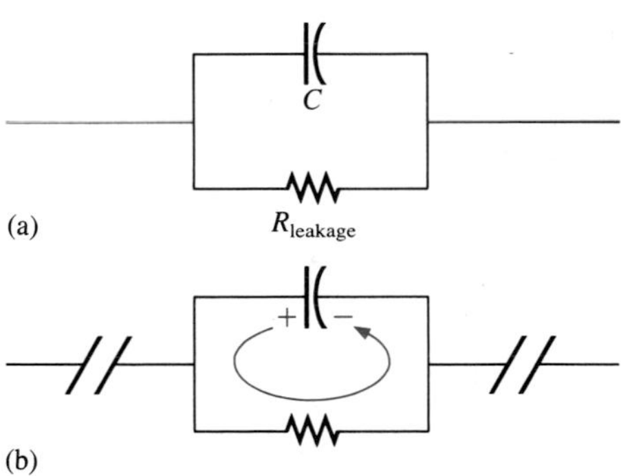

FIG. 10.9
Demonstrating the effect of the leakage current.

10.6 TYPES OF CAPACITORS

Like resistors, all capacitors can be included under either of two general headings: fixed or variable. The symbol for a fixed capacitor is ⊥ and for a variable capacitor, ⊥̸. The curved line represents the plate that is usually connected to the point of lower potential.

Fixed Capacitors

Many types of fixed capacitors are available today. Some of the most common are the mica, ceramic, electrolytic, tantalum, and polyester-film capacitors. The typical *mica capacitor* consists basically of mica sheets separated by sheets of metal foil. The plates are connected to two electrodes, as shown in Fig. 10.10. The total area is the area of one sheet

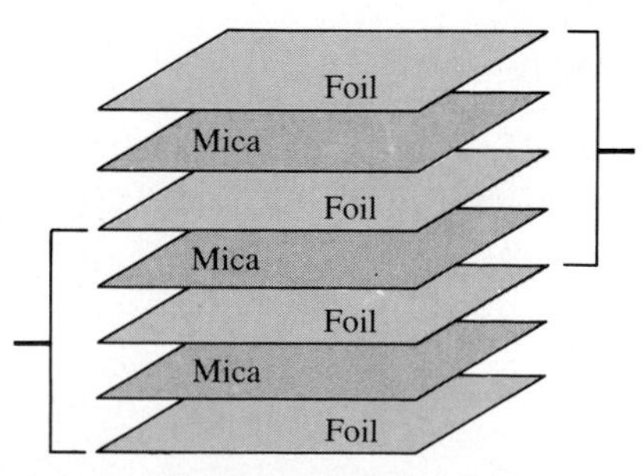

FIG. 10.10
Basic structure of a stacked mica capacitor.

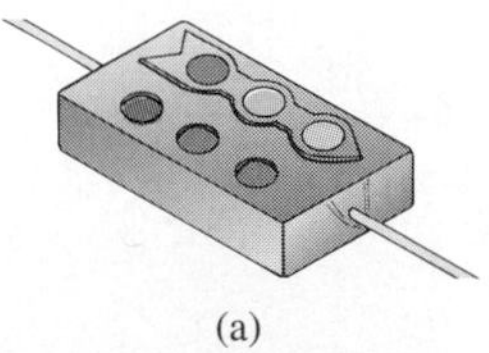

(a)

(b)

FIG. 10.11
Mica capacitors. (Part (b) courtesy of Custom Electronics Inc.)

times the number of dielectric sheets. The entire system is encased in a plastic insulating material as shown in Fig. 10.11(a). The mica capacitor exhibits excellent characteristics under stress of temperature variations and high voltage applications (its dielectric strength is 5000 V/mil). Its leakage current is also very small (R_{leakage} about 1000 MΩ).

Mica capacitors are typically between a few picofarads and 0.2 μF, with voltages of 100 V or more. The color code for the mica capacitors of Fig. 10.11(a) can be found in Appendix D.

A second type of mica capacitor appears in Fig. 10.11(b). Note in particular the cylindrical unit in the bottom left-hand corner of the figure. The ability to "roll" the mica to form the cylindrical shape is due to a process whereby the soluble contaminants in natural mica are removed, leaving a paperlike structure due to the cohesive forces in natural mica. It is commonly referred to as *reconstituted mica,* although the terminology does not mean "recycled" or "second-hand" mica. For some of the units in the photograph, different levels of capacitance are available between different sets of terminals.

The *ceramic capacitor* is made in many shapes and sizes, some of which are shown in Fig. 10.12. The basic construction, however, is about the same for each, as shown in Fig. 10.13. A ceramic base is coated on two sides with a metal, such as copper or silver, to act as the two plates. The leads are then attached through electrodes to the plates. An insulating coating of ceramic or plastic is then applied over the plates and dielectric. Ceramic capacitors also have a very low leakage current (R_{leakage} about 1000 MΩ) and can be used in both dc and ac networks. They can be found in values ranging from a few picofarads to perhaps 2 μF, with very high working voltages such as 5000 V or more.

In recent years there has been increasing interest in monolithic (single-structure) chip capacitors such as those appearing in Fig. 10.14(a) due to their application on hybrid circuitry [networks using both discrete and integrated circuit (IC) components]. There has also been increasing use of microstrip (strip-line) circuitry such as the one in Fig. 10.14(b). Note the small chips in this cutaway section. The *L* and *H* of Fig. 10.14(a) indicate the level of capacitance. For example, the letter *H* in black letters represents 16 units of capacitance (in picofarads), or 16 pF. If blue ink is used, a multiplier of 100 is applied, resulting in 1600 pF. Although the size is similar, the type of ceramic material controls the capacitance level.

The *electrolytic capacitor* is used most commonly in situations where capacitances of the order of one to several thousand microfarads

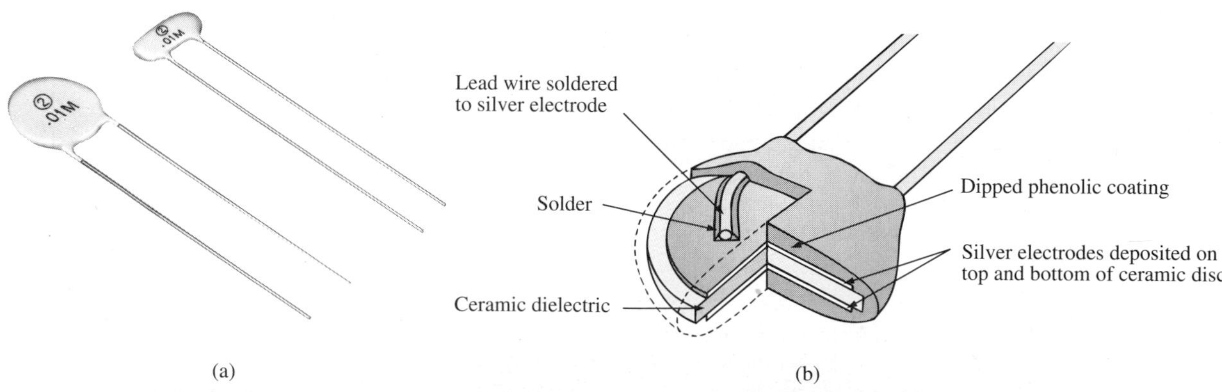

(a) (b)

FIG. 10.12
Ceramic disc capacitors: (a) photograph; (b) construction.

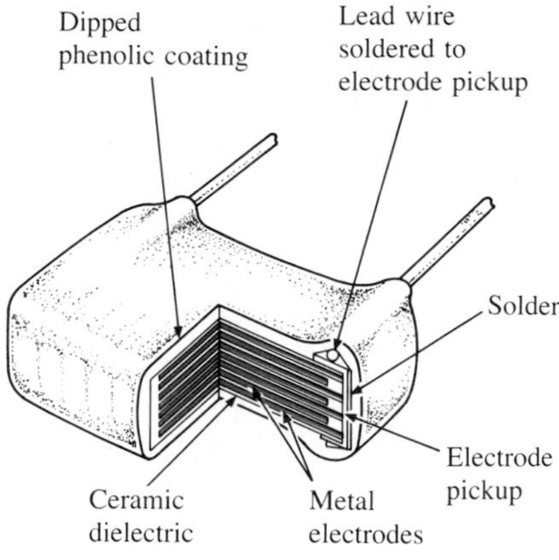

FIG. 10.13
Multilayer, radial-lead ceramic capacitor.

(a)

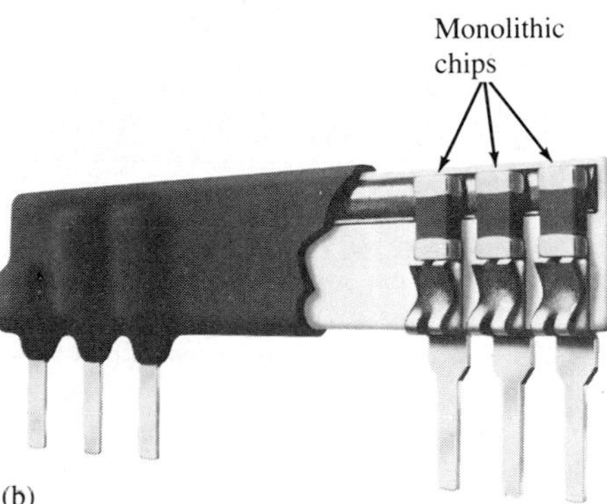

(b)

FIG. 10.14
Monolithic chip capacitors. (Courtesy of Vitramon, Inc.)

are required. They are designed primarily for use in networks where only dc voltages will be applied across the capacitor because they have good insulating characteristics (high leakage current) between the plates in one direction but take on the characteristics of a conductor in the other direction. There are electrolytic capacitors available that can be used in ac circuits (for starting motors) and in cases where the polarity of the dc voltage will reverse across the capacitor for short periods of time.

The basic construction of the electrolytic capacitor consists of a roll of aluminum foil coated on one side with an aluminum oxide, the aluminum being the positive plate and the oxide the dielectric. A layer of paper or gauze saturated with an electrolyte is placed over the aluminum oxide on the positive plate. Another layer of aluminum without the oxide coating is then placed over this layer to assume the role of the negative plate. In most cases the negative plate is connected directly to the aluminum container, which then serves as the negative terminal for external connections. Because of the size of the roll of aluminum foil, the overall area of this capacitor is large; and due to the use of an oxide as the dielectric, the distance between the plates is extremely small. The negative terminal of the electrolytic capacitor is usually the one with no visible identification on the casing. The positive is usually indicated by such designs as +, △, □, and so on. Due to the polarity requirement, the symbol for an electrolytic capacitor will normally appear as ⊤⁺.

Associated with each electrolytic capacitor are the dc working voltage and the surge voltage. The *working voltage* is the voltage that can be applied across the capacitor for long periods of time without breakdown. The *surge voltage* is the maximum dc voltage that can be applied for a short period of time. Electrolytic capacitors are characterized as having low breakdown voltages and high leakage currents (R_{leakage} about 1 MΩ). Various types of electrolytic capacitors are shown in Fig. 10.15. They can be found in values extending from a few microfarads to several thousand microfarads and working voltages as high as 500 V. However, increased levels of voltage are normally associated with lower values of available capacitance.

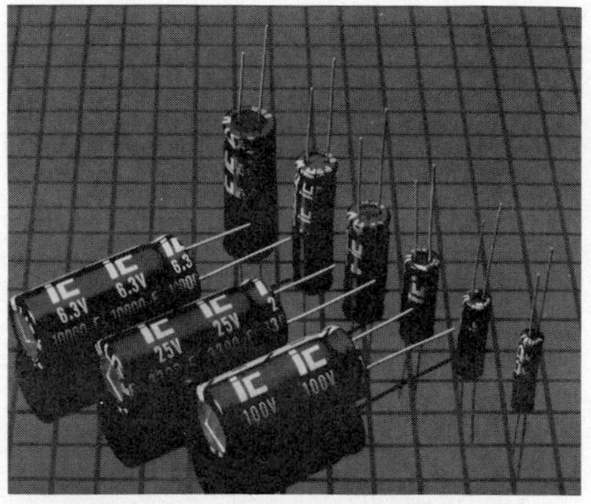

(a)

(b)

FIG. 10.15
Electrolytic capacitors: (a) Radial lead with extended endurance rating of 2000 hours at 85°C. Capacitance range: 0.1–15,000 μF with a voltage range of 6.3 to 250 WVDC (Courtesy of Illinois Capacitor, Inc.) (b) Solid aluminum electrolytic capacitors available in axial, resin-dipped, and surface mount configurations to withstand harsh environmental conditions. (Courtesy of Philips Components, Inc.)

There are fundamentally two types of *tantalum capacitors:* the *solid* and the *wet-slug*. In each case, tantalum powder of high purity is pressed into a rectangular or cylindrical shape, as shown in Fig. 10.16. The anode (+) connection is then simply pressed into the resulting structures, as shown in the figure. The resulting unit is then sintered (baked) in a vacuum at very high temperatures to establish a very porous material. The result is a structure with a very large surface area in a limited volume. Through immersion in an acid solution, a very thin manganese dioxide (MnO_2) coating is established on the large, porous surface area. An electrolyte is then added to establish contact between the surface area and the cathode, producing a solid tantalum capacitor. If an appropriate "wet" acid is introduced, it is called a *wet-slug* tantalum capacitor.

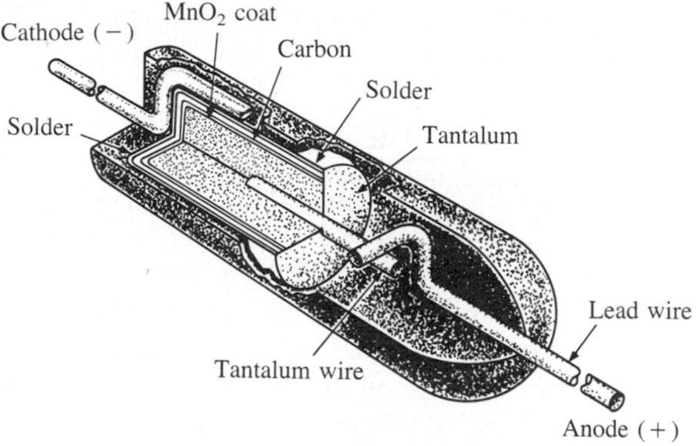

FIG. 10.16
Tantalum capacitor. (Courtesy of Union Carbide Corp.)

The last type of fixed capacitor to be introduced is the *polyester-film capacitor,* the basic construction of which is shown in Fig. 10.17. It consists simply of two metal foils separated by a strip of polyester material such as Mylar®. The outside layer of polyester is applied to act as an insulating jacket. Each metal foil is connected to a lead that extends either axially or radially from the capacitor. The rolled construction results in a large surface area, and the use of the plastic dielectric results in a very thin layer between the conducting surfaces.

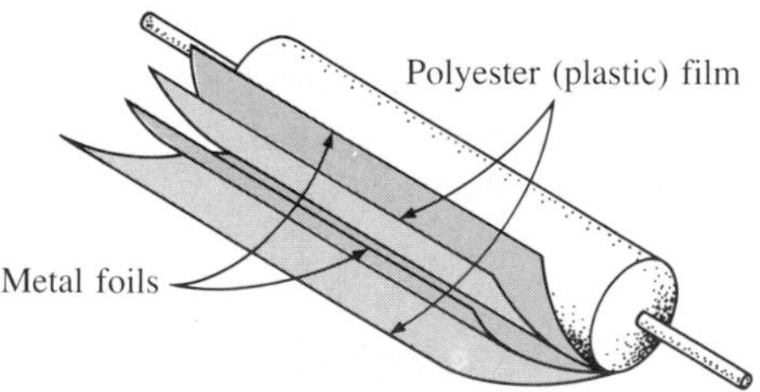

Polyester-film capacitor.

Data such as capacitance and working voltage are printed on the outer wrapping if the polyester capacitor is large enough. Color coding is used on smaller devices (see Appendix E). A band (usually black) is sometimes printed near the lead that is connected to the outer metal foil. The lead nearest this band should always be connected to the point of

lower potential. This capacitor can be used for both dc and ac networks. Its leakage resistance is of the order of 100 MΩ. An axial lead and radial lead polyester-film capacitor appear in Fig. 10.18. The axial lead variety is available with capacitance levels of 0.1 μF to 18 μF, with working voltages extending to 630 V. The radial lead variety has a capacitance range of 0.01 μF to 10 μF, with working voltages extending to 1000 V.

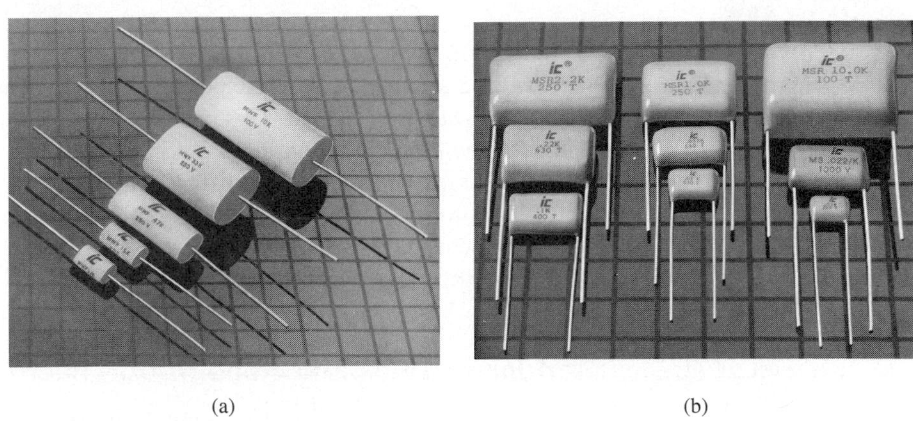

(a) (b)

FIG. 10.18
Polyester-film capacitors: (a) axial lead; (b) radial lead. (Courtesy of Illinois Capacitor, Inc.)

Variable Capacitors

The most common of the variable-type capacitors is shown in Fig. 10.19. The dielectric for each capacitor is air. The capacitance in Fig. 10.19(a) is changed by turning the shaft at one end to vary the common area of the movable and fixed plates. The greater the common area, the larger the capacitance, as determined by Eq. (10.10). The capacitance of the trimmer capacitor in Fig. 10.19(b) is changed by turning the screw, which will vary the distance between the plates and thereby the capacitance.

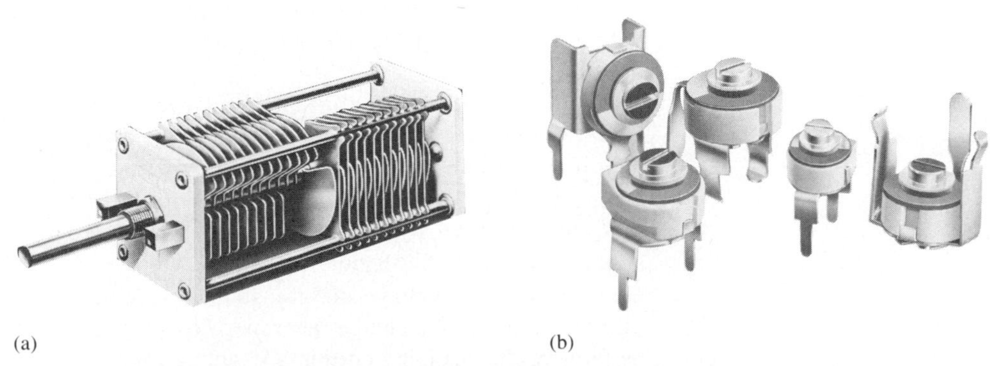

(a) (b)

FIG. 10.19
Variable air capacitors. (Part (a) courtesy of James Millen Manufacturing Co.; part (b) courtesy of Johnson Manufacturing Co.)

FIG. 10.20
Digital reading capacitance meter. (Courtesy of BK PRECISION, Maxtec International Corp.)

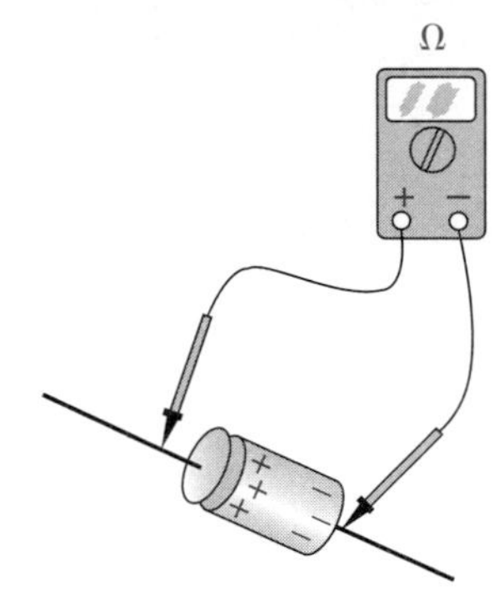

FIG. 10.21
Checking the dielectric of an electrolytic capacitor.

Measurement and Testing

A digital reading capacitance meter appears in Fig. 10.20; it displays the level of capacitance by simply placing the capacitor between the provided clips with the proper polarity.

The best check of a capacitor is to use a meter designed to perform the necessary tests. However, an ohmmeter can identify those in which the dielectric has deteriorated (especially in paper and electrolytic capacitors). As the dielectric breaks down, the insulating qualities decrease to a point where the resistance between the plates drops to a relatively low level. After ensuring that the capacitor is fully discharged, place an ohmmeter across the capacitor, as shown in Fig. 10.21. In a polarized capacitor, the polarities of the meter should match those of the capacitor. A low resistance reading (zero ohms to a few hundred ohms) normally indicates a defective capacitor.

The above test of leakage is not all-inclusive, since some capacitors will break down only when higher voltages are applied. The test, however, does identify those capacitors that have lost the insulating quality of the dielectric between the plates.

Standard Values and Recognition Factor

The standard values for capacitors employ the same numerical multipliers encountered for resistors. The most common have the same numerical multipliers as the most common resistors, that is, those available with the full range of tolerances (5%, 10%, and 20%) as shown in Table 3.8. They include **0.1** μF, **0.15** μF, **0.22** μF, **0.33** μF, **0.47** μF, and **0.68** μF and then **1** μF, **1.5** μF, **2.2** μF, **3.3** μF, **4.7** μF, and so on.

Figure 10.22 was developed to establish a recognition factor when it comes to the various types of capacitors. In other words, it will help you to develop the skills to identify types of capacitors, their typical range of values, and some of the most common applications. The figure is certainly not inclusive, but it does offer a first step in establishing a sense for what to expect for various applications.

10.7 TRANSIENTS IN CAPACITIVE NETWORKS: CHARGING PHASE

Section 10.3 described how a capacitor acquires its charge. Let us now extend this discussion to include the potentials and current developed within the network of Fig. 10.23 following the closing of the switch (to position 1).

You will recall that the instant the switch is closed, electrons are drawn from the top plate and deposited on the bottom plate by the battery, resulting in a net positive charge on the top plate, and a negative charge on the bottom plate. The transfer of electrons is very rapid at first, slowing down as the potential across the capacitor approaches the applied voltage of the battery. When the voltage across the capacitor equals the battery voltage, the transfer of electrons will cease and the plates will have a net charge determined by $Q = CV_C = CE$.

Plots of the changing current and voltage appear in Figs. 10.24 and 10.25, respectively. When the switch is closed at $t = 0$ s, the current jumps to a value limited only by the resistance of the network and then decays to zero as the plates are charged. Note the rapid decay in current

Type: Miniature Axial Electrolytic
Typical Values: 0.1 μF to 15,000 μF
Typical Voltage Range: 5 V to 450 V
Capacitor tolerance: ±20%
Description: Polarized, used in DC power supplies, bypass filters, DC blocking.

Type: Miniature Radial Electrolyte
Typical Values: 0.1 μF to 15,000 μF
Typical Voltage Range: 5 V to 450 V
Capacitor tolerance: ±20%
Description: Polarized, used in DC power supplies, bypass filters, DC blocking.

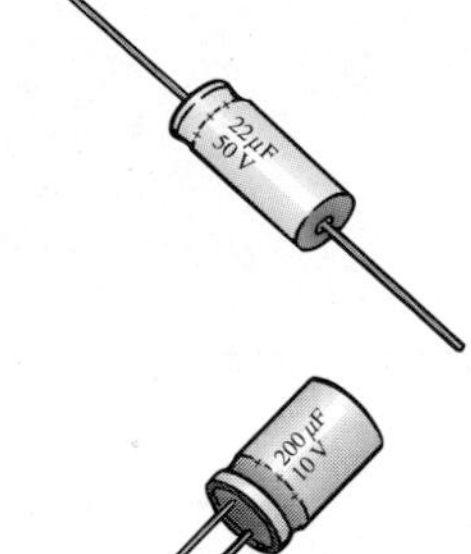

Type: Ceramic Disc
Typical Values: 10 pF to 0.047 μF
Typical Voltage Range: 100 V to 6 kV
Capacitor tolerance: ±5%, ±10%
Description: Non-polarized, NPO type, stable for a wide range of temperatures. Used in oscillators, noise filters, circuit coupling, tank circuits.

Type: Dipped Tantalum
Typical Values: 0.047 μF to 470 μF
Typical Voltage Range: 6.3 V to 50 V
Capacitor tolerance: ±10%, ±20%
Description: Polarized, low leakage current, used in power supplies, high frequency noise filters, bypass filter.

Type: Surface Mount Type (SMT)
Typical Values: 10 pF to 10 μF
Typical Voltage Range: 6.3 V to 16 V
Capacitor tolerance: ±10%
Description: Polarized and non-polarized, used in all types of circuits, requires a minimum amount of PC board real estate.

Type: Silver Mica
Typical Value: 10 pF to 0.001 μF
Typical Voltage Range: 50 V to 500 V
Capacitor tolerance: ±5%
Description: Non-polarized, used in oscillators, in circuits that require a stable component over a range of temperatures and voltages.

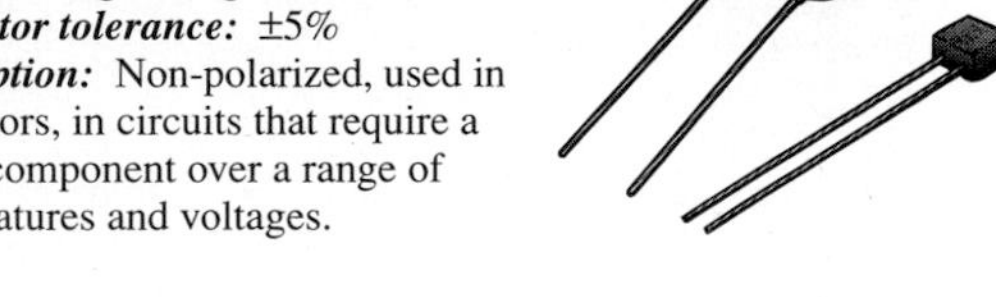

Type: Mylar Paper
Typical Value: 0.001 μF to 0.68 μF
Typical Voltage Range: 50 V to 600 V
Capacitor tolerance: ±22%
Description: Non-polarized, used in all types of circuits, moisture resistant.

Type: AC/DC Motor Run
Typical Value: 0.25 μF to 1200 μF
Typical Voltage Range: 240 V to 660 V
Capacitor tolerance: ±10%
Description: Non-polarized, used in motor run - start, high intensity lighting supplies, AC noise filtering.

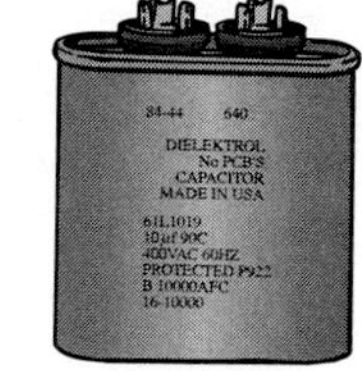

Type: Trimmer Variable
Typical Value: 1.5 pF to 600 pF
Typical Voltage Range: 5 V to 100 V
Capacitor tolerance: ±10%
Description: Non-polarized, used in oscillators, tuning circuits, AC filters.

Type: Tuning variable
Typical Value: 10 pF to 600 pF
Typical Voltage Range: 5 V to 100 V
Capacitor tolerance: ±10%
Description: Non-polarized, used in oscillators, radio tuning circuit.

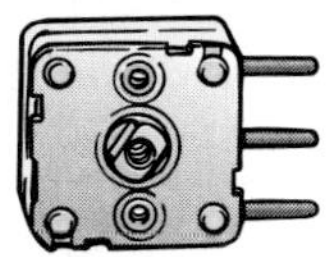

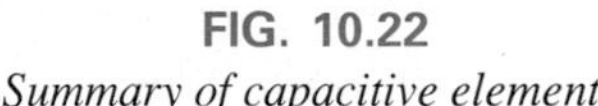

FIG. 10.22
Summary of capacitive elements.

level, revealing that the amount of charge deposited on the plates per unit time is rapidly decaying also. Since the voltage across the plates is directly related to the charge on the plates by $v_C = q/C$, the rapid rate with which charge is initially deposited on the plates will result in a rapid increase in v_C. Obviously, as the rate of flow of charge (I)

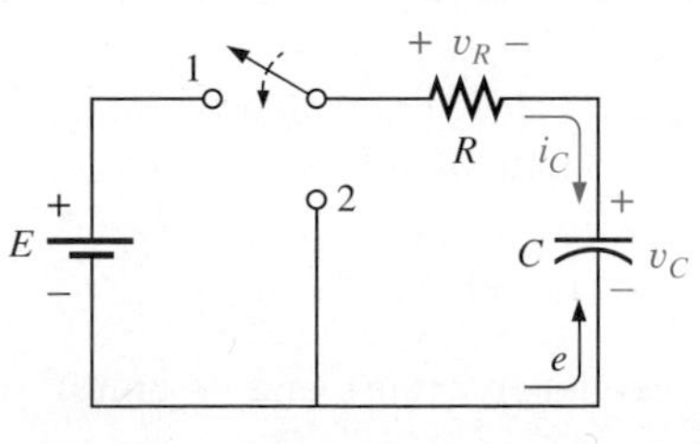

FIG. 10.23
Basic charging network.

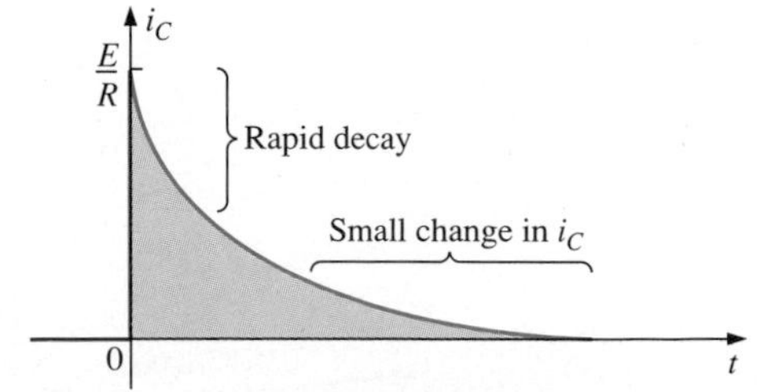

FIG. 10.24
i_C during the charging phase.

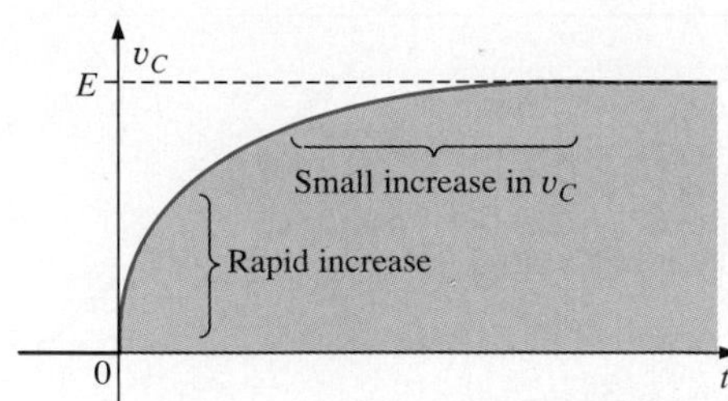

FIG. 10.25
v_C during the charging phase.

decreases, the rate of change in voltage will follow suit. Eventually, the flow of charge will stop, the current I will be zero, and the voltage will cease to change in magnitude—the *charging phase* has passed. At this point the capacitor takes on the characteristics of an open circuit: a voltage drop across the plates without a flow of charge "between" the plates. As demonstrated in Fig. 10.26, the voltage across the capacitor is the source voltage since $i = i_C = i_R = 0$ A and $v_R = i_R R = (0)R = 0$ V. For all future analysis:

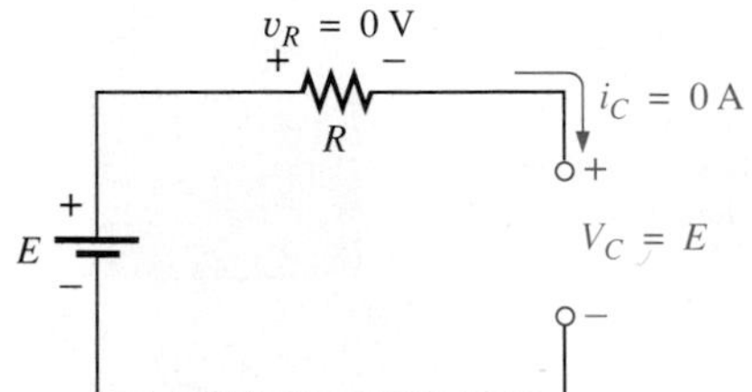

FIG. 10.26
Open-circuit equivalent for a capacitor following the charging phase.

A capacitor can be replaced by an open-circuit equivalent once the charging phase in a dc network has passed.

Looking back at the instant the switch is closed, we can also surmise that a capacitor behaves like a short circuit the moment the switch is closed in a dc charging network, as shown in Fig. 10.27. The current $i = i_C = i_R = E/R$, and the voltage $v_C = E - v_R = E - i_R R = E - (E/R)R = E - E = 0$ V at $t = 0$ s.

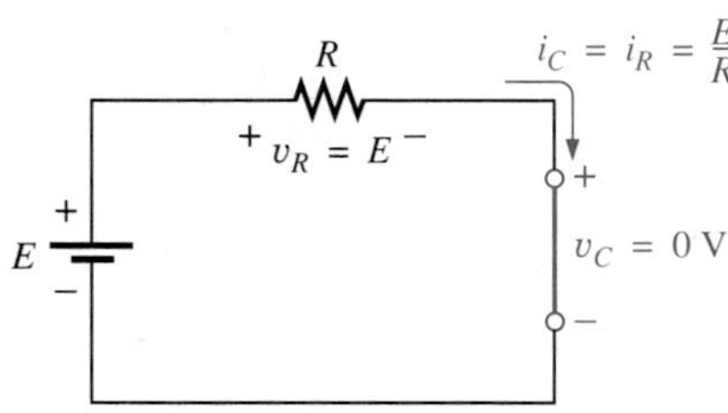

FIG. 10.27
Short-circuit equivalent for a capacitor (switch closed, $t = 0$).

Through the use of calculus, the following mathematical equation for the charging current i_C can be obtained:

$$i_C = \frac{E}{R}e^{-t/RC} \qquad \textbf{(10.13)}$$

The factor $e^{-t/RC}$ is an exponential function of the form e^{-x}, where $x = -t/RC$ and $e = 2.71828 \ldots$. A plot of e^{-x} for $x \geq 0$ appears in Fig. 10.28. Exponentials are mathematical functions that all students of electrical, electronic, or computer systems must become very familiar with. They will appear throughout the analysis to follow in this course, and in succeeding courses.

Our current interest in the function e^{-x} is limited to values of x greater than zero, as noted by the curve of Fig. 10.24. All modern-day scientific calculators have the function e^x. To obtain e^{-x}, the sign of x must be changed using the sign key before the exponential function is keyed in. The magnitude of e^{-x} has been listed in Table 10.3 for a range of values of x. Note the rapidly decreasing magnitude of e^{-x} with increasing value of x.

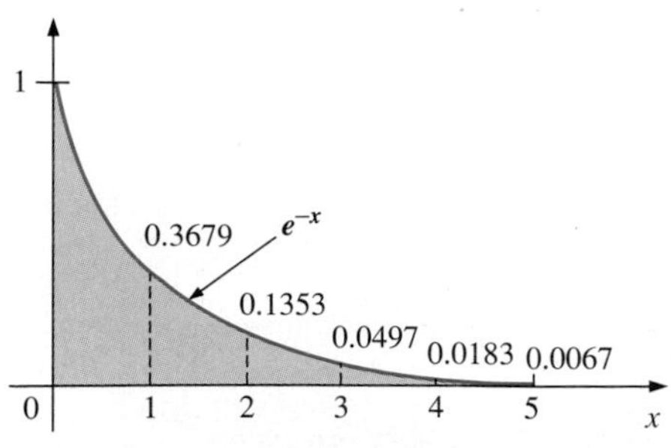

FIG. 10.28
The e^{-x} function ($x \geq 0$).

The factor RC in Eq. (10.13) is called the *time constant* of the system and has the units of time as follows:

$$RC = \left(\frac{V}{I}\right)\left(\frac{Q}{V}\right) = \left(\frac{\cancel{V}}{\cancel{Q}/t}\right)\left(\frac{\cancel{Q}}{\cancel{V}}\right) = t$$

Its symbol is the Greek letter τ (tau), and its unit of measure is the second. Thus,

$$\tau = RC \qquad \text{(seconds, s)} \qquad \textbf{(10.14)}$$

TABLE 10.3
Selected values of e^{-x}.

$x = 0$	$e^{-x} = e^{-0} = \frac{1}{e^0} = \frac{1}{1} = 1$
$x = 1$	$e^{-1} = \frac{1}{e} = \frac{1}{2.71828\ldots} = 0.3679$
$x = 2$	$e^{-2} = \frac{1}{e^2} = 0.1353$
$x = 5$	$e^{-5} = \frac{1}{e^5} = 0.00674$
$x = 10$	$e^{-10} = \frac{1}{e^{10}} = 0.0000454$
$x = 100$	$e^{-100} = \frac{1}{e^{100}} = 3.72 \times 10^{-44}$

If we substitute $\tau = RC$ into the exponential function $e^{-t/RC}$, we obtain $e^{-t/\tau}$. In one time constant, $e^{-t/\tau} = e^{-\tau/\tau} = e^{-1} = 0.3679$, or the function equals 36.79% of its maximum value of 1. At $t = 2\tau$, $e^{-t/\tau} = e^{-2\tau/\tau} = e^{-2} = 0.1353$, and the function has decayed to only 13.53% of its maximum value.

The magnitude of $e^{-t/\tau}$ and the percentage change between time constants have been tabulated in Tables 10.4 and 10.5, respectively. Note that the current has dropped 63.2% (100% − 36.8%) in the first time constant but only 0.4% between the fifth and sixth time constants. The rate of change of i_C is therefore quite sensitive to the time constant determined by the network parameters R and C. For this reason, the universal time constant chart of Fig. 10.29 is provided to permit a more

TABLE 10.4
i_C versus τ (charging phase).

t	Magnitude	
0	100%	
1τ	36.8%	
2τ	13.5%	
3τ	5.0%	
4τ	1.8%	
5τ	**0.67%** ←	Less than 1% of maximum
6τ	0.24%	

TABLE 10.5
Change in i_C between time constants.

$(0 \rightarrow 1)\tau$	63.2%
$(1 \rightarrow 2)\tau$	23.3%
$(2 \rightarrow 3)\tau$	8.6%
$(3 \rightarrow 4)\tau$	3.0%
$(4 \rightarrow 5)\tau$	**1.2%**
$(5 \rightarrow 6)\tau$	0.4% ← Less than 1%

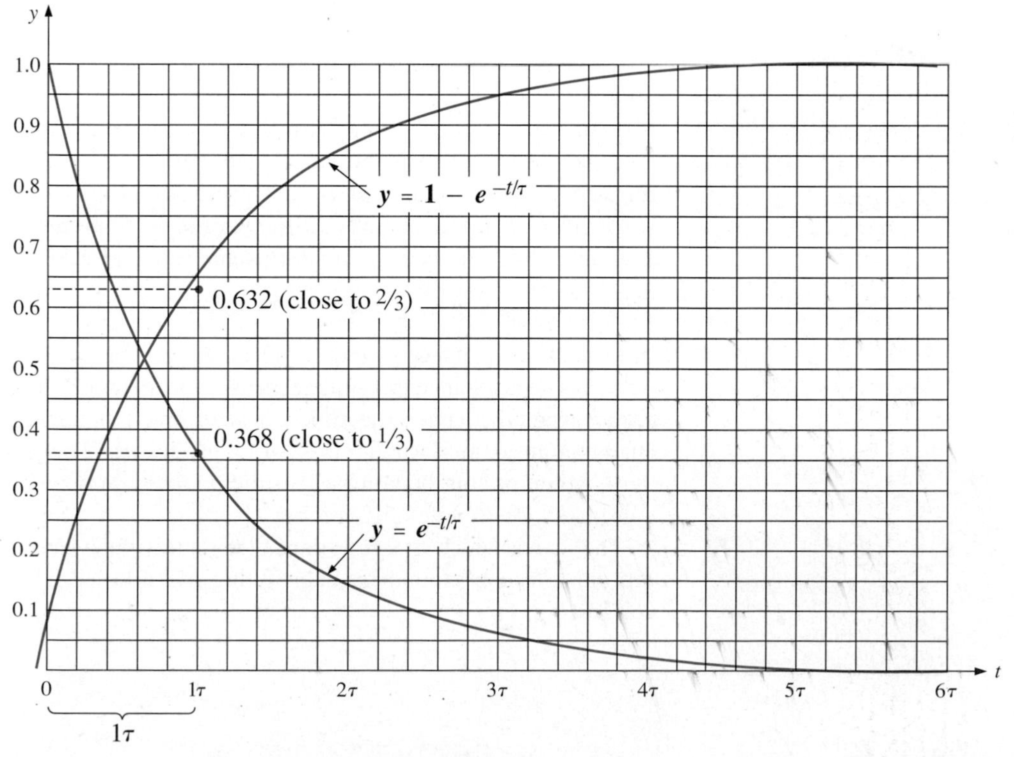

FIG. 10.29
Universal time constant chart.

accurate estimate of the value of the function e^{-x} for specific time intervals related to the time constant. The term *universal* is used because the axes are not scaled to specific values.

Returning to Eq. (10.13), we find that the multiplying factor E/R is the maximum value the current i_C can attain, as shown in Fig. 10.24. Substituting $t = 0$ s into Eq. (10.13) yields

$$i_C = \frac{E}{R}e^{-t/RC} = \frac{E}{R}e^{-0} = \frac{E}{R}$$

verifying our earlier conclusion.

For increasing values of t, the magnitude of $e^{-t/\tau}$, and therefore the value of i_C, will decrease, as shown in Fig. 10.30. Since the magnitude of i_C is less than 1% of its maximum after five time constants, we will assume for future analysis that:

The current i_C of a capacitive network is essentially zero after five time constants of the charging phase have passed in a dc network.

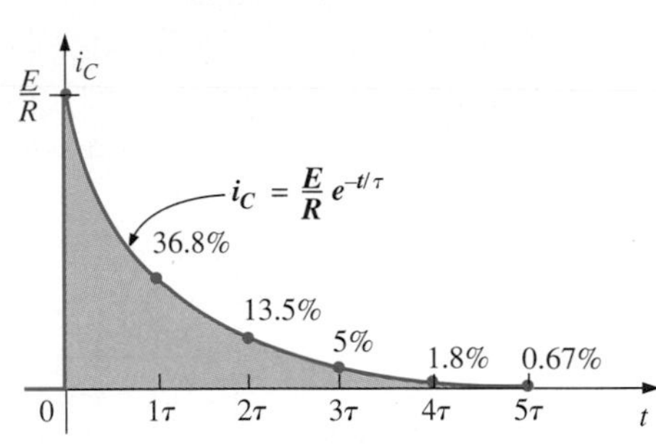

FIG. 10.30
i_C versus t during the charging phase.

Since C is usually found in microfarads or picofarads, the time constant $\tau = RC$ will never be greater than a few seconds unless R is very large.

Let us now turn our attention to the charging voltage across the capacitor. Through further mathematical analysis, the following equation for the voltage across the capacitor can be determined:

$$v_C = E(1 - e^{-t/RC}) \quad \textbf{(10.15)}$$

Note the presence of the same factor $e^{-t/RC}$ and the function $(1 - e^{-t/RC})$ appearing in Fig. 10.29. Since $e^{-t/\tau}$ is a decaying function, the factor $(1 - e^{-t/\tau})$ will grow toward a maximum value of 1 with time, as shown in Fig. 10.29. In addition, since E is the multiplying factor, we can conclude that, for all practical purposes, the voltage v_C is E volts after five time constants of the charging phase. A plot of v_C versus t is provided in Fig. 10.31.

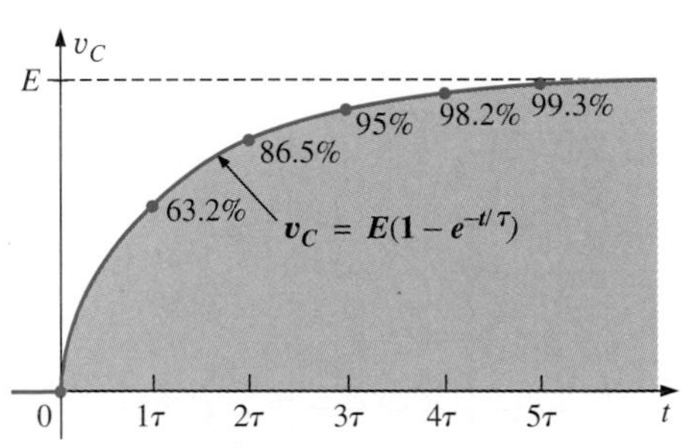

FIG. 10.31
v_C versus t during the charging phase.

If we keep R constant and reduce C, the product RC will decrease, and the rise time of five time constants will decrease. The change in transient behavior of the voltage v_C is plotted in Fig. 10.32 for various values of C. The product RC will always have some numerical value, even though it may be very small in some cases. For this reason:

The voltage across a capacitor cannot change instantaneously.

In fact, the capacitance of a network is also a measure of how much it will oppose a change in voltage across the network. The larger the capacitance, the larger the time constant and the longer it takes to charge up to its final value (curve of C_3 in Fig. 10.32). A lesser capacitance would permit the voltage to build up more quickly since the time constant is less (curve of C_1 in Fig. 10.32).

The rate at which charge is deposited on the plates during the charging phase can be found by substituting the following for v_C in Eq. (10.15):

$$v_C = \frac{q}{C}$$

and

$$q = Cv_C = CE(1 - e^{-t/\tau}) \quad \text{charging} \quad \textbf{(10.16)}$$

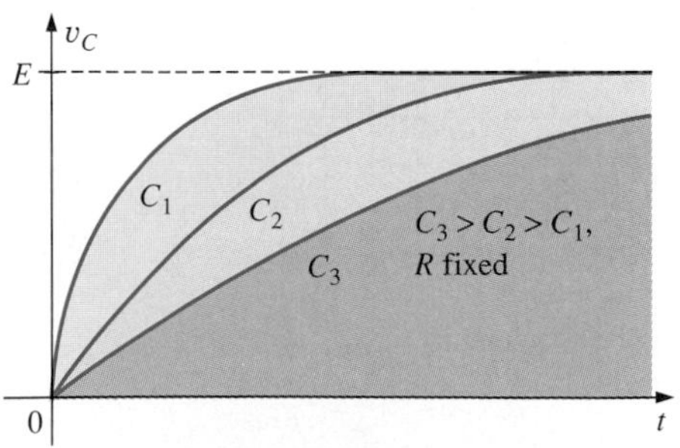

FIG. 10.32
Effect of C on the charging phase.

indicating that the charging rate is very high during the first few time constants and less than 1% after five time constants.

The voltage across the resistor is determined by Ohm's law:

$$v_R = i_R R = Ri_C = R\frac{E}{R}e^{-t/\tau}$$

or

$$\boxed{v_R = Ee^{-t/\tau}} \qquad \textbf{(10.17)}$$

A plot of v_R appears in Fig. 10.33.

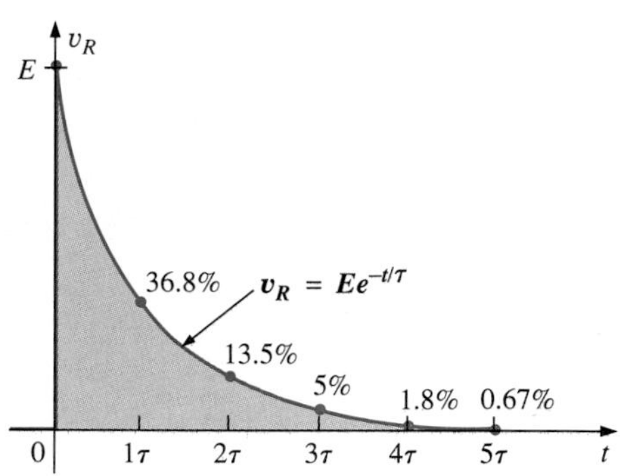

FIG. 10.33
v_R versus t during the charging phase.

Applying Kirchhoff's voltage law to the circuit of Fig. 10.23 will result in

$$v_C = E - v_R$$

Substituting Eq. (10.17):

$$v_C = E - Ee^{-t/\tau}$$

Factoring gives $v_C = E(1 - e^{-t/\tau})$, as obtained earlier.

EXAMPLE 10.5

a. Find the mathematical expressions for the transient behavior of v_C, i_C, and v_R for the circuit of Fig. 10.34 when the switch is moved to position 1. Plot the curves of v_C, i_C, and v_R.
b. How much time must pass before it can be assumed, for all practical purposes, that $i_C \cong 0$ A and $v_C \cong E$ volts?

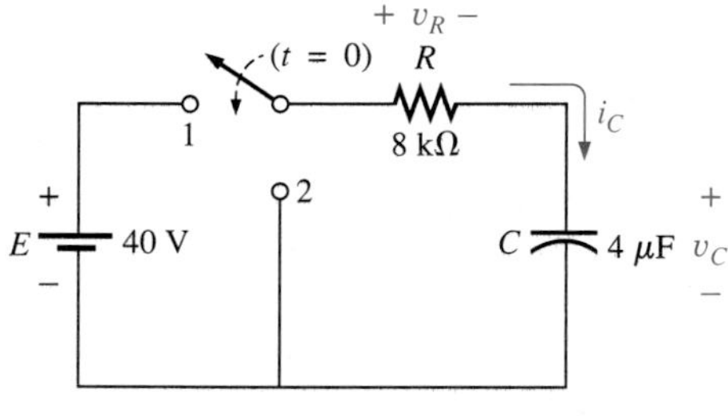

FIG. 10.34
Example 10.5.

Solutions:

a. $\tau = RC = (8 \times 10^3\ \Omega)(4 \times 10^{-6}\ \text{F}) = 32 \times 10^{-3}\ \text{s} = \mathbf{32\ ms}$
By Eq. (10.15),

$$v_C = E(1 - e^{-t/\tau}) = \mathbf{40(1 - e^{-t/(32\times10^{-3})})}$$

By Eq. (10.13),

$$i_C = \frac{E}{R}e^{-t/\tau} = \frac{40\ \text{V}}{8\ \text{k}\Omega}e^{-t/(32\times10^{-3})}$$
$$= \mathbf{(5 \times 10^{-3})e^{-t/(32\times10^{-3})}}$$

By Eq. (10.17),

$$v_R = Ee^{-t/\tau} = \mathbf{40e^{-t/(32\times10^{-3})}}$$

The curves appear in Fig. 10.35.

b. $5\tau = 5(32\ \text{ms}) = \mathbf{160\ ms}$

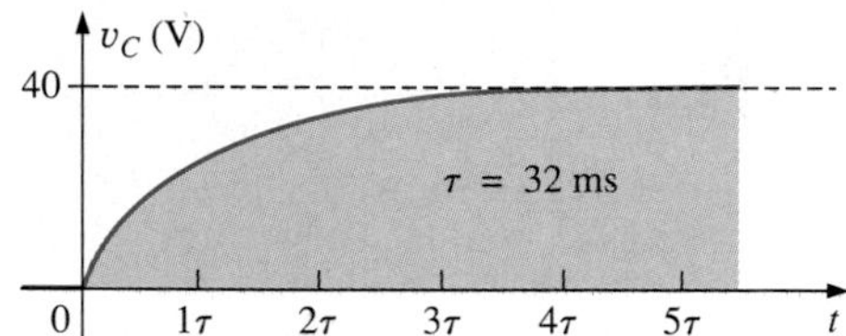

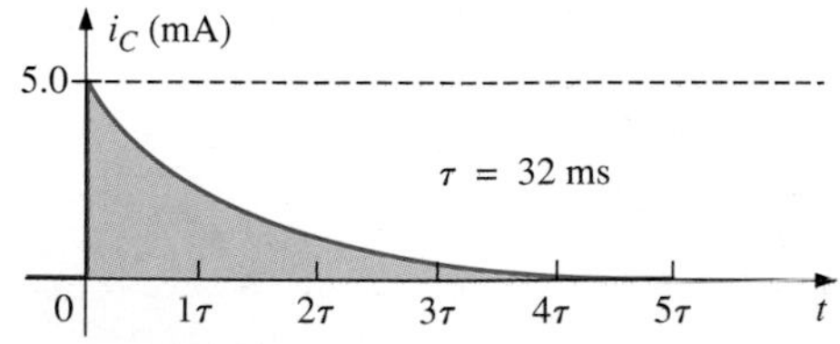

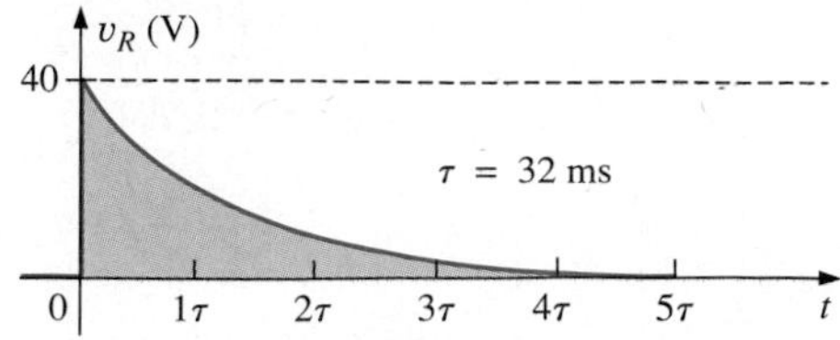

FIG. 10.35
Waveforms for the network of Fig. 10.34.

Once the voltage across the capacitor has reached the input voltage *E*, the capacitor is fully charged and will remain in this state if no further changes are made in the circuit.

If the switch of Fig. 10.23 is opened, as shown in Fig. 10.36(a), the capacitor will retain its charge for a period of time determined by its leakage current. For capacitors such as the mica and ceramic, the leakage current ($i_{\text{leakage}} = v_C/R_{\text{leakage}}$) is very small, enabling the capacitor to retain its charge, and hence the potential difference across its plates, for a long time. For electrolytic capacitors, which have very high leakage currents, the capacitor will discharge more rapidly, as shown in Fig. 10.36(b). In any event, to ensure that they are completely discharged,

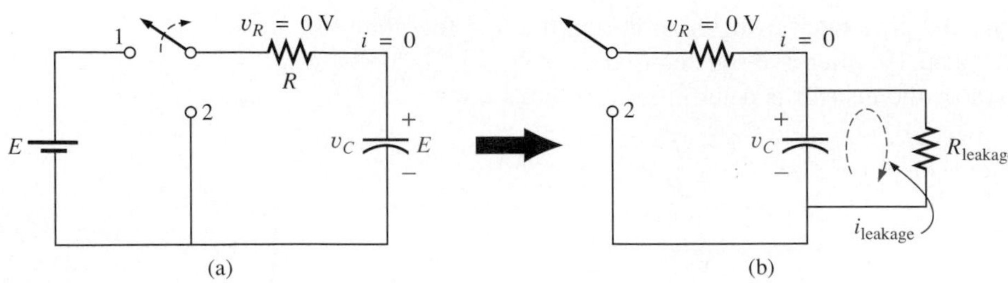

FIG. 10.36
Effect of the leakage current on the steady-state behavior of a capacitor.

capacitors should be shorted by a lead or a screwdriver before they are handled.

10.8 DISCHARGE PHASE

The network of Fig. 10.23 is designed to both charge and discharge the capacitor. When the switch is placed in position 1, the capacitor will charge toward the supply voltage, as described in the last section. At any point in the charging process, if the switch is moved to position 2, the capacitor will begin to discharge at a rate sensitive to the same time constant $\tau = RC$. The established voltage across the capacitor will create a flow of charge in the closed path that will eventually discharge the capacitor completely. In essence, the capacitor functions like a battery with a decreasing terminal voltage. Note in particular that the current i_C has reversed direction, changing the polarity of the voltage across R.

If the capacitor had charged to the full battery voltage as indicated in Fig. 10.37, the equation for the decaying voltage across the capacitor would be the following:

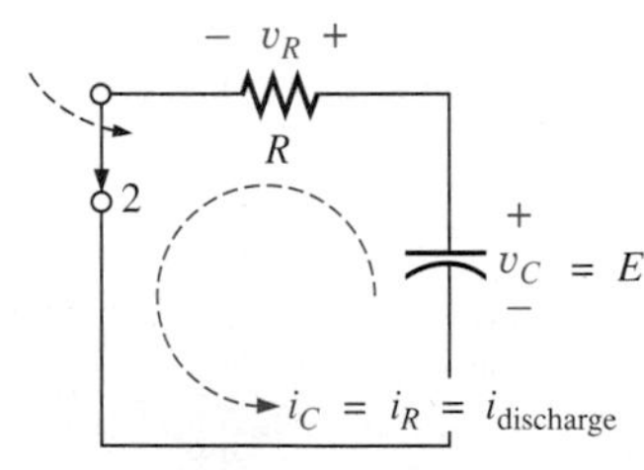

FIG. 10.37
Demonstrating the discharge behavior of a capacitive network.

$$v_C = Ee^{-t/RC} \quad \textit{discharging} \tag{10.18}$$

which employs the function e^{-x} and the same time constant used above. The resulting curve will have the same shape as the curve for i_C and v_R in the last section. During the discharge phase, the current i_C will also decrease with time, as defined by the following equation:

$$i_C = \frac{E}{R}e^{-t/RC} \quad \textit{discharging} \tag{10.19}$$

The voltage $v_R = v_C$, and

$$v_R = Ee^{-t/RC} \quad \textit{discharging} \tag{10.20}$$

The complete discharge will occur, for all practical purposes, in five time constants. If the switch is moved between terminals 1 and 2 every five time constants, the waveshapes of Fig. 10.38 will result for v_C, i_C, and v_R. For each curve, the current direction and voltage polarities were defined by Fig. 10.23. Since the polarity of v_C is the same for both the charging and discharging phases, the entire curve lies above the axis. The current i_C reverses direction during the charging and discharging phases, producing a negative pulse for both the current and the voltage

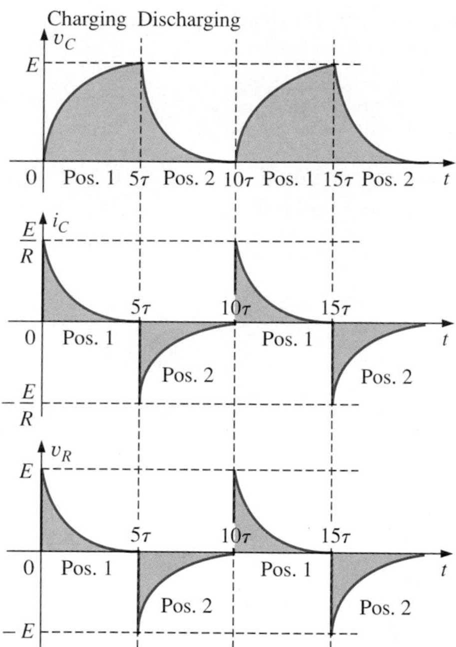

FIG. 10.38
The charging and discharging cycles for the network of Fig. 10.23.

v_R. Note that the voltage v_C never changes magnitude instantaneously but that the current i_C has the ability to change instantaneously, as demonstrated by its vertical rises and drops to maximum values.

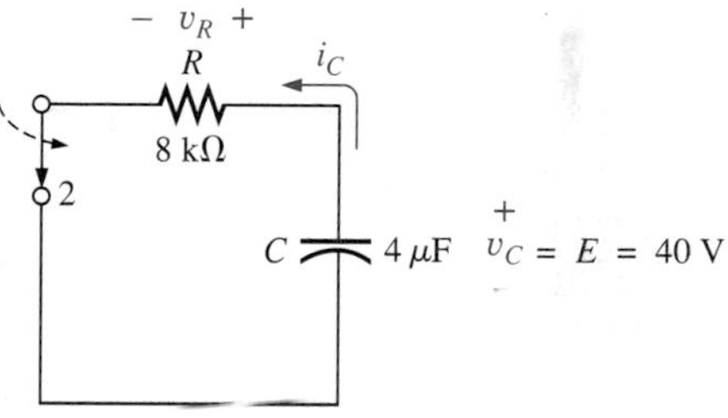

FIG. 10.39
Example 10.6.

EXAMPLE 10.6 After v_C in Example 10.5 has reached its final value of 40 V, the switch is thrown into position 2, as shown in Fig. 10.39. Find the mathematical expressions for the transient behavior of v_C, i_C, and v_R after the closing of the switch. Plot the curves for v_C, i_C, and v_R using the defined directions and polarities of Fig. 10.34. Assume $t = 0$ when the switch is moved to position 2.

Solution:

$$\tau = 32 \text{ ms}$$

By Eq. (10.18),

$$v_C = Ee^{-t/\tau} = \mathbf{40}e^{-t/(32\times10^{-3})}$$

By Eq. (10.19),

$$i_C = -\frac{E}{R}e^{-t/\tau} = -(\mathbf{5\times10^{-3}})e^{-t/(32\times10^{-3})}$$

By Eq. (10.20),

$$v_R = -Ee^{-t/\tau} = -\mathbf{40}e^{-t/(32\times10^{-3})}$$

The curves appear in Fig. 10.40.

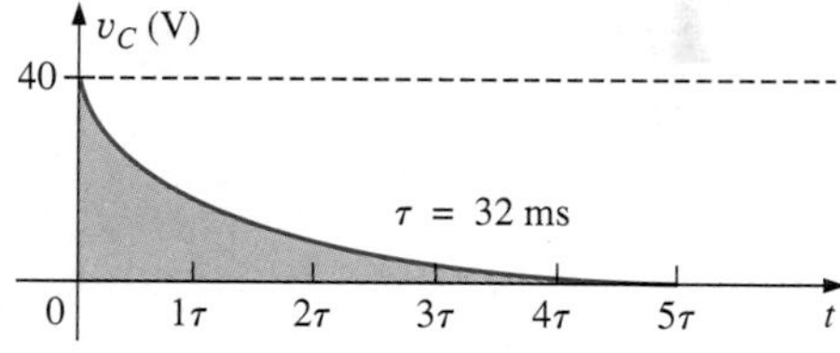

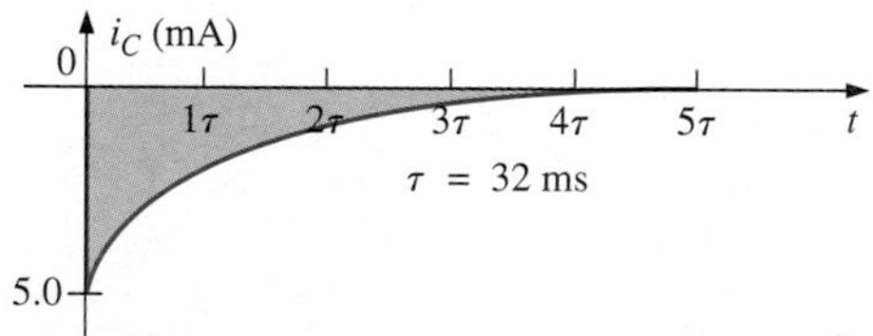

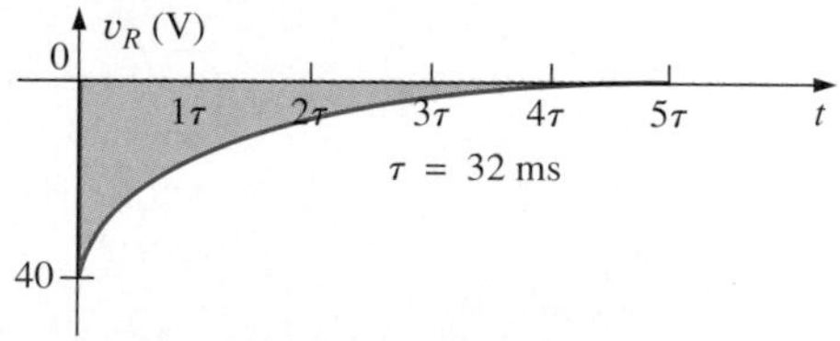

FIG. 10.40
The waveforms for the network of Fig. 10.39.

The preceding discussion and examples apply to situations in which the capacitor charges to the battery voltage. If the charging phase is disrupted before reaching the supply voltage, the capacitive voltage will be

less, and the equation for the discharging voltage v_C will take on the form

$$v_C = V_i e^{-t/RC} \tag{10.21}$$

where V_i is the starting or initial voltage for the discharge phase. The equation for the decaying cusrrent is also modified by simply substituting V_i for E; that is,

$$i_C = \frac{V_i}{R} e^{-t/\tau} = I_i e^{-t/\tau} \tag{10.22}$$

Use of the above equations will be demonstrated in Examples 10.7 and 10.8.

EXAMPLE 10.7

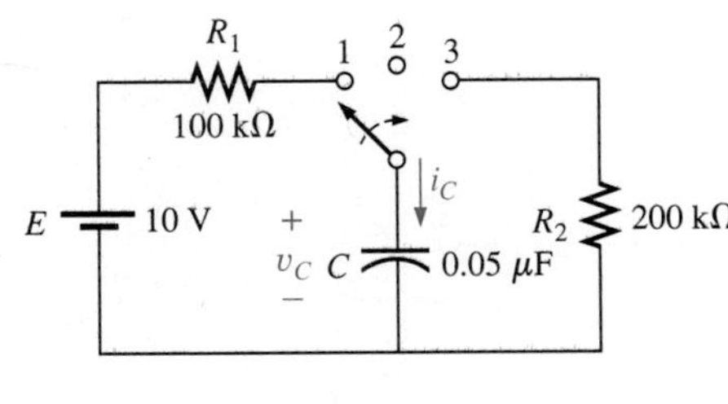

FIG. 10.41
Example 10.7.

a. Find the mathematical expression for the transient behavior of the voltage across the capacitor of Fig. 10.41 if the switch is thrown into position 1 at $t = 0$ s.
b. Repeat part (a) for i_C.
c. Find the mathematical expressions for the response of v_C and i_C if the switch is thrown into position 2 at 30 ms (assuming the leakage resistance of the capacitor is infinite ohms).
d. Find the mathematical expressions for the voltage v_C and current i_C if the switch is thrown into position 3 at $t = 48$ ms.
e. Plot the waveforms obtained in parts (a) through (d) on the same time axis for the voltage v_C and the current i_C using the defined polarity and current direction of Fig. 10.41.

a. Charging phase:

$$v_C = E(1 - e^{-t/\tau})$$
$$\tau = R_1 C = (100 \times 10^3\ \Omega)(0.05 \times 10^{-6}\ \text{F}) = 5 \times 10^{-3}\ \text{s} = 5\ \text{ms}$$
$$v_C = \mathbf{10(1 - e^{-t/(5\times10^{-3})})}$$

b. $$i_C = \frac{E}{R_1} e^{-t/\tau}$$
$$= \frac{10\ \text{V}}{100 \times 10^3\ \Omega} e^{-t/(5\times10^{-3})}$$
$$i_C = \mathbf{(0.1 \times 10^{-3})e^{-t/(5\times10^{-3})}}$$

c. Storage phase:

$$v_C = E = \mathbf{10\ V}$$
$$i_C = \mathbf{0\ A}$$

d. Discharge phase (starting at 48 ms with $t = 0$ s for the following equations):

$$v_C = Ee^{-t/\tau'}$$
$$\tau' = R_2 C = (200 \times 10^3\ \Omega)(0.05 \times 10^{-6}\ \text{F}) = 10 \times 10^{-3}\ \text{s} = 10\ \text{ms}$$
$$v_C = \mathbf{10e^{-t/(10\times10^{-3})}}$$

$$i_C = -\frac{E}{R_2}e^{-t/\tau'}$$

$$= -\frac{10\text{ V}}{200 \times 10^3\ \Omega}e^{-t/(10\times10^{-3})}$$

$$i_C = \mathbf{-(0.05 \times 10^{-3})e^{-t/(10\times10^{-3})}}$$

e. See Fig. 10.42.

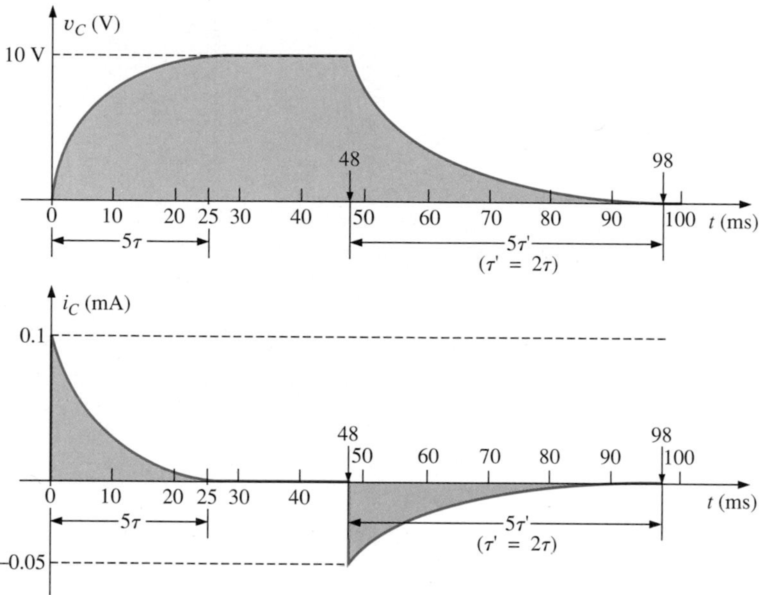

FIG. 10.42
The waveforms for the network of Fig. 10.41.

EXAMPLE 10.8

a. Find the mathematical expression for the transient behavior of the voltage across the capacitor of Fig. 10.43 if the switch is thrown into position 1 at $t = 0$ s.

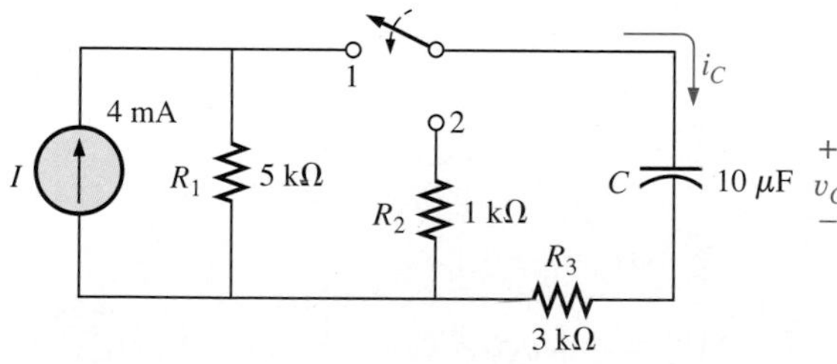

FIG. 10.43
Example 10.8.

b. Repeat part (a) for i_C.

c. Find the mathematical expression for the response of v_C and i_C if the switch is thrown into position 2 at $t = 1\tau$ of the charging phase.

d. Plot the waveforms obtained in parts (a) through (c) on the same time axis for the voltage v_C and the current i_C using the defined polarity and current direction of Fig. 10.43.

Solutions:

a. *Charging phase:* Converting the current source to a voltage source will result in the network of Fig. 10.44.

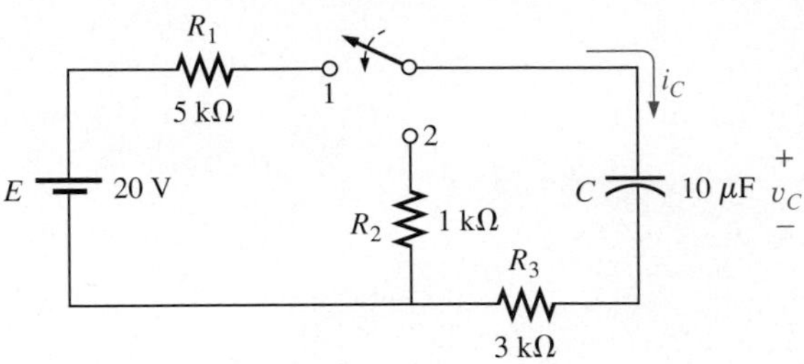

FIG. 10.44
The charging phase for the network of Fig. 10.43.

$$v_C = E(1 - e^{-t/\tau_1})$$
$$\tau_1 = (R_1 + R_3)C = (5\text{ k}\Omega + 3\text{ k}\Omega)(10 \times 10^{-6}\text{ F}) = 80\text{ ms}$$
$$v_C = \mathbf{20(1 - e^{-t/(80\times 10^{-3})})}$$

b. $$i_C = \frac{E}{R_1 + R_3}e^{-t/\tau_1} = \frac{20\text{ V}}{8\text{ k}\Omega}e^{-t/(80\times 10^{-3})}$$
$$i_C = \mathbf{(2.5 \times 10^{-3})e^{-t/(80\times 10^{-3})}}$$

c. At $t = 1\tau_1$, $v_C = 0.632E = 0.632(20\text{ V}) = 12.64\text{ V}$, resulting in the network of Fig. 10.45. Then $v_C = V_i e^{-t/\tau_2}$

with $$\tau_2 = (R_2 + R_3)C = (1\text{ k}\Omega + 3\text{ k}\Omega)(10 \times 10^{-6}\text{ F}) = 40\text{ ms}$$

and $$v_C = \mathbf{12.64e^{-t/(40\times 10^{-3})}}$$

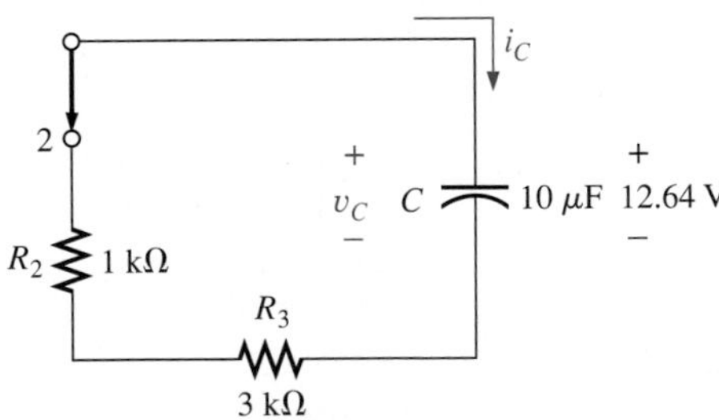

FIG. 10.45
Network of Fig. 10.44 when the switch is moved to position 2 at $t = 1\tau_1$.

At $t = 1\tau_1$, i_C drops to $(0.368)(2.5\text{ mA}) = 0.92\text{ mA}$. Then it switches to

$$i_C = -I_i e^{-t/\tau_2}$$

$$= -\frac{V_i}{R_2 + R_3} e^{-t/\tau_2} = -\frac{12.64\ \text{V}}{1\ \text{k}\Omega + 3\ \text{k}\Omega} e^{-t/(40\times 10^{-3})}$$

$$i_C = \mathbf{-3.16 \times 10^{-3} e^{-t/(40\times 10^{-3})}}$$

d. See Fig. 10.46.

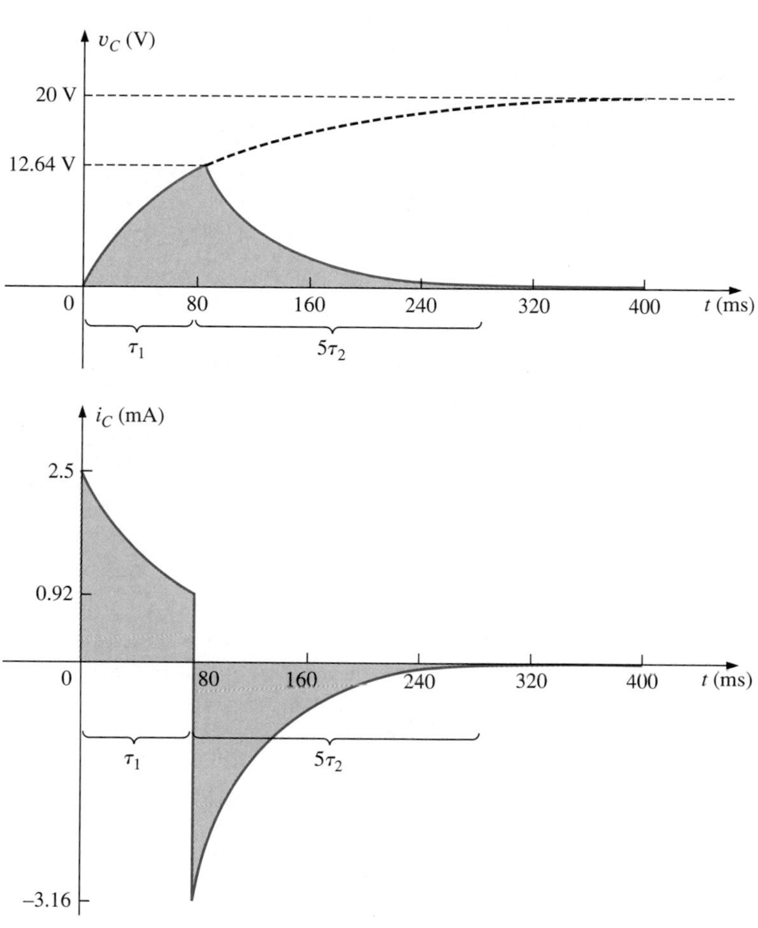

FIG. 10.46
The waveforms for the network of Fig. 10.43.

10.9 INITIAL VALUES

In all the examples examined in the previous sections, the capacitor was uncharged before the switch was thrown. We will now examine the effect of a charge, and therefore a voltage ($V = Q/C$), on the plates at the instant the switching action takes place. The voltage across the capacitor at this instant is called the *initial* value, as shown for the general waveform of Fig. 10.47. Once the switch is thrown, the transient phase will commence until a leveling off occurs after 5 time constants. This region of relatively fixed value that follows the transient response is called the *steady-state* region, and the resulting value is called the *steady-state* or *final* value. The steady-state value is found by simply substituting the open-circuit equivalent for the capacitor and finding the voltage across the plates. Using the transient equation developed in the

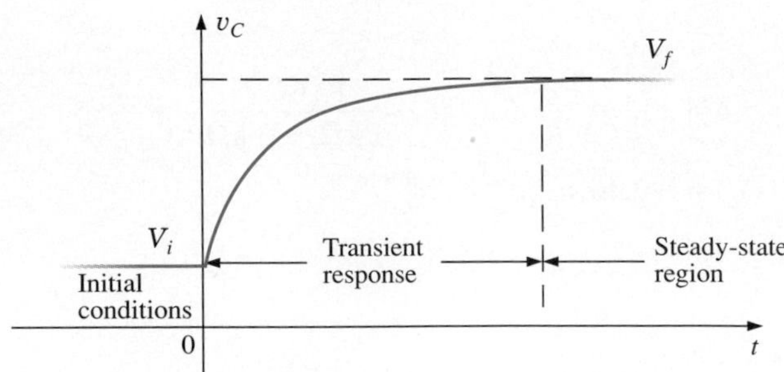

FIG. 10.47
Defining the regions associated with a transient response.

previous section, an equation for the voltage v_C can be written for the entire time interval of Fig. 10.47; that is,

$$v_C = V_i + (V_f - V_i)(1 - e^{-t/\tau})$$

However, by multiplying through and rearranging terms:

$$\begin{aligned} v_C &= V_i + V_f - V_f e^{-t/\tau} - V_i + V_i e^{-t/\tau} \\ &= V_f - V_f e^{-t/\tau} + V_i e^{-t/\tau} \end{aligned}$$

we find

$$v_C = V_f + (V_i - V_f)e^{-t/\tau} \tag{10.23}$$

If you are required to draw the waveform for the voltage v_C from the initial value to the final value, start by drawing a line at the initial and steady-state levels and then add the transient response (sensitive to the time constant) between the two levels. The example to follow will clarify the procedure.

EXAMPLE 10.9 The capacitor of Fig. 10.48 has an initial voltage of 4 volts.

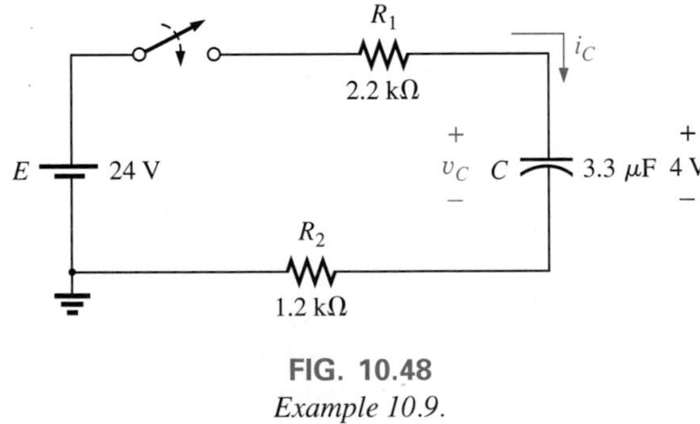

FIG. 10.48
Example 10.9.

a. Find the mathematical expression for the voltage across the capacitor once the switch is closed.
b. Find the mathematical expression for the current during the transient period.
c. Sketch the waveform for each from initial value to final value.

Solutions:

a. Substituting the open-circuit equivalent for the capacitor will result in a final or steady-state voltage υ_C of 24 V.

The time constant is determined by

$$\tau = (R_1 + R_2)C$$
$$= (2.2\text{ k}\Omega + 1.2\text{ k}\Omega)(3.3\ \mu\text{F})$$
$$= 11.22\text{ ms}$$

with $5\tau = 56.1\text{ ms}$

Applying Eq. (10.23):

$$\upsilon_C = V_f + (V_i - V_f)e^{-t/\tau}$$
$$= 24\text{ V} + (4\text{ V} - 24\text{ V})e^{-t/11.22\text{ms}}$$

and $\upsilon_C = \mathbf{24\ V - 20\ V}e^{\mathbf{-t/11.22ms}}$

b. Since the voltage across the capacitor is constant at 4 V prior to the closing of the switch, the current (whose level is sensitive only to changes in voltage across the capacitor) must have an initial value of 0 mA. At the instant the switch is closed, the voltage across the capacitor cannot change instantaneously, so the voltage across the resistive elements at this instant is the applied voltage less the initial voltage across the capacitor. The resulting peak current is

$$I_m = \frac{E - V_C}{R_1 + R_2} = \frac{24\text{ V} - 4\text{ V}}{2.2\text{ k}\Omega + 1.2\text{ k}\Omega} = \frac{20\text{ V}}{3.4\text{ k}\Omega} = 5.88\text{ mA}$$

The current will then decay (with the same time constant as the voltage υ_C) to zero because the capacitor is approaching its open-circuit equivalence.

The equation for i_C is therefore:

$$i_C = \mathbf{(5.88\ mA)}e^{\mathbf{-t/11.22ms}}$$

c. See Fig. 10.49.

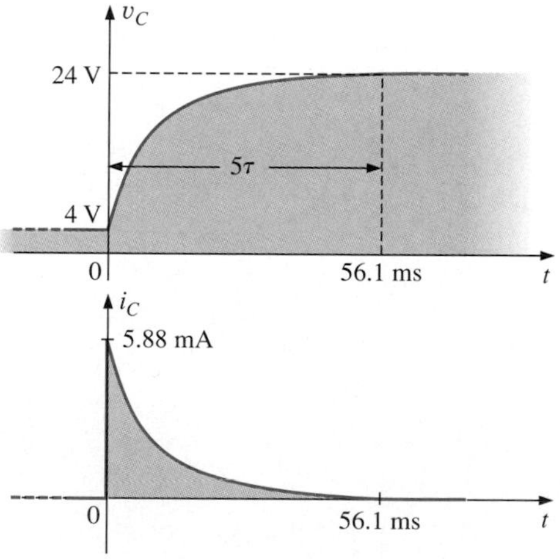

FIG. 10.49
υ_C and i_C for the network of Fig. 10.48.

The initial and final values of the voltage were drawn first and then the transient response was included between these levels. For the current, the waveform begins and ends at zero, with the peak

value having a sign sensitive to the defined direction of i_C in Fig. 10.48.

Let us now test the validity of the equation for v_C by substituting $t = 0$ s to reflect the instant the switch is closed.

$$e^{-t/\tau} = e^{-0} = 1$$

and $\quad v_C = 24\text{ V} - 20\text{ V}e^{-t/\tau} = 24\text{ V} - 20\text{ V} = 4\text{ V}$

When $\quad t > 5\tau,\ e^{-t/\tau} \cong 0$

and $\quad v_C = 24\text{ V} - 20\text{ V}e^{-t/\tau} = 24\text{ V} - 0\text{ V} = 24\text{ V}$

10.10 INSTANTANEOUS VALUES

On occasion it will be necessary to determine the voltage or current at a particular instant of time that is not an integral multiple of τ, as in the previous sections. For example, if

$$v_C = 20(1 - e^{-t/(2\times10^{-3})})$$

the voltage v_C may be required at $t = 5$ ms, which does not correspond to a particular value of τ. Figure 10.29 reveals that $(1 - e^{-t/\tau})$ is approximately 0.93 at $t = 5\text{ ms} = 2.5\tau$, resulting in $v_C = 20(0.93) = 18.6$ V. Additional accuracy can be obtained simply by substituting $t = 5$ ms into the equation and solving for v_C using a calculator or table to determine $e^{-2.5}$. Thus,

$$\begin{aligned} v_C &= 20(1 - e^{-5\text{ms}/2\text{ms}}) \\ &= 20(1 - e^{-2.5}) \\ &= 20(1 - 0.082) \\ &= 20(0.918) \\ &= \mathbf{18.36\ V} \end{aligned}$$

The results are close, but accuracy beyond the tenths' place is suspect using Fig. 10.29. The above procedure can also be applied to any other equation introduced in this chapter for currents or other voltages.

There are also occasions when the time to reach a particular voltage or current is required. The procedure is complicated somewhat by the use of natural logs ($\log_e$, or ln), but today's calculators are equipped to handle the operation with ease. There are two forms that require some development. First, consider the following sequence:

$$v_C = E(1 - e^{-t/\tau})$$

$$\frac{v_C}{E} = 1 - e^{-t/\tau}$$

$$1 - \frac{v_C}{E} = e^{-t/\tau}$$

$$\log_e\left(1 - \frac{v_C}{E}\right) = \log_e e^{-t/\tau}$$

$$\log_e\left(1 - \frac{v_C}{E}\right) = -\frac{t}{\tau}$$

and $$t = -\tau \log_e\left(1 - \frac{v_C}{E}\right)$$

but $$-\log_e\frac{x}{y} = +\log_e\frac{y}{x}$$

Therefore,

$$t = \tau \log_e\left(\frac{E}{E - v_C}\right) \tag{10.24}$$

The second form is as follows:

$$v_C = Ee^{-t/\tau}$$

$$\frac{v_C}{E} = e^{-t/\tau}$$

$$\log_e\frac{v_C}{E} = \log_e e^{-t/\tau}$$

$$\log_e\frac{v_C}{E} = -\frac{t}{\tau}$$

and

$$t = -\tau \log_e \frac{v_C}{E}$$

or

$$t = \tau \log_e \frac{E}{v_C} \tag{10.25}$$

For $i_C = (E/R)e^{-t/\tau}$,

$$t = \tau \log_e \frac{E}{i_C R} \tag{10.26}$$

For example, suppose

$$v_C = 20(1 - e^{-t/(2\times10^{-3})})$$

and the time to reach 10 V is required. Substituting into Eq. (10.24), we have

$$\begin{aligned} t &= (2 \text{ ms})\log_e\left(\frac{20 \text{ V}}{20 \text{ V} - 10 \text{ V}}\right) \\ &= (2 \text{ ms})\log_e 2 \\ &= (2 \text{ ms})(0.693) \quad \leftarrow \boxed{\text{ln}} \text{ key on calculator} \\ &= \mathbf{1.386 \text{ ms}} \end{aligned}$$

Using Fig. 10.29, we find at $(1 - e^{-t/\tau}) = v_C/E = 0.5$ that $t \cong 0.7\tau = 0.7(2 \text{ ms}) = 1.4$ ms, which is relatively close to the above.

10.11 $\tau = R_{Th}C$

Occasions will arise in which the network does not have the simple series form of Fig. 10.23. It will then be necessary first to find the Thévenin equivalent circuit for the network external to the capacitive element. E_{Th} will then be the source voltage E of Eqs. (10.15) through (10.20) and R_{Th} will be the resistance R. The time constant is then $\tau = R_{Th}C$.

EXAMPLE 10.10 For the network of Fig. 10.50:

a. Find the mathematical expression for the transient behavior of the voltage v_C and the current i_C following the closing of the switch (position 1 at $t = 0$ s).

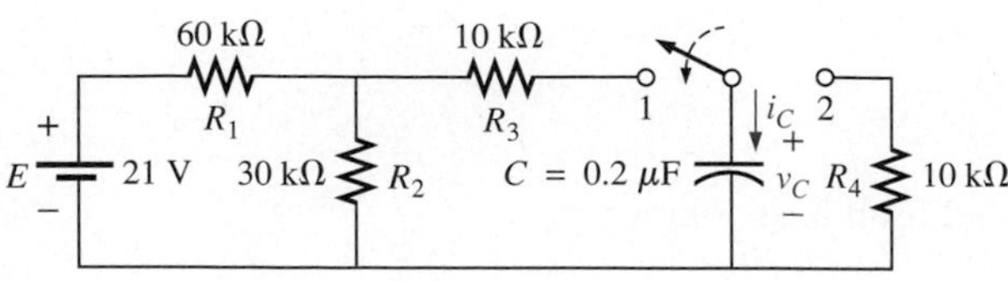

FIG. 10.50
Example 10.10.

b. Find the mathematical expression for the voltage v_C and current i_C as a function of time if the switch is thrown into position 2 at t = 9 ms.

c. Draw the resultant waveforms of parts (a) and (b) on the same time axis.

Solutions:

a. Applying Thévenin's theorem to the 0.2-μF capacitor, we obtain Fig. 10.51:

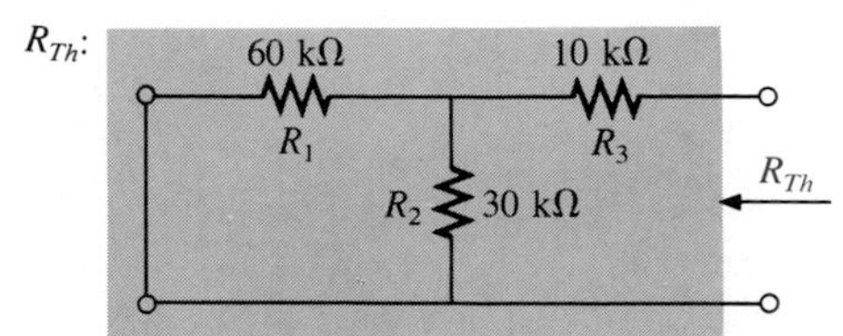

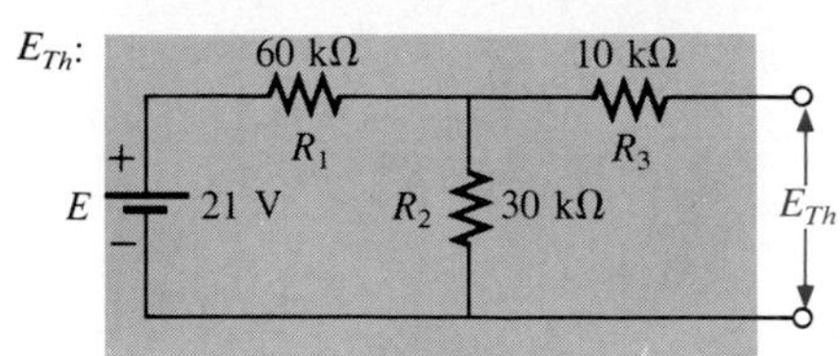

FIG. 10.51
Applying Thévenin's theorem to the network of Fig. 10.50.

$$R_{Th} = R_1 \| R_2 + R_3 = \frac{(60\text{ k}\Omega)(30\text{ k}\Omega)}{90\text{ k}\Omega} + 10\text{ k}\Omega$$

$$= 20\text{ k}\Omega + 10\text{ k}\Omega$$

$$R_{Th} = 30\text{ k}\Omega$$

$$E_{Th} = \frac{R_2 E}{R_2 + R_1} = \frac{(30\text{ k}\Omega)(21\text{ V})}{30\text{ k}\Omega + 60\text{ k}\Omega} = \frac{1}{3}(21\text{ V}) = 7\text{ V}$$

The resultant Thévenin equivalent circuit with the capacitor replaced is shown in Fig. 10.52:

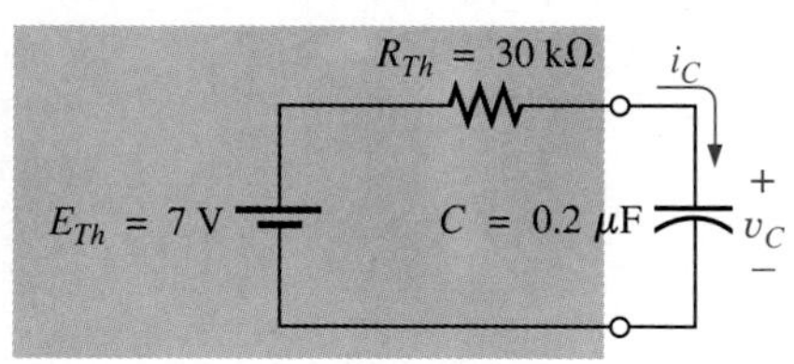

FIG. 10.52
Substituting the Thévenin equivalent for the network of Fig. 10.50.

$$v_C = E_{Th}(1 - e^{-t/\tau})$$

$$\tau = R_{Th}C = (30\text{ k}\Omega)(0.2\ \mu\text{F})$$

$$= (30 \times 10^3\ \Omega)(0.2 \times 10^{-6}\text{ F}) = 6 \times 10^{-3}\text{ s}$$

$$\tau = 6\text{ ms}$$

$$v_C = \mathbf{7(1 - e^{-t/(6 \times 10^{-3})})}$$

and

$$i_C = \frac{E_{Th}}{R}e^{-t/RC}$$

$$= \frac{7\text{ V}}{30\text{ k}\Omega}e^{-t/(6\times10^{-3})}$$

$$i_C = \mathbf{(0.233 \times 10^{-3})e^{-t/(6\times10^{-3})}}$$

b. At t = 9 ms,

$$v_C = E_{Th}(1 - e^{-t/\tau}) = 7(1 - e^{-(9\times10^{-3})/(6\times10^{-3})})$$

$$= 7(1 - e^{-1.5}) = 7(1 - 0.223)$$

$$v_C = 7(0.777) = 5.44\text{ V}$$

$$i_C = \frac{E_{Th}}{R}e^{-t/\tau} = (0.233 \times 10^{-3})e^{-1.5}$$

$$= (0.233 \times 10^{-3})(0.223)$$

$$i_C = 0.052 \times 10^{-3} = 0.052\text{ mA}$$

By Eq. (10.21),

$$v_C = V_i e^{-t/\tau'}$$

with

$$\tau' = R_4 C = (10 \times 10^3\ \Omega)(0.2 \times 10^{-6}\text{ F}) = 2 \times 10^{-3}\text{ s}$$

$$= 2\text{ ms}$$

and $\quad v_C = \mathbf{5.44}e^{-t/(2\times10^{-3})}$

By Eq. (10.22),

$$I_i = \frac{5.44 \text{ V}}{10 \text{ k}\Omega} = 0.054 \text{ mA}$$

and $\quad i_C = I_i e^{-t/\tau} = \mathbf{-(0.54 \times 10^{-3})}e^{-t/(2\times10^{-3})}$

c. See Fig. 10.53.

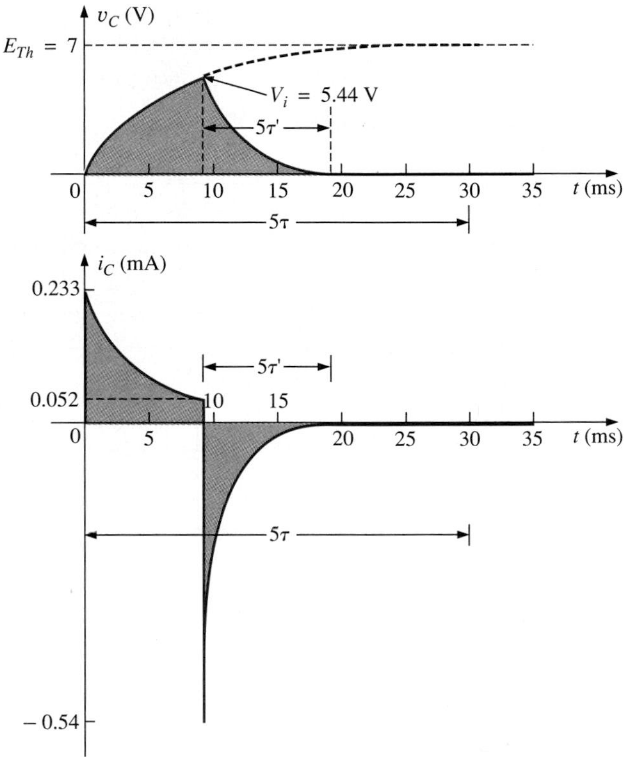

FIG. 10.53
The resulting waveforms for the network of Fig. 10.50.

EXAMPLE 10.11 The capacitor of Fig. 10.54 is initially charged to 40 V. Find the mathematical expression for v_C after the closing of the switch.

Solution: The network is redrawn in Fig. 10.55:

E_{Th}:

$$E_{Th} = \frac{R_3E}{R_3 + R_1 + R_4} = \frac{18 \text{ k}\Omega(120 \text{ V})}{18 \text{ k}\Omega + 7 \text{ k}\Omega + 2 \text{ k}\Omega}$$
$$= 80 \text{ V}$$

R_{Th}:

$$R_{Th} = 5 \text{ k}\Omega + 18 \text{ k}\Omega \parallel (7 \text{ k}\Omega + 2 \text{ k}\Omega)$$
$$= 5 \text{ k}\Omega + 6 \text{ k}\Omega$$
$$= 11 \text{ k}\Omega$$

Therefore, $\quad V_i = 40 \text{ V}, V_f = 80 \text{ V}$

and $\quad \tau = R_{Th}C = (11 \text{ k}\Omega)(40 \, \mu\text{F}) = 0.44 \text{ s}$

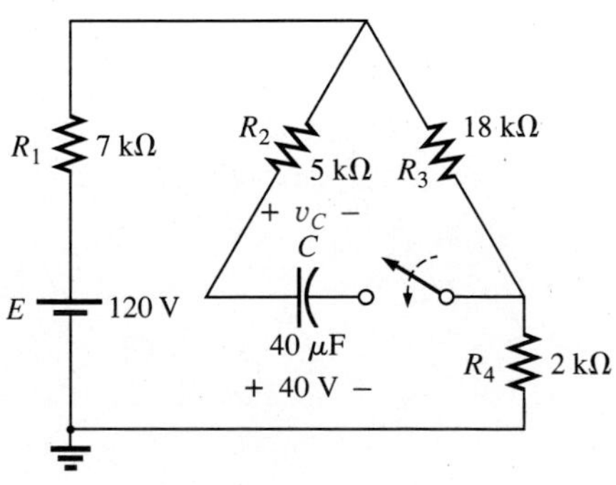

FIG. 10.54
Example 10.11.

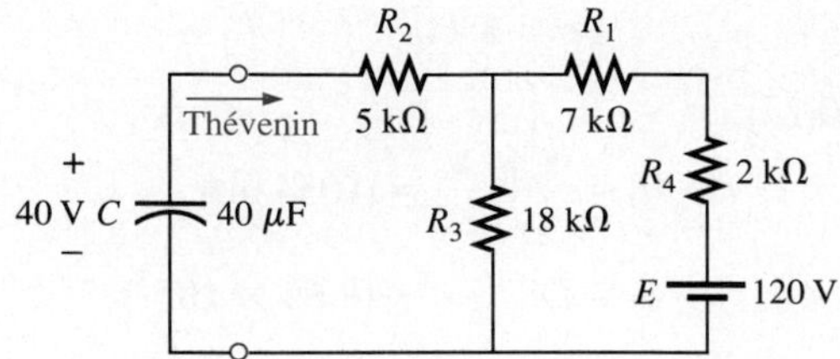

FIG. 10.55
Network of Fig. 10.54 redrawn.

Eq. (10.23): $\upsilon_C = V_f + (V_i - V_f)e^{-t/\tau}$

$$= 80\text{ V} + (40\text{ V} - 80\text{ V})e^{-t/0.44\text{s}}$$

and $\boldsymbol{\upsilon_C = 80\text{ V} - 40\text{ V}e^{-t/0.44\text{s}}}$

EXAMPLE 10.12 For the network of Fig. 10.56, find the mathematical expression for the voltage υ_C after the closing of the switch (at $t = 0$).

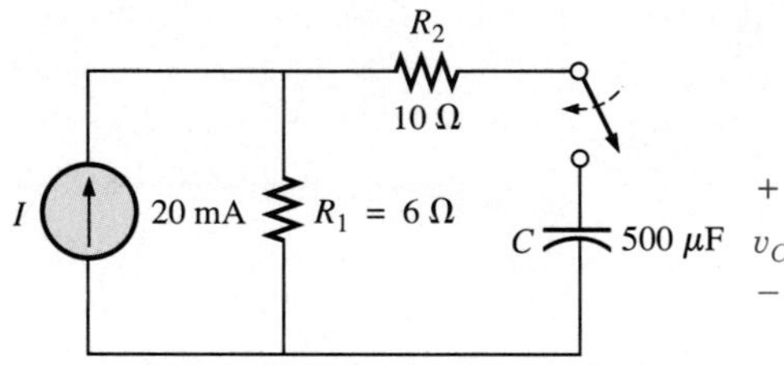

FIG. 10.56
Example 10.12.

Solution:

$$R_{Th} = R_1 + R_2 = 6\ \Omega + 10\ \Omega = 16\ \Omega$$

$$E_{Th} = V_1 + V_2 = IR_1 + 0$$

$$= (20 \times 10^{-3}\text{ A})(6\ \Omega) = 120 \times 10^{-3}\text{ V} = 0.12\text{ V}$$

and $\tau = R_{Th}C = (16\ \Omega)(500 \times 10^{-6}\text{ F}) = 8\text{ ms}$

so that $\boldsymbol{\upsilon_C = 0.12(1 - e^{-t/(8\times10^{-3})})}$

10.12 THE CURRENT i_C

The current i_C associated with a capacitance C is related to the voltage across the capacitor by

$$i_C = C\frac{d\upsilon_C}{dt} \tag{10.27}$$

where $d\upsilon_C/dt$ is a measure of the change in υ_C in a vanishingly small period of time. The function $d\upsilon_C/dt$ is called the *derivative* of the voltage υ_C with respect to time t.

If the voltage fails to change at a particular instant, then

$$d\upsilon_C = 0$$

and $$i_C = C\frac{d\upsilon_C}{dt} = 0$$

In other words, if the voltage across a capacitor fails to change with time, the current i_C associated with the capacitor is zero. To take this a step further, the equation also states that the more rapid the change in voltage across the capacitor, the greater the resulting current.

In an effort to develop a clearer understanding of Eq. (10.27), let us calculate the average current associated with a capacitor for various voltages impressed across the capacitor. The average current is defined by the equation

$$i_{C_{av}} = C\frac{\Delta v_C}{\Delta t} \tag{10.28}$$

where Δ indicates a finite (measurable) change in charge, voltage, or time. The instantaneous current can be derived from Eq. (10.28) by letting Δt become vanishingly small; that is,

$$i_{C_{inst}} = \lim_{\Delta t \to 0} C\frac{\Delta v_C}{\Delta t} = C\frac{dv_C}{dt}$$

In the following example, the change in voltage Δv_C will be considered for each slope of the voltage waveform. If the voltage increases with time, the average current is the change in voltage divided by the change in time, with a positive sign. If the voltage decreases with time, the average current is again the change in voltage divided by the change in time, but with a negative sign.

EXAMPLE 10.13 Find the waveform for the average current if the voltage across a 2-μF capacitor is as shown in Fig. 10.57.

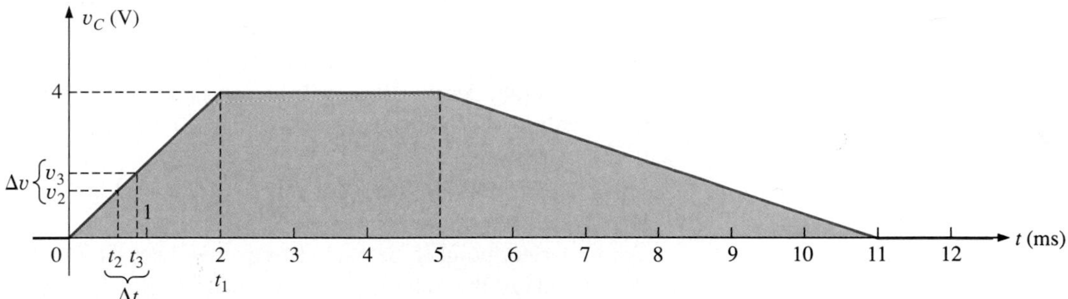

FIG. 10.57
Example 10.13.

Solution:

a. From 0 to 2 ms, the voltage increases linearly from 0 to 4 V, the change in voltage $\Delta v = 4\text{ V} - 0 = 4\text{ V}$ (with a positive sign since the voltage increases with time). The change in time $\Delta t = 2\text{ ms} - 0 = 2\text{ ms}$, and

$$i_{C_{av}} = C\frac{\Delta v_C}{\Delta t} = (2 \times 10^{-6}\text{ F})\left(\frac{4\text{ V}}{2 \times 10^{-3}\text{ s}}\right)$$
$$= 4 \times 10^{-3}\text{ A} = 4\text{ mA}$$

b. From 2 to 5 ms, the voltage remains constant at 4 V; the change in voltage $\Delta v = 0$. The change in time $\Delta t = 3$ ms, and

$$i_{C_{av}} = C\frac{\Delta v_C}{\Delta t} = C\frac{0}{\Delta t} = 0$$

c. From 5 to 11 ms, the voltage decreases from 4 to 0 V. The change in voltage Δv is, therefore, $4 \text{ V} - 0 = 4 \text{ V}$ (with a negative sign since the voltage is decreasing with time). The change in time $\Delta t = 11 \text{ ms} - 5 \text{ ms} = 6 \text{ ms}$, and

$$i_{C_{av}} = C\frac{\Delta v_C}{\Delta t} = -(2 \times 10^{-6} \text{ F})\left(\frac{4 \text{ V}}{6 \times 10^{-3} \text{ s}}\right)$$
$$= -1.33 \times 10^{-3} \text{ A} = -1.33 \text{ mA}$$

d. From 11 ms on, the voltage remains constant at 0 and $\Delta v = 0$, so $i_{C_{av}} = 0$. The waveform for the average current for the impressed voltage is as shown in Fig. 10.58.

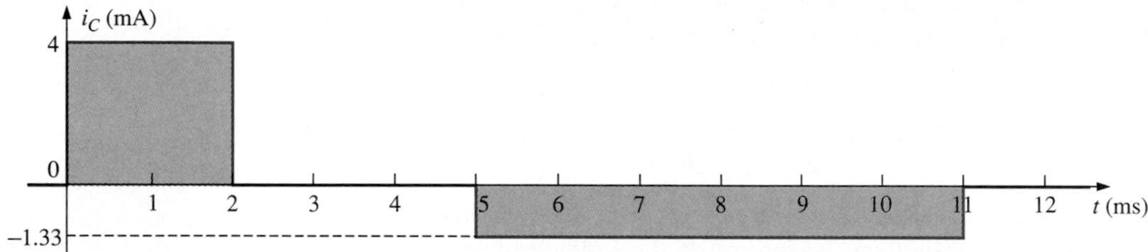

FIG. 10.58
The resulting current i_C for the applied voltage of Fig. 10.57.

Note in Example 10.13 that, in general, the steeper the slope, the greater the current, and when the voltage fails to change, the current is zero. In addition, the average value is the same as the instantaneous value at any point along the slope over which the average value was found. For example, if the interval Δt is reduced from $0 \rightarrow t_1$ to $t_2 - t_3$, as noted in Fig. 10.57, $\Delta v/\Delta t$ is still the same. In fact, no matter how small the interval Δt, the slope will be the same, and therefore the current $i_{C_{av}}$ will be the same. If we consider the limit as $\Delta t \rightarrow 0$, the slope will still remain the same, and therefore $i_{C_{av}} = i_{C_{inst}}$ at any instant of time between 0 and t_1. The same can be said about any portion of the voltage waveform that has a constant slope.

An important point to be gained from this discussion is that it is not the magnitude of the voltage across a capacitor that determines the current but rather how quickly the voltage *changes* across the capacitor. An applied steady dc voltage of 10,000 V would (ideally) not create any flow of charge (current), but a change in voltage of 1 V in a very brief period of time could create a significant current.

The method described above is only for waveforms with straight-line (linear) segments. For nonlinear (curved) waveforms, a method of calculus (differentiation) must be employed.

10.13 CAPACITORS IN SERIES AND PARALLEL

Capacitors, like resistors, can be placed in series and in parallel. Increasing levels of capacitance can be obtained by placing capacitors in parallel, while decreasing levels can be obtained by placing capacitors in series.

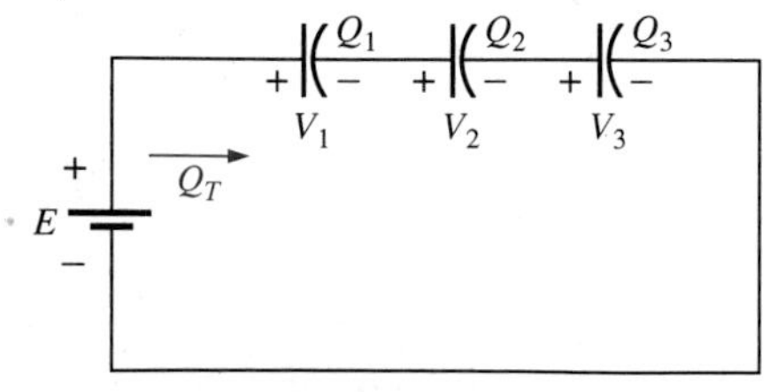

FIG. 10.59
Series capacitors.

For capacitors in series, the charge is the same on each capacitor (Fig. 10.59):

$$Q_T = Q_1 = Q_2 = Q_3 \tag{10.29}$$

Applying Kirchhoff's voltage law around the closed loop gives

$$E = V_1 + V_2 + V_3$$

However,
$$V = \frac{Q}{C}$$

so that
$$\frac{Q_T}{C_T} = \frac{Q_1}{C_1} + \frac{Q_2}{C_2} + \frac{Q_3}{C_3}$$

Using Eq. (10.29) and dividing both sides by Q yields

$$\frac{1}{C_T} = \frac{1}{C_1} + \frac{1}{C_2} + \frac{1}{C_3} \tag{10.30}$$

which is similar to the manner in which we found the total resistance of a parallel resistive circuit. The total capacitance of two capacitors in series is

$$C_T = \frac{C_1C_2}{C_1 + C_2} \tag{10.31}$$

The voltage across each capacitor of Fig. 10.59 can be found by first recognizing that

$$Q_T = Q_1$$

or
$$C_TE = C_1V_1$$

Solving for V_1:

$$V_1 = \frac{C_TE}{C_1}$$

and substituting for C_T:

$$V_1 = \frac{\dfrac{1}{C_1}(E)}{\dfrac{1}{C_1} + \dfrac{1}{C_2} + \dfrac{1}{C_3}} \tag{10.32}$$

A similar equation will result for each capacitor of the network.

For capacitors in parallel, as shown in Fig. 10.60, the voltage is the same across each capacitor, and the total charge is the sum of that on each capacitor:

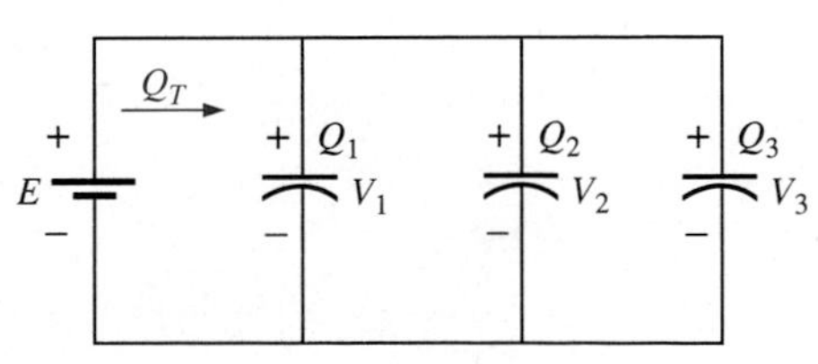

FIG. 10.60
Parallel capacitors.

$$Q_T = Q_1 + Q_2 + Q_3 \tag{10.33}$$

However,
$$Q = CV$$

Therefore,
$$C_TE = C_1V_1 + C_2V_2 + C_3V_3$$

but
$$E = V_1 = V_2 = V_3$$

Thus,
$$C_T = C_1 + C_2 + C_3 \tag{10.34}$$

which is similar to the manner in which the total resistance of a series circuit is found.

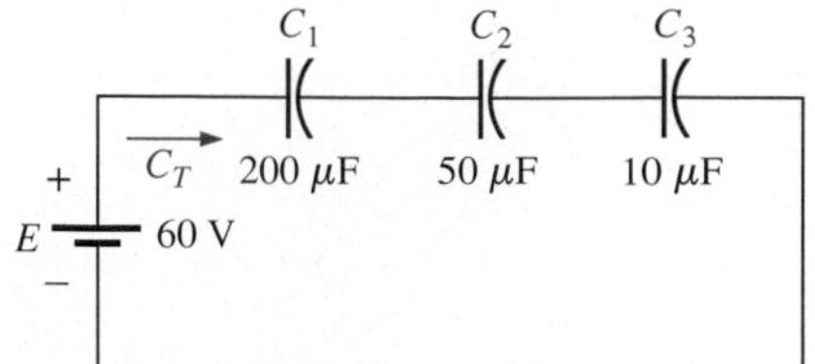

FIG. 10.61
Example 10.14.

EXAMPLE 10.14 For the circuit of Fig. 10.61:

a. Find the total capacitance.
b. Determine the charge on each plate.
c. Find the voltage across each capacitor.

Solutions:

a. $$\frac{1}{C_T} = \frac{1}{C_1} + \frac{1}{C_2} + \frac{1}{C_3}$$
$$= \frac{1}{200 \times 10^{-6}\,\text{F}} + \frac{1}{50 \times 10^{-6}\,\text{F}} + \frac{1}{10 \times 10^{-6}\,\text{F}}$$
$$= 0.005 \times 10^6 + 0.02 \times 10^6 + 0.1 \times 10^6$$
$$= 0.125 \times 10^6$$

and $$C_T = \frac{1}{0.125 \times 10^6} = \mathbf{8\ \mu F}$$

b. $$Q_T = Q_1 = Q_2 = Q_3$$
$$Q_T = C_T E = (8 \times 10^{-6}\,\text{F})(60\,\text{V}) = \mathbf{480\ \mu C}$$

c. $$V_1 = \frac{Q_1}{C_1} = \frac{480 \times 10^{-6}\,\text{C}}{200 \times 10^{-6}\,\text{F}} = \mathbf{2.4\ V}$$

$$V_2 = \frac{Q_2}{C_2} = \frac{480 \times 10^{-6}\,\text{C}}{50 \times 10^{-6}\,\text{F}} = \mathbf{9.6\ V}$$

$$V_3 = \frac{Q_3}{C_3} = \frac{480 \times 10^{-6}\,\text{C}}{10 \times 10^{-6}\,\text{F}} = \mathbf{48.0\ V}$$

and $$E = V_1 + V_2 + V_3 = 2.4\,\text{V} + 9.6\,\text{V} + 48\,\text{V}$$
$$= \mathbf{60\ V} \qquad \text{(checks)}$$

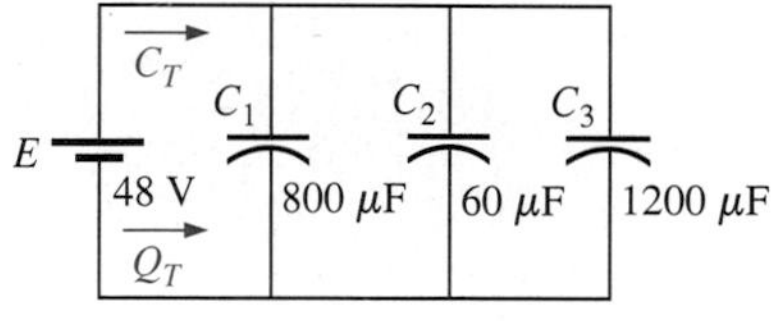

FIG. 10.62
Example 10.15.

EXAMPLE 10.15 For the network of Fig. 10.62:

a. Find the total capacitance.
b. Determine the charge on each plate.
c. Find the total charge.

Solutions:

a. $$C_T = C_1 + C_2 + C_3 = 800\,\mu\text{F} + 60\,\mu\text{F} + 1200\,\mu\text{F}$$
$$= \mathbf{2060\ \mu F}$$

b. $$Q_1 = C_1E = (800 \times 10^{-6}\,\text{F})(48\,\text{V}) = \mathbf{38.4\ mC}$$
$$Q_2 = C_2E = (60 \times 10^{-6}\,\text{F})(48\,\text{V}) = \mathbf{2.88\ mC}$$
$$Q_3 = C_3E = (1200 \times 10^{-6}\,\text{F})(48\,\text{V}) = \mathbf{57.6\ mC}$$

c. $$Q_T = Q_1 + Q_2 + Q_3 = 38.4\,\text{mC} + 2.88\,\text{mC} + 57.6\,\text{mC}$$
$$= \mathbf{98.88\ mC}$$

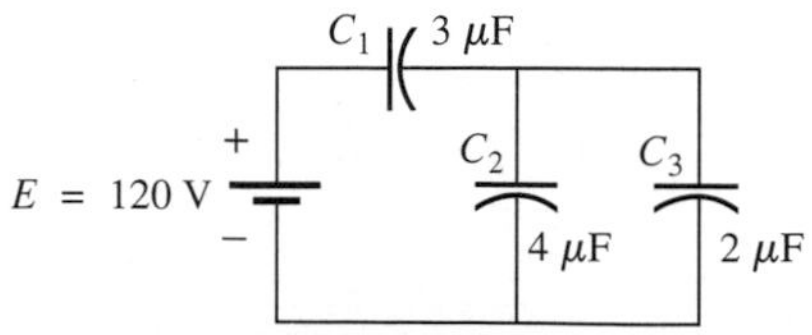

FIG. 10.63
Example 10.16.

EXAMPLE 10.16 Find the voltage across and charge on each capacitor for the network of Fig. 10.63.

Solution:

$$C'_T = C_2 + C_3 = 4\,\mu\text{F} + 2\,\mu\text{F} = 6\,\mu\text{F}$$
$$C_T = \frac{C_1C'_T}{C_1 + C'_T} = \frac{(3\,\mu\text{F})(6\,\mu\text{F})}{3\,\mu\text{F} + 6\,\mu\text{F}} = 2\,\mu\text{F}$$

$$Q_T = C_T E = (2 \times 10^{-6}\text{ F})(120\text{ V})$$
$$= \mathbf{240\ \mu C}$$

An equivalent circuit (Fig. 10.64) has

$$Q_T = Q_1 = Q'_T$$

and, therefore,
$$Q_1 = \mathbf{240\ \mu C}$$

and
$$V_1 = \frac{Q_1}{C_1} = \frac{240 \times 10^{-6}\text{ C}}{3 \times 10^{-6}\text{ F}} = \mathbf{80\ V}$$

$$Q'_T = 240\ \mu\text{C}$$

and, therefore,

$$V'_T = \frac{Q'_T}{C'_T} = \frac{240 \times 10^{-6}\text{ C}}{6 \times 10^{-6}\text{ F}} = \mathbf{40\ V}$$

and
$$Q_2 = C_2 V'_T = (4 \times 10^{-6}\text{ F})(40\text{ V}) = \mathbf{160\ \mu C}$$
$$Q_3 = C_3 V'_T = (2 \times 10^{-6}\text{ F})(40\text{ V}) = \mathbf{80\ \mu C}$$

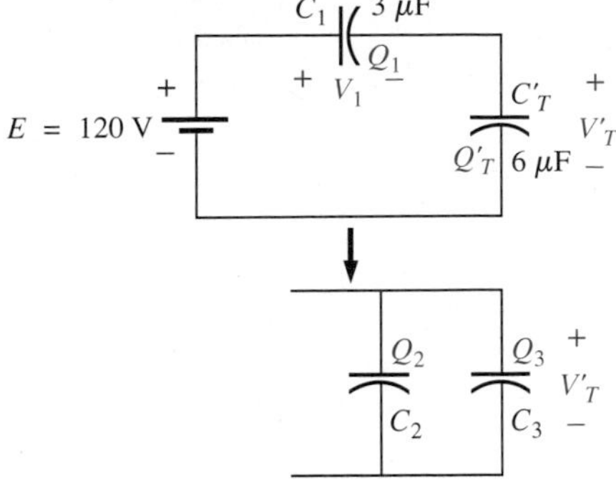

FIG. 10.64
Reduced equivalent for the network of Fig. 10.63.

EXAMPLE 10.17 Find the voltage across and charge on capacitor C_1 of Fig. 10.65 after it has charged up to its final value.

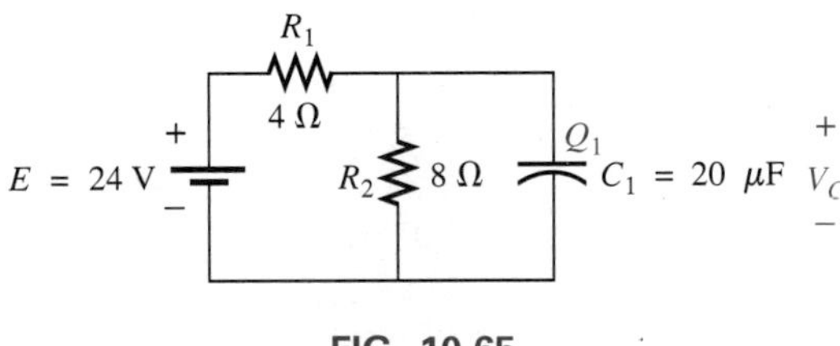

FIG. 10.65
Example 10.17.

Solution: As previously discussed, the capacitor is effectively an open circuit for dc after charging up to its final value (Fig. 10.66). Therefore,

$$V_C = \frac{(8\ \Omega)(24\text{ V})}{4\ \Omega + 8\ \Omega} = \mathbf{16\ V}$$
$$Q_1 = C_1 V_C = (20 \times 10^{-6}\text{ F})(16\text{ V})$$
$$= \mathbf{320\ \mu C}$$

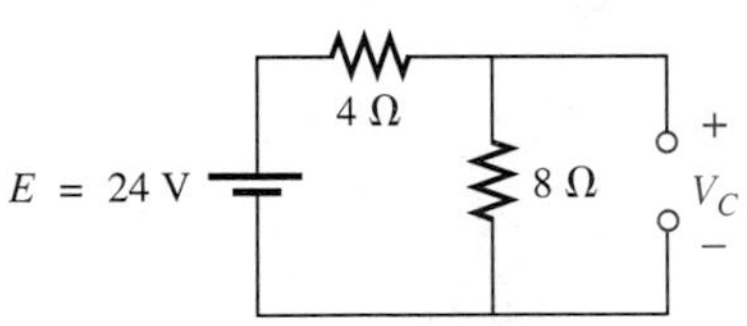

FIG. 10.66
Determining the final (steady-state) value for v_C.

EXAMPLE 10.18 Find the voltage across and charge on each capacitor of the network of Fig. 10.67 after each has charged up to its final value.

Solution:

$$V_{C_2} = \frac{(7\ \Omega)(72\text{ V})}{7\ \Omega + 2\ \Omega} = \mathbf{56\ V}$$

$$V_{C_1} = \frac{(2\ \Omega)(72\text{ V})}{2\ \Omega + 7\ \Omega} = \mathbf{16\ V}$$

$$Q_1 = C_1 V_{C_1} = (2 \times 10^{-6}\text{ F})(16\text{ V}) = \mathbf{32\ \mu C}$$
$$Q_2 = C_2 V_{C_2} = (3 \times 10^{-6}\text{ F})(56\text{ V}) = \mathbf{168\ \mu C}$$

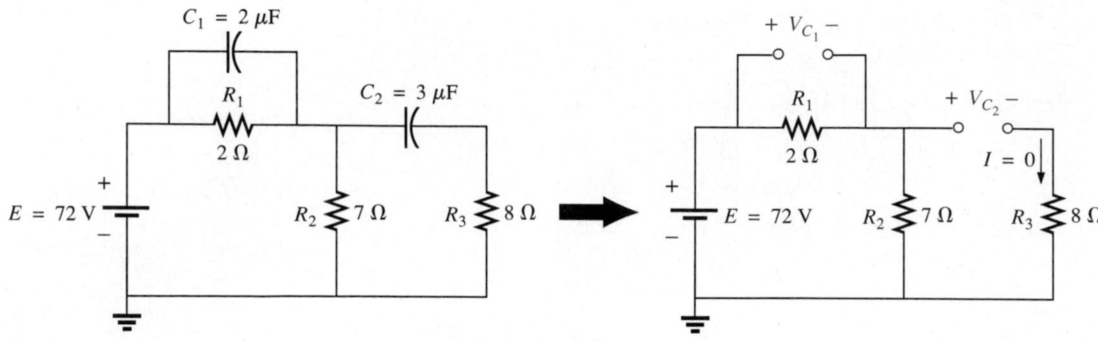

FIG. 10.67
Example 10.18.

10.14 ENERGY STORED BY A CAPACITOR

The ideal capacitor does not dissipate any of the energy supplied to it. It stores the energy in the form of an electric field between the conducting surfaces. A plot of the voltage, current, and power to a capacitor during the charging phase is shown in Fig. 10.68. The power curve can be obtained by finding the product of the voltage and current at selected instants of time and connecting the points obtained. The energy stored is represented by the shaded area under the power curve. Using calculus, we can determine the area under the curve:

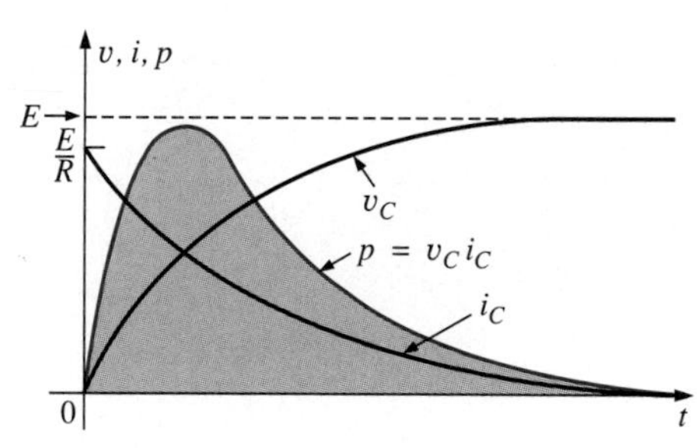

FIG. 10.68
Plotting the power to a capacitive element during the transient phase.

$$W_C = \frac{1}{2}CE^2$$

In general,

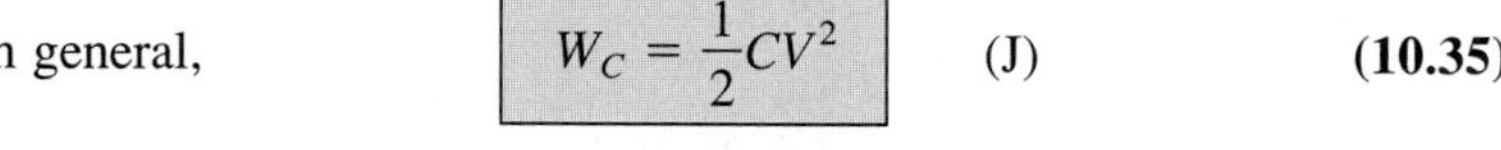

$$W_C = \frac{1}{2}CV^2 \quad \text{(J)} \qquad \textbf{(10.35)}$$

where V is the steady-state voltage across the capacitor. In terms of Q and C,

$$W_C = \frac{1}{2}C\left(\frac{Q}{C}\right)^2$$

or

$$W_C = \frac{Q^2}{2C} \quad \text{(J)} \qquad \textbf{(10.36)}$$

EXAMPLE 10.19 For the network of Fig. 10.67, determine the energy stored by each capacitor.

Solution:

For C_1, $W_C = \frac{1}{2}CV^2$

$$= \frac{1}{2}(2 \times 10^{-6}\ \text{F})(16\ \text{V})^2 = (1 \times 10^{-6})(256)$$

$$= \mathbf{256\ \mu J}$$

For C_2, $W_C = \frac{1}{2}CV^2$

$$= \frac{1}{2}(3 \times 10^{-6}\ \text{F})(56\ \text{V})^2 = (1.5 \times 10^{-6})(3136)$$

$$= \mathbf{4704\ \mu J}$$

Due to the squared term, note the difference in energy stored because of a higher voltage.

10.15 STRAY CAPACITANCES

In addition to the capacitors discussed so far in this chapter, there are stray capacitances that exist not through design but simply because two conducting surfaces are relatively close to each other. Two conducting wires in the same network will have a capacitive effect between them, as shown in Fig. 10.69(a). In electronic circuits, capacitance levels exist between conducting surfaces of the transistor, as shown in Fig. 10.69(b). In Chapter 12 we will discuss another element called the *inductor,* which will have capacitive effects between the windings [Fig. 10.69(c)]. Stray capacitances can often lead to serious errors in system design if they are not considered carefully.

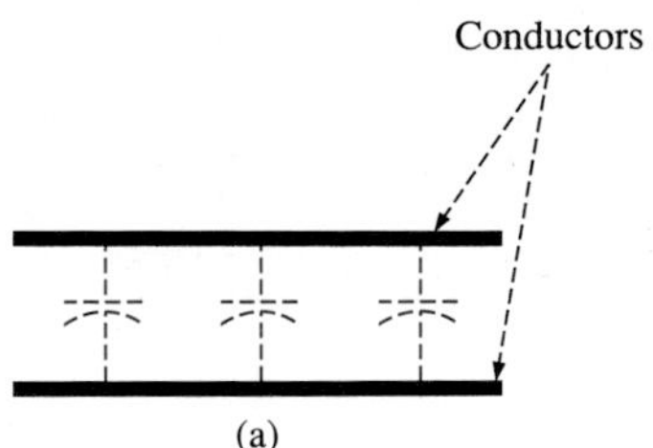

(a)

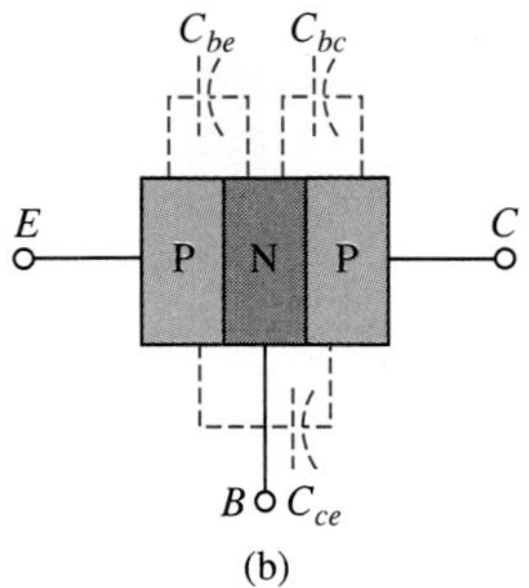

(b)

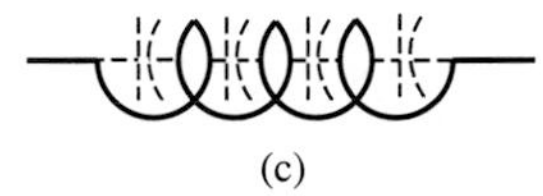

(c)

FIG. 10.69
Examples of stray capacitance.

10.16 COMPUTER ANALYSIS

PSpice (DOS)

PSpice has a specific command (.TRAN) that will determine the transient response over time. A plot of the response can then be obtained using the .PLOT or .PROBE commands. The .TRAN command determines the transient response of a network from $t = 0$ s to some final value. The format of the command is the following:

```
.TRAN    TSTEP    TSTOP
```

TSTEP: Time increment between determined values

TSTOP: Final value of t for calculations

Example:
```
.TRAN    5U    100U
```

TSTEP and TSTOP are specified by the user. TSTEP is the time interval between determined values, and TSTOP is the final value of time (t) for which a magnitude is to be determined. The example specifies that the magnitude will be determined every 5 μs from 0 to 100 μs, with the first level determined at $t = 5$ μs.

Capacitors are entered in much the same manner as resistors, as demonstrated by the format below:

```
CBYPASS    6    7    100U    IC=2V
```

CBYPASS: Name; 6: + Node; 7: − Node; 100U: Value; IC=2V: Initial value

The user fills in those quantities with brackets underneath, although the initial condition entry (the voltage across the capacitor before the switching action occurs) can be omitted if the initial value is zero volts. The preceding entry is for a 100-μF bypass capacitor between nodes 6 and 7, with node 6 as the defined higher potential. In PSpice, capacitors have the additional limitation that they cannot be entered as series elements. However, this problem can be circumvented by placing a very large resistance (equivalent to an open circuit for the network of interest) in parallel with one of the capacitors.

Let us now write the input file for the network of Fig. 10.70. Rather than use a switch to establish the 20 V across the network at $t = 0$ s, a pulse is applied that starts at $t = 0$ s, switching to 20 V for a period of time to exceed 5τ of the network.

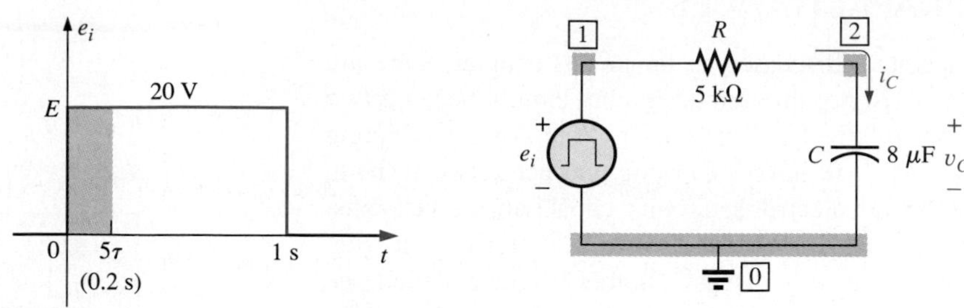

FIG. 10.70
Using PSpice (DOS) to apply a pulse waveform to an R-C network.

In general, the format for an applied pulse has the following format:

VPULSE	3	2	PULSE (2	8	1U	10N	20N	50U	100U)
Name	+ Node	− Node	Initial level of pulse	Pulsed level	Delay time (DT)	Rise time (RT)	Fall time (FT)	Pulse width (PW)	Period of pulse (PER)

All the quantities with a bracket below the entry are controlled by the user. Don't forget the parentheses at each end of the PULSE statement! The 2 V is the initial level of the pulse and the 8 V is the pulsed level. The delay time (1 μs) is the time interval following $t = 0$ s before the pulse changes levels. If desired, it can be zero seconds. The rise and fall times are the time intervals required to change levels. The rise and fall times must be specified (other than zero seconds), or PSpice will fall back to TSTEP of the .TRAN command for RT and FT. For the preceding example, the rise time is 10 ns and the fall time is 20 ns. The pulse width (50 μs) is the time interval at the pulsed level, and the period (100 μs) is the time between successive pulses. If the pulse width and period are unspecified, TSTOP of the .TRAN command will be employed for each.

For the input of Fig. 10.70, the statement for the pulse appears in the input file of Fig. 10.71. The pulse starts at $t = 0$ s (TD = 0) and has a rise and fall time of 1 ns, which are both significantly less than the time scale of the input or time constant of the network. The pulse width is 1 s, which is 5 times the period associated with 5τ (200 ms) for this network and beyond the range of interest as defined by the .TRAN command. In other words, the 20-V level will be present for the period of interest defined by the 0–5τ interval.

The .PRINT statement of the input file specifies that the magnitude of the pulse waveform, v_C and i_C, be tabulated for the .TRAN period of 10 ms to 200 ms at intervals of 10 ms. Note in the output file that the first value of each quantity is not determined until 10 ms have passed. Note also, however, that once past the 10-ms interval, 20 V are available at the input and that v_C is very close to the 20-V level after 5τ (or 200 ms). The current i_C drops from 4 mA to 3.541 mA in the first 10 ms and then continues to drop to a very low level during the rest of the defined range of interest.

Replacing the .PRINT command with the .PROBE command will result in the beautiful plots of Fig. 10.72 for the input voltage v_C and i_C. Note how .PROBE assigns a notation for each of the plots and deter-

```
Chapter 10 - R-C Circuit Transient Analysis

 ****      CIRCUIT DESCRIPTION

******************************************************************************


VE 1 0 PULSE(0 20V 0 1N 1N 1)
R  1 2 5K
C  2 0 8U
.TRAN 10M 200M
.PRINT TRAN V(1) V(2) I(C)
.OPTIONS NOPAGE
.END

****      TRANSIENT ANALYSIS                      TEMPERATURE =    27.000 DEG C


 TIME          V(1)          V(2)          I(C)

  0.000E+00    0.000E+00     0.000E+00     0.000E+00
  1.000E-02    2.000E+01     4.421E+00     3.116E-03
  2.000E-02    2.000E+01     7.858E+00     2.428E-03
  3.000E-02    2.000E+01     1.055E+01     1.889E-03
  4.000E-02    2.000E+01     1.264E+01     1.472E-03
  5.000E-02    2.000E+01     1.427E+01     1.145E-03
  6.000E-02    2.000E+01     1.554E+01     8.926E-04
  7.000E-02    2.000E+01     1.653E+01     6.944E-04
  8.000E-02    2.000E+01     1.729E+01     5.412E-04
  9.000E-02    2.000E+01     1.790E+01     4.210E-04
  1.000E-01    2.000E+01     1.836E+01     3.281E-04
  1.100E-01    2.000E+01     1.872E+01     2.552E-04
  1.200E-01    2.000E+01     1.901E+01     1.989E-04
  1.300E-01    2.000E+01     1.923E+01     1.547E-04
  1.400E-01    2.000E+01     1.940E+01     1.206E-04
  1.500E-01    2.000E+01     1.953E+01     9.382E-05
  1.600E-01    2.000E+01     1.963E+01     7.311E-05
  1.700E-01    2.000E+01     1.972E+01     5.688E-05
  1.800E-01    2.000E+01     1.978E+01     4.433E-05
  1.900E-01    2.000E+01     1.983E+01     3.448E-05
  2.000E-01    2.000E+01     1.987E+01     2.684E-05
```

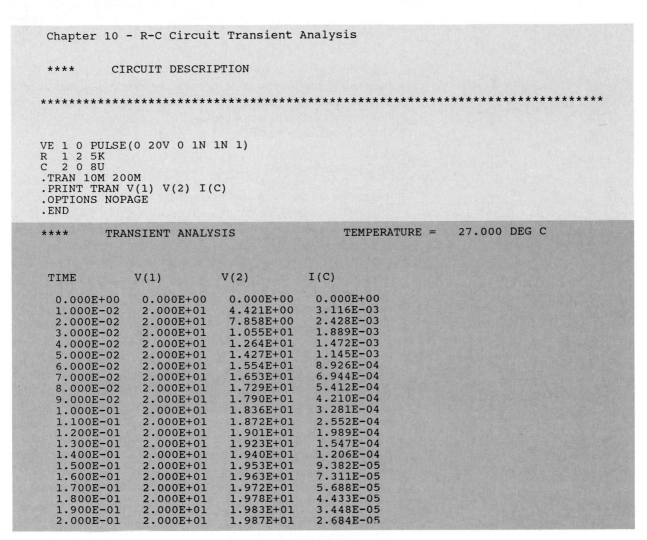

FIG. 10.71
The transient response for the voltages and current of Fig. 10.70.

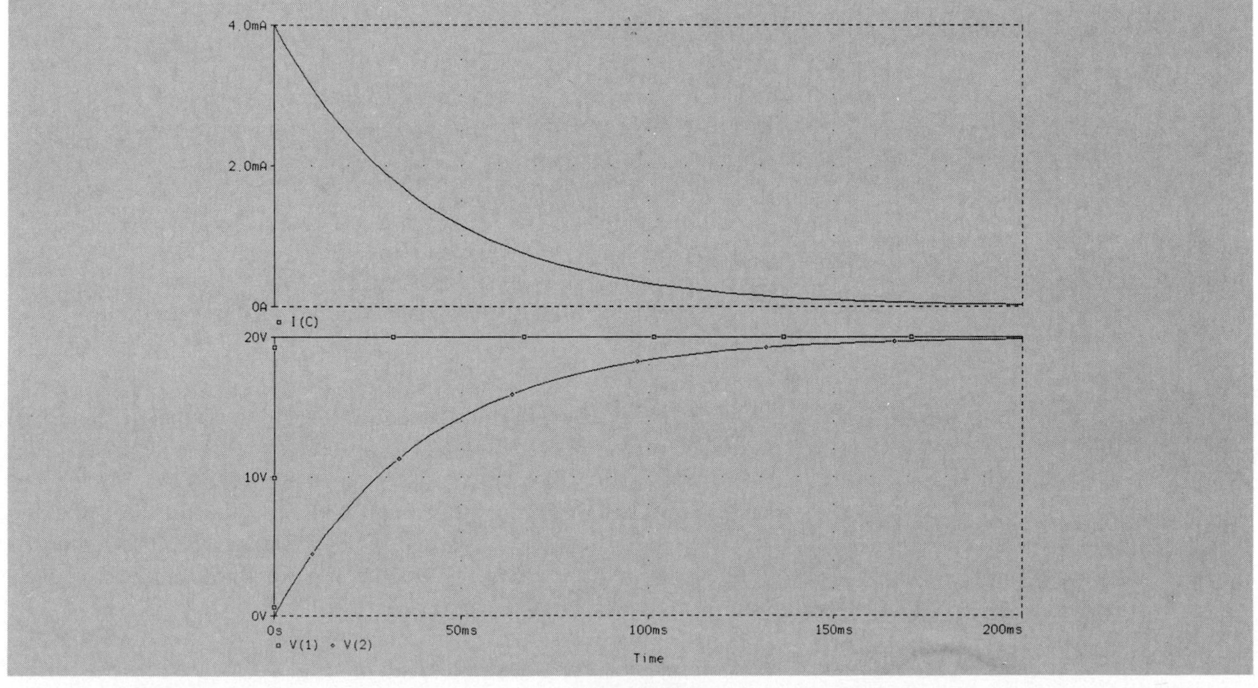

FIG. 10.72
The .PROBE response for i_C and v_C for the network of Fig. 10.70.

mines the appropriate scaling for the horizontal and vertical axes. As expected, v_C rises toward 20 V in 5τ and the input V(1) jumps to 20 V at zero milliseconds and continues at that level for the rest of the plot. The current i_C starts at 4 mA and then decays exponentially to almost zero milliampere in the 5τ period.

PSpice (Windows)

Rather than use the schematics to obtain the same type of response just described for PSpice (DOS), this session with PSpice (Windows) will expand on the use of the pulse function with a verification of the results of Example 10.13. The outcome will be a better understanding of the impact of the various parameters of the pulse function and an indication of how the use of some functions can be extended beyond the immediate, most obvious applications.

The pulse source is obtained through the sequence **Draw-Get New Part-Browse-source.slb-VPULSE-OK.** Once located, a double-click on the symbol will result in the dialog box that controls the various attributes. The parameters to be specified are actually the same as encountered for PSpice (DOS). **V1** is the initial level and **V2** is the pulse level, with **TD,** the delay time; **TR,** the rise time; **TF,** the fall time; **PW,** the pulse width; and **PER** the period of the waveform. Be aware that **PW** is the width of the pulse at the level **V2.** It is not the pulse width at the 50% level, as often defined in pulse systems. For each parameter, the **Save Attr-Change Display-Both name and value** was chosen to list the parameters on the display. The reference designator for the source was changed by simply double-clicking on the current reference **V1** and then entering **VPulse.** Then **Draw-Get New Part-Browse-analog.slb-C-OK** will result in the capacitor needed to complete the simple network of Fig. 10.73. Its value is then set at 2 μF and the label is left as C1.

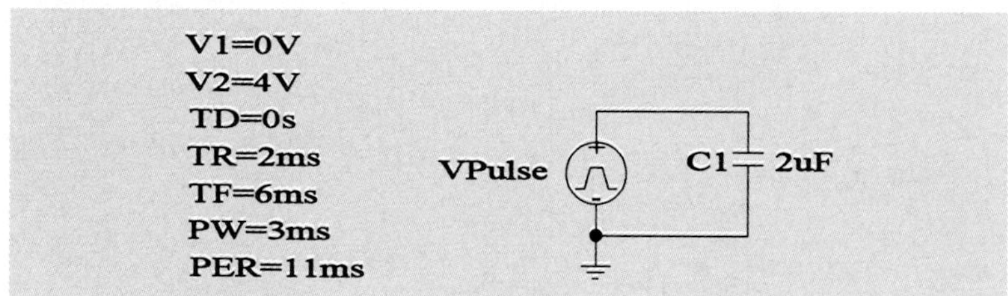

FIG. 10.73
Verifying the results of Example 10.13 using PSpice (Windows).

Under **Analysis, Setup** is chosen, followed by **Transient.** Since the time interval is about 10 ms and 1000 points would give a complete, clear display, the **Print Step** was chosen as 10 ms/1000 = 10 μs. The **Final Time** is 11 ms, the **No-Print Delay** 0 s, and the **Step Ceiling** 10 μs to ensure at least 1000 data points. After **OK-Close-Analysis-Probe Setup-Automatically Run Probe After Simulation,** the analysis was initiated by **Analysis-Simulation.**

When completed and the Probe screen appears, **Trace-Add-V(C1:1)** was chosen, resulting in the display of Fig. 10.74, which is a perfect match with Fig. 10.57. Note that the rise and fall times are linear

(straight-line) functions and the pulse width is as specified. The resulting horizontal axis extends for 12 ms, which can be reduced to the interval of interest by choosing **Plot-X-Axis Settings-User Defined 0 ms-11 ms-OK.** A plot of the capacitive current can also be displayed on the same screen by choosing **Plot-Add Plot-Trace-Add-I(C1:1)-OK.** The resulting display, as shown in Fig. 10.74, verifies the results of Example 10.13. The horizontal line at 0 mA was added by using **Tools-Label-Line** and then the pencil as described for drawing networks. The labels V(C) and I(C) were added by **Tools-Label-Text Enter text label-V(C)-OK.** The negative level of the current I(C) is difficult to read, but the **Tools-Cursor-Display** sequence will result in two intersecting lines on the plot defining a specific point. The horizontal and vertical positions of this point appear in a dialog box at the bottom of Fig. 10.74. Placing the vertical line at 10 ms by using the mouse will result in a value of -1.3333 mA, which also matches the solution of Example 10.13. Be aware that since two plots now appear on the screen, the display under immediate control for operations such as **Tools** is defined by the **SEL≫** notation on one of the displays, as noted for I(C) in Fig. 10.74. It can be changed to the other display by simple clicking the mouse on the chosen plot.

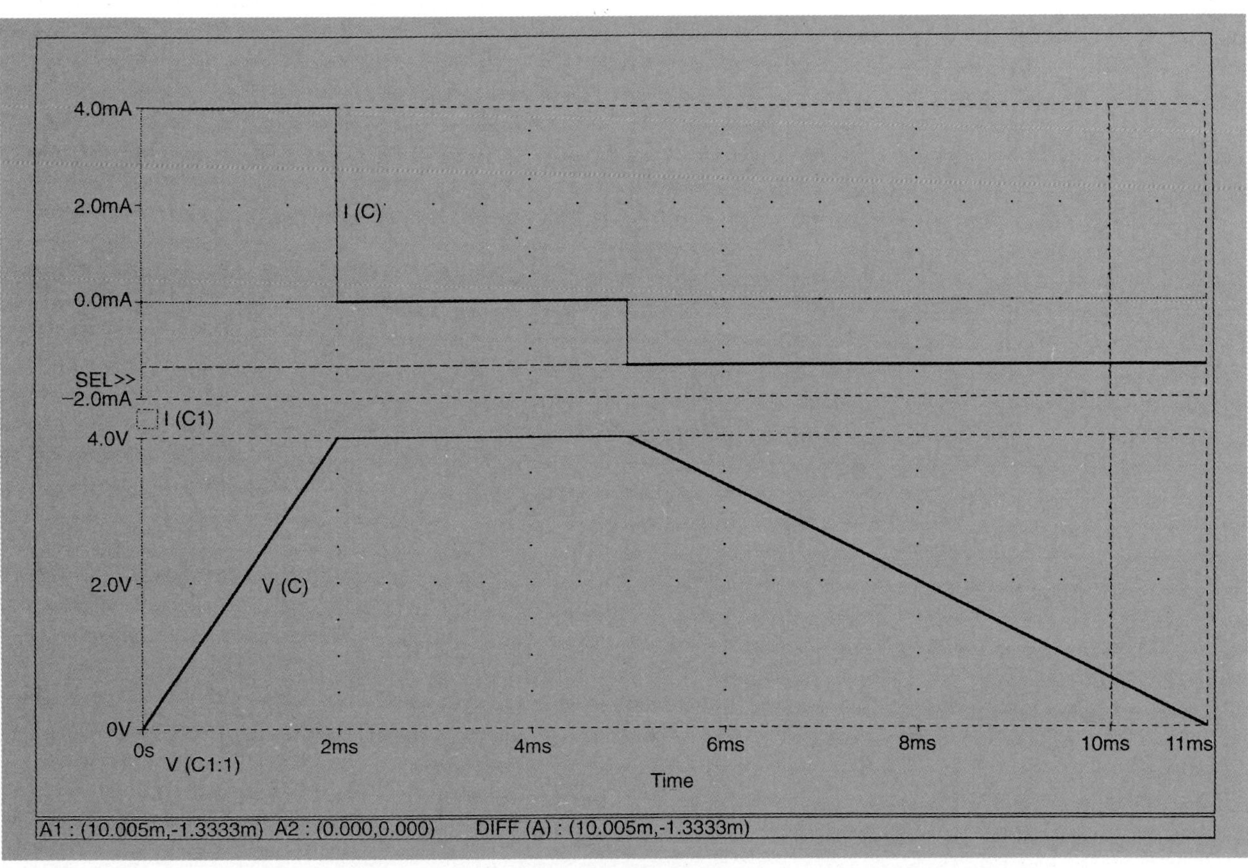

FIG. 10.74
The waveforms for the voltage and current of Example 10.13.

PROBLEMS

SECTION 10.2 The Electric Field

1. Find the electric field strength at a point 2 m from a charge of 4 μC.
2. The electric field strength is 36 newtons/coulomb (N/C) at a point r meters from a charge of 0.064 μC. Find the distance r.

SECTION 10.3 Capacitance

3. Find the capacitance of a parallel plate capacitor if 1400 μC of charge are deposited on its plates when 20 V are applied across the plates.
4. How much charge is deposited on the plates of a 0.05-μF capacitor if 45 V are applied across the capacitor?
5. Find the electric field strength between the plates of a parallel plate capacitor if 100 mV are applied across the plates and the plates are 2 mm apart.
6. Repeat Problem 5 if the plates are separated by 4 mils.
7. A 4-μF parallel plate capacitor has 160 μC of charge on its plates. If the plates are 5 mm apart, find the electric field strength between the plates.
8. Find the capacitance of a parallel plate capacitor if the area of each plate is 0.075 m^2 and the distance between the plates is 1.77 mm. The dielectric is air.
9. Repeat Problem 8 if the dielectric is paraffin-coated paper.
10. Find the distance in mils between the plates of a 2-μF capacitor if the area of each plate is 0.09 m^2 and the dielectric is transformer oil.
11. The capacitance of a capacitor with a dielectric of air is 1200 pF. When a dielectric is inserted between the plates, the capacitance increases to 0.006 μF. Of what material is the dielectric made?
12. The plates of a parallel plate air capacitor are 0.2 mm apart and have an area of 0.08 m^2, and 200 V are applied across the plates.
    a. Determine the capacitance.
    b. Find the electric field intensity between the plates.
    c. Find the charge on each plate if the dielectric is air.
13. A sheet of Bakelite 0.2 mm thick having an area of 0.08 m^2 is inserted between the plates of Problem 12.
    a. Find the electric field strength between the plates.
    b. Determine the charge on each plate.
    c. Determine the capacitance.

SECTION 10.4 Dielectric Strength

14. Find the maximum voltage ratings of the capacitors of Problems 12 and 13 assuming a linear relationship between the breakdown voltage and the thickness of the dielectric.
15. Find the maximum voltage that can be applied across a parallel plate capacitor of 0.006 μF. The area of one plate is 0.02 m^2 and the dielectric is mica. Assume a linear relationship between the dielectric strength and the thickness of the dielectric.

16. Find the distance in millimeters between the plates of a parallel plate capacitor if the maximum voltage that can be applied across the capacitor is 1250 V. The dielectric is mica. Assume a linear relationship between the breakdown strength and the thickness of the dielectric.

SECTION 10.7 Transients in Capacitive Networks: Charging Phase

17. For the circuit of Fig. 10.75:
    a. Determine the time constant of the circuit.
    b. Write the mathematical equation for the voltage v_C following the closing of the switch.
    c. Determine the voltage v_C after one, three, and five time constants.
    d. Write the equations for the current i_C and the voltage v_R.
    e. Sketch the waveforms for v_C and i_C.

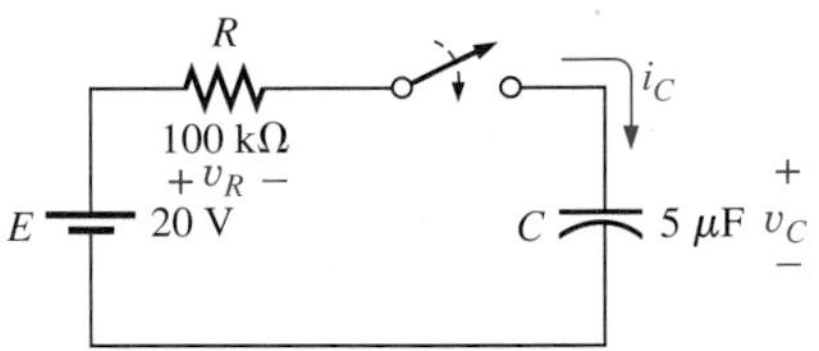

FIG. 10.75
Problems 17 and 18.

18. Repeat Problem 17 for $R = 1\ \text{M}\Omega$ and compare the results.

19. For the circuit of Fig. 10.76:
    a. Determine the time constant of the circuit.
    b. Write the mathematical equation for the voltage v_C following the closing of the switch.
    c. Determine v_C after one, three, and five time constants.
    d. Write the equations for the current i_C and the voltage v_R.
    e. Sketch the waveforms for v_C and i_C.

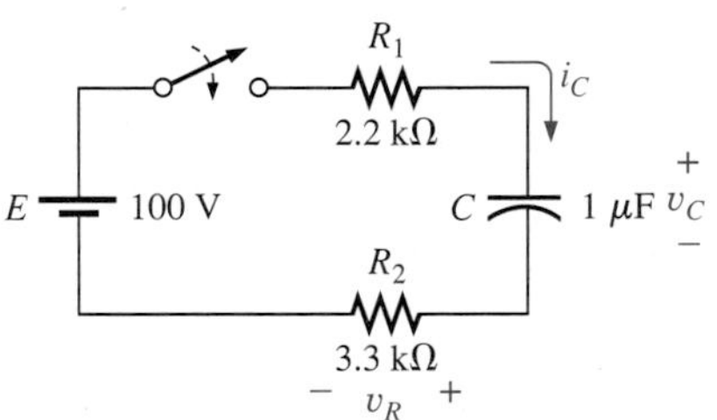

FIG. 10.76
Problem 19.

20. For the circuit of Fig. 10.77:
    a. Determine the time constant of the circuit.
    b. Write the mathematical equation for the voltage v_C following the closing of the switch.
    c. Write the mathematical expression for the current i_C following the closing of the switch.
    d. Sketch the waveforms of v_C and i_C.

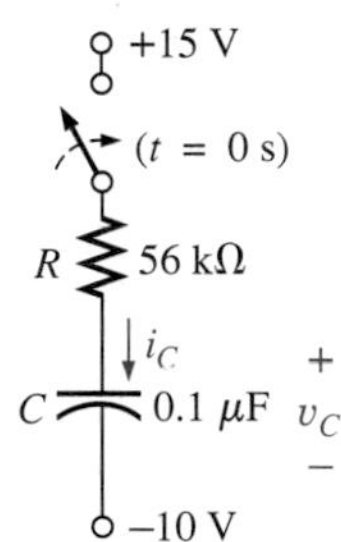

FIG. 10.77
Problem 20.

SECTION 10.8 Discharge Phase

21. For the circuit of Fig. 10.78:
    a. Determine the time constant of the circuit when the switch is thrown into position 1.
    b. Find the mathematical expression for the voltage across the capacitor after the switch is thrown into position 1.
    c. Determine the mathematical expression for the current following the closing of the switch (position 1).
    d. Determine the voltage v_C and the current i_C if the switch is thrown into position 2 at $t = 100$ ms.
    e. Determine the mathematical expressions for the voltage v_C and the current i_C if the switch is thrown into position 3 at $t = 200$ ms.
    f. Plot the waveforms of v_C and i_C for a period of time extending from $t = 0$ to $t = 300$ ms.

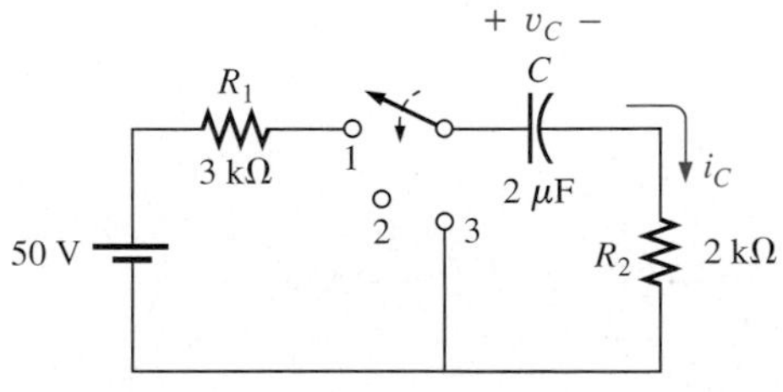

FIG. 10.78
Problems 21 and 22.

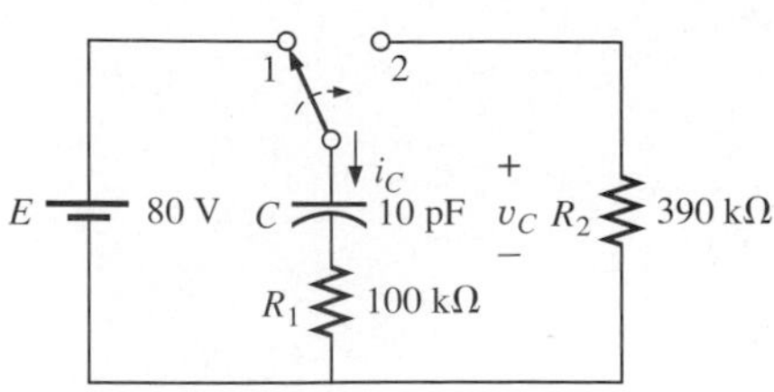

FIG. 10.79
Problem 23.

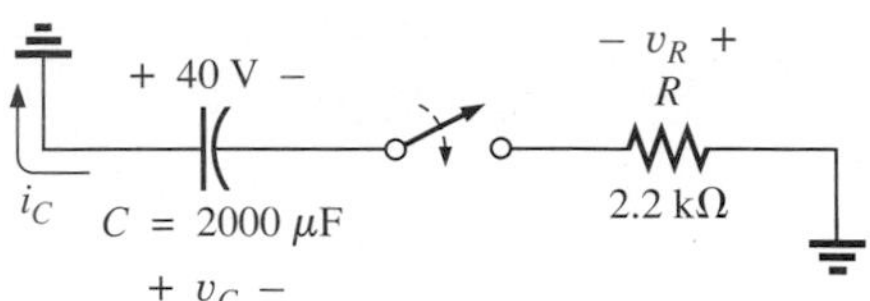

FIG. 10.80
Problem 24.

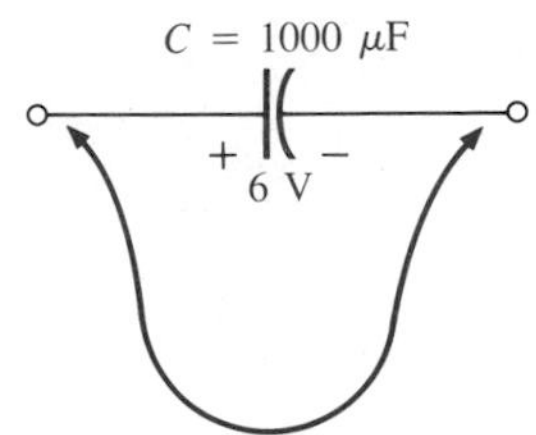

FIG. 10.81
Problems 25 and 29.

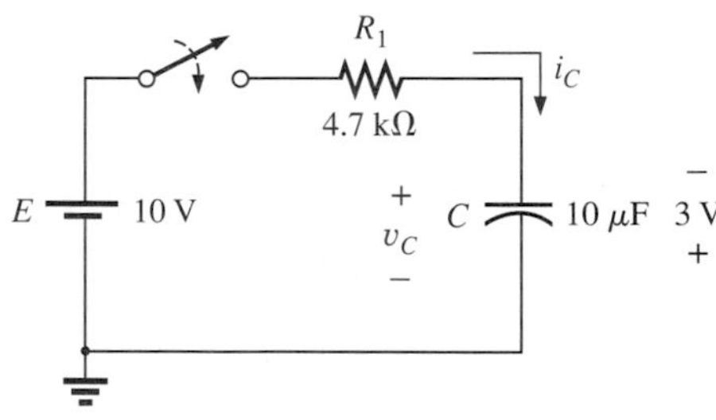

FIG. 10.82
Problem 26.

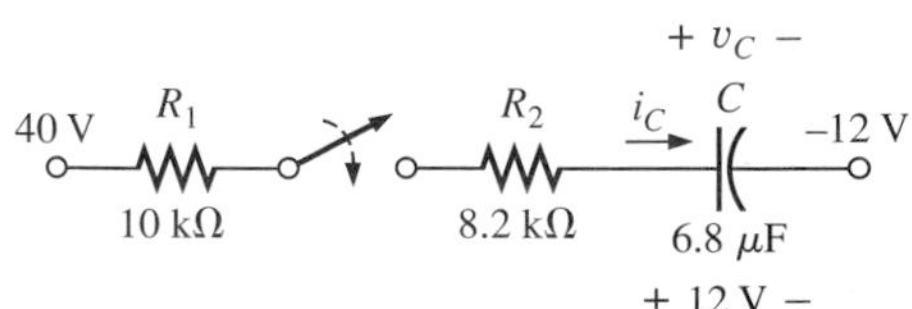

FIG. 10.83
Problem 27.

22. Repeat Problem 21 for a capacitance of 20 μF.

***23.** For the network of Fig. 10.79:

a. Find the mathematical expression for the voltage across the capacitor after the switch is thrown into position 1.

b. Repeat part (a) for the current i_C.

c. Find the mathematical expressions for the voltage v_C and current i_C if the switch is thrown into position 2 at a time equal to five time constants of the charging circuit.

d. Plot the waveforms of v_C and i_C for a period of time extending from $t = 0$ to $t = 30\ \mu s$.

24. The capacitor of Fig. 10.80 is initially charged to 40 V before the switch is closed. Write the expressions for the voltages v_C and v_R and the current i_C for the decay phase.

25. The 1000-μF capacitor of Fig. 10.81 is charged to 6 V. To discharge the capacitor before further use, a wire with a resistance of 0.002 Ω is placed across the capacitor.

a. How long will it take to discharge the capacitor?

b. What is the peak value of the current?

c. Based on the answer to part (b), is a spark expected when contact is made with both ends of the capacitor?

SECTION 10.9 Initial Values

26. The capacitor in Fig. 10.82 is initially charged to 3 V with the polarity shown.

a. Find the mathematical expressions for the voltage v_C and the current i_C when the switch is closed.

b. Sketch the waveforms for v_C and i_C.

***27.** The capacitor of Fig. 10.83 is initially charged to 12 V with the polarity shown.

a. Find the mathematical expressions for the voltage v_C and the current i_C when the switch is closed.

b. Sketch the waveforms for v_C and i_C.

SECTION 10.10 Instantaneous Values

28. Given the expression $v_C = 8(1 - e^{-t/(20\times10^{-6})})$:

a. Determine v_C after five time constants.

b. Determine v_C after 10 time constants.

c. Determine v_C at $t = 5\ \mu s$.

29. For the situation of Problem 25, determine when the discharge current is one-half its maximum value if contact is made at $t = 0$ s.

30. For the network of Fig. 10.84, V_L must be 8 V before the system is activated. If the switch is closed at $t = 0$ s, how long will it take for the system to be activated?

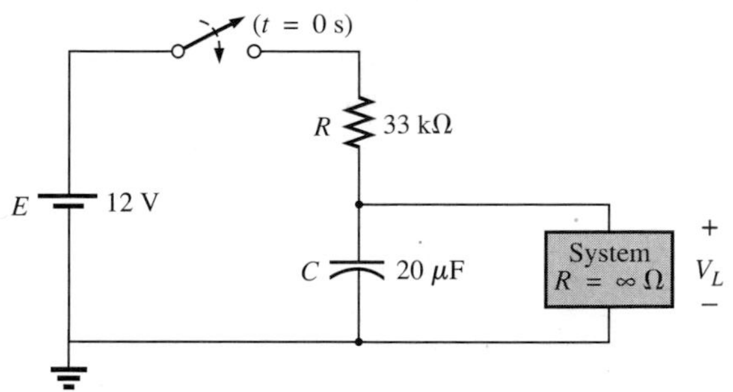

FIG. 10.84
Problem 30.

*31. Design the network of Fig. 10.85 such that the system will turn on 10 s after the switch is closed.

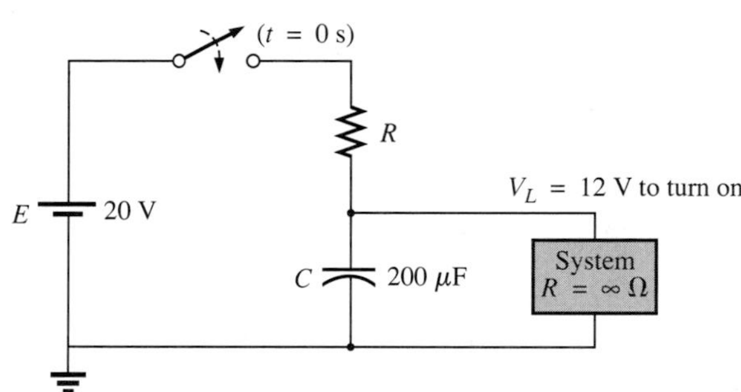

FIG. 10.85
Problem 31.

32. For the circuit of Fig. 10.86:
   a. Find the time required for v_C to reach 60 V following the closing of the switch.
   b. Calculate the current i_C at the instant $v_C = 60$ V.
   c. Determine the power delivered by the source at the instant $t = 2\tau$.

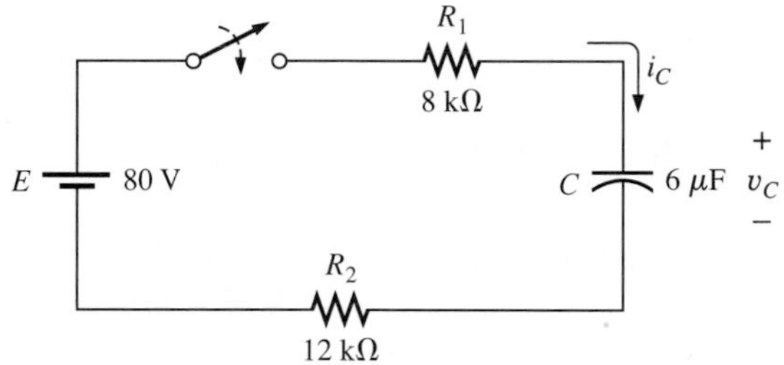

FIG. 10.86
Problem 32.

*33. For the network of Fig. 10.87:
   a. Calculate v_C, i_C, and v_{R_1} at 0.5 s and 1 s after the switch makes contact with position 1.
   b. The network sits in position 1 10 min before the switch is moved to position 2. How long after making contact with position 2 will it take for the current i_C to drop to 8 μA? How much *longer* will it take for v_C to drop to 10 V?

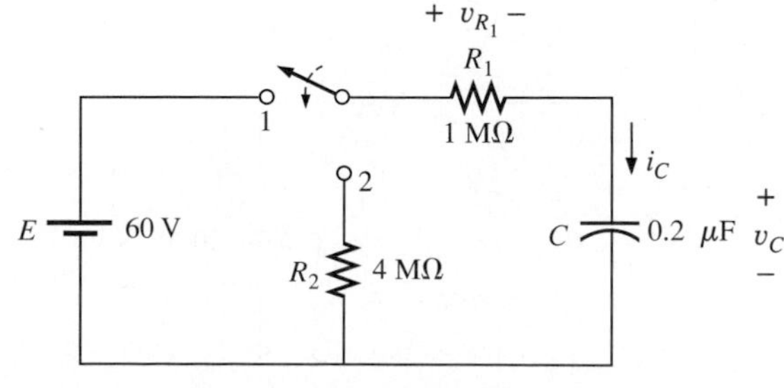

FIG. 10.87
Problem 33.

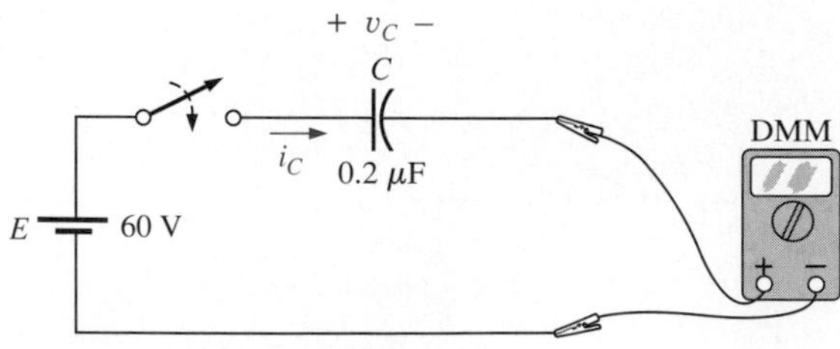

FIG. 10.88
Problem 34.

34. For the system of Fig. 10.88, using a DMM with a 10-MΩ internal resistance in the voltmeter mode:

a. Determine the voltmeter reading 1 time constant after the switch is closed.

b. Find the current i_C 2 time constants after the switch is closed.

c. Calculate the time that must pass after the closing of the switch for the voltage v_C to be 50 V.

SECTION 10.11 $\tau = R_{Th}C$

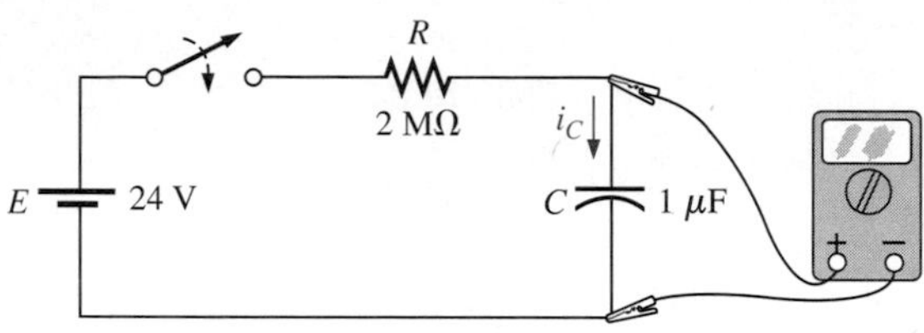

FIG. 10.89
Problem 35.

35. For the system of Fig. 10.89, using a DMM with a 10-MΩ internal resistance in the voltmeter mode:

a. Determine the voltmeter reading 4 time constants after the switch is closed.

b. Find the time that must pass before i_C drops to 3 μA.

c. Find the time that must pass after the closing of the switch for the voltage across the meter to reach 10 V.

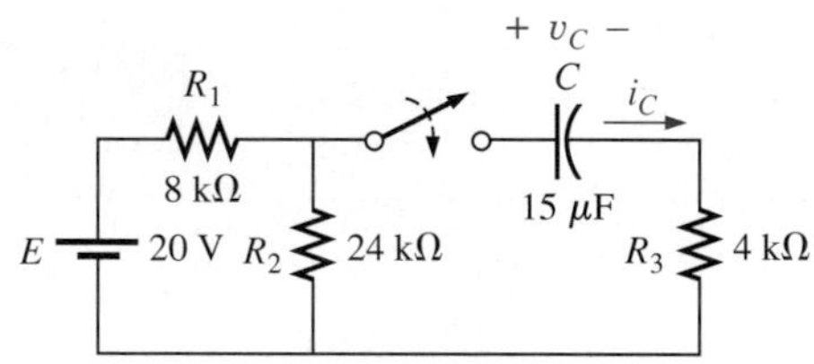

FIG. 10.90
Problem 36.

36. For the circuit of Fig. 10.90:

a. Find the mathematical expressions for the transient behavior of the voltage v_C and the current i_C following the closing of the switch.

b. Sketch the waveforms of v_C and i_C.

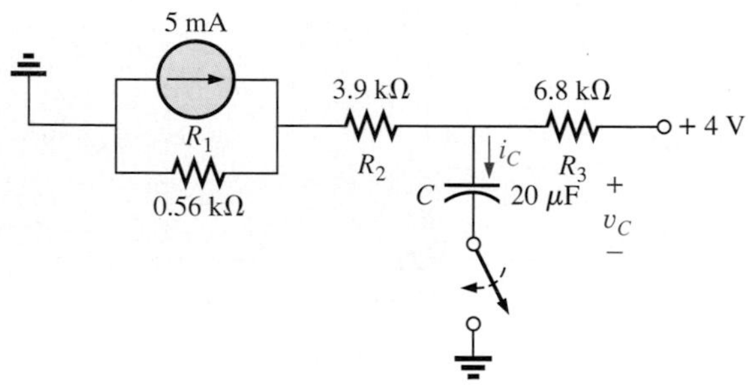

FIG. 10.91
Problems 37 and 58.

***37.** Repeat Problem 36 for the circuit of Fig. 10.91.

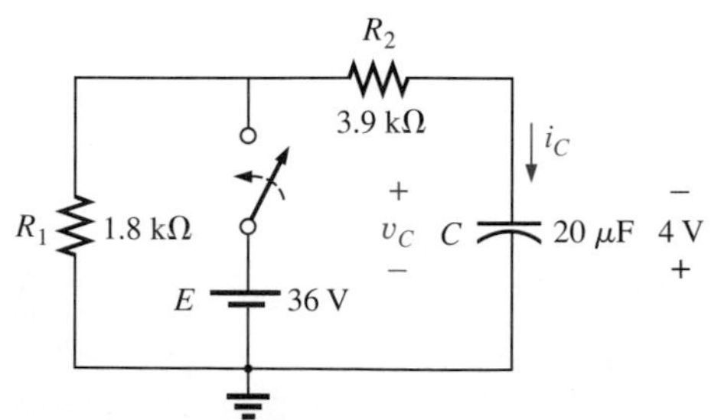

FIG. 10.92
Problem 38.

38. The capacitor of Fig. 10.92 is initially charged to 4 V with the polarity shown.

a. Write the mathematical expressions for the voltage v_C and the current i_C when the switch is closed.

b. Sketch the waveforms of v_C and i_C.

39. The capacitor of Fig. 10.93 is initially charged to 2 V with the polarity shown.

a. Write the mathematical expressions for the voltage v_C and the current i_C when the switch is closed.

b. Sketch the waveforms of v_C and i_C.

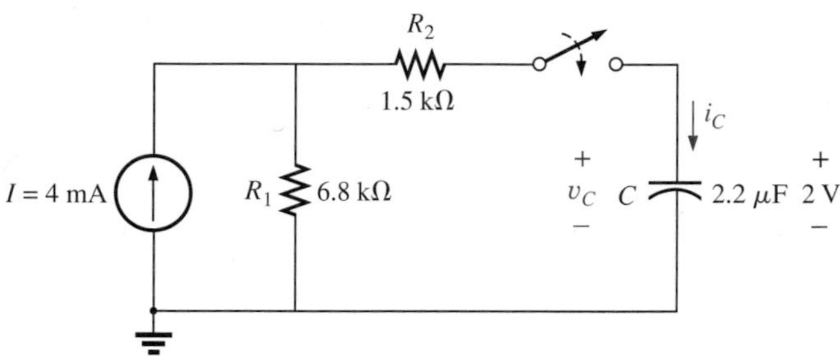

FIG. 10.93
Problem 39.

***40.** The capacitor of Fig. 10.94 is initially charged to 3 V with the polarity shown.

a. Write the mathematical expressions for the voltage v_C and the current i_C when the switch is closed.

b. Sketch the waveforms of v_C and i_C.

+10 V 2 kΩ 6.8 kΩ + v_C − C −20 V i_C 39 μF + 3 V −

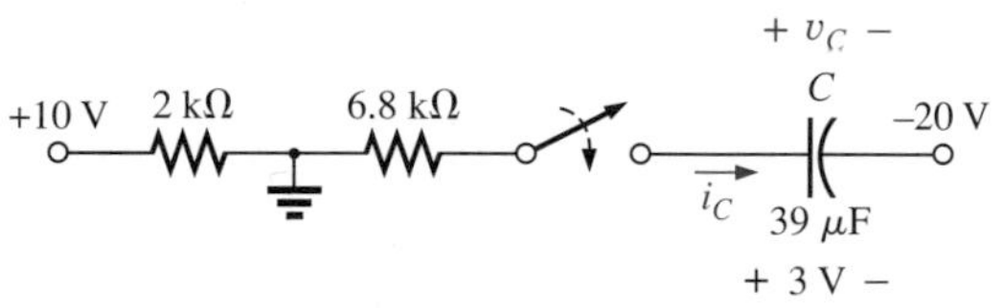

FIG. 10.94
Problem 40.

SECTION 10.12 The Current i_C

41. Find the waveform for the average current if the voltage across a 0.06-μF capacitor is as shown in Fig. 10.95.

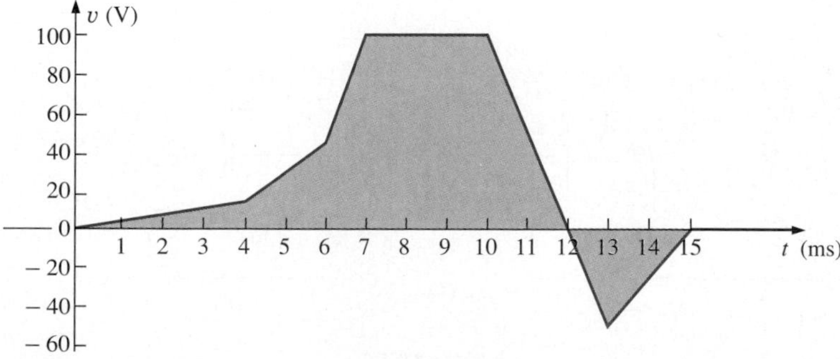

FIG. 10.95
Problem 41.

42. Repeat Problem 41 for the waveform of Fig. 10.96.

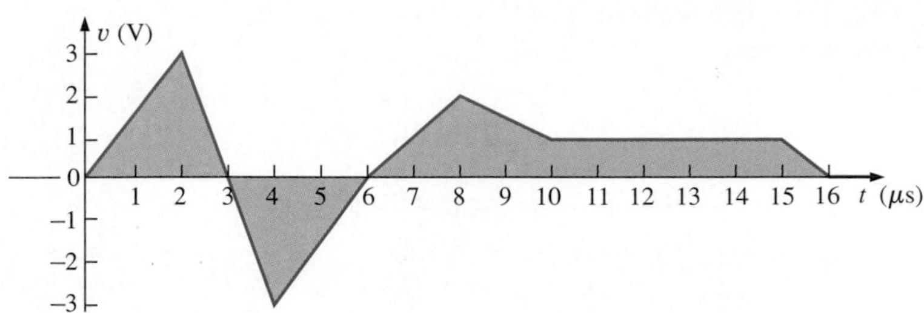

FIG. 10.96
Problem 42.

***43.** Given the following waveform (Fig. 10.97) for the current of a 20-μF capacitor, sketch the waveform of the voltage v_C across the capacitor if $v_C = 0$ V at $t = 0$ s.

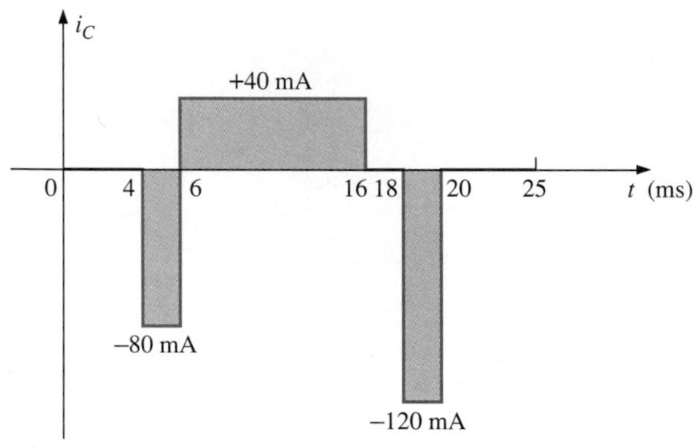

FIG. 10.97
Problem 43.

SECTION 10.13 Capacitors in Series and Parallel

44. Find the total capacitance C_T between points a and b of the circuits of Fig. 10.98.

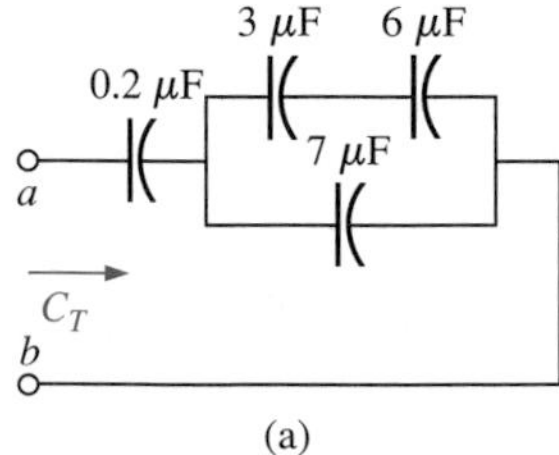

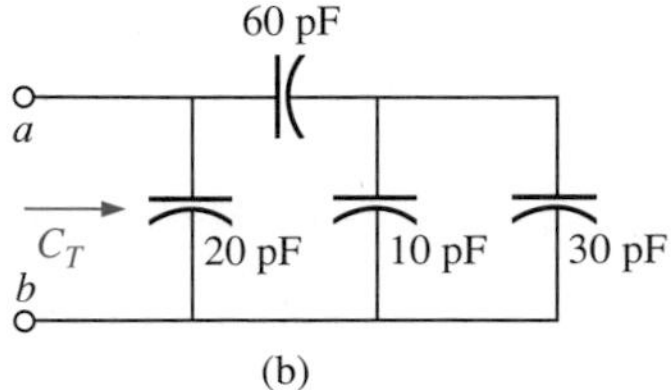

FIG. 10.98
Problem 44.

45. Find the voltage across and charge on each capacitor for the circuits of Fig. 10.99.

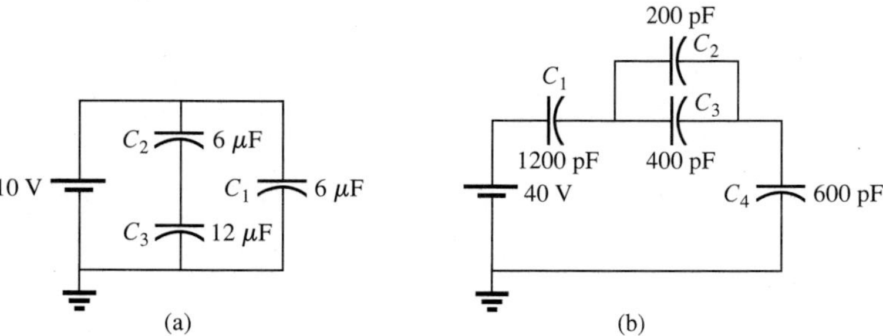

FIG. 10.99
Problem 45.

***46.** For each configuration of Fig. 10.100, determine the voltage across each capacitor and the charge on each capacitor.

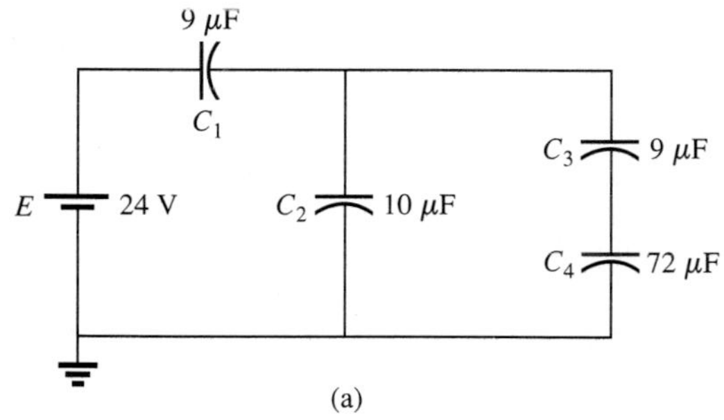

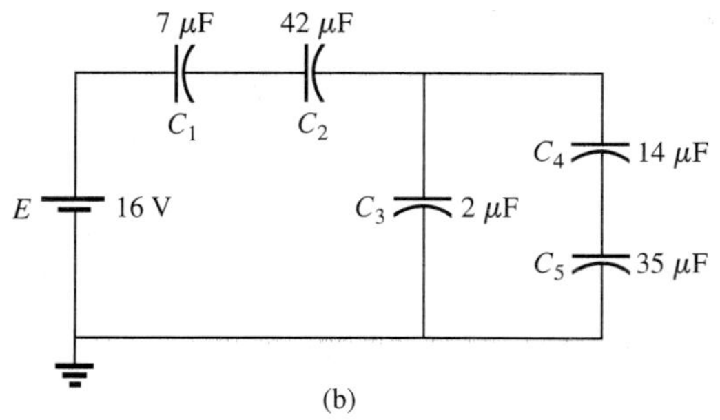

FIG. 10.100
Problem 46.

***47.** For the network of Fig. 10.101, determine the following 100 ms after the switch is closed.

a. V_{ab}
b. V_{ac}
c. V_{cb}
d. V_{da}
e. If the switch is moved to position 2 one hour later, find the time required for v_{R_2} to drop to 20 V.

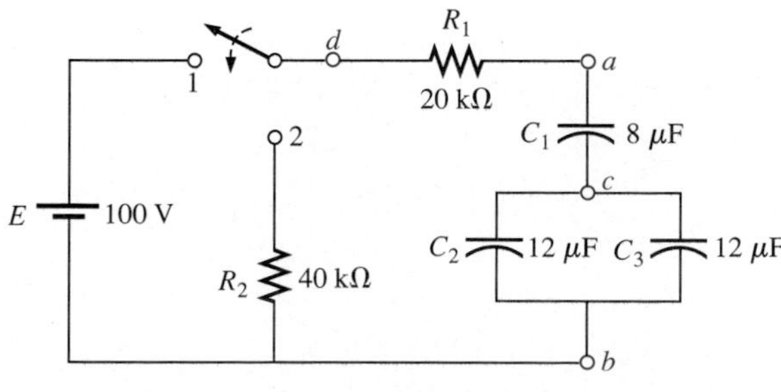

FIG. 10.101
Problem 47.

48. For the circuits of Fig. 10.102, find the voltage across and charge on each capacitor after each capacitor has charged to its final value.

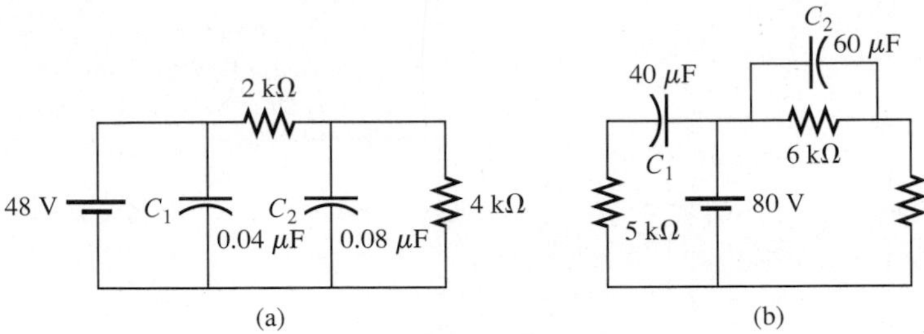

FIG. 10.102
Problem 48.

SECTION 10.14 Energy Stored by a Capacitor

49. Find the energy stored by a 120-pF capacitor with 12 V across its plates.

50. If the energy stored by a 6-μF capacitor is 1200 J, find the charge Q on each plate of the capacitor.

*51. An electronic flashgun has a 1000-μF capacitor that is charged to 100 V.
   a. How much energy is stored by the capacitor?
   b. What is the charge on the capacitor?
   c. When the photographer takes a picture, the flash fires for 1/2000 s. What is the average current through the flashtube?
   d. Find the power delivered to the flashtube.
   e. After a picture is taken, the capacitor has to be recharged by a power supply that delivers a maximum current of 10 mA. How long will it take to charge the capacitor?

52. For the network of Fig. 10.103:
   a. Determine the energy stored by each capacitor under steady-state conditions.
   b. Repeat part (a) if the capacitors are in series.

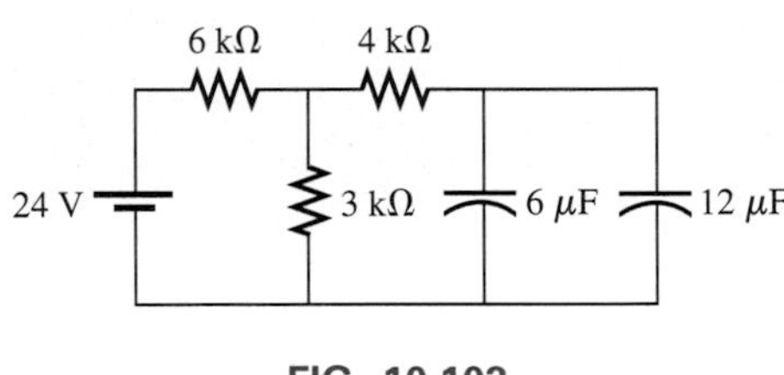

FIG. 10.103
Problem 52

SECTION 10.16 Computer Analysis

PSpice (DOS)

53. Write the input file to obtain the waveforms of Fig. 10.35 for the network of Fig. 10.34.

54. Write the input file to obtain the waveforms of v_C and i_C for the network of Fig. 10.43 if the switch is moved to position 1 at $t = 0$ s.

*55. Write the input file to obtain the waveforms of v_C and i_C for the network of Fig. 10.50 if the switch is moved to position 1 at $t = 0$ s.

PSpice (Windows)

56. Using schematics:
   a. Obtain the waveforms for v_C and i_C versus time for the network of Fig. 10.34.
   b. Obtain the power curve (representing the energy stored by the capacitor over the same time interval) and compare it to the plot of Fig. 10.68.

***57.** Using schematics, obtain the waveforms of v_C and i_C versus time for the network of Fig. 10.48 using the IC option.

58. Verify your solution to Problem 37 using schematics.

Programming Language (C++, BASIC, PASCAL, etc.)

59. Write a BASIC program to tabulate the voltage v_C and current i_C for the network of Fig. 10.43 for five time constants after the switch is moved to position 1 at $t = 0$ s. Use an increment of $(1/5)\tau$.

***60.** Write a program to write the mathematical expression for the voltage v_C for the network of Fig. 10.50 for any element values when the switch is moved to position 1.

***61.** Given three capacitors in any series-parallel arrangement, write a program to determine the total capacitance; that is, determine the total number of possibilities and ask the user to identify the configuration and provide the capacitor values. Then calculate the total capacitance.

GLOSSARY

Breakdown voltage Another term for *dielectric strength,* listed below.

Capacitance A measure of a capacitor's ability to store charge; measured in farads (F).

Capacitive time constant The product of resistance and capacitance that establishes the required time for the charging and discharging phases of a capacitive transient.

Capacitive transient The waveforms for the voltage and current of a capacitor that result during the charging and discharging phases.

Capacitor A fundamental electrical element having two conducting surfaces separated by an insulating material and having the capacity to store charge on its plates.

Coulomb's law An equation relating the force between two like or unlike charges.

Dielectric The insulating material between the plates of a capacitor that can have a pronounced effect on the charge stored on the plates of a capacitor.

Dielectric constant Another term for *relative permittivity,* listed below.

Dielectric strength An indication of the voltage required for unit length to establish conduction in a dielectric.

Electric field strength The force acting on a unit positive charge in the region of interest.

Electric flux lines Lines drawn to indicate the strength and direction of an electric field in a particular region.

Fringing An effect established by flux lines that do not pass directly from one conducting surface to another.

Leakage current The current that will result in the total discharge of a capacitor if the capacitor is disconnected from the charging network for a sufficient length of time.

Permittivity A measure of how well a dielectric will *permit* the establishment of flux lines within the dielectric.

Relative permittivity The permittivity of a material compared to that of air.

Stray capacitance Capacitances that exist not through design but simply because two conducting surfaces are relatively close to each other.

Surge voltage The maximum voltage that can be applied across the capacitor for very short periods of time.

.TRAN PSpice command to specify a transient analysis.

TSTEP PSpice entry to specify the interval between values to be determined.

TSTOP PSpice entry to specify the final value for which function is to be determined.

Working voltage The voltage that can be applied across a capacitor for long periods of time without concern for dielectric breakdown.

11

Magnetic Circuits

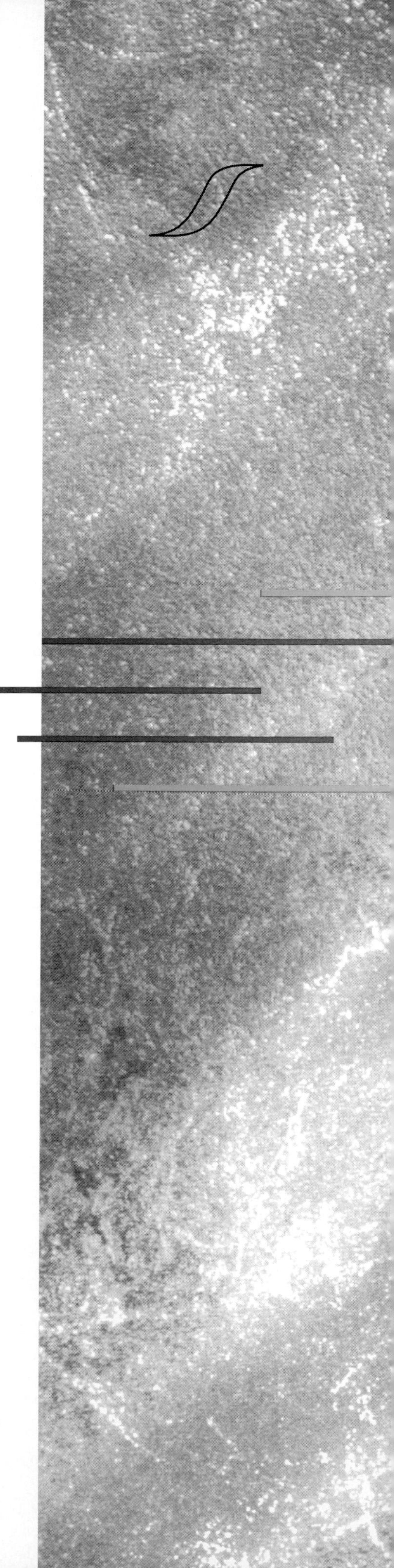

11.1 INTRODUCTION

Magnetism plays an integral part in almost every electrical device used today in industry, research, or the home. Generators, motors, transformers, circuit breakers, televisions, computers, tape recorders, and telephones all employ magnetic effects to perform a variety of important tasks.

The compass, used by Chinese sailors as early as the second century A.D., relies on a *permanent magnet* for indicating direction. The permanent magnet is made of a material, such as steel or iron, that will remain magnetized for long periods of time without the need for an external source of energy.

In 1820, the Danish physicist Hans Christian Oersted discovered that the needle of a compass would deflect if brought near a current-carrying conductor. For the first time it was demonstrated that electricity and magnetism were related, and in the same year the French physicist André-Marie Ampère performed experiments in this area and developed what is presently known as *Ampère's circuital law.* In subsequent years, men such as Michael Faraday, Karl Friedrich Gauss, and James Clerk Maxwell continued to experiment in this area and developed many of the basic concepts of *electromagnetism*—magnetic effects induced by the flow of charge, or current.

There is a great deal of similarity between the analyses of electric circuits and magnetic circuits. This will be demonstrated later in this chapter when we compare the basic equations and methods used to solve magnetic circuits with those used for electric circuits.

Difficulty in understanding methods used with magnetic circuits will often arise in simply learning to use the proper set of units, not because of the equations themselves. The problem exists because three different systems of units are still used in the industry. To the extent practical, SI will be used throughout this chapter. For the CGS and English systems, a conversion table is provided in Appendix G.

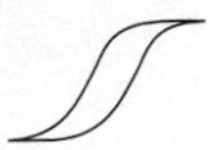

11.2 MAGNETIC FIELDS

In the region surrounding a permanent magnet there exists a magnetic field, which can be represented by magnetic flux lines similar to electric flux lines. Magnetic flux lines, however, do not have origins or terminating points like electric flux lines but exist in continuous loops, as shown in Fig. 11.1. The symbol for magnetic flux is the Greek letter Φ (phi).

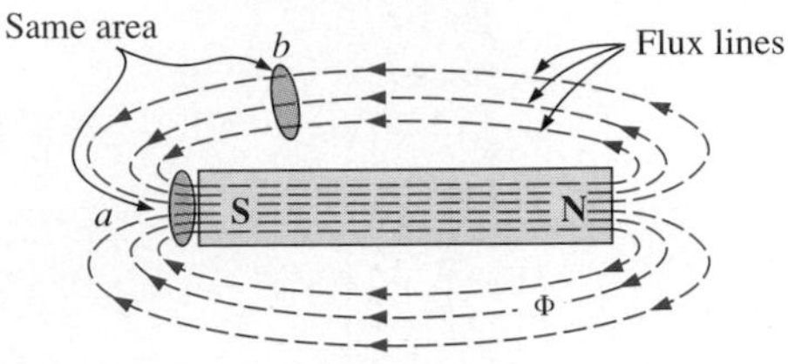

FIG. 11.1
Flux distribution for a permanent magnet.

The magnetic flux lines radiate from the north pole to the south pole, returning to the north pole through the metallic bar. Note the equal spacing between the flux lines within the core and the symmetric distribution outside the magnetic material. These are additional properties of magnetic flux lines in homogeneous materials (that is, materials having uniform structure or composition throughout). It is also important to realize that the continuous magnetic flux line will strive to occupy as small an area as possible. This will result in magnetic flux lines of minimum length between the like poles, as shown in Fig. 11.2. The strength of a magnetic field in a particular region is directly related to the density of flux lines in that region. In Fig. 11.1, for example, the magnetic field strength at *a* is twice that at *b* since there are twice as many magnetic flux lines associated with the perpendicular plane at *a* than at *b*. Recall from childhood experiments how the strength of permanent magnets was always stronger near the poles.

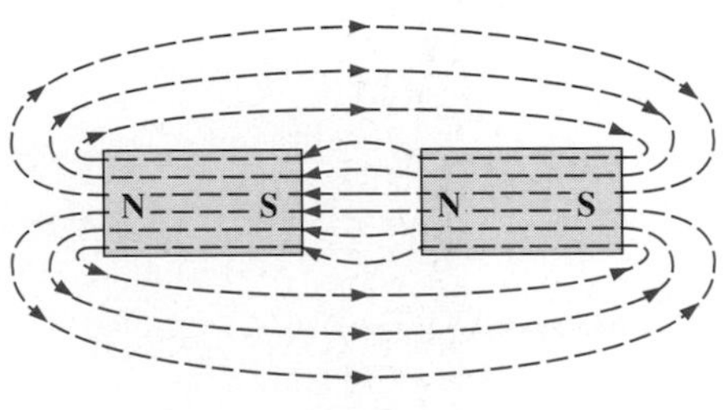

FIG. 11.2
Flux distribution for two adjacent, opposite poles.

If unlike poles of two permanent magnets are brought together, the magnets will attract, and the flux distribution will be as shown in Fig. 11.2. If like poles are brought together, the magnets will repel, and the flux distribution will be as shown in Fig. 11.3.

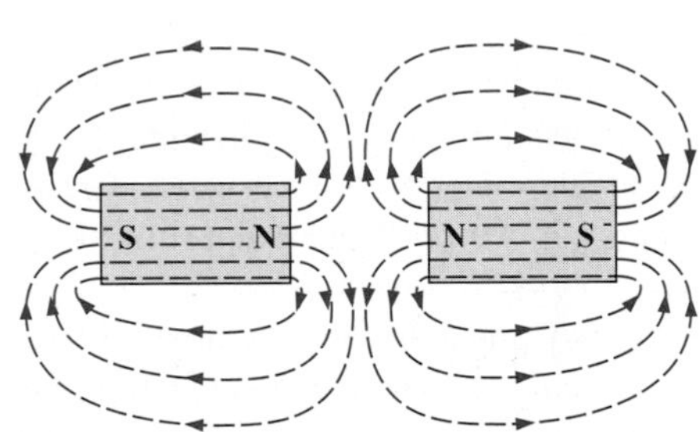

FIG. 11.3
Flux distribution for two adjacent, like poles.

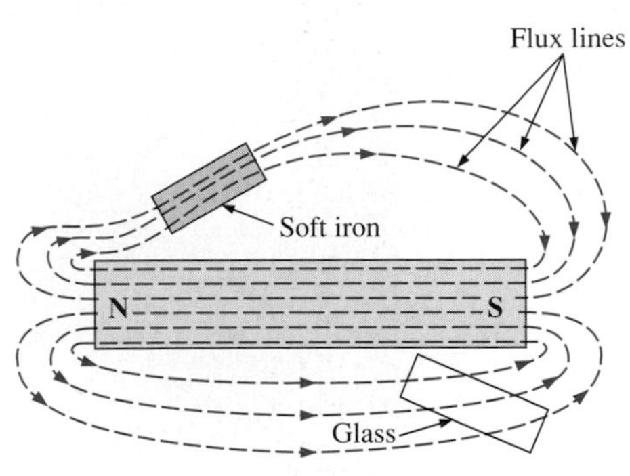

FIG. 11.4
Effect of a ferromagnetic sample on the flux distribution of a permanent magnet.

If a nonmagnetic material, such as glass or copper, is placed in the flux paths surrounding a permanent magnet, there will be an almost unnoticeable change in the flux distribution (Fig. 11.4). However, if a magnetic material, such as soft iron, is placed in the flux path, the flux lines will pass through the soft iron rather than the surrounding air because flux lines pass with greater ease through magnetic materials than through air. This principle is put to use in the shielding of sensitive electrical elements and instruments that can be affected by stray magnetic fields (Fig. 11.5).

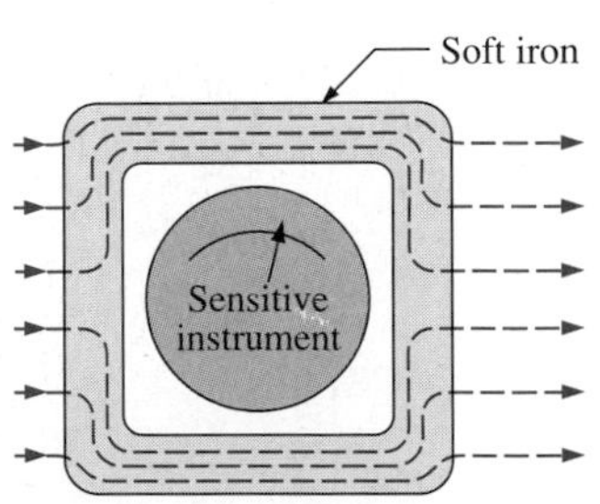

FIG. 11.5
Effect of a magnetic shield on the flux distribution.

As indicated in the introduction, a magnetic field (represented by concentric magnetic flux lines, as in Fig. 11.6) is present around every wire that carries an electric current. The direction of the magnetic flux lines can be found simply by placing the thumb of the *right* hand in the direction of *conventional* current flow and noting the direction of the

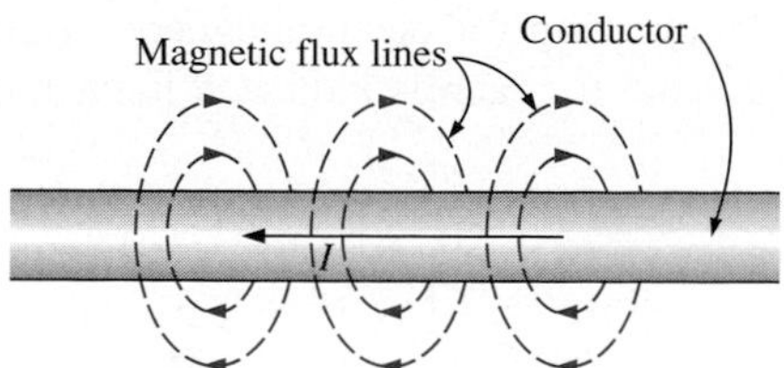

FIG. 11.6
Magnetic flux lines around a current-carrying conductor.

fingers. (This method is commonly called the *right-hand rule.*) If the conductor is wound in a single-turn coil (Fig. 11.7), the resulting flux will flow in a common direction through the center of the coil. A coil of more than one turn would produce a magnetic field that would exist in a continuous path through and around the coil (Fig. 11.8).

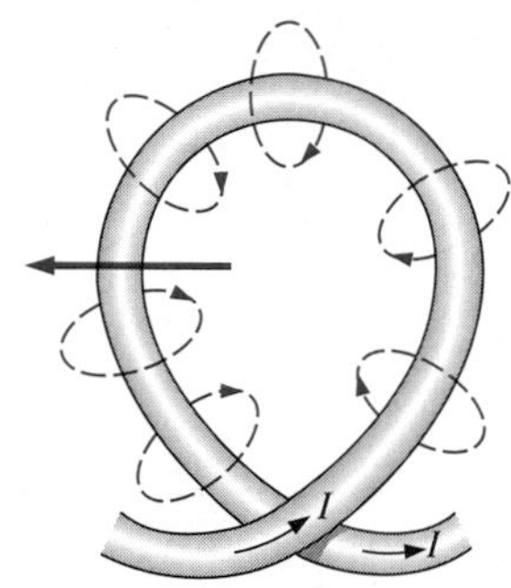

FIG. 11.7
Flux distribution of a single-turn coil.

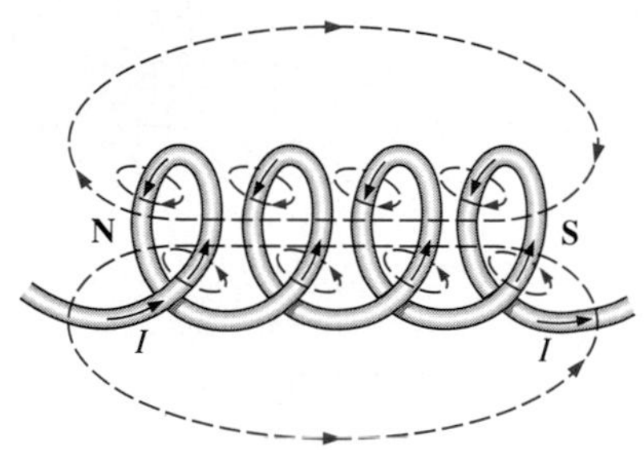

FIG. 11.8
Flux distribution of a current-carrying coil.

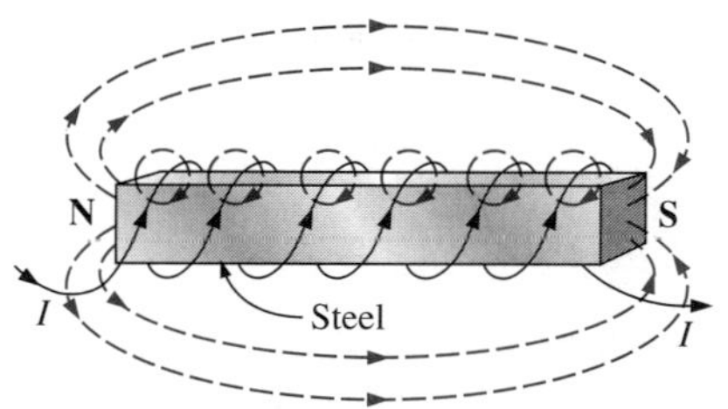

FIG. 11.9
Electromagnet.

The flux distribution of the coil is quite similar to that of the permanent magnet. The flux lines leaving the coil from the left and entering to the right simulate a north and south pole, respectively. The principal difference between the two flux distributions is that the flux lines are more concentrated for the permanent magnet than for the coil. Also, since the strength of a magnetic field is determined by the density of the flux lines, the coil has a weaker field strength. The field strength of the coil can be effectively increased by placing certain materials, such as iron, steel, or cobalt, within the coil to increase the flux density within the coil. By increasing the field strength with the addition of the core, we have devised an *electromagnet* (Fig. 11.9) that, in addition to having all the properties of a permanent magnet, also has a field strength that can be varied by changing one of the component values (current, turns, and so on). Of course, current must pass through the coil of the electromagnet in order for magnetic flux to be developed, whereas there is no need for the coil or current in the permanent magnet. The direction of flux lines can be determined for the electromagnet (or in any core with a wrapping of turns) by placing the fingers of the right hand in the direction of current flow around the core. The thumb will then point in the direction of the north pole of the induced magnetic flux, as demonstrated in Fig. 11.10(a). A cross section of the same electromagnet is included as Fig. 11.10(b) to introduce the convention for directions perpendicular to the page. The cross and dot refer to the tail and head of the arrow, respectively.

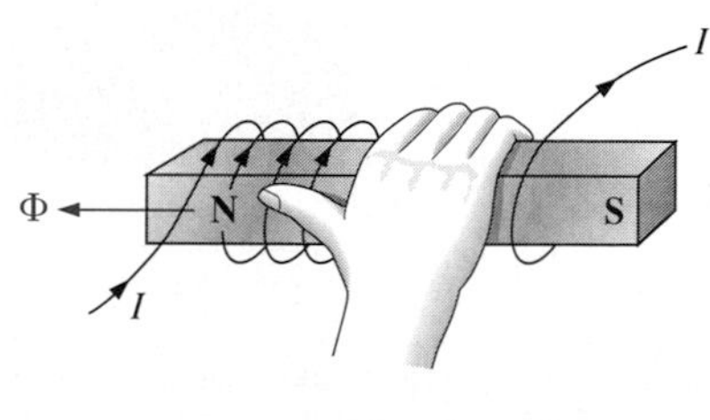

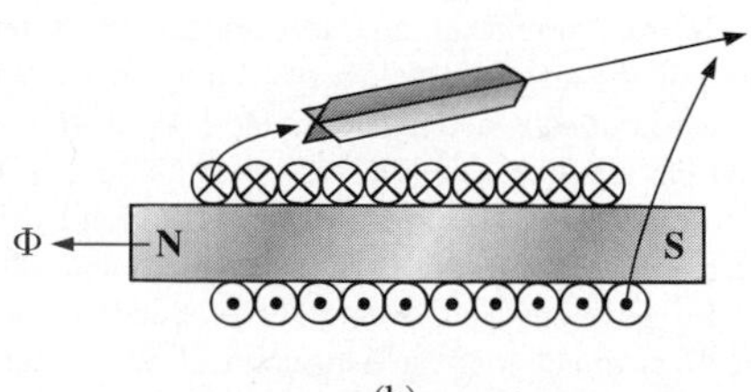

FIG. 11.10
Determining the direction of flux for an electromagnet: (a) method, (b) notation.

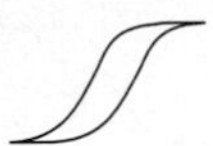

Other areas of application for electromagnetic effects are shown in Fig. 11.11. The flux path for each is indicated in each figure.

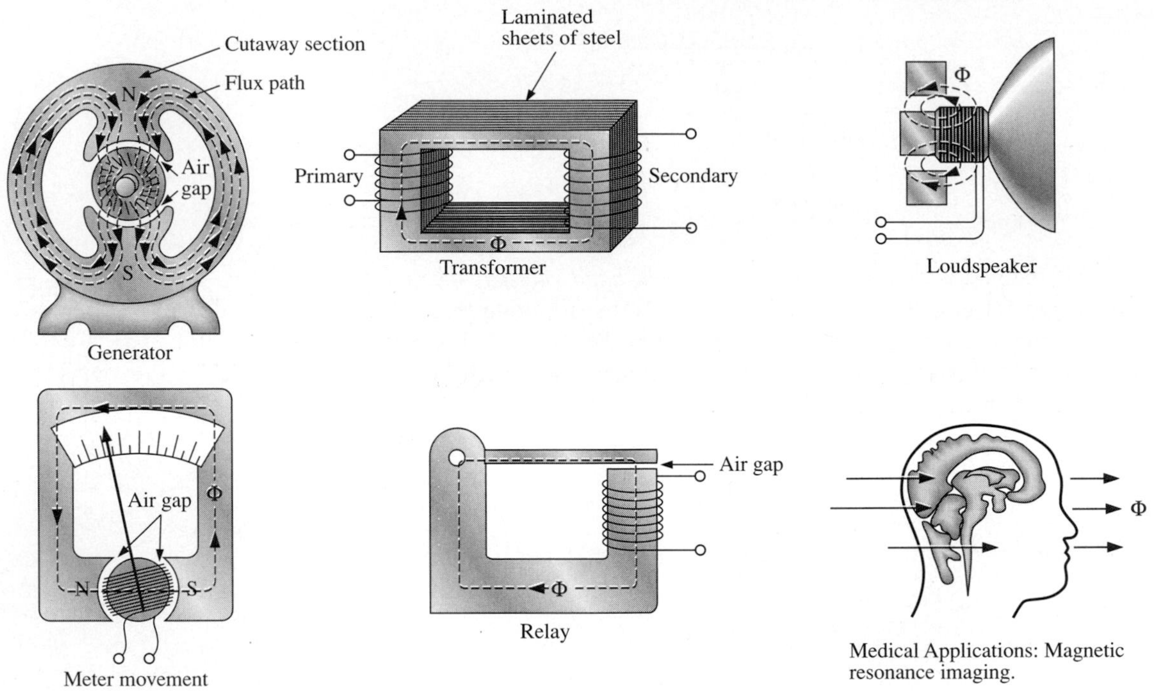

FIG. 11.11
Some areas of application of magnetic effects.

German (Wittenberg, Göttingen)
(1804–1891)
Physicist
Professor of Physics, University of Göttingen

Courtesy of the Smithsonian Institution Photo No. 52,604

An important contributor to the establishment of a system of *absolute units* for the electrical sciences, which was beginning to become a very active area of research and development. Established a definition of electric current in an electromagnetic system based on the magnetic field produced by the current. He was politically active and, in fact, was dismissed from the faculty of the University of Göttingen for protesting the suppression of the constitution by the King of Hanover in 1837. However, he found other faculty positions and eventually returned to Göttingen as director of the astronomical observatory. Received honors from England, France, and Germany, including the Copley Medal of the Royal Society.

FIG. 11.12
Wilhelm Eduard Weber.

11.3 FLUX DENSITY

In the SI system of units, magnetic flux is measured in *webers* (note Fig. 11.12) and has the symbol Φ. The number of flux lines per unit area is called the *flux density,* is denoted by the capital letter B, and is measured in *teslas* (note Fig. 11.15). Its magnitude is determined by the following equation:

$$B = \frac{\Phi}{A} \qquad \begin{aligned} B &= \text{teslas (T)} \\ \Phi &= \text{webers (Wb)} \\ A &= \text{square meters (m}^2\text{)} \end{aligned} \tag{11.1}$$

where Φ is the number of flux lines passing through the area A (Fig. 11.13). The flux density at position a in Fig. 11.1 is twice that at b because twice as many flux lines are passing through the same area.

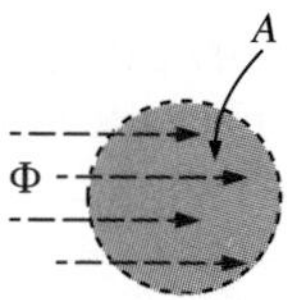

FIG. 11.13
Defining the flux density B.

By definition,

$$1\ \text{T} = 1\ \text{Wb/m}^2$$

EXAMPLE 11.1 For the core of Fig. 11.14, determine the flux density B in teslas.

Solution:

$$B = \frac{\Phi}{A} = \frac{6 \times 10^{-5}\text{ Wb}}{1.2 \times 10^{-3}\text{ m}^2} = \mathbf{5 \times 10^{-2}\text{ T}}$$

EXAMPLE 11.2 In Fig. 11.14, if the flux density is 1.2 T and the area is 0.25 in.^{2}, determine the flux through the core.

Solution: By Eq. (11.1),

$$\Phi = BA$$

However, converting 0.25 in.^{2} to metric units,

$$A = 0.25\text{ in.}^2\left(\frac{1\text{ m}}{39.37\text{ in.}}\right)\left(\frac{1\text{ m}}{39.37\text{ in.}}\right) = 1.613 \times 10^{-4}\text{ m}^2$$

and

$$\Phi = (1.2\text{ T})(1.613 \times 10^{-4}\text{ m}^2) = \mathbf{1.936 \times 10^{-4}\text{ Wb}}$$

An instrument designed to measure flux density in gauss (CGS system) appears in Fig. 11.16. Appendix G reveals that 1 T = 10^4 gauss. The magnitude of the reading appearing on the face of the meter in Fig. 11.16 is therefore

$$1.964\text{ gauss}\left(\frac{1\text{ T}}{10^4\text{ gauss}}\right) = 1.964 \times 10^{-4}\text{ T}$$

11.4 PERMEABILITY

If cores of different materials with the same physical dimensions are used in the electromagnet described in Section 11.2, the strength of the magnet will vary in accordance with the core used. This variation in strength is due to the greater or lesser number of flux lines passing through the core. Materials in which flux lines can readily be set up are said to be *magnetic* and to have *high permeability.* The permeability (μ) of a material, therefore, is a measure of the ease with which magnetic flux lines can be established in the material. It is similar in many respects to conductivity in electric circuits. The permeability of free space μ_o (vacuum) is

$$\mu_o = 4\pi \times 10^{-7}\,\frac{\text{Wb}}{\text{A}\cdot\text{m}}$$

As indicated above, μ has the units of Wb/A·m. Practically speaking, the permeability of all nonmagnetic materials, such as copper, aluminum, wood, glass, and air, is the same as that for free space. Materials that have permeabilities slightly less than that of free space are said to be *diamagnetic,* and those with permeabilities slightly greater than that of free space are said to be *paramagnetic*. Magnetic materials, such as iron, nickel, steel, cobalt, and alloys of these metals, have permeabilities hundreds and even thousands of times that of free space. Materials with these very high permeabilities are referred to as *ferromagnetic.*

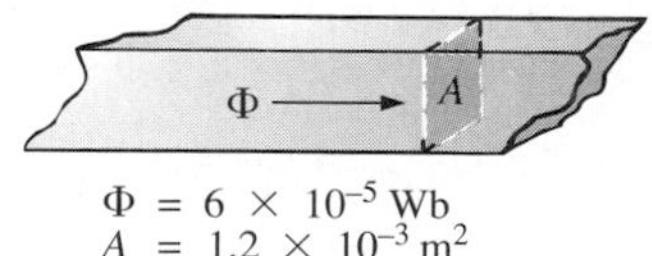

$\Phi = 6 \times 10^{-5}$ Wb
$A = 1.2 \times 10^{-3}$ m^2

FIG. 11.14
Example 11.1.

Croatian-American
(Smiljan, Paris, Colorado Springs, New York City)
(1856–1943)
Electrical Engineer and Inventor
Recipient of the Edison Medal in 1917

Courtesy of the Smithsonian Institution Photo No. 52,223

Nikola Tesla is often regarded as one of the most innovative and inventive individuals in the history of the sciences. He was the first to introduce the *alternating-current machine,* removing the need for commutator bars of dc machines. After emigrating to the United States in 1884, he sold a number of his patents on *ac machines, transformers,* and *induction coils* (including the *Tesla coil* as we know it today) to the Westinghouse Electric Company. Some say that his most important discovery was made at his laboratory in Colorado Springs, where in 1900 he discovered *terrestrial stationary waves.* The range of his discoveries and inventions is too extensive to list here but extends from lighting systems to *polyphase power systems* to a *wireless world broadcasting system.*

FIG. 11.15
Nikola Tesla.

FIG. 11.16
Digital display gaussmeter. (Courtesy of LDJ Electronics, Inc.)

The ratio of the permeability of a material to that of free space is called its *relative permeability*; that is,

$$\mu_r = \frac{\mu}{\mu_o} \tag{11.2}$$

In general, for ferromagnetic materials, $\mu_r \geq 100$, and for nonmagnetic materials, $\mu_r = 1$.

Since μ_r is a variable, dependent on other quantities of the magnetic circuit, values of μ_r are not tabulated. Methods of calculating μ_r from the data supplied by manufacturers will be considered in a later section.

11.5 RELUCTANCE

The resistance of a material to the flow of charge (current) is determined for electric circuits by the equation

$$R = \rho\frac{l}{A} \qquad \text{(ohms, } \Omega\text{)}$$

The *reluctance* of a material to the setting up of magnetic flux lines in the material is determined by the following equation:

$$\mathcal{R} = \frac{l}{\mu A} \qquad \text{(rels, or At/Wb)} \tag{11.3}$$

where $\mathcal{R}$ is the reluctance, l is the length of the magnetic path, and A is its cross-sectional area. The t in the units At/Wb is the number of turns of the applied winding. More is said about ampere-turns (At) in the next section. Note that the resistance and reluctance are inversely proportional to the area, indicating that an increase in area will result in a reduction in each and an *increase* in the desired result: current and flux. For an increase in length the opposite is true, and the desired effect is reduced. The reluctance, however, is inversely proportional to the permeability, while the resistance is directly proportional to the resistivity. The larger the μ or the smaller the ρ, the smaller the reluctance and resistance, respectively. Obviously, therefore, materials with high permeability, such as the ferromagnetics, have very small reluctances and will result in an increased measure of flux through the core. There is no widely accepted unit for reluctance, although the *rel* and the At/Wb are usually applied.

11.6 OHM'S LAW FOR MAGNETIC CIRCUITS

Recall the equation

$$\text{Effect} = \frac{\text{cause}}{\text{opposition}}$$

appearing in Chapter 4 to introduce Ohm's law for electric circuits. For magnetic circuits, the effect desired is the flux Φ. The cause is the *magnetomotive force* (mmf) $\mathcal{F}$, which is the external force (or "pressure") required to set up the magnetic flux lines within the magnetic material. The opposition to the setting up of the flux Φ is the reluctance $\mathcal{R}$.

Substituting, we have

$$\Phi = \frac{\mathscr{F}}{\mathscr{R}} \tag{11.4}$$

The magnetomotive force $\mathscr{F}$ is proportional to the product of the number of turns around the core (in which the flux is to be established) and the current through the turns of wire (Fig. 11.17). In equation form,

$$\mathscr{F} = NI \qquad \text{(ampere-turns, At)} \tag{11.5}$$

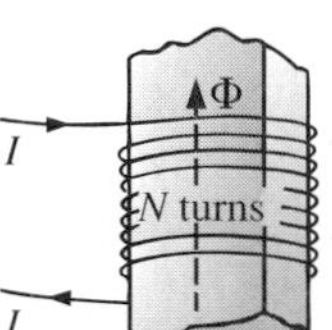

FIG. 11.17
Defining the components of a magnetomotive force.

This equation clearly indicates that an increase in the number of turns or the current through the wire will result in an increased "pressure" on the system to establish flux lines through the core.

Although there is a great deal of similarity between electric and magnetic circuits, one must continue to realize that the flux Φ is not a "flow" variable such as current in an electric circuit. Magnetic flux is established in the core through the alteration of the atomic structure of the core due to external pressure and is not a measure of the flow of some charged particles through the core.

11.7 MAGNETIZING FORCE

The magnetomotive force per unit length is called the *magnetizing force* (H). In equation form,

$$H = \frac{\mathscr{F}}{l} \qquad \text{(At/m)} \tag{11.6}$$

Substituting for the magnetomotive force will result in

$$H = \frac{NI}{l} \qquad \text{(At/m)} \tag{11.7}$$

For the magnetic circuit of Fig. 11.18, if $NI = 40$ At and $l = 0.2$ m, then

$$H = \frac{NI}{l} = \frac{40\text{ At}}{0.2\text{ m}} = 200\text{ At/m}$$

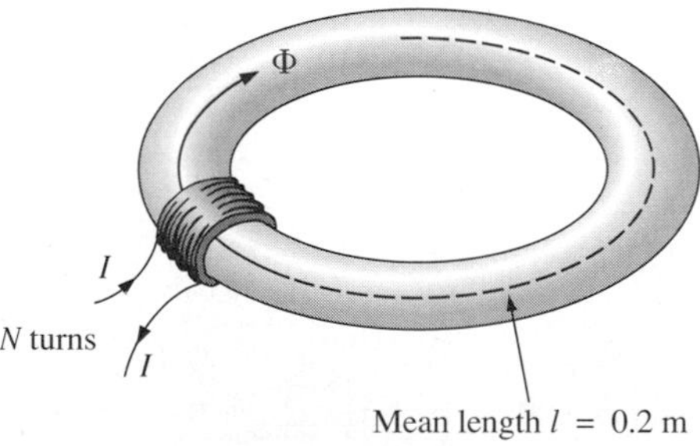

FIG. 11.18
Defining the magnetizing force of a magnetic circuit.

In words, the result indicates that there are 200 At of "pressure" per meter to establish flux in the core.

Note in Fig. 11.18 that the direction of the flux Φ can be determined by placing the fingers of the right hand in the direction of current around the core and noting the direction of the thumb. It is interesting to realize that *the magnetizing force is independent of the type of core material*—it is determined solely by the number of turns, the current, and the length of the core.

The applied magnetizing force has a pronounced effect on the resulting permeability of a magnetic material. As the magnetizing force increases, the permeability rises to a maximum and then drops to a minimum, as shown in Fig. 11.19 for three commonly employed magnetic materials.

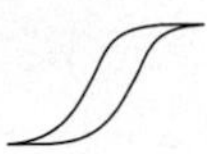

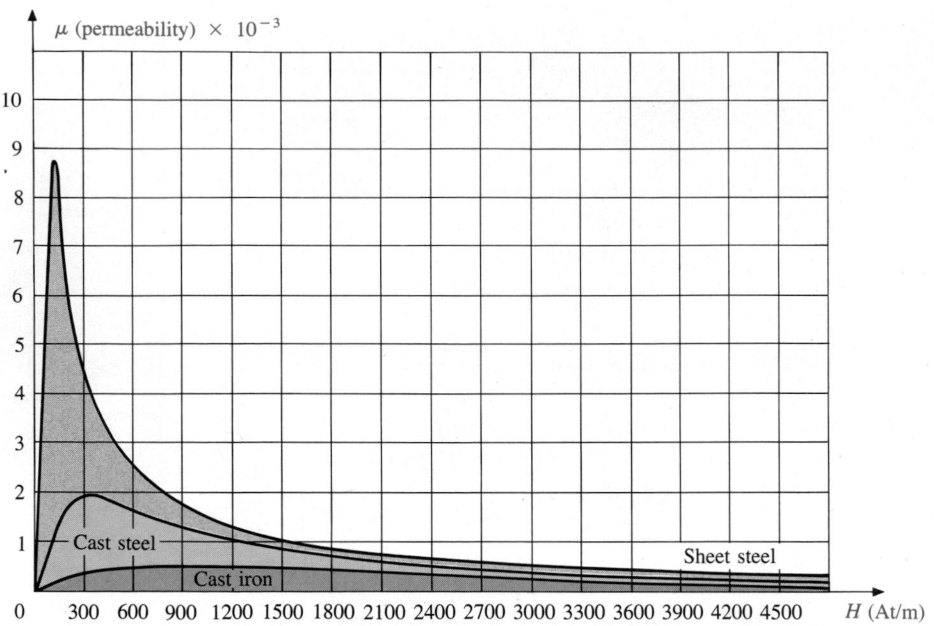

FIG. 11.19
Variation of μ with the magnetizing force.

The flux density and the magnetizing force are related by the following equation:

$$B = \mu H \quad (11.8)$$

This equation indicates that for a particular magnetizing force, the greater the permeability, the greater will be the induced flux density.

Now that henries (H) and the magnetizing force (H) use the same capital letter, it must be pointed out that all units of measurement in the text, such as henries, use roman letters, such as H, whereas variables such as the magnetizing force use italic letters, such as H.

11.8 HYSTERESIS

A curve of the flux density B versus the magnetizing force H of a material is of particular importance to the engineer. Curves of this type can usually be found in manuals, descriptive pamphlets, and brochures published by manufacturers of magnetic materials. A typical B-H curve for a ferromagnetic material such as steel can be derived using the setup of Fig. 11.20.

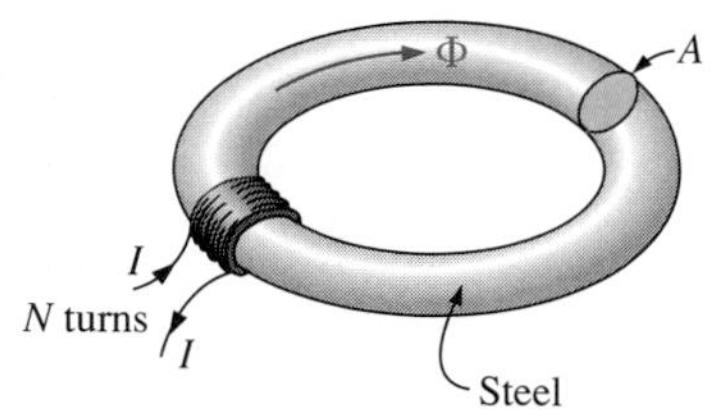

FIG. 11.20
Series magnetic circuit used to define the hysteresis curve.

The core is initially unmagnetized and the current $I = 0$. If the current I is increased to some value above zero, the magnetizing force H will increase to a value determined by

$$H\uparrow = \frac{NI\uparrow}{l}$$

The flux Φ and the flux density B ($B = \Phi/A$) will also increase with the current I (or H). If the material has no residual magnetism and the magnetizing force H is increased from zero to some value H_a, the B-H curve will follow the path shown in Fig. 11.21 between o and a. If the mag-

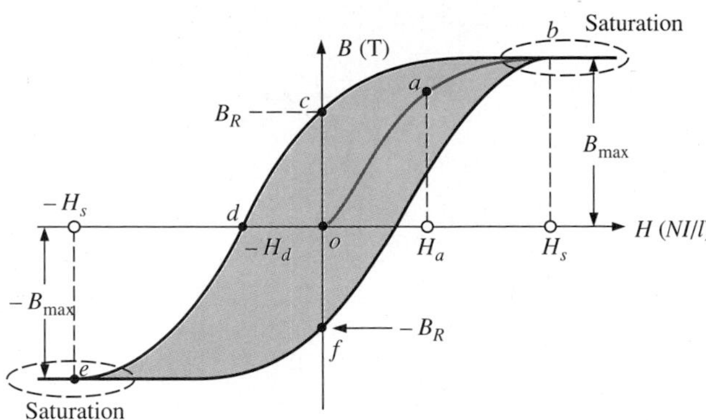

FIG. 11.21
Hysteresis curve.

netizing force H is increased until saturation (H_s) occurs, the curve will continue as shown in the figure to point b. When saturation occurs, the flux density has, *for all practical purposes,* reached its maximum value. Any further increase in current through the coil increasing $H = NI/l$ will result in a very small increase in flux density B.

If the magnetizing force is reduced to zero by letting I decrease to zero, the curve will follow the path of the curve between b and c. The flux density B_R, which remains when the magnetizing force is zero, is called the *residual flux density.* It is this residual flux density that makes it possible to create permanent magnets. If the coil is now removed from the core of Fig. 11.20, the core will still have the magnetic properties determined by the residual flux density, a measure of its "retentivity." If the current I is reversed, developing a magnetizing force, $-H$, the flux density B will decrease with an increase in I. Eventually, the flux density will be zero when $-H_d$ (the portion of curve from c to d) is reached. The magnetizing force $-H_d$ required to "coerce" the flux density to reduce its level to zero is called the *coercive force,* a measure of the coercivity of the magnetic sample. As the force $-H$ is increased until saturation again occurs and is then reversed and brought back to zero, the path *def* will result. If the magnetizing force is increased in the positive direction ($+H$), the curve will trace the path shown from f to b. The entire curve represented by *bcdefb* is called the *hysteresis curve* for the ferromagnetic material, from the Greek *hysterein,* meaning "to lag behind." The flux density B *lagged* behind the magnetizing force H during the entire plotting of the curve. When H was zero at c, B was not zero but had only begun to decline. Long after H had passed through zero and had become equal to $-H_d$ did the flux density B finally become equal to zero.

If the entire cycle is repeated, the curve obtained for the same core will be determined by the maximum H applied. Three hysteresis loops for the same material for maximum values of H less than the saturation value are shown in Fig. 11.22. In addition, the saturation curve is repeated for comparison purposes.

Note from the various curves that for a particular value of H, say, H_x, the value of B can vary widely, as determined by the history of the core. In an effort to assign a particular value of B to each value of H, we compromise by connecting the tips of the hysteresis loops. The resulting curve, shown by the heavy, solid line in Fig. 11.22 and for various materials in Fig. 11.23, is called the *normal magnetization curve.* An expanded view of one region appears in Fig. 11.24.

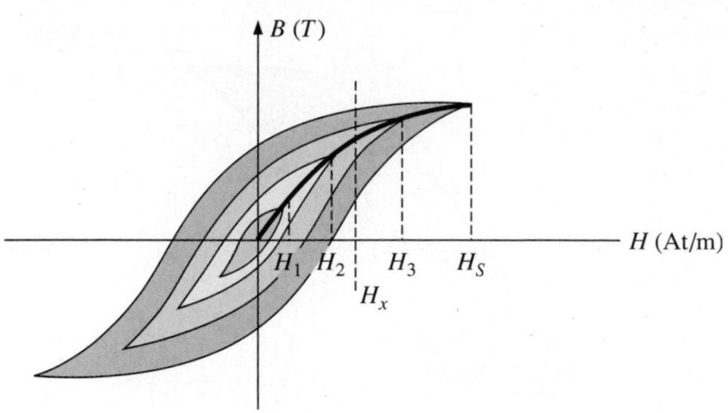

FIG. 11.22
Defining the normal magnetization curve.

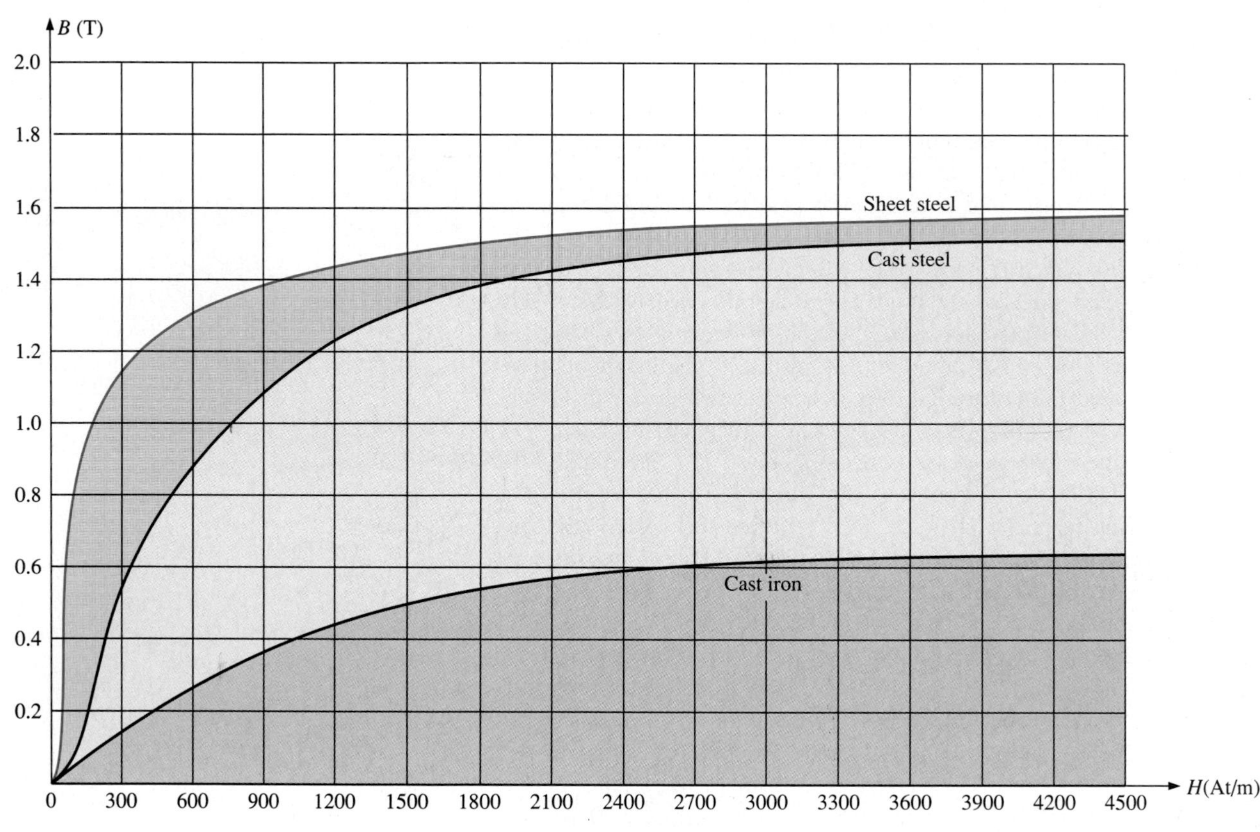

FIG. 11.23
Normal magnetization curve for three ferromagnetic materials.

A comparison of Figs. 11.19 and 11.23 shows that for the same value of H, the value of B is higher in Fig. 11.23 for the materials with the higher μ in Fig. 11.19. This is particularly obvious for low values of H. This correspondence between the two figures must exist since $B = \mu H$. In fact, if in Fig. 11.23 we find μ for each value of H using the equation $\mu = B/H$, we will obtain the curves of Fig. 11.19. An instrument that will provide a plot of the B-H curve for a magnetic sample appears in Fig. 11.25.

It is interesting to note that the hysteresis curves of Fig. 11.22 have a *point symmetry* about the origin; that is, the inverted pattern to the left of the vertical axis is the same as that appearing to the right of the vertical axis. In addition, you will find that a further application of the

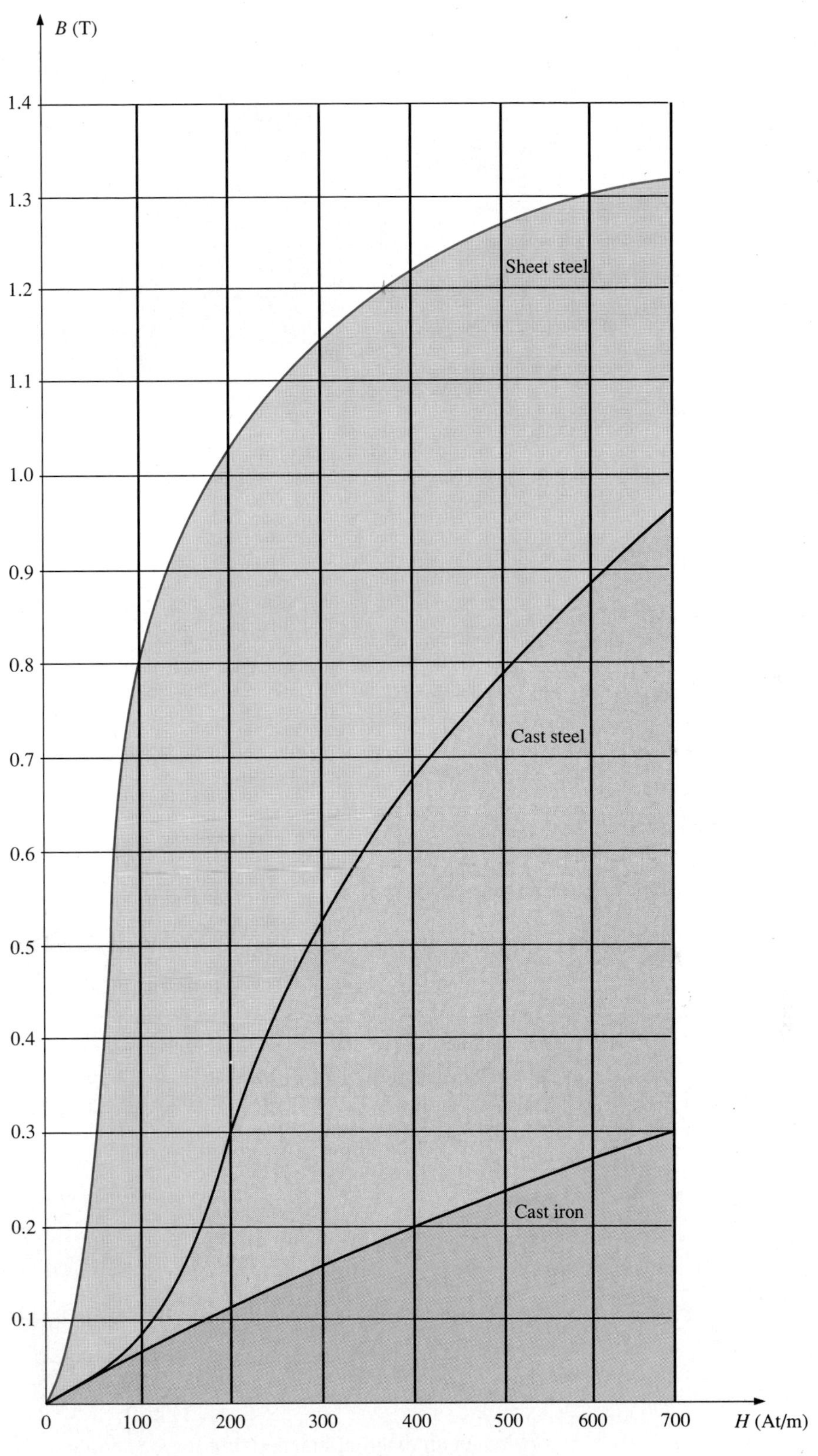

FIG. 11.24
Expanded view of Fig. 11.23 for the low magnetizing force region.

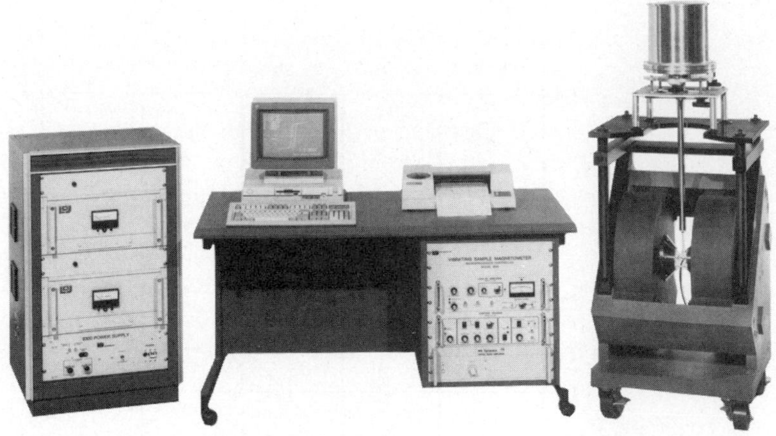

FIG. 11.25
Model 9600 vibrating sample magnetometer. (Courtesy of LDJ Electronics, Inc.)

same magnetizing forces to the sample will result in the same plot. For a current I in $H = NI/l$ that will move between positive and negative maximums at a fixed rate, the same B-H curve will result during each cycle. Such will be the case when we examine ac (sinusoidal) networks in the later chapters. The reversal of the field (Φ) due to the changing current direction will result in a loss of energy that can best be described by first introducing the *domain theory of magnetism.*

Within each atom, the orbiting electrons (described in Chapter 2) are also spinning as they revolve around the nucleus. The atom, due to its spinning electrons, has a magnetic field associated with it. In nonmagnetic materials, the net magnetic field is effectively zero since the magnetic fields due to the atoms of the material oppose each other. In magnetic materials such as iron and steel, however, the magnetic fields of groups of atoms numbering in the order of 10^{12} are aligned, forming very small bar magnets. This group of magnetically aligned atoms is called a *domain.* Each domain is a separate entity; that is, each domain is independent of the surrounding domains. For an unmagnetized sample of magnetic material, these domains appear in a random manner, such as shown in Fig. 11.26(a). The net magnetic field in any one direction is zero.

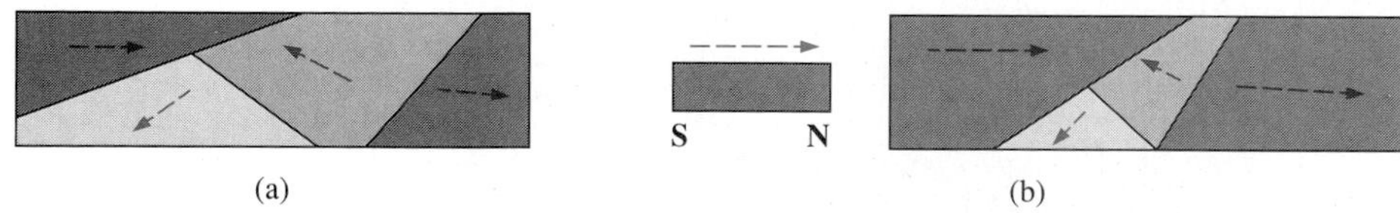

FIG. 11.26
Demonstrating the domain theory of magnetism.

When an external magnetizing force is applied, the domains that are nearly aligned with the applied field will grow at the expense of the less favorably oriented domains, such as shown in Fig. 11.26(b). Eventually, if a sufficiently strong field is applied, all of the domains will have the orientation of the applied magnetizing force, and any further increase in external field will not increase the strength of the magnetic flux through the core—a condition referred to as *saturation.* The elasticity of the

above is evidenced by the fact that when the magnetizing force is removed, the alignment will be lost to some measure and the flux density will drop to B_R. In other words, the removal of the magnetizing force will result in the return of a number of misaligned domains within the core. The continued alignment of a number of the domains, however, accounts for our ability to create permanent magnets.

At a point just before saturation, the opposing unaligned domains are reduced to small cylinders of various shapes referred to as *bubbles.* These bubbles can be moved within the magnetic sample through the application of a *controlling* magnetic field. These magnetic bubbles form the basis of the recently designed bubble memory system for computers.

11.9 AMPÈRE'S CIRCUITAL LAW

It was mentioned in the introduction to this chapter that there is a broad similarity between the analyses of electric and magnetic circuits. This has already been demonstrated to some extent for the quantities in Table 11.1.

TABLE 11.1

	Electric Circuits	Magnetic Circuits
Cause	E	$\mathscr{F}$
Effect	I	Φ
Opposition	R	$\mathscr{R}$

If we apply the "cause" analogy to Kirchhoff's voltage law ($\Sigma_{\circlearrowright} V = 0$), we obtain the following:

$$\Sigma_{\circlearrowright} \mathscr{F} = 0 \qquad \text{(for magnetic circuits)} \qquad \textbf{(11.9)}$$

which, in words, states that the algebraic sum of the rises and drops of the mmf around a closed loop of a magnetic circuit is equal to zero; that is, the sum of the mmf rises equals the sum of the mmf drops around a closed loop.

Equation (11.9) is referred to as *Ampère's circuital law.* When it is applied to magnetic circuits, sources of mmf are expressed by the equation

$$\mathscr{F} = NI \qquad \text{(At)} \qquad \textbf{(11.10)}$$

The equation for the mmf drop across a portion of a magnetic circuit can be found by applying the relationships listed in Table 11.1; that is, for electric circuits,

$$V = IR$$

resulting in the following for magnetic circuits:

$$\mathscr{F} = \Phi\mathscr{R} \qquad \text{(At)} \qquad \textbf{(11.11)}$$

where Φ is the flux passing through a section of the magnetic circuit and $\mathscr{R}$ is the reluctance of that section. The reluctance, however, is sel-

dom calculated in the analysis of magnetic circuits. A more practical equation for the mmf drop is

$$\mathscr{F} = Hl \qquad \text{(At)} \qquad \textbf{(11.12)}$$

as derived from Eq. (11.6), where H is the magnetizing force on a section of a magnetic circuit and l is the length of the section. As an example of Eq. (11.9), consider the magnetic circuit appearing in Fig. 11.27 constructed of three different ferromagnetic materials.

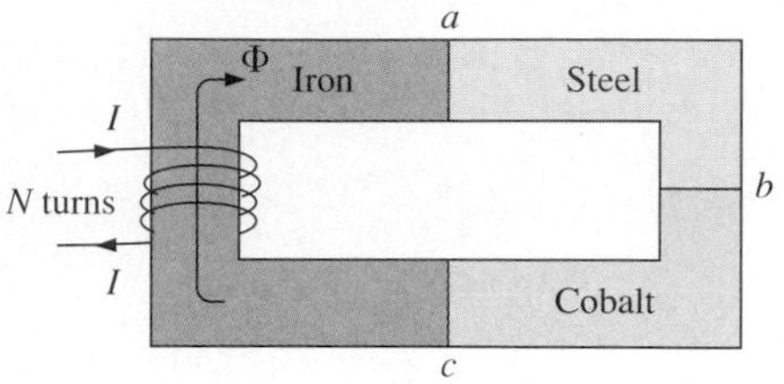

FIG. 11.27
Series magnetic circuit of three different materials.

Applying Ampère's circuital law, we have

$$\Sigma_{\circlearrowleft} \mathscr{F} = 0$$

$$\underbrace{+NI}_{\text{Rise}} - \underbrace{H_{ab}l_{ab}}_{\text{Drop}} - \underbrace{H_{bc}l_{bc}}_{\text{Drop}} - \underbrace{H_{ca}l_{ca}}_{\text{Drop}} = 0$$

$$\underbrace{NI}_{\substack{\text{Impressed}\\ \text{mmf}}} = \underbrace{H_{ab}l_{ab} + H_{bc}l_{bc} + H_{ca}l_{ca}}_{\text{mmf drops}}$$

All the terms of the equation are known except the magnetizing force for each portion of the magnetic circuit, which can be found by using the B-H curve if the flux density B is known.

11.10 THE FLUX Φ

If we continue to apply the relationships described in the previous section to Kirchhoff's current law, we will find that the sum of the fluxes entering a junction is equal to the sum of the fluxes leaving a junction; that is, for the circuit of Fig. 11.28,

$$\Phi_a = \Phi_b + \Phi_c \qquad \text{(at junction } a\text{)}$$

or

$$\Phi_b + \Phi_c = \Phi_a \qquad \text{(at junction } b\text{)}$$

both of which are equivalent.

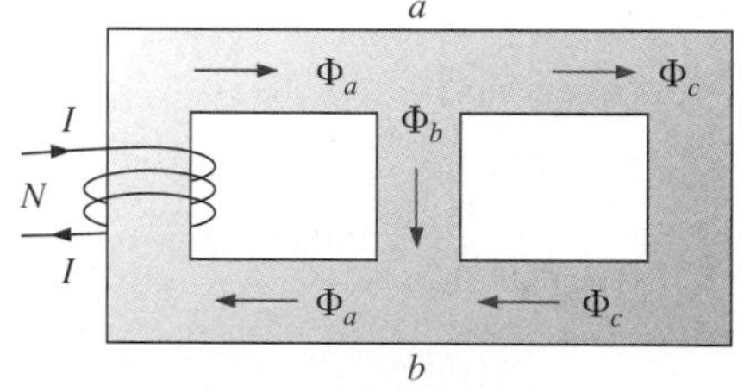

FIG. 11.28
Flux distribution of a series-parallel magnetic network.

11.11 SERIES MAGNETIC CIRCUITS: DETERMINING *NI*

We are now in a position to solve a few magnetic circuit problems, which are basically of two types. In one type, Φ is given, and the impressed mmf NI must be computed. This is the type of problem encountered in the design of motors, generators, and transformers. In the other type, NI is given, and the flux Φ of the magnetic circuit must be found. This type of problem is encountered primarily in the design of magnetic amplifiers and is more difficult since the approach is "hit or miss."

As indicated in earlier discussions, the value of μ will vary from point to point along the magnetization curve. This eliminates the possibility of finding the reluctance of each "branch" or the "total reluctance" of a network, as was done for electric circuits where ρ had a fixed value for any applied current or voltage. If the total reluctance could be determined, Φ could then be determined using the Ohm's law analogy for magnetic circuits.

For magnetic circuits, the level of B or H is determined from the other using the B-H curve, and μ is seldom calculated unless asked for.

An approach frequently employed in the analysis of magnetic circuits is the *table* method. Before a problem is analyzed in detail, a table is prepared listing in the extreme left-hand column the various sections of the magnetic circuit. The columns on the right are reserved for the quantities to be found for each section. In this way, the individual doing the problem can keep track of what is required to complete the problem and also of what the next step should be. After a few examples, the usefulness of this method should become clear.

This section will consider only *series* magnetic circuits in which the flux Φ is the same throughout. In each example, the magnitude of the magnetomotive force is to be determined.

EXAMPLE 11.3 For the series magnetic circuit of Fig. 11.29:

a. Find the value of I required to develop a magnetic flux of $\Phi = 4 \times 10^{-4}$ Wb.
b. Determine μ and μ_r for the material under these conditions.

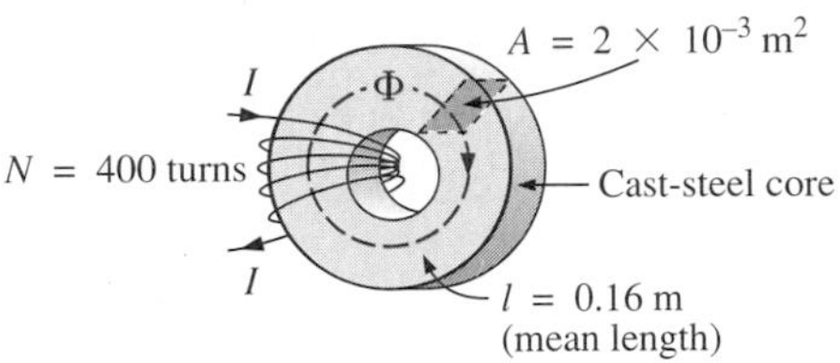

FIG. 11.29
Example 11.3.

Solutions: The magnetic circuit can be represented by the system shown in Fig. 11.30(a). The electric circuit analogy is shown in Fig. 11.30(b). Analogies of this type can be very helpful in the solution of magnetic circuits. Table 11.2 is for part (a) of this problem. The table is fairly trivial for this example but it does define the quantities to be found.

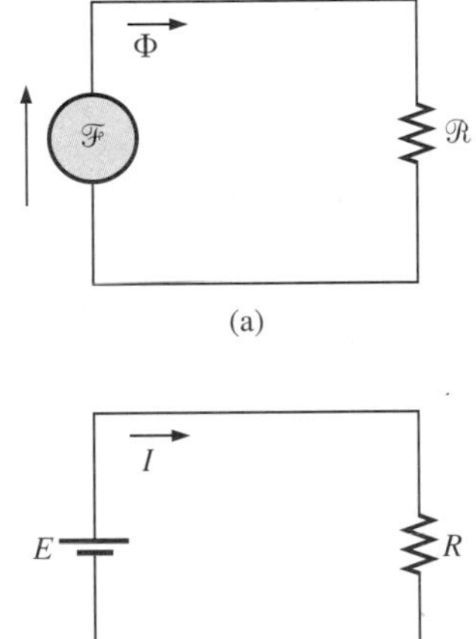

FIG. 11.30
(a) Magnetic circuit equivalent and (b) electric circuit analogy.

TABLE 11.2

Section	Φ (Wb)	A (m²)	B (T)	H (At/m)	l (m)	Hl (At)
One continuous section	4×10^{-4}	2×10^{-3}			0.16	

a. The flux density B is

$$B = \frac{\Phi}{A} = \frac{4 \times 10^{-4}\ \text{Wb}}{2 \times 10^{-3}\ \text{m}^2} = 2 \times 10^{-1}\ \text{T} = 0.2\ \text{T}$$

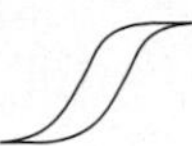

Using the B-H curves of Fig. 11.24, we can determine the magnetizing force H:

$$H \text{ (cast steel)} = 170 \text{ At/m}$$

Applying Ampère's circuital law yields

$$NI = Hl$$

and
$$I = \frac{Hl}{N} = \frac{(170 \text{ At/m})(0.16 \text{ m})}{400 \text{ t}} = \mathbf{68 \text{ mA}}$$

(Recall that t represents turns.)

b. The permeability of the material can be found using Eq. (11.8):

$$\mu = \frac{B}{H} = \frac{0.2 \text{ T}}{170 \text{ At/m}} = \mathbf{1.176 \times 10^{-3} \text{ Wb/A}\cdot\text{m}}$$

and the relative permeability is

$$\mu_r = \frac{\mu}{\mu_o} = \frac{1.176 \times 10^{-3}}{4\pi \times 10^{-7}} = \mathbf{935.83}$$

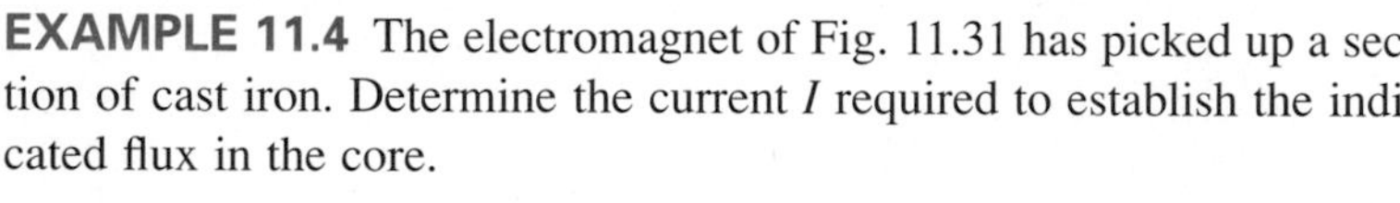

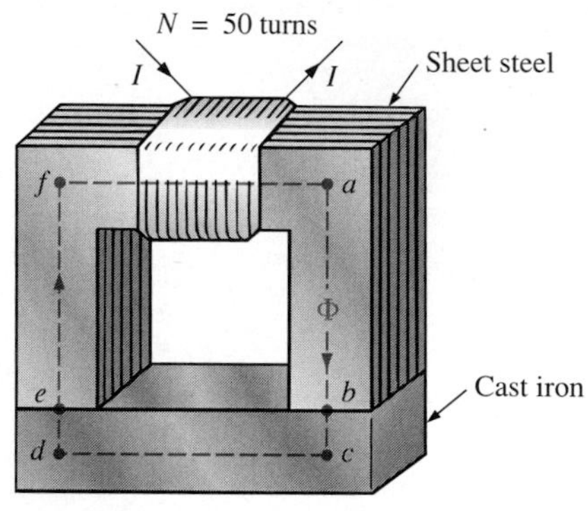

$l_{ab} = l_{cd} = l_{ef} = l_{fa} = 4$ in.
$l_{bc} = l_{de} = 0.5$ in.
Area (throughout) $= 1$ in.^{2}
$\Phi = 3.5 \times 10^{-4}$ Wb

FIG. 11.31
Electromagnet for Example 11.4.

EXAMPLE 11.4 The electromagnet of Fig. 11.31 has picked up a section of cast iron. Determine the current I required to establish the indicated flux in the core.

Solution: To be able to use Figs. 11.23 and 11.24, the dimensions must first be converted to the metric system. However, since the area is the same throughout, we can determine the length for each material rather than work with the individual sections:

$$l_{efab} = 4 \text{ in.} + 4 \text{ in.} + 4 \text{ in.} = 12 \text{ in.}$$
$$l_{bcde} = 0.5 \text{ in.} + 4 \text{ in.} + 0.5 \text{ in.} = 5 \text{ in.}$$
$$12 \cancel{\text{in.}}\left(\frac{1 \text{ m}}{39.37 \cancel{\text{in.}}}\right) = 304.8 \times 10^{-3} \text{ m}$$
$$5 \cancel{\text{in.}}\left(\frac{1 \text{ m}}{39.37 \cancel{\text{in.}}}\right) = 127 \times 10^{-3} \text{ m}$$
$$1 \cancel{\text{in.}}^2\left(\frac{1 \text{ m}}{39.37 \cancel{\text{in.}}}\right)\left(\frac{1 \text{ m}}{39.37 \cancel{\text{in.}}}\right) = 6.452 \times 10^{-4} \text{ m}^2$$

The information available from the specifications of the problem has been inserted in Table 11.3. When the problem has been completed, each space will contain some information. Sufficient data to complete the problem can be found if we fill in each column from left to right. As the various quantities are calculated, they will be placed in a similar table found at the end of the example.

TABLE 11.3

Section	Φ (Wb)	A (m^2)	B (T)	H (At/m)	l (m)	Hl (At)
efab	3.5×10^{-4}	6.452×10^{-4}			304.8×10^{-3}	
bcde	3.5×10^{-4}	6.452×10^{-4}			127×10^{-3}	

The flux density for each section is

$$B = \frac{\Phi}{A} = \frac{3.5 \times 10^{-4} \text{ Wb}}{6.452 \times 10^{-4} \text{ m}^2} = 0.542 \text{ T}$$

and the magnetizing force is

$$H \text{ (sheet steel, Fig. 11.24)} \cong 70 \text{ At/m}$$
$$H \text{ (cast iron, Fig. 11.23)} \cong 1600 \text{ At/m}$$

Note the extreme difference in magnetizing force for each material for the required flux density. In fact, when we apply Ampère's circuital law, we will find that the sheet steel section could be ignored with a minimal error in the solution.

Determining *Hl* for each section yields

$$H_{efab}l_{efab} = (70 \text{ At/m})(304.8 \times 10^{-3} \text{ m}) = 21.34 \text{ At}$$
$$H_{bcde}l_{bcde} = (1600 \text{ At/m})(127 \times 10^{-3} \text{ m}) = 203.2 \text{ At}$$

Inserting the above data in Table 11.3 will result in Table 11.4.

TABLE 11.4

Section	Φ (Wb)	A (m²)	B (T)	H (At/m)	l (m)	Hl (At)
efab	3.5×10^{-4}	6.452×10^{-4}	0.542	60	304.8×10^{-3}	21.34
bcde	3.5×10^{-4}	6.452×10^{-4}	0.542	1600	127×10^{-3}	203.2

The magnetic circuit equivalent and the electric circuit analogy for the system of Fig. 11.31 appear in Fig. 11.32.

Applying Ampère's circuital law,

$$NI = H_{efab}l_{efab} + H_{bcde}l_{bcde}$$
$$= 21.34 \text{ At} + 203.2 \text{ At} = 224.54 \text{ At}$$

and

$$(50 \text{ t})I = 224.54 \text{ At}$$

so that

$$I = \frac{224.54 \text{ At}}{50 \text{ t}} = \mathbf{4.49 \text{ A}}$$

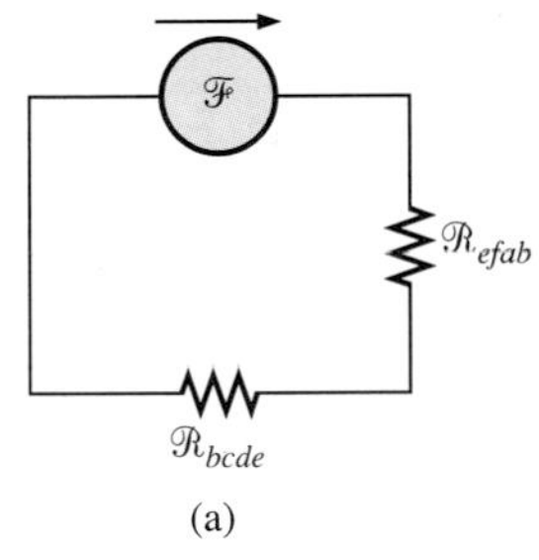

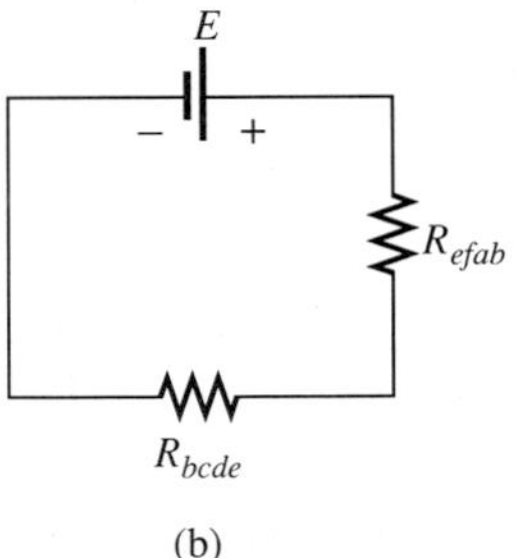

FIG. 11.32
(a) Magnetic circuit equivalent and (b) electric circuit analogy for the electromagnet of Fig. 11.31.

EXAMPLE 11.5 Determine the secondary current I_2 for the transformer of Fig. 11.33 if the resultant clockwise flux in the core is 1.5×10^{-5} Wb.

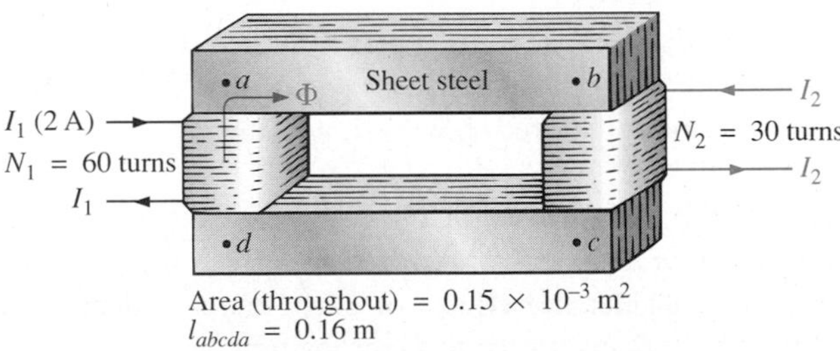

FIG. 11.33
Transformer for Example 11.5.

Solution: This is the first example with two magnetizing forces to consider. In the analogies of Fig. 11.34 you will note that the resulting flux of each is opposing, just as the two sources of voltage are opposing in the electric circuit analogy.

The structural data appear in Table 11.5.

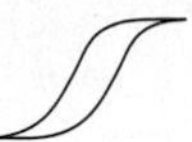

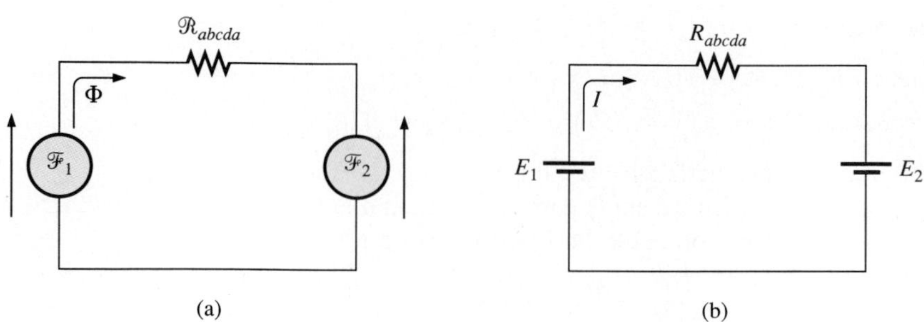

FIG. 11.34
(a) Magnetic circuit equivalent and (b) electric circuit analogy for the transformer of Fig. 11.33.

TABLE 11.5

Section	Φ (Wb)	A (m²)	B (T)	H (At/m)	l (m)	Hl (At)
abcda	1.5×10^{-5}	0.15×10^{-3}			0.16	

The flux density throughout is

$$B = \frac{\Phi}{A} = \frac{1.5 \times 10^{-5}\ \text{Wb}}{0.15 \times 10^{-3}\ \text{m}^2} = 10 \times 10^{-2}\ \text{T} = 0.10\ \text{T}$$

and

$$H\ (\text{from Fig. 11.24}) \cong \frac{1}{5}(100\ \text{At/m}) = 20\ \text{At/m}$$

Applying Ampère's circuital law,

$$N_1I_1 - N_2I_2 = H_{abcda}l_{abcda}$$
$$(60\ \text{t})(2\ \text{A}) - (30\ \text{t})(I_2) = (20\ \text{At/m})(0.16\ \text{m})$$
$$120\ \text{At} - (30\ \text{t})I_2 = 3.2\ \text{At}$$

and

$$(30\ \text{t})I_2 = 120\ \text{At} - 3.2\ \text{At}$$

or

$$I_2 = \frac{116.8\ \text{At}}{30\ \text{t}} = \mathbf{3.89\ A}$$

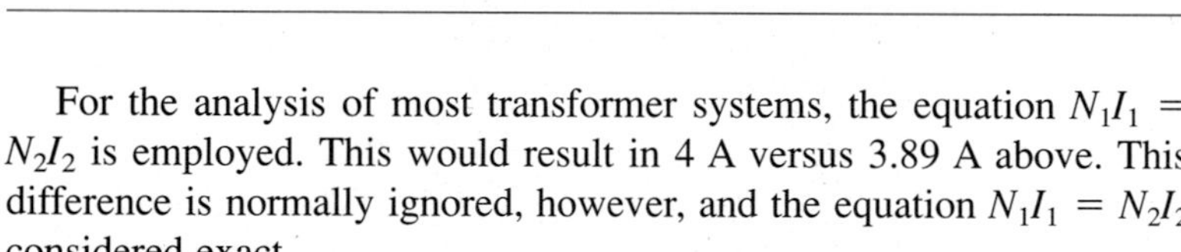

For the analysis of most transformer systems, the equation $N_1I_1 = N_2I_2$ is employed. This would result in 4 A versus 3.89 A above. This difference is normally ignored, however, and the equation $N_1I_1 = N_2I_2$ considered exact.

Because of the nonlinearity of the *B-H* curve, *it is not possible to apply superposition to magnetic circuits*; that is, in the previous example, we cannot consider the effects of each source independently and then find the total effects by using superposition.

11.12 AIR GAPS

Before continuing with the illustrative examples, let us consider the effects an air gap has on a magnetic circuit. Note the presence of air gaps in the magnetic circuits of the motor and meter of Fig. 11.11. The spreading of the flux lines outside the common area of the core for the air gap in Fig. 11.35(a) is known as *fringing*. For our purposes, we shall neglect this effect and assume the flux distribution to be as in Fig. 11.35(b).

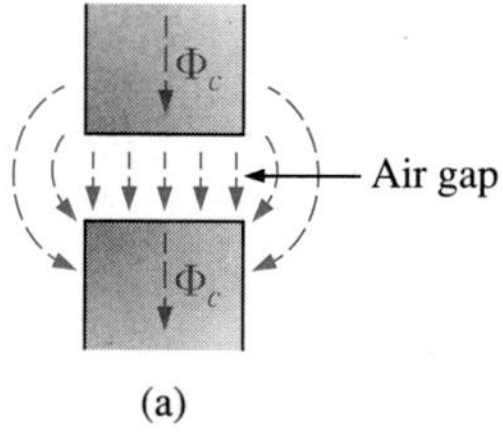

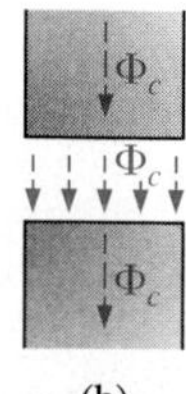

FIG. 11.35
Air gaps: (a) with fringing, (b) ideal.

The flux density of the air gap in Fig. 11.35(b) is given by

$$B_g = \frac{\Phi_g}{A_g} \quad \textbf{(11.13)}$$

where, for our purposes,

$$\Phi_g = \Phi_{\text{core}}$$

and

$$A_g = A_{\text{core}}$$

For most practical applications, the permeability of air is taken to be equal to that of free space. The magnetizing force of the air gap is then determined by

$$H_g = \frac{B_g}{\mu_o} \quad \textbf{(11.14)}$$

and the mmf drop across the air gap is equal to $H_g l_g$. An equation for H_g is as follows:

$$H_g = \frac{B_g}{\mu o} = \frac{B_g}{4\pi \times 10^{-7}}$$

and

$$H_g = (7.96 \times 10^5)B_g \quad \text{(At/m)} \quad \textbf{(11.15)}$$

EXAMPLE 11.6 Find the value of I required to establish a magnetic flux of $\Phi = 0.75 \times 10^{-4}$ Wb in the series magnetic circuit of Fig. 11.36.

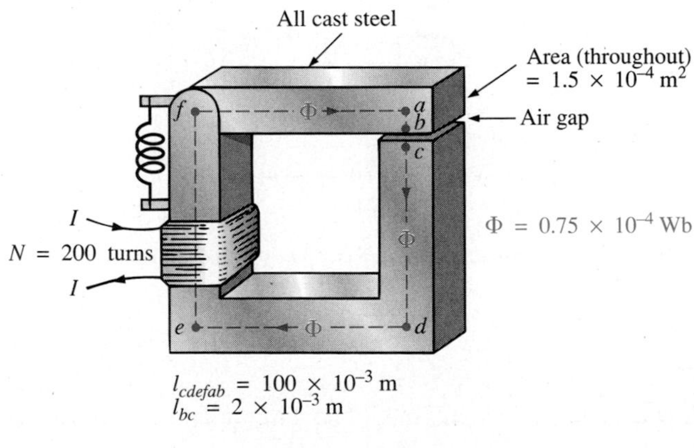

FIG. 11.36
Relay for Example 11.6.

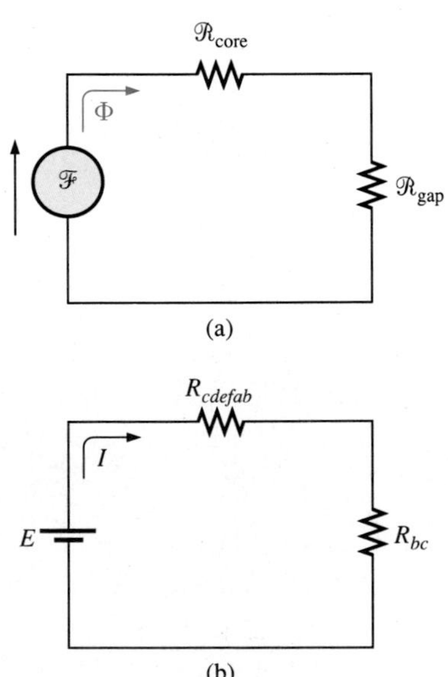

FIG. 11.37
(a) Magnetic circuit equivalent and (b) electric circuit analogy for the relay of Fig. 11.36.

Solution: An equivalent magnetic circuit and its electric circuit analogy are shown in Fig. 11.37.

The flux density for each section is

$$B = \frac{\Phi}{A} = \frac{0.75 \times 10^{-4}\ \text{Wb}}{1.5 \times 10^{-4}\ \text{m}^2} = 0.5\ \text{T}$$

From the B-H curves of Fig. 11.24,

$$H \text{ (cast steel)} \cong 280 \text{ At/m}$$

Applying Eq. (11.15),

$$H_g = (7.96 \times 10^5)B_g = (7.96 \times 10^5)(0.5 \text{ T}) = 3.98 \times 10^5 \text{ At/m}$$

The mmf drops are

$$H_{core}l_{core} = (280 \text{ At/m})(100 \times 10^{-3} \text{ m}) = 28 \text{ At}$$
$$H_g l_g = (3.98 \times 10^5 \text{ At/m})(2 \times 10^{-3} \text{ m}) = 796 \text{ At}$$

Applying Ampère's circuital law,

$$NI = H_{core}l_{core} + H_g l_g$$
$$= 28 \text{ At} + 796 \text{ At}$$
$$(200 \text{ t})I = 824 \text{ At}$$
$$I = \mathbf{4.12\ A}$$

Note from the above that the air gap requires the biggest share (by far) of the impressed NI due to the fact that air is nonmagnetic.

11.13 SERIES-PARALLEL MAGNETIC CIRCUITS

As one might expect, the close analogies between electric and magnetic circuits will eventually lead to series-parallel magnetic circuits similar in many respects to those encountered in Chapter 7. In fact, the electric circuit analogy will prove helpful in defining the procedure to follow toward a solution.

EXAMPLE 11.7 Determine the current I required to establish a flux of 1.5×10^{-4} Wb in the section of the core indicated in Fig. 11.38.

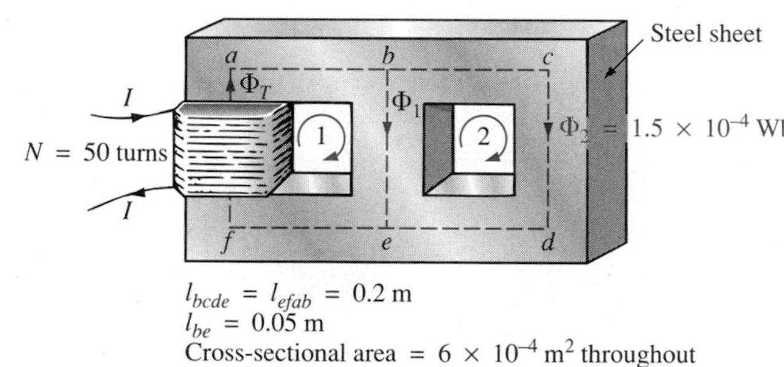

FIG. 11.38
Example 11.7.

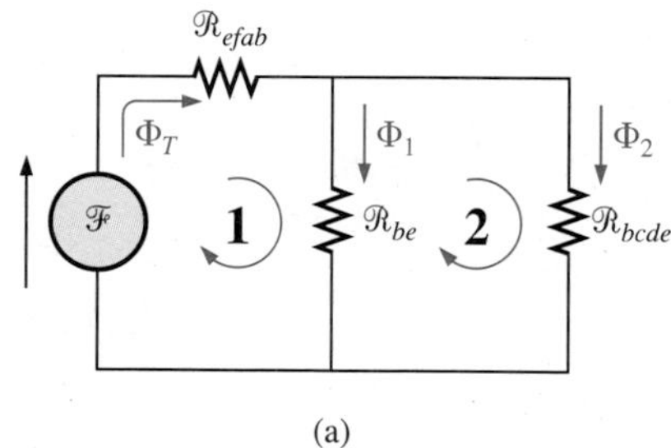

(a)

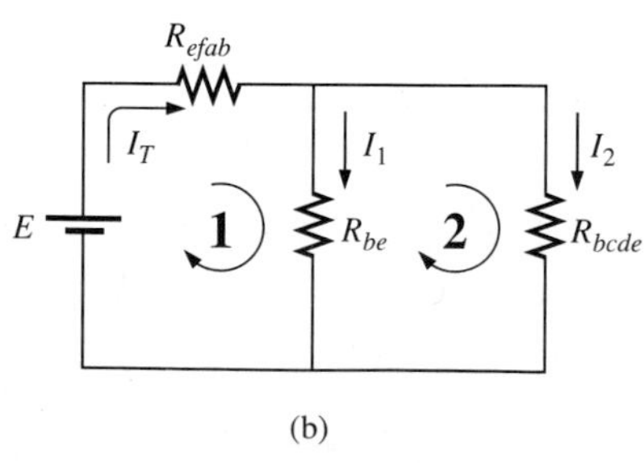

(b)

FIG. 11.39
(a) Magnetic circuit equivalent and (b) electric circuit analogy for the series-parallel system of Fig. 11.38.

Solution: The equivalent magnetic circuit and the electric circuit analogy appear in Fig. 11.39. We have

$$B_2 = \frac{\Phi_2}{A} = \frac{1.5 \times 10^{-4} \text{ Wb}}{6 \times 10^{-4} \text{ m}^2} = 0.25 \text{ T}$$

From Fig. 11.24,

$$H_{bcde} \cong 40 \text{ At/m}$$

Applying Ampère's circuital law around loop 2 of Figs. 11.38 and 11.39,

$$\Sigma_{\circlearrowright} \mathscr{F} = 0$$
$$H_{be}l_{be} - H_{bcde}l_{bcde} = 0$$
$$H_{be}(0.05 \text{ m}) - (40 \text{ At/m})(0.2 \text{ m}) = 0$$
$$H_{be} = \frac{8 \text{ At}}{0.05 \text{ m}} = 160 \text{ At/m}$$

From Fig. 11.24,

$$B_1 \cong 0.97 \text{ T}$$

and

$$\Phi_1 = B_1A = (0.97 \text{ T})(6 \times 10^{-4} \text{ m}^2) = 5.82 \times 10^{-4} \text{ Wb}$$

The results are then entered in Table 11.6.

TABLE 11.6

Section	Φ (Wb)	A (m²)	B (T)	H (At/m)	l (m)	Hl (At)
bcde	1.5×10^{-4}	6×10^{-4}	0.25	40	0.2	8
be	5.82×10^{-4}	6×10^{-4}	0.97	160	0.05	8
efab		6×10^{-4}			0.2	

The table reveals that we must now turn our attention to section *efab*:

$$\Phi_T = \Phi_1 + \Phi_2 = 5.82 \times 10^{-4} \text{ Wb} + 1.5 \times 10^{-4} \text{ Wb}$$
$$= 7.32 \times 10^{-4} \text{ Wb}$$
$$B = \frac{\Phi_T}{A} = \frac{7.32 \times 10^{-4} \text{ Wb}}{6 \times 10^{-4} \text{ m}^2}$$
$$= 1.22 \text{ T}$$

From Fig. 11.23,

$$H_{efab} \cong 400 \text{ At}$$

Applying Ampère's circuital law,

$$+NI - H_{efab}l_{efab} - H_{be}l_{be} = 0$$
$$NI = (400 \text{ At/m})(0.2 \text{ m}) + (160 \text{ At/m})(0.05 \text{ m})$$
$$(50 \text{ t})I = 80 \text{ At} + 8 \text{ At}$$
$$I = \frac{88 \text{ At}}{50 \text{ t}} = \mathbf{1.76 \text{ A}}$$

To demonstrate that μ is sensitive to the magnetizing force H, the permeability of each section is determined as follows. For section *bcde*,

$$\mu = \frac{B}{H} = \frac{0.25 \text{ T}}{40 \text{ At/m}} = 6.25 \times 10^{-3}$$

and

$$\mu_r = \frac{\mu}{\mu_o} = \frac{6.25 \times 10^{-3}}{12.57 \times 10^{-7}} = \mathbf{4972.2}$$

For section *be*,

$$\mu = \frac{B}{H} = \frac{0.97 \text{ T}}{160 \text{ At/m}} = 6.06 \times 10^{-3}$$

and $$\mu_r = \frac{\mu}{\mu_o} = \frac{6.06 \times 10^{-3}}{12.57 \times 10^{-7}} = \mathbf{4821}$$

For section *efab*,

$$\mu = \frac{B}{H} = \frac{1.22\ \text{T}}{400\ \text{At/m}} = 3.05 \times 10^{-3}$$

and $$\mu_r = \frac{\mu}{\mu_o} = \frac{3.05 \times 10^{-3}}{12.57 \times 10^{-7}} = \mathbf{2426.41}$$

11.14 DETERMINING Φ

The examples of this section are of the second type, where NI is given and the flux Φ must be found. This is a relatively straightforward problem if only one magnetic section is involved. Then

$$H = \frac{NI}{l} \qquad H \rightarrow B\ (B\text{-}H\ \text{curve})$$

and $$\Phi = BA$$

For magnetic circuits with more than one section, there is no set order of steps that will lead to an exact solution for every problem on the first attempt. In general, however, we proceed as follows. We must find the impressed mmf for a *calculated guess* of the flux Φ and then compare this with the specified value of mmf. We can then make adjustments on our guess to bring it closer to the actual value. For most applications, a value within ±5% of the actual Φ or specified NI is acceptable.

We can make a reasonable guess at the value of Φ if we realize that the maximum mmf drop appears across the material with the smallest permeability if the length and area of each material are the same. As shown in Example 11.6, if there is an air gap in the magnetic circuit, there will be a considerable drop in mmf across the gap. As a starting point for problems of this type, therefore, we shall assume that the total mmf (NI) is across the section with the lowest μ or greatest $\mathcal{R}$ (if the other physical dimensions are relatively similar). This assumption gives a value of Φ that will produce a calculated NI greater than the specified value. Then, after considering the results of our original assumption very carefully, we shall *cut* Φ and NI by introducing the effects (reluctance) of the other portions of the magnetic circuit and *try* the new solution. For obvious reasons, this approach is frequently called the *cut and try* method.

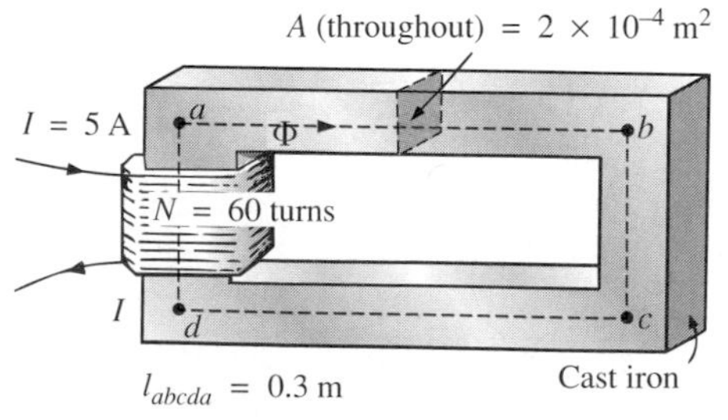

FIG. 11.40
Example 11.8.

EXAMPLE 11.8 Calculate the magnetic flux Φ for the magnetic circuit of Fig. 11.40.

Solution: By Ampère's circuital law,

$$NI = H_{abcda} l_{abcda}$$

or $$H_{abcda} = \frac{NI}{l_{abcda}} = \frac{(60\ \text{t})(5\ \text{A})}{0.3\ \text{m}} = \frac{300\ \text{At}}{0.3\ \text{m}} = 1000\ \text{At/m}$$

and $$B_{abcda}\ (\text{from Fig. 11.23}) \cong 0.39\ \text{T}$$

Since $B = \Phi/A$, we have

$$\Phi = BA = (0.39\ \text{T})(2 \times 10^{-4}\ \text{m}^2) = \mathbf{0.78 \times 10^{-4}\ Wb}$$

EXAMPLE 11.9 Find the magnetic flux Φ for the series magnetic circuit of Fig. 11.41 for the specified impressed mmf.

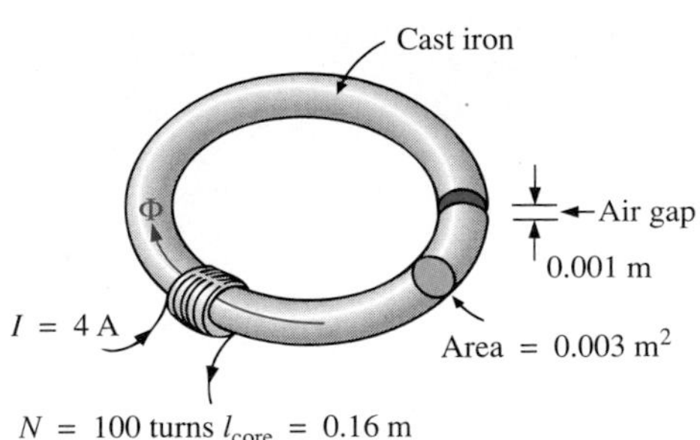

FIG. 11.41
Example 11.9.

Solution: Assuming that the total impressed mmf *NI* is across the air gap,

$$NI = H_g l_g$$

or
$$H_g = \frac{NI}{l_g} = \frac{400 \text{ At}}{0.001 \text{ m}} = 4 \times 10^5 \text{ At/m}$$

and
$$B_g = \mu_o H_g = (4\pi \times 10^{-7})(4 \times 10^5 \text{ At/m}) = 0.503 \text{ T}$$

The flux

$$\Phi_g = \Phi_{\text{core}} = B_g A = (0.503 \text{ T})(0.003 \text{ m}^2)$$
$$\Phi_{\text{core}} = 1.51 \times 10^{-3} \text{ Wb}$$

Using this value of Φ, we can find *NI*. The data are inserted in Table 11.7.

TABLE 11.7

Section	Φ (Wb)	A (m²)	B (T)	H (At/m)	l (m)	Hl (At)
Core	1.51×10^{-3}	0.003	0.503	1500 (*B-H* curve)	0.16	
Gap	1.51×10^{-3}	0.003	0.503	4×10^5	0.001	400

$$H_{\text{core}} l_{\text{core}} = (1500 \text{ At/m})(0.16 \text{ m}) = 240 \text{ At}$$

Applying Ampère's circuital law results in

$$NI = H_{\text{core}} l_{\text{core}} + H_g l_g = 240 \text{ At} + 400 \text{ At}$$
$$NI = 640 \text{ At} > 400 \text{ At}$$

Since we neglected the reluctance of all the magnetic paths but the air gap, the calculated value is greater than the specified value. We must therefore reduce this value by including the effect of these reluctances. Since approximately (640 At − 400 At)/640 At = 240 At/640 At ≅ 37.5% of our calculated value is above the desired value, let us reduce Φ by 30% and see how close we come to the impressed mmf of 400 At:

$$\Phi = (1 - 0.3)(1.51 \times 10^{-3} \text{ Wb}) = 1.057 \times 10^{-3} \text{ Wb}$$

See Table 11.8.

TABLE 11.8

Section	Φ (Wb)	A (m²)	B (T)	H (At/m)	l (m)	Hl (At)
Core	1.057×10^{-3}	0.003			0.16	
Gap	1.057×10^{-3}	0.003			0.001	

$$B = \frac{\Phi}{A} = \frac{1.057 \times 10^{-3}\text{ Wb}}{0.003\text{ m}^2} \cong 0.352\text{ T}$$

$$\begin{aligned} H_g l_g &= (7.96 \times 10^5) B_g l_g \\ &= (7.96 \times 10^5)(0.352\text{ T})(0.001\text{ m}) \\ &\cong 280.19\text{ At} \end{aligned}$$

From *B-H* curves,

$$H_{\text{core}} \cong 850\text{ At/m}$$

$$H_{\text{core}} l_{\text{core}} = (850\text{ At/m})(0.16\text{ m}) = 136\text{ At}$$

Applying Ampère's circuital law yields

$$\begin{aligned} NI &= H_{\text{core}} l_{\text{core}} + H_g l_g \\ &= 136\text{ At} + 280.19\text{ At} \\ NI &= \mathbf{416.19\text{ At}} > 400\text{ At} \end{aligned}$$

(but within $\pm 5\%$ and therefore acceptable)

The solution is, therefore,

$$\Phi \cong \mathbf{1.057 \times 10^{-3}\text{ Wb}}$$

11.15 AREAS OF APPLICATION

Recording Systems

The most common application of magnetic material is probably in the increasing number of recording instruments used every day in the office and the home. For instance, the eight-track tape and cassette of Fig. 11.42 are used almost daily by every family with a VCR or cassette player. The basic recording process is not that difficult to understand and will be described in detail in the section to follow on computer hard disks. Simply remember for the discussion to follow that the typical VCR tape has eight tracks.

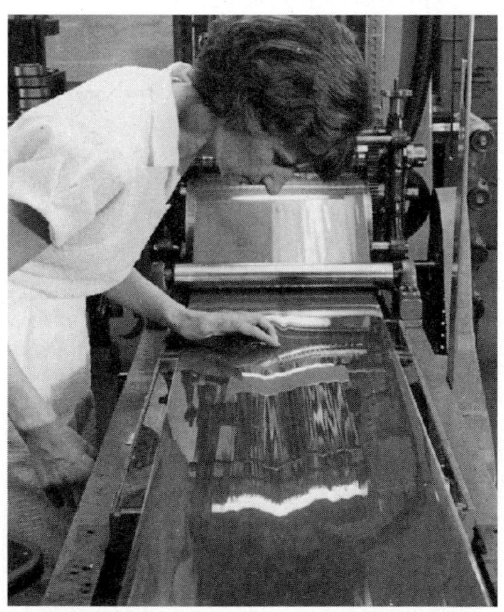

FIG. 11.42
Magnetic tape: (a) videocassette and recording cassette (Courtesy of Maxell Corporation of America), (b) manufacturing process (Courtesy of Ampex Corporation).

Speakers and Microphones

Electromagnetic effects are the moving force in the design of speakers such as that shown in Fig. 11.43. The shape of the pulsating waveform of the input current is determined by the sound to be reproduced by the speaker at a high audio level. As the current peaks and returns to the valleys of the sound pattern, the strength of the electromagnet varies in exactly the same manner. This causes the cone of the speaker to vibrate at a frequency directly proportional to the pulsating input. The higher the pitch of the sound pattern, the higher the oscillating frequency between the peaks and valleys and the higher the frequency of vibration of the cone.

A second design used more frequently in more expensive speaker systems appears in Fig. 11.44. In this case the permanent magnet is fixed and the input is applied to a movable core within the magnet, as shown in the figure. High peaking currents at the input produce a strong flux pattern in the voice coil, causing it to be drawn well into the flux pattern of the permanent magnet. As occurred for the speaker of Fig. 11.43, the core then vibrates at a rate determined by the input and provides the audible sound.

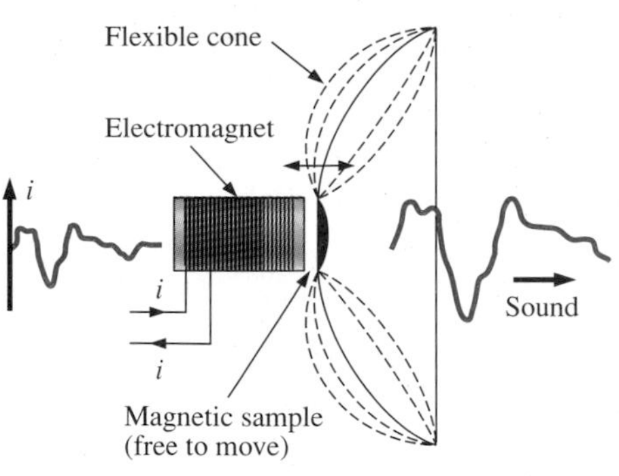

FIG. 11.43
Speaker.

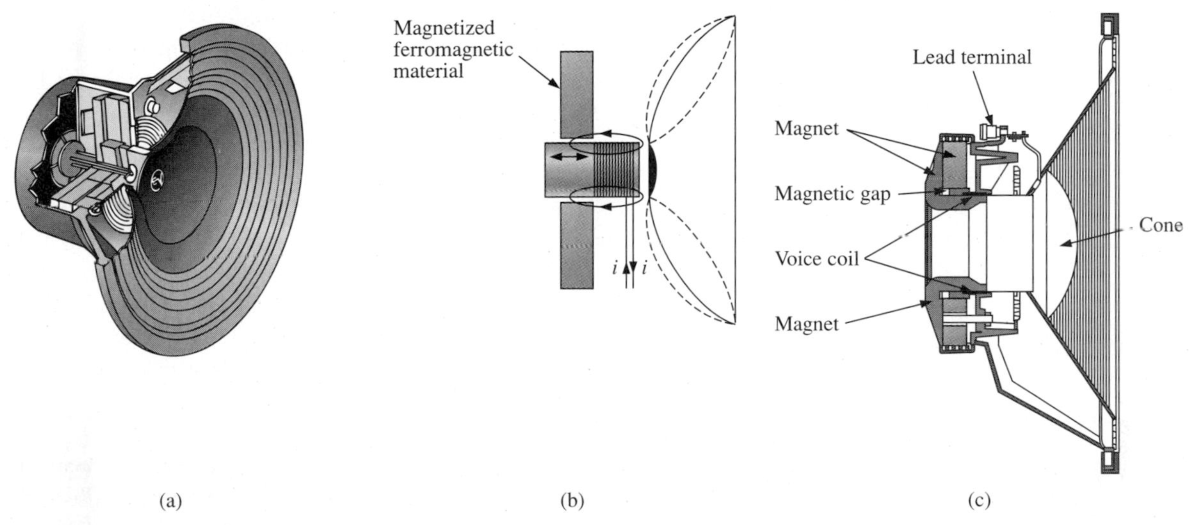

FIG. 11.44
Coaxial high-fidelity loudspeaker: (a) photograph, (b) basic operation, (c) cross section of actual unit. (Courtesy of Electro-Voice, Inc.)

Microphones such as those in Fig. 11.45 also employ electromagnetic effects. The sound to be reproduced at a higher audio level causes the core and attached moving coil to move within the magnetic field of the permanent magnet. Through Faraday's law ($e = N\, d\phi/dt$) a voltage is induced across the movable coil proportional to the speed with which it is moving through the magnetic field. The resulting induced voltage pattern can then be amplified and reproduced at a much higher audio level through the use of speakers, as described earlier. Microphones of this type are the most frequently employed, although other types that use capacitive, carbon granular, and piezoelectric* effects are available. This particular design is commercially referred to as a *dynamic* microphone.

*Piezoelectricity is the generation of a small voltage by exerting pressure across certain crystals.

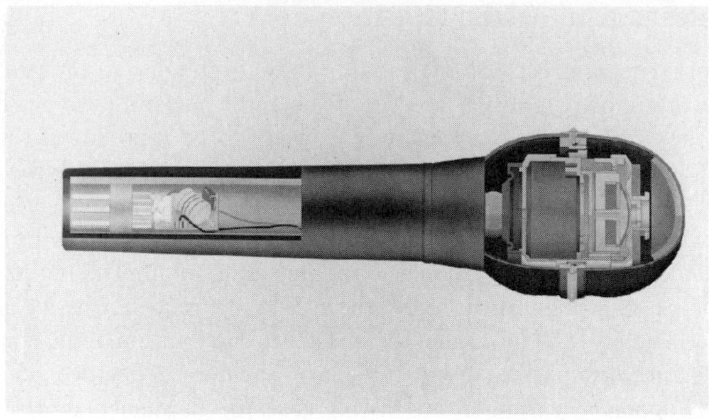

FIG. 11.45
Dynamic microphone. (Courtesy of Electro-Voice, Inc.)

Computer Hard Disks

The computer *hard disk* is a sealed unit in a computer that stores data on a magnetic coating applied to the surface of circular platters that spin like a record. The platters are constructed on a base of aluminum or glass (both nonferromagnetic), which makes them rigid, hence the term *hard disk.* Since the unit is sealed, the internal platters and components are inaccessible, and a "crash" (a term applied to the loss of data from a disk or the malfunction thereof) usually requires that the entire unit be replaced. Hard disks are currently available with diameters from 1⅓ in. to 5¼ in., with the 3⅕ in. the most popular for today's desk- and laptop units. All hard disk drives are often referred to as *Winchester drives,* a term first applied in the 1960s to an IBM drive that had 30 MB [a byte is a series of binary bits (0s and 1s) representing a number, letter, or symbol] of fixed (nonaccessible) data storage and 30 MB of accessible data storage. The term *Winchester* was applied because the 30-30 data capacity matched the name of the popular 30-30 Winchester rifle.

The magnetic coating on the platters is called the *media* and is of either the *oxide* or the *thin-film* variety. The oxide coating is formed by first coating the platter with a gel containing iron-oxide (ferromagnetic) particles. The disk is then spun at a very high speed to spread the material evenly across the surface of the platter. The resulting surface is then covered with a protective coating that is made as smooth as possible. The thin-film coating is very thin, but durable, with a surface that is smooth and consistent throughout the disk area. In recent years the trend has been toward the thin-film coating because the read/write heads (to be described shortly) must travel closer to the surface of the platter, requiring a consistent coating thickness. Recent techniques have resulted in thin-film magnetic coatings as thin as one-millionth of an inch.

The information on a disk is stored around the disk in circular paths called *tracks,* with each track containing so many bits of information per inch. The product of the number of bits per inch and the number of tracks per inch is the *Areal* density of the disk, which provides an excellent quantity for comparison with early systems and reveals how far the field has progressed in recent years. In the 1950s the first drives had an Areal density of about 2 kbits/in.^{2} as compared to today's typical 100 Mbits/in.^{2} Prototypes have now reached levels of 1–2 Gbits/in.^{2}, an

incredible achievement; consider 1,000,000,000,000 bits of information on an area the size of the face of your watch. *Electromagnetism* is the key element in the *writing* of information on the disk and the *reading* of information off the disk. In its simplest form the *write/read head* of a hard disk (or floppy disk) is a U-shaped electromagnet with an air gap that rides just above the surface of the disk, as shown in Fig. 11.46. As the disk rotates, information in the form of a voltage with changing polarities is applied to the winding of the electromagnet. For our purposes we will associate a positive voltage level with a 1 level (of binary arithmetic) and a negative voltage level with a 0 level. Combinations of these 0 and 1 levels can be used to represent letters, numbers, or symbols. If energized as shown in Fig. 11.46 with a 1 level (positive voltage), the resulting magnetic flux pattern will have the direction shown in the core. When the flux pattern encounters the air gap of the core, it jumps to the magnetic material (since magnetic flux always seeks the path of least reluctance and air has a high reluctance) and establishes a flux pattern, as shown on the disk, until it reaches the other end of the core air gap, where it returns to the electromagnet and completes the path. As the head then moves to the next bit sector, it leaves behind the magnetic flux pattern just established from the left to the right. The next bit sector has a 0 level input (negative voltage) that reverses the polarity of the applied voltage and the direction of the magnetic flux in the core of the head. The result is a flux pattern in the disk opposite that associated with a 1 level. The next bit of information is also a 0 level, resulting in the same pattern just generated. In total, therefore, information is stored on the disk in the form of small magnets whose polarity defines whether they are representing a 0 or a 1. Now that the data have been stored, we have to have some method to retrieve the information when desired. The first few hard disks used the same head for both the write and read functions. In Fig. 11.47(a) the U-shaped electromagnet in the read mode simply picks up the flux pattern of the current bit of information. *Faraday's law of electromagnet induction* states that a voltage is induced across a coil if exposed to a changing magnetic field. The change in flux for the core in Fig. 11.47(a) is minimal as it passes over the induced bar magnet on the surface of the disk. A flux pattern is established in the core because of the bar magnet on the disk, but the

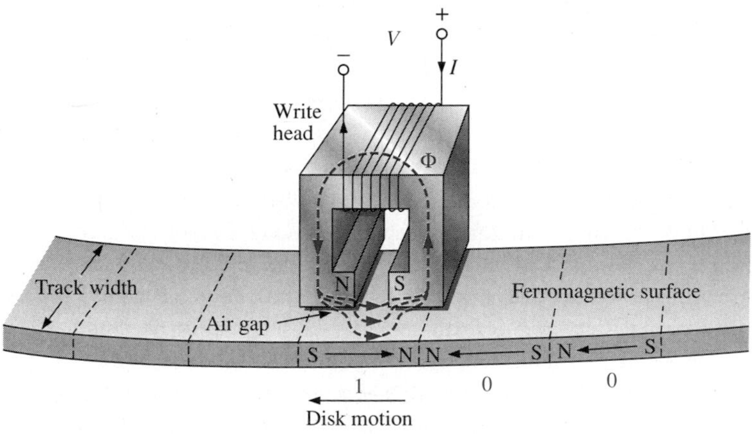

FIG. 11.46
Hard disk storage using a U-shaped electromagnet write head.

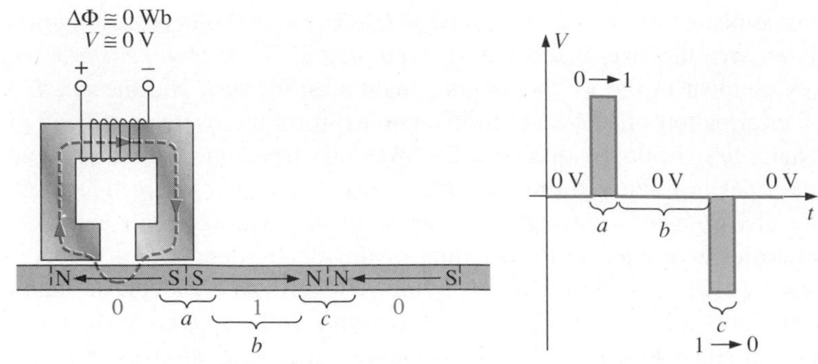

FIG. 11.47
Reading the information off a hard disk using a U-shaped electromagnet.

lack of a significant change in flux level results in an induced voltage at the output terminals of the pickup of approximately 0 V, as shown in Fig. 11.47(b) for the read-out waveform. A significant change in flux occurs when the head passes over the transition region so marked in Fig. 11.47(a). In region *a* the flux pattern changes from one direction to the other—a significant change in flux occurs in the core as it reverses direction, causing a measurable voltage to be generated across the terminals of the pickup coil as dictated by Faraday's law and indicated in Fig. 11.47(b). In region *b* there is no significant change in the flux pattern from one bit area to the next, and a voltage is not generated, as also revealed in Fig. 11.47(b). However, when region *c* is reached, the change in flux is significant but opposite that occurring in region *a*, resulting in another pulse but of opposite polarity. In total, therefore, the output bits of information are in the form of pulses that have a shape totally different from the read signals but are certainly representative of the information being stored. In addition, note that the output is generated at the transition regions and not in the constant flux region of the bit storage.

In the early years the use of the same head for the read and write functions was acceptable, but as the tracks became narrower and the seek time (the average time required to move from one track to another a random distance away) had to be reduced, it became increasingly difficult to construct the coil or core configuration in a manner that was sufficiently thin with minimum weight. In the late 1970s IBM introduced the *thin-film inductive head,* which was manufactured in much the same way as the small integrated circuits of today. The result is a head having a length typically less than 1⁄10 in., a height less than 1⁄50 in., with minimum mass and high durability. The average seek time has dropped from the few hundred milliseconds for the first units to about 8 ms for current models. In addition, production methods have improved to the point that the head can "float" above the surface (to minimize damage to the disk) at a height of only 5 microinches or 0.000005 in. Using a typical hard speed of 3600 rpm (some are as high as 7200 rpm) and an average diameter of 1.75 in. for a 3.5-in. disk, the speed of the head over the track is about 38 mph. Scaling the floating height up to 1⁄4 in. (multiplying by a factor of 50,000) the speed would increase to about 1.9×10^6 mph. In other words, the speed of the head over the surface of the platter is analogous to a mass traveling 1⁄4 in. above a surface at 1.9 million miles per hour—quite a technical achievement and amazingly enough one that perhaps will be improved

by a factor of 10 in the next decade. Incidentally, the speed of rotation of floppy disks is about 1/10 that of the hard disk, or 360 rpm. In addition, the head touches the magnetic surface of the floppy disk, limiting the storage life of the unit. The typical magnetizing force needed to lay down the magnetic orientation is 400 mA-turn (peak-to-peak). The result is a write current of only 40 mA for a 10-turn thin-film inductive head.

Although the thin-film inductive head could be used as a read head also, the *magnetoresistive* (MR) head has improved reading characteristics. The MR head depends on the fact that the resistance of a soft ferromagnetic conductor such as permolloy is sensitive to changes in external magnetic fields. As the hard disk rotates, the changes in magnetic flux from the induced magnetized regions of the platter change the terminal resistance of the head. A constant current passed through the sensor displays a terminal voltage sensitive to the magnitude of the resistance. The result is output voltages with peak values in excess of 300 V, which exceeds that of typical inductive read heads by a factor of 2 or 3:1.

Further investigation will reveal that the best write head is of the thin-film inductive variety and that the optimum read head is of the MR variety. Each has particular design criteria for maximum performance, resulting in the increasingly common *dual-element head,* with each head containing separate conductive paths and different gap widths. The Areal density of the new hard disks will essentially require the dual-head assembly for optimum performance.

The above is clear evidence of the importance of magnetic effects in today's growing industrial, computer-oriented society. Although research continues to maximize the Areal density, it appears certain that the storage will remain magnetic for the write/read process and not be replaced by any of the growing alternatives such as the optic laser variety used so commonly in CD-ROMS.

A 3.5-in. full-height disk drive, which is manufactured by the Micropolis Corporation and has a formatted capacity of 1.75 gigabytes (GB) with an average search time of 10 ms, appears in Fig. 11.48.

FIG. 11.48
A 3.5-in. hard-disk drive with a capacity of 1.75 GB and an average search time of 10 ms. (Courtesy of Micropolis Corporation.)

Hall Effect Sensor

The Hall effect sensor is a semiconductor device that generates an output voltage when exposed to a magnetic field. The basic construction consists of a slab of semiconductor material through which a current is passed, as shown in Fig. 11.49(a). If a magnetic field is applied as shown in the figure perpendicular to the direction of the current, a voltage V_H will be generated between the two terminals, as indicated in Fig. 11.49(a). The difference in potential is due to the separation of charge established by the Lorentz force first studied by Professor Hendrick Lorentz in the early eighteenth century. He found that electrons in a magnetic field are subjected to a force proportional to the velocity of the electrons through the field and the strength of the magnetic field. The direction of the force is determined by the left-hand rule. Simply place the index finger of the left hand in the direction of the magnetic field with the second finger at right angles to the index finger in the direction of conventional current through the semiconductor material, as shown in Fig. 11.49(b). The thumb, if placed at right angles to the index finger, will indicate the direction of the force on

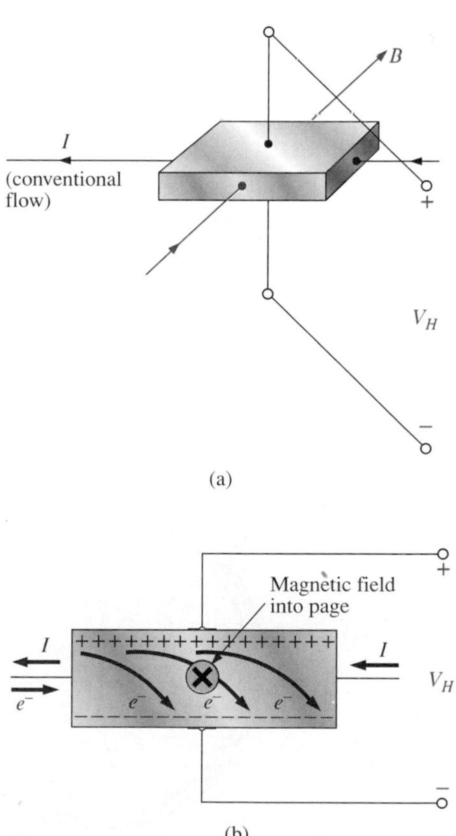

FIG. 11.49
Hall effect sensor: (a) orientation of controlling parameters, (b) effect on electron flow.

the electrons. In Fig. 11.49(b) the force causes the electrons to accumulate in the bottom region of the semiconductor (connected to the negative terminal of the voltage V_H), leaving a net positive charge in the upper region of the material (connected to the positive terminal of V_H). The stronger the current or strength of the magnetic field, the greater the induced voltage V_H.

In essence, therefore, the Hall effect sensor can reveal the strength of a magnetic field or the level of current through a device if the other determining factor is held fixed. Two applications of the sensor are therefore apparent—to measure the strength of a magnetic field in the vicinity of a sensor (for an applied fixed current) and to measure the level of current through a sensor (with knowledge of the strength of the magnetic field linking the sensor). The gaussmeter in Fig. 11.16 employs a Hall effect sensor. Internal to the meter, a fixed current is passed through the sensor with the voltage V_H indicating the relative strength of the field. Through amplification, calibration, and proper scaling the meter can display the relative strength in gauss.

The Hall effect sensor has a broad range of applications that are often quite interesting and innovative. The most widespread is as a trigger for an alarm system in large department stores, where theft is often a difficult problem. A magnetic strip is attached to the merchandise that sounds an alarm when a customer passes through the exit gates without paying for the product. The sensor, control current, and monitoring system are housed in the exit fence and react to the presence of the magnetic field as the product leaves the store. When the product is paid for, the cashier removes the strip or demagnetizes the strip by applying a magnetizing force that reduces the residual magnetism in the strip to essentially zero.

The Hall effect sensor is also used to indicate the speed of a bicycle on a digital display conveniently mounted on the handlebars. As shown in Fig. 11.50(a), the sensor is mounted on the frame of the bike, and a small permanent magnet is mounted on a spoke of the front wheel. The magnet must be carefully mounted to be sure that it passes over the proper region of the sensor. When the magnet passes over the sensor,

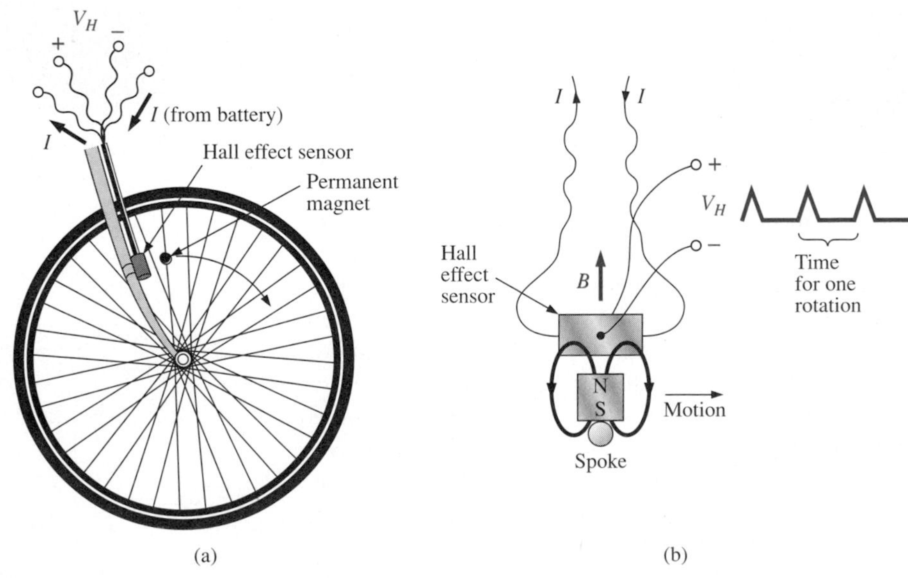

FIG. 11.50

Obtaining a speed indication for a bicycle using a Hall effect sensor: (a) mounting the components, (b) Hall effect response.

the flux pattern in Fig. 11.50(b) results, and a voltage with a sharp peak is developed by the sensor. Assuming a bicycle with a 26-in. diameter wheel, the circumference will be about 82 in. Over 1 mi the number of rotations is

$$5280\ \cancel{\text{ft}}\left(\frac{12\ \cancel{\text{in.}}}{1\ \cancel{\text{ft}}}\right)\left(\frac{1\ \text{rotation}}{82\ \cancel{\text{in.}}}\right) \cong 773\ \text{rotations}$$

If the bicycle is traveling at 20 mph, an output pulse will occur at a rate of 4.29 per second. It is interesting to note that at a speed of 20 mph, the wheel is rotating at more than 4 revolutions per second and the total number of rotations over 20 mi is 15,460.

Magnetic Reed Switch

One of the most frequently employed switches in alarm systems is the *magnetic reed switch* in Fig. 11.51. As shown by the figure, there are two components of the reed switch—a permanent magnet embedded in one unit that is normally connected to the movable element (door, window, and so on) and a reed switch in the other unit that is connected to the electrical control circuit. The reed switch is constructed of two iron alloy (ferromagnetic) reeds in a hermetically sealed capsule. The cantilevered ends of the two reeds do not touch but are in very close proximity to one another. In the absence of a magnetic field the reeds remain separated. However, if a magnetic field is introduced, the reeds will be drawn to each other because flux lines seek the path of least reluctance and, if possible, exercise every alternative to establish the path of least reluctance. It is similar to placing a ferromagnetic bar close to the ends of a U-shaped magnet. The bar is drawn to the poles of the magnet, establishing a magnetic flux path without air gaps and minimum reluctance. In the open-circuit state the resistance between reeds is in excess of 100 MΩ, while in the on state it drops to less than 1 Ω.

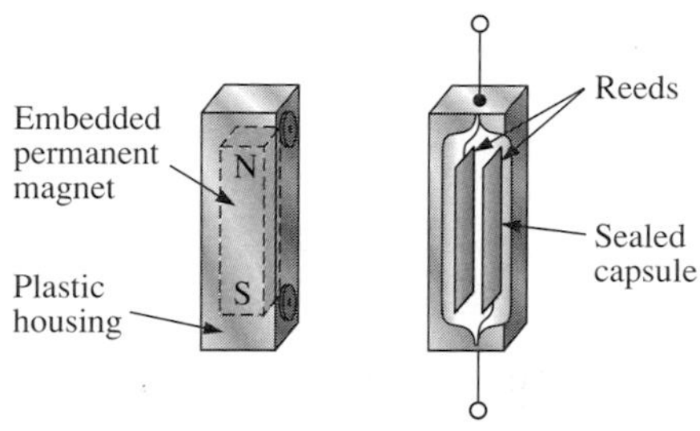

FIG. 11.51
Magnetic reed switch.

In Fig. 11.52 a reed switch has been placed on the fixed frame of a window and a magnet on the movable window unit. When the window is closed as shown in Fig. 11.52, the magnet and reed switch are sufficiently close to establish contact between the reeds, and a current is established through the reed switch to the control panel. In the armed state the alarm system accepts the resulting current flow as a normal secure response. If the window is opened, the magnet will leave the vicinity of the reed switch and the switch will open. The current through the switch will be interrupted, and the alarm will react appropriately.

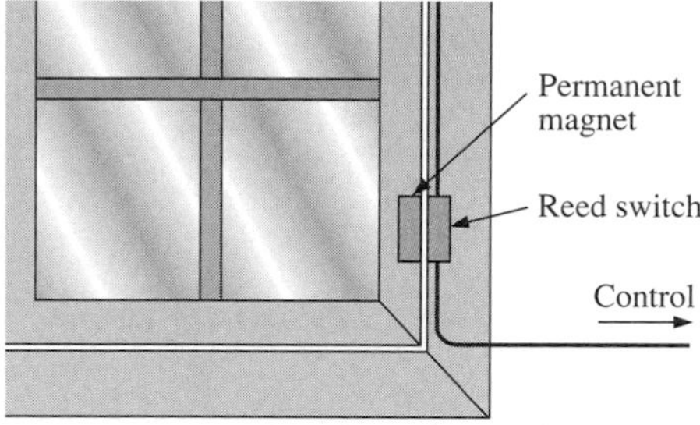

FIG. 11.52
Using a magnetic reed switch to monitor the state of a window.

One of the distinct advantages of the magnetic reed switch is that the proper operation of any switch can be checked with a portable magnetic element. Simply bring the magnet to the switch and note the output response. There is no need to continually open and close windows and doors. In addition, the reed switch is hermetically enclosed so that oxidation and foreign objects cannot damage it, and the result is a unit that can last indefinitely. Magnetic reed switches are also available in other shapes and sizes, allowing them to be concealed from obvious view. One is a circular variety that can be set into the edge of a door and door jam, resulting in only two small visible disks when the door is open.

Magnetic Resonance Imaging

Magnetic resonance imaging [MRI, also called nuclear magnetic resonance (NMR)] is receiving more and more attention as we strive to

improve the quality of the cross-sectional images of the body so useful in medical diagnosis and treatment. MRI does not expose the patient to potentially hazardous X-rays or injected contrast materials such as those employed to obtain computerized axial tomography (CAT) scans.

The three major components of an MRI system are a huge magnet that can weigh up to 100 tons, a table for transporting the patient into the circular hole in the magnet, and a control center, as shown in Fig. 11.53. The image is obtained by placing the patient in the tube to a precise depth depending on the cross section to be obtained and applying

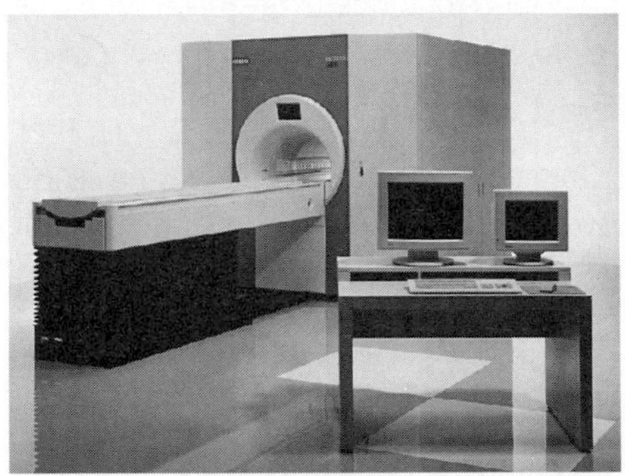

FIG. 11.53
Magnetic resonance imaging equipment. (Courtesy of Siemens Medical Systems, Inc.)

a strong magnetic field that causes the nuclei of certain atoms in the body to line up. Radio waves of different frequencies are then applied to the patient in the region of interest, and if the frequency of the wave matches the natural frequency of the atom, the nuclei will be set into a state of resonance and absorb energy from the applied signal. When the signal is removed, the nuclei release the acquired energy in the form of weak but detectable signals. The strength and duration of the energy emission vary from one tissue of the body to another. The weak signals are then amplified, digitized, and translated to provide a cross-sectional image such as the one shown in Fig. 11.54.

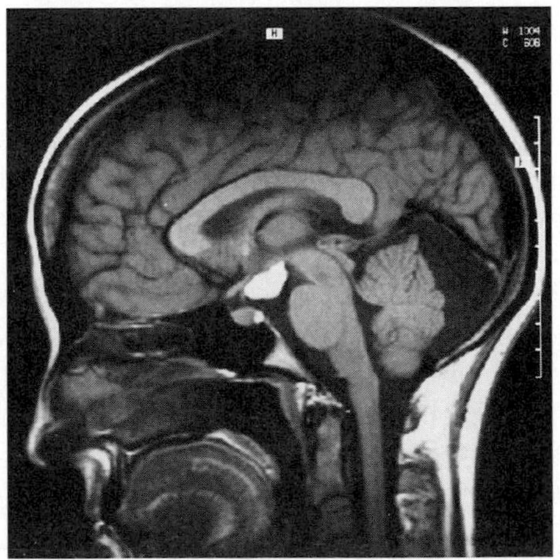

FIG. 11.54
Magnetic resonance image. (Courtesy of Siemens Medical Systems, Inc.)

MRI units are very expensive and therefore are not available at all locations. In recent years, however, their numbers have grown and one is available in almost every major community. For some patients the claustrophobic feeling they experience while in the circular tube is difficult to contend with. Today, however, a more open unit has been developed, as shown in Fig. 11.55, that has removed most of this discomfort.

Patients who have metallic implants or pacemakers or those who have worked in industrial environments where minute ferromagnetic particles may have become lodged in open, sensitive areas such as the eyes, nose, and so on, may have to use a CAT scan system because it does not employ magnetic effects. The attending physician is well trained in such areas of concern and will remove any unfounded fears or suggest alternative methods.

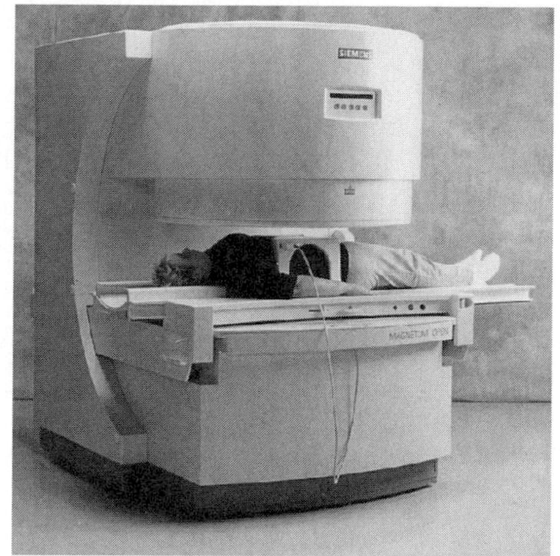

FIG. 11.55
Magnetic resonance imaging equipment (open variety). (Courtesy of Siemens Medical Systems, Inc.)

PROBLEMS

SECTION 11.3 Flux Density

1. Using Appendix G, fill in the blanks in the following table. Indicate the units for each quantity.

	Φ	B
SI	5×10^{-4} Wb	8×10^{-4} T
CGS	___	___
English	___	___

2. Repeat Problem 1 for the following table if area = 2 in.^{2}:

	Φ	B
SI	___	___
CGS	60,000 maxwells	___
English	___	___

3. For the electromagnet of Fig. 11.56:
  a. Find the flux density in the core.
  b. Sketch the magnetic flux lines and indicate their direction.
  c. Indicate the north and south poles of the magnet.

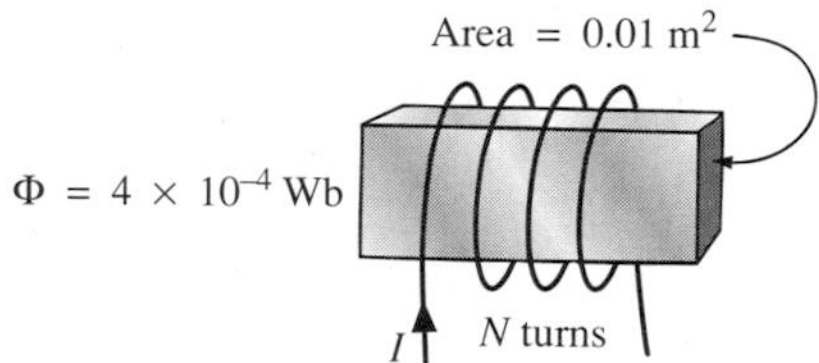

FIG. 11.56
Problem 3.

SECTION 11.5 Reluctance

4. Which section of Fig. 11.57 [(a), (b), or (c)] has the largest reluctance to the setting up of flux lines through its longest dimension?

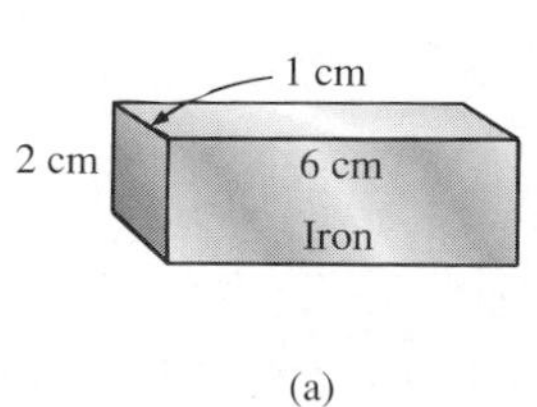

(a)

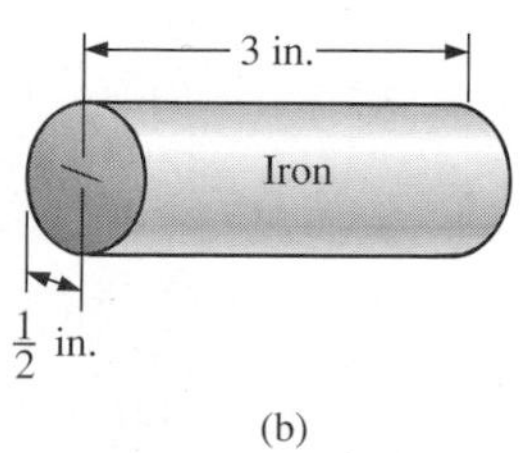

(b)

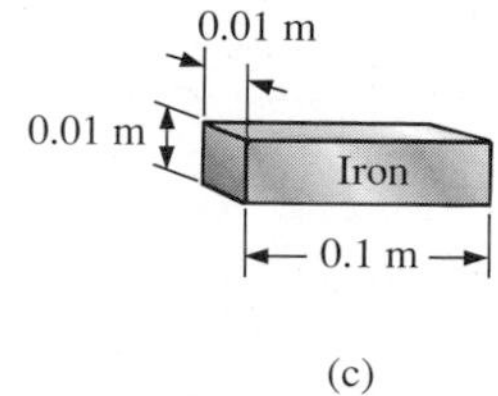

(c)

FIG. 11.57
Problem 4.

SECTION 11.6 Ohm's Law for Magnetic Circuits

5. Find the reluctance of a magnetic circuit if a magnetic flux $\Phi = 4.2 \times 10^{-4}$ Wb is established by an impressed mmf of 400 At.

6. Repeat Problem 5 for $\Phi = 72{,}000$ maxwells and an impressed mmf of 120 gilberts.

SECTION 11.7 Magnetizing Force

7. Find the magnetizing force H for Problem 5 in SI units if the magnetic circuit is 6 in. long.

8. If a magnetizing force H of 600 At/m is applied to a magnetic circuit, a flux density B of 1200×10^{-4} Wb/m^2 is established. Find the permeability μ of a material that will produce twice the original flux density for the same magnetizing force.

SECTION 11.8 Hysteresis

9. For the series magnetic circuit of Fig. 11.58, determine the current I necessary to establish the indicated flux.

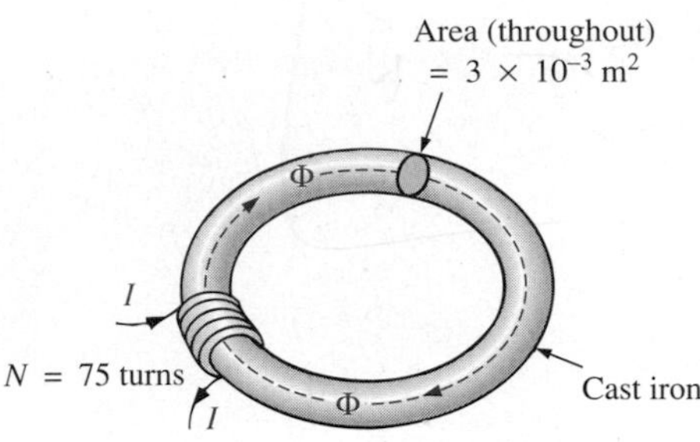

FIG. 11.58
Problem 9.

10. Find the current necessary to establish a flux of $\Phi = 3 \times 10^{-4}$ Wb in the series magnetic circuit of Fig. 11.59.

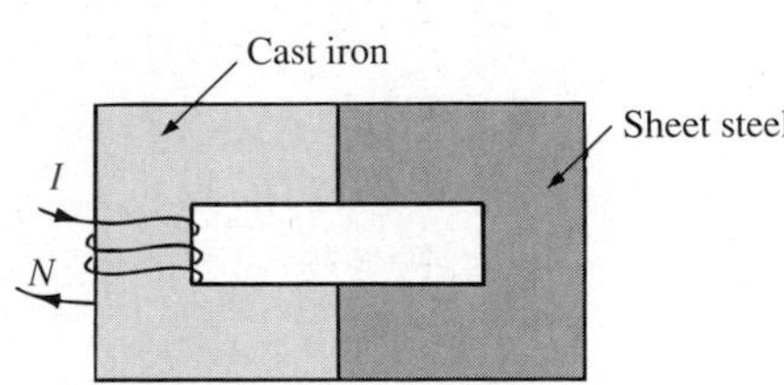

$l_{iron\ core} = l_{steel\ core} = 0.3$ m
Area (throughout) $= 5 \times 10^{-4}$ m^2
$N = 100$ turns

FIG. 11.59
Problem 10.

11. a. Find the number of turns N_1 required to establish a flux $\Phi = 12 \times 10^{-4}$ Wb in the magnetic circuit of Fig. 11.60.
    b. Find the permeability μ of the material.

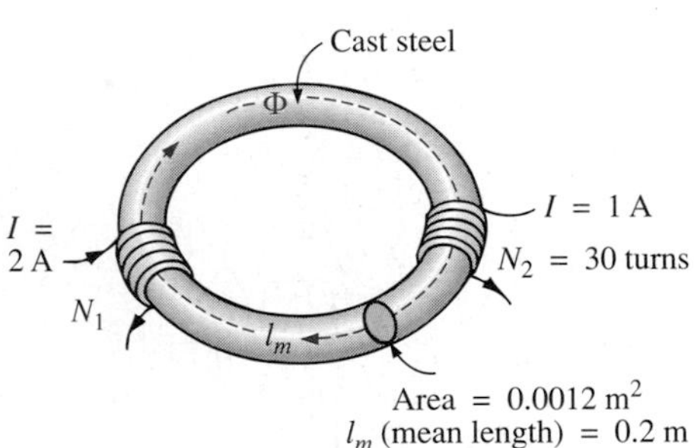

Area = 0.0012 m^2
l_m (mean length) = 0.2 m

FIG. 11.60
Problem 11.

12. a. Find the mmf (NI) required to establish a flux $\Phi = 80{,}000$ lines in the magnetic circuit of Fig. 11.61.
    b. Find the permeability of each material.

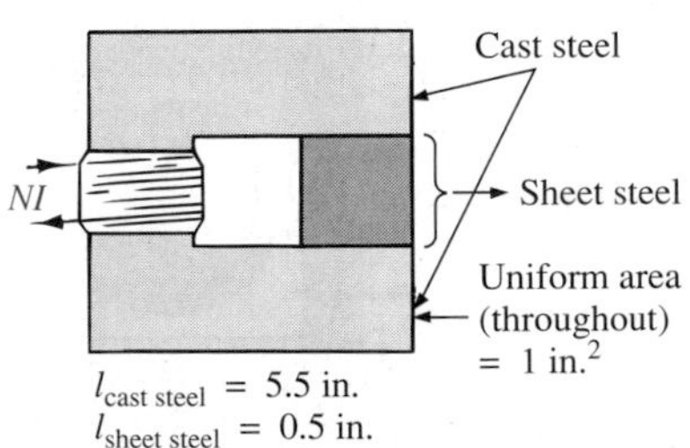

$l_{cast\ steel} = 5.5$ in.
$l_{sheet\ steel} = 0.5$ in.

FIG. 11.61
Problem 12.

*13. For the series magnetic circuit of Fig. 11.62 with two impressed sources of magnetic "pressure," determine the current I. Each applied mmf establishes a flux pattern in the clockwise direction.

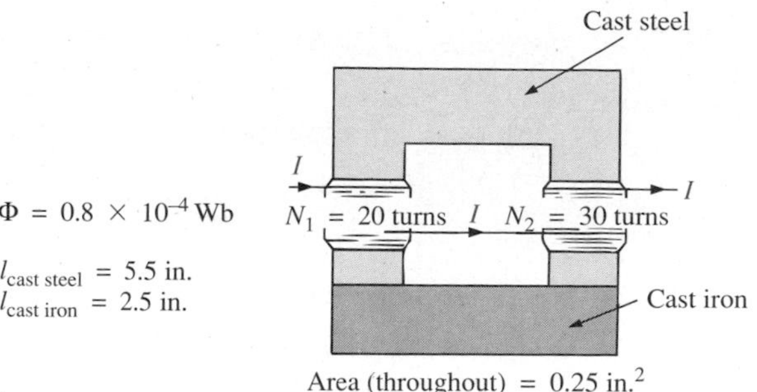

FIG. 11.62
Problem 13.

SECTION 11.12 Air Gaps

14. **a.** Find the current I required to establish a flux $\Phi = 2.4 \times 10^{-4}$ Wb in the magnetic circuit of Fig. 11.63.
    b. Compare the mmf drop across the air gap to that across the rest of the magnetic circuit. Discuss your results using the value of μ for each material.

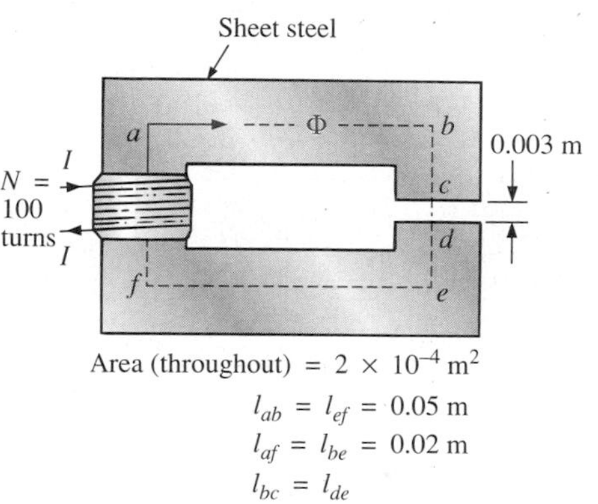

FIG. 11.63
Problem 14.

*15. The force carried by the plunger of the door chime of Fig. 11.64 is determined by

$$f = \frac{1}{2}NI\frac{d\phi}{dx} \qquad \text{(newtons)}$$

where $d\phi/dx$ is the rate of change of flux linking the coil as the core is drawn into the coil. The greatest rate of change of flux will occur when the core is ¼ to ¾ the way through. In this region, if Φ changes from 0.5×10^{-4} Wb to 8×10^{-4} Wb, what is the force carried by the plunger?

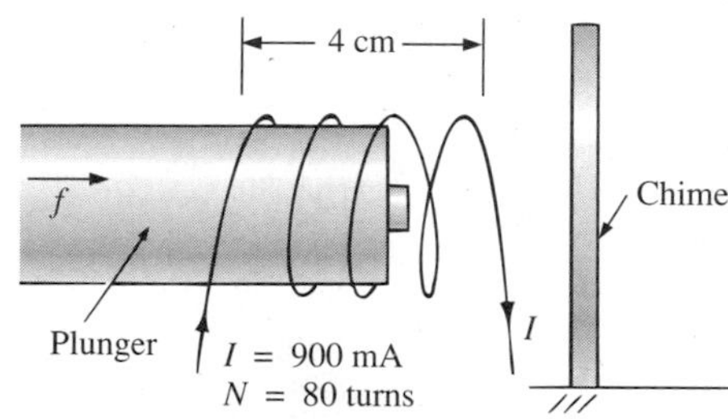

FIG. 11.64
Door chime for Problem 15.

16. Determine the current I_1 required to establish a flux of $\Phi = 2 \times 10^{-4}$ Wb in the magnetic circuit of Fig. 11.65.

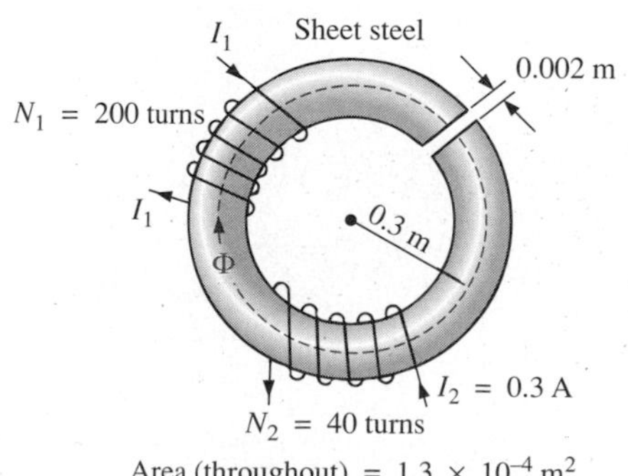

FIG. 11.65
Problem 16.

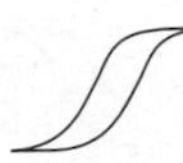

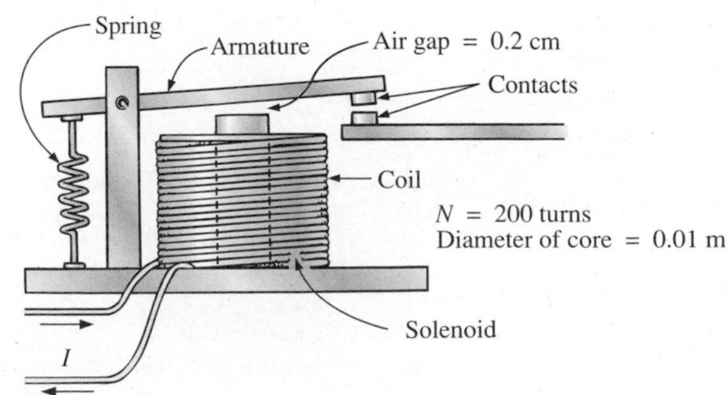

FIG. 11.66
Relay for Problem 17.

*17. a. A flux of 0.2×10^{-4} Wb will establish sufficient attractive force for the armature of the relay of Fig. 11.66 to close the contacts. Determine the required current to establish this flux level if we assume the total mmf drop is across the air gap.

b. The force exerted on the armature is determined by the equation

$$F\text{ (newtons)} = \frac{1}{2} \cdot \frac{B_g^2 A}{\mu_o}$$

where B_g is the flux density within the air gap and A is the common area of the air gap. Find the force in newtons exerted when the flux Φ specified in part (a) is established.

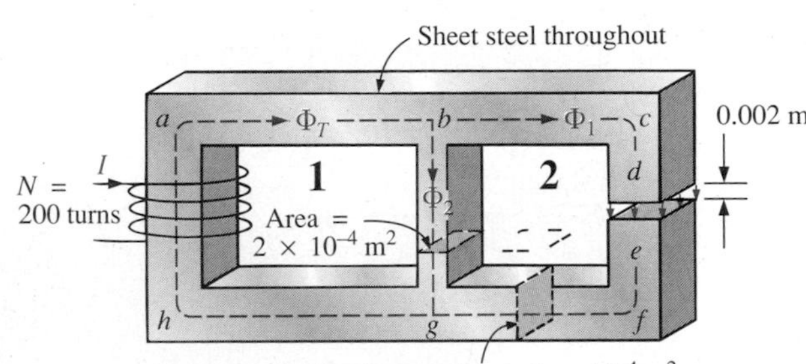

Area for sections other than $bg = 5 \times 10^{-4}\ \text{m}^2$
$l_{ab} = l_{bg} = l_{gh} = l_{ha} = 0.2$ m
$l_{bc} = l_{fg} = 0.1$ m, $l_{cd} = l_{ef} = 0.099$ m

FIG. 11.67
Problem 18.

*18. For the series-parallel magnetic circuit of Fig. 11.67, find the value of I required to establish a flux in the gap $\Phi_g = 2 \times 10^{-4}$ Wb.

SECTION 11.14 Determining Φ

19. Find the magnetic flux Φ established in the series magnetic circuit of Fig. 11.68.

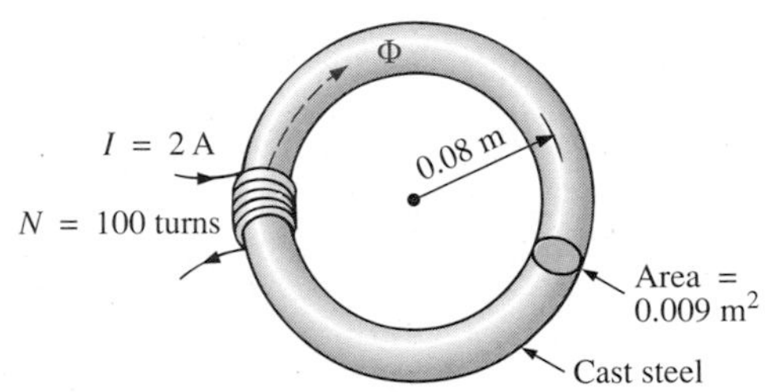

FIG. 11.68
Problem 19.

*20. Determine the magnetic flux Φ established in the series magnetic circuit of Fig. 11.69.

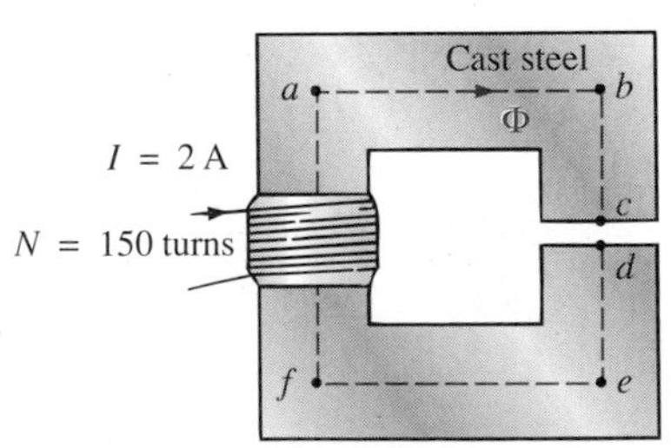

$l_{cd} = 8 \times 10^{-4}$ m
$l_{ab} = l_{be} = l_{ef} = l_{fa} = 0.2$ m
Area (throughout) $= 2 \times 10^{-4}\ \text{m}^2$
$l_{bc} = l_{de}$

FIG. 11.69
Problem 20.

***21.** Note how closely the *B-H* curve of cast steel in Fig. 11.23 matches the curve for the voltage across a capacitor as it charges from zero volts to its final value.

a. Using the equation for the charging voltage as a guide, write an equation for B as a function of H ($B = f(H)$) for cast steel.

b. Test the resulting equation at $H = 900$ At/m, 1800 At/m, and 2700 At/m.

c. Using the equation of part (a), derive an equation for H in terms of B ($H = f(B)$).

d. Test the resulting equation at $B = 1$ T and $B = 1.4$ T.

e. Using the result of part (c), perform the analysis of Example 11.3 and compare the results for the current I.

COMPUTER PROBLEMS Programming Language (C++, BASIC, PASCAL, etc.)

***22.** Using the results of Problem 21, write a program to perform the analysis of a core such as that shown in Example 11.3; that is, let the dimensions of the core and the applied turns be input variables requested by the program.

***23.** Using the results of Problem 21, develop a program to perform the analysis appearing in Example 11.9 for cast steel. A test routine will have to be developed to determine whether the results obtained are sufficiently close to the applied ampere-turns.

GLOSSARY

Ampère's circuital law A law establishing the fact that the algebraic sum of the rises and drops of the mmf around a closed loop of a magnetic circuit is equal to zero.

Diamagnetic materials Materials that have permeabilities slightly less than that of free space.

Domain A group of magnetically aligned atoms.

Electromagnetism Magnetic effects introduced by the flow of charge or current.

Ferromagnetic materials Materials having permeabilities hundreds and thousands of times greater than that of free space.

Flux density (B) A measure of the flux per unit area perpendicular to a magnetic flux path. It is measured in teslas (T) or webers per square meter (Wb/m^2).

Hysteresis The lagging effect between the flux density of a material and the magnetizing force applied.

Magnetic flux lines Lines of a continuous nature that reveal the strength and direction of a magnetic field.

Magnetizing force (H) A measure of the magnetomotive force per unit length of a magnetic circuit.

Magnetomotive force ($\mathscr{F}$) The "pressure" required to establish magnetic flux in a ferromagnetic material. It is measured in ampere-turns (At).

Paramagnetic materials Materials that have permeabilities slightly greater than that of free space.

Permanent magnet A material such as steel or iron that will remain magnetized for long periods of time without the aid of external means.

Permeability (μ) A measure of the ease with which magnetic flux can be established in a material. It is measured in Wb/Am.

Relative permeability (μ_r) The ratio of the permeability of a material to that of free space.

Reluctance ($\mathscr{R}$) A quantity determined by the physical characteristics of a material that will provide an indication of the "reluctance" of that material to the setting up of magnetic flux lines in the material. It is measured in rels or At/Wb.

12

Inductors

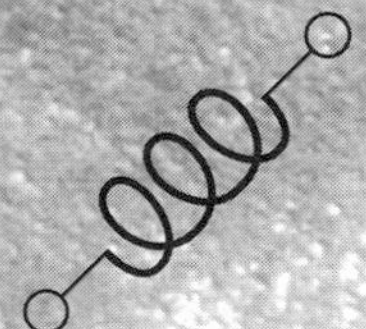

12.1 INTRODUCTION

We have examined the resistor and the capacitor in detail. In this chapter we shall consider a third element, the *inductor,* which has a number of response characteristics similar in many respects to those of the capacitor. In fact, some sections of this chapter will proceed parallel to those for the capacitor to emphasize the similarity that exists between the two elements.

12.2 FARADAY'S LAW OF ELECTROMAGNETIC INDUCTION

If a conductor is moved through a magnetic field so that it cuts magnetic lines of flux, a voltage will be induced across the conductor, as shown in Fig. 12.1. The greater the number of flux lines cut per unit time (by increasing the speed with which the conductor passes through the field), or the stronger the magnetic field strength (for the same tra-

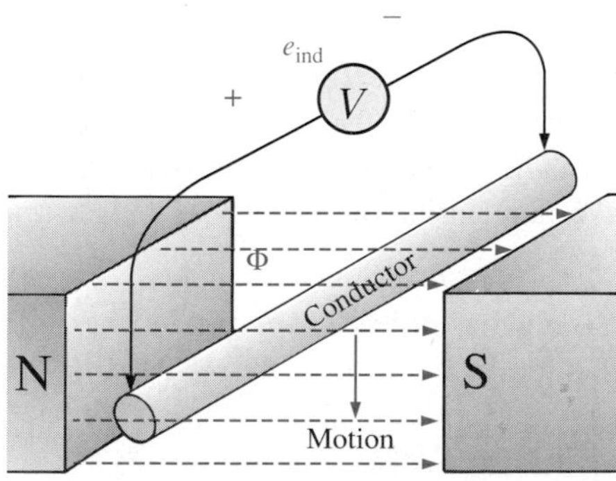

FIG. 12.1
Generating an induced voltage by moving a conductor through a magnetic field.

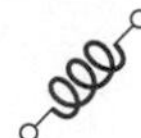

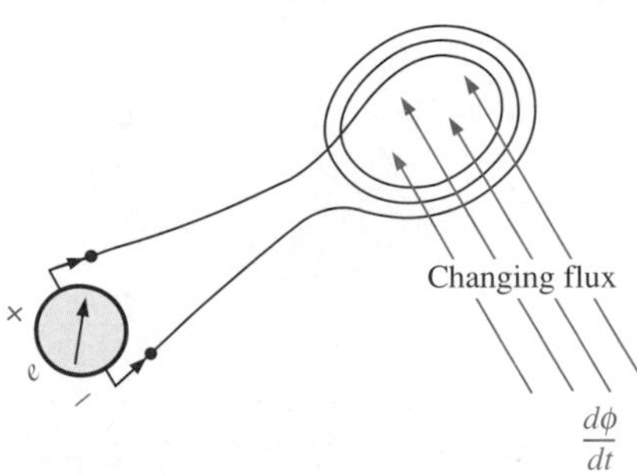

FIG. 12.2
Demonstrating Faraday's law.

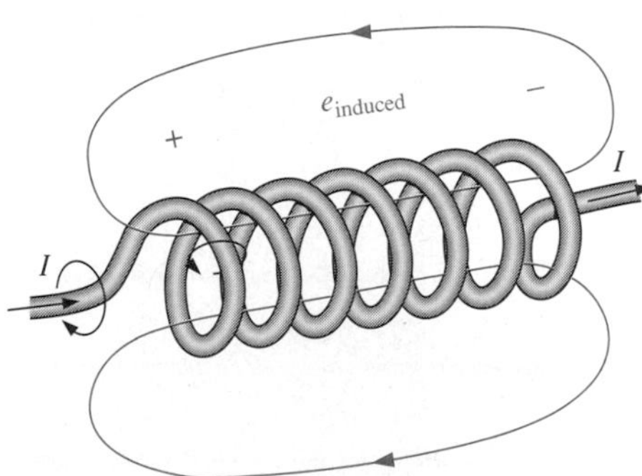

FIG. 12.3
Demonstrating the effect of Lenz's law.

American (Albany, Princeton)
(1797–1878)
Physicist and Mathematician
Professor of Natural Philosophy,
Princeton University

Courtesy of the Smithsonian Institution
Photo No. 59,054

In the early 1800s the title Professor of Natural Philosophy was applied to educators in the sciences. As a student and teacher at the Albany Academy, he performed extensive research in the area of electromagnetism. He improved the design of *electromagnets* by insulating the coil wire to permit a tighter wrap on the core. One of his earlier designs was capable of lifting 3600 pounds. In 1832 he discovered and delivered a paper on *self-induction.* This was followed by the construction of an effective *electric telegraph transmitter and receiver* and extensive research on the oscillatory nature of lightning and discharges from a *Leyden jar.* In 1845 he was appointed the first Secretary of the Smithsonian.

FIG. 12.4
Joseph Henry.

versing speed), the greater will be the induced voltage across the conductor. If the conductor is held fixed and the magnetic field is moved so that its flux lines cut the conductor, the same effect will be produced.

If a coil of N turns is placed in the region of a changing flux, as in Fig. 12.2, a voltage will be induced across the coil as determined by *Faraday's law:*

$$\boxed{e = N\frac{d\phi}{dt}} \qquad \text{(volts, V)} \tag{12.1}$$

where N represents the number of turns of the coil and $d\phi/dt$ is the instantaneous change in flux (in webers) linking the coil. The term *linking* refers to the flux within the turns of wire. The term *changing* simply indicates that either the strength of the field linking the coil changes in magnitude or the coil is moved through the field in such a way that the number of flux lines through the coil changes with time.

If the flux linking the coil ceases to change, such as when the coil simply sits still in a magnetic field of fixed strength, $d\phi/dt = 0$, and the induced voltage $e = N(d\phi/dt) = N(0) = 0$.

12.3 LENZ'S LAW

In Section 11.2 it was shown that the magnetic flux linking a coil of N turns with a current I has the distribution of Fig. 12.3.

If the current increases in magnitude, the flux linking the coil also increases. It was shown in Section 12.2, however, that a changing flux linking a coil induces a voltage across the coil. For this coil, therefore, an induced voltage is developed *across* the coil due to the change in current *through* the coil. The polarity of this induced voltage tends to establish a current in the coil that produces a flux that will oppose any change in the original flux. In other words, the induced effect (e_{ind}) is a result of the increasing current through the coil. However, the resulting induced voltage will tend to establish a current that will oppose the increasing change in current through the coil. Keep in mind that this is all occurring simultaneously. The instant the current begins to increase in magnitude, there will be an opposing effect trying to limit the change. It is "choking" the change in current through the coil. Hence, the term *choke* is often applied to the inductor or coil. In fact, we will find shortly that the current through a coil cannot change instantaneously. A period of time determined by the coil and the resistance of the circuit is required before the inductor discontinues its opposition to a momentary change in current. Recall a similar situation for the voltage across a capacitor in Chapter 10. The reaction above is true for increasing or decreasing levels of current through the coil. This effect is an example of a general principle known as *Lenz's law*, which states that

an induced effect is always such as to oppose the cause that produced it.

12.4 SELF-INDUCTANCE

The ability of a coil to oppose any change in current is a measure of the *self-inductance L* of the coil. For brevity, the prefix *self* is usually dropped. Inductance is measured in henries (H), after the American physicist Joseph Henry (Fig. 12.4).

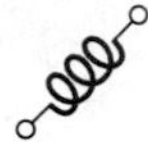

Inductors are coils of various dimensions designed to introduce specified amounts of inductance into a circuit. The inductance of a coil varies directly with the magnetic properties of the coil. Ferromagnetic materials, therefore, are frequently employed to increase the inductance by increasing the flux linking the coil.

A close approximation, in terms of physical dimensions, for the inductance of the coils of Fig. 12.5 can be found using the following equation:

$$L = \frac{N^2 \mu A}{l} \quad \text{(henries, H)} \tag{12.2}$$

where N represents the number of turns; μ, the permeability of the core (as introduced in Section 11.4; recall that μ is not a constant but depends on the level of B and H since $\mu = B/H$); A, the area of the core in square meters; and l, the mean length of the core in meters.

Substituting $\mu = \mu_r \mu_o$ into Eq. (12.2) yields

$$L = \frac{N^2 \mu_r \mu_o A}{l} = \mu_r \frac{N^2 \mu_o A}{l}$$

and

$$L = \mu_r L_o \tag{12.3}$$

where L_o is the inductance of the coil with an air core. In other words, the inductance of a coil with a ferromagnetic core is the relative permeability of the core times the inductance achieved with an air core.

Equations for the inductance of coils different from those shown above can be found in reference handbooks. Most of the equations are more complex than those just described.

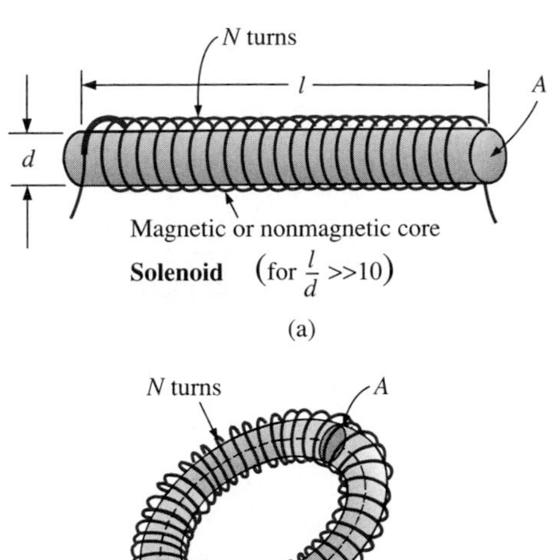

FIG. 12.5
Inductor configurations for which Eq. (12.2) is appropriate.

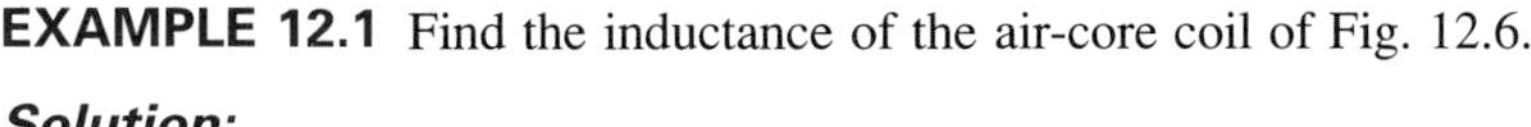

EXAMPLE 12.1 Find the inductance of the air-core coil of Fig. 12.6.

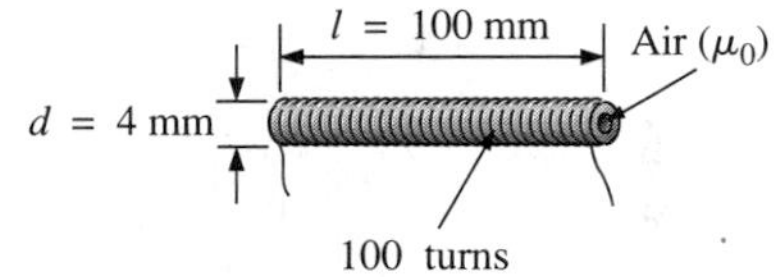

FIG. 12.6
Example 12.1.

Solution:

$$\mu = \mu_r \mu_o = (1)(\mu_o) = \mu_o$$

$$A = \frac{\pi d^2}{4} = \frac{(\pi)(4 \times 10^{-3}\ \text{m})^2}{4} = 12.57 \times 10^{-6}\ \text{m}^2$$

$$L_o = \frac{N^2 \mu_o A}{l} = \frac{(100\ \text{t})^2 (4\pi \times 10^{-7}\ \text{Wb/A·m})(12.57 \times 10^{-6}\ \text{m}^2)}{0.1\ \text{m}}$$

$$= \mathbf{1.58\ \mu H}$$

EXAMPLE 12.2 Repeat Example 12.1, but with an iron core and conditions such that $\mu_r = 2000$.

Solution: By Eq. (12.3),

$$L = \mu_r L_o = (2000)(1.58 \times 10^{-6}\ \text{H}) = \mathbf{3.16\ mH}$$

12.5 TYPES OF INDUCTORS

Practical Equivalent

Inductors, like capacitors, are not ideal. Associated with every inductor are a resistance equal to the resistance of the turns and a stray capaci-

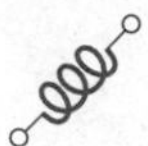

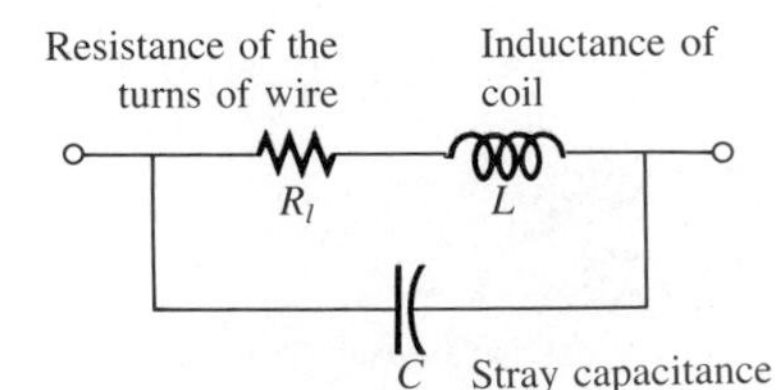

FIG. 12.7
Complete equivalent model for an inductor.

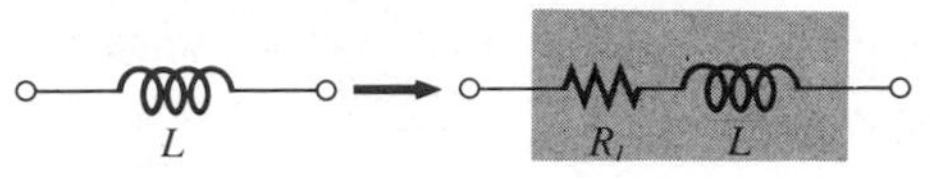

FIG. 12.8
Practical equivalent model for an inductor.

tance due to the capacitance between the turns of the coil. To include these effects, the equivalent circuit for the inductor is as shown in Fig. 12.7. However, for most applications considered in this text, the stray capacitance appearing in Fig. 12.7 can be ignored, resulting in the equivalent model of Fig. 12.8. The resistance R_l can play an important role in the analysis of networks with inductive elements. For most applications, we have been able to treat the capacitor as an ideal element and maintain a high degree of accuracy. For the inductor, however, R_l must often be included in the analysis and can have a pronounced effect on the response of a system (see Chapter 20, "Resonance"). The level of R_l can extend from a few ohms to a few hundred ohms. Keep in mind that the longer or thinner the wire used in the construction of the inductor, the greater will be the dc resistance as determined by $R = \rho l/A$. Our initial analysis will treat the inductor as an ideal element. Once a general feeling for the response of the element is established, the effects of R_l will be included.

Symbols

The primary function of the inductor, however, is to introduce inductance—not resistance or capacitance—into the network. For this reason, the symbols employed for inductance are as shown in Fig. 12.9.

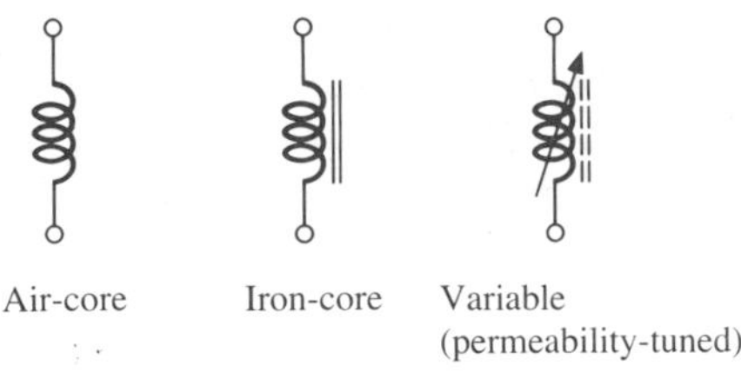

FIG. 12.9
Inductor symbols.

Appearance

All inductors, like capacitors, can be listed under two general headings: *fixed* and *variable*. The fixed air-core and iron-core inductors were described in the last section. The permeability-tuned variable coil has a ferromagnetic shaft that can be moved within the coil to vary the flux linkages of the coil and thereby its inductance. Several fixed and variable inductors appear in Fig. 12.10.

Testing

The primary reasons for inductor failure are shorts that develop between the windings and open circuits in the windings due to factors such as excessive currents, overheating, and age. The open-circuit condition can be checked easily with an ohmmeter (∞ ohms indication), but the short-circuit condition is harder because the resistance of many good inductors is relatively small and the shorting of a few windings will not adversely affect the total resistance. Of course, if one is aware of the typical resistance of the coil, it can be compared to the measured

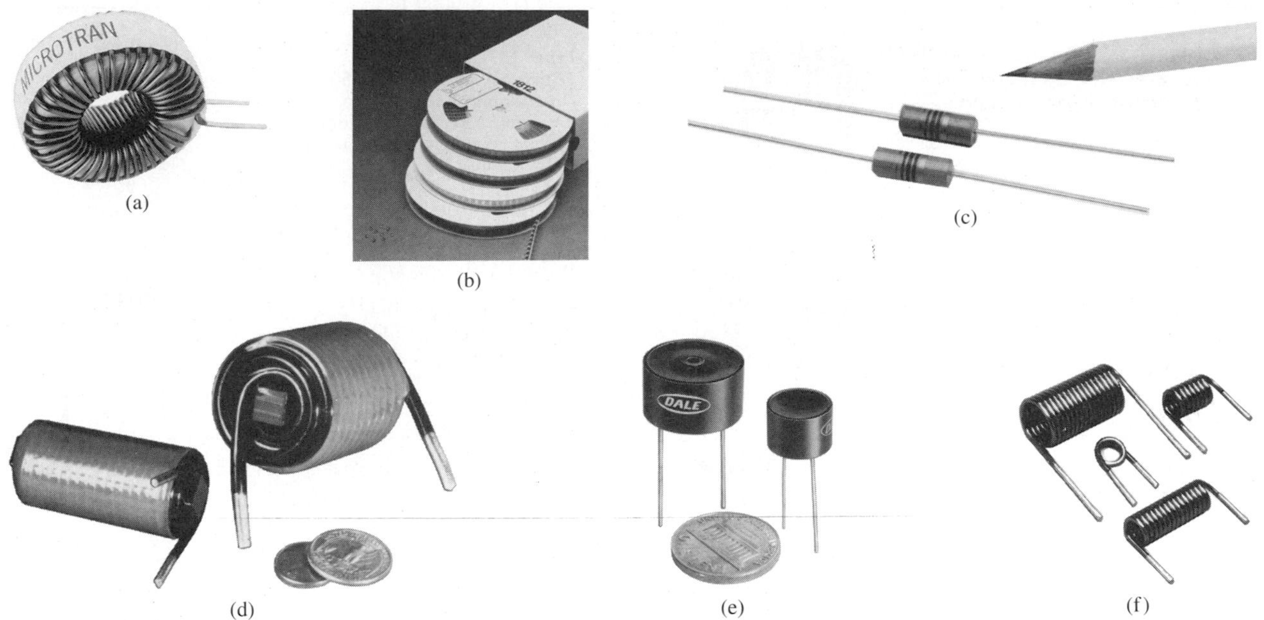

FIG. 12.10

Various types of inductors: (a) toroidal power inductor (1.4 μH to 5.6 mH) (courtesy of Microtan Co., Inc.); (b) surface mount inductors on reels (0.1 μH through 1000 μH on 500-piece reels in 46 values) (courtesy of Bell Industries); (c) molded inductors (0.1 μH to 10 μH); (d) high current filter inductors (24 μH at 60 A to 500 μH at 15 A); (e) toroid filter inductors (40 μH to 5 H); (f) air-core inductors (1 to 32 turns) for high-frequency applications. (Parts (c) through (f) courtesy of Dale Electronics, Inc.)

value. A short between the windings and the core can be checked by simply placing one lead of the meter on one wire (terminal) and the other on the core itself. An indication of zero ohms reflects a short between the two because the wire that makes up the winding has an insulation jacket throughout. The universal LCR meter of Fig. 10.20 can be used to check the inductance level.

Standard Values and Recognition Factor

The standard values for inductors employ the same numerical multipliers used with resistors and inductors. Like the capacitor, the most common employ the same numerical multipliers as the most common resistors, that is, those with the full range of tolerances (5%, 10%, and 20%), as appearing in Table 3.8. However, inductors are also readily available with the multipliers associated with the 5% and 10% resistors of Table 3.8. In general, therefore, expect to find inductors with the following multipliers: **0.1** μH, **0.12** μH, **0.15** μH, **0.18** μH, **0.22** μH, **0.27** μH, **0.33** μH, **0.39** μH, **0.47** μH, **0.56** μH, **0.68** μH, and **0.82** μH, and then **1** mH, **1.2** mH, **1.5** mH, **1.8** mH, **2.2** mH, **2.7** mH, and so on.

Fig. 12.11 was developed to establish a recognition factor when it comes to the various types and uses for inductors. In other words, develop the skills to identify types of inductors, their typical range of values, and some of the most common applications. Figure 12.11 is certainly noninclusive, but it does offer a first step in establishing a sense for what to expect for various applications.

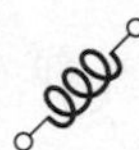

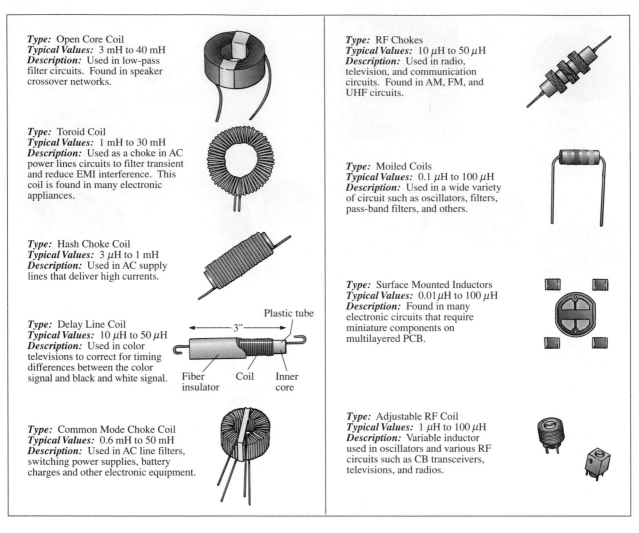

Type: Open Core Coil
Typical Values: 3 mH to 40 mH
Description: Used in low-pass filter circuits. Found in speaker crossover networks.

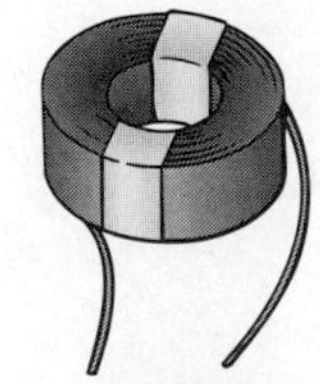

Type: Toroid Coil
Typical Values: 1 mH to 30 mH
Description: Used as a choke in AC power lines circuits to filter transient and reduce EMI interference. This coil is found in many electronic appliances.

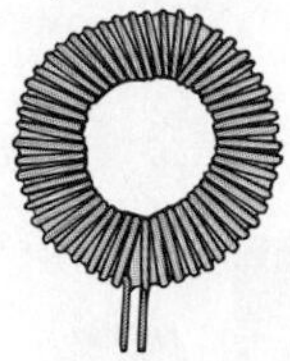

Type: Hash Choke Coil
Typical Values: 3 μH to 1 mH
Description: Used in AC supply lines that deliver high currents.

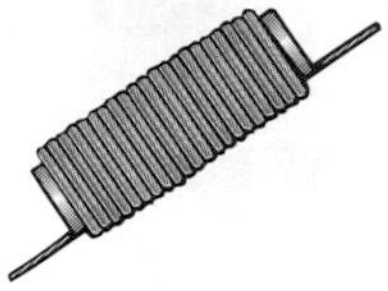

Type: Delay Line Coil
Typical Values: 10 μH to 50 μH
Description: Used in color televisions to correct for timing differences between the color signal and black and white signal.

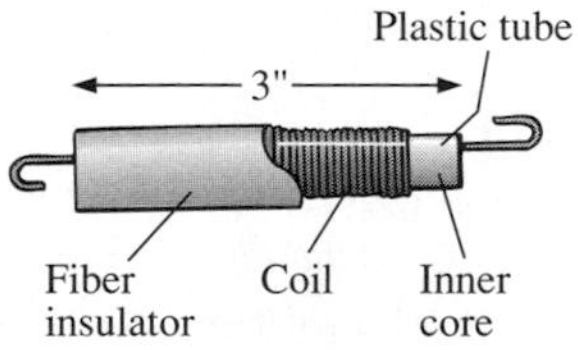

Plastic tube
3"
Fiber insulator
Coil
Inner core

Type: Common Mode Choke Coil
Typical Values: 0.6 mH to 50 mH
Description: Used in AC line filters, switching power supplies, battery charges and other electronic equipment.

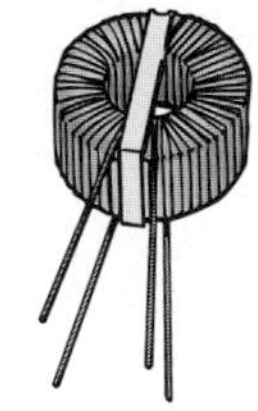

Type: RF Chokes
Typical Values: 10 μH to 50 μH
Description: Used in radio, television, and communication circuits. Found in AM, FM, and UHF circuits.

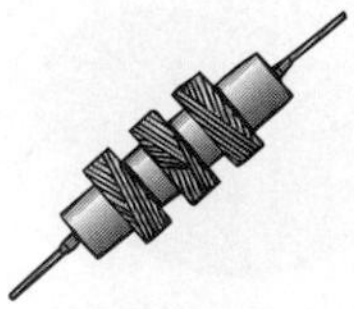

Type: Moiled Coils
Typical Values: 0.1 μH to 100 μH
Description: Used in a wide variety of circuit such as oscillators, filters, pass-band filters, and others.

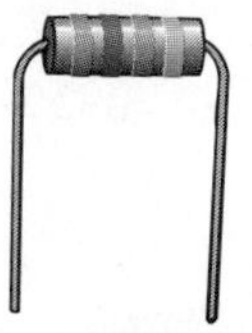

Type: Surface Mounted Inductors
Typical Values: 0.01 μH to 100 μH
Description: Found in many electronic circuits that require miniature components on multilayered PCB.

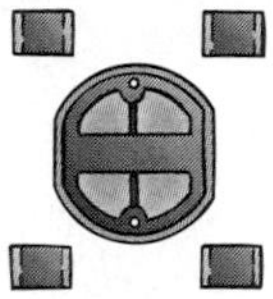

Type: Adjustable RF Coil
Typical Values: 1 μH to 100 μH
Description: Variable inductor used in oscillators and various RF circuits such as CB transceivers, televisions, and radios.

FIG. 12.11
Typical areas of application for inductive elements.

12.6 INDUCED VOLTAGE

The inductance of a coil is also a measure of the change in flux linking a coil due to a change in current through the coil; that is,

$$L = N\frac{d\phi}{di} \qquad \text{(H)} \qquad \textbf{(12.4)}$$

where N is the number of turns, ϕ is the flux in webers, and i is the current through the coil. If a change in current through the coil fails to result in a significant change in the flux linking the coil through its center, the resulting inductance level will be relatively small. For this reason the inductance of a coil is sensitive to the point of operation on the hysteresis curve (described in detail in Section 11.8). If operating on the steep slope, the change in flux will be relatively high for a change in current through the coil. If operating near or in saturation, the change in flux will be relatively small for the same change in current, resulting in a reduced level of inductance. This effect is particularly important when we examine ac circuits since a dc level associated with the applied ac

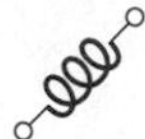

signal may put the coil at or near saturation, and the resulting inductance level for the applied ac signal will be significantly less than expected. You will find that the maximum dc current is normally provided in supply manuals and data sheets to ensure avoidance of the saturation region.

Equation (12.4) also reveals that the larger the inductance of a coil (with N fixed), the larger will be the instantaneous change in flux linking the coil due to an instantaneous change in current through the coil.

If we write Eq. (12.1) as

$$e_L = N\frac{d\phi}{dt} = \left(N\frac{d\phi}{di}\right)\left(\frac{di}{dt}\right)$$

and substitute Eq. (12.4), we then have

$$\boxed{e_L = L\frac{di}{dt}} \quad \text{(V)} \qquad \textbf{(12.5)}$$

revealing that the magnitude of the voltage across an inductor is directly related to the inductance L and the instantaneous rate of change of current through the coil. Obviously, therefore, the greater the *rate* of change of current through the coil, the greater will be the induced voltage. This certainly agrees with our earlier discussion of Lenz's law.

When induced effects are employed in the generation of voltages such as those available from dc or ac generators, the symbol e is appropriate for the induced voltage. However, in network analysis the voltage across an inductor will always have a polarity such as to oppose the source that produced it, and therefore the following notation will be used throughout the analysis to come:

$$\boxed{v_L = L\frac{di}{dt}} \qquad \textbf{(12.6)}$$

If the current through the coil fails to change at a particular instant, the induced voltage across the coil will be zero. For dc applications, after the transient effect has passed, $di/dt = 0$, and the induced voltage is

$$v_L = L\frac{di}{dt} = L(0) = 0 \text{ V}$$

Recall that the equation for the current of a capacitor is the following:

$$i_C = C\frac{dv_C}{dt}$$

Note the similarity between this equation and Eq. (12.6). In fact, if we apply the duality $v \leftrightarrows i$ (that is, interchange the two) and $L \leftrightarrows C$ for capacitance and inductance, each equation can be derived from the other.

The average voltage across the coil is defined by the equation

$$\boxed{v_{L_{\text{av}}} = L\frac{\Delta i}{\Delta t}} \quad \text{(V)} \qquad \textbf{(12.7)}$$

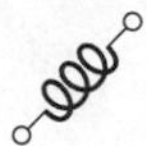

where Δ signifies finite change (a measurable change). Compare this to $i_C = C(\Delta v/\Delta t)$, and the meaning of Δ and application of this equation should be clarified from Chapter 10. An example follows.

EXAMPLE 12.3 Find the waveform for the average voltage across the coil if the current through a 4-mH coil is as shown in Fig. 12.12.

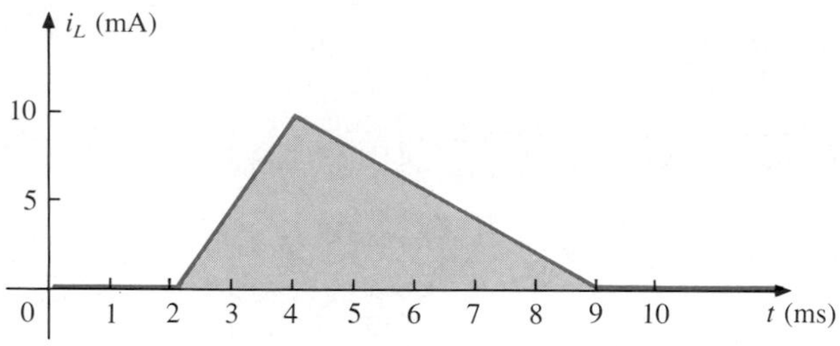

FIG. 12.12
Example 12.3.

Solution:

a. 0 to 2 ms: Since there is no change in current through the coil, there is no voltage induced across the coil; that is,

$$v_L = L\frac{\Delta i}{\Delta t} = L\frac{0}{\Delta t} = \mathbf{0}$$

b. 2 ms to 4 ms:

$$v_L = L\frac{\Delta i}{\Delta t} = (4 \times 10^{-3}\ \text{H})\left(\frac{10 \times 10^{-3}\ \text{A}}{2 \times 10^{-3}\ \text{s}}\right) = 20 \times 10^{-3}\ \text{V}$$
$$= \mathbf{20\ mV}$$

c. 4 ms to 9 ms:

$$v_L = L\frac{\Delta i}{\Delta t} = (-4 \times 10^{-3}\ \text{H})\left(\frac{10 \times 10^{-3}\ \text{A}}{5 \times 10^{-3}\ \text{s}}\right) = -8 \times 10^{-3}\ \text{V}$$
$$= \mathbf{-8\ mV}$$

d. 9 ms to ∞:

$$v_L = L\frac{\Delta i}{\Delta t} = L\frac{0}{\Delta t} = \mathbf{0}$$

The waveform for the average voltage across the coil is shown in Fig. 12.13. Note from the curve that

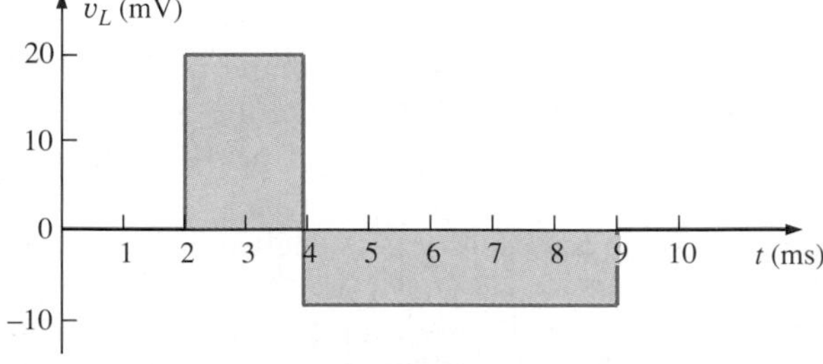

FIG. 12.13
Voltage across a 4-mH coil due to the current of Fig. 12.12.

the voltage across the coil is not determined solely by the magnitude of the change in current through the coil (Δi), but by the rate of change of current through the coil ($\Delta i/\Delta t$).

A similar statement was made for the current of a capacitor due to a change in voltage across the capacitor.

A careful examination of Fig. 12.13 will also reveal that the area under the positive pulse from 2 ms to 4 ms equals the area under the negative pulse from 4 ms to 9 ms. In Section 12.13, we will find that the area under the curves represents the energy stored or released by the inductor. From 2 ms to 4 ms, the inductor is storing energy, whereas from 4 ms to 9 ms, the inductor is releasing the energy stored. For the full period zero to 10 ms, energy has simply been stored and released; there has been no dissipation as experienced for the resistive elements. Over a full cycle, both the ideal capacitor and inductor do not consume energy but simply store and release it in their respective forms.

12.7 *R-L* TRANSIENTS: STORAGE CYCLE

The changing voltages and current that result during the storing of energy in the form of a magnetic field by an inductor in a dc circuit can best be described using the circuit of Fig. 12.14. At the instant the switch is closed, the inductance of the coil will prevent an instantaneous change in current through the coil. The potential drop across the coil, υ_L, will equal the impressed voltage E as determined by Kirchhoff's voltage law since $\upsilon_R = iR = (0)R = 0$ V. The current i_L will then build up from zero, establishing a voltage drop across the resistor and a corresponding drop in υ_L. The current will continue to increase until the voltage across the inductor drops to zero volts and the full impressed voltage appears across the resistor. Initially, the current i_L increases quite rapidly, followed by a continually decreasing rate until it reaches its maximum value of E/R.

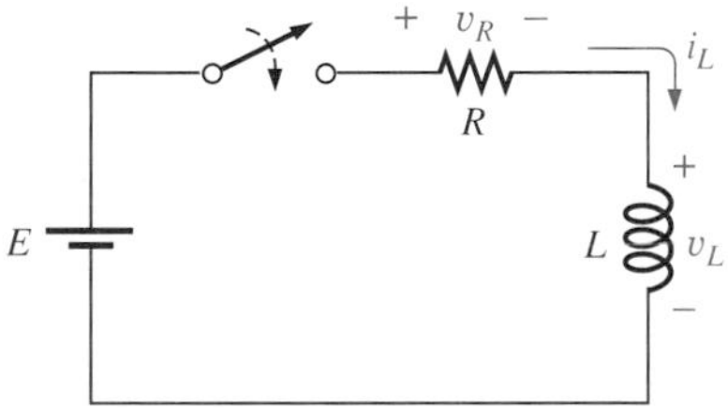

FIG. 12.14
Basic R-L transient network.

You will recall from the discussion of capacitors that a capacitor has a short-circuit equivalent when the switch is first closed and an open-circuit equivalent when steady-state conditions are established. The inductor assumes the opposite equivalents for each stage. The instant the switch of Fig. 12.14 is closed, the equivalent network will appear as shown in Fig. 12.15. Note the correspondence with the earlier comments regarding the levels of voltage and current. The inductor obviously meets all the requirements for an open-circuit equivalent—$\upsilon_L = E$ volts, $i_L = 0$ A.

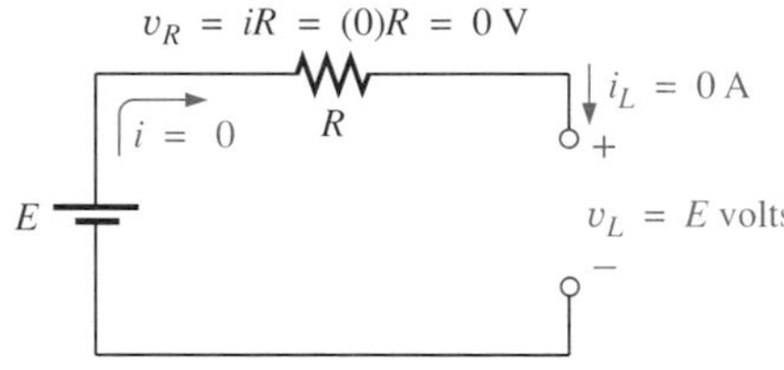

FIG. 12.15
Circuit of Fig. 12.14 the instant the switch is closed.

When steady-state conditions have been established and the storage phase is complete, the "equivalent" network will appear as shown in Fig. 12.16. The network clearly reveals that:

An ideal inductor ($R_l = 0\ \Omega$) assumes a short-circuit equivalent in a dc network once steady-state conditions have been established.

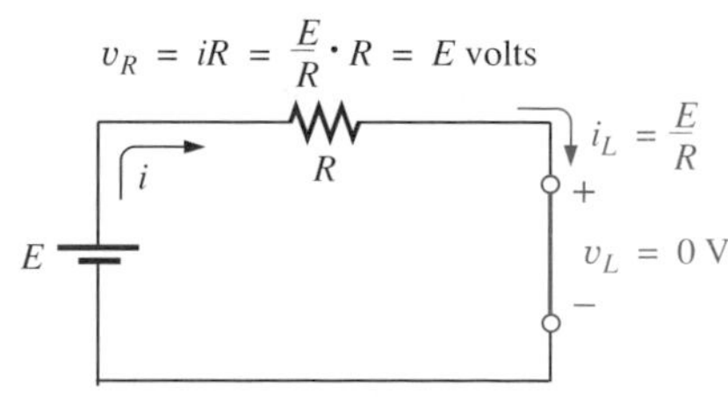

FIG. 12.16
Circuit of Fig. 12.14 under steady-state conditions.

Fortunately, the mathematical equations for the voltages and current for the storage phase are similar in many respects to those encountered for the *R-C* network. The experience gained with these equations in Chapter 10 will undoubtedly make the analysis of *R-L* networks somewhat easier to understand.

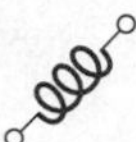

The equation for the current i_L during the storage phase is the following:

$$i_L = I_m(1 - e^{-t/\tau}) = \frac{E}{R}(1 - e^{-t/(L/R)}) \qquad (12.8)$$

Note the factor $(1 - e^{-t/\tau})$, which also appeared for the voltage υ_C of a capacitor during the charging phase. A plot of the equation is given in Fig. 12.17, clearly indicating that the maximum steady-state value of i_L is E/R, and that the rate of change in current decreases as time passes. The abscissa is scaled in time constants, with τ for inductive circuits defined by the following:

$$\tau = \frac{L}{R} \qquad \text{(seconds, s)} \qquad (12.9)$$

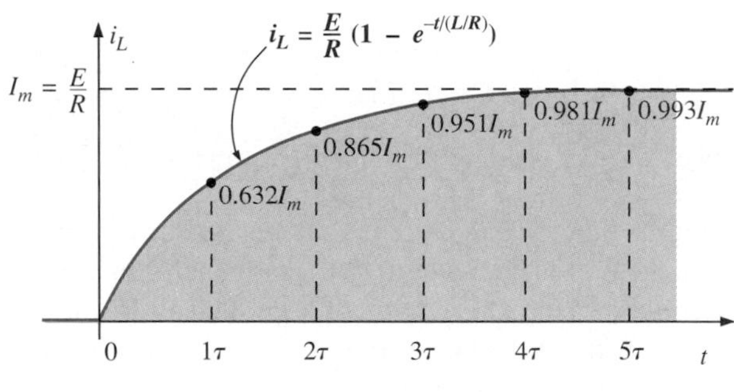

FIG. 12.17
Plotting the waveform for i_L during the storage cycle.

The fact that τ has the units of time can be verified by taking the equation for the induced voltage

$$\upsilon_L = L\frac{di}{dt}$$

and solving for L:

$$L = \frac{\upsilon_L}{di/dt}$$

which leads to the ratio

$$\tau = \frac{L}{R} = \frac{\dfrac{\upsilon_L}{di/dt}}{R} = \frac{\upsilon_L}{\dfrac{di}{dt}R} \longrightarrow \frac{V}{\dfrac{IR}{t}} = \frac{V}{\dfrac{V}{t}} = t \quad \text{(s)}$$

Our experience with the factor $(1 - e^{-t/\tau})$ verifies the level of 63.2% after one time constant, 86.5% after two time constants, and so on. For convenience, Fig. 10.28 is repeated as Fig. 12.18 to evaluate the functions $(1 - e^{-t/\tau})$ and $e^{-t/\tau}$ at various values of τ.

If we keep R constant and increase L, the ratio L/R increases and the rise time increases. The change in transient behavior for the current i_L is plotted in Fig. 12.19 for various values of L. Note again the duality between these curves and those obtained for the R-C network in Fig. 10.31.

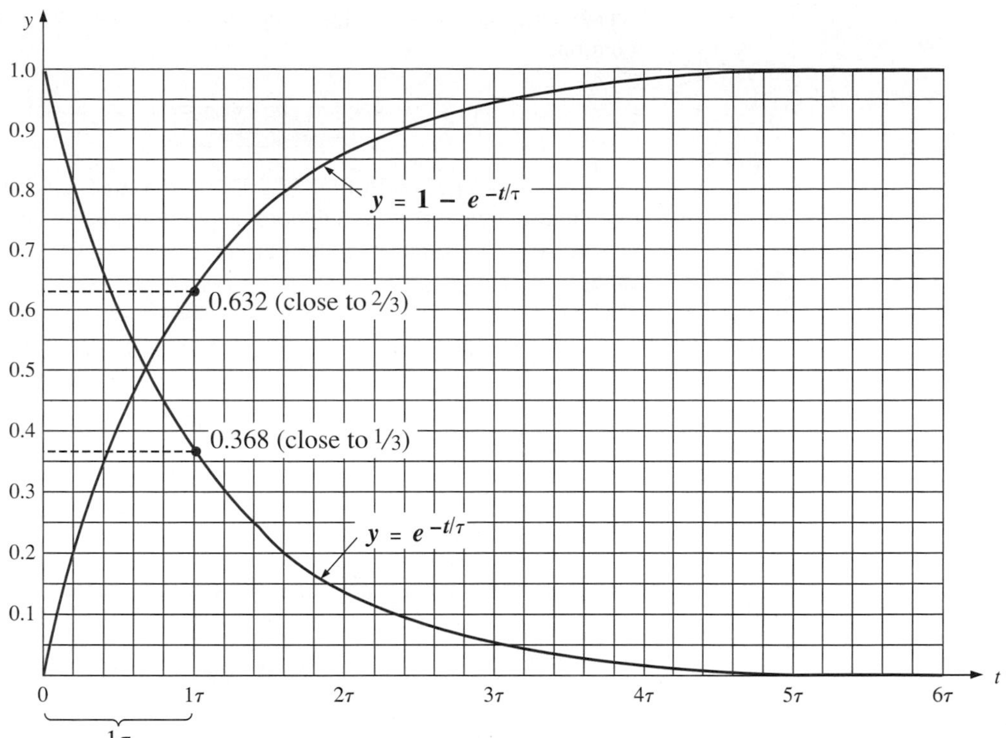

FIG. 12.18
Plotting the functions $y = 1 - e^{-t/\tau}$ and $y = e^{-t/\tau}$.

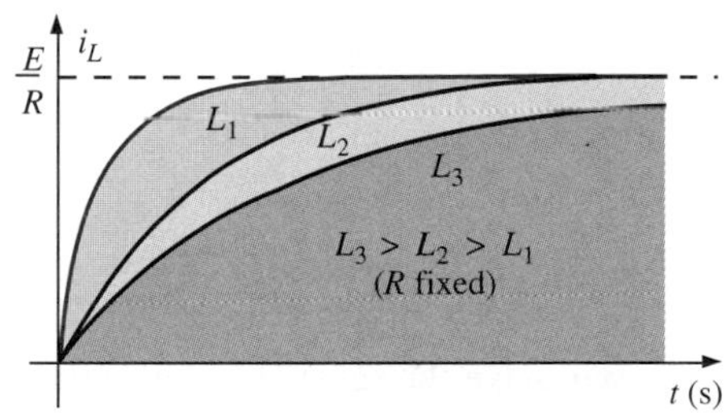

FIG. 12.19
Effect of L on the shape of the i_L storage waveform.

For most practical applications, we will assume that:

The storage phase has passed and steady-state conditions have been established once a period of time equal to five time constants has occurred.

In addition, since L/R will always have some numerical value, even though it may be very small, the period 5τ will always be greater than zero, confirming the fact that

the current cannot change instantaneously in an inductive network.

In fact, the larger the inductance, the more the circuit will oppose a rapid buildup in current level.

Figures 12.16 and 12.17 clearly reveal that the voltage across the coil jumps to E volts when the switch is closed and decays to zero volts with time. The decay occurs in an exponential manner, and v_L during

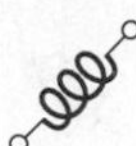

the storage phase can be described mathematically by the following equation:

$$v_L = Ee^{-t/\tau} \tag{12.10}$$

A plot of v_L appears in Fig. 12.20 with the time axis again divided into equal increments of τ. Obviously, the voltage v_L will decrease to zero volts at the same rate the current presses toward its maximum value.

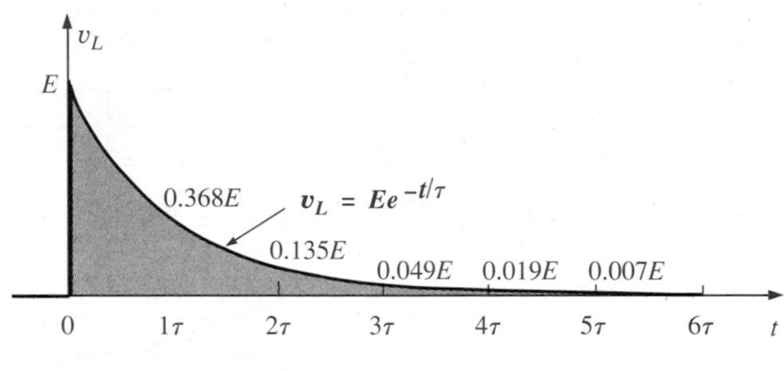

FIG. 12.20
Plotting the voltage v_R versus time for the network of Fig. 12.14.

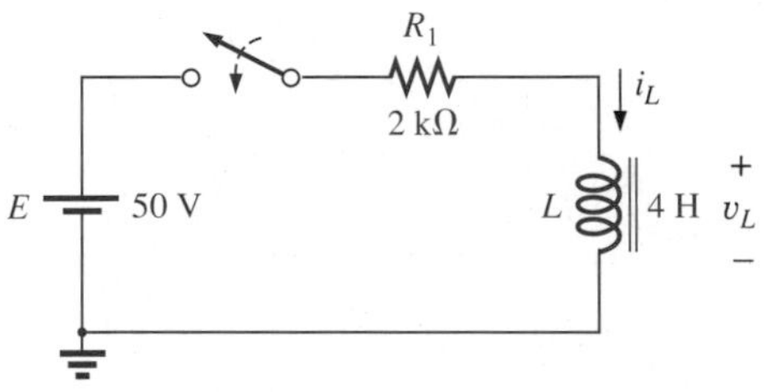

FIG. 12.21
Example 12.4.

In five time constants, $i_L = E/R$, $v_L = 0$ V, and the inductor can be replaced by its short-circuit equivalent.

Since $$v_R = i_R R = i_L R$$

then $$v_R = \left[\frac{E}{R}(1 - e^{-t/\tau})\right]R$$

and $$v_R = E(1 - e^{-t/\tau}) \tag{12.11}$$

and the curve for v_R will have the same shape as obtained for i_L.

EXAMPLE 12.4 Find the mathematical expressions for the transient behavior of i_L and v_L for the circuit of Fig. 12.21 after the closing of the switch. Sketch the resulting curves.

Solution:

$$\tau = \frac{L}{R_1} = \frac{4\text{ H}}{2\text{ k}\Omega} = 2\text{ ms}$$

By Eq (12.8),

$$I_m = \frac{E}{R_1} = \frac{50}{2\text{ k}\Omega} = 25 \times 10^{-3}\text{ A} = 25\text{ mA}$$

and $$i_L = \mathbf{(25 \times 10^{-3})(1 - e^{-t/(2\times 10^{-3})})}$$

By Eq. (12.10),

$$v_L = \mathbf{50e^{-t/(2\times 10^{-3})}}$$

Both waveforms appear in Fig. 12.22.

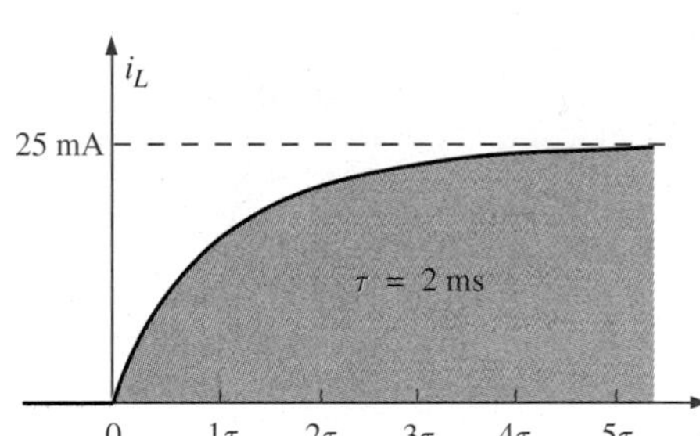

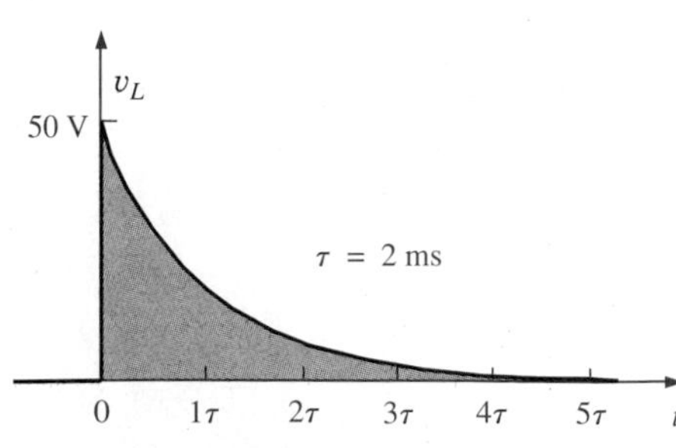

FIG. 12.22
i_L and v_L for the network of Fig. 12.21.

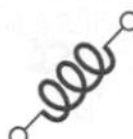

12.8 R-L TRANSIENTS: DECAY PHASE

In the analysis of R-C circuits, we found that the capacitor could hold its charge and store energy in the form of an electric field for a period of time determined by the leakage factors. In R-L circuits, the energy is stored in the form of a magnetic field established by the current through the coil. Unlike the capacitor, however, an isolated inductor cannot continue to store energy since the absence of a closed path would cause the current to drop to zero, releasing the energy stored in the form of a magnetic field. If the switch of Fig. 12.14 were opened quickly, a spark would probably occur across the contacts due to the rapid change in current from a maximum of E/R to zero amperes. The change in current di/dt of the equation $v_L = L(di/dt)$ would establish a high voltage v_L across the coil that would discharge across the points of the switch. This is the same mechanism as applied in the ignition system of a car to ignite the fuel in the cylinder. Some 25,000 V are generated by the rapid decrease in ignition coil current that occurs when the switch in the system is opened. (In older systems, the "points" in the distributor served as the switch.) This inductive reaction is significant when you consider that the only independent source in a car is a 12-V battery.

If opening the switch to move it to another position will cause such a rapid discharge in stored energy, how can the decay phase of an R-L circuit be analyzed in much the same manner as for the R-C circuit? The solution is to use a network such as that appearing in Fig. 12.23. When the switch is closed, the voltage across the resistor R_2 is E volts and the R-L branch will respond in the same manner as described above, with the same waveforms and levels. A Thévenin network of E in parallel with R_2 would simply result in the source since R_2 would be shorted out by the short-circuit replacement of the voltage source E when the Thévenin resistance is determined.

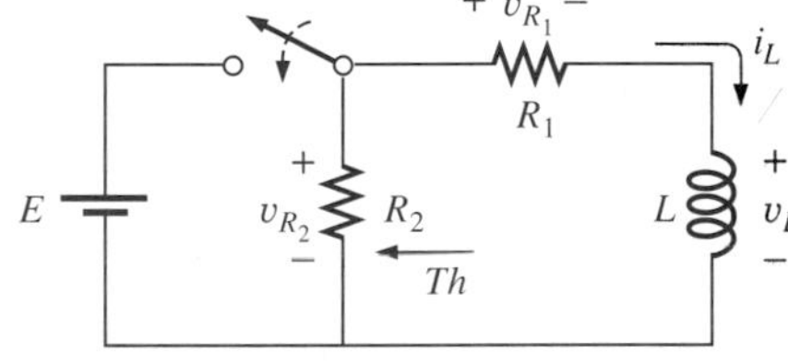

FIG. 12.23
Initiating the storage phase for the inductor L by closing the switch.

After the storage phase has passed and steady-state conditions are established, the switch can be opened without the sparking effect or rapid discharge due to the resistor R_2, which provides a complete path for the current i_L. In fact, for clarity the discharge path is isolated in Fig. 12.24. The voltage v_L across the inductor will reverse polarity and have a magnitude determined by

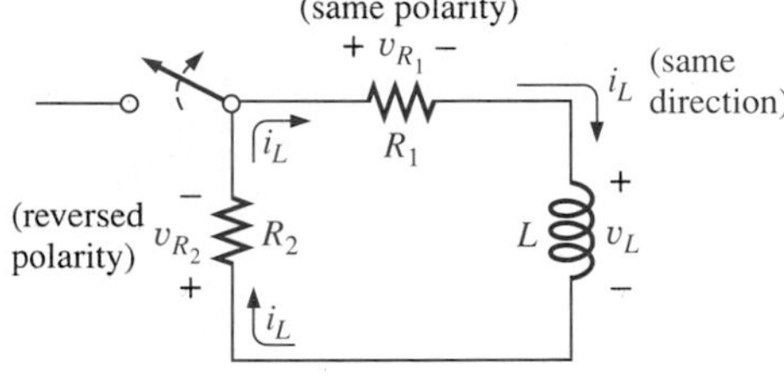

FIG. 12.24
Network of Fig. 12.23 the instant the switch is opened.

$$v_L = v_{R_1} + v_{R_2} \tag{12.12}$$

Recall that the voltage across an inductor can change instantaneously but the current cannot. The result is that the current i_L must maintain the same direction and magnitude as shown in Fig. 12.24. Therefore, the instant after the switch is opened, i_L is still $I_m = E/R_1$, and

$$\begin{aligned} v_L &= v_{R_1} + v_{R_2} = i_1R_1 + i_2R_2 \\ &= i_L(R_1 + R_2) = \frac{E}{R_1}(R_1 + R_2) = \left(\frac{R_1}{R_1} + \frac{R_2}{R_1}\right)E \end{aligned}$$

and

$$v_L = \left(1 + \frac{R_2}{R_1}\right)E \tag{12.13}$$

which is bigger than E volts by the ratio R_2/R_1. In other words, when the switch is opened, the voltage across the inductor will reverse polar-

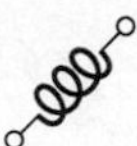

ity and drop instantaneously from E to $-[1 + (R_2/R_1)]E$ volts. The minus sign is a result of v_L having a polarity opposite to the defined polarity of Fig. 12.24.

As an inductor releases its stored energy, the voltage across the coil will decay to zero in the following manner:

$$v_L = V_i e^{-t/\tau'} \tag{12.14}$$

with
$$V_i = \left(1 + \frac{R_2}{R_1}\right)E$$

and
$$\tau' = \frac{L}{R_T} = \frac{L}{R_1 + R_2}$$

The current will decay from a maximum of $I_m = E/R_1$ to zero, in the following manner:

$$i_L = I_m e^{-t/\tau'} \tag{12.15}$$

with
$$I_m = \frac{E}{R_1} \quad \text{and} \quad \tau' = \frac{L}{R_1 + R_2}$$

The mathematical expression for the voltage across either resistor can then be determined using Ohm's law:

$$\begin{aligned} v_{R_1} &= i_{R_1}R_1 = i_L R_1 \\ &= I_m e^{-t/\tau'} R_1 \\ &= \frac{E}{R_1} R_1 e^{-t/\tau'} \end{aligned}$$

and
$$v_{R_1} = Ee^{-t/\tau'} \tag{12.16}$$

The voltage v_{R_1} has the same polarity as during the storage phase since the current i_L has the same direction. The voltage v_{R_2} is expressed as follows:

$$\begin{aligned} v_{R_2} &= i_{R_2}R_2 = i_L R_2 \\ &= I_m e^{-t/\tau'} R_2 \\ &= \frac{E}{R_1} R_2 e^{-t/\tau'} \end{aligned}$$

and
$$v_{R_2} = \frac{R_2}{R_1} Ee^{-t/\tau'} \tag{12.17}$$

with the polarity indicated in Fig. 12.24.

EXAMPLE 12.5 The resistor R_2 was added to the network of Fig. 12.22, as shown in Fig. 12.25.

a. Find the mathematical expressions for i_L, v_L, v_{R_1}, and v_{R_2} after the storage phase has been completed and the switch is opened.
b. Sketch the waveforms for each voltage and current for both phases covered by this example and Example 12.4 if five time constants pass between phases. Use the defined polarities of Fig. 12.23.

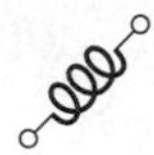

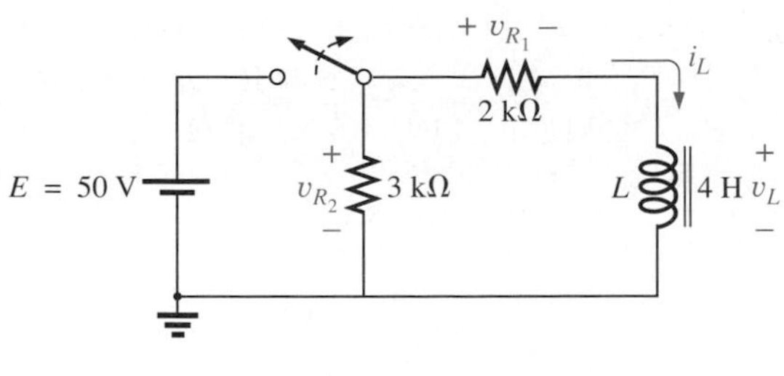

FIG. 12.25
Example 12.5.

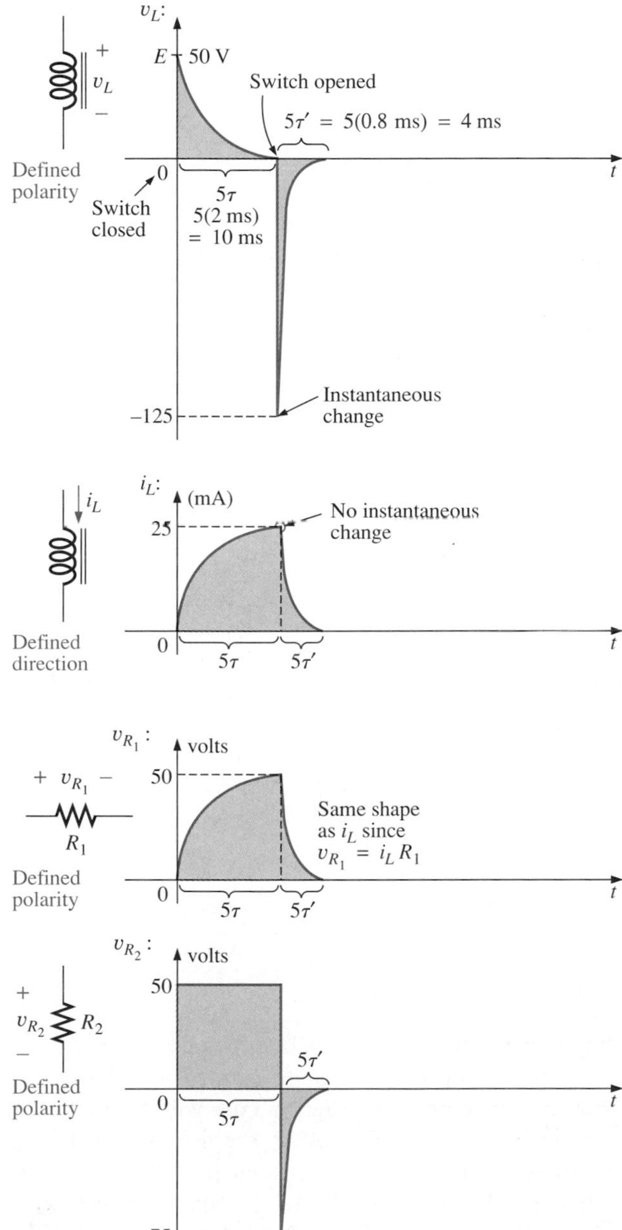

FIG. 12.26
The various voltages and the current for the network of Fig. 12.25.

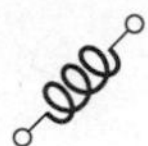

Solutions:

a. $$\tau' = \frac{L}{R_1 + R_2} = \frac{4\text{ H}}{2\text{ k}\Omega + 3\text{ k}\Omega} = \frac{4\text{ H}}{5 \times 10^3\ \Omega} = 0.8 \times 10^{-3}\text{ s}$$
$$= 0.8\text{ ms}$$

By Eq. (12.14),

$$V_i = \left(1 + \frac{R_2}{R_1}\right)E = \left(1 + \frac{3\text{ k}\Omega}{2\text{ k}\Omega}\right)(50\text{ V}) = 125\text{ V}$$

and $$v_L = -V_i e^{-t/\tau'} = \mathbf{-125}\boldsymbol{e}^{\mathbf{-t/(0.8\times10^{-3})}}$$

By Eq. (12.15),

$$I_m = \frac{E}{R_1} = \frac{50\text{ V}}{2\text{ k}\Omega} = 25\text{ mA}$$

and $$i_L = I_m e^{-t/\tau'} = \mathbf{(25 \times 10^{-3})}\boldsymbol{e}^{\mathbf{-t/(0.8\times10^{-3})}}$$

By Eq. (12.16),

$$v_{R_1} = Ee^{-t/\tau'} = \mathbf{50}\boldsymbol{e}^{\mathbf{-t/(0.8\times10^{-3})}}$$

By Eq. (12.17),

$$v_{R_2} = -\frac{R_2}{R_1}Ee^{-t/\tau'} = -\frac{3\text{ k}\Omega}{2\text{ k}\Omega}(50\text{ V})e^{-t/\tau'} = \mathbf{-75}\boldsymbol{e}^{\mathbf{-t/(0.8\times10^{-3})}}$$

b. See Fig. 12.26.

In the preceding analysis, it was assumed that steady-state conditions were established during the charging phase and $I_m = E/R_1$, with $v_L = 0$ V. However, if the switch of Fig. 12.24 is opened before i_L reaches its maximum value, the equation for the decaying current of Fig. 12.24 must change to

$$\boxed{i_L = I_i e^{-t/\tau'}} \qquad \textbf{(12.18)}$$

where I_i is the starting or initial current. Equation (12.14) would be modified as follows:

$$\boxed{v_L = V_i e^{-t/\tau'}} \qquad \textbf{(12.19)}$$

with $$V_i = I_i(R_1 + R_2)$$

12.9 INITIAL VALUES

This section will parallel Section 10.9 on the effect of *initial values* on the transient phase. Since the current through a coil cannot change instantaneously, the current through a coil will begin the *transient phase* at the *initial value* established by the network (note Fig. 12.27) before the switch was closed. It will then pass through the transient phase until it reaches the *steady-state* (or *final*) level after about 5 time constants. The steady-state level of the inductor current can be found by simply substituting its short-circuit equivalent (or R_l for the practical equivalent) and finding the resulting current through the element.

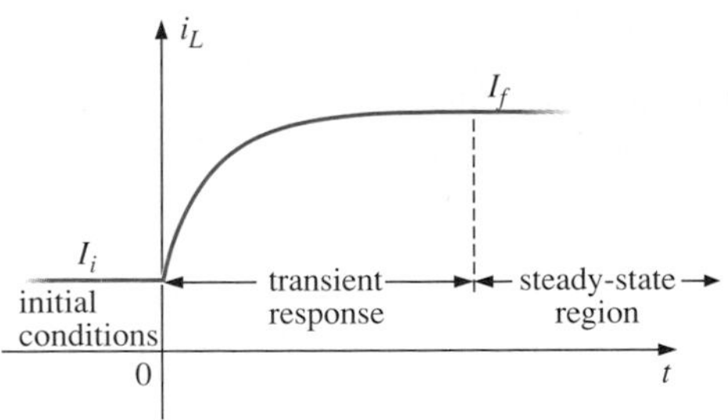

FIG. 12.27
Defining the three phases of a transient waveform.

Using the transient equation developed in the previous section, an equation for the current i_L can be written for the entire time interval of Fig. 12.27; that is,

$$i_L = I_i + (I_f - I_i)(1 - e^{-t/\tau})$$

with $(I_f - I_i)$ representing the total change during the transient phase. However, by multiplying through and rearranging terms:

$$\begin{aligned} i_L &= I_i + I_f - I_f e^{-t/\tau} - I_i + I_i e^{-t/\tau} \\ &= I_f - I_f e^{-t/\tau} + I_i e^{-t/\tau} \end{aligned}$$

we find

$$\boxed{i_L = I_f + (I_i - I_f)e^{-t/\tau}} \qquad \textbf{(12.20)}$$

If you are required to draw the waveform for the current i_L from initial value to final value, start by drawing a line at the initial value and steady-state levels, and then add the transient response (sensitive to the time constant) between the two levels. The following example will clarify the procedure.

EXAMPLE 12.6 The inductor of Fig. 12.28 has an initial current level of 4 mA in the direction shown. (Specific methods to establish the initial current will be presented in the sections and problems to follow.)

a. Find the mathematical expression for the current through the coil once the switch is closed.
b. Find the mathematical expression for the voltage across the coil during the same transient period.
c. Sketch the waveform for each from initial value to final value.

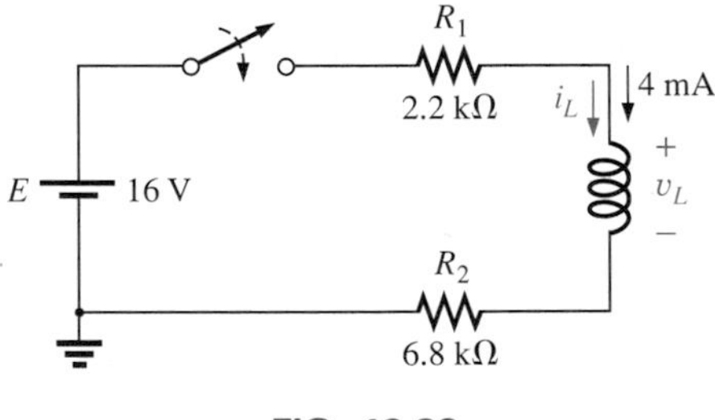

FIG. 12.28
Example 12.6.

Solutions:

a. Substituting the short-circuit equivalent for the inductor will result in a final or steady-state current determined by Ohm's law:

$$I_f = \frac{E}{R_1 + R_2} = \frac{16\text{ V}}{2.2\text{ k}\Omega + 6.8\text{ k}\Omega} = \frac{16\text{ V}}{9\text{ k}\Omega} = 1.78\text{ mA}$$

The time constant is determined by:

$$\tau = \frac{L}{R_T} = \frac{100\text{ mH}}{2.2\text{ k}\Omega + 6.8\text{ k}\Omega} = \frac{100\text{ mH}}{9\text{ k}\Omega} = 11.11\ \mu\text{s}$$

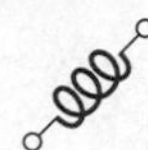

Applying Eq. (12.20):

$$
\begin{aligned}
i_L &= I_f + (I_i - I_f)e^{-t/\tau} \\
&= 1.78 \text{ mA} + (4 \text{ mA} - 1.78 \text{ mA})e^{-t/11.11\ \mu s} \\
&= \mathbf{1.78\ mA + 2.22\ mA}\ e^{-t/11.11\ \mu s}
\end{aligned}
$$

b. Since the current through the inductor is constant at 4 mA prior to the closing of the switch, the voltage (whose level is sensitive only to changes in current through the coil) must have an initial value of 0 V. At the instant the switch is closed, the current through the coil cannot change instantaneously, so the current through the resistive elements will be 4 mA. The resulting peak voltage at $t = 0$ s can then be found using Kirchhoff's voltage law as follows:

$$
\begin{aligned}
V_m &= E - V_{R_1} - V_{R_2} \\
&= 16 \text{ V} - (4 \text{ mA})(2.2 \text{ k}\Omega) - (4 \text{ mA})(6.8 \text{ k}\Omega) \\
&= 16 \text{ V} - 8.8 \text{ V} - 27.2 \text{ V} = 16 \text{ V} - 36 \text{ V} \\
&= -20 \text{ V}
\end{aligned}
$$

Note the minus sign to indicate that the polarity of the voltage v_L is opposite to the defined polarity of Fig. 12.28.

The voltage will then decay (with the same time constant as the current i_L) to zero because the inductor is approaching its short-circuit equivalence.

The equation for v_L is therefore:

$$v_L = \mathbf{-20}e^{-t/11.11\ \mu s}$$

c. See Fig. 12.29. The initial and final values of the current were drawn first and then the transient response was included between these levels. For the voltage, the waveform begins and ends at zero, with the peak value having a sign sensitive to the defined polarity of v_L in Fig. 12.28.

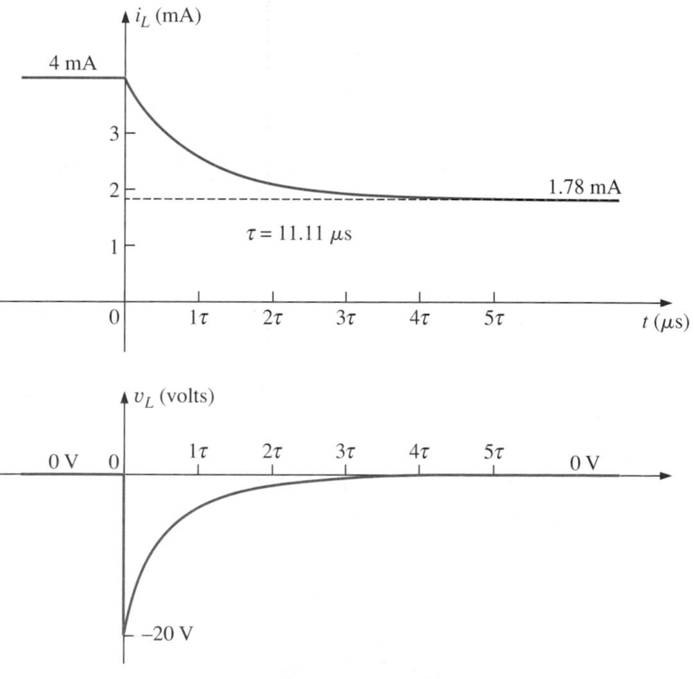

FIG. 12.29
i_L and v_L for the network of Fig. 12.28.

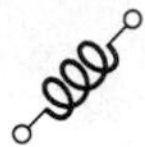

Let us now test the validity of the equation for i_L by substituting $t = 0$ s to reflect the instant the switch is closed.

$$e^{-t/\tau} = e^{-0} = 1$$

and $i_L = 1.78 \text{ mA} + 2.22 \text{ mA } e^{-t/\tau} = 1.78 \text{ mA} + 2.22 \text{ mA}$
$= 4 \text{ mA}$

When $t > 5\tau,\ e^{-t/\tau} \cong 0$

and $i_L = 1.78 \text{ mA} + 2.22 \text{ mA } e^{-t/\tau} = 1.78 \text{ mA}$

12.10 INSTANTANEOUS VALUES

The development presented in Section 10.10 for capacitive networks can also be applied to *R-L* networks to determine instantaneous voltages, currents, and time. The instantaneous values of any voltage or current can be determined by simply inserting t into the equation and using a calculator or table to determine the magnitude of the exponential term.

The similarity between the equations $v_C = E(1 - e^{-t/\tau})$ and $i_L = I_m(1 - e^{-t/\tau})$ results in a derivation of the following for t that is identical to that used to obtain Eq. (10.24).

$$t = \tau \log_e \left(\frac{I_m}{I_m - i_L}\right) \tag{12.21}$$

For the other form, the equation $v_C = Ee^{-t/\tau}$ is a close match with $v_L = Ee^{-t/\tau}$, permitting a derivation similar to that employed for Eq. (10.25):

$$t = \tau \log_e \frac{E}{v_L} \tag{12.22}$$

The similarities between the above and the equations in Chapter 10 should make the equation for t fairly easy to obtain.

12.11 $\tau = L/R_{Th}$

In Chapter 10 ("Capacitors"), we found that there are occasions when the circuit does not have the basic form of Fig. 12.14. The same is true for inductive networks. Again, it is necessary to find the Thévenin equivalent circuit before proceeding in the manner described in this chapter. Consider the following example.

EXAMPLE 12.7 For the network of Fig. 12.30:

a. Find the mathematical expression for the transient behavior of the current i_L and the voltage v_L after the closing of the switch ($I_i = 0$ mA).
b. Draw the resultant waveform for each.

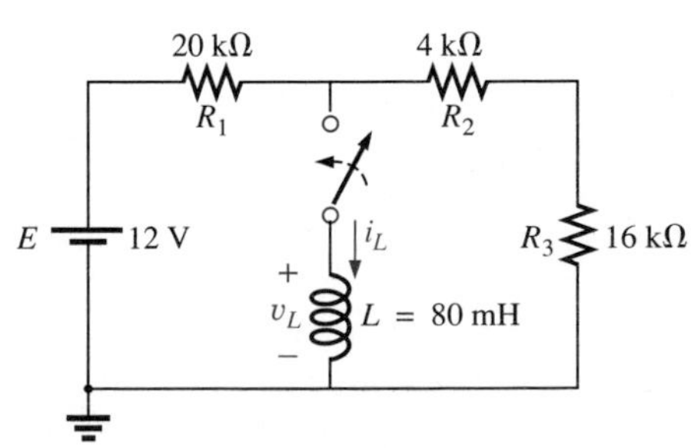

FIG. 12.30
Example 12.7.

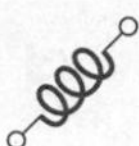

Solutions:

a. Applying Thévenin's theorem to the 80-mH inductor (Fig. 12.31) yields

$$R_{Th} = \frac{R}{N} = \frac{20\text{ k}\Omega}{2} = 10\text{ k}\Omega$$

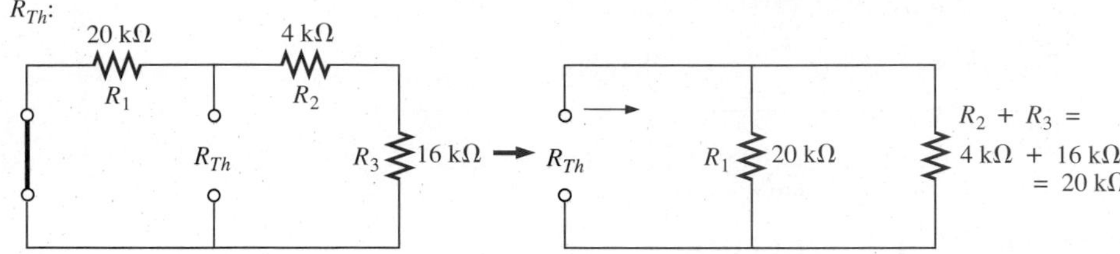

FIG. 12.31
Determining R_{Th} for the network of Fig. 12.30.

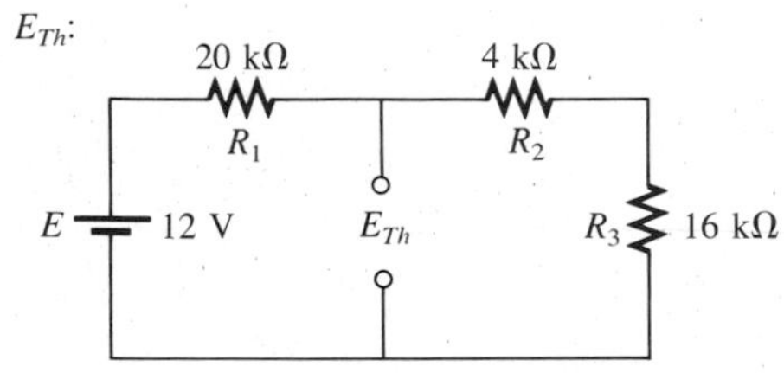

FIG. 12.32
Determining E_{Th} for the network of Fig. 12.30.

Applying the voltage divider rule (Fig. 12.32),

$$E_{Th} = \frac{(R_2 + R_3)E}{R_1 + R_2 + R_3}$$

$$= \frac{(4\text{ k}\Omega + 16\text{ k}\Omega)(12\text{ V})}{20\text{ k}\Omega + 4\text{ k}\Omega + 16\text{ k}\Omega} = \frac{(20\text{ k}\Omega)(12\text{ V})}{40\text{ k}\Omega} = 6\text{ V}$$

The Thévenin equivalent circuit is shown in Fig. 12.33. Using Eq. (12.8),

$$i_L = \frac{E_{Th}}{R}(1 - e^{-t/\tau})$$

$$\tau = \frac{L}{R_{Th}} = \frac{80 \times 10^{-3}\text{ H}}{10 \times 10^3\ \Omega} = 8 \times 10^{-6}\text{ s}$$

$$I_m = \frac{E_{Th}}{R_{Th}} = \frac{6\text{ V}}{10 \times 10^3\ \Omega} = 0.6 \times 10^{-3}\text{ A}$$

and $$\boldsymbol{i_L = (0.6 \times 10^{-3})(1 - e^{-t/(8\times10^{-6})})}$$

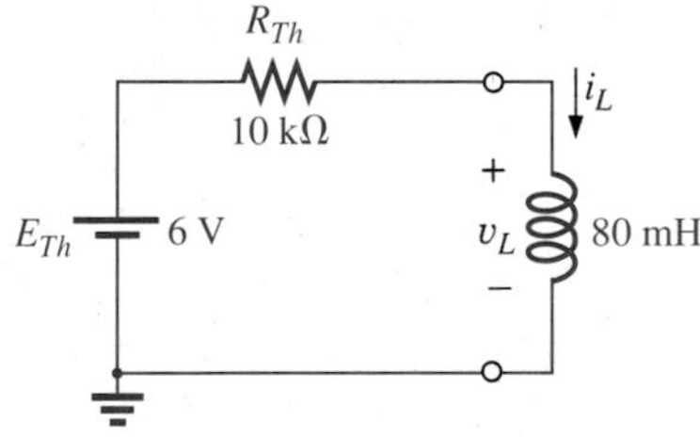

FIG. 12.33
The resulting Thévenin equivalent circuit for the network of Fig. 12.30.

Using Eq. (12.10),

$$v_L = E_{Th}e^{-t/\tau}$$

so that $$\boldsymbol{v_L = 6e^{-t/(8\times10^{-6})}}$$

b. See Fig. 12.34.

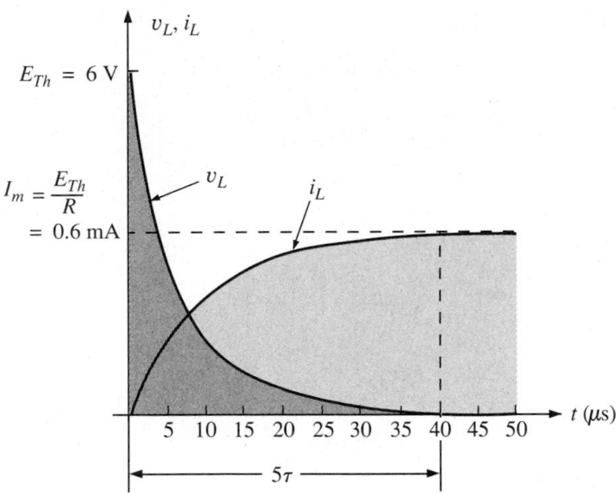

FIG. 12.34
The resulting waveforms for i_L and v_L for the network of Fig. 12.30.

EXAMPLE 12.8 The switch S_1 of Fig. 12.35 has been closed for a long time. At $t = 0$ s, S_1 is opened at the same instant S_2 is closed to avoid an interruption in current through the coil.

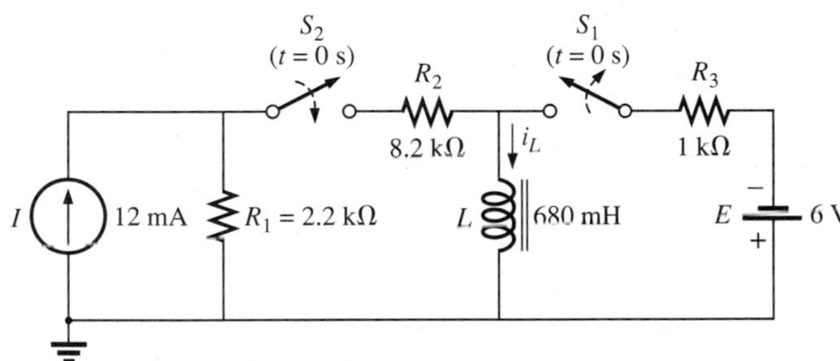

FIG. 12.35
Example 12.8.

a. Find the initial current through the coil. Pay particular attention to its direction.
b. Find the mathematical expression for the current i_L following the closing of the switch S_2.
c. Sketch the waveform for i_L.

Solutions:

a. Using Ohm's law the initial current through the coil is determined by:

$$I_i = -\frac{E}{R_3} = -\frac{6\text{ V}}{1\text{ k}\Omega} = -6\text{ mA}$$

b. Applying Thévenin's theorem:

$$R_{Th} = R_1 + R_2 = 2.2\text{ k}\Omega + 8.2\text{ k}\Omega = 10.4\text{ k}\Omega$$
$$E_{Th} = IR_1 = (12\text{ mA})(2.2\text{ k}\Omega) = 26.4\text{ V}$$

The Thévenin equivalent network appears in Fig. 12.36.

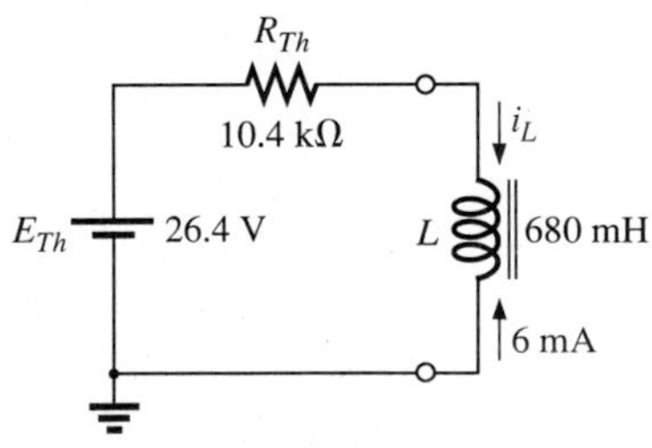

FIG. 12.36
Thévenin equivalent circuit for the network of Fig. 12.35 for $t \geq 0$ s.

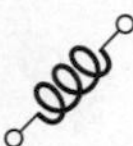

The steady-state current can then be determined by substituting the short-circuit equivalent for the inductor:

$$I_f = \frac{E}{R_{Th}} = \frac{26.4 \text{ V}}{10.4 \text{ k}\Omega} = 2.54 \text{ mA}$$

The time constant:

$$\tau = \frac{L}{R_{Th}} = \frac{680 \text{ mH}}{10.4 \text{ k}\Omega} = 65.39 \; \mu\text{s}$$

Applying Eq. (12.20):

$$\begin{aligned} i_L &= I_f + (I_i - I_f)e^{-t/\tau} \\ &= 2.54 \text{ mA} + (-6 \text{ mA} - 2.54 \text{ mA})e^{-t/65.39 \; \mu\text{s}} \\ &= \mathbf{2.54 \text{ mA} - 8.54 \text{ mA} \; e^{-t/(65.39 \; \mu s)}} \end{aligned}$$

c. Note Fig. 12.37.

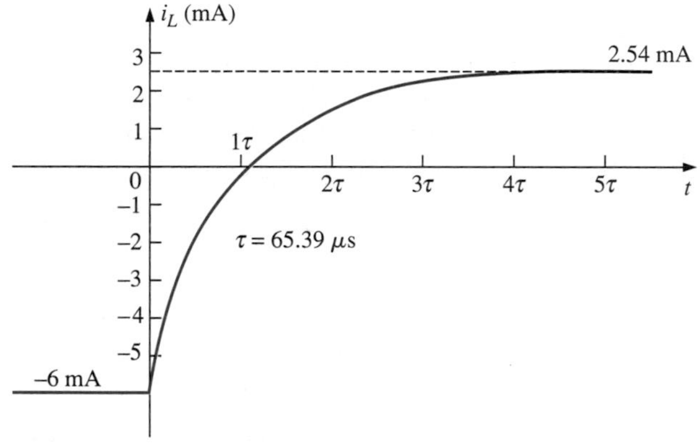

FIG. 12.37
The current i_L for the network of Fig. 12.36.

12.12 INDUCTORS IN SERIES AND PARALLEL

Inductors, like resistors and capacitors, can be placed in series or parallel. Increasing levels of inductance can be obtained by placing inductors in series, while decreasing levels can be obtained by placing inductors in parallel.

For inductors in series, the total inductance is found in the same manner as the total resistance of resistors in series (Fig. 12.38):

$$L_T = L_1 + L_2 + L_3 + \cdots + L_N \tag{12.23}$$

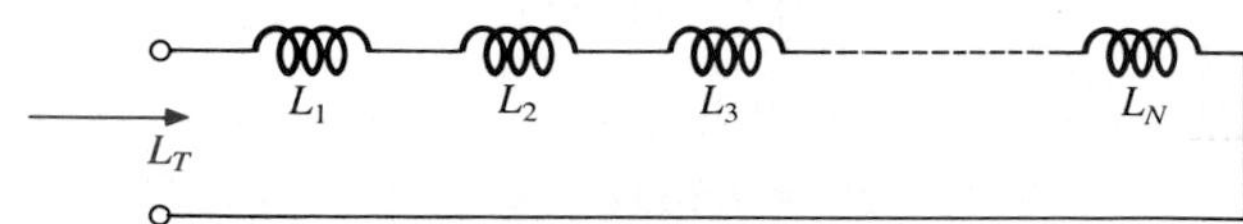

FIG. 12.38
Inductors in series.

For inductors in parallel, the total inductance is found in the same manner as the total resistance of resistors in parallel (Fig. 12.39):

$$\frac{1}{L_T} = \frac{1}{L_1} + \frac{1}{L_2} + \frac{1}{L_3} + \cdots + \frac{1}{L_N} \tag{12.24}$$

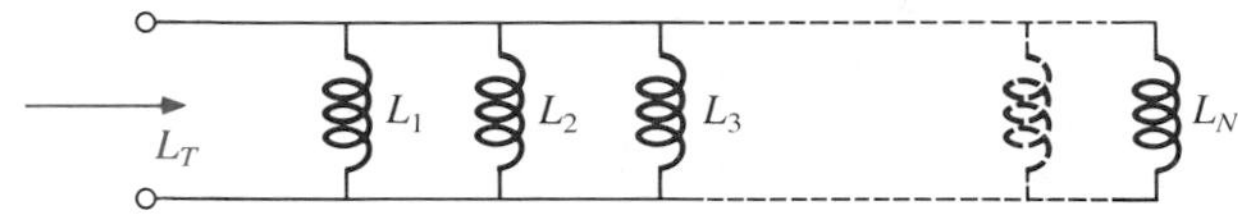

FIG. 12.39
Inductors in parallel.

For two inductors in parallel,

$$L_T = \frac{L_1 L_2}{L_1 + L_2} \tag{12.25}$$

EXAMPLE 12.9 Reduce the network of Fig. 12.40 to its simplest form.

Solution: The inductors L_2 and L_3 are equal in value and they are in parallel, resulting in an equivalent parallel value of

$$L'_T = \frac{L}{N} = \frac{1.2\text{ H}}{2} = 0.6\text{ H}$$

The resulting 0.6 H is then in parallel with the 1.8-H inductor, and

$$L''_T = \frac{(L'_T)(L_4)}{L'_T + L_4} = \frac{(0.6\text{ H})(1.8\text{ H})}{0.6\text{ H} + 1.8\text{ H}}$$
$$= 0.45\text{ H}$$

The inductor L_1 is then in series with the equivalent parallel value, and

$$L_T = L_1 + L''_T = 0.56\text{ H} + 0.45\text{ H}$$
$$= 1.01\text{ H}$$

The reduced equivalent network appears in Fig. 12.41.

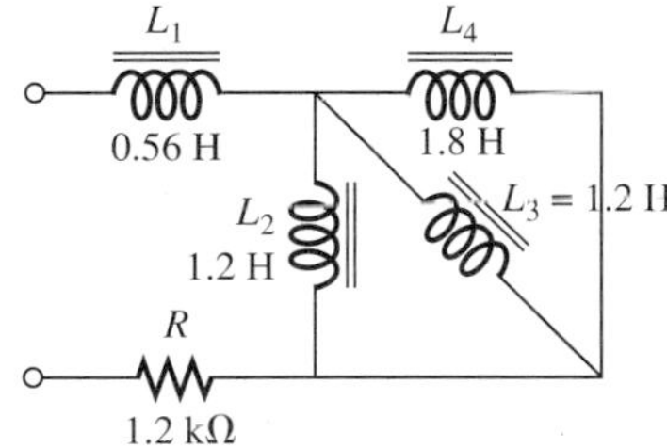

FIG. 12.40
Example 12.9.

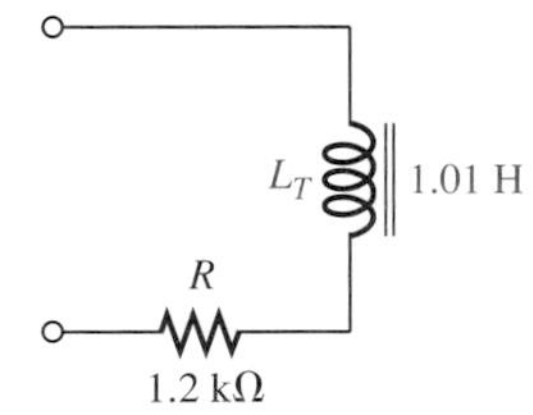

FIG. 12.41
Terminal equivalent of the network of Fig. 12.40.

12.13 *R-L* AND *R-L-C* CIRCUITS WITH dc INPUTS

We found in Section 12.7 that, for all practical purposes, an inductor can be replaced by a short circuit in a dc circuit after a period of time greater than five time constants has passed. If in the following circuits we assume that all of the currents and voltages have reached their final values, the current through each inductor can be found by replacing each inductor with a short circuit. For the circuit of Fig. 12.42, for example,

$$I_1 = \frac{E}{R_1} = \frac{10\text{ V}}{2\ \Omega} = \mathbf{5\ A}$$

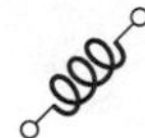

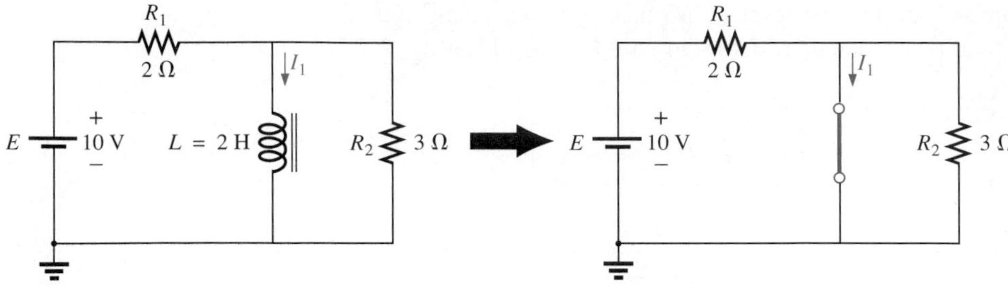

FIG. 12.42
Substituting the short-circuit equivalent for the inductor for $t > 5\tau$.

For the circuit of Fig. 12.43,

$$I = \frac{E}{R_2 \parallel R_3} = \frac{21\text{ V}}{2\ \Omega} = \mathbf{10.5\ A}$$

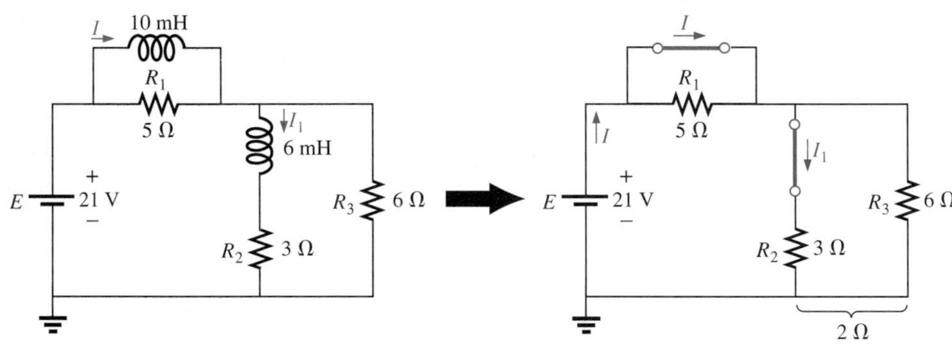

FIG. 12.43
Establishing the equivalent network for $t > 5\tau$.

Applying the current divider rule,

$$I_1 = \frac{R_3 I}{R_3 + R_2} = \frac{(6\ \Omega)(10.5\text{ A})}{6\ \Omega + 3\ \Omega} = \frac{63\text{ A}}{9} = \mathbf{7\ A}$$

In the following examples we will assume that the voltage across the capacitors and the current through the inductors have reached their final values. Under these conditions, the inductors can be replaced with short circuits, and the capacitors can be replaced with open circuits.

EXAMPLE 12.10 Find the current I_L and the voltage V_C for the network of Fig. 12.44.

Solution:

$$I_L = \frac{E}{R_1 + R_2} = \frac{10\text{ V}}{5\ \Omega} = \mathbf{2\ A}$$

$$V_C = \frac{R_2 E}{R_2 + R_1} = \frac{(3\ \Omega)(10\text{ V})}{3\ \Omega + 2\ \Omega} = \mathbf{6\ V}$$

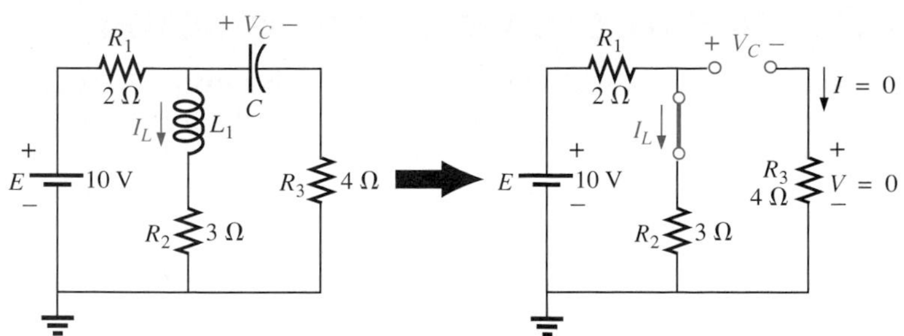

FIG. 12.44
Example 12.10.

EXAMPLE 12.11 Find the currents I_1 and I_2 and the voltages V_1 and V_2 for the network of Fig. 12.45.

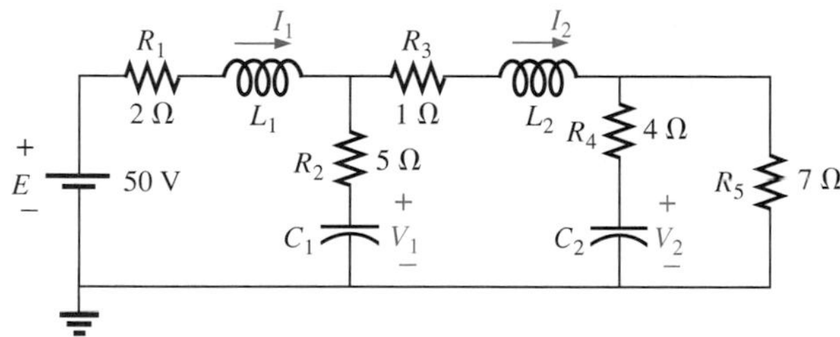

FIG. 12.45
Example 12.11.

Solution: Note Fig. 12.46:

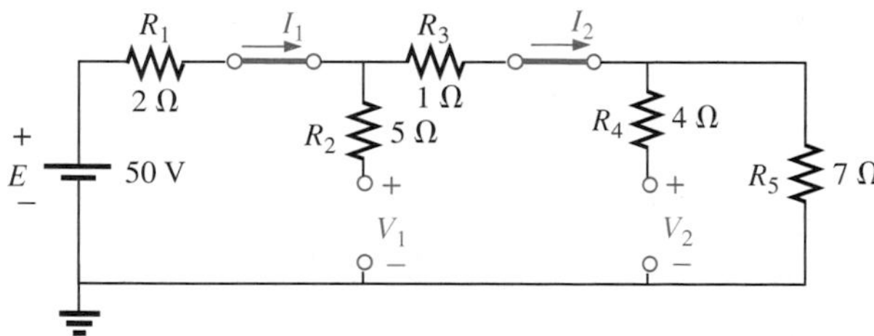

FIG. 12.46
Substituting the short-circuit equivalents for the inductors and open-circuit equivalents for the capacitor for $t > 5\tau$.

$$I_1 = I_2$$

$$I_1 = \frac{E}{R_1 + R_3 + R_5} = \frac{50\ \text{V}}{2\ \Omega + 1\ \Omega + 7\ \Omega} = \frac{50\ \text{V}}{10\ \Omega} = \mathbf{5\ A}$$

$$V_2 = I_2 R_5 = (5\ \text{A})(7\ \Omega) = \mathbf{35\ V}$$

Applying the voltage divider rule,

$$V_1 = \frac{(R_3 + R_5)E}{R_1 + R_3 + R_5} = \frac{(1\ \Omega + 7\ \Omega)(50\ \text{V})}{2\ \Omega + 1\ \Omega + 7\ \Omega} = \frac{(8\ \Omega)(50\ \text{V})}{10\ \Omega} = \mathbf{40\ V}$$

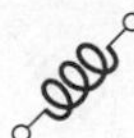

12.14 ENERGY STORED BY AN INDUCTOR

The ideal inductor, like the ideal capacitor, does not dissipate the electrical energy supplied to it. It stores the energy in the form of a magnetic field. A plot of the voltage, current, and power to an inductor is shown in Fig. 12.47 during the buildup of the magnetic field surrounding the inductor. The energy stored is represented by the shaded area under the power curve. Using calculus, we can show that the evaluation of the area under the curve yields

$$W_{\text{stored}} = \frac{1}{2} L I_m^2 \qquad \text{(joules, J)} \qquad \textbf{(12.26)}$$

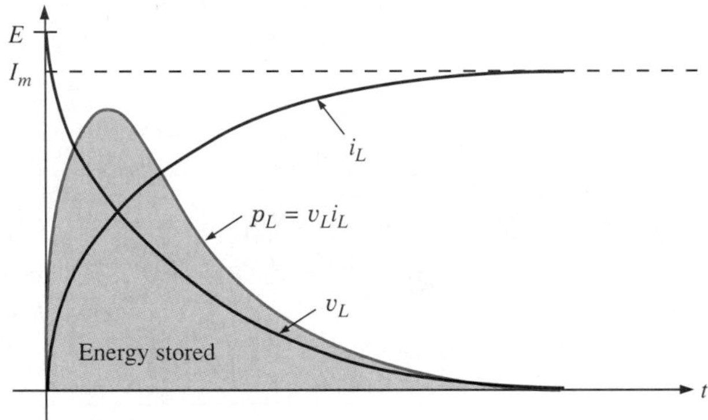

FIG. 12.47
The power curve for an inductive element under transient conditions.

EXAMPLE 12.12 Find the energy stored by the inductor in the circuit of Fig. 12.48 when the current through it has reached its final value.

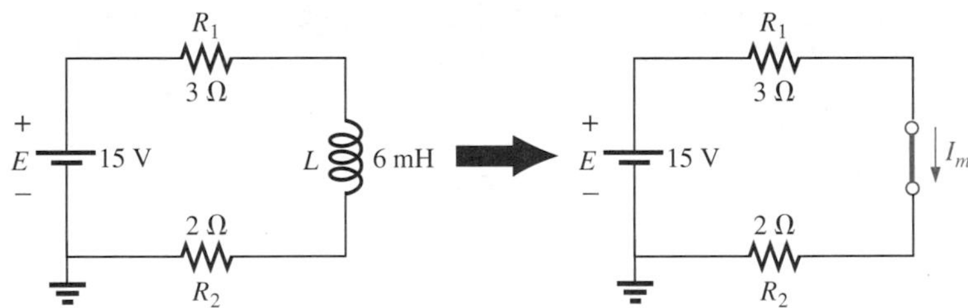

FIG. 12.48
Example 12.12.

Solution:

$$I_m = \frac{E}{R_1 + R_2} = \frac{15\text{ V}}{3\ \Omega + 2\ \Omega} = \frac{15\text{ V}}{5\ \Omega} = 3\text{ A}$$

$$W_{\text{stored}} = \frac{1}{2}LI_m^2 = \frac{1}{2}(6 \times 10^{-3}\text{ H})(3\text{ A})^2 = \frac{54}{2} \times 10^{-3}\text{ J}$$

$$= \mathbf{27\ mJ}$$

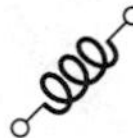

12.15 COMPUTER ANALYSIS

Both PSpice and a programming language can provide the transient response for an *R-L* circuit. When using a programming language, the appropriate equations are employed to determine the voltages and currents as they change with time. A table of values can then be generated or a plot obtained using a plotting routine.

PSpice (DOS)

Inductors are entered in much the same manner as are resistors and capacitors, as demonstrated by the format below:

LTOROID	3	4	5M	IC=2M
	+	–		
Name	Node	Node	Value	Initial value

The user fills in those quantities with braces underneath, although the initial condition entry (the current through the inductor before the switching action occurs) can be omitted if the initial value is zero amperes. The above entry is for a 5-mH coil between nodes 3 and 4, with node 3 as the defined higher potential, and an initial current of 2 mA. In PSpice, inductors have the additional limitation that they cannot form a closed loop (like parallel inductors). However, this limitation can be circumvented by placing a small resistor (negligible compared to the parameters of the network) in series with one of the inductors (as in the example to follow).

The network to be analyzed appears in Fig. 12.49, with the input file in Fig. 12.50. Since two parallel inductors form a closed loop, a 1-mΩ resistor was placed in series with one of the coils before the nodes were defined. The input pulse is as defined for *R-C* circuits to simulate the closing of a switch at $t = 0$ s and to establish 50 V across the network. The parallel combination of the inductors is 4 H $\|$ 12 H = 3 H and $\tau = L/R = 3$ H/2 kΩ = 1.5 ms. The .TRAN command is therefore established from 0.5 ms to 10 ms to provide at least three data points in each time-constant interval. The .PROBE command then permits a request for V(3), which is v_L, and I(R), which is the total current i_L through the parallel coils, as shown in the output file of Fig. 12.51. Note how v_L approaches zero volts after $5\tau = 7.5$ ms and i_L approaches its final value of $E/R = 50$ V/2 kΩ = 25 mA in the same time interval. It is satisfying to watch the PSpice and .PROBE combination present the

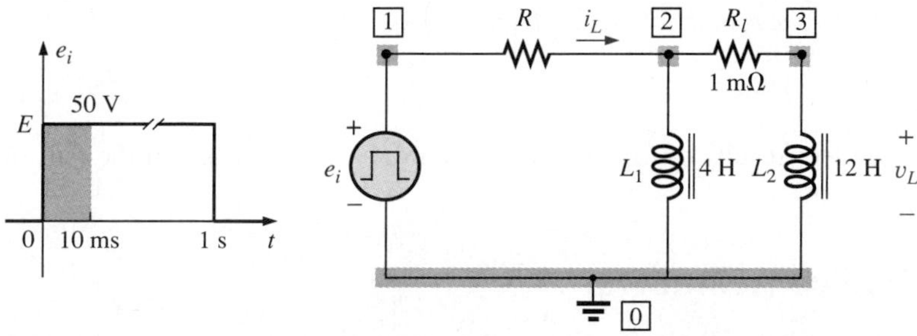

FIG. 12.49
R-L network to be analyzed using PSpice (DOS).

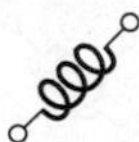

```
Chapter 12 - R-L Circuit Transient Analysis

****     CIRCUIT DESCRIPTION

******************************************************************************

VE 1 0 PULSE(0 50 0 1N 1N 1)
R  1 2 2K
RL 2 3 1M
L1 2 0 4H
L2 3 0 12H
.TRAN 0.5M 10M
.PROBE
.OPTIONS NOPAGE
.END
```

FIG. 12.50
Input file for the network of Fig. 12.49.

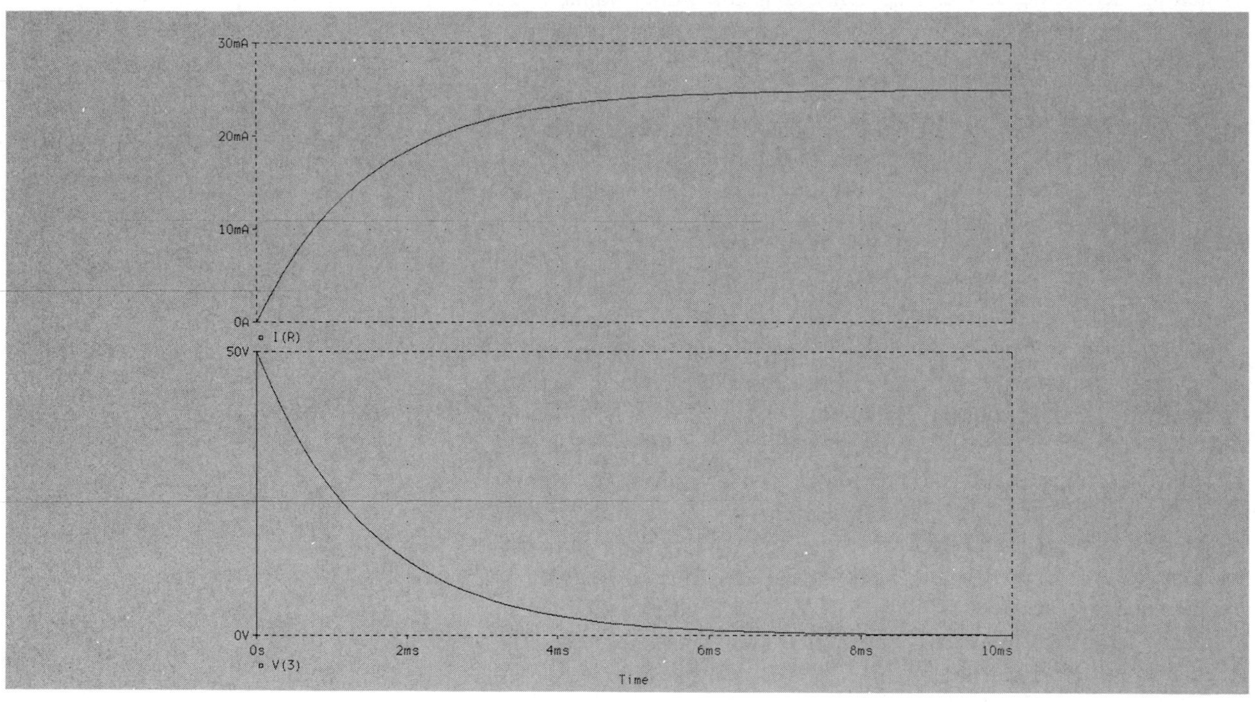

FIG. 12.51
Output file for the network of Fig. 12.49.

excellent results of Fig. 12.51 with a minimum of effort on the part of the user.

PSpice (Windows)

The results of Example 12.6 will now be verified using schematics. A pulse source will be employed, as introduced in Chapter 10, to simulate the switching action at $t = 0$ s. Following **DRAW-Get New Part-**

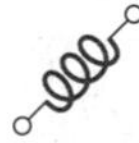

Browse-source.slb-VPULSE the pulse parameters are set as indicated on Fig. 12.52. For each parameter the sequence **Save Attr-Change-Display Both name and value** was chosen to generate the listing. The pulse width and period were set at 60 μs to exceed five constants ($5\tau = 55.55$ μs) of the network. The rise time and fall time were chosen as sufficiently small compared to a time constant, and the delay time was set at 0 s to reflect a closing of the switch at $t = 0$ s. By double-clicking the inductor symbol, the value can be set at 100 mH and the initial current can be set at 4 mA. For each, **Save Attr-Change Display-Both name and value** was chosen. Once the network is drawn using procedures described in earlier chapters, **Analysis-Setup-Transient** is chosen followed by **Print Step** = 100 ns, **Final Time** = 60 μs, **No-Print Delay** = 0 s, and **Step Ceiling** = 100 ns-**Close.** Then **Probe Setup-Automatically Run Probe After Simulation** is chosen and the analysis initiated by **Analysis-Simulate.**

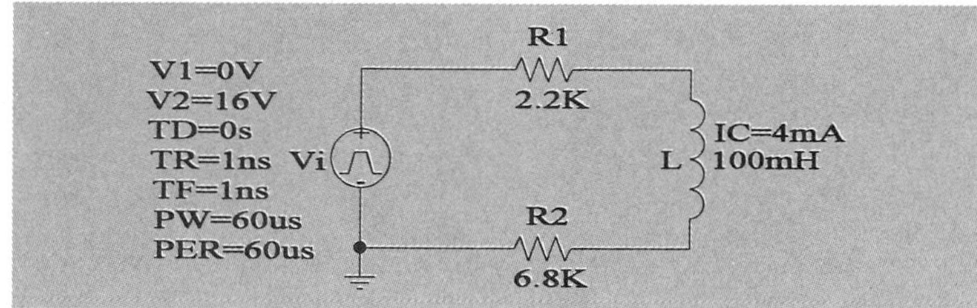

FIG. 12.52

R-L network with initial conditions to be analyzed using PSpice (Windows).

Once the **Probe** heading is obtained **Trace-Add-I(L)** will return a plot of i_L versus time, as shown in Fig. 12.53. Note that it falls off from the initial value of 4 mA to a steady-state level of about 1.79 mA in five time constants. The **Tools-Cursor-Display** option was employed to read the current level at 55.604 μs (about 5 time constants), as indicated in the narrow dialog box at the bottom of the figure. At about one time constant the other cursor reveals a current of 2.5962 mA. The **I(L)** label on the figure was added through **Tools-Label-Text.** A second plot of the voltage across the inductor can be obtained by **Plot-Add-Plot-V(L:1)-V(L:2),** as appearing in the same figure. **V(L)** does not appear as a single entity in the **Trace** listing because it is not available with respect to ground. The plot again verifies the solution of Example 12.6, dropping down to −20 V and eventually rising to 0 V with the same time constant as the inductor current. The scale was changed to −20 V to 0 V using **Plot-Y-Axis Settings-User Defined** and the label added as described above. The final plot is the power delivered to the source during the transient phase, with the area under the curve reflecting the energy stored by the inductor. The curve does not start at 0 W like Fig. 12.47 because of the initial value of inductor current. In fact note that the peak value of the power curve is the product (4 mA)(20 V) = 80 mW. The vertical scale was changed and the label added as described above. The minus sign preceding the mathematical expression for the power is included to obtain a positive plot for the power absorbed.

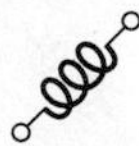

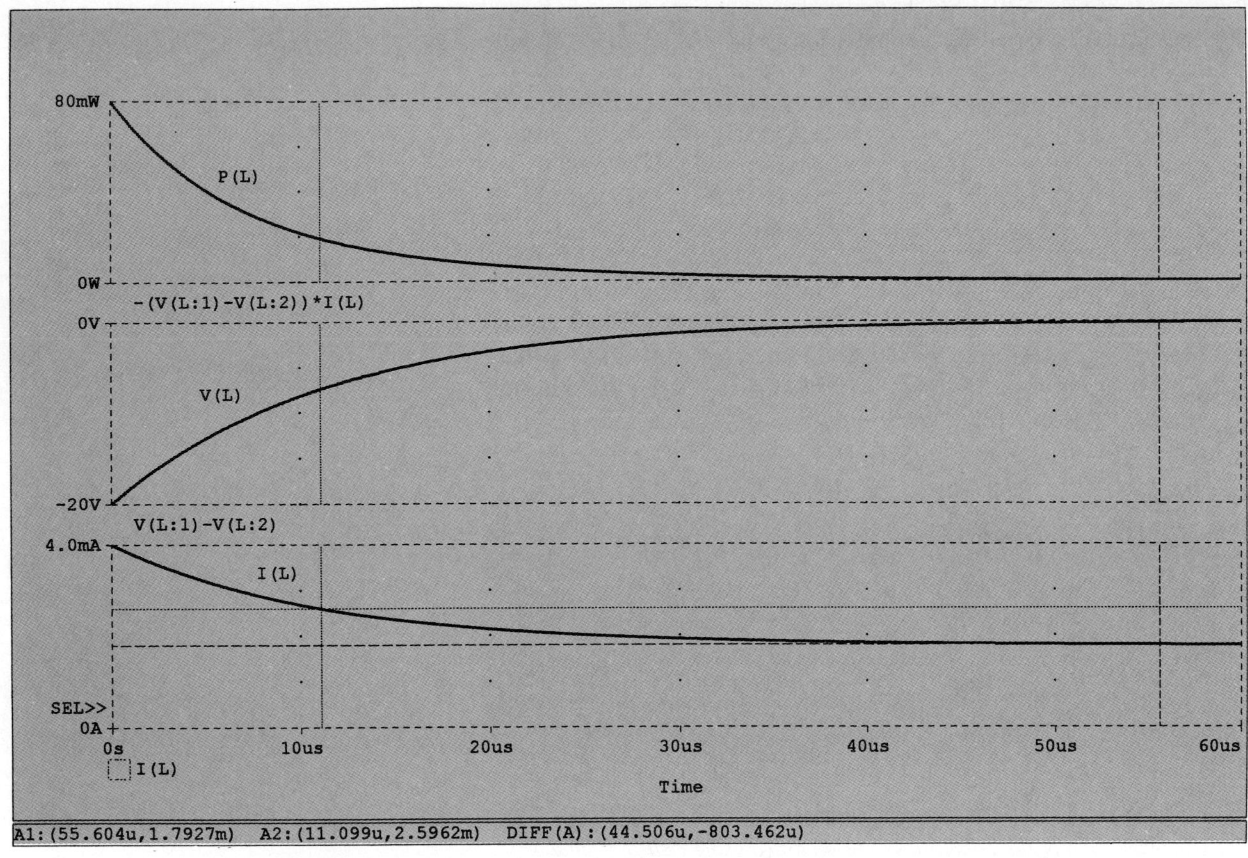

FIG. 12.53
Probe response for i_L, v_L, and p_L for the network of Fig. 12.52.

PROBLEMS

SECTION 12.2 Faraday's Law of Electromagnetic Induction

1. If the flux linking a coil of 50 turns changes at a rate of 0.085 Wb/s, what is the induced voltage across the coil?
2. Determine the rate of change of flux linking a coil if 20 V are induced across a coil of 40 turns.
3. How many turns does a coil have if 42 mV are induced across the coil by a change of flux of 0.003 Wb/s?

SECTION 12.4 Self-Inductance

4. Find the inductance L in henries of the inductor of Fig. 12.54.

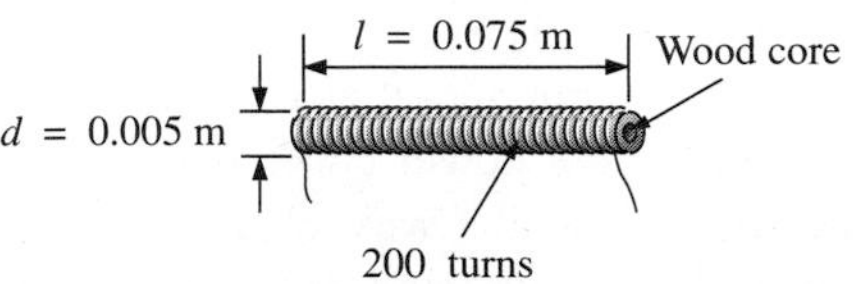

FIG. 12.54
Problems 4 and 5.

5. Repeat Problem 4 with $l = 4$ in. and $d = 0.25$ in.
6. a. Find the inductance L in henries of the inductor of Fig. 12.55.
   b. Repeat part (a) if a ferromagnetic core is added having a μ_r of 2000.

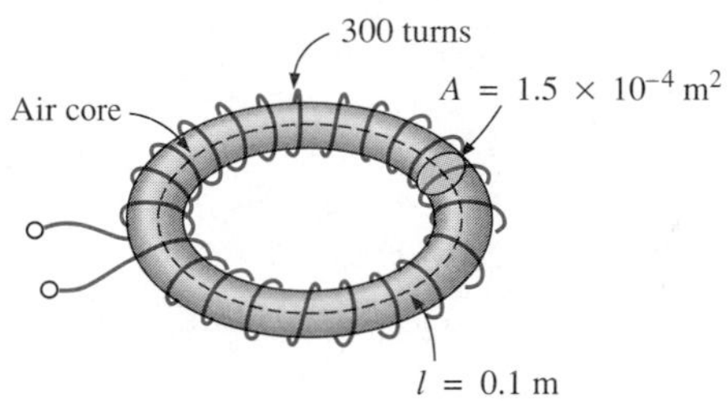

FIG. 12.55
Problem 6.

SECTION 12.6 Induced Voltage

7. Find the voltage induced across a coil of 5 H if the rate of change of current through the coil is
   a. 0.5 A/s
   b. 60 mA/s
   c. 0.04 A/ms
8. Find the induced voltage across a 50-mH inductor if the current through the coil changes at a rate of 0.1 mA/μs.
9. Find the waveform for the voltage induced across a 200-mH coil if the current through the coil is as shown in Fig. 12.56.

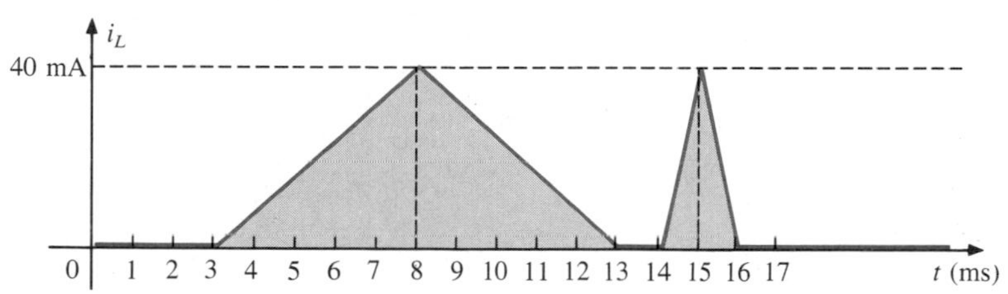

FIG. 12.56
Problem 9.

10. Sketch the waveform for the voltage induced across a 0.2-H coil if the current through the coil is as shown in Fig. 12.57.

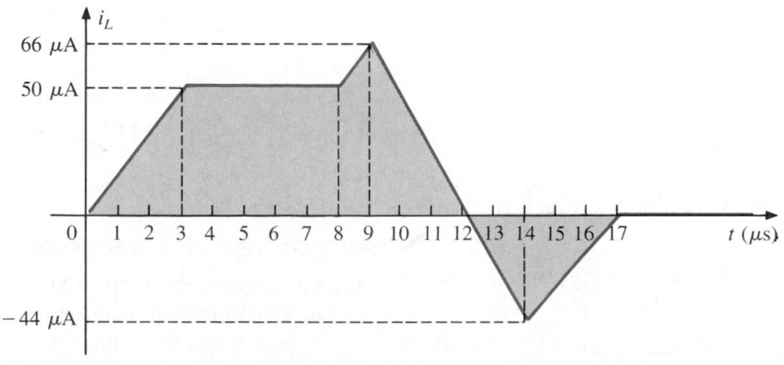

FIG. 12.57
Problem 10.

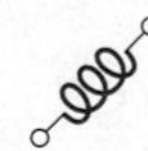

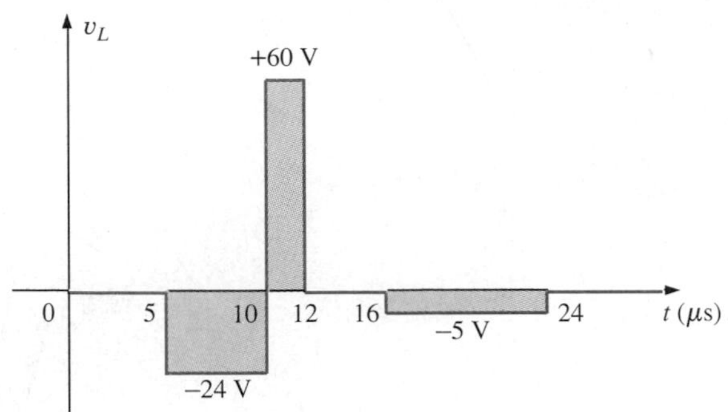

FIG. 12.58
Problem 11.

***11.** Find the waveform for the current of a 10-mH coil if the voltage across the coil follows the pattern of Fig. 12.58. The current i_L is 4 mA at $t = 0$ s.

SECTION 12.7 *R-L* Transients: Storage Cycle

12. For the circuit of Fig. 12.59:
   a. Determine the time constant.
   b. Write the mathematical expression for the current i_L after the switch is closed.
   c. Repeat part (b) for v_L and v_R.
   d. Determine i_L and v_L at one, three, and five time constants.
   e. Sketch the waveforms of i_L, v_L, and v_R.

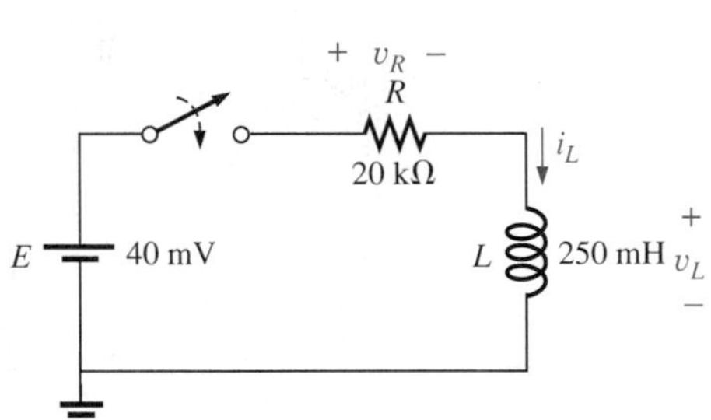

FIG. 12.59
Problem 12.

13. For the circuit of Fig. 12.60:
   a. Determine τ.
   b. Write the mathematical expression for the current i_L after the switch is closed at $t = 0$ s.
   c. Write the mathematical expressions for v_L and v_R after the switch is closed at $t = 0$ s.
   d. Determine i_L and v_L at $t = 1\tau$, 3τ, and 5τ.
   e. Sketch the waveforms of i_L, v_L, and v_R for the storage phase.

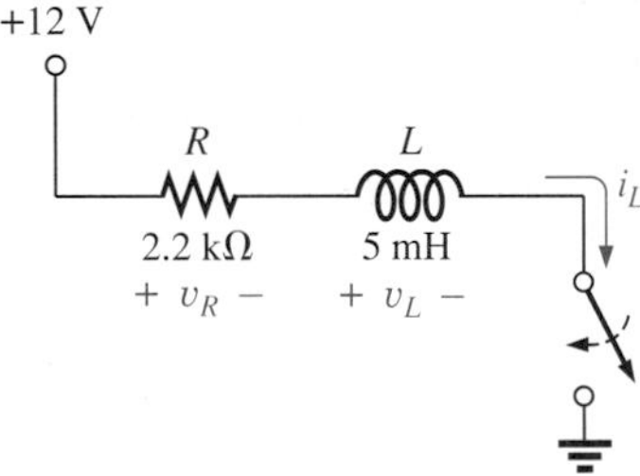

FIG. 12.60
Problem 13.

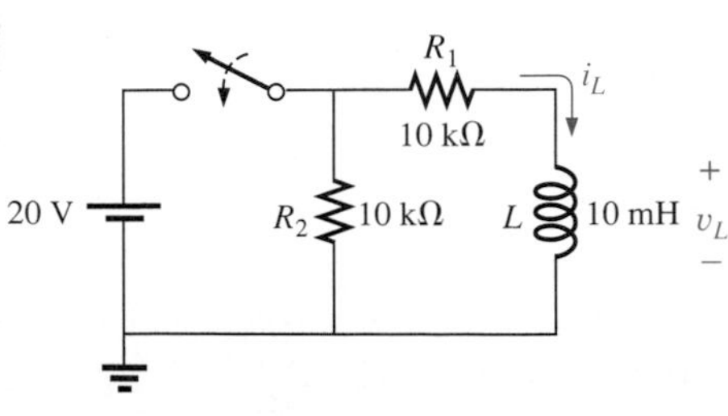

FIG. 12.61
Problems 14, 48, and 49.

SECTION 12.8 *R-L* Transients: Decay Phase

14. For the network of Fig. 12.61:
   a. Determine the mathematical expressions for the current i_L and the voltage v_L when the switch is closed.
   b. Repeat part (a) if the switch is opened after a period of five time constants has passed.
   c. Sketch the waveforms of parts (a) and (b) on the same axis.

***15.** For the network of Fig. 12.62:

a. Write the mathematical expression for the current i_L and the voltage v_L following the closing of the switch.

b. Determine the mathematical expressions for i_L and v_L if the switch is opened after a period of five time constants has passed.

c. Sketch the waveforms of i_L and v_L for the time periods defined by parts (a) and (b).

d. Sketch the waveform for the voltage across R_2 for the same period of time encompassed by i_L and v_L. Take careful note of the defined polarities and directions of Fig. 12.62.

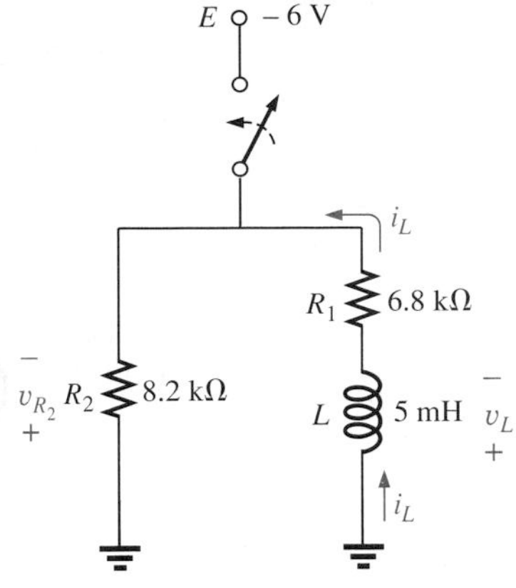

FIG. 12.62
Problem 15.

***16.** For the network of Fig. 12.63:

a. Determine the mathematical expressions for the current i_L and the voltage v_L following the closing of the switch.

b. Repeat part (a) if the switch is opened at $t = 1\ \mu s$.

c. Sketch the waveforms of parts (a) and (b) on the same axis.

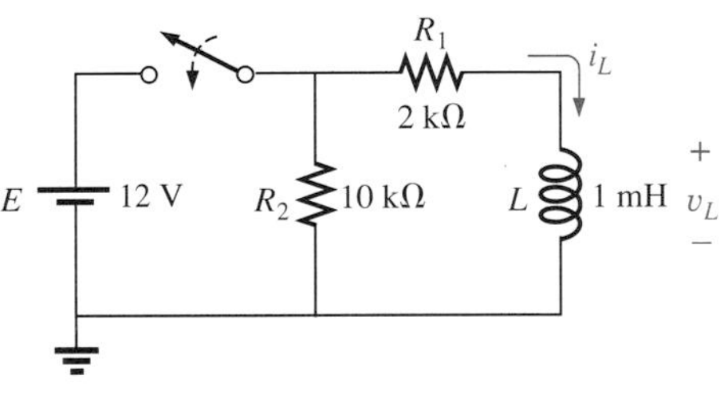

FIG. 12.63
Problem 16.

SECTION 12.9 Initial Values

17. For the network of Fig. 12.64:

a. Write the mathematical expressions for the current i_L and the voltage v_L following the closing of the switch. Note the magnitude and direction of the initial current.

b. Sketch the waveform of i_L and v_L for the entire period from initial value to steady-state level.

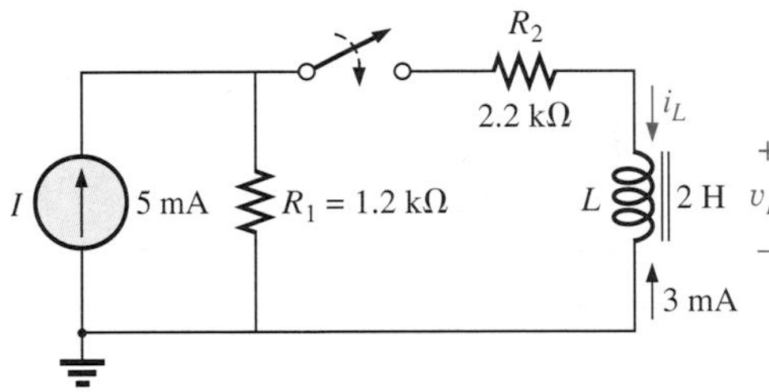

FIG. 12.64
Problem 17.

18. For the network of Fig. 12.65:

a. Write the mathematical expressions for the current i_L and the voltage v_L following the closing of the switch. Note the magnitude and direction of the initial current.

b. Sketch the waveform of i_L and v_L for the entire period from initial value to steady-state level.

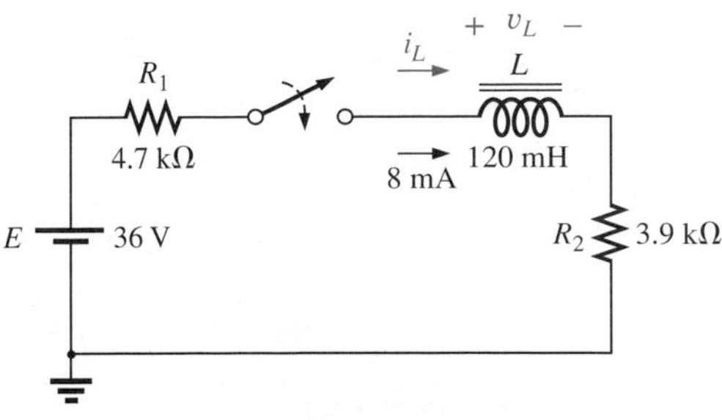

FIG. 12.65
Problem 18.

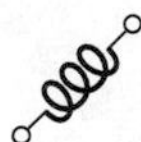

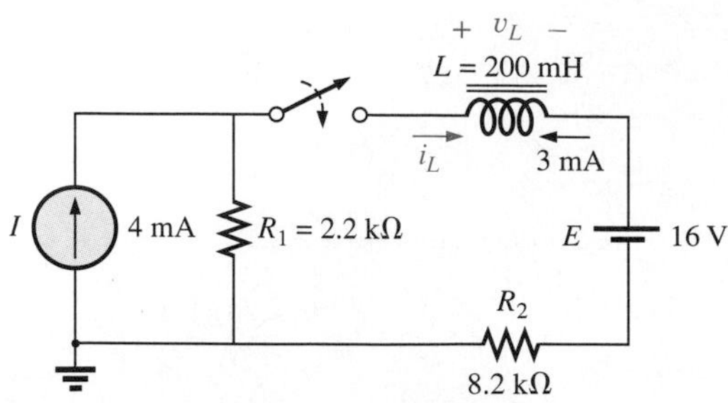

FIG. 12.66
Problem 19.

***19.** For the network of Fig. 12.66:

a. Write the mathematical expressions for the current i_L and the voltage v_L following the closing of the switch. Note the magnitude and direction of the initial current.

b. Sketch the waveform of i_L and v_L for the entire period from initial value to steady-state level.

SECTION 12.10 Instantaneous Values

20. Referring to the solution to Example 12.4, determine the time when the current i_L reaches a level of 10 mA. Then determine the time when the voltage drops to a level of 10 V.

21. Referring to the solution to Example 12.6, determine the time when the current i_L drops to 2 mA.

SECTION 12.11 $\tau = L/R_{Th}$

22. a. Determine the mathematical expressions for i_L and v_L following the closing of the switch in Fig. 12.67.

b. Determine i_L and v_L at t = 100 ns.

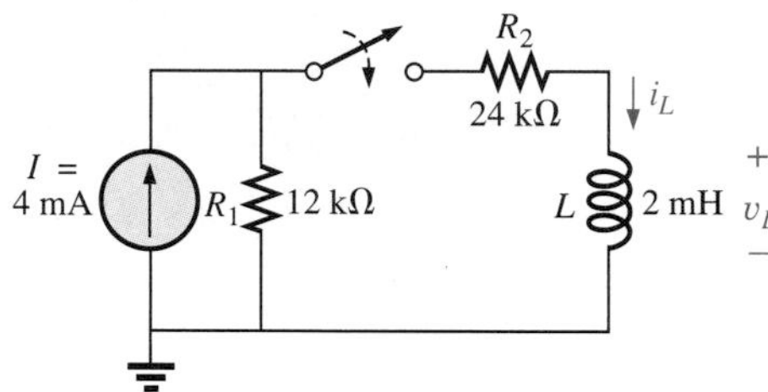

FIG. 12.67
Problems 22 and 41.

***23. a.** Determine the mathematical expressions for i_L and v_L following the closing of the switch in Fig. 12.68.

b. Calculate i_L and v_L at t = 10 μs.

c. Write the mathematical expressions for the current i_L and the voltage v_L if the switch is opened at t = 10 μs.

d. Sketch the waveforms of i_L and v_L for parts (a) and (c).

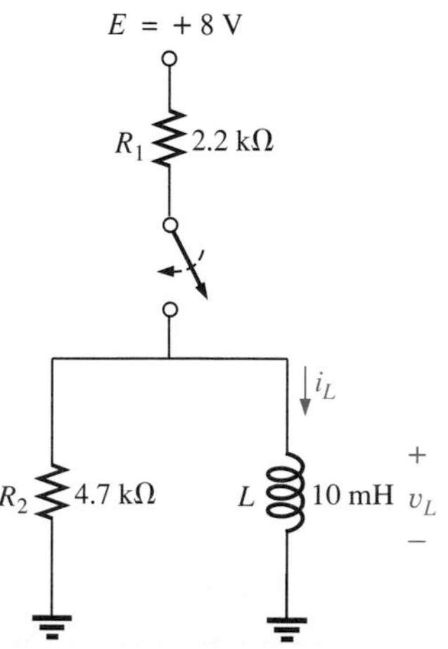

FIG. 12.68
Problem 23.

*24. **a.** Determine the mathematical expressions for i_L and v_L following the closing of the switch in Fig. 12.69.

b. Determine i_L and v_L after two time constants of the storage phase.

c. Write the mathematical expressions for the current i_L and the voltage v_L if the switch is opened at the instant defined by part (b).

d. Sketch the waveforms of i_L and v_L for parts (a) and (c).

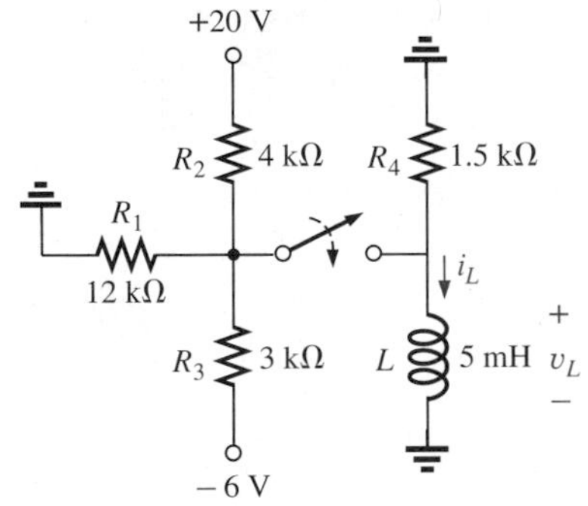

FIG. 12.69
Problem 24.

*25. For the network of Fig. 12.70, the switch is closed at $t = 0$ s.

a. Determine v_L at $t = 25$ ms.

b. Find v_L at $t = 1$ ms.

c. Calculate v_{R_1} at $t = 1\tau$.

d. Find the time required for the current i_L to reach 100 mA.

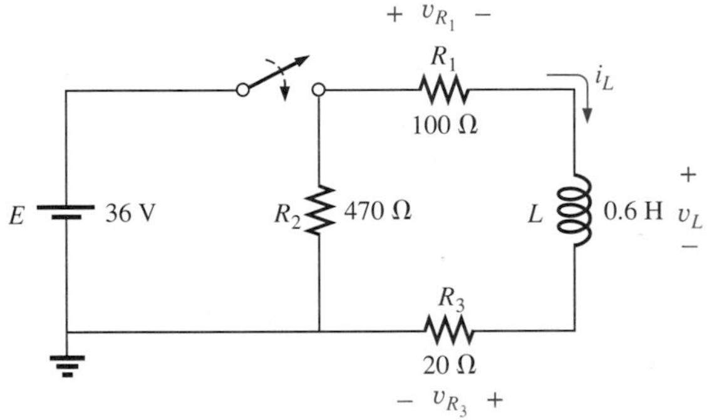

FIG. 12.70
Problem 25.

*26. The switch for the network of Fig. 12.71 has been closed for about 1 h. It is then opened at the time defined as $t =$ 0 s.

a. Determine the time required for the current i_R to drop to 1 mA.

b. Find the voltage v_L at $t = 1$ ms.

c. Calculate v_{R_3} at $t = 5\tau$.

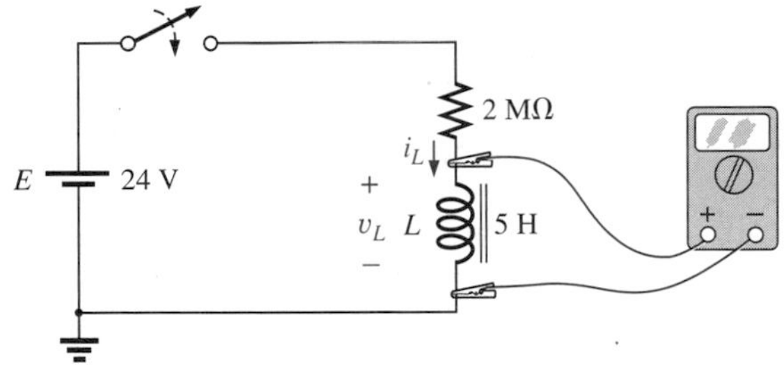

FIG. 12.71
Problem 26.

27. The network of Fig. 12.71 employs a DMM with an internal resistance of 10 MΩ in the voltmeter mode. The switch is closed at $t = 0$ s.

a. Find the voltage across the coil the instant after the switch is closed.

b. What is the final value of the current i_L?

c. How much time must pass before i_L reaches 10 μA?

d. What is the voltmeter reading at $t = 12$ μs?

*28. The switch in Fig. 12.72 has been open for a long time. It is then closed at $t = 0$ s.

a. Write the mathematical expression for the current i_L and the voltage v_L after the switch is closed.

b. Sketch the waveform of i_L and v_L from the initial value to the steady-state level.

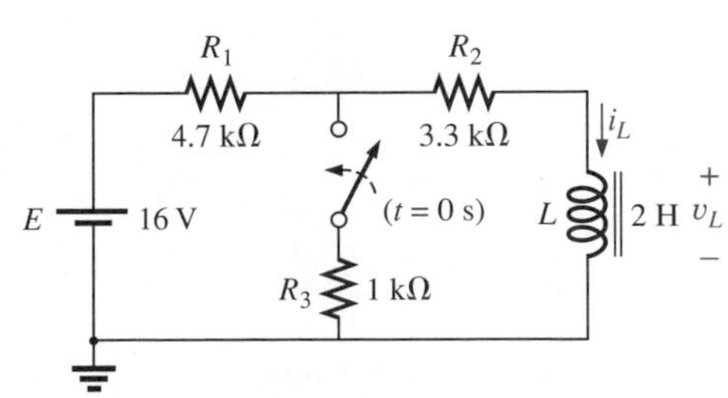

FIG. 12.72
Problems 28 and 42.

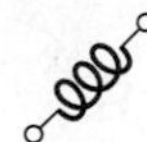

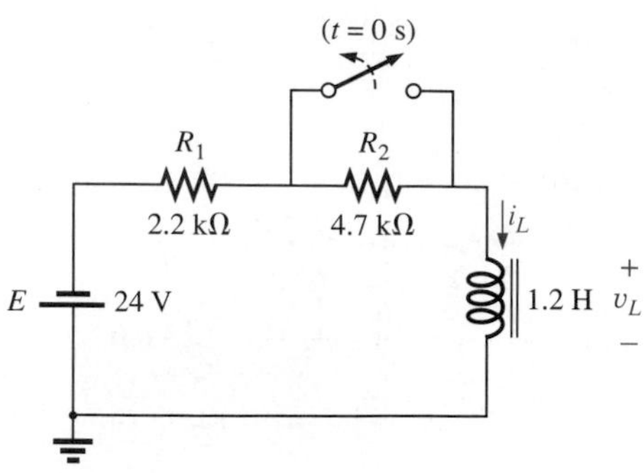

FIG. 12.73
Problem 29.

***29.** The switch of Fig. 12.73 has been closed for a long time. It is then opened at $t = 0$ s.

a. Write the mathematical expression for the current i_L and the voltage v_L after the switch is opened.

b. Sketch the waveform of i_L and v_L from initial value to the steady-state level.

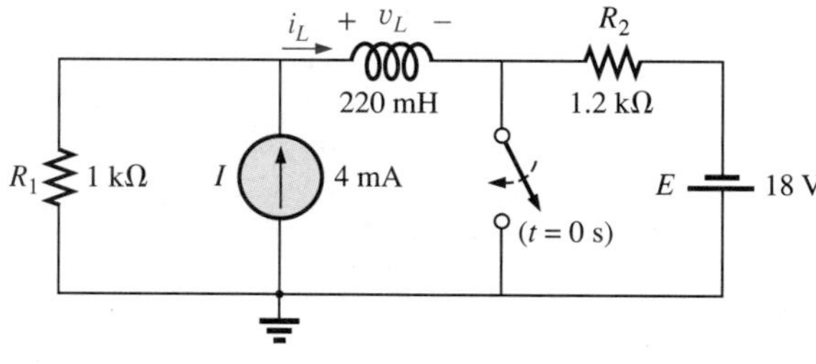

FIG. 12.74
Problems 30 and 46.

***30.** The switch of Fig. 12.74 has been open for a long time. It is then closed at $t = 0$ s.

a. Write the mathematical expression for the current i_L and the voltage v_L after the switch is closed.

b. Sketch the waveform of i_L and v_L from initial value to the steady-state level.

SECTION 12.12 Inductors in Series and Parallel

31. Find the total inductance of the circuits of Fig. 12.75.

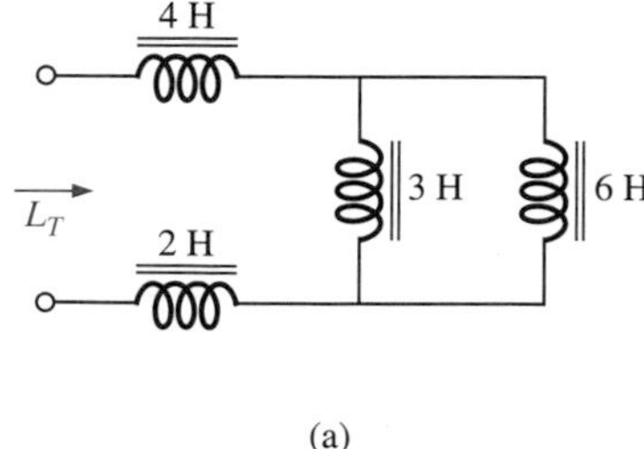

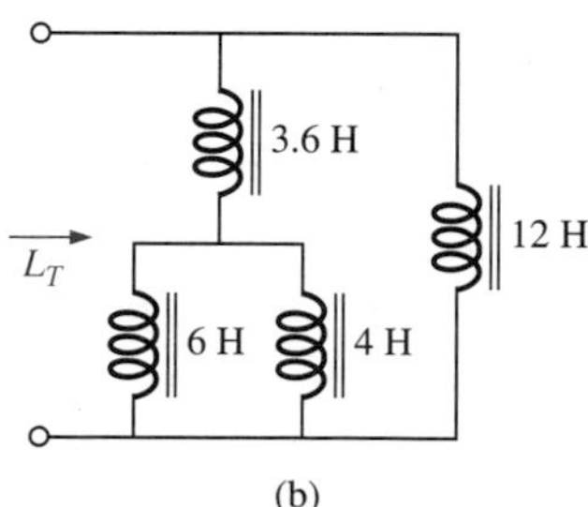

FIG. 12.75
Problem 31.

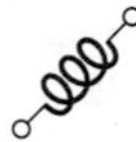

32. Reduce the networks of Fig. 12.76 to the fewest elements.

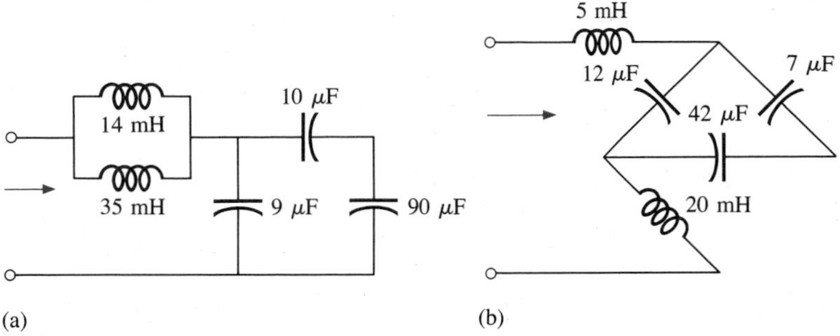

FIG. 12.76
Problem 32.

33. Reduce the network of Fig. 12.77 to the fewest number of components.

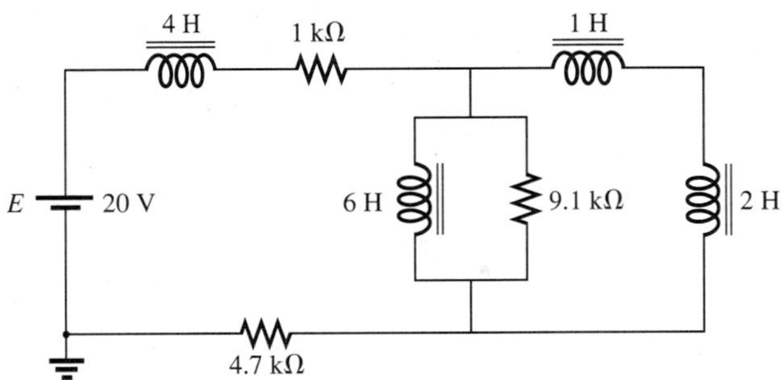

FIG. 12.77
Problem 33.

***34.** For the network of Fig. 12.78:

a. Find the mathematical expressions for the voltage v_L and the current i_L following the closing of the switch.

b. Sketch the waveforms of v_L and i_L obtained in part (a).

c. Determine the mathematical expression for the voltage v_{L_3} following the closing of the switch, and sketch the waveform.

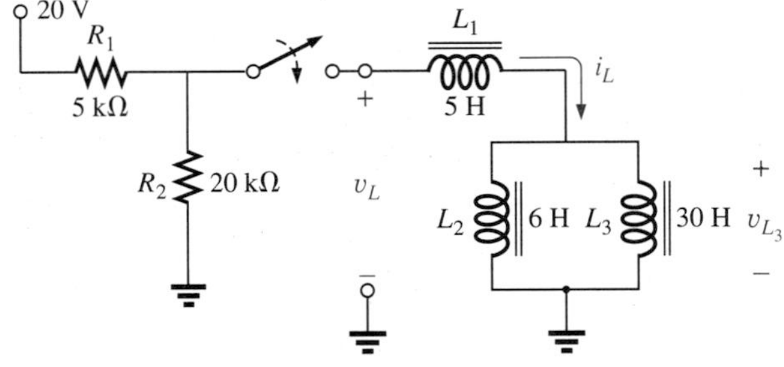

FIG. 12.78
Problems 34 and 43.

SECTION 12.13 *R-L* and *R-L-C* Circuits with dc Inputs

For Problems 35 through 37, assume that the voltage across each capacitor and the current through each inductor have reached their final values.

35. Find the voltages V_1 and V_2 and the current I_1 for the circuit of Fig. 12.79.

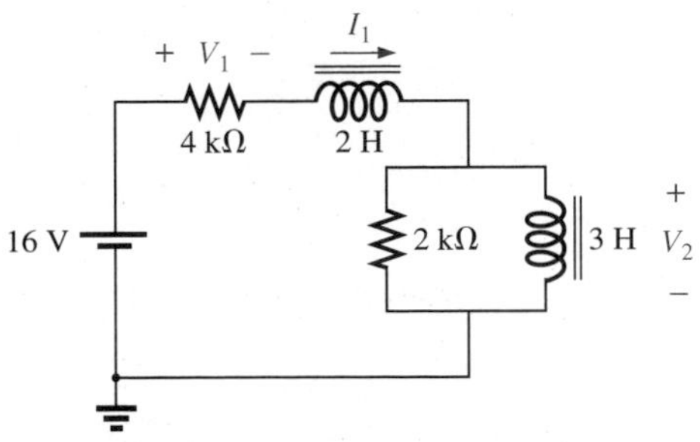

FIG. 12.79
Problems 35 and 38.

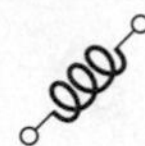

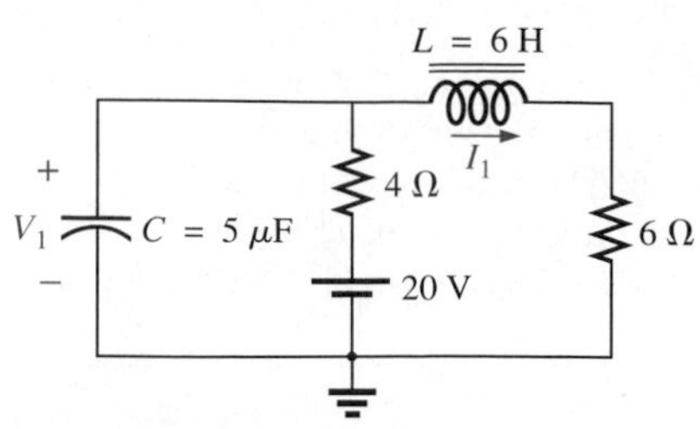

FIG. 12.80
Problems 36 and 39.

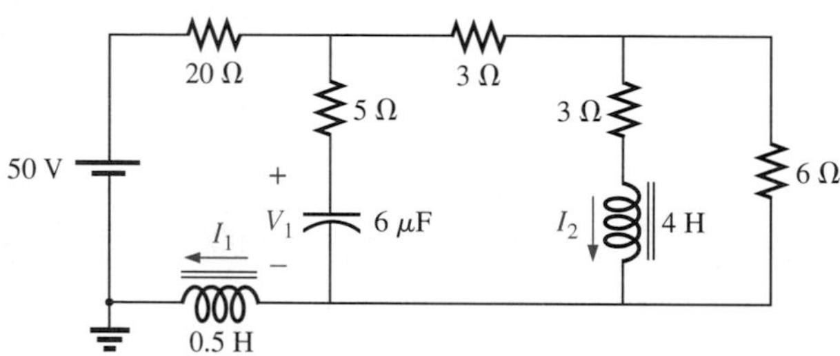

FIG. 12.81
Problems 37 and 40.

36. Find the current I_1 and the voltage V_1 for the circuit of Fig. 12.80.

37. Find the voltage V_1 and the current through each inductor in the circuit of Fig. 12.81.

SECTION 12.14 Energy Stored by an Inductor

38. Find the energy stored in each inductor of Problem 35.

39. Find the energy stored in the capacitor and inductor of Problem 36.

40. Find the energy stored in each inductor of Problem 37.

SECTION 12.15 Computer Analysis

PSpice (DOS)

41. Write the input file to obtain a plot of v_L and i_L for the network of Fig. 12.67 following the closing of the switch.

***42.** Write the input file to obtain a plot of v_L and i_L for the network of Fig. 12.72 following the closing of the switch.

***43.** Write the input file to obtain a plot of v_{L_3}, i_L, and v_L, for the network of Fig. 12.78 following the closing of the switch.

PSpice (Windows)

***44.** Verify the results of Example 12.5 using the VPULSE function and a PW equal to five time constants of the charging network.

***45.** Verify the results of Example 12.3 using the VPULSE function and a PW equal to 1 ns.

***46.** Verify the results of Problem 30 using the VPULSE function and the appropriate initial current.

Programming Language (C++, BASIC, PASCAL, etc.)

47. Write a program to provide a general solution for the circuit of Fig. 12.14; that is, given the network parameters, generate the equations for i_L, v_L, and v_R.

48. Write a program that will provide a general solution for the storage and decay phase of the network of Fig. 12.61; that is, given the network values, generate the equations for i_L and v_L for each phase. In this case, assume that the storage phase has passed through five time constants before the decay phase begins.

49. Repeat Problem 48, but assume that the storage phase was not completed, requiring that the instantaneous values of i_L and v_L be determined when the switch is opened.

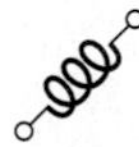

GLOSSARY

Choke A term often applied to an inductor, due to the ability of an inductor to resist a change in current through it.

Faraday's law A law relating the voltage induced across a coil to the number of turns in the coil and the rate at which the flux linking the coil is changing.

Inductor A fundamental element of electrical systems constructed of numerous turns of wire around a ferromagnetic or air core.

Lenz's law A law stating that an induced effect is always such as to oppose the cause that produced it.

Self-inductance A measure of the ability of a coil to oppose any change in current through the coil and to store energy in the form of a magnetic field in the region surrounding the coil.

13

Sinusoidal Alternating Waveforms

13.1 INTRODUCTION

The analysis thus far has been limited to dc networks, networks in which the currents or voltages are fixed in magnitude except for transient effects. We will now turn our attention to the analysis of networks in which the magnitude of the source varies in a set manner. Of particular interest is the time-varying voltage that is commercially available in large quantities and is commonly called the *ac voltage.* (The letters *ac* are an abbreviation for *alternating current.*) To be absolutely rigorous, the terminology *ac voltage* or *ac current* is not sufficient to describe the type of signal we will be analyzing. Each waveform of Fig. 13.1 is an alternating waveform available from commercial supplies. The term *alternating* indicates only that the waveform alternates between two prescribed levels in a set time sequence (Fig. 13.1). To be

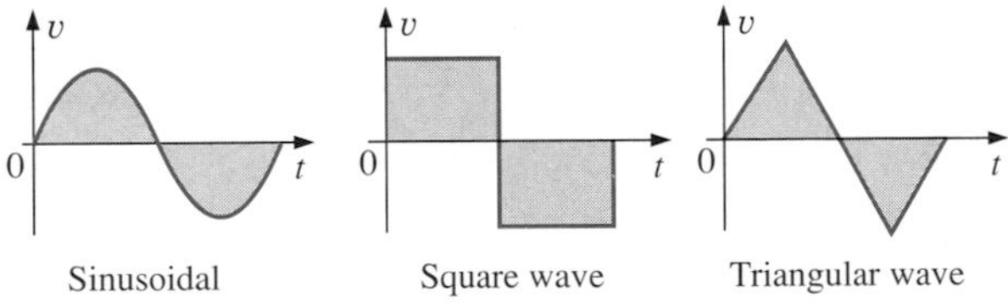

FIG. 13.1
Alternating waveforms.

absolutely correct, the term *sinusoidal, square wave,* or *triangular* must also be applied. The pattern of particular interest is the *sinusoidal* ac voltage of Fig. 13.1. Since this type of signal is encountered in the vast majority of instances, the abbreviated phrases *ac voltage* and *ac current* are commonly applied without confusion. For the other patterns of Fig. 13.1, the descriptive term is always present, but frequently the *ac* abbreviation is dropped, resulting in the designation *square-wave* or *triangular* waveforms.

One of the important reasons for concentrating on the sinusoidal ac voltage is that it is the voltage generated by utilities throughout the world. Other reasons include its application throughout electrical, electronic, communication, and industrial systems. In addition, the chapters to follow will reveal that the waveform itself has a number of characteristics that will result in a unique response when it is applied to the basic electrical elements. The wide range of theorems and methods introduced for dc networks will also be applied to sinusoidal ac systems. Although the application of sinusoidal signals will raise the required math level, once the notation given in Chapter 14 is understood, most of the concepts introduced in the dc chapters can be applied to ac networks with a minimum of added difficulty.

The increasing number of computer systems used in the industrial community requires, at the very least, a brief introduction to the terminology employed with pulse waveforms and the response of some fundamental configurations to the application of such signals. Chapter 22 will serve such a purpose.

13.2 SINUSOIDAL ac VOLTAGE CHARACTERISTICS AND DEFINITIONS

Generation

Sinusoidal ac voltages are available from a variety of sources. The most common source is the typical home outlet, which provides an ac voltage that originates at a power plant; such a power plant is most commonly fueled by water power, oil, gas, or nuclear fusion. In each case an *ac generator* (also called *alternator*), as in Fig. 13.2(a), is the

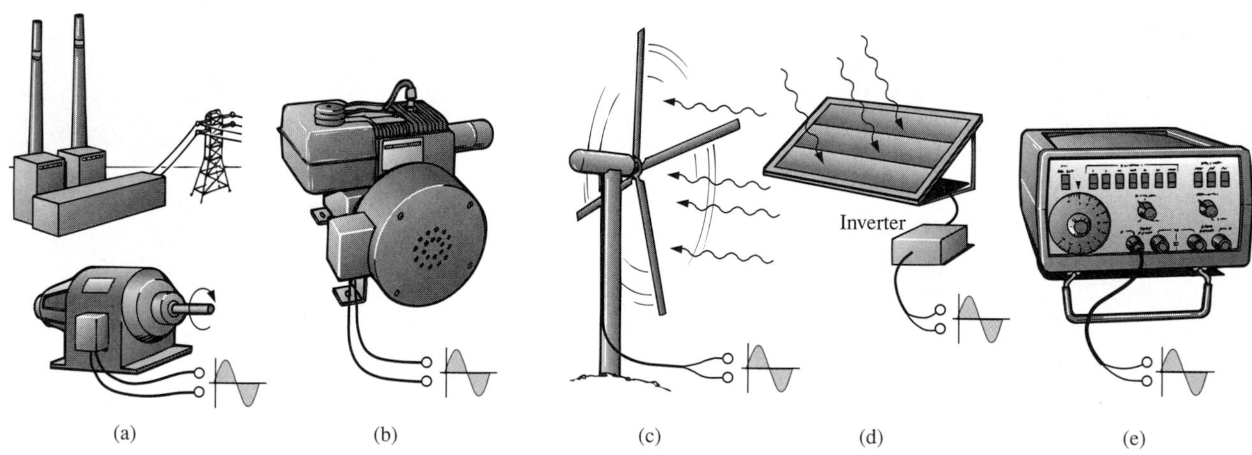

FIG. 13.2
Various sources of ac power: (a) generating plant; (b) portable ac generator; (c) wind-power station; (d) solar panel; (e) function generator.

primary component in the energy-conversion process. The power to the shaft developed by one of the energy sources listed will turn a *rotor* (constructed of alternating magnetic poles) inside a set of windings housed in the *stator* (the stationary part of the dynamo) and induce a voltage across the windings of the stator, as defined by Faraday's law,

$$e = N\frac{d\phi}{dt}$$

Through proper design of the generator, a sinusoidal ac voltage is developed that can be transformed to higher levels for distribution through the power lines to the consumer. For isolated locations where power lines have not been installed, portable ac generators [Fig. 13.2(b)] are available that run on gasoline. As in the larger power plants, however, an ac generator is an integral part of the design.

In an effort to conserve our natural resources, wind power and solar energy are receiving increasing interest from various districts of the world that have such energy sources available in level and duration that make the conversion process viable. The turning propellers of the windpower station [Fig. 13.2(c)] are connected directly to the shaft of an ac generator to provide the ac voltage described above. Through light energy absorbed in the form of *photons,* solar cells [Fig. 13.2(d)] can generate dc voltages. Through an electronic package called an *inverter,* the dc voltage can be converted to one of a sinusoidal nature. Boats, recreational vehicles (RVs), etc., make frequent use of the inversion process in isolated areas.

Sinusoidal ac voltages with characteristics that can be controlled by the user are available from *function generators,* such as the one in Fig. 13.2(e). By setting the various switches and controlling the position of the knobs on the face of the instrument, sinusoidal voltages of different peak values and different repetition rates can be made available. The function generator plays an integral role in the investigation of the variety of theorems, methods of analysis, and topics to be introduced in the chapters that follow.

Definitions

The sinusoidal waveform of Fig. 13.3 with its additional notation will now be used as a model in defining a few basic terms. These terms can, however, be applied to any alternating waveform. It is important to remember as you proceed through the various definitions that the vertical scaling is in volts or amperes and the horizontal scaling is *always* in units of time.

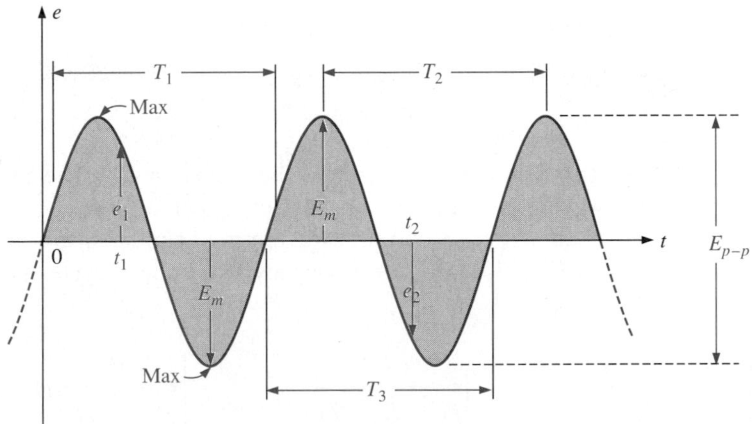

FIG. 13.3
Important parameters for a sinusoidal voltage.

Waveform: The path traced by a quantity, such as the voltage in Fig. 13.3, plotted as a function of some variable such as time (as above), position, degrees, radians, temperature, and so on.

Instantaneous value: The magnitude of a waveform at any instant of time; denoted by lowercase letters (e_1, e_2).

Peak amplitude: The maximum value of a waveform as measured from its *average,* or *mean,* value, denoted by uppercase letters (such as E_m for sources of voltage and V_m for the voltage drop across a load). For the waveform of Fig. 13.3 the average value is zero volts and E_m is as defined by the figure.

Peak value: The maximum instantaneous value of a function as measured from the zero-volt level. For the waveform of Fig. 13.3 the peak amplitude and peak value are the same, since the average value of the function is zero volts.

Peak-to-peak value: Denoted by $E_{p\text{-}p}$ or $V_{p\text{-}p}$, the full voltage between positive and negative peaks of the waveform, that is, the sum of the magnitude of the positive and negative peaks.

Periodic waveform: A waveform that continually repeats itself after the same time interval. The waveform of Fig. 13.3 is a periodic waveform.

Period (T): The time interval between successive repetitions of a periodic waveform (the period $T_1 = T_2 = T_3$ in Fig. 13.3), so long as successive *similar points* of the periodic waveform are used in determining *T.*

Cycle: The portion of a waveform contained in *one period* of time. The cycles within T_1, T_2, and T_3 of Fig. 13.3 may appear different in Fig. 13.4, but they are all bounded by one period of time and therefore satisfy the definition of a cycle.

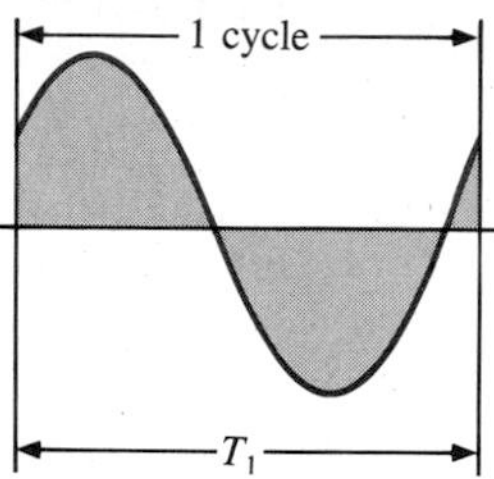

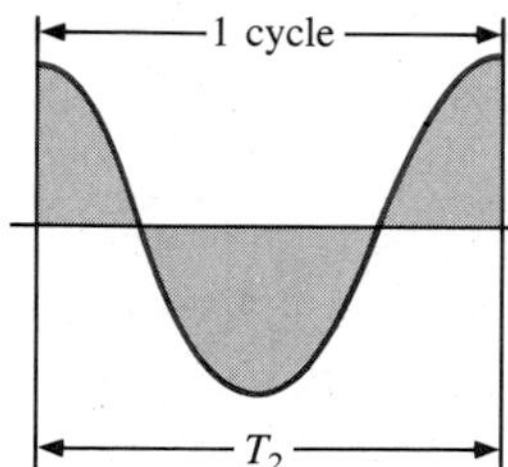

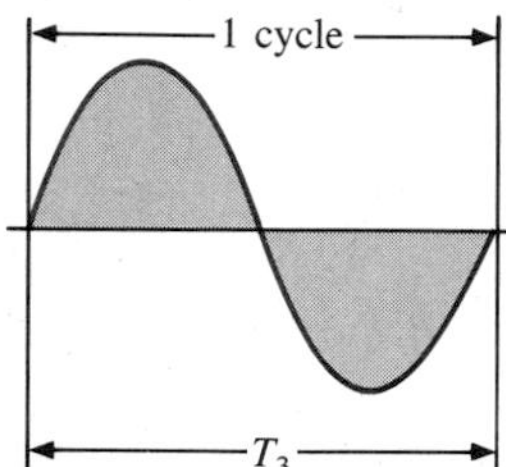

FIG. 13.4
Defining the cycle and period of a sinusoidal waveform.

Frequency (f): The number of cycles that occur in 1 s. The frequency of the waveform of Fig. 13.5(a) is 1 cycle per second, and for Fig. 13.5(b), 2½ cycles per second. If a waveform of similar shape had a period of 0.5 s [Fig. 13.5(c)], the frequency would be 2 cycles per second.

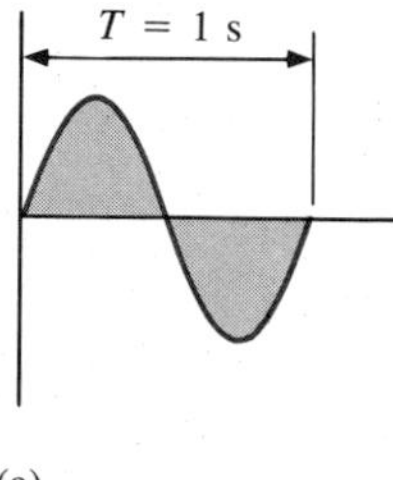

(a)

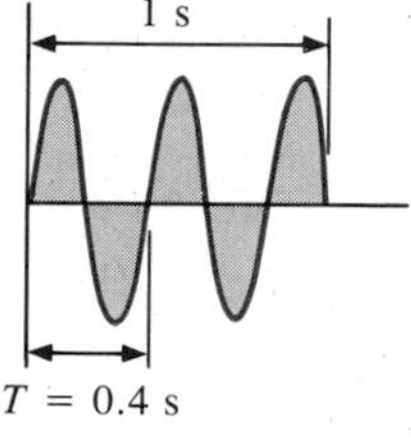

(b)

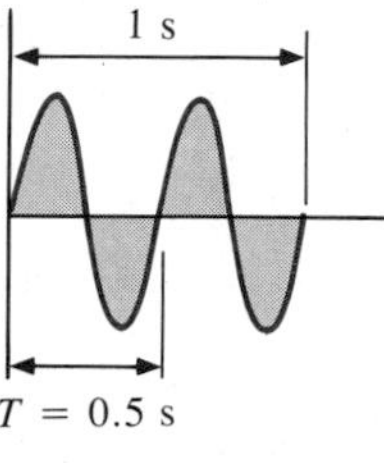

(c)

FIG. 13.5
Demonstrating the effect of a changing frequency on the period of a sinusoidal waveform.

The unit of measure for frequency is the *hertz* (Hz), where

$$1 \text{ hertz (Hz)} = 1 \text{ cycle per second (c/s)} \tag{13.1}$$

The unit hertz is derived from the surname of Heinrich Rudolph Hertz (Fig. 13.6), who did original research in the area of alternating currents and voltages and their effect on the basic R, L, and C elements. The frequency standard for North America is 60 Hz, whereas for Europe it is predominantly 50 Hz.

As with all standards, any variation from the norm will cause difficulties. In 1993, Berlin, Germany, received all its power from eastern plants, whose output frequency was varying between 50.03 and 51 Hz. The result was that clocks were gaining as much as 4 minutes a day. Alarms went off too soon, VCRs clicked off before the end of the program, etc., requiring that clocks be continually reset. In 1994, however, when power was linked with the rest of Europe, the precise standard of 50 Hz was reestablished and everyone was on time again.

Using a log scale (described in detail in Chapter 21) a frequency spectrum from 1 to 1000 GHz can be scaled off on the same axis, as shown in Fig. 13.7. A number of terms in the various spectrums are probably familiar to the reader from everyday experiences. Note that the audio range (human ear) extends from only 15 Hz to 20 kHz, but the transmission of radio signals can occur between 3 kHz and 300 GHz. The uniform process of defining the intervals of the radio frequency spectrum from VLF to EHF is quite evident from the length of the bars in the figure (although keep in mind that it is a log scale, so the frequencies encompassed within each segment are quite different). Other frequencies of particular interest (TV, CB, microwave, etc.) are also included for reference purposes. Although it is numerically easy to talk about frequencies in the megahertz and gigahertz range, keep in mind that a frequency of 100 MHz, for instance, represents a sinusoidal waveform that passes through 100,000,000 cycles in only 1 s—an incredible number when we compare it to the 60 Hz of our conventional power sources The new Pentium Pro chip manufactured by Intel can run at speeds up to 200 MHz. Imagine a product able to handle 200,000,000 instructions per second—an incredible achievement.

Since the frequency is inversely related to the period—that is, as one increases, the other decreases by an equal amount—the two can be related by the following equation:

$$f = \frac{1}{T} \qquad \begin{aligned} f &= \text{Hz} \\ T &= \text{seconds (s)} \end{aligned} \tag{13.2}$$

or

$$T = \frac{1}{f} \tag{13.3}$$

German (Hamburg, Berlin, Karlsruhe)
(1857–1894)
Physicist
Professor of Physics, Karlsruhe Polytechnic and University of Bonn

Courtesy of the Smithsonian Institution Photo No. 66,606

Spurred on by the earlier predictions of the English physicist James Clerk Maxwell, Heinrich Hertz produced *electromagnetic waves* in his laboratory at the Karlsruhe Polytechnic while in his early 30s. The rudimentary *transmitter* and *receiver* were in essence the first to broadcast and receive radio waves. He was able to measure the *wavelength* of the electromagnetic waves and confirmed that the *velocity of propagation* is in the same order of magnitude as light. In addition, he demonstrated that the *reflective* and *refractive* properties of electromagnetic waves are the same as those for heat and light waves. It was indeed unfortunate that such an ingenious, industrious individual should pass away at the very early age of 37 due to a bone disease.

FIG. 13.6
Heinrich Rudolph Hertz.

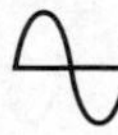

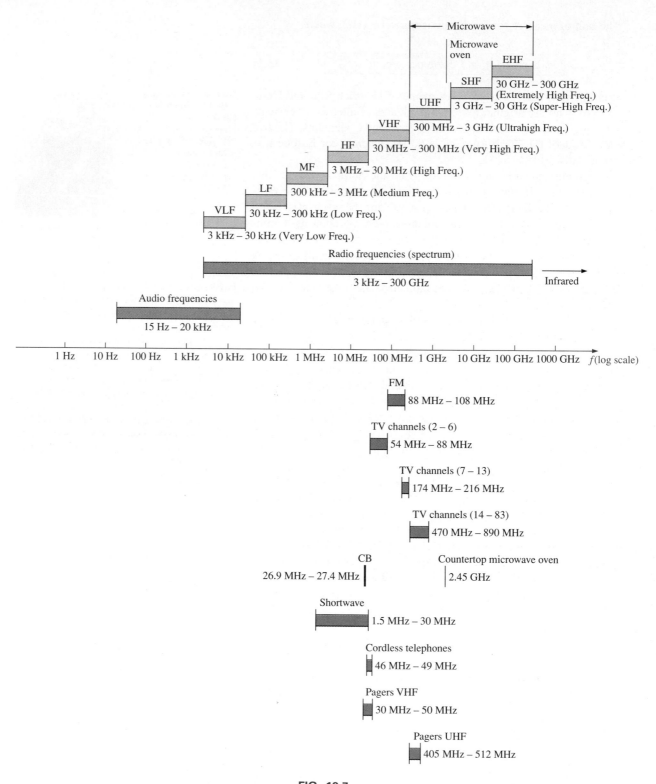

FIG. 13.7
Areas of application for specific frequency bands.

EXAMPLE 13.1 Find the period of a periodic waveform with a frequency of
a. 60 Hz
b. 1000 Hz

Solutions:

a. $T = \dfrac{1}{f} = \dfrac{1}{60\text{ Hz}} \cong 0.01667\text{ s, or } \mathbf{16.67\ ms}$

(a recurring value since 60 Hz is so prevalent)

b. $T = \dfrac{1}{f} = \dfrac{1}{1000\text{ Hz}} = 10^{-3}\text{ s} = \mathbf{1\ ms}$

EXAMPLE 13.2 Determine the frequency of the waveform of Fig. 13.8.

Solution: From the figure, $T = (25\text{ ms} - 5\text{ ms}) = 20\text{ ms}$ and

$$f = \frac{1}{T} = \frac{1}{20 \times 10^{-3}\text{ s}} = \mathbf{50\ Hz}$$

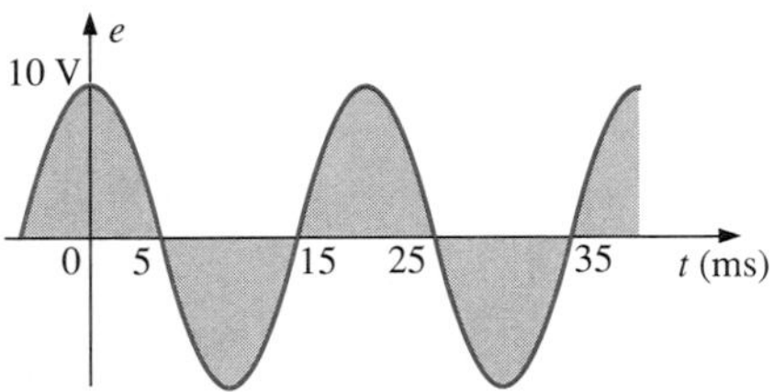

FIG. 13.8
Example 13.2.

EXAMPLE 13.3 The oscilloscope is an instrument that will display alternating waveforms such as those described above. A sinusoidal pattern appears on the oscilloscope of Fig. 13.9 with the indicated vertical and horizontal sensitivities. The vertical sensitivity defines the voltage associated with each vertical division of the display. Virtually all oscilloscope screens are cut into a crosshatch pattern of lines separated by 1 cm in the vertical and horizontal directions. The horizontal sensitivity defines the time period associated with each horizontal division of the display.

For the pattern of Fig. 13.9 and the indicated sensitivities, determine the period, frequency, and peak value of the waveform.

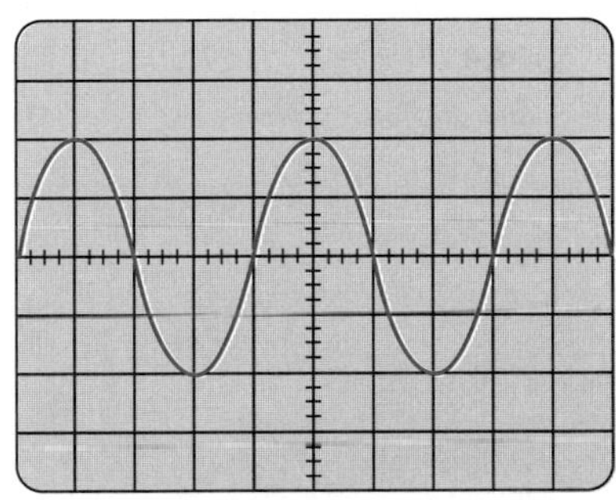

Vertical sensitivity = 0.1 V/div.
Horizontal sensitivity = 50 μs/div.

FIG. 13.9
Example 13.3.

Solution: One cycle spans 4 divisions. The period is therefore

$$T = 4\ \cancel{\text{div.}}\left(\frac{50\ \mu\text{s}}{\cancel{\text{div.}}}\right) = \mathbf{200\ \mu s}$$

and the frequency is

$$f = \frac{1}{T} = \frac{1}{200 \times 10^{-6}\text{ s}} = \mathbf{5\ kHz}$$

The vertical height above the horizontal axis encompasses 2 divisions. Therefore,

$$V_m = 2\ \cancel{\text{div.}}\left(\frac{0.1\text{ V}}{\cancel{\text{div.}}}\right) = \mathbf{0.2\ V}$$

Defined Polarities and Direction

In the following analysis, we will find it necessary to establish a set of polarities for the sinusoidal ac voltage and a direction for the sinusoidal ac current. In each case, the polarity and current direction will be for an instant of time in the positive portion of the sinusoidal waveform. This is shown in Fig. 13.10 with the symbols for the sinusoidal ac voltage and current. A lowercase letter is employed for each to indicate that the quantity is time dependent; that is, its magnitude will change with time.

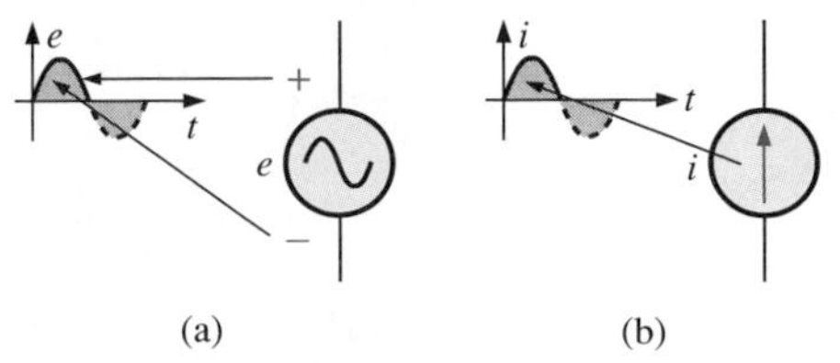

FIG. 13.10
(a) Sinusoidal ac voltage sources; (b) sinusoidal current sources.

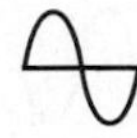

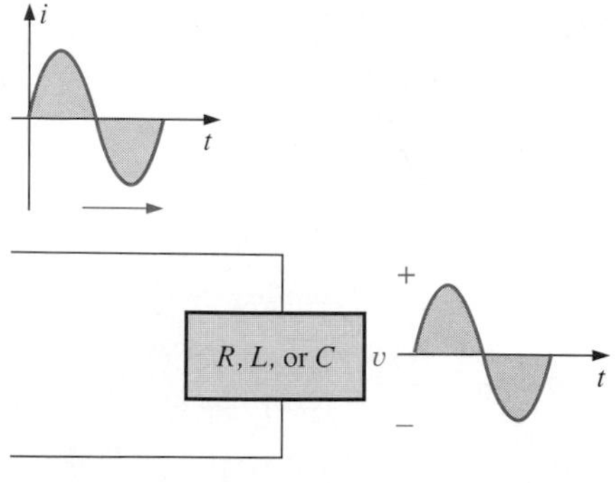

FIG. 13.11
The sine wave is the only alternating waveform whose shape is not altered by the response characteristics of a pure resistor, inductor, or capacitor.

The need for defining polarities and current direction will become quite obvious when we consider multisource ac networks. Note in the last sentence the absence of the term *sinusoidal* before the phrase *ac networks.* This will occur to an increasing degree as we progress; *sinusoidal* is to be understood unless otherwise indicated.

13.3 THE SINE WAVE

The terms defined in the previous section can be applied to any type of periodic waveform, whether smooth or discontinuous. The sinusoidal waveform is of particular importance, however, since it lends itself readily to the mathematics and the physical phenomena associated with electric circuits. Consider the power of the following statement:

The sine wave is the only alternating waveform whose shape is unaffected by the response characteristics of R, L, and C elements.

In other words, if the voltage across (or current through) a resistor, coil, or capacitor is sinusoidal in nature, the resulting current (or voltage, respectively) for each will also have sinusoidal characteristics, as shown in Fig. 13.11. If a square wave or a triangular wave were applied, such would not be the case. It must be pointed out that the above statement is also applicable to the cosine wave, since the waves differ only by a 90° shift on the horizontal axis, as shown in Fig. 13.12.

The unit of measurement for the horizontal axis of Fig. 13.12 is the *degree.* A second unit of measurement frequently used is the *radian* (rad). It is defined by a quadrant of a circle such as in Fig. 13.13 where the distance subtended on the circumference equals the radius of the circle.

If we define x as the number of intervals of r (the radius) around the circumference of the circle, then

$$C = 2\pi r = x \cdot r$$

and we find

$$x = 2\pi$$

Therefore, there are 2π rad around a 360° circle, as shown in Fig. 13.14, and

$$\boxed{2\pi \text{ rad} = 360°} \qquad \textbf{(13.4)}$$

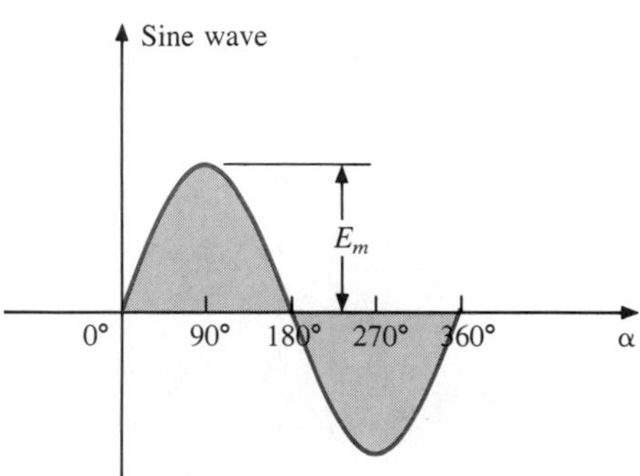

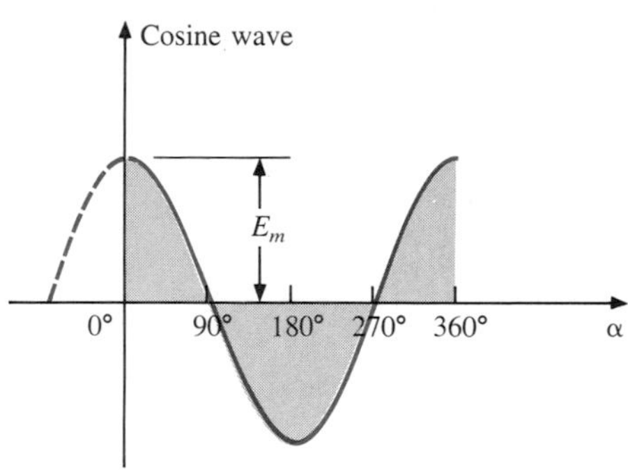

FIG. 13.12
Sine wave and cosine wave with the horizontal axis in degrees.

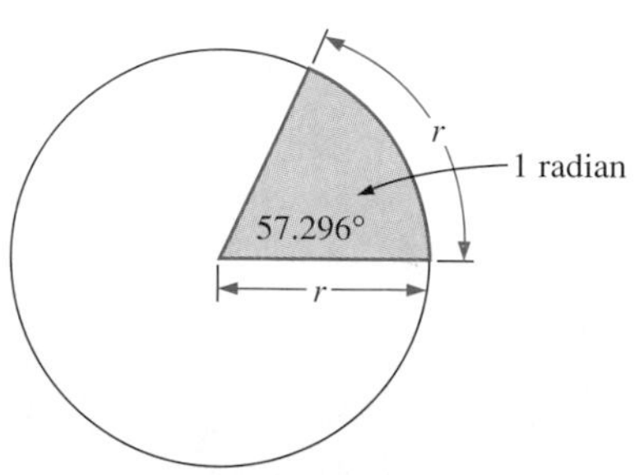

FIG. 13.13
Defining the radian.

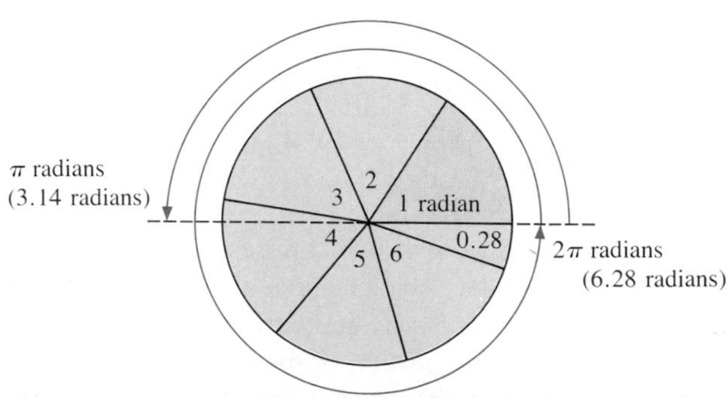

FIG. 13.14
There are 2π radians in one full circle of 360°.

with

$$1 \text{ rad} = 57.296^\circ \cong 57.3^\circ \tag{13.5}$$

A number of electrical formulas contain a multiplier of π. This is one reason it is sometimes preferable to measure angles in radians rather than in degrees.

The quantity π is the ratio of the circumference of a circle to its diameter.

π has been determined to an extended number of places primarily in an attempt to see if a repetitive sequence of numbers appears. It does not. A sampling of the effort appears below:

$$\pi = 3.14159\ 26535\ 89793\ 23846\ 26433\ldots$$

Although the approximation $\pi \cong 3.14$ is often applied, all the calculations in this text will use the π function as provided on all scientific calculators.

For 180° and 360°, the two units of measurement are related as shown in Fig. 13.14. The conversion equations between the two are the following:

$$\text{Radians} = \left(\frac{\pi}{180^\circ}\right) \times (\text{degrees}) \tag{13.6}$$

$$\text{Degrees} = \left(\frac{180^\circ}{\pi}\right) \times (\text{radians}) \tag{13.7}$$

Applying these equations, we find

$$\mathbf{90^\circ:} \quad \text{Radians} = \frac{\pi}{180^\circ}(90^\circ) = \frac{\pi}{2}\ \textbf{rad}$$

$$\mathbf{30^\circ:} \quad \text{Radians} = \frac{\pi}{180^\circ}(30^\circ) = \frac{\pi}{6}\ \textbf{rad}$$

$$\frac{\pi}{3}\ \textbf{rad:} \quad \text{Degrees} = \frac{180^\circ}{\pi}\left(\frac{\pi}{3}\right) = \mathbf{60^\circ}$$

$$\frac{3\pi}{2}\ \textbf{rad:} \quad \text{Degrees} = \frac{180^\circ}{\pi}\left(\frac{3\pi}{2}\right) = \mathbf{270^\circ}$$

Using the radian as the unit of measurement for the abscissa, we would obtain a sine wave, as shown in Fig. 13.15.

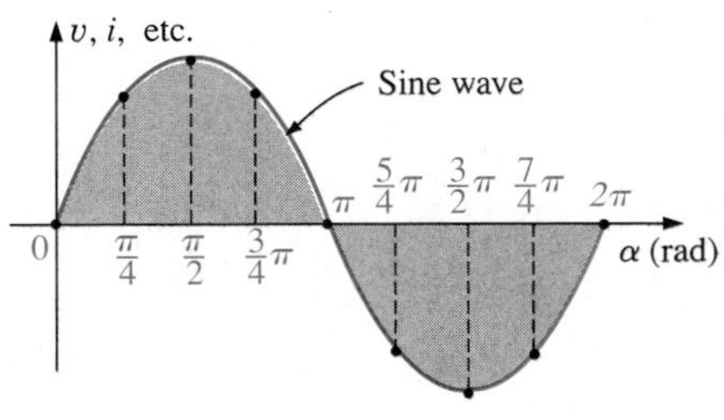

FIG. 13.15
Plotting a sine wave versus radians.

It is of particular interest that the sinusoidal waveform can be derived from the length of the *vertical projection* of a radius vector rotating in a uniform circular motion about a fixed point. Starting as shown in Fig. 13.16(a) and plotting the amplitude (above and below zero) on the coordinates drawn to the right [Figs. 13.16(b) through (i)], we will trace a complete sinusoidal waveform after the radius vector has completed a 360° rotation about the center.

The velocity with which the radius vector rotates about the center, called the *angular velocity,* can be determined from the following equation:

$$\text{Angular velocity} = \frac{\text{distance (degrees or radians)}}{\text{time (seconds)}} \tag{13.8}$$

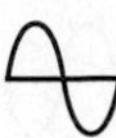

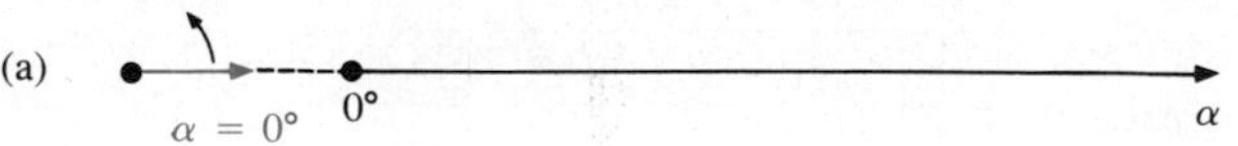

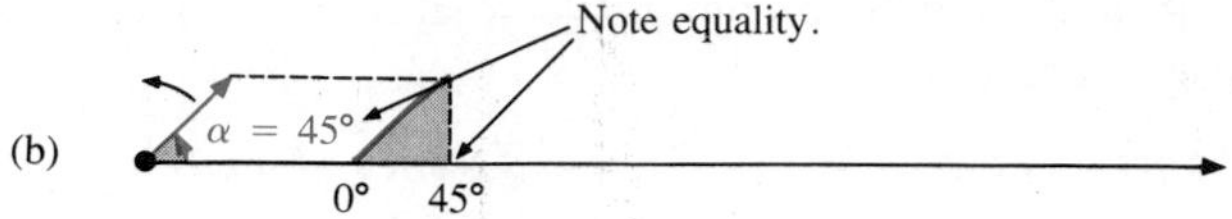

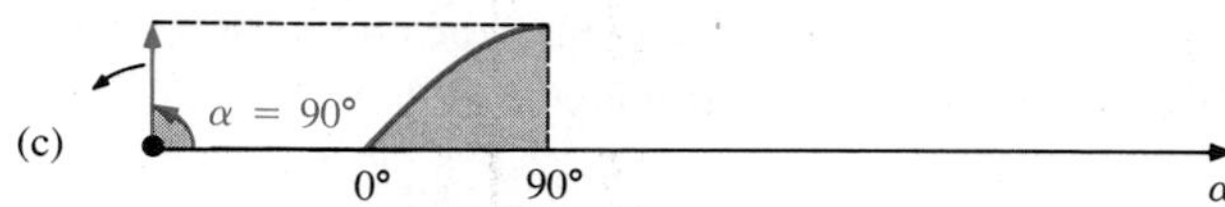

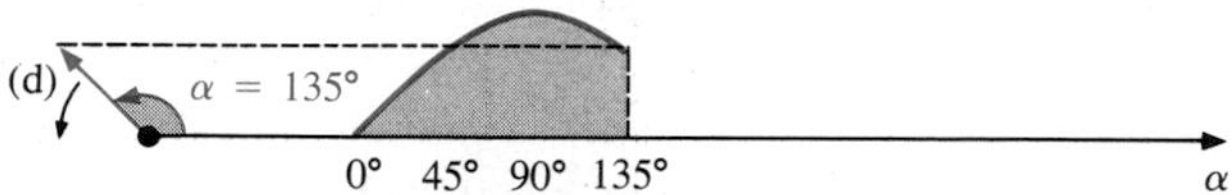

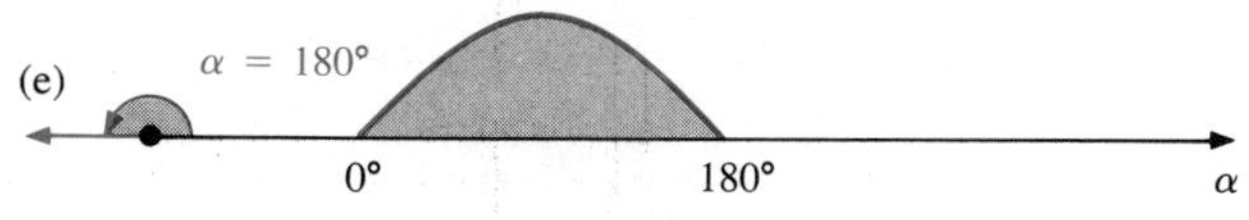

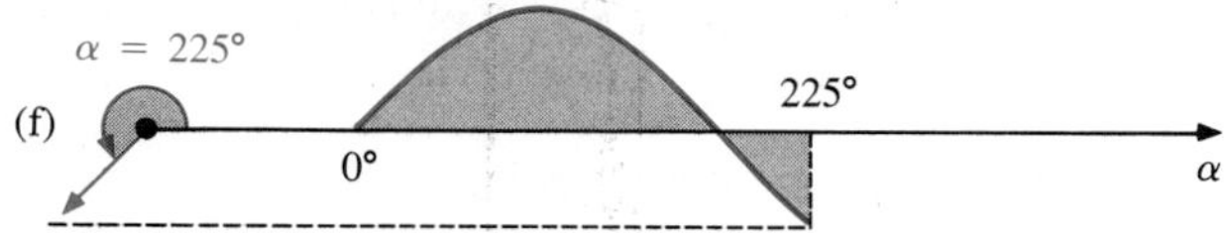

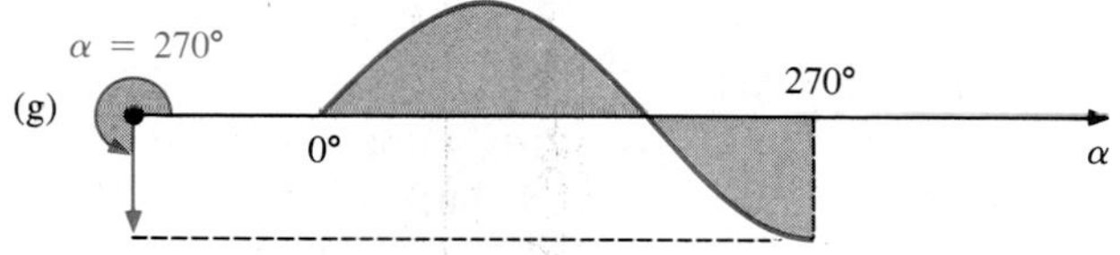

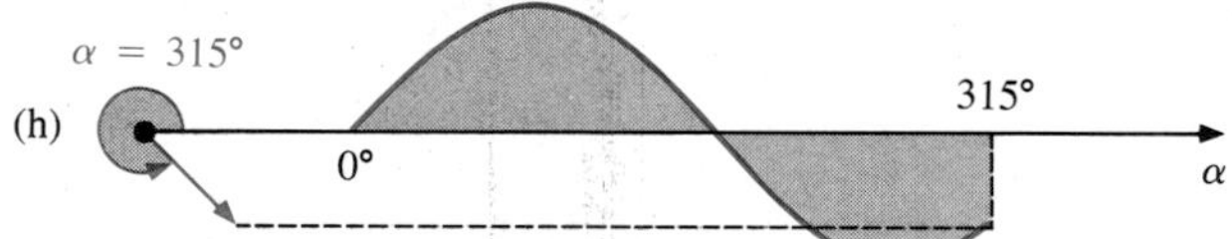

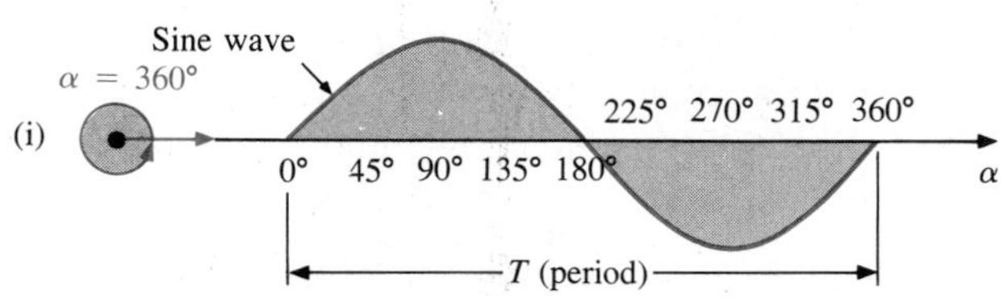

FIG. 13.16

Generating a sinusoidal waveform through the vertical projection of a rotating vector.

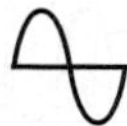

Substituting into Eq. (13.8) and assigning the Greek letter omega (ω) to the angular velocity, we have

$$\omega = \frac{\alpha}{t} \tag{13.9}$$

and

$$\alpha = \omega t \tag{13.10}$$

Since ω is typically provided in radians per second, the angle α obtained using Eq. (13.10) is usually in radians. If α is required in degrees, Eq. (13.7) must be applied. The importance of remembering the above will become obvious in the examples to follow.

In Fig. 13.16, the time required to complete one revolution is equal to the period (T) of the sinusoidal waveform of Fig. 13.16(i). The radians subtended in this time interval are 2π. Substituting, we have

$$\omega = \frac{2\pi}{T} \quad \text{(rad/s)} \tag{13.11}$$

In words, this equation states that the smaller the period of the sinusoidal waveform of Fig. 13.16(i), or the smaller the time interval before one complete cycle is generated, the greater must be the angular velocity of the rotating radius vector. Certainly this statement agrees with what we have learned thus far. We can now go one step further and apply the fact that the frequency of the generated waveform is inversely related to the period of the waveform; that is, $f = 1/T$. Thus,

$$\omega = 2\pi f \quad \text{(rad/s)} \tag{13.12}$$

This equation states that the higher the frequency of the generated sinusoidal waveform, the higher must be the angular velocity. Equations (13.11) and (13.12) are verified somewhat by Fig. 13.17, where for the same radius vector, ω = 100 rad/s and 500 rad/s.

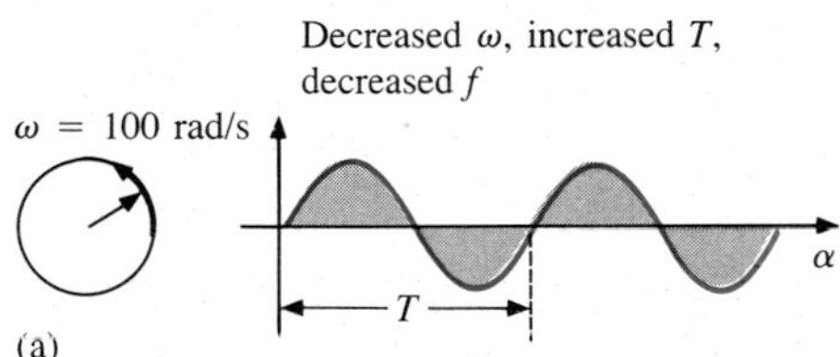

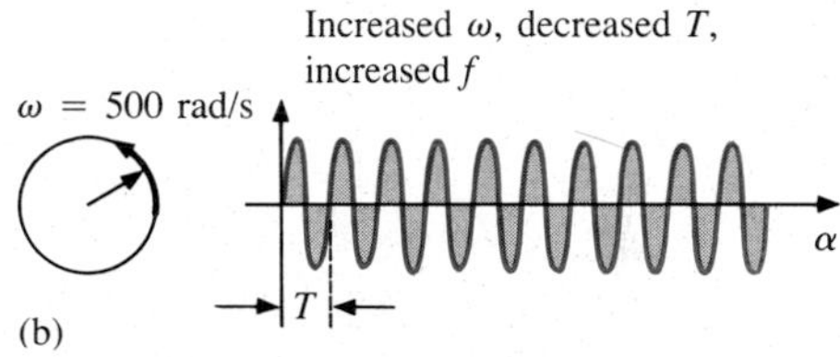

FIG. 13.17
Demonstrating the effect of ω on the frequency and period.

EXAMPLE 13.4 Determine the angular velocity of a sine wave having a frequency of 60 Hz.

Solution:

$$\omega = 2\pi f = (2\pi)(60\text{ Hz}) \cong \mathbf{377\ rad/s}$$

(a recurring value due to 60-Hz predominance)

EXAMPLE 13.5 Determine the frequency and period of the sine wave of Fig. 13.17(b).

Solution: Since $\omega = 2\pi/T$,

$$T = \frac{2\pi}{\omega} = \frac{2\pi\text{ rad}}{500\text{ rad/s}} = \frac{2\pi\text{ rad}}{500\text{ rad/s}} = \mathbf{12.57\ ms}$$

and

$$f = \frac{1}{T} = \frac{1}{12.57 \times 10^{-3}\text{ s}} = \mathbf{79.58\ Hz}$$

EXAMPLE 13.6 Given $\omega = 200$ rad/s, determine how long it will take the sinusoidal waveform to pass through an angle of 90°.

Solution: Eq. (13.10): $\alpha = \omega t$, and

$$t = \frac{\alpha}{\omega}$$

However, α must be substituted as $\pi/2$ (= 90°) since ω is in radians per second:

$$t = \frac{\alpha}{\omega} = \frac{\pi/2 \text{ rad}}{200 \text{ rad/s}} = \frac{\pi}{400 \text{ s}} = \mathbf{7.85\ ms}$$

EXAMPLE 13.7 Find the angle through which a sinusoidal waveform of 60 Hz will pass in a period of 5 ms.

Solution: Eq. (13.11): $\alpha = \omega t$, or

$$\alpha = 2\pi f t = (2\pi)(60 \text{ Hz})(5 \times 10^{-3} \text{ s}) = \mathbf{1.885\ rad}$$

If not careful, one might be tempted to interpret the answer as 1.885°. However,

$$\alpha\,(°) = \frac{180°}{\pi \text{ rad}}(1.885 \text{ rad}) = \mathbf{108°}$$

13.4 GENERAL FORMAT FOR THE SINUSOIDAL VOLTAGE OR CURRENT

The basic mathematical format for the sinusoidal waveform is

$$\boxed{A_m \sin \alpha} \qquad \textbf{(13.13)}$$

where A_m is the peak value of the waveform and α is the unit of measure for the horizontal axis, as shown in Fig. 13.18.

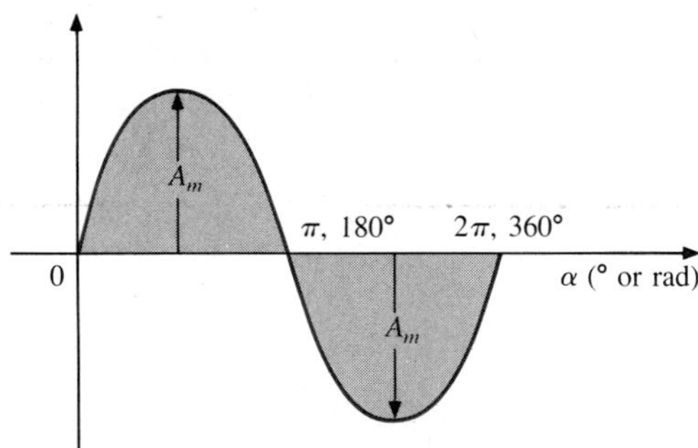

FIG. 13.18
Basic sinusoidal function.

The equation $\alpha = \omega t$ states that the angle α through which the rotating vector of Fig. 13.16 will pass is determined by the angular velocity of the rotating vector and the length of time the vector rotates. For example, for a particular angular velocity (fixed ω), the longer the radius vector is permitted to rotate (that is, the greater the value of t), the greater will be the number of degrees or radians through which the vector will pass. Relating this statement to the sinusoidal waveform, for a particular angular velocity, the longer the time, the greater the num-

ber of cycles shown. For a fixed time interval, the greater the angular velocity, the greater the number of cycles generated.

Due to Eq. (13.10), the general format of a sine wave can also be written

$$\boxed{A_m \sin \omega t} \tag{13.14}$$

with ωt as the horizontal unit of measure.

For electrical quantities such as current and voltage, the general format is

$$i = I_m \sin \omega t = I_m \sin \alpha$$
$$e = E_m \sin \omega t = E_m \sin \alpha$$

where the capital letters with the subscript m represent the amplitude and the lowercase letters i and e represent the instantaneous value of current or voltage, respectively, at any time t. This format is particularly important since it presents the sinusoidal voltage or current as a function of time, which is the horizontal scale for the oscilloscope. Recall that the horizontal sensitivity of a scope is in time per division and not degrees per centimeter.

EXAMPLE 13.8 Given $e = 5 \sin \alpha$, determine e at $\alpha = 40°$ and $\alpha = 0.8\pi$.

Solution: For $\alpha = 40°$,

$$e = 5 \sin 40° = 5(0.6428) = \mathbf{3.214\ V}$$

For $\alpha = 0.8\pi$,

$$\alpha\ (°) = \frac{180°}{\pi}(0.8\pi) = 144°$$

and

$$e = 5 \sin 144° = 5(0.5878) = \mathbf{2.939\ V}$$

The conversion to degrees will not be required for most modern-day scientific calculators since they can perform the function directly. First, be sure the calculator is in the RAD mode and then simply enter the radian measure and use the appropriate trigonometric key (sin, cos, tan, etc.).

The angle at which a particular voltage level is attained can be determined by rearranging the equation

$$e = E_m \sin \alpha$$

in the following manner:

$$\sin \alpha = \frac{e}{E_m}$$

which can be written

$$\boxed{\alpha = \sin^{-1} \frac{e}{E_m}} \tag{13.15}$$

Similarly, for a particular current level,

$$\boxed{\alpha = \sin^{-1} \frac{i}{I_m}} \tag{13.16}$$

The function $\sin^{-1}$ is available on all scientific calculators.

EXAMPLE 13.9

a. Determine the angle at which the magnitude of the sinusoidal function $v = 10 \sin 377t$ is 4 V.
b. Determine the time at which the magnitude is attained.

Solutions:

a. Eq. (13.15):

$$\alpha_1 = \sin^{-1}\frac{v}{E_m} = \sin^{-1}\frac{4\text{ V}}{10\text{ V}} = \sin^{-1} 0.4 = \mathbf{23.578^\circ}$$

However, Fig. 13.19 reveals that the magnitude of 4 V (positive) will be attained at two points between 0° and 180°. The second intersection is determined by

$$\alpha_2 = 180^\circ - 23.578^\circ = \mathbf{156.422^\circ}$$

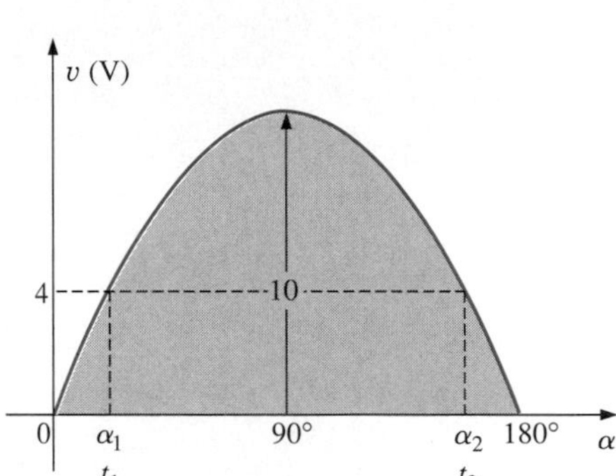

FIG. 13.19
Example 13.9.

In general, therefore, keep in mind that Eqs. (13.15) and (13.16) will provide an angle with a magnitude between 0° and 90°.

b. Eq. (13.10): $\alpha = \omega t$, and so $t = \alpha/\omega$. However, α must be in radians. Thus,

$$\alpha\text{ (rad)} = \frac{\pi}{180^\circ}(23.578^\circ) = 0.411\text{ rad}$$

and
$$t_1 = \frac{\alpha}{\omega} = \frac{0.411\text{ rad}}{377\text{ rad/s}} = \mathbf{1.09\text{ ms}}$$

For the second intersection,

$$\alpha\text{ (rad)} = \frac{\pi}{180^\circ}(156.422^\circ) = 2.73\text{ rad}$$

$$t_2 = \frac{\alpha}{\omega} = \frac{2.73\text{ rad}}{377\text{ rad/s}} = \mathbf{7.24\text{ ms}}$$

The sine wave can also be plotted against *time* on the horizontal axis. The time period for each interval can be determined from $t = \alpha/\omega$, but the most direct route is simply to find the period T from $T = 1/f$ and break it up into the required intervals. This latter technique will be demonstrated in Example 13.10.

Before reviewing the example, take special note of the relative simplicity of the mathematical equation that can represent a sinusoidal waveform. Any alternating waveform whose characteristics differ from those of the sine wave cannot be represented by a single term, but may require two, four, six, or perhaps an infinite number of terms to be represented accurately. Additional description of nonsinusoidal waveforms can be found in Chapter 24.

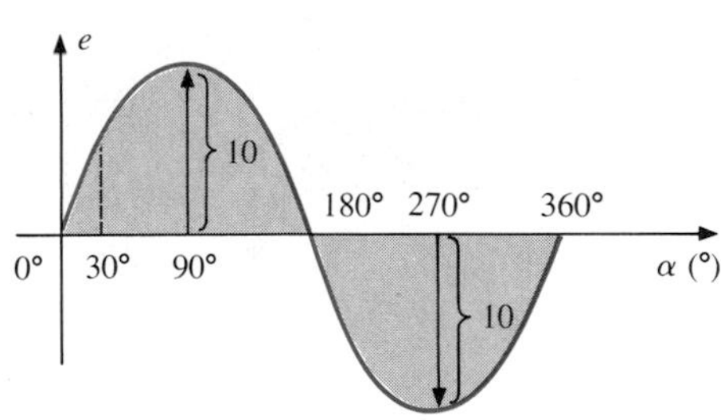

FIG. 13.20
Example 13.10, horizontal axis in degrees.

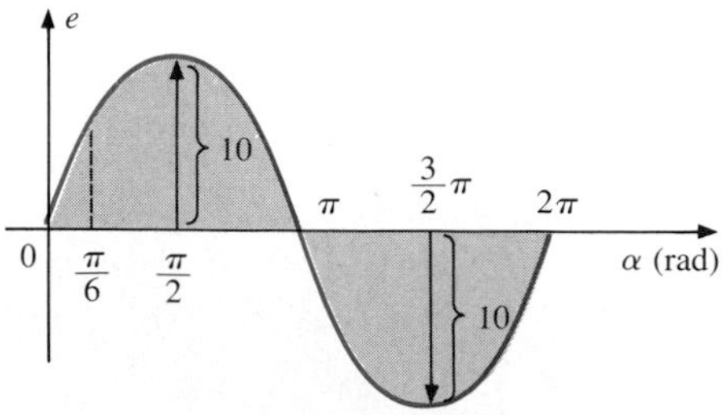

FIG. 13.21
Example 13.10, horizontal axis in radians.

EXAMPLE 13.10 Sketch $e = 10 \sin 314t$ with the abscissa

a. angle (α) in degrees.
b. angle (α) in radians.
c. time (t) in seconds.

Solutions:

a. See Fig 13.20. (Note that no calculations are required.)
b. See Fig. 13.21. (Once the relationship between degrees and radians is understood, there is again no need for calculations.)

c. 360°: $T = \dfrac{2\pi}{\omega} = \dfrac{2\pi}{314} = 20 \text{ ms}$

180°: $\dfrac{T}{2} = \dfrac{20 \text{ ms}}{2} = 10 \text{ ms}$

90°: $\dfrac{T}{4} = \dfrac{20 \text{ ms}}{4} = 5 \text{ ms}$

30°: $\dfrac{T}{12} = \dfrac{20 \text{ ms}}{12} = 1.67 \text{ ms}$

See Fig. 13.22.

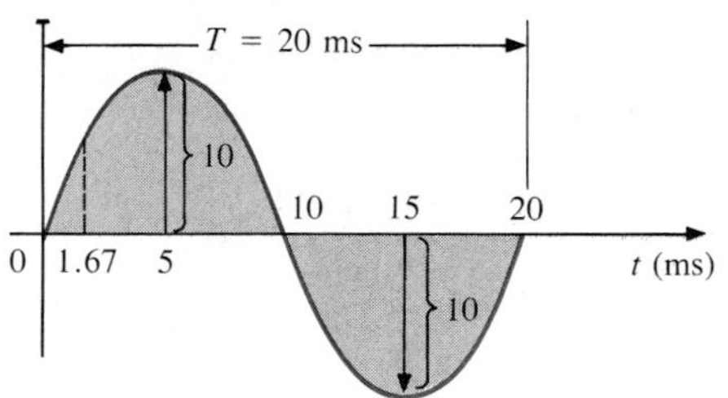

FIG. 13.22
Example 13.10, horizontal axis in milliseconds.

EXAMPLE 13.11 Given $i = 6 \times 10^{-3} \sin 1000t$, determine i at $t =$ 2 ms.

Solution:

$$\alpha = \omega t = 1000t = (1000 \text{ rad/s})(2 \times 10^{-3} \text{ s}) = 2 \text{ rad}$$

$$\alpha\ (°) = \frac{180°}{\pi \text{ rad}}(2 \text{ rad}) = 114.59°$$

$$i = (6 \times 10^{-3})(\sin 114.59°)$$
$$= (6 \text{ mA})(0.9093) = \mathbf{5.46\ mA}$$

13.5 PHASE RELATIONS

Thus far, we have considered only sine waves that have maxima at $\pi/2$ and $3\pi/2$, with a zero value at 0, π, and 2π, as shown in Fig. 13.21. If the waveform is shifted to the right or left of 0°, the expression becomes

$$A_m \sin(\omega t \pm \theta) \qquad \textbf{(13.17)}$$

where θ is the angle in degrees or radians that the waveform has been shifted.

If the waveform passes through the horizontal axis with a *positive-going* (increasing with time) slope *before* 0°, as shown in Fig. 13.23, the expression is

$$A_m \sin(\omega t + \theta) \qquad \textbf{(13.18)}$$

At $\omega t = \alpha = 0°$, the magnitude is determined by $A_m \sin\theta$. If the waveform passes through the horizontal axis with a positive-going slope *after* 0°, as shown in Fig. 13.24, the expression is

$$A_m \sin(\omega t - \theta) \qquad \textbf{(13.19)}$$

And at $\omega t = \alpha = 0°$, the magnitude is $A_m \sin(-\theta)$, which, by a trigonometric identity, is $-A_m \sin\theta$.

If the waveform crosses the horizontal axis with a positive-going slope 90° ($\pi/2$) sooner, as shown in Fig. 13.25, it is called a *cosine wave;* that is,

$$\sin(\omega t + 90°) = \sin\left(\omega t + \frac{\pi}{2}\right) = \cos\omega t \qquad \textbf{(13.20)}$$

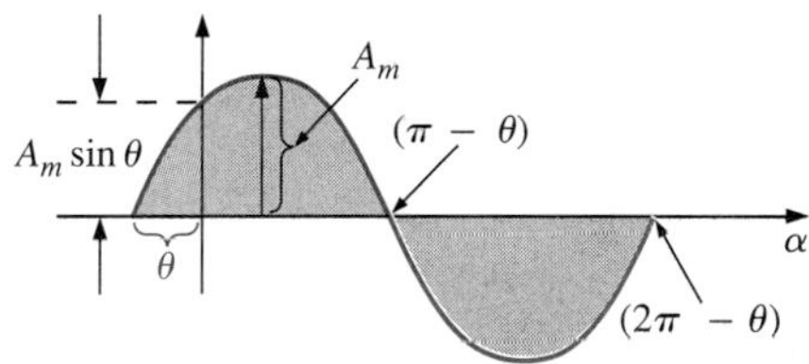

FIG. 13.23
Defining the phase shift for a sinusoidal function that crosses the horizontal axis with a positive slope before 0°.

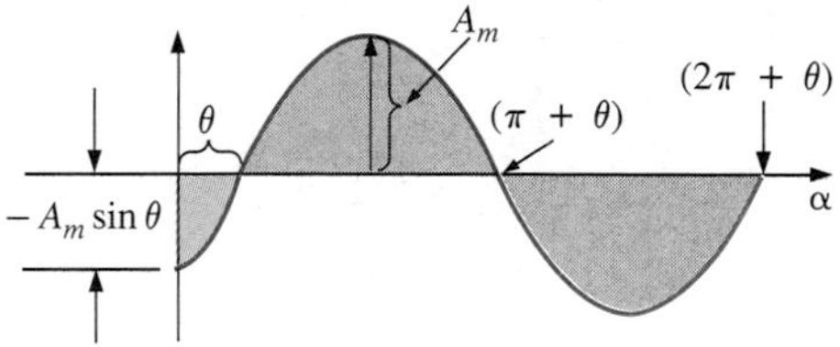

FIG. 13.24
Defining the phase shift for a sinusoidal function that crosses the horizontal axis with a positive slope after 0°.

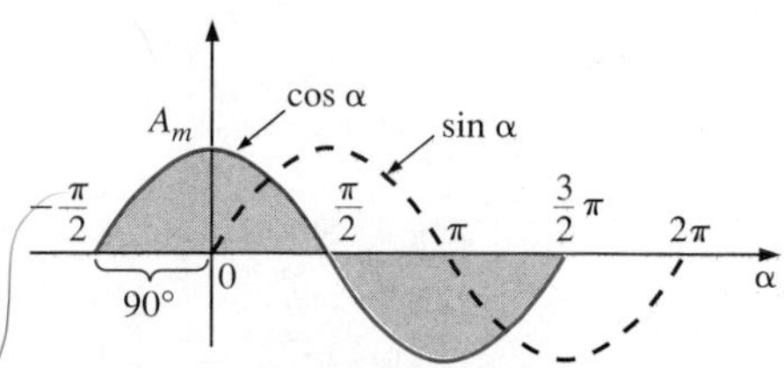

FIG. 13.25
Phase relationship between a sine wave and a cosine wave.

or

$$\sin \omega t = \cos(\omega t - 90°) = \cos\left(\omega t - \frac{\pi}{2}\right) \tag{13.21}$$

The terms *lead* and *lag* are used to indicate the relationship between two sinusoidal waveforms of the *same frequency* plotted on the same set of axes. In Fig. 13.25, the cosine curve is said to *lead* the sine curve by 90°, and the sine curve is said to *lag* the cosine curve by 90°. The 90° is referred to as the phase angle between the two waveforms. In language commonly applied, the waveforms are *out of phase* by 90°. Note that the phase angle between the two waveforms is measured between those two points on the horizontal axis through which each passes with the *same slope*. If both waveforms cross the axis at the same point with the same slope, they are *in phase*.

The geometric relationship between various forms of the sine and cosine functions can be derived from Fig. 13.26. For instance, starting at the $\sin \alpha$ position, we find that $\cos \alpha$ is an additional 90° in the counterclockwise direction. Therefore, $\cos \alpha = \sin(\alpha + 90°)$. For $-\sin \alpha$ we must travel 180° in the counterclockwise (or clockwise) direction so that $-\sin \alpha = \sin(\alpha \pm 180°)$, and so on, as listed below:

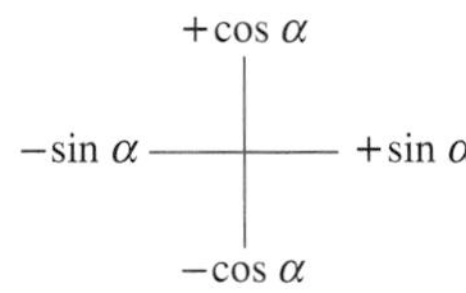

FIG. 13.26
Graphic tool for finding the relationship between specific sine and cosine functions.

$$\begin{aligned} \cos \alpha &= \sin(\alpha + 90°) \\ \sin \alpha &= \cos(\alpha - 90°) \\ -\sin \alpha &= \sin(\alpha \pm 180°) \\ -\cos \alpha &= \sin(\alpha + 270°) = \sin(\alpha - 90°) \\ &\text{etc.} \end{aligned} \tag{13.22}$$

In addition one should be aware that

$$\begin{aligned} \sin(-\alpha) &= -\sin \alpha \\ \cos(-\alpha) &= \cos \alpha \end{aligned} \tag{13.23}$$

If a sinusoidal expression should appear as

$$e = -E_m \sin \omega t$$

the negative sign is associated with the sine portion of the expression, not the peak value E_m. In other words, the expression, if not for convenience, would be written

$$e = E_m(-\sin \omega t)$$

Since

$$-\sin \omega t = \sin(\omega t \pm 180°)$$

the expression can also be written

$$e = E_m \sin(\omega t \pm 180°)$$

revealing that a negative sign can be replaced by a 180° change in phase angle (+ or −); that is,

$$\begin{aligned} e = E_m \sin \omega t &= E_m \sin(\omega t + 180°) \\ &= E_m \sin(\omega t - 180°) \end{aligned}$$

A plot of each will clearly show their equivalence. There are, therefore, two correct mathematical representations for the functions.

The *phase relationship* between two waveforms indicates which one leads or lags, and by how many degrees or radians.

EXAMPLE 13.12 What is the phase relationship between the sinusoidal waveforms of each of the following sets?

a. $v = 10 \sin(\omega t + 30°)$
   $i = 5 \sin(\omega t + 70°)$
b. $i = 15 \sin(\omega t + 60°)$
   $v = 10 \sin(\omega t - 20°)$
c. $i = 2 \cos(\omega t + 10°)$
   $v = 3 \sin(\omega t - 10°)$
d. $i = -\sin(\omega t + 30°)$
   $v = 2 \sin(\omega t + 10°)$
e. $i = -2 \cos(\omega t - 60°)$
   $v = 3 \sin(\omega t - 150°)$

Solutions:

a. See Fig. 13.27.
   ***i* leads *v* by 40°, or *v* lags *i* by 40°.**

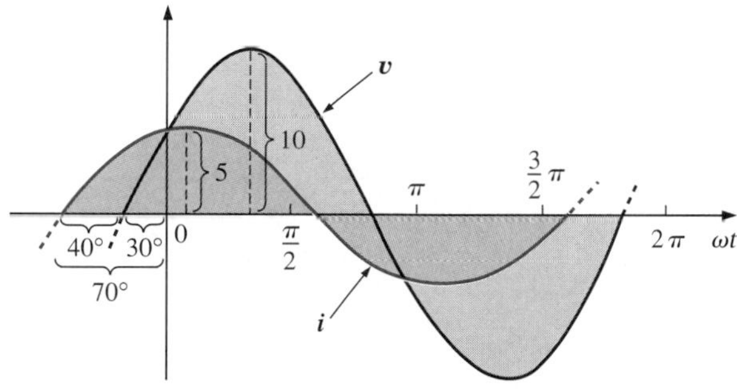

FIG. 13.27
Example 13.12, i leads v by 40°.

b. See Fig. 13.28.
   ***i* leads *v* by 80°, or *v* lags *i* by 80°.**

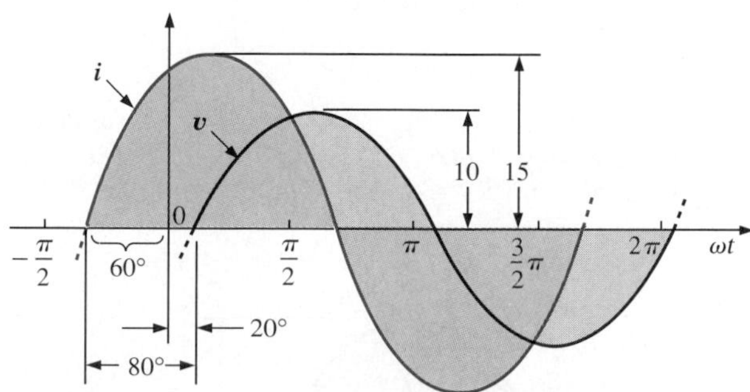

FIG. 13.28
Example 13.12, i leads v by 80°.

c. See Fig. 13.29.

$$i = 2\cos(\omega t + 10°) = 2\sin(\omega t + 10° + 90°)$$
$$= 2\sin(\omega t + 100°)$$

***i* leads *v* by 110°, or *v* lags *i* by 110°.**

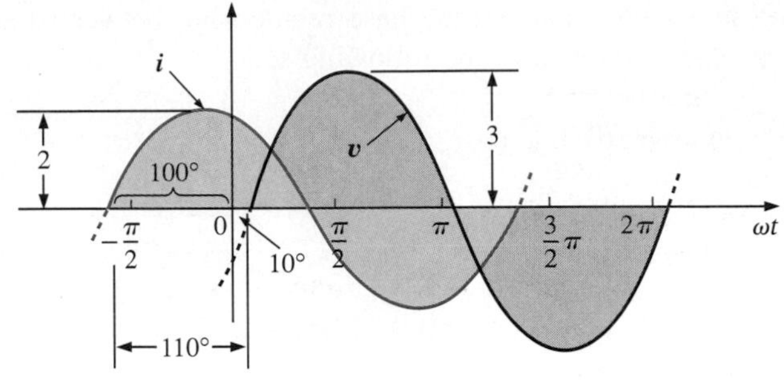

FIG. 13.29
Example 13.12, i leads v by 110°.

d. See Fig. 13.30.

$$-\sin(\omega t + 30°) = \sin(\omega t + 30° - 180°) \quad \text{(Note)}$$
$$= \sin(\omega t - 150°)$$

***v* leads *i* by 160°, or *i* lags *v* by 160°.**

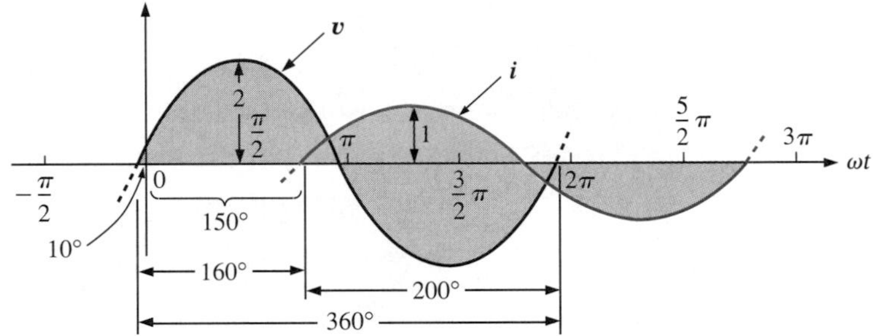

FIG. 13.30
Example 13.12, v leads i by 160°.

Or using

$$-\sin(\omega t + 30°) = \sin(\omega t + 30° + 180°) \quad \text{(Note)}$$
$$= \sin(\omega t + 210°)$$

***i* leads *v* by 200°, or *v* lags *i* by 200°.**

e. See Fig. 13.31.

$$i = -2\cos(\omega t - 60°) = 2\cos(\omega t - 60° - 180°) \quad \text{(By choice)}$$
$$= 2\cos(\omega t - 240°)$$

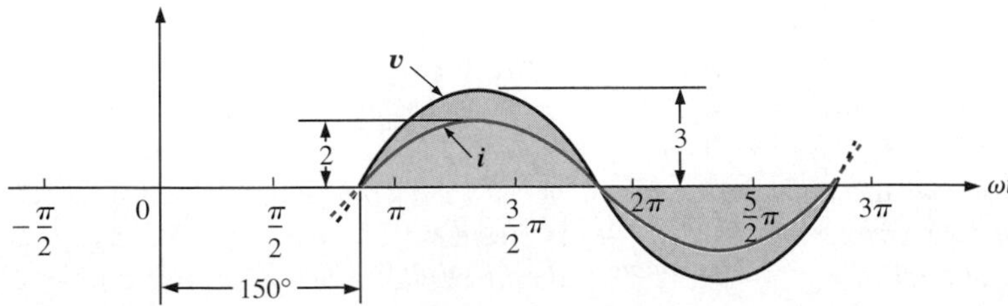

FIG. 13.31
Example 13.12, v and i are in phase.

However, $\cos\alpha = \sin(\alpha + 90°)$

so that $2\cos(\omega t - 240°) = 2\sin(\omega t - 240° + 90°)$
$= 2\sin(\omega t - 150°)$

***v* and *i* are in phase.**

Phase Measurements

The hookup procedure for using an oscilloscope to measure phase angles is covered in detail in Section 15.13. However, the equation for determining the phase angle can be introduced using Fig. 13.32. First, note that each sinusoidal function has the same frequency, permitting the use of either waveform to determine the period. For the waveform chosen in Fig. 13.32 the period encompasses 5 divisions at 0.2 ms/div. The phase shift between the waveforms (irrespective of which is leading or lagging) is 2 divisions. Since the full period represents a cycle of 360°, the following ratio [from which Eq. (13.24) can be derived] can be formed:

$$\frac{360°}{T\text{ (no. of div.)}} = \frac{\theta}{\text{phase shift (no. of div.)}}$$

and

$$\theta = \frac{\text{phase shift (no. of div.)}}{T\text{ (no. of div.)}} \times 360° \qquad (13.24)$$

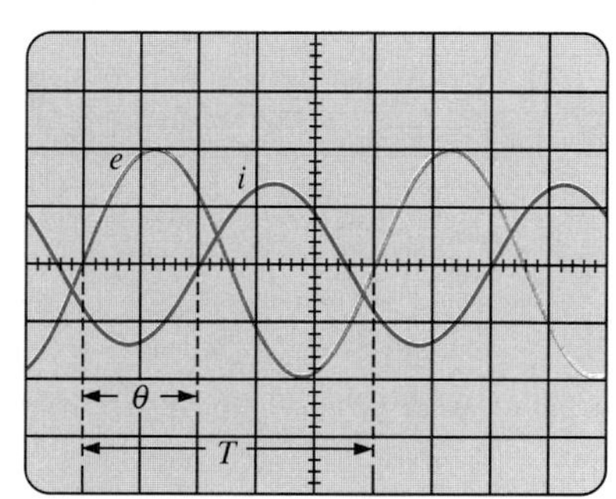

Vertical sensitivity = 2 V/div.
Horizontal sensitivity = 0.2 ms/div.

FIG. 13.32
Finding the phase angle between waveforms using a dual-trace oscilloscope.

Substituting into Eq. (13.24) will result in

$$\theta = \frac{(2\text{ div.})}{(5\text{ div.})} \times 360° = 144°$$

and *e* leads *i* by 144°.

13.6 AVERAGE VALUE

Even though the concept of the *average value* is an important one in most technical fields, its true meaning is often misunderstood. In Fig. 13.33(a), for example, the average height of the sand may be required to determine the volume of sand available. The average height of the sand is that height obtained if the distance from one end to the other is maintained while the sand is leveled off, as shown in Fig. 13.33(b). The area under the mound of Fig. 13.33(a) will then equal the area under the rectangular shape of Fig. 13.33(b) as determined by $A = b \times h$. Of course, the depth (into the page) of the sand must be the same for Fig. 13.33(a) and 13.33(b) for the preceding conclusions to have any meaning.

In Fig. 13.33 the distance was measured from one end to the other. In Fig. 13.34(a) the distance extends beyond the end of the original pile of Fig. 13.33. The situation could be one where a landscaper would like to know the average height of the sand if spread out over a distance such as defined in Fig. 13.34(a). The result of an increased distance is as shown in Fig. 13.34(b). The average height has decreased compared to Fig. 13.33. Quite obviously, therefore, the longer the distance, the lower is the average value.

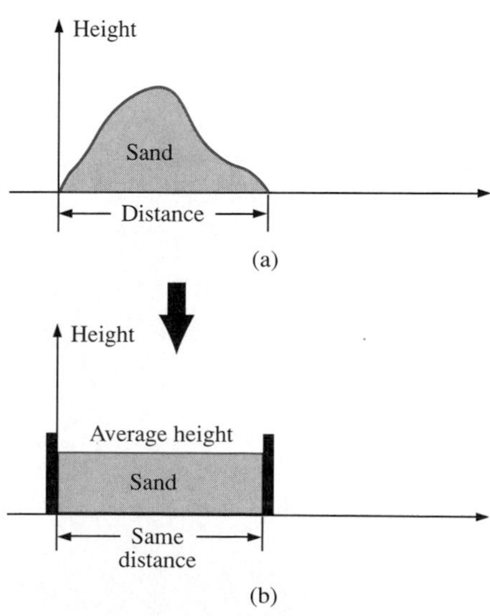

FIG. 13.33
Defining average value.

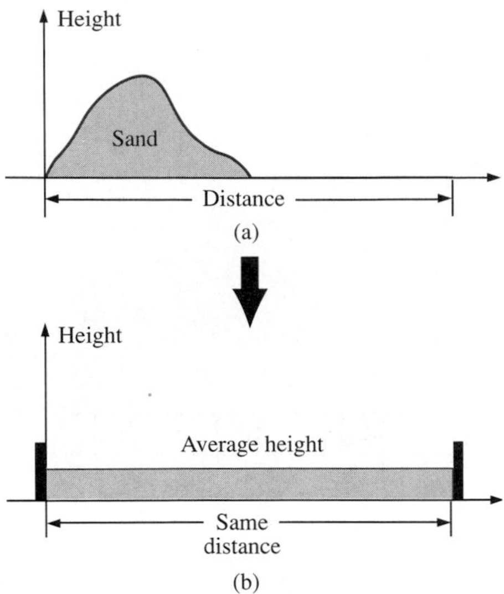

FIG. 13.34
Effect of distance (length) on average value.

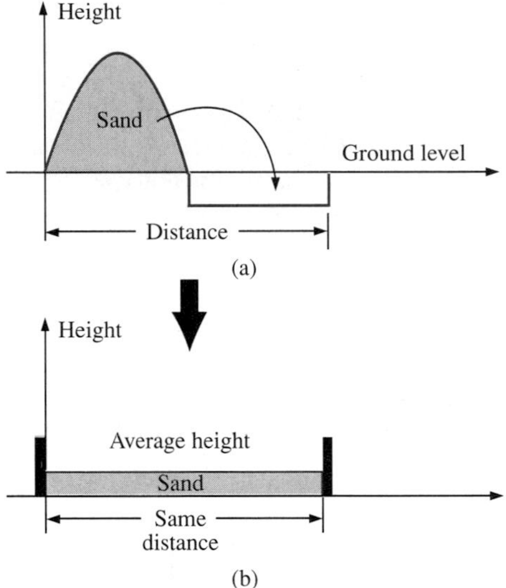

FIG. 13.35
Effect of depressions (negative excursions) on average value.

If the distance parameter includes a depression, as shown in Fig. 13.35(a), some of the sand will be used to fill the depression, resulting in an even lower average value for the landscaper, as shown in Fig. 13.35(b). For a sinusoidal waveform, the depression would have the same shape as the mound of sand (over one full cycle), resulting in an average value at ground level (or zero volts for a sinusoidal voltage over one full period).

After traveling a considerable distance by car, some drivers like to calculate their average speed for the entire trip. This is usually done by dividing the miles traveled by the hours required to drive that distance. For example, if a person traveled 180 mi in 5 h, the average speed was 180 mi/5 h, or 36 mi/h. This same distance may have been traveled at various speeds for various intervals of time, as shown in Fig. 13.36.

By finding the total area under the curve for the 5 h and then dividing the area by 5 h (the total time for the trip), we obtain the same result of 36 mi/h; that is,

$$\text{Average speed} = \frac{\text{area under curve}}{\text{length of curve}} \tag{13.25}$$

$$= \frac{A_1 + A_2}{5\text{ h}}$$

$$= \frac{(40\text{ mi/h})(2\text{ h}) + (50\text{ mi/h})(2\text{ h})}{5\text{ h}}$$

$$= \frac{180}{5}\text{ mi/h}$$

$$= \mathbf{36\ mi/h}$$

Equation (13.25) can be extended to include any variable quantity, such as current or voltage, if we let G denote the average value, as follows:

$$G\text{ (average value)} = \frac{\text{algebraic sum of areas}}{\text{length of curve}} \tag{13.26}$$

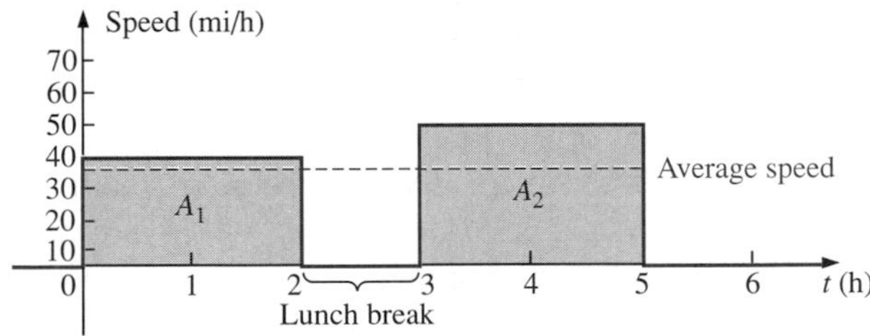

FIG. 13.36
Plotting speed versus time for an automobile excursion.

The *algebraic* sum of the areas must be determined, since some area contributions will be from below the horizontal axis. Areas above the axis will be assigned a positive sign, and those below, a negative sign. A positive average value will then be above the axis, and a negative value, below.

The average value of *any* current or voltage is the value indicated on a dc meter. In other words, over a complete cycle, the average value is

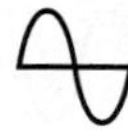

the equivalent dc value. In the analysis of electronic circuits to be considered in a later course, both dc and ac sources of voltage will be applied to the same network. It will then be necessary to know or determine the dc (or average value) and ac components of the voltage or current in various parts of the system.

EXAMPLE 13.13 Determine the average value of the waveforms of Fig. 13.37.

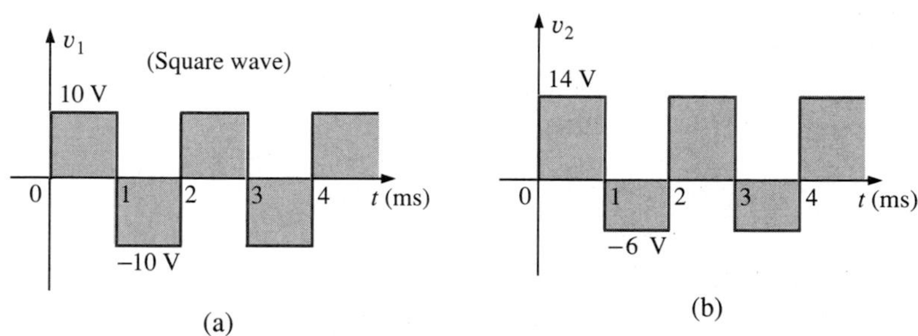

FIG. 13.37
Example 13.13.

Solutions:

a. By inspection, the area above the axis equals the area below over one cycle, resulting in an average value of zero volts. Using Eq. (13.26):

$$G = \frac{(10\text{ V})(1\text{ ms}) - (10\text{ V})(1\text{ ms})}{2\text{ ms}}$$

$$= \frac{0}{2\text{ ms}} = \mathbf{0\ V}$$

b. Using Eq. (13.26):

$$G = \frac{(14\text{ V})(1\text{ ms}) - (6\text{ V})(1\text{ ms})}{2\text{ ms}}$$

$$= \frac{14\text{ V} - 6\text{ V}}{2} = \frac{8\text{ V}}{2} = \mathbf{4\ V}$$

as shown in Fig. 13.38.

In reality, the waveform of Fig. 13.37(b) is simply the square wave of Fig. 13.37(a) with a dc shift of 4 V; that is,

$$v_2 = v_1 + 4\text{ V}$$

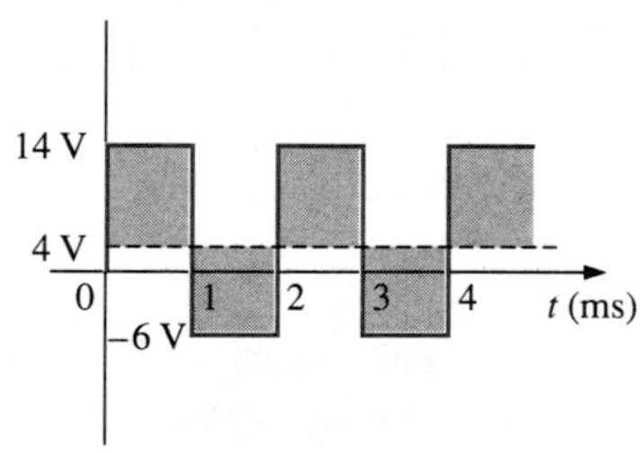

FIG. 13.38
Defining the average value for the waveform of Fig. 13.37(b).

EXAMPLE 13.14 Find the average values of the following waveforms over one full cycle:

a. Figure 13.39.
b. Figure 13.40.

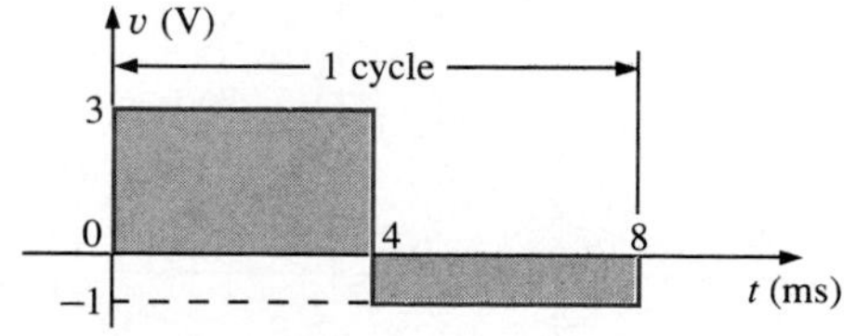

FIG. 13.39
Example 13.14, part (a).

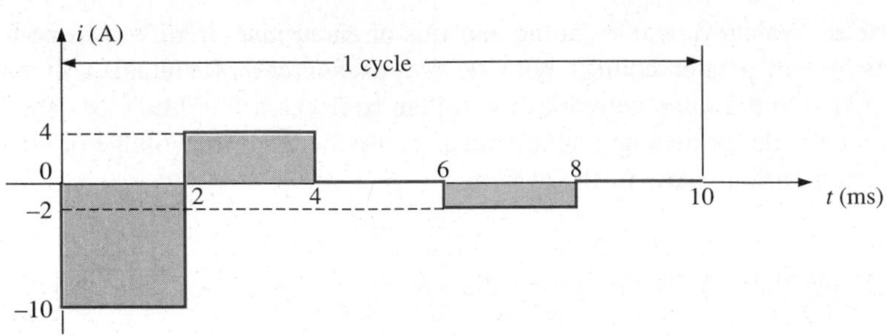

FIG. 13.40
Example 13.14, part (b).

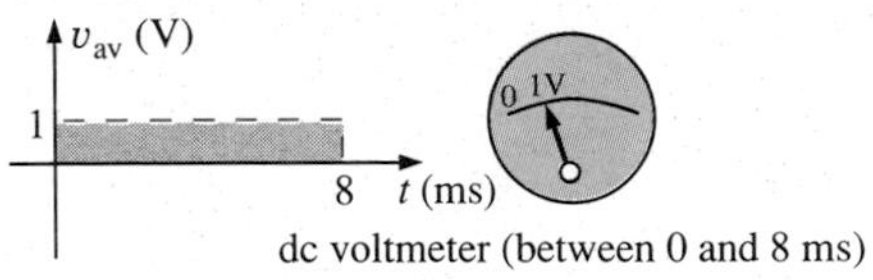

FIG. 13.41
The response of a dc meter to the waveform of Fig. 13.39.

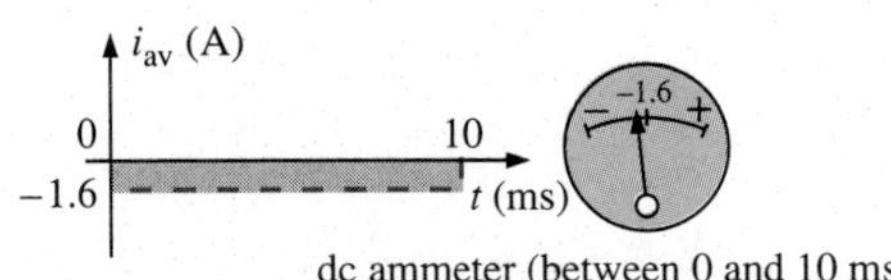

FIG. 13.42
The response of a dc meter to the waveform of Fig. 13.40.

Solutions:

a. $G = \dfrac{+(3\text{ V})(4\text{ ms}) - (1\text{ V})(4\text{ ms})}{8\text{ ms}} = \dfrac{12\text{ V} - 4\text{ V}}{8} = \mathbf{1\text{ V}}$

Note Fig. 13.41.

b. $G = \dfrac{-(10\text{ V})(2\text{ ms}) + (4\text{ V})(2\text{ ms}) - (2\text{ V})(2\text{ ms})}{10\text{ ms}}$

$= \dfrac{-20\text{ V} + 8\text{ V} - 4\text{ V}}{10} = -\dfrac{16\text{ V}}{10} = \mathbf{-1.6\text{ V}}$

Note Fig. 13.42.

We found the areas under the curves in the preceding example by using a simple geometric formula. If we should encounter a sine wave or any other unusual shape, however, we must find the area by some other means. We can obtain a good approximation of the area by attempting to reproduce the original wave shape using a number of small rectangles or other familiar shapes, the area of which we already know through simple geometric formulas. For example,

the area of the positive (or negative) pulse of a sine wave is $2A_m$.

Approximating this waveform by two triangles (Fig. 13.43), we obtain (using *area* = 1/2 *base* × *height* for the area of a triangle) a rough idea of the actual area:

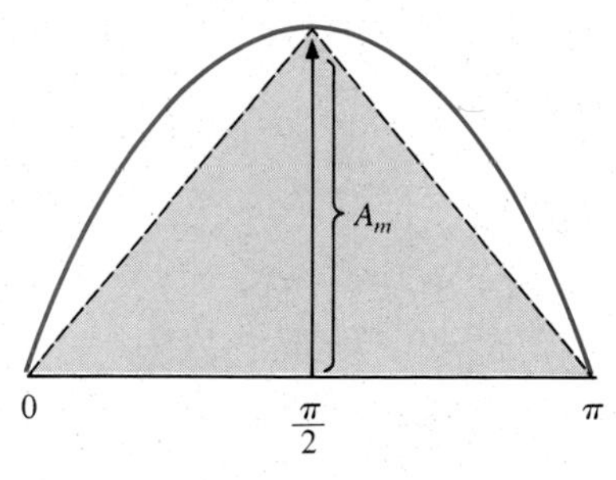

FIG. 13.43
Approximating the shape of the positive pulse of a sinusoidal waveform with two right triangles.

$$\text{Area shaded} = 2\left(\frac{1}{2}bh\right) = 2\left[\left(\frac{1}{2}\right)\overbrace{\left(\frac{\pi}{2}\right)}^{b}\overbrace{(A_m)}^{h}\right] = \frac{\pi}{2}A_m$$
$$\cong 1.58A_m$$

A closer approximation might be a rectangle with two similar triangles (Fig. 13.44):

$$\text{Area} = A_m\frac{\pi}{3} + 2\left(\frac{1}{2}bh\right) = A_m\frac{\pi}{3} + \frac{\pi}{3}A_m = \frac{2}{3}\pi A_m$$
$$= 2.094A_m$$

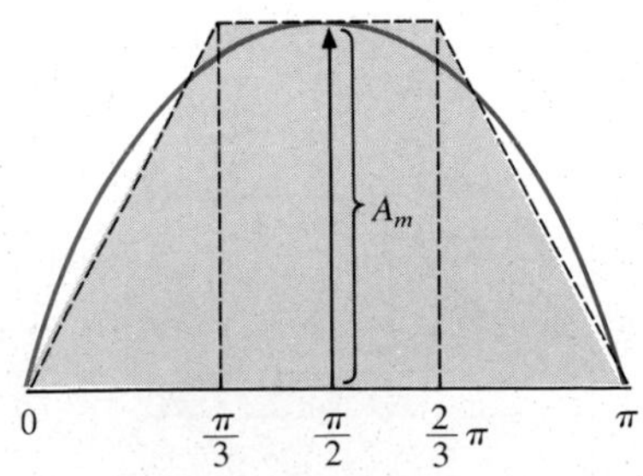

FIG. 13.44
A better approximation for the shape of the positive pulse of a sinusoidal waveform.

which is certainly close to the actual area. If an infinite number of forms were used, an exact answer of $2A_m$ could be obtained. For irregular waveforms, this method can be especially useful if data such as the average value are desired.

The procedure of calculus that gives the exact solution $2A_m$ is known as *integration*. Integration is presented here only to make the

method recognizable to the reader; it is not necessary to be proficient in its use to continue with this text. It is a useful mathematical tool, however, and should be learned. Finding the area under the positive pulse of a sine wave using integration, we have

$$\text{Area} = \int_0^{\pi} A_m \sin \alpha \, d\alpha$$

where $\int$ is the sign of integration, 0 and π are the limits of integration, $A_m \sin \alpha$ is the function to be integrated, and $d\alpha$ indicates that we are integrating with respect to α.

Integrating, we obtain

$$\begin{aligned} \text{Area} &= A_m[-\cos \alpha]_0^{\pi} \\ &= -A_m(\cos \pi - \cos 0°) \\ &= -A_m[-1 - (+1)] = -A_m(-2) \end{aligned}$$

$$\boxed{\text{Area} = 2A_m} \qquad (13.27)$$

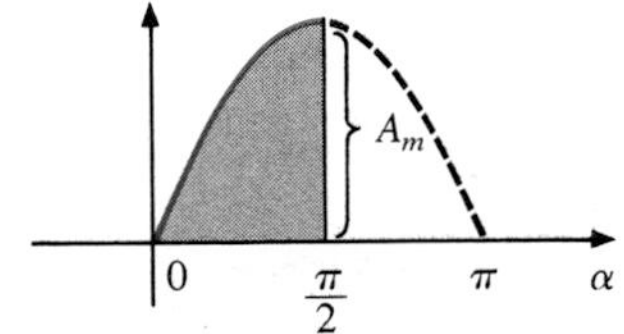

FIG. 13.45
Finding the average value of one-half the positive pulse of a sinusoidal waveform.

Since we know the area under the positive (or negative) pulse, we can easily determine the average value of the positive (or negative) region of a sine wave pulse by applying Eq. (13.26):

$$G = \frac{2A_m}{\pi}$$

and

$$\boxed{G = 0.637A_m} \qquad (13.28)$$

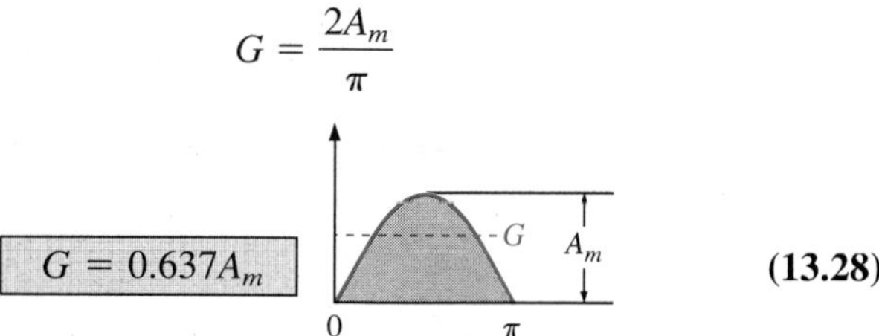

For the waveform of Fig. 13.45,

$$G = \frac{(2A_m/2)}{\pi/2} = \frac{2A_m}{\pi} \qquad \text{(average the same as for a full pulse)}$$

EXAMPLE 13.15 Determine the average value of the sinusoidal waveform of Fig. 13.46.

Solution: By inspection it is fairly obvious that

the average value of a pure sinusoidal waveform over one full cycle is zero.

Equation (13.26):

$$G = \frac{+2A_m - 2A_m}{2\pi} = \mathbf{0\ V}$$

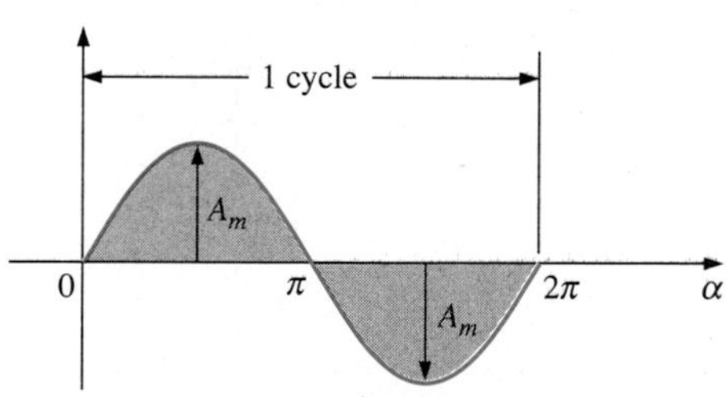

FIG. 13.46
Example 13.15.

EXAMPLE 13.16 Determine the average value of the waveform of Fig. 13.47.

Solution: The peak-to-peak value of the sinusoidal function is 16 mV + 2 mV = 18 mV. The peak amplitude of the sinusoidal waveform is, therefore, 18 mV/2 = 9 mV. Counting down 9 mV from 2 mV (or 9 mV up from −16 mV) results in an average or dc level of −7 mV, as noted by the dashed line of Fig. 13.47.

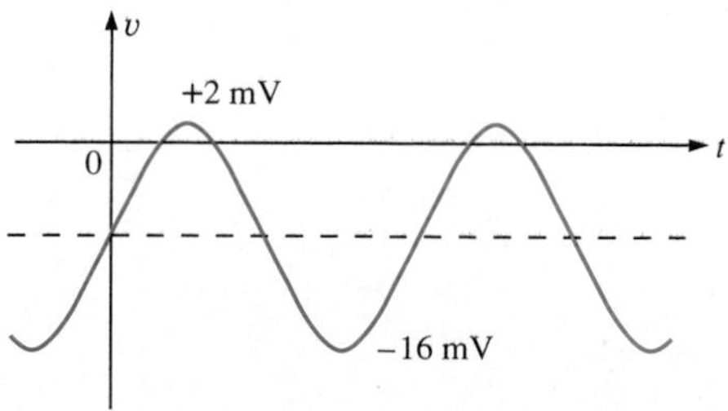

FIG. 13.47
Example 13.16.

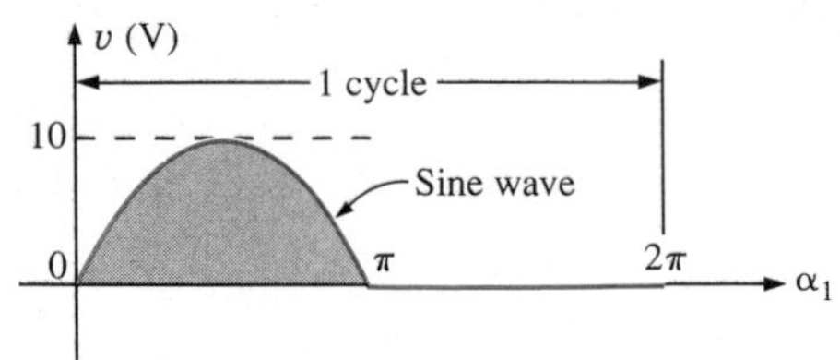

FIG. 13.48
Example 13.17.

EXAMPLE 13.17 Determine the average value of the waveform of Fig. 13.48.

Solution:

$$G = \frac{2A_m + 0}{2\pi} = \frac{2(10\text{ V})}{2\pi} \cong \mathbf{3.18\ V}$$

EXAMPLE 13.18 For the waveform of Fig. 13.49 determine whether the average value is positive or negative and determine its approximate value.

Solution: From the appearance of the waveform, the average value is positive and in the vicinity of 2 mV. Occasionally, judgments of this type will have to be made.

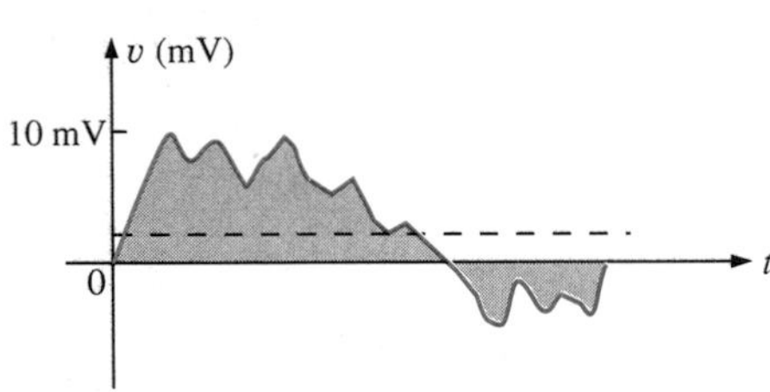

FIG. 13.49
Example 13.18.

Instrumentation

The dc level or average value of any waveform can be found using a digital multimeter (DMM) or an oscilloscope. For purely dc circuits, simply set the DMM on dc and read the voltage or current levels. Oscilloscopes are limited to voltage levels using the sequence of steps listed below:

1. First choose GND from the DC-GND-AC option list associated with each vertical channel. The GND option blocks any signal to which the oscilloscope probe may be connected from entering the oscilloscope and responds with just a horizontal line. Set the resulting line in the middle of the vertical axis on the horizontal axis, as shown in Fig. 13.50(a).

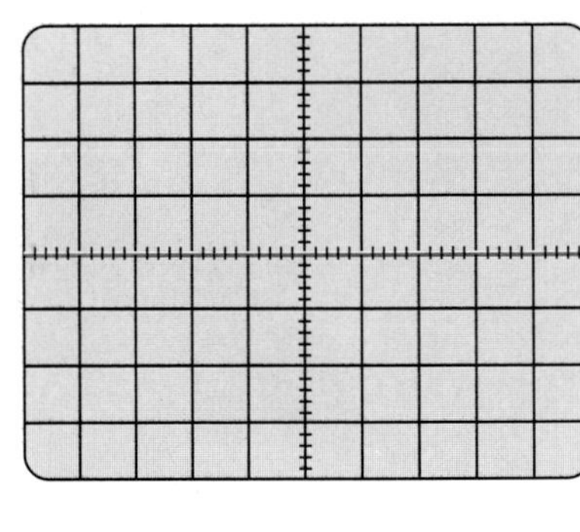

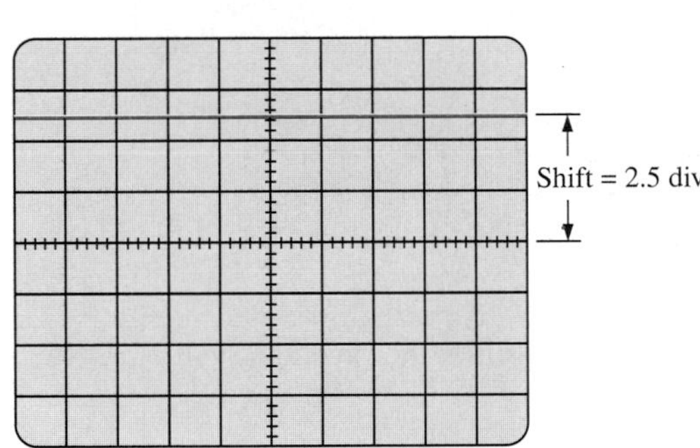

FIG. 13.50
Using the oscilloscope to measure dc voltages: (a) setting the GND condition; (b) the vertical shift resulting from a dc voltage when shifted to the DC option.

2. Apply the oscilloscope probe to the voltage to be measured (if not already connected) and switch to the DC option. If a dc voltage is present, the horizontal line will shift up or down, as demonstrated in Fig. 13.50(b). Multiplying the shift by the vertical sensitivity will result in the dc voltage. An upward shift is a positive voltage (higher potential at the red or positive lead of the oscilloscope), while a downward shift is a negative voltage (lower potential at the red or positive lead of the oscilloscope).

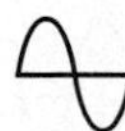

In general,

$$V_{dc} = (\text{vertical shift in div.}) \times (\text{vertical sensitivity in V/div.}) \qquad (13.29)$$

For the waveform of Fig. 13.50(b),

$$V_{dc} = (2.5 \text{ div.})(50 \text{ mV/div.}) = \mathbf{125 \text{ mV}}$$

The oscilloscope can also be used to measure the dc or average level of any waveform using the following sequence:

1. Using the GND option, reset the horizontal line to the middle of the screen.
2. Switch to AC (all dc components of the signal to which the probe is connected will be blocked from entering the oscilloscope—only the alternating, or changing, components will be displayed) and note the location of some definitive point on the waveform, such as the bottom of the half-wave rectified waveform of Fig. 13.51(a); that is, note its position on the vertical scale. For the future, whenever you use the AC option, keep in mind that the computer will distribute the waveform above and below the horizontal axis such that the average value is zero; that is, the area above the axis will equal the area below.
3. Then switch to DC (to permit both the dc and ac components of the waveform to enter the oscilloscope) and note the shift in the chosen level of part 2, as shown in Fig. 13.51(b). Equation (13.29) can then be used to determine the dc or average value of the waveform. For the waveform of Fig. 13.51(b), the average value is about

$$V_{average} = V_{dc} = (0.9 \text{ div.})(5 \text{ V/div.}) = \mathbf{4.5 \text{ V}}$$

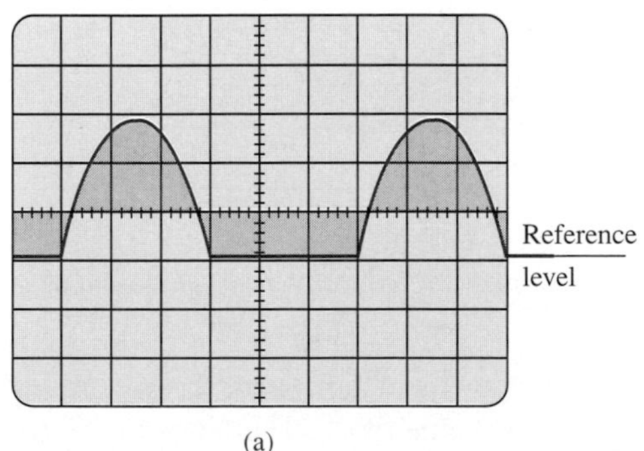

(a)

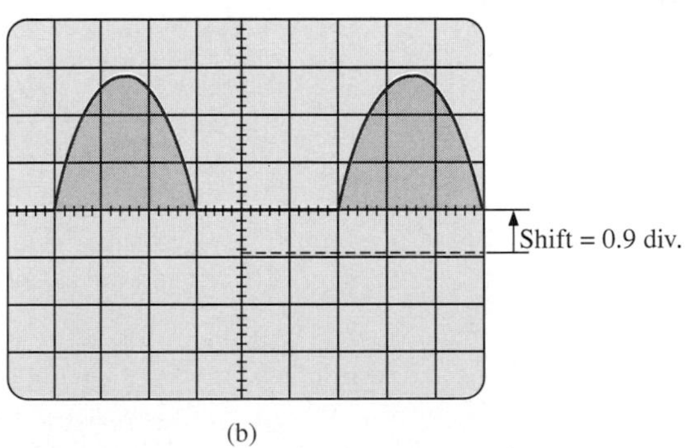

(b)

FIG. 13.51
Determining the average value of a nonsinusoidal waveform using the oscilloscope: (a) vertical channel on the ac mode; (b) vertical channel on the dc mode.

The procedure outlined above can be applied to any alternating waveform such as the one in Fig. 13.49. In some cases the average value may require moving the starting position of the waveform under the AC option to a different region of the screen or choosing a higher voltage scale. DMMs can read the average or dc level of any waveform by simply choosing the appropriate scale.

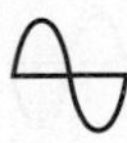

13.7 EFFECTIVE VALUES

This section will begin to relate dc and ac quantities with respect to the power delivered to a load. It will help us determine the amplitude of a sinusoidal ac current required to deliver the same power as a particular dc current. The question frequently arises, How is it possible for a sinusoidal ac quantity to deliver a net power if, over a full cycle, the net current in any one direction is zero (average value = 0)? It would almost appear that the power delivered during the positive portion of the sinusoidal waveform is withdrawn during the negative portion, and since the two are equal in magnitude, the net power delivered is zero. However, understand that *irrespective of direction,* current of any magnitude through a resistor will deliver power *to that resistor.* In other words, during the positive or negative portions of a sinusoidal ac current, power is being delivered at *each instant of time* to the resistor. The power delivered at each instant will, of course, vary with the magnitude of the sinusoidal ac current, but there will be a net flow during either the positive or negative pulses with a net flow over the full cycle. The net power flow will equal twice that delivered by either the positive or negative regions of sinusoidal quantity.

A fixed relationship between ac and dc voltages and currents can be derived from the experimental setup shown in Fig. 13.52. A resistor in a water bath is connected by switches to a dc and an ac supply. If switch 1 is closed, a dc current I, determined by the resistance R and battery voltage E, will be established through the resistor R. The temperature reached by the water is determined by the dc power dissipated in the form of heat by the resistor.

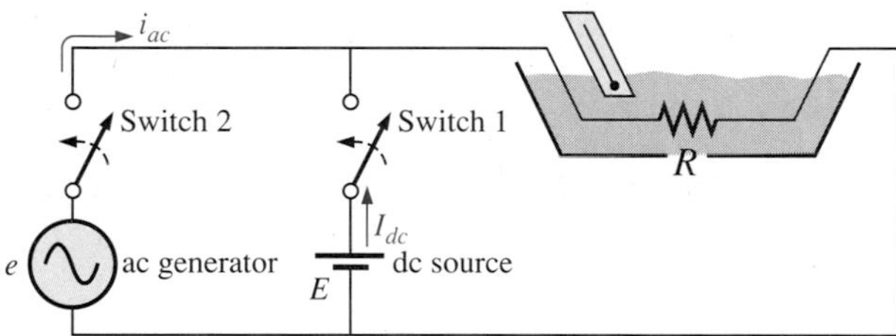

FIG. 13.52
An experimental setup to establish a relationship between dc and ac quantities.

If switch 2 is closed and switch 1 left open, the ac current through the resistor will have a peak value of I_m. The temperature reached by the water is now determined by the ac power dissipated in the form of heat by the resistor. The ac input is varied until the temperature is the same as that reached with the dc input. When this is accomplished, the average electrical power delivered to the resistor R by the ac source is the same as that delivered by the dc source.

The power delivered by the ac supply at any instant of time is

$$P_{\text{ac}} = (i_{\text{ac}})^2R = (I_m \sin \omega t)^2R = (I_m^2 \sin^2\omega t)R$$

but

$$\sin^2\omega t = \frac{1}{2}(1 - \cos 2\omega t) \qquad \text{(trigonometric identity)}$$

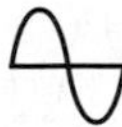

Therefore,

$$P_{ac} = I_m^2\left[\frac{1}{2}(1 - \cos 2\omega t)\right]R$$

and

$$P_{ac} = \frac{I_m^2 R}{2} - \frac{I_m^2 R}{2}\cos 2\omega t \tag{13.30}$$

The *average power* delivered by the ac source is just the first term, since the average value of a cosine wave is zero even though the wave may have twice the frequency of the original input current waveform. Equating the average power delivered by the ac generator to that delivered by the dc source,

$$P_{av(ac)} = P_{dc}$$

$$\frac{I_m^2 R}{2} = I_{dc}^2 R \quad \text{and} \quad I_m = \sqrt{2}I_{dc}$$

or

$$I_{dc} = \frac{I_m}{\sqrt{2}} = 0.707 I_m$$

which, in words, states that

the equivalent dc value of a sinusoidal current or voltage is 1/$\sqrt{2}$ or 0.707 of its maximum value.

The equivalent dc value is called the effective value of the sinusoidal quantity.

In summary,

$$I_{eq\ dc} = I_{eff} = 0.707 I_m \tag{13.31}$$

or

$$I_m = \sqrt{2}I_{eff} = 1.414 I_{eff} \tag{13.32}$$

and

$$E_{eff} = 0.707 E_m \tag{13.33}$$

or

$$E_m = \sqrt{2}E_{eff} = 1.414 E_{eff} \tag{13.34}$$

As a simple numerical example, it would require an ac current with a peak value of $\sqrt{2}(10) = 14.14$ A to deliver the same power to the resistor in Fig. 13.52 as a dc current of 10 A. The effective value of any quantity plotted as a function of time can be found by using the following equation derived from the experiment just described:

$$I_{eff} = \sqrt{\frac{\int_0^T i^2(t)\,dt}{T}} \tag{13.35}$$

or

$$I_{eff} = \sqrt{\frac{\text{area}\,(i^2(t))}{T}} \tag{13.36}$$

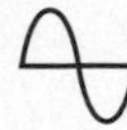

which, in words, states that to find the effective value, the function $i(t)$ must first be squared. After $i(t)$ is squared, the area under the curve is found by integration. It is then divided by T, the length of the cycle or period of the waveform, to obtain the average or *mean* value of the squared waveform. The final step is to take the *square root* of the mean value. This procedure gives us another designation for the effective value, the *root-mean-square* (rms) value.

EXAMPLE 13.19 Find the effective values of the sinusoidal waveform in each part of Fig. 13.53.

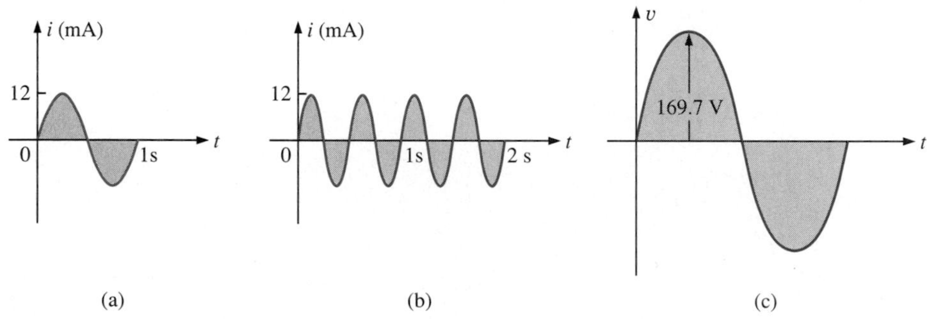

FIG. 13.53
Example 13.19.

Solution: For part (a), $I_{\text{eff}} = 0.707(12 \times 10^{-3}\text{ A}) = \mathbf{8.484\text{ mA}}$. For part (b), again $I_{\text{eff}} = \mathbf{8.484\text{ mA}}$. Note that frequency did not change the effective value in (b) above as compared to (a). For part (c), $V_{\text{eff}} = 0.707(169.73\text{ V}) \cong \mathbf{120\text{ V}}$, the same as available from a home outlet.

EXAMPLE 13.20 The 120-V dc source of Fig. 13.54(a) delivers 3.6 W to the load. Determine the peak value of the applied voltage (E_m) and the current (I_m) if the ac source (Fig. 13.54(b)) is to deliver the same power to the load.

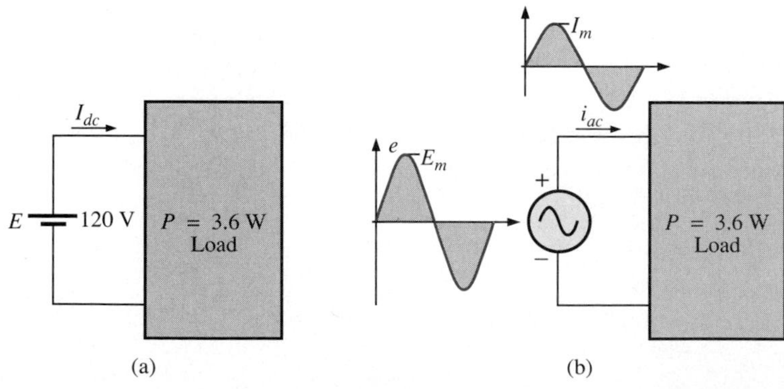

FIG. 13.54
Example 13.20.

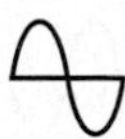

Solution:

$$P_{dc} = V_{dc}I_{dc}$$

and
$$I_{dc} = \frac{P_{dc}}{V_{dc}} = \frac{3.6\text{ W}}{120\text{ V}} = 30\text{ mA}$$
$$I_m = \sqrt{2}I_{dc} = (1.414)(30\text{ mA}) = \mathbf{42.42\text{ mA}}$$
$$E_m = \sqrt{2}E_{dc} = (1.414)(120\text{ V}) = \mathbf{169.68\text{ V}}$$

EXAMPLE 13.21 Find the effective or rms value of the waveform of Fig. 13.55.

Solution:
v^2 (Fig. 13.56):

$$V_{eff} = \sqrt{\frac{(9)(4) + (1)(4)}{8}} = \sqrt{\frac{40}{8}} = \mathbf{2.236\text{ V}}$$

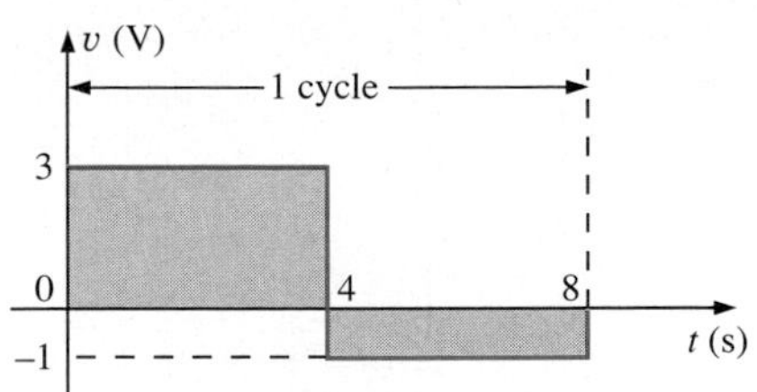

FIG. 13.55
Example 13.21.

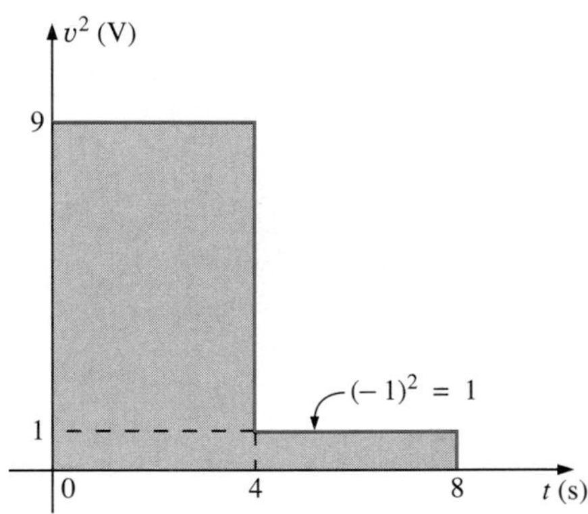

FIG. 13.56
The squared waveform of Fig. 13.55.

EXAMPLE 13.22 Calculate the effective value of the voltage of Fig. 13.57.

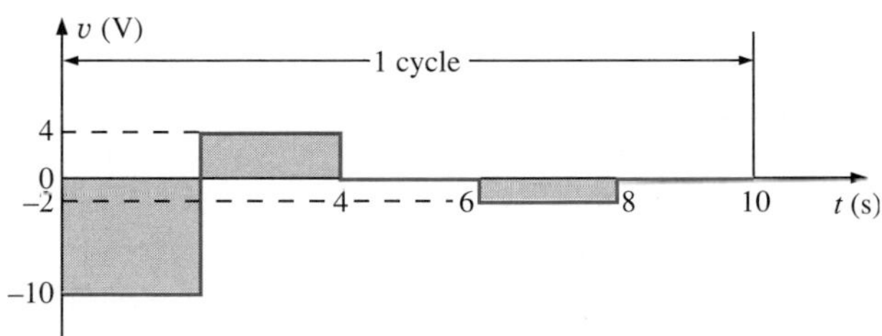

FIG. 13.57
Example 13.22.

Solution:
v^2 (Fig. 13.58):

$$V_{eff} = \sqrt{\frac{(100)(2) + (16)(2) + (4)(2)}{10}} = \sqrt{\frac{240}{10}}$$
$$= \mathbf{4.899\text{ V}}$$

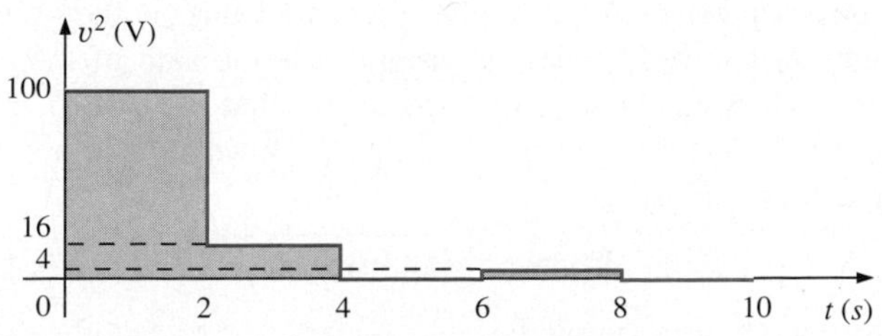

FIG. 13.58
The squared waveform of Fig. 13.57.

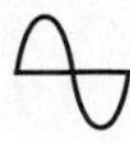

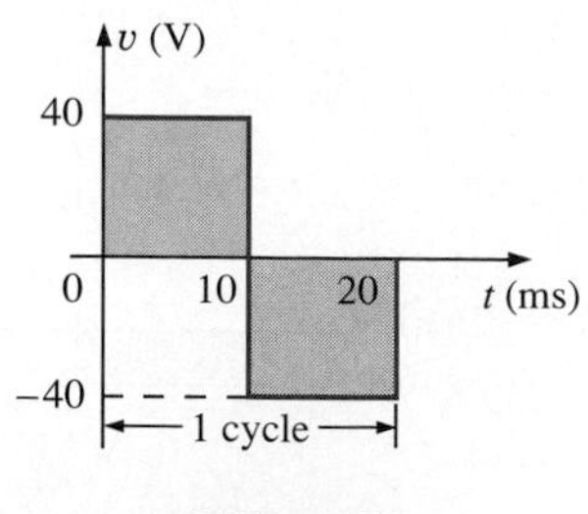

FIG. 13.59
Example 13.23.

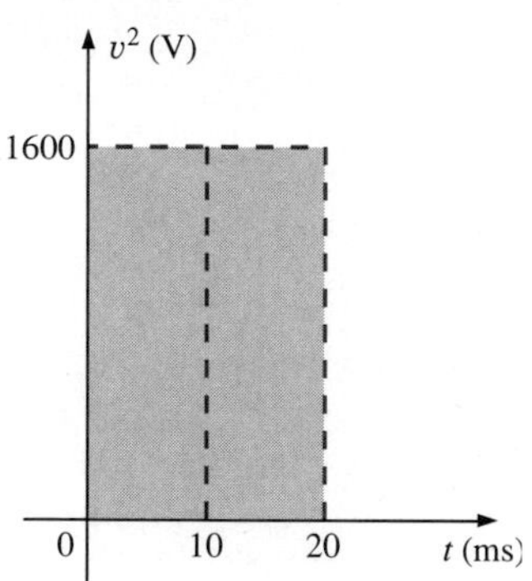

FIG. 13.60
The squared waveform of Fig. 13.59.

EXAMPLE 13.23 Determine the average and effective values of the square wave of Fig. 13.59.

Solution: By inspection, the average value is zero.

v^2 (Fig. 13.60):

$$V_{eff} = \sqrt{\frac{(1600)(10 \times 10^{-3}) + (1600)(10 \times 10^{-3})}{20 \times 10^{-3}}}$$

$$= \sqrt{\frac{32{,}000 \times 10^{-3}}{20 \times 10^{-3}}} = \sqrt{1600}$$

$$V_{eff} = \mathbf{40\ V}$$

(the maximum value of the waveform of Fig. 13.60)

The waveforms appearing in these examples are the same as those used in the examples on the average value. It might prove interesting to compare the effective and average values of these waveforms.

The effective values of sinusoidal quantities such as voltage or current will be represented by E and I. These symbols are the same as those used for dc voltages and currents. To avoid confusion, the peak value of a waveform will always have a subscript m associated with it: $I_m \sin \omega t$. *Caution:* When finding the effective value of the positive pulse of a sine wave, note that the squared area is *not* simply $(2A_m)^2 = 4A^2{}_m$; it must be found by a completely new integration. This will always be the case for any waveform that is not rectangular.

A unique situation arises if a waveform has both a dc and an ac component that may be due to a source such as the one in Fig. 13.61. The combination appears frequently in the analysis of electronic networks where both dc and ac levels are present in the same system.

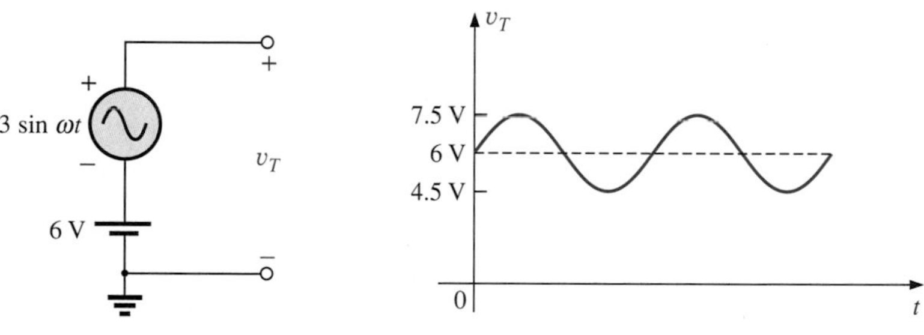

FIG. 13.61
Generation and display of a waveform having a dc and an ac component.

The question arises, What is the effective value of the voltage v_T? One might be tempted to simply assume it is the sum of the effective values of each component of the waveform; that is, $V_T(\text{eff}) = 0.7071$ (1.5 V) + 6 V = 1.06 V + 6 V = 7.06 V. However, the rms value is actually determined by

$$V_{eff} = \sqrt{V_{dc}^2 + V_{ac\ rms}^2} \tag{13.37}$$

which for the above example is

$$V_{eff} = \sqrt{(6\ \text{V})^2 + (1.06\ \text{V})^2}$$
$$= \sqrt{37.124}\ \text{V}$$
$$\cong \mathbf{6.1\ V}$$

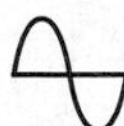

This result is noticeably less than the above solution. The development of Eq. (13.37) can be found in Chapter 24.

Instrumentation

It is important to note whether the DMM in use is a *true rms* meter or simply a meter where the average value is calibrated (as described in the next section) to indicate the rms level. A *true rms* meter will read the effective value of any waveform (such as Figs. 13.49 and 13.61) and is not limited to only sinusoidal waveforms. Since the label *true rms* is normally not placed on the face of the meter, it is prudent to check the manual if waveforms other than purely sinusoidal are to be encountered.

13.8 ac METERS AND INSTRUMENTS

The d'Arsonval movement employed in dc meters can also be used to measure sinusoidal voltages and currents if the *bridge rectifier* of Fig. 13.62 is placed between the signal to be measured and the average reading movement.

The bridge rectifier, composed of four diodes (electronic switches), will convert the input signal of zero average value to one having an average value sensitive to the peak value of the input signal. The conversion process is well described in most basic electronics texts. Fundamentally, conduction is permitted through the diodes in such a manner as to convert the sinusoidal input of Fig. 13.63(a) to one having the appearance of Fig. 13.63(b). The negative portion of the input has been effectively "flipped over" by the bridge configuration. The resulting waveform of Fig. 13.63(b) is called a *full-wave rectified waveform.*

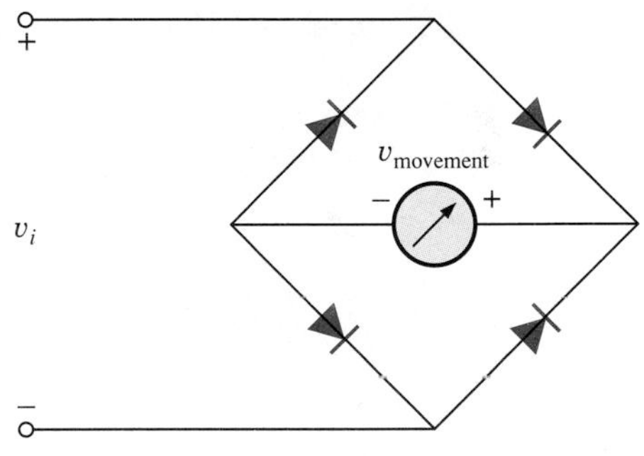

FIG. 13.62
Full-wave bridge rectifier.

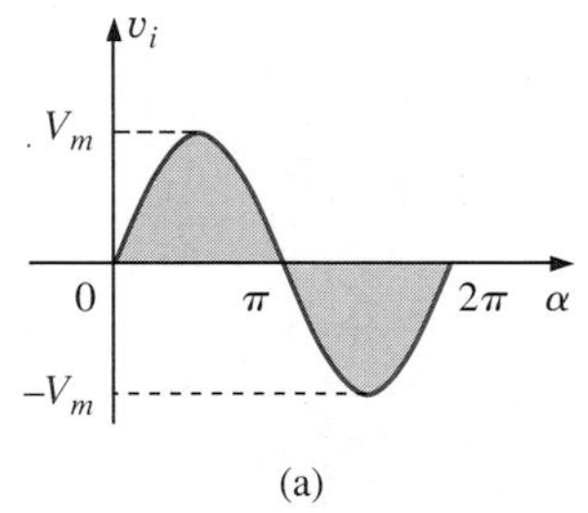

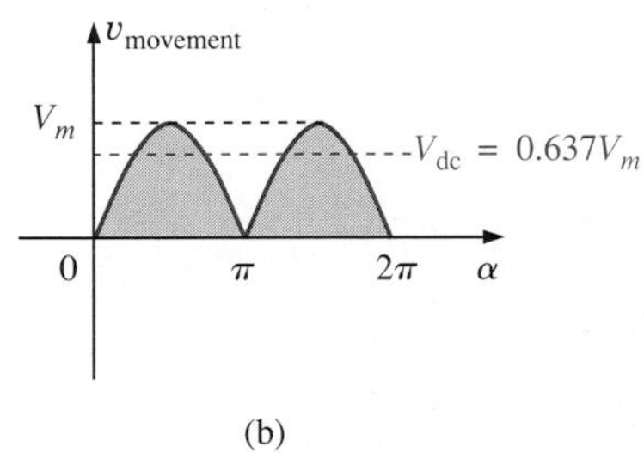

FIG. 13.63
(a) Sinusoidal input; (b) full-wave rectified signal.

The zero average value of Fig. 13.63(a) has been replaced by a pattern having an average value determined by

$$G = \frac{2V_m + 2V_m}{2\pi} = \frac{4V_m}{2\pi} = \frac{2V_m}{\pi} = 0.637V_m$$

The movement of the pointer will therefore be directly related to the peak value of the signal by the factor 0.637.

Forming the ratio between the rms and dc levels will result in

$$\frac{V_{rms}}{V_{dc}} = \frac{0.707V_m}{0.637V_m} \cong 1.11$$

revealing that the scale indication is 1.11 times the dc level measured by the movement; that is,

$$\text{Meter indication} = 1.11 \text{ (dc or average value)} \quad \textit{full-wave} \tag{13.38}$$

Some ac meters use a half-wave rectifier arrangement that results in the waveform of Fig. 13.64, which has half the average value of Fig. 13.63(b) over one full cycle. The result is

$$\text{Meter indication} = 2.22 \text{ (dc or average value)} \quad \textit{half-wave} \tag{13.39}$$

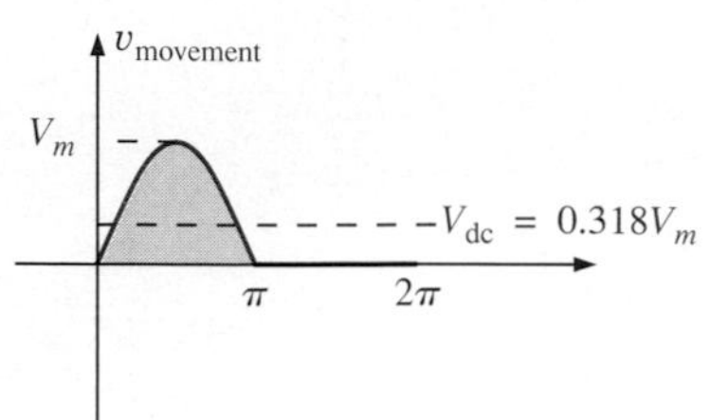

FIG. 13.64
Half-wave rectified signal.

A second movement, called the electrodynamometer movement (Fig. 13.65), can measure both ac and dc quantities without a change in internal circuitry. The movement can, in fact, read the effective value of any periodic or nonperiodic waveform because a reversal in current direction reverses the fields of both the stationary and the movable coils, so the deflection of the pointer is always up-scale.

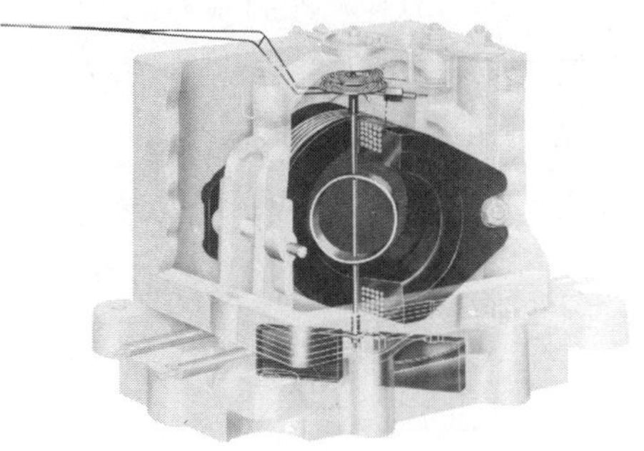

FIG. 13.65
Electrodynamometer movement. (Courtesy of Weston Instruments, Inc.)

The VOM, introduced in Chapter 2, can be used to measure both dc and ac voltages using a d'Arsonval movement and the proper switching networks. That is, when the meter is used for dc measurements, the dial setting will establish the proper series resistance for the chosen scale and permit the appropriate dc level to pass directly to the movement. For ac measurements, the dial setting will introduce a network that employs a full- or half-wave rectifier to establish a dc level. As discussed above, each setting is properly calibrated to indicate the desired quantity on the face of the instrument.

EXAMPLE 13.24 Determine the reading of each meter for each situation of Fig. 13.66.

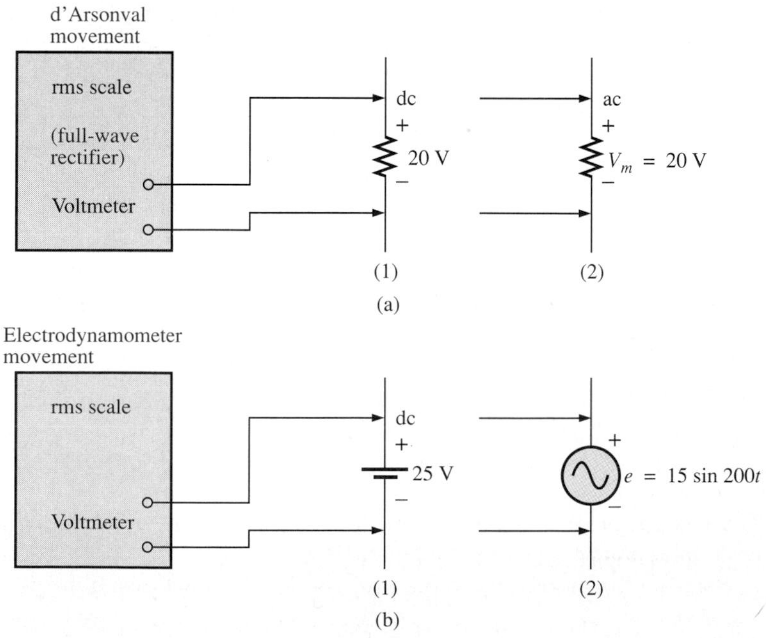

FIG. 13.66
Example 13.24.

Solution: For part (a), situation (1): By Eq. (13.38),

$$\text{Meter indication} = 1.11(20\ \text{V}) = \mathbf{22.2\ V}$$

For part (a), situation (2):

$$V_{\text{rms}} = 0.707V_m = 0.707(20\ \text{V}) = \mathbf{14.14\ V}$$

For part (b), situation (1):

$$V_{\text{rms}} = V_{\text{dc}} = \mathbf{25\ V}$$

For part (b), situation (2):

$$V_{\text{rms}} = 0.707V_m = 0.707(15\ \text{V}) \cong \mathbf{10.6\ V}$$

Most DMMs employ a full-wave rectification system to convert the input ac signal to one with an average value. In fact, for the DMM of Fig. 2.27, the same scale factor of Eq. (13.38) is employed; that is, the average value is scaled up by a factor of 1.11 to obtain the rms value. In the digital meters, however, there are no moving parts such as in the d'Arsonval or electrodynamometer movements to display the signal level. Rather, the average value is sensed by a multiprocessor integrated circuit (IC), which in turn determines which digits should appear on the digital display.

Digital meters can also be used to measure nonsinusoidal signals, but the scale factor of each input waveform must first be known (normally provided by the manufacturer in the operator's manual). For instance, the scale factor for an average responding DMM on the ac rms scale will produce an indication for a square-wave input that is 1.11 times the peak value. For a triangular input, the response is 0.555 times the peak value. Obviously, for a sine wave input, the response is 0.707 times the peak value.

For any instrument, it is always good practice to read (if only briefly) the operator's manual if it appears you will use the instrument on a regular basis.

For frequency measurements, the frequency counter of Fig. 13.67 provides a digital readout of sine, square, and triangular waves from 5 Hz to 100 MHz at input levels from 30 mV to 42 V. Note the relative simplicity of the panel and the high degree of accuracy available.

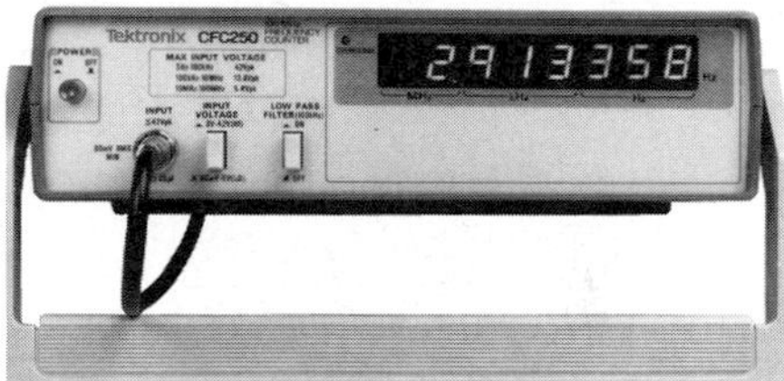

FIG. 13.67
Frequency counter. (Courtesy of Tektronix, Inc.)

The Amp-Clamp® of Fig. 13.68 is an instrument that can measure alternating current in the ampere range without having to open the circuit. The loop is opened by squeezing the "trigger"; then it is placed around the current-carrying conductor. Through transformer action, the level of current in rms units will appear on the appropriate scale. The accuracy of this instrument is $\pm 3\%$ of full scale at 60 Hz, and its scales have maximum values ranging from 6 A to 300 A. The addition of two leads, as indicated in the figure, permits its use as both a voltmeter and an ohmmeter.

FIG. 13.68
Amp-Clamp®. (Courtesy of Simpson Instruments, Inc.)

One of the most versatile and important instruments in the electronics industry is the oscilloscope, which has already been introduced in this chapter. It provides a display of the waveform on a cathode-ray tube to permit the detection of irregularities and the determination of quantities such as magnitude, frequency, period, dc component, and so on. The analog oscilloscope of Fig. 13.69 can display two waveforms at the same time (dual-channel) using an innovative interface (front panel). It employs menu buttons to set the vertical and horizontal scales by choosing from selections appearing on the screen. One can also store up to four measurement setups for future use.

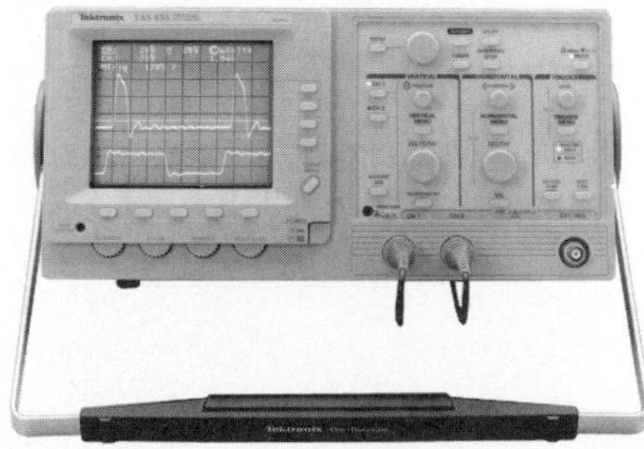

FIG. 13.69
Dual-channel oscilloscope. (Courtesy of Tektronix, Inc.)

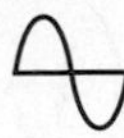

A student accustomed to watching TV might be confused when first introduced to an oscilloscope. There is, at least initially, an assumption that the oscilloscope is generating the waveform on the screen—much like a TV broadcast. However, it is important to clearly understand that

an oscilloscope displays only those signals generated elsewhere and connected to the input terminals of the oscilloscope. The absence of an external signal will simply result in a horizontal line on the screen of the scope.

On most modern-day oscilloscopes, there is a switch or knob with the choice DC/GND/AC, as shown in Fig. 13.70(a), that is often ignored or treated too lightly in the early stages of scope utilization. The effect of each position is fundamentally as shown in Fig. 13.70(b). In the DC mode the dc and ac components of the input signal can pass directly to the display. In the AC position the dc input is blocked by the capacitor, but the ac portion of the signal can pass through to the screen. In the GND position the input signal is prevented from reaching the scope display by a direct ground connection, which reduces the scope display to a single horizontal line.

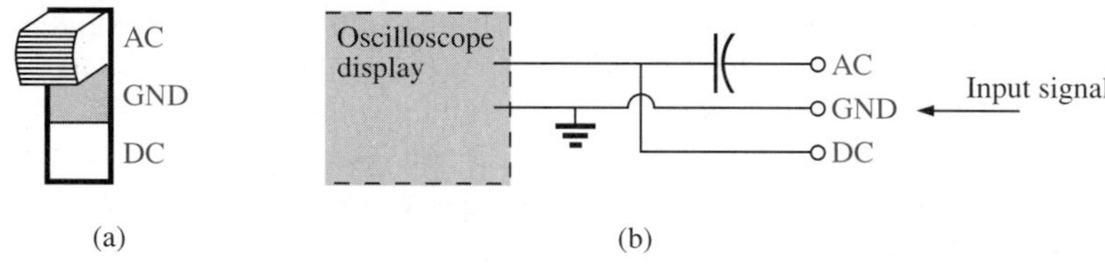

FIG. 13.70
AC-GND-DC switch for the vertical channel of an oscilloscope.

13.9 COMPUTER ANALYSIS

PSpice (DOS)

As with the analysis just described, there is a great deal of similarity between the application of PSpice to dc and ac systems. The basic format of the input file is similar with passive elements (*R*, *L*, and *C*) entered in exactly the same manner. The two elements that have the most impact on the command statements are the frequency or frequencies of interest and the phase angles applied and obtained, as demonstrated by the following commands.

Independent ac Sources The term *independent* (as described more than once in the text) simply means that the magnitude of the source is independent of any other parameters of the network. Since effective values are the most frequently applied and measured, the magnitude associated with all sinusoidal voltages and currents will be the effective value. The format for an independent ac voltage source is the following:

VSOURCE	1	0	AC	20V	0
Name	+ Node	− Node		Magnitude ACMAG	Phase angle ACPHASE

As with the commands for dc networks, the quantities with brackets are those controlled by the user. The others must appear as shown above in the same relative positions.

For a current source $\mathbf{I}_L = 10 \text{ mA } \angle 30°$ (whose arrow within the graphic symbol points from node 3 to 4), the following command would be included in the input file:

```
I2   3   4   AC   10MA   30
```

If the ACMAG entry is left blank, a default value of 1 V is applied, whereas the absence of an ACPHASE entry will result in a default value of zero degrees.

Note that the command statement for an independent ac source does not include the applied frequency. The effect of frequency is relegated to the next command.

PSpice (Windows)

Schematics offers a variety of ac voltage and current sources. However, for the purposes of this text, the voltage source **VSIN** and the current source **ISIN** are the most appropriate because the attributes to be defined include all of those of normal interest for sinusoidal networks. On occasion the **ISRC** source will be used because it has an arrow symbol like that appearing in this text. The symbol for **ISIN** is a sine wave with the direction of supply indicated by a plus or minus sign. The sources **VAC, VSRC,** and **ISRC** are fine if the magnitude and phase of a specific quantity are desired or if a transient plot is against frequency. However, they will not provide a transient response against time even if the frequency and transient information is provided under **Analysis.**

For all the sinusoidal sources, the magnitude provided is the peak value of the waveform and not its rms value. This will become clear when a plot of a quantity is desired and the magnitude calculated is the peak value in a transient response. However, for the **AC** response, the magnitude provided can be assumed to be the rms value and the output read as an rms value. Only when a plot is desired will the value define the peak value of the waveform. The phase angle is the same whether it is related to the peak value or rms value of the voltage or current. A number of default values are set by the program if values are not entered. If not specified, **DC** and **AC** values default to 0, and **transient** values default to the **DC** value. When using **VSIN** always specify **VOFF** at 0 V (unless a specific value is appropriate) and provide both the **AC** and the **VAMPL** values (at the same level) and provide the **PHASE** angle associated with the source. The **TD** (time delay), **DF** (damping factor), and **DC** value will all default to 0 if not specified. Similar statements apply to **ISIN.** Additional information about the various types of sources available can be found in MicroSim Circuit Analysis User's Guide.

Each source can be obtained by the sequence **Draw-Get New Part-Add Part-Browse-source.slb,** followed by identification of the source desired with a click of the left side of the mouse. An **OK** and the source will appear on the screen. A left click will place the source, and a right click will end the process. To set the attributes of the source, double-click on the source and set the desired levels. For a source such as **VSIN** note that **TD, DF,** and **PHASE** are already set on their default values of 0. To change the assigned name on the

schematic, simply enable **Include System defined Attributes** and set **PKGREF** to the desired name. For each entry for the source, be sure to **Save Attr,** and if the attribute is to be displayed, choose **Change Display.** Note at the top of the dialog box how the name and value are defined. If both are to be displayed, choose **Both name and value.**

C++

The absence of any network configurations to analyze in this chapter severely limits the content with respect to packaged computer programs. However, the door is still wide open for the application of a language to write programs that can be helpful in the application of some of the concepts introduced in the chapter. In particular, let us examine the C++ program of Fig. 13.71, designed to calculate the average value of a pulse waveform having up to 5 different levels.

```
Heading [ //C++ Average Waveform Voltage Calculation
Preprocessor directive [ #include <iostream.h>          //needed for input/output

Body of program:
          main()
          {
Define form and name of variables:
                float Vave;                  //average value of waveform
                float Vlevel;                //voltage level during time Tlevel
                float VTsum = 0;             //used for adding voltage-time products
                float T = 0;                 //total waveform time
                float Tlevel;                //time duration of Vlevel
                int levels;                  //the number of levels in the waveform
                int count;                   //loop counter
Obtain # of levels:
                cout << "How many levels do you wish to enter (1..5) ? ";
                cin >> levels;               //get number of levels from user
Iterative for statement:
                for(count = 1; count <= levels; count++)     //begin loop
                {
                      cout << "\n";
                      cout << "Enter voltage level " << count << ": ";
                      cin >> Vlevel;       //get voltage from user
                      cout << "Enter time for level " << count << ": ";
                      cin >> Tlevel;       //get time from user
                      VTsum += Vlevel * Tlevel;      //add product to VTsum
                      T += Tlevel;         //add Tlevel to total waveform time
                }
Calculate Vave:
                Vave = VTsum / T;          //calculate average value
Display results:
                cout << "\n";
                cout << "The average value of the waveform is ";
                cout << Vave << " volts.\n";
          }
```

FIG. 13.71
C++ program designed to calculate the average value of a waveform with up to five positive or negative pulses.

The program begins with a heading and preprocessor directive. Recall that the *iostream.h* header file sets up the input–output path between the program and the disk operating system. Note that the *main* ( ) part of the program extends all the way down to the bottom, as identified by the braces { }. Within this region all the calculations will be performed, and the results will be displayed.

Within the *main* ( ) part of the program, all the variables to be employed in the calculations are defined as floating point (decimal values) or integer (whole numbers). The comments on the right

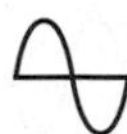

identify the identity of each variable. This is followed by a display of the question about how many levels will be encountered in the waveform using *cout* (comment out). The *cin* (comment in) statement permits a response from the user. Next, the loop statement *for* is employed to establish a fixed number of repetitions of the sequence appearing within the parentheses ( ) for a number of loops defined by the variable *levels*. The format of this for statement is such that the first entry within the parentheses ( ) is the initial value of the variable *count* (1 in this case), followed by a semicolon and then a test expression determining how many times the sequence to follow will be repeated. In other words, if *levels* is 5, then the first pass through the *for* statement will result in 1 being compared to 5, and the test expression will be satisfied because 5 is greater than or equal to 1(< =). On the next pass, *count* will be increased to 2 and the same test will be performed. Eventually *count* will equal 5, the test expression will not be satisfied, and the program will move to its next statement, which is *Vave* = *VT sum*/*T*. The last entry *count*++ of the *for* statement simply increments the variable *count* after each iteration. The first line within the *for* statement calls for a line to be skipped, followed by a question on the display about the level of voltage for the first time interval. The question will include the current state of the *count* variable followed by a colon. In C++ all character outputs must be displayed in quotes (not required for numerical values). However, note the absence of the quotes for *count* since it will be a numerical value. Next the user enters the first voltage level through *cin,* followed by a request for the time interval. In this case units are not provided but simply measured as an increment of the whole; that is, if the total period is 5 μs and the first interval is 2 μs, then just a 2 is entered.

The area under the pulse is then calculated to establish the variable *VTsum,* which was initially set at 0. On the next pass the value of *VTsum* will be the value obtained by the first run plus the new area. In other words, *VTsum* is a storage for the total accumulated area. Similarly, *T* is the accumulated sum of the time intervals.

Following a FALSE response from the test expression of the *for* statement, the program will move to calculate the average value of the waveform using the accumulated values of the area and time. A line is then skipped and the average value is displayed with the remaining *cout* statements. Brackets have been added along the edge of the program to help identify the various components of the program.

A program is now available that can find the average value of any pulse waveform having up to 5 positive or negative pulses. It can be placed in storage and simply called for when needed. Operations such as the above are not available in either form of PSpice or in any commercially available software package. It took the knowledge of a language and a few minutes of time to generate a short program of lifetime value.

Two runs will clearly reveal what will be displayed and how the output will appear. The waveform of Fig. 13.72 has 5 levels, entered as shown in the output file of Fig. 13.73. As indicated the average value is 1.6 volts. The waveform of Fig. 13.74 has only three pulses, and the time interval for each is different. Note the manner in which the time intervals were entered. Each is entered as a multiplier of the standard unit of measure for the horizontal axis. The variable *levels* will be only 3 requiring only three iterations of the *for* statement. The result is a negative value of −0.933 volt, as shown in the output file of Fig. 13.75.

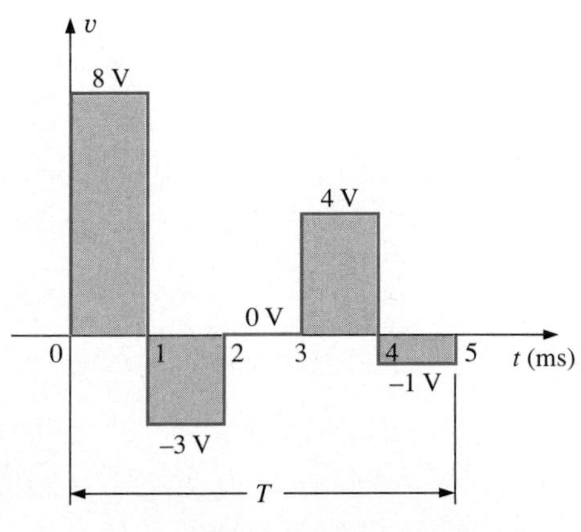

FIG. 13.72
Waveform with five pulses to be analyzed by the C++ program of Fig. 13.71.

```
How many levels do you wish to enter (1..5) ? 5

Enter voltage level 1: 8
Enter time for level 1: 1

Enter voltage level 2: -3
Enter time for level 2: 1

Enter voltage level 3: 0
Enter time for level 3: 1

Enter voltage level 4: 4
Enter time for level 4: 1

Enter voltage level 5: -1
Enter time for level 5: 1

The average value of the waveform is 1.6 volts.
```

FIG. 13.73
Output results for the waveform of Fig. 13.72.

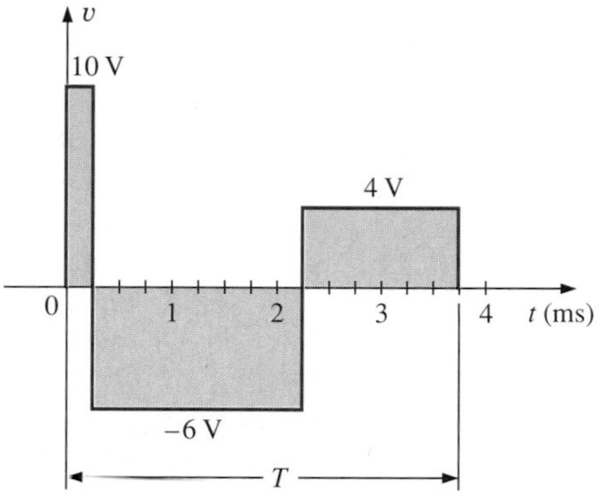

FIG. 13.74
Waveform with three pulses to be analyzed by the C++ program of Fig. 13.71.

```
How many levels do you wish to enter (1..5) ? 3

Enter voltage level 1: 10
Enter time for level 1: .25

Enter voltage level 2: -6
Enter time for level 2: 2

Enter voltage level 3: 4
Enter time for level 3: 1.5

The average value of the waveform is -0.933333 volts.
```

FIG. 13.75
Output results for the waveform of Fig. 13.74.

PROBLEMS

SECTION 13.2 Sinusoidal ac Voltage Characteristics and Definitions

1. For the periodic waveform of Fig. 13.76:
   a. Find the period T.
   b. How many cycles are shown?
   c. What is the frequency?
   ***d.** Determine the positive amplitude and peak-to-peak value (think!).

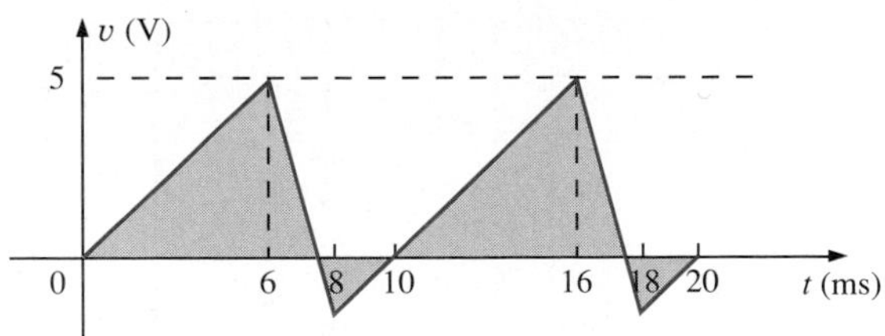

FIG. 13.76
Problem 1.

2. Repeat Problem 1 for the periodic waveform of Fig. 13.77.

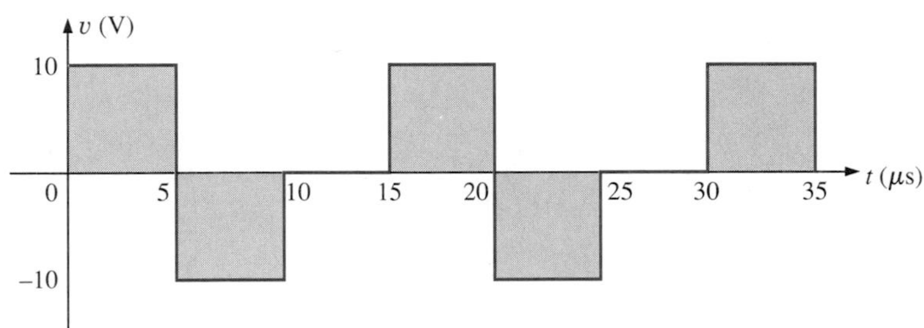

FIG. 13.77
Problems 2 and 47.

3. Determine the period and frequency of the sawtooth waveform of Fig. 13.78.

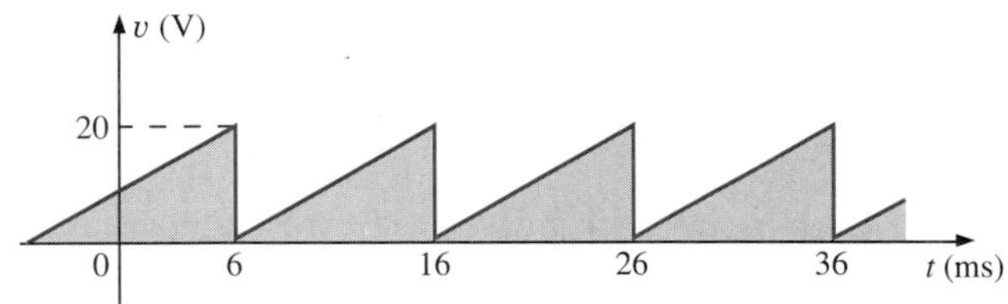

FIG. 13.78
Problems 3 and 48.

4. Find the period of a periodic waveform whose frequency is
   a. 25 Hz
   b. 35 MHz
   c. 55 kHz
   d. 1 Hz

5. Find the frequency of a repeating waveform whose period is
   a. 1/60 s
   b. 0.01 s
   c. 34 ms
   d. 25 μs

6. Find the period of a sinusoidal waveform that completes 80 cycles in 24 ms.

7. If a periodic waveform has a frequency of 20 Hz, how long (in seconds) will it take to complete 5 cycles?

8. What is the frequency of a periodic waveform that completes 42 cycles in 6 s?

9. Sketch a periodic square wave like that appearing in Fig. 13.77 with a frequency of 20,000 Hz and a peak value of 10 mV.

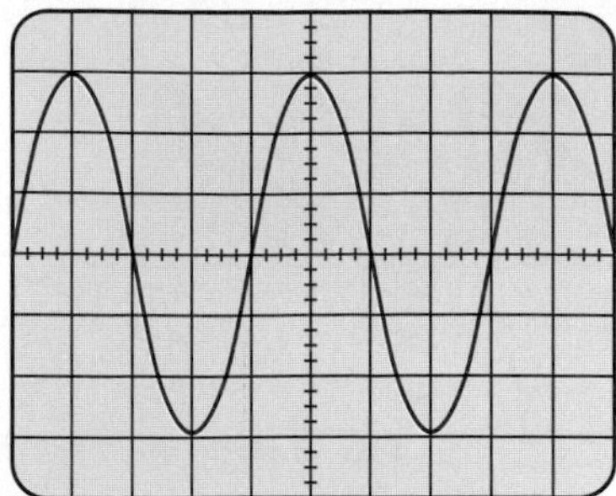

Vertical sensitivity = 50 mV/div.
Horizontal sensitivity = 10 μs/div.

FIG. 13.79
Problem 10.

10. For the oscilloscope pattern of Fig. 13.79:
  a. Determine the peak amplitude.
  b. Find the period.
  c. Calculate the frequency.
  Redraw the oscilloscope pattern if a +25-mV dc level were added to the input waveform.

SECTION 13.3 The Sine Wave

11. Convert the following degrees to radians:
  a. 45° **b.** 60°
  c. 120° **d.** 270°
  e. 178° **f.** 221°

12. Convert the following radians to degrees:
  a. $\pi/4$ **b.** $\pi/6$
  c. $\frac{1}{10}\pi$ **d.** $\frac{7}{6}\pi$
  e. 3π **f.** 0.55π

13. Find the angular velocity of a waveform with a period of
  a. 2 s **b.** 0.3 ms
  c. 4 μs **d.** $\frac{1}{25}$ s

14. Find the angular velocity of a waveform with a frequency of
  a. 50 Hz **b.** 600 Hz
  c. 2 kHz **d.** 0.004 MHz

15. Find the frequency and period of sine waves having an angular velocity of
  a. 754 rad/s **b.** 8.4 rad/s
  c. 6000 rad/s **d.** $\frac{1}{16}$ rad/s

16. Given $f = 60$ Hz, determine how long it will take the sinusoidal waveform to pass through an angle of 45°.

17. If a sinusoidal waveform passes through an angle of 30° in 5 ms, determine the angular velocity of the waveform.

SECTION 13.4 General Format for the Sinusoidal Voltage or Current

18. Find the amplitude and frequency of the following waves:
  a. $20 \sin 377t$ **b.** $5 \sin 754t$
  c. $10^6 \sin 10{,}000t$ **d.** $0.001 \sin 942t$
  e. $-7.6 \sin 43.6t$ **f.** $(\frac{1}{42}) \sin 6.283t$

19. Sketch $5 \sin 754t$ with the abscissa
  a. angle in degrees.
  b. angle in radians.
  c. time in seconds.

20. Sketch $10^6 \sin 10{,}000t$ with the abscissa
  a. angle in degrees.
  b. angle in radians.
  c. time in seconds.

21. Sketch $-7.6 \sin 43.6t$ with the abscissa
  a. angle in degrees.
  b. angle in radians.
  c. time in seconds.

22. If $e = 300 \sin 157t$, how long (in seconds) does it take this waveform to complete 1/2 cycle?

23. Given $i = 0.5 \sin \alpha$, determine i at $\alpha = 72°$.

24. Given $v = 20 \sin \alpha$, determine v at $\alpha = 1.2\pi$.

***25.** Given $v = 30 \times 10^{-3} \sin \alpha$, determine the angles at which v will be 6 mV.

***26.** If $v = 40$ V at $\alpha = 30°$ and $t = 1$ ms, determine the mathematical expression for the sinusoidal voltage.

SECTION 13.5 Phase Relations

27. Sketch $\sin(377t + 60°)$ with the abscissa
  a. angle in degrees.
  b. angle in radians.
  c. time in seconds.

28. Sketch the following waveforms:
  a. $50 \sin(\omega t + 0°)$ **b.** $-20 \sin(\omega t + 2°)$
  c. $5 \sin(\omega t + 60°)$ **d.** $4 \cos \omega t$
  e. $2 \cos(\omega t + 10°)$ **f.** $-5 \cos(\omega t + 20°)$

29. Find the phase relationship between the waveforms of each set:
  a. $v = 4 \sin(\omega t + 50°)$
  $i = 6 \sin(\omega t + 40°)$
  b. $v = 25 \sin(\omega t - 80°)$
  $i = 5 \times 10^{-3} \sin(\omega t - 10°)$
  c. $v = 0.2 \sin(\omega t - 60°)$
  $i = 0.1 \sin(\omega t + 20°)$
  d. $v = 200 \sin(\omega t - 210°)$
  $i = 25 \sin(\omega t - 60°)$

***30.** Repeat Problem 29 for the following sets:
  a. $v = 2 \cos(\omega t - 30°)$
  $i = 5 \sin(\omega t + 60°)$
  b. $v = -1 \sin(\omega t + 20°)$
  $i = 10 \sin(\omega t - 70°)$
  c. $v = -4 \cos(\omega t + 90°)$
  $i = -2 \sin(\omega t + 10°)$

31. Write the analytical expression for the waveforms of Fig. 13.80 with the phase angle in degrees.

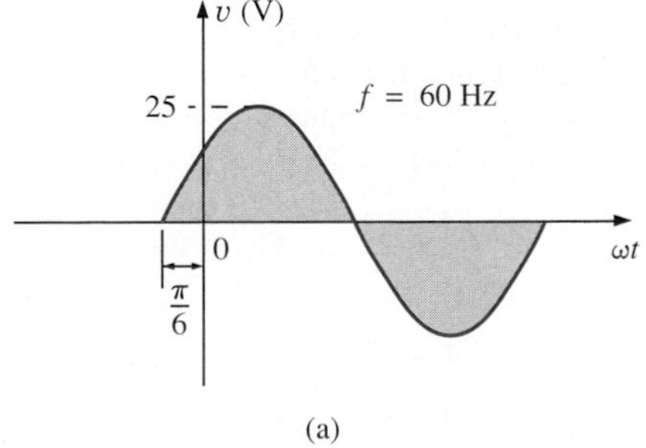

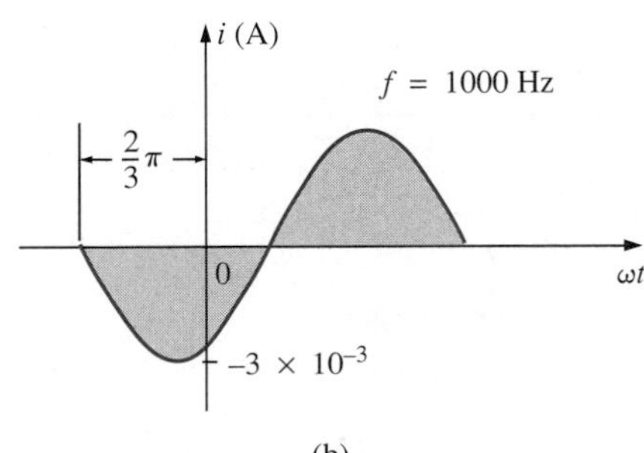

FIG. 13.80
Problem 31.

32. Repeat Problem 31 for the waveforms of Fig. 13.81.

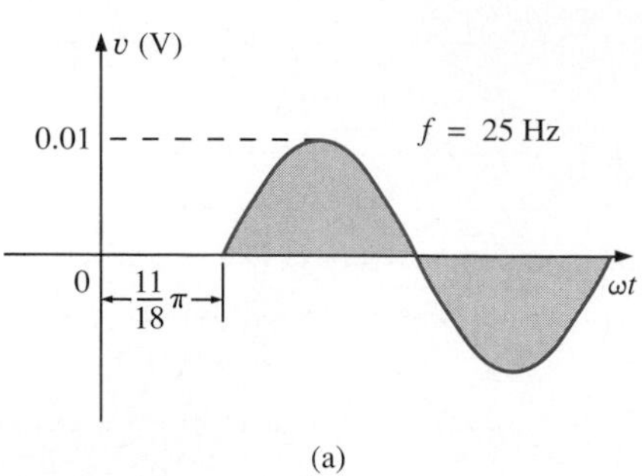

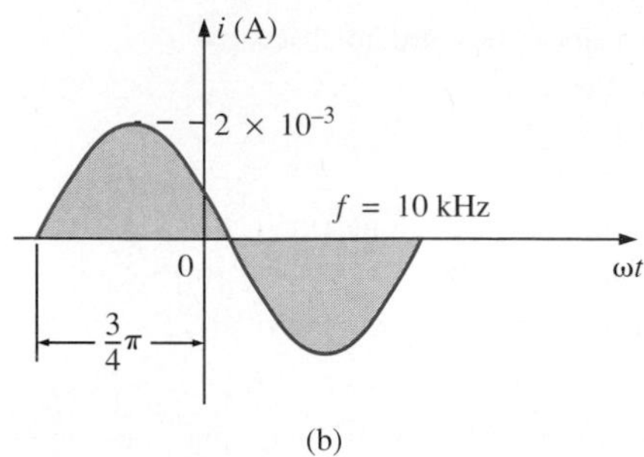

FIG. 13.81
Problem 32.

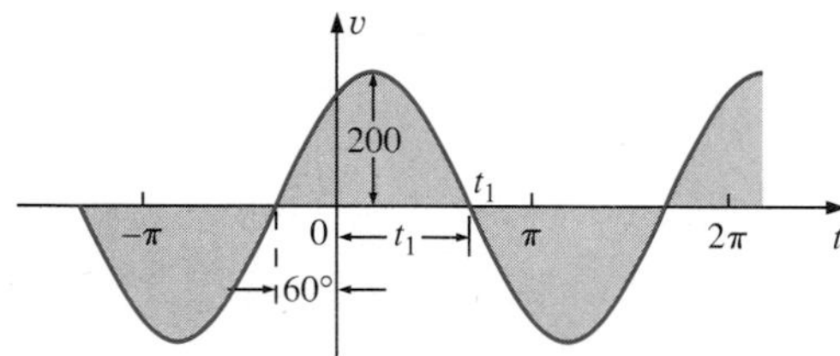

FIG. 13.82
Problem 33.

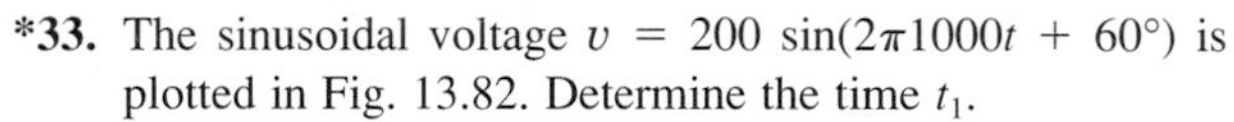

***33.** The sinusoidal voltage $v = 200 \sin(2\pi 1000t + 60°)$ is plotted in Fig. 13.82. Determine the time t_1.

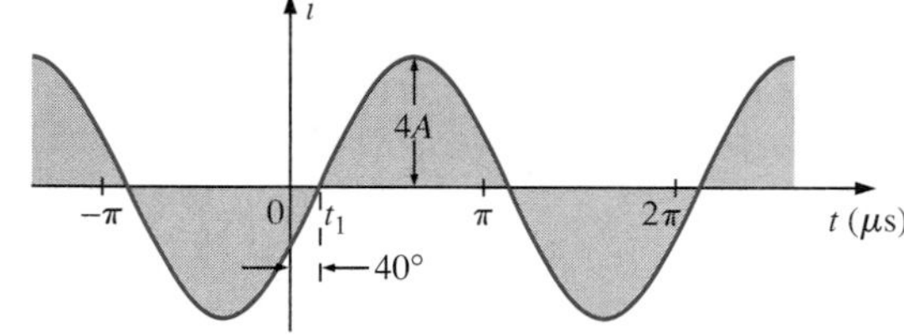

FIG. 13.83
Problem 34.

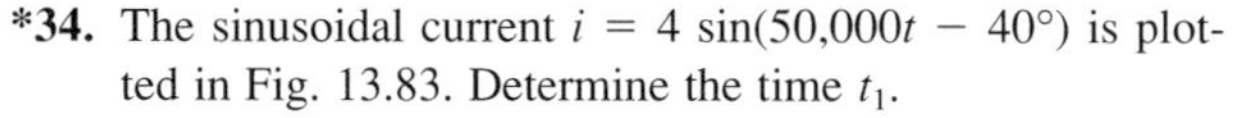

***34.** The sinusoidal current $i = 4 \sin(50{,}000t - 40°)$ is plotted in Fig. 13.83. Determine the time t_1.

***35.** Determine the phase delay in milliseconds between the following two waveforms:

$$v = 60 \sin(1800t + 20°)$$
$$i = 1.2 \sin(1800t - 20°)$$

36. For the oscilloscope display of Fig. 13.84:
a. Determine the period of each waveform.
b. Determine the frequency of each waveform.
c. Find the rms value of each waveform.
d. Determine the phase shift between the two waveforms and which leads or lags.

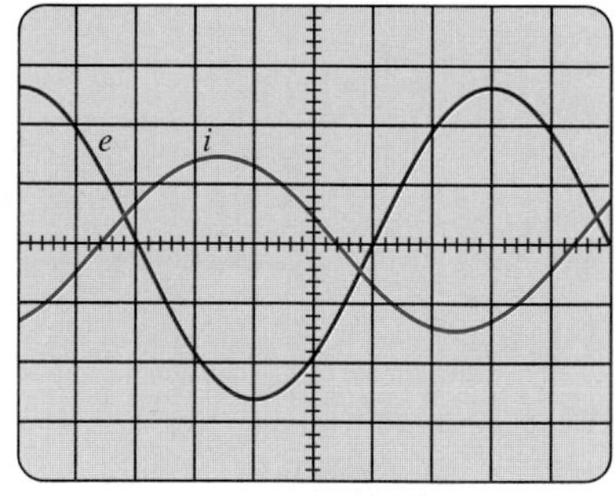

Vertical sensitivity = 0.5 V/div.
Horizontal sensitivity = 1 ms/div.

FIG. 13.84
Problem 36.

SECTION 13.6 Average Value

37. For the waveform of Fig. 13.85:

a. Determine the period.

b. Find the frequency.

c. Determine the average value.

d. Sketch the resulting oscilloscope display if the vertical channel is switched from DC to AC.

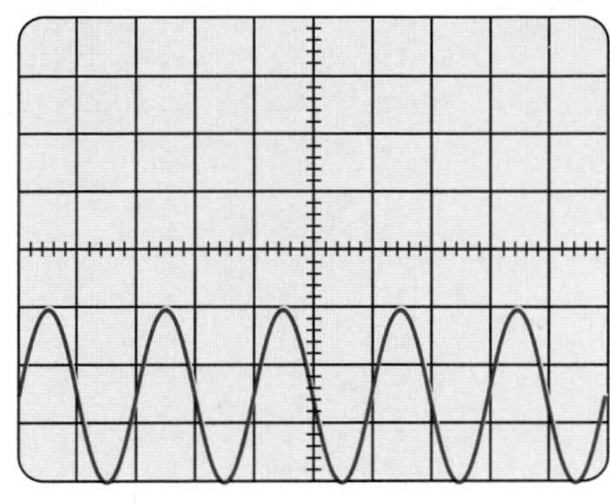

Vertical sensitivity = 10 mV/div.
Horizontal sensitivity = 0.2 ms/div.

FIG. 13.85
Problem 37.

38. Find the average value of the periodic waveforms of Fig. 13.86 over one full cycle.

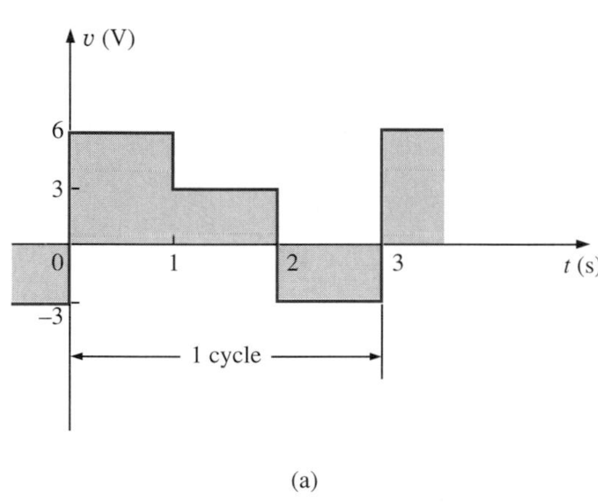

(a)

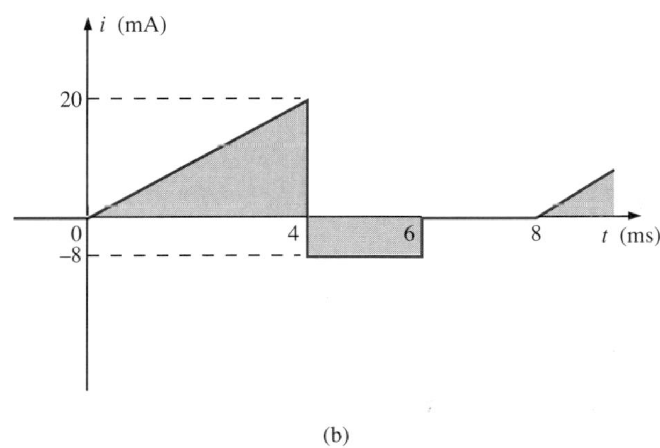

(b)

FIG. 13.86
Problem 38.

39. Find the average value of the periodic waveforms of Fig. 13.87 over one full cycle.

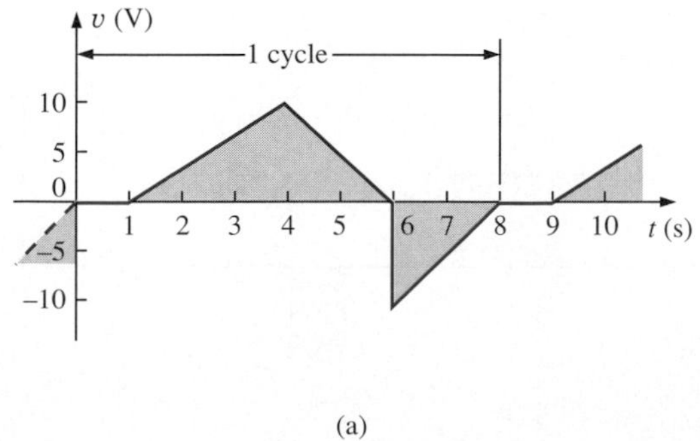

(a)

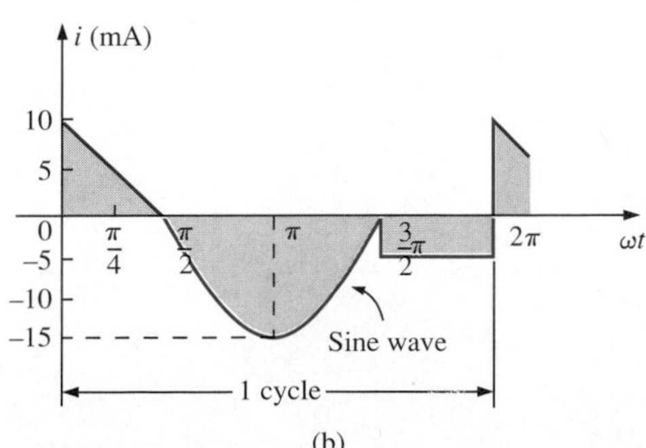

(b)

FIG. 13.87
Problem 39.

***40. a.** By the method of approximation, using familiar geometric shapes, find the area under the curve of Fig. 13.88 from zero to 10 s. Compare your solution with the actual area of 5 volt-seconds (V·s).

b. Find the average value of the waveform from zero to 10 s.

***41.** For the waveform of Fig. 13.89:

a. Determine the period.

b. Find the frequency.

c. Determine the average value.

d. Sketch the resulting oscilloscope display if the vertical channel is switched from DC to AC.

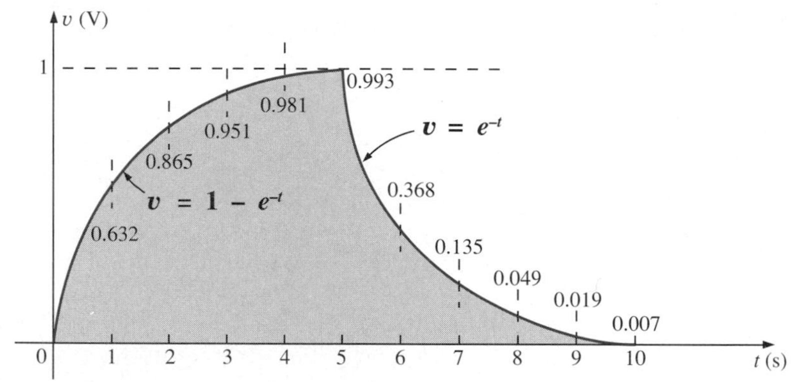

FIG. 13.88
Problem 40.

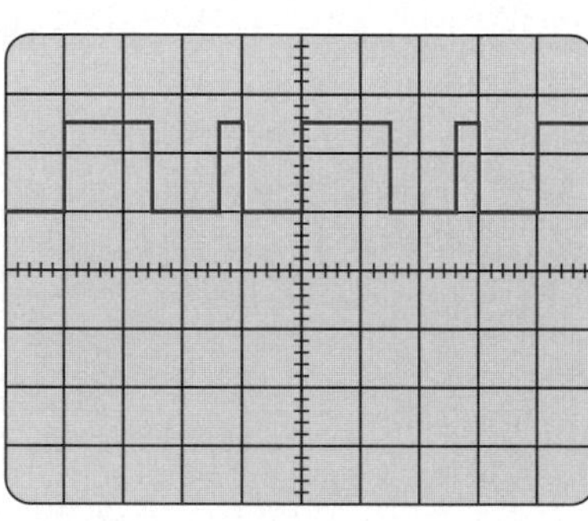

FIG. 13.89
Problem 41.

SECTION 13.7 Effective Values

42. Find the effective values of the following sinusoidal waveforms:

a. $v = 20 \sin 754t$

b. $v = 7.07 \sin 377t$

c. $i = 0.006 \sin(400t + 20°)$

d. $i = 16 \times 10^{-3} \sin(377t - 10°)$

43. Write the sinusoidal expressions for voltages and currents having the following effective values at a frequency of 60 Hz with zero phase shift:

a. 1.414 V **b.** 70.7 V

c. 0.06 A **d.** 24 μA

44. Find the effective value of the periodic waveform of Fig. 13.90 over one full cycle.

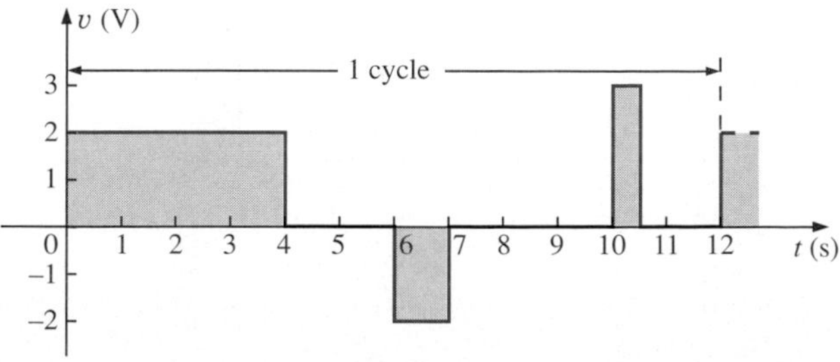

FIG. 13.90
Problem 44.

45. Find the effective value of the periodic waveform of Fig. 13.91 over one full cycle.

46. What are the average and effective values of the square wave of Fig. 13.92?

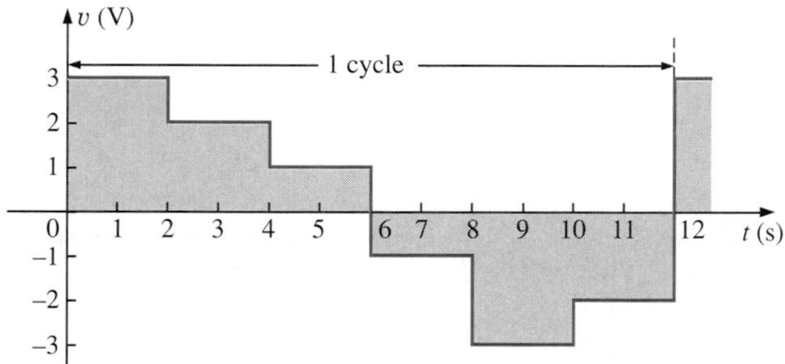

FIG. 13.91
Problem 45.

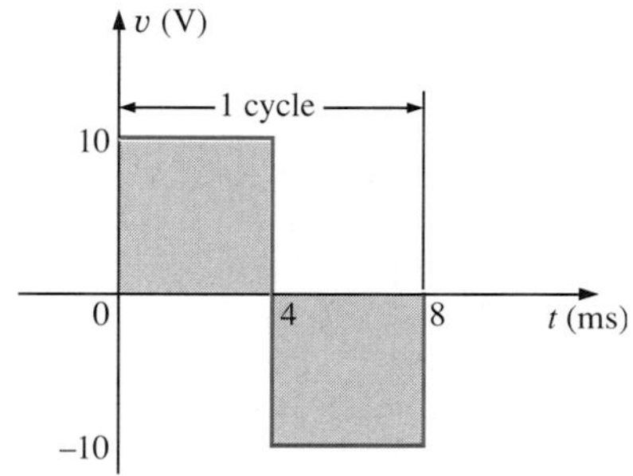

FIG. 13.92
Problem 46.

47. What are the average and effective values of the waveform of Fig. 13.77?

48. What is the average value of the waveform of Fig. 13.78?

49. For each waveform of Fig. 13.93 determine the period, frequency, average value, and effective value.

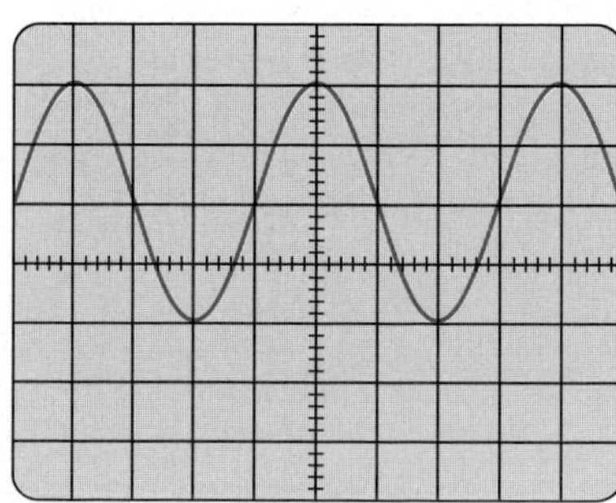

Vertical sensitivity = 20 mV/div.
Horizontal sensitivity = 10 μs/div.
(a)

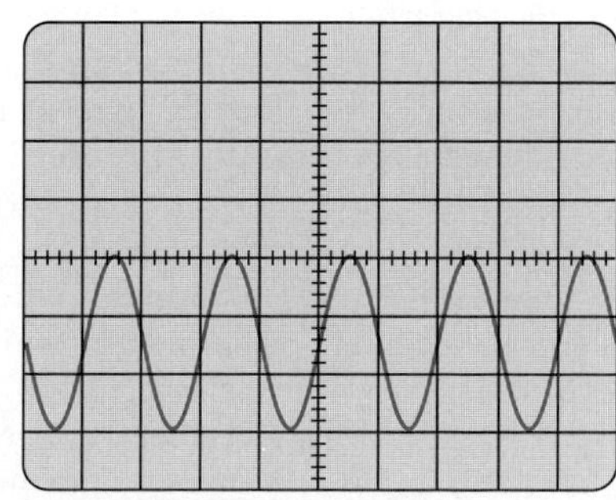

Vertical sensitivity = 0.2 V/div.
Horizontal sensitivity = 50 μs/div.
(b)

FIG. 13.93
Problem 49.

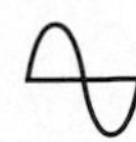

SECTION 13.8 ac Meters and Instruments

50. Determine the reading of the meter for each situation of Fig. 13.94.

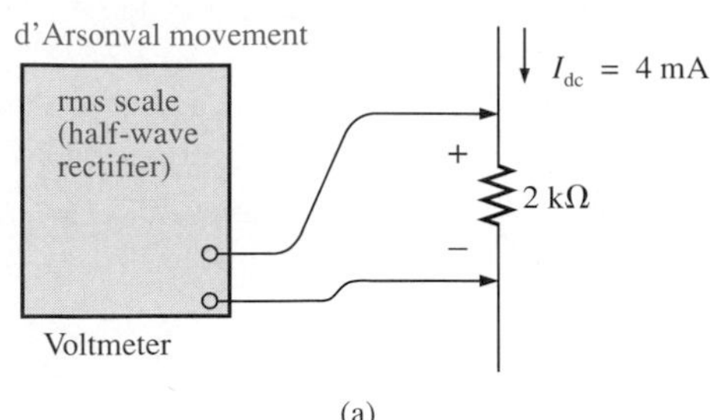

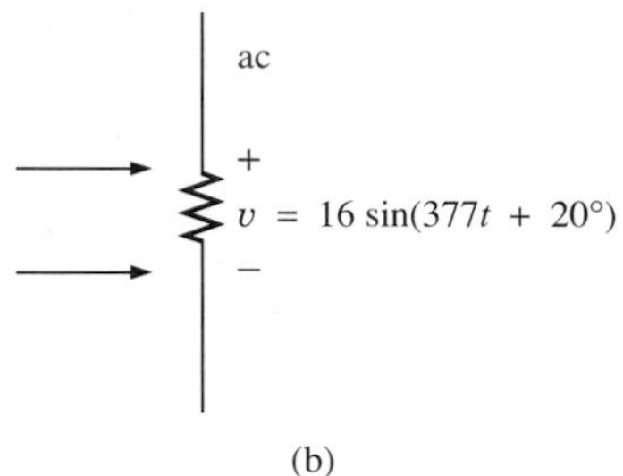

FIG. 13.94
Problem 50.

COMPUTER PROBLEMS Programming Language (C++, BASIC, PASCAL, etc.)

51. Given a sinusoidal function, write a program to determine the effective value, frequency, and period.

52. Given two sinusoidal functions, write a program to determine the phase shift between the two waveforms, and indicate which is leading or lagging.

53. Given an alternating pulse waveform, write a program to determine the average and effective value of the waveform over one complete cycle.

GLOSSARY

Alternating waveform A waveform that oscillates above and below a defined reference level.

Amp-Clamp® A clamp-type instrument that will permit noninvasive current measurements and that can be used as a conventional voltmeter or ohmmeter.

Angular velocity The velocity with which a radius vector projecting a sinusoidal function rotates about its center.

Average value The level of a waveform defined by the condition that the area enclosed by the curve above this level is exactly equal to the area enclosed by the curve below this level.

Cycle A portion of a waveform contained in one period of time.

Effective value The equivalent dc value of any alternating voltage or current.

Electrodynamometer meters Instruments that can measure both ac and dc quantities without a change in internal circuitry.

Frequency (f) The number of cycles of a periodic waveform that occur in one second.

Frequency counter An instrument that will provide a digital display of the frequency or period of a periodic time-varying signal.

Instantaneous value The magnitude of a waveform at any instant of time, denoted by lowercase letters.

Oscilloscope An instrument that will display, through the use of a cathode-ray tube, the characteristics of a time-varying signal.

Peak-to-peak value The magnitude of the total swing of a signal from positive to negative peaks. The sum of the absolute values of the positive and negative peak values.

Peak value The maximum value of a waveform, denoted by uppercase letters.

Period (T) The time interval between successive repetitions of a periodic waveform.

Periodic waveform A waveform that continually repeats itself after a defined time interval.

Phase relationship An indication of which of two waveforms leads or lags the other, and by how many degrees or radians.

Radian A unit of measure used to define a particular segment of a circle. One radian is approximately equal to 57.3°; 2π rad are equal to 360°.

Rectifier-type ac meter An instrument calibrated to indicate the effective value of a current or voltage through the use of a rectifier network and d'Arsonval-type movement.

rms value The root-mean-square or effective value of a waveform.

Sinusoidal ac waveform An alternating waveform of unique characteristics that oscillates with equal amplitude above and below a given axis.

VOM A multimeter with the capability to measure resistance and both ac and dc levels of current and voltage.

Waveform The path traced by a quantity, plotted as a function of some variable such as position, time, degrees, temperature, and so on.

14

The Basic Elements and Phasors

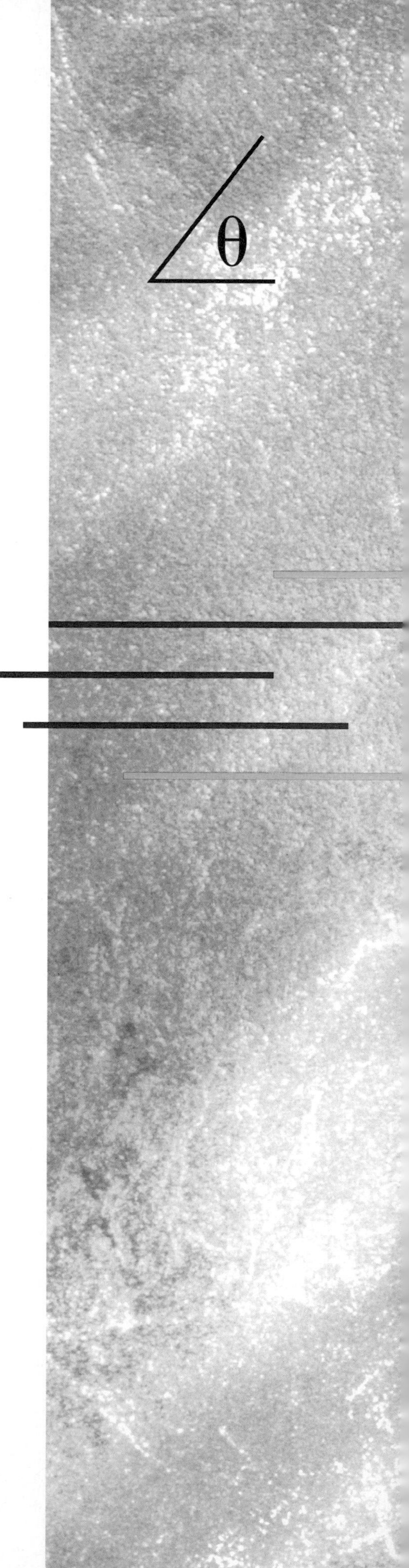

14.1 INTRODUCTION

The response of the basic R, L, and C elements to a sinusoidal voltage and current will be examined in this chapter, with special note of how frequency will affect the "opposing" characteristic of each element. Phasor notation will then be introduced to establish a method of analysis that permits a direct correspondence with a number of the methods, theorems, and concepts introduced in the dc chapters.

14.2 THE DERIVATIVE

It is fundamental to the understanding of the response of the basic R, L, and C elements to a sinusoidal signal that the concept of the *derivative* be examined in some detail. It will not be necessary that you become proficient in the mathematical technique, but simply that you understand the impact of a relationship defined by a derivative.

Recall from Section 10.11 that the derivative dx/dt is defined as the rate of change of x with respect to time. If x fails to change at a particular instant, $dx = 0$, and the derivative is zero. For the sinusoidal waveform, dx/dt is zero only at the positive and negative peaks ($\omega t = \pi/2$ and $\frac{3}{2}\pi$ in Fig. 14.1), since x fails to change at these instants of time. The derivative dx/dt is actually the slope of the graph at any instant of time.

A close examination of the sinusoidal waveform will also indicate that the greatest change in x will occur at the instants $\omega t = 0$, π, and 2π. The derivative is therefore a maximum at these points. At 0 and 2π, x increases at its greatest rate, and the derivative is given a positive sign since x increases with time. At π, dx/dt decreases at the same rate as it increases at 0 and 2π, but the derivative is given a negative sign since x decreases with time. Since the rate of change at 0, π, and 2π is the same, the magnitude of the derivative at these points is the same also. For various values of ωt between these maxima and minima, the derivative will exist and have values from the minimum to the maximum inclusive. A plot of the derivative in Fig. 14.2 shows that

the derivative of a sine wave is a cosine wave.

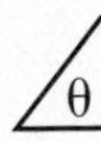

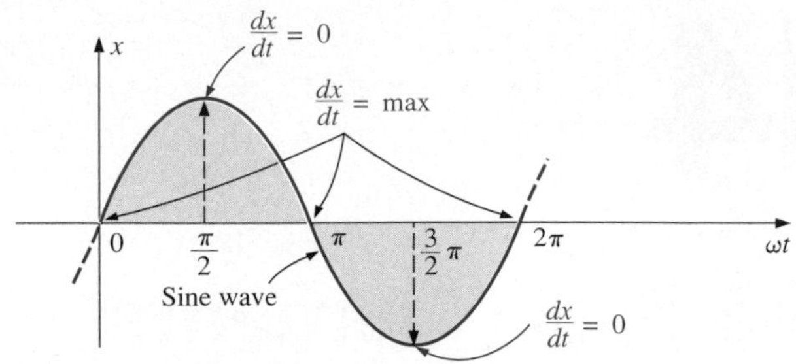

FIG. 14.1
Defining those points in a sinusoidal waveform that have maximum and minimum derivatives.

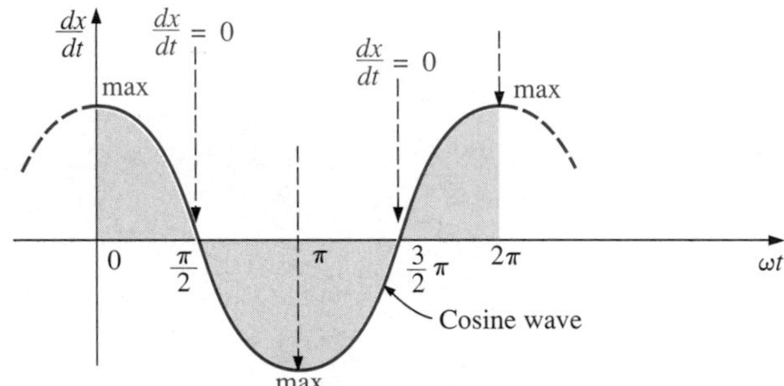

FIG. 14.2
Derivative of the sine wave of Fig. 14.1.

The peak value of the cosine wave is directly related to the frequency of the original waveform. The higher the frequency, the steeper the slope at the horizontal axis and the greater the value of dx/dt, as shown in Fig. 14.3 for two different frequencies.

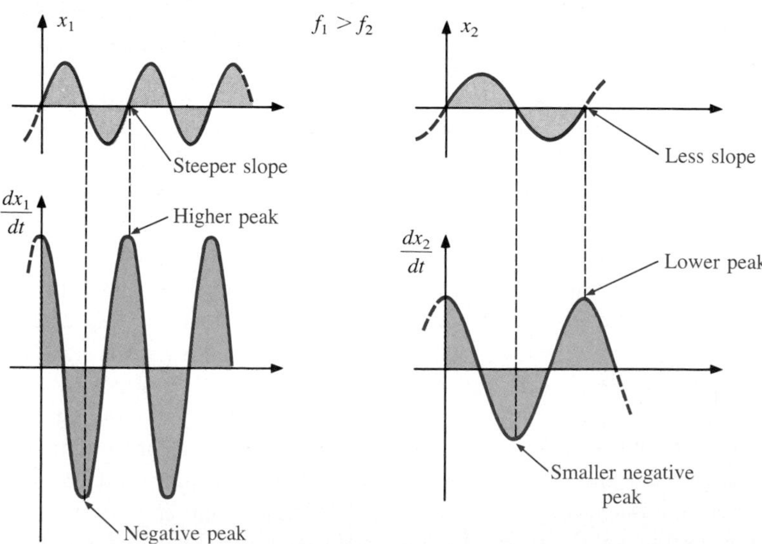

FIG. 14.3
Effect of frequency on the peak value of the derivative.

Note in Fig. 14.3 that even though both waveforms (x_1 and x_2) have the same peak value, the sinusoidal function with the higher frequency produces the larger peak value for the derivative. In addition, note that

the derivative of a sine wave has the same period and frequency as the original sinusoidal waveform.

For the sinusoidal voltage

$$e(t) = E_m \sin(\omega t \pm \theta)$$

the derivative can be found directly by differentiation (calculus) to produce the following:

$$\begin{aligned} \frac{d}{dt} e(t) &= \omega E_m \cos(\omega t \pm \theta) \\ &= 2\pi f E_m \cos(\omega t \pm \theta) \end{aligned} \tag{14.1}$$

The mechanics of the differentiation process will not be discussed or investigated here, nor will they be required to continue with the text. Note, however, that the peak value of the derivative, $2\pi f E_m$, is a function of the frequency of $e(t)$ and the derivative of a sine wave is a cosine wave.

14.3 RESPONSE OF BASIC *R, L,* AND *C* ELEMENTS TO A SINUSOIDAL VOLTAGE OR CURRENT

Now that we are familiar with the characteristics of the derivative of a sinusoidal function, we can investigate the response of the basic elements R, L, and C to a sinusoidal voltage or current.

Resistor

For power-line frequencies and frequencies up to a few hundred kilohertz, resistance is, for all practical purposes, unaffected by the frequency of the applied sinusoidal voltage or current. For this frequency region, the resistor R of Fig. 14.4 can be treated as a constant, and Ohm's law can be applied as follows. For $\upsilon = V_m \sin \omega t$,

$$i = \frac{\upsilon}{R} = \frac{V_m \sin \omega t}{R} = \frac{V_m}{R} \sin \omega t = I_m \sin \omega t$$

where

$$I_m = \frac{V_m}{R} \tag{14.2}$$

In addition, for a given i,

$$\upsilon = iR = (I_m \sin \omega t)R = I_m R \sin \omega t = V_m \sin \omega t$$

where

$$V_m = I_m R \tag{14.3}$$

A plot of υ and i in Fig. 14.5 reveals that

for a purely resistive element, the voltage across and the current through the element are in phase with their peak values related by Ohm's law.

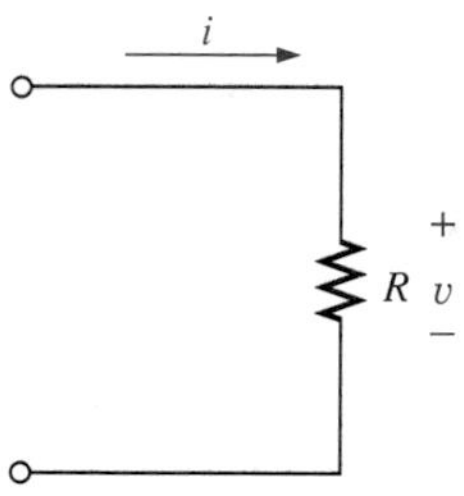

FIG. 14.4
Determining the sinusoidal response for a resistive element.

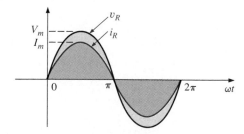

FIG. 14.5
The voltage and current of a resistive element are in phase.

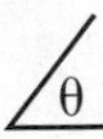

Inductor

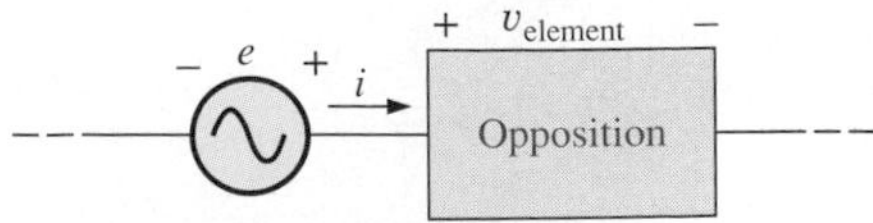

FIG. 14.6
Defining the opposition of an element to the flow of charge through the element.

For the series configuration of Fig. 14.6, the voltage v_{element} of the boxed-in element opposes the source e and thereby reduces the magnitude of the current i. The magnitude of the voltage across the element is determined by the opposition of the element to the flow of charge, or current i. For a resistive element, we have found that the opposition is its resistance and that v_{element} and i are determined by $v_{\text{element}} = iR$.

For the inductor, we found in Chapter 12 that the voltage across an inductor is directly related to the rate of change of current through the coil. Consequently, the higher the frequency, the greater will be the rate of change of current through the coil, and the greater the magnitude of the voltage. In addition, we found in the same chapter that the inductance of a coil will determine the rate of change of the flux linking a coil for a particular change in current through the coil. The higher the inductance, the greater the rate of change of the flux linkages, and the greater the resulting voltage across the coil.

The inductive voltage, therefore, is directly related to the frequency (or, more specifically, the angular velocity of the sinusoidal ac current through the coil) and the inductance of the coil. For increasing values of f and L in Fig. 14.7, the magnitude of v_L will increase as described above.

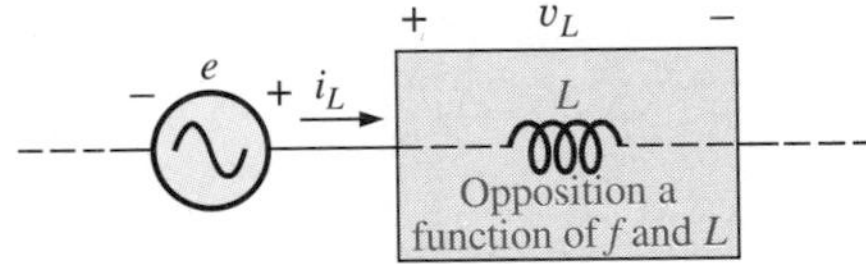

FIG. 14.7
Defining the parameters that determine the opposition of an inductive element to the flow of charge.

Utilizing the similarities between Figs. 14.6 and 14.7, we find that increasing levels of v_L are directly related to increasing levels of opposition in Fig. 14.6. Since v_L will increase with both ω $(= 2\pi f)$ and L, the opposition of an inductive element is as defined in Fig. 14.7.

We will now verify some of the preceding conclusions using a more mathematical approach and then define a few important quantities to be employed in the sections and chapters to follow.

For the inductor of Fig. 14.8, we recall from Chapter 12 that

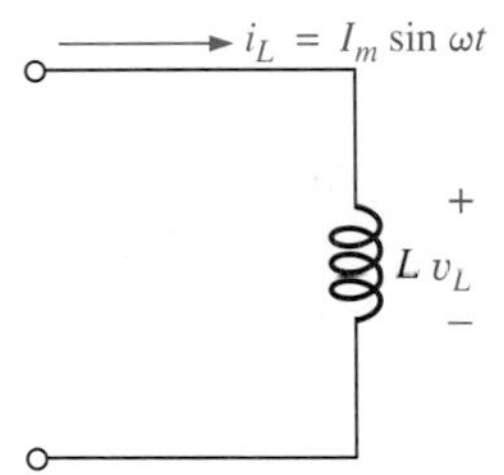

FIG. 14.8
Investigating the sinusoidal response of an inductive element.

$$v_L = L\frac{di_L}{dt}$$

and, applying differentiation,

$$\frac{di_L}{dt} = \frac{d}{dt}(I_m \sin \omega t) = \omega I_m \cos \omega t$$

Therefore, $$v_L = L\frac{di_L}{dt} = L(\omega I_m \cos \omega t) = \omega L I_m \cos \omega t$$

or $$v_L = V_m \sin(\omega t + 90^\circ)$$

where $$V_m = \omega L I_m$$

Note that the peak value of v_L is directly related to ω $(= 2\pi f)$ and L as predicted in the discussion above.

A plot of v_L and i_L in Fig. 14.9 reveals that

for an inductor, v_L leads i_L by 90°, or i_L lags v_L by 90°.

If a phase angle is included in the sinusoidal expression for i_L, such as

$$i_L = I_m \sin(\omega t \pm \theta)$$

then $$v_L = \omega L I_m \sin(\omega t \pm \theta + 90^\circ)$$

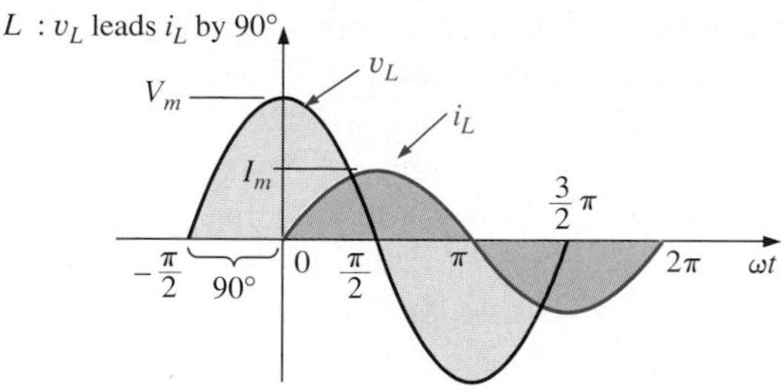

FIG. 14.9
For a pure inductor, the voltage across the coil leads the current through the coil by 90°.

The opposition established by an inductor in a sinusoidal ac network can now be found by applying Eq. (4.1):

$$\text{Effect} = \frac{\text{cause}}{\text{opposition}}$$

which, for our purposes, can be written

$$\text{Opposition} = \frac{\text{cause}}{\text{effect}}$$

Substituting values, we have

$$\text{Opposition} = \frac{V_m}{I_m} = \frac{\omega L I_m}{I_m} = \omega L$$

revealing that the opposition established by an inductor in an ac sinusoidal network is directly related to the product of the angular velocity ($\omega = 2\pi f$) and the inductance, verifying our earlier conclusions.

The quantity ωL, called the *reactance* (from the word *reaction*) of an inductor, is symbolically represented by X_L and is measured in ohms; that is,

$$\boxed{X_L = \omega L} \qquad \text{(ohms, } \Omega) \qquad \textbf{(14.4)}$$

In an Ohm's law format, its magnitude can be determined from

$$\boxed{X_L = \frac{V_m}{I_m}} \qquad \text{(ohms, } \Omega) \qquad \textbf{(14.5)}$$

Inductive reactance is the opposition to the flow of current, which results in the continual interchange of energy between the source and the magnetic field of the inductor. In other words, inductive reactance, unlike resistance (which dissipates energy in the form of heat), does not dissipate electrical energy (ignoring the effects of the internal resistance of the inductor).

Capacitor

Let us now return to the series configuration of Fig. 14.6 and insert the capacitor as the element of interest. For the capacitor, however, we will determine i for a particular voltage across the element. When this approach reaches its conclusion, the relationship between the voltage and current will be known, and the opposing voltage (v_{element}) can be determined for any sinusoidal current i.

Our investigation of the inductor revealed that the inductive voltage across a coil opposes the instantaneous change in current through the coil. For capacitive networks, the voltage across the capacitor is limited by the rate at which charge can be deposited on, or released by, the plates of the capacitor during the charging and discharging phases, respectively. In other words, an instantaneous change in voltage across a capacitor is opposed by the fact that there is an element of time required to deposit charge on (or release charge from) the plates of a capacitor, and $V = Q/C$.

Since capacitance is a measure of the rate at which a capacitor will store charge on its plates,

for a particular change in voltage across the capacitor, the greater the value of capacitance, the greater will be the resulting capacitive current.

In addition, the fundamental equation relating the voltage across a capacitor to the current of a capacitor [$i = C(dv/dt)$] indicates that

for a particular capacitance, the greater the rate of change of voltage across the capacitor, the greater the capacitive current.

Certainly, an increase in frequency corresponds to an increase in the rate of change of voltage across the capacitor and to an increase in the current of the capacitor.

The current of a capacitor is therefore directly related to the frequency (or, again more specifically, the angular velocity) and the capacitance of the capacitor. An increase in either quantity will result in an increase in the current of the capacitor. For the basic configuration of Fig. 14.10, however, we are interested in determining the opposition of the capacitor as related to the resistance of a resistor and ωL for the inductor. Since an increase in current corresponds to a decrease in opposition, and i_C is proportional to ω and C, the opposition of a capacitor is inversely related to ω (= $2\pi f$) and C.

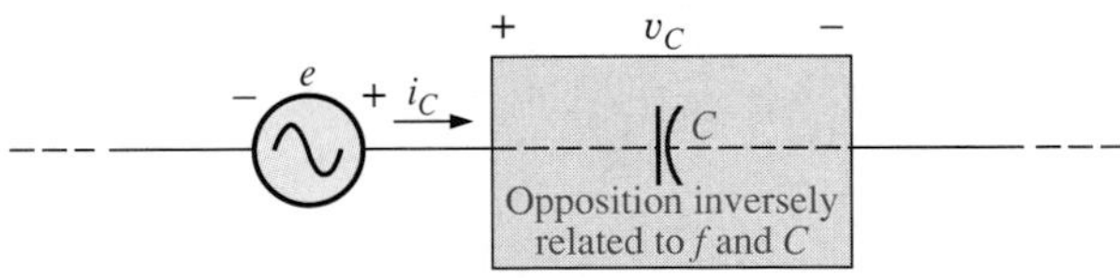

FIG. 14.10
Defining the parameters that determine the opposition of a capacitive element to the flow of the charge.

We will now verify, as we did for the inductor, some of the above conclusions using a more mathematical approach.

For the capacitor of Fig. 14.11, we recall from Chapter 10 that

$$i_C = C\,\frac{dv_C}{dt}$$

and, applying differentiation,

$$\frac{dv_C}{dt} = \frac{d}{dt}(V_m \sin \omega t) = \omega V_m \cos \omega t$$

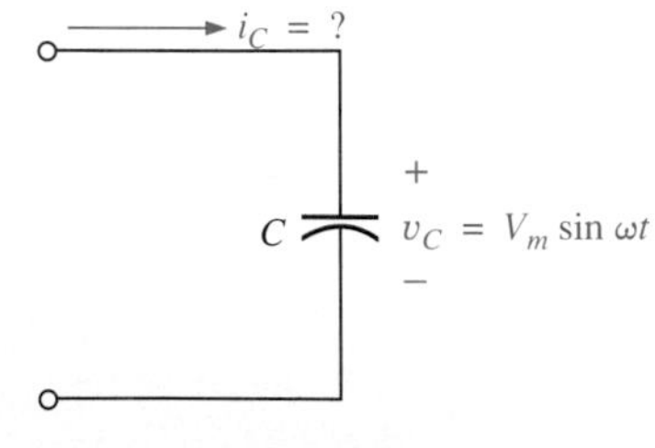

FIG. 14.11
Investigating the sinusoidal response of a capacitive element.

Therefore,

$$i_C = C\frac{dv_C}{dt} = C(\omega V_m \cos \omega t) = \omega C V_m \cos \omega t$$

or

$$i_C = I_m \sin(\omega t + 90°)$$

where

$$I_m = \omega C V_m$$

Note that the peak value of i_C is directly related to ω $(= 2\pi f)$ and C, as predicted in the discussion above.

A plot of v_C and i_C in Fig. 14.12 reveals that

for a capacitor, i_C leads v_C by 90°, or v_C lags i_C by 90°.*

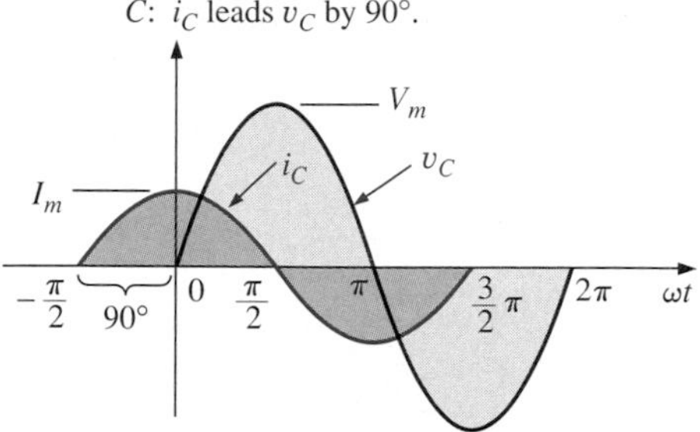

FIG. 14.12
The current of a purely capacitive element leads the voltage across the element by 90°.

If a phase angle is included in the sinusoidal expression for v_C, such as

$$v_C = V_m \sin(\omega t \pm \theta)$$

then

$$i_C = \omega C V_m \sin(\omega t \pm \theta + 90°)$$

Applying

$$\text{Opposition} = \frac{\text{cause}}{\text{effect}}$$

and substituting values, we obtain

$$\text{Opposition} = \frac{V_m}{I_m} = \frac{V_m}{\omega C V_m} = \frac{1}{\omega C}$$

which agrees with the results obtained above.

The quantity $1/\omega C$, called the *reactance* of a capacitor, is symbolically represented by X_C and is measured in ohms; that is,

$$X_C = \frac{1}{\omega C} \quad \text{(ohms, } \Omega) \qquad \textbf{(14.6)}$$

In an Ohm's law format, its magnitude can be determined from

$$X_C = \frac{V_m}{I_m} \quad \text{(ohms, } \Omega) \qquad \textbf{(14.7)}$$

Capacitive reactance is the opposition to the flow of charge, which results in the continual interchange of energy between the source and the electric field of the capacitor. Like the inductor, the capacitor does *not* dissipate energy in any form (ignoring the effects of the leakage resistance).

In the circuits just considered, the current was given in the inductive circuit, and the voltage in the capacitive circuit. This was done to avoid the use of integration in finding the unknown quantities. In the inductive circuit,

$$v_L = L\frac{di_L}{dt}$$

---

*A mnemonic phrase sometimes used to remember the phase relationship between the voltage and current of a coil and capacitor is "*ELI* the *ICE* man." Note that the *L* (inductor) has the *E* before the *I* (*e* leads *i* by 90°), and the *C* (capacitor) has the *I* before the *E* (*i* leads *e* by 90°).

but $$i_L = \frac{1}{L}\int v_L\,dt \tag{14.8}$$

In the capacitive circuit,

$$i_C = C\frac{dv_C}{dt}$$

but $$v_C = \frac{1}{C}\int i_C\,dt \tag{14.9}$$

Shortly, we shall consider a method of analyzing ac circuits that will permit us to solve for an unknown quantity with sinusoidal input without having to use direct integration or differentiation.

It is possible to determine whether a network with one or more elements is predominantly capacitive or inductive by noting the phase relationship between the input voltage and current.

If the source current leads the applied voltage, the network is predominantly capacitive, and if the applied voltage leads the source current, it is predominantly inductive.

Since we now have an equation for the reactance of an inductor or capacitor, we do not need to use derivatives or integration in the examples to be considered. Simply applying Ohm's law, $I_m = E_m/X_L$ (or X_C), and keeping in mind the phase relationship between the voltage and current for each element, will be sufficient to complete the examples.

EXAMPLE 14.1 The voltage across a resistor is indicated. Find the sinusoidal expression for the current if the resistor is 10 Ω. Sketch the curves for v and i.

a. $v = 100 \sin 377t$
b. $v = 25 \sin(377t + 60°)$

Solutions:

a. Eq. (14.2): $I_m = \dfrac{V_m}{R} = \dfrac{100\text{ V}}{10\ \Omega} = 10\text{ A}$

(v and i are in phase), resulting in

$$i = \mathbf{10 \sin 377}\boldsymbol{t}$$

The curves are sketched in Fig. 14.13.

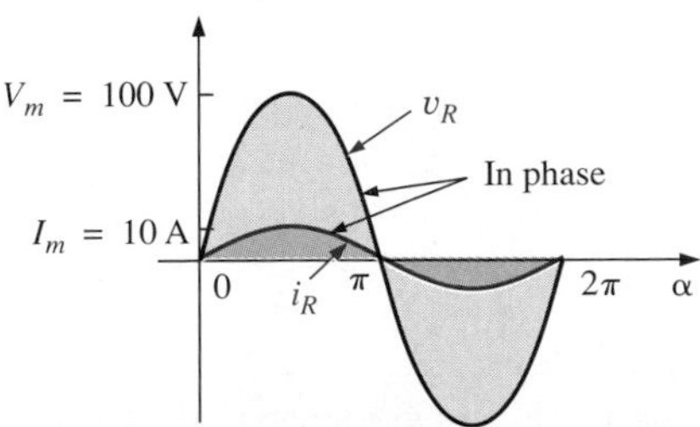

FIG. 14.13
Example 14.1(a).

b. Eq. (14.2): $I_m = \dfrac{V_m}{R} = \dfrac{25\text{ V}}{10\ \Omega} = 2.5\text{ A}$

(v and i are in phase), resulting in

$$i = \mathbf{2.5\ sin(377}t\ \mathbf{+ 60°)}$$

The curves are sketched in Fig. 14.14.

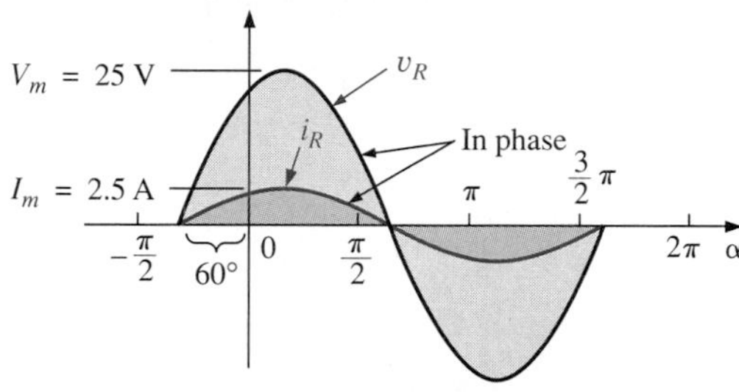

FIG. 14.14
Example 14.1(b).

EXAMPLE 14.2 The current through a 5-Ω resistor is given. Find the sinusoidal expression for the voltage across the resistor for $i = 40\sin(377t + 30°)$.

Solution: Eq. (14.3): $V_m = I_mR = (40\text{ A})(5\ \Omega) = 200\text{ V}$

(v and i are in phase), resulting in

$$v = \mathbf{200\ sin(377}t\ \mathbf{+ 30°)}$$

EXAMPLE 14.3 The current through a 0.1-H coil is provided. Find the sinusoidal expression for the voltage across the coil. Sketch the v and i curves.

a. $i = 10\sin 377t$
b. $i = 7\sin(377t - 70°)$

Solutions:

a. Eq. (14.4): $X_L = \omega L = (377\text{ rad/s})(0.1\text{ H}) = 37.7\ \Omega$
   Eq. (14.5): $V_m = I_mX_L = (10\text{ A})(37.7\ \Omega) = 377\text{ V}$

   and we know for a coil that v leads i by 90°. Therefore,

$$v = \mathbf{377\ sin(377}t\ \mathbf{+ 90°)}$$

   The curves are sketched in Fig. 14.15.

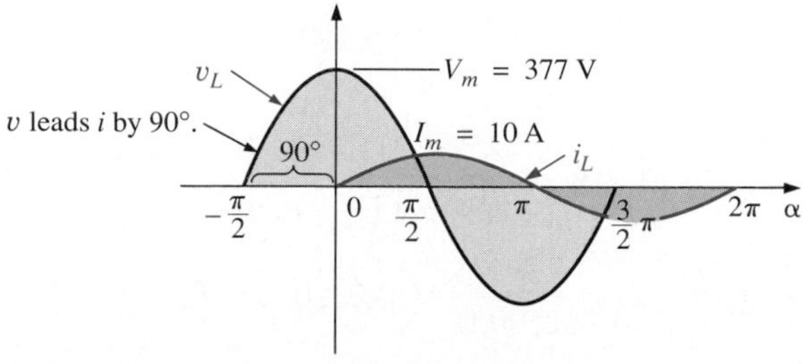

FIG. 14.15
Example 14.3(a).

b. X_L remains at 37.7 Ω.

$$V_m = I_m X_L = (7 \text{ A})(37.7\ \Omega) = 263.9 \text{ V}$$

and we know for a coil that v leads i by 90°. Therefore,

$$v = 263.9 \sin(377t - 70° + 90°)$$

and

$$\mathbf{v = 263.9 \sin(377t + 20°)}$$

The curves are sketched in Fig. 14.16.

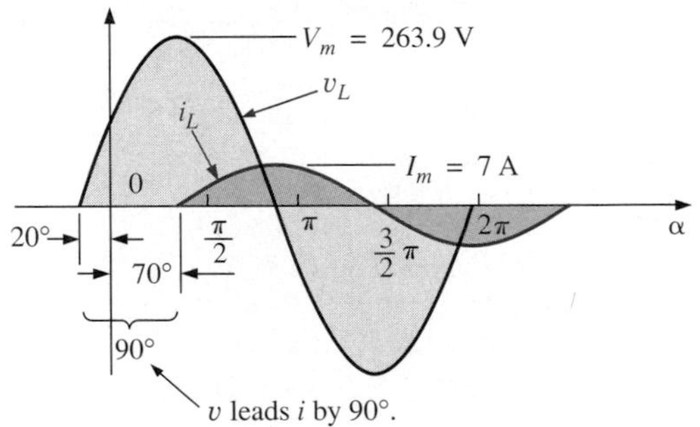

FIG. 14.16
Example 14.3(b).

EXAMPLE 14.4 The voltage across a 0.5-H coil is provided below. What is the sinusoidal expression for the current?

$$v = 100 \sin 20t$$

Solution:

$$X_L = \omega L = (20 \text{ rad/s})(0.5 \text{ H}) = 10\ \Omega$$

$$I_m = \frac{V_m}{X_L} = \frac{100 \text{ V}}{10\ \Omega} = 10 \text{ A}$$

and we know that i lags v by 90°. Therefore,

$$\mathbf{i = 10 \sin(20t - 90°)}$$

EXAMPLE 14.5 The voltage across a 1-μF capacitor is provided below. What is the sinusoidal expression for the current? Sketch the v and i curves.

$$v = 30 \sin 400t$$

Solution:

$$\text{Eq. (14.6):}\quad X_C = \frac{1}{\omega C} = \frac{1}{(400 \text{ rad/s})(1 \times 10^{-6} \text{ F})} = \frac{10^6\ \Omega}{400} = 2500\ \Omega$$

$$\text{Eq. (14.7):}\quad I_m = \frac{V_m}{X_C} = \frac{30 \text{ V}}{2500\ \Omega} = 0.0120 \text{ A} = 12 \text{ mA}$$

and we know for a capacitor that i leads v by 90°. Therefore,

$$\mathbf{i = 12 \times 10^{-3} \sin(400t + 90°)}$$

The curves are sketched in Fig. 14.17.

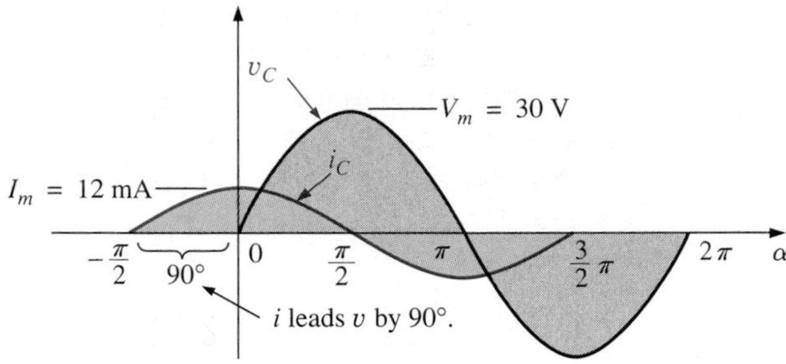

FIG. 14.17
Example 14.5.

EXAMPLE 14.6 The current through a 100-μF capacitor is given. Find the sinusoidal expression for the voltage across the capacitor.

$$i = 40 \sin(500t + 60°)$$

Solution:

$$X_C = \frac{1}{\omega C} = \frac{1}{(500 \text{ rad/s})(100 \times 10^{-6} \text{ F})} = \frac{10^6 \ \Omega}{5 \times 10^4} = \frac{10^2 \ \Omega}{5} = 20 \ \Omega$$

$$V_m = I_m X_C = (40 \text{ A})(20 \ \Omega) = 800 \text{ V}$$

and we know for a capacitor that v lags i by 90°. Therefore,

$$v = 800 \sin(500t + 60° - 90°)$$

and $$v = \mathbf{800 \sin(500}\boldsymbol{t}\mathbf{ - 30°)}$$

EXAMPLE 14.7 For the following pairs of voltages and currents, determine whether the element involved is a capacitor, inductor, or resistor, and determine the value of C, L, or R, if sufficient data are provided (Fig. 14.18):

a. $v = 100 \sin(\omega t + 40°)$
   $i = 20 \sin(\omega t + 40°)$
b. $v = 1000 \sin(377t + 10°)$
   $i = 5 \sin(377t - 80°)$
c. $v = 500 \sin(157t + 30°)$
   $i = 1 \sin(157t + 120°)$
d. $v = 50 \cos(\omega t + 20°)$
   $i = 5 \sin(\omega t + 110°)$

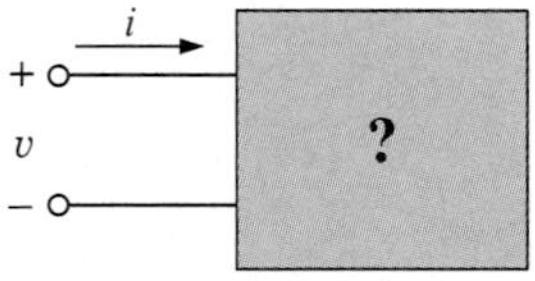

FIG. 14.18
Example 14.7.

Solutions:

a. Since v and i are *in phase*, the element is a *resistor*, and

$$R = \frac{V_m}{I_m} = \frac{100 \text{ V}}{20 \text{ A}} = \mathbf{5 \ \Omega}$$

b. Since v *leads* i by 90°, the element is an *inductor*, and

$$X_L = \frac{V_m}{I_m} = \frac{1000 \text{ V}}{5 \text{ A}} = 200 \ \Omega$$

so that $X_L = \omega L = 200 \ \Omega$

$$\text{or } L = \frac{200\ \Omega}{\omega} = \frac{200\ \Omega}{377\ \text{rad/s}} = \mathbf{0.531\ H}$$

c. Since i *leads* v by 90°, the element is a *capacitor,* and

$$X_C = \frac{V_m}{I_m} = \frac{500\ \text{V}}{1\ \text{A}} = 500\ \Omega$$

$$\text{so that } X_C = \frac{1}{\omega C} = 500\ \Omega$$

$$\text{or } C = \frac{1}{\omega 500\ \Omega} = \frac{1}{(157\ \text{rad/s})(500\ \Omega)} = \mathbf{12.74\ \mu F}$$

d. $v = 50\cos(\omega t + 20°) = 50\sin(\omega t + 20° + 90°)$
$= 50\sin(\omega t + 110°)$

Since v and i are *in phase,* the element is a *resistor,* and

$$R = \frac{V_m}{I_m} = \frac{50\ \text{V}}{5\ \text{A}} = \mathbf{10\ \Omega}$$

dc, High-, and Low-Frequency Effects on *L* and *C*

For dc circuits, the frequency is zero, and the reactance of a coil is

$$X_L = 2\pi f L = 2\pi(0)L = 0\ \Omega$$

The use of the short-circuit equivalence for the inductor in dc circuits (Chapter 12) is now validated. At very high frequencies, $X_L\uparrow = 2\pi f\uparrow L$ is very large, and for some practical applications the inductor can be replaced by an open circuit. In equation form

$$\boxed{X_L = 0\ \Omega} \quad \text{dc, } f = 0\ \text{Hz} \tag{14.10}$$

and

$$\boxed{X_L \cong \infty\ \Omega} \quad f = \text{very high frequencies} \tag{14.11}$$

The capacitor can be replaced by an open-circuit equivalence in dc circuits since $f = 0$, and

$$X_C = \frac{1}{2\pi f C} = \frac{1}{2\pi(0)C} = \infty\ \Omega$$

once again substantiating our previous action (Chapter 10). At very high frequencies, for finite capacitances,

$$X_C\downarrow = \frac{1}{2\pi f\uparrow C}$$

is very small, and for some practical applications the capacitor can be replaced by a short circuit. In equation form

$$\boxed{X_C \cong \infty\ \Omega} \quad \text{dc, } f = 0\ \text{Hz} \tag{14.12}$$

and

$$\boxed{X_C \cong 0\ \Omega} \quad f = \text{very high frequencies} \tag{14.13}$$

Table 14.1 reviews the preceding conclusions.

TABLE 14.1
Effect of high and low frequencies on the circuit model of an inductor and capacitor.

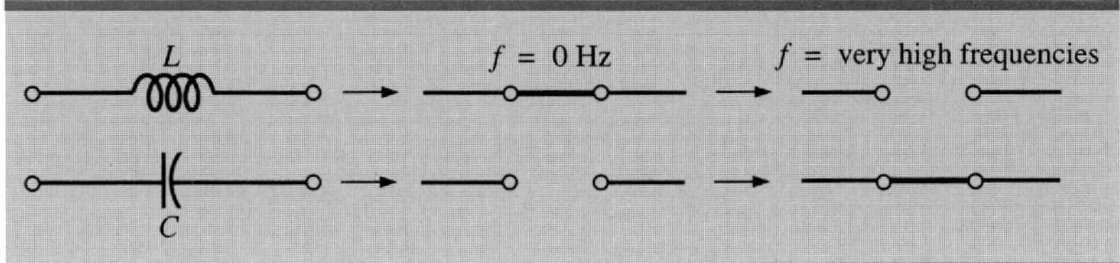

Phase Angle Measurements Between the Applied Voltage and Source Current

Now that we are familiar with phase relationships and understand how the elements affect the phase relationship between the applied voltage and resulting current, the use of the oscilloscope to measure the phase angle can be introduced. Recall from past discussions that the oscilloscope can be used only to display voltage levels versus time. However, now that we realize that the voltage across a resistor is in phase with the current through a resistor, we can consider the phase angle associated with the voltage across any resistor actually to be the phase angle of the current. For example, suppose we want to find the phase angle introduced by the unknown system of Fig. 14.19(a). In Fig. 14.19(b) a resistor was added to the input leads, and the two channels of a dual trace (most modern-day oscilloscopes can display two signals at the same time) were connected as shown. One channel will display the input voltage v_i, whereas the other will display v_R, as shown in Fig. 14.19(c). However, as noted before, since v_R and i_R are in phase, the phase angle appearing in Fig. 14.19(c) is also the phase angle between v_i and i_i. The addition of a "sensing" resistor (a resistor of a magnitude that will not adversely affect the input characteristics of the system), therefore, can be used to determine the phase angle introduced by the system and can be used to determine the magnitude of the resulting current. The details of the connections that must be made and how the actual phase angle is determined will be left for the laboratory experience.

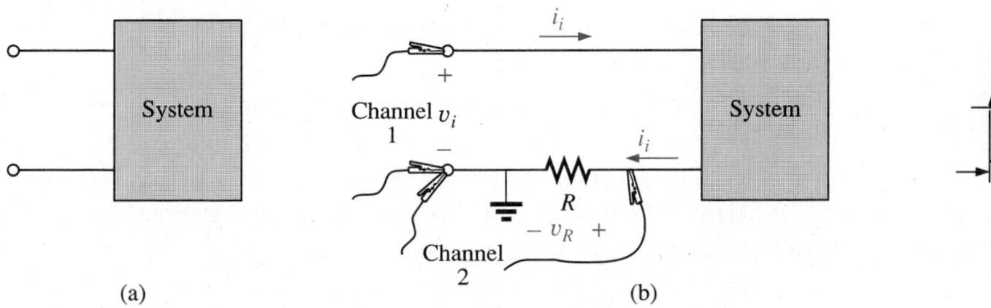

v_i
v_R (i_i)
θ
v_i leads v_R (i_i) by θ (inductive network)
(c)

FIG. 14.19
Using an oscilloscope to determine the phase angle between the applied voltage and the source current.

14.4 FREQUENCY RESPONSE OF THE BASIC ELEMENTS

The analysis of Section 14.3 was limited to a particular applied frequency. What is the effect of varying the frequency on the level of opposition offered by a resistive, inductive, or capacitive element? We are aware from the last section that the inductive reactance increases with frequency while the capacitive reactance decreases. However, what is the pattern to this increase or decrease in opposition? Does it continue indefinitely on the same path? Since applied signals may have frequencies extending from a few hertz to megahertz, it is important to be aware of the effect of frequency on the opposition level.

R

Thus far we have assumed that the resistance of a resistor is independent of the applied frequency. However, in the real world each resistive element has stray capacitance levels and lead inductance that are sensitive to the applied frequency. However, the capacitive and inductive levels involved are usually so small that their real effect is not noticed until the megahertz range. The resistance-versus-frequency curves for a number of carbon composition resistors are provided in Fig. 14.20. Note that the lower resistance levels seem to be less affected by the frequency level. The 100-Ω resistor is essentially stable up to about 300 MHz, whereas the 100-kΩ resistor starts its radical decline at about 15 MHz.

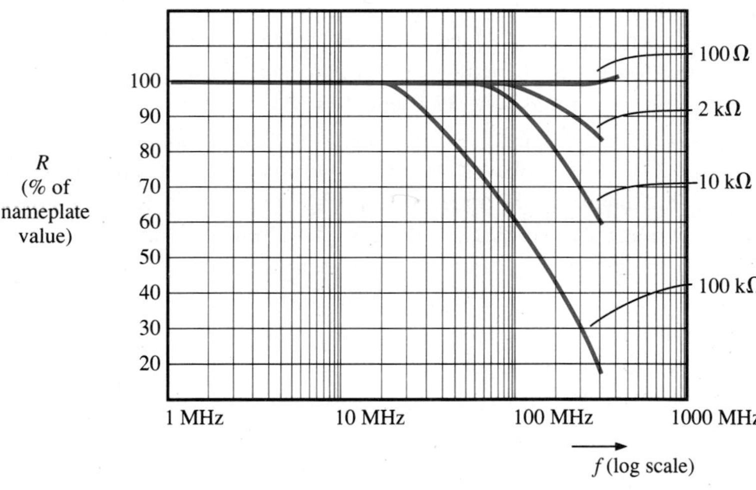

FIG. 14.20
Typical resistance-versus-frequency curves for carbon compound resistors.

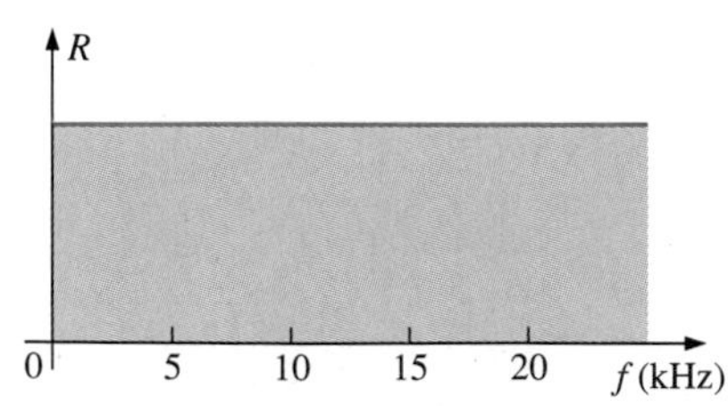

FIG. 14.21
R versus f for the range of interest.

Frequency, therefore, does have impact on the resistance of an element, but for our current frequency range of interest, we will assume the resistance-versus-frequency plot of Fig. 14.21 (like Fig. 14.20 up to 15 MHz), which essentially specifies that the resistance level of a resistor is independent of frequency.

L

For inductors, the equation

$$X_L = \omega L = 2\pi fL = 2\pi Lf$$

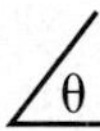

is directly related to the straight-line equation

$$y = mx + b = (2\pi L)f + 0$$

with a slope (m) of $2\pi L$ and a y-intercept (b) of zero. X_L is the y variable and f is the x variable, as shown in Fig. 14.22.

The larger the inductance, the greater the slope ($m = 2\pi L$) for the same frequency range, as shown in Fig. 14.22. Keep in mind, as reemphasized by Fig. 14.22, that the opposition of an inductor at very low frequencies approaches that of a short circuit, while at high frequencies the reactance approaches that of an open circuit.

For the capacitor, the reactance equation

$$X_C = \frac{1}{2\pi fC}$$

can be written

$$X_C f = \frac{1}{2\pi C}$$

which matches the basic format of a hyperbola,

$$yx = k$$

with $y = X_C$, $x = f$, and the constant $k = 1/(2\pi C)$.

At $f = 0$ Hz, the reactance of the capacitor is so large, as shown in Fig. 14.23, that it can be replaced by an open-circuit equivalent. As the frequency increases, the reactance decreases, until eventually a short-circuit equivalent would be appropriate. Note how an increase in capacitance causes the reactance to drop off more rapidly with frequency.

In summary, therefore, as the applied frequency increases, the resistance of a resistor remains constant, the reactance of an inductor increases linearly, and the reactance of a capacitor decreases nonlinearly.

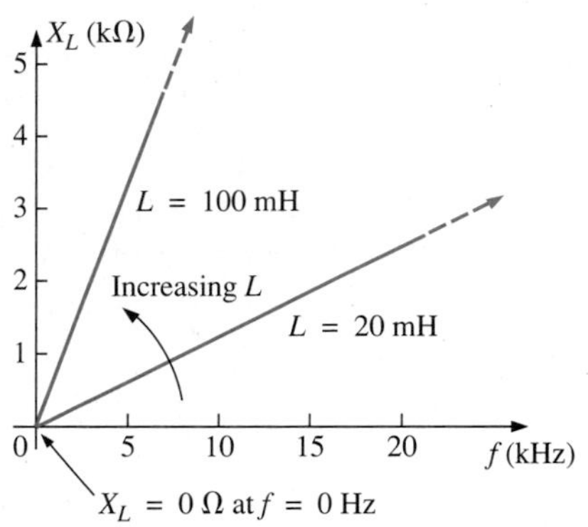

FIG. 14.22
X_L versus frequency.

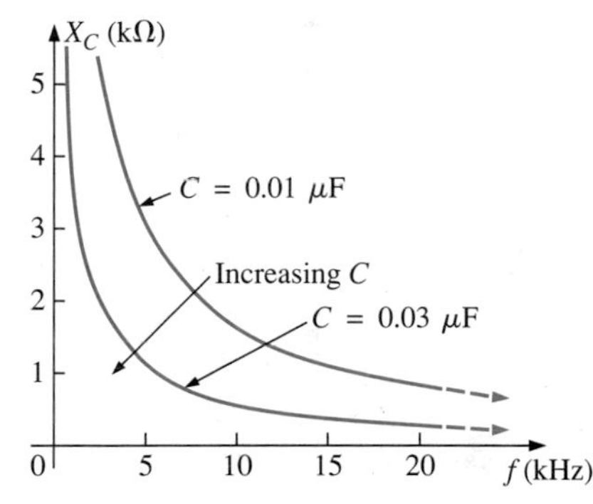

FIG. 14.23
X_C versus frequency.

EXAMPLE 14.8 At what frequency will the reactance of a 200-mH inductor match the resistance level of a 5-kΩ resistor?

Solution: The resistance remains constant at 5 kΩ for the frequency range of the inductor. Therefore,

$$R = 5000\ \Omega = X_L = 2\pi fL = 2\pi Lf$$
$$= 2\pi(200 \times 10^{-3}\ \text{H})f = 1.257f$$

and

$$f = \frac{5000\ \text{Hz}}{1.257} \cong \mathbf{3.98\ kHz}$$

EXAMPLE 14.9 At what frequency will an inductor of 5 mH have the same reactance as a capacitor of 0.1 μF?

Solution:

$$X_L = X_C$$

$$2\pi fL = \frac{1}{2\pi fC}$$

$$f^2 = \frac{1}{4\pi^2 LC}$$

and

$$f = \frac{1}{2\pi\sqrt{LC}} = \frac{1}{2\pi\sqrt{(5 \times 10^{-3}\ \text{H})(0.1 \times 10^{-6}\ \text{F})}}$$

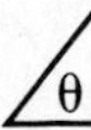

$$= \frac{1}{2\pi\sqrt{5 \times 10^{-10}}} = \frac{1}{(2\pi)(2.236 \times 10^{-5})}$$

$$f = \frac{10^5 \text{ Hz}}{14.05} \cong \mathbf{7.12\ kHz}$$

One must also be aware that commercial inductors are not ideal elements. In other words, the terminal characteristics of an inductance will vary with several factors, such as frequency, temperature, and current. A true equivalent for an inductor appears in Fig. 14.24. The series resistance R_s represents the copper losses (resistance of the many turns of thin copper wire), the eddy current losses (to be described in Chapter 19; losses due to small circular currents in the core when an ac voltage is applied), and hysteresis losses (also to be described in Chapter 19; losses due to core losses created by the rapidly reversing field in the core). The capacitance C_p is the stray capacitance that exists between the windings of the inductor. For most inductors, the construction is usually such that the larger the inductance, the lower the frequency at which the parasitic elements become important. That is, for inductors in the milli-henry range (which is very typical), frequencies approaching 100 kHz can have an effect on the ideal characteristics of the element. For inductors in the micro-henry range, a frequency of 1 MHz may introduce negative effects. This is not to suggest that the inductors lose their effect at these frequencies but more that they can no longer be considered ideal (purely inductive elements). Figure 14.25 is a plot of

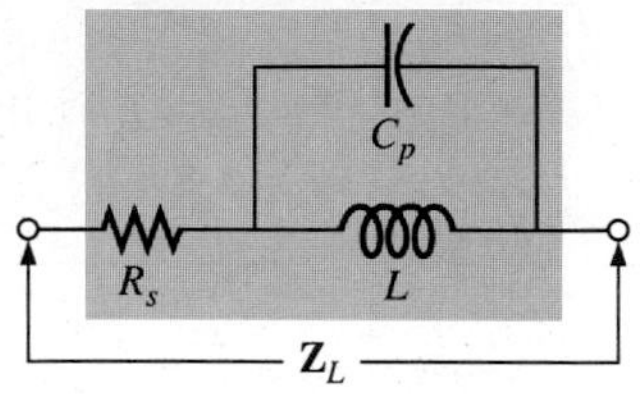

FIG. 14.24
Practical equivalent for an inductor.

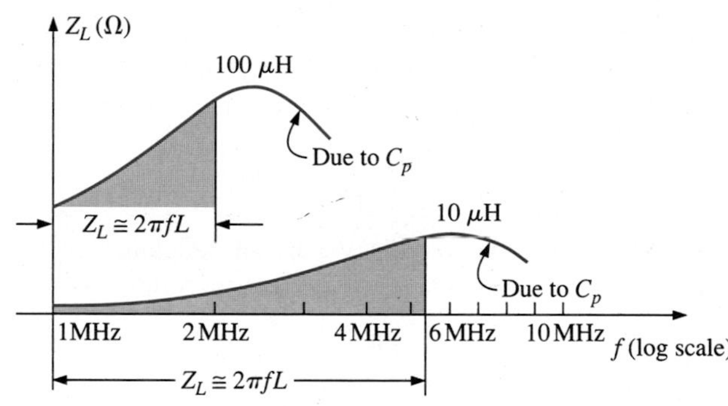

FIG. 14.25
Z_L versus frequency for the practical inductor equivalent of Fig. 14.24.

the magnitude of the impedance Z_L of Fig. 14.24 versus frequency. Note that up to about 2 MHz the impedance increases almost linearly with frequency, clearly suggesting that the 100 μH inductor is essentially ideal. However, above 2 MHz all the factors contributing to R_s will start to increase, while the reactance due to the capacitive element C_p will be more pronounced. The dropping level of capacitive reactance will begin to have a shorting effect across the windings of the inductor and reduce the overall inductive effect. Eventually, if the frequency continues to increase, the capacitive effects will overcome the inductive effects, and the element will actually begin to behave in a capacitive fashion. Note the similarities of this region with the curves of Fig. 14.23. Also note that decreasing levels of inductance (available with fewer turns and therefore lower levels of C_p) will not demonstrate the degrading effect until higher frequencies are applied. In general, therefore, the frequency of application for a coil becomes important at increasing frequencies. Inductors lose their ideal characteristics and in fact begin to act as capacitive elements with increasing losses at very high frequencies.

The capacitor, like the inductor, is not ideal at higher frequencies. In fact, a transition point can be defined where the characteristics of the capacitor will actually be inductive. The complete equivalent model for a capacitor is provided in Fig. 14.26. The resistance R_s, defined by the resistivity of the dielectric (typically 10^{12} $\Omega \cdot$m or better) and the case resistance, will determine the level of leakage current to expect during the discharge cycle. In other words, a charged capacitor can discharge both through the case and through the dielectric at a rate determined by the resistance of each path. Depending on the capacitor, the discharge time can extend from a few seconds for some electrolytic capacitors to hours (paper) or perhaps days (polystyrene). Inversely, therefore, electrolytics obviously have much lower levels of R_s than paper or polystyrene. The resistance R_p reflects the energy lost as the atoms continually realign themselves in the dielectric due to the applied alternating ac voltage. There is molecular friction present due to the motion of the atoms as they respond to the alternating applied electric field. Interestingly enough, however, the relative permittivity will decrease with increasing frequencies but eventually take a complete turnaround and begin to increase at very high frequencies. The inductance L_s includes the inductance of the capacitor leads and any inductive effects introduced by the design of the capacitor. Be aware that the inductance of the leads is about 0.05 μH per centimeter or 0.2 μH for a capacitor with two 2-cm leads—a level that can be important at high frequencies. As for the inductor, the capacitor will behave quite ideally for the low- and mid-frequency range, as shown by the plot of Fig. 14.27 for a 0.01-μF metalized film capacitor with 2-cm leads. As the frequency increases, however, and the reactance X_s becomes larger, a frequency will eventually be reached where the reactance of the coil equals that of the capacitor (a resonant condition to be described in Chapter 20). Any additional increase in frequency will simply result in X_s being greater than X_C, and the element will behave like an inductor. In general, therefore, the frequency of application is important for capacitive elements because

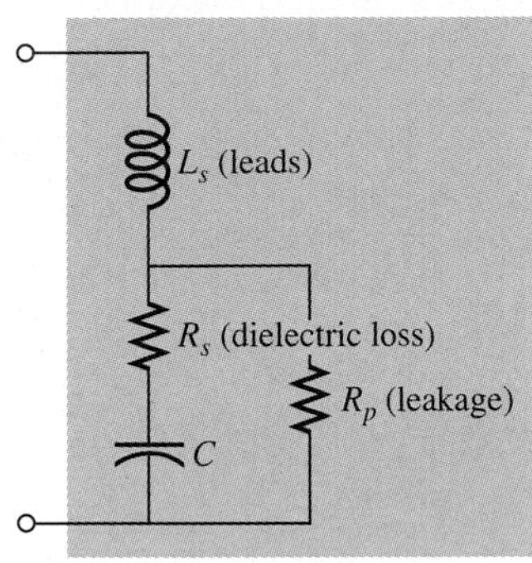

FIG. 14.26
Practical equivalent for a capacitor.

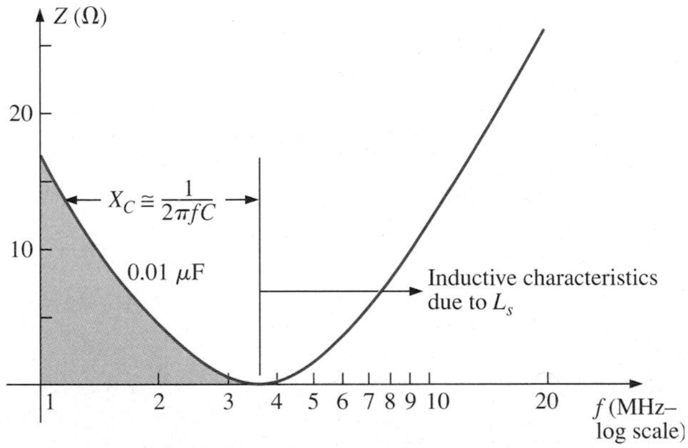

FIG. 14.27
Impedance characteristics of a 0.01-μF metalized film capacitor versus frequency.

there comes a point with increasing frequency when the element will take on inductive characteristics. It also points out that the frequency of application defines the type of capacitor (or inductor) that would be applied: Electrolytics are limited to frequencies up to perhaps 10 kHz, while ceramic or mica can handle frequencies beyond 10 MHz.

The expected temperature range of operation can have an important impact on the type of capacitor chosen for a particular application. Electrolytics, tantalum, and some high-*k* ceramic capacitors are very sensitive to colder temperatures. In fact, most electrolytics lose 20% of their room temperature capacitance at 0°C (freezing). Higher temperatures (up to 100°C or 212°F) seem to have less of an impact in general than colder temperatures, but high-*k* ceramics can lose up to 30% of their capacitance level at 100°C compared to room temperature. With exposure and experience, you will learn the type of capacitor employed for each application, and concern will arise only when very high frequencies, extreme temperatures, or very high currents or voltages are encountered.

14.5 AVERAGE POWER AND POWER FACTOR

For any load in a sinusoidal ac network, the voltage across the load and the current through the load will vary in a sinusoidal nature. The question then arises, How does the power to the load determined by the product $v \cdot i$ vary and what fixed value can be assigned to the power since it will vary with time?

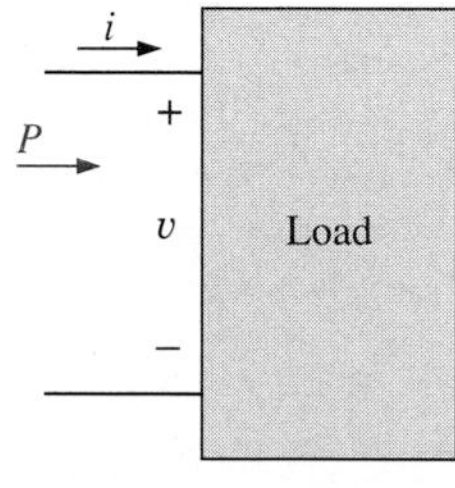

FIG. 14.28
Determining the power delivered in a sinusoidal ac network.

If we take the general case depicted in Fig. 14.28 and use the following for v and i:

$$v = V_m \sin(\omega t + \theta_v)$$
$$i = I_m \sin(\omega t + \theta_i)$$

then the power is defined by

$$\begin{aligned} p = vi &= V_m \sin(\omega t + \theta_v) I_m \sin(\omega t + \theta_i) \\ &= V_m I_m \sin(\omega t + \theta_v) \sin(\omega t + \theta_i) \end{aligned}$$

Using the trigonometric identity

$$\sin A \sin B = \frac{\cos(A - B) - \cos(A + B)}{2}$$

the function $\sin(\omega t + \theta_v) \sin(\omega t + \theta_i)$ becomes

$$\begin{aligned} \sin(\omega t + \theta_v) \sin(\omega t + \theta_i) \\ &= \frac{\cos[(\omega t + \theta_v) - (\omega t + \theta_i)] - \cos[(\omega t + \theta_v) + (\omega t + \theta_i)]}{2} \\ &= \frac{\cos(\theta_v - \theta_i) - \cos(2\omega t + \theta_v + \theta_i)}{2} \end{aligned}$$

so that

$$p = \overbrace{\left[\frac{V_m I_m}{2} \cos(\theta_v - \theta_i)\right]}^{\text{Fixed value}} - \overbrace{\left[\frac{V_m I_m}{2} \cos(2\omega t + \theta_v + \theta_i)\right]}^{\text{Time-varying (function of } t\text{)}}$$

A plot of v, i, and p on the same set of axes is shown in Fig. 14.29.

Note that the second factor in the preceding equation is a cosine wave with an amplitude of $V_m I_m/2$, and a frequency twice that of the

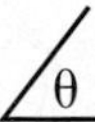

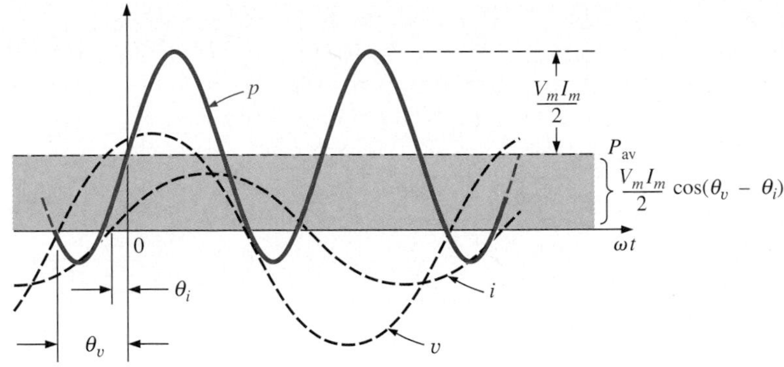

FIG. 14.29
Defining the average power for a sinusoidal ac network.

voltage or current. The average value of this term is zero over one cycle, producing no net transfer of energy in any one direction.

The first term in the preceding equation, however, has a constant magnitude (no time dependence) and therefore provides some net transfer of energy. This term is referred to as the *average power,* the reason for which is obvious from Fig. 14.29. The average power, or *real* power as it is sometimes called, is the power delivered to and dissipated by the load. It corresponds to the power calculations performed for dc networks. The angle $(\theta_v - \theta_i)$ is the phase angle between v and i. Since $\cos(-\alpha) = \cos\alpha$,

the magnitude of average power delivered is independent of whether v leads i or i leads v.

Defining θ as equal to $|\theta_v - \theta_i|$, where | | indicates that only the magnitude is important and the sign is immaterial, we have

$$P = \frac{V_m I_m}{2}\cos\theta \qquad \text{(watts, W)} \qquad \textbf{(14.14)}$$

where P is the average power in watts. This equation can also be written

$$P = \left(\frac{V_m}{\sqrt{2}}\right)\left(\frac{I_m}{\sqrt{2}}\right)\cos\theta$$

or, since

$$V_{\text{eff}} = \frac{V_m}{\sqrt{2}} \quad \text{and} \quad I_{\text{eff}} = \frac{I_m}{\sqrt{2}}$$

Eq. (14.14) becomes

$$P = V_{\text{eff}} I_{\text{eff}} \cos\theta \qquad \textbf{(14.15)}$$

Let us now apply Eqs. (14.14) and (14.15) to the basic R, L, and C elements.

Resistor

In a purely resistive circuit, since v and i are in phase, $|\theta_v - \theta_i| = \theta = 0°$, and $\cos\theta = \cos 0° = 1$, so that

$$P = \frac{V_m I_m}{2} = V_{\text{eff}} I_{\text{eff}} \qquad \text{(W)} \qquad \textbf{(14.16)}$$

Or, since $$I_{\text{eff}} = \frac{V_{\text{eff}}}{R}$$

then $$P = \frac{V^2_{\text{eff}}}{R} = I^2_{\text{eff}} R \qquad \text{(W)} \qquad \textbf{(14.17)}$$

Inductor

In a purely inductive circuit, since v leads i by 90°, $|\theta_v - \theta_i| = \theta = |-90°| = 90°$. Therefore,

$$P = \frac{V_m I_m}{2} \cos 90° = \frac{V_m I_m}{2}(0) = \mathbf{0\ W}$$

The average power or power dissipated by the ideal inductor (no associated resistance) is zero watts.

Capacitor

In a purely capacitive circuit, since i leads v by 90°, $|\theta_v - \theta_i| = \theta = |-90°| = 90°$. Therefore,

$$P = \frac{V_m I_m}{2} \cos(90°) = \frac{V_m I_m}{2}(0) = \mathbf{0\ W}$$

The average power or power dissipated by the ideal capacitor (no associated resistance) is zero watts.

EXAMPLE 14.10 Find the average power dissipated in a network whose input current and voltage are the following:

$$i = 5 \sin(\omega t + 40°)$$
$$v = 10 \sin(\omega t + 40°)$$

Solution: Since v and i are in phase, the circuit appears to be purely resistive at the input terminals. Therefore,

$$P = \frac{V_m I_m}{2} = \frac{(10\text{ V})(5\text{ A})}{2} = \mathbf{25\ W}$$

or $$R = \frac{V_m}{I_m} = \frac{10\text{ V}}{5\text{ A}} = 2\ \Omega$$

and $$P = \frac{V^2_{\text{eff}}}{R} = \frac{[(0.707)(10\text{ V})]^2}{2} = \mathbf{25\ W}$$

or $$P = I^2_{\text{eff}} R = [(0.707)(5\text{ A})]^2(2) = \mathbf{25\ W}$$

For the following example, the circuit consists of a combination of resistances and reactances producing phase angles between the input current and voltage different from 0° or 90°.

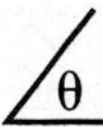

EXAMPLE 14.11 Determine the average power delivered to networks having the following input voltage and current:

a. $v = 100 \sin(\omega t + 40°)$
   $i = 20 \sin(\omega t + 70°)$
b. $v = 150 \sin(\omega t - 70°)$
   $i = 3 \sin(\omega t - 50°)$

Solutions:

a. $V_m = 100,\ \theta_v = 40°$
   $I_m = 20,\ \theta_i = 70°$
   $\theta = |\theta_v - \theta_i| = |40° - 70°| = |-30°| = 30°$

and

$$P = \frac{V_m I_m}{2} \cos\theta = \frac{(100\text{ V})(20\text{ A})}{2} \cos(30°) = (1000\text{ W})(0.866)$$
$$= \mathbf{866\ W}$$

b. $V_m = 150\text{ V},\ \theta_v = -70°$
   $I_m = 3\text{ A},\ \theta_i = -50°$
   $\theta = |\theta_v - \theta_i| = |-70° - (-50°)|$
   $= |-70° + 50°| = |-20°| = 20°$

and

$$P = \frac{V_m I_m}{2} \cos\theta = \frac{(150\text{ V})(3\text{ A})}{2} \cos(20°) = (225\text{ W})(0.9397)$$
$$= \mathbf{211.43\ W}$$

Power Factor

In the equation $P = (V_m I_m/2)\cos\theta$ the factor that has significant control on the delivered power level is the $\cos\theta$. No matter how large the voltage or current, if $\cos\theta = 0$, the power is zero; if $\cos\theta = 1$, the power delivered is a maximum. Since it has such control, the expression was given the name *power factor* and is defined by

$$\text{Power factor} = F_p = \cos\theta \qquad \textbf{(14.18)}$$

For a purely resistive load such as the one shown in Fig. 14.30, the phase angle between v and i is 0° and $F_p = \cos\theta = \cos 0° = 1$. The power delivered is a maximum of $(V_m I_m/2)\cos\theta = ((100\text{ V})(5\text{ A})/2)\ (1) = 250\text{ W}$.

For a purely reactive load (inductive or capacitive) such as the one shown in Fig. 14.31, the phase angle between v and i is 90° and $F_p = \cos\theta = \cos 90° = 0$. The power delivered is then the minimum value of 0 watts, **even though the current has the same peak value** as that encountered in Fig. 14.30.

For situations where the load is a combination of resistive and reactive elements, the power factor will vary between 0 and 1. The more resistive the total impedance, the closer the power factor is to 1; the more reactive the total impedance, the closer the power factor is to 0.

In terms of the average power and the terminal voltage and current,

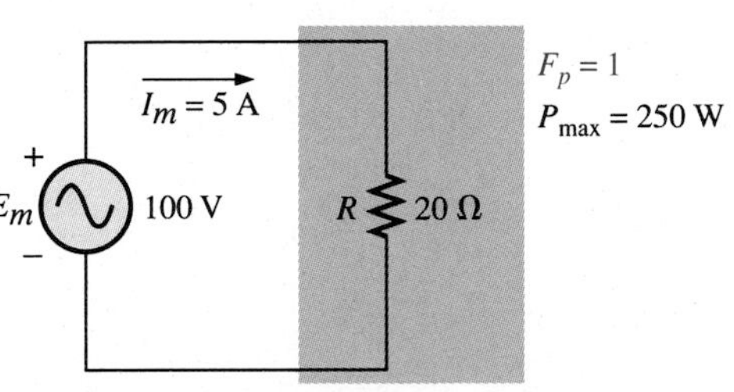

FIG. 14.30
Purely resistive load with $F_p = 1$.

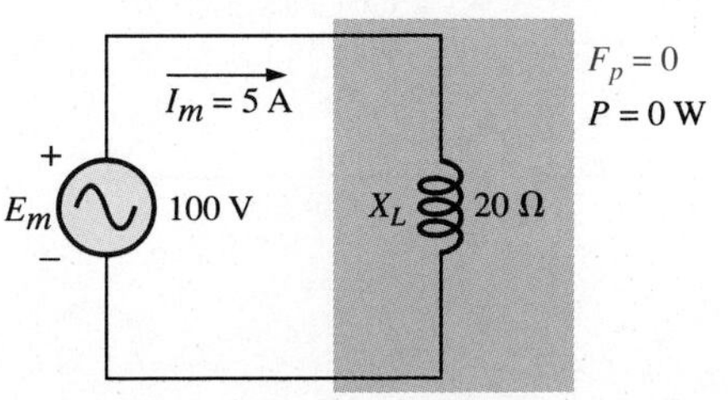

FIG. 14.31
Purely inductive load with $F_p = 0$.

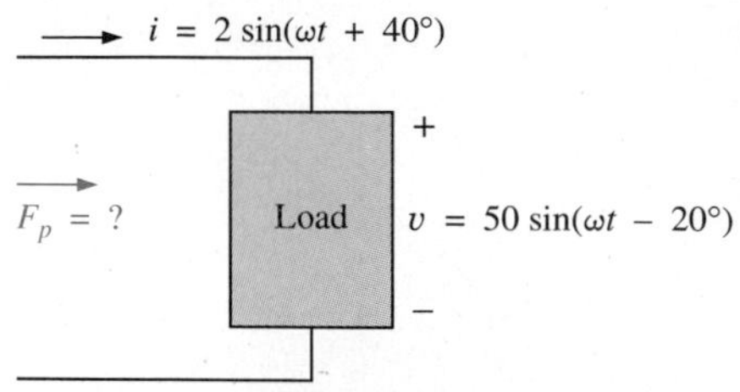

FIG. 14.32
Example 14.12(a).

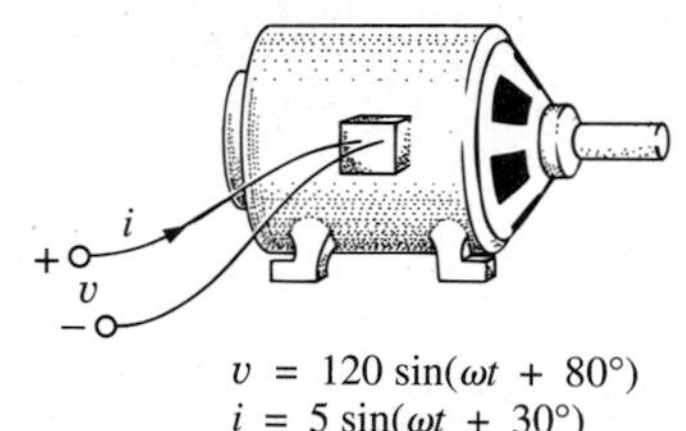

FIG. 14.33
Example 14.12(b).

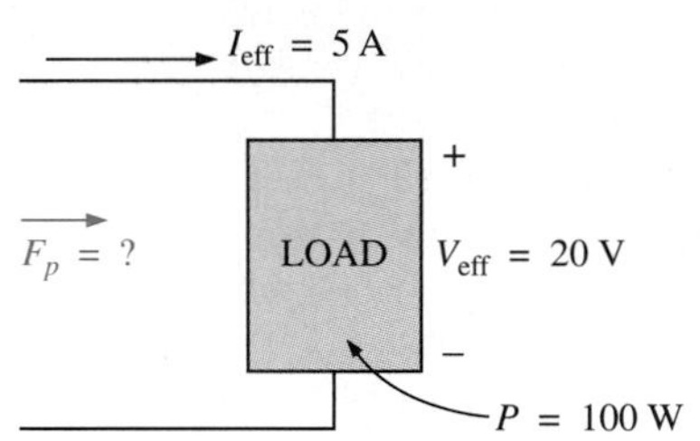

FIG. 14.34
Example 14.12(c).

$$F_p = \cos\theta = \frac{P}{V_{eff}I_{eff}} \tag{14.19}$$

The terms *leading* and *lagging* are often written in conjunction with the power factor. *They are defined by the current through the load.* If the current leads the voltage across a load, the load has a leading power factor. If the current lags the voltage across the load, the load has a lagging power factor. In other words,

capacitive networks have leading power factors, and inductive networks have lagging power factors.

The importance of the power factor on power distribution systems is examined in Chapter 19. In fact, one section is devoted to power factor correction.

EXAMPLE 14.12 Determine the power factors of the following loads, and indicate whether they are leading or lagging:

a. Fig. 14.32
b. Fig. 14.33
c. Fig. 14.34

Solutions:

a. $F_p = \cos\theta = \cos|40° - (-20°)| = \cos 60° = \mathbf{0.5\ leading}$
b. $F_p = \cos\theta\ |80° - 30°| = \cos 50° = \mathbf{0.6428\ lagging}$
c. $F_p = \cos\theta = \dfrac{P}{V_{eff}I_{eff}} = \dfrac{100\text{ W}}{(20\text{ V})(5\text{ A})} = \dfrac{100\text{ W}}{100\text{ W}} = \mathbf{1}$

The load is resistive, and F_p is neither leading nor lagging.

14.6 COMPLEX NUMBERS

In our analysis of dc networks, we found it necessary to determine the algebraic sum of voltages and currents. Since the same will also be true for ac networks, the question arises, How do we determine the algebraic sum of two or more voltages (or currents) that are varying sinusoidally? Although one solution would be to find the algebraic sum on a point-to-point basis (as shown in Section 14.12), this would be a long and tedious process in which accuracy would be directly related to the scale employed.

It is the purpose of this chapter to introduce a system of *complex numbers* that, when related to the sinusoidal ac waveform, will result in a technique for finding the algebraic sum of sinusoidal waveforms that is quick, direct, and accurate. In the following chapters, the technique will be extended to permit the analysis of sinusoidal ac networks in a manner very similar to that applied to dc networks. The methods and theorems as described for dc networks can then be applied to sinusoidal ac networks with little difficulty.

A *complex number* represents a point in a two-dimensional plane located with reference to two distinct axes. This point can also determine a radius vector drawn from the origin to the point. The horizontal axis is called the *real* axis, while the vertical axis is called the *imaginary* axis. Both are labeled in Fig. 14.35. For reasons that will be obvious later, the real axis is sometimes called the *resistance* axis, and the imaginary axis,

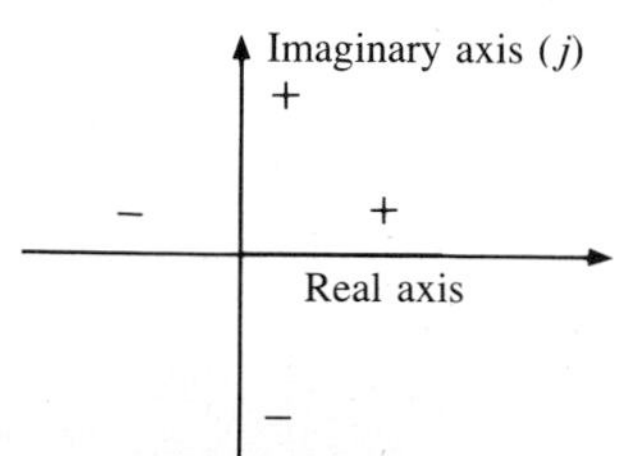

FIG. 14.35
Defining the real and imaginary axes of a complex plane.

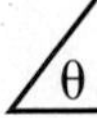

the *reactance* axis. Every number from zero to $\pm\infty$ can be represented by some point along the real axis. Prior to the development of this system of complex numbers, it was believed that any number not on the real axis would not exist—hence the term *imaginary* for the vertical axis.

In the complex plane, the horizontal or real axis represents all positive numbers to the right of the imaginary axis and all negative numbers to the left of the imaginary axis. All positive imaginary numbers are represented above the real axis, and all negative imaginary numbers, below the real axis. The symbol j (or sometimes i) is used to denote the imaginary component.

There are two forms used to represent a complex number: *rectangular* and *polar*. Each can represent a point in the plane or a radius vector drawn from the origin to that point.

14.7 RECTANGULAR FORM

The format for the rectangular form is

$$\mathbf{C} = A + jB \tag{14.20}$$

as shown in Fig. 14.36.

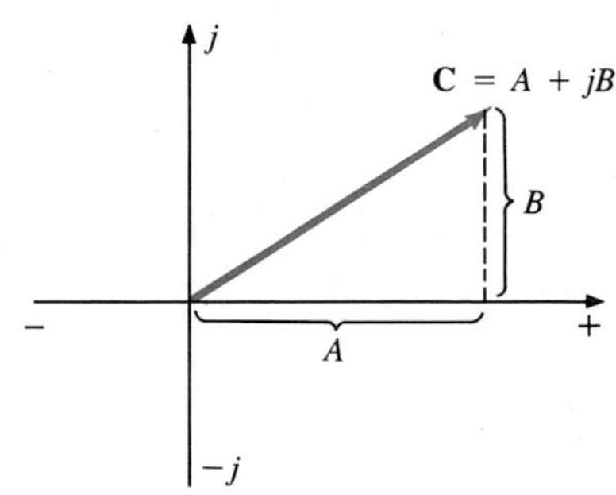

FIG. 14.36
Defining the rectangular form.

EXAMPLE 14.13 Sketch the following complex numbers in the complex plane:
a. $\mathbf{C} = 3 + j4$
b. $\mathbf{C} - 0 - j6$
c. $\mathbf{C} = -10 - j20$

Solutions:
a. See Fig. 14.37.
b. See Fig. 14.38.
c. See Fig. 14.39.

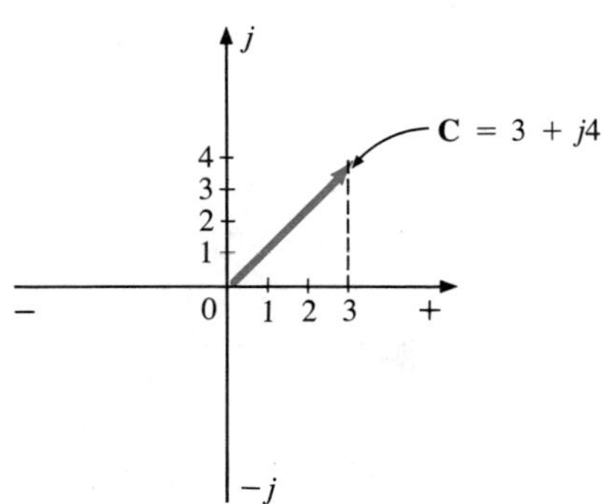

FIG. 14.37.
Example 14.13(a).

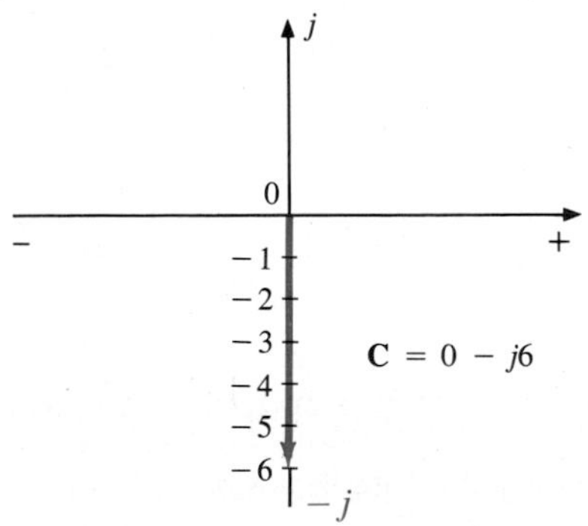

FIG. 14.38
Example 14.13(b).

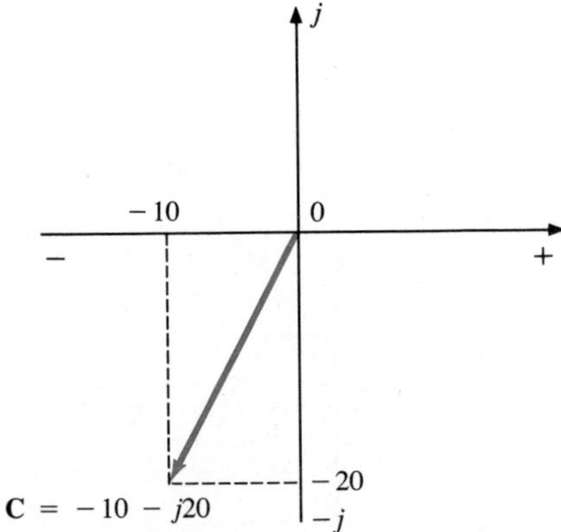

FIG. 14.39
Example 14.13(c).

14.8 POLAR FORM

The format for the polar form is

$$\mathbf{C} = C \angle\theta \tag{14.21}$$

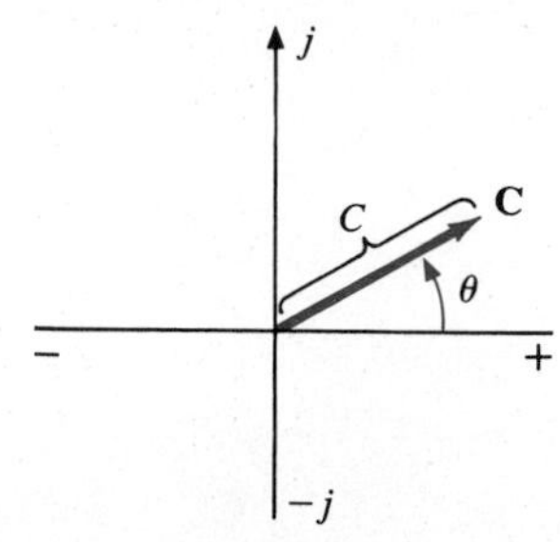

FIG. 14.40
Defining the polar form.

where C indicates magnitude only and θ is always measured counterclockwise (CCW) from the *positive real axis,* as shown in Fig. 14.40. Angles measured in the clockwise direction from the positive real axis must have a negative sign associated with them.

A negative sign in front of the polar form has the effect shown in Fig. 14.41. Note that it results in a complex number directly opposite the complex number with a positive sign.

$$-\mathbf{C} = -C \angle\theta = C \angle\theta \pm \pi \tag{14.22}$$

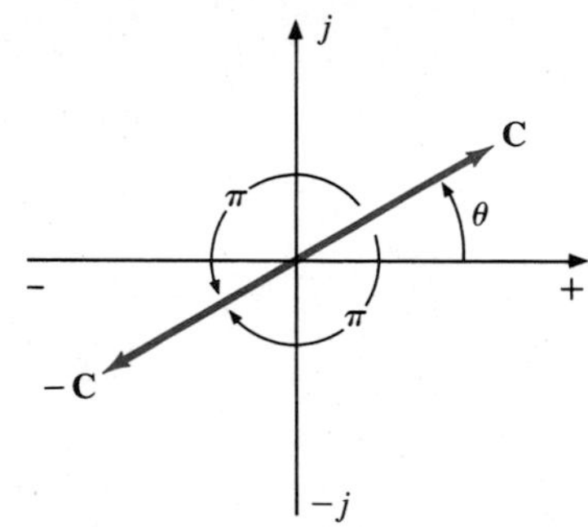

FIG. 14.41
Demonstrating the effect of a negative sign on the polar form.

EXAMPLE 14.14 Sketch the following complex numbers in the complex plane:

a. $\mathbf{C} = 5 \angle 30°$
b. $\mathbf{C} = 7 \angle -120°$
c. $\mathbf{C} = -4.2 \angle 60°$

Solutions:

a. See Fig. 14.42.
b. See Fig. 14.43.
c. See Fig. 14.44.

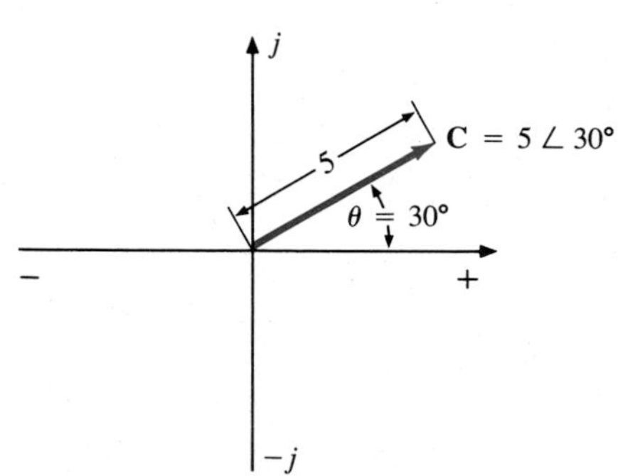

FIG. 14.42
Example 14.14(a).

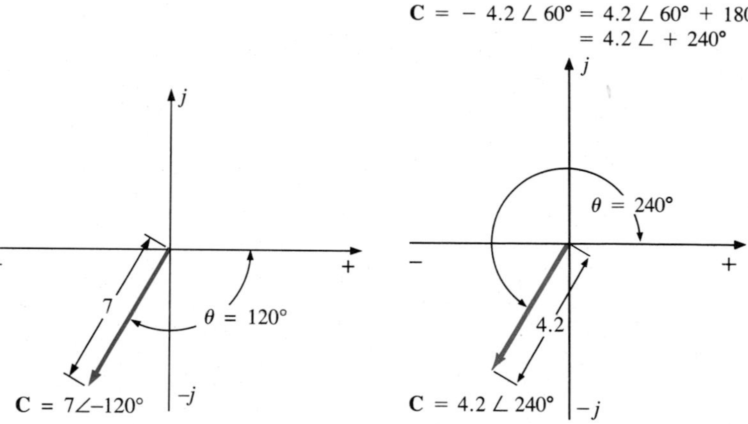

FIG. 14.43
Example 14.14(b).

FIG. 14.44
Example 14.14(c).

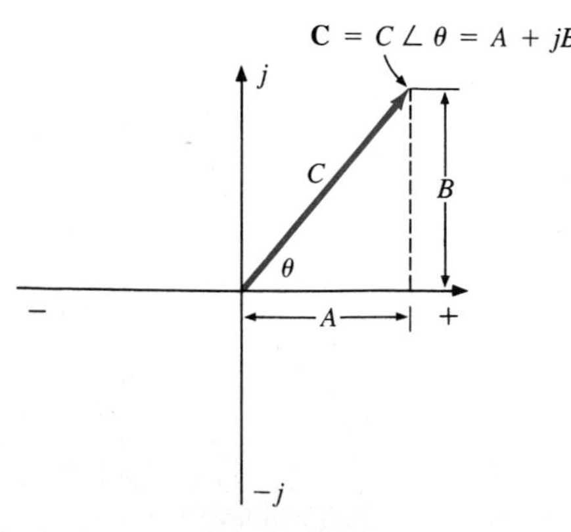

FIG. 14.45
Conversion between forms.

14.9 CONVERSION BETWEEN FORMS

The two forms are related by the following equations.

Rectangular to Polar

$$C = \sqrt{A^2 + B^2} \tag{14.23}$$

$$\theta = \tan^{-1} \frac{B}{A} \tag{14.24}$$

Note Fig. 14.45.

Polar to Rectangular

$$A = C\cos\theta \qquad (14.25)$$

$$B = C\sin\theta \qquad (14.26)$$

EXAMPLE 14.15 Convert the following from rectangular to polar form:

$$\mathbf{C} = 3 + j4 \qquad \text{(Fig. 14.46)}$$

Solution:

$$C = \sqrt{(3)^2 + (4)^2} = \sqrt{25} = 5$$

Example 14.18 $\theta = \tan^{-1}\left(\frac{4}{3}\right) = 53.13°$

and $$\mathbf{C = 5 \angle 53.13°}$$

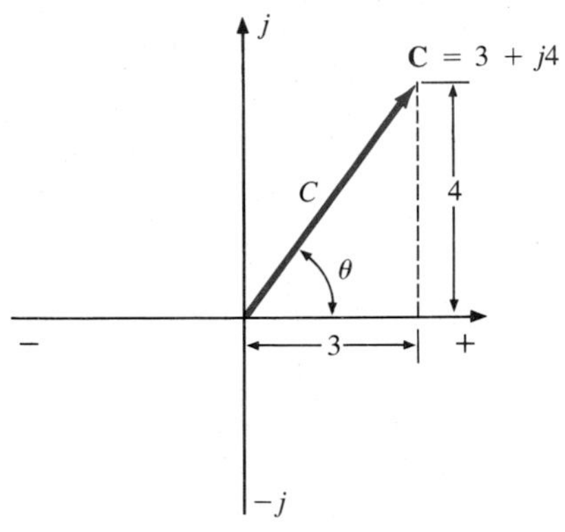

FIG. 14.46
Example 14.15.

EXAMPLE 14.16 Convert the following from polar to rectangular form:

$$\mathbf{C} = 10\angle 45° \qquad \text{(Fig. 14.47)}$$

Solution:

$$A = 10\cos 45° = (10))(0.707) = 7.07$$
$$B = 10\sin 45° = (10)(0.707) = 7.07$$

and $$\mathbf{C = 7.07 + j\,7.07}$$

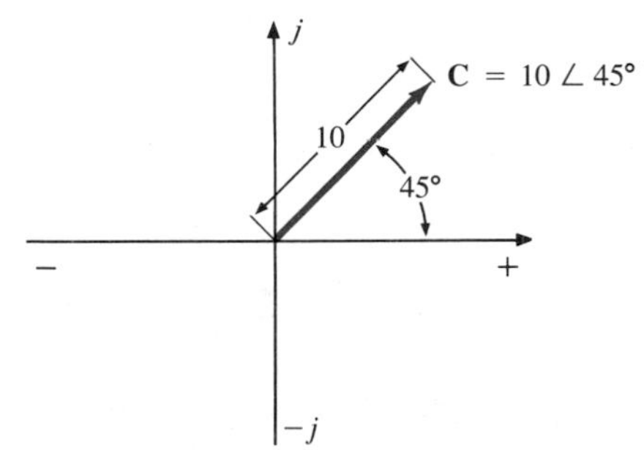

FIG. 14.47
Example 14.16.

If the complex number should appear in the second, third, or fourth quadrant, simply convert it in that quadrant, and carefully determine the proper angle to be associated with the magnitude of the vector.

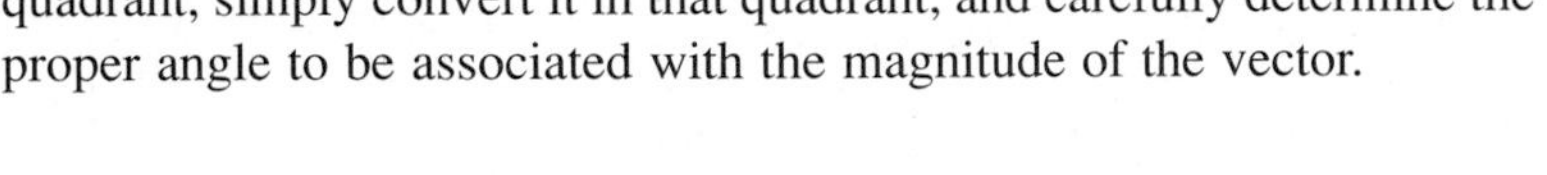

EXAMPLE 14.17 Convert the following from rectangular to polar form:

$$\mathbf{C} = -6 + j3 \qquad \text{(Fig. 14.48)}$$

Solution:

$$C = \sqrt{(6)^2 + (3)^2} = \sqrt{45} = 6.71$$
$$\beta = \tan^{-1}\left(\frac{3}{6}\right) = 26.57°$$
$$\theta = 180 - 26.57° = 153.43°$$

and $$\mathbf{C = 6.71 \angle 153.43°}$$

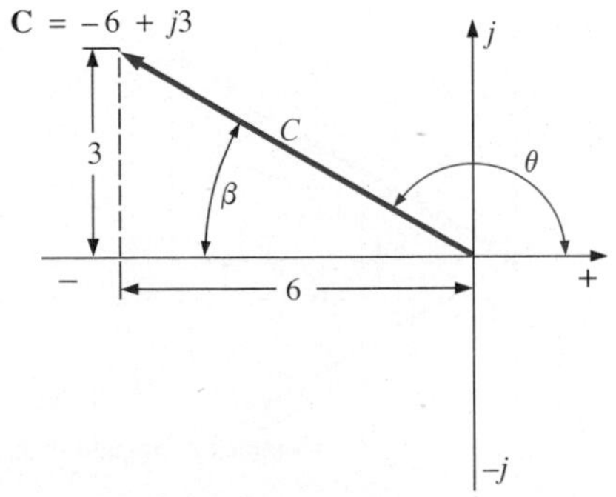

FIG. 14.48
Example 14.17.

EXAMPLE 14.18 Convert the following from polar to rectangular form:

$$\mathbf{C} = 10\angle 230° \qquad \text{(Fig. 14.49)}$$

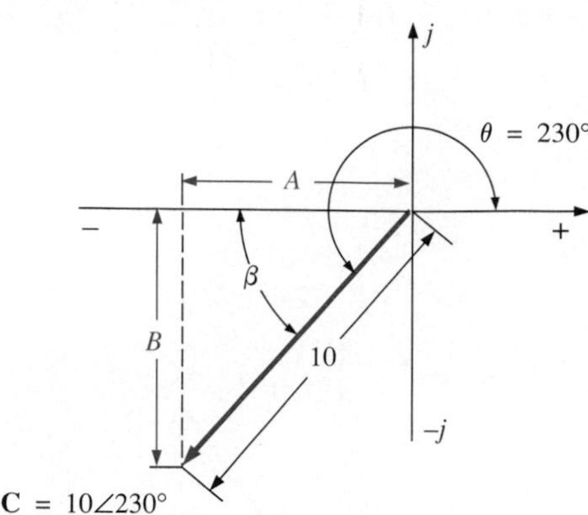

FIG. 14.49
Example 14.18.

Solution:

$$A = C\cos\beta = 10\cos(230° - 180°) = 10\cos 50°$$
$$= (10)(0.6428) = 6.428$$
$$B = C\sin\beta = 10\sin 50° = (10)(0.7660) = 7.660$$

and $\mathbf{C} = \mathbf{-6.428 - \mathit{j}7.660}$

14.10 MATHEMATICAL OPERATIONS WITH COMPLEX NUMBERS

Complex numbers lend themselves readily to the basic mathematical operations of addition, subtraction, multiplication, and division. A few basic rules and definitions must be understood before considering these operations.

Let us first examine the symbol j associated with imaginary numbers. By definition,

$$j = \sqrt{-1} \tag{14.27}$$

Thus,

$$j^2 = -1 \tag{14.28}$$

and

$$j^3 = j^2 j = -1j = -j$$

with

$$j^4 = j^2 j^2 = (-1)(-1) = +1$$
$$j^5 = j$$

and so on. Further,

$$\frac{1}{j} = (1)\left(\frac{1}{j}\right) = \left(\frac{j}{j}\right)\left(\frac{1}{j}\right) = \frac{j}{j^2} = \frac{j}{-1}$$

and

$$\frac{1}{j} = -j \tag{14.29}$$

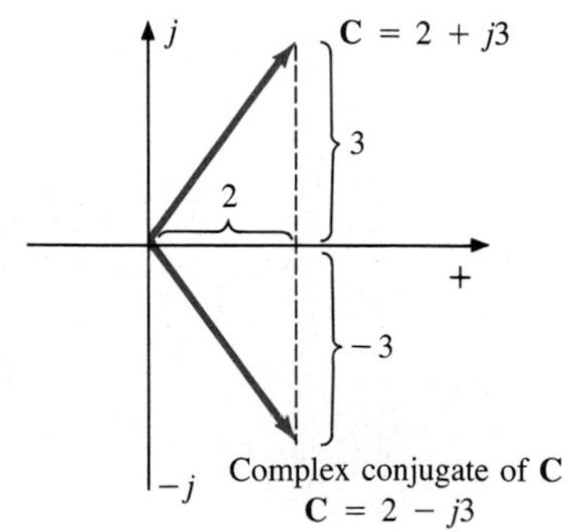

FIG. 14.50
Defining the complex conjugate of a complex number in rectangular form.

Complex Conjugate

The *conjugate* or *complex conjugate* of a complex number can be found by simply changing the sign of the imaginary part in the rectangular form or by using the negative of the angle of the polar form. For example, the conjugate of

$$\mathbf{C} = 2 + j3$$

is

$$2 - j3$$

as shown in Fig. 14.50. The conjugate of

$$\mathbf{C} = 2\angle 30°$$

is

$$2\angle -30°$$

as shown in Fig. 14.51.

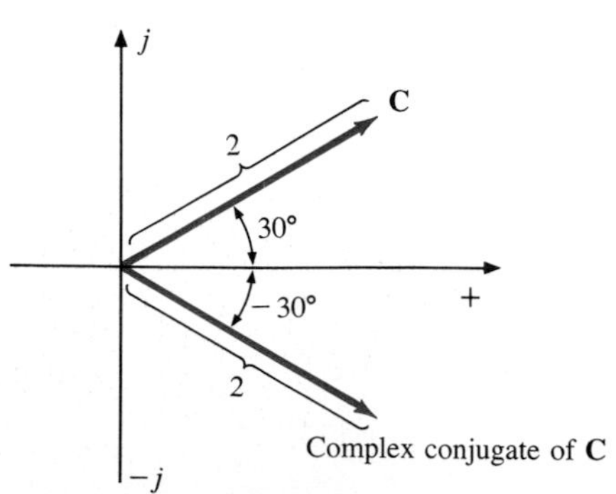

FIG. 14.51
Defining the complex conjugate of a complex number in polar form.

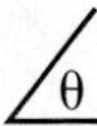

Reciprocal

The *reciprocal* of a complex number is 1 divided by the complex number. For example, the reciprocal of

$$\mathbf{C} = A + jB$$

is

$$\frac{1}{A + jB}$$

and of $C \angle\theta$,

$$\frac{1}{C \angle\theta}$$

We are now prepared to consider the four basic operations of *addition, subtraction, multiplication,* and *division* with complex numbers.

Addition

To add two or more complex numbers, simply add the real and imaginary parts separately. For example, if

$$\mathbf{C}_1 = \pm A_1 \pm jB_1 \quad \text{and} \quad \mathbf{C}_2 = \pm A_2 \pm jB_2$$

then

$$\mathbf{C}_1 + \mathbf{C}_2 = (\pm A_1 \pm A_2) + j(\pm B_1 \pm B_2) \tag{14.30}$$

There is really no need to memorize the equation. Simply set one above the other and consider the real and imaginary parts separately, as shown in Example 14.19.

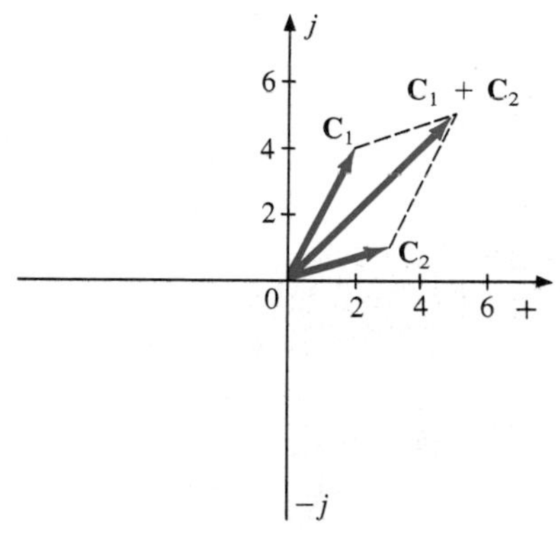

FIG. 14.52
Example 14.19(a).

EXAMPLE 14.19

a. Add $\mathbf{C}_1 = 2 + j4$ and $\mathbf{C}_2 = 3 + j1$.
b. Add $\mathbf{C}_1 = 3 + j6$ and $\mathbf{C}_2 = -6 + j3$.

Solutions:

a. By Eq. (14.30),

$$\mathbf{C}_1 + \mathbf{C}_2 = (2 + 3) + j(4 + 1) = \mathbf{5 + j5}$$

Note Fig. 14.52. An alternative method is

$$\begin{array}{c} 2 + j4 \\ \underline{3 + j1} \\ \downarrow \quad \downarrow \\ \mathbf{5 + j5} \end{array}$$

b. By Eq. (14.30),

$$\mathbf{C}_1 + \mathbf{C}_2 = (3 - 6) + j(6 + 3) = \mathbf{-3 + j9}$$

Note Fig. 14.53. An alternative method is

$$\begin{array}{c} 3 + j6 \\ \underline{-6 + j3} \\ \downarrow \quad \downarrow \\ \mathbf{-3 + j9} \end{array}$$

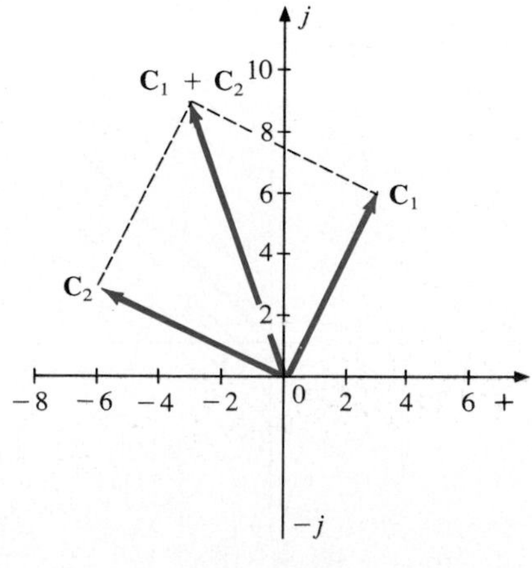

FIG. 14.53
Example 14.19(b).

Subtraction

In subtraction, the real and imaginary parts are again considered separately. For example, if

$$\mathbf{C}_1 = \pm A_1 \pm jB_1 \quad \text{and} \quad \mathbf{C}_2 = \pm A_2 \pm jB_2$$

then

$$\mathbf{C}_1 - \mathbf{C}_2 = [\pm A_2 - (\pm A_2)] + j\,[\pm B_1 - (\pm B_2)] \quad \textbf{(14.31)}$$

Again, there is no need to memorize the equation if the alternative method of Example 14.20 is employed.

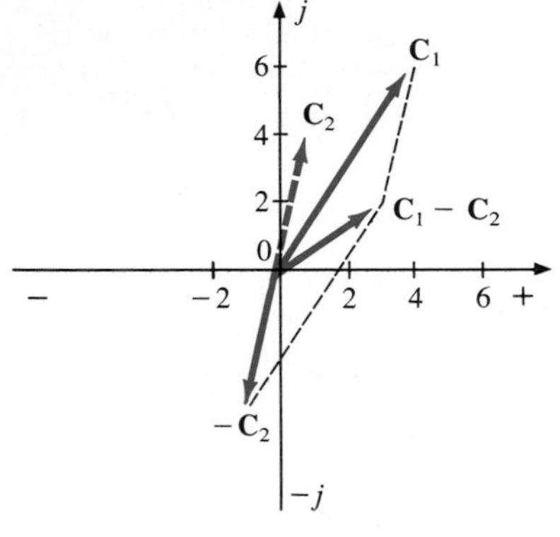

FIG. 14.54
Example 14.20(a).

EXAMPLE 14.20

a. Subtract $\mathbf{C}_2 = 1 + j4$ from $\mathbf{C}_1 = 4 + j6$.
b. Subtract $\mathbf{C}_2 = -2 + j5$ from $\mathbf{C}_1 = +3 + j3$.

Solutions:

a. By Eq. (14.31),

$$\mathbf{C}_1 - \mathbf{C}_2 = (4 - 1) + j(6 - 4) = \mathbf{3 + j2}$$

Note Fig. 14.54. An alternative method is

$$\begin{array}{c} 4 + j6 \\ \underline{-(1 + j4)} \\ \downarrow \quad \downarrow \\ \mathbf{3 + j2} \end{array}$$

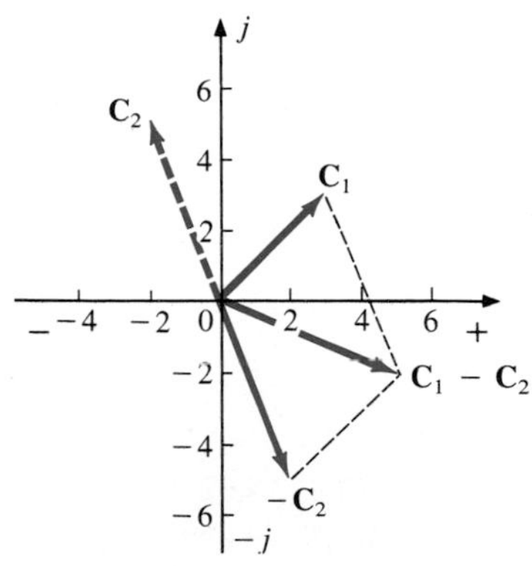

FIG. 14.55
Example 14.20(b).

b. By Eq. (14.31),

$$\mathbf{C}_1 - \mathbf{C}_2 = [3 - (-2)] + j(3 - 5) = \mathbf{5 - j2}$$

Note Fig. 14.55. An alternative method is

$$\begin{array}{c} 3 + j3 \\ \underline{-(-2 + j5)} \\ \downarrow \quad \downarrow \\ \mathbf{5 - j2} \end{array}$$

Addition or subtraction cannot be performed in polar form unless the complex numbers have the same angle θ or differ only by multiples of 180°.

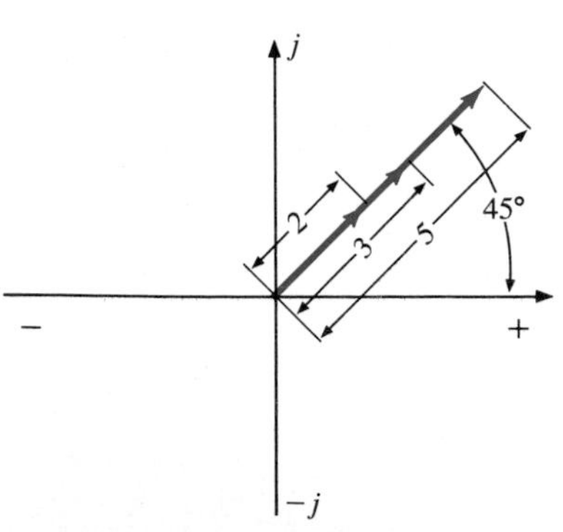

FIG. 14.56
Example 14.21(a).

EXAMPLE 14.21

a. $2\,\angle 45° + 3\,\angle 45° = \mathbf{5\,\angle 45°}$

Note Fig. 14.56. Or

b. $2\,\angle 0° - 4\,\angle 180° = \mathbf{6\,\angle 0°}$

Note Fig. 14.57.

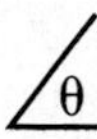

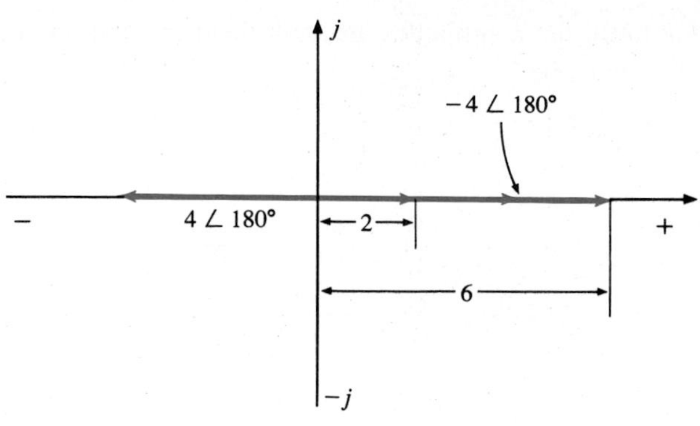

FIG. 14.57
Example 14.21(b).

Multiplication

To multiply two complex numbers in *rectangular* form, multiply the real and imaginary parts of one in turn by the real and imaginary parts of the other. For example, if

$$\mathbf{C}_1 = A_1 + jB_1 \quad \text{and} \quad \mathbf{C}_2 = A_2 + jB_2$$

then $\mathbf{C}_1 \cdot \mathbf{C}_2$:

$$\begin{array}{l} A_1 + jB_1 \\ A_2 + jB_2 \\ \hline A_1A_2 + jB_1A_2 \\ \quad + jA_1B_2 + j^2B_1B_2 \\ \hline A_1A_2 + j(B_1A_2 + A_1B_2) + B_1B_2(-1) \end{array}$$

and

$$\boxed{\mathbf{C}_1 \cdot \mathbf{C}_2 = (A_1A_2 - B_1B_2) + j\,(B_1A_2 + A_1B_2)} \qquad \textbf{(14.32)}$$

In Example 14.22(b), we obtain a solution without resorting to memorizing Eq. (14.32). Simply carry along the j factor when multiplying each part of one vector with the real and imaginary parts of the other.

EXAMPLE 14.22

a. Find $\mathbf{C}_1 \cdot \mathbf{C}_2$ if

$$\mathbf{C}_1 = 2 + j3 \quad \text{and} \quad \mathbf{C}_2 = 5 + j10$$

b. Find $\mathbf{C}_1 \cdot \mathbf{C}_2$ if

$$\mathbf{C}_1 = -2 - j3 \quad \text{and} \quad \mathbf{C}_2 = +4 - j6$$

Solutions:

a. Using the format above, we have

$$\begin{aligned}\mathbf{C}_1 \cdot \mathbf{C}_2 &= [(2)(5) - (3)(10)] + j\,[(3)(5) + (2)(10)] \\ &= \mathbf{-20 + j35}\end{aligned}$$

b. Without using the format, we obtain

$$\begin{array}{l} -2 - j3 \\ +4 - j6 \\ \hline -8 - j12 \\ \quad + j12 + j^2 18 \\ \hline -8 + j(-12 + 12) - 18 \end{array}$$

and

$$\mathbf{C}_1 \cdot \mathbf{C}_2 = \mathbf{-26 = 26\,\angle 180°}$$

In *polar* form, the magnitudes are multiplied and the angles added algebraically. For example, for

$$\mathbf{C}_1 = C_1 \angle\theta_1 \quad \text{and} \quad \mathbf{C}_2 = C_2 \angle\theta_2$$

we write

$$\boxed{\mathbf{C}_1 \cdot \mathbf{C}_2 = C_1C_2 \,\underline{/\theta_1 + \theta_2}} \tag{14.33}$$

EXAMPLE 14.23

a. Find $\mathbf{C}_1 \cdot \mathbf{C}_2$ if

$$\mathbf{C}_1 = 5 \angle 20° \quad \text{and} \quad \mathbf{C}_2 = 10 \angle 30°$$

b. Find $\mathbf{C}_1 \cdot \mathbf{C}_2$ if

$$\mathbf{C}_1 = 2 \angle -40° \quad \text{and} \quad \mathbf{C}_2 = 7 \angle +120°$$

Solutions:

a. $\mathbf{C}_1 \cdot \mathbf{C}_2 = (5 \angle 20°)(10 \angle 30°) = (5)(10) \,\underline{/20° + 30°} = \mathbf{50 \angle 50°}$

b. $\mathbf{C}_1 \cdot \mathbf{C}_2 = (2 \angle -40°)(7 \angle +120°) = (2)(7) \,\underline{/-40° + 120°}$
$= \mathbf{14 \angle +80°}$

To multiply a complex number in rectangular form by a real number requires that both the real part and the imaginary part be multiplied by the real number. For example,

$$(10)(2 + j3) = 20 + j30$$

and $$50 \angle 0°(0 + j6) = j300 = 300 \angle 90°$$

Division

To divide two complex numbers in *rectangular* form, multiply the numerator and denominator by the conjugate of the denominator and the resulting real and imaginary parts collected. That is, if

$$\mathbf{C}_1 = A_1 + jB_1 \quad \text{and} \quad \mathbf{C}_2 = A_2 + jB_2$$

then

$$\frac{\mathbf{C}_1}{\mathbf{C}_2} = \frac{(A_1 + jB_1)(A_2 - jB_2)}{(A_2 + jB_2)(A_2 - jB_2)}$$

$$= \frac{(A_1A_2 + B_1B_2) + j(A_2B_1 - A_1B_2)}{A_2^2 + B_2^2}$$

and

$$\boxed{\frac{\mathbf{C}_1}{\mathbf{C}_2} = \frac{A_1A_2 + B_1B_2}{A_2^2 + B_2^2} + j\frac{A_2B_1 - A_1B_2}{A_2^2 + B_2^2}} \tag{14.34}$$

The equation does not have to be memorized if the steps above used to obtain it are employed. That is, first multiply the numerator by the complex conjugate of the denominator and separate the real and imaginary terms. Then divide each term by the sum of each term of the denominator squared.

EXAMPLE 14.24

a. Find $\mathbf{C}_1/\mathbf{C}_2$ if $\mathbf{C}_1 = 1 + j4$ and $\mathbf{C}_2 = 4 + j5$.

b. Find $\mathbf{C}_1/\mathbf{C}_2$ if $\mathbf{C}_1 = -4 - j8$ and $\mathbf{C}_2 = +6 - j1$.

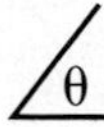

Solutions:

a. By Eq. (14.34),

$$\frac{\mathbf{C}_1}{\mathbf{C}_2} = \frac{(1)(4) + (4)(5)}{4^2 + 5^2} + j\frac{(4)(4) - (1)(5)}{4^2 + 5^2}$$

$$= \frac{24}{41} + \frac{j11}{41} \cong \mathbf{0.585 + j\,0.268}$$

b. Using an alternative method, we obtain

$$\begin{array}{l} -4 - j8 \\ \underline{+6 + j1} \\ -24 - j48 \\ \underline{\quad - j4 - j^2 8} \\ -24 - j52 + 8 = -16 - j52 \end{array}$$

$$\begin{array}{l} +6 - j1 \\ \underline{+6 + j1} \\ 36 + j6 \\ \underline{\quad - j6 - j^2 1} \\ 36 + 0 + \quad 1 = 37 \end{array}$$

and $$\frac{\mathbf{C}_1}{\mathbf{C}_2} = \frac{-16}{37} - \frac{j52}{37} = \mathbf{-0.432 - j1.405}$$

To divide a complex number in rectangular form by a real number, both the real part and the imaginary part must be divided by the real number. For example,

$$\frac{8 + j10}{2} = 4 + j5$$

and $$\frac{6.8 - j0}{2} = 3.4 - j0 = 3.4\angle 0^\circ$$

In *polar* form, division is accomplished by simply dividing the magnitude of the numerator by the magnitude of the denominator and subtracting the angle of the denominator from that of the numerator. That is, for

$$\mathbf{C}_1 = C_1\angle\theta_1 \quad \text{and} \quad \mathbf{C}_2 = C_2\angle\theta_2$$

we write

$$\boxed{\frac{\mathbf{C}_1}{\mathbf{C}_2} = \frac{C_1}{C_2}\underline{/\theta_1 - \theta_2}} \qquad \textbf{(14.35)}$$

EXAMPLE 14.25

a. Find $\mathbf{C}_1/\mathbf{C}_2$ if $\mathbf{C}_1 = 15\angle 10^\circ$ and $\mathbf{C}_2 = 2\angle 7^\circ$.

b. Find $\mathbf{C}_1/\mathbf{C}_2$ if $\mathbf{C}_1 = 8\angle 120^\circ$ and $\mathbf{C}_2 = 16\angle -50^\circ$.

Solutions:

a. $$\frac{\mathbf{C}_1}{\mathbf{C}_2} = \frac{15\angle 10^\circ}{2\angle 7^\circ} = \frac{15}{2}\underline{/10^\circ - 7^\circ} = \mathbf{7.5\angle 3^\circ}$$

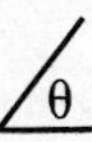

b. $\dfrac{\mathbf{C}_1}{\mathbf{C}_2} = \dfrac{8 \angle 120^\circ}{16 \angle -50^\circ} = \dfrac{8}{16} \underline{/120^\circ - (-50^\circ)} = \mathbf{0.5 \angle 170^\circ}$

We obtain the *reciprocal* in the rectangular form by multiplying the numerator and denominator by the complex conjugate of the denominator:

$$\frac{1}{A + jB} = \left(\frac{1}{A + jB}\right)\left(\frac{A - jB}{A - jB}\right) = \frac{A - jB}{A^2 + B^2}$$

and

$$\frac{1}{A + jB} = \frac{A}{A^2 + B^2} - j\frac{B}{A^2 + B^2} \qquad (14.36)$$

In the polar form, the reciprocal is

$$\frac{1}{C \angle \theta} = \frac{1}{C} \angle -\theta \qquad (14.37)$$

A concluding example using the four basic operations follows.

EXAMPLE 14.26 Perform the following operations, leaving the answer in polar or rectangular form:

a. $\dfrac{(2 + j3) + (4 + j6)}{(7 + j7) - (3 - j3)} = \dfrac{(2 + 4) + j(3 + 6)}{(7 - 3) + j(7 + 3)}$

$$= \frac{(6 + j9)(4 - j10)}{(4 + j10)(4 - j10)}$$

$$= \frac{[(6)(4) + (9)(10)] + j[(4)(9) - (6)(10)]}{4^2 + 10^2}$$

$$= \frac{114 - j24}{116} = \mathbf{0.983 - j0.207}$$

b. $\dfrac{(50 \angle 30^\circ)(5 + j5)}{10 \angle -20^\circ} = \dfrac{(50 \angle 30^\circ)(7.07 \angle 45^\circ)}{10 \angle -20^\circ} = \dfrac{353.5 \angle 75^\circ}{10 \angle -20^\circ}$

$$= 35.35 \underline{/75^\circ - (-20^\circ)} = \mathbf{35.35 \angle 95^\circ}$$

c. $\dfrac{(2 \angle 20^\circ)^2(3 + j4)}{8 - j6} = \dfrac{(2 \angle 20^\circ)(2 \angle 20^\circ)(5 \angle 53.13^\circ)}{10 \angle -36.87^\circ}$

$$= \frac{(4 \angle 40^\circ)(5 \angle 53.13^\circ)}{10 \angle -36.87^\circ} = \frac{20 \angle 93.13^\circ}{10 \angle -36.87^\circ}$$

$$= 2 \underline{/93.13^\circ - (-36.87^\circ)} = \mathbf{2.0 \angle 130^\circ}$$

d. $3 \angle 27^\circ - 6 \angle -40^\circ = (2.673 + j1.362) - (4.596 - j3.857)$

$$= (2.673 - 4.596) + j(1.362 + 3.857)$$

$$= \mathbf{-1.923 + j5.219}$$

14.11 CALCULATOR AND COMPUTER METHODS WITH COMPLEX NUMBERS

The process of converting from one form to another or working through lengthy operations with complex numbers can be time-consuming and often frustrating if one lost minus sign or decimal point invalidates the solution. Fortunately, technologists of today have calculators and computer methods that make the process measurably easier with higher degrees of reliability and accuracy.

Calculators

The TI-85 calculator of Fig. 14.58 is only one of numerous calculators that can convert from one form to another and perform lengthy calculations with complex numbers in a concise, neat form. All the details of using a specific calculator will not be included here because each has its own format and sequence of steps. However, the basic operations with the TI-85 will be included primarily to demonstrate the ease with which the conversions can be made and the format for more complex operations.

For the TI-85 calculator, one must first call up the 2nd function CPLX from the keyboard, which results in a menu at the bottom of the display including conj, real, imag, abs, and angle. By choosing the key MORE, ► Rec and ► Pol will appear as options (for the conversion process). To convert from one form to another, simply enter the current form in brackets with a comma between components for the rectangular form and an angle symbol for the polar form. Follow this form with the operation to be performed and press the ENTER key—the result will appear on the screen in the desired format.

FIG. 14.58
TI-85 scientific calculator. (Courtesy of Texas Instruments, Inc.)

EXAMPLE 14.27 This example is for demonstration purposes only. It is not expected that all the readers will have a TI-85 calculator. The sole purpose of the example is to demonstrate the power of today's calculators.

Using the TI-85 calculator, perform the following conversions:

a. $3 - j4$ to polar form.
b. $0.006 \angle 20.6^\circ$ to rectangular form.

Solutions:

a. The TI-85 display for part (a) is the following:

CALC. 14.1

b. The TI-85 display for part (b) is the following:

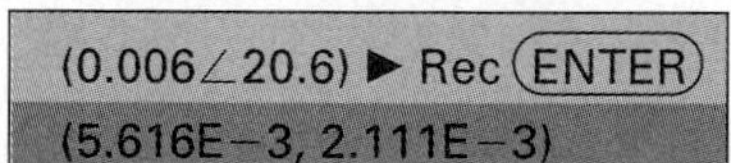

CALC. 14.2

EXAMPLE 14.28 Using the TI-85 calculator, perform the desired operations required in part (c) of Example 14.26 and compare solutions.

Solution: One must now be aware of the hierarchy of mathematical operations. In other words, in which sequence will the calculator perform the desired operations? In most cases, it is in the same order as is

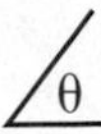

done in long-hand, although one must become adept at setting up the parentheses to ensure the correct order of operations. For this example, the TI-85 display is the following:

```
((2∠20)²*(3,4))/(8,−6)► Pol (ENTER)
(2.000E0∠130.000E0)
```

CALC. 14.3

which is a perfect match with the earlier solution.

Computer Methods

MathCad is not limited to the dc analysis performed in the earlier chapters but can perform the same operations on complex numbers. The format is slightly different because MathCad employs the letter i to represent the imaginary component and it is placed after the magnitude. The j can be employed instead of the i, however, simply by using <Esc>, entering FORMAT, and responding as prompted.

EXAMPLE 14.29 Use MathCad to perform the operations required in part (a) of Example 14.26.

Solution: The printout for the solution appears below:

$$\frac{(2 + 3j) + (4 + 6j)}{(7 + 7j) - (3 - 3j)} = 0.983 - 0.207j$$

MATHCAD 14.1

German-American
(Breslau, Germany; Yonkers and Schenectady, New York, USA)
(1865–1923)
Mathematician, Scientist, Engineer, Inventor, Professor of Electrical Engineering and Electrophysics
Union College
Department Head:
General Electric Co.

Courtesy of the Hall of History Foundation, Schenectady, New York

Although the holder of some 200 patents and recognized worldwide for his contributions to the study of hysteresis losses and electrical transients, Charles Proteus Steinmetz is best recognized for his contribution to the study of ac networks. His "Symbolic Method of Alternating-current Calculations" provided an approach to the analysis of ac networks that removed a great deal of the confusion and frustration experienced by engineers of that day as they made the transition from dc to ac systems. His approach (from which the phasor notation of this text is premised) permitted a direct analysis of ac systems using many of the theorems and methods of analysis developed for dc systems. In 1897 he authored the epic work *Theory and Calculation of Alternating Current Phenomena*, which became the "bible" for practicing engineers. Dr. Steinmetz was fondly referred to as "The Doctor" at General Electric Company where he worked for some 30 years in a number of important capacities. His recognition as a "multigifted genius" is supported by the fact that he maintained active friendships with such individuals as Albert Einstein, Guglielmo Marconi (radio), and Thomas A. Edison, to name just a few. He was President of the American Institute of Electrical Engineers (AIEE) and the National Association of Corporation Schools and actively supported his local community (Schenectady) as president of the Board of Education and the Commission on Parks and City Planning.

CHARLES PROTEUS STEINMETZ

14.12 PHASORS

As noted earlier in this chapter, the addition of sinusoidal voltages and currents will frequently be required in the analysis of ac circuits. One lengthy but valid method of performing this operation is to place both sinusoidal waveforms on the same set of axes and add algebraically the magnitudes of each at every point along the abscissa, as shown for $c = a + b$ in Fig. 14.59. This, however, can be a long and tedious process with limited accuracy. A shorter method uses the rotating radius vector first appearing in Fig. 13.16. This *radius vector,* having a *constant magnitude* (length) with *one end fixed at the origin,* is called a *phasor* when applied to electric circuits. During its rotational development of the sine wave, the phasor will, at the instant $t = 0$, have the positions shown in Fig. 14.60(a) for each waveform in Fig. 14.60(b).

Note in Fig. 14.60(b) that v_2 passes through the horizontal axis at $t = 0$ s, requiring that the radius vector in Fig. 14.60(a) be on the horizontal axis to ensure a vertical projection of zero volts at $t = 0$ s. Its length in Fig. 14.60(a) is equal to the peak value of the sinusoid as required by the radius vector of Fig. 13.16. The other sinusoid has passed through 90° of its rotation by the time $t = 0$ s is reached and

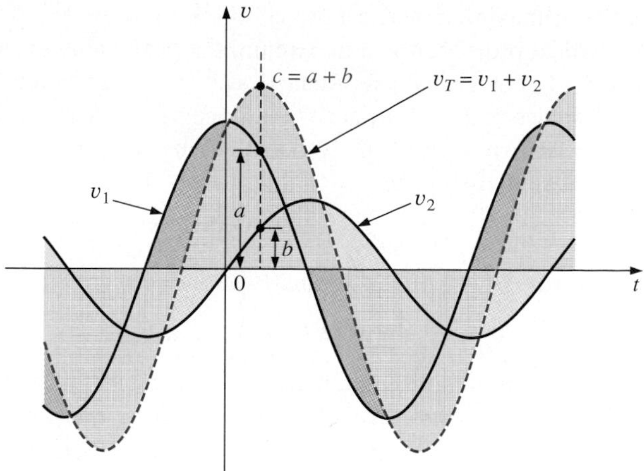

FIG. 14.59
Adding two sinusoidal waveforms on a point-by-point basis.

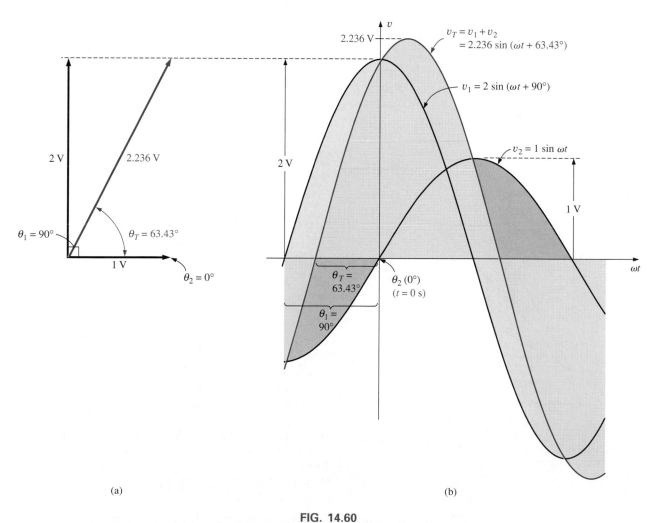

FIG. 14.60
(a) The phasor representation of the sinusoidal waveforms of Fig. 14.60(b); (b) finding the sum of two sinusoidal waveforms of v_1 and v_2.

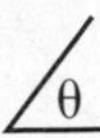

therefore has its maximum vertical projection as shown in Fig. 14.60(a). Since the vertical projection is a maximum, the peak value of the sinusoid that it will generate is also attained at $t = 0$ s, as shown in Fig. 14.60(b). Note also that $v_T = v_1$ at $t = 0$ s since $v_2 = 0$ V at this instant.

It can be shown [see Fig. 14.60(a)] using the vector algebra described in Section 14.10 that

$$1\text{ V}\angle 0^\circ + 2\text{ V}\angle 90^\circ = 2.236\text{ V}\angle 63.43^\circ$$

In other words, if we convert v_1 and v_2 to the phasor form using

$$v = V_m \sin(\omega t \pm \theta) \Rightarrow V_m \angle \pm \theta$$

and add them using complex number algebra, we can find the phasor form for v_T with very little difficulty. It can then be converted to the time domain and plotted on the same set of axes, as shown in Fig. 14.60(b). Figure 14.60(a), showing the magnitudes and relative positions of the various phasors, is called a *phasor diagram*. It is actually a "snapshot" of the rotating radius vectors at $t = 0$ s.

In the future, therefore, if the addition of two sinusoids is required, they should first be converted to the phasor domain and the sum found using complex algebra. The result can then be converted to the time domain.

The case of two sinusoidal functions having phase angles different from 0° and 90° appears in Fig. 14.61. Note again that the vertical height of the functions in Fig. 14.61(b) at $t = 0$ s is determined by the rotational positions of the radius vectors in Fig. 14.61(a).

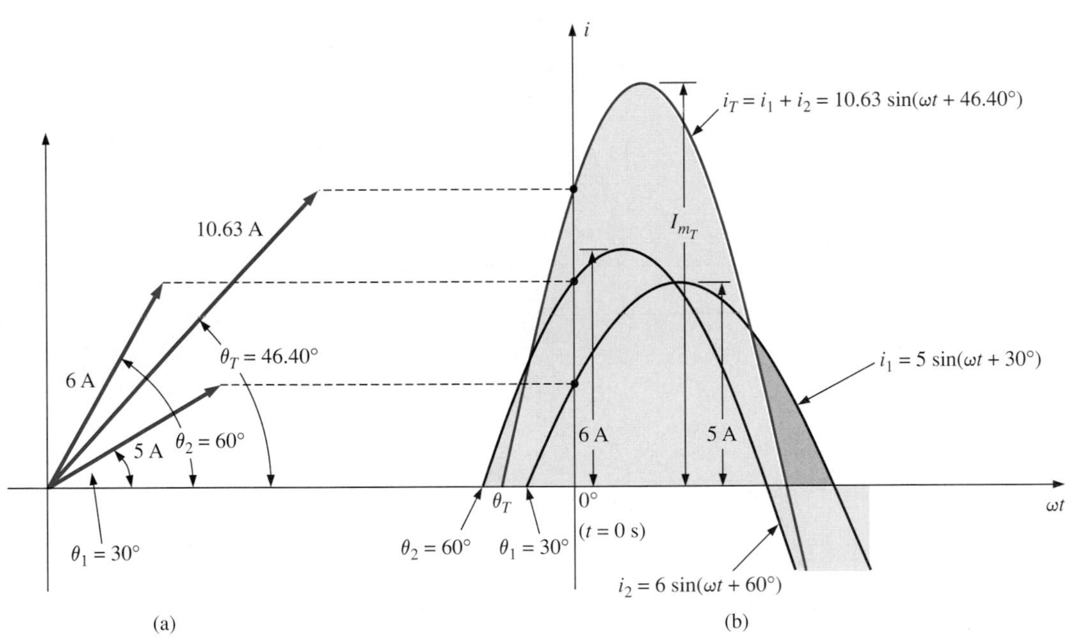

FIG. 14.61
Adding two sinusoidal currents with phase angles other than 90°.

Since the effective, rather than the peak, values are used almost exclusively in the analysis of ac circuits, the phasor will now be redefined for the purposes of practicality and uniformity as having a magnitude equal to the *effective value* of the sine wave it represents. The angle associated with the phasor will remain as previously described—the phase angle.

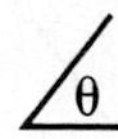

In general, for all of the analyses to follow, the phasor form of a sinusoidal voltage or current will be

$$\mathbf{V} = V \angle\theta \quad \text{and} \quad \mathbf{I} = I \angle\theta$$

where V and I are effective values and θ is the phase angle. It should be pointed out that in phasor notation, the sine wave is always the reference, and the frequency is not represented.

Phasor algebra for sinusoidal quantities is applicable only for waveforms having the same frequency.

EXAMPLE 14.30 Convert the following from the time to the phasor domain:

Time Domain	Phasor Domain
a. $\sqrt{2}(50) \sin \omega t$	$\mathbf{50 \angle 0°}$
b. $69.6 \sin(\omega t + 72°)$	$(0.707)(69.6) \angle 72° = \mathbf{49.21 \angle 72°}$
c. $45 \cos \omega t$	$(0.707)(45) \angle 90° = \mathbf{31.82 \angle 90°}$

EXAMPLE 14.31 Write the sinusoidal expression for the following phasors if the frequency is 60 Hz:

Phasor Domain	Time Domain
a. $\mathbf{I} = 10 \angle 30°$	$i = \sqrt{2}(10) \sin(2\pi 60t + 30°)$ and $i = \mathbf{14.14 \sin(377t + 30°)}$
b. $\mathbf{V} = 115 \angle -70°$	$v = \sqrt{2}(115) \sin(377t - 70°)$ and $v = \mathbf{162.6 \sin(377t - 70°)}$

EXAMPLE 14.32 Find the input voltage of the circuit of Fig. 14.62 if

$$\left.\begin{aligned} v_a &= 50 \sin(377t + 30°) \\ v_b &= 30 \sin(377t + 60°) \end{aligned}\right\} f = 60 \text{ Hz}$$

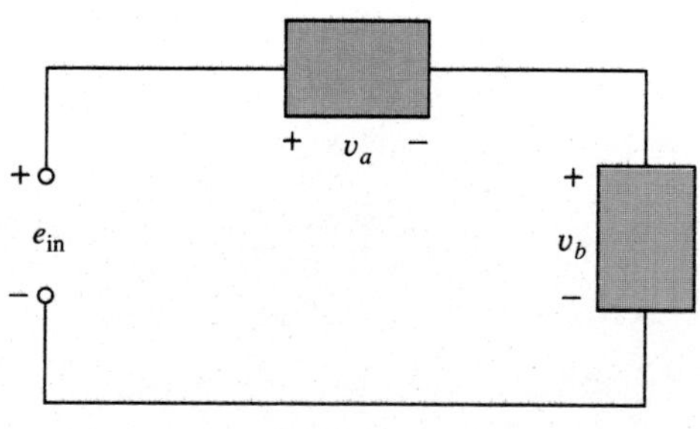

FIG. 14.62
Example 14.32.

Solution: Applying Kirchhoff's voltage law, we have

$$e_{in} = v_a + v_b$$

Converting from the time to the phasor domain yields

$$v_a = 50 \sin(377t + 30°) \Rightarrow \mathbf{V}_a = 35.35 \text{ V} \angle 30°$$
$$v_b = 30 \sin(377t + 60°) \Rightarrow \mathbf{V}_b = 21.21 \text{ V} \angle 60°$$

Converting from polar to rectangular form for addition yields

$$\mathbf{V}_a = 35.35 \text{ V} \angle 30° = 30.61 \text{ V} + j17.68 \text{ V}$$
$$\mathbf{V}_b = 21.21 \text{ V} \angle 60° = 10.61 \text{ V} + j18.37 \text{ V}$$

Then

$$\begin{aligned} \mathbf{E}_{in} = \mathbf{V}_a + \mathbf{V}_b &= (30.61 \text{ V} + j17.68 \text{ V}) + (10.61 \text{ V} + j18.37 \text{ V}) \\ &= 41.22 \text{ V} + j36.05 \text{ V} \end{aligned}$$

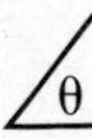

Converting from rectangular to polar form, we have

$$\mathbf{E}_{\text{in}} = 41.22\text{ V} + j36.05\text{ V} = 54.76\text{ V}\angle 41.17°$$

Converting from the phasor to the time domain, we obtain

$$\mathbf{E}_{\text{in}} = 54.76\text{ V}\angle 41.17° \Rightarrow e_{\text{in}} = \sqrt{2}(54.76)\sin(377t + 41.17°)$$

and $$e_{\text{in}} = \mathbf{77.43\ sin(377}\boldsymbol{t}\ \mathbf{+ 41.17°)}$$

A plot of the three waveforms is shown in Fig. 14.63. Note that at each instant of time, the sum of the two waveforms does in fact add up to e_{in}. At $t = 0$ ($\omega t = 0$), e_{in} is the sum of the two positive values, while at a value of ωt, almost midway between $\pi/2$ and π, the sum of the positive value of v_a and the negative value of v_b results in $e_{\text{in}} = 0$.

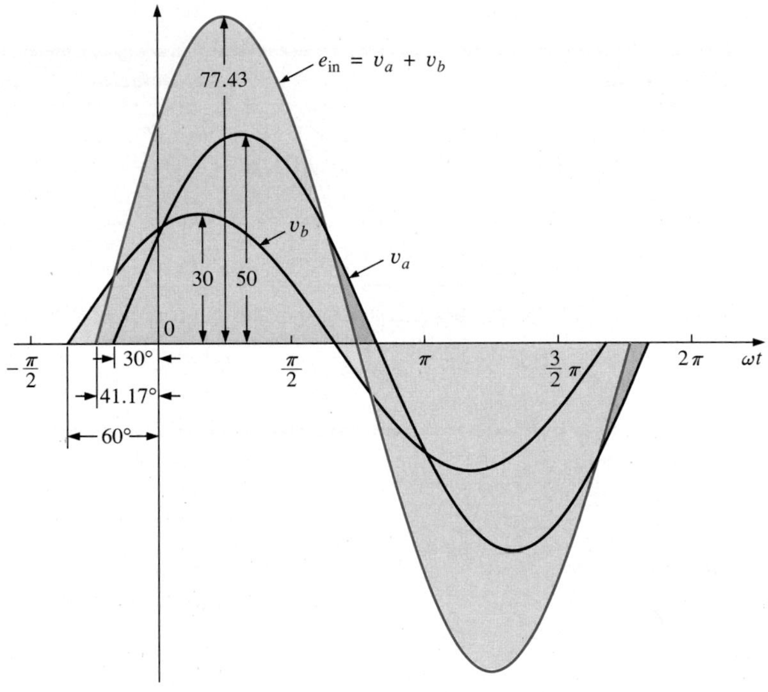

FIG. 14.63
Solution to Example 14.32.

EXAMPLE 14.33 Determine the current i_2 for the network of Fig. 14.64.

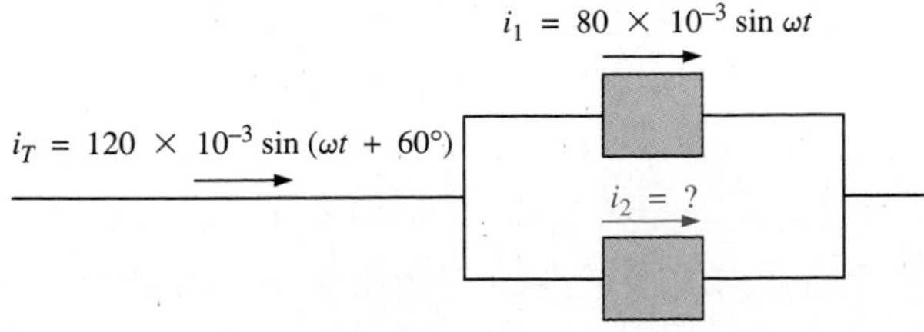

FIG. 14.64
Example 14.33.

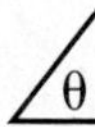

Solution: Applying Kirchhoff's current law, we obtain

$$i_T = i_1 + i_2 \quad \text{or} \quad i_2 = i_T - i_1$$

Converting from the time to the phasor domain yields

$$i_T = 120 \times 10^{-3} \sin(\omega t + 60°) \Rightarrow 84.84 \text{ mA } \angle 60°$$
$$i_1 = 80 \times 10^{-3} \sin \omega t \Rightarrow 56.56 \text{ mA } \angle 0°$$

Converting from polar to rectangular form for subtraction yields

$$\mathbf{I}_T = 84.84 \text{ mA } \angle 60° = 42.42 \text{ mA} + j73.47 \text{ mA}$$
$$\mathbf{I}_1 = 56.56 \text{ mA } \angle 0° = 56.56 \text{ mA} + j0$$

Then

$$\begin{aligned}\mathbf{I}_2 &= \mathbf{I}_T - \mathbf{I}_1 \\ &= (42.42 \text{ mA} + j73.47 \text{ mA}) - (56.56 \text{ mA} + j0)\end{aligned}$$

and
$$\mathbf{I}_2 = -14.14 \text{ mA} + j73.47 \text{ mA}$$

Converting from rectangular to polar form, we have

$$\mathbf{I}_2 = 74.82 \text{ mA } \angle 100.89°$$

Converting from the phasor to the time domain, we have

$$\mathbf{I}_2 = 74.82 \text{ mA } \angle 100.89° \Rightarrow$$
$$i_2 = \sqrt{2}(74.82 \times 10^{-3}) \sin(\omega t + 100.89°)$$

and
$$\mathbf{i_2 = 105.8 \times 10^{-3} \sin(\omega t + 100.89°)}$$

A plot of the three waveforms appears in Fig. 14.65. The waveforms clearly indicate that $i_T = i_1 + i_2$.

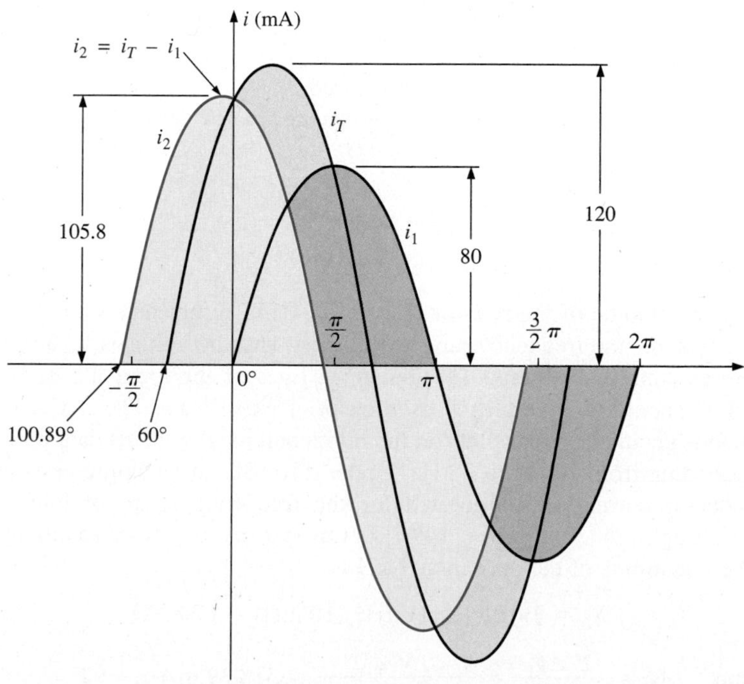

FIG. 14.65
Solution to Example 14.33.

14.13 COMPUTER ANALYSIS

PSpice (DOS)

The first ac system to be examined using the PSpice package is the isolated inductor of Fig. 14.66(a). However, due to a restriction of PSpice that does not permit isolated inductors (no internal resistance), a resistor must be added in series with the inductor, as shown in Fig. 14.66(b).

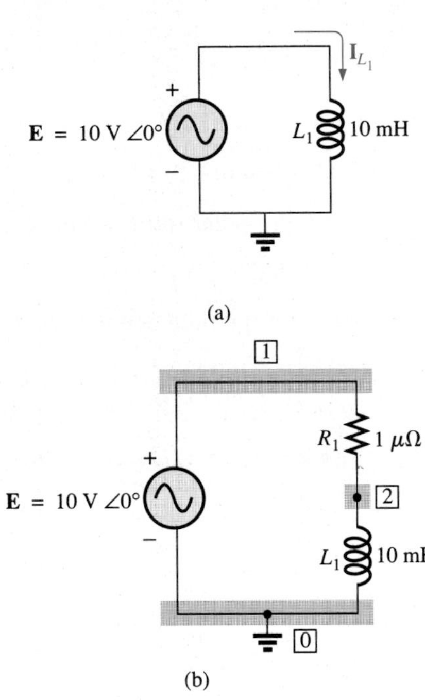

FIG. 14.66
Inductive network to be analyzed using PSpice (DOS).

The magnitude of the resistor (1×10^{-6} Ω) is sufficiently small to be ignored for the frequency range of interest, and the voltages V_1 and V_2 are essentially the same. The first three lines of the input file of Fig. 14.67 should now be clear, based on prior experience. The .AC command specifies a linear plot (on the horizontal axis) with 40 data points extending from 50 Hz to 2 kHz. Under .PROBE, the magnitude of the inductor current was requested for the frequency range of interest, resulting in the plot of Fig. 14.67. Let us test the results by calculating the magnitude of the current at 1 kHz:

$$X_L = 2\pi fL = 2\pi(1\text{ kHz})(10\text{ mH}) = 62.83\ \Omega$$

and $$I_L = \frac{E\angle\theta}{X_L\angle 90^\circ} = \frac{20\text{ V}\angle 0^\circ}{62.83\ \Omega\angle 90^\circ} = \mathbf{318.32\text{ mA}\angle -90^\circ}$$

which is a match with the results of Fig. 14.67.

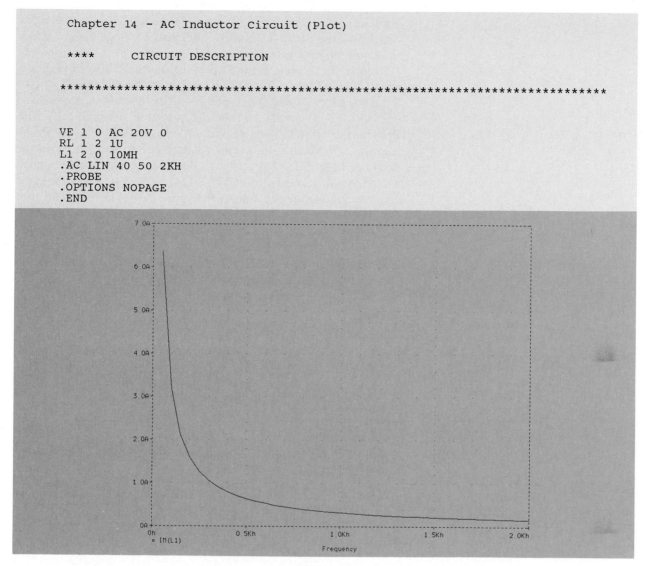

```
Chapter 14 - AC Inductor Circuit (Plot)

 ****     CIRCUIT DESCRIPTION

******************************************************************************

VE 1 0 AC 20V 0
RL 1 2 1U
L1 2 0 10MH
.AC LIN 40 50 2KH
.PROBE
.OPTIONS NOPAGE
.END
```

7.0A
6.0A
5.0A
4.0A
3.0A
2.0A
1.0A
0A
0h
0.5Kh
1.0Kh
1.5Kh
2.0Kh
▫ IM(L1)
Frequency

FIG. 14.67
Input and output files for the network of Fig. 14.66.

PSpice (Windows)

A capacitive network will now be analyzed using schematics. The **VSIN** source was chosen from **source.slb** and placed on the schematic, as shown in Fig. 14.68. Double-clicking the source will permit setting the attributes as follows: **DC** = 0V, **AC** = 5V, **VOFF** = 0V, **VAMPL** = 5V, **FREQ** = 1kHz, and **PHASE** = 0. In each case be sure to **Save Attr** and **Change Display** if the attribute is to appear on the schematic. For **AC, FREQ,** and **PHASE, Both name and value** was chosen. After ensuring that **Include System defined Attributes** is enabled in the **VSIN** dialog box, drop down to **PKGREF** and set equal to *E*, as shown in Fig. 14.68. In this latter case only the **Value** was displayed under **Change Display.**

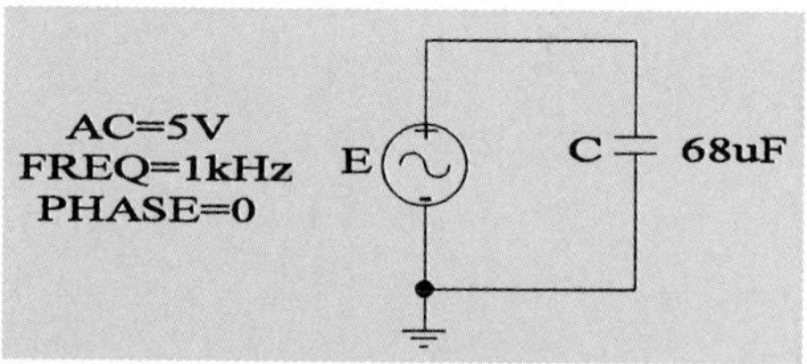

FIG. 14.68
Capacitive network to be analyzed using PSpice (Windows).

The remaining elements were chosen and placed as in earlier schematics. Under **Analysis-Setup-AC Sweep, Total Pts.** = 1, **Start Freq.** = 1 kHz and **End Freq.** = 1 kHz. Then under **Transient,** to permit the display of a function versus time, **Print Step** = 10 μs (time between print elements), **Final Time** = 3 ms (for three cycles), **No-Print Delay** = 0 s (print immediately), and **Step Ceiling** = 10 μs (ensure values calculated every 10 μs). Then under **Probe Setup** choose **Automatically Run Probe After Simulation** to avoid a delay in obtaining the **Probe** response after **Simulation.** Return to **Analysis, Simulate,** and when the **Analysis Type** dialog box appears, choose **Transient** (the **AC** option will be considered next). Then choose **Trace-Add-V(E:+)** to obtain the waveform for the applied voltage *E,* as appearing in the bottom half of Fig. 14.69. Repeating **Trace-Add-I(C)** will result in the waveform for the current, as shown in the same plot. Note that the current of a capacitor does in fact lead the voltage across the capacitor by 90°. The period of the applied voltage at 1 kHz is 1 ms. Note also that three cycles of each appear in the figure since the time scale extends to 3 ms.

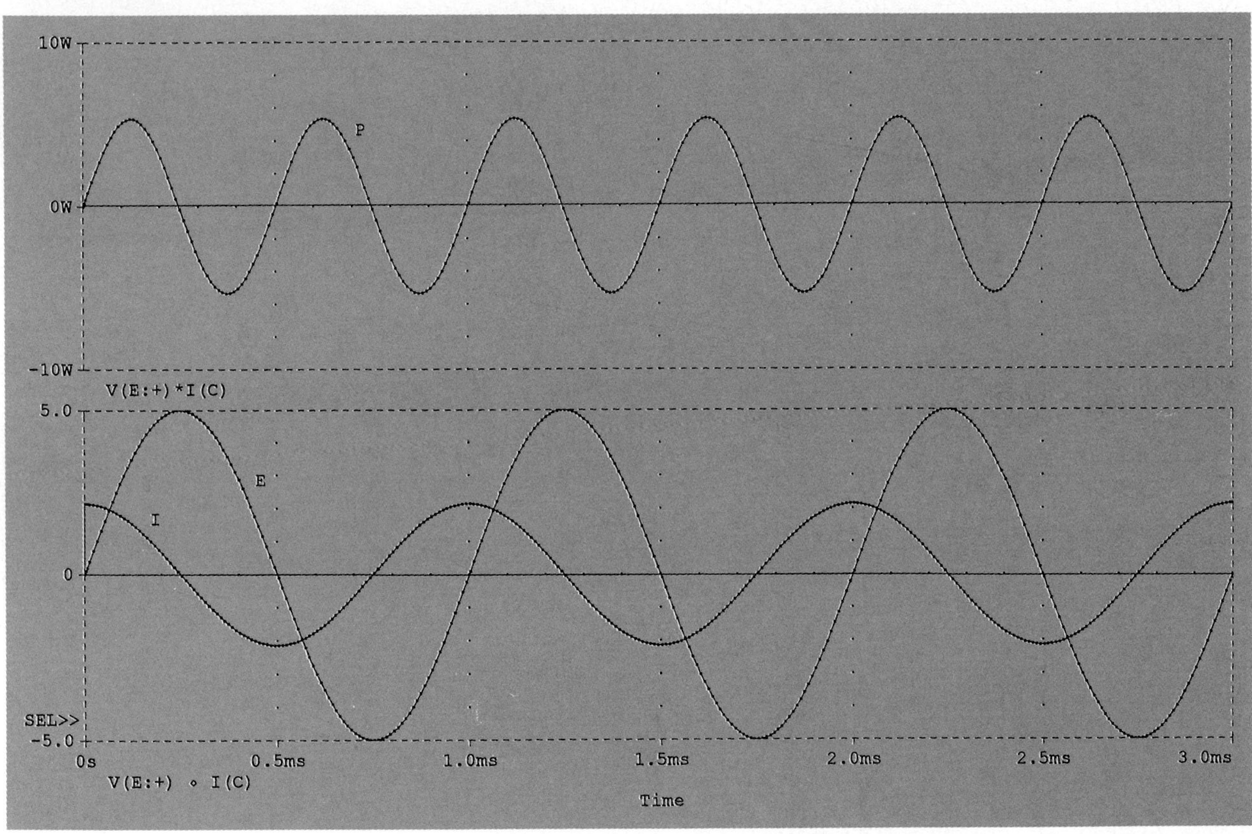

FIG. 14.69
Plots of e, i_C, and p for the capacitive network of Fig. 14.68.

Additional plots can be obtained by choosing **Plot-Add Plot** and then **Trace-Add.** In this case let us get the power curve for the power to the capacitive element as determined by the product of the voltage across the capacitor and the current of the capacitor. This is accomplished under the **Add Trace** dialog box by simply typing in **V(E:+)*I(C),** as appearing at the bottom of the plot in Fig. 14.69. Note

in this case that the average value of the power curve is zero (average or real power delivered = 0 W), as it must be for a purely reactive element. The lines at 0 W and 0 V were added for each plot by first selecting a plot (by simply clicking on the plot) and noting the position of **SEL>>,** which will define the plot of current interest. Then choose **Tools-Label-Line,** and a pencil that is used similar to line under **Draw** will appear. The labels **P, E,** and **I** were added using **Tools-Label-Text.**

The **AC** solutions option can now be chosen by returning to **Plot** and selecting **AC.** The screen will be cleared and a new horizontal scale will appear with 1 kHz as the only indicated intersection, as shown in Fig. 14.70. Then **Trace-Add-I(C)** will result in a mark in a vertical line at the 1-kHz point. Looking to the left of the current scale will reveal the magnitude of the current. Since it is difficult to read, return to the **Tools** option and select **Cursor-Display.** A horizontal line will appear through the above intersection that extends across the full length of the plot. The dialog box at the bottom of the plot will reveal the location of the cursor as 1 kHz and 2.1363 A. Other quantities for the network,

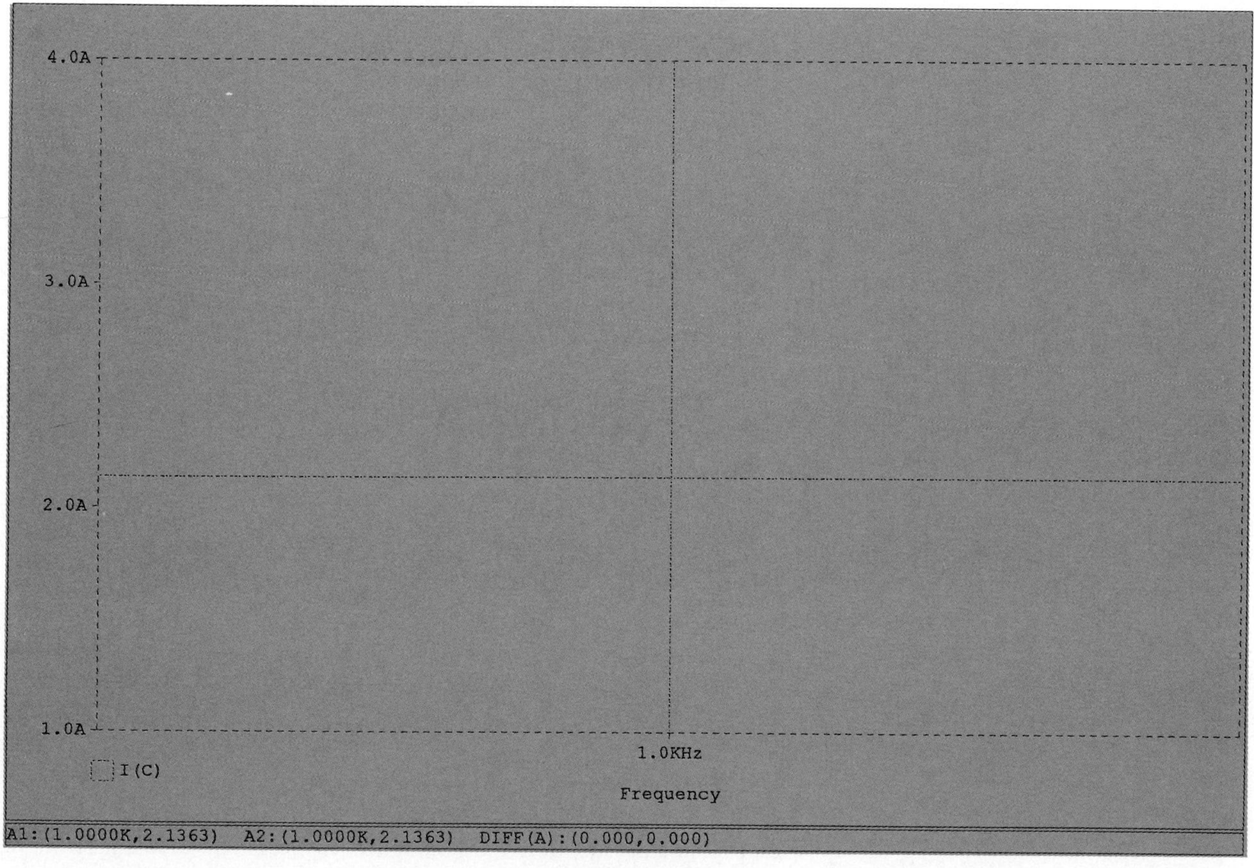

FIG. 14.70
Determining the magnitude of I_C using the cursor option.

such as the phase angle of $I(C)$ and the peak value of the power curve, can be found in the same manner. In fact, the peak value of the power curve of Fig. 14.69 is 5.341 W, which is an exact match with that obtained using the equation

$$\frac{V_m I_m}{2} = \frac{(5\text{ V})(2.1363\text{ A})}{2} = 5.341\text{ W}$$

appearing on Fig. 14.29. Note also that the frequency of the power curve is twice that of the applied voltage shown in Fig. 14.29. The **Cursor** option can also be used on the curves of Fig. 14.68, but additional uses of this option will be left for the chapters to follow.

C++

The versatility of the C++ programming language is clearly demonstrated by the following program designed to perform conversions between the polar and rectangular forms. Comments are provided on the right side of the program to help identify the function of specific lines or sections of the program. Recall that any comments to the right of the parallel slash bars // are ignored by the compiler. In this case the file *math.h* must be added to the preprocessor directive list, as shown in Fig. 14.71, to provide the mathematical functions to be employed in the program. A complete list of operations can be found in the compiler reference manual. The *#define* directive defines the level of *PI* to be employed when called for in the program and specifies the operations to be performed when *SQR*(*N*) and *SGN*(*N*) appear. The *?* associated with the *SGN*(*N*) directive is a *conditional operator* that specifies +1 if *N* is greater than or equal to 0 and −1 if not.

Next the variables are introduced and defined as floating points. The next entry includes the term *void* to indicate that the variable *to _polar* will not return a specific numerical value when part of an execution but rather may identify a subroutine or string of words or characters. The *void* within the parentheses reveals that the variable does not have a list of parameters associated with it for possible use in an application.

As described in earlier programs the *main* ( ) defines the point at which execution will begin, with the body of *main* defined by the opening and closing braces { }. Within *main*, an integer variable *choice* is introduced to handle the integer number (1 or 2) the user will choose in response to the question posed under *cout*. Through *cin* the user will respond with a 1 or 2, which will define the variable *choice*. The *switch* is a conditional response that will follow a path defined by the variable *choice*. The possible paths for the program to follow under *switch* are enclosed in the braces { }. Since a numerical value will determine the path, the options must begin with the word *case*. In this case a 1 will follow the *to_polar* structured variable, and a 2 will follow the *to_rectangular* structured variable. The *break* simply marks the end of the selection process.

On a *to_polar* choice the program will move to the subroutine *void to_polar* and convert the number to the polar form. The first six lines simply create line shifts and ask for the values of *X* and *Y*. The next line calculates the magnitude of the polar form (*Z*) using *SQR*(*N*), defined above, and the *sqrt* from the *math.h* header file. An *if* statement sensitive to the value of *X* and *Y* will then delineate which line will determine the phase angle of the polar form. The *SGN*(*N*), as introduced in the preprocessor listing, will determine the sign to be employed in the equation. The *a* preceding the tan function indicates arc tan or $\tan^{-1}$, while *PI* is as defined above in the preprocessor section. Note also that the angles must first be converted to radians by multiplying by the ratio $180°/\pi$. Once determined, the polar form is printed out using the *cout* statements.

Choosing the *to_rectangular* structured variable will cause the program to bypass the above subroutine and move directly to the polar-to-rectangular-conversion sequence. Again, the first six lines simply ask

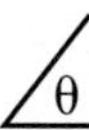


```
//C++ Rectangular/Polar Conversion

#include <iostream.h>                   //needed for input/output
#include <math.h>                       //needed for sqrt() and fabs()

#define PI 3.14159265
#define SQR(N) N * N                     //calculates N squared
#define SGN(N) ((N >= 0) ? 1 : -1) //calculates sign of number

       float X, Y, Z, TH; //define system variables

void to_polar(void);        //convert into polar function
void to_rectangular(void); //convert into rectangular function

main()
{
     int choice;   //needed for user choice input

     cout << "Enter (1) for rectangular to polar conversion\n";
     cout << "      (2) for polar to rectangular conversion\n";
     cout << "                  Choice=? ";
     cin >> choice;          //get choice from user
     switch(choice)          //match choice with case value
     {
             case 1 : to_polar(); break;
             case 2 : to_rectangular(); break;
     }
}

void to_polar()          //convert from rectangular to polar
{
     cout << "\n";
     cout << "Enter rectangular data:\n\n";
     cout << "X=? ";
     cin >> X;              //get X value from user
     cout << "Y=? ";
     cin >> Y;              //get Y value from user
     Z = sqrt(SQR(X) + SQR(Y));    //calculate magnitude
     //now do all the angle tests
     if (X > 0) TH = atan(Y / X) * 180.0 / PI;
     if (X < 0) TH = 180.0 * SGN(Y) + atan(Y / X) * 180.0 / PI;
     if (X == 0) TH = 90.0 * SGN(Y);
     if ((Y == 0) && (X < 0)) TH = 180.0;
     cout << "\n";
     cout << "Polar form is " << Z;
     cout << " at an angle of " << TH << " degrees\n";
}

void to_rectangular()   //convert from polar into rectangular
{
     cout << "\n";
     cout << "Enter polar data:\n\n";
     cout << "Z=? ";
     cin >> Z;              //get magnitude from user
     cout << "Angle(degrees). TH=? ";
     cin >> TH;             //get angle from user
     X = Z * cos(TH * PI / 180.0);    //calculate X value
     Y = Z * sin(TH * PI / 180.0);    //calculate Y value
     cout << "\n";
     cout << "Rectangular form is " << X;

       if (Y >= 0)        //test Y value
             cout << " +j ";
       else
             cout << " -j ";
       cout << fabs(Y) << "\n";     //display absolute value of Y value
}
```

FIG. 14.71
C++ program for complex number conversions.

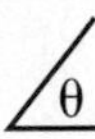

for the components of the polar form. The real and imaginary parts are then calculated and the results printed out. Note the *if-else* statement required to associate the proper signed *j* with the imaginary part.

In an effort to clearly identify the major components of the program, brackets have been added at the edge of the program with a short description of the function performed. As mentioned earlier, do not be concerned if a number of questions arise about the program structure or specific commands or statements. The purpose here is simply to introduce the basic format of the C++ programming language and not to provide all the details required to write your own programs.

Two runs of the program have been provided in Figs. 14.72 and 14.73, one for a polar-to-rectangular conversion and the other for a rectangular-to-polar conversion. Note in each case the result of the *cout* and *cin* statements and in general the clean, clear, and direct format of the resulting output.

```
Enter (1) for rectangular to polar conversion
      (2) for polar to rectangular conversion
                   Choice=? 2

Enter polar data:

Z=? 12
Angle(degrees). TH=? 35

Rectangular form is 9.829824 +j 6.882917
```

FIG. 14.72
Polar-to-rectangular conversion using the C++ program of Fig. 14.71.

```
Enter (1) for rectangular to polar conversion
      (2) for polar to rectangular conversion
                   Choice=? 1

Enter rectangular data:
X=? -10
Y=? 20

Polar form is 22.36068 at an angle of 116.565048 degrees
```

FIG. 14.73
Rectangular-to-polar conversion using the C++ program of Fig. 14.71.

BASIC

The BASIC program of Fig. 14.74 will perform the same conversion process just described for C++ to permit a direct comparison between the resulting programs for each. The first obvious difference is the need for line numbers in BASIC that do not appear in C++. Keep in mind that the line numbers not only identify each line but also define the sequence in which the operations are to be performed and actually

```
10 REM ***** PROGRAM 14-1 *****
20 REM *******************************************
30 REM Program to perform selected conversions
40 REM *******************************************
50 REM
100 PRINT
110 PRINT "Enter (1) for rectangular to polar conversion"
120 PRINT "      (2) for polar to rectangular conversion"
130 PRINT TAB(20);
140 INPUT "Choice="; C: REM C is choice 1 or 2
150 IF C < 0 OR C > 2 THEN GOTO 110
160 ON C GOSUB 200, 300
170 PRINT : INPUT "More(YES or NO)"; A$
180 IF A$ = "YES" THEN GOTO 100
190 END
200 REM Use rectangular to polar conversion module
210 PRINT : PRINT : PRINT "Enter rectangular data:"
220 INPUT "X="; X: INPUT "Y="; Y
230 GOSUB 2000
240 PRINT : PRINT "Polar form is"; Z; "at an angle of"; TH; "degrees"
250 RETURN
300 REM Use polar to rectangular conversion
310 PRINT : PRINT "Enter polar data:": PRINT : INPUT "Z="; Z
320 INPUT "Angle(degrees). TH="; TH
330 GOSUB 2100
340 PRINT : PRINT "Rectangular form is"; X;
350 IF Y >= 0 THEN PRINT "+j"; Y
360 IF Y < 0 THEN PRINT "-j"; ABS(Y)
370 RETURN
2000 REM Module to convert from rectangular to polar form.
2010 REM Enter with X,Y - Return with Z, TH(eta)
2020 Z = SQR(X ^ 2 + Y ^ 2)
2030 IF X > 0 THEN TH = (180 / 3.14159) * ATN(Y / X)
2040 IF X < 0 THEN TH = 180 * SGN(Y) + (180 / 3.14159) * ATN(Y / X)
2050 IF X = 0 THEN TH = 90 * SGN(Y)
2060 IF Y = 0 THEN IF X < 0 THEN TH = 180
2070 RETURN
2100 REM Module to convert from polar to rectangular form.
2110 REM Enter with Z, TH(eta) - return with X,Y
2120 X = Z * COS(TH * 3.14159 / 180)
2130 Y = Z * SIN(TH * 3.14159 / 180)
2140 RETURN
```

Margin annotations: Input-(Rect.) [line 220]; Output-(Polar) [line 240]; Input (Polar) [lines 310–320]; Output (Rect.) [lines 340–360]; Rect. ↓ Polar [lines 2000–2070]; Polar ↓ Rect. [lines 2100–2140]

FIG. 14.74
BASIC program for complex number conversions.

appear in the program as target points for a GOTO or GOSUB command. In C++ the operating sequence is line by line, with statements such as loop altering the eventual path. In general, a reader with no experience in BASIC or C++ will probably find the BASIC language easier to follow. However, keep in mind that C++ is a more efficient link with the computer hardware, resulting in faster running programs and in design advantages. In total, the C++ program seems longer, but be aware that the format of the C++ program was chosen for clarity. More efficient formats can be chosen to make the two equal in length or in fact make the C++ program shorter.

For the BASIC program the input parameters of the rectangular form are entered on line 220, and the polar form on lines 310 and 320. Line 240 outputs the polar form, and lines 340 through 360 output the rectangular form. The rectangular-to-polar conversion routine appears on lines 2000 through 2070, whereas the polar-to-rectangular conversion appears on lines 2100 through 2140. Note on line 2020 the equation for the magnitude of the polar form ($Z = \sqrt{X^2 + Y^2}$) and the testing of X and Y to determine the correct value of θ on lines 2030 through 2060. Lines 2120 and 2130 determine X and Y using the equations $X = Z \cos \theta$ and $Y = Z \sin \theta$, respectively. Note the need to convert the input angle in degrees (TH) to radians before the BASIC language can act on the SIN and COS functions.

Two runs have been provided in Fig. 14.75. The first converts the polar form $5 \angle -53.13°$ to $3 - j4$, and the second, the rectangular form $-10 + j20$ to $22.3607 \angle 116.565°$.

```
Enter (1) for rectangular to polar conversion
      (2) for polar to rectangular conversion
                 Choice=? 2

Enter polar data:

Z=? 5
Angle(degrees). TH=? -53.13

Rectangular form is 3.00001 -j 3.999993

More(YES or NO)? YES

Enter (1) for rectangular to polar conversion
      (2) for polar to rectangular conversion
                 Choice=? 1

Enter rectangular data:
X=? -10
Y=? 20

Polar form is 22.36068 at an angle of 116.565 degrees

More(YES or NO)? NO
```

FIG. 14.75

Complex number conversions using the BASIC program of Fig. 14.74.

PROBLEMS

SECTION 14.2 The Derivative

1. Plot the following waveform versus time showing one clear, complete cycle. Then determine the derivative of the waveform using Eq. (14.1), and sketch one complete cycle of the derivative directly under the original waveform. Compare the magnitude of the derivative at various points versus the slope of the original sinusoidal function.

$$v = 1 \sin 3.14t$$

2. Repeat Problem 1 for the following sinusoidal function and compare results. In particular, determine the frequency of the waveforms of Problems 1 and 2 and compare the magnitude of the derivative.

$$v = 1 \sin 15.71t$$

3. What is the derivative of each of the following sinusoidal expressions?

a. $10 \sin 377t$ **b.** $0.6 \sin(754t + 20°)$
c. $\sqrt{2}\, 20 \sin(157t - 20°)$ **d.** $-200 \sin(t + 180°)$

SECTION 14.3 Response of Basic *R*, *L*, and *C* Elements to a Sinusoidal Voltage or Current

4. The voltage across a 5-Ω resistor is as indicated. Find the sinusoidal expression for the current. In addition, sketch the v and i sinusoidal waveforms on the same axis.

a. $150 \sin 377t$ **b.** $30 \sin(377t + 20°)$
c. $40 \cos(\omega t + 10°)$ **d.** $-80 \sin(\omega t + 40°)$

5. The current through a 7-kΩ resistor is as indicated. Find the sinusoidal expression for the voltage. In addition, sketch the v and i sinusoidal waveforms on the same axis.

a. $0.03 \sin 754t$
b. $2 \times 10^{-3} \sin(400t - 120°)$
c. $6 \times 10^{-6} \cos(\omega t - 2°)$
d. $-0.004 \cos(\omega t - 90°)$

6. Determine the inductive reactance (in ohms) of a 2-H coil for

a. dc

and for the following frequencies:

b. 25 Hz **c.** 60 Hz
d. 2000 Hz **e.** 100,000 Hz

7. Determine the inductance of a coil that has a reactance of
   a. 20 Ω at $f = 2$ Hz.
   b. 1000 Ω at $f = 60$ Hz.
   c. 5280 Ω at $f = 1000$ Hz.

8. Determine the frequency at which a 10-H inductance has the following inductive reactances:
   a. 50 Ω b. 3770 Ω
   c. 15.7 kΩ d. 243 Ω

9. The current through a 20-Ω inductive reactance is given. What is the sinusoidal expression for the voltage? Sketch the v and i sinusoidal waveforms on the same axis.
   a. $i = 5 \sin \omega t$ b. $i = 0.4 \sin(\omega t + 60°)$
   c. $i = -6 \sin(\omega t - 30°)$ d. $i = 3 \cos(\omega t + 10°)$

10. The current through a 0.1-H coil is given. What is the sinusoidal expression for the voltage?
    a. $30 \sin 30t$
    b. $0.006 \sin 377t$
    c. $5 \times 10^{-6} \sin(400t + 20°)$
    d. $-4 \cos(20t - 70°)$

11. The voltage across a 50-Ω inductive reactance is given. What is the sinusoidal expression for the current? Sketch the v and i sinusoidal waveforms on the same axis.
    a. $50 \sin \omega t$ b. $30 \sin(\omega t + 20°)$
    c. $40 \cos(\omega t + 10°)$ d. $-80 \sin(377t + 40°)$

12. The voltage across a 0.2-H coil is given. What is the sinusoidal expression for the current?
    a. $1.5 \sin 60t$
    b. $0.016 \sin(t + 4°)$
    c. $-4.8 \sin(0.05t + 50°)$
    d. $9 \times 10^{-3} \cos(377t + 360°)$

13. Determine the capacitive reactance (in ohms) of a 5-μF capacitor for
    a. dc

    and for the following frequencies:
    b. 60 Hz c. 120 Hz
    d. 1800 Hz e. 24,000 Hz

14. Determine the capacitance in microfarads if a capacitor has a reactance of
    a. 250 Ω at $f = 60$ Hz.
    b. 55 Ω at $f = 312$ Hz.
    c. 10 Ω at $f = 25$ Hz.

15. Determine the frequency at which a 50-μF capacitor has the following capacitive reactances:
    a. 342 Ω b. 684 Ω
    c. 171 Ω d. 2000 Ω

16. The voltage across a 2.5-Ω capacitive reactance is given. What is the sinusoidal expression for the current? Sketch the v and i sinusoidal waveforms on the same axis.
    a. $100 \sin \omega t$ b. $0.4 \sin(\omega t + 20°)$
    c. $8 \cos(\omega t + 10°)$ d. $-70 \sin(\omega t + 40°)$

17. The voltage across a 1-μF capacitor is given. What is the sinusoidal expression for the current?
    a. $30 \sin 200t$ b. $90 \sin 377t$
    c. $-120 \sin(374t + 30°)$ d. $70 \cos(800t - 20°)$

18. The current through a 10-Ω capacitive reactance is given. Write the sinusoidal expression for the voltage. Sketch the v and i sinusoidal waveforms on the same axis.
    a. $i = 50 \sin \omega t$ b. $i = 40 \sin(\omega t + 60°)$
    c. $i = -6 \sin(\omega t - 30°)$ d. $i = 3 \cos(\omega t + 10°)$

19. The current through a 0.5-μF capacitor is given. What is the sinusoidal expression for the voltage?
    a. $0.20 \sin 300t$ b. $0.007 \sin 377t$
    c. $0.048 \cos 754t$ d. $0.08 \sin(1600t - 80°)$

*20. For the following pairs of voltages and currents, indicate whether the element involved is a capacitor, inductor, or resistor, and the value of C, L, or R if sufficient data are given:
    a. $v = 550 \sin(377t + 40°)$
       $i = 11 \sin(377t - 50°)$
    b. $v = 36 \sin(754t + 80°)$
       $i = 4 \sin(754t + 170°)$
    c. $v = 10.5 \sin(\omega t + 13°)$
       $i = 1.5 \sin(\omega t + 13°)$

*21. Repeat Problem 20 for the following pairs of voltages and currents:
    a. $v = 2000 \sin \omega t$
       $i = 5 \cos \omega t$
    b. $v = 80 \sin(157t + 150°)$
       $i = 2 \sin(157t + 60°)$
    c. $v = 35 \sin(\omega t - 20°)$
       $i = 7 \cos(\omega t - 110°)$

SECTION 14.4 Frequency Response of the Basic Elements

22. Plot X_L versus frequency for a 5-mH coil using a frequency range of zero to 100 kHz on a linear scale.

23. Plot X_C versus frequency for a 1-μF capacitor using a frequency range of zero to 10 kHz on a linear scale.

24. At what frequency will the reactance of a 1-μF capacitor equal the resistance of a 2-kΩ resistor?

25. The reactance of a coil equals the resistance of a 10-kΩ resistor at a frequency of 5 kHz. Determine the inductance of the coil.

26. Determine the frequency at which a 1-μF capacitor and a 10-mH inductor will have the same reactance.

27. Determine the capacitance required to establish a capacitive reactance that will match that of a 2-mH coil at a frequency of 50 kHz.

SECTION 14.5 Average Power and Power Factor

28. Find the average power loss in watts for each set in Problem 20.

29. Find the average power loss in watts for each set in Problem 21.

***30.** Find the average power loss and power factor for each of the circuits whose input current and voltage are as follows:

a. $v = 60 \sin(\omega t + 30°)$
$i = 15 \sin(\omega t + 60°)$

b. $v = -50 \sin(\omega t - 20°)$
$i = -2 \sin(\omega t + 40°)$

c. $v = 50 \sin(\omega t + 80°)$
$i = 3 \cos(\omega t + 20°)$

d. $v = 75 \sin(\omega t - 5°)$
$i = 0.08 \sin(\omega t - 35°)$

31. If the current through and voltage across an element are $i = 8 \sin(\omega t + 40°)$ and $v = 48 \sin(\omega t + 40°)$, respectively, compute the power by I^2R, $(V_mI_m/2) \cos \theta$, and $VI \cos \theta$, and compare answers.

32. A circuit dissipates 100 W (average power) at 150 V (effective input voltage) and 2 A (effective input current). What is the power factor? Repeat if the power is 0 W; 300 W.

***33.** The power factor of a circuit is 0.5 lagging. The power delivered in watts is 500. If the input voltage is $50 \sin(\omega t + 10°)$, find the sinusoidal expression for the input current.

34. In Fig. 14.76, $e = 30 \sin(377t + 20°)$.

a. What is the sinusoidal expression for the current?

b. Find the power loss in the circuit.

c. How long (in seconds) does it take the current to complete 6 cycles?

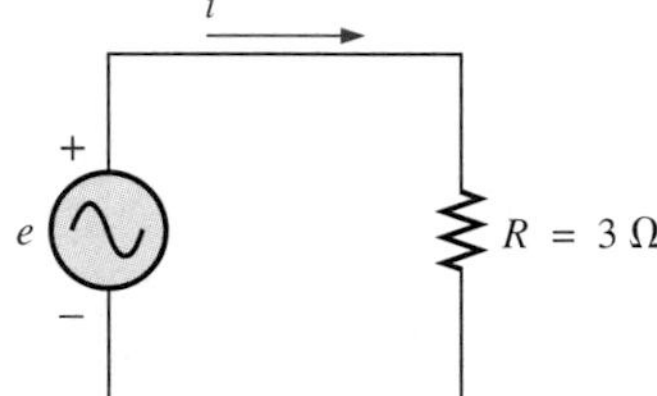

FIG. 14.76
Problem 34.

35. In Fig. 14.77, $e = 100 \sin(157t + 30°)$.

a. Find the sinusoidal expression for i.

b. Find the value of the inductance L.

c. Find the average power loss by the inductor.

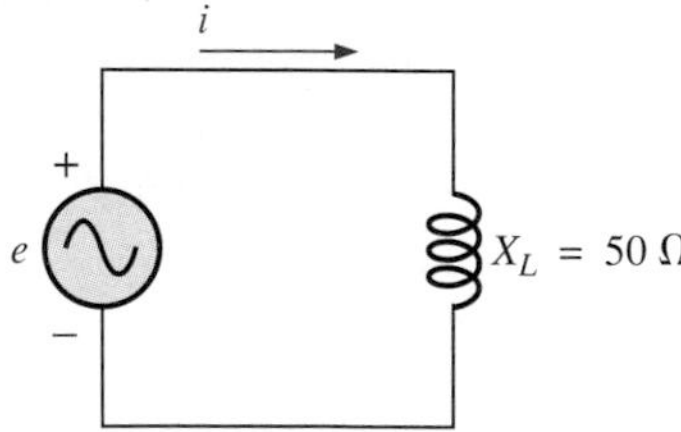
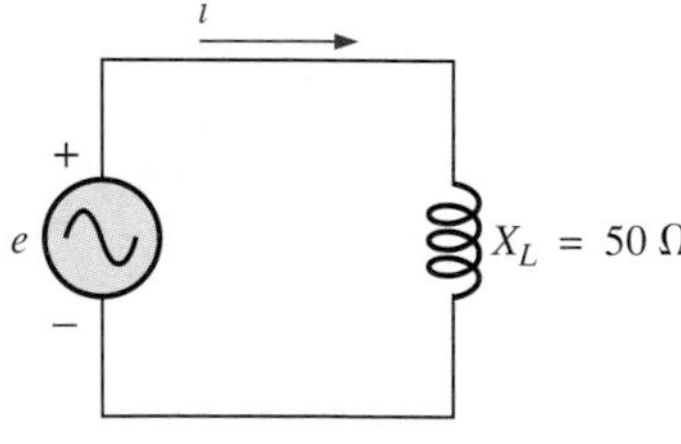

FIG. 14.77
Problem 35.

36. In Fig. 14.78, $i = 3 \sin(377t - 20°)$.

a. Find the sinusoidal expression for e.

b. Find the value of the capacitance C in microfarads.

c. Find the average power loss in the capacitor.

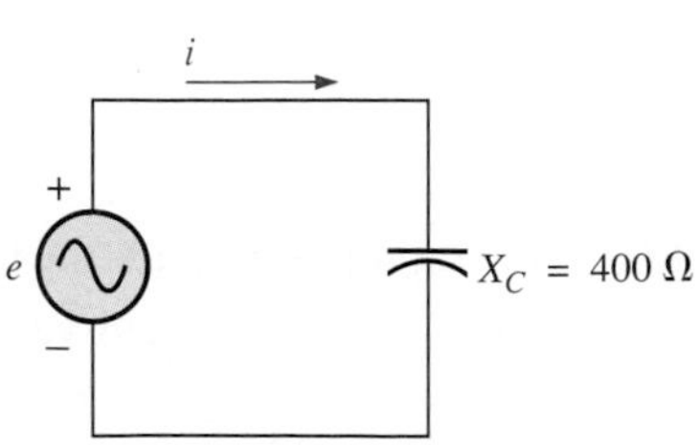

FIG. 14.78
Problem 36.

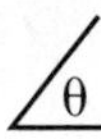

***37.** For the network of Fig. 14.79 and the applied signal:

a. Determine i_1 and i_2.

b. Find i_s.

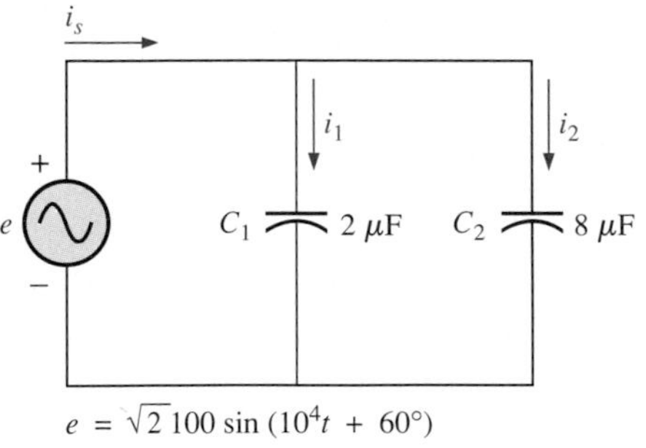

FIG. 14.79
Problem 37.

***38.** For the network of Fig. 14.80 and the applied source:

a. Determine the source voltage v_s.

b. Find the currents i_1 and i_2.

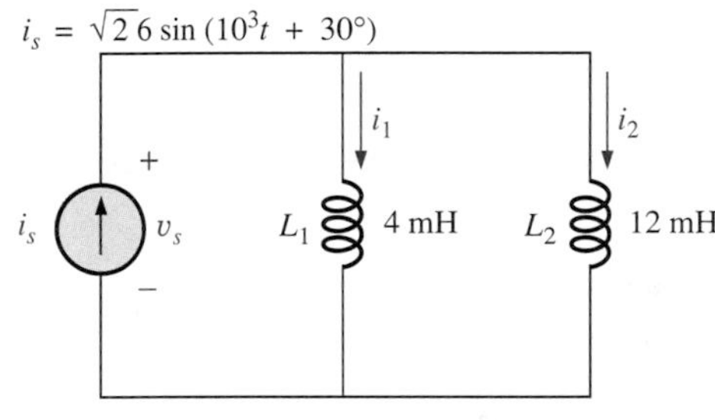

FIG. 14.80
Problem 38.

SECTION 14.9 Conversion Between Forms

39. Convert the following from rectangular to polar form:

a. $4 + j3$ **b.** $2 + j2$
c. $3.5 + j16$ **d.** $100 + j800$
e. $1000 + j400$ **f.** $0.001 + j0.0065$
g. $7.6 - j9$ **h.** $-8 + j4$
i. $-15 - j60$ **j.** $+78 - j65$
k. $-2400 + j3600$
l. $5 \times 10^{-3} - j25 \times 10^{-3}$

40. Convert the following from polar to rectangular form:

a. $6 \angle 30°$ **b.** $40 \angle 80°$
c. $7400 \angle 70°$ **d.** $4 \times 10^{-4} \angle 8°$
e. $0.04 \angle 80°$ **f.** $0.0093 \angle 23°$
g. $65 \angle 150°$ **h.** $1.2 \angle 135°$
i. $500 \angle 200°$ **j.** $6320 \angle -35°$
k. $7.52 \angle -125°$ **l.** $0.008 \angle 310°$

41. Convert the following from rectangular to polar form:

a. $1 + j15$ **b.** $60 + j5$
c. $0.01 + j0.3$ **d.** $100 - j2000$
e. $-5.6 + j86$ **f.** $-2.7 - j38.6$

42. Convert the following from polar to rectangular form:

a. $13 \angle 5°$ **b.** $160 \angle 87°$
c. $7 \times 10^{-6} \angle 2°$ **d.** $8.7 \angle 177°$
e. $76 \angle -4°$ **f.** $396 \angle +265°$

SECTION 14.10 Mathematical Operations with Complex Numbers

Perform the following operations.

43. Addition and subtraction (express your answers in rectangular form):

a. $(4.2 + j6.8) + (7.6 + j0.2)$

b. $(142 + j7) + (9.8 + j42) + (0.1 + j0.9)$

c. $(4 \times 10^{-6} + j76) + (7.2 \times 10^{-7} - j5)$
d. $(9.8 + j6.2) - (4.6 + j4.6)$
e. $(167 + j243) - (-42.3 - j68)$
f. $(-36.0 + j78) - (-4 - j6) + (10.8 - j72)$
g. $6 \angle 20° + 8 \angle 80°$
h. $42 \angle 45° + 62 \angle 60° - 70 \angle 120°$

44. Multiplication [express your answers in rectangular form for parts (a) through (d), and in polar form for parts (e) through (h)]:
a. $(2 + j3)(6 + j8)$
b. $(7.8 + j1)(4 + j2)(7 + j6)$
c. $(0.002 + j0.006)(-2 + j2)$
d. $(400 - j200)(-0.01 - j0.5)(-1 + j3)$
e. $(2 \angle 60°)(4 \angle 22°)$
f. $(6.9 \angle 8°)(7.2 \angle -72°)$
g. $0.002 \angle 120°)(0.5 \angle 200°)(40 \angle -60°)$
h. $(540 \angle -20°)(-5 \angle 180°)(6.2 \angle 0°)$

45. Division (express your answers in polar form):
a. $(42 \angle 10°)/(7 \angle 60°)$
b. $(0.006 \angle 120°)/(30 \angle -20°)$
c. $(4360 \angle -20°)/(40 \angle 210°)$
d. $(650 \angle -80°)/(8.5 \angle 360°)$
e. $(8 + j8)/(2 + j2)$
f. $(8 + j42)/(-6 + j60)$
g. $(0.05 + j0.25)/(8 - j60)$
h. $(-4.5 - j6)/(0.1 - j0.4)$

***46.** Perform the following operations (express your answers in rectangular form):
a. $\dfrac{(4 + j3) + (6 - j8)}{(3 + j3) - (2 + j3)}$
b. $\dfrac{8 \angle 60°}{(2 \angle 0°) + (100 + j100)}$
c. $\dfrac{(6 \angle 20°)(120 \angle -40°)(3 + j4)}{2 \angle -30°}$
d. $\dfrac{(0.4 \angle 60°)^2(300 \angle 40°)}{3 + j9}$
e. $\left(\dfrac{1}{(0.02 \angle 10°)^2}\right)\left(\dfrac{2}{j}\right)^3\left(\dfrac{1}{6^2 - j\sqrt{900}}\right)$

***47. a.** Determine a solution for x and y if

$$(x + j4) + (3x + jy) - j7 = 16 \angle 0°$$

b. Determine x if

$$(10 \angle 20°)(x \angle -60°) = 30.64 - j25.72$$

c. Determine a solution for x and y if

$$(5x + j10)(2 - jy) = 90 - j70$$

d. Determine θ if

$$\frac{80 \angle 0°}{20 \angle \theta} = 3.464 - j2$$

SECTION 14.12 Phasors

48. Express the following in phasor form:
a. $\sqrt{2}(100) \sin(\omega t + 30°)$
b. $\sqrt{2}(0.25) \sin(157t - 40°)$
c. $100 \sin(\omega t - 90°)$
d. $42 \sin(377t + 0°)$

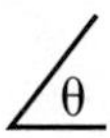

e. $6 \times 10^{-6} \cos \omega t$
f. $3.6 \times 10^{-6} \cos(754t - 20°)$

49. Express the following phasor currents and voltages as sine waves if the frequency is 60 Hz:

a. $\mathbf{I} = 40 \text{ A} \angle 20°$ **b.** $\mathbf{V} = 120 \text{ V} \angle 0°$
c. $\mathbf{I} = 8 \times 10^{-3} \text{ A} \angle 120°$ **d.** $\mathbf{V} = 5 \text{ V} \angle 90°$
e. $\mathbf{I} = 1200 \text{ A} \angle -120°$ **f.** $\mathbf{V} = \dfrac{6000}{\sqrt{2}} \text{ V} \angle -180°$

50. For the system of Fig. 14.81, find the sinusoidal expression for the unknown voltage v_a if

$$e_{in} = 60 \sin(377t + 20°)$$
$$v_b = 20 \sin 377t$$

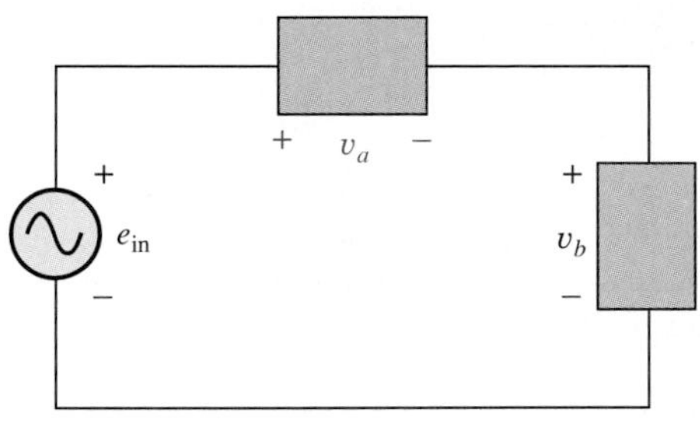

FIG. 14.81
Problem 50.

51. For the system of Fig. 14.82, find the sinusoidal expression for the unknown current i_1 if

$$i_s = 20 \times 10^{-6} \sin(\omega t + 90°)$$
$$i_2 = 6 \times 10^{-6} \sin(\omega t - 60°)$$

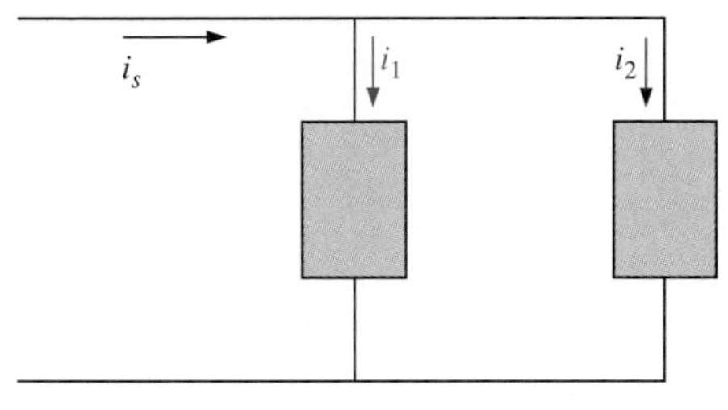

FIG. 14.82
Problem 51.

52. Find the sinusoidal expression for the applied voltage e for the system of Fig. 14.83 if

$$v_a = 60 \sin(\omega t + 30°)$$
$$v_b = 30 \sin(\omega t - 30°)$$
$$v_c = 40 \sin(\omega t + 120°)$$

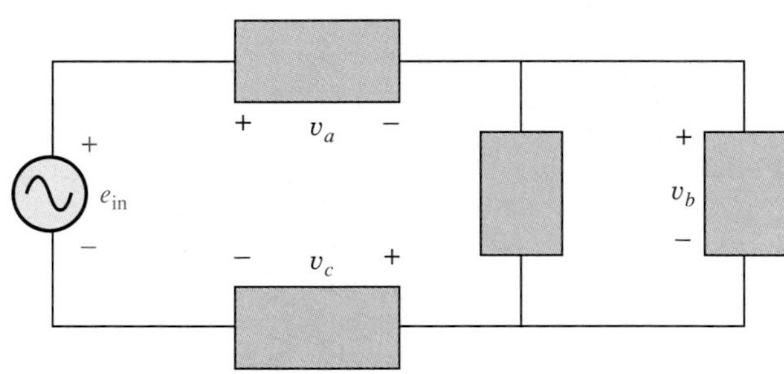

FIG. 14.83
Problem 52.

53. Find the sinusoidal expression for the current i_s for the system of Fig. 14.84 if

$$i_1 = 6 \times 10^{-3} \sin(377t + 180°)$$
$$i_2 = 8 \times 10^{-3} \sin 377t$$
$$i_3 = 2i_2$$

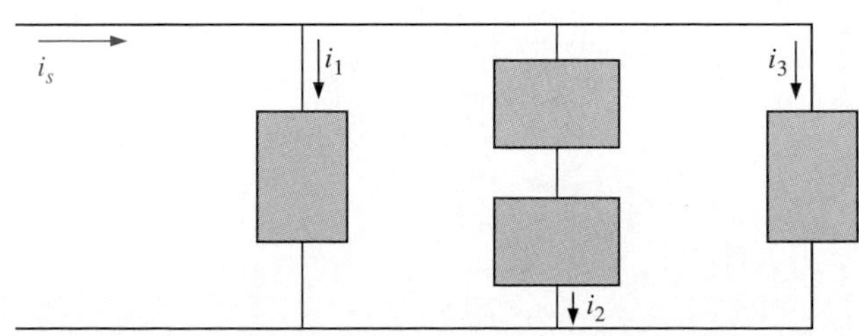

FIG. 14.84
Problem 53.

SECTION 14.13 Computer Analysis

PSpice (DOS)

54. Plot the source current versus frequency (100 Hz to 10 kHz) for the network of Fig. 14.68.

55. Obtain a listing of the magnitude and phase angle for the current of the capacitor of Fig. 14.68 versus frequency. Obtain at least 20 points separated by the same increment.

56. Plot Z_L (magnitude only) versus frequency (100 kHz to 100 MHz) for the inductor equivalent of Fig. 14.24 if $L = 1$ mH, $C_p = 1.6$ pF, and $R_s = 20$ mΩ. For what frequency range is the inductor "inductive"?

PSpice (Windows)

57. Plot i_L and v_L versus time for the network of Fig. 14.66 for two cycles if the frequency is 0.2 kHz.

58. Plot the magnitude and phase angle of the current i_C versus frequency (100 Hz to 100 kHz) for the network of Fig. 14.68.

59. Plot the total impedance of the configuration of Fig. 14.26 versus frequency (100 kHz to 100 MHz) for the following parameter values: $C = 0.1$ μF, $L_s = 0.2$ μH, $R_s = 2$ MΩ, and $R_p = 100$ MΩ. For what frequency range is the capacitor "capacitive"?

Programming Language (C++, BASIC, PASCAL, etc.)

60. Given a sinusoidal function, write a program to print out the derivative.

61. Given the sinusoidal expression for the current, determine the expression for the voltage across a resistor, capacitor, or inductor, depending on the element involved. In other words, the program will ask which element is to be investigated and will then request the pertinent data to obtain the mathematical expression for the sinusoidal voltage.

62. Write a program to tabulate the reactance versus frequency for an inductor or capacitor for a specified frequency range.

63. Given the sinusoidal expression for the voltage and current of a load, write a program to determine the average power and power factor.

64. Given two sinusoidal functions, write a program to convert each to the phasor domain, add the two, and print out the sum in the phasor and time domains.

GLOSSARY

Average or **real power** The power delivered to and dissipated by the load over a full cycle.

Complex conjugate A complex number defined by simply changing the sign of an imaginary component of a complex number in the rectangular form.

Complex number A number that represents a point in a two-dimensional plane located with reference to two distinct axes. It defines a vector drawn from the origin to that point.

Derivative The instantaneous rate of change of a function with respect to time or another variable.

Leading and **lagging power factors** An indication of whether a network is primarily capacitive or inductive in nature. Leading power factors are associated with capacitive networks, and lagging power factors with inductive networks.

Phasor A radius vector that has a constant magnitude at a fixed angle from the positive real axis and that represents a sinusoidal voltage or current in the vector domain.

Phasor diagram A "snapshot" of the phasors that represent a number of sinusoidal waveforms at $t = 0$.

Polar form A method of defining a point in a complex plane that includes a single magnitude to represent the distance from the origin, and an angle to reflect the counterclockwise distance from the positive real axis.

Power factor (F_p) An indication of how reactive or resistive an electrical system is. The higher the power factor, the greater the resistive component.

Reactance The opposition of an inductor or capacitor to the flow of charge that results in the continual exchange of energy between the circuit and magnetic field of an inductor or the electric field of a capacitor.

Reciprocal A format defined by 1 divided by the complex number.

Rectangular form A method of defining a point in a complex plane that includes the magnitude of the real component and the magnitude of the imaginary component, the latter component being defined by an associated letter *j*.

15

Series and Parallel ac Circuits

15.1 INTRODUCTION

In this chapter, phasor algebra will be used to develop a quick, direct method for solving both the series and the parallel ac circuits. The close relationship that exists between this method for solving for unknown quantities and the approach used for dc circuits will become apparent after a few simple examples are considered. Once this association is established, many of the rules (current divider rule, voltage divider rule, and so on) for dc circuits can be readily applied to ac circuits.

SERIES ac CIRCUITS

15.2 IMPEDANCE AND THE PHASOR DIAGRAM

Resistive Elements

In Chapter 14, we found, for the purely resistive circuit of Fig. 15.1, that v and i were in phase, and the magnitude

$$I_m = \frac{V_m}{R} \qquad \text{or} \qquad V_m = I_m R$$

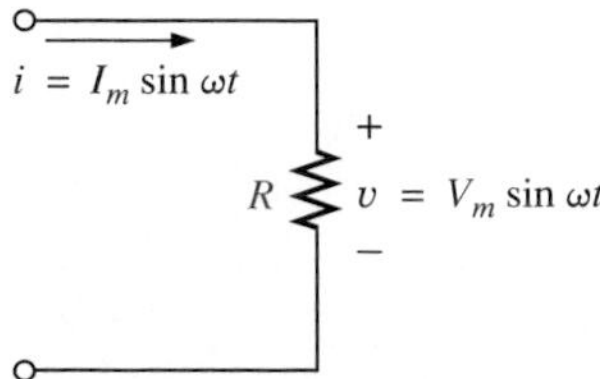

FIG. 15.1
Resistive ac circuit.

In phasor form,

$$v = V_m \sin \omega t \Rightarrow \mathbf{V} = V \angle 0^\circ$$

where $V = 0.707V_m$.

Applying Ohm's law and using phasor algebra, we have

$$\mathbf{I} = \frac{V \angle 0^\circ}{R \angle \theta_R} = \frac{V}{R} \underline{/0^\circ - \theta_R}$$

Since i and v are in phase, the angle associated with i also must be 0°. To satisfy this condition, θ_R must equal 0°. Substituting $\theta_R = 0°$, we find

$$\mathbf{I} = \frac{V \angle 0^\circ}{R \angle 0^\circ} = \frac{V}{R} \underline{/0^\circ - 0^\circ} = \frac{V}{R} \angle 0^\circ$$

so that in the time domain,

$$i = \sqrt{2}\left(\frac{V}{R}\right) \sin \omega t$$

The fact that $\theta_R = 0°$ will now be employed in the following polar format to ensure the proper phase relationship between the voltage and current of a resistor.

$$\boxed{\mathbf{Z}_R = R \underline{/0^\circ}} \qquad \textbf{(15.1)}$$

The boldface roman quantity $\mathbf{Z}_R$, having both magnitude and an associated angle, is referred to as the *impedance* of a resistive element. It is measured in ohms and is a measure of how much the element will "impede" the flow of charge through the network. The above format will prove to be a useful "tool" when the networks become more complex and phase relationships become less obvious. It is important to realize, however, that $\mathbf{Z}_R$ is *not a phasor,* even though the format $R \angle 0°$ is very similar to the phasor notation for sinusoidal currents and voltages. The term *phasor* is reserved for quantities that vary with time, and R and its associated angle of 0° are fixed, nonvarying quantities.

EXAMPLE 15.1 Using complex algebra, find the current i for the circuit of Fig. 15.2. Sketch the waveforms of v and i.

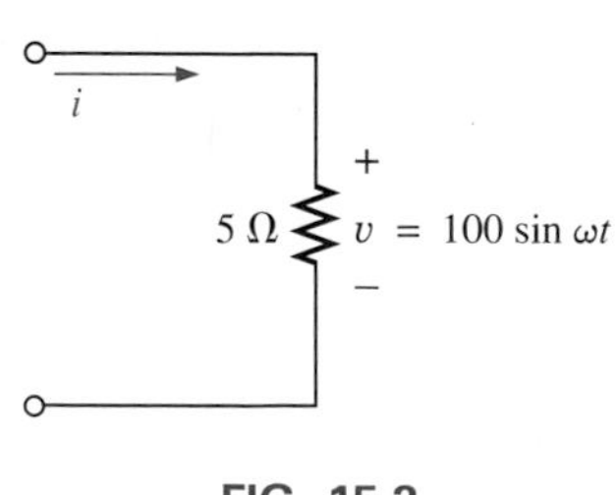

FIG. 15.2
Example 15.1.

Solution: Note Fig. 15.3:

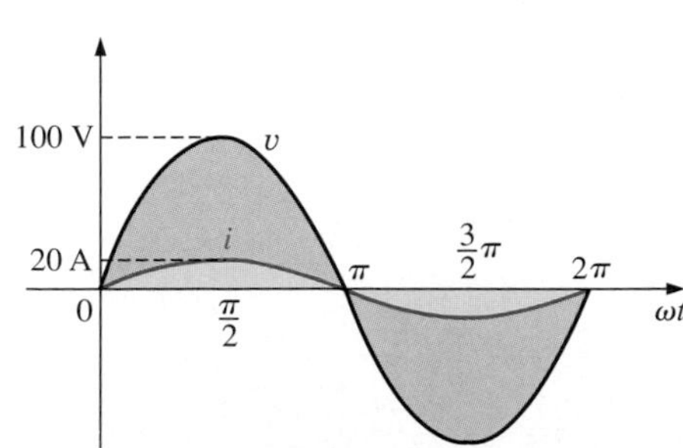

FIG. 15.3
Waveforms for Example 15.1.

$$v = 100 \sin \omega t \Rightarrow \text{phasor form } \mathbf{V} = 70.71 \text{ V} \angle 0°$$

$$\mathbf{I} = \frac{\mathbf{V}}{\mathbf{Z}_R} = \frac{V \angle \theta}{R \angle 0°} = \frac{70.71 \text{ V} \angle 0°}{5 \ \Omega \angle 0°} = 14.14 \text{ A} \angle 0°$$

and $$i = \sqrt{2}(14.14) \sin \omega t = \mathbf{20 \sin \omega t}$$

EXAMPLE 15.2 Using complex algebra, find the voltage v for the circuit of Fig. 15.4. Sketch the waveforms of v and i.

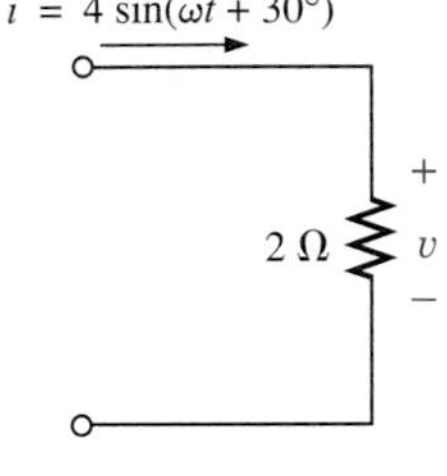

FIG. 15.4
Example 15.2.

Solution: Note Fig. 15.5:

$$i = 4 \sin(\omega t + 30°) \Rightarrow \text{phasor form } \mathbf{I} = 2.828 \text{ A} \angle 30°$$

$$\mathbf{V} = \mathbf{I}\mathbf{Z}_R = (I \angle \theta)(R \angle 0°) = (2.828 \text{ A} \angle 30°)(2 \ \Omega \angle 0°) = 5.656 \text{ V} \angle 30°$$

and $$v = \sqrt{2}(5.656) \sin(\omega t + 30°) = \mathbf{8.0 \sin(\omega t + 30°)}$$

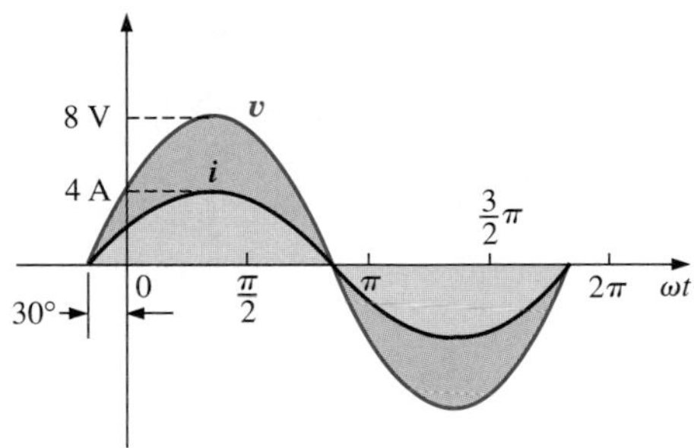

FIG. 15.5
Waveforms for Example 15.2.

It is often helpful in the analysis of networks to have a *phasor diagram*, which shows at a glance the *magnitudes* and *phase relations* among the various quantities within the network. For example, the phasor diagrams of the circuits considered in the preceding examples would be as shown in Fig. 15.6. In both cases, it is immediately obvious that v and i are in phase since they both have the same phase angle.

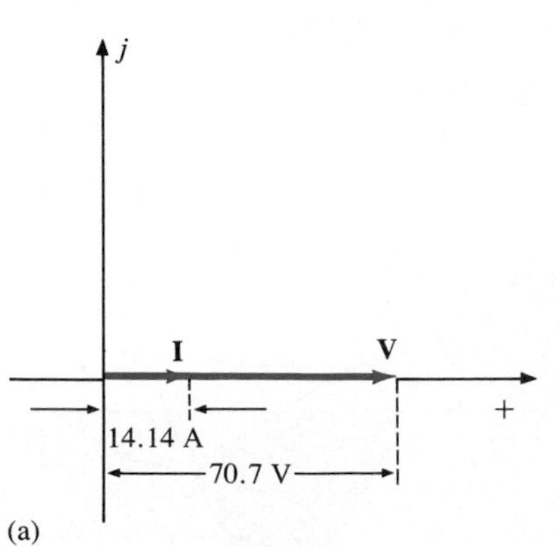

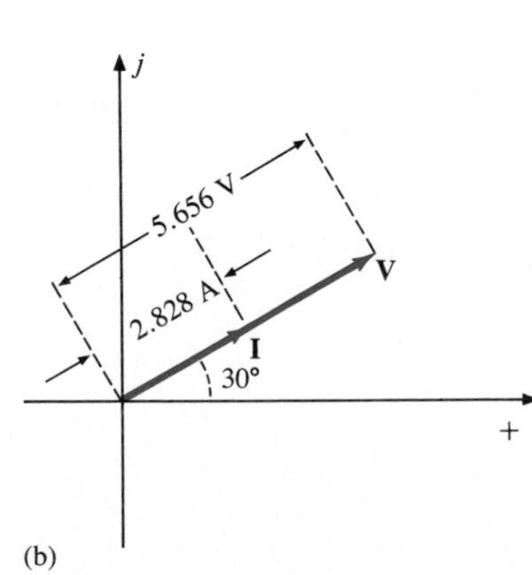

FIG. 15.6
Phasor diagrams for Examples 15.1 and 15.2.

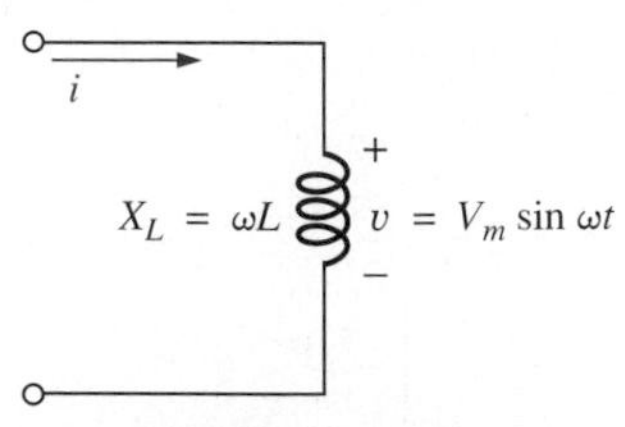

FIG. 15.7
Inductive ac circuit.

Inductive Reactance

For the pure inductor of Fig. 15.7, it was learned in Chapter 13 that the voltage leads the current by 90°, and that the reactance of the coil X_L is determined by ωL.

$$v = V_m \sin \omega t \Rightarrow \text{phasor form } \mathbf{V} = V \angle 0^\circ$$

By Ohm's law,

$$\mathbf{I} = \frac{V \angle 0^\circ}{X_L \angle \theta_L} = \frac{V}{X_L} \angle 0^\circ - \theta_L$$

Since v leads i by 90°, i must have an angle of $-90°$ associated with it. To satisfy this condition, θ_L must equal $+90°$. Substituting $\theta_L = 90°$, we obtain

$$\mathbf{I} = \frac{V \angle 0^\circ}{X_L \angle 90^\circ} = \frac{V}{X_L} \angle 0^\circ - 90^\circ = \frac{V}{X_L} \angle -90^\circ$$

so that in the time domain,

$$i = \sqrt{2}\left(\frac{V}{X_L}\right)\sin(\omega t - 90^\circ)$$

The fact that $\theta_L = 90°$ will now be employed in the following polar format for inductive reactance to ensure the proper phase relationship between the voltage and current of an inductor.

$$\boxed{\mathbf{Z}_L = X_L \angle 90^\circ} \qquad (15.2)$$

The boldface roman quantity $\mathbf{Z}_L$, having both magnitude and an associated angle, is referred to as the *impedance* of an inductive element. It is measured in ohms and is a measure of how much the inductive element will "control or impede" the level of current through the network (always keep in mind that inductive elements are storage devices and do not dissipate like resistors). The above format, like that defined for the resistive element, will prove to be a useful "tool" in the analysis of ac networks. Again, be aware that $\mathbf{Z}_L$ is not a phasor quantity, for the same reasons indicated for a resistive element.

EXAMPLE 15.3 Using complex algebra, find the current i for the circuit of Fig. 15.8. Sketch the v and i curves.

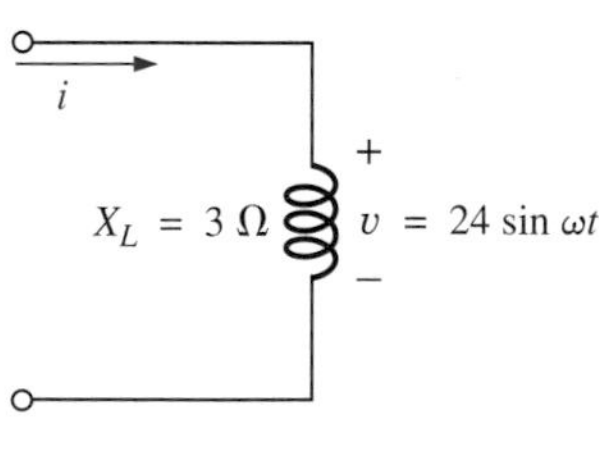

FIG. 15.8
Example 15.3.

Solution: Note Fig. 15.9:

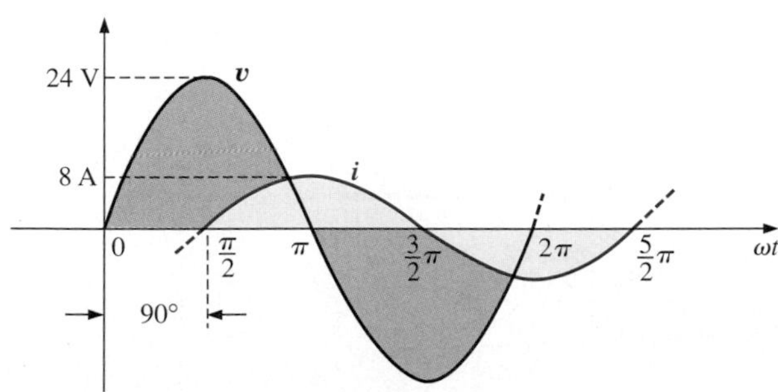

FIG. 15.9
Waveforms for Example 15.3.

$$v = 24 \sin \omega t \Rightarrow \text{phasor form } \mathbf{V} = 16.968 \text{ V } \angle 0°$$

$$\mathbf{I} = \frac{\mathbf{V}}{\mathbf{Z}_L} = \frac{V \angle \theta}{X_L \angle 90°} = \frac{16.968 \text{ V } \angle 0°}{3 \text{ } \Omega \angle 90°} = 5.656 \text{ A } \angle -90°$$

and $$i = \sqrt{2}(5.656) \sin(\omega t - 90°) = \mathbf{8.0 \sin(\omega t - 90°)}$$

EXAMPLE 15.4 Using complex algebra, find the voltage v for the circuit of Fig. 15.10. Sketch the v and i curves.

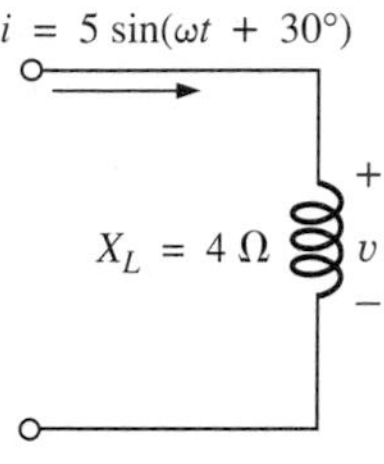

FIG. 15.10
Example 15.4.

Solution: Note Fig. 15.11:

$$i = 5 \sin(\omega t + 30°) \Rightarrow \text{phasor form } \mathbf{I} = 3.535 \text{ A } \angle 30°$$

$$\mathbf{V} = \mathbf{I}\mathbf{Z}_L = (I \angle \theta)(X_L \angle 90°) = (3.535 \text{ A } \angle 30°)(4 \text{ } \Omega \angle +90°)$$
$$= 14.140 \text{ V } \angle 120°$$

and $$v = \sqrt{2}(14.140) \sin(\omega t + 120°) = \mathbf{20 \sin(\omega t + 120°)}$$

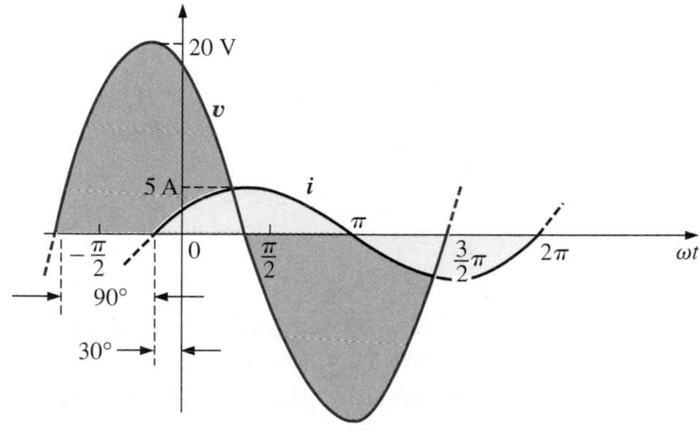

FIG. 15.11
Waveforms for Example 15.4.

The phasor diagrams for the two circuits of the preceding examples are shown in Fig. 15.12. Both indicate quite clearly that the voltage leads the current by 90°.

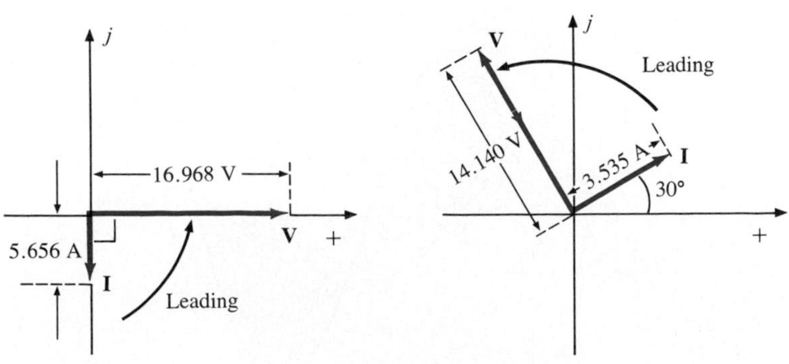

FIG. 15.12
Phasor diagrams for Examples 15.3 and 15.4.

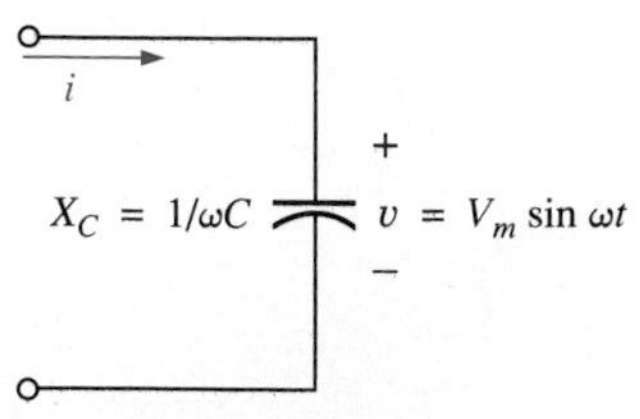

FIG. 15.13
Capacitive ac circuit.

Capacitive Reactance

For the pure capacitor of Fig. 15.13, it was learned in Chapter 13 that the current leads the voltage by 90°, and that the reactance of the capacitor X_C is determined by $1/\omega C$.

$$v = V_m \sin \omega t \Rightarrow \text{phasor form } \mathbf{V} = V \angle 0^\circ$$

Applying Ohm's law and using phasor algebra, we find

$$\mathbf{I} = \frac{V \angle 0^\circ}{X_C \angle \theta_C} = \frac{V}{X_C} \underline{/0^\circ - \theta_C}$$

Since we know i leads v by 90°, i must have an angle of +90° associated with it. To satisfy this condition, θ_C must equal −90°. Substituting $\theta_C = -90^\circ$ yields

$$\mathbf{I} = \frac{V \angle 0^\circ}{X_C \angle -90^\circ} = \frac{V}{X_C} \underline{/0^\circ - (-90^\circ)} = \frac{V}{X_C} \angle 90^\circ$$

so, in the time domain,

$$i = \sqrt{2}\left(\frac{V}{X_C}\right) \sin(\omega t + 90^\circ)$$

The fact that $\theta_C = -90^\circ$ will now be employed in the following polar format for capacitive reactance to ensure the proper phase relationship between the voltage and current of a capacitor.

$$\boxed{\mathbf{Z}_C = X_C \angle -90^\circ} \qquad (15.3)$$

The boldface roman quantity $\mathbf{Z}_C$, having both magnitude and an associated angle, is referred to as the *impedance* of a capacitive element. It is measured in ohms and is a measure of how much the capacitive element will "control or impede" the level of current through the network (always keep in mind that capacitive elements are storage devices and do not dissipate like resistors). The above format, like that defined for the resistive element, will prove a very useful "tool" in the analysis of ac networks. Again, be aware that $\mathbf{Z}_C$ is not a phasor quantity, for the same reasons indicated for a resistive element.

EXAMPLE 15.5 Using complex algebra, find the current i for the circuit of Fig. 15.14. Sketch the v and i curves.

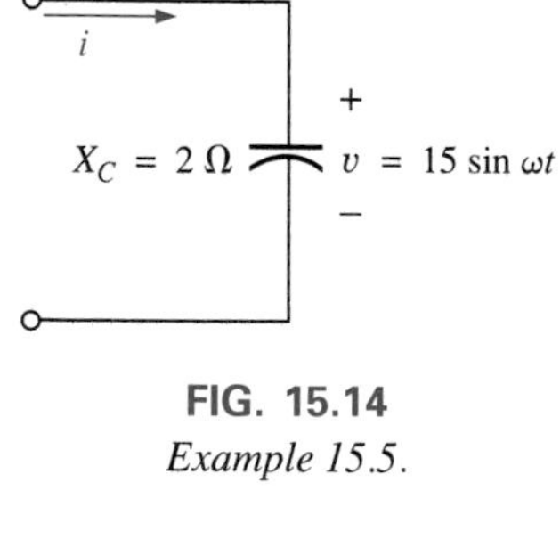

FIG. 15.14
Example 15.5.

Solution: Note Fig. 15.15:

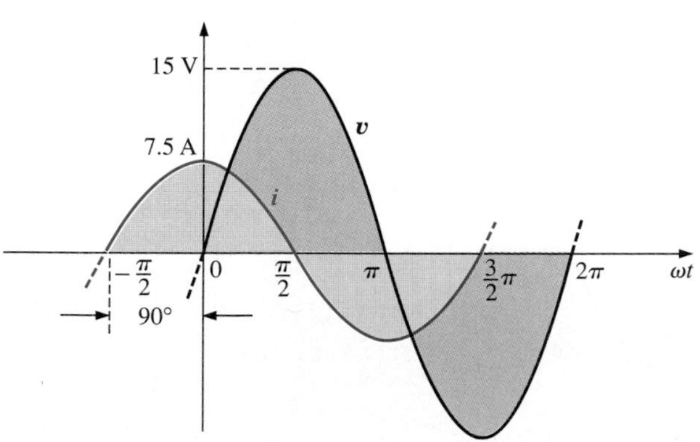

FIG. 15.15
Waveforms for Example 15.5.

$$v = 15 \sin \omega t \Rightarrow \text{phasor notation } \mathbf{V} = 10.605 \text{ V } \angle 0°$$

$$\mathbf{I} = \frac{\mathbf{V}}{\mathbf{Z}_C} = \frac{V \angle \theta}{X_C \angle -90°} = \frac{10.605 \text{ V } \angle 0°}{2\ \Omega \angle -90°} = 5.303 \text{ A } \angle 90°$$

and $$i = \sqrt{2}(5.303) \sin(\omega t + 90°) = \mathbf{7.5 \sin(\omega t + 90°)}$$

EXAMPLE 15.6 Using complex algebra, find the voltage v for the circuit of Fig. 15.16. Sketch the v and i curves.

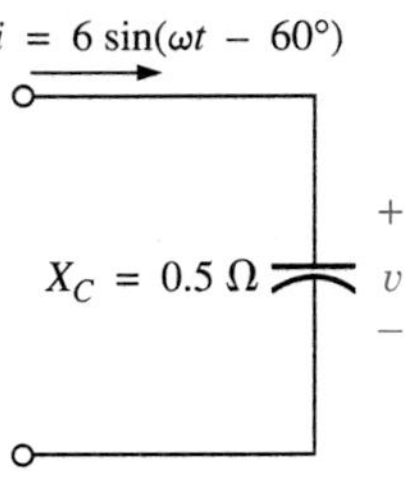

FIG. 15.16
Example 15.6.

Solution: Note Fig. 15.17:

$$i = 6 \sin(\omega t - 60°) \Rightarrow \text{phasor notation } \mathbf{I} = 4.242 \text{ A } \angle -60°$$

$$\mathbf{V} = \mathbf{I}\mathbf{Z}_C = (I \angle \theta)(X_C \angle -90°) = (4.242 \text{ A } \angle -60°)(0.5\ \Omega \angle -90°)$$
$$= 2.121 \text{ V } \angle -150°$$

and $$v = \sqrt{2}(2.121) \sin(\omega t - 150°) = \mathbf{3.0 \sin(\omega t - 150°)}$$

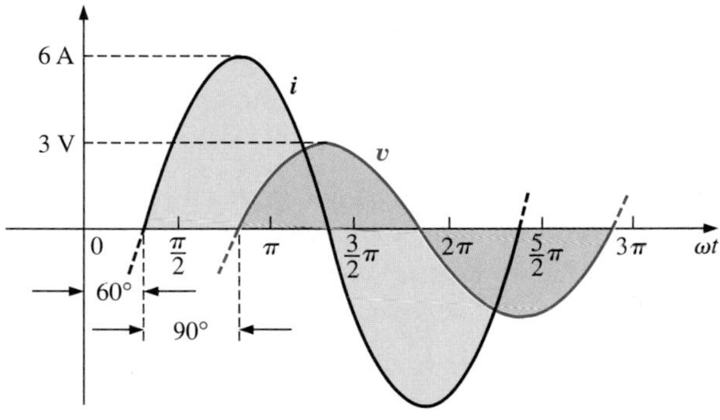

FIG. 15.17
Waveforms for Example 15.6.

The phasor diagrams for the two circuits of the preceding examples are shown in Fig. 15.18. Both indicate quite clearly that the current i leads the voltage v by 90°.

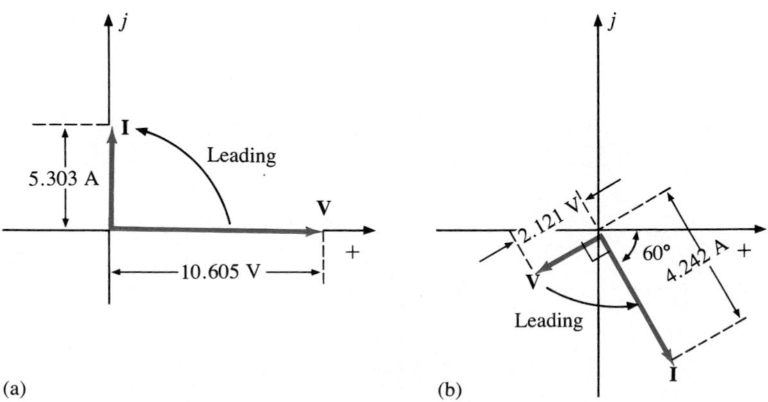

FIG. 15.18
Phasor diagrams for Examples 15.5 and 15.6.

Impedance Diagram

Now that an angle is associated with resistance, inductive reactance, and capacitive reactance, each can be placed on a complex plane dia-

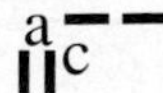

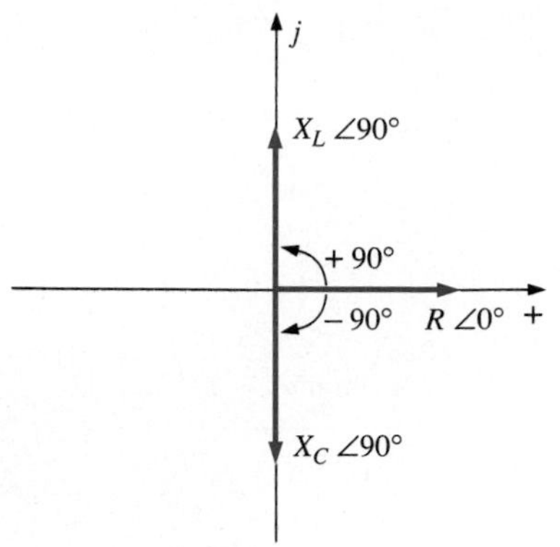

FIG. 15.19
Impedance diagram.

gram, as shown in Fig. 15.19. For any network, the resistance will *always* appear on the positive real axis, the inductive reactance on the positive imaginary axis, and the capacitive reactance on the negative imaginary axis. The result is an *impedance diagram* that can reflect the individual and total impedance levels of an ac network.

We will find in the sections and chapters to follow that networks combining different types of elements will have total impedances that extend from $-90°$ to $+90°$. If the total impedance has an angle of 0°, it is said to be resistive in nature. If it is closer to 90°, it is inductive in nature; and if it is closer to $-90°$, it is capacitive in nature.

Of course, for single-element networks the angle associated with the impedance will be the same as that of the resistive or reactive element, as revealed by Eqs. (15.1) to (15.3). It is important to stay aware that impedance, like resistance or reactance, is not a phasor quantity representing a time-varying function with a particular phase shift. It is simply an operating "tool" that is extremely useful in determining the magnitude and angle of quantities in a sinusoidal ac network.

Once the total impedance of a network is determined, its magnitude will define the resulting current level (through Ohm's law), whereas its angle will reveal whether the network is primarily inductive or capacitive or simply resistive.

For any configuration (series, parallel, series-parallel, etc.), the angle associated with the total impedance is the angle by which the applied voltage leads the source current. For inductive networks, θ_T will be positive, whereas for capacitive networks, θ_T will be negative.

15.3 SERIES CONFIGURATION

The overall properties of series ac circuits (Fig. 15.20) are the same as those for dc circuits. For instance, the total impedance of a system is the sum of the individual impedances:

$$\mathbf{Z}_T = \mathbf{Z}_1 + \mathbf{Z}_2 + \mathbf{Z}_3 + \cdots + \mathbf{Z}_N \qquad \textbf{(15.4)}$$

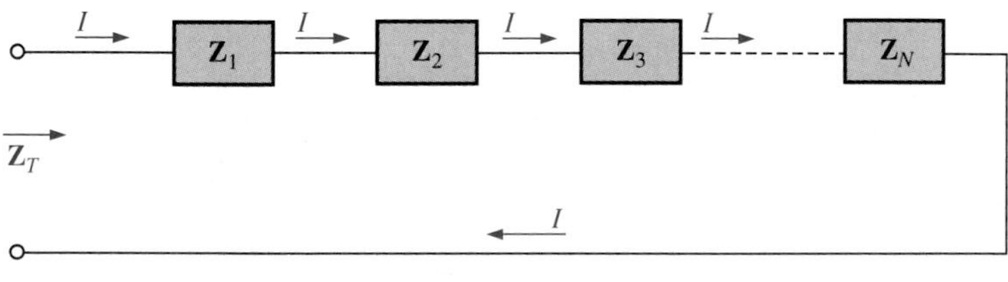

FIG. 15.20
Series impedances.

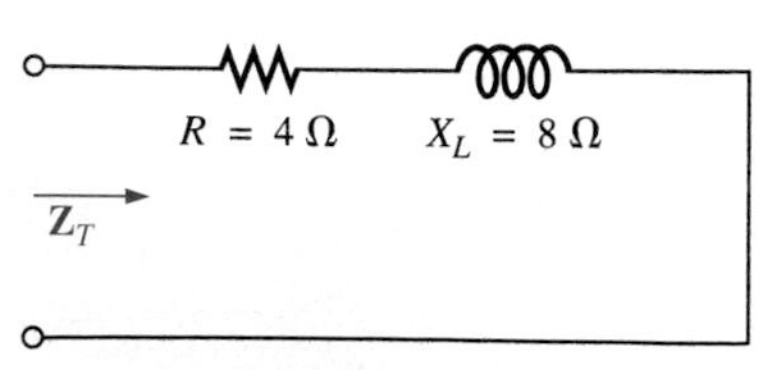

FIG. 15.21
Example 15.7.

EXAMPLE 15.7 Draw the impedance diagram for the circuit of Fig. 15.21 and find the total impedance.

Solution: As indicated by Fig. 15.22, the input impedance can be found graphically from the impedance diagram by properly scaling the real and imaginary axes and finding the length of the resultant vector Z_T and angle θ_T. Or, by using vector algebra, we obtain

$$\begin{aligned}\mathbf{Z}_T &= \mathbf{Z}_1 + \mathbf{Z}_2 \\ &= R \angle 0^\circ + X_L \angle 90^\circ \\ &= R + jX_L = 4\ \Omega + j8\ \Omega \\ \mathbf{Z}_T &= \mathbf{8.944\ \Omega \angle 63.43^\circ}\end{aligned}$$

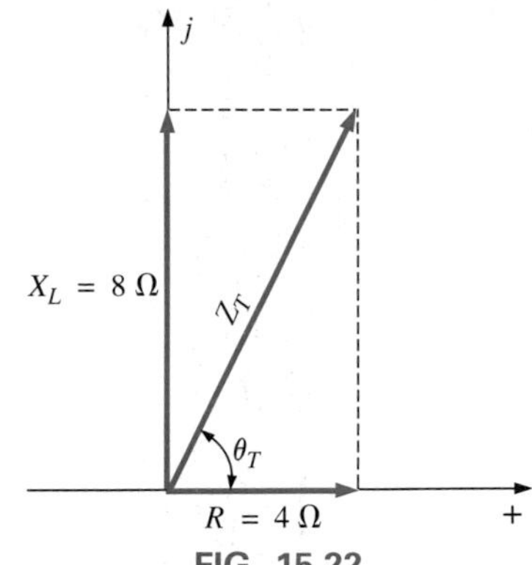

FIG. 15.22
Impedance diagram for Example 15.7.

EXAMPLE 15.8 Determine the input impedance to the series network of Fig. 15.23. Draw the impedance diagram.

Solution:

$$\begin{aligned}\mathbf{Z}_T &= \mathbf{Z}_1 + \mathbf{Z}_2 + \mathbf{Z}_3 \\ &= R \angle 0^\circ + X_L \angle 90^\circ + X_C \angle -90^\circ \\ &= R + jX_L - jX_C \\ &= R + j(X_L - X_C) = 6\ \Omega + j(10\ \Omega - 12\ \Omega) = 6\ \Omega - j2\ \Omega \\ \mathbf{Z}_T &= \mathbf{6.325\ \Omega \angle -18.43^\circ}\end{aligned}$$

The impedance diagram appears in Fig. 15.24. Note that in this example, series inductive and capacitive reactances are in direct opposition. For the circuit of Fig. 15.23, if the inductive reactance were equal to the capacitive reactance, the input impedance would be purely resistive. We will have more to say about this particular condition in a later chapter.

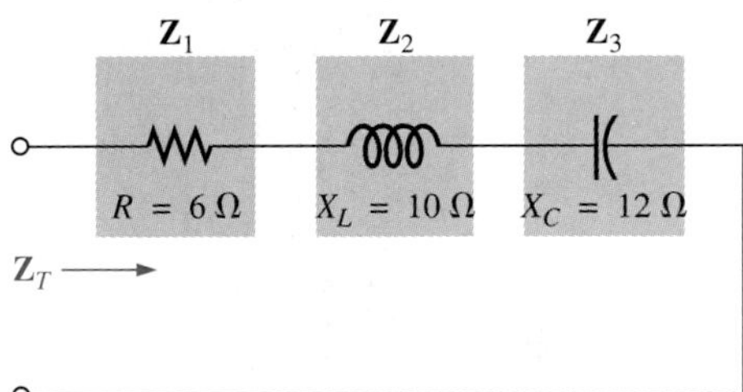

FIG. 15.23
Example 15.8

For the representative series ac network of Fig. 15.25 having two impedances, *the current is the same through each element* (as it was for the series dc circuits) and is determined by Ohm's law:

$$\mathbf{Z}_T = \mathbf{Z}_1 + \mathbf{Z}_2$$

and

$$\mathbf{I} = \frac{\mathbf{E}}{\mathbf{Z}_T} \tag{15.5}$$

The voltage across each element can then be found by another application of Ohm's law:

$$\mathbf{V}_1 = \mathbf{I}\mathbf{Z}_1 \tag{15.6a}$$

$$\mathbf{V}_2 = \mathbf{I}\mathbf{Z}_2 \tag{15.6b}$$

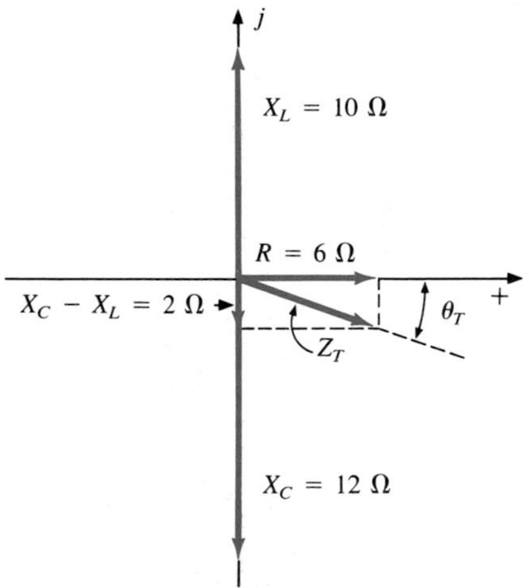

FIG. 15.24
Impedance diagram for Example 15.8.

Kirchhoff's voltage law can then be applied in the same manner as it is employed for dc circuits. However, keep in mind that we are now dealing with the algebraic manipulation of quantities that have both magnitude and direction.

$$\mathbf{E} - \mathbf{V}_1 - \mathbf{V}_2 = 0$$

or

$$\mathbf{E} = \mathbf{V}_1 + \mathbf{V}_2 \tag{15.7}$$

The power to the circuit can be determined by

$$P = EI \cos \theta_T \tag{15.8}$$

where θ_T is the phase angle between **E** and **I**.

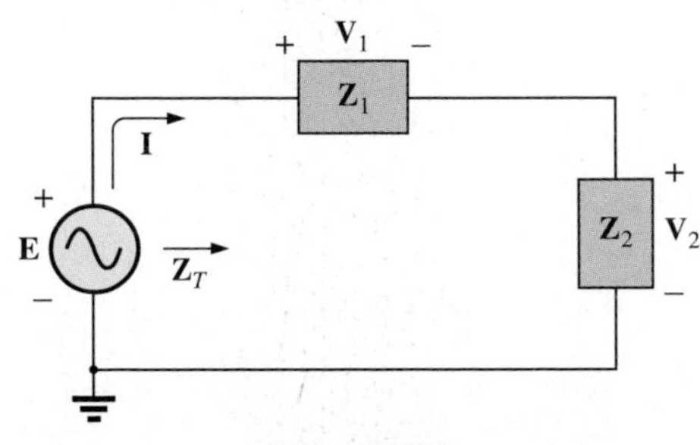

FIG. 15.25
Series ac circuit.

Now that a general approach has been introduced, the simplest of series configurations will be investigated in detail to further emphasize the similarities in the analysis of dc circuits. In many of the circuits to be considered, $3 + j4 = 5 \angle 53.13°$ and $4 + j3 = 5 \angle 36.87°$ will be used quite frequently to ensure that the approach is as clear as possible and not lost in mathematical complexity. Of course, the problems at the end of the chapter will provide plenty of experience with random values.

R-L

$R = 3\ \Omega$ $X_L = 4\ \Omega$

$+\ v_R\ -$ $+\ v_L\ -$

$e = 141.4 \sin \omega t$ i

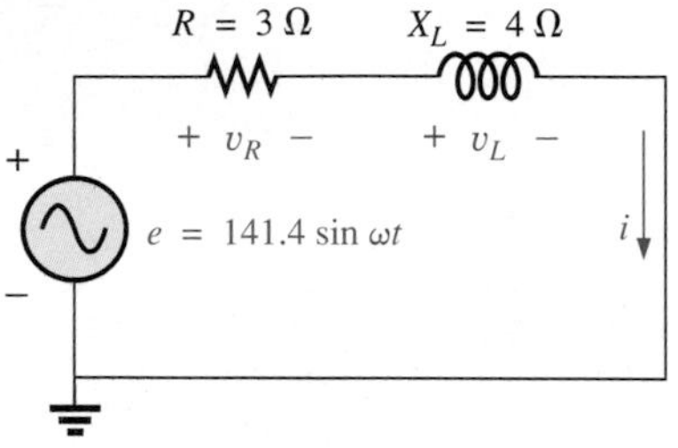

FIG. 15.26
Series R-L circuit.

Refer to Fig. 15.26.

Phasor notation:

$$e = 141.4 \sin \omega t \Rightarrow \mathbf{E} = 100\text{ V} \angle 0°$$

Note Fig. 15.27.

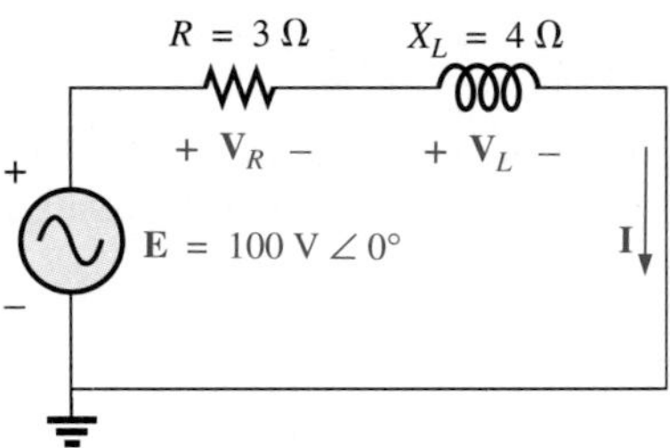

FIG. 15.27
Applying phasor notation to the network of Fig. 15.26.

$\mathbf{Z}_T$:

$$\mathbf{Z}_T = \mathbf{Z}_1 + \mathbf{Z}_2 = 3\ \Omega \angle 0° + 4\ \Omega \angle 90° = 3\ \Omega + j4\ \Omega$$

and $$\mathbf{Z}_T = \mathbf{5\ \Omega \angle 53.13°}$$

Impedance diagram: See Fig. 15.28.

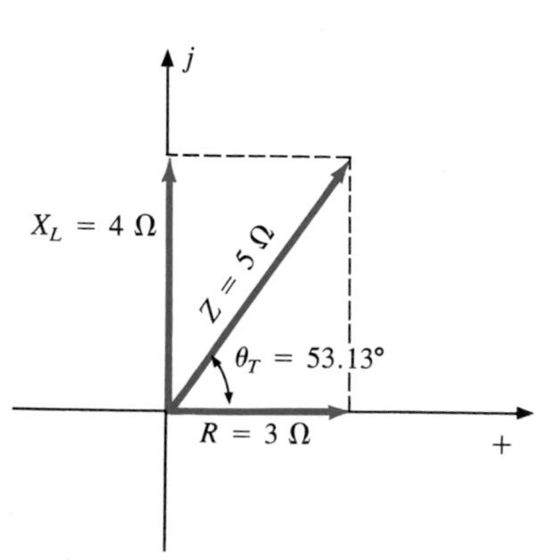

FIG. 15.28
Impedance diagram for the series R-L circuit of Fig. 15.26.

I:

$$\mathbf{I} = \frac{\mathbf{E}}{\mathbf{Z}_T} = \frac{100\text{ V} \angle 0°}{5\ \Omega \angle 53.13°} = \mathbf{20\ A \angle -53.13°}$$

$\mathbf{V}_R$, $\mathbf{V}_L$:

Ohm's law:

$$\mathbf{V}_R = \mathbf{I}\mathbf{Z}_R = (20\text{ A} \angle -53.13°)(3\ \Omega \angle 0°) = \mathbf{60\ V \angle -53.13°}$$

$$\mathbf{V}_L = \mathbf{I}\mathbf{Z}_L = (20\text{ A} \angle -53.13°)(4\ \Omega \angle 90°) = \mathbf{80\ V \angle 36.87°}$$

Kirchhoff's voltage law:

$$\Sigma_{\circlearrowright} \mathbf{V} = \mathbf{E} - \mathbf{V}_R - \mathbf{V}_L = 0$$

or $$\mathbf{E} = \mathbf{V}_R + \mathbf{V}_L$$

In rectangular form,

$$\mathbf{V}_R = 60 \text{ V} \angle -53.13^\circ = 36 \text{ V} - j48 \text{ V}$$
$$\mathbf{V}_L = 80 \text{ V} \angle +36.87^\circ = 64 \text{ V} + j48 \text{ V}$$

and

$$\mathbf{E} = \mathbf{V}_R + \mathbf{V}_L = (36 \text{ V} - j48 \text{ V}) + (64 \text{ V} + j48 \text{ V}) = 100 \text{ V} + j0 = 100 \text{ V} \angle 0^\circ$$

as applied.

Phasor diagram: Note that for the phasor diagram of Fig. 15.29, **I** is in phase with the voltage across the resistor and lags the voltage across the inductor by 90°.

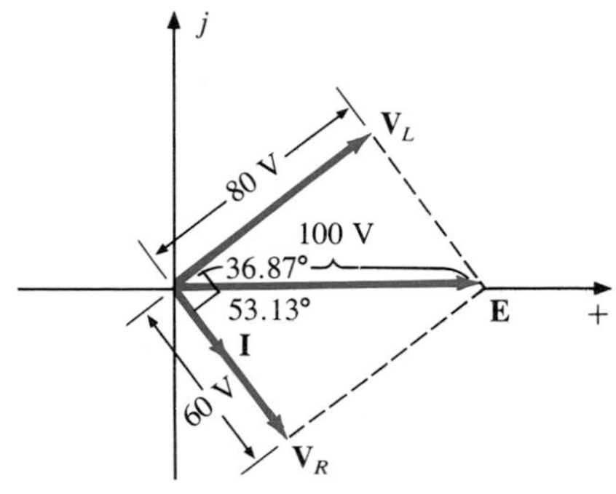

FIG. 15.29
Phasor diagram for the series R-L circuit of Fig. 15.26.

Power: The total power in watts delivered to the circuit is

$$\begin{aligned} P_T &= EI \cos \theta_T \\ &= (100 \text{ V})(20 \text{ A}) \cos 53.13^\circ = (2000 \text{ W})(0.6) \\ &= \mathbf{1200 \text{ W}} \end{aligned}$$

where E and I are effective values and θ_T is the phase angle between E and I, or

$$\begin{aligned} P_T &= I^2R \\ &= (20 \text{ A})^2(3 \ \Omega) = (400)(3) \\ &= \mathbf{1200 \text{ W}} \end{aligned}$$

where I is the effective value, or, finally,

$$\begin{aligned} P_T = P_R + P_L &= V_R I \cos \theta_R + V_L I \cos \theta_L \\ &= (60 \text{ V})(20 \text{ A}) \cos 0^\circ + (80 \text{ V})(20 \text{ A}) \cos 90^\circ \\ &= 1200 \text{ W} + 0 \\ &= \mathbf{1200 \text{ W}} \end{aligned}$$

where θ_R is the phase angle between $\mathbf{V}_R$ and **I**, and θ_L is the phase angle between $\mathbf{V}_L$ and **I**.

Power factor: The power factor F_p of the circuit is cos 53.13° = **0.6 lagging,** where 53.13° is the phase angle between **E** and **I.**

If we write the basic power equation $P = EI \cos \theta$ as follows:

$$\cos \theta = \frac{P}{EI}$$

where E and I are the input quantities and P is the power delivered to the network, and then perform the following substitutions from the basic series ac circuit:

$$\cos \theta = \frac{P}{EI} = \frac{I^2R}{EI} = \frac{IR}{E} = \frac{R}{E/I} = \frac{R}{Z_T}$$

we find

$$F_p = \cos \theta_T = \frac{R}{Z_T} \qquad \textbf{(15.9)}$$

Reference to Fig. 15.28 also indicates that θ is the impedance angle θ_T as written in Eq. (15.9), further supporting the fact that the impedance angle θ_T is also the phase angle between the input voltage and current for a series ac circuit. To determine the power factor, it is necessary only to form the ratio of the total resistance to the magnitude of the input impedance. For the case at hand,

$$F_p = \cos\theta = \frac{R}{Z_T} = \frac{3\ \Omega}{5\ \Omega} = \mathbf{0.6\ lagging}$$

as found above.

R-C

Refer to Fig. 15.30.

FIG. 15.30
Series R-C ac circuit.

Phasor notation:

$$i = 7.07\sin(\omega t + 53.13^\circ) \Rightarrow \mathbf{I} = 5\text{ A}\angle 53.13^\circ$$

Note Fig. 15.31.

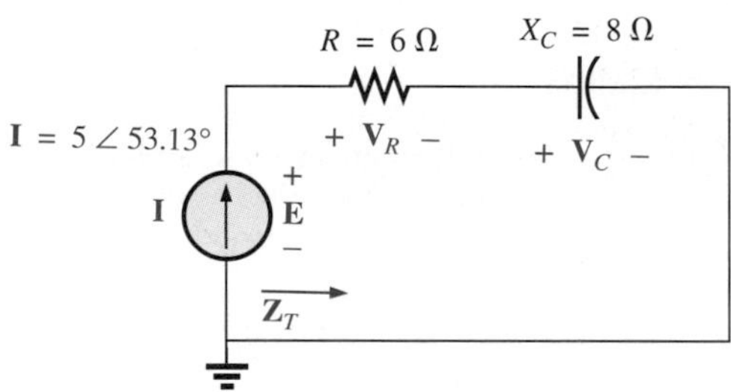

FIG. 15.31
Applying phasor notation to the circuit of Fig. 15.30.

$\mathbf{Z}_T$:

$$\mathbf{Z}_T = \mathbf{Z}_1 + \mathbf{Z}_2 = 6\ \Omega\angle 0^\circ + 8\ \Omega\angle -90^\circ = 6\ \Omega - j8\ \Omega$$

and $$\mathbf{Z}_T = \mathbf{10\ \Omega\angle -53.13^\circ}$$

Impedance diagram: As shown in Fig. 15.32.

FIG. 15.32
Impedance diagram for the series R-C circuit of Fig. 15.30.

E:

$$\mathbf{E} = \mathbf{I}\mathbf{Z}_T = (5\text{ A}\angle 53.13^\circ)(10\ \Omega\angle -53.13^\circ) = \mathbf{50\ V\angle 0^\circ}$$

$\mathbf{V}_R$, $\mathbf{V}_C$:

$$\begin{aligned}\mathbf{V}_R &= \mathbf{I}\mathbf{Z}_R = (I\angle\theta)(R\angle 0^\circ) = (5\text{ A}\angle 53.13^\circ)(6\ \Omega\angle 0^\circ)\\ &= \mathbf{30\ V\angle 53.13^\circ}\end{aligned}$$

$$\begin{aligned}\mathbf{V}_C &= \mathbf{I}\mathbf{Z}_C = (I\angle\theta)(X_C\angle -90^\circ) = (5\text{ A}\angle 53.13^\circ)(8\ \Omega\angle -90^\circ)\\ &= \mathbf{40\ V\angle -36.87^\circ}\end{aligned}$$

Kirchhoff's voltage law:

$$\Sigma_{\circlearrowright} \mathbf{V} = \mathbf{E} - \mathbf{V}_R - \mathbf{V}_C = 0$$

or $$\mathbf{E} = \mathbf{V}_R + \mathbf{V}_C$$

which can be verified by vector algebra as demonstrated for the *R-L* circuit.

Phasor diagram: Note on the phasor diagram of Fig. 15.33 that the current **I** is in phase with the voltage across the resistor and leads the voltage across the capacitor by 90°.

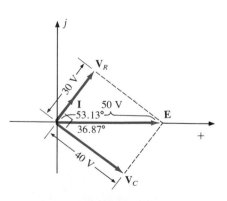

FIG. 15.33
Phasor diagram for the series R-C circuit of Fig. 15.30.

Time domain: In the time domain,

$$e = \sqrt{2}(50) \sin \omega t = \mathbf{70.70 \sin \omega t}$$
$$v_R = \sqrt{2}(30) \sin(\omega t + 53.13°) = \mathbf{42.42 \sin(\omega t + 53.13°)}$$
$$v_C = \sqrt{2}(40) \sin(\omega t - 36.87°) = \mathbf{56.56 \sin(\omega t - 36.87°)}$$

A plot of all of the voltages and the current of the circuit appears in Fig. 15.34. Note again that i and v_R are in phase and that v_C lags i by 90°.

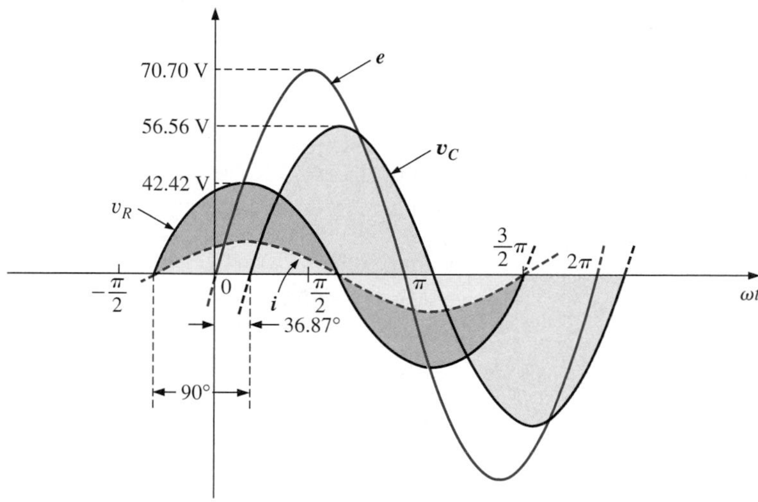

FIG. 15.34
Waveforms for the series R-C circuit of Fig. 15.30.

Power: The total power in watts delivered to the circuit is

$$P_T = EI \cos \theta_T = (50 \text{ V})(5 \text{ A}) \cos 53.13°$$
$$= (250)(0.6) = \mathbf{150 \text{ W}}$$

or

$$P_T = I^2R = (5 \text{ A})^2(6 \ \Omega) = (25)(6)$$
$$= \mathbf{150 \text{ W}}$$

or, finally,

$$P_T = P_R + P_C = V_RI \cos \theta_R + V_CI \cos \theta_C$$
$$= (30 \text{ V})(5 \text{ A}) \cos 0° + (40 \text{ V})(5 \text{ A}) \cos 90°$$
$$= 150 \text{ W} + 0$$
$$= \mathbf{150 \text{ W}}$$

Power factor: The power factor of the circuit is

$$F_p = \cos \theta = \cos 53.13° = \mathbf{0.6 \text{ leading}}$$

Using Eq. (15.9), we obtain

$$F_p = \cos \theta = \frac{R}{Z_T} = \frac{6 \ \Omega}{10 \ \Omega}$$
$$= \mathbf{0.6 \text{ leading}}$$

as determined above.

R-L-C

Refer to Fig. 15.35.

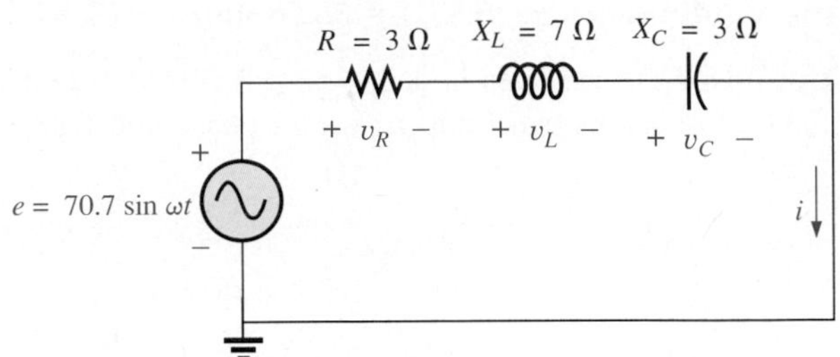

FIG. 15.35
Series R-L-C ac circuit.

Phasor notation: As shown in Fig. 15.36.

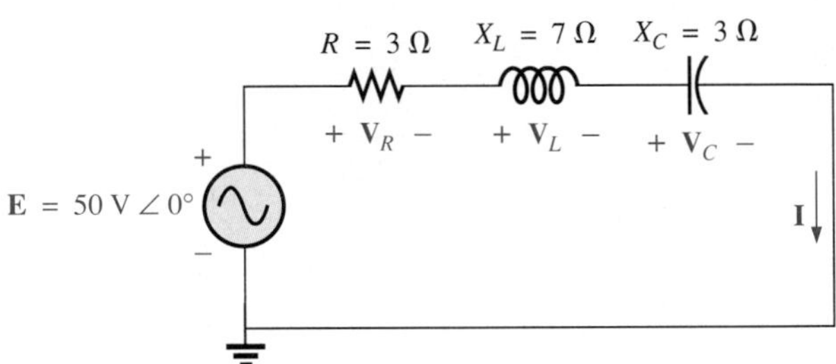

FIG. 15.36
Applying phasor notation to the circuit of Fig. 15.35.

$\mathbf{Z}_T$:

$$\mathbf{Z}_T = \mathbf{Z}_1 + \mathbf{Z}_2 + \mathbf{Z}_3 = R\angle 0° + X_L\angle 90° + X_C\angle -90°$$
$$= 3\ \Omega + j\,7\ \Omega - j\,3\ \Omega = 3\ \Omega + j\,4\ \Omega$$

and $$\mathbf{Z}_T = \mathbf{5\ \Omega\angle 53.13°}$$

Impedance diagram: As shown in Fig. 15.37.

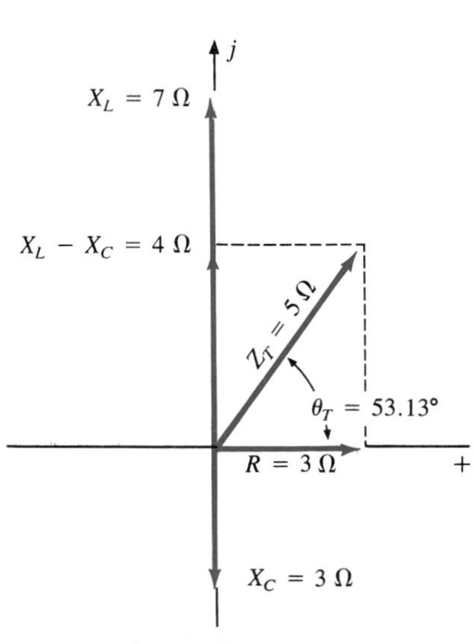

FIG. 15.37
Impedance diagram for the series R-L-C circuit of Fig. 15.35.

I:

$$\mathbf{I} = \frac{\mathbf{E}}{\mathbf{Z}_T} = \frac{50\text{ V}\angle 0°}{5\ \Omega\angle 53.13°} = \mathbf{10\ A\angle -53.13°}$$

$\mathbf{V}_R$, $\mathbf{V}_L$, $\mathbf{V}_C$:

$$\mathbf{V}_R = \mathbf{IZ}_R = (I\angle\theta)(R\angle 0°) = (10\text{ A}\angle -53.13°)(3\ \Omega\angle 0°)$$
$$= \mathbf{30\ V\angle -53.13°}$$

$$\mathbf{V}_L = \mathbf{IZ}_L = (I\angle\theta)(X_L\angle 90°) = (10\text{ A}\angle -53.13°)(7\ \Omega\angle 90°)$$
$$= \mathbf{70\ V\angle 36.87°}$$

$$\mathbf{V}_C = \mathbf{IZ}_C = (I\angle\theta)(X_C\angle -90°) = (10\text{ A}\angle -53.13°)(3\ \Omega\angle -90°)$$
$$= \mathbf{30\ V\angle -143.13°}$$

Kirchhoff's voltage law:

$$\Sigma_{\circlearrowright} \mathbf{V} = \mathbf{E} - \mathbf{V}_R - \mathbf{V}_L - \mathbf{V}_C = 0$$

or $$\mathbf{E} = \mathbf{V}_R + \mathbf{V}_L + \mathbf{V}_C$$

which can also be verified through vector algebra.

Phasor diagram: The phasor diagram of Fig. 15.38 indicates that the current **I** is in phase with the voltage across the resistor, lags the voltage across the inductor by 90°, and leads the voltage across the capacitor by 90°.

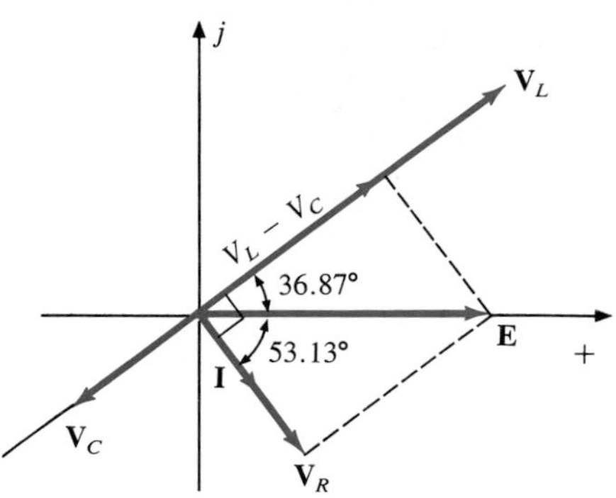

FIG. 15.38
Phasor diagram for the series R-L-C circuit of Fig. 15.35.

Time domain:

$$i = \sqrt{2}(10)\sin(\omega t - 53.13°) = \mathbf{14.14\,sin(\omega t - 53.13°)}$$
$$v_R = \sqrt{2}(30)\sin(\omega t - 53.13°) = \mathbf{42.42\,sin(\omega t - 53.13°)}$$
$$v_L = \sqrt{2}(70)\sin(\omega t + 36.87°) = \mathbf{98.98\,sin(\omega t + 36.87°)}$$
$$v_C = \sqrt{2}(30)\sin(\omega t - 143.13°) = \mathbf{42.42\,sin(\omega t - 143.13°)}$$

A plot of all the voltages and the current of the circuit appears in Fig. 15.39.

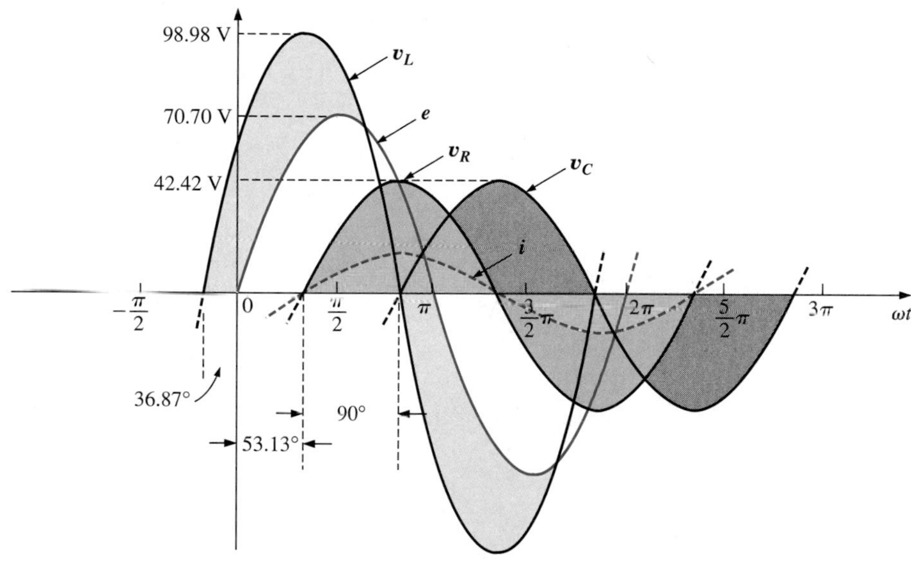

FIG. 15.39
Waveforms for the series R-L circuit of Fig. 15.35.

Power: The total power in watts delivered to the circuit is

$$P_T = EI\cos\theta_T = (50\text{ V})(10\text{ A})\cos 53.13° = (500)(0.6) = \mathbf{300\ W}$$

or $$P_T = I^2R = (10\text{ A})^2(3\ \Omega) = (100)(3) = \mathbf{300\ W}$$

or
$$\begin{aligned}P_T &= P_R + P_L + P_C\\ &= V_RI\cos\theta_R + V_LI\cos\theta_L + V_CI\cos\theta_C\\ &= (30\text{ V})(10\text{ A})\cos 0° + (70\text{ V})(10\text{ A})\cos 90°\\ &\qquad + (30\text{ V})(10\text{ A})\cos 90°\\ &= (30\text{ V})(10\text{ A}) + 0 + 0 = \mathbf{300\ W}\end{aligned}$$

Power factor: The power factor of the circuit is

$$F_p = \cos\theta_T = \cos 53.13° = \mathbf{0.6\ lagging}$$

Using Eq. (15.9), we obtain

$$F_p = \cos\theta = \frac{R}{Z_T} = \frac{3\ \Omega}{5\ \Omega} = \mathbf{0.6\ lagging}$$

15.4 VOLTAGE DIVIDER RULE

The basic format for the voltage divider rule in ac circuits is exactly the same as that for dc circuits:

$$\mathbf{V}_x = \frac{\mathbf{Z}_x\mathbf{E}}{\mathbf{Z}_T} \quad (15.10)$$

where $\mathbf{V}_x$ is the voltage across one or more elements in series that have total impedance $\mathbf{Z}_x$, $\mathbf{E}$ is the total voltage appearing across the series circuit, and $\mathbf{Z}_T$ is the total impedance of the series circuit.

EXAMPLE 15.9 Using the voltage divider rule, find the voltage across each element of the circuit of Fig. 15.40.

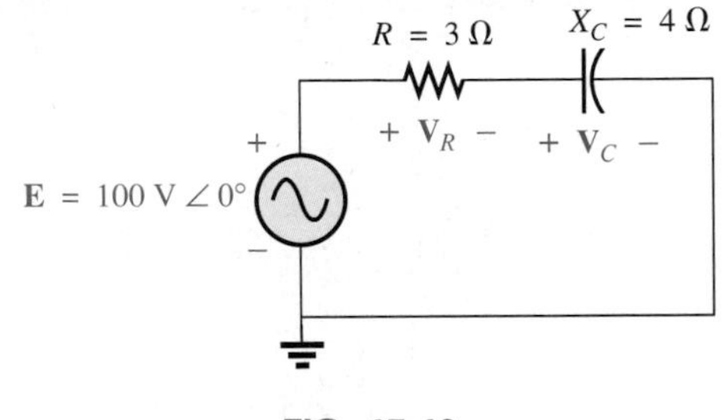

FIG. 15.40
Example 15.9.

Solution:

$$\mathbf{V}_C = \frac{\mathbf{Z}_C\mathbf{E}}{\mathbf{Z}_C + \mathbf{Z}_R} = \frac{(4\ \Omega\ \angle -90^\circ)(100\ \text{V}\ \angle 0^\circ)}{4\ \Omega\ \angle -90^\circ + 3\ \Omega\ \angle 0^\circ} = \frac{400\ \angle -90^\circ}{3 - j\,4}$$
$$= \frac{400\ \angle -90^\circ}{5\ \angle -53.13^\circ} = \mathbf{80\ V\ \angle -36.87^\circ}$$

$$\mathbf{V}_R = \frac{\mathbf{Z}_R\mathbf{E}}{\mathbf{Z}_C + \mathbf{Z}_R} = \frac{(3\ \Omega\ \angle 0^\circ)(100\ \text{V}\ \angle 0^\circ)}{5\ \Omega\ \angle -53.13^\circ} = \frac{300\ \angle 0^\circ}{5\ \angle -53.13^\circ}$$
$$= \mathbf{60\ V\ \angle +53.13^\circ}$$

EXAMPLE 15.10 Using the voltage divider rule, find the unknown voltages $\mathbf{V}_R$, $\mathbf{V}_L$, $\mathbf{V}_C$, and $\mathbf{V}_1$ for the circuit of Fig. 15.41.

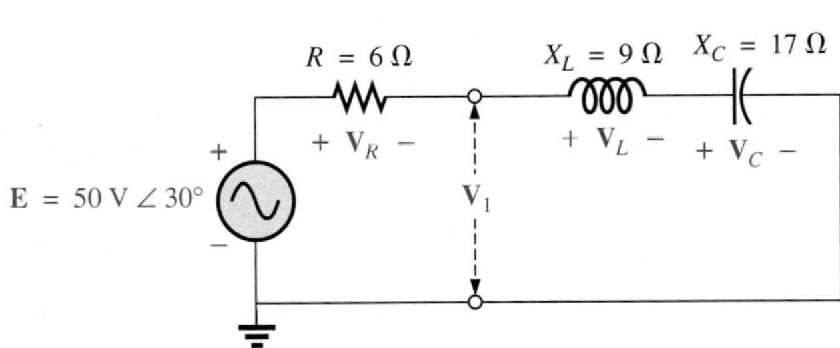

FIG. 15.41
Example 15.10.

Solution:

$$\mathbf{V}_R = \frac{\mathbf{Z}_R\mathbf{E}}{\mathbf{Z}_R + \mathbf{Z}_L + \mathbf{Z}_C} = \frac{(6\ \Omega\ \angle 0^\circ)(50\ \text{V}\ \angle 30^\circ)}{6\ \Omega\ \angle 0^\circ + 9\ \Omega\ \angle 90^\circ + 17\ \Omega\ \angle -90^\circ}$$
$$= \frac{300\ \angle 30^\circ}{6 + j\,9 - j\,17} = \frac{300\ \angle 30^\circ}{6 - j\,8}$$
$$= \frac{300\ \angle 30^\circ}{10\ \angle -53.13^\circ} = \mathbf{30\ V\ \angle 83.13^\circ}$$

Calculator The above calculation provides an excellent opportunity to demonstrate the power of today's calculators. Using the notation of the TI-85 calculator, the above calculation and the result are:

```
(6∠0)*(50∠30)/((6∠0)+(9∠90)+(17∠-90))
    (3.588E0,29.785E0)
Ans ▶ Pol
    (30.000E0∠83.130E0)
```

CALC. 15.1

$$\mathbf{V}_L = \frac{\mathbf{Z}_L\mathbf{E}}{\mathbf{Z}_T} = \frac{(9\ \Omega\ \angle 90^\circ)(50\ \text{V}\ \angle 30^\circ)}{10\ \Omega\ \angle -53.13^\circ} = \frac{450\ \text{V}\angle 120^\circ}{10\ \angle -53.13^\circ}$$
$$= \mathbf{45\ V\ \angle 173.13^\circ}$$

$$\mathbf{V}_C = \frac{\mathbf{Z}_C\mathbf{E}}{\mathbf{Z}_T} = \frac{(17\ \Omega\ \angle -90^\circ)(50\ \text{V}\ \angle 30^\circ)}{10\ \Omega\ \angle -53.13^\circ} = \frac{850\ \text{V}\angle -60^\circ}{10\ \angle -53^\circ}$$
$$= \mathbf{85\ V\ \angle -6.87^\circ}$$

$$\mathbf{V}_1 = \frac{(\mathbf{Z}_L + \mathbf{Z}_C)\mathbf{E}}{\mathbf{Z}_T} = \frac{(9\ \Omega\ \angle 90^\circ + 17\ \Omega\ \angle -90^\circ)(50\ \text{V}\ \angle 30^\circ)}{10\ \Omega\ \angle -53.13^\circ}$$
$$= \frac{(8\ \angle -90^\circ)(50\ \angle 30^\circ)}{10\ \angle -53.13^\circ}$$
$$= \frac{400\ \angle -60^\circ}{10\ \angle -53.13^\circ} = \mathbf{40\ V\ \angle -6.87^\circ}$$

EXAMPLE 15.11 For the circuit of Fig. 15.42:

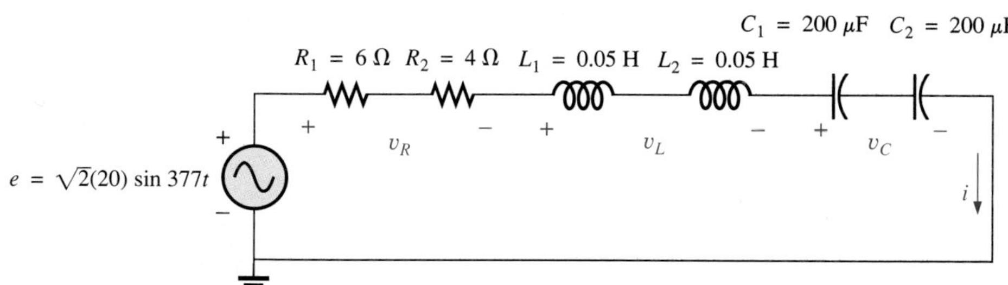

FIG. 15.42
Example 15.11.

a. Calculate **I**, $\mathbf{V}_R$, $\mathbf{V}_L$, and $\mathbf{V}_C$ in phasor form.
b. Calculate the total power factor.
c. Calculate the average power delivered to the circuit.
d. Draw the phasor diagram.
e. Obtain the phasor sum of $\mathbf{V}_R$, $\mathbf{V}_L$, and $\mathbf{V}_C$, and show that it equals the input voltage **E**.
f. Find $\mathbf{V}_R$ and $\mathbf{V}_C$ using the voltage divider rule.

Solutions:

a. Combining common elements and finding the reactance of the inductor and capacitor, we obtain

$$R_T = 6\ \Omega + 4\ \Omega = 10\ \Omega$$
$$L_T = 0.05\ \text{H} + 0.05\ \text{H} = 0.1\ \text{H}$$
$$C_T = \frac{200\ \mu\text{F}}{2} = 100\ \mu\text{F}$$

$$X_L = \omega L = (377 \text{ rad/s})(0.1 \text{ H}) = 37.70 \ \Omega$$

$$X_C = \frac{1}{\omega C} = \frac{1}{(377 \text{ rad/s})(100 \times 10^{-6} \text{ F})} = \frac{10^6 \ \Omega}{37{,}700} = 26.53 \ \Omega$$

Redrawing the circuit using phasor notation results in Fig. 15.43.

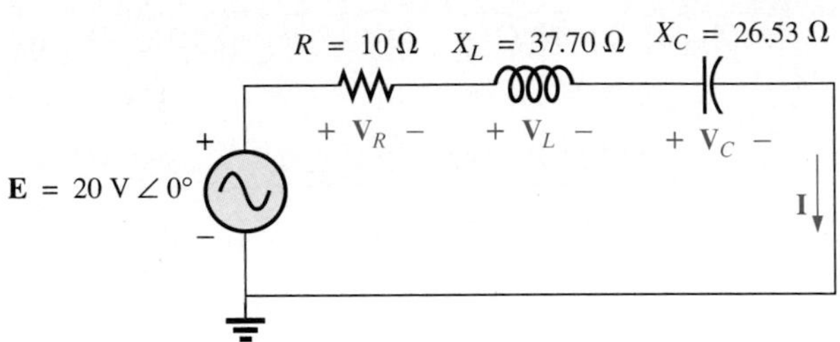

FIG. 15.43
Applying phasor notation to the circuit of Fig. 15.42.

For the circuit of Fig. 15.43,

$$\begin{aligned}\mathbf{Z}_T &= R \angle 0^\circ + X_L \angle 90^\circ + X_C \angle -90^\circ \\ &= 10 \ \Omega + j\,37.70 \ \Omega - j\,26.53 \ \Omega \\ &= 10 \ \Omega + j\,11.17 \ \Omega = \mathbf{15 \ \Omega \angle 48.16^\circ}\end{aligned}$$

The current **I** is

$$\mathbf{I} = \frac{\mathbf{E}}{\mathbf{Z}_T} = \frac{20 \text{ V} \angle 0^\circ}{15 \ \Omega \angle 48.16^\circ} = \mathbf{1.33 \text{ A} \angle -48.16^\circ}\text{ss}$$

The voltage across the resistor, inductor, and capacitor can be found using Ohm's law:

$$\begin{aligned}\mathbf{V}_R = \mathbf{I}\mathbf{Z}_R &= (I \angle \theta)(R \angle 0^\circ) = (1.33 \text{ A} \angle -48.16^\circ)(10 \ \Omega \angle 0^\circ) \\ &= \mathbf{13.30 \text{ V} \angle -48.16^\circ}\end{aligned}$$

$$\begin{aligned}\mathbf{V}_L = \mathbf{I}\mathbf{Z}_L &= (I \angle \theta)(X_L \angle 90^\circ) = (1.33 \text{ A} \angle -48.16^\circ)(37.70 \ \Omega \angle 90^\circ) \\ &= \mathbf{50.14 \text{ V} \angle 41.84^\circ}\end{aligned}$$

$$\begin{aligned}\mathbf{V}_C = \mathbf{I}\mathbf{Z}_C &= (I \angle \theta)(X_C \angle -90^\circ) = (1.33 \text{ A} \angle -48.16^\circ)(26.53 \ \Omega \angle -90^\circ) \\ &= \mathbf{35.28 \text{ V} \angle -138.16^\circ}\end{aligned}$$

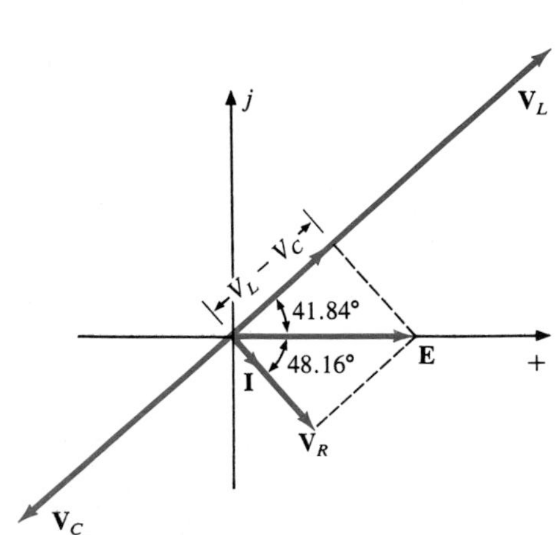

FIG. 15.44
Phasor diagram for the circuit of Fig. 15.42.

b. The total power factor, determined by the angle between the applied voltage **E** and the resulting current **I**, is 48.16°:

$$F_p = \cos \theta = \cos 48.16^\circ = \mathbf{0.667 \text{ lagging}}$$

or
$$F_p = \cos \theta = \frac{R}{Z_T} = \frac{10 \ \Omega}{15 \ \Omega} = \mathbf{0.667 \text{ lagging}}$$

c. The total power in watts delivered to the circuit is

$$P_T = EI \cos \theta = (20 \text{ V})(1.33 \text{ A})(0.667) = \mathbf{17.74 \text{ W}}$$

d. The phasor diagram appears in Fig. 15.44.
e. The phasor sum of $\mathbf{V}_R$, $\mathbf{V}_L$, and $\mathbf{V}_C$ is

$$\begin{aligned}\mathbf{E} &= \mathbf{V}_R + \mathbf{V}_L + \mathbf{V}_C \\ &= 13.30 \text{ V} \angle -48.16^\circ + 50.14 \text{ V} \angle 41.84^\circ + 35.28 \text{ V} \angle -138.16^\circ \\ \mathbf{E} &= 13.30 \text{ V} \angle -48.16^\circ + 14.86 \text{ V} \angle 41.84^\circ\end{aligned}$$

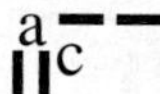

Therefore,

$$E = \sqrt{(13.30\text{ V})^2 + (14.86\text{ V})^2} = \mathbf{20\text{ V}}$$

and $\theta_E = \mathbf{0°}$ (from phasor diagram)

and $\mathbf{E} = 20\ \angle 0°$

f. $$\mathbf{V}_R = \frac{\mathbf{Z}_R\mathbf{E}}{\mathbf{Z}_T} = \frac{(10\ \Omega\ \angle 0°)(20\text{ V}\ \angle 0°)}{15\ \Omega\ \angle 48.16°} = \frac{200\text{ V}\angle 0°}{15\ \angle 48.16°}$$

$$= \mathbf{13.3\text{ V}\ \angle -48.16°}$$

$$\mathbf{V}_C = \frac{\mathbf{Z}_C\mathbf{E}}{\mathbf{Z}_T} = \frac{(26.5\ \Omega\ \angle -90°)(20\text{ V}\ \angle 0°)}{15\ \Omega\ \angle 48.16°} = \frac{530.6\text{ V}\angle -90°}{15\ \angle 48.16°}$$

$$= \mathbf{35.37\text{ V}\ \angle -138.16°}$$

15.5 FREQUENCY RESPONSE OF THE *R-C* CIRCUIT

Thus far, the analysis of series circuits has been limited to a particular frequency. We will now examine the effect of frequency on the response of an *R-C* series configuration such as that in Fig. 15.45. The magnitude of the source is fixed at 10 V, but the frequency range of analysis will extend from zero to 20 kHz.

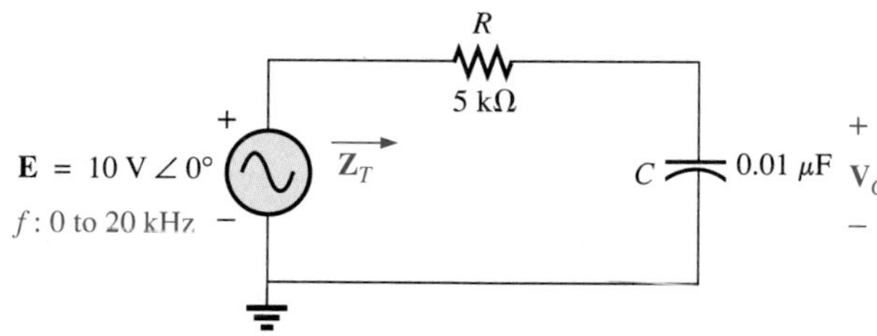

FIG. 15.45
Determining the frequency response of a series R-C circuit.

$\mathbf{Z}_T$:

Let us first determine how the impedance of the circuit $\mathbf{Z}_T$ will vary with frequency for the specified frequency range of interest. Before getting into specifics, however, let us first develop a sense for what we should expect by noting the impedance-versus-frequency curve of each element, as drawn in Fig. 15.46.

At low frequencies the reactance of the capacitor will be quite high and considerably more than the level of the resistance R, suggesting that the total impedance will be primarily capacitive in nature. At high frequencies the reactance X_C will drop below the $R = 5\text{-k}\Omega$ level and the network will start to shift toward one of a purely resistive nature (at 5 kΩ). The frequency at which $X_C = R$ can be determined in the following manner:

$$X_C = \frac{1}{2\pi f_1 C} = R$$

and

$$\boxed{f_1 = \frac{1}{2\pi RC}} \quad X_C = R \qquad \textbf{(15.11)}$$

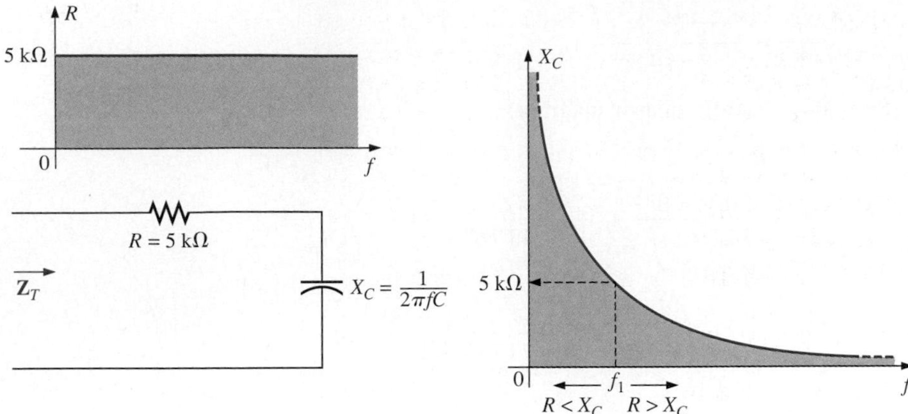

FIG. 15.46
The frequency response of the individual elements of a series R-C circuit.

which for the network of interest is

$$f_1 = \frac{1}{2\pi(5\text{ k}\Omega)(0.01\ \mu\text{F})} \cong \mathbf{3183.1\ Hz}$$

For frequencies less than f_1, $X_C > R$, and for frequencies greater than f_1, $R > X_C$, as shown in Fig. 15.46.

Now for the details. The total impedance is determined by the following equation:

$$\mathbf{Z}_T = R - jX_C$$

and

$$\mathbf{Z}_T = Z_T \angle \theta_T = \sqrt{R^2 + X_C^2} \angle -\tan^{-1}\frac{X_C}{R} \qquad \textbf{(15.12)}$$

The magnitude and angle of the total impedance can now be found at any frequency of interest by simply substituting into Eq. (15.12). The presence of the capacitor suggests that we start from a low frequency (100 Hz) and then open the spacing until we reach the upper limit of interest (20 kHz).

***f* = 100 Hz:**

$$X_C = \frac{1}{2\pi fC} = \frac{1}{2\pi(100\text{ Hz})(0.01\ \mu\text{F})} = 159.16\text{ k}\Omega$$

and $Z_T = \sqrt{R^2 + X_C^2} = \sqrt{(5\text{ k}\Omega)^2 + (159.16\text{ k}\Omega)^2} = 159.24\text{ k}\Omega$

with $\theta_T = -\tan^{-1}\frac{X_C}{R} = -\tan^{-1}\frac{159.16\text{ k}\Omega}{5\text{ k}\Omega} = -\tan^{-1} 31.83$
$= -88.2°$

and

$$\mathbf{Z}_T = \mathbf{159.24\ k\Omega \angle -88.2°}$$

which compares very closely with $\mathbf{Z}_C = 159.16\text{ k}\Omega \angle -90°$ if the circuit were purely capacitive ($R = 0\ \Omega$). Our assumption that the circuit is primarily capacitive at low frequencies is therefore confirmed.

***f* = 1 kHz:**

$$X_C = \frac{1}{2\pi fC} = \frac{1}{2\pi(1\text{ kHz})(0.01\ \mu\text{F})} = 15.92\text{ k}\Omega$$

and $Z_T = \sqrt{R^2 + X_C^2} = \sqrt{(5\text{ k}\Omega)^2 + (15.92\text{ k}\Omega)^2} = 16.69\text{ k}\Omega$

with $\theta_T = -\tan^{-1}\dfrac{X_C}{R} = -\tan^{-1}\dfrac{15.92\text{ k}\Omega}{5\text{ k}\Omega} = -\tan^{-1} 3.18 = -72.54°$

and $$\mathbf{Z}_T = \mathbf{16.69\ k\Omega \angle -72.54°}$$

A noticeable drop in the magnitude has occurred and the impedance angle has dropped almost 17° from the purely capacitive level.

Continuing:

$$f = 5\text{ kHz:}\quad \mathbf{Z}_T = \mathbf{5.93\ k\Omega \angle -32.48°}$$
$$f = 10\text{ kHz:}\quad \mathbf{Z}_T = \mathbf{5.25\ k\Omega \angle -17.66°}$$
$$f = 15\text{ kHz:}\quad \mathbf{Z}_T = \mathbf{5.11\ k\Omega \angle -11.98°}$$
$$f = 20\text{ kHz:}\quad \mathbf{Z}_T = \mathbf{5.06\ k\Omega \angle -9.04°}$$

Note how close the magnitude of Z_T at $f = 20$ kHz is to the resistance level of 5 kΩ. In addition, note how the phase angle is approaching that associated with a pure resistive network (0°).

A plot of Z_T versus frequency in Fig. 15.47 completely supports our assumption based on the curves of Fig. 15.46. The plot of θ_T versus frequency in Fig. 15.48 further suggests the fact that the total impedance made a transition from one of a capacitive nature ($\theta_T = -90°$) to one with resistive characteristics ($\theta_T = 0°$).

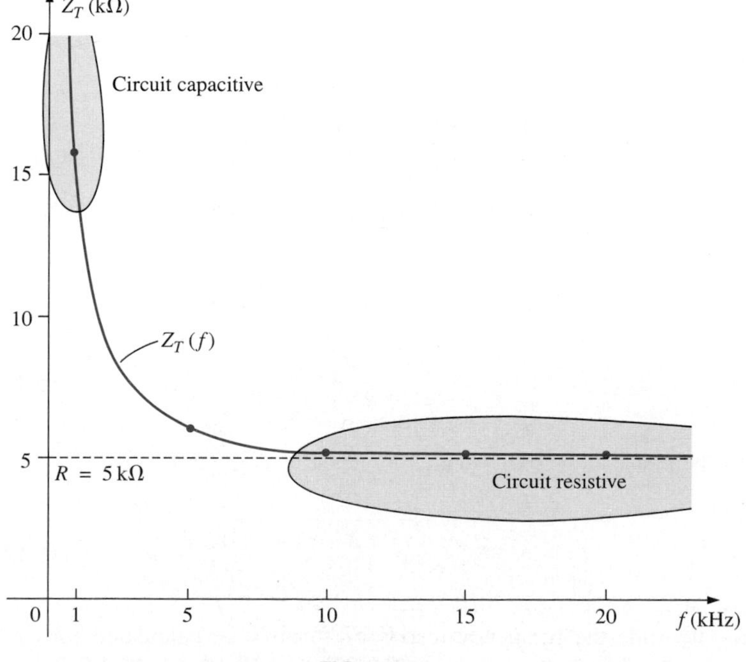

FIG. 15.47
The magnitude of the input impedance versus frequency for the circuit of Fig. 15.45.

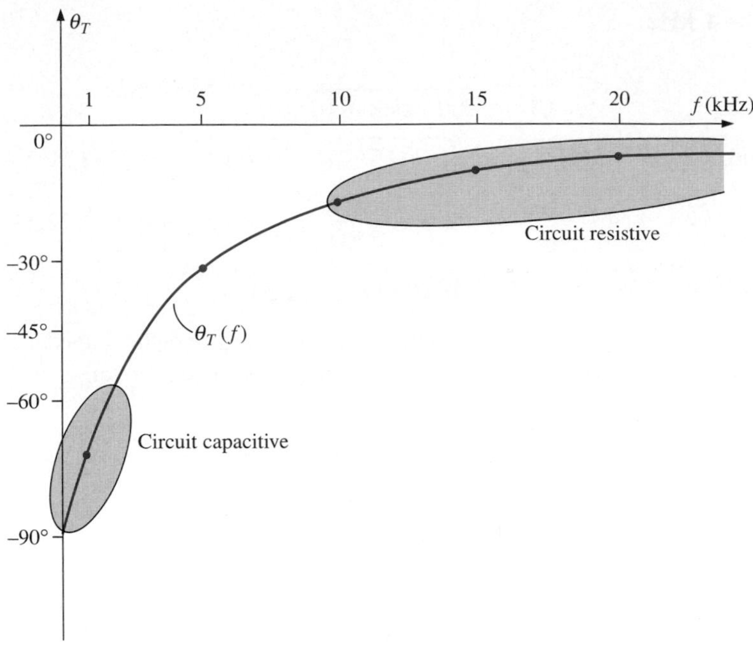

FIG. 15.48
The phase angle of the input impedance versus frequency for the circuit of Fig. 15.45.

Applying the voltage divider rule to determine the voltage across the capacitor in phasor form yields

$$\begin{aligned}\mathbf{V}_C &= \frac{\mathbf{Z}_C\mathbf{E}}{\mathbf{Z}_R + \mathbf{Z}_C}\\ &= \frac{(X_C \angle -90^\circ)(E \angle 0^\circ)}{R - jX_C} = \frac{X_C E \angle -90^\circ}{R - jX_C}\\ &= \frac{X_C E \angle -90^\circ}{\sqrt{R^2 + X_C^2} \angle -\tan^{-1} X_C/R}\end{aligned}$$

or $\mathbf{V}_C = V_C \angle \theta_C = \dfrac{X_C E}{\sqrt{R^2 + X_C^2}} \angle -90^\circ + \tan^{-1}(X_C/R)$

The magnitude of $\mathbf{V}_C$ is therefore determined by

$$V_C = \frac{X_C E}{\sqrt{R^2 + X_C^2}} \tag{15.13}$$

and the phase angle θ_C by which $\mathbf{V}_C$ leads $\mathbf{E}$ is given by

$$\theta_C = -90^\circ + \tan^{-1}\frac{X_C}{R} = -\tan^{-1}\frac{R}{X_C} \tag{15.14}$$

To determine the frequency response, X_C must be calculated for each frequency of interest and inserted into Eqs. (15.13) and (15.14).

To begin our analysis, it makes good sense to consider the case of $f = 0$ Hz (dc conditions).

***f* = 0 Hz:**

$$X_C = \frac{1}{2\pi(0)C} = \frac{1}{0} \Rightarrow \text{very large value}$$

Applying the open-circuit equivalent for the capacitor based on the above calculation will result in the following:

$$\mathbf{V}_C = \mathbf{E} = 10\text{ V}\angle 0^\circ$$

If we apply Eq. (15.13), we find

$$X_C^2 >> R^2$$

and
$$\sqrt{R^2 + X_C^2} \cong \sqrt{X_C^2} = X_C$$

and
$$V_C = \frac{X_C E}{\sqrt{R^2 + X_C^2}} = \frac{X_C E}{X_C} = E$$

with
$$\theta_C = -\tan^{-1}\frac{R}{X_C} = -\tan^{-1} 0 = 0^\circ$$

verifying the above conclusions.

***f* = 1 kHz:**

Applying Eq. (15.13):

$$X_C = \frac{1}{2\pi f C} = \frac{1}{(2\pi)(1 \times 10^3\text{ Hz})(0.01 \times 10^{-6}\text{ F})} \cong \mathbf{15.92\ k\Omega}$$

$$\sqrt{R^2 + X_C^2} = \sqrt{(5\text{ k}\Omega)^2 + (15.92\text{ k}\Omega)^2} \cong 16.69\text{ k}\Omega$$

and
$$V_C = \frac{X_C E}{\sqrt{R^2 + X_C^2}} = \frac{(15.92\text{ k}\Omega)(10)}{16.69\text{ k}\Omega} = \mathbf{9.54\ V}$$

Applying Eq. (15.14):

$$\theta_C = -\tan^{-1}\frac{R}{X_C} = -\tan^{-1}\frac{5\text{ k}\Omega}{15.9\text{ k}\Omega}$$

$$= -\tan^{-1} 0.314 = \mathbf{-17.46^\circ}$$

and
$$\mathbf{V}_C = \mathbf{9.53\ V\angle -17.46^\circ}$$

As expected, the high reactance of the capacitor at low frequencies has resulted in the major part of the applied voltage appearing across the capacitor.

If we plot the phasor diagrams for $f = 0$ Hz and $f = 1$ kHz, as shown in Fig. 15.49, we find that $\mathbf{V}_C$ is beginning a clockwise rotation with an increase in frequency that will increase the angle θ_C and decrease the phase angle between **I** and **E.** Recall that for a purely capacitive net-

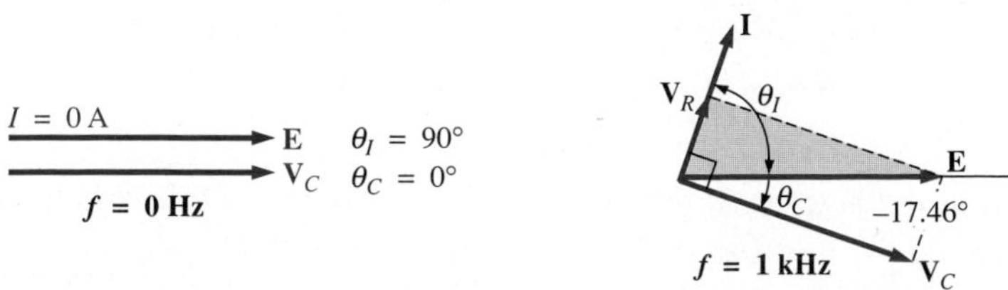

FIG. 15.49

The phasor diagram for the circuit of Fig. 15.45 for f = 0 Hz and 1 kHz.

work, **I** leads **E** by 90°. As the frequency increases, therefore, the capacitive reactance is decreasing, and eventually $R >> X_C$ with $\theta_C = -90°$, and the angle between **I** and **E** will approach 0°. Keep in mind as we proceed through the other frequencies that θ_C is the phase angle between $\mathbf{V}_C$ and **E** and that the magnitude of the angle by which **I** leads **E** is determined by

$$|\theta_I| = 90° - |\theta_C| \tag{15.15}$$

***f* = 5 kHz:**

Applying Eq. (15.13):

$$X_C = \frac{1}{2\pi fC} = \frac{1}{(2\pi)(5 \times 10^3\ \text{Hz})(0.01 \times 10^{-6}\ \text{F})} \cong \mathbf{3.18\ k\Omega}$$

Note the dramatic drop in X_C from 1 kHz to 5 kHz. In fact, X_C is now less than the resistance R of the network, and the phase angle determined by $\tan^{-1}(X_C/R)$ must be less than 45°. Here,

$$V_C = \frac{X_C E}{\sqrt{R^2 + X_C^2}} = \frac{(3.18\ \text{k}\Omega)(10\ \text{V})}{\sqrt{(5\ \text{k}\Omega)^2 + (3.18\ \text{k}\Omega)^2}} = \mathbf{5.37\ V}$$

with $$\theta_C = -\tan^{-1}\frac{R}{X_C} = -\tan^{-1}\frac{5\ \text{k}\Omega}{3.2\ \text{k}\Omega}$$

$$= -\tan^{-1} 1.56 = \mathbf{-57.38°}$$

***f* = 10 kHz:**

$$X_C \cong \mathbf{1.59\ k\Omega},\ V_C = \mathbf{3.03\ V},\ \theta_C = \mathbf{-72.34°}$$

***f* = 15 kHz:**

$$X_C \cong \mathbf{1.06\ k\Omega},\ V_C = \mathbf{2.07\ V},\ \theta_C = \mathbf{-78.02°}$$

***f* = 20 kHz:**

$$X_C \cong \mathbf{795.78\ \Omega},\ V_C = \mathbf{1.57\ V},\ \theta_C = \mathbf{-80.96°}$$

The phasor diagrams for f = 5 kHz and f = 20 kHz appear in Fig. 15.50 to show the continuing rotation of the $\mathbf{V}_C$ vector.

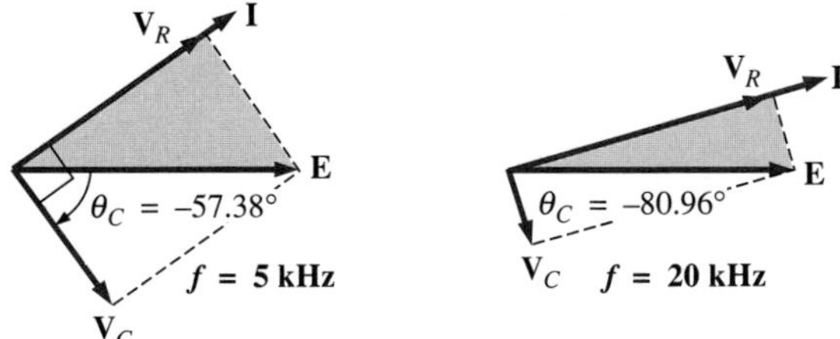

FIG. 15.50
The phasor diagram for the circuit of Fig. 15.45 for f = 5 kHz and 20 kHz.

Note also from Figs. 15.49 and 15.50 that the vector $\mathbf{V}_R$ and the current **I** have grown in magnitude with the reduction in the capacitive reactance. Eventually, at very high frequencies X_C will approach zero ohms and the short-circuit equivalent can be applied, resulting in $V_C \cong$ 0 V and $\theta_C \cong -90°$, and producing the phasor diagram of Fig. 15.51. The network is then resistive and the phase angle between **I** and **E** is essentially zero degrees, and V_R and I are their maximum values.

A plot of V_C versus frequency appears in Fig. 15.52. At low frequencies $X_C >> R$ and V_C is very close to E in magnitude. As the applied frequency increases, X_C decreases in magnitude along with V_C as V_R captures more of the applied voltage. A plot of θ_C versus frequency is provided in Fig. 15.53. At low frequencies the phase angle

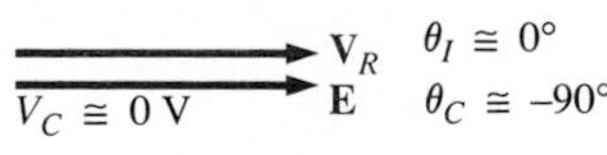

FIG. 15.51
The phasor diagram for the circuit of Fig. 15.45 at very high frequencies.

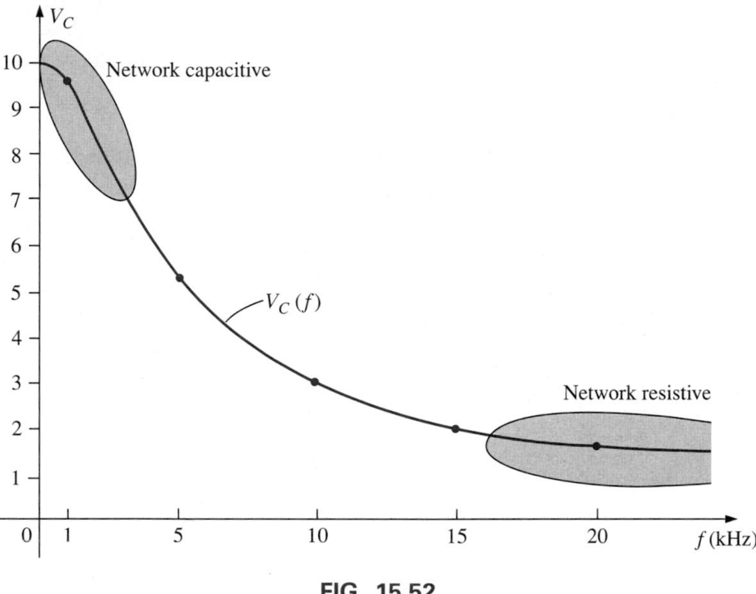

FIG. 15.52
The magnitude of the voltage V_C versus frequency for the circuit of Fig. 15.45.

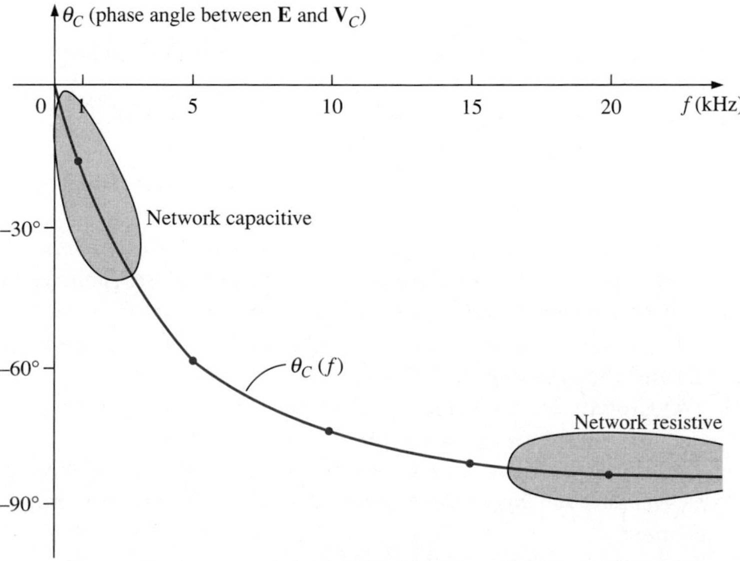

FIG. 15.53
*The phase angle between **E** and $\mathbf{V}_C$ versus frequency for the circuit of Fig. 15.45.*

between $\mathbf{V}_C$ and $\mathbf{E}$ is very small since $\mathbf{V}_C \cong \mathbf{E}$. Recall that if two phasors are equal, they must have the same angle. As the applied frequency increases, the network becomes more resistive and the phase angle between $\mathbf{V}_C$ and $\mathbf{E}$ approaches 90°. Keep in mind that, at high frequencies, $\mathbf{I}$ and $\mathbf{E}$ are approaching an in-phase situation and the angle between $\mathbf{V}_C$ and $\mathbf{E}$ will approach that between $\mathbf{V}_C$ and $\mathbf{I}$, which we know must be 90° ($\mathbf{I}_C$ leading $\mathbf{V}_C$).

A plot of V_R versus frequency would approach E volts from zero volts with an increase in frequency, but remember $V_R \neq E - V_C$ due to the vector relationship. The phase angle between $\mathbf{I}$ and $\mathbf{E}$ could be plotted directly from Fig. 15.53 using Eq. (15.15).

In Chapter 21, the analysis of this section will be extended to a much wider frequency range using a log axis for frequency. It will be demonstrated that an R-C circuit such as that in Fig. 15.45 can be used as a filter to determine which frequencies will have the greatest impact on the stage to follow. From our current analysis, it is obvious that any network connected across the capacitor will receive the greatest potential level at low frequencies and be effectively "shorted out" at very high frequencies.

The analysis of a series R-L circuit would proceed in much the same manner, except that X_L and V_L would increase with frequency and the angle between $\mathbf{I}$ and $\mathbf{E}$ would approach 90° (voltage leading the current) rather than 0°. If $\mathbf{V}_L$ were plotted versus frequency, $\mathbf{V}_L$ would approach $\mathbf{E}$, and X_L would eventually attain a level at which the open-circuit equivalent would be appropriate.

15.6 SUMMARY—SERIES ac CIRCUITS

The following is a review of important conclusions that can be derived from the discussion and examples of the previous sections. The list is not all-inclusive, but it does emphasize some of the conclusions that should be carried forward in the future analysis of ac systems.

For series ac circuits with reactive elements:

1. ***The total impedance will be frequency dependent.***
2. ***The impedance of any one element can be greater than the total impedance of the network.***
3. ***The inductive and capacitive reactances are always in direct opposition on an impedance diagram.***
4. ***Depending on the frequency applied, the same circuit can be either predominantly inductive or capacitive.***
5. ***At lower frequencies the capacitive elements will usually have the most impact on the total impedance, while at high frequencies the inductive elements will usually have the most impact.***
6. ***The magnitude of the voltage across any one element can be greater than the applied voltage.***
7. ***The magnitude of the voltage across an element as compared to the other elements of the circuit is directly related to the magnitude of its impedance; that is, the larger the impedance of an element, the larger the magnitude of the voltage across the element.***
8. ***The voltages across a coil or capacitor are always in direct opposition on a phasor diagram.***
9. ***The current is always in phase with the voltage across the resistive elements, lags the voltage across all the inductive***

elements by 90°, and leads the voltage across all the capacitive elements by 90°.

10. The larger the resistive element of a circuit compared to the net reactive impedance, the closer the power factor is to unity.

PARALLEL ac CIRCUITS

15.7 ADMITTANCE AND SUSCEPTANCE

The discussion for parallel ac circuits will be very similar to that for dc circuits. In dc circuits, *conductance* (G) was defined as being equal to $1/R$. The total conductance of a parallel circuit was then found by adding the conductance of each branch. The total resistance R_T is simply $1/G_T$.

In ac circuits, we define *admittance* ($\mathbf{Y}$) as being equal to $1/\mathbf{Z}$. The unit of measure for admittance as defined by the SI system is *siemens*, which has the symbol S. Admittance is a measure of how well an ac circuit will *admit*, or allow, current to flow in the circuit. The larger its value, therefore, the heavier the current flow for the same applied potential. The total admittance of a circuit can also be found by finding the sum of the parallel admittances. The total impedance $\mathbf{Z}_T$ of the circuit is then $1/\mathbf{Y}_T$; that is, for the network of Fig. 15.54:

$$\mathbf{Y}_T = \mathbf{Y}_1 + \mathbf{Y}_2 + \mathbf{Y}_3 + \cdots + \mathbf{Y}_N \qquad (15.16)$$

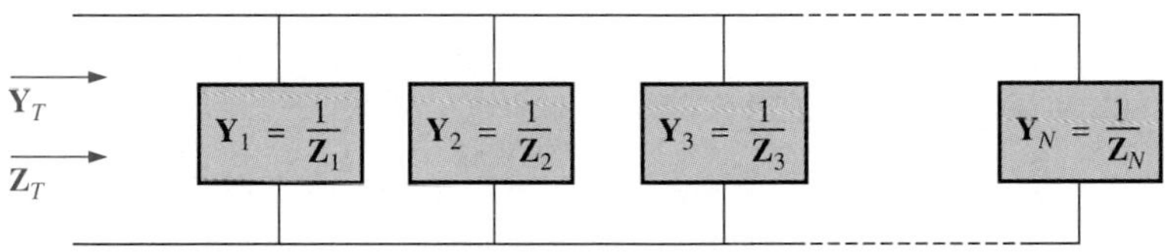

FIG. 15.54
Parallel ac network.

or, since $\mathbf{Z} = 1/\mathbf{Y}$,

$$\frac{1}{\mathbf{Z}_T} = \frac{1}{\mathbf{Z}_1} + \frac{1}{\mathbf{Z}_2} + \frac{1}{\mathbf{Z}_3} + \cdots + \frac{1}{\mathbf{Z}_N} \qquad (15.17)$$

For two impedances in parallel,

$$\frac{1}{\mathbf{Z}_T} = \frac{1}{\mathbf{Z}_1} + \frac{1}{\mathbf{Z}_2}$$

If the manipulations used in Chapter 6 to find the total resistance of two parallel resistors are now applied, the following similar equation will result:

$$\mathbf{Z}_T = \frac{\mathbf{Z}_1\mathbf{Z}_2}{\mathbf{Z}_1 + \mathbf{Z}_2} \qquad (15.18)$$

For three parallel impedances,

$$\mathbf{Z}_T = \frac{\mathbf{Z}_1\mathbf{Z}_2\mathbf{Z}_3}{\mathbf{Z}_1\mathbf{Z}_2 + \mathbf{Z}_2\mathbf{Z}_3 + \mathbf{Z}_1\mathbf{Z}_3} \tag{15.19}$$

As pointed out in the introduction to this section, conductance is the reciprocal of resistance, and

$$\mathbf{Y}_R = \frac{1}{\mathbf{Z}_R} = \frac{1}{R \angle 0^\circ} = G \angle 0^\circ \tag{15.20}$$

The reciprocal of reactance ($1/X$) is called *susceptance* and is a measure of how *susceptible* an element is to the passage of current through it. Susceptance is also measured in *siemens* and is represented by the capital letter B.

For the inductor,

$$\mathbf{Y}_L = \frac{1}{\mathbf{Z}_L} = \frac{1}{X_L \angle 90^\circ} = \frac{1}{X_L} \angle -90^\circ \tag{15.21}$$

Defining

$$B_L = \frac{1}{X_L} \quad \text{(siemens, S)} \tag{15.22}$$

we have

$$\mathbf{Y}_L = B_L \angle -90^\circ \tag{15.23}$$

Note that for inductance, an increase in frequency or inductance will result in a decrease in susceptance or, correspondingly, in admittance.

For the capacitor,

$$\mathbf{Y}_C = \frac{1}{\mathbf{Z}_C} = \frac{1}{X_C \angle -90^\circ} = \frac{1}{X_C} \angle 90^\circ \tag{15.24}$$

Defining

$$B_C = \frac{1}{X_C} \quad \text{(siemens, S)} \tag{15.25}$$

we have

$$\mathbf{Y}_C = B_C \angle 90^\circ \tag{15.26}$$

For the capacitor, therefore, an increase in frequency or capacitance will result in an increase in its susceptibility.

For parallel ac circuits, the *admittance diagram* is used with the three admittances, represented as shown in Fig. 15.55.

Note in Fig. 15.55 that the conductance (like resistance) is on the positive real axis, whereas inductive and capacitive susceptances are in direct opposition on the imaginary axis.

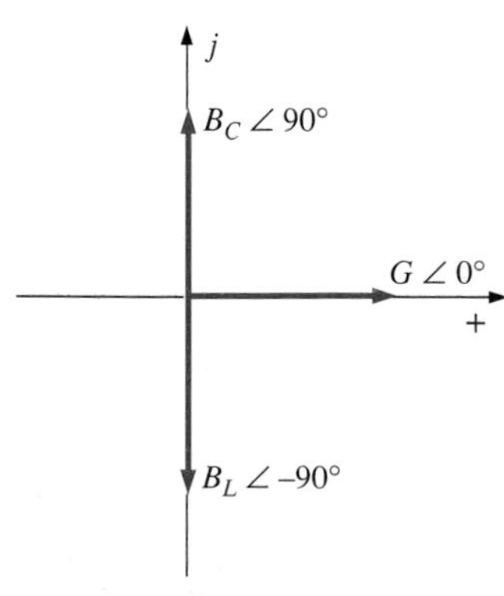

FIG. 15.55
Admittance diagram.

For any configuration (series, parallel, series-parallel, etc.), the angle associated with the total admittance is the angle by which the source current leads the applied voltage. For inductive networks, θ_T is negative, whereas for capacitive networks, θ_T is positive.

EXAMPLE 15.12 For the network of Fig. 15.56:

a. Find the admittance of each parallel branch.
b. Determine the input admittance.
c. Calculate the input impedance.
d. Draw the admittance diagram.

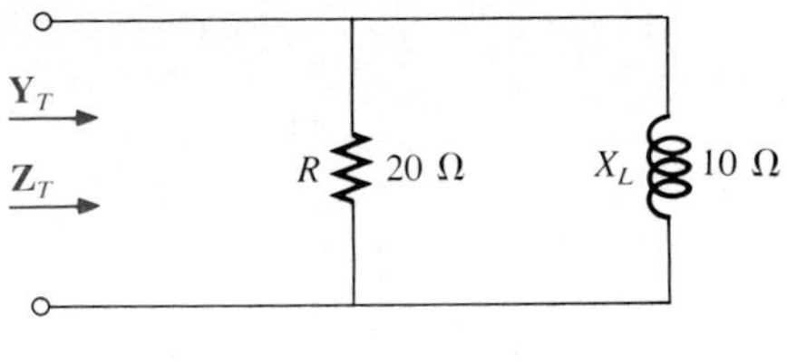

FIG. 15.56
Example 15.12.

Solutions:

a. $\mathbf{Y}_R = G\,\angle 0° = \dfrac{1}{R}\,\angle 0° = \dfrac{1}{20\,\Omega}\,\angle 0°$

$= \mathbf{0.05\ S\,\angle 0° = 0.05\ S} + \boldsymbol{j}\,\mathbf{0}$

$\mathbf{Y}_L = B_L\,\angle -90° = \dfrac{1}{X_L}\,\angle -90° = \dfrac{1}{10\,\Omega}\,\angle -90°$

$= \mathbf{0.1\ S\,\angle -90° = 0} - \boldsymbol{j}\,\mathbf{0.1\ S}$

b. $\mathbf{Y}_T = \mathbf{Y}_R + \mathbf{Y}_L = (0.05\text{ S} + j\,0) + (0 - j\,0.1\text{ S})$

$= \mathbf{0.05\ S} - \boldsymbol{j}\,\mathbf{0.1\ S} = G - jB_L$

c. $\mathbf{Z}_T = \dfrac{1}{\mathbf{Y}_T} = \dfrac{1}{0.05\text{ S} - j\,0.1\text{ S}} = \dfrac{1}{0.112\text{ S}\,\angle -63.43°}$

$= \mathbf{8.93\ \Omega\,\angle 63.43°}$

or Eq. (15.17):

$$\mathbf{Z}_T = \frac{\mathbf{Z}_R\mathbf{Z}_L}{\mathbf{Z}_R + \mathbf{Z}_L} = \frac{(20\,\Omega\,\angle 0°)(10\,\Omega\,\angle 90°)}{20\,\Omega + j\,10\,\Omega}$$

$$= \frac{200\,\Omega\,\angle 90°}{22.36\,\angle 26.57°} = \mathbf{8.93\ \Omega\,\angle 63.43°}$$

d. The admittance diagram appears in Fig. 15.57.

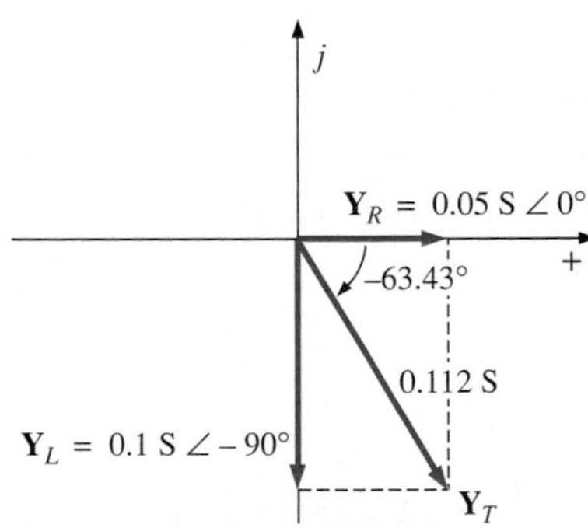

FIG. 15.57
Admittance diagram for the network of Fig. 15.56.

EXAMPLE 15.13 Repeat Example 15.12 for the parallel network of Fig. 15.58.

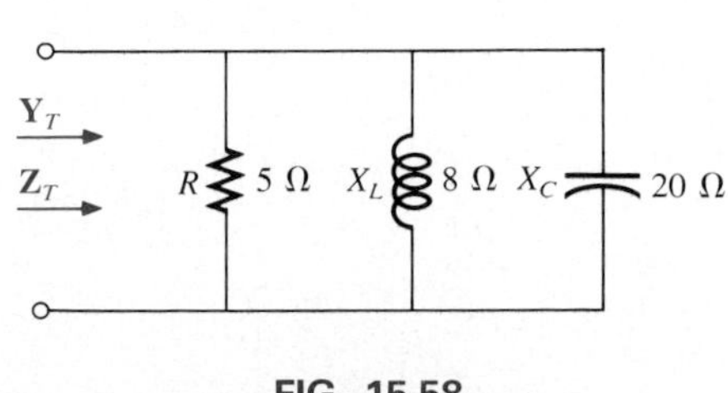

FIG. 15.58
Example 15.13.

Solutions:

a. $\mathbf{Y}_R = G\,\angle 0° = \dfrac{1}{R}\,\angle 0° = \dfrac{1}{5\,\Omega}\,\angle 0°$

$= \mathbf{0.2\ S\,\angle 0° = 0.2\ S} + \boldsymbol{j}\,\mathbf{0}$

$$\mathbf{Y}_L = B_L \angle -90° = \frac{1}{X_L} \angle -90° = \frac{1}{8\ \Omega} \angle -90°$$
$$= \mathbf{0.125\ S} \angle \mathbf{-90°} = \mathbf{0 - j\,0.125\ S}$$
$$\mathbf{Y}_C = B_C \angle 90° = \frac{1}{X_C} \angle 90° = \frac{1}{20\ \Omega} \angle 90°$$
$$= \mathbf{0.050\ S} \angle \mathbf{+90°} = \mathbf{0 + j\,0.050\ S}$$

b. $\mathbf{Y}_T = \mathbf{Y}_R + \mathbf{Y}_L + \mathbf{Y}_C$
$$= (0.2\ \text{S} + j\,0) + (0 - j\,0.125\ \text{S}) + (0 + j\,0.050\ \text{S})$$
$$= 0.2\ \text{S} - j\,0.075\ \text{S} = \mathbf{0.2136\ S} \angle \mathbf{-20.56°}$$

c. $\mathbf{Z}_T = \dfrac{1}{0.2136\ \text{S} \angle -20.56°} = \mathbf{4.68\ \Omega} \angle \mathbf{20.56°}$

or

$$\mathbf{Z}_T = \frac{\mathbf{Z}_R\mathbf{Z}_L\mathbf{Z}_C}{\mathbf{Z}_R\mathbf{Z}_L + \mathbf{Z}_L\mathbf{Z}_C + \mathbf{Z}_R\mathbf{Z}_C}$$
$$= \frac{(5\ \Omega \angle 0°)(8\ \Omega \angle 90°)(20\ \Omega \angle -90°)}{(5\ \Omega \angle 0°)(8\ \Omega \angle 90°) + (8\ \Omega \angle 90°)(20\ \Omega \angle -90°) + (5\ \Omega \angle 0°)(20\ \Omega \angle -90°)}$$
$$= \frac{800\ \Omega \angle 0°}{40 \angle 90° + 160 \angle 0° + 100 \angle -90°}$$
$$= \frac{800\ \Omega}{160 + j\,40 - j\,100} = \frac{800\ \Omega}{160 - j\,60}$$
$$= \frac{800\ \Omega}{170.88 \angle -20.56°}$$
$$= \mathbf{4.68\ \Omega} \angle \mathbf{20.56°}$$

d. The admittance diagram appears in Fig. 15.59.

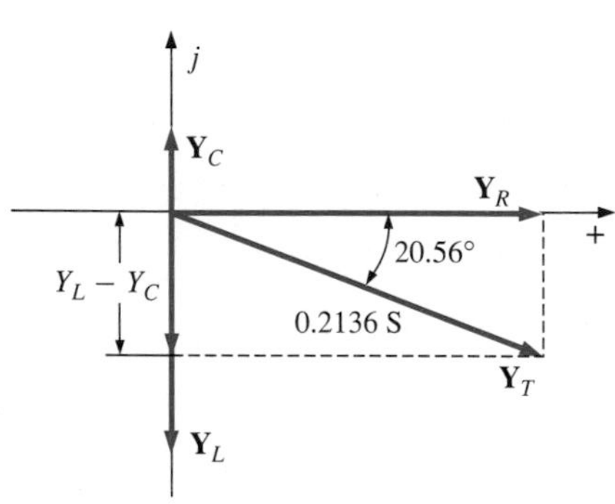

FIG. 15.59
Admittance diagram for the network of Fig. 15.58.

On many occasions, the inverse relationship $\mathbf{Y}_T = 1/\mathbf{Z}_T$ or $\mathbf{Z}_T = 1/\mathbf{Y}_T$ will require that we divide the number 1 by a complex number having a real and an imaginary part. This division, if not performed in the polar form, requires that we multiply the numerator and denominator by the conjugate of the denominator, as follows:

$$\mathbf{Y}_T = \frac{1}{\mathbf{Z}_T} = \frac{1}{4\ \Omega + j\,6\ \Omega} = \left(\frac{1}{4\ \Omega + j\,6\ \Omega}\right)\left(\frac{(4\ \Omega - j\,6\ \Omega)}{(4\ \Omega - j\,6\ \Omega)}\right) = \frac{4 - j\,6}{4^2 + 6^2}$$

and
$$\mathbf{Y}_T = \frac{4}{52}\ \text{S} - j\,\frac{6}{52}\ \text{S}$$

To avoid this laborious task each time we want to find the reciprocal of a complex number in rectangular form, a format can be developed using the following complex number, which is symbolic of any impedance or admittance in the first or fourth quadrant:

$$\frac{1}{a_1 \pm j\,b_1} = \left(\frac{1}{a_1 \pm j\,b_1}\right)\left(\frac{a_1 \mp j\,b_1}{a_1 \mp j\,b_1}\right) = \frac{a_1 \mp j\,b_1}{a_1^2 + b_1^2}$$

or

$$\frac{1}{a_1 \pm j\,b_1} = \frac{a_1}{a_1^2 + b_1^2} \mp j\,\frac{b_1}{a_1^2 + b_1^2} \qquad \textbf{(15.27)}$$

Note that the denominator is simply the sum of the squares of each term. The sign is inverted between the real and imaginary parts. A few examples will develop some familiarity with the use of this equation.

EXAMPLE 15.14 Find the admittance of each set of series elements in Fig. 15.60.

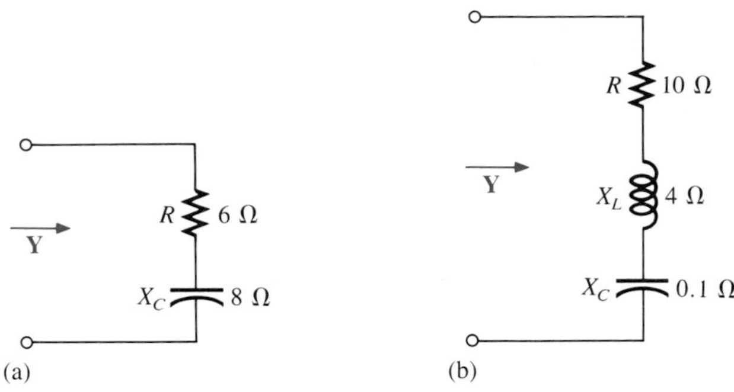

FIG. 15.60
Example 15.14.

Solutions:

a. $\mathbf{Z} = R - j X_C = 6\ \Omega - j\ 8\ \Omega$

$$\text{Eq. (15.27): } \mathbf{Y} = \frac{1}{6\ \Omega - j\ 8\ \Omega} = \frac{6}{(6)^2 + (8)^2} + j\frac{8}{(6)^2 + (8)^2}$$

$$= \mathbf{\frac{6}{100}\ S + j\frac{8}{100}\ S}$$

b. $\mathbf{Z} = 10\ \Omega + j\ 4\ \Omega + (-j\ 0.1\ \Omega) = 10\ \Omega + j\ 3.9\ \Omega$

Eq. (15.27):

$$\mathbf{Y} = \frac{1}{\mathbf{Z}} = \frac{1}{10\ \Omega + j\ 3.9\ \Omega} = \frac{10}{(10)^2 + (3.9)^2} - j\frac{3.9}{(10)^2 + (3.9)^2}$$

$$= \frac{10}{115.21} - j\frac{3.9}{115.21} = \mathbf{0.087\ S - j\ 0.034\ S}$$

15.8 PARALLEL ac NETWORKS

For the representative parallel ac network of Fig. 15.61, the total impedance or admittance is determined as described in the previous section, and the source current is determined by Ohm's law as follows:

$$\mathbf{I} = \frac{\mathbf{E}}{\mathbf{Z}_T} = \mathbf{E}\mathbf{Y}_T \qquad \textbf{(15.28)}$$

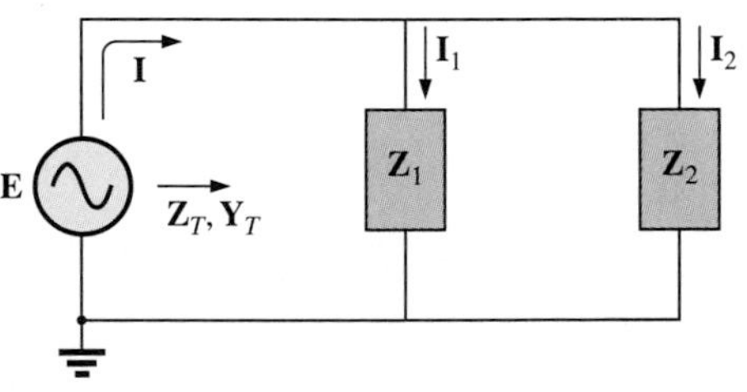

FIG. 15.61
Parallel ac network.

Since the voltage is the same across parallel elements, the current through each branch can then be found through another application of Ohm's law:

$$\mathbf{I}_1 = \frac{\mathbf{E}}{\mathbf{Z}_1} = \mathbf{E}\mathbf{Y}_1 \qquad \textbf{(15.29a)}$$

$$\mathbf{I}_2 = \frac{\mathbf{E}}{\mathbf{Z}_2} = \mathbf{E}\mathbf{Y}_2 \qquad \textbf{(15.29b)}$$

Kirchhoff's current law can then be applied in the same manner as employed for dc networks. However, keep in mind that we are now dealing with the algebraic manipulation of quantities that have both magnitude and direction.

$$\mathbf{I} - \mathbf{I}_1 - \mathbf{I}_2 = 0$$

or

$$\boxed{\mathbf{I} = \mathbf{I}_1 + \mathbf{I}_2} \tag{15.30}$$

The power to the network can be determined by

$$\boxed{P = EI \cos \theta_T} \tag{15.31}$$

where θ_T is the phase angle between **E** and **I**.

Let us now look at a few examples carried out in great detail for the first exposure.

R-L

Refer to Fig. 15.62.

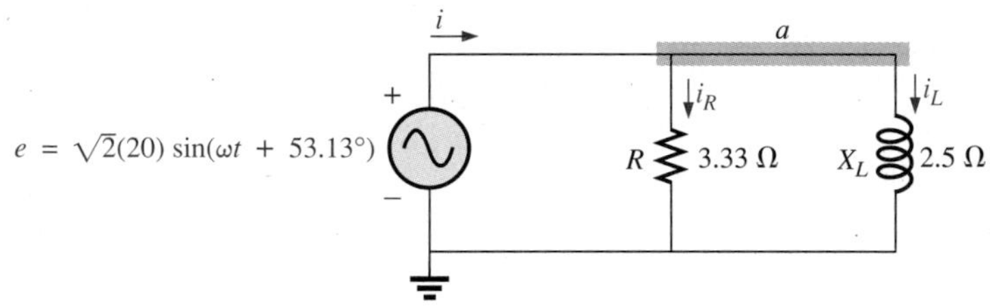

FIG. 15.62
Parallel R-L network.

Phasor notation: As shown in Fig. 15.63.

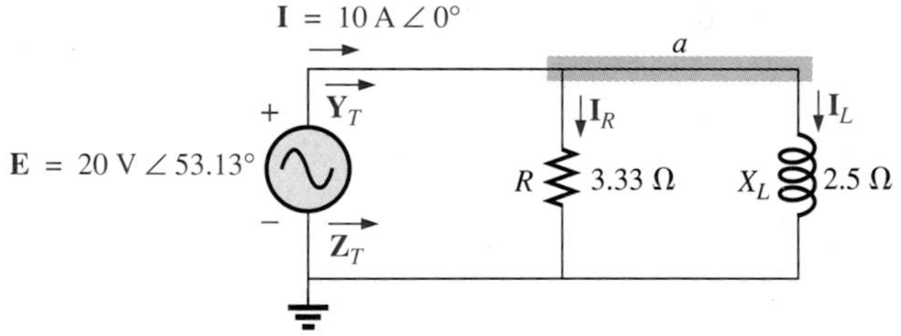

FIG. 15.63
Applying phasor notation to the network of Fig. 15.62.

$\mathbf{Y}_T$, $\mathbf{Z}_T$:

$$\begin{aligned}
\mathbf{Y}_T &= \mathbf{Y}_R + \mathbf{Y}_L \\
&= G \angle 0^\circ + B_L \angle -90^\circ = \frac{1}{3.33\ \Omega} \angle 0^\circ + \frac{1}{2.5\ \Omega} \angle -90^\circ \\
&= 0.3\ \text{S} \angle 0^\circ + 0.4\ \text{S} \angle -90^\circ = 0.3\ \text{S} - j\,0.4\ \text{S} \\
&= \mathbf{0.5\ S \angle -53.13^\circ}
\end{aligned}$$

$$\mathbf{Z}_T = \frac{1}{\mathbf{Y}_T} = \frac{1}{0.5\ \text{S} \angle -53.13^\circ} = \mathbf{2\ \Omega \angle 53.13^\circ}$$

Admittance diagram: As shown in Fig. 15.64.

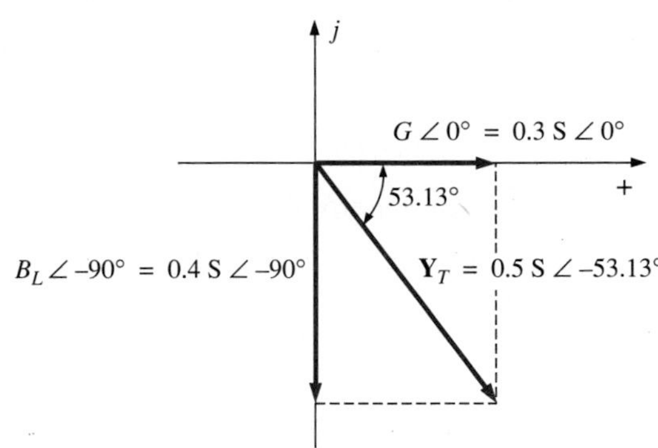

FIG. 15.64
Admittance diagram for the parallel R-L network of Fig. 15.62.

I:

$$\mathbf{I} = \frac{\mathbf{E}}{\mathbf{Z}_T} = \mathbf{E}\mathbf{Y}_T = (20 \text{ V} \angle 53.13°)(0.5 \text{ S} \angle -53.13°) = \mathbf{10 \text{ A} \angle 0°}$$

$\mathbf{I}_R$, $\mathbf{I}_L$:

$$\begin{aligned}\mathbf{I}_R &= \frac{E \angle \theta}{R \angle 0°} = (E \angle \theta)(G \angle 0°) \\ &= (20 \text{ V} \angle 53.13°)(0.3 \text{ S} \angle 0°) = \mathbf{6 \text{ A} \angle 53.13°} \\ \mathbf{I}_L &= \frac{E \angle \theta}{X_L \angle 90°} = (E \angle \theta)(B_L \angle -90°) \\ &= (20 \text{ V} \angle 53.13°)(0.4 \text{ S} \angle -90°) \\ &= \mathbf{8 \text{ A} \angle -36.87°}\end{aligned}$$

Kirchhoff's current law: At node a,

$$\mathbf{I} - \mathbf{I}_R - \mathbf{I}_L = 0$$

or

$$\begin{aligned}\mathbf{I} &= \mathbf{I}_R + \mathbf{I}_L \\ 10 \text{ A} \angle 0° &= 6 \text{ A} \angle 53.13° + 8 \text{ A} \angle -36.87° \\ 10 \text{ A} \angle 0° &= (3.60 \text{ A} + j\,4.80 \text{ A}) + (6.40 \text{ A} - j\,4.80 \text{ A}) = 10 \text{ A} + j\,0\end{aligned}$$

and $\mathbf{10 \text{ A} \angle 0° = 10 \text{ A} \angle 0°}$ (checks)

Phasor diagram: The phasor diagram of Fig. 15.65 indicates that the applied voltage $\mathbf{E}$ is in phase with the current $\mathbf{I}_R$ and leads the current $\mathbf{I}_L$ by 90°.

Power: The total power in watts delivered to the circuit is

$$\begin{aligned}P_T &= EI \cos \theta_T \\ &= (20 \text{ V})(10 \text{ A}) \cos 53.13° = (200 \text{ W})(0.6) \\ &= \mathbf{120 \text{ W}}\end{aligned}$$

or $$P_T = I^2R = \frac{V_R^2}{R} = V_R^2 G = (20 \text{ V})^2(0.3 \text{ S}) = \mathbf{120 \text{ W}}$$

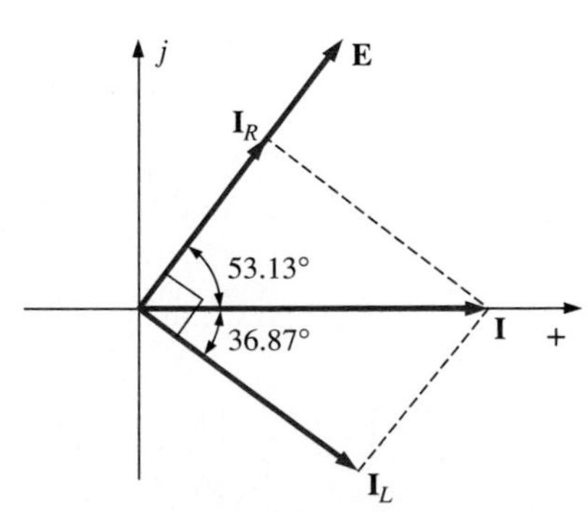

FIG. 15.65
Phasor diagram for the parallel R-L network of Fig. 15.62.

or, finally,

$$\begin{aligned} P_T &= P_R + P_L = EI_R \cos\theta_R + EI_L \cos\theta_L \\ &= (20\text{ V})(6\text{ A})\cos 0° + (20\text{ V})(8\text{ A})\cos 90° = 120\text{ W} + 0 \\ &= \mathbf{120\ W} \end{aligned}$$

Power factor: The power factor of the circuit is

$$F_p = \cos\theta_T = \cos 53.13° = \mathbf{0.6\ lagging}$$

or, through an analysis similar to that employed for a series ac circuit,

$$\cos\theta_T = \frac{P}{EI} = \frac{E^2/R}{EI} = \frac{EG}{I} = \frac{G}{I/V} = \frac{G}{Y_T}$$

and

$$F_p = \cos\theta_T = \frac{G}{Y_T} \tag{15.32}$$

where G and Y_T are the magnitudes of the total conductance and admittance of the parallel network. For this case,

$$F_p = \cos\theta_T = \frac{0.3\text{ S}}{0.5\text{ S}} = \mathbf{0.6\ lagging}$$

Impedance approach: The current **I** can also be found by first finding the total impedance of the network:

$$\begin{aligned} \mathbf{Z}_T &= \frac{\mathbf{Z}_R\mathbf{Z}_L}{\mathbf{Z}_R + \mathbf{Z}_L} = \frac{(3.33\ \Omega\ \angle 0°)(2.5\ \Omega\ \angle 90°)}{3.33\ \Omega\ \angle 0° + 2.5\ \Omega\ \angle 90°} \\ &= \frac{8.325\ \angle 90°}{4.164\ \angle 36.87°} = \mathbf{2\ \Omega\ \angle 53.13°} \end{aligned}$$

And then, using Ohm's law, we obtain

$$\mathbf{I} = \frac{\mathbf{E}}{\mathbf{Z}_T} = \frac{20\text{ V}\ \angle 53.13°}{2\ \Omega\ \angle 53.13°} = \mathbf{10\ A\ \angle 0°}$$

R-C

Refer to Fig. 15.66.

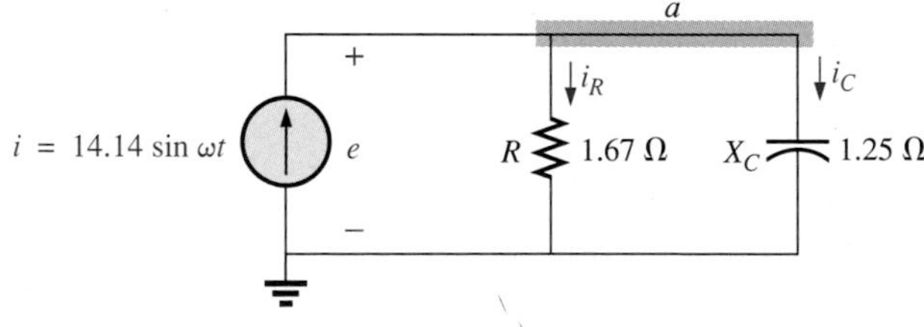

FIG. 15.66
Parallel R-C network.

Phasor notation: As shown in Fig. 15.67.

$\mathbf{Y}_T$, $\mathbf{Z}_T$:

$$\begin{aligned} \mathbf{Y}_T &= \mathbf{Y}_R + \mathbf{Y}_C = G\ \angle 0° + B_C\ \angle 90° = \frac{1}{1.67\ \Omega}\ \angle 0° + \frac{1}{1.25\ \Omega}\ \angle 90° \\ &= 0.6\text{ S}\ \angle 0° + 0.8\text{ S}\ \angle 90° = 0.6\text{ S} + j\,0.8\text{ S} = \mathbf{1.0\ S\ \angle 53.13°} \end{aligned}$$

$$\mathbf{Z}_T = \frac{1}{\mathbf{Y}_T} = \frac{1}{1.0\text{ S}\ \angle 53.13°} = \mathbf{1\ \Omega\ \angle -53.13°}$$

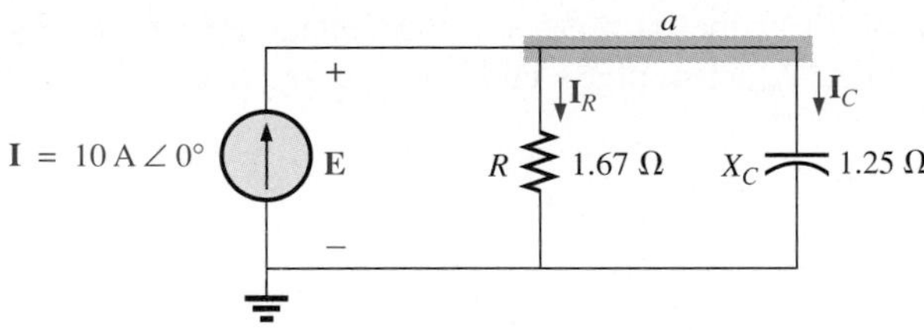

FIG. 15.67
Applying phasor notation to the network of Fig. 15.66.

Admittance diagram: As shown in Fig. 15.68.

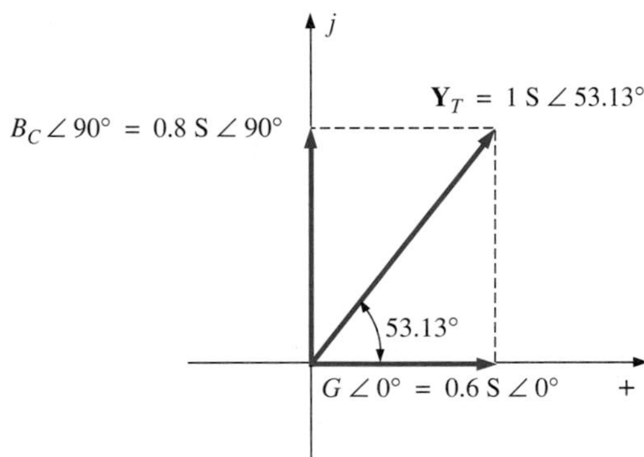

FIG. 15.68
Admittance diagram for the parallel R-C network of Fig. 15.66.

E:

$$\mathbf{E} = \mathbf{I}\mathbf{Z}_T = \frac{\mathbf{I}}{\mathbf{Y}_T} = \frac{10\text{ A }\angle 0^\circ}{1\text{ S }\angle 53.13^\circ} = \mathbf{10\ V\ \angle -53.13^\circ}$$

$\mathbf{I}_R$, $\mathbf{I}_C$:

$$\begin{aligned}\mathbf{I}_R &= (E\ \angle\theta)(G\ \angle 0^\circ)\\ &= (10\text{ V }\angle -53.13^\circ)(0.6\text{ S }\angle 0^\circ) = \mathbf{6\ A\ \angle -53.13^\circ}\\ \mathbf{I}_C &= (E\ \angle\theta)(B_C\ \angle 90^\circ)\\ &= (10\text{ V }\angle -53.13^\circ)(0.8\text{ S }\angle 90^\circ) = \mathbf{8\ A\ \angle 36.87^\circ}\end{aligned}$$

Kirchhoff's current law: At node *a*,

$$\mathbf{I} - \mathbf{I}_R - \mathbf{I}_C = 0$$

or

$$\mathbf{I} = \mathbf{I}_R + \mathbf{I}_C$$

which can also be verified (as for the *R-L* network) through vector algebra.

Phasor diagram: The phasor diagram of Fig. 15.69 indicates that **E** is in phase with the current through the resistor $\mathbf{I}_R$ and lags the capacitive current $\mathbf{I}_C$ by 90°.

Time domain:

$$\begin{aligned}e &= \sqrt{2}(10)\sin(\omega t - 53.13^\circ) = \mathbf{14.14\ sin(\omega t - 53.13^\circ)}\\ i_R &= \sqrt{2}(6)\sin(\omega t - 53.13^\circ) = \mathbf{8.48\ sin(\omega t - 53.13^\circ)}\\ i_C &= \sqrt{2}(8)\sin(\omega t + 36.87^\circ) = \mathbf{11.31\ sin(\omega t + 36.87^\circ)}\end{aligned}$$

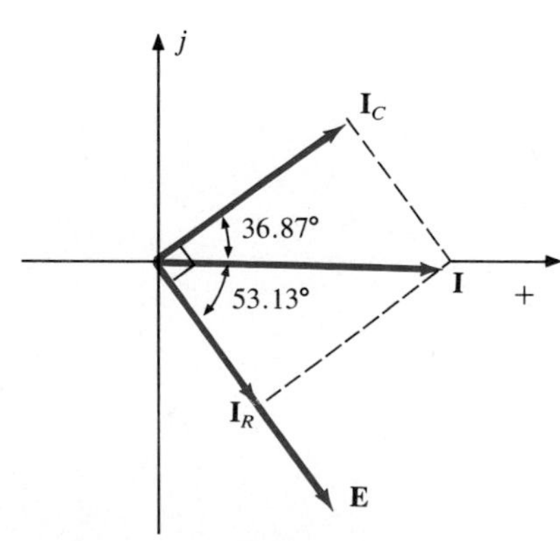

FIG. 15.69
Phasor diagram for the parallel R-C network of Fig. 15.66.

A plot of all of the currents and the voltage appears in Fig. 15.70. Note that e and i_R are in phase and e lags i_C by 90°.

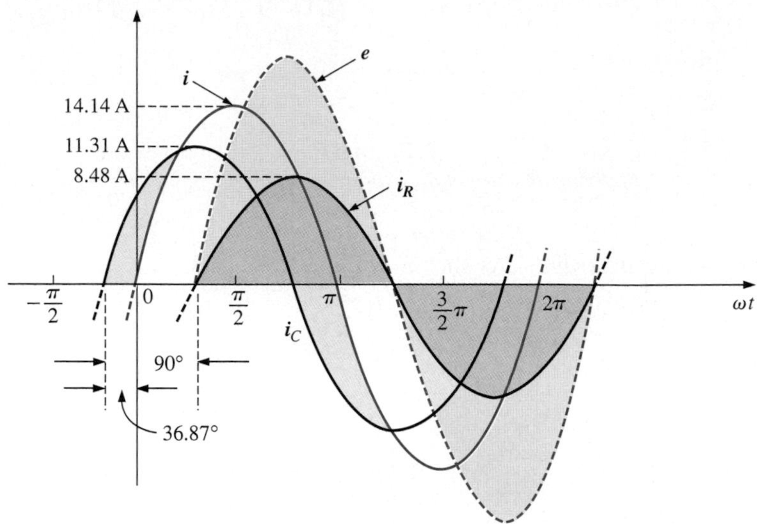

FIG. 15.70
Waveforms for the parallel R-C network of Fig. 15.66.

Power:

$$P_T = EI \cos\theta = (10 \text{ V})(10 \text{ A}) \cos 53.13° = (10)^2(0.6) = \mathbf{60\ W}$$

or

$$P_T = E^2G = (10 \text{ V})^2(0.6 \text{ S}) = \mathbf{60\ W}$$

or, finally,

$$\begin{aligned} P_T = P_R + P_C &= EI_R \cos\theta_R + EI_C \cos\theta_C \\ &= (10 \text{ V})(6 \text{ A}) \cos 0° + (10 \text{ V})(8 \text{ A}) \cos 90° \\ &= \mathbf{60\ W} \end{aligned}$$

Power factor: The power factor of the circuit is

$$F_p = \cos 53.13° = \mathbf{0.6\ leading}$$

Using Eq. (15.32), we have

$$F_p = \cos\theta_T = \frac{G}{Y_T} = \frac{0.6 \text{ S}}{1.0 \text{ S}} = \mathbf{0.6\ leading}$$

Impedance approach: The voltage **E** can also be found by first finding the total impedance of the circuit:

$$\begin{aligned} \mathbf{Z}_T = \frac{\mathbf{Z}_R\mathbf{Z}_C}{\mathbf{Z}_R + \mathbf{Z}_C} &= \frac{(1.67\ \Omega \angle 0°)(1.25\ \Omega \angle -90°)}{1.67\ \Omega \angle 0° + 1.25\ \Omega \angle -90°} \\ &= \frac{2.09 \angle -90°}{2.09 \angle -36.81°} = \mathbf{1\ \Omega \angle -53.19°} \end{aligned}$$

and then, using Ohm's law, we find

$$\mathbf{E} = \mathbf{IZ}_T = (10 \text{ A} \angle 0°)(1\ \Omega \angle -53.19°) = \mathbf{10\ V \angle -53.19°}$$

R-L-C

Refer to Fig. 15.71.

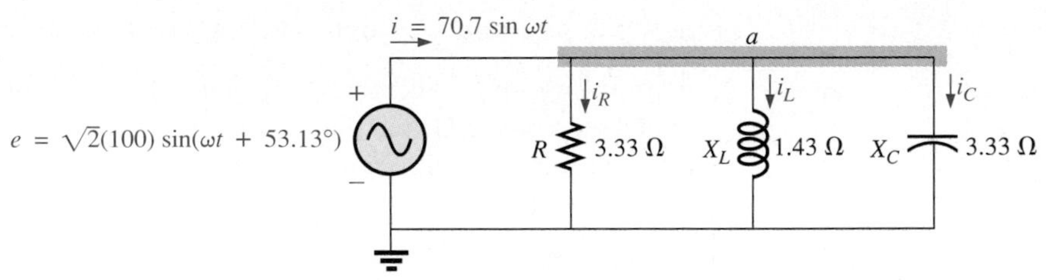

FIG. 15.71
Parallel R-L-C ac network.

Phasor notation: As shown in Fig. 15.72.

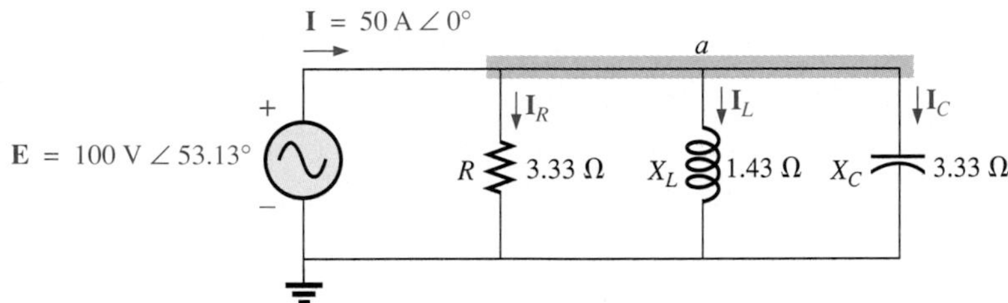

FIG. 15.72
Applying phasor notation to the network of Fig. 15.71.

$\mathbf{Y}_T$, $\mathbf{Z}_T$:

$$\begin{aligned}\mathbf{Y}_T &= \mathbf{Y}_R + \mathbf{Y}_L + \mathbf{Y}_C = G\angle 0° + B_L\angle -90° + B_C\angle 90°\\ &= \frac{1}{3.33\ \Omega}\angle 0° + \frac{1}{1.43\ \Omega}\angle -90° + \frac{1}{3.33\ \Omega}\angle 90°\\ &= 0.3\text{ S}\angle 0° + 0.7\text{ S}\angle -90° + 0.3\text{ S}\angle 90°\\ &= 0.3\text{ S} - j\,0.7\text{ S} + j\,0.3\text{ S}\\ &= 0.3\text{ S} - j\,0.4\text{ S} = \mathbf{0.5\ S\angle -53.13°}\end{aligned}$$

$$\mathbf{Z}_T = \frac{1}{\mathbf{Y}_T} = \frac{1}{0.5\text{ S}\angle -53.13°} = \mathbf{2\ \Omega\angle 53.13°}$$

Admittance diagram: As shown in Fig. 15.73.

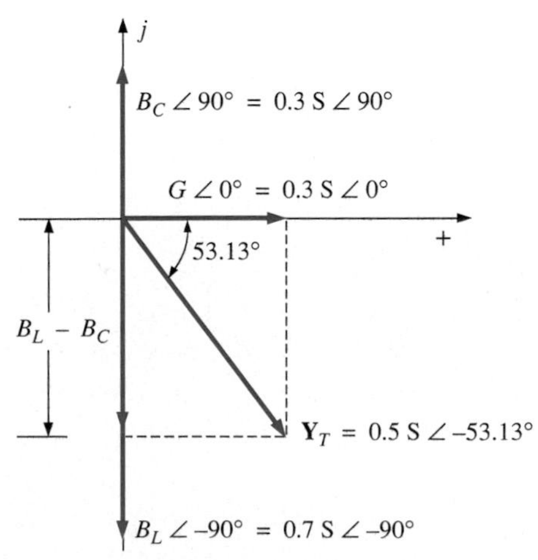

FIG. 15.73
Admittance diagram for the parallel R-L-C network of Fig. 15.71.

I:

$$\mathbf{I} = \frac{\mathbf{E}}{\mathbf{Z}_T} = \mathbf{E}\mathbf{Y}_T = (100\text{ V}\angle 53.13°)(0.5\text{ S}\angle -53.13°) = \mathbf{50\ A\angle 0°}$$

$\mathbf{I}_R$, $\mathbf{I}_L$, $\mathbf{I}_C$:

$$\begin{aligned}\mathbf{I}_R &= (E\angle\theta)(G\angle 0°)\\ &= (100\text{ V}\angle 53.13°)(0.3\text{ S}\angle 0°) = \mathbf{30\ A\angle 53.13°}\\ \mathbf{I}_L &= (E\angle\theta)(B_L\angle -90°)\\ &= (100\text{ V}\angle 53.13°)(0.7\text{ S}\angle -90°) = \mathbf{70\ A\angle -36.87°}\\ \mathbf{I}_C &= (E\angle\theta)(B_C\angle 90°)\\ &= (100\text{ V}\angle 53.13°)(0.3\text{ S}\angle +90°) = \mathbf{30\ A\angle 143.13°}\end{aligned}$$

Kirchhoff's current law: At node *a*,

$$\mathbf{I} - \mathbf{I}_R - \mathbf{I}_L - \mathbf{I}_C = 0$$

or $$\mathbf{I} = \mathbf{I}_R + \mathbf{I}_L + \mathbf{I}_C$$

Phasor diagram: The phasor diagram of Fig. 15.74 indicates that the impressed voltage **E** is in phase with the current $\mathbf{I}_R$ through the resistor, leads the current $\mathbf{I}_L$ through the inductor by 90°, and lags the current $\mathbf{I}_C$ of the capacitor by 90°.

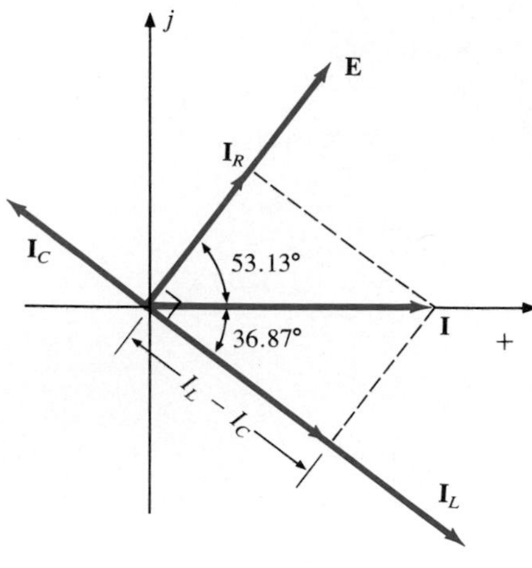

FIG. 15.74
Phasor diagram for the parallel R-L-C network of Fig. 15.71.

Time domain:

$$i = \sqrt{2}(50)\sin\omega t = \mathbf{70.70\ sin\ \omega t}$$
$$i_R = \sqrt{2}(30)\sin(\omega t + 53.13°) = \mathbf{42.42\ sin(\omega t + 53.13°)}$$
$$i_L = \sqrt{2}(70)\sin(\omega t - 36.87°) = \mathbf{98.98\ sin(\omega t - 36.87°)}$$
$$i_C = \sqrt{2}(30)\sin(\omega t + 143.13°) = \mathbf{42.42\ sin(\omega t + 143.13°)}$$

A plot of all of the currents and the impressed voltage appears in Fig. 15.75.

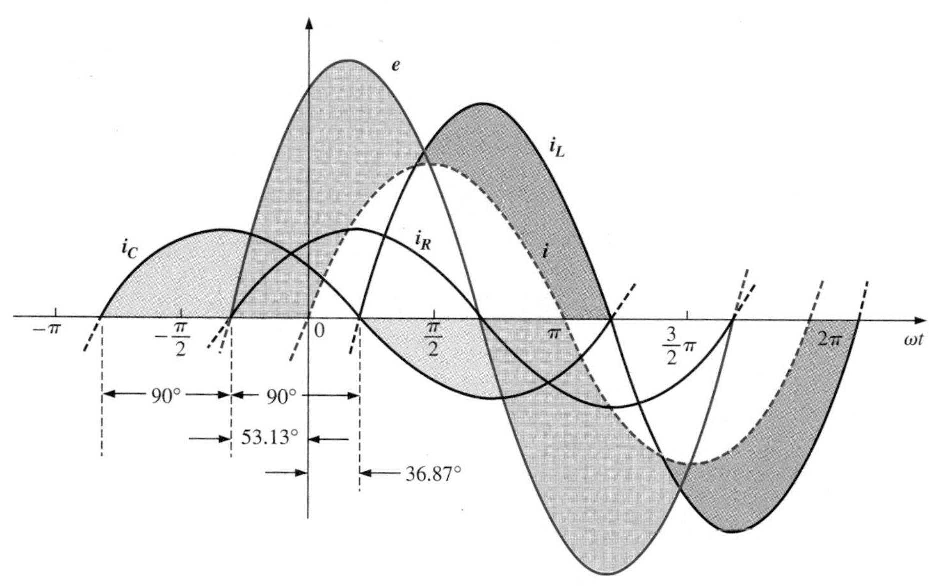

FIG. 15.75
Waveforms for the parallel R-L-C network of Fig. 15.71.

Power: The total power in watts delivered to the circuit is

$$P_T = EI\cos\theta = (100\ \text{V})(50\ \text{A})\cos 53.13° = (5000)(0.6)$$
$$= \mathbf{3000\ W}$$

or $$P_T = E^2G = (100\ \text{V})^2(0.3\ \text{S}) = \mathbf{3000\ W}$$

or, finally,

$$\begin{aligned}P_T &= P_R + P_L + P_C \\ &= EI_R\cos\theta_R + EI_L\cos\theta_L + EL_C\cos\theta_C \\ &= (100\ \text{V})(30\ \text{A})\cos 0° + (100\ \text{V})(70\ \text{A})\cos 90° \\ &\qquad + (100\ \text{V})(30\ \text{A})\cos 90° \\ &= 3000\ \text{W} + 0 + 0 \\ &= \mathbf{3000\ W}\end{aligned}$$

Power factor: The power factor of the circuit is

$$F_p = \cos\theta_T = \cos 53.13° = \mathbf{0.6\ lagging}$$

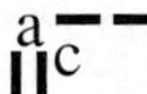

Using Eq. (15.32), we obtain

$$F_p = \cos\theta_T = \frac{G}{Y_T} = \frac{0.3\text{ S}}{0.5\text{ S}} = \mathbf{0.6\ lagging}$$

Impedance approach: The input current **I** can also be determined by first finding the total impedance in the following manner:

$$\mathbf{Z}_T = \frac{\mathbf{Z}_R\mathbf{Z}_L\mathbf{Z}_C}{\mathbf{Z}_R\mathbf{Z}_L + \mathbf{Z}_L\mathbf{Z}_C + \mathbf{Z}_R\mathbf{Z}_C} = \mathbf{2\ \Omega\ \angle 53.13^\circ}$$

and, applying Ohm's law, we obtain

$$\mathbf{I} = \frac{\mathbf{E}}{\mathbf{Z}_T} = \frac{100\text{ V}\ \angle 53.13^\circ}{2\ \Omega\ \angle 53.13^\circ} = \mathbf{50\ A\ \angle 0^\circ}$$

15.9 CURRENT DIVIDER RULE

The basic format for the current divider rule in ac circuits is exactly the same as that for dc circuits; that is, for two parallel branches with impedances $\mathbf{Z}_1$ and $\mathbf{Z}_2$ as shown in Fig. 15.76,

$$\mathbf{I}_1 = \frac{\mathbf{Z}_2\mathbf{I}_T}{\mathbf{Z}_1 + \mathbf{Z}_2} \quad \text{or} \quad \mathbf{I}_2 = \frac{\mathbf{Z}_1\mathbf{I}_T}{\mathbf{Z}_1 + \mathbf{Z}_2} \tag{15.33}$$

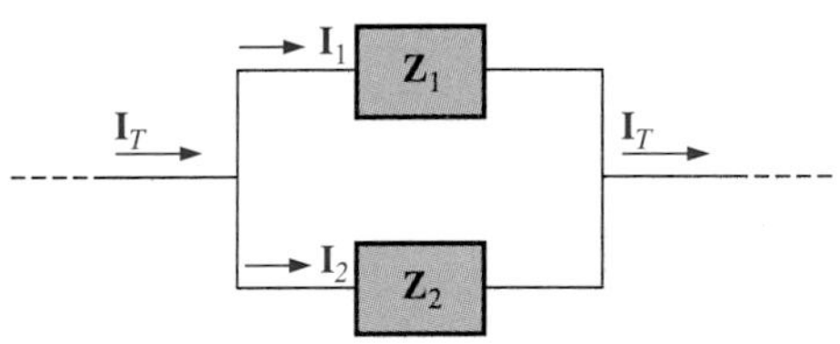

FIG. 15.76
Applying the current divider rule.

EXAMPLE 15.15 Using the current divider rule, find the current through each impedance of Fig. 15.77.

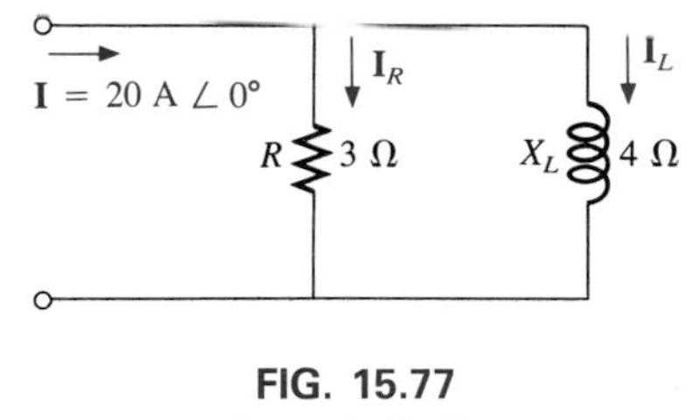

FIG. 15.77
Example 15.15.

Solution:

$$\mathbf{I}_R = \frac{\mathbf{Z}_L\mathbf{I}_T}{\mathbf{Z}_R + \mathbf{Z}_L} = \frac{(4\ \Omega\ \angle 90^\circ)(20\text{ A}\ \angle 0^\circ)}{3\ \Omega\ \angle 0^\circ + 4\ \Omega\ \angle 90^\circ} = \frac{80\text{ A}\angle 90^\circ}{5\ \angle 53.13^\circ}$$
$$= \mathbf{16\ A\ \angle 36.87^\circ}$$

$$\mathbf{I}_L = \frac{\mathbf{Z}_R\mathbf{I}_T}{\mathbf{Z}_R + \mathbf{Z}_L} = \frac{(3\ \Omega\ \angle 0^\circ)(20\text{ A}\ \angle 0^\circ)}{5\ \Omega\ \angle 53.13^\circ} = \frac{60\text{ A}\angle 0^\circ}{5\ \angle 53.13^\circ}$$
$$= \mathbf{12\ A\ \angle -53.13^\circ}$$

EXAMPLE 15.16 Using the current divider rule, find the current through each parallel branch of Fig. 15.78.

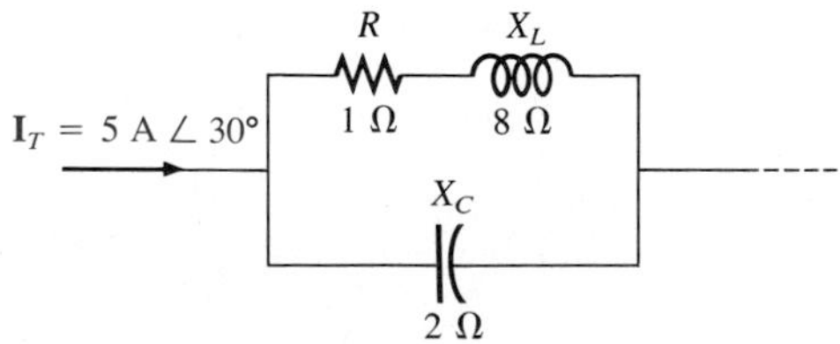

FIG. 15.78
Example 15.16.

Solution:

$$\mathbf{I}_{R\text{-}L} = \frac{\mathbf{Z}_C\mathbf{I}_T}{\mathbf{Z}_C + \mathbf{Z}_{R\text{-}L}} = \frac{(2\ \Omega\ \angle -90^\circ)(5\text{ A}\ \angle 30^\circ)}{-j\,2\ \Omega + 1\ \Omega + j\,8\ \Omega} = \frac{10\text{ A}\angle -60^\circ}{1 + j\,6}$$
$$= \frac{10\text{ A}\angle -60^\circ}{6.083\ \angle 80.54^\circ} \cong \mathbf{1.644\ A\ \angle -140.54^\circ}$$

$$\mathbf{I}_C = \frac{\mathbf{Z}_{R\text{-}L}\mathbf{I}_T}{\mathbf{Z}_{R\text{-}L} + \mathbf{Z}_C} = \frac{(1\ \Omega + j\,8\ \Omega)(5\ \text{A} \angle 30^\circ)}{6.08\ \Omega \angle 80.54^\circ}$$

$$= \frac{(8.06 \angle 82.87^\circ)(5\ \text{A} \angle 30^\circ)}{6.08 \angle 80.54^\circ} = \frac{40.30\ \text{A} \angle 112.87^\circ}{6.083 \angle 80.54^\circ}$$

$$= \mathbf{6.625\ A \angle 32.33^\circ}$$

15.10 FREQUENCY RESPONSE OF THE PARALLEL *R-L* NETWORK

In Section 15.5 the frequency response of a series *R-C* circuit was analyzed. Let us now note the impact of frequency on the total impedance and inductive current for the parallel *R-L* network of Fig. 15.79 for a frequency range of 0 through 40 kHz.

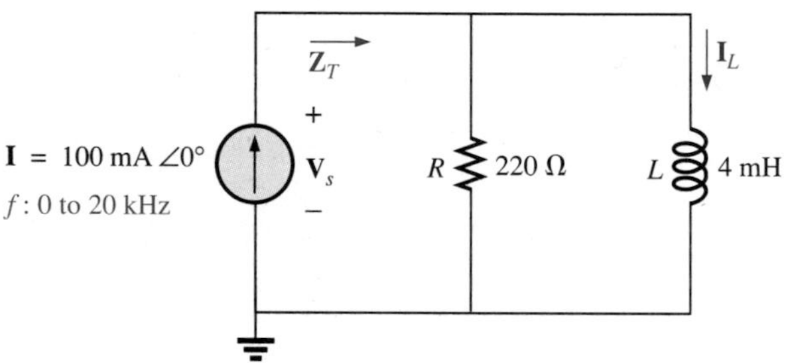

FIG. 15.79
Determining the frequency response of a parallel R-L network.

$\mathbf{Z}_T$:

Before getting into specifics, let us first develop a "sense" for the impact of frequency on the network of Fig. 15.79 by noting the impedance-versus-frequency curves of the individual elements, as shown in Fig. 15.80. The fact that the elements are now in parallel requires that we consider their characteristics in a different manner than occurred for the series *R-C* circuit of Section 15.5. Recall that for parallel elements, the element with the smallest impedance will have the greatest impact

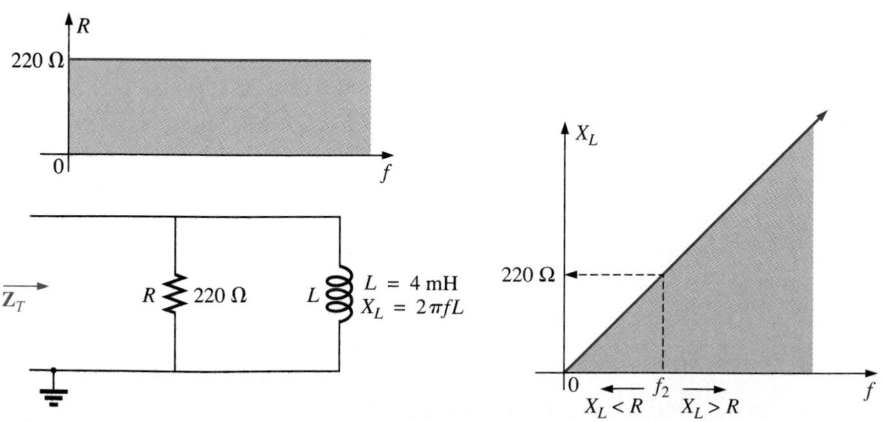

FIG. 15.80
The frequency response of the individual elements of a parallel R-L network.

on the total impedance at that frequency. In Fig. 15.80, for example, X_L is very small at low frequencies compared to R, establishing X_L as the predominant factor in this frequency range. In other words, at low frequencies the network will be primarily inductive and the angle associated with the total impedance will be close to 90°, as with a pure inductor. As the frequency increases, X_L will increase until it equals the impedance of the resistor (220 Ω). The frequency at which this situation occurs can be determined in the following manner:

$$X_L = 2\pi f_2 L = R$$

and

$$f_2 = \frac{R}{2\pi L} \tag{15.34}$$

which for the network of Fig. 15.79 is

$$f_2 = \frac{R}{2\pi L} = \frac{220\ \Omega}{2\pi(4 \times 10^{-3}\ \text{H})}$$
$$\cong \mathbf{8.75\ kHz}$$

which falls within the frequency range of interest.

For frequencies less than f_2, $X_L < R$, and for frequencies greater than f_2, $X_L > R$, as shown in Fig. 15.80. A general equation for the total impedance in vector form can be developed in the following manner:

$$\mathbf{Z}_T = \frac{\mathbf{Z}_R\mathbf{Z}_L}{\mathbf{Z}_R + \mathbf{Z}_L}$$
$$= \frac{(R \angle 0°)(X_L \angle 90°)}{R + jX_L} = \frac{RX_L \angle 90°}{\sqrt{R^2 + X_L^2} \angle \tan^{-1} X_L/R}$$

and

$$\mathbf{Z}_T = \frac{RX_L}{\sqrt{R^2 + X_L^2}} \angle 90° - \tan^{-1} X_L/R$$

so that

$$\mathbf{Z}_T = \frac{RX_L}{\sqrt{R^2 + X_L^2}} \tag{15.35}$$

and

$$\theta_T = 90° - \tan^{-1}\frac{X_L}{R} = \tan^{-1}\frac{R}{X_L} \tag{15.36}$$

The magnitude and angle of the total impedance can now be found at any frequency of interest simply by substituting Eqs. (15.35) and (15.36).

***f* = 1 kHz:**

$$X_L = 2\pi f L = 2\pi(1\ \text{kHz})(4 \times 10^{-3}\ \text{H}) = 25.12\ \Omega$$

and

$$Z_T = \frac{RX_L}{\sqrt{R^2 + X_L^2}} = \frac{(220\ \Omega)(25.12\ \Omega)}{\sqrt{(220\ \Omega)^2 + (25.12\Omega)^2}} = \mathbf{24.96\ \Omega}$$

with

$$\theta_T = \tan^{-1}\frac{R}{X_L} = \tan^{-1}\frac{220\ \Omega}{25.12\ \Omega}$$
$$= \tan^{-1} 8.76 = 83.49°$$

and $$\mathbf{Z}_T = \mathbf{24.96\ \Omega\ \angle 83.49^\circ}$$

This value compares very closely with $X_L = 25.12\ \Omega\ \angle 90^\circ$, which it would be if the network were purely inductive ($R = \infty\ \Omega$). Our assumption that the network is primarily inductive at low frequencies is therefore confirmed.

Continuing:

$$f = 5\text{ kHz:}\quad \mathbf{Z}_T = \mathbf{109.1\ \Omega\ \angle 60.23^\circ}$$
$$f = 10\text{ kHz:}\quad \mathbf{Z}_T = \mathbf{165.5\ \Omega\ \angle 41.21^\circ}$$
$$f = 15\text{ kHz:}\quad \mathbf{Z}_T = \mathbf{189.99\ \Omega\ \angle 30.28^\circ}$$
$$f = 20\text{ kHz:}\quad \mathbf{Z}_T = \mathbf{201.53\ \Omega\ \angle 23.65^\circ}$$
$$f = 30\text{ kHz:}\quad \mathbf{Z}_T = \mathbf{211.19\ \Omega\ \angle 16.27^\circ}$$
$$f = 40\text{ kHz:}\quad \mathbf{Z}_T = \mathbf{214.91\ \Omega\ \angle 12.35^\circ}$$

At $f = 40$ kHz, note how closely the magnitude of Z_T has approached the resistance level of 220 Ω and how the associated angle with the total impedance is approaching zero degrees. The result is a network with terminal characteristics that are becoming more and more resistive as the frequency increases, which further confirms the earlier conclusions developed by the curves of Fig. 15.80.

Plots of Z_T versus frequency in Fig. 15.81 and θ_T in Fig. 15.82 clearly reveal the transition from an inductive network to one that has resistive characteristics. Note that the transition frequency of 8.75 kHz occurs right in the middle of the knee of the curves for both Z_T and θ_T.

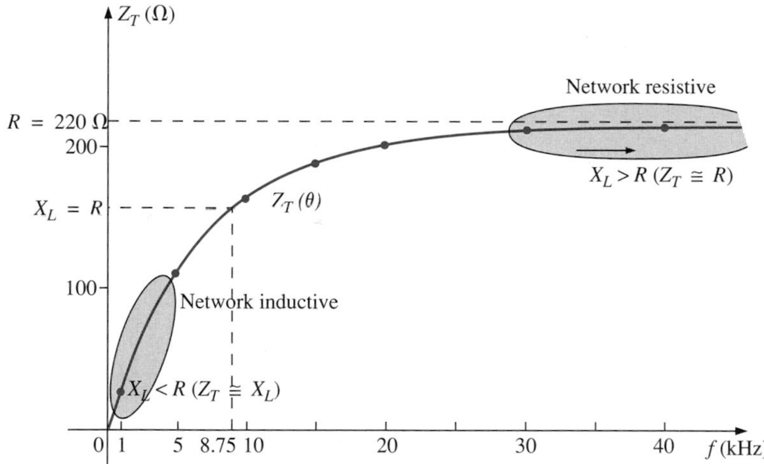

FIG. 15.81
The magnitude of the input impedance versus frequency for the network of Fig. 15.79.

A review of Figs. 15.47 and 15.81 will reveal that a series *R-C* and a parallel *R-L* network will have an impedance level that approaches the resistance of the network at high frequencies. The capacitive circuit approaches the level from above, whereas the inductive network does the same from below. For the series *R-L* circuit and the parallel *R-C* network, the total impedance will begin at the resistance level and then display the characteristics of the reactive elements at high frequencies.

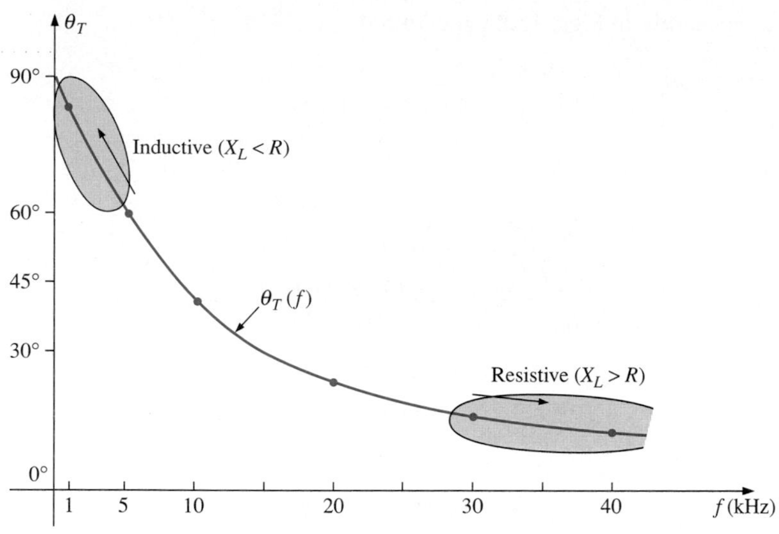

FIG. 15.82
The phase angle of the input impedance versus frequency for the network of Fig. 15.79.

I_L:
Applying the current divider rule to the network of Fig. 15.79 will result in the following:

$$\mathbf{I}_L = \frac{\mathbf{Z}_R\mathbf{I}}{\mathbf{Z}_R + \mathbf{Z}_L}$$

$$= \frac{(R\angle 0°)(I\angle 0°)}{R + jX_L} = \frac{RI\angle 0°}{\sqrt{R^2 + X_L^2}\ \angle \tan^{-1} X_L/R}$$

and
$$\mathbf{I}_L = I_L\angle\theta_L = \frac{RI}{\sqrt{R^2 + X_L^2}}\ \angle -\tan^{-1} X_L/R$$

The magnitude of I_L is therefore determined by

$$I_L = \frac{RI}{\sqrt{R^2 + X_L^2}} \qquad (15.37)$$

and the phase angle θ_L, by which $\mathbf{I}_L$ leads $\mathbf{I}$, is given by

$$\theta_L = -\tan^{-1}\frac{X_L}{R} \qquad (15.38)$$

Because θ_L is always negative, the magnitude of θ_L is, in actuality, the angle by which $\mathbf{I}_L$ lags $\mathbf{I}$.

To begin our analysis, let us first consider the case of $f = 0$ Hz (dc conditions).

***f* = 0 Hz:**

$$X_L = 2\pi fL = 2\pi(0\text{ Hz})L = 0\ \Omega$$

Applying the short-circuit equivalent for the inductor in Fig. 15.79 would result in

$$\mathbf{I}_L = \mathbf{I} = 100\text{ mA}\angle 0°$$

as appearing in Figs. 15.83 and 15.84.

***f* = 1 kHz:**
Applying Eq. (15.37):

$$X_L = 2\pi fL = 2\pi(1\text{ kHz})(4\text{ mH}) = 25.12\ \Omega$$

and $$\sqrt{R^2 + X_L^2} = \sqrt{(220\ \Omega)^2 + (25.12\ \Omega)^2} = 221.43\ \Omega$$

and $$I_L = \frac{RI}{\sqrt{R^2 + X_L^2}} = \frac{(220\ \Omega)(100\text{ mA})}{221.43\ \Omega} = \mathbf{99.35\ mA}$$

with

$$\theta_L = \tan^{-1}\frac{X_L}{R} = -\tan^{-1}\frac{25.12\ \Omega}{220\ \Omega} = -\tan^{-1}0.114 = \mathbf{-6.51°}$$

and $$\mathbf{I}_L = \mathbf{99.35\ mA\ \angle -6.51°}$$

The result is a current $\mathbf{I}_L$ that is still very close to the source current $\mathbf{I}$ in both magnitude and phase.

Continuing:

$$f = 5\text{ kHz:}\quad \mathbf{I}_L = \mathbf{86.84\ mA\ \angle -29.72°}$$
$$f = 10\text{ kHz:}\quad \mathbf{I}_L = \mathbf{65.88\ mA\ \angle -48.79°}$$
$$f = 15\text{ kHz:}\quad \mathbf{I}_L = \mathbf{50.43\ mA\ \angle -59.72°}$$
$$f = 20\text{ kHz:}\quad \mathbf{I}_L = \mathbf{40.11\ mA\ \angle -66.35°}$$
$$f = 30\text{ kHz:}\quad \mathbf{I}_L = \mathbf{28.02\ mA\ \angle -73.73°}$$
$$f = 40\text{ kHz:}\quad \mathbf{I}_L = \mathbf{21.38\ mA\ \angle -77.65°}$$

The plot of the magnitude of I_L versus frequency is provided in Fig. 15.83 and reveals that the current through the coil dropped from its maximum of 100 mA to almost 20 mA at 40 kHz. As the reactance of the coil increased with frequency, more of the source current chose the

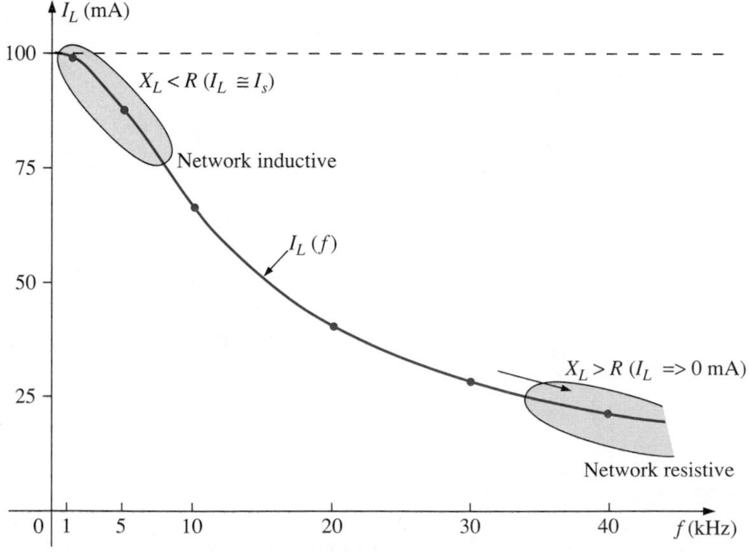

FIG. 15.83
The magnitude of the current $\mathbf{I}_L$ versus frequency for the parallel R-L network of Fig. 15.79.

lower-resistance path of the resistor. The magnitude of the phase angle between $\mathbf{I}_L$ and $\mathbf{I}$ is approaching 90° with an increase in frequency, as shown in Fig. 15.84, leaving its initial value of zero degrees at $f = 0$ Hz far behind.

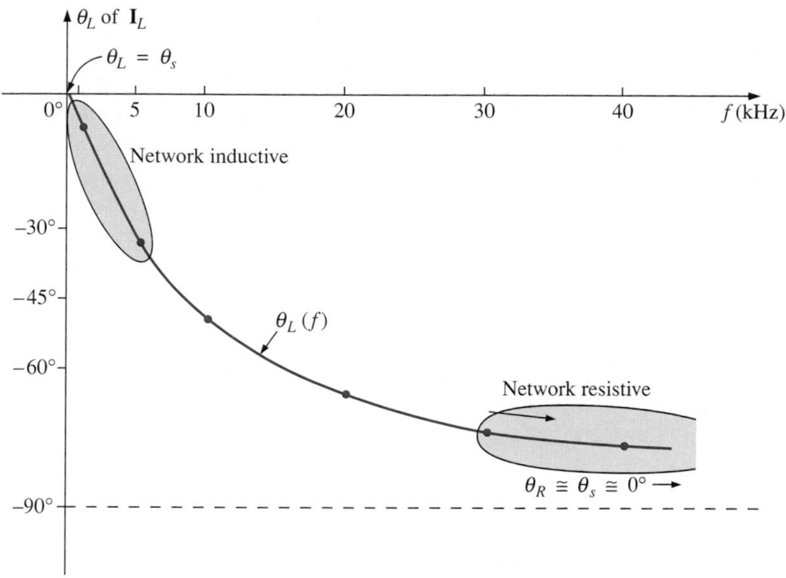

FIG. 15.84
The phase angle of the current $\mathbf{I}_L$ versus frequency for the parallel R-L network of Fig. 15.79.

At $f = 1$ kHz the phasor diagram of the network appears as shown in Fig. 15.85. First note that the magnitude and phase angle of $\mathbf{I}_L$ are very close to those of $\mathbf{I}$. Since the voltage across a coil must lead the current through a coil by 90°, the voltage $\mathbf{V}_s$ appears as shown. The voltage across a resistor is in phase with the current through the resistor, resulting in the direction of $\mathbf{I}_R$ shown in Fig. 15.85. Of course, at this frequency $R > X_L$, and the current I_R is relatively small in magnitude.

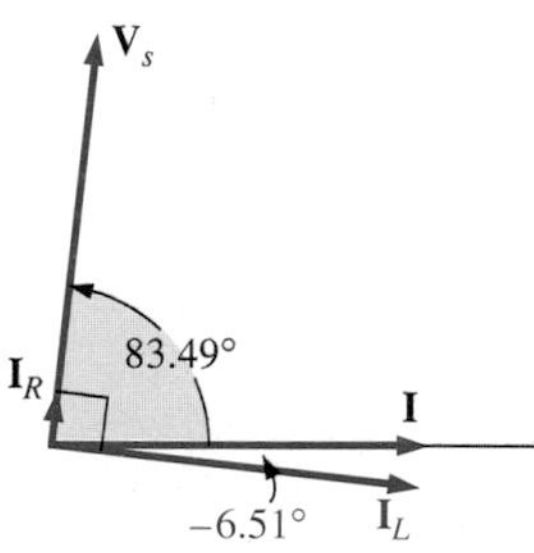

FIG. 15.85
The phasor diagram for the parallel R-L network of Fig. 15.79 at $f = 1$ kHz.

At $f = 40$ kHz the phasor diagram changes to that appearing in Fig. 15.86. Note that now $\mathbf{I}_R$ and $\mathbf{I}$ are close in magnitude and phase because $X_L > R$. The magnitude of $\mathbf{I}_L$ has dropped to very low levels and the phase angle associated with $\mathbf{I}_L$ is approaching $-90°$. The network is now more "resistive" as compared to its "inductive" characteristics at low frequencies.

The analysis of a parallel *R-C* or *R-L-C* network would proceed in much the same manner, with the inductive impedance predominating at low frequencies and the capacitive reactance predominating at high frequencies.

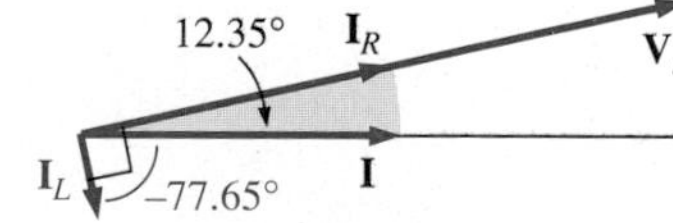

FIG. 15.86
The phasor diagram for the parallel R-L network of Fig. 15.79 at $f = 40$ kHz.

15.11 SUMMARY—PARALLEL ac NETWORKS

The following is a review of important conclusions that can be derived from the discussion and examples of the previous sections. The list is not all-inclusive, but it does emphasize some of the conclusions that should be carried forward in the future analysis of ac systems.

For parallel ac networks with reactive elements:

1. ***The total admittance (impedance) will be frequency dependent.***
2. ***The impedance of any one element can be less than the total impedance (for dc circuits recall that the total resistance must always be less than the smallest parallel resistor).***
3. ***The inductive and capacitive susceptances are in direct opposition on an admittance diagram.***
4. ***Depending on the frequency applied, the same network can be either predominantly inductive or capacitive.***
5. ***At lower frequencies the inductive elements will usually have the most impact on the total impedance, while at high frequencies the capacitive elements will usually have the most impact.***
6. ***The magnitude of the current through any one branch can be greater than the source current.***
7. ***The magnitude of the current through an element, as compared to the other elements of the network, is directly related to the magnitude of its impedance; that is, the smaller the impedance of an element, the larger the magnitude of the current through the element.***
8. ***The current through a coil is always in direct opposition with the current through a capacitor on a phasor diagram.***
9. ***The applied voltage is always in phase with the current through the resistive elements, leads the voltage across all the inductive elements by 90°, and lags the current through all capacitive elements by 90°.***
10. ***The smaller the resistive element of a network compared to the net reactive susceptance, the closer the power factor is to unity.***

15.12 EQUIVALENT CIRCUITS

In a series ac circuit, the total impedance of two or more elements in series is often equivalent to an impedance that can be achieved with fewer elements of different values, the elements and their values being determined by the frequency applied. This is also true for parallel circuits. For the circuit of Fig. 15.87(a),

$$\mathbf{Z}_T = \frac{\mathbf{Z}_C\mathbf{Z}_L}{\mathbf{Z}_C + \mathbf{Z}_L} = \frac{(5\ \Omega\ \angle -90^\circ)(10\ \Omega\ \angle 90^\circ)}{5\ \Omega\ \angle -90^\circ + 10\ \Omega\ \angle 90^\circ} = \frac{50\ \angle 0^\circ}{5\ \angle 90^\circ}$$
$$= 10\ \Omega\ \angle -90^\circ$$

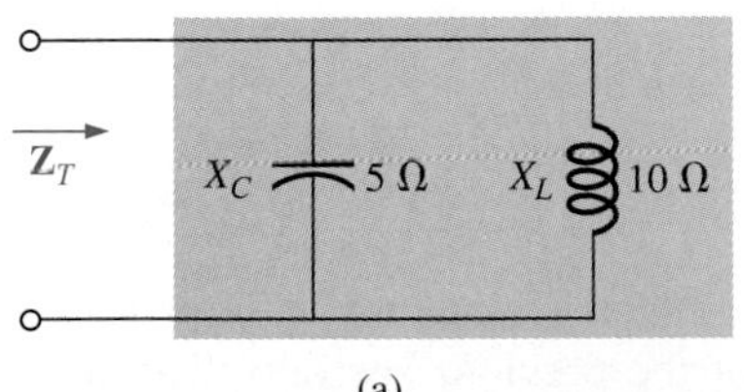

(a)

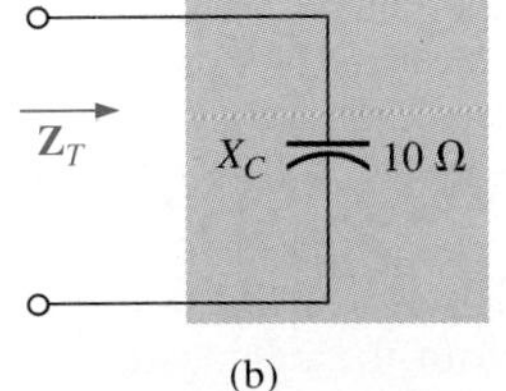

(b)

FIG. 15.87
Defining the equivalence between two networks at a specific frequency.

The total impedance at the frequency applied is equivalent to a capacitor with a reactance of 10 Ω, as shown in Fig. 15.87(b). Always keep in mind that this equivalence is true only at the applied frequency. If the frequency changes, the reactance of each element changes, and the equivalent circuit will change—perhaps from capacitive to inductive in the above example.

Another interesting development appears if the impedance of a parallel circuit, such as the one of Fig. 15.88(a), is found in rectangular form. In this case,

$$\mathbf{Z}_T = \frac{\mathbf{Z}_L\mathbf{Z}_R}{\mathbf{Z}_L + \mathbf{Z}_R} = \frac{(4\ \Omega\ \angle 90^\circ)(3\ \Omega\ \angle 0^\circ)}{4\ \Omega\ \angle 90^\circ + 3\ \Omega\ \angle 0^\circ}$$

$$= \frac{12\ \angle 90^\circ}{5\ \angle 53.13^\circ} = 2.40\ \Omega\ \angle 36.87^\circ$$

$$= 1.920\ \Omega + j\,1.440\ \Omega$$

which is the impedance of a series circuit with a resistor of 1.92 Ω and an inductive reactance of 1.44 Ω, as shown in Fig. 15.88(b).

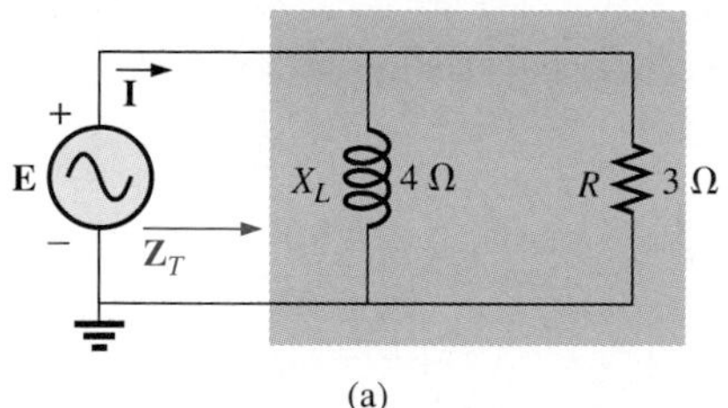

(a)

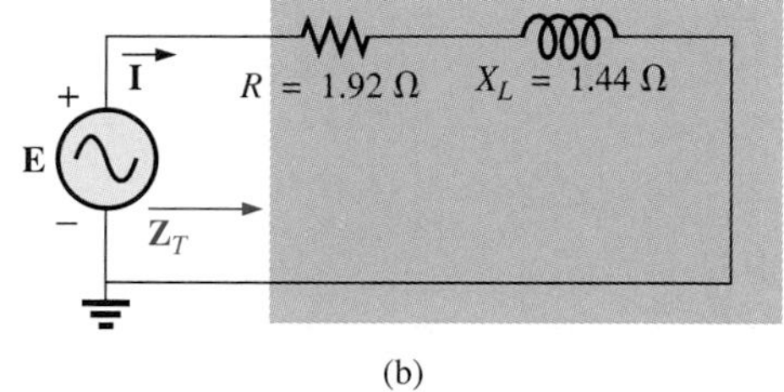

(b)

FIG. 15.88
Finding the series equivalent circuit for a parallel R-L network.

The current **I** will be the same in each circuit of Fig. 15.87 or Fig. 15.88 if the same input voltage **E** is applied. For a parallel circuit of one resistive element and one reactive element, the series circuit with the same input impedance will always be composed of one resistive and one reactive element. The impedance of each element of the series circuit will be different from that of the parallel circuit, but the reactive elements will always be of the same type; that is, an *R-L* circuit and an *R-C* parallel circuit will have an equivalent *R-L* and *R-C* series circuit, respectively. The same is true when converting from a series to a parallel circuit. In the discussion to follow, keep in mind that

the term equivalent refers only to the fact that for the same applied potential, the same impedance and input current will result.

To formulate the equivalence between the series and parallel circuits, the equivalent series circuit for a resistor and reactance in parallel can be found by determining the total impedance of the circuit in rectangular form; that is, for the circuit of Fig. 15.89(a),

$$\mathbf{Y}_p = \frac{1}{R_p} + \frac{1}{\pm j\,X_p} = \frac{1}{R_P} \mp j\frac{1}{X_p}$$

and

$$\mathbf{Z}_p = \frac{1}{\mathbf{Y}_p} = \frac{1}{(1/R_p) \pm j\,(1/X_p)}$$

$$= \frac{1/R_p}{(1/R_p)^2 + (1/X_p)^2} \pm j\frac{1/X_p}{(1/R_p)^2 + (1/X_p)^2}$$

Multiplying the numerator and denominator of each term by $R_p^2X_p^2$ results in

$$\mathbf{Z}_p = \frac{R_pX_p^2}{X_p^2 + R_p^2} \pm j\frac{R_p^2X_p}{X_p^2 + R_p^2}$$

$$= R_s \pm j\,X_s \qquad \text{[Fig. 15.89(b)]}$$

and

$$R_s = \frac{R_pX_p^2}{X_p^2 + R_p^2} \qquad \textbf{(15.39)}$$

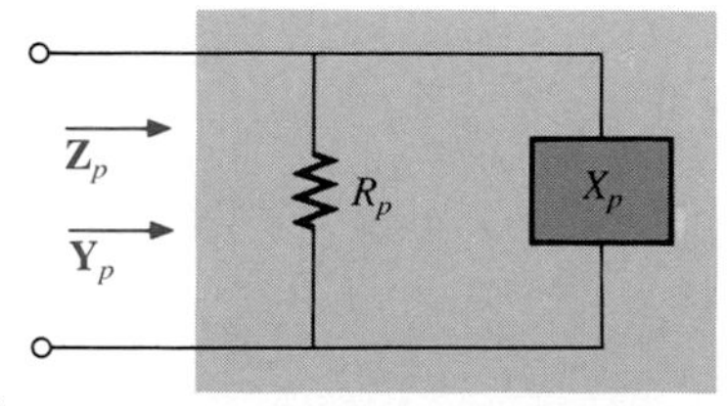

(a)

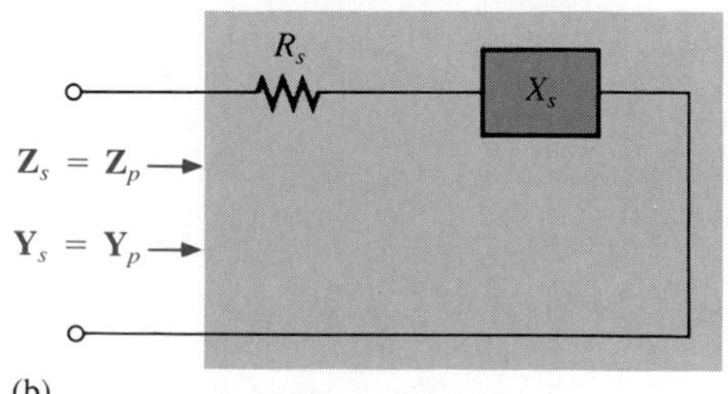

(b)

FIG. 15.89
Defining the parameters of equivalent series and parallel networks.

with

$$X_s = \frac{R_p^2 X_p}{X_p^2 + R_p^2} \tag{15.40}$$

For the network of Fig. 15.88,

$$R_s = \frac{R_p X_p^2}{X_p^2 + R_p^2} = \frac{(3\ \Omega)(4\ \Omega)^2}{(4\ \Omega)^2 + (3\ \Omega)^2} = \frac{48\ \Omega}{25} = \mathbf{1.920\ \Omega}$$

and

$$X_s = \frac{R_p^2 X_p}{X_p^2 + R_p^2} = \frac{(3\ \Omega)^2(4\ \Omega)}{(4\ \Omega)^2 + (3\ \Omega)^2} = \frac{36\ \Omega}{25} = \mathbf{1.440\ \Omega}$$

which agrees with the previous result.

The equivalent parallel circuit for a circuit with a resistor and reactance in series can be found by simply finding the total admittance of the system in rectangular form; that is, for the circuit of Fig. 15.89(b),

$$\mathbf{Z}_s = R_s \pm j X_s$$

$$\mathbf{Y}_s = \frac{1}{\mathbf{Z}_s} = \frac{1}{R_s \pm j X_s} = \frac{R_s}{R_s^2 + X_s^2} \mp j \frac{X_s}{R_s^2 + X_s^2}$$

$$= G_p \mp j B_p = \frac{1}{R_p} \mp j \frac{1}{X_p} \qquad \text{[Fig. 15.89(a)]}$$

or

$$R_p = \frac{R_s^2 + X_s^2}{R_s} \tag{15.41}$$

with

$$X_p = \frac{R_s^2 + X_s^2}{X_s} \tag{15.42}$$

For the above example,

$$R_p = \frac{R_s^2 + X_s^2}{R_s} = \frac{(1.92\ \Omega)^2 + (1.44\ \Omega)^2}{1.92\ \Omega} = \frac{5.76\ \Omega}{1.92} = \mathbf{3.0\ \Omega}$$

and

$$X_p = \frac{R_s^2 + X_s^2}{X_s} = \frac{5.76\ \Omega}{1.44} = \mathbf{4.0\ \Omega}$$

as shown in Fig. 15.88(a).

EXAMPLE 15.17 Determine the series equivalent circuit for the network of Fig. 15.90.

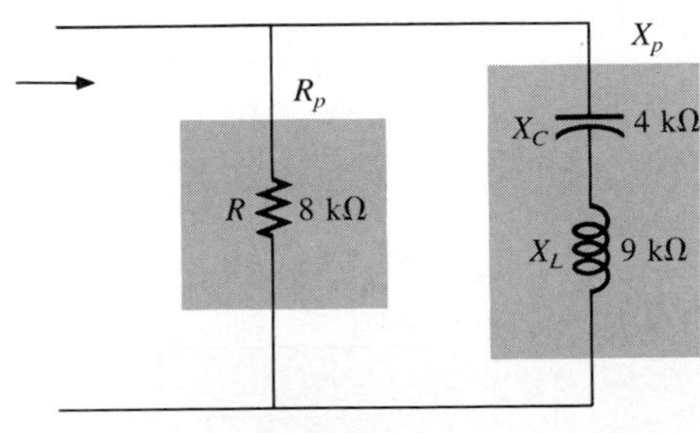

FIG. 15.90
Example 15.17.

Solution:

$$R_p = 8\text{ k}\Omega$$
$$X_p\text{ (resultant)} = |X_L - X_C| = |9\text{ k}\Omega - 4\text{ k}\Omega|$$
$$= 5\text{ k}\Omega$$

and

$$R_s = \frac{R_p X_p^2}{X_p^2 + R_p^2} = \frac{(8\text{ k}\Omega)(5\text{ k}\Omega)^2}{(5\text{ k}\Omega)^2 + (8\text{ k}\Omega)^2} = \frac{200\text{ k}\Omega}{89} = \mathbf{2.247\text{ k}\Omega}$$

with

$$X_s = \frac{R_p^2 X_p}{X_p^2 + R_p^2} = \frac{(8\text{ k}\Omega)^2(5\text{ k}\Omega)}{(5\text{ k}\Omega)^2 + (8\text{ k}\Omega)^2} = \frac{320\text{ k}\Omega}{89}$$
$$= \mathbf{3.596\text{ k}\Omega} \quad \textbf{(inductive)}$$

The equivalent series circuit appears in Fig. 15.91.

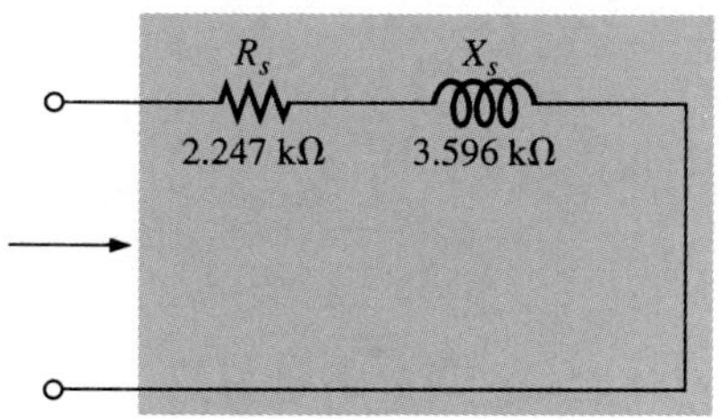

FIG. 15.91
The equivalent series circuit for the parallel network of Fig. 15.90.

EXAMPLE 15.18 For the network of Fig. 15.92:

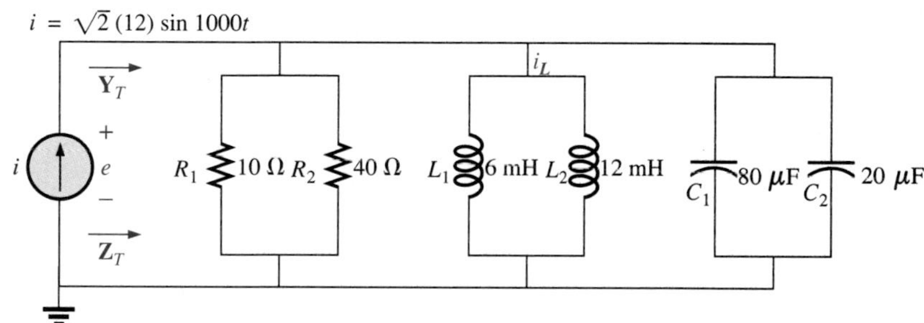

FIG. 15.92
Example 15.18.

a. Determine $\mathbf{Y}_T$.
b. Sketch the admittance diagram.
c. Find $\mathbf{E}$ and $\mathbf{I}_L$.
d. Compute the power factor of the network and the power delivered to the network.
e. Determine the equivalent series circuit as far as the terminal characteristics of the network are concerned.
f. Using the equivalent circuit developed in part (e), calculate $\mathbf{E}$ and compare it with the result of part (c).
g. Determine the power delivered to the network and compare it with the solution of part (d).
h. Determine the equivalent parallel network from the equivalent series circuit and calculate the total admittance $\mathbf{Y}_T$. Compare the result with the solution of part (a).

Solutions:

a. Combining common elements and finding the reactance of the inductor and capacitor, we obtain

$$R_T = 10\ \Omega \,\|\, 40\ \Omega = 8\ \Omega$$
$$L_T = 6\text{ mH} \,\|\, 12\text{ mH} = 4\text{ mH}$$
$$C_T = 80\ \mu\text{F} + 20\ \mu\text{F} = 100\ \mu\text{F}$$

$$X_L = \omega L = (1000 \text{ rad/s})(4 \text{ mH}) = 4 \ \Omega$$

$$X_C = \frac{1}{\omega C} = \frac{1}{(1000 \text{ rad/s})(100 \ \mu\text{F})} = 10 \ \Omega$$

The network is redrawn in Fig. 15.93 with phasor notation. The total admittance is

$$\begin{aligned}\mathbf{Y}_T &= \mathbf{Y}_R + \mathbf{Y}_L + \mathbf{Y}_C \\ &= G \angle 0^\circ + B_L \angle -90^\circ + B_C \angle +90^\circ \\ &= \frac{1}{8 \ \Omega} \angle 0^\circ + \frac{1}{4 \ \Omega} \angle -90^\circ + \frac{1}{10 \ \Omega} \angle +90^\circ \\ &= 0.125 \text{ S} \angle 0^\circ + 0.25 \text{ S} \angle -90^\circ + 0.1 \text{ S} \angle +90^\circ \\ &= 0.125 \text{ S} - j\,0.25 \text{ S} + j\,0.1 \text{ S} \\ &= 0.125 \text{ S} - j\,0.15 \text{ S} = \mathbf{0.195 \ S \angle -50.194^\circ}\end{aligned}$$

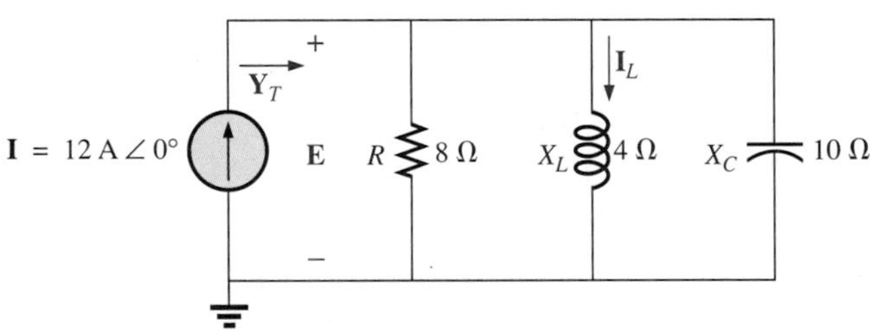

FIG. 15.93
Applying phasor notation to the network of Fig. 15.92.

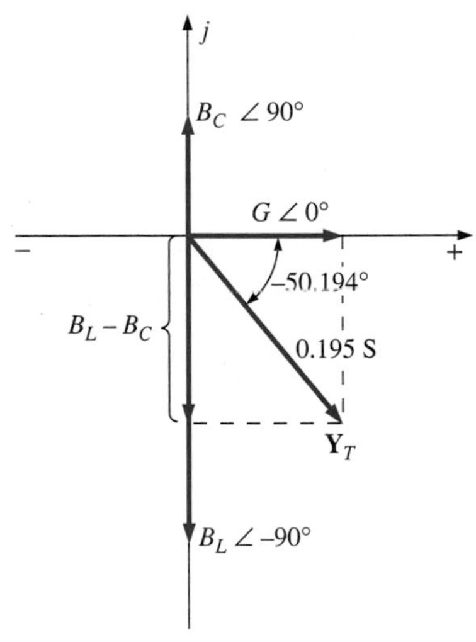

FIG. 15.94
Admittance diagram for the parallel R-L-C network of Fig. 15.92.

b. See Fig. 15.94.

c. $\mathbf{E} = \mathbf{I}\mathbf{Z}_T = \dfrac{\mathbf{I}}{\mathbf{Y}_T} = \dfrac{12 \text{ A} \angle 0^\circ}{0.195 \text{ S} \angle -50.194^\circ} = \mathbf{61.538 \ V \angle 50.194^\circ}$

$\mathbf{I}_L = \dfrac{\mathbf{V}_L}{\mathbf{Z}_L} = \dfrac{\mathbf{E}}{\mathbf{Z}_L} = \dfrac{61.538 \text{ V} \angle 50.194^\circ}{4 \ \Omega \angle 90^\circ} = \mathbf{15.385 \ A \angle -39.81^\circ}$

d. $F_p = \cos\theta = \dfrac{G}{Y_T} = \dfrac{0.125 \text{ S}}{0.195 \text{ S}} = \mathbf{0.641}$ lagging (**E** leads **I**)

$P = EI\cos\theta = (61.538 \text{ V})(12 \text{ A}) \cos 50.194^\circ$
$= \mathbf{472.75 \ W}$

e. $\mathbf{Z}_T = \dfrac{1}{\mathbf{Y}_T} = \dfrac{1}{0.195 \text{ S} \angle -50.194^\circ} = 5.128 \ \Omega \angle +50.194^\circ$
$= 3.283 \ \Omega + j3.939 \ \Omega$
$= R + j X_L$

$X_L = 3.939 \ \Omega = \omega L$

and $L = \dfrac{3.939 \ \Omega}{\omega} = \dfrac{3.939 \ \Omega}{1000 \text{ rad/s}} = \mathbf{3.939 \ mH}$

The series equivalent circuit appears in Fig. 15.95.

f. $\mathbf{E} = \mathbf{I}\mathbf{Z}_T = (12 \text{ A} \angle 0^\circ)(5.128 \ \Omega \angle 50.194^\circ)$
$= \mathbf{61.536 \ V \angle 50.194^\circ}$, as above

g. $P = I^2 R = (12 \text{ A})^2 (3.283 \ \Omega) = \mathbf{472.75 \ W}$ (as above)

h. $R_p = \dfrac{R_s^2 + X_s^2}{R_s} = \dfrac{(3.283 \ \Omega)^2 + (3.939 \ \Omega)^2}{3.283 \ \Omega} = \mathbf{8 \ \Omega}$

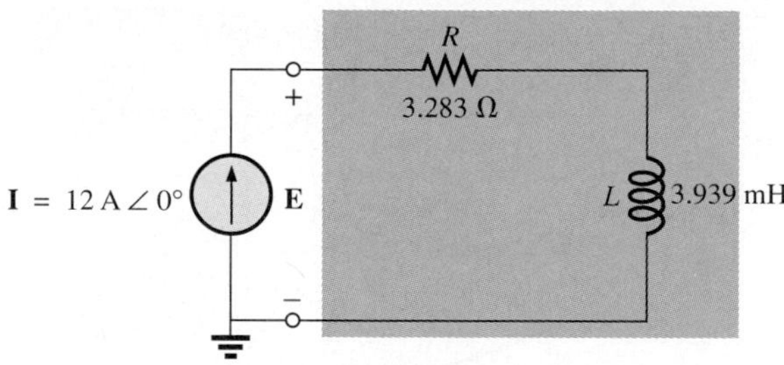

FIG. 15.95
Series equivalent circuit for the parallel R-L-C network of Fig. 15.92 with $\omega = 1000$ rad/s.

$$X_p = \frac{R_s^2 + X_s^2}{X_s} = \frac{(3.283\ \Omega)^2 + (3.939\ \Omega)^2}{3.939\ \Omega} = \mathbf{6.675\ \Omega}$$

The parallel equivalent circuit appears in Fig. 15.96.

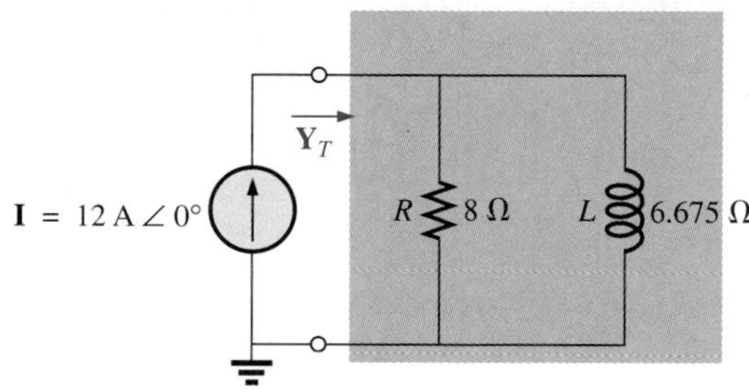

FIG. 15.96
Parallel equivalent of the circuit of Fig. 15.95.

$$\begin{aligned}\mathbf{Y}_T = G \angle 0° + B_L \angle -90° &= \frac{1}{8\ \Omega} \angle 0° + \frac{1}{6.675\ \Omega} \angle -90° \\ &= 0.125\ \text{S} \angle 0° + 0.15\ \text{S} \angle -90° \\ &= 0.125\ \text{S} - j\,0.15\ \text{S} = \mathbf{0.195\ S \angle -50.194°} \qquad \text{(as above)}\end{aligned}$$

15.13 PHASE MEASUREMENTS DUAL-TRACE OSCILLOSCOPE

The phase shift between the voltages of a network or between the voltages and currents of a network can be found using a dual-trace (two signals displayed at the same time) oscilloscope. Phase-shift measurements can also be performed using a single-trace oscilloscope by properly interpreting the resulting Lissajous patterns obtained on the screen. This latter approach, however, will be left for the laboratory experience.

In Fig. 15.97 channel 1 of the dual-trace oscilloscope is hooked up to display the applied voltage e. Channel 2 is connected to display the voltage across the inductor v_L. Of particular importance is the fact that the ground of the scope is connected to the ground of the oscilloscope for both channels. In other words, there is only one common ground for the circuit and oscilloscope. The resulting waveforms may appear as shown in Fig. 15.98.

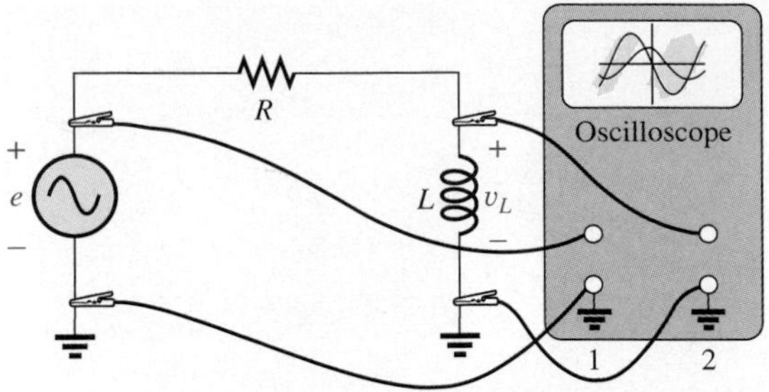

FIG. 15.97
Determining the phase relationship between e and v_L.

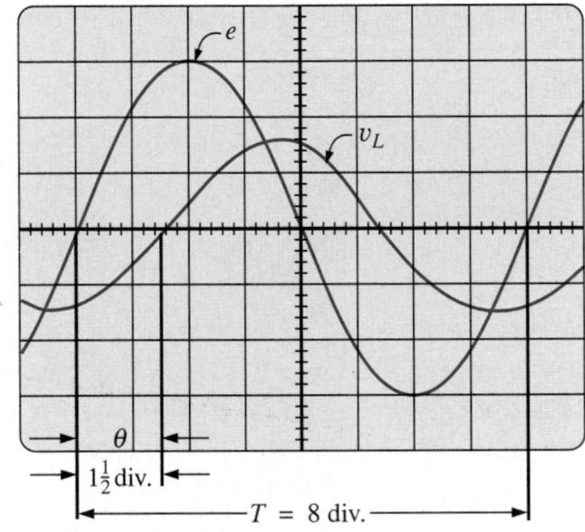

FIG. 15.98
Determining the phase angle between e and v_L.

For the chosen horizontal sensitivity, each waveform of Fig. 15.98 has a period T defined by eight horizontal divisions, and the phase angle between the two waveforms is defined by $1\frac{1}{2}$ divisions. Using the fact that each period of a sinusoidal waveform encompasses 360°, the following ratios can be set up to determine the phase angle θ.

$$\frac{8 \text{ divisions}}{360^\circ} = \frac{1.5 \text{ divisions}}{\theta}$$

and
$$\theta = \left(\frac{1.5}{8}\right)360^\circ = \mathbf{67.5^\circ}$$

In general,

$$\boxed{\theta = \frac{(\text{Div. for } \theta)}{(\text{Div. for } T)} \times 360^\circ} \qquad \mathbf{(15.43)}$$

If the phase relationship between e and v_R is required, the oscilloscope *must not* be hooked up as shown in Fig. 15.99. Points a and b have a common ground that will establish a zero-volt drop between the two points; this drop will have the same effect as a short-circuit connection between a and b. The resulting short circuit will "short out" the inductive element, and the current will increase due to the drop in impedance for the circuit. A dangerous situation can arise if the inductive element has a high impedance and the resistor has a relatively low

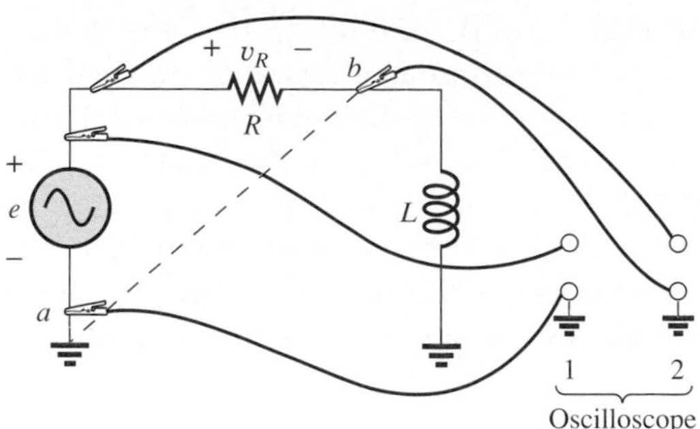

FIG. 15.99
An improper phase-measurement connection.

impedance. The current, controlled solely by the resistance R, could jump to dangerous levels and damage the equipment.

The phase relationship between e and v_R can be determined by simply interchanging the positions of the coil and resistor or by introducing a sensing resistor, as shown in Fig. 15.100. A sensing resistor is exactly that, introduced to "sense" a quantity without adversely affecting the behavior of the network. In other words, the sensing resistor must be small enough compared to the other impedances of the network not to cause a significant change in the voltage and current levels or phase relationships. Note that the sensing resistor is introduced in a way that will result in one end being connected to the common ground of the network. In Fig. 15.100, channel 2 will display the voltage v_{R_s}, which is in phase with the current i. However, the current i is also in phase with the voltage v_R across the resistor R. The net result is that the voltages v_{R_s} and v_R are in phase and the phase relationship between e and v_R can be determined from the waveforms e and v_{R_s}. Since v_{R_s} and i are in phase, the above procedure will also determine the phase angle between the applied voltage e and the source current i. If the magnitude of R_s is sufficiently small compared to R or X_L, the phase measurements of Fig. 15.97 can be performed with R_s in place. That is, channel 2 can be connected to the top of the inductor and to ground, and the effect of R_s can be ignored. In the above application, the sensing resistor will not reveal the magnitude of the voltage v_R but simply the phase relationship between e and v_R.

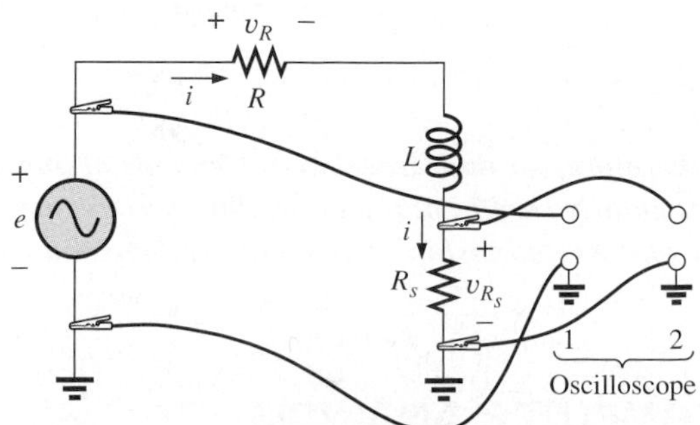

FIG. 15.100
Determining the phase relationship between e and v_R or e and i using a sensing resistor.

For the parallel network of Fig. 15.101, the phase relationship between two of the branch currents, i_R and i_L, can be determined using a sensing resistor, as shown in the figure. Channel 1 will display the voltage v_R and channel 2 will display the voltage v_{R_s}. Since v_R is in phase with i_R and v_{R_s} is in phase with the current i_L, the phase relationship between v_R and v_{R_s} will be the same as that between i_R and i_L. In this case, the magnitudes of the current levels can be determined using Ohm's law and the resistance levels R and R_s, respectively.

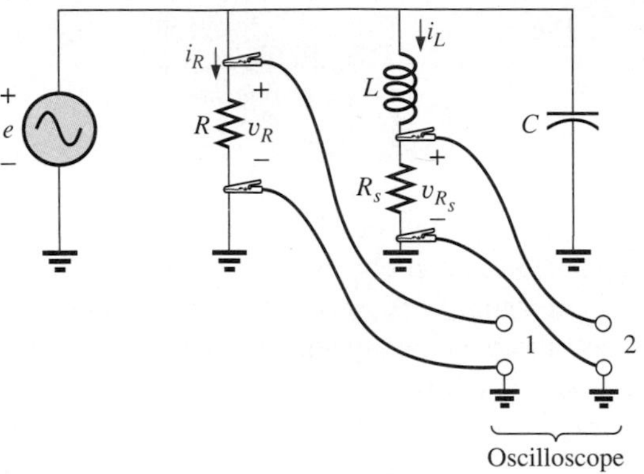

FIG. 15.101
Determining the phase relationship between i_R and i_L.

If the phase relationship between e and i_s of Fig. 15.101 is required, a sensing resistor can be employed, as shown in Fig. 15.102.

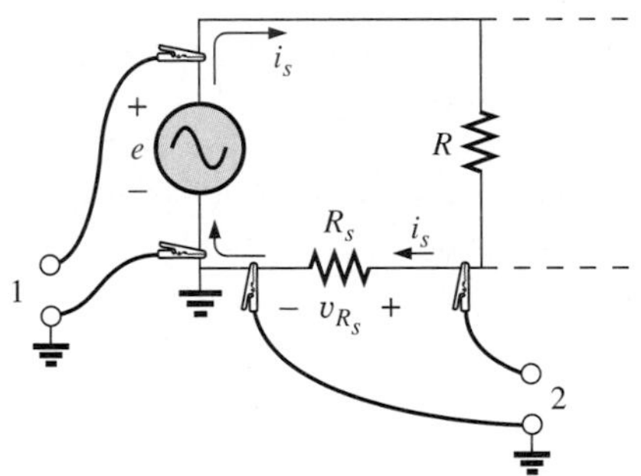

FIG. 15.102
Determining the phase relationship between e and i_s.

In general, therefore, for dual-trace measurements of phase relationships, be particularly careful of the grounding arrangement and fully utilize the in-phase relationship between the voltage and current of a resistor.

15.14 COMPUTER ANALYSIS

We have now covered sufficient material to make effective use of the PSpice software package. Although all the examples to appear in this section can be solved using C++ or BASIC, this section will be limited

to PSpice and the commands required to apply the package to ac systems.

PSpice (DOS)

As with the analysis just described, there is a great deal of similarity between the application of PSpice to dc and ac systems. The basic format of the input file is similar, with passive elements (R, L, and C) entered in exactly the same manner. The format for independent ac sources was introduced in Chapter 13. The two elements that have the most impact on the command statements are the frequency or frequencies of interest and the phase angles applied and obtained, as demonstrated by the following commands.

.AC Command

Recall the importance of the .DC statement for specifying sweep values and permitting control of the output format. The .AC command provides the same control for ac networks, specifying the frequency or frequencies of interest. The basic format of the command is the following:

```
.AC    LIN    11        1KH        2KH
```

		Number of data points	Starting frequency	Final frequency

As before, the quantities with brackets are controlled by the user. The LIN notation is an abbreviation for the word *linear*, which specifies that the increment between frequencies to be applied should be a fixed value. In other words, the spacing between frequencies must be the same from the initial to final values. The choice of the number of data points is an important one if the output file is to contain the desired results. In this light, let us take a moment to be absolutely sure we understand how to determine the number of data points. In Fig. 15.103, the initial (f_I) and final (f_F) frequencies have been specified. Let us define N as the total number of data points to be determined between f_I and f_F, *including* f_I and f_F. Defining INC as the increment between frequencies will result in $N - 1$ increments to progress from f_I to f_F. That is,

$$f_I + (N - 1)(\text{INC}) = f_F$$

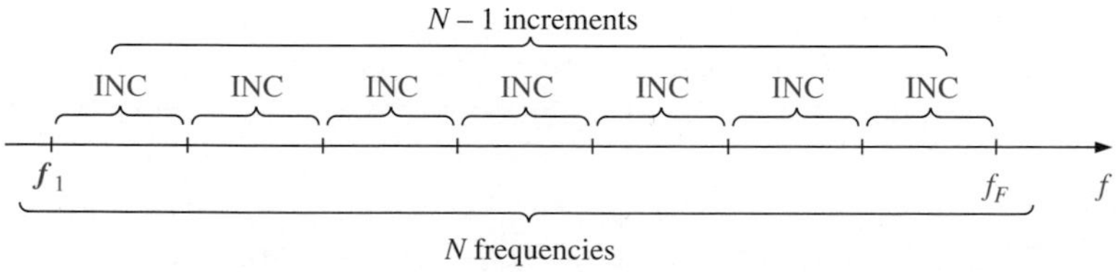

FIG. 15.103
Defining INC for a frequency range.

Solving for the resulting increment:

$$\boxed{\text{INC (in Hz)} = \frac{f_F - f_I}{N - 1}} \qquad \textbf{(15.44)}$$

For the preceding command statement,

$$\text{INC} = \frac{2000\text{ Hz} - 1000\text{ Hz}}{11 - 1} = \frac{1000\text{ Hz}}{10} = 100\text{ Hz}$$

Fig. 15.104 provides a list of the frequencies to be applied for the preceding command. Note that there are 11 distinct frequencies, and the increment between each is 100 Hz.

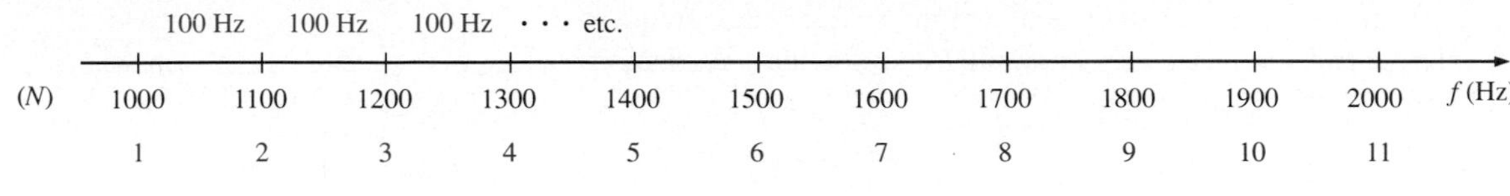

FIG. 15.104
The resulting INC for 11 distinct frequencies between 1 kHz and 2 kHz.

If the analysis is to be performed at a fixed frequency such as 1 kHz, the command statement will be entered as

```
.AC   LIN   1   1KH   1KH
```

.PRINT Command

The last command required to perform a basic ac analysis using PSpice is the .PRINT statement. It has the following general format:

```
.PRINT   AC   VM(2)   VP(2)
```

VM(2): Magnitude of the voltage V(2)
VP(2): Phase angle of the voltage V(2)

The quantities within the brackets are controlled by the user and can define a voltage or current. The uppercase letter M specifies that the magnitude of the voltage V(2) be printed, and the uppercase letter P specifies that the phase angle of the same voltage appear in the output file.

Series *R-C* Circuit

The series *R-C* circuit of Fig. 15.45, analyzed in detail in Section 15.5, will now be examined using PSpice. The input file of Fig. 15.105 includes a source $\mathbf{E} = 10\text{ V} \angle 0°$, a resistor of 5 kΩ, and a capacitor of

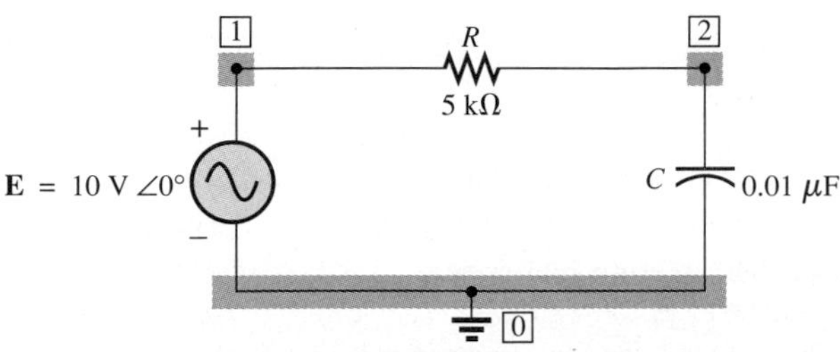

FIG. 15.105
Series R-C circuit to be analyzed using PSpice (DOS).

0.01 μF. The analysis will take place from f = 1 kHz to 20 kHz, in increments of (20 kHz − 1 kHz)/(20 − 1) = 1 kHz, as appearing in Fig. 15.106. The .PRINT statement requests the magnitude and phase angle of $\mathbf{V}_C$ as the applied frequency moves through the specified range.

```
Chapter 15 - Series RC Circuit - AC Analysis (Table)

 ****     CIRCUIT DESCRIPTION

******************************************************************************

VE 1 0 AC 10V 0
R 1 2 5K
C 2 0 0.01UF
.AC LIN 20 1KH 20KH
.PRINT AC VM(C) VP(C)
.OPTIONS NOPAGE
.END

****     AC ANALYSIS                          TEMPERATURE =   27.000 DEG C


FREQ          VM(C)        VP(C)

1.000E+03    9.540E+00   -1.744E+01
2.000E+03    8.467E+00   -3.214E+01
3.000E+03    7.277E+00   -4.330E+01
4.000E+03    6.227E+00   -5.149E+01
5.000E+03    5.370E+00   -5.752E+01
6.000E+03    4.687E+00   -6.205E+01
7.000E+03    4.139E+00   -6.555E+01
8.000E+03    3.697E+00   -6.830E+01
9.000E+03    3.334E+00   -7.052E+01
1.000E+04    3.033E+00   -7.234E+01
1.100E+04    2.780E+00   -7.386E+01
1.200E+04    2.564E+00   -7.514E+01
1.300E+04    2.378E+00   -7.624E+01
1.400E+04    2.217E+00   -7.719E+01
1.500E+04    2.076E+00   -7.802E+01
1.600E+04    1.951E+00   -7.875E+01
1.700E+04    1.840E+00   -7.939E+01
1.800E+04    1.741E+00   -7.997E+01
1.900E+04    1.652E+00   -8.049E+01
2.000E+04    1.572E+00   -8.096E+01
```

FIG. 15.106

Input and output files for the magnitude and phase of the voltage across the capacitor of the circuit of Fig. 15.105.

Note the similarities between the tabulated values for the magnitude and phase angle and the plots of Figs. 15.52 and 15.53. In addition, note the correspondence between the magnitudes and angles in the output file compared to the calculated values of Section 15.5. Imagine the time saved and the accuracy maintained using the PSpice software package versus the longhand approach—a very definite plus for the package approach, providing, of course, the theory is already clearly and correctly understood!

The next input file (Fig. 15.107) provides a plot of V_C and θ_C for the circuit of Fig. 15.105 using the .PROBE command in place of the

```
Chapter 15 - Series RC Circuit - AC Analysis (Plot)

 ****     CIRCUIT DESCRIPTION

*****************************************************************************

VE 1 0 AC 10V 0
R 1 2 5K
C 2 0 0.01UF
.AC LIN 100 100H 20KH
.PROBE
.OPTIONS NOPAGE
.END
```

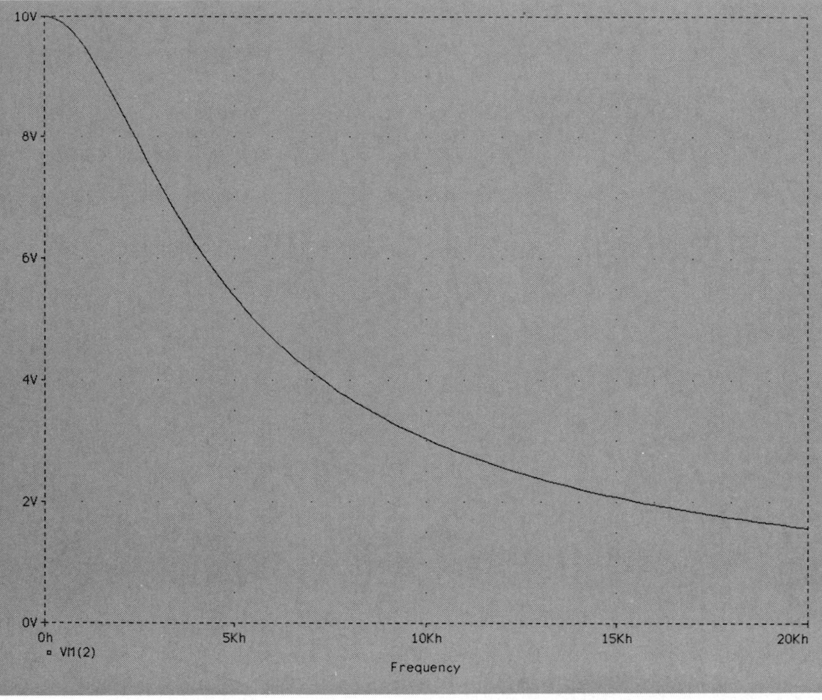

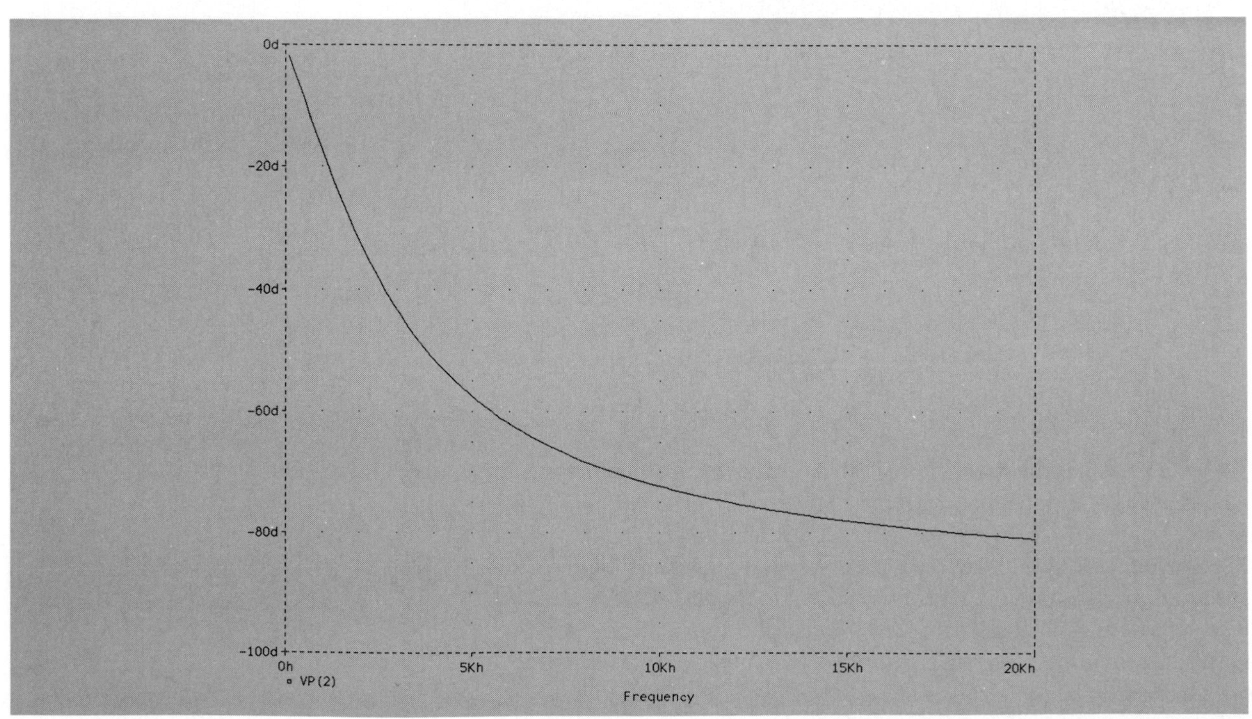

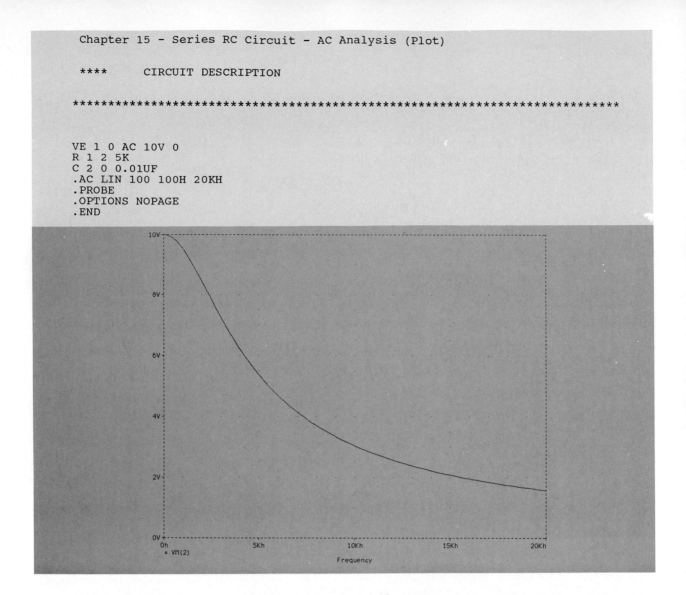

FIG. 15.107

Probe response for the magnitude and phase angle of the voltage across the capacitor of the network of Fig. 15.105.

.PRINT command. Compare the resulting curves with those obtained in Figs. 15.52 and 15.53.

PSpice (Windows)

Applying PSpice (Windows) to the network of Fig. 15.35 will result in the schematic of Fig. 15.108. In Fig. 15.35 the inductive and capacitive reactances were provided but cannot be entered on the schematic. Rather a frequency, 1 kHz in this case, must be chosen, and the inductive and capacitive values determined as follows.

$$X_L = 2\pi f L \Rightarrow L = \frac{X_L}{2\pi f} = \frac{7\ \Omega}{2\pi(1\text{ kHz})} = 1.114\text{ mH}$$

$$X_C = \frac{1}{2\pi f C} \Rightarrow C = \frac{1}{2\pi f X_C} = \frac{1}{2\pi(1\text{ kHz})3\ \Omega} = 53.05\ \mu\text{F}$$

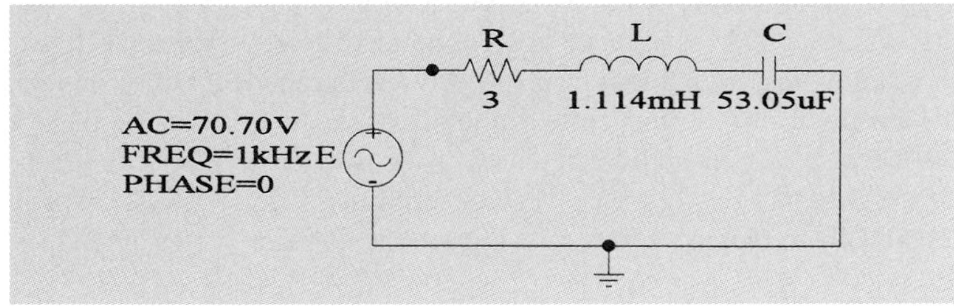

FIG. 15.108
Schematic for the PSpice (Windows) analysis of the circuit of Fig. 15.35.

The ac source chosen was **VSIN** with the following settings: **AC** = 70.70 V, **VOFF** = 0 V, **FREQ** = 1 kHz, and **PHASE** = 0 degrees. By choosing **Include System defined Attributes** under the **VSIN** dialog box, the package reference **PKGREF** can be set to **E**, as shown on the schematic. For each quantity, be sure to **Save Attribute**, or the value will not be entered into memory for the simulation. Then **Change Display** must be chosen to display specific quantities on the schematic. For most of the attributes **Both name and value** were chosen, but for **PKGREF** only **value** was displayed.

Using the sequence **Analysis-Setup-AC Sweep,** the following was then entered: **Total Pts** = 1, **Start Freq** = 1 kHz and **End Freq** = 1 kHz (because the analysis will take place at only one frequency). Since we will also want to plot the current and various voltages, we must also choose **Transient** and enter the following: **Print Step** = 5 μs, **Final Time** = 5 ms, **No-Print Delay** = 0 s, and **Step Ceiling** = 5 μs. The choices were made because the period of the applied sine wave is 1 ms and 5 cycles will be displayed. We do not want any hesitancy in the printout, so the **Delay time** is 0 s. In addition we want to limit the **Step Ceiling** to 5 μs to ensure a data point every 5 μs. The **Print Step** of 5 μs was chosen to provide 1000 data points for the full 5-ms period (5 ms/5 μs = 1000). Both **AC Sweep** and **Transient** should be enabled before **Close** is chosen. Turning to the **Probe Setup** under **Analysis**, the choice was **Automatically Run Probe** after **Simulation** to make the response go directly to the **Probe** screen once simulation is called for.

When **simulation** is chosen, a display will appear asking us whether we want to perform an **AC** or **Transient** solution since both were enabled under the **Analysis Setup.** By choosing **AC** we are telling the **Probe** response that we want to limit the display to the frequency of interest, which is 1 kHz for this example. When chosen, the display will simply have one frequency indicated on the horizontal axis, as shown in Fig. 15.109. The sequence **Trace-Add** will then result in the **Add Traces** dialog box, where the choice **Alias Names** will be picked to provide a complete display of available references for the network under investigation. Choosing **I(R)** (the current through the resistor) as the current of the circuit will make **I(R)** appear in the **Trace Command** box and it can then be displayed with an **OK.**

The short horizontal mark at 1 kHz reveals that the peak value of the current is near 14 A. To limit the plot to the region around 14 A, choose **Plot** and then **Y-Axis Settings**, followed by **User Defined**, and set the range to 10 A to 20 A. An **OK** adjusts the display. Next the actual value of the display can be obtained by using the cursor feature as follows. **Tools-Cursor-Display** and the **Probe Cursor** dialog box will reveal that **A1** is at 1 kHz (horizontal position) and 14.14 A (vertical position). In other words, the **Cursor** can be used to read the level of points on the **Probe** display. Using **Tools-Label-Text** the 14.14 A can be placed on the display. An additional plot can be added by **Plot-Add Plot-Trace-Add** and typing in the **Trace Command V(L:1)-V(L:2)**, which tells **Probe** to plot the difference between the voltages across the induc-

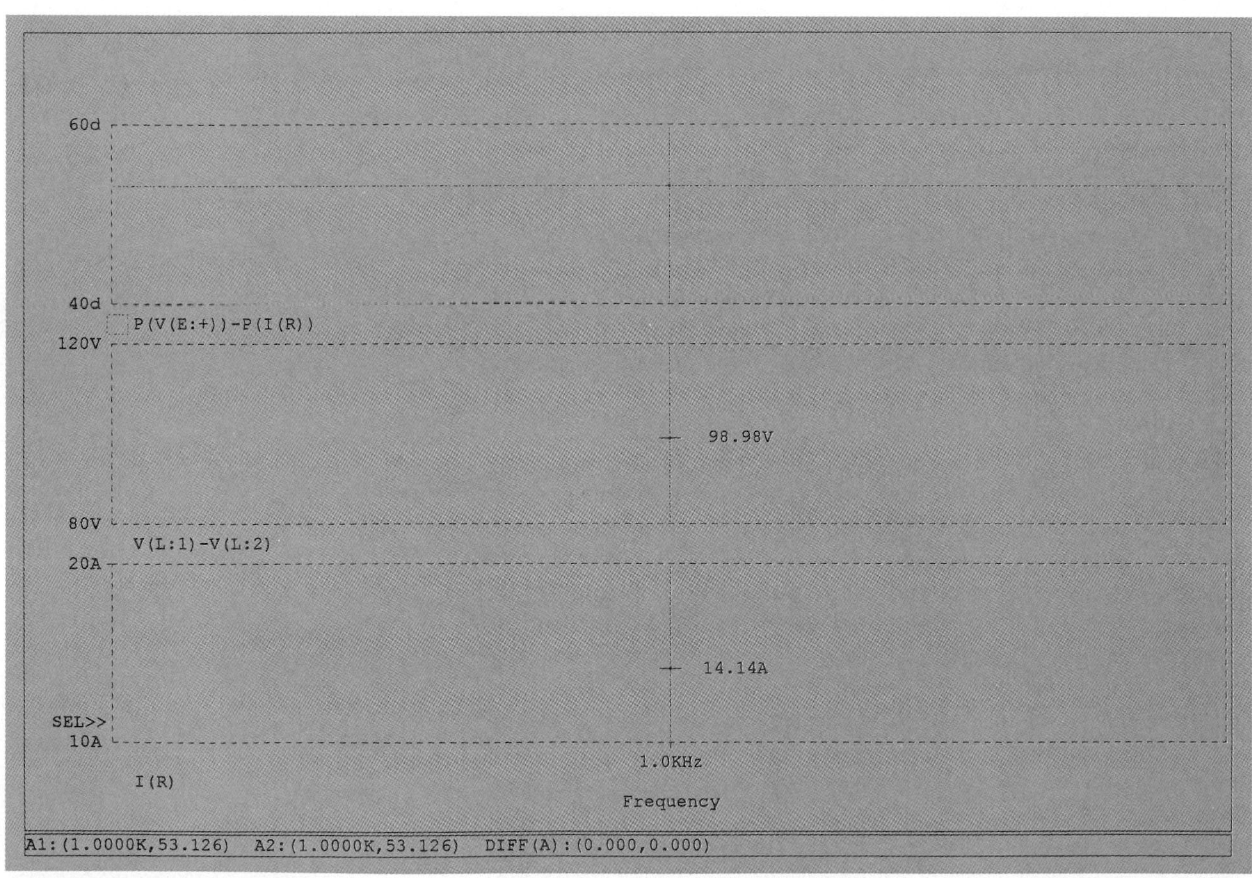

FIG. 15.109

Determining the peak value of the current i_R and the voltage v_L, and finding the phase angle between the applied voltage and the resulting current.

tor. Always keep in mind that the voltage listing in **Probe** is for the nodal voltages, and, unless the element is connected to ground, the above format must be used to find the voltage across an element. After an **OK** the second display will appear with a value near 100 V. Using the **Cursor** option we find the actual level to be 98.98 V, so the vertical scale was changed to 80 V to 120 V using the procedure described above. The phase angle between the applied voltage and the circuit current can be found by the **Add Plot** sequence and the **Trace Command P(V(E:+))-P(V(I(R))**, which in words says to display the difference between the phase angles of the applied voltage and the circuit current. The letter **P** is defined as the phase angle in a long list of permissible **Probe** arithmetic operations in the Schematics manual. The result is exactly 53.126°, as appears in the **Cursor** dialog box at the bottom of Fig. 15.108. The actual **Cursor**, with its vertical and horizontal lines, appears in the resulting plot. The vertical scale was also changed to 40° to 60° to better reflect the area of interest.

The above has demonstrated how **Probe** can be used to find specific quantities of an ac network at a specific frequency. Using the available arithmetic operations and some ingenuity, most quantities of interest can be determined.

A transient response can now be obtained by turning to **Plot** and choosing **Transient** at the bottom of the listing. A display will now appear with a horizontal axis range of 0 to 5 ms, as set in the **Transient** dialog box. Choosing **Trace-Add** and **V(E:+)** followed by **OK** makes the applied voltage appear on the screen as shown in Fig. 15.110. The

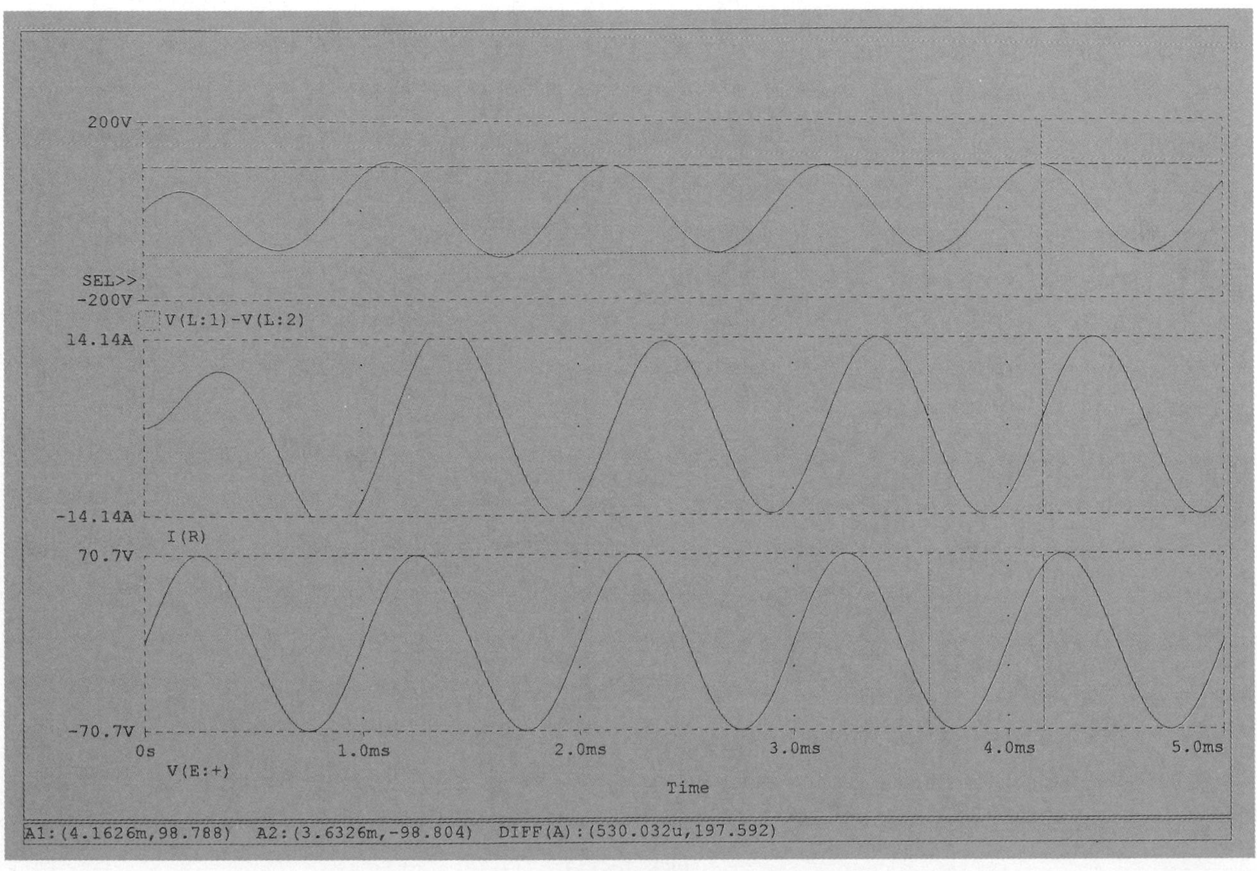

FIG. 15.110
Plotting e, i_R, and v_L versus time for the circuit of Fig. 15.108.

peak value can then be checked using the **Cursor** option, and the vertical scale set to the peak value as described above. The current of the network can then be displayed using **Plot-Add Plot-Trace-Add** and choosing **I(R)** from the available list. Again, the peak value was determined using the **Cursor** and the **Y-Axis Setting** under **Plot** set to the peak values. The distortion in the first few milliseconds is only a reflection of the effect of the reactive elements going through their transient phase before reaching a steady-state condition at about 3 ms. The last plot to be added is the voltage across the coil, which is done using the **Trace Command** described for the **AC** response. The **Cursors** were left in position for the voltage across the coil, showing that the positive peak value is at 98.788 V and the negative peak at −98.904 V. The difference is 197.592 V, which is a very close match with the difference obtained using the longhand approach (2 × 98.98 V = 197.96 V).

A plot to match Fig. 15.39 can be obtained by first clearing the screen using the sequence **Plot-AC-Plot-Transient** and choosing the various voltages and the current, as listed at the bottom of Fig. 15.111. In this case simply choose **Trace-Add** to add each waveform. New plots are not being introduced, just new traces on the same plot. The distortion in the period 0–3 ms can be eliminated and the display limited to just one cycle by choosing **Plot-X-Axis Setting** and **User Defining** 4 ms to 5 ms as the period of interest. The resulting plot is now an exact match with Fig. 15.39. The labels for the various waveforms were added with the **Tools-Label-Text** sequence, and the line at 0 was added with **Tools-Label-Line.**

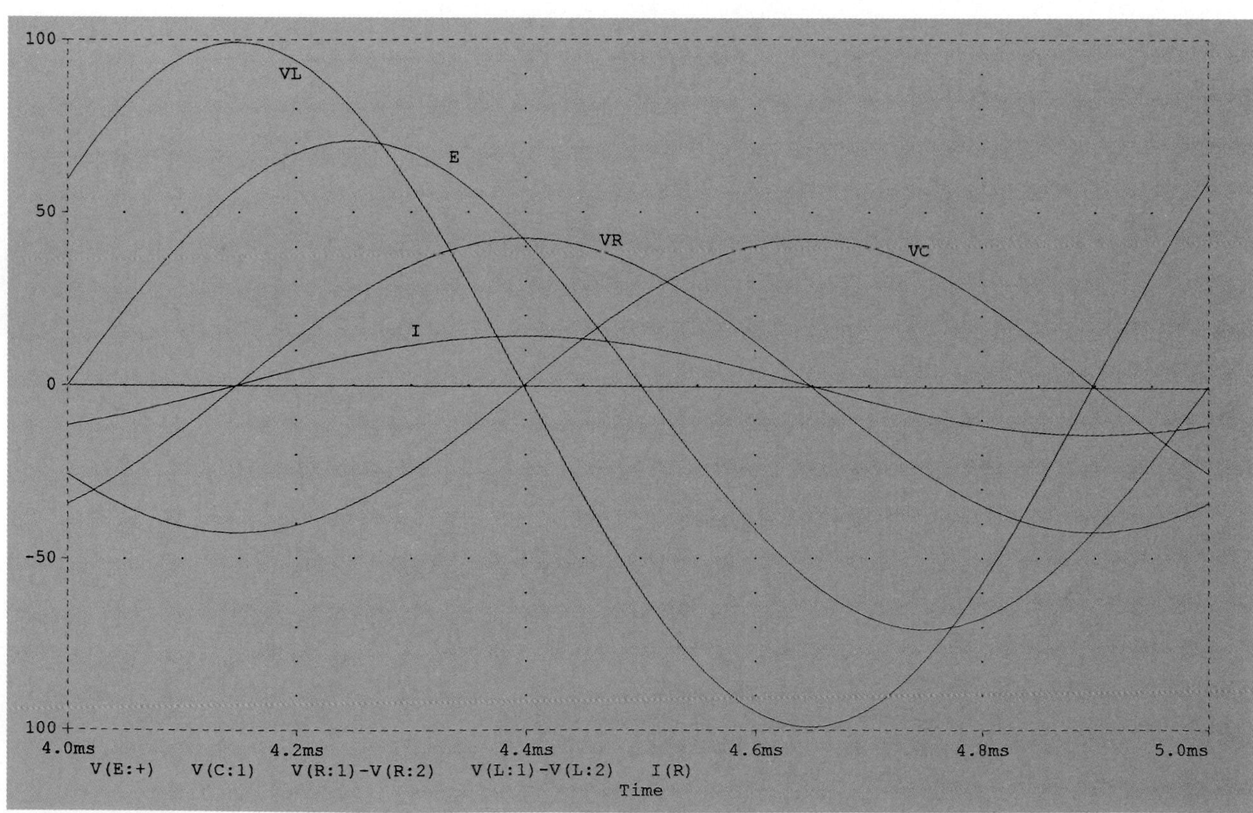

FIG. 15.111

Generating a plot of e, v_C, v_R, v_L, and i to compare against Fig. 15.39.

The frequency response of the parallel ac network of Fig. 15.79 will now be investigated using PSpice (Windows). The schematic appears in Fig. 15.112 with the addition of a 1-$\mu\Omega$ resistor in series with the inductor, to satisfy the condition that inductive elements must have a resistor in series. The current source chosen was **ISRC** from the **source.slb**, which results in a symbol that had the arrow. This will limit the list of attributes that can be assigned to the **DC** and **AC** level and **PKGREF.** Both the **AC** level and new **PKGREF** were displayed on the schematic. For such situations a phase angle of 0° is assumed, and the frequencies of interest will be assigned by the **AC Sweep.** The 1-A amplitude was chosen to permit plotting the magnitude of the impedance of the network versus frequency using the relationship $Z_T = V/I = V/1 = V$. Since the voltage across the network is also of interest, a **VPRINT1** (from the **special.slb** library) was placed as shown with the **PKGREF** set as **VL.** The **AC, MAG,** and **PHASE** were assigned the **ok** status for the printout.

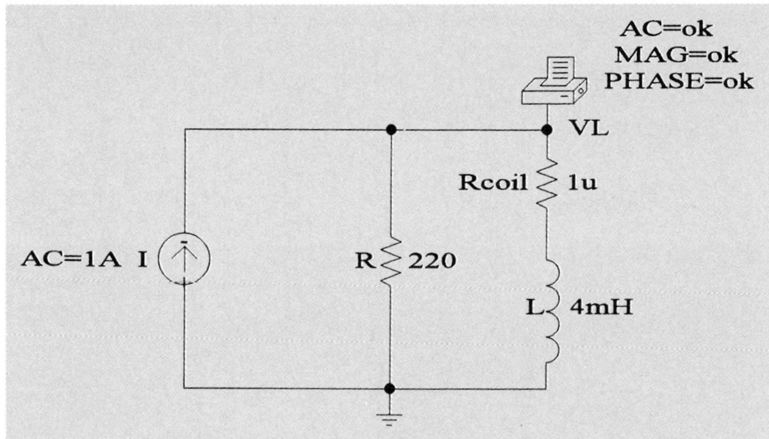

FIG. 15.112
Schematic representation of the network of Fig. 15.79.

The first analysis will provide a table of values of Z_T versus specific frequencies to match the listing appearing in Section 15.10. This is accomplished by enabling only the **AC Sweep** of the **Analysis** dialog box and setting the following values: **Total Pts** = 8, **Start Freq** = 5 kHz, and **End Freq** = 40 kHz. Under **Probe Setup** the **Do Not Auto-Run Probe** was chosen and **simulation** initiated. Once completed, click on **File** and then **Examine Output**, which will result in the output file where the listing of Fig. 15.113 can be found. Note the perfect match with the longhand solution, and think of the time saved in obtaining such a list.

A plot of the magnitude and phase of the total impedance versus frequency can then be obtained to compare with the results of Figs. 15.81 and 15.82 using the following procedure: **AC Sweep: Total Pts** = 1000, **Start Freq** = 100 Hz (remember the beginning frequency cannot be zero for an **AC Sweep**), and **End Freq** = 40 kHz. The result will be a plot versus frequency that makes the **Transient** option inappropriate since it provides a plot versus time. Choosing **Automatically Run Probe After Simulation** will result in the Probe screen, from which **Trace-Add-V(I:-)** can be chosen. The resulting plot of Fig. 15.114 has a log scale for the horizontal axis that is different from that

```
     ****       AC ANALYSIS                          TEMPERATURE =    27.000
     DEG C

     ****************************************************************************
     ************


          FREQ          VM($N_0001) VP($N_0001)


          5.000E+03    1.091E+02    6.027E+01
          1.000E+04    1.655E+02    4.120E+01
          1.500E+04    1.900E+02    3.027E+01
          2.000E+04    2.015E+02    2.364E+01
          2.500E+04    2.076E+02    1.930E+01
          3.000E+04    2.112E+02    1.627E+01
          3.500E+04    2.134E+02    1.404E+01
          4.000E+04    2.149E+02    1.234E+01
```

FIG. 15.113
Listing the magnitude and phase angle for the total impedance $\mathbf{Z}_T$ *versus frequency for the parallel R-L network of Fig. 15.112.*

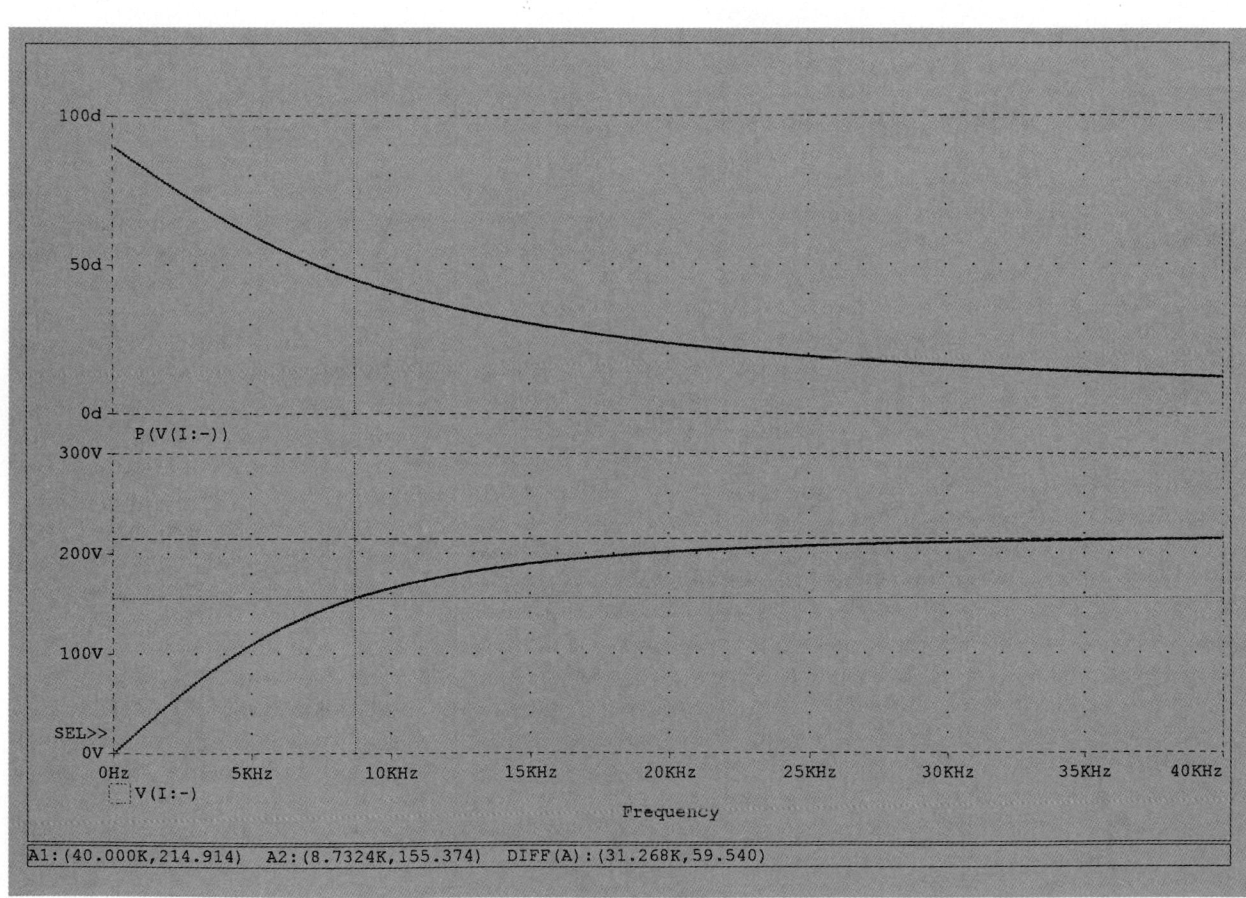

FIG. 15.114
A plot of the magnitude and phase angle of the total impedance $\mathbf{Z}_T$ *versus frequency for the network of Fig. 15.112.*

of Figs. 15.81 and 15.82. A linear scale can be obtained by going to **Plot-X Axis Setting-Scale-Linear-OK.** The **Cursor** option was then employed to show that the peak value of **V(I:-)** (actually Z_T) is approaching the 220 Ω of the resistor (214.914 Ω at 40 kHz) at very high frequencies, and the frequency at which $R = X_L$ is 8.7324 kHz, comparing well with the longhand calculated value of 8.75 kHz. Note that the phase angle of the total impedance is 90° at 0 Hz and drops to 45° when $R = X_L$. Eventually, at very high frequencies (beyond 40 kHz), the phase angle will approach the zero degrees associated with a purely resistive network.

PROBLEMS

SECTION 15.2 Impedance and the Phasor Diagram

1. Express the impedances of Fig. 15.115 in both polar and rectangular form.

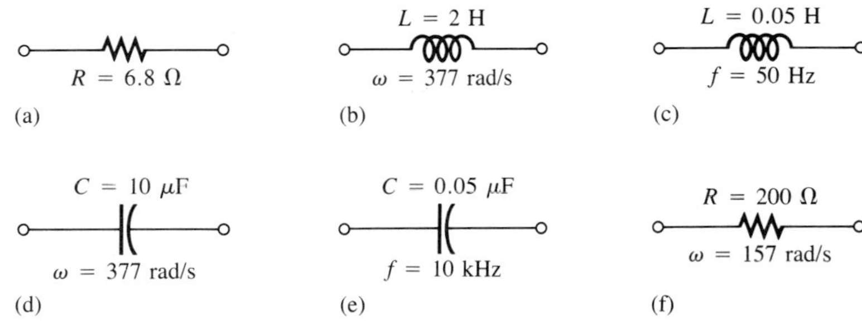

FIG. 15.115
Problem 1.

2. Find the current i for the elements of Fig. 15.116 using complex algebra. Sketch the waveforms for v and i on the same set of axes.

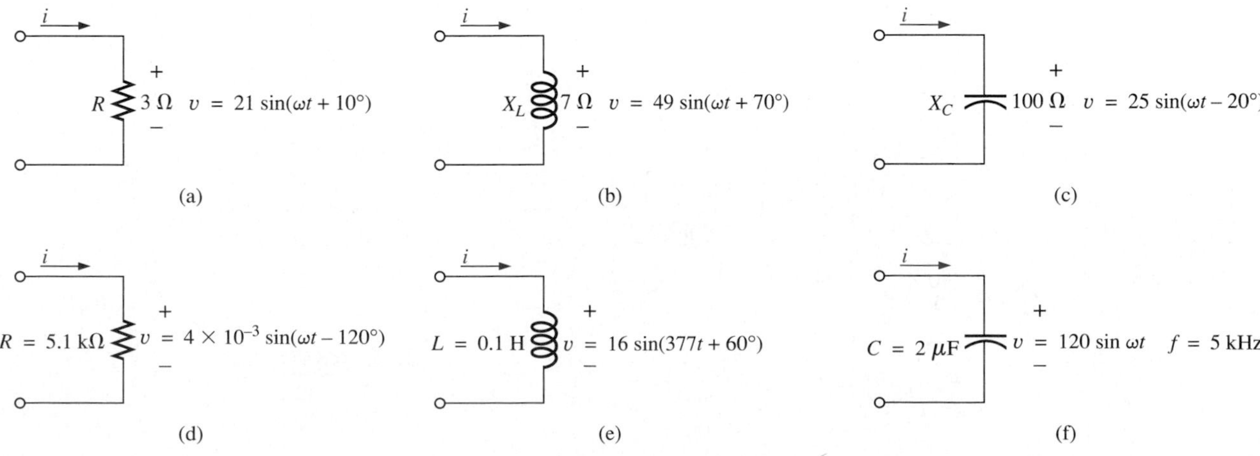

FIG. 15.116
Problem 2.

3. Find the voltage v for the elements of Fig. 15.117 using complex algebra. Sketch the waveforms of v and i on the same set of axes.

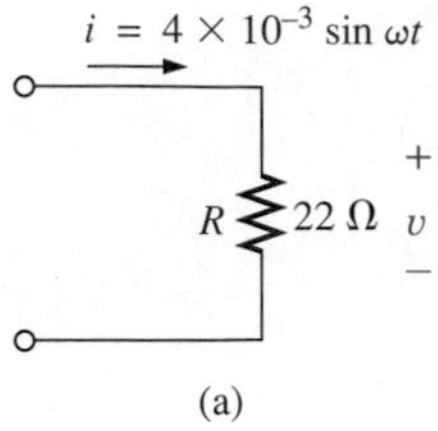

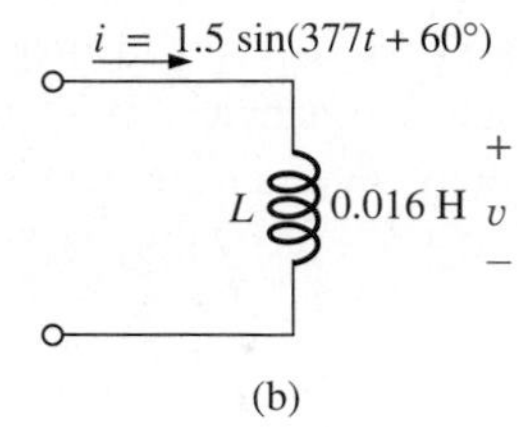

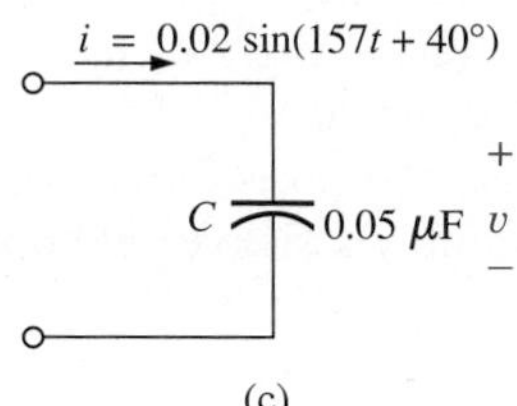

FIG. 15.117
Problem 3.

SECTION 15.3 Series Configuration

4. Calculate the total impedance of the circuits of Fig. 15.118. Express your answer in rectangular and polar form, and draw the impedance diagram.

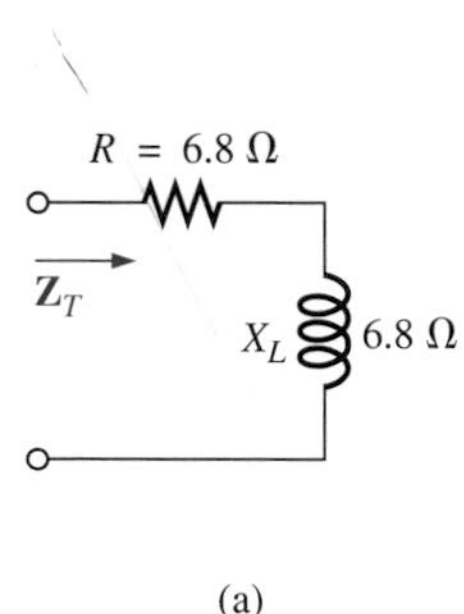

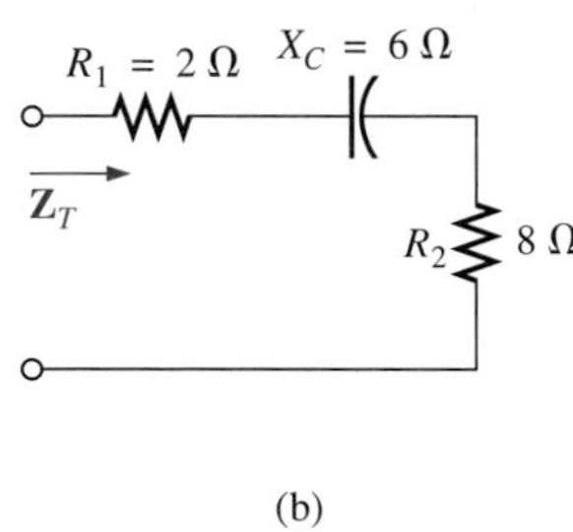

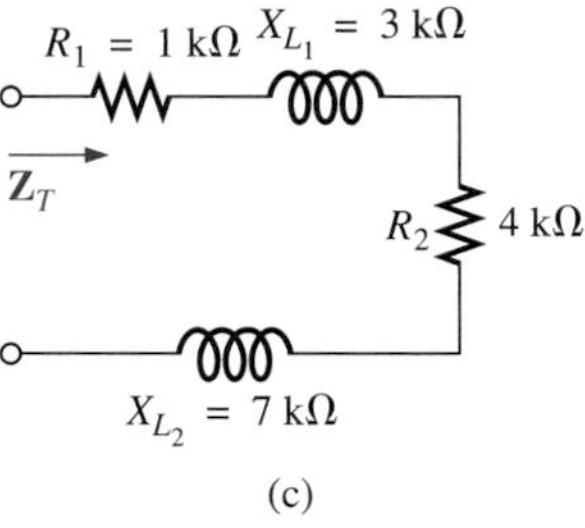

FIG. 15.118
Problem 4.

5. Calculate the total impedance of the circuits of Fig. 15.119. Express your answer in rectangular and polar form, and draw the impedance diagram.

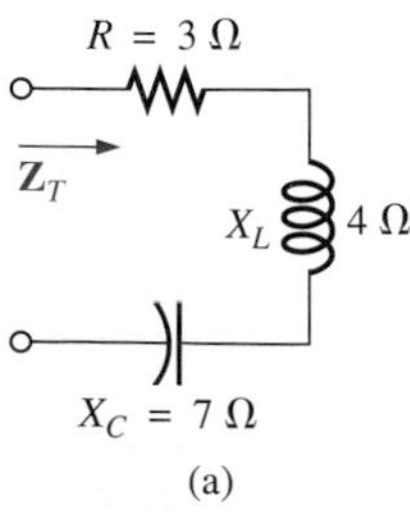

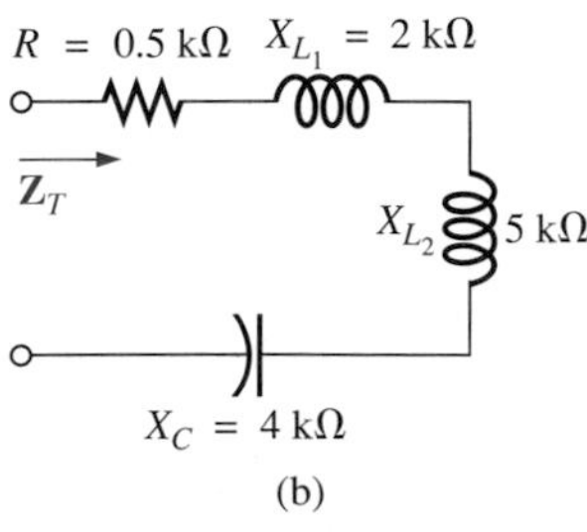

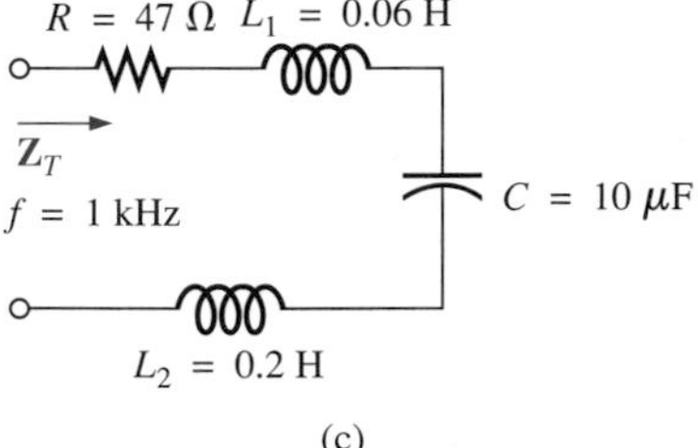

FIG. 15.119
Problem 5.

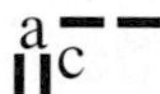

6. Find the type and impedance in ohms of the series circuit elements that must be in the closed container of Fig. 15.120 for the indicated voltages and currents to exist at the input terminals. (Find the simplest series circuit that will satisfy the indicated conditions.)

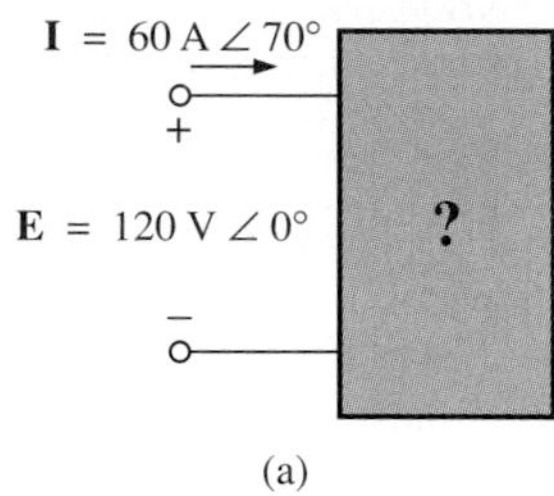

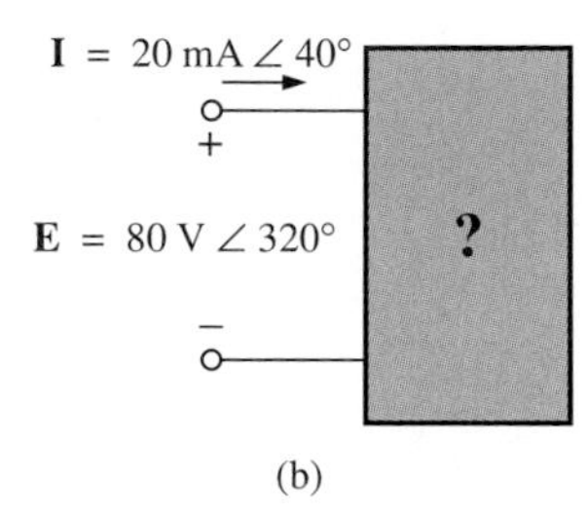

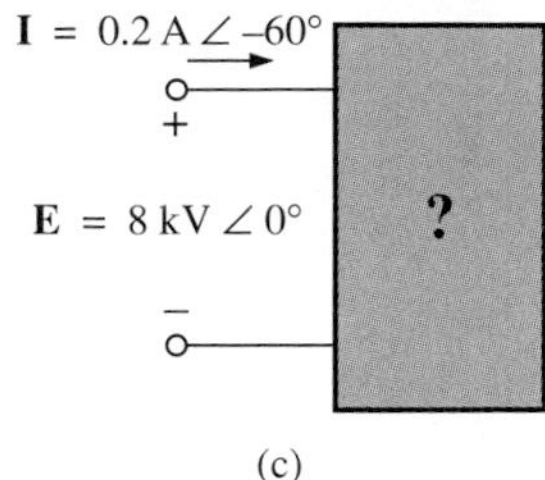

FIG. 15.120
Problems 6 and 26.

7. For the circuit of Fig. 15.121:
   a. Find the total impedance $\mathbf{Z}_T$ in polar form.
   b. Draw the impedance diagram.
   c. Find the current **I** and the voltages $\mathbf{V}_R$ and $\mathbf{V}_L$ in phasor form.
   d. Draw the phasor diagram of the voltages **E**, $\mathbf{V}_R$, and $\mathbf{V}_L$, and the current **I**.
   e. Verify Kirchhoff's voltage law around the closed loop.
   f. Find the average power delivered to the circuit.
   g. Find the power factor of the circuit and indicate whether it is leading or lagging.
   h. Find the sinusoidal expressions for the voltages and current if the frequency is 60 Hz.
   i. Plot the waveforms for the voltages and current on the same set of axes.

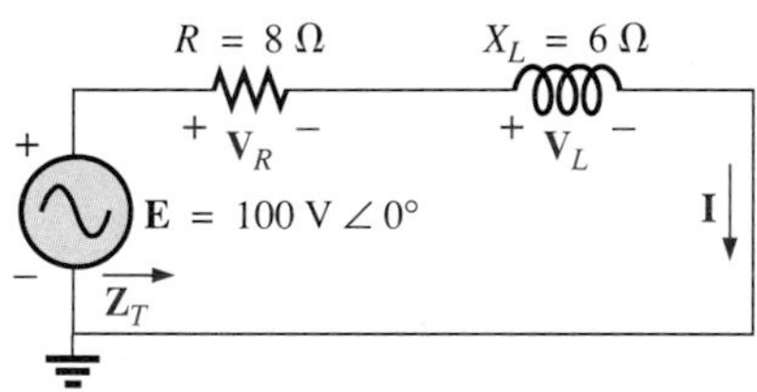

FIG. 15.121
Problems 7, 47, and 51.

8. Repeat Problem 7 for the circuit of Fig. 15.122, replacing $\mathbf{V}_L$ with $\mathbf{V}_C$ in parts (c) and (d).

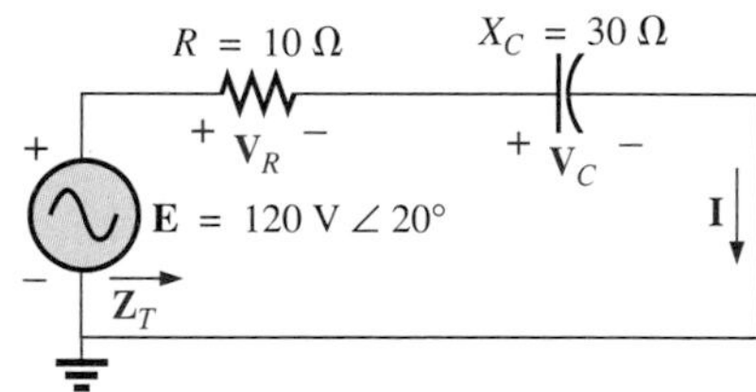

FIG. 15.122
Problem 8.

9. Given the network of Fig. 15.123:
   a. Determine $\mathbf{Z}_T$.
   b. Find **I**.
   c. Calculate $\mathbf{V}_R$ and $\mathbf{V}_L$.
   d. Find P and F_p.

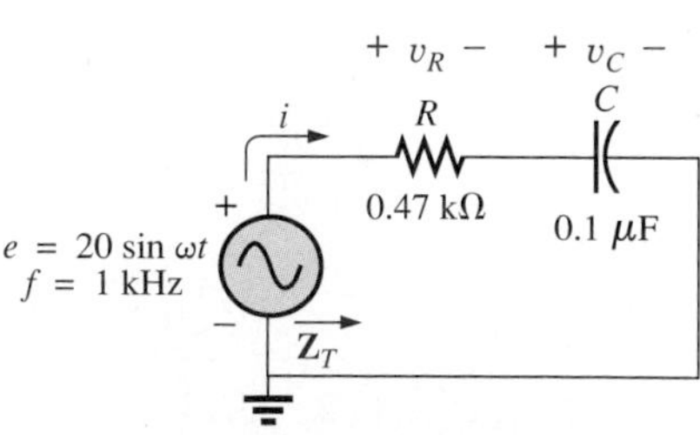

FIG. 15.123
Problems 9 and 53.

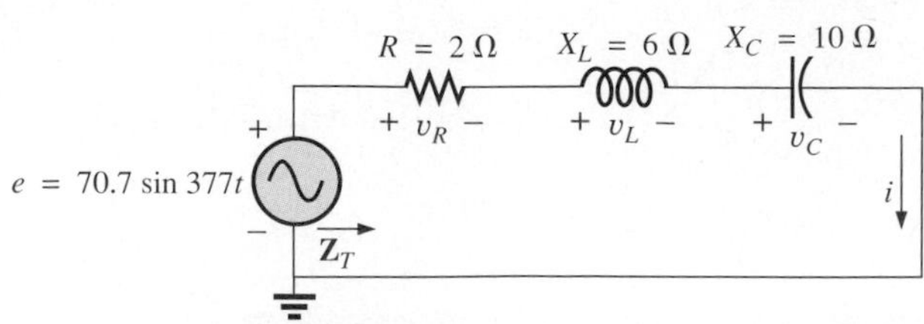

FIG. 15.124
Problem 10.

10. For the circuit of Fig. 15.124:
    a. Find the total impedance $\mathbf{Z}_T$ in polar form.
    b. Draw the impedance diagram.
    c. Find the value of C in microfarads and L in henries.
    d. Find the current **I** and the voltages $\mathbf{V}_R$, $\mathbf{V}_L$, and $\mathbf{V}_C$ in phasor form.
    e. Draw the phasor diagram of the voltages **E**, $\mathbf{V}_R$, $\mathbf{V}_L$, and $\mathbf{V}_C$, and the current **I**.
    f. Verify Kirchhoff's voltage law around the closed loop.
    g. Find the average power delivered to the circuit.
    h. Find the power factor of the circuit and indicate whether it is leading or lagging.
    i. Find the sinusoidal expressions for the voltages and current.
    j. Plot the waveforms for the voltages and current on the same set of axes.

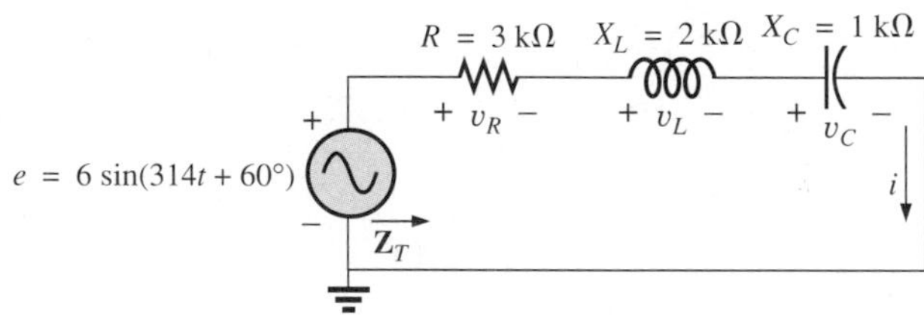

FIG. 15.125
Problem 11.

11. Repeat Problem 10 for the circuit of Fig. 15.125.

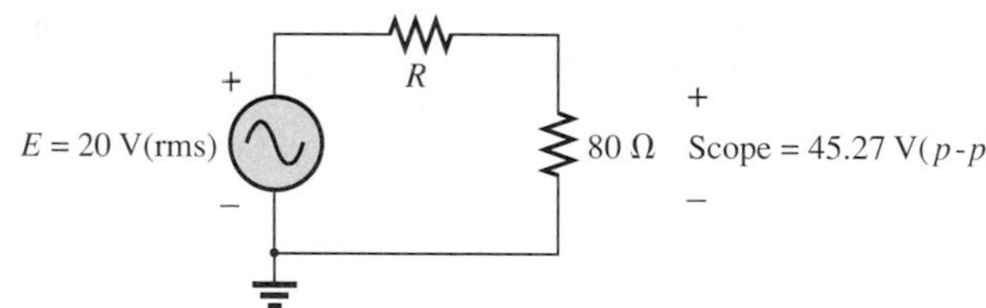

FIG. 15.126
Problem 12.

12. Using the oscilloscope reading of Fig. 15.126, determine the resistance R.

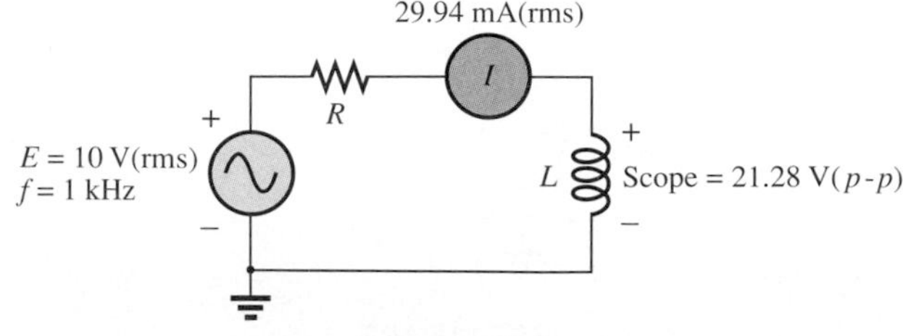

FIG. 15.127
Problem 13.

*13. Using the DMM current reading and the oscilloscope measurement of Fig. 15.127:
    a. Determine the inductance L.
    b. Find the resistance R.

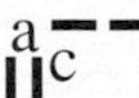

*14. Using the oscilloscope reading of Fig. 15.128, determine the capacitance C.

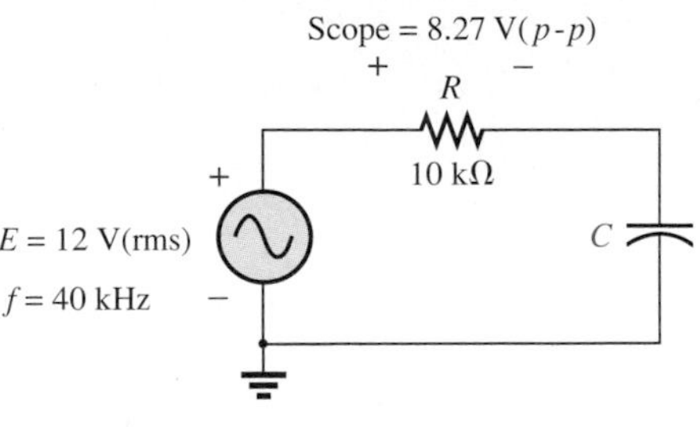

FIG. 15.128
Problem 14.

SECTION 15.4 Voltage Divider Rule

15. Calculate the voltages $\mathbf{V}_1$ and $\mathbf{V}_2$ for the circuit of Fig. 15.129 in phasor form using the voltage divider rule.

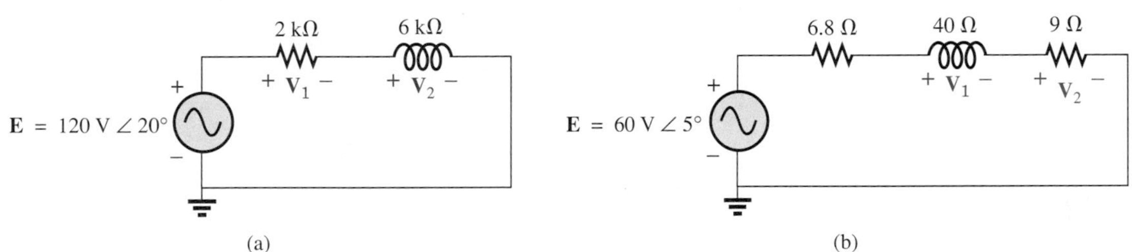

FIG. 15.129
Problem 15.

16. Calculate the voltages $\mathbf{V}_1$ and $\mathbf{V}_2$ for the circuit of Fig. 15.130 in phasor form using the voltage divider rule.

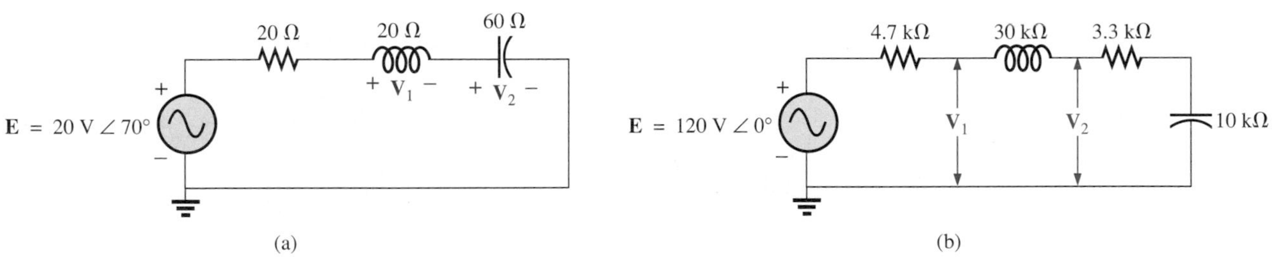

FIG. 15.130
Problem 16.

*17. For the circuit of Fig. 15.131:
a. Determine $\mathbf{I}$, $\mathbf{V}_R$, and $\mathbf{V}_C$ in phasor form.
b. Calculate the total power factor and indicate whether it is leading or lagging.

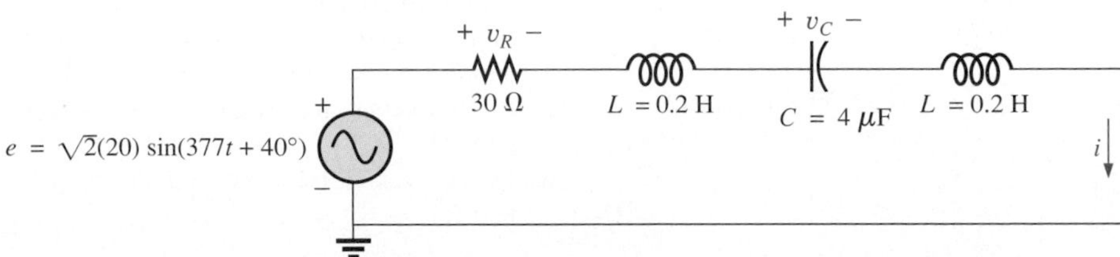

FIG. 15.131
Problems 17, 18, and 54.

c. Calculate the average power delivered to the circuit.
d. Draw the impedance diagram.
e. Draw the phasor diagram of the voltages **E**, $\mathbf{V}_R$, and $\mathbf{V}_C$, and the current **I**.
f. Find the voltages $\mathbf{V}_R$ and $\mathbf{V}_C$ using the voltage divider rule, and compare them with the results of part (a) above.
g. Draw the equivalent series circuit of the above as far as the total impedance and the current i are concerned.

***18.** Repeat Problem 17 if the capacitance is changed to 1000 μF.

19. An electrical load has a power factor of 0.8 lagging. It dissipates 8 kW at a voltage of 200 V. Calculate the impedance of this load in rectangular coordinates.

***20.** Find the series element or elements that must be in the enclosed container of Fig. 15.132 to satisfy the following conditions:
a. Average power to circuit = 300 W.
b. Circuit has a lagging power factor.

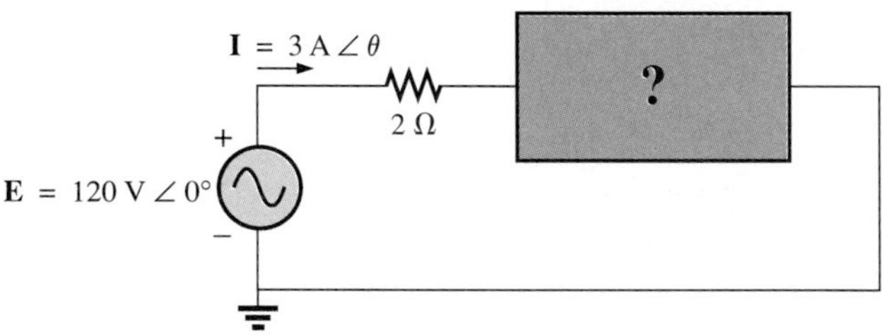

FIG. 15.132
Problem 20.

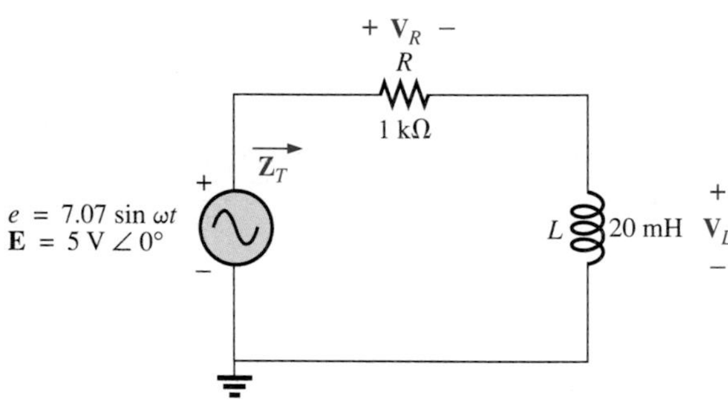

FIG. 15.133
Problem 21.

SECTION 15.5 Frequency Response of the *R-C* Circuit

***21.** For the circuit of Fig. 15.133:
a. Plot Z_T and θ_T versus frequency for a frequency range of zero to 20 kHz.
b. Plot V_L versus frequency for the frequency range of part (a).
c. Plot θ_L versus frequency for the frequency range of part (a).
d. Plot V_R versus frequency for the frequency range of part (a).

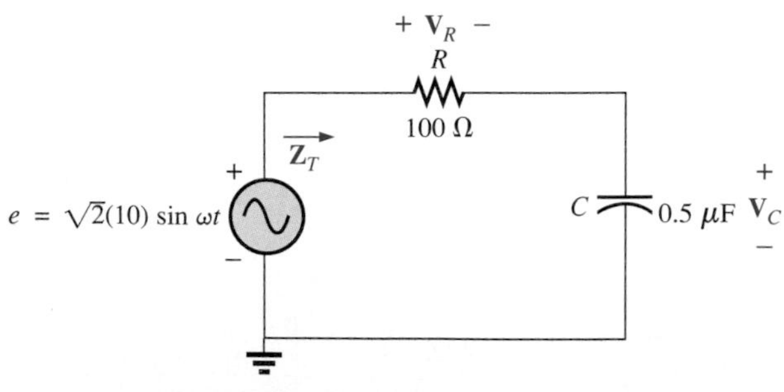

FIG. 15.134
Problem 22.

***22.** For the circuit of Fig. 15.134:
a. Plot Z_T and θ_T versus frequency for a frequency range of zero to 10 kHz.
b. Plot V_C versus frequency for the frequency range of part (a).
c. Plot θ_C versus frequency for the frequency range of part (a).
d. Plot V_R versus frequency for the frequency range of part (a).

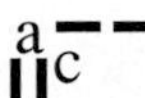

*23. For the series R-L-C circuit of Fig. 15.135:
   a. Plot Z_T and θ_T versus frequency for a frequency range of zero to 20 kHz in increments of 1 kHz.
   b. Plot V_C (magnitude only) versus frequency for the same frequency range of part (a).
   c. Plot I (magnitude only) versus frequency for the same frequency range of part (a).

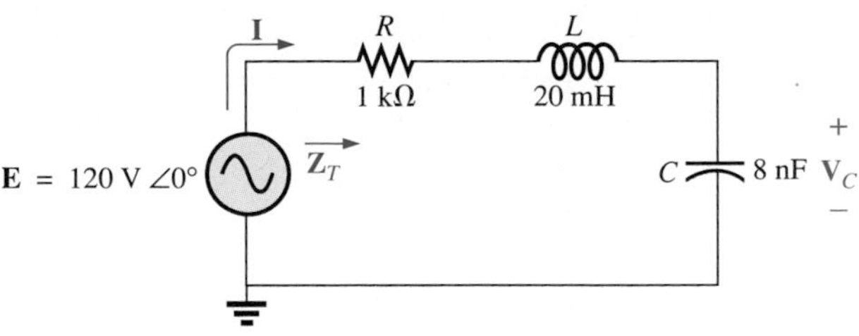

FIG. 15.135
Problems 23 and 48.

SECTION 15.7 Admittance and Susceptance

24. Find the total admittance and impedance of the circuits of Fig. 15.136. Identify the values of conductance and susceptance, and draw the admittance diagram.

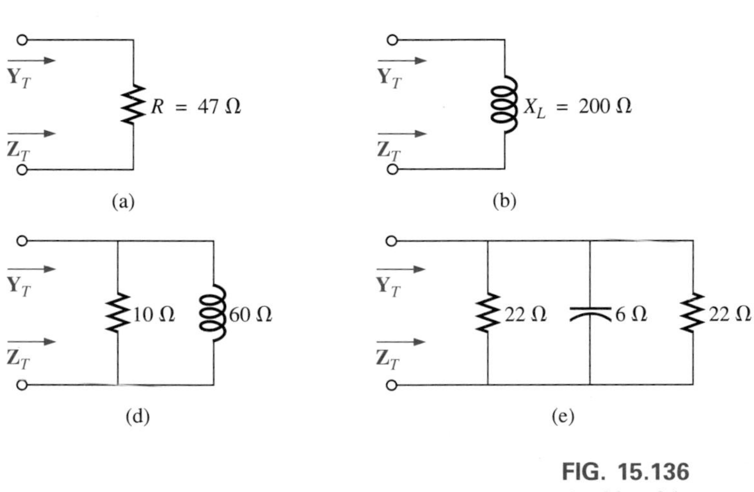

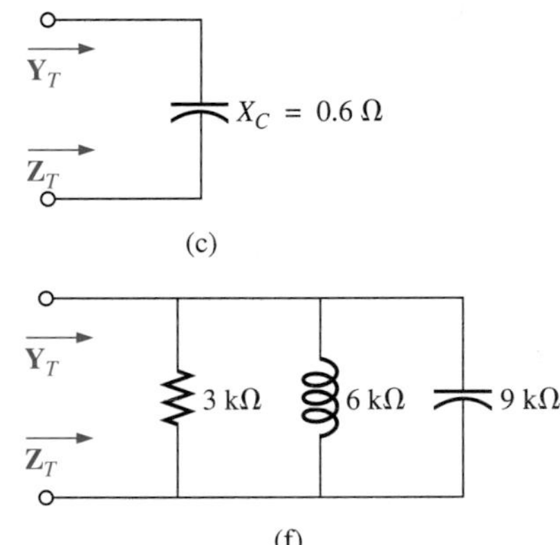

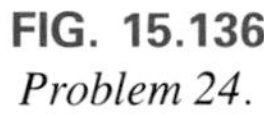

FIG. 15.136
Problem 24.

25. Find the total admittance and impedance of the circuits of Fig. 15.137. Identify the values of conductance and susceptance, and draw the admittance diagram.

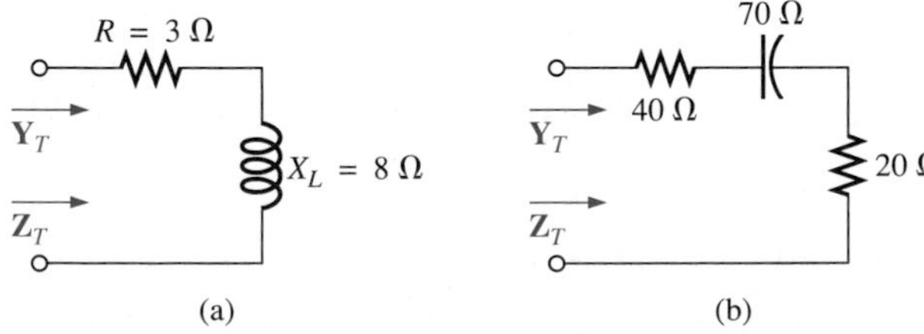

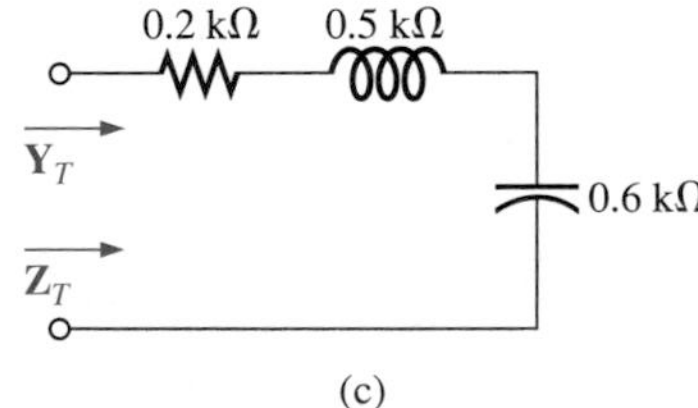

FIG. 15.137
Problem 25.

26. Repeat Problem 6 for the parallel circuit elements that must be in the closed container for the same voltage and current to exist at the input terminals. (Find the simplest parallel circuit that will satisfy the conditions indicated.)

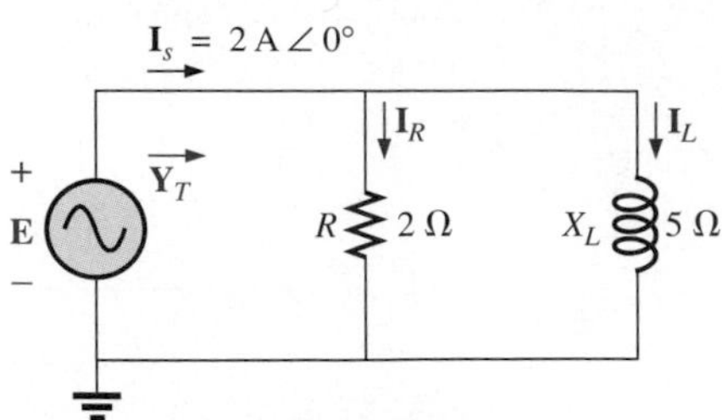

FIG. 15.138
Problem 27.

SECTION 15.8 Parallel ac Networks

27. For the circuit of Fig. 15.138:

a. Find the total admittance $\mathbf{Y}_T$ in polar form.
b. Draw the admittance diagram.
c. Find the voltage **E** and the currents $\mathbf{I}_R$ and $\mathbf{I}_L$ in phasor form.
d. Draw the phasor diagram of the currents $\mathbf{I}_s$, $\mathbf{I}_R$, and $\mathbf{I}_L$ and the voltage **E**.
e. Verify Kirchhoff's current law at one node.
f. Find the average power delivered to the circuit.
g. Find the power factor of the circuit and indicate whether it is leading or lagging.
h. Find the sinusoidal expressions for the currents and voltage if the frequency is 60 Hz.
i. Plot the waveforms for the currents and voltage on the same set of axes.

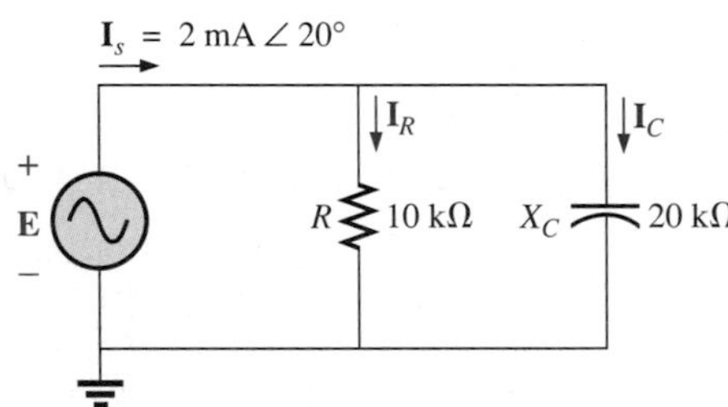

FIG. 15.139
Problem 28.

28. Repeat Problem 27 for the circuit of Fig. 15.139, replacing $\mathbf{I}_L$ with $\mathbf{I}_C$ in parts (c) and (d).

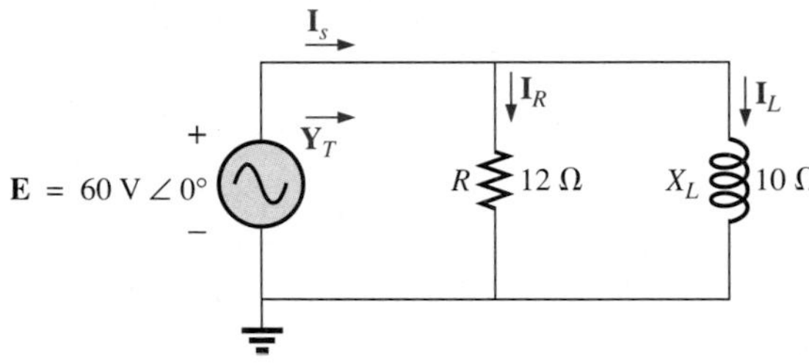

FIG. 15.140
Problems 29 and 52.

29. Repeat Problem 27 for the circuit of Fig. 15.140, replacing **E** with $\mathbf{I}_s$ in part (c).

30. For the circuit of Fig. 15.141:

a. Find the total admittance $\mathbf{Y}_T$ in polar form.
b. Draw the admittance diagram.
c. Find the value of C in microfarads and L in henries.
d. Find the voltage **E** and currents $\mathbf{I}_R$, $\mathbf{I}_L$, and $\mathbf{I}_C$ in phasor form.

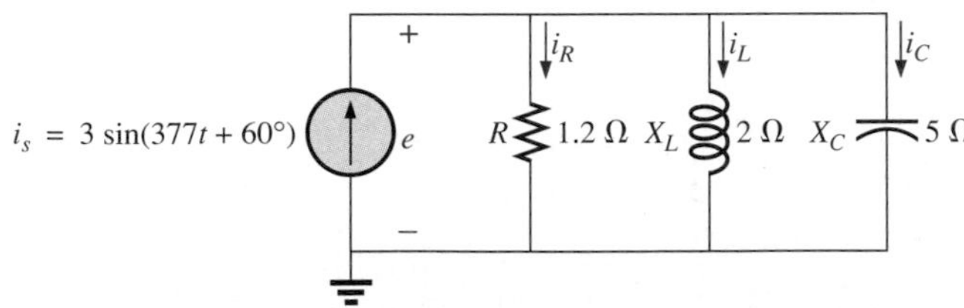

FIG. 15.141
Problems 30 and 49.

e. Draw the phasor diagram of the currents $\mathbf{I}_s$, $\mathbf{I}_R$, $\mathbf{I}_L$, and $\mathbf{I}_C$, and the voltage $\mathbf{E}$.

f. Verify Kirchhoff's current law at one node.

g. Find the average power delivered to the circuit.

h. Find the power factor of the circuit and indicate whether it is leading or lagging.

i. Find the sinusoidal expressions for the currents and voltage.

j. Plot the waveforms for the currents and voltage on the same set of axes.

31. Repeat Problem 30 for the circuit of Fig. 15.142.

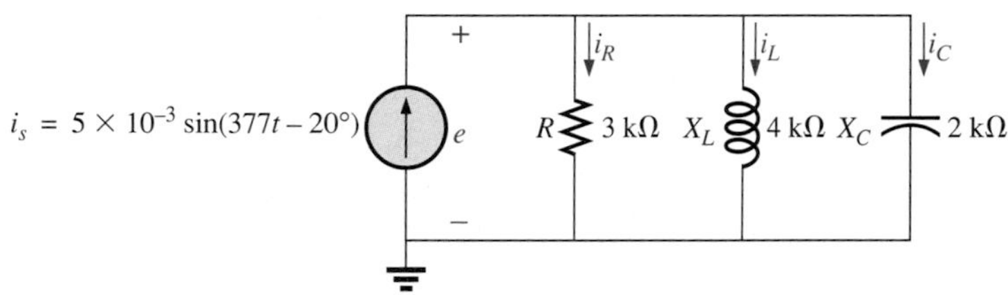

FIG. 15.142
Problem 31.

32. Repeat Problem 30 for the circuit of Fig. 15.143, replacing e with i_s in part (d).

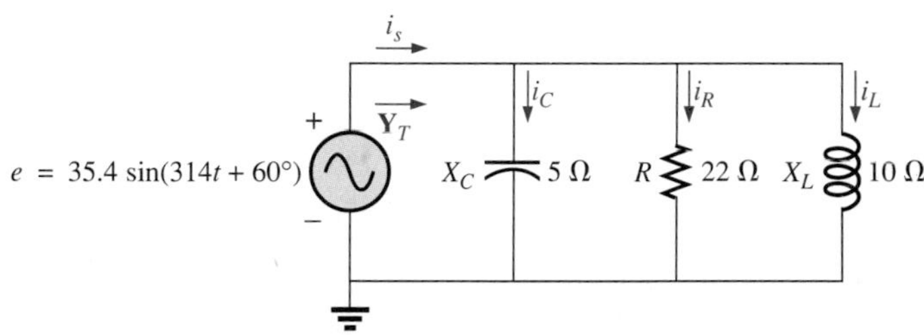

FIG. 15.143
Problem 32.

SECTION 15.9 Current Divider Rule

33. Calculate the currents $\mathbf{I}_1$ and $\mathbf{I}_2$ of Fig. 15.144 in phasor form using the current divider rule.

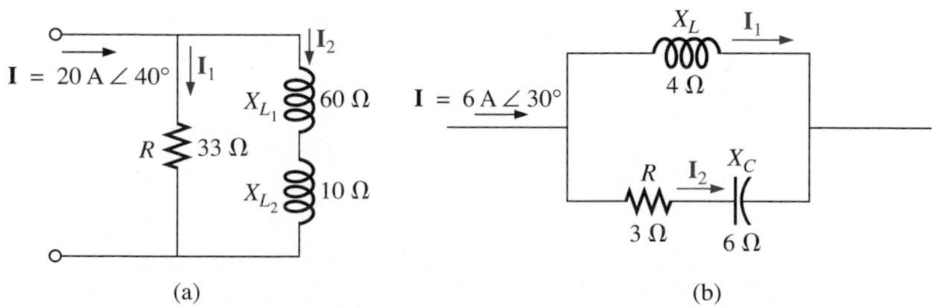

FIG. 15.144
Problem 33.

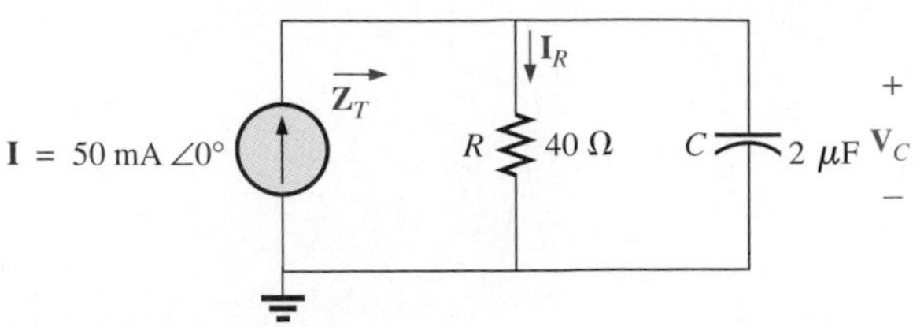

FIG. 15.145
Problems 34 and 36.

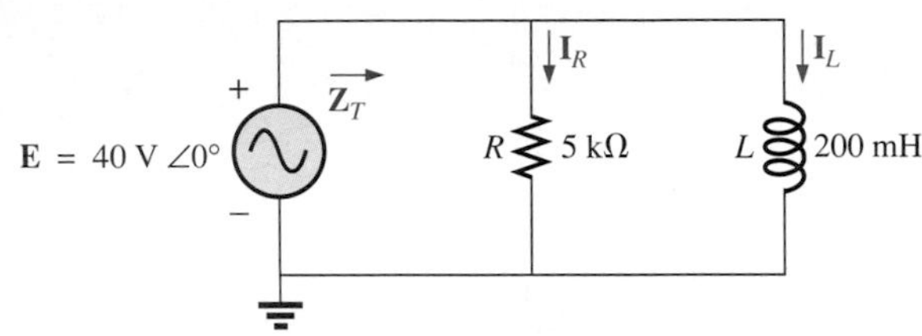

FIG. 15.146
Problems 35, 37, and 50.

SECTION 15.10 Frequency Response of the Parallel *R-L* Network

***34.** For the parallel *R-C* network of Fig. 15.145:
   a. Plot Z_T and θ_T versus frequency for a frequency range of zero to 20 kHz.
   b. Plot V_C versus frequency for the frequency range of part (a).
   c. Plot I_R versus frequency for the frequency range of part (a).

***35.** For the parallel *R-L* network of Fig. 15.146:
   a. Plot Z_T and θ_T versus frequency for a frequency range of zero to 10 kHz.
   b. Plot I_L versus frequency for the frequency range of part (a).
   c. Plot I_R versus frequency for the frequency range of part (a).

36. Plot Y_T and θ_T (of $\mathbf{Y}_T = Y_T \angle \theta_T$) for a frequency range of zero to 20 kHz for the network of Fig. 15.145.

37. Plot Y_T and θ_T (of $\mathbf{Y}_T = Y_T \angle \theta_T$) for a frequency range of zero to 10 kHz for the network of Fig. 15.146.

38. For the parallel *R-L-C* network of Fig. 15.147:
   a. Plot Y_T and θ_T (of $\mathbf{Y}_T = Y_T \angle \theta_T$) for a frequency range of zero to 20 kHz.
   b. Repeat part (a) for Z_T and θ_T (of $\mathbf{Z}_T = Z_T \angle \theta_T$).
   c. Plot V_C versus frequency for the frequency range of part (a).
   d. Plot I_L versus frequency for the frequency range of part (a).

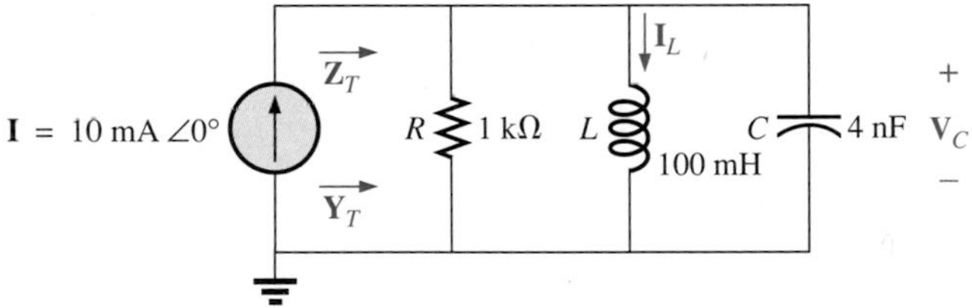

FIG. 15.147
Problem 38.

SECTION 15.12 Equivalent Circuits

39. For the series circuits of Fig. 15.148, find a parallel circuit that will have the same total impedance ($\mathbf{Z}_T$).

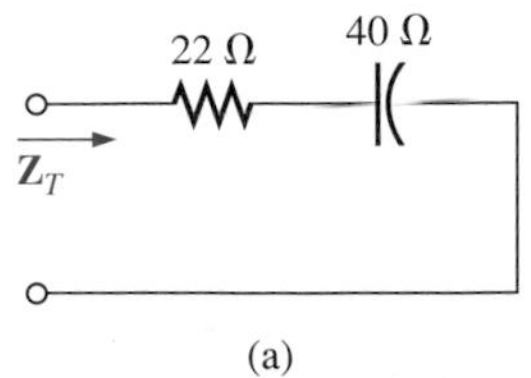

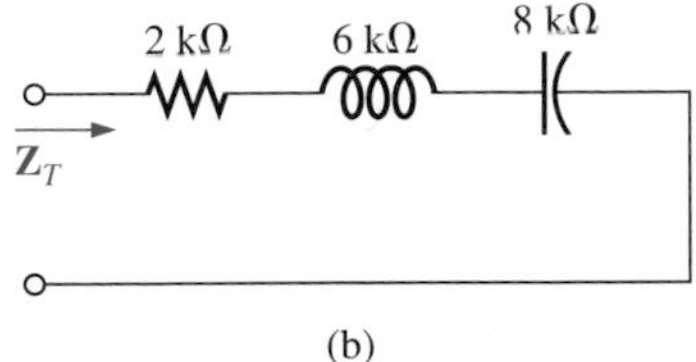

FIG. 15.148
Problem 39.

40. For the parallel circuits of Fig. 15.149, find a series circuit that will have the same total impedance.

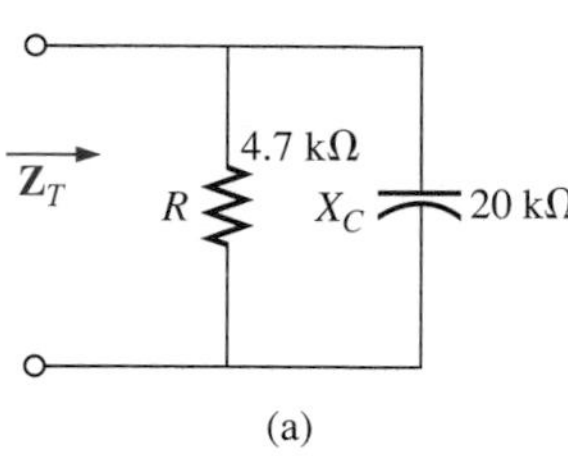

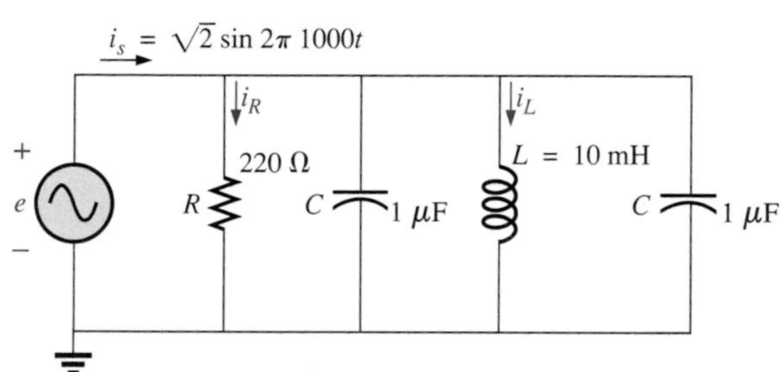

FIG. 15.149
Problem 40.

41. For the network of Fig. 15.150:

a. Calculate $\mathbf{E}$, $\mathbf{I}_R$, and $\mathbf{I}_L$ in phasor form.
b. Calculate the total power factor and indicate whether it is leading or lagging.
c. Calculate the average power delivered to the circuit.
d. Draw the admittance diagram.
e. Draw the phasor diagram of the currents $\mathbf{I}_s$, $\mathbf{I}_R$, and $\mathbf{I}_L$, and the voltage $\mathbf{E}$.
f. Find the current $\mathbf{I}_C$ for each capacitor using only Kirchhoff's current law.
g. Find the series circuit of one resistive and reactive element that will have the same impedance as the original circuit.

***42.** Repeat Problem 41 if the inductance is changed to 1 H.

FIG. 15.150
Problems 41 and 42.

43. Find the element or elements that must be in the closed container of Fig. 15.151 to satisfy the following conditions. (Find the simplest parallel circuit that will satisfy the indicated conditions.)

a. Average power to the circuit = 3000 W.
b. Circuit has a lagging power factor.

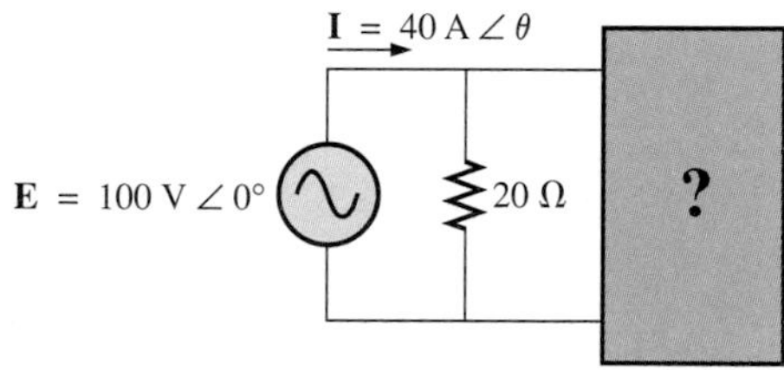

FIG. 15.151
Problem 43.

SECTION 15.13 Phase Measurements (Dual-Trace Oscilloscope)

44. For the circuit of Fig. 15.152, determine the phase relationship between the following using a dual-trace oscilloscope. The circuit can be reconstructed differently for each part but do not use sensing resistors. Show all connections on a redrawn diagram.

a. e and v_C
b. e and i_s
c. e and v_L

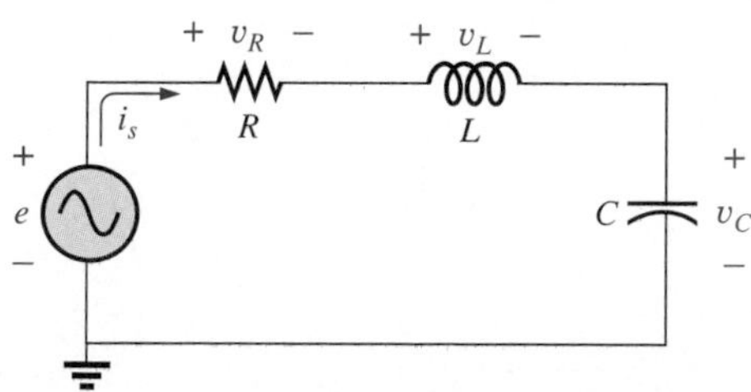

FIG. 15.152
Problem 44.

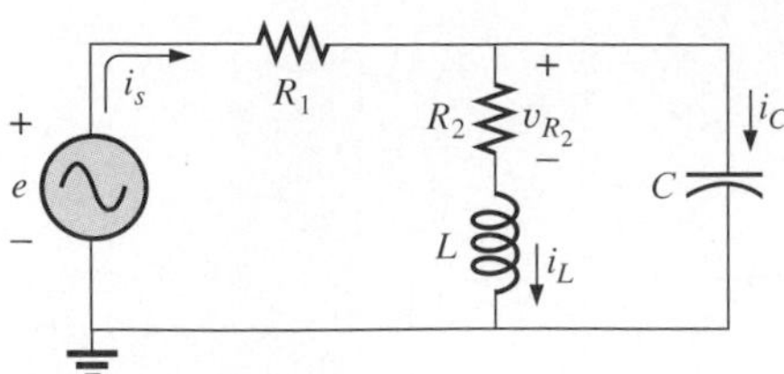

FIG. 15.153
Problem 45.

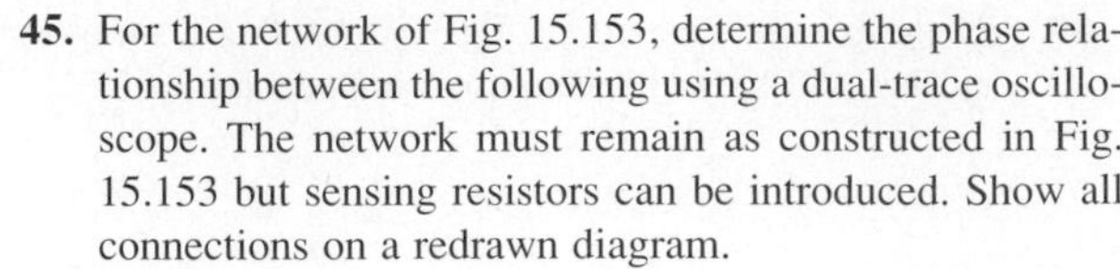

45. For the network of Fig. 15.153, determine the phase relationship between the following using a dual-trace oscilloscope. The network must remain as constructed in Fig. 15.153 but sensing resistors can be introduced. Show all connections on a redrawn diagram.

a. e and v_{R_2}
b. e and i_s
c. i_L and i_C

46. For the oscilloscope traces of Fig. 15.154:

a. Determine the phase relationship between the waveforms and indicate which one leads or lags.
b. Determine the peak-to-peak and rms values of each waveform.
c. Find the frequency of each waveform.

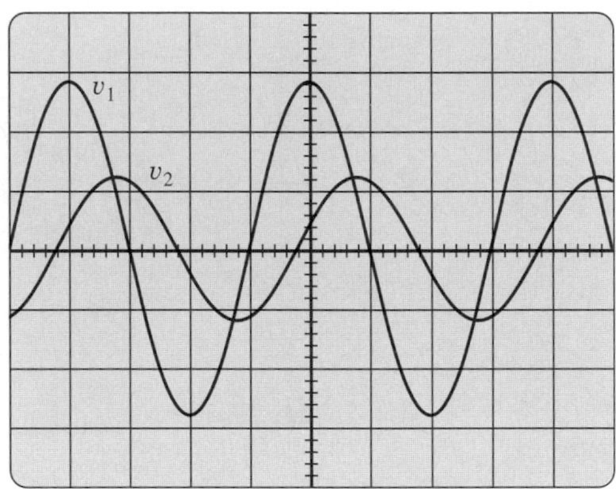

Vertical sensitivity = 0.5 V/div.
Horizontal sensitivity = 0.2 ms/div.

(I)

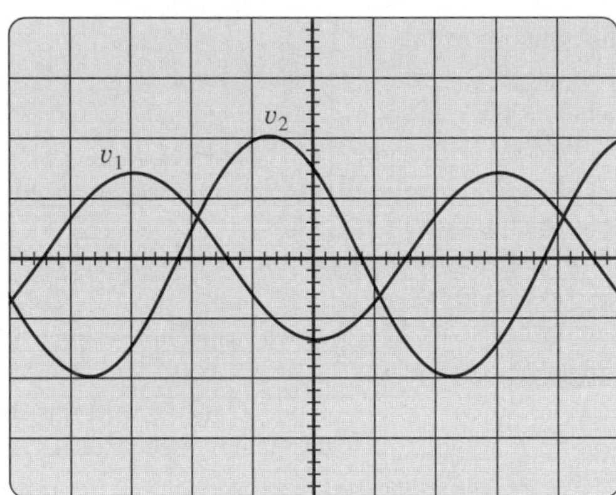

Vertical sensitivity = 2 V/div.
Horizontal sensitivity = 10 μs/div.

(II)

FIG. 15.154
Problem 46.

SECTION 15.14 Computer Analysis

PSpice (DOS)

47. Write the input file to calculate the magnitude and phase angle of $\mathbf{V}_R$, $\mathbf{V}_L$, and $\mathbf{I}$ for the network of Fig. 15.121. Use a frequency of 1 kHz to determine the appropriate value of L for the analysis.

***48.** For the network of Fig. 15.135:

a. Write the input file to print the magnitude and angle of $\mathbf{Z}_T$ for a frequency range of zero to 20 kHz in increments of 1 kHz.
b. Repeat part (a) for $\mathbf{I}_C$.

49. Write the input file to calculate the magnitude and phase angle of $\mathbf{I}_R$, $\mathbf{I}_L$, $\mathbf{I}_C$, and $\mathbf{E}$ for the network of Fig. 15.141 using any appropriate frequency to determine the value of L and C for the calculations.

***50.** For the network of Fig. 15.146:

a. Write the input file to print out the magnitude and angle of $\mathbf{Y}_T$ for a frequency range of zero to 20 kHz in increments of 500 Hz.
b. Repeat part (a) for $\mathbf{I}_R$.

PSpice (Windows)

51. For the network of Fig. 15.121 (use $f = 1$ kHz):
  a. Determine the rms values of the voltages $\mathbf{V}_R$ and $\mathbf{V}_L$ and the current **I** using the Probe AC response.
  b. Plot v_R, v_L, and i versus time on separate plots.
  c. Place e, v_R, v_L, and i on the same plot and label accordingly.

52. For the network of Fig. 15.140:
  a. Determine the rms values of the currents $\mathbf{I}_s$, $\mathbf{I}_R$, and $\mathbf{I}_L$ using the Probe AC response.
  b. Plot i_s, i_R, and i_L versus time on separate plots.
  c. Place e, i_s, i_R, and i_L on the same plot and label accordingly.

53. For the network of Fig. 15.123:
  a. Tabulate the impedance of the network at frequencies of 500 Hz to 10 kHz in increments of 500 Hz.
  b. Plot the impedance versus frequency for the frequency range 0 to 10 kHz.
  c. Plot the current i versus frequency for the frequency range 0 to 10 kHz.

***54.** For the network of Fig. 15.131:
  a. Find the rms values of the voltages v_R and v_C at a frequency of 1 kHz.
  b. Plot v_C versus frequency for the frequency range 0 to 10 kHz.
  c. Plot the phase angle between e and i for the frequency range 0 Hz to 10 kHz.

Programming Language (C++, BASIC, PASCAL, etc.)

55. Write a program to generate the sinusoidal expression for the current of a resistor, inductor, or capacitor given the value of R, L, or C and the applied voltage in sinusoidal form.

56. Given the impedance of each element in rectangular form, write a program to determine the total impedance in rectangular form of any number of series elements.

57. Given two phasors in polar form in the first quadrant, write a program to generate the sum of the two phasors in polar form.

GLOSSARY

Admittance A measure of how easily a network will "admit" the passage of current through that system. It is measured in siemens, abbreviated S, and is represented by the capital letter Y.

Admittance diagram A vector display that clearly depicts the magnitude of the admittance of the conductance, capacitive susceptance, and inductive susceptance, and the magnitude and angle of the total admittance of the system.

Current divider rule A method by which the current through either of two parallel branches can be determined in an ac network without first finding the voltage across the parallel branches.

Equivalent circuits For every series ac network there is a parallel ac network (and vice versa) that will be "equivalent" in the sense that the input current and impedance are the same.

Impedance diagram A vector display that clearly depicts the magnitude of the impedance of the resistive, reactive, and capacitive components of a network, and the magnitude and angle of the total impedance of the system.

Parallel ac circuits A connection of elements in an ac network in which all the elements have two points in common. The voltage is the same across each element.

Phasor diagram A vector display that provides at a glance the magnitude and phase relationships among the various voltages and currents of a network.

Series ac configuration A connection of elements in an ac network in which no two impedances have more than one terminal in common and the current is the same through each element.

Susceptance A measure of how "susceptible" an element is to the passage of current through it. It is measured in siemens, abbreviated S, and is represented by the capital letter B.

Voltage divider rule A method through which the voltage across one element of a series of elements in an ac network can be determined without first having to find the current through the elements.

16

Series-Parallel ac Networks

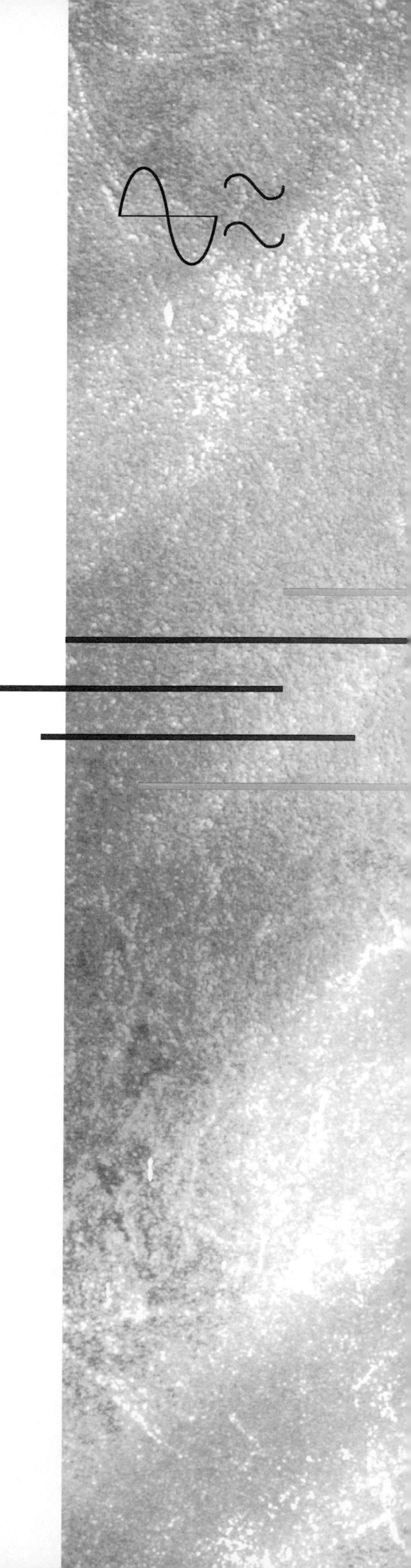

16.1 INTRODUCTION

In this chapter, we shall utilize the fundamental concepts of the previous chapter to develop a technique for solving series-parallel ac networks. A brief review of Chapter 7 may be helpful before considering these networks since the approach here will be quite similar to that undertaken earlier. The circuits to be discussed will have only one source of energy, either potential or current. Networks with two or more sources will be considered in Chapters 17 and 18, using methods previously described for dc circuits.

In general, when working with series-parallel ac networks, consider the following approach:

1. ***Redraw the network, employing block impedances to combine obvious series and parallel elements, which will reduce the network to one that clearly reveals the fundamental structure of the system.***
2. ***Study the problem and make a brief mental sketch of the overall approach you plan to use. Doing this may result in time- and energy-saving shortcuts. In some cases a lengthy, drawn out analysis may not be necessary. A single application of a fundamental law of circuit analysis may result in the desired solution.***
3. ***After the overall approach has been determined, it is usually best to consider each branch involved in your method independently before tying them together in series-parallel combinations. In most cases, work back from the obvious series and parallel combinations to the source to determine the total impedance of the network. The source current can then be determined, and the path back to specific unknowns can be defined. As you progress back to the source, continually define those unknowns that have not been lost in the reduction process. It will save time when you have to work back through the network to find specific quantities.***

4. ***When you have arrived at a solution, check to see that it is reasonable by considering the magnitudes of the energy source and the elements in the circuit. If not, either solve the network using another approach, or check over your work very carefully. At this point a computer solution can be an invaluable asset in the validation process.***

16.2 ILLUSTRATIVE EXAMPLES

EXAMPLE 16.1 For the network of Fig. 16.1:

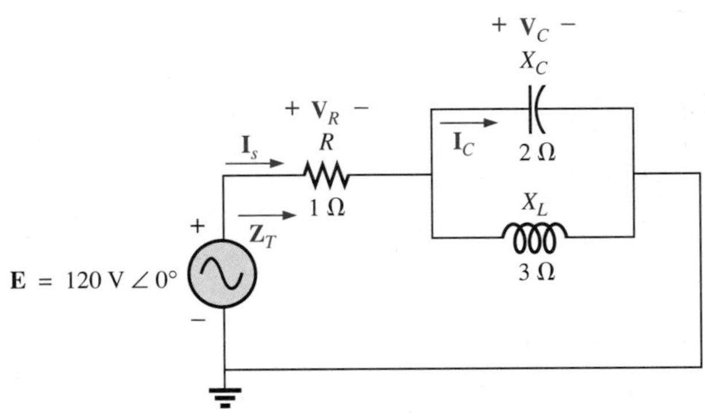

FIG. 16.1
Example 16.1.

a. Calculate $\mathbf{Z}_T$.
b. Determine $\mathbf{I}_s$.
c. Calculate $\mathbf{V}_R$ and $\mathbf{V}_C$.
d. Find $\mathbf{I}_C$.
e. Compute the power delivered.
f. Find F_p of the network.

Solutions:

a. As suggested in the introduction, the network has been redrawn with block impedances, as shown in Fig. 16.2. The impedance $\mathbf{Z}_1$ is simply the resistor R of 1 Ω, and $\mathbf{Z}_2$ is the parallel combination of X_C and X_L. The network now clearly reveals that it is fundamentally a series circuit suggesting a direct path toward the total impedance and the source current. As noted in the introduction, for many such problems you must first work back to the source to find the total impedance and then the source current. When the unknown quantities are found in terms of these subscripted impedances, the numerical values can then be substituted to find the magnitude and phase angle of the unknown. In other words, try to find the desired solution solely in terms of the subscripted impedances before substituting numbers. This approach will usually enhance the clarity of the chosen path toward a solution while saving time and preventing careless calculation errors. Note also in Fig. 16.2 that all the unknown quantities except $\mathbf{I}_C$ have been preserved, meaning that Fig. 16.2 can be used to determine these quantities rather than having to return to the more complex network of Fig. 16.1.

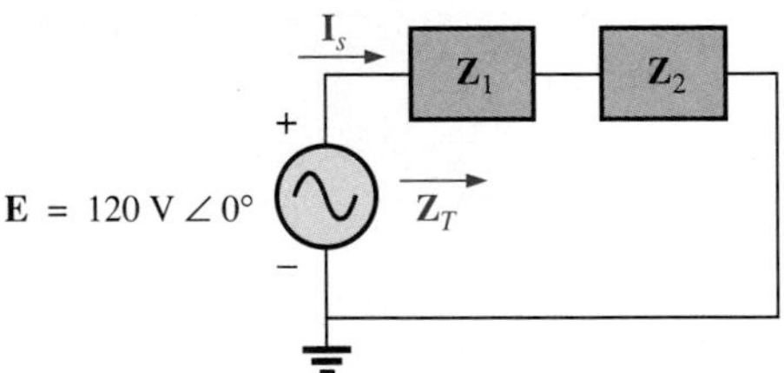

FIG. 16.2
Network of Fig. 16.1 after assigning the block impedances.

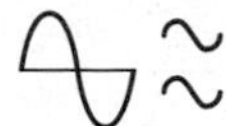

The total impedance is defined by

$$\mathbf{Z}_T = \mathbf{Z}_1 + \mathbf{Z}_2$$

with

$$\mathbf{Z}_1 = R \angle 0° = 1\ \Omega \angle 0°$$

$$\mathbf{Z}_2 = \mathbf{Z}_C \parallel \mathbf{Z}_L = \frac{(X_C \angle -90°)(X_L \angle 90°)}{-jX_C + jX_L} = \frac{(2\ \Omega \angle -90°)(3\ \Omega \angle 90°)}{-j2\ \Omega + j3\ \Omega}$$

$$= \frac{6\ \Omega \angle 0°}{j1} = \frac{6\ \Omega \angle 0°}{1 \angle 90°} = 6\ \Omega \angle -90°$$

and

$$\mathbf{Z}_T = \mathbf{Z}_1 + \mathbf{Z}_2 = 1\ \Omega - j6\ \Omega = \mathbf{6.08\ \Omega \angle -80.54°}$$

b. $\mathbf{I}_s = \dfrac{\mathbf{E}}{\mathbf{Z}_T} = \dfrac{120\text{ V} \angle 0°}{6.08\ \Omega \angle -80.5°} = \mathbf{19.74\ A \angle 80.54°}$

c. Referring to Fig. 16.2, we find that $\mathbf{V}_R$ and $\mathbf{V}_C$ can be found by a direct application of Ohm's law:

$$\mathbf{V}_R = \mathbf{I}_s\mathbf{Z}_1 = (19.74\text{ A} \angle 80.54°)(1\ \Omega \angle 0°) = \mathbf{19.74\ V \angle 80.54°}$$

$$\mathbf{V}_C = \mathbf{I}_s\mathbf{Z}_2 = (19.74\text{ A} \angle 80.54°)(6\ \Omega \angle -90°) = \mathbf{118.44\ V \angle -9.46°}$$

d. Now that $\mathbf{V}_C$ is known, the current $\mathbf{I}_C$ can also be found using Ohm's law.

$$\mathbf{I}_C = \frac{\mathbf{V}_C}{\mathbf{Z}_C} = \frac{118.44\text{ V} \angle -9.46°}{2\ \Omega \angle -90°} = \mathbf{59.22\ A \angle 80.54°}$$

e. $P_{\text{del}} = I_s^2R = (19.74\text{ A})^2(1\ \Omega) = \mathbf{389.67\ W}$

f. $F_p = \cos\theta = \cos 80.54° = \mathbf{0.164\ leading}$

The fact that the total impedance has a negative phase angle (revealing that $\mathbf{I}_s$ leads $\mathbf{E}$) is a clear indication that the network is capacitive in nature and therefore has a leading power factor. The fact that the network is capacitive can be determined from the original network by first realizing that, for the parallel *L-C* elements, the smaller impedance predominates and results in an *R-C* network.

EXAMPLE 16.2 For the network of Fig. 16.3:

a. If **I** is 50 A $\angle 30°$, calculate $\mathbf{I}_1$ using the current divider rule.
b. Repeat part (a) for $\mathbf{I}_2$.
c. Verify Kirchhoff's current law at one node.

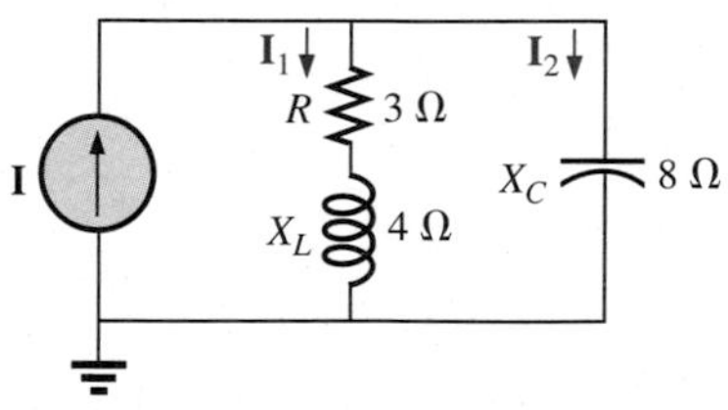

FIG. 16.3
Example 16.2.

Solutions:

a. Redrawing the circuit as in Fig. 16.4, we have

$$\mathbf{Z}_1 = R + jX_L = 3\ \Omega + j4\ \Omega = 5\ \Omega \angle 53.13°$$

$$\mathbf{Z}_2 = -jX_C = -j8\ \Omega = 8\ \Omega \angle -90°$$

Using the current divider rule yields

$$\mathbf{I}_1 = \frac{\mathbf{Z}_2\mathbf{I}}{\mathbf{Z}_2 + \mathbf{Z}_1} = \frac{(8\ \Omega \angle -90°)(50\text{ A} \angle 30°)}{(-j8\ \Omega) + (3\ \Omega + j4\ \Omega)} = \frac{400 \angle -60°}{3 - j4}$$

$$= \frac{400 \angle -60°}{5 \angle -53.13°} = \mathbf{80\ A \angle -6.87°}$$

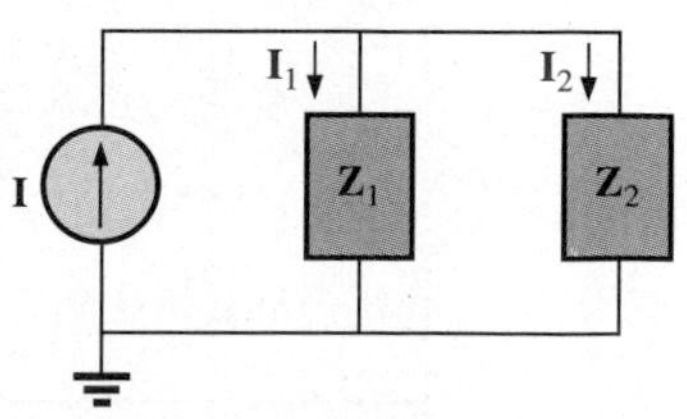

FIG. 16.4
Network of Fig. 16.3 after assigning the block impedances.

b. $\mathbf{I}_2 = \dfrac{\mathbf{Z}_1\mathbf{I}}{\mathbf{Z}_2 + \mathbf{Z}_1} = \dfrac{(5\ \Omega \angle 53.13°)(50\ \text{A} \angle 30°)}{5\ \Omega \angle -53.13°} = \dfrac{250 \angle 83.13°}{5 \angle -53.13°}$

$$= \mathbf{50\ A \angle 136.26°}$$

c.

$$\mathbf{I} = \mathbf{I}_1 + \mathbf{I}_2$$
$$50\ \text{A} \angle 30° = 80\ \text{A} \angle -6.87° + 50\ \text{A} \angle 136.26°$$
$$= (79.43 - j\,9.57) + (-36.12 + j\,34.57)$$
$$= 43.31 + j\,25.0$$
$$50\ \text{A} \angle 30° = 50\ \text{A} \angle 30° \qquad \text{(checks)}$$

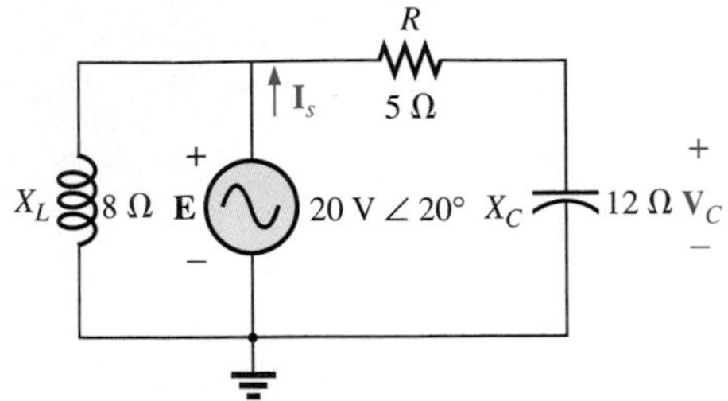

FIG. 16.5
Example 16.3.

EXAMPLE 16.3 For the network of Fig. 16.5:

a. Calculate the voltage $\mathbf{V}_C$ using the voltage divider rule.
b. Calculate the current $\mathbf{I}_s$.

Solutions:

a. The network is redrawn as shown in Fig. 16.6, with

$$\mathbf{Z}_1 = 5\ \Omega = 5\ \Omega \angle 0°$$
$$\mathbf{Z}_2 = -j\,12\ \Omega = 12\ \Omega \angle -90°$$
$$\mathbf{Z}_3 = +j\,8\ \Omega = 8\ \Omega \angle 90°$$

Since $\mathbf{V}_C$ is desired, we will not combine R and X_C into a single block impedance. Note also how Fig. 16.6 clearly reveals that $\mathbf{E}$ is the total voltage across the series combination of $\mathbf{Z}_1$ and $\mathbf{Z}_2$, permitting the use of the voltage divider rule to calculate $\mathbf{V}_C$. In addition, note that all the currents necessary to determine $\mathbf{I}_s$ have been preserved in Fig. 16.6, revealing that there is no need to ever return to the network of Fig. 16.5—everything is defined by Fig. 16.6.

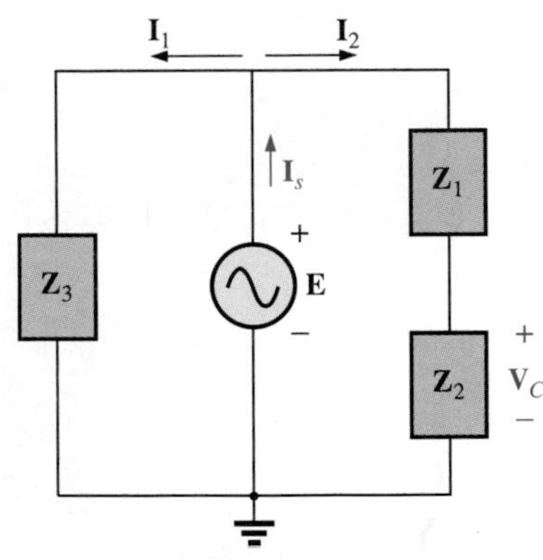

FIG. 16.6
Network of Fig. 16.5 after assigning the block impedances.

$$\mathbf{V}_C = \frac{\mathbf{Z}_2\mathbf{E}}{\mathbf{Z}_1 + \mathbf{Z}_2} = \frac{(12\ \Omega \angle -90°)(20\ \text{V} \angle 20°)}{5\ \Omega - j\,12\ \Omega} = \frac{240\ \text{V} \angle -70°}{13 \angle -67.38°}$$
$$= \mathbf{18.46\ V \angle -2.62°}$$

b. $\mathbf{I}_1 = \dfrac{\mathbf{E}}{\mathbf{Z}_3} = \dfrac{20\ \text{V} \angle 20°}{8\ \Omega \angle 90°} = 2.5\ \text{A} \angle -70°$

$$\mathbf{I}_2 = \frac{\mathbf{E}}{\mathbf{Z}_1 + \mathbf{Z}_2} = \frac{20\ \text{V} \angle 20°}{13\ \Omega \angle -67.38°} = 1.54\ \text{A} \angle 87.38°$$

and

$$\mathbf{I}_s = \mathbf{I}_1 + \mathbf{I}_2$$
$$= 2.5\ \text{A} \angle -70° + 1.54\ \text{A} \angle 87.38°$$
$$= (0.86 - j\,2.35) + (0.07 + j\,1.54)$$
$$\mathbf{I}_s = 0.93 - j\,0.81 = \mathbf{1.23\ A \angle -41.05°}$$

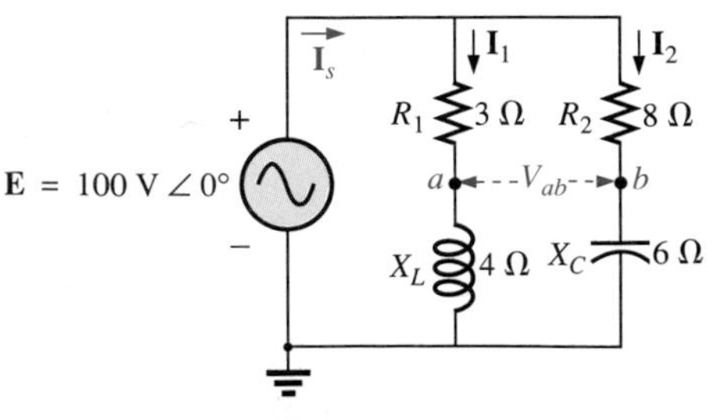

FIG. 16.7
Example 16.4.

EXAMPLE 16.4 For Fig. 16.7:

a. Calculate the current $\mathbf{I}_s$.
b. Find the voltage $\mathbf{V}_{ab}$.

Solutions:

a. Redrawing the circuit as in Fig. 16.8, we obtain

$$\mathbf{Z}_1 = R_1 + j\,X_L = 3\ \Omega + j\,4\ \Omega = 5\ \Omega \angle 53.13°$$
$$\mathbf{Z}_2 = R_2 - j\,X_C = 8\ \Omega - j\,6\ \Omega = 10\ \Omega \angle -36.87°$$

In this case the voltage $\mathbf{V}_{ab}$ is lost in the redrawn network, but the currents $\mathbf{I}_1$ and $\mathbf{I}_2$ remain defined for future calculations neces-

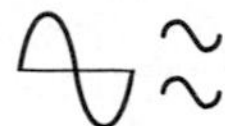

sary to determine $\mathbf{V}_{ab}$. Figure 16.8 clearly reveals that the total impedance can be found using the equation for two parallel impedances:

$$\mathbf{Z}_T = \frac{\mathbf{Z}_1\mathbf{Z}_2}{\mathbf{Z}_1 + \mathbf{Z}_2} = \frac{(5\ \Omega\ \angle 53.13°)(10\ \Omega\ \angle -36.87°)}{3\ \Omega + j\,4\ \Omega) + (8\ \Omega - j\,6\ \Omega)}$$

$$= \frac{50\ \Omega\ \angle 16.26°}{11 - j\,2} = \frac{50\ \Omega\ \angle 16.26°}{11.18\ \angle -10.30°}$$

$$= \mathbf{4.472\ \Omega\ \angle 26.56°}$$

and $\mathbf{I}_s = \dfrac{\mathbf{E}}{\mathbf{Z}_T} = \dfrac{100\ \text{V}\ \angle 0°}{4.472\ \Omega\ \angle 26.56°} = \mathbf{22.36\ A\ \angle -26.56°}$

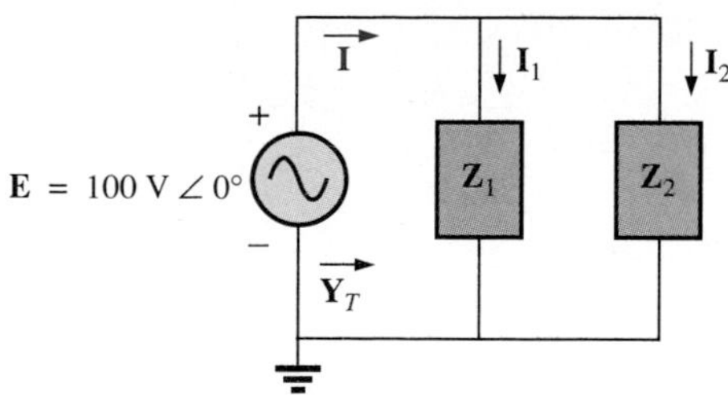

FIG. 16.8
Network of Fig. 16.7 after assigning the block impedances.

b. By Ohm's law,

$$\mathbf{I}_1 = \frac{\mathbf{E}}{\mathbf{Z}_1} = \frac{100\ \text{V}\ \angle 0°}{5\ \Omega\ \angle 53.13°} = \mathbf{20\ A\ \angle -53.13°}$$

$$\mathbf{I}_2 = \frac{\mathbf{E}}{\mathbf{Z}_2} = \frac{100\ \text{V}\ \angle 0°}{10\ \Omega\ \angle -36.87°} = \mathbf{10\ A\ \angle 36.87°}$$

Returning to Fig. 16.7, we have

$$\mathbf{V}_{R_1} = \mathbf{I}_1\mathbf{Z}_{R_1} = (20\ \text{A}\ \angle -53.13°)(3\ \Omega\ \angle 0°) = \mathbf{60\ V\ \angle -53.13°}$$

$$\mathbf{V}_{R_2} = \mathbf{I}_1\mathbf{Z}_{R_2} = (10\ \text{A}\ \angle +36.87°)(8\ \Omega\ \angle 0°) = \mathbf{80\ V\ \angle +36.87°}$$

Instead of using the two steps just shown, $\mathbf{V}_{R_1}$ or $\mathbf{V}_{R_2}$ could have been determined in one step using the voltage divider rule:

$$\mathbf{V}_{R_1} = \frac{(3\ \Omega\ \angle 0°)(100\ \text{V}\ \angle 0°)}{3\ \Omega\ \angle 0° + 4\ \Omega\ \angle 90°} = \frac{300\ \text{V}\ \angle 0°}{5\ \angle 53.13°} = \mathbf{60\ V\ \angle -53.13°}$$

To find $\mathbf{V}_{ab}$, Kirchhoff's voltage law must be applied around the loop (Fig.16.9) consisting of the 3-Ω and 8-Ω resistors. By Kirchhoff's voltage law,

$$\mathbf{V}_{ab} + \mathbf{V}_{R_1} - \mathbf{V}_{R_2} = 0$$

or

$$\begin{aligned}\mathbf{V}_{ab} &= \mathbf{V}_{R_2} - \mathbf{V}_{R_1}\\ &= 80\ \text{V}\ \angle 36.87° - 60\ \text{V}\ \angle -53.13°\\ &= (64 + j\,48) - (36 - j\,48)\\ &= 28 + j\,96\\ \mathbf{V}_{ab} &= \mathbf{100\ V\ \angle 73.74°}\end{aligned}$$

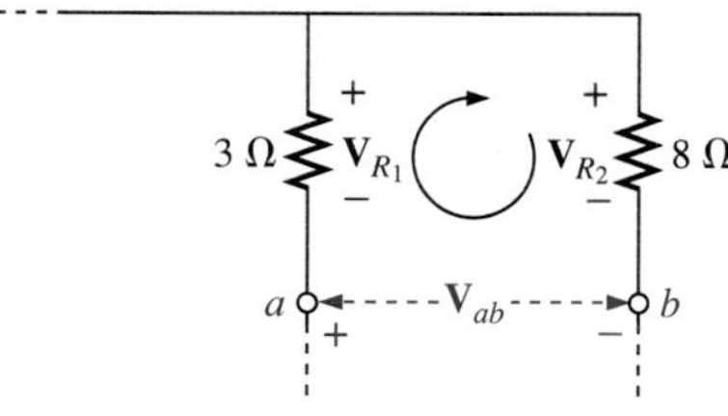

FIG. 16.9
Determining the voltage V_{ab} for the network of Fig. 16.7.

EXAMPLE 16.5 The network of Fig. 16.10 is frequently encountered in the analysis of transistor networks. The transistor equivalent circuit includes a current source **I** and an output impedance R_o. The resistor R_C is a biasing resistor to establish specific dc conditions, and the resistor R_i represents the loading of the next stage. The coupling capacitor is designed to be an open circuit for dc and as low an impedance as possible for the frequencies of interest to ensure that $\mathbf{V}_L$ is a maximum value. The frequency range of the example includes the entire audio (hearing) spectrum from 100 Hz to 20 kHz. The purpose of the example is to demonstrate that, for the full audio range, the effect of the capacitor can be ignored. It performs its function as a dc blocking agent but permits the ac to pass through with little disturbance.

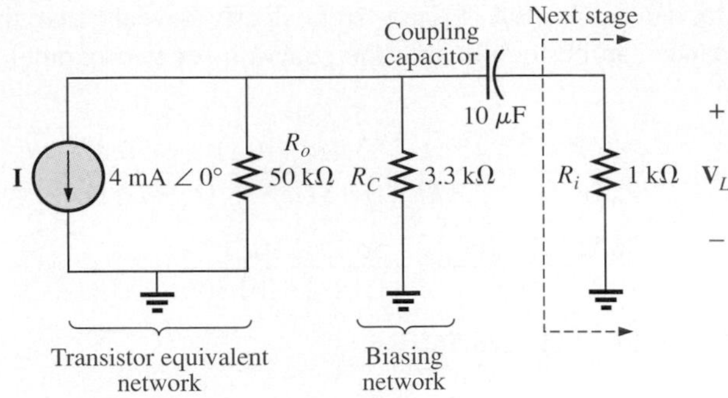

FIG. 16.10
Basic transistor amplifier.

a. Determine $\mathbf{V}_L$ for the network of Fig. 16.10 at a frequency of 100 Hz.
b. Repeat part (a) at a frequency of 20 kHz.
c. Compare the results of parts (a) and (b).

Solutions:

a. The network is redrawn with subscripted impedances in Fig. 16.11.

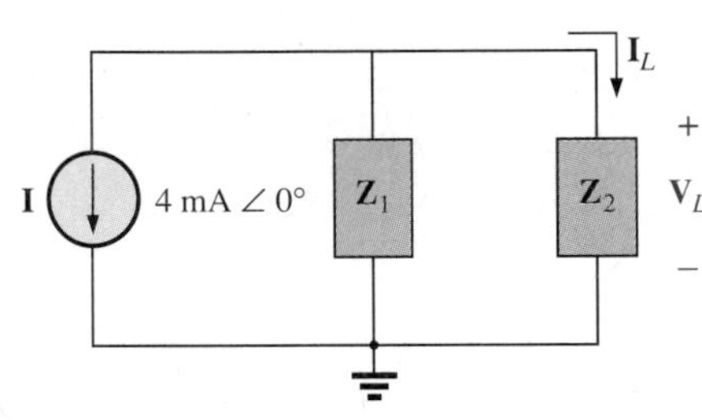

FIG. 16.11
Network of Fig. 16.10 following the assignment of the block impedances.

$$\mathbf{Z}_1 = 50\ \text{k}\Omega\ \angle 0° \parallel 3.3\ \text{k}\Omega\ \angle 0° = 3.096\ \text{k}\Omega\ \angle 0°$$
$$\mathbf{Z}_2 = R_i - j\,X_C$$

At $f = 100$ Hz: $X_C = \dfrac{1}{2\pi fC} = \dfrac{1}{2\pi(100\ \text{Hz})(10\ \mu\text{F})} = 159.16\ \Omega$

and $\mathbf{Z}_2 = 1\ \text{k}\Omega - j\,159.16\ \Omega$

Current divider rule:

$$\begin{aligned}\mathbf{I}_L &= \frac{\mathbf{Z}_1\mathbf{I}}{\mathbf{Z}_1 + \mathbf{Z}_2} = \frac{(3.096\ \text{k}\Omega\ \angle 0°)(4\ \text{mA}\ \angle 0°)}{3.096\ \text{k}\Omega + 1\ \text{k}\Omega - j\,159.16\ \Omega}\\ &= \frac{12.384\ \text{A}\ \angle 0°}{4096 - j\,159.16} = \frac{12.384\ \text{A}\ \angle 0°}{4099\ \angle -2.225°}\\ &= 3.021\ \text{mA}\ \angle 2.225°\end{aligned}$$

and
$$\begin{aligned}\mathbf{V}_L &= \mathbf{I}_L\mathbf{Z}_R\\ &= (3.021\ \text{mA}\ \angle 2.225°)(1\ \text{k}\Omega\ \angle 0°)\\ &= \mathbf{3.021\ V\ \angle 2.225°}\end{aligned}$$

b. At $f = 20$ kHz: $X_C = \dfrac{1}{2\pi fC} = \dfrac{1}{2\pi(20\ \text{kHz})(10\ \mu\text{F})} = 0.796\ \Omega$

Note the dramatic change in X_C with frequency. Obviously, the higher the frequency, the better the short-circuit approximation for X_C for ac conditions.

$$\mathbf{Z}_2 = 1\ \text{k}\Omega - j\,0.796\ \Omega$$

Current divider rule:

$$\begin{aligned}\mathbf{I}_L &= \frac{\mathbf{Z}_1\mathbf{I}}{\mathbf{Z}_1 + \mathbf{Z}_2} = \frac{(3.096\ \text{k}\Omega\ \angle 0°)(4\ \text{mA}\ \angle 0°)}{3.096\ \text{k}\Omega + 1\ \text{k}\Omega - j\,0.796\ \Omega}\\ &= \frac{12.384\ \text{A}\ \angle 0°}{4096 - j\,0.796\ \Omega} = \frac{12.384\ \text{A}\ \angle 0°}{4096\ \angle -0.011°}\\ &= 3.023\ \text{mA}\ \angle 0.011°\end{aligned}$$

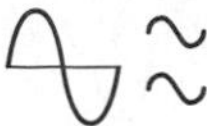

and $$\mathbf{V}_L = \mathbf{I}_L\mathbf{Z}_R$$
$$= (3.023 \text{ mA} \angle 0.011°)(1 \text{ k}\Omega \angle 0°)$$
$$= \mathbf{3.023 \text{ V} \angle 0.011°}$$

c. The results clearly indicate that the capacitor had little effect on the frequencies of interest. In addition, note that most of the supply current reached the load for the typical parameters employed.

EXAMPLE 16.6 For the network of Fig. 16.12:

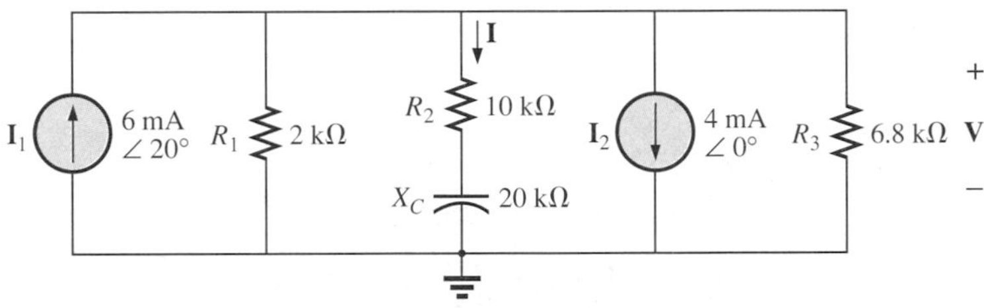

FIG. 16.12
Example 16.6.

a. Determine the current **I**.
b. Find the voltage **V**.

Solutions:

a. The rules for parallel current sources are the same for dc and ac networks. That is, the equivalent current source is their sum or difference (as phasors). Therefore,

$$\mathbf{I}_T = 6 \text{ mA} \angle 20° - 4 \text{ mA} \angle 0°$$
$$= 5.638 \text{ mA} + j\,2.052 \text{ mA} - 4 \text{ mA}$$
$$= 1.638 \text{ mA} + j\,2.052 \text{ mA}$$
$$= 2.626 \text{ mA} \angle 51.402°$$

Redrawing the network using block impedances will result in the network of Fig. 16.13 where

$$\mathbf{Z}_1 = 2 \text{ k}\Omega \angle 0° \parallel 6.8 \text{ k}\Omega \angle 0° = 1.545 \text{ k}\Omega \angle 0°$$

and $$\mathbf{Z}_2 = 10 \text{ k}\Omega - j\,20 \text{ k}\Omega = 22.361 \text{ k}\Omega \angle -63.435°$$

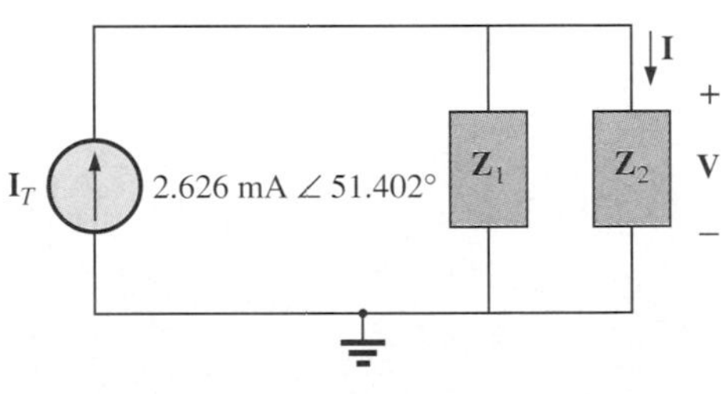

FIG. 16.13
Network of Fig. 16.12 following the assignment of the subscripted impedances.

Note that **I** and **V** are still defined in Fig. 16.13.

Current divider rule:

$$\mathbf{I} = \frac{\mathbf{Z}_1\mathbf{I}_T}{\mathbf{Z}_1 + \mathbf{Z}_2} = \frac{(1.545 \text{ k}\Omega \angle 0°)(2.626 \text{ mA} \angle 51.402°)}{1.545 \text{ k}\Omega + 10 \text{ k}\Omega - j\,20 \text{ k}\Omega}$$
$$= \frac{4.057 \text{ A} \angle 51.402°}{11.545 \times 10^3 - j\,20 \times 10^3} = \frac{4.057 \text{ A} \angle 51.402°}{23.093 \times 10^3 \angle -60.004°}$$
$$= \mathbf{0.176 \text{ mA} \angle 111.406°}$$

b. $$\mathbf{V} = \mathbf{I}\mathbf{Z}_2$$
$$= (0.176 \text{ mA} \angle 111.406°)(22.36 \text{ k}\Omega \angle -63.435°)$$
$$= \mathbf{3.936 \text{ V} \angle 47.971°}$$

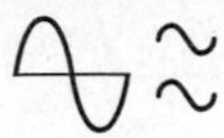

EXAMPLE 16.7 For the network of Fig. 16.14:

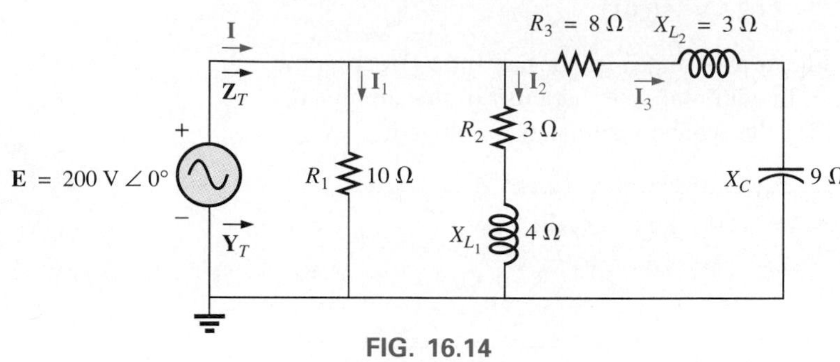

FIG. 16.14
Example 16.7.

a. Compute **I**.
b. Find $\mathbf{I}_1$, $\mathbf{I}_2$, and $\mathbf{I}_3$.
c. Verify Kirchhoff's current law by showing that

$$\mathbf{I} = \mathbf{I}_1 + \mathbf{I}_2 + \mathbf{I}_3$$

d. Find the total impedance of the circuit.

Solutions:

a. Redrawing the circuit as in Fig. 16.15 reveals a strictly parallel network where

$$\mathbf{Z}_1 = R_1 = 10\ \Omega \angle 0°$$
$$\mathbf{Z}_2 = R_2 + jX_{L_1} = 3\ \Omega + j\,4\ \Omega$$
$$\mathbf{Z}_3 = R_3 + jX_{L_2} - jX_C = 8\ \Omega + j\,3\ \Omega - j\,9\ \Omega = 8\ \Omega - j\,6\ \Omega$$

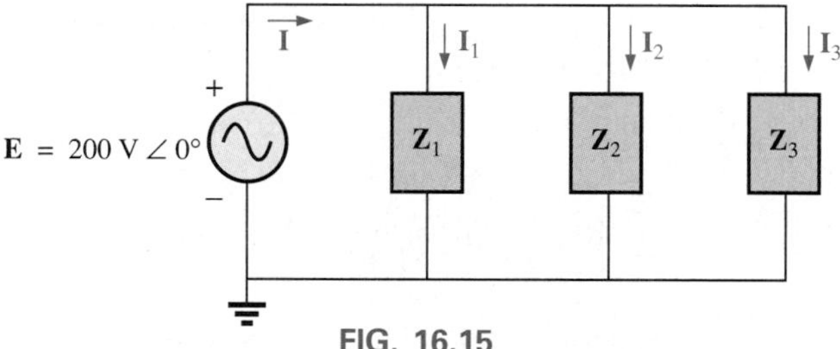

FIG. 16.15
Network of Fig. 16.14 following the assignment of the subscripted impedances.

The total admittance is

$$\begin{aligned}\mathbf{Y}_T &= \mathbf{Y}_1 + \mathbf{Y}_2 + \mathbf{Y}_3\\ &= \frac{1}{\mathbf{Z}_1} + \frac{1}{\mathbf{Z}_2} + \frac{1}{\mathbf{Z}_3} = \frac{1}{10\ \Omega} + \frac{1}{3\ \Omega + j\,4\ \Omega} + \frac{1}{8\ \Omega - j\,6\ \Omega}\\ &= 0.1\ \text{S} + \frac{1}{5\ \Omega \angle 53.13°} + \frac{1}{10\ \Omega \angle -36.87°}\\ &= 0.1\ \text{S} + 0.2\ \text{S} \angle -53.13° + 0.1\ \text{S} \angle 36.87°\\ &= 0.1\ \text{S} + 0.12\ \text{S} - j\,0.16\ \text{S} + 0.08\ \text{S} + j\,0.06\ \text{S}\\ &= 0.3\ \text{S} - j\,0.1\ \text{S} = 0.316\ \text{S} \angle -18.435°\end{aligned}$$

Calculator The above mathematical exercise presents an excellent opportunity to demonstrate the power of some of today's calculators. Using the TI-85, the above operation would appear as follows on the display:

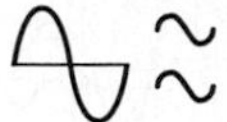


```
1/(10,0)+1/(3,4)+1/(8,−6)
```

with the result:

```
(300.000E−3,−100.000E−3)
```

Converting to polar form:

```
Ans ▶ Pol
(316.228E−3∠−18.435E0)
```

The current **I**:

$$\mathbf{I} = \mathbf{E}\mathbf{Y}_T = (200 \text{ V} \angle 0°)(0.316 \text{ S} \angle -18.435°)$$
$$= \mathbf{63.2\ A \angle -18.435°}$$

b. Since the voltage is the same across parallel branches,

$$\mathbf{I}_1 = \frac{\mathbf{E}}{\mathbf{Z}_1} = \frac{200 \text{ V} \angle 0°}{10\ \Omega \angle 0°} = \mathbf{20\ A \angle 0°}$$

$$\mathbf{I}_2 = \frac{\mathbf{E}}{\mathbf{Z}_2} = \frac{200 \text{ V} \angle 0°}{5\ \Omega \angle 53.13°} = \mathbf{40\ A \angle -53.13°}$$

$$\mathbf{I}_3 = \frac{\mathbf{E}}{\mathbf{Z}_3} = \frac{200 \text{ V} \angle 0°}{10\ \Omega \angle -36.87°} = \mathbf{20\ A \angle +36.87°}$$

c. $\mathbf{I} = \mathbf{I}_1 + \mathbf{I}_2 + \mathbf{I}_3$

$$60 - j\,20 = 20 \angle 0° + 40 \angle -53.13° + 20 \angle +36.87°$$
$$= (20 + j\,0) + (24 - j\,32) + (16 + j\,12)$$
$$60 - j\,20 = 60 - j\,20 \qquad \text{(checks)}$$

d. $\mathbf{Z}_T = \dfrac{1}{\mathbf{Y}_T} = \dfrac{1}{0.316 \text{ S} \angle -18.435°}$

$$= \mathbf{3.165\ \Omega \angle 18.435°}$$

EXAMPLE 16.8 For the network of Fig. 16.16:

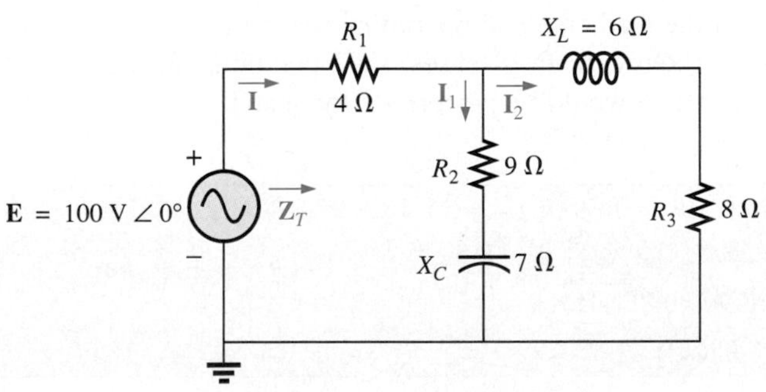

FIG. 16.16
Example 16.8.

a. Calculate the total impedance $\mathbf{Z}_T$.
b. Compute **I**.
c. Find the total power factor.
d. Calculate $\mathbf{I}_1$ and $\mathbf{I}_2$.
e. Find the average power delivered to the circuit.

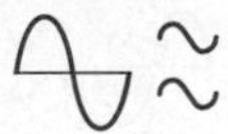

Solutions:

a. Redrawing the circuit as in Fig. 16.17, we have

$$\mathbf{Z}_1 = R_1 = 4\ \Omega \angle 0°$$
$$\mathbf{Z}_2 = R_2 - jX_C = 9\ \Omega - j7\ \Omega = 11.40\ \Omega \angle -37.87°$$
$$\mathbf{Z}_3 = R_3 + jX_L = 8\ \Omega + j6\ \Omega = 10\ \Omega \angle +36.87°$$

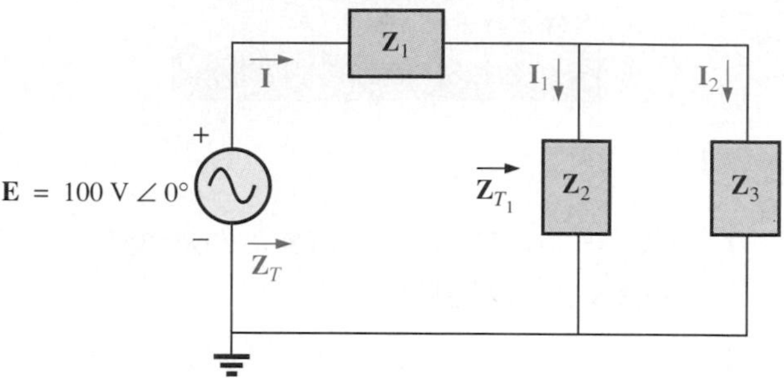

FIG. 16.17
Network of Fig. 16.16 following the assignment of the subscripted impedances.

Notice that all the desired quantities were conserved in the redrawn network. The total impedance:

$$\mathbf{Z}_T = \mathbf{Z}_1 + \mathbf{Z}_{T_1}$$
$$= \mathbf{Z}_1 + \frac{\mathbf{Z}_2\mathbf{Z}_3}{\mathbf{Z}_2 + \mathbf{Z}_3}$$
$$= \frac{4\ \Omega + (11.4\ \Omega \angle -37.87°)(10\ \Omega \angle 36.87°)}{(9\ \Omega - j7\ \Omega) + (8\ \Omega + j6\ \Omega)}$$
$$= 4\ \Omega + \frac{114\ \Omega \angle -1.00°}{17.03 \angle -3.37°} = 4\ \Omega + 6.69\ \Omega \angle 2.37°$$
$$= 4\ \Omega + 6.68\ \Omega + j0.28\ \Omega = 10.68\ \Omega + j0.28\ \Omega$$
$$\mathbf{Z}_T = \mathbf{10.684\ \Omega \angle 1.5°}$$

Calculator Another opportunity to demonstrate the versatility of the calculator! For the above operation, however, one must be aware of the priority of the mathematical operations, as demonstrated in the calculator display below. In most cases, the operations are performed in the same order they would be performed longhand.

```
(4,0)+((9,−7)+(8,6))^−1* (11.4∠−37.87)(10∠36.87) (ENTER)
(10.689E0,276.413E−3)
Ans ► Pol(ENTER)
(10.692E0∠1.481E0)
```

b. $\mathbf{I} = \frac{\mathbf{E}}{\mathbf{Z}_T} = \frac{100\ \text{V} \angle 0°}{10.684\ \Omega \angle 1.5°} = \mathbf{9.36\ A \angle -1.5°}$

c. $F_p = \cos\theta_T = \frac{R}{Z_T} = \frac{10.68\ \Omega}{10.684\ \Omega} \cong \mathbf{1}$

(essentially resistive, which is interesting, considering the complexity of the network)

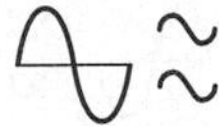

d. Current divider rule:

$$\mathbf{I}_2 = \frac{\mathbf{Z}_2\mathbf{I}}{\mathbf{Z}_2 + \mathbf{Z}_3} = \frac{(11.40\ \Omega \angle -37.87°)(9.36\ \text{A} \angle -1.5°)}{(9\ \Omega - j\,7\ \Omega) + (8\ \Omega + j\,6\ \Omega)}$$

$$= \frac{106.7\ \text{A} \angle -39.37°}{17 - j\,1} = \frac{106.7\ \text{A} \angle -39.37°}{17.03 \angle -3.37°}$$

$$\mathbf{I}_2 = \mathbf{6.27\ A \angle -36°}$$

Applying Kirchhoff's current law (rather than another application of the current divider rule) yields

$$\mathbf{I}_1 = \mathbf{I} - \mathbf{I}_2$$

or

$$\begin{aligned} \mathbf{I} &= \mathbf{I}_1 - \mathbf{I}_2 \\ &= (9.36\ \text{A} \angle -1.5°) - (6.27\ \text{A} \angle -36°) \\ &= (9.36\ \text{A} - j\,0.25\ \text{A}) - (5.07\ \text{A} - j\,3.69\ \text{A}) \\ \mathbf{I}_1 &= 4.29\ \text{A} + j\,3.44\ \text{A} = \mathbf{5.5\ A \angle 38.72°} \end{aligned}$$

e.
$$\begin{aligned} P_T &= EI \cos \theta_T \\ &= (100\ \text{V})(9.36\ \text{A}) \cos 1.5° \\ &= (936)(0.99966) \\ P_T &= \mathbf{935.68\ W} \end{aligned}$$

16.3 LADDER NETWORKS

Ladder networks were discussed in some detail in Chapter 7. This section will simply apply the first method described in Section 7.3 to the general sinusoidal ac ladder network of Fig. 16.18. The current $\mathbf{I}_6$ is desired.

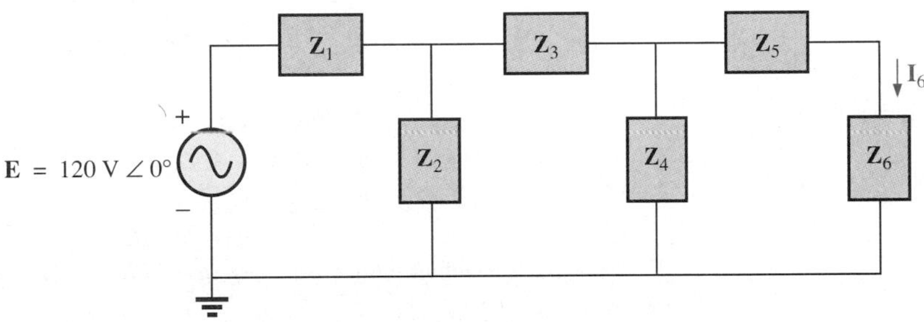

FIG. 16.18
Ladder network.

Impedances $\mathbf{Z}_T$, $\mathbf{Z}'_T$, and $\mathbf{Z}''_T$ and currents $\mathbf{I}_1$ and $\mathbf{I}_3$ are defined in Fig. 16.19:

$$\mathbf{Z}''_T = \mathbf{Z}_5 + \mathbf{Z}_6$$

and

$$\mathbf{Z}'_T = \mathbf{Z}_3 + \mathbf{Z}_4 \parallel \mathbf{Z}''_T$$

with

$$\mathbf{Z}_T = \mathbf{Z}_1 + \mathbf{Z}_2 \parallel \mathbf{Z}'_T$$

Then

$$\mathbf{I} = \frac{\mathbf{E}}{\mathbf{Z}_T}$$

and

$$\mathbf{I}_3 = \frac{\mathbf{Z}_2\mathbf{I}}{\mathbf{Z}_2 + \mathbf{Z}'_T}$$

with

$$\mathbf{I}_6 = \frac{\mathbf{Z}_4\mathbf{I}_3}{\mathbf{Z}_4 + \mathbf{Z}''_T}$$

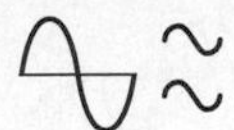

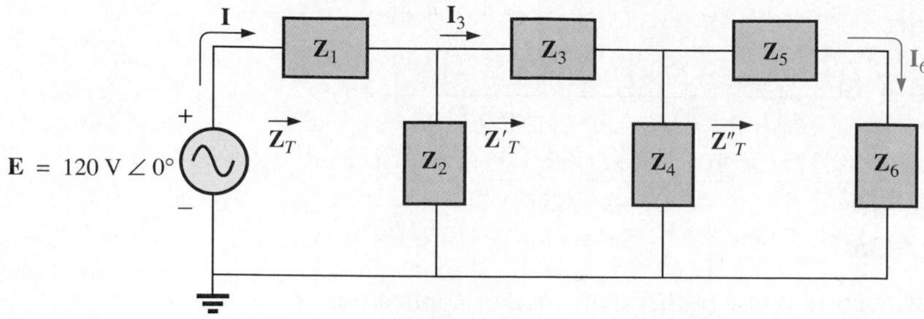

FIG. 16.19
Defining an approach to the analysis of ladder networks.

16.4 COMPUTER ANALYSIS

PSpice (DOS)

The first run is for the network of Fig. 16.20, which appeared as Example 16.4. Review the number of steps required to determine the requested unknown quantities in the example versus the time required to obtain a solution with PSpice—a total of 9.88 s once the input file was entered!

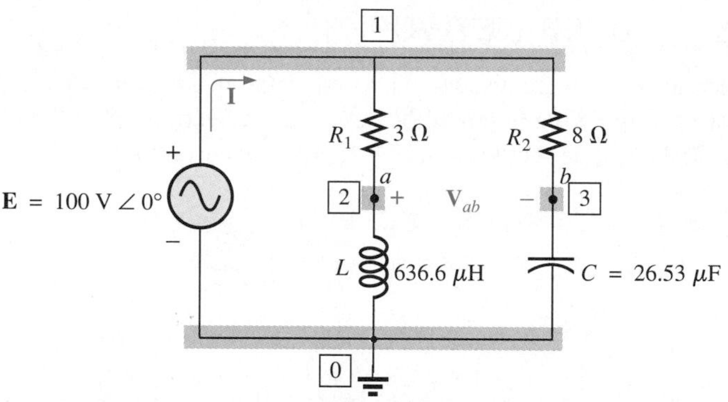

FIG. 16.20
Defining the parameters for a PSpice (DOS) analysis.

There are a few factors associated with the use of packaged programs that have not been touched on in the earlier chapters and that deserve discussion at this point. First, we have to assume the package was properly written and can handle any nuances that a circuit configuration may introduce. In other words, no matter what elements or frequency are entered, and in whatever configuration, the package will provide a *correct* solution. Any package can provide a solution—but more important, is it the correct solution? For the PSpice software package, the background of its development and the extensive use it is enjoying in recent years suggest that it is of the highest caliber and, for the analysis of the type we are currently interested in, fully trustworthy. Let us assume PSpice is absolutely bug-free. What are the other concerns? How often have you used a calculator, obtained a result, and assumed the answer was correct for future calculations—only to find way down the line that the result was incorrect due to a wrong entry in the calculator sequence. The same thing can happen with PSpice. An incorrect entry in the input file can result in a solution that is totally erroneous, but unless we have a sense for what to

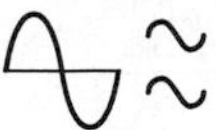

expect and some idea of the maneuvers the program is going through, we will have no way to check the results. For the future, therefore, be skeptical of all results—check and recheck the input file, compare the solutions to the magnitudes involved, and evaluate the results based on your own experience with similar systems. This is not to downgrade PSpice but to simply ensure that the types of errors often encountered using a calculator are not repeated using the package approach.

The input file for Fig. 16.20 is provided in Fig. 16.21. If you compare the network of Fig. 16.7 with that of Fig. 16.20, you will find that the reactances of the elements were provided in ohms in Fig. 16.7 and in their individual units of measurement in Fig. 16.20. This maneuver was required because PSpice calculates the reactance level at the applied frequency and does not permit the entry of reactance levels in ohms. To duplicate the analysis of Example 16.4, a frequency of 1 kHz was chosen, and the required inductor ($L = X_L/2\pi f$) and capacitor ($C = 1/2\pi f X_C$) were determined using the appropriate equation.

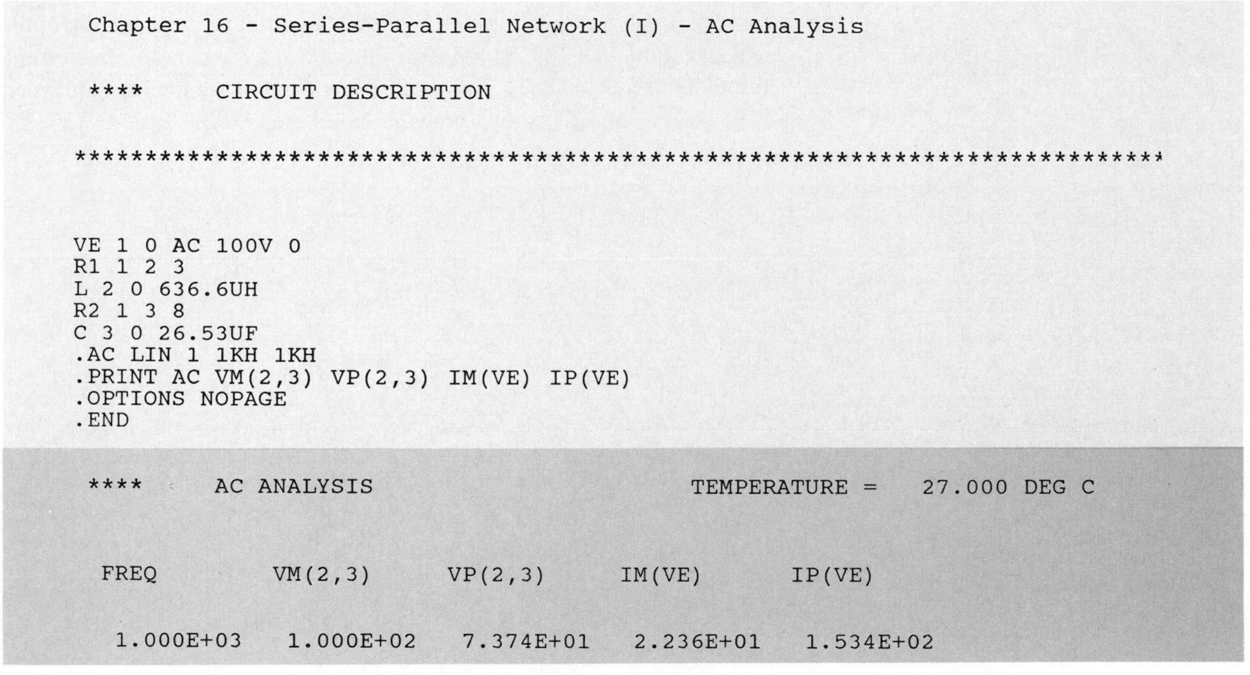

```
Chapter 16 - Series-Parallel Network (I) - AC Analysis

****     CIRCUIT DESCRIPTION

*****************************************************************************

VE 1 0 AC 100V 0
R1 1 2 3
L 2 0 636.6UH
R2 1 3 8
C 3 0 26.53UF
.AC LIN 1 1KH 1KH
.PRINT AC VM(2,3) VP(2,3) IM(VE) IP(VE)
.OPTIONS NOPAGE
.END

****     AC ANALYSIS                        TEMPERATURE =   27.000 DEG C

  FREQ        VM(2,3)      VP(2,3)      IM(VE)       IP(VE)

  1.000E+03   1.000E+02    7.374E+01    2.236E+01    1.534E+02
```

FIG. 16.21
Input and output files for the network of Fig. 16.20.

In Example 16.4 $\mathbf{V}_{ab}$ was found to be 100 V $\angle 73.74°$, matching the result for V(2, 3) in the output file of Fig. 16.21. For the source current, Example 16.4 provided a result of $\mathbf{I} = 22.36\text{ A } \angle -26.56°$, whereas PSpice produced a result of 22.36 A $\angle 153.4°$. The magnitude is the same, but the phase angle appears to be totally different. This difference in phase angle is due to the fact that current through an independent source (dc or ac) is defined by PSpice to flow from the positive to the negative node. In the example, the currents **I** of Figs. 16.7 and 16.20 have the opposite direction, requiring that a 180° phase shift be introduced to the PSpice solution, as shown in Fig. 16.22. Now note the close correlation with the result of Example 16.4, where $\mathbf{I} = 22.36\text{ A } \angle -26.56°$.

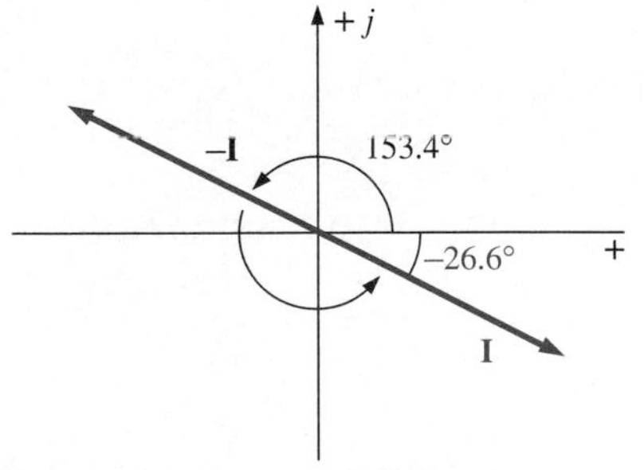

FIG. 16.22
Demonstrating the equivalence between solutions.

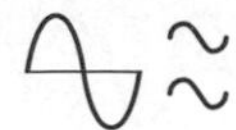

The next application of PSpice is to the series-parallel network of Fig. 16.23, which is Problem 10 in the text. Although the solutions are provided here, they should provide added incentive to obtain the same results longhand.

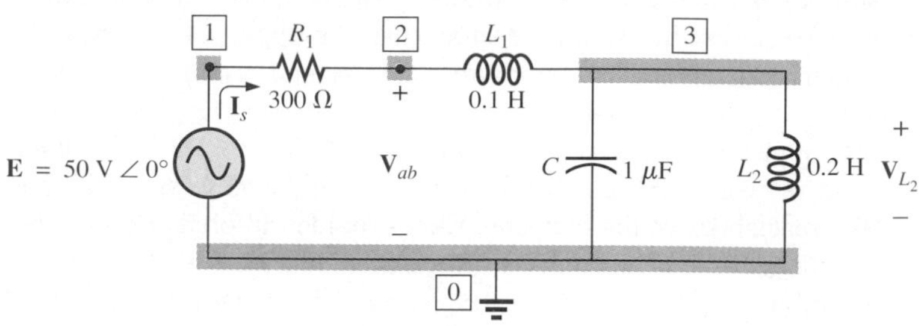

FIG. 16.23
Performing a PSpice (DOS) analysis of the network of Problem 10.

The input file is provided in Fig. 16.24 with a request for $\mathbf{I}_s$, $\mathbf{V}_{L_2}$, and $\mathbf{V}_2$ (= $\mathbf{V}_{ab}$). The current $\mathbf{I}_s$ is requested as I(R1), since I(R1) will be the same as $\mathbf{I}_s$ (series configuration) and I(R1) will have the same direction as $\mathbf{I}_s$, avoiding the 180° phase shift introduced by asking for the current through the source. The solution for $\mathbf{I}_s$ obtained in the longhand manner is 93 mA ∠−56.07°, comparing very closely to the 93.01 mA ∠−56.08° obtained with PSpice. The longhand solution for $\mathbf{V}_{L_2}$ is 16.931 V

```
Chapter 16 - Series-Parallel Network (II) - AC Analysis

 ****     CIRCUIT DESCRIPTION

******************************************************************************

VE 1 0 AC 50V 0
R1 1 2 300
L1 2 3 0.1H
C  3 0 1UF
L2 3 0 0.2H
.AC LIN 1 1KH 1KH
.PRINT AC IM(R1) IP(R1) VM(L2) VP(L2) VM(2) VP(2)
.OPTIONS NOPAGE
.END

 ****     AC ANALYSIS                           TEMPERATURE =   27.000 DEG C

  FREQ        IM(R1)        IP(R1)       VM(L2)       VP(L2)       VM(2)

   1.000E+03   9.301E-02   -5.608E+01   1.695E+01   -1.461E+02   4.149E+01

 ****     AC ANALYSIS                           TEMPERATURE =   27.000 DEG C

  FREQ        VP(2)

   1.000E+03   3.392E+01
```

FIG. 16.24
Input and output files for the network of Fig. 16.23.

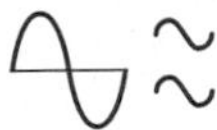

∠213.93° versus 16.95 V ∠−143.1° for PSpice. The results are very close when one changes the phase angle for the PSpice solution to a positive number. That is, $\mathbf{V}_{L_2}$ = 16.95 V ∠+360° − 143.1° = 16.95 V ∠213.9°. The result for $\mathbf{V}_2 = \mathbf{V}_{ab}$ using PSpice is exactly the same as obtained in the longhand manner: $\mathbf{V}_2 = \mathbf{V}_{ab}$ = 41.49 V ∠33.92°.

PSpice (Windows)

The schematics analysis will determine the voltage and current for the last element of the ladder network of Fig. 16.25. The mathematical content of this chapter would certainly suggest that this analysis will be a lengthy exercise in complex algebra, with one mistake invalidating the entire effort. However, we will find that when we simulate this schematic, the computer operating time will be only 5.05 seconds—simply amazing!

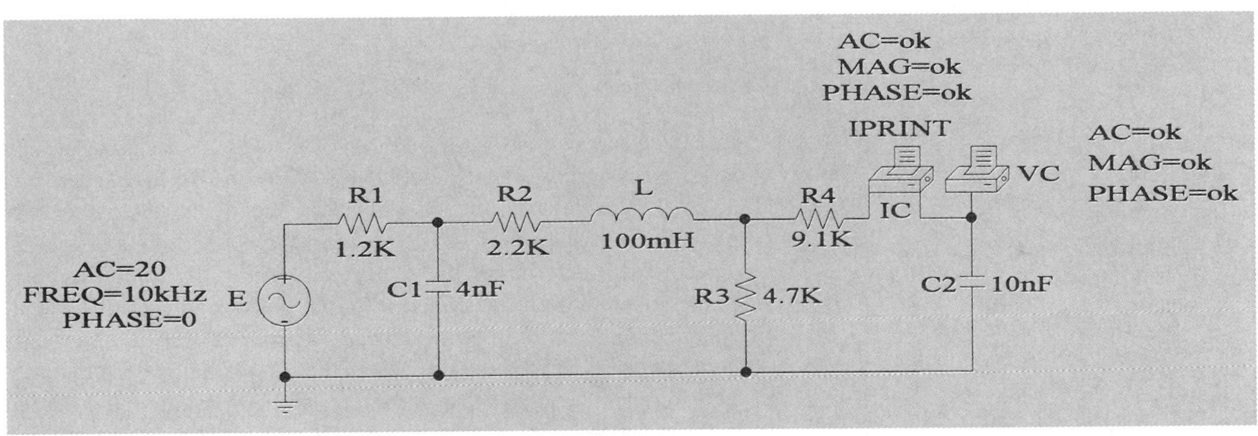

FIG. 16.25
Applying PSpice (Windows) to a ladder network.

The components are entered as described in earlier chapters, with the **IPRINT** and **VPRINT** introduced to define the desired output quantities. The **AC Sweep** is set for one point at 10 kHz, and the **Automatically Run Probe** is disabled before the simulation. The results of Fig. 16.26 reveal that

$$\mathbf{I}_C = \mathbf{0.7506\ mA\ \angle{-52.15°}}$$

with $$\mathbf{V}_C = \mathbf{1.195\ V\ \angle{-142.1°}}$$

Keep in mind that since the applied signal had a peak value of 20 V, all the numerical results are the peak value and not the effective value.

The **Probe** option under **AC analysis** at a fixed frequency can provide many important quantities that are not available directly from the schematic. Since it is a fixed frequency application, the only frequency appearing on the horizontal axis of the plot is the frequency of interest. The magnitude of the quantity of interest as chosen under the **Add Trace** dialog box will simply be indicated by a short horizontal hash mark. For instance, the magnitude (actually, the peak value) of the source current is provided by the lower plot of Fig. 16.27. The cursor was employed to determine the magnitude of current, as shown in the narrow horizontal box at the bottom of the figure. Note that the frequency is 10 kHz and the peak value of I(E) is 3.1935 mA. To set the

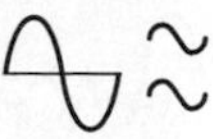

```
 ****       AC ANALYSIS                              TEMPERATURE =    27.000
DEG C

****************************************************************************
************

  FREQ          IM(V_IC)     IP(V_IC)

   1.000E+04   7.506E-04   -5.215E+01

  FREQ          VM($N_0006) VP($N_0006)

   1.000E+04   1.195E+00   -1.421E+02
```

FIG. 16.26
Output file for the schematics analysis of Fig. 16.25.

cursor, simply follow the sequence **Tools-Cursor-Display.** When you click on the display, the cursor will automatically move to the only point on the plot. The phase angle between the applied voltage and the source current is determined in the other plot of Fig. 16.27. In this case, since the source current is defined from the positive to negative potential of the supply, a minus sign appears in front of I(E) to change its defined direction. The prefix **P** specifies the phase angle of the quantity in the parentheses and produces the expression appearing on the bottom of the plot for the phase angle between **E** and $\mathbf{I}_s$. The angle turns out negative because **E** is at zero degrees, and zero minus anything is a negative number. For this example, the phase angle is about $-53°$, as noted by the very light, short dashed line.

If the power to the network is desired, the general equation has the following form:

$$P = \frac{V_{\text{peak}} I_{\text{peak}}}{2} \cos\theta = V_{\text{rms}} I_{\text{rms}} \cos\theta$$

However, the cosine function under Probe finds the cosine of an angle in radians. Since the conversion between degrees and radians is the following:

$$\theta_{\text{radians}} = \left(\frac{\pi}{180°}\right)\theta° = (17.453 \times 10^{-3})\theta°$$

the factor 17.452E−3 must precede the angle in degrees, as shown in the lower part of Fig. 16.28.

The **M** preceding both **E** and $\mathbf{I}_s$ specifies the magnitude of the quantity, which in this case will give the magnitude of the peak value of each quantity. The resulting input power, as noted by the cursor, is 19.211 mW for the network of Fig. 16.25. If the power to the resistor R_1 were desired, it could be obtained by the following equation:

$$P_R = \frac{V_{R_1(\text{peak})} \cdot I_{R_1(\text{peak})}}{2} = V_{R_1(\text{rms})} \cdot I_{R_1(\text{rms})}$$

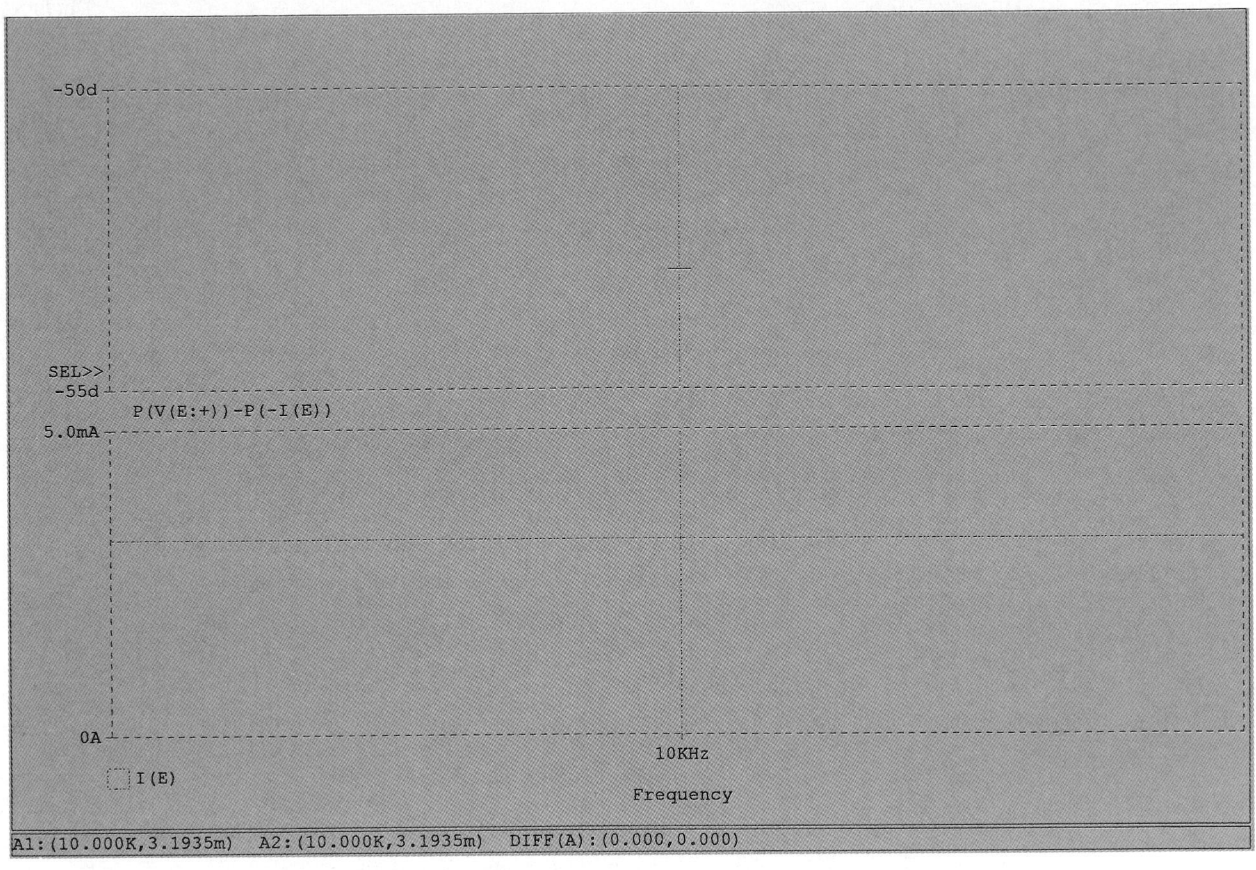

FIG. 16.27
The peak value of the source current and the phase angle between **E** *and* $\mathbf{I}_s$ *for the network of Fig. 16.25.*

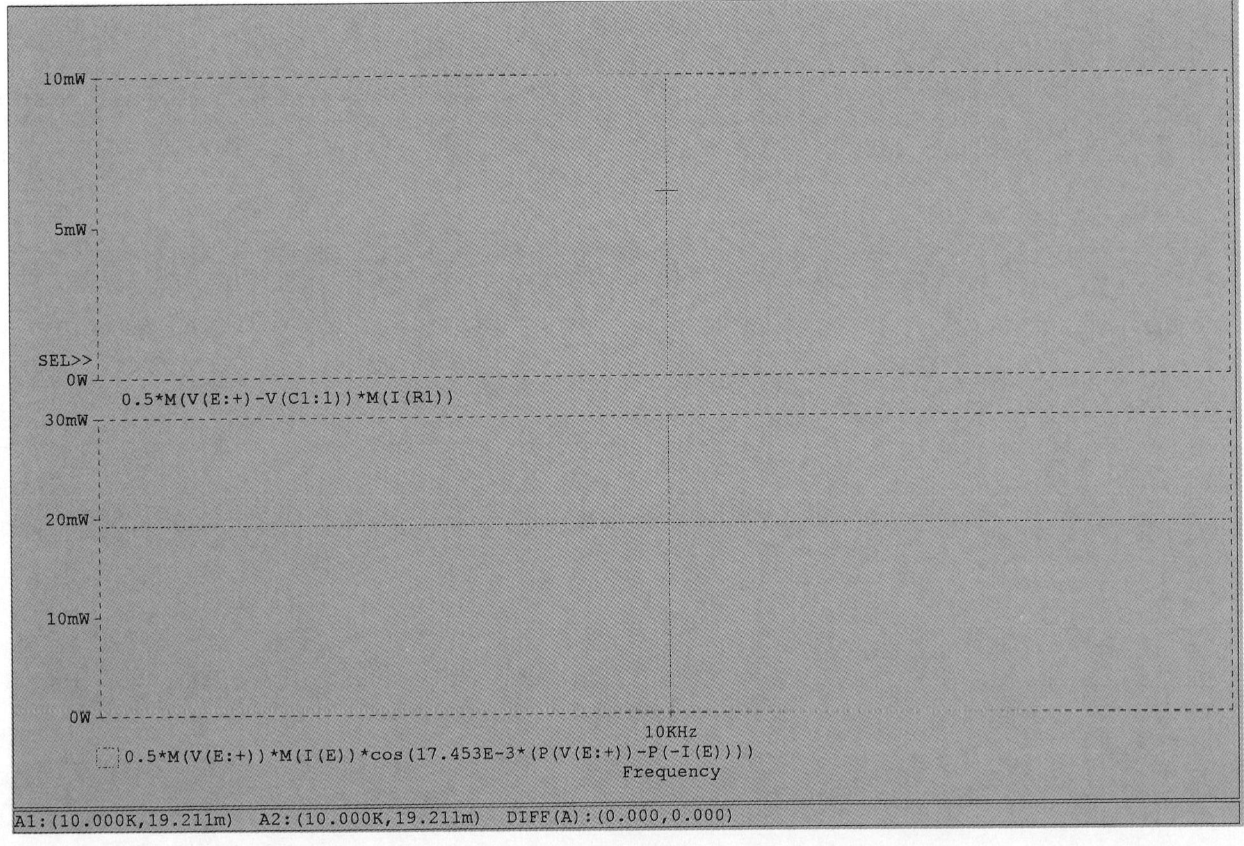

FIG. 16.28
The power delivered to the network and the resistor R_1 of Fig. 16.25.

Note the format for the above equation in the top part of Fig. 16.28. The voltage across the resistor must be found first, followed by a determination of the magnitude of the resulting peak value. The result is then multiplied by the magnitude of the peak value of the current and 0.5 to convert the peak value into effective values. The resulting power is about 6.1 mW.

The **MicroSim PSpice A/D Circuit Analysis User's Guide with Schematics** has a long list of valid Probe arithmetic operations that can be quite powerful once their full impact is understood.

PROBLEMS

SECTION 16.2 Illustrative Examples

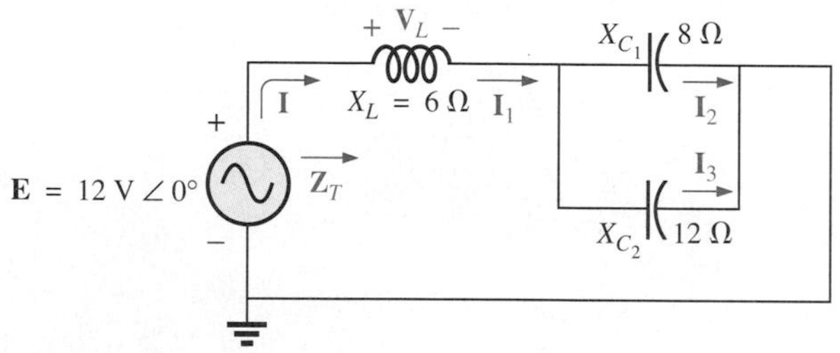

FIG. 16.29
Problems 1 and 19.

1. For the series-parallel network of Fig. 16.29:
   a. Calculate $\mathbf{Z}_T$.
   b. Determine **I**.
   c. Determine $\mathbf{I}_1$.
   d. Find $\mathbf{I}_2$ and $\mathbf{I}_3$.
   e. Find $\mathbf{V}_L$.

2. Find the network of Fig. 16.30:
   a. Find the total impedance $\mathbf{Z}_T$.
   b. Determine the current $\mathbf{I}_s$.
   c. Calculate $\mathbf{I}_C$ using the current divider rule.
   d. Calculate $\mathbf{V}_L$ using the voltage divider rule.

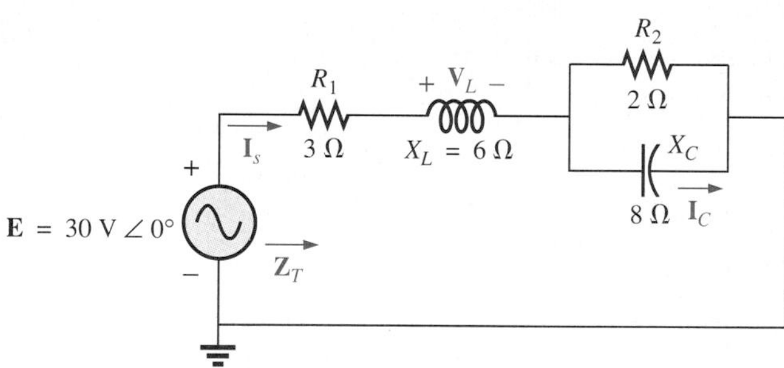

FIG. 16.30
Problems 2 and 15.

3. For the network of Fig. 16.31:
   a. Find the total impedance $\mathbf{Z}_T$ and the total admittance $\mathbf{Y}_T$.
   b. Find the current $\mathbf{I}_s$.
   c. Calculate $\mathbf{I}_2$ using the current divider rule.
   d. Calculate $\mathbf{V}_C$.
   e. Calculate the average power delivered to the network.

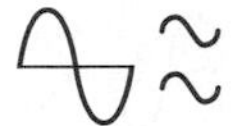

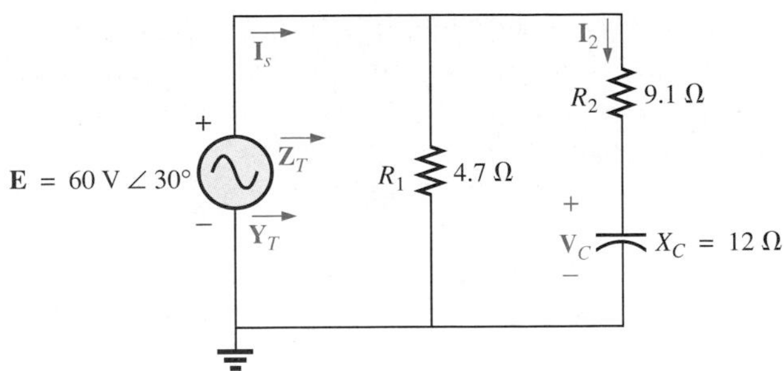

FIG. 16.31
Problems 3 and 20.

4. For the network of Fig. 16.32:
   a. Find the total impedance $\mathbf{Z}_T$.
   b. Calculate the voltage $\mathbf{V}_2$ and the current $\mathbf{I}_L$.
   c. Find the power factor of the network.

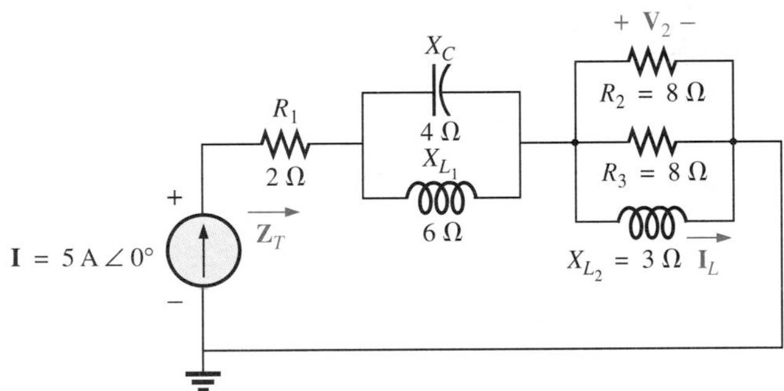

FIG. 16.32
Problem 4.

5. For the network of Fig. 16.33:
   a. Find the current **I.**
   b. Find the voltage $\mathbf{V}_C$.
   c. Find the average power delivered to the network.

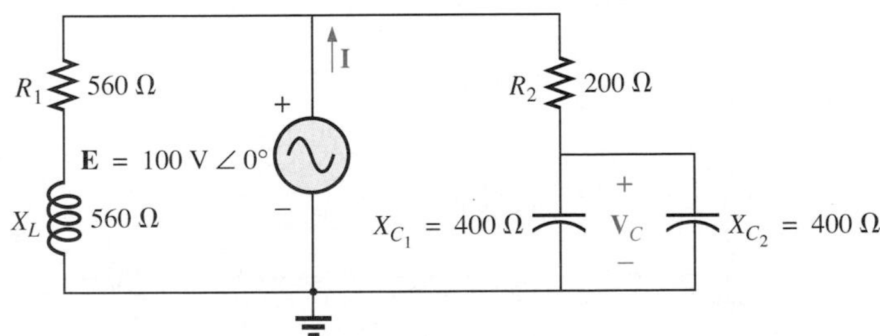

FIG. 16.33
Problems 5 and 21.

***6.** For the network of Fig. 16.34:
   a. Find the current $\mathbf{I}_1$.
   b. Calculate the voltage $\mathbf{V}_C$ using the voltage divider rule.
   c. Find the voltage $\mathbf{V}_{ab}$.

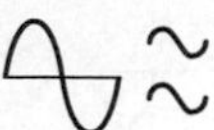

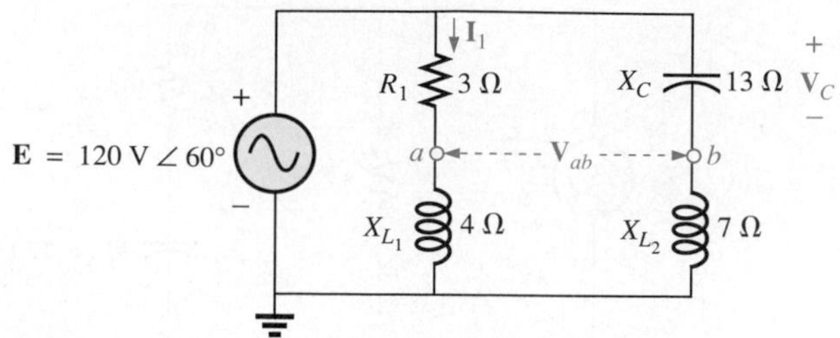

FIG. 16.34
Problem 6.

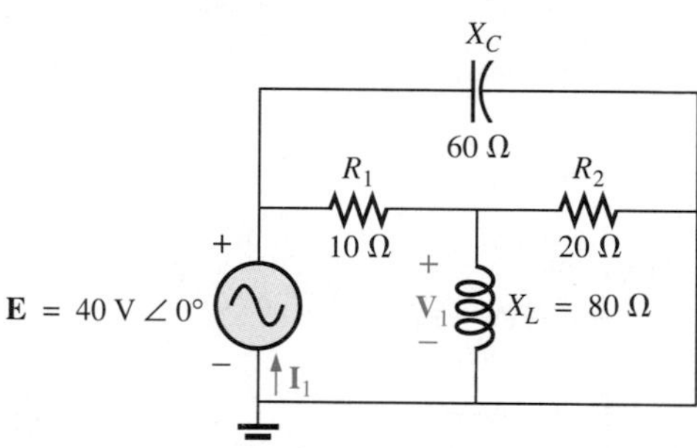

FIG. 16.35
Problems 7 and 16.

***7.** For the network of Fig. 16.35:

a. Find the current $\mathbf{I}_1$.
b. Find the voltage $\mathbf{V}_1$.
c. Calculate the average power delivered to the network.

8. For the network of Fig. 16.36:

a. Find the total impedance $\mathbf{Z}_T$ and the admittance $\mathbf{Y}_T$.
b. Find the currents $\mathbf{I}_1$, $\mathbf{I}_2$, and $\mathbf{I}_3$.
c. Verify Kirchhoff's current law by showing that $\mathbf{I}_s = \mathbf{I}_1 + \mathbf{I}_2 + \mathbf{I}_3$.
d. Find the power factor of the network and indicate whether it is leading or lagging.

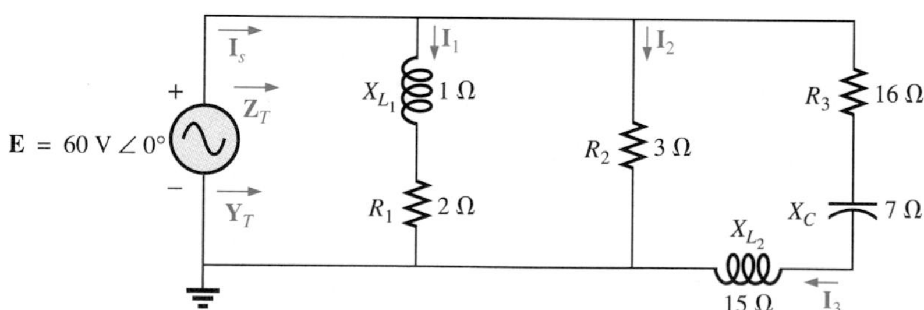

FIG. 16.36
Problem 8.

***9.** For the network of Fig. 16.37:

a. Find the total admittance $\mathbf{Y}_T$.
b. Find the voltages $\mathbf{V}_1$ and $\mathbf{V}_2$.
c. Find the current $\mathbf{I}_3$.

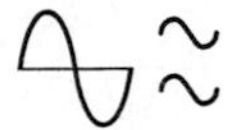

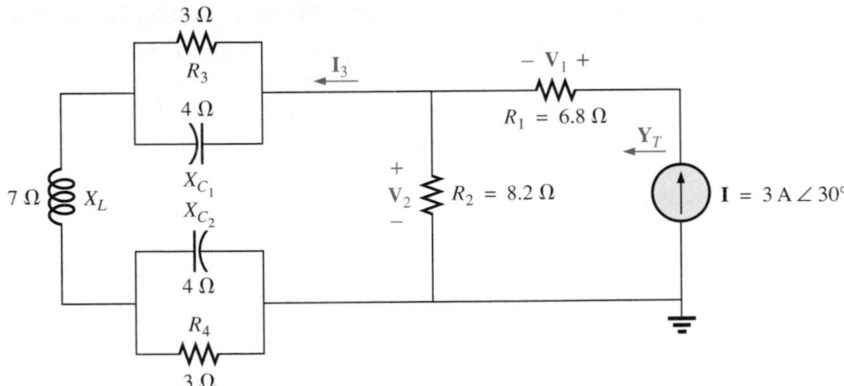

FIG. 16.37
Problem 9.

***10.** For the network of Fig. 16.38:
  a. Find the total impedance $\mathbf{Z}_T$ and the admittance $\mathbf{Y}_T$.
  b. Find the source current $\mathbf{I}_s$ in phasor form.
  c. Find the currents $\mathbf{I}_1$ and $\mathbf{I}_2$ in phasor form.
  d. Find the voltages $\mathbf{V}_1$ and $\mathbf{V}_{ab}$ in phasor form.
  e. Find the average power delivered to the network.
  f. Find the power factor of the network and indicate whether it is leading or lagging.

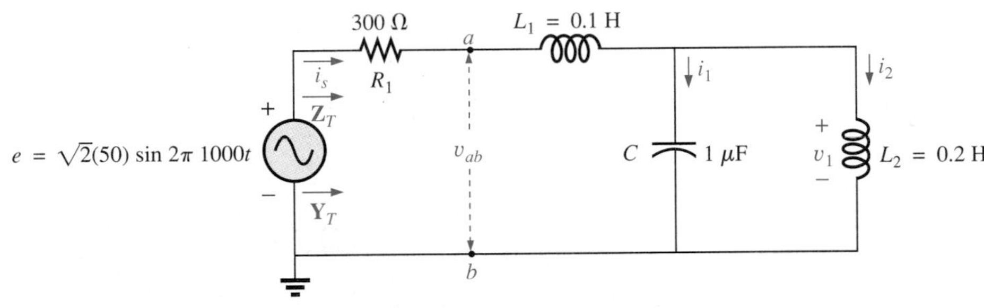

FIG. 16.38
Problem 10.

***11.** Find the current **I** for the network of Fig. 16.39.

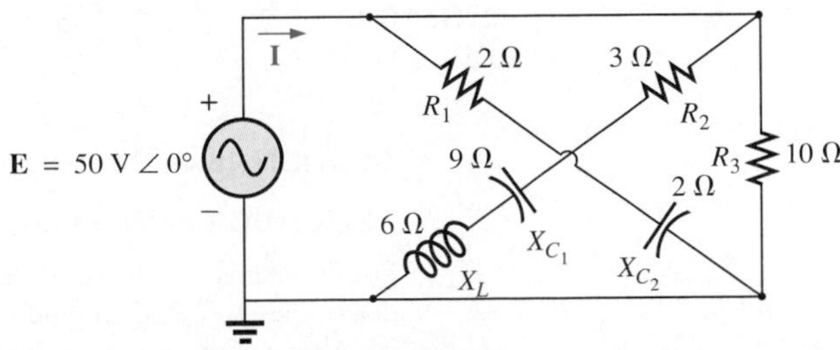

FIG. 16.39
Problems 11 and 17.

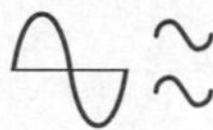

SECTION 16.3 Ladder Networks

12. Find the current $\mathbf{I}_5$ for the network of Fig. 16.40. Note the effect of one reactive element on the resulting calculations.

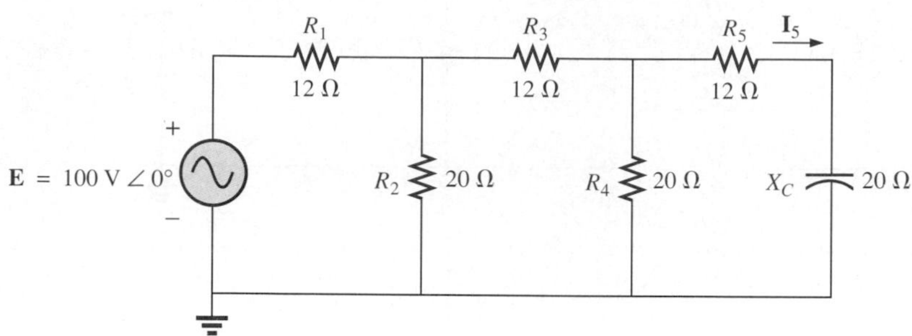

FIG. 16.40
Problem 12.

13. Find the average power delivered to R_4 in Fig. 16.41.

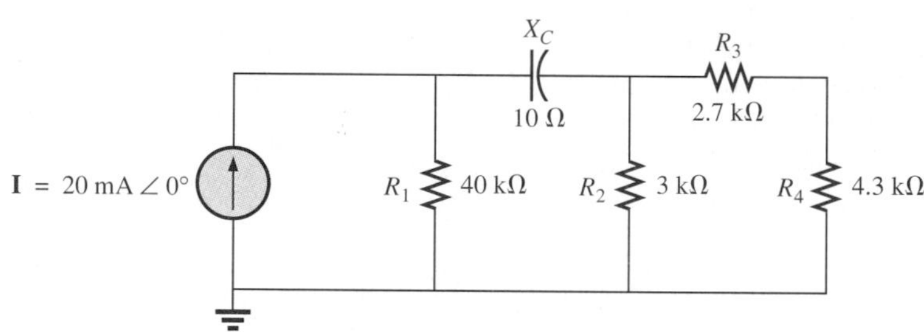

FIG. 16.41
Problem 13.

14. Find the current $\mathbf{I}_1$ for the network of Fig. 16.42.

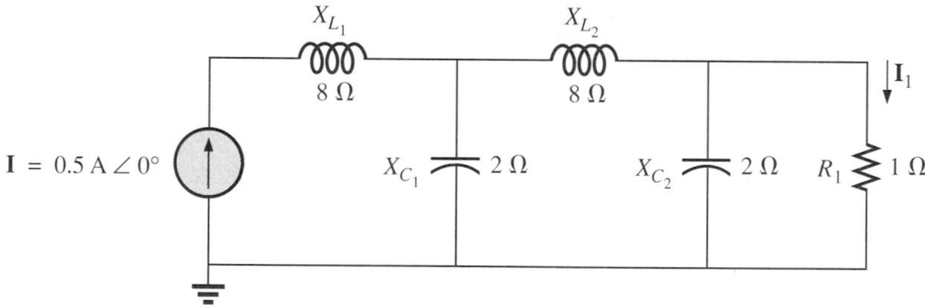

FIG. 16.42
Problems 14 and 18.

SECTION 16.4 Computer Analysis

PSpice (DOS or Windows)

For Problems 15 through 18, use a frequency of 1 kHz to determine the inductive and capacitive levels required for the input files. In each case write the required input file.

***15.** Repeat Problem 2 using PSpice. A separate input file must be written for part (a).

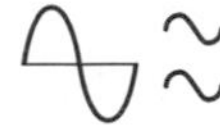

*16. Repeat Problem 7, parts (a) and (b), using PSpice.

*17. Repeat Problem 11 using PSpice.

*18. Repeat Problem 14 using PSpice.

Programming Language (C++, BASIC, PASCAL, etc.)

19. Write a program to provide a general solution to Problem 1; that is, given the reactance of each element, generate a solution for parts (a) through (e).

20. Given the network of Fig. 16.31, write a program to generate a solution for parts (a) and (b) of Problem 2. Use the values given.

21. Generate a program to obtain a general solution for the network of Fig. 16.33 for the questions asked in parts (a) through (c) of Problem 2. That is, given the resistance and reactance of the elements, determine the requested current, voltage, and power.

GLOSSARY

Ladder network A repetitive combination of series and parallel branches that has the appearance of a ladder.

Series-parallel ac network A combination of series and parallel branches in the same network configuration. Each branch may contain any number of elements whose impedance is dependent on the applied frequency.

17

Methods of Analysis and Selected Topics (ac)

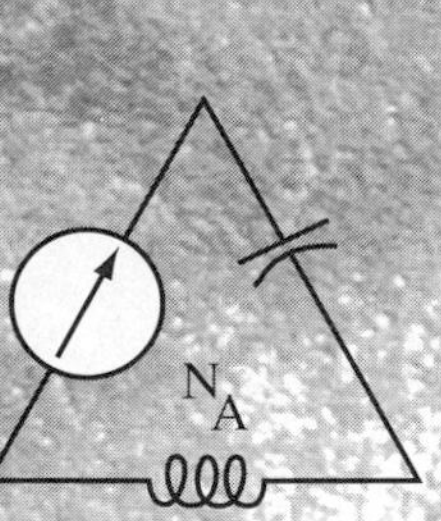

17.1 INTRODUCTION

For networks with two or more sources that are not in series or parallel, the methods described in the last two chapters cannot be applied. Rather, methods such as mesh analysis or nodal analysis must be employed. Since these methods were discussed in detail for dc circuits in Chapter 8, this chapter will consider the variations required to apply these methods to ac circuits. Dependent sources will also be introduced for both mesh and modal analysis.

The branch-current method will not be discussed again because it falls within the framework of mesh analysis. In addition to the methods mentioned above, the bridge network and Δ-Y, Y-Δ conversions will also be discussed for ac circuits.

Before we examine these topics, however, we must consider the subject of independent and controlled sources.

17.2 INDEPENDENT VERSUS DEPENDENT (CONTROLLED) SOURCES

In the previous chapters, each source appearing in the analysis of dc or ac networks was an *independent source,* such as E and I (or $\mathbf{E}$ and $\mathbf{I}$) in Fig. 17.1.

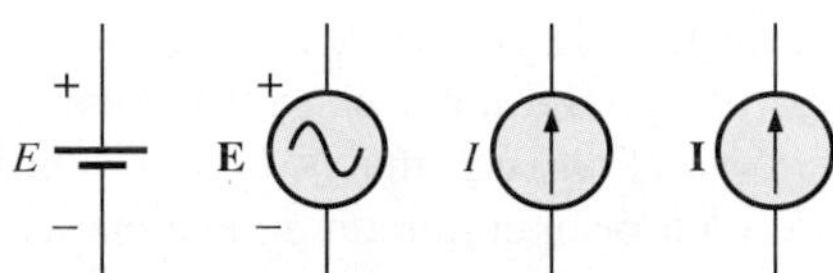

FIG. 17.1
Independent sources.

The term independent specifies that the magnitude of the source is independent of the network to which it is applied and that it displays its terminal characteristics even if completely isolated.

A dependent or controlled source is one whose magnitude is determined (or controlled) by a current or voltage of the system in which it appears.

There are currently two symbols used for controlled sources. One simply uses the independent symbol with an indication of the controlling element, as shown in Fig. 17.2. In Fig. 17.2(a), the magnitude and phase of the voltage are controlled by a voltage **V** elsewhere in the system, with the magnitude further controlled by the constant k_1. In Fig.

(a) (b)

FIG. 17.2
Controlled or dependent sources.

17.2(b), the magnitude and phase of the current source are controlled by a current **I** elsewhere in the system, with the magnitude further controlled by the constant k_2. To distinguish between the dependent and independent sources, the notation of Fig. 17.3 was introduced. In recent years many respected publications on circuit analysis have accepted the notation of Fig. 17.3, although a number of excellent publications in the area of electronics continue to use the symbol of Fig. 17.2, especially in the circuit modeling for a variety of electronic devices such as the transistor and FET. This text will employ the symbols of Fig. 17.3.

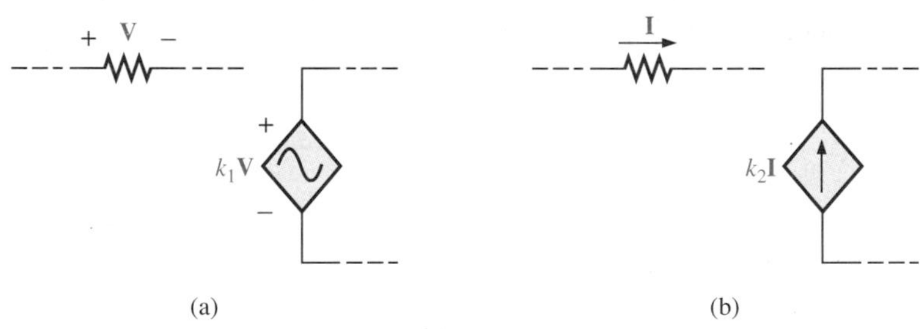

FIG. 17.3
Special notation for controlled or dependent sources.

Possible combinations for controlled sources are indicated in Fig. 17.4. Note that the magnitude of current sources or voltage sources can be controlled by a voltage and a current, respectively. Unlike with the independent source, isolation such that **V** or **I** = 0 in Fig. 17.4(a) will result in the short-circuit or open-circuit equivalent as indicated in Fig. 17.4(b). Note that the type of representation under these conditions is controlled by whether it is a current source or a voltage source, not by the controlling agent (**V** or **I**).

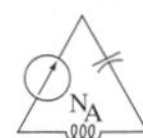

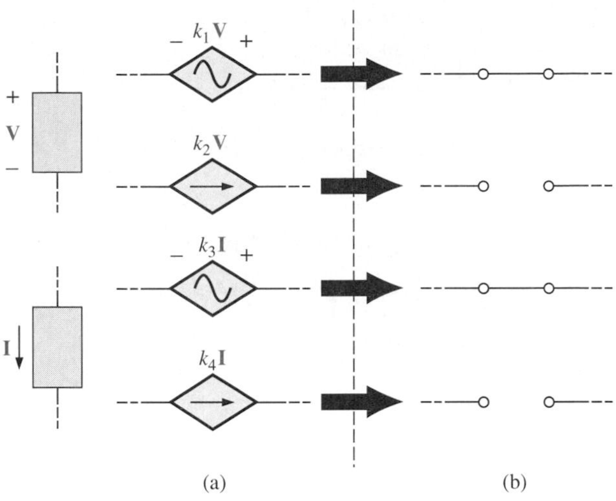

FIG. 17.4
Conditions of V = 0 V and I = 0 A for a controlled source.

17.3 SOURCE CONVERSIONS

When applying the methods to be discussed, it may be necessary to convert a current source to a voltage source, or a voltage source to a current source. This can be accomplished in much the same manner as for dc circuits, except now we shall be dealing with phasors and impedances instead of just real numbers and resistors.

Independent Sources

In general, the format for converting one type of independent source to another is as shown in Fig. 17.5.

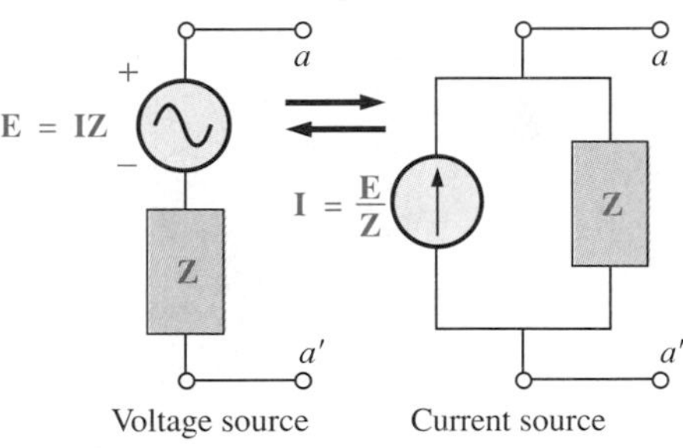

FIG. 17.5
Source conversion.

EXAMPLE 17.1 Convert the voltage source of Fig. 17.6(a) to a current source.

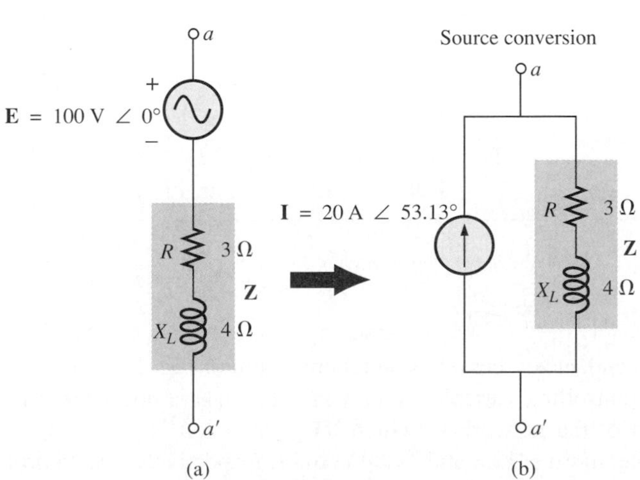

FIG. 17.6
Example 17.1.

Solution:

$$\mathbf{I} = \frac{\mathbf{E}}{\mathbf{Z}} = \frac{100 \text{ V} \angle 0°}{5 \ \Omega \angle 53.13°}$$
$$= \mathbf{20 \ A \angle -53.13°} \qquad \text{[Fig. 17.6(b)]}$$

EXAMPLE 17.2 Convert the current source of Fig. 17.7(a) to a voltage source.

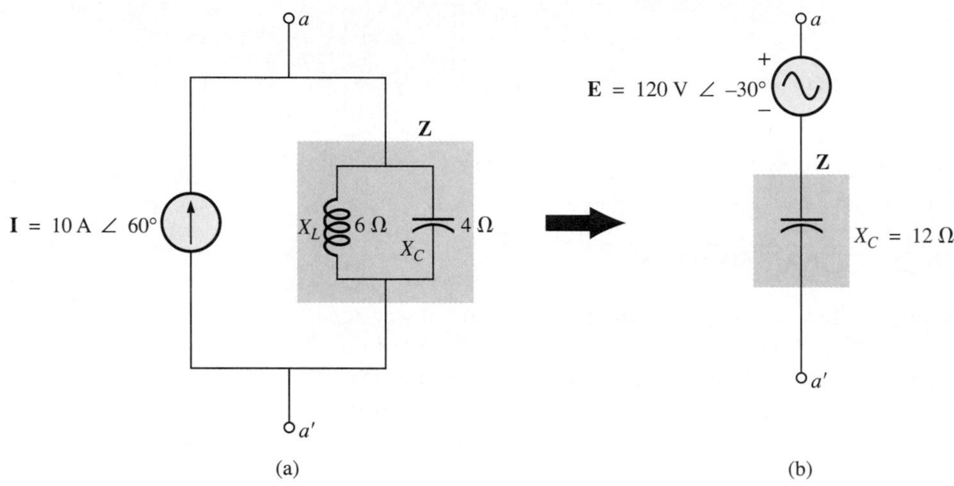

FIG. 17.7
Example 17.2.

Solution:

$$\mathbf{Z} = \frac{\mathbf{Z}_C\mathbf{Z}_L}{\mathbf{Z}_C + \mathbf{Z}_L} = \frac{(X_C \angle -90°)(X_L \angle 90°)}{-j X_C + j X_L}$$
$$= \frac{(4 \ \Omega \angle -90°)(6 \ \Omega \angle 90°)}{-j 4 \ \Omega + j 6 \ \Omega} = \frac{24 \ \Omega \angle 0°}{2 \angle 90°}$$
$$= \mathbf{12 \ \Omega \angle -90°} \qquad \text{[Fig. 17.7(b)]}$$
$$\mathbf{E} = \mathbf{IZ} = (10 \text{ A} \angle 60°)(12 \ \Omega \angle -90°)$$
$$= \mathbf{120 \ V \angle -30°} \qquad \text{[Fig. 17.7(b)]}$$

Dependent Sources

For dependent sources, the direct conversion of Fig. 17.5 can be applied if the controlling variable (**V** or **I** in Fig. 17.4) is not determined by a portion of the network to which the conversion is to be applied. For example, in Figs. 17.8 and 17.9, **V** and **I**, respectively, are controlled by an external portion of the network. Conversions of the other kind, where **V** and **I** are controlled by a portion of the network to be converted, will be considered in Sections 18.3 and 18.4.

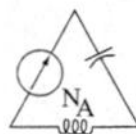

EXAMPLE 17.3 Convert the voltage source of Fig. 17.8(a) to a current source.

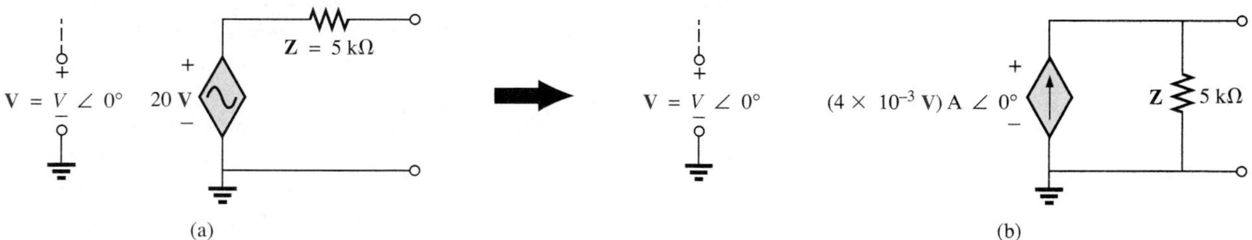

FIG. 17.8
Source conversion with a voltage-controlled voltage source.

Solution:

$$\mathbf{I} = \frac{\mathbf{E}}{\mathbf{Z}} = \frac{(20\mathbf{V})\ \mathrm{V}\ \angle 0^\circ}{5\ \mathrm{k}\Omega\ \angle 0^\circ}$$
$$= \mathbf{(4 \times 10^{-3}\ V)\ A\ \angle 0^\circ} \qquad \text{[Fig. 17.8(b)]}$$

EXAMPLE 17.4 Convert the current source of Fig. 17.9(a) to a voltage source.

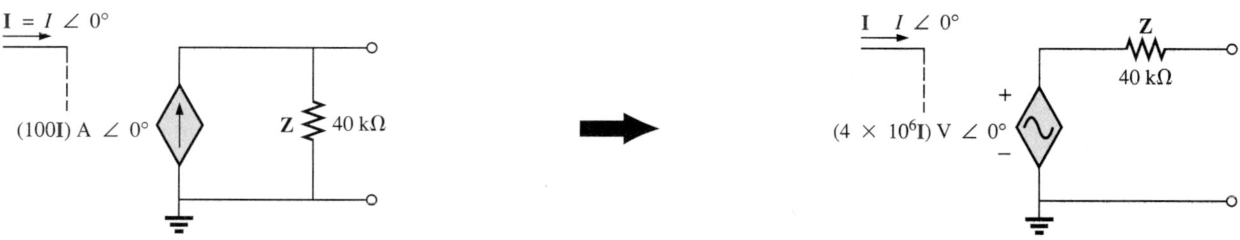

FIG. 17.9
Source conversion with a current-controlled current source.

Solution:

$$\mathbf{E} = \mathbf{IZ} = [(100\mathbf{I})\ \mathrm{A}\ \angle 0^\circ][40\ \mathrm{k}\Omega\ \angle 0^\circ]$$
$$= \mathbf{(4 \times 10^{6} I)\ V\ \angle 0^\circ} \qquad \text{[Fig. 17.9(b)]}$$

17.4 MESH ANALYSIS

General Approach

Independent Voltage Sources Before examining the application of the method to ac networks, the student should first review the appropriate sections on mesh analysis in Chapter 8 since the content of this section will be limited to the general conclusions of Chapter 8.

The general approach to mesh analysis for independent sources includes the same sequence of steps appearing in Chapter 8. In fact, throughout this section the only change from the dc coverage will be to substitute impedance for resistance and admittance for conductance in the general procedure.

1. ***Assign a distinct current in the clockwise direction to each independent closed loop of the network. It is not absolutely necessary to choose the clockwise direction for each loop current. However, it eliminates the need to have to choose a direction for each application. Any direction can be chosen for each loop current with no loss in accuracy as long as the remaining steps are followed properly.***
2. ***Indicate the polarities within each loop for each impedance as determined by the assumed direction of loop current for that loop.***
3. ***Apply Kirchhoff's voltage law around each closed loop in the clockwise direction. Again, the clockwise direction was chosen to establish uniformity and prepare us for the format approach to follow.***
   a. ***If an impedance has two or more assumed currents through it, the total current through the impedance is the assumed current of the loop in which Kirchhoff's voltage law is being applied, plus the assumed currents of the other loops passing through in the same direction, minus the assumed currents passing through in the opposite direction.***
   b. ***The polarity of a voltage source is unaffected by the direction of the assigned loop currents.***
4. ***Solve the resulting simultaneous linear equations for the assumed loop currents.***

The technique is applied as above for all networks with independent sources or networks with *dependent sources where the controlling variable is not a part of the network under investigation.* If the controlling variable is part of the network being examined, a method to be described shortly must be applied.

EXAMPLE 17.5 Using the general approach to mesh analysis, find the current $\mathbf{I}_1$ in Fig. 17.10.

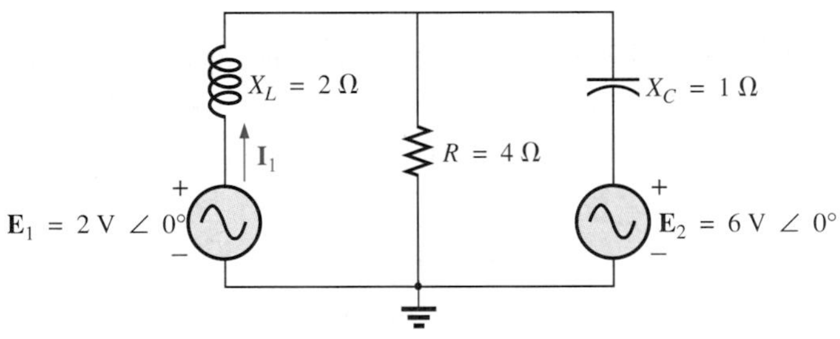

FIG. 17.10
Example 17.5.

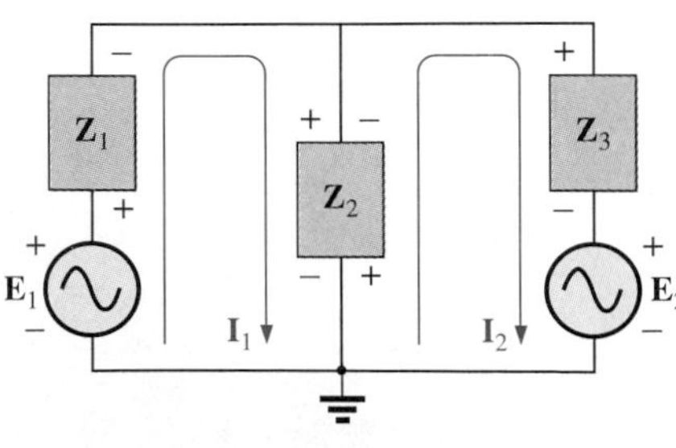

FIG. 17.11
Assigning the mesh currents and subscripted impedances for the network of Fig. 17.10.

Solution: When applying these methods to ac circuits, it is good practice to represent the resistors and reactances (or combinations thereof) by subscripted impedances. When the total solution is found in terms of these subscripted impedances, the numerical values can be substituted to find the unknown quantities.

The network is redrawn in Fig. 17.11 with subscripted impedances:

$$\mathbf{Z}_1 = +j\,X_L = +j\,2\ \Omega \qquad \mathbf{E}_1 = 2\ \text{V} \angle 0°$$
$$\mathbf{Z}_2 = R = 4\ \Omega \qquad \mathbf{E}_2 = 6\ \text{V} \angle 0°$$
$$\mathbf{Z}_3 = -j\,X_C = -j\,1\ \Omega$$

Steps 1 and 2 are as indicated in Fig. 17.11.

Step 3:

$$+\mathbf{E}_1 - \mathbf{I}_1\mathbf{Z}_1 - \mathbf{Z}_2(\mathbf{I}_1 - \mathbf{I}_2) = 0$$
$$-\mathbf{Z}_2(\mathbf{I}_2 - \mathbf{I}_1) - \mathbf{I}_2\mathbf{Z}_3 - \mathbf{E}_2 = 0$$

or

$$\mathbf{E}_1 - \mathbf{I}_1\mathbf{Z}_1 - \mathbf{I}_1\mathbf{Z}_2 + \mathbf{I}_2\mathbf{Z}_2 = 0$$
$$-\mathbf{I}_2\mathbf{Z}_2 + \mathbf{I}_1\mathbf{Z}_2 - \mathbf{I}_2\mathbf{Z}_3 - \mathbf{E}_2 = 0$$

so that

$$\mathbf{I}_1(\mathbf{Z}_1 + \mathbf{Z}_2) - \mathbf{I}_2\mathbf{Z}_2 = \mathbf{E}_1$$
$$\mathbf{I}_2(\mathbf{Z}_2 + \mathbf{Z}_3) - \mathbf{I}_1\mathbf{Z}_2 = -\mathbf{E}_2$$

which are rewritten as

$$\begin{aligned} \mathbf{I}_1(\mathbf{Z}_1 + \mathbf{Z}_2) - \mathbf{I}_2\mathbf{Z}_2 \qquad\qquad &= \mathbf{E}_1 \\ -\mathbf{I}_1\mathbf{Z}_2 \qquad + \mathbf{I}_2(\mathbf{Z}_2 + \mathbf{Z}_3) &= -\mathbf{E}_2 \end{aligned}$$

Step 4: Using determinants, we obtain

$$\mathbf{I}_1 = \frac{\begin{vmatrix} \mathbf{E}_1 & -\mathbf{Z}_2 \\ -\mathbf{E}_2 & \mathbf{Z}_2 + \mathbf{Z}_3 \end{vmatrix}}{\begin{vmatrix} \mathbf{Z}_1 + \mathbf{Z}_2 & -\mathbf{Z}_2 \\ -\mathbf{Z}_2 & \mathbf{Z}_2 + \mathbf{Z}_3 \end{vmatrix}} = \frac{\mathbf{E}_1(\mathbf{Z}_2 + \mathbf{Z}_3) - \mathbf{E}_2(\mathbf{Z}_2)}{(\mathbf{Z}_1 + \mathbf{Z}_2)(\mathbf{Z}_2 + \mathbf{Z}_3) - (\mathbf{Z}_2)^2} = \frac{(\mathbf{E}_1 - \mathbf{E}_2)\mathbf{Z}_2 + \mathbf{E}_1\mathbf{Z}_3}{\mathbf{Z}_1\mathbf{Z}_2 + \mathbf{Z}_1\mathbf{Z}_3 + \mathbf{Z}_2\mathbf{Z}_3}$$

Substituting numerical values yields

$$\mathbf{I}_1 = \frac{(2\text{ V} - 6\text{ V})(4\ \Omega) + (2\text{ V})(-j\,1\ \Omega)}{(+j\,2\ \Omega)(4\ \Omega) + (+j\,2\ \Omega)(-j\,2\ \Omega) + (4\ \Omega)(-j\,2\ \Omega)}$$
$$= \frac{-16 - j\,2}{j\,8 - j^2 2 - j\,4} = \frac{-16 - j\,2}{2 + j\,4} = \frac{16.12\text{ A}\angle -172.87^\circ}{4.47\angle 63.43^\circ}$$
$$= \mathbf{3.61\ A\ \angle -236.30^\circ} \quad \text{or} \quad \mathbf{3.61\ A\ \angle 123.70^\circ}$$

Dependent Voltage Sources For dependent voltage sources, the procedure is modified as follows:

1. Steps 1 and 2 are the same as those applied for independent voltage sources.
2. Step 3 is modified as follows: Treat each dependent source like an independent source when Kirchhoff's voltage law is applied to each independent loop. However, once the equation is written, substitute the equation for the controlling quantity to ensure that the unknowns are limited solely to the chosen mesh currents.
3. Step 4 is as before.

EXAMPLE 17.6 Write the mesh currents for the network of Fig. 17.12 having a dependent voltage source.

Solution:

Steps 1 and 2 are defined on Fig. 17.12.

Step 3:
$$\mathbf{E}_1 - \mathbf{I}_1 R_1 - R_2(\mathbf{I}_1 - \mathbf{I}_2) = 0$$
$$R_2(\mathbf{I}_2 - \mathbf{I}_1) + \mu\mathbf{V}_x - \mathbf{I}_2 R_3 = 0$$

with $\mathbf{V}_x = (\mathbf{I}_1 - \mathbf{I}_2)R_2$

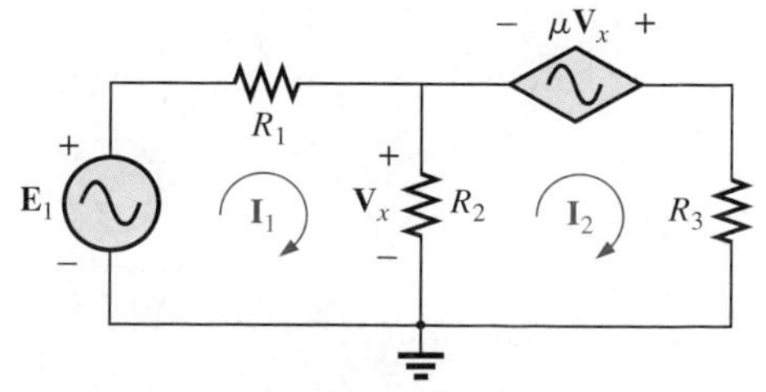

FIG. 17.12
Applying mesh analysis to a network with a voltage-controlled voltage source.

The result is two equations and two unknowns.

$$\mathbf{E}_1 - \mathbf{I}_1 R_1 - R_2(\mathbf{I} - \mathbf{I}_2) = 0$$
$$R_2(\mathbf{I}_2 - \mathbf{I}_1) + \mu R_2(\mathbf{I}_1 - \mathbf{I}_2) - \mathbf{I}_2 R_3 = 0$$

Independent Current Sources For independent current sources, the procedure is modified as follows:

1. Steps 1 and 2 are the same as those applied for independent sources.
2. Step 3 is modified as follows: Treat each current source as an open circuit (recall the *supermesh* designation of Chapter 8), and write the mesh equations for each remaining independent path. Then relate the chosen mesh currents to the dependent sources to ensure that the unknowns of the final equations are limited simply to the mesh currents.
3. Step 4 is as before.

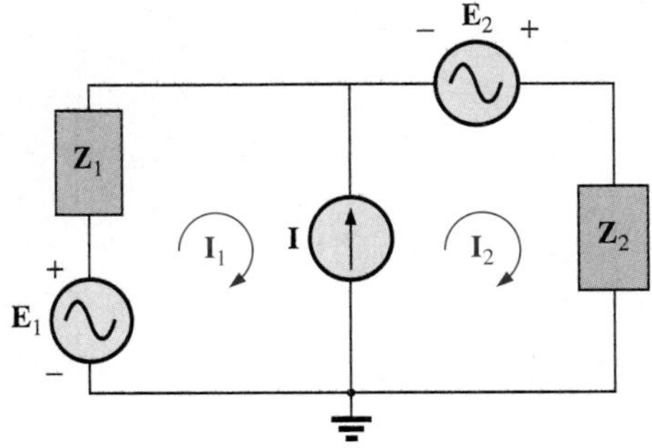

FIG. 17.13
Applying mesh analysis to a network with an independent current source.

EXAMPLE 17.7 Write the mesh currents for the network of Fig. 17.13 having an independent current source.

Solution:

Steps 1 and 2 are defined on Fig. 17.13.

Step 3: $\mathbf{E}_1 - \mathbf{I}_1\mathbf{Z}_1 + \mathbf{E}_2 - \mathbf{I}_2\mathbf{Z}_2 = 0$ (only remaining independent path)

with $\mathbf{I}_1 + \mathbf{I} = \mathbf{I}_2$

The result is two equations and two unknowns.

Dependent Current Sources For dependent current sources, the procedure is modified as follows:

1. Steps 1 and 2 are the same as those applied for independent sources.
2. Step 3 is modified as follows: The procedure is essentially the same as that applied for independent current sources, except now the dependent sources have to be defined in terms of the chosen mesh currents to ensure that the final equations have only mesh currents as the unknown quantities.
3. Step 4 is as before.

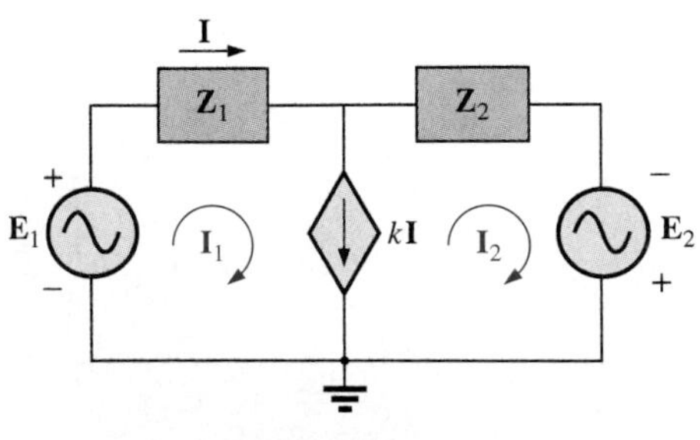

FIG. 17.14
Applying mesh analysis to a network with a current-controlled current source.

EXAMPLE 17.8 Write the mesh currents for the network of Fig. 17.14 having a dependent current source.

Solution:

Steps 1 and 2 are defined on Fig. 17.14.

Step 3: $\mathbf{E}_1 - \mathbf{I}_1\mathbf{Z}_1 - \mathbf{I}_2\mathbf{Z}_2 + \mathbf{E}_2 = 0$

with $kR\mathbf{I} = \mathbf{I}_1 - \mathbf{I}_2$ and $\mathbf{I} = \mathbf{I}_1$

so that $k\mathbf{I}_1 = \mathbf{I}_1 - \mathbf{I}_2$ or $\mathbf{I}_2 = \mathbf{I}_1(1 - k)$

The result is two equations and two unknowns.

Format Approach

The format approach was introduced in Section 8.9. The steps for applying this method are repeated here with changes for its use in ac circuits:

1. *Assign a loop current to each independent closed loop (as in the previous section) in a clockwise direction.*
2. *The number of required equations is equal to the number of chosen independent closed loops. Column 1 of each equation is formed by simply summing the impedance values of those impedances through which the loop current of interest passes and multiplying the result by that loop current.*
3. *We must now consider the mutual terms that are always subtracted from the terms in the first column. It is possible to have more than one mutual term if the loop current of interest has an element in common with more than one other loop current. Each mutual term is the product of the mutual impedance and the other loop current passing through the same element.*
4. *The column to the right of the equality sign is the algebraic sum of the voltage sources through which the loop current of interest passes. Positive signs are assigned to those sources of voltage having a polarity such that the loop current passes from the negative to the positive terminal. A negative sign is assigned to those potentials for which the reverse is true.*
5. *Solve resulting simultaneous equations for the desired loop currents.*

The technique is applied as above for all networks with independent sources or networks with dependent sources where the controlling variable is not a part of the network under investigation. If the controlling variable is part of the network being examined, additional care must be taken when applying the above steps.

EXAMPLE 17.9 Using the format approach to mesh analysis, find the current $\mathbf{I}_2$ in Fig. 17.15.

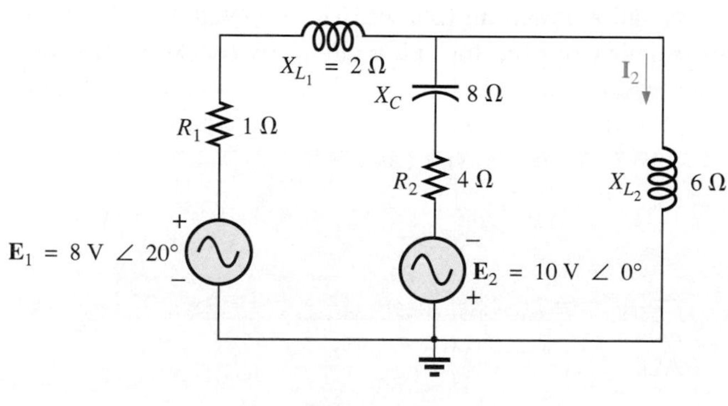

FIG. 17.15
Example 17.9.

Solution: The network is redrawn in Fig. 17.16:

$$\mathbf{Z}_1 = R_1 + jX_{L_1} = 1\ \Omega + j\,2\ \Omega \qquad \mathbf{E}_1 = 8\ \text{V}\ \angle 20°$$
$$\mathbf{Z}_2 = R_2 - jX_C = 4\ \Omega - j\,8\ \Omega \qquad \mathbf{E}_2 = 10\ \text{V}\ \angle 0°$$
$$\mathbf{Z}_3 = +jX_{L_2} = +j\,6\ \Omega$$

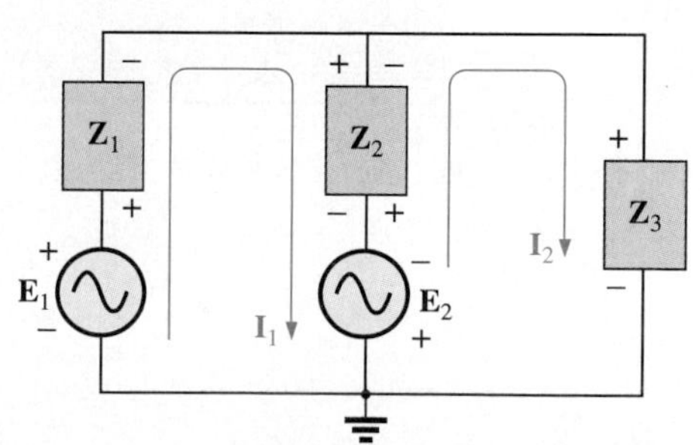

FIG. 17.16
Assigning the mesh currents and subscripted impedances for the network of Fig. 17.15.

Note the reduction in complexity of the problem with the substitution of the subscripted impedances.

Step 1 is as indicated in Fig. 17.16.

Steps 2 to 4:

$$\begin{aligned}\mathbf{I}_1(\mathbf{Z}_1 + \mathbf{Z}_2) - \mathbf{I}_2\mathbf{Z}_2 &= \mathbf{E}_1 + \mathbf{E}_2\\ \mathbf{I}_2(\mathbf{Z}_2 + \mathbf{Z}_3) - \mathbf{I}_1\mathbf{Z}_2 &= -\mathbf{E}_2\end{aligned}$$

which are rewritten as

$$\begin{aligned}\mathbf{I}_1(\mathbf{Z}_1 + \mathbf{Z}_2) - \mathbf{I}_2\mathbf{Z}_2 \qquad\qquad &= \mathbf{E}_1 + \mathbf{E}_2\\ -\mathbf{I}_1\mathbf{Z}_2 \qquad + \mathbf{I}_2(\mathbf{Z}_2 + \mathbf{Z}_3) &= -\mathbf{E}_2\end{aligned}$$

Step 5: Using determinants, we have

$$\mathbf{I}_2 = \frac{\begin{vmatrix}\mathbf{Z}_1 + \mathbf{Z}_2 & \mathbf{E}_1 + \mathbf{E}_2\\ -\mathbf{Z}_2 & -\mathbf{E}_2\end{vmatrix}}{\begin{vmatrix}\mathbf{Z}_1 + \mathbf{Z}_2 & -\mathbf{Z}_2\\ -\mathbf{Z}_2 & \mathbf{Z}_2 + \mathbf{Z}_3\end{vmatrix}}$$

$$= \frac{-(\mathbf{Z}_1 + \mathbf{Z}_2)\mathbf{E}_2 + \mathbf{Z}_2(\mathbf{E}_1 + \mathbf{E}_2)}{(\mathbf{Z}_1 + \mathbf{Z}_2)(\mathbf{Z}_2 + \mathbf{Z}_3) - \mathbf{Z}_2^2}$$

$$= \frac{\mathbf{Z}_2\mathbf{E}_1 - \mathbf{Z}_1\mathbf{E}_2}{\mathbf{Z}_1\mathbf{Z}_2 + \mathbf{Z}_1\mathbf{Z}_3 + \mathbf{Z}_2\mathbf{Z}_3}$$

Substituting numerical values yields

$$\mathbf{I}_2 = \frac{(4\ \Omega - j\,8\ \Omega)(8\ \text{V}\ \angle\ 20^\circ) - (1\ \Omega + j\,2\ \Omega)(10\ \text{V}\ \angle 0^\circ)}{(1\ \Omega + j\,2\ \Omega)(4\ \Omega - j\,8\ \Omega) + (1\ \Omega + j\,2\ \Omega)(+j\,6\ \Omega) + (4\ \Omega - j\,8\ \Omega)(+j\,6\ \Omega)}$$

$$= \frac{(4 - j\,8)(7.52 + j\,2.74) - (10 + j\,20)}{20 + (j\,6 - 12) + (j\,24 + 48)}$$

$$= \frac{(52.0 - j\,49.20) - (10 + j\,20)}{56 + j\,30} = \frac{42.0 - j\,69.20}{56 + j\,30} = \frac{80.95\ \text{A}\ \angle -58.74^\circ}{63.53\ \angle 28.18^\circ}$$

$$= \mathbf{1.27\ A\ \angle -86.92^\circ}$$

Calculator The calculator (TI-85 or equivalent) can also be an effective tool in performing the long, laborious calculations involved with the final equation appearing above. However, you must be very careful with the number of brackets and defining by brackets the order of the arithmetic operations.

```
((4,−8)*8(∠20)−(1,2)*(10∠0))/((1,2)*(4,−8)+(1,2)*(0,6)+(4,−8)*(0,6)) (ENTER)
(67.854E−3,−1.272E0)
Ans ▶ Pol
(1.274E0∠−86.956E0)
```

CALC. 17.1

MathCad The above example provides an excellent opportunity to demonstrate the power of MathCad. Note in MathCad 17.1 that the magnitude and angle of the source $\mathbf{E}_1$ are entered as E1a and E1b, respectively (similarly for $\mathbf{E}_2$). In addition, the angle must be converted to degrees, as shown by the $\pi/180$ multiplying factor. Otherwise, everything is entered as shown and the result for $\mathbf{I}_2$ appears as shown, with the magnitude in amperes and the angle in degrees.

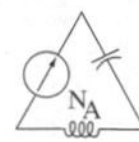


```
E1a := 8          E2a := 10          rad := 1          deg := π/180 · rad

E1b := 20         E2b := 0

E1 := E1a·cos(E1b·deg) + j ·E1a·sin(E1b·deg)     Enter "j" as "1j" even
                                                  though the "1" is not
E2 := E2a·cos(E2b·deg) + j ·E2a·sin(E2b·deg)     displayed.

Z1 := 1 + 2j          Z2 := 4 - 8j          Z3 := 6j

A := [Z1 + Z2   E1 + E2]          B := [Z1 + Z2    -Z2   ]
     [  -Z2       -E2  ]               [  -Z2    Z2 + Z3]

I2 := |A|/|B|     I2 = 0.068 - 1.272j     |I2| = 1.274     arg(I2) = -86.946·deg

When arg(I2) is entered, the answer appears is radians, but
changes to degrees when "deg" is entered at the cursor and
followed by a recalculation.
```

MATHCAD 17.1

EXAMPLE 17.10 Write the mesh equations for the network of Fig. 17.17. Do not solve.

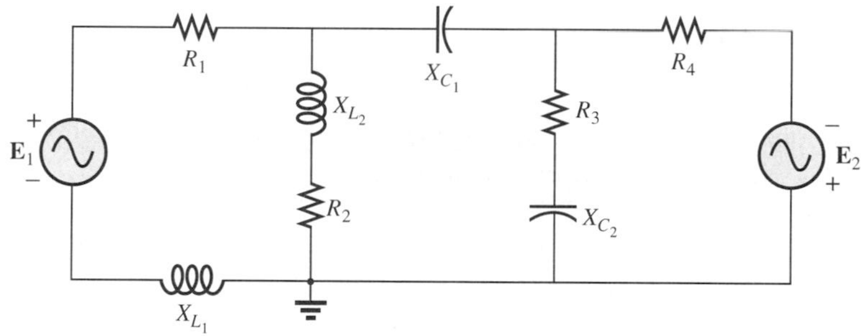

FIG. 17.17
Example 17.10.

Solution: The network is redrawn in Fig. 17.18. Again note the reduced complexity and increased clarity by the use of subscripted impedances:

$$\mathbf{Z}_1 = R_1 + jX_{L_1} \qquad \mathbf{Z}_4 = R_3 - jX_{C_2}$$
$$\mathbf{Z}_2 = R_2 + jX_{L_2} \qquad \mathbf{Z}_5 = R_4$$
$$\mathbf{Z}_3 = jX_{C_1}$$

and

$$\mathbf{I}_1(\mathbf{Z}_1 + \mathbf{Z}_2) - \mathbf{I}_2\mathbf{Z}_2 = \mathbf{E}_1$$
$$\mathbf{I}_2(\mathbf{Z}_2 + \mathbf{Z}_3 + \mathbf{Z}_4) - \mathbf{I}_1\mathbf{Z}_2 - \mathbf{I}_3\mathbf{Z}_4 = 0$$
$$\mathbf{I}_3(\mathbf{Z}_4 + \mathbf{Z}_5) - \mathbf{I}_2\mathbf{Z}_4 = \mathbf{E}_2$$

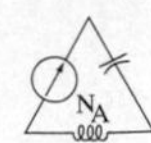

$$
\begin{array}{llll}
\text{or} & \mathbf{I}_1(\mathbf{Z}_1 + \mathbf{Z}_2) - \mathbf{I}_2(\mathbf{Z}_2) & + 0 & = \mathbf{E}_1 \\
& \mathbf{I}_1\mathbf{Z}_2 \quad - \mathbf{I}_2(\mathbf{Z}_2 + \mathbf{Z}_3 + \mathbf{Z}_4) & + \mathbf{I}_3(\mathbf{Z}_4) & = 0 \\
& 0 \quad - \mathbf{I}_2(\mathbf{Z}_4) & + \mathbf{I}_3(\mathbf{Z}_4 + \mathbf{Z}_5) & = \mathbf{E}_2
\end{array}
$$

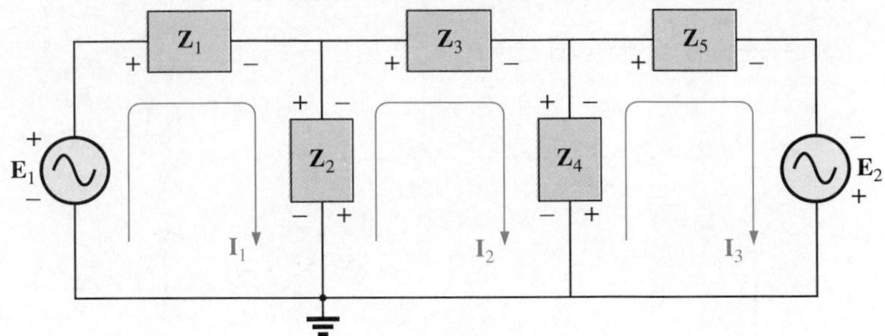

FIG. 17.18
Assigning the mesh currents and subscripted impedances for the network of Fig. 17.17.

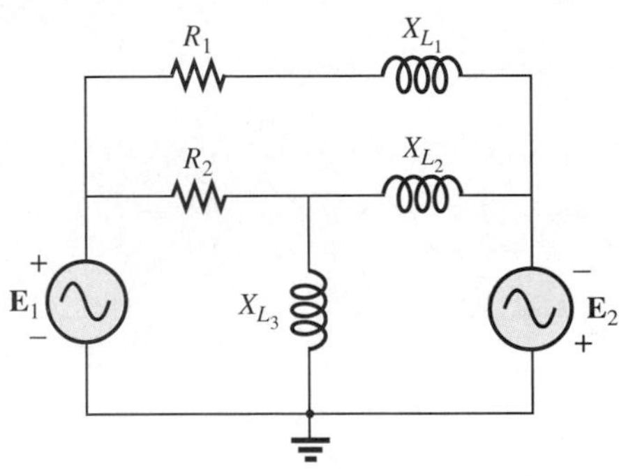

FIG. 17.19
Example 17.11

EXAMPLE 17.11 Using the format approach, write the mesh equations for the network of Fig. 17.19.

Solution: The network is redrawn as shown in Fig. 17.20, where

$$
\mathbf{Z}_1 = R_1 + j X_{L_1} \qquad \mathbf{Z}_3 = j X_{L_2}
$$
$$
\mathbf{Z}_2 = R_2 \qquad \mathbf{Z}_4 = j X_{L_3}
$$

and

$$
\mathbf{I}_1(\mathbf{Z}_2 + \mathbf{Z}_4) - \mathbf{I}_2\mathbf{Z}_2 - \mathbf{I}_3\mathbf{Z}_4 = \mathbf{E}_1
$$
$$
\mathbf{I}_2(\mathbf{Z}_1 + \mathbf{Z}_2 + \mathbf{Z}_3) - \mathbf{I}_1\mathbf{Z}_2 - \mathbf{I}_3\mathbf{Z}_3 = 0
$$
$$
\mathbf{I}_3(\mathbf{Z}_3 + \mathbf{Z}_4) - \mathbf{I}_2\mathbf{Z}_3 - \mathbf{I}_1\mathbf{Z}_4 = \mathbf{E}_2
$$

or

$$
\begin{array}{llll}
\mathbf{I}_1(\mathbf{Z}_2 + \mathbf{Z}_4) & - \mathbf{I}_2\mathbf{Z}_2 & - \mathbf{I}_3\mathbf{Z}_4 & = \mathbf{E}_1 \\
-\mathbf{I}_1\mathbf{Z}_2 & + \mathbf{I}_2(\mathbf{Z}_1 + \mathbf{Z}_2 + \mathbf{Z}_3) & - \mathbf{I}_3\mathbf{Z}_3 & = 0 \\
-\mathbf{I}_1\mathbf{Z}_4 & - \mathbf{I}_2\mathbf{Z}_3 & + \mathbf{I}_3(\mathbf{Z}_3 + \mathbf{Z}_4) & = \mathbf{E}_2
\end{array}
$$

Note the symmetry *about* the diagonal axis; that is, note the location of $-\mathbf{Z}_2$, $-\mathbf{Z}_4$, and $-\mathbf{Z}_3$ off the diagonal.

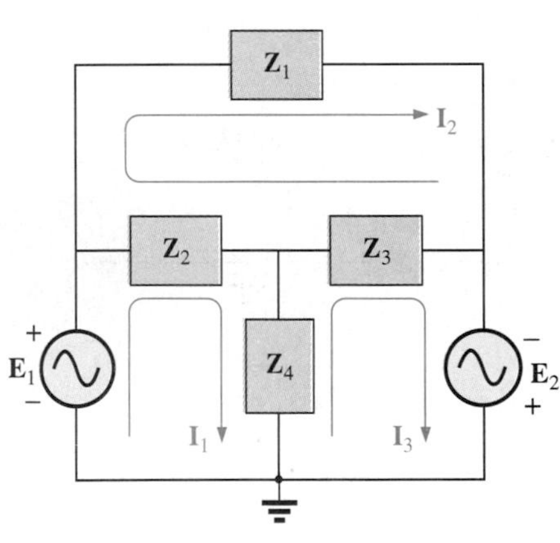

FIG. 17.20
Assigning the mesh currents and subscripted impedances for the network of Fig. 17.19.

17.5 NODAL ANALYSIS

General Approach

Independent Sources Before examining the application of the method to ac networks, a review of the appropriate sections on nodal analysis in Chapter 8 is suggested since the content of this section will be limited to the general conclusions of Chapter 8.

The fundamental steps are the following:

1. *Determine the number of nodes within the network.*
2. *Pick a reference node and label each remaining node with a subscripted value of voltage:* $\mathbf{V}_1$, $\mathbf{V}_2$, *and so on.*
3. *Apply Kirchhoff's current law at each node except the reference. Assume that all unknown currents leave the node for each application of Kirchhoff's current law.*
4. *Solve the resulting equations for the nodal voltages.*

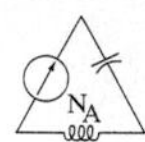

A few examples will refresh your memory about the content of Chapter 8 and the general approach to a nodal analysis solution.

EXAMPLE 17.12 Determine the voltage across the inductor for the network of Fig. 17.21.

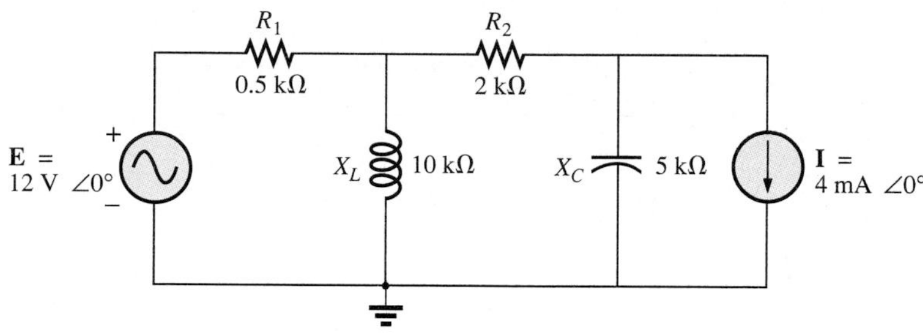

FIG. 17.21
Example 17.12.

Solution:

Steps 1 and 2 are as indicated in Fig. 17.22.

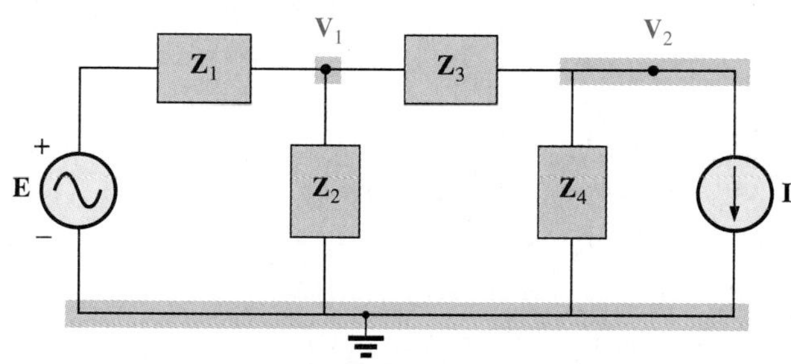

FIG. 17.22
Assigning the nodal voltages and subscripted impedances to the network of Fig. 17.21.

Step 3: Note Fig. 17.23 for the application of Kirchhoff's current law to node $\mathbf{V}_1$.

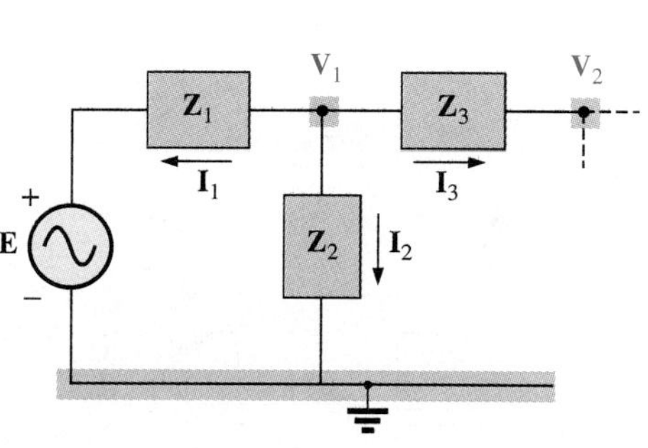

FIG. 17.23
Applying Kirchhoff's current law to the node $\mathbf{V}_1$ of Fig. 17.22.

$$\Sigma\mathbf{I}_i = \Sigma\mathbf{I}_o$$

$$0 = \mathbf{I}_1 + \mathbf{I}_2 + \mathbf{I}_3$$

$$\frac{\mathbf{V}_1 - \mathbf{E}}{\mathbf{Z}_1} + \frac{\mathbf{V}_1}{\mathbf{Z}_2} + \frac{\mathbf{V}_1 - \mathbf{V}_2}{\mathbf{Z}_3} = 0$$

Rearranging terms:

$$\mathbf{V}_1\left[\frac{1}{\mathbf{Z}_1} + \frac{1}{\mathbf{Z}_2} + \frac{1}{\mathbf{Z}_3}\right] - \mathbf{V}_2\left[\frac{1}{\mathbf{Z}_3}\right] = \frac{\mathbf{E}_1}{\mathbf{Z}_1} \qquad \textbf{(17.1)}$$

Note Fig. 17.24 for the application of Kirchhoff's current law to node $\mathbf{V}_2$.

$$0 = \mathbf{I}_3 + \mathbf{I}_4 + \mathbf{I}$$

$$\frac{\mathbf{V}_2 - \mathbf{V}_1}{\mathbf{Z}_3} + \frac{\mathbf{V}_2}{\mathbf{Z}_4} + \mathbf{I} = 0$$

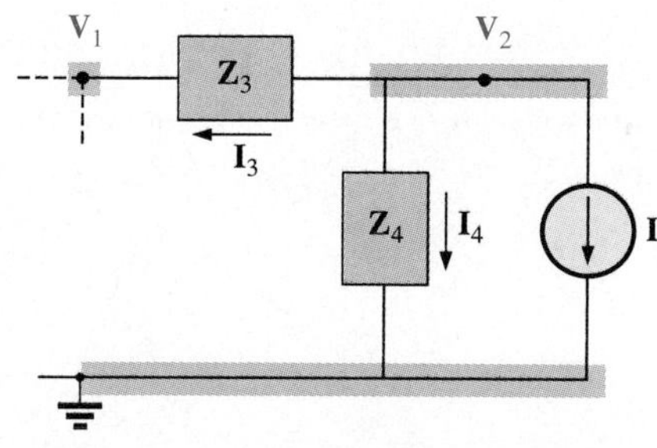

FIG. 17.24
Applying Kirchhoff's current law to the node $\mathbf{V}_2$ of Fig. 17.22.

Rearranging terms:

$$\mathbf{V}_2\left[\frac{1}{\mathbf{Z}_3}+\frac{1}{\mathbf{Z}_4}\right]-\mathbf{V}_1\left[\frac{1}{\mathbf{Z}_3}\right]=-\mathbf{I} \qquad \textbf{(17.2)}$$

Grouping equations:

$$\mathbf{V}_1\left[\frac{1}{\mathbf{Z}_1}+\frac{1}{\mathbf{Z}_2}+\frac{1}{\mathbf{Z}_3}\right]-\mathbf{V}_2\left[\frac{1}{\mathbf{Z}_3}\right]=\frac{\mathbf{E}}{\mathbf{Z}_1}$$

$$\mathbf{V}_1\left[\frac{1}{\mathbf{Z}_3}\right]-\mathbf{V}_2\left[\frac{1}{\mathbf{Z}_3}+\frac{1}{\mathbf{Z}_4}\right]=\mathbf{I}$$

$$\frac{1}{\mathbf{Z}_1}+\frac{1}{\mathbf{Z}_2}+\frac{1}{\mathbf{Z}_3}=\frac{1}{0.5\text{ k}\Omega}+\frac{1}{j\,10\text{ k}\Omega}+\frac{1}{2\text{ k}\Omega}=2.5\text{ mS}\angle-2.29^\circ$$

$$\frac{1}{\mathbf{Z}_3}+\frac{1}{\mathbf{Z}_4}=\frac{1}{2\text{ k}\Omega}+\frac{1}{-j\,5\text{ k}\Omega}=0.539\text{ mS}\angle21.80^\circ$$

and

$$\mathbf{V}_1[2.5\text{ mS}\angle-2.29^\circ]-\mathbf{V}_2[0.5\text{ mS}\angle0^\circ]=24\text{ mA}\angle0^\circ$$
$$\mathbf{V}_1[0.5\text{ mS}\angle0^\circ]-\mathbf{V}_2[0.539\text{ mS}\angle21.80^\circ]=4\text{ mA}\angle0^\circ$$

with

$$\mathbf{V}_1=\frac{\begin{vmatrix}24\text{ mA}\angle0^\circ & -0.5\text{ mS}\angle0^\circ\\ 4\text{ mA}\angle0^\circ & -0.539\text{ mS}\angle21.80^\circ\end{vmatrix}}{\begin{vmatrix}2.5\text{ mS}\angle-2.29^\circ & -0.5\text{ mS}\angle0^\circ\\ 0.5\text{ mS}\angle0^\circ & -0.539\text{ mS}\angle21.80^\circ\end{vmatrix}}$$

$$=\frac{(24\text{ mA}\angle0^\circ)(-0.539\text{ mS}\angle21.80^\circ)+(0.5\text{ mS}\angle0^\circ)(4\text{ mA}\angle0^\circ)}{(2.5\text{ mS}\angle-2.29^\circ)(-0.539\text{ mS}\angle21.80^\circ)+(0.5\text{ mS}\angle0^\circ)(0.5\text{ mS}\angle0^\circ)}$$

$$=\frac{-12.94\times10^{-6}\text{ V}\angle21.80^\circ+2\times10^{-6}\text{ V}\angle0^\circ}{-1.348\times10^{-6}\angle19.51^\circ+0.25\times10^{-6}\angle0^\circ}$$

$$=\frac{-(12.01+j\,4.81)\times10^{-6}\text{ V}+2\times10^{-6}\text{ V}}{-(1.271+j\,0.45)\times10^{-6}+0.25\times10^{-6}}$$

$$=\frac{-10.01\text{ V}-j\,4.81\text{ V}}{-1.021-j\,0.45}=\frac{11.106\text{ V}\angle-154.33^\circ}{1.116\angle-156.21^\circ}$$

$$\mathbf{V}_1=\mathbf{9.95\text{ V}\angle1.88^\circ}$$

MathCad The length and complexity of the above mathematical development strongly suggest the use of an alternative approach such as MathCad. Note in MathCad 17.2 that the equations are entered in the same format as Eqs. (17.1) and (17.2). Both $\mathbf{V}_1$ and $\mathbf{V}_2$ were generated, but because only $\mathbf{V}_1$ was asked for, it was the only solution converted to the polar form. In the lower solution the complexity was significantly reduced by simply recognizing that the current is in milliamperes and the impedances in kilohms. The result will then be in volts.

$$K := 10^{3} \qquad m := 10^{-3} \qquad rad := 1$$

$$V1 := 1 + j \qquad V2 := 1 + j \qquad deg := \frac{\pi}{180}$$

Given

$$V1 \cdot \left[\frac{1}{.5 \cdot K} + \frac{1}{10j \cdot K} + \frac{1}{2 \cdot K}\right] - V2 \cdot \frac{1}{2 \cdot K} \approx 24 \cdot m$$

$$V1 \cdot \left[\frac{1}{2 \cdot K}\right] - V2 \cdot \left[\frac{1}{2 \cdot K} + \frac{1}{-5j \cdot K}\right] \approx 4 \cdot m$$

$$Find(V1, V2) = \begin{bmatrix} 9.944 + 0.319j \\ 1.786 - 0.396j \end{bmatrix} \begin{matrix} Volts \\ Volts \end{matrix}$$

$$V1 := 9.944 + .319j \qquad |V1| = 9.949 \qquad arg(V1) = 1.837 \cdot deg$$

Recognizing that current in mA results when Z is in kilohms, an alternative format follows:

Given

$$V1 \cdot \left[0.5^{-1} + (10j)^{-1} + 2^{-1}\right] - V2 \cdot 2^{-1} \approx 24$$

$$V1 \cdot 2^{-1} - V2 \cdot \left[2^{-1} + (-5j)^{-1}\right] \approx 4$$

$$Find(V1, V2) = \begin{bmatrix} 9.944 + 0.319j \\ 1.786 - 0.396j \end{bmatrix} \begin{matrix} Volts \\ Volts \end{matrix}$$

$$V1 := 9.944 + .319j \qquad |V1| = 9.949 \qquad arg(V1) = 1.837 \cdot deg$$

MATHCAD 17.2

Dependent Current Sources For dependent current sources, the procedure is modified as follows:

1. Steps 1 and 2 are the same as those applied for independent sources.
2. Step 3 is modified as follows: Treat each dependent current source like an independent source when Kirchhoff's current law is applied to each defined node. However, once the equations are established, substitute the equation for the controlling quantity to ensure that the unknowns are limited solely to the chosen nodal voltages.
3. Step 4 is as before.

EXAMPLE 17.13 Write the nodal equations for the network of Fig. 17.25 having a dependent current source.

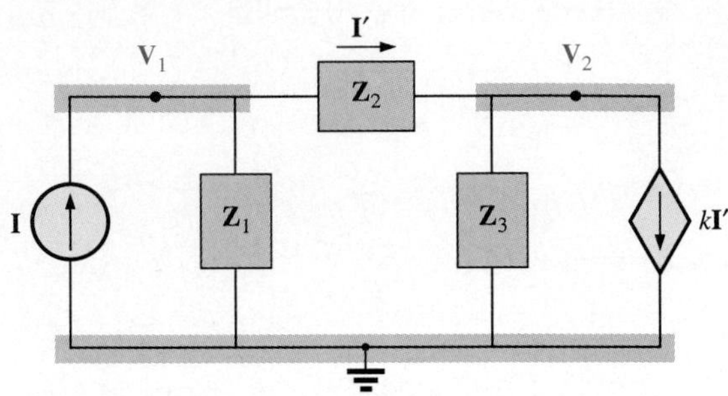

FIG. 17.25
Applying nodal analysis to a network with a current-controlled current source.

Solution:

Steps 1 and 2 are as defined in Fig. 17.25.

Step 3: At node $\mathbf{V}_1$,

$$\mathbf{I} = \mathbf{I}_1 + \mathbf{I}_2$$

$$\frac{\mathbf{V}_1}{\mathbf{Z}_1} + \frac{\mathbf{V}_1 - \mathbf{V}_2}{\mathbf{Z}_2} - \mathbf{I} = 0$$

and

$$\mathbf{V}_1\left[\frac{1}{\mathbf{Z}_1} + \frac{1}{\mathbf{Z}_2}\right] - \mathbf{V}_2\left[\frac{1}{\mathbf{Z}_2}\right] = \mathbf{I}$$

At node $\mathbf{V}_2$,

$$0 = \mathbf{I}_2 + \mathbf{I}_3 + k\mathbf{I}'$$

$$\frac{\mathbf{V}_2 - \mathbf{V}_1}{\mathbf{Z}_2} + \frac{\mathbf{V}_2}{\mathbf{Z}_3} + k\left[\frac{\mathbf{V}_1 - \mathbf{V}_2}{\mathbf{Z}_2}\right] = 0$$

and

$$\mathbf{V}_1\left[\frac{1-k}{\mathbf{Z}_2}\right] - \mathbf{V}_2\left[\frac{1-k}{\mathbf{Z}_2} + \frac{1}{\mathbf{Z}_3}\right] = 0$$

resulting in two equations and two unknowns.

Independent Voltage Sources Between Assigned Nodes

For independent voltage sources between assigned nodes, the procedure is modified as follows:

1. Steps 1 and 2 are the same as those applied for independent sources.
2. Step 3 is modified as follows: Treat each source between defined nodes as a short circuit (recall the *supernode* classification of Chapter 8), and write the nodal equations for each remaining independent node. Then relate the chosen nodal voltages to the independent voltage source to ensure that the unknowns of the final equations are limited solely to the nodal voltages.
3. Step 4 is as before.

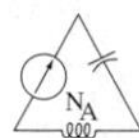

EXAMPLE 17.14 Write the nodal equations for the network of Fig. 17.26 having an independent source between two assigned nodes.

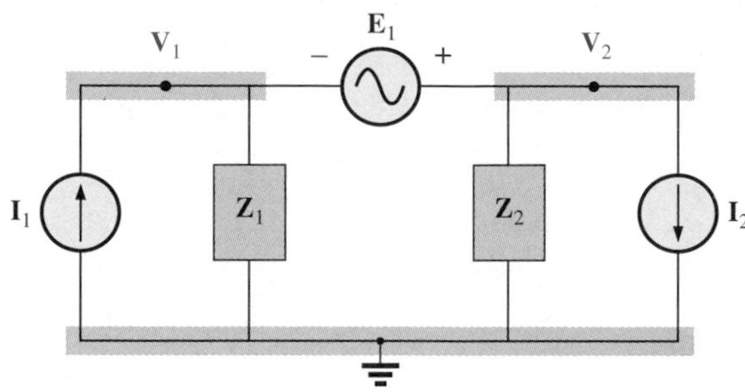

FIG. 17.26
Applying nodal analysis to a network with an independent voltage source between defined nodes.

Solution:

Steps 1 and 2 are defined in Fig. 17.26.

Step 3: Replacing the independent source **E** with a short-circuit equivalent results in a supernode that will generate the following equation when Kirchhoff's current law is applied to node $\mathbf{V}_1$:

$$\mathbf{I}_1 = \frac{\mathbf{V}_1}{\mathbf{Z}_1} + \frac{\mathbf{V}_2}{\mathbf{Z}_2} + \mathbf{I}_2$$

with
$$\mathbf{V}_2 - \mathbf{V}_1 = \mathbf{E}$$

and we have two equations and two unknowns.

Dependent Voltage Sources Between Defined Nodes For dependent voltage sources between defined nodes, the procedure is modified as follows:

1. Steps 1 and 2 are the same as those applied for independent voltage sources.
2. Step 3 is modified as follows: The procedure is essentially the same as that applied for independent voltage sources, except now the dependent sources have to be defined in terms of the chosen nodal voltages to ensure that the final equations have only nodal voltages as their unknown quantities.
3. Step 4 is as before.

EXAMPLE 17.15 Write the nodal equations for the network of Fig. 17.27 having a dependent voltage source between two defined nodes.

Solution:

Steps 1 and 2 are defined in Fig. 17.27.

Step 3: Replacing the dependent source $\mu\mathbf{V}_x$ with a short-circuit equivalent will result in the following equation when Kirchhoff's current law is applied at node $\mathbf{V}_1$:

$$\mathbf{I} = \mathbf{I}_1 + \mathbf{I}_2$$

$$\frac{\mathbf{V}_1}{\mathbf{Z}_1} + \frac{(\mathbf{V}_1 - \mathbf{V}_2)}{\mathbf{Z}_2} - \mathbf{I} = 0$$

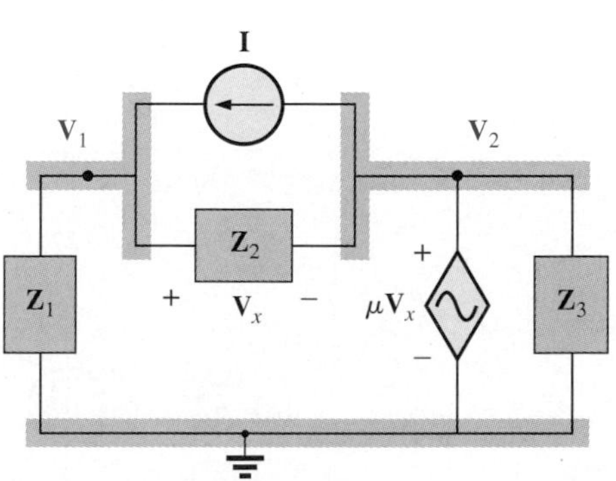

FIG. 17.27
Applying nodal analysis to a network with a voltage-controlled voltage source.

and $$\mathbf{V}_2 = \mu\mathbf{V}_x = \mu[\mathbf{V}_1 - \mathbf{V}_2]$$

or $$\mathbf{V}_2 = \frac{\mu}{1 + \mu}\mathbf{V}_1$$

resulting in two equations and two unknowns. Note that because the impedance $\mathbf{Z}_3$ is in parallel with a voltage source it does not appear in the analysis. It will, however, affect the current through the dependent voltage source.

Format Approach

A close examination of Eqs. (17.1) and (17.2) in Example 17.11 will reveal that they are the same equations that would have been obtained using the format approach introduced in Chapter 8. Recall that the approach required that the voltage source first be converted to a current source, but the writing of the equations was quite direct and minimized any chances of an error due to a lost sign or missing term.

The sequence of steps required to apply the format approach is the following:

1. ***Choose a reference node and assign a subscripted voltage label to the (N − 1) remaining independent nodes of the network.***
2. ***The number of equations required for a complete solution is equal to the number of subscripted voltages (N − 1). Column 1 of each equation is formed by summing the admittances tied to the node of interest and multiplying the result by that subscripted nodal voltage.***
3. ***The mutual terms are always subtracted from the terms of the first column. It is possible to have more than one mutual term if the nodal voltage of interest has an element in common with more than one other nodal voltage. Each mutual term is the product of the mutual admittance and the other nodal voltage tied to that admittance.***
4. ***The column to the right of the equality sign is the algebraic sum of the current sources tied to the node of interest. A current source is assigned a positive sign if it supplies current to a node, and a negative sign if it draws current from the node.***
5. ***Solve resulting simultaneous equations for the desired nodal voltages. The comments offered for mesh analysis regarding independent and dependent sources apply here also.***

EXAMPLE 17.16 Using the format approach to nodal analysis, find the voltage across the 4-Ω resistor in Fig. 17.28.

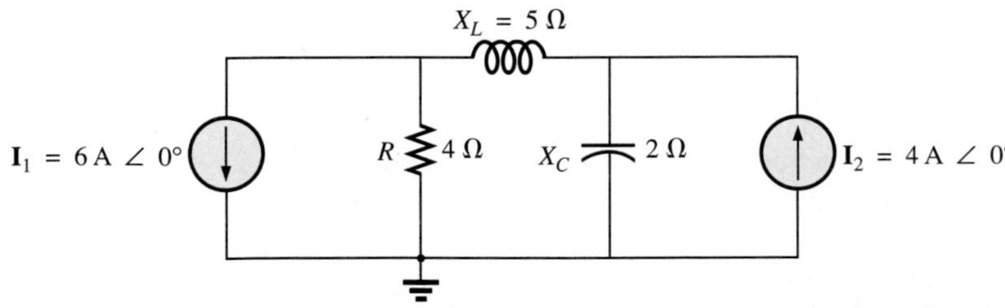

FIG. 17.28
Example 17.16.

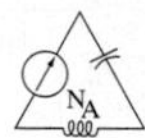

Solution: Choosing nodes (Fig. 17.29) and writing the nodal equations, we have

$$\mathbf{Z}_1 = R = 4\,\Omega \qquad \mathbf{Z}_2 = jX_L = j\,5\,\Omega \quad \mathbf{Z}_3 = -jX_C = -j\,2\,\Omega$$

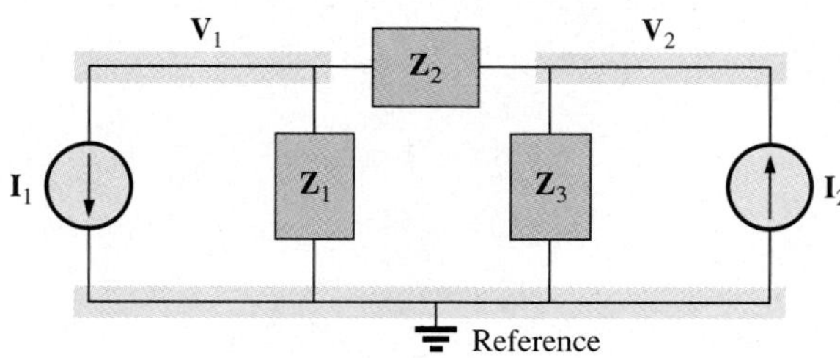

FIG. 17.29
Assigning the nodal voltages and subscripted impedances for the network of Fig. 17.28.

$$\mathbf{V}_1(\mathbf{Y}_1 + \mathbf{Y}_2) - \mathbf{V}_2(\mathbf{Y}_2) = -\mathbf{I}_1$$
$$\mathbf{V}_2(\mathbf{Y}_3 + \mathbf{Y}_2) - \mathbf{V}_1(\mathbf{Y}_2) = +\mathbf{I}_2$$

or

$$\begin{aligned} \mathbf{V}_1(\mathbf{Y}_1 + \mathbf{Y}_2) - \mathbf{V}_2(\mathbf{Y}_2) \qquad &= -\mathbf{I}_1 \\ -\mathbf{V}_1(\mathbf{Y}_2) \qquad + \mathbf{V}_2(\mathbf{Y}_3 + \mathbf{Y}_2) &= +\mathbf{I}_2 \end{aligned}$$

$$\mathbf{Y}_1 = \frac{1}{\mathbf{Z}_1} \qquad \mathbf{Y}_2 = \frac{1}{\mathbf{Z}_2} \qquad \mathbf{Y}_3 = \frac{1}{\mathbf{Z}_3}$$

Using determinants yields

$$\begin{aligned} \mathbf{V}_1 &= \frac{\begin{vmatrix} -\mathbf{I}_1 & -\mathbf{Y}_2 \\ +\mathbf{I}_2 & \mathbf{Y}_3 + \mathbf{Y}_2 \end{vmatrix}}{\begin{vmatrix} \mathbf{Y}_1 + \mathbf{Y}_2 & -\mathbf{Y}_2 \\ -\mathbf{Y}_2 & \mathbf{Y}_3 + \mathbf{Y}_2 \end{vmatrix}} \\ &= \frac{-(\mathbf{Y}_3 + \mathbf{Y}_2)\mathbf{I}_1 + \mathbf{I}_2\mathbf{Y}_2}{(\mathbf{Y}_1 + \mathbf{Y}_2)(\mathbf{Y}_3 + \mathbf{Y}_2) - \mathbf{Y}_2^2} \\ &= \frac{-(\mathbf{Y}_3 + \mathbf{Y}_2)\mathbf{I}_1 + \mathbf{I}_2\mathbf{Y}_2}{\mathbf{Y}_1\mathbf{Y}_3 + \mathbf{Y}_2\mathbf{Y}_3 + \mathbf{Y}_1\mathbf{Y}_2} \end{aligned}$$

Substituting numerical values, we have

$$\begin{aligned} \mathbf{V}_1 &= \frac{-[(1/-j\,2\,\Omega) + (1/j\,5\,\Omega)]6\text{ A}\angle 0^\circ + 4\text{ A}\angle 0^\circ(1/j\,5\,\Omega)}{(1/4\,\Omega)(1/-j\,2\,\Omega) + (1/j\,5\,\Omega)(1/-j\,2\,\Omega) + (1/4\,\Omega)(1/j\,5\,\Omega)} \\ &= \frac{-(+j\,0.5 - j\,0.2)6\angle 0^\circ + 4\angle 0^\circ(-j\,0.2)}{(1/-j\,8) + (1/10) + (1/j\,20)} \\ &= \frac{(-0.3\angle 90^\circ)(6\angle 0^\circ) + (4\angle 0^\circ)(0.2\angle -90^\circ)}{j\,0.125 + 0.1 - j\,0.05} \\ &= \frac{-1.8\angle 90^\circ + 0.8\angle -90^\circ}{0.1 + j\,0.075} \\ &= \frac{2.6\text{ V}\angle -90^\circ}{0.125\angle 36.87^\circ} \\ &= \mathbf{20.80\text{ V}\angle -126.87^\circ} \end{aligned}$$

MathCad Using MathCad and the matrix format with the admittance parameters will quickly provide a solution for $\mathbf{V}_1$ in Example 17.16.

```
Z1 := 4        Z2 := 5j       Z3 := -2j       rad := 1        deg := π/180

        ⎡⎡  -1      -1⎤          -1          ⎤
        ⎢⎣Z1   + Z2   ⎦       -Z2            ⎥             ⎡-6⎤
Y :=    ⎢       -1         ⎡  -1      -1⎤    ⎥      I :=   ⎣ 4⎦
        ⎣   -Z2            ⎣Z2   + Z3   ⎦    ⎦

 -1         ⎡-12.48 - 16.64j⎤     Volts
Y  · I  =   ⎣  8.32 - 2.24j ⎦     Volts

V1 := -12.48 - 16.64j       |V1| = 20.8        arg(V1) = -126.87·deg

V2 := 8.32 - 2.24j          |V2| = 8.616       arg(V2) = -15.068·deg
```

MATHCAD 17.3

EXAMPLE 17.17 Using the format approach, write the nodal equations for the network of Fig. 17.30.

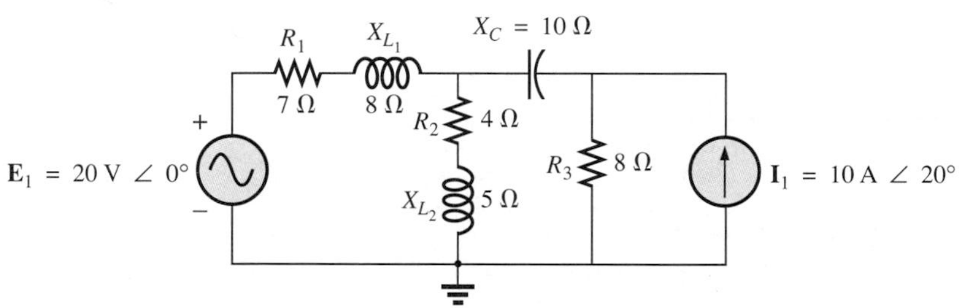

FIG. 17.30
Example 17.17.

Solution: The circuit is redrawn in Fig. 17.31, where

$$\mathbf{Z}_1 = R_1 + jX_{L_1} = 7\,\Omega + j\,8\,\Omega \qquad \mathbf{E}_1 = 20\text{ V}\angle 0^\circ$$
$$\mathbf{Z}_2 = R_2 + jX_{L_2} = 4\,\Omega + j\,5\,\Omega \qquad \mathbf{I}_1 = 10\text{ A}\angle 20^\circ$$
$$\mathbf{Z}_3 = -jX_C = -j\,10\,\Omega$$
$$\mathbf{Z}_4 = R_3 = 8\,\Omega$$

Converting the voltage source to a current source and choosing nodes, we obtain Fig. 17.32. Note the "neat" appearance of the network using the subscripted impedances. Working directly with Fig. 17.30 would be more difficult and may produce errors.

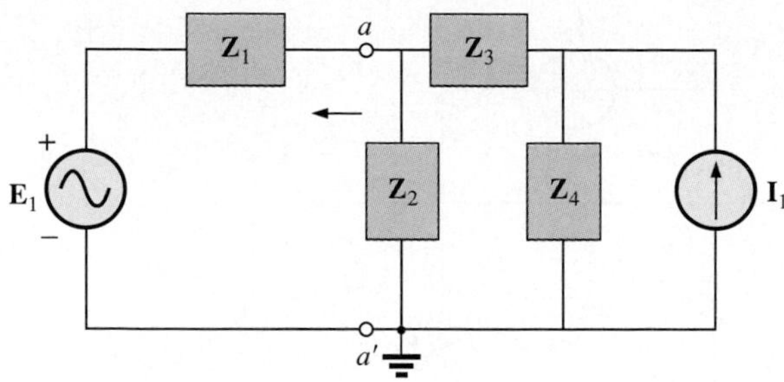

FIG. 17.31
Assigning the subscripted impedances for the network of Fig. 17.30.

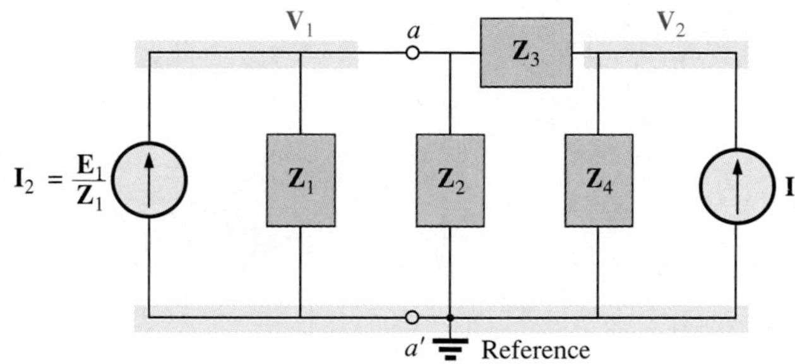

FIG. 17.32
Converting the voltage source of Fig. 17.31 to a current source and defining the nodal voltages.

Write the nodal equations:

$$\mathbf{V}_1(\mathbf{Y}_1 + \mathbf{Y}_2 + \mathbf{Y}_3) - \mathbf{V}_2(\mathbf{Y}_3) = +\mathbf{I}_2$$
$$\mathbf{V}_2(\mathbf{Y}_3 + \mathbf{Y}_4) - \mathbf{V}_1(\mathbf{Y}_3) = +\mathbf{I}_1$$

$$\mathbf{Y}_1 = \frac{1}{\mathbf{Z}_1} \quad \mathbf{Y}_2 = \frac{1}{\mathbf{Z}_2} \quad \mathbf{Y}_3 = \frac{1}{\mathbf{Z}_3} \quad \mathbf{Y}_4 = \frac{1}{\mathbf{Z}_4}$$

which are rewritten as

$$\begin{aligned} \mathbf{V}_1(\mathbf{Y}_1 + \mathbf{Y}_2 + \mathbf{Y}_3) - \mathbf{V}_2(\mathbf{Y}_3) \qquad &= +\mathbf{I}_2 \\ -\mathbf{V}_1(\mathbf{Y}_3) \qquad + \mathbf{V}_2(\mathbf{Y}_3 + \mathbf{Y}_4) &= +\mathbf{I}_1 \end{aligned}$$

EXAMPLE 17.18 Write the nodal equations for the network of Fig. 17.33. Do not solve.

Solution: Choose nodes (Fig. 17.34):

$$\mathbf{Z}_1 = R_1 \qquad \mathbf{Z}_2 = jX_{L_1} \qquad \mathbf{Z}_3 = R_2 - jX_{C_2}$$
$$\mathbf{Z}_4 = -jX_{C_1} \qquad \mathbf{Z}_5 = R_3 \qquad \mathbf{Z}_6 = jX_{L_2}$$

and write the nodal equations:

$$\mathbf{V}_1(\mathbf{Y}_1 + \mathbf{Y}_2) - \mathbf{V}_2(\mathbf{Y}_2) = +\mathbf{I}_1$$
$$\mathbf{V}_2(\mathbf{Y}_2 + \mathbf{Y}_3 + \mathbf{Y}_4) - \mathbf{V}_1(\mathbf{Y}_2) - \mathbf{V}_3(\mathbf{Y}_4) = -\mathbf{I}_2$$
$$\mathbf{V}_3(\mathbf{Y}_4 + \mathbf{Y}_5 + \mathbf{Y}_6) - \mathbf{V}_2(\mathbf{Y}_4) = +\mathbf{I}_2$$

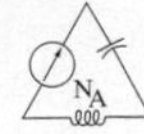

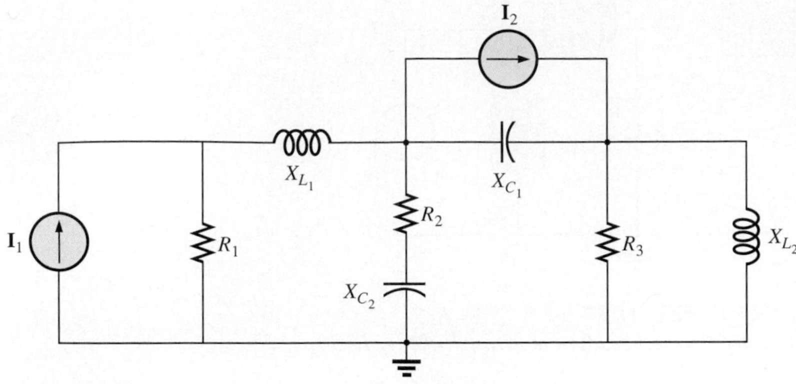

FIG. 17.33
Example 17.18.

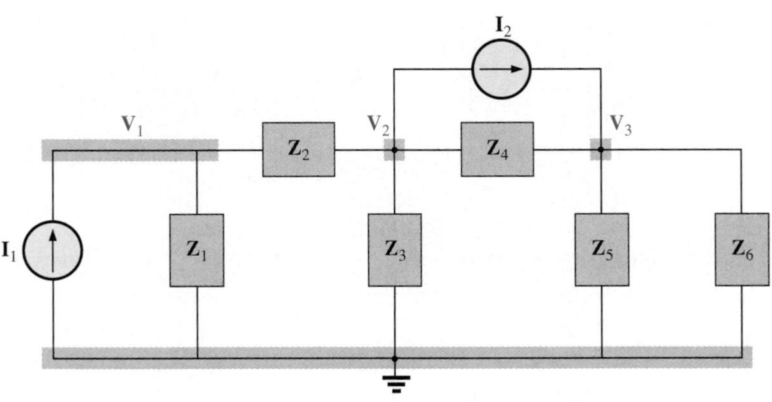

FIG. 17.34
Assigning the nodal voltages and subscripted impedances for the network of Fig. 17.33.

which are rewritten as

$$\begin{array}{llll}
\mathbf{V}_1(\mathbf{Y}_1 + \mathbf{Y}_2) - \mathbf{V}_2(\mathbf{Y}_2) & & + 0 & = +\mathbf{I}_1 \\
-\mathbf{V}_1(\mathbf{Y}_2) & + \mathbf{V}_2(\mathbf{Y}_2 + \mathbf{Y}_3 + \mathbf{Y}_4) & - \mathbf{V}_3(\mathbf{Y}_4) & = -\mathbf{I}_2 \\
0 & - \mathbf{V}_2(\mathbf{Y}_4) & + \mathbf{V}_3(\mathbf{Y}_4 + \mathbf{Y}_5 + \mathbf{Y}_6) & = +\mathbf{I}_2
\end{array}$$

$$\mathbf{Y}_1 = \frac{1}{R_1} \qquad \mathbf{Y}_2 = \frac{1}{j X_{L_1}} \qquad \mathbf{Y}_3 = \frac{1}{R_2 - j X_{C_2}}$$

$$\mathbf{Y}_4 = \frac{1}{-j X_{C_1}} \qquad \mathbf{Y}_5 = \frac{1}{R_3} \qquad \mathbf{Y}_6 = \frac{1}{j X_{L_2}}$$

Note the symmetry about the diagonal for this example and those preceding it in this section.

EXAMPLE 17.19 Apply nodal analysis to the network of Fig. 17.35. Determine the voltage $\mathbf{V}_L$.

Solution: In this case there is no need for a source conversion. The network is redrawn in Fig. 17.36 with the chosen node voltage and subscripted impedances.

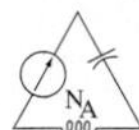

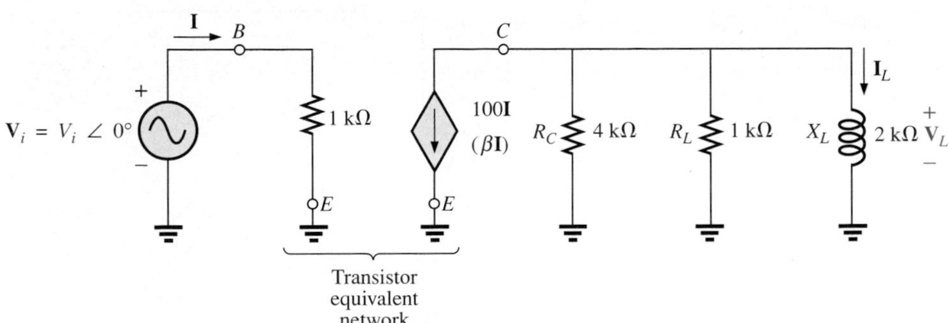

FIG. 17.35
Example 17.19.

Apply the format approach:

$$\mathbf{Y}_1 = \frac{1}{\mathbf{Z}_1} = \frac{1}{4\text{ k}\Omega} = 0.25\text{ mS}\angle 0° = G_1\angle 0°$$

$$\mathbf{Y}_2 = \frac{1}{\mathbf{Z}_2} = \frac{1}{1\text{ k}\Omega} = 1\text{ mS}\angle 0° = G_2\angle 0°$$

$$\mathbf{Y}_3 = \frac{1}{\mathbf{Z}_3} = \frac{1}{2\text{ k}\Omega\angle 90°} = 0.5\text{ mS}\angle -90°$$

$$= -j\,0.5\text{ mS} = -j\,B_L$$

$$\mathbf{V}_1\!: (\mathbf{Y}_1 + \mathbf{Y}_2 + \mathbf{Y}_3)\mathbf{V}_1 = -100\mathbf{I}$$

and
$$\mathbf{V}_1 = \frac{-100\mathbf{I}}{\mathbf{Y}_1 + \mathbf{Y}_2 + \mathbf{Y}_3}$$

$$= \frac{-100\mathbf{I}}{0.25\text{ mS} + 1\text{ mS} - j\,0.5\text{ mS}}$$

$$= \frac{-100\times 10^3\mathbf{I}}{1.25 - j\,0.5} = \frac{-100\times 10^3\mathbf{I}}{1.3463\angle -21.80°}$$

$$= -74.28\times 10^3\mathbf{I}\angle 21.80°$$

$$= -74.28\times 10^3\left(\frac{\mathbf{V}_i}{1\text{ k}\Omega}\right)\angle 21.80°$$

$$\mathbf{V}_1 = \mathbf{V}_L = \mathbf{-(74.28V}_i\mathbf{)\ V\angle 21.80°}$$

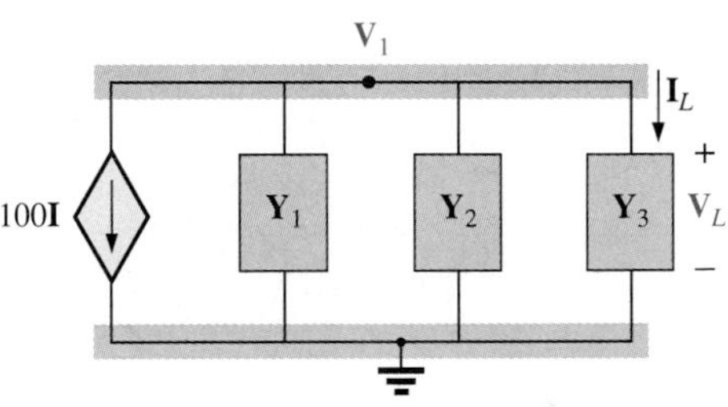

FIG. 17.36
Assigning the nodal voltage and subscripted impedances for the network of Fig. 17.35.

17.6 BRIDGE NETWORKS (ac)

The basic bridge configuration was discussed in some detail in Section 8.11 for dc networks. We now continue to examine bridge networks by considering those that have reactive components and a sinusoidal ac voltage or current applied.

We will first analyze various familiar forms of the bridge network using mesh analysis and nodal analysis (the format approach). The balance conditions will be investigated throughout the section.

Apply *mesh analysis* to the network of Fig. 17.37. The network is redrawn in Fig. 17.38, where

$$\mathbf{Z}_1 = \frac{1}{\mathbf{Y}_1} = \frac{1}{G_1 + j\,B_C} = \frac{G_1}{G_1^2 + B_C^2} - j\,\frac{B_C}{G_1^2 + B_C^2}$$

$$\mathbf{Z}_2 = R_2 \qquad \mathbf{Z}_3 = R_3 \qquad \mathbf{Z}_4 = R_4 + j\,X_L \qquad \mathbf{Z}_5 = R_5$$

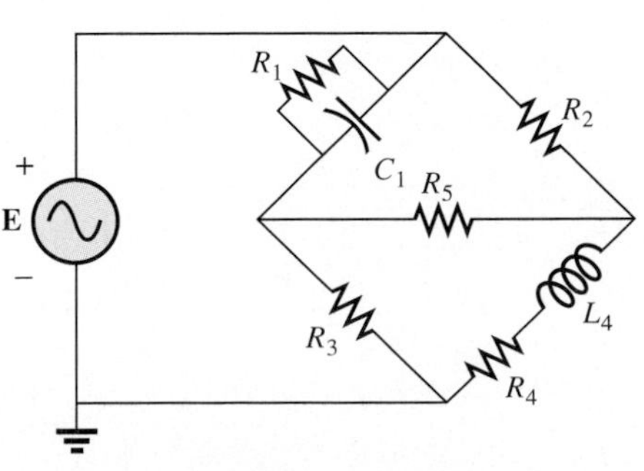

FIG. 17.37
Maxwell bridge.

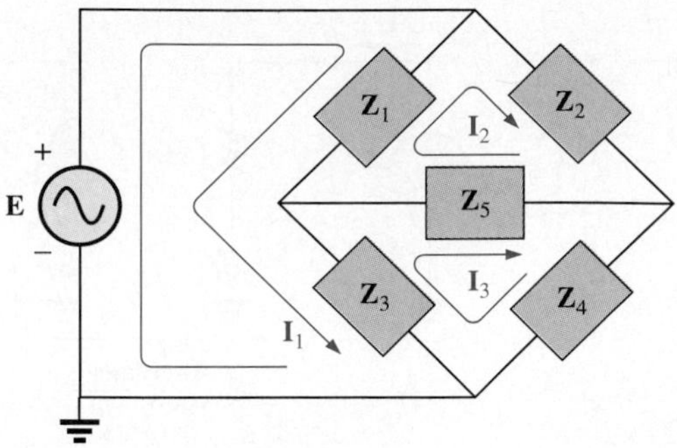

FIG. 17.38
Assigning the mesh currents and subscripted impedances for the network of Fig. 17.37.

Applying the format approach:

$$\begin{aligned}(\mathbf{Z}_1 + \mathbf{Z}_3)\mathbf{I}_1 - (\mathbf{Z}_1)\mathbf{I}_2 - (\mathbf{Z}_3)\mathbf{I}_3 &= \mathbf{E}\\ (\mathbf{Z}_1 + \mathbf{Z}_2 + \mathbf{Z}_5)\mathbf{I}_2 - (\mathbf{Z}_1)\mathbf{I}_1 - (\mathbf{Z}_5)\mathbf{I}_3 &= 0\\ (\mathbf{Z}_3 + \mathbf{Z}_4 + \mathbf{Z}_5)\mathbf{I}_3 - (\mathbf{Z}_3)\mathbf{I}_1 - (\mathbf{Z}_5)\mathbf{I}_2 &= 0\end{aligned}$$

which are rewritten as

$$\begin{array}{llll}\mathbf{I}_1(\mathbf{Z}_1 + \mathbf{Z}_3) & -\ \mathbf{I}_2\mathbf{Z}_1 & -\ \mathbf{I}_3\mathbf{Z}_3 & = \mathbf{E}\\ -\mathbf{I}_1\mathbf{Z}_1 & +\ \mathbf{I}_2(\mathbf{Z}_1 + \mathbf{Z}_2 + \mathbf{Z}_5) & -\ \mathbf{I}_3\mathbf{Z}_5 & = 0\\ -\mathbf{I}_1\mathbf{Z}_3 & -\ \mathbf{I}_2\mathbf{Z}_5 & +\ \mathbf{I}_3(\mathbf{Z}_3 + \mathbf{Z}_4 + \mathbf{Z}_5) & = 0\end{array}$$

Note the symmetry about the diagonal of the above equations. For balance, $\mathbf{I}_{\mathbf{Z}_5} = 0$ A, and

$$\mathbf{I}_{\mathbf{Z}_5} = \mathbf{I}_2 - \mathbf{I}_3 = 0$$

From the above equations,

$$\mathbf{I}_2 = \frac{\begin{vmatrix}\mathbf{Z}_1 + \mathbf{Z}_3 & \mathbf{E} & -\mathbf{Z}_3\\ -\mathbf{Z}_1 & 0 & -\mathbf{Z}_5\\ -\mathbf{Z}_3 & 0 & (\mathbf{Z}_3 + \mathbf{Z}_4 + \mathbf{Z}_5)\end{vmatrix}}{\begin{vmatrix}\mathbf{Z}_1 + \mathbf{Z}_3 & -\mathbf{Z}_1 & -\mathbf{Z}_3\\ -\mathbf{Z}_1 & (\mathbf{Z}_1 + \mathbf{Z}_2 + \mathbf{Z}_5) & -\mathbf{Z}_5\\ -\mathbf{Z}_3 & -\mathbf{Z}_5 & (\mathbf{Z}_3 + \mathbf{Z}_4 + \mathbf{Z}_5)\end{vmatrix}}$$

$$= \frac{\mathbf{E}(\mathbf{Z}_1\mathbf{Z}_3 + \mathbf{Z}_1\mathbf{Z}_4 + \mathbf{Z}_1\mathbf{Z}_5 + \mathbf{Z}_3\mathbf{Z}_5)}{\Delta}$$

where Δ signifies the determinant of the denominator (or coefficients). Similarly,

$$\mathbf{I}_3 = \frac{\mathbf{E}(\mathbf{Z}_1\mathbf{Z}_3 + \mathbf{Z}_3\mathbf{Z}_2 + \mathbf{Z}_1\mathbf{Z}_5 + \mathbf{Z}_3\mathbf{Z}_5)}{\Delta}$$

and
$$\mathbf{I}_{\mathbf{Z}_5} = \mathbf{I}_2 - \mathbf{I}_3 = \frac{\mathbf{E}(\mathbf{Z}_1\mathbf{Z}_4 - \mathbf{Z}_3\mathbf{Z}_2)}{\Delta}$$

For $\mathbf{I}_{\mathbf{Z}_5} = 0$, the following must be satisfied (for a finite Δ not equal to zero):

$$\boxed{\mathbf{Z}_1\mathbf{Z}_4 = \mathbf{Z}_3\mathbf{Z}_2} \qquad \mathbf{I}_{\mathbf{Z}_5} = 0 \tag{17.3}$$

This condition will be analyzed in greater depth later in this section.

Applying *nodal analysis* to the network of Fig. 17.39 will result in the configuration of Fig. 17.40, where

$$\mathbf{Y}_1 = \frac{1}{\mathbf{Z}_1} = \frac{1}{R_1 - jX_C} \qquad \mathbf{Y}_2 = \frac{1}{\mathbf{Z}_2} = \frac{1}{R_2}$$

$$\mathbf{Y}_3 = \frac{1}{\mathbf{Z}_3} = \frac{1}{R_3} \qquad \mathbf{Y}_4 = \frac{1}{\mathbf{Z}_4} = \frac{1}{R_4 + jX_L} \qquad \mathbf{Y}_5 = \frac{1}{R_5}$$

and

$$\begin{aligned}(\mathbf{Y}_1 + \mathbf{Y}_2)\mathbf{V}_1 - (\mathbf{Y}_1)\mathbf{V}_2 - (\mathbf{Y}_2)\mathbf{V}_3 &= \mathbf{I}\\ (\mathbf{Y}_1 + \mathbf{Y}_3 + \mathbf{Y}_5)\mathbf{V}_2 - (\mathbf{Y}_1)\mathbf{V}_1 - (\mathbf{Y}_5)\mathbf{V}_3 &= 0\\ (\mathbf{Y}_2 + \mathbf{Y}_4 + \mathbf{Y}_5)\mathbf{V}_3 - (\mathbf{Y}_2)\mathbf{V}_1 - (\mathbf{Y}_5)\mathbf{V}_2 &= 0\end{aligned}$$

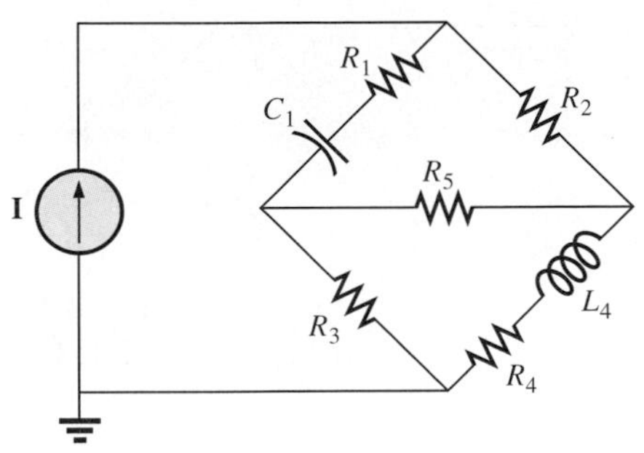

FIG. 17.39
Hay bridge.

which are rewritten as

$$\begin{aligned}\mathbf{V}_1(\mathbf{Y}_1 + \mathbf{Y}_2) - \mathbf{V}_2\mathbf{Y}_1 \qquad\qquad - \mathbf{V}_3\mathbf{Y}_2 &= \mathbf{I}\\ -\mathbf{V}_1\mathbf{Y}_1 + \mathbf{V}_2(\mathbf{Y}_1 + \mathbf{Y}_3 + \mathbf{Y}_5) - \mathbf{V}_3\mathbf{Y}_5 &= 0\\ -\mathbf{V}_1\mathbf{Y}_2 - \mathbf{V}_2\mathbf{Y}_5 + \mathbf{V}_3(\mathbf{Y}_2 + \mathbf{Y}_4 + \mathbf{Y}_5) &= 0\end{aligned}$$

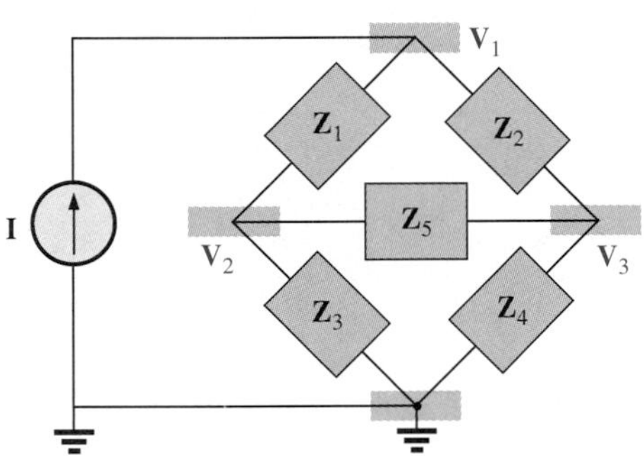

FIG. 17.40
Assigning the nodal voltages and subscripted impedances for the network of Fig. 17.39.

Again, note the symmetry about the diagonal axis. For balance, $\mathbf{V}_{\mathbf{Z}_5} = 0$ V, and

$$\mathbf{V}_{\mathbf{Z}_5} = \mathbf{V}_2 - \mathbf{V}_3 = 0$$

From the above equations,

$$\mathbf{V}_2 = \frac{\begin{vmatrix}\mathbf{Y}_1 + \mathbf{Y}_2 & \mathbf{I} & -\mathbf{Y}_2\\ -\mathbf{Y}_1 & 0 & -\mathbf{Y}_5\\ -\mathbf{Y}_2 & 0 & (\mathbf{Y}_2 + \mathbf{Y}_4 + \mathbf{Y}_5)\end{vmatrix}}{\begin{vmatrix}\mathbf{Y}_1 + \mathbf{Y}_2 & -\mathbf{Y}_1 & -\mathbf{Y}_2\\ -\mathbf{Y}_1 & (\mathbf{Y}_1 + \mathbf{Y}_3 + \mathbf{Y}_5) & -\mathbf{Y}_5\\ -\mathbf{Y}_2 & -\mathbf{Y}_5 & (\mathbf{Y}_2 + \mathbf{Y}_4 + \mathbf{Y}_5\end{vmatrix}}$$

$$= \frac{\mathbf{I}(\mathbf{Y}_1\mathbf{Y}_3 + \mathbf{Y}_1\mathbf{Y}_4 + \mathbf{Y}_1\mathbf{Y}_5 + \mathbf{Y}_3\mathbf{Y}_5)}{\Delta}$$

Similarly,

$$\mathbf{V}_3 = \frac{\mathbf{I}(\mathbf{Y}_1\mathbf{Y}_3 + \mathbf{Y}_3\mathbf{Y}_2 + \mathbf{Y}_1\mathbf{Y}_5 + \mathbf{Y}_3\mathbf{Y}_5)}{\Delta}$$

Note the similarities between the above equations and those obtained for mesh analysis. Then

$$\mathbf{V}_{\mathbf{Z}_5} = \mathbf{V}_2 - \mathbf{V}_3 = \frac{\mathbf{I}(\mathbf{Y}_1\mathbf{Y}_4 - \mathbf{Y}_3\mathbf{Y}_2)}{\Delta}$$

For $\mathbf{V}_{\mathbf{Z}_5} = 0$, the following must be satisfied for a finite Δ not equal to zero:

$$\boxed{\mathbf{Y}_1\mathbf{Y}_4 = \mathbf{Y}_3\mathbf{Y}_2} \qquad \mathbf{V}_{\mathbf{Z}_5} = 0 \qquad \textbf{(17.4)}$$

However, substituting $\mathbf{Y}_1 = 1/\mathbf{Z}_1$, $\mathbf{Y}_2 = 1/\mathbf{Z}_2$, $\mathbf{Y}_3 = 1/\mathbf{Z}_3$, and $\mathbf{Y}_4 = 1/\mathbf{Z}_4$, we have

$$\frac{1}{\mathbf{Z}_1\mathbf{Z}_4} = \frac{1}{\mathbf{Z}_3\mathbf{Z}_2}$$

or

$$\boxed{\mathbf{Z}_1\mathbf{Z}_4 = \mathbf{Z}_3\mathbf{Z}_2} \qquad \mathbf{V}_{\mathbf{Z}_5} = 0$$

corresponding with Eq. (17.3) obtained earlier.

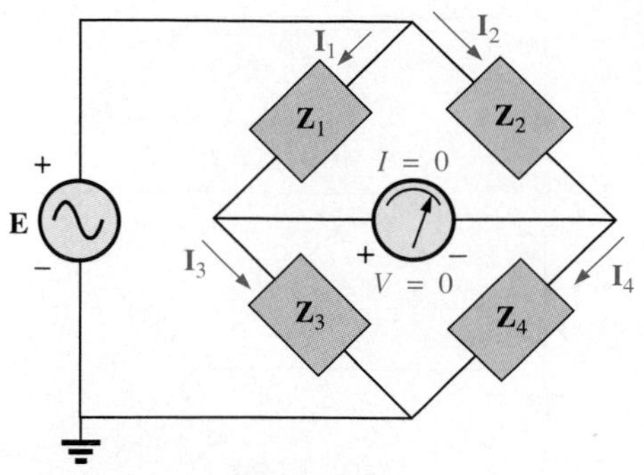

FIG. 17.41
Investigating the balance criteria for an ac bridge configuration.

Let us now investigate the balance criteria in more detail by considering the network of Fig. 17.41, where it is specified that **I**, **V** = 0.

Since **I** = 0,

$$\mathbf{I}_1 = \mathbf{I}_3 \tag{17.5a}$$

and

$$\mathbf{I}_2 = \mathbf{I}_4 \tag{17.5b}$$

In addition, for **V** = 0,

$$\mathbf{I}_1\mathbf{Z}_1 = \mathbf{I}_2\mathbf{Z}_2 \tag{17.5c}$$

and

$$\mathbf{I}_3\mathbf{Z}_3 = \mathbf{I}_4\mathbf{Z}_4 \tag{17.5d}$$

Substituting the preceding current relations into Eq. (17.5d), we have

$$\mathbf{I}_1\mathbf{Z}_3 = \mathbf{I}_2\mathbf{Z}_4$$

and

$$\mathbf{I}_2 = \frac{\mathbf{Z}_3}{\mathbf{Z}_4}\mathbf{I}_1$$

Substituting this relationship for $\mathbf{I}_2$ into Eq. (17.5c) yields

$$\mathbf{I}_1\mathbf{Z}_1 = \left(\frac{\mathbf{Z}_3}{\mathbf{Z}_4}\mathbf{I}_1\right)\mathbf{Z}_2$$

and

$$\mathbf{Z}_1\mathbf{Z}_4 = \mathbf{Z}_2\mathbf{Z}_3$$

as obtained earlier. Rearranging, we have

$$\frac{\mathbf{Z}_1}{\mathbf{Z}_3} = \frac{\mathbf{Z}_2}{\mathbf{Z}_4} \tag{17.6}$$

corresponding with Eq. (8.4) for dc resistive networks.

For the network of Fig. 17.39, which is referred to as a *Hay bridge* when $\mathbf{Z}_5$ is replaced by a sensitive galvanometer,

$$\mathbf{Z}_1 = R_1 - jX_C$$
$$\mathbf{Z}_2 = R_2$$
$$\mathbf{Z}_3 = R_3$$
$$\mathbf{Z}_4 = R_4 + jX_L$$

This particular network is used for measuring the resistance and inductance of coils in which the resistance is a small fraction of the reactance X_L.

Substitute into Eq. (17.6) in the following form:

$$\mathbf{Z}_2\mathbf{Z}_3 = \mathbf{Z}_4\mathbf{Z}_1$$
$$R_2R_3 = (R_4 + jX_L)(R_1 - jX_C)$$

or

$$R_2R_3 = R_1R_4 + j(R_1X_L - R_4X_C) + X_CX_L$$

so that

$$R_2R_3 + j0 = (R_1R_4 + X_CX_L) + j(R_1X_L - R_4X_C)$$

For the equations to be equal, *the real and imaginary parts must be equal.* Therefore, for a balanced Hay bridge,

$$\boxed{R_2R_3 = R_1R_4 + X_CX_L} \tag{17.7a}$$

and

$$\boxed{0 = R_1X_L - R_4X_C} \tag{17.7b}$$

or substituting $X_L = \omega L$ and $X_C = \dfrac{1}{\omega C}$

we have

$$X_CX_L = \left(\frac{1}{\omega C}\right)(\omega L) = \frac{L}{C}$$

and

$$R_2R_3 = R_1R_4 + \frac{L}{C}$$

with

$$R_1\omega L = \frac{R_4}{\omega C}$$

Solving for R_4 in the last equation yields

$$R_4 = \omega^2LCR_1$$

and substituting into the previous equation, we have

$$R_2R_3 = R_1(\omega^2LCR_1) + \frac{L}{C}$$

Multiply through by C and factor:

$$CR_2R_3 = L(\omega^2C^2R_1^2 + 1)$$

and

$$\boxed{L = \frac{CR_2R_3}{1 + \omega^2C^2R_1^2}} \tag{17.8a}$$

With additional algebra this yields:

$$\boxed{R_4 = \frac{\omega^2C^2R_1R_2R_3}{1 + \omega^2C^2R^2{}_1}} \tag{17.8b}$$

Equations (17.7) and (17.8) are the balance conditions for the Hay bridge. Note that each is frequency dependent. For different frequencies, the resistive and capacitive elements must vary for a particular coil to achieve balance. For a coil placed in the Hay bridge as shown in Fig. 17.39, the resistance and inductance of the coil can be determined by Eqs. (17.8a) and (17.8b) when balance is achieved.

The bridge of Fig. 17.37 is referred to as a *Maxwell bridge* when $\mathbf{Z}_5$ is replaced by a sensitive galvanometer. This setup is used for inductance measurements when the resistance of the coil is large enough not to require a Hay bridge.

Application of Eq. (17.6) in the form:

$$\mathbf{Z}_2\mathbf{Z}_3 = \mathbf{Z}_4\mathbf{Z}_1$$

and substituting

$$\mathbf{Z}_1 = R_1 \angle 0° \parallel X_{C_1} \angle -90° = \frac{(R_1 \angle 0°)(X_{C_1} \angle -90°)}{R_1 - jX_{C_1}}$$

$$= \frac{R_1X_{C_1} \angle -90°}{R_1 - jX_{C_1}} = \frac{-jR_1X_{C_1}}{R_1 - jX_{C_1}}$$

$$\mathbf{Z}_2 = R_2$$
$$\mathbf{Z}_3 = R_3$$
and $$\mathbf{Z}_4 = R_4 + jX_{L_4}$$

we have $$(R_2)(R_3) = (R_4 + jX_{L_4})\left(\frac{-jR_1X_{C_1}}{R_1 - jX_{C_1}}\right)$$

$$R_2R_3 = \frac{-jR_1R_4X_{C_1} + R_1X_{C_1}X_{L_4}}{R_1 - jX_{C_1}}$$

or $$(R_2R_3)(R_1 - jX_{C_1}) = R_1X_{C_1}X_{L_4} - jR_1R_4X_{C_1}$$

and $$R_1R_2R_3 - jR_2R_3X_{C_1} = R_1X_{C_1}X_{L_4} - jR_1R_4X_{C_1}$$

so that for balance

$$\not{R}_1R_2R_3 = \not{R}_1X_{C_1}X_{L_4}$$
$$R_2R_3 = \left(\frac{1}{2\not{\pi}fC_1}\right)(2\not{\pi}fL_4)$$

and $$\boxed{L_4 = C_1R_2R_3} \tag{17.9}$$

and $$R_2R_3\not{X}_{C_1} = R_1R_4\not{X}_{C_1}$$

so that $$\boxed{R_4 = \frac{R_2R_3}{R_1}} \tag{17.10}$$

Note the absence of frequency in Eqs. (17.9) and (17.10).

One remaining popular bridge is the *capacitance comparison bridge* of Fig. 17.42. An unknown capacitance and its associated resistance can be determined using this bridge. Application of Eq. (17.6) will yield the following results:

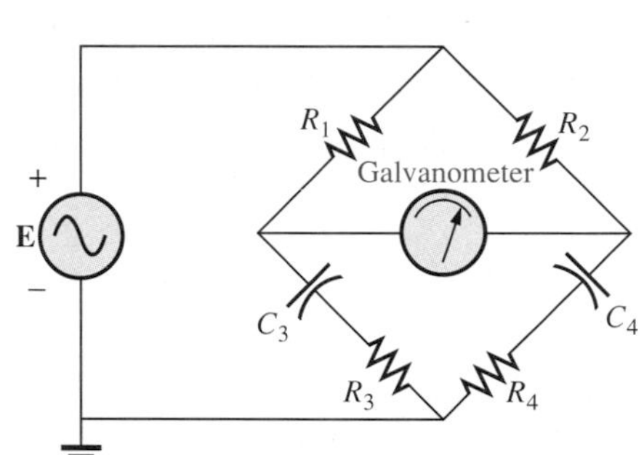

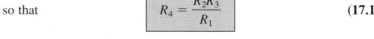

FIG. 17.42
Capacitance comparison bridge.

$$\boxed{C_4 = C_3\frac{R_1}{R_2}} \tag{17.11}$$

$$\boxed{R_4 = \frac{R_2R_3}{R_1}} \tag{17.12}$$

The derivation of these equations will appear as a problem at the end of the chapter.

17.7 Δ-Y, Y-Δ CONVERSIONS

The Δ-Y, Y-Δ (or π-T, T-π as defined in Section 8.12) conversions for ac circuits will not be derived here since the development corresponds exactly with that for dc circuits. Taking the Δ-Y configuration shown in Fig. 17.43, we find the general equations for the impedances of the Y in terms of those for the Δ:

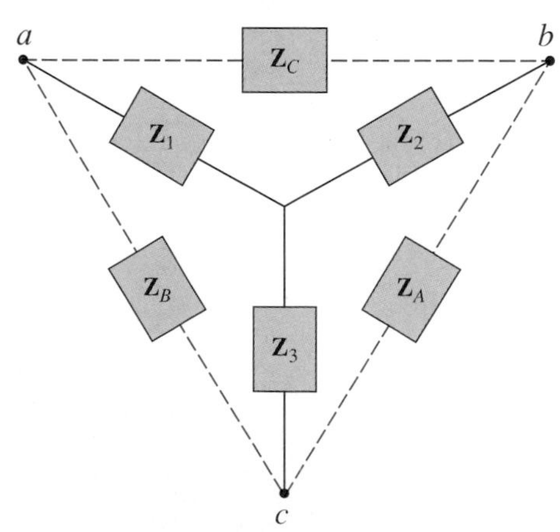

FIG. 17.43
Δ-Y configuration.

$$\boxed{\mathbf{Z}_1 = \frac{\mathbf{Z}_B\mathbf{Z}_C}{\mathbf{Z}_A + \mathbf{Z}_B + \mathbf{Z}_C}} \tag{17.13}$$

$$\mathbf{Z}_2 = \frac{\mathbf{Z}_A\mathbf{Z}_C}{\mathbf{Z}_A + \mathbf{Z}_B + \mathbf{Z}_C} \quad \textbf{(17.14)}$$

$$\mathbf{Z}_3 = \frac{\mathbf{Z}_A\mathbf{Z}_B}{\mathbf{Z}_A + \mathbf{Z}_B + \mathbf{Z}_C} \quad \textbf{(17.15)}$$

For the impedances of the Δ in terms of those for the Y, the equations are

$$\mathbf{Z}_B = \frac{\mathbf{Z}_1\mathbf{Z}_2 + \mathbf{Z}_1\mathbf{Z}_3 + \mathbf{Z}_2\mathbf{Z}_3}{\mathbf{Z}_2} \quad \textbf{(17.16)}$$

$$\mathbf{Z}_A = \frac{\mathbf{Z}_1\mathbf{Z}_2 + \mathbf{Z}_1\mathbf{Z}_3 + \mathbf{Z}_2\mathbf{Z}_3}{\mathbf{Z}_1} \quad \textbf{(17.17)}$$

$$\mathbf{Z}_C = \frac{\mathbf{Z}_1\mathbf{Z}_2 + \mathbf{Z}_1\mathbf{Z}_3 + \mathbf{Z}_2\mathbf{Z}_3}{\mathbf{Z}_3} \quad \textbf{(17.18)}$$

Note that each impedance of the Y is equal to the product of the impedances in the two closest branches of the Δ, divided by the sum of the impedances in the Δ.

Further, the value of each impedance of the Δ is equal to the sum of the possible product combinations of the impedances of the Y, divided by the impedances of the Y farthest from the impedance to be determined.

Drawn in different forms (Fig. 17.44), they are also referred to as the T and π configurations.

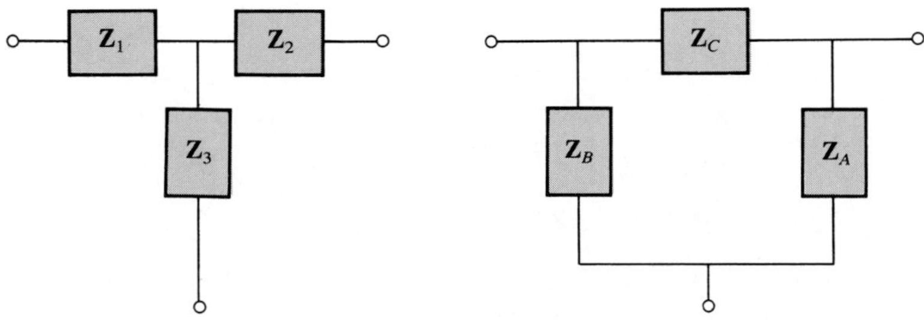

FIG. 17.44
The T and π configurations.

In the study of dc networks, we found that if all of the resistors of the Δ or Y were the same, the conversion from one to the other could be accomplished using the equation

$$R_\Delta = 3R_Y \quad \text{or} \quad R_Y = \frac{R_\Delta}{3}$$

For ac networks,

$$\mathbf{Z}_\Delta = 3\mathbf{Z}_Y \quad \text{or} \quad \mathbf{Z}_Y = \frac{\mathbf{Z}_\Delta}{3} \quad \textbf{(17.19)}$$

Be careful when using this simplified form. It is not sufficient for all the impedances of the Δ or Y to be of the same magnitude: *The angle associated with each must also be the same.*

EXAMPLE 17.20 Find the total impedance $\mathbf{Z}_T$ of the network of Fig. 17.45.

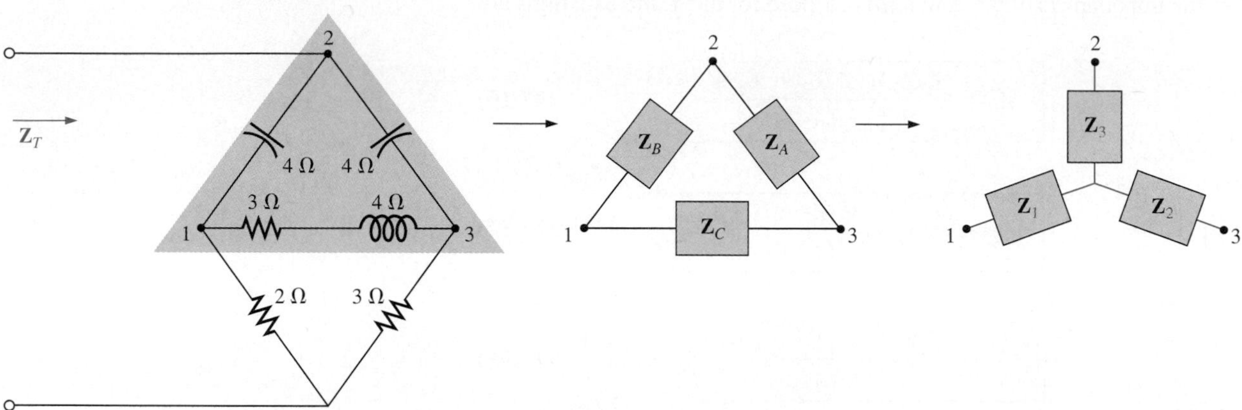

FIG. 17.45
Converting the upper Δ of a bridge configuration to a Y.

Solution:

$$\mathbf{Z}_B = -j\,4 \qquad \mathbf{Z}_A = -j\,4 \qquad \mathbf{Z}_C = 3 + j\,4$$

$$\mathbf{Z}_1 = \frac{\mathbf{Z}_B\mathbf{Z}_C}{\mathbf{Z}_A + \mathbf{Z}_B + \mathbf{Z}_C} = \frac{(-j\,4\,\Omega)(3\,\Omega + j\,4\,\Omega)}{(-j\,4\,\Omega) + (-j\,4\,\Omega) + (3\,\Omega + j\,4\,\Omega)}$$

$$= \frac{(4\,\angle -90^\circ)(5\,\angle 53.13^\circ)}{3 - j\,4} = \frac{20\,\angle -36.87^\circ}{5\,\angle -53.13^\circ}$$

$$= 4\,\Omega\,\angle 16.13^\circ = 3.84\,\Omega + j\,1.11\,\Omega$$

$$\mathbf{Z}_2 = \frac{\mathbf{Z}_A\mathbf{Z}_C}{\mathbf{Z}_A + \mathbf{Z}_B + \mathbf{Z}_C} = \frac{(-j\,4\,\Omega)(3\,\Omega + j\,4\,\Omega)}{5\,\Omega\,\angle -53.13^\circ}$$

$$= 4\,\Omega\,\angle 16.13^\circ = 3.84\,\Omega + j\,1.11\,\Omega$$

Recall from the study of dc circuits that if two branches of the Y or Δ were the same, the corresponding Δ or Y, respectively, would also have two similar branches. In this example, $\mathbf{Z}_A = \mathbf{Z}_B$. Therefore, $\mathbf{Z}_1 = \mathbf{Z}_2$, and

$$\mathbf{Z}_3 = \frac{\mathbf{Z}_A\mathbf{Z}_B}{\mathbf{Z}_A + \mathbf{Z}_B + \mathbf{Z}_C} = \frac{(-j\,4\,\Omega)(-j\,4\,\Omega)}{5\,\Omega\,\angle -53.13^\circ}$$

$$= \frac{16\,\Omega\,\angle -180^\circ}{5\,\angle -53.13^\circ} = 3.2\,\Omega\,\angle -126.87^\circ = -1.92\,\Omega - j\,2.56\,\Omega$$

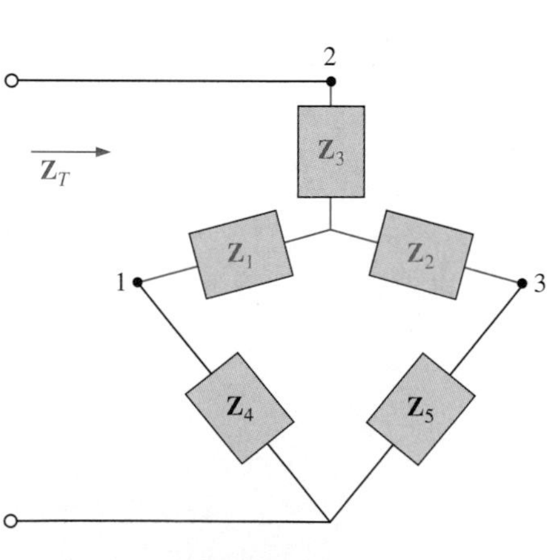

FIG. 17.46
The network of Fig. 17.45 following the substitution of the Y configuration.

Replace the Δ by the Y (Fig. 17.46):

$$\mathbf{Z}_1 = 3.84\,\Omega + j\,1.11\,\Omega \qquad \mathbf{Z}_2 = 3.84\,\Omega + j\,1.11\,\Omega$$
$$\mathbf{Z}_3 = -1.92\,\Omega - j\,2.56\,\Omega \quad \mathbf{Z}_4 = 2\,\Omega \qquad \mathbf{Z}_5 = 3\,\Omega$$

Impedances $\mathbf{Z}_1$ and $\mathbf{Z}_4$ are in series:

$$\mathbf{Z}_{T_1} = \mathbf{Z}_1 + \mathbf{Z}_4 = 3.84\,\Omega + j\,1.11\,\Omega + 2\,\Omega = 5.84\,\Omega + j\,1.11\,\Omega$$
$$= 5.94\,\Omega\,\angle 10.76^\circ$$

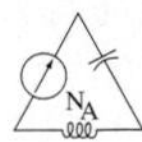

Impedances $\mathbf{Z}_2$ and $\mathbf{Z}_5$ are in series:

$$\begin{aligned}\mathbf{Z}_{T_2} &= \mathbf{Z}_2 + \mathbf{Z}_5 = 3.84\ \Omega + j\,1.11\ \Omega + 3\ \Omega = 6.84\ \Omega + j\,1.11\ \Omega \\ &= 6.93\ \Omega\ \angle 9.22^\circ\end{aligned}$$

Impedances $\mathbf{Z}_{T_1}$ and $\mathbf{Z}_{T_2}$ are in parallel:

$$\begin{aligned}\mathbf{Z}_{T_3} &= \frac{\mathbf{Z}_{T_1}\mathbf{Z}_{T_2}}{\mathbf{Z}_{T_1} + \mathbf{Z}_{T_2}} = \frac{(5.94\ \Omega\ \angle 10.76^\circ)(6.93\ \Omega\ \angle 9.22^\circ)}{5.84\ \Omega + j\,1.11\ \Omega + 6.84\ \Omega + j\,1.11\ \Omega} \\ &= \frac{41.16\ \Omega\ \angle 19.98^\circ}{12.68 + j\,2.22} = \frac{41.16\ \Omega\ \angle 19.98^\circ}{12.87\ \angle 9.93^\circ} = 3.198\ \Omega\ \angle 10.05^\circ \\ &= 3.15\ \Omega + j\,0.56\ \Omega\end{aligned}$$

Impedances $\mathbf{Z}_3$ and $\mathbf{Z}_{T_3}$ are in series. Therefore,

$$\begin{aligned}\mathbf{Z}_T = \mathbf{Z}_3 + \mathbf{Z}_{T_3} &= -1.92\ \Omega - j\,2.56\ \Omega + 3.15\ \Omega + j\,0.56\ \Omega \\ &= 1.23\ \Omega - j\,2.0\ \Omega = \mathbf{2.35\ \Omega\ \angle -58.41^\circ}\end{aligned}$$

EXAMPLE 17.21 Using both the Δ-Y and Y-Δ transformations, find the total impedance $\mathbf{Z}_T$ for the network of Fig. 17.47.

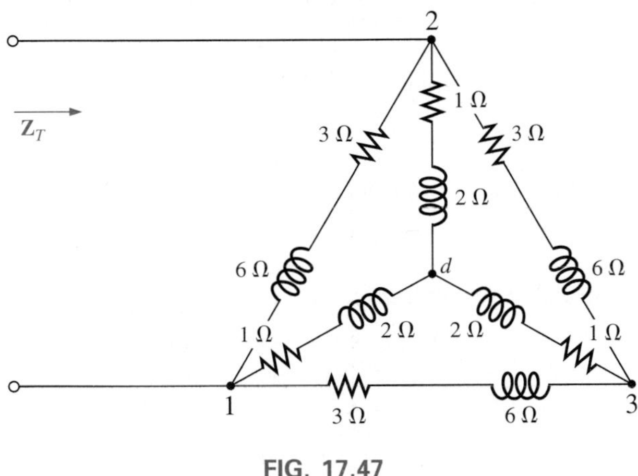

FIG. 17.47
Example 17.21.

Solution: *Using the Δ-Y transformation,* we obtain Fig. 17.48. In this case, since both systems are balanced (same impedance in each

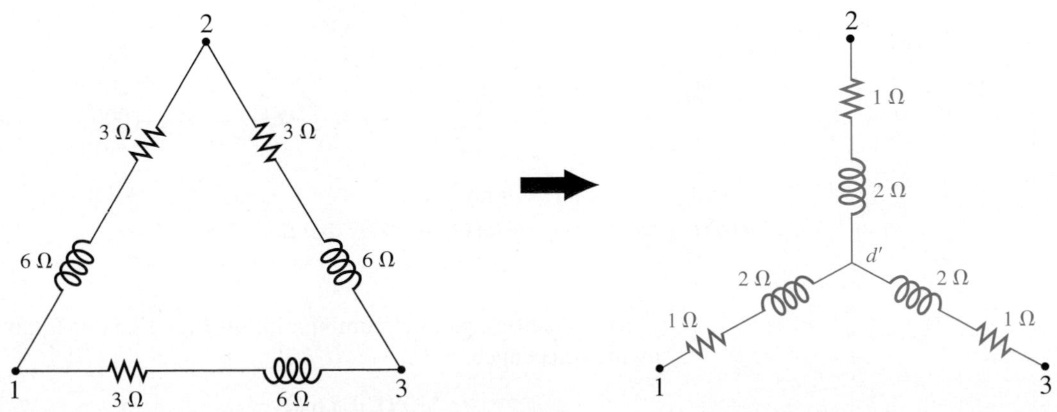

FIG. 17.48
Converting a Δ configuration to a Y configuration.

branch), the center point d' of the transformed Δ will be the same as point d of the original Y:

$$\mathbf{Z}_{Y} = \frac{\mathbf{Z}_{\Delta}}{3} = \frac{3\,\Omega + j\,6\,\Omega}{3} = 1\,\Omega + j\,2\,\Omega$$

and (Fig. 17.49)

$$\mathbf{Z}_{T} = 2\left(\frac{1\,\Omega + j\,2\,\Omega}{2}\right) = \mathbf{1\,\Omega + j\,2\,\Omega}$$

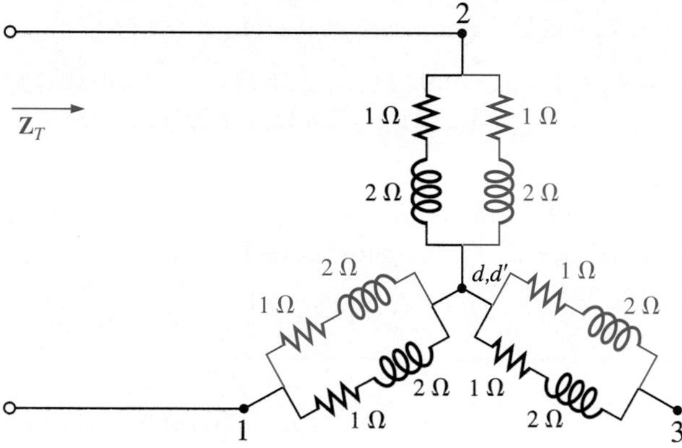

FIG. 17.49
Substituting the Y configuration of Fig. 17.18 into the network of Fig. 17.47.

Using the Y-Δ transformation (Fig. 17.50), we obtain

$$\mathbf{Z}_{\Delta} = 3\mathbf{Z}_{Y} = 3(1\,\Omega + j\,2\,\Omega) = 3\,\Omega + j\,6\,\Omega$$

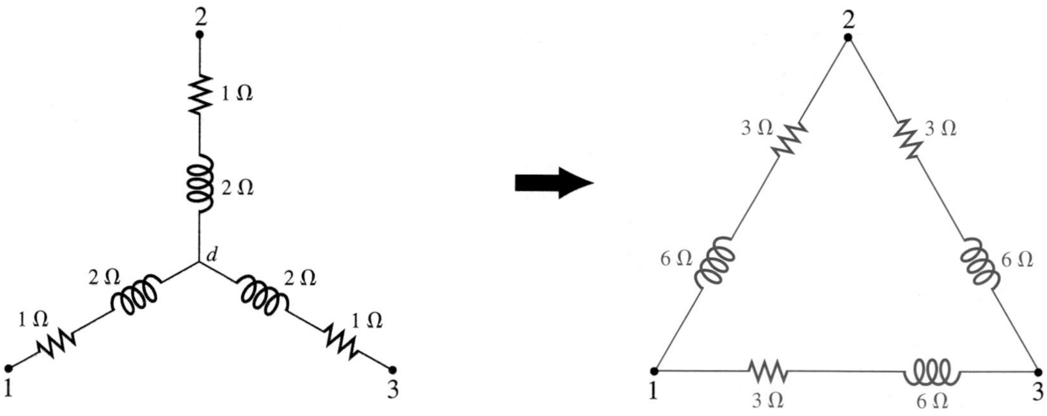

FIG. 17.50
Converting the Y configuration of Fig. 17.47 to a Δ.

Each resulting parallel combination in Fig. 17.51 will have the following impedance:

$$\mathbf{Z}' = \frac{3\,\Omega + j\,6\,\Omega}{2} = 1.5\,\Omega + j\,3\,\Omega$$

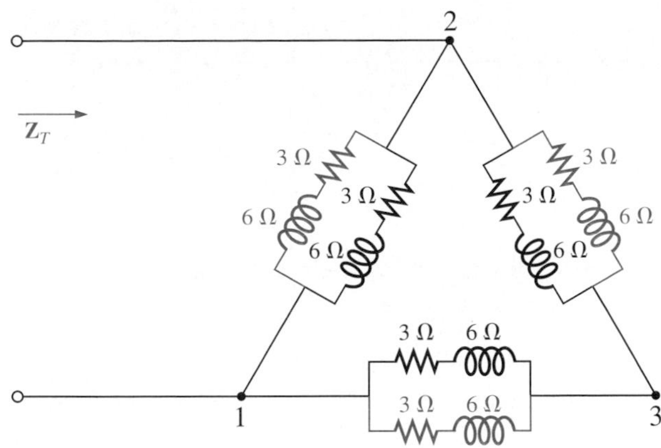

FIG. 17.51
Substituting the Δ configuration of Fig. 17.50 into the network of Fig. 17.47.

and
$$\mathbf{Z}_T = \frac{\mathbf{Z}'(2\mathbf{Z}')}{\mathbf{Z}' + 2\mathbf{Z}'} = \frac{2(\mathbf{Z}')^2}{3\mathbf{Z}'} = \frac{2\mathbf{Z}'}{3}$$
$$= \frac{2(1.5\ \Omega + j\,3\ \Omega)}{3} = \mathbf{1\ \Omega + j\,2\ \Omega}$$

which compares with the above result.

17.8 COMPUTER ANALYSIS

Programming Language

The nodal voltages and mesh currents for any network appearing in this chapter can be obtained using any computer language to determine the solution to the equations that result from each configuration. Once established, the parameters of a network can be changed and the solution quickly determined by the program. The sequence, therefore, is to find a general solution for the network in terms of the network impedances and applied sources and then apply the language to determine the solution using vector algebra. Of course, any change in the original configuration will result in a different set of equations and require the writing of a new program.

PSpice (DOS)

In this case, the obvious advantage of PSpice is that a general solution does not have to be obtained before the software package can be applied. Through proper input of the network parameters, the software package can determine the relationships among parameters and print out the desired quantities. There is no need specifically to convey to the package which elements are in series or parallel and assist in the generation of a general solution. The input file is sufficient to provide this information to the package, from which it generates and prints out a solution for the user.

The first application of PSpice in this section is to determine the nodal voltages for the network of Example 17.16 and compare solutions. The network is redrawn in Fig. 17.52 with the defined nodes and parameter values as determined at a frequency of 1 kHz from the reactance levels of Fig. 17.28. The input file of Fig. 17.53 requires five lines for the network

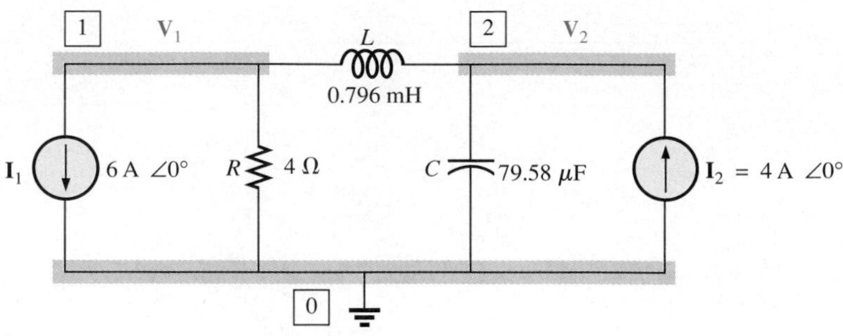

FIG. 17.52
Applying PSpice (DOS) to the network of Example 17.16.

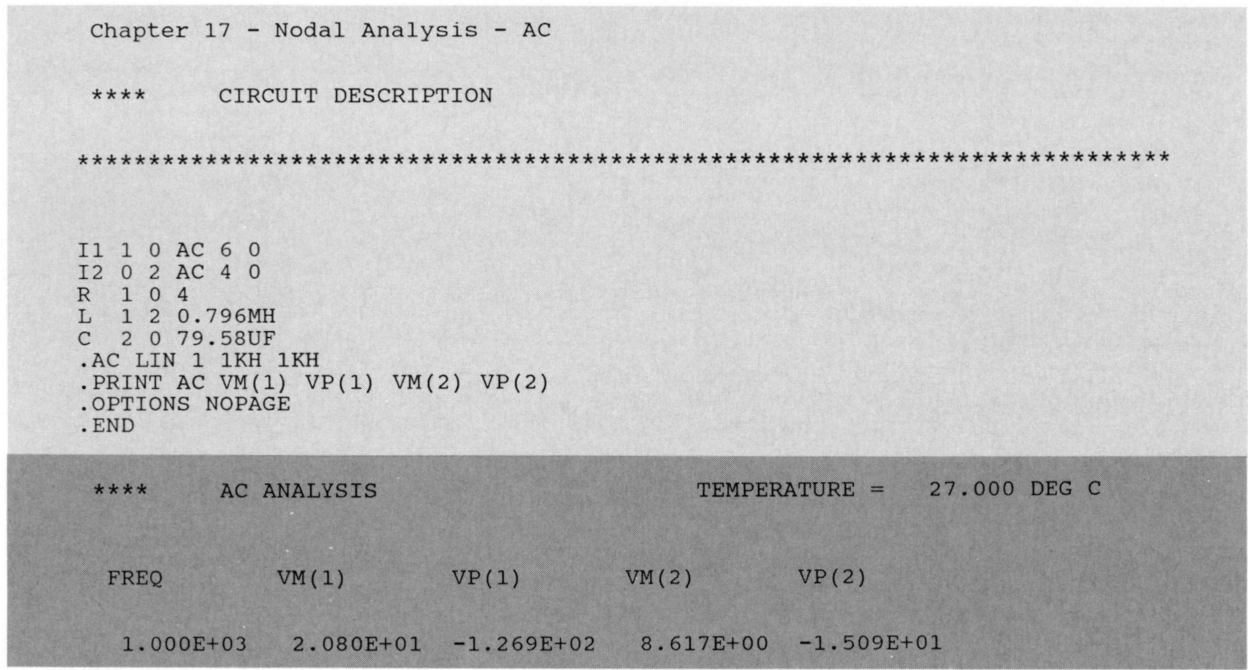

```
Chapter 17 - Nodal Analysis - AC

****      CIRCUIT DESCRIPTION

******************************************************************************

I1 1 0 AC 6 0
I2 0 2 AC 4 0
R  1 0 4
L  1 2 0.796MH
C  2 0 79.58UF
.AC LIN 1 1KH 1KH
.PRINT AC VM(1) VP(1) VM(2) VP(2)
.OPTIONS NOPAGE
.END

****      AC ANALYSIS                        TEMPERATURE =   27.000 DEG C

  FREQ        VM(1)        VP(1)        VM(2)        VP(2)

  1.000E+03   2.080E+01   -1.269E+02   8.617E+00   -1.509E+01
```

FIG. 17.53
The input and output files for the network of Fig. 17.52.

parameters, an .AC line to define the analysis and frequency, and a .PRINT statement to output the desired results. Note that $\mathbf{V}_1 = 20.80\text{ V}\angle -126.9°$ is exactly the same as that obtained in the longhand fashion of Example 17.16. In addition, $\mathbf{V}_2 = 8.617\text{ V}\angle -150.9°$.

The second application is to the bridge network of Fig. 17.54, which is Problem 29. The voltage across the bridge arm was requested to determine if a balance condition exists, as defined by $|\mathrm{V}| = 0$ V. The input file of Fig. 17.55 is quite lengthy but follows the same format as previously defined. The frequency applied was determined from the equation

$$f = \frac{\omega}{2\pi} = \frac{1000\text{ rad/s}}{2\pi\text{ rad}} \cong 159.15\text{ Hz}$$

The result of 18.31×10^{-15} V $\cong 0$ V certainly satisfies the balance condition, and the system is in a balanced state. There is a phase angle present, but it is immaterial when associated with such a small current level. The rms-reading galvanometer will obviously reflect a current of zero amperes.

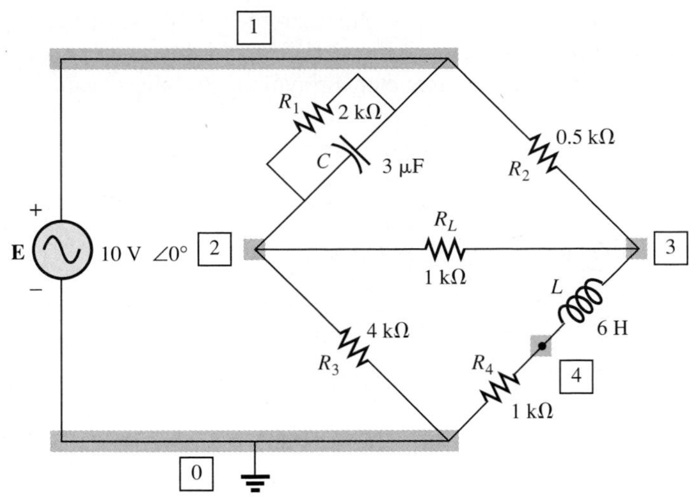

FIG. 17.54
Applying PSpice (DOS) to the bridge configuration of Problem 29.

```
Chapter 17 - Bridge Network - AC

****      CIRCUIT DESCRIPTION

******************************************************************************


VE 1 0 AC 10V 0
R1 1 2 2K
R2 1 3 0.5K
R3 2 0 4K
R4 4 0 1K
RL 2 3 1K
L  3 4 6H
C  1 2 3UF
.AC LIN 1 159.15H 159.15H
.PRINT AC VM(2,3) VP(2,3)
.OPTIONS NOPAGE
.END

****      AC ANALYSIS                          TEMPERATURE =    27.000 DEG C


  FREQ          VM(2,3)       VP(2,3)


   1.592E+02    1.831E-15     1.400E+01
```

FIG. 17.55
Input and output files for the network of Fig. 17.54.

PSpice (Windows)

The emphasis in this chapter will be on the application of controlled sources in a Windows environment. Controlled sources are not particularly difficult to apply once a few important characteristics of the device are understood along with methods of defining the important parameters.

The network of Fig. 17.14 had a current-controlled current source in the center leg of the configuration whose magnitude was controlled by the current through the impedance R_1. The **current-controlled current source** (CCCS) is called up from the **analog.slb** library under the listing

F. Once chosen it will appear as shown in the schematic of Fig. 17.56. The source can be turned using the **Ctrl-R** combination. The controlled current appears within the circle of the symbol, and the sensing current

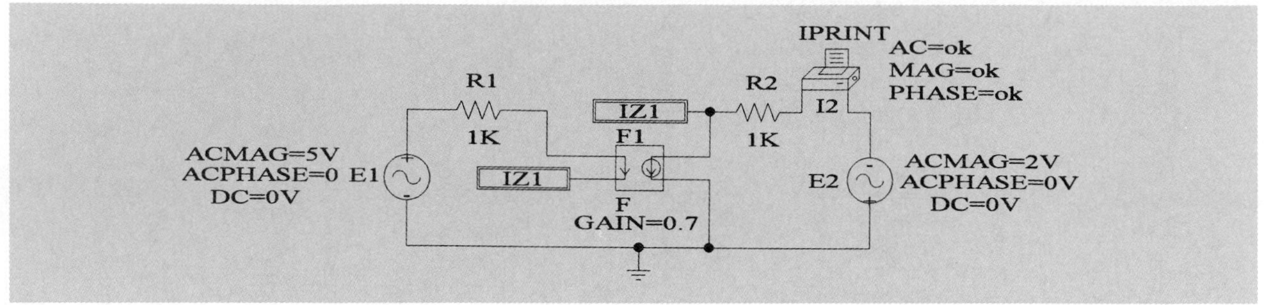

FIG. 17.56
Schematic representation of the network of Fig. 17.14 with a current-controlled current source.

passes through the source in the direction indicated by the other arrow. On most occasions it can be difficult to make direct wire connections on the schematic for the sensing and controlled currents. This problem is overcome using the **GLOBAL** choice under the **port.slb** library. Once chosen and displayed on the screen, it can be positioned and its attribute defined. In Fig. 17.56 two points are defined as IZ1 (representing the current through the impedance $Z_1 = R_1$). For the simulation, both points will be connected. In fact, any other **GLOBAL** location with the same attribute will be considered connected to the same point. By clicking on the CCCS, the gain can be set at 0.7 as shown on the schematic. Note that IZ1 is in fact sensing the current through R_1, and the controlled source is connected from the junction of R_1 and R_2. The other elements of the network have been defined in earlier chapters. A simulation of the network will result in the output file of Fig. 17.57, which reveals that the current $\mathbf{I}_2$ is 1.615 mA at an angle of zero degrees.

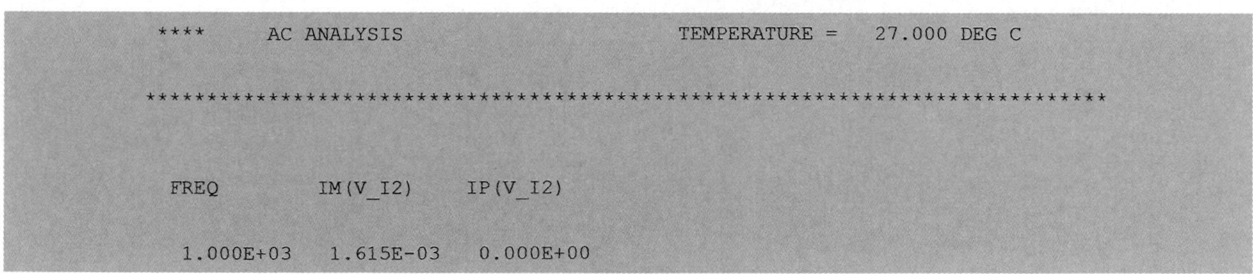

```
****     AC ANALYSIS                         TEMPERATURE =   27.000 DEG C

*****************************************************************************

  FREQ        IM(V_I2)     IP(V_I2)

  1.000E+03   1.615E-03    0.000E+00
```

FIG. 17.57
Output file for the PSpice analysis of the network of Fig. 17.56.

The impedances and sources chosen in this example were chosen to permit a quick, longhand analysis of the network to check the results. The earlier equations obtained using the supermesh approach were

$$\mathbf{E} - \mathbf{I}_1\mathbf{Z}_1 - \mathbf{I}_2\mathbf{Z}_2 + \mathbf{E}_2 = 0 \qquad \text{or} \qquad \mathbf{I}_1\mathbf{Z}_1 + \mathbf{I}_2\mathbf{Z}_2 = \mathbf{E}_1 + \mathbf{E}_2$$

$$\text{and} \qquad k\mathbf{I} = k\mathbf{I}_1 = \mathbf{I}_1 - \mathbf{I}_2 \qquad \text{resulting in} \qquad \mathbf{I}_1 = \frac{\mathbf{I}_2}{1-k}$$

Substituting numerical values (all have an associated angle of zero degrees):

$$\mathbf{I}_1(1\ \text{k}\Omega) + \mathbf{I}_2(1\ \text{k}\Omega) = 7\ \text{V}$$

and
$$\mathbf{I}_1 = \frac{\mathbf{I}_2}{1 - 0.7} = \frac{\mathbf{I}_2}{0.3} = 3.333\mathbf{I}_2$$

so that
$$(3.333\mathbf{I}_2)1\ \text{k}\Omega + \mathbf{I}_2(1\ \text{k}\Omega) = 7\ \text{V}$$

or
$$(4.333\ \text{k}\Omega)\mathbf{I}_2 = 7\ \text{V}$$

and
$$\mathbf{I}_2 = \frac{7\ \text{V}}{4.333\ \text{k}\Omega} = \mathbf{1.615\ mA\ \angle 0°}$$

confirming the computer solution.

The next example investigates the **voltage-controlled voltage source** (VCVS) listed as **E** under the analog.slb library. The network chosen is that of Fig. 17.27, appearing in schematic form in Fig. 17.58.

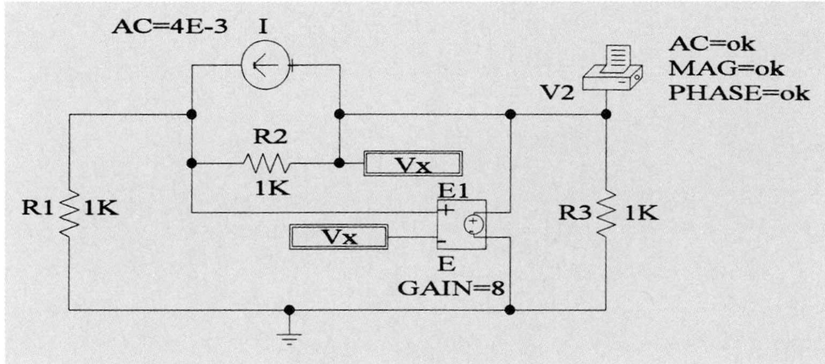

FIG. 17.58
Schematic representation of the network of Fig. 17.27 with a voltage-controlled voltage source.

Note that the controlled voltage is denoted by the circle in the symbol, while the sensing voltage is defined between the positive and negative polarities at the input. The **GLOBAL** connections were employed again to define the sensing voltage across the impedance Z2. The gain is set at 8, and the **VPRINT** is set across the output voltage V2. There are two choices for the current source since the phase angle of the source is zero degrees. The **ISRC** choice under source.slb will provide a symbol with an arrow, but the only attributes that can be set are the **AC, DC,** and **TRAN** levels. The phase angle or frequency cannot be set as attributes for the source. If the additional quantities need to be specified, then the **ISIN** choice should be made, although the symbol is the same as that used for ac voltage sources, with the direction indicated by the plus and minus signs on the schematics symbol. For this run the results are indicated in Fig. 17.59, which reveals that the voltage V2 is 3.2 volts.

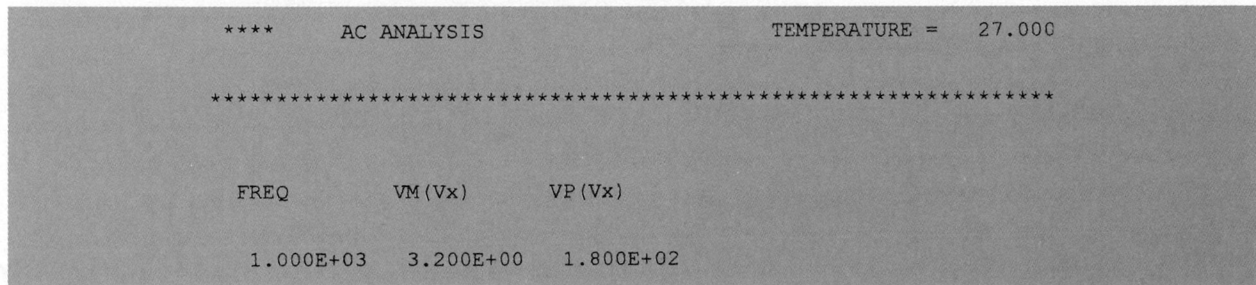

**** AC ANALYSIS TEMPERATURE = 27.000
****
FREQ VM(Vx) VP(Vx)
1.000E+03 3.200E+00 1.800E+02

FIG. 17.59
Output file for the PSpice analysis of the network of Fig. 17.58.

Again, the elements were chosen to permit a quick determination of the solution using the longhand approach. The equations obtained earlier were

$$\frac{\mathbf{V}_1}{\mathbf{Z}_1} + \frac{(\mathbf{V}_1 - \mathbf{V}_2)}{\mathbf{Z}_2} - \mathbf{I} = 0 \quad \text{or} \quad \mathbf{V}_1\left[\frac{1}{\mathbf{Z}_1} + \frac{1}{\mathbf{Z}_2}\right] - \frac{\mathbf{V}_2}{\mathbf{Z}_2} = \mathbf{I}$$

with $\mathbf{V}_2 = \mu \mathbf{V}_x = \mu[\mathbf{V}_1 - \mathbf{V}_2]$ so that $\mathbf{V}_1 = \left(\frac{\mu + 1}{\mu}\right)\mathbf{V}_2$

Substituting numerical values (all have an associated angle of zero degrees):

$$\mathbf{V}_1\left[\frac{1}{1\text{ k}\Omega} + \frac{1}{1\text{ k}\Omega}\right] - \frac{\mathbf{V}_2}{1\text{ k}\Omega} = 4\text{ mA}$$

and $$\mathbf{V}_1 = \left(\frac{8+1}{8}\right)\mathbf{V}_2 = \frac{9}{8}\mathbf{V}_2$$

so that $$\left(\frac{9}{8}\mathbf{V}_2\right)\left(\frac{2}{1\text{ k}\Omega}\right) - \frac{\mathbf{V}_2}{1\text{ k}\Omega} = 4\text{ mA}$$

or $$\mathbf{V}_2\left[\frac{18}{8\text{ k}\Omega} - \frac{1}{1\text{ k}\Omega}\right] = 4\text{ mA}$$

and $$\mathbf{V}_2[2.25\text{ mS} - 1\text{ mS}] = 4\text{ mA}$$

with $$\mathbf{V}_2[1.25\text{ mS}] = 4\text{ mA}$$

and $$\mathbf{V}_2 = \frac{4\text{ mA}}{1.25\text{ mS}} = \mathbf{3.2\text{ V} \angle 0°}$$

confirming the computer solution.

PROBLEMS

SECTION 17.2 Independent Versus Dependent (Controlled) Sources

1. Discuss, in your own words, the difference between a controlled and an independent source.

SECTION 17.3 Source Conversions

2. Convert the voltage sources of Fig. 17.60 to current sources.

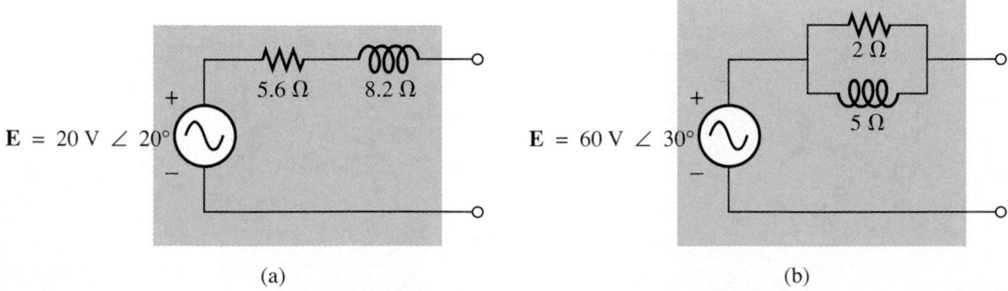

FIG. 17.60
Problem 2.

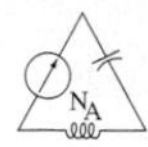

3. Convert the current sources of Fig. 17.61 to voltage sources.

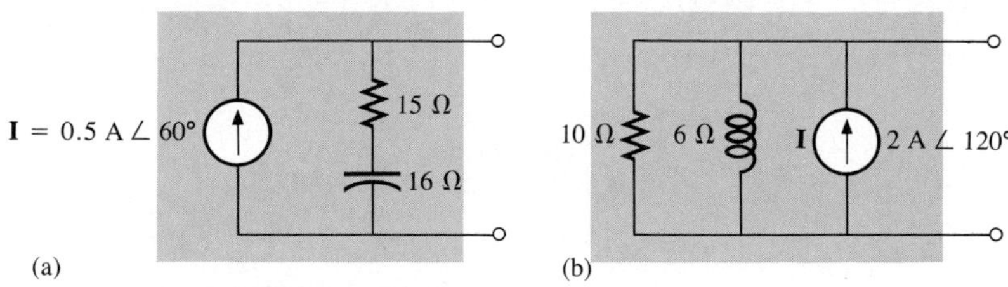

FIG. 17.61
Problem 3.

4. Convert the voltage source of Fig. 17.62(a) to a current source and the current source of Fig. 17.62(b) to a voltage source.

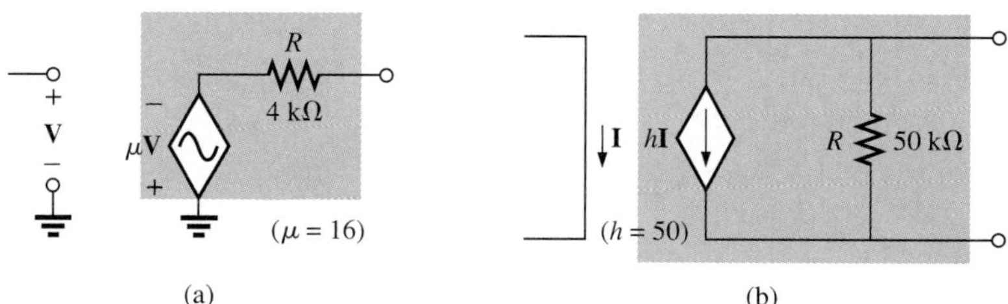

FIG. 17.62
Problem 4.

SECTION 17.4 Mesh Analysis

5. Write the mesh equations for the networks of Fig. 17.63. Determine the current through the resistor R.

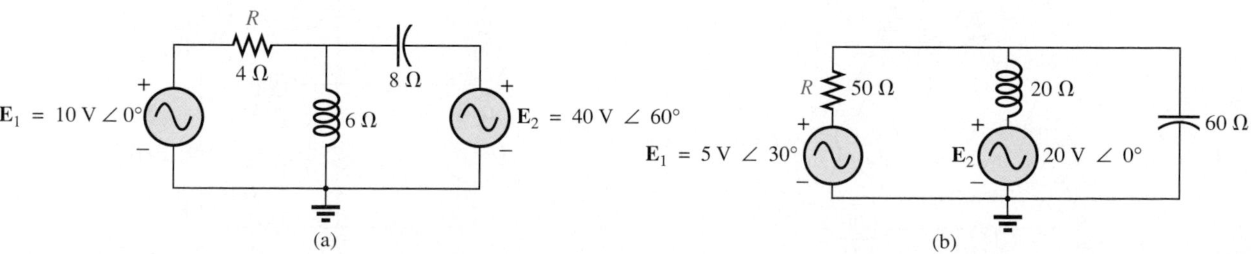

FIG. 17.63
Problems 5 and 34.

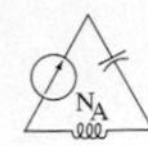

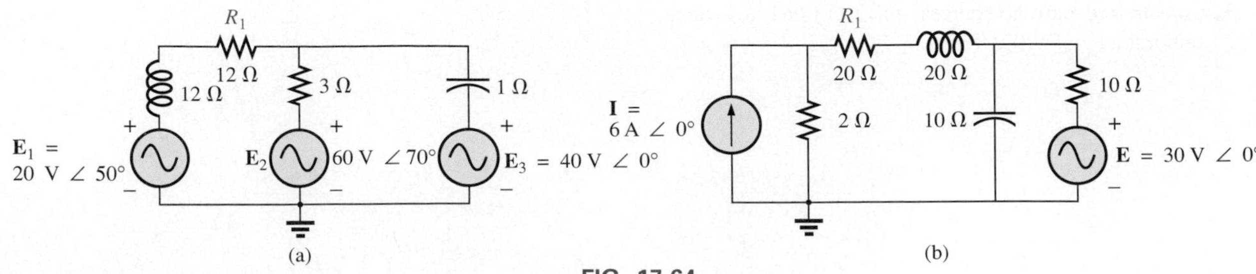

FIG. 17.64
Problems 6 and 16.

6. Write the mesh equations for the networks of Fig. 17.64. Determine the current through the resistor R_1.

***7.** Write the mesh equations for the networks of Fig. 17.65. Determine the current through the resistor R_1.

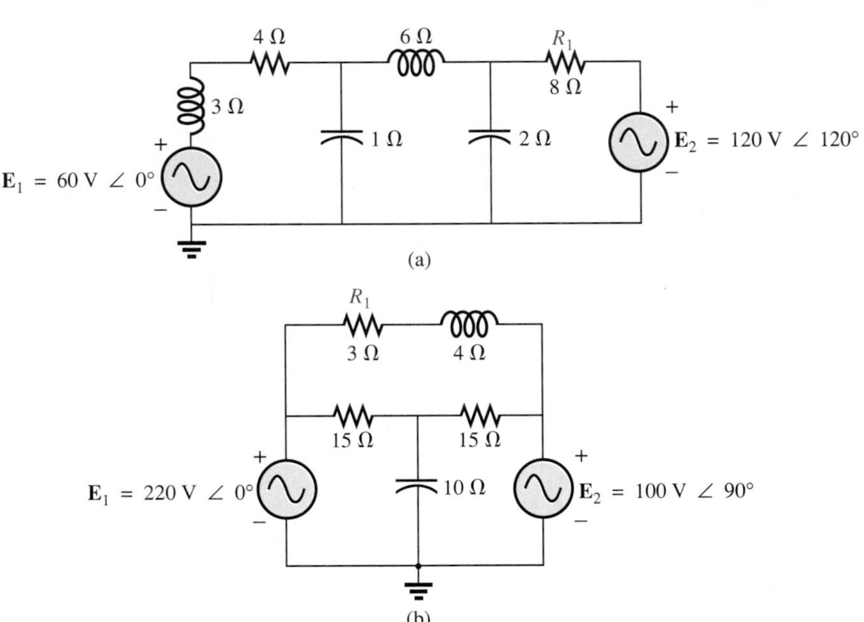

FIG. 17.65
Problems 7, 17, and 35.

***8.** Write the mesh equations for the networks of Fig. 17.66. Determine the current through the resistor R_1.

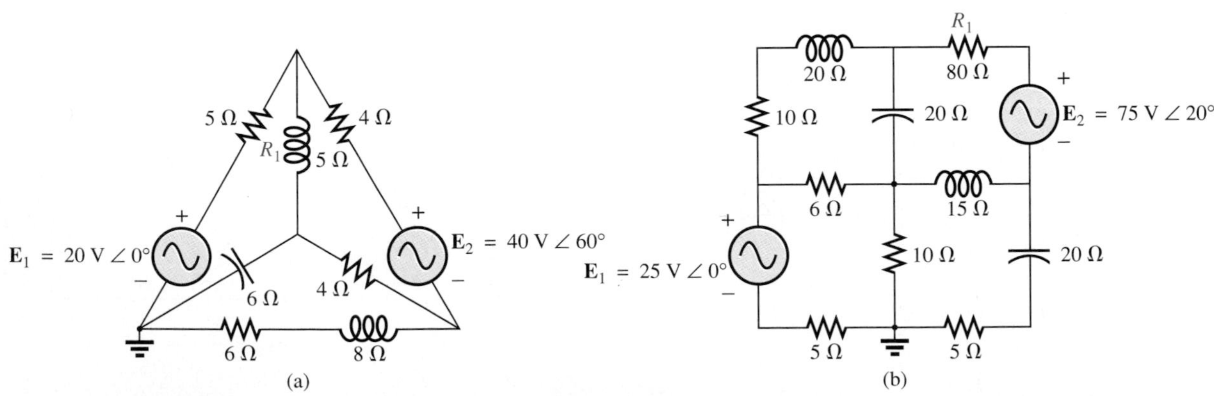

FIG. 17.66
Problems 8, 18, and 19.

9. Using mesh analysis, determine the current $\mathbf{I}_L$ (in terms of **V**) for the network of Fig. 17.67.

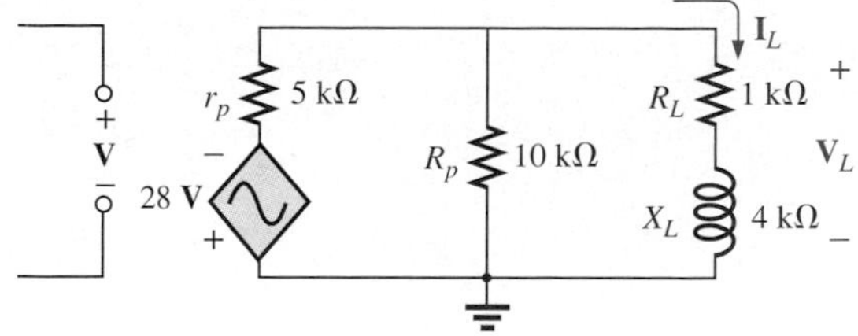

FIG. 17.67
Problem 9.

*10. Using mesh analysis, determine the current $\mathbf{I}_L$ (in terms of **I**) for the network of Fig. 17.68.

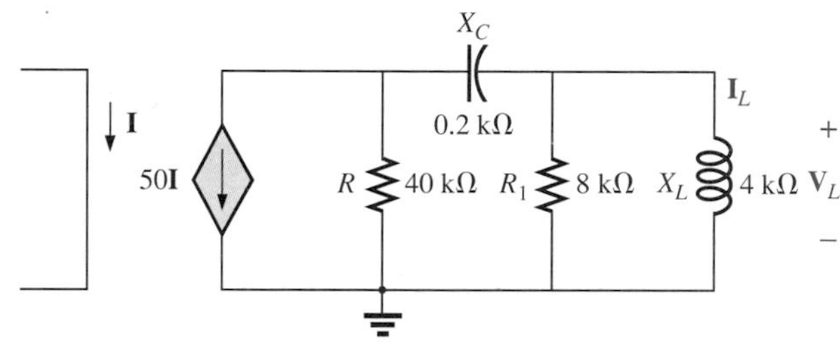

FIG. 17.68
Problem 10.

*11. Write the mesh equations for the network of Fig. 17.69 and determine the current through the 1-kΩ and 2-kΩ resistors.

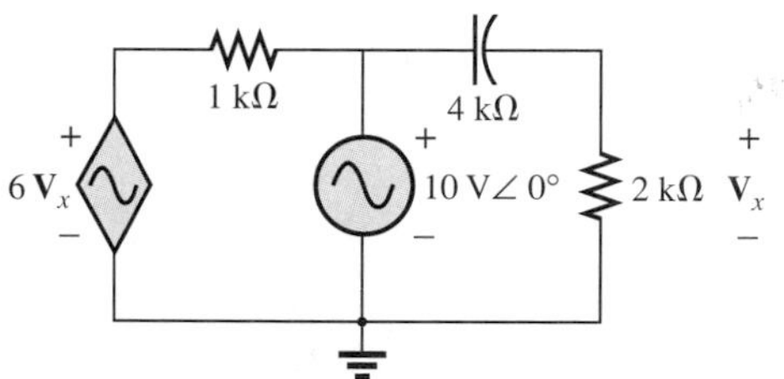

FIG. 17.69
Problems 11 and 36.

*12. Write the mesh equations for the network of Fig. 17.70 and determine the current through the 10-kΩ resistor.

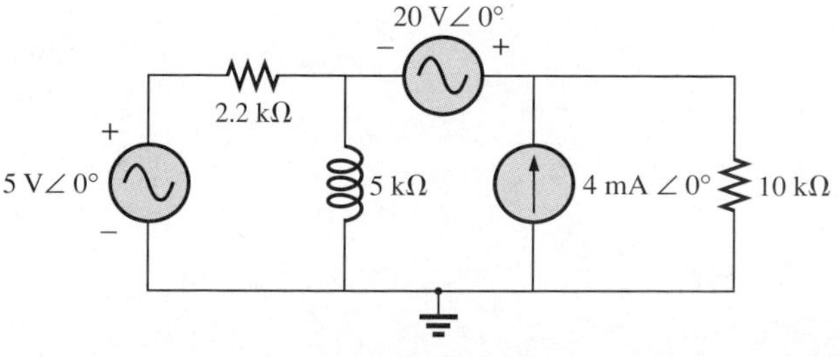

FIG. 17.70
Problems 12 and 37.

*13. Write the mesh equations for the network of Fig. 17.71 and determine the current through the inductive element.

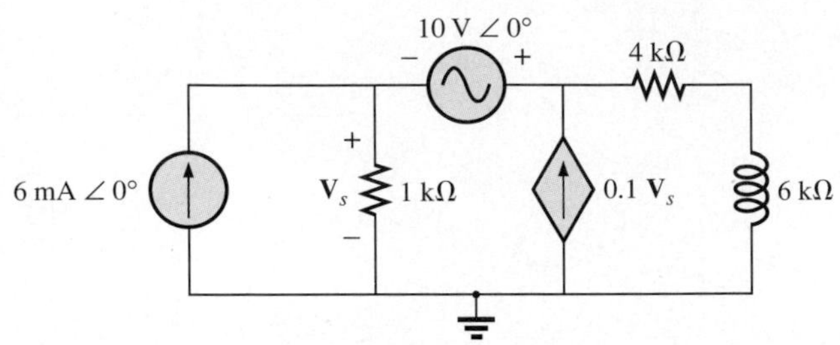

FIG. 17.71
Problems 13 and 38.

SECTION 17.5 Nodal Analysis

14. Determine the nodal voltages for the networks of Fig. 17.72.

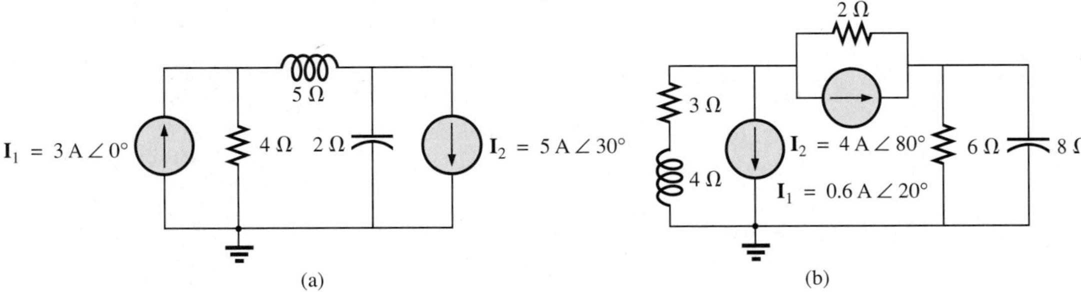

FIG. 17.72
Problems 14 and 39.

15. Determine the nodal voltages for the networks of Fig. 17.73.

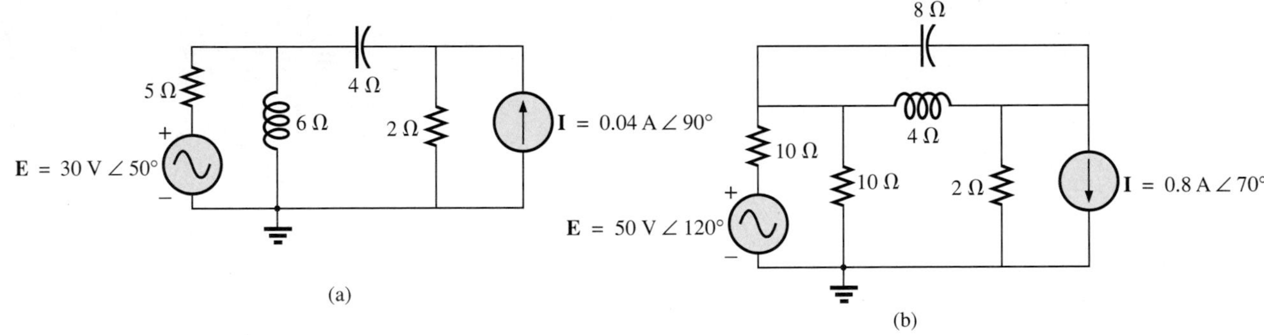

FIG. 17.73
Problem 15.

16. Determine the nodal voltages for the network of Fig. 17.64(b).

17. Determine the nodal voltages for the network of Fig. 17.65(b).

*18. Determine the nodal voltages for the network of Fig. 17.66(a).

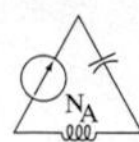

***19.** Determine the nodal voltages for the network of Fig. 17.66(b).

***20.** Determine the nodal voltages for the networks of Fig. 17.74.

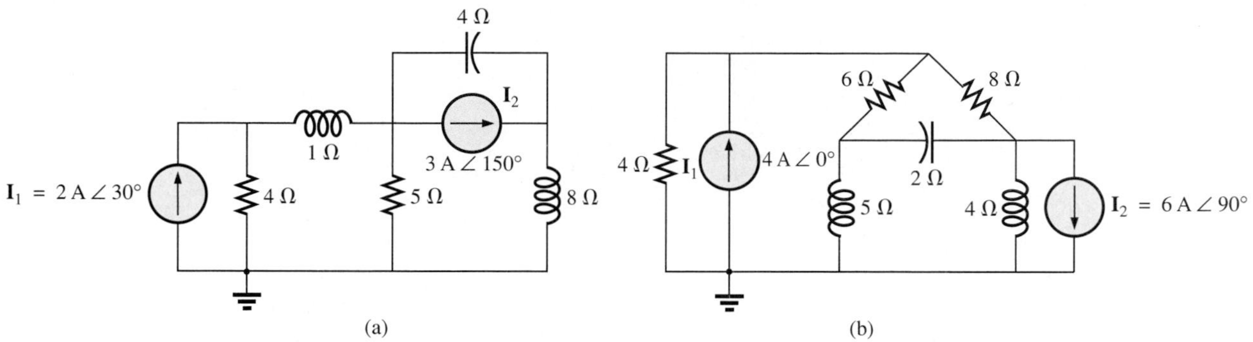

FIG. 17.74
Problem 20.

***21.** Write the nodal equations for the network of Fig. 17.75 and find the voltage across the 1-kΩ resistor.

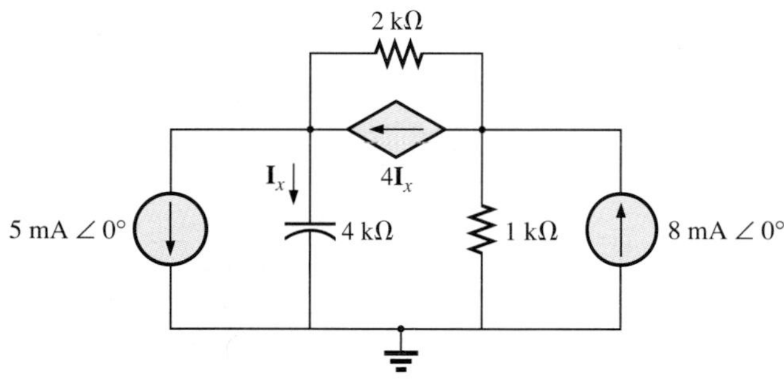

FIG. 17.75
Problems 21 and 40.

***22.** Write the nodal equations for the network of Fig. 17.76 and find the voltage across the capacitive element.

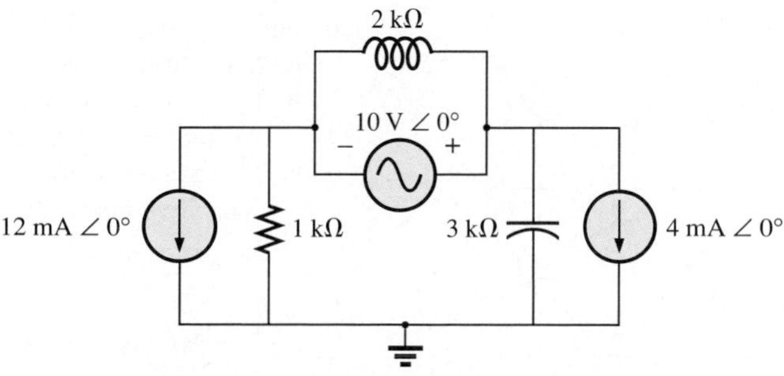

FIG. 17.76
Problems 22 and 41.

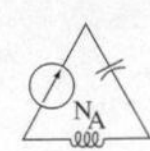

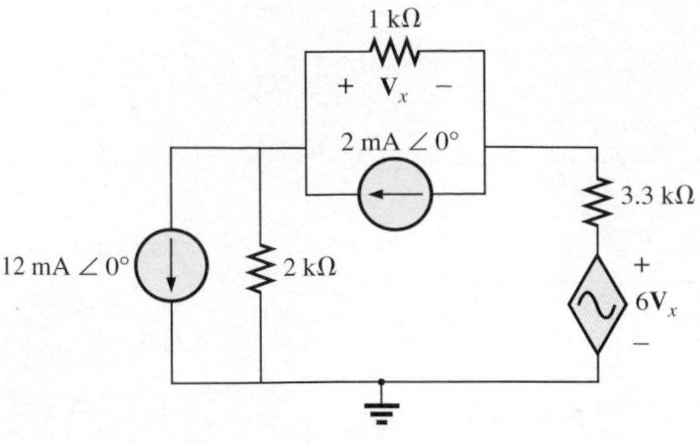

FIG. 17.77
Problems 23 and 42.

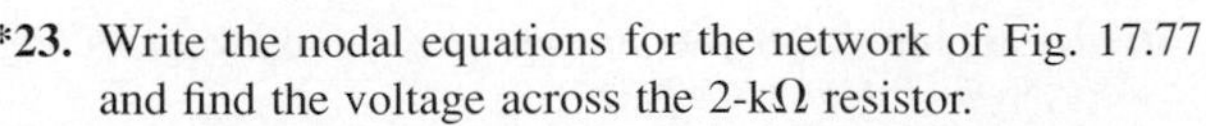

*23. Write the nodal equations for the network of Fig. 17.77 and find the voltage across the 2-kΩ resistor.

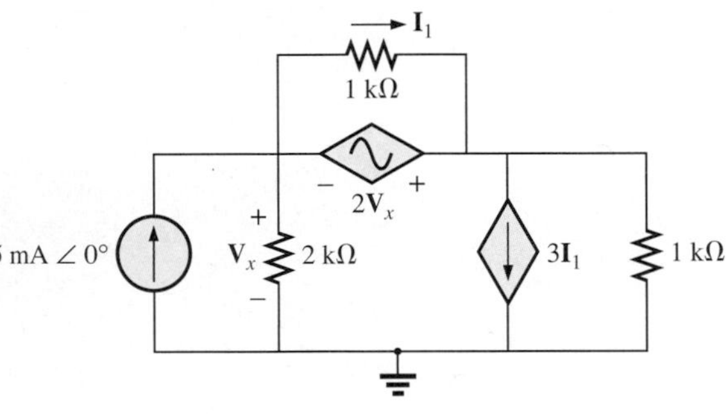

FIG. 17.78
Problems 24 and 43.

*24. Write the nodal equations for the network of Fig. 17.78 and find the voltage across the 2-kΩ resistor.

*25. For the network of Fig. 17.79, determine the voltage $\mathbf{V}_L$ in terms of the voltage $\mathbf{E}_i$.

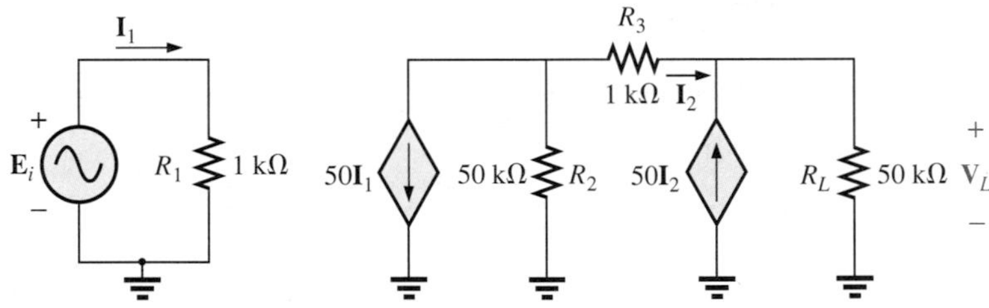

FIG. 17.79
Problem 25.

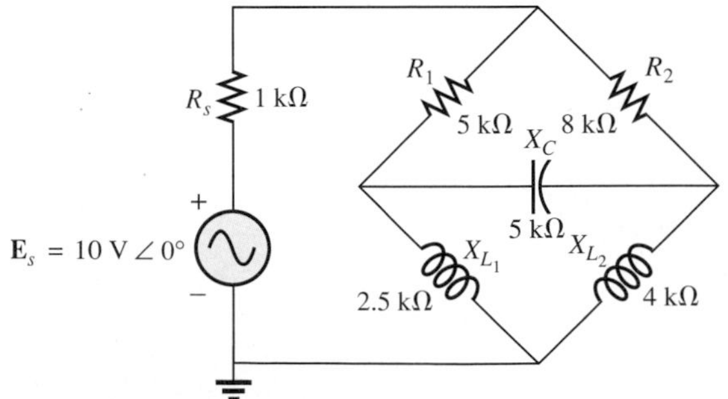

SECTION 17.6 Bridge Networks (ac)

26. For the bridge network of Fig. 17.80:
   a. Is the bridge balanced?
   b. Using mesh analysis, determine the current through the capacitive reactance.
   c. Using nodal analysis, determine the voltage across the capacitive reactance.

FIG. 17.80
Problem 26.

27. For the bridge network of Fig. 17.81:
    a. Is the bridge balanced?
    b. Using mesh analysis, determine the current through the capacitive reactance.
    c. Using nodal analysis, determine the voltage across the capacitive reactance.

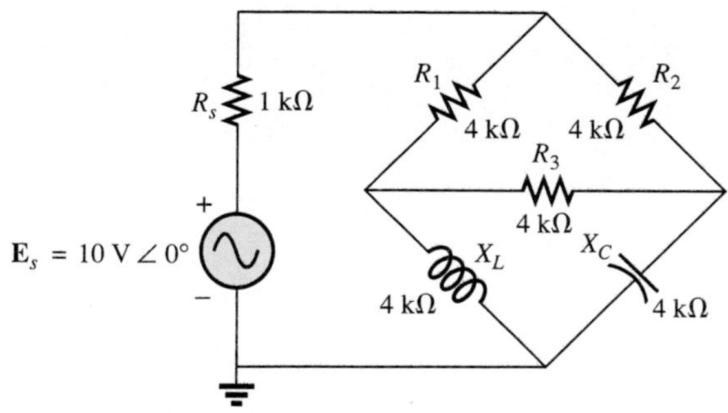

FIG. 17.81
Problem 27.

28. The Hay bridge of Fig. 17.82 is balanced. Using Eq. (17.3), determine the unknown inductance L_x and resistance R_x.

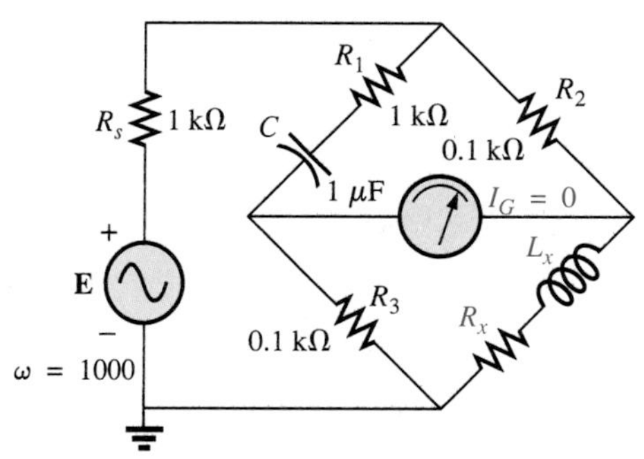

FIG. 17.82
Problem 28.

29. Determine whether the Maxwell bridge of Fig. 17.83 is balanced (ω = 1000 rad/s).

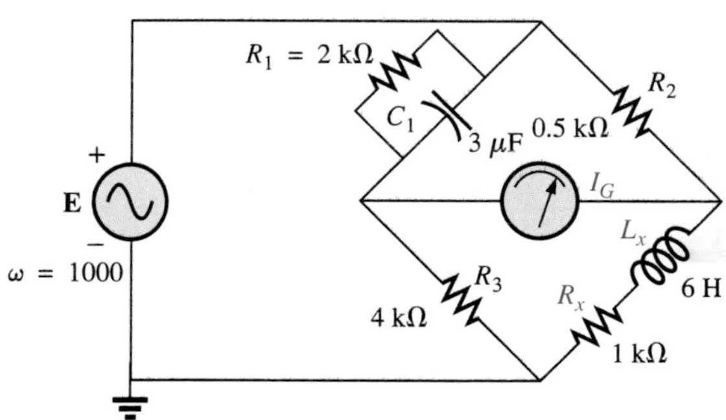

FIG. 17.83
Problem 29.

30. Derive the balance equations (17.11) and (17.12) for the capacitance comparison bridge.

31. Determine the balance equations for the inductance bridge of Fig. 17.84.

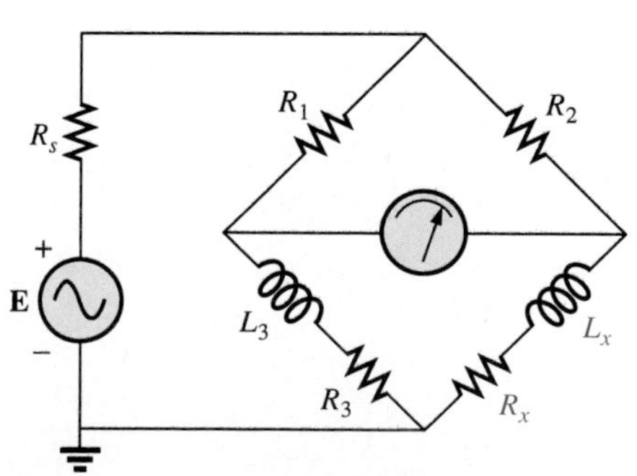

FIG. 17.84
Problem 31.

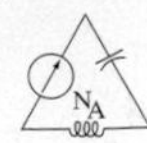

SECTION 17.7 Δ-Y, Y-Δ Conversions

32. Using the Δ-Y or Y-Δ conversion, determine the current **I** for the networks of Fig. 17.85.

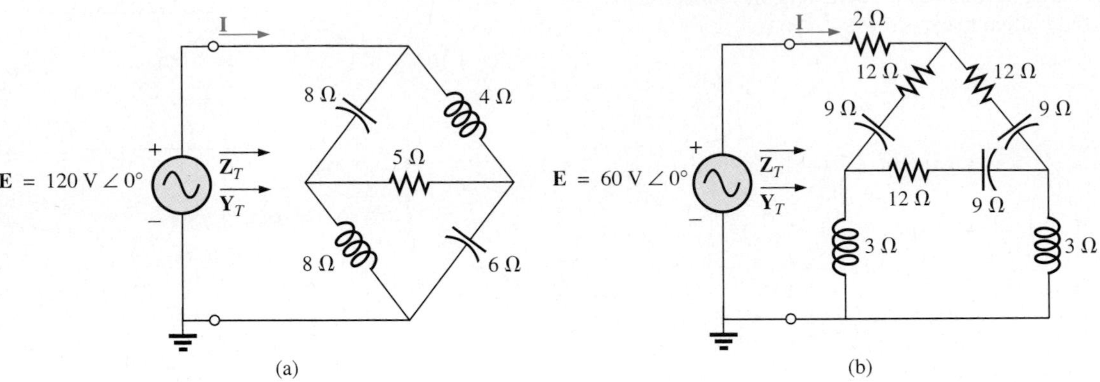

FIG. 17.85
Problem 32.

33. Using the Δ-Y or Y-Δ conversion, determine the current **I** for the networks of Fig. 17.86. (**E** = 100 V ∠0° in each case.)

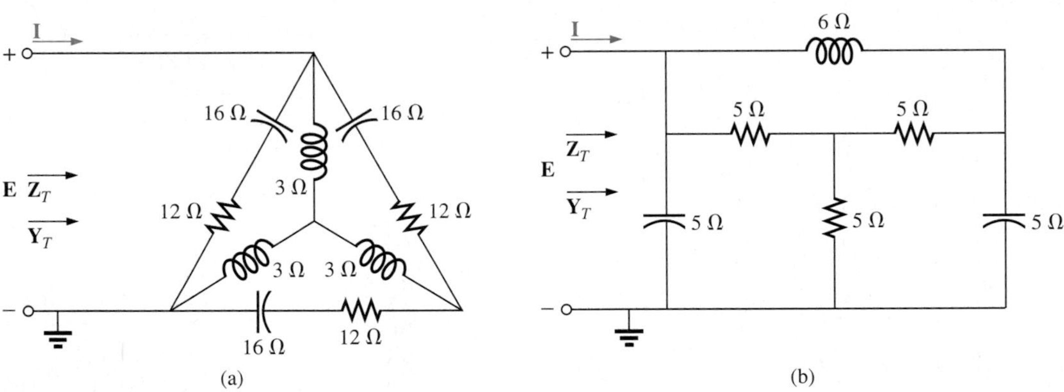

FIG. 17.86
Problem 33.

SECTION 17.8 Computer Analysis

PSpice (DOS or Windows)

34. Determine the mesh currents for the network of Fig. 17.63(a).

35. Determine the mesh currents for the network of Fig. 17.65(a).

***36.** Determine the mesh currents for the network of Fig. 17.69.

***37.** Determine the mesh currents for the network of Fig. 17.70.

***38.** Determine the mesh currents for the network of Fig. 17.71.

39. Determine the nodal voltages for the network of Fig. 17.72(b).

***40.** Determine the nodal voltages for the network of Fig. 17.75.

***41.** Determine the nodal voltages for the network of Fig. 17.76.

***42.** Determine the nodal voltages for the network of Fig. 17.77.

***43.** Determine the nodal voltages for the network of Fig. 17.78.

Programming Language (C++, BASIC, PASCAL, etc.)

44. Write a computer program that will provide a general solution for the network of Fig. 17.10. That is, given the reactance of each element and the parameters of the source voltages, generate a solution in phasor form for both mesh currents.

45. Repeat Problem 35 for the nodal voltages of Fig. 17.28.

46. Given a bridge composed of series impedances in each branch, write a program to test the balance condition as defined by Eq. (17.6).

GLOSSARY

Bridge network A network configuration having the appearance of a diamond in which no two branches are in series or parallel.

Capacitance comparison bridge A bridge configuration having a galvanometer in the bridge arm that is used to determine an unknown capacitance and associated resistance.

Delta (Δ) configuration A network configuration having the appearance of the capital Greek letter delta.

Dependent (controlled) source A source whose magnitude and/or phase angle is determined (controlled) by a current or voltage of the system in which it appears.

Hay bridge A bridge configuration used for measuring the resistance and inductance of coils in those cases where the resistance is a small fraction of the reactance of the coil.

Independent source A source whose magnitude is independent of the network to which it is applied. It displays its terminal characteristics even if completely isolated.

Maxwell bridge A bridge configuration used for inductance measurements when the resistance of the coil is large enough not to require a Hay bridge.

Mesh analysis A method through which the loop (or mesh) currents of a network can be determined. The branch currents of the network can then be determined directly from the loop currents.

Nodal analysis A method through which the node voltages of a network can be determined. The voltage across each element can then be determined through application of Kirchhoff's voltage law.

Source conversion The changing of a voltage source to a current source, or vice versa, which will result in the same terminal behavior of the source. In other words, the external network is unaware of the change in sources.

Wye (Y) configuration A network configuration having the appearance of the capital letter Y.

18

Network Theorems (ac)

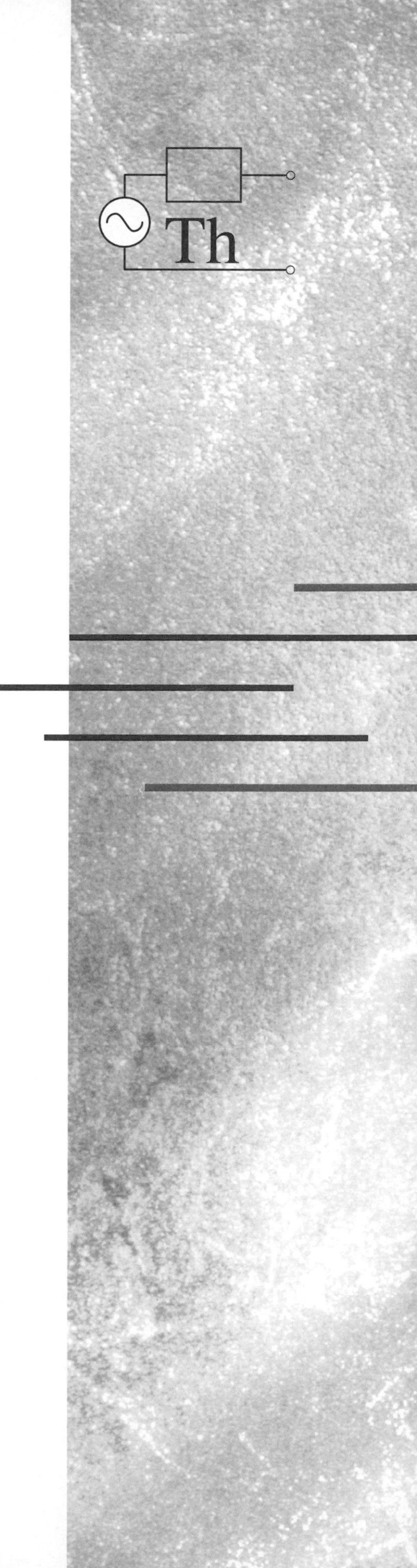

18.1 INTRODUCTION

This chapter will parallel Chapter 9, which dealt with network theorems as applied to dc networks. It would be time well spent to review each theorem in Chapter 9 before beginning this chapter because many of the comments offered there will not be repeated.

Due to the need for developing confidence in the application of the various theorems to networks with controlled (dependent) sources, some sections have been divided into two parts: independent sources and dependent sources.

Theorems to be considered in detail include the superposition theorem, Thévenin and Norton theorems, and the maximum power theorem. The substitution and reciprocity theorems and Millman's theorem are not discussed in detail here because a review of Chapter 9 will enable you to apply them to sinusoidal ac networks with little difficulty.

18.2 SUPERPOSITION THEOREM

You will recall from Chapter 9 that the superposition theorem eliminated the need for solving simultaneous linear equations by considering the effects of each source independently. To consider the effects of each source, we had to remove the remaining sources. This was accomplished by setting voltage sources to zero (short-circuit representation) and current sources to zero (open-circuit representation). The current through, or voltage across, a portion of the network produced by each source was then added algebraically to find the total solution for the current or voltage.

The only variation in applying this method to ac networks with independent sources is that we will now be working with impedances and phasors instead of just resistors and real numbers.

The superposition theorem is not applicable to power effects in ac networks since we are still dealing with a nonlinear relationship. It can be applied to networks with sources of different frequencies only if

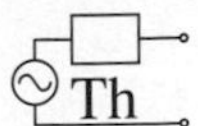

the total response for *each* frequency is found independently and the results are expanded in a nonsinusoidal expression, as appearing in Chapter 24.

One of the most frequent applications of the superposition theorem is to electronic systems in which the dc and ac analysis are treated separately and the total solution is the sum of the two. It is an important application of the theorem because the impact of the reactive elements changes dramatically in response to the two types of independent sources. In addition, the dc analysis of an electronic system can often define important parameters for the ac analysis. The fourth example will demonstrate the impact of the applied source on the general configuration of the network.

We will first consider networks with only independent sources to provide a close association with the analysis of Chapter 9.

Independent Sources

EXAMPLE 18.1 Using the superposition theorem, find the current **I** through the 4-Ω reactance (X_{L_2}) of Fig. 18.1.

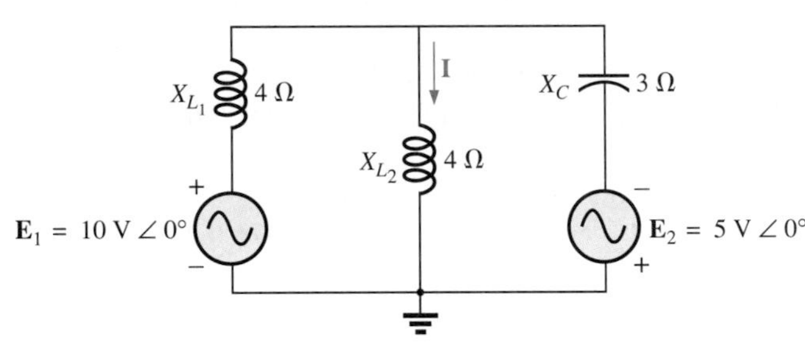

FIG. 18.1
Example 18.1.

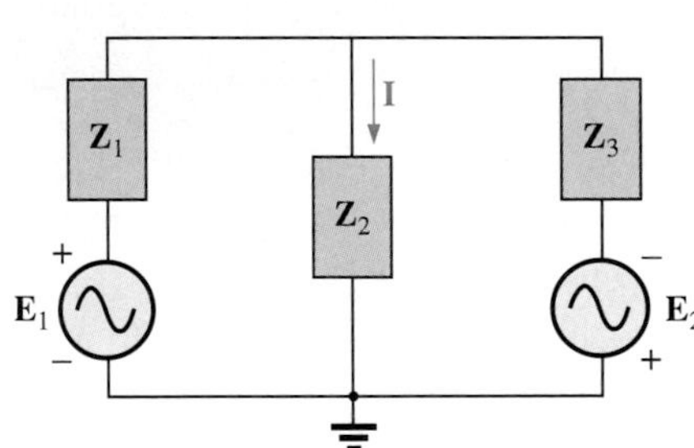

FIG. 18.2
Assigning the subscripted impedances to the network of Fig. 18.1.

Solution: For the redrawn circuit (Fig. 18.2),

$$\mathbf{Z}_1 = +jX_{L_1} = j\,4\ \Omega$$
$$\mathbf{Z}_2 = +jX_{L_2} = j\,4\ \Omega$$
$$\mathbf{Z}_3 = -jX_C = -j\,3\ \Omega$$

Considering the effects of the voltage source $\mathbf{E}_1$ (Fig. 18.3), we have

$$\mathbf{Z}_{2\|3} = \frac{\mathbf{Z}_2\mathbf{Z}_3}{\mathbf{Z}_2 + \mathbf{Z}_3} = \frac{(j\,4\ \Omega)(-j\,3\ \Omega)}{j\,4\ \Omega - j\,3\ \Omega} = \frac{12\ \Omega}{j} = -j\,12\ \Omega$$
$$= 12\ \Omega\ \angle -90°$$

$$I_{s_1} = \frac{\mathbf{E}_1}{\mathbf{Z}_{2\|3} + \mathbf{Z}_1} = \frac{10\ \text{V}\ \angle 0°}{-j\,12\ \Omega + j\,4\ \Omega} = \frac{10\ \text{V}\ \angle 0°}{8\ \Omega\ \angle -90°}$$
$$= 1.25\ \text{A}\ \angle 90°$$

and

$$\mathbf{I}' = \frac{\mathbf{Z}_3\mathbf{I}_{s_1}}{\mathbf{Z}_2 + \mathbf{Z}_3} \qquad \text{(current divider rule)}$$
$$= \frac{(-j\,3\ \Omega)(j\,1.25\ \text{A})}{j\,4\ \Omega - j\,3\ \Omega} = \frac{3.75\ \text{A}}{j\,1} = 3.75\ \text{A}\ \angle -90°$$

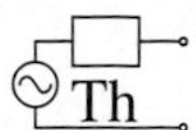

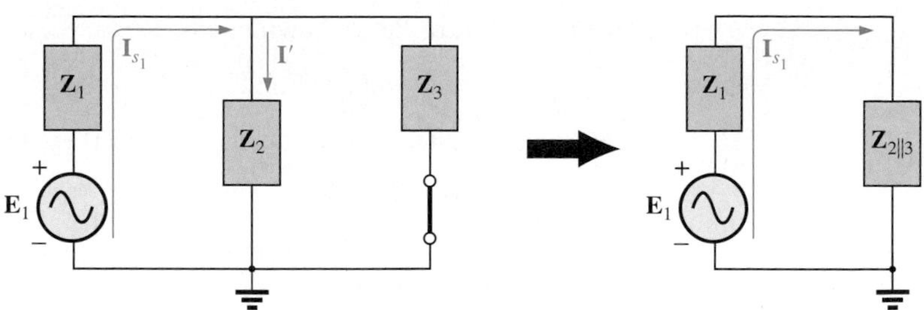

FIG. 18.3
Determining the effect of the voltage source $\mathbf{E}_1$ *on the current* **I** *of the network of Fig. 18.1.*

Considering the effects of the voltage source $\mathbf{E}_2$ (Fig. 18.4), we have

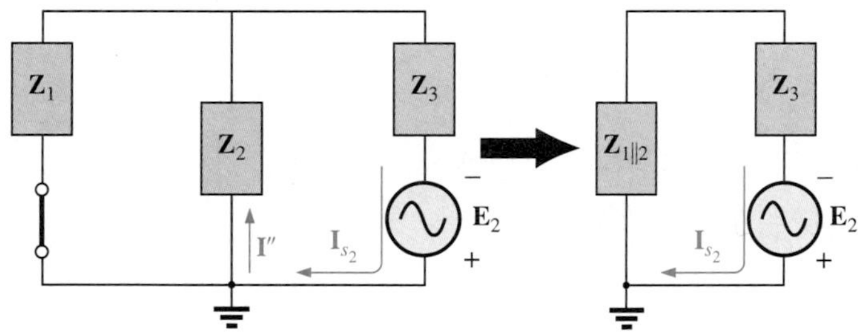

FIG. 18.4
Determining the effect of the voltage source $\mathbf{E}_2$ *on the current* **I** *of the network of Fig. 18.1.*

$$\mathbf{Z}_{1\|2} = \frac{\mathbf{Z}_1}{N} = \frac{j\,4\,\Omega}{2} = j\,2\,\Omega$$

$$\mathbf{I}_{s_2} = \frac{\mathbf{E}_2}{\mathbf{Z}_{1\|2} + \mathbf{Z}_3} = \frac{5\text{ V}\angle 0^\circ}{j\,2\,\Omega - j\,3\,\Omega} = \frac{5\text{ V}\angle 0^\circ}{1\,\Omega\angle -90^\circ} = 5\text{ A}\angle 90^\circ$$

and
$$\mathbf{I}'' = \frac{\mathbf{I}_{s_2}}{2} = 2.5\text{ A}\angle 90^\circ$$

The resultant current through the 4-Ω reactance X_{L_2} (Fig. 18.5) is

$$\begin{aligned}\mathbf{I} &= \mathbf{I}' - \mathbf{I}''\\ &= 3.75\text{ A}\angle -90^\circ - 2.50\text{ A}\angle 90^\circ = -j\,3.75\text{ A} - j\,2.50\text{ A}\\ &= -j\,6.25\text{ A}\\ \mathbf{I} &= \mathbf{6.25\text{ A}\angle -90^\circ}\end{aligned}$$

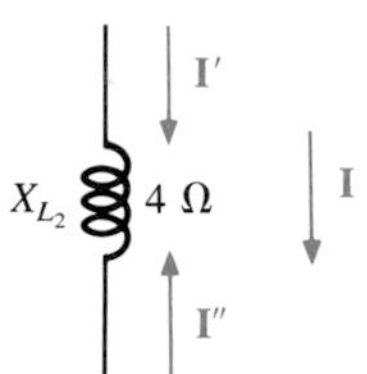

FIG. 18.5
Determining the resultant current for the network of Fig. 18.1.

EXAMPLE 18.2 Using superposition, find the current **I** through the 6-Ω resistor of Fig. 18.6.

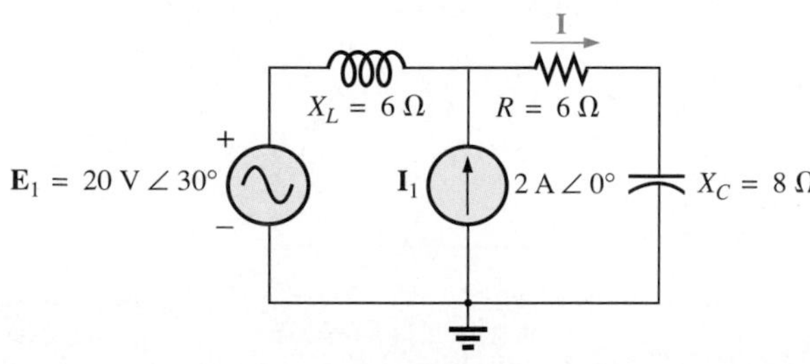

FIG. 18.6
Example 18.2.

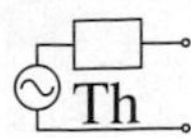

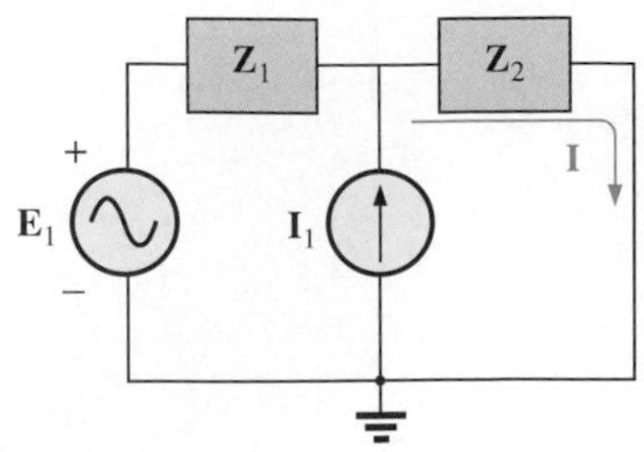

FIG. 18.7
Assigning the subscripted impedances to the network of Fig. 18.6.

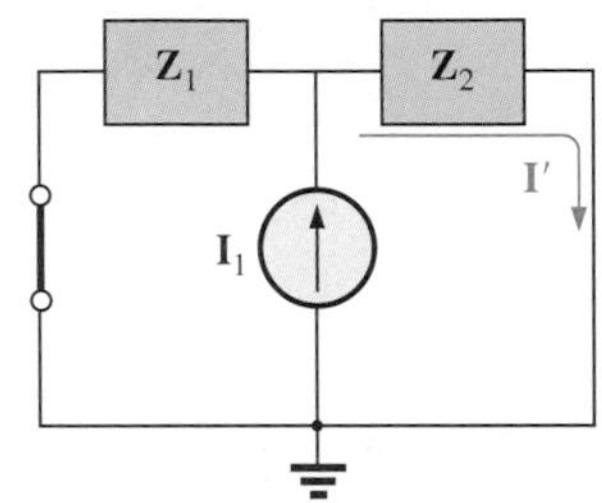

FIG. 18.8
Determining the effect of the current source $\mathbf{I}_1$ *on the current* **I** *of the network of Fig. 18.6.*

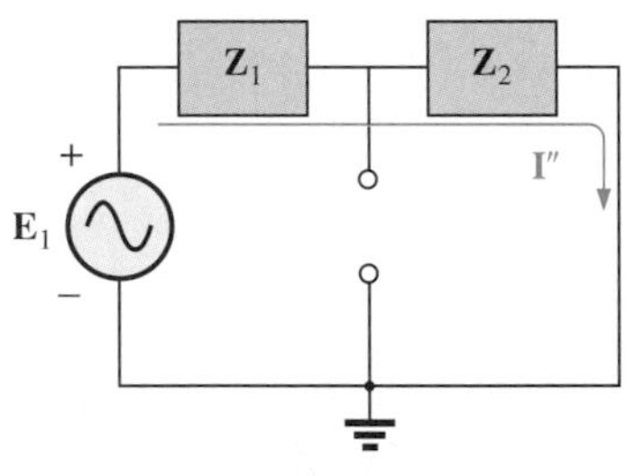

Determining the effect of the voltage source $\mathbf{E}_1$ *on the current* **I** *of the network of Fig. 18.6.*

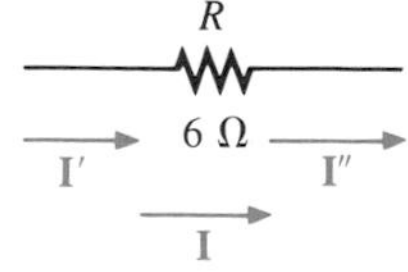

FIG. 18.10
Determining the resultant current **I** *for the network of Fig. 18.6.*

Solution: For the redrawn circuit (Fig. 18.7),

$$\mathbf{Z}_1 = j\,6\,\Omega \qquad \mathbf{Z}_2 = 6 - j\,8\,\Omega$$

Consider the effects of the current source (Fig. 18.8). Applying the current divider rule, we have

$$\mathbf{I}' = \frac{\mathbf{Z}_1\mathbf{I}_1}{\mathbf{Z}_1 + \mathbf{Z}_2} = \frac{(j\,6\,\Omega)(2\text{ A})}{j\,6\,\Omega + 6\,\Omega - j\,8\,\Omega} = \frac{j\,12\text{ A}}{6 - j\,2}$$
$$= \frac{12\text{ A}\,\angle 90^\circ}{6.32\,\angle -18.43^\circ}$$
$$\mathbf{I}' = 1.9\text{ A}\,\angle 108.43^\circ$$

Consider the effects of the voltage source (Fig. 18.9). Applying Ohm's law gives us

$$\mathbf{I}'' = \frac{\mathbf{E}_1}{\mathbf{Z}_T} = \frac{\mathbf{E}_1}{\mathbf{Z}_1 + \mathbf{Z}_2} = \frac{20\text{ V}\,\angle 30^\circ}{6.32\,\Omega\,\angle -18.43^\circ}$$
$$= 3.16\text{ A}\,\angle 48.43^\circ$$

The total current through the 6-Ω resistor (Fig. 18.10) is

$$\mathbf{I} = \mathbf{I}' + \mathbf{I}''$$
$$= 1.9\text{ A}\,\angle 108.43^\circ + 3.16\text{ A}\,\angle 48.43^\circ$$
$$= (-0.60\text{ A} + j\,1.80\text{ A}) + (2.10\text{ A} + j\,2.36\text{ A})$$
$$= 1.50\text{ A} + j\,4.16\text{ A}$$
$$\mathbf{I} = \mathbf{4.42\text{ A}\,\angle 70.2^\circ}$$

EXAMPLE 18.3 Using superposition, find the voltage across the 6-Ω resistor in Fig. 18.6. Check the results against $\mathbf{V}_{6\Omega} = \mathbf{I}(6\,\Omega)$, where **I** is the current found through the 6-Ω resistor in the previous example.

Solution: For the current source,

$$\mathbf{V}'_{6\Omega} = \mathbf{I}'(6\,\Omega) = (1.9\text{ A}\,\angle 108.43^\circ)(6\,\Omega) = 11.4\text{ V}\,\angle 108.43^\circ$$

For the voltage source,

$$\mathbf{V}''_{6\Omega} = \mathbf{I}''(6) = (3.16\text{ A}\,\angle 48.43^\circ)(6\,\Omega) = 18.96\text{ V}\,\angle 48.43^\circ$$

The total voltage across the 6-Ω resistor (Fig. 18.11) is

$$\mathbf{V}_{6\Omega} = \mathbf{V}'_{6\Omega} + \mathbf{V}''_{6\Omega}$$
$$= 11.4\text{ V}\,\angle 108.43^\circ + 18.96\text{ V}\,\angle 48.43^\circ$$
$$= (-3.60\text{ V} + j\,10.82\text{ V}) + (12.58\text{ V} + j\,14.18\text{ V})$$
$$= 8.98\text{ V} + j\,25.0\text{ V}$$
$$\mathbf{V}_{6\Omega} = \mathbf{26.5\text{ V}\,\angle 70.2^\circ}$$

Checking the result, we have

$$\mathbf{V}_{6\Omega} = \mathbf{I}(6\,\Omega) = (4.42\text{ A}\,\angle 70.2^\circ)(6\,\Omega)$$
$$= \mathbf{26.5\text{ V}\,\angle 70.2^\circ} \qquad \text{(checks)}$$

\+ $\mathbf{V}'_{6\Omega}$ –

\+ $\mathbf{V}''_{6\Omega}$ –

R

6 Ω

\+ $\mathbf{V}_{6\Omega}$ –

FIG. 18.11
Determining the resultant voltage $\mathbf{V}_{6\Omega}$ *for the network of Fig. 18.6.*

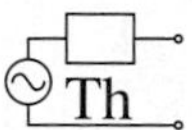

EXAMPLE 18.4 For the network of Fig. 18.12, determine the sinusoidal expression for the voltage v_3 using superposition.

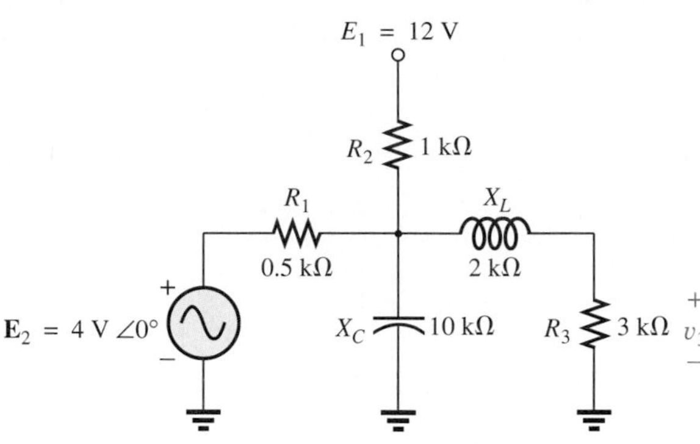

FIG. 18.12
Example 18.4.

Solution: For the dc source, recall that for dc analysis, in the steady state the capacitor can be replaced by an open-circuit equivalent and the inductor by a short-circuit equivalent. The result is the network of Fig. 18.13.

The resistors R_1 and R_3 are then in parallel and the voltage V_3 can be determined using the voltage divider rule:

$$R' = R_1 \parallel R_3 = 0.5\text{ k}\Omega \parallel 3\text{ k}\Omega = 0.429\text{ k}\Omega$$

and

$$V_3 = \frac{R'E_1}{R' + R_2}$$

$$= \frac{(0.429\text{ k}\Omega)(12\text{ V})}{0.429\text{ k}\Omega + 1\text{ k}\Omega} = \frac{5.148\text{ V}}{1.429}$$

$$V_3 \cong \mathbf{3.6\text{ V}}$$

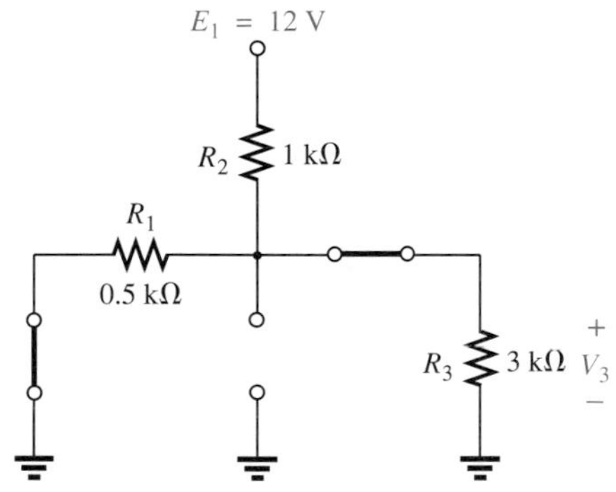

FIG. 18.13
Determining the effect of the dc voltage source E_1 on the voltage v_3 of the network of Fig. 18.12.

For ac analysis, the dc source is set to zero and the network is redrawn, as shown in Fig. 18.14.

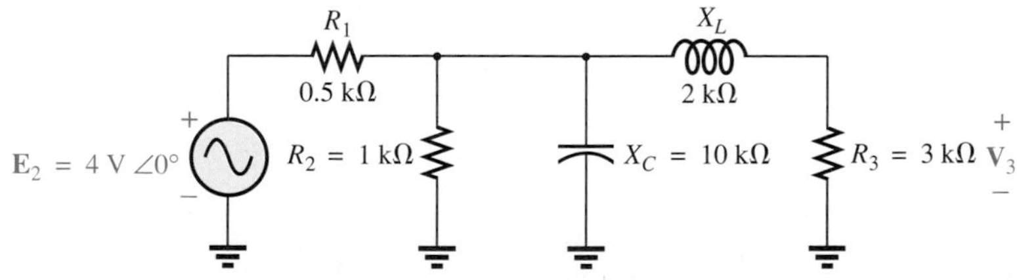

FIG. 18.14
Redrawing the network of Fig. 18.12 to determine the effect of the ac voltage source $\mathbf{E}_2$.

The block impedances are then defined as in Fig. 18.15 and series-parallel techniques are applied as follows:

$$\mathbf{Z}_1 = 0.5\text{ k}\Omega \angle 0°$$

$$\mathbf{Z}_2 = (R_2 \angle 0° \parallel (X_C \angle -90°)$$

$$= \frac{(1\text{ k}\Omega \angle 0°)(10\text{ k}\Omega \angle -90°)}{1\text{ k}\Omega - j\,10\text{ k}\Omega} = \frac{10\text{ k}\Omega \angle -90°}{10.05 \angle -84.29°}$$

$$= 0.995\text{ k}\Omega \angle -5.71°$$

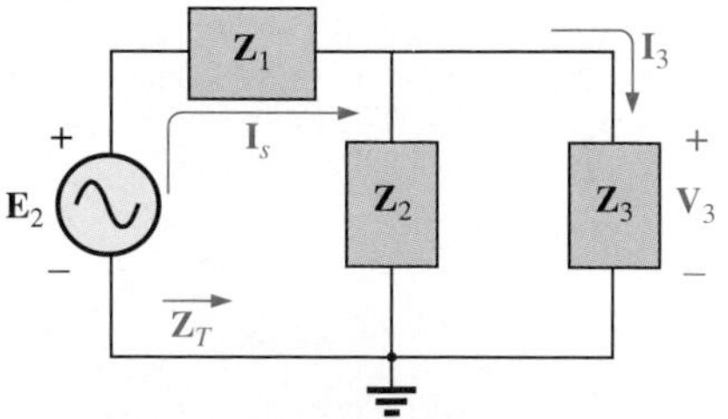

FIG. 18.15
Assigning the subscripted impedances to the network of Fig. 18.14.

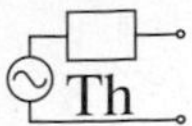

$$\mathbf{Z}_3 = R_3 + jX_L = 3\text{ k}\Omega + j\,2\text{ k}\Omega = 3.61\text{ k}\Omega\ \angle 33.69°$$

and

$$\begin{aligned}\mathbf{Z}_T &= \mathbf{Z}_1 + \mathbf{Z}_2 \parallel \mathbf{Z}_3\\ &= 0.5\text{ k}\Omega + (0.995\text{ k}\Omega\ \angle -5.71°) \parallel (3.61\text{ k}\Omega\ \angle 33.69°)\\ &= 1.312\text{ k}\Omega\ \angle 1.57°\end{aligned}$$

Calculator Performing the above on the TI-85 calculator:

```
(0.5,0)+((0.995∠−5.71)*(3.61∠33.69))/((0.995∠−5.71)+(3.61∠33.69)) (Enter)
  (1.311E0,35.373E−3)
Ans ▶ Pol
  (1.312E0∠1.545E0)
```

CALC. 18.1

$$\mathbf{I}_s = \frac{\mathbf{E}_2}{\mathbf{Z}_T} = \frac{4\text{ V}\ \angle 0°}{1.312\text{ k}\Omega\ \angle 1.57°} = 3.05\text{ mA}\ \angle -1.57°$$

Current divider rule:

$$\begin{aligned}\mathbf{I}_3 &= \frac{\mathbf{Z}_2\mathbf{I}_s}{\mathbf{Z}_2 + \mathbf{Z}_3} = \frac{(0.995\text{ k}\Omega\ \angle -5.71°)(3.05\text{ mA}\ \angle -1.57°)}{0.995\text{ k}\Omega\ \angle -5.71° + 3.61\text{ k}\Omega\ \angle 33.69°}\\ &= 0.686\text{ mA}\ \angle -32.74°\end{aligned}$$

with

$$\begin{aligned}\mathbf{V}_3 &= (I_3\ \angle\theta)(R_3\ \angle 0°)\\ &= (0.686\text{ mA}\ \angle -32.74°)(3\text{ k}\Omega\ \angle 0°)\\ &= \mathbf{2.06\ V\ \angle -32.74°}\end{aligned}$$

The total solution:

$$\begin{aligned}\upsilon_3 &= \upsilon_3\text{ (dc)} + \upsilon_3\text{ (ac)}\\ &= 3.6\text{ V} + 2.06\text{ V}\ \angle -32.74°\\ \upsilon_3 &= \mathbf{3.6 + 2.91\sin(\omega t - 32.74°)}\end{aligned}$$

The result is a sinusoidal voltage having a peak value of 2.91 V riding on an average value of 3.6 V, as shown in Fig. 18.16.

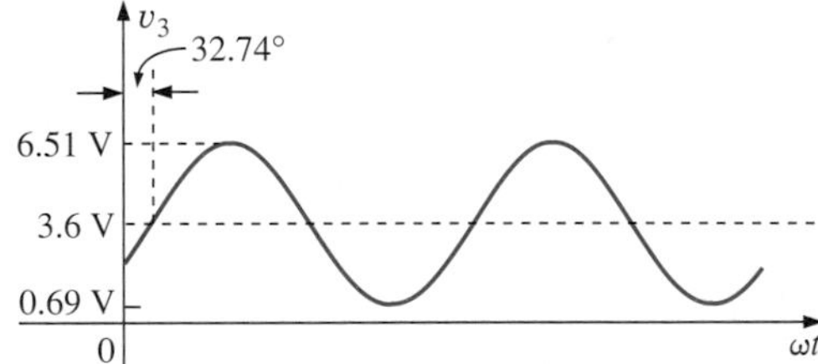

FIG. 18.16
The resultant voltage υ_3 for the network of Fig.18.12.

Dependent Sources

For dependent sources in which *the controlling variable is not determined by the network to which the superposition theorem is to be applied,* the application of the theorem is basically the same as for inde-

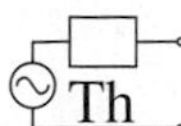

pendent sources. The solution obtained will simply be in terms of the controlling variables.

EXAMPLE 18.5 Using the superposition theorem, determine the current $\mathbf{I}_2$ for the network of Fig. 18.17. The quantities μ and h are constants.

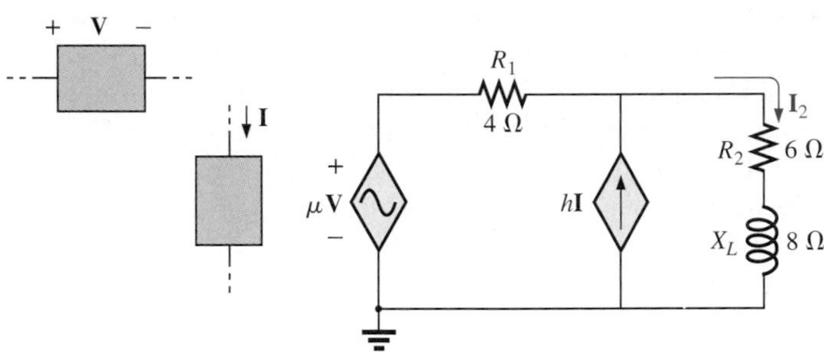

FIG. 18.17
Example 18.5.

Solution: With a portion of the system redrawn (Fig. 18.18),

$$\mathbf{Z}_1 = R_1 = 4\ \Omega \qquad \mathbf{Z}_2 = R_2 + j X_L = 6 + j\,8\ \Omega$$

For the voltage source (Fig. 18.19),

$$\mathbf{I}' = \frac{\mu\mathbf{V}}{\mathbf{Z}_1 + \mathbf{Z}_2} = \frac{\mu\mathbf{V}}{4\ \Omega + 6\ \Omega + j\,8\ \Omega} = \frac{\mu\mathbf{V}}{10\ \Omega + j\,8\ \Omega}$$

$$= \frac{\mu\mathbf{V}}{12.8\ \Omega\ \angle 38.66^\circ} = 0.078\ \mu\mathbf{V}/\Omega\ \angle -38.66^\circ$$

For the current source (Fig. 18.20),

$$\mathbf{I}'' = \frac{\mathbf{Z}_1(h\mathbf{I})}{\mathbf{Z}_1 + \mathbf{Z}_2} = \frac{(4\ \Omega)(h\mathbf{I})}{12.8\ \Omega\ \angle 38.66^\circ} = 4(0.078)h\mathbf{I}\ \angle -38.66^\circ$$

$$= 0.312h\mathbf{I}\ \angle -38.66^\circ$$

The current $\mathbf{I}_2$ is

$$\mathbf{I}_2 = \mathbf{I}' + \mathbf{I}''$$
$$= 0.078\ \mu\mathbf{V}/\Omega\ \angle -38.66^\circ + 0.312h\mathbf{I}\ \angle \mathbf{-38.66^\circ}$$

For $\mathbf{V} = 10\ \text{V}\ \angle 0^\circ$, $\mathbf{I} = 20\ \text{mA}\ \angle 0^\circ$, $\mu = 20$, $h = 100$,

$$\mathbf{I}_2 = 0.078(20)(10\ \text{V}\ \angle 0^\circ)/\Omega\ \angle -38.66^\circ + 0.312(100)(20\ \text{mA}\ \angle 0^\circ)\ \angle -38.66^\circ$$
$$= 15.60\ \text{A}\ \angle -38.66^\circ + 0.62\ \text{A}\ \angle -38.66^\circ$$
$$\mathbf{I}_2 = \mathbf{16.22\ A\ \angle -38.66^\circ}$$

For dependent sources in which *the controlling variable is determined by the network to which the theorem is to be applied,* the dependent source cannot be set to zero unless the controlling variable is also zero. For networks containing dependent sources such as indicated in Example 18.5 and dependent sources of the type just introduced above, the superposition theorem is applied for each independent source and each dependent source not having a controlling variable in the portions of the network under investigation. It must be reemphasized that depen-

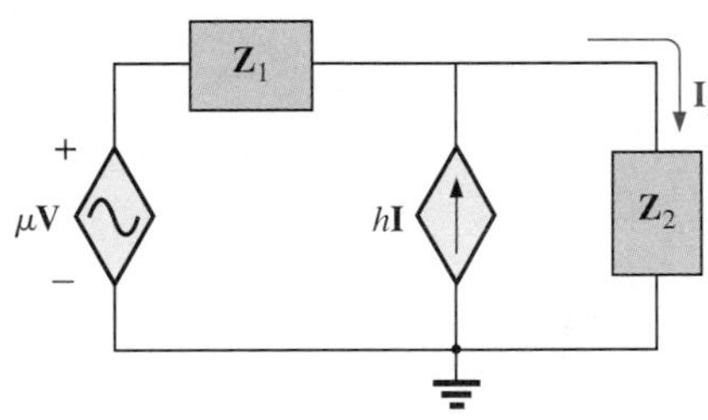

FIG. 18.18
Assigning the subscripted impedances to the network of Fig. 18.17.

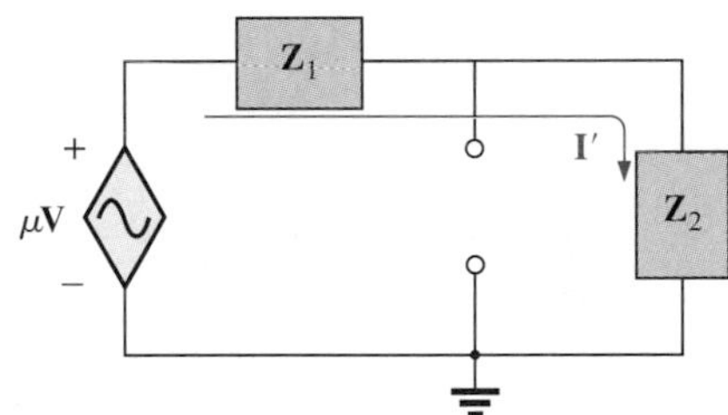

FIG. 18.19
Determining the effect of the voltage-controlled voltage source on the current $\mathbf{I}_2$ *for the network of Fig. 18.17.*

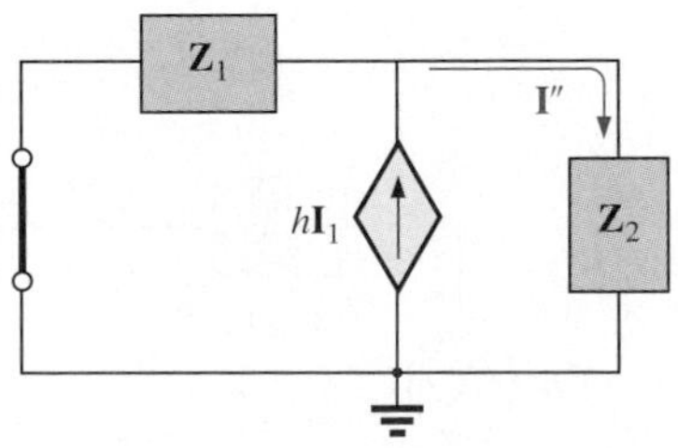

FIG. 18.20
Determining the effect of the current-controlled current source on the current $\mathbf{I}_2$ *for the network of Fig. 18.17.*

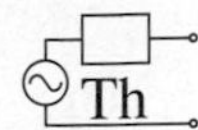

dent sources are not sources of energy in the sense that, if all independent sources are removed from a system, all currents and voltages must be zero.

EXAMPLE 18.6 Determine the current $\mathbf{I}_L$ through the resistor R_L of Fig. 18.21.

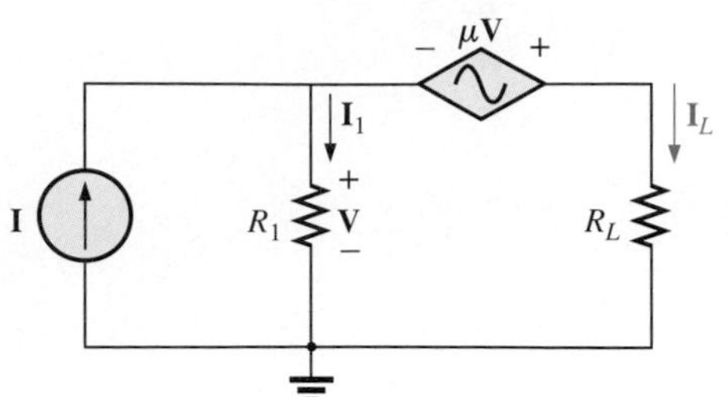

FIG. 18.21
Example 18.6.

Solution: Note that the controlling variable $\mathbf{V}$ is determined by the network to be analyzed. From the above discussions, it is understood that the dependent source cannot be set to zero unless $\mathbf{V}$ is zero. If we set $\mathbf{I}$ to zero, the network lacks a source of voltage, and $\mathbf{V} = 0$ with $\mu\mathbf{V} = 0$. The resulting $\mathbf{I}_L$ under this condition is zero. Obviously, therefore, the network must be analyzed as it appears in Fig. 18.21, with the result that neither source can be eliminated, as is normally done using the superposition theorem.

Applying Kirchhoff's voltage law, we have

$$\mathbf{V}_L = \mathbf{V} + \mu\mathbf{V} = (1 + \mu)\mathbf{V}$$

and

$$\mathbf{I}_L = \frac{\mathbf{V}_L}{R_L} = \frac{(1 + \mu)\mathbf{V}}{R_L}$$

The result, however, must be found in terms of $\mathbf{I}$ since $\mathbf{V}$ and $\mu\mathbf{V}$ are only dependent variables.

Applying Kirchhoff's current law gives us

$$\mathbf{I} = \mathbf{I}_1 + \mathbf{I}_L = \frac{\mathbf{V}}{R_1} + \frac{(1 + \mu)\mathbf{V}}{R_L}$$

and

$$\mathbf{I} = \mathbf{V}\left(\frac{1}{R_1} + \frac{1 + \mu}{R_L}\right)$$

or

$$\mathbf{V} = \frac{\mathbf{I}}{(1/R_1) + [(1 + \mu)/R_L]}$$

Substituting into the above yields

$$\mathbf{I}_L = \frac{(1 + \mu)\mathbf{V}}{R_L} = \frac{(1 + \mu)}{R_L}\left(\frac{\mathbf{I}}{(1/R_1) + [(1 + \mu)/R_L]}\right)$$

Therefore,

$$\mathbf{I}_L = \frac{\mathbf{(1 + \mu)}\boldsymbol{R}_1\mathbf{I}}{\boldsymbol{R}_L + \mathbf{(1 + \mu)}\boldsymbol{R}_1}$$

18.3 THÉVENIN'S THEOREM

Thévenin's theorem, as stated for sinusoidal ac circuits, is changed only to include the term *impedance* instead of *resistance*; that is,

any two-terminal linear ac network can be replaced with an equivalent circuit consisting of a voltage source and an impedance in series, as shown in Fig. 18.22.

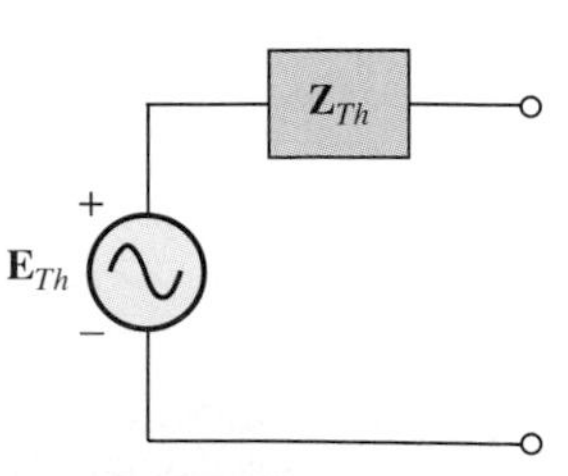

FIG. 18.22
Thévenin equivalent circuit for ac networks.

Since the reactances of a circuit are frequency dependent, the Thévenin circuit found for a particular network is applicable only at *one* frequency.

The steps required to apply this method to dc circuits are repeated here with changes for sinusoidal ac circuits. As before, the only change

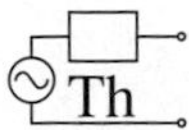

is the replacement of the term *resistance* with *impedance*. Again, dependent and independent sources will be treated separately.

The last example of the independent source section will include a network with dc and ac sources to establish the groundwork for possible use in the electronics area.

Independent Sources

1. ***Remove that portion of the network across which the Thévenin equivalent circuit is to be found.***
2. ***Mark (○, •, and so on) the terminals of the remaining two-terminal network.***
3. ***Calculate Z_{Th} by first setting all voltage and current sources to zero (short circuit and open circuit, respectively) and then finding the resulting impedance between the two marked terminals.***
4. ***Calculate E_{Th} by first replacing the voltage and current sources and then finding the open-circuit voltage between the marked terminals.***
5. ***Draw the Thévenin equivalent circuit with the portion of the circuit previously removed replaced between the terminals of the Thévenin equivalent circuit.***

EXAMPLE 18.7 Find the Thévenin equivalent circuit for the network external to resistor R in Fig. 18.23.

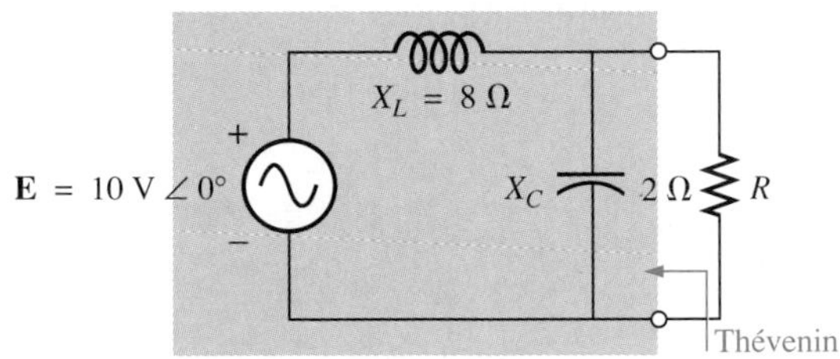

FIG. 18.23
Example 18.7.

Solution:

Steps 1 and 2 (Fig. 18.24):

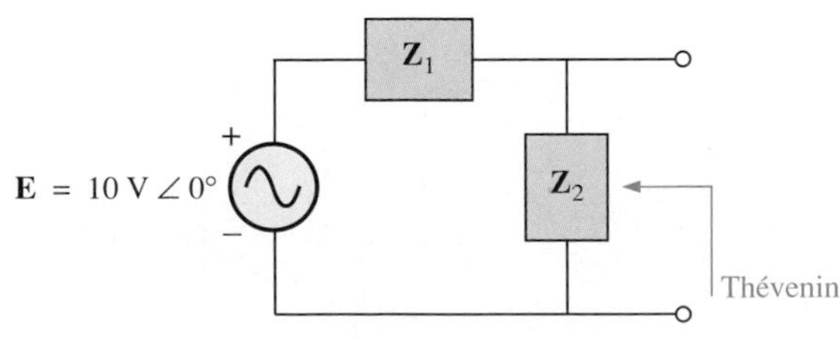

FIG. 18.24
Assigning the subscripted impedances to the network of Fig. 18.23.

$$\mathbf{Z}_1 = jX_L = j\,8\ \Omega \qquad \mathbf{Z}_2 = -jX_C = -j\,2\ \Omega$$

Step 3 (Fig. 18.25):

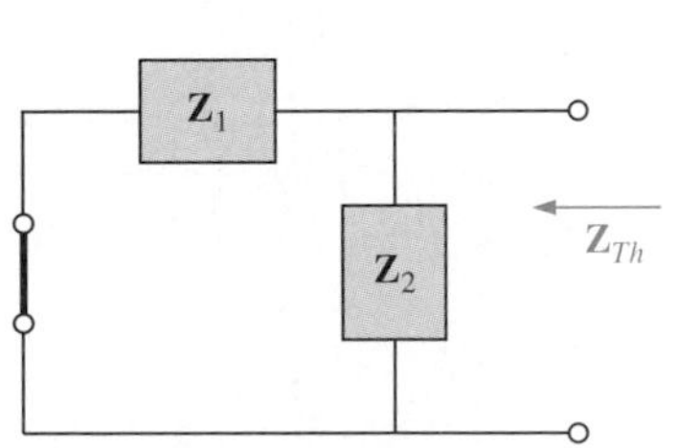

FIG. 18.25
Determining the Thévenin impedance for the network of Fig. 18.23.

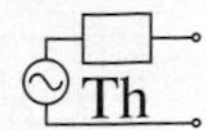

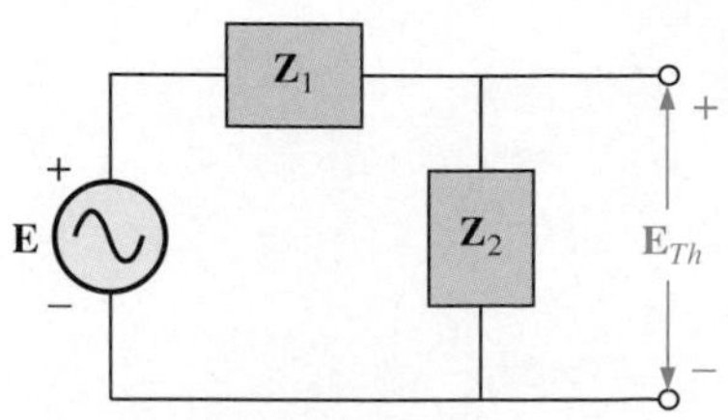

FIG. 18.26
Determining the open-circuit Thévenin voltage for the network of Fig. 18.23.

$$\mathbf{Z}_{Th} = \frac{\mathbf{Z}_1\mathbf{Z}_2}{\mathbf{Z}_1 + \mathbf{Z}_2} = \frac{(j\,8\,\Omega)(-j\,2\,\Omega)}{j\,8\,\Omega - j\,2\,\Omega} = \frac{-j^2 16\,\Omega}{j\,6} = \frac{16\,\Omega}{6\,\angle 90^\circ}$$
$$= \mathbf{2.67\,\Omega\,\angle -90^\circ}$$

Step 4 (Fig. 18.26):

$$\mathbf{E}_{Th} = \frac{\mathbf{Z}_2\mathbf{E}}{\mathbf{Z}_1 + \mathbf{Z}_2} \qquad \text{(voltage divider rule)}$$
$$= \frac{(-j\,2\,\Omega)(10\text{ V})}{j\,8\,\Omega - j\,2\,\Omega} = \frac{-j\,20\text{ V}}{j\,6} = \mathbf{3.33\text{ V}\,\angle -180^\circ}$$

Step 5: The Thévenin equivalent circuit is shown in Fig. 18.27.

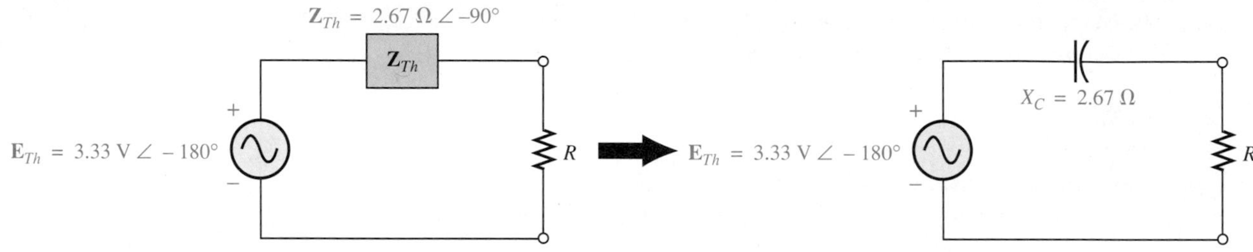

FIG. 18.27
The Thévenin equivalent circuit for the network of Fig. 18.23.

EXAMPLE 18.8 Find the Thévenin equivalent circuit for the network external to branch a-a' in Fig. 18.28.

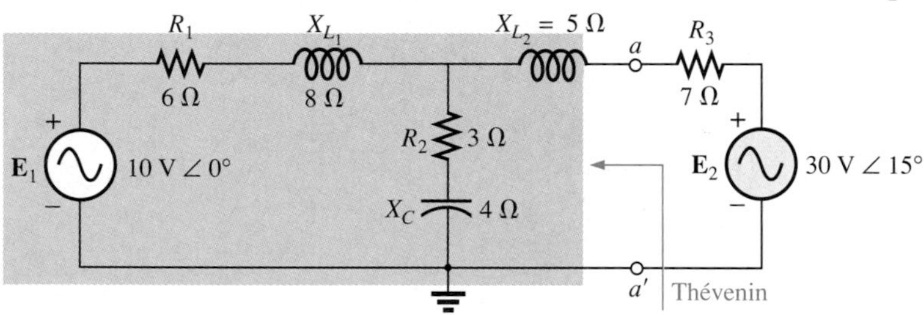

FIG. 18.28
Example 18.8.

Solution:

Steps 1 and 2 (Fig. 18.29): Note the reduced complexity with subscripted impedances:

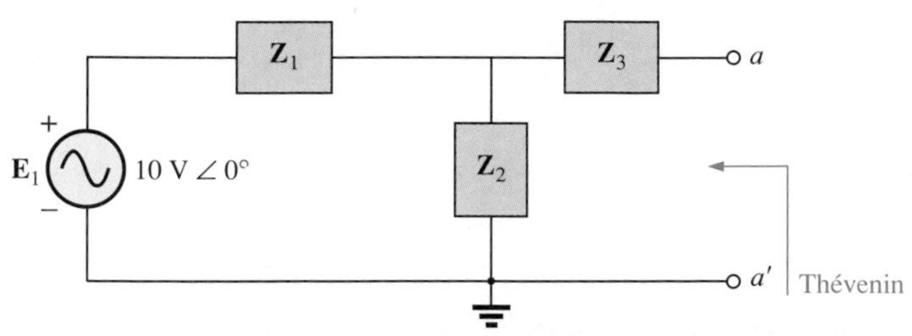

FIG. 18.29
Assigning the subscripted impedances to the network of Fig. 18.28.

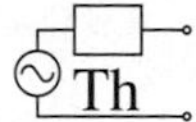

$$\mathbf{Z}_1 = R_1 + jX_{L_1} = 6\ \Omega + j\,8\ \Omega$$
$$\mathbf{Z}_2 = R_2 - jX_C = 3\ \Omega - j\,4\ \Omega$$
$$\mathbf{Z}_3 = +jX_{L_2} = j\,5\ \Omega$$

Step 3 (Fig. 18.30):

$$\mathbf{Z}_{Th} = \mathbf{Z}_3 + \frac{\mathbf{Z}_1\mathbf{Z}_2}{\mathbf{Z}_1 + \mathbf{Z}_2} = j\,5\ \Omega + \frac{(10\ \Omega \angle 53.13^\circ)(5\ \Omega \angle -53.13^\circ)}{(6\ \Omega + j\,8\ \Omega) + (3\ \Omega - j\,4\ \Omega)}$$
$$= j\,5 + \frac{50 \angle 0^\circ}{9 + j\,4} = j\,5 + \frac{50 \angle 0^\circ}{9.85 \angle 23.96^\circ}$$
$$= j\,5 + 5.08 \angle -23.96^\circ = j\,5 + 4.64 - j\,2.06$$
$$\mathbf{Z}_{Th} = \mathbf{4.64\ \Omega + j\,2.94\ \Omega = 5.49\ \Omega \angle 32.36^\circ}$$

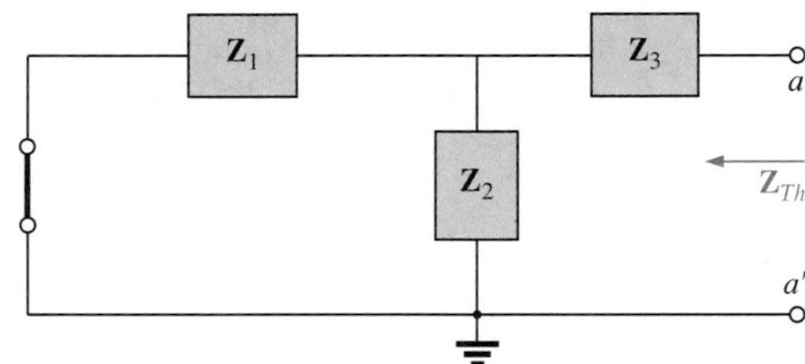

FIG. 18.30
Determining the Thévenin impedance for the network of Fig. 18.28.

Step 4 (Fig. 18.31): Since *a-a′* is an open circuit, $\mathbf{I}_{\mathbf{Z}_3} = 0$. Then $\mathbf{E}_{Th}$ is the voltage drop across $\mathbf{Z}_2$:

$$\mathbf{E}_{Th} = \frac{\mathbf{Z}_2\mathbf{E}}{\mathbf{Z}_2 + \mathbf{Z}_1} \qquad \text{(voltage divider rule)}$$
$$= \frac{(5\ \Omega \angle -53.13^\circ)(10\ \text{V} \angle 0^\circ)}{9.85\ \Omega \angle 23.96^\circ}$$
$$\mathbf{E}_{Th} = \frac{50\ \text{V} \angle -53.13^\circ}{9.85 \angle 23.96^\circ} = \mathbf{5.08\ V \angle -77.09^\circ}$$

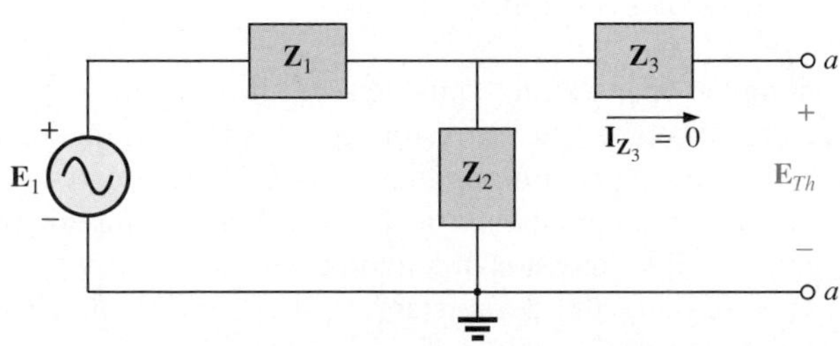

FIG. 18.31
Determining the open-circuit Thévenin voltage for the network of Fig. 18.28.

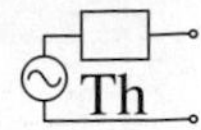

Step 5: The Thévenin equivalent circuit is shown in Fig. 18.32.

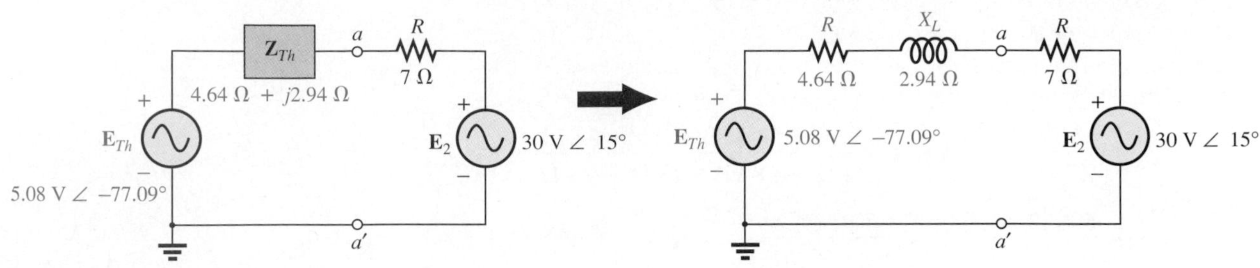

FIG. 18.32
The Thévenin equivalent circuit for the network of Fig. 18.28.

The next example demonstrates how superposition is applied to electronic circuits to permit *a separation of the dc and ac analysis.* The fact that the controlling variable in this analysis is not in the portion of the network connected directly to the terminals of interest permits an analysis of the network in the same manner as applied above for independent sources.

EXAMPLE 18.9 Determine the Thévenin equivalent circuit for the transistor network external to the resistor R_L in the following network (Fig. 18.33) and then determine $\mathbf{V}_L$.

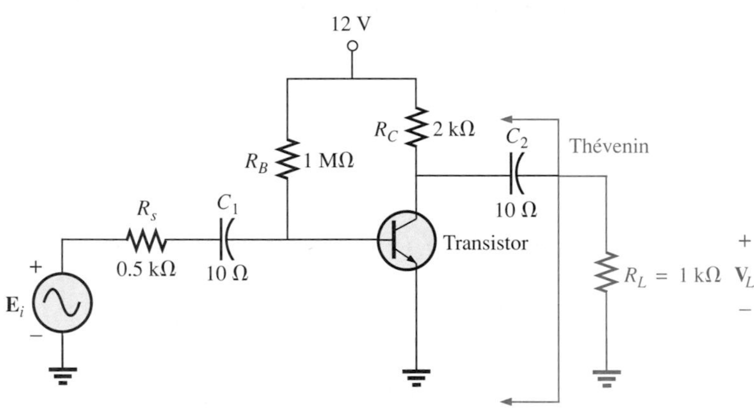

FIG. 18.33
Example 18.9.

Solution: Applying superposition.

dc conditions:

Substituting the open-circuit equivalent for the coupling capacitor C_2 will isolate the dc source and the resulting currents from the load resistor. The result is for dc conditions that $V_L = 0$ V. Although the output dc voltage is zero, the application of the dc voltage is important to the basic operation of the transistor in a number of important ways, one of which is to determine the parameters of the "equivalent circuit" to appear in the ac analysis to follow.

ac conditions:

For the ac analysis an equivalent circuit is substituted for the transistor, as established by the dc conditions above, that will behave like

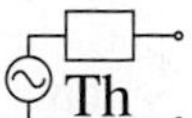

the actual transistor. A great deal more will be said about equivalent circuits and the operations performed to obtain the network of Fig. 18.34, but for now let us limit our attention to the manner in which the Thévenin equivalent circuit is obtained. Note in Fig. 18.34 that the equivalent circuit includes a resistor of 2.3 kΩ and a controlled current source whose magnitude is determined by the product of a factor of 100 and the current I_1 in another part of the network.

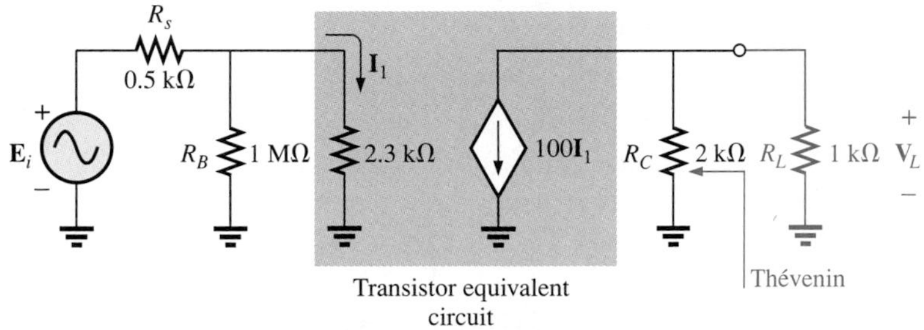

FIG. 18.34
The ac equivalent network for the transistor amplifier of Fig. 18.33.

Note in Fig. 18.34 the absence of the coupling capacitors for the ac analysis. In general, coupling capacitors are designed to be open circuits for dc and short circuits for ac analysis. The short-circuit equivalent is valid because the other impedances in series with the coupling capacitors are so much larger in magnitude that the effect of the coupling capacitors can be ignored. Both R_B and R_C are now tied to ground because the dc source was set to zero volts (superposition) and replaced by a short-circuit equivalent to ground.

For the analysis to follow, the effect of the resistor R_B will be ignored since it is so much larger than the parallel 2.3-kΩ resistor.

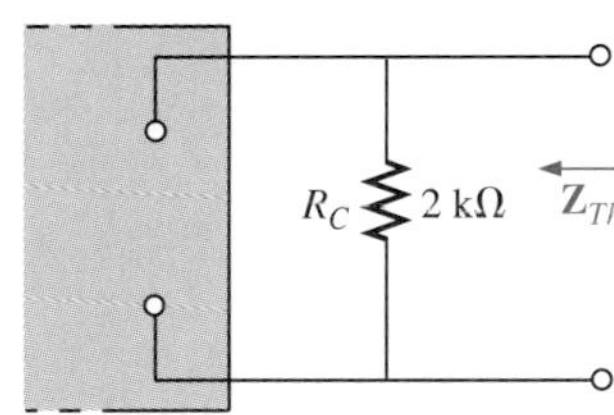

FIG. 18.35
Determining the Thévenin impedance for the network of Fig. 18.34.

$\mathbf{Z}_{Th}$:

When $\mathbf{E}_i$ is set to zero volts, the current $\mathbf{I}_1$ will be zero amperes, and the controlled source $100\mathbf{I}_1$ will be zero amperes also. The result is an open-circuit equivalent for the source, as appearing in Fig. 18.35.

It is fairly obvious from Fig. 18.35 that

$$\mathbf{Z}_{Th} = \mathbf{2\ k\Omega}$$

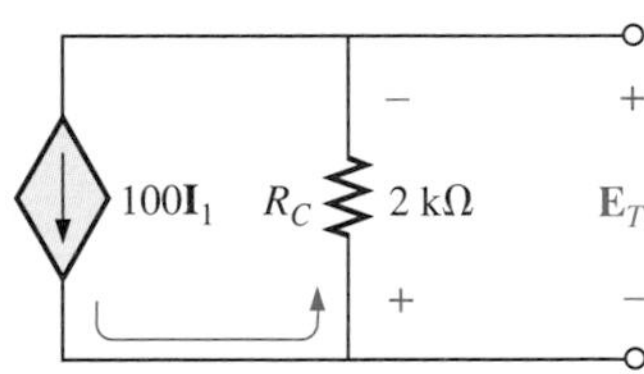

FIG. 18.36
Determining the Thévenin voltage for the network of Fig. 18.34.

$\mathbf{E}_{Th}$:

For $\mathbf{E}_{Th}$ the current $\mathbf{I}_1$ of Fig. 18.34 will be

$$\mathbf{I}_1 = \frac{\mathbf{E}_i}{R_s + 2.3\ \text{k}\Omega} = \frac{\mathbf{E}_i}{0.5\ \text{k}\Omega + 2.3\ \text{k}\Omega} = \frac{\mathbf{E}_i}{2.8\ \text{k}\Omega}$$

and $$100\mathbf{I}_1 = (100)\left(\frac{\mathbf{E}_i}{2.8\ \text{k}\Omega}\right) = 35.71 \times 10^{-3}/\Omega\ \mathbf{E}_i$$

Referring to Fig. 18.36, we find that

$$\begin{aligned}\mathbf{E}_{Th} &= -(100\mathbf{I}_1)R_C\\ &= -(35.71 \times 10^{-3}/\Omega\ \mathbf{E}_i)(2 \times 10^3\ \Omega)\\ \mathbf{E}_{Th} &= -71.42\mathbf{E}_i\end{aligned}$$

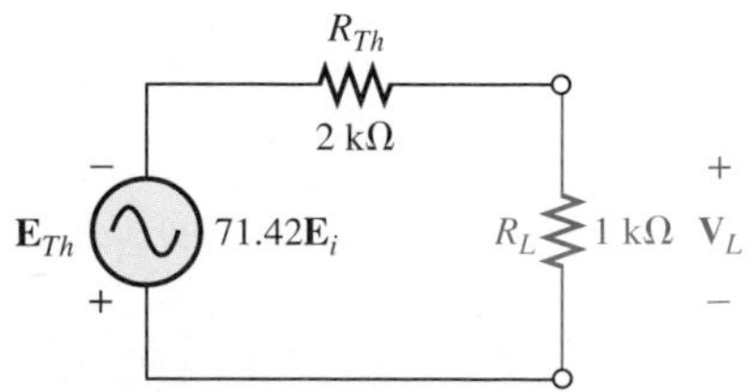

FIG. 18.37
The Thévenin equivalent circuit for the network of Fig. 18.34.

The Thévenin equivalent circuit appears in Fig. 18.37 with the original load R_L.

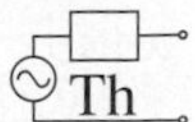

The output voltage $\mathbf{V}_L$:

$$\mathbf{V}_L = \frac{-R_L\mathbf{E}_{Th}}{R_L + \mathbf{Z}_{Th}} = \frac{-(1\text{ k}\Omega)(71.42\mathbf{E}_i)}{1\text{ k}\Omega + 2\text{ k}\Omega}$$

and $$\mathbf{V}_L = \mathbf{-23.81E}_i$$

revealing that the output voltage is 23.81 times the applied voltage with a phase shift of 180° due to the minus sign.

Dependent Sources

For dependent sources with a *controlling variable not in the network under investigation,* the procedure indicated above can be applied. However, for dependent sources of the other type, where the *controlling variable is part of the network to which the theorem is to be applied,* another approach must be employed. The necessity for a different approach will be demonstrated in an example to follow. The method is *not limited to dependent sources* of the latter type. It can also be applied to any dc or sinusoidal ac network. However, for networks of independent sources, the method of application employed in Chapter 9 and the first portion of this section is generally more direct, with the usual savings in time and errors.

The new approach to Thévenin's theorem can best be introduced at this stage in the development by considering the Thévenin equivalent circuit of Fig. 18.38(a). As indicated in Fig. 18.38(b), the open-circuit terminal voltage ($\mathbf{E}_{oc}$) of the Thévenin equivalent circuit is the Thévenin equivalent voltage; that is,

$$\mathbf{E}_{oc} = \mathbf{E}_{Th} \tag{18.1}$$

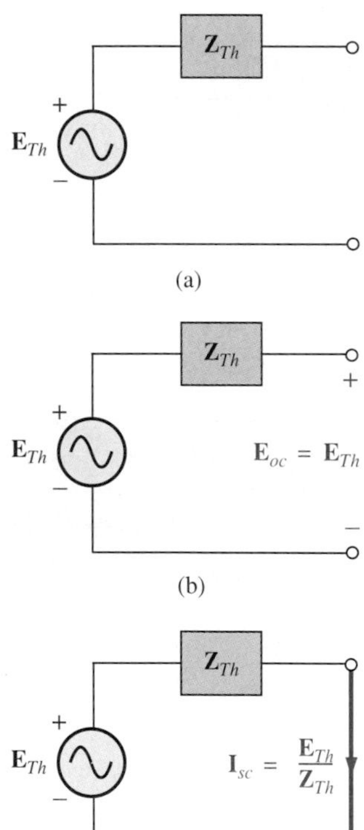

FIG. 18.38
Defining an alternative approach for determining the Thévenin impedance.

If the external terminals are short circuited as in Fig. 18.38(c), the resulting short-circuit current is determined by

$$\mathbf{I}_{sc} = \frac{\mathbf{E}_{Th}}{\mathbf{Z}_{Th}} \tag{18.2}$$

or, rearranged,

$$\mathbf{Z}_{Th} = \frac{\mathbf{E}_{Th}}{\mathbf{I}_{sc}}$$

and $$\mathbf{Z}_{Th} = \frac{\mathbf{E}_{oc}}{\mathbf{I}_{sc}} \tag{18.3}$$

Equations (18.1) and (18.3) indicate that for any linear bilateral dc or ac network with or without dependent sources of any type, if the open-circuit terminal voltage of a portion of a network can be determined along with the short-circuit current between the same two terminals, the Thévenin equivalent circuit is effectively known. A few examples will make the method quite clear. The advantage of the method, which was stressed earlier in this section for independent sources, should now be more obvious. The current $\mathbf{I}_{sc}$, which is necessary to find $\mathbf{Z}_{Th}$, is in general more difficult to obtain since all of the sources are present.

There is a third approach to the Thévenin equivalent circuit that is also useful from a practical viewpoint. The Thévenin voltage is found as in the two previous methods. However, the Thévenin impedance is

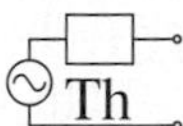

obtained by applying a source of voltage to the terminals of interest and determining the source current as indicated in Fig. 18.39. For this method, the source voltage of the original network is set to zero. The Thévenin impedance is then determined by the following equation:

$$\mathbf{Z}_{Th} = \frac{\mathbf{E}_g}{\mathbf{I}_g} \tag{18.4}$$

Note that for each technique, $\mathbf{E}_{Th} = \mathbf{E}_{oc}$, but the Thévenin impedance is found in different ways.

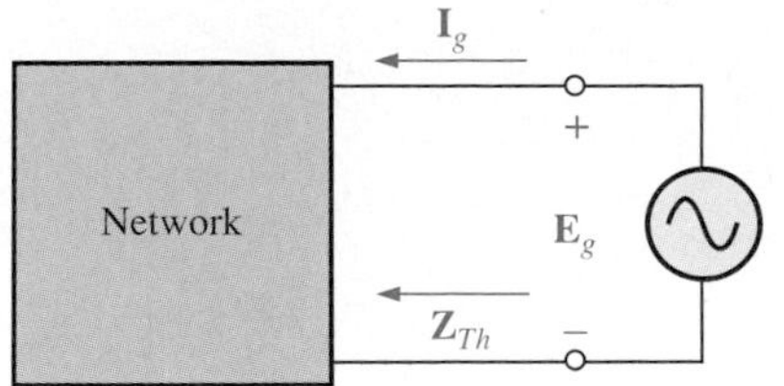

FIG. 18.39
Determining $\mathbf{Z}_{Th}$ using the approach $\mathbf{Z}_{Th} = \mathbf{E}_g / \mathbf{I}_g$.

EXAMPLE 18.10 Using each of the three techniques described in this section, determine the Thévenin equivalent circuit for the network of Fig. 18.40.

Solution: Since for each approach the Thévenin voltage is found in exactly the same manner, it will be determined first. From Fig. 18.40, where $\mathbf{I}_{X_C} = 0$,

Due to the polarity for **V** and defined terminal polarities

$$\mathbf{V}_{R_1} = \mathbf{E}_{Th} = \mathbf{E}_{oc} = \overset{\downarrow}{-}\frac{R_2(\mu\mathbf{V})}{R_1 + R_2} = -\frac{\mu R_2 \mathbf{V}}{R_1 + R_2}$$

The following three methods for determining the Thévenin impedance appear in the order in which they were introduced in this section.

Method 1 (Fig. 18.41):

$$\mathbf{Z}_{Th} = R_1 \parallel R_2 - j X_C$$

Method 2 (Fig. 18.42): Converting the voltage source to a current source (Fig. 18.43), we have (current divider rule)

$$\mathbf{I}_{sc} = \frac{-(R_1 \parallel R_2)\dfrac{\mu\mathbf{V}}{R_1}}{(R_1 \parallel R_2) - jX_C} = \frac{-\dfrac{R_1R_2}{R_1 + R_2}\left(\dfrac{\mu\mathbf{V}}{R_1}\right)}{(R_1 \parallel R_2) - jX_C}$$

$$= \frac{\dfrac{-\mu R_2\mathbf{V}}{R_1 + R_2}}{(R_1 \parallel R_2) - jX_C}$$

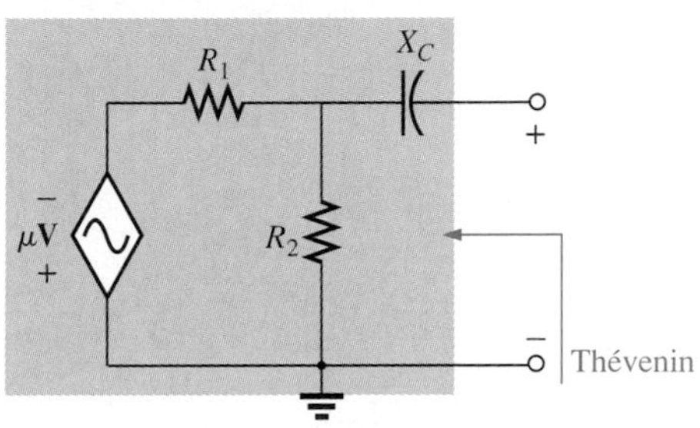

FIG. 18.40
Example 18.10.

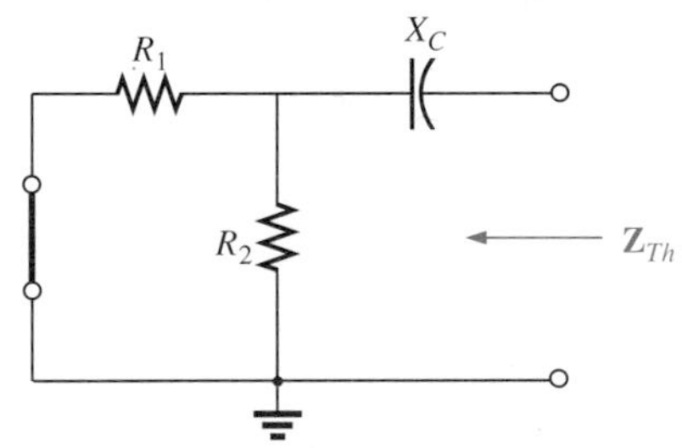

FIG. 18.41
Determining the Thévenin impedance for the network of Fig. 18.40.

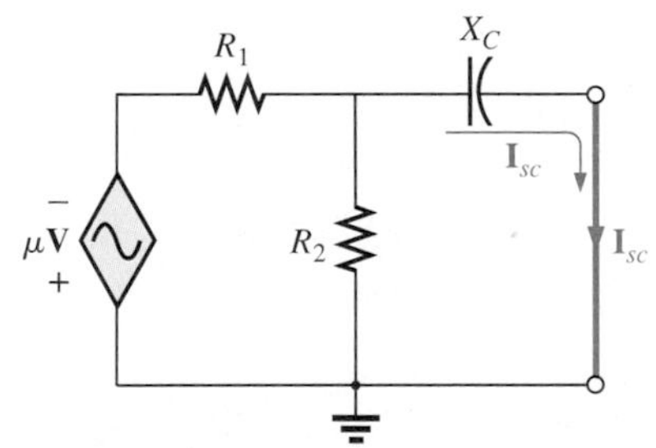

FIG. 18.42
Determining the short-circuit current for the network of Fig. 18.40.

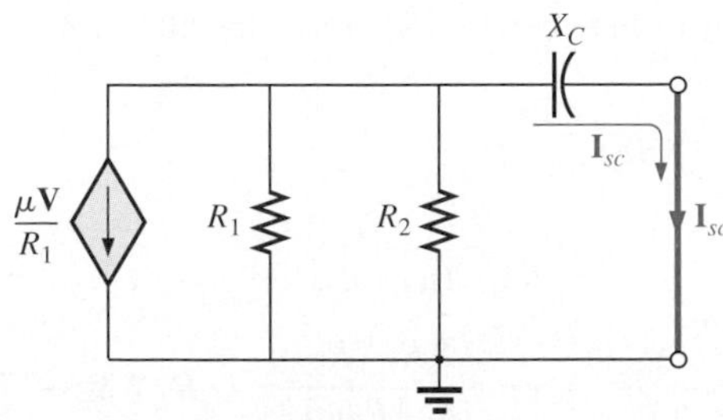

FIG. 18.43
Converting the voltage source of Fig. 18.42 to a current source.

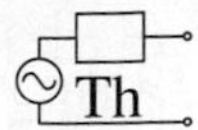

and

$$\mathbf{Z}_{Th} = \frac{\mathbf{E}_{oc}}{\mathbf{I}_{sc}} = \frac{\dfrac{-\mu R_2 \mathbf{V}}{R_1 + R_2}}{\dfrac{\dfrac{-\mu R_2 \mathbf{V}}{R_1 + R_2}}{(R_1 \parallel R_2) - j X_C}} = \frac{1}{\dfrac{1}{(R_1 \parallel R_2) - j X_C}}$$

$$= \boldsymbol{R_1 \parallel R_2 - j X_C}$$

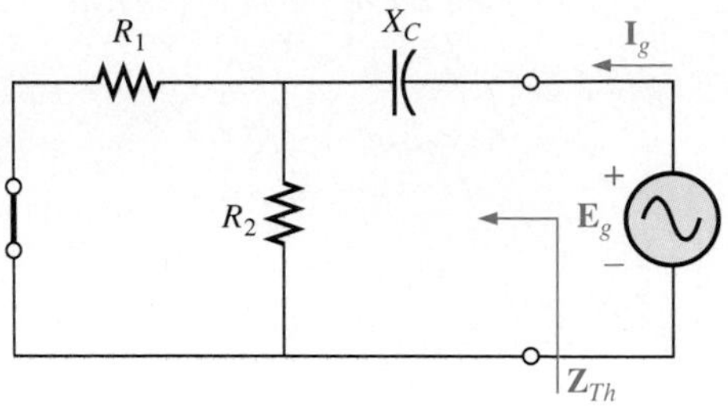

Method 3 (Fig. 18.44):

$$\mathbf{I}_g = \frac{\mathbf{E}_g}{(R_1 \parallel R_2) - j X_C}$$

and
$$\mathbf{Z}_{Th} = \frac{\mathbf{E}_g}{\mathbf{I}_g} = \boldsymbol{R_1 \parallel R_2 - j X_C}$$

FIG. 18.44
Determining the Thévenin impedance for the network of Fig. 18.40 using the approach $\mathbf{Z}_{Th} = \mathbf{E}_g / \mathbf{I}_g$.

In each case, the Thévenin impedance is the same. The resulting Thévenin equivalent circuit is shown in Fig. 18.45.

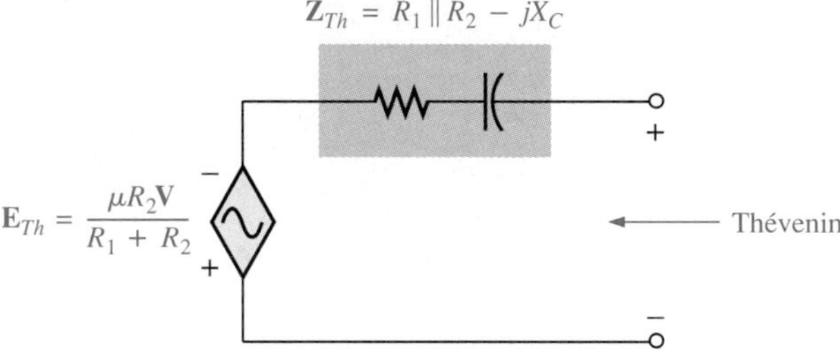

FIG. 18.45
The Thévenin equivalent circuit for the network of Fig. 18.40.

EXAMPLE 18.11 Repeat Example 18.10 for the network of Fig. 18.46.

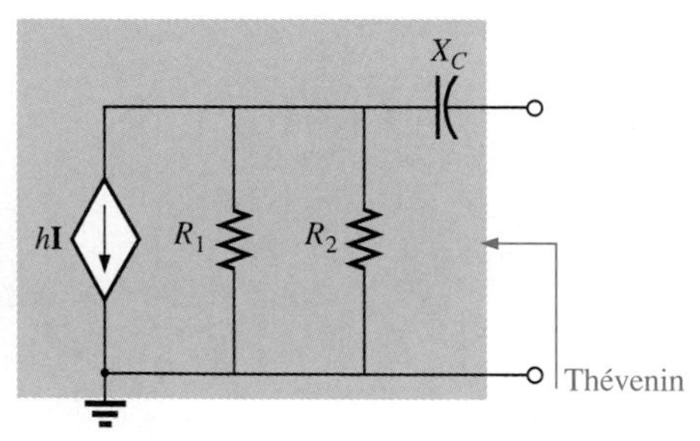

FIG. 18.46
Example 18.11.

Solution: From Fig. 18.46, $\mathbf{E}_{Th}$ is

$$\mathbf{E}_{Th} = \mathbf{E}_{oc} = -h\mathbf{I}(R_1 \parallel R_2) = -\frac{\boldsymbol{hR_1R_2\mathbf{I}}}{\boldsymbol{R_1 + R_2}}$$

Method 1 (Fig. 18.47):

$$\mathbf{Z}_{Th} = \boldsymbol{R_1 \parallel R_2 - j X_C}$$

Note the similarity between this solution and that obtained for the previous example.

Method 2 (Fig. 18.48):

$$\mathbf{I}_{sc} = \frac{-(R_1 \parallel R_2)h\mathbf{I}}{(R_1 \parallel R_2) - j X_C}$$

and
$$\mathbf{Z}_{Th} = \frac{\mathbf{E}_{oc}}{\mathbf{I}_{sc}} = \frac{-h\mathbf{I}(R_1 \parallel R_2)}{\dfrac{-(R_1 \parallel R_2)h\mathbf{I}}{(R_1 \parallel R_2) - j X_C}} = \boldsymbol{R_1 \parallel R_2 - j X_C}$$

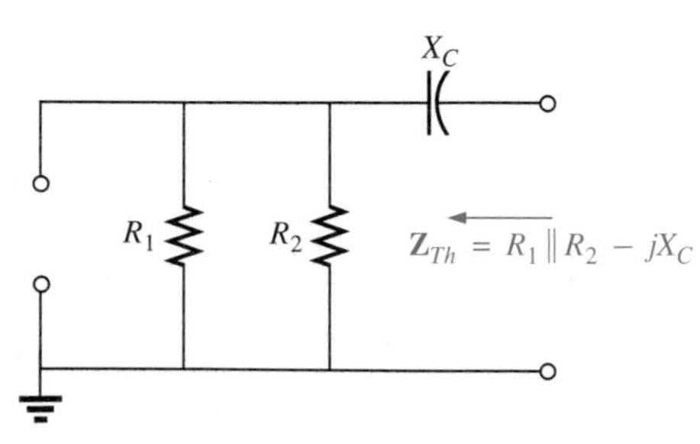

FIG. 18.47
Determining the Thévenin impedance for the network of Fig. 18.46.

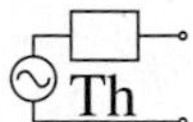

Method 3 (Fig. 18.49):

$$\mathbf{I}_g = \frac{\mathbf{E}_g}{(R_1 \parallel R_2) - jX_C}$$

and
$$\mathbf{Z}_{Th} = \frac{\mathbf{E}_g}{\mathbf{I}_g} = \boldsymbol{R_1 \parallel R_2 - jX_C}$$

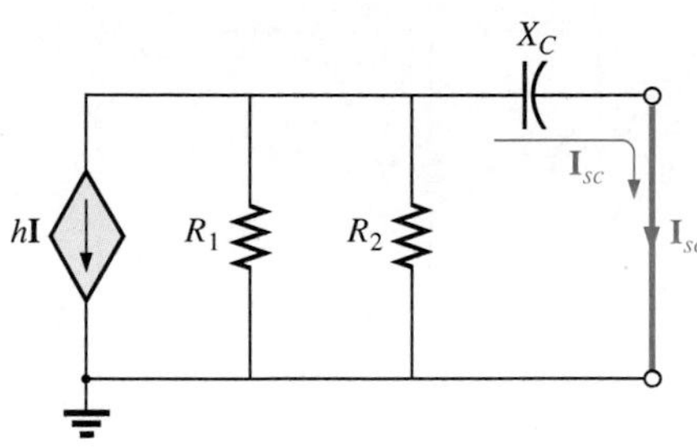

FIG. 18.48
Determining the short-circuit current for the network of Fig. 18.46.

The following example has a dependent source that will not permit the use of the method described at the beginning of this section for independent sources. All three methods will be applied, however, so that the results can be compared.

EXAMPLE 18.12 For the network of Fig. 18.50 (introduced in Example 18.6), determine the Thévenin equivalent circuit between the indicated terminals using each method described in this section. Compare your results.

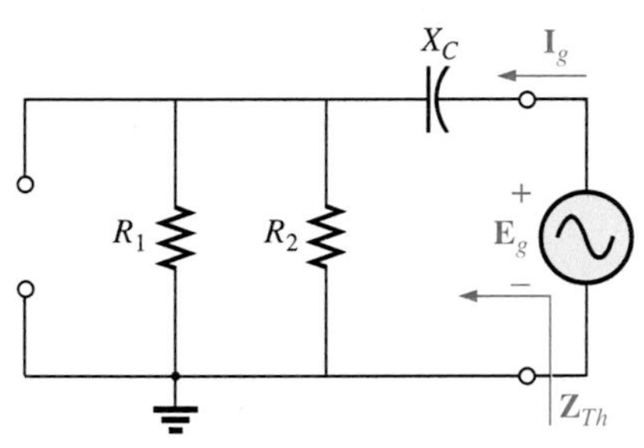

FIG. 18.49
Determining the Thévenin impedance using the approach $\mathbf{Z}_{Th} = \mathbf{E}_g / \mathbf{I}_g$.

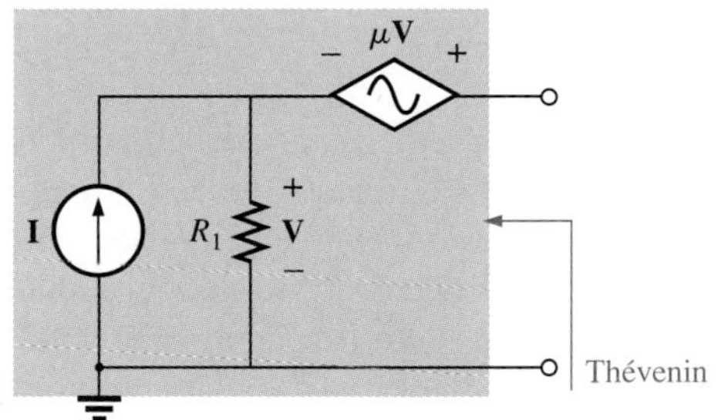

FIG. 18.50
Example 18.12.

Solution: First, using Kirchhoff's voltage law, $\mathbf{E}_{Th}$ (which is the same for each method) is written

$$\mathbf{E}_{Th} = \mathbf{V} + \mu\mathbf{V} = (1 + \mu)\mathbf{V}$$

However,
$$\mathbf{V} = \mathbf{I}R_1$$

so
$$\mathbf{E}_{Th} = \boldsymbol{(1 + \mu)\mathbf{I}R_1}$$

$\mathbf{Z}_{Th}$:

Method 1 (Fig. 18.51): Since $\mathbf{I} = 0$, $\mathbf{V}$ and $\mu\mathbf{V} = 0$, and

$$\cancel{Z_{Th} = R_1} \qquad \text{(incorrect)}$$

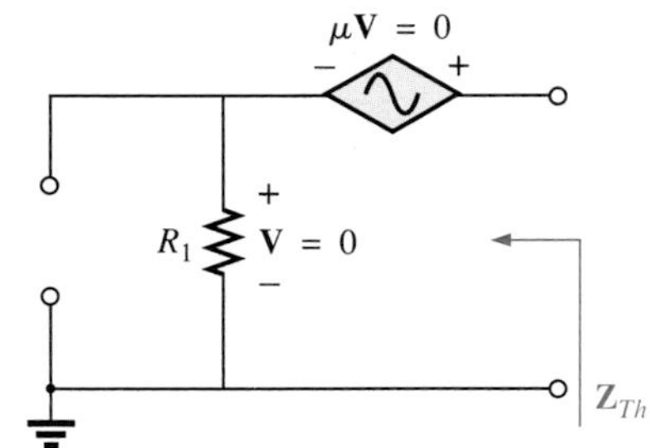

FIG. 18.51
Determining $\mathbf{Z}_{Th}$ incorrectly.

Method 2 (Fig. 18.52): Kirchhoff's voltage law around the indicated loop gives us

$$\mathbf{V} + \mu\mathbf{V} = 0$$

and
$$\mathbf{V}(1 + \mu) = 0$$

Since μ is a positive constant, the above equation can be satisfied only when $\mathbf{V} = 0$. Substitution of this result into Fig. 18.52 will yield the configuration of Fig. 18.53, and

$$\mathbf{I}_{sc} = \mathbf{I}$$

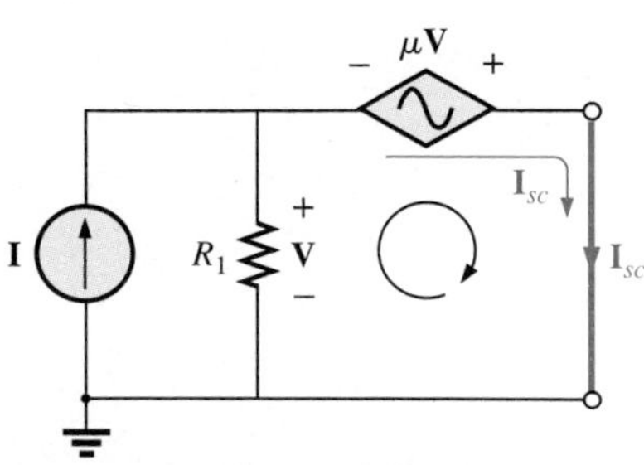

FIG. 18.52
Determining $\mathbf{I}_{sc}$ for the network of Fig. 18.50.

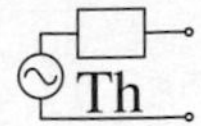

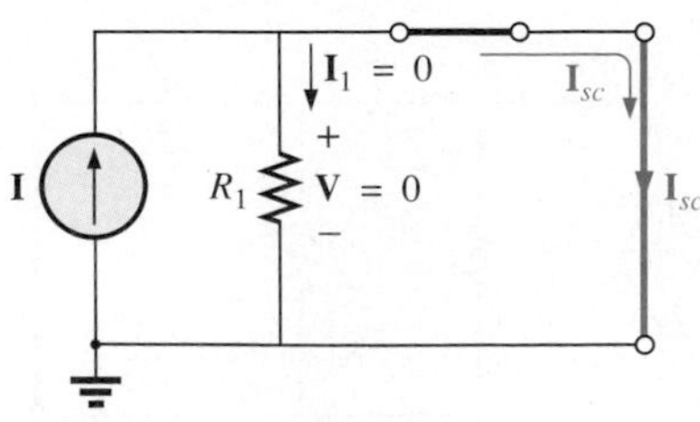

FIG. 18.53
Substituting **V** *= 0 into the network of Fig. 18.52.*

with

$$\mathbf{Z}_{Th} = \frac{\mathbf{E}_{oc}}{\mathbf{I}_{sc}} = \frac{(1 + \mu)\mathbf{I}R_1}{\mathbf{I}} = (\mathbf{1} + \boldsymbol{\mu})\boldsymbol{R}_1 \qquad \text{(correct)}$$

Method 3 (Fig. 18.54):

$$\mathbf{E}_g = \mathbf{V} + \mu\mathbf{V} = (1 + \mu)\mathbf{V}$$

or

$$\mathbf{V} = \frac{\mathbf{E}_g}{1 + \mu}$$

and

$$\mathbf{I}_g = \frac{\mathbf{V}}{R_1} = \frac{\mathbf{E}_g}{(1 + \mu)R_1}$$

and

$$\mathbf{Z}_{Th} = \frac{\mathbf{E}_g}{\mathbf{I}_g} = (\mathbf{1} + \boldsymbol{\mu})\boldsymbol{R}_1 \qquad \text{(correct)}$$

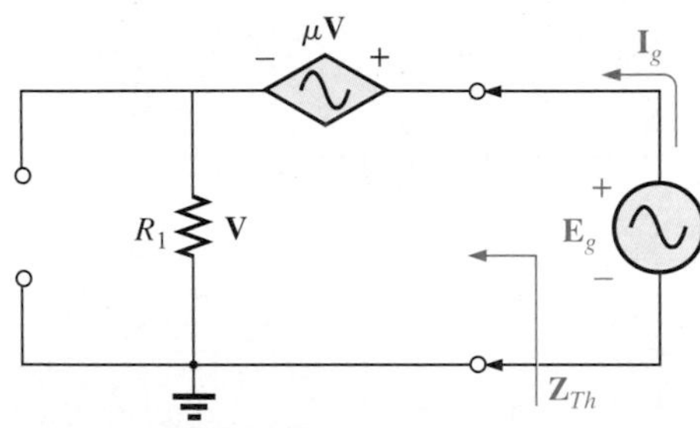

FIG. 18.54
Determining $\mathbf{Z}_{Th}$ *using the approach* $\mathbf{Z}_{Th} = \mathbf{E}_g / \mathbf{I}_g$.

The Thévenin equivalent circuit appears in Fig. 18.55, and

$$\mathbf{I}_L = \frac{(\mathbf{1} + \boldsymbol{\mu})\boldsymbol{R}_1\mathbf{I}}{\boldsymbol{R}_L + (\mathbf{1} + \boldsymbol{\mu})\boldsymbol{R}_1}$$

which compares with the result of Example 18.6.

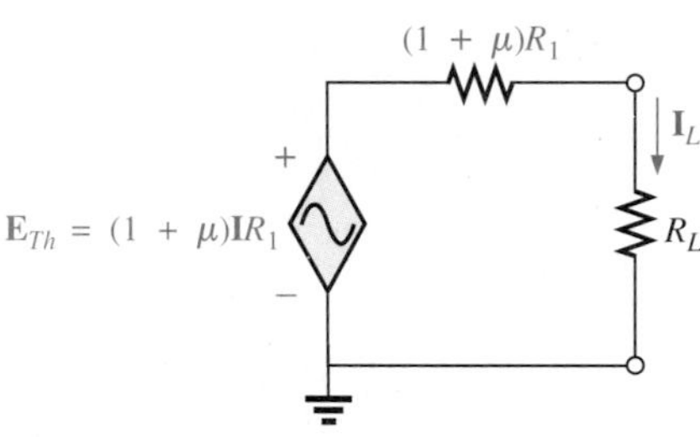

FIG. 18.55
The Thévenin equivalent circuit for the network of Fig. 18.50.

The network of Fig. 18.56 is the basic configuration of the transistor equivalent circuit applied most frequently today (although most texts in electronics will use the circle rather than the diamond outline for the source). Needless to say, it is necessary to know its characteristics and be adept in its use. Note that there is a controlled voltage and current source, each controlled by variables in the configuration.

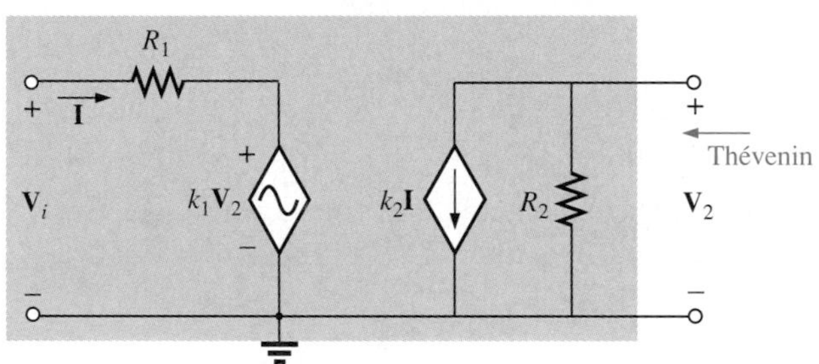

FIG. 18.56
Example 18.13: Transistor equivalent network.

EXAMPLE 18.13 Determine the Thévenin equivalent circuit for the indicated terminals of the network of Fig. 18.56.

Solution: Apply the second method introduced in this section.

$\mathbf{E}_{Th}$:

$$\mathbf{E}_{oc} = \mathbf{V}_2$$

$$\mathbf{I} = \frac{\mathbf{V}_i - k_1\mathbf{V}_2}{R_1} = \frac{\mathbf{V}_i - k_1\mathbf{E}_{oc}}{R_1}$$

and

$$\mathbf{E}_{oc} = -k_2\mathbf{I}R_2 = -k_2R_2\left(\frac{\mathbf{V}_i - k_1\mathbf{E}_{oc}}{R_1}\right)$$

$$= \frac{-k_2R_2\mathbf{V}_i}{R_1} + \frac{k_1k_2R_2\mathbf{E}_{oc}}{R_1}$$

or
$$\mathbf{E}_{oc}\left(1 - \frac{k_1 k_2 R_2}{R_1}\right) = \frac{-k_2 R_2 \mathbf{V}_i}{R_1}$$

and
$$\mathbf{E}_{oc}\left(\frac{R_1 - k_1 k_2 R_2}{\cancel{R_1}}\right) = \frac{-k_2 R_2 \mathbf{V}_i}{\cancel{R_1}}$$

so
$$\boxed{\mathbf{E}_{oc} = \frac{-k_2 R_2 \mathbf{V}_i}{R_1 - k_1 k_2 R_2} = \mathbf{E}_{Th}} \tag{18.5}$$

$\mathbf{I}_{sc}$:

For the network of Fig. 18.57, where
$$\mathbf{V}_2 = 0 \qquad k_1\mathbf{V}_2 = 0 \qquad \mathbf{I} = \frac{\mathbf{V}_i}{R_1}$$

and
$$\mathbf{I}_{sc} = -k_2\mathbf{I} = \frac{-k_2\mathbf{V}_i}{R_1}$$

so
$$\mathbf{Z}_{Th} = \frac{\mathbf{E}_{oc}}{\mathbf{I}_{sc}} = \frac{\dfrac{-k_2 R_2 \mathbf{V}_i}{R_1 - k_1 k_2 R_2}}{\dfrac{-k_2 \mathbf{V}_i}{R_1}} = \frac{R_1 R_2}{R_1 - k_1 k_2 R_2}$$

and
$$\boxed{\mathbf{Z}_{Th} = \frac{R_2}{1 - \dfrac{k_1 k_2 R_2}{R_1}}} \tag{18.6}$$

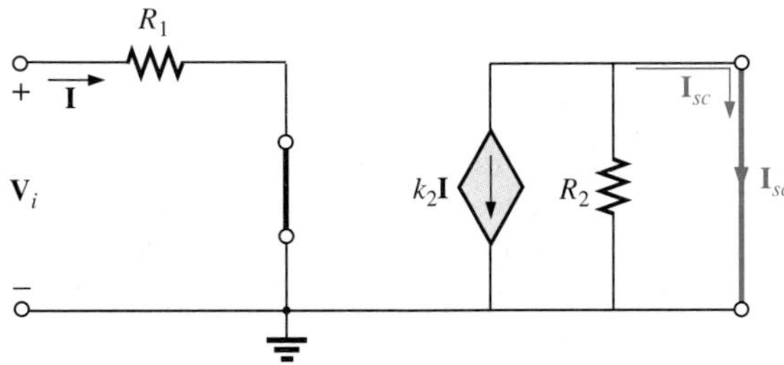

FIG. 18.57
Determining $\mathbf{I}_{sc}$ for the network of Fig. 18.56.

Frequently, the approximation $k_1 \cong 0$ is applied. Then the Thévenin voltage and impedance are

$$\boxed{\mathbf{E}_{Th} = \frac{-k_2 R_2 \mathbf{V}_i}{R_1}} \qquad k_1 = 0 \tag{18.7}$$

$$\boxed{\mathbf{Z}_{Th} = R_2} \qquad k_1 = 0 \tag{18.8}$$

Apply $\mathbf{Z}_{Th} = \mathbf{E}_g/\mathbf{I}_g$ to the network of Fig. 18.58, where

$$\mathbf{I} = \frac{-k_1\mathbf{V}_2}{R_1}$$

But
$$\mathbf{V}_2 = \mathbf{E}_g$$

so
$$\mathbf{I} = \frac{-k_1\mathbf{E}_g}{R_1}$$

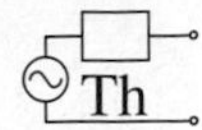

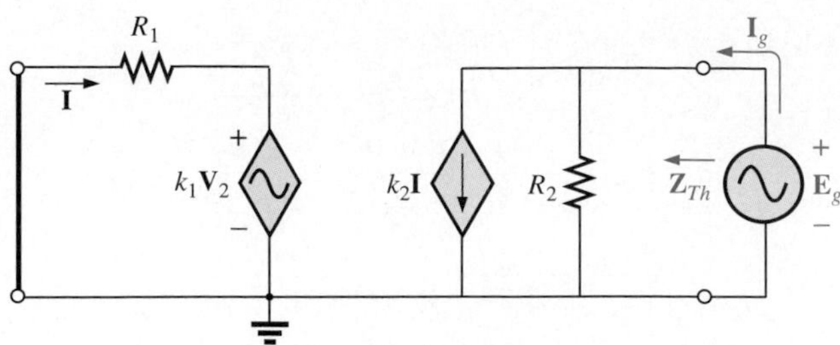

FIG. 18.58
Determining $\mathbf{Z}_{Th}$ using the procedure $\mathbf{Z}_{Th} = \mathbf{E}_g / \mathbf{I}_g$.

Applying Kirchhoff's current law, we have

$$\mathbf{I}_g = k_2\mathbf{I} + \frac{\mathbf{E}_g}{R_2} = k_2\left(-\frac{k_1\mathbf{E}_g}{R_1}\right) + \frac{\mathbf{E}_g}{R_2}$$

$$= \mathbf{E}_g\left(\frac{1}{R_2} - \frac{k_1k_2}{R_1}\right)$$

and

$$\frac{\mathbf{I}_g}{\mathbf{E}_g} = \frac{R_1 - k_1k_2R_2}{R_1R_2}$$

or

$$\mathbf{Z}_{Th} = \frac{\mathbf{E}_g}{\mathbf{I}_g} = \frac{\boldsymbol{R_1R_2}}{\boldsymbol{R_1 - k_1k_2R_2}}$$

as obtained above.

The last two methods presented in this section were applied only to networks in which the magnitudes of the controlled sources were dependent on a variable within the network for which the Thévenin equivalent circuit was to be obtained. Understand that both of these methods can also be applied to any dc or sinusoidal ac network containing only independent sources or dependent sources of the other kind.

18.4 NORTON'S THEOREM

The three methods described for Thévenin's theorem will each be altered to permit their use with Norton's theorem. Since the Thévenin and Norton impedances are the same for a particular network, certain portions of the discussion will be quite similar to those encountered in the previous section. We will first consider independent sources and the approach developed in Chapter 9, followed by dependent sources and the new techniques developed for Thévenin's theorem.

You will recall from Chapter 9 that Norton's theorem allows us to replace any two-terminal linear bilateral ac network with an equivalent circuit consisting of a current source and impedance, as in Fig. 18.59.

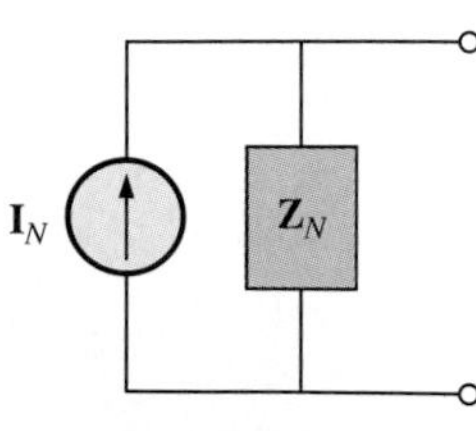

FIG. 18.59
The Norton equivalent circuit for ac networks.

The Norton equivalent circuit, like the Thévenin equivalent circuit, is applicable at only one frequency since the reactances are frequency dependent.

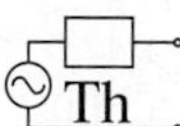

Independent Sources

The procedure outlined below to find the Norton equivalent of a sinusoidal ac network is changed (from that in Chapter 9) in only one respect: the replacement of the term *resistance* with the term *impedance*.

1. ***Remove that portion of the network across which the Norton equivalent circuit is to be found.***
2. ***Mark (○, •, and so on) the terminals of the remaining two-terminal network.***
3. ***Calculate $\mathbf{Z}_N$ by first setting all voltage and current sources to zero (short circuit and open circuit, respectively) and then finding the resulting impedance between the two marked terminals.***
4. ***Calculate $\mathbf{I}_N$ by first replacing the voltage and current sources and then finding the short-circuit current between the marked terminals.***
5. ***Draw the Norton equivalent circuit with the portion of the circuit previously removed replaced between the terminals of the Norton equivalent circuit.***

The Norton and Thévenin equivalent circuits can be found from each other by using the source transformation shown in Fig. 18.60. The source transformation is applicable for any Thévenin or Norton equivalent circuit determined from a network with any combination of independent or dependent sources.

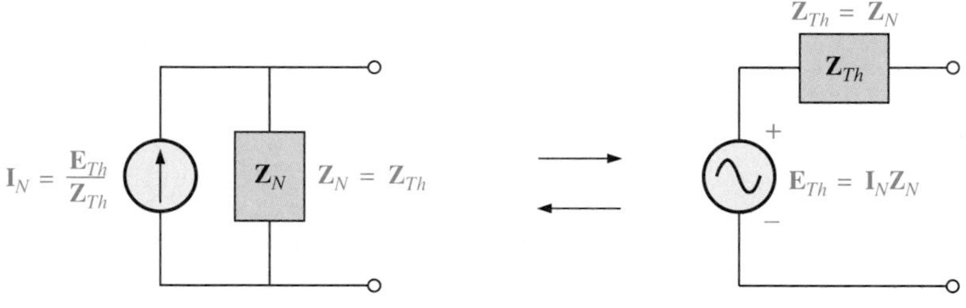

FIG. 18.60
Conversion between the Thévenin and Norton equivalent circuits.

EXAMPLE 18.14 Determine the Norton equivalent circuit for the network external to the 6-Ω resistor of Fig. 18.61.

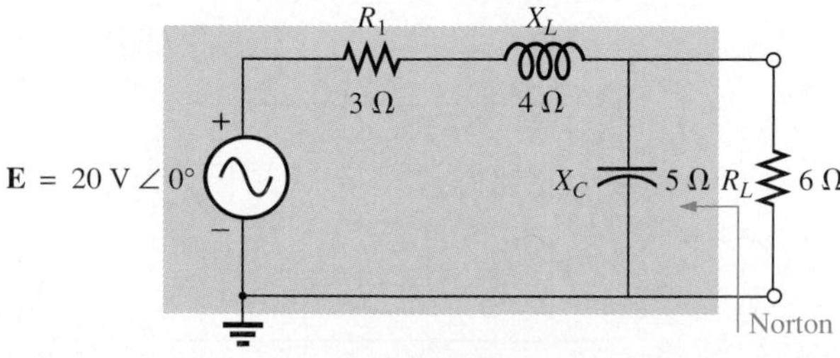

FIG. 18.61
Example 18.14.

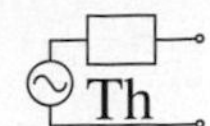

Solution:

Steps 1 and 2 (Fig. 18.62):

$$\mathbf{Z}_1 = R_1 + j\,X_L = 3\,\Omega + j\,4\,\Omega = 5\,\Omega\,\angle 53.13^\circ$$
$$\mathbf{Z}_2 = -j\,X_C = -j\,5\,\Omega$$

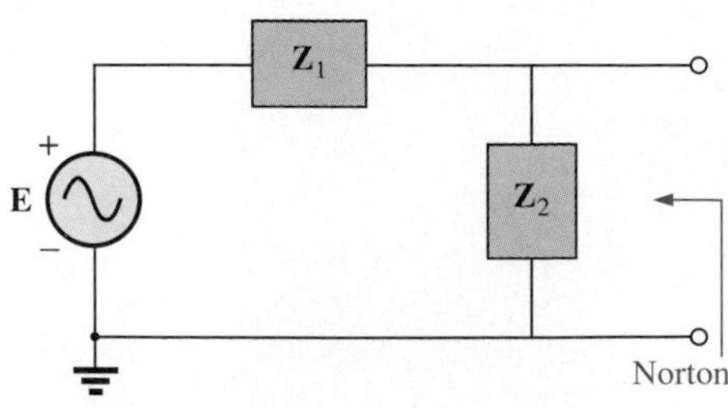

FIG. 18.62
Assigning the subscripted impedances to the network of Fig. 18.61.

Step 3 (Fig. 18.63):

$$\mathbf{Z}_N = \frac{\mathbf{Z}_1\mathbf{Z}_2}{\mathbf{Z}_1 + \mathbf{Z}_2} = \frac{(5\,\Omega\,\angle 53.13^\circ)(5\,\Omega\,\angle -90^\circ)}{3\,\Omega + j\,4\,\Omega - j\,5\,\Omega} = \frac{25\,\Omega\,\angle -36.87^\circ}{3 - j\,1}$$
$$= \frac{25\,\Omega\,\angle -36.87^\circ}{3.16\,\angle -18.43^\circ} = 7.91\,\Omega\,\angle -18.44^\circ = \mathbf{7.50\,\Omega - j\,2.50\,\Omega}$$

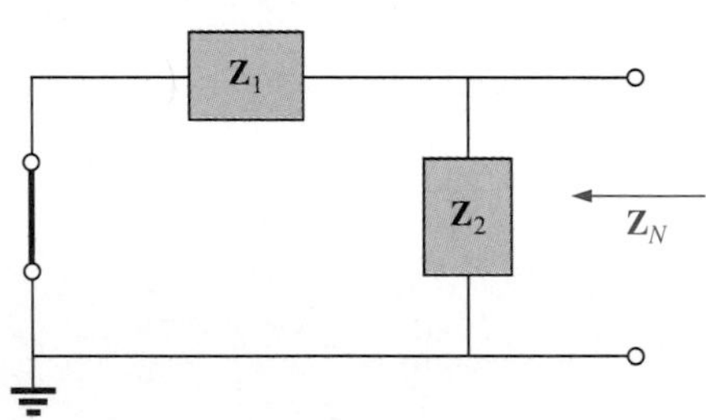

FIG. 18.63
Determining the Norton impedance for the network of Fig. 18.61.

Step 4 (Fig. 18.64):

$$\mathbf{I}_N = \mathbf{I}_1 = \frac{\mathbf{E}}{\mathbf{Z}_1} = \frac{20\text{ V}\,\angle 0^\circ}{5\,\Omega\,\angle 53.13^\circ} = \mathbf{4\text{ A}\,\angle -53.13^\circ}$$

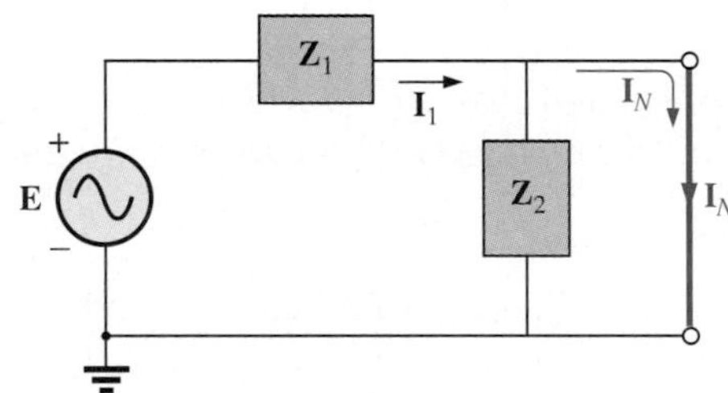

FIG. 18.64
Determining $\mathbf{I}_N$ for the network of Fig. 18.61.

Step 5: The Norton equivalent circuit is shown in Fig. 18.65.

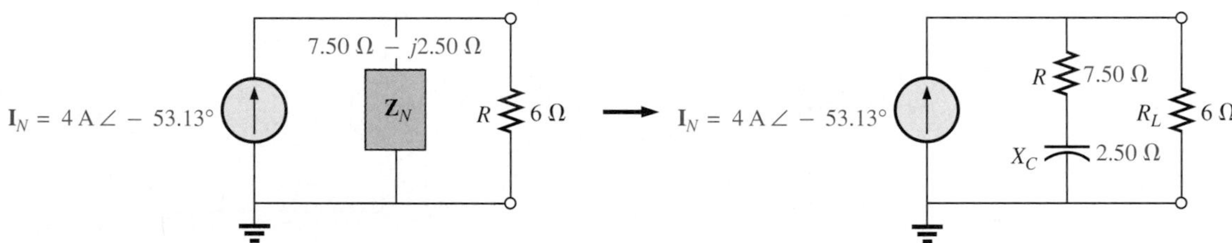

FIG. 18.65
The Norton equivalent circuit for the network of Fig. 18.61.

EXAMPLE 18.15 Find the Norton equivalent circuit for the network external to the 7-Ω capacitive reactance in Fig. 18.66.

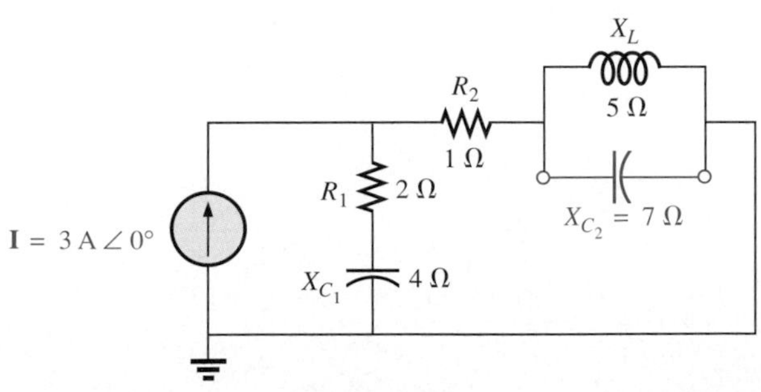

FIG. 18.66
Example 18.15.

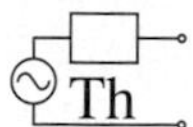

Solution:

Steps 1 and 2 (Fig. 18.67):

$$\mathbf{Z}_1 = R_1 - j X_{C_1} = 2\ \Omega - j\,4\ \Omega$$
$$\mathbf{Z}_2 = R_2 = 1\ \Omega$$
$$\mathbf{Z}_3 = +j X_L = j\,5\ \Omega$$

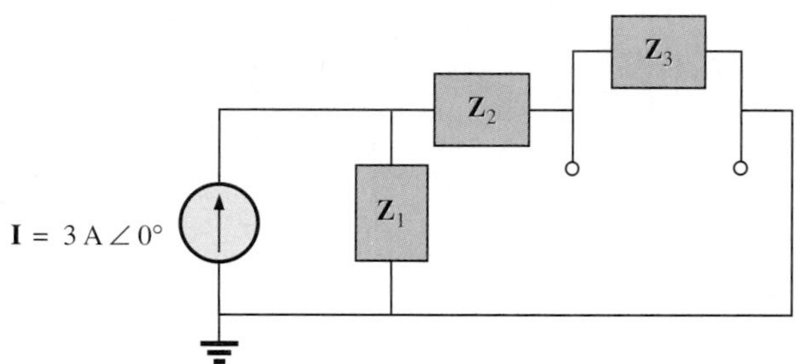

FIG. 18.67
Assigning the subscripted impedances to the network of Fig. 18.66.

Step 3 (Fig. 18.68):

$$\mathbf{Z}_N = \frac{\mathbf{Z}_3(\mathbf{Z}_1 + \mathbf{Z}_2)}{\mathbf{Z}_3 + (\mathbf{Z}_1 + \mathbf{Z}_2)}$$

$$\mathbf{Z}_1 + \mathbf{Z}_2 = 2\ \Omega - j\,4\ \Omega + 1\ \Omega = 3\ \Omega - j\,4\ \Omega = 5\ \Omega\ \angle -53.13^\circ$$

$$\mathbf{Z}_N = \frac{(5\ \Omega\ \angle 90^\circ)(5\ \Omega\ \angle -53.13^\circ)}{j\,5\ \Omega + 3\ \Omega - j\,4\ \Omega} = \frac{25\ \Omega\ \angle 36.87^\circ}{3 + j\,1}$$

$$= \frac{25\ \Omega\ \angle 36.87^\circ}{3.16\ \angle +18.43^\circ}$$

$$\mathbf{Z}_N = 7.91\ \Omega\ \angle 18.44^\circ = \mathbf{7.50\ \Omega + j\,2.50\ \Omega}$$

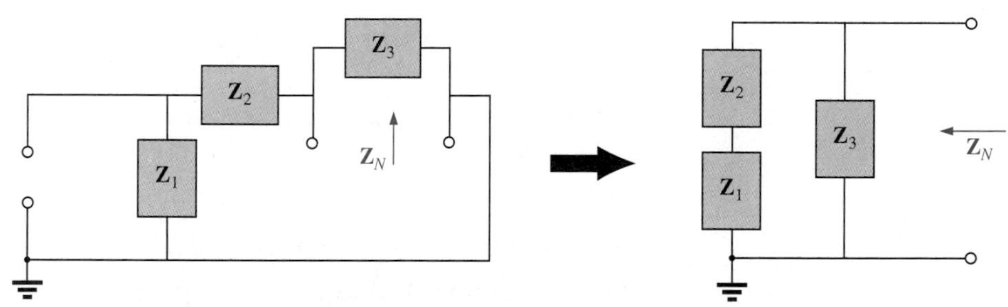

FIG. 18.68
Finding the Norton impedance for the network of Fig. 18.66.

Calculator Performing the above on the TI-85 calculator:

```
((0,5)*((2,−4)+(1,0)))/((0,5)+((2,−4)+(1,0)))
    (7.500E0,2.500E0)
Ans ▶ Pol
    (7.906E0∠18.435E0)
```

CALC. 18.1

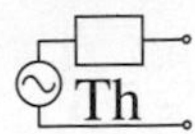

Step 4 (Fig. 18.69):

$$\mathbf{I}_N = \mathbf{I}_1 = \frac{\mathbf{Z}_1\mathbf{I}}{\mathbf{Z}_1 + \mathbf{Z}_2} \qquad \text{(current divider rule)}$$

$$= \frac{(2\,\Omega - j\,4\,\Omega)(3\text{ A})}{3\,\Omega - j\,4\,\Omega} = \frac{6\text{ A} - j\,12\text{ A}}{5\,\angle -53.13^\circ} = \frac{13.4\text{ A}\,\angle -63.43^\circ}{5\,\angle -53.13^\circ}$$

$$\mathbf{I}_N = \mathbf{2.68\ A\ \angle -10.3^\circ}$$

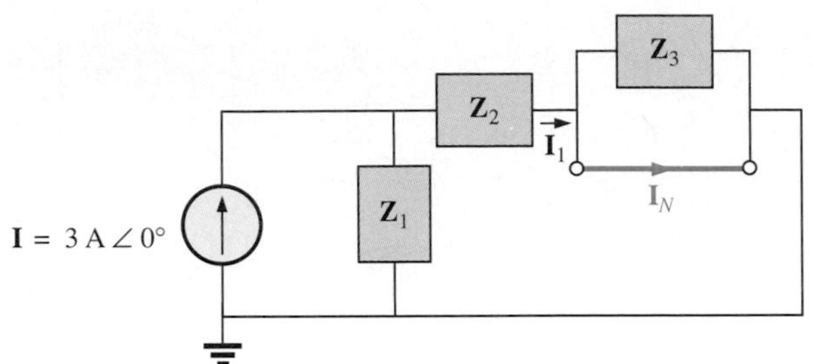

FIG. 18.69
Determining $\mathbf{I}_N$ for the network of Fig. 18.66.

Step 5: The Norton equivalent circuit is shown in Fig. 18.70.

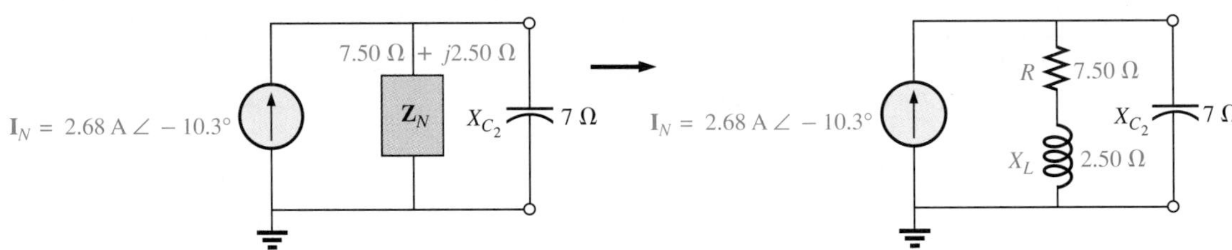

FIG. 18.70
The Norton equivalent circuit for the network of Fig. 18.66.

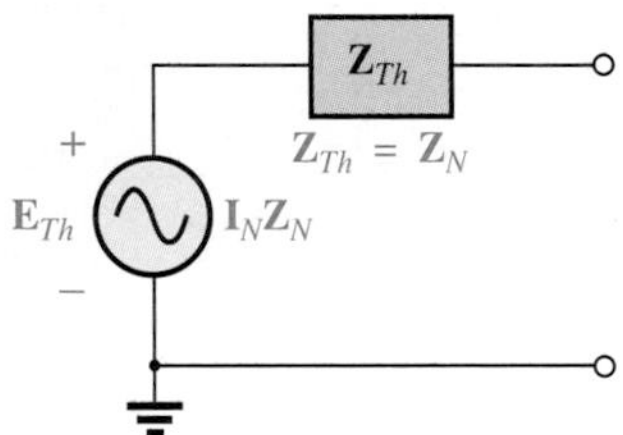

FIG. 18.71
Determining the Thévenin equivalent circuit for the Norton equivalent of Fig. 18.70.

EXAMPLE 18.16 Find the Thévenin equivalent circuit for the network external to the 7-Ω capacitive reactance in Fig. 18.66.

Solution: Using the conversion between sources (Fig. 18.71), we obtain

$$\mathbf{Z}_{Th} = \mathbf{Z}_N = \mathbf{7.50\ \Omega + \mathit{j}\,2.50\ \Omega}$$

$$\mathbf{E}_{Th} = \mathbf{I}_N\mathbf{Z}_N = (2.68\text{ A}\,\angle -10.3^\circ)(7.91\,\Omega\,\angle 18.44^\circ)$$

$$= \mathbf{21.2\ V\ \angle 8.14^\circ}$$

The Thévenin equivalent circuit is shown in Fig. 18.72.

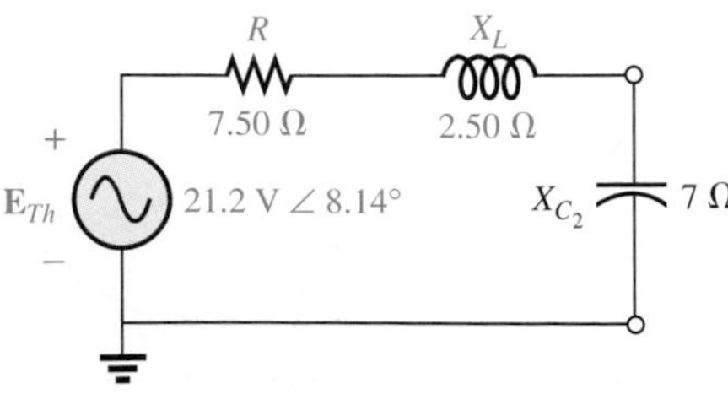

FIG. 18.72
The Thévenin equivalent circuit for the network of Fig. 18.66.

Dependent Sources

As stated for Thévenin's theorem, *dependent sources in which the controlling variable is not determined by the network* for which the Norton equivalent circuit is to be found do not alter the procedure outlined above.

For dependent sources of the other kind, one of the following procedures must be applied. Both of these procedures can also be applied to networks with any combination of independent sources and dependent sources not controlled by the network under investigation.

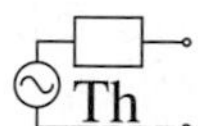

The Norton equivalent circuit appears in Fig. 18.73(a). In Fig. 18.73(b), we find that

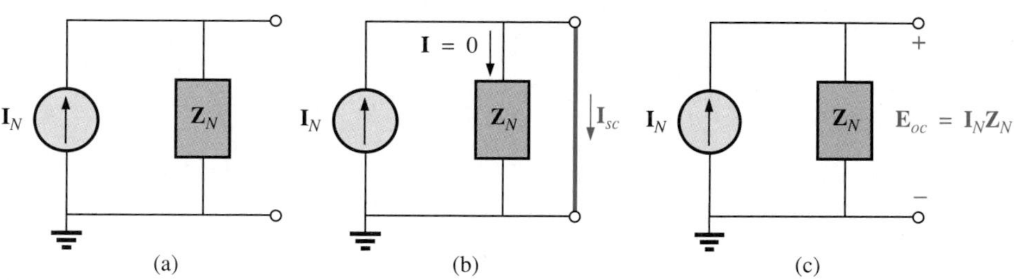

FIG. 18.73
Defining an alternative approach for determining $\mathbf{Z}_N$.

$$\boxed{\mathbf{I}_{sc} = \mathbf{I}_N} \qquad \textbf{(18.9)}$$

and in Fig. 18.73(c) that

$$\mathbf{E}_{oc} = \mathbf{I}_N \mathbf{Z}_N$$

Or, rearranging, we have

$$\mathbf{Z}_N = \frac{\mathbf{E}_{oc}}{\mathbf{I}_N}$$

and

$$\boxed{\mathbf{Z}_N = \frac{\mathbf{E}_{oc}}{\mathbf{I}_{sc}}} \qquad \textbf{(18.10)}$$

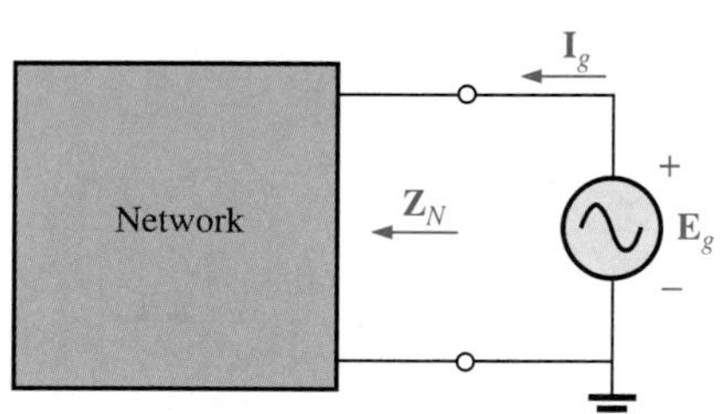

FIG. 18.74
Determining the Norton impedance using the approach $\mathbf{Z}_N = \mathbf{E}_g / \mathbf{I}_g$.

The Norton impedance can also be determined by applying a source of voltage $\mathbf{E}_g$ to the terminals of interest and finding the resulting $\mathbf{I}_g$, as shown in Fig. 18.74. All independent sources and dependent sources not controlled by a variable in the network of interest are set to zero, and

$$\boxed{\mathbf{Z}_N = \frac{\mathbf{E}_g}{\mathbf{I}_g}} \qquad \textbf{(18.11)}$$

For this latter approach, the Norton current is still determined by the short-circuit current.

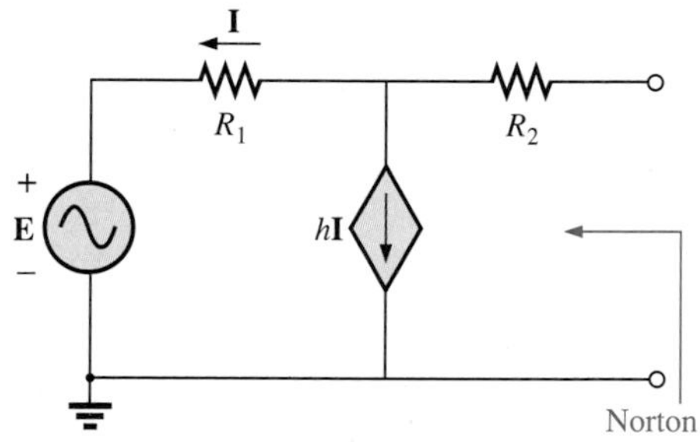

FIG. 18.75
Example 18.17.

EXAMPLE 18.17 Using each method described for dependent sources, find the Norton equivalent circuit for the network of Fig. 18.75.

Solution:

$\mathbf{I}_N$:

For each method, $\mathbf{I}_N$ is determined in the same manner. From Fig. 18.76, using Kirchhoff's current law, we have

$$0 = \mathbf{I} + h\mathbf{I} + \mathbf{I}_{sc}$$

or

$$\mathbf{I}_{sc} = -(1 + h)\mathbf{I}$$

Applying Kirchhoff's voltage law gives us

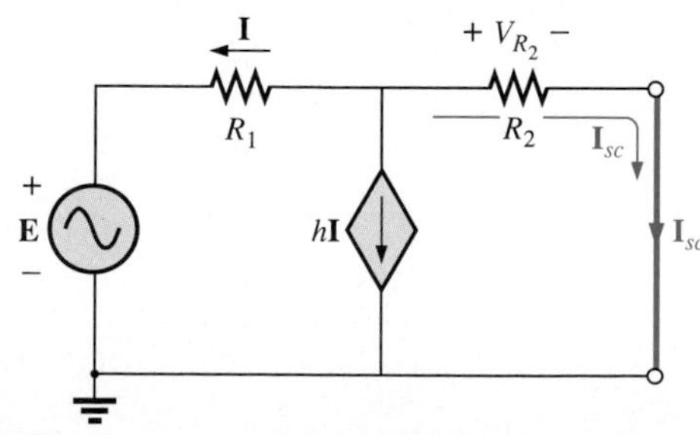

FIG. 18.76
Determining $\mathbf{I}_{sc}$ *for the network of Fig. 18.75.*

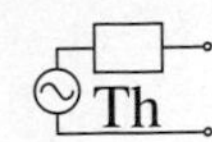

$$\mathbf{E} + \mathbf{I}R_1 - \mathbf{I}_{sc}R_2 = 0$$

and
$$\mathbf{I}R_1 = \mathbf{I}_{sc}R_2 - \mathbf{E}$$

or
$$\mathbf{I} = \frac{\mathbf{I}_{sc}R_2 - \mathbf{E}}{R_1}$$

so
$$\mathbf{I}_{sc} = -(1 + h)\mathbf{I} = -(1 + h)\left(\frac{\mathbf{I}_{sc}R_2 - \mathbf{E}}{R_1}\right)$$

or
$$R_1\mathbf{I}_{sc} = -(1 + h)\mathbf{I}_{sc}R_2 + (1 + h)\mathbf{E}$$
$$\mathbf{I}_{sc}[R_1 + (1 + h)R_2] = (1 + h)\mathbf{E}$$
$$\mathbf{I}_{sc} = \frac{\mathbf{(1 + \mathit{h})E}}{\mathbf{\mathit{R}_1 + (1 + \mathit{h})\mathit{R}_2}} = \mathbf{I}_N$$

$\mathbf{Z}_N$:

Method 1: $\mathbf{E}_{oc}$ is determined from the network of Fig. 18.77. By Kirchhoff's current law,

$$0 = \mathbf{I} + h\mathbf{I} \quad \text{or} \quad \mathbf{I}(h + 1) = 0$$

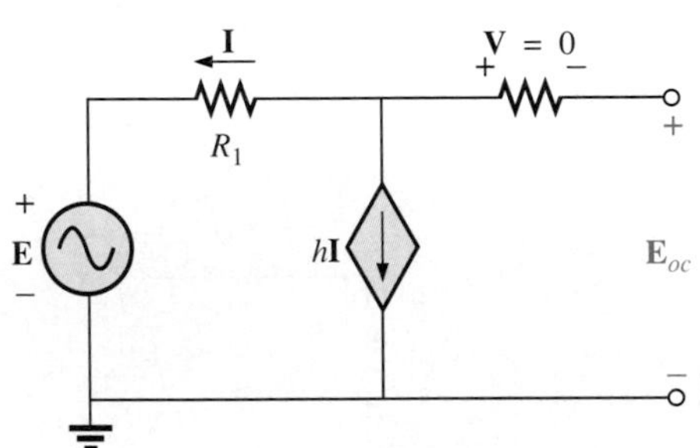

FIG. 18.77
Determining $\mathbf{E}_{oc}$ for the network of Fig. 18.75.

For h, a positive constant $\mathbf{I}$ must equal zero to satisfy the above. Therefore,

$$\mathbf{I} = 0 \quad \text{and} \quad h\mathbf{I} = 0$$

and
$$\mathbf{E}_{oc} = \mathbf{E}$$

with
$$\mathbf{Z}_N = \frac{\mathbf{E}_{oc}}{\mathbf{I}_{sc}} = \frac{\mathbf{E}}{\dfrac{(1 + h)\mathbf{E}}{R_1 + (1 + h)R_2}} = \frac{\boldsymbol{R_1 + (1 + h)R_2}}{\boldsymbol{(1 + h)}}$$

Method 2: Note Fig. 18.78. By Kirchhoff's current law,

$$\mathbf{I}_g = \mathbf{I} + h\mathbf{I} = (1 + h)\mathbf{I}$$

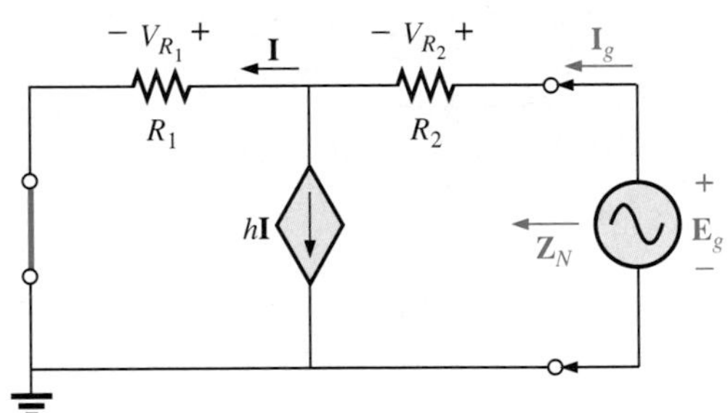

FIG. 18.78
Determining the Norton impedance using the approach $\mathbf{Z}_N = \mathbf{E}_g/\mathbf{E}_g$.

By Kirchhoff's voltage law,

$$\mathbf{E}_g - \mathbf{I}_gR_2 - \mathbf{I}R_1 = 0$$

or
$$\mathbf{I} = \frac{\mathbf{E}_g - \mathbf{I}_gR_2}{R_1}$$

Substituting, we have

$$\mathbf{I}_g = (1 + h)\mathbf{I} = (1 + h)\left(\frac{\mathbf{E}_g - \mathbf{I}_gR_2}{R_1}\right)$$

and
$$\mathbf{I}_gR_1 = (1 + h)\mathbf{E}_g - (1 + h)\mathbf{I}_gR_2$$

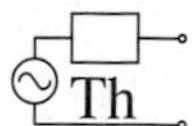

so $$\mathbf{E}_g(1+h) = \mathbf{I}_g[R_1 + (1+h)R_2]$$

or $$\mathbf{Z}_N = \frac{\mathbf{E}_g}{\mathbf{I}_g} = \frac{R_1 + (1+h)R_2}{1+h}$$

which agrees with the above.

EXAMPLE 18.18 Find the Norton equivalent circuit for the network configuration of Fig. 18.56.

Solution: By source conversion,

$$\mathbf{I}_N = \frac{\mathbf{E}_{Th}}{\mathbf{Z}_{Th}} = \frac{\dfrac{-k_2 R_2 \mathbf{V}_i}{R_1 - k_1 k_2 R_2}}{\dfrac{R_1 R_2}{R_1 - k_1 k_2 R_2}}$$

and $$\boxed{\mathbf{I}_N = \frac{-k_2 \mathbf{V}_i}{R_1}} \qquad (18.12)$$

which is $\mathbf{I}_{sc}$ as determined in Example 18.13, and

$$\boxed{\mathbf{Z}_N = \mathbf{Z}_{Th} = \frac{R_2}{1 - \dfrac{k_1 k_2 R_2}{R_1}}} \qquad (18.13)$$

For $k_1 \cong 0$, we have

$$\boxed{\mathbf{I}_N = \frac{-k_2 \mathbf{V}_i}{R_1}} \qquad k_1 = 0 \qquad (18.14)$$

$$\boxed{\mathbf{Z}_N = R_2} \qquad k_1 = 0 \qquad (18.15)$$

18.5 MAXIMUM POWER TRANSFER THEOREM

When applied to ac circuits, the maximum power transfer theorem states that

maximum power will be delivered to a load when the load impedance is the conj ugate of the Thévenin impedance across its terminals.

That is, for Fig. 18.79, for maximum power transfer to the load,

$$\boxed{Z_L = Z_{Th} \quad \text{and} \quad \theta_L = -\theta_{Th_Z}} \qquad (18.16)$$

or, in rectangular form,

$$\boxed{R_L = R_{Th} \quad \text{and} \quad \pm j\,X_{\text{load}} = \mp j\,X_{Th}} \qquad (18.17)$$

The conditions just mentioned will make the total impedance of the circuit appear purely resistive, as indicated in Fig. 18.80:

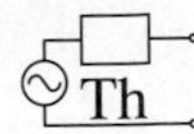

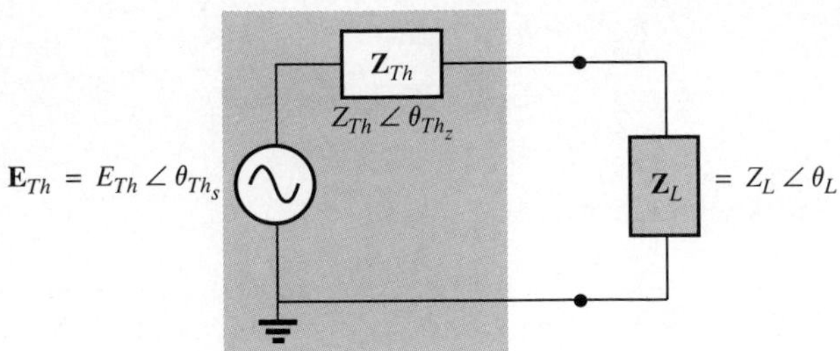

FIG. 18.79
Defining the conditions for maximum power transfer to a load.

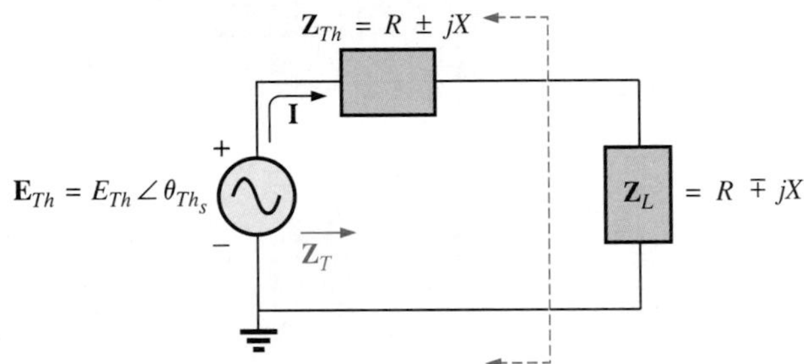

FIG. 18.80
Conditions for maximum power transfer to $\mathbf{Z}_L$.

$$\mathbf{Z}_T = (R \pm jX) + (R \mp jX)$$

and

$$\boxed{\mathbf{Z}_T = 2R} \tag{18.18}$$

Since the circuit is purely resistive, the power factor of the circuit under maximum power conditions is 1; that is,

$$\boxed{F_p = 1} \quad \text{(maximum power transfer)} \tag{18.19}$$

The magnitude of the current **I** of Fig. 18.80 is

$$I = \frac{E_{Th}}{Z_T} = \frac{E_{Th}}{2R}$$

The maximum power to the load is

$$P_{\max} = I^2R = \left(\frac{E_{Th}}{2R}\right)^2 R$$

and

$$\boxed{P_{\max} = \frac{E_{Th}^2}{4R}} \tag{18.20}$$

EXAMPLE 18.19 Find the load impedance in Fig. 18.81 for maximum power to the load, and find the maximum power.

Solution: Determine $\mathbf{Z}_{Th}$ [Fig. 18.82(a)]:

$$\mathbf{Z}_1 = R - jX_C = 6\ \Omega - j\,8\ \Omega = 10\ \Omega \angle -53.13^\circ$$
$$\mathbf{Z}_2 = +jX_L = j\,8\ \Omega$$

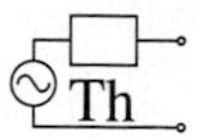

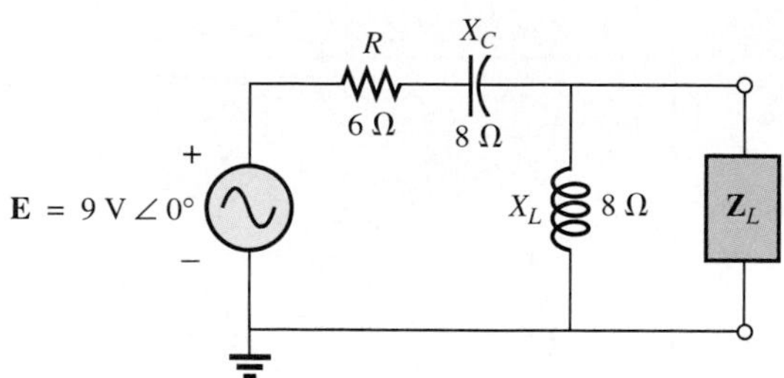

FIG. 18.81
Example 18.19.

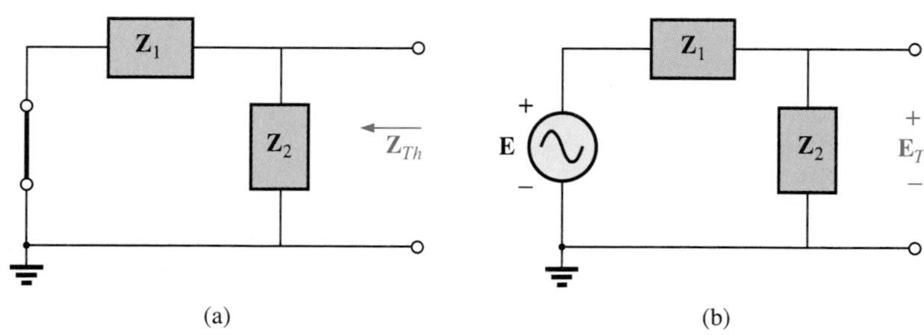

FIG. 18.82
Determining (a) $\mathbf{Z}_{Th}$ *and (b)* $\mathbf{E}_{Th}$ *for the network external to the load in Fig. 18.81.*

$$\mathbf{Z}_{Th} = \frac{\mathbf{Z}_1\mathbf{Z}_2}{\mathbf{Z}_1 + \mathbf{Z}_2} = \frac{(10\ \Omega \angle -53.13°)(8\ \Omega \angle 90°)}{6\ \Omega - j\,8\ \Omega + j\,8\ \Omega} = \frac{80\ \Omega \angle 36.87°}{6 \angle 0°}$$
$$= 13.33\ \Omega \angle 36.87° = 10.66\ \Omega + j\,8\ \Omega$$

and $\mathbf{Z}_L = 13.3\ \Omega \angle -36.87° = \mathbf{10.66\ \Omega - j\,8\ \Omega}$

To find the maximum power, we must first find $\mathbf{E}_{Th}$ [Fig. 18.82(b)], as follows:

$$\mathbf{E}_{Th} = \frac{\mathbf{Z}_2\mathbf{E}}{\mathbf{Z}_2 + \mathbf{Z}_1} \quad \text{(voltage divider rule)}$$
$$= \frac{(8\ \Omega \angle 90°)(9\ \text{V} \angle 0°)}{j\,8\ \Omega + 6\ \Omega - j\,8\ \Omega} = \frac{72\ \text{V} \angle 90°}{6 \angle 0°} = 12\ \text{V} \angle 90°$$

Then $$P_{\max} = \frac{E_{Th}^2}{4R} = \frac{(12\ \text{V})^2}{4(10.66\ \Omega)} = \frac{144}{42.64} = \mathbf{3.38\ W}$$

EXAMPLE 18.20 Find the load impedance in Fig. 18.83 for maximum power to the load, and find the maximum power.

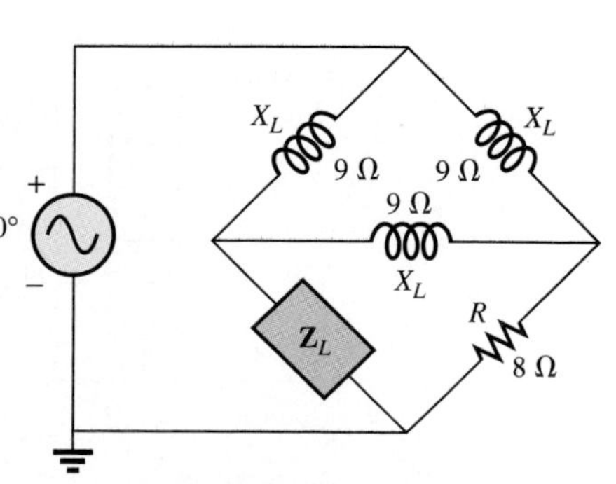

FIG. 18.83
Example 18.20.

Solution: First we must find $\mathbf{Z}_{Th}$ (Fig. 18.84).

$$\mathbf{Z}_1 = +j\,X_L = j\,9\ \Omega \qquad \mathbf{Z}_2 = R = 8\ \Omega$$

Converting from a Δ to a Y (Fig. 18.85), we have

$$\mathbf{Z}'_1 = \frac{\mathbf{Z}_1}{3} = j\,3\ \Omega \qquad \mathbf{Z}_2 = 8\ \Omega$$

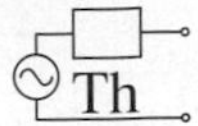

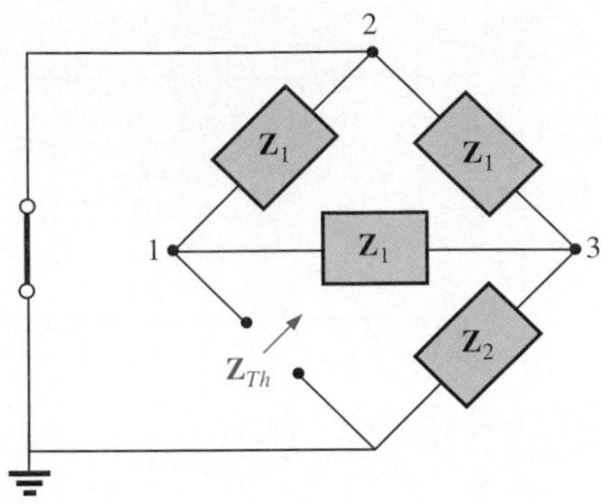

FIG. 18.84
Defining the subscripted impedances for the network of Fig. 18.83.

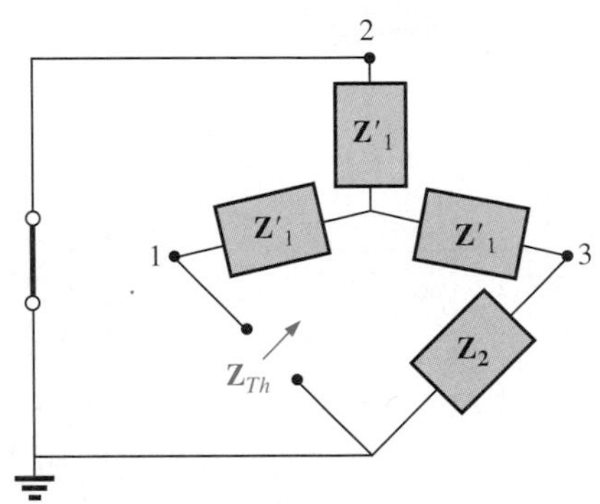

FIG. 18.85
Substituting the Y equivalent for the upper Δ configuration of Fig. 18.84.

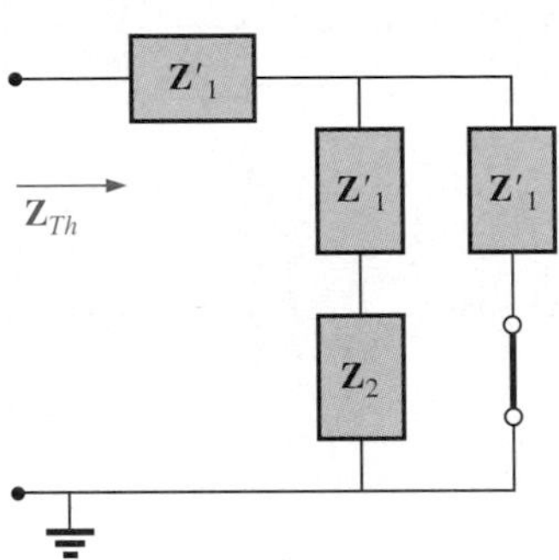

FIG. 18.86
Determining $\mathbf{Z}_{Th}$ for the network of Fig. 18.33.

The redrawn circuit (Fig. 18.86) shows

$$\mathbf{Z}_{Th} = \mathbf{Z'}_1 + \frac{\mathbf{Z'}_1(\mathbf{Z'}_1 + \mathbf{Z}_2)}{\mathbf{Z'}_1 + (\mathbf{Z'}_1 + \mathbf{Z}_2)}$$

$$= j\,3\ \Omega + \frac{3\ \Omega\ \angle 90°(j\,3\ \Omega + 8\ \Omega)}{j\,6\ \Omega + 8\ \Omega}$$

$$= j\,3 + \frac{(3\ \angle 90°)(8.54\ \angle 20.56°)}{10\ \angle 36.87°}$$

$$= j\,3 + \frac{25.62\ \angle 110.56°}{10\ \angle 36.87°} = j\,3 + 2.56\ \angle 73.69°$$

$$= j\,3 + 0.72 + j\,2.46$$

$$\mathbf{Z}_{Th} = 0.72\ \Omega + j\,5.46\ \Omega$$

and

$$\mathbf{Z}_L = \mathbf{0.72\ \Omega - j\,5.46\ \Omega}$$

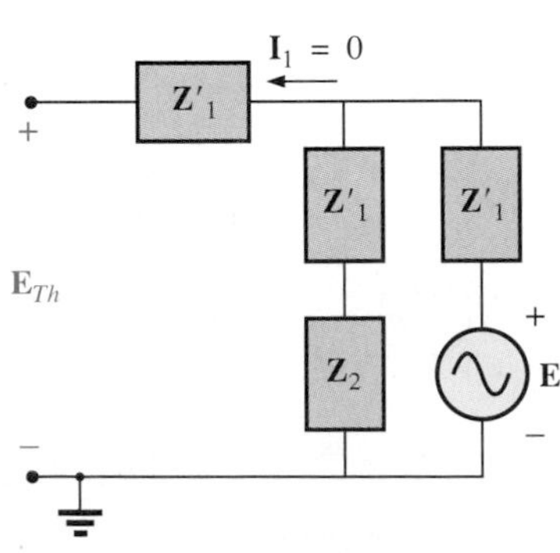

FIG. 18.87
Finding the Thévenin voltage for the network of Fig. 18.83.

For $\mathbf{E}_{Th}$, use the modified circuit of Fig. 18.87 with the voltage source replaced in its original position. Since $I_1 = 0$, $\mathbf{E}_{Th}$ is the voltage across the series impedance of $\mathbf{Z'}_1$ and $\mathbf{Z}_2$. Using the voltage divider rule gives us

$$\mathbf{E}_{Th} = \frac{(\mathbf{Z'}_1 + \mathbf{Z}_2)\mathbf{E}}{\mathbf{Z'}_1 + \mathbf{Z}_2 + \mathbf{Z'}_1} = \frac{(j\,3\ \Omega + 8\ \Omega)(10\ \text{V}\ \angle 0°)}{8\ \Omega + j\,6\ \Omega}$$

$$= \frac{(8.54\ \angle 20.56°)(10\ \text{V}\ \angle 0°)}{10\ \angle 36.87°}$$

$$\mathbf{E}_{Th} = 8.54\ \text{V}\ \angle -16.31°$$

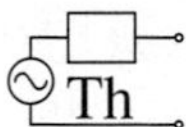

and $$P_{max} = \frac{E_{Th}^2}{4R} = \frac{(8.54\text{ V})^2}{4(0.72\ \Omega)} = \frac{72.93}{2.88}\text{ W}$$
$$= \mathbf{25.32\ W}$$

If the load resistance is adjustable but the magnitude of the load reactance cannot be set equal to the magnitude of the Thévenin reactance, then the maximum power *that can be delivered* to the load will occur when the load reactance is made as close to the Thévenin reactance as possible and the load resistance is set to the following value:

$$R_L = \sqrt{R_{Th}^2 + (X_{Th} + X_{load})^2} \qquad \textbf{(18.21)}$$

where each reactance carries a positive sign if inductive and a negative sign if capacitive.

The power delivered will be determined by

$$P = E_{Th}^2/4R_{av} \qquad \textbf{(18.22)}$$

where $$R_{av} = \frac{R_{Th} + R_L}{2} \qquad \textbf{(18.23)}$$

The derivation of the above equations is given in Appendix H of the text. The following example demonstrates the use of the above.

EXAMPLE 18.21 For the network of Fig. 18.88:

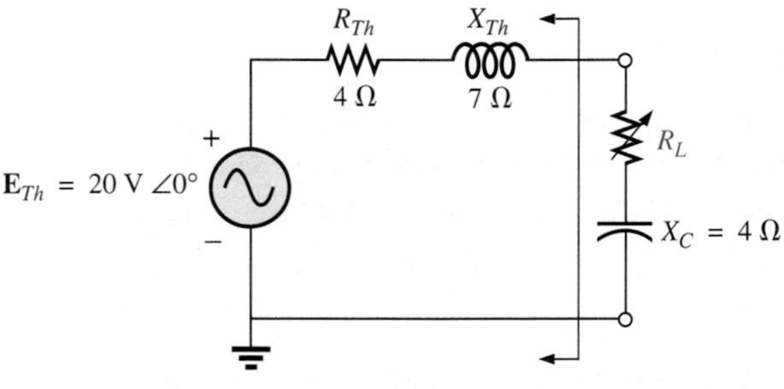

FIG. 18.88
Example 18.21.

a. Determine the value of R_L for maximum power to the load if the load reactance is fixed at 4 Ω.
b. Find the power delivered to the load under the conditions of part (a).
c. Find the maximum power to the load if the load reactance is made adjustable to any value, and compare the result to part (b) above.

Solutions:

a. Eq. (18.21): $$R_L = \sqrt{R_{Th}^2 + (X_{Th} + X_{load})^2}$$
$$= \sqrt{(4\ \Omega)^2 + (7\ \Omega - 4\ \Omega)^2}$$

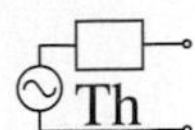

$$= \sqrt{16 + 9} = \sqrt{25}$$
$$R_L = \mathbf{5\ \Omega}$$

b. Eq. (18.23): $R_{av} = \dfrac{R_{Th} + R_L}{2} = \dfrac{4\ \Omega + 5\ \Omega}{2}$

$$= \mathbf{4.5\ \Omega}$$

Eq. (18.22): $P = \dfrac{E_{Th}^2}{4R_{av}}$

$$= \frac{(20\ \text{V})^2}{4(4.5\ \Omega)} = \frac{400}{18}\ \text{W}$$
$$\cong \mathbf{22.22\ W}$$

c. For $\mathbf{Z}_L = 4\ \Omega - j\,7\ \Omega$,

$$P_{max} = \frac{E_{Th}^2}{4R_{Th}} = \frac{(20\ \text{V})^2}{4(4\ \Omega)}$$
$$= \mathbf{25\ W}$$

exceeding the result of part (b) by 2.78 W.

18.6 SUBSTITUTION, RECIPROCITY, AND MILLMAN'S THEOREMS

As indicated in the introduction to this chapter, the substitution and reciprocity theorems and Millman's theorem will not be considered here in detail. A careful review of Chapter 9 will enable you to apply these theorems to sinusoidal ac networks with little difficulty. A number of problems in the use of these theorems appear in the problems section.

18.7 COMPUTER ANALYSIS

The computer analysis of this chapter will include a detailed discussion of controlled sources. The extended use of controlled sources in the analysis of electronic systems places a high priority on this subject. There are essentially four types of controlled sources in electronic systems with which a student should become familiar: current-controlled current sources (CCCS), voltage-controlled current sources (VCCS), current-controlled voltage sources (CCVS), and voltage-controlled voltage sources (VCVS). Each has a controlling variable that must be listed properly in the input file to ensure the proper magnitude and phase for the controlled source. In BASIC an analysis of the network as appearing in this chapter would have to be applied—simply an extension of the longhand approach to a specific configuration. PSpice permits a fairly standard entry of network parameters, with the software package able to perform the proper analysis and provide the desired solution. This section will be limited to the PSpice approach to permit increased coverage of the controlled source input format.

PSpice (DOS)

Thévenin's Theorem The network to be analyzed using Thévenin's theorem is the same as shown in Fig. 18.28 of the text. To enter the proper inductive and capacitive levels for the network of Fig. 18.89(a), a

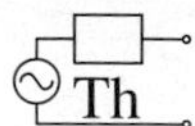

frequency of 1 kHz was chosen and the nameplate values were calculated for the reactance levels of Fig. 18.28.

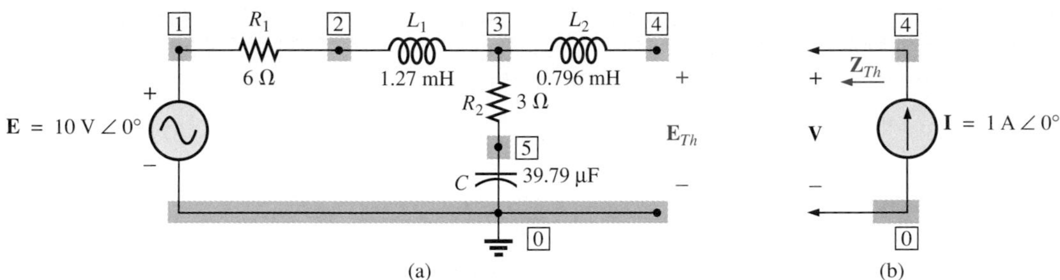

FIG. 18.89
(a) Using PSpice (DOS) to determine the Thévenin voltage for the network of Fig. 18.28. (b) Applying a 1-A source (at 0°) to determine the Thévenin impedance.

$\mathbf{E}_{Th}$:

For $\mathbf{E}_{Th}$ the input file appears in Fig. 18.90 with the values defined in Fig. 18.89(a) and an applied frequency of 1 kHz. Both the magnitude and phase angle of the Thévenin voltage between nodes 4 and 0 were requested. The result of $\mathbf{E}_{Th} = 5.081\text{ V} \angle -76.98°$ is a close match with the result of Example 18.8 ($\mathbf{E}_{Th} = 5.08\text{ V} \angle -77.09°$).

$\mathbf{Z}_{Th}$:

For $\mathbf{Z}_{Th}$ the technique applied to dc systems was applied with the source set to zero volts and current source of 1 A applied between nodes 4 and 0, as shown in Fig. 18.89(b). Since $\mathbf{Z}_T = \mathbf{V}/\mathbf{I}$ and $\mathbf{I} = 1\text{ A} \angle 0°$, the

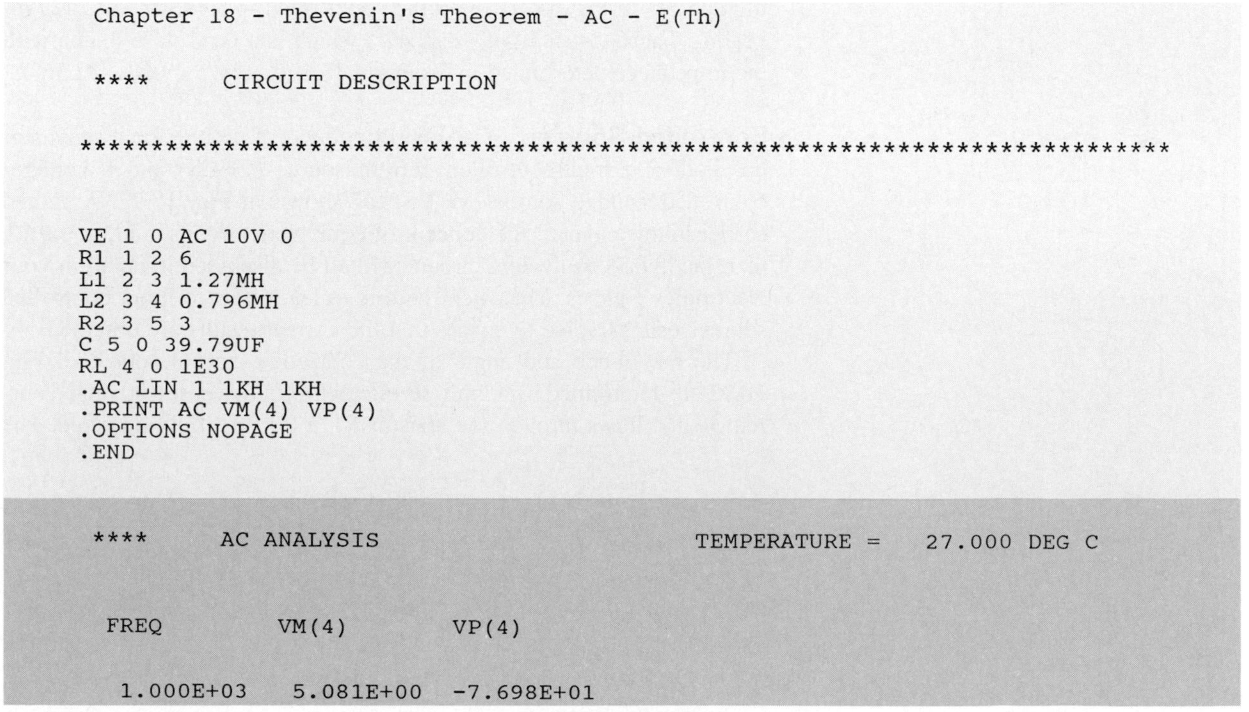

```
Chapter 18 - Thevenin's Theorem - AC - E(Th)

****     CIRCUIT DESCRIPTION

******************************************************************************

VE 1 0 AC 10V 0
R1 1 2 6
L1 2 3 1.27MH
L2 3 4 0.796MH
R2 3 5 3
C 5 0 39.79UF
RL 4 0 1E30
.AC LIN 1 1KH 1KH
.PRINT AC VM(4) VP(4)
.OPTIONS NOPAGE
.END

****     AC ANALYSIS                          TEMPERATURE =   27.000 DEG C

  FREQ          VM(4)         VP(4)

  1.000E+03    5.081E+00   -7.698E+01
```

FIG. 18.90
Input and output files for the network of Fig. 18.89(a).

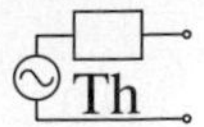


```
Chapter 18 - Thevenin's Theorem - AC - Z(Th)

****      CIRCUIT DESCRIPTION

******************************************************************************

VE 1 0 AC 0V 0
R1 1 2 6
L1 2 3 1.27MH
L2 3 4 0.796MH
R2 3 5 3
C 5 0 39.79UF
RL 4 0 1E30
II 0 4 AC 1 0
.AC LIN 1 1KH 1KH
.PRINT AC VM(4) VP(4)
.OPTIONS NOPAGE
.END

****      AC ANALYSIS                         TEMPERATURE =   27.000 DEG C

  FREQ          VM(4)         VP(4)

   1.000E+03    5.493E+00    3.242E+01
```

FIG. 18.91
Input and output files for the network of Fig. 18.89(a), with the current source of Fig. 18.89(b) applied.

magnitude and angle of **V** will be the same as those for $\mathbf{Z}_T$. The input file appears in Fig. 18.91, with the result of $\mathbf{V} = 5.493 \text{ V} \angle 32.42°$ revealing that $\mathbf{Z}_{Th} = 5.493 \ \Omega \angle 32.42°$, which is a very close match with the impedance determined in Example 18.8 ($\mathbf{Z}_{Th} = 5.49 \ \Omega \angle 32.36°$).

Controlled Sources The hybrid equivalent circuit for a transistor has both a current-controlled current source (CCCS) and a voltage-controlled voltage source (VCVS), as shown in Fig. 18.92. As noted earlier in this chapter, the concept of equivalent circuits, and, in particular, the hybrid equivalent circuit, will all be discussed in detail in your electronics courses. The intent here is to learn how to input controlled sources using PSpice to ensure that the correct results are obtained.

The magnitude and angle of the controlled current source of Fig. 18.92 are determined by a multiplying factor (in this case 100) and a current $\mathbf{I}_1$ that flows through the resistor R_1 in the direction indicated. The

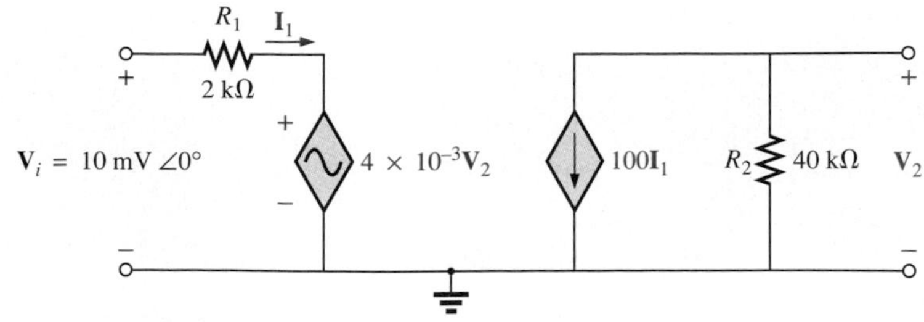

FIG. 18.92
Applying PSpice (DOS) to networks with controlled sources.

magnitude and angle of the controlled voltage source are determined by a multiplying factor of 4×10^{-3}, and the output voltage $\mathbf{V}_2$ is defined by nodes 4 and 0. Note in passing that the controlled current source is part of the output circuit, whereas the controlling variable is part of the input circuit. The reverse is true for the controlled voltage source, which is actually a "feedback" of the output voltage to the input circuit.

The formats for CCCS and VCVS controlled sources are as follows.

Current-Controlled Current Source (CCCS) For CCCS sources it is particularly important that the direction of the controlling current be properly entered. In Fig. 18.92, if the controlling current $\mathbf{I}_1$ were entered with a direction opposite to that of Fig. 18.92, the result would be meaningless.

In PSpice a controlling current has a direction defined by an independent source in series with the branch in which the controlling current is defined. Keep in mind, however, that in PSpice the current through an independent source has the direction shown in Fig. 18.93—from the + to − potential.

In Fig. 18.92 this would result in a wrong direction for the controlling current $\mathbf{I}_1$. A correct direction can be obtained in PSpice by introducing an independent sensing source of zero volts in the branch in which $\mathbf{I}_1$ is defined, as shown in Fig. 18.94. Note that the direction of $\mathbf{I}_1$ corresponds with the current direction defined for the "dummy" source VSENSE. The fact that the magnitude of the dummy source is zero volts removes any concern about it affecting the overall behavior of the network. Of course, if $\mathbf{I}_1$ had the opposite direction, the voltage source $\mathbf{V}_i$ could have been used to define the direction of $\mathbf{I}_1$ for the input file.

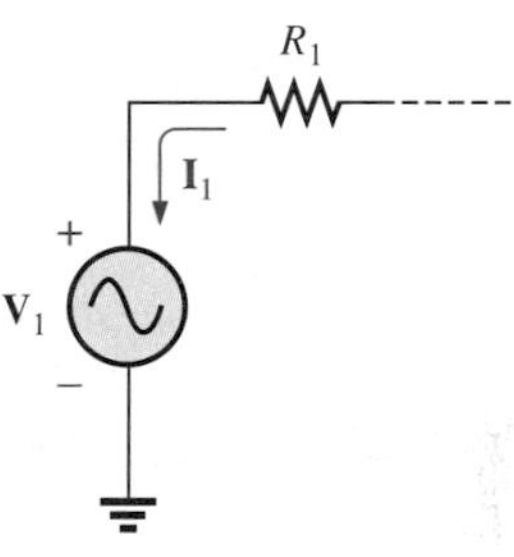

FIG. 18.93
Defining the relationship between the polarity of a voltage source and the direction of its source current when applying PSpice.

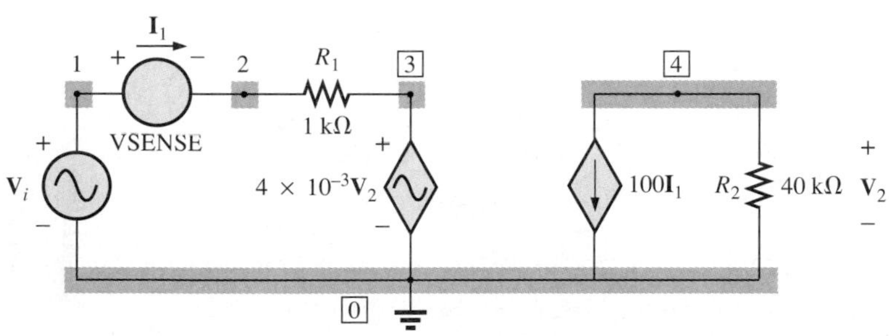

FIG. 18.94
Applying PSpice (DOS) to a network with a CCCS and VCVS.

In review:

The direction of a controlling current is defined by an independent source in series with the branch in which the controlling current is defined. The absence of an independent source or the presence of a source with the wrong polarity requires the insertion of an independent source with the correct polarity and a magnitude of zero volts.

The general format for a CCCS is as follows.

```
FXISTOR    4    0    VSENSE    100
```

FXISTOR	4	0	VSENSE	100
Name	+ node	− node	Name of controlling voltage for direction of controlling current	Multiplying factor
	Supply current from node 4 to 0			

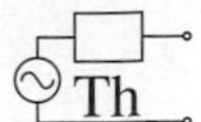

Note in the preceding format that a single letter F defines a CCCS and that a voltage is defining the direction of $\mathbf{I}_1$, as discussed earlier. Of course, the use of VSENSE in the CCCS command will require that VSENSE appear as an independent source in the input file.

Voltage-Controlled Voltage Source (VCVS) The input format for a VCVS is easier to grasp because the polarity of the controlling voltage source is entered as defined in Fig. 18.92. The general format is the following:

EFBACK	3	0	4	0	4M
Name	+ node	− node	+ node	− node	Multiplying factor
	For controlled source		Of controlling voltage		

Again note the use of a single letter E to define a VCVS and the relative simplicity of writing the input line compared to the CCCS.

Before looking at the entire input file, let us take a moment to analyze the network of Fig. 18.92 longhand (as we would in BASIC) to be sure we understand the impact of a controlled source and have an answer to compare with the output of the PSpice run.

For the controlling variables:

$$\mathbf{I}_1 = \frac{\mathbf{V}_i - 4 \times 10^{-3}\mathbf{V}_2}{1\ \text{k}\Omega}$$

and
$$\mathbf{V}_2 = -(100\mathbf{I}_1)(40\ \text{k}\Omega)$$

Substituting:

$$\mathbf{V}_2 = -(100)\left(\frac{\mathbf{V}_i - 4 \times 10^{-3}\mathbf{V}_2}{1\ \text{k}\Omega}\right)(40\ \text{k}\Omega)$$

$$\mathbf{V}_2 = -(4000)[\mathbf{V}_i - 4 \times 10^{-3}\mathbf{V}_2] = -4000\mathbf{V}_i + 16\mathbf{V}_2$$

and
$$-15\mathbf{V}_2 = -4000\mathbf{V}_i$$

or
$$\mathbf{V}_2 = \frac{-4000\mathbf{V}_i}{15} = -266.67\mathbf{V}_i$$

but
$$\mathbf{V}_i = 10\ \text{mV}\ \angle 0^\circ$$

and
$$\mathbf{V}_2 = -(266.67)(10 \times 10^{-3}\ \text{V}\ \angle 0^\circ)$$
$$\mathbf{V}_2 = \mathbf{-2.67\ V\ \angle 0^\circ = 2.67\ V\ \angle 180^\circ}$$

The input file for Fig. 18.94 appears in Fig. 18.95, with the two controlled sources ER and FF and the independent source VSENSE to define the correct direction for $\mathbf{I}_1$. Even though there are no reactive elements in the network, a frequency is defined to establish the .AC command. Note how closely the result for $\mathbf{V}_2$ in the output file corresponds with the preceding solution.

Although not appearing in Fig. 18.92, the format of the remaining two types of controlled sources are now defined.

Voltage-Controlled Current Source (VCCS)

GFET	2	3	5	8	4M
Name	+ node	− node	+ node	− node	Multiplying factor (units: siemens)
	Source current from node 2 to node 3		Defined by polarity of controlling voltage		

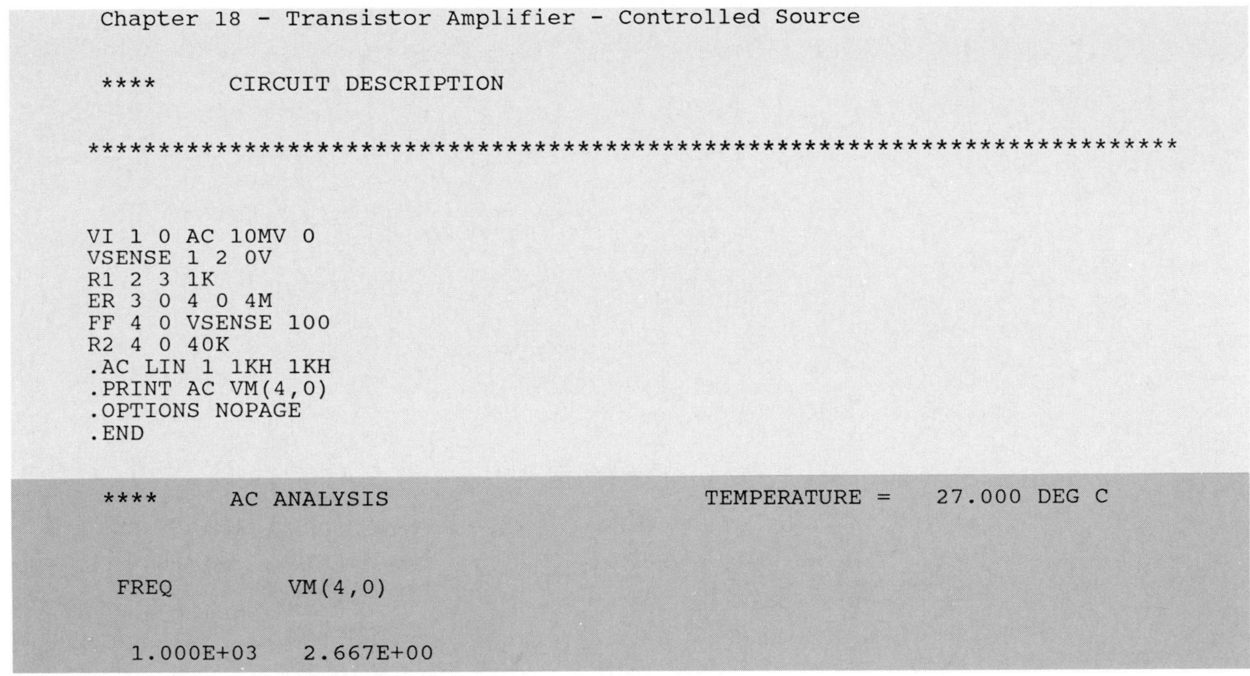

```
Chapter 18 - Transistor Amplifier - Controlled Source

****     CIRCUIT DESCRIPTION

******************************************************************************

VI 1 0 AC 10MV 0
VSENSE 1 2 0V
R1 2 3 1K
ER 3 0 4 0 4M
FF 4 0 VSENSE 100
R2 4 0 40K
.AC LIN 1 1KH 1KH
.PRINT AC VM(4,0)
.OPTIONS NOPAGE
.END

****     AC ANALYSIS                         TEMPERATURE =   27.000 DEG C

  FREQ        VM(4,0)

   1.000E+03   2.667E+00
```

FIG. 18.95
Input and output files for the network of Fig. 18.94.

Current-Controlled Voltage Source (CCVS)

HCCVS	6	5	VSENSE	4
Name	+ node	− node	Defines direction of controlling variable	Multiplying factor (units: ohms)
	Polarity of controlled source			

PSpice (Windows)

The application of the schematics approach will begin with an analysis of the network of Fig 18.12 from Example 18.4. The network was chosen because it has both dc and ac sources. The following will demonstrate that the analysis can include both the dc and ac in one simulation, and **Probe** can include dc levels in the waveforms generated under the **Trace** command. The schematic for the network appears in Fig. 18.96, with inductive and capacitive parameters determined at a frequency of 1 kHz—the frequency chosen in the **AC Sweep** of the **Analysis Setup** dialog box. The label V3 for **VPRINT** was chosen by double-clicking on the symbol and choosing **Include System-defined Attributes** and changing **PKGREF** to V3. In the **Analysis Setup** dialog box, the **AC Sweep** is set at a frequency of 1 kHz, and the **Bias Point Detail** option is enabled to obtain the dc solution.

After simulation, the output file appearing as Fig. 18.97 can be reviewed by selecting **File,** followed by **Examine Output.** Note under the **SMALL SIGNAL BIAS SOLUTION** that the voltage at node 2 and at node 3 is the same dc level of 3.6 volts. This is a result of the inductor of Fig. 18.96 acting like a short circuit to dc. Nodes 2 and 3 are at the top of the capacitor and the 3-kΩ resistor, respectively. Of course, node 3 is our output voltage (with respect to ground), and the

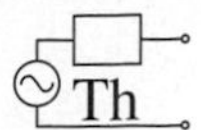

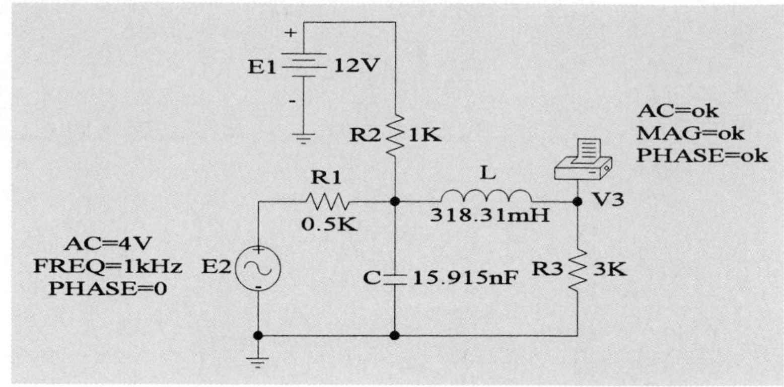

FIG. 18.96
Applying PSpice (Windows) to the network of Fig. 18.12.

```
 ****      SMALL SIGNAL BIAS SOLUTION          TEMPERATURE =   27.000 DEG C

***************************************************************************

 NODE   VOLTAGE      NODE   VOLTAGE      NODE   VOLTAGE      NODE   VOLTAGE

($N_0001)     0.0000                     ($N_0002)     3.6000

($N_0003)     3.6000                     ($N_0004)    12.0000
 ****      AC ANALYSIS                         TEMPERATURE =   27.000 DEG C

***************************************************************************

  FREQ          VM($N_0003) VP($N_0003)

   1.000E+03    2.060E+00   -3.274E+01
```

FIG. 18.97
The output file for the analysis performed on the network of Fig. 18.96.

3.6 volts is an exact match with the longhand solution. The **AC ANALYSIS** reveals that the voltage $\mathbf{V}_3$ is 2.06 V $\angle -32.74°$, which also matches our longhand solution.

A display of the voltage v_3 can be obtained by setting up a **Transient** response under the **Analysis Setup** dialog box and using a **Print Step** of 10 μs, a **Final Time** of 5 ms (for five full periods of the applied signal), a **No Print Delay** of 0 s, and a **Step Ceiling** of 10 μs (to be sure the output results are crisp and clear). Before running **Probe** the input voltage must be changed to 5.657 V (1.414 $\times$ 4 V since it is the peak value of the applied rms voltage of 4 V). It is okay to leave the 4 V as the input (even though it is the peak value and not the rms value) for a general analysis because the results obtained in the output file will simply be regarded as rms values. However, for a plot, the waveform will treat the 4 V as a peak value and will plot the resulting waveforms accordingly. Once **Probe** is run, simply click the sequence **Trace-Add-Alias Names** and choose **V(R3:1)** to produce the waveform of Fig. 18.98. The horizontal line at the dc level is obtained by choosing **Tools,**

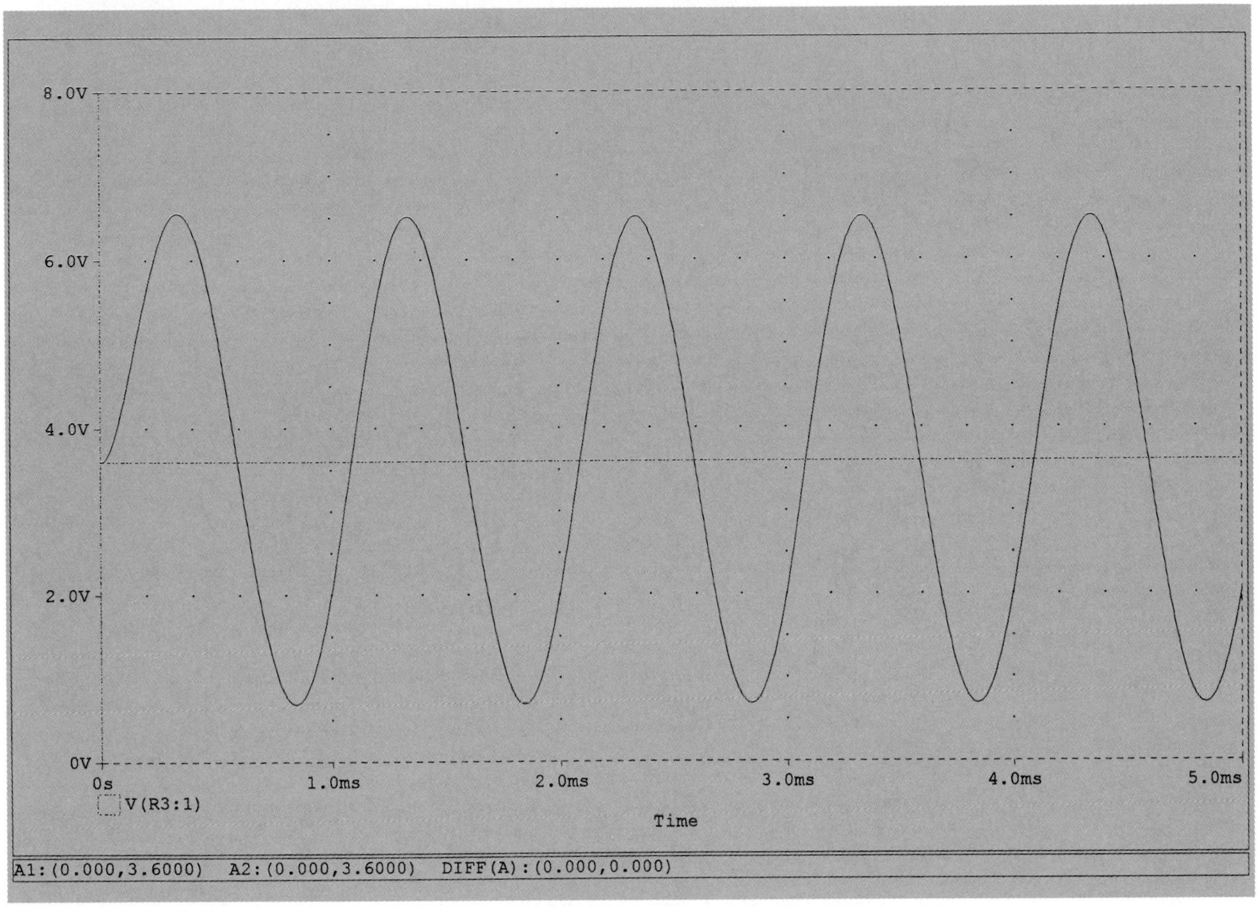

FIG. 18.98
The Probe response for v_3 for the network of Fig. 18.96.

followed by **Cursor** and **Display.** The cursor will automatically appear at the dc level and will reveal its level in the narrow horizontal box at the bottom of the graph. Note, as indicated above, that the waveform includes the dc shift of 3.6 V and has a peak value equal to the longhand solution.

The next application of schematics will reinforce the analysis done earlier on the network of Fig. 18.28, reappearing as shown in Fig. 18.99. Note the very large resistor at the point where the open-circuit Thévenin voltage is determined. The resistor is chosen sufficiently large compared to the other elements of the network to assume there is essentially an open circuit between the two points of interest. The connection (also for DOS) is necessary for a complete **Netlist.** The results of the **simulation** appear in Fig. 18.100 and confirm our earlier results; that is, $\mathbf{E}_{Th} = 5.081 \text{ V} \angle -76.98$.

Next the short-circuit current was determined, as shown in Fig. 18.101, to permit a determination of the Thévenin impedance. The resistance

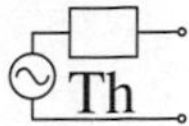

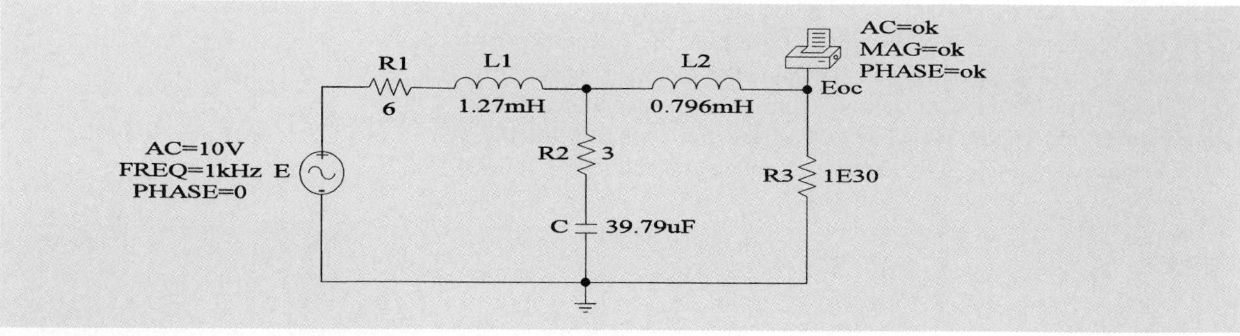

FIG. 18.99
Determining the Thévenin voltage for the network of Fig. 18.28 using PSpice (Windows).

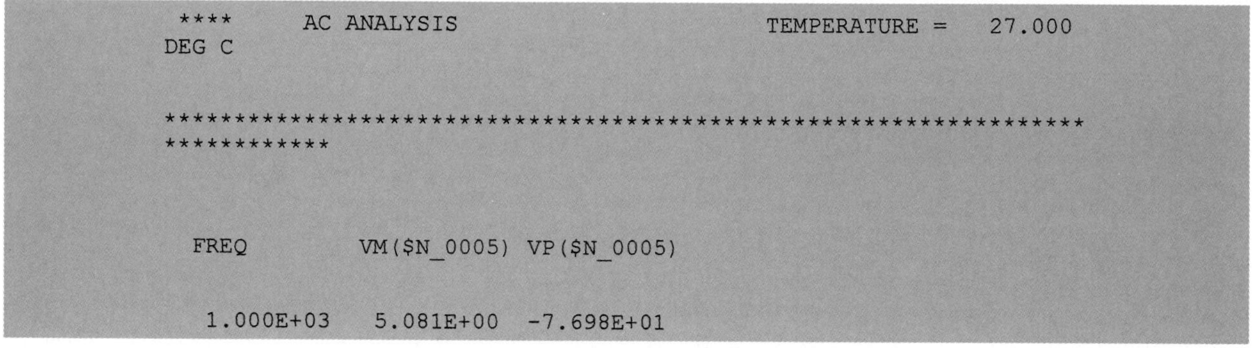

```
****     AC ANALYSIS                          TEMPERATURE =   27.000
DEG C

**********************************************************************
************

  FREQ          VM($N_0005) VP($N_0005)

   1.000E+03    5.081E+00   -7.698E+01
```

FIG. 18.100
The output file for the analysis performed on the schematic of Fig. 18.99.

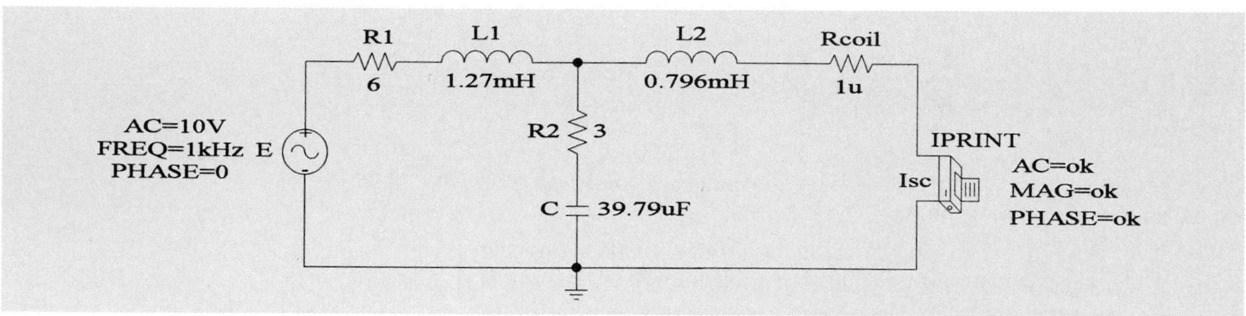

FIG. 18.101
Determining the short-circuit current for the network of Fig. 18.28.

Rcoil of 1 $\mu\Omega$ had to be introduced because inductors cannot sit alone as ideal elements in PSpice. The printout of Fig. 18.102 reveals that $\mathbf{I}_{sc} =$ 0.925 mA $\angle -10.94$. The Thévenin impedance is then defined by:

$$\mathbf{Z}_{Th} = \frac{\mathbf{E}_{Th}}{\mathbf{I}_{sc}} = \frac{5.081 \text{ V} \angle -76.98^\circ}{0.925 \text{ A} \angle -109.4^\circ}$$

$$= \mathbf{5.493 \ \Omega \angle 32.42^\circ}$$

which matches the result of Example 18.8.

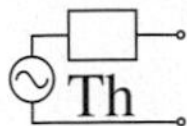

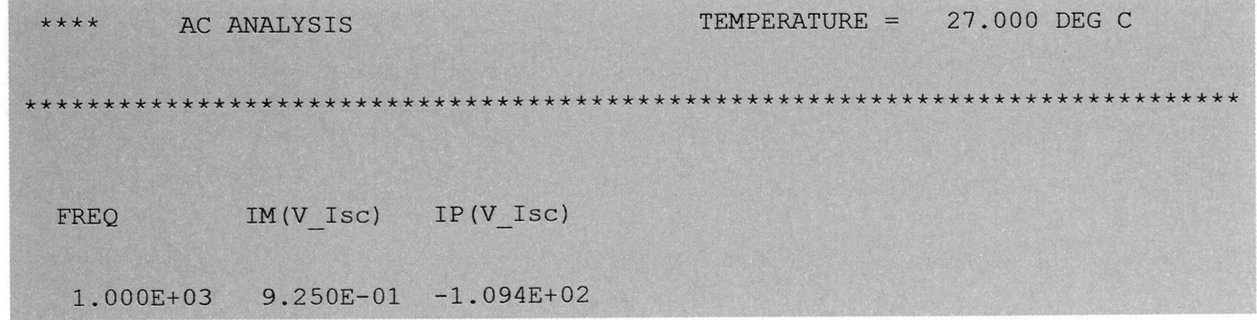

```
****     AC ANALYSIS                          TEMPERATURE =    27.000 DEG C

******************************************************************************

  FREQ          IM(V_Isc)     IP(V_Isc)

   1.000E+03    9.250E-01    -1.094E+02
```

FIG. 18.102
The output file for the analysis performed on the schematic of Fig. 18.101.

The last application of schematics will be to verify the results of Example 18.12 and to get some practice using controlled (dependent) sources. The network of Fig. 18.50, with its voltage-controlled voltage source, will have the schematic appearing in Fig. 18.103. The VCVS appears as **E** in the **analog.slb** library. Note the use of two ground symbols to establish a connection between the negative side of

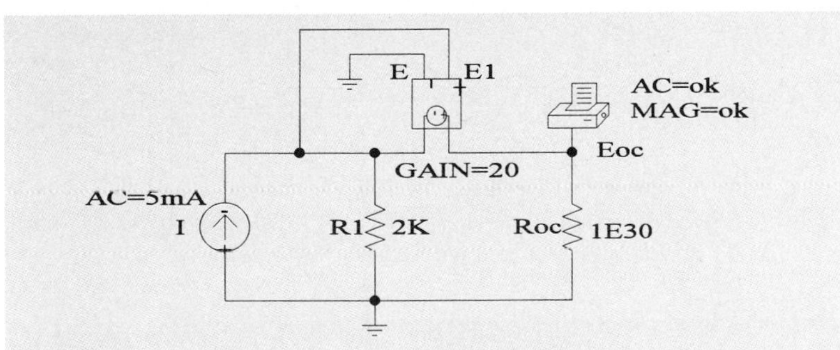

FIG. 18.103
Applying PSpice (Windows) to the network of Fig. 18.50 to determine the open-circuit voltage across the terminals of interest.

the controlled-source-sensing voltage and ground. The positive side of the sensing voltage was connected directly to the network as called for without having to resort to a **GLOBAL** application. Again, a large resistor of 1E30 was chosen to establish the open-circuit condition at the point where the Thévenin voltage is measured. A **simulation** resulted in the output of Fig. 18.104 and established the Thévenin voltage at 210 V. Substituting the numerical values into the equation obtained in Example 18.12 confirms the result; that is,

$$\mathbf{E}_{Th} = (1 + \mu)\mathbf{I}R_1 = (1 + 20)(5\text{ mA} \angle 0°)(2\text{ k}\Omega)$$
$$= \mathbf{210\ V} \angle \mathbf{0°}$$

Next, the short-circuit current was determined using the schematic of Fig. 18.105, where **IPRINT** was placed right across the terminals of interest. The result is the output file of Fig. 18.106, where the short-circuit current is the source current of 5 mA, as was obtained in Example 18.12. The ratio of the two determined quantities will result in the following Thévenin impedance:

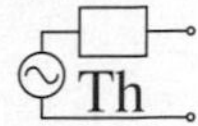

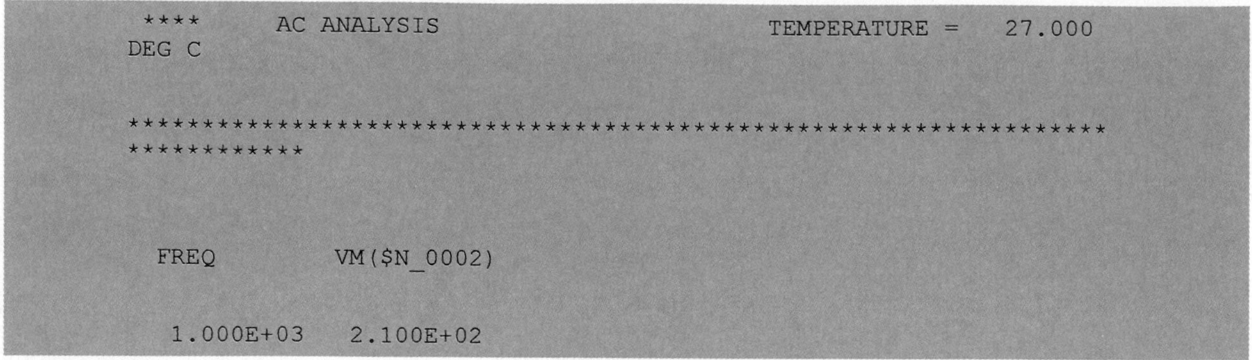

```
****        AC ANALYSIS                        TEMPERATURE =    27.000
DEG C

**********************************************************************
************

 FREQ          VM($N_0002)

  1.000E+03    2.100E+02
```

FIG. 18.104
The output file for the analysis of Fig. 18.103.

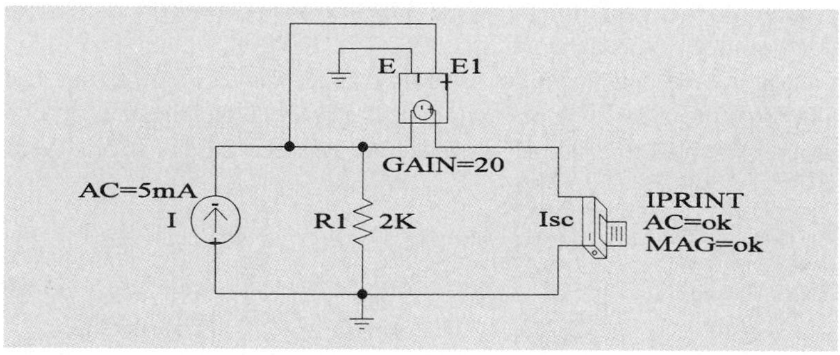

FIG. 18.105
Applying PSpice (Windows) to the network of Fig. 18.50 to determine the short-circuit current between the terminals of interest.

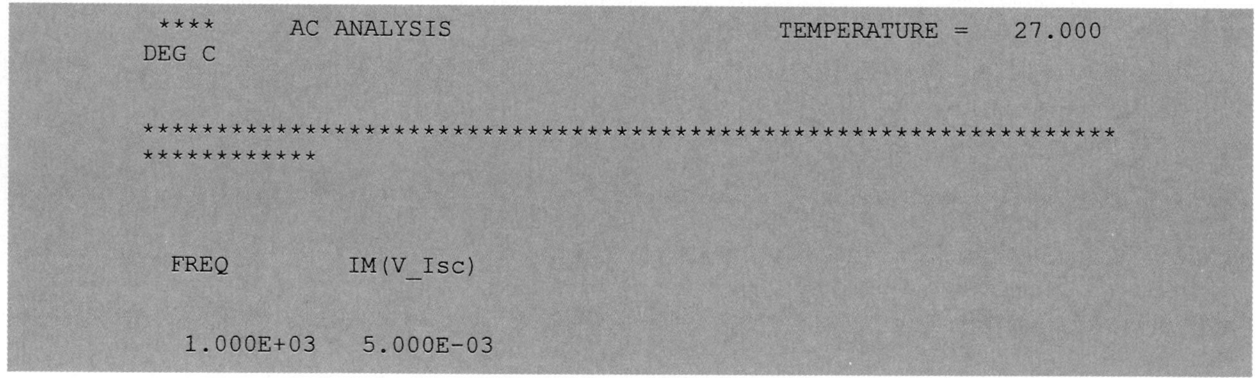

```
****        AC ANALYSIS                        TEMPERATURE =    27.000
DEG C

**********************************************************************
************

 FREQ          IM(V_Isc)

  1.000E+03    5.000E-03
```

FIG. 18.106
The output file for the analysis of Fig. 18.105.

$$\mathbf{Z}_{Th} = \frac{\mathbf{E}_{oc}}{\mathbf{I}_{sc}} = \frac{\mathbf{E}_{Th}}{\mathbf{I}_{sc}} = \frac{210\text{ V}\angle 0^\circ}{5\text{ mA}\angle 0^\circ}$$
$$= 42\text{ k}\Omega$$

which also matches the longhand solution of Example 18.12: $\mathbf{Z}_{Th} = (1 + \mu)R_1 = (21)2\text{ k}\Omega = 42\text{ k}\Omega$.

An analysis of the full transistor equivalent network of Fig. 18.56 with two controlled sources can be found in the PSpice (Windows) section of Chapter 26.

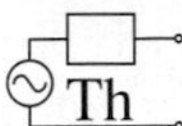

PROBLEMS

SECTION 18.2 Superposition Theorem

1. Using superposition, determine the current through the inductance X_L for each network of Fig. 18.107.

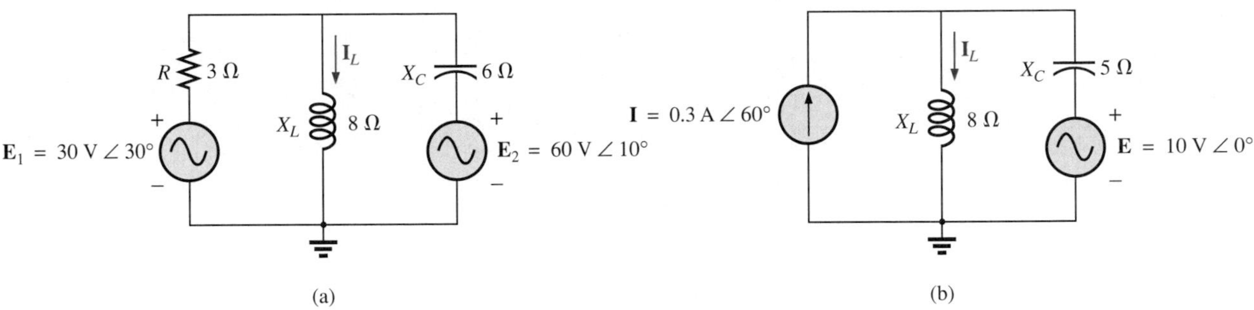

FIG. 18.107
Problem 1.

*2. Using superposition, determine the current $\mathbf{I}_L$ for each network of Fig. 18.108.

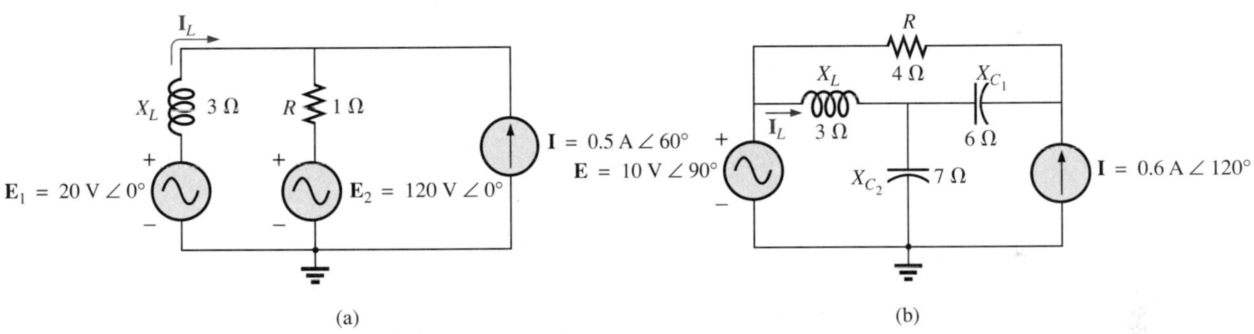

FIG. 18.108
Problem 2.

*3. Using superposition, find the sinusoidal expression for the current i for the network of Fig. 18.109.

4. Using superposition, find the sinusoidal expression for the voltage v_C for the network of Fig. 18.110.

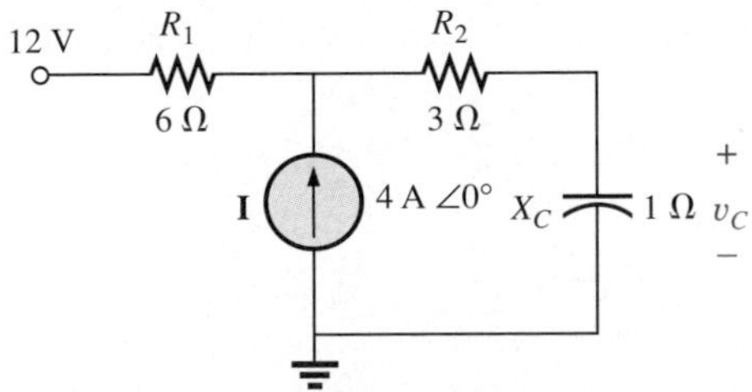

FIG. 18.110
Problems 4, 16, 31, and 43.

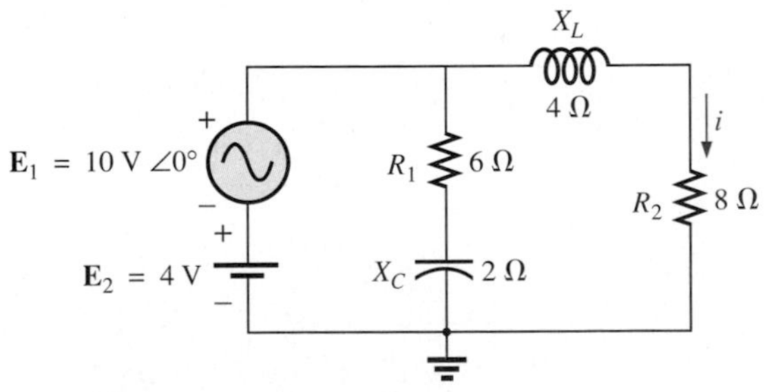

FIG. 18.109
Problems 3, 30, and 42.

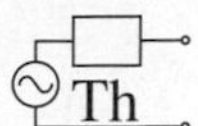

*5. Using superposition, find the current **I** for the network of Fig. 18.111.

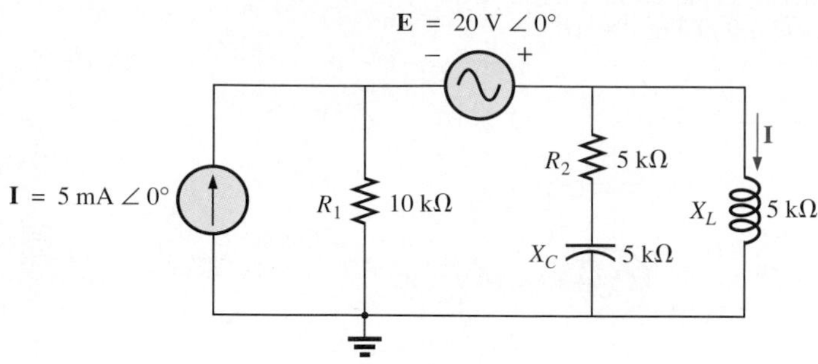

FIG. 18.111
Problems 5, 17, 32, and 44.

6. Using superposition, determine the current $\mathbf{I}_L$ ($h = 100$) for the network of Fig. 18.112.

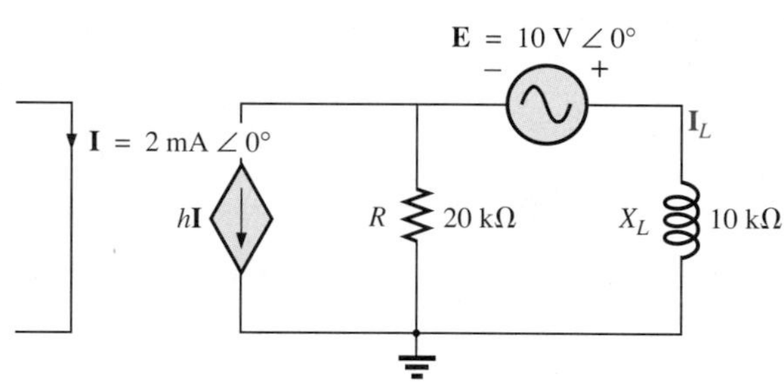

FIG. 18.112
Problems 6 and 20.

7. Using superposition, for the network of Fig. 18.113, determine the voltage $\mathbf{V}_L$ ($\mu = 20$).

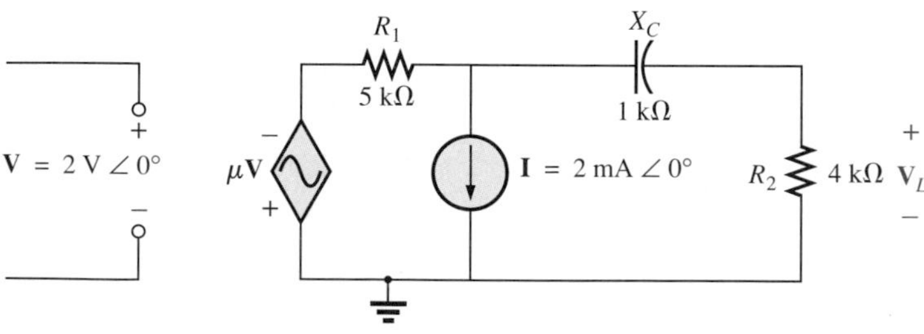

FIG. 18.113
Problems 7, 21, and 35.

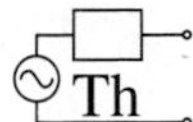

***8.** Using superposition, determine the current $\mathbf{I}_L$ for the network of Fig. 18.114 ($\mu = 20$; $h = 100$).

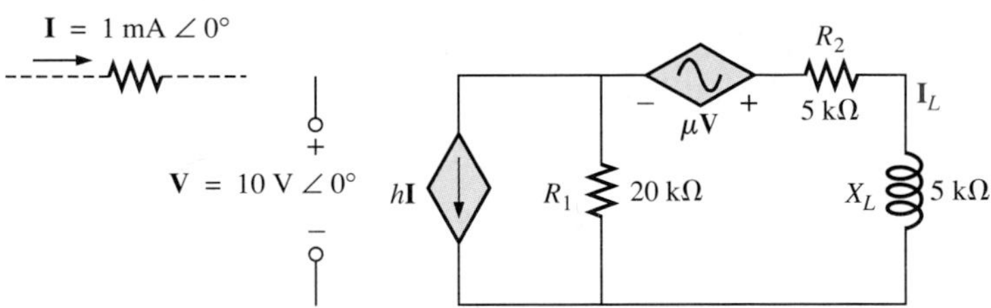

FIG. 18.114
Problems 8, 22, and 36.

***9.** Determine $\mathbf{V}_L$ for the network of Fig. 18.115 ($h = 50$).

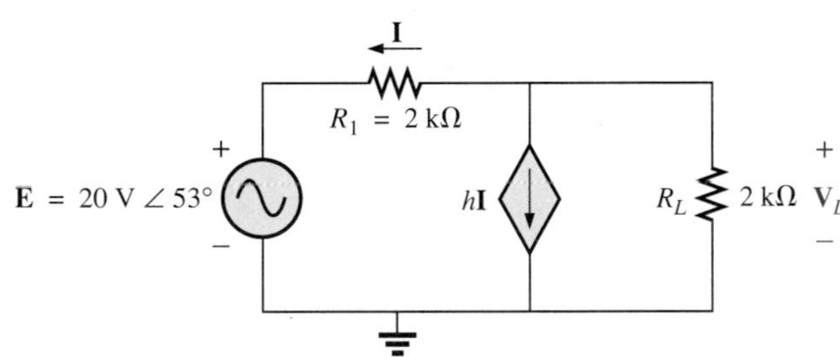

FIG. 18.115
Problems 9 and 23.

***10.** Calculate the current **I** for the network of Fig. 18.116.

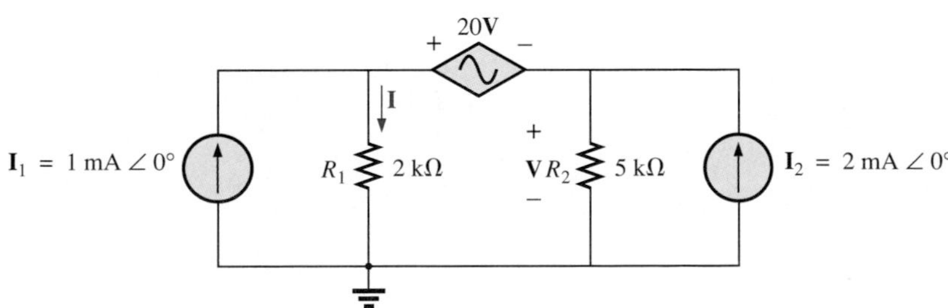

FIG. 18.116
Problems 10, 24, and 38.

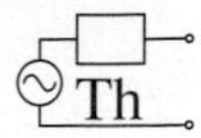

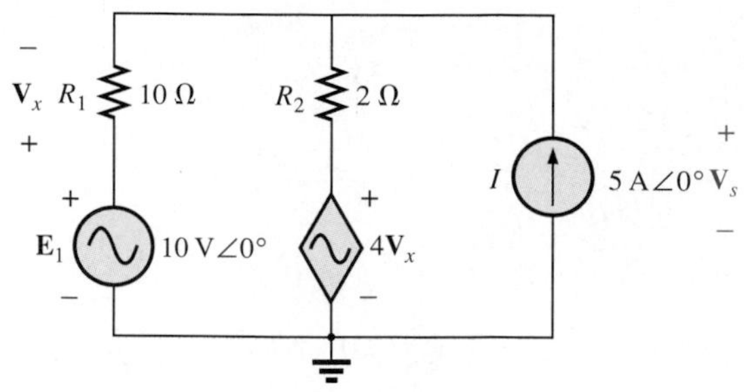

FIG. 18.117
Problem 11.

11. Find the voltage $\mathbf{V}_s$ for the network of Fig. 18.117.

SECTION 18.3 Thévenin's Theorem

12. Find the Thévenin equivalent circuit for the portions of the networks of Fig. 18.118 external to the elements between points *a* and *b*.

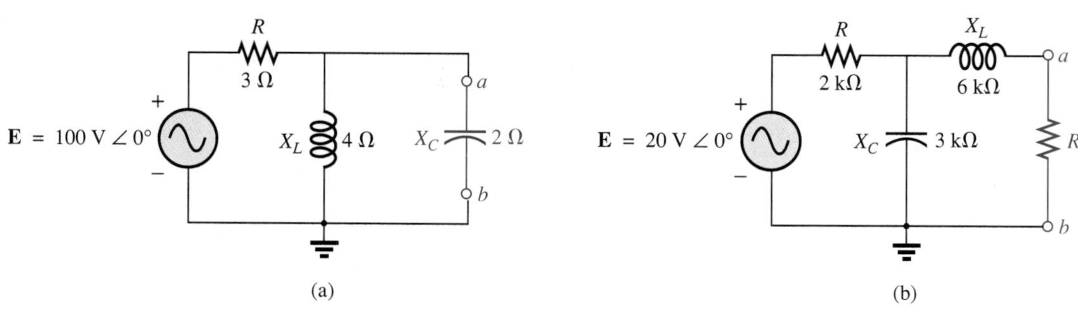

FIG. 18.118
Problems 12 and 26.

***13.** Find the Thévenin equivalent circuit for the portions of the networks of Fig. 18.119 external to the elements between points *a* and *b*.

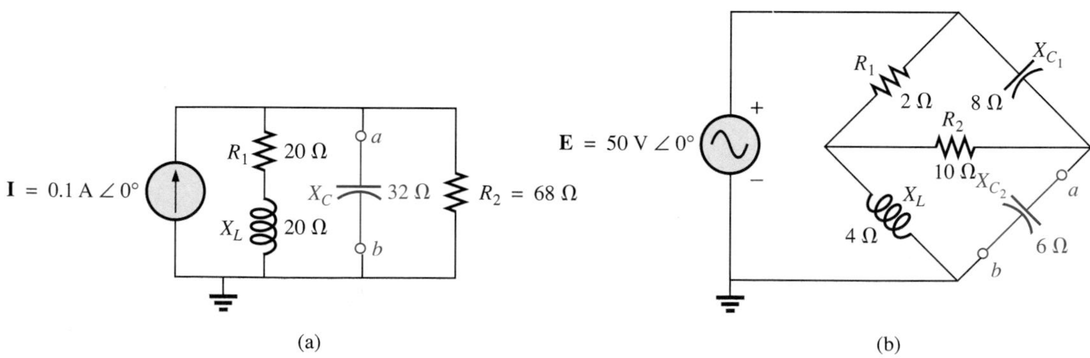

FIG. 18.119
Problems 13, 27, and 51.

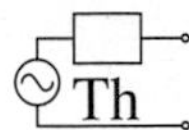

*14. Find the Thévenin equivalent circuit for the portions of the networks of Fig. 18.120 external to the elements between points a and b.

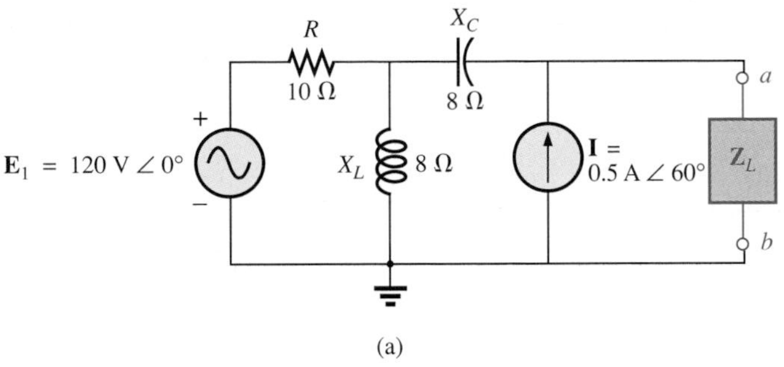

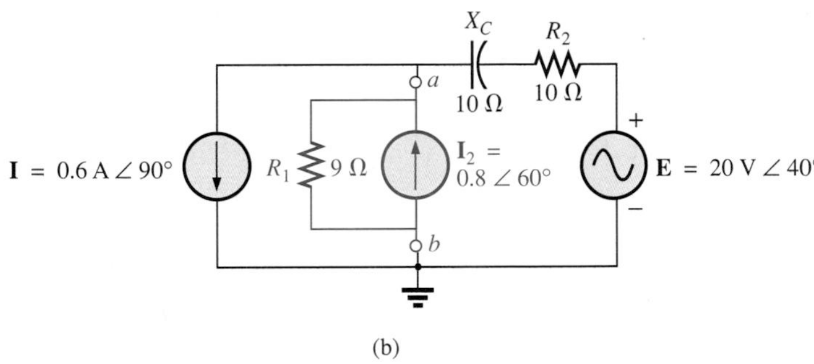

FIG. 18.120
Problems 14 and 28.

*15. **a.** Find the Thévenin equivalent circuit for the network external to the resistor R_2 in Fig. 18.109.
b. Using the results of part (a), determine the current i of the same figure.

16. **a.** Find the Thévenin equivalent circuit for the network external to the capacitor of Fig. 18.110.
b. Using the results of part (a), determine the voltage $\mathbf{V}_C$ for the same figure.

*17. **a.** Find the Thévenin equivalent circuit for the network external to the inductor of Fig. 18.111.
b. Using the results of part (a), determine the current **I** of the same figure.

18. Determine the Thévenin equivalent circuit for the network external to the 5-kΩ inductive reactance of Fig. 18.121 (in terms of **V**).

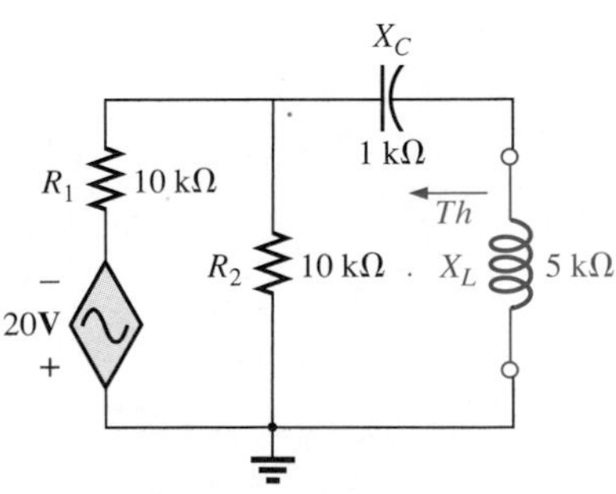

FIG. 18.121
Problems 18 and 33.

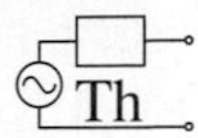

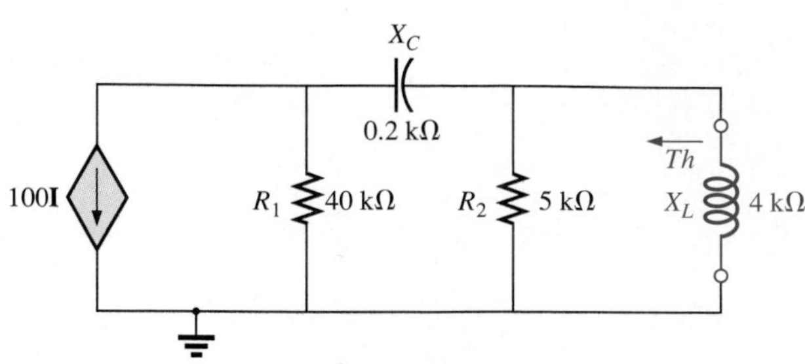

FIG. 18.122
Problems 19 and 34.

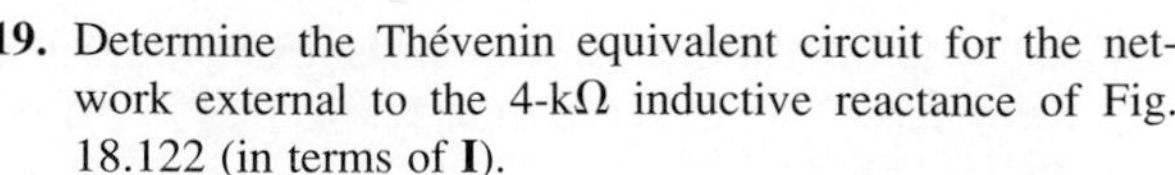

19. Determine the Thévenin equivalent circuit for the network external to the 4-kΩ inductive reactance of Fig. 18.122 (in terms of **I**).

20. Find the Thévenin equivalent circuit for the network external to the 10-kΩ inductive reactance of Fig. 18.112.

21. Determine the Thévenin equivalent circuit for the network external to the 4-kΩ resistor of Fig. 18.113.

***22.** Find the Thévenin equivalent circuit for the network external to the 5-kΩ inductive reactance of Fig. 18.114.

***23.** Determine the Thévenin equivalent circuit for the network external to the 2-kΩ resistor of Fig. 18.115.

***24.** Find the Thévenin equivalent circuit for the network external to the resistor R_1 of Fig. 18.116.

***25.** Find the Thévenin equivalent circuit for the network to the left of terminals *a-a′* of Fig. 18.123.

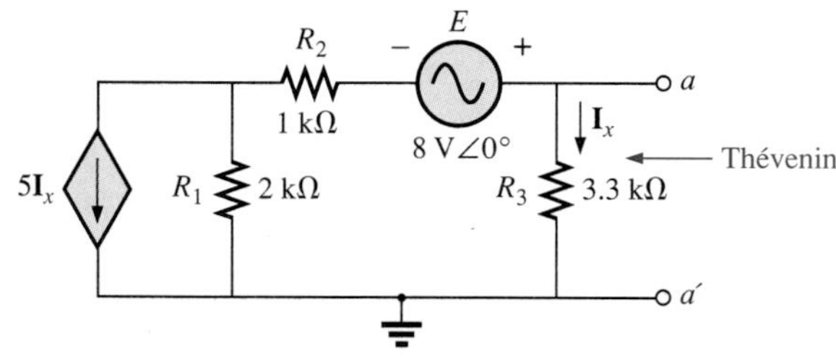

FIG. 18.123
Problem 25.

SECTION 18.4 Norton's Theorem

26. Find the Norton equivalent circuit for the network external to the elements between *a* and *b* for the networks of Fig. 18.118.

27. Find the Norton equivalent circuit for the network external to the elements between *a* and *b* for the networks of Fig. 18.119.

28. Find the Norton equivalent circuit for the network external to the elements between *a* and *b* for the networks of Fig. 18.120.

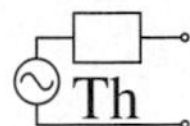

***29.** Find the Norton equivalent circuit for the portions of the networks of Fig. 18.124 external to the elements between points *a* and *b*.

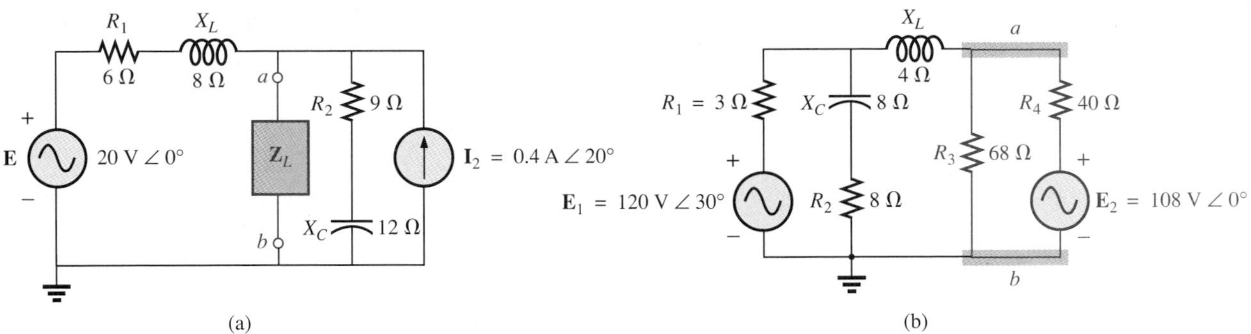

FIG. 18.124
Problem 29.

***30. a.** Find the Norton equivalent circuit for the network external to the resistor R_2 in Fig. 18.109.

b. Using the results of part (a), determine the current **I** of the same figure.

***31. a.** Find the Norton equivalent circuit for the network external to the capacitor of Fig. 18.110.

b. Using the results of part (a), determine the voltage $\mathbf{V}_C$ for the same figure.

***32. a.** Find the Norton equivalent circuit for the network external to the inductor of Fig. 18.111.

b. Using the results of part (a), determine the current **I** of the same figure.

33. Determine the Norton equivalent circuit for the network external to the 5-kΩ inductive reactance of Fig. 18.121.

34. Determine the Norton equivalent circuit for the network external to the 4-kΩ inductive reactance of Fig. 18.122.

35. Find the Norton equivalent circuit for the network external to the 4-kΩ resistor of Fig. 18.113.

***36.** Find the Norton equivalent circuit for the network external to the 5-kΩ inductive reactance of Fig. 18.114.

***37.** For the network of Fig. 18.125, find the Norton equivalent circuit for the network external to the 2-kΩ resistor.

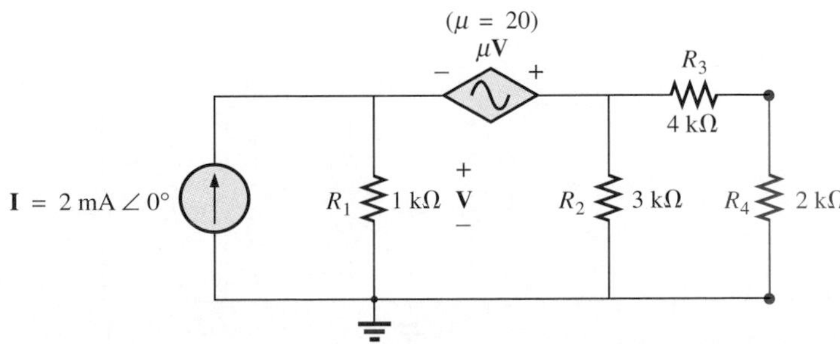

FIG. 18.125
Problem 37.

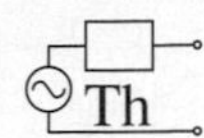

*38. Find the Norton equivalent circuit for the network external to the $\mathbf{I}_1$ current source of Fig. 18.116.

SECTION 18.5 Maximum Power Transfer Theorem

39. Find the load impedance $\mathbf{Z}_L$ for the networks of Fig. 18.126 for maximum power to the load, and find the maximum power to the load.

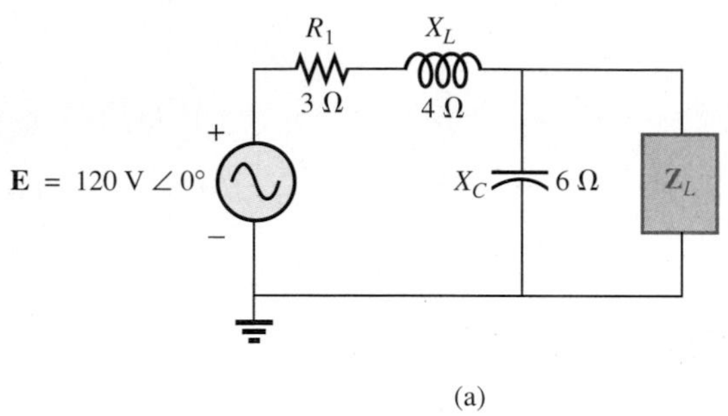

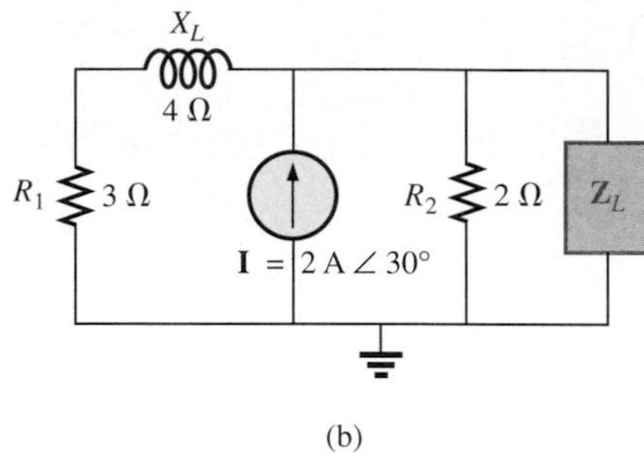

FIG. 18.126
Problem 39.

*40. Find the load impedance $\mathbf{Z}_L$ for the networks of Fig. 18.127 for maximum power to the load, and find the maximum power to the load.

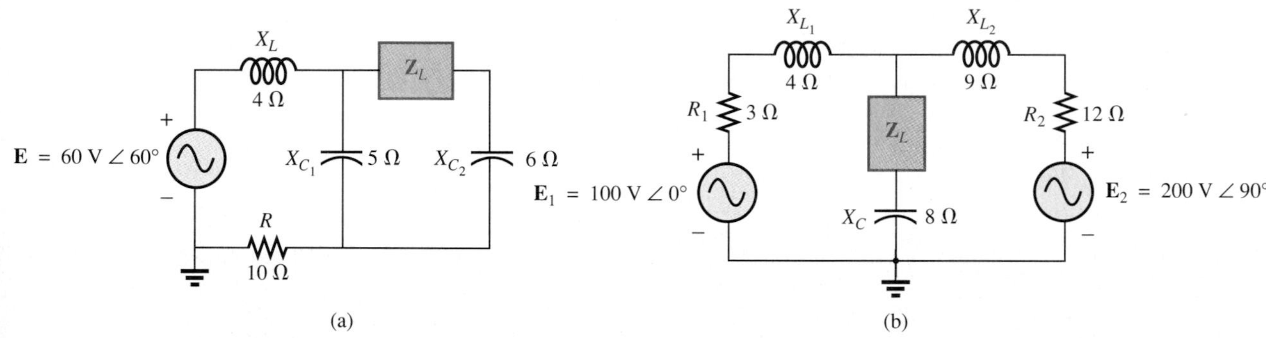

FIG. 18.127
Problem 40.

41. Find the load impedance R_L for the network of Fig. 18.128 for maximum power to the load, and find the maximum power to the load.

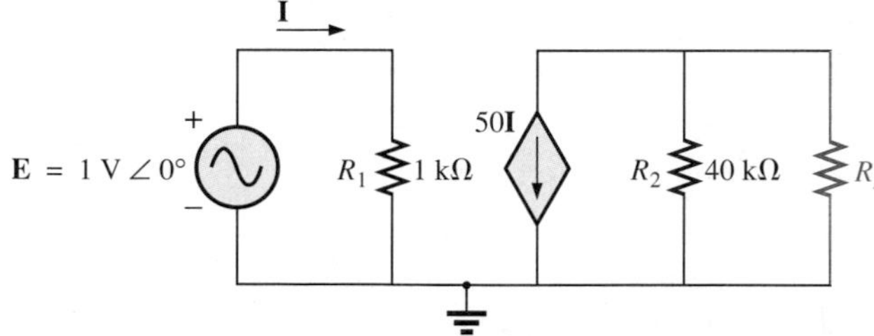

FIG. 18.128
Problem 41.

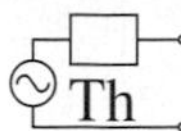

***42. a.** Determine the load impedance to replace the resistor R_2 of Fig. 18.109 to ensure maximum power to the load.

b. Using the results of part (a), determine the maximum power to the load.

***43. a.** Determine the load impedance to replace the capacitor X_C of Fig. 18.110 to ensure maximum power to the load.

b. Using the results of part (a), determine the maximum power to the load.

***44. a.** Determine the load impedance to replace the inductor X_L of Fig. 18.111 to ensure maximum power to the load.

b. Using the results of part (a), determine the maximum power to the load.

45. a. For the network of Fig. 18.129, determine the value of R_L that will result in maximum power to the load.

b. Using the results of part (a), determine the maximum power delivered.

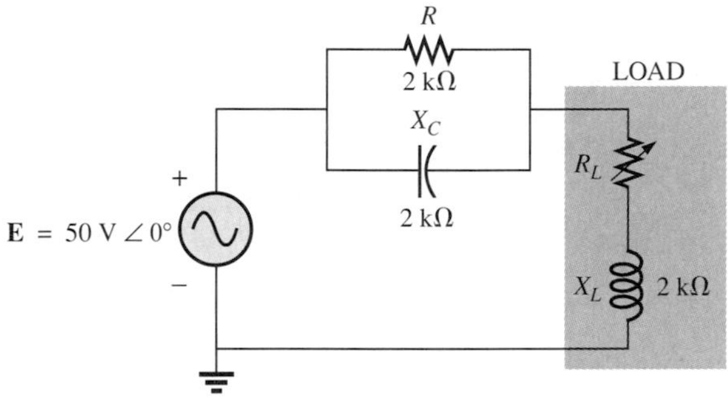

FIG. 18.129
Problem 45.

***46. a.** For the network of Fig. 18.130, determine the level of capacitance that will ensure maximum power to the load if the range of capacitance is limited to 1 to 5 nF.

b. Using the results of part (a), determine the value of R_L that will ensure maximum power to the load.

c. Using the results of parts (a) and (b), determine the maximum power to the load.

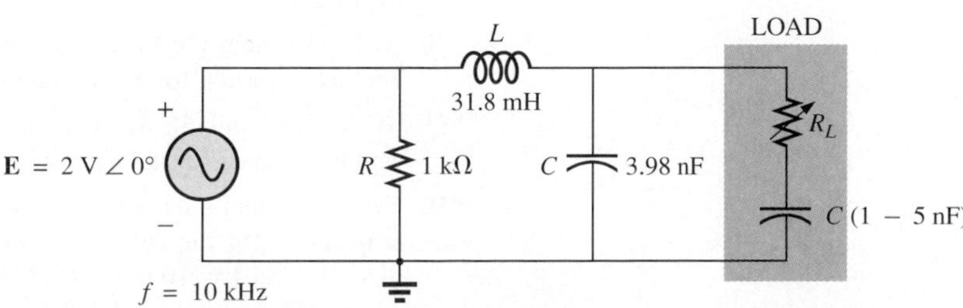

FIG. 18.130
Problem 46.

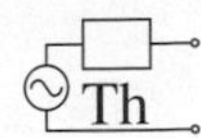

SECTION 18.6 Substitution, Reciprocity, and Millman's Theorems

47. For the network of Fig. 18.131, determine two equivalent branches through the substitution theorem for the branch *a-b*.

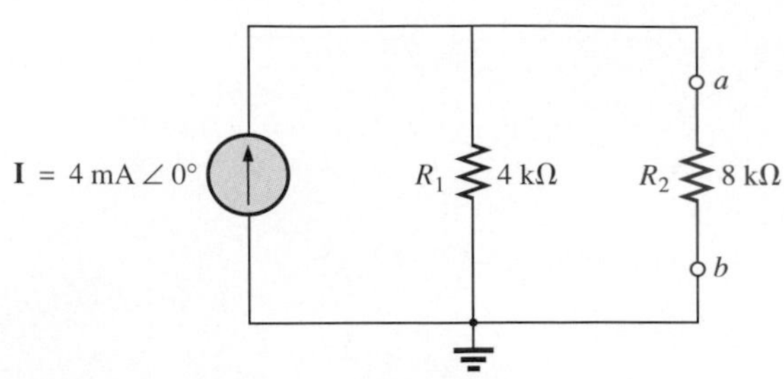

FIG. 18.131
Problem 47.

48. a. For the network of Fig. 18.132(a), find the current **I**.
b. Repeat part (a) for the network of Fig. 18.132(b).
c. Do the results of parts (a) and (b) compare?

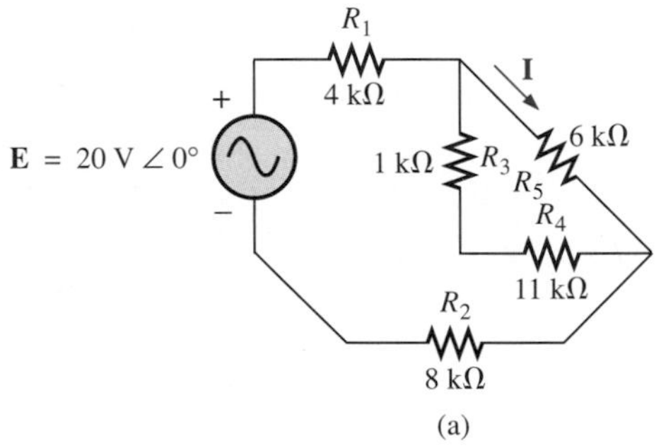

FIG. 18.132
Problem 48.

49. Using Millman's theorem, determine the current through the 4-kΩ capacitive reactance of Fig. 18.133.

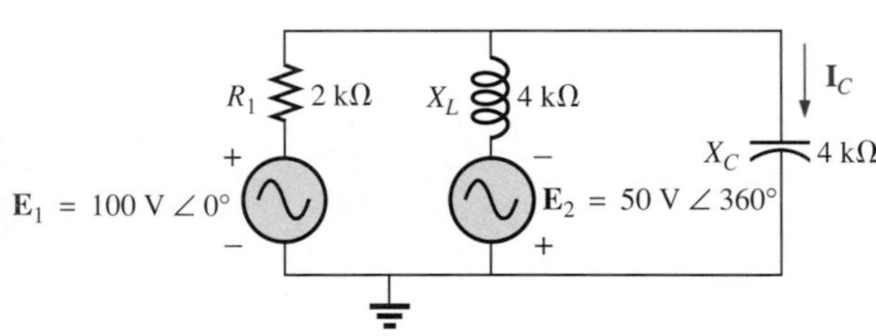

FIG. 18.133
Problem 49.

SECTION 18.7 Computer Analysis

PSpice (DOS)

50. Write the input file for the network of Fig. 18.1 that will provide a format for a solution using superposition.

***51.** Write the input file to determine the Thévenin voltage and impedance for the network of Fig. 18.119(b).

***52.** Write the input file to determine the Thévenin voltage and impedance for the network of Fig. 18.50 if $R_1 = 2\ \text{k}\Omega$, $\mu = 50$, and $\mathbf{I} = 10\ \text{mA}\ \angle 0°$. Compare with the solution obtained using the results of Example 18.12.

***53.** Write the input file to determine the Norton equivalent circuit for the network of Example 18.15.

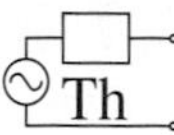

PSpice (Windows)

54. Apply superposition to the network of Fig. 18.6. That is, determine the current **I** due to each source and then find the resultant current.

***55.** Determine the current $\mathbf{I}_L$ for the network of Fig. 18.21 using schematics.

***56.** Using schematics, determine $\mathbf{V}_2$ for the network of Fig. 18.56 if $\mathbf{V}_i = 1\text{ V}\angle 0°$, $R_1 = 0.5\text{ k}\Omega$, $k_1 = 3 \times 10^{-4}$, $k_2 = 50$, and $R_2 = 20\text{ k}\Omega$.

***57.** Find the Norton equivalent circuit for the network of Fig. 18.75 using schematics.

***58.** Using schematics, plot the power to the *R-C* load of Fig. 18.88 for values of R_L from 1 Ω to 10 Ω.

Programming Language (C++, BASIC, PASCAL, etc.)

59. Given the network of Fig. 18.1, write a program to determine a general solution for the current **I** using superposition. That is, given the reactance of the same network elements, determine **I** for voltage sources of any magnitude but the same angle.

60. Given the network of Fig. 18.23, write a program to determine the Thévenin voltage and impedance for any level of reactance for each element and any magnitude of voltage for the voltage source. The angle of the voltage source should remain at zero degrees.

61. Given the configuration of Fig. 18.134, demonstrate that maximum power is delivered to the load when $X_C = X_L$ by tabulating the power to the load for X_C varying from 0.1 kΩ to 2 kΩ in increments of 0.1 kΩ.

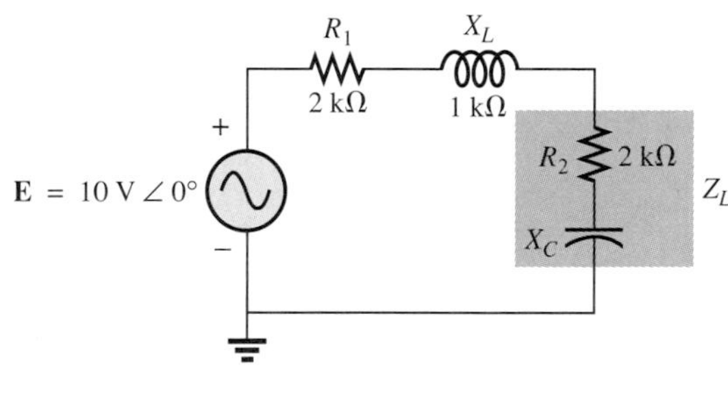

FIG. 18.134
Problem 61.

GLOSSARY

Current-controlled current source (CCCS) A current source whose parameters are controlled by a current elsewhere in the system.

Current-controlled voltage source (CCVS) A voltage source whose parameters are controlled by a voltage elsewhere in the system.

Maximum power transfer theorem A theorem used to determine the load impedance necessary to ensure maximum power to the load.

Millman's theorem A method employing voltage-to-current source conversions that will permit the determination of unknown variables in a multiloop network.

Norton's theorem A theorem that permits the reduction of any two-terminal linear ac network to one having a single current source and parallel impedance. The resulting configuration can then be employed to determine a particular current or voltage in the original network or to examine the effects of a specific portion of the network on a particular variable.

Reciprocity theorem A theorem stating that for single-source networks, the magnitude of the current in any branch of a network, due to a single voltage source anywhere else in the network, will equal the magnitude of the current through the branch in which the source was originally located if the source is placed in the branch in which the current was originally measured.

Substitution theorem A theorem stating that if the voltage across and current through any branch of an ac bilateral network are known, the branch can be replaced by any combination of elements that will maintain the same voltage across and current through the chosen branch.

Superposition theorem A method of network analysis that permits considering the effects of each source independently. The resulting current and/or voltage is the phasor sum of the currents and/or voltages developed by each source independently.

Thévenin's theorem A theorem that permits the reduction of any two-terminal linear ac network to one having a single voltage source and series impedance. The resulting configuration can then be employed to determine a particular current or voltage in the original network or to examine the effects of a specific portion of the network on a particular variable.

Voltage-controlled current source (VCCS) A current source whose parameters are controlled by a voltage elsewhere in the system.

Voltage-controlled voltage source (VCVS) A voltage source whose parameters are controlled by a voltage elsewhere in the system.

19

Power (ac)

P_S^q

19.1 INTRODUCTION

The discussion of power in Chapter 14 included only the average power delivered to an ac network. We will now examine the total power equation in a slightly different form and introduce two additional types of power: *apparent* and *reactive*.

For any system such as Fig. 19.1, the power delivered to a load at any instant is defined by the product of the applied voltage and the resulting current; that is,

$$p = vi$$

In this case, since v and i are sinusoidal quantities, let us establish a general case where

$$v = V_m \sin(\omega t + \theta)$$

and

$$i = I_m \sin \omega t$$

The chosen v and i include all possibilities because, if the load is purely resistive, $\theta = 0°$. If the load is purely inductive or capacitive, $\theta = 90°$ or $\theta = -90°$, respectively. For a network that is primarily inductive, θ is positive (v leads i), and for a network that is primarily capacitive θ is negative (i leads v).

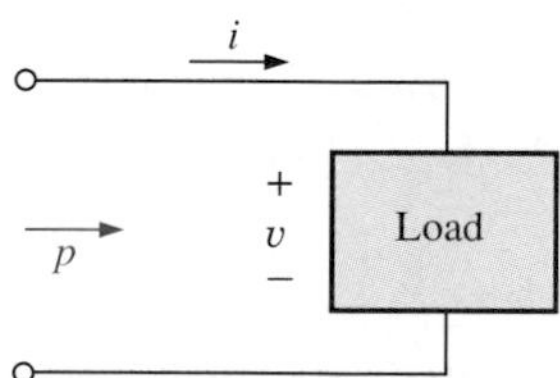

FIG. 19.1
Defining the power delivered to a load.

Substituting the above equations for v and i into the power equation will result in

$$p = V_m I_m \sin \omega t \sin(\omega t + \theta)$$

If we now apply a number of trigonometric identities, the following form for the power equation will result:

$$\boxed{p = VI \cos \theta(1 - \cos 2\omega t) + VI \sin \theta(\sin 2\omega t)} \qquad \textbf{(19.1)}$$

where V and I are now effective values. The conversion from peak values V_m and I_m to effective values resulted from the operations performed using the trigonometric identities.

It would appear initially that nothing has been gained by putting the equation in this form. However, the usefulness of the form of Eq. (19.1) will be demonstrated in the following sections. The derivation of Eq. (19.1) from the initial form will appear as an assignment at the end of the chapter.

If Eq. (19.1) is expanded to the form

$$p = \underbrace{VI \cos \theta}_{\text{Average}} - \underbrace{VI \cos \theta}_{\text{Peak}} \cos \underbrace{2\omega t}_{2x} + \underbrace{VI \sin \theta}_{\text{Peak}} \sin \underbrace{2\omega t}_{2x}$$

there are two obvious points that can be made. First, the average power still appears as an isolated term that is time independent. Second, both terms that follow vary at a frequency twice that of the applied voltage or current with peak values having a very similar format.

In an effort to ensure completeness and order in presentation, each basic element (R, L, and C) will be treated separately.

19.2 RESISTIVE CIRCUIT

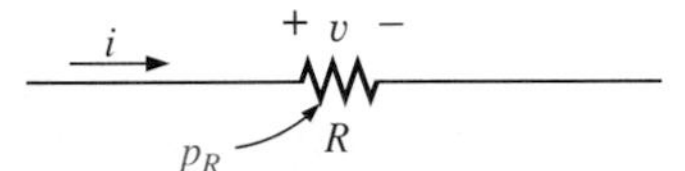

FIG. 19.2
Determining the power delivered to a purely resistive load.

For a purely resistive circuit (such as that in Fig. 19.2), v and i are in phase, and $\theta = 0°$, as appearing in Fig. 19.3. Substituting $\theta = 0°$ into Eq. (19.1), we obtain

$$\begin{aligned} p_R &= VI \cos(0°)(1 - \cos 2\omega t) + VI \sin(0°) \sin 2\omega t \\ &= VI(1 - \cos 2\omega t) + 0 \end{aligned}$$

or

$$\boxed{p_R = VI - VI \cos 2\omega t} \qquad \textbf{(19.2)}$$

where VI is the average or dc term and $-VI \cos 2\omega t$ is a negative cosine wave with twice the frequency of either input quantity (v or i) and a peak value of VI.

Plotting the waveform for p_R (Fig. 19.3), we see that

$$T_1 = \text{period of input quantities}$$
$$T_2 = \text{period of power curve } p_R$$

Note that in Fig. 19.3 the power curve passes through two cycles about its average value of VI for each cycle of either v or i ($T_1 = 2T_2$ or $f_2 = 2f_1$). Consider also that since the peak and average values of the power curve are the same, the curve is always above the horizontal axis. This indicates that

the total power delivered to a resistor will be dissipated in the form of heat.

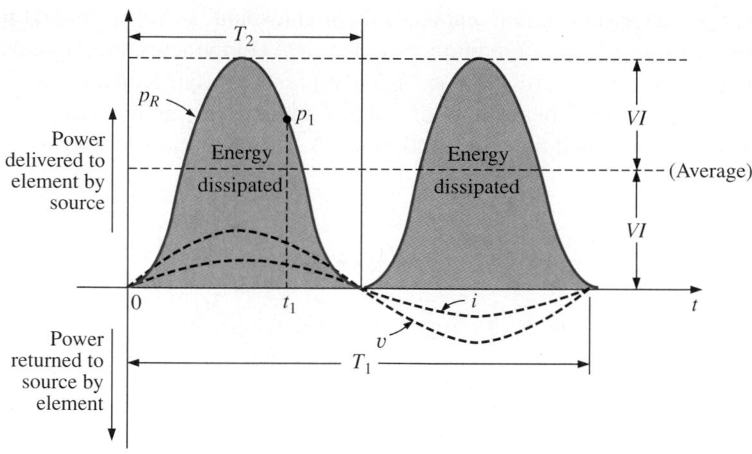

FIG. 19.3
Power versus time for a purely resistive load.

The power returned to the source is represented by the portion of the curve below the axis, which is zero in this case. The power dissipated at any instant of time t_1 by the resistor can be found by simply substituting the time t_1 into Eq. (19.2) to find p_1, as indicated in Fig. 19.3. The average power from Eq. (19.2), or Fig. 19.3, is VI; or, as a summary,

$$P = VI = \frac{V_m I_m}{2} = I^2R = \frac{V^2}{R} \quad \text{(watts, W)} \qquad \textbf{(19.3)}$$

as derived in Chapter 14.

The energy dissipated by the resistor (W_R) over one full cycle of the applied voltage (Fig. 19.3) can be found using the following equation:

$$W = Pt$$

where P is the average value and t is the period of the applied voltage; that is,

$$W_R = VIT_1 \quad \text{(joules, J)} \qquad \textbf{(19.4)}$$

or, since $T_1 = 1/f_1$,

$$W_R = \frac{VI}{f_1} \quad \text{(joules, J)} \qquad \textbf{(19.5)}$$

19.3 APPARENT POWER

From our analysis of dc networks (and resistive elements above), it would seem *apparent* that the power delivered to the load of Fig. 19.4 is simply determined by the product of the applied voltage and current, with no concern for the components of the load; that is, $P = VI$. However, we found in Chapter 14 that the power factor ($\cos\theta$) of the load will have a pronounced effect on the power dissipated, less pronounced for more reactive loads. Although the product of the voltage and current is not always the power delivered, it is a power rating of significant use-

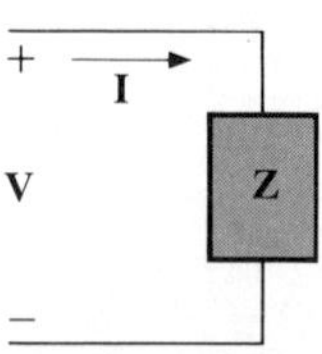

FIG. 19.4
Defining the apparent power to a load.

P_S^q

fulness in the description and analysis of sinusoidal ac networks and in the maximum rating of a number of electrical components and systems. It is called the *apparent power* and is represented symbolically by S.* Since it is simply the product of voltage and current, its units are *volt-amperes*, for which the abbreviation is VA. Its magnitude is determined by

$$S = VI \quad \text{(volt-amperes, VA)} \tag{19.6}$$

or, since $V = IZ$ and $I = \dfrac{V}{Z}$

then

$$S = I^2 Z \quad \text{(VA)} \tag{19.7}$$

and

$$S = \frac{V^2}{Z} \quad \text{(VA)} \tag{19.8}$$

The average power to the load of Fig. 19.4 is

$$P = VI \cos \theta$$

However, $S = VI$

Therefore,

$$P = S \cos \theta \quad \text{(W)} \tag{19.9}$$

and the power factor of a system F_p is

$$F_p = \cos \theta = \frac{P}{S} \quad \text{(unitless)} \tag{19.10}$$

The power factor of a circuit, therefore, is the ratio of the average power to the apparent power. For a purely resistive circuit, we have

$$P = VI = S$$

and

$$F_p = \cos \theta = \frac{P}{S} = 1$$

In general, power equipment is rated in volt-amperes (VA) or in kilovolt-amperes (kVA) and not in watts. By knowing the volt-ampere rating and the rated voltage of a device, we can readily determine the *maximum* current rating. For example, a device rated at 10 kVA at 200 V has a maximum current rating of I = 10,000 VA/200 V = 50 A when operated under rated conditions. The volt-ampere rating of a piece of equipment is equal to the wattage rating only when the F_p is 1. It is therefore a maximum power dissipation rating. This condition exists only when the total impedance of a system $Z \angle \theta$ is such that $\theta = 0°$.

The exact current demand of a device, when used under normal operating conditions, could be determined if the wattage rating and power factor were given instead of the volt-ampere rating. However, the power factor is sometimes not available, or it may vary with the load.

---

*Prior to 1968, the symbol for apparent power was the more descriptive P_a.

The reason for rating some electrical equipment in kilovolt-amperes rather than in kilowatts can be described using the configuration of Fig. 19.5. The load has an apparent power rating of 10 kVA and a current rating of 50 A at the applied voltage, 200 V. As indicated, the current demand of 70 A is above the rated value and could damage the load element, yet the reading on the wattmeter is relatively low since the load is highly reactive. In other words, the wattmeter reading is an indication of the watts dissipated and may not reflect the magnitude of the current drawn. Theoretically, if the load were purely reactive, the wattmeter reading would be zero even if the load was being damaged by a high current level.

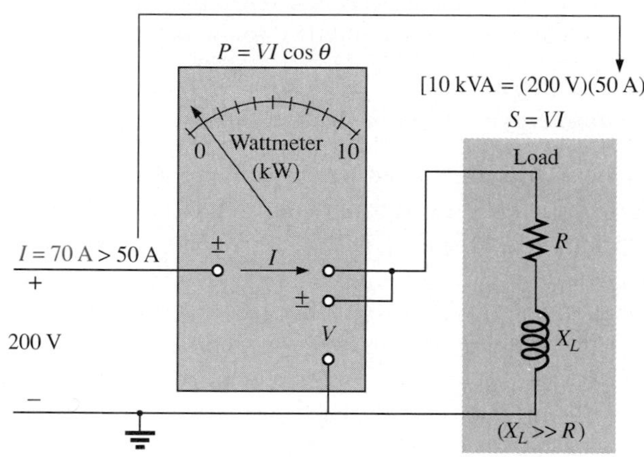

FIG. 19.5
Demonstrating the reason for rating a load in kVA rather than kW.

19.4 INDUCTIVE CIRCUIT AND REACTIVE POWER

For a purely inductive circuit (such as that in Fig. 19.6), v leads i by 90°, as shown in Fig. 19.7. Therefore, in Eq. (19.1), $\theta = 90°$. Substituting $\theta = 90°$ into Eq. (19.1) yields

$$p_L = VI\cos(90°)(1 - \cos 2\omega t) + VI\sin(90°)(\sin 2\omega t)$$
$$= 0 + VI \sin 2\omega t$$

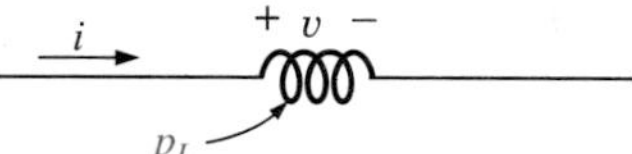

FIG. 19.6
Defining the power level for a purely inductive load.

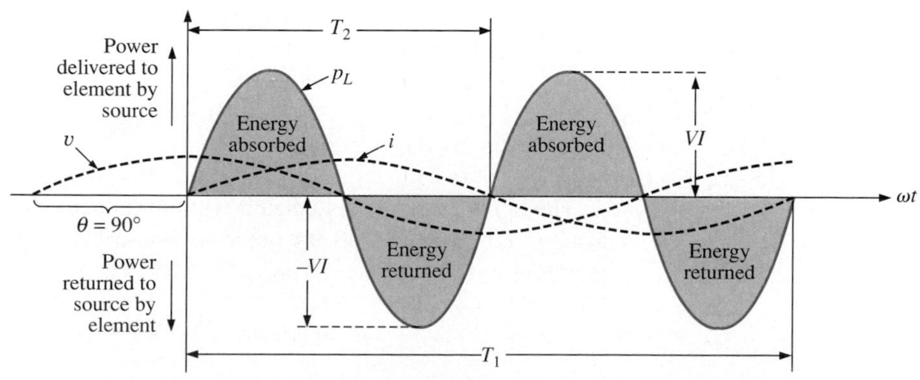

FIG. 19.7
The power curve for a purely inductive load.

or

$$p_L = VI \sin 2\omega t \qquad \textbf{(19.11)}$$

where $VI \sin 2\omega t$ is a sine wave with twice the frequency of either input quantity (v or i) and a peak value of VI. Note the absence of an average or constant term in the equation.

Plotting the waveform for p_L (Fig. 19.7), we obtain

$$T_1 = \text{period of either input quantity}$$
$$T_2 = \text{period of } p_L \text{ curve}$$

Note that over one full cycle of p_L (T_2), the area above the horizontal axis in Fig. 19.7 is exactly equal to that below the axis. This indicates that over a full cycle of p_L, the power delivered by the source to the inductor is exactly equal to that returned to the source by the inductor.

The net flow of power to the pure (ideal) inductor is zero over a full cycle, and no energy is lost in the transaction.

The power absorbed or returned by the inductor at any instant of time t_1 can be found simply by substituting t_1 into Eq. (19.11). The peak value of the curve VI is defined as the *reactive power* associated with a pure inductor.

In general, the reactive power associated with any circuit is defined to be $VI \sin \theta$, a factor appearing in the second term of Eq. (19.1). Note that it is the peak value of that term of the total power equation that produces no net transfer of energy. The symbol for reactive power is Q, and its unit of measure is the *volt-ampere reactive* (VAR).* The Q is derived from the quadrature (90°) relationship between the various powers, to be discussed in detail in a later section. Therefore,

$$Q = VI \sin \theta \qquad \text{(volt-ampere-reactive, VAR)} \qquad \textbf{(19.12)}$$

where θ is the phase angle between V and I.

For the inductor,

$$Q_L = VI \qquad \text{(VAR)} \qquad \textbf{(19.13)}$$

or, since $V = IX_L$ or $I = V/X_L$,

$$Q_L = I^2X_L \qquad \text{(VAR)} \qquad \textbf{(19.14)}$$

or

$$Q_L = \frac{V^2}{X_L} \qquad \text{(VAR)} \qquad \textbf{(19.15)}$$

The apparent power associated with an inductor is $S = VI$, and the average power is $P = 0$, as noted in Fig. 19.7. The power factor is therefore

$$F_p = \cos \theta = \frac{P}{S} = \frac{0}{VI} = 0$$

---

*Prior to 1968, the symbol for reactive power was the more descriptive P_q.

If the average power is zero, and the energy supplied is returned within one cycle, why is reactive power of any significance? The reason is not obvious but can be explained using the curve of Fig. 19.7. At every instant of time along the power curve that the curve is above the axis (positive), energy must be supplied to the inductor, even though it will be returned during the negative portion of the cycle. This power requirement during the positive portion of the cycle requires that the generating plant provide this energy during that interval. Therefore, the effect of reactive elements such as the inductor can be to raise the power requirement of the generating plant, even though the reactive power is not dissipated but simply "borrowed." The increased power demand during these intervals is a cost factor that must be passed on to the industrial consumer. In fact, most larger users of electrical energy pay for the apparent power demand rather than the watts dissipated since the volt-amperes used are sensitive to the reactive power requirement (see Section 19.6). In other words, the closer the power factor of an industrial outfit is to 1, the more efficient is the plant's operation since it is limiting its use of "borrowed" power.

The energy stored by the inductor during the positive portion of the cycle (Fig. 19.7) is equal to that returned during the negative portion and can be determined using the equation:

$$W = Pt$$

where P is the average value for the interval and t is the associated interval of time.

Recall from Chapter 14 that the average value of the positive portion of a sinusoid equals 2(peak value/π) and $t = T_2/2$. Therefore,

$$W_L = \left(\frac{2VI}{\pi}\right) \times \left(\frac{T_2}{2}\right)$$

and

$$W_L = \frac{VIT_2}{\pi} \quad \text{(J)} \qquad \textbf{(19.16)}$$

or, since $T_2 = 1/f_2$, where f_2 is the frequency of the p_L curve, we have

$$W_L = \frac{VI}{\pi f_2} \quad \text{(J)} \qquad \textbf{(19.17)}$$

Since the frequency f_2 of the power curve is twice that of the input quantity, if we substitute the frequency f_1 of the input voltage or current, Eq. (19.17) becomes

$$W_L = \frac{VI}{\pi(2f_1)} = \frac{VI}{\omega_1}$$

However,

$$V = IX_L = I\omega_1 L$$

so that

$$W_L = \frac{(I\omega_1 L)I}{\omega_1}$$

and

$$W_L = LI^2 \quad \text{(J)} \qquad \textbf{(19.18)}$$

providing an equation for the energy stored or released by the inductor in one half-cycle of the applied voltage in terms of the inductance and effective value of the current squared.

19.5 CAPACITIVE CIRCUIT

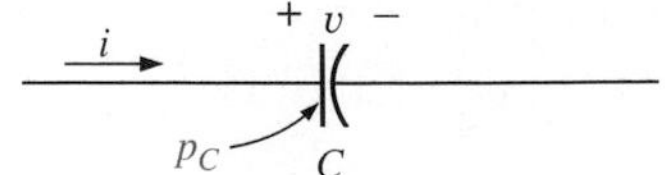

FIG. 19.8
Defining the power level for a purely capacitive load.

For a purely capacitive circuit (such as that in Fig. 19.8), i leads v by 90°, as shown in Fig. 19.9. Therefore, in Eq. (19.1), $\theta = -90°$. Substituting $\theta = -90°$ into Eq. (19.1), we obtain

$$\begin{aligned} p_C &= VI\cos(-90°)(1 - \cos 2\omega t) + VI\sin(-90°)(\sin 2\omega t) \\ &= 0 - VI\sin 2\omega t \end{aligned}$$

or

$$\boxed{p_C = -VI\sin 2\omega t} \qquad \textbf{(19.19)}$$

where $-VI\sin 2\omega t$ is a negative sine wave with twice the frequency of either input (v or i) and a peak value of VI. Again, note the absence of an average or constant term.

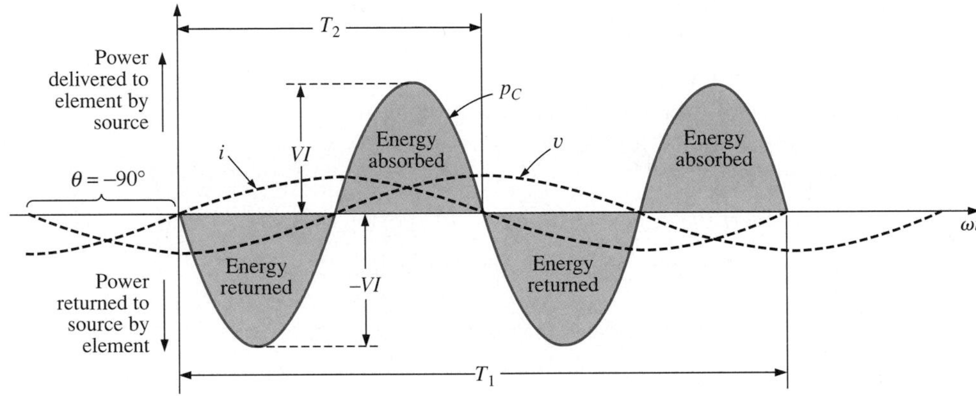

FIG. 19.9
The power curve for a purely capacitive load.

Plotting the waveform for p_C (Fig. 19.9) gives us

T_1 = period of either input quantity
T_2 = period of p_C curve

Note that the same situation exists here for the p_C curve as existed for the p_L curve. The power delivered by the source to the capacitor is exactly equal to that returned to the source by the capacitor over one full cycle.

The net flow of power to the pure (ideal) capacitor is zero over a full cycle,

and no energy is lost in the transaction. The power absorbed or returned by the capacitor at any instant of time t_1 can be found by substituting t_1 into Eq. (19.19).

The reactive power associated with the capacitor is equal to the peak value of the p_C curve, as follows:

$$\boxed{Q_C = VI} \quad \text{(VAR)} \qquad \textbf{(19.20)}$$

but, since $V = IX_C$ and $I = V/X_C$, the reactive power to the capacitor can also be written

$$\boxed{Q_C = I^2X_C} \quad \text{(VAR)} \qquad \textbf{(19.21)}$$

and
$$Q_C = \frac{V^2}{X_C} \qquad \text{(VAR)} \tag{19.22}$$

The apparent power associated with the capacitor is

$$S = VI \qquad \text{(VA)} \tag{19.23}$$

and the average power is $P = 0$, as noted from Eq. (19.19) or Fig. 19.9. The power factor is, therefore,

$$F_p = \cos\theta = \frac{P}{S} = \frac{0}{VI} = 0$$

The energy stored by the capacitor during the positive portion of the cycle (Fig. 19.9) is equal to that returned during the negative portion and can be determined using the equation $W = Pt$.

Proceeding in a manner similar to that used for the inductor, we can show that

$$W_C = \frac{VIT_2}{\pi} \qquad \text{(J)} \tag{19.24}$$

or, since $T_2 = 1/f_2$, where f_2 is the frequency of the p_C curve,

$$W_C = \frac{VI}{\pi f_2} \qquad \text{(J)} \tag{19.25}$$

In terms of the frequency f_1 of the input quantities v and i,

$$W_C = \frac{VI}{\pi(2f_1)} = \frac{VI}{\omega_1} = \frac{V(V\omega_1 C)}{\omega_1}$$

and
$$W_C = CV^2 \qquad \text{(J)} \tag{19.26}$$

providing an equation for the energy stored or released by the capacitor in one half-cycle of the applied voltage in terms of the capacitance and effective value of the voltage squared.

19.6 THE POWER TRIANGLE

The three quantities average power, apparent power, and reactive power can be related in the vector domain by

$$\mathbf{S} = \mathbf{P} + \mathbf{Q} \tag{19.27}$$

with

$$\mathbf{P} = P\angle 0^\circ \qquad \mathbf{Q}_L = Q_L\angle 90^\circ \qquad \mathbf{Q}_C = Q_C\angle -90^\circ$$

For an inductive load, the *phasor power* **S**, as it is often called, is defined by

$$\mathbf{S} = P + jQ_L$$

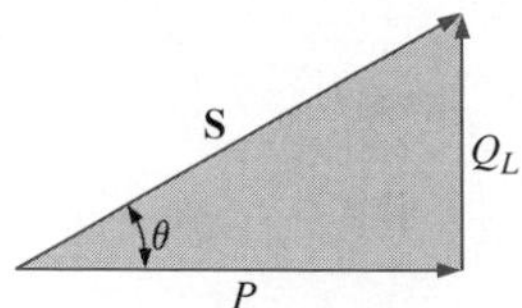

FIG. 19.10
Power diagram for inductive loads.

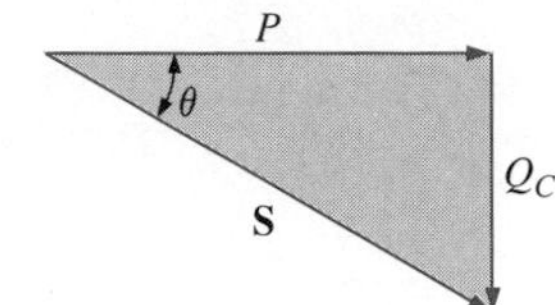

FIG. 19.11
Power diagram for capacitive loads.

as shown in Fig. 19.10.

The 90° shift in Q_L from P is the source of another term for reactive power: *quadrature power*.

For a capacitive load, the phasor power **S** is defined by

$$\mathbf{S} = P - j\,Q_C$$

as shown in Fig. 19.11.

If a network has both capacitive and inductive elements, the reactive component of the power triangle will be determined by the *difference* between the reactive power delivered to each. If $Q_L > Q_C$, the resultant power triangle will be similar to Fig. 19.10. If $Q_C > Q_L$, the resultant power triangle will be similar to Fig. 19.11.

That the total reactive power is the difference between the reactive powers of the inductive and capacitive elements can be demonstrated by considering Eqs. (19.11) and (19.19). Using these equations, the reactive power delivered to each reactive element has been plotted for a series *L-C* circuit on the same set of axes in Fig. 19.12. The reactive elements were chosen such that $X_L > X_C$. Note that the power curve for each is exactly 180° out of phase. The curve for the resultant reactive power is therefore determined by the algebraic resultant of the two at each instant of time. Since the reactive power is defined as the peak value, the reactive component of the power triangle is as indicated in the figure: $I^2(X_L - X_C)$.

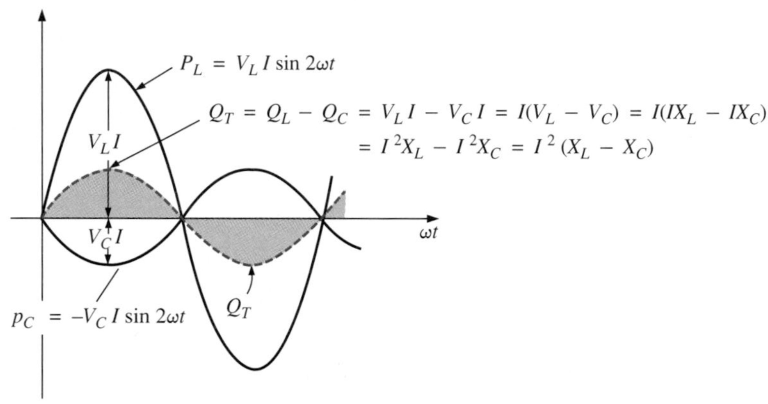

FIG. 19.12
Demonstrating why the net reactive power is the difference between that delivered to inductive and capacitive elements.

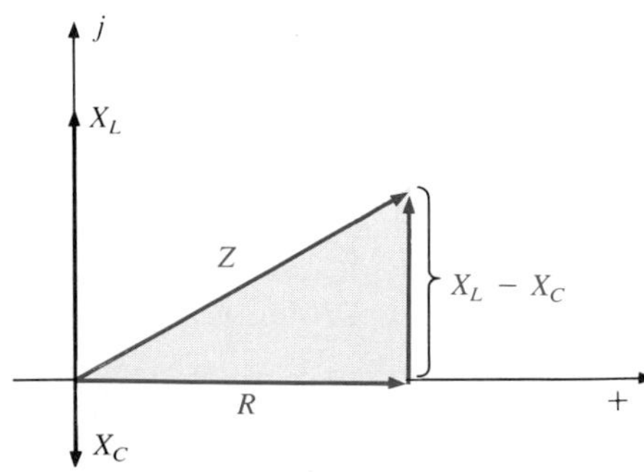

FIG. 19.13
Impedance diagram for a series R-L-C circuit.

An additional verification can be derived by first considering the impedance diagram of a series *R-L-C* circuit (Fig. 19.13). If we multiply each radius vector by the current squared (I^2), we obtain the results shown in Fig. 19.14, which is the power triangle for a predominantly inductive circuit.

Since the reactive power and average power are always angled 90° to each other, the three powers are related by the Pythagorean theorem; that is,

$$S^2 = P^2 + Q^2 \tag{19.28}$$

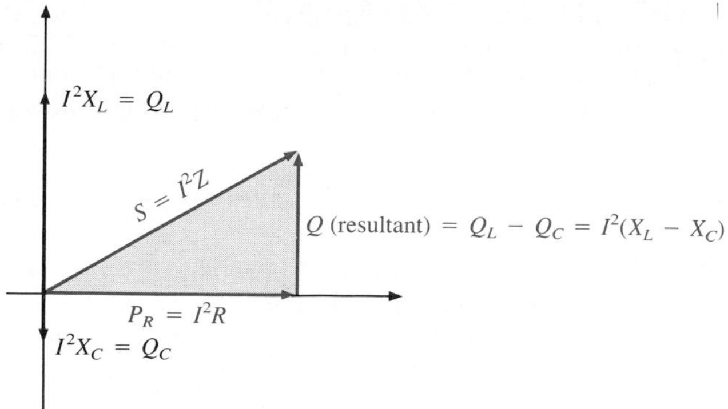

FIG. 19.14
The result of multiplying each vector of Fig. 19.13 by I^2 for a series R-L-C circuit.

Therefore, the third power can always be found if the other two are known.

It is particularly interesting that the equation

$$\mathbf{S} = \mathbf{VI}^* \tag{19.29}$$

will provide the vector form of the apparent power of a system. Here, **V** is the voltage across the system and $\mathbf{I}^*$ is the complex conjugate of the current.

Consider, for example, the simple *R-L* circuit of Fig. 19.15, where

$$\mathbf{I} = \frac{\mathbf{V}}{\mathbf{Z}_T} = \frac{10\text{ V}\angle 0°}{3\ \Omega + j\,4\ \Omega} = \frac{10\text{ V}\angle 0°}{5\ \Omega\angle 53.13°} = 2\text{ A}\angle -53.13°$$

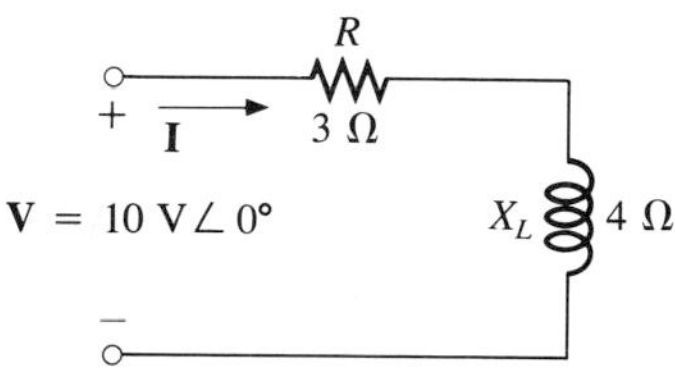

FIG. 19.15
Demonstrating the validity of Eq. (19.29).

The real power (the term *real* being derived from the positive real axis of the complex plane) is

$$P = I^2R = (2\text{ A})^2(3\ \Omega) = 12\text{ W}$$

and the reactive power is

$$Q_L = I^2X_L = (2\text{ A})^2(4\ \Omega) = 16\text{ VAR}$$

with $\quad \mathbf{S} = P + j\,Q_L = 12\text{ W} + j\,16\text{ VAR} = 20\text{ VA}\angle 53.13°$

as shown in Fig. 19.16. Applying Eq. (19.29) yields

$$\mathbf{S} = \mathbf{VI}^* = (10\text{ V}\angle 0°)(2\text{ A}\angle +53.13°) = 20\text{ VA}\angle 53.13°$$

as obtained above.

The angle θ associated with **S** and appearing in Figs. 19.10, 19.11, and 19.16 is the power-factor angle of the network. Since

$$P = VI\cos\theta$$

or

$$P = S\cos\theta$$

then

$$F_p = \cos\theta = \frac{P}{S} \tag{19.30}$$

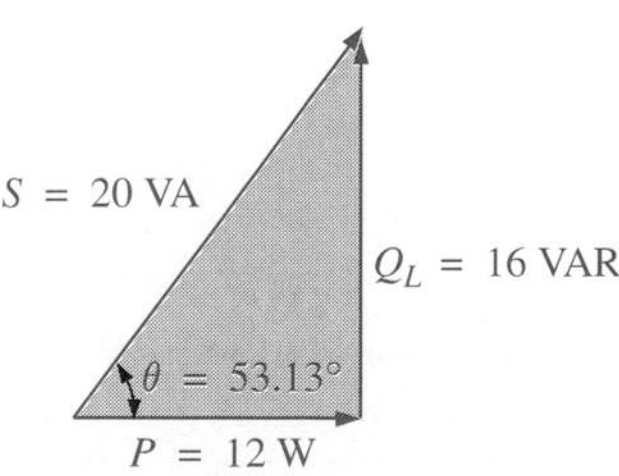

FIG. 19.16
The power triangle for the circuit of Fig. 19.15.

19.7 THE TOTAL *P, Q,* AND *S*

The total number of watts, volt-amperes reactive, and volt-amperes, and the power factor of any system can be found using the following procedure:

1. ***Find the real power and reactive power for each branch of the circuit.***
2. ***The total real power of the system (P_T) is then the sum of the average power delivered to each branch.***
3. ***The total reactive power (Q_T) is the difference between the reactive power of the inductive loads and that of the capacitive loads.***
4. ***The total apparent power is $S_T = \sqrt{P_T^2 + Q_T^2}$.***
5. ***The total power factor is P_T/S_T.***

There are two important points in the above tabulation. First, the total apparent power must be determined from the total average and reactive powers and *cannot* be determined from the apparent powers of each branch. Second, and more important, it is *not necessary* to consider the series-parallel arrangement of branches. In other words, the total real, reactive, or apparent power is independent of whether the loads are in series, parallel, or series-parallel. The following examples will demonstrate the relative ease with which all the quantities of interest can be found.

EXAMPLE 19.1 Find the total number of watts, volt-amperes reactive, and volt-amperes, and the power factor F_p of the network in Fig. 19.17. Draw the power triangle and find the current in phasor form.

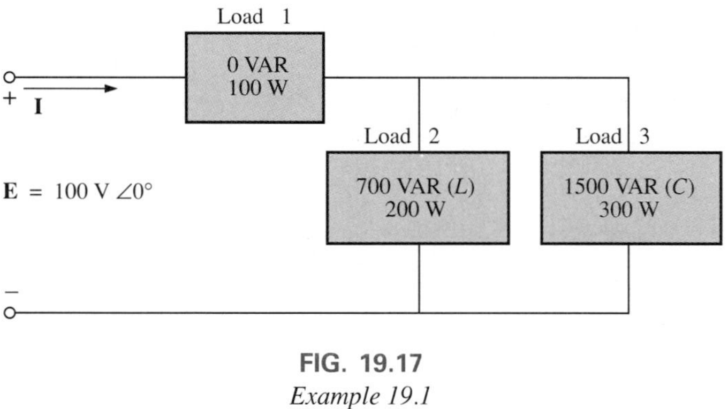

FIG. 19.17
Example 19.1

Solution: Use a table:

Load	W	VAR	VA
1	100	0	100
2	200	700 (L)	$\sqrt{(200)^2 + (700)^2} = 728.0$
3	300	1500 (C)	$\sqrt{(300)^2 + (1500)^2} = 1529.71$
	$P_T = \mathbf{600}$ Total power dissipated	$Q_T = \mathbf{800\ (C)}$ Resultant reactive power of network	$S_T = \sqrt{(600)^2 + (800)^2} = \mathbf{1000}$ (Note that $S_T \neq$ sum of each branch: $1000 \neq 100 + 728 + 1529.71$)

$$F_p = \frac{P_T}{S_T} = \frac{600\text{ W}}{1000\text{ VA}} = \mathbf{0.6\ leading\ (C)}$$

The power triangle is shown in Fig. 19.18.

Since $S_T = VI = 1000$ VA, $I = 1000$ VA/100 V = 10 A; and since θ of $\cos\theta = F_p$ is the angle between the input voltage and current,

$$\mathbf{I} = \mathbf{10\ A\ \angle{+53.13°}}$$

The plus sign is associated with the phase angle since the circuit is predominantly capacitive.

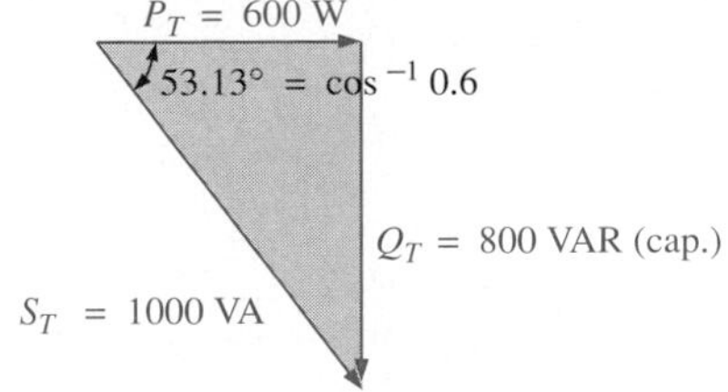

FIG. 19.18
Power triangle for Example 19.1.

EXAMPLE 19.2

a. Find the total number of watts, volt-amperes reactive, and volt-amperes, and the power factor F_p for the network of Fig. 19.19.

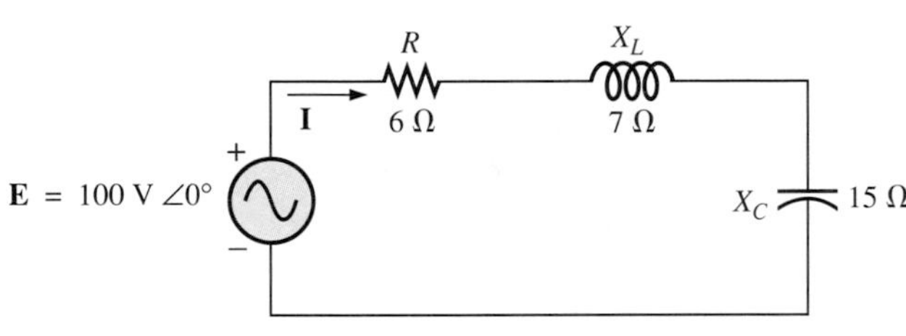

FIG. 19.19
Example 19.2.

b. Sketch the power triangle.
c. Find the energy dissipated by the resistor over one full cycle of the input voltage if the frequency of the input quantities is 60 Hz.
d. Find the energy stored in, or returned by, the capacitor or inductor over one half-cycle of the power curve for each if the frequency of the input quantities is 60 Hz.

Solutions:

a. $$\mathbf{I} = \frac{\mathbf{E}}{\mathbf{Z}_T} = \frac{100\text{ V}\angle 0°}{6\ \Omega + j\,7\ \Omega - j\,15\ \Omega} = \frac{100\text{ V}\angle 0°}{10\ \Omega\angle -53.13°}$$
$$= 10\text{ A}\angle 53.13°$$
$$\mathbf{V}_R = (10\text{ A}\angle 53.13°)(6\ \Omega\angle 0°) = 60\text{ V}\angle 53.13°$$
$$\mathbf{V}_L = (10\text{ A}\angle 53.13°)(7\ \Omega\angle 90°) = 70\text{ V}\angle 143.13°$$
$$\mathbf{V}_C = (10\text{ A}\angle 53.13°)(15\ \Omega\angle -90°) = 150\text{ V}\angle -36.87°$$
$$P_T = EI\cos\theta = (100\text{ V})(10\text{ A})\cos 53.13° = \mathbf{600\ W}$$
$$= I^2R = (10\text{ A})^2(6\ \Omega) = \mathbf{600\ W}$$
$$= \frac{V_R^2}{R} = \frac{(60\text{ V})^2}{6} = \mathbf{600\ W}$$
$$S_T = EI = (100\text{ V})(10\text{ A}) = \mathbf{1000\ VA}$$
$$= I^2Z_T = (10\text{ A})^2(10\ \Omega) = \mathbf{1000\ VA}$$
$$= \frac{E^2}{Z_T} = \frac{(100\text{ V})^2}{10\ \Omega} = \mathbf{1000\ VA}$$
$$Q_T = EI\sin\theta = (100\text{ V})(10\text{ A})\sin 53.13° = \mathbf{800\ VAR}$$
$$= Q_C - Q_L$$
$$= I^2(X_C - X_L) = (10\text{ A})^2(15\ \Omega - 7\ \Omega) = \mathbf{800\ VAR}$$

$$Q_T = \frac{V_C^2}{X_C} - \frac{V_L^2}{X_L} = \frac{(150\text{ V})^2}{15\ \Omega} - \frac{(70\text{ V})^2}{7\ \Omega}$$
$$= 1500\text{ VAR} - 700\text{ VAR} = \mathbf{800\ VAR}$$
$$F_p = \frac{P_T}{S_T} = \frac{600\text{ W}}{1000\text{ VA}} = \mathbf{0.6\ leading\ (cap.)}$$

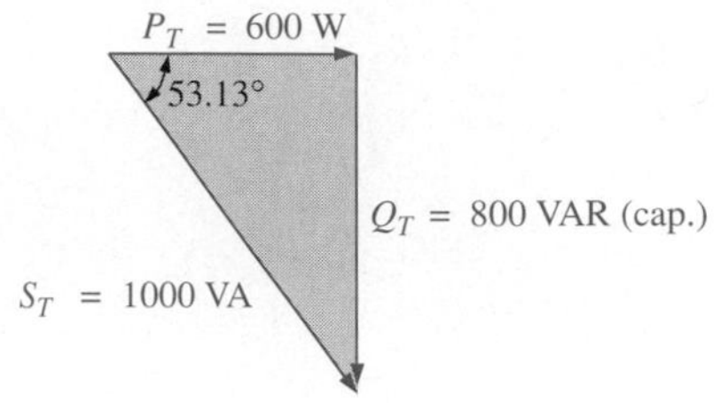

FIG. 19.20
Power triangle for Example 19.2.

b. The power triangle is as shown in Fig. 19.20.

c. $W_R = \dfrac{V_R I}{f_1} = \dfrac{(60\text{ V})(10\text{ A})}{60\text{ Hz}} = \mathbf{10\ J}$

d. $W_L = \dfrac{V_L I}{\omega_1} = \dfrac{(70\text{ V})(10\text{ A})}{(2\pi)(60\text{ Hz})} = \dfrac{700\text{ J}}{377} = \mathbf{1.86\ J}$

$$W_C = \frac{V_C I}{\omega_1} = \frac{(150\text{ V})(10\text{ A})}{377\text{ rad/s}} = \frac{1500\text{ J}}{377} = \mathbf{3.98\ J}$$

EXAMPLE 19.3 For the system of Fig. 19.21,

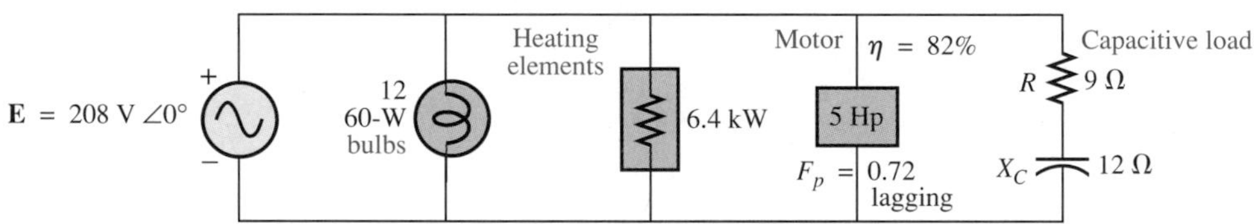

FIG. 19.21
Example 19.3.

a. Find the average power, apparent power, reactive power, and F_p for each branch.
b. Find the total number of watts, volt-amperes reactive, and volt-amperes, and the power factor of the system. Sketch the power triangle.
c. Find the source current I.

Solutions:

a. *Bulbs:*

Total dissipation of applied power

$$P_1 = 12(60\text{ W}) = \mathbf{720\ W}$$
$$Q_1 = \mathbf{0\ VAR}$$
$$S_1 = P_1 = \mathbf{720\ VA}$$
$$F_{p_1} = \mathbf{1}$$

Heating elements:

Total dissipation of applied power

$$P_2 = \mathbf{6.4\ kW}$$
$$Q_2 = \mathbf{0\ VAR}$$
$$S_2 = P_2 = \mathbf{6.4\ kVA}$$
$$F_{p_2} = \mathbf{1}$$

Motor:

$$\eta = \frac{P_o}{P_i} \longrightarrow P_i = \frac{P_o}{\eta} = \frac{5(746\text{ W})}{0.82} = \mathbf{4548.78\text{ W}} = P_3$$

$$F_p = \mathbf{0.72\text{ lagging}}$$

$$P_3 = S_3\cos\theta \longrightarrow S_3 = \frac{P_3}{\cos\theta} = \frac{4548.78\text{ W}}{0.72} = \mathbf{6317.75\text{ VA}}$$

Also, $\theta = \cos^{-1} 0.72 = 43.95°$, so that

$$Q_3 = S_3\sin\theta = (6317.75\text{ VA})(\sin 43.95°)$$
$$= (6317.75\text{ VA})(0.694) = \mathbf{4384.71\text{ VAR}}$$

Capacitive load:

$$\mathbf{I} = \frac{\mathbf{E}}{\mathbf{Z}} = \frac{208\text{ V}\angle 0°}{9\ \Omega - j\,12\ \Omega} = \frac{208\text{ V}\angle 0°}{15\ \Omega\angle -53.13°} = 13.87\text{ A}\angle 53.13°$$

$$P_4 = I^2R = (13.87\text{ A})^2\cdot 9\ \Omega = \mathbf{1731.39\text{ W}}$$

$$Q_4 = I^2X_C = (13.87\text{ A})^2\cdot 12\ \Omega = \mathbf{2308.52\text{ VAR}}$$

$$S_4 = \sqrt{P_4^2 + Q_4^2} = \sqrt{(1731.39\text{ W})^2 + (2308.52\text{ VAR})^2}$$
$$= \mathbf{2885.65\text{ VA}}$$

$$F_p = \frac{P_4}{S_4} = \frac{1731.39\text{ W}}{2885.65\text{ VA}} = \mathbf{0.6\text{ leading}}$$

b. $P_T = P_1 + P_2 + P_3 + P_4$

$$= 720\text{ W} + 6400\text{ W} + 4548.78\text{ W} + 1731.39\text{ W}$$
$$= \mathbf{13{,}400.17\text{ W}}$$

$$Q_T = \pm Q_1 \pm Q_2 \pm Q_3 \pm Q_4$$
$$= 0 + 0 + 4384.71\text{ VAR} - 2308.52\text{ VAR}$$
$$= \mathbf{2076.19\text{ VAR (ind.)}}$$

$$S_T = \sqrt{P_T^2 + Q_T^2} = \sqrt{(13{,}400.17\text{ W})^2 + (2076.19\text{ VAR})^2}$$
$$= 13{,}560.06\text{ VA}$$

$$F_p = \frac{P_T}{S_T} = \frac{13.4\text{ kW}}{13{,}560.56\text{ VA}} = \mathbf{0.988\text{ lagging}}$$

$$\theta = \cos^{-1} 0.988 = 8.89°$$

Note Fig. 19.22.

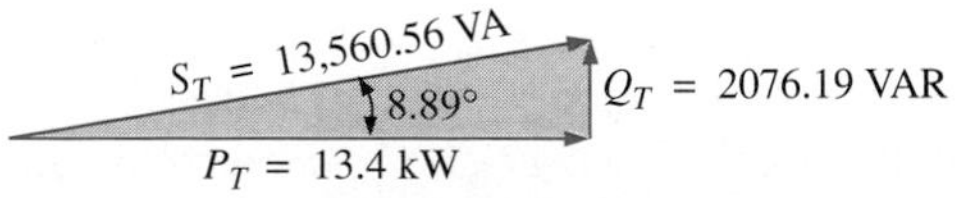

S_T = 13,560.56 VA
8.89°
Q_T = 2076.19 VAR
P_T = 13.4 kW

FIG. 19.22
Power triangle for Example 19.3.

c. $S_T = EI \longrightarrow I = \dfrac{S_T}{E} = \dfrac{13{,}559.89\text{ VA}}{208\text{ V}} = 65.19\text{ A}$

Lagging power factor: **E** leads **I** by 8.89° and

$$\mathbf{I = 65.19\text{ A}\angle -8.89°}$$

EXAMPLE 19.4 An electrical device is rated 5 kVA, 100 V at a 0.6 power-factor lag. What is the impedance of the device in rectangular coordinates?

Solution:

$$S = EI = 5000 \text{ VA}$$

Therefore,

$$I = \frac{5000 \text{ VA}}{100 \text{ V}} = 50 \text{ A}$$

For $F_p = 0.6$, we have

$$\theta = \cos^{-1} 0.6 = 53.13°$$

Since the power factor is lagging, the circuit is predominantly inductive, and **I** lags **E**. Or, for $\mathbf{E} = 100 \text{ V} \angle 0°$,

$$\mathbf{I} = 50 \text{ A} \angle -53.13°$$

However,

$$\mathbf{Z}_T = \frac{\mathbf{E}}{\mathbf{I}} = \frac{100 \text{ V} \angle 0°}{50 \text{ A} \angle -53.13°} = 2 \,\Omega \angle 53.13° = \mathbf{1.2 \,\Omega \; + j\,1.6 \,\Omega}$$

which is the impedance of the circuit of Fig. 19.23.

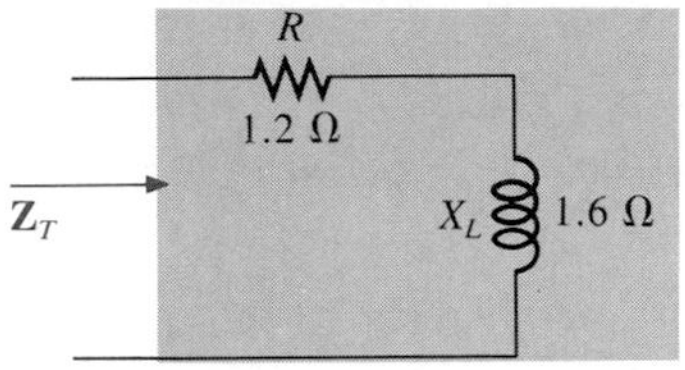

FIG. 19.23
Example 19.4.

19.8 POWER-FACTOR CORRECTION

The design of any power transmission system is very sensitive to the magnitude of the current in the lines as determined by the applied loads. Increased currents result in increased power losses (by a squared factor since $P = I^2R$) in the transmission lines due to the resistance of the lines. Heavier currents also require larger conductors, increasing the amount of copper needed for the system, and, quite obviously, they require increased generating capacities by the utility company.

Every effort must therefore be made to keep current levels at a minimum. Since the line voltage of a transmission system is fixed, the apparent power is directly related to the current level. In turn, the smaller the net apparent power, the smaller the current drawn from the supply. Minimum current is therefore drawn from a supply when $S = P$ and $Q_T = 0$. Note the effect of decreasing levels of Q_T on the length (and magnitude) of S in Fig. 19.24 for the same real power. Note also that the power-factor angle approaches zero degrees and F_p approaches 1, revealing that the network is appearing more and more resistive at the input terminals.

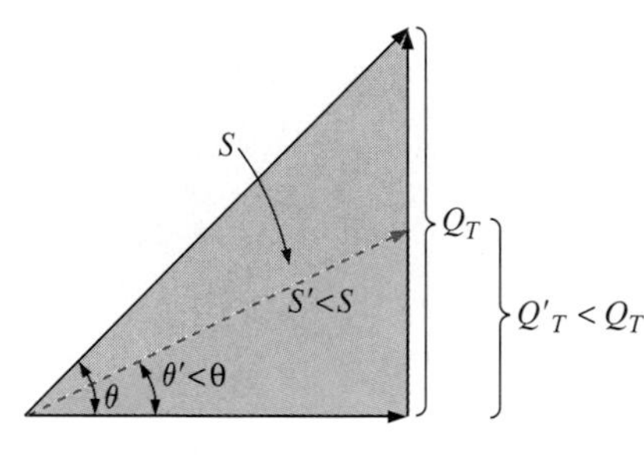

FIG. 19.24
Demonstrating the impact of power-factor correction on the power triangle of a network.

The process of introducing reactive elements to bring the power factor closer to unity is called *power-factor correction*. Since most loads are inductive, the process normally involves introducing elements with capacitive terminal characteristics having the sole purpose of improving the power factor.

In Fig. 19.25(a), for instance, an inductive load is drawing a current I_L that has a real and an imaginary component. In Fig. 19.25(b) a capacitive load was added in parallel with the original load to raise the power factor of the total system to the unity power-factor level. Note that by placing all the elements in parallel, the load still receives the same terminal voltage and draws the same current I_L. In other words, the load is unaware of and unconcerned about whether it is hooked up as shown in Fig. 19.25(a) or Fig. 19.25(b).

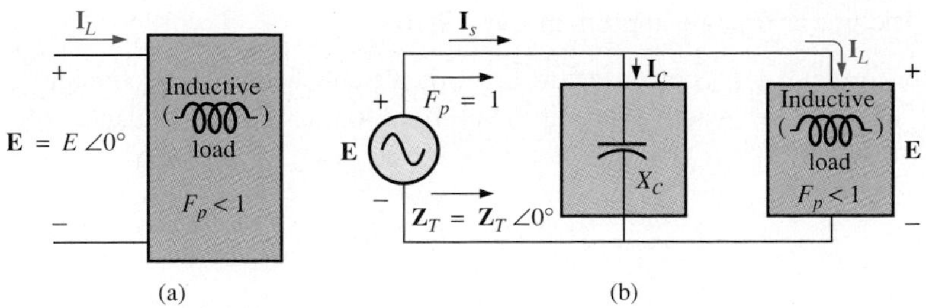

FIG. 19.25
Demonstrating the impact of a capacitive element on the power factor of a network.

Solving for the source current in Fig. 19.25(b):

$$
\begin{aligned}
\mathbf{I}_s &= \mathbf{I}_C + \mathbf{I}_L \\
&= -jI_C + (I_L + jI'_L) \\
&= I_L + j(I'_L - I_C)
\end{aligned}
$$

If X_C is chosen such that $I_C = I'_L$,

$$\mathbf{I}_s = I_L + j(0) = I_L \angle 0°$$

The result is a source current whose magnitude is simply equal to the real part of the load current, which can be considerably less than the magnitude of the load current of Fig. 19.25(a). In addition, since the phase angle associated with both the applied voltage and the source current is the same, the system appears "resistive" at the input terminals and all the power supplied is absorbed, creating maximum efficiency for a generating utility.

EXAMPLE 19.5 A 5-hp motor with a 0.6 lagging power factor and an efficiency of 92% is connected to a 208-V, 60-Hz supply.

a. Establish the power triangle for the load.
b. Determine the power factor capacitor that must be placed in parallel with the load to raise the power factor to unity.
c. Determine the change in supply current from the uncompensated to the compensated system.
d. Find the network equivalent of the above and verify the conclusions.

Solutions:

a. Since 1 hp = 746 W,

$$P_o = 5 \text{ hp} = 5(746 \text{ W}) = 3730 \text{ W}$$

and $\quad P_i \text{ (drawn from the line)} = \dfrac{P_o}{\eta} = \dfrac{3730 \text{ W}}{0.92} = 4054.35 \text{ W}$

Also, $\quad F_P = \cos\theta = 0.6$

and $\quad \theta = \cos^{-1} 0.6 = 53.13°$

Applying $\quad \tan\theta = \dfrac{Q_L}{P_i}$

we obtain $\quad Q_L = P_i \tan\theta = (4054.35 \text{ W}) \tan 53.13° = 5405.8 \text{ VAR}$

and

$$
\begin{aligned}
S &= \sqrt{P_i^2 + Q_L^2} = \sqrt{(4054.35 \text{ W})^2 + (5405.8 \text{ VAR})^2} \\
&= 6757.25 \text{ VA}
\end{aligned}
$$

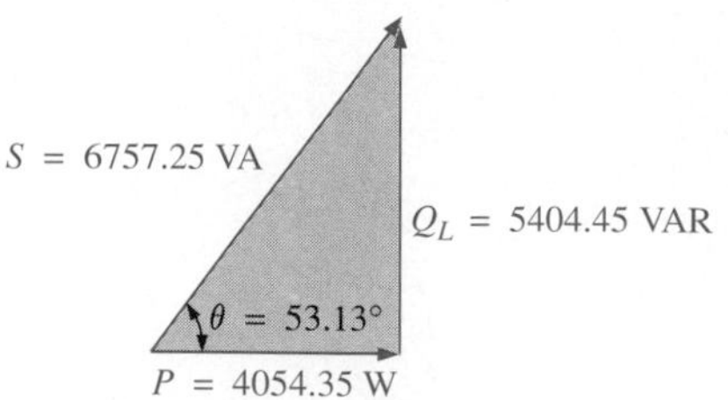

FIG. 19.26
Initial power triangle for the load of Example 19.5.

The power triangle appears in Fig. 19.26.

b. A net unity power-factor level is established by introducing a capacitive reactive power level of 5405.8 VAR to balance Q_L. Since

$$Q_C = \frac{V^2}{X_C}$$

then
$$X_C = \frac{V^2}{Q_C} = \frac{(208 \text{ V})^2}{5405.8 \text{ VAR}} = 8 \ \Omega$$

and
$$C = \frac{1}{2\pi f X_C} = \frac{1}{(2\pi)(60 \text{ Hz})(8 \ \Omega)} = \mathbf{331.6 \ \mu F}$$

c. **At 0.6F_p,**

$$S = VI = 6757.25 \text{ VA}$$

and
$$I = \frac{S}{V} = \frac{6757.25 \text{ VA}}{208 \text{ V}} = \mathbf{32.49 \ A}$$

At unity F_p,

$$S = VI = 4054.35 \text{ VA}$$

and
$$I = \frac{S}{V} = \frac{4054.35 \text{ VA}}{208 \text{ V}} = \mathbf{19.49 \ A}$$

producing a 40% reduction in supply current.

d. For the motor, the angle by which the applied voltage leads the current is

$$\theta = \cos^{-1} 0.6 = 53.13°$$

and $P = EI_m \cos\theta = 4054.35$ W, from above, so that

$$I_m = \frac{P}{E\cos\theta} = \frac{4054.35 \text{ W}}{(208 \text{ V})(0.6)} = \mathbf{32.49 \ A} \qquad \text{(as above)}$$

resulting in

$$\mathbf{I}_m = 32.49 \text{ A} \angle -53.13°$$

Therefore,

$$\mathbf{Z}_m = \frac{\mathbf{E}}{\mathbf{I}_m} = \frac{208 \text{ V} \angle 0°}{32.49 \text{ A} \angle -53.13°} = 6.4 \ \Omega \angle 53.13°$$
$$= 3.84 \ \Omega + j\,5.12 \ \Omega$$

as shown in Fig. 19.27(a).

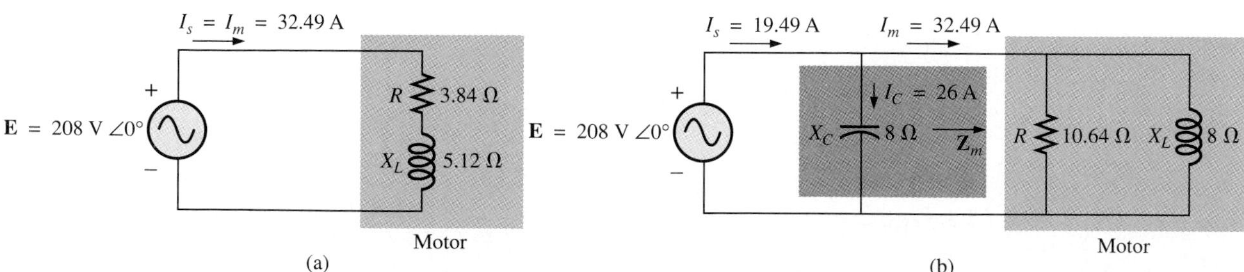

FIG. 19.27
Demonstrating the impact of power-factor corrections on the source current.

The equivalent parallel load is determined from

$$\mathbf{Y} = \frac{1}{\mathbf{Z}} = \frac{1}{6.4\ \Omega\ \angle 53.13^\circ}$$
$$= 0.156\ \text{S}\ \angle -53.13^\circ = 0.094\ \text{S} - j\,0.125\ \text{S}$$
$$= \frac{1}{10.64\ \Omega} + \frac{1}{j\,8\ \Omega}$$

as shown in Fig. 19.27(b).

It is now clear that the effect of the 8-Ω inductive reactance can be compensated for by a parallel capacitive reactance of 8 Ω using a power factor correction capacitor of 332 μF.

Since

$$\mathbf{Y}_T = \frac{1}{-jX_C} + \frac{1}{R} + \frac{1}{+jX_L} = \frac{1}{R}$$

$$I_s = EY_T = E\left(\frac{1}{R}\right) = (208\ \text{V})\left(\frac{1}{10.64\ \Omega}\right) = \mathbf{19.54\ A} \qquad \text{as above}$$

In addition, the magnitude of the capacitive current can be determined as follows:

$$I_C = \frac{E}{X_C} = \frac{208\ \text{V}}{8\ \Omega} = \mathbf{26\ A}$$

EXAMPLE 19.6 A small industrial plant has a 10-kW heating load and a 20-kVA inductive load due to a bank of induction motors. The heating elements are considered purely resistive ($F_p = 1$) and the induction motors have a lagging power factor of 0.7. If the supply is 1000 V at 60 Hz, determine the capacitive element required to raise the power factor to 0.95.

Solution: For the induction motors,

$$S = VI = 20\ \text{kVA}$$
$$P = VI\cos\theta = (20 \times 10^3\ \text{VA})(0.7) = 14 \times 10^3\ \text{W}$$
$$\theta = \cos^{-1} 0.7 \cong 45.6^\circ$$

and

$$Q_L = VI\sin\theta = (20 \times 10^3\ \text{VA})(0.714) = 14.28 \times 10^3\ \text{VAR}$$

The power triangle for the total system appears in Fig. 19.28. Note the addition of real powers and the resulting S_T:

$$S_T = \sqrt{(24\ \text{kW})^2 + (14.28\ \text{kVAR})^2} = 27.93\ \text{kVA}$$

with
$$I = \frac{S_T}{V} = \frac{27.93\ \text{kVA}}{1000\ \text{V}} = 27.93\ \text{A}$$

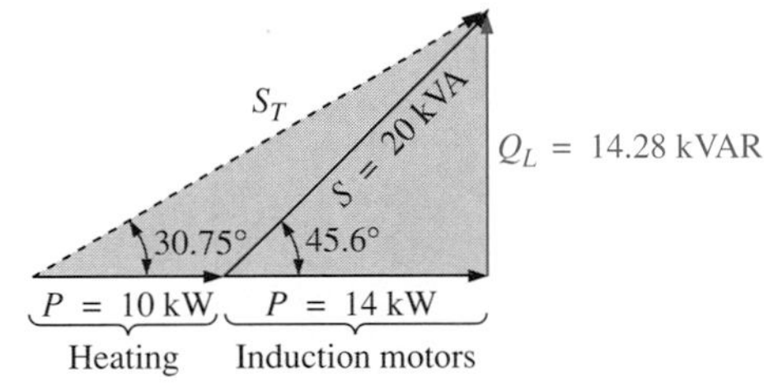

FIG. 19.28
Initial power triangle for the load of Example 19.6.

The desired power factor of 0.95 results in an angle between S and P of

$$\theta = \cos^{-1} 0.95 = 18.19^\circ$$

changing the power triangle to that of Fig. 19.29:

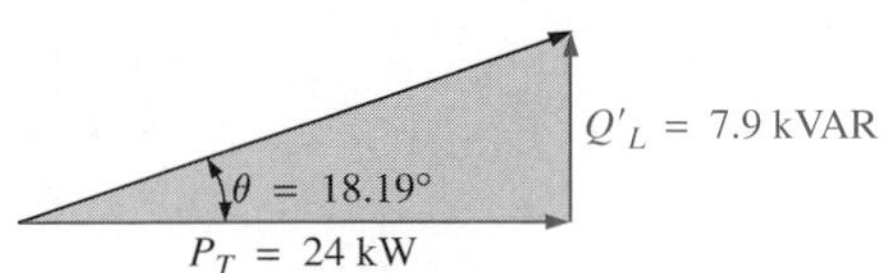

FIG. 19.29
Power triangle for the load of Example 19.6 after raising the power factor to 0.95.

$$\text{with } \tan\theta = \frac{Q'_L}{P_T} \longrightarrow Q'_L = P_T\tan\theta = (24 \times 10^3\ \text{W})(\tan 18.19^\circ)$$
$$= (24 \times 10^3\ \text{W})(0.329) = 7.9\ \text{kVAR}$$

The inductive reactive power must therefore be reduced by

$$Q_L - Q'_L = 14.28\text{ kVAR} - 7.9\text{ kVAR} = 6.38\text{ kVAR}$$

Therefore, $Q_C = 6.38$ kVAR, and using

$$Q_C = \frac{V^2}{X_C}$$

we obtain

$$X_C = \frac{V^2}{Q_C} = \frac{(10^3\text{ V})^2}{6.38 \times 10^3\text{ VAR}} = 156.74\ \Omega$$

and $$C = \frac{1}{2\pi f X_C} = \frac{1}{(2\pi)(60\text{ Hz})(156.74\ \Omega)} = \mathbf{16.93\ \mu F}$$

19.9 WATTMETERS AND POWER-FACTOR METERS

The electrodynamometer wattmeter was introduced in Section 4.4 along with its movement and terminal connections. The same meter can be used to measure the power in a dc or ac network using the same connection strategy; in fact, it can be used to measure the wattage of any network with a periodic or nonperiodic input.

FIG. 19.30
Digital wattmeter. (Courtesy of Yokogawa Corporation of America)

The digital display wattmeter of Fig. 19.30 employs a sophisticated electronic package to sense the voltage and current levels and, through the use of an analog-to-digital conversion unit, display the proper digits on the display. It is capable of providing a digital readout for distorted nonsinusoidal waveforms and can provide the phase power, total power, apparent power, reactive power, and power factor.

When using a wattmeter, the operator must take care not to exceed the current, voltage, or wattage rating. The product of the voltage and current ratings may or may not equal the wattage rating. In the high-power-factor wattmeter, the product of the voltage and current ratings is usually equal to the wattage rating, or at least 80% of it. For a low-power-factor wattmeter, the product of the current and voltage ratings is much greater than the wattage rating. For obvious reasons, the low-power-factor meter is used only in circuits with low power factors (total impedance highly reactive). Typical ratings for high-power-factor (HPF) and low-power-factor (LPF) meters are shown in Table 19.1. Meters of both high and low power factors have an accuracy of 0.5% to 1% of full scale.

TABLE 19.1

Meter	Current Ratings	Voltage Ratings	Wattage Ratings
HPF	2.5 A	150 V	1500/750/375
	5.0 A	300 V	
LPF	2.5 A	150 V	300/150/75
	5.0 A	300 V	

As the name implies, power-factor meters are designed to read the power factor of a load under operating conditions. Most are designed to be used on single- or three-phase systems. Both the voltage and current are typically measured using nonintrusive methods; that is, connections are made directly to the terminals for the voltage measurements, whereas clamp-on current transformers are used to sense the current level, as shown for the power-factor meter of Fig. 19.31.

Once the power factor is known, most power-factor meters come with a set of tables that will help define the power-factor capacitor that should be used to improve the power factor. Power-factor capacitors are typically rated in kVARs, with typical ratings extending from 1 to 25 kVARs at 240 V and 1 to 50 kVARs at 480 V or 600 V.

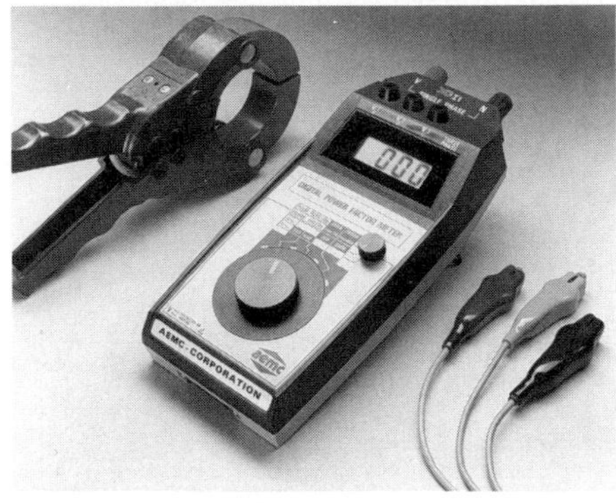

FIG. 19.31
Clamp-on power-factor meter. (Courtesy of the AEMC Corporation.)

19.10 EFFECTIVE RESISTANCE

The resistance of a conductor as determined by the equation $R = \rho(l/A)$ is often called the *dc*, *ohmic*, or *geometric* resistance. It is a constant quantity determined only by the material used and its physical dimensions. In ac circuits, the actual resistance of a conductor (called the *effective* resistance) differs from the dc resistance because of the varying currents and voltages that introduce effects not present in dc circuits.

These effects include radiation losses, skin effect, eddy currents, and hysteresis losses. The first two effects apply to any network, while the latter two are concerned with the additional losses introduced by the presence of ferromagnetic materials in a changing magnetic field.

Experimental Procedure

The effective resistance of an ac circuit cannot be measured by the ratio V/I since this ratio is now the impedance of a circuit that may have both resistance and reactance. The effective resistance can be found, however, by using the power equation $P = I^2R$, where

$$R_{eff} = \frac{P}{I^2} \tag{19.31}$$

A wattmeter and an ammeter are therefore necessary for measuring the effective resistance of an ac circuit.

Radiation Losses

Let us now examine the various losses in greater detail. The radiation loss is the loss of energy in the form of electromagnetic waves during the transfer of energy from one element to another. This loss in energy requires that the input power be larger to establish the same current I, causing R to increase as determined by Eq. (19.31). At a frequency of 60 Hz, the effects of radiation losses can be completely ignored. However, at radio frequencies, this is an important effect and may in fact become the main effect in an electromagnetic device such as an antenna.

Skin Effect

The explanation of skin effect requires the use of some basic concepts previously described. Remember from Chapter 11 that a magnetic field

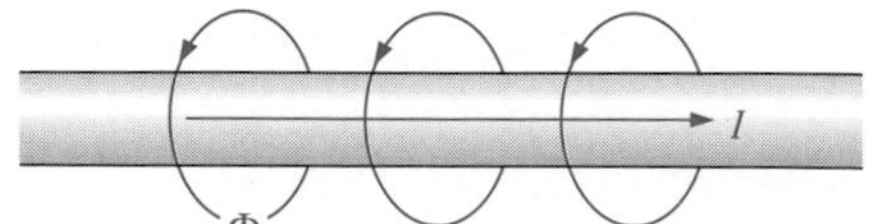

FIG. 19.32
Demonstrating the skin effect on the effective resistance of a conductor.

exists around every current-carrying conductor (Fig. 19.32). Since the amount of charge flowing in ac circuits changes with time, the magnetic field surrounding the moving charge (current) also changes. Recall also that a wire placed in a changing magnetic field will have an induced voltage across its terminals as determined by Faraday's law $e = N \times (d\phi/dt)$. The higher the frequency of the changing flux as determined by an alternating current, the greater the induced voltage will be.

For a conductor carrying alternating current, the changing magnetic field surrounding the wire links the wire itself, thus developing within the wire an induced voltage that opposes the original flow of charge or current. These effects are more pronounced at the center of the conductor than at the surface because the center is linked by the changing flux inside the wire as well as that outside the wire. As the frequency of the applied signal increases, the flux linking the wire will change at a greater rate. An increase in frequency will therefore increase the counter-induced voltage at the center of the wire to the point where the current will, for all practical purposes, flow on the surface of the conductor. At 60 Hz, the skin effect is almost noticeable. However, at radio frequencies the skin effect is so pronounced that conductors are frequently made hollow because the center part is relatively ineffective. The skin effect, therefore, reduces the effective area through which the current can flow, and causes the resistance of the conductor, given by the equation $R\uparrow = \rho(l/A\downarrow)$, to increase.

Hysteresis and Eddy Current Losses

As mentioned earlier, hysteresis and eddy current losses will appear when a ferromagnetic material is placed in the region of a changing magnetic field. To describe eddy current losses in greater detail, we will consider the effects of an alternating current passing through a coil wrapped around a ferromagnetic core. As the alternating current passes through the coil, it will develop a changing magnetic flux Φ linking both the coil and the core that will develop an induced voltage within the core as determined by Faraday's law. This induced voltage and the geometric resistance of the core $R_C = \rho(l/A)$ cause currents to be developed within the core, $i_{core} = (e_{ind}/R_C)$, called *eddy currents*. The currents flow in circular paths, as shown in Fig. 19.33, changing direction with the applied ac potential.

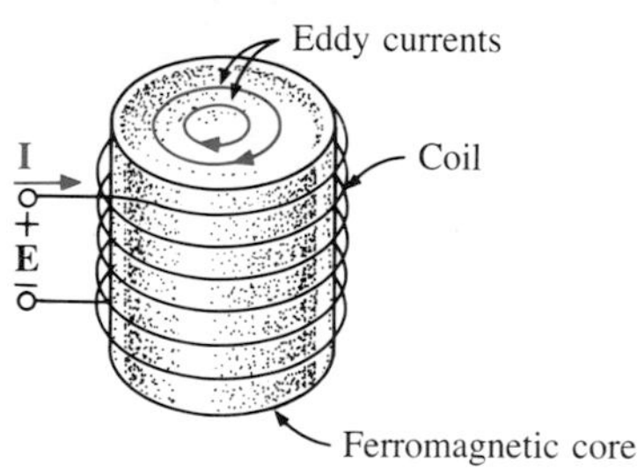

FIG. 19.33
Defining the eddy current losses of a ferromagnetic core.

The eddy current losses are determined by

$$P_{eddy} = i_{eddy}^2 R_{core}$$

The magnitude of these losses is determined primarily by the type of core used. If the core is nonferromagnetic—and has a high resistivity like wood or air—the eddy current losses can be neglected. In terms of the frequency of the applied signal and the magnetic field strength produced, the eddy current loss is proportional to the square of the frequency times the square of the magnetic field strength:

$$P_{eddy} \propto f^2B^2$$

Eddy current losses can be reduced if the core is constructed of thin, laminated sheets of ferromagnetic material insulated from one another and aligned parallel to the magnetic flux. Such construction reduces the magnitude of the eddy currents by placing more resistance in their path.

Hysteresis losses were described in Section 11.8. You will recall that in terms of the frequency of the applied signal and the magnetic field strength produced, the hysteresis loss is proportional to the frequency to the 1st power times the magnetic field strength to the nth power:

$$P_{\text{hys}} \propto f^1 B^n$$

where n can vary from 1.4 to 2.6, depending on the material under consideration.

Hysteresis losses can be effectively reduced by the injection of small amounts of silicon into the magnetic core, constituting some 2% or 3% of the total composition of the core. This must be done carefully, however, because too much silicon makes the core brittle and difficult to machine into the shape desired.

EXAMPLE 19.7

a. An air-core coil is connected to a 120-V, 60-Hz source as shown in Fig. 19.34. The current is found to be 5 A, and a wattmeter reading of 75 W is observed. Find the effective resistance and the inductance of the coil.

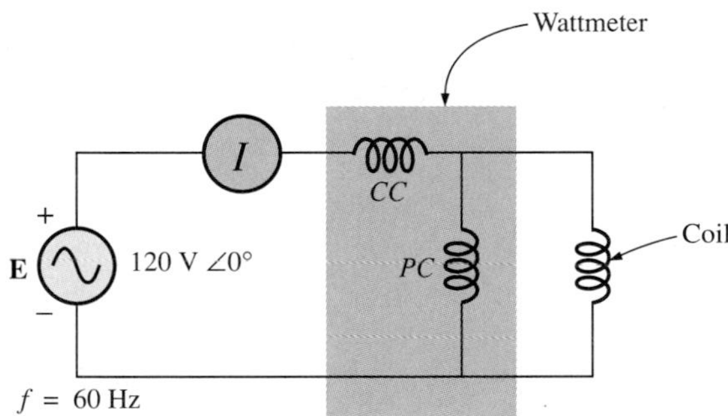

FIG. 19.34
The basic components required to determine the effective resistance and inductance of the coil.

b. A brass core is then inserted in the coil, and the ammeter reads 4 A, and the wattmeter reads 80 W. Calculate the effective resistance of the core. To what do you attribute the increase in value over that of part (a)?

c. If a solid iron core is inserted in the coil, the current is found to be 2 A, and the wattmeter reads 52 W. Calculate the resistance and the inductance of the coil. Compare these values to those of part (a), and account for the changes.

Solutions:

a. $R = \dfrac{P}{I^2} = \dfrac{75 \text{ W}}{(5 \text{ A})^2} = \mathbf{3\ \Omega}$

$$Z_T = \frac{E}{I} = \frac{120 \text{ V}}{5 \text{ A}} = 24\ \Omega$$

$$X_L = \sqrt{Z_T^2 - R^2} = \sqrt{(24\ \Omega)^2 - (3\ \Omega)^2} = 23.81\ \Omega$$

and $X_L = 2\pi f L$

or $L = \dfrac{X_L}{2\pi f} = \dfrac{23.81\ \Omega}{377 \text{ rad/s}} = \mathbf{63.16\ mH}$

b. $R = \dfrac{P}{I^2} = \dfrac{80 \text{ W}}{(4 \text{ A})^2} = \dfrac{80\ \Omega}{16} = \mathbf{5\ \Omega}$

The brass core has less reluctance than the air core. Therefore, a greater magnetic flux density B will be created in it. Since $P_{\text{eddy}} \propto f^2B^2$, and $P_{\text{hys}} \propto f^1B^n$, as the flux density increases, the core losses and the effective resistance increase.

c. $R = \dfrac{P}{I^2} = \dfrac{52 \text{ W}}{(2 \text{ A})^2} = \dfrac{52\ \Omega}{4} = \mathbf{13\ \Omega}$

$$Z_T = \frac{E}{I} = \frac{120 \text{ V}}{2 \text{ A}} = 60\ \Omega$$

$$X_L = \sqrt{Z_T^2 - R^2} = \sqrt{(60\ \Omega)^2 - (13\ \Omega)^2} = 58.57\ \Omega$$

$$L = \frac{X_L}{2\pi f} = \frac{58.57\ \Omega}{377 \text{ rad/s}} = \mathbf{155.36\ mH}$$

The iron core has less reluctance than the air or brass cores. Therefore, a greater magnetic flux density B will be developed in the core. Again, since $P_{\text{eddy}} \propto f^2B^2$, and $P_{\text{hys}} \propto f^1B^n$, the increased flux density will cause the core losses and the effective resistance to increase.

Since the inductance L is related to the change in flux by the equation $L = N(d\phi/di)$, the inductance will be greater for the iron core because the changing flux linking the core will increase.

19.11 COMPUTER ANALYSIS

PSpice (Windows)

The computer analysis will begin with a verification of the curves of Fig. 19.3. The network of Fig. 19.35 is established first, followed by a Probe analysis using a transient solution from 0 to 1 ms (the period of the applied sinusoidal signal). The response of Fig. 19.36 clearly reveals that v_R = **V(R)** = **V(R:1)** is in phase with i_R = **I(R)** and the power curve defined by $p_R = v_R i_R$ = **V(R:1)*I(R)** has a frequency twice that of v_R or i_R. The average value of the power curve occurs at 25 W, as shown by the horizontal line introduced through the use of the **Line** option under **Tools.** An analysis of the circuit reveals that $P = V_R^2/R = (10 \text{ V})^2/4\ \Omega = 25$ W, as obtained above. In particular, note that the power curve obtained from the product of two sinusoidal functions is indeed sinusoidal, with a peak value equal to the average power delivered to the resistive load. Note also that for a purely resistive network, the power curve is always above the horizontal axis, revealing that the network is always absorbing power—just at different rates.

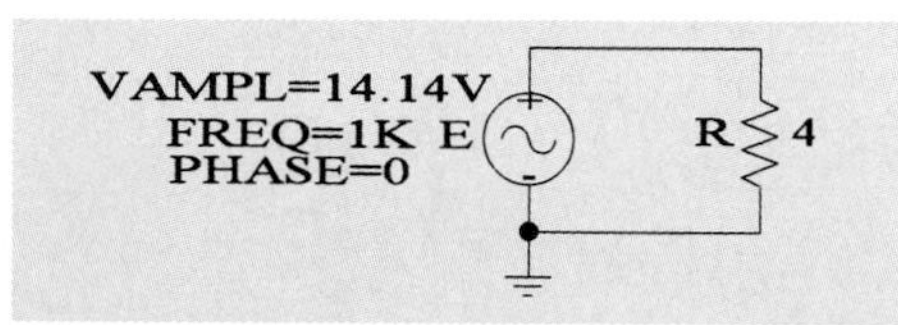

FIG. 19.35
Applying ac power to a purely resistive load using PSpice schematics.

The network of Fig. 19.37, with its combination of elements, will now be used to demonstrate that, no matter what the physical makeup of the network, the average value of the power curve established by the product of the input voltage and source current will reveal the total average power dissipated by the circuit. At a frequency of 1 kHz, $X_L = 8\ \Omega$ and $X_C = 4\ \Omega$, resulting in a lagging network, as noted by the fact that e leads i in the resulting waveforms of Fig. 19.38. A line was again drawn at the average value of the power curve intersecting the vertical axis at a real power level of 12.5 W. An analysis of the network will result in $\mathbf{Z}_T = 4\ \Omega + j\,4\ \Omega = 5.657\ \Omega \angle 45°$ with $\mathbf{I} = \mathbf{E/R} = 10 \text{ V} \angle 0°/5.657\ \Omega \angle 45° = 1.768 \text{ A} \angle -45°$ and $P = I^2R = (1.768 \text{ A})^2 4$

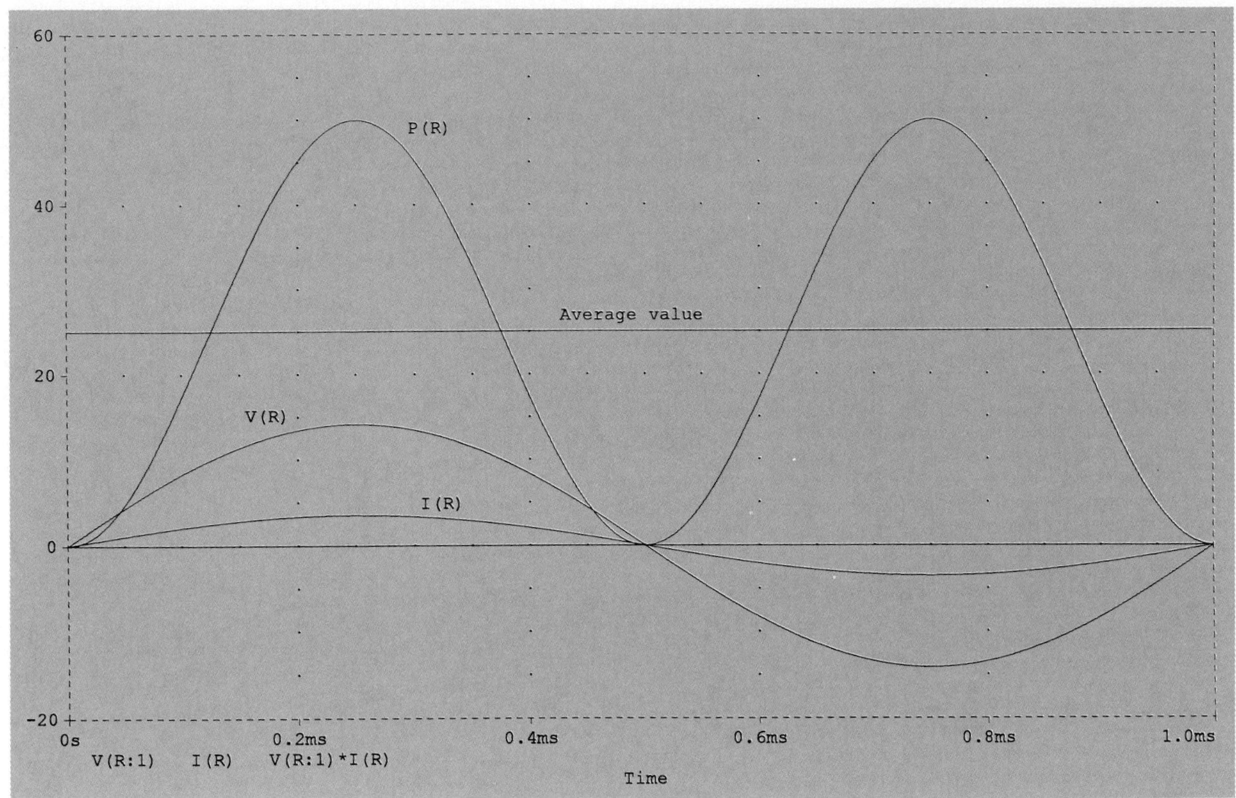

FIG. 19.36
Power versus time for the resistive load of Fig. 19.35.

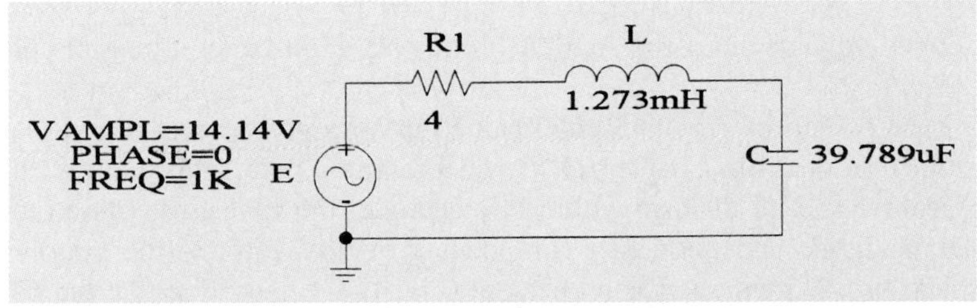

FIG. 19.37
Average power investigation for a series R-L-C circuit.

Ω = 12.5 W. There is confirmation therefore that *the average value of the power curve is the power absorbed by the system.* The phase angle of 45° between **E** and **I** is verified by Fig. 19.38. Note the use of a **No-Print Delay** of 19 ms assigned within the **Transient** Option of the **Analysis Setup** to permit the passing of the transient period and a response reflecting the steady-state response. The **Final Time** is 20 ms, and a 1 μs **Print Step** was employed along with a 20 μs **Step Ceiling.** A careful review of the plotted response curves under the **Add Trace** option reveals the use of a negative sign preceding **I(E)** in two cases. This was done because the current of an element is defined as flowing from the positive to negative potential level, and this would be opposite to the actual situation for the applied source. Note again that the power curve does have twice the frequency of the applied signal and has a purely sinusoidal waveshape, even though the phase angle between the

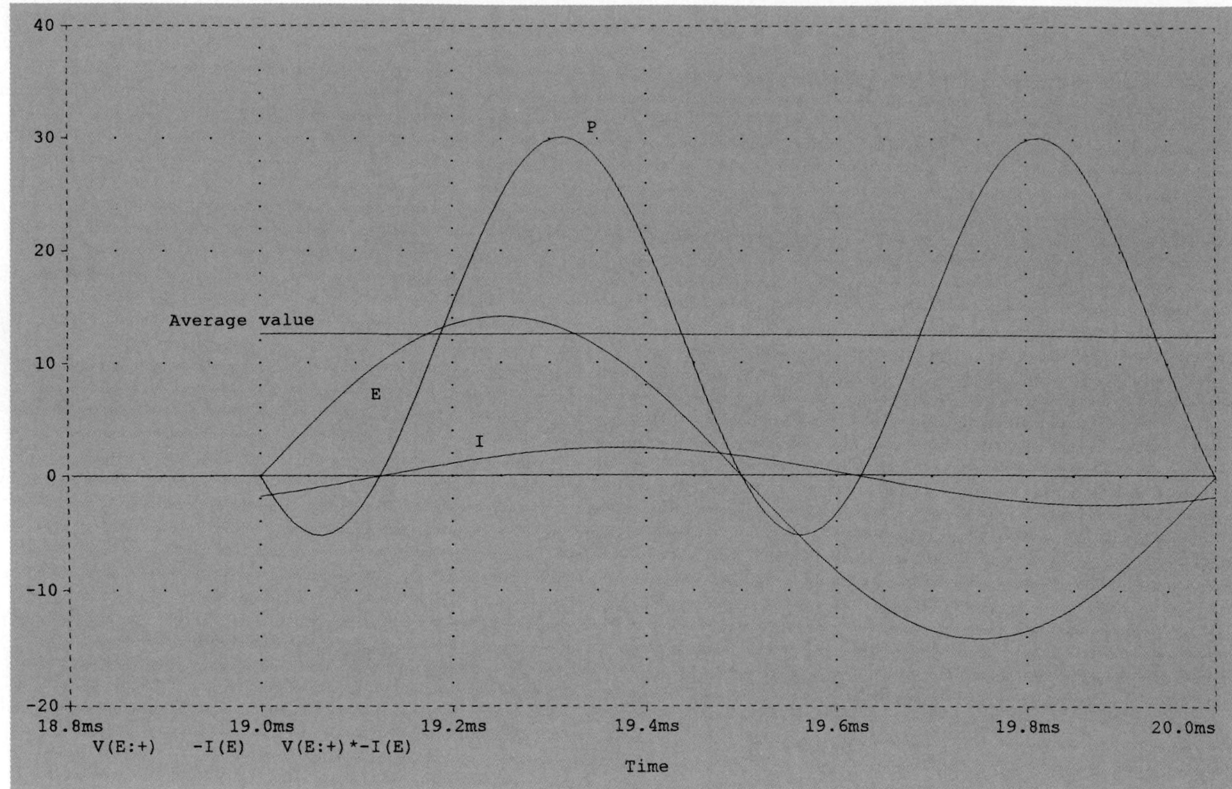

FIG. 19.38
Plot of the power delivered to the series R-L-C circuit of Fig. 19.37.

applied voltage and source current in this situation is not a multiple of 90°.

The pattern of Fig. 19.39 is obtained using a procedure that will be defined in detail in Chapter 25. It was presented here simply to reveal that there is a mechanism within PSpice under the **Plot** menu option to obtain the dc and frequency components of a waveform (the Fourier spectrum). Note the left-hand peak at 0 Hz (DC), revealing that the dc value of the power curve defined by **V(E:+)*−I(E)** is 12.5 W. The second peak at 2 kHz (twice the applied frequency) is at about 18 W, representing the peak value of the 2-kHz component and defined by the product $V \cdot I = (10\text{ V})(1.768\text{ A}) = 17.68\text{ W}$. The sum of dc and the 2-kHz component is 12.5 W + 17.68 W = 30.2 W, which matches the peak value of the power curve in Fig. 19.39.

BASIC

Although PSpice can calculate power levels using the .PROBE option, it cannot tabulate real and reactive power levels using simply the .PRINT and .PLOT commands. However, with C++ or BASIC, a convenient program can be developed and used to solve most of the networks appearing in this chapter. The BASIC program of Fig. 19.40 can determine the total real, reactive, and apparent power levels, in addition to the power factor and input current. It is designed to handle up to five different loads of varying power-factor angles as defined by lines 110 through 130.

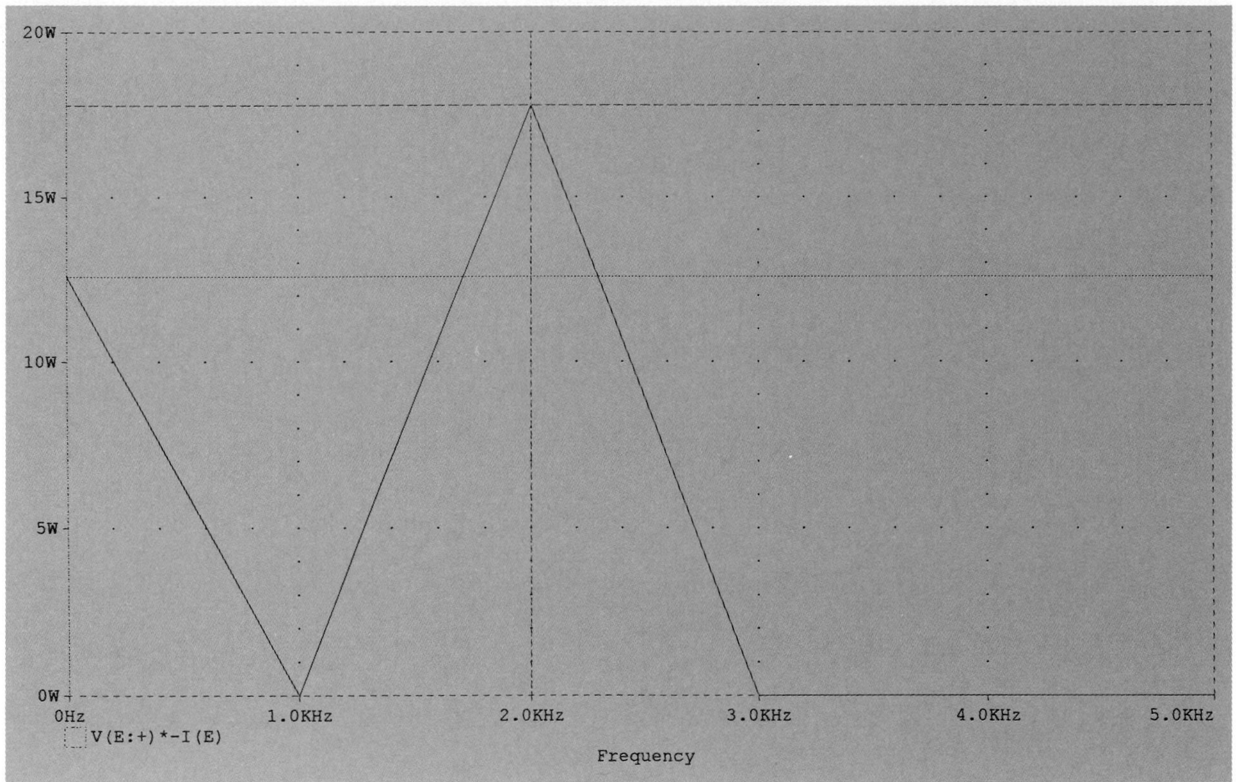

FIG. 19.39
Spectrum distribution for the power curve of Fig. 19.38.

Lines 260 and 270 determine the total real and reactive power using a loop routine that begins on line 210 and ends on line 280. The apparent power for each load is determined on line 340, and the total apparent power is determined on line 370. The results are printed out by lines 390 through 430, and the power factor is determined by lines 440 through 530. The input current is then determined by lines 540 through 560.

A run of the program for four loads is provided with parameter values that permit a relatively easy check of the program through review of the results.

```
      10 REM ***** PROGRAM 19-1 *****
      20 REM *****************************************
      30 REM This program calculates the total real,
      40 REM reactive and apparent power of a network
      50 REM with 5 individual loads.
      60 REM *****************************************
      70 REM
      100 DIM P(5), Q(5), S(5)
      110 PRINT "This program calculates the total real,"
      120 PRINT "reactive and apparent power of a network"
      130 PRINT "with five individual loads."
      140 PRINT
      150 PRINT "Input the following data:"
      160 PRINT "(use negative sign for capacitive vars)"
      170 PRINT
Input [ 180 INPUT "E="; E
      [ 190 INPUT "at an angle="; EA
      200 PRINT
```

FIG. 19.40
Program 19.1.

```
210 FOR I = 1 TO 5
220 PRINT "For"; I; "  ";
230 INPUT "P(watts)="; P(I)
240 PRINT TAB(8);
250 INPUT "Q(vars)="; Q(I)
260 PT = PT + P(I)
270 QT = QT + Q(I)
280 NEXT I
290 PRINT
300 PRINT "The apparent power associated with each load"
310 PRINT "is the following:"
320 PRINT
330 FOR I = 1 TO 5
340 S(I) = SQR(P(I) ^ 2 + Q(I) ^ 2)
350 PRINT "S"; I; "="; S(I)
360 NEXT I
370 ST = SQR(PT ^ 2 + QT ^ 2)
380 PRINT : PRINT
390 PRINT "Total real power, PT="; PT; "watts"
400 PRINT
410 PRINT "Total reactive power, QT="; QT; "vars"
420 PRINT
430 PRINT "Total apparent power, ST="; ST; "VA"
440 FP = PT / ST
450 TH = -57.296 * ATN(QT / PT)
460 IF QT > 0 THEN IA = EA - TH
470 IF QT < 0 THEN IA = EA + TH
480 PRINT
490 PRINT "Power factor angle="; IA; "degrees"
500 PRINT
510 PRINT "Power factor="; FP;
520 IF QT > 0 THEN PRINT "(lagging)"
530 IF QT < 0 THEN PRINT "(leading)"
540 I = ST / E
550 PRINT
560 PRINT "Input current:"; I; "at an angle of"; IA; "degrees"
570 END
This program calculates the total real,
reactive and apparent power of a network
with five individual loads.

Input the following data:
(use negative sign for capacitive vars)

E=? 50
at an angle=? 60

For 1   P(watts)=? 200
        Q(vars)=? 100
For 2   P(watts)=? 200
        Q(vars)=? 100
For 3   P(watts)=? 100
        Q(vars)=? -200
For 4   P(watts)=? 100
        Q(vars)=? -200
For 5   P(watts)=? 0
        Q(vars)=? 0

The apparent power associated with each load
is the following:

S 1 = 223.6068
S 2 = 223.6068
S 3 = 223.6068
S 4 = 223.6068
S 5 = 0


Total real power, PT= 600 watts

Total reactive power, QT=-200 vars

Total apparent power, ST= 632.4555 VA

Power factor angle= 78.43502 degrees

Power factor= .9486833 (leading)

Input current: 12.64911 at an angle of 78.43502 degrees
```

FIG. 19.40 continued

PROBLEMS

SECTIONS 19.1 THROUGH 19.7

1. For the battery of bulbs (purely resistive) appearing in Fig. 19.41:
   a. Determine the total power dissipation.
   b. Calculate the total reactive and apparent power.
   c. Find the source current I_s.
   d. Calculate the resistance of each bulb for the specified operating conditions.
   e. Determine the currents I_1 and I_2.

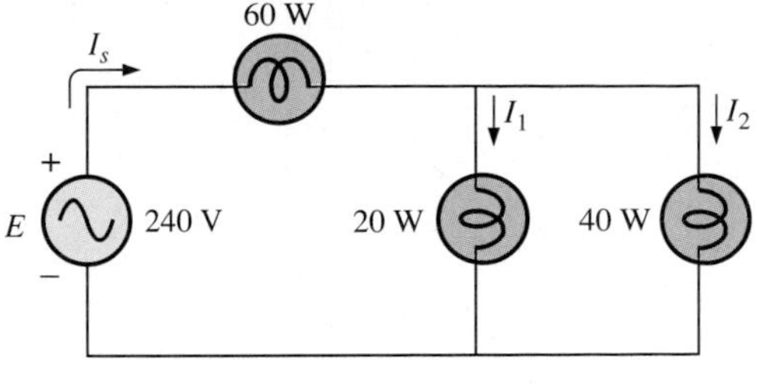

FIG. 19.41
Problem 1.

2. For the network of Fig. 19.42:
   a. Find the average power delivered to each element.
   b. Find the reactive power for each element.
   c. Find the apparent power for each element.
   d. Find the total number of watts, volt-amperes reactive, and volt-amperes, and the power factor F_p of the circuit.
   e. Sketch the power triangle.
   f. Find the energy dissipated by the resistor over one full cycle of the input voltage.
   g. Find the energy stored or returned by the capacitor and the inductor over one half-cycle of the power curve for each.

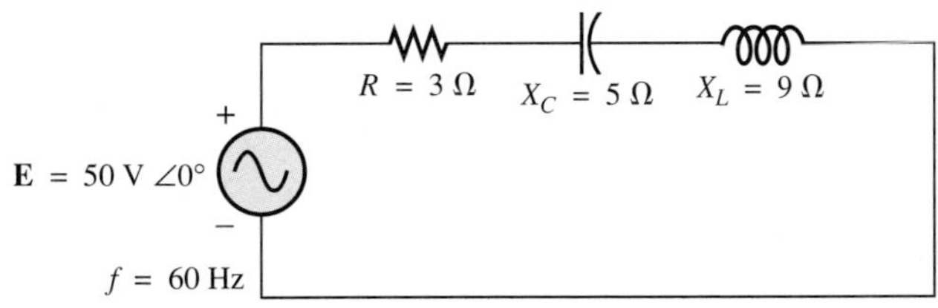

FIG. 19.42
Problem 2.

3. For the system of Fig. 19.43:
   a. Find the total number of watts, volt-amperes reactive, and volt-amperes, and the power factor F_p.
   b. Draw the power triangle.
   c. Find the current $\mathbf{I}_s$.

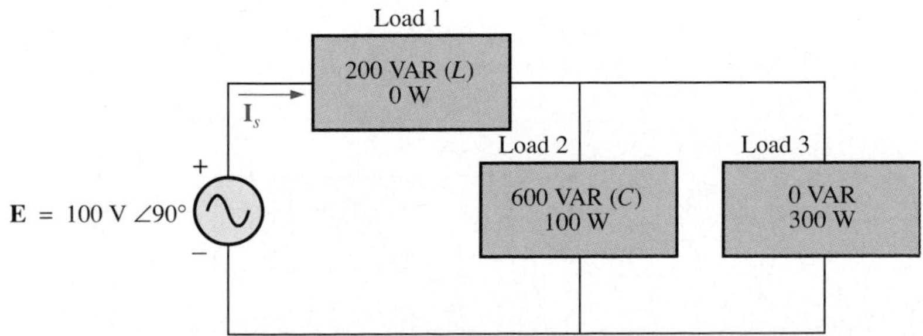

FIG. 19.43
Problem 3.

4. For the system of Fig. 19.44:
   a. Find P_T, Q_T, and S_T.
   b. Determine the power factor F_p.
   c. Draw the power triangle.
   d. Find $\mathbf{I}_s$.

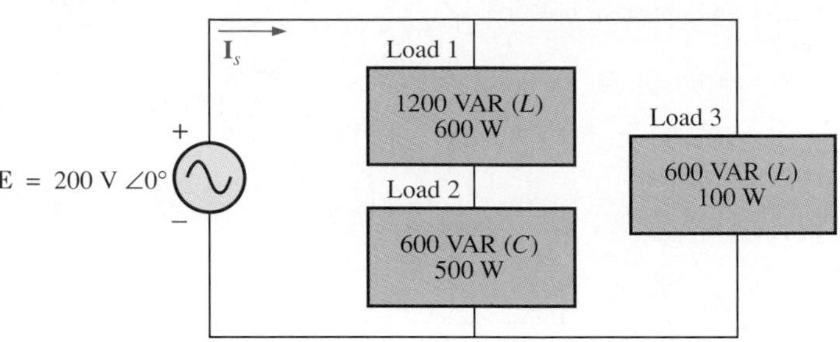

FIG. 19.44
Problem 4.

5. For the system of Fig. 19.45:
   a. Find P_T, Q_T, and S_T.
   b. Find the power factor F_p.
   c. Draw the power triangle.
   d. Find $\mathbf{I}_s$.

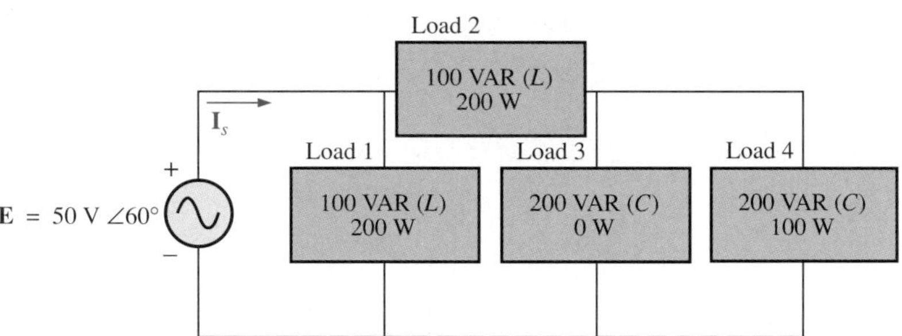

FIG. 19.45
Problem 5.

6. For the circuit of Fig. 19.46:
   a. Find the average, reactive, and apparent power for the 20-Ω resistor.
   b. Repeat part (a) for the 10-Ω inductive reactance.
   c. Find the total number of watts, volt-amperes reactive, and volt-amperes, and the power factor F_p.
   d. Find the current $\mathbf{I}_s$.

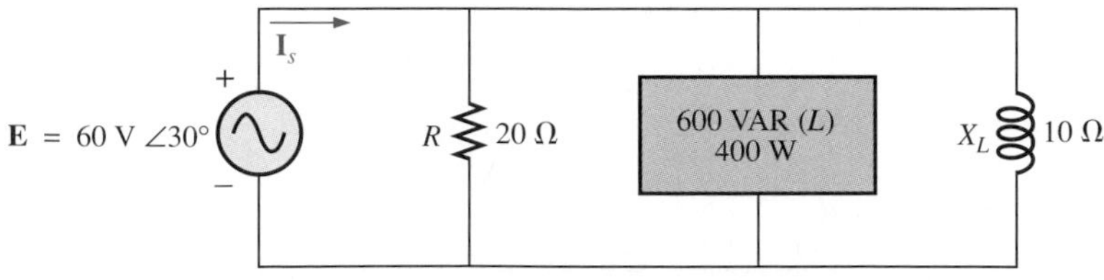

FIG. 19.46
Problem 6.

7. For the network of Fig. 19.47:
   a. Find the average power delivered to each element.
   b. Find the reactive power for each element.
   c. Find the apparent power for each element.
   d. Find P_T, Q_T, S_T, and F_p for the system.
   e. Sketch the power triangle.
   f. Find $\mathbf{I}_s$.

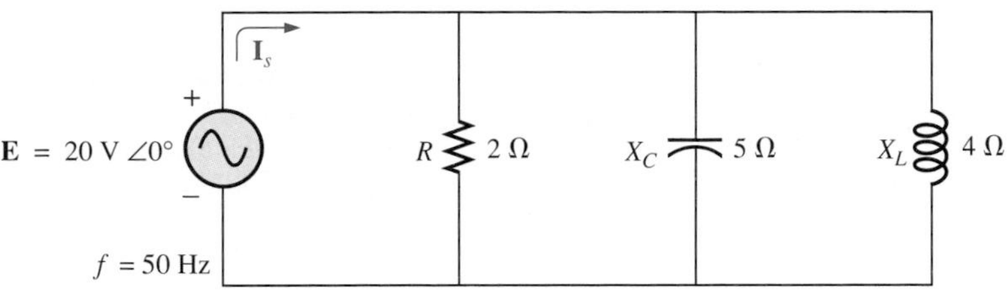

FIG. 19.47
Problem 7.

8. Repeat Problem 7 for the circuit of Fig. 19.48.

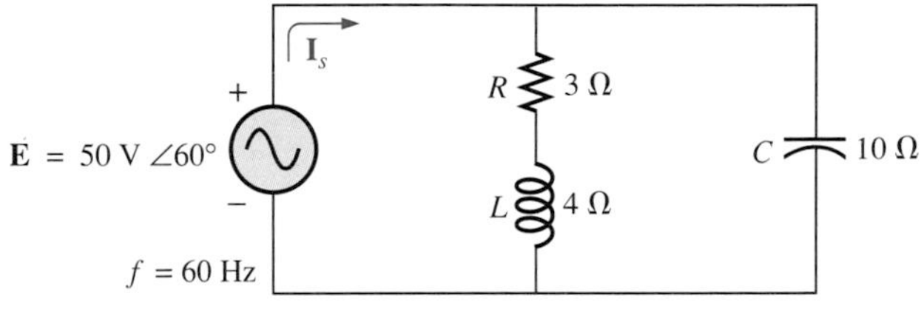

FIG. 19.48
Problem 8.

*9. For the network of Fig. 19.49:
   a. Find the average power delivered to each element.
   b. Find the reactive power for each element.
   c. Find the apparent power for each element.
   d. Find the total number of watts, volt-amperes reactive, and volt-amperes, and the power factor F_p of the circuit.
   e. Sketch the power triangle.
   f. Find the energy dissipated by the resistor over one full cycle of the input voltage.
   g. Find the energy stored or returned by the capacitor and the inductor over one half-cycle of the power curve for each.

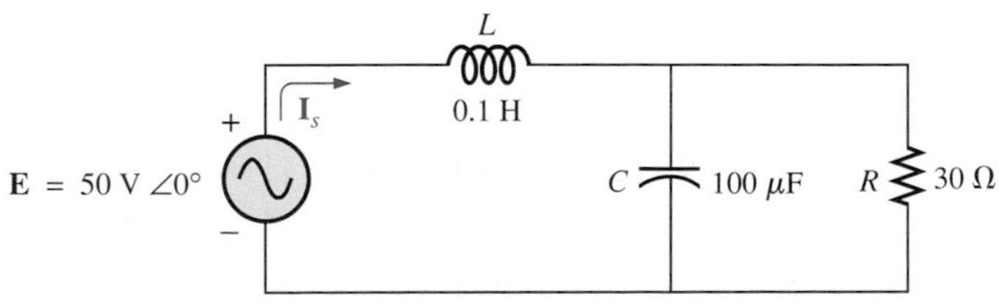

FIG. 19.49
Problem 9.

10. An electrical system is rated 10 kVA, 200 V at a 0.5 leading power factor.
   a. Determine the impedance of the system in rectangular coordinates.
   b. Find the average power delivered to the system.

11. An electrical system is rated 5 kVA, 120 V, at a 0.8 lagging power factor.

a. Determine the impedance of the system in rectangular coordinates.

b. Find the average power delivered to the system.

***12.** For the system of Fig. 19.50:

a. Find the total number of watts, volt-amperes reactive, and volt-amperes, and F_p.

b. Find the current $\mathbf{I}_s$.

c. Draw the power triangle.

d. Find the type elements and their impedance in ohms within each electrical box. (Assume that all elements of a load are in series.)

e. Verify that the result of part (b) is correct by finding the current $\mathbf{I}_s$ using only the input voltage **E** and the results of part (d). Compare the value of $\mathbf{I}_s$ with that obtained for part (b).

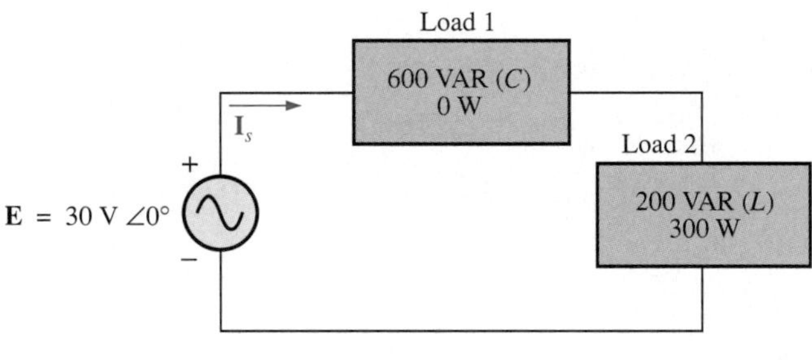

FIG. 19.50
Problem 12.

***13.** Repeat Problem 12 for the system of Fig. 19.51:

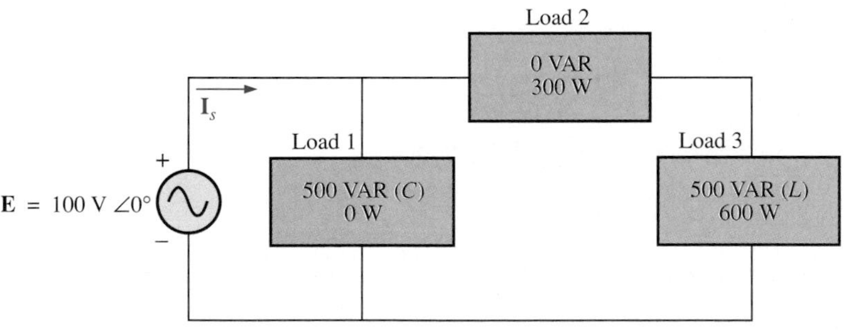

FIG. 19.51
Problem 13.

*14. For the circuit of Fig. 19.52:

a. Find the total number of watts, volt-amperes reactive, and volt-amperes, and F_p.

b. Find the current $\mathbf{I}_s$.

c. Find the type elements and their impedance in each box. (Assume that the elements within each box are in series.)

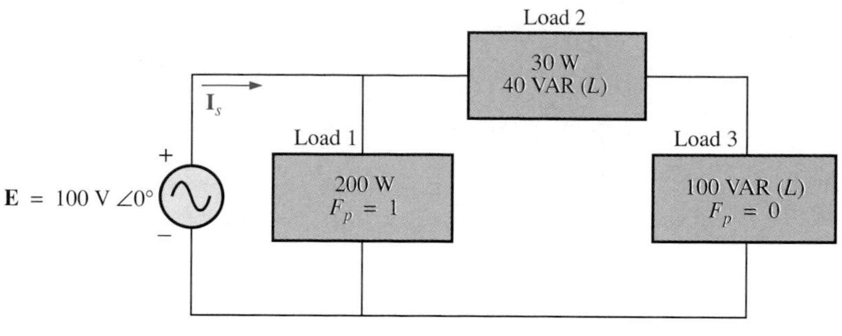

FIG. 19.52
Problem 14.

15. For the circuit of Fig. 19.53:

a. Find the total number of watts, volt-amperes reactive, and volt-amperes, and F_p.

b. Find the voltage **E**.

c. Find the type elements and their impedance in each box. (Assume that the elements within each box are in series.)

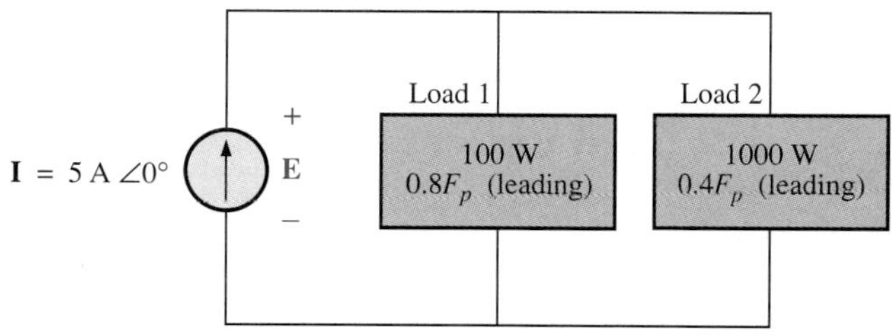

FIG. 19.53
Problem 15.

SECTION 19.8 Power-Factor Correction

*16. The lighting and motor loads of a small factory establish a 10-kVA power demand at a 0.7 lagging power factor on a 208-V, 60-Hz supply.

a. Establish the power triangle for the load.

b. Determine the power-factor capacitor that must be placed in parallel with the load to raise the power factor to unity.

c. Determine the change in supply current from the uncompensated to compensated system.

d. Repeat parts (b) and (c) if the power factor is increased to 0.9.

17. The load on a 120-V, 60-Hz supply is 5 kW (resistive), 8 kVAR (inductive), and 2 kVAR (capacitive).

a. Find the total kilovolt-amperes.

b. Determine the F_p of the combined loads.

c. Find the current drawn from the supply.
d. Calculate the capacitance necessary to establish a unity power factor.
e. Find the current drawn from the supply at unity power factor and compare it to the uncompensated level.

18. The loading of a factory on a 1000-V, 60-Hz system includes:

20-kW heating (unity power factor)
10-kW (P_i) induction motors (0.7 lagging power factor)
5-kW lighting (0.85 lagging power factor)

a. Establish the power triangle for the total loading on the supply.
b. Determine the power factor capacitor required to raise the power factor to unity.
c. Determine the change in supply current from the uncompensated to the compensated system.

SECTION 19.9 Wattmeters and Power-Factor Meters

19. a. A wattmeter is connected with its current coil as shown in Fig. 19.54 and the potential coil across points *f-g*. What does the wattmeter read?
b. Repeat part (a) with the potential coil (*PC*) across *a-b, b-c, a-c, a-d, c-d, d-e,* and *f-e*.

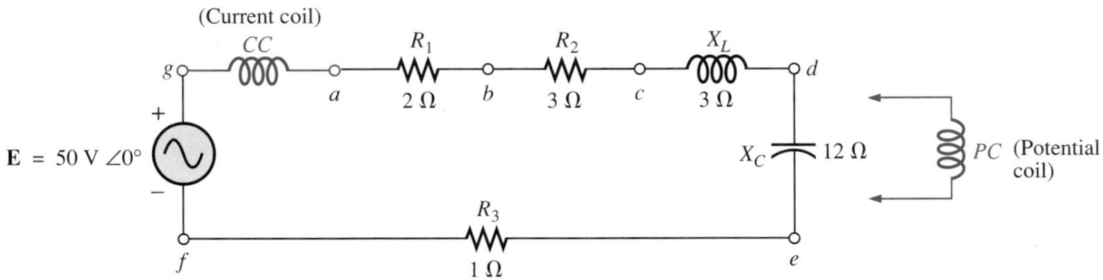

FIG. 19.54
Problem 19.

20. The voltage source of Fig. 19.55 delivers 660 VA at 120 V, with a supply current that lags the voltage by a power factor of 0.6.
a. Determine the voltmeter, ammeter, and wattmeter readings.
b. Find the load impedance in rectangular form.

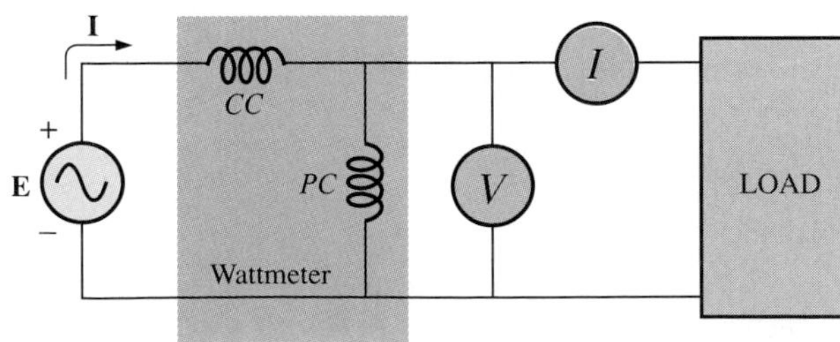

FIG. 19.55
Problem 20.

P_S^q

SECTION 19.10 Effective Resistance

21. a. An air-core coil is connected to a 200-V, 60-Hz source. The current is found to be 4 A, and a wattmeter reading of 80 W is observed. Find the effective resistance and the inductance of the coil.

b. A brass core is inserted in the coil. The ammeter reads 3 A, and the wattmeter reads 90 W. Calculate the effective resistance of the core. Explain the increase over the value of part (a).

c. If a solid iron core is inserted in the coil, the current is found to be 2 A, and the wattmeter reads 60 W. Calculate the resistance and inductance of the coil. Compare these values to the values of part (a), and account for the changes.

22. a. The inductance of an air-core coil is 0.08 H, and the effective resistance is 4 Ω when a 60-V, 50-Hz source is connected across the coil. Find the current passing through the coil and the reading of a wattmeter across the coil.

b. If a brass core is inserted in the coil, the effective resistance increases to 7 Ω, and the wattmeter reads 30 W. Find the current passing through the coil and the inductance of the coil.

c. If a solid iron core is inserted in the coil, the effective resistance of the coil increases to 10 Ω, and the current decreases to 1.7 A. Find the wattmeter reading and the inductance of the coil.

SECTION 19.11 Computer Analysis

PSpice (Windows)

23. Using PSpice, obtain a plot of reactive power for a pure capacitor of 636.62 μF at a frequency of 1 kHz for one cycle of the input voltage using an applied voltage $\mathbf{E} = 10\text{ V} \angle 0°$. On the same graph, plot both the applied voltage and resulting current. Apply appropriate labels to the resulting curves to generate results similar to Fig. 19.36.

24. Repeat the analysis of Fig. 19.37 for a parallel *R-L-C* network of the same values and frequency.

25. Plot both the applied voltage and the source current on the same axis for the network of Fig. 19.27(b) and show that they are both in phase due to the resulting unity power factor.

Programming Language (C++, BASIC, PASCAL, etc.)

26. Write a program that provides a general solution for the network of Fig. 19.19. That is, given the resistance or reactance of each element and the source voltage at zero degrees, calculate the real, reactive, and apparent power of the system.

27. Write a program that will demonstrate the effect of increasing reactive power on the power factor of a system. Tabulate the real power, reactive power, and power factor of the system for a fixed real power and a reactive power that starts at 10% of the real power and continues through to five times the real power in increments of 10% of the real power.

GLOSSARY

Apparent power The power delivered to a load without consideration of the effects of a power-factor angle of the load. It is determined solely by the product of the terminal voltage and current of the load.

Average (real) power The delivered power dissipated in the form of heat by a network or system.

Eddy currents Small, circular currents in a paramagnetic core causing an increase in the power losses and the effective resistance of the material.

Effective resistance The resistance value that includes the effects of radiation losses, skin effect, eddy currents, and hysteresis losses.

Hysteresis losses Losses in a magnetic material introduced by changes in the direction of the magnetic flux within the material.

Power-factor correction The addition of reactive components (typically capacitive) to establish a system power factor closer to unity.

Radiation losses The loss of energy in the form of electromagnetic waves during the transfer of energy from one element to another.

Reactive power The power associated with reactive elements that provides a measure of the energy associated with setting up the magnetic and electric fields of inductive and capacitive elements, respectively.

Skin effect At high frequencies, a counter-induced voltage builds up at the center of a conductor, resulting in an increased flow near the surface (skin) of the conductor and a sharp reduction near the center. As a result, the effective area of conduction decreases and the resistance increases as defined by the basic equation for the geometric resistance of a conductor.

20

Resonance

20.1 INTRODUCTION

This chapter will introduce the very important resonant (or tuned) circuit, which is fundamental to the operation of a wide variety of electrical and electronic systems in use today. The resonant circuit is a combination of R, L, and C elements having a frequency response characteristic similar to the one appearing in Fig. 20.1. Note in the figure that the response is a

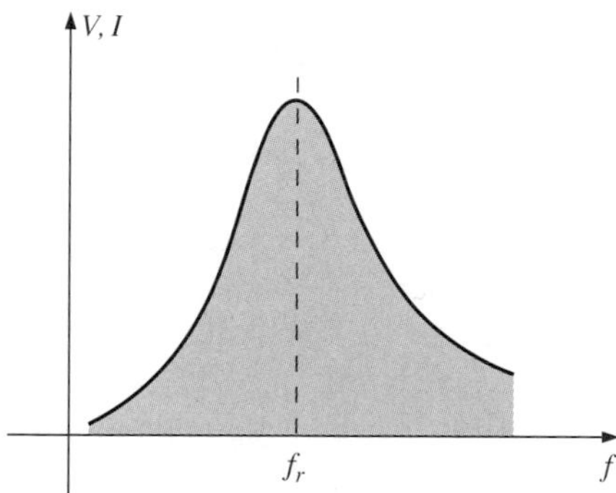

FIG. 20.1
Resonance curve.

maximum for the frequency f_r, decreasing to the right and left of this frequency. In other words, for a particular range of frequencies the response will be near or equal to the maximum. The frequencies to the far left or right have very low voltage or current levels and, for all practical purposes, have little effect on the system's response. The radio or television receiver has a response curve for each broadcast station of the type indicated in Fig. 20.1. When the receiver is set (or tuned) to a particular station, it is set on or near the frequency f_r of Fig. 20.1. Stations transmitting at frequencies to the far right or left of this resonant frequency are not carried through with significant power to affect the program of interest. The tuning process (setting the dial to f_r) as described above is the reason for

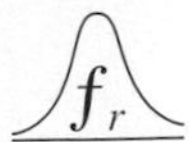

the terminology *tuned circuit*. When the response is at or near the maximum, the circuit is said to be in a state of *resonance*.

The concept of resonance is not limited to electrical or electronic systems. If mechanical impulses are applied to a mechanical system at the proper frequency, the system will enter a state of resonance in which sustained vibrations of very large amplitude will develop. The frequency at which this occurs is called the *natural frequency* of the system. The classic example of this effect was the Tacoma Narrows Bridge built in 1940 over Puget Sound in Washington State. Four months after the bridge, with its suspended span of 2800 ft, was completed, a 42-mi/h pulsating gale set the bridge into oscillations at its natural frequency. The amplitude of the oscillations increased to the point where the main span broke up and fell into the water below. It has since been replaced by the new Tacoma Narrows Bridge, completed in 1950.

The resonant electrical circuit *must* have both inductance and capacitance. In addition, resistance will always be present due either to the lack of ideal elements or to the control offered on the shape of the resonance curve. When resonance occurs due to the application of the proper frequency (f_r), the energy absorbed by one reactive element is the same as that released by another reactive element within the system. In other words, energy pulsates from one reactive element to the other. Therefore, once an ideal (pure C, L) system has reached a state of resonance, it requires no further reactive power since it is self-sustaining. In a practical circuit, there is some resistance associated with the reactive elements that will result in the eventual "damping" of the oscillations between reactive elements.

There are two types of resonant circuits: *series* and *parallel*. Each will be considered in some detail in this chapter.

SERIES RESONANCE

20.2 SERIES RESONANT CIRCUIT

A resonant circuit (series or parallel) must have an inductive and capacitive element. A resistive element will always be present due to the internal resistance of the source (R_s), the internal resistance of the inductor (R_l), and any added resistance to control the shape of the response curve (R_{design}). The basic configuration for the series resonant circuit appears in Fig. 20.2(a) with the resistive elements listed above. The "cleaner" appearance of Fig. 20.2(b) is a result of combining the series resistive elements into one total value. That is,

$$R = R_s + R_l + R_d \tag{20.1}$$

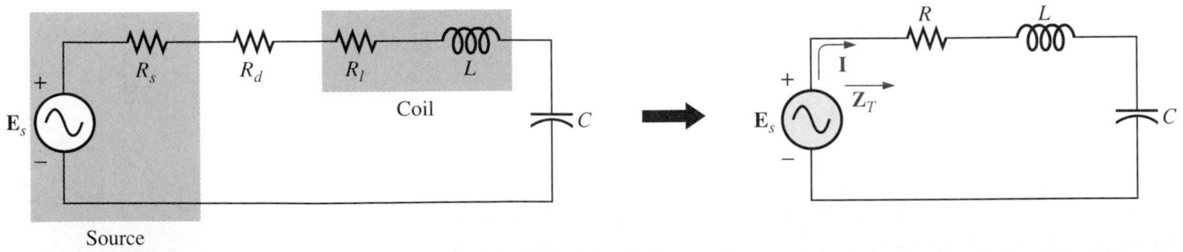

FIG. 20.2
Series resonant circuit.

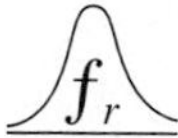

The total impedance of this network at any frequency is determined by

$$\mathbf{Z}_T = R + jX_L - jX_C = R + j(X_L - X_C)$$

The resonant conditions described in the introduction will occur when

$$X_L = X_C \tag{20.2}$$

removing the reactive component from the total impedance equation. The total impedance at resonance is then simply

$$\mathbf{Z}_{T_s} = R \tag{20.3}$$

representing the minimum value of $\mathbf{Z}_T$ at any frequency. The subscript s will be employed to indicate series resonant conditions.

The resonant frequency can be determined in terms of the inductance and capacitance by examining the defining equation for resonance [Eq. (20.2)]:

$$X_L = X_C$$

Substituting yields

$$\omega L = \frac{1}{\omega C} \quad \text{and} \quad \omega^2 = \frac{1}{LC}$$

and

$$\omega_s = \frac{1}{\sqrt{LC}} \tag{20.4}$$

or

$$f_s = \frac{1}{2\pi\sqrt{LC}} \qquad \begin{aligned} f &= \text{hertz (Hz)} \\ L &= \text{henries (H)} \\ C &= \text{farads (F)} \end{aligned} \tag{20.5}$$

The current through the circuit at resonance is

$$\mathbf{I} = \frac{E \angle 0^\circ}{R \angle 0^\circ} = \frac{E}{R} \angle 0^\circ$$

which you will note is the maximum current for the circuit of Fig. 20.2 for an applied voltage **E** since $\mathbf{Z}_T$ is a minimum value. Consider also that *the input voltage and current are in phase at resonance.*

Since the current is the same through the capacitor and inductor, the voltage across each is equal in magnitude but 180° out of phase at resonance:

$$\left.\begin{aligned} \mathbf{V}_L &= (I \angle 0^\circ)(X_L \angle 90^\circ) = IX_L \angle 90^\circ \\ \mathbf{V}_C &= (I \angle 0^\circ)(X_C \angle -90^\circ) = IX_C \angle -90^\circ \end{aligned}\right\} \begin{array}{l} 180^\circ \\ \text{out of} \\ \text{phase} \end{array}$$

and, since $X_L = X_C$, the magnitude of V_L equals V_C at resonance; that is,

$$V_{L_s} = V_{C_s} \tag{20.6}$$

Figure 20.3, a phasor diagram of the voltages and current, clearly indicates that the voltage across the resistor at resonance is the input voltage and **E,** and **I,** and $\mathbf{V}_R$ are in phase at resonance.

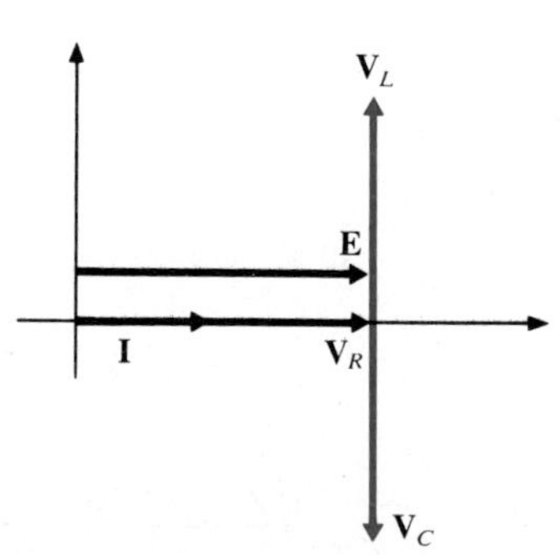

FIG. 20.3
Phasor diagram for the series resonant circuit at resonance.

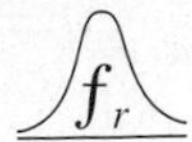

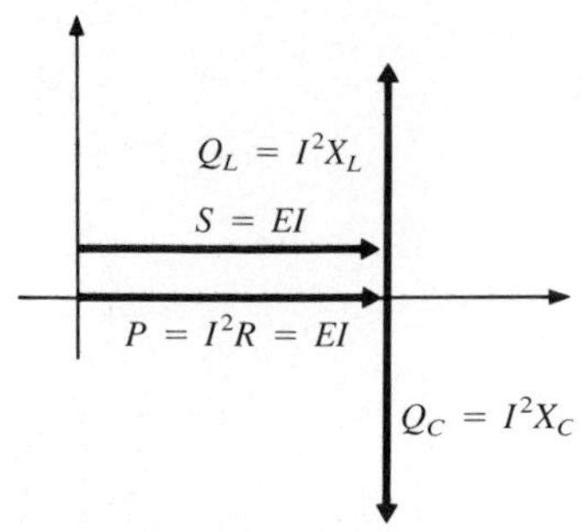

FIG. 20.4
Power triangle for the series resonant circuit at resonance.

The average power to the resistor at resonance is equal to I^2R, and the reactive power to the capacitor and inductor are I^2X_C and I^2X_L, respectively.

The power triangle at resonance (Fig. 20.4) shows that the total apparent power is equal to the average power dissipated by the resistor since $Q_L = Q_C$. The power factor of the circuit at resonance is

$$F_p = \cos\theta = \frac{P}{S}$$

and

$$\boxed{F_{p_s} = 1} \tag{20.7}$$

Plotting the power curves of each element on the same set of axes (Fig. 20.5), we note that, even though the total reactive power at any instant is equal to zero (note $t = t'$), energy is still being absorbed and released by the inductor and capacitor at resonance.

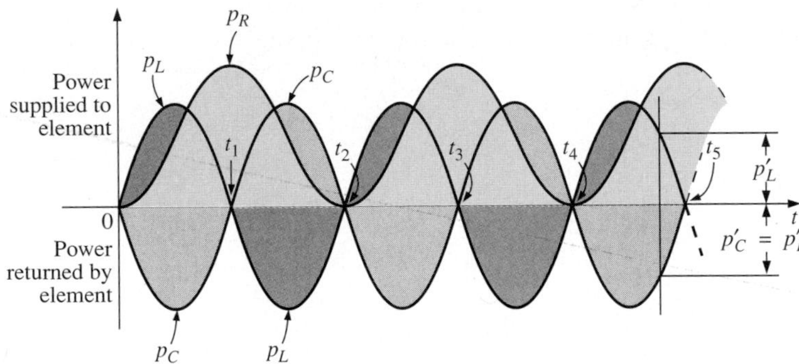

FIG. 20.5
Power curves at resonance for the series resonant circuit.

A closer examination reveals that the energy absorbed by the inductor from time 0 to t_1 is the same as the energy released by the capacitor from 0 to t_1. The reverse occurs from t_1 to t_2, and so on. Therefore, the total apparent power continues to be equal to the average power, even though the inductor and capacitor are absorbing and releasing energy. This condition occurs only at resonance. The slightest change in frequency introduces a reactive component into the power triangle, which will increase the apparent power of the system above the average power dissipation, and resonance will no longer exist.

20.3 THE QUALITY FACTOR (Q)

The *quality factor* Q of a series resonant circuit is defined as the ratio of the reactive power of either the inductor or the capacitor to the average power of the resistor at resonance; that is,

$$\boxed{Q_s = \frac{\text{reactive power}}{\text{average power}}} \tag{20.8}$$

The quality factor is also an indication of how much energy is placed in storage (continual transfer from one reactive element to the other) as compared to that dissipated. The lower the level of dissipation for the

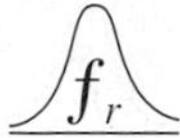

same reactive power, the larger the Q_s factor and the more concentrated and intense the region of resonance.

Substituting for an inductive reactance in Eq. (20.8) at resonance gives us

$$Q_s = \frac{I^2 X_L}{I^2 R}$$

and

$$Q_s = \frac{X_L}{R} = \frac{\omega_s L}{R} \qquad \textbf{(20.9)}$$

If the resistance R is just the resistance of the coil (R_l), we can speak of the Q of the coil, where

$$Q_{\text{coil}} = Q_l = \frac{X_L}{R_l} \qquad R = R_l \qquad \textbf{(20.10)}$$

Since the quality factor of a coil is typically the information provided by manufacturers of inductors, it is often given the symbol Q without an associated subscript. It would appear from Eq. (20.10) that Q_l will increase linearly with frequency since $X_L = 2\pi fL$. That is, if the frequency doubles, then Q_l will also increase by a factor of 2. This is approximately true for the low range to the midrange of frequencies such as shown for the coils of Fig. 20.6. Unfortunately, however, as the frequency increases, the effective resistance of the coil will also increase, due primarily to skin effect phenomena, and the resulting Q_l will decrease. In addition, the capacitive effects between the windings will increase, further reducing the Q_l of the coil. For this reason, Q_l must be specified for a particular frequency or frequency range. For wide frequency applications, a plot of Q_l versus frequency is often provided. The maximum Q_l for most commercially available coils is less than 200, with most having a maximum near 100. Note in Fig. 20.6 that for coils of the same type, Q_l drops off more quickly for higher levels of inductance.

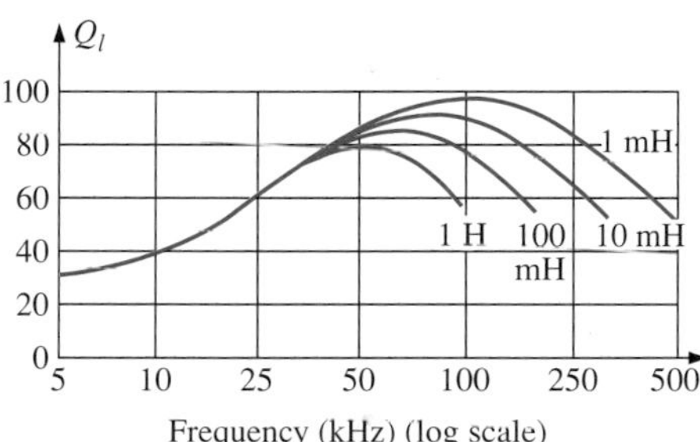

FIG. 20.6

Q_l versus frequency for a series of inductors of similar construction.

If we substitute

$$\omega_s = 2\pi f_s$$

and then

$$f_s = \frac{1}{2\pi\sqrt{LC}}$$

into Eq. (20.9), we have

$$Q_s = \frac{\omega_s L}{R} = \frac{2\pi f_s L}{R} = \frac{2\pi}{R}\left(\frac{1}{2\pi\sqrt{LC}}\right)L$$

$$= \frac{L}{R}\left(\frac{1}{\sqrt{LC}}\right) = \left(\frac{\sqrt{L}}{\sqrt{L}}\right)\frac{L}{R\sqrt{LC}}$$

and

$$Q_s = \frac{1}{R}\sqrt{\frac{L}{C}} \qquad \textbf{(20.11)}$$

providing Q_s in terms of the circuit parameters.

For series resonant circuits used in communication systems, Q_s is usually greater than 1. By applying the voltage divider rule to the circuit of Fig. 20.2, we obtain

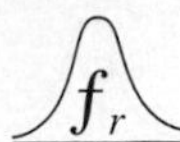

$$V_L = \frac{X_L E}{Z_T} = \frac{X_L E}{R} \quad \text{(at resonance)}$$

and

$$V_{L_s} = Q_s E \tag{20.12}$$

or

$$V_C = \frac{X_C E}{Z_T} = \frac{X_C E}{R}$$

and

$$V_{C_s} = Q_s E \tag{20.13}$$

Since Q_s is usually greater than 1, the voltage across the capacitor or inductor of a series resonant circuit can be significantly greater than the input voltage. In fact, in many cases the Q_s is so high that careful design and handling (including adequate insulation) are mandatory with respect to the voltage across the capacitor and inductor.

In the circuit of Fig. 20.7, for example, which is in the state of resonance,

$$Q_s = \frac{X_L}{R} = \frac{480\ \Omega}{6\ \Omega} = 80$$

and $$V_L = V_C = Q_s E = (80)(10\text{ V}) = \mathbf{800\ V}$$

which is certainly a potential of significant magnitude.

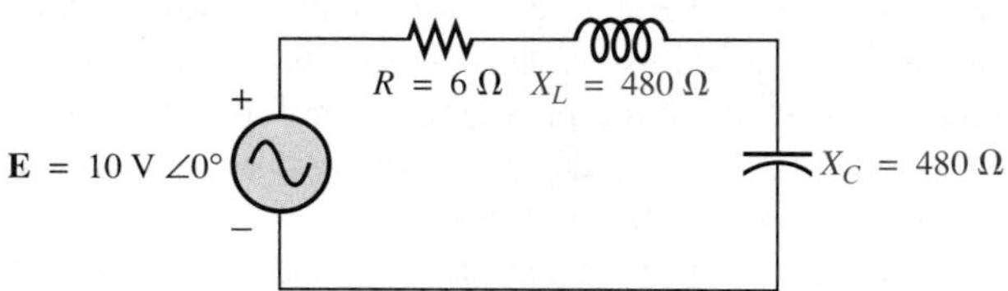

FIG. 20.7
High-Q series resonant circuit.

20.4 Z_T VERSUS FREQUENCY

The total impedance of the series *R-L-C* circuit of Fig. 20.2 at any frequency is determined by

$$\mathbf{Z}_T = R + jX_L - jX_C \quad \text{or} \quad \mathbf{Z}_T = R + j(X_L - X_C)$$

The magnitude of the impedance $\mathbf{Z}_T$ versus frequency is determined by

$$Z_T = \sqrt{R^2 + (X_L - X_C)^2}$$

The total-impedance-versus-frequency curve for the series resonant circuit of Fig. 20.2 can be found by applying the impedance-versus-frequency curve for each element of the equation just derived, written in the following form:

$$Z_T(f) = \sqrt{[R(f)]^2 + [X_L(f) - X_C(f)]^2} \tag{20.14}$$

where $Z_T(f)$ "means" the total impedance as a *function* of frequency. For the frequency range of interest, we will assume that the resistance R does not change with frequency, resulting in the plot of Fig. 20.8. The curve for the inductance, as determined by the reactance equation, is a

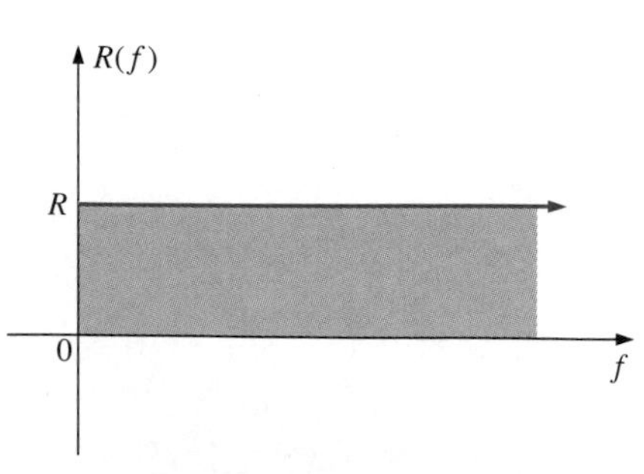

FIG. 20.8
Resistance versus frequency.

straight line intersecting the origin with a slope equal to the inductance of the coil. The mathematical expression for any straight line in a two-dimensional plane is given by

$$y = mx + b$$

Thus, for the coil,

$$\begin{array}{ccccccc} X_L = 2\pi fL + 0 = & (2\pi L) & (f) & + & 0 \\ \downarrow & \downarrow & \downarrow & & \downarrow \\ y = & m & \cdot\ x & + & b \end{array}$$

(where $2\pi L$ is the slope), producing the results shown in Fig. 20.9.

For the capacitor,

$$X_C = \frac{1}{2\pi fC} \quad \text{or} \quad X_C f = \frac{1}{2\pi C}$$

which becomes $yx = k$, the equation for a hyperbola, where

$$y \text{ (variable)} = X_C$$
$$x \text{ (variable)} = f$$
$$k \text{ (constant)} = \frac{1}{2\pi C}$$

The hyperbolic curve for $X_C(f)$ is plotted in Fig. 20.10. In particular, note its very large magnitude at low frequencies and its rapid drop-off as the frequency increases.

If we place Figs. 20.9 and 20.10 on the same axis, we obtain the curves of Fig. 20.11. The condition of resonance is now clearly defined by the point of intersection, where $X_L = X_C$. For frequencies less than f_s, it is also quite clear that the network is primarily capacitive ($X_C > X_L$). For frequencies above the resonant condition, $X_L > X_C$ and the network is inductive.

Applying

$$Z_T(f) = \sqrt{[R(f)]^2 + [X_L(f) - X_C(f)]^2}$$
$$= \sqrt{[R(f)]^2 + [X(f)]^2}$$

to the curves of Fig. 20.11, where $X(f) = X_L(f) - X_C(f)$, we obtain the curve for $Z_T(f)$ as shown in Fig. 20.12. The minimum impedance occurs at the resonant frequency and is equal to the resistance R. Note that the curve is not symmetrical about the resonant frequency (especially at higher values of Z_T).

The phase angle associated with the total impedance is

$$\theta = \tan^{-1}\frac{(X_L - X_C)}{R} \qquad \textbf{(20.15)}$$

For the $\tan^{-1}x$ function (resulting when $X_L > X_C$), the larger x is, the larger the angle θ (closer to 90°). However, for regions where $X_C > X_L$ one must also be aware that

$$\tan^{-1}(-x) = -\tan^{-1}x \qquad \textbf{(20.16)}$$

At low frequencies, $X_C > X_L$ and θ will approach $-90°$ (capacitive), as shown in Fig. 20.13, whereas at high frequencies, $X_L > X_C$ and θ will approach 90°. In general, therefore, for a series resonant circuit:

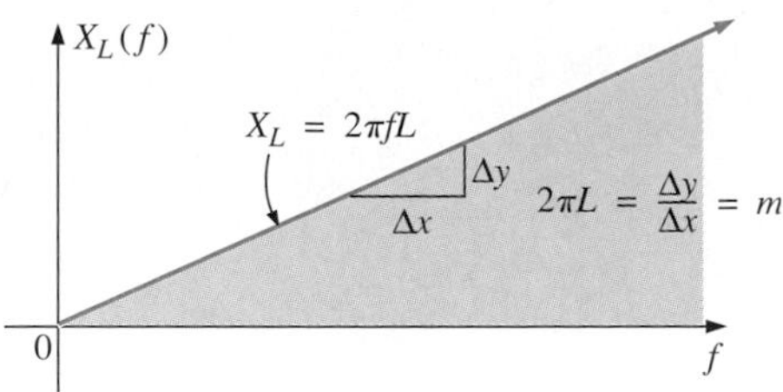

FIG. 20.9
Inductive reactance versus frequency.

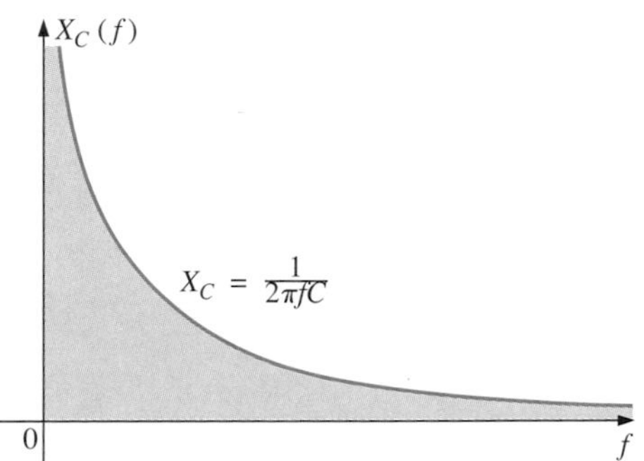

FIG. 20.10
Capacitive reactance versus frequency.

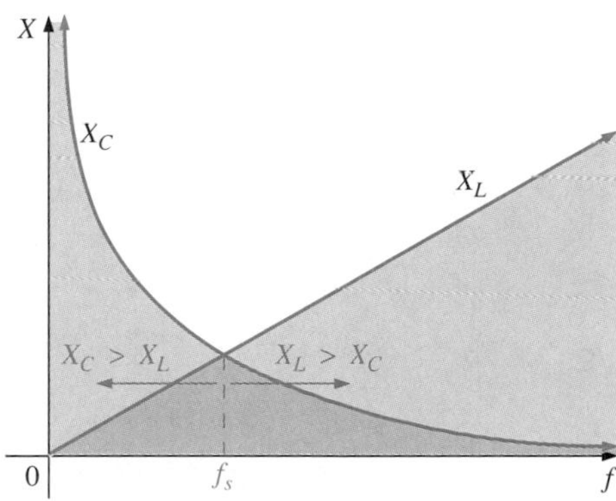

FIG. 20.11
Placing the frequency response of the inductive and capacitive reactance of a series R-L-C circuit on the same axis.

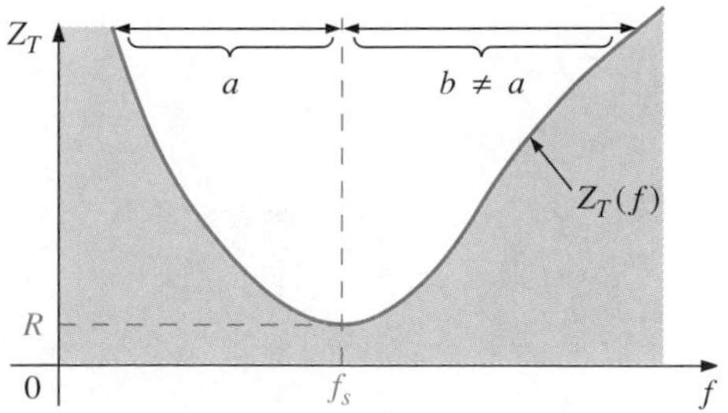

FIG. 20.12
Z_T versus frequency for the series resonant circuit.

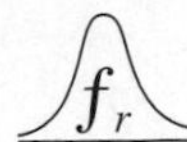

$f < f_s$:	network capacitive, **I** leads **E**
$f > f_s$:	network inductive, **E** leads **I**
$f = f_s$:	network resistive, **E** and **I** are in phase

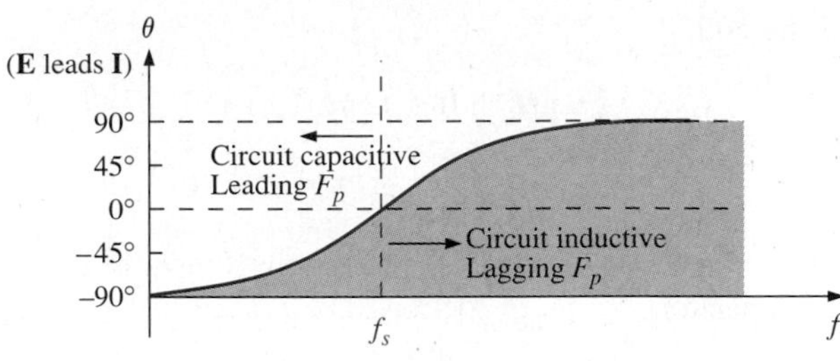

FIG. 20.13
Phase plot for the series resonant circuit.

20.5 SELECTIVITY

If we now plot the magnitude of the current $I = E/Z_T$ versus frequency for a *fixed* applied voltage E, we obtain the curve shown in Fig. 20.14, which rises from zero to a maximum value of E/R (where Z_T is a minimum) and then drops toward zero (as Z_T increases) at a slower rate than it rose to its peak value. The curve is actually the inverse of the impedance-versus-frequency curve. Since the Z_T curve is not absolutely symmetrical about the resonant frequency, the curve of the current versus frequency has the same property.

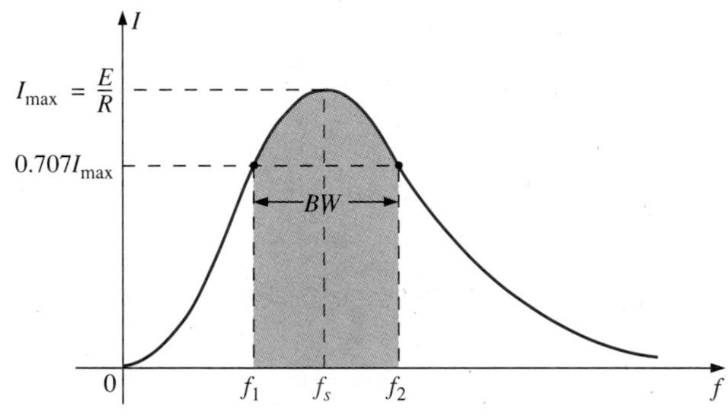

FIG. 20.14
I versus frequency for the series resonant circuit.

There is a definite range of frequencies at which the current is near its maximum value and the impedance is at a minimum. Those frequencies corresponding to 0.707 of the maximum current are called the *band frequencies, cutoff frequencies,* or *half-power frequencies.* They are indicated by f_1 and f_2 in Fig. 20.14. The range of frequencies between the two is referred to as the *bandwidth* (abbreviated BW) of the resonant circuit.

Half-power frequencies are those frequencies at which the power delivered is one-half that delivered at the resonant frequency; that is,

$$P_{HPF} = \frac{1}{2} P_{max} \tag{20.17}$$

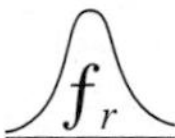

The above condition is derived using the fact that

$$P_{max} = I^2_{max}R$$

and $\quad P_{HPF} = I^2R = (0.707I_{max})^2R = (0.5)(I^2_{max}R) = \frac{1}{2}P_{max}$

Since the resonant circuit is adjusted to select a band of frequencies, the curve of Fig. 20.14 is called the *selectivity curve*. The term is derived from the fact that one must be *selective* in choosing the frequency to ensure that it is in the bandwidth. The smaller the bandwidth, the higher the selectivity. The shape of the curve, as shown in Fig. 20.15, depends on each element of the series *R-L-C* circuit. If the resistance is made smaller with a fixed inductance and capacitance, the bandwidth decreases and the selectivity increases. Similarly, if the ratio L/C increases with fixed resistance, the bandwidth again decreases with an increase in selectivity.

In terms of Q_s, if R is larger for the same X_L, then Q_s is less, as determined by the equation $Q_s = \omega_s L/R$.

A small Q_s, therefore, is associated with a resonant curve having a large bandwidth and a small selectivity, while a large Q_s indicates the opposite.

For circuits where $Q_s \geq 10$, a widely accepted approximation is that the resonant frequency bisects the bandwidth and that the resonant curve is symmetrical about the resonant frequency.

These conditions are shown in Fig. 20.16, indicating that the cutoff frequencies are then equidistant from the resonant frequency.

For any Q_s, the preceding is not true. The cutoff frequencies f_1 and f_2 can be found for the general case (any Q_s) by first employing the fact that a drop in current to 0.707 of its resonant value corresponds to an increase in impedance equal to $1/0.707 = \sqrt{2}$ times the resonant value, which is R.

Substituting $\sqrt{2}R$ into the equation for the magnitude of Z_T, we find that

$$Z_T = \sqrt{R^2 + (X_L - X_C)^2}$$

becomes

$$\sqrt{2}R = \sqrt{R^2 + (X_L - X_C)^2}$$

or, squaring both sides, that

$$2R^2 = R^2 + (X_L - X_C)^2$$

and

$$R^2 = (X_L - X_C)^2$$

Taking the square root of both sides gives us

$$R = X_L - X_C \quad \text{or} \quad R - X_L + X_C = 0$$

Let us first consider the case where $X_L > X_C$, which relates to f_2 or ω_2. Substituting $\omega_2 L$ for X_L and $1/\omega_2 C$ for X_C and bringing both quantities to the left of the equal sign, we have

$$R - \omega_2 L + \frac{1}{\omega_2 C} = 0 \quad \text{or} \quad R\omega_2 - \omega_2^2 L + \frac{1}{C} = 0$$

which can be written

$$\omega_2^2 - \frac{R}{L}\omega_2 - \frac{1}{LC} = 0$$

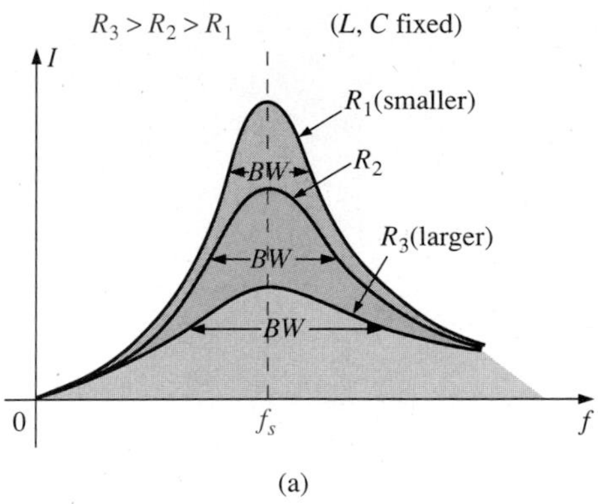

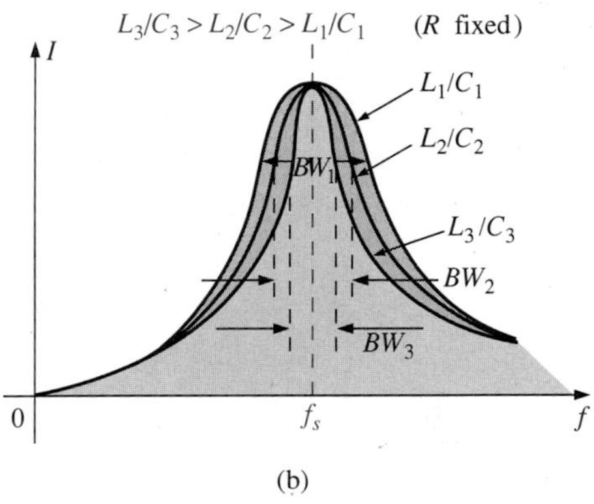

FIG. 20.15
Effect of R, L, and C on the selectivity curve for the series resonant circuit.

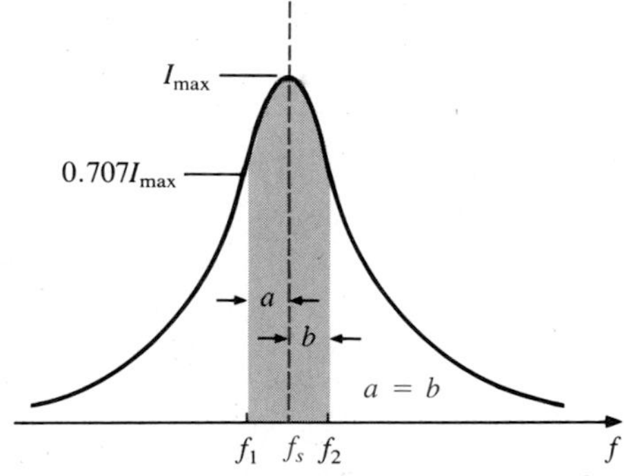

FIG. 20.16
Approximate series resonance curve for $Q_s \geq 10$.

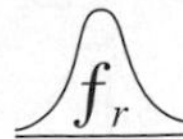

Solving the quadratic, we have

$$\omega_2 = \frac{-(-R/L) \pm \sqrt{[-(R/L)]^2 - [-(4/LC)]}}{2}$$

and
$$\omega_2 = +\frac{R}{2L} \pm \frac{1}{2}\sqrt{\frac{R^2}{L^2} + \frac{4}{LC}}$$

with
$$f_2 = \frac{1}{2\pi}\left[\frac{R}{2L} + \frac{1}{2}\sqrt{\left(\frac{R}{L}\right)^2 + \frac{4}{LC}}\right] \quad \text{(Hz)} \qquad \textbf{(20.18)}$$

The negative sign in front of the second factor was dropped because $(1/2)\sqrt{(R/L)^2 + 4/LC}$ is always greater than $R/(2L)$. If it were not dropped, there would be a negative solution for the radian frequency ω.

If we repeat the same procedure for $X_C > X_L$, which relates to ω_1 or f_1 such that $Z_T = \sqrt{R^2 + (X_C - X_L)^2}$, the solution f_1 becomes

$$f_1 = \frac{1}{2\pi}\left[-\frac{R}{2L} + \frac{1}{2}\sqrt{\left(\frac{R}{L}\right)^2 + \frac{4}{LC}}\right] \quad \text{(Hz)} \qquad \textbf{(20.19)}$$

The bandwidth (BW) is

$$BW = f_2 - f_1 = \text{Eq. (20.18)} - \text{Eq. (20.19)}$$

and
$$BW = f_2 - f_1 = \frac{R}{2\pi L} \qquad \textbf{(20.20)}$$

Substituting $R/L = \omega_s/Q_s$ from $Q_s = \omega_s L/R$ and $1/2\pi = f_s/\omega_s$ from $\omega_s = 2\pi f_s$ gives us

$$BW = \frac{R}{2\pi L} = \left(\frac{1}{2\pi}\right)\left(\frac{R}{L}\right) = \left(\frac{f_s}{\omega_s}\right)\left(\frac{\omega_s}{Q_s}\right)$$

or
$$BW = \frac{f_s}{Q_s} \qquad \textbf{(20.21)}$$

which is a very convenient form since it relates the bandwidth to the Q_s of the circuit. As mentioned earlier, Eq. (20.21) verifies that the larger the Q_s, the smaller the bandwidth, and vice versa.

Written in a slightly different form, Eq. (20.21) becomes

$$\frac{f_2 - f_1}{f_s} = \frac{1}{Q_s} \qquad \textbf{(20.22)}$$

The ratio $(f_2 - f_1)/f_s$ is sometimes called the *fractional bandwidth,* providing an indication of the width of the bandwidth as compared to the resonant frequency.

It can also be shown through mathematical manipulations of the pertinent equations that the resonant frequency is related to the geometric mean of the band frequencies; that is,

$$f_s = \sqrt{f_1 f_2} \tag{20.23}$$

20.6 V_R, V_L, AND V_C

Plotting the magnitude (effective value) of the voltages $\mathbf{V}_R$, $\mathbf{V}_L$, and $\mathbf{V}_C$ and the current $\mathbf{I}$ versus frequency for the series resonant circuit on the same set of axes, we obtain the curves shown in Fig. 20.17. Note that the V_R curve has the same shape as the I curve and a peak value equal to the magnitude of the input voltage E. The V_C curve builds up slowly at first from a value equal to the input voltage since the reactance of the capacitor is infinite (open circuit) at zero frequency and the reactance of the inductor is zero (short circuit) at this frequency. As the frequency increases, $1/\omega C$ of the equation

$$V_C = IX_C = (I)\left(\frac{1}{\omega C}\right)$$

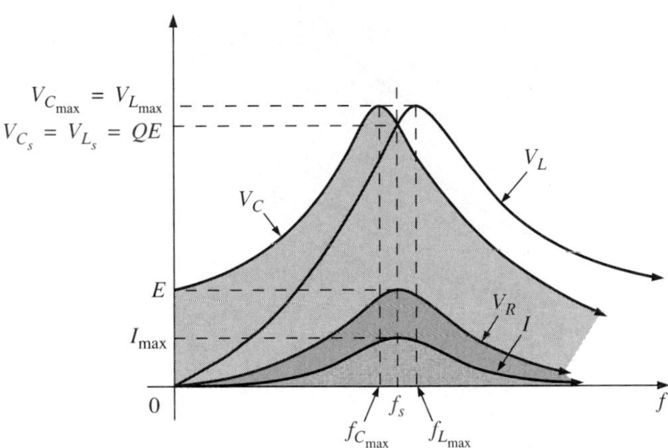

FIG. 20.17
V_R, V_L, V_C, and I versus frequency for a series resonant circuit.

becomes smaller, but I increases at a rate faster than that at which $1/\omega C$ drops. Therefore, V_C rises and will continue to rise due to the quickly rising current, until the frequency nears resonance. As it approaches the resonant condition, the rate of change of I decreases. When this occurs, the factor $1/\omega C$, which decreased as the frequency rose, will overcome the rate of change of I, and V_C will start to drop. The peak value will occur at a frequency just before resonance. After resonance, both V_C and I drop in magnitude, and V_C approaches zero.

The higher the Q_s of the circuit, the closer $f_{C_{max}}$ will be to f_s, and the closer $V_{C_{max}}$ will be to Q_sE. For circuits with $Q_s \geq 10$, $f_{C_{max}} \cong f_s$, and $V_{C_{max}} \cong Q_s E$.

The curve for V_L increases steadily from zero to the resonant frequency since both quantities ωL and I of the equation $V_L = IX_L = (I)(\omega L)$ increase over this frequency range. At resonance, I has reached its maximum value, but ωL is still rising. Therefore, V_L will reach its maximum value after resonance. After reaching its peak value, the voltage V_L will drop toward E since the drop in I will overcome the rise in ωL. It approaches E because X_L will eventually be infinite, and X_C will be zero.

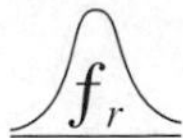

As Q_s of the circuit increases, the frequency $f_{L_{max}}$ drops toward f_s, and $V_{L_{max}}$ approaches $Q_s E$. For circuits with $Q_s \geq 10$, $f_{L_{max}} \cong f_s$, and $V_{L_{max}} \cong Q_s E$.

The V_L curve has a greater magnitude than the V_C curve for any frequency above resonance, and the V_C curve has a greater magnitude than the V_L curve for any frequency below resonance. This again verifies that the series R-L-C circuit is predominantly capacitive from zero to the resonant frequency and predominantly inductive for any frequency above resonance.

For the condition $Q_s \geq 10$, the curves of Fig. 20.17 will appear as shown in Fig. 20.18. Note that they each peak (on an approximate basis) at the resonant frequency and have a similar shape.

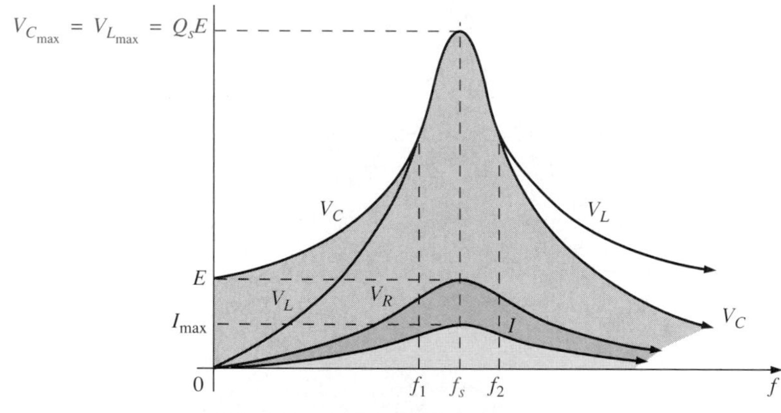

FIG. 20.18
V_R, V_L, V_C, and I for a series resonant circuit where $Q_s \geq 10$.

In review,

1. V_C and V_L are at their maximum values at or near resonance (depending on Q_s).
2. At very low frequencies, V_C is very close to the source voltage and V_L is very close to zero volts, whereas at very high frequencies, V_L approaches the source voltage and V_C approaches zero volts.
3. Both V_R and I peak at the resonant frequency and have the same shape.

20.7 EXAMPLES (SERIES RESONANCE)

EXAMPLE 20.1

a. For the series resonant circuit of Fig. 20.19, find $\mathbf{I}$, $\mathbf{V}_R$, $\mathbf{V}_L$, and $\mathbf{V}_C$ at resonance.
b. What is the Q_s of the circuit?
c. If the resonant frequency is 5000 Hz, find the bandwidth.
d. What is the power dissipated in the circuit at the half-power frequencies?

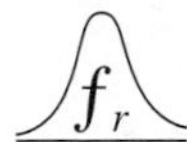

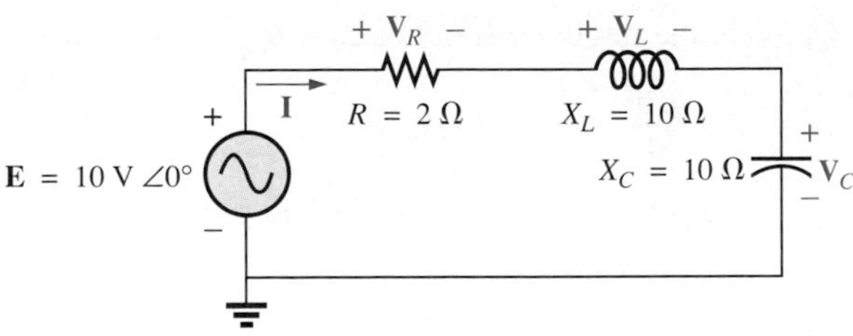

FIG. 20.19
Example 20.1.

Solutions:

a. $\mathbf{Z}_{T_s} = R = 2\ \Omega$

$$\mathbf{I} = \frac{\mathbf{E}}{\mathbf{Z}_{T_s}} = \frac{10\ \text{V}\ \angle 0^\circ}{2\ \Omega\ \angle 0^\circ} = \mathbf{5\ A\ \angle 0^\circ}$$

$$\mathbf{V}_R = \mathbf{E} = 10\ \text{V}\ \angle 0^\circ$$

$$\mathbf{V}_L = (I\ \angle 0^\circ)(X_L\ \angle 90^\circ) = (5\ \text{A}\ \angle 0^\circ)(10\ \Omega\ \angle 90^\circ) = \mathbf{50\ V\ \angle 90^\circ}$$

$$\mathbf{V}_C = (I\ \angle 0^\circ)(X_C\ \angle -90^\circ) = (5\ \text{A}\ \angle 0^\circ)(10\ \Omega\ \angle -90^\circ) = \mathbf{50\ V\ \angle -90^\circ}$$

b. $Q_s = \dfrac{X_L}{R} = \dfrac{10\ \Omega}{2\ \Omega} = \mathbf{5}$

c. $BW = f_2 - f_1 = \dfrac{f_s}{Q_s} = \dfrac{5000\ \text{Hz}}{5} = \mathbf{1000\ Hz}$

d. $P_{\text{HPF}} = \dfrac{1}{2}P_{\text{max}} = \dfrac{1}{2}I^2_{\text{max}}R = \left(\dfrac{1}{2}\right)(5\ \text{A})^2(2\ \Omega) = \mathbf{25\ W}$

EXAMPLE 20.2 The bandwidth of a series resonant circuit is 400 Hz.

a. If the resonant frequency is 4000 Hz, what is the value of Q_s?
b. If $R = 10\ \Omega$, what is the value of X_L at resonance?
c. Find the inductance L and capacitance C of the circuit.

Solutions:

a. $BW = \dfrac{f_s}{Q_s}$ or $Q_s = \dfrac{f_s}{BW} = \dfrac{4000\ \text{Hz}}{400\ \text{Hz}} = \mathbf{10}$

b. $Q_s = \dfrac{X_L}{R}$ or $X_L = Q_sR = (10)(10\ \Omega) = \mathbf{100\ \Omega}$

c. $X_L = 2\pi f_sL$ or $L = \dfrac{X_L}{2\pi f_s} = \dfrac{100\ \Omega}{2\pi(4000\ \text{Hz})} = \mathbf{3.98\ mH}$

$$X_C = \frac{1}{2\pi f_sC} \quad \text{or} \quad C = \frac{1}{2\pi f_sX_C} = \frac{1}{2\pi(4000\ \text{Hz})(100\ \Omega)}$$

$$= \mathbf{0.398\ \mu F}$$

EXAMPLE 20.3 A series *R-L-C* circuit has a series resonant frequency of 12,000 Hz.

a. If $R = 5\ \Omega$ and X_L at resonance is 300 Ω, find the bandwidth.
b. Find the cutoff frequencies.

Solutions:

a. $Q_s = \dfrac{X_L}{R} = \dfrac{300\ \Omega}{5\ \Omega} = 60$

$$BW = \frac{f_s}{Q_s} = \frac{12{,}000\ \text{Hz}}{60} = \mathbf{200\ Hz}$$

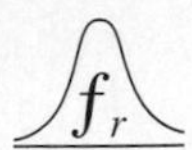

b. Since $Q_s \geq 10$, the bandwidth is bisected by f_s. Therefore,

$$f_2 = f_s + \frac{BW}{2} = 12{,}000 \text{ Hz} + 100 \text{ Hz} = \mathbf{12{,}100 \text{ Hz}}$$

and $f_1 = 12{,}000 \text{ Hz} - 100 \text{ Hz} = \mathbf{11{,}900 \text{ Hz}}$

EXAMPLE 20.4

a. Determine the Q_s and bandwidth for the response curve of Fig. 20.20.
b. For $C = 101.5$ nF, determine L and R for the series resonant circuit.
c. Determine the applied voltage.

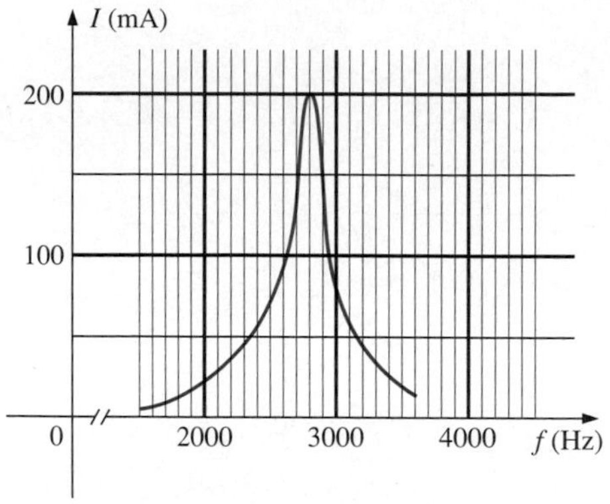

FIG. 20.20
Example 20.4.

Solutions:

a. The resonant frequency is 2800 Hz. At 0.707 times the peak value,

$$BW = \mathbf{200 \text{ Hz}}$$

and
$$Q_s = \frac{f_s}{BW} = \frac{2800 \text{ Hz}}{200 \text{ Hz}} = \mathbf{14}$$

b. $f_s = \dfrac{1}{2\pi\sqrt{LC}}$ or $L = \dfrac{1}{4\pi^2 f_s^2 C}$

$$= \frac{1}{4\pi^2(2.8 \times 10^3 \text{ Hz})^2(101.5 \times 10^{-9} \text{ F})}$$

$$= \mathbf{31.832 \text{ mH}}$$

$$Q_s = \frac{X_L}{R} \quad \text{or} \quad R = \frac{X_L}{Q_s} = \frac{2\pi(2800 \text{ Hz})(31.832 \times 10^{-3} \text{ H})}{14}$$

$$= \mathbf{40 \ \Omega}$$

c. $I_{\max} = \dfrac{E}{R}$ or $E = I_{\max}R$

and $E = (200 \text{ mA})(40 \ \Omega) = \mathbf{8 \text{ V}}$

EXAMPLE 20.5 A series *R-L-C* circuit is designed to resonant at $\omega_s = 10^5$ rad/s, have a bandwidth of $0.15\omega_s$, and draw 16 W from a 120-V source at resonance.

a. Determine the value of R.
b. Find the bandwidth in hertz.
c. Find the nameplate values of L and C.
d. What is the Q_s of the circuit?
e. Determine the fractional bandwidth.

Solutions:

a. $P = \dfrac{E^2}{R}$ and $R = \dfrac{E^2}{P} = \dfrac{(120 \text{ V})^2}{16 \text{ W}} = \mathbf{900 \ \Omega}$

b. $f_s = \dfrac{\omega_s}{2\pi} = \dfrac{10^5 \text{ rad/s}}{2\pi} = 15{,}915.49 \text{ Hz}$

$$BW = 0.15f_s = 0.15(15{,}915.49 \text{ Hz}) = \mathbf{2387.32 \text{ Hz}}$$

c. Eq. (20.20):

$$BW = \frac{R}{2\pi L} \quad \text{and} \quad L = \frac{R}{2\pi BW} = \frac{900 \ \Omega}{2\pi(2387.32 \text{ Hz})} = \mathbf{60 \text{ mH}}$$

$$f_s = \frac{1}{2\pi\sqrt{LC}} \quad \text{and} \quad C = \frac{1}{4\pi^2 f_s^2 L} = \frac{1}{4\pi^2(15{,}915.49 \text{ Hz})^2(60 \times 10^{-3} \text{ H})}$$

$$= \mathbf{1.67 \text{ nF}}$$

d. $Q_s = \frac{X_L}{R} = \frac{2\pi f_s L}{R} = \frac{2\pi(15{,}915.49 \text{ Hz})(60 \text{ mH})}{900\ \Omega} = \mathbf{6.67}$

e. $\frac{f_2 - f_1}{f_s} = \frac{BW}{f_s} = \frac{1}{Q_s} = \frac{1}{6.67} = \mathbf{0.15}$

PARALLEL RESONANCE

20.8 PARALLEL RESONANT CIRCUIT

The basic format of the series resonant circuit is a series *R-L-C* combination in series with an applied voltage source. The parallel resonant circuit has the basic configuration of Fig. 20.21, a parallel *R-L-C* combination in parallel with an applied current source.

For the series circuit, the impedance was a minimum at resonance, producing a significant current that resulted in a high output voltage for $\mathbf{V}_C$ and $\mathbf{V}_L$. For the parallel resonant circuit, the impedance is relatively high at resonance, producing a significant voltage for $\mathbf{V}_C$ and $\mathbf{V}_L$ through the Ohm's law relationship ($\mathbf{V}_C = \mathbf{IZ}_T$). For the network of Fig. 20.21, resonance will occur when $X_L = X_C$, and the resonant frequency will have the same format obtained for series resonance.

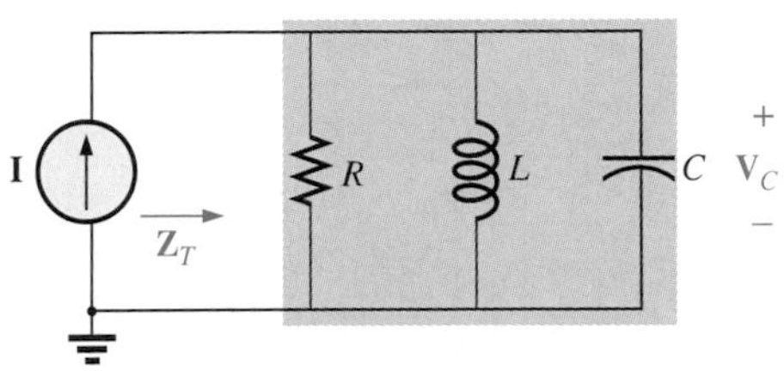

FIG. 20.21
Ideal parallel resonant network.

If the practical equivalent of Fig. 20.21 had the format of Fig. 20.21, the analysis would be as direct and lucid as that experienced for series resonance. However, in the practical world, the internal resistance of the coil must be placed in series with the inductor, as shown in Fig. 20.22. The resistance R_l can no longer be included in a simple series or parallel combination with the source resistance and any other resistance added for design purposes. Even though R_l is usually relatively small in magnitude compared with other resistance and reactance levels of the network, it does have an important impact on the parallel resonant condition, as will be demonstrated in the sections to follow. In other words, the network of Fig. 20.21 is an ideal situation that can be assumed only for specific network conditions.

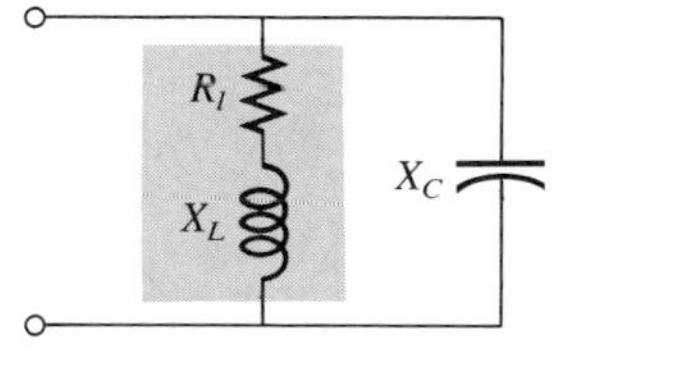

FIG. 20.22
Practical parallel L-C network.

Our first effort will be to find a parallel network equivalent (at the terminals) for the series *R-L* branch of Fig. 20.22 using the technique introduced in Section 15.10. That is,

$$\mathbf{Z}_{R\text{-}L} = R_l + j X_L$$

and
$$\mathbf{Y}_{R\text{-}L} = \frac{1}{\mathbf{Z}_{R\text{-}L}} = \frac{1}{R_l + j X_L} = \frac{R_l}{R_l^2 + X_L^2} - j \frac{X_L}{R_l^2 + X_L^2}$$

$$= \frac{1}{\dfrac{R_l^2 + X_L^2}{R_l}} + \frac{1}{j\left(\dfrac{R_l^2 + X_L^2}{X_L}\right)} = \frac{1}{R_p} + \frac{1}{j X_{L_p}}$$

with
$$R_p = \frac{R_l^2 + X_L^2}{R_l} \tag{20.24}$$

and
$$X_{L_p} = \frac{R_l^2 + X_L^2}{X_L} \tag{20.25}$$

as shown in Fig. 20.23.

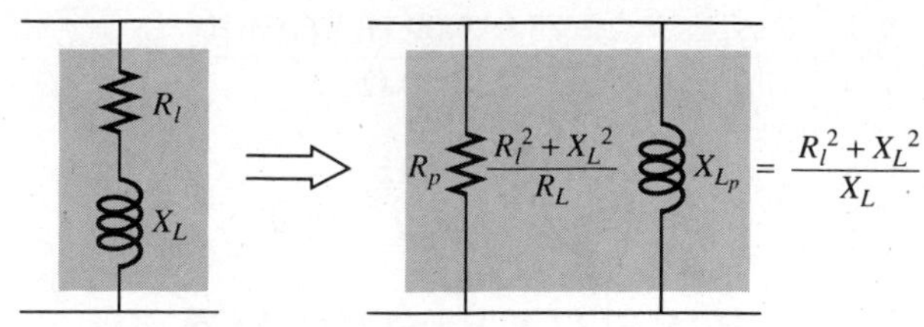

FIG. 20.23
Equivalent parallel network for a series R-L combination.

Redrawing the network of Fig. 20.22 with the equivalent of Fig. 20.23 and a practical current source having an internal resistance R_s will result in the network of Fig. 20.24.

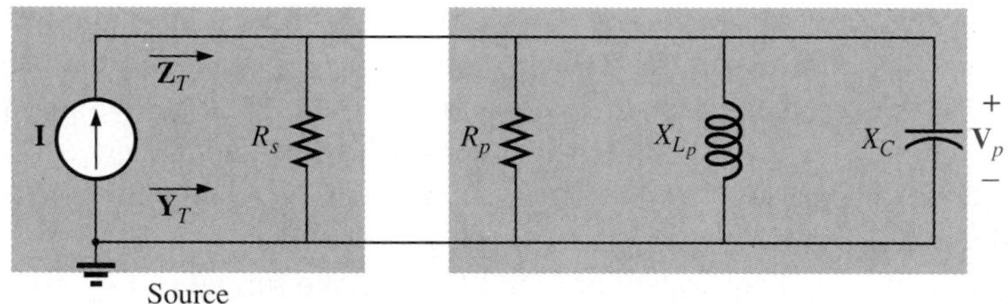

FIG. 20.24
Substituting the equivalent parallel network for the series R-L combination of Fig. 20.22.

If we define the parallel combination of R_s and R_p by the notation

$$R = R_s \| R_p \qquad \textbf{(20.26)}$$

the network of Fig. 20.25 will result. It has the same format as the ideal configuration of Fig. 20.21.

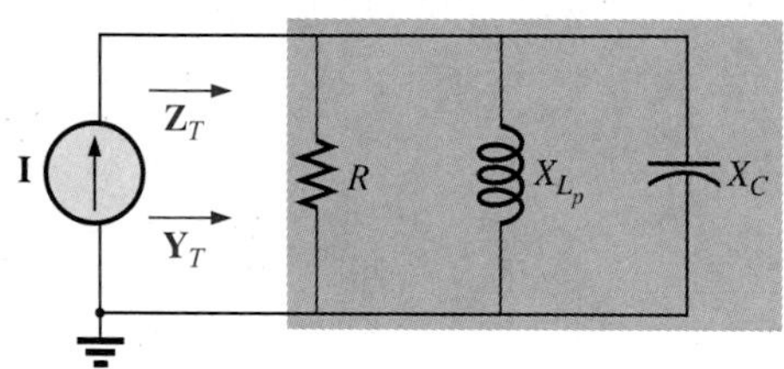

FIG. 20.25
Substituting $R = R_s \| R_p$ for the network of Fig. 20.24.

We are now at a point where we can define the resonance conditions for the practical parallel resonant configuration. Recall for series resonance that the resonant frequency was the frequency at which the impedance was a minimum, the current a maximum, and the input impedance purely resistive, and the network had a unity power factor. For parallel networks, since the resistance R_p in our equivalent model is frequency dependent, the frequency at which maximum V_C is obtained is not the same as required for the unity power factor characteristic. Since both conditions are often used to define the resonant state, the frequency at which each occurs will be designated by different subscripts.

Unity Power Factor, f_p

For the network of Fig. 20.25,

$$\mathbf{Y}_T = \frac{1}{\mathbf{Z}_1} + \frac{1}{\mathbf{Z}_2} + \frac{1}{\mathbf{Z}_3} = \frac{1}{R} + \frac{1}{jX_{L_p}} + \frac{1}{-jX_C}$$

$$= \frac{1}{R} - j\left(\frac{1}{X_{L_p}}\right) + j\left(\frac{1}{X_C}\right)$$

and
$$\mathbf{Y}_T = \frac{1}{R} + j\left(\frac{1}{X_C} - \frac{1}{X_{L_p}}\right) \quad \textbf{(20.27)}$$

For unity power factor, the reactive component must be zero as defined by

$$\frac{1}{X_C} - \frac{1}{X_{L_p}} = 0$$

Therefore,
$$\frac{1}{X_C} = \frac{1}{X_{L_p}}$$

and
$$X_{L_p} = X_C \quad \textbf{(20.28)}$$

Substituting for X_{L_p} yields

$$\frac{R_l^2 + X_L^2}{X_L} = X_C \quad \textbf{(20.29)}$$

The resonant frequency, f_p, can now be determined from Eq. (20.29) as follows:

$$R_l^2 + X_L^2 = X_C X_L = \left(\frac{1}{\omega C}\right)\omega L = \frac{L}{C}$$

or
$$X_L^2 = \frac{L}{C} - R_l^2$$

with
$$2\pi f_p L = \sqrt{\frac{L}{C} - R_l^2}$$

and
$$f_p = \frac{1}{2\pi L}\sqrt{\frac{L}{C} - R_l^2}$$

Multiplying the top and bottom of the factor within the square-root sign by C/L produces

$$f_p = \frac{1}{2\pi L}\sqrt{\frac{1 - R_l^2(C/L)}{C/L}} = \frac{1}{2\pi L\sqrt{C/L}}\sqrt{1 - \frac{R_l^2 C}{L}}$$

and
$$f_p = \frac{1}{2\pi\sqrt{LC}}\sqrt{1 - \frac{R_l^2 C}{L}} \quad \textbf{(20.30)}$$

or
$$f_p = f_s\sqrt{1 - \frac{R_l^2 C}{L}} \quad \textbf{(20.31)}$$

where f_p is the resonant frequency of a parallel resonant circuit (for $F_p = 1$) and f_s is the resonant frequency as determined by $X_L = X_C$ for series resonance. Note that unlike a series resonant circuit, the resonant

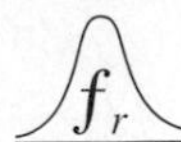

frequency f_p is a function of resistance (in this case R_l). Note also, however, the absence of the source resistance R_s in Eqs. (20.30) and (20.31). Since the factor $\sqrt{1 - (R_l^2C/L)}$ is less than one, f_p is less than f_s. Recognize also that as the magnitude of R_l approaches zero, f_p rapidly approaches f_s.

Maximum Impedance, f_m

At $f = f_p$ the input impedance of a parallel resonant circuit will be near its maximum value but not quite its maximum value due to the frequency dependence of R_p. The frequency at which maximum impedance will occur is defined by f_m and is slightly more than f_p, as demonstrated in Fig. 20.26. The frequency f_m is determined by differentiating (calculus) the general equation for Z_T with respect to frequency and then determining the frequency at which the resulting equation is equal to zero. The algebra is quite extensive and cumbersome and will not be included here. The resulting equation, however, is the following:

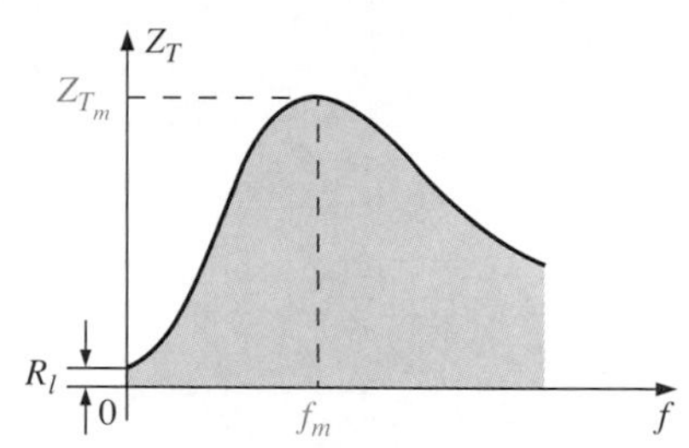

FIG. 20.26
Z_T versus frequency for the parallel resonant circuit.

$$f_m = f_s\sqrt{1 - \frac{1}{4}\left(\frac{R_l^2C}{L}\right)} \qquad \textbf{(20.32)}$$

Note the similarities with Eq. (20.31). Since the square root factor of Eq. (20.32) is always more than the similar factor of Eq. (20.31), f_m is always closer to f_s and more than f_p. In general,

$$f_s > f_m > f_p \qquad \textbf{(20.33)}$$

Once f_m is determined, the network of Fig. 20.25 can be used to determine the magnitude and phase angle of the total impedance at the resonance condition simply by substituting $f = f_m$ and performing the required calculations. That is,

$$Z_{T_m} = R \parallel X_{L_p} \parallel X_C \Big|_{f = f_m} \qquad \textbf{(20.34)}$$

20.9 SELECTIVITY CURVE FOR PARALLEL RESONANT CIRCUITS

The Z_T-versus-frequency curve of Fig. 20.26 clearly reveals that a parallel resonant circuit exhibits maximum impedance at resonance (f_m), unlike the series resonant circuit, which experiences minimum resistance levels at resonance. Note also that Z_T is approximately R_l at $f =$ 0 Hz since $Z_T = R_s \parallel R_l \cong R_l$.

Since the current I of the current source is constant for any value of Z_T or frequency, the voltage across the parallel circuit will have the same shape as the total impedance Z_T, as shown in Fig. 20.27.

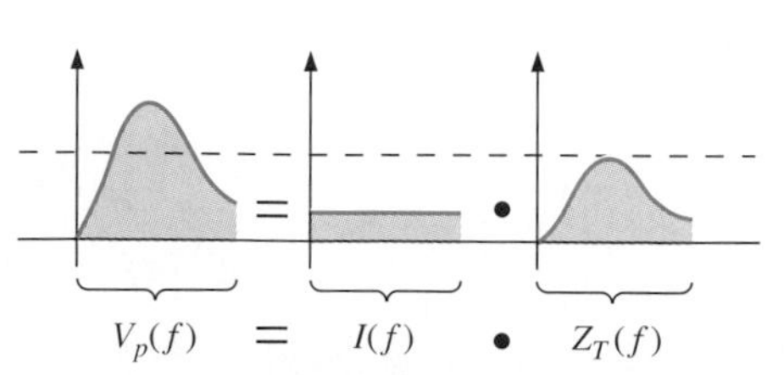

FIG. 20.27
Defining the shape of the $V_p(f)$ curve.

For the parallel circuit, the resonance curve of interest is that of the voltage V_C across the capacitor. The reason for this interest in V_C derives from electronic considerations that often place the capacitor at the input to another stage of a network.

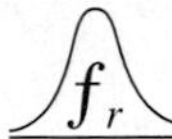

Since the voltage across parallel elements is the same,

$$V_C = V_p = IZ_T \tag{20.35}$$

The resonant value of V_C is therefore determined by the value of Z_{T_m} and the magnitude of the current source I.

The quality factor of the parallel resonant circuit continues to be determined by the ratio of the reactive power to the real power. That is,

$$Q_p = \frac{V_p^2/X_{L_p}}{V_p^2/R}$$

where $R = R_s \| R_p$, and V_p is the voltage across the parallel branches. The result is

$$Q_p = \frac{R}{X_{L_p}} = \frac{R_s \| R_p}{X_{L_p}} \tag{20.36a}$$

or since $X_{L_p} = X_C$ at resonance,

$$Q_p = \frac{R_s \| R_p}{X_C} \tag{20.36b}$$

For the ideal current source ($R_s = \infty\ \Omega$) or when R_s is sufficiently large compared to R_p, we can make the following approximation:

$$R = R_s \| R_p \cong R_p$$

and

$$Q_p = \frac{R_s \| R_p}{X_{L_p}} = \frac{R_p}{X_{L_p}} = \frac{(R_l^2 + X_L^2)/R_l}{(R_l^2 + X_L^2)/X_L}$$

so that

$$Q_p = \frac{X_L}{R_l} = Q_l \qquad R_s \gg R_p \tag{20.37}$$

which is simply the quality factor Q_l of the coil.

In general, the bandwidth is still related to the resonant frequency and the quality factor by

$$BW = f_2 - f_1 = \frac{f_r}{Q_p} \tag{20.38}$$

The cutoff frequencies f_1 and f_2 can be determined using the equivalent network of Fig. 20.25 and the unity power condition for resonance. The half-power frequencies are defined by the condition that the output voltage is 0.707 times the maximum value. However, for parallel resonance with a current source driving the network, the frequency response for the driving point impedance is the same as that for the output voltage. This similarity permits defining each cutoff frequency as the frequency at which the input impedance is 0.707 times its maximum value. Since the maximum value is the equivalent resistance R of Fig. 20.25, the cutoff frequencies will be associated with an impedance equal to $0.707R$ or $(1/\sqrt{2})R$.

Setting the input impedance for the network of Fig. 20.25 equal to this value will result in the following relationship:

$$\mathbf{Z} = \frac{1}{\dfrac{1}{R} + j\left(\omega C - \dfrac{1}{\omega L}\right)} = 0.707R$$

which can be written as

$$\mathbf{Z} = \frac{1}{\dfrac{1}{R}\left[1 + jR\left(\omega C - \dfrac{1}{\omega L}\right)\right]} = \frac{R}{\sqrt{2}}$$

or

$$\frac{R}{1 + jR\left(\omega C - \dfrac{1}{\omega L}\right)} = \frac{R}{\sqrt{2}}$$

and finally

$$\frac{1}{1 + jR\left(\omega C - \dfrac{1}{\omega L}\right)} = \frac{1}{\sqrt{2}}$$

The only way the equality can be satisfied is if the magnitude of the imaginary term on the bottom left is equal to 1 because the magnitude of $1 + j\,1$ must be equal to $\sqrt{2}$.

The following relationship, therefore, defines the cutoff frequencies for the system:

$$R\left(\omega C - \frac{1}{\omega L}\right) = 1$$

Substituting $\omega = 2\pi f$ and rearranging will result in the following quadratic equation:

$$f^2 - \frac{f}{2\pi RC} - \frac{1}{4\pi^2 LC} = 0$$

having the form

$$af^2 + bf + c = 0$$

with

$$a = 1,\ b = -\frac{1}{2\pi RC},\ \text{and } c = -\frac{1}{4\pi^2 LC}$$

Substituting into the equation:

$$f = \frac{-b \pm \sqrt{b^2 - 4ac}}{2a}$$

will result in the following after a series of careful mathematical manipulations:

$$f_1 = \frac{1}{4\pi C}\left[\frac{1}{R} - \sqrt{\frac{1}{R^2} + \frac{4C}{L}}\right] \tag{20.39a}$$

$$f_2 = \frac{1}{4\pi C}\left[\frac{1}{R} + \sqrt{\frac{1}{R^2} + \frac{4C}{L}}\right] \tag{20.39b}$$

Since the term in the brackets of Eq. (20.39a) will always be negative, simply associate f_1 with the magnitude of the result.

The effect of R_l, L, and C on the shape of the parallel resonance curve, as shown in Fig. 20.28 for the input impedance, is quite similar

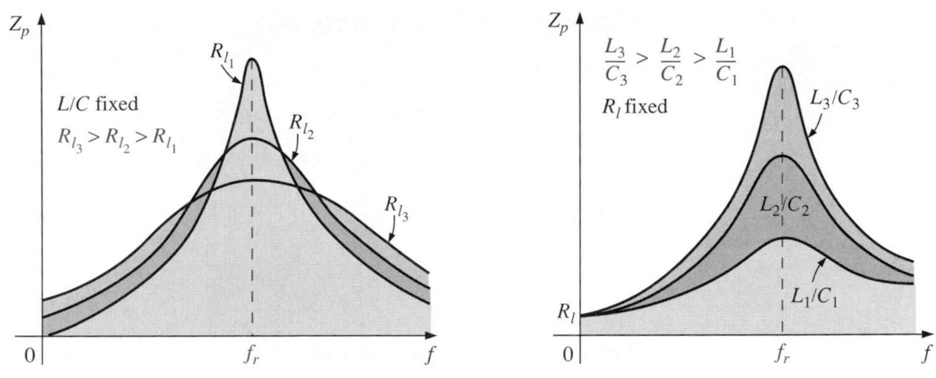

FIG. 20.28
Effect of R_l, L, and C on the parallel resonance curve.

to their effect on the series resonance curve. Whether or not R_l is zero, the parallel resonant circuit will frequently appear in a network schematic as shown in Fig. 20.28.

At resonance, an increase in R_l or a decrease in the ratio L/C will result in a decrease in the resonant impedance, with a corresponding increase in the current. The bandwidth of the resonance curves is given by Eq. (20.38). For increasing R_l or decreasing L (or L/C for constant C), the bandwidth will increase as shown in Fig. 20.28.

At low frequencies, the capacitive reactance is quite high, and the inductive reactance is low. Since the elements are in parallel, the total impedance at low frequencies will therefore be inductive. At high frequencies, the reverse is true, and the network is capacitive. At resonance (f_p), the network appears resistive. These facts lead to the phase plot of Fig. 20.29. Note that it is the inverse of that appearing for the series resonant circuit because at low frequencies the series resonant circuit was capacitive and at high frequencies it was inductive.

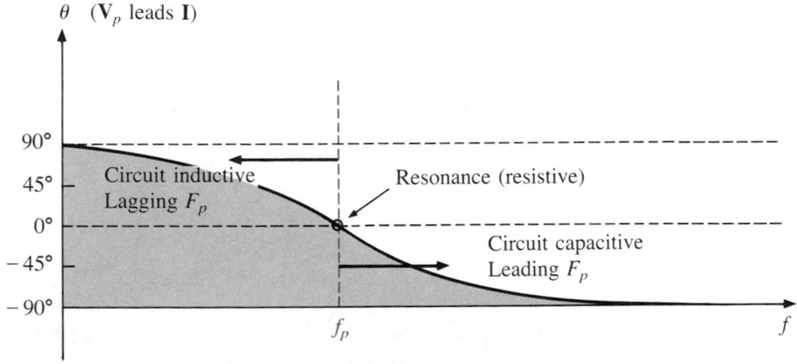

FIG. 20.29
Phase plot for the parallel resonant circuit.

20.10 EFFECT OF $Q_l \geq 10$

The content of the previous section may suggest that the analysis of parallel resonant circuits is significantly more complex than encountered for series resonant circuits. Fortunately, however, this is not the case since, for the majority of parallel resonant circuits, the quality factor of the coil Q_l is sufficiently large to permit a number of approximations that simplify the required analysis.

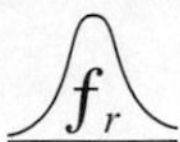

Inductive Reactance, X_{L_p}

If we expand X_{L_p} as

$$X_{L_p} = \frac{R_l^2 + X_L^2}{X_L} = \frac{R_l^2(X_L)}{X_L(X_L)} + X_L = \frac{X_L}{Q_l^2} + X_L$$

then, for $Q_l \geq 10$, $X_L/Q_l^2 \cong 0$ compared to X_L and

$$X_{L_p} \cong X_L \qquad Q_l \geq 10 \tag{20.40}$$

and since resonance is defined by $X_{L_p} = X_C$, the resulting condition for resonance is reduced to:

$$X_L \cong X_C \qquad Q_l \geq 10 \tag{20.41}$$

Resonant Frequency, f_p (Unity Power Factor)

We can rewrite the factor R_l^2C/L of Eq. (20.31) as

$$\frac{R_l^2C}{L} = \frac{1}{\dfrac{L}{R_l^2C}} = \frac{1}{\dfrac{(\omega)}{(\omega)}\dfrac{L}{R_l^2C}} = \frac{1}{\dfrac{\omega L}{R_l^2\omega C}} = \frac{1}{\dfrac{X_LX_C}{R_l^2}}$$

and substitute Eq. (20.40) ($X_L = X_C$):

$$\frac{1}{\dfrac{X_LX_C}{R_l^2}} = \frac{1}{\dfrac{X_L^2}{R_l^2}} = \frac{1}{Q_l^2}$$

Equation (20.31) then becomes

$$f_p = f_s\sqrt{1 - \frac{1}{Q_l^2}} \qquad Q_l \geq 10 \tag{20.42}$$

clearly revealing that as Q_l increases, f_p becomes closer and closer to f_s.

For $Q_l \geq 10$,

$$1 - \frac{1}{Q_l^2} \cong 1$$

and

$$f_p \cong f_s = \frac{1}{2\pi\sqrt{LC}} \qquad Q_l \geq 10 \tag{20.43}$$

Resonant Frequency, f_m (Maximum V_C)

Using the equivalency $R_l^2C/L = 1/Q_l^2$ derived for Eq. (20.42), Eq. (20.32) will take on the following form:

$$f_m \cong f_s\sqrt{1 - \frac{1}{4}\left(\frac{1}{Q_l^2}\right)} \qquad Q_l \geq 10 \tag{20.44}$$

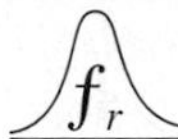

The fact that the negative term under the square root will always be less than that appearing in the equation for f_p reveals that f_m will always be closer to f_s than f_p.

For $Q_l \geq 10$ the negative term becomes very small and can be dropped from consideration, leaving:

$$f_m \cong f_s = \frac{1}{2\pi\sqrt{LC}} \quad Q_l \geq 10 \tag{20.45}$$

In total, therefore, for $Q_l \geq 10$,

$$f_p \cong f_m \cong f_s \quad Q_l \geq 10 \tag{20.46}$$

R_p

$$R_p = \frac{R_l^2 + X_L^2}{R_l} = R_l + \frac{X_L^2}{R_l}\left(\frac{R_l}{R_l}\right) = R_l + \frac{X_L^2}{R_l^2}R_l$$

$$= R_l + Q_l^2 R_l = (1 + Q_l^2)R_l$$

For $Q_l \geq 10$, $1 + Q_l^2 \cong Q_l^2$ and

$$R_p \cong Q_l^2 R_l \quad Q_l \geq 10 \tag{20.47}$$

Applying the approximations just derived to the network of Fig. 20.24 will result in the approximate equivalent network for $Q_l \geq 10$ of Fig. 20.30, which is certainly a lot "cleaner" in general appearance.

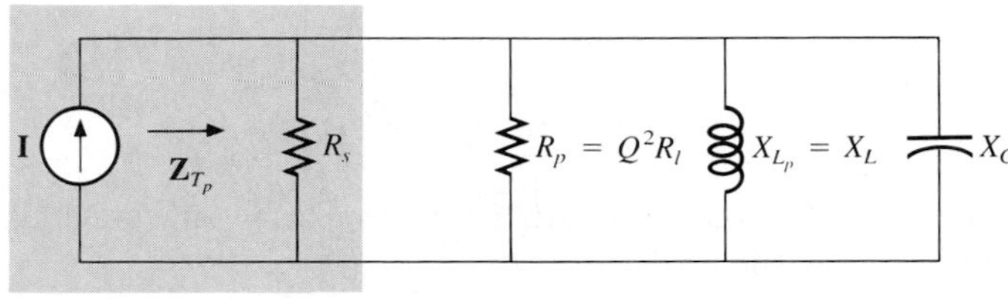

FIG. 20.30
Approximate equivalent circuit for $Q_l \geq 10$.

Substituting $Q_l = \dfrac{X_L}{R_l}$ into Eq. (20.47)

$$R_p \cong Q_l^2 R_l = \frac{X_L^2}{R_l^2}R_l = \frac{X_L^2}{R_l} = \frac{X_L X_C}{R_l} = \frac{2\pi f L}{R_l(2\pi f C)}$$

and

$$R_p \cong \frac{L}{R_l C} \quad Q_l \geq 10 \tag{20.48}$$

Z_{T_p}

The total impedance at resonance is now defined by:

$$Z_{T_p} \cong R_s \parallel R_p = R_s \parallel Q_l^2 R_l \quad Q_l \geq 10 \tag{20.49}$$

For an ideal current source ($R_s = \infty\ \Omega$) or if $R_s \gg R_p$, the equation reduces to

$$Z_{T_p} \cong Q_l^2 R_l \qquad Q_l \geq 10,\ R_s \gg R_p \tag{20.50}$$

Q_p

The quality factor is now defined by

$$Q_p = \frac{R}{X_{L_p}} \cong \frac{R_s \parallel Q_l^2 R_l}{X_L} \tag{20.51}$$

Quite obviously, therefore, R_s does have an impact on the quality factor of the network and the shape of the resonant curves.

If an ideal current source ($R_s = \infty\ \Omega$) is employed or if $R_s \gg R_p$,

$$Q_p \cong \frac{R_s \parallel Q_l^2 R_l}{X_L} = \frac{Q_l^2 R_l}{X_L} = \frac{Q_l^2}{X_L/R_l} = \frac{Q_l^2}{Q_l}$$

and

$$Q_p \cong Q_l \qquad Q_l \geq 10,\ R_s \gg R_p \tag{20.52}$$

BW

The bandwidth defined by f_p is

$$BW = f_2 - f_1 = \frac{f_p}{Q_p} \tag{20.53}$$

By substituting Q_p from above and performing a few algebraic manipulations, it can be shown that

$$BW = f_2 - f_1 \cong \frac{1}{2\pi}\left[\frac{R_l}{L} + \frac{1}{R_s C}\right] \tag{20.54}$$

clearly revealing the impact of R_s on the resulting bandwidth. Of course, if $R_s = \infty\ \Omega$ (ideal current source):

$$BW = f_2 - f_1 \cong \frac{R_l}{2\pi L} \qquad R_s = \infty\ \Omega \tag{20.55}$$

I_L and I_C

A portion of Fig. 20.30 is reproduced in Fig. 20.31, with I_T defined as shown.

As indicated, Z_{T_p} at resonance is $Q_l^2 R_l$. The voltage across the parallel network is, therefore,

$$V_C = V_L = V_R = I_T Z_{T_p} = I_T Q_l^2 R_l$$

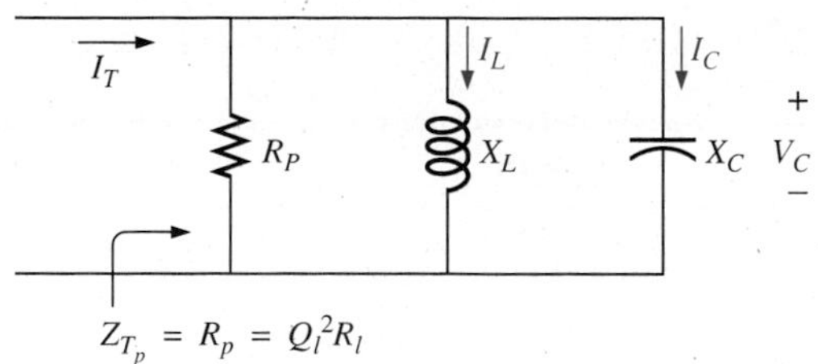

FIG. 20.31
Establishing the relationship between I_C and I_L and the current I_T.

The magnitude of the current I_C can then be determined using Ohm's law, as follows:

$$I_C = \frac{V_C}{X_C} = \frac{I_T Q_l^2 R_l}{X_C}$$

Substituting $X_C = X_L$ when $Q_l \geq 10$,

$$I_C = \frac{I_T Q_l^2 R_l}{X_L} = I_T \frac{Q_l^2}{\frac{X_L}{R_l}} = I_T \frac{Q_l^2}{Q_l}$$

and

$$\boxed{I_C \cong Q_l I_T} \qquad Q_l \geq 10 \qquad \textbf{(20.56)}$$

revealing that the capacitive current is Q_l times the magnitude of the current entering the parallel resonant circuit. For large Q_l, the current I_C can be significant.

A similar derivation results in

$$\boxed{I_L \cong Q_l I_T} \qquad Q_l \geq 10 \qquad \textbf{(20.57)}$$

Conclusions

The equations resulting from the application of the condition $Q_l \geq 10$ are obviously a great deal easier to apply than those obtained earlier. It is, therefore, a condition that should be checked early in an analysis to determine which approach must be applied. Although the condition $Q_l \geq 10$ was applied throughout, many of the equations are still good approximations for $Q_l < 10$. For instance, if $Q_l = 5$, $X_{L_p} = (X_L/Q_l^2) + X_l = (X_L/25) + X_L = 1.04X_L$, which is very close to X_L. In fact, for $Q_l = 2$, $X_{L_p} = (X_L/4) + X_L = 1.25X_L$, which agreeably is not X_L, but it is only 25% off. In general, be aware that the approximate equations can be applied with good accuracy with $Q_l < 10$. The smaller the level of Q_l, however, the less valid the approximation. The approximate equations are certainly valid for a range of values of $Q_l < 10$ if a rough approximation to the actual response is desired rather than one accurate to the hundredths place.

20.11 SUMMARY TABLE

In an effort to limit any confusion resulting from the introduction of f_p and f_m and an approximate approach dependent on Q_l, summary Table 20.1 was developed. One can always use the equations for any Q_l, but

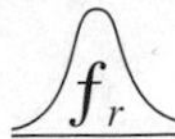

TABLE 20.1
Parallel resonant circuit ($f_s = 1/(2\pi\sqrt{LC})$).

	Any Q_l	$Q_l \geq 10$	$Q_l \geq 10, R_s \gg Q_l^2 R_l$
f_p	$f_s\sqrt{1 - \frac{R_l^2 C}{L}}$	f_s	f_s
f_m	$f_s\sqrt{1 - \frac{1}{4}\left[\frac{R_l^2 C}{L}\right]}$	f_s	f_s
Z_{T_p}	$R_s \parallel R_p = R_s \parallel \left(\frac{R_l^2 + X_L^2}{R_l}\right)$	$R_s \parallel Q_l^2 R_l$	$Q_l^2 R_l$
Z_{T_m}	$R_s \parallel \mathbf{Z}_{R-L} \parallel \mathbf{Z}_C$	$R_s \parallel Q_l^2 R_l$	$Q_l^2 R_l$
Q_p	$\frac{Z_{T_p}}{X_{L_p}} = \frac{Z_{T_p}}{X_C}$	$\frac{Z_{T_p}}{X_L} = \frac{Z_{T_p}}{X_C}$	Q_l
BW	$\frac{f_p}{Q}$ or $\frac{f_m}{Q}$	$\frac{f_p}{Q_p} = \frac{f_s}{Q_p}$	$\frac{f_p}{Q_l} = \frac{f_s}{Q_l}$
I_L, I_C	Network analysis	$I_L = I_C = Q_l I_T$	$I_L = I_C = Q_l I_T$

a proficiency in applying the approximate equations defined by Q_l will pay dividends in the long run.

For the future, the analysis of a parallel resonant network might proceed as follows:

1. Determine f_s to obtain some idea of the resonant frequency. Recall that for most situations, f_s, f_m, and f_p will be relatively close to each other.
2. Calculate an approximate Q_l using f_s from above and compare to the condition $Q_l \geq 10$. If satisfied, the approximate approach should be the chosen path unless a high degree of accuracy is required.
3. If Q_l is less than 10, the approximate approach can be applied, but it must be understood that the smaller the level of Q_l, the less accurate the solution. However, considering the typical variations from nameplate values for many of our components and that a resonant frequency to the tenths place is seldom required, the use of the approximate approach for many practical situations is usually quite valid.

20.12 EXAMPLES (PARALLEL RESONANCE)

EXAMPLE 20.6 Given the parallel network of Fig. 20.32 composed of "ideal" elements:

a. Determine the resonant frequency f_p.
b. Find the total impedance at resonance.
c. Calculate the quality factor, bandwidth, and cutoff frequencies f_1 and f_2 of the system.

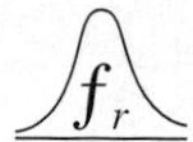

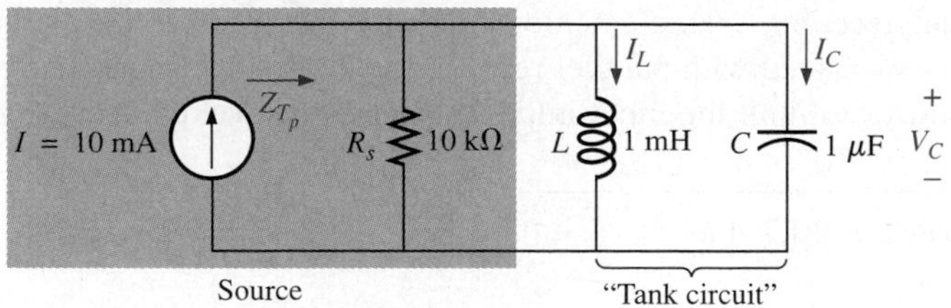

FIG. 20.32
Example 20.6.

d. Find the voltage V_C at resonance.
e. Determine the currents I_L and I_C at resonance.

Solutions:

a. The fact that R_l is zero ohms results in a very high Q_l ($= X_L/R_l$), permitting the use of the following equation for f_p.

$$f_p = f_s = \frac{1}{2\pi\sqrt{LC}} = \frac{1}{2\pi\sqrt{(1\text{ mH})(1\ \mu\text{F})}}$$
$$= \mathbf{5.03\ kHz}$$

b. For the parallel reactive elements:

$$\mathbf{Z}_L \parallel \mathbf{Z}_C = \frac{(X_L \angle 90°)(X_C \angle -90°)}{+j(X_L - X_C)}$$

but $X_L = X_C$ at resonance, resulting in a zero in the denominator of the equation and a very high impedance that can be approximated by an open circuit. Therefore,

$$Z_{T_p} = R_s \parallel \mathbf{Z}_L \parallel \mathbf{Z}_C = R_s = \mathbf{10\ k\Omega}$$

c. $$Q_p = \frac{R_s}{X_{L_p}} = \frac{R_s}{2\pi f_p L} = \frac{10\text{ k}\Omega}{2\pi(5.03\text{ kHz})(1\text{ mH})} = \mathbf{316.41}$$

$$BW = \frac{f_p}{Q_p} = \frac{5.03\text{ kHz}}{316.41} = \mathbf{15.90\ Hz}$$

Eq. (20.39a):

$$f_1 = \frac{1}{4\pi C}\left[\frac{1}{R} - \sqrt{\frac{1}{R^2} + \frac{4C}{L}}\right]$$
$$= \frac{1}{4\pi(1\ \mu\text{F})}\left[\frac{1}{10\text{ k}\Omega} - \sqrt{\frac{1}{(10\text{ k}\Omega)^2} + \frac{4(1\ \mu\text{F})}{1\text{ mH}}}\right]$$
$$= \mathbf{5.025\ kHz}$$

Eq. (20.39b):

$$f_2 = \frac{1}{4\pi C}\left[\frac{1}{R} + \sqrt{\frac{1}{R^2} + \frac{4C}{L}}\right]$$
$$= \mathbf{5.041\ kHz}$$

d. $V_C = IZ_{T_p} = (10\text{ mA})(10\text{ k}\Omega) = \mathbf{100\ V}$

e. $$I_L = \frac{V_L}{X_L} = \frac{V_C}{2\pi f_p L} = \frac{100\text{ V}}{2\pi(5.03\text{ kHz})(1\text{ mH})} = \frac{100\text{ V}}{31.6\ \Omega} = \mathbf{3.16\ A}$$

$$I_C = \frac{V_C}{X_C} = \frac{100\text{ V}}{31.6\ \Omega} = \mathbf{3.16\ A}\ (= Q_p I)$$

The preceding example demonstrates the impact of R_s on the calculations associated with parallel resonance. The source impedance is the only factor to limit the input impedance and the level of V_C.

EXAMPLE 20.7 For the parallel resonant circuit of Fig. 20.33 with $R_s = \infty\ \Omega$:

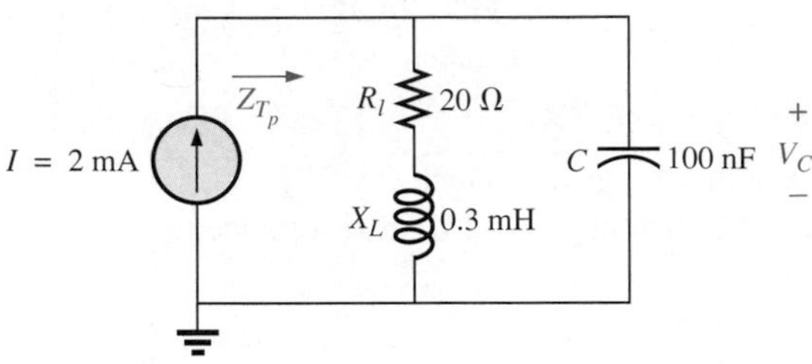

FIG. 20.33
Example 20.7.

a. Determine f_s, f_m, and f_p and compare their levels.
b. Calculate the maximum impedance and the magnitude of the voltage V_C at f_m.
c. Determine the quality factor Q_p.
d. Calculate the bandwidth.
e. Compare the above results with those obtained using the equations associated with $Q_l \geq 10$.

Solutions:

a. $$f_s = \frac{1}{2\pi\sqrt{LC}} = \frac{1}{2\pi\ \sqrt{(0.3\text{ mH})(100\text{ nF})}} = \mathbf{29{,}057.58\ Hz}$$

$$f_m = f_s\sqrt{1 - \frac{1}{4}\left[\frac{R_l^2C}{L}\right]}$$

$$= (29{,}057.58\text{ Hz})\sqrt{1 - \frac{1}{4}\left[\frac{(20\ \Omega)^2(100\text{ nF})}{0.3\text{ mH}}\right]}$$

$$= \mathbf{28{,}569.19\ Hz}$$

$$f_p = f_s\sqrt{1 - \frac{R_l^2C}{L}} = (29{,}057.58\text{ Hz})\sqrt{1 - \left[\frac{(20\ \Omega)^2(100\text{ nF})}{0.3\text{ mH}}\right]}$$

$$= \mathbf{27{,}051.14\ Hz}$$

Both f_m and f_p are less than f_s, as predicted. In addition, f_m is closer to f_s than f_p, as forecast. f_m is about 0.5 kHz less than f_s, whereas f_p is about 2 kHz less. The differences among f_s, f_m, and f_p suggest a low-Q network.

b. $\mathbf{Z}_{T_m} = (R_l + j\,X_L) \parallel -j\,X_C$ at $f = f_m$

$$X_L = 2\pi f_m L = 2\pi(28{,}569.19\text{ Hz})(0.3\text{ mH}) = 53{,}852\ \Omega$$

$$X_C = \frac{1}{2\pi f_m C} = \frac{1}{2\pi(28{,}569.19\text{ Hz})(100\text{ nF})} = 55{,}709\ \Omega$$

$$R_l + j\,X_L = 20\ \Omega + j\,53.852\ \Omega = 57.446\ \Omega\ \angle 69.626^\circ$$

$$\mathbf{Z}_{T_m} = \frac{(57.446\ \Omega\ \angle 69.626^\circ)(55.709\ \Omega\ \angle -90^\circ)}{20\ \Omega + j\,53.852\ \Omega - j\,55.709\ \Omega}$$

$$= \mathbf{159.34\ \Omega\ \angle -15.069^\circ}$$

$$V_{C_{\max}} = IZ_{T_m} = (2\text{ mA})(159.34\ \Omega) = \mathbf{318.68\ mV}$$

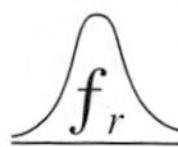

c. $R_s = \infty\ \Omega$; therefore,

$$Q_p = \frac{R_s \| R_p}{X_{L_p}} = \frac{R_p}{X_{L_p}} = Q_l = \frac{X_L}{R_l}$$

$$= \frac{2\pi(27{,}051.14\text{ Hz})(0.3\text{ mH})}{20\ \Omega} = \frac{50.990\ \Omega}{20\ \Omega} = \mathbf{2.55}$$

The low Q confirms our conclusion of part (a). The differences among f_s, f_m, and f_p will be significantly less for higher-Q networks.

d. $BW = \dfrac{f_p}{Q_p} = \dfrac{27{,}051.14\text{ Hz}}{2.55} = \mathbf{10{,}608.29\ Hz}$

e. For $Q_l \geq 10$, $f_m = f_p = f_s = \mathbf{29{,}057.28\ Hz}$

$$Q_p = Q_l = \frac{2\pi f_s L}{R_l} = \frac{2\pi(29{,}057.58\text{ Hz})(0.3\text{ mH})}{20\ \Omega} = \mathbf{2.739}$$

(versus 2.55 above)

$$Z_{T_p} = Q_l^2 R_l = (2.739)^2 \cdot 20\ \Omega = \mathbf{150.04\ \Omega \angle 0°}$$

(versus 159.34 Ω ∠−15.069° above)

$$V_{C_{max}} = IZ_{T_p} = (2\text{ mA})(150.04\ \Omega) = \mathbf{300.08\ mV}$$

(versus 318.68 mV above)

$$BW = \frac{f_p}{Q_p} = \frac{29{,}057.58\text{ Hz}}{2.739} = \mathbf{10{,}608.83\ Hz}$$

(versus 10,608.29 Hz above)

The results reveal that, even for a relatively low Q system, the approximate solutions are still in the ball park compared to those obtained using the full equations. The primary difference is between f_s and f_p (about 7%), with the difference between f_s and f_m at less than 2%. For the future, using f_s to determine Q_l will certainly provide a measure of Q_l that can be used to determine whether the approximate approach is appropriate.

EXAMPLE 20.8 For the network of Fig. 20.34 with f_p provided:

a. Determine Q_l.
b. Determine R_p.
c. Calculate Z_{T_p}.
d. Find C at resonance.
e. Find Q_p.
f. Calculate the BW and cutoff frequencies.

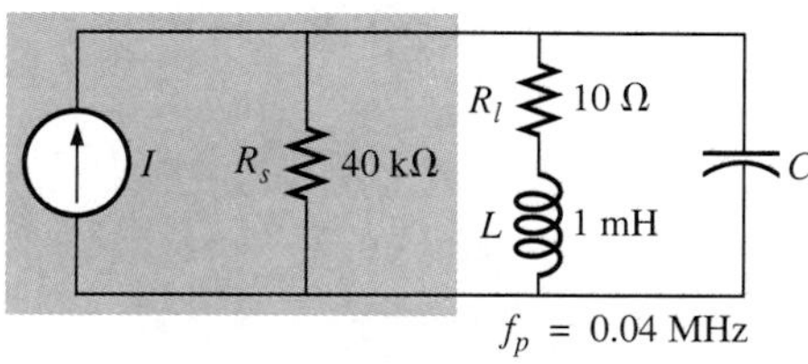

FIG. 20.34
Example 20.8.

Solutions:

a. $Q_l = \dfrac{X_L}{R_l} = \dfrac{2\pi f_p L}{R_l} = \dfrac{2\pi(0.04\text{ MHz})(1\text{ mH})}{10\ \Omega} = \mathbf{25.12}$

b. $Q_l \geq 10$. Therefore,

$$R_p \cong Q_l^2 R_l = (25.12)^2(10\ \Omega) = \mathbf{6.31\ k\Omega}$$

c. $Z_{T_p} = R_s \| R_p = 40\text{ k}\Omega \| 6.31\text{ k}\Omega = \mathbf{5.45\ k\Omega}$

d. $Q_l \geq 10$. Therefore,

$$f_p \cong \frac{1}{2\pi\sqrt{LC}}$$

and $C = \dfrac{1}{4\pi^2 f^2 L} = \dfrac{1}{4\pi^2(0.04\text{ MHz})^2(1\text{ mH})} = \mathbf{15.83\ nF}$

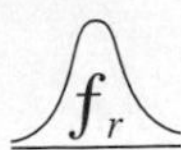

e. $Q_l \geq 10$. Therefore,

$$Q_p = \frac{Z_{T_p}}{X_L} = \frac{R_s \parallel Q_l^2 R_l}{2\pi f_p L} = \frac{5.45\text{ k}\Omega}{2\pi(0.04\text{ MHz})(1\text{ mH})} = \mathbf{21.68}$$

f. $$BW = \frac{f_p}{Q_p} = \frac{0.04\text{ MHz}}{21.68} = \mathbf{1.85\text{ kHz}}$$

$$f_1 = \frac{1}{4\pi C}\left[\frac{1}{R} - \sqrt{\frac{1}{R^2} + \frac{4C}{L}}\right]$$

$$= \frac{1}{4\pi(15.9\text{ mF})}\left[\frac{1}{5.45\text{ k}\Omega} - \sqrt{\frac{1}{(5.45\text{ k}\Omega)^2} + \frac{4(15.9\text{ mF})}{1\text{ mH}}}\right]$$

$$= 5.005 \times 10^6[183.486 \times 10^{-6} - 7.977 \times 10^{-3}]$$

$$= 5.005 \times 10^6[-7.794 \times 10^{-3}]$$

$$= \mathbf{39.009\text{ kHz}} \quad \text{(ignoring the negative sign)}$$

$$f_2 = \frac{1}{4\pi C}\left[\frac{1}{R} + \sqrt{\frac{1}{R^2} + \frac{4C}{L}}\right]$$

$$= 5.005 \times 10^6[183.486 \times 10^{-6} + 7.977 \times 10^{-3}]$$

$$= 5.005 \times 10^6[8.160 \times 10^{-3}]$$

$$= \mathbf{40.843\text{ kHz}}$$

Note that $f_2 - f_1 = 40.843\text{ kHz} - 39.009\text{ kHz} = 1.834\text{ kHz}$, confirming our solution for the bandwidth above. Note also that the bandwidth is not symmetrical about the resonant frequency, with 991 Hz below and 843 Hz above.

EXAMPLE 20.9 The equivalent network for the transistor configuration of Fig. 20.35 is provided in Fig. 20.36.

a. Find f_p.
b. Determine Q_p.
c. Calculate the BW.
d. Determine V_p at resonance.
e. Sketch the curve of V_C versus frequency.

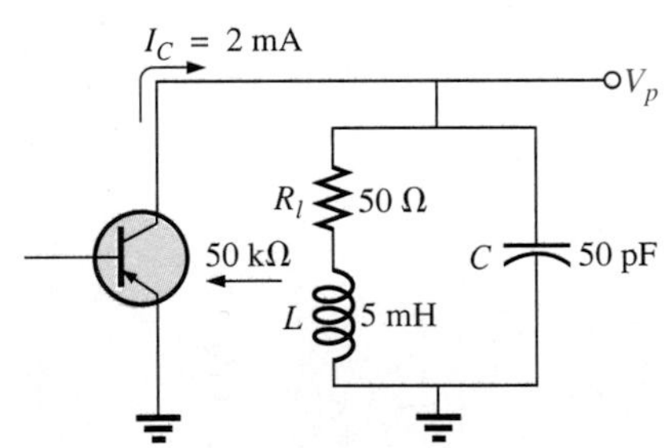

FIG. 20.35
Example 20.9.

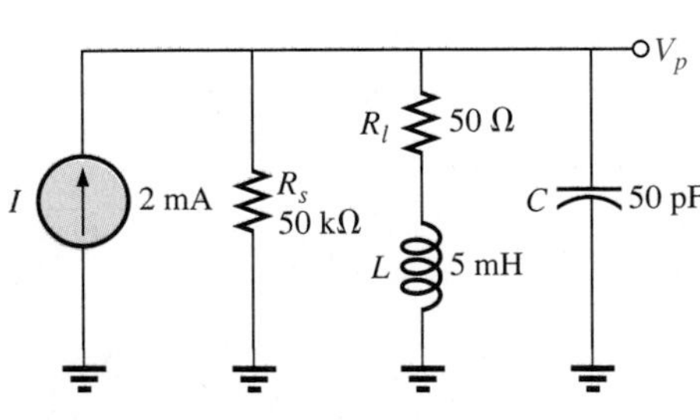

FIG. 20.36
Equivalent network for the transistor configuration of Fig. 20.35.

Solutions:

a. $$f_s = \frac{1}{2\pi\sqrt{LC}} = \frac{1}{2\pi\sqrt{(5\text{ mH})(50\text{ pF})}} = 318.31\text{ kHz}$$

$$X_L = 2\pi f_s L = 2\pi(318.31\text{ kHz})(5\text{ mH}) = 10\text{ k}\Omega$$

$$Q_l = \frac{X_L}{R_l} = \frac{10\text{ k}\Omega}{50\ \Omega} = 200 > 10$$

Therefore, $f_p = f_s = \mathbf{318.31\text{ kHz}}$. Using Eq. (20.31) would result in $\cong 318.5$ kHz.

b. $$Q_p = \frac{R_s \parallel R_p}{X_L}$$

$$R_p = Q_l^2 R_l = (200)^2 50\ \Omega = 2\text{ M}\Omega$$

$$Q_p = \frac{50\text{ k}\Omega \parallel 2\text{ M}\Omega}{10\text{ k}\Omega} = \frac{48.78\text{ k}\Omega}{10\text{ k}\Omega} = \mathbf{4.88}$$

Note the drop in Q from $Q_l = 200$ to $Q_p = 4.88$ due to R_s.

c. $$BW = \frac{f_p}{Q_p} = \frac{318.31\text{ kHz}}{4.88} = \mathbf{65.23\text{ kHz}}$$

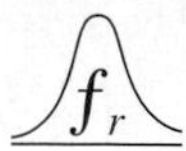

On the other hand,

$$BW = \frac{1}{2\pi}\left(\frac{R_l}{L} + \frac{1}{R_s C}\right) = \frac{1}{2\pi}\left[\frac{50\ \Omega}{5\text{ mH}} + \frac{1}{(50\text{ k}\Omega)(50\text{ pF})}\right]$$
$$= \mathbf{65.25\ kHz}$$

compares very favorably with the above solution.

d. $V_p = IZ_{T_p} = (2\text{ mA})(R_s \parallel R_p) = (2\text{ mA})(48.78\text{ k}\Omega) = \mathbf{97.56\ V}$

e. See Fig. 20.37.

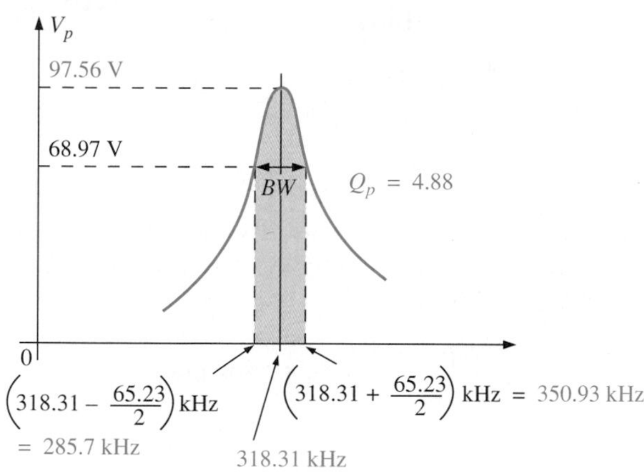

FIG. 20.37
Resonance curve for the network of Fig. 20.36.

EXAMPLE 20.10 Repeat Example 20.9, but ignore the effects of R_s and compare results.

Solutions:

a. f_p is the same, **318.31 kHz.**

b. For $R_s = \infty\ \Omega$

$$Q_p = Q_l = \mathbf{200} \qquad \text{(versus 4.88)}$$

c. $BW = \dfrac{f_p}{Q_p} = \dfrac{318.31\text{ kHz}}{200} = \mathbf{1.592\ kHz} \qquad$ (versus 65.23 kHz)

d. $Z_{T_p} = R_p = 2\text{ M}\Omega \qquad$ (versus 48.78 kΩ)
$V_p = IZ_{T_p} = (2\text{ mA})(2\text{ M}\Omega) = \mathbf{4000\ V} \qquad$ (versus 97.56 V)

The results obtained clearly reveal that the source resistance can have a significant impact on the response characteristics of a parallel resonant circuit.

EXAMPLE 20.11 Design a parallel resonant circuit to have the response curve of Fig. 20.38 using a 1-mH, 10-Ω inductor, and a current source with an internal resistance of 40 kΩ.

Solution:

$$BW = \frac{f_p}{Q_p}$$

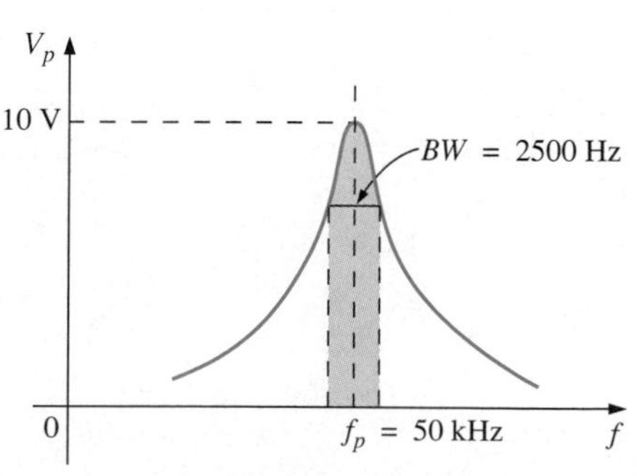

FIG. 20.38
Example 20.11.

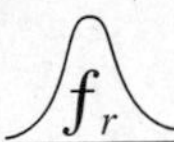

Therefore,

$$Q_p = \frac{f_p}{BW} = \frac{50{,}000\text{ Hz}}{2500\text{ Hz}} = \mathbf{20}$$

$$X_L = 2\pi f_p L = 2\pi(50\text{ kHz})(1\text{ mH}) = 314\ \Omega$$

and
$$Q_l = \frac{X_L}{R_l} = \frac{314\ \Omega}{10\ \Omega} = \mathbf{31.4}$$

$$R_p = Q_l^2 R = (31.4)^2(10\ \Omega) = \mathbf{9859.6\ \Omega}$$

$$Q_p = \frac{R}{X_L} = \frac{R_s \parallel 9859.6\ \Omega}{314\ \Omega} = 20 \qquad \text{(from above)}$$

so that
$$\frac{(R_s)(9859.6)}{R_s + 9859.6} = 6280$$

resulting in
$$R_s = 17.298\text{ k}\Omega$$

However, the source resistance was given as 40 kΩ. We must therefore add a parallel resistor (R') that will reduce the 40 kΩ to approximately 17.298 kΩ; that is,

$$\frac{(40\text{ k}\Omega)(R')}{40\text{ k}\Omega + R'} = 17.298\text{ k}\Omega$$

Solving for R':

$$R' = \mathbf{30.481\text{ k}\Omega}$$

The closest commercial value is **30 kΩ**. At resonance, $X_L = X_C$, and

$$X_C = \frac{1}{2\pi f_p C}$$

$$C = \frac{1}{2\pi f_p X_C} = \frac{1}{2\pi(50\text{ kHz})(314\ \Omega)}$$

and
$$C \cong \mathbf{0.01\ \mu F} \qquad \text{(commercially available)}$$

$$\begin{aligned} Z_{T_p} &= R_s \parallel Q_l^2 R_l \\ &= 17.298\text{ k}\Omega \parallel 9859.6\ \Omega \\ &= 6.28\text{ k}\Omega \end{aligned}$$

with
$$V_p = IZ_{T_p}$$

and
$$I = \frac{V_p}{Z_{T_p}} = \frac{10\text{ V}}{6.28\text{ k}\Omega} \cong \mathbf{1.6\text{ mA}}$$

The network appears in Fig. 20.39.

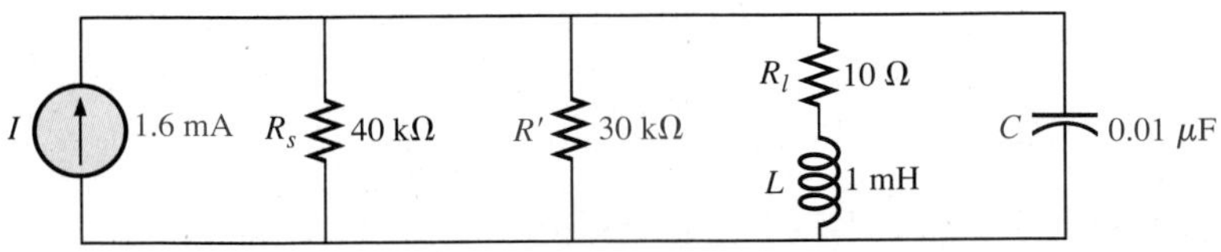

FIG. 20.39
Network designed to meet the criteria of Fig. 20.38.

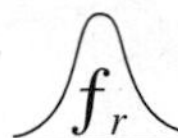

20.13 COMPUTER ANALYSIS

PSpice (DOS)

The beauty of the .PROBE response can be fully appreciated in the plots that result for series and parallel resonant circuits. The insertion of a few network parameters into the PSpice file will result in a curve that immediately reveals the resonant frequency, the bandwidth of the response, and the relative quality factor. Also immediately available is the range of frequencies that will not receive sufficient power to significantly affect the response of the succeeding stage of the system.

The series resonant circuit of Fig. 20.40 is very similar to the circuit employed in Example 20.4 of the text with the addition of a voltage source. The magnitude of the source was chosen to produce a maximum current of $I = 400\text{ mV}/40\ \Omega = 10\text{ mA}$ at resonance, and the reactive elements were chosen to have a resonant frequency of

$$f_s = \frac{1}{2\pi\sqrt{LC}} = \frac{1}{2\pi\sqrt{(30\text{ mH})(0.1\ \mu\text{F})}} \cong \mathbf{2.9\ kHz}$$

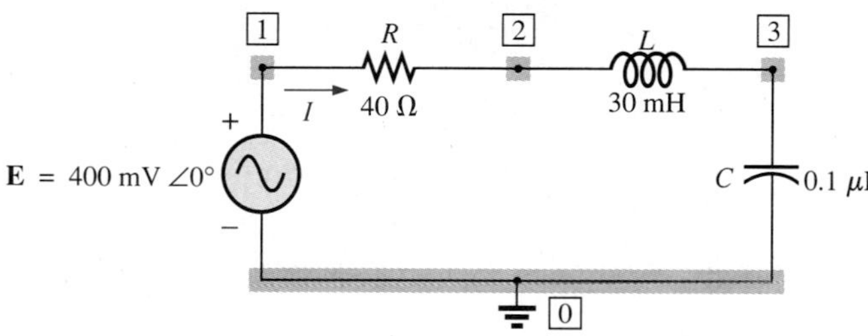

FIG. 20.40
Defining the parameters for a PSpice (DOS) analysis.

The quality factor is

$$Q_l = \frac{X_L}{R_l} = \frac{546.64\ \Omega}{40\ \Omega} = \mathbf{13.7}$$

and the bandwidth is

$$BW = \frac{f_s}{Q_l} = \frac{2.9\text{ kHz}}{13.7} \cong \mathbf{212\ Hz}$$

The input file of Fig. 20.41 requests 91 data points between 1 kHz and 10 kHz for an increment of

$$\text{Increment} = \frac{10\text{ kHz} - 1\text{ kHz}}{91 - 1} = \frac{9\text{ kHz}}{90} = \mathbf{100\ Hz}$$

The output file of Fig. 20.42 clearly substantiates the above results, with a resonant frequency of 2.9 kHz, a maximum current of 10 mA, and a bandwidth at 7.07 mA of about 200 Hz. The shape of the curve also suggests a quality factor greater than 10. In addition, note that the current level at 1 kHz and 8 kHz is about the same, and any frequency less than 1 kHz or greater than 8 kHz has a lower value.

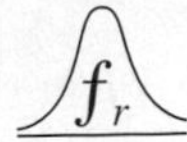

```
Chapter 20 -Series Resonance

 ****     CIRCUIT DESCRIPTION

*****************************************************************************

VE 1 0 AC 400MV 0
R  1 2 40
L  2 3 30MH
C  3 0 0.1UF
.AC LIN 91 1KH 10KH
.PROBE
.OPTIONS NOPAGE
.END
```

FIG. 20.41
Input file for the network of Fig. 20.40.

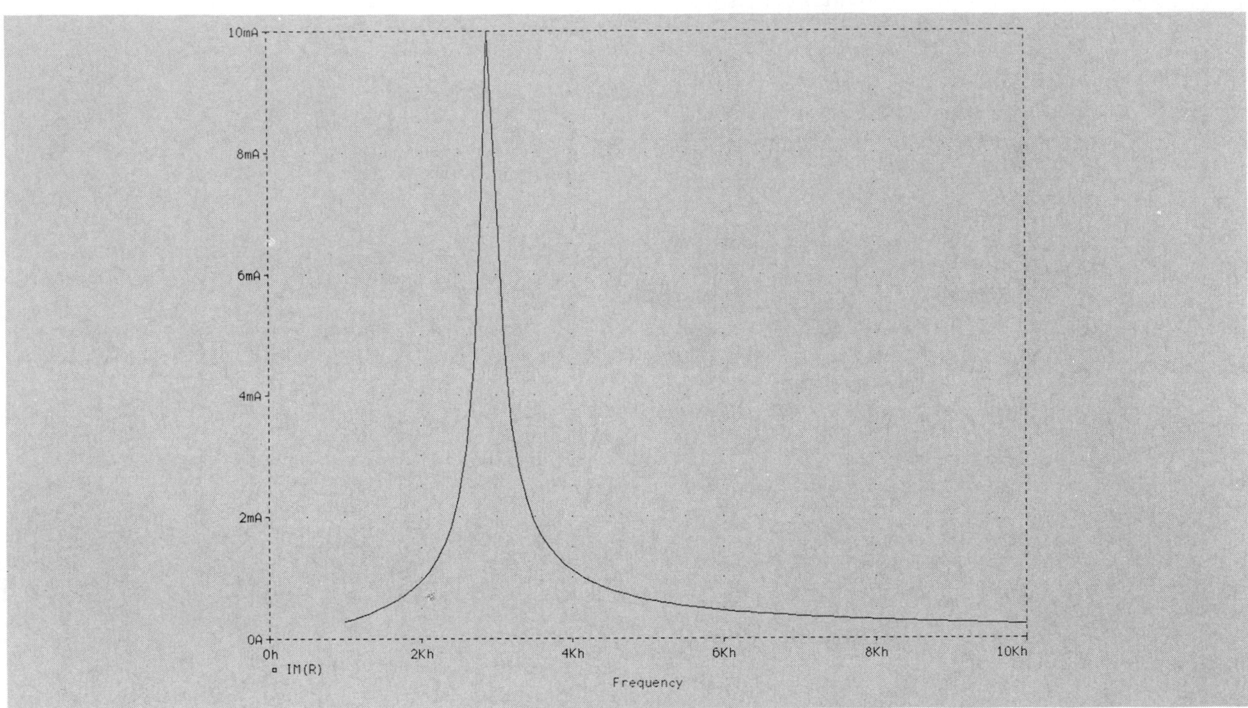

FIG. 20.42
Output file for the network of Fig. 20.40.

PSpice (Windows)

The schematic for the network of Example 20.9 appears in Fig. 20.43. The current source chosen was **ISRC** primarily because of the arrow symbol. Attributes chosen were **DC** = 0 A, **AC** = 2 mA, and **PKGREF** = I. Under **Analysis**, the **Setup** included the **AC Swee**p with only the following settings: **Linear, Total Pts.:** 1000, **Start Freq.:** 10 Hz, **End Freq.:** 500 kHz. Under **Probe Setup** the **Automatically Run Probe After Simulation** was chosen. After **Simulation**, the following sequence was applied to obtain a linear scale from 10 Hz to 500 kHz: **Plot-X-Axis Settings-Linear-User Defined-10 Hz to 500 kHz**. Next,

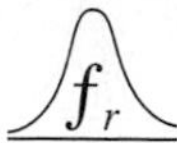

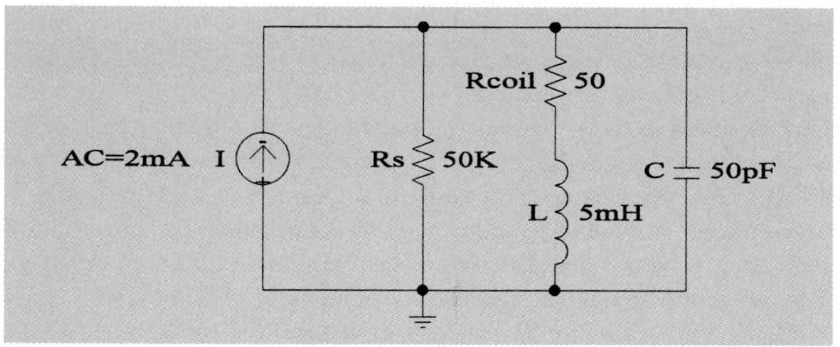

FIG. 20.43
Schematic representation of the parallel resonant network of Fig. 20.36.

the output voltage across the capacitor was displayed with **Trace-Add-Alias Names-V(C:1)-OK**, resulting in the response of Fig. 20.44. The **Cursor** was then used through **Tools-Cursor-Display**-Click left to find the peak value, which occurred at 317.822 kHz, as appearing in the **Cursor** dialog box at the bottom left-hand corner of the plot. The peak value of 97.543 V is very close to the 97.56 V obtained in Example 20.9. By clicking right, the second **Cursor** was used to define the low cutoff frequency. Multiplying the peak value by 0.707 to define the

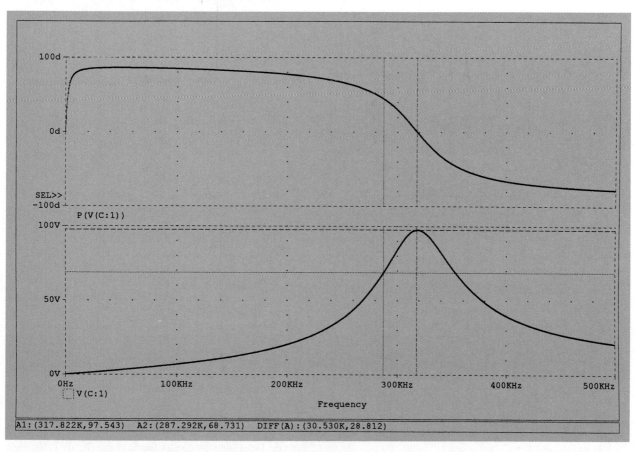

FIG. 20.44
The frequency response for V_C and θ_C for the parallel resonant network of Fig. 20.43.

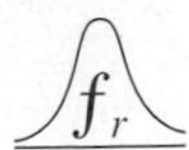

level of V_C at the cutoff frequencies resulted in 0.707(97.543 V) = 68.96 V. Setting the **Cursor** as close to this value as possible resulted in an intersection at a frequency of 287.292 kHz, as also shown in the **Cursor** dialog box. By moving the same **Cursor** to the right side of the resonant frequency to the same voltage level, a high cutoff frequency of 353.35 kHz was obtained, revealing that the upper cutoff frequency is about 5 kHz more above the resonant frequency than the lower cutoff frequency is below—evidence of a relatively low-Q system (about 5). The two cutoff frequencies establish a bandwidth of 66.06 kHz.

Using **Plot-Add Plot-Trace-Add-P(V(C:1))** the phase angle of the output voltage can be plotted versus frequency, as shown in Fig. 20.44. Note how close the phase angle is to 0 degrees, as it should be if the total impedance is to be purely resistive. Using the **Cursor**, we find that the frequency closest to 0 degrees is 318.322 kHz, which is slightly more than the frequency obtained for the **V(C:1)** plot. According to the text, however, we should expect that the frequency obtained at 0 degrees (referred to as f_p) should be less than that obtained for maximum voltage (referred to as f_m). However, one must remember that the two are usually quite close to each other and the level of accuracy for the reading must be considered. By looking at each plot individually (full screen) to increase the accuracy, the following results are obtained, verifying the general conclusion of the text: f_m = 318.822 kHz and f_p = 318.322 kHz. The slight distortion at low frequencies for the phase plot is simply due to the rapid change in capacitive reactance at these frequencies.

PROBLEMS

SECTIONS 20.2 THROUGH 20.7 Series Resonance

1. Find the resonant ω_s and f_s for the series circuit with the following parameters:
   a. $R = 10\ \Omega, L = 1\ \text{H}, C = 16\ \mu\text{F}$
   b. $R = 300\ \Omega, L = 0.5\ \text{H}, C = 0.16\ \mu\text{F}$
   c. $R = 20\ \Omega, L = 0.28\ \text{mH}, C = 7.46\ \mu\text{F}$

2. For the series circuit of Fig. 20.45:
   a. Find the value of X_C for resonance.
   b. Determine the total impedance of the circuit at resonance.
   c. Find the magnitude of the current I.
   d. Calculate the voltages V_R, V_L, and V_C at resonance. How are V_L and V_C related? How does V_R compare to the applied voltage E?
   e. What is the quality factor of the circuit? Is it a high- or low-Q circuit?
   f. What is the power dissipated by the circuit at resonance?

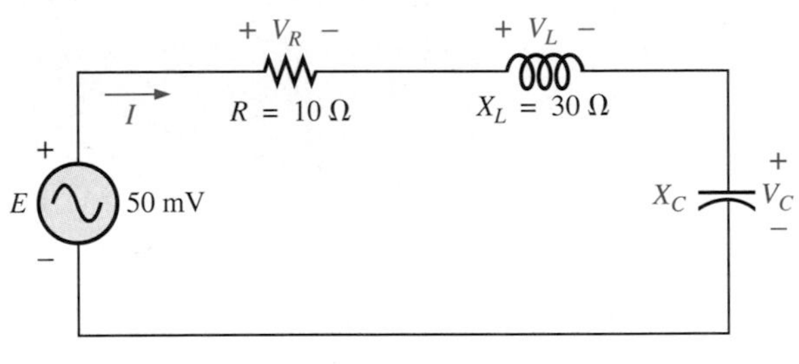

FIG. 20.45
Problem 2.

3. For the series circuit of Fig. 20.46:
   a. Find the value of X_L for resonance.
   b. Determine the magnitude of the current I at resonance.
   c. Find the voltages V_R, V_L, and V_C at resonance and compare their magnitudes.
   d. Determine the quality factor of the circuit. Is it a high- or low-Q circuit?
   e. If the resonant frequency is 5 kHz, determine the value of L and C.

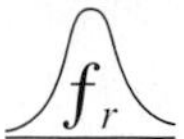

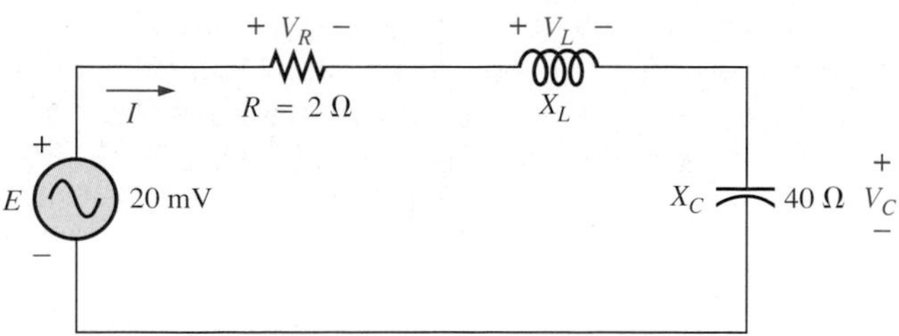

FIG. 20.46
Problem 3.

f. Find the bandwidth of the response if the resonant frequency is 5 kHz.
g. What are the low and high cutoff frequencies?

4. For the circuit of Fig. 20.47:
   a. Find the value of L in millihenries if the resonant frequency is 1800 Hz.
   b. Calculate X_L and X_C. How do they compare?
   c. Find the magnitude of the current I_{rms} at resonance.
   d. Find the power dissipated by the circuit at resonance.
   e. What is the apparent power delivered to the system at resonance?
   f. What is the power factor of the circuit at resonance?
   g. Calculate the Q of the circuit and the resulting bandwidth.
   h. Find the cutoff frequencies and calculate the power dissipated by the circuit at these frequencies.

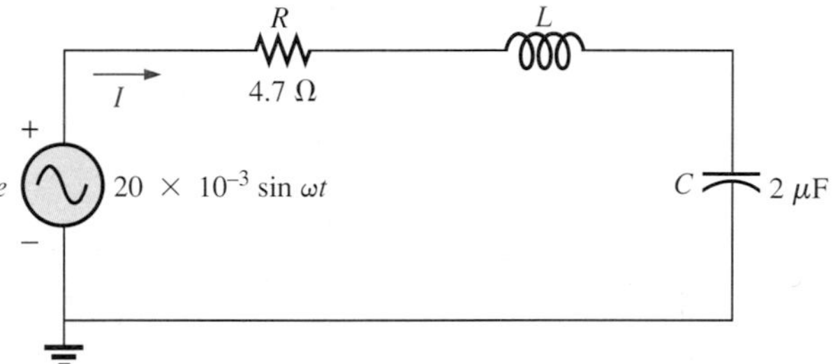

FIG. 20.47
Problem 4.

5. a. Find the bandwidth of a series resonant circuit having a resonant frequency of 6000 Hz and a Q_s of 15.
   b. Find the cutoff frequencies.
   c. If the resistance of the circuit at resonance is 3 Ω, what are the values of X_L and X_C in ohms?
   d. What is the power dissipated at the half-power frequencies if the maximum current flowing through the circuit is 0.5 A?

6. A series circuit has a resonant frequency of 10 kHz. The resistance of the circuit is 5 Ω, and X_C at resonance is 200 Ω.
   a. Find the bandwidth.
   b. Find the cutoff frequencies.
   c. Find Q_s.
   d. If the input voltage is $30\ \text{V} \angle 0°$, find the voltage across the coil and capacitor in phasor form.
   e. Find the power dissipated at resonance.

7. a. The bandwidth of a series resonant circuit is 200 Hz. If the resonant frequency is 2000 Hz, what is the value of Q_s for the circuit?
   b. If $R = 2\ \Omega$, what is the value of X_L at resonance?
   c. Find the value of L and C at resonance.
   d. Find the cutoff frequencies.

8. The cutoff frequencies of a series resonant circuit are 5400 Hz and 6000 Hz.
   a. Find the bandwidth of the circuit.
   b. If Q_s is 9.5, find the resonant frequency of the circuit.
   c. If the resistance of the circuit is 2 Ω, find the value of X_L and X_C at resonance.
   d. Find the value of L and C at resonance.

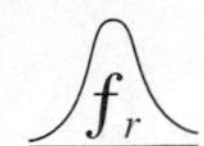

***9.** Design a series resonant circuit with an input voltage of 5 V $\angle 0°$ to have the following specifications:
  a. A peak current of 500 mA at resonance
  b. A bandwidth of 120 Hz
  c. A resonant frequency of 8400 Hz
  Find the value of L and C and the cutoff frequencies.

***10.** Design a series resonant circuit to have a bandwidth of 400 Hz using a coil with a Q_l of 20 and a resistance of 2 Ω. Find the value of L and C and the cutoff frequencies.

***11.** A series resonant circuit is to resonate at $\omega_s = 2\pi \times 10^6$ rad/s and draw 20 W from a 120-V source at resonance. If the fractional bandwidth is 0.16,
  a. Determine the resonant frequency in hertz.
  b. Calculate the bandwidth in hertz.
  c. Determine the values of R, L, and C.
  d. Find the resistance of the coil if $Q_l = 80$.

***12.** A series resonant circuit will resonate at a frequency of 1 MHz with a fractional bandwidth of 0.2. If the quality factor of the coil at resonance is 12.5 and its inductance is 100 μH, determine
  a. The resistance of the coil.
  b. The additional resistance required to establish the indicated fractional bandwidth.
  c. The required value of capacitance.

SECTIONS 20.8 THROUGH 20.12
Parallel Resonance

13. For the "ideal" parallel resonant circuit of Fig. 20.48:
  a. Determine the resonant frequency (f_p).
  b. Find the voltage V_C at resonance.
  c. Determine the currents I_L and I_C at resonance.
  d. Find Q_p.

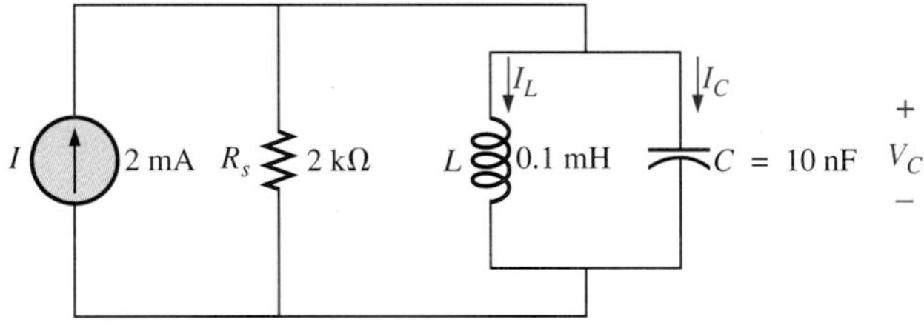

FIG. 20.48
Problem 13.

14. For the parallel resonant network of Fig. 20.49:
  a. Calculate f_s.
  b. Determine Q_l using $f = f_s$. Can the approximate approach be applied?
  c. Determine f_p and f_m.
  d. Calculate X_L and X_C using f_p. How do they compare?
  e. Find the total impedance at resonance (f_p).
  f. Calculate V_C at resonance (f_p).
  g. Determine Q_p and the BW using f_p.
  h. Calculate I_L and I_C at f_p.

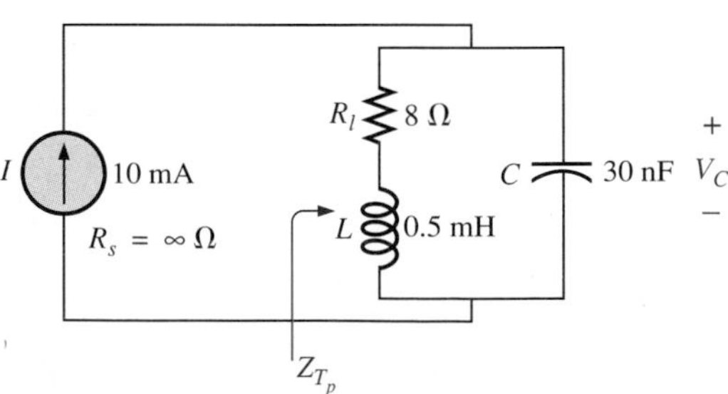

FIG. 20.49
Problem 14.

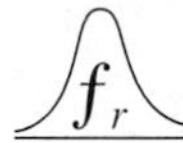

15. Repeat Problem 14 for the network of Fig. 20.50.

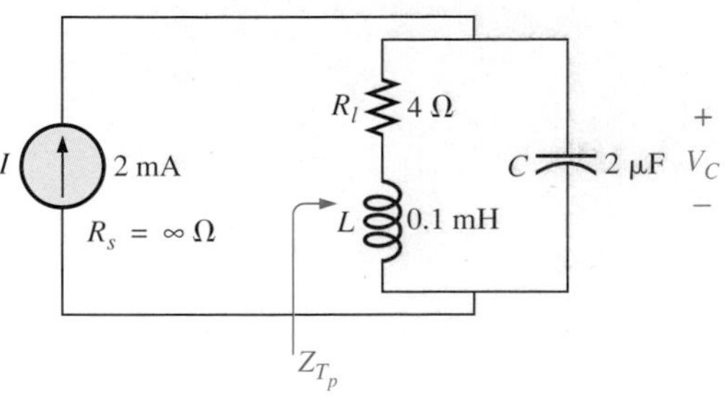

FIG. 20.50
Problem 15.

16. For the network of Fig. 20.51:

a. Find the value of X_C at resonance (f_p).

b. Find the total impedance Z_{T_p} at resonance (f_p).

c. Find the currents I_L and I_C at resonance (f_p).

d. If the resonant frequency is 20,000 Hz, find the value of L and C at resonance.

e. Find Q_p and BW.

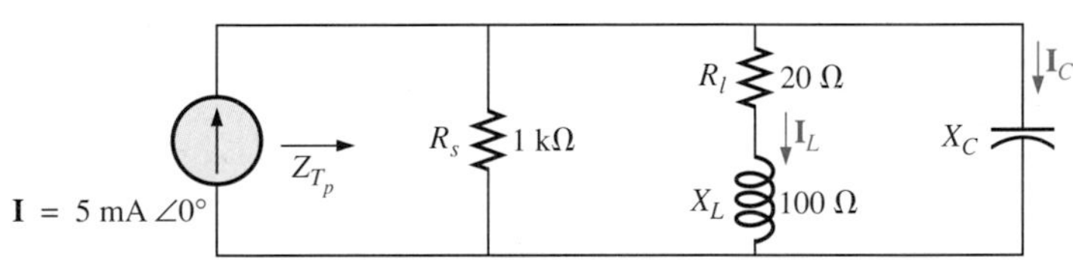

FIG. 20.51
Problem 16.

17. Repeat Problem 16 for the network of Fig. 20.52.

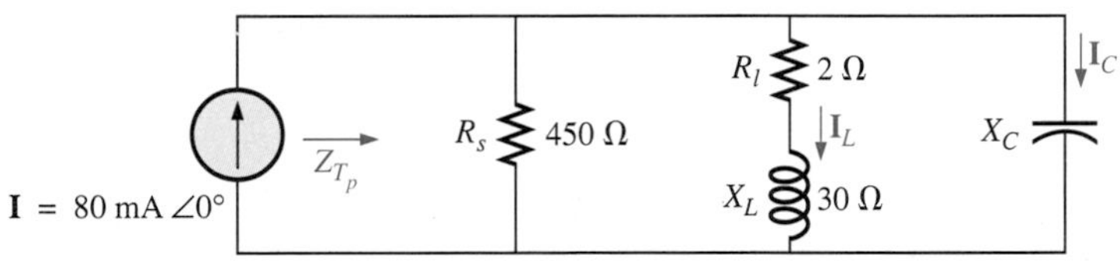

FIG. 20.52
Problem 17.

18. For the network of Fig. 20.53:

a. Find the resonant frequencies f_s, f_p, and f_m. What do the results suggest about the Q_p of the network?

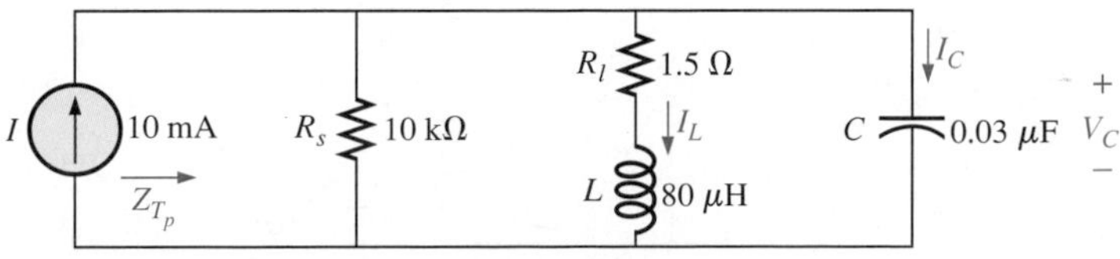

FIG. 20.53
Problem 18.

b. Find the value of X_L and X_C at resonance (f_p). How do they compare?
c. Find the impedance Z_{T_p} at resonance (f_p).
d. Calculate Q_p and the BW.
e. Find the magnitude of currents I_L and I_C at resonance (f_p).
f. Calculate the voltage V_C at resonance (f_p).

***19.** Repeat Problem 18 for the network of Fig. 20.54.

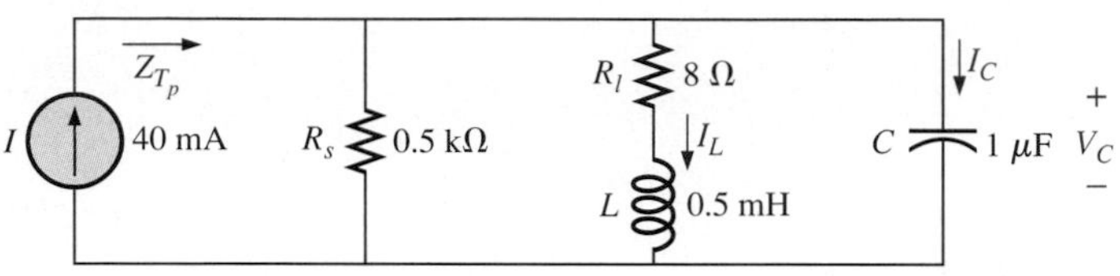

FIG. 20.54
Problems 19, 29, and 31.

20. It is desired that the impedance $\mathbf{Z}_T$ of the high-Q circuit of Fig. 20.55 be 50 kΩ ∠0° at resonance (f_p).
a. Find the value of X_L.
b. Compute X_C.
c. Find the resonant frequency (f_p) if L = 16 mH.
d. Find the value of C.

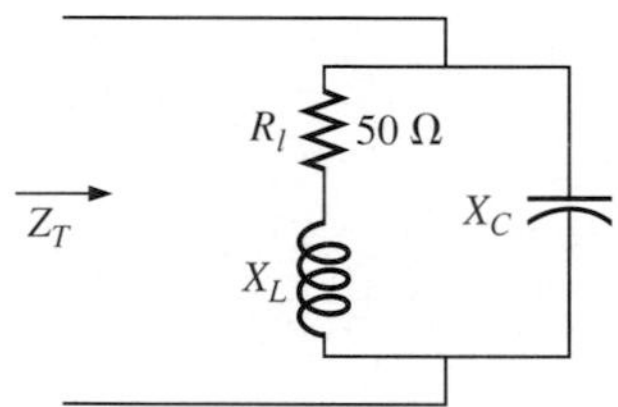

FIG. 20.55
Problem 20.

21. For the network of Fig. 20.56:
a. Find f_p.
b. Calculate the magnitude of V_C at resonance (f_p).
c. Determine the power absorbed at resonance.
d. Find the BW.

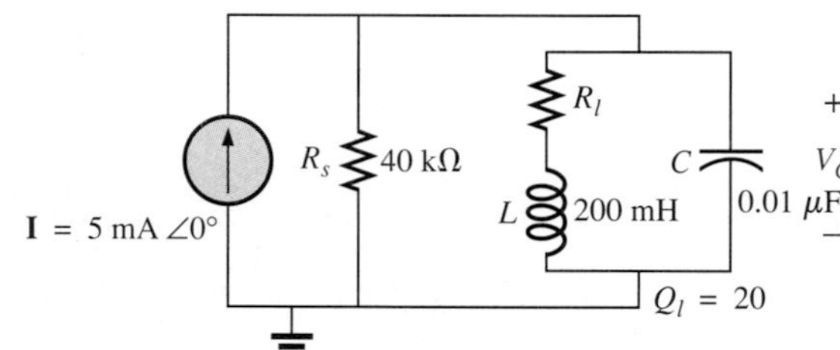

FIG. 20.56
Problem 21.

***22.** For the network of Fig. 20.57:
a. Find the value of X_L for resonance.
b. Find Q_l.

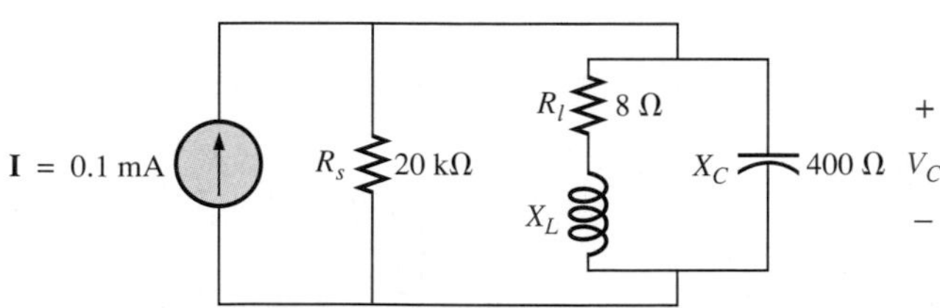

FIG. 20.57
Problem 22.

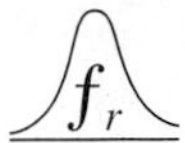

c. Find the resonant frequency (f_p) if the bandwidth is 1 kHz.

d. Find the maximum value of the voltage V_C.

e. Sketch the curve of V_C versus frequency. Indicate its peak value, resonant frequency, and band frequencies.

***23.** Repeat Problem 22 for the network of Fig. 20.58.

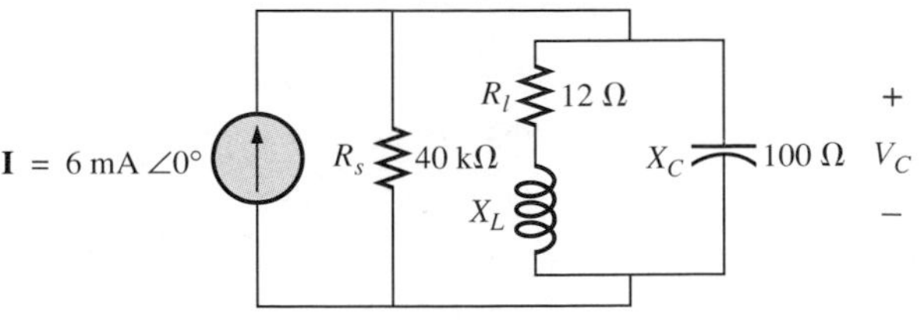

FIG. 20.58
Problem 23.

***24.** For the network of Fig. 20.59:

a. Find f_s, f_p, and f_m.

b. Determine Q_l and Q_p at f_p after a source conversion is performed.

c. Find the input impedance Z_{T_p}.

d. Find the magnitude of the voltage V_C.

e. Calculate the bandwidth using f_p.

f. Determine the magnitude of the currents I_C and I_L.

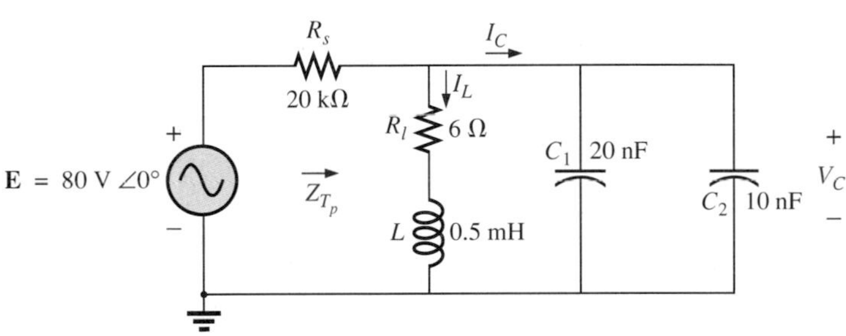

FIG. 20.59
Problem 24.

***25.** For the network of Fig. 20.60, the following are specified:

$$f_p = 20 \text{ kHz}$$
$$BW = 1.8 \text{ kHz}$$
$$L = 2 \text{ mH}$$
$$Q_l = 80$$

Find R_s and C.

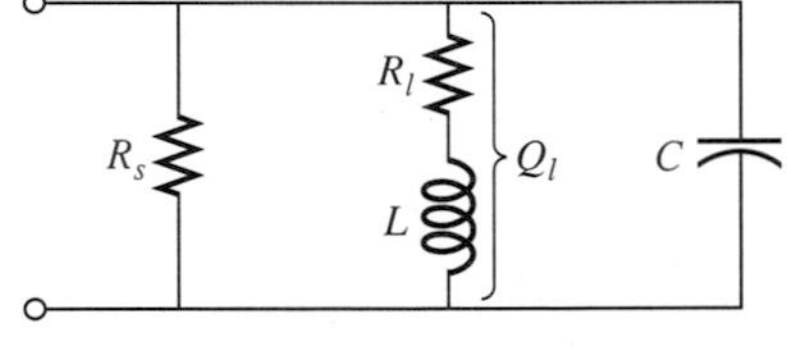

FIG. 20.60
Problem 25.

***26.** Design the network of Fig. 20.61 to have the following characteristics:

a. a bandwidth of 500 Hz

b. $Q_p = 30$

c. $V_{C_{max}} = 1.8$ V

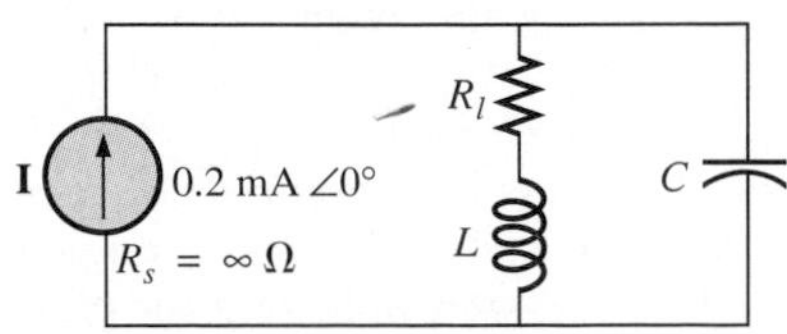

FIG. 20.61
Problem 26.

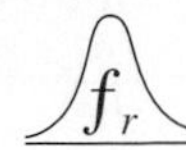

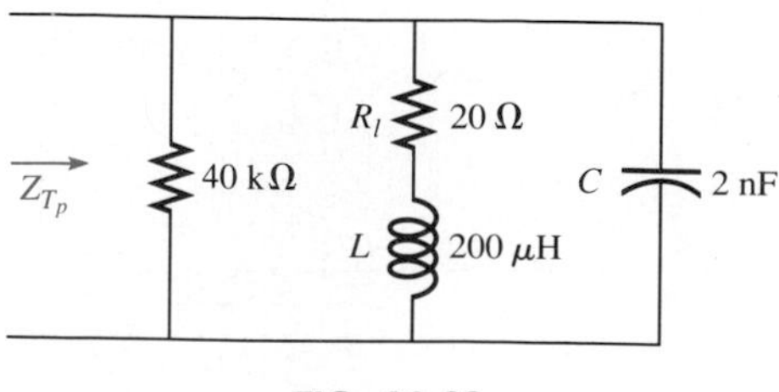

FIG. 20.62

***27.** For the parallel resonant circuit of Fig. 20.62:
- **a.** Determine the resonant frequency.
- **b.** Find the total impedance at resonance.
- **c.** Find Q_p.
- **d.** Calculate the bandwidth.
- **e.** Repeat parts (a) through (d) for $L = 20\ \mu\text{H}$ and $C = 20\ \text{nF}$.
- **f.** Repeat parts (a) through (d) for $L = 0.4\ \text{mH}$ and $C = 1\ \text{nF}$.
- **g.** For the network of Fig. 20.62 and the parameters of parts (e) and (f), determine the ratio L/C.
- **h.** Do your results confirm the conclusions of Fig. 20.28 for changes in the L/C ratio?

SECTION 20.13 Computer Analysis

PSpice (DOS)

28. Given a series R-L-C circuit with $R = 10\ \Omega$, $L = 3.98\ \text{mH}$, and $C = 0.398\ \mu\text{F}$, write an input file to determine the voltages V_R, V_L, and V_C and the current I at resonance ($f_s = 2800\ \text{Hz}$).

29. Write the input file to obtain a .PROBE response for the voltage V_C for the network of Fig. 20.54.

PSpice (Windows)

30. Verify the results of Example 20.8 using PSpice Windows; that is, show that the resonant frequency is in fact 40 kHz, the cutoff frequencies are as calculated, and the bandwidth is 1.85 kHz.

31. Find f_p and f_m for the parallel resonant network of Fig. 20.54 and comment on the resulting bandwidth as it relates to the quality factor of the network.

Programming Language (C++, BASIC, PASCAL, etc.)

32. Write a program to tabulate the impedance and current of the network of Fig. 20.2 versus frequency for a frequency range extending from $0.1f_s$ to $2f_s$ in increments of $0.1f_s$. For the first run, use the parameters defined by Example 20.1.

33. Write a program to provide a general solution for the network of Fig. 20.36; that is, determine the parameters requested in parts (a) through (e) of Example 20.9.

GLOSSARY

Band (cutoff, half-power, corner) frequencies Frequencies that define the points on the resonance curve that are 0.707 of the peak current or voltage value. In addition, they define the frequencies at which the power transfer to the resonant circuit will be half the maximum power level.

Bandwidth The range of frequencies between the band, cutoff, or half-power frequencies.

Quality factor (Q) A ratio that provides an immediate indication of the sharpness of the peak of a resonance curve. The higher the Q, the sharper the peak and the more quickly it drops off to the right and left of the resonant frequency.

Resonance A condition established by the application of a particular frequency (the resonant frequency) to a series or parallel R-L-C network. The transfer of power to the system is a maximum, and, for frequencies above and below, the power transfer drops off to significantly lower levels.

Selectivity A characteristic of resonant networks directly related to the bandwidth of the resonant system. High selectivity is associated with small bandwidth (high Q's), and low selectivity with larger bandwidths (low Q's).

21

dB

Decibels, Filters, and Bode Plots

21.1 LOGARITHMS

The use of logarithms in industry is so extensive that a clear understanding of their purpose and use is an absolute necessity. At first exposure, logarithms often appear vague and mysterious due to the mathematical operations required to find the logarithm and antilogarithm using the longhand table approach that is typically taught in mathematics courses. However, almost all of today's scientific calculators have the common and natural log functions, eliminating the complexity of applying logarithms and allowing us to concentrate on the positive characteristics of the function.

Basic Relationships

Let us first examine the relationship between the variables of the logarithmic function. The mathematical expression

$$N = (b)^x$$

states that the number N is equal to the base b taken to the power x. A few examples:

$$100 = (10)^2$$
$$27 = (3)^3$$
$$54.6 = (e)^4 \qquad \text{where } e = 2.7183$$

If the question were to find the power x to satisfy the equation

$$1200 = (10)^x$$

the value of x could be determined using logarithms in the following manner:

$$x = \log_{10} 1200 = \mathbf{3.079}$$

revealing that

$$10^{3.079} = 1200$$

Note that the logarithm was taken to the base 10—the number to be taken to the power of x. There is no limitation to the numerical value of the base except that tables and calculators are designed to handle either a base of 10 (common logarithm, log) or base $e = 2.7183$ (natural logarithm, ln). In review, therefore,

$$\text{If } N = (b)^x, \text{ then } x = \log_b N. \tag{21.1}$$

The base to be employed is a function of the area of application. If a conversion from one base to the other is required, the following equation can be applied:

$$\log_e x = 2.3 \log_{10} x \tag{21.2}$$

The content of this chapter is such that we will concentrate solely on the common logarithm. However, a number of the conclusions are also applicable to natural logarithms.

Some Areas of Application

The following is a short list of the most common applications of the logarithmic function.

1. This chapter will demonstrate that the use of logarithms permits plotting the response of a system for a range of values that may otherwise be impossible or unwieldy with a linear scale.
2. Levels of power, voltage, and the like, can be compared without dealing with very large or small numbers that often cloud the true impact of the difference in magnitudes.
3. There are a number of systems that respond to outside stimuli in a nonlinear logarithmic manner. The result is a mathematical model that permits a direct calculation of the response of the system to a particular input signal.
4. The response of a cascaded or compound system can be rapidly determined using logarithms if the gain of each stage is known on a logarithmic basis. This characteristic will be demonstrated in an example to follow.

Graphs

Graph paper is available in the *semilog* and *log-log* varieties. Semilog paper has only one log scale, with the other a linear scale. Both scales of log-log paper are log scales. A section of semilog paper appears in Fig. 21.1. Note the linear (even-spaced-interval) vertical scaling and the repeating intervals of the log scale at multiples of 10.

The spacing of the log scale is determined by taking the common log (base 10) of the number. The scaling starts with 1, since $\log_{10} 1 = 0$. The distance between 1 and 2 is determined by $\log_{10} 2 = 0.3010$, or approximately 30% of the full distance of a log interval, as shown on the graph. The distance between 1 and 3 is determined by $\log_{10} 3 = 0.4771$, or about 48% of the full width. For future reference, keep in mind that almost 50% of the width of one log interval is represented by a 3 rather than by the 5 of a linear scale. In addition, note that the number 5 is about 70% of the full width, and 8 is about 90%. Remembering the percentage of full width of the

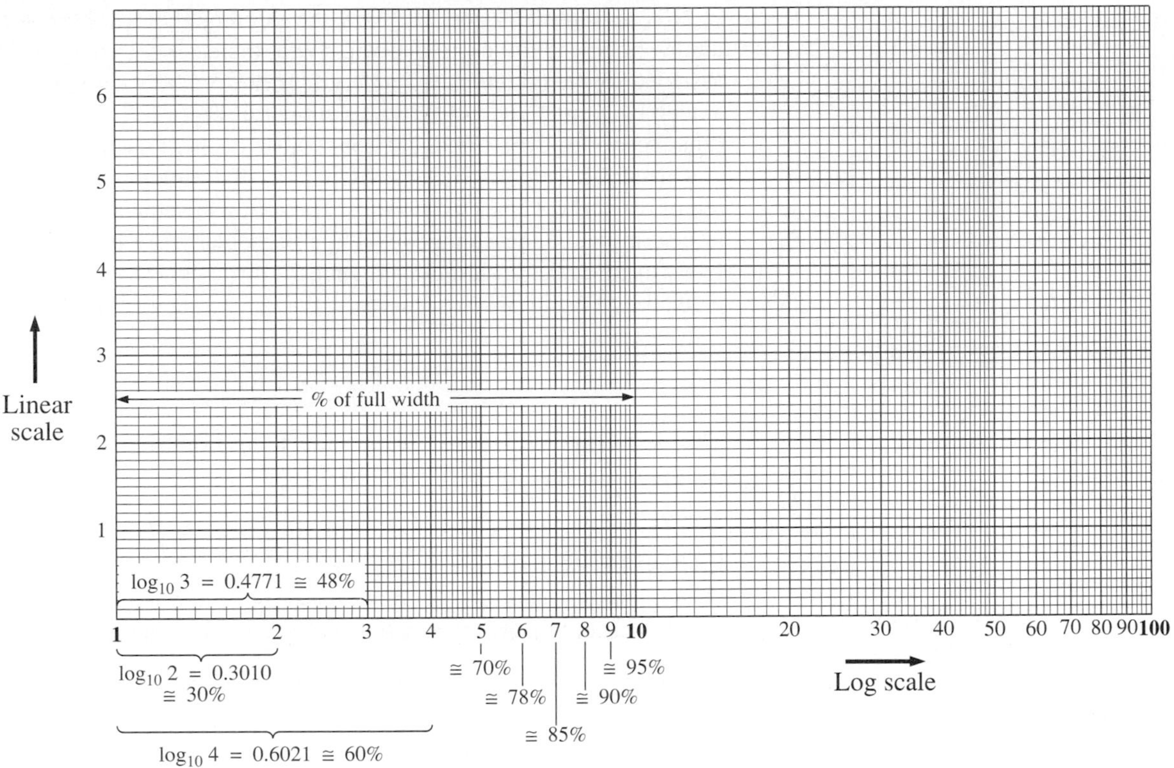

FIG. 21.1
Semilog graph paper.

lines 2, 3, 5, and 8 will be particularly useful when the various lines of a log plot are left unnumbered.

Since

$$\log_{10} 1 = 0$$
$$\log_{10} 10 = 1$$
$$\log_{10} 100 = 2$$
$$\log_{10} 1000 = 3$$
$$\vdots$$

the spacing between 1 and 10, 10 and 100, 100 and 1000, and so on, will be the same as shown in Figs. 21.1 and 21.2.

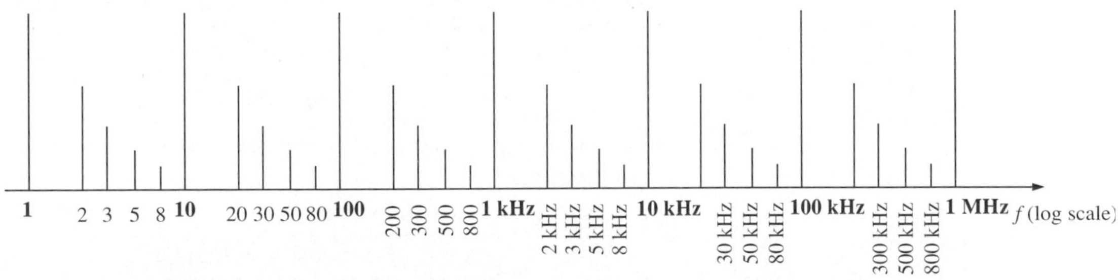

FIG. 21.2
Frequency log scale.

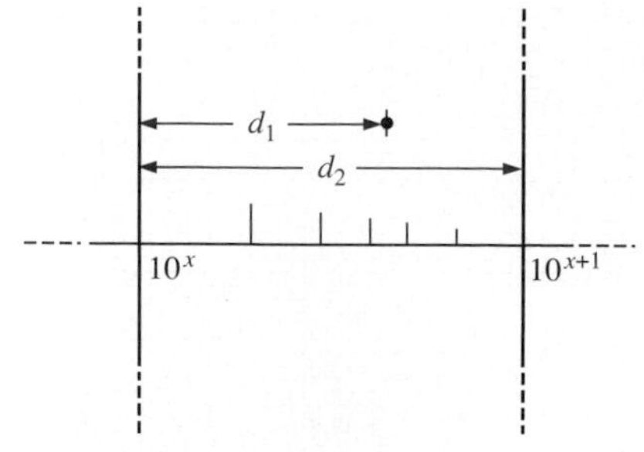

FIG. 21.3
Finding a value on a log plot.

Note in Figs. 21.1 and 21.2 how the log scale becomes compressed at the high end of each interval. With increasing frequency levels assigned to each interval, a single graph can provide a frequency plot extending from 1 Hz to 1 MHz, as shown in Fig. 21.2, with particular reference to the 30%, 50%, 70%, and 90% levels of each interval.

On many log plots the tick marks for most of the intermediate levels are left off because of space constraints. The following equation can be used to determine the logarithmic level at a particular point between known levels using a ruler or simply estimating the distances. The parameters are defined by Fig. 21.3.

$$\boxed{\text{Value} = 10^x \times 10^{d_1/d_2}} \tag{21.3}$$

The derivation of Eq. (21.3) is simply an extension of the details regarding distance appearing on Fig. 21.1.

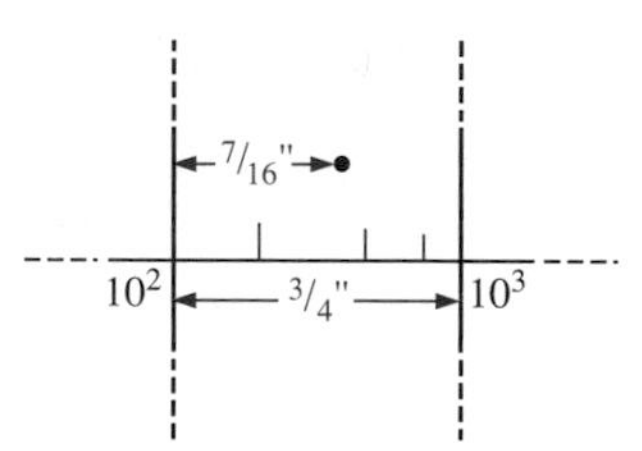

FIG. 21.4
Example 21.1.

EXAMPLE 21.1

Determine the value of the point appearing on the logarithmic plot of Fig. 21.4 using the measurements made by a ruler (linear).

Solution:

$$\frac{d_1}{d_2} = \frac{7/16''}{3/4''} = \frac{0.438''}{0.750''} = 0.584$$

Using a calculator:

$$10^{d_1/d_2} = 10^{0.584} = 3.837$$

Applying Eq. (21.3):

$$\text{Value} = 10^x \times 10^{d_1/d_2} = 10^2 \times 3.837$$
$$= \mathbf{383.7}$$

21.2 PROPERTIES OF LOGARITHMS

There are a few characteristics of logarithms that should be emphasized.

1. ***The common or natural logarithm of the number 1 is 0.***

$$\boxed{\log_{10} 1 = 0} \tag{21.4}$$

just as $10^x = 1$ requires that $x = 0$.

2. ***The log of any number less than 1 is a negative number.***

$$\log_{10} \tfrac{1}{2} = \log_{10} 0.5 = -0.3$$
$$\log_{10} \tfrac{1}{10} = \log_{10} 0.1 = -1$$

3. ***The log of the product of two numbers is the sum of the logs of the numbers.***

$$\boxed{\log_{10} ab = \log_{10} a + \log_{10} b} \tag{21.5}$$

4. ***The log of the quotient of two numbers is the log of the numerator minus the log of the denominator.***

$$\log_{10} \frac{a}{b} = \log_{10} a - \log_{10} b \quad (21.6)$$

5. *The log of a number taken to a power is equal to the product of the power and the log of the number.*

$$\log_{10} a^n = n \log_{10} a \quad (21.7)$$

Calculator Functions

On most calculators the log of a number is found by simply entering the number and pressing the **log** or **ln** key.

For example,

$$\log_{10} 80 = \boxed{8}\ \boxed{0}\ \boxed{\text{log}}$$

with a display of **1.903.**

For the reverse process, where N, or the antilogarithm, is desired, the function 10^x is employed. On most calculators 10^x appears as a second function above the **log** key. For the case of

$$0.6 = \log_{10} N$$

the following keys are employed:

$$\boxed{\cdot}\ \boxed{6}\ \boxed{\text{2ndF}}\ \boxed{10^x}$$

with a display of **3.981.** Checking: $\log_{10} 3.981 = 0.6$.

EXAMPLE 21.2 Evaluate each of the following logarithmic expressions.

a. $\log_{10} 0.004$
b. $\log_{10} 250{,}000$
c. $\log_{10}(0.08)(240)$
d. $\log_{10} \dfrac{1 \times 10^4}{1 \times 10^{-4}}$
e. $\log_{10}(10)^4$

Solutions:

a. **−2.398**
b. **+5.398**
c. $\log_{10}(0.08)(240) = \log_{10} 0.08 + \log_{10} 240 = -1.097 + 2.380 = \mathbf{1.283}$
d. $\log_{10} \dfrac{1 \times 10^4}{1 \times 10^{-4}} = \log_{10} 1 \times 10^4 - \log_{10} 1 \times 10^{-4} = 4 - (-4) = \mathbf{8}$
e. $\log_{10} 10^4 = 4 \log_{10} 10 = 4(1) = \mathbf{4}$

21.3 DECIBELS

Power Gain

Two levels of power can be compared using a unit of measure called the *bel,* which is defined by the following equation:

$$\boxed{B = \log_{10} \frac{P_2}{P_1}} \qquad \text{(bels)} \qquad \textbf{(21.8)}$$

However, to provide a unit of measure of *less* magnitude, a *decibel* is defined, where

$$\boxed{1 \text{ bel} = 10 \text{ decibels (dB)}} \qquad \textbf{(21.9)}$$

The result is the following important equation, which compares power levels P_2 and P_1 in decibels.

$$\boxed{\text{dB} = 10 \log_{10} \frac{P_2}{P_1}} \qquad \text{(decibels, dB)} \qquad \textbf{(21.10)}$$

If the power levels are equal ($P_2 = P_1$), there is no change in power level, and dB $= 0$. If there is an increase in power level ($P_2 > P_1$), the resulting decibel level is positive. If there is a decrease in power level ($P_2 < P_1$), the resulting decibel level will be negative.

For the special case of $P_2 = 2P_1$, the gain in decibels is

$$\text{dB} = 10 \log_{10} \frac{P_2}{P_1} = 10 \log_{10} 2 = \mathbf{3\ dB}$$

Therefore, for a speaker system, a 3-dB increase in output would require that the power level be doubled. In the audio industry, it is a generally accepted rule that an increase in sound level is accomplished with 3-dB increments in the output level. In other words, a 1-dB increase is barely detectable and a 2-dB increase, just discernible. A 3-dB increase normally results in a readily detectable increase in sound level. An additional increase in the sound level is normally accomplished by simply increasing the output level another 3 dB. If an 8-W system were in use, a 3-dB increase would require a 16-W output, whereas an additional increase of 3 dB (a total of 6 dB) would require a 32-W system, as demonstrated by the calculations below:

$$\text{dB} = 10 \log_{10} \frac{P_2}{P_1} = 10 \log_{10} \frac{16}{8} = 10 \log_{10} 2 = \mathbf{3\ dB}$$

$$\text{dB} = 10 \log_{10} \frac{P_2}{P_1} = 10 \log_{10} \frac{32}{8} = 10 \log_{10} 4 = \mathbf{6\ dB}$$

For $P_2 = 10P_1$,

$$\text{dB} = 10 \log_{10} \frac{P_2}{P_1} = 10 \log_{10} 10 = 10(1) = \mathbf{10\ dB}$$

resulting in the unique situation where the power gain has the same magnitude as the decibel level.

For some applications, a reference level is established to permit a comparison of decibel levels from one situation to another. For communication systems a commonly applied reference level is

$$P_{ref} = 1 \text{ mW} \quad \text{(across a 600-}\Omega\text{ load)}$$

Equation (21.10) is then typically written as

$$\text{dB}_m = 10 \log_{10} \frac{P}{1\text{ mW}}\bigg|_{600\ \Omega} \qquad \textbf{(21.11)}$$

Note the subscript m to denote that the decibel level is determined with a reference level of 1 mW.

In particular, for $P = 40$ mW,

$$\text{dB}_m = 10 \log_{10} \frac{40\text{ mW}}{1\text{ mW}} = 10 \log_{10} 40 = 10(1.6) = \mathbf{16\ dB_m}$$

whereas for $P = 4$ W,

$$\text{dB}_m = 10 \log_{10} \frac{4000\text{ mW}}{1\text{ mW}} = 10 \log_{10} 4000 = 10(3.6) = \mathbf{36\ dB_m}$$

Even though the power level has increased by a factor of 4000 mW/40 mW = 100, the dB_m increase is limited to 20 dB_m. In time, the significance of dB_m levels of 16 dB_m and 36 dB_m will generate an immediate appreciation regarding the power levels involved. An increase of 20 dB_m will also be associated with a significant gain in power levels.

Voltage Gain

Decibels are also used to provide a comparison between voltage levels. Substituting the basic power equations $P_2 = V_2^2/R_2$ and $P_1 = V_1^2/R_1$ into Eq. (21.10) will result in

$$\begin{aligned}\text{dB} &= 10 \log_{10} \frac{P_2}{P_1} = 10 \log_{10} \frac{V_2^2/R_2}{V_1^2/R_1} \\ &= 10 \log_{10} \frac{V_2^2/V_1^2}{R_2/R_1} = 10 \log_{10}\left(\frac{V_2}{V_1}\right)^2 - 10 \log_{10}\left(\frac{R_2}{R_1}\right)\end{aligned}$$

and

$$\text{dB} = 20 \log_{10} \frac{V_2}{V_1} - 10 \log_{10} \frac{R_2}{R_1}$$

For the situation where $R_2 = R_1$, a condition normally assumed when comparing voltage levels on a decibel basis, the second term of the preceding equation will drop out ($\log_{10} 1 = 0$) and

$$\text{dB}_v = 20 \log_{10} \frac{V_2}{V_1} \quad \text{(dB)} \qquad \textbf{(21.12)}$$

Note the subscript v to define the decibel level obtained.

EXAMPLE 21.3 Find the voltage gain in dB of a system where the applied signal is 2 mV and the output voltage is 1.2 V.

Solution:

$$\text{dB}_v = 20 \log_{10} \frac{V_o}{V_i} = 20 \log_{10} \frac{1.2\text{ V}}{2\text{ mV}} = 20 \log_{10} 600 = \mathbf{55.56\ dB}$$

for a voltage gain $A_v = V_o/V_i$ of 600.

EXAMPLE 21.4 If a system has a voltage gain of 36 dB, find the applied voltage if the output voltage is 6.8 V.

Solution:

$$\text{dB}_v = 10 \log_{10} \frac{V_o}{V_i}$$

$$36 = 20 \log_{10} \frac{V_o}{V_i}$$

$$1.8 = \log_{10} \frac{V_o}{V_i}$$

From the antilogarithm: $\dfrac{V_o}{V_i} = 63.096$

and $$V_i = \frac{V_o}{63.096} = \frac{6.8 \text{ V}}{63.096} = \mathbf{107.77 \ mV}$$

TABLE 21.1

V_o/V_i	$\text{dB} = 20 \log_{10}(V_o/V_i)$
1	0 dB
2	6 dB
10	20 dB
20	26 dB
100	40 dB
1,000	60 dB
100,000	100 dB

Table 21.1 compares the magnitude of specific gains to the resulting decibel level. In particular, note that when voltage levels are compared, a doubling of the level results in a change of 6 dB rather than 3 dB as obtained for power levels.

In addition, note that an increase in gain from 1 to 100,000 results in a change in decibels that can easily be plotted on a single graph. Also note that doubling the gain (from 1 to 2 and 10 to 20) results in a 6-dB increase in the decibel level, while a change of 10 to 1 (from 1 to 10, 10 to 100, and so on) always results in a 20-dB decrease in the decibel level.

The Human Auditory Response

One of the most frequent applications of the decibel scale is in the communication and entertainment industries. The human ear does not respond in a linear fashion to changes in source power level; that is, a doubling of the audio power level from 1/2 W to 1 W does not result in a doubling of the loudness level for the human ear. In addition, a change from 5 W to 10 W will be received by the ear as the same change in sound intensity as experienced from 1/2 W to 1 W. In other words, the ratio between levels is the same in each case (1 W/0.5 W = 10 W/5 W = 2), resulting in the same decibel or logarithmic change defined by Eq. (21.7). The ear, therefore, responds in a logarithmic fashion to changes in audio power levels.

To establish a basis for comparison between audio levels, a reference level of 0.0002 microbar (μbar) was chosen, where 1 μbar is equal to the sound pressure of 1 dyne per square centimeter, or about 1 millionth of the normal atmospheric pressure at sea level. The 0.0002-μbar level is the threshold level of hearing. Using this reference level, the sound pressure level in decibels is defined by the following equation:

$$\text{dB}_s = 20 \log_{10} \frac{P}{0.0002 \ \mu\text{bar}} \tag{21.13}$$

where P is the sound pressure in microbars.

The decibel levels of Fig. 21.5 are defined by Eq. (21.13). Meters designed to measure audio levels are calibrated to the levels defined by Eq. (21.13) and shown in Fig. 21.5.

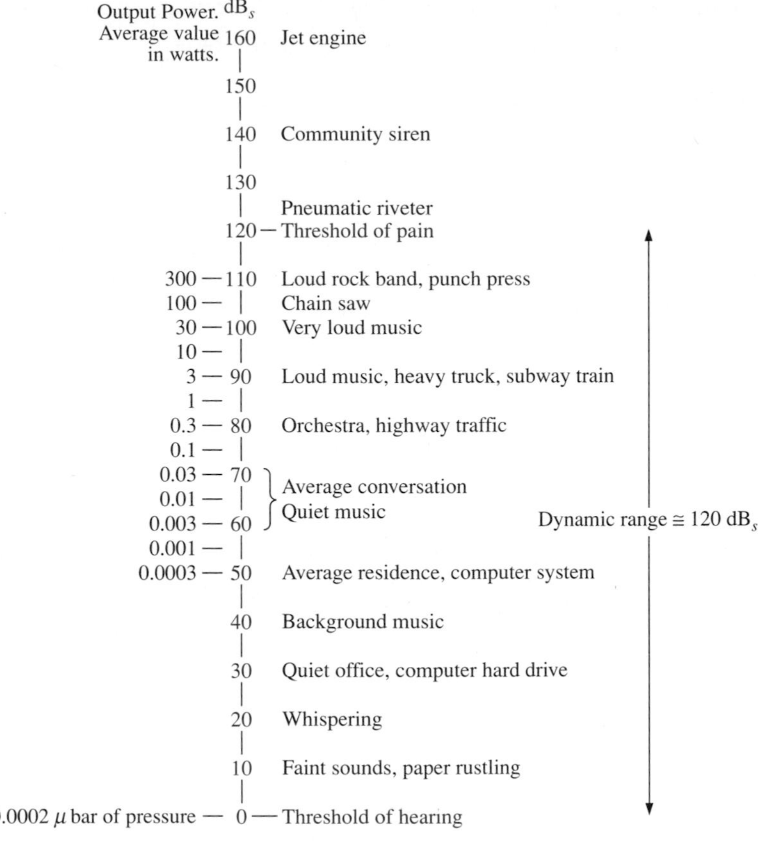

FIG. 21.5
Typical sound levels and their decibel levels.

A common question regarding audio levels is how much the power level of an acoustical source must be increased to double the sound level received by the human ear. The question is not as simple as it first seems due to considerations such as the frequency content of the sound, the acoustical conditions of the surrounding area, the physical characteristics of the surrounding medium, and—of course—the unique characteristics of the human ear. However, a general conclusion can be formulated that has practical value if we note the power levels of an acoustical source appearing to the left of Fig. 21.5. Each power level is associated with a particular decibel level, and a change of 10 dB in the scale corresponds with an increase or decrease in power by a factor of 10. For instance, a change from 90 dB to 100 dB is associated with a change in wattage from 3 W to 30 W. Through experimentation it has been found on an average basis that the loudness level will double for every 10-dB change in audio level—a conclusion somewhat verified by the examples to the right Fig. 21.5. Using the fact that a 10-dB change corresponds with a tenfold increase in power level supports the following conclusion (on an approximate basis): Through experimentation it has been found on an average basis that the loudness level will double for every 10-dB change in audio level.

To double the sound level received by the human ear, the power rating of the acoustical source (in watts) must be increased by a factor of 10.

In other words, doubling the sound level available from a 1-W acoustical source would require moving up to a 10-W source.

Instrumentation

A number of modern VOMs and DMMs have a dB scale designed to provide an indication of power ratios referenced to a standard level of 1 mW at 600 Ω. Since the reading is accurate only if the load has a characteristic impedance of 600 Ω, the 1-mW, 600 reference level is normally printed somewhere on the face of the meter, as shown in Fig. 21.6. The dB scale is usually calibrated to the lowest ac scale of the meter. In other

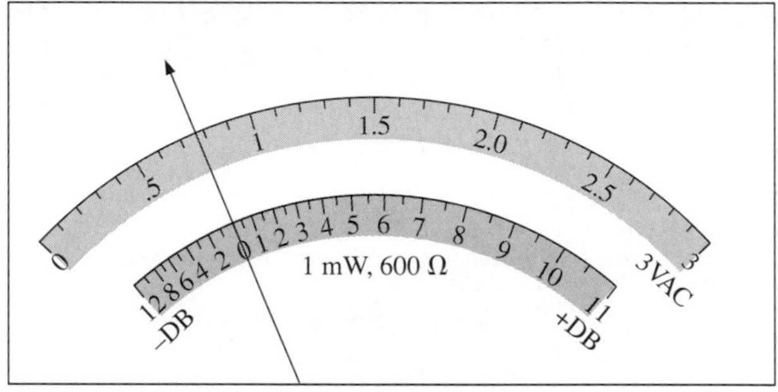

FIG. 21.6
Defining the relationship between a dB scale referenced to 1 mW, 600 Ω and a 3 V rms voltage scale.

words, when making the dB measurement, choose the lowest ac voltage scale, but read the dB scale. If a higher voltage scale is chosen, a correction factor must be employed that is sometimes printed on the face of the meter but always available in the meter manual. If the impedance is other than 600 Ω or not purely resistive, other correction factors must be used that are normally included in the meter manual. Using the basic power equation $P = V^2/R$ will reveal that 1 mW across a 600-Ω load is the same as applying 0.775 V rms across a 600-Ω load; that is, $V = \sqrt{PR} = \sqrt{(1\text{ mW})(600\ \Omega)} = 0.775$ V. The result is that an analog display will have 0 dB [defining the reference point of 1 mW, $\text{dB} = 10 \log_{10} P_2/P_1 = 10 \log_{10} (1\text{ mW}/1\text{ mW(ref)}] = 0\text{ dB}$] and 0.775 V rms on the same pointer projection, as shown in Fig. 21.6. A voltage of 2.5 V across a 600-Ω load would result in a dB level of $\text{dB} = 20 \log_{10} V_2/V_1 = 20 \log_{10} 2.5\text{ V}/0.775 = 10.17$ dB, resulting in 2.5 V and 10.17 dB appearing along the same pointer projection. A voltage of less than 0.775 V, such as 0.5 V, will result in a dB level of $\text{dB} = 20 \log_{10} V_2/V_1 = 20 \log_{10} 0.5\text{ V}/0.775\text{ V} = -3.8$ dB, as is also shown on the scale of Fig. 21.6. Although a reading of 10 dB will reveal that the power level is 10 times the reference, don't assume that a reading of 5 dB means the output level is 5 mW. The 10:1 ratio is a special one in logarithmic circles. For the 5-dB level, the power level must be found using the antilogarithm (3.126), which reveals that the power level associated with 5 dB is about 3.1 times the reference, or 3.1 mW. A conversion table is usually provided in the manual for such conversions.

21.4 FILTERS

Any combination of passive (*R*, *L*, and *C*) and/or active (transistors or operational amplifiers) elements designed to select or reject a band of frequencies is called a *filter.* In communication systems, filters are employed to pass those frequencies containing the desired information and reject the remaining frequencies. In stereo systems, filters can be used to isolate particular bands of frequencies for increased or decreased emphasis by the output acoustical system (amplifier, speaker, etc.). Filters are employed to filter out any unwanted frequencies, commonly called *noise,* due to the nonlinear characteristics of some electronic devices or signals picked up from the surrounding medium. In general, there are two classifications of filters:

1. *Passive filters* are those filters composed of series or parallel combinations of R, L, and C elements.
2. *Active filters* are filters that employ active devices such as transistors and operational amplifiers in combination with R, L, and C elements.

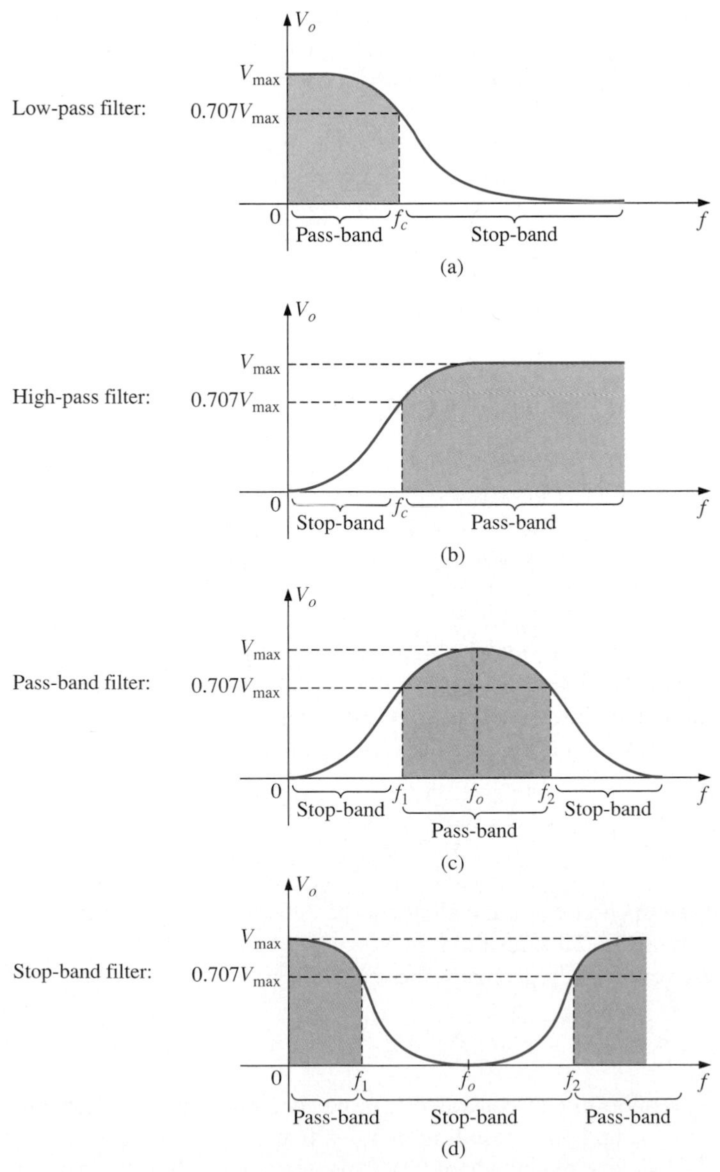

FIG. 21.7
Defining the four broad categories of filters.

Since this text is limited to passive devices, the analysis of this chapter will be limited to passive filters. In addition, only the most fundamental forms will be examined in the next few sections. The subject of filters is a very broad one that continues to receive extensive research support from industry and the government as new communication systems are developed to meet the demands of increased volume and speed. There are courses and texts devoted solely to the analysis and design of filter systems that can become quite complex and sophisticated. In general, however, all filters belong to the four broad categories of *low-pass, high-pass, pass-band,* and *stop-band,* as depicted in Fig. 21.7. For each form there are critical frequencies that define the regions of pass- and stop-bands (often called *reject* bands). Any frequency in the pass-band will pass through to the next stage with at least 70.7% of the maximum output voltage. Recall the use of the 0.707 level to define the bandwidth of a series or parallel resonant circuit (both with the general shape of the pass-band filter).

For some stop-band filters, the stop-band is defined by conditions other than the 0.707 level. In fact, for many stop-band filters, the condition that $V_o = 1/1000V_{\max}$ (corresponding with -60 dB in the discussion to follow) is used to define the stop-band region, with the pass-band continuing to be defined by the 0.707-V level. The resulting frequencies between the two regions are then called the *transition frequencies* and establish the *transition region.*

At least one example of each filter of Fig. 21.7 will be discussed in some detail in the sections to follow. Take particular note of the relative simplicity of some of the designs.

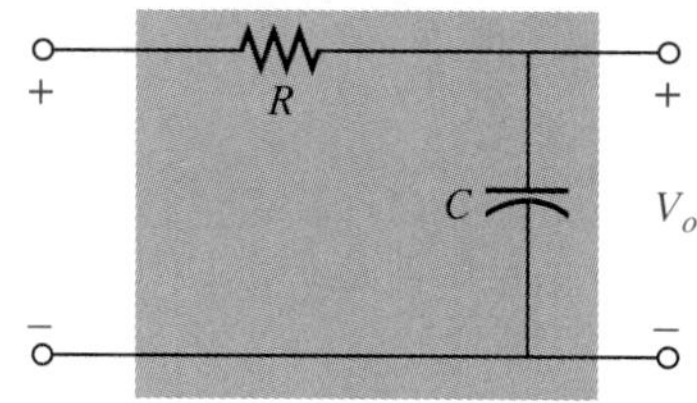

FIG. 21.8
Low-pass filter.

21.5 *R-C* LOW-PASS FILTER

The *R-C* filter, incredibly simple in design, can be used as a low-pass or high-pass filter. If the output is taken off the capacitor, as shown in Fig. 21.8, it will respond as a low-pass filter. If the positions of the resistor and capacitor are interchanged and the output is taken off the resistor, the response will be that of a high-pass filter.

A glance at Fig. 21.7(a) reveals that the circuit should behave in a manner that will result in a high-level output for low frequencies and a declining level for frequencies above the critical value. Let us first examine the network at the frequency extremes of $f = 0$ Hz and very high frequencies to test the response of the circuit.

At $f = 0$ Hz,

$$X_C = \frac{1}{2\pi fC} = \infty\ \Omega$$

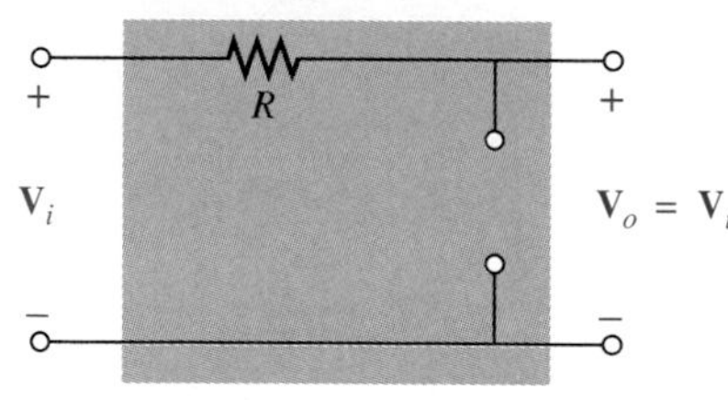

FIG. 21.9
R-C low-pass filter at low frequencies.

and the open-circuit equivalent can be substituted for the capacitor, as shown in Fig. 21.9, resulting in $\mathbf{V}_o = \mathbf{V}_i$.

At very high frequencies, the reactance is

$$X_C = \frac{1}{2\pi fC} \cong 0\ \Omega$$

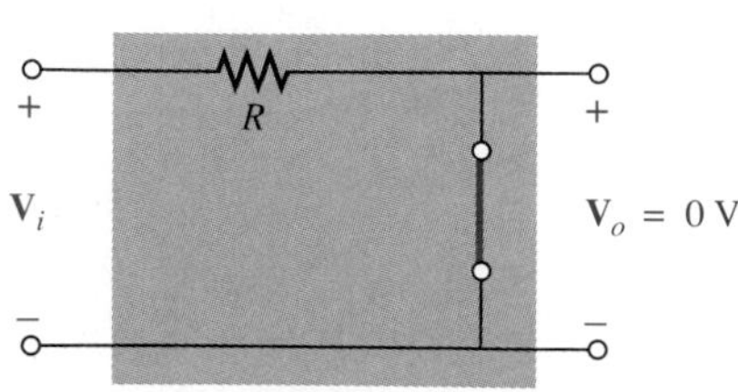

FIG. 21.10
R-C low-pass filter at high frequencies.

and the short-circuit equivalent can be substituted for the capacitor, as shown in Fig. 21.10, resulting in $\mathbf{V}_o = 0$ V.

A plot of the magnitude of V_o versus frequency will result in the curve of Fig. 21.11. Our next goal is now clearly defined: Find the frequency at which the transition takes place from a pass-band to a stop-band.

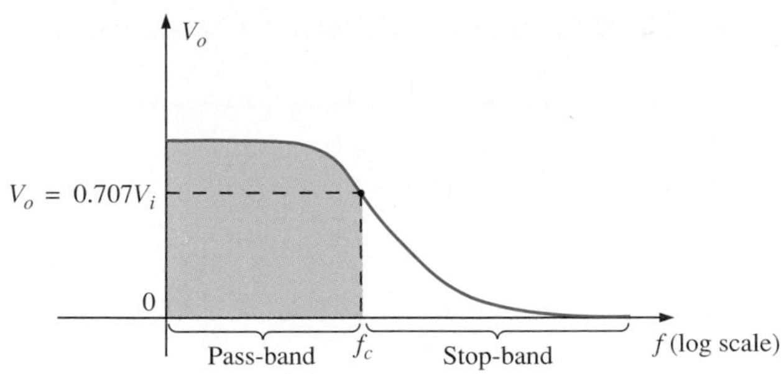

FIG. 21.11
V_o versus frequency for a low-pass R-C filter.

For filters, a normalized plot is employed more often than the plot of V_o versus frequency of Fig. 21.11.

Normalization is a process whereby a quantity such as voltage, current, or impedance is divided by a quantity of the same unit of measure to establish a dimensionless level of a specific value or range.

A normalized plot in the filter domain can be obtained by dividing the plotted quantity such as V_o of Fig. 21.11 with the applied voltage V_i for the frequency range of interest. Since the maximum value of V_o for the low-pass filter of Fig. 21.8 is V_i, each level of V_o in Fig. 21.11 is divided by the level of V_i. The result is the plot of $A_v = V_o/V_i$ of Fig. 21.12. Note that the maximum value is 1 and the cutoff frequency is defined at the 0.707 level.

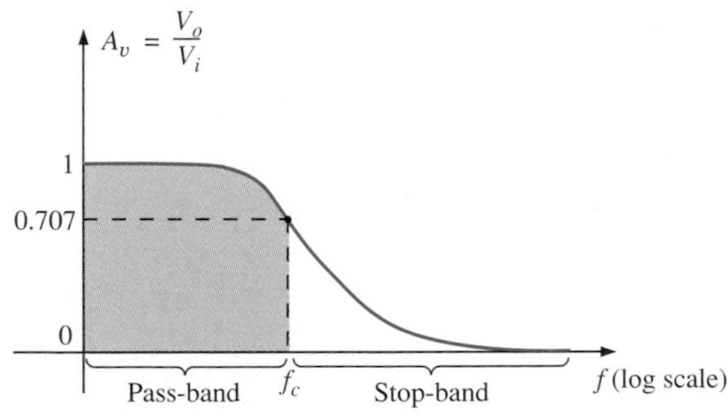

FIG. 21.12
Normalized plot of Fig. 21.11.

At any intermediate frequency, the output voltage $\mathbf{V}_o$ of Fig. 21.8 can be determined using the voltage divider rule:

$$\mathbf{V}_o = \frac{X_C \angle -90° \mathbf{V}_i}{R - jX_C}$$

or

$$\mathbf{A}_v = \frac{\mathbf{V}_o}{\mathbf{V}_i} = \frac{X_C \angle -90°}{R - jX_C} = \frac{X_C \angle -90°}{\sqrt{R^2 + X_C^2} \angle -\tan^{-1}(X_C/R)}$$

and

$$\mathbf{A}_v = \frac{\mathbf{V}_o}{\mathbf{V}_i} = \frac{X_C}{\sqrt{R^2 + X_C^2}} \angle -90° + \tan^{-1}\left(\frac{X_C}{R}\right)$$

The magnitude of the ratio V_o/V_i is therefore determined by

$$A_v = \frac{V_o}{V_i} = \frac{X_C}{\sqrt{R^2 + X_C^2}} = \frac{1}{\sqrt{\left(\frac{R}{X_C}\right)^2 + 1}} \tag{21.14}$$

and the phase angle is determined by

$$\theta = -90° + \tan^{-1}\frac{X_C}{R} = -\tan^{-1}\frac{R}{X_C} \tag{21.15}$$

For the special frequency at which $X_C = R$, the magnitude becomes

$$A_v = \frac{V_o}{V_i} = \frac{1}{\sqrt{\left(\frac{R}{X_C}\right)^2 + 1}} = \frac{1}{\sqrt{1+1}} = \frac{1}{\sqrt{2}} = 0.707$$

which defines the critical or cutoff frequency of Fig. 21.12.

The frequency at which $X_C = R$ is determined by

$$\frac{1}{2\pi f_c C} = R$$

and

$$f_c = \frac{1}{2\pi RC} \tag{21.16}$$

The impact of Eq. (21.16) extends beyond its relative simplicity. For any low-pass filter, the application of any frequency less than f_c will result in an output voltage V_o that is at least 70.7% of the maximum. For any frequency above f_c, the output is less than 70.7% of the applied signal.

Solving for $\mathbf{V}_o$ and substituting $\mathbf{V}_i = V_i \angle 0°$ gives

$$\mathbf{V}_o = \left[\frac{X_C}{\sqrt{R^2 + X_C^2}} \angle\theta\right]\mathbf{V}_i = \left[\frac{X_C}{\sqrt{R^2 + X_C^2}} \angle\theta\right] V_i \angle 0°$$

and

$$\mathbf{V}_o = \frac{X_C V_i}{\sqrt{R^2 + X_C^2}} \angle\theta$$

The angle θ is, therefore, the angle by which $\mathbf{V}_o$ leads $\mathbf{V}_i$. Since $\theta = -\tan^{-1} R/X_C$ is always negative (except at $f = 0$ Hz), it is clear that $\mathbf{V}_o$ will always lag $\mathbf{V}_i$, leading to the label *lagging network* for the network of Fig. 21.8.

At high frequencies, X_C is very small and R/X_C is quite large, resulting in $\theta = -\tan^{-1} R/X_C$ approaching $-90°$.

At low frequencies, X_C is quite large and R/X_C is very small, resulting in θ approaching 0°.

At $X_C = R$, or $f = f_c$, $-\tan^{-1} R/X_C = -\tan^{-1} 1 = -45°$.

A plot of θ versus frequency results in the phase plot of Fig. 21.13.

The plot is of $\mathbf{V}_o$ leading $\mathbf{V}_i$, but since the phase angle is always negative, the phase plot of Fig. 21.14 ($\mathbf{V}_o$ lagging $\mathbf{V}_i$) is more appropriate. Note that a change in sign requires that the vertical axis be changed to the angle by which $\mathbf{V}_o$ lags $\mathbf{V}_i$. In particular, note that the phase angle between $\mathbf{V}_o$ and $\mathbf{V}_i$ is less than 45° in the pass-band and approaches 0°

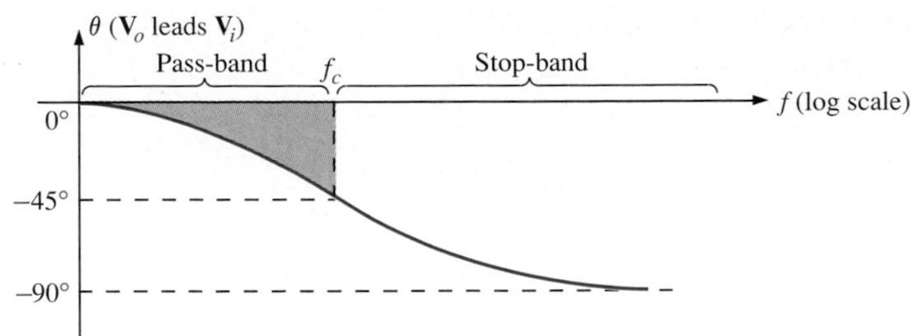

FIG. 21.13
Angle by which $\mathbf{V}_o$ *leads* $\mathbf{V}_i$.

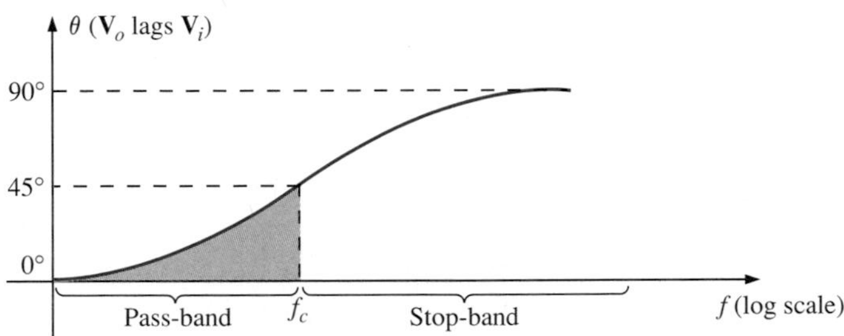

FIG. 21.14
Angle by which $\mathbf{V}_o$ *lags* $\mathbf{V}_i$.

at lower frequencies. In summary, for the low-pass *R-C* filter of Fig. 21.8:

$$f_c = \frac{1}{2\pi RC}$$

For	$f < f_c$,	$V_o > 0.707V_i$
whereas for	$f > f_c$,	$V_o < 0.707V_i$
At f_c,	$\mathbf{V}_o$ lags $\mathbf{V}_i$ by 45°.	

The low-pass filter response of Fig. 21.7(a) can also be obtained using the *R-L* combination of Fig. 21.15 with

$$f_c = \frac{R}{2\pi L} \qquad \textbf{(21.17)}$$

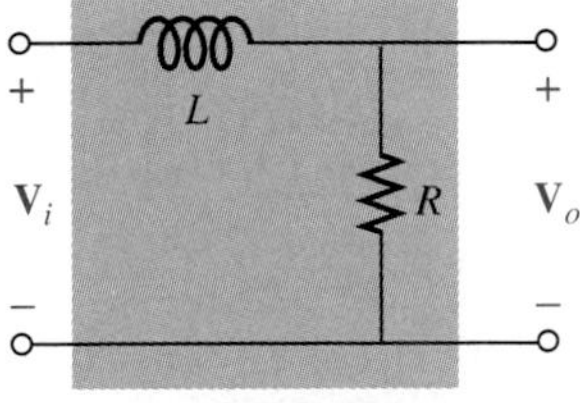

FIG. 21.15
Low-pass R-L filter.

In general, however, the *R-C* combination is more popular due to the smaller size of capacitive elements and the nonlinearities associated with inductive elements. The details of the analysis of the low-pass *R-L* will be left as an exercise for the reader.

EXAMPLE 21.5

a. Sketch the output voltage V_o versus frequency for the low-pass *R-C* filter of Fig. 21.16.
b. Determine the voltage V_o at f = 100 kHz and 1 MHz and compare the results to the results obtained from the curve of part (a).
c. Sketch the normalized gain $A_v = V_o/V_i$.

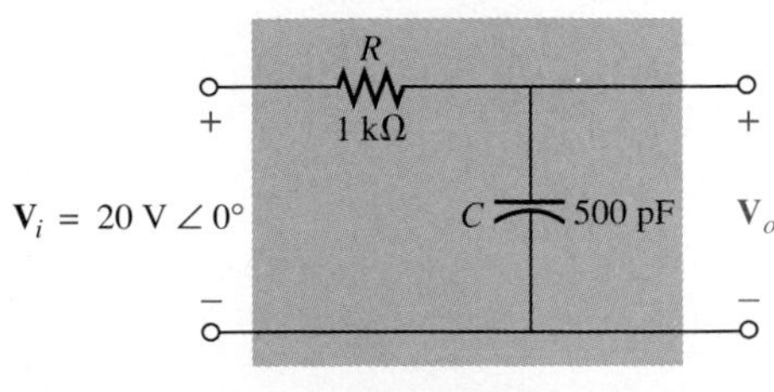

FIG. 21.16
Example 21.5.

Solutions:

a. Equation (21.16):

$$f_c = \frac{1}{2\pi RC} = \frac{1}{2\pi(1\text{ k}\Omega)(500\text{ pF})} = \mathbf{318.31\text{ kHz}}$$

At f_c, $V_o = 0.707(20\text{ V}) = 14.14\text{ V}$. See Fig. 21.17.

b. Eq. (21.14):

$$V_o = \frac{V_i}{\sqrt{\left(\frac{R}{X_C}\right)^2 + 1}}$$

At $f = 100$ kHz:

$$X_C = \frac{1}{2\pi fC} = \frac{1}{2\pi(100\text{ kHz})(500\text{ pF})} = 3.18\text{ k}\Omega$$

and
$$V_o = \frac{20\text{ V}}{\sqrt{\left(\frac{1\text{ k}\Omega}{3.18\text{ k}\Omega}\right)^2 + 1}} = \mathbf{19.08\text{ V}}$$

At $f = 1$ MHz:

$$X_C = \frac{1}{2\pi fC} = \frac{1}{2\pi(1\text{ MHz})(500\text{ pF})} = 0.32\text{ k}\Omega$$

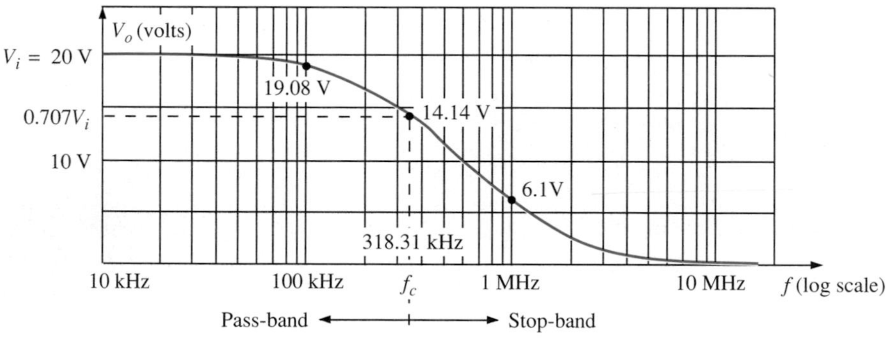

FIG. 21.17

Frequency response for the low-pass R-C network of Fig. 21.16.

and
$$V_o = \frac{20\text{ V}}{\sqrt{\left(\frac{1\text{ k}\Omega}{0.32\text{ k}\Omega}\right)^2 + 1}} = \mathbf{6.1\text{ V}}$$

Both levels are verified by Fig. 21.17.

c. Dividing every level of Fig. 21.17 by $V_i = 20$ V will result in the normalized plot of Fig. 21.18.

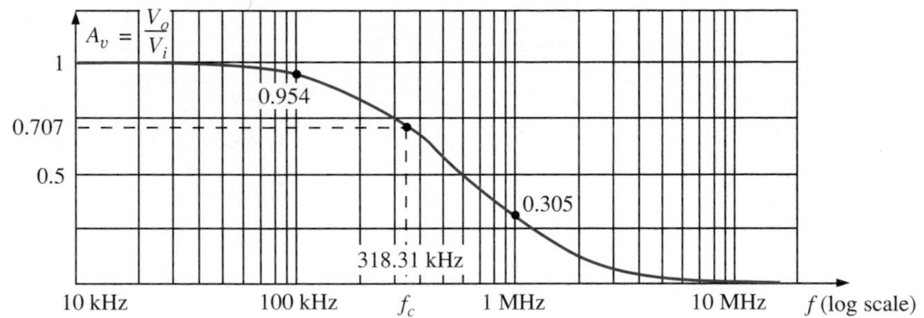

FIG. 21.18
Normalized plot of Fig. 21.17.

21.6 *R-C* HIGH-PASS FILTER

As noted early in Section 21.5, a high-pass *R-C* filter can be constructed by simply reversing the positions of the capacitor and resistor, as shown in Fig. 21.19.

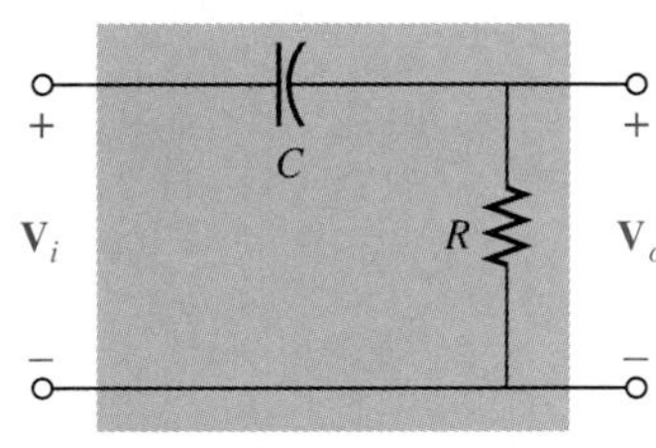

FIG. 21.19
High-pass filter.

At very high frequencies the reactance of the capacitor is very small and the short-circuit equivalent can be substituted, as shown in Fig. 21.20. The result is $\mathbf{V}_o = \mathbf{V}_i$.

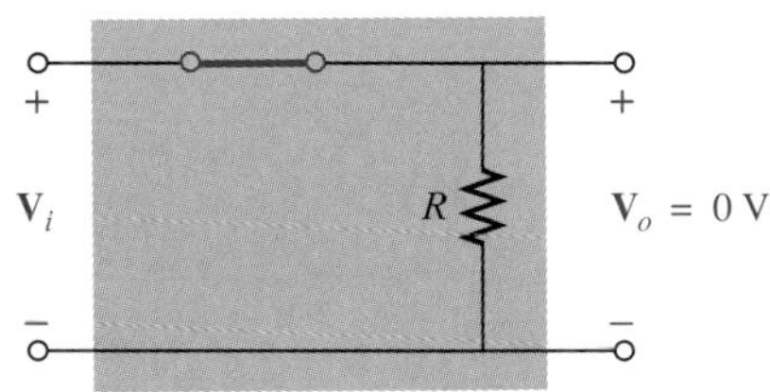

FIG. 21.20
R-C high-pass filter at very high frequencies.

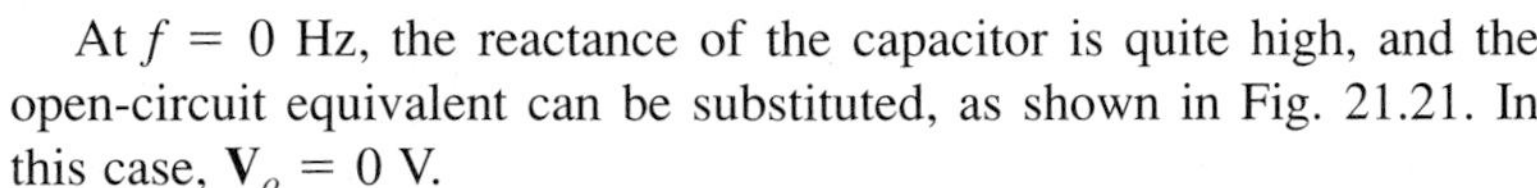

At $f = 0$ Hz, the reactance of the capacitor is quite high, and the open-circuit equivalent can be substituted, as shown in Fig. 21.21. In this case, $\mathbf{V}_o = 0$ V.

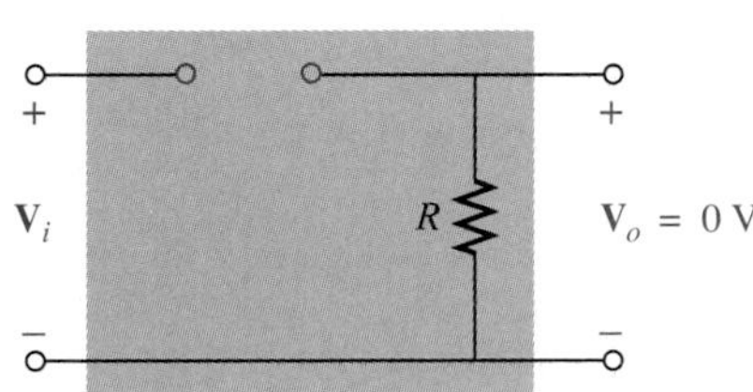

FIG. 21.21
R-C high-pass filter at $f = 0$ Hz.

A plot of the magnitude versus frequency is provided in Fig. 21.22, with the normalized plot in Fig. 21.23.

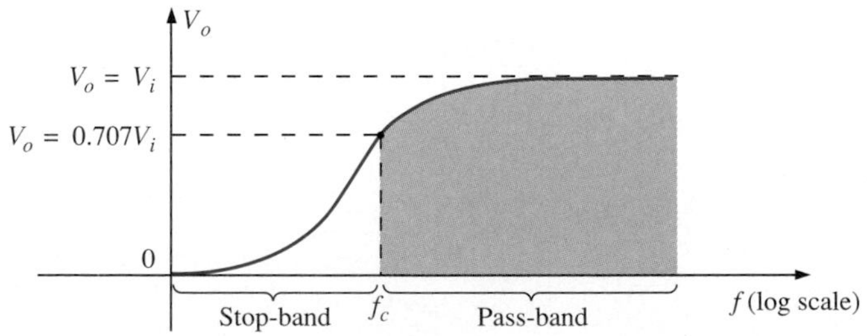

FIG. 21.22
V_o versus frequency for a high-pass R-C filter.

At any intermediate frequency, the output voltage can be determined using the voltage divider rule:

$$\mathbf{V}_o = \frac{R \angle 0^\circ \, \mathbf{V}_i}{R - jX_C}$$

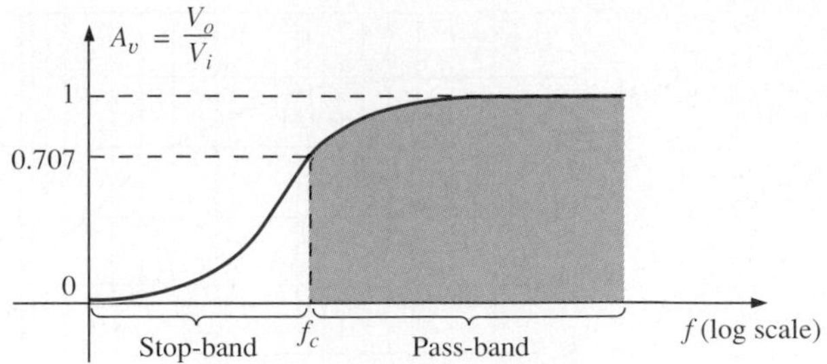

FIG. 21.23
Normalized plot of Fig. 21.22.

or

$$\frac{\mathbf{V}_o}{\mathbf{V}_i} = \frac{R \angle 0^\circ}{R - jX_C} = \frac{R \angle 0^\circ}{\sqrt{R^2 + X_C^2} \angle -\tan^{-1}(X_C/R)}$$

and

$$\frac{\mathbf{V}_o}{\mathbf{V}_i} = \frac{R}{\sqrt{R^2 + X_C^2}} \angle \tan^{-1}(X_C/R)$$

The magnitude of the ratio $\mathbf{V}_o/\mathbf{V}_i$ is therefore determined by

$$A_v = \frac{V_o}{V_i} = \frac{X_C}{\sqrt{R^2 + X_C^2}} = \frac{1}{\sqrt{\left(\frac{R}{X_C}\right)^2 + 1}} \quad \textbf{(21.18)}$$

and the phase angle θ, by

$$\theta = \tan^{-1}\frac{X_C}{R} \quad \textbf{(21.19)}$$

For the frequency at which $X_C = R$, the magnitude becomes

$$\frac{V_o}{V_i} = \frac{1}{\sqrt{1 + \left(\frac{X_C}{R}\right)^2}} = \frac{1}{\sqrt{1 + 1}} = \frac{1}{\sqrt{2}} = 0.707$$

as shown in Fig. 21.23.

The frequency at which $X_C = R$ is determined by

$$X_C = \frac{1}{2\pi f_c C} = R$$

and

$$f_c = \frac{1}{2\pi RC} \quad \textbf{(21.20)}$$

For the high-pass R-C filter, the application of any frequency greater than f_c will result in an output voltage V_o that is at least 70.7% of the magnitude of the input signal. For any frequency below f_c, the output is less than 70.7% of the applied signal.

For the phase angle, high frequencies result in small values of X_C, and the ratio X_C/R will approach zero with $\tan^{-1}(X_C/R)$ approaching 0°, as shown in Fig. 21.24. At low frequencies, the ratio X_C/R becomes

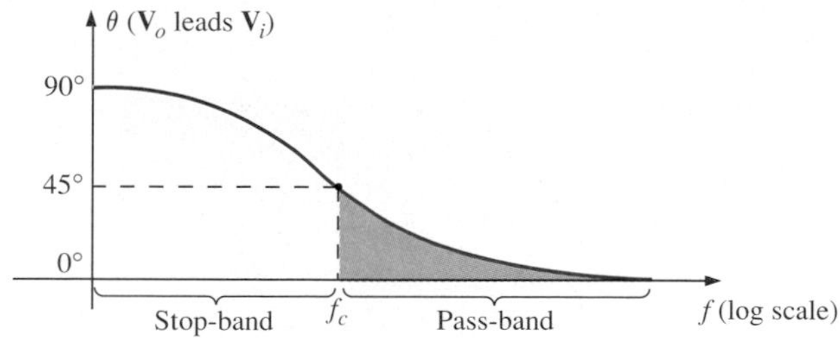

FIG. 21.24
Phase-angle response for the high-pass R-C filter.

quite large and $\tan^{-1}(X_C/R)$ approaches 90°. For the case $X_C = R$, $\tan^{-1}(X_C/R) = \tan^{-1} 1 = 45°$. Assigning a phase angle of 0° to $\mathbf{V}_i$ such that $\mathbf{V}_i = V_i \angle 0°$, the phase angle associated with $\mathbf{V}_o$ is θ, resulting in $\mathbf{V}_o = V_o \angle \theta$ and revealing that θ is the angle by which $\mathbf{V}_o$ leads $\mathbf{V}_i$. Since the angle θ is the angle by which $\mathbf{V}_o$ leads $\mathbf{V}_i$ throughout the frequency range of Fig. 21.24, the high-pass *R-C* filter is referred to as a *leading network.*

In summary, for the high-pass *R-C* filter:

$$f_c = \frac{1}{2\pi RC}$$

For	$f < f_c$,	$V_o < 0.707V_i$
whereas for	$f > f_c$,	$V_o > 0.707V_i$
At f_c,	$\mathbf{V}_o$ leads $\mathbf{V}_i$ by 45°.	

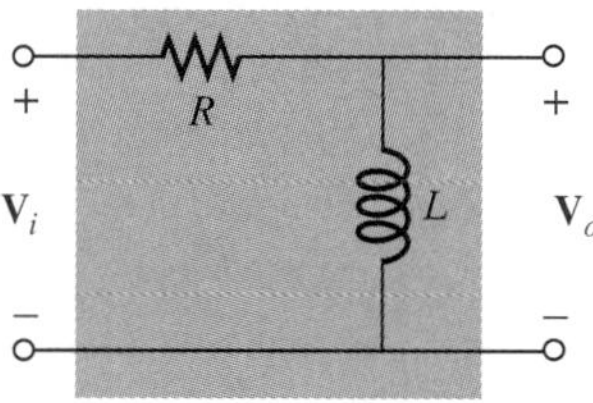

FIG. 21.25
High-pass R-L filter.

The high-pass filter response of Fig. 21.23 can also be obtained using the same elements of Fig. 21.15 but interchanging their positions, as shown in Fig. 21.25.

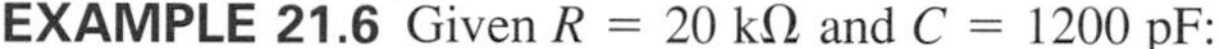

EXAMPLE 21.6 Given $R = 20\text{ k}\Omega$ and $C = 1200\text{ pF}$:

a. Sketch the normalized plot if the filter is used as both a high- and a low-pass filter.
b. Sketch the phase plot for both filters of part (a).
c. Determine the magnitude and phase of $\mathbf{A}_v = \mathbf{V}_o/\mathbf{V}_i$ at $f = \frac{1}{2}f_c$ for the high-pass filter.

Solutions:

a. $f_c = \dfrac{1}{2\pi RC} = \dfrac{1}{(2\pi)(20\text{ k}\Omega)(1200\text{ pF})}$

$= \mathbf{6631.46\ Hz}$

The normalized plots appear in Fig. 21.26.

b. The phase plots appear in Fig. 21.27.

c. $f = \dfrac{1}{2}f_c = \dfrac{1}{2}(6631.46\text{ Hz}) = 3315.73\text{ Hz}$

$$X_C = \frac{1}{2\pi fC} = \frac{1}{(2\pi)(3315.73\text{ Hz})(1200\text{ pF})}$$

$$\cong 40\text{ k}\Omega$$

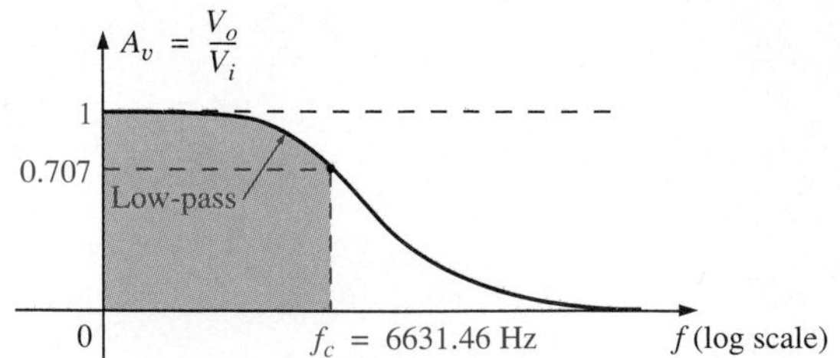

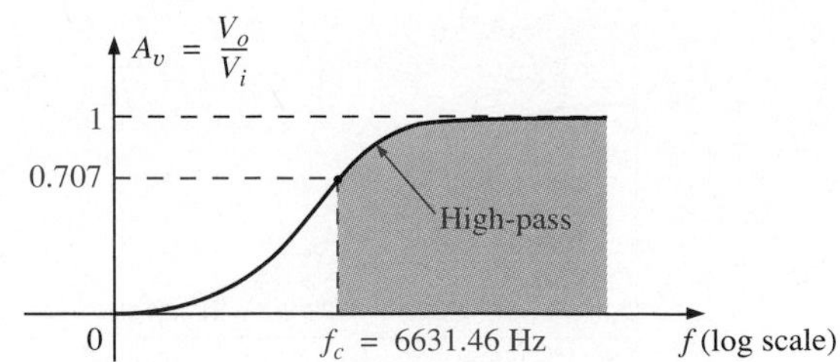

FIG. 21.26
Normalized plots for a low-pass and high-pass filter using the same elements.

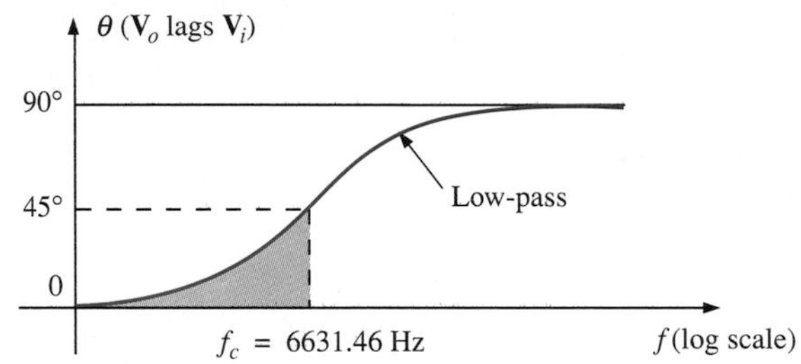

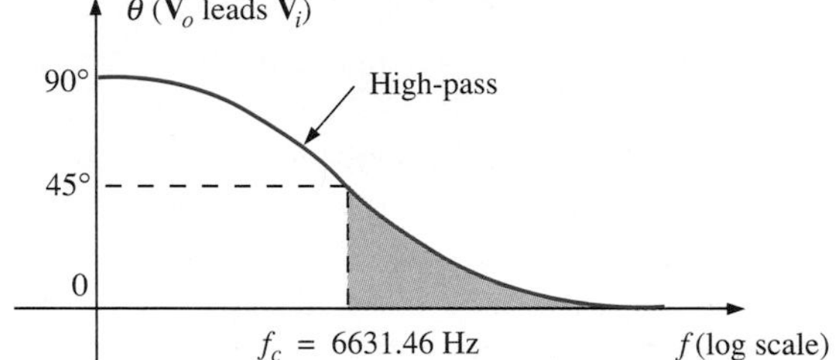

FIG. 21.27
Phase plots for a low-pass and high-pass filter using the same elements.

$$A_v = \frac{V_o}{V_i} = \frac{1}{\sqrt{1 + \left(\frac{X_C}{R}\right)^2}} = \frac{1}{\sqrt{1 + \left(\frac{40\text{ k}\Omega}{20\text{ k}\Omega}\right)^2}} = \frac{1}{\sqrt{1 + (2)^2}}$$

$$= \frac{1}{\sqrt{5}} = 0.4472$$

$$\theta = \tan^{-1}\frac{X_C}{R} = \tan^{-1}\frac{40\text{ k}\Omega}{20\text{ k}\Omega} = \tan^{-1} 2 = 63.43°$$

and $$\mathbf{A}_v = \frac{\mathbf{V}_o}{\mathbf{V}_i} = \mathbf{0.4472 \angle 63.43°}$$

21.7 PASS-BAND FILTERS

There are a number of methods to establish the pass-band characteristic of Fig. 21.7(c). One method employs both a low-pass and high-pass filter in cascade, as shown in Fig. 21.28.

The components are chosen to establish a cutoff frequency for the high-pass filter that is lower than the critical frequency of the low-pass filter, as shown in Fig. 21.29. A frequency f_1 may pass through the low-

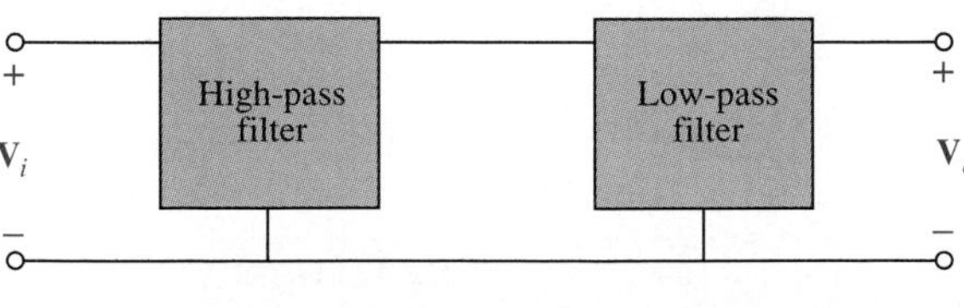

FIG. 21.28
Pass-band filter.

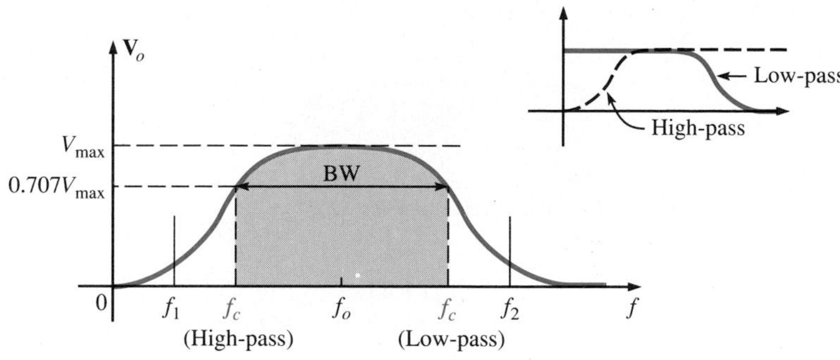

FIG. 21.29
Pass-band characteristics.

pass filter but have little effect on V_o due to the reject characteristics of the high-pass filter. A frequency f_2 may pass through the high-pass filter unmolested but be prohibited from reaching the high-pass filter by the low-pass characteristics. A frequency f_o near the center of the pass-band will pass through both filters with very little degeneration.

The network of Example 21.7 will generate the characteristics of Fig. 21.29. However, for a circuit such as the one shown in Fig. 21.30, there is a loading between stages at each frequency that will affect the level of $\mathbf{V}_o$. Through proper design, the level of $\mathbf{V}_o$ may be very near $\mathbf{V}_i$ in the pass-band, but it will never be exactly equal. In addition, as the critical frequencies of each filter get closer and closer together to increase the quality factor of the response curve, the peak values within the pass-band will continue to drop.

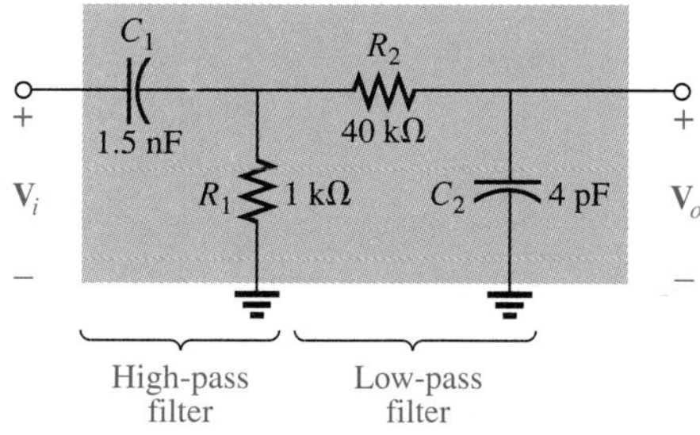

FIG. 21.30
Pass-band filter.

EXAMPLE 21.7 For the pass-band filter of Fig. 21.30:

a. Determine the critical frequencies for the low- and high-pass filters.
b. Using only the critical frequencies, sketch the ideal response characteristics and determine the bandwidth of the pass-band.
c. Determine the actual value of V_o at the high-pass critical frequency and compare to the ideal value of $0.707V_i$.

Solutions:

a. High-pass filter:

$$f_c = \frac{1}{2\pi R_1 C_1} = \frac{1}{2\pi(1\text{ k}\Omega)(1.5\text{ nF})} = \mathbf{106.1\ kHz}$$

Low-pass filter:

$$f_c = \frac{1}{2\pi R_2 C_2} = \frac{1}{2\pi(40\text{ k}\Omega)(4\text{ pF})} = \mathbf{994.72\ kHz}$$

b. See Fig. 21.31.
c. At $f = 994.72$ kHz,

$$X_{C_1} = \frac{1}{2\pi f C_1} \cong 107\ \Omega$$

and

$$X_{C_2} = \frac{1}{2\pi f C_2} = R_2 = 40\text{ k}\Omega$$

resulting in the network of Fig. 21.32.

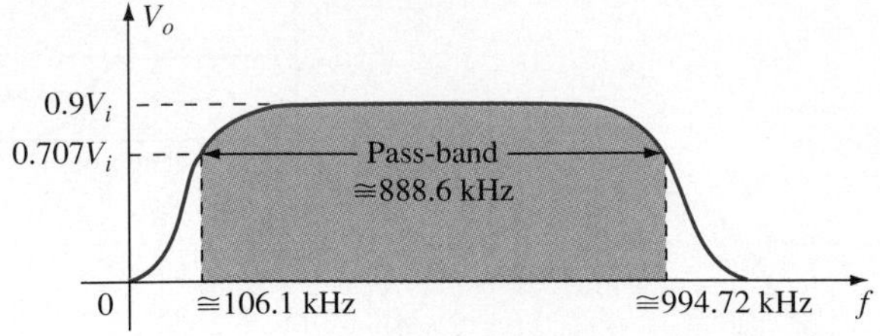

FIG. 21.31
Pass-band characteristics for the filter of Fig. 21.30.

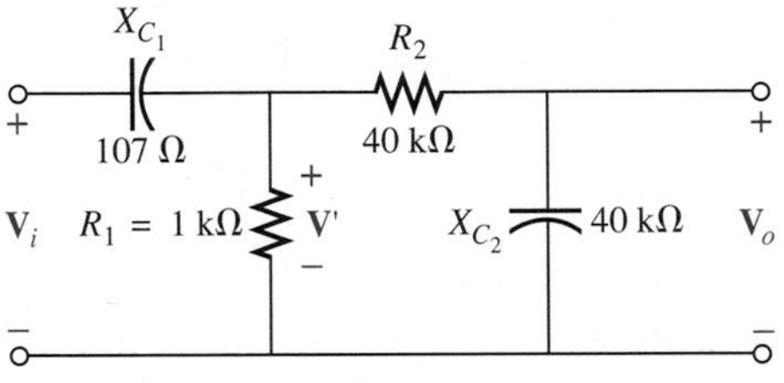

FIG. 21.32
Network of Fig. 21.30 at f = 994.72 kHz.

The magnitude of the series R_2-X_{C_2} combination is so large compared to the parallel resistor R_1 that the loading effect on R_1 can be ignored as a good approximation.

The result is

$$\mathbf{V}' = \frac{R_1 \angle 0^\circ \, \mathbf{V}_i}{R_1 - jX_{C_1}} = \frac{(1\text{ k}\Omega \angle 0^\circ)\mathbf{V}_i}{1\text{ k}\Omega - j\,0.107\text{ k}\Omega} = 0.994V_i \angle 6.11^\circ$$

At $f = f_c$,

$$V_o = 0.707V' = 0.707(0.994V_i)$$

and

$$A_v = \frac{V_o}{V_i} = \mathbf{0.703}$$

which is very close to the 0.707 level.

A PSpice analysis using schematics will result in a level of 0.707 at a frequency of 107.5 kHz, and 0.710 at 970.98 kHz, verifying our design. The frequency response, however, did not reach a maximum in the pass-band region but maximizes at 0.901 at a frequency of 320.33 kHz.

The pass-band response can also be obtained using the series and parallel resonant circuits discussed in Chapter 20. In each case, however, V_o will not be equal to V_i in the pass-band, but a frequency range in which V_o will be equal to or greater than $0.707V_{max}$ can be defined.

For the series resonant circuit of Fig. 21.33, $X_L = X_C$ at resonance and

$$V_{o_{max}} = \frac{R}{R + R_l} V_i \qquad f = f_s \tag{21.21}$$

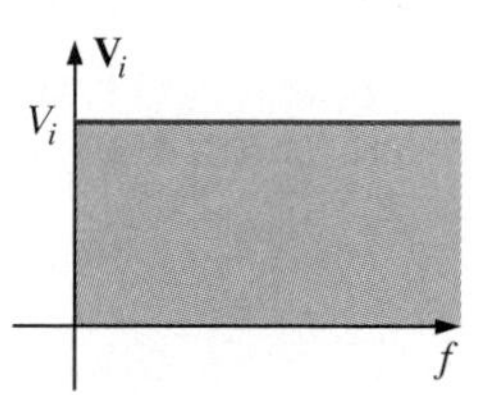

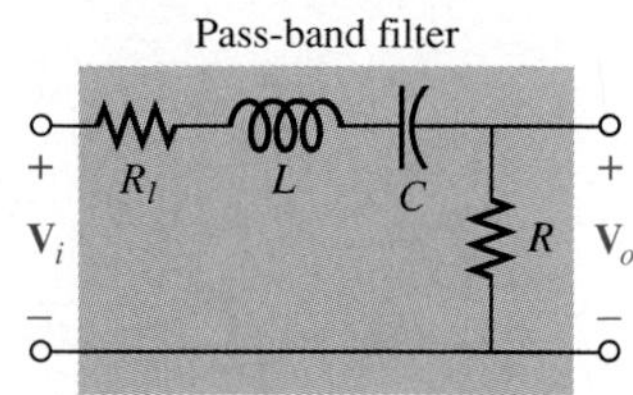

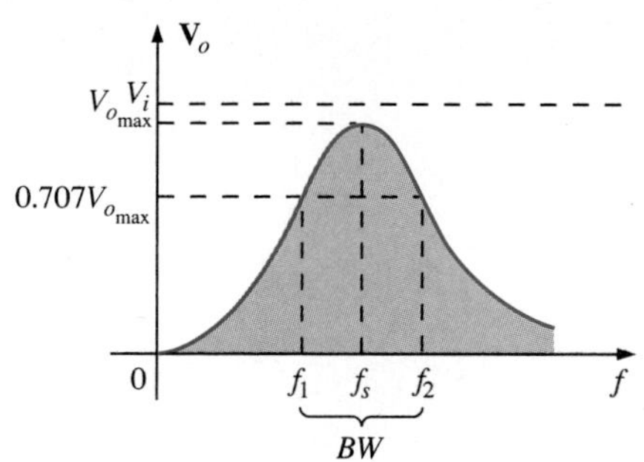

FIG. 21.33
Series resonant pass-band filter.

and
$$f_s = \frac{1}{2\pi\sqrt{LC}} \tag{21.22}$$

with
$$Q_s = \frac{X_L}{R + R_l} \tag{21.23}$$

and
$$BW = \frac{f_s}{Q_s} \tag{21.24}$$

$$\frac{Z_{T_p}V_i}{Z_{T_p} + R}$$

For the parallel resonant circuit of Fig. 21.34, Z_{T_p} is a maximum value at resonance and

$$V_{o_{max}} = \frac{Z_{T_p}V_i}{Z_{T_p} + R} \Bigg|_{f = f_p} \tag{21.25}$$

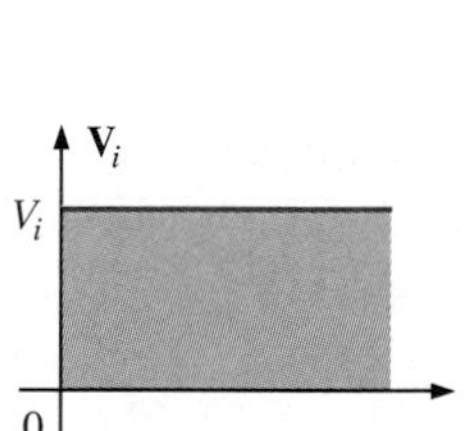

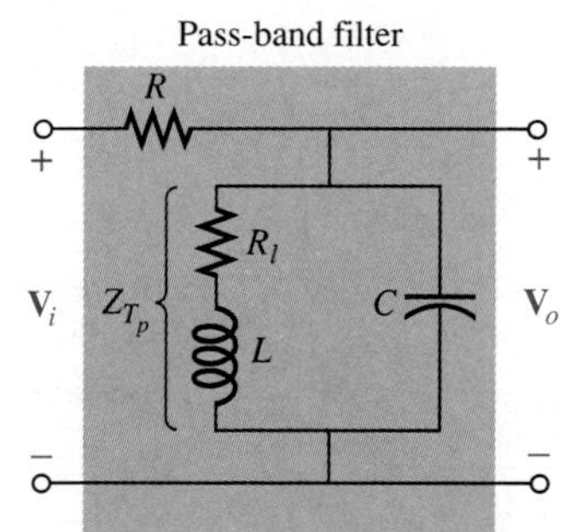

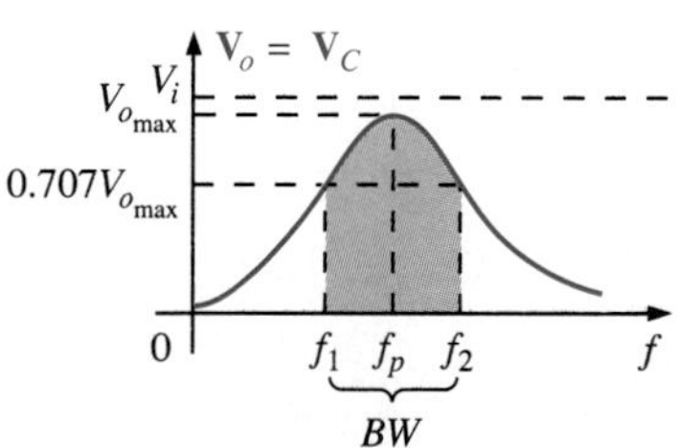

FIG. 21.34
Parallel resonant pass-band filter.

with
$$Z_{T_p} = Q_l^2 R_l \qquad Q_l \geq 10 \tag{21.26}$$

and
$$f_p = \frac{1}{2\pi\sqrt{LC}} \qquad Q_l \geq 10 \tag{21.27}$$

For the parallel resonant circuit

$$Q_p = \frac{X_L}{R_l} \tag{21.28}$$

and

$$BW = \frac{f_p}{Q_p} \tag{21.29}$$

As a first approximation that is acceptable for most practical applications, it can be assumed that the resonant frequency bisects the bandwidth.

EXAMPLE 21.8

a. Determine the frequency response for the voltage V_o for the series circuit of Fig. 21.35.
b. Plot the normalized response $A_v = V_o/V_i$.
c. Plot a normalized response defined by $A'_v = \dfrac{A_v}{A_{v_{max}}}$.

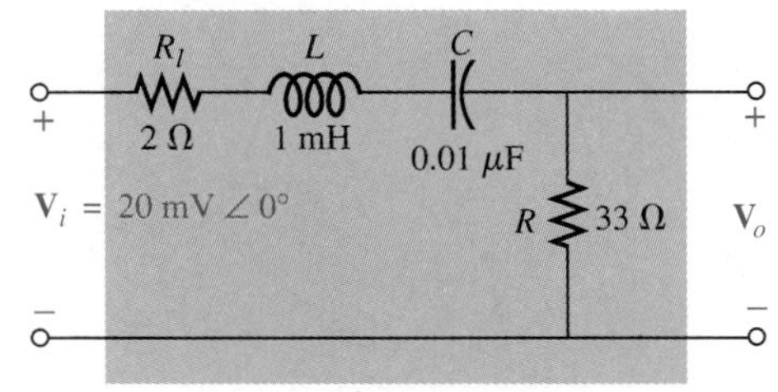

FIG. 21.35
Series resonant pass-band filter for Example 21.8.

Solutions:

a. $$f_s = \frac{1}{2\pi\sqrt{LC}} = \frac{1}{2\pi\sqrt{(1\text{ mH})(0.01\ \mu\text{F})}} = \mathbf{50{,}329.21\ Hz}$$

$$Q_s = \frac{X_L}{R + R_l} = \frac{2\pi(50{,}329.21\text{ Hz})(1\text{ mH})}{33\ \Omega + 2\ \Omega} = \mathbf{9.04}$$

$$BW = \frac{f_s}{Q_s} = \frac{50{,}329.21\text{ Hz}}{9.04} = \mathbf{5.57\ kHz}$$

At resonance:

$$V_{o_{max}} = \frac{RV_i}{R + R_l} = \frac{33\ \Omega(V_i)}{33\ \Omega + 2\ \Omega} = 0.943V_i = 0.943(20\text{ mV}) = \mathbf{18.86\ mV}$$

At the cutoff frequencies:

$$V_o = (0.707)(0.943V_i) = 0.667V_i = 0.667(20\text{ mV}) = \mathbf{13.34\ mV}$$

Note Fig. 21.36.

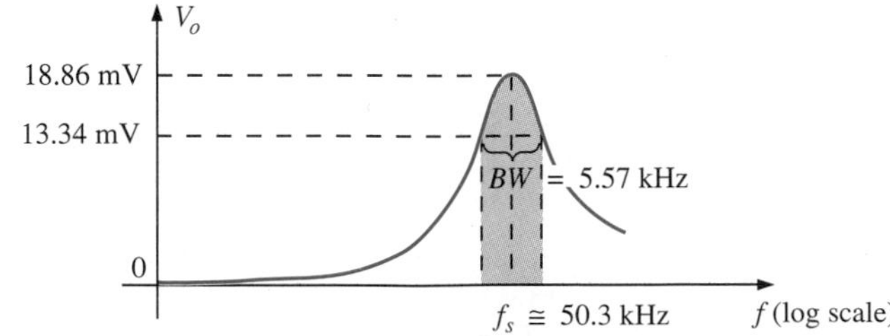

FIG. 21.36
Pass-band response for the network.

b. Dividing all levels of Fig. 21.36 by V_i = 20 mV will result in the normalized plot of Fig. 21.37(a).

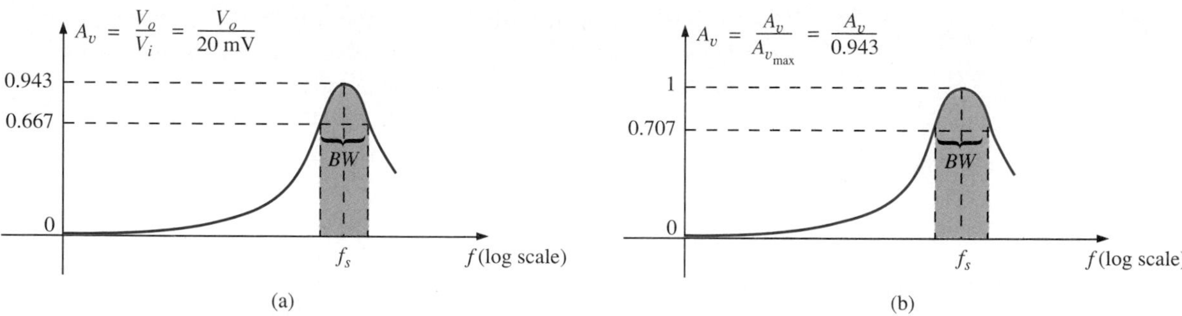

FIG. 21.37
Normalized plots for the pass-band filter of Fig. 21.35.

c. Dividing all levels of Fig. 21.37(a) by $A_{v_{max}} = 0.943$ will result in the normalized plot of Fig. 21.37(b).

21.8 STOP-BAND FILTERS

Stop-band filters can also be constructed using a low-pass and a high-pass filter. However, rather than the cascaded configuration used for the pass-band filter, a parallel arrangement is required, as shown in Fig. 21.38. A low frequency f_1 can pass through the low-pass filter and a higher frequency f_2 can use the parallel path, as shown in Figs. 21.38 and 21.39. However, a frequency such as f_o in the reject-band is higher than the low-pass critical frequency and lower than the high-pass critical frequency, and is therefore prevented from contributing to the levels of V_o above $0.707V_{max}$.

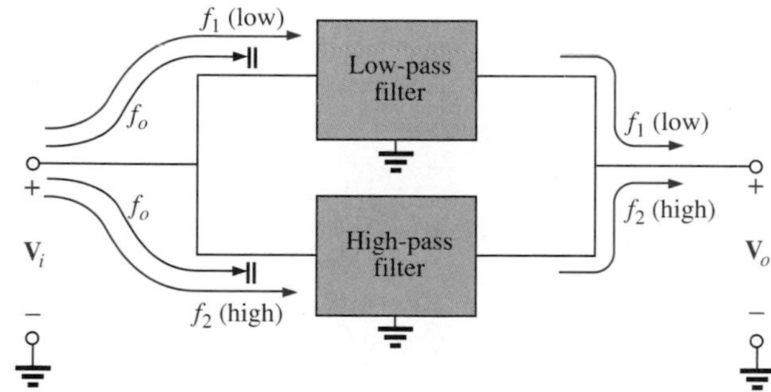

FIG. 21.38
Stop-band filter.

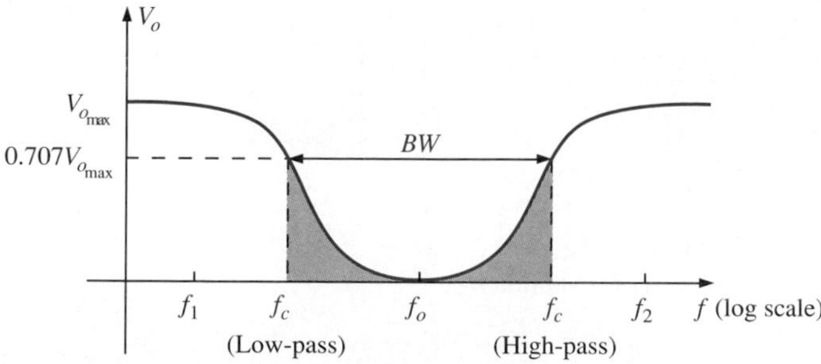

FIG. 21.39
Stop-band characteristics.

Since the characteristics of a stop-band filter are the inverse of the pattern obtained for the pass-band filters, we can employ the fact that at any frequency the sum of the magnitudes of the two waveforms to the right of the equals sign in Fig. 21.40 will equal the applied voltage V_i.

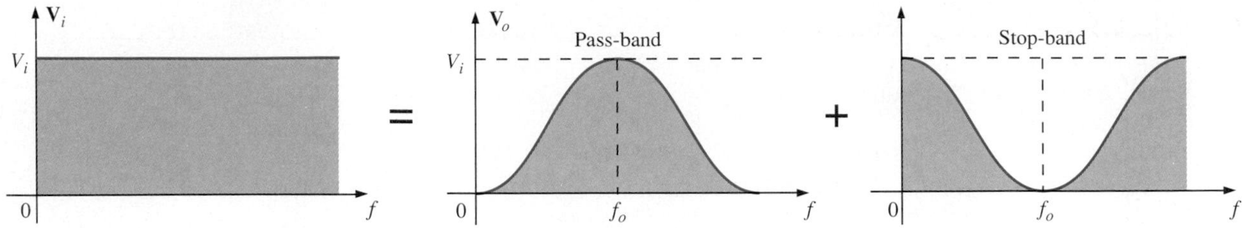

FIG. 21.40
Demonstrating how an applied signal of fixed magnitude can be broken down into a pass-band and stop-band response curve.

For the pass-band filters of Figs. 21.33 and 21.34, therefore, if we take the output off the other series elements as shown in Figs. 21.41 and 21.42, a stop-band characteristic will be obtained, as required by Kirchhoff's voltage law.

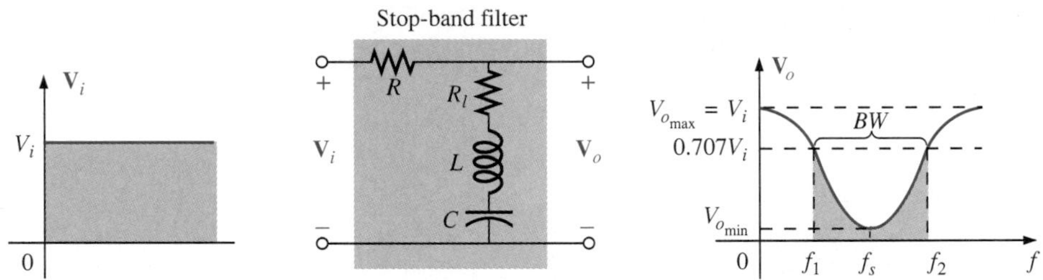

FIG. 21.41
Stop-band filter using a series resonant circuit.

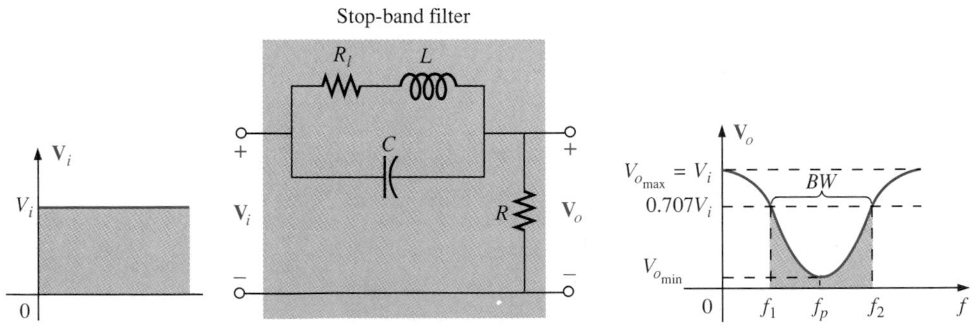

FIG. 21.42
Stop-band filter using a parallel resonant network.

For the series resonant circuit of Fig. 21.41, Eqs. (21.22) to (21.24) still apply, but now at resonance,

$$V_{o_{\min}} = \frac{R_l V_i}{R_l + R} \qquad \textbf{(21.30)}$$

For the parallel resonant circuit of Fig. 21.42, Eqs. (21.26) to (21.29) are still applicable, but now at resonance,

$$V_{o_{\min}} = \frac{RV_i}{R + Z_{T_p}} \qquad \textbf{(21.31)}$$

The maximum value of V_o for the series resonant circuit is V_i at the low end due to the open-circuit equivalent for the capacitor and V_i at the high end due to the high impedance of the inductive element.

For the parallel resonant circuit, at $f = 0$ Hz, the coil can be replaced by a short-circuit equivalent and the capacitor can be replaced by its open circuit and $V_o = RV_i/(R + R_l)$. At the high-frequency end, the capacitor approaches a short-circuit equivalent, and V_o increases toward V_i.

21.9 DOUBLE-TUNED FILTER

There are some network configurations that display both a pass-band and a stop-band characteristic, such as shown in Fig. 21.43. For the network of Fig. 21.43(a), the parallel resonant circuit will establish a stop-band for the range of frequencies not permitted to establish a significant V_L. The greater part of the applied voltage will appear across the parallel resonant circuit for this frequency range due to its very high impedance compared with R_L. For the pass-band, the parallel resonant circuit is designed to be capacitive (inductive if L_s is replaced by C_s). The inductance L_s is chosen to cancel the effects of the resulting net capacitive reactance at the resonant pass-band frequency of the tank circuit, thereby acting as a series resonant circuit. The applied voltage will then appear across R_L at this frequency.

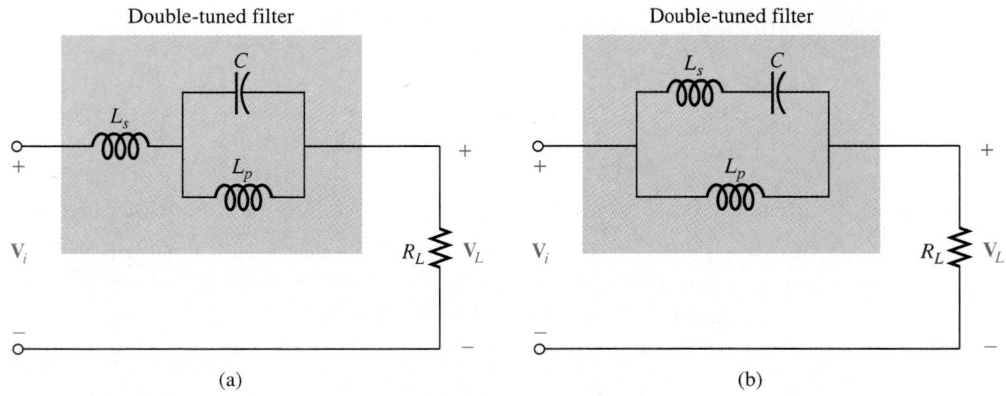

FIG. 21.43
Double-tuned networks.

For the network of Fig. 21.43(b), the series resonant circuit will still determine the pass-band, acting as a very low impedance across the parallel inductor at resonance. At the desired stop-band resonant frequency, the series resonant circuit is capacitive. The inductance L_p is chosen to establish parallel resonance at the resonant stop-band frequency. The high impedance of the parallel resonant circuit will result in a very low load voltage V_L.

For rejected frequencies below the pass-band, the networks should appear as shown in Fig. 21.43. For the reverse situation, L_s in Fig. 21.43(a) and L_p in Fig. 21.43(b) are replaced by capacitors.

EXAMPLE 21.9 For the network of Fig. 21.43(b), determine L_s and L_p for a capacitance C of 500 pF if a frequency of 200 kHz is to be rejected and a frequency of 600 kHz accepted.

Solution: For series resonance, we have

$$f_s = \frac{1}{2\pi\sqrt{LC}}$$

and $$L_s = \frac{1}{4\pi^2 f_s^2 C} = \frac{1}{4\pi^2(600\text{ kHz})^2(500\text{ pF})} = \mathbf{140.7\ \mu H}$$

At 200 kHz,

$$X_{L_s} = \omega L = 2\pi f_s L_s = (2\pi)(200\text{ kHz})(140.7\ \mu\text{H}) = 176.8\ \Omega$$

and $$X_C = \frac{1}{\omega C} = \frac{1}{(2\pi)(200\text{ kHz})(500\text{ pF})} = 1591.5\ \Omega$$

For the series elements,

$$j(X_{L_s} - X_C) = j(176.8\ \Omega - 1591.5\ \Omega) = -j\,1414.7\ \Omega = -jX'_C$$

At parallel resonance ($Q_l \geq 10$ assumed),

$$X_{L_p} = X'_C$$

and $$L_p = \frac{X_{L_p}}{\omega} = \frac{1414.7\ \Omega}{(2\pi)(200\text{ kHz})} = \mathbf{1.13\ mH}$$

The frequency response for the preceding network appears as one of the examples of PSpice in the last section of the chapter.

American (Madison, Wis.; Summit, N.J.; Cambridge, Mass.)
(1905–1981)
V.P. at Bell Laboratories
Professor of Systems Engineering, Harvard University

Courtesy of AT&T Archives

In his early years at Bell Laboratories, Hendrik Bode was involved with *electric filter* and *equalizer design.* He then transferred to the Mathematics Research Group, where he specialized in research pertaining to electrical networks theory and its application to long distance communication facilities. In 1946 he was awarded the Presidential Certificate of Merit for his work in electronic fire control devices. In addition to the publication of the book *Network Analysis and Feedback Amplifier Design* in 1945, which is considered a classic in its field, he has been granted 25 patents in electrical engineering and systems design. Upon retirement, Bode was elected Gordon McKay Professor of Systems Engineering at Harvard University. He was a fellow of the IEEE and American Academy of Arts and Sciences.

FIG. 21.44
Hendrik Wade Bode

21.10 BODE PLOTS

There is a technique for sketching the frequency response of such factors as filters, amplifiers, and systems on a decibel scale that can save a great deal of time and effort and provide an excellent way to compare decibel levels at different frequencies.

The curves obtained for the magnitude and/or phase angle versus frequency are called Bode plots (Fig. 21.44). Through the use of straight-line segments called idealized Bode plots, the frequency response of a system can be found efficiently and accurately.

To ensure that the derivation of the method is correctly and clearly understood, the first network to be analyzed will be examined in some detail. The second network will be treated in a shorthand manner, and finally a method for quickly determining the response will be introduced.

High-Pass *R-C* Filter

Let us start by reexamining the high-pass filter of Fig. 21.45. The high-pass filter was chosen as our starting point since the frequencies of primary interest are at the low end of the frequency spectrum.

The voltage gain of the system is given by:

$$\mathbf{A}_v = \frac{\mathbf{V}_o}{\mathbf{V}_i} = \frac{R}{R - jX_C} = \frac{1}{1 - j\dfrac{X_C}{R}} = \frac{1}{1 - j\dfrac{1}{2\pi fCR}}$$

$$= \frac{1}{1 - j\left(\dfrac{1}{2\pi RC}\right)\dfrac{1}{f}}$$

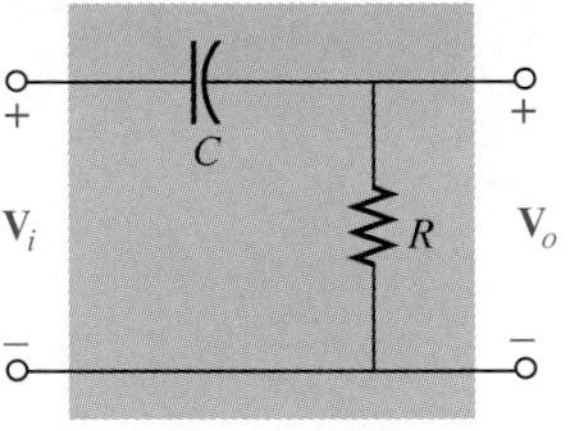

FIG. 21.45
High-pass filter.

If we substitute

$$f_c = \frac{1}{2\pi RC} \qquad \textbf{(21.32)}$$

which we recognize as the cutoff frequency of earlier sections, we obtain

$$\mathbf{A}_v = \frac{1}{1 - j\,(f_c/f)} \qquad \textbf{(21.33)}$$

We will find in the analysis to follow that the ability to reformat the gain to one having the general characteristics of Eq. (21.33) is critical to the application of the Bode technique. Different configurations will result in variations of the format of Eq. (21.33), but the desired similarities will become obvious as we progress through the material.

In magnitude and phase form:

$$\mathbf{A}_v = \frac{\mathbf{V}_o}{\mathbf{V}_i} = A_v \angle\theta = \frac{1}{\sqrt{1 + (f_c/f)^2}} \angle \tan^{-1}(f_c/f) \qquad \textbf{(21.34)}$$

providing an equation for the magnitude and phase of the high-pass filter in terms of the frequency levels.

Using Eq. (21.12),

$$A_{v_{\text{dB}}} = 20 \log_{10} A_v$$

and, substituting the magnitude component of Eq. (21.34),

$$A_{v_{\text{dB}}} = 20 \log_{10} \frac{1}{\sqrt{1 + (f_c/f)^2}} = \underbrace{20 \log_{10} 1}_{0} - 20 \log_{10} \sqrt{1 + (f_c/f)^2}$$

and

$$\mathbf{A}_{v_{\text{dB}}} = -20 \log_{10} \sqrt{1 + \left(\frac{f_c}{f}\right)^2}$$

Recognizing that $\log_{10}\sqrt{x} = \log_{10} x^{1/2} = \frac{1}{2}\log_{10} x$,

we have

$$A_{v_{\text{dB}}} = -\frac{1}{2}(20)\log_{10}\left[1 + \left(\frac{f_c}{f}\right)^2\right]$$

$$= -10 \log_{10}\left[1 + \left(\frac{f_c}{f}\right)^2\right]$$

For frequencies where $f \ll f_c$ or $(f_c/f)^2 \gg 1$,

$$1 + \left(\frac{f_c}{f}\right)^2 \cong \left(\frac{f_c}{f}\right)^2$$

and
$$A_{v_{\text{dB}}} = -10 \log_{10}\left(\frac{f_c}{f}\right)^2$$

but
$$\log_{10}x^2 = 2 \log_{10}x$$

resulting in
$$A_{v_{\text{dB}}} = -20 \log_{10} \frac{f_c}{f}$$

However, logarithms are such that

$$-\log_{10} b = +\log_{10} \frac{1}{b}$$

and substituting $b = f_c/f$ we have

$$\boxed{A_{v_{\text{dB}}} = +20 \log_{10} \frac{f}{f_c}} \quad f \ll f_c \tag{21.35}$$

First note the similarities between Eq. (21.35) and the basic equation for gain in decibels: $G_{\text{dB}} = 20 \log_{10} V_o/V_i$. The comments regarding changes in decibel levels due to changes in V_o/V_i can therefore be applied here also, except now a change in frequency by a 2:1 ratio will result in a 6-dB change in gain. A change in frequency by a 10:1 ratio will result in a 20-dB change in gain.

Two frequencies separated by a 2:1 ratio are said to be an octave apart.

For Bode plots, a change in frequency by one octave will result in a 6-dB change in gain.

Two frequencies separated by a 10:1 ratio are said to be a decade apart.

For Bode plots, a change in frequency by one decade will result in a 20-dB change in gain.

One may wonder about all the mathematical development to obtain an equation that initially appears confusing and of limited value. As specified, Eq. (21.35) is accurate only for frequency levels much less than f_c.

First, realize that the mathematical development of Eq. (21.35) will not have to be repeated for each configuration encountered. Second, the equation itself is seldom applied but simply used in a manner to be described to define a straight line on a log plot that permits a sketch of the frequency response of a system with a minimum of effort and a high degree of accuracy.

To plot Eq. (21.35), consider the following levels of increasing frequency:

For $f = f_c/10$, $f/f_c = 0.1$ and $+20 \log_{10} 0.1 = -20$ dB

For $f = f_c/4$, $f/f_c = 0.25$ and $+20 \log_{10} 0.25 = -12$ dB

For $f = f_c/2$, $f/f_c = 0.5$ and $+20 \log_{10} 0.5 = -6$ dB

For $f = f_c$, $f/f_c = 1$ and $+20 \log_{10} 1 = 0$ dB

Note from the above equations that as the frequency of interest approaches f_c, the dB gain becomes less negative and approaches the final normalized value of 0 dB. The positive sign in front of Eq. (21.35) can therefore be interpreted as an indication that the dB gain will have a positive slope with an increase in frequency. A plot of these points on a log scale will result in the straight-line segment of Fig. 21.46 to the left of f_c.

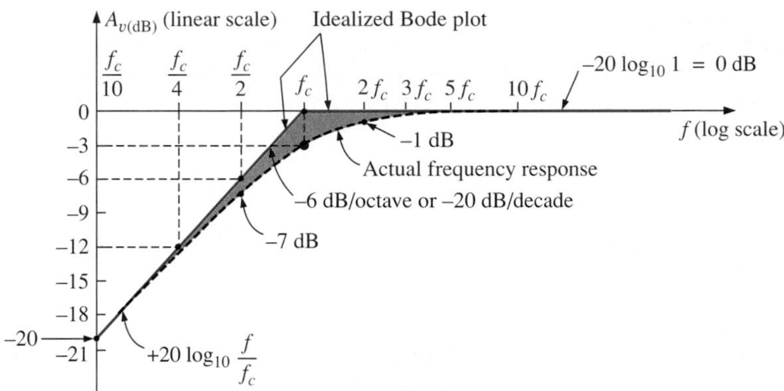

FIG. 21.46
Idealized Bode plot for the low-frequency region.

For the future, note that the resulting plot is a straight line intersecting the 0-dB line at f_c. It increases to the right at a rate of +6 dB per octave or +20 dB per decade. In other words, once f_c is determined, find $f_c/2$ and a plot point exists at −6 dB (or find $f_c/10$ and a plot point exists at −20 dB).

Bode plots are straight-line segments because the dB change per decade or octave is constant.

The actual response will approach an asymptote (straight-line segment) defined by $A_{v_{dB}} = 0$ dB since at high frequencies

$$f \gg f_c \quad \text{and} \quad f_c/f \cong 0$$

with
$$A_{v_{dB}} = 20 \log_{10} \frac{1}{\sqrt{1 + (f_c/f)^2}} = 20 \log_{10} \frac{1}{\sqrt{1 + 0}}$$
$$= 20 \log_{10} 1 = 0 \text{ dB}$$

The two asymptotes defined above will intersect at f_c, as shown in Fig. 21.46, forming an envelope for the actual frequency response.

At $f = f_c$, the cutoff frequency,

$$A_{v_{dB}} = 20 \log_{10} \frac{1}{\sqrt{1 + (f_c/f)^2}} = 20 \log_{10} \frac{1}{\sqrt{1 + 1}} = 20 \log_{10} \frac{1}{\sqrt{2}}$$
$$= \mathbf{-3\ dB}$$

At $f = 2f_c$,

$$A_{v_{dB}} = -20 \log_{10} \sqrt{1 + \left(\frac{f_c}{2f_c}\right)^2} = -20 \log_{10} \sqrt{1 + \left(\frac{1}{2}\right)^2}$$
$$= -20 \log_{10} \sqrt{1.25} = \mathbf{-1\ dB}$$

as shown in Fig. 21.46.

At $f = f_c/2$,

$$A_{v_{dB}} = -20 \log_{10} \sqrt{1 + \left(\frac{f_c}{f_c/2}\right)^2} = -20 \log_{10} \sqrt{1 + (2)^2}$$
$$= -20 \log_{10} \sqrt{5}$$
$$= \mathbf{-7\ dB}$$

separating the idealized Bode plot from the actual response by 7 dB − 6 dB = 1 dB, as shown in Fig. 21.46.

Reviewing the above,

at $f = f_c$ the actual response curve is 3 dB down from the idealized Bode plot, whereas at $f = 2f_c$ and $f_c/2$, the actual response curve is 1 dB down from the asymptotic response.

The phase response can also be sketched using straight-line asymptotes by considering a few critical points in the frequency spectrum.

Equation (21.34) specifies the phase response (the angle by which $\mathbf{V}_o$ leads $\mathbf{V}_i$) by

$$\theta = \tan^{-1}\frac{f_c}{f} \qquad \textbf{(21.36)}$$

For frequencies well below $f_c (f \ll f_c)$, $\theta = \tan^{-1}(f_c/f)$ approaches 90° and for frequencies well above $f_c (f \gg f_c)$, $\theta = \tan^{-1}(f_c/f)$ will approach 0°, as discovered in earlier sections of the chapter. At $f = f_c$, $\theta = \tan^{-1}(f_c/f) = \tan^{-1} 1 = 45°$.

Defining $f \ll f_c$ for $f = f_c/10$ (and less) and $f \gg f_c$ for $f = 10f_c$ (and more), we can define

an asymptote at $\theta = 90°$ for $f \ll f_c/10$, an asymptote at $\theta = 0°$ for $f \gg 10f_c$, and an asymptote from $f_c/10$ to $10f_c$ that passes through $\theta = 45°$ at $f = f_c$.

The asymptotes defined above all appear in Fig. 21.47. Again, the Bode plot for Eq.(21.36) is a straight line because the change in phase angle will be 45° for every tenfold change in frequency.

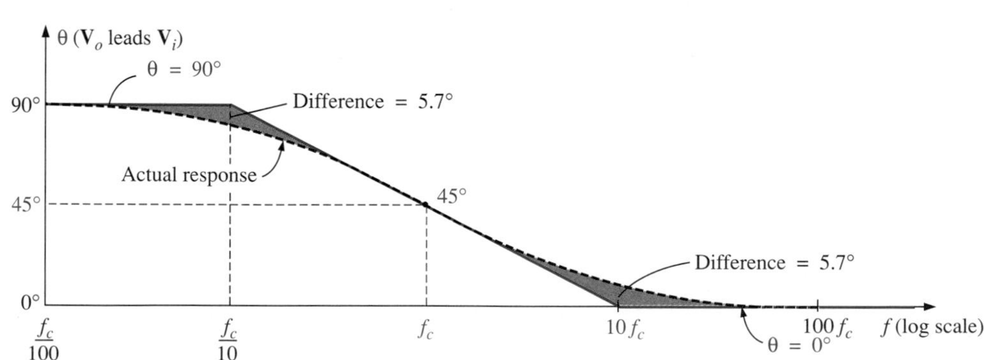

FIG. 21.47
Phase response for a high-pass R-C filter.

Substituting $f = f_c/10$ into Eq. (21.36),

$$\theta = \tan^{-1}\left(\frac{f_c}{f_c/10}\right) = \tan^{-1} 10 = \mathbf{84.29°}$$

for a difference of $90^\circ - 84.29^\circ \cong 5.7^\circ$ from the idealized response.

Substituting $f = 10f_c$

$$\theta = \tan^{-1}\left(\frac{f_c}{10f_c}\right) = \tan^{-1}\frac{1}{10} \cong \mathbf{5.7^\circ}$$

In summary, therefore,

at $f = f_c$, $\theta = 45^\circ$, whereas at $f = f_c/10$ and $10f_c$, the difference between the actual phase response and the asymptotic plot is 5.7°.

EXAMPLE 21.10

a. Sketch $A_{v_{\text{dB}}}$ versus frequency for the high-pass *R-C* filter of Fig. 21.48.
b. Determine the decibel level at $f = 1$ kHz.
c. Sketch the phase response versus frequency on a log scale.

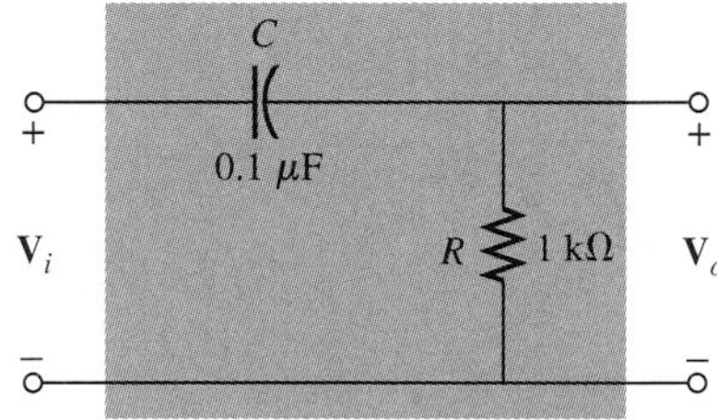

FIG. 21.48
Example 21.10.

Solutions:

a. $f_c = \dfrac{1}{2\pi RC} = \dfrac{1}{(2\pi)(1\text{ k}\Omega)(0.1\ \mu\text{F})} = 1591.55\text{ Hz}$

The frequency f_c is identified on the log scale as shown in Fig. 21.49. A straight line is then drawn from f_c with a slope that will intersect -20 dB at $f_c/10 = 159.15$ Hz or -6 dB at $f_c/2 = 795.77$ Hz. A second asymptote is drawn from f_c to higher frequencies at 0 dB. The actual response curve can then be drawn through the -3-dB level at f_c approaching the two asymptotes of Fig. 21.49. Note the 1-dB difference between the actual response and the idealized Bode plot at $f = 2f_c$ and $0.5f_c$.

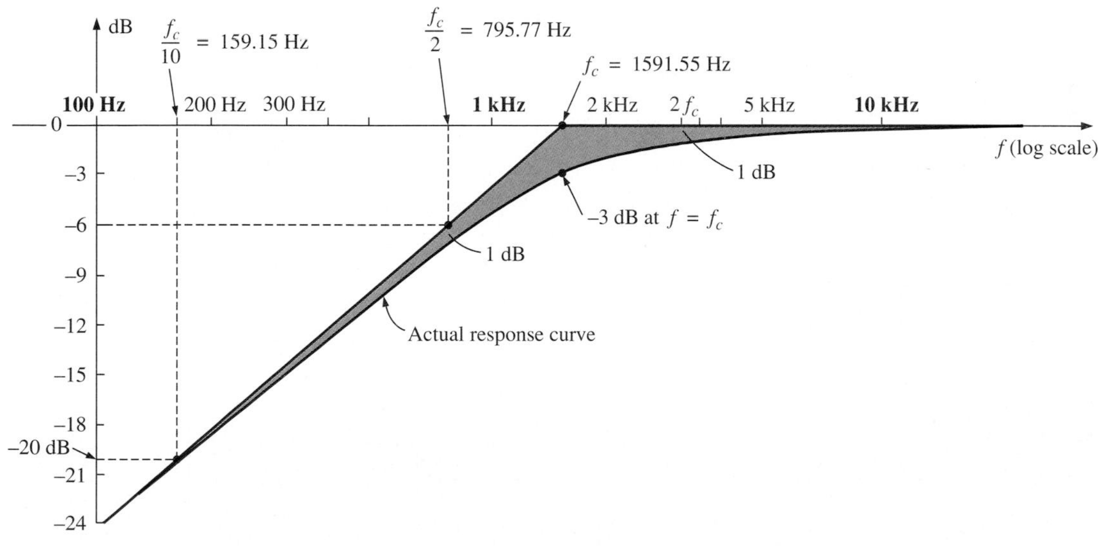

FIG. 21.49
Frequency response for the high-pass filter of Fig. 21.48.

Note in the preceding solution that there was no need to employ Eq. (21.35) or to perform any extensive mathematical manipulations.

b. Eq. (21.33):

$$|A_{v\text{dB}}| = 20\log_{10}\frac{1}{\sqrt{1+\left(\dfrac{f_c}{f}\right)^2}} = 20\log_{10}\frac{1}{\sqrt{1+\left(\dfrac{1591.55\text{ Hz}}{1000}\right)^2}}$$

$$= 20\log_{10}\frac{1}{\sqrt{1+(1.592)^2}} = 20\log_{10}0.5318 = \mathbf{-5.49\ dB}$$

as verified by Fig. 21.49.

c. See Fig. 21.50. Note that $\theta = 45°$ at $f = f_c = 1591.55$ Hz, and the difference between the straight-line segment and the actual response is 5.7° at $f = f_c/10 = 159.2$ Hz and $f = 10f_c = 15{,}923.6$ Hz.

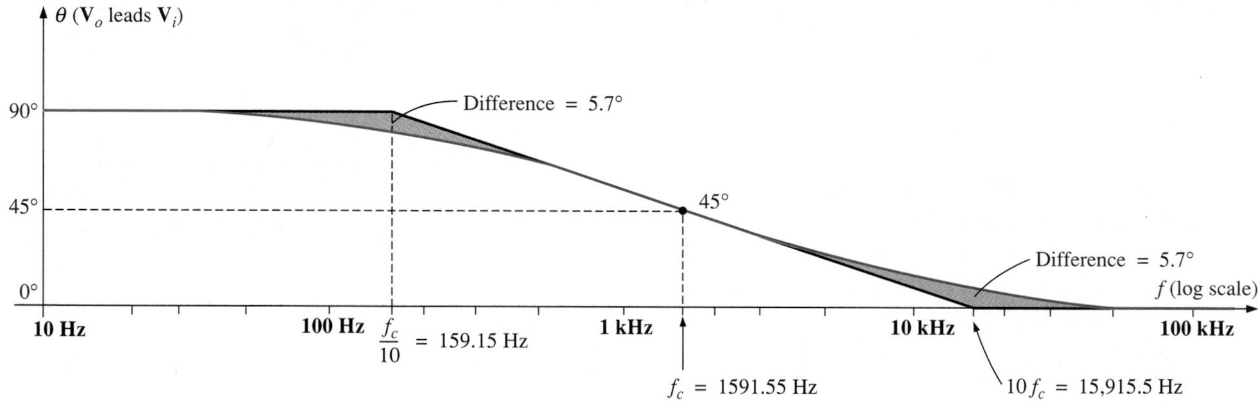

FIG. 21.50
Phase plot for the high-pass R-C filter.

Low-Pass *R-C* Filter

For the low-pass filter of Fig. 21.51,

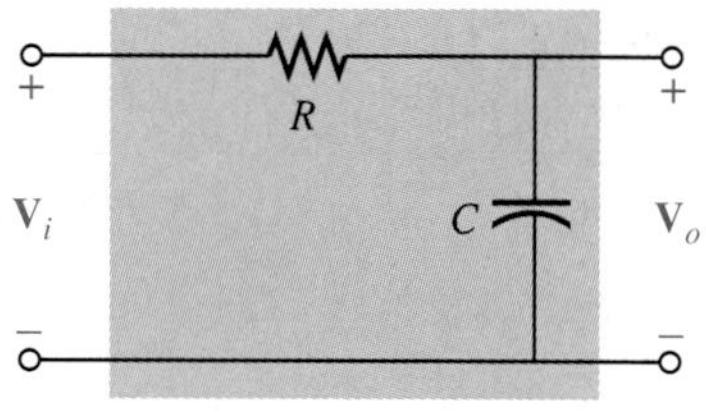

$$\mathbf{A}_v = \frac{\mathbf{V}_o}{\mathbf{V}_i} = \frac{-jX_C}{R - jX_C} = \frac{1}{\dfrac{R}{-jX_C}} + 1$$

$$= \frac{1}{1 + j\dfrac{R}{X_C}} = \frac{1}{1 + j\dfrac{R}{\dfrac{1}{2\pi fC}}} = \frac{1}{1 + j\dfrac{f}{\dfrac{1}{2\pi RC}}}$$

FIG. 21.51
Low-pass filter.

and

$$\mathbf{A}_v = \frac{1}{1 + j(f/f_c)} \qquad (21.37)$$

with

$$f_c = \frac{1}{2\pi RC} \qquad (21.38)$$

as defined earlier.

Note that now the sign of the imaginary component in the denominator is positive and f_c appears in the denominator of the frequency ratio rather than in the numerator, as in the case of f_c for the high-pass filter.

In terms of magnitude and phase,

$$\mathbf{A}_v = \frac{\mathbf{V}_o}{\mathbf{V}_i} = A_v \angle\theta = \frac{1}{\sqrt{1 + (f/f_c)^2}} \angle -\tan^{-1}(f/f_2) \qquad \textbf{(21.39)}$$

An analysis similar to that performed for the high-pass filter will result in

$$A_{v_{dB}} = -20 \log_{10} \frac{f}{f_c} \qquad f \gg f_c \qquad \textbf{(21.40)}$$

Note in particular that the equation is exact only for frequencies much greater than f_c, but a plot of Eq. (21.40) does provide an asymptote that performs the same function as the asymptote derived for the high-pass filter. In addition, note that it is exactly the same as Eq. (21.35), except for the minus sign, which suggests that the resulting Bode plot will have a negative slope [recall the positive slope for Eq. (21.35)] for increasing frequencies beyond f_c.

A plot of Eq. (21.40) appears in Fig. 21.52 for $f_c = 1$ kHz. Note the 6-dB drop at $f = 2f_c$ and the 20-dB drop at $f = 10f_c$.

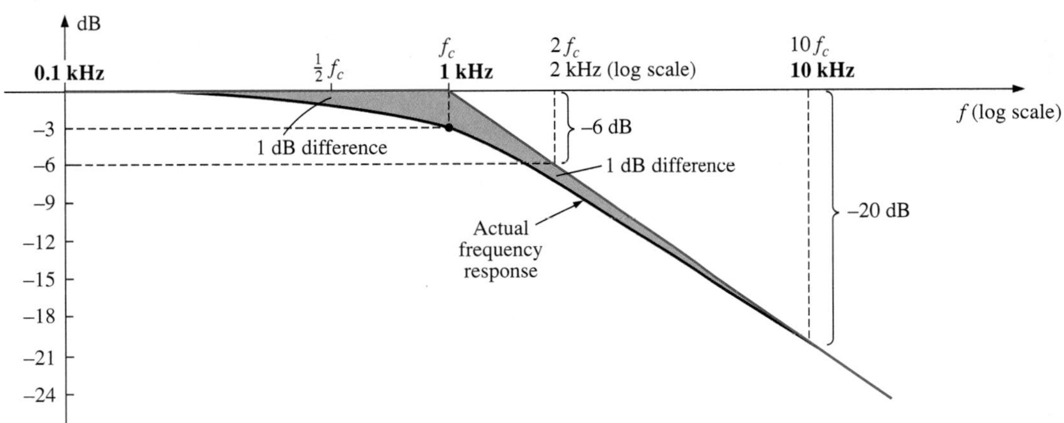

FIG. 21.52
Bode plot for the high-frequency region of a low-pass R-C filter.

At $f \gg f_c$, the phase angle $\theta = -\tan^{-1}(f/f_c)$ approaches $-90°$, whereas for frequencies $f \ll f_c$, $\theta = -\tan^{-1}(f/f_c)$ approaches $0°$. At $f = f_c$, $\theta = -\tan^{-1} 1 = -45°$, establishing the plot of Fig. 21.53. Note again the 45° change in phase angle for each tenfold increase in frequency.

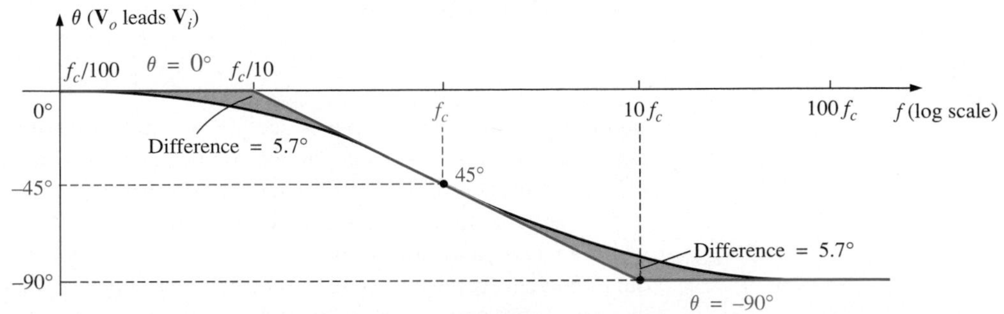

FIG. 21.53
Phase plot for a low-pass R-C filter.

Even though the preceding analysis has been limited solely to the *R*-*C* combination, the results obtained will have an impact on networks a great deal more complicated. One good example is the high- and low-frequency response of a standard transistor configuration. There are some capacitive elements in a practical transistor network that will affect the low-frequency response and others that will affect the high-frequency response. In the absence of the capacitive elements, the frequency response of a transistor would ideally stay level at the midband value. However, the coupling capacitors at low frequencies and the bypass and parasitic capacitors at high frequencies will define a bandwidth for numerous transistor configurations. In the low-frequency region, specific capacitors and resistors will form an *R*-*C* combination that will define a low cutoff frequency. There are then other elements and capacitors forming a second *R*-*C* combination that will define a high cutoff frequency. Once the cutoff frequencies are known, the −3-dB points are set and the bandwidth of the system can be determined.

21.11 SKETCHING THE BODE RESPONSE

In the previous section we found that normalized functions of the form appearing in Fig. 21.54 had the Bode envelope and dB response indicated in the same figure. In this section we introduce additional functions and their responses that can be used in conjunction with those of Fig. 21.54 to determine the dB response of more sophisticated systems in a systematic, time-saving, and accurate manner.

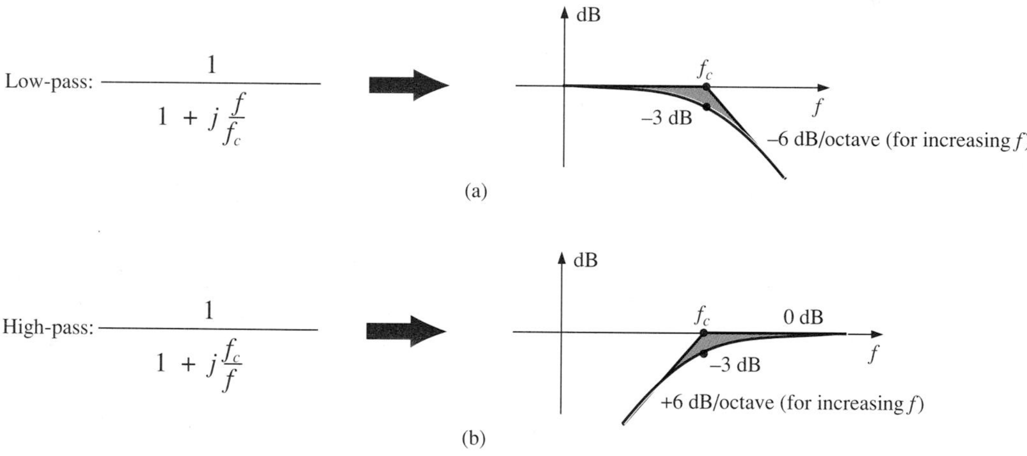

FIG. 21.54
dB response of (a) low-pass filter and (b) high-pass filter.

As an avenue toward introducing an additional function that appears quite frequently, let us examine the high-pass filter of Fig. 21.55 that has a high-frequency output less than the full applied voltage.

Before developing a mathematical expression for $\mathbf{A}_v = \mathbf{V}_o/\mathbf{V}_i$, let us first make a rough sketch of the expected response.

At $f = 0$ Hz, the capacitor will assume its open-circuit equivalence and $V_o = 0$ V. At very high frequencies, the capacitor can assume its short-circuit equivalence and

$$V_o = \frac{R_2}{R_1 + R_2} V_i = \frac{4\text{ k}\Omega}{1\text{ k}\Omega + 4\text{ k}\Omega} V_i = 0.8V_i$$

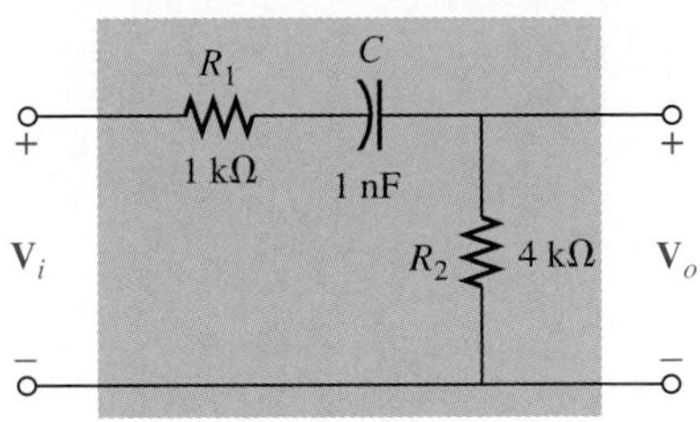

FIG. 21.55
High-pass filter with attenuated output.

The resistance to be employed in the cutoff frequency equation can be determined by simply determining the Thévenin resistance "seen" by the capacitor. Setting $V_i = 0$ V and solving for R_{Th} (for the capacitor C) will result in the network of Fig. 21.56, where it is quite clear that

$$R_{Th} = R_1 + R_2 = 1\text{ k}\Omega + 4\text{ k}\Omega = 5\text{ k}\Omega$$

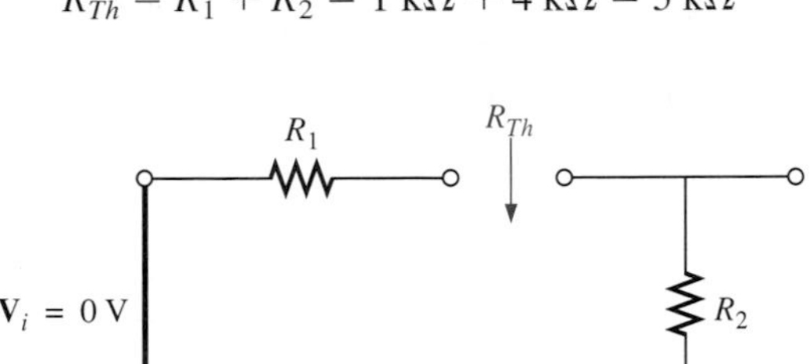

FIG. 21.56
Determining R_{Th} for the cutoff frequency equation.

Therefore,

$$f_c = \frac{1}{2\pi R_{Th}C} = \frac{1}{2\pi(5\text{ k}\Omega)(1\text{ nF})} = 31.83\text{ kHz}$$

A sketch of V_o versus frequency is provided in Fig. 21.57(a). A normalized plot using V_i as the normalizing quantity will result in the response of Fig. 21.57(b). If the maximum value of A_v is used in the normalization process, the response of Fig. 21.57(c) will be obtained.

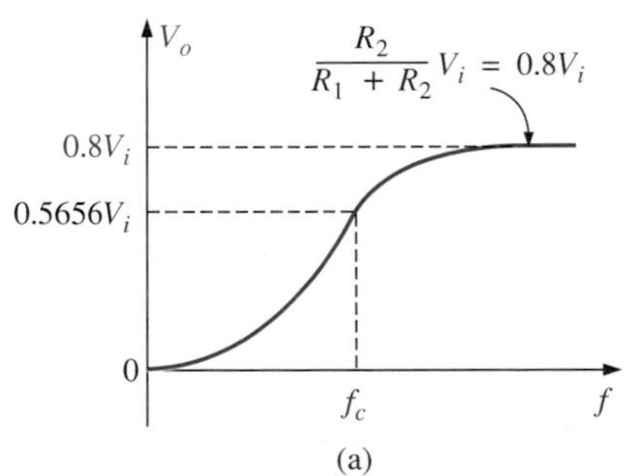

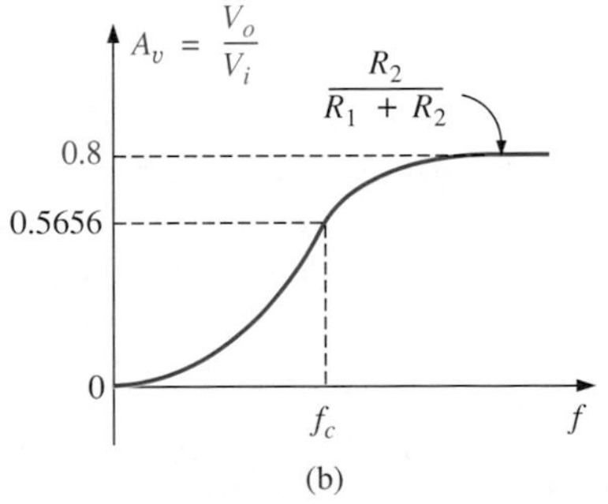

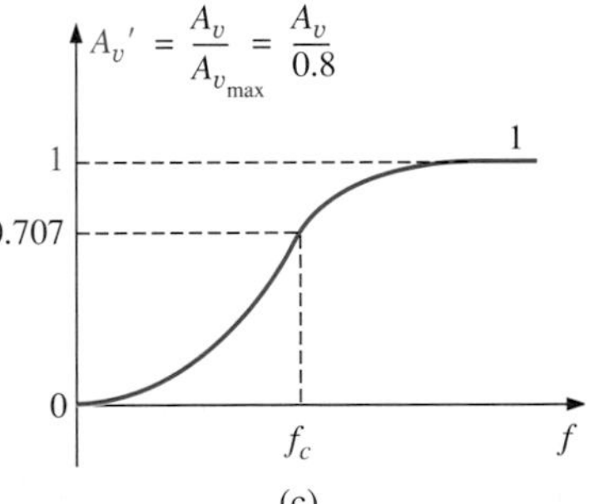

FIG. 21.57
Finding the normalized plot for the gain of the high-pass filter of Fig. 21.55 with attenuated output.

For all the plots obtained in the previous section, V_i was the maximum value and the ratio V_o/V_i had a maximum value of 1. For many situations, this will not be the case, and we must be aware of which ratio is being plotted versus frequency. The dB response curves for the plots of Figs. 21.57(b) and 21.57(c) can both be obtained quite directly using the foundation established by the conclusions depicted in Fig. 21.54, but we must be aware of what to expect and how they will differ. In Fig. 21.57(b) we are comparing the output level to the input voltage. In

Fig. 21.57(c) we are plotting A_v versus the maximum value of A_v. On most data sheets and for the majority of the investigative techniques commonly employed, the normalized plot of Fig. 21.57(c) is used because it establishes 0 dB as an asymptote for the dB plot. To ensure that the impact of using either Fig. 21.57(b) or Fig. 21.57(c) in a frequency plot is understood, the analysis of the filter of Fig. 21.55 will include the resulting dB plot for both normalized curves.

For the network of Fig. 21.55:

$$\mathbf{V}_o = \frac{R_2\mathbf{V}_i}{R_1 + R_2 - jX_C} = R_2\left[\frac{1}{R_1 + R_2 - jX_C}\right]\mathbf{V}_i$$

Dividing the top and bottom of the equation by $R_1 + R_2$ results in

$$\mathbf{V}_o = \frac{R_2}{R_1 + R_2}\left[\frac{1}{1 - j\dfrac{X_C}{R_1 + R_2}}\right]$$

but $$-j\frac{X_C}{R_1 + R_2} = -j\frac{1}{\omega(R_1 + R_2)C} = -j\frac{1}{2\pi f(R_1 + R_2)C}$$

$$= -j\frac{f_c}{f} \quad \text{with} \quad f_c = \frac{1}{2\pi R_{Th}C} \quad \text{and} \quad R_{Th} = R_1 + R_2$$

so that $$\mathbf{V}_o = \frac{R_2}{R_1 + R_2}\left[\frac{1}{1 - j(f_c/f)}\right]\mathbf{V}_i$$

If we divide both sides by $\mathbf{V}_i$, we obtain

$$\mathbf{A}_v = \frac{\mathbf{V}_o}{\mathbf{V}_i} = \frac{R_2}{R_1 + R_2}\left[\frac{1}{1 - j(f_c/f)}\right] \tag{21.41}$$

from which the magnitude plot of Fig. 21.57(b) can be obtained. If we divide both sides by $\mathbf{A}_{v_{\max}} = R_2/(R_1 + R_2)$, we have

$$\mathbf{A}'_v = \frac{\mathbf{A}_v}{\mathbf{A}_{v_{\max}}} = \frac{1}{1 - j(f_c/f)} \tag{21.42}$$

from which the magnitude plot of Fig. 21.57(c) can be obtained.

Based on the past section, a dB plot of the magnitude of $A'_v = A_v/A_{v_{\max}}$ is now quite direct using Fig. 12.54(b). The plot appears in Fig. 21.58.

For the gain $A_v = V_o/V_i$ we can apply Eq. (21.5):

$$20 \log_{10} ab = 20 \log_{10} a + 20 \log_{10} b$$

where

$$20 \log_{10}\left\{\frac{R_2}{R_1 + R_2}\left[\frac{1}{1 - j(f_c/f)}\right]\right\} = 20 \log_{10}\frac{R_2}{R_1 + R_2} + 20 \log_{10}\frac{1}{\sqrt{1 + (f_c/f)^2}}$$

The second term will result in the same plot of Fig. 21.58, but the first term must be added to the second to obtain the total dB response.

Since $R_2/(R_1 + R_2)$ must always be less than 1, we can rewrite the first term as

$$20 \log_{10}\frac{R_2}{R_1 + R_2} = 20 \log_{10}\frac{1}{\dfrac{R_1 + R_2}{R_2}} = \underbrace{20 \log_{10} 1}_{0} - 20 \log_{10}\frac{R_1 + R_2}{R_2}$$

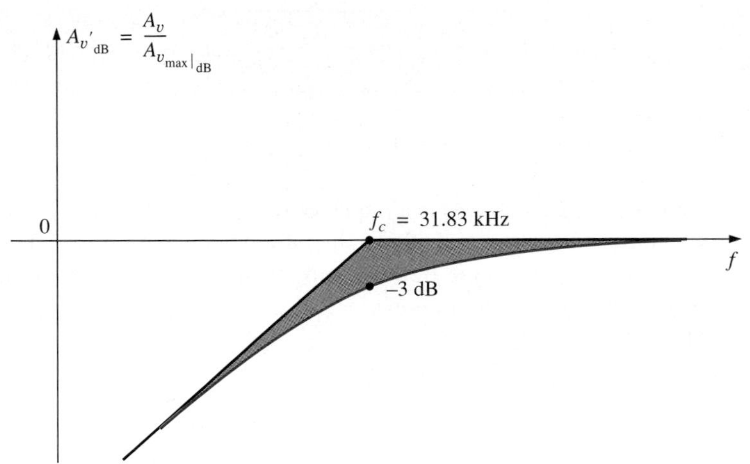

FIG. 21.58
dB plot for A'_v for the high-pass filter of Fig. 21.55.

and

$$20 \log_{10} \frac{R_2}{R_1 + R_2} = -20 \log_{10} \frac{R_1 + R_2}{R_2} \tag{21.43}$$

providing the drop in dB from the 0-dB level for the plot. Adding one log plot to the other *at each frequency,* as permitted by Eq. (21.5), will result in the plot of Fig. 21.59.

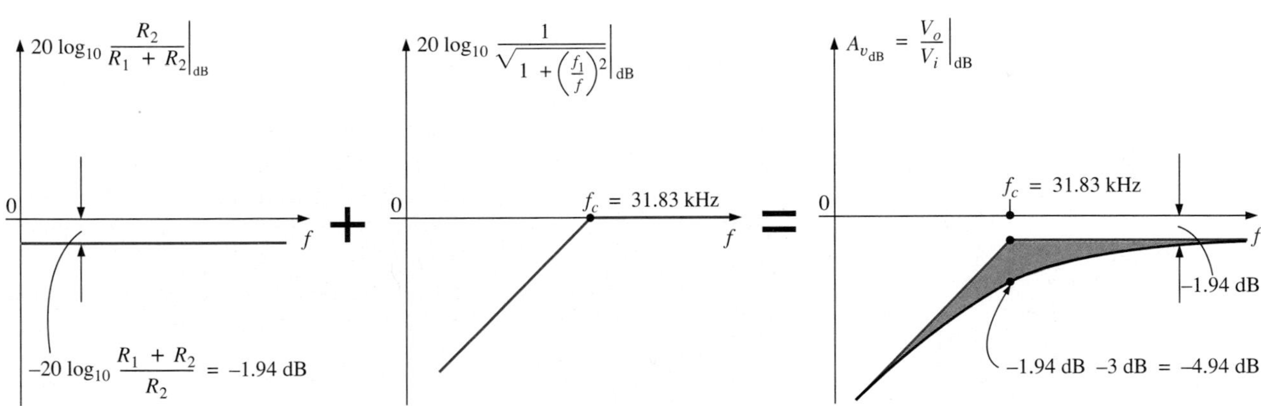

FIG. 21.59
Obtaining a dB plot of $A_{v_{dB}} = \frac{V_o}{V_i}\Big|_{dB}$.

For the network of Fig. 21.55, the gain $\mathbf{A}_v = \mathbf{V}_o/\mathbf{V}_i$ can also be found in the following manner:

$$\mathbf{V}_o = \frac{R_2\mathbf{V}_i}{R_1 + R_2 - jX_C}$$

$$\mathbf{A}_v = \frac{\mathbf{V}_o}{\mathbf{V}_i} = \frac{R_2}{R_1 + R_2 - jX_C} = \frac{jR_2}{j(R_1 + R_2) + X_C} = \frac{jR_2/X_C}{j(R_1 + R_2)/X_C + 1}$$

$$= \frac{j\,\omega R_2C}{1 + j\,\omega(R_1 + R_2)C} = \frac{j\,2\pi f R_2 C}{1 + j\,2\pi f(R_1 + R_2)C}$$

and
$$\mathbf{A}_v = \frac{\mathbf{V}_o}{\mathbf{V}_i} = \frac{j\,(f/f_1)}{1 + j\,(f/f_c)} \tag{21.44}$$

with
$$f_1 = \frac{1}{2\pi R_2 C} \quad \text{and} \quad f_c = \frac{1}{2\pi(R_1 + R_2)C}$$

The bottom of Eq. (21.44) is a match of the denominator of the low-pass function of Fig. 21.54(a). The numerator, however, is a new function that will define a unique Bode asymptote that will prove useful for a variety of network configurations.

Applying Eq. (21.5):

$$20 \log_{10}\frac{V_o}{V_i} = 20 \log_{10}\left[\frac{f}{f_1}\right]\left[\frac{1}{\sqrt{1 + (f/f_c)^2}}\right]$$
$$= 20 \log_{10}(f/f_1) + 20 \log_{10}\frac{1}{\sqrt{1 + (f/f_c)^2}}$$

Let us now consider specific frequencies for the first term.

At $f = f_1$:

$$20 \log_{10}\frac{f}{f_1} = 20 \log_{10} 1 = 0\text{ dB}$$

At $f = 2f_1$:

$$20 \log_{10}\frac{f}{f_1} = 20 \log_{10} 2 = +6\text{ dB}$$

At $f = \frac{1}{2}f_1$:

$$20 \log_{10}\frac{f}{f_1} = 20 \log_{10} 0.5 = -6\text{ dB}$$

A plot of $20 \log_{10}(f/f_1)$ is provided in Fig. 21.60.

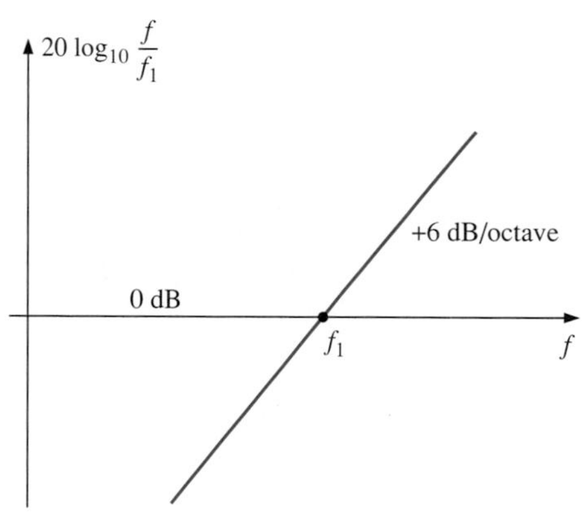

FIG. 21.60
dB plot of f/f_1.

Note that the asymptote passes through the 0-dB line at $f = f_1$ and has a positive slope of +6 dB/octave (or 20 dB/decade) for frequencies above and below f_1 for increasing values of f.

If we examine the original function A_v, we find that the phase angle associated with $jf/f_1 = f/f_1 \angle 90°$ is fixed at 90°, resulting in a phase angle for $\mathbf{A}_v$ of $90° - \tan^{-1}(f/f_c) = +\tan^{-1}(f_c/f)$.

Now that we have a plot of the dB response for the magnitude of the function f/f_1, we can plot the dB response of the magnitude of $\mathbf{A}_v$ using a procedure outlined by Fig. 21.61.

Solving for f_1 and f_c:

$$f_1 = \frac{1}{2\pi R_2 C} = \frac{1}{2\pi(4\text{ k}\Omega)(1\text{ nF})} = 39.79\text{ kHz}$$

with
$$f_c = \frac{1}{2\pi(R_1 + R_2)C} = \frac{1}{2\pi(5\text{ k}\Omega)(1\text{ nF})} = 31.83\text{ kHz}$$

For this development the straight-line asymptotes for each term resulting from the application of Eq. (21.5) will be drawn on the same frequency axis to permit an examination of the impact of one line section on the other. For clarity, the frequency spectrum of Fig. 21.61 has been divided into two regions.

In region 1 we have a 0-dB asymptote and one increasing at 6 dB/octave for increasing frequencies. The sum of the two as defined by Eq. (21.5) is simply the 6-dB/octave asymptote shown in the figure.

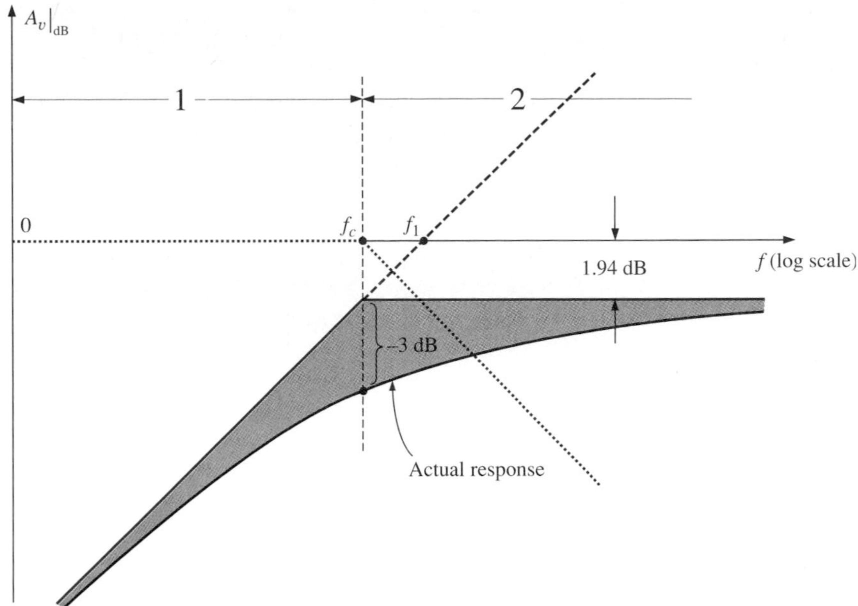

FIG. 21.61
Plot of $A_v|_{dB}$ for the network of Fig. 21.55.

In region 2 one asymptote is increasing at 6 dB, and the other is decreasing at −6 dB/octave for increasing frequencies. The net effect is for one to cancel the other for the region greater than $f = f_c$, leaving a horizontal asymptote beginning at $f = f_c$. A careful sketch of the asymptotes on a log scale will reveal that the horizontal asymptote is at −1.94 dB, as obtained earlier for the same function. The horizontal level can also be determined by simply plugging $f = f_c$ into the Bode plot defined by f/f_1; that is,

$$20 \log \frac{f}{f_1} = 20 \log_{10} \frac{f_c}{f_1} = 20 \log_{10} \frac{31.83 \text{ kHz}}{39.79 \text{ kHz}}$$

$$= 20 \log_{10} 0.799 = -1.94 \text{ dB}$$

The actual response can then be drawn using the asymptotes and the known differences at $f = f_c$ (−3 dB) and at $f = 0.5f_c$ or $2f_c$ (−1 dB).

In summary, therefore, the same dB response for $\mathbf{A}_v = \mathbf{V}_o/\mathbf{V}_i$ can be obtained by isolating the maximum value or defining the gain in a different form. The latter approach permitted the introduction of a new function for our catalog of idealized Bode plots that will prove useful in the future.

21.12 LOW-PASS FILTER WITH LIMITED ATTENUATION

Our analysis will now continue with the low-pass filter of Fig. 21.62, which has limited attentuation at the high-frequency end. That is, the output will not drop to zero as the frequency becomes relatively high. The filter is similar in construction to that of Fig. 21.55, but note that now $\mathbf{V}_o$ includes the capacitive element.

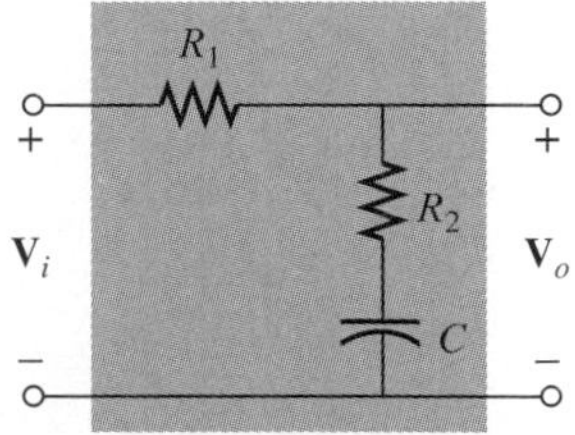

FIG. 21.62
Low-pass filter with limited attenuation.

At $f = 0$ Hz, the capacitor can assume its open-circuit equivalence and $\mathbf{V}_o = \mathbf{V}_i$. At high frequencies the capacitor can be approximated by a short-circuit equivalence and

$$\mathbf{V}_o = \frac{R_2}{R_1 + R_2}\mathbf{V}_i$$

A plot of V_o versus frequency is provided in Fig. 21.63(a). A sketch of $A_v = V_o/V_i$ will appear as shown in Fig. 21.63(b).

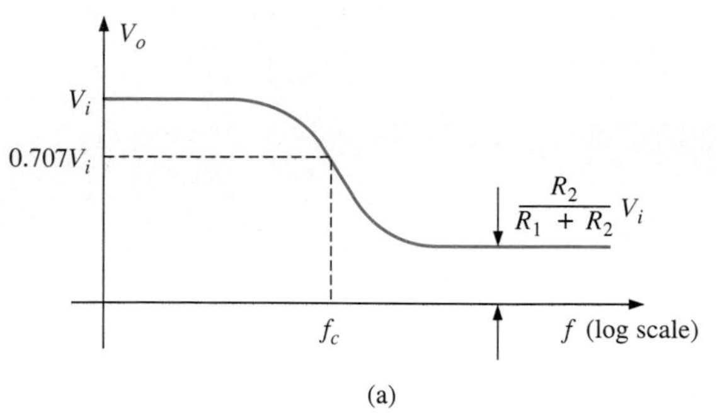

(a)

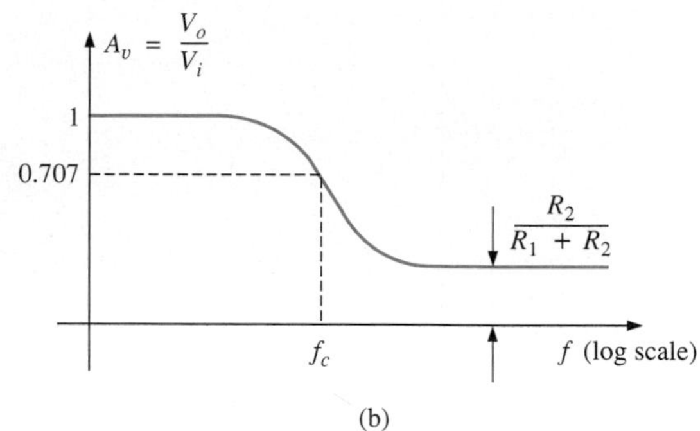

(b)

FIG. 21.63
Low-pass filter with limited attenuation.

An equation for $\mathbf{V}_o$ in terms of $\mathbf{V}_i$ can be derived by first applying the voltage divider rule:

$$\mathbf{V}_o = \frac{(R_2 - jX_C)\mathbf{V}_i}{R_1 + R_2 - jX_C}$$

and
$$\mathbf{A}_v = \frac{\mathbf{V}_o}{\mathbf{V}_i} = \frac{R_2 - jX_C}{R_1 + R_2 - jX_C} = \frac{R_2/X_C - j}{(R_1 + R_2)/X_C - j}$$
$$= \frac{(j)(R_2X_C - j)}{(j)((R_1 + R_2)/X_C - j)}$$
$$= \frac{j(R_2/X_C) + 1}{j((R_1 + R_2)/X_C) + 1} = \frac{1 + j\,2\pi f R_2 C}{1 + j\,2\pi f(R_1 + R_2)C}$$

so that
$$\boxed{\mathbf{A}_v = \frac{\mathbf{V}_o}{\mathbf{V}_i} = \frac{1 + j\,(f/f_1)}{1 + j\,(f/f_c)}} \qquad \textbf{(21.45)}$$

with
$$f_1 = \frac{1}{2\pi R_2 C}, \qquad f_c = \frac{1}{2\pi(R_1 + R_2)C}$$

The denominator of Eq. (21.45) is simply the denominator of the low-pass function of Fig. 21.54(a). The numerator, however, is new and must be investigated.

Applying Eq. (21.5):

$$A_{v_{\text{dB}}} = 20\log_{10}\frac{V_o}{V_i} = 20\log_{10}\sqrt{1 + (f/f_1)^2} + 20\log_{10}\frac{1}{\sqrt{1 + (f/f_c)^2}}$$

For $f \gg f_1$, $(f/f_1)^2 \gg 1$, and the first term becomes

$$20\log_{10}\sqrt{(f/f_1)^2} = 20\log_{10}((f/f_1)^2)^{1/2} = 20\log_{10}(f/f_1)\Big|_{f \gg f_1}$$

which defines the idealized Bode asymptote for the numerator of Eq. (21.45).

At $f = f_1$, $20\log_{10} 1 = 0$ dB and at $f = 2f_1$, $20\log_{10} 2 = 6$ dB. For frequencies much less than f_1, $(f/f_1)^2 \ll 1$, and the first term of the Eq. (21.5) expansion becomes $20\log_{10}\sqrt{1} = 20\log_{10} 1 = 0$ dB, which establishes the low-frequency asymptote.

The full idealized Bode response for the numerator of Eq. (21.45) is provided in Fig. 21.64.

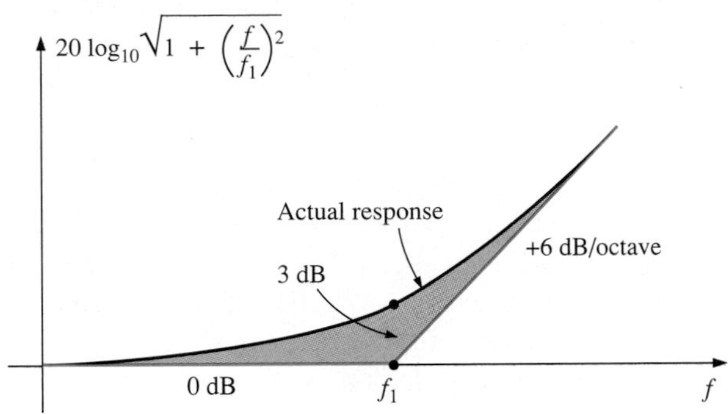

FIG. 21.64
Idealized and actual Bode response for the magnitude of $(1 + j(f/f_1))$.

We are now in a position to determine $A_v|_{dB}$ by plotting the asymptote for each function of Eq. (21.45) on the same frequency axis, as shown in Fig. 21.65. Note that f_c must be less than f_1 since the denominator of f_1 includes only R_2, whereas the denominator of f_c includes both R_2 and R_1.

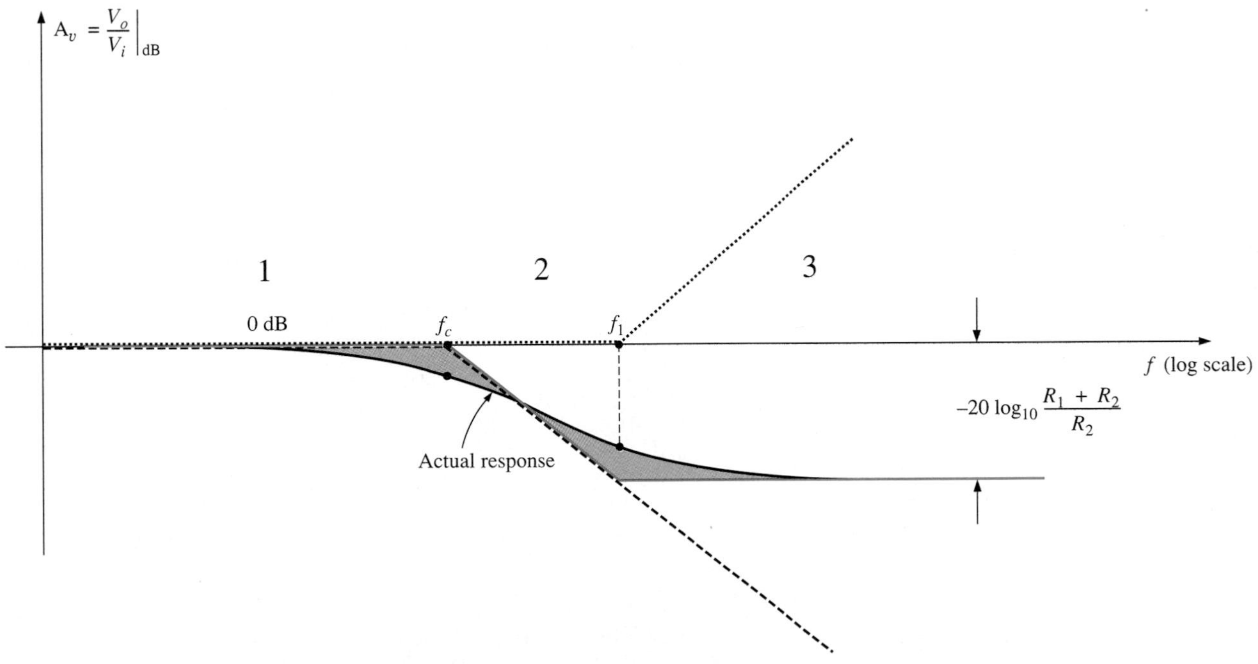

FIG. 21.65
$A_{v_{dB}}$ versus frequency for the low-pass filter with limited attenuation of Fig. 21.62.

Since $R_2/(R_1 + R_2)$ will always be less than 1, we can use an earlier development to obtain an equation for the drop in dB below the 0-dB axis at high frequencies. That is,

$$\begin{aligned} 20 \log_{10} R_2/(R_1 + R_2) &= 20 \log_{10} 1/((R_1 + R_2)/R_2) \\ &= \underbrace{20 \log_{10} 1}_{0} - 20 \log_{10}((R_1 + R_2)/R_2) \end{aligned}$$

and
$$20 \log_{10} \frac{R_2}{R_1 + R_2} = -20 \log_{10} \frac{R_1 + R_2}{R_2} \quad \textbf{(21.46)}$$

as shown in Fig. 21.65.

In region 1 of Fig. 21.65, both asymptotes are at 0 dB, resulting in a net Bode asymptote at 0 dB for the region. At $f = f_c$ one asymptote maintains its 0-dB level, whereas the other is dropping by 6 dB/octave. The sum of the two is the 6-dB drop per octave shown for the region. In region 3 the −6-dB/octave asymptote is balanced by the +6-dB/octave asymptote, establishing a level asymptote at the negative dB level attained by the f_c asymptote at $f = f_1$. The dB level of the horizontal asymptote in region 3 can be determined using Eq. (21.46) or by simply substituting $f = f_1$ into the asymptotic expression defined by f_c.

The full idealized Bode envelope is now defined, permitting a sketch of the actual response by simply shifting 3 dB in the right direction at each corner frequency, as shown in Fig. 21.65.

The phase angle associated with $\mathbf{A}_v$ can be determined directly from Eq. (21.45). That is,

$$\theta = \tan^{-1} f/f_1 - \tan^{-1} f/f_c \quad \textbf{(21.47)}$$

A full plot of θ versus frequency can be obtained by simply substituting various key frequencies into Eq. (21.47) and plotting the result on a log scale.

The first term of Eq. (21.47) defines the phase angle established by the numerator of Eq. (21.45). The asymptotic plot established by the numerator is provided in Fig. 21.66. Note the phase angle of 45° at $f = f_1$ and the straight-line asymptote between $f_1/10$ and $10f_1$.

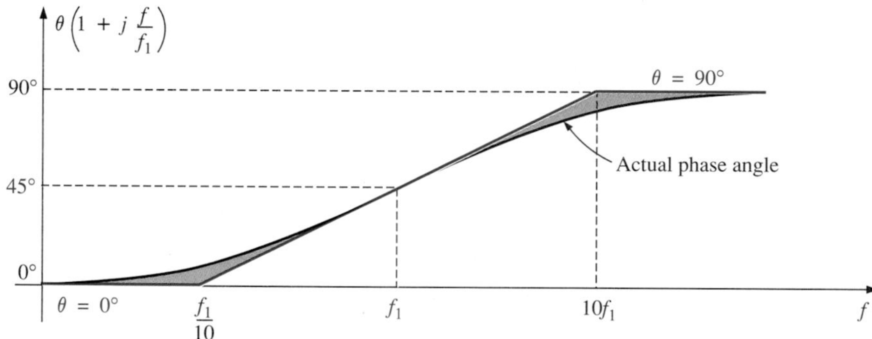

FIG. 21.66
Phase angle for $(1 + j(f/f_1))$.

Now that we have an asymptotic plot for the phase angle of the numerator, we can plot the full phase response by sketching the asymptotes for both functions of Eq. (21.45) on the same graph, as shown in Fig. 21.67.

The asymptotes of Fig. 21.67 clearly indicate that the phase angle will be 0° in the low-frequency range and 0° (90° − 90° = 0°) in the high-frequency range. In region 2 the phase plot drops below 0° due to the impact of the f_c asymptote. In region 4 the phase angle increases since the asymptote due to f_c remains fixed at −90°, whereas that due to f_1 is increasing. In the midrange the plot due to f_1 is balancing the

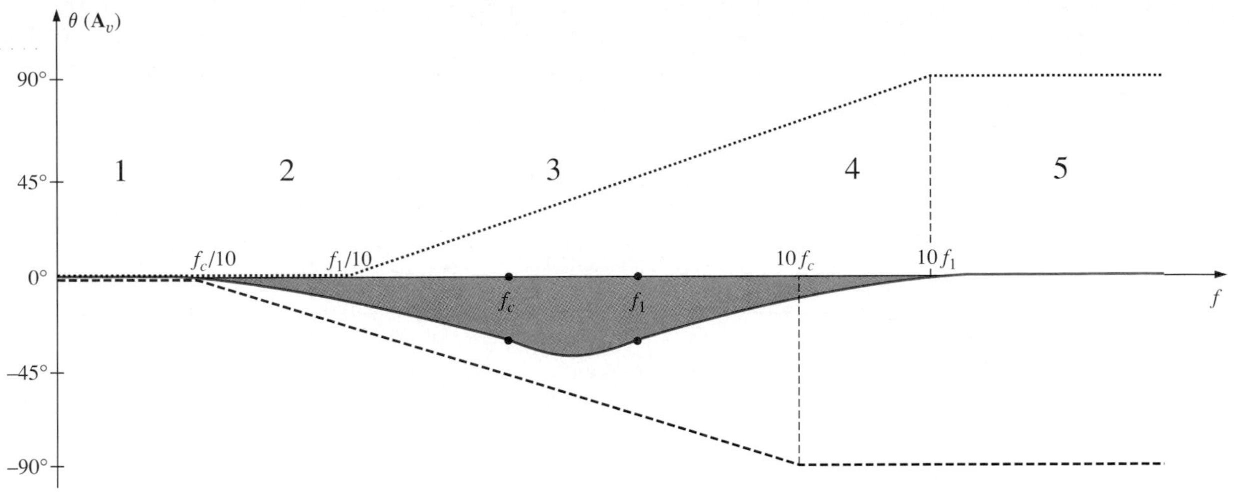

FIG. 21.67
Phase angle for the low-pass filter of Fig. 21.62.

continued negative drop due to the f_c asymptote, resulting in the leveling response indicated. Due to the equal and opposite slopes of the asymptotes in the midregion, the angles of f_1 and f_c will be the same, but note that they are less than 45°. The maximum negative angle will occur between f_1 and f_c. The remaining points on the curve of Fig. 21.67 can be determined by simply substituting specific frequencies into Eq. (21.45). However, it is also useful to know that the most dramatic (the quickest) changes in the phase angle occur when the dB plot of the magnitude also goes through its greatest changes (such as at f_1 and f_c).

21.13 HIGH-PASS FILTER WITH LIMITED ATTENUATION

The filter of Fig. 21.68 is designed to limit the low-frequency attenuation in much the same manner as described for the low-pass filter of the previous section.

At $f = 0$ Hz the capacitor can assume its open-circuit equivalence and $\mathbf{V}_o = (R_2/(R_1 + R_2))\mathbf{V}_i$. At high frequencies the capacitor can be approximated by a short-circuit equivalence and $\mathbf{V}_o = \mathbf{V}_i$.

The resistance to be employed when determining f_c can be found by finding the Thévenin resistance for the capacitor C, as shown in Fig. 21.69. A careful examination of the resulting configuration will reveal that $R_{Th} = R_1 \parallel R_2$ and $f_c = 1/2\pi(R_1 \parallel R_2)C$.

A plot of V_o versus frequency is provided in Fig. 21.70(a) and a sketch of $A_v = V_o/V_i$ appears in Fig. 21.70(b).

An equation for $\mathbf{A}_v = \mathbf{V}_o/\mathbf{V}_i$ can be derived by first applying the voltage divider rule:

$$\mathbf{V}_o = \frac{R_2\mathbf{V}_i}{R_2 + R_1\| -jX_C}$$

and

$$\mathbf{A}_v = \frac{\mathbf{V}_o}{\mathbf{V}_i} = \frac{R_2}{R_2 + R_1\| -jX_C} = \frac{R_2}{R_2 + \dfrac{R_1(-jX_C)}{R_1 - jX_C}}$$

$$= \frac{R_2(R_1 - jX_C)}{R_2(R_1 - jX_C) - jR_1X_C} = \frac{R_1R_2 - jR_2X_C}{R_1R_2 - jR_2X_C - jR_1X_C}$$

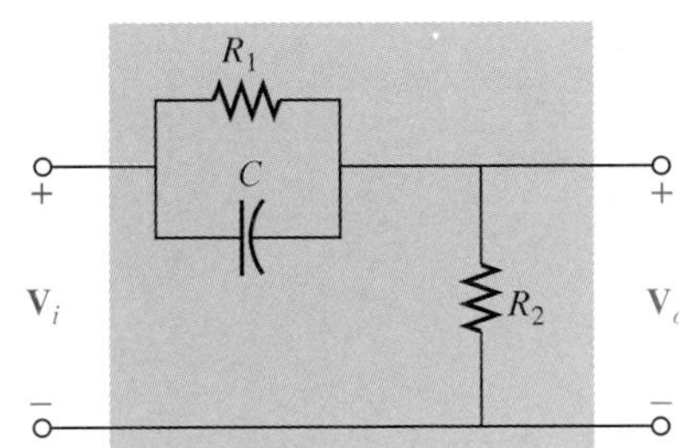

FIG. 21.68
High-pass filter with limited attenuation.

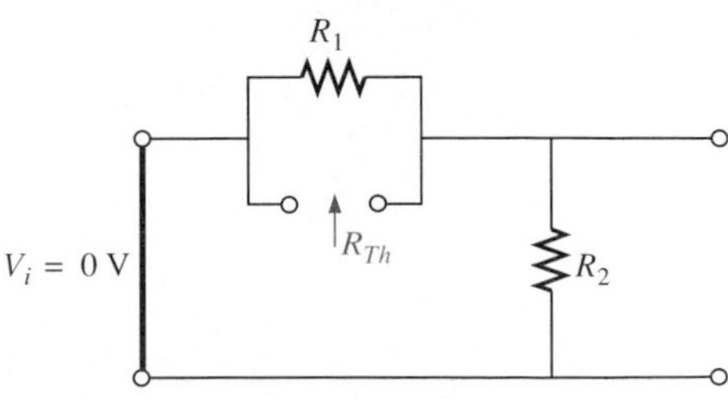

FIG. 21.69
Determining R for the f_c calculation for the filter of Fig. 21.68.

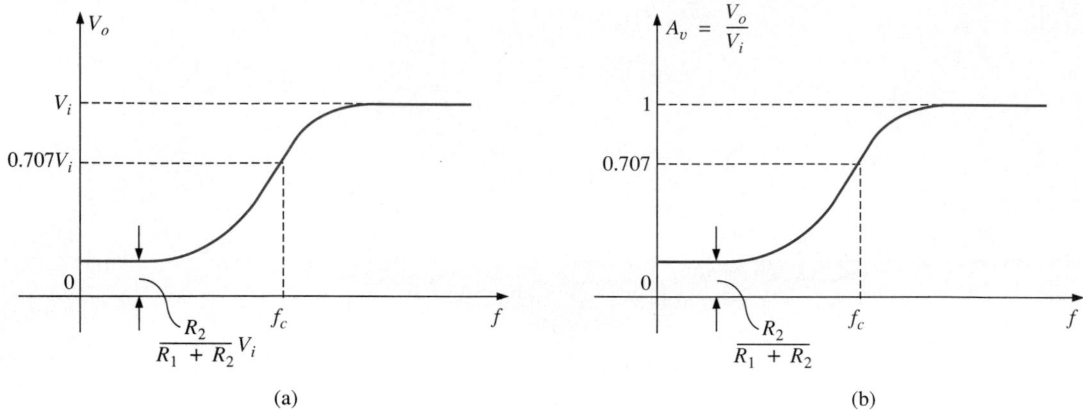

FIG. 21.70
High-pass filter with limited attenuation.

$$= \frac{R_1R_2 - jR_2X_C}{R_1R_2 - j(R_1 + R_2)X_C} = \frac{1 - j\dfrac{R_2X_C}{R_1R_2}}{1 - j\dfrac{(R_1 + R_2)}{R_1R_2}X_C}$$

$$= \frac{1 - j\dfrac{X_C}{R_1}}{1 - j\dfrac{X_C}{\dfrac{R_1R_2}{R_1 + R_2}}} = \frac{1 - j\dfrac{X_C}{R_1}}{1 - j\dfrac{X_C}{R_1 \| R_2}} = \frac{1 - j\dfrac{1}{2\pi fR_1C}}{1 - j\dfrac{1}{2\pi f(R_1 \| R_2)C}}$$

so that

$$\mathbf{A}_v = \frac{\mathbf{V}_o}{\mathbf{V}_i} = \frac{1 - j(f_1/f)}{1 - j(f_c/f)} \tag{21.48}$$

with

$$f_1 = \frac{1}{2\pi R_1C} \quad \text{and} \quad f_c = \frac{1}{2\pi(R_1 \| R_2)C}$$

The denominator of Eq. (21.48) is simply the denominator of the high-pass function of Fig. 21.54(b). The numerator, however, is new and must be investigated.

Applying Eq. (21.5):

$$A_{v_{dB}} = 20\log_{10}\frac{V_o}{V_i} = 20\log_{10}\sqrt{1 + (f_1/f)^2} + 20\log_{10}\frac{1}{\sqrt{1 + (f_c/f)^2}}$$

For $f \ll f_1$, $\left(\dfrac{f_1}{f}\right)^2 \gg 1$ and the first term becomes

$$20\log_{10}\sqrt{(f_1/f)^2} = 20\log_{10}(f_1/f)\Big|_{f \ll f_1}$$

which defines the idealized Bode asymptote for the numerator of Eq. (21.48).

At $f = f_1$, $20\log_{10} 1 = 0$ dB
At $f = 0.5f_1$, $20\log_{10} 2 = 6$ dB
At $f = 0.1f_1$, $20\log_{10} 10 = 20$ dB

For frequencies greater than f_1, $f_1/f \ll 1$ and $20\log_{10} 1 = 0$ dB, which establishes the high-frequency asymptote. The full idealized Bode plot for the numerator of Eq. (21.48) is provided in Fig. 21.71.

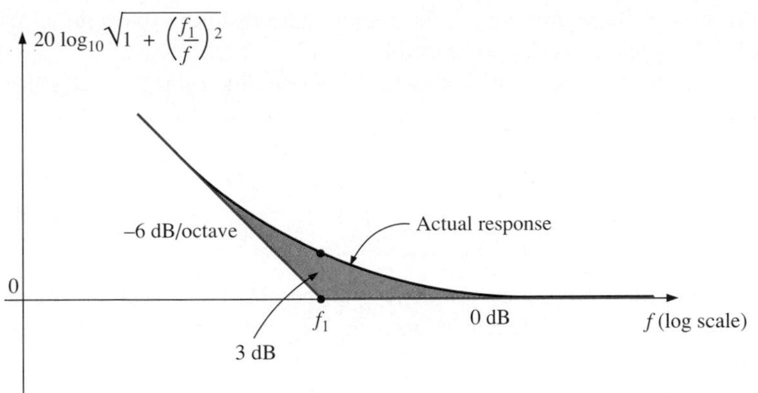

FIG. 21.71
Idealized and actual Bode response for the magnitude of $(1 - j(f_1/f))$.

We are now in a position to determine $A_{v_{\text{dB}}}$ by plotting the asymptotes for each function of Eq. (21.48) on the same frequency axis, as shown in Fig. 21.72. Note that f_c must be more than f_1 since $R_1 \parallel R_2$ must be less than R_1.

When determining the linearized Bode response, let us first examine region 2, where one function is 0 dB and the other is dropping at 6 dB/octave for decreasing frequencies. The result is a decreasing asymptote from f_c to f_1. At the intersection of the resultant of region 2 with f_1, we enter region 1, where the asymptotes have opposite slopes and cancel the effect of each other. The resulting level at f_1 is determined by $-20 \log_{10}(R_1 + R_2)/R_2$, as found in earlier sections. The drop can also be determined by simply substituting $f = f_1$ into the asymptotic equation defined for f_c. In region 3 both are at 0 dB, result-

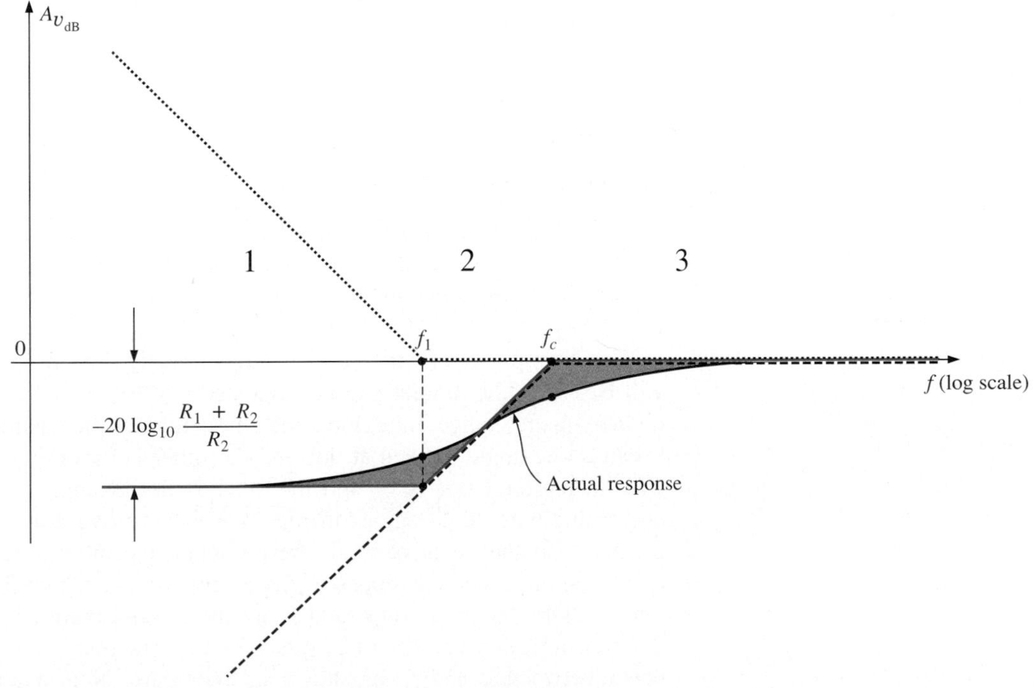

FIG. 21.72
$A_{v_{\text{dB}}}$ versus frequency for the high-pass filter with limited attenuation of Fig. 21.68.

ing in a 0-dB asymptote for the region. The resulting asymptotic and actual responses both appear in Fig. 21.72.

The phase angle associated with A_v can be determined directly from Eq. (21.48); that is,

$$\theta = -\tan^{-1}\frac{f_1}{f} + \tan^{-1}\frac{f_c}{f} \tag{21.49}$$

A full plot of θ versus frequency can be obtained by simply substituting various key frequencies into Eq. (21.49) and plotting the result on a log scale.

The first term of Eq. (21.49) defines the phase angle established by the numerator of Eq. (21.48). The asymptotic plot resulting from the numerator is provided in Fig. 21.73. Note the leading phase angle of 45° at $f = f_1$ and the straight-line asymptote from $f_1/10$ to $10f_1$.

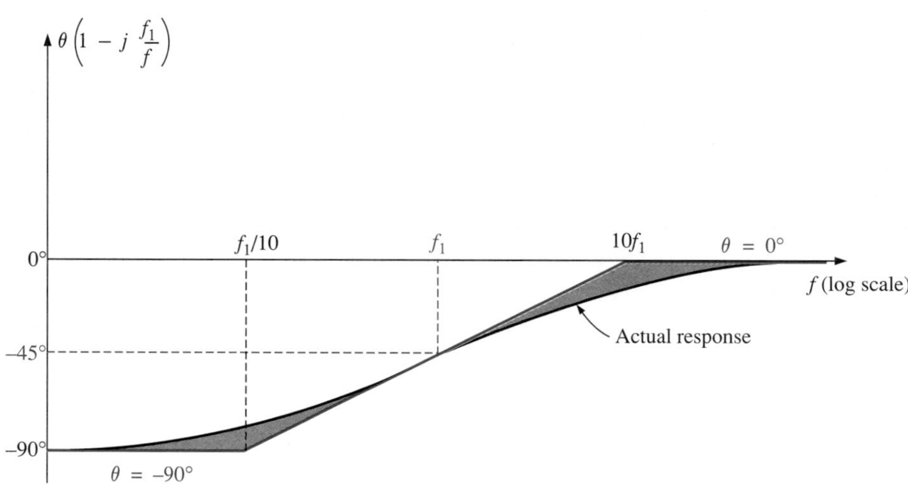

FIG. 21.73
Phase angle for (1 − j (f₁/f)).

Now that we have an asymptotic plot for the phase angle of the numerator, we can plot the full phase response by sketching the asymptotes for both functions of Eq. (21.48) on the same graph, as shown in Fig. 21.74.

The asymptotes of Fig. 21.74 clearly indicate that the phase angle will be 90° in the low-frequency range and 0° (90° − 90° = 0°) in the high-frequency range. In region 2 the phase angle is increasing above 0° because one angle is fixed at 90° and the other is becoming less negative. In region 4 one is 0° and the other is decreasing, resulting in a decreasing θ for this region. In region 3 the positive angle is always greater than the negative angle, resulting in a positive angle for the entire region. Since the slopes of the asymptotes in region 3 are equal but opposite, the angles at f_c and f_1 are the same. Figure 21.74 reveals that the angle at f_c and f_1 will be less than 45°. The maximum angle will occur between f_c and f_1, as shown in the figure. Note again that the greatest change in θ occurs at the corner frequencies, matching the regions of greatest change in the dB plot.

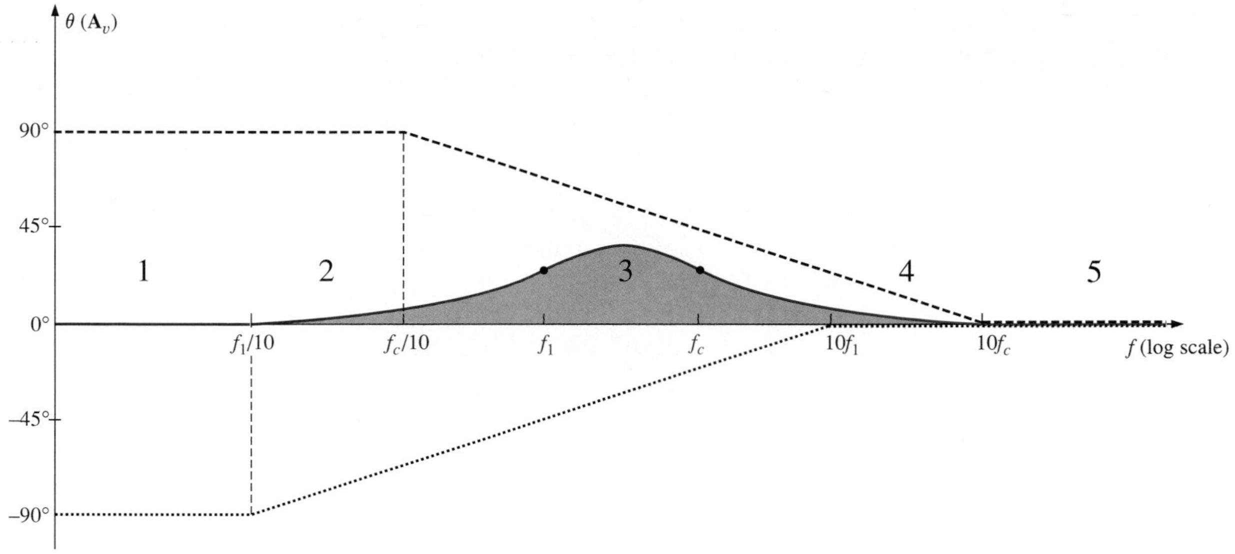

FIG. 21.74
Phase response for the high-pass filter of Fig. 21.68.

EXAMPLE 21.11 For the filter of Fig. 21.75:

a. Sketch the curve of $A_{v_{dB}}$ versus frequency using a log scale.
b. Sketch the curve of θ versus frequency using a log scale.

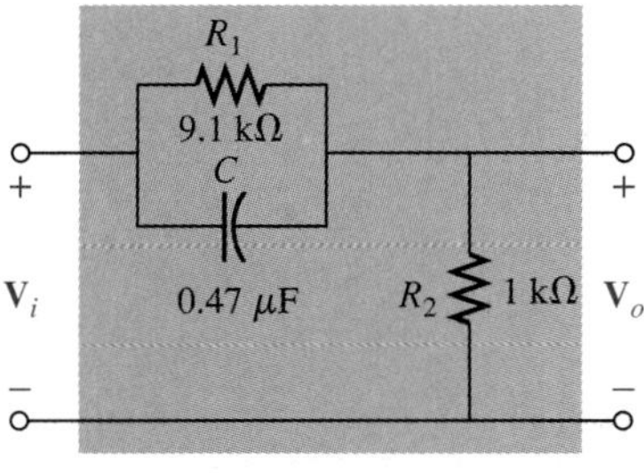

FIG. 21.75
Example 21.11.

Solutions:

a. For the break frequencies:

$$f_1 = \frac{1}{2\pi R_1 C} = \frac{1}{2\pi(9.1\ \text{k}\Omega)(0.47\ \mu\text{F})} = \mathbf{37.2\ Hz}$$

$$f_c = \frac{1}{2\pi\left(\frac{R_1 R_2}{R_1 + R_2}\right)C} = \frac{1}{2\pi(0.9\ \text{k}\Omega)(0.47\ \mu\text{F})} = \mathbf{376.25\ Hz}$$

The maximum low-level attentuation is

$$-20\log_{10}\frac{R_1 + R_2}{R_2} = -20\log_{10}\frac{9.1\ \text{k}\Omega + 1\ \text{k}\Omega}{1\ \text{k}\Omega}$$
$$= -20\log_{10} 10.1 = \mathbf{-20.09\ dB}$$

The resulting plot appears in Fig. 21.76.

b. For the break frequencies:

$f = f_1 = 37.2$ Hz:

$$\theta = -\tan^{-1}\frac{f_1}{f} + \tan^{-1}\frac{f_c}{f}$$
$$= -\tan^{-1} 1 + \tan^{-1}\frac{376.25\ \text{Hz}}{37.2\ \text{Hz}}$$
$$= -45° + 84.35°$$
$$= \mathbf{39.35°}$$

$f = f_c = 376.26$ Hz:

$$\theta = -\tan^{-1}\frac{37.2\ \text{Hz}}{376.26\ \text{Hz}} + \tan^{-1} 1$$

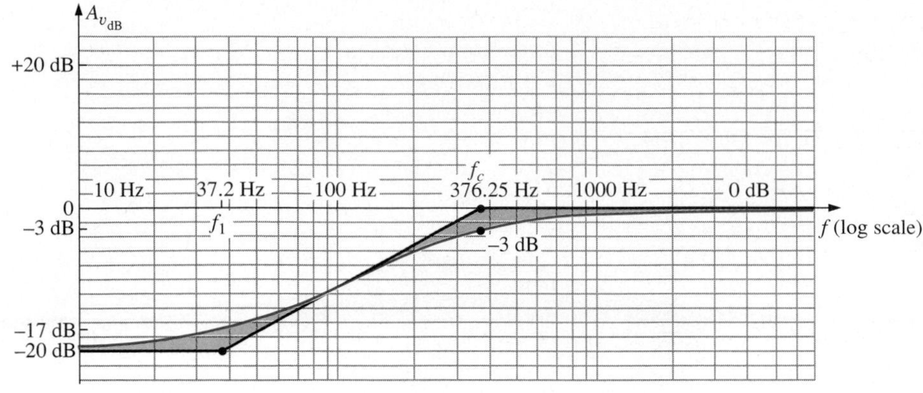

FIG. 21.76
$A_{v_{dB}}$ versus frequency for the filter of Fig. 21.75.

$$= -5.65° + 45°$$
$$= \mathbf{39.35°}$$

At a frequency midway between f_c and f_1 on a log scale, for example, 120 Hz:

$$\theta = -\tan^{-1}\frac{37.2\text{ Hz}}{120\text{ Hz}} + \tan^{-1}\frac{376.26\text{ Hz}}{120\text{ Hz}}$$
$$= -17.22° + 72.31°$$
$$= \mathbf{55.09°}$$

The resulting phase plot appears in Fig. 21.77.

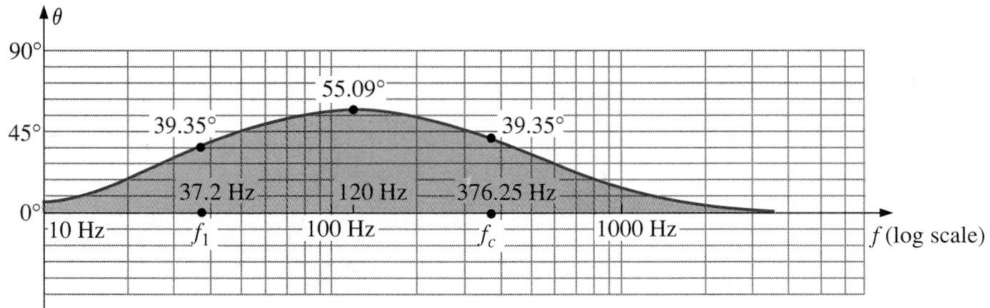

FIG. 21.77
θ (the phase angle associated with A_v) versus frequency for the filter of Fig. 21.75.

21.14 OTHER PROPERTIES AND A SUMMARY TABLE

Bode plots are not limited to filters but can be applied to any system for which a dB-versus-frequency plot is desired. Although the previous sections did not cover all the functions that lend themselves to the idealized linear asymptotes, many of those most commonly encountered have been introduced.

We now examine some of the special situations that can develop that will further demonstrate the adaptability and usefulness of the linear Bode approach to frequency analysis.

In all the situations described in this chapter, there was only one term in the numerator or denominator. For situations where there is more than one term, there will be an interaction between functions that must be examined and understood. In many cases the use of Eq. (21.5) will prove useful. For example, if $\mathbf{A}_v$ should have the format

$$\mathbf{A}_v = \frac{200(1 - jf_2/f)(jf/f_1)}{(1 - jf_1/f)(1 + jf/f_2)} = \frac{(a)(b)(c)}{(d)(e)} \tag{21.50}$$

we can expand the function in the following manner:

$$A_{v_{\text{dB}}} = 20 \log_{10} \frac{(a)(b)(c)}{(d)(e)} = 20 \log_{10} a + 20 \log_{10} b$$
$$+20 \log_{10} c - 20 \log_{10} d - 20 \log_{10} e$$

revealing that the net or resultant dB level is equal to the algebraic sum of the contributions from all the terms of the original function. We will, therefore, be able to add algebraically the linearized Bode plots of all the terms in each frequency interval to determine the idealized Bode plot for the full function.

If two terms happen to have the same format and corner frequency, as in the function

$$\mathbf{A}_v = \frac{1}{(1 - jf_1/f)(1 - jf_1/f)}$$

the function can be rewritten as

$$\mathbf{A}_v = \frac{1}{(1 - jf_1/f)^2}$$

so that $A_{v_{\text{dB}}} = 20 \log_{10} \dfrac{1}{(\sqrt{1 + (f_1/f)^2})^2} = -20 \log_{10}(1 + (f_1/f)^2)$

for $f \ll f_1$, $(f_1/f)^2 \gg 1$ and

$$A_{v_{\text{dB}}} = -20 \log_{10}(f_1/f)^2 = -40 \log_{10} f_1/f$$

versus the $-20 \log_{10}(f_1/f)$ obtained for a single term in the denominator. The resulting dB asymptote will drop, therefore, at a rate of -12 dB/octave (-40 dB/decade) for decreasing frequencies rather than -6 dB/octave. The corner frequency is the same and the high-frequency asymptote is still at 0 dB. The idealized Bode plot for the above function is provided in Fig. 21.78.

Note the steeper slope of the asymptote and the fact that the actual curve will now pass -6 dB below the corner frequency rather than -3 dB, as for a single term.

If the corner frequencies of the two terms in the numerator or denominator are close but not exactly equal, keep in mind that the total dB drop is the algebraic sum of the contributing terms of the expansion. For instance, consider the linearized Bode plot of Fig. 21.79 with corner frequencies f_1 and f_2.

In region 3 both asymptotes are 0 dB, resulting in an asymptote at 0 dB for frequencies greater than f_2. For region 2, one asymptote is at 0 dB, whereas the other drops at -6 dB/octave for decreasing frequencies. The net result for this region is an asymptote dropping at -6 dB, as shown in the same figure. At f_1, we find two asymptotes dropping off at -6 dB for decreasing frequencies. The result is an asymptote dropping off at -12 dB/octave for this region.

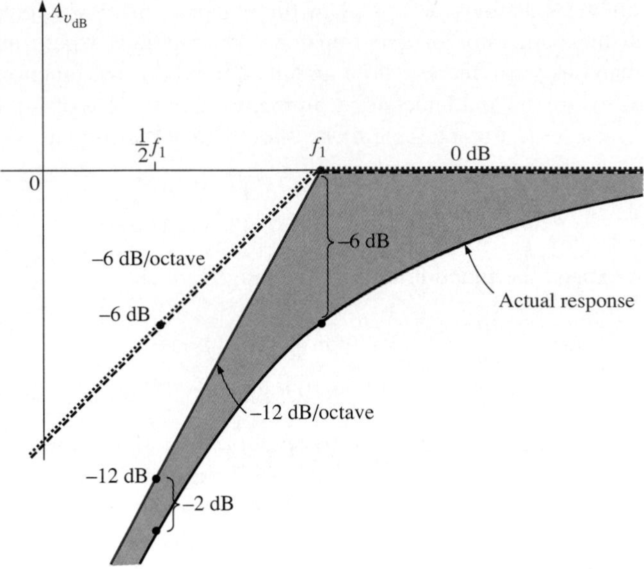

FIG. 21.78
Plotting the linearized Bode plot of $\frac{1}{(1 - j(f_1/f))^2}$.

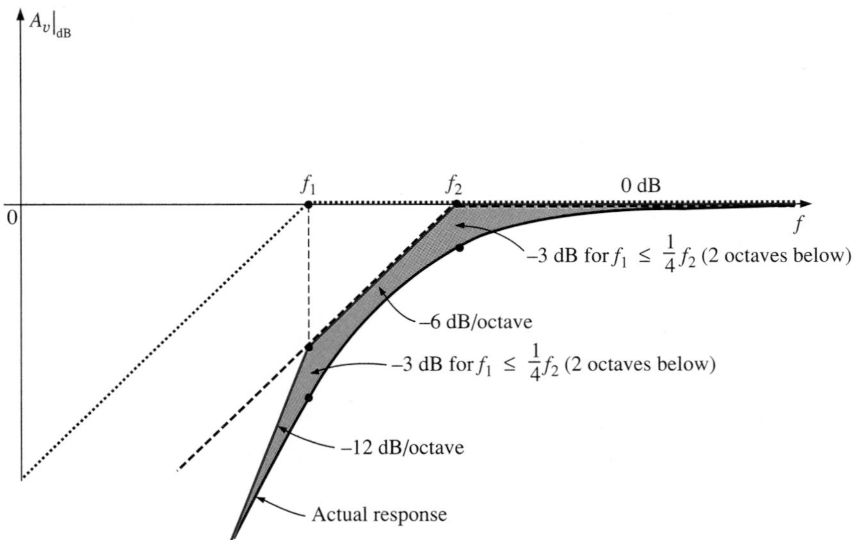

FIG. 21.79
Plot of $A_v|_{dB}$ *for* $\frac{1}{(1 - j(f_1/f))(1 - j(f_2/f))}$ *with* $f_1 < f_2$.

If f_1 and f_2 are at least two octaves apart, the effect of one on the plotting of the actual response for the other can just about be ignored. In other words, for this example, if $f_1 < \frac{1}{4}f_2$, then the actual response will be down -3 dB at $f = f_2$ and f_1.

The above discussion can be expanded for any number of terms at the same frequency or in the same region. For three equal terms in the denominator, the asymptote will drop at -18 dB/octave, and so on. In time the procedure will be somewhat self-evident and relatively straightforward to apply. In many cases the hardest part of finding a solution is to put the original function in the desired form.

EXAMPLE 21.12 A transistor amplifier has the following gain:

$$\mathbf{A}_v = \frac{100}{\left(1 - j\,\dfrac{50\text{ Hz}}{f}\right)\left(1 - j\,\dfrac{200\text{ Hz}}{f}\right)\left(1 + j\,\dfrac{f}{10\text{ kHz}}\right)\left(1 + j\,\dfrac{f}{20\text{ kHz}}\right)}$$

a. Sketch the normalized response $A'_v = A_v/A_{v_{\max}}$ and determine the bandwidth of the amplifier.
b. Sketch the phase response and determine a frequency where the phase angle is close to 0°.

Solutions:

a. $$A'_v = \frac{A_v}{A_{v_{\max}}} = \frac{A_v}{100}$$

$$= \frac{1}{\left(1 - j\,\dfrac{50\text{ Hz}}{f}\right)\left(1 - j\,\dfrac{200\text{ Hz}}{f}\right)\left(1 + j\,\dfrac{f}{10\text{ kHz}}\right)\left(1 + j\,\dfrac{f}{20\text{ kHz}}\right)}$$

$$= \frac{1}{(a)(b)(c)(d)} = \left(\frac{1}{a}\right)\left(\frac{1}{b}\right)\left(\frac{1}{c}\right)\left(\frac{1}{d}\right)$$

and

$$A'_{v_{\text{dB}}} = -20\log_{10} a - 20\log_{10} b - 20\log_{10} c - 20\log_{10} d$$

clearly substantiating the fact that the total number of decibels is equal to the algebraic sum of the contributing terms.

A careful examination of the original function will reveal that the first two terms in the denominator are high-pass filter functions, whereas the last two are low-pass functions. Figure 21.80 demonstrates how the combination of the two types of functions defines a bandwidth for the amplifier. The high-frequency filter functions have defined the low-cutoff frequency, and the low-frequency filter functions have defined the high-cutoff frequency.

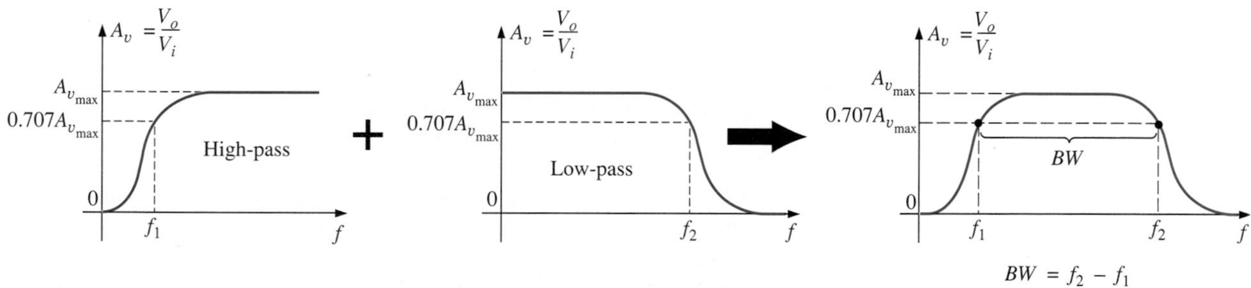

FIG. 21.80
Finding the overall gain versus frequency for Example 21.12.

Plotting all the idealized Bode plots on the same axis will result in the plot of Fig. 21.81. Note for frequencies less than 50 Hz that the resulting asymptote drops off at −12 dB/octave. In addition, since 50 Hz and 200 Hz are separated by two octaves, the actual response will be down by only about −3 dB at the corner frequencies of 50 Hz and 200 Hz.

For the high-frequency region, the corner frequencies are not separated by two octaves, and the difference between the idealized plot

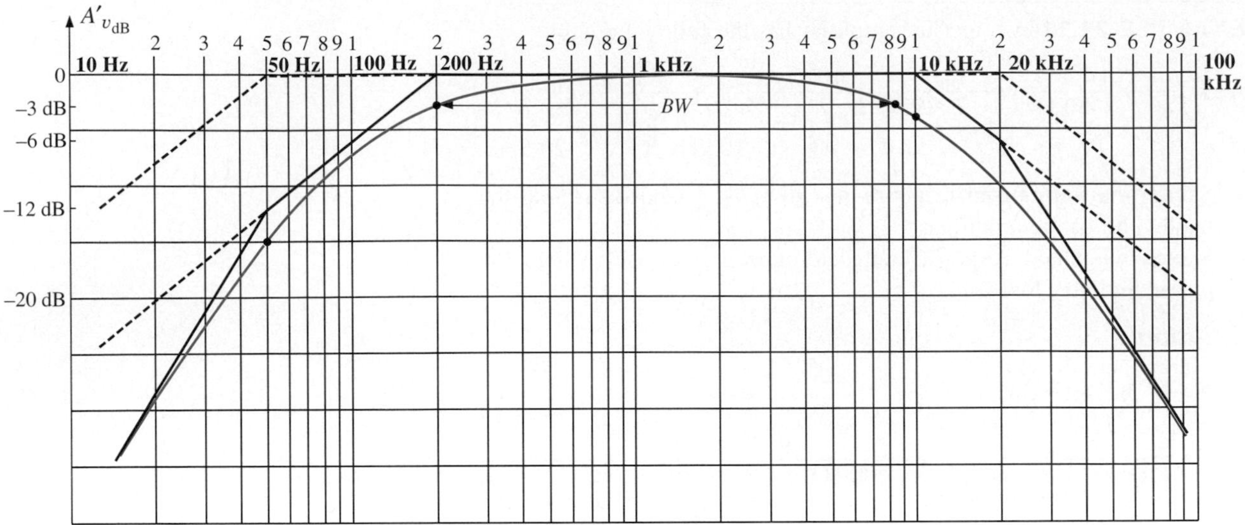

FIG. 21.81
$A'_{v_{dB}}$ versus frequency for Example 21.12.

and the actual Bode response must be examined more carefully. Since 10 kHz is one octave below 20 kHz, we can use the fact that the difference between the idealized response and the actual response for a single corner frequency is 1 dB. If we add an additional −1-dB drop due to the 20-kHz corner frequency to the −3-dB drop at $f = 10$ kHz, we can conclude that the drop at 10 kHz will be −4 dB, as shown on the plot. To check the conclusion, let us write the full expression for the dB level at 10 kHz and find the actual level for comparison purposes.

$$A'_{v_{dB}} = -20 \log_{10} \sqrt{1 + \left(\frac{50 \text{ Hz}}{10 \text{ kHz}}\right)^2} - 20 \log_{10} \sqrt{1 + \left(\frac{200 \text{ Hz}}{10 \text{ kHz}}\right)^2}$$
$$-20 \log_{10} \sqrt{1 + \left(\frac{10 \text{ kHz}}{10 \text{ kHz}}\right)^2} - 20 \log_{10} \sqrt{1 + \left(\frac{10 \text{ kHz}}{20 \text{ kHz}}\right)^2}$$
$$= -0.00011 \text{ dB} - 0.0017 \text{ dB} - 3.01 \text{ dB} - 0.969 \text{ dB}$$
$$= -3.98 \text{ dB} \cong \mathbf{-4\ dB}, \qquad \text{as before}$$

An examination of the above calculations clearly reveals that the last two terms predominate in the high-frequency region and essentially eliminate the need to consider the first two terms in that region. For the low-frequency region an examination of the first two terms is sufficient.

Proceeding in a similar fashion, we find a −4-dB difference at $f = 20$ kHz, resulting in the actual response appearing in Fig. 21.81. Since the bandwidth is defined at the −3-dB level, a judgment must be made as to where the actual response crosses the −3-dB level in the high-frequency region. A rough sketch suggests that it is near 8.5 kHz. Plugging this frequency into the high-frequency terms results in:

$$A'_{v_{dB}} = -20 \log_{10} \sqrt{1 + \left(\frac{8.5 \text{ kHz}}{10 \text{ kHz}}\right)^2} - 20 \log_{10} \sqrt{1 + \left(\frac{8.5 \text{ kHz}}{20 \text{ kHz}}\right)^2}$$
$$= -2.148 \text{ dB} - 0.645 \text{ dB} \cong \mathbf{-2.8\ dB}$$

which is relatively close to the −3-dB level and

$$BW = f_{high} - f_{low} = 8.5\text{ kHz} - 200\text{ Hz} = \mathbf{8.3\text{ kHz}}$$

In the midrange of the bandwidth, $A'_{v_{dB}}$ will approach 0 dB. At $f =$ 1 kHz:

$$A'_{v_{dB}} = -20\log_{10}\sqrt{1+\left(\frac{50\text{ Hz}}{1\text{ kHz}}\right)^2} - 20\log_{10}\sqrt{1+\left(\frac{200\text{ Hz}}{1\text{ kHz}}\right)^2}$$

$$-20\log_{10}\sqrt{1+\left(\frac{1\text{ kHz}}{10\text{ kHz}}\right)^2} - 20\log_{10}\sqrt{1+\left(\frac{1\text{ kHz}}{20\text{ kHz}}\right)^2}$$

$$= -0.0108\text{ dB} - 0.1703\text{ dB} - 0.0432\text{ dB} - 0.0108\text{ dB}$$

$$= \mathbf{-0.2351\text{ dB}} \cong -\frac{1}{5}\text{ dB}$$

which is certainly close to the 0-dB level, as shown on the plot.

b. The phase response can be determined by simply substituting a number of key frequencies into the following equation, derived directly from the original function $\mathbf{A}_v$:

$$\theta = \tan^{-1}\frac{50\text{ Hz}}{f} + \tan^{-1}\frac{200\text{ Hz}}{f} - \tan^{-1}\frac{f}{10\text{ kHz}} - \tan^{-1}\frac{f}{20\text{ kHz}}$$

However, let us make full use of the asymptotes defined by each term of $\mathbf{A}_v$ and sketch the response by finding the resulting phase angle at critical points on the frequency axis. The resulting asymptotes and phase plot are provided in Fig. 21.82. Note at $f = 50$ Hz that the sum of the two angles determined by the straight-line asymptotes is 45° + 75° = 120° (actual = 121°). At $f = 1$ kHz, if we subtract 5.7° for one corner frequency, we obtain a net angle of 14° − 5.7° ≅ 8.3° (actual = 5.6°).

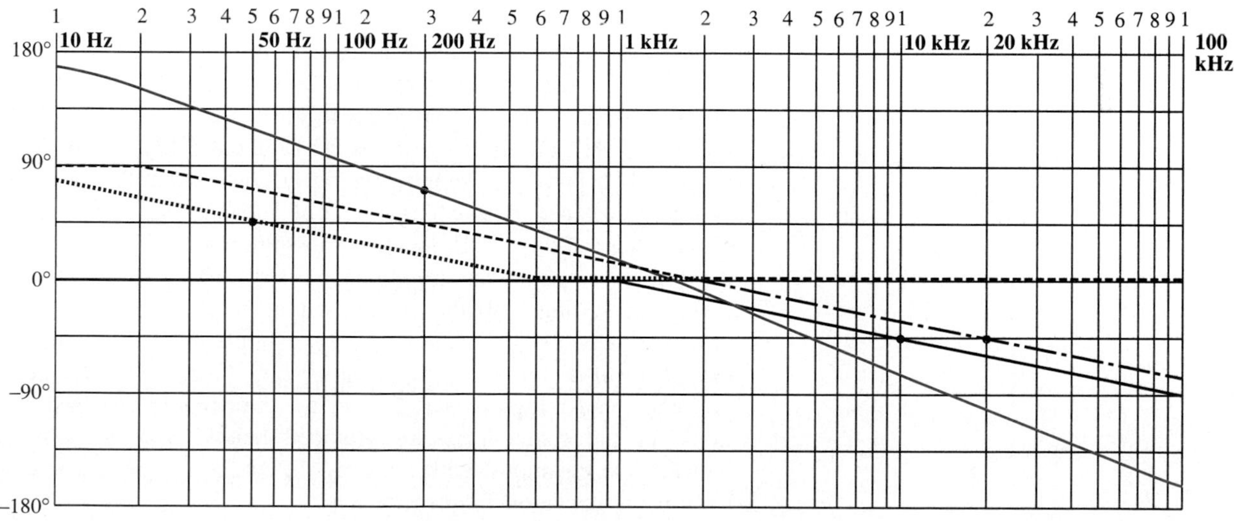

FIG. 21.82
Phase response for Example 21.12.

At 10 kHz the asymptotes leave us with $\theta \cong -45° - 32° = -77°$ (actual = −71.56°). The net phase plot appears to be close to 0° at about 1300 Hz. As a check on our assumptions and the use of the asymptotic approach, let us plug in $f = 1300$ Hz into the equation for θ:

$$\theta = \tan^{-1}\frac{50\text{ Hz}}{1300\text{ Hz}} + \tan^{-1}\frac{200\text{ Hz}}{1300\text{ Hz}} - \tan^{-1}\frac{1300\text{ Hz}}{10\text{ kHz}} - \tan^{-1}\frac{1300\text{ Hz}}{20\text{ kHz}}$$
$$= 2.2° + 8.75° - 7.41° - 3.72°$$
$$= -0.18° \cong 0° \quad \text{as predicted}$$

In total, the phase plot appears to shift from a positive angle of 180° ($\mathbf{V}_o$ leading $\mathbf{V}_i$) to a negative angle of 180° as the frequency spectrum extends from very low frequencies to high frequencies. In the midregion the phase plot is close to 0° ($\mathbf{V}_o$ in phase with $\mathbf{V}_i$), much like the response to a common-base transistor amplifier.

In an effort to consolidate some of the material introduced in this chapter and provide a reference for future investigations, Table 21.2 was developed; it includes the linearized dB and phase plots for the functions appearing in the first column. These are by no means all the functions encountered, but they do provide a foundation to which additional functions can be added.

Reviewing the development of the filters of Sections 21.12 and 21.13, it is probably evident that establishing the function $\mathbf{A}_v$ in the proper form is the most difficult part of the analysis. However, with practice and an awareness of the desired format, methods will surface that will significantly reduce the effort involved.

21.15 CROSSOVER NETWORKS

The topic of *crossover networks* is included primarily to present an excellent demonstration of filter operation without a high level of complexity. Crossover networks are employed in audio systems to ensure that the proper frequencies are channeled to the appropriate speaker. Although less expensive audio systems have to rely on one speaker to cover the full audio range from about 20 Hz to 20 kHz, better systems will employ at least three speakers to cover the low range (20 Hz to about 500 Hz), the midrange (500 Hz to about 5 kHz), and the high range (5 kHz and up). The term *crossover* comes from the fact that the system is designed to have a crossover of frequency spectrums for adjacent speakers at the −3-dB level, as shown in Fig. 21.83. Depending on the design, each filter can drop off at 6 dB, 12 dB, or 18 dB, with complexity increasing with the desired dB drop-off rate. The three-way crossover network of Fig. 21.83 is quite simple in design, with a low-pass *R-L* filter for the *woofer,* an *R-L-C* pass-band filter for the midrange, and a high-pass *R-C* filter for the *tweeter.* The basic equations for the components are provided below. Note the similarity between the equations, with the only difference for each type of element being the cutoff frequency.

$$L_{\text{low}} = \frac{R}{2\pi f_1} \qquad L_{\text{mid}} = \frac{R}{2\pi f_2} \tag{21.51}$$

$$C_{\text{mid}} = \frac{1}{2\pi f_1 R} \qquad C_{\text{high}} = \frac{1}{2\pi f_2 R} \tag{21.52}$$

TABLE 21.2

Idealized Bode plots for various functions.

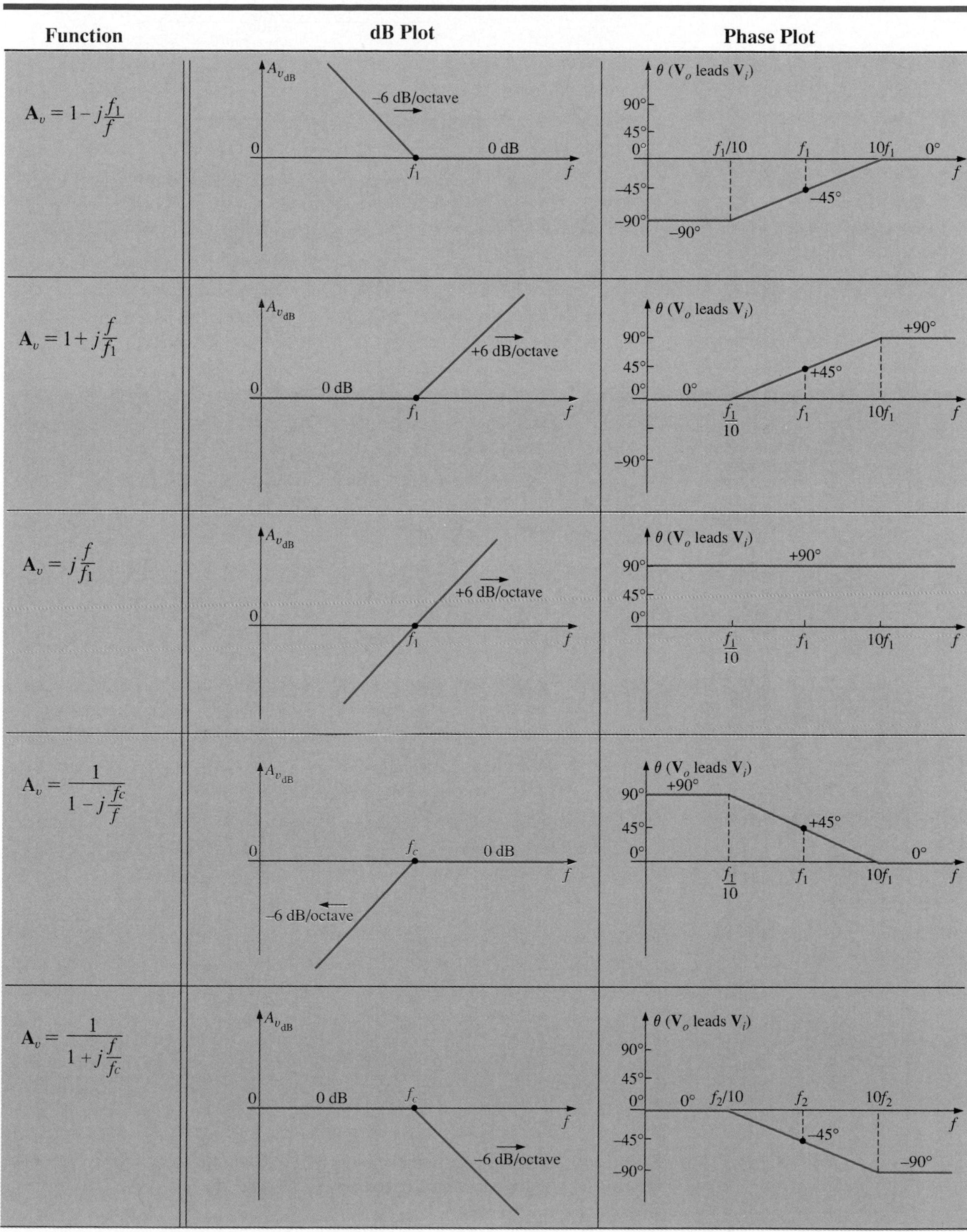

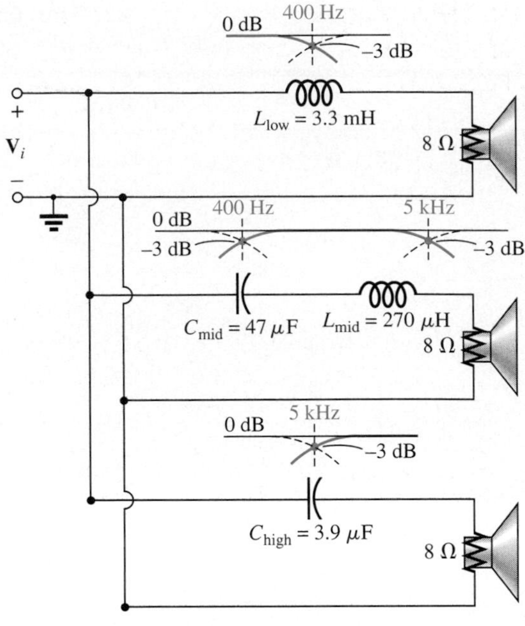

FIG. 21.83
Three-way, 6-dB-per-octave, crossover network.

For the crossover network of Fig. 21.83 with three 8-Ω speakers, the resulting values are

$$L_{low} = \frac{R}{2\pi f_1} = \frac{8\ \Omega}{2\pi(400\text{ Hz})} = 3.183\text{ mH} \longrightarrow 3.3\text{ mH}$$
(commercial value)

$$L_{mid} = \frac{R}{2\pi f_2} = \frac{8\ \Omega}{2\pi(5\text{ kHz})} = 254.65\ \mu\text{H} \longrightarrow 270\ \mu\text{H}$$
(commercial value)

$$C_{mid} = \frac{1}{2\pi f_1 R} = \frac{1}{2\pi(400\text{ Hz})(8\ \Omega)} = 49.736\ \mu\text{F} \longrightarrow 47\ \mu\text{F}$$
(commercial value)

$$C_{high} = \frac{1}{2\pi f_2 R} = \frac{1}{2\pi(5\text{ kHz})(8\ \Omega)} = 3.979\ \mu\text{F} \longrightarrow 3.9\ \mu\text{F}$$
(commercial value)

as appearing on Fig. 21.83.

For each filter, a rough sketch of the frequency response is included to show the crossover at the specific frequencies of interest. Because all three speakers are in parallel, the source voltage and impedance for each are the same. The total loading on the source is obviously a function of the frequency applied, but the total delivered is determined solely by the speakers since they are essentially resistive in nature.

To test the system let us apply a 4-V signal at a frequency of 1 kHz (a predominant frequency of the typical human auditory response curve) and see which speaker will have the highest power level.

f = 1 kHz: $X_{L_{low}} = 2\pi f L_{low} = 2\pi(1\text{ kHz})(3.3\text{ mH}) = 20.74\ \Omega$

$$\mathbf{V}_o = \frac{(\mathbf{Z}_R \angle 0°)(V_i \angle 0°)}{\mathbf{Z}_T} = \frac{(8\ \Omega \angle 0°)(4\text{ V}\angle 0°)}{8\ \Omega + j\,20.74\ \Omega}$$
$$= 1.44\text{ V}\angle -68.90°$$

$$X_{L_{\text{mid}}} = 2\pi f L_{\text{mid}} = 2\pi(1\text{ kHz})(270\ \mu\text{H}) = 1.696\ \Omega$$

$$X_{C_{\text{mid}}} = \frac{1}{2\pi f C_{\text{mid}}} = \frac{1}{2\pi(1\text{ kHz})(47\ \mu\text{F})} = 3.386\ \Omega$$

$$\mathbf{V}_o = \frac{(\mathbf{Z}_R \angle 0^\circ)(V_i \angle 0^\circ)}{\mathbf{Z}_T} = \frac{(8\ \Omega \angle 0^\circ)(4\text{ V} \angle 0^\circ)}{8\ \Omega + j\,1.696\ \Omega - j\,3.386\ \Omega}$$
$$= 3.94\text{ V}\angle 11.93^\circ$$

$$X_{C_{\text{high}}} = \frac{1}{2\pi f C_{\text{high}}} = \frac{1}{2\pi(1\text{ kHz})(3.9\ \mu\text{F})} = 40.81\ \Omega$$

$$\mathbf{V}_o = \frac{(Z_R \angle 0^\circ)(V_i \angle 0^\circ)}{\mathbf{Z}_T} = \frac{(8\ \Omega \angle 0^\circ)(4\text{ V} \angle 0^\circ)}{8\ \Omega - j\,40.81\ \Omega}$$
$$= 0.77\text{ V} \angle 78.91^\circ$$

Using the basic power equation $P = V^2/R$, the power to the woofer is:

$$P_{\text{low}} = \frac{V^2}{R} = \frac{(1.44\text{ V})^2}{8\ \Omega} = \mathbf{0.259\ W}$$

to the midrange speaker:

$$P_{\text{mid}} = \frac{V^2}{R} = \frac{(3.94\text{ V})^2}{8\ \Omega} = \mathbf{1.94\ W}$$

and to the tweeter:

$$P_{\text{high}} = \frac{V^2}{R} = \frac{(0.77\text{ V})^2}{8\ \Omega} = \mathbf{0.074\ W}$$

resulting in a power ratio of 7.5:1 between the midrange and the woofer and 26:1 between the midrange and the tweeter. Obviously, the response of the midrange speaker will totally overshadow the other two.

The same system is analyzed using PSpice Windows in Section 21.16 to permit a glimpse at the dB response of each filter and whether our chosen crossover frequencies are, in fact, at the −3-dB level.

21.16 COMPUTER ANALYSIS

PSpice (DOS)

This chapter provides an excellent opportunity to demonstrate the versatility of the PROBE option. By simply indicating the preference for each axis, the vertical and horizontal scales can be linear or log scales; in fact, a decibel scale can be chosen if desired.

The first input file of this section will investigate the impact of using a log-log scale to plot the current versus frequency for the inductor of Fig. 21.84, initially investigated in Chapter 15 with linear scales. By choosing a log-log scale, the range of current and frequency values can go far beyond the plot of Chapter 15. The input file appears in Fig. 21.85, with the PROBE response appearing in Fig. 21.86.

The vertical scale of Fig. 21.86 must be read with particular care since intermediate log values are not included. For instance, at $f =$ 1 Hz, the interval between 10 A and 1000 A must first be broken down, as shown in Fig. 21.87, to include the 100-A level. The intersection of the plot is then about 300 A on a log scale rather than 500 A if it were on a linear scale. Checking our conclusion, at $f = 1$ Hz,

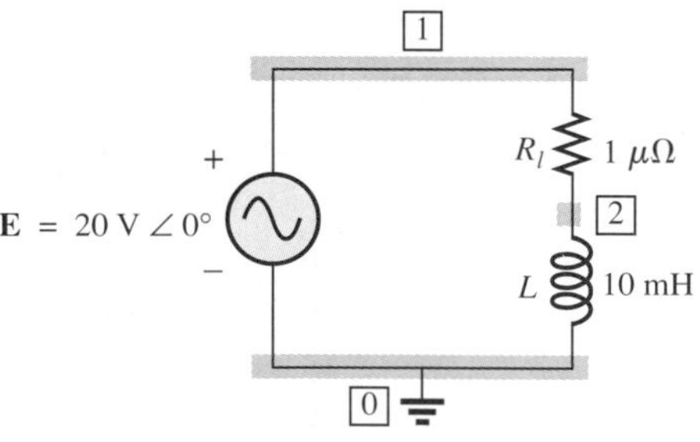

FIG. 21.84
Demonstrating the effect of a log-log plot on the frequency response of an inductive element.

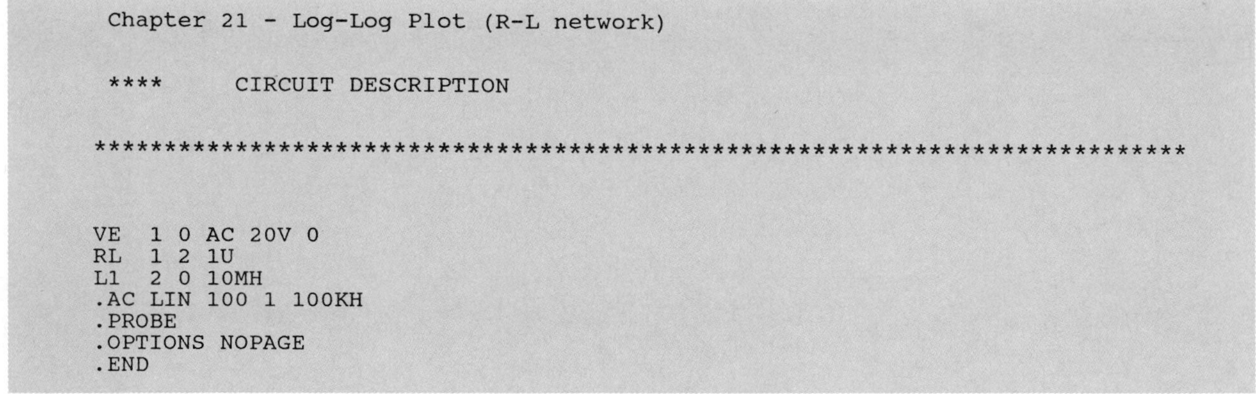

```
Chapter 21 - Log-Log Plot (R-L network)

****      CIRCUIT DESCRIPTION

******************************************************************************

VE  1 0 AC 20V 0
RL  1 2 1U
L1  2 0 10MH
.AC LIN 100 1 100KH
.PROBE
.OPTIONS NOPAGE
.END
```

FIG. 21.85
Input file for the network of Fig. 21.84.

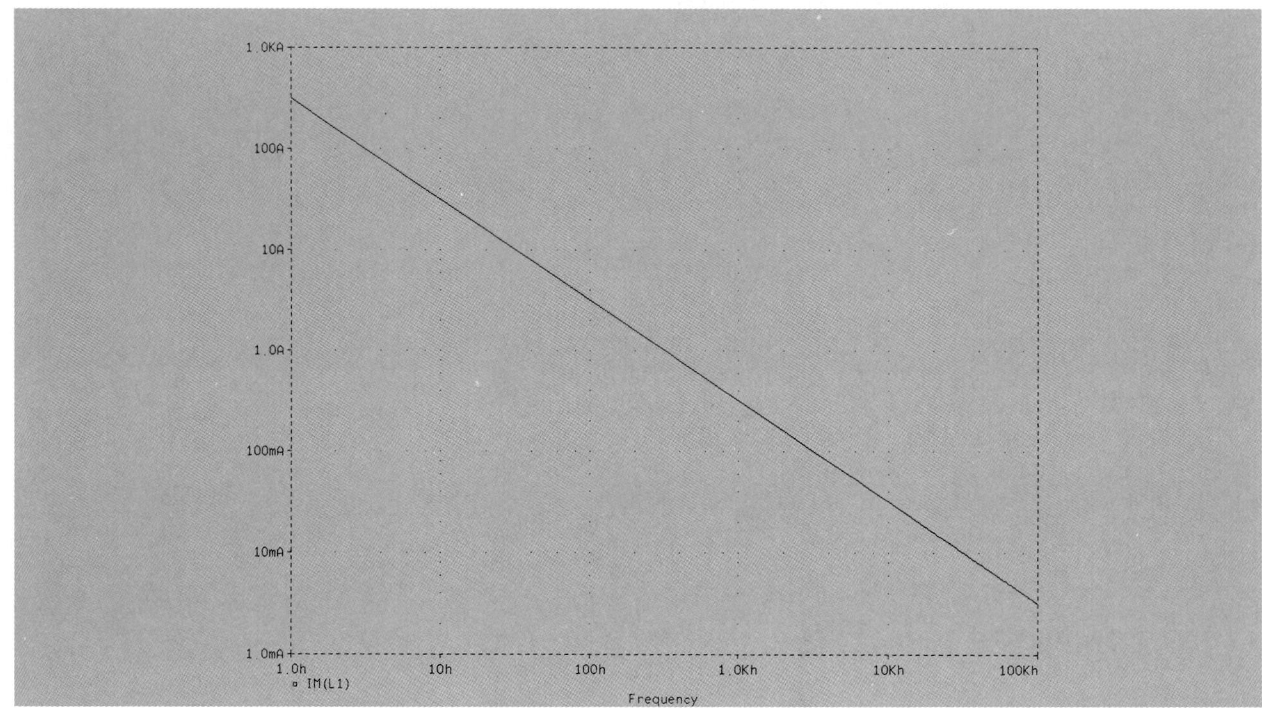

FIG. 21.86
Output PROBE response for the network of Fig. 21.84.

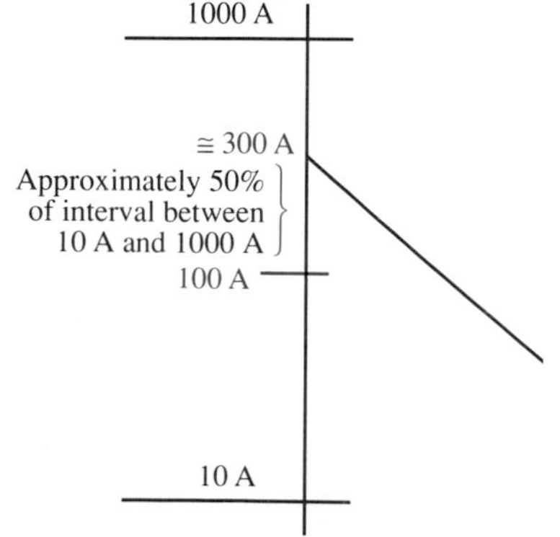

FIG. 21.87
Reading the vertical log scale.

$$X_L = 2\pi fL = 2\pi(1\text{ Hz})(10\text{ mH}) = 62.8\text{ m}\Omega$$

and
$$I = \frac{E}{X_L} = \frac{20\text{ V}}{62.8\text{ m}\Omega} = 318.47\text{ A}$$

At f = 100 kHz, the 10-mA level must be added between 1 mA and 100 mA before a level of about 3 mA can be estimated.

In particular, note the enormous range of current and frequency that can appear on the same plot. The resulting curve does not reflect the exponential relationship between the current and applied frequency, but it does reveal the level of current for a wide range of frequencies. Just imagine trying to plot the preceding graph using linear scales—especially if we use the interval designated for 1 mA to 100 mA for the full range of current values.

The frequency response of the *R-C* network of Fig. 21.88 will now be determined using the input file of Fig. 21.89. The PROBE response of Fig. 21.90 requested the magnitude of the output voltage versus frequency on a log scale. Quite obviously, as the frequency increased, the magnitude of V_o increased also, as required for a high-pass filter. The critical frequency is defined by 0.707 V, but reading the resulting frequency on the horizontal axis requires a bit of care due to the log scale. A careful breakdown of the region between 1 kHz and 3 kHz will result in a frequency of about 1.6 kHz versus the actual value of 1591.55 Hz. The phase plot of Fig. 21.91 reveals a phase shift of about 83° at f = 100 Hz, 45° at f_c = 1.6 kHz, and about 10° at 10 kHz, supporting the plot of Fig. 21.24 of the text.

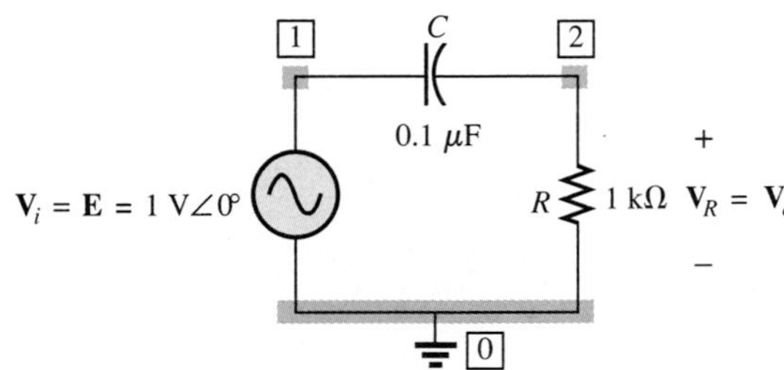

FIG. 21.88
Applying PSpice (DOS) to a high-pass R-C filter.

```
Chapter 21 - High Pass R-C Filter

****      CIRCUIT DESCRIPTION

******************************************************************************

V 1 0 AC 1 0
C 1 2 0.1UF
R 2 0 1K
.AC LIN 101 100H 10KH
.PROBE
.OPTIONS NOPAGE
.END
```

FIG. 21.89
Input file for the network of Fig. 21.88.

PSpice (Windows)

Using schematics, the response to the double-tuned filter of Example 21.9 will be verified. The schematic appears in Fig. 21.92, with **VSIN** chosen as the source and **AC** = **VAMPL** = 1 V and **DC** = **VOFF** = 0 V and **FREQ.** = 100 Hz. In addition **PKGREF** = E. The **AC Sweep** was set at 1000 Pts. for a clear continuous plot with the frequency range from 100 Hz to 1E6. **Automatically Run Probe After Simulation** was chosen and **Simulation** was activated. Using the sequence **Trace-Add-V(RL:1),** the bottom plot of Fig. 21.93 was obtained with a log scale for the horizontal axis. Choosing **Plot-X, Axis-Settings-Linear**

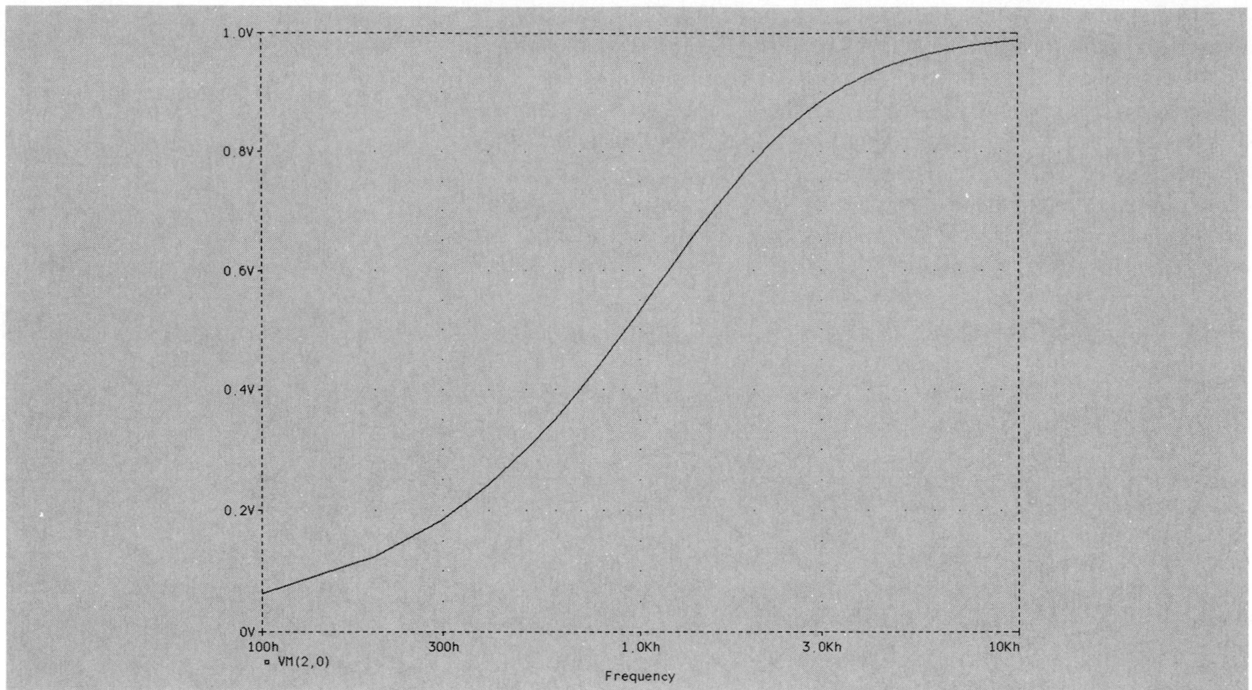

FIG. 21.90
Output voltage versus frequency for the high-pass filter of Fig. 21.88.

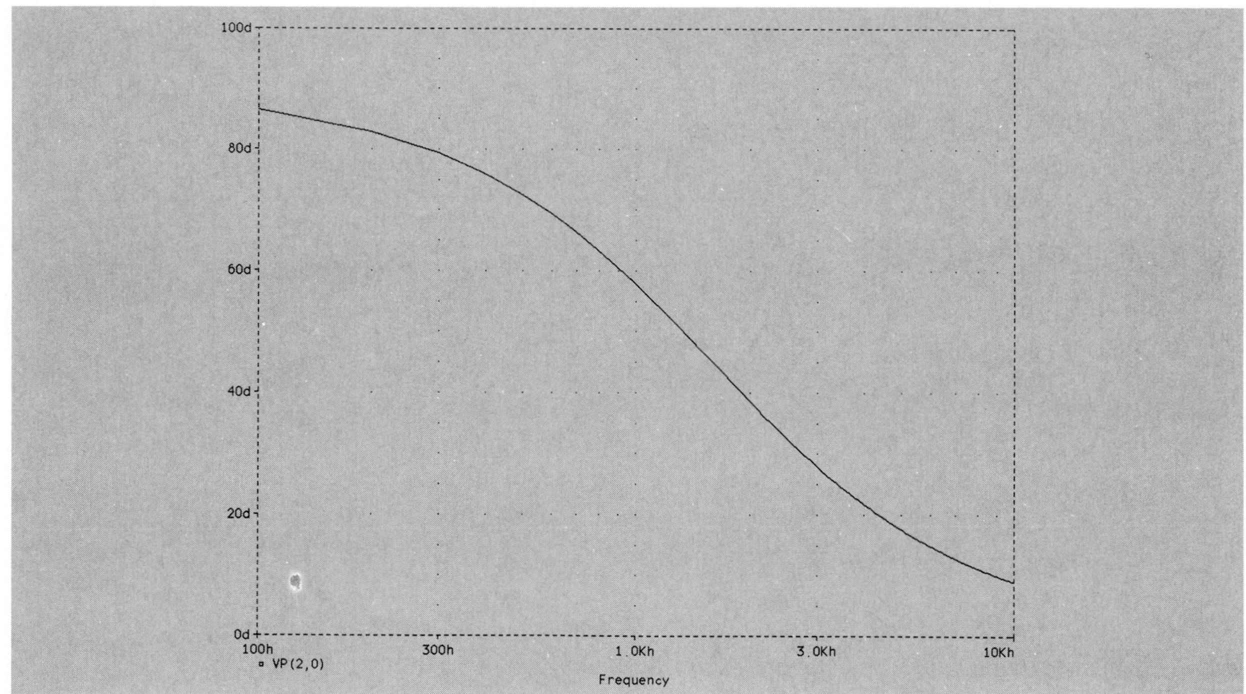

FIG. 21.91
Phase angle versus frequency for the output voltage of Fig. 21.88.

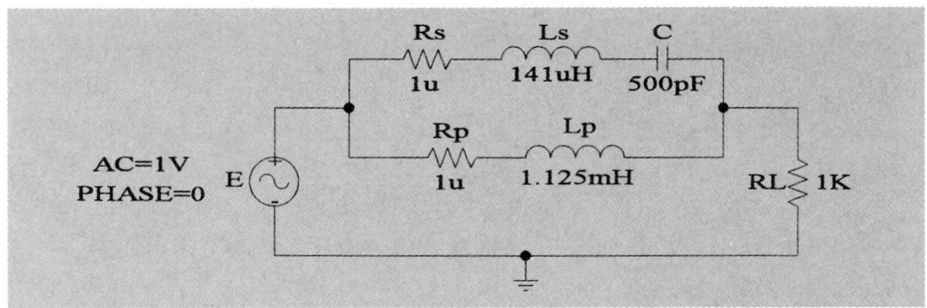

FIG. 21.92
Schematic for the double-tuned filter of Example 21.9.

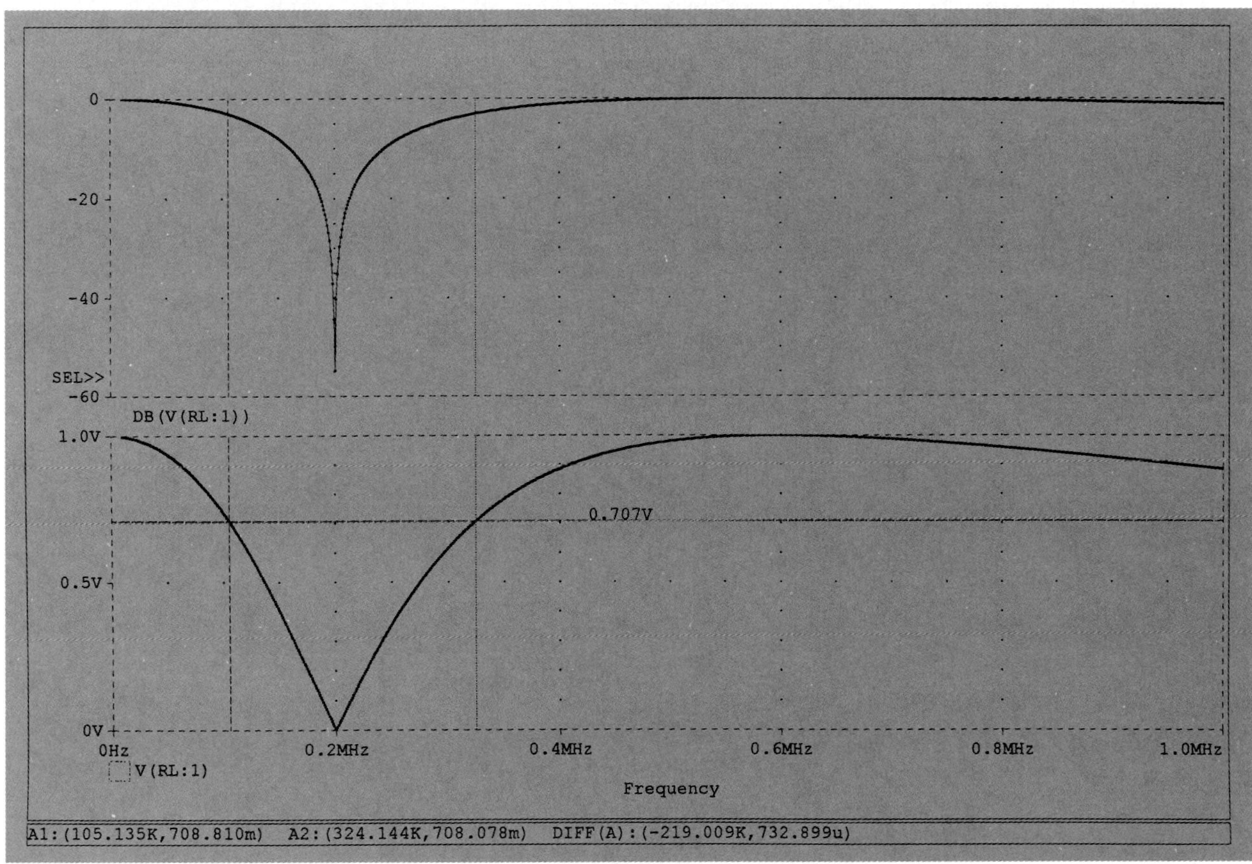

FIG. 21.93
Frequency response for the output voltage of the double-tuned network of Fig. 21.92.

resulted in the desired linear scale. The results clearly validate the results of Example 21.9, with a rejection frequency of 200 kHz and an acceptance frequency of 600 kHz. Using the **Cursor** option set at 0.707, the bandwidth of rejection frequencies and the cutoff frequencies can be determined. The dialog box in the bottom left of the plot reveals that the lower cutoff frequency is 105.14 kHz and the upper cutoff frequency is 324.14 kHz with a bandwidth of 219 kHz. The 0.708 level was the closest available to 0.707 for the number of data points chosen.

A dB plot can be obtained by simply following the sequence **Plot-Add Plot-Trace-Add-DB(V(RL:1)).** The vertical axis was changed to a minimum of 60 dB with **Plot-Y-Axis-Settings-User Defined −60.**

The result is the plot appearing in the top of Fig. 21.93 with about −55 dB at 200 kHz. Note also how the curve is at −3 dB at the cutoff frequencies defined by the **Cursors** in the plot below.

The schematic of Fig. 21.94 will verify the conclusions of the crossover network appearing as Fig. 21.83 in the text. Again the **VSIN** source was chosen, with **FREQ** = 10 Hz and the other attributes as above. The **GLOBAL** option under **port.slb** established the indicated points as connected together. Any **GLOBAL** point will be connected with the E (in this case, all three points). The **AC Sweep** was set with **Total Pts.** at 10,000 to maximize the accuracy of the plot, with the **Start Freq.** at 10 Hz and the **End Freq.** at 100 kHz. Note in this case

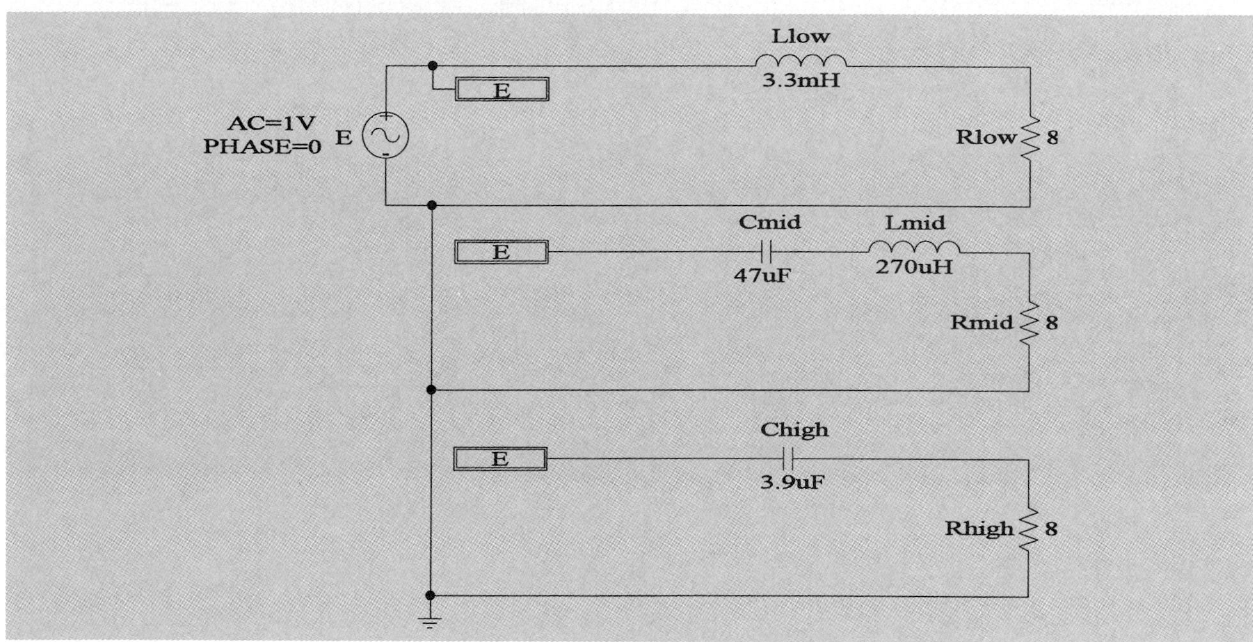

FIG. 21.94
Verifying the response of the crossover network of Fig. 21.83 using PSpice (Windows).

the absence of any 1-$\mu\Omega$ resistors in series with the coil because of the presence of the series 8-Ω load. Inductive elements simply require a resistance in series at some point in the network. Choosing **Automatically Run Probe After Simulation** followed by **Simulation-Trace-Add-Alias Names-V(Rmid:1)** will result in the plot appearing at the bottom of Fig. 21.95. The plots of **V(Rhigh:1)** and **V(Rlow:1)** were then added with the **Trace-Add** sequence. Note that the curves have the filter response expected, with crossover frequencies defined at 392.83 Hz and 5.14 kHz, which compare well with the desired 400 Hz and 5 kHz. (Recall that commercial values for the elements were chosen, which will certainly have an effect on the resulting crossover frequencies.) A dB plot for the midband region was obtained with **Plot-Add Plot-Trace-Add-DB(V(Rmid:1).** The vertical scale was changed to −20 dB to 0 using the **Plot-Y-Axis Settings** following the addition of the two other traces. Note from the added artwork that the waveforms do drop off at −6 dB per octave, as called for in the design.

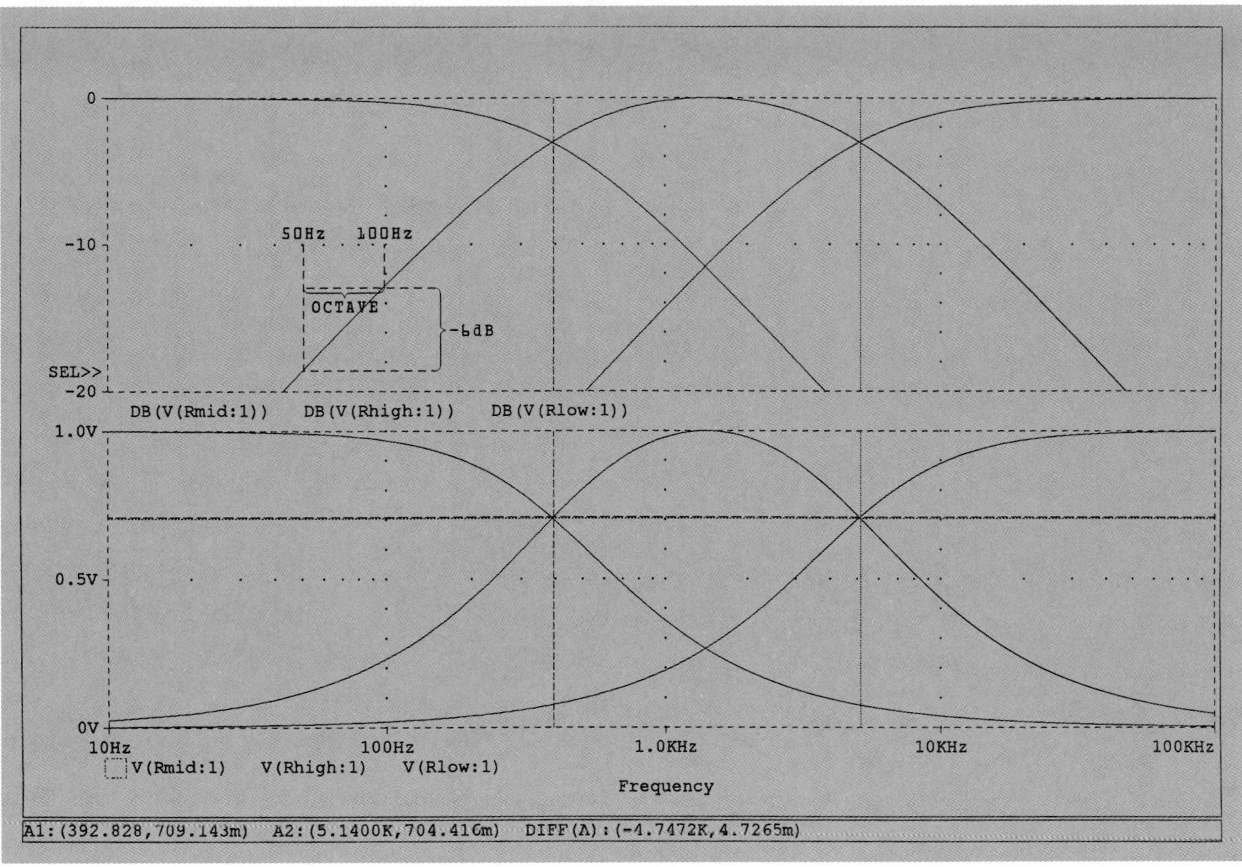

FIG. 21.95
Frequency response for the crossover network of Fig. 21.94.

PROBLEMS

SECTION 21.1 Logarithms

1. **a.** Determine the frequencies (in kHz) at the points indicated on the plot of Fig. 21.96(a).
   b. Determine the voltages (in mV) at the points indicated on the plot of Fig. 21.96(b).

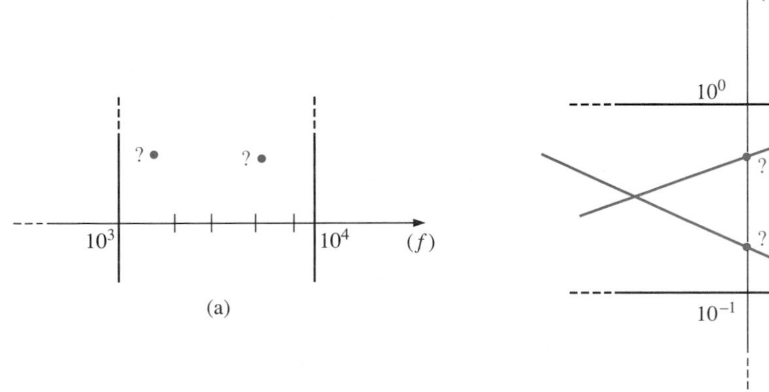

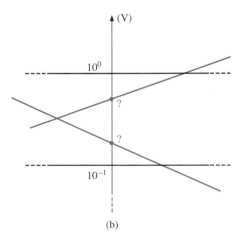

FIG. 21.96
Problem 1.

SECTION 21.2 Properties of Logarithms

2. Determine $\log_{10} x$ for each value of x.

a. 100,000 **b.** 0.0001
c. 10^8 **d.** 10^{-6}
e. 20 **f.** 8643.4
g. 56,000 **h.** 0.318

3. Given $N = \log_{10} x$, determine x for each value of N.

a. 3 **b.** 12
c. 0.2 **d.** 0.04
e. 10 **f.** 3.18
g. 1.001 **h.** 6.1

4. Determine $\log_e x$ for each value of x.

a. 100,000 **b.** 0.0001
c. 20 **d.** 8643.4

Compare with the solutions to Problem 2.

5. Determine $\log_{10} 48 = \log_{10}(8)(6)$ and compare to $\log_{10} 8 + \log_{10} 6$.

6. Determine $\log_{10} 0.2 = \log_{10} 18/90$ and compare to $\log_{10} 18 - \log_{10} 90$.

7. Verify that $\log_{10} 0.5$ is equal to $-\log_{10} 1/0.5 = -\log_{10} 2$.

8. Find $\log_{10}(3)^3$ and compare with $3 \log_{10} 3$.

SECTION 21.3 Decibels

9. **a.** Determine the number of bels that relate power levels of $P_2 = 280$ mW and $P_1 = 4$ mW.

b. Determine the number of decibels for the power levels of part (a) and compare results.

10. A power level of 100 W is 6 dB above what power level?

11. If a 2-W speaker is replaced by one with a 40-W output, what is the increase in decibel level?

12. Determine the dB_m level for an output power of 120 mW.

13. Find the dB_v gain of an amplifier that raises the voltage level from 0.1 mV to 8.4 V.

14. Find the output voltage of an amplifier if the applied voltage is 20 mV and a dB_v gain of 22 dB is attained.

15. If the sound pressure level is increased from 0.001 μbar to 0.016 μbar, what is the increase in dB_s level?

16. What is the required increase in acoustical power to raise a sound level from that of quiet music to very loud music? Use Fig. 21.5.

17. **a.** Using semilog paper, plot X_L versus frequency for a 10-mH coil and a frequency range of 100 Hz to 1 MHz. Choose the best vertical scaling for the range of X_L.

b. Repeat part (a) using log-log graph paper. Compare to the results of part (a). Which plot is more informative?

c. Using semilog paper, plot X_C versus frequency for a 1-μF capacitor and a frequency range of 10 Hz to 100 kHz. Again choose the best vertical scaling for the range of X_C.

d. Repeat part (a) using log-log graph paper. Compare to the results of part (c). Which plot is more informative?

18. **a.** For the meter of Fig. 21.6, find the power delivered to a load for an 8-dB reading.

b. Repeat part (a) for a −5-dB reading.

SECTION 21.5 *R-C* Low-Pass Filter

19. For the *R-C* low-pass filter of Fig. 21.97:

a. Sketch $A_v = V_o/V_i$ versus frequency using a log scale for the frequency axis. Determine $A_v = V_o/V_i$ at $0.1f_c$, $0.5f_c$, f_c, $2f_c$, and $10f_c$.

b. Sketch the phase plot of θ versus frequency, where θ is the angle by which $\mathbf{V}_o$ leads $\mathbf{V}_i$. Determine θ at $f = 0.1f_c$, $0.5f_c$, f_c, $2f_c$, and $10f_c$.

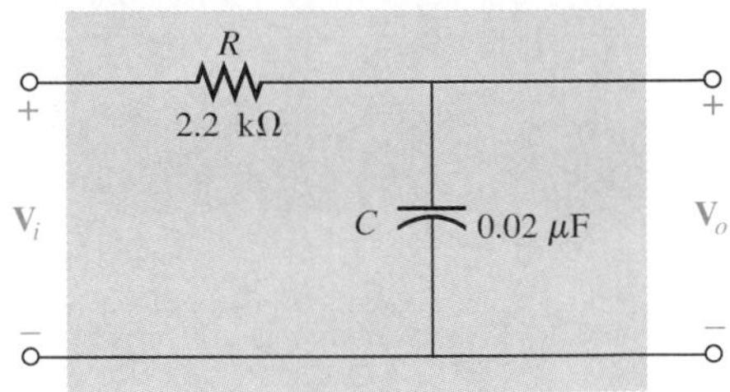

FIG. 21.97
Problems 19 and 54.

***20.** For the network of Fig. 21.98:

a. Determine V_o at a frequency one octave above the critical frequency.

b. Determine V_o at a frequency one decade below the critical frequency.

c. Do the levels of parts (a) and (b) verify the expected frequency plot of V_o versus frequency for the filter?

21. Design an *R-C* low-pass filter to have a cutoff frequency of 500 Hz using a resistor of 1.2 kΩ. Then sketch the resulting magnitude and phase plot for a frequency range of $0.1f_c$ to $10f_c$.

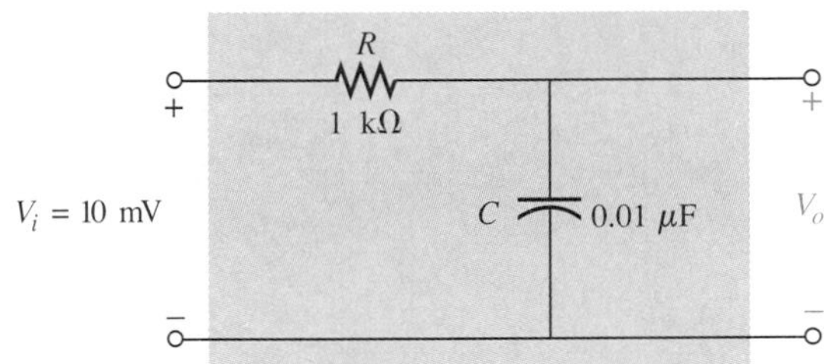

FIG. 21.98
Problem 20.

22. For the low-pass filter of Fig. 21.99:

a. Determine f_c.

b. Find $A_v = V_o/V_i$ at $f = 0.1f_c$ and compare to the maximum value of 1 for the low-frequency range.

c. Find $A_v = V_o/V_i$ at $f = 10f_c$ and compare to the minimum value of 0 for the high-frequency range.

d. Determine the frequency at which $A_v = 0.01$ or $V_o = \frac{1}{100} V_i$.

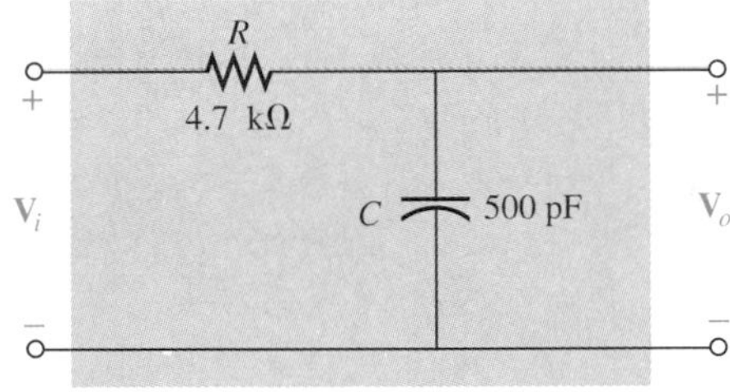

FIG. 21.99
Problem 22.

SECTION 21.6 *R-C* High-Pass Filter

23. For the *R-C* high-pass filter of Fig. 21.100:

a. Sketch $A_v = V_o/V_i$ versus frequency using a log scale for the frequency axis. Determine $A_v = V_o/V_i$ at f_c, one octave above and below f_c, and one decade above and below f_c.

b. Sketch the phase plot of θ versus frequency, where θ is the angle by which $\mathbf{V}_o$ leads $\mathbf{V}_i$. Determine θ at the same frequencies noted in part (a).

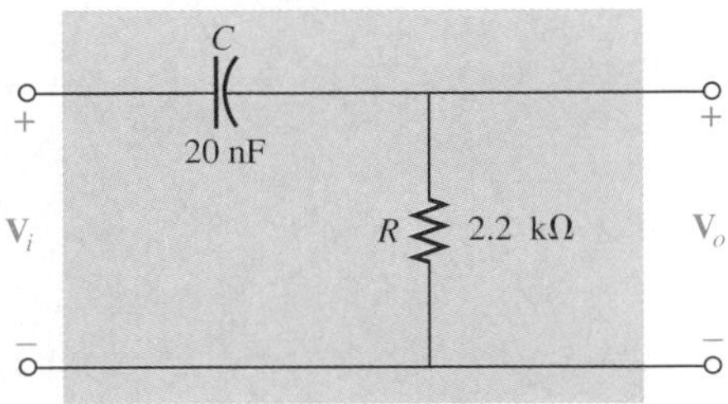

FIG. 21.100
Problem 23.

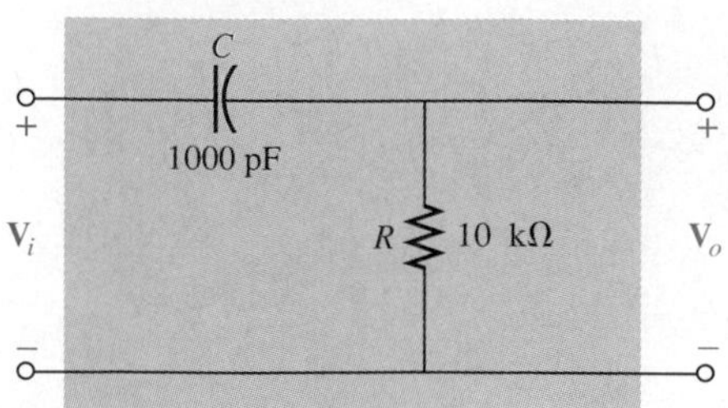

FIG. 21.101
Problem 24.

24. For the network of Fig. 21.101:
   a. Determine $A_v = V_o/V_i$ at $f = f_c$ for the high-pass filter.
   b. Determine $A_v = V_o/V_i$ at two octaves above f_c. Is the rise in V_o significant from the $f = f_c$ level?
   c. Determine $A_v = V_o/V_i$ at two decades above f_c. Is the rise in V_o significant from the $f = f_c$ level?
   d. If $V_i = 10$ mV, what is the power delivered to R at the critical frequency?

25. Design a high-pass R-C filter to have a cutoff or corner frequency of 2 kHz, given a capacitor of 0.1 μF. Choose the closest commercial value for R and then recalculate the resulting corner frequency. Sketch the normalized gain $A_v = V_o/V_i$ for a frequency range of $0.1f_c$ to $10f_c$.

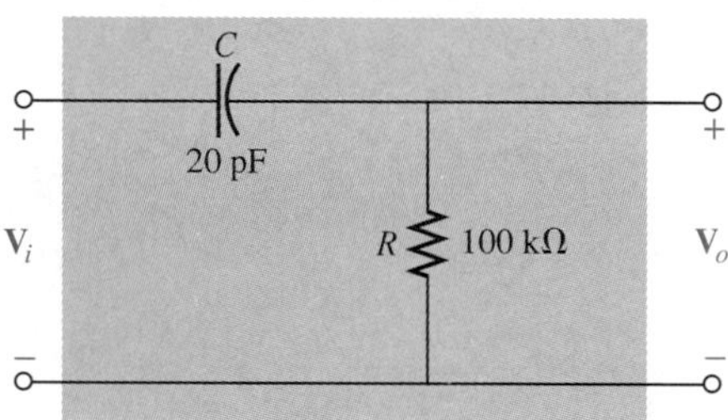

FIG. 21.102
Problems 26 and 57.

26. For the high-pass filter of Fig. 21.102:
   a. Determine f_c.
   b. Find $A_v = V_o/V_i$ at $f = 0.01f_c$ and compare to the minimum level of 0 for the low-frequency region.
   c. Find $A_v = V_o/V_i$ at $f = 100f_c$ and compare to the maximum level of 1 for the high-frequency region.
   d. Determine the frequency at which $V_o = \frac{1}{2}V_i$.

SECTION 21.7 Pass-Band Filters

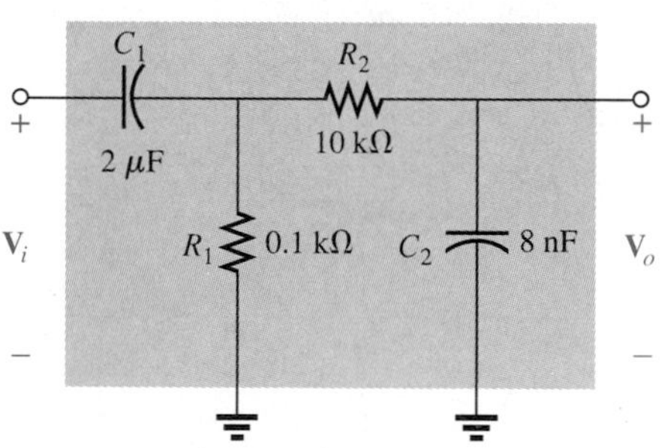

FIG. 21.103
Problems 27, 28, and 55.

27. For the pass-band filter of Fig. 21.103:
   a. Sketch the frequency response of $A_v = V_o/V_i$ against a log scale extending from 10 Hz to 10 kHz.
   b. What are the bandwidth and the center frequency?

*28. Design a pass-band filter such as the one appearing in Fig. 21.103 to have a low cutoff frequency of 4 kHz and a high cutoff frequency of 80 kHz.

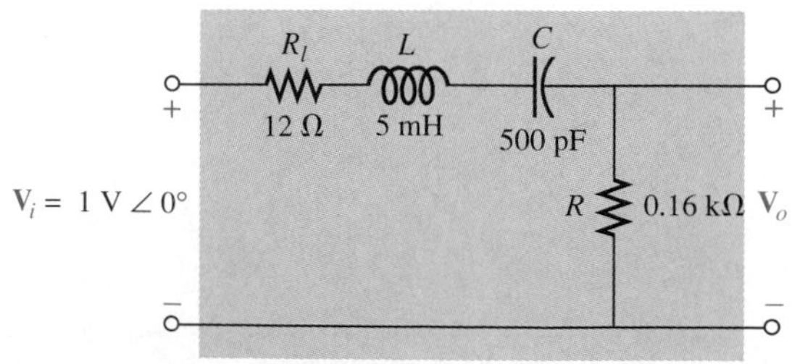

FIG. 21.104
Problem 29.

29. For the pass-band filter of Fig. 21.104:
   a. Determine f_s.
   b. Calculate Q_s and the bandwidth for $\mathbf{V}_o$.
   c. Sketch $A_v = V_o/V_i$ for a frequency range of 1 kHz to 1 MHz.
   d. Find the magnitude of V_o at $f = f_s$ and the cutoff frequencies.

30. For the pass-band filter of Fig. 21.105:

a. Determine the frequency response of $A_v = V_o/V_i$ for a frequency range of 100 Hz to 1 MHz.

b. Find the quality factor Q_p and the bandwidth of the response.

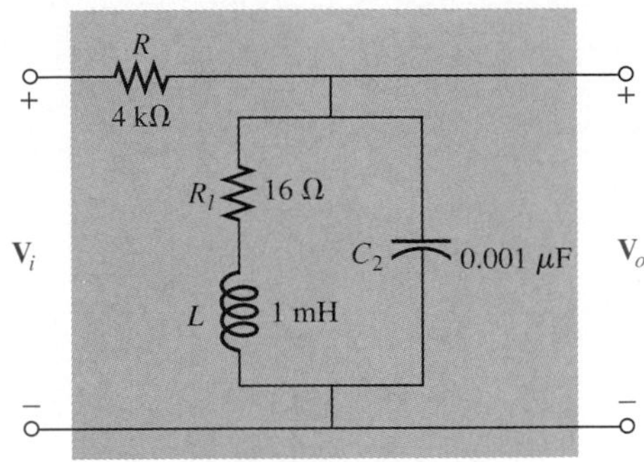

FIG. 21.105
Problems 30 and 58.

SECTION 21.8 Stop-Band Filters

***31.** For the stop-band filter of Fig. 21.106:

a. Determine Q_s.

b. Find the bandwidth and half-power frequencies.

c. Sketch the frequency characteristics of $A_v = V_o/V_i$.

d. What is the effect on the curve of part (c) if a load of 2 kΩ is applied?

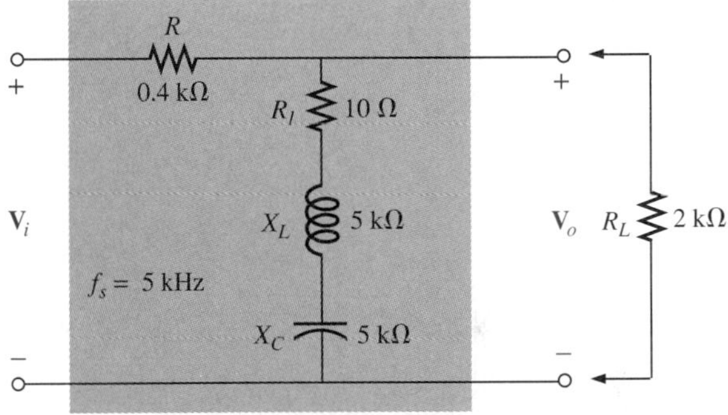

FIG. 21.106
Problem 31.

***32.** For the pass-band filter of Fig. 21.107:

a. Determine Q_p ($R_L = \infty$ Ω, an open circuit).

b. Sketch the frequency characteristics of $A_v = V_o/V_i$.

c. Find Q_p (loaded) for R_L = 100 kΩ, and indicate the effect of R_L on the characteristics of part (b).

d. Repeat part (c) for R_L = 20 kΩ.

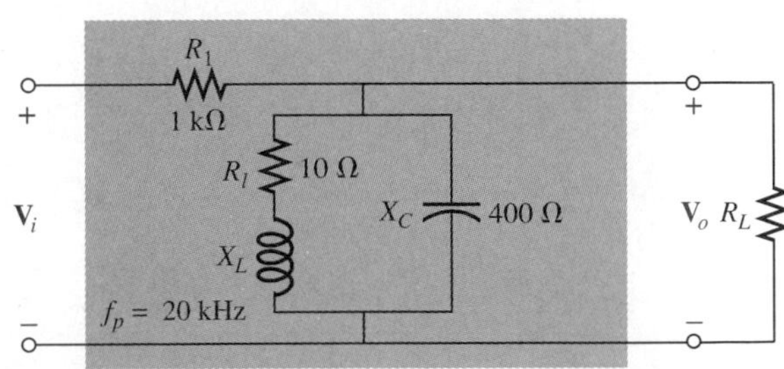

FIG. 21.107
Problem 32.

SECTION 21.9 Double-Tuned Filter

33. a. For the network of Fig. 21.43(a), if $L_p = 400\ \mu\text{H}$ ($Q > 10$), $L_s = 60\ \mu\text{H}$, and $C = 120$ pF, determine the rejected and accepted frequencies.

b. Sketch the response curve for part (a).

34. a. For the network of Fig. 21.43(b), if the rejected frequency is 30 kHz and the accepted is 100 kHz, determine the values of L_s and L_p ($Q > 10$) for a capacitance of 200 pF.

b. Sketch the response curve for part (a).

SECTION 21.10 Bode Plots

35. a. Sketch the idealized Bode plot for $A_v = V_o/V_i$ for the high-pass filter of Fig. 21.108.

b. Using the results of part (a), sketch the actual frequency response for the same frequency range.

c. Determine the decibel level at f_c, $\frac{1}{2}f_c$, $2f_c$, $\frac{1}{10}f_c$, and $10f_c$.

d. Determine the gain $A_v = V_o/V_i$ as $f = f_c$, $\frac{1}{2}f_c$, and $2f_c$.

e. Sketch the phase response for the same frequency range.

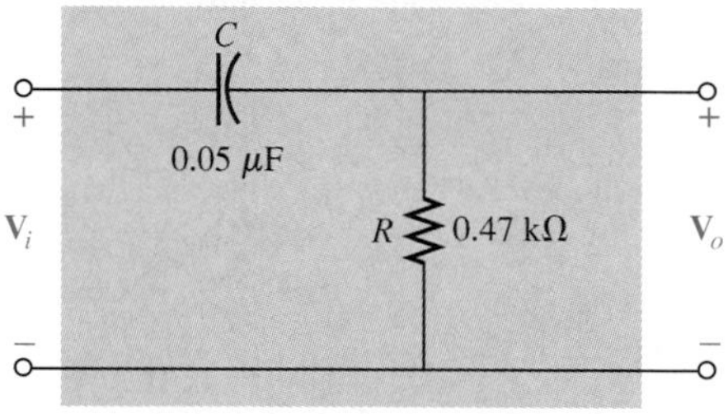

FIG. 21.108
Problem 35.

***36. a.** Sketch the response of the magnitude of V_o (in terms of V_i) versus frequency for the high-pass filter of Fig. 21.109.

b. Using the results of part (a), sketch the response $A_v = V_o/V_i$ for the same frequency range.

c. Sketch the idealized Bode plot.

d. Sketch the actual response, indicating the dB difference between the idealized and actual response at $f = f_c$, $0.5f_c$, and $2f_c$.

e. Determine $A_{v_{\text{dB}}}$ at $f = 1.5f_c$ from the plot of part (d) and then determine the corresponding magnitude of $A_v = V_o/V_i$.

f. Sketch the phase response for the same frequency range (the angle by which $\mathbf{V}_o$ leads $\mathbf{V}_i$).

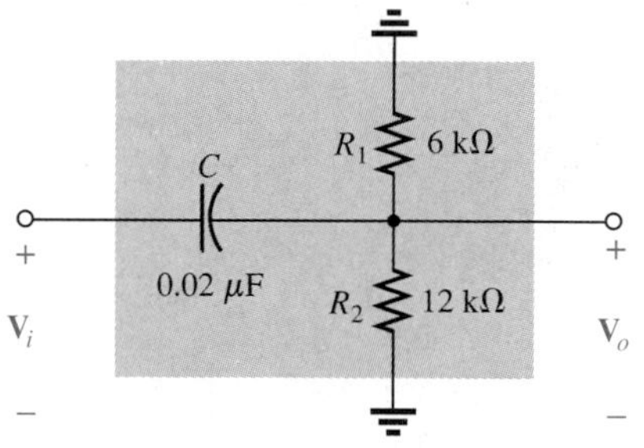

FIG. 21.109
Problem 36.

37. a. Sketch the idealized Bode plot for $A_v = V_o/V_i$ for the low-pass filter of Fig. 21.110.

b. Using the results of part (a), sketch the actual frequency response for the same frequency range.

c. Determine the decibel level at f_c, $\frac{1}{2}f_c$, $2f_c$, $\frac{1}{10}f_c$, and $10f_c$.

d. Determine the gain $A_v = V_o/V_i$ at $f = f_c$, $\frac{1}{2}f_c$, and $2f_c$.

e. Sketch the phase response for the same frequency range.

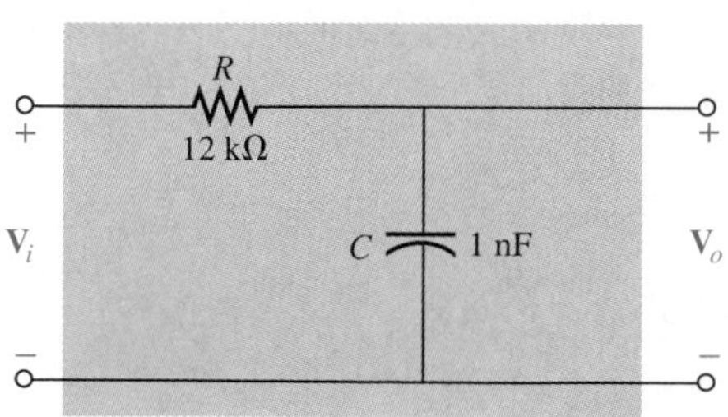

FIG. 21.110
Problem 37.

*38. a. Sketch the response of the magnitude of V_o (in terms of V_i) versus frequency for the low-pass filter of Fig. 21.111.

b. Using the results of part (a), sketch the response $A_v = V_o/V_i$ for the same frequency range.

c. Sketch the idealized Bode plot.

d. Sketch the actual response indicating the dB difference between the idealized and actual response at $f = f_c$, $0.5f_c$, and $2f_c$.

e. Determine $A_{v_{dB}}$ at $f = 0.25f_c$ from the plot of part (d) and then determine the corresponding magnitude of $A_v = V_o/V_i$.

f. Sketch the phase response for the same frequency range (the angle by which $\mathbf{V}_o$ leads $\mathbf{V}_i$).

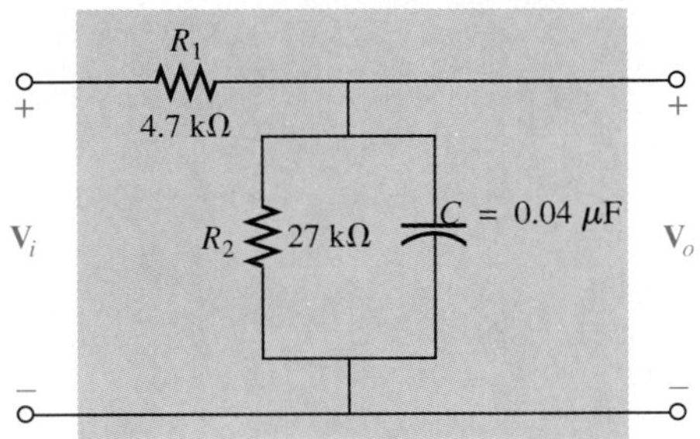

FIG. 21.111
Problem 38.

SECTION 21.11 Sketching the Bode Response

39. For the filter of Fig. 21.112:

a. Sketch the curve of $A_{v_{dB}}$ versus frequency using a log scale.

b. Sketch the curve of θ versus frequency for the same frequency range as in part (a).

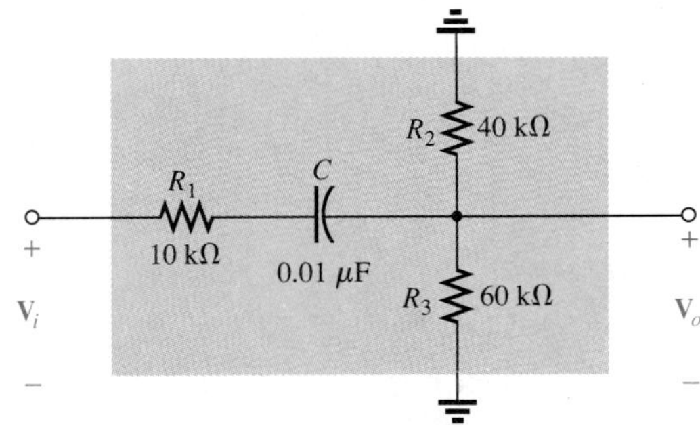

FIG. 21.112
Problem 39.

*40. For the filter of Fig. 21.113:

a. Sketch the curve of $A_{v_{dB}}$ versus frequency using a log scale.

b. Sketch the curve of θ versus frequency for the same frequency range as in part (a).

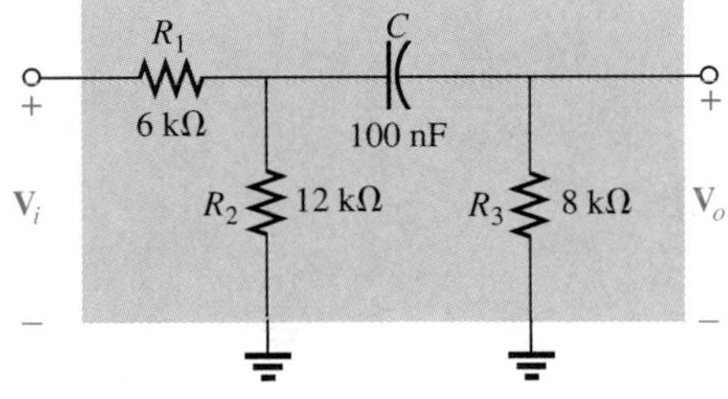

FIG. 21.113
Problem 40.

SECTION 21.12 Low-Pass Filter with Limited Attenuation

41. For the filter of Fig. 21.114:

a. Sketch the curve of $A_{v_{dB}}$ versus frequency using the idealized Bode plots as a guide.

b. Sketch the curve of θ versus frequency.

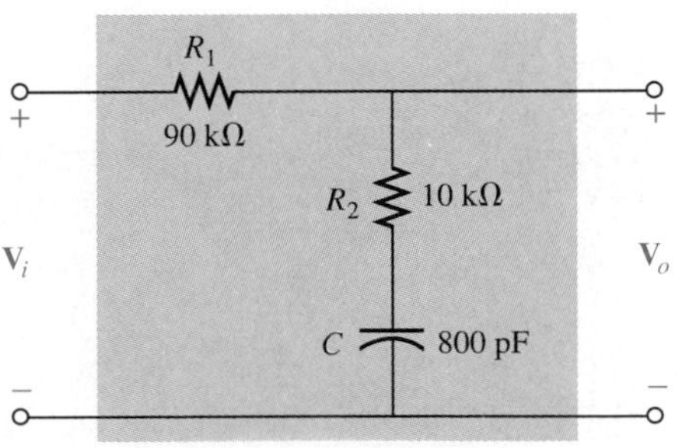

FIG. 21.114
Problem 41.

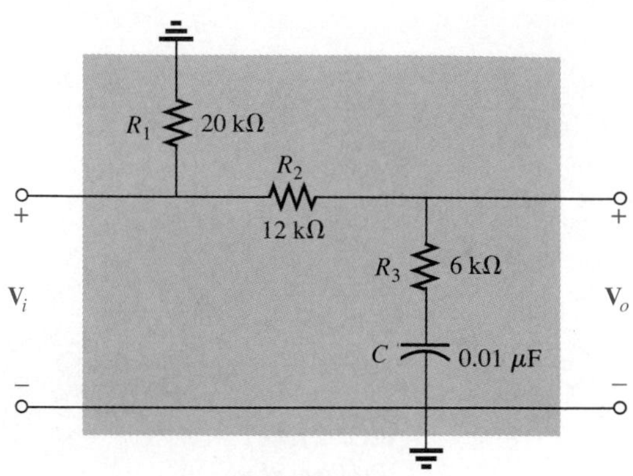

FIG. 21.115
Problem 42.

***42.** For the filter of Fig. 21.115:

a. Sketch the curve of $A_{v_{dB}}$ versus frequency using the idealized Bode plots as a guide.

b. Sketch the curve of θ versus frequency.

SECTION 21.13 High-Pass Filter with Limited Attenuation

43. For the filter of Fig. 21.116:

a. Sketch the curve of $A_{v_{dB}}$ versus frequency using the idealized Bode plots as an envelope for the actual response.

b. Sketch the curve of θ (the angle by which $\mathbf{V}_o$ leads $\mathbf{V}_i$) versus frequency.

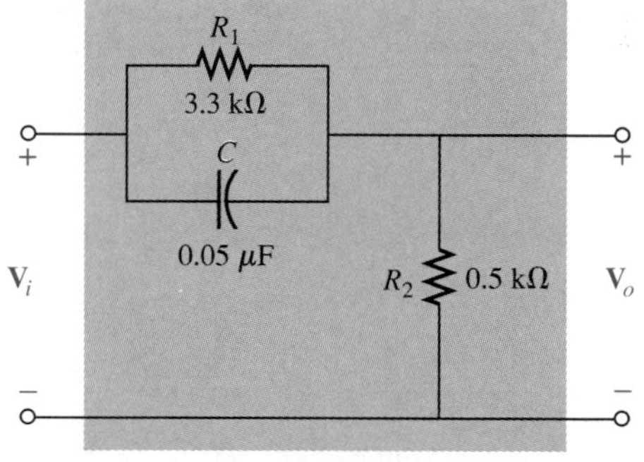

FIG. 21.116
Problem 43.

***44.** For the filter of Fig. 21.117:

a. Sketch the curve of $A_{v_{dB}}$ versus frequency using the idealized Bode plots as an envelope for the actual response.

b. Sketch the curve of θ (the angle by which $\mathbf{V}_o$ leads $\mathbf{V}_i$) versus frequency.

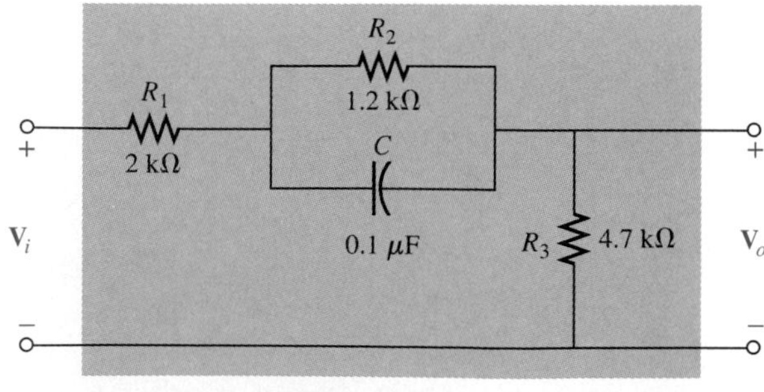

FIG. 21.117
Problems 44 and 56.

SECTION 21.14 Other Properties and a Summary Table

45. A bipolar transistor amplifier has the following gain:

$$\mathbf{A}_v = \frac{160}{\left(1 - j\dfrac{100\text{ Hz}}{f}\right)\left(1 - j\dfrac{130\text{ Hz}}{f}\right)\left(1 + j\dfrac{f}{20\text{kHz}}\right)\left(1 + j\dfrac{f}{50\text{ kHz}}\right)}$$

a. Sketch the normalized Bode response $A'_{v_{\text{dB}}} = (A_v/A_{v_{\max}})|_{\text{dB}}$ and determine the bandwidth of the amplifier. Be sure to note the corner frequencies.

b. Sketch the phase response and determine a frequency where the phase angle is relatively close to 45°.

46. A JFET transistor amplifier has the following gain:

$$\mathbf{A}_v = \frac{-5.6}{\left(1 - j\dfrac{10\text{ Hz}}{f}\right)\left(1 - j\dfrac{45\text{ Hz}}{f}\right)\left(1 - j\dfrac{68\text{ Hz}}{f}\right)\left(1 + j\dfrac{f}{23\text{ kHz}}\right)\left(1 + j\dfrac{f}{50\text{ kHz}}\right)}$$

a. Sketch the normalized Bode response $A'_{v_{\text{dB}}} = (A_v/A_{v_{\max}}|\text{dB})$ and determine the bandwidth of the amplifier. When you normalize, be sure the maximum value of A'_v is +1. Clearly indicate the cutoff frequencies on the plot.

b. Sketch the phase response and note the regions of greatest change in phase angle. How do the regions correspond to the frequencies appearing in the function $\mathbf{A}_v$?

47. A transistor amplifier has a midband gain of −120, a high cutoff frequency of 36 kHz, and a bandwidth of 35.8 kHz. In addition, the actual response is also about −15 dB at $f = 50$ Hz. Write the transfer function $\mathbf{A}_v$ for the amplifier.

48. Sketch the Bode plot of the following function:

$$\mathbf{A}_v = \frac{0.05}{0.05 - j\,100/f}$$

49. Sketch the Bode plot of the following function:

$$\mathbf{A}_v = \frac{200}{200 + j\,0.1f}$$

50. Sketch the Bode plot of the following function:

$$\mathbf{A}_v = \frac{jf/1000}{(1 + jf/1000)(1 + jf/10{,}000)}$$

***51.** Sketch the Bode plot of the following function:

$$\mathbf{A}_v = \frac{(1 + jf/1000)(1 + jf/2000)}{(1 + jf/3000)^2}$$

***52.** Sketch the Bode plot of the following function (note the presence of ω rather than f):

$$\mathbf{A}_v = \frac{40(1 + j\,0.001\omega)}{(j\,0.001\omega)(1 + j\,0.0002\omega)}$$

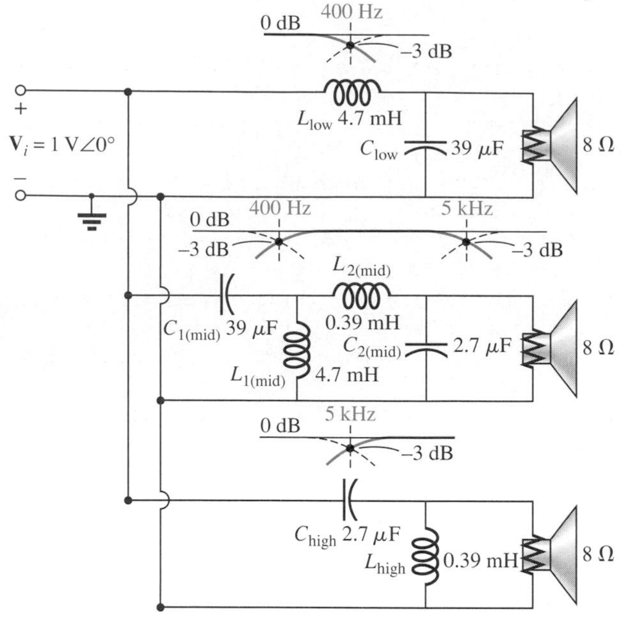

FIG. 21.118
Problems 53 and 60.

SECTION 21.15 Crossover Networks

***53.** The three-way crossover network of Fig. 21.118 has a 12-dB rolloff at the cutoff frequencies.

a. Determine the ratio V_o/V_i for the woofer and tweeter at the cutoff frequencies of 400 Hz and 5 kHz, respectively, and compare to the desired level of 0.707.

b. Calculate the ratio V_o/V_i for the woofer and tweeter at a frequency of 3 kHz, where the mid-range speaker is designed to predominate.

c. Determine the ratio V_o/V_i for the mid-range speaker at a frequency of 3 kHz and compare to the desired level of 1.

SECTION 21.16 Computer Analysis

PSpice (DOS)

54. Setting $\mathbf{V}_i = 1\text{ V} \angle 0°$, write the input file to determine the frequency response of the magnitude of $\mathbf{V}_o$ for the network of Fig. 21.97.

55. Setting $\mathbf{V}_i = 1\text{ V} \angle 0°$, write the input file to obtain the frequency response of $\mathbf{V}_o$ for the magnitude of the network of Fig. 21.103.

56. Setting $\mathbf{V}_i = 1\text{ V} \angle 0°$, write the input file to obtain the frequency response of the magnitude of the output voltage $\mathbf{V}_o$ for the network of Fig. 21.117.

PSpice (Windows)

57. Using schematics, obtain the magnitude and phase response versus frequency for the network of Fig. 21.102.

58. Using schematics, obtain the magnitude and phase response versus frequency for the network of Fig. 21.105.

***59.** Obtain the dB and phase plots for the network of Fig. 21.75 and compare with the plots of Figs. 21.76 and 21.77.

***60.** Using schematics, obtain the magnitude and dB plot versus frequency for each filter of Fig. 21.118 and verify that the curves drop off at 12 dB per octave.

Programming Language (C++, BASIC, PASCAL, etc.)

61. Write a program that will tabulate the gain of Eq. (21.14) versus frequency for a frequency range extending from $0.1f_1$ to $2f_1$ in increments of $0.1f_1$. Note whether $f = f_1$ when $V_o/V_i = 0.707$. Use $R = 1\text{ k}\Omega$ and $C = 500\text{ pF}$.

62. Write a program to tabulate $A_{v_{dB}}$ as determined from Eq. (21.34) and $A_{v_{dB}}$ as calculated by Eq. (21.35). For a frequency range extending from $0.01f_1$ to f_1 in increments of $0.01f_1$, compare the magnitudes, and note whether the values are closer when $f \ll f_1$ and whether $A_{v_{dB}} = -3\text{ dB}$ at $f = f_1$ for Eq. (21.34) and zero for Eq. (21.35).

GLOSSARY

Active filter A filter that employs active devices such as transistors or operational amplifiers in combination with *R*, *L*, and *C* elements.

Bode plot A plot of the frequency response of a system using straight-line segments called asymptotes.

Decibel A unit of measurement used to compare power levels.

Double-tuned filter A network having both a pass-band and a stop-band region.

Filter Networks designed to either pass or reject the transfer of signals at certain frequencies to a load.

High-pass filter A filter designed to pass high frequencies and reject low frequencies.

Log-log paper Graph paper with vertical and horizontal log scales.

Low-pass filter A filter designed to pass low frequencies and reject high frequencies.

Microbar A unit of measurement for sound pressure levels that permits comparing audio levels on a dB scale.

Pass-band (band-pass) filter A network designed to pass signals within a particular frequency range.

Passive filter A filter constructed of series, parallel, or series-parallel *R*, *L*, and *C* elements.

Semilog paper Graph paper with one log scale and one linear scale.

Stop-band filter A network designed to reject (block) signals within a particular frequency range.

22

Pulse Waveforms and the *R-C* Response

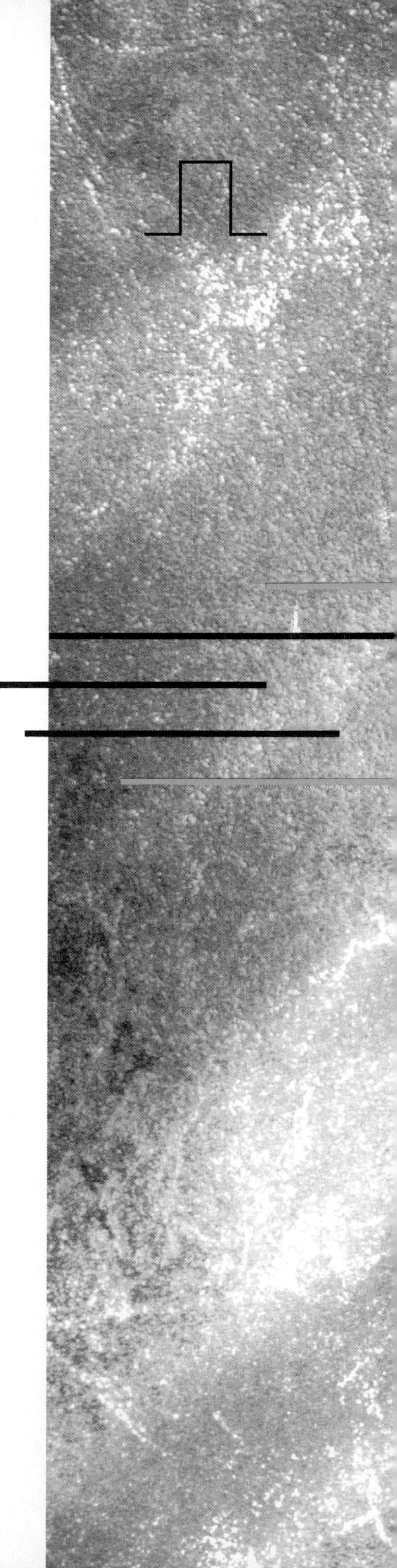

22.1 INTRODUCTION

Our analysis thus far has been limited to alternating waveforms that vary in a sinusoidal manner. This chapter will introduce the basic terminology associated with the pulse waveform and examine the response of an *R-C* circuit to a square-wave input. The importance of the pulse waveform to the electrical/electronics industry cannot be overstated. A vast array of instrumentation, communication systems, computers, radar systems, and so on, all employ pulse signals to control operation, transmit data, and display information in a variety of formats.

The response of the networks described thus far to a pulse signal is quite different from that obtained for sinusoidal signals. In fact, we will be returning to the dc chapter on capacitors to retrieve a few fundamental concepts and equations that will help us in the analysis to follow. The content of this chapter is quite introductory in nature, designed simply to provide the fundamentals that will be helpful when the pulse waveform is encountered in specific areas of application.

22.2 IDEAL VERSUS ACTUAL

The *ideal* pulse of Fig. 22.1 has vertical sides, sharp corners, and a flat peak characteristic, and starts instantaneously at t_1 and ends just as abruptly at t_2.

The waveform of Fig. 22.1 will be applied in the analysis to follow in this chapter and will probably appear in the initial investigation of areas of application beyond the scope of this text. Once the fundamental operation of a device, package, or system is clearly understood using ideal characteristics, the effect of a *true, actual,* or *practical* pulse must be considered. If an attempt were made to introduce all the differences between an ideal and actual pulse in a single figure, the result would probably be complex and confusing. A number of waveforms will therefore be used to define the critical parameters.

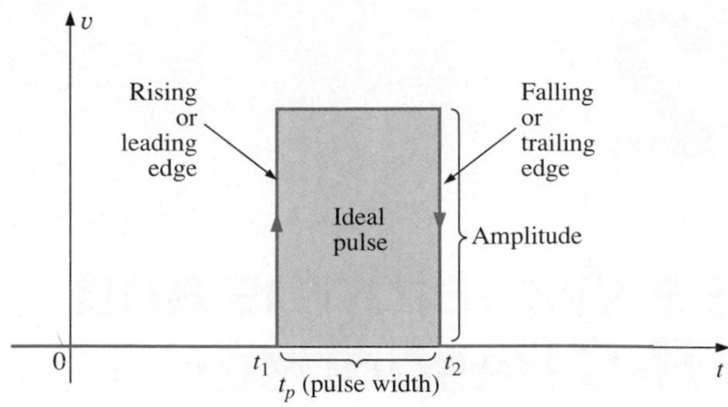

FIG. 22.1
Ideal pulse waveform.

The reactive elements of a network, in their effort to prevent instantaneous changes in voltage (capacitor) and current (inductor), establish a slope to both edges of the pulse waveform, as shown in Fig. 22.2. The *rising* edge of the waveform of Fig. 22.2 is defined as the edge that increases from a lower to a higher level.

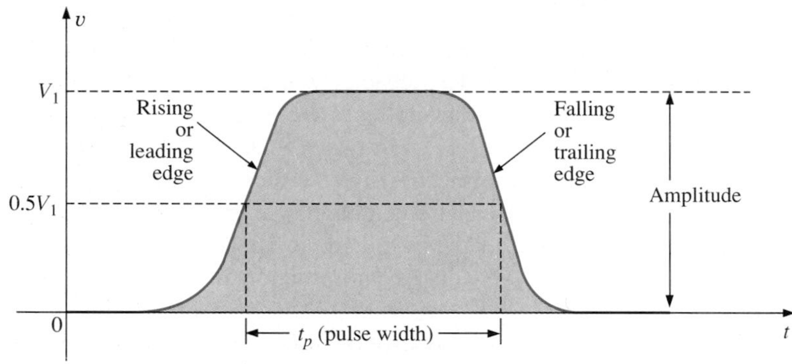

FIG. 22.2
Actual pulse waveform.

The falling edge is defined by the region or edge where the waveform decreases from a higher to a lower level. Since the rising edge is the first to be encountered (closest to t = 0 s), it is also called the leading edge. The falling edge always follows the leading edge and is therefore often called the trailing edge.

Both regions are defined in Figs. 22.1 and 22.2.

Amplitude

For most applications, the amplitude of a pulse waveform is defined as the peak-to-peak value. Of course, if the waveforms all start and return to the zero-volt level, then the peak and peak-to-peak values are synonymous.

For the purposes of this text, the amplitude of a pulse waveform is the peak-to-peak value, as illustrated in Figs. 22.1 and 22.2.

Pulse Width

The pulse width (t_p), or pulse duration, is defined by a pulse level equal to 50% of the peak value.

For the ideal pulse of Fig. 22.1, the pulse width is the same at any level, whereas t_p for the waveform of Fig. 22.2 is a very specific value.

Base-Line Voltage

The base-line voltage (V_b) is the voltage level from which the pulse is initiated.

The waveforms of Figs. 22.1 and 22.2 both have a 0-V base-line voltage. In Fig. 22.3(a) the base-line voltage is 1 V, whereas in Fig. 22.3(b) the base-line voltage is −4 V.

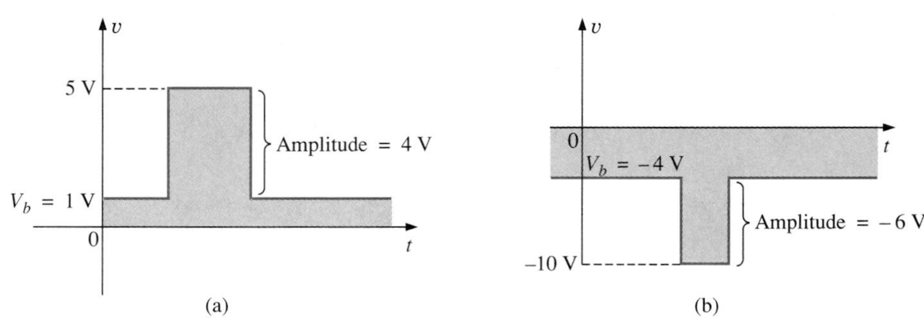

FIG. 22.3
Defining the base-line voltage.

Positive-Going and Negative-Going Pulses

A positive-going pulse increases positively from the base-line voltage, whereas a negative-going pulse increases in the negative direction from the base-line voltage.

The waveform of Fig. 22.3(a) is a positive-going pulse, whereas the waveform of Fig. 22.3(b) is a negative-going pulse.

Even though the base-line voltage of Fig. 22.4 is negative, the waveform is positive-going (with an amplitude of 10 V) since the voltage increased in the positive direction from the base-line voltage.

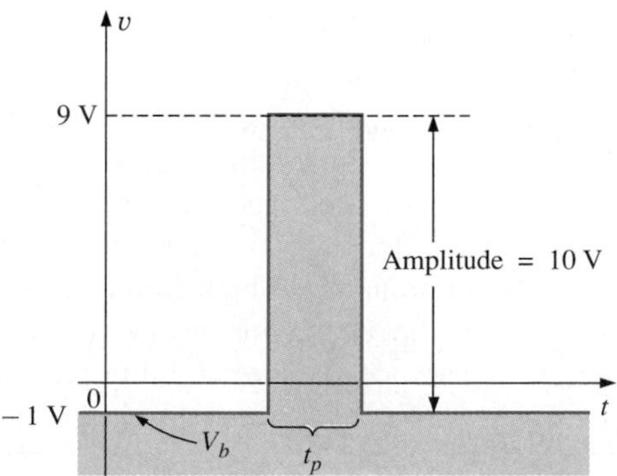

FIG. 22.4
Positive-going pulse.

Rise Time (t_r) and Fall Time (t_f)

Of particular importance is the time required for the pulse to shift from one level to another. The *rounding* (defined in Fig. 22.5) that occurs at the beginning and end of each transition makes it difficult to define the exact point at which the rise time should be initiated and terminated. For this reason,

the rise and fall times are defined by the 10% and 90% levels, as indicated in Fig. 22.5.

Note that there is no requirement that t_r equal t_f.

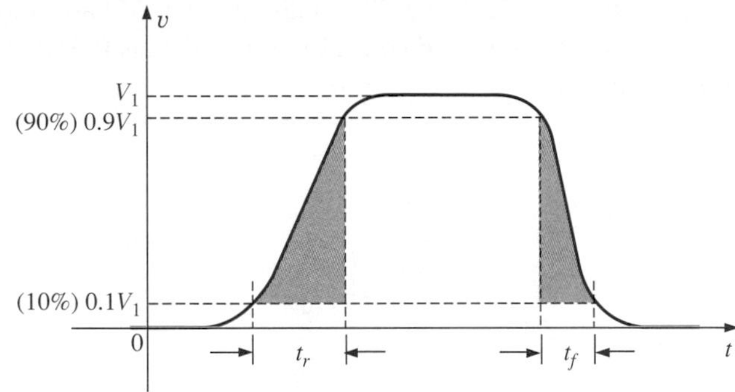

FIG. 22.5
Defining t_r and t_f.

Tilt

An undesirable but common distortion normally occurring due to a poor low-frequency response characteristic of the system through which a pulse has passed appears in Fig. 22.6. The drop in peak value is called *tilt, droop,* or *sag*. The percentage tilt is defined by

$$\% \text{ tilt} = \frac{V_1 - V_2}{V} \times 100\% \qquad \textbf{(22.1)}$$

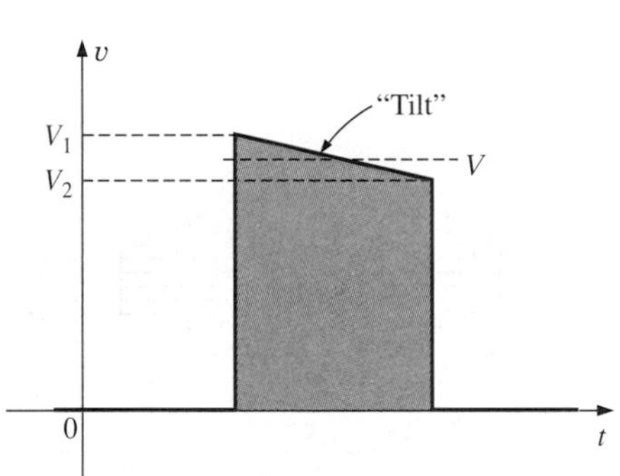

FIG. 22.6
Defining tilt.

where V is the average value of the peak amplitude as determined by

$$V = \frac{V_1 + V_2}{2} \qquad \textbf{(22.2)}$$

Naturally, the less the percentage tilt or sag, the more ideal the pulse. Due to rounding, it may be difficult to define the values of V_1 and V_2. It is then necessary only to approximate the sloping region by a straight-line approximation and use the resulting values of V_1 and V_2.

Other distortions include the *preshoot* and *overshoot* appearing in Fig. 22.7, normally due to pronounced high-frequency effects of a system, and *ringing,* due to the interaction between the capacitive and inductive elements of a network at their natural or resonant frequency.

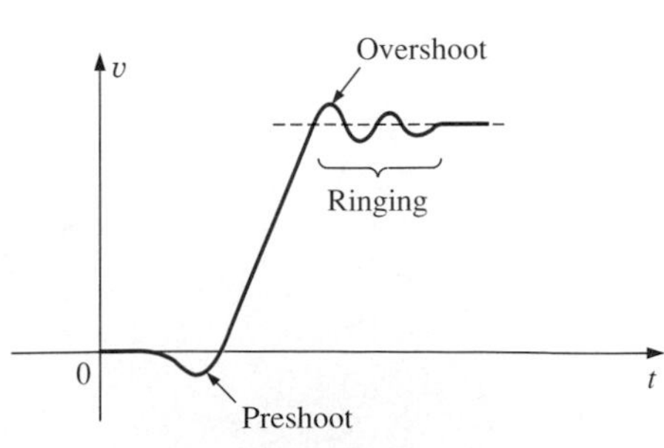

FIG. 22.7
Defining preshoot, overshoot, and ringing.

EXAMPLE 22.1 Determine the following for the pulse waveform of Fig. 22.8:

a. positive- or negative-going?
b. base-line voltage

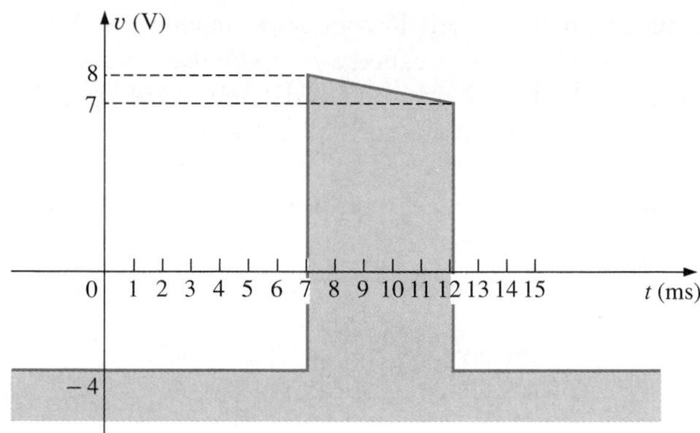

FIG. 22.8
Example 22.1.

c. pulse width
d. maximum amplitude
e. tilt

Solutions:

a. **positive-going**
b. $V_b = \mathbf{-4\ V}$
c. $t_p = (12 - 7)\ \text{ms} = \mathbf{5\ ms}$
d. $V_{max} = 8\ \text{V} + 4\ \text{V} = \mathbf{12\ V}$
e. $V = \dfrac{V_1 + V_2}{2} = \dfrac{12\ \text{V} + 11\ \text{V}}{2} = \dfrac{23\ \text{V}}{2} = 11.5\ \text{V}$

$$\%\ \text{tilt} = \frac{V_1 - V_2}{V} \times 100\% = \frac{12\ \text{V} - 11\ \text{V}}{11.5\ \text{V}} \times 100\% = \mathbf{8.696\%}$$

(Remember, V is defined by the average value of the peak amplitude.)

EXAMPLE 22.2 Determine the following for the pulse waveform of Fig. 22.9:

a. positive- or negative-going?
b. base-line voltage
c. tilt
d. amplitude
e. t_p
f. t_r and t_f

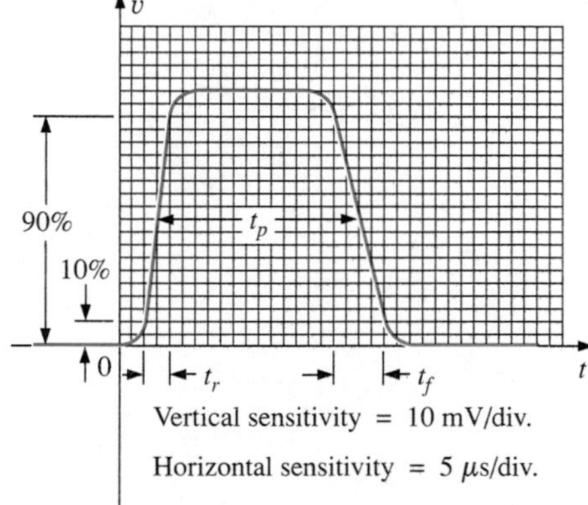

FIG. 22.9
Example 22.2.

Solutions:

a. **positive-going**
b. $V_b = \mathbf{0\ V}$
c. $\%\ \text{tilt} = \mathbf{0\%}$
d. amplitude = (4 div.)(10 mV/div.) = **40 mV**
e. $t_p = (3.2\ \text{div.})(5\ \mu\text{s/div.}) = \mathbf{16\ \mu s}$
f. $t_r = (0.4\ \text{div.})(5\ \mu\text{s/div.}) = \mathbf{2\ \mu s}$
   $t_f = (0.8\ \text{div.})(5\ \mu\text{s/div.}) = \mathbf{4\ \mu s}$

22.3 PULSE REPETITION RATE AND DUTY CYCLE

A series of pulses such as those appearing in Fig. 22.10 is called a *pulse train*. The varying widths and heights may contain information that can be decoded at the receiving end.

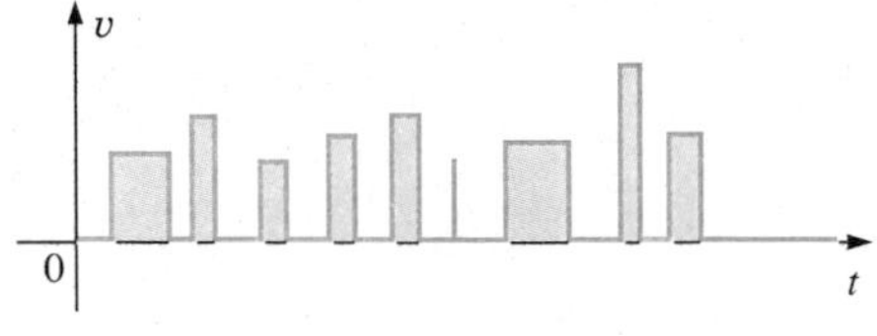

FIG. 22.10
Pulse train.

If the pattern repeats itself in a periodic manner as shown in Fig. 22.11(a) and (b), the result is called a *periodic pulse train.*

The *period* (T) of the pulse train is defined as the time differential between any two similar points on the pulse train, as shown in Figs. 22.11(a) and (b).

The *pulse repetition frequency* (prf), or *pulse repetition rate* (prr), is defined by

$$\text{prf (or prr)} = \frac{1}{T} \qquad \text{(Hz or pulses/s)} \tag{22.3}$$

Applying Eq. (22.3) to each waveform of Fig. 22.11 will result in the same pulse repetition frequency since the periods are the same. The result clearly reveals that

the shape of the periodic pulse does not affect the determination of the pulse repetition frequency.

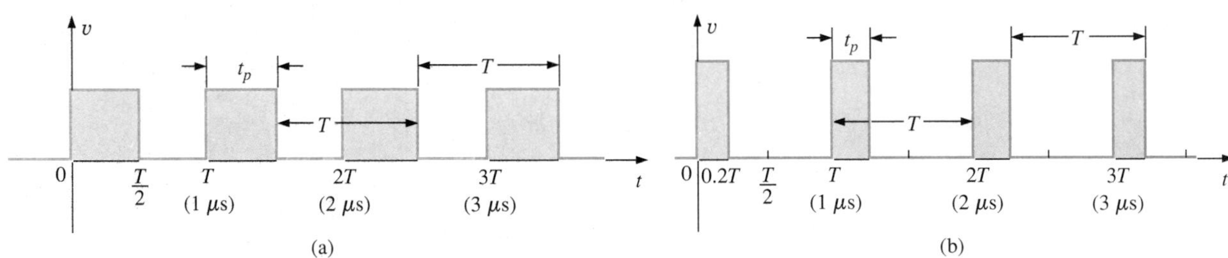

FIG. 22.11
Periodic pulse trains.

The pulse repetition frequency is determined solely by the period of the repeating pulse. The factor that will reveal how much of the period is encompassed by the pulse is called the *duty cycle,* defined as follows:

$$\text{Duty cycle} = \frac{\text{pulse width}}{\text{period}} \times 100\%$$

or

$$\text{Duty cycle} = \frac{t_p}{T} \times 100\% \tag{22.4}$$

For Fig. 22.11(a) (a square-wave pattern),

$$\text{Duty cycle} = \frac{0.5T}{T} \times 100\% = \mathbf{50\%}$$

and for Fig. 22.11(b),

$$\text{Duty cycle} = \frac{0.2T}{T} \times 100\% = \mathbf{20\%}$$

The above results clearly reveal that

the duty cycle provides a percentage indication of the portion of the total period encompassed by the pulse waveform.

EXAMPLE 22.3 Determine the pulse repetition frequency and the duty cycle for the periodic pulse waveform of Fig. 22.12.

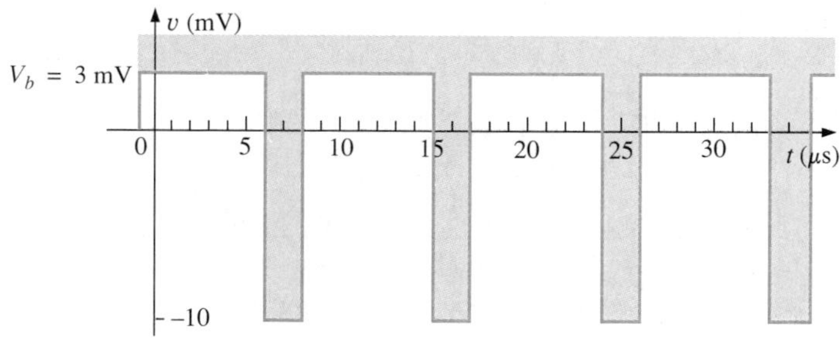

FIG. 22.12
Example 22.3.

Solution:

$$T = (15 - 6)\ \mu\text{s} = 9\ \mu\text{s}$$

$$\text{prf} = \frac{1}{T} = \frac{1}{9\ \mu\text{s}} \cong \mathbf{111.11\ kHz}$$

$$\text{Duty cycle} = \frac{t_p}{T} \times 100\% = \frac{(8 - 6)\ \mu\text{s}}{9\ \mu\text{s}} \times 100\%$$

$$= \frac{2}{9} \times 100\% \cong \mathbf{22.22\%}$$

EXAMPLE 22.4 Determine the pulse repetition frequency and the duty cycle for the oscilloscope pattern of Fig. 22.13 having the indicated sensitivities.

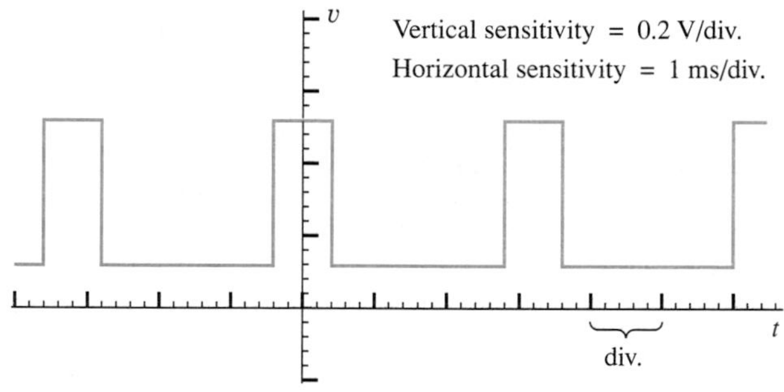

FIG. 22.13
Example 22.4.

Solution:

$$T = (3.2\ \text{div.})(1\ \text{ms/div.}) = 3.2\ \text{ms}$$

$$t_p = (0.8\ \text{div.})(1\ \text{ms/div.}) = 0.8\ \text{ms}$$

$$\text{prf} = \frac{1}{T} = \frac{1}{3.2\ \text{ms}} = \mathbf{312.5\ Hz}$$

$$\text{Duty cycle} = \frac{t_p}{T} \times 100\% = \frac{0.8\ \text{ms}}{3.2\ \text{ms}} \times 100\% = \mathbf{25\%}$$

EXAMPLE 22.5 Determine the pulse repetition rate and duty cycle for the trigger waveform of Fig. 22.14.

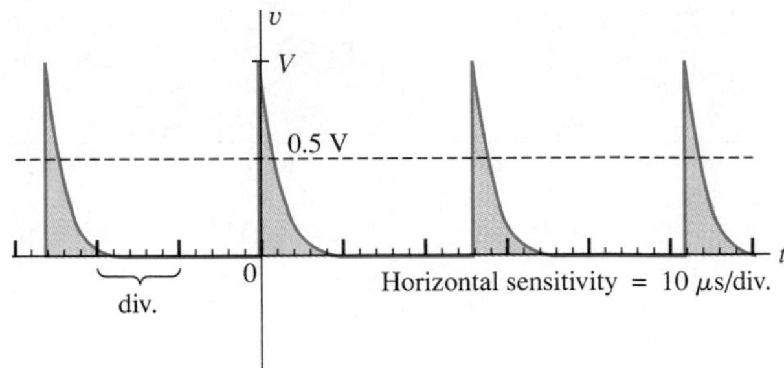

FIG. 22.14
Example 22.5.

Solution:

$$T = (2.6 \text{ div.})(10\ \mu\text{s/div.}) = 26\ \mu\text{s}$$

$$\text{prf} = \frac{1}{T} = \frac{1}{26\ \mu\text{s}} = \mathbf{38{,}462\ kHz}$$

$$t_p \cong (0.2 \text{ div.})(10\ \mu\text{s/div.}) = 2\ \mu\text{s}$$

$$\text{Duty cycle} = \frac{t_p}{T} \times 100\% = \frac{2\ \mu\text{s}}{26\ \mu\text{s}} \times 100\% = \mathbf{7.69\%}$$

22.4 AVERAGE VALUE

The average value of a pulse waveform can be determined using one of two methods. The first is the procedure outlined in Section 13.6, which can be applied to any alternating waveform. The second can be applied only to pulse waveforms since it utilizes terms specifically related to pulse waveforms; that is,

$$V_{av} = (\text{duty cycle})(\text{peak value}) + (1 - \text{duty cycle})(V_b) \qquad \textbf{(22.5)}$$

In Eq. (22.5), the peak value is the maximum deviation from the reference or zero-volt level, and the duty cycle is in decimal form. Equation (22.5) does not include the effect of any tilt pulse waveforms with sloping sides.

EXAMPLE 22.6 Determine the average value for the periodic pulse waveform of Fig. 22.15.

Solution: By the method of Section 13.6,

$$G = \frac{\text{area under curve}}{T}$$

$$T = (12 - 2)\ \mu\text{s} = 10\ \mu\text{s}$$

$$G = \frac{(8\text{ mV})(4\ \mu\text{s}) + (2\text{ mV})(6\ \mu\text{s})}{10\ \mu\text{s}} = \frac{32 \times 10^{-9} + 12 \times 10^{-9}}{10 \times 10^{-6}}$$

$$= \frac{44 \times 10^{-9}}{10 \times 10^{-6}} = \mathbf{4.4\ mV}$$

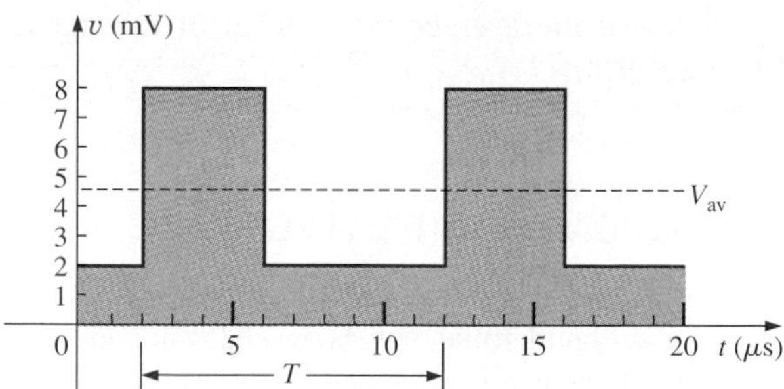

FIG. 22.15
Example 22.6.

By Eq. (22.5),

$$V_b = +2 \text{ mV}$$

$$\text{Duty cycle} = \frac{t_p}{T} = \frac{(6 - 2)\ \mu\text{s}}{10\ \mu\text{s}} = \frac{4}{10} = 0.4 \text{ (decimal form)}$$

$$\text{Peak value (from 0-V reference)} = 8 \text{ mV}$$

$$\begin{aligned} V_{av} &= (\text{duty cycle})(\text{peak value}) + (1 - \text{duty cycle})(V_b) \\ &= (0.4)(8 \text{ mV}) + (1 - 0.4)(2 \text{ mV}) \\ &= 3.2 \text{ mV} + 1.2 \text{ mV} = \mathbf{4.4\ mV} \end{aligned}$$

as obtained above.

EXAMPLE 22.7 Given a periodic pulse waveform with a duty cycle of 28%, a peak value of 7 V, and a base-line voltage of −3 V:

a. Determine the average value.
b. Sketch the waveform.
c. Verify the result of part (a) using the method of Section 13.6.

Solutions:

a. By Eq. (22.5),

$$\begin{aligned} V_{av} &= (\text{duty cycle})(\text{peak value}) + (1 - \text{duty cycle})(V_b) \\ &= (0.28)(7 \text{ V}) + (1 - 0.28)(-3 \text{ V}) = 1.96 \text{ V} + (-2.16 \text{ V}) \\ &= \mathbf{-0.2\ V} \end{aligned}$$

b. See Fig. 22.16.

c. $$G = \frac{(7 \text{ V})(0.28T) - (3 \text{ V})(0.72T)}{T} = 1.96 \text{ V} - 2.16 \text{ V}$$

$$= \mathbf{-0.2\ V}$$

as obtained above.

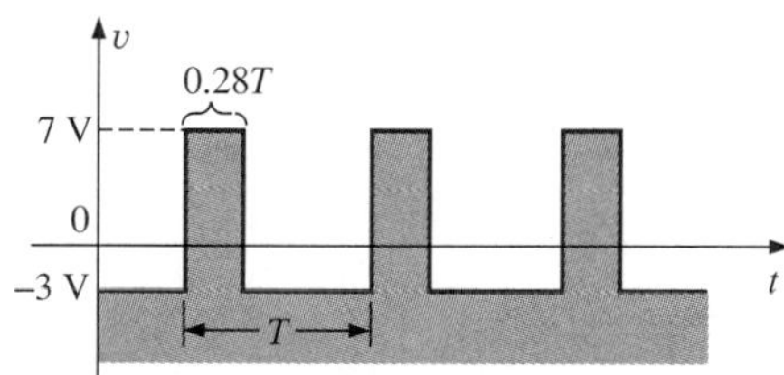

FIG. 22.16
Solution to part (b) of Example 22.7.

Instrumentation

The average value (dc value) of any waveform can be easily determined using the oscilloscope. If the mode switch of the scope is set in the ac position, the average or dc component of the applied waveform will be blocked by an internal capacitor from reaching the screen. The pattern can be adjusted to establish the display of Fig. 22.17(a). If the mode switch is then placed in the dc position, the vertical shift (positive or

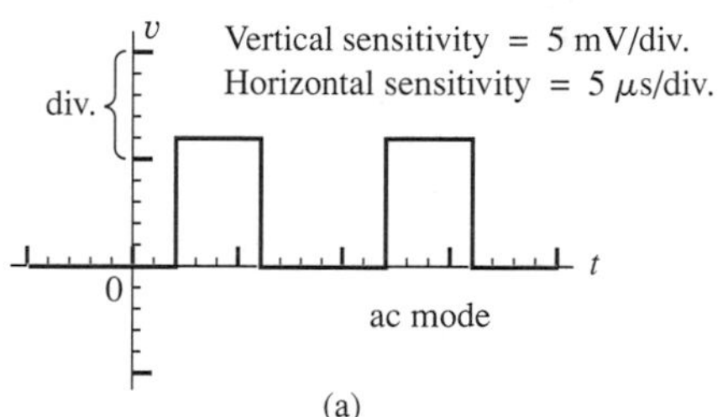

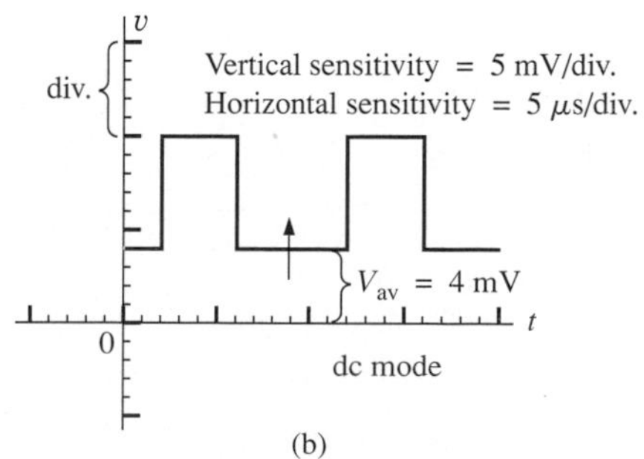

FIG. 22.17
Determining the average value of a pulse waveform using an oscilloscope.

negative) will reveal the average or dc level of the input signal, as shown in Fig. 22.17(b).

22.5 TRANSIENT R-C NETWORKS

In Chapter 10 the general solution for the transient behavior of an R-C network with or without initial values was developed. The resulting equation for the voltage across a capacitor is repeated below for convenience.

$$\boxed{\upsilon_C = V_i + (V_f - V_i)(1 - e^{-t/RC})} \tag{22.6}$$

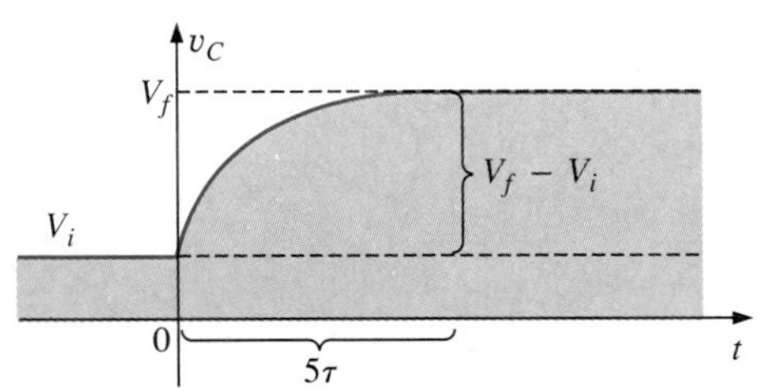

FIG. 22.18
Defining the parameters of Eq. (22.6).

Recall that V_i is the initial voltage across the capacitor when the transient phase is initiated as shown in Fig. 22.18. The voltage V_f is the steady-state (resting) value of the voltage across the capacitor when the transient phase has ended. The transient period is approximated as 5τ, where τ is the time constant of the network and equal to the product RC.

For the situation where the initial voltage is zero volts, the equation reduces to the following familiar form, where V_f is often the applied voltage:

$$\boxed{\upsilon_C = V_f(1 - e^{-t/RC})} \quad V_i = 0\text{ V} \tag{22.7}$$

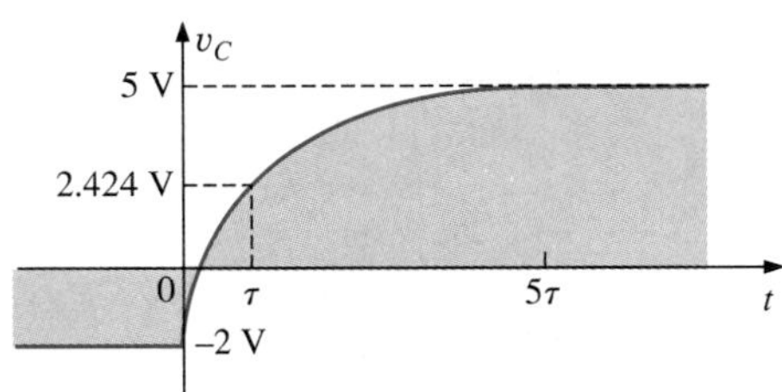

FIG. 22.19
Example of the use of Eq. (22.6).

For the case of Fig. 22.19, $V_i = -2$ V, $V_f = +5$ V, and

$$\begin{aligned}\upsilon_C &= V_i + (V_f - V_i)(1 - e^{-t/RC})\\ &= -2\text{ V} + [5\text{ V} - (-2\text{ V})](1 - e^{-t/RC})\\ \upsilon_C &= -2\text{ V} + 7\text{ V}(1 - e^{-t/RC})\end{aligned}$$

For the case where $t = \tau = RC$,

$$\begin{aligned}\upsilon_C &= -2\text{ V} + 7\text{ V}(1 - e^{-t/\tau}) = -2\text{ V} + 7\text{ V}(1 - e^{-1})\\ &= -2\text{ V} + 7\text{ V}(1 - 0.368) = -2\text{ V} + 7\text{ V}(0.632)\\ \upsilon_C &= 2.424\text{ V}\end{aligned}$$

as verified by Fig. 22.19.

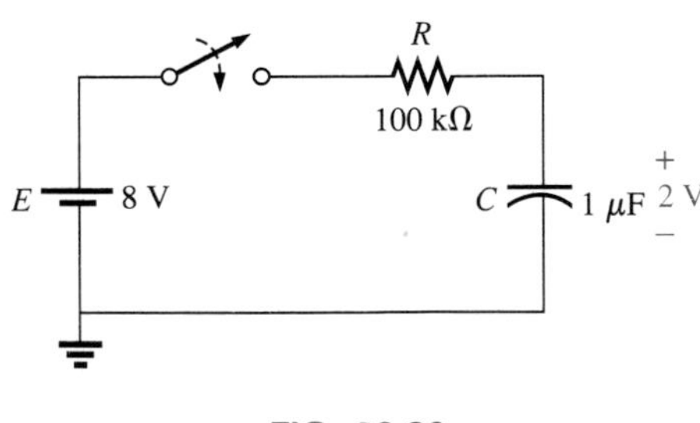

FIG. 22.20
Example 22.8.

EXAMPLE 22.8 The capacitor of Fig. 22.20 is initially charged to 2 V before the switch is closed. The switch is then closed.

a. Determine the mathematical expression for υ_C.
b. Determine the mathematical expression for i_C.
c. Sketch the waveforms of υ_C and i_C.

Solutions:

a. $V_i = 2$ V
$V_f\text{ (after }5\tau) = E = 8$ V
$\tau = RC = (100\text{ k}\Omega)(1\ \mu\text{F}) = 100$ ms

By Eq. (22.6),

$$\begin{aligned}\upsilon_C &= V_i + (V_f - V_i)(1 - e^{-t/\tau})\\ &= 2\text{ V} + (8\text{ V} - 2\text{ V})(1 - e^{-t/\tau})\end{aligned}$$

and $\qquad \upsilon_C = \mathbf{2\text{ V} + 6\text{ V}(1 - e^{-t/\tau})}$

b. When the switch is first closed, the voltage across the capacitor cannot change instantaneously, and $V_R = E - V_i = 8\text{ V} - 2\text{ V} = 6\text{ V}$. The current therefore jumps to a level determined by Ohm's law:

$$I_{R_{\max}} = \frac{V_R}{R} = \frac{6\text{ V}}{100\text{ k}\Omega} = 0.06\text{ mA}$$

The current will then decay to zero amperes with the same time constant calculated in part (a), and

$$i_C = \mathbf{0.06\ mA}e^{-t/\tau}$$

c. See Fig. 22.21.

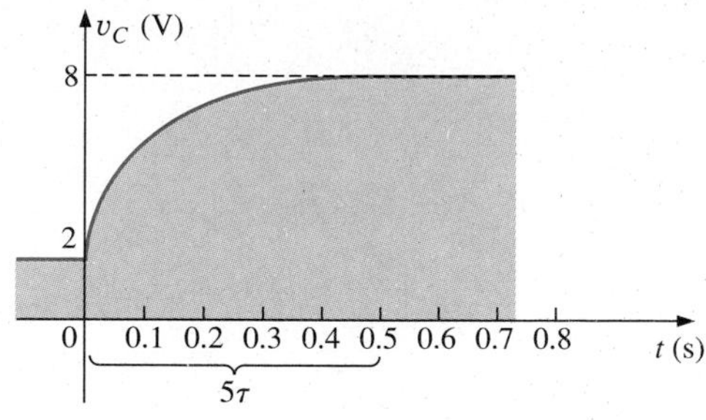

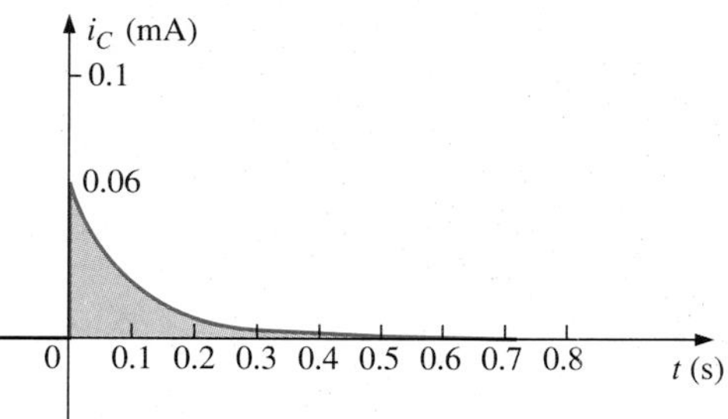

FIG. 22.21
v_C and i_C for the network of Fig. 22.20.

EXAMPLE 22.9 Sketch v_C for the step input shown in Fig. 22.22. Assume the −4 mV has been present for a period of time in excess of five time constants of the network. Then determine when $v_C = 0$ V if the step changes levels at $t = 0$ s.

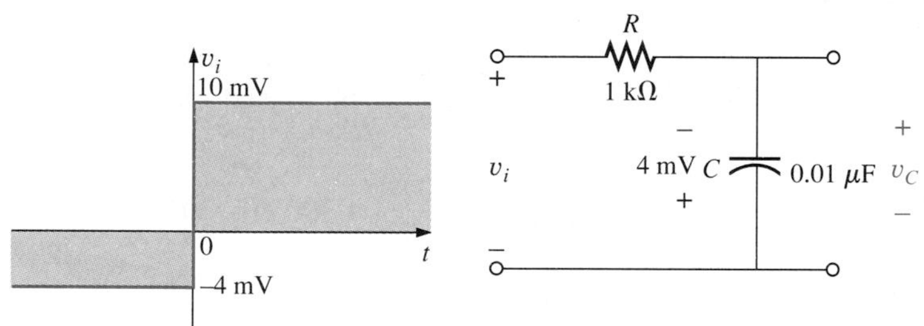

FIG. 22.22
Example 22.9.

Solution:

$$V_i = -4\text{ mV} \qquad V_f = 10\text{ mV}$$
$$\tau = RC = (1\text{ k}\Omega)(0.01\ \mu\text{F}) = 10\ \mu\text{s}$$

By Eq. (22.6),

$$\begin{aligned} v_C &= V_i + (V_f - V_i)(1 - e^{-t/\tau}) \\ &= -4\text{ mV} + [10\text{ mV} - (-4\text{ mV})](1 - e^{-t/\tau}) \end{aligned}$$

and $$\boldsymbol{v_C = -4\ \mathrm{mV} + 14\ \mathrm{mV}(1 - e^{-t/(10\ \mu s)})}$$

The waveform appears in Fig. 22.23.

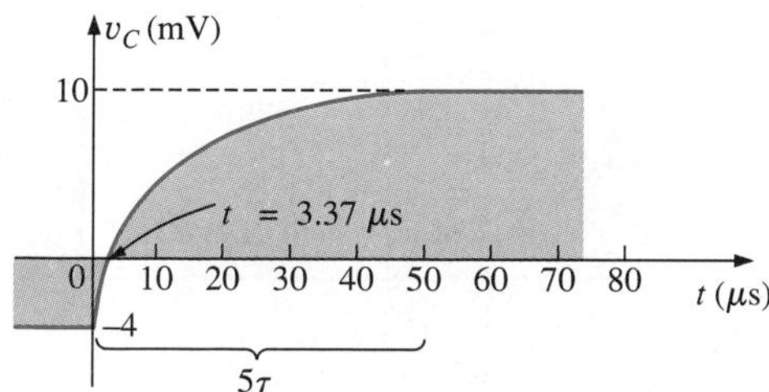

FIG. 22.23
v_C for the network of Fig. 22.22.

Substituting $v_C = 0$ V into the above equation yields

$$v_C = 0 = -4 \text{ mV} + 14 \text{ mV}(1 - e^{-t/\tau})$$

and
$$\frac{4 \text{ mV}}{14 \text{ mV}} = 1 - e^{-t/\tau}$$

or
$$0.286 - 1 = -e^{-t/\tau}$$

and
$$0.714 = e^{-t/\tau}$$

but
$$\log_e 0.714 = \log_e(e^{-t/\tau}) = \frac{-t}{\tau}$$

and
$$t = -\tau \log_e 0.714 = -(10 \,\mu\text{s})(-0.377) = \mathbf{3.37 \,\mu s}$$

as indicated in Fig. 22.23.

22.6 *R-C* RESPONSE TO SQUARE-WAVE INPUTS

The square wave of Fig. 22.24 is a particular form of pulse waveform. It has a duty cycle of 50% and an average value of zero volts, as calculated below:

$$\text{Duty cycle} = \frac{t_p}{T} \times 100\% = \frac{T/2}{T} \times 100\% = \mathbf{50\%}$$

$$V_{av} = \frac{(V_1)(T/2) + (-V_1)(T/2)}{T} = \frac{0}{T} = \mathbf{0 \text{ V}}$$

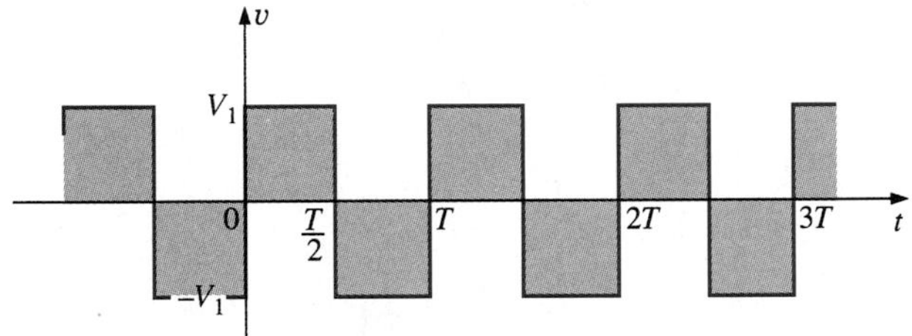

FIG. 22.24
Periodic square wave.

The application of a dc voltage V_1 in series with the square wave of Fig. 22.24 can raise the base-line voltage from $-V_1$ to zero volts and the average value to V_1 volts.

If a square wave such as developed in Fig. 22.25 is applied to an

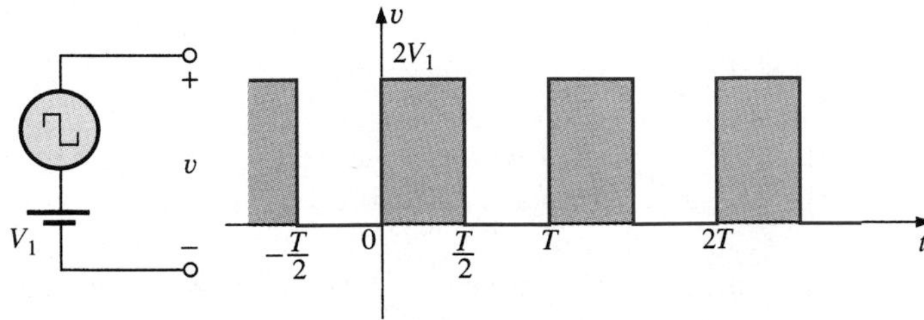

FIG. 22.25
Raising the base-line voltage of a square wave to zero volts.

R-C circuit as shown in Fig. 22.26, the period of the square wave can have a pronounced effect on the resulting waveform for v_C.

For the analysis to follow, we will assume that steady-state conditions will be established after a period of five time constants has passed. The types of waveforms developed across the capacitor can then be separated into three fundamental types.

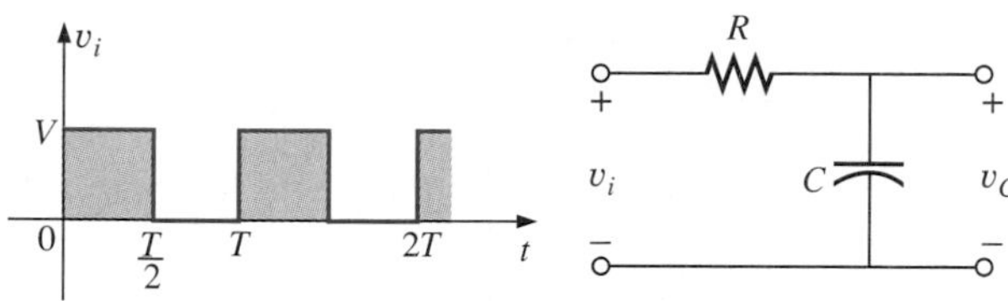

FIG. 22.26
Applying a periodic square-wave pulse train to an R-C network.

$T/2 > 5\tau$

The condition $T/2 > 5\tau$, or $T > 10\tau$, establishes a situation where the capacitor can charge to its steady-state value in advance of $t = T/2$. The resulting waveforms for v_C and i_C will appear as shown in Fig. 22.27. Note how closely the voltage v_C shadows the applied waveform and how i_C is nothing more than a series of very sharp spikes. Note also that the change of V_i from V to zero volts during the trailing edge simply results in a rapid discharge of v_C to zero volts. In essence, when $V_i = 0$ the capacitor and resistor are in parallel and the capacitor simply discharges through R with a time constant equal to that encountered during the charging phase but with a direction of charge flow (current) opposite to that established during the charging phase.

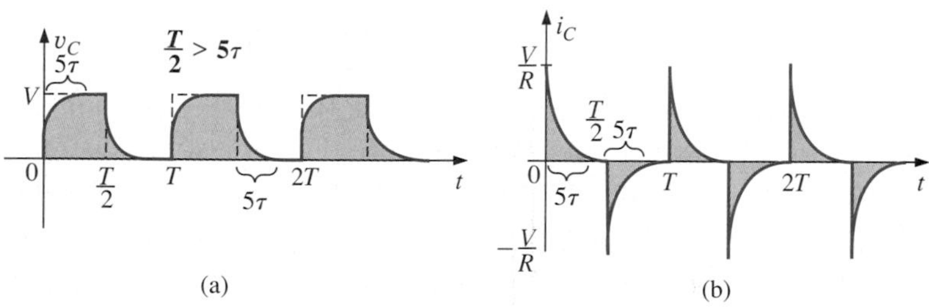

FIG. 22.27
v_C and i_C for $T/2 > 5\tau$.

$T/2 = 5\tau$

If the frequency of the square wave is chosen such that $T/2 = 5\tau$ or $T = 10\tau$, the voltage v_C will reach its final value just before beginning its discharge phase, as shown in Fig. 22.28. The voltage v_C no longer resembles the square-wave input and, in fact, has some of the characteristics of a triangular waveform. The increased time constant has resulted in a more rounded v_C, and i_C has increased substantially in width to reveal the longer charging period.

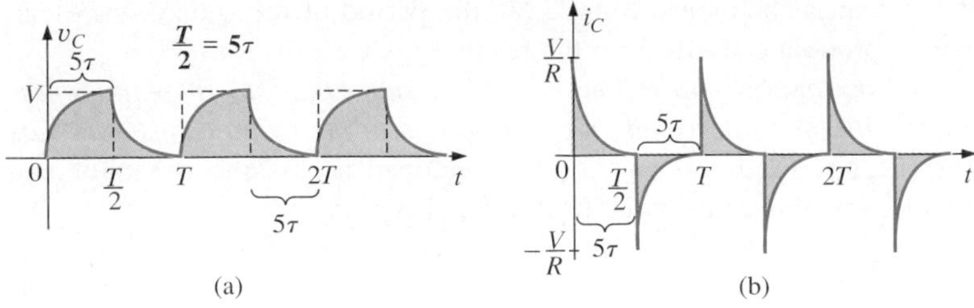

FIG. 22.28
v_C and i_C for $T/2 = 5\tau$.

$T/2 < 5\tau$

If $T/2 < 5\tau$ or $T < 10\tau$, the voltage v_C will not reach its final value during the first pulse (Fig. 22.29), and the discharge cycle will not return to zero volts. In fact, the initial value for each succeeding pulse will change until steady-state conditions are reached. In most instances, it is a good approximation to assume that steady-state conditions have been established in 5 cycles of the applied waveform.

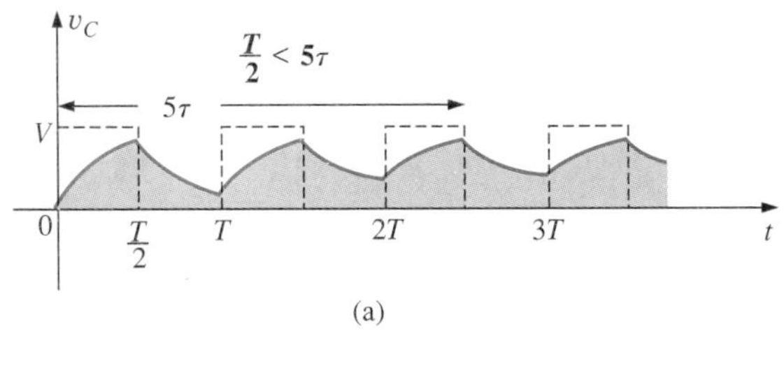

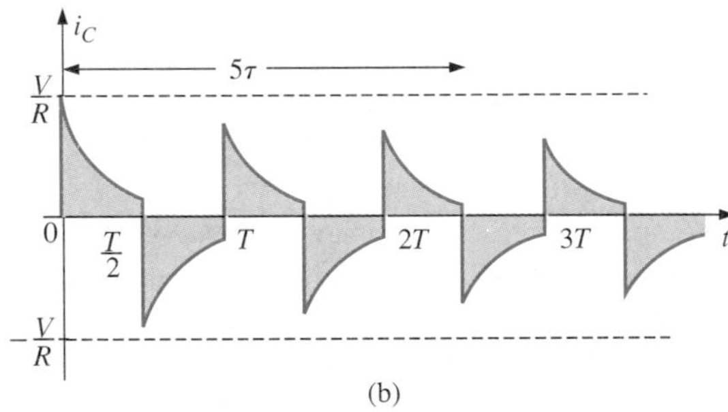

FIG. 22.29
v_C and i_C for $T/2 < 5\tau$.

As the frequency increases and the period decreases, there will be a flattening of the response for v_C until a pattern like that in Fig. 22.30 results. Figure 22.30 begins to reveal an important conclusion regarding the response curve for v_C:

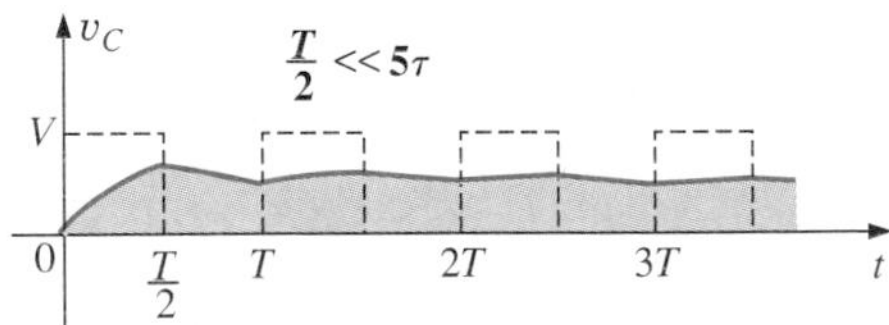

FIG. 22.30
v_C for $T/2 \ll 5\tau$ or $T \ll 10\tau$.

Under steady-state conditions, the average value of v_C will equal the average value of the applied square wave.

Note in Figs. 22.29 and 22.30 how the waveform for v_C approaches an average value of $V/2$.

EXAMPLE 22.10 The 1000-Hz square wave of Fig. 22.31 is applied to the *R-C* circuit of the same figure.

a. Compare the pulse width of the square wave to the time constant of the circuit.
b. Sketch v_C.
c. Sketch i_C.

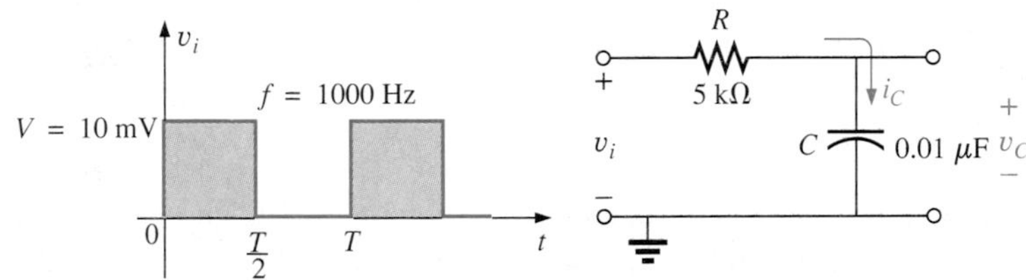

FIG. 22.31
Example 22.10.

Solutions:

a. $T = \dfrac{1}{f} = \dfrac{1}{1000} = 1\text{ ms}$

$$t_p = \frac{T}{2} = 0.5\text{ ms}$$

$$\tau = RC = (5 \times 10^3\ \Omega)(0.01 \times 10^{-6}\text{ F}) = 0.05\text{ ms}$$

$$\frac{t_p}{\tau} = \frac{0.5\text{ ms}}{0.05\text{ ms}} = 10 \text{ and } \boldsymbol{t_p = 10\tau = \frac{T}{2}}$$

The result reveals that v_C will charge to its final value in half the pulse width.

b. For the charging phase, $V_i = 0$ V and $V_f = 10$ mV, and

$$v_C = V_i + (V_f - V_i)(1 - e^{-t/\tau})$$
$$= 0 + (10\text{ mV} - 0)(1 - e^{-t/\tau})$$

and $$v_C = \mathbf{10\ mV(1 - e^{-t/\tau})}$$

For the discharge phase, $V_i = 10$ mV and $V_f = 0$ V, and

$$v_C = V_i + (V_f - V_i)(1 - e^{-t/\tau})$$
$$= 10\text{ mV} + (0 - 10\text{ mV})(1 - e^{-t/\tau})$$
$$v_C = 10\text{ mV} - 10\text{ mV} + 10\text{ mV}(e^{-t/\tau})$$

and $$v_C = \mathbf{10\ mVe^{-t/\tau}}$$

The waveform for v_C appears in Fig. 22.32.

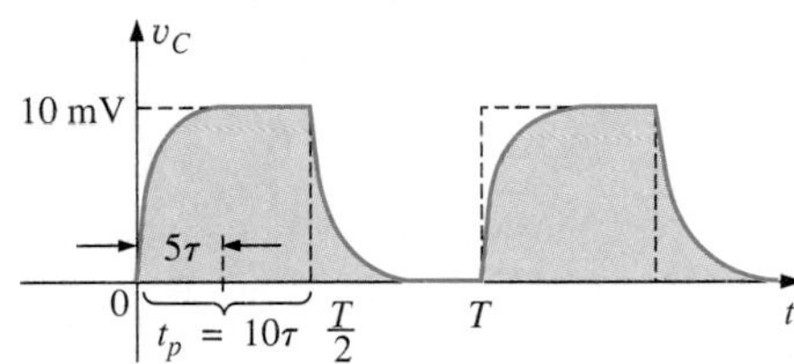

FIG. 22.32
v_C for the R-C network of Fig. 22.31.

c. For the charging phase at $t = 0$ s, $V_R = V$ and $I_{R_{max}} = V/R = 10\text{ mV}/5\text{ k}\Omega = 2\ \mu\text{A}$, and

$$i_C = I_{max}e^{-t/\tau} = \mathbf{2\ \mu Ae^{-t/\tau}}$$

For the discharge phase, the current will have the same mathematical formulation but the opposite direction, as shown in Fig. 22.33.

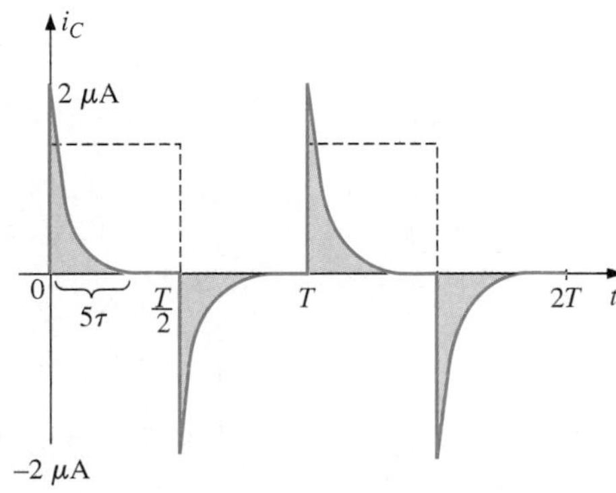

FIG. 22.33
i_C for the R-C network of Fig. 22.31.

EXAMPLE 22.11 Repeat Example 22.10 for $f = 10$ kHz.

Solution:

$$T = \frac{1}{f} = \frac{1}{10\text{ kHz}} = 0.1\text{ ms}$$

and
$$\frac{T}{2} = 0.05\text{ ms}$$

with
$$\tau = t_p = \frac{T}{2} = 0.05\text{ ms}$$

In other words, the pulse width is exactly equal to the time constant of the network. The voltage υ_C will not reach the final value before the first pulse of the square-wave input returns to zero volts.

For t in the range $t = 0$ to $t = T/2$, $V_i = 0$ V and $V_f = 10$ mV, and

$$\upsilon_C = 10\text{ mV}(1 - e^{-t/\tau})$$

At $t = \tau$, we recall from Chapter 10 that $\upsilon_C = 63.2\%$ of the final value. Substituting $t = \tau$ into the equation above yields

$$\begin{aligned}\upsilon_C &= (10\text{ mV})(1 - e^{-1}) = (10\text{ mV})(1 - 0.368)\\ &= (10\text{ mV})(0.632) = 6.32\text{ mV}\end{aligned}$$

as shown in Fig. 22.34.

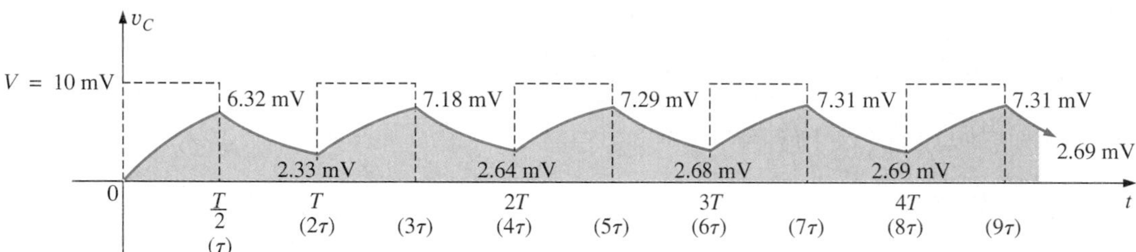

FIG. 22.34
υ_C response for $t_p = \tau = T/2$.

For the discharge phase between $t = T/2$ and T, $V_i = 6.32$ mV and $V_f = 0$ V, and

$$\begin{aligned}\upsilon_C &= V_i + (V_f - V_i)(1 - e^{-t/\tau})\\ &= 6.32\text{ mV} + (0 - 6.32\text{ mV})(1 - e^{-t/\tau})\\ \upsilon_C &= 6.32\text{ mV}e^{-t/\tau}\end{aligned}$$

with t now being measured from $t = T/2$ in Fig. 22.34. In other words, for each interval of Fig. 22.34, the beginning of the transient waveform is defined as $t = 0$ s. The value of υ_C at $t = T$ is therefore determined by substituting $t = \tau$ into the above equation, and not 2τ as defined by Fig. 22.34.

Substituting $t = \tau$,

$$\begin{aligned}\upsilon_C &= (6.32\text{ mV})(e^{-1}) = (6.32\text{ mV})(0.368)\\ &= 2.33\text{ mV}\end{aligned}$$

as shown in Fig. 22.34.

For the next interval, $V_i = 2.33$ mV and $V_f = 10$ mV, and

$$\begin{aligned} v_C &= V_i + (V_f - V_i)(1 - e^{-t/\tau}) \\ &= 2.33\text{ mV} + (10\text{ mV} - 2.33\text{ mV})(1 - e^{-t/\tau}) \\ v_C &= 2.33\text{ mV} + 7.67\text{ mV}(1 - e^{-t/\tau}) \end{aligned}$$

At $t = \tau$ (since $t = T = 2\tau$ is now $t = 0$ s for this interval),

$$\begin{aligned} v_C &= 2.33\text{ mV} + 7.67\text{ mV}(1 - e^{-1}) \\ &= 2.33\text{ mV} + 4.85\text{ mV} \\ v_C &= 7.18\text{ mV} \end{aligned}$$

as shown in Fig. 22.34.

For the discharge interval, $V_i = 7.18$ mV and $V_f = 0$ V, and

$$\begin{aligned} v_C &= V_i + (V_f - V_i)(1 - e^{-t/\tau}) \\ &= 7.18\text{ mV} + (0 - 7.18\text{ mV})(1 - e^{-t/\tau}) \\ v_C &= 7.18\text{ mV}e^{-t/\tau} \end{aligned}$$

At $t = \tau$ (measured from 3τ of Fig. 22.34),

$$\begin{aligned} v_C &= (7.18\text{ mV})(e^{-1}) = (7.18\text{ mV})(0.368) \\ &= 2.64\text{ mV} \end{aligned}$$

as shown in Fig. 22.34.

Continuing in the same manner, the remaining waveform for v_C will be generated as depicted in Fig. 22.34. Note that repetition occurs after $t = 8\tau$ and the waveform has essentially reached steady-state conditions in a period of time less than 10τ, or 5 cycles of the applied square wave.

A closer look will reveal that both the peak and lower levels continued to increase until steady-state conditions were established. Since the exponential waveforms between $t = 4T$ and $t = 5T$ have the same time constant, the average value of v_C can be determined from the steady-state 7.31-mV and 2.69-mV levels as follows:

$$V_{\text{av}} = \frac{7.31\text{ mV} + 2.69\text{ mV}}{2} = \frac{10\text{ mV}}{2} = 5\text{ mV}$$

which equals the average value of the applied signal as stated earlier in this section.

We can use the results of Fig. 22.34 to plot i_C. At any instant of time,

$$v_i = v_R + v_C \quad \text{or} \quad v_R = v_i - v_C$$

and

$$i_R = i_C = \frac{v_i - v_C}{R}$$

At $t = 0^+$, $v_C = 0$ V, and

$$i_R = \frac{v_i - v_C}{R} = \frac{10\text{ mV} - 0}{5\text{ k}\Omega} = 2\ \mu\text{A}$$

as shown in Fig. 22.35.

As the charging process proceeds, the current i_C will decay at a rate determined by

$$i_C = 2\ \mu\text{A}e^{-t/\tau}$$

At $t = \tau$,

$$\begin{aligned} i_C &= (2\ \mu\text{A})(e^{-\tau/\tau}) = (2\ \mu\text{A})(e^{-1}) = (2\ \mu\text{A})(0.368) \\ &= 0.736\ \mu\text{A} \end{aligned}$$

as shown in Fig. 22.35.

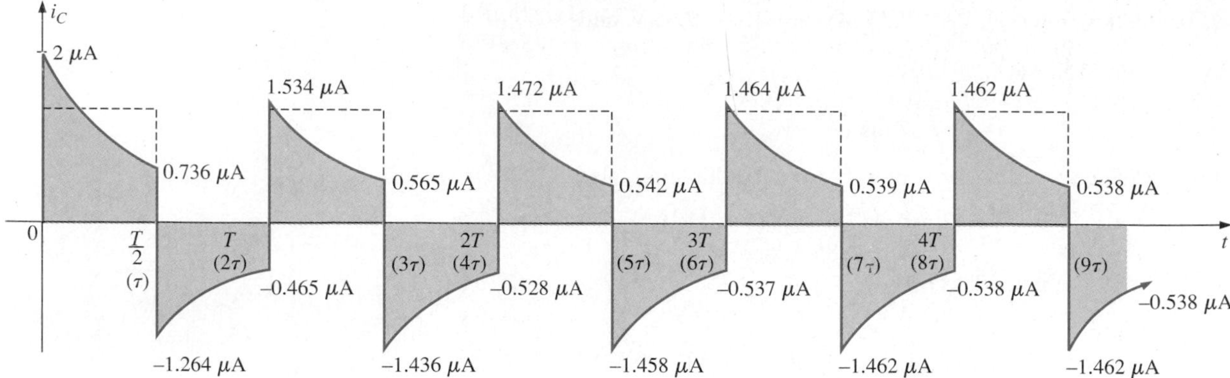

FIG. 22.35
i_C response for $t_p = \tau = T/2$.

For the trailing edge of the first pulse, the voltage across the capacitor cannot change instantaneously, resulting in the following when v_i drops to zero volts:

$$i_C = i_R = \frac{v_i - v_C}{R} = \frac{0 - 6.32\text{ mV}}{5\text{ k}\Omega} = -1.264\ \mu\text{A}$$

as illustrated in Fig. 22.35. The current will then decay as determined by

$$i_C = -1.264\ \mu\text{A}e^{-t/\tau}$$

and at $t = \tau$ (actually $t = 2\tau$ in Fig. 22.35),

$$\begin{aligned} i_C &= (-1.264\ \mu\text{A})(e^{-\tau/\tau}) = (-1.264\ \mu\text{A})(e^{-1}) \\ &= (-1.264\ \mu\text{A})(0.368) = -0.465\ \mu\text{A} \end{aligned}$$

as shown in Fig. 22.35.

At $t = T$ ($t = 2\tau$), $v_C = 2.33$ mV and v_i returns to 10 mV, resulting in

$$i_C = i_R = \frac{v_i - v_C}{R} = \frac{10\text{ mV} - 2.33\text{ mV}}{5\text{ k}\Omega} = 1.534\ \mu\text{A}$$

The equation for the decaying current is now

$$i_C = 1.534\ \mu\text{A}e^{-t/\tau}$$

and at $t = \tau$ (actually $t = 3\tau$ in Fig. 22.35),

$$i_C = (1.534\ \mu\text{A})(0.368) = 0.565\ \mu\text{A}$$

The process will continue until steady-state conditions are reached at the same time they were attained for v_C. Note in Fig. 22.35 that the positive peak current decreased toward steady-state conditions while the negative peak became more negative. It is also interesting and important to realize that the current waveform becomes symmetrical about the axis when steady-state conditions are established. The result is that the net average current over one cycle is zero, as it should be in a series *R-C* circuit. Recall from Chapter 10 that the capacitor under dc steady-state conditions can be replaced by an open-circuit equivalent, resulting in $I_C = 0$ A.

Although both examples provided above started with an uncharged capacitor, there is no reason that the same approach cannot be used effectively for initial conditions. Simply substitute the initial voltage on the capacitor as V_i in Eq. (22.6) and proceed as above.

22.7 OSCILLOSCOPE ATTENUATOR AND COMPENSATING PROBE

The ×10 attenuator probe employed with oscilloscopes is designed to reduce the magnitude of the input voltage by a factor of 10. If the input impedance to a scope is 1 MΩ, the ×10 attenuator probe would have an internal resistance of 9 MΩ, as shown in Fig. 22.36.

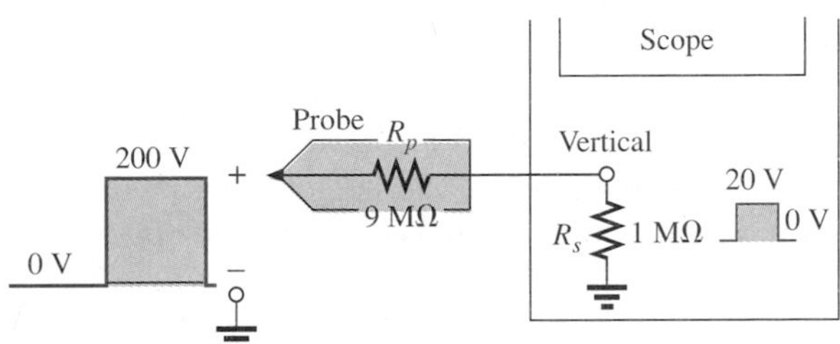

FIG. 22.36
×10 attenuator probe.

Applying the voltage divider rule,

$$V_{\text{scope}} = \frac{(1\text{ M}\Omega)(V_i)}{1\text{ M}\Omega + 9\text{ M}\Omega} = \frac{1}{10}V_i$$

In addition to the input resistance, oscilloscopes have some internal input capacitance, and the probe will add an additional capacitance in parallel with the oscilloscope capacitance, as shown in Fig. 22.37. The probe capacitance is typically about 10 pF for a 1-m (3.3-ft) long cable, reaching about 15 pF for a 3-m (9.9-ft) cable. The total input capacitance is therefore the sum of the two capacitive elements, resulting in the equivalent network of Fig. 22.38.

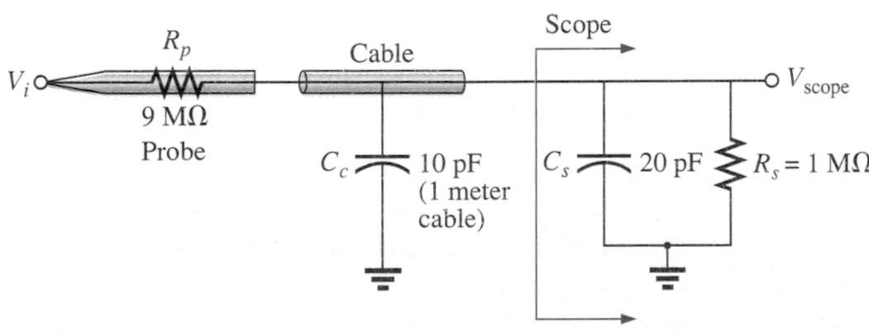

FIG. 22.37
Capacitive elements present in an attenuator probe arrangement.

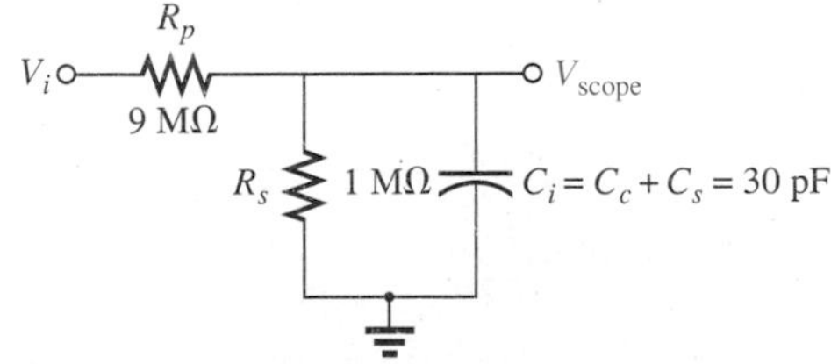

FIG. 22.38
Equivalent network of Fig. 22.37.

For the analysis to follow, let us determine the Thévenin equivalent circuit for the capacitor C_i:

$$E_{Th} = \frac{(1\text{ M}\Omega)(V_i)}{1\text{ M}\Omega + 9\text{ M}\Omega} = \frac{1}{10}V_i$$

and

$$R_{Th} = 9\text{ M}\Omega \parallel 1\text{ M}\Omega = 0.9\text{ M}\Omega$$

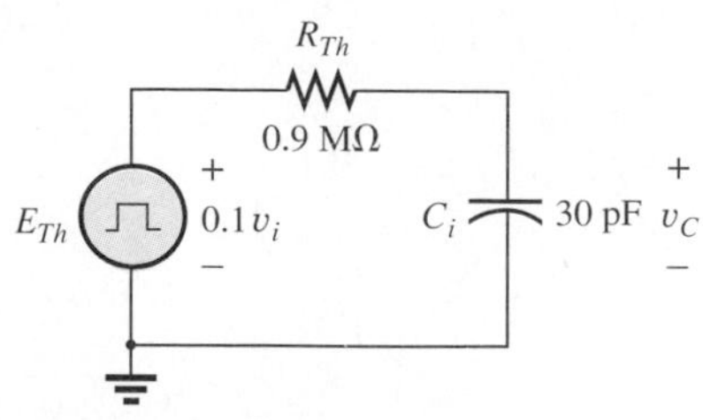

FIG. 22.39
Thévenin equivalent for C_i of Fig. 22.38.

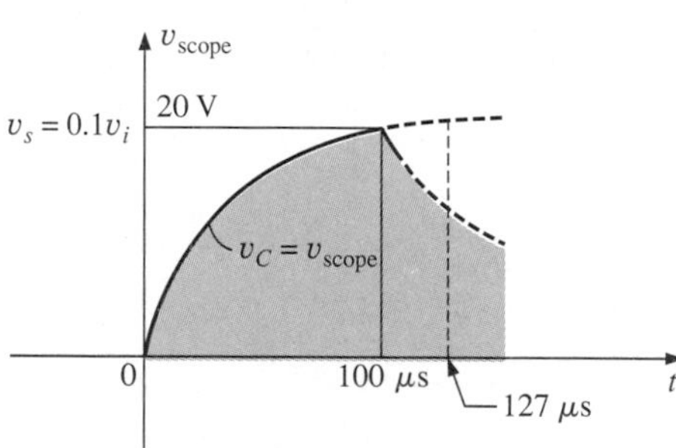

FIG. 22.40
The scope pattern for the conditions of Fig. 22.38 with v_i = 200 V peak.

FIG. 22.41
Commercial compensated 10:1 attenuator probe. (Courtesy of Tektronix, Inc.)

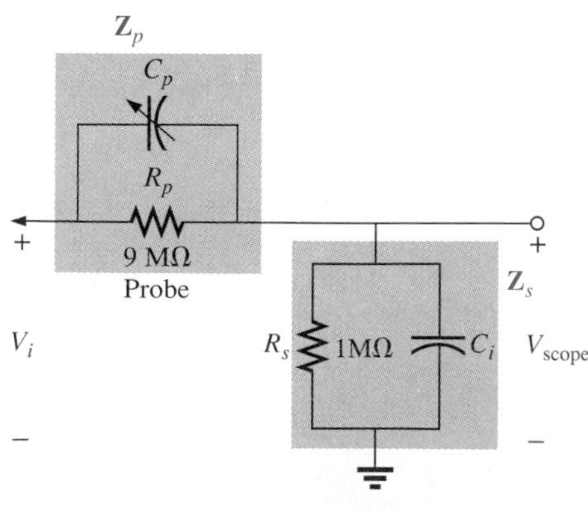

FIG. 22.42
Compensated attenuator and input impedance to a scope, including the cable capacitance.

The Thévenin network is shown in Fig. 22.39.

For v_i = 200 V (peak),

$$E_{Th} = 0.1v_i = 20 \text{ V (peak)}$$

and for v_C, V_f = 20 V and V_i = 0 V, with

$$\tau = RC = (0.9 \times 10^6 \ \Omega)(30 \times 10^{-12} \text{ F}) = 27 \ \mu\text{s}$$

For an applied frequency of 5 kHz,

$$T = \frac{1}{f} = 0.2 \text{ ms} \quad \text{and} \quad \frac{T}{2} = 0.1 \text{ ms} = 100 \ \mu\text{s}$$

with 5τ = 135 μs > 100 μs, as shown in Fig. 22.40, clearly producing a severe rounding distortion of the square wave and a poor representation of the applied signal.

To improve matters, a variable capacitor is often added in parallel with the resistance of the attenuator, resulting in a *compensated attenuator* such as the one shown in Fig. 22.41. In Chapter 25, it will be demonstrated that a square wave can be generated by a summation of sinusoidal signals of particular frequency and amplitude. If we therefore design a network such as the one shown in Fig. 22.42 that will ensure that V_{scope} is $0.1v_i$ for any frequency, then the rounding distortion will be removed and V_{scope} will have the same appearance as v_i.

Applying the voltage divider rule to the network of Fig. 22.42,

$$\mathbf{V}_{scope} = \frac{\mathbf{Z}_s\mathbf{V}_i}{\mathbf{Z}_s + \mathbf{Z}_p} \qquad \textbf{(22.8)}$$

If the parameters are chosen or adjusted such that

$$R_pC_p = R_sC_s \qquad \textbf{(22.9)}$$

the phase angle of $\mathbf{Z}_s$ and $\mathbf{Z}_p$ will be the same, and Eq. (22.8) will reduce to

$$\mathbf{V}_{scope} = \frac{R_s\mathbf{V}_i}{R_s + R_p} \qquad \textbf{(22.10)}$$

which is insensitive to frequency since the capacitive elements have dropped out of the relationship.

In the laboratory, simply adjust the probe capacitance using a standard or known square-wave signal until the desired sharp corners of the square wave are obtained. If you avoid the calibration step, you may make a rounded signal look square since you assumed a square wave at the point of measurement.

Too much capacitance will result in an overshoot effect, whereas too little will continue to show the rounding effect.

22.8 COMPUTER ANALYSIS

PSpice (DOS)

In this section we will make full use of the .PROBE command in PSpice to generate some of the waveforms appearing in this chapter. In particular, let us obtain the pulse response to the network of Fig. 22.43, which also appears as Fig. 22.31.

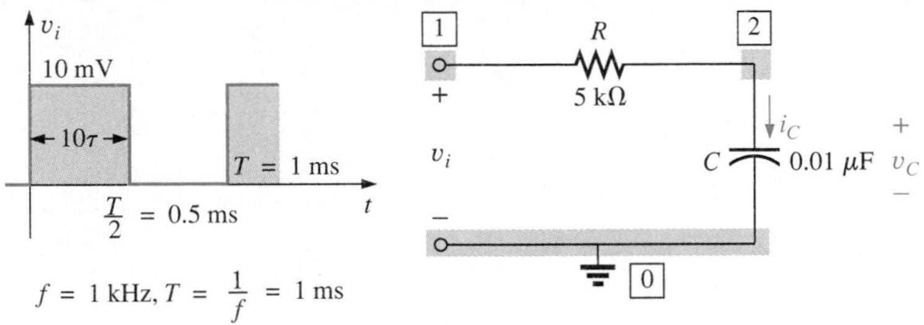

FIG. 22.43
Defining the parameters for a PSpice (DOS) analysis of an R-C network.

For the applied frequency, the pulse is 10 mV for 0.5 ms and 0 V for the succeeding 0.5 ms. The first line of the input file of Fig. 22.44 specifies a 0-V to 10-mV transition with a 0-s delay time. The next two 1-ns entries represent the rise and fall times, which—you recall—must be specified or they default back to TSTEP of the .TRAN command. The 0.5-ms entry is the length of the pulse at the 10-mV level, and 1 ms specifies the period of the waveform. The .TRAN command will generate a data point every 0.05 ms, or 50 μs, up to 1 ms (1 M).

```
    Chapter 22 - R-C Pulse Response

    ****      CIRCUIT DESCRIPTION

******************************************************************************


VI 1 0 PULSE(0V 10MV 0 1NS 1NS 0.5MS 1MS)
R  1 2 5K
C  2 0 0.01UF
.TRAN 0.05M 1M
.PROBE
.OPTIONS NOPAGE
.END
```

FIG. 22.44
Input file for the network of Fig. 22.43.

The .PROBE command was then used to obtain a plot of v_i (nodal voltage V_1 of Fig. 22.43(b)) and v_C (nodal voltage V_2 of Fig. 22.43(b)), as shown in Fig. 22.45. Note how the results for each plot match the curves appearing in Figs. 22.32 and 22.33. Note also how .PROBE specifies at the bottom of the plot the notation for each voltage, □ for v_i and ■ for v_C. The plot of i_C reveals a sensitivity of .PROBE to negative and positive values as it sets the horizontal axis in the middle of the page.

The second input file (Fig. 22.46) is for the applied pulse of Fig. 22.47, which has a frequency of 10 kHz (as applied in Example 22.11). In this case, the 10-mV pulse is present for a period of time matching the time constant of the network—a period insufficient (less than 5τ) for the voltage v_C to climb to its final value. The result will be a level of v_C at $t = 0.05$ ms less than 10 mV, as shown in the figure. The only

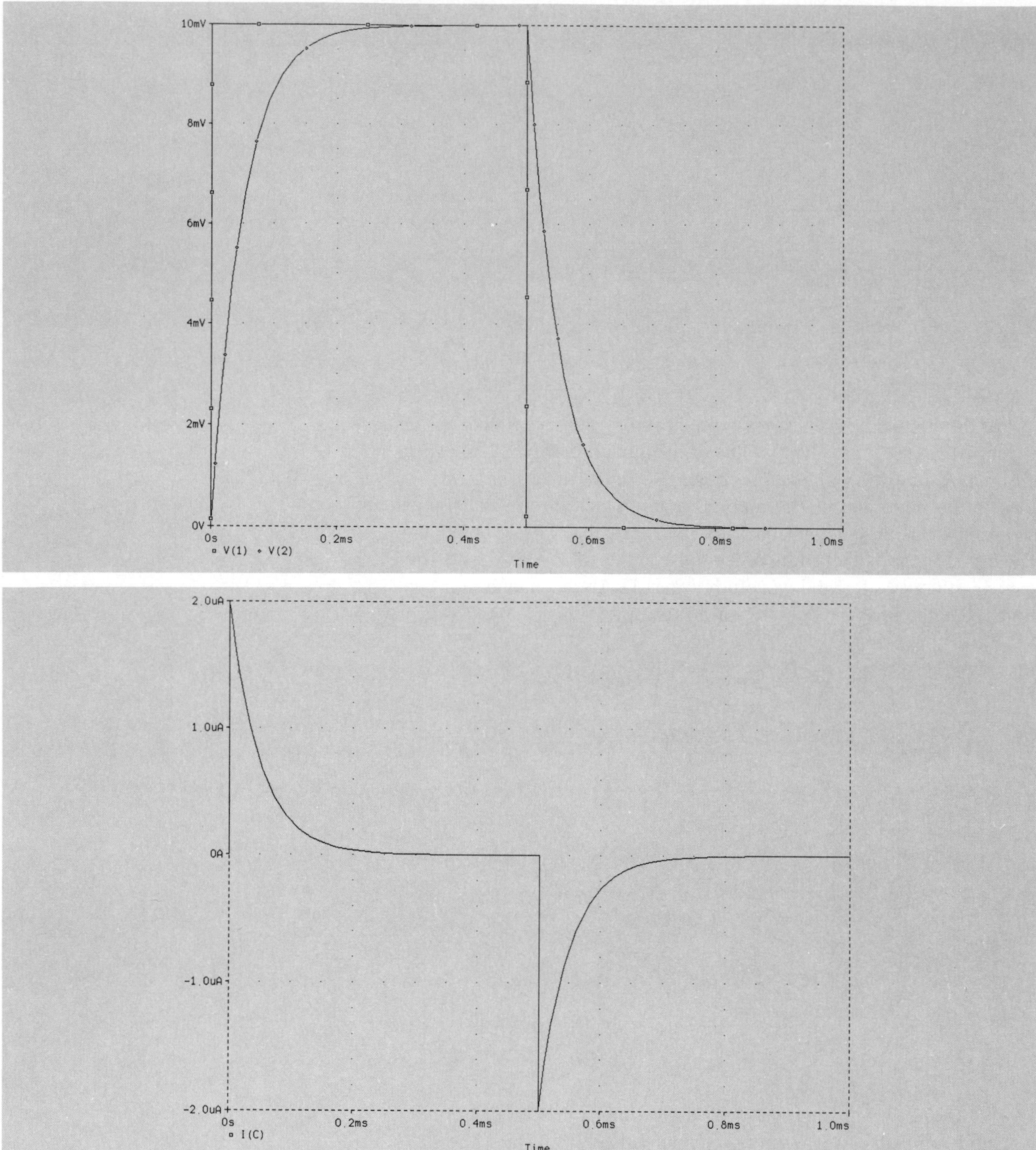

FIG. 22.45
Output file for the network of Fig. 22.43.

changes to the input file of Fig. 22.44 are to change the pulse period to 0.1 ms (2 × 0.05 ms) and to modify the .TRAN command with a data point every 0.005 ms = 5 μs for a period extending to 0.1 ms = 100 μs. A request for a plot of v_i and v_C resulted in the curves of Fig. 22.48, which compare very favorably with the curves of Fig. 22.34 for the same conditions. The value of v_C has risen to about 6.3 V in 50 μs and drops to about 2.3 V at t = 100 μs. Accuracy beyond the tenths

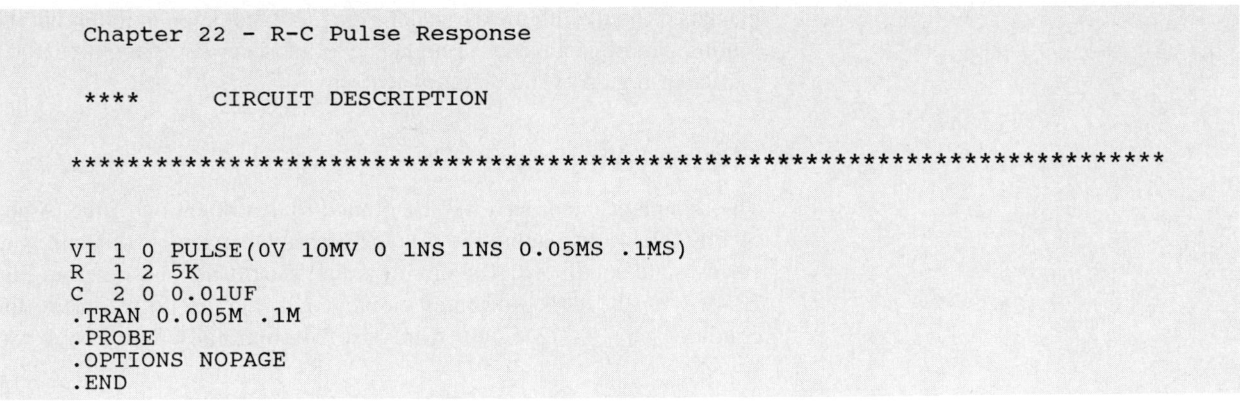

```
Chapter 22 - R-C Pulse Response

 ****     CIRCUIT DESCRIPTION

****************************************************************************

VI 1 0 PULSE(0V 10MV 0 1NS 1NS 0.05MS .1MS)
R  1 2 5K
C  2 0 0.01UF
.TRAN 0.005M .1M
.PROBE
.OPTIONS NOPAGE
.END
```

FIG. 22.46
Input file for the conditions of Fig. 22.47.

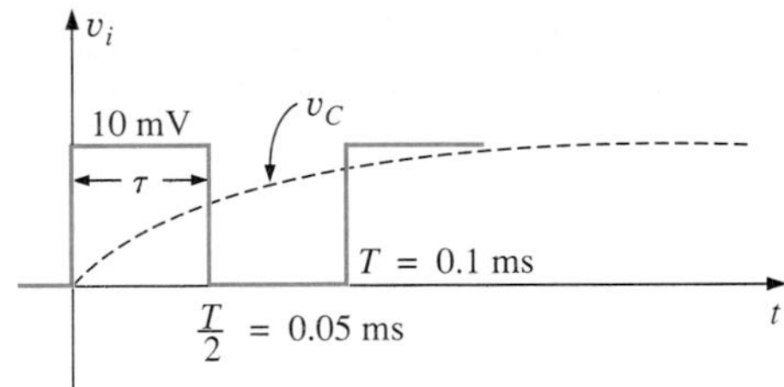

FIG. 22.47
v_C response for $5\tau > T$.

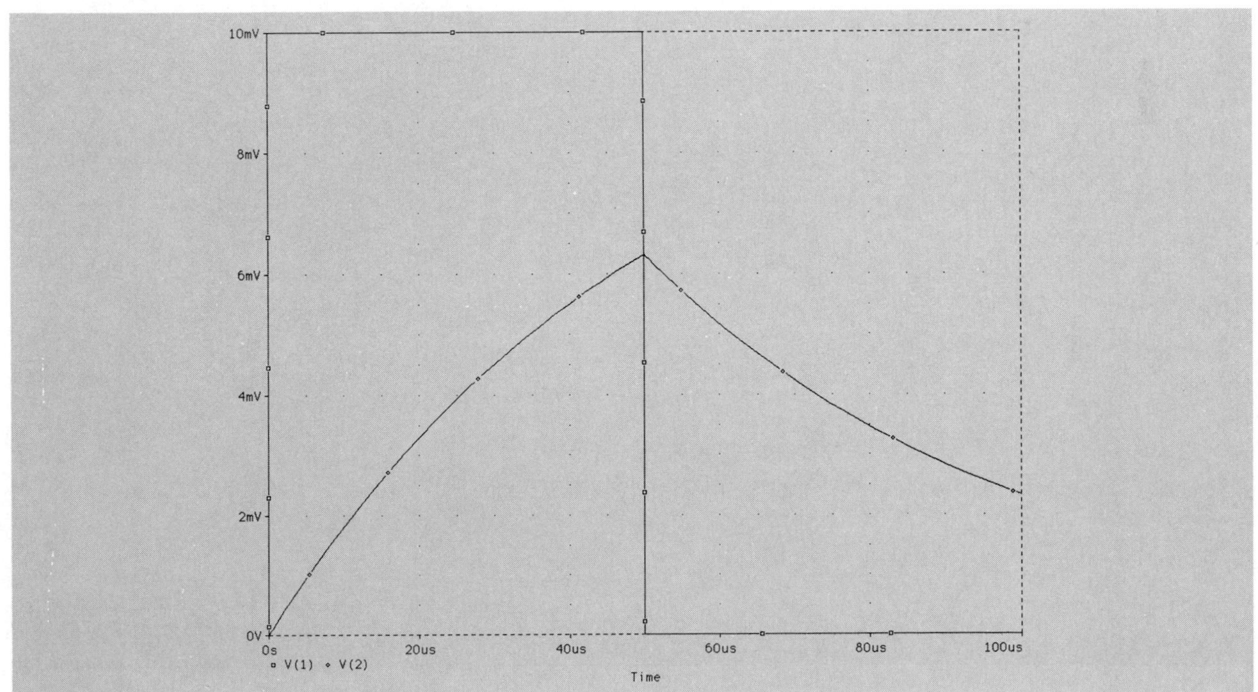

FIG. 22.48
Output response for the conditions defined by Fig. 22.47.

place is difficult with the curves of Fig. 22.48, but keep in mind that the scaling can be changed if a higher level of accuracy is required for a particular region of the output waveforms.

PSpice (Windows)

The schematics approach will be limited to a verification of the results of Fig. 22.34, displaying the voltage across a capacitor as it builds to steady-state conditions. The circuit is first established as shown in Fig. 22.49 with the same attributes employed in Fig. 22.46 to ensure the condition of $\tau = T/2$. Under **Analysis, Automatically Run Probe** was

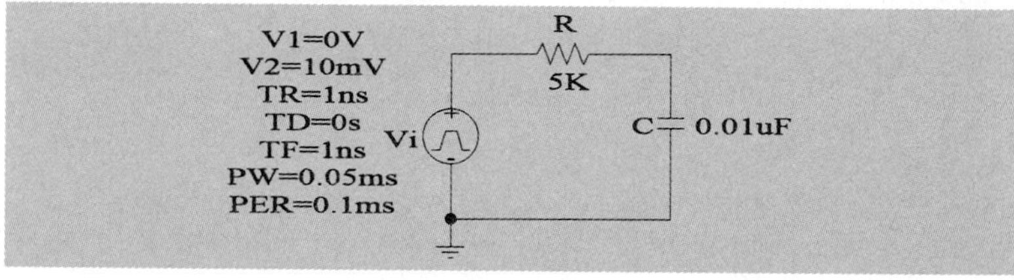

FIG. 22.49
Schematics version of the circuit of Fig. 22.31 with f = 10 kHz.

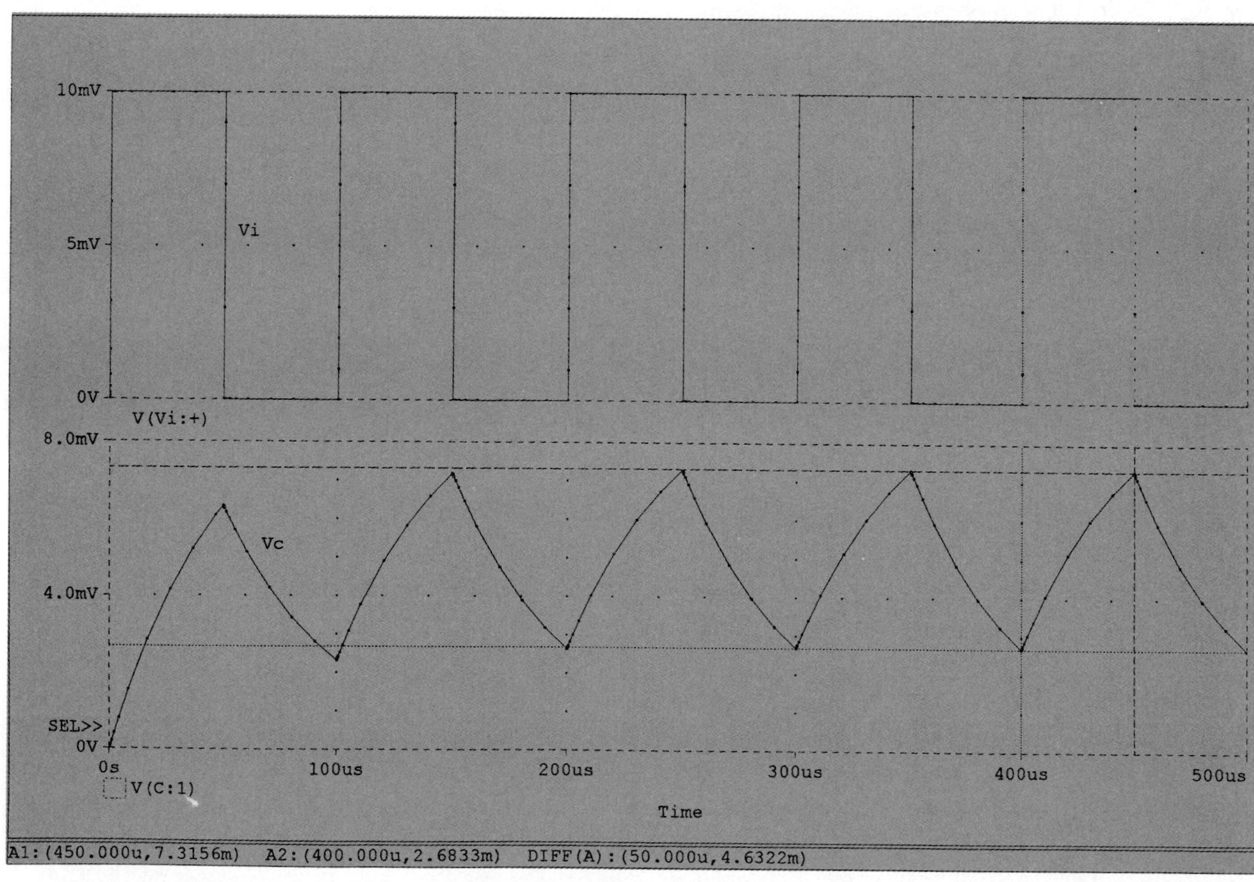

FIG. 22.50
Confirmation of the results of Fig. 22.34.

chosen, and after simulation, **Trace, Add,** and **Add Trace** led to the choice of **V(C:1)** = V_C, which will then appear on the screen as reproduced as Fig. 22.50. Then **Tools-Cursor-Display** was followed by a click of the left side of the mouse to establish a cursor line. By returning to **Tools-Cursor-Max,** the cursor will place itself at the maximum value of 7.31 mV, which occurred at 450 μs, as shown at the bottom of Fig. 22.50. Checking Fig. 22.34 we find it is a perfect match with the longhand solution. A second cursor was added with a right click of the mouse and was then moved with the right clicker down until we find the minimum point of 2.68 mV at 400 μs. Again, the result obtained is a very close match with the longhand solution. The only difference is that PSpice provided a solution in 8.73 s (with a little additional time to obtain the desired display) compared to the very time-consuming, longhand analysis in Section 22.6.

A second plot of the applied voltage V_i was then added by the sequence **Plot-Add Plot-Trace-Add-V(Vi:+).** Labels can then be added to each waveform by first using the left clicker to identify which network is to receive the label. Just click the left side anywhere on the plot, and a **SEL≫** will appear at the bottom left of the selected waveform. Then use **Tools-LABEL-Text** to enter V_i and V_C.

PROBLEMS

SECTION 22.2 Ideal Versus Actual

1. Determine the following for the pulse waveform of Fig. 22.51:
   a. positive- or negative-going?
   b. base-line voltage
   c. pulse width
   d. amplitude
   e. % tilt

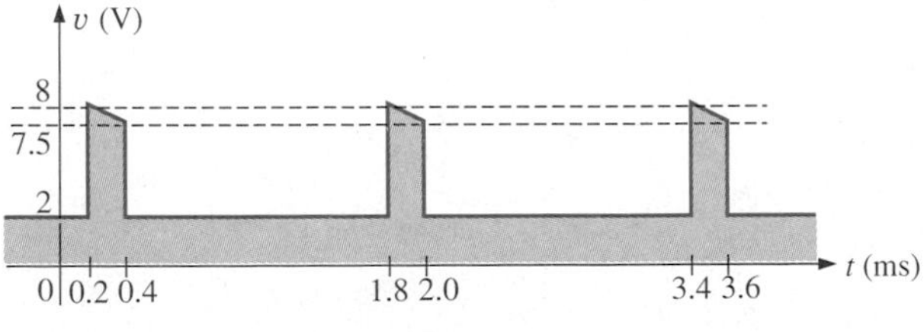

FIG. 22.51
Problems 1, 8, and 12.

2. Repeat Problem 1 for the pulse waveform of Fig. 22.52.

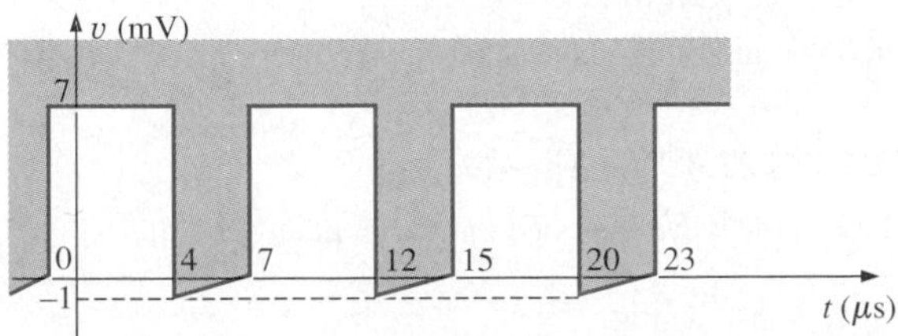

FIG. 22.52
Problems 2 and 9.

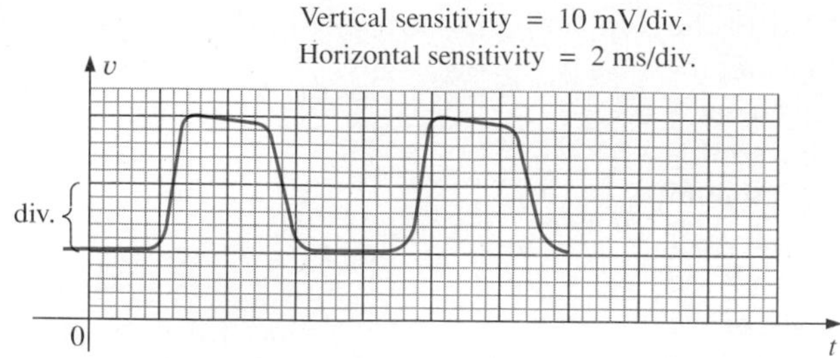

FIG. 22.53
Problems 3, 4, 10, and 13.

3. Repeat Problem 1 for the pulse waveform of Fig. 22.53.

4. Determine the rise and fall time for the waveform of Fig. 22.53.

5. Sketch a pulse waveform that has a base-line voltage of −5 mV, a pulse width of 2 μs, an amplitude of 15 mV, a 10% tilt, a period of 10 μs, and vertical sides, and is positive-going.

6. For the waveform of Fig. 22.54, established by straight-line approximations of the original waveform:
   a. Determine the rise time.
   b. Find the fall time.
   c. Find the pulse width.
   d. Calculate the frequency.

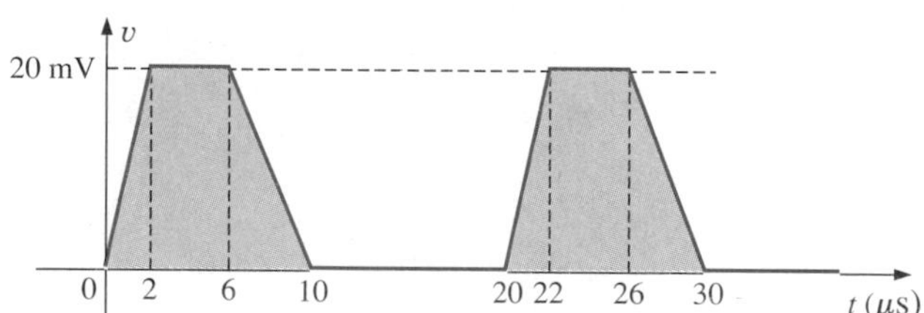

FIG. 22.54
Problems 6 and 14.

7. For the waveform of Fig. 22.55:
   a. Determine the period.
   b. Find the frequency.
   c. Find the maximum and minimum amplitude.

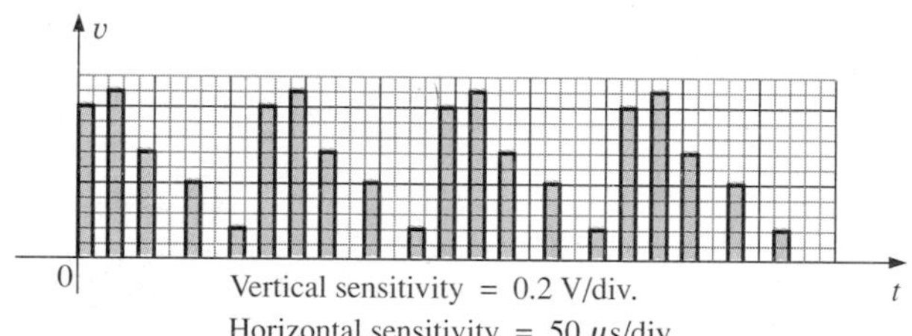

FIG. 22.55
Problems 7 and 15.

SECTION 22.3 Pulse Repetition Rate and Duty Cycle

8. Determine the pulse repetition frequency and duty cycle for the waveform of Fig. 22.51.

9. Determine the pulse repetition frequency and duty cycle for the waveform of Fig. 22.52.

10. Determine the pulse repetition frequency and duty cycle for the waveform of Fig. 22.53.

SECTION 22.4 Average Value

11. For the waveform of Fig. 22.56, determine the
    a. period.
    b. pulse width.
    c. pulse repetition frequency.
    d. average value.
    e. effective value.

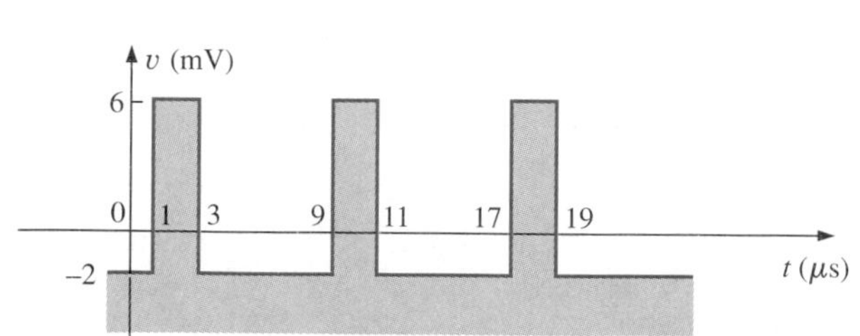

FIG. 22.56
Problem 11.

12. Determine the average value of the periodic pulse waveform of Fig. 22.51.

13. To the best accuracy possible, determine the average value of the waveform of Fig. 22.53.

14. Determine the average value of the waveform of Fig. 22.54.

15. Determine the average value of the periodic pulse train of Fig. 22.55.

SECTION 22.5 Transient *R-C* Networks

16. The capacitor of Fig. 22.57 is initially charged to 5 V, with the polarity indicated in the figure. The switch is then closed at $t = 0$ s.
   a. What is the mathematical expression for the voltage v_C?
   b. Sketch v_C versus t.
   c. What is the mathematical expression for the current i_C?
   d. Sketch i_C versus t.

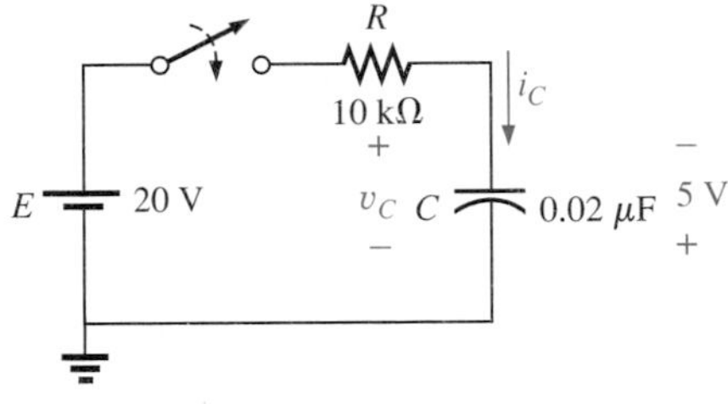

FIG. 22.57
Problem 16.

17. For the input voltage v_i appearing in Fig. 22.58, sketch the waveform for v_o. Assume that steady-state conditions were established with $v_i = 8$ V.

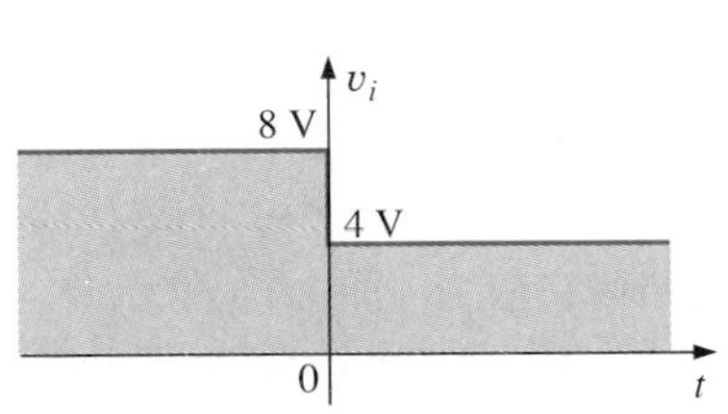

FIG. 22.58
Problem 17.

18. The switch of Fig. 22.59 is in position 1 until steady-state conditions are established. Then the switch is moved (at $t = 0$ s) to position 2. Sketch the waveform for the voltage v_C.

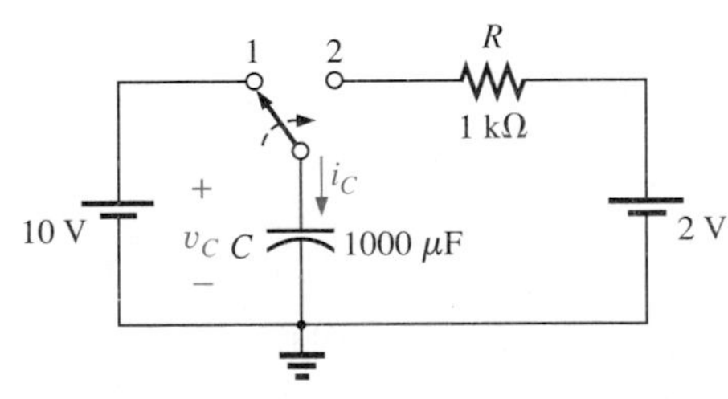

FIG. 22.59
Problems 18 and 19.

19. Sketch the waveform for i_C for Problem 18.

SECTION 22.6 *R-C* Response to Square-Wave Inputs

20. Sketch the voltage v_C for the network of Fig. 22.60 due to the square-wave input of the same figure with a frequency of
 a. 500 Hz.
 b. 100 Hz.
 c. 5000 Hz.

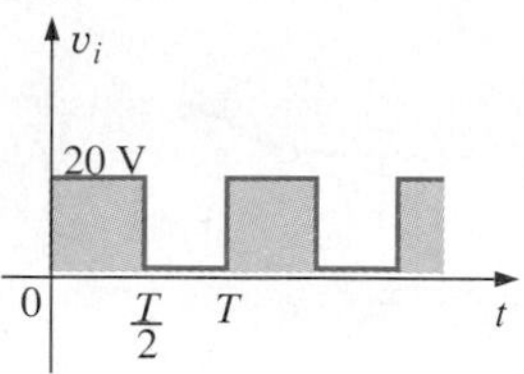

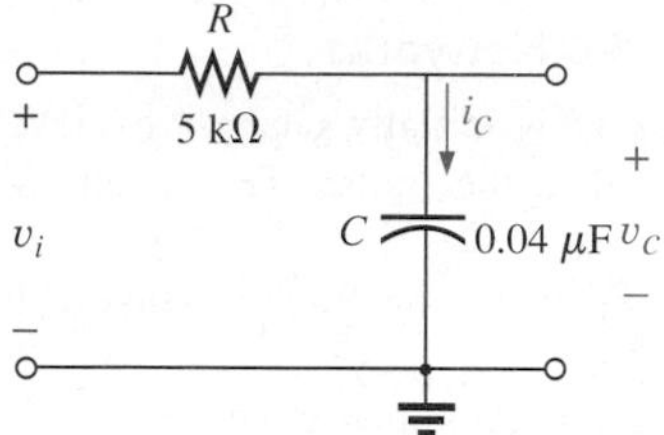

FIG. 22.60
Problems 20, 21, 23, 24, 29, 30, and 31.

21. Sketch the current i_C for each frequency of Problem 20.
22. Sketch the response (v_C) of the network of Fig. 22.60 to the square-wave input of Fig. 22.61.

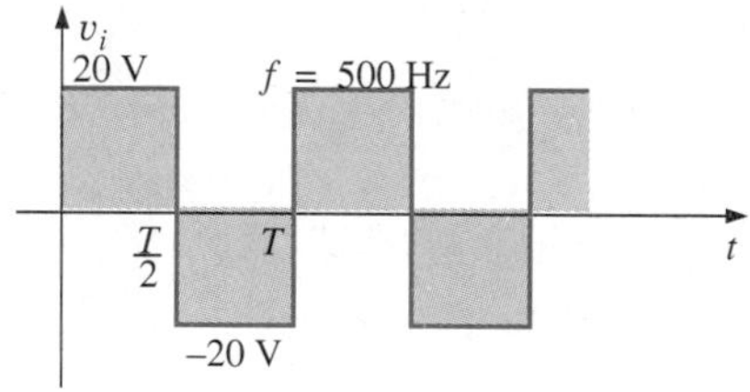

FIG. 22.61
Problems 22 and 28.

23. If the capacitor of Fig. 22.60 is initially charged to 20 V, sketch the response (v_C) to the same input signal (of Fig. 22.60) at a frequency of 500 Hz.
24. Repeat Problem 23 if the capacitor is initially charged to −10 V.

SECTION 22.7 Oscilloscope Attenuator and Compensating Probe

25. Given the network of Fig. 22.42 with $R_p = 9\text{ M}\Omega$, and $R_s = 1\text{ M}\Omega$, find $\mathbf{V}_{\text{scope}}$ in polar form if $C_p = 3$ pF, $C_s = 18$ pF, $C_c = 9$ pF, and $v_i = \sqrt{2}(100)\sin 2\pi 10{,}000t$. That is, determine $\mathbf{Z}_s$ and $\mathbf{Z}_p$, substitute into Eq. (22.8), and compare the results obtained with Eq. (22.10). Is it verified that the phase angle of $\mathbf{Z}_s$ and $\mathbf{Z}_p$ is the same under the condition $R_pC_p = R_sC_s$?
26. Repeat Problem 25 at $\omega = 10^5$ rad/s.

SECTION 22.8 Computer Analysis

PSpice (DOS)

27. Write the input file to obtain the waveforms of v_C and i_C for the network of Fig. 22.31 for a frequency of 1 kHz.
28. Write the input file to print out the plot v_C versus time as requested in Problem 22. Use an interval of $1/5\tau$ for data points from 0 to 5τ.

*29. Write the input file to print out and plot i_C versus time for the conditions of Problem 23. Consult your PSpice manual to determine how to handle initial conditions.

PSpice (Windows)

30. Using schematics, obtain the waveforms for v_C and i_C for the network of Fig. 22.60 for a frequency of 1 kHz.

*31. Using schematics, place the waveforms of v_i, v_C, and i_C on the same printout for the network of Fig. 22.60 at a frequency of 2 kHz.

*32. Using schematics, obtain the waveform appearing on the scope of Fig. 22.37 with a 20-V pulse input at a frequency of 5 kHz.

*33. Place a capacitor in parallel with R_p in Fig. 22.37 that will establish an in-phase relationship between v_{scope} and v_i. Using schematics, obtain the waveform appearing on the scope of Fig. 22.37 with a 20-V pulse input at a frequency of 5 kHz.

Programming Language (C++, BASIC, PASCAL, etc.)

34. Given a periodic pulse train such as that in Fig. 22.11, write a program to determine the average value, given the base-line voltage, peak value, and duty cycle.

35. Given the initial and final values and the network parameters (R and C), write a program to tabulate the values of v_C at each time constant (of the first five) of the transient phase.

36. For the case of $T/2 < 5\tau$, as defined by Fig. 22.29, write a program to determine the values of v_C at each half-period of the applied square wave. Test the solution by entering the conditions of Example 22.10.

GLOSSARY

Actual (true, practical) pulse A pulse waveform having a leading and trailing edge that are not vertical, along with other distortion effects such as tilt, ringing, or overshoot.

Attenuator probe A scope probe that will reduce the strength of the signal applied to the vertical channel of a scope.

Base-line voltage The voltage level from which a pulse is initiated.

Compensated attenuator probe A scope probe that can reduce the applied signal and balance the effects of the input capacitance of a scope on the signal to be displayed.

Duty cycle Reveals how much of a period is encompassed by the pulse waveform.

Fall time (t_f) The time required for the trailing edge of a pulse waveform to drop from the 90% to the 10% level.

Ideal pulse A pulse waveform characterized as having vertical sides, sharp corners, and a flat peak response.

Negative-going pulse A pulse that increases in the negative direction from the base-line voltage.

Periodic pulse train A sequence of pulses that repeats itself after a specific period of time.

Positive-going pulse A pulse that increases in the positive direction from the base-line voltage.

Pulse amplitude The peak-to-peak value of a pulse waveform.

Pulse repetition frequency The frequency of a periodic pulse train.

Pulse train A series of pulses that may have varying heights and widths.

Pulse width (t_p) The pulse width defined by the 50% voltage level.

Rise time (t_r) The time required for the leading edge of a pulse waveform to travel from the 10% to the 90% level.

Square wave A periodic pulse waveform with a 50% duty cycle.

Step function A waveform that abruptly changes from one level to another.

Tilt The drop in peak value across the pulse width of a pulse waveform.

23

Polyphase Systems

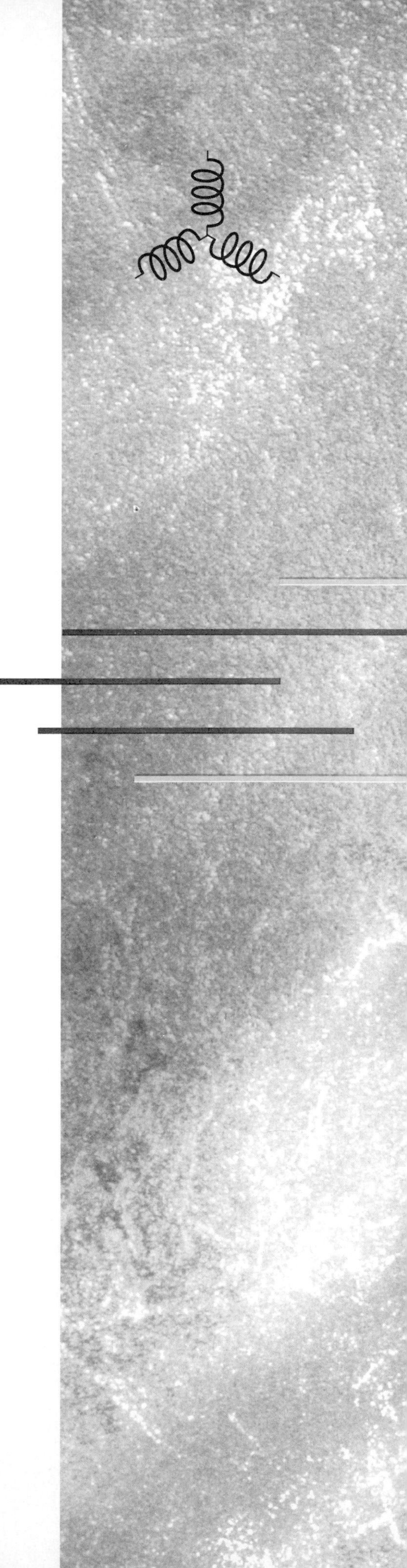

23.1 INTRODUCTION

An ac generator designed to develop a single sinusoidal voltage for each rotation of the shaft (rotor) is referred to as a *single-phase generator.* If the number of coils on the rotor is increased in a specified manner, the result is a *polyphase generator,* which develops more than one ac phase voltage per rotation of the rotor. In this chapter, the three-phase system will be discussed in detail since it is the most frequently used for power transmission.

In general, three-phase systems are preferred over single-phase systems for the transmission of power for many reasons, including the following:

1. Thinner conductors can be used to transmit the same kVA at the same voltage, which reduces the amount of copper required (typically about 25% less) and in turn reduces construction and maintenance costs.
2. The lighter lines are easier to install, and the supporting structures can be less massive and farther apart.
3. Three-phase equipment and motors have preferred running and starting characteristics compared to single-phase systems because of a more even flow of power to the transducer than can be delivered with a single-phase supply.
4. In general, most larger motors are three phase because they are essentially self-starting and do not require a special design or additional starting circuitry.

The frequency generated is determined by the number of poles on the *rotor* (the rotating part of the generator) and the speed with which the shaft is turned. Throughout the United States the line frequency is 60 Hz, whereas in Europe the chosen standard is 50 Hz. Both frequencies were chosen primarily because they can be generated by a relatively efficient and stable mechanical design that is sensitive to the size of the generating systems and the demand that must be met during peak

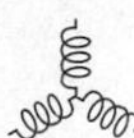

periods. On aircraft and ships the demand levels permit the use of a 400-Hz line frequency.

The three-phase system is used by almost all commercial electric generators. This does not mean that single-phase and two-phase generating systems are obsolete. Most small emergency generators, such as the gasoline type, are one-phase generating systems. The two-phase system is commonly used in servomechanisms, which are self-correcting control systems capable of detecting and adjusting their own operation. Servomechanisms are used in ships and aircraft to keep them on course automatically, or, in simpler devices such as a thermostatic circuit, to regulate heat output. In many cases, however, where single-phase and two-phase inputs are required, they are supplied by one and two phases of a three-phase generating system rather than generated independently.

The number of phase voltages that can be produced by a polyphase generator is not limited to three. Any number of phases can be obtained by spacing the windings for each phase at the proper angular position around the rotor. Some electrical systems operate more efficiently if more than three phases are used. One such system involves the process of rectification, which is used to convert an alternating output to one having an average, or dc, value. The greater the number of phases, the smoother the dc output of the system.

23.2 THE THREE-PHASE GENERATOR

The three-phase generator of Fig. 23.1(a) has three induction coils placed 120° apart on the rotor (armature), as shown symbolically by Fig. 23.1(b). Since the three coils have an equal number of turns, and each coil rotates with the same angular velocity, the voltage induced across each coil will have the same peak value, shape, and frequency. As the shaft of the generator is turned by some external means, the induced voltages e_{AN}, e_{BN}, and e_{CN} will be generated simultaneously, as shown in Fig. 23.2. Note the 120° phase shift between waveforms and the similarities in appearance of the three sinusoidal functions.

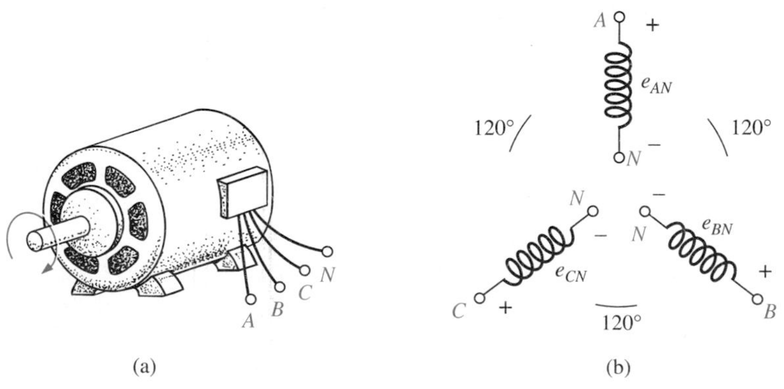

FIG. 23.1
(a) Three-phase generator; (b) induced voltages of a three-phase generator.

In particular, note that

at any instant of time, the algebraic sum of the three phase voltages of a three-phase generator is zero.

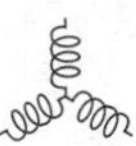

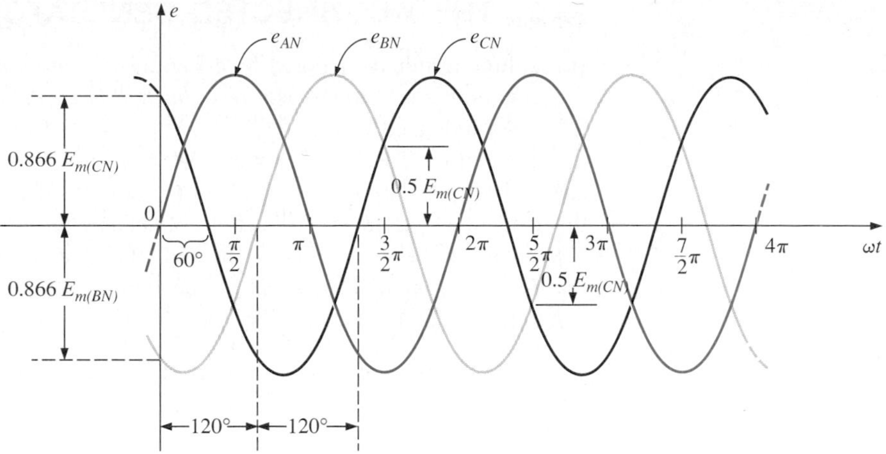

FIG. 23.2
Phase voltages of a three-phase generator.

This is shown at $\omega t = 0$ in Fig. 23.2, where it is also evident that *when one induced voltage is zero, the other two are 86.6% of their positive or negative maximums. In addition, when any two are equal in magnitude and sign (at $0.5E_m$), the remaining induced voltage has the opposite polarity and a peak value.*

The sinusoidal expression for each of the induced voltages of Fig. 23.2 is

$$\begin{aligned} e_{AN} &= E_{m(AN)} \sin \omega t \\ e_{BN} &= E_{m(BN)} \sin(\omega t - 120°) \\ e_{CN} &= E_{m(CN)} \sin(\omega t - 240°) = E_{m(CN)} \sin(\omega t + 120°) \end{aligned} \tag{23.1}$$

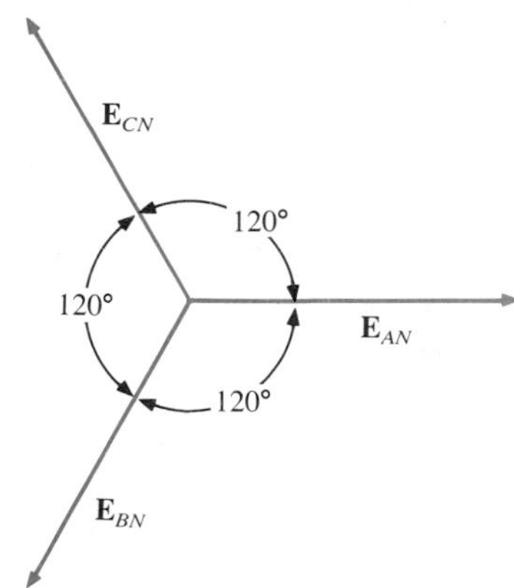

FIG. 23.3
Phasor diagram for the phase voltages of a three-phase generator.

The phasor diagram of the induced voltages is shown in Fig. 23.3, where the effective value of each is determined by

$$E_{AN} = 0.707E_{m(AN)}$$
$$E_{BN} = 0.707E_{m(BN)}$$
$$E_{CN} = 0.707E_{m(CN)}$$

and

$$\mathbf{E}_{AN} = E_{AN} \angle 0°$$
$$\mathbf{E}_{BN} = E_{BN} \angle -120°$$
$$\mathbf{E}_{CN} = E_{CN} \angle +120°$$

By rearranging the phasors as shown in Fig. 23.4 and applying a law of vectors which states that *the vector sum of any number of vectors drawn such that the "head" of one is connected to the "tail" of the next, and that the head of the last vector is connected to the tail of the first is zero,* we can conclude that the phasor sum of the phase voltages in a three-phase system is zero. That is,

$$\Sigma\, \mathbf{E}_{AN} + \mathbf{E}_{BN} + \mathbf{E}_{CN} = 0 \tag{23.2}$$

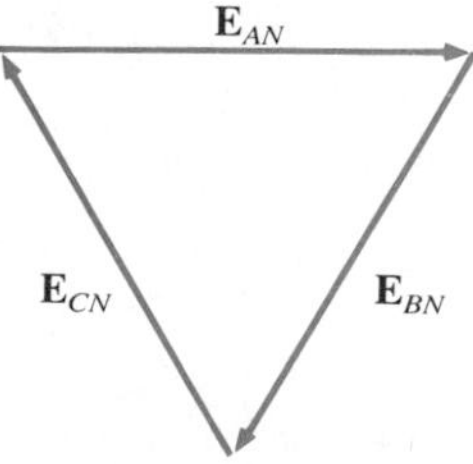

FIG. 23.4
Demonstrating that the vector sum of the phase voltages of a three-phase generator is zero.

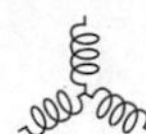

23.3 THE Y-CONNECTED GENERATOR

If the three terminals denoted N of Fig. 23.1(b) are connected together, the generator is referred to as a *Y-connected three-phase generator* (Fig. 23.5). As indicated in Fig. 23.5, the Y is inverted for ease of notation and for clarity. The point at which all the terminals are connected is called the *neutral point*. If a conductor is not attached from this point to the load, the system is called a *Y-connected, three-phase, three-wire generator*. If the neutral is connected, the system is a *Y-connected, three-phase, four-wire generator*. The function of the neutral will be discussed in detail when we consider the load circuit.

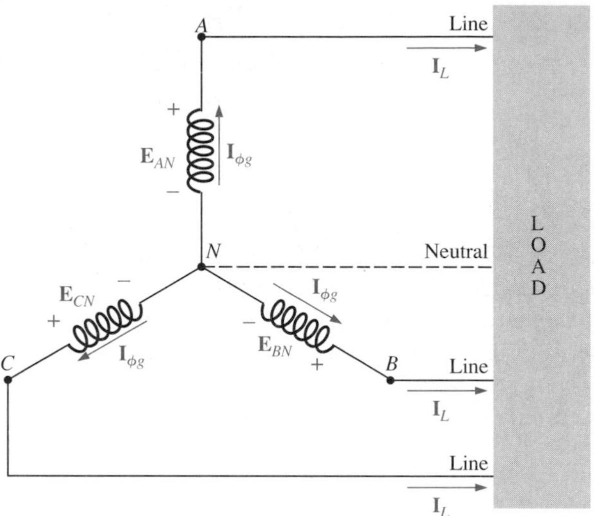

FIG. 23.5
Y-connected generator.

The three conductors connected from A, B, and C to the load are called *lines*. For the Y-connected system, it should be obvious from Fig. 23.5 that the line current equals the phase current for each phase; that is,

$$\mathbf{I}_L = \mathbf{I}_{\phi g} \tag{23.3}$$

where ϕ is used to denote a phase quantity and g is a generator parameter.

The voltage from one line to another is called a *line voltage*. On the phasor diagram (Fig. 23.6) it is the phasor drawn from the end of one phase to another in the counterclockwise direction.

Applying Kirchhoff's voltage law around the indicated loop of Fig. 23.6, we obtain

$$\mathbf{E}_{AB} - \mathbf{E}_{AN} + \mathbf{E}_{BN} = 0$$

or

$$\mathbf{E}_{AB} = \mathbf{E}_{AN} - \mathbf{E}_{BN} = \mathbf{E}_{AN} + \mathbf{E}_{NB}$$

The phasor diagram is redrawn to find $\mathbf{E}_{AB}$ as shown in Fig. 23.7. Since each phase voltage, when reversed ($\mathbf{E}_{NB}$), will bisect the other two, $\alpha = 60°$. The angle β is 30° since a line drawn from opposite ends of a rhombus will divide in half both the angle of origin and the opposite angle. Lines drawn between opposite corners of a rhombus will also bisect each other at right angles.

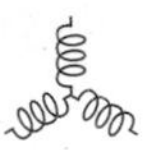

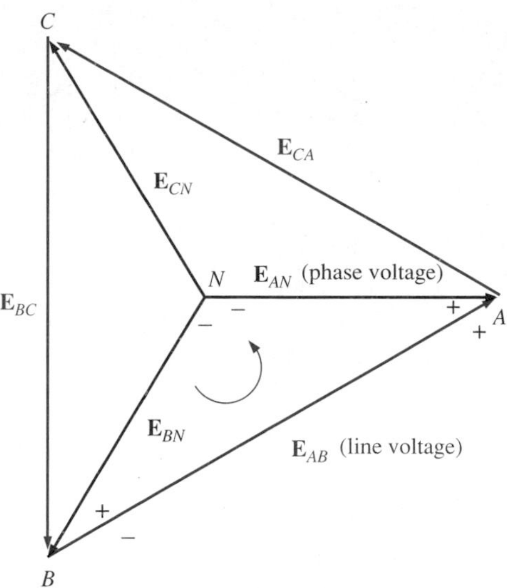

FIG. 23.6
Line and phase voltages of the Y-connected three-phase generator.

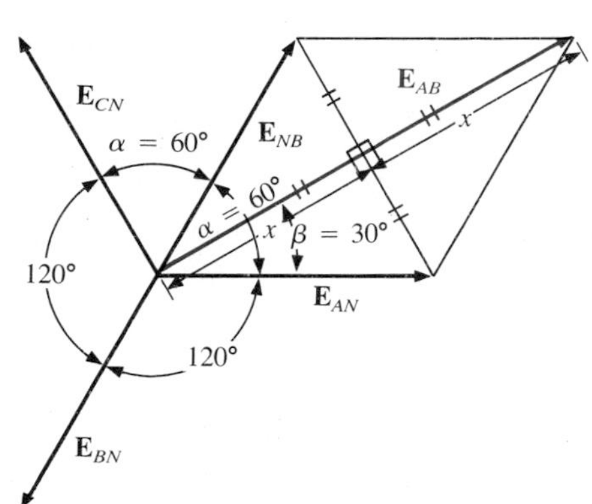

FIG. 23.7
Determining a line voltage for a three-phase generator.

The length x is

$$x = E_{AN} \cos 30^\circ = \frac{\sqrt{3}}{2} E_{AN}$$

and
$$E_{AB} = 2x = (2)\frac{\sqrt{3}}{2} E_{AN} = \sqrt{3}E_{AN}$$

Noting from the phasor diagram that θ of $\mathbf{E}_{AB} = \beta = 30^\circ$, the result is

$$\mathbf{E}_{AB} = E_{AB} \angle 30^\circ = \sqrt{3}E_{AN} \angle 30^\circ$$

and
$$\mathbf{E}_{CA} = \sqrt{3}E_{CA} \angle 150^\circ$$
$$\mathbf{E}_{BC} = \sqrt{3}E_{BC} \angle 270^\circ$$

In words, the magnitude of the line voltage of a Y-connected generator is $\sqrt{3}$ times the phase voltage:

$$\boxed{E_L = \sqrt{3}E_\phi} \qquad \textbf{(23.4)}$$

with the phase angle between any line voltage and the nearest phase voltage at 30°.

In sinusoidal notation,

$$e_{AB} = \sqrt{2}E_{AB} \sin(\omega t + 30^\circ)$$
$$e_{CA} = \sqrt{2}E_{CA} \sin(\omega t + 150^\circ)$$

and
$$e_{BC} = \sqrt{2}E_{BC} \sin(\omega t + 270^\circ)$$

The phasor diagram of the line and phase voltages is shown in Fig. 23.8. If the phasors representing the line voltages in Fig. 23.8(a) are rearranged slightly, they will form a closed loop [Fig. 23.8(b)]. Therefore, we can conclude that the sum of the line voltages is also zero; that is,

$$\boxed{\Sigma\, \mathbf{E}_{AB} + \mathbf{E}_{CA} + \mathbf{E}_{BC} = 0} \qquad \textbf{(23.5)}$$

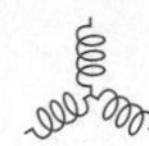

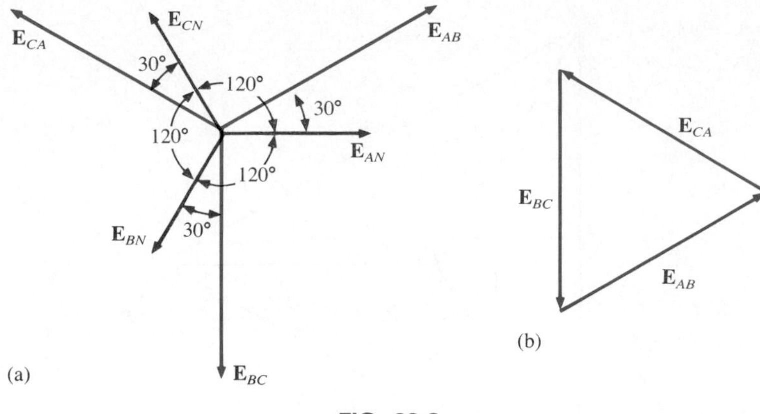

FIG. 23.8
(a) Phasor diagram of the line and phase voltages of a three-phase generator; (b) demonstrating that the vector sum of the line voltages of a three-phase system is zero.

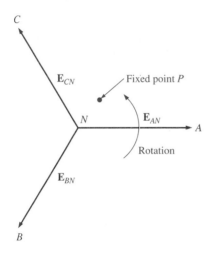

FIG. 23.9
Determining the phase sequence from the phase voltages of a three-phase generator.

23.4 PHASE SEQUENCE (Y-CONNECTED GENERATOR)

The phase sequence can be determined by the order in which the phasors representing the phase voltages pass through a fixed point on the phasor diagram if the phasors are rotated in a counterclockwise direction. For example, in Fig. 23.9 the phase sequence is *ABC*. However, since the fixed point can be chosen anywhere on the phasor diagram, the sequence can also be written as *BCA* or *CAB*. The phase sequence is quite important in the three-phase distribution of power. In a three-phase motor, for example, if two phase voltages are interchanged, the sequence will change and the direction of rotation of the motor will be reversed. Other effects will be described when we consider the loaded three-phase system.

The phase sequence can also be described in terms of the line voltages. Drawing the line voltages on a phasor diagram in Fig. 23.10, we are able to determine the phase sequence by again rotating the phasors in the counterclockwise direction. In this case, however, the sequence can be determined by noting the order of the passing first or second subscripts. In the system of Fig. 23.10, for example, the phase sequence of the first subscripts passing point *P* is *ABC*, and the phase sequence of the second subscripts is *BCA*. But we know that *BCA* is equivalent to *ABC*, so the sequence is the same for each. Note that the phase sequence is the same as that of the phase voltages described in Fig. 23.9.

If the sequence is given, the phasor diagram can be drawn by simply picking a reference voltage, placing it on the reference axis, and then drawing the other voltages at the proper angular position. For a sequence of *ACB*, for example, we might choose E_{AB} to be the reference [Fig. 23.11(a)] if we wanted the phasor diagram of the line voltages, or E_{NA} for the phase voltages [Fig. 23.11(b)]. For the sequence indicated, the phasor diagrams would be as in Fig. 23.11. In phasor notation,

$$\text{Line voltages} \begin{cases} \mathbf{E}_{AB} = E_{AB}\angle 0^\circ & \text{(reference)} \\ \mathbf{E}_{CA} = E_{CA}\angle -120^\circ \\ \mathbf{E}_{BC} = E_{BC}\angle +120^\circ \end{cases}$$

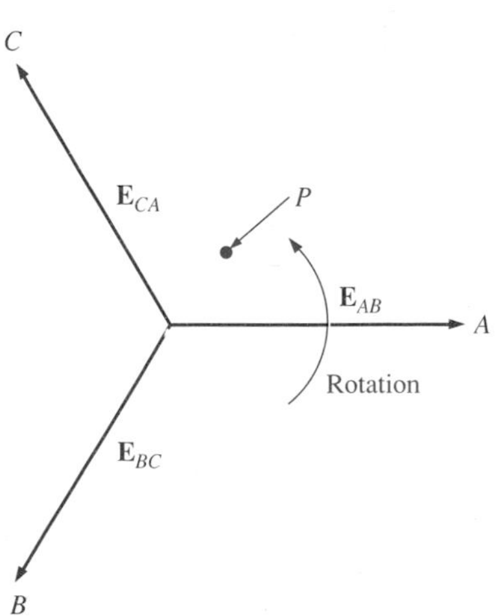

FIG. 23.10
Determining the phase sequence from the line voltages of a three-phase generator.

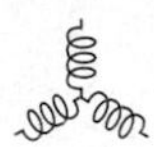

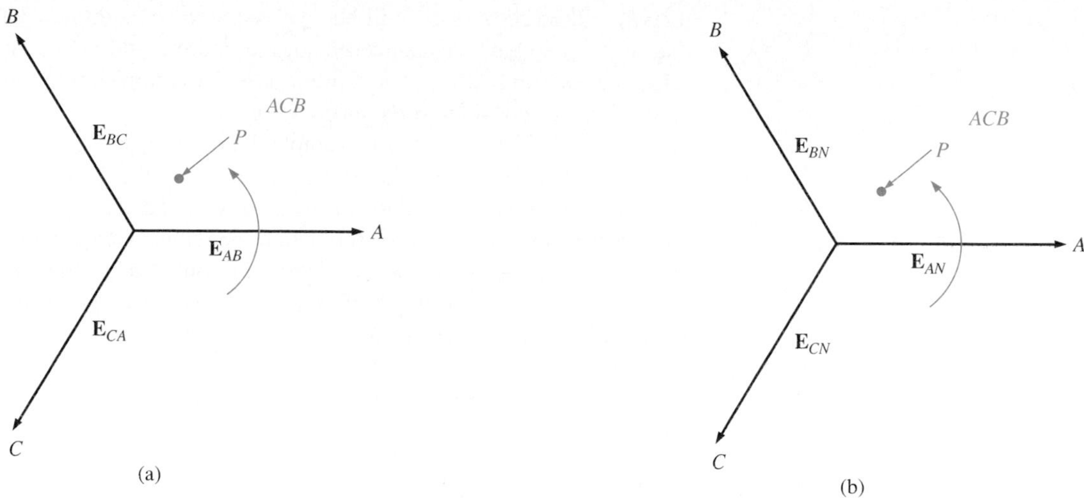

FIG. 23.11
Drawing the phasor diagram from the phase sequence.

Phase voltages

$$\begin{cases} \mathbf{E}_{AN} = E_{AN} \angle 0^\circ & \text{(reference)} \\ \mathbf{E}_{CN} = E_{CN} \angle -120^\circ \\ \mathbf{E}_{BN} = E_{BN} \angle +120^\circ \end{cases}$$

23.5 THE Y-CONNECTED GENERATOR WITH A Y-CONNECTED LOAD

Loads connected to three-phase supplies are of two types: the Y and the Δ. If a Y-connected load is connected to a Y-connected generator, the system is symbolically represented by Y-Y. The physical setup of such a system is shown in Fig. 23.12.

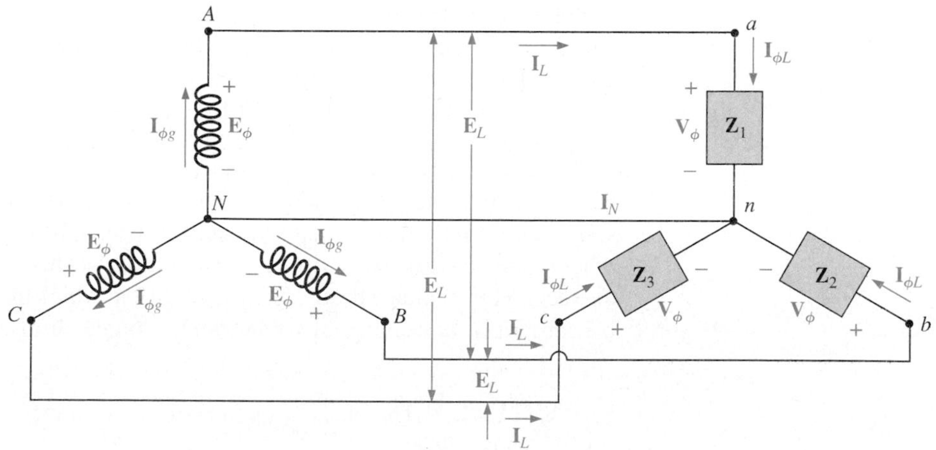

FIG. 23.12
Y-connected generator with a Y-connected load.

If the load is balanced, the neutral can be removed without affecting the circuit in any manner; that is, if

$$\mathbf{Z}_1 = \mathbf{Z}_2 = \mathbf{Z}_3$$

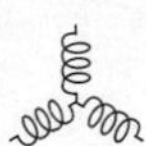

then I_N will be zero. (This will be demonstrated in Example 23.1.) Note that in order to have a balanced load, the phase angle must also be the same for each impedance—a condition that was unnecessary in dc circuits when we considered balanced systems.

In practice, if a factory, for example, had only balanced, three-phase loads, the absence of the neutral would have no effect since ideally, the system would always be balanced. The cost would therefore be less since the number of required conductors would be reduced. However, lighting and most other electrical equipment will use only one of the phase voltages, and even if the loading is designed to be balanced (as it should be), there will never be perfect continuous balancing since lights and other electrical equipment will be turned on and off, upsetting the balanced condition. The neutral is therefore necessary to carry the resulting current away from the load and back to the Y-connected generator. This will be demonstrated when we consider unbalanced Y-connected systems.

We shall now examine the *four-wire Y-Y-connected system*. The current passing through each phase of the generator is the same as its corresponding line current, which in turn for a Y-connected load is equal to the current in the phase of the load to which it is attached:

$$\boxed{\mathbf{I}_{\phi g} = \mathbf{I}_L = \mathbf{I}_{\phi L}} \qquad \textbf{(23.6)}$$

For a balanced or unbalanced load, since the generator and load have a common neutral point, then

$$\boxed{\mathbf{V}_\phi = \mathbf{E}_\phi} \qquad \textbf{(23.7)}$$

In addition, since $\mathbf{I}_{\phi L} = \mathbf{V}_\phi/\mathbf{Z}_\phi$, the magnitude of the current in each phase will be equal for a balanced load and unequal for an unbalanced load. You will recall that for the Y-connected generator, the magnitude of the line voltage is equal to $\sqrt{3}$ times the phase voltage. This same relationship can be applied to a balanced or unbalanced four-wire Y-connected load:

$$\boxed{E_L = \sqrt{3}V_\phi} \qquad \textbf{(23.8)}$$

For a voltage drop across a load element, the first subscript refers to that terminal through which the current enters the load element, and the second subscript refers to the terminal from which the current leaves. In other words, the first subscript is, by definition, positive with respect to the second for a voltage drop. Note Fig. 23.13, in which the standard double subscripts for a source of voltage and a voltage drop are indicated.

EXAMPLE 23.1 The phase sequence of the Y-connected generator in Fig. 23.13 is *ABC*.

a. Find the phase angles θ_2 and θ_3.
b. Find the magnitude of the line voltages.
c. Find the line currents.
d. Verify that, since the load is balanced, $\mathbf{I}_N = 0$.

Solutions:

a. For an *ABC* phase sequence,

$$\theta_2 = \mathbf{-120^\circ} \quad \text{and} \quad \theta_3 = \mathbf{+120^\circ}$$

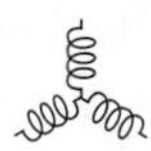

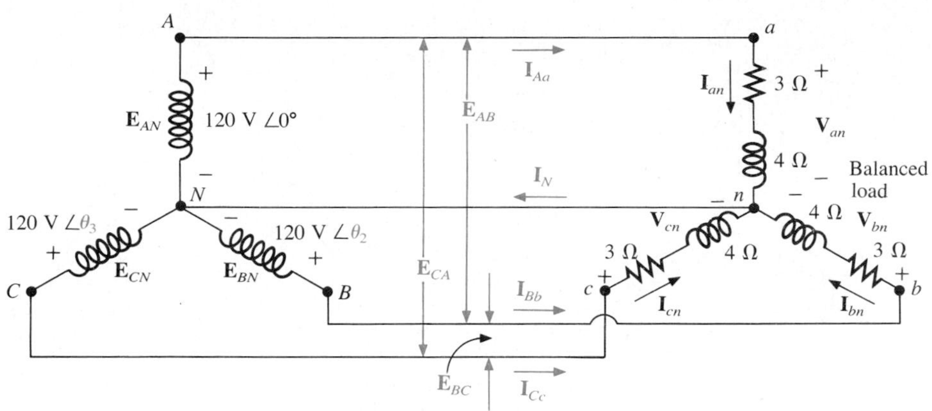

FIG. 23.13
Example 23.1.

b. $E_L = \sqrt{3}E_\phi = (1.73)(120\text{ V}) = 208\text{ V}$. Therefore,

$$E_{AB} = E_{BC} = E_{CA} = \mathbf{208\ V}$$

c. $\mathbf{V}_\phi = \mathbf{E}_\phi$. Therefore,

$$\mathbf{V}_{an} = \mathbf{E}_{AN} \qquad \mathbf{V}_{bn} = \mathbf{E}_{BN} \qquad \mathbf{V}_{cn} = \mathbf{E}_{CN}$$

$$\mathbf{I}_{\phi L} = \mathbf{I}_{an} = \frac{\mathbf{V}_{an}}{\mathbf{Z}_{an}} = \frac{120\text{ V}\angle 0^\circ}{3\ \Omega + j\,4\ \Omega} = \frac{120\text{ V}\angle 0^\circ}{5\ \Omega\angle 53.13^\circ}$$

$$= 24\text{ A}\angle -53.13^\circ$$

$$\mathbf{I}_{bn} = \frac{\mathbf{V}_{bn}}{\mathbf{Z}_{bn}} = \frac{120\text{ V}\angle -120^\circ}{5\ \Omega\angle 53.13^\circ} = 24\text{ A}\angle -173.13^\circ$$

$$\mathbf{I}_{cn} = \frac{\mathbf{V}_{cn}}{\mathbf{Z}_{cn}} = \frac{120\text{ V}\angle +120^\circ}{5\ \Omega\angle 53.13^\circ} = 24\text{ A}\angle 66.87^\circ$$

and, since $\mathbf{I}_L = \mathbf{I}_{\phi L}$,

$$\mathbf{I}_{Aa} = \mathbf{I}_{an} = \mathbf{24\ A\angle -53.13^\circ}$$
$$\mathbf{I}_{Bb} = \mathbf{I}_{bn} = \mathbf{24\ A\angle -173.13^\circ}$$
$$\mathbf{I}_{Cc} = \mathbf{I}_{cn} = \mathbf{24\ A\angle 66.87^\circ}$$

d. Applying Kirchhoff's current law, we have

$$\mathbf{I}_N = \mathbf{I}_{Aa} + \mathbf{I}_{Bb} + \mathbf{I}_{Cc}$$

In rectangular form,

$$\begin{aligned} \mathbf{I}_{Aa} &= 24\text{ A}\angle -53.13^\circ &&= 14.40\text{ A} - j\,19.20\text{ A} \\ \mathbf{I}_{Bb} &= 24\text{ A}\angle -173.13^\circ &&= -23.83\text{ A} - j\,2.87\text{ A} \\ \mathbf{I}_{Cc} &= 24\text{ A}\angle 66.87^\circ &&= 9.43\text{ A} + j\,22.07\text{ A} \\ \Sigma(\mathbf{I}_{Aa} + \mathbf{I}_{Bb} + \mathbf{I}_{Cc}) & &&= 0 + j\,0 \end{aligned}$$

and $\mathbf{I}_N$ is in fact equal to **zero,** as required for a balanced load.

23.6 THE Y-Δ SYSTEM

There is no neutral connection for the Y-Δ system of Fig. 23.14. Any variation in the impedance of a phase that produces an unbalanced system will simply vary the line and phase currents of the system.

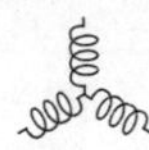

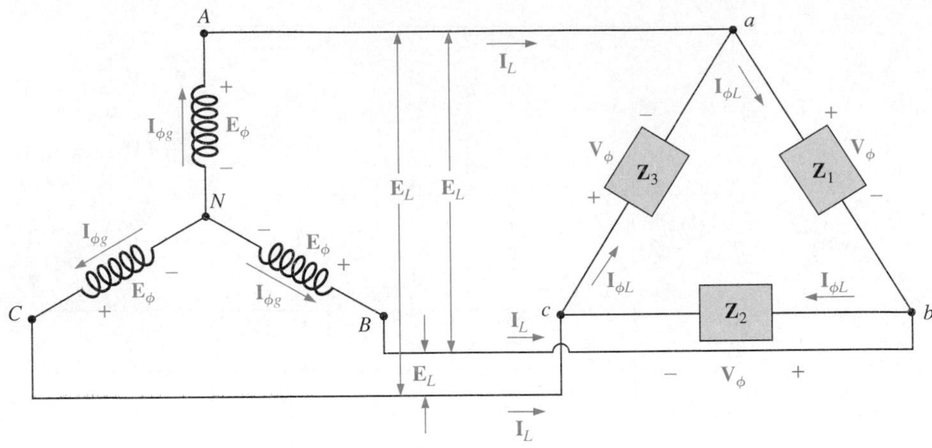

FIG. 23.14
Y-connected generator with a Δ-connected load.

For a balanced load,

$$\mathbf{Z}_1 = \mathbf{Z}_2 = \mathbf{Z}_3 \tag{23.9}$$

The voltage across each phase of the load is equal to the line voltage of the generator for a balanced or unbalanced load:

$$\mathbf{V}_\phi = \mathbf{E}_L \tag{23.10}$$

The relationship between the line currents and phase currents of a balanced Δ load can be found using an approach very similar to that used in Section 23.3 to find the relationship between the line voltages and phase voltages of a Y-connected generator. For this case, however, Kirchhoff's current law is employed instead of Kirchhoff's voltage law.

The results obtained are

$$I_L = \sqrt{3}I_\phi \tag{23.11}$$

and the phase angle between a line current and the nearest phase current is 30°. A more detailed discussion of this relationship between the line and phase currents of a Δ-connected system can be found in Section 23.7.

For a balanced load, the line currents will be equal in magnitude, as will the phase currents.

EXAMPLE 23.2 For the three-phase system of Fig. 23.15:

a. Find the phase angles θ_2 and θ_3.
b. Find the current in each phase of the load.
c. Find the magnitude of the line currents.

Solutions:

a. For an *ABC* sequence,

$$\theta_2 = \mathbf{-120°} \quad \text{and} \quad \theta_3 = \mathbf{+120°}$$

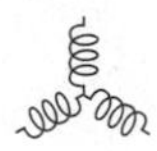

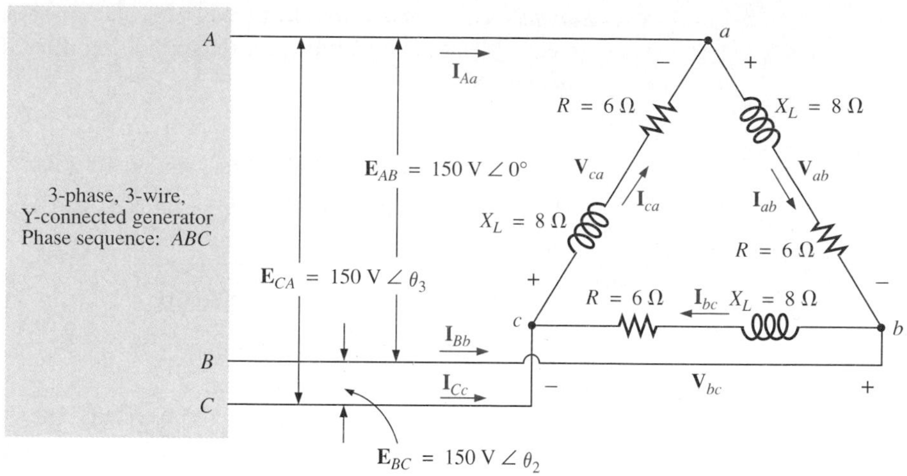

FIG. 23.15
Example 23.2.

b. $\mathbf{V}_\phi = \mathbf{E}_L$. Therefore,

$$\mathbf{V}_{ab} = \mathbf{E}_{AB} \qquad \mathbf{V}_{ca} = \mathbf{E}_{CA} \qquad \mathbf{V}_{bc} = \mathbf{E}_{BC}$$

The phase currents are

$$\mathbf{I}_{ab} = \frac{\mathbf{V}_{ab}}{\mathbf{Z}_{ab}} = \frac{150 \text{ V} \angle 0°}{6\ \Omega + j\,8\ \Omega} = \frac{150 \text{ V} \angle 0°}{10\ \Omega \angle 53.13°} = \mathbf{15\ A \angle -53.13°}$$

$$\mathbf{I}_{bc} = \frac{\mathbf{V}_{bc}}{\mathbf{Z}_{bc}} = \frac{150 \text{ V} \angle -120°}{10\ \Omega \angle 53.13°} = \mathbf{15\ A \angle -173.13°}$$

$$\mathbf{I}_{ca} = \frac{\mathbf{V}_{ca}}{\mathbf{Z}_{ca}} = \frac{150 \text{ V} \angle +120°}{10\ \Omega \angle 53.13°} = \mathbf{15\ A \angle 66.87°}$$

c. $I_L = \sqrt{3} I_\phi = (1.73)(15 \text{ A}) = 25.95 \text{ A}$. Therefore,

$$I_{Aa} = I_{Bb} = I_{Cc} = \mathbf{25.95\ A}$$

23.7 THE Δ-CONNECTED GENERATOR

If we rearrange the coils of the generator in Fig. 23.16(a) as shown in Fig. 23.16(b), the system is referred to as a *three-phase, three-wire,*

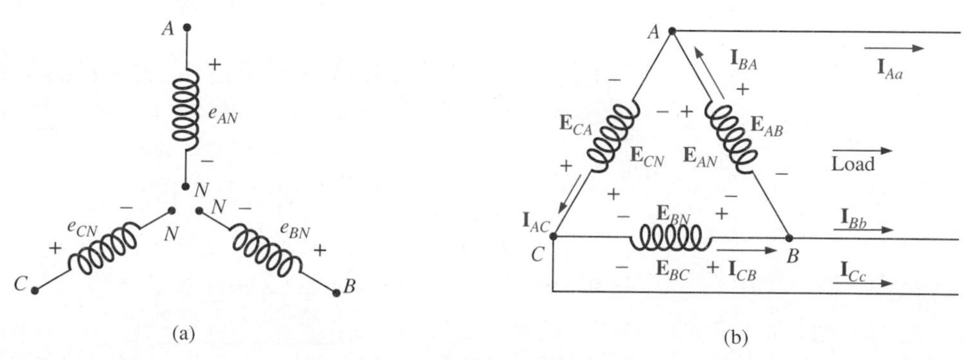

FIG. 23.16
Δ-connected generator.

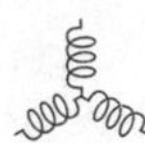

Δ-*connected ac generator*. In this system, the phase and line voltages are equivalent and equal to the voltage induced across each coil of the generator; that is,

$$\left.\begin{array}{lll} \mathbf{E}_{AB} = \mathbf{E}_{AN} & \text{and} & e_{AN} = \sqrt{2}E_{AN}\sin\omega t \\ \mathbf{E}_{BC} = \mathbf{E}_{BN} & \text{and} & e_{BN} = \sqrt{2}E_{BN}\sin(\omega t - 120^\circ) \\ \mathbf{E}_{CA} = \mathbf{E}_{CN} & \text{and} & e_{CN} = \sqrt{2}E_{CN}\sin(\omega t + 120^\circ) \end{array}\right\} \begin{array}{l}\text{Phase}\\ \text{sequence}\\ ABC\end{array}$$

or

$$\boxed{\mathbf{E}_L = \mathbf{E}_{\phi g}} \tag{23.12}$$

Note that only one voltage (magnitude) is available instead of the two available in the Y-connected system.

Unlike the line current for the Y-connected generator, the line current for the Δ-connected system is not equal to the phase current. The relationship between the two can be found by applying Kirchhoff's current law at one of the nodes and solving for the line current in terms of the phase currents; that is, at node A,

$$\mathbf{I}_{BA} = \mathbf{I}_{Aa} + \mathbf{I}_{AC}$$

or

$$\mathbf{I}_{Aa} = \mathbf{I}_{BA} - \mathbf{I}_{AC} = \mathbf{I}_{BA} + \mathbf{I}_{CA}$$

The phasor diagram is shown in Fig. 23.17 for a balanced load.

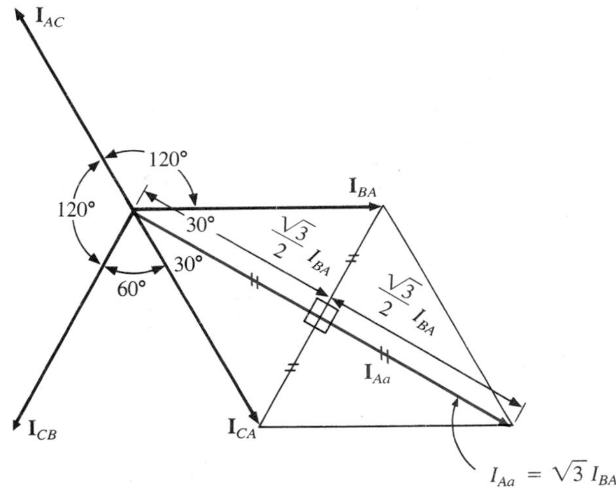

FIG. 23.17
Determining a line current from the phase currents of a Δ-connected, three-phase generator.

Using the same procedure to find the line current as was used to find the line voltage of a Y-connected generator produces the following:

$$I_{Aa} = \sqrt{3}I_{BA}\angle -30^\circ$$
$$I_{Bb} = \sqrt{3}I_{CB}\angle -150^\circ$$
$$I_{Cc} = \sqrt{3}I_{AC}\angle 90^\circ$$

In general:

$$\boxed{I_L = \sqrt{3}I_{\phi g}} \tag{23.13}$$

with the phase angle between a line current and the nearest phase current at 30°. The phasor diagram of the currents is shown in Fig. 23.18.

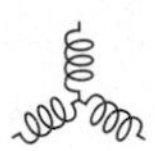

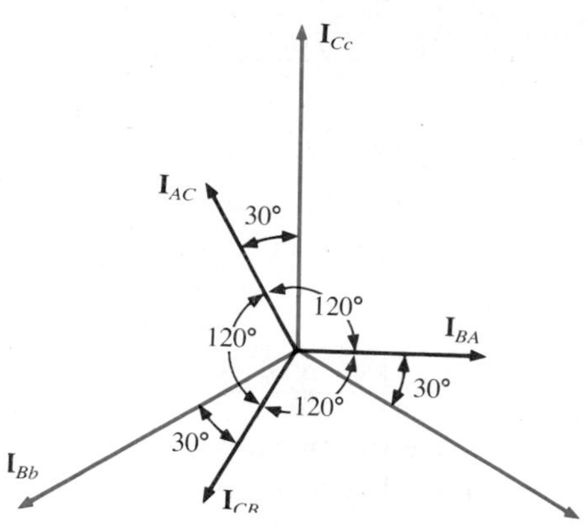

FIG. 23.18
The phasor diagram of the currents of a three-phase, Δ-connected generator.

It can be shown in the same manner employed for the voltages of a Y-connected generator that the phasor sum of the line currents or phase currents for Δ-connected systems with balanced loads is zero.

23.8 PHASE SEQUENCE (Δ-CONNECTED GENERATOR)

Even though the line and phase voltages of a Δ-connected system are the same, it is standard practice to describe the phase sequence in terms of the line voltages. The method used is the same as that described for the line voltages of the Y-connected generator. For example, the phasor diagram of the line voltages for a phase sequence *ABC* is shown in Fig. 23.19. In drawing such a diagram, one must take care to have the sequence of the first and second subscripts the same. In phasor notation,

$$\mathbf{E}_{AB} = E_{AB} \angle 0^\circ$$
$$\mathbf{E}_{BC} = E_{BC} \angle -120^\circ$$
$$\mathbf{E}_{CA} = E_{CA} \angle 120^\circ$$

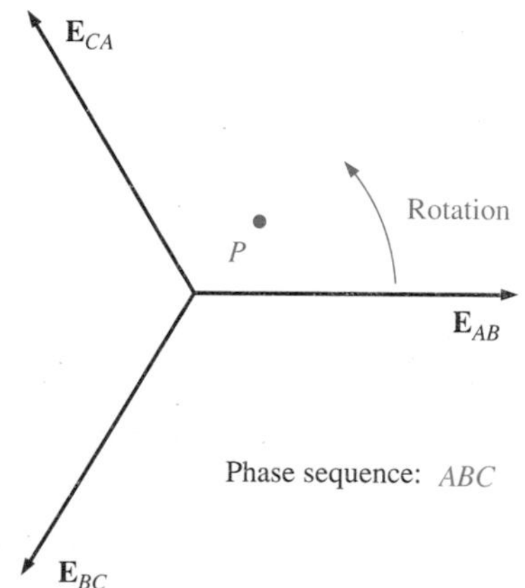

FIG. 23.19
Determining the phase sequence for a Δ-connected, three-phase generator.

23.9 THE Δ-Δ, Δ-Y THREE-PHASE SYSTEMS

The basic equations necessary to analyze either of the two systems (Δ-Δ, Δ-Y) have been presented at least once in this chapter. We will therefore proceed directly to two descriptive examples, one with a Δ-connected load and one with a Y-connected load.

EXAMPLE 23.3 For the Δ-Δ system shown in Fig. 23.20:

a. Find the phase angles θ_2 and θ_3 for the specified phase sequence.
b. Find the current in each phase of the load.
c. Find the magnitude of the line currents.

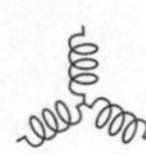

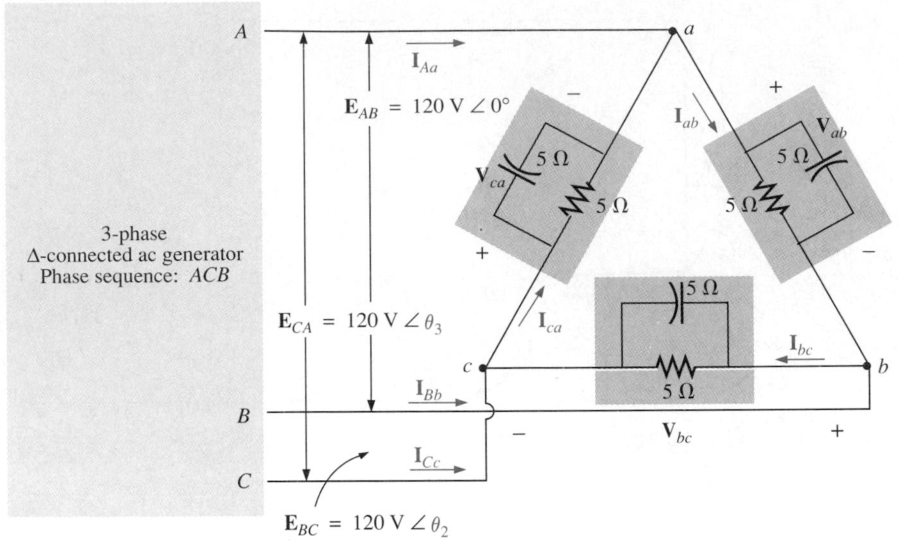

FIG. 23.20
Example 23.3, Δ-Δ system.

Solutions:

a. For an *ACB* phase sequence,

$$\theta_2 = \mathbf{120°} \quad \text{and} \quad \theta_3 = \mathbf{-120°}$$

b. $\mathbf{V}_\phi = \mathbf{E}_L$. Therefore,

$$\mathbf{V}_{ab} = \mathbf{E}_{AB} \qquad \mathbf{V}_{ca} = \mathbf{E}_{CA} \qquad \mathbf{V}_{bc} = \mathbf{E}_{BC}$$

The phase currents are

$$\mathbf{I}_{ab} = \frac{\mathbf{V}_{ab}}{\mathbf{Z}_{ab}} = \frac{120\text{ V}\angle 0°}{\dfrac{(5\ \Omega\ \angle 0°)(5\ \Omega\ \angle -90°)}{5\ \Omega - j\,5\ \Omega}} = \frac{120\text{ V}\angle 0°}{\dfrac{25\ \Omega\ \angle -90°}{7.071\ \angle -45°}}$$

$$= \frac{120\text{ V}\angle 0°}{3.54\ \Omega\ \angle -45°} = \mathbf{33.9\ A\ \angle 45°}$$

$$\mathbf{I}_{bc} = \frac{\mathbf{V}_{bc}}{\mathbf{Z}_{bc}} = \frac{120\text{ V}\angle 120°}{3.54\ \Omega\ \angle -45°} = \mathbf{33.9\ A\ \angle 165°}$$

$$\mathbf{I}_{ca} = \frac{\mathbf{V}_{ca}}{\mathbf{Z}_{ca}} = \frac{120\text{ V}\angle -120°}{3.54\ \Omega\ \angle -45°} = \mathbf{33.9\ A\ \angle -75°}$$

c. $I_L = \sqrt{3}I_\phi = (1.73)(34\text{ A}) = 58.82\text{ A}$. Therefore,

$$I_{Aa} = I_{Bb} = I_{Cc} = \mathbf{58.82\ A}$$

EXAMPLE 23.4 For the Δ-Y system shown in Fig. 23.21:

a. Find the voltage across each phase of the load.
b. Find the magnitude of the line voltages.

Solutions:

a. $\mathbf{I}_{\phi L} = \mathbf{I}_L$. Therefore,

$$\mathbf{I}_{an} = \mathbf{I}_{Aa} = 2\text{ A}\angle 0°$$
$$\mathbf{I}_{bn} = \mathbf{I}_{Bb} = 2\text{ A}\angle -120°$$
$$\mathbf{I}_{cn} = \mathbf{I}_{Cc} = 2\text{ A}\angle 120°$$

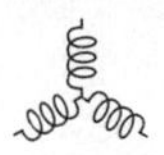

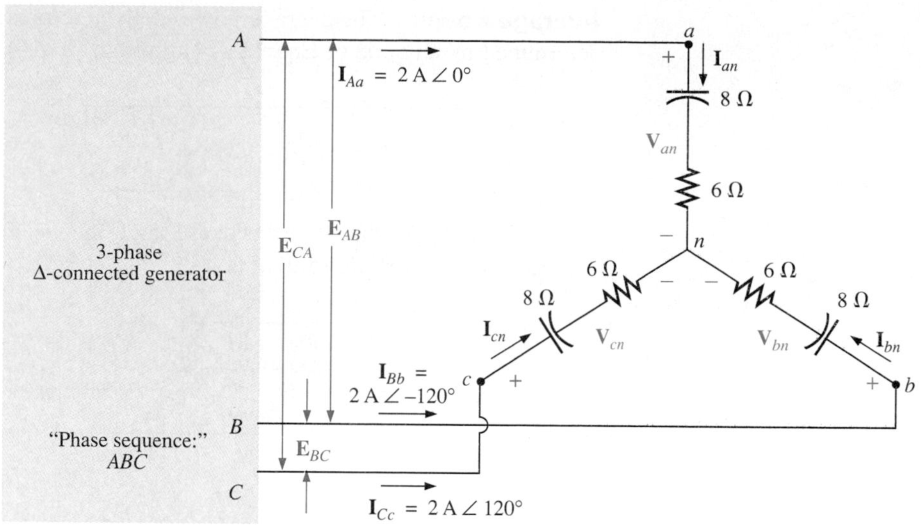

FIG. 23.21
Example 23.4, Δ-Y system.

The phase voltages are

$$\mathbf{V}_{an} = \mathbf{I}_{an}\mathbf{Z}_{an} = (2\text{ A}\angle 0°)(10\ \Omega\angle -53.13°) = \mathbf{20\ V\angle -53.13°}$$
$$\mathbf{V}_{bn} = \mathbf{I}_{bn}\mathbf{Z}_{bn} = (2\text{ A}\angle -120°)(10\ \Omega\angle -53.13°) = \mathbf{20\ V\angle -173.13°}$$
$$\mathbf{V}_{cn} = \mathbf{I}_{cn}\mathbf{Z}_{cn} = (2\text{ A}\angle 120°)(10\ \Omega\angle -53.13°) = \mathbf{20\ V\angle 66.87°}$$

b. $E_L = \sqrt{3}V_\phi = (1.73)(20\text{ V}) = 34.6\text{ V}$. Therefore,

$$E_{BA} = E_{CB} = E_{AC} = \mathbf{34.6\ V}$$

23.10 POWER

Y-Connected Balanced Load

Please refer to Fig. 23.22.

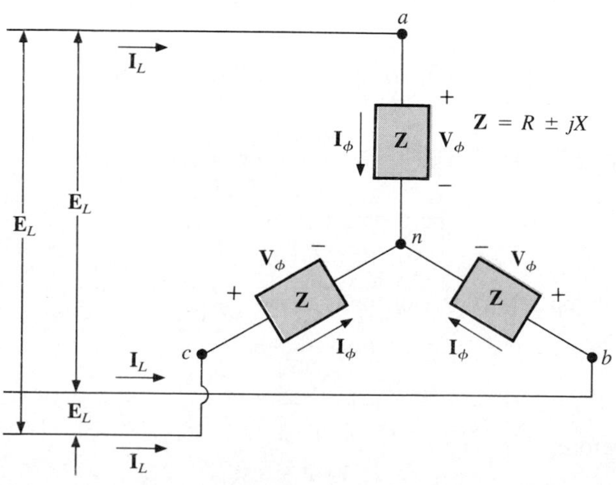

FIG. 23.22
Y-connected balanced load.

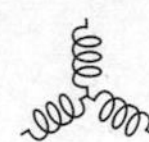

Average Power The average power delivered to each phase can be determined by any one of Eqs. (23.14) through (23.16).

$$P_\phi = V_\phi I_\phi \cos\theta_{I_\phi}^{V_\phi} = I_\phi^2 R_\phi = \frac{V_R^2}{R_\phi} \quad \text{(watts, W)} \tag{23.14}$$

where $\theta_{I_\phi}^{V_\phi}$ indicates that θ is the phase angle between V_ϕ and I_ϕ. The total power to the balanced load is

$$P_T = 3P_\phi \quad \text{(W)} \tag{23.15}$$

or, since $$V_\phi = \frac{E_L}{\sqrt{3}} \quad \text{and} \quad I_\phi = I_L$$

then $$P_T = 3\frac{E_L}{\sqrt{3}} I_L \cos\theta_{I_\phi}^{V_\phi}$$

But $$\left(\frac{3}{\sqrt{3}}\right)(1) = \left(\frac{3}{\sqrt{3}}\right)\left(\frac{\sqrt{3}}{\sqrt{3}}\right) = \frac{3\sqrt{3}}{3} = \sqrt{3}$$

Therefore,

$$P_T = \sqrt{3}E_L I_L \cos\theta_{I_\phi}^{V_\phi} = 3I_L^2 R_\phi \quad \text{(W)} \tag{23.16}$$

Reactive Power The reactive power of each phase (in volt-amperes reactive) is

$$Q_\phi = V_\phi I_\phi \sin\theta_{I_\phi}^{V_\phi} = I_\phi^2 X_\phi = \frac{V_X^2}{X_\phi} \quad \text{(VAR)} \tag{23.17}$$

The total reactive power of the load is

$$Q_T = 3Q_\phi \quad \text{(VAR)} \tag{23.18}$$

or, proceeding in the same manner as above, we have

$$Q_T = \sqrt{3}E_L I_L \sin\theta_{I_\phi}^{V_\phi} = 3I_L^2 X_\phi \quad \text{(VAR)} \tag{23.19}$$

Apparent Power The apparent power of each phase is

$$S_\phi = V_\phi I_\phi \quad \text{(VA)} \tag{23.20}$$

The total apparent power of the load is

$$S_T = 3S_\phi \quad \text{(VA)} \tag{23.21}$$

or, as before,

$$S_T = \sqrt{3}E_L I_L \quad \text{(VA)} \tag{23.22}$$

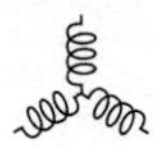

Power Factor The power factor of the system is given by

$$F_p = \frac{P_T}{S_T} = \cos\theta \quad \text{(leading or lagging)} \tag{23.23}$$

EXAMPLE 23.5 For the Y-connected load of Fig. 23.23:

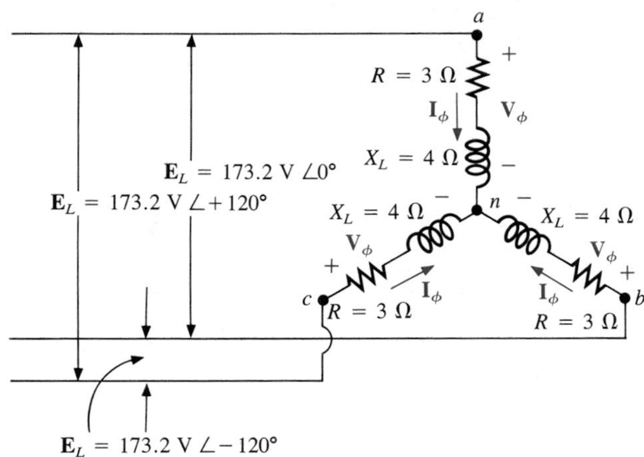

FIG. 23.23
Example 23.5.

a. Find the average power to each phase and the total load.
b. Determine the reactive power to each phase and the total reactive power.
c. Find the apparent power to each phase and the total apparent power.
d. Find the power factor of the load.

Solutions:

a. The *average power* is

$$P_\phi = V_\phi I_\phi \cos\theta_{I_\phi}^{V_\phi} = (100\text{ V})(20\text{ A})\cos 53.13° = (2000)(0.6)$$
$$= \mathbf{1200\ W}$$

$$P_\phi = I_\phi^2 R_\phi = (20\text{ A})^2(3\ \Omega) = (400)(3) = \mathbf{1200\ W}$$

$$P_\phi = \frac{V_R^2}{R_\phi} = \frac{(60\text{ V})^2}{3\ \Omega} = \frac{3600}{3} = \mathbf{1200\ W}$$

$$P_T = 3P_\phi = (3)(1200\text{ W}) = \mathbf{3600\ W}$$

or

$$P_T = \sqrt{3}E_L I_L \cos\theta_{I_\phi}^{V_\phi} = (1.732)(173.2\text{ V})(20\text{ A})(0.6) = \mathbf{3600\ W}$$

b. The *reactive power* is

$$Q_\phi = V_\phi I_\phi \sin\theta_{I_\phi}^{V_\phi} = (100\text{ V})(20\text{ A})\sin 53.13° = (2000)(0.8)$$
$$= \mathbf{1600\ VAR}$$

or $$Q_\phi = I_\phi^2 X_\phi = (20\text{ A})^2(4\ \Omega) = (400)(4) = \mathbf{1600\ VAR}$$
$$Q_T = 3Q_\phi = (3)(1600\text{ VAR}) = \mathbf{4800\ VAR}$$

or

$$Q_T = \sqrt{3}E_L I_L \sin\theta_{I_\phi}^{V_\phi} = (1.732)(173.2\text{ V})(20\text{ A})(0.8) = \mathbf{4800\ VAR}$$

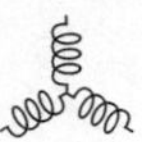

c. The *apparent power* is

$$S_\phi = V_\phi I_\phi = (100\text{ V})(20\text{ A}) = \mathbf{2000\ VA}$$
$$S_T = 3S_\phi = (3)(2000\text{ VA}) = \mathbf{6000\ VA}$$

or $$S_T = \sqrt{3}E_L I_L = (1.732)(173.2\text{ V})(20\text{ A}) = \mathbf{6000\ VA}$$

d. The *power factor* is

$$F_p = \frac{P_T}{S_T} = \frac{3600\text{ W}}{6000\text{ VA}} = \mathbf{0.6\ lagging}$$

Δ-Connected Balanced Load

Please refer to Fig. 23.24.

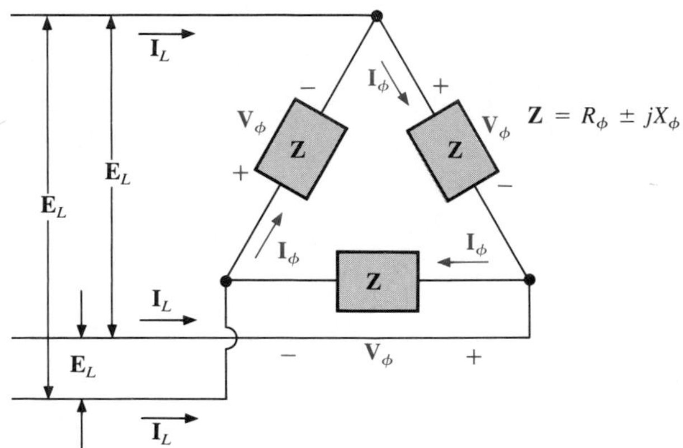

FIG. 23.24
Δ-connected balanced load.

Average Power

$$P_\phi = V_\phi I_\phi \cos \theta^{V_\phi}_{I_\phi} = I^2_\phi R_\phi = \frac{V^2_R}{R_\phi} \qquad \text{(W)} \qquad \textbf{(23.24)}$$

$$P_T = 3P_\phi \qquad \text{(W)} \qquad \textbf{(23.25)}$$

Reactive Power

$$Q_\phi = V_\phi I_\phi \sin \theta^{V_\phi}_{I_\phi} = I^2_\phi X_\phi = \frac{V^2{}_X}{X_\phi} \qquad \text{(VAR)} \qquad \textbf{(23.26)}$$

$$Q_T = 3Q_\phi \qquad \text{(VAR)} \qquad \textbf{(23.27)}$$

Apparent power

$$S_\phi = V_\phi I_\phi \qquad \text{(VA)} \qquad \textbf{(23.28)}$$

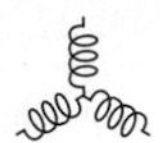

$$\boxed{S_T = 3S_\phi = \sqrt{3}E_L I_L} \qquad \text{(VA)} \qquad \textbf{(23.29)}$$

Power Factor

$$\boxed{F_p = \frac{P_T}{S_T}} \qquad \textbf{(23.30)}$$

EXAMPLE 23.6 For the Δ-Y connected load of Fig. 23.25, find the total average, reactive, and apparent power. In addition, find the power factor of the load.

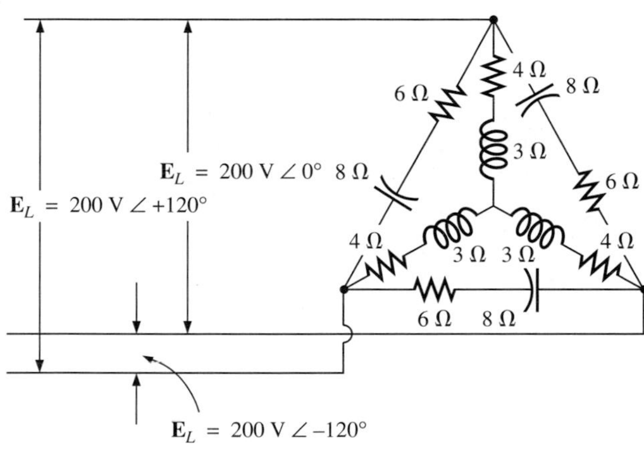

FIG. 23.25
Example 23.6.

Solutions: Consider the Δ and Y separately.

For the Δ:

$$\mathbf{Z}_\Delta = 6\ \Omega - j\,8\ \Omega = 10\ \Omega\ \angle -53.13^\circ$$

$$I_\phi = \frac{E_L}{Z_\Delta} = \frac{200\ \text{V}}{10\ \Omega} = 20\ \text{A}$$

$$P_{T_\Delta} = 3I_\phi^2 R_\phi = (3)(20\ \text{A})^2(6\ \Omega) = \mathbf{7200\ W}$$

$$Q_{T_\Delta} = 3I_\phi^2 X_\phi = (3)(20\ \text{A})^2(8\ \Omega) = \mathbf{9600\ VAR\ (cap.)}$$

$$S_{T_\Delta} = 3V_\phi I_\phi = (3)(200\ \text{A})(20\ \Omega) = \mathbf{12{,}000\ VA}$$

For the Y:

$$\mathbf{Z}_Y = 4\ \Omega + j\,3\ \Omega = 5\ \Omega\ \angle 36.87^\circ$$

$$I_\phi = \frac{E_L/\sqrt{3}}{Z_Y} = \frac{200\ \text{V}/\sqrt{3}}{5\ \Omega} = \frac{116\ \text{V}}{5\ \Omega} = 23.12\ \text{A}$$

$$P_{T_Y} = 3I_\phi^2 R_\phi = (3)(23.12\ \text{A})^2(4\ \Omega) = \mathbf{6414.41\ W}$$

$$Q_{T_Y} = 3I_\phi^2 X_\phi = (3)(23.12\ \text{A})^2(3\ \Omega) = \mathbf{4810.81\ VAR\ (ind.)}$$

$$S_{T_Y} = 3V_\phi I_\phi = (3)(116\ \text{V})(23.12\ \text{A}) = \mathbf{8045.76\ VA}$$

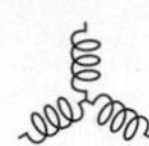

For the total load:

$$P_T = P_{T_\Delta} + P_{T_Y} = 7200 \text{ W} + 6414.41 \text{ W} = \mathbf{13{,}614.41 \text{ W}}$$

$$Q_T = Q_{T_\Delta} - Q_{T_Y} = 9600 \text{ VAR (cap.)} - 4810.81 \text{ VAR (ind.)}$$
$$= \mathbf{4789.19 \text{ VAR (cap.)}}$$

$$S_T = \sqrt{P_T^2 + Q_T^2} = \sqrt{(13{,}614.41 \text{ W})^2 + (4789.19 \text{ VAR})^2}$$
$$= \mathbf{14{,}432.2 \text{ VA}}$$

$$F_p = \frac{P_T}{S_T} = \frac{13{,}614.41 \text{ W}}{14{,}432.20 \text{ VA}} = \mathbf{0.943 \text{ leading}}$$

EXAMPLE 23.7 Each transmission line of the three-wire, three-phase system of Fig. 23.26 has an impedance of 15 Ω + *j* 20 Ω. The system delivers a total power of 160 kW at 12,000 V to a balanced three-phase load with a lagging power factor of 0.86.

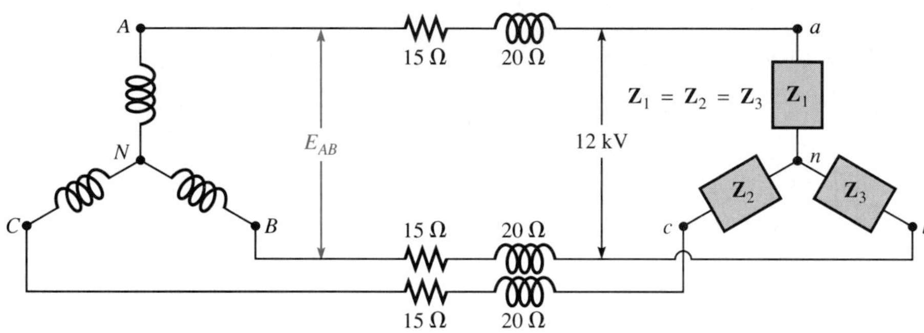

FIG. 23.26
Example 23.7.

a. Determine the magnitude of the line voltage E_{AB} of the generator.
b. Find the power factor of the total load applied to the generator.
c. What is the efficiency of the system?

Solutions:

a. $$V_\phi \text{ (load)} = \frac{V_L}{\sqrt{3}} = \frac{12{,}000 \text{ V}}{1.73} = 6936.42 \text{ V}$$

$$P_T \text{ (load)} = 3V_\phi I_\phi \cos\theta$$

and $$I_\phi = \frac{P_T}{3V_\phi \cos\theta} = \frac{160{,}000 \text{ W}}{3(6936.42 \text{ V})(0.86)}$$
$$= \mathbf{8.94 \text{ A}}$$

Since $\theta = \cos^{-1} 0.86 = 30.68°$, assigning $\mathbf{V}_\phi$ an angle of 0° or $\mathbf{V}_\phi = V_\phi \angle 0°$, a lagging power factor results in

$$\mathbf{I}_\phi = 8.94 \text{ A} \angle -30.68°$$

For each phase, the system will appear as shown in Fig. 23.27, where

$$\mathbf{E}_{AN} - \mathbf{I}_\phi \mathbf{Z}_{\text{line}} - \mathbf{V}_\phi = 0$$

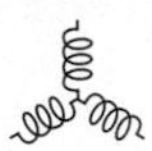

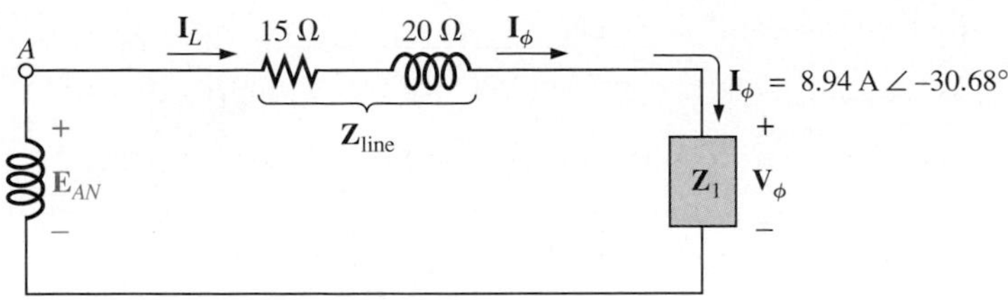

FIG. 23.27
The loading on each phase of the system of Fig. 23.26.

or

$$\begin{aligned}\mathbf{E}_{AN} &= \mathbf{I}_\phi \mathbf{Z}_{\text{line}} + \mathbf{V}_\phi \\ &= (8.94 \text{ A} \angle -30.68^\circ)(25 \ \Omega \angle 53.13^\circ) + 6936.42 \text{ V} \angle 0^\circ \\ &= 223.5 \text{ V} \angle 22.45^\circ + 6936.42 \text{ V} \angle 0^\circ \\ &= 206.56 \text{ V} + j\,85.35 \text{ V} + 6936.42 \text{ V} \\ &= 7142.98 \text{ V} + j\,85.35 \text{ V} \\ &= 7143.5 \text{ V} \angle 0.68^\circ\end{aligned}$$

then
$$\begin{aligned}E_{AB} &= \sqrt{3}E_{\phi g} = (1.73)(7143.5 \text{ V}) \\ &= \mathbf{12{,}358.26 \text{ V}}\end{aligned}$$

b.
$$\begin{aligned}P_T &= P_{\text{load}} + P_{\text{lines}} \\ &= 160 \text{ kW} + 3(I_L)^2 R_{\text{line}} \\ &= 160 \text{ kW} + 3(8.94 \text{ A})^2 15 \ \Omega \\ &= 160{,}000 \text{ W} + 3596.55 \text{ W} \\ &= 163{,}596.55 \text{ W}\end{aligned}$$

and $P_T = \sqrt{3}V_L I_L \cos\theta_T$

or
$$\cos\theta_T = \frac{P_T}{\sqrt{3}V_L I_L} = \frac{163{,}596.55 \text{ W}}{(1.73)(12{,}358.26 \text{ V})(8.94 \text{ A})}$$

and
$$F_p = \mathbf{0.856} < 0.86 \text{ of load}$$

c.
$$\eta = \frac{P_o}{P_i} = \frac{P_o}{P_o + P_{\text{losses}}} = \frac{160 \text{ kW}}{160 \text{ kW} + 3596.55 \text{ W}} = 0.978$$
$$= \mathbf{97.8\%}$$

23.11 THE THREE-WATTMETER METHOD

The power delivered to a balanced or an unbalanced four-wire, Y-connected load can be found using three wattmeters in the manner shown in Fig. 23.28. Each wattmeter measures the power delivered to each phase. The potential coil of each wattmeter is connected parallel with the load, while the current coil is in series with the load. The total average power of the system can be found by summing the three wattmeter readings; that is,

$$P_{T_Y} = P_1 + P_2 + P_3 \qquad \textbf{(23.31)}$$

For the load (balanced or unbalanced), the wattmeters are connected as shown in Fig. 23.29. The total power is again the sum of the three wattmeter readings:

$$P_{T_\Delta} = P_1 + P_2 + P_3 \qquad \textbf{(23.32)}$$

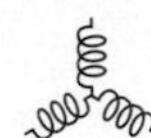

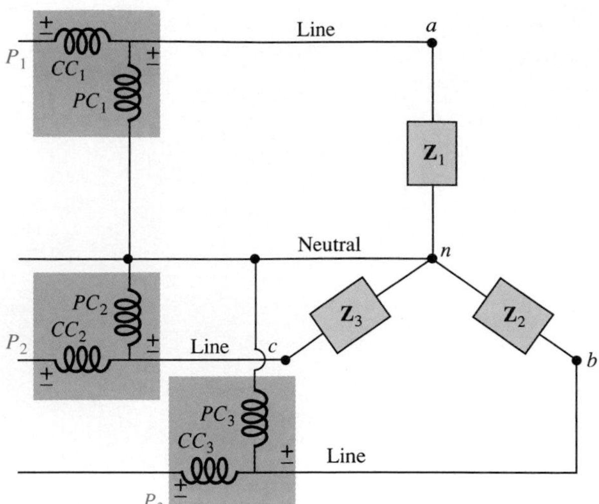

FIG. 23.28
Three-wattmeter method for a Y-connected load.

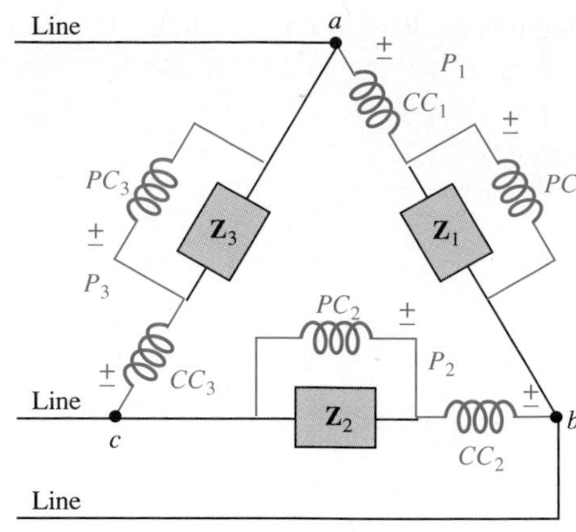

FIG. 23.29
Three-wattmeter method for a Δ-connected load.

If in either of the cases just described the load is balanced, the power delivered to each phase will be the same. The total power is then just three times any one wattmeter reading.

23.12 THE TWO-WATTMETER METHOD

The power delivered to a three-phase, three-wire, Δ- or Y-connected, balanced or unbalanced load can be found using only two wattmeters if the proper connection is employed and if the wattmeter readings are interpreted properly. The basic connections are shown in Fig. 23.30. One end of each potential coil is connected to the same line. The current coils are then placed in the remaining lines.

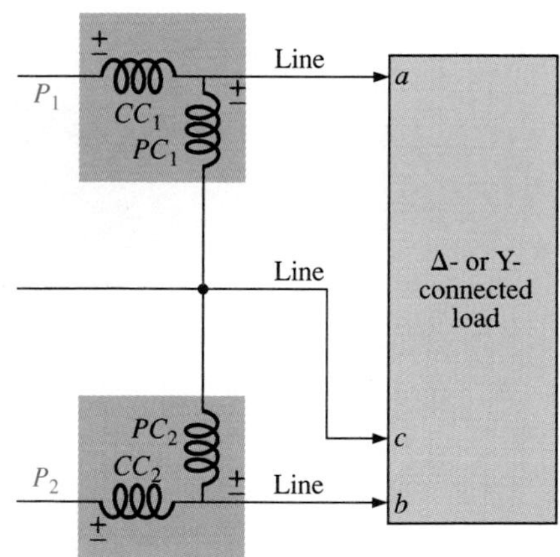

FIG. 23.30
Two-wattmeter method for a Δ- or a Y-connected load.

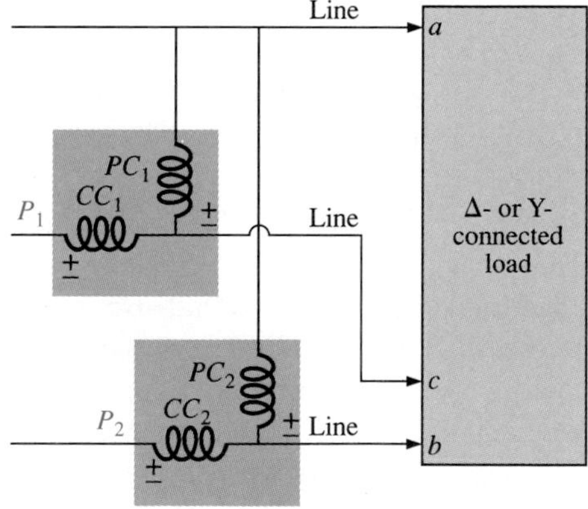

FIG. 23.31
Alternative hookup for the two-wattmeter method.

The connection shown in Fig. 23.31 will also satisfy the requirements. A third hookup is also possible, but this is left to the reader as an exercise.

The total power delivered to the load is the algebraic sum of the two wattmeter readings. For a *balanced* load, we will now consider two methods of determining whether the total power is the sum or the difference of the two wattmeter readings. The first method to be described requires that we know or be able to find the power factor (leading or lagging) of any one phase of the load. When this information has been obtained, it can be applied directly to the curve of Fig. 23.32.

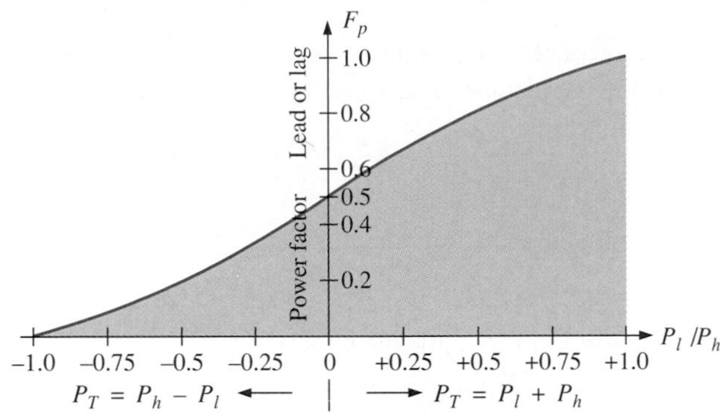

FIG. 23.32
Determining whether the readings obtained using the two-wattmeter method should be added or subtracted.

The curve in Fig. 23.32 is a plot of the power factor of the load (phase) versus the ratio P_l/P_h, where P_l and P_h are the magnitudes of the lower- and higher-reading wattmeters, respectively. Note that for a power factor (leading or lagging) greater than 0.5, the ratio has a positive value. This indicates that both wattmeters are reading positive, and the total power is the sum of the two wattmeter readings; that is, $P_T = P_l + P_h$. For a power factor less than 0.5 (leading or lagging), the ratio has a negative value. This indicates that the smaller-reading wattmeter is reading negative, and the total power is the difference of the two wattmeter readings; that is, $P_T = P_h - P_l$.

A closer examination will reveal that, when the power factor is 1 (cos 0° = 1), corresponding to a purely resistive load, $P_l/P_h = 1$ or $P_l = P_h$, and both wattmeters will have the same wattage indication. At a power factor equal to 0 (cos 90° = 0), corresponding to a purely reactive load, $P_l/P_h = -1$ or $P_l = -P_h$, and both wattmeters will again have the same wattage indication but with opposite signs. The transition from a negative to a positive ratio occurs when the power factor of the load is 0.5 or $\theta = \cos^{-1} 0.5 = 60°$. At this power factor, $P_l/P_h = 0$, so that $P_l = 0$, while P_h will read the total power delivered to the load.

The second method for determining whether the total power is the sum or difference of the two wattmeter readings involves a simple laboratory test. For the test to be applied, both wattmeters must first have an up-scale deflection. If one of the wattmeters has a below-zero indication, an up-scale deflection can be obtained by simply reversing the leads of the current coil of the wattmeter. To perform the test:

1. Take notice of which line does not have a current coil sensing the line current.
2. For the lower-reading wattmeter, disconnect the lead of the potential coil connected to the line without the current coil.
3. Take the disconnected lead of the lower-reading wattmeter's potential coil, and touch a connection point on the line that has the current coil of the higher-reading wattmeter.

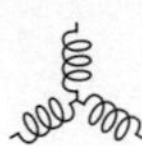

4. If the pointer deflects downward (below zero watts), the wattage reading of the lower-reading wattmeter should be subtracted from that of the higher-reading wattmeter. Otherwise the readings should be added.

For a *balanced system,* since

$$P_T = P_h \pm P_l = \sqrt{3}E_L I_L \cos\theta_{I\phi}^{V\phi}$$

the power factor of the load (phase) can be found from the wattmeter readings and the magnitude of the line voltage and current:

$$F_p = \cos\theta_{I\phi}^{V\phi} = \frac{P_h \pm P_l}{\sqrt{3}E_L I_L} \qquad \textbf{(23.33)}$$

EXAMPLE 23.8 For the unbalanced Δ-connected load of Fig. 23.33 with two properly connected wattmeters,

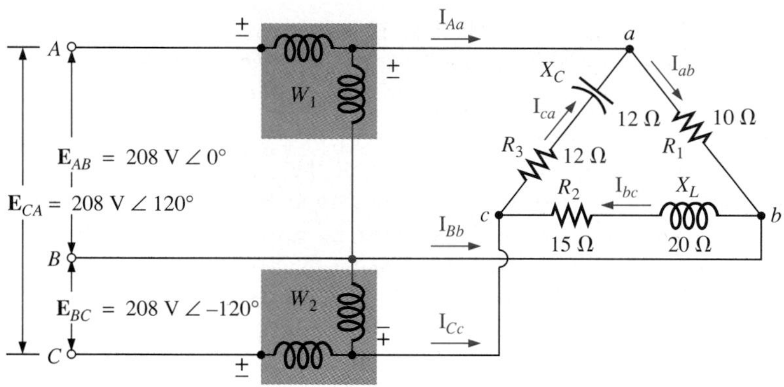

FIG. 23.33
Example 23.8.

a. Determine the magnitude and angle of the phase currents.
b. Calculate the magnitude and angle of the line currents.
c. Determine the power reading of each wattmeter.
d. Calculate the total power absorbed by the load.
e. Compare the result of part (d) with the total power calculated using the phase currents and the resistive elements.

Solutions:

a. $\mathbf{I}_{ab} = \dfrac{\mathbf{V}_{ab}}{\mathbf{Z}_{ab}} = \dfrac{\mathbf{E}_{AB}}{\mathbf{Z}_{ab}} = \dfrac{208\text{ V}\angle 0^\circ}{10\ \Omega\angle 0^\circ} = \mathbf{20.8\ A\angle 0^\circ}$

$\mathbf{I}_{bc} = \dfrac{\mathbf{V}_{bc}}{\mathbf{Z}_{bc}} = \dfrac{\mathbf{E}_{BC}}{\mathbf{Z}_{bc}} = \dfrac{208\text{ V}\angle -120^\circ}{15\ \Omega + j\,20\ \Omega} = \dfrac{208\text{ V}\angle -120^\circ}{25\ \Omega\angle 53.13^\circ}$
$= \mathbf{8.32\ A\angle -173.13^\circ}$

$\mathbf{I}_{ca} = \dfrac{\mathbf{V}_{ca}}{\mathbf{Z}_{ca}} = \dfrac{\mathbf{E}_{CA}}{\mathbf{Z}_{ca}} = \dfrac{208\text{ V}\angle +120^\circ}{12\ \Omega + j\,12\ \Omega} = \dfrac{208\text{ V}\angle +120^\circ}{16.97\ \Omega\angle -45^\circ}$
$= \mathbf{12.26\ A\angle 165^\circ}$

b. $\mathbf{I}_{Aa} = \mathbf{I}_{ab} - \mathbf{I}_{ca}$
$= 20.8\text{ A}\angle 0^\circ - 12.26\text{ A}\angle 165^\circ$
$= 20.8\text{ A} - (-11.84\text{ A} + j\,3.17\text{ A})$
$= 20.8\text{ A} + 11.84\text{ A} - j\,3.17\text{ A} = 32.64\text{ A} - j\,3.17\text{ A}$
$= \mathbf{32.79\ A\angle -5.55^\circ}$

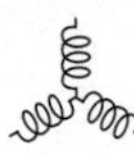

$$\begin{aligned}\mathbf{I}_{Bb} &= \mathbf{I}_{bc} - \mathbf{I}_{ab}\\ &= 8.32 \text{ A} \angle -173.13° - 20.8 \text{ A} \angle 0°\\ &= (-8.26 \text{ A} - j\,1 \text{ A}) - 20.8 \text{ A}\\ &= -8.26 \text{ A} - 20.8 \text{ A} - j\,1 \text{ A} = -29.06 \text{ A} - j\,1 \text{ A}\\ &= \mathbf{29.08 \text{ A} \angle -178.03°}\end{aligned}$$

$$\begin{aligned}\mathbf{I}_{Cc} &= \mathbf{I}_{ca} - \mathbf{I}_{bc}\\ &= 12.26 \text{ A} \angle 165° - 8.32 \text{ A} \angle -173.13°\\ &= (-11.84 \text{ A} + j\,3.17 \text{ A}) - (-8.26 \text{ A} - j\,1 \text{ A})\\ &= -11.84 \text{ A} + 8.26 \text{ A} + j(3.17 \text{ A} + 1 \text{ A}) = -3.58 \text{ A} + j\,4.17 \text{ A}\\ &= \mathbf{5.5 \text{ A} \angle 130.65°}\end{aligned}$$

c. $P_1 = V_{ab} I_{Aa} \cos \theta_{\mathbf{I}_{Aa}}^{\mathbf{V}_{ab}}$, $\mathbf{V}_{ab} = 208 \text{ V} \angle 0°$, $\mathbf{I}_{Aa} = 32.79 \text{ A} \angle -5.55°$
$= (208 \text{ V})(32.79 \text{ A}) \cos 5.55°$
$= \mathbf{6788.35 \text{ W}}$

$\mathbf{V}_{bc} = \mathbf{E}_{BC} = 208 \text{ V} \angle -120°$
but $\mathbf{V}_{cb} = \mathbf{E}_{CB} = 208 \text{ V} \angle -120° + 180°$
$= 208 \text{ V} \angle 60°$
with $\mathbf{I}_{Cc} = 5.5 \text{ A} \angle 130.65°$

$$\begin{aligned}P_2 &= V_{cb} I_{Cc} \cos \theta_{\mathbf{I}_{Cc}}^{\mathbf{V}_{cb}}\\ &= (208 \text{ V})(5.5 \text{ A}) \cos 70.65°\\ &= \mathbf{379.1 \text{ W}}\end{aligned}$$

d. $P_T = P_1 + P_2 = 6788.35 \text{ W} + 379.1 \text{ W}$
$= \mathbf{7167.45 \text{ W}}$

e. $P_T = (I_{ab})^2 R_1 + (I_{bc})^2 R_2 + (I_{ca})^2 R_3$
$= (20.8 \text{ A})^2 10\ \Omega + (8.32 \text{ A})^2 15\ \Omega + (12.26 \text{ A})^2 12\ \Omega$
$= 4326.4 \text{ W} + 1038.34 \text{ W} + 1803.69 \text{ W}$
$= \mathbf{7168.43 \text{ W}}$

(The slight difference is due to the level of accuracy carried through the calculations.)

23.13 UNBALANCED, THREE-PHASE, FOUR-WIRE, Y-CONNECTED LOAD

For the three-phase, four-wire, Y-connected load of Fig. 23.34, conditions are such that *none* of the load impedances are equal. Since the neutral is a common point between the load and source, no matter what the impedance of each phase of the load and source, the voltage across each phase is the phase voltage of the generator:

$$\mathbf{V}_\phi = \mathbf{E}_\phi \tag{23.34}$$

The phase currents can therefore be determined by Ohm's law:

$$\mathbf{I}_{\phi_1} = \frac{\mathbf{V}_{\phi_1}}{\mathbf{Z}_1} = \frac{\mathbf{E}_{\phi_1}}{\mathbf{Z}_1}, \text{ and so on} \tag{23.35}$$

The current in the neutral for any unbalanced system can then be found by applying Kirchhoff's current law at the common point n:

$$\mathbf{I}_N = \mathbf{I}_{\phi_1} + \mathbf{I}_{\phi_2} + \mathbf{I}_{\phi_3} = \mathbf{I}_{L_1} + \mathbf{I}_{L_2} + \mathbf{I}_{L_3} \tag{23.36}$$

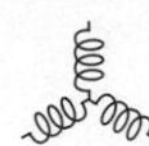

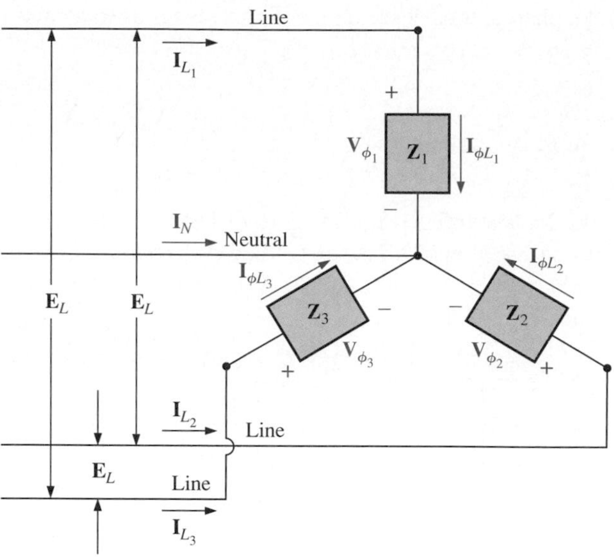

FIG. 23.34
Unbalanced Y-connected load.

Because of the variety of equipment found in an industrial environment, both three-phase and single-phase power are usually provided with the single-phase obtained off the three-phase system. In addition, since the load on each phase is continually changing, a four-wire system (with a neutral) is normally employed to ensure steady voltage levels and to provide a path for the current resulting from an unbalanced load. The system of Fig. 23.35 has a three-phase transformer dropping the line voltage from 13,800 V to 208 V. All the lower-power-demand loads such as lighting, wall outlets, security, etc., use the single-phase, 120-V line to neutral voltage. Higher power loads such as air conditioners, electric ovens or dryers, etc., use the single-phase, 208 V available from line to line. For larger motors and special high-demand equipment, the full three-phase power can be taken directly off the system, as shown in Fig. 23.35. In the design and construction of a commercial establishment, the National Electric Code requires that every effort be made to ensure that the expected loads, whether they be single- or multiphase, result in a total load that is as balanced as possible

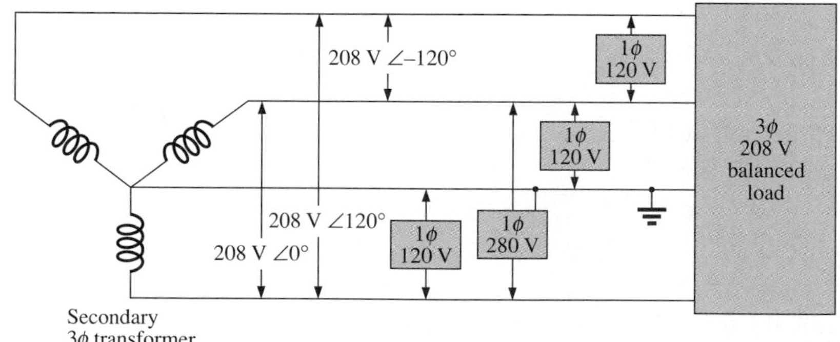

FIG. 23.35
3ϕ/1ϕ, 208 V/120 V industrial supply.

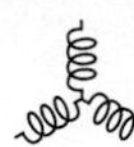

between the phases, thus ensuring the highest level of transmission efficiency.

23.14 UNBALANCED, THREE-PHASE, THREE-WIRE, Y-CONNECTED LOAD

For the system shown in Fig. 23.36, the required equations can be derived by first applying Kirchhoff's voltage law around each closed loop to produce

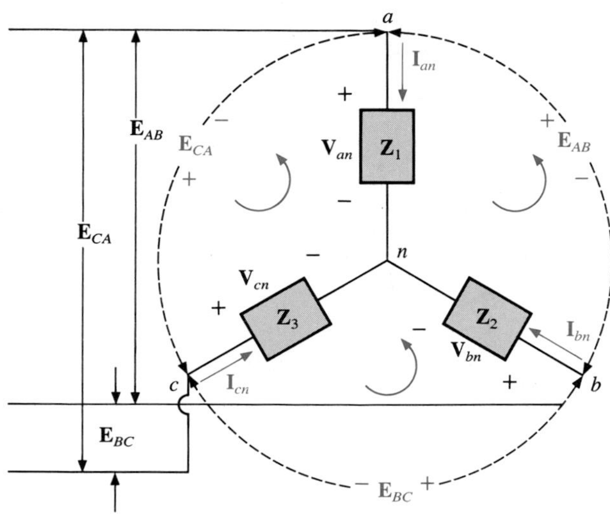

FIG. 23.36
Unbalanced, three-phase, three-wire, Y-connected load.

$$\mathbf{E}_{AB} - \mathbf{V}_{an} + \mathbf{V}_{bn} = 0$$
$$\mathbf{E}_{BC} - \mathbf{V}_{bn} + \mathbf{V}_{cn} = 0$$
$$\mathbf{E}_{CA} - \mathbf{V}_{cn} + \mathbf{V}_{an} = 0$$

Substituting, we have

$$\mathbf{V}_{an} = \mathbf{I}_{an}\mathbf{Z}_1 \qquad \mathbf{V}_{bn} = \mathbf{I}_{bn}\mathbf{Z}_2 \qquad \mathbf{V}_{cn} = \mathbf{I}_{cn}\mathbf{Z}_3$$

$$\mathbf{E}_{AB} = \mathbf{I}_{an}\mathbf{Z}_1 - \mathbf{I}_{bn}\mathbf{Z}_2 \tag{23.37a}$$
$$\mathbf{E}_{BC} = \mathbf{I}_{bn}\mathbf{Z}_2 - \mathbf{I}_{cn}\mathbf{Z}_3 \tag{23.37b}$$
$$\mathbf{E}_{CA} = \mathbf{I}_{cn}\mathbf{Z}_3 - \mathbf{I}_{an}\mathbf{Z}_1 \tag{23.37c}$$

Applying Kirchhoff's current law at node n results in

$$\mathbf{I}_{an} + \mathbf{I}_{bn} + \mathbf{I}_{cn} = 0 \quad \text{and} \quad \mathbf{I}_{bn} = -\mathbf{I}_{an} - \mathbf{I}_{cn}$$

Substituting for $\mathbf{I}_{bn}$ in Eqs. (23.37a) and (23.37b) yields

$$\mathbf{E}_{AB} = \mathbf{I}_{an}\mathbf{Z}_1 - [-(\mathbf{I}_{an} + \mathbf{I}_{cn})]\mathbf{Z}_2$$
$$\mathbf{E}_{BC} = -(\mathbf{I}_{an} + \mathbf{I}_{cn})\mathbf{Z}_2 - \mathbf{I}_{cn}\mathbf{Z}_3$$

which are rewritten as

$$\mathbf{E}_{AB} = \mathbf{I}_{an}(\mathbf{Z}_1 + \mathbf{Z}_2) + \mathbf{I}_{cn}\mathbf{Z}_2$$
$$\mathbf{E}_{BC} = \mathbf{I}_{an}(-\mathbf{Z}_2) + \mathbf{I}_{cn}[-(\mathbf{Z}_2 + \mathbf{Z}_3)]$$

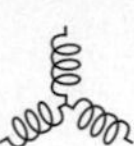

Using determinants, we have

$$\mathbf{I}_{an} = \frac{\begin{vmatrix} \mathbf{E}_{AB} & \mathbf{Z}_2 \\ \mathbf{E}_{BC} & -(\mathbf{Z}_2 + \mathbf{Z}_3) \end{vmatrix}}{\begin{vmatrix} \mathbf{Z}_1 + \mathbf{Z}_2 & \mathbf{Z}_2 \\ -\mathbf{Z}_2 & -(\mathbf{Z}_2 + \mathbf{Z}_3) \end{vmatrix}}$$

$$= \frac{-(\mathbf{Z}_2 + \mathbf{Z}_3)\mathbf{E}_{AB} - \mathbf{E}_{BC}\mathbf{Z}_2}{-\mathbf{Z}_1\mathbf{Z}_2 - \mathbf{Z}_1\mathbf{Z}_3 - \mathbf{Z}_2\mathbf{Z}_3 - \mathbf{Z}_2^2 + \mathbf{Z}_2^2}$$

$$\mathbf{I}_{an} = \frac{-\mathbf{Z}_2(\mathbf{E}_{AB} + \mathbf{E}_{BC}) - \mathbf{Z}_3\mathbf{E}_{AB}}{-\mathbf{Z}_1\mathbf{Z}_2 - \mathbf{Z}_1\mathbf{Z}_3 - \mathbf{Z}_2\mathbf{Z}_3}$$

Applying Kirchhoff's voltage law to the line voltages:

$$\mathbf{E}_{AB} + \mathbf{E}_{CA} + \mathbf{E}_{BC} = 0 \quad \text{or} \quad \mathbf{E}_{AB} + \mathbf{E}_{BC} = -\mathbf{E}_{CA}$$

Substituting for ($\mathbf{E}_{AB} + \mathbf{E}_{CB}$) in the above equation for $\mathbf{I}_{an}$:

$$\mathbf{I}_{an} = \frac{-\mathbf{Z}_2(-\mathbf{E}_{CA}) - \mathbf{Z}_3\mathbf{E}_{AB}}{-\mathbf{Z}_1\mathbf{Z}_2 - \mathbf{Z}_1\mathbf{Z}_3 - \mathbf{Z}_2\mathbf{Z}_3}$$

and

$$\mathbf{I}_{an} = \frac{\mathbf{E}_{AB}\mathbf{Z}_3 - \mathbf{E}_{CA}\mathbf{Z}_2}{\mathbf{Z}_1\mathbf{Z}_2 + \mathbf{Z}_1\mathbf{Z}_3 + \mathbf{Z}_2\mathbf{Z}_3} \quad \textbf{(23.38)}$$

In the same manner, it can be shown that

$$\mathbf{I}_{cn} = \frac{\mathbf{E}_{CA}\mathbf{Z}_2 - \mathbf{E}_{BC}\mathbf{Z}_1}{\mathbf{Z}_1\mathbf{Z}_2 + \mathbf{Z}_1\mathbf{Z}_3 + \mathbf{Z}_2\mathbf{Z}_3} \quad \textbf{(23.39)}$$

Substituting Eq. (23.39) for $\mathbf{I}_{cn}$ in the right-hand side of Eq. (23.37b), we obtain

$$\mathbf{I}_{bn} = \frac{\mathbf{E}_{BC}\mathbf{Z}_1 - \mathbf{E}_{AB}\mathbf{Z}_3}{\mathbf{Z}_1\mathbf{Z}_2 + \mathbf{Z}_1\mathbf{Z}_3 + \mathbf{Z}_2\mathbf{Z}_3} \quad \textbf{(23.40)}$$

EXAMPLE 23.9 A *phase-sequence indicator* is an instrument that can display the phase sequence of a polyphase circuit. A network that will perform this function appears in Fig. 23.37. The applied phase sequence is *ABC*. The bulb corresponding to this phase sequence will burn more brightly than the bulb indicating the *ACB* sequence because a greater current is passing through the *ABC* bulb. Calculating the phase currents will demonstrate that this situation does in fact exist:

$$Z_1 = X_C = \frac{1}{\omega C} = \frac{1}{(377\text{ rad/s})(16 \times 10^{-6}\text{ F})} = 166\ \Omega$$

By Eq. (23.39),

$$\mathbf{I}_{cn} = \frac{\mathbf{E}_{CA}\mathbf{Z}_2 - \mathbf{E}_{BC}\mathbf{Z}_1}{\mathbf{Z}_1\mathbf{Z}_2 + \mathbf{Z}_1\mathbf{Z}_3 + \mathbf{Z}_2\mathbf{Z}_3}$$

$$= \frac{(200\text{ V}\angle 120°)(200\ \Omega\angle 0°) - (200\text{ V}\angle -120°)(166\ \Omega\angle -90°)}{(166\ \Omega\angle -90°)(200\ \Omega\angle 0°) + (166\ \Omega\angle -90°)(200\ \Omega\angle 0°) + (200\ \Omega\angle 0°)(200\ \Omega\angle 0°)}$$

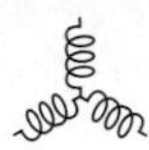

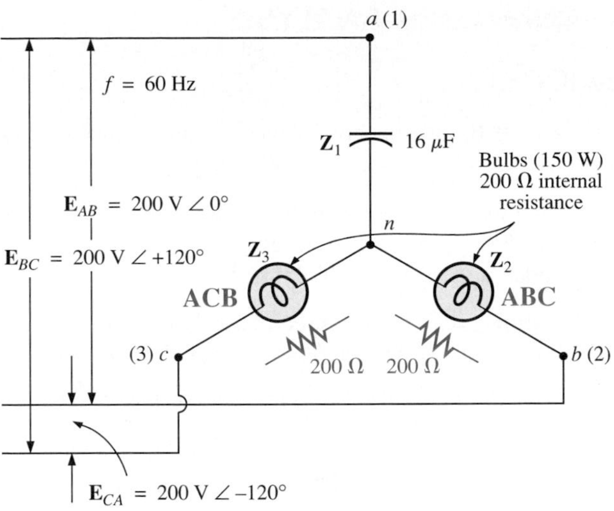

FIG. 23.37
Example 23.9.

$$\mathbf{I}_{cn} = \frac{40{,}000 \text{ V} \angle 120° + 33{,}200 \text{ V} \angle -30°}{33{,}200 \ \Omega \angle -90° + 33{,}200 \ \Omega \angle -90° + 40{,}000 \ \Omega \angle 0°}$$

Dividing the numerator and denominator by 1000 and converting both to the rectangular domain yields

$$\mathbf{I}_{cn} = \frac{(-20 + j\,34.64) + (28.75 - j\,16.60)}{40 - j\,66.4}$$

$$= \frac{8.75 + j\,18.04}{77.52 \angle -58.93°} = \frac{20.05 \angle 64.13°}{77.52 \angle -58.93°}$$

$$\mathbf{I}_{cn} = \mathbf{0.259 \text{ A} \angle 123.06°}$$

By Eq. (23.40),

$$\mathbf{I}_{bn} = \frac{\mathbf{E}_{BC}\mathbf{Z}_1 - \mathbf{E}_{AB}\mathbf{Z}_3}{\mathbf{Z}_1\mathbf{Z}_2 + \mathbf{Z}_1\mathbf{Z}_3 + \mathbf{Z}_2\mathbf{Z}_3}$$

$$= \frac{(200 \text{ V} \angle -120°)(166 \angle -90°) - (200 \text{ V} \angle 0°)(200 \angle 0°)}{77.52 \times 10^3 \ \Omega \angle -58.93°}$$

$$\mathbf{I}_{bn} = \frac{33{,}200 \text{ V} \angle -210° - 40{,}000 \text{ V} \angle 0°}{77.52 \times 10^3 \ \Omega \angle -58.93°}$$

Dividing by 1000 and converting to the rectangular domain yields

$$\mathbf{I}_{bn} = \frac{-28.75 + j\,16.60 - 40.0}{77.52 \angle -58.93°} = \frac{-68.75 + j\,16.60}{77.52 \angle -58.93°}$$

$$= \frac{70.73 \angle 166.43°}{77.52 \angle -58.93°} = \mathbf{0.91 \text{ A} \angle 225.36°}$$

and $I_{bn} > I_{cn}$ by a factor of more than 3:1. Therefore, the bulb indicating an *ABC* sequence will burn more brightly due to the greater current. If the phase sequence were *ACB*, the reverse would be true.

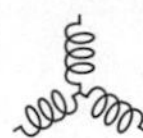

23.15 COMPUTER ANALYSIS

PSpice (DOS)

The application of PSpice to three-phase systems requires that the user be particularly careful to ensure the following:

1. A continuous loop of independent voltage sources is avoided.
2. All nodes have a dc path to ground.
3. A reference node is properly defined.

Writing an input file for the three-phase balanced network of Fig. 23.38 requires that a resistor of relatively negligible value be placed between nodes 1 and 2 to void the continuous loop of independent voltage sources $\mathbf{V}_{AB}$, $\mathbf{V}_{BC}$, and $\mathbf{V}_{CA}$. Note the choice of reference node at the center of the balanced load to establish a reference node for each branch voltage of the Y and each node of the Δ-connected load. To ensure a dc path to ground for each node, a resistor of $10^{30}\ \Omega$ is placed in parallel with each capacitor, as shown for R7, R8, and R9 in the input file of Fig. 23.39.

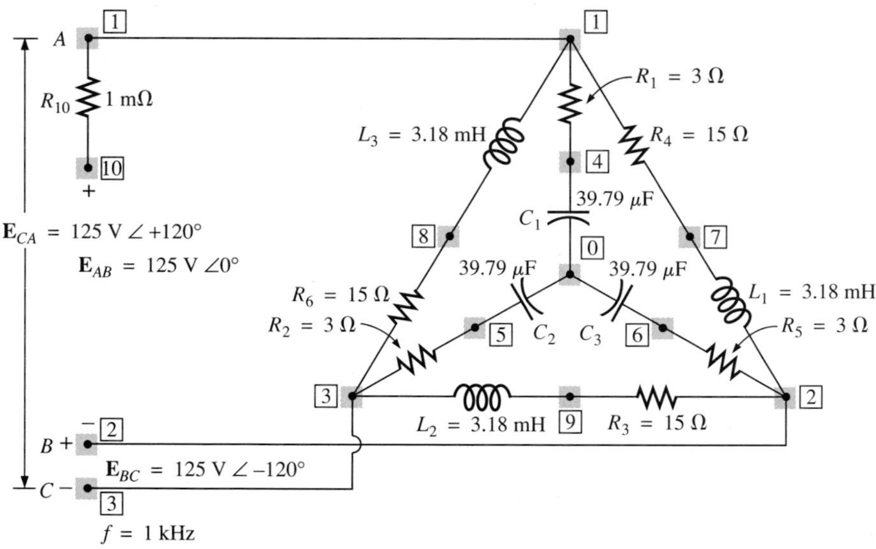

FIG. 23.38
Defining the parameters for a PSpice (DOS) analysis.

The network of Fig. 23.38 is the same as that appearing in Problem 42 of this chapter, with the inductor and capacitor values determined at a frequency of 1 kHz. Note in the printout that the magnitude of the branch current of each leg of the balanced Y is 1.443 A, with phase angles that differ by 120°. The magnitude of each phase current of the delta is 5 A, with phase angles that also differ by 120°. The capacitive Y load establishes phase currents that lead the line voltages, and the inductive delta establishes phase currents that lag the applied line voltages.

```
Chapter 23 - Three Phase System (balanced)

****     CIRCUIT DESCRIPTION

******************************************************************************


V1 10 2 AC 125V 0DEG
R10 10 2 1M
V2 2 3 AC 125V -120DEG
V3 3 1 AC 125V +120DEG
R1 1 4 3
R2 3 5 3
R3 2 6 3
C1 4 0 39.79UF
C2 5 0 39.79UF
C3 6 0 39.79UF
R7 4 0 1E30
R8 6 0 1E30
R9 5 0 1E30
R4 1 7 15
R5 2 9 15
R6 3 8 15
L1 7 2 3.18MH
L2 9 3 3.18MH
L3 8 1 3.18MH
.AC LIN 1 1KH 1KH
.PRINT AC IM(R1) IM(R2) IM(R3)
.PRINT AC IP(R1) IP(R2) IP(R3)
.PRINT AC IM(R4) IM(R5) IM(R6)
.PRINT AC IP(R4) IP(R5) IP(R6)
.OPTIONS NOPAGE
.END
```

```
****     AC ANALYSIS                        TEMPERATURE =   27.000 DEG C


 FREQ        IM(R1)       IM(R2)       IM(R3)

  1.000E+03   1.443E+01    1.443E+01    1.443E+01


****     AC ANALYSIS                        TEMPERATURE =   27.000 DEG C


 FREQ        IP(R1)       IP(R2)       IP(R3)

  1.000E+03   2.313E+01    1.431E+02   -9.687E+01

****     AC ANALYSIS                        TEMPERATURE =   27.000 DEG C


 FREQ        IM(R4)       IM(R5)       IM(R6)

  1.000E+03   5.003E+00    5.003E+00    5.003E+00


****     AC ANALYSIS                        TEMPERATURE =   27.000 DEG C


 FREQ        IP(R4)       IP(R5)       IP(R6)

  1.000E+03  -5.310E+01   -1.731E+02    6.690E+01
```

FIG. 23.39

Input and output files for the PSpice analysis of the three-phase load of Fig. 23.38.

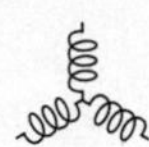

The second three-phase system to be analyzed by PSpice is the unbalanced load of Fig. 23.40. Again, the input file of Fig. 23.41 requires a relatively small resistance between nodes 6 and 1 to eliminate the continuous loop of independent sources, and a resistor of $10^{30}\ \Omega$ must be placed across the capacitor to provide a dc path to ground for node 1.

The results are:

$$\mathbf{I}_1 = 10.71\ \text{A}\ \angle 29.58^\circ$$
$$\mathbf{I}_2 = 6.512\ \text{A}\ \angle 42.3^\circ$$
$$\mathbf{I}_3 = 17.12\ \text{A}\ \angle -145.6^\circ$$

Note the different magnitudes for each phase current due to the unbalanced conditions and the fact that the phase angles are not out of phase by 120°, as in the balanced situation.

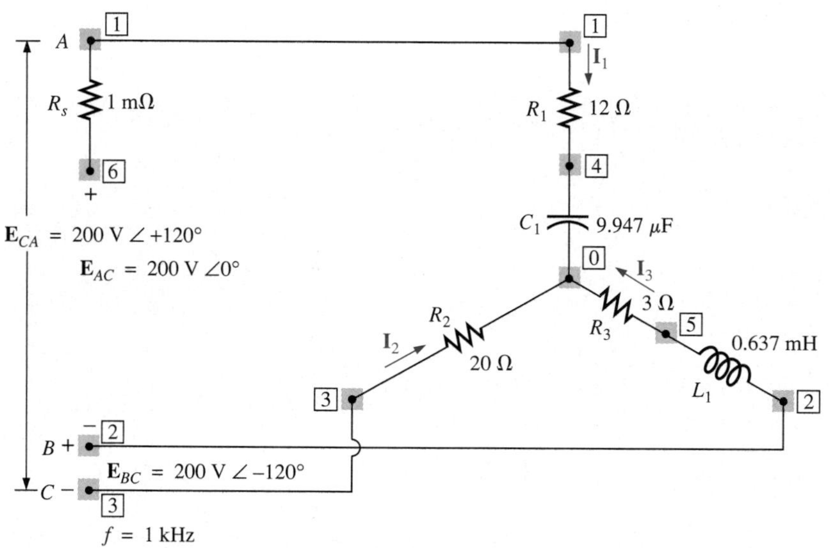

FIG. 23.40
Defining the parameters for a PSpice (DOS) analysis.

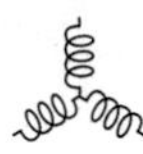


```
Chapter 23 - Three Phase Systems (unbalanced)

****     CIRCUIT DESCRIPTION

******************************************************************************


V1 6 2 AC 200V 0
R21 6 1 1M
V2 2 3 AC 200V -120
V3 3 1 AC 200V -240
R1 1 4 12
C1 4 0 9.947UF
R2 3 0 20
R3 5 0 3
L3 2 5 0.637MH
.AC LIN 1 1KH 1KH
.PRINT AC IM(R1) IP(R1)
.PRINT AC IM(R2) IP(R2)
.PRINT AC IM(R3) IP(R3)
.OPTIONS NOPAGE
.END

****     AC ANALYSIS                          TEMPERATURE =    27.000 DEG C


 FREQ         IM(R1)       IP(R1)

  1.000E+03   1.071E+01   2.958E+01

****     AC ANALYSIS                          TEMPERATURE =    27.000 DEG C


 FREQ         IM(R2)       IP(R2)

  1.000E+03   6.512E+00   4.230E+01

****     AC ANALYSIS                          TEMPERATURE =    27.000 DEG C


 FREQ         IM(R3)       IP(R3)

  1.000E+03   1.712E+01  -1.456E+02
```

FIG. 23.41

Input and output files for the PSpice analysis of the three-phase load of Fig. 23.40.

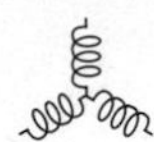

PROBLEMS

SECTION 23.5 The Y-Connected Generator with a Y-Connected Load

1. A balanced Y load having a 10-Ω resistance in each leg is connected to a three-phase, four-wire, Y-connected generator having a line voltage of 208 V. Calculate the magnitude of
   a. the phase voltage of the generator.
   b. the phase voltage of the load.
   c. the phase current of the load.
   d. the line current.

2. Repeat Problem 1 if each phase impedance is changed to a 12-Ω resistor in series with a 16-Ω capacitive reactance.

3. Repeat Problem 1 if each phase impedance is changed to a 10-Ω resistor in parallel with a 10-Ω capacitive reactance.

4. The phase sequence for the Y-Y system of Fig. 23.42 is *ABC*.
   a. Find the angles θ_2 and θ_3 for the specified phase sequence.
   b. Find the voltage across each phase impedance in phasor form.
   c. Find the current through each phase impedance in phasor form.
   d. Draw the phasor diagram of the currents found in part (c) and show that their phasor sum is zero.
   e. Find the magnitude of the line currents.
   f. Find the magnitude of the line voltages.

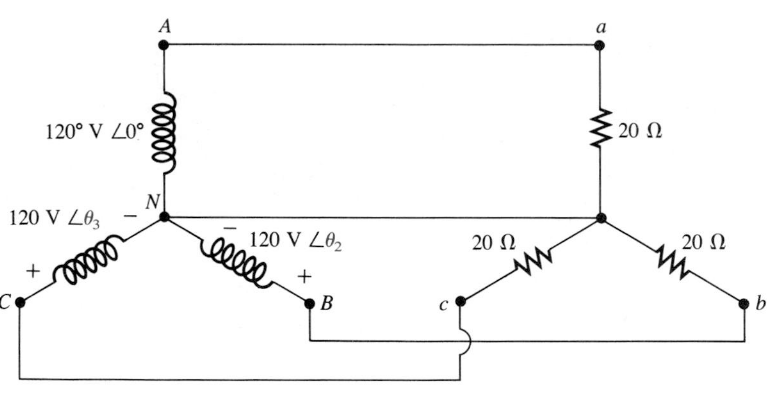

FIG. 23.42
Problem 4.

5. Repeat Problem 4 if the phase impedances are changed to a 9-Ω resistor in series with a 12-Ω inductive reactance.

6. Repeat Problem 4 if the phase impedances are changed to a 6-Ω resistance in parallel with an 8-Ω capacitive reactance.

7. For the system of Fig. 23.43, find the magnitude of the unknown voltages and currents.

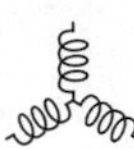

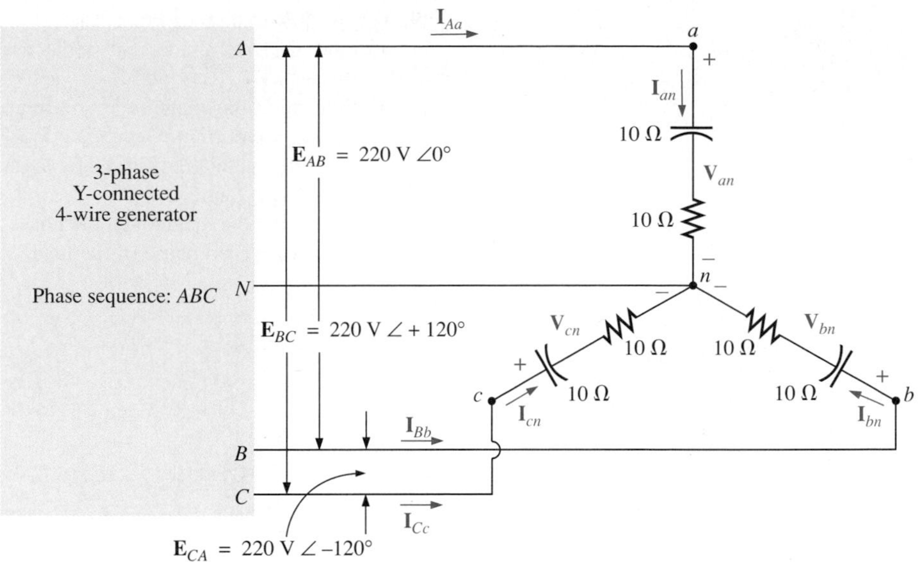

FIG. 23.43
Problems 7, 44, and 51.

***8.** Compute the magnitude of the voltage E_{AB} for the balanced three-phase system of Fig. 23.44.

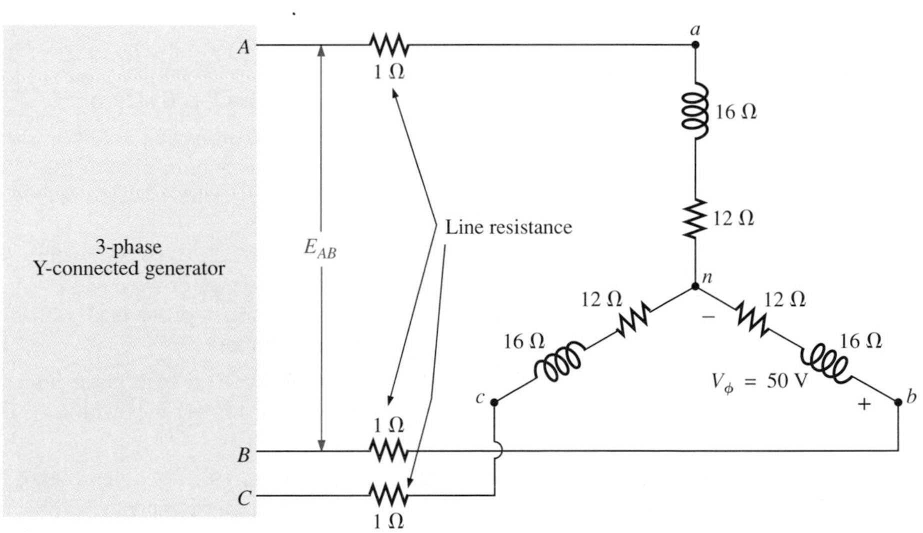

FIG. 23.44
Problem 8.

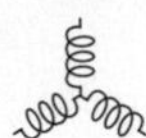

***9.** For the Y-Y system of Fig. 23.45:

a. Find the magnitude and angle associated with the voltages $\mathbf{E}_{AN}$, $\mathbf{E}_{BN}$, and $\mathbf{E}_{CN}$.

b. Determine the magnitude and angle associated with each phase current of the load: $\mathbf{I}_{an}$, $\mathbf{I}_{bn}$, and $\mathbf{I}_{cn}$.

c. Find the magnitude and phase angle of each line current: $\mathbf{I}_{Aa}$, $\mathbf{I}_{Bb}$, and $\mathbf{I}_{Cc}$.

d. Determine the magnitude and phase angle of the voltage across each phase of the load: $\mathbf{V}_{an}$, $\mathbf{V}_{bn}$, and $\mathbf{V}_{cn}$.

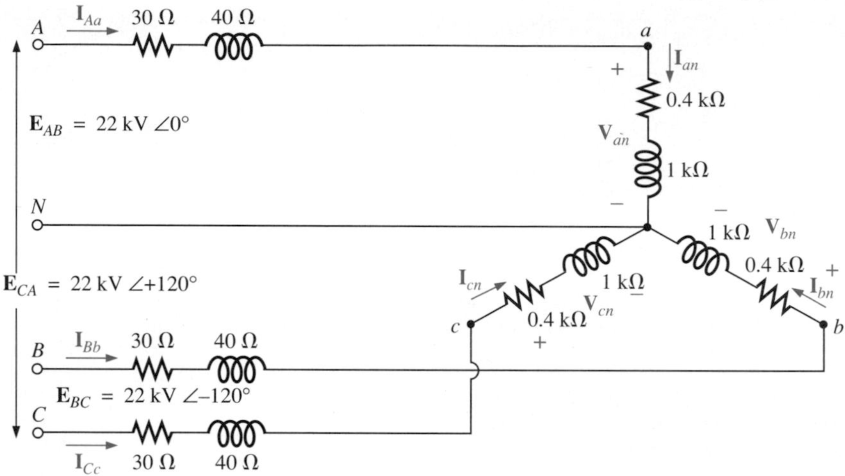

FIG. 23.45
Problems 9 and 53.

SECTION 23.6 The Y-Δ System

10. A balanced Δ load having a 20-Ω resistance in each leg is connected to a three-phase, three-wire, Y-connected generator having a line voltage of 208 V. Calculate the magnitude of

a. the phase voltage of the generator.

b. the phase voltage of the load.

c. the phase current of the load.

d. the line current.

11. Repeat Problem 10 if each phase impedance is changed to a 6.8-Ω resistor in series with a 14-Ω inductive reactance.

12. Repeat Problem 10 if each phase impedance is changed to an 18-Ω resistance in parallel with an 18-Ω capacitive reactance.

13. The phase sequence for the Y-Δ system of Fig. 23.46 is *ABC*.

a. Find the angles θ_2 and θ_3 for the specified phase sequence.

b. Find the voltage across each phase impedance in phasor form.

c. Draw the phasor diagram of the voltages found in part (b) and show that their sum is zero around the closed loop of the Δ load.

d. Find the current through each phase impedance in phasor form.

e. Find the magnitude of the line currents.

f. Find the magnitude of the generator phase voltages.

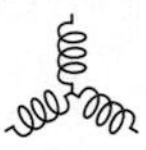

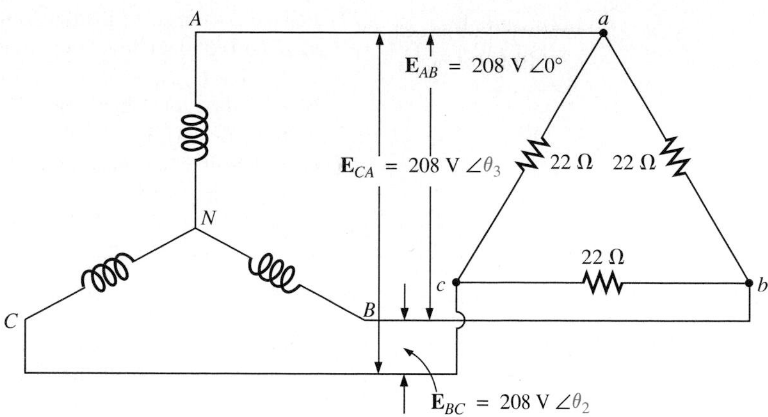

FIG. 23.46
Problems 13 and 45.

14. Repeat the Problem 13 if the phase impedances are changed to a 100-Ω resistor in series with a capacitive reactance of 100 Ω.

15. Repeat Problem 14 if the phase impedances are changed to a 3-Ω resistor in parallel with an inductive reactance of 4 Ω.

16. For the system of Fig. 23.47, find the magnitude of the unknown voltages and currents.

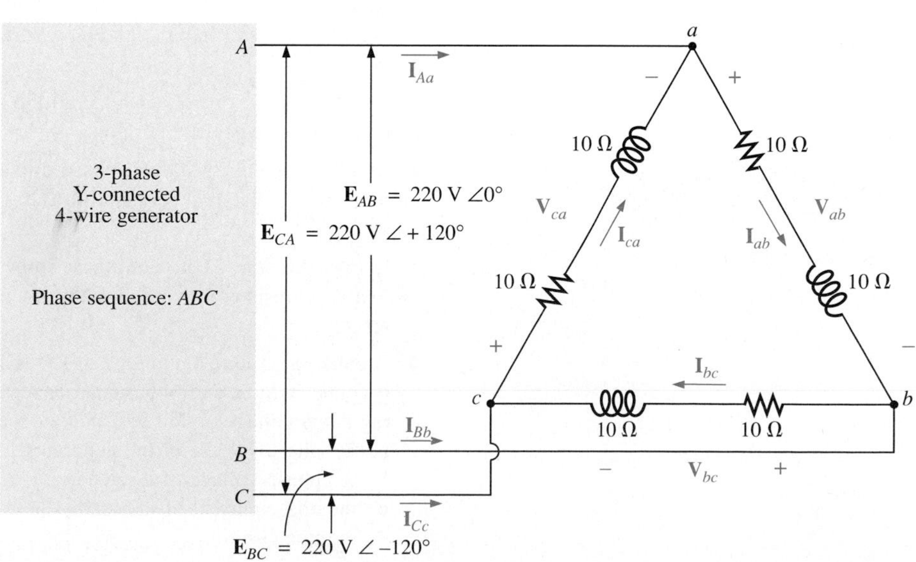

FIG. 23.47
Problems 16 and 52.

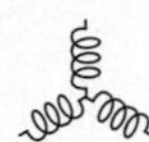

***17.** For the Δ-connected load of Fig. 23.48,

a. Find the magnitude and angle of each phase current: $\mathbf{I}_{ab}$, $\mathbf{I}_{bc}$, $\mathbf{I}_{ca}$.

b. Calculate the magnitude and angle of each line current: $\mathbf{I}_{Aa}$, $\mathbf{I}_{Bb}$, $\mathbf{I}_{Cc}$.

c. Determine the magnitude and angle of the voltages $\mathbf{E}_{AB}$, $\mathbf{E}_{BC}$, and $\mathbf{E}_{CA}$.

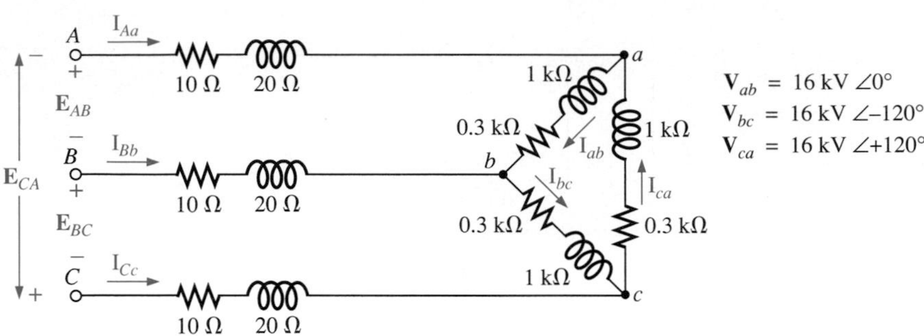

FIG. 23.48
Problem 17.

SECTION 23.9 The Δ-Δ, Δ-Y Three-Phase Systems

18. A balanced Y load having a 30-Ω resistance in each leg is connected to a three-phase, Δ-connected generator having a line voltage of 208 V. Calculate the magnitude of

a. the phase voltage of the generator.

b. the phase voltage of the load.

c. the phase current of the load.

d. the line current.

19. Repeat Problem 18 if each phase impedance is changed to a 12-Ω resistor in series with a 12-Ω inductive reactance.

20. Repeat Problem 18 if each phase impedance is changed to a 15-Ω resistor in parallel with a 20-Ω capacitive reactance.

***21.** For the system of Fig. 23.49, find the magnitude of the unknown voltages and currents.

22. Repeat Problem 21 if each phase impedance is changed to a 10-Ω resistor in series with a 20-Ω inductive reactance.

23. Repeat Problem 21 if each phase impedance is changed to a 20-Ω resistor in parallel with a 15-Ω capacitive reactance.

24. A balanced Δ load having a 220-Ω resistance in each leg is connected to a three-phase, Δ-connected generator having a line voltage of 440 V. Calculate the magnitude of

a. the phase voltage of the generator.

b. the phase voltage of the load.

c. the phase current of the load.

d. the line current.

25. Repeat Problem 24 if each phase impedance is changed to a 12-Ω resistor in series with a 9-Ω capacitive reactance.

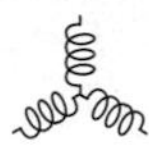

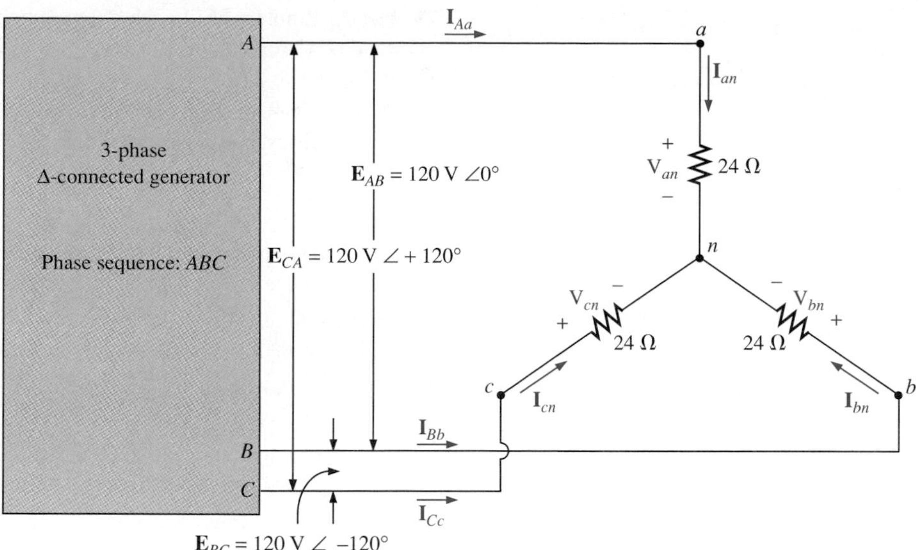

FIG. 23.49
Problem 21.

26. Repeat Problem 24 if each phase impedance is changed to a 22-Ω resistor in parallel with a 22-Ω inductive reactance.

27. The phase sequence for the Δ-Δ system of Fig. 23.50 is *ABC*.

a. Find the angles θ_2 and θ_3 for the specified phase sequence.

b. Find the voltage across each phase impedance in phasor form.

c. Draw the phasor diagram of the voltages found in part (b) and show that their phasor sum is zero around the closed loop of the Δ load.

d. Find the current through each phase impedance in phasor form.

e. Find the magnitude of the line currents.

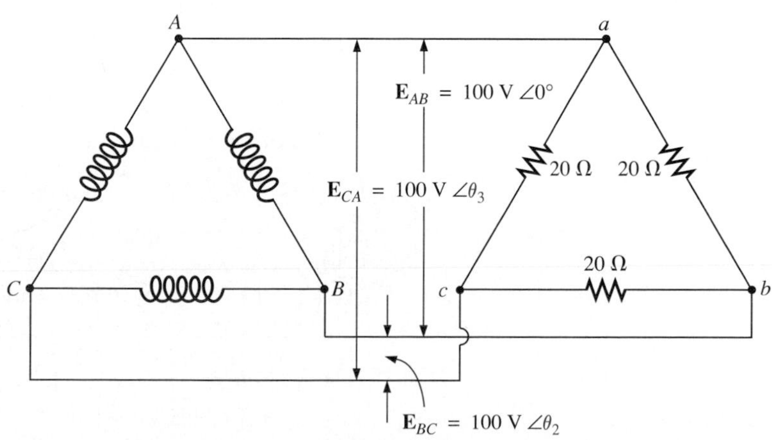

FIG. 23.50
Problem 27.

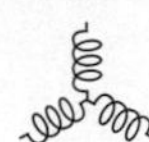

28. Repeat Problem 25 if each phase impedance is changed to a 12-Ω resistor in series with a 16-Ω inductive reactance.

29. Repeat Problem 25 if each phase impedance is changed to a 20-Ω resistor in parallel with a 20-Ω capacitive reactance.

SECTION 23.10 Power

30. Find the total watts, volt-amperes reactive, volt-amperes, and F_p of the three-phase system of Problem 2.

31. Find the total watts, volt-amperes reactive, volt-amperes, and F_p of the three-phase system of Problem 4.

32. Find the total watts, volt-amperes reactive, volt-amperes, and F_p of the three-phase system of Problem 7.

33. Find the total watts, volt-amperes reactive, volt-amperes, and F_p of the three-phase system of Problem 12.

34. Find the total watts, volt-amperes reactive, volt-amperes, and F_p of the three-phase system of Problem 14.

35. Find the total watts, volt-amperes reactive, volt-amperes, and F_p of the three-phase system of Problem 16.

36. Find the total watts, volt-amperes reactive, volt-amperes, and F_p of the three-phase system of Problem 20.

37. Find the total watts, volt-amperes reactive, volt-amperes, and F_p of the three-phase system of Problem 22.

38. Find the total watts, volt-amperes reactive, volt-amperes, and F_p of the three-phase system of Problem 26.

39. Find the total watts, volt-amperes reactive, volt-amperes, and F_p of the three-phase system of Problem 28.

40. A balanced, three-phase, Δ-connected load has a line voltage of 200 and a total power consumption of 4800 W at a lagging power factor of 0.8. Find the impedance of each phase in rectangular coordinates.

41. A balanced, three-phase, Y-connected load has a line voltage of 208 and a total power consumption of 1200 W at a leading power factor of 0.6. Find the impedance of each phase in rectangular coordinates.

***42.** Find the total watts, volt-amperes reactive, volt-amperes, and F_p of the system of Fig. 23.51.

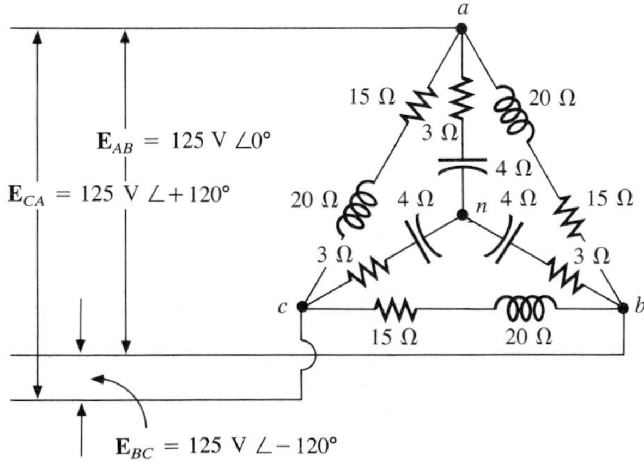

FIG. 23.51
Problem 42.

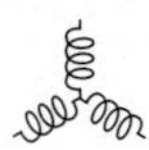

*43. The Y-Y system of Fig. 23.52 has a balanced load and a line impedance $\mathbf{Z}_{\text{line}} = 4\ \Omega + j\ 20\ \Omega$. If the line voltage at the generator is 16,000 V and the total power delivered to the load is 1200 kW at 80 A, determine each of the following:

   a. The magnitude of each phase voltage of the generator.
   b. The magnitude of the line currents.
   c. The total power delivered by the source.
   d. The power factor angle of the entire load "seen" by the source.
   e. The magnitude and angle of the current $\mathbf{I}_{Aa}$ if $\mathbf{E}_{AN} = E_{AN} \angle 0°$.
   f. The magnitude and angle of the phase voltage $\mathbf{V}_{an}$.
   g. The impedance of the load of each phase in rectangular coordinates.
   h. The difference between the power factor of the load and the power factor of the entire system (including $\mathbf{Z}_{\text{line}}$).
   i. The efficiency of the system.

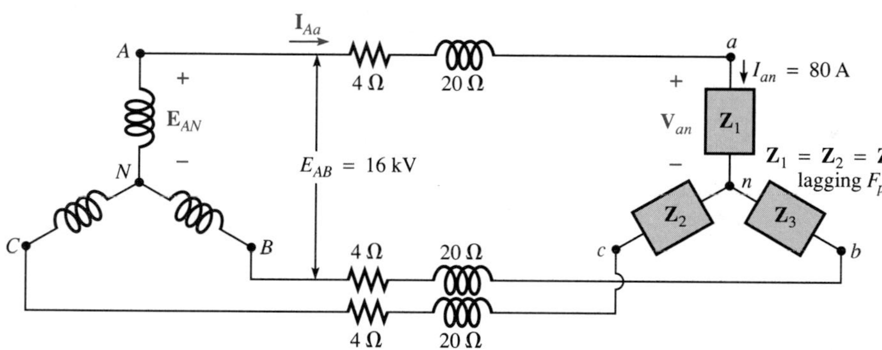

FIG. 23.52
Problem 43.

SECTION 23.11 The Three-Wattmeter Method

44. a. Sketch the connections required to measure the total watts delivered to the load of Fig. 23.43 using three wattmeters.
    b. Determine the total wattage dissipation and the reading of each wattmeter.

45. Repeat Problem 44 for the network of Fig. 23.46.

SECTION 23.12 The Two-Wattmeter Method

46. a. For the three-wire system of Fig. 23.53, properly connect a second wattmeter so that the two will measure the total power delivered to the load.
    b. If one wattmeter has a reading of 200 W and the other a reading of 85 W, what is the total dissipation in watts if the total power factor is 0.8 leading?
    c. Repeat part (b) if the total power factor is 0.2 lagging and $P_l = 100$ W.

47. Sketch three different ways that two wattmeters can be connected to measure the total power delivered to the load of Problem 16.

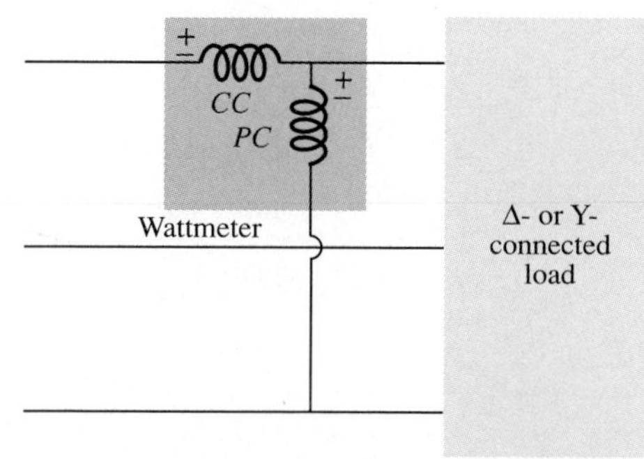

FIG. 23.53
Problem 46.

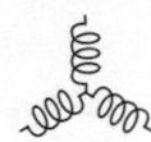

***48.** For the Y-Δ system of Fig. 23.54:

a. Determine the magnitude and angle of the phase currents.

b. Find the magnitude and angle of the line currents.

c. Determine the reading of each wattmeter.

d. Find the total power delivered to the load.

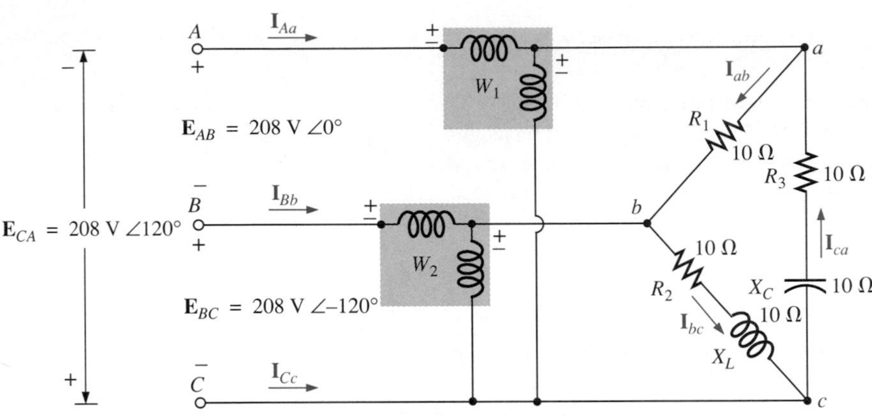

FIG. 23.54
Problems 48 and 54.

SECTION 23.13 Unbalanced, Three-Phase, Four-Wire, Y-Connected Load

***49.** For the system of Fig. 23.55:

a. Calculate the magnitude of the voltage across each phase of the load.

b. Find the magnitude of the current through each phase of the load.

c. Find the total watts, volt-amperes reactive, volt-amperes, and F_p of the system.

d. Find the phase currents in phasor form.

e. Using the results of part (c), determine the current $\mathbf{I}_N$.

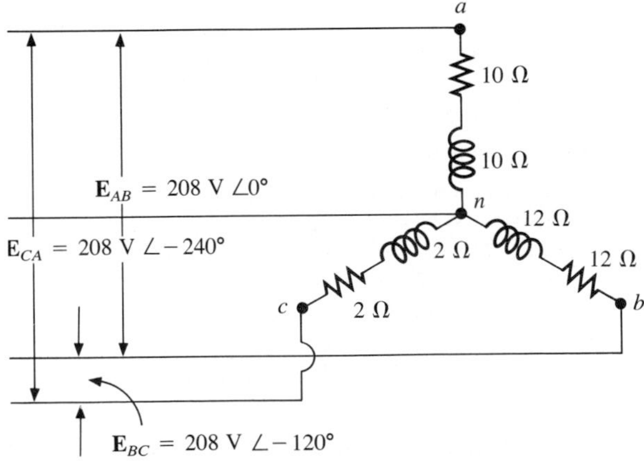

FIG. 23.55
Problem 49.

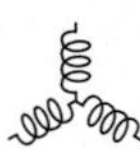

SECTION 23.14 Unbalanced, Three-Phase, Three-Wire, Y-Connected Load

***50.** For the three-phase, three-wire system of Fig. 23.56, find the magnitude of the current through each phase of the load and the total watts, volt-amperes reactive, volt-amperes, and F_p of the load.

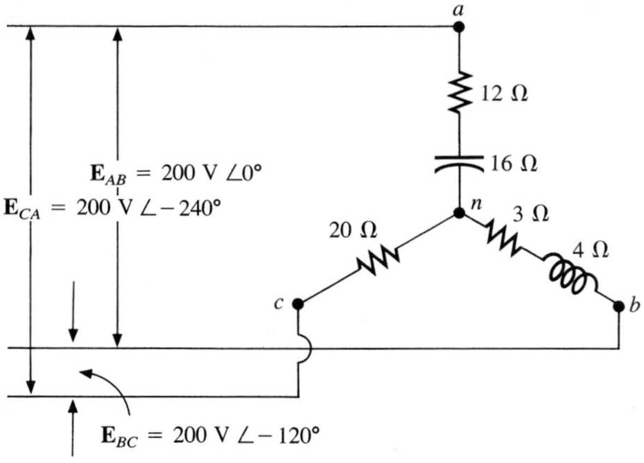

FIG. 23.56
Problem 50.

SECTION 23.15 Computer Analysis

PSpice (DOS)

51. Write the input file to determine the phase voltages and currents for the network of Fig. 23.43.

52. Write the input file to determine the phase currents for the network of Fig. 23.47.

53. Write the input file to determine the phase voltages and currents for the load of Fig. 23.45.

54. Write the input file to determine the phase currents for the network of Fig. 23.54.

Programming Language (C++, BASIC, PASCAL, etc.)

55. Given the magnitude of the line voltages and the impedance of each phase (in series or parallel), write a program to determine the magnitude of all the voltages and currents of a balanced Y-connected load.

56. Repeat Problem 55 for a Δ-connected load.

57. For a balanced Y-connected load, write a program to determine
  a. the magnitude of the load currents and voltages.
  b. the real, reactive, and apparent power to each phase.
  c. the total real, reactive, and apparent power to the load.
  d. the load power factor.

58. Repeat Problem 57 for a balanced Δ-connected load.

GLOSSARY

Delta (Δ)-connected generator A three-phase generator having the three phases connected in the shape of the capital Greek letter delta (Δ).

Line current The current that flows from the generator to the load of a single-phase or polyphase system.

Line voltage The potential difference that exists between the lines of a single-phase or polyphase system.

Neutral connection The connection between the generator and the load that, under balanced conditions, will have zero current associated with it.

Phase current The current that flows through each phase of a single-phase or polyphase generator load.

Phase sequence The order in which the generated sinusoidal voltages of a polyphase generator will affect the load to which they are applied.

Phase voltage The voltage that appears between the line and neutral of a Y-connected generator and from line to line in a Δ-connected generator.

Polyphase ac generator An electromechanical source of ac power that generates more than one sinusoidal voltage per rotation of the rotor. The frequency generated is determined by the speed of rotation and the number of poles of the rotor.

Single-phase ac generator An electromechanical source of ac power that generates a single sinusoidal voltage having a frequency determined by the speed of rotation and the number of poles of the rotor.

Three-wattmeter method A method for determining the total power delivered to a three-phase load using three wattmeters.

Two-wattmeter method A method for determining the total power delivered to a Δ- or Y-connected three-phase load using only two wattmeters and considering the power factor of the load.

Unbalanced polyphase load A load not having the same impedance in each phase.

Wye (Y)-connected generator A three-phase source of ac power in which the three phases are connected in the shape of the letter Y.

24

Nonsinusoidal Circuits

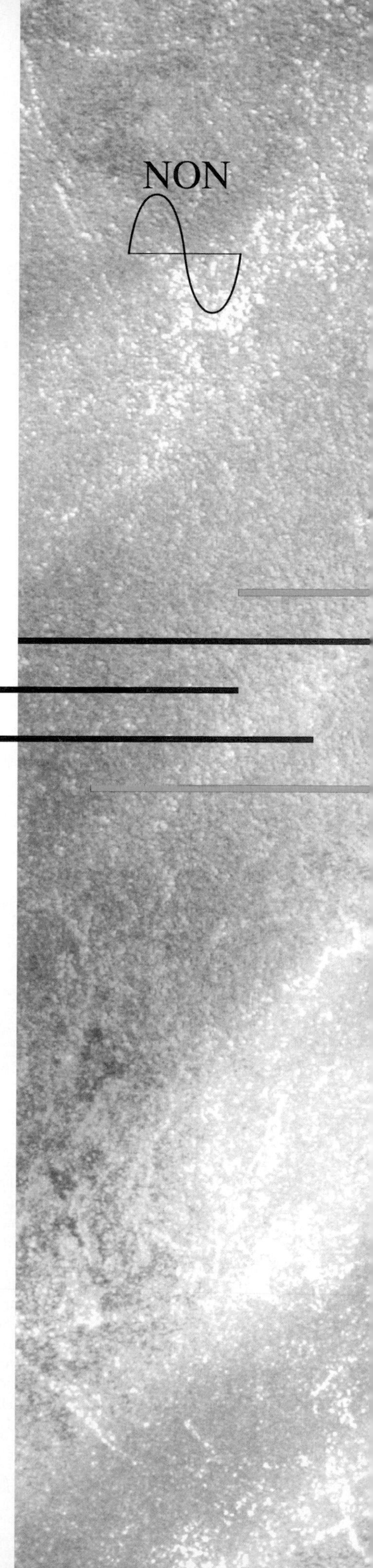

24.1 INTRODUCTION

Any waveform that differs from the basic description of the sinusoidal waveform is referred to as *nonsinusoidal*. The most obvious and familiar are the dc, square-wave, triangular, sawtooth, and rectified waveforms of Fig. 24.1.

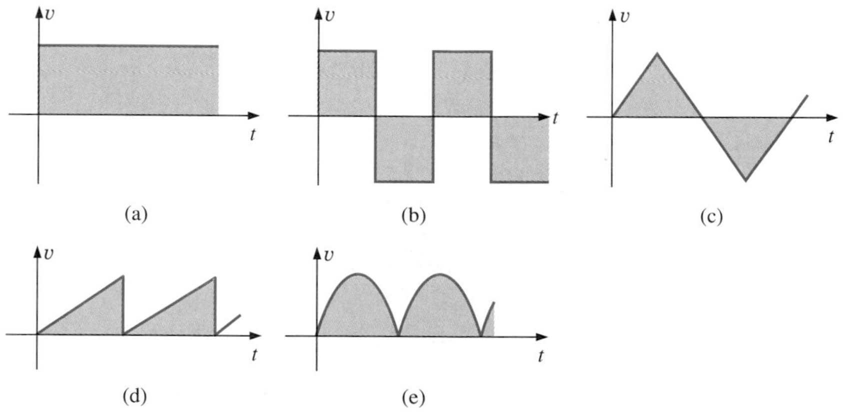

FIG. 24.1
Common nonsinusoidal waveforms: (a) dc, (b) square, (c) triangular, (d) sawtooth, (e) rectified.

The output of many electrical and electronic devices will be nonsinusoidal, even though the applied signal may be purely sinusoidal. For example, the network of Fig. 24.2 employs a diode to clip off the negative portion of the applied signal in a process called half-wave rectification, which is used in the development of dc levels from a sinusoidal input. You will find in your electronics courses that the diode is similar to a mechanical switch, but it is different because it can conduct current in only one direction. The output waveform is definitely nonsinusoidal, but note that it has the same period as the applied signal and matches the input for half the period.

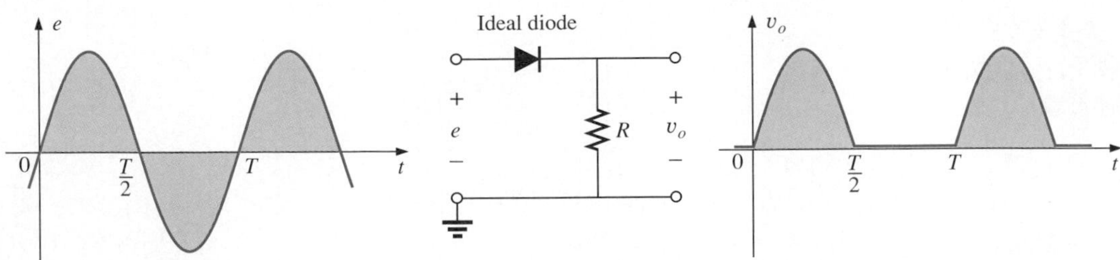

FIG. 24.2
Half-wave rectifier producing a nonsinusoidal waveform.

This chapter will demonstrate how a nonsinusoidal waveform like the output of Fig. 24.2 can be represented by a series of terms. You will also learn how to determine the response of a network to such an input.

24.2 FOURIER SERIES

Fourier series refers to a series of terms, developed in 1826 by Baron Jean Fourier (Fig. 24.3), that can be used to represent a nonsinusoidal periodic waveform. In the analysis of these waveforms, we solve for each term in the Fourier series:

$$f(t) = \underbrace{A_0}_{\text{dc or average value}} + \underbrace{A_1 \sin \omega t + A_2 \sin 2\omega t + A_3 \sin 3\omega t + \cdots + A_n \sin n\omega t}_{\text{sine terms}} + \underbrace{B_1 \cos \omega t + B_2 \cos 2\omega t + B_3 \cos 3\omega t + \cdots + B_n \cos n\omega t}_{\text{cosine terms}} \tag{24.1}$$

French (Auxerre, Grenoble, Paris)
(1768–1830)
Mathematician, Egyptologist, and **Administrator**
Professor of Mathematics, École Polytechnique

Courtesy of the Smithsonian Institution Photo No. 56,822

He is best known for an infinite mathematical series of sine and cosine terms called the *Fourier series* that he used to show how the conduction of heat in solids can be analyzed and defined. Although primarily a mathematician, a great deal of Fourier's work revolved around real-world physical occurrences such as heat transfer, sun spots, and the weather. He joined the École Polytechnic in Paris as a faculty member when the institute first opened. Napoleon requested his aid in the research of Egyptian antiquities, resulting in a three-year stay in Egypt as Secretary of the Institut d'Égypte. Napoleon made him a baron in 1809 and he was elected to the Académie des Sciences in 1817.

FIG. 24.3
Baron Jean Fourier.

Depending on the waveform, a large number of these terms may be required to approximate the waveform closely for the purpose of circuit analysis.

The Fourier series has three basic parts. The first is the dc term A_0, which is the average value of the waveform over one full cycle. The second is a series of sine terms. There are no restrictions on the values or relative values of the amplitudes of these sine terms, but each will have a frequency that is an integer multiple of the frequency of the first sine term of the series. The third part is a series of cosine terms. There are again *no* restrictions on the values or relative values of the amplitudes of these cosine terms, but each will have a frequency that is an integer multiple of the frequency of the first cosine term of the series. For a particular waveform, it is quite possible that all of the sine *or* cosine terms are zero. Characteristics of this type can be determined by simply examining the nonsinusoidal waveform and its position on the horizontal axis.

The first term of the sine and cosine series is called the *fundamental component*. It represents the minimum frequency term required to represent a particular waveform, and it also has the same frequency as the waveform being represented. A fundamental term, therefore, must be present in any Fourier series representation. The other terms with higher-order frequencies (integer multiples of the fundamental) are called the *harmonic terms*. A term that has a frequency equal to twice the fundamental is the second harmonic; three times, the third harmonic; and so on.

Average Value: A_0

The dc term of the Fourier series is the average value of the waveform over one full cycle. If the net area above the horizontal axis equals that below in one full period, $A_0 = 0$, and the dc term does not appear in the expansion. If the area above the axis is greater than that below over one full cycle, A_0 is positive and will appear in the Fourier series representation. If the area below the axis is greater, A_0 is negative and will appear with the negative sign in the expansion.

Odd Function (Point Symmetry)

If a waveform is such that its value for $+t$ is the negative of that for $-t$, it is called an odd function or is said to have point symmetry.

Figure 24.4(a) is an example of a waveform with point symmetry. Note that the waveform has a peak value at t_1 that matches the magnitude (with the opposite sign) of the peak value at $-t_1$. For waveforms of this type, all the parameters $B_{1\to\infty}$ of Eq. (24.1) will be zero. In fact,

waveforms with point symmetry can be fully described by just the dc and sine terms of the Fourier series.

Note in Fig. 24.4(b) that a sine wave is an odd function with point symmetry.

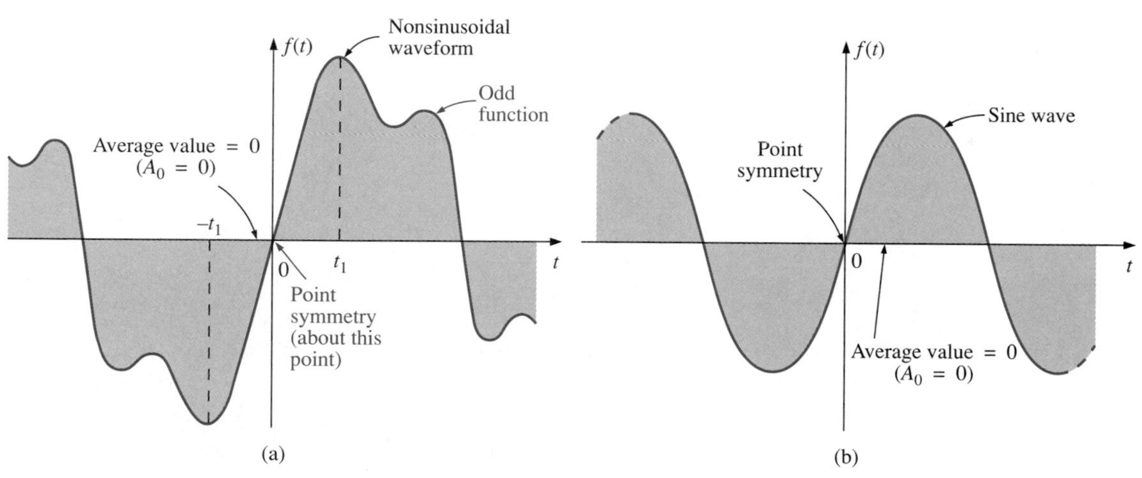

FIG. 24.4
Point symmetry.

For both waveforms of Fig. 24.4, the following mathematical relationship is true:

$$f(t) = -f(-t) \qquad \text{(odd function)} \qquad \textbf{(24.2)}$$

In words, it states that the magnitude of the function at $+t$ is equal to the negative of the magnitude at $-t$ [t_1 in Fig. 24.4(a)].

Even Function (Axis Symmetry)

If a waveform is symmetric about the vertical axis, it is called an even function or is said to have axis symmetry.

Figure 24.5(a) is an example of such a waveform. Note that the value of the function at t_1 is equal to the value at $-t_1$. For waveforms of this type, all the parameters $A_{1\rightarrow\infty}$ will be zero. In fact,

waveforms with axis symmetry can be fully described by just the dc and cosine terms of the Fourier series.

Note in Fig. 24.5(b) that a cosine wave is an even function with axis symmetry.

For both waveforms of Fig. 24.5, the following mathematical relationship is true:

$$f(t) = f(-t) \qquad \text{(even function)} \qquad \textbf{(24.3)}$$

In words, it states that the magnitude of the function is the same at $+t_1$ as at $-t$ [t_1 in Fig. 24.5(a)].

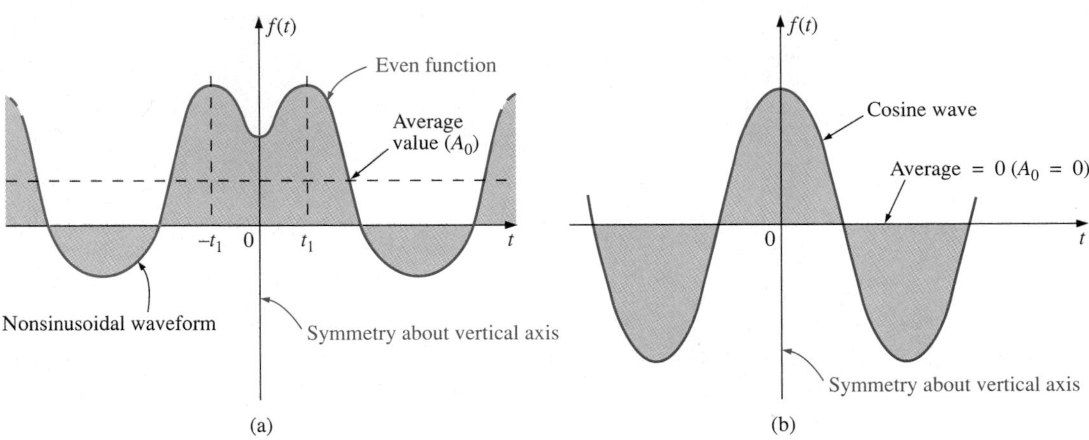

FIG. 24.5
Axis symmetry.

Mirror or Half-Wave Symmetry

If a waveform has half-wave or mirror symmetry as demonstrated by the waveform of Fig. 24.6, the even harmonics of the series of sine and cosine terms will be zero.

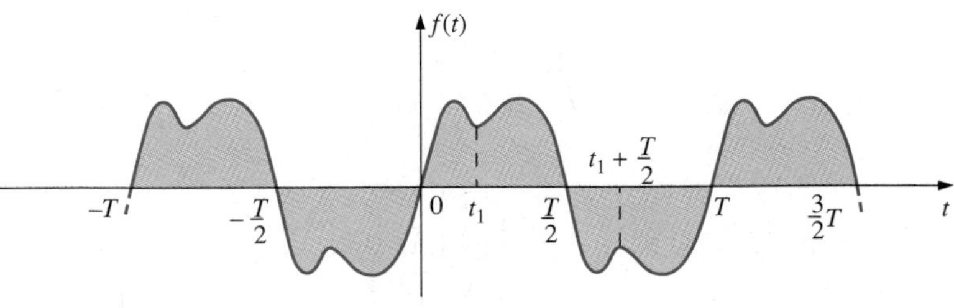

FIG. 24.6
Mirror symmetry.

In functional form the waveform must satisfy the following relationship:

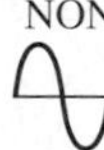

$$f(t) = -f\left(t + \frac{T}{2}\right) \tag{24.4}$$

Equation (24.4) states that the waveform encompassed in one time interval $T/2$ will repeat itself in the next $T/2$ time interval, but in the negative sense (t_1 in Fig. 24.6). For example, the waveform of Fig. 24.6 from zero to $T/2$ will repeat itself in the time interval $T/2$ to T, but below the horizontal axis.

Repetitive on the Half-Cycle

The repetitive nature of a waveform can determine whether specific harmonics will be present in the Fourier series expansion. In particular,

if a waveform is repetitive on the half-cycle as demonstrated by the waveform of Fig. 24.7, the odd harmonics of the series of sine and cosine terms are zero.

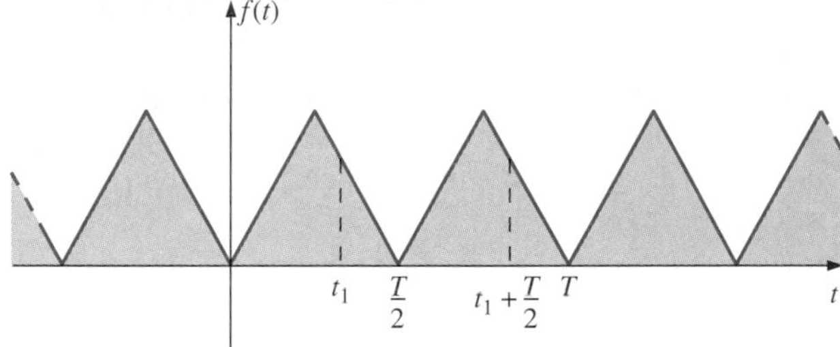

FIG. 24.7
A waveform repetitive on the half-cycle.

In functional form the waveform must satisfy the following relationship:

$$f(t) = f\left(t + \frac{T}{2}\right) \tag{24.5}$$

Equation (24.5) states that the function repeats itself after each $T/2$ time interval (t_1 in Fig. 24.7). The waveform, however, will also repeat itself after each period T. In general, therefore, for a function of this type, if the period T of the waveform is chosen to be twice that of the minimum period ($T/2$), the odd harmonics will all be zero.

Mathematical Approach

The constants A_0, $A_{1\rightarrow n}$, $B_{1\rightarrow n}$ can be determined by using the following integral formulas:

$$A_0 = \frac{1}{T}\int_0^T f(t)\,dt \tag{24.6}$$

$$A_n = \frac{2}{T}\int_0^T f(t)\sin n\omega t\,dt \tag{24.7}$$

$$B_n = \frac{2}{T}\int_0^T f(t)\cos n\omega t\,dt \tag{24.8}$$

These equations have been presented for recognition purposes only; they will not be used in the following analysis.

Instrumentation

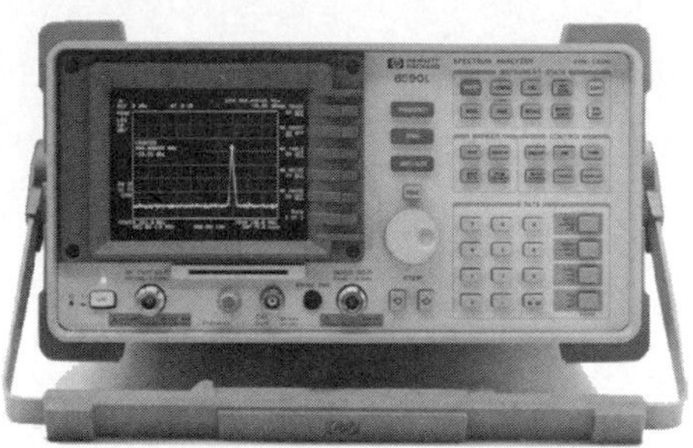

FIG. 24.8
Spectrum analyzer. (Courtesy of Hewlett Packard)

There are three types of instrumentation available that will reveal the dc, fundamental, and harmonic content of a waveform: *the spectrum analyzer, wave analyzer,* and *Fourier analyzer.* The purpose of such instrumentation is not solely to determine the composition of a particular waveform but also to reveal the level of distortion that may have been introduced by a system. For instance, an amplifier may be increasing the applied signal by a factor of 50, but in the process it may have distorted the waveform in a way that is quite unnoticeable from the oscilloscope display. The amount of distortion would appear in the form of harmonics at frequencies that are multiples of the applied frequency. Each of the above instruments would reveal which frequencies are having the most impact on the distortion, permitting their removal with properly designed filters.

The *spectrum analyzer* has the appearance of an oscilloscope, as shown in Fig. 24.8, but rather than display a waveform that is voltage (vertical axis) versus time (horizontal axis), it generates a display scaled off in dB (vertical axis) versus frequency (horizontal axis). Such a display is said to be in the *frequency domain* versus the *time domain* of the standard oscilloscope. The height of the vertical line in the display of Fig. 24.8 reveals the impact of that frequency on the shape of the waveform. Spectrum analyzers are unable to provide the phase angle associated with each component.

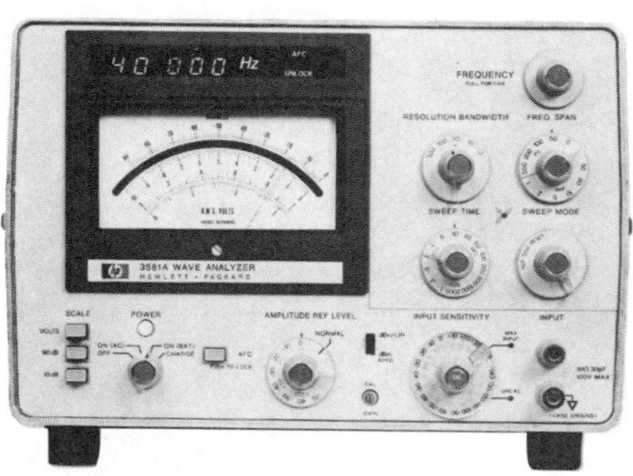

FIG. 24.9
Wave analyzer. (Courtesy of Hewlett Packard)

The *wave analyzer* of Fig. 24.9 is a true rms voltmeter whose frequency of measurement can be changed manually. In other words, the operator works through the frequencies of interest, and the analog display will indicate the rms value of each harmonic component present. Of course, once the fundamental component is determined, the operator can quickly move through the possible harmonic levels. The wave analyzer, like the spectrum analyzer, is unable to provide the angle associated with the various components.

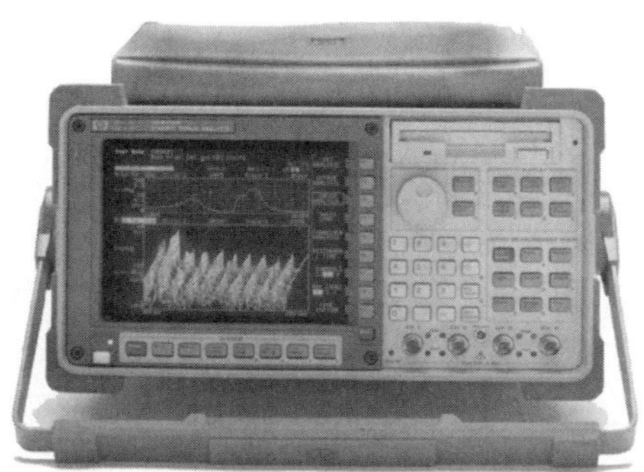

FIG. 24.10
Fourier analyzer. (Courtesy of Hewlett Packard)

The *Fourier analyzer* of Fig. 24.10 is similar in many respects to the spectrum analyzer except for its ability to investigate all the frequencies of interest at one time. The spectrum analyzer must review the signal one frequency at a time. The Fourier analyzer has the distinct advantage of being able to determine the phase angle of each component.

The following examples will demonstrate the use of the equations and concepts introduced thus far in this chapter.

EXAMPLE 24.1 Determine which components of the Fourier series are present in the waveforms of Fig. 24.11.

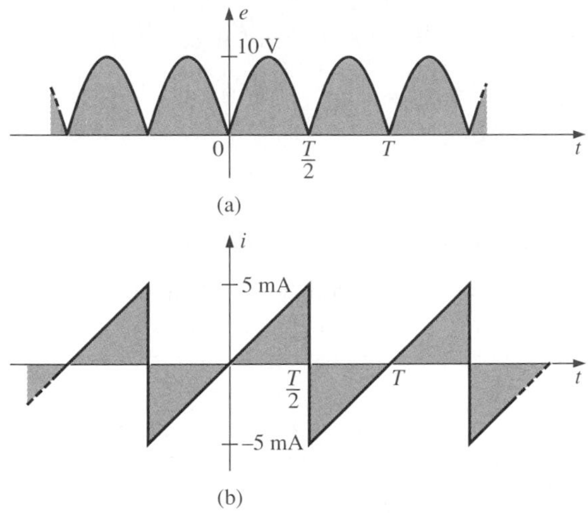

FIG. 24.11
Example 24.1.

Solutions:

Fig. 24.11(a): the waveform has a net area above the horizontal axis and therefore will have **a positive dc term A_0**.

The waveform has axis symmetry, resulting in **only cosine terms** in the expansion.

The waveform has half-cycle symmetry, resulting in **only even terms** in the cosine series.

Fig. 24.11(b): the waveform has the same area above and below the horizontal axis within each period, resulting in $\boldsymbol{A_0 = 0}$.

The waveform has point symmetry, resulting in **only sine terms** in the expansion.

EXAMPLE 24.2 Write the Fourier series expansion for the waveforms of Fig. 24.12.

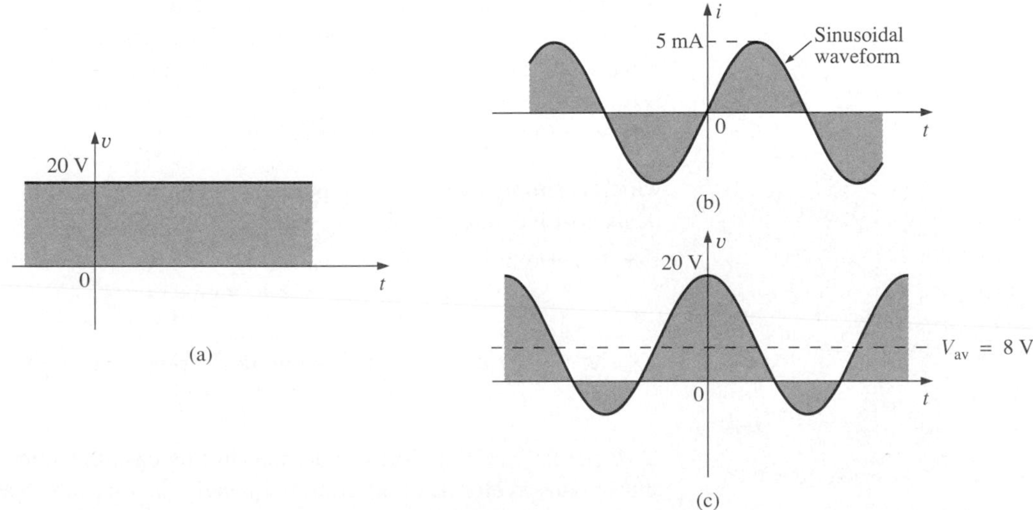

FIG. 24.12
Example 24.2.

Solutions:

a. $A_0 = 20, A_{1\to n} = 0, B_{1\to n} = 0$
   $\boldsymbol{v = 20}$

b. $A_0 = 0, A_1 = 5 \times 10^{-3}, A_{2\to n} = 0, B_{1\to n} = 0$
   $\boldsymbol{i = 5 \times 10^{-3} \sin \omega t}$

c. $A_0 = 8, A_{1-n} = 0, B_1 = 12, B_{2\to n} = 0$
   $\boldsymbol{v = 8 + 12 \cos \omega t}$

EXAMPLE 24.3 Sketch the following Fourier series expansion:

$$v = 2 + 1 \cos \alpha + 2 \sin \alpha$$

Solution: Note Fig. 24.13.

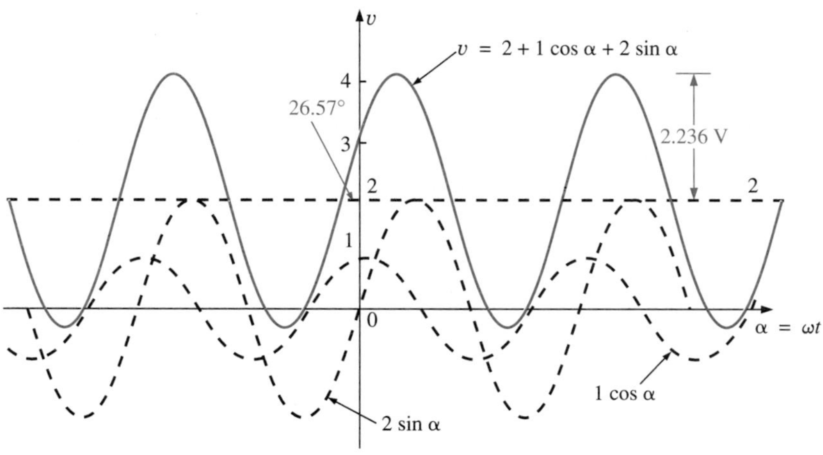

FIG. 24.13
Example 24.3.

The solution could be obtained graphically by first plotting all of the functions and considering a sufficient number of points on the horizontal axis; or phasor algebra could be employed as follows:

$$\begin{aligned} 1 \cos \alpha + 2 \sin \alpha &= 1 \text{ V} \angle 90^\circ + 2 \text{ V} \angle 0^\circ = j\,1 \text{ V} + 2 \text{ V} \\ &= 2 \text{ V} + j\,1 \text{ V} = 2.236 \text{ V} \angle 26.57^\circ \\ &= 2.236 \sin(\alpha + 26.57^\circ) \end{aligned}$$

and
$$v = 2 + 2.236 \sin(\alpha + 26.57^\circ)$$

which is simply the sine wave portion riding on a dc level of 2 V. That is, its positive maximum is 2 V + 2.236 V = 4.236 V, and its minimum is 2 V − 2.236 V = −0.236 V.

EXAMPLE 24.4 Sketch the following Fourier series expansion:

$$i = 1 \sin \omega t + 1 \sin 2\omega t$$

Solution: See Fig. 24.14. Note that in this case the sum of the two sinusoidal waveforms of different frequencies is *not* a sine wave. Recall that complex algebra can be applied only to waveforms having the *same* frequency. In this case the solution is obtained graphically point by point, as shown for $t = t_1$.

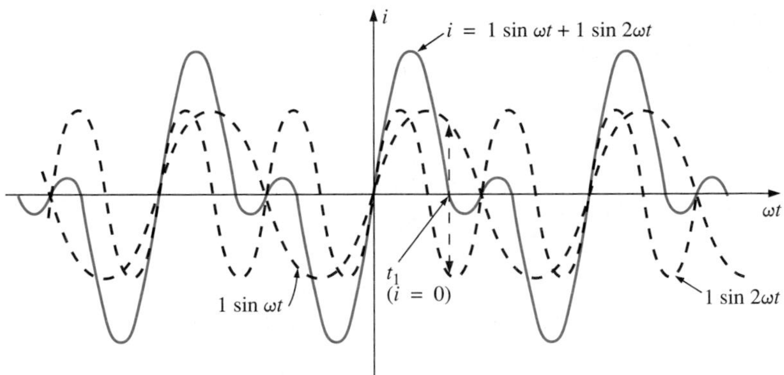

FIG. 24.14
Example 24.4.

As an additional example in the use of the Fourier series approach, consider the square wave shown in Fig. 24.15. The average value is zero, so $A_0 = 0$. It is an odd function, so all the constants $B_{1\rightarrow n}$ equal zero; only sine terms will be present in the series expansion. Since the waveform satisfies the criteria for $f(t) = -f(t + T/2)$, the even harmonics will also be zero.

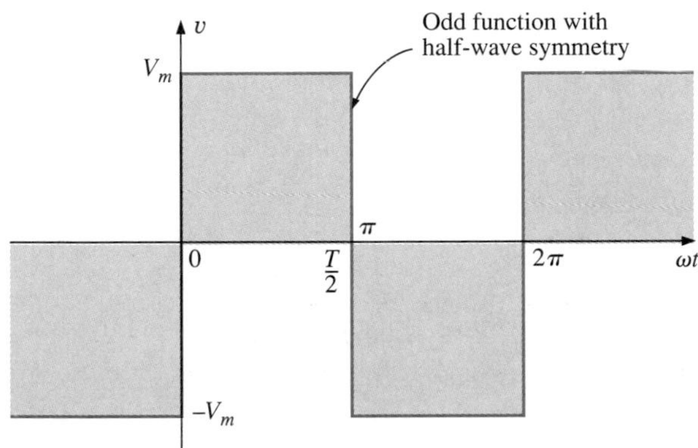

FIG. 24.15
Square wave.

The expression obtained after evaluating the various coefficients using Eq. (24.8) is

$$v = \frac{4}{\pi} V_m\left(\sin \omega t + \frac{1}{3} \sin 3\omega t + \frac{1}{5} \sin 5\omega t + \frac{1}{7} \sin 7\omega t + \cdots + \frac{1}{n} \sin n\omega t\right) \quad \textbf{(24.9)}$$

Note that the fundamental does indeed have the same frequency as that of the square wave. If we add the fundamental and third harmonics, we obtain the results shown in Fig. 24.16.

Even with only the first two terms, a few characteristics of the square wave are beginning to appear. If we add the next two terms (Fig. 24.17), the width of the pulse increases, and the number of peaks increases.

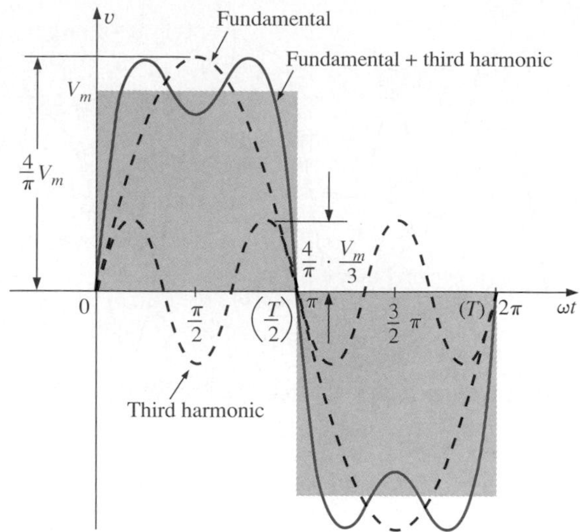

FIG. 24.16
Fundamental plus third harmonic.

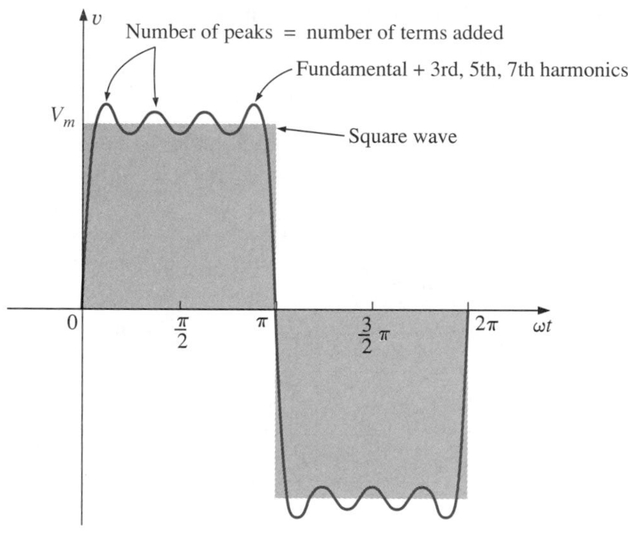

FIG. 24.17
Fundamental plus third, fifth, and seventh harmonics.

As we continue to add terms, the series will better approximate the square wave. Note, however, that the amplitude of each succeeding term diminishes to the point at which it will be negligible compared with those of the first few terms. A good approximation would be to assume that the waveform is composed of the harmonics up to and including the ninth. Any higher harmonics would be less than one-tenth the fundamental. If the waveform just described were shifted above or below the horizontal axis, the Fourier series would be altered only by a change in the dc term. Figure 24.18(c), for example, is the sum of Fig. 24.18(a) and (b). The Fourier series for the complete waveform is, therefore,

$$\begin{aligned} v &= v_1 + v_2 = V_m + \text{Eq. (24.9)} \\ &= V_m + \frac{4}{\pi}V_m\left(\sin \omega t + \frac{1}{3}\sin 3\omega t + \frac{1}{5}\sin 5\omega t + \frac{1}{7}\sin 7\omega t + \cdots\right) \end{aligned}$$

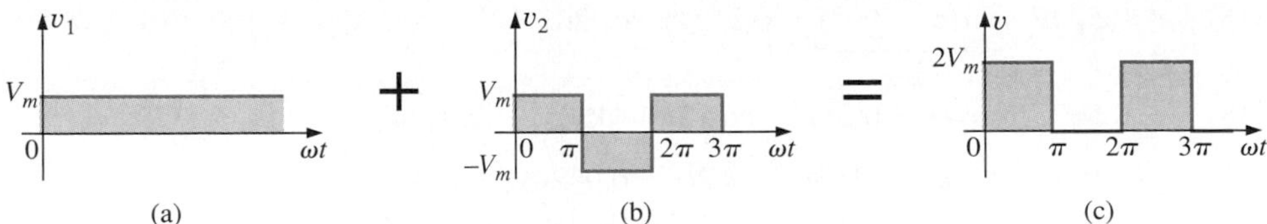

FIG. 24.18
Shifting a waveform vertically with the addition of a dc term.

and
$$v = V_m\left[1 + \frac{4}{\pi}\left(\sin \omega t + \frac{1}{3}\sin 3\omega t + \frac{1}{5}\sin 5\omega t + \frac{1}{7}\sin 7\omega t + \cdots\right)\right]$$

The equation for the half-wave rectified pulsating waveform of Fig. 24.19(c) is

$$\boxed{v_2 = 0.318V_m + 0.500V_m \sin \alpha - 0.212V_m \cos 2\alpha - 0.0424V_m \cos 4\alpha - \cdots} \qquad \textbf{(24.10)}$$

The waveform in Fig. 24.19(a) is the sum of the two in Fig. 24.19(b) and (c). The Fourier series for the waveform of Fig. 24.19(a) is, therefore,

$$\begin{aligned} v_T = v_1 + v_2 &= -\frac{V_m}{2} + \text{Eq. (24.10)} \\ &= -0.500V_m + 0.318V_m + 0.500V_m \sin \alpha - 0.212V_m \cos 2\alpha - 0.0424V_m \cos 4\alpha + \cdots \end{aligned}$$

and
$$v_T = -0.182V_m + 0.5V_m \sin \alpha - 0.212V_m \cos 2\alpha - 0.0424V_m \cos 4\alpha + \cdots$$

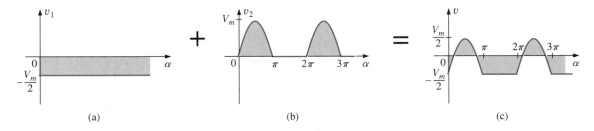

FIG. 24.19
Lowering a waveform with the addition of a negative dc component.

If either waveform were shifted to the right or left, the phase shift would be subtracted or added, respectively, from the sine and cosine terms. The dc term would not change with a shift to the right or left.

If the half-wave rectified signal is shifted 90° to the left, as in Fig. 24.20, the Fourier series becomes

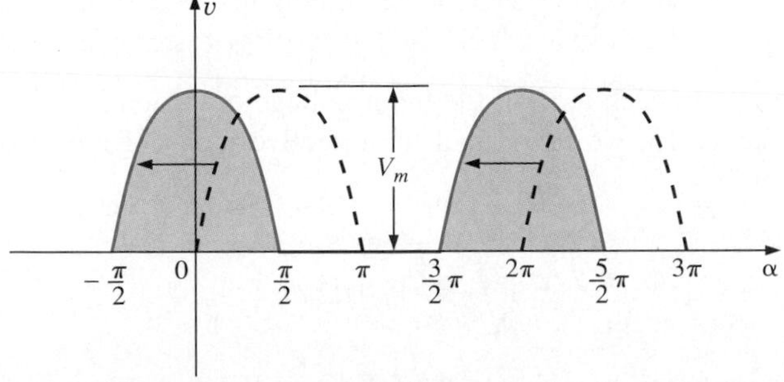

FIG. 24.20
Changing the phase angle of a waveform.

$$v = 0.318V_m + 0.500V_m \underbrace{\sin(\alpha + 90°)}_{\cos\alpha} - 0.212V_m \cos 2(\alpha + 90°) - 0.0424V_m \cos 4(\alpha + 90°) + \cdots$$

$$= 0.318V_m + 0.500V_m \cos\alpha - 0.212V_m \cos(2\alpha + 180°) - 0.0424V_m \cos(4\alpha + 360°) + \cdots$$

and $$v = 0.318V_m + 0.500V_m \cos\alpha + 0.212V_m \cos 2\alpha - 0.0424V_m \cos 4\alpha + \cdots$$

24.3 CIRCUIT RESPONSE TO A NONSINUSOIDAL INPUT

The Fourier series representation of a nonsinusoidal input can be applied to a linear network using the principle of superposition. Recall that this theorem allowed us to consider the effects of each source of a circuit independently. If we replace the nonsinusoidal input with the terms of the Fourier series deemed necessary for practical considerations, we can use superposition to find the response of the network to each term (Fig. 24.21).

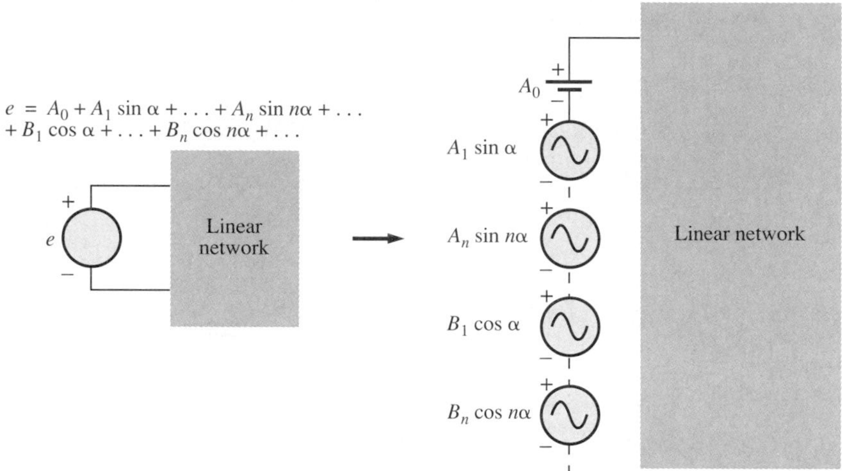

FIG. 24.21
Setting up the application of a Fourier series of terms to a linear network.

The total response of the system is then the algebraic sum of the values obtained for each term. The major change between using this theorem for nonsinusoidal circuits and using it for the circuits previously described is that the frequency will be different for each term in the nonsinusoidal application. Therefore, the reactances

$$X_L = 2\pi fL \quad \text{and} \quad X_C = \frac{1}{2\pi fC}$$

will change for each term of the input voltage or current.

In Chapter 13, we found that the effective value of any waveform was given by

$$\sqrt{\frac{1}{T}\int_0^T [f(t)]2\,dt}$$

If we apply this equation to the following Fourier series:

$$v(\alpha) = V_0 + V_{m_1} \sin \alpha + \cdots + V_{m_n} \sin n\alpha + V'_{m_1} \cos \alpha + \cdots + V'_{m_n} \cos n\alpha$$

then

$$V_{\text{eff}} = \sqrt{V_0^2 + \frac{V_{m_1}^2 + \cdots + V_{m_n}^2 + V'^2_{m_1} + \cdots + V'^2_{m_n}}{2}} \qquad \textbf{(24.11)}$$

However, since

$$\frac{V_{m_1}^2}{2} = \left(\frac{V_{m_1}}{\sqrt{2}}\right)\left(\frac{V_{m_1}}{\sqrt{2}}\right) = (V_{1_{\text{eff}}})(V_{1_{\text{eff}}}) = V_{1_{\text{eff}}}^2$$

then

$$V_{\text{eff}} = \sqrt{V_0^2 + V_{1_{\text{eff}}}^2 + \cdots + V_{n_{\text{eff}}}^2 + V'^2_{1_{\text{eff}}} + \cdots + V'^2_{n_{\text{eff}}}} \qquad \textbf{(24.12)}$$

Similarly, for

$$i(\alpha) = I_0 + I_{m_1} \sin \alpha + \cdots + I_{m_n} \sin n\alpha + I'_{m_1} \cos \alpha + \cdots + I'_{m_n} \cos n\alpha$$

we have

$$I_{\text{eff}} = \sqrt{I_0^2 + \frac{I_{m_1}^2 + \cdots + I_{m_n}^2 + I'^2_{m_1} + \cdots + I'^2_{m_n}}{2}} \qquad \textbf{(24.13)}$$

and

$$I_{\text{eff}} = \sqrt{I_0^2 + I_{1_{\text{eff}}}^2 + \cdots + I_{n_{\text{eff}}}^2 + I'^2_{1_{\text{eff}}} + \cdots + I'^2_{n_{\text{eff}}}} \qquad \textbf{(24.14)}$$

The total power delivered is the sum of that delivered by the corresponding terms of the voltage and current. In the following equations, all voltages and currents are effective values:

$$P_T = V_0 I_0 + V_1 I_1 \cos \theta_1 + \cdots + V_n I_n \cos \theta_n + \cdots \qquad \textbf{(24.15)}$$

$$P_T = I_0^2 R + I_1^2 R + \cdots + I_n^2 R + \cdots \qquad \textbf{(24.16)}$$

or

$$P_T = I_{\text{eff}}^2 R \qquad \textbf{(24.17)}$$

with I_{eff} as defined by Eq. (24.13), and, similarly,

$$P_T = \frac{V^2_{\text{eff}}}{R} \qquad \textbf{(24.18)}$$

with V_{eff} as defined by Eq. (24.11).

EXAMPLE 24.5

a. Sketch the input resulting from the combination of sources in Fig. 24.22.

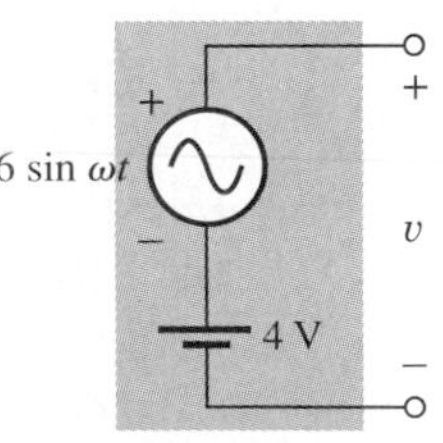

FIG. 24.22
Example 24.5.

b. Determine the effective value of the input of Fig. 24.22.

Solutions:

a. Note Fig. 24.23.

b. Eq. (24.12):

$$V_{\text{eff}} = \sqrt{V_0^2 + \frac{V_m^2}{2}}$$

$$= \sqrt{(4\text{ V})^2 + \frac{(6\text{ V})^2}{2}} = \sqrt{16 + \frac{36}{2}}\text{ V} = \sqrt{34}\text{ V}$$

$$= \mathbf{5.831\ V}$$

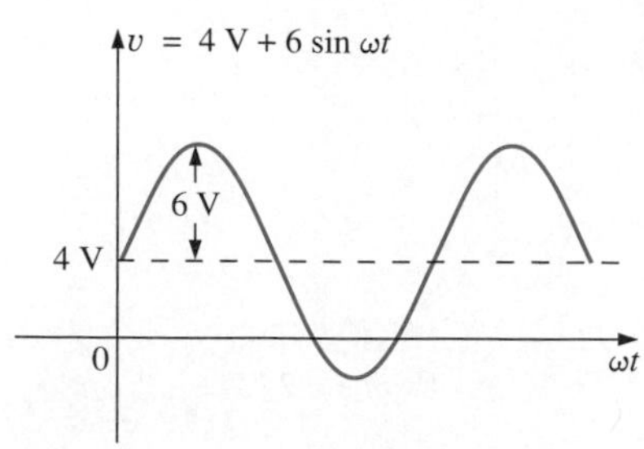

FIG. 24.23
Wave pattern generated by the source of Fig. 24.22.

It is particularly interesting to note from the preceding example that the rms value of a waveform having both dc and ac components is not simply the sum of the effective values of each. In other words, there is a temptation in the absence of Eq. (24.12) to state that $V_{\text{eff}} = 4\text{ V} + 0.707(6\text{ V}) = 8.242\text{ V}$, which is incorrect and, in fact, exceeds the correct level by some 41%.

Instrumentation

It is important to realize that not every DMM will read the rms value of nonsinusoidal waveforms such as the one appearing in Fig. 24.23. Many are designed to read the rms value of sinusoidal waveforms only. It is important to read the manual provided with the meter to see if it is a *true rms* meter that can read the rms value of any waveform.

EXAMPLE 24.6 We learned in Chapter 13 that the effective value of a square wave is the peak value of the waveform. Let us test this result using the Fourier expansion and Eq. (24.11).

Determine the effective value of the square wave of Fig. 24.15 with $V_m = 20$ V using the first six terms of the Fourier expansion and compare to the actual effective value of 20 V.

Solution:

$$\upsilon = \frac{4}{\pi}(20\text{ V})\sin\omega t + \frac{4}{\pi}\left(\frac{1}{3}\right)(20\text{ V})\sin 3\omega t + \frac{4}{\pi}\left(\frac{1}{5}\right)(20\text{ V})\sin 5\omega t + \frac{4}{\pi}\left(\frac{1}{7}\right)(20\text{ V})\sin 7\omega t$$
$$+ \frac{4}{\pi}\left(\frac{1}{9}\right)(20\text{ V})\sin 9\omega t + \frac{4}{\pi}\left(\frac{1}{11}\right)(20\text{ V})\sin 11\omega t$$

$$\upsilon = 25.465\sin\omega t + 8.488\sin 3\omega t + 5.093\sin 5\omega t + 3.638\sin 7\omega t + 2.829\sin 9\omega t + 2.315\sin 11\,\omega t$$

Eq. (24.11):

$$V_{\text{eff}} = \sqrt{V_0^2 + \frac{V_{m_1}^2 + V_{m_2}^2 + V_{m_3}^2 + V_{m_4}^2 + V_{m_5}^2 + V_{m_6}^2}{2}}$$

$$= \sqrt{(0\text{ V})^2 + \frac{(25.465\text{ V})^2 + (8.488\text{ V})^2 + (5.093\text{ V})^2 + (3.638\text{ V})^2 + (2.829\text{ V})^2 + (2.315\text{ V})^2}{2}}$$

$$= 19.66\text{ V}$$

The solution is less than 0.4 V from the correct answer of 20 V. However, each additional term in the Fourier series will bring the result closer to the 20-V level. An infinite number would result in an exact solution of 20 V.

EXAMPLE 24.7 The input to the circuit of Fig. 24.24 is the following:

$$e = 12 + 10 \sin 2t$$

a. Find the current i and the voltages v_R and v_C.
b. Find the effective values of i, v_R, and v_C.
c. Find the power delivered to the circuit.

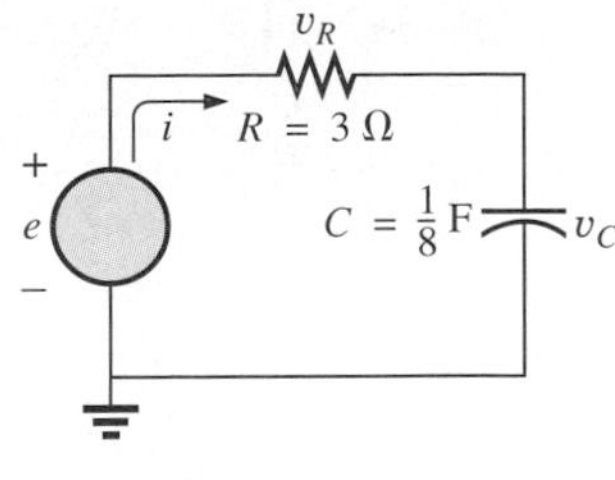

FIG. 24.24
Example 24.7.

Solutions:

a. Redraw the original circuit as shown in Fig. 24.25. Then apply superposition:

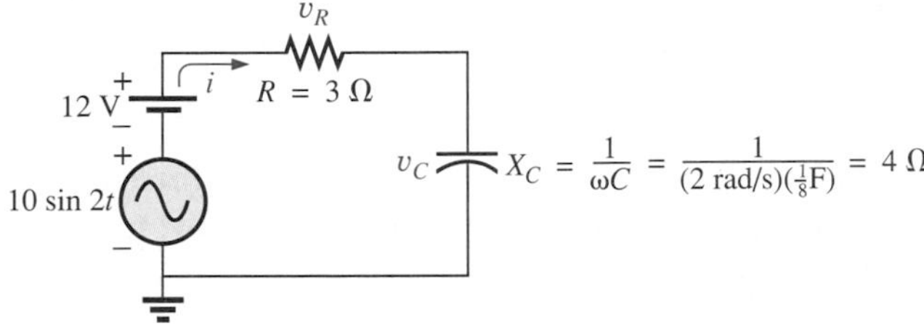

FIG. 24.25
Circuit of Fig. 24.24 with the components of the Fourier series input.

1. *For the 12-V dc supply portion of the input,* $I = 0$ since the capacitor is an open circuit to dc when v_C has reached its final (steady-state) value. Therefore,

$$V_R = IR = 0\text{ V} \quad \text{and} \quad V_C = 12\text{ V}$$

2. *For the ac supply,*

$$\mathbf{Z} = 3\ \Omega - j\,4\ \Omega = 5\ \Omega \angle -53.13^\circ$$

$$\text{and} \quad \mathbf{I} = \frac{\mathbf{E}}{\mathbf{Z}} = \frac{\dfrac{10}{\sqrt{2}}\text{ V} \angle 0^\circ}{5\ \Omega \angle -53.13^\circ} = \frac{2}{\sqrt{2}}\text{ A} \angle +53.13^\circ$$

$$\mathbf{V}_R = (I \angle \theta)(R \angle 0^\circ) = \left(\frac{2}{\sqrt{2}}\text{ A} \angle +53.13^\circ\right)(3\ \Omega \angle 0^\circ) = \frac{6}{\sqrt{2}}\text{ V} \angle +53.13^\circ$$

$$\text{and} \quad \mathbf{V}_C = (I \angle \theta)(X_C \angle -90^\circ) = \left(\frac{2}{\sqrt{2}}\text{ A} \angle +53.13^\circ\right)(4\ \Omega \angle -90^\circ) = \frac{8}{\sqrt{2}}\text{ V} \angle -36.87^\circ$$

In the time domain,

$$i = \mathbf{0 + 2 \sin(2t + 53.13^\circ)}$$

Note that even though the dc term was present in the expression for the input voltage, the dc term for the current in this circuit is zero:

$$v_R = \mathbf{0 + 6 \sin(2t + 53.13^\circ)}$$

$$\text{and} \quad v_C = \mathbf{12 + 8 \sin(2t - 36.87^\circ)}$$

b. Eq. (24.14): $I_{\text{eff}} = \sqrt{(0)^2 + \frac{(2\text{ A})^2}{2}} = \sqrt{2}\text{ V} = \mathbf{1.414\text{ A}}$

Eq. (24.12): $V_{R_{\text{eff}}} = \sqrt{(0)^2 + \frac{(6\text{ V})^2}{2}} = \sqrt{18}\text{ V} = \mathbf{4.243\text{ V}}$

Eq. (24.12): $V_{C_{\text{eff}}} = \sqrt{(12\text{ V})^2 + \frac{(8\text{ V})^2}{2}} = \sqrt{176}\text{ V} = \mathbf{13.267\text{ V}}$

c. $P = I_{\text{eff}}^2 R = \left(\frac{2}{\sqrt{2}}\text{ A}\right)^2 (3\ \Omega) = \mathbf{6\text{ W}}$

EXAMPLE 24.8 Find the response of the circuit of Fig. 24.26 to the input shown.

$$e = 0.318E_m + 0.500E_m \sin \omega t - 0.212E_m \cos 2\omega t - 0.0424E_m \cos 4\omega t + \cdots$$

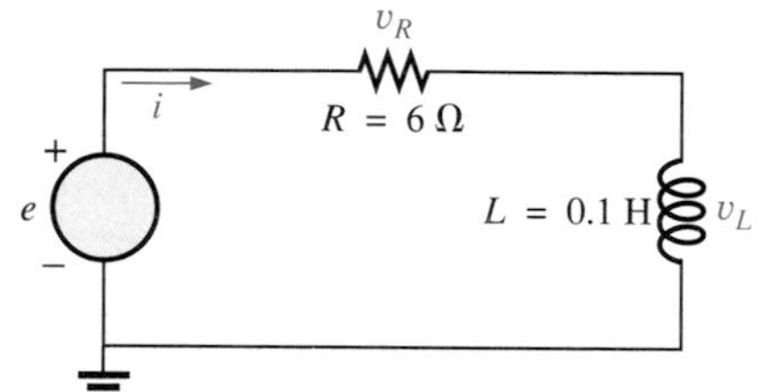

(a)

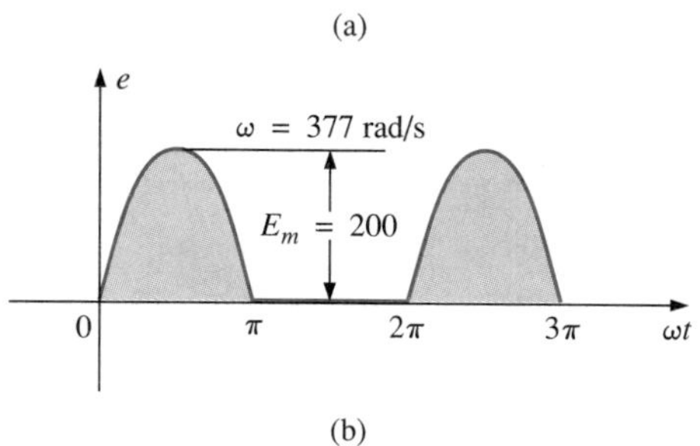

(b)

FIG. 24.26
Example 24.8.

Solution: For discussion purposes, only the first three terms will be used to represent e. Converting the cosine terms to sine terms and substituting for E_m gives us

$$e = 63.60 + 100.0 \sin \omega t - 42.40 \sin(2\omega t + 90°)$$

Using phasor notation, the original circuit becomes like the one shown in Fig. 24.27. Applying superposition:

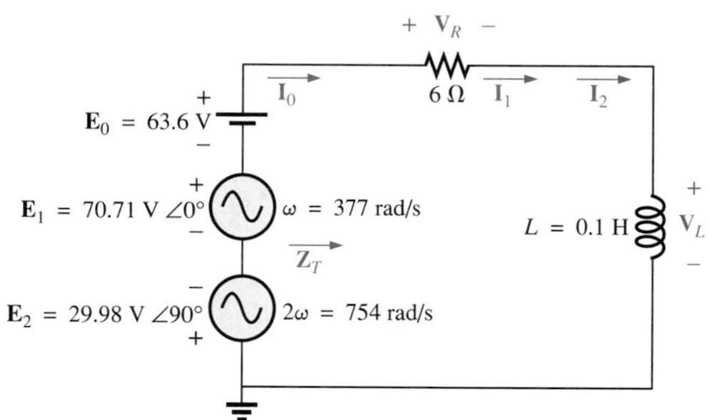

FIG. 24.27
Circuit of Fig. 24.26 with the components of the Fourier series input.

For the dc term ($E_0 = 63.6$ V),

$$X_L = 0 \quad \text{(short for dc)}$$
$$\mathbf{Z}_T = R\ \angle 0° = 6\ \Omega\ \angle 0°$$
$$I_0 = \frac{E_0}{R} = \frac{63.6\text{ V}}{6\ \Omega} = 10.60\text{ A}$$
$$V_{R_0} = I_0 R = E_0 = 63.60\text{ V}$$
$$V_{L_0} = 0$$

The average power is

$$P_0 = I_0^2 R = (10.60\text{ A})^2(6\ \Omega) = 674.2\text{ W}$$

For the fundamental term ($\mathbf{E}_1 = 70.71\text{ V}\ \angle 0°,\ \omega = 377$),

$$X_{L_1} = \omega L = (377\text{ rad/s})(0.1\text{ H}) = 37.7\ \Omega$$

$$\mathbf{Z}_{T_1} = 6\ \Omega + j\,37.7\ \Omega = 38.17\ \Omega\ \angle 80.96°$$

$$\mathbf{I}_1 = \frac{\mathbf{E}_1}{\mathbf{Z}_{T_1}} = \frac{70.71\ \text{V}\ \angle 0°}{38.17\ \Omega\ \angle 80.96°} = 1.85\ \text{A}\ \angle -80.96°$$

$$\mathbf{V}_{R_1} = (I_1 \angle \theta)(R \angle 0°) = (1.85\ \text{A}\ \angle -80.96°)(6\ \Omega\ \angle 0°)$$
$$= 11.10\ \text{V}\ \angle -80.96°$$

$$\mathbf{V}_{L_1} = (I_1 \angle \theta)(X_{L_1} \angle 90°) = (1.85\ \text{A}\ \angle -80.96°)(37.7\ \Omega\ \angle 90°)$$
$$= 69.75\ \text{V}\ \angle 9.04°$$

The average power is

$$P_1 = I_1^2 R = (1.85\ \text{A})^2(6\ \Omega) = 20.54\ \text{W}$$

For the second harmonic ($\mathbf{E}_2 = 29.98\ \text{V}\ \angle -90°$, $\omega = 754$). The phase angle of $\mathbf{E}_2$ was changed to $-90°$ to give it the same polarity as the input voltages $\mathbf{E}_0$ and $\mathbf{E}_1$.

$$X_{L_2} = \omega L = (754\ \text{rad/s})(0.1\ \text{H}) = 75.4\ \Omega$$

$$\mathbf{Z}_{T_2} = 6\ \Omega + j\,75.4\ \Omega = 75.64\ \Omega\ \angle 85.45°$$

$$\mathbf{I}_2 = \frac{\mathbf{E}_2}{\mathbf{Z}_{T_2}} = \frac{29.98\ \text{V}\ \angle -90°}{75.64\ \Omega\ \angle 85.45} = 0.396\ \text{A}\ \angle -174.45°$$

$$\mathbf{V}_{R_2} = (I_2 \angle \theta)(R \angle 0°) = (0.396\ \text{A}\ \angle -174.45°)(6\ \Omega\ \angle 0°)$$
$$= 2.38\ \text{V}\ \angle -174.45°$$

$$\mathbf{V}_{L_2} = (I_2 \angle \theta)(X_{L_2} \angle 90°) = (0.396\ \text{A}\ \angle -174.45°)(75.4\ \Omega\ \angle 90°)$$
$$= 29.9\ \text{V}\ \angle -84.45°$$

The average power is

$$P_2 = I_2^2 R = (0.396\ \text{A})^2(6\ \Omega) = 0.941\ \text{W}$$

The Fourier series expansion for i is

$$i = \mathbf{10.6 + \sqrt{2}(1.85)\ sin(377}t\ \mathbf{- 80.96°) + \sqrt{2}(0.396)\ sin(754}t\ \mathbf{- 174.45°)}$$

and

$$I_{\text{eff}} = \sqrt{(10.6\ \text{A})^2 + (1.85\ \text{A})^2 + (0.396\ \text{A})^2} = \mathbf{10.77\ A}$$

The Fourier series expansion for v_R is

$$v_R = \mathbf{63.6 + \sqrt{2}(11.10)\ sin(377}t\ \mathbf{- 80.96°) + \sqrt{2}(2.38)\ sin(754}t\ \mathbf{- 174.45°)}$$

and

$$V_{R_{\text{eff}}} = \sqrt{(63.6\ \text{V})^2 + (11.10\ \text{V})^2 + (2.38\ \text{V})^2} = \mathbf{64.61\ V}$$

The Fourier series expansion for v_L is

$$v_L = \mathbf{\sqrt{2}(69.75)\ sin(377}t\ \mathbf{+ 9.04°) + \sqrt{2}(29.93)\ sin(754}t\ \mathbf{- 84.45°)}$$

and $$V_{L_{\text{eff}}} = \sqrt{(69.75\ \text{V})^2 + (29.93\ \text{V})^2} = \mathbf{75.90\ V}$$

The total average power is

$$P_T = I_{\text{eff}}^2 R = (10.77\ \text{A})^2(6\ \Omega) = \mathbf{695.96\ W} = P_0 + P_1 + P_2$$

24.4 ADDITION AND SUBTRACTION OF NONSINUSOIDAL WAVEFORMS

The Fourier series expression for the waveform resulting from the addition or subtraction of two nonsinusoidal waveforms can be found using

phasor algebra if the terms having the same frequency are considered separately.

For example, the sum of the following two nonsinusoidal waveforms is found using this method:

$$v_1 = 30 + 20 \sin 20t + \cdots + 5 \sin(60t + 30°)$$
$$v_2 = 60 + 30 \sin 20t + 20 \sin 40t + 10 \cos 60t$$

1. dc terms:

$$V_{T_0} = 30 \text{ V} + 60 \text{ V} = 90 \text{ V}$$

2. $\omega = 20$:

$$V_{T_{1(\max)}} = 30 \text{ V} + 20 \text{ V} = 50 \text{ V}$$

and $$v_{T_1} = 50 \sin 20t$$

3. $\omega = 40$:

$$v_{T_2} = 20 \sin 40t$$

4. $\omega = 60$:

$$5 \sin(60t + 30°) = (0.707)(5) \text{ V} \angle 30° = 3.54 \text{ V} \angle 30°$$
$$10 \cos 60t = 10 \sin(60t + 90°) \Rightarrow (0.707)(10) \text{ V} \angle 90° = 7.07 \text{ V} \angle 90°$$
$$\mathbf{V}_{T_3} = 3.54 \text{ V} \angle 30° + 7.07 \text{ V} \angle 90° = 3.07 \text{ V} + j\,1.77 \text{ V} + j\,7.07 \text{ V} = 3.07 \text{ V} + j\,8.84 \text{ V}$$
$$\mathbf{V}_{T_3} = 9.36 \text{ V} \angle 70.85°$$

and $$v_{T_3} = 13.24 \sin(60t + 70.85°)$$

with

$$v_T = v_1 + v_2 = \mathbf{90 + 50 \sin 20t + 20 \sin 40t + 13.24 \sin(60t + 70.85°)}$$

24.5 COMPUTER ANALYSIS

PSpice (Windows)

The computer analysis will begin with a verification of the waveform of Fig. 24.17, demonstrating that only four terms of a Fourier series can generate a waveform that has a number of the characteristics of a square wave. The square wave has a positive peak of 10 V, resulting in the following Fourier series using Eq. (24.9):

$$v = \frac{4}{\pi}(10 \text{ V})(\sin \omega t + \frac{1}{3} \sin 3\omega t + \frac{1}{5} \sin 5\,\omega t + \frac{1}{7} \sin 7\omega t)$$
$$v = 12.732 \sin \omega t + 4.244 \sin 3\omega t + 2.546 \sin 5\omega t + 1.819 \sin 7\omega t$$

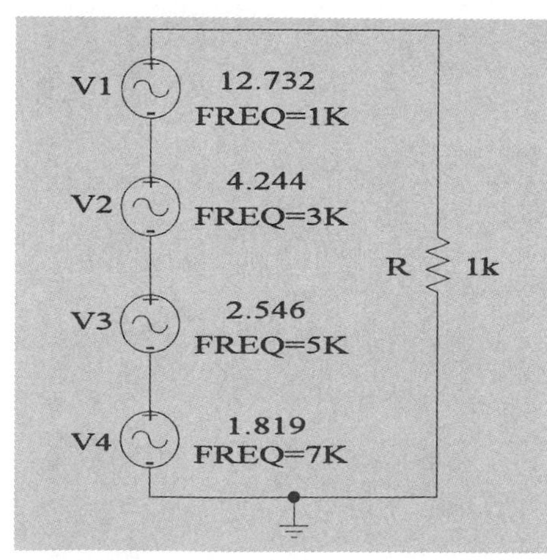

FIG. 24.28
Applying four terms of the Fourier expansion of a 10-V square wave to a load resistor of 1 kΩ.

Each term of the Fourier series is treated as an independent ac source, as shown in Fig. 24.28, with its peak value and applicable frequency. The sum of the source voltages will appear across the resistor *R* and will generate the waveform of Fig. 24.29, which matches that of Fig. 24.17. The vertical scaling was chosen as −20 V to 20 V using the **User Defined** option under **Y-Axis Setting** of **Plot** to show clearly how the waveform oscillates about the level of 10 V. The **X-Axis Setting** under **Plot** was also **User Defined** as 0 s to 2.0 ms to limit the waveform to 2 cycles.

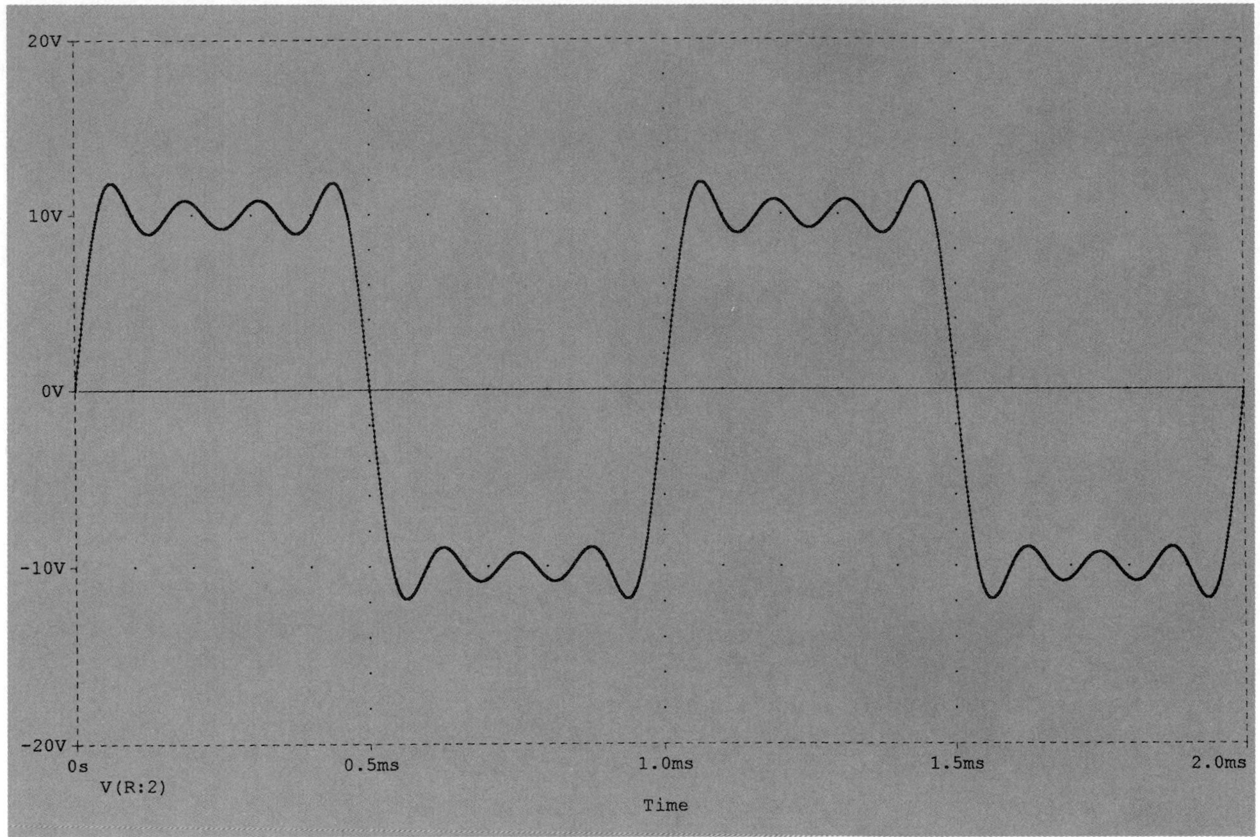

FIG. 24.29
The waveform across the load resistor of Fig. 24.28.

A frequency spectrum plot revealing the magnitude of each component at its frequency can be obtained by returning to **Plot,** choosing the **X-Axis Setting,** and enabling the **Fourier** option in the dialog box. For our purposes the **User defined** frequency spectrum was limited to 0 Hz to 10 kHz, resulting in the plot of Fig. 24.30. Note that a peak occurs at 1 kHz, 3 kHz, 5 kHz, and 7 kHz, with the magnitude of the peak dropping as the frequency increases. The maximum value of the first peak can be identified using the **Cursor** option under the **Tools** heading. After choosing **Display** and clicking the cursor on to the screen, we can return to the sequence **Tools-Cursor-Max,** and A1 = 1 kHz at a peak value of 12.66 will be obtained, as shown on the bottom of Fig. 24.30. The 12.66 V is not an exact match of the first maximum value of 12.732 V, but it certainly is an excellent result for the peak value of the 1-kHz contribution. The other peaks can also be obtained by simply moving the cursor to the peak at each frequency.

In this case we knew the components of the waveform beforehand, but consider the power of this option when it is used to determine the Fourier coefficients of any other waveform for which the components are unknown. For any nonlinear device it can reveal the harmonic content of the distortion introduced to permit proper filtering that will restore the shape of the applied signal.

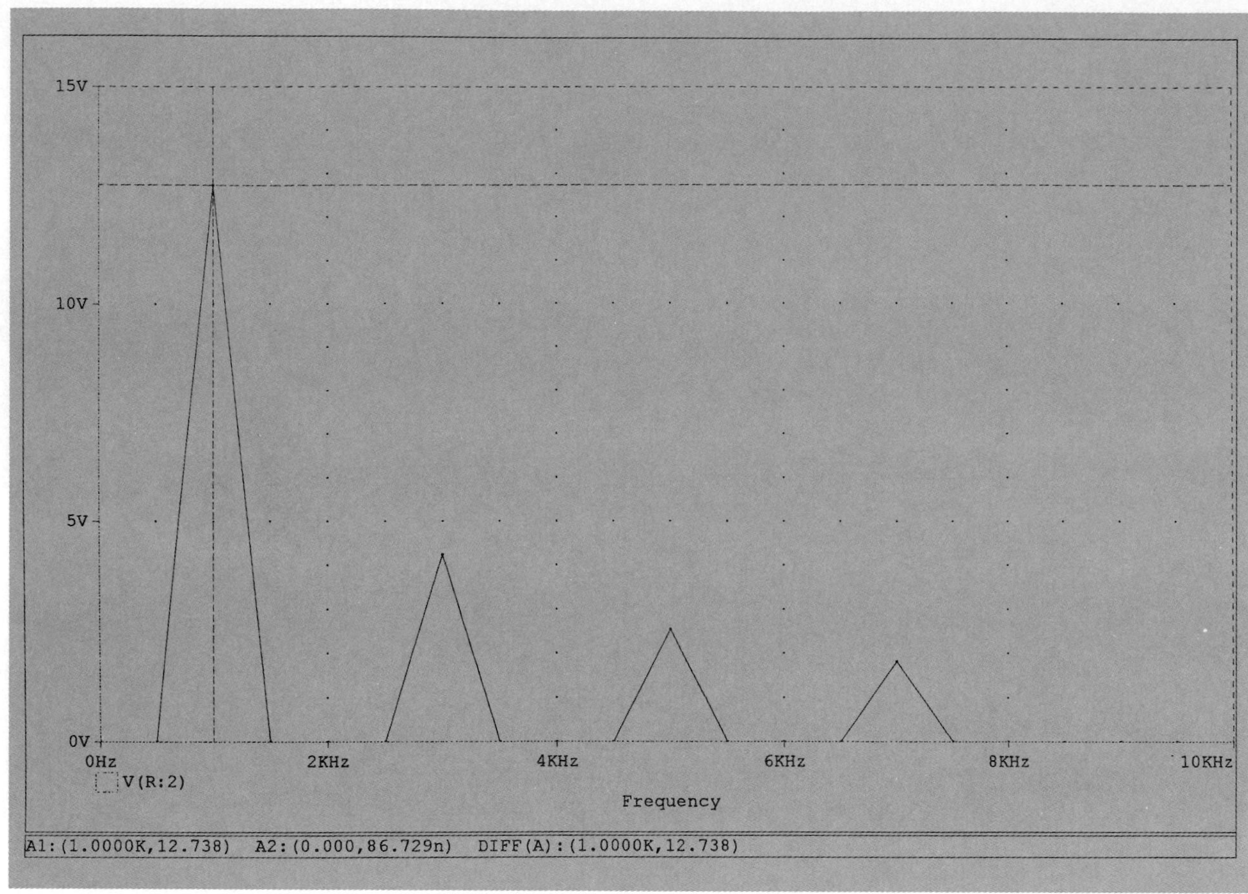

FIG. 24.30
The Fourier components of the waveform of Fig. 24.28.

PROBLEMS

SECTION 24.2 Fourier Series

1. For the waveforms of Fig. 24.31, determine whether the following will be present in the Fourier series representation:
   a. dc term
   b. cosine terms
   c. sine terms
   d. even-ordered harmonics
   e. odd-ordered harmonics

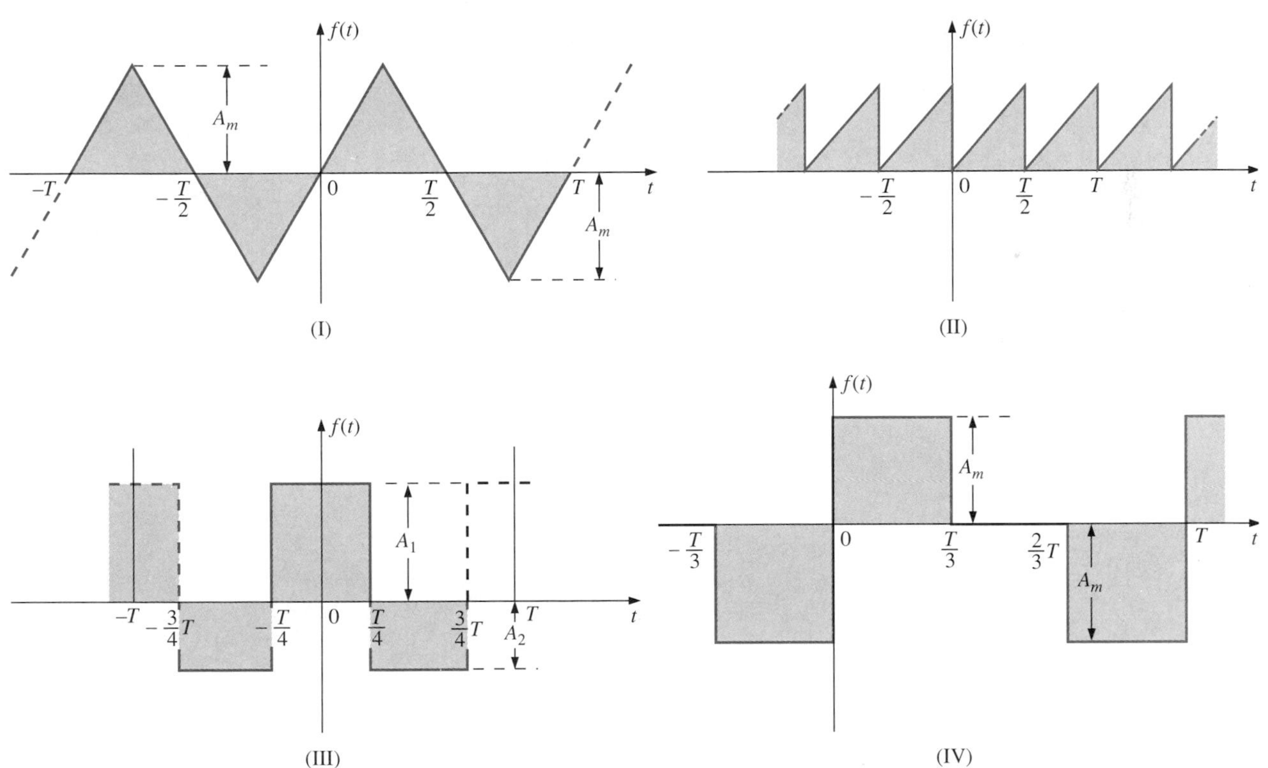

FIG. 24.31
Problem 1.

2. If the Fourier series for the waveform of Fig. 24.32(a) is

$$i = \frac{2I_m}{\pi}\left(1 + \frac{2}{3}\cos 2\omega t - \frac{2}{15}\cos 4\omega t + \frac{2}{35}\cos 6\omega t + \cdots\right)$$

find the Fourier series representation for waveforms (b), (c), and (d).

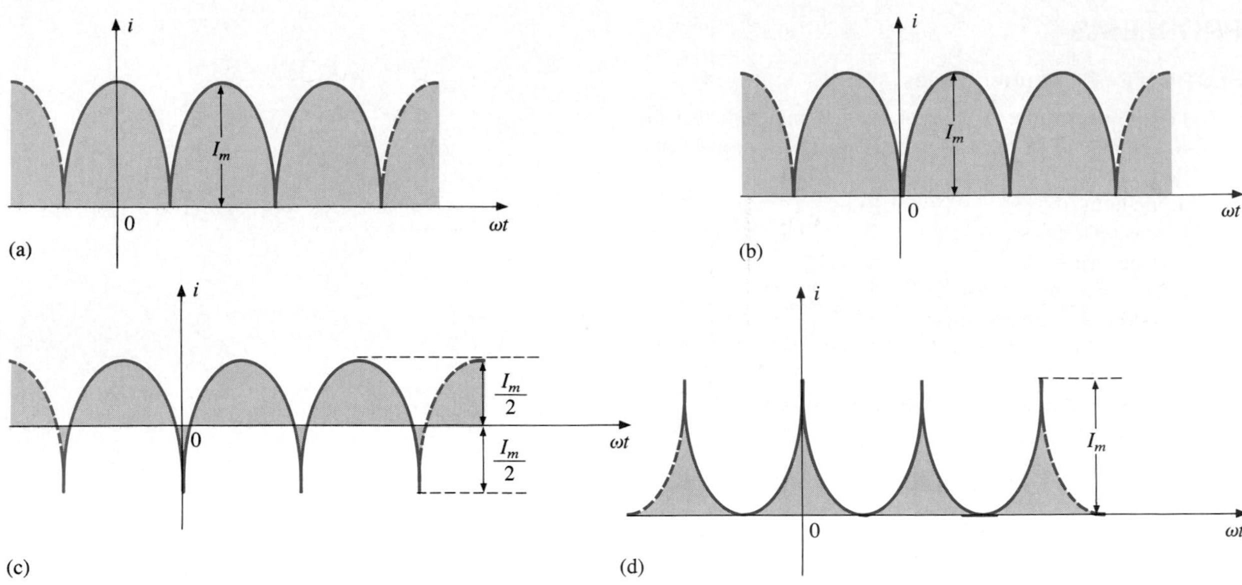

FIG. 24.32
Problem 2.

3. Sketch the following nonsinusoidal waveforms with $\alpha = \omega t$ as the abscissa:
a. $v = -4 + 2 \sin \alpha$ **b.** $v = (\sin \alpha)^2$
c. $i = 2 - 2 \cos \alpha$

4. Sketch the following nonsinusoidal waveforms with α as the abscissa:
a. $i = 3 \sin \alpha - 6 \sin 2\alpha$
b. $v = 2 \cos 2\alpha + \sin \alpha$

5. Sketch the following nonsinusoidal waveforms with ωt as the abscissa:
a. $i = 50 \sin \omega t + 25 \sin 3\omega t$
b. $i = 50 \sin \alpha - 25 \sin 3\alpha$
c. $i = 4 + 3 \sin \omega t + 2 \sin 2\omega t - 1 \sin 3\omega t$

SECTION 24.3 Circuit Response to a Nonsinusoidal Input

6. Find the average and effective values of the following nonsinusoidal waves:
a. $v = 100 + 50 \sin \omega t + 25 \sin 2\omega t$
b. $i = 3 + 2 \sin(\omega t - 53^\circ) + 0.8 \sin(2\omega t - 70^\circ)$

7. Find the effective value of the following nonsinusoidal waves:
a. $v = 20 \sin \omega t + 15 \sin 2\omega t - 10 \sin 3\omega t$
b. $i = 6 \sin(\omega t + 20^\circ) + 2 \sin(2\omega t + 30^\circ) - 1 \sin(3\omega t + 60^\circ)$

8. Find the total average power to a circuit whose voltage and current are as indicated in Problem 6.

9. Find the total average power to a circuit whose voltage and current are as indicated in Problem 7.

10. The Fourier series representation for the input voltage to the circuit of Fig. 24.33 is

$$e = 18 + 30 \sin 400t$$

a. Find the nonsinusoidal expression for the current i.
b. Calculate the effective value of the current.
c. Find the expression for the voltage across the resistor.
d. Calculate the effective value of the voltage across the resistor.
e. Find the expression for the voltage across the reactive element.
f. Calculate the effective value of the voltage across the reactive element.
g. Find the average power delivered to the resistor.

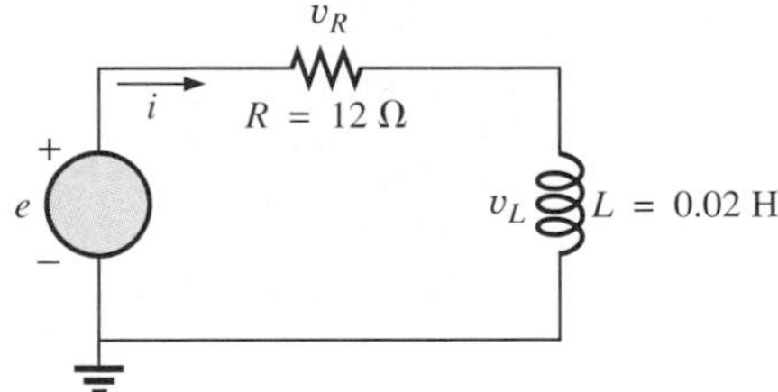

FIG. 24.33
Problem 10.

11. Repeat Problem 10 for

$$e = 24 + 30 \sin 400t + 10 \sin 800t$$

12. Repeat Problem 10 for the following input voltage:

$$e = -60 + 20 \sin 300t - 10 \sin 600t$$

13. Repeat Problem 10 for the circuit of Fig. 24.34.

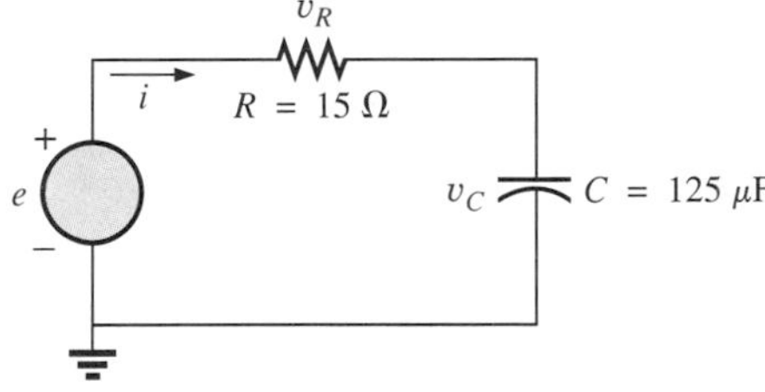

FIG. 24.34
Problem 13.

***14.** The input voltage [Fig. 24.35(a)] to the circuit of Fig. 24.35(b) is a full-wave rectified signal having the following Fourier series expansion:

$$e = \frac{(2)(100\text{ V})}{\pi}\left(1 + \frac{2}{3}\cos 2\omega t - \frac{2}{15}\cos 4\omega t + \frac{2}{53}\cos 6\omega t + \cdots\right)$$

where $\omega = 377$.

a. Find the Fourier series expression for the voltage v_o using only the first three terms of the expression.
b. Find the effective value of v_o.
c. Find the average power delivered to the 1-kΩ resistor.

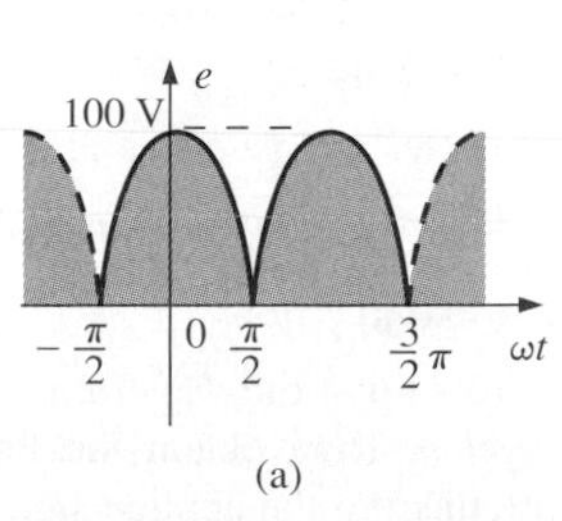

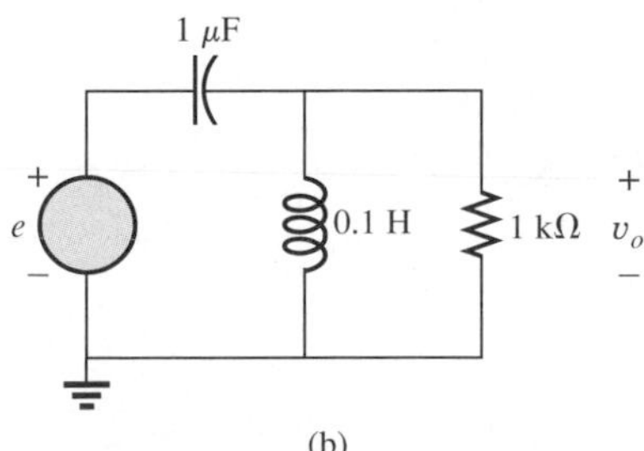

FIG. 24.35
Problem 14.

***15.** Find the Fourier series expression for the voltage v_o of Fig. 24.36.

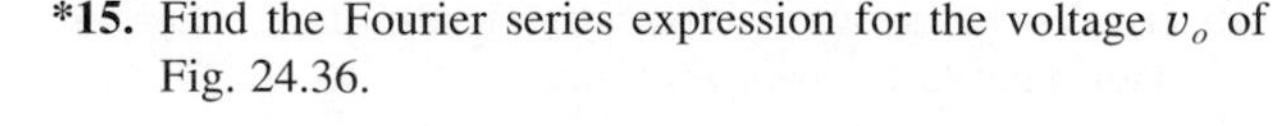

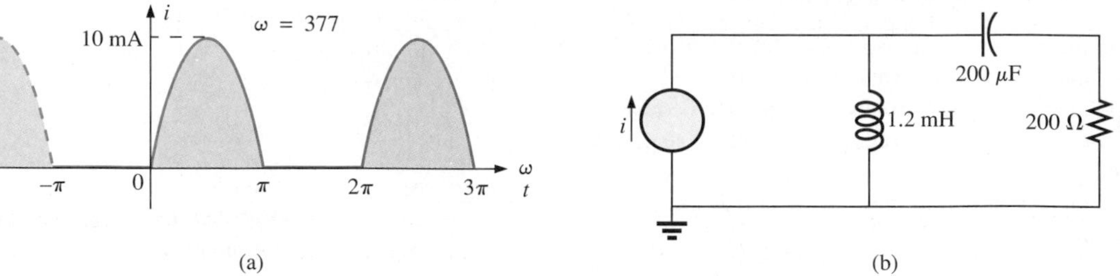

FIG. 24.36
Problem 15.

SECTION 24.4 Addition and Subtraction of Nonsinusoidal Waveforms

16. Perform the indicated operations on the following nonsinusoidal waveforms:

a.

$$[60 + 70 \sin \omega t + 20 \sin(2\omega t + 90°) + 10 \sin(3\omega t + 60°)] + [20 + 30 \sin \omega t - 20 \cos 2\omega t + 5 \cos 3\omega t]$$

b.

$$[20 + 60 \sin \alpha + 10 \sin(2\alpha - 180°) + 5 \cos(3\alpha + 90°)] - [5 - 10 \sin \alpha + 4 \sin(3\alpha - 30°)]$$

17. Find the nonsinusoidal expression for the current i_s of the diagram of Fig. 24.37.

$$i_2 = 10 + 30 \sin 20t - 0.5 \sin(40t + 90°)$$
$$i_1 = 20 + 4 \sin(20t + 90°) + 0.5 \sin(40t + 30°)$$

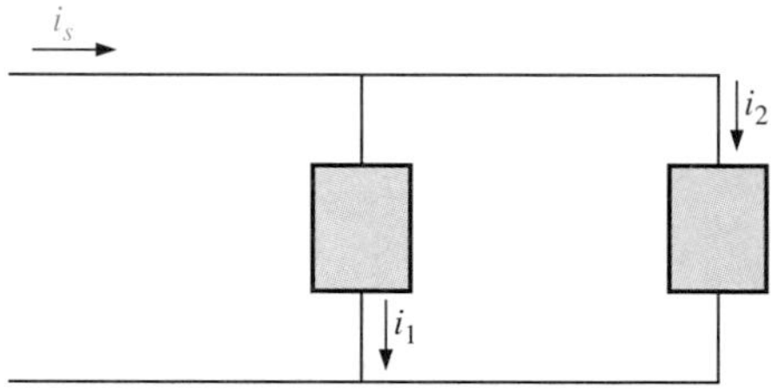

i_s i_2 i_1

FIG. 24.37
Problem 17.

18. Find the nonsinusoidal expression for the voltage e of the diagram of Fig. 24.38.

$$v_1 = 20 - 200 \sin 600t + 100 \cos 1200t + 75 \sin 1800t$$
$$v_2 = -10 + 150 \sin(600t + 30°) + 50 \sin(1800t + 60°)$$

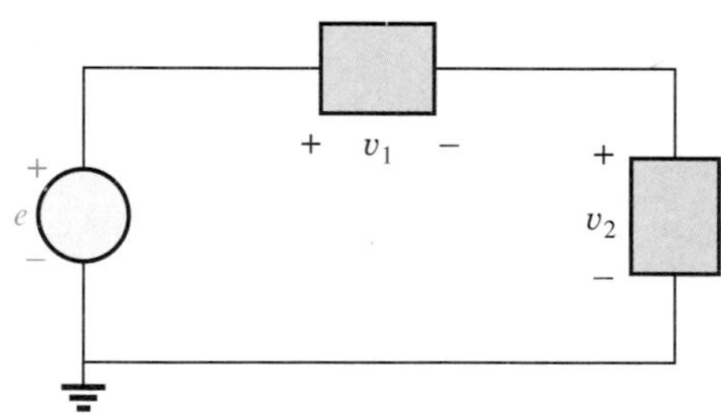

FIG. 24.38
Problem 18.

SECTION 24.5 Computer Analysis

PSpice (Windows)

19. Using Probe, plot the waveform of Fig. 24.13 for two or three cycles. Then obtain the Fourier components and compare them to the applied signal.

20. Plot a half-rectified waveform with a peak value of 20 V using Eq. (24.10). Use the dc term, the fundamental term, and four harmonics. Compare the resulting waveform to the ideal half-rectified waveform.

21. Demonstrate the effect of adding two more terms to the waveform of Fig. 25.29 and generate the Fourier spectrum.

Computer Language (C++, BASIC, PASCAL, etc.)

22. Write a program to obtain the Fourier expansion resulting from the addition of two nonsinusoidal waveforms.

23. Write a program to determine the sum of the first 10 terms of Eq. (24.9) at $\omega t = \pi/2$, π, and $(3/2)\pi$, and compare to the values determined by Fig. 24.15. That is, enter Eq. (24.9) into memory and calculate the sum of the terms at the points listed above.

24. Given any nonsinusoidal function, write a program that will determine the average and effective values of the waveform. The program should request the data required from the nonsinusoidal function.

25. Write a program that will provide a general solution for the network of Fig. 24.24 for a single dc and ac term in the applied voltage. In other words, the parameter values are given along with the particulars regarding the applied signal, and the nonsinusoidal expression for the current and each voltage is generated by the program.

GLOSSARY

Axis symmetry A sinusoidal or nonsinusoidal function that has symmetry about the vertical axis.

Even harmonics The terms of the Fourier series expansion that have frequencies that are even multiples of the fundamental component.

Fourier series A series of terms, developed in 1826 by Baron Jean Fourier, that can be used to represent a nonsinusoidal function.

Fundamental component The minimum frequency term required to represent a particular waveform in the Fourier series expansion.

Half-wave (mirror) symmetry A sinusoidal or nonsinusoidal function that satisfies the relationship $f(t) = -f\left(\frac{T}{2} + t\right)$.

Harmonic The terms of the Fourier series expansion that have frequencies that are integer multiples of the fundamental component.

Nonsinusoidal waveform Any waveform that differs from the fundamental sinusoidal function.

Odd harmonics The terms of the Fourier series expansion that have frequencies that are odd multiples of the fundamental component.

Point symmetry A sinusoidal or nonsinusoidal function that satisfies the relationship $f(\alpha) = -f(-\alpha)$.

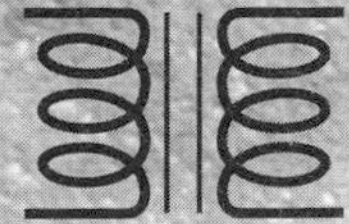

25 Transformers

25.1 INTRODUCTION

Chapter 12 discussed the *self-inductance* of a coil. We shall now examine the *mutual inductance* that exists between coils of the same or different dimensions. Mutual inductance is a phenomenon basic to the operation of the transformer, an electrical device used today in almost every field of electrical engineering. This device plays an integral part in power distribution systems and can be found in many electronic circuits and measuring instruments. In this chapter, we will discuss three of the basic applications of a transformer: to build up or step down the voltage or current, to act as an impedance matching device, and to isolate (no physical connection) one portion of a circuit from another. In addition, we will introduce the dot convention and consider the transformer equivalent circuit. The chapter will conclude with a word about writing mesh equations for a network with mutual inductance.

25.2 MUTUAL INDUCTANCE

The transformer is constructed of two coils placed so that the changing flux developed by one will link the other, as shown in Fig. 25.1. This

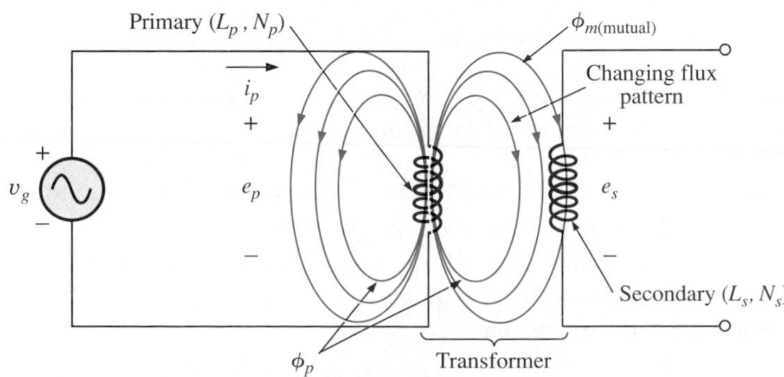

FIG. 25.1
Defining the components of a transformer.

will result in an induced voltage across each coil. To distinguish between the coils, we will apply the transformer convention that

the coil to which the source is applied is called the primary, and the coil to which the load is applied is called the secondary.

For the primary of the transformer of Fig. 25.1, an application of Faraday's law [Eq. (12.1)] will result in

$$e_p = N_p \frac{d\phi_p}{dt} \qquad \text{(volts, V)} \tag{25.1}$$

revealing that the voltage induced across the primary is directly related to the number of turns in the primary and the rate of change of magnetic flux linking the primary coil. Or, from Eq. (12.5),

$$e_p = L_p \frac{di_p}{dt} \qquad \text{(volts, V)} \tag{25.2}$$

revealing that the induced voltage across the primary is also directly related to the self-inductance of the primary and the rate of change of current through the primary winding.

The magnitude of e_s, the voltage induced across the secondary, is determined by

$$e_s = N_s \frac{d\phi_m}{dt} \qquad \text{(volts, V)} \tag{25.3}$$

where N_s is the number of turns in the secondary winding and ϕ_m is the portion of the primary flux ϕ_p that links the secondary winding.

If all of the flux linking the primary links the secondary, then

$$\phi_m = \phi_p$$

and

$$e_s = N_s \frac{d\phi_p}{dt} \qquad \text{(volts, V)} \tag{25.4}$$

The *coefficient of coupling* between two coils is determined by

$$k \text{ (coefficient of coupling)} = \frac{\phi_m}{\phi_p} \tag{25.5}$$

Since the maximum level of ϕ_m is ϕ_p, the coefficient of coupling between two coils can never be greater than 1.

The coefficient of coupling between various coils is indicated in Fig. 25.2. Note that for the iron core, k approaches 1, whereas for the air core, k is considerably less. Those coils with low coefficients of coupling are said to be *loosely coupled.*

For the secondary, we have

$$e_s = N_s \frac{d\phi_m}{dt} = N_s \frac{dk\phi_p}{dt}$$

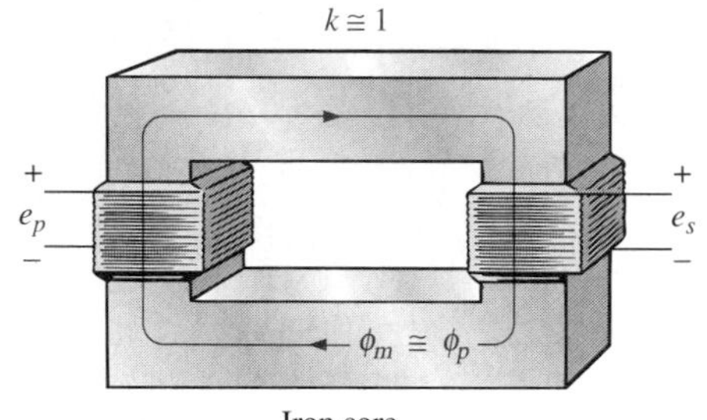

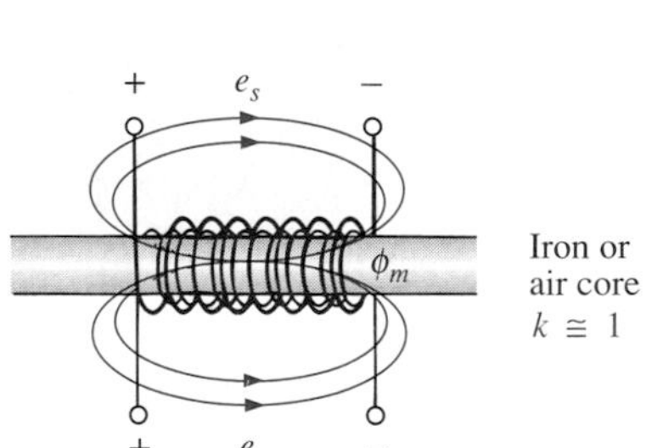

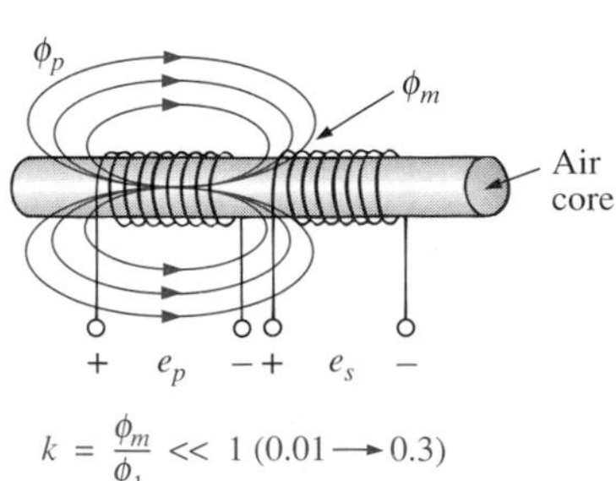

FIG. 25.2
Windings having different coefficients of coupling.

and
$$e_s = kN_s \frac{d\phi_p}{dt} \quad \text{(volts, V)} \tag{25.6}$$

The mutual inductance between the two coils of Fig. 25.1 is determined by

$$M = N_s \frac{d\phi_m}{di_p} \quad \text{(henries, H)} \tag{25.7}$$

or
$$M = N_p \frac{d\phi_p}{di_s} \quad \text{(henries, H)} \tag{25.8}$$

Note in the above equations that the symbol for mutual inductance is the capital letter *M*, and that its unit of measurement, like that of self-inductance, is the *henry.* In words, Eqs. (25.7) and (25.8) state that the

mutual inductance between two coils is proportional to the instantaneous change in flux linking one coil due to an instantaneous change in current through the other coil.

In terms of the inductance of each coil and the coefficient of coupling, the mutual inductance is determined by

$$M = k\sqrt{L_pL_s} \quad \text{(henries, H)} \tag{25.9}$$

The greater the coefficient of coupling (greater flux linkages), or the greater the inductance of either coil, the higher the mutual inductance between the coils. Relate this fact to the configurations of Fig. 25.2.

The secondary voltage e_s can also be found in terms of the mutual inductance if we rewrite Eq. (25.3) as

$$e_s = N_s\left(\frac{d\phi_m}{di_p}\right)\left(\frac{di_p}{dt}\right)$$

and, since $M = N_s(d\phi_m/di_p)$, it can also be written

$$e_s = M\frac{di_p}{dt} \quad \text{(volts, V)} \tag{25.10}$$

Similarly,
$$e_p = M\frac{di_s}{dt} \quad \text{(volts, V)} \tag{25.11}$$

EXAMPLE 25.1 For the transformer in Fig. 25.3:

a. Find the mutual inductance *M*.
b. Find the induced voltage e_p if the flux ϕ_p changes at the rate of 450 mWb/s.
c. Find the induced voltage e_s for the same rate of change indicated in part (b).
d. Find the induced voltages e_p and e_s if the current i_p changes at the rate of 0.2 A/ms.

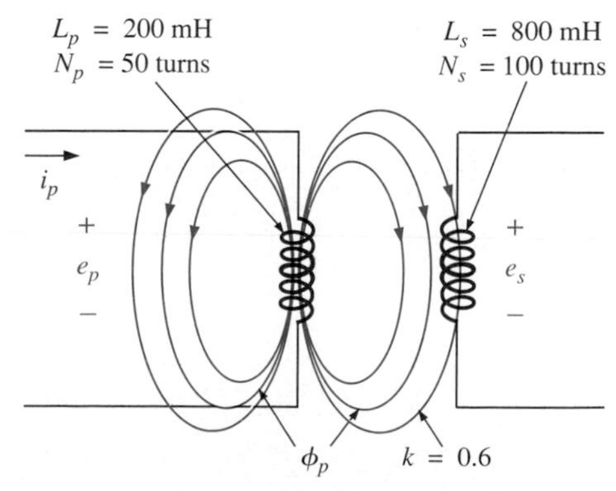

FIG. 25.3
Example 25.1.

Solutions:

a. $M = k\sqrt{L_pL_s} = 0.6\sqrt{(200\text{ mH})(800\text{ mH})}$
$= 0.6\sqrt{16 \times 10^{-2}} = (0.6)(400 \times 10^{-3}) = \mathbf{240\text{ mH}}$

b. $e_p = N_p \dfrac{d\phi_p}{dt} = (50)(450\text{ mWb/s}) = \mathbf{22.5\text{ V}}$

c. $e_s = kN_s \dfrac{d\phi_p}{dt} = (0.6)(100)(450\text{ mWb/s}) = \mathbf{27\text{ V}}$

d. $e_p = L_p \dfrac{di_p}{dt} = (200\text{ mH})(0.2\text{ A/ms}) = (200\text{ mH})(200\text{ A/s}) = \mathbf{40\text{ V}}$

$e_s = M \dfrac{di_p}{dt} = (240\text{ mH})(200\text{ A/s}) = \mathbf{48\text{ V}}$

25.3 SERIES CONNECTION OF MUTUALLY COUPLED COILS

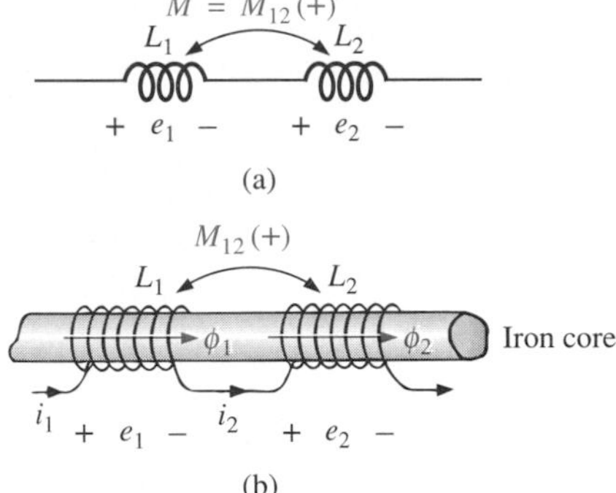

FIG. 25.4
Mutually coupled coils connected in series.

In Chapter 12, we found that the total inductance of series isolated coils was determined simply by the sum of the inductances. For two coils that are connected in series but also share the same flux linkages, such as those in Fig. 25.4(a), a mutual term is introduced that will alter the total inductance of the series combination. The physical picture of how the coils are connected is indicated in Fig. 25.4(b). An iron core is included, although the equations to be developed are for any two mutually coupled coils with any value of coefficient of coupling k. When referring to the voltage induced across the inductance L_1 (or L_2) due to the change in flux linkages of the inductance L_2 (or L_1, respectively), the mutual inductance is represented by M_{12}. This type of subscript notation is particularly important when there are two or more mutual terms.

Due to the presence of the mutual term, the induced voltage e_1 is composed of that due to the self-inductance L_1 and that due to the mutual inductance M_{12}. That is,

$$e_1 = L_1\frac{di_1}{dt} + M_{12}\frac{di_2}{dt}$$

However, since $i_1 = i_2 = i$,

$$e_1 = L_1\frac{di}{dt} + M_{12}\frac{di}{dt}$$

or

$$e_1 = (L_1 + M_{12})\frac{di}{dt} \qquad \text{(volts, V)} \qquad \mathbf{(25.12)}$$

and, similarly,

$$e_2 = (L_2 + M_{12})\frac{di}{dt} \qquad \text{(volts, V)} \qquad \mathbf{(25.13)}$$

For the series connection, the total induced voltage across the series coils, represented by e_T, is

$$e_T = e_1 + e_2 = (L_1 + M_{12})\frac{di}{dt} + (L_2 + M_{12})\frac{di}{dt}$$

or

$$e_T = (L_1 + L_2 + M_{12} + M_{12})\frac{di}{dt}$$

and the total effective inductance is

$$\boxed{L_{T(+)} = L_1 + L_2 + 2M_{12}} \quad \text{(henries, H)} \qquad \textbf{(25.14)}$$

The subscript (+) was included to indicate that the mutual terms have a positive sign and are added to the self-inductance values to determine the total inductance. If the coils were wound such as shown in Fig. 25.5, where ϕ_1 and ϕ_2 are in opposition, the induced voltages due to the mutual terms would oppose that due to the self-inductance, and the total inductance would be determined by

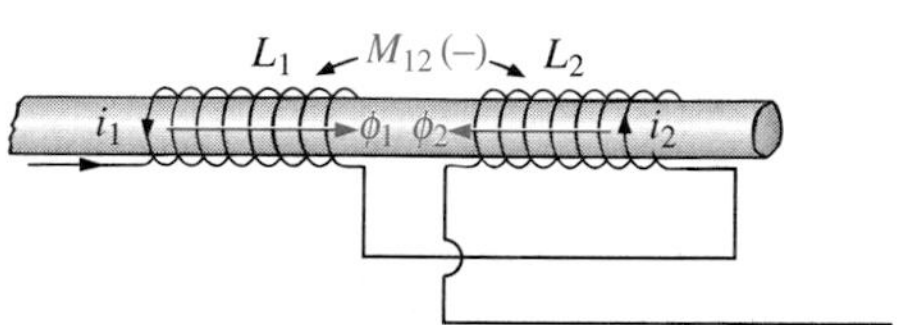

FIG. 25.5
Mutually coupled coils connected in series with negative mutual inductance.

$$\boxed{L_{T(-)} = L_1 + L_2 - 2M_{12}} \quad \text{(henries, H)} \qquad \textbf{(25.15)}$$

Through Eqs. (25.14) and (25.15), the mutual inductance can be determined by

$$\boxed{M_{12} = \frac{1}{4}(L_{T(+)} - L_{T(-)})} \qquad \textbf{(25.16)}$$

Equation (25.16) is very effective in determining the mutual inductance between two coils. It states that the mutual inductance is equal to one-quarter the difference between the total inductance with a positive and negative mutual effect.

From the preceding, it should be clear that the mutual inductance will directly affect the magnitude of the voltage induced across a coil since it will determine the net inductance of the coil. Additional examination reveals that the sign of the mutual term for each coil of a coupled pair is the same. For $L_{T(+)}$ they were both positive, and for $L_{T(-)}$ they were both negative. On a network schematic where it is inconvenient to indicate the windings and the flux path, a system of dots is employed that will determine whether the mutual terms are to be positive or negative. The dot convention is shown in Fig. 25.6 for the series coils of Figs. 25.4 and 25.5.

If the current through *each* of the mutually coupled coils is going away from (or toward) the dot as it *passes through the coil,* the mutual term will be positive, as shown for the case in Fig. 25.6(a). If the arrow indicating current direction through the coil is leaving the dot for one coil and entering the dot for the other, the mutual term is negative.

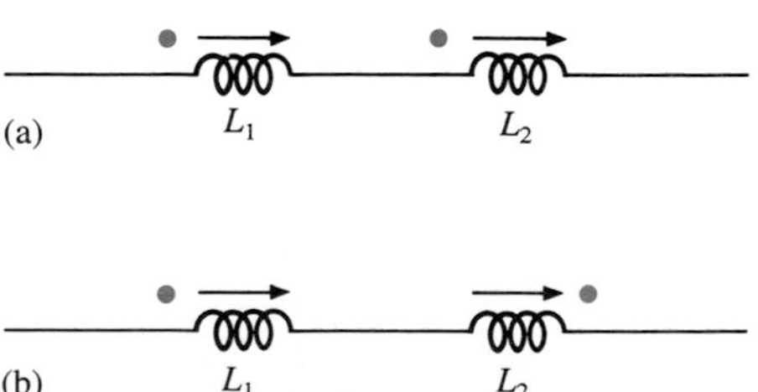

Dot convention for the series coils of (a) Fig. 25.4 and (b) Fig. 25.5.

A few possibilities for mutually coupled transformer coils are indicated in Fig. 25.7(a). The sign of M is indicated for each. When determining the sign, be sure to examine the current direction within the coil itself. In Fig. 25.7(b), one direction was indicated outside for one coil and through for the other. It initially might appear that the sign should be positive since both currents enter the dot, but the current *through* coil 1 is leaving the dot; hence a negative sign is in order.

The dot convention also reveals the polarity of the *induced* voltage across the mutually coupled coil. If the reference direction for the current *in* a coil leaves the dot, the polarity at the dot for the induced volt-

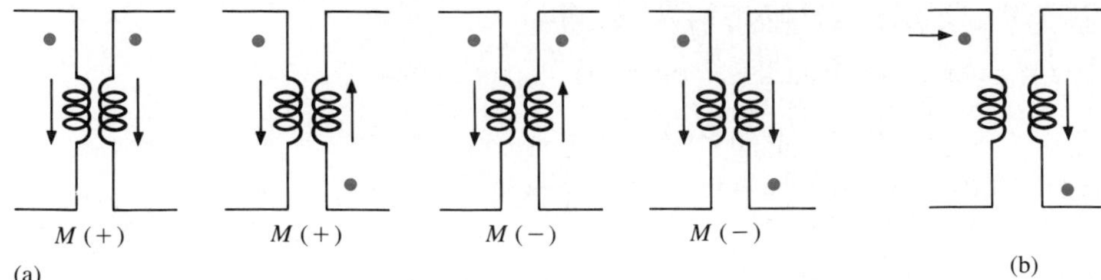

FIG. 25.7
Defining the sign of M for mutually coupled transformer coils.

age of the mutually coupled coil is positive. In the first two figures of Fig. 25.7(a), the polarity at the dots of the induced voltages is positive. In the third figure of Fig. 25.7(a), the polarity at the dot of the right-hand coil is negative, while the polarity at the dot of the left-hand coil is positive, since the current enters the dot (within the coil) of the right-hand coil. The comments for the third figure of Fig. 25.7(a) can also be applied to the last figure of Fig. 25.7(a).

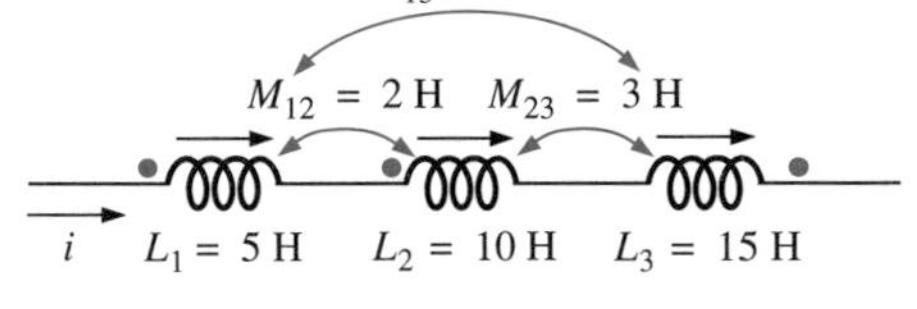

FIG. 25.8
Example 25.2.

EXAMPLE 25.2 Find the total inductance of the series coils of Fig. 25.8.

Solution:

Current vectors leave dot.

$$\text{Coil 1: } L_1 + M_{12} - M_{13}$$

One current vector enters dot, while one leaves.

$$\text{Coil 2: } L_2 + M_{12} - M_{23}$$
$$\text{Coil 3: } L_3 - M_{23} - M_{13}$$

and

$$\begin{aligned} L_T &= (L_1 + M_{12} - M_{13}) + (L_2 + M_{12} - M_{23}) + (L_3 - M_{23} - M_{13}) \\ &= L_1 + L_2 + L_3 + 2M_{12} - 2M_{23} - 2M_{13} \end{aligned}$$

Substituting values, we find

$$\begin{aligned} L_T &= 5\text{ H} + 10\text{ H} + 15\text{ H} + 2(2\text{ H}) - 2(3\text{ H}) - 2(1\text{ H}) \\ &= 34\text{ H} - 8\text{ H} = \mathbf{26\text{ H}} \end{aligned}$$

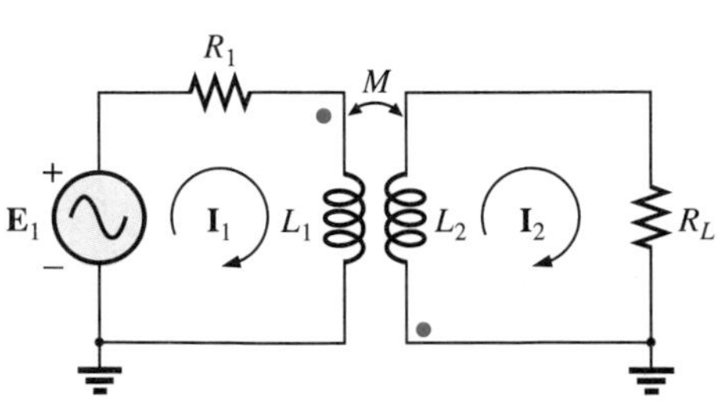

FIG. 25.9
Example 25.3.

EXAMPLE 25.3 Write the mesh equations for the transformer network in Fig. 25.9.

Solution: For each coil, the mutual term is positive, and the sign of M in $\mathbf{X}_m = \omega M \angle 90°$ is positive, as determined by the direction of $\mathbf{I}_1$ and $\mathbf{I}_2$. Thus,

$$\mathbf{E}_1 - \mathbf{I}_1 R_1 - \mathbf{I}_1 X_{L_1} \angle 90° - \mathbf{I}_2 X_m \angle 90° = 0$$

or $$\mathbf{E}_1 - \mathbf{I}_1(R_1 + j X_{L_1}) - \mathbf{I}_2 X_m \angle 90° = 0$$

For the other loop,

$$-\mathbf{I}_2 X_{L_2} \angle 90° - \mathbf{I}_1 X_m \angle 90° - \mathbf{I}_2 R_L = 0$$

or $$\mathbf{I}_2(R_L + j X_{L_2}) + \mathbf{I}_1 X_m \angle 90° = 0$$

25.4 THE IRON-CORE TRANSFORMER

An iron-core transformer under loaded conditions is shown in Fig. 25.10. The iron core will serve to increase the coefficient of coupling between the coils by increasing the mutual flux ϕ_m. Recall from Chapter 11 that magnetic flux lines will always take the path of least reluctance, which in this case is the iron core.

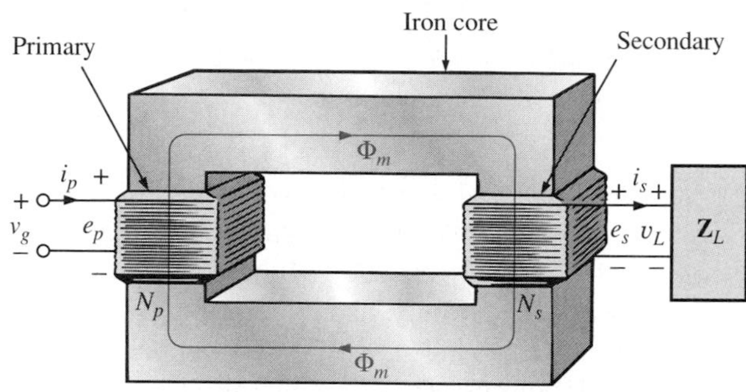

FIG. 25.10
Iron-core transformer.

We will assume in the analyses to follow in this chapter that all of the flux linking coil 1 will link coil 2. In other words, the coefficient of coupling is its maximum value, 1, and $\phi_m = \phi_p = \phi_s$. In addition, we will first analyze the transformer from an ideal viewpoint; that is, we will neglect losses such as the geometric or dc resistance of the coils, the leakage reactance due to the flux linking either coil that forms no part of ϕ_m, and the hysteresis and eddy current losses. This is not to convey the impression, however, that we will be far from the actual operation of a transformer. Most transformers manufactured today can be considered almost ideal. The equations we will develop under ideal conditions will be, in general, a first approximation to the actual response, which will never be off by more than a few percentage points. The losses will be considered in greater detail in Section 25.6.

When the current i_p through the primary circuit of the iron-core transformer is a maximum, the flux ϕ_m linking both coils is also a maximum. In fact, the magnitude of the flux is directly proportional to the current through the primary windings. Therefore, the two are in phase, and for sinusoidal inputs, the magnitude of the flux will vary as a sinusoid also. That is, if

$$i_p = \sqrt{2}I_p \sin \omega t$$

then

$$\phi_m = \Phi_m \sin \omega t$$

The induced voltage across the primary due to a sinusoidal input can be determined by Faraday's law:

$$e_p = N_p \frac{d\phi_p}{dt} = N_p \frac{d\phi_m}{dt}$$

Substituting for ϕ_m gives us

$$e_p = N_p \frac{d}{dt}(\Phi_m \sin \omega t)$$

and differentiating, we obtain

$$e_p = \omega N_p \Phi_m \cos \omega t$$

or

$$e_p = \omega N_p \Phi_m \sin(\omega t + 90°)$$

indicating that the induced voltage e_p leads the current through the primary coil by 90°.

The effective value of e_p is

$$E_p = \frac{\omega N_p \Phi_m}{\sqrt{2}} = \frac{2\pi f N_p \Phi_m}{\sqrt{2}}$$

and

$$\boxed{E_p = 4.44 f N_p \Phi_m} \qquad \textbf{(25.17)}$$

which is an equation for the effective value of the voltage across the primary coil in terms of the frequency of the input current or voltage, the number of turns of the primary, and the maximum value of the magnetic flux linking the primary.

For the case under discussion, where the flux linking the secondary equals that of the primary, if we repeat the procedure just described for the induced voltage across the secondary, we get

$$\boxed{E_s = 4.44 f N_s \Phi_m} \qquad \textbf{(25.18)}$$

Dividing Eq. (25.17) by Eq. (25.18), as follows:

$$\frac{E_p}{E_s} = \frac{4.44 f N_p \Phi_m}{4.44 f N_s \Phi_m}$$

we obtain

$$\boxed{\frac{E_p}{E_s} = \frac{N_p}{N_s}} \qquad \textbf{(25.19)}$$

revealing an important relationship for transformers:

The ratio of the magnitudes of the induced voltages is the same as the ratio of the corresponding turns.

If we consider that

$$e_p = N_p \frac{d\phi_m}{dt} \quad \text{and} \quad e_s = N_s \frac{d\phi_m}{dt}$$

and divide one by the other; that is,

$$\frac{e_p}{e_s} = \frac{N_p(d\phi_m/dt)}{N_s(d\phi_m/dt)}$$

then

$$\frac{e_p}{e_s} = \frac{N_p}{N_s}$$

The *instantaneous* values of e_1 and e_2 are therefore related by a constant determined by the turns ratio. Since their instantaneous magnitudes are related by a constant, the induced voltages are in phase, and Eq. (25.19) can be changed to include phasor notation; that is,

$$\frac{\mathbf{E}_p}{\mathbf{E}_s} = \frac{N_p}{N_s} \tag{25.20}$$

or, since $\mathbf{V}_g = \mathbf{E}_1$ and $\mathbf{V}_L = \mathbf{E}_2$ for the ideal situation,

$$\frac{\mathbf{V}_g}{\mathbf{V}_L} = \frac{N_p}{N_s} \tag{25.21}$$

The ratio N_p/N_s, usually represented by the lowercase letter a, is referred to as the *transformation ratio:*

$$a = \frac{N_p}{N_s} \tag{25.22}$$

If $a < 1$, the transformer is called a *step-up transformer* since the voltage $E_s > E_p$; that is,

$$\frac{E_p}{E_s} = \frac{N_p}{N_s} = a \quad \text{or} \quad E_s = \frac{E_p}{a}$$

and, if $a < 1$,

$$E_s > E_p$$

If $a > 1$, the transformer is called a *step-down transformer* since $E_s < E_p$; that is,

$$E_p = aE_s$$

and, if $a > 1$, then

$$E_p > E_s$$

EXAMPLE 25.4 For the iron-core transformer of Fig. 25.11:

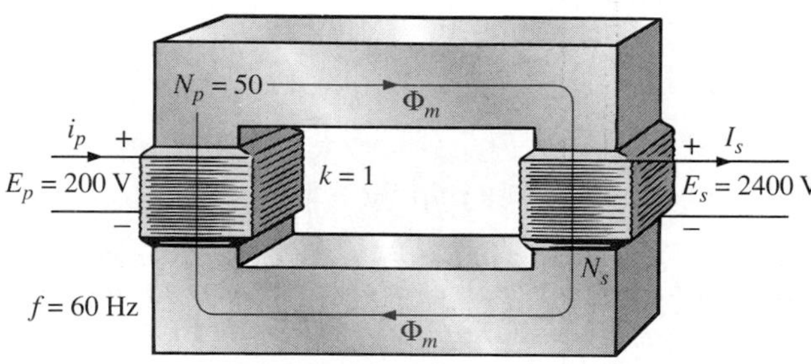

FIG. 25.11
Example 25.4.

a. Find the maximum flux Φ_m.
b. Find the secondary turns N_s.

Solutions:

a. $E_p = 4.44N_pf\Phi_m$

Therefore, $\Phi_m = \dfrac{E_p}{4.44N_pf} = \dfrac{200 \text{ V}}{(4.44)(50 \text{ t})(60 \text{ Hz})}$

and $\Phi_m = \mathbf{15.02 \text{ mWb}}$

b. $\dfrac{E_p}{E_s} = \dfrac{N_p}{N_s}$

$$\text{Therefore, } N_s = \frac{N_p E_s}{E_p} = \frac{(50\text{ t})(2400\text{ V})}{200\text{ V}}$$
$$= \mathbf{600\ turns}$$

The induced voltage across the secondary of the transformer of Fig. 25.10 will establish a current i_s through the load Z_L and the secondary windings. This current and the turns N_s will develop an mmf $N_s i_s$ that would not be present under no-load conditions since $i_s = 0$ and $N_s i_s = 0$. Under loaded or unloaded conditions, however, the net ampere-turns on the core produced by both the primary and the secondary must remain unchanged for the same flux ϕ_m to be established in the core. The flux ϕ_m must remain the same to have the same induced voltage across the primary and to balance the voltage impressed across the primary. In order to counteract the mmf of the secondary, which is tending to change ϕ_m, an additional current must flow in the primary. This current is called the *load component of the primary current* and is represented by the notation i'_p.

For the balanced or equilibrium condition,

$$N_p i'_p = N_s i_s$$

The total current in the primary under loaded conditions is

$$i_p = i'_p + i_{\phi_m}$$

where i_{ϕ_m} is the current in the primary necessary to establish the flux ϕ_m. For most practical applications, $i'_p > i_{\phi_m}$. For our analysis, we will assume $i_p \cong i'_p$, so

$$N_p i_p = N_s i_s$$

Since the instantaneous values of i_p and i_s are related by the turns ratio, the phasor quantities $\mathbf{I}_p$ and $\mathbf{I}_s$ are also related by the same ratio:

$$N_p \mathbf{I}_p = N_s \mathbf{I}_s$$

or

$$\boxed{\frac{\mathbf{I}_p}{\mathbf{I}_s} = \frac{N_s}{N_p}} \qquad \mathbf{(25.23)}$$

The primary and secondary currents of a transformer are therefore related by the inverse ratios of the turns.

Keep in mind that Eq. (25.23) holds true only if we neglect the effects of i_{ϕ_m}. Otherwise, the magnitudes of $\mathbf{I}_p$ and $\mathbf{I}_s$ are not related by the turns ratio, and $\mathbf{I}_p$ and $\mathbf{I}_s$ are not in phase.

For the step-up transformer, $a < 1$, and the current in the secondary, $I_s = aI_p$, is less in magnitude than that in the primary. For a step-down transformer, the reverse is true.

25.5 REFLECTED IMPEDANCE AND POWER

In the previous section we found that

$$\frac{\mathbf{V}_g}{\mathbf{V}_L} = \frac{N_p}{N_s} = a \quad \text{and} \quad \frac{\mathbf{I}_p}{\mathbf{I}_s} = \frac{N_s}{N_p} = \frac{1}{a}$$

Dividing the first by the second, we have

$$\frac{\mathbf{V}_g/\mathbf{V}_L}{\mathbf{I}_p/\mathbf{I}_s} = \frac{a}{1/a}$$

or

$$\frac{\mathbf{V}_g/\mathbf{I}_p}{\mathbf{V}_L/\mathbf{I}_s} = a^2 \quad \text{and} \quad \frac{\mathbf{V}_g}{\mathbf{I}_p} = a^2\frac{\mathbf{V}_L}{\mathbf{I}_s}$$

However, since

$$\mathbf{Z}_p = \frac{\mathbf{V}_g}{\mathbf{I}_p} \quad \text{and} \quad \mathbf{Z}_L = \frac{\mathbf{V}_L}{\mathbf{I}_s}$$

then

$$\mathbf{Z}_p = a^2\mathbf{Z}_L \tag{25.24}$$

which in words states that the impedance of the primary circuit of an ideal transformer is the transformation ratio squared times the impedance of the load. If a transformer is used, therefore, an impedance can be made to appear larger or smaller at the primary by placing it in the secondary of a step-down ($a > 1$) or step-up ($a < 1$) transformer, respectively. Note that if the load is capacitive or inductive, the reflected impedance will also be capacitive or inductive.

For the ideal iron-core transformer,

$$\frac{E_p}{E_s} = a = \frac{I_s}{I_p}$$

or

$$E_pI_p = E_sI_s \tag{25.25}$$

and

$$P_{\text{in}} = P_{\text{out}} \quad \text{(ideal conditions)} \tag{25.26}$$

EXAMPLE 25.5 For the iron-core transformer of Fig. 25.12:

a. Find the magnitude of the current in the primary and the impressed voltage across the primary.
b. Find the input resistance of the transformer.

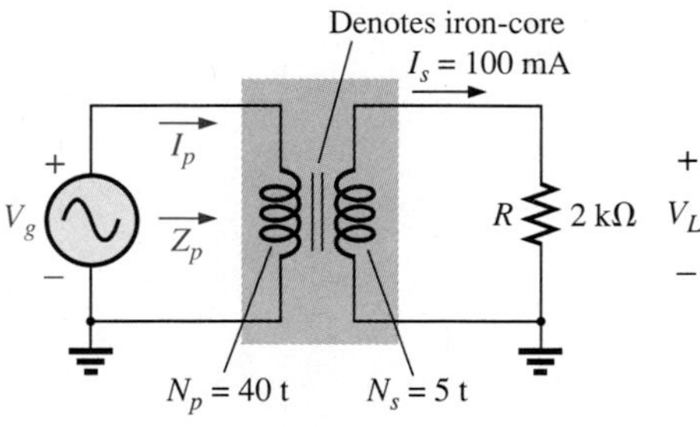

FIG. 25.12
Example 25.5.

Solutions:

a. $\dfrac{I_p}{I_s} = \dfrac{N_s}{N_p}$

$$I_p = \frac{N_s}{N_p}I_s = \left(\frac{5\text{ t}}{40\text{ t}}\right)(0.1\text{ A}) = \mathbf{12.5\text{ mA}}$$

$$V_L = I_sZ_L = (0.1\text{ A})(2\text{ k}\Omega) = 200\text{ V}$$

Also, $\dfrac{V_g}{V_L} = \dfrac{N_p}{N_s}$

$$V_g = \frac{N_p}{N_s}V_L = \left(\frac{40\text{ t}}{5\text{ t}}\right)(200\text{ V}) = \mathbf{1600\text{ V}}$$

b. $Z_p = a^2Z_L$

$$a = \frac{N_p}{N_s} = 8$$

$$Z_p = (8)^2(2\text{ k}\Omega) = R_p = \mathbf{128\text{ k}\Omega}$$

EXAMPLE 25.6 For the residential supply appearing in Fig. 25.13, determine (assuming a totally resistive load) the following:

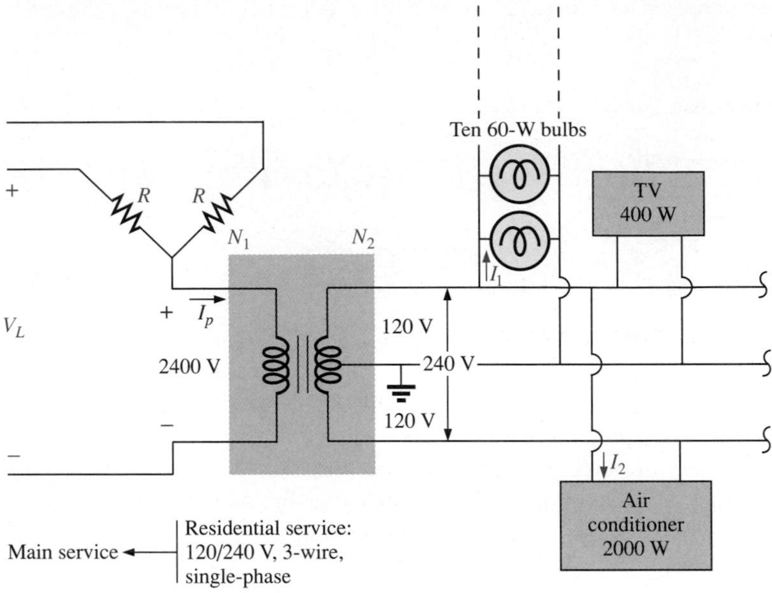

FIG. 25.13
Single-phase residential supply.

a. the value of R to ensure a balanced load
b. the magnitude of I_1 and I_2
c. the line voltage V_L
d. the total power delivered
e. the turns ratio $a = N_p/N_s$

Solutions:

a. $P_T = (10)(60\text{ W}) + 400\text{ W} + 2000\text{ W}$
$= 600\text{ W} + 400\text{ W} + 2000\text{ W} = 3000\text{ W}$
$P_{\text{in}} = P_{\text{out}}$
$V_pI_p = V_sI_s = 3000\text{ W}$ (purely resistive load)
$(2400\text{ V})I_p = 3000\text{ W}$ and $I_p = 1.25\text{ A}$

$$R = \frac{V_\phi}{I_p} = \frac{2400\text{ V}}{1.25\text{ A}} = \mathbf{1920\ \Omega}$$

b. $P_1 = 600\text{ W} = VI_1 = (120\text{ V})I_1$

and $I_1 = \mathbf{5\ A}$

$P_2 = 2000\text{ W} = VI_2 = (240\text{ V})I_2$

and $I_2 = \mathbf{8.33\ A}$

c. $V_L = \sqrt{3}V_\phi = 1.73(2400\text{ V}) = \mathbf{4152\ V}$

d. $P_T = 3P_\phi = 3(3000\text{ W}) = \mathbf{9\ kW}$

e. $$a = \frac{N_p}{N_s} = \frac{V_p}{V_s} = \frac{2400\text{ V}}{240\text{ V}} = \mathbf{10}$$

25.6 EQUIVALENT CIRCUIT (IRON-CORE TRANSFORMER)

For the nonideal or practical iron-core transformer, the equivalent circuit appears as in Fig. 25.14. As indicated, part of this equivalent circuit

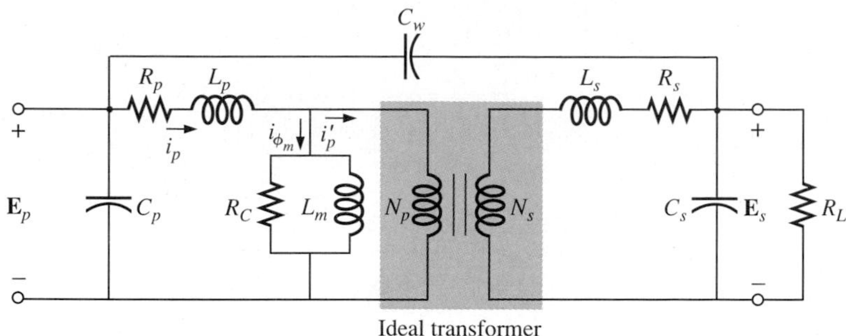

FIG. 25.14
Equivalent circuit for the practical iron-core transformer.

includes an ideal transformer. The remaining elements of Fig. 25.14 are those elements that contribute to the nonideal characteristics of the device. The resistances R_p and R_s are simply the dc or geometric resistance of the primary and secondary windings, respectively. For the primary and secondary coils of a transformer, there is a small amount of flux that links each coil but does not pass through the core, as shown in Fig. 25.15 for the primary winding. This *leakage* flux, representing a definite loss in the system, is represented by an inductance L_p in the primary circuit and an inductance L_s in the secondary.

The resistance R_c represents the hysteresis and eddy current losses (core losses) within the core due to an ac flux through the core. The inductance L_m (magnetizing inductance) is the inductance associated with the magnetization of the core, that is, the establishing of the flux Φ_m in the core. The capacitances C_p and C_s are the lumped capacitances of the primary and secondary circuits, respectively, and C_w represents the equivalent lumped capacitances between the windings of the transformer.

Since i'_p is normally considerably larger than i_{Φ_m} (the magnetizing current), we will ignore i_{Φ_m} for the moment (set it equal to zero), resulting in the absence of R_c and L_m in the reduced equivalent circuit of Fig. 25.16. The capacitances C_p, C_w, and C_s do not appear in the equivalent circuit of Fig. 25.16 since their reactance at typical operating frequencies will not appreciably affect the transfer characteristics of the transformer.

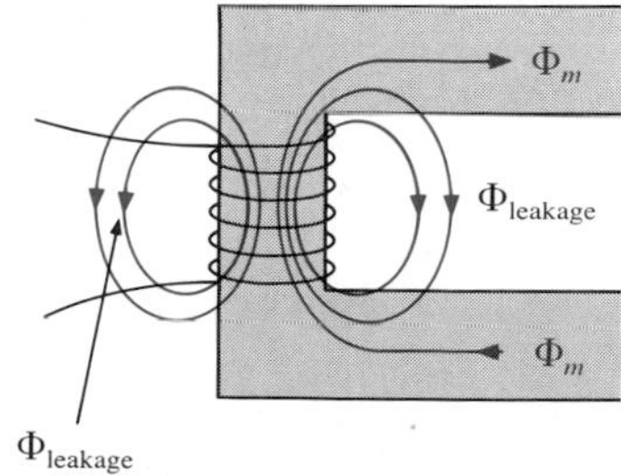

FIG. 25.15
Identifying the leakage flux of the primary.

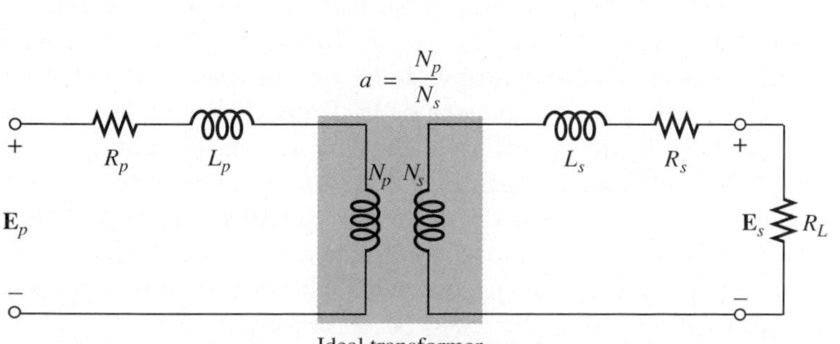

FIG. 25.16
Reduced equivalent circuit for the nonideal iron-core transformer.

If we now reflect the secondary circuit through the ideal transformer using Eq. (25.24), as shown in Fig. 25.17(a), we will have the load and generator voltage in the same continuous circuit. The total resistance and inductive reactance of the primary circuit are determined by

$$R_{\text{equivalent}} = R_e = R_p + a^2R_s \tag{25.27}$$

and

$$X_{\text{equivalent}} = X_e = X_p + a^2X_s \tag{25.28}$$

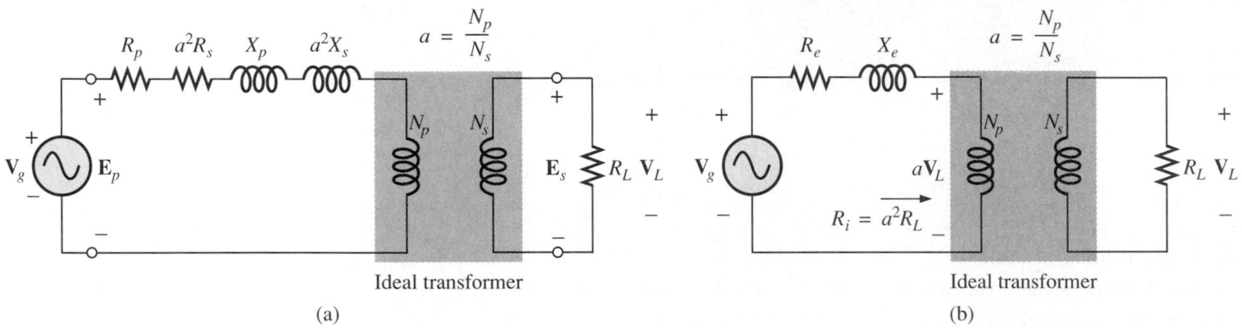

FIG. 25.17
Reflecting the secondary circuit into the primary side of the iron-core transformer.

which result in the useful equivalent circuit of Fig. 25.17(b). The load voltage can be obtained directly from the circuit of Fig. 25.17(b) through the voltage divider rule:

$$a\mathbf{V}_L = \frac{(R_i)\mathbf{V}_g}{(R_e + R_i) + jX_e}$$

and

$$\mathbf{V}_L = \frac{a^2R_L\mathbf{V}_g}{(R_e + a^2R_L) + jX_e} \tag{25.29}$$

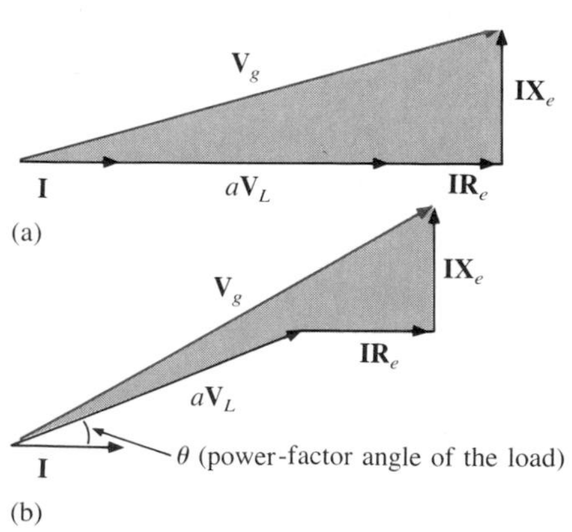

FIG. 25.18
Phasor diagram for the iron-core transformer with (a) unity power-factor load (resistive) and (b) lagging power-factor load (inductive).

The network of Fig. 25.17(b) will also allow us to calculate the generator voltage necessary to establish a particular load voltage. The voltages across the elements of Fig. 25.17(b) have the phasor relationship indicated in Fig. 25.18(a). Note that the current is the reference phasor for drawing the phasor diagram. That is, the voltages across the resistive elements are *in phase* with the current phasor, while the voltage across the equivalent inductance leads the current by 90°. The primary voltage, by Kirchhoff's voltage law, is then the phasor sum of these voltages, as indicated in Fig. 25.18(a). For an inductive load, the phasor diagram appears in Fig. 25.18(b). Note that $a\mathbf{V}_L$ leads $\mathbf{I}$ by the power-factor angle of the load. The remainder of the diagram is then similar to that for a resistive load. (The phasor diagram for a capacitive load will be left to the reader as an exercise.)

The effect of R_e and X_e on the magnitude of $\mathbf{V}_g$ for a particular $\mathbf{V}_L$ is obvious from Eq. (25.29) or Fig. 25.18. For increased values of R_e or X_e, an increase in $\mathbf{V}_g$ is required for the same load voltage. For R_e and $X_e = 0$, $\mathbf{V}_L$ and $\mathbf{V}_g$ are simply related by the turns ratio.

EXAMPLE 25.7 For a transformer having the equivalent circuit of Fig. 25.19:

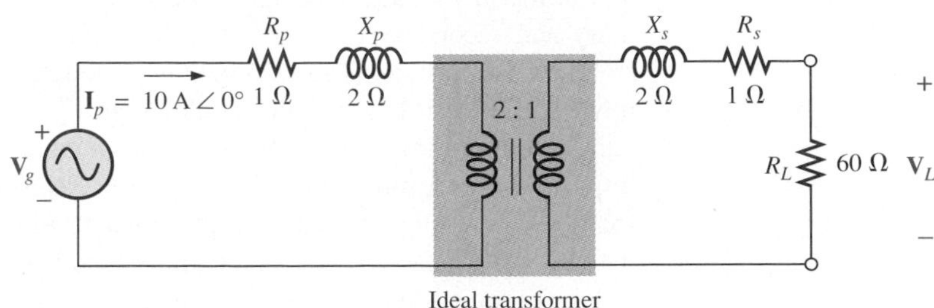

FIG. 25.19
Example 25.7.

a. Determine R_e and X_e.
b. Determine the magnitude of the voltages V_L and V_g.
c. Determine the magnitude of the voltage V_g to establish the same load voltage in part (b) if R_e and $X_e = 0\ \Omega$. Compare with the result of part (b).

Solutions:

a. $R_e = R_p + a^2R_s = 1\ \Omega + (2)^2(1\ \Omega) = \mathbf{5\ \Omega}$

$X_e = X_p + a^2X_s = 2\ \Omega + (2)^2(2\ \Omega) = \mathbf{10\ \Omega}$

b. The transformed equivalent circuit appears in Fig. 25.20.

$$aV_L = (I_p)(a^2R_L) = 2400\text{ V}$$

Thus, $V_L = \dfrac{2400\text{ V}}{a} = \dfrac{2400\text{ V}}{2} = \mathbf{1200\ V}$ and

$$\mathbf{V}_g = \mathbf{I}_p(R_e + a^2R_L + j\,X_e)$$
$$= 10\text{ A}(5\ \Omega + 240\ \Omega + j\,10\ \Omega) = 10\text{ A}(245\ \Omega + j\,10\ \Omega)$$
$$\mathbf{V}_g = 2450\text{ V} + j\,100\text{ V} = 2452.04\text{ V}\angle 2.34° = \mathbf{2452.04\ V\angle 2.34°}$$

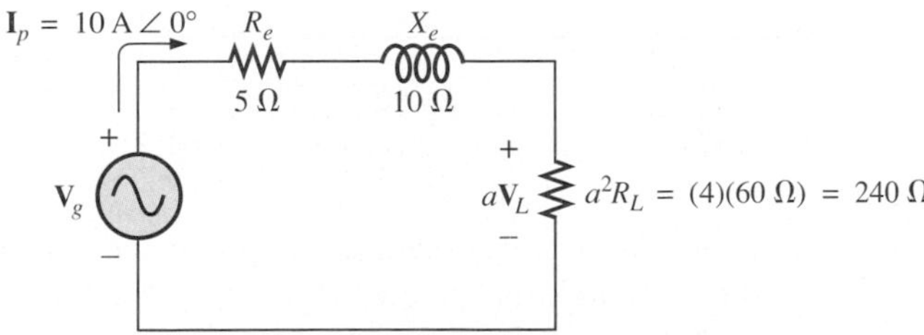

FIG. 25.20
Transformed equivalent circuit of Fig. 25.19.

c. For R_e and $X_e = 0$, $V_g = aV_L = (2)(1200\text{ V}) = 2400\text{ V}$.

Therefore, it is necessary to increase the generator voltage by 52.04 V (due to R_e and X_e) to obtain the same load voltage.

25.7 FREQUENCY CONSIDERATIONS

For certain frequency ranges, the effect of some parameters in the equivalent circuit of the iron-core transformer of Fig. 25.14 should not be ignored. Since it is convenient to consider a low-, mid-, and high-frequency region, the equivalent circuits for each will now be introduced and briefly examined.

For the low-frequency region, the series reactance ($2\pi fL$) of the primary and secondary leakage reactances can be ignored since they are small in magnitude. The magnetizing inductance must be included, however, since it appears in parallel with the secondary reflected circuit, and small impedances in a parallel network can have a dramatic impact on the terminal characteristics. The resulting equivalent network for the low-frequency region is provided in Fig. 25.21(a). As the frequency decreases, the reactance of the magnetizing inductance will reduce in magnitude, causing a reduction in the voltage across the secondary circuit. For $f = 0$ Hz, L_m is ideally a short circuit, and $V_L = 0$. As the frequency increases, the reactance of L_m will eventually be sufficiently large compared with the reflected secondary impedance to be neglected. The mid-frequency reflected equivalent circuit will then appear as shown in Fig. 25.21(b). Note the absence of reactive elements, resulting in an *in-phase* relationship between load and generator voltages.

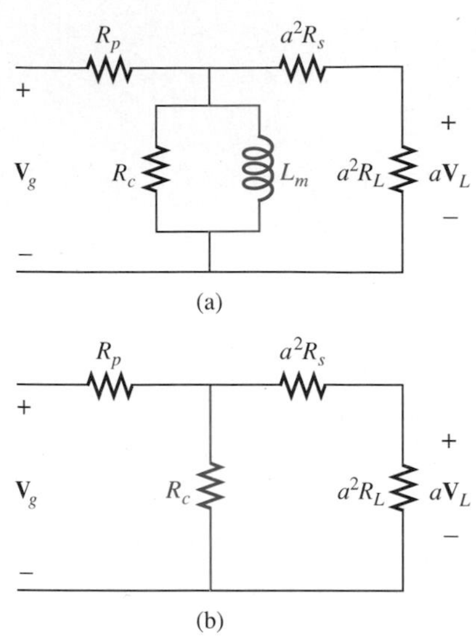

FIG. 25.21
(a) Low-frequency reflected equivalent circuit; (b) mid-frequency reflected circuit.

For higher frequencies, the capacitive elements and primary and secondary leakage reactances must be considered, as shown in Fig. 25.22. For discussion purposes, the effects of C_w and C_s appear as a lumped capacitor C in the reflected network of Fig. 25.22; C_p does not appear since the effect of C will predominate. As the frequency of interest increases, the capacitive reactance ($X_C = 1/2\pi fC$) will decrease to the point that it will have a shorting effect across the secondary circuit of the transformer, causing V_L to decrease in magnitude.

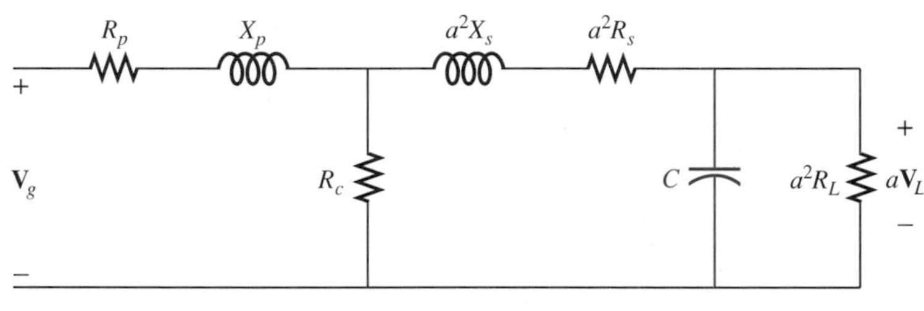

FIG. 25.22
High-frequency reflected equivalent circuit.

A typical iron-core transformer-frequency response curve appears in Fig. 25.23. For the low- and high-frequency regions, the primary ele-

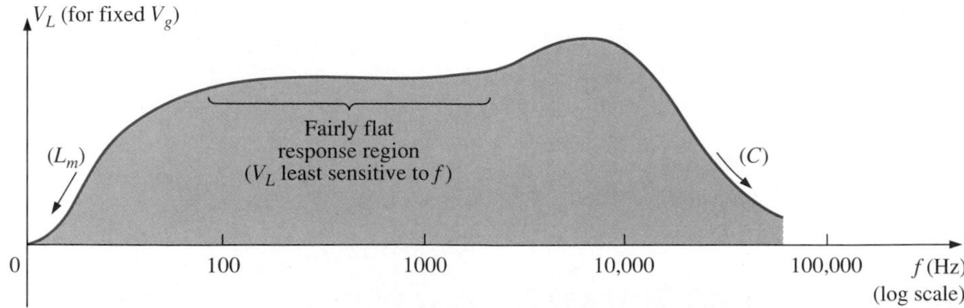

FIG. 25.23
Transformer-frequency response curve.

ment responsible for the drop-off is indicated. The peaking that occurs in the high-frequency region is due to the series resonant circuit established by the inductive and capacitive elements of the equivalent circuit. In the peaking region, the series resonant circuit is in, or near, its resonant or tuned state.

25.8 AIR-CORE TRANSFORMER

As the name implies, the air-core transformer does not have a ferromagnetic core to link the primary and secondary coils. Rather, the coils are placed sufficiently close to have a mutual inductance that will establish the desired transformer action. In Fig. 25.24, current direction and polarities have been defined for the air-core transformer. Note the presence of a mutual inductance term M, which will be positive in this case, as determined by the dot convention.

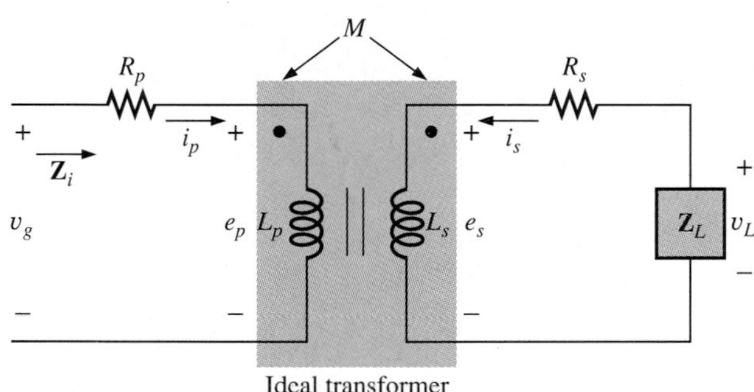

FIG. 25.24
Air-core transformer equivalent circuit.

From past analysis in this chapter, we now know that

$$e_p = L_p\frac{di_p}{dt} + M\frac{di_s}{dt} \tag{25.30}$$

for the primary circuit.

We found in Chapter 12 that for the pure inductor, with no mutual inductance present, the mathematical relationship

$$v_1 = L\frac{di_1}{dt}$$

resulted in the following useful form of the voltage across an inductor:

$$\mathbf{V}_1 = \mathbf{I}_1 X_L \angle 90°, \quad \text{where } X_L = \omega L$$

Similarly, it can be shown, for a mutual inductance, that

$$v_1 = M\frac{di_2}{dt}$$

will result in

$$\mathbf{V}_1 = \mathbf{I}_2 X_m \angle 90°, \quad \text{where } X_m = \omega M \tag{25.31}$$

Equation (25.30) can then be written (using phasor notation) as

$$\mathbf{E}_p = \mathbf{I}_p X_{L_p} \angle 90^\circ + \mathbf{I}_s X_m \angle 90^\circ \tag{25.32}$$

and $$\mathbf{V}_g = \mathbf{I}_p R_p \angle 0^\circ + \mathbf{I}_p X_{L_p} \angle 90^\circ + \mathbf{I}_s X_m \angle 90^\circ$$

or $$\mathbf{V}_g = \mathbf{I}_p(R_p + j X_{L_p}) + \mathbf{I}_s X_m \angle 90^\circ \tag{25.33}$$

For the secondary circuit,

$$\mathbf{E}_s = \mathbf{I}_s X_{L_s} \angle 90^\circ + \mathbf{I}_p X_m \angle 90^\circ \tag{25.34}$$

and $$\mathbf{V}_L = \mathbf{I}_s R_s \angle 0^\circ + \mathbf{I}_s X_{L_s} \angle 90^\circ + \mathbf{I}_p X_m \angle 90^\circ$$

or $$\mathbf{V}_L = \mathbf{I}(R_s + j X_{L_s}) + \mathbf{I}_p X_m \angle 90^\circ \tag{25.35}$$

Substituting $$\mathbf{V}_L = -\mathbf{I}_s \mathbf{Z}_L$$

into Eq. (25.35) results in

$$0 = \mathbf{I}_s(R_s + j X_{L_s} + \mathbf{Z}_L) + \mathbf{I}_p X_m \angle 90^\circ$$

Solving for $\mathbf{I}_s$, we have

$$\mathbf{I}_s = \frac{-\mathbf{I}_p X_m \angle 90^\circ}{R_s + j X_{L_s} + \mathbf{Z}_L}$$

and, substituting into Eq. (25.33), we obtain

$$\mathbf{V}_g = \mathbf{I}_p(R_p + j X_{L_p}) + \left(\frac{-\mathbf{I}_p X_m \angle 90^\circ}{R_s + j X_{L_s} + \mathbf{Z}_L}\right) X_m \angle 90^\circ$$

Thus, the input impedance is

$$\mathbf{Z}_i = \frac{\mathbf{V}_g}{\mathbf{I}_p} = R_p + j X_{L_p} - \frac{(X_m \angle 90^\circ)^2}{R_s + j X_{L_s} + \mathbf{Z}_L}$$

or, defining

$$\mathbf{Z}_p = R_p + j X_{L_p}, \quad \mathbf{Z}_s = R_s + j X_{L_s}, \quad \text{and} \quad X_m \angle 90^\circ = +j\,\omega M$$

we have

$$\mathbf{Z}_i = \mathbf{Z}_p - \frac{(+j\,\omega M)^2}{\mathbf{Z}_s + \mathbf{Z}_L}$$

and $$\mathbf{Z}_i = \mathbf{Z}_p - \frac{(\omega M)^2}{\mathbf{Z}_s + \mathbf{Z}_L} \tag{25.36}$$

The term $(\omega M)^2/(\mathbf{Z}_s + \mathbf{Z}_L)$ is called the *coupled impedance,* and it is independent of the sign of M since it is squared in the equation. Consider also that since $(\omega M)^2$ is a constant with 0° phase angle, if the load $\mathbf{Z}_L$ is resistive, the resulting coupled impedance term will appear capacitive due to division of $(\mathbf{Z}_s + R_L)$ into $(\omega M)^2$. This resulting capacitive reactance will oppose the series primary inductance L_p, causing a reduction in $\mathbf{Z}_i$. Including the effect of the mutual term, the input impedance to the network will appear as shown in Fig. 25.25.

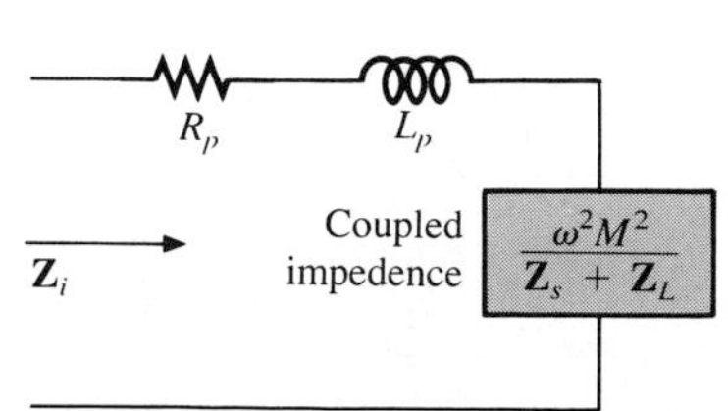

FIG. 25.25
Input characteristics for the air-core transformer.

EXAMPLE 25.8 Determine the input impedance to the air-core transformer in Fig. 25.26.

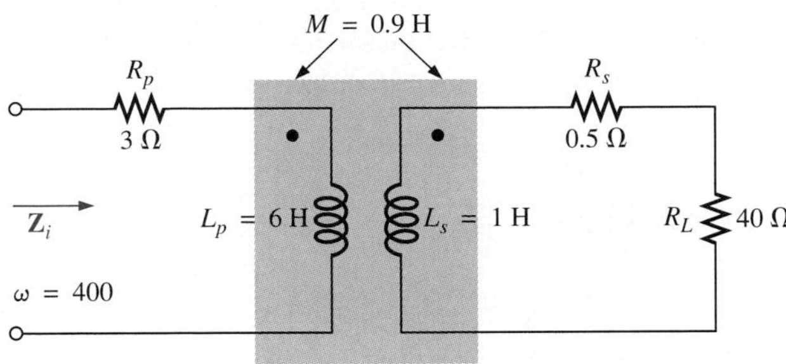

FIG. 25.26
Example 25.8.

Solution:

$$\mathbf{Z}_i = \mathbf{Z}_p + \frac{(\omega M)^2}{\mathbf{Z}_s + \mathbf{Z}_L}$$

$$= R_p + jX_{L_p} + \frac{(\omega M)^2}{R_s + jX_{L_s} + R_L}$$

$$= 3\ \Omega + j\,2.4\ \text{k}\Omega + \frac{((400\ \text{rad/s})(0.9\ \text{H}))^2}{0.5\ \Omega + j\,400\ \Omega + 40\ \Omega}$$

$$\cong j\,2.4\ \text{k}\Omega + \frac{129.6 \times 10^3\ \Omega}{40.5 + j\,400}$$

$$= j\,2.4\ \text{k}\Omega + \underbrace{322.4\ \Omega\ \angle -84.22^\circ}_{\text{capacitive}}$$

$$= j\,2.4\ \text{k}\Omega + (0.0325\ \text{k}\Omega - j\,0.3208\ \text{k}\Omega)$$

$$= 0.0325\ \text{k}\Omega + j\,(2.40 - 0.3208)\ \text{k}\Omega$$

and $\mathbf{Z}_i = R_i + jX_{L_i} = \mathbf{32.5\ \Omega + j\,2079\ \Omega = 2079.25\ \Omega\ \angle 89.10^\circ}$

25.9 IMPEDANCE MATCHING, ISOLATION, AND DISPLACEMENT

Transformers can be particularly useful when you are trying to ensure that a load receives maximum power from a source. Recall that maximum power is transferred to a load when its impedance is a match with the internal resistance of the supply. Even if a perfect match is unattainable, the closer the load matches the internal resistance, the greater the power to the load and the more efficient the system. Unfortunately, unless it is planned as part of the design, most loads are not a close match with the internal impedance of the supply. However, transformers have a unique relationship between their primary and secondary impedances that can be put to good use in the impedance matching process. Example 25.9 will demonstrate the significant difference in the power delivered to the load with and without an impedance matching transformer.

EXAMPLE 25.9

a. The source impedance for the supply of Fig. 25.27(a) is 512 Ω, which is a poor match with the 8-Ω input impedance of the speaker. One can only expect that the power delivered to the speaker will be significantly less than the maximum possible level. Determine the power to the speaker under the conditions of Fig. 25.27(a).

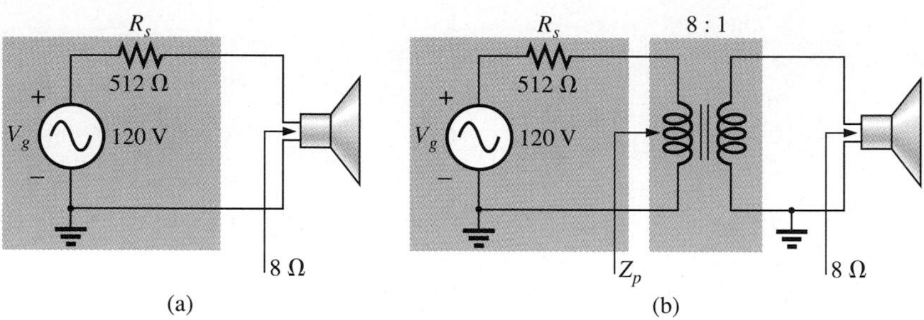

FIG. 25.27
Example 25.9.

b. In Fig. 25.27(b), an audio impedance matching transformer was introduced between the speaker and the source, and it was designed to ensure maximum power to the 8-Ω speaker. Determine the input impedance of the transformer and the power delivered to the speaker.

c. Compare the power delivered to the speaker under the conditions of parts (a) and (b).

Solutions:

a. The source current:

$$I_s = \frac{E}{R_T} = \frac{120\text{ V}}{512\ \Omega + 8\ \Omega} = \frac{120\text{ V}}{520\ \Omega} = 230.8\text{ mA}$$

The power to the speaker:

$$P = I^2R = (230.8\text{ mA})^2 \cdot 8\ \Omega = \mathbf{426.15\ mW} \cong \mathbf{0.43\ W}$$

or less than half a watt.

b.
$$Z_p = a^2Z_L \qquad a = \frac{N_p}{N_s} = \frac{8}{1} = 8$$

and
$$Z_p = (8)^2 8\ \Omega = \mathbf{512\ \Omega}$$

which matches that of the source. Maximum power transfer conditions have been established and the source current is now determined by:

$$I_s = \frac{E}{R_T} = \frac{120\text{ V}}{512\ \Omega + 512\ \Omega} = \frac{120\text{ V}}{1024\ \Omega} = 117.19\text{ mA}$$

The power to the primary (which equals that to the secondary for the ideal transformer) is:

$$P = I^2R = (117.19\text{ mA})^2\ 512\ \Omega = \mathbf{7.032\ W}$$

The result is not in milliwatts, as obtained above, and exceeds 7 W—a significant improvement.

c. Comparing levels, 7.032 W/426.15 mW = 16.5, or more than 16 times the power delivered to the speaker using the impedance matching transformer.

Another important application of the impedance matching capabilities of a transformer is the matching of the 300-Ω twin line transmission line from a television antenna to the 75-Ω input impedance of today's televisions (ready-made for the 75-Ω coaxial cable), as shown in Fig. 25.28. A match must be made to ensure the strongest signal to the television receiver.

Using the equation $Z_p = a^2 Z_L$ we find:

$$300\ \Omega = a^2 75\ \Omega$$

and
$$a = \sqrt{\frac{300\ \Omega}{75\ \Omega}} = \sqrt{4} = \mathbf{2}$$

with $N_p{:}N_s = 2{:}1$ (a step-down transformer)

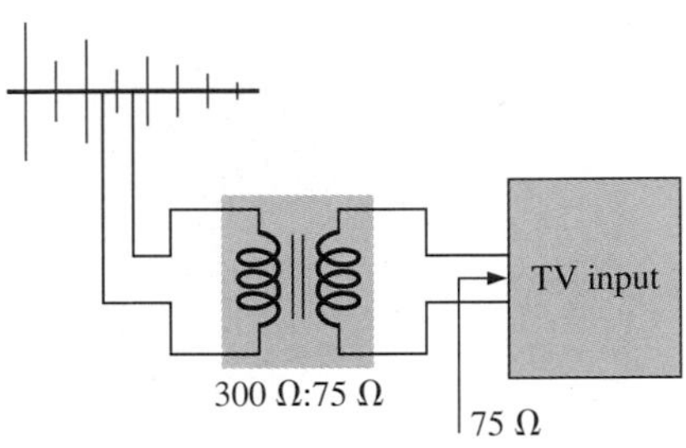

FIG. 25.28
Television impedance matching transformer.

EXAMPLE 25.10 Impedance matching transformers are also quite evident in public address systems, such as the one appearing in the 70.7-V system of Fig. 25.29. Although the system has only one set of output terminals, up to four speakers can be connected to this system (the number is a function of the chosen system). Each 8-Ω speaker is connected to the 70.7-V line through a 10-W audio-matching transformer (defining the frequency range of linear operation).

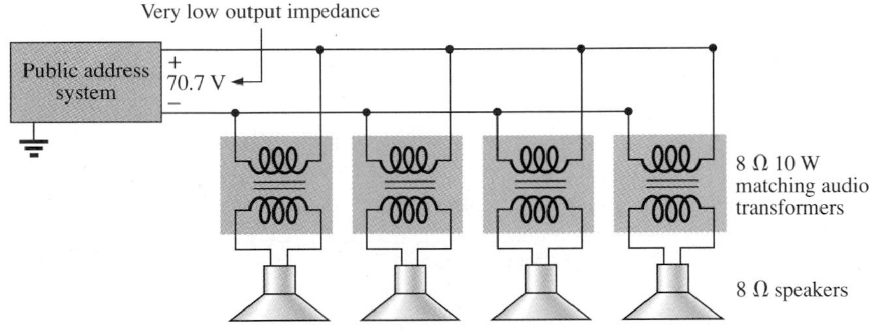

FIG. 25.29
Public address system.

a. If each speaker of Fig. 25.29 can receive 10 W of power, what is the maximum power drain on the source?
b. For each speaker, determine the impedance seen at the input side of the transformer if each is operating under its full 10 W of power.
c. Determine the turns ratio of the transformers.
d. At 10 W, what is the speaker voltage and current?
e. What is the load "seen" by the source with 1, 2, 3, or 4 speakers connected?

Solutions:

a. Ideally, the primary power equals the power delivered to the load, resulting in a maximum of 40 W from the supply.

b. The power at the primary:

$$P_p = V_p I_p = (70.7\text{ V})\,I_p = 10\text{ W}$$

and $$I_p = \frac{10\text{ W}}{70.7\text{ V}} = 141.4\text{ mA}$$

so that $$Z_p = \frac{V_p}{I_p} = \frac{70.7\text{ V}}{141.4\text{ mA}} = \mathbf{500\ \Omega}$$

c. $$Z_p = a^2 Z_L \Rightarrow a = \sqrt{\frac{Z_p}{Z_L}} = \sqrt{\frac{500\ \Omega}{8\ \Omega}} = \sqrt{62.5} = \mathbf{7.91} \cong 8{:}1$$

d. $$V_s = V_L = \frac{V_p}{a} = \frac{70.7\text{ V}}{7.91} = \mathbf{8.94\text{ V} \cong 9\text{ V}}$$

e. All the speakers are in parallel. Therefore,

1 speaker: $R_T = \mathbf{500\ \Omega}$

2 speakers: $R_T = \dfrac{500\ \Omega}{2} = \mathbf{250\ \Omega}$

3 speakers: $R_T = \dfrac{500\ \Omega}{3} = \mathbf{167\ \Omega}$

4 speakers: $R_T = \dfrac{500\ \Omega}{4} = \mathbf{125\ \Omega}$

Even though the load "seen" by the source will vary with the number of speakers connected, the source impedance is so low (compared to the lowest load of 125 Ω) that the terminal voltage of 70.7 V is essentially constant. This is not the case where the desired result is to match the load to the input impedance; rather, it was to ensure 70.7 V at each primary, no matter how many speakers were connected, and to limit the current drawn from the supply.

The transformer is frequently used to isolate one portion of an electrical system from another. *Isolation* implies the absence of any direct physical connection. As a first example of its use as an isolation device, consider the measurement of line voltages on the order of 40,000 V (Fig. 25.30).

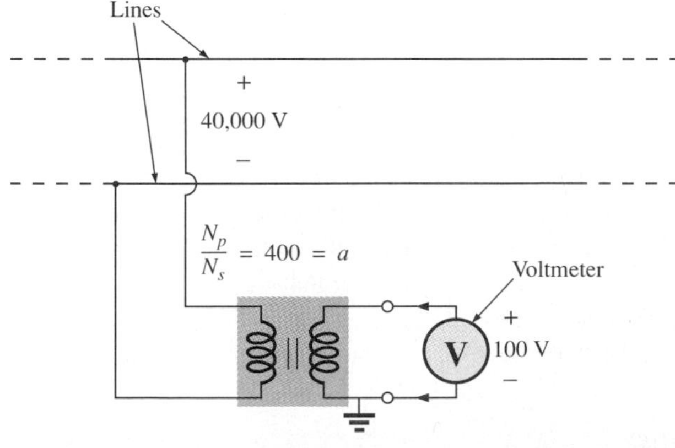

FIG. 25.30
Isolating a high-voltage line from the point of measurement.

To apply a voltmeter across 40,000 V would obviously be a dangerous task due to the possibility of physical contact with the lines when making the necessary connections. By including a transformer in the transmission system as original equipment, one can bring the potential down to a safe level for measurement purposes and can determine the line voltage using the turns ratio. Therefore, the transformer will serve both to isolate and to step down the voltage.

As a second example, consider the application of the voltage v_x to the vertical input of the oscilloscope (a measuring instrument) in Fig. 25.31. If the connections are made as shown and the generator and oscilloscope have a common ground, the impedance $\mathbf{Z}_2$ has been effectively shorted out of the circuit by the ground connection of the oscilloscope. The input voltage to the oscilloscope will therefore be meaningless as far as the voltage v_x is concerned. In addition, if $\mathbf{Z}_2$ is the current-limiting impedance in the circuit, the current in the circuit may rise to a level that will cause severe damage to the circuit. If a transformer is used as shown in Fig. 25.32, this problem will be eliminated, and the input voltage to the oscilloscope will be v_x.

The linear variable differential transformer (LVDT) is a sensor that can reveal displacement using transformer effects. In its simplest form, the LVDT has a central winding and two secondary windings, as shown in Fig. 25.33(a). A ferromagnetic core inside the windings is free to move as dictated by some external force. A constant, low-level ac voltage is applied to the primary, and the output voltage is the difference between the voltages induced in the secondaries. If the core is in the position shown in Fig. 25.33(b), a relatively large voltage will be induced across the secondary winding labeled coil 1, and a relatively small voltage will be induced across the secondary winding labeled coil 2 (essentially an air-core transformer for this position). The result is a relatively large secondary output voltage. If the core is in the position shown in Fig. 25.33(c), the flux linking each coil is the same, and the output voltage (being the difference) will be quite small. In total, therefore, the position of the core can be related to the secondary voltage, and a position-versus-voltage graph can be developed

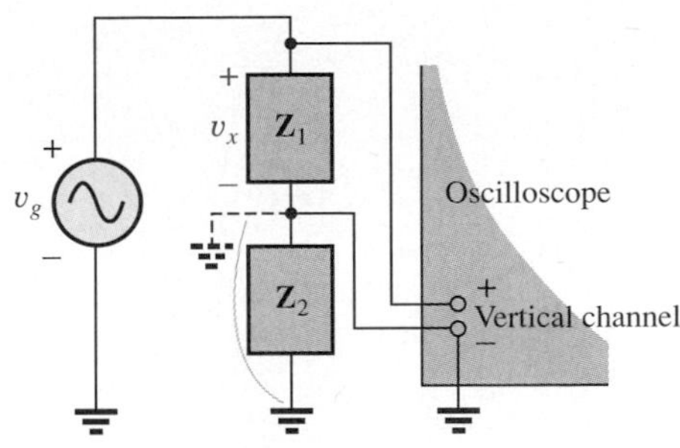

FIG. 25.31
Demonstrating the shorting effect introduced by the grounded side of the vertical channel of an oscilloscope.

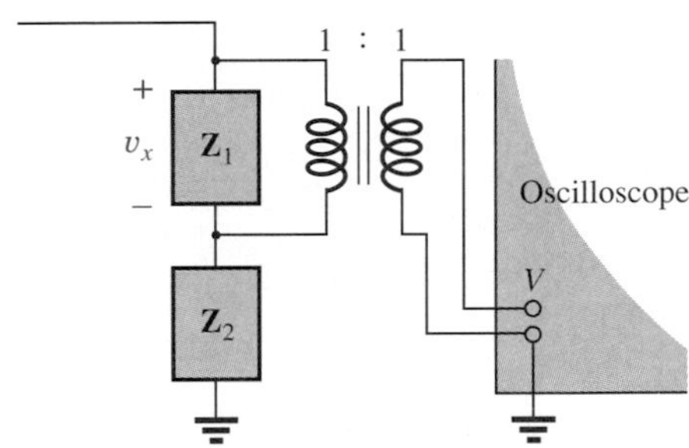

FIG. 25.32
Correcting the situation of Fig. 25.31 using an isolation transformer.

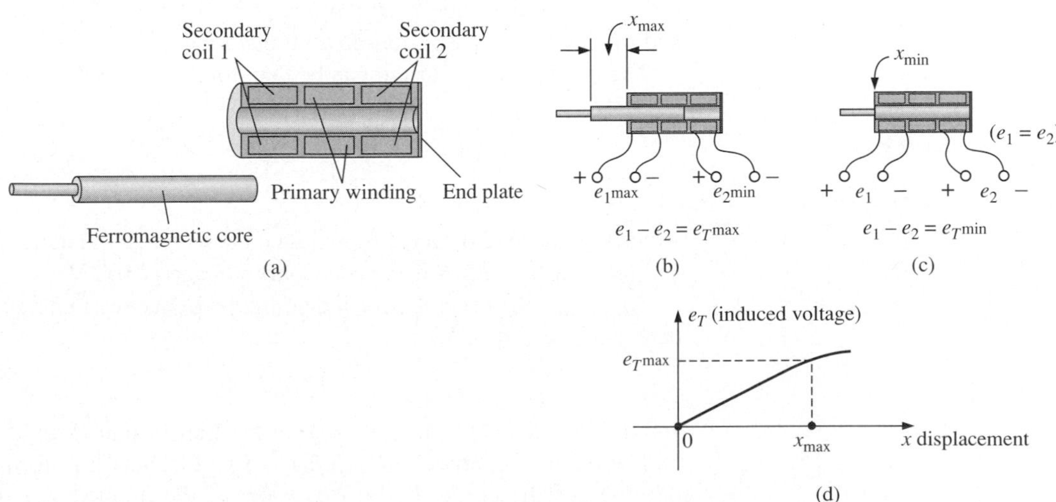

FIG. 25.33
LVDT transformer: (a) construction, (b) maximum displacement, (c) minimum displacement, (d) induced voltage versus displacement.

as shown in Fig. 25.33(d). Due to the nonlinearity of the *B-H* curve, the curve becomes somewhat nonlinear if the core is moved too far out of the unit.

25.10 NAMEPLATE DATA

A typical iron-core power transformer rating might be the following:

5 kVA, 2000/100 V, 60 Hz

The 2000 V or the 100 V can be either the primary or the secondary voltage; that is, if 2000 V is the primary voltage, then 100 V is the secondary voltage, and vice versa. The 5 kVA is the apparent power ($S = VI$) rating of the transformer. If the secondary voltage is 100 V, then the maximum load current is

$$I_L = \frac{S}{V_L} = \frac{5000 \text{ VA}}{100 \text{ V}} = 50 \text{ A}$$

and if the secondary voltage is 2000 V, then the maximum load current is

$$I_L = \frac{S}{V_L} = \frac{5000 \text{ VA}}{2000 \text{ V}} = 2.5 \text{ A}$$

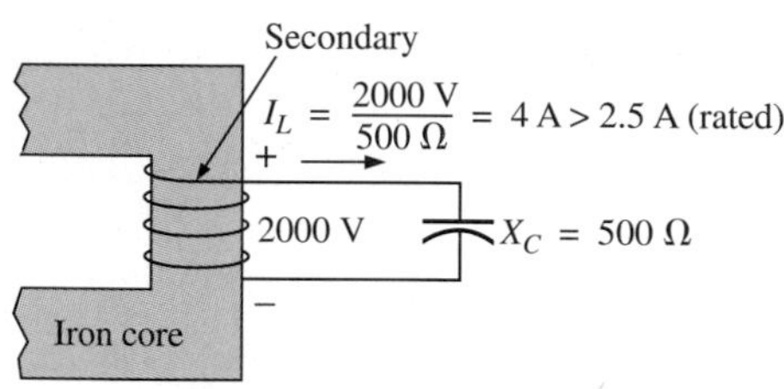

FIG. 25.34
Demonstrating why transformers are rated in kVA rather than kW.

The transformer is rated in terms of the apparent power rather than the average, or real, power for the reason demonstrated by the circuit of Fig. 25.34. Since the current through the load is greater than that determined by the apparent power rating, the transformer may be permanently damaged. Note, however, that since the load is purely capacitive, the average power to the load is zero. The wattage rating would therefore be meaningless regarding the ability of this load to damage the transformer.

The transformation ratio of the transformer under discussion can be either of two values. If the secondary voltage is 2000 V, the transformation ratio is $a = N_p/N_s = V_g/V_L = 100 \text{ V}/2000 \text{ V} = 1/20$, and the transformer is a step-up transformer. If the secondary voltage is 100 V, the transformation ratio is $a = N_p/N_s = V_g/V_L = 2000 \text{ V}/100 \text{ V} = 20$, and the transformer is a step-down transformer.

The rated primary current can be determined simply by applying Eq. (25.23):

$$I_p = \frac{I_s}{a}$$

which is equal to [2.5 A/(1/20)] = 50 A if the secondary voltage is 2000 V, and (50 A/20) = 2.5 A if the secondary voltage is 100 V.

To explain the necessity for including the frequency in the nameplate data, consider Eq. (25.7):

$$E_p = 4.44 f_p N_p \Phi_m$$

and the *B-H* curve for the iron core of the transformer (Fig. 25.35).

The point of operation on the *B-H* curve for most transformers is at the knee of the curve. If the frequency of the applied signal should drop, and N_p and E_p remain the same, then Φ_m must increase in magnitude, as determined by Eq. (25.7):

$$\Phi_m\uparrow = \frac{E_p}{4.44 f_p\downarrow N_p}$$

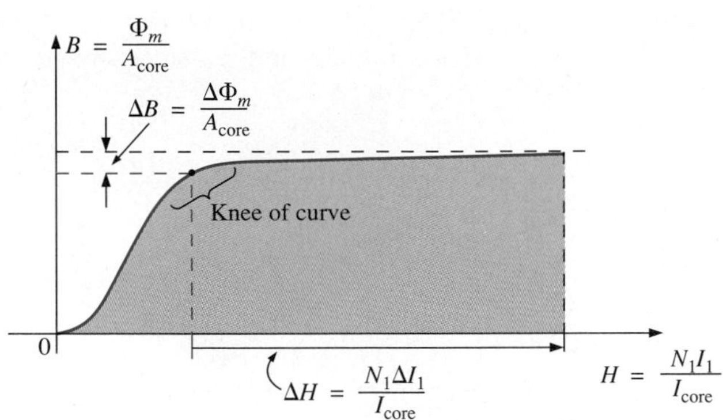

FIG. 25.35
Demonstrating why the frequency of application is important for transformers.

The result is that B will increase, as shown in Fig. 25.35, causing H to increase also. The resulting ΔI could cause a very high current in the primary, resulting in possible damage to the transformer.

25.11 TYPES OF TRANSFORMERS

Transformers are available in many different shapes and sizes. Some of the more common types include the power transformer, audio transformer, I-F (intermediate-frequency) transformer, and R-F (radio-frequency) transformer. Each is designed to fulfill a particular requirement in a specific area of application. The symbols for some of the basic types of transformers are shown in Fig. 25.36.

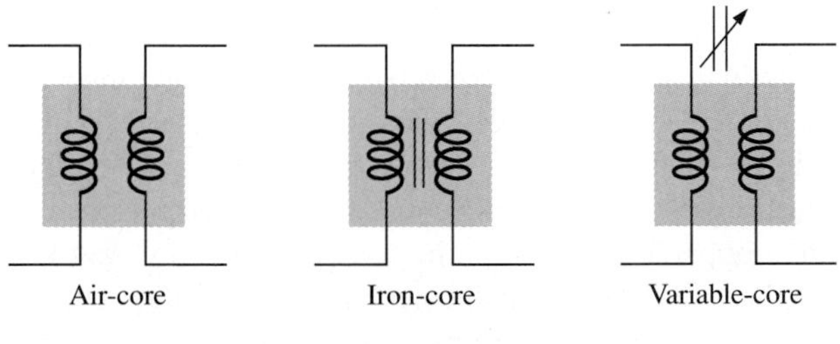

FIG. 25.36
Transformer symbols.

The method of construction varies from one transformer to another. Two of the many different ways in which the primary and secondary coils can be wound around an iron core are shown in Fig. 25.37. In either case, the core is made of laminated sheets of ferromagnetic material separated by an insulator to reduce the eddy current losses. The sheets themselves will also contain a small percentage of silicon to increase the electrical resistivity of the material and further reduce the eddy current losses.

A variation of the core-type transformer appears in Fig. 25.38. This transformer is designed for low-profile (2.5 VA size has a maximum height of only 0.65 in.) applications in power, control, and instrumentation applications. There are actually two transformers on the same core, with the primary and secondary of each wound side by side. The schematic representation appears in the same figure. Each set of terminals on the left can accept 115 V at 50 or 60 Hz, whereas each side of

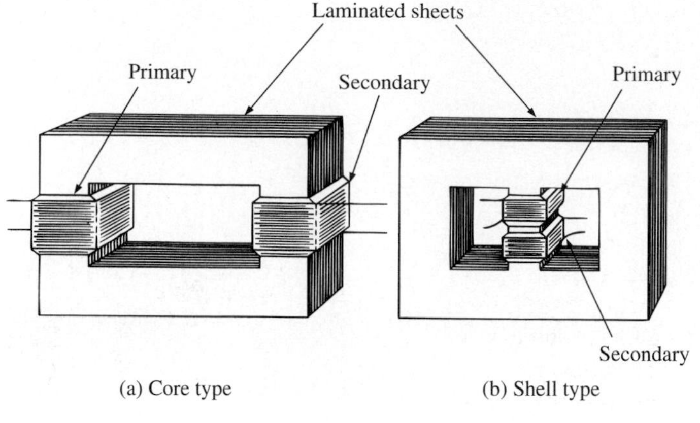

FIG. 25.37
Types of ferromagnetic core construction.

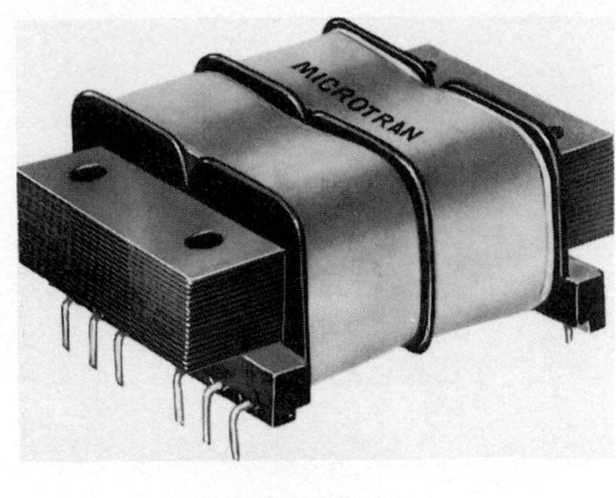

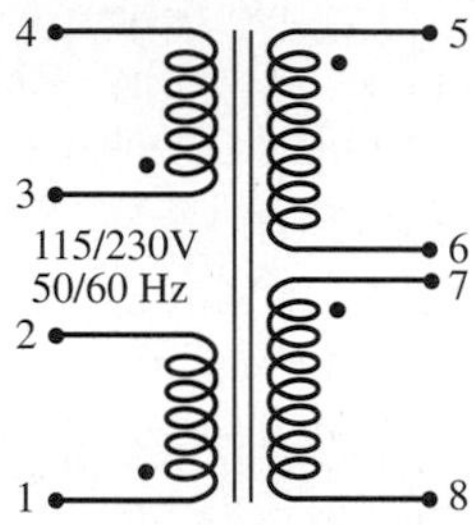

FIG. 25.38
Split bobbin, low-profile power transformer. (Courtesy of Microtran Company, Inc.)

the output will provide 230 V at the same frequency. Note the dot convention, as described earlier in the chapter.

The *autotransformer* [Fig. 25.39(b)] is a type of power transformer that, instead of employing the two-circuit principle (complete isolation between coils), has one winding common to both the input and output circuits. The induced voltages are related to the turns ratio in the same manner as that described for the two-circuit transformer. If the proper connection is used, a two-circuit power transformer can be employed as an autotransformer. The advantage of using it as an autotransformer is that a larger apparent power can be transformed. This can be demonstrated by the two-circuit transformer of Fig. 25.39(a). It is shown in Fig. 25.39(b) as an autotransformer.

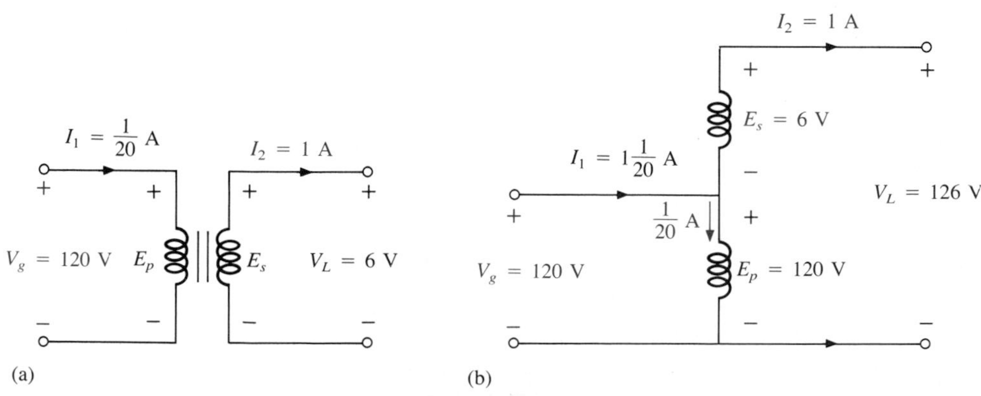

FIG. 25.39
(a) Two-circuit transformer; (b) autotransformer.

For the two-circuit transformer, note that $S = (\frac{1}{20}\ \text{A})(120\ \text{V}) = 6\ \text{VA}$, whereas for the autotransformer, $S = (1\frac{1}{20}\ \text{A})(120\ \text{V}) = 126\ \text{VA}$, which is many times that of the two-circuit transformer. Note also that the current and voltage of each coil are the same as those for the two-circuit configuration. The disadvantage of the autotransformer is obvious: loss of the isolation between the primary and secondary circuits.

A pulse transformer designed for printed-circuit applications where high-amplitude, long-duration pulses must be transferred without saturation appears in Fig. 25.40. Turns ratios are available from 1:1 to 5:1 at maximum line voltages of 240 V rms at 60 Hz. The upper unit is for printed-circuit applications with isolated dual primaries, whereas the lower unit is the bobbin variety with a single primary winding.

Two miniature ($\frac{1}{4}$ in. by $\frac{1}{4}$ in.) transformers with plug-in or insulated leads appear in Fig. 25.41, along with their schematic representations. Power ratings of 100 mW or 125 mW are available with a variety of turns ratios, such as 1:1, 5:1, 9.68:1, and 25:1.

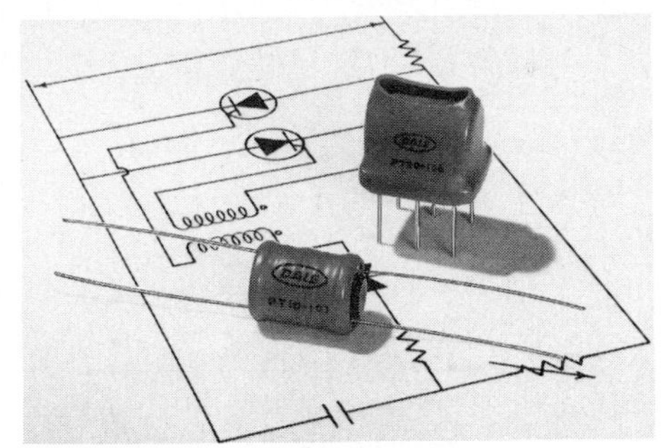

FIG. 25.40
Pulse transformers. (Courtesy of DALE Electronics, Inc.)

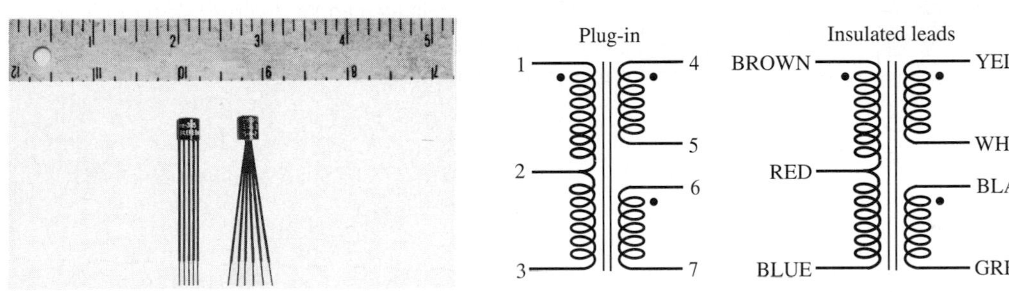

FIG. 25.41
Miniature transformers. (Courtesy of PICO Electronics, Inc.)

25.12 TAPPED AND MULTIPLE-LOAD TRANSFORMERS

For the center-tapped (primary) transformer of Fig. 25.42, where the voltage from the center tap to either outside lead is defined as $E_p/2$, the relationship between E_p and E_s is

$$\boxed{\frac{\mathbf{E}_p}{\mathbf{E}_s} = \frac{N_p}{N_s}} \qquad (25.37)$$

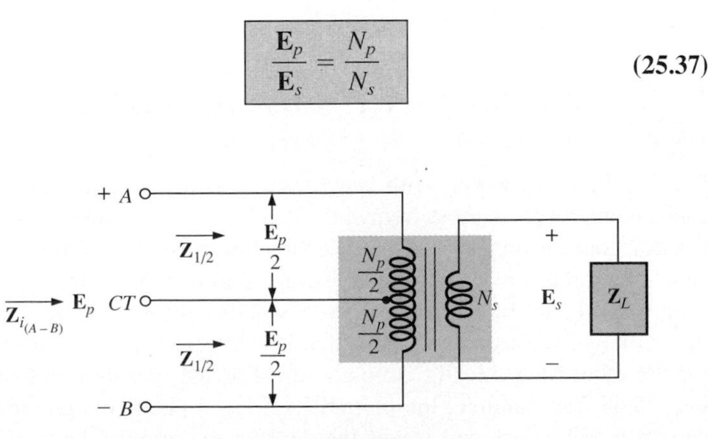

FIG. 25.42
Ideal transformer with a center-tapped primary.

For each half-section of the primary,

$$\mathbf{Z}_{1/2} = \left(\frac{N_p/2}{N_s}\right)^2 \mathbf{Z}_L = \frac{1}{4}\left(\frac{N_p}{N_s}\right)^2 \mathbf{Z}_L$$

with $$\mathbf{Z}_{i_{(A-B)}} = \left(\frac{N_p}{N_s}\right)^2 \mathbf{Z}_L$$

Therefore, $$\boxed{\mathbf{Z}_{1/2} = \frac{1}{4}\mathbf{Z}_i} \quad (25.38)$$

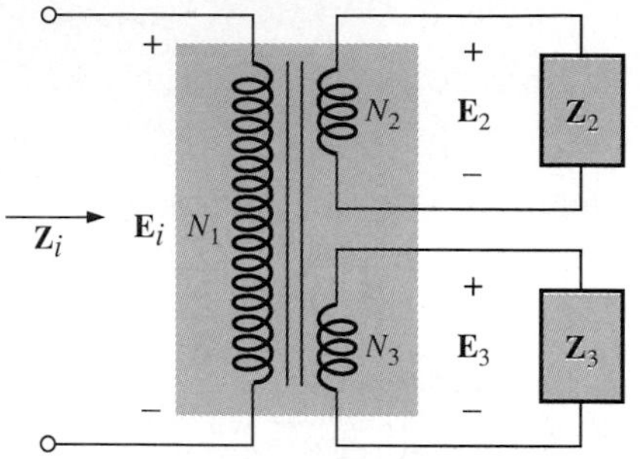

FIG. 25.43
Ideal transformer with multiple loads.

For the multiple-load transformer of Fig. 25.43, the following equations apply:

$$\boxed{\frac{\mathbf{E}_i}{\mathbf{E}_2} = \frac{N_1}{N_2} \qquad \frac{\mathbf{E}_1}{\mathbf{E}_3} = \frac{N_1}{N_3} \qquad \frac{\mathbf{E}_2}{\mathbf{E}_3} = \frac{N_2}{N_3}} \quad (25.39)$$

The total input impedance can be determined by first noting that, for the ideal transformer, the power delivered to the primary is equal to the power dissipated by the load; that is,

$$P_1 = P_{L_2} + P_{L_3}$$

and, for resistive loads ($\mathbf{Z}_i = R_i$, $\mathbf{Z}_2 = R_2$, and $\mathbf{Z}_3 = R_3$),

$$\frac{E_i^2}{R_i} = \frac{E_2^2}{R_2} + \frac{E_3^2}{R_3}$$

or, since $$E_2 = \frac{N_2}{N_1}E_i \quad \text{and} \quad E_3 = \frac{N_3}{N_1}E_1$$

then $$\frac{E_i^2}{R_i} = \frac{[(N_2/N_1)E_i]^2}{R_2} + \frac{[(N_3/N_1)E_i]^2}{R_3}$$

and $$\frac{E_i^2}{R_i} = \frac{E_i^2}{(N_1/N_2)^2R_2} + \frac{E_i^2}{(N_1/N_3)^2R_3}$$

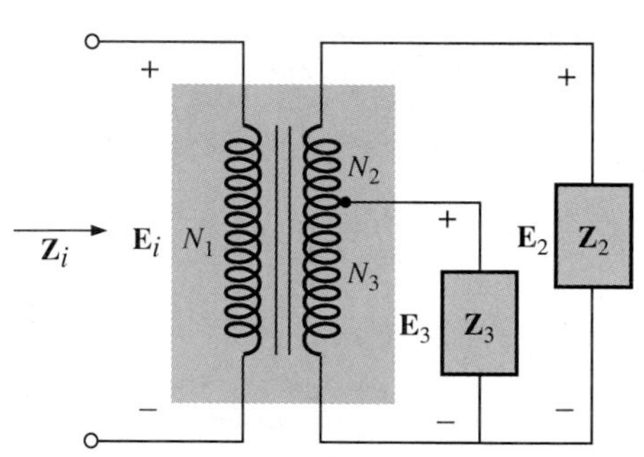

FIG. 25.44
Ideal transformer with a tapped secondary and multiple loads.

Thus,

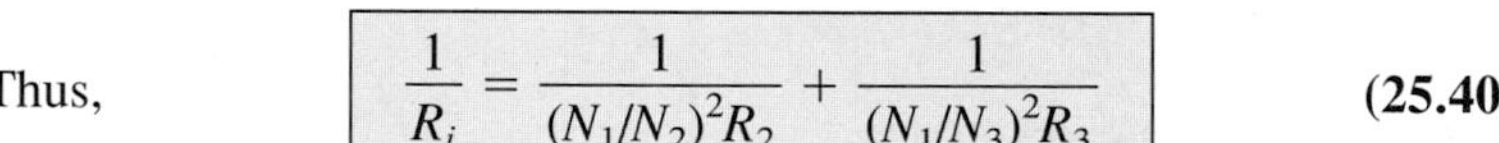

$$\boxed{\frac{1}{R_i} = \frac{1}{(N_1/N_2)^2R_2} + \frac{1}{(N_1/N_3)^2R_3}} \quad (25.40)$$

indicating that the load resistances are reflected in parallel.

For the configuration of Fig. 25.44, with E_2 and E_3 defined as shown, Eqs. (25.39) and (25.40) are applicable.

25.13 NETWORKS WITH MAGNETICALLY COUPLED COILS

For multiloop networks with magnetically coupled coils, the mesh-analysis approach is most frequently applied. A firm understanding of the dot convention discussed earlier should make the writing of the equations quite direct and free of errors. Before writing the equations for any particular loop, first determine whether the mutual term is positive or negative, keeping in mind that it will have the same sign as that for the other magnetically coupled coil. For the two-loop network of Fig. 25.45, for example, the mutual term has a positive sign since the current through each coil leaves the dot. For the primary loop,

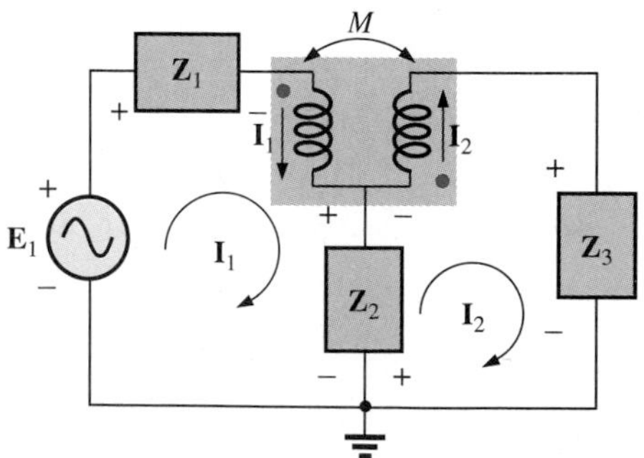

FIG. 25.45
Applying mesh analysis to magnetically coupled coils.

$$\mathbf{E}_1 - \mathbf{I}_1\mathbf{Z}_1 - \mathbf{I}_1\mathbf{Z}_{L_1} - \mathbf{I}_2\mathbf{Z}_m - \mathbf{Z}_2(\mathbf{I}_1 - \mathbf{I}_2) = 0$$

where M of $\mathbf{Z}_m = \omega M \angle 90°$ is positive and

$$\mathbf{I}_1(\mathbf{Z}_1 + \mathbf{Z}_{L_1} + \mathbf{Z}_2) - \mathbf{I}_2(\mathbf{Z}_2 - \mathbf{Z}_m) = \mathbf{E}_1$$

Note in the above that the mutual impedance was treated as if it were an additional inductance in series with the inductance L_1 having a sign determined by the dot convention and the voltage across which is determined by the current in the magnetically coupled loop.

For the secondary loop,

$$-\mathbf{Z}_2(\mathbf{I}_2 - \mathbf{I}_1) - \mathbf{I}_2\mathbf{Z}_{L_2} - \mathbf{I}_1\mathbf{Z}_m - \mathbf{I}_2\mathbf{Z}_3 = 0$$

or

$$\mathbf{I}_2(\mathbf{Z}_2 + \mathbf{Z}_{L_2} + \mathbf{Z}_3) - \mathbf{I}_1(\mathbf{Z}_2 - \mathbf{Z}_m) = 0$$

For the network of Fig. 25.46, we find a mutual term between L_1 and L_2 and L_1 and L_3 labeled M_{12} and M_{13}, respectively.

For the coils with the dots (L_1 and L_3), since each current through the coils leaves the dot, M_{13} is positive for the chosen direction of I_1 and I_3. However, since the current I_1 leaves the dot through L_1, and I_2 enters the dot through coil L_2, M_{12} is negative. Consequently, for the input circuit,

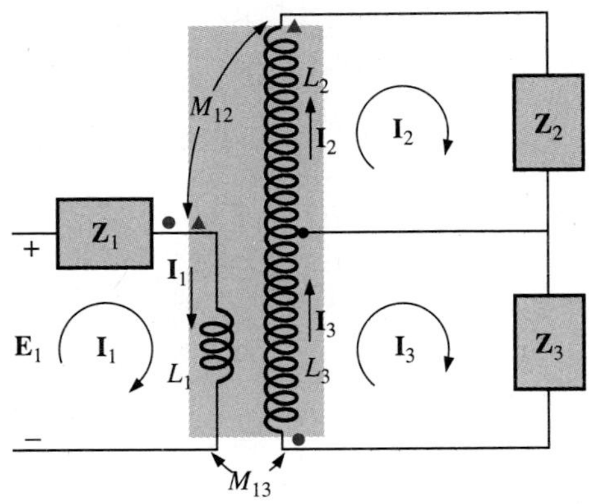

FIG. 25.46
Applying mesh analysis to a network with two magnetically coupled coils.

$$\mathbf{E}_1 - \mathbf{I}_1\mathbf{Z}_1 - \mathbf{I}_1\mathbf{Z}_{L_1} - \mathbf{I}_2(-\mathbf{Z}_{m_{12}}) - \mathbf{I}_3\mathbf{Z}_{m_{13}} = 0$$

or

$$\mathbf{E}_1 - \mathbf{I}_1(\mathbf{Z}_1 + \mathbf{Z}_{L_1}) + \mathbf{I}_2\mathbf{Z}_{m_{12}} - \mathbf{I}_3\mathbf{Z}_{m_{13}} = 0$$

For loop 2,

$$-\mathbf{I}_2\mathbf{Z}_2 - \mathbf{I}_2\mathbf{Z}_{L_2} - \mathbf{I}_1(-\mathbf{Z}_{m_{12}}) = 0$$
$$-\mathbf{I}_1\mathbf{Z}_{m_{12}} + \mathbf{I}_2(\mathbf{Z}_2 + \mathbf{Z}_{L_2}) = 0$$

and for loop 3,

$$-\mathbf{I}_3\mathbf{Z}_3 - \mathbf{I}_3\mathbf{Z}_{L_3} - \mathbf{I}_1\mathbf{Z}_{m_{13}} = 0$$

or

$$\mathbf{I}_1\mathbf{Z}_{m_{13}} + \mathbf{I}_3(\mathbf{Z}_3 + \mathbf{Z}_{L_3}) = 0$$

In determinant form,

$$\begin{array}{llll} \mathbf{I}_1(\mathbf{Z}_1 + \mathbf{Z}_{L_1}) - \mathbf{I}_2\mathbf{Z}_{m_{12}} & & + \mathbf{I}_3\mathbf{Z}_{m_{13}} & = \mathbf{E}_1 \\ -\mathbf{I}_1\mathbf{Z}_{m_{12}} & + \mathbf{I}_2(\mathbf{Z}_2 + \mathbf{Z}_{L_{12}}) & + 0 & = 0 \\ \mathbf{I}_1\mathbf{Z}_{m_{13}} & + 0 & + \mathbf{I}_3(\mathbf{Z}_3 + \mathbf{Z}_{13}) & = 0 \end{array}$$

25.14 COMPUTER ANALYSIS

PSpice (DOS)

In PSpice, the correct modeling of a device is a very important element of the input file if the results are to have any meaning whatsoever. In fact, to ensure that the model is correct, it must first be tested against known results before being applied to the system to be analyzed. In other words, do not become too confident in your use of certain commands in new environments—there may be some very special limitations on their application.

For transformers, a model for the ideal device requires the use of two controlled sources and an independent sensing source, as shown in Fig. 25.47.

The multiplying factor for both controlling sources is one over the turns ratio ($1/a$). Figure 25.47 reveals that the secondary voltage is $1/a$ times the primary voltage, and the primary current is $1/a$ times the secondary current. The use of both a controlled current and voltage source includes the

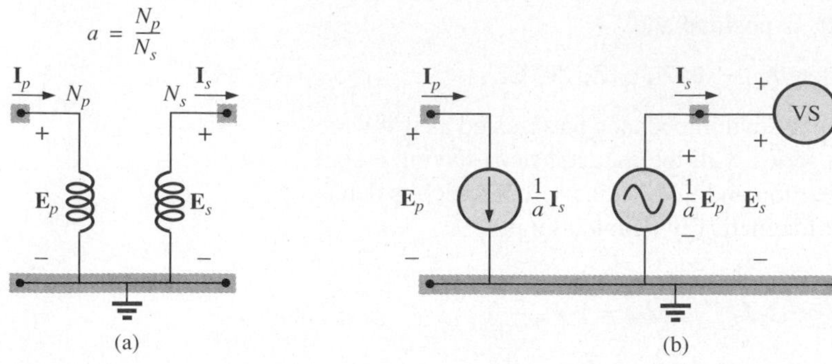

FIG. 25.47
(a) Ideal transformer; (b) PSpice model.

important voltage and current relationships for a transformer ($E_p = aE_s$, $I_s = aI_p$). The independent source VS is required to define the direction of I_s for the CCCS. The controlling voltage for the VCVS is simply the primary voltage of the transformer. There is no need to include the impedance relationships ($Z_i = a^2Z_L$, etc.) since they are already incorporated in the voltage and current relationships of the transformer.

As a first example let us analyze the basic transformer configuration of Fig. 25.48 and determine the load voltage V_L. The transformation ratio is

$$a = \frac{N_p}{N_s} = \frac{1}{4} < 1$$

revealing that the configuration is a step-up transformer.

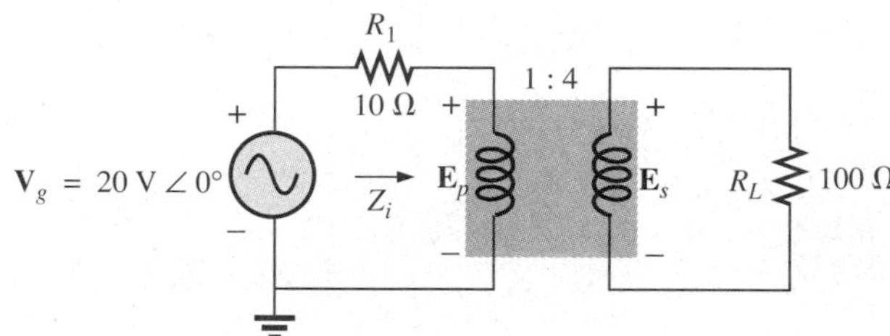

FIG. 25.48
Applying PSpice (DOS) to a step-up transformer.

Before assigning nodes and writing the input file, let us first analyze the network in the longhand fashion to provide a result for comparison with the output file.

$$Z_i = a^2Z_L$$
$$= \left(\frac{1}{4}\right)^2 100\ \Omega$$
$$= 6.25\ \Omega$$

and
$$E_p = \frac{(6.25\ \Omega)(20\ \text{V})}{6.25\ \Omega + 10\ \Omega} = 7.692\ \text{V}$$

with
$$E_s = \frac{1}{a}E_p = \frac{1}{\left(\frac{1}{4}\right)}(7.692\ \text{V}) = 4(7.692\ \text{V}) = 30.77\ \text{V}$$

and $\qquad V_L = E_s = \mathbf{30.77\ V}$

The nodes are defined in Fig. 25.49. The reference node is carried through to the output circuit to establish a common ground. If preferred, isolation between the input and output circuits can be maintained by assigning a different node label for the base of the output circuit and connecting the two bases by a large impedance of perhaps 1E30 Ω to simulate an open circuit. In any event, a common ground is required for the entire system.

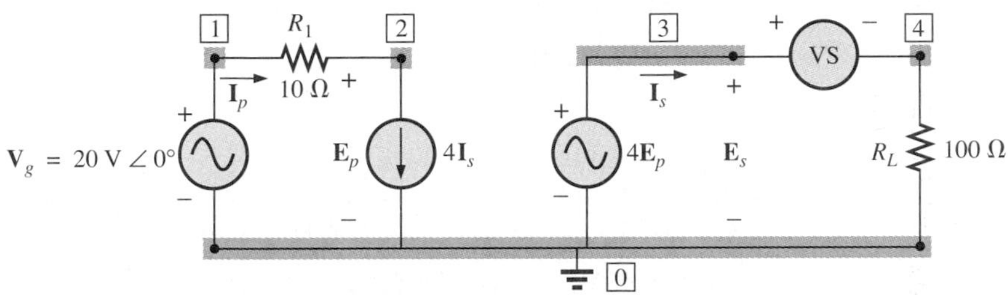

FIG. 25.49
Defining the parameters for a PSpice (DOS) analysis of the network of Fig. 25.48.

The input file of Fig. 25.50 employed the "names" PRI and SEC to distinguish between sources, and it requested the magnitude of the voltage across the load resistor, as calculated before. Note for both controlled sources that the multiplying factor is $1/a = 1/(\frac{1}{4}) = 4$ and that a fre-

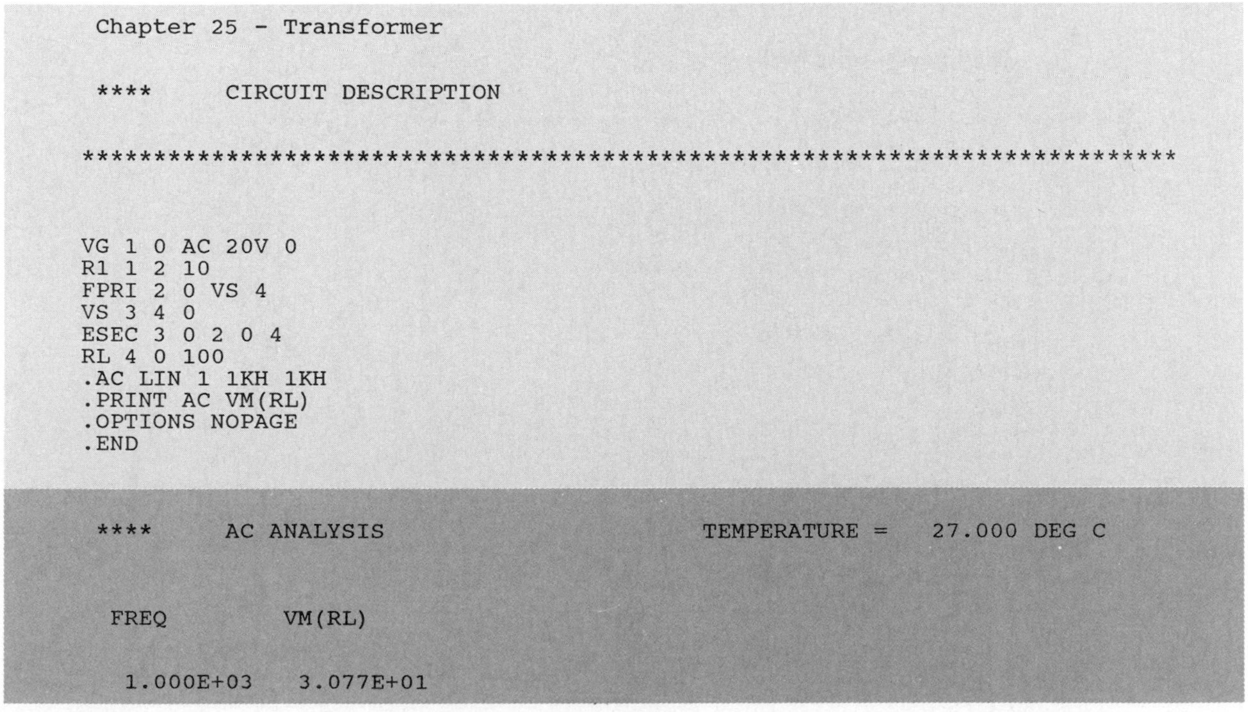

```
Chapter 25 - Transformer

****     CIRCUIT DESCRIPTION

*****************************************************************************

VG 1 0 AC 20V 0
R1 1 2 10
FPRI 2 0 VS 4
VS 3 4 0
ESEC 3 0 2 0 4
RL 4 0 100
.AC LIN 1 1KH 1KH
.PRINT AC VM(RL)
.OPTIONS NOPAGE
.END

****     AC ANALYSIS                        TEMPERATURE =   27.000 DEG C

  FREQ          VM(RL)

   1.000E+03    3.077E+01
```

FIG. 25.50
Input and output files for the PSpice (DOS) analysis of the network of Fig. 25.48.

quency of 1 kHz was chosen to establish an .AC analysis. The output file reveals an exact match between the longhand and computer solutions.

The second transformer network to be analyzed appears in Fig. 25.19. In Example 25.7 the primary current established a load voltage of 1200 V, with the requirement that $\mathbf{V}_g = 2452.04\text{ V} \angle 2.34°$. In this exercise we will apply a source voltage $\mathbf{V}_g = 2452.04\text{ V} \angle 2.34°$ and note whether a load voltage of 1200 V results, as in the example.

The nodes are chosen as in Fig. 25.51, with the inductor values determined at a frequency of 60 Hz. In this case, $a = N_p/N_s = 2/1 = 2$, resulting in a multiplying factor of $1/a = 1/2 = 0.5$, as appearing in the input file of Fig. 25.52. The output file reveals an exact match of 1200 V.

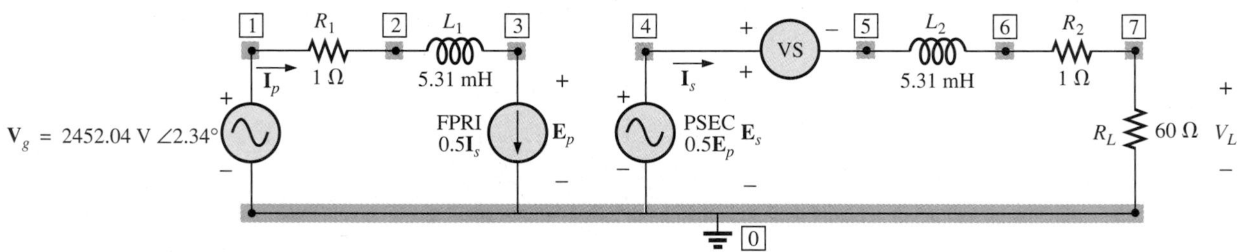

FIG. 25.51
Defining the parameters for a PSpice (DOS) analysis of the network of Example 25.7.

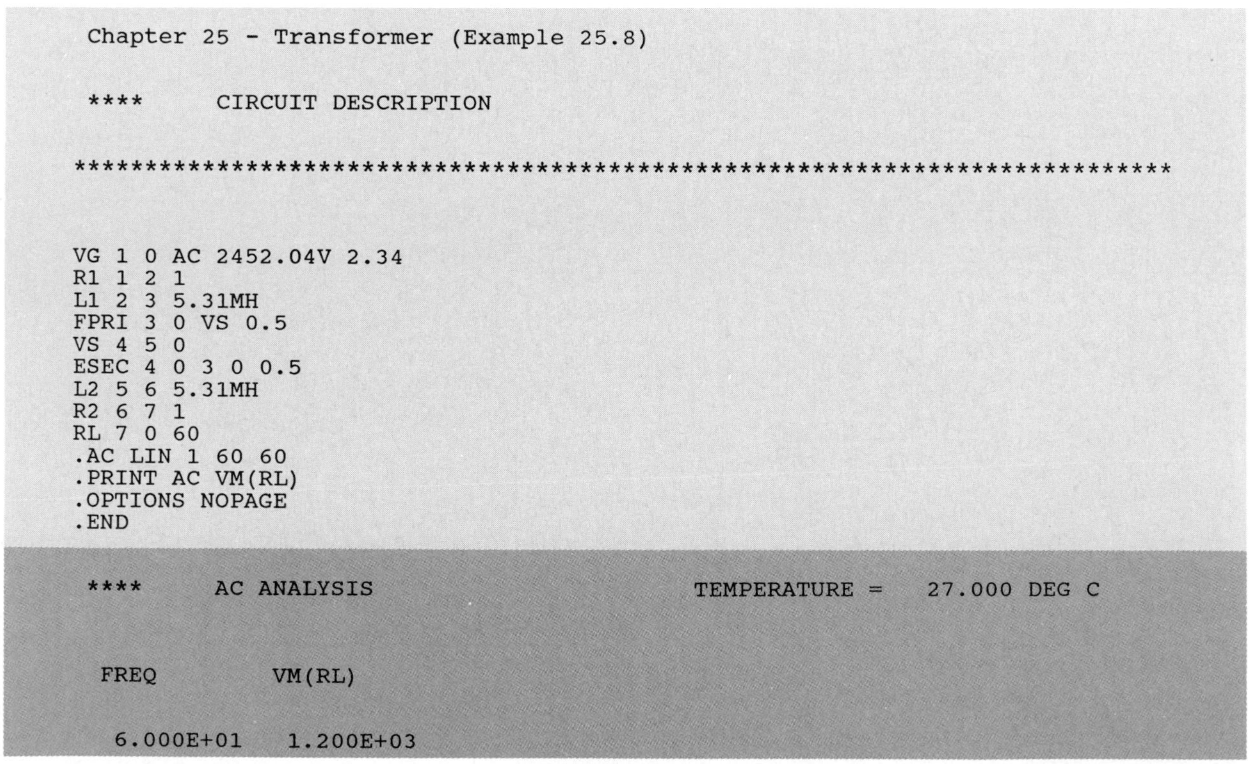

```
Chapter 25 - Transformer (Example 25.8)

****      CIRCUIT DESCRIPTION

******************************************************************************

VG 1 0 AC 2452.04V 2.34
R1 1 2 1
L1 2 3 5.31MH
FPRI 3 0 VS 0.5
VS 4 5 0
ESEC 4 0 3 0 0.5
L2 5 6 5.31MH
R2 6 7 1
RL 7 0 60
.AC LIN 1 60 60
.PRINT AC VM(RL)
.OPTIONS NOPAGE
.END

****      AC ANALYSIS                    TEMPERATURE =   27.000 DEG C

 FREQ          VM(RL)

  6.000E+01    1.200E+03
```

FIG. 25.52
Input and output files for the PSpice (DOS) analysis of the network of Example 25.7.

PSpice (Windows)

A schematics analysis of the network of Fig. 25.48 will result in the network of Fig. 25.53 with two controlled sources. The current-controlled current source **(CCCS) F1** has the symbol **F** under the **analog.slb** library file. The voltage-controlled voltage source **(VCVS) E1** has the symbol **E** under the same library file. For each, the magnitude of the gain is set by simply double-clicking the symbol on the schematic and entering the gain for the proper attribute. Be sure to save the attribute! The network introduced is complex because it has two controlled sources, so the **GLOBAL** option was chosen from the **port.slb** library file to provide the proper connections for the sensing current of **F1**. The label **Is** must be applied to each by choosing **GLOBAL** again and giving it the proper **LABEL.** All points labeled **Is** will now be connected directly without the necessity of drawing a physical wire. Note that the current directions of the current-controlled source and the voltage polarities of the voltage-controlled source match those of the original network under investigation. The output file of Fig. 25.54 provides the same results as those obtained using DOS and a longhand solution.

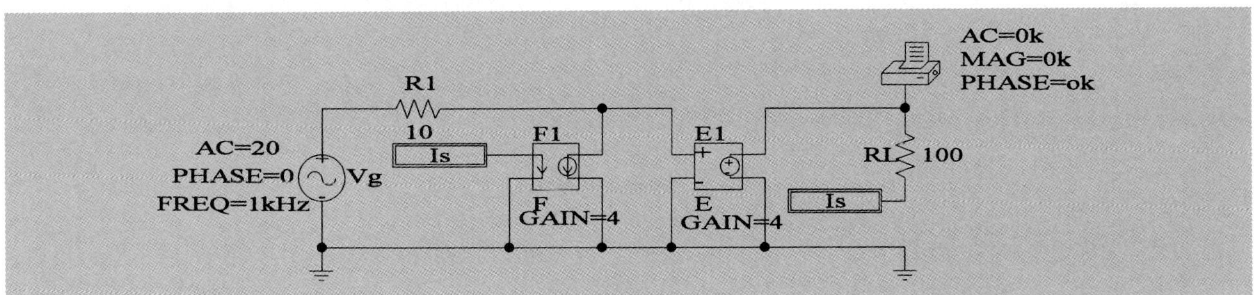

FIG. 25.53
Determining the magnitude and phase of the load voltage for the network of Fig. 25.48.

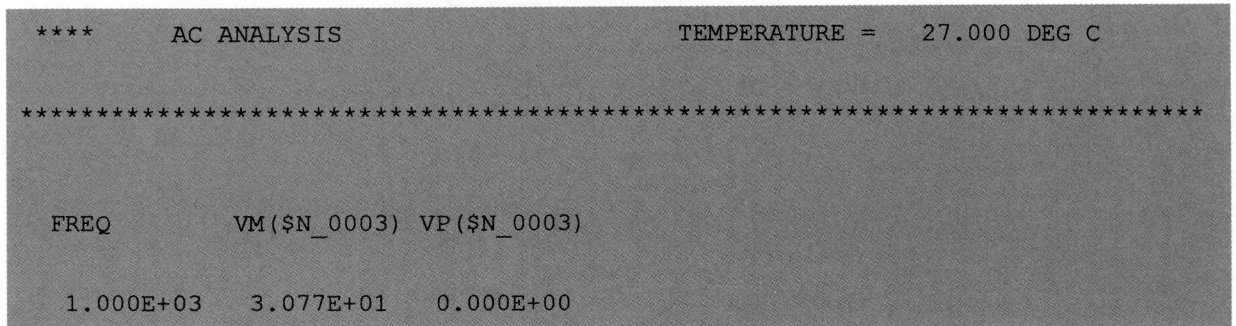

```
****     AC ANALYSIS                              TEMPERATURE =   27.000 DEG C

******************************************************************************

 FREQ        VM($N_0003) VP($N_0003)

  1.000E+03   3.077E+01   0.000E+00
```

FIG. 25.54
Output results for the analysis indicated in Fig. 25.48.

The same network can be analyzed by choosing one of the transformers from the **eval.slb** library file, as shown in Fig. 25.55. The first listed transformer labeled **K3019PL_3C8** was chosen and the proper attributes were set by double-clicking the device symbol on the

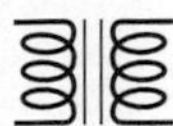

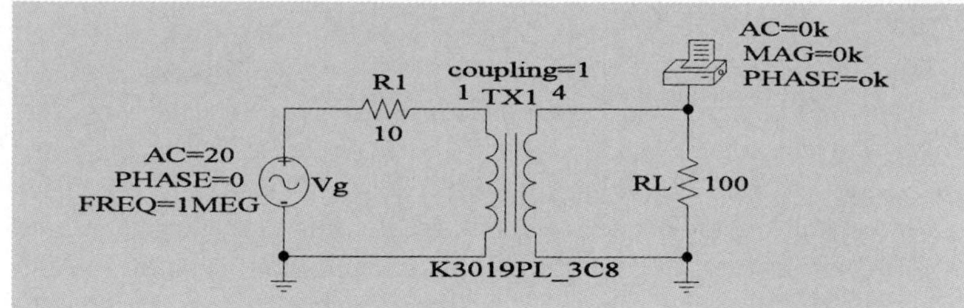

FIG. 25.55
Using a provided transformer of eval.slb to analyze the network of Fig. 25.48.

schematic. The **coupling** was set at 1 with **L1 TURNS** at 1 and **L2 TURNS** at 4. The choice was made using **Change Display** to list all three values on the schematic. At a frequency of 1 MHz, the transformer acted like an ideal transformer and provided the results of Fig. 25.56.

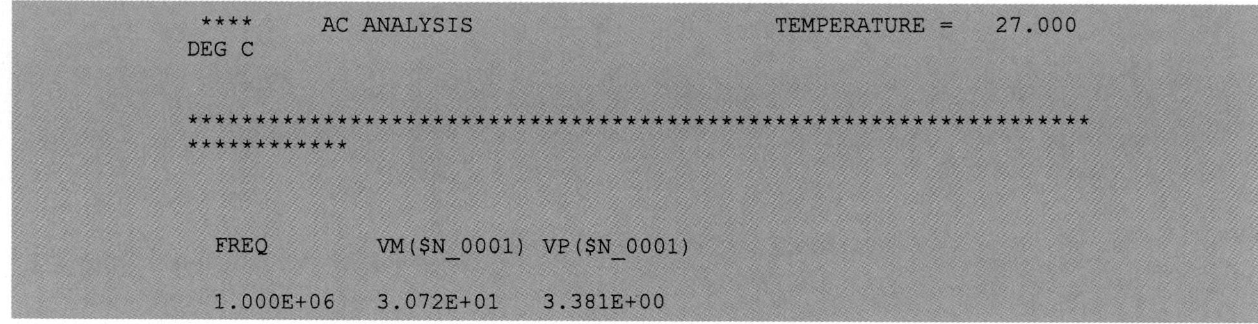

```
****      AC ANALYSIS                        TEMPERATURE =   27.000
DEG C

**********************************************************************
************

  FREQ          VM($N_0001) VP($N_0001)

  1.000E+06   3.072E+01   3.381E+00
```

FIG. 25.56
Output file for the analysis indicated in Fig. 25.48.

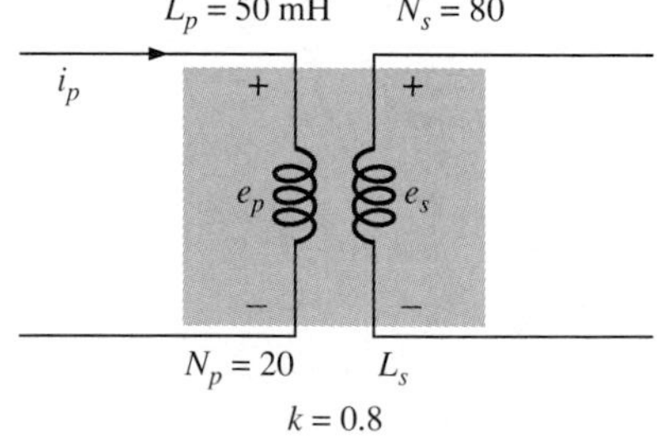

FIG. 25.57
Problems 1, 2, and 3.

PROBLEMS

SECTION 25.2 Mutual Inductance

1. For the air-core transformer of Fig. 25.57:
   a. Find the value of L_s if the mutual inductance M is equal to 80 mH.
   b. Find the induced voltages e_p and e_s if the flux linking the primary coil changes at the rate of 0.08 Wb/s.
   c. Find the induced voltages e_p and e_s if the current i_p changes at the rate of 0.3 A/ms.
2. a. Repeat Problem 1 if k is changed to 1.
   b. Repeat Problem 1 if k is changed to 0.2.
   c. Compare the results of parts (a) and (b).
3. Repeat Problem 1 for $k = 0.9$, $N_p = 300$ turns, and $N_s = 25$ turns.

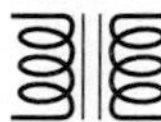

SECTION 25.3 Series Connection of Mutually Coupled Coils

4. Determine the total inductance of the series coils of Fig. 25.58.

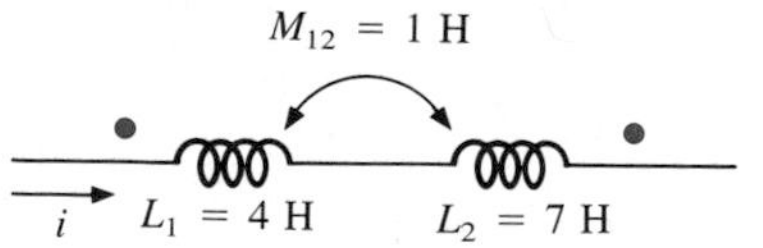

FIG. 25.58
Problem 4.

5. Determine the total inductance of the series coils of Fig. 25.59.

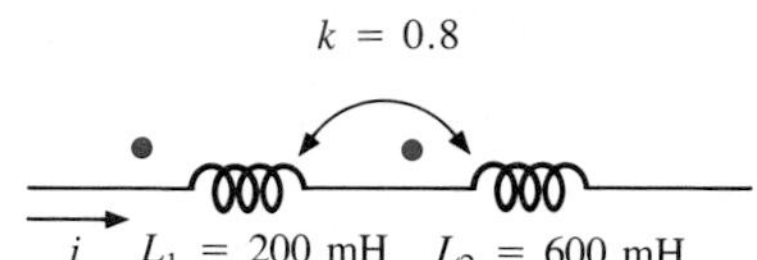

FIG. 25.59
Problem 5.

6. Determine the total inductance of the series coils of Fig. 25.60.

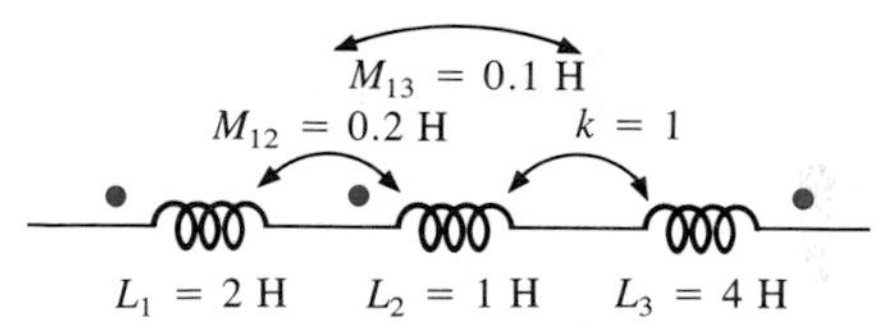

FIG. 25.60
Problem 6.

7. Write the mesh equations for the network of Fig. 25.61.

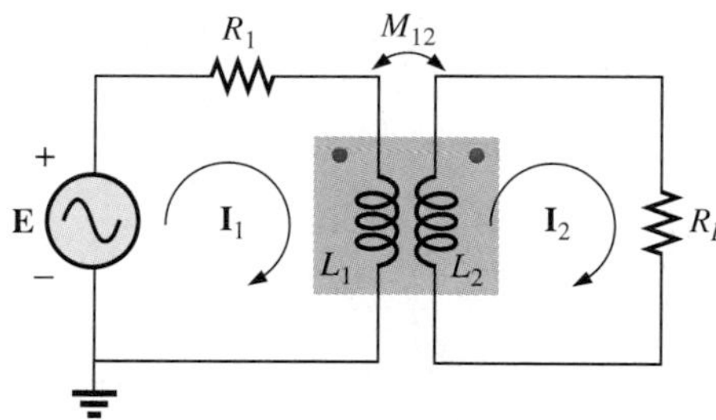

FIG. 25.61
Problem 7.

SECTION 25.4 The Iron-Core Transformer

8. For the iron-core transformer ($k = 1$) of Fig. 25.62:

a. Find the magnitude of the induced voltage E_s.

b. Find the maximum flux Φ_m.

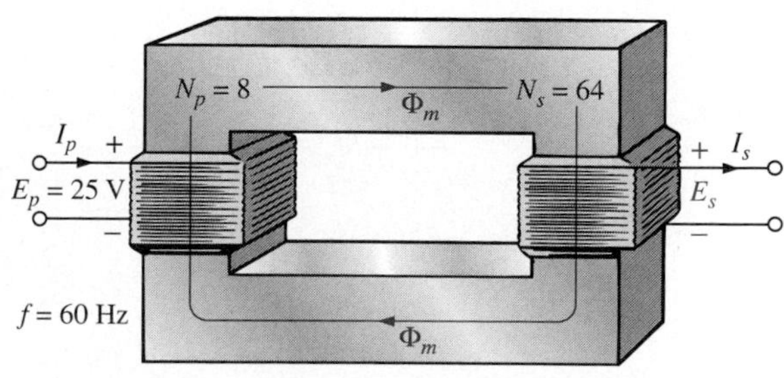

FIG. 25.62
Problems 8, 9, and 11.

9. Repeat Problem 8 for $N_p = 240$ and $N_s = 30$.

10. Find the applied voltage of an iron-core transformer with a secondary voltage of 240 V, and $N_p = 60$ with $N_s = 720$.

11. If the maximum flux passing through the core of Problem 8 is 12.5 mWb, find the frequency of the input voltage.

SECTION 25.5 Reflected Impedance and Power

12. For the iron-core transformer of Fig. 25.63:
    a. Find the magnitude of the current I_L and the voltage V_L if $a = 1/5$, $I_p = 2$ A, and $Z_L = 2\text{-}\Omega$ resistor.
    b. Find the input resistance for the data specified in part (a).

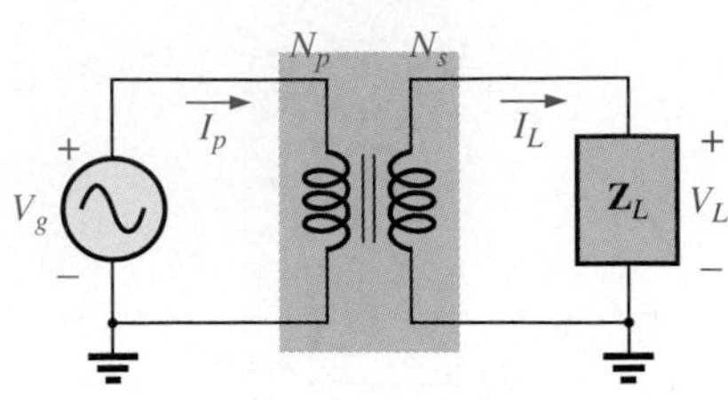

FIG. 25.63
Problems 12 to 16.

13. Find the input impedance for the iron-core transformer of Fig. 25.63 if $a = 2$, $I_p = 4$ A, and $V_g = 1600$ V.

14. Find the voltage V_g and the current I_p if the input impedance of the iron-core transformer of Fig. 25.63 is 4 Ω, and $V_L = 1200$ V and $a = 1/4$.

15. If $V_L = 240$ V, $Z_L = 20\text{-}\Omega$ resistor, $I_p = 0.05$ A, and $N_s = 50$, find the number of turns in the primary circuit of the iron-core transformer of Fig. 25.63.

16. a. If $N_p = 400$, $N_s = 1200$, and $V_g = 100$ V, find the magnitude of I_p for the iron-core transformer of Fig. 25.63 if $Z_L = 9\ \Omega + j\ 12\ \Omega$.
    b. Find the magnitude of the voltage V_L and the current I_L for the conditions of part (a).

SECTION 25.6 Equivalent Circuit (Iron-Core Transformer)

17. For the transformer of Fig. 25.64, determine
    a. the equivalent resistance R_e.
    b. the equivalent reactance X_e.
    c. the equivalent circuit reflected to the primary.
    d. the primary current for $\mathbf{V}_g = 50\text{ V}\ \angle 0°$.
    e. the load voltage V_L.
    f. the phasor diagram of the reflected primary circuit.
    g. the new load voltage if we assume the transformer to be ideal with a 4:1 turns ratio. Compare the result with that of part (e).

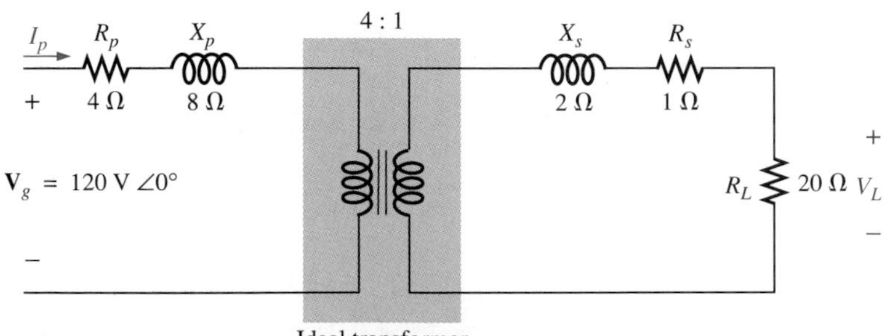

FIG. 25.64
Problems 17, 30, 33, and 34.

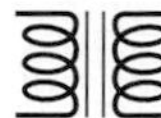

18. For the transformer of Fig. 25.65, if the resistive load is replaced by an inductive reactance of 20 Ω:
    a. Determine the total reflected primary impedance.
    b. Calculate the primary current.
    c. Determine the voltage across R_e and X_e, and the reflected load.
    d. Draw the phasor diagram.

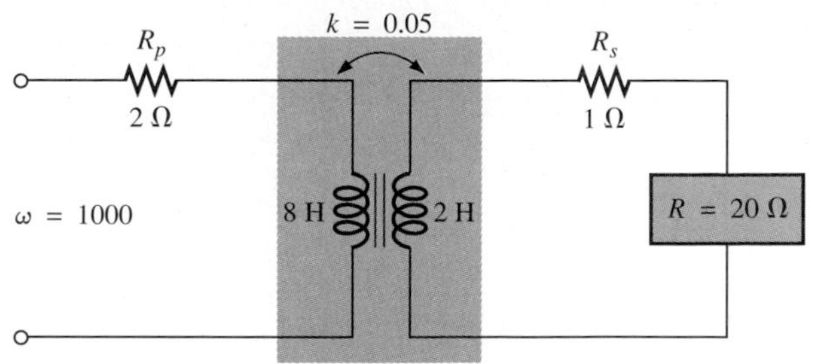

FIG. 25.65
Problems 18, 19, and 35.

19. Repeat Problem 18 for a capacitive load having a reactance of 20 Ω.

SECTION 25.7 Frequency Considerations

20. Discuss in your own words the frequency characteristics of the transformer. Employ the applicable equivalent circuit and frequency characteristics appearing in this chapter.

SECTION 25.8 Air-Core Transformer

21. Determine the input impedance to the air-core transformer of Fig. 25.66. Sketch the reflected primary network.

SECTION 25.9 Impedance Matching, Isolation, and Displacement

22. a. For the circuit of Fig. 25.66, find the transformation ratio required to deliver maximum power to the speaker.
    b. Find the maximum power delivered to the speaker.

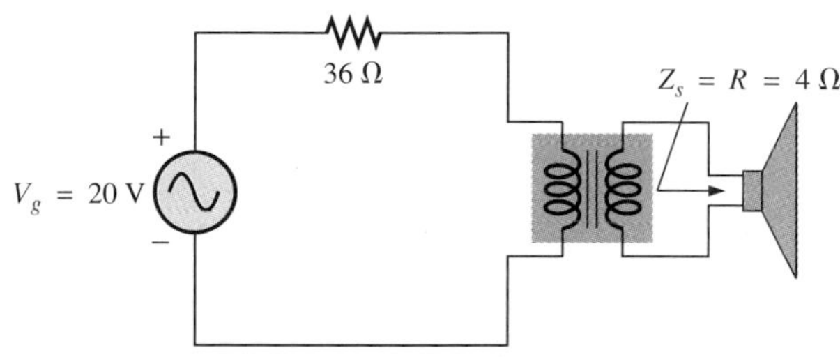

FIG. 25.66
Problems 21 and 22.

SECTION 25.10 Nameplate Data

23. An ideal transformer is rated 10 kVA, 2400/120 V, 60 Hz.
    a. Find the transformation ratio if the 120 V is the secondary voltage.
    b. Find the current rating of the secondary if the 120 V is the secondary voltage.
    c. Find the current rating of the primary if the 120 V is the secondary voltage.
    d. Repeat parts (a) through (c) if the 2400 V is the secondary voltage.

SECTION 25.11 Types of Transformers

24. Determine the primary and secondary voltages and currents for the autotransformer of Fig. 25.67.

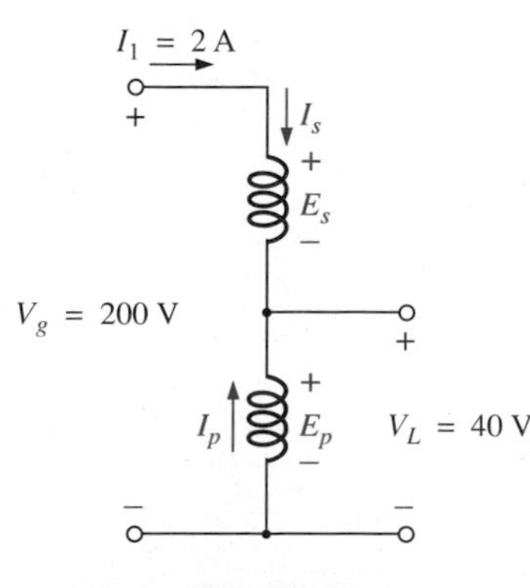

FIG. 25.67
Problem 24.

SECTION 25.12 Tapped and Multiple-Load Transformers

25. For the center-tapped transformer of Fig. 25.42 where $N_p = 100$, $N_s = 25$, $Z_L = R \angle 0° = 5\ \Omega \angle 0°$, and $\mathbf{E}_p = 100\ \text{V} \angle 0°$:
- **a.** Determine the load voltage and current.
- **b.** Find the impedance $\mathbf{Z}_i$.
- **c.** Calculate the impedance $\mathbf{Z}_{1/2}$.

26. For the multiple-load transformer of Fig. 25.43 where $N_1 = 90$, $N_2 = 15$, $N_3 = 45$, $\mathbf{Z}_2 = R_2 \angle 0° = 8\ \Omega \angle 0°$, $\mathbf{Z}_3 = R_L \angle 0° = 5\ \Omega \angle 0°$, and $\mathbf{E}_i = 60\ \text{V} \angle 0°$:
- **a.** Determine the load voltages and currents.
- **b.** Calculate $\mathbf{Z}_1$.

27. For the multiple-load transformer of Fig. 25.44 where $N_1 = 120$, $N_2 = 40$, $N_3 = 30$, $\mathbf{Z}_2 = R_2 \angle 0° = 12\ \Omega \angle 0°$, $\mathbf{Z}_3 = R_3 \angle 0° = 10\ \Omega \angle 0°$, and $\mathbf{E}_1 = 120\ \text{V} \angle 60°$:
- **a.** Determine the load voltages and currents.
- **b.** Calculate $\mathbf{Z}_1$.

SECTION 25.13 Networks with Magnetically Coupled Coils

28. Write the mesh equations for the network of Fig. 25.68.

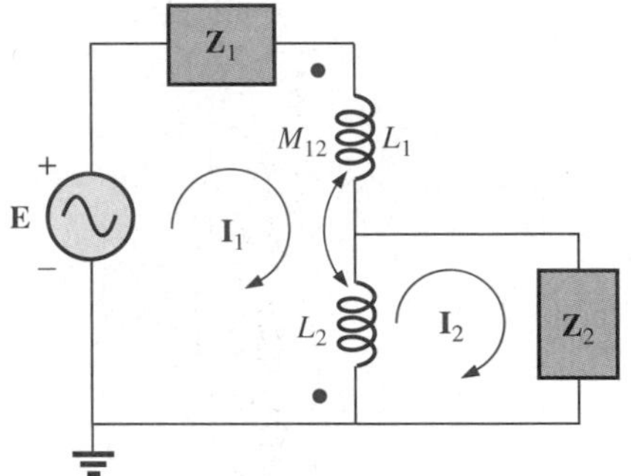

FIG. 25.68
Problem 28.

29. Write the mesh equations for the network of Fig. 25.69.

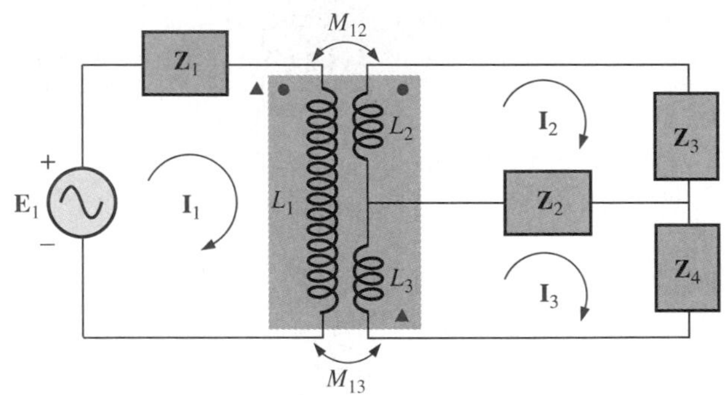

FIG. 25.69
Problem 29.

SECTION 25.14 Computer Analysis

PSpice (DOS)

***30.** Write the input file to determine the magnitude and phase of the voltage $\mathbf{V}_L$ for the network of Fig. 25.64.

***31.** Write the input file for the network of Fig. 25.9 using the equations developed in the example to establish a network model. Print out the voltage $\mathbf{V}_L$ for $\mathbf{E}_1 = 100\ \text{V}\ \angle 0°$, $R_L = 100\ \Omega$, $R_1 = 10\ \Omega$, $L_1 = 2\ \text{H}$, $L_2 = 5\ \text{H}$, $M = 0.9\ \text{H}$, and $f = 60\ \text{Hz}$.

***32.** Write the input file to determine the input impedance $\mathbf{Z}_i$ for the air-core transformer of Fig. 25.26. Use the technique of earlier chapters to find the impedance level, and use Eqs. (25.33) and (25.35) to establish a network model for the system.

PSpice (Windows)

***33.** Generate the schematic for the network of Fig. 25.64 and find the voltage V_L.

***34.** Develop a technique using Windows PSpice to find the input impedance at the source for the network of Fig. 25.64.

***35.** Use a transformer from the eval.slb library and find the load voltage for the network of Fig. 25.65 for an applied voltage of 40 V $\angle 0°$.

Programming Language (C++, BASIC, PASCAL, etc.)

36. Write a program to provide a general solution to the problem of impedance matching as defined by Example 25.9. That is, given the speaker impedance and the internal resistance of the source, determine the required turns ratio and the power delivered to the speaker. In addition, calculate the load and source current and the primary and secondary voltages. The source voltage will have to be provided with the other parameters of the network.

37. Given the equivalent model of an iron-core transformer appearing in Fig. 25.19, write a program to calculate the magnitude of the voltage $\mathbf{V}_g$.

38. Given the parameters of Example 25.7, write a program to calculate the input impedance in polar form.

GLOSSARY

Autotransformer A transformer with one winding common to both the primary and the secondary circuits. A loss in isolation is balanced by the increase in its kilovolt-ampere rating.

Coefficient of coupling (k) A measure of the magnetic coupling of two coils that ranges from a minimum of zero to a maximum of 1.

Dot convention A technique for labeling the effect of the mutual inductance on a net inductance of a network or system.

Leakage flux The flux linking the coil that does not pass through the ferromagnetic path of the magnetic circuit.

Loosely coupled A term applied to two coils that have a low coefficient of coupling.

Multiple-load transformers Transformers having more than a single load connected to the secondary winding or windings.

Mutual inductance The inductance that exists between magnetically coupled coils of the same or different dimensions.

Nameplate data Information such as the kilovolt-ampere rating, voltage transformation ratio, and frequency of application that is of primary importance in choosing the proper transformer for a particular application.

Primary The coil or winding to which the source of electrical energy is normally applied.

Reflected impedance The impedance appearing at the primary of a transformer due to a load connected to the secondary. Its magnitude is controlled directly by the transformation ratio.

Secondary The coil or winding to which the load is normally applied.

Step-down transformer A transformer whose secondary voltage is less than its primary voltage. The transformation ratio a is greater than 1.

Step-up transformer A transformer whose secondary voltage is greater than its primary voltage. The magnitude of the transformation ratio a is less than 1.

Tapped transformer A transformer having an additional connection between the terminals of the primary or secondary windings.

Transformation ratio (a) The ratio of primary to secondary turns of a transformer.

26

System Analysis—An Introduction

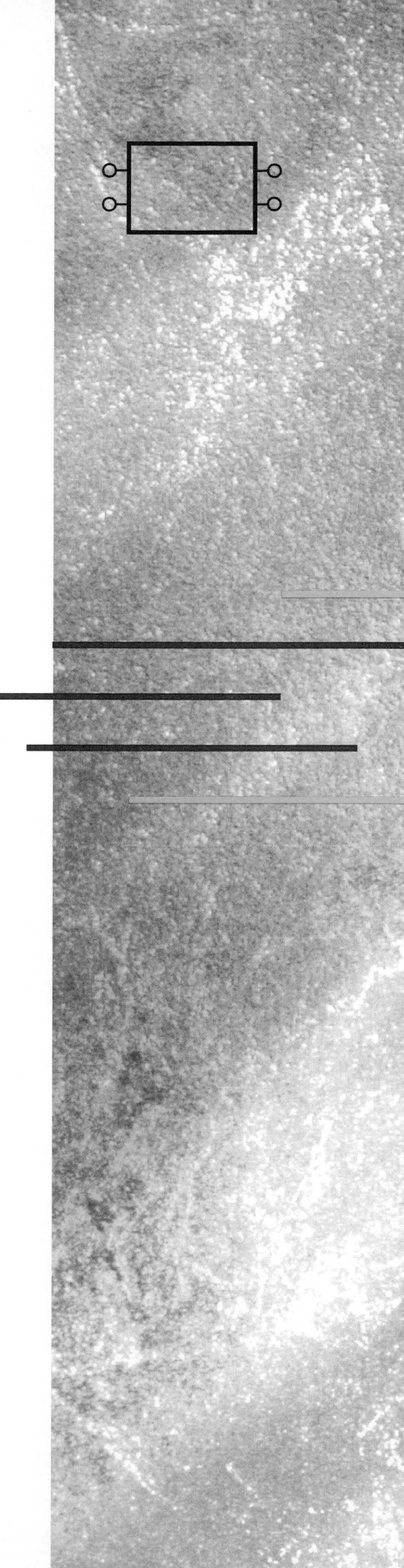

26.1 INTRODUCTION

The growing number of packaged systems in the electrical, electronic, and computer fields now requires that some form of system analysis appear in the syllabus of the introductory course. Although the content of this chapter will be a surface treatment at best, the material will introduce a number of important terms and techniques employed in the system-analysis approach. The increasing use of packaged systems is quite understandable when we consider the advantages associated with such structures: reduced size, sophisticated and tested design, reduced construction time, reduced cost compared to discrete designs, and so forth. The integrated circuit of Fig. 1.1 is an excellent example of the current state of the art due to its 1.2 million transistors in a structure with dimensions less than one-half inch. The MC68040 is a microprocessor that is the heart of a number of micro- and minicomputers. The use of any packaged system is limited solely to the proper utilization of the provided terminals of the system. Entry into the internal structure is not permitted, which also eliminates the possibility of repair to such systems.

System analysis includes the development of two-, three-, or multiport models of devices, systems, or structures. The emphasis in this chapter will be on the configuration most frequently subject to modeling techniques: the two-port system of Fig. 26.1.

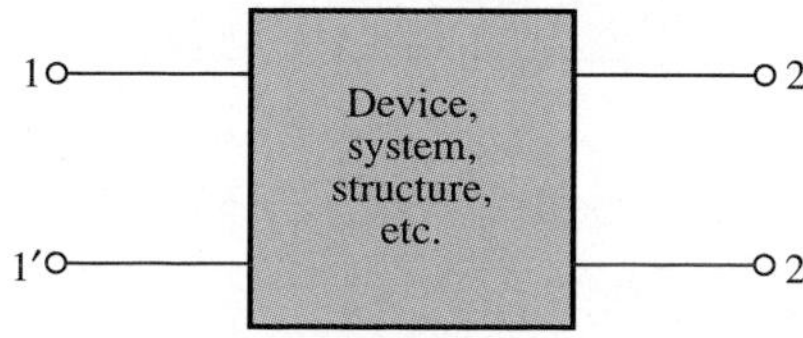

FIG. 26.1
Two-port system.

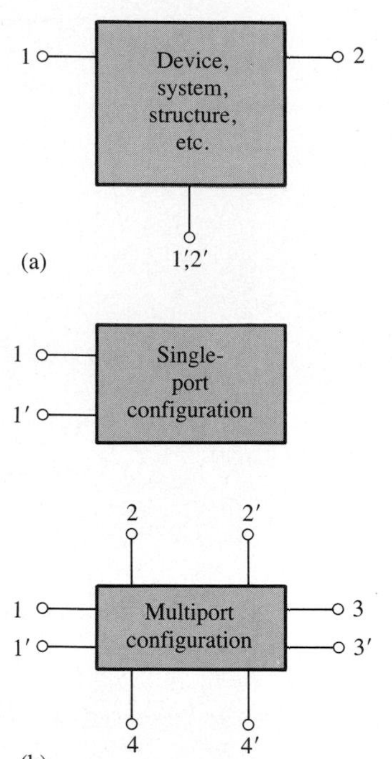

FIG. 26.2
(a) Two-port system, (b) single-port system, (c) multiport system.

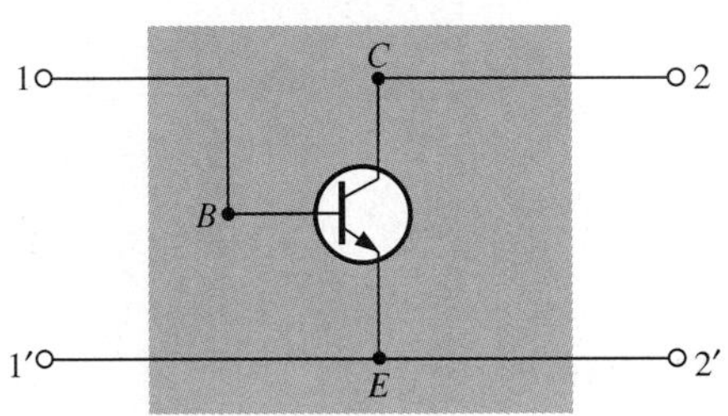

FIG. 26.3
Two-port transistor configuration.

Note that in Fig. 26.1 there are two ports of entry or interest, each having a pair of terminals. For some devices, the two-port configuration of Fig. 26.1 may appear as shown in Fig. 26.2(a). The block diagram of Fig. 26.2(a) simply indicates that terminals 1′ and 2′ are in common—a particular case of the general two-port network. A single-port and a multiport network appear in Fig. 26.2(b). The former has been analyzed throughout the text, while the characteristics of the latter will be touched on in this chapter, with a more extensive coverage left for a more advanced course.

The latter part of this chapter introduces a set of equations (and, subsequently, networks) that will allow us to model the device or system appearing within the enclosed structure of Fig. 26.1. That is, we will be able to establish a network that will display the same terminal characteristics as those of the original system, device, and so on. In Fig. 26.3, for example, a transistor appears between the four external terminals. Through the analysis to follow, we will find a combination of network elements that will allow us to replace the transistor with a network that will behave very much like the original device for a specific set of operating conditions. Methods such as mesh and nodal analysis can then be applied to determine any unknown quantities. The models, when reduced to their simplest forms as determined by the operating conditions, can also provide very quick estimates of network behavior without a lengthy mathematical derivation. In other words, someone well-versed in the use of models can analyze the operation of large, complex systems in short order. The results may be only approximate in most cases, but this quick return for a minimum of effort is often worthwhile.

The analysis of this chapter is limited to linear (fixed-value) systems with bilateral elements. Three sets of parameters are developed for the two-port configuration, referred to as the *impedance* (**z**), *admittance* (**y**), and *hybrid* (**h**) parameters. A table is provided at the end of the chapter relating the three sets of parameters.

26.2 THE IMPEDANCE PARAMETERS Z_i AND Z_o

For the two-port system of Fig. 26.4, $\mathbf{Z}_i$ is the input impedance between terminals 1 and 1′, and $\mathbf{Z}_o$ is the output impedance between terminals 2 and 2′. For multiport networks an impedance level can be defined between any two (adjacent or not) terminals of the network.

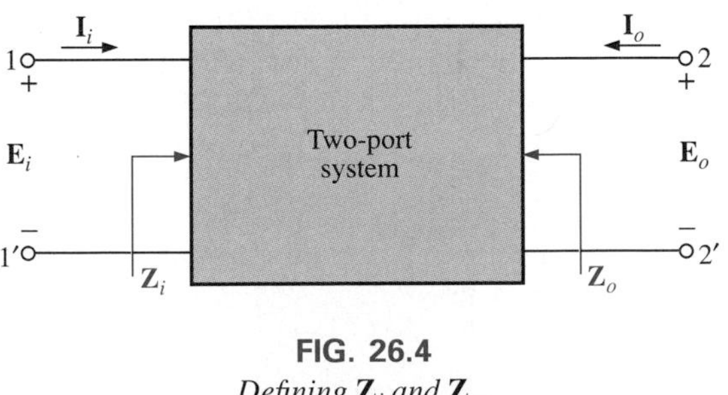

FIG. 26.4
Defining $\mathbf{Z}_i$ *and* $\mathbf{Z}_o$.

The input impedance is defined by Ohm's law in the following form:

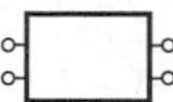

$$\mathbf{Z}_i = \frac{\mathbf{E}_i}{\mathbf{I}_i} \qquad \text{(ohms, } \Omega) \qquad \textbf{(26.1)}$$

with $\mathbf{I}_i$ the current resulting from the application of a voltage $\mathbf{E}_i$.

The output impedance $\mathbf{Z}_o$ is defined by

$$\mathbf{Z}_o = \frac{\mathbf{E}_o}{\mathbf{I}_o}\bigg|_{\mathbf{E}_i = 0\text{ V}} \qquad \text{(ohms, } \Omega) \qquad \textbf{(26.2)}$$

with $\mathbf{I}_o$ the current resulting from the application of a voltage $\mathbf{E}_o$ to the output terminals, with $\mathbf{E}_i$ set to zero.

Note that both $\mathbf{I}_i$ and $\mathbf{I}_o$ are defined as entering the package. This is common practice for a number of system analysis methods to avoid concern about the actual direction for each current and also to define $\mathbf{Z}_i$ and $\mathbf{Z}_o$ as positive quantities in Eqs. (26.1) and (26.2), respectively. If $\mathbf{I}_o$ were chosen to be leaving the system, $\mathbf{Z}_o$ as defined in Eq. (26.2) would have to have a negative sign.

An experimental setup for determining $\mathbf{Z}_i$ for any two input terminals is provided in Fig. 26.5. The sensing resistor R_s is chosen small enough not to disturb the basic operation of the system or require too large a voltage $\mathbf{E}_g$ to establish the desired level of $\mathbf{E}_i$. Under operating conditions, the voltage across R_s is $\mathbf{E}_g - \mathbf{E}_i$, and the current through the sensing resistor is

$$\mathbf{I}_{R_s} = \frac{\mathbf{V}_{R_s}}{R_s} = \frac{\mathbf{E}_g - \mathbf{E}_i}{R_s}$$

but

$$\mathbf{I}_i = \mathbf{I}_{R_s} \quad \text{and} \quad \mathbf{Z}_i = \frac{\mathbf{E}_i}{\mathbf{I}_i} = \frac{\mathbf{E}_i}{\mathbf{I}_{R_s}}$$

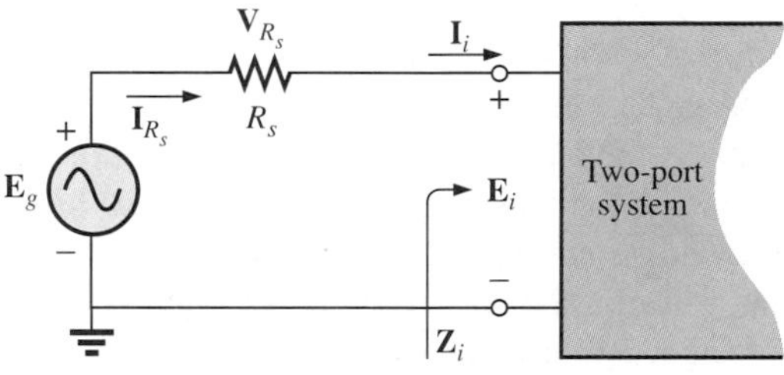

FIG. 26.5
Determining $\mathbf{Z}_i$.

The sole purpose of the sensing resistor, therefore, was to determine $\mathbf{I}_i$ using purely voltage measurements.

As we progress through this chapter, keep in mind that we cannot use an ohmmeter to measure $\mathbf{Z}_i$ or $\mathbf{Z}_o$ since we are dealing with ac systems whose impedance may be sensitive to the applied frequency. Ohmmeters can be used to measure resistance in a dc or ac network, but recall that ohmmeters are employed only on a de-energized network, and their internal source is a dc battery.

The output impedance $\mathbf{Z}_o$ can be determined experimentally using the setup of Fig. 26.6. Note that a sensing resistor is introduced again, with $\mathbf{E}_g$ being an applied voltage to establish typical operating conditions. In addition, note that the input signal must be set to zero, as defined by Eq.

(26.2). The voltage across the sensing resistor is $\mathbf{E}_g - \mathbf{E}_o$, and the current through the sensing resistor is

$$\mathbf{I}_{R_s} = \frac{\mathbf{V}_{R_s}}{R_s} = \frac{\mathbf{E}_g - \mathbf{E}_o}{R_s}$$

but $$\mathbf{I}_o = \mathbf{I}_{R_s} \quad \text{and} \quad \mathbf{Z}_o = \frac{\mathbf{E}_o}{\mathbf{I}_o} = \frac{\mathbf{E}_o}{\mathbf{I}_{R_s}}$$

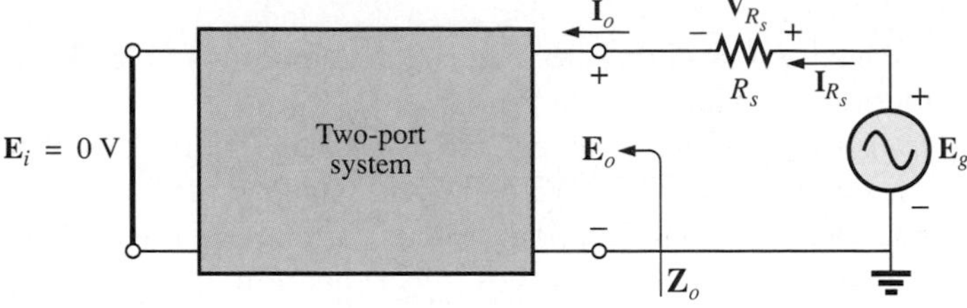

FIG. 26.6
Determining $\mathbf{Z}_o$.

For the majority of situations $\mathbf{Z}_i$ and $\mathbf{Z}_o$ will be purely resistive, resulting in an angle of zero degrees for each impedance. The result is that either a DMM or a scope can be used to find the required magnitude of the desired quantity. For instance, for both $\mathbf{Z}_i$ and $\mathbf{Z}_o$, $\mathbf{V}_{R_s}$ can be measured directly with the DMM, as can the required levels of $\mathbf{E}_g$, $\mathbf{E}_i$, or $\mathbf{E}_o$. The current for each case can then be determined using Ohm's law, and the impedance level can be determined using either Eq. (26.1) or Eq. (26.2).

If we use an oscilloscope, we must be more sensitive to the common ground requirement. For instance, in Fig. 26.4, $\mathbf{E}_g$ and $\mathbf{E}_i$ can be measured with the oscilloscope since they have a common ground. Trying to measure $\mathbf{V}_{R_s}$ directly with the ground of the oscilloscope at the top input terminal of $\mathbf{E}_i$ would result in a shorting effect across the input terminals of the system due to the common ground between the supply and oscilloscope. If the input impedance of the system is "shorted out," the current $\mathbf{I}_i$ can rise to dangerous levels because the only resistance in the input circuit is the relatively small sensing resistor R_s. If we use the DMM to avoid concern about the grounding situation, we must be sure the meter is designed to operate properly at the frequency of interest. Many commercial units are limited to a few kilohertz.

If the input impedance has an angle other than zero degrees (purely resistive), a DMM cannot be used to find the reactive component at the input terminals. The magnitude of the total impedance will be correct if measured as described above, but the angle from which the resistive and reactive components can be determined will not be provided. If an oscilloscope is used, the network must be hooked up as shown in Fig. 26.7. Both the voltage $\mathbf{E}_g$ and $\mathbf{V}_{R_s}$ can be displayed on the oscilloscope at the same time, and the phase angle between $\mathbf{E}_g$ and $\mathbf{V}_{R_s}$ can be determined. Since $\mathbf{V}_{R_s}$ and $\mathbf{I}_i$ are in phase, the angle determined will also be the angle between $\mathbf{E}_g$ and $\mathbf{I}_i$. The angle we are looking for is between $\mathbf{E}_i$ and $\mathbf{I}_i$, not between $\mathbf{E}_g$ and $\mathbf{I}_i$, but since R_s is usually chosen small enough, we can assume that the voltage drop across R_s is so small compared to $\mathbf{E}_g$ that $\mathbf{E}_i \cong \mathbf{E}_g$. Substituting the peak, peak-to-peak, or rms values from the oscilloscope measurements, along with the angle just determined, will permit a determination of the magnitude and angle for

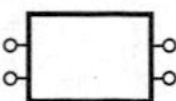

$\mathbf{Z}_i$, from which the resistive and reactive components can be determined using a few basic geometric relationships. The reactive nature (inductive or capacitive) of the input impedance can be determined when the angle between $\mathbf{E}_i$ and $\mathbf{I}_i$ is computed. For a dual-trace oscilloscope, if $\mathbf{E}_g$ leads $\mathbf{V}_{R_s}$ ($\mathbf{E}_i$ leads $\mathbf{I}_i$), the network is inductive; if the reverse is true, the network is capacitive.

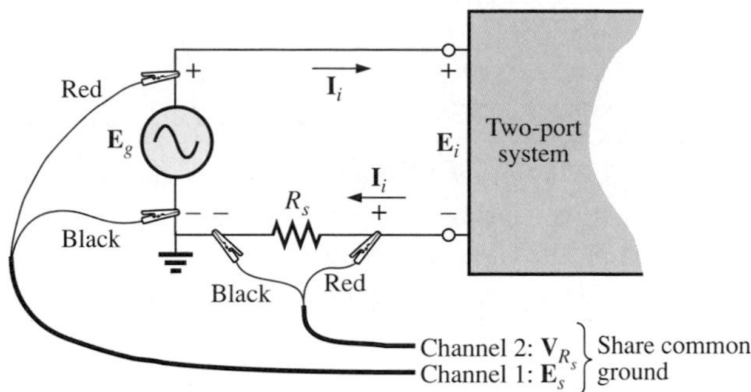

FIG. 26.7
Determining $\mathbf{Z}_i$ using an oscilloscope.

To determine the angle associated with $\mathbf{Z}_o$, the sensing resistor must again be moved to the bottom to form a common ground with the supply $\mathbf{E}_g$. Then, using the approximation $\mathbf{E}_g \cong \mathbf{E}_o$, the magnitude and angle of $\mathbf{Z}_o$ can be determined.

EXAMPLE 26.1 Given the DMM measurements appearing in Fig. 26.8, determine the input impedance $\mathbf{Z}_i$ for the system if the input impedance is known to be purely resistive.

Solution:

$$V_{R_s} = E_g - E_i = 100 \text{ mV} - 96 \text{ mV} = 4 \text{ mV}$$

$$I_i = I_{R_s} = \frac{V_{R_s}}{R_s} = \frac{4 \text{ mV}}{100 \ \Omega} = 40 \ \mu\text{A}$$

$$Z_i = R_i = \frac{E_i}{I_i} = \frac{96 \text{ mV}}{40 \ \mu\text{A}} = \mathbf{2.4 \ k\Omega}$$

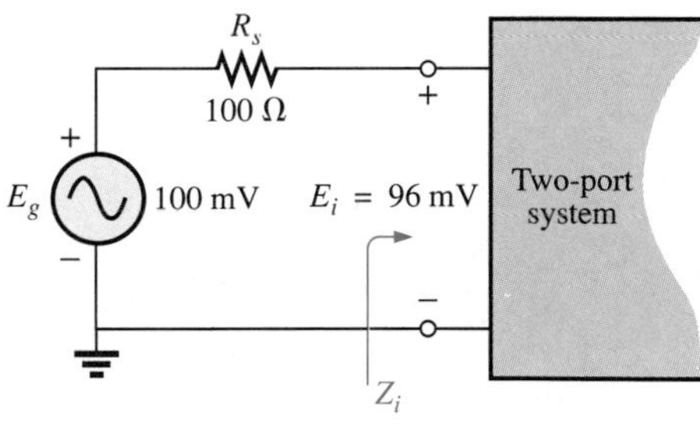

FIG. 26.8
Example 26.1.

EXAMPLE 26.2 Using the provided DMM measurements of Fig. 26.9, determine the output impedance Z_o for the system if the output impedance is known to be purely resistive.

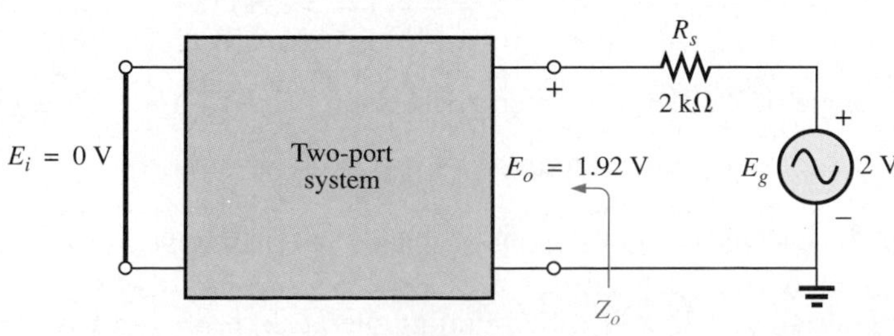

FIG. 26.9
Example 26.2.

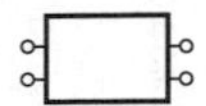

Solution:

$$V_{R_s} = E_g - E_o = 2\text{ V} - 1.92\text{ V} = 0.08\text{ V} = 80\text{ mV}$$

$$I_o = I_{R_s} = \frac{V_{R_s}}{R_s} = \frac{80\text{ mV}}{2\text{ k}\Omega} = 40\ \mu\text{A}$$

$$Z_o = \frac{E_o}{I_o} = \frac{1.92\text{ V}}{40\ \mu\text{A}} = \mathbf{48\ k\Omega}$$

EXAMPLE 26.3 The input characteristics for the system of Fig. 26.10(a) are unknown. Using the oscilloscope measurements of Fig. 26.10(b), determine the input impedance for the system. If a reactive component exists, determine its magnitude and whether it is inductive or capacitive.

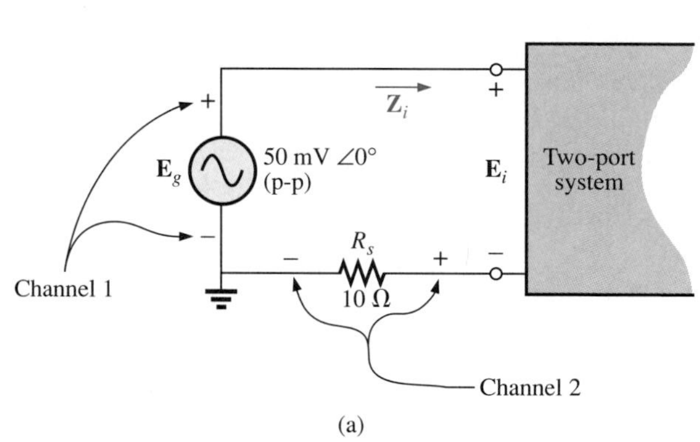

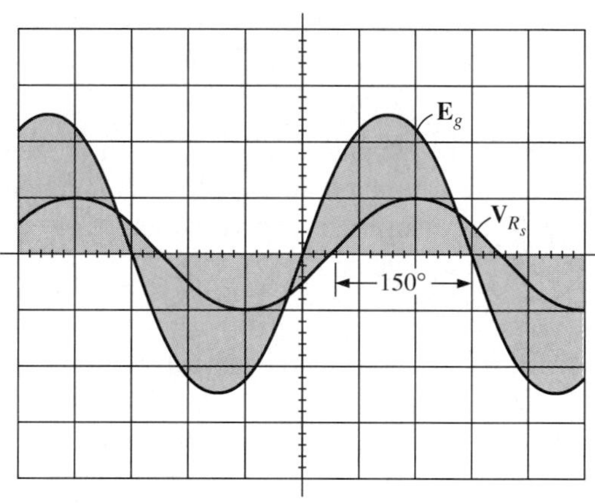

FIG. 26.10
Example 26.3.

Solution: The magnitude of $\mathbf{Z}_i$:

$$I_{i(p\text{-}p)} = I_{R_{s(p\text{-}p)}} = \frac{V_{R_{s(p\text{-}p)}}}{R_s} = \frac{2\text{ mV}}{10\ \Omega} = 200\ \mu\text{A}$$

$$Z_i = \frac{E_i}{I_i} \cong \frac{E_g}{I_i} = \frac{50\text{ mV}}{200\ \mu\text{A}} = 250\ \Omega$$

The angle of $\mathbf{Z}_i$: The phase angle between $\mathbf{E}_g$ and $\mathbf{V}_{R_s}$ (or $\mathbf{I}_{R_s} = \mathbf{I}_i$) is

$$180^\circ - 150^\circ = 30^\circ$$

with $\mathbf{E}_g$ leading $\mathbf{I}_i$, so the system is inductive. Therefore,

$$\mathbf{Z}_i = 250\ \Omega\ \angle 30^\circ$$
$$= \mathbf{216.51\ \Omega + \mathit{j}\,125\ \Omega} = R + jX_L$$

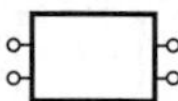

26.3 THE VOLTAGE GAINS $\mathbf{A}_{v_{NL}}$, $\mathbf{A}_v$, AND $\mathbf{A}_{v_T}$

The voltage gain for the two-port system of Fig. 26.11 is defined by

$$\mathbf{A}_{v_{NL}} = \frac{\mathbf{E}_o}{\mathbf{E}_i} \tag{26.3}$$

FIG. 26.11
Defining the no-load gain $\mathbf{A}_{v_{NL}}$.

The capital letter **A** in the notation was chosen from the term *a*mplification factor, with the subscript v selected to specify that voltage levels are involved. The subscript *NL* reveals that the ratio was determined under *no-load* conditions; that is, a load was not applied to the output terminals when the gain was determined. The no-load voltage gain is the gain typically provided with packaged systems since the applied load is a function of the application.

The magnitude of the ratio can be determined using a DMM or an oscilloscope. The oscilloscope, however, must be used to determine the phase shift between the two voltages.

In Fig. 26.12 a load has been introduced to establish a loaded gain that will be denoted simply as $\mathbf{A}_v$ and defined by

$$\mathbf{A}_v = \left.\frac{\mathbf{E}_o}{\mathbf{E}_i}\right|_{\text{with } R_L} \tag{26.4}$$

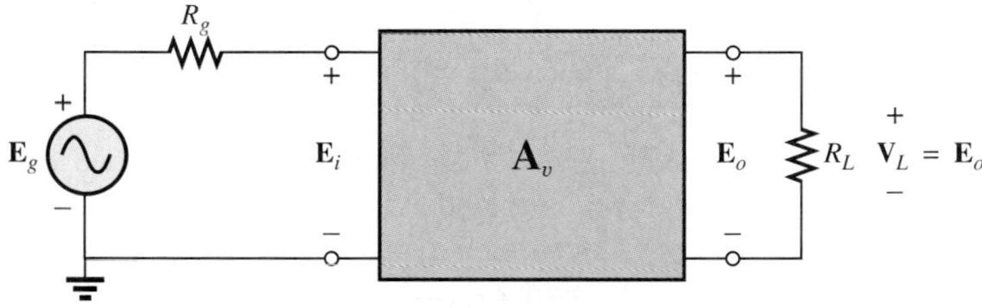

FIG. 26.12
Defining the loaded voltage gain $\mathbf{A}_v$ (and $\mathbf{A}_{v_T}$).

For all two-port systems the loaded gain $\mathbf{A}_v$ will always be less than the no-load gain.

In other words, the application of a load will always reduce the gain below the no-load level.

There is a third voltage gain that can be defined using Fig. 26.12 since it has an applied voltage source with an associated internal resistance—a situation often encountered in electronic systems. The total voltage gain of the system is represented by $\mathbf{A}_{v_T}$ and is defined by

$$\mathbf{A}_{v_T} = \frac{\mathbf{E}_o}{\mathbf{E}_g} \tag{26.5}$$

It is the voltage gain from the source $\mathbf{E}_g$ to the output terminals $\mathbf{E}_o$. Due to loss of signal voltage across the source resistance,

the voltage gain $\mathbf{A}_{v_T}$ is always less than the loaded voltage gain $\mathbf{A}_v$ or unloaded gain $\mathbf{A}_{v_{NL}}$.

If we expand Eq. (26.5) as follows:

$$\mathbf{A}_{v_T} = \frac{\mathbf{E}_o}{\mathbf{E}_g} = \frac{\mathbf{E}_o}{\mathbf{E}_g}(1) = \frac{\mathbf{E}_o}{\mathbf{E}_g}\left(\frac{\mathbf{E}_i}{\mathbf{E}_i}\right) = \frac{\mathbf{E}_o}{\mathbf{E}_i} \cdot \frac{\mathbf{E}_i}{\mathbf{E}_g}$$

then $$\mathbf{A}_{v_T} = \mathbf{A}_v \frac{\mathbf{E}_i}{\mathbf{E}_g} \quad \text{(if loaded)}$$

or $$\mathbf{A}_{v_T} = \mathbf{A}_{v_{NL}} \frac{\mathbf{E}_i}{\mathbf{E}_g} \quad \text{(if unloaded)}$$

The relationship between $\mathbf{E}_i$ and $\mathbf{E}_g$ can be determined from Fig. 26.12 if we recognize that $\mathbf{E}_i$ is across the input impedance $\mathbf{Z}_i$ and thus apply the voltage divider rule as follows:

$$\mathbf{E}_i = \frac{\mathbf{Z}_i(\mathbf{E}_g)}{\mathbf{Z}_i + R_g}$$

or $$\frac{\mathbf{E}_i}{\mathbf{E}_g} = \frac{\mathbf{Z}_i}{\mathbf{Z}_i + R_g}$$

Substituting into the above relationships will result in

$$\mathbf{A}_{v_T} = \mathbf{A}_v \frac{\mathbf{Z}_i}{\mathbf{Z}_i + R_g} \quad \text{(if loaded)} \qquad \textbf{(26.6)}$$

$$\mathbf{A}_{v_T} = \mathbf{A}_{v_{NL}} \frac{\mathbf{Z}_i}{\mathbf{Z}_i + R_g} \quad \text{(if unloaded)} \qquad \textbf{(26.7)}$$

A two-port equivalent model for an unloaded system based on the definitions of $\mathbf{Z}_i$, $\mathbf{Z}_o$, and $\mathbf{A}_{v_{NL}}$ is provided in Fig. 26.13. Both $\mathbf{Z}_i$ and $\mathbf{Z}_o$ appear as resistive values since this is typically the case for most electronic amplifiers. However, both $\mathbf{Z}_i$ and $\mathbf{Z}_o$ can have reactive components and not invalidate the equivalency of the model.

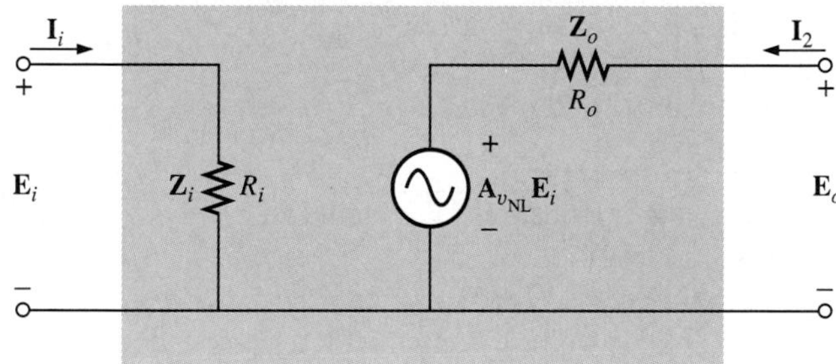

FIG. 26.13
Equivalent model for two-port amplifier.

The input impedance is defined by $\mathbf{Z}_i = \mathbf{E}_i/\mathbf{I}_i$ and the voltage $\mathbf{E}_o = \mathbf{A}_{v_{NL}}\mathbf{E}_i$ in the absence of a load, resulting in $\mathbf{A}_{v_{NL}} = \mathbf{E}_o/\mathbf{E}_i$ as defined. The output impedance is defined with $\mathbf{E}_i$ set to zero volts, resulting in $\mathbf{A}_{v_{NL}}\mathbf{E}_i = 0$ V, which permits the use of a short-circuit equivalent for the controlled source. The result is $\mathbf{Z}_o = \mathbf{E}_o/\mathbf{I}_o$, as defined, and the parameters and structure of the equivalent model are validated.

If a load is applied as in Fig. 26.14, an application of the voltage divider rule will result in

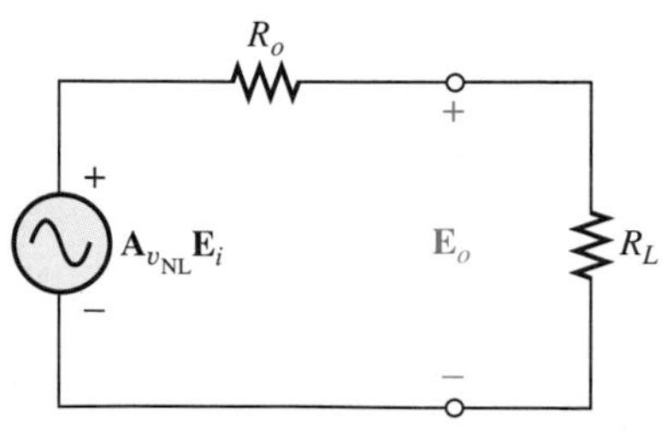

FIG. 26.14
Applying load to the output of Fig. 26.13.

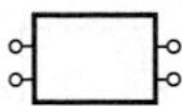

$$\mathbf{E}_o = \frac{R_L(\mathbf{A}_{v_{NL}}\mathbf{E}_i)}{R_L + R_o}$$

and

$$\boxed{\mathbf{A}_v = \frac{\mathbf{E}_o}{\mathbf{E}_i} = A_{v_{NL}}\frac{R_L}{R_L + R_o}} \tag{26.8}$$

For any value of R_L or R_o, the ratio $R_L/(R_L + R_o)$ must be less than 1, mandating that $\mathbf{A}_v$ is always less than $\mathbf{A}_{v_{NL}}$ as stated earlier. Further,

for a fixed output impedance (R_o), the larger the load resistance (R_L), the closer the loaded gain to the no-load level.

An experimental procedure for determining R_o can be developed if we solve Eq. (26.8) for the output impedance R_o:

$$\mathbf{A}_v = \frac{R_L}{R_L + R_o}\mathbf{A}_{v_{NL}}$$

or

$$\mathbf{A}_v(R_L + R_o) = R_L\mathbf{A}_{v_{NL}}$$

$$\mathbf{A}_vR_L + \mathbf{A}_vR_o = R_L\mathbf{A}_{v_{NL}}$$

and

$$\mathbf{A}_vR_o = R_L\mathbf{A}_{v_{NL}} - \mathbf{A}_vR_L$$

with

$$R_o = \frac{R_L(\mathbf{A}_{v_{NL}} - \mathbf{A}_v)}{\mathbf{A}_v}$$

or

$$\boxed{R_o = R_L\left(\frac{\mathbf{A}_{v_{NL}}}{\mathbf{A}_v} - 1\right)} \tag{26.9}$$

Equation (26.9) reveals that the output impedance R_o of an amplifier can be determined by first measuring the voltage gain $\mathbf{E}_o/\mathbf{E}_i$ without a load in place to find $\mathbf{A}_{v_{NL}}$ and then measuring the gain with a load R_L to find $\mathbf{A}_v$. Substitution of $\mathbf{A}_{v_{NL}}$, $\mathbf{A}_v$, and R_L into Eq. (26.9) will then provide the value for R_o.

EXAMPLE 26.4 For the system of Fig. 26.15(a) employed in the loaded amplifier of Fig. 26.15(b):

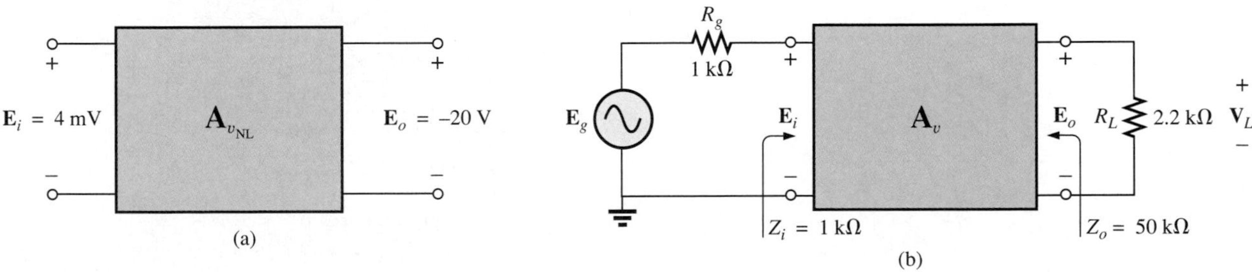

FIG. 26.15
Example 26.4.

a. Determine the no-load voltage gain $\mathbf{A}_{v_{NL}}$.
b. Find the loaded voltage gain $\mathbf{A}_v$.
c. Calculate the loaded voltage gain $\mathbf{A}_{v_T}$.
d. Determine R_o from Eq. (26.9) and compare it to the specified value of Fig. 26.15.

Solutions:

a. $\mathbf{A}_{v_{NL}} = \dfrac{\mathbf{E}_o}{\mathbf{E}_i} = \dfrac{-20\text{ V}}{4\text{ mV}} = \mathbf{-5000}$

b. $\mathbf{A}_v = \mathbf{A}_{v_{NL}} \dfrac{R_L}{R_L + R_o} = (-5000)\left(\dfrac{2.2\text{ k}\Omega}{2.2\text{ k}\Omega + 50\text{ k}\Omega}\right)$

$= (-5000)(0.0421) = \mathbf{-210.73}$

c. $\mathbf{A}_{v_T} = \mathbf{A}_v \dfrac{\mathbf{Z}_i}{\mathbf{Z}_i + R_g} = (-210.73)\left(\dfrac{1\text{ k}\Omega}{1\text{ k}\Omega + 1\text{ k}\Omega}\right)$

$= (-210.73)\left(\dfrac{1}{2}\right) = \mathbf{-105.36}$

d. $R_o = R_L\left(\dfrac{\mathbf{A}_{v_{NL}}}{\mathbf{A}_v} - 1\right) = 2.2\text{ k}\Omega\left(\dfrac{-5000}{-210.73} - 1\right)$

$= 2.2\text{ k}\Omega(23.727 - 1) = 2.2\text{ k}\Omega(22.727)$

$= \mathbf{50\text{ k}\Omega}$, as specified

26.4 THE CURRENT GAINS A_i AND A_{i_T}, AND THE POWER GAIN A_G

The current gain of two-port systems is typically calculated from voltage levels. A no-load gain is not defined for current gain since the absence of R_L requires that $\mathbf{I}_o = \mathbf{E}_o/R_L = 0$ A and $\mathbf{A}_i = \mathbf{I}_o/\mathbf{I}_i = 0$.

For the system of Fig. 26.16, however, a load has been applied and

$$\mathbf{I}_o = -\frac{\mathbf{E}_o}{R_L}$$

with

$$\mathbf{I}_i = \frac{\mathbf{E}_i}{\mathbf{Z}_i}$$

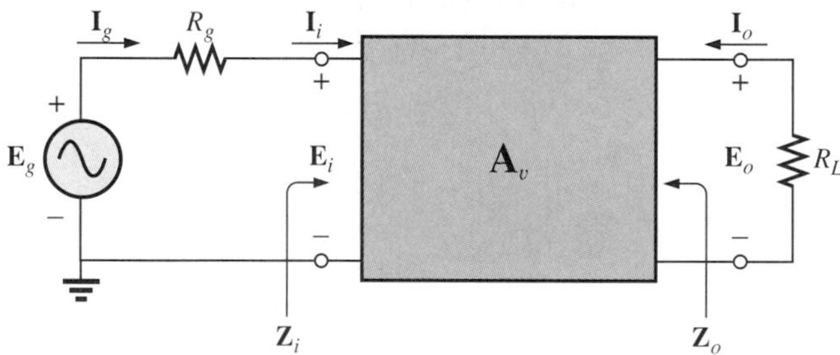

FIG. 26.16
Defining A_i and A_{i_T}.

Note the need for a minus sign when $\mathbf{I}_o$ is defined, because the defined polarity of $\mathbf{E}_o$ would establish the opposite direction for $\mathbf{I}_o$ through R_L.

The loaded current gain is

$$\mathbf{A}_i = \frac{\mathbf{I}_o}{\mathbf{I}_i} = \frac{-\mathbf{E}_o/R_L}{\mathbf{E}_i/\mathbf{Z}_i} = -\frac{\mathbf{E}_o}{\mathbf{E}_i}\left(\frac{\mathbf{Z}_i}{R_L}\right)$$

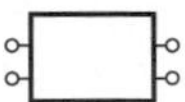

and

$$\boxed{\mathbf{A}_i = -\mathbf{A}_v \frac{\mathbf{Z}_i}{R_L}} \qquad \textbf{(26.10)}$$

In general, therefore, the loaded current gain can be obtained directly from the loaded voltage gain and the ratio of impedance levels, $\mathbf{Z}_i$ over R_L.

If the ratio $\mathbf{A}_{i_T} = \mathbf{I}_o/\mathbf{I}_g$ were required, we would proceed as follows:

$$\mathbf{I}_o = -\frac{\mathbf{E}_o}{R_L}$$

with

$$\mathbf{I}_i = \frac{\mathbf{E}_g}{R_g + \mathbf{Z}_i}$$

and

$$\mathbf{A}_{i_T} = \frac{\mathbf{I}_o}{\mathbf{I}_g} = \frac{-\mathbf{E}_o/R_L}{\mathbf{E}_g/(R_g + \mathbf{Z}_i)} = -\left(\frac{\mathbf{E}_o}{\mathbf{E}_g}\right)\left(\frac{R_g + \mathbf{Z}_i}{R_L}\right)$$

or

$$\boxed{\mathbf{A}_{i_T} = \frac{\mathbf{I}_o}{\mathbf{I}_g} = -\mathbf{A}_{v_T}\left(\frac{R_g + \mathbf{Z}_i}{R_L}\right)} \qquad \textbf{(26.11)}$$

The result obtained with Eq. (26.10) or (26.11) will be the same since $\mathbf{I}_g = \mathbf{I}_i$, but the option of which gain is available or which you choose to use is now available.

Returning to Fig. 26.13 (repeated in Fig. 26.17), an equation for the current gain can be determined in terms of the no-load voltage gain.

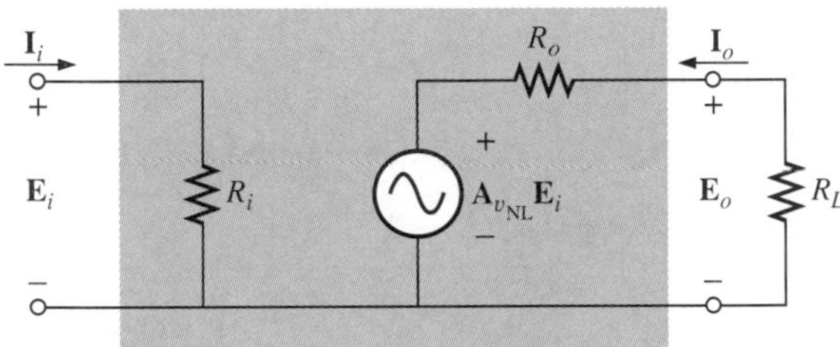

FIG. 26.17
Developing an equation for A_i in terms of $A_{v_{NL}}$.

Through Ohm's law:

$$\mathbf{I}_o = -\frac{\mathbf{A}_{v_{NL}}\mathbf{E}_i}{R_L + R_o}$$

but

$$\mathbf{E}_i = \mathbf{I}_i R_i$$

and

$$\mathbf{I}_o = -\frac{\mathbf{A}_{v_{NL}}(\mathbf{I}_i R_i)}{R_L + R_o}$$

so that

$$\boxed{\mathbf{A}_i = \frac{\mathbf{I}_o}{\mathbf{I}_i} = -\mathbf{A}_{v_{NL}} \frac{R_i}{R_L + R_o}} \qquad \textbf{(26.12)}$$

The result is an equation for the loaded current gain of an amplifier in terms of the nameplate no-load voltage gain and the resistive elements of the network.

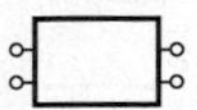

Recall an earlier conclusion that the larger the value of R_L, the larger the loaded voltage gain. For current levels, Eq. (26.12) reveals that

the larger the level of R_L, the less the current gain of a loaded amplifier.

In the design of an amplifier, therefore, one must balance the desired voltage gain with the current gain and resulting ac output power level.

For the system of Fig. 26.17, the power delivered to the load is determined by E_o^2/R_L, whereas the power delivered at the input terminals is E_i^2/R_i. The power gain is therefore defined by

$$A_G = \frac{P_o}{P_i} = \frac{E_o^2/R_L}{E_i^2/R_i} = \frac{E_o^2}{E_i^2}\frac{R_i}{R_L} = \left(\frac{E_o}{E_i}\right)^2 \frac{R_i}{R_L}$$

and

$$A_G = A_v^2 \frac{R_i}{R_L} \qquad \textbf{(26.13)}$$

Expanding the conclusion,

$$A_G = (A_v)\left(A_v \frac{R_i}{R_L}\right) = (A_v)(-A_i)$$

so

$$A_G = -A_v A_i \qquad \textbf{(26.14)}$$

Don't be concerned about the minus sign. A_v or A_i will be negative to ensure that the power gain is positive, as obtained from Eq. (26.13).

If we substitute $A_v = -A_i R_L/R_i$ [from Eq. (26.10)] into Eq. (26.14), we will find

$$A_G = -A_v A_i = -\left(\frac{-A_i R_L}{R_i}\right) A_i$$

or

$$A_G = A_i^2 \frac{R_L}{R_i} \qquad \textbf{(26.15)}$$

which has a format similar to Eq. (26.13), but now A_G is given in terms of the current gain of the system.

The last power gain to be defined is the following:

$$A_{G_T} = \frac{P_L}{P_g} = \frac{E_o^2/R_L}{E_g I_g} = \frac{E_o^2/R_L}{E_g^2/(R_g + R_i)} = \left(\frac{E_o}{E_g}\right)^2 \left(\frac{R_g + R_i}{R_L}\right)$$

or

$$A_{G_T} = A_{v_T}^2 \left(\frac{R_g + R_i}{R_L}\right) \qquad \textbf{(26.16)}$$

Expanding:

$$A_{G_T} = A_{v_T}\left(A_{v_T} \frac{R_g + R_i}{R_L}\right)$$

and

$$A_{G_T} = -A_{v_T} A_{i_T} \qquad \textbf{(26.17)}$$

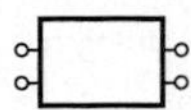

EXAMPLE 26.5 Given the system of Fig. 26.18 with its nameplate data:

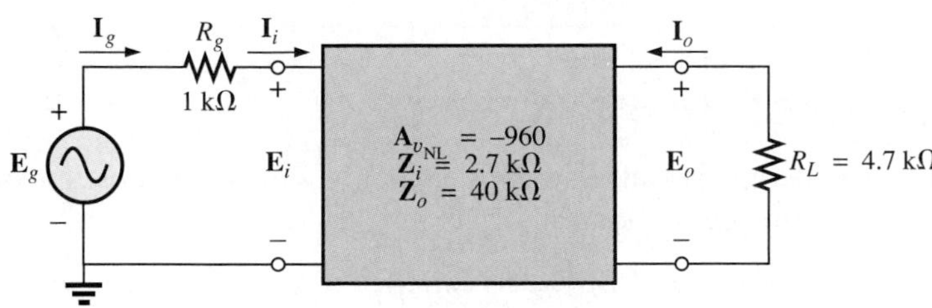

FIG. 26.18
Example 26.5.

a. Determine $\mathbf{A}_v$.
b. Calculate $\mathbf{A}_i$.
c. Increase R_L to double its current value and note the effect on $\mathbf{A}_v$ and $\mathbf{A}_i$.
d. Find $\mathbf{A}_{i_T}$.
e. Calculate A_G.
f. Determine A_i from Eq. (26.1) and compare it to the value obtained in part (b).

Solutions:

a. $$\mathbf{A}_v = \mathbf{A}_{v_{NL}}\frac{R_L}{R_L + R_o} = (-960)\left(\frac{4.7\text{ k}\Omega}{4.7\text{ k}\Omega + 40\text{ k}\Omega}\right) = \mathbf{-100.94}$$

b. $$\mathbf{A}_i = -\mathbf{A}_{v_{NL}}\frac{R_i}{R_L + R_o} = -(-960)\left(\frac{2.7\text{ k}\Omega}{4.7\text{ k}\Omega + 40\text{ k}\Omega}\right) = \mathbf{57.99}$$

c. $R_L = 2(4.7\text{ k}\Omega) = 9.4\text{ k}\Omega$

$$\mathbf{A}_v = \mathbf{A}_{v_{NL}}\left(\frac{R_L}{R_L + R_o}\right) = (-960)\left(\frac{9.4\text{ k}\Omega}{9.4\text{ k}\Omega + 40\text{ k}\Omega}\right)$$

$= \mathbf{-182.67}$ versus -100.94, which is a significant increase.

$$\mathbf{A}_i = -\mathbf{A}_{v_{NL}}\left(\frac{R_i}{R_L + R_o}\right) = -(-960)\left(\frac{2.7\text{ k}\Omega}{40\text{ k}\Omega + 9.4\text{ k}\Omega}\right)$$

$= \mathbf{52.47}$ versus 57.99

which is a drop in level but not as significant as the change in $\mathbf{A}_v$.

d. $\mathbf{A}_{i_T} = \mathbf{A}_i = \mathbf{57.99}$ as obtained in part (b)

However, $$\mathbf{A}_{i_T} = -\mathbf{A}_{v_T}\left(\frac{R_g + R_i}{R_L}\right)$$

$$= -\left[\mathbf{A}_v\frac{R_i}{(R_i + R_g)}\right]\left[\frac{(R_g + R_i)}{R_L}\right]$$

$$= -\mathbf{A}_v\frac{R_i}{R_L} = -(-100.94)\left(\frac{2.7\text{ k}\Omega}{4.7\text{ k}\Omega}\right)$$

$= \mathbf{57.99}$ as well

e. $$A_G = A_v^2\frac{R_i}{R_L} = (100.94)^2\left(\frac{2.7\text{ k}\Omega}{4.7\text{ k}\Omega}\right) = \mathbf{5853.19}$$

f. $A_G = -A_v A_i$

$$\text{or } A_i = \frac{A_G}{A_v} = -\frac{(5853.19)}{(-100.94)}$$

$$= \mathbf{57.99} \text{ as found in part (b)}$$

26.5 CASCADED SYSTEMS

When considering cascaded systems, as in Fig. 26.19, the most important fact to remember is that

the equations for cascaded systems employ the loaded voltage and current gains for each stage and not the nameplate unloaded levels.

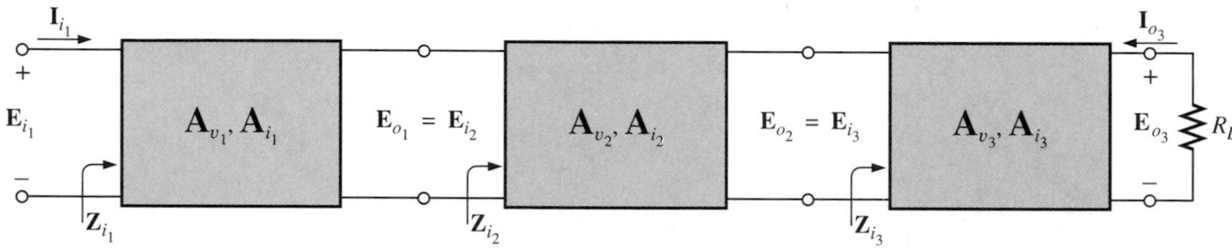

FIG. 26.19
Cascaded system.

Too often the labeled no-load gains are employed, resulting in enormous overall gains and unreasonably high expectations for the system. In addition, bear in mind that the input impedance of stage 3 may affect the input impedance of stage 2 and, therefore, the load on stage 1.

In general, therefore, the equations for cascaded systems initially appear to offer a high level of simplicity to the analysis. Simply be aware, however, that each term of the overall equations must be carefully evaluated before using the equation.

The total voltage gain for the system of Fig. 26.19 is

$$\mathbf{A}_{v_T} = \mathbf{A}_{v_1} \cdot \mathbf{A}_{v_2} \cdot \mathbf{A}_{v_3} \qquad \textbf{(26.18)}$$

where, as noted above, the amplification factor of each stage is determined under loaded conditions.

The total current gain for the system of Fig. 26.19 is

$$\mathbf{A}_{i_T} = \mathbf{A}_{i_1} \cdot \mathbf{A}_{i_2} \cdot \mathbf{A}_{i_3} \qquad \textbf{(26.19)}$$

where, again, the gain of each stage is determined under loaded (connected) conditions.

The current gain between any two stages can also be determined using an equation developed earlier in the chapter. For cascaded systems, the equation has the following general format:

$$\mathbf{A}_i = \mathbf{A}_v \frac{Z_i}{R_L} \qquad \textbf{(26.20)}$$

where $\mathbf{A}_v$ is the loaded voltage gain corresponding to the desired loaded current gain. That is, if the gain is from the first to the third stages, then

the voltage gain substituted is also from the first to third stages. The input impedance $\mathbf{Z}_i$ is for the first stage of interest, and R_L is the loading on the last stage of interest.

For example, for the three-stage amplifier of Fig. 26.19,

$$\mathbf{A}_{i_T} = \mathbf{A}_{v_T}\frac{Z_{i_1}}{R_L}$$

whereas for the first two stages

$$\mathbf{A}_i' = \mathbf{A}_v'\frac{Z_{i_1}}{Z_{i_3}}$$

where $\mathbf{A}_i' = \dfrac{\mathbf{I}_{o_2}}{\mathbf{I}_{i_1}}$ and $\mathbf{A}_v' = \dfrac{\mathbf{E}_{o_2}}{\mathbf{E}_{i_1}}$

The total power gain is determined by

$$\boxed{A_{G_T} = A_{v_T}A_{i_T}} \qquad \textbf{(26.21)}$$

whereas the gain between specific stages is simply the product of the voltage and current gains for each section. For example, for the first two stages of Fig. 26.19,

$$A_G' = A_{v_2}' \cdot A_{i_2}'$$

where $A_{v_2}' = A_{v_1} \cdot A_{v_2}$ and $A_{i_2}' = A_{i_1} \cdot A_{i_2}$

EXAMPLE 26.6 For the cascaded system of Fig. 26.20, with its nameplate no-load parameters:

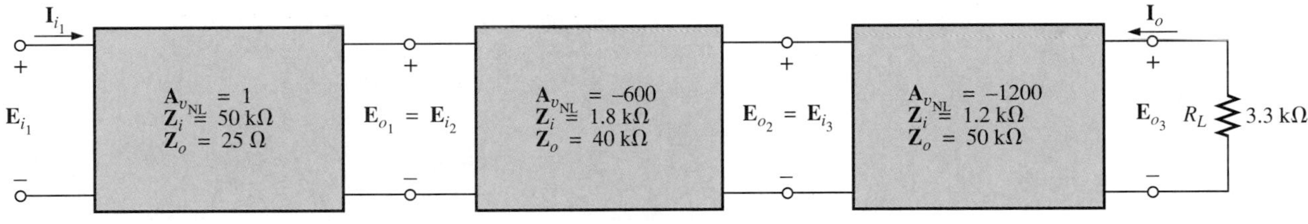

FIG. 26.20
Example 26.6.

a. Determine the load voltage and current gain for each stage and redraw the system of Fig. 26.20 with the loaded parameters.
b. Calculate the total voltage and current gain.
c. Find the total power gain of the system using Eq. (26.21).
d. Calculate the voltage and current gain for the first two stages using Eqs. (26.18) and (26.19).
e. Determine the current gain for the first two stages using Eq. (26.20) and compare with the result of part (d).
f. Calculate the power gain for the first two stages using Eq. (26.21).
g. Determine the power gain for the first two stages using Eq. (26.13). Compare this answer with the result of part (f).
h. Calculate the incorrect voltage gain for the entire system using Eq. (26.18) and the no-load nameplate level for each stage. Compare this answer to the result of part (b).

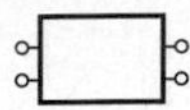

Solutions:

a. $$\mathbf{A}_{v_1} = \mathbf{A}_{v_{NL1}}\frac{R_L}{R_L + R_o} = \mathbf{A}_{v_{NL1}}\frac{Z_{i_2}}{Z_{i_2} + R_{o_1}} = (1)\frac{1.8\text{ k}\Omega}{1.8\text{ k}\Omega + 25\ \Omega}$$
$$= \mathbf{0.986}$$

$$\mathbf{A}_{v_2} = \mathbf{A}_{v_{NL2}}\frac{\mathbf{Z}_{i_3}}{Z_{i_3} + R_{o_2}} = (-600)\frac{1.2\text{ k}\Omega}{1.2\text{ k}\Omega + 40\text{ k}\Omega} = \mathbf{-17.476}$$

$$\mathbf{A}_{v_3} = \mathbf{A}_{v_{NL3}}\frac{R_L}{R_L + R_{o_3}} = (-1200)\frac{3.3\text{ k}\Omega}{3.3\text{ k}\Omega + 50\text{ k}\Omega} = \mathbf{-74.296}$$

$$\mathbf{A}_{i_1} = -\mathbf{A}_{v_{NL}}\frac{R_i}{R_L + R_o} = -\mathbf{A}_{v_{NL1}}\frac{Z_{i_1}}{Z_{i_2} + R_{o_1}} = -(1)\frac{50\text{ k}\Omega}{1.8\text{ k}\Omega + 25\ \Omega}$$
$$= \mathbf{-27.397}$$

$$\mathbf{A}_{i_2} = -\mathbf{A}_{v_{NL2}}\frac{Z_{i_2}}{Z_{i_3} + R_{o_2}} = -(-600)\frac{1.8\text{ k}\Omega}{1.2\text{ k}\Omega + 40\text{ k}\Omega} = \mathbf{26.214}$$

$$\mathbf{A}_{i_3} = -\mathbf{A}_{v_{NL3}}\frac{Z_{i_3}}{R_L + R_{o_3}} = -(-1200)\frac{1.2\text{ k}\Omega}{3.3\text{ k}\Omega + 50\text{ k}\Omega} = \mathbf{27.017}$$

Note Fig. 26.21.

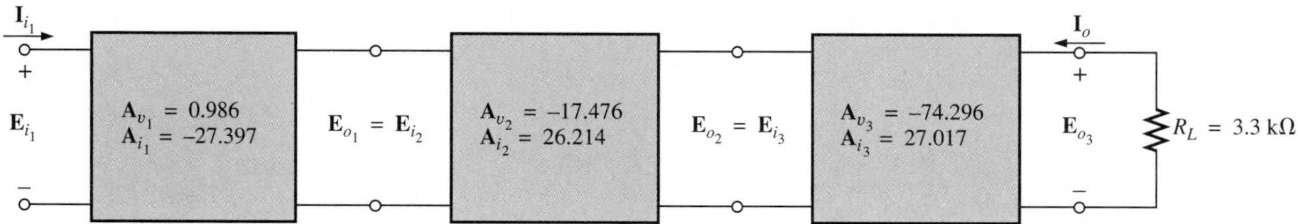

FIG. 26.21
Solution to Example 26.6.

b. $$\mathbf{A}_{v_T} = \frac{\mathbf{E}_{o_3}}{\mathbf{E}_{i_1}} = \mathbf{A}_{v_1}\cdot\mathbf{A}_{v_2}\cdot\mathbf{A}_{v_3} = (0.986)(-17.476)(-74.296)$$
$$= \mathbf{1280.22}$$

$$\mathbf{A}_{i_T} = \frac{\mathbf{I}_{o_3}}{\mathbf{I}_{i_1}} = \mathbf{A}_{i_1}\cdot\mathbf{A}_{i_2}\cdot\mathbf{A}_{i_3} = (-27.397)(26.214)(27.017)$$
$$= \mathbf{-19{,}403.20}$$

c. $$A_{G_T} = -A_{v_T}\cdot A_{i_T} = -(1280.22)(-19{,}403.20) = \mathbf{24.84 \times 10^6}$$

d. $$A'_{v_2} = \mathbf{A}_{v_1}\cdot\mathbf{A}_{v_2} = (0.986)(-17.476) = \mathbf{-17.231}$$
$$A'_{i_2} = \mathbf{A}_{i_1}\cdot\mathbf{A}_{i_2} = (-27.397)(26.214) = \mathbf{-718.185}$$

e. $$\mathbf{A}'_{i_2} = \mathbf{A}_v\frac{\mathbf{Z}_i}{R_L} = \mathbf{A}'_{v_2}\frac{\mathbf{Z}_{i_1}}{\mathbf{Z}_{i_3}} = (-17.231)\frac{50\text{ k}\Omega}{1.2\text{ k}\Omega}$$
$$= \mathbf{-717.958}\text{ versus }-718.185$$

with the difference due to the level of accuracy carried through the calculations.

f. $$A'_{G_2} = A'_{v_2}\cdot A'_{i_2} = (-17.231)(-718.185) = \mathbf{12{,}375.05}$$

g. $$A'_{G_2} = A_v^2\frac{R_i}{R_L} = (A'_{v_2})^2\frac{R_{i_1}}{Z_{i_3}} = (-17.231)^2\frac{50\text{ k}\Omega}{1.2\text{ k}\Omega} = \mathbf{12{,}371.14}$$

h. $\mathbf{A}_{v_T} = \mathbf{A}_{v_1} \cdot \mathbf{A}_{v_2} \cdot \mathbf{A}_{v_3} = (1)(-600)(-1200) = 7.2 \times 10^5$
$720{,}000{:}1280.22 = 562.40{:}1$
which is certainly a significant difference in results.

26.6 IMPEDANCE (z) PARAMETERS

For the two-port configuration of Fig. 26.22, four variables are specified. For most situations, if any two are specified, the remaining two variables can be determined. The four variables can be related by the following equations:

$$\mathbf{E}_1 = \mathbf{z}_{11}\mathbf{I}_1 + \mathbf{z}_{12}\mathbf{I}_2 \qquad \textbf{(26.22a)}$$

$$\mathbf{E}_2 = \mathbf{z}_{21}\mathbf{I}_1 + \mathbf{z}_{22}\mathbf{I}_2 \qquad \textbf{(26.22b)}$$

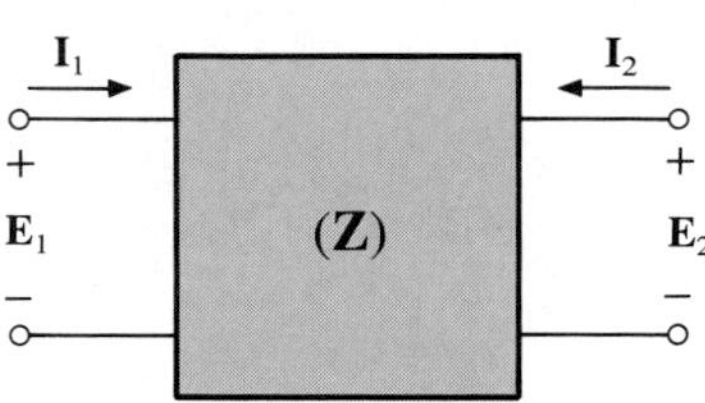

FIG. 26.22
Two-port impedance parameter configuration.

The *impedance parameters* $\mathbf{z}_{11}$, $\mathbf{z}_{12}$, and $\mathbf{z}_{22}$ are measured in ohms.

To model the system, each impedance parameter must be determined by setting a particular variable to zero.

z_{11}

For $\mathbf{z}_{11}$, if $\mathbf{I}_2$ is set to zero, as shown in Fig. 26.23, Eq. (26.22a) becomes

$$\mathbf{E}_1 = \mathbf{z}_{11}\mathbf{I}_1 + \mathbf{z}_{12}(0)$$

and

$$\mathbf{z}_{11} = \left.\frac{\mathbf{E}_1}{\mathbf{I}_1}\right|_{\mathbf{I}_2 = 0} \quad \text{(ohms, } \Omega\text{)} \qquad \textbf{(26.23)}$$

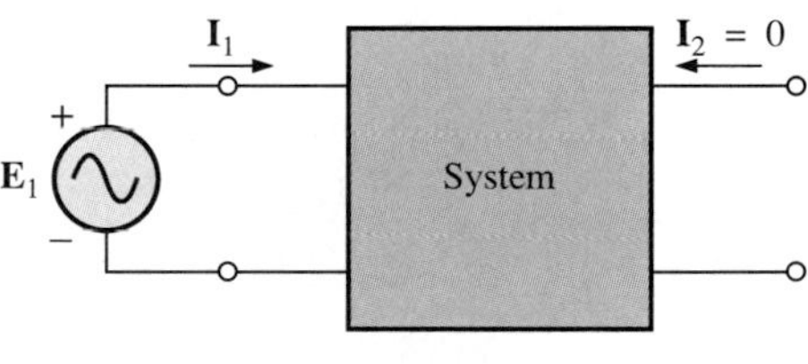

FIG. 26.23
Determining $\mathbf{z}_{11}$.

Equation (26.23) reveals that with $\mathbf{I}_2$ set to zero, the impedance parameter is determined by the resulting ratio of $\mathbf{E}_1$ to $\mathbf{I}_1$. Since $\mathbf{E}_1$ and $\mathbf{I}_1$ are both input quantities, with $\mathbf{I}_2$ set to zero, the parameter $\mathbf{z}_{11}$ is formally referred to in the following manner:

$\mathbf{z}_{11}$ = ***open-circuit, input-impedance parameter***

z_{12}

For $\mathbf{z}_{12}$, $\mathbf{I}_1$ is set to zero, and Eq. (26.22a) results in

$$\mathbf{z}_{12} = \left.\frac{\mathbf{E}_1}{\mathbf{I}_2}\right|_{\mathbf{I}_1 = 0} \quad \text{(ohms, } \Omega\text{)} \qquad \textbf{(26.24)}$$

For most systems where input and output quantities are to be compared, the ratio of interest is usually that of the output quantity divided by the input quantity. In this case, the *reverse* is true, resulting in the following:

$\mathbf{z}_{12}$ = ***open-circuit, reverse-transfer impedance parameter***

The term *transfer* is included to indicate that $\mathbf{z}_{12}$ will relate an input and output quantity (for the condition $\mathbf{I}_1 = 0$). The network configuration for determining $\mathbf{z}_{12}$ is shown in Fig. 26.24.

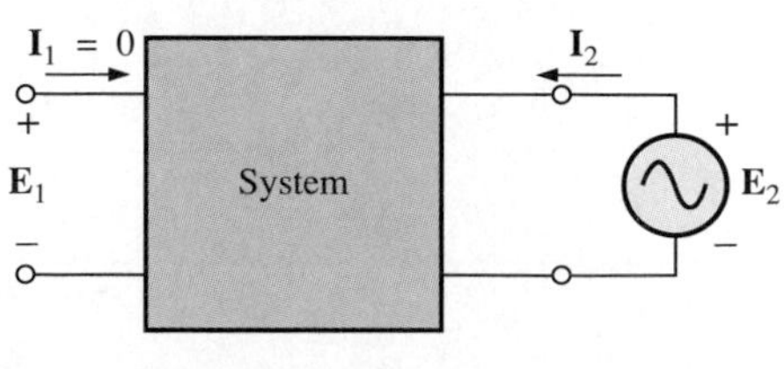

FIG. 26.24
Determining $\mathbf{z}_{12}$.

For an applied source $\mathbf{E}_2$, the ratio $\mathbf{E}_1/\mathbf{I}_2$ will determine $\mathbf{z}_{12}$ with $\mathbf{I}_1$ set to zero.

$\mathbf{z}_{21}$

To determine $\mathbf{z}_{21}$, set $\mathbf{I}_2$ to zero and find the ratio $\mathbf{E}_2/\mathbf{I}_1$ as determined by Eq. (26.22b). That is,

$$\boxed{\mathbf{z}_{21} = \frac{\mathbf{E}_2}{\mathbf{I}_1}\bigg|_{\mathbf{I}_2 = 0}} \quad \text{(ohms, } \Omega) \tag{26.25}$$

In this case, input and output quantities are again the determining variables, requiring the term *transfer* in the nomenclature. However, the ratio is that of an output to an input quantity, so the descriptive term *forward* is applied, and

$\mathbf{z}_{21}$ = ***open-circuit, forward-transfer impedance parameter***

The determining network is shown in Fig. 26.25. For an applied voltage $\mathbf{E}_1$, it is determined by the ratio $\mathbf{E}_2/\mathbf{I}_1$ with $\mathbf{I}_2$ set to zero.

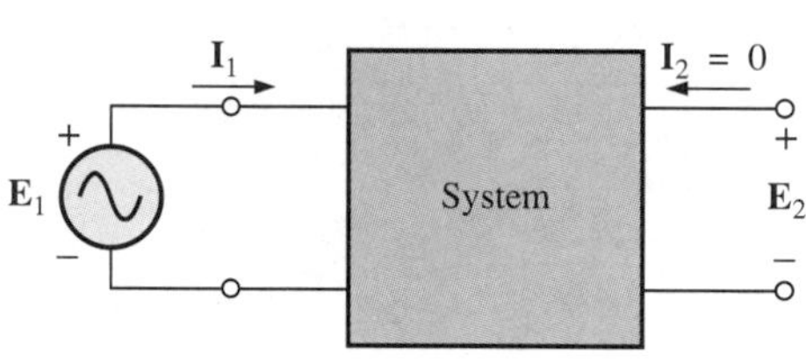

FIG. 26.25
Determining $\mathbf{z}_{21}$.

$\mathbf{z}_{22}$

The remaining parameter, $\mathbf{z}_{22}$, is determined by

$$\boxed{\mathbf{z}_{22} = \frac{\mathbf{E}_2}{\mathbf{I}_2}\bigg|_{\mathbf{I}_1 = 0}} \quad \text{(ohms, } \Omega) \tag{26.26}$$

as derived from Eq. (26.22b) with $\mathbf{I}_1$ set to zero. Since it is the ratio of the output voltage to the output current with $\mathbf{I}_1$ set to zero, it has the terminology

$\mathbf{z}_{22}$ = ***open-circuit, output-impedance parameter***

The required network is shown in Fig. 26.26. For an applied voltage $\mathbf{E}_2$, it is determined by the resulting ratio $\mathbf{E}_2/\mathbf{I}_2$ with $\mathbf{I}_1 = 0$.

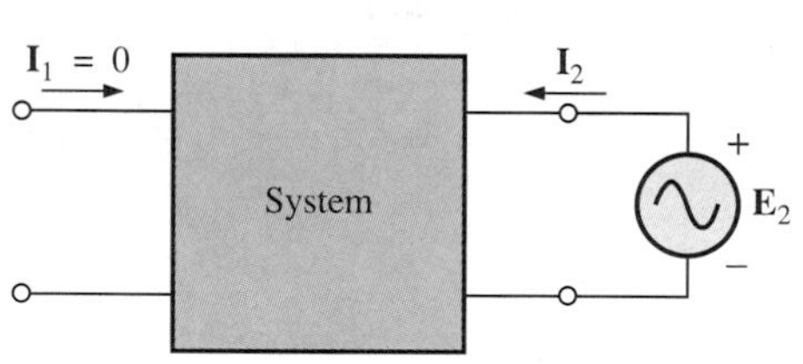

FIG. 26.26
Determining $\mathbf{z}_{22}$.

EXAMPLE 26.7 Determine the impedance (**z**) parameters for the T network of Fig. 26.27.

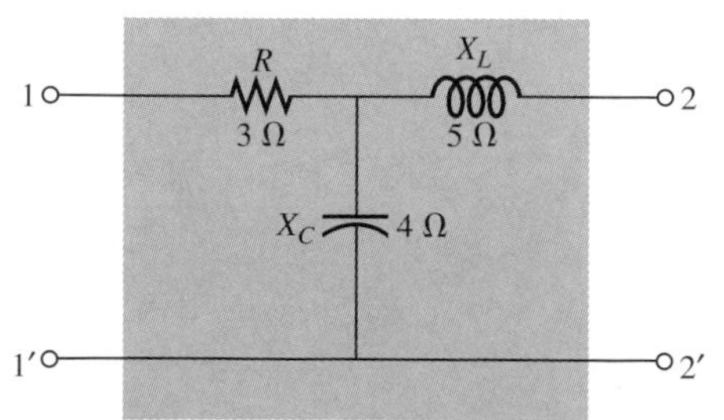

FIG. 26.27
T configuration.

Solution: For $\mathbf{z}_{11}$, the network will appear as shown in Fig. 26.28, with $\mathbf{Z}_1 = 3\ \Omega\ \angle 0°$, $\mathbf{Z}_2 = 5\ \Omega\ \angle 90°$, and $\mathbf{Z}_3 = 4\ \Omega\ \angle -90°$:

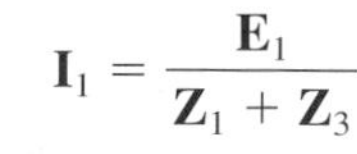

$$\mathbf{I}_1 = \frac{\mathbf{E}_1}{\mathbf{Z}_1 + \mathbf{Z}_3}$$

Thus
$$\mathbf{z}_{11} = \frac{\mathbf{E}_1}{\mathbf{I}_1}\bigg|_{\mathbf{I}_2 = 0}$$

and
$$\boxed{\mathbf{z}_{11} = \mathbf{Z}_1 + \mathbf{Z}_3} \tag{26.27}$$

For $\mathbf{z}_{12}$, the network will appear as shown in Fig. 26.29:

$$\mathbf{E}_1 = \mathbf{I}_2\mathbf{Z}_3$$

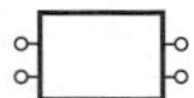

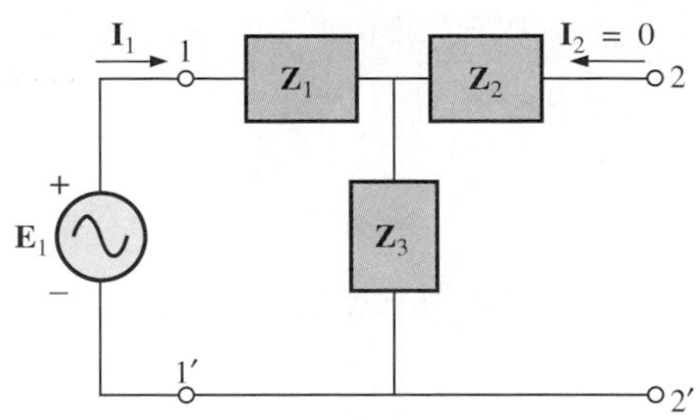

FIG. 26.28
Determining $\mathbf{z}_{11}$.

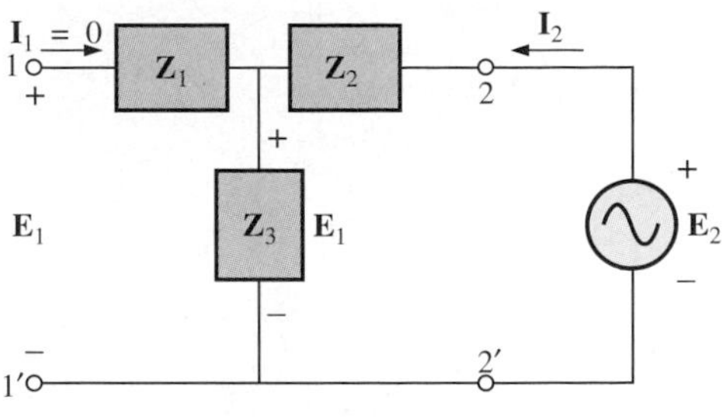

FIG. 26.29
Determining $\mathbf{z}_{12}$.

Thus
$$\mathbf{z}_{12} = \frac{\mathbf{E}_1}{\mathbf{I}_2}\bigg|_{\mathbf{I}_1 = 0} = \frac{\mathbf{I}_2\mathbf{Z}_3}{\mathbf{I}_2}$$

and
$$\boxed{\mathbf{z}_{12} = \mathbf{Z}_3} \tag{26.28}$$

For $\mathbf{z}_{21}$, the required network appears in Fig. 26.30:

$$\mathbf{E}_2 = \mathbf{I}_1\mathbf{Z}_3$$

Thus
$$\mathbf{z}_{21} = \frac{\mathbf{E}_2}{\mathbf{I}_1}\bigg|_{\mathbf{I}_2 = 0} = \frac{\mathbf{I}_1\mathbf{Z}_3}{\mathbf{I}_1}$$

and
$$\boxed{\mathbf{z}_{21} = \mathbf{Z}_3} \tag{26.29}$$

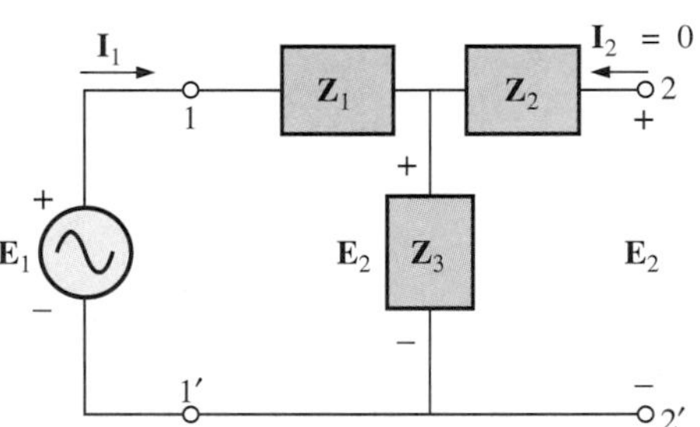

FIG. 26.30
Determining $\mathbf{z}_{21}$.

For $\mathbf{z}_{22}$, the determining configuration is shown in Fig. 26.31:

$$\mathbf{I}_2 = \frac{\mathbf{E}_2}{\mathbf{Z}_2 + \mathbf{Z}_3}$$

Thus
$$\mathbf{z}_{22} = \frac{\mathbf{E}_2}{\mathbf{I}_2}\bigg|_{\mathbf{I}_1 = 0} = \frac{\mathbf{I}_2(\mathbf{Z}_2 + \mathbf{Z}_3)}{\mathbf{I}_2}$$

and
$$\boxed{\mathbf{z}_{22} = \mathbf{Z}_2 + \mathbf{Z}_3} \tag{26.30}$$

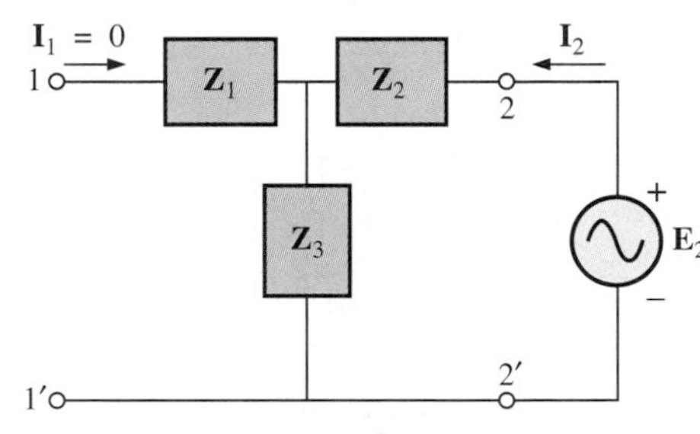

FIG. 26.31
Determining $\mathbf{z}_{22}$.

Note that for the T configuration, $\mathbf{z}_{12} = \mathbf{z}_{21}$. For $\mathbf{Z}_1 = 3\ \Omega\ \angle 0°$, $\mathbf{Z}_2 = 5\ \Omega\ \angle 90°$, and $\mathbf{Z}_3 = 4\ \Omega\ \angle -90°$, we have

$$\mathbf{z}_{11} = \mathbf{Z}_1 + \mathbf{Z}_3 = \mathbf{3\ \Omega - \mathit{j}\,4\ \Omega}$$
$$\mathbf{z}_{12} = \mathbf{z}_{21} = \mathbf{Z}_3 = 4\ \Omega\ \angle -90° = \mathbf{-\mathit{j}\,4\ \Omega}$$
$$\mathbf{z}_{22} = \mathbf{Z}_2 + \mathbf{Z}_3 = 5\ \Omega\ \angle 90° + 4\ \Omega\ \angle -90° = 1\ \Omega\ \angle 90° = \mathbf{\mathit{j}\,1\ \Omega}$$

For a set of impedance parameters, the terminal (external) behavior of the device or network within the configuration of Fig. 26.22 is determined. An *equivalent circuit* for the system can be developed using the impedance parameters and Eqs. (26.22a) and (26.22b). Two possibilities for the impedance parameters appear in Fig. 26.32.

Applying Kirchhoff's voltage law to the input and output loops of the network of Fig. 26.32(a) results in

$$\mathbf{E}_1 - \mathbf{z}_{11}\mathbf{I}_1 - \mathbf{z}_{12}\mathbf{I}_2 = 0$$

and
$$\mathbf{E}_2 - \mathbf{z}_{22}\mathbf{I}_2 - \mathbf{z}_{21}\mathbf{I}_1 = 0$$

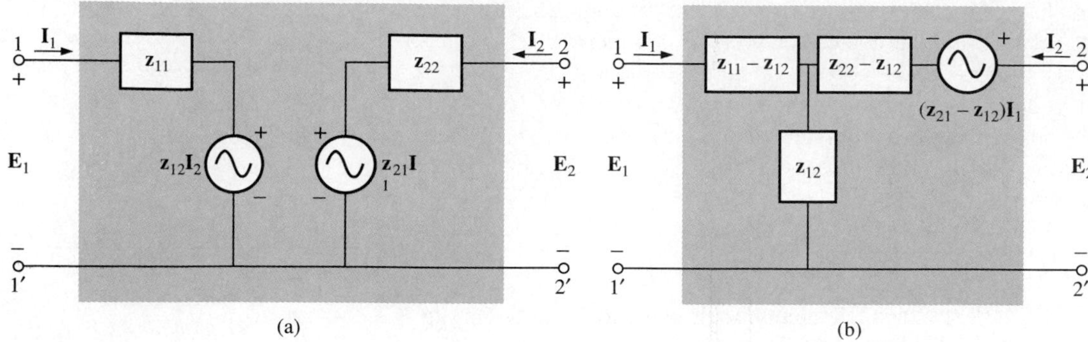

FIG. 26.32
*Two possible two-port, **z**-parameter equivalent networks.*

which, when rearranged, become

$$\mathbf{E}_1 = \mathbf{z}_{11}\mathbf{I}_1 + \mathbf{z}_{12}\mathbf{I}_2 \qquad \mathbf{E}_2 = \mathbf{z}_{21}\mathbf{I}_1 + \mathbf{z}_{22}\mathbf{I}_2$$

matching Eqs. (26.22a) and (26.22b).

For the network of Fig. 26.32(b),

$$\mathbf{E}_1 - \mathbf{I}_1(\mathbf{z}_{11} - \mathbf{z}_{12}) - \mathbf{z}_{12}(\mathbf{I}_1 + \mathbf{I}_2) = 0$$

and $$\mathbf{E}_2 - \mathbf{I}_1(\mathbf{z}_{21} - \mathbf{z}_{12}) - \mathbf{I}_2(\mathbf{z}_{22} - \mathbf{z}_{12}) - \mathbf{z}_{12}(\mathbf{I}_1 + \mathbf{I}_2) = 0$$

which, when rearranged, are

$$\mathbf{E}_1 = \mathbf{I}_1(\mathbf{z}_{11} - \mathbf{z}_{12} + \mathbf{z}_{12}) + \mathbf{I}_2\mathbf{z}_{12}$$
$$\mathbf{E}_2 = \mathbf{I}_1(\mathbf{z}_{21} - \mathbf{z}_{12} + \mathbf{z}_{12}) + \mathbf{I}_2(\mathbf{z}_{22} - \mathbf{z}_{12} + \mathbf{z}_{12})$$

and
$$\mathbf{E}_1 = \mathbf{z}_{11}\mathbf{I}_1 + \mathbf{z}_{12}\mathbf{I}_2$$
$$\mathbf{E}_2 = \mathbf{z}_{21}\mathbf{I}_1 + \mathbf{z}_{22}\mathbf{I}_2$$

Note in each network the necessity for a current-controlled voltage source, that is, a voltage source the magnitude of which is determined by a particular current of the network.

The usefulness of the impedance parameters and the resulting equivalent networks can best be described by considering the system of Fig. 26.33(a), which contains a device (or system) for which the impedance

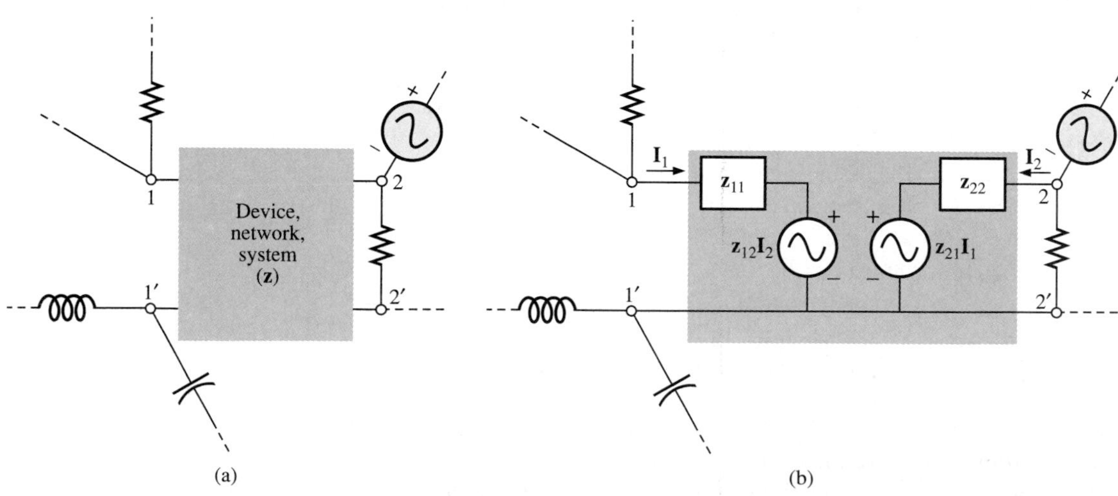

FIG. 26.33
*Substitution of the **z**-parameter equivalent network into a complex system.*

parameters have been determined. As shown in Fig. 26.33(b), the equivalent network for the device (or system) can then be substituted, and methods such as mesh analysis, nodal analysis, and so on, can be employed to determine required unknown quantities. The device itself can then be replaced with an equivalent circuit and the desired solutions obtained more directly and with less effort than is required using only the characteristics of the device.

EXAMPLE 26.8 Draw the equivalent circuit in the form shown in Fig. 26.32(b) using the impedance parameters determined in Example 26.7.

Solution: The circuit appears in Fig. 26.34.

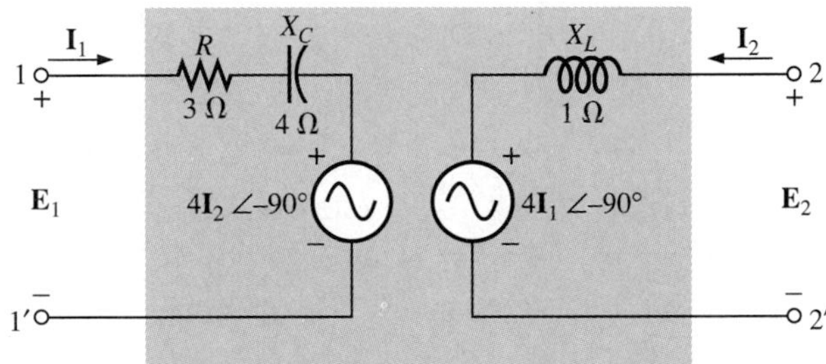

FIG. 26.34
Example 26.8.

26.7 ADMITTANCE (y) PARAMETERS

The equations relating the four terminal variables of Fig. 26.22 can also be written in the following form:

$$\mathbf{I}_1 = \mathbf{y}_{11}\mathbf{E}_1 + \mathbf{y}_{12}\mathbf{E}_2 \qquad \textbf{(26.31a)}$$

$$\mathbf{I}_2 = \mathbf{y}_{21}\mathbf{E}_1 + \mathbf{y}_{22}\mathbf{E}_2 \qquad \textbf{(26.31b)}$$

Note that in this case each term of each equation has the units of current, as compared to voltage for each term of Eqs. (26.22a) and (26.22b). In addition, the unit of each coefficient is siemens, compared with the ohm for the impedance parameters.

The impedance parameters were determined by setting a particular current to zero through an open-circuit condition. For the *admittance parameters* of Eqs. (26.31a) and (26.31b), a voltage is set to zero through a short-circuit condition.

The terminology applied to each of the admittance parameters follows directly from the descriptive terms applied to each of the impedance parameters. The equations for each are determined directly from Eqs. (26.31a) and (26.31b) by setting a particular voltage to zero.

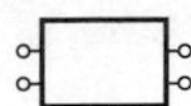

$\mathbf{y}_{11}$

$$\mathbf{y}_{11} = \frac{\mathbf{I}_1}{\mathbf{E}_1}\Bigg|_{\mathbf{E}_2 = 0} \quad \text{(siemens, S)} \tag{26.32}$$

$\mathbf{y}_{11}$ = ***short-circuit, input-admittance parameter***

The determining network appears in Fig. 26.35.

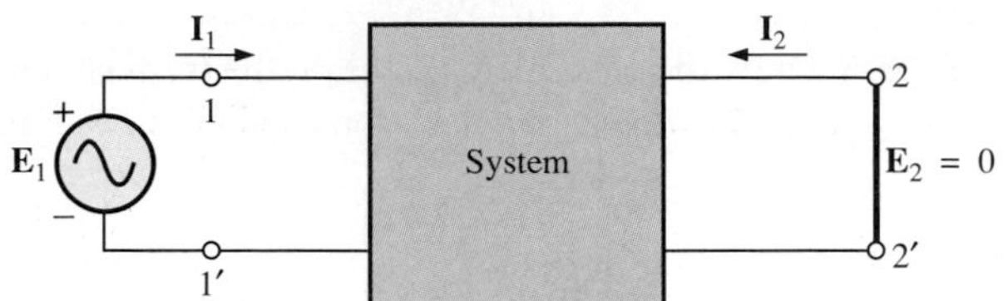

FIG. 26.35
$\mathbf{y}_{11}$ *determination.*

$\mathbf{y}_{12}$

$$\mathbf{y}_{12} = \frac{\mathbf{I}_1}{\mathbf{E}_2}\Bigg|_{\mathbf{E}_1 = 0} \quad \text{(siemens, S)} \tag{26.33}$$

$\mathbf{y}_{12}$ = ***short-circuit, reverse-transfer admittance parameter***

The network for determining $\mathbf{y}_{12}$ appears in Fig. 26.36.

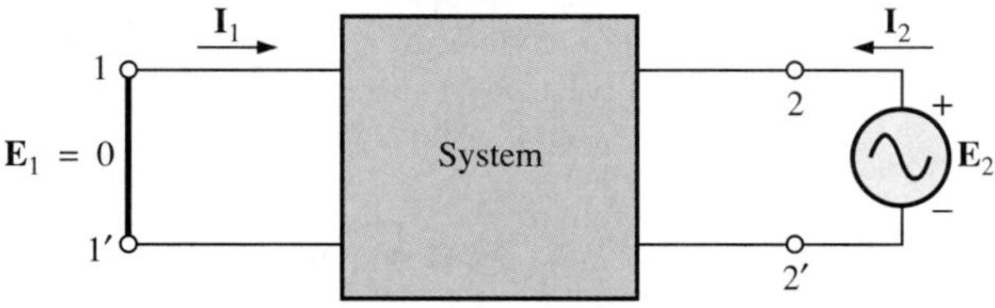

FIG. 26.36
$\mathbf{y}_{12}$ *determination.*

$\mathbf{y}_{21}$

$$\mathbf{y}_{21} = \frac{\mathbf{I}_2}{\mathbf{E}_1}\Bigg|_{\mathbf{E}_2 = 0} \quad \text{(siemens, S)} \tag{26.34}$$

$\mathbf{y}_{21}$ = ***short-circuit, forward-transfer admittance parameter***

The network for determining $\mathbf{y}_{21}$ appears in Fig. 26.37.

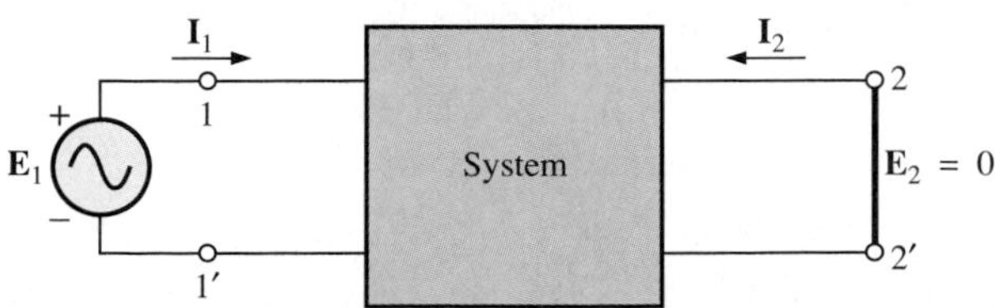

FIG. 26.37
$\mathbf{y}_{21}$ *determination.*

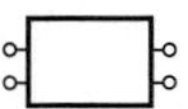

$\mathbf{y}_{22}$

$$\mathbf{y}_{22} = \left.\frac{\mathbf{I}_2}{\mathbf{E}_2}\right|_{\mathbf{E}_1 = 0} \quad \text{(siemens, S)} \tag{26.35}$$

$\mathbf{y}_{22}$ = ***short-circuit, output-admittance parameter***

The required network appears in Fig. 26.38.

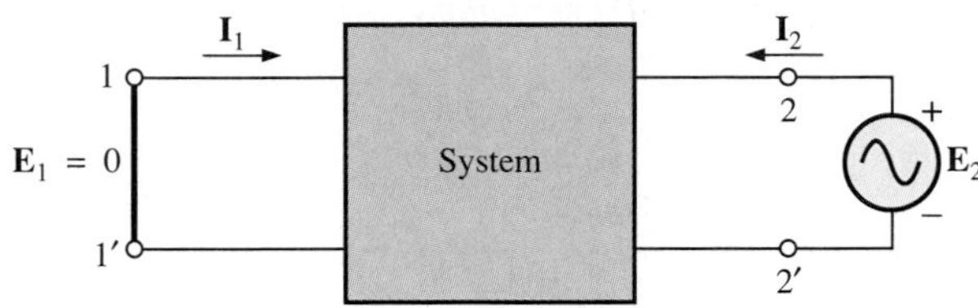

FIG. 26.38
$\mathbf{y}_{22}$ *determination.*

EXAMPLE 26.9 Determine the admittance parameters for the π network of Fig. 26.39.

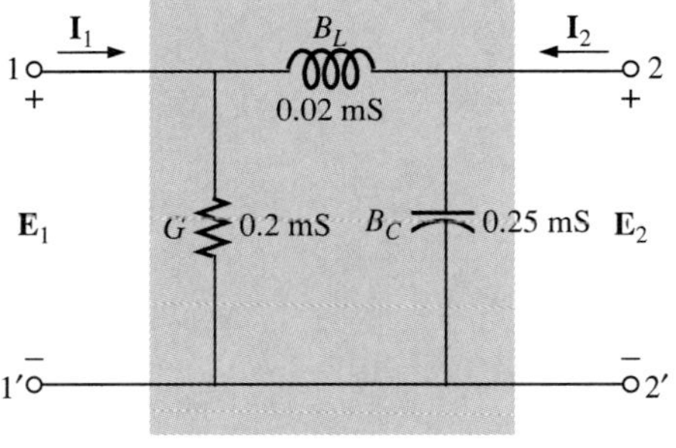

FIG. 26.39
π *network.*

Solution: The network for $\mathbf{y}_{11}$ will appear as shown in Fig. 26.40, with

$\mathbf{Y}_1 = 0.2 \text{ mS} \angle 0^\circ \qquad \mathbf{Y}_2 = 0.02 \text{ mS} \angle -90^\circ \qquad \mathbf{Y}_3 = 0.25 \text{ mS} \angle 90^\circ$

We use $\qquad \mathbf{I}_1 = \mathbf{E}_1\mathbf{Y}_T = \mathbf{E}_1(\mathbf{Y}_1 + \mathbf{Y}_2)$

with $\qquad \mathbf{y}_{11} = \left.\dfrac{\mathbf{I}_1}{\mathbf{E}_1}\right|_{\mathbf{E}_2 = 0}$

and

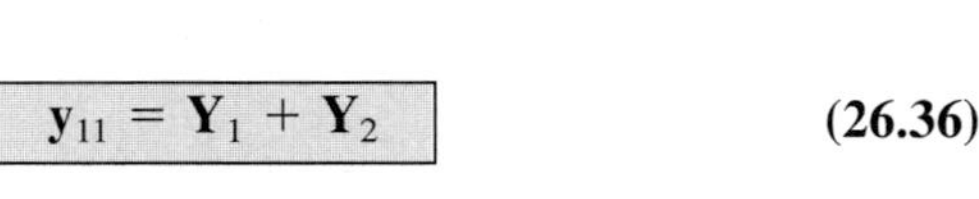

$$\mathbf{y}_{11} = \mathbf{Y}_1 + \mathbf{Y}_2 \tag{26.36}$$

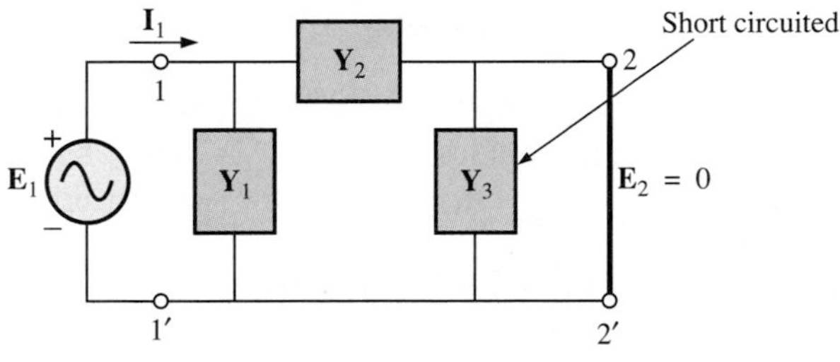

FIG. 26.40
Determining $\mathbf{y}_{11}$.

The determining network for $\mathbf{y}_{12}$ appears in Fig. 26.41. $\mathbf{Y}_1$ is short circuited; so $\mathbf{I}_{\mathbf{Y}_2} = \mathbf{I}_1$, and

$$\mathbf{I}_{\mathbf{Y}_2} = \mathbf{I}_1 = -\mathbf{E}_2\mathbf{Y}_2$$

The minus sign results because the defined direction of $\mathbf{I}_1$ in Fig. 26.41 is opposite to the actual flow direction due to the applied source $\mathbf{E}_2$;

$$\mathbf{y}_{12} = \left.\frac{\mathbf{I}_1}{\mathbf{E}_2}\right|_{\mathbf{E}_1 = 0}$$

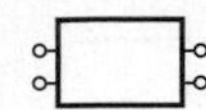

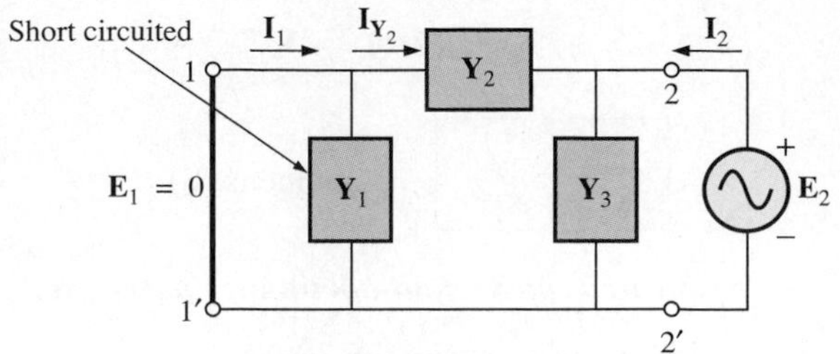

FIG. 26.41
Determining $\mathbf{y}_{12}$.

and $$\mathbf{y}_{12} = -\mathbf{Y}_2 \tag{26.37}$$

The network employed for $\mathbf{y}_{21}$ appears in Fig. 26.42. In this case, $\mathbf{Y}_3$ is short circuited, resulting in

$$\mathbf{I}_{Y_2} = \mathbf{I}_2 \quad \text{and} \quad \mathbf{I}_{Y_2} = \mathbf{I}_2 = -\mathbf{E}_1\mathbf{Y}_2$$

with $$\mathbf{y}_{21} = \frac{\mathbf{I}_2}{\mathbf{E}_1}\bigg|_{\mathbf{E}_2 = 0}$$

and $$\mathbf{y}_{21} = -\mathbf{Y}_2 \tag{26.38}$$

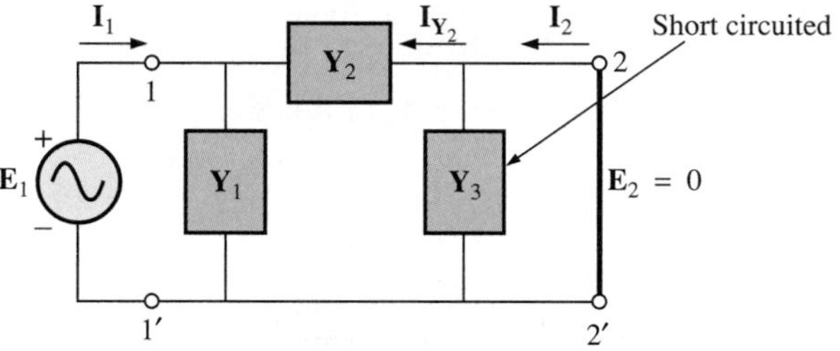

FIG. 26.42
Determining $\mathbf{y}_{21}$.

Note that for the π configuration, $\mathbf{y}_{12} = \mathbf{y}_{21}$, which was expected since the impedance parameters for the T network were such that $\mathbf{z}_{12} = \mathbf{z}_{21}$. A T network can be converted directly to a π network using the Y-Δ transformation.

The determining network for $\mathbf{y}_{22}$ appears in Fig. 26.43:

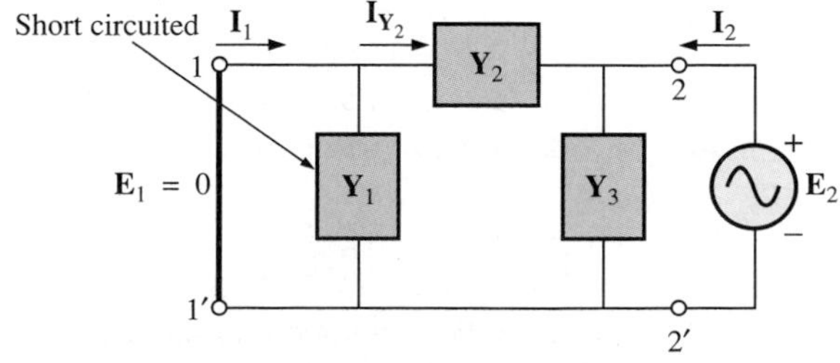

FIG. 26.43
Determining $\mathbf{y}_{22}$.

$$\mathbf{Y}_T = \mathbf{Y}_2 + \mathbf{Y}_3 \quad \text{and} \quad \mathbf{I}_2 = \mathbf{E}_2(\mathbf{Y}_2 + \mathbf{Y}_3)$$

Thus

$$\mathbf{y}_{22} = \left.\frac{\mathbf{I}_2}{\mathbf{E}_2}\right|_{\mathbf{E}_1 = 0}$$

and

$$\boxed{\mathbf{y}_{22} = \mathbf{Y}_2 + \mathbf{Y}_3} \qquad \textbf{(26.39)}$$

Substituting values, we have

$$\mathbf{Y}_1 = 0.2 \text{ mS } \angle 0^\circ$$
$$\mathbf{Y}_2 = 0.02 \text{ mS } \angle -90^\circ$$
$$\mathbf{Y}_3 = 0.25 \text{ mS } \angle 90^\circ$$
$$\mathbf{y}_{11} = \mathbf{Y}_1 + \mathbf{Y}_2$$
$$= \mathbf{0.2\ mS} - \boldsymbol{j}\,\mathbf{0.02\ mS\ (ind.)}$$
$$\mathbf{y}_{12} = \mathbf{y}_{21} = -\mathbf{Y}_2 = -(-j\,0.02 \text{ mS})$$
$$= \boldsymbol{j}\,\mathbf{0.02\ mS\ (cap.)}$$
$$\mathbf{y}_{22} = \mathbf{Y}_2 + \mathbf{Y}_3 = -j\,0.02 \text{ mS} + j\,0.25 \text{ mS}$$
$$= \boldsymbol{j}\,\mathbf{0.23\ mS\ (cap.)}$$

Note the similarities between the results for $\mathbf{y}_{11}$ and $\mathbf{y}_{22}$ for the π network compared with $\mathbf{z}_{11}$ and $\mathbf{z}_{22}$ for the T network.

Two networks satisfying the terminal relationships of Eqs. (26.31a) and (26.31b) are shown in Fig. 26.44. Note the use of parallel branches

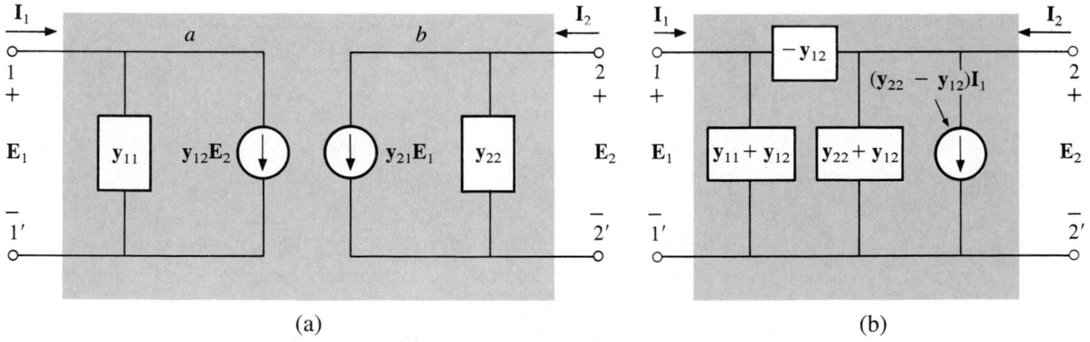

FIG. 26.44
Two possible two-port, **y***-parameter equivalent networks.*

since each term of Eqs. (26.31a) and (26.31b) has the units of current, and the most direct route to the equivalent circuit is an application of Kirchhoff's current law in reverse. That is, find the network that satisfies Kirchhoff's current law relationship. For the impedance parameters, each term had the units of volts, so Kirchhoff's voltage law was applied in reverse to determine the series combination of elements in the equivalent circuit of Fig. 26.44(a).

Applying Kirchhoff's current law to the network of Fig. 26.44(a), we have

$$\text{Node } a\text{: } \mathbf{I}_1 = \overbrace{\mathbf{y}_{11}\mathbf{E}_1}^{\text{Entering}} + \overbrace{\mathbf{y}_{12}\mathbf{E}_2}^{\text{Leaving}}$$
$$\text{Node } b\text{: } \mathbf{I}_2 = \mathbf{y}_{22}\mathbf{E}_2 + \mathbf{y}_{21}\mathbf{E}_1$$

which, when rearranged, are Eqs. (26.31a) and (26.31b).

For the results of Example 26.9, the network of Fig. 26.45 will result if the equivalent network of Fig. 26.44(a) is employed.

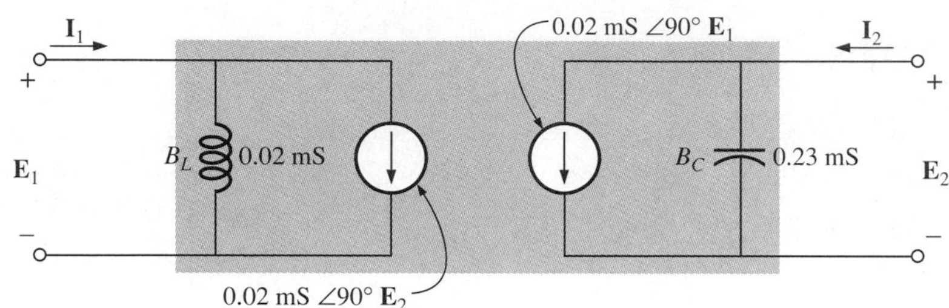

FIG. 26.45
Equivalent network for the results of Example 26.9.

26.8 HYBRID (h) PARAMETERS

The *hybrid* (**h**) *parameters* are employed extensively in the analysis of transistor networks. The term *hybrid* is derived from the fact that the parameters have a mixture of units (a hybrid set) rather than a single unit of measurement such as ohms or siemens, used for the **z** and **y** parameters, respectively. The defining hybrid equations have a mixture of current *and* voltage variables on one side, as follows:

$$\boxed{\mathbf{E}_1 = \mathbf{h}_{11}\mathbf{I}_1 + \mathbf{h}_{12}\mathbf{E}_2} \tag{26.40a}$$

$$\boxed{\mathbf{I}_2 = \mathbf{h}_{21}\mathbf{I}_1 + \mathbf{h}_{22}\mathbf{E}_2} \tag{26.40b}$$

To determine the hybrid parameters, it will be necessary to establish both the short-circuit and the open-circuit condition, depending on the parameter desired.

$\mathbf{h}_{11}$

$$\boxed{\mathbf{h}_{11} = \frac{\mathbf{E}_1}{\mathbf{I}_1}}\Bigg|_{\mathbf{E}_2 = 0} \quad \text{(ohms, }\Omega\text{)} \tag{26.41}$$

$\mathbf{h}_{11}$ = ***short-circuit, input-impedance parameter***

The determining network is shown in Fig. 26.46.

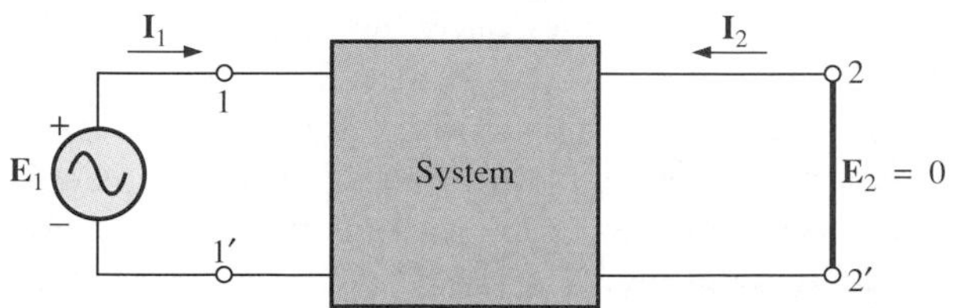

FIG. 26.46
$\mathbf{h}_{11}$ *determination.*

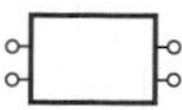

$\mathbf{h}_{12}$

$$\mathbf{h}_{12} = \frac{\mathbf{E}_1}{\mathbf{E}_2}\bigg|_{\mathbf{I}_1 = 0} \quad \text{(dimensionless)} \tag{26.42}$$

$\mathbf{h}_{12}$ = ***open-circuit, reverse-transfer voltage ratio parameter***

The network employed in determining $\mathbf{h}_{12}$ is shown in Fig. 26.47.

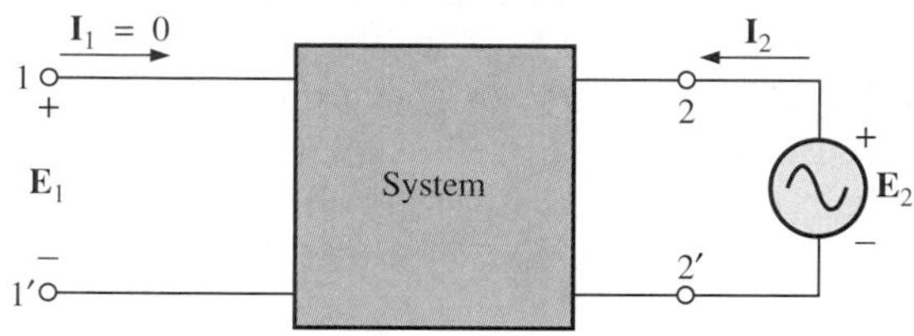

FIG. 26.47
$\mathbf{h}_{12}$ *determination.*

$\mathbf{h}_{21}$

$$\mathbf{h}_{21} = \frac{\mathbf{I}_2}{\mathbf{I}_1}\bigg|_{\mathbf{E}_2 = 0} \quad \text{(dimensionless)} \tag{26.43}$$

$\mathbf{h}_{21}$ = ***short-circuit, forward-transfer current ratio parameter***

The determining network appears in Fig. 26.48.

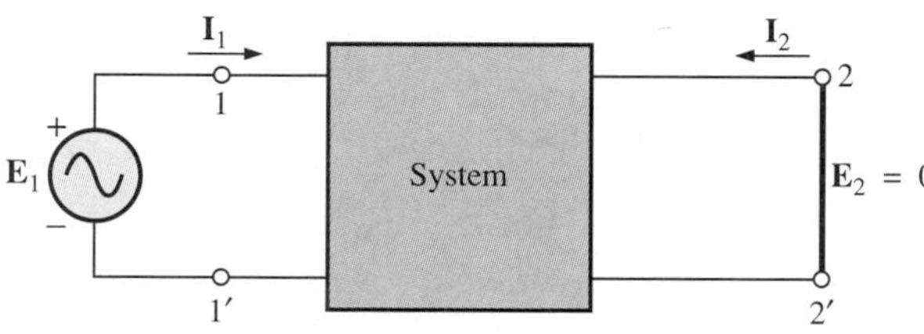

FIG. 26.48
$\mathbf{h}_{21}$ *determination.*

$\mathbf{h}_{22}$

$$\mathbf{h}_{22} = \frac{\mathbf{I}_2}{\mathbf{E}_2}\bigg|_{\mathbf{I}_1 = 0} \quad \text{(siemens, S)} \tag{26.44}$$

$\mathbf{h}_{22}$ = ***open-circuit, output admittance parameter***

The network employed to determine $\mathbf{h}_{22}$ is shown in Fig. 26.49.

The subscript notation for the hybrid parameters is reduced to the following for most applications. The letter chosen is that letter appearing in boldface in the preceding description of each parameter:

$$\mathbf{h}_{11} = \mathbf{h}_i \qquad \mathbf{h}_{12} = \mathbf{h}_r \qquad \mathbf{h}_{21} = \mathbf{h}_f \qquad \mathbf{h}_{22} = \mathbf{h}_o$$

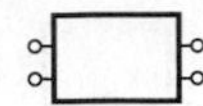

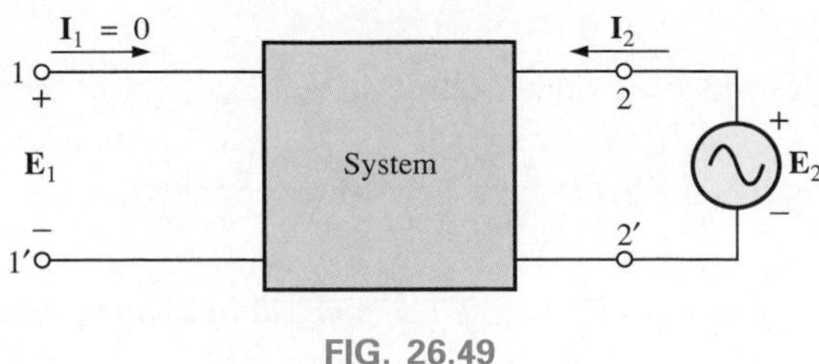

FIG. 26.49
$\mathbf{h}_{22}$ *determination.*

The hybrid equivalent circuit appears in Fig. 26.50. Since the unit of measurement for each term of Eq. (26.40a) is the volt, Kirchhoff's voltage law was applied in reverse to obtain the series input circuit indicated. The unit of measurement of each term of Eq. (26.40b) has the units of current, resulting in the parallel elements of the output circuit as obtained by applying Kirchhoff's current law in reverse.

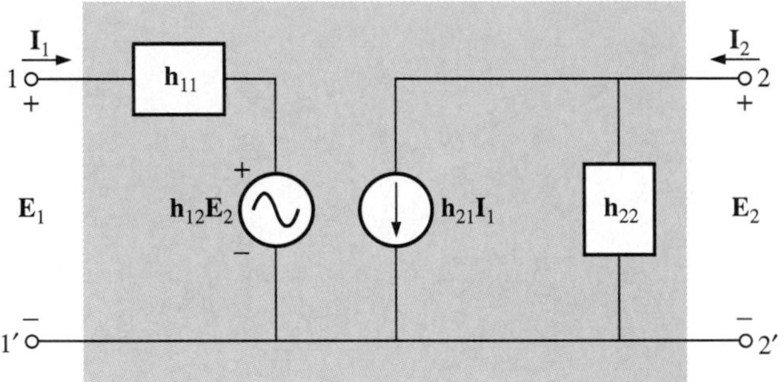

FIG. 26.50
Two-port, hybrid-parameter equivalent network.

Note that the input circuit has a voltage-controlled voltage source whose controlling voltage is the output terminal voltage, while the output circuit has a current-controlled current source whose controlling current is the current of the input circuit.

EXAMPLE 26.10 For the hybrid equivalent circuit of Fig. 26.51:

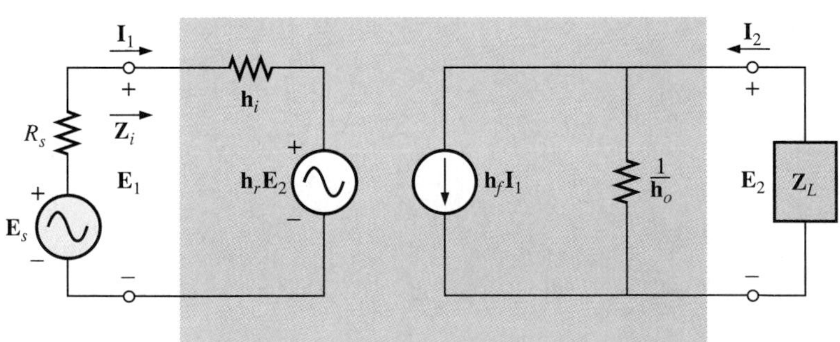

FIG. 26.51
Example 26.10.

a. Determine the current ratio (gain) $\mathbf{A}_i = \mathbf{I}_2/\mathbf{I}_1$.
b. Determine the voltage ratio (gain) $\mathbf{A}_v = \mathbf{E}_2/\mathbf{E}_1$.

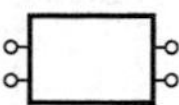

Solutions:

a. Using the current divider rule, we have

$$\mathbf{I}_2 = \frac{(1/\mathbf{h}_o)\mathbf{h}_f\mathbf{I}_1}{(1/\mathbf{h}_o) + \mathbf{Z}_L} = \frac{\mathbf{h}_f\mathbf{I}_1}{1 + \mathbf{h}_o\mathbf{Z}_L}$$

and

$$\boxed{\mathbf{A}_i = \frac{\mathbf{I}_2}{\mathbf{I}_1} = \frac{\mathbf{h}_f}{1 + \mathbf{h}_o\mathbf{Z}_L}} \qquad \textbf{(26.45)}$$

b. Applying Kirchhoff's voltage law to the input circuit gives us

$$\mathbf{E}_1 - \mathbf{h}_i\mathbf{I}_1 - \mathbf{h}_r\mathbf{E}_2 = 0 \quad \text{and} \quad \mathbf{I}_1 = \frac{\mathbf{E}_1 - \mathbf{h}_r\mathbf{E}_2}{\mathbf{h}_i}$$

Apply Kirchhoff's current law to the output circuit:

$$\mathbf{I}_2 = \mathbf{h}_f\mathbf{I}_1 + \mathbf{h}_o\mathbf{E}_2$$

However,

$$\mathbf{I}_2 = -\frac{\mathbf{E}_2}{\mathbf{Z}_L}$$

so

$$-\frac{\mathbf{E}_2}{\mathbf{Z}_L} = \mathbf{h}_f\mathbf{I}_1 + \mathbf{h}_o\mathbf{E}_2$$

Substituting for $\mathbf{I}_1$ gives us

$$-\frac{\mathbf{E}_2}{\mathbf{Z}_L} = \mathbf{h}_f\left(\frac{\mathbf{E}_1 - \mathbf{h}_r\mathbf{E}_2}{\mathbf{h}_i}\right) + \mathbf{h}_o\mathbf{E}_2$$

or

$$\mathbf{h}_i\mathbf{E}_2 = -\mathbf{h}_f\mathbf{Z}_L\mathbf{E}_1 + \mathbf{h}_r\mathbf{h}_f\mathbf{Z}_L\mathbf{E}_2 - \mathbf{h}_i\mathbf{h}_o\mathbf{Z}_L\mathbf{E}_2$$

and

$$\mathbf{E}_2(\mathbf{h}_i - \mathbf{h}_r\mathbf{h}_f\mathbf{Z}_L + \mathbf{h}_i\mathbf{h}_o\mathbf{Z}_L) = -\mathbf{h}_f\mathbf{Z}_L\mathbf{E}_1$$

with the result that

$$\boxed{\mathbf{A}_v = \frac{\mathbf{E}_2}{\mathbf{E}_1} = \frac{-\mathbf{h}_f\mathbf{Z}_L}{\mathbf{h}_i(1 + \mathbf{h}_o\mathbf{Z}_L) - \mathbf{h}_r\mathbf{h}_f\mathbf{Z}_L}} \qquad \textbf{(26.46)}$$

EXAMPLE 26.11 For a particular transistor, $\mathbf{h}_i = 1\text{ k}\Omega$, $\mathbf{h}_r = 4 \times 10^{-4}$, $\mathbf{h}_f = 50$, and $\mathbf{h}_o = 25\ \mu\text{s}$. Determine the current and the voltage gain if $\mathbf{Z}_L$ is a 2-kΩ resistive load.

Solution:

$$\mathbf{A}_i = \frac{\mathbf{h}_f}{1 + \mathbf{h}_o\mathbf{Z}_L} = \frac{50}{1 + (25\ \mu\text{S})(2\text{ k}\Omega)}$$

$$= \frac{50}{1 + (50 \times 10^{-3})} = \frac{50}{1.050} = \mathbf{47.62}$$

$$\mathbf{A}_v = \frac{-\mathbf{h}_f\mathbf{Z}_L}{\mathbf{h}_i(1 + \mathbf{h}_o\mathbf{Z}_L) - \mathbf{h}_r\mathbf{h}_f\mathbf{Z}_L}$$

$$= \frac{-(50)(2\text{ k}\Omega)}{(1\text{ k}\Omega)(1.050) - (4 \times 10^{-4})(50)(2\text{ k}\Omega)}$$

$$= \frac{-100 \times 10^3}{(1.050 \times 10^3) - (0.04 \times 10^3)} = -\frac{100}{1.01} = \mathbf{-99}$$

The minus sign simply indicates a phase shift of 180° between $\mathbf{E}_2$ and $\mathbf{E}_1$ for the defined polarities in Fig. 26.51.

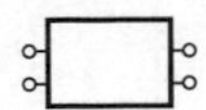

26.9 INPUT AND OUTPUT IMPEDANCES

The input and output impedances will now be determined for the hybrid equivalent circuit and a **z** parameter equivalent circuit. The input impedance can always be determined by the ratio of the input voltage to the input current with or without a load applied. The output impedance is always determined with the source voltage or current set to zero. We found in the previous section that for the hybrid equivalent circuit of Fig. 26.51,

$$\mathbf{E}_1 = \mathbf{h}_i\mathbf{I}_1 + \mathbf{h}_r\mathbf{E}_2$$

$$\mathbf{E}_2 = -\mathbf{I}_2\mathbf{Z}_L$$

and

$$\frac{\mathbf{I}_2}{\mathbf{I}_1} = \frac{\mathbf{h}_f}{1 + \mathbf{h}_o\mathbf{Z}_L}$$

By substituting for $\mathbf{I}_2$ in the second equation (using the relationship of the last equation), we have

$$\mathbf{E}_2 = -\left(\frac{\mathbf{h}_f\mathbf{I}_1}{1 + \mathbf{h}_o\mathbf{Z}_L}\right)\mathbf{Z}_L$$

so the first equation becomes

$$\mathbf{E}_1 = \mathbf{h}_i\mathbf{I}_1 + \mathbf{h}_r\left(-\frac{\mathbf{h}_f\mathbf{I}_1\mathbf{Z}_L}{1 + \mathbf{h}_o\mathbf{Z}_L}\right)$$

and

$$\mathbf{E}_1 = \mathbf{I}_1\left(\mathbf{h}_i - \frac{\mathbf{h}_r\mathbf{h}_f\mathbf{Z}_L}{1 + \mathbf{h}_o\mathbf{Z}_L}\right)$$

Thus,

$$\mathbf{Z}_i = \frac{\mathbf{E}_1}{\mathbf{I}_1} = \mathbf{h}_i - \frac{\mathbf{h}_r\mathbf{h}_f\mathbf{Z}_L}{1 + \mathbf{h}_o\mathbf{Z}_L} \qquad \textbf{(26.47)}$$

For the output impedance, we will set the source voltage to zero but preserve its internal resistance R_s as shown in Fig. 26.52.

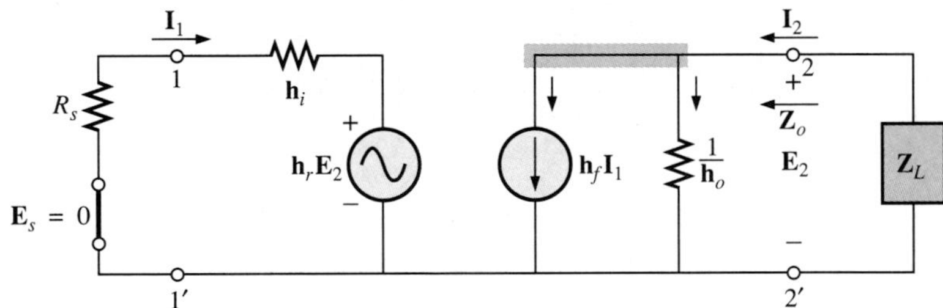

FIG. 26.52
Determining $\mathbf{Z}_o$ for the hybrid equivalent network.

Since

$$\mathbf{E}_s = 0$$

then

$$\mathbf{I}_1 = -\frac{\mathbf{h}_r\mathbf{E}_2}{\mathbf{h}_i + R_s}$$

From the output circuit,

$$\mathbf{I}_2 = \mathbf{h}_f\mathbf{I}_1 + \mathbf{h}_o\mathbf{E}_2$$

or

$$\mathbf{I}_2 = \mathbf{h}_f\left(-\frac{\mathbf{h}_r\mathbf{E}_2}{\mathbf{h}_i + R_s}\right) + \mathbf{h}_o\mathbf{E}_2$$

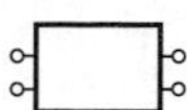

and
$$\mathbf{I}_2 = \left(-\frac{\mathbf{h}_r\mathbf{h}_f}{\mathbf{h}_i + R_s} + \mathbf{h}_o\right)\mathbf{E}_2$$

Thus,
$$\mathbf{Z}_o = \frac{\mathbf{E}_2}{\mathbf{I}_2} = \frac{1}{\mathbf{h}_o - \dfrac{\mathbf{h}_r\mathbf{h}_f}{\mathbf{h}_i + R_s}} \tag{26.48}$$

EXAMPLE 26.12 Determine $\mathbf{Z}_i$ and $\mathbf{Z}_o$ for the transistor having the parameters of Example 26.11 if $R_s = 1\ \text{k}\Omega$.

Solution:

$$\mathbf{Z}_i = \mathbf{h}_i - \frac{\mathbf{h}_r\mathbf{h}_f\mathbf{Z}_L}{1 + \mathbf{h}_o\mathbf{Z}_L} = 1\ \text{k}\Omega - \frac{0.04\ \text{k}\Omega}{1.050}$$
$$= 1 \times 10^3 - 0.0381 \times 10^3 = \mathbf{961.9\ \Omega}$$

$$\mathbf{Z}_o = \frac{1}{\mathbf{h}_o - \dfrac{\mathbf{h}_r\mathbf{h}_f}{\mathbf{h}_i + R_s}} = \frac{1}{25\ \mu\text{S} - \dfrac{(4 \times 10^{-4})(50)}{1\ \text{k}\Omega + 1\ \text{k}\Omega}}$$
$$= \frac{1}{25 \times 10^{-6} - 10 \times 10^{-6}} = \frac{1}{15 \times 10^{-6}}$$
$$\mathbf{Z}_o = \mathbf{66.67\ k\Omega}$$

For the **z** parameter equivalent circuit of Fig. 26.53,

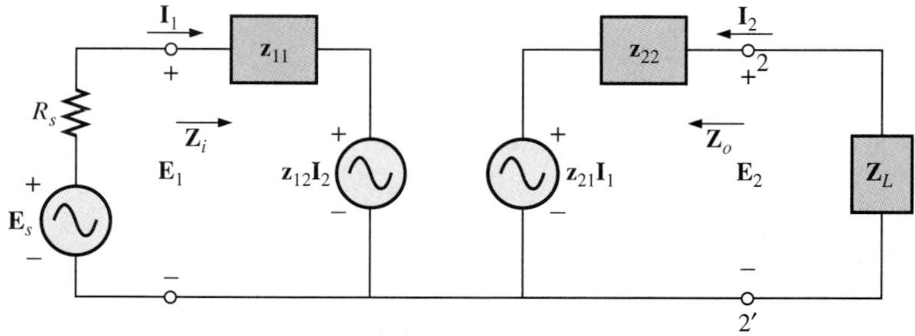

FIG. 26.53
Determining $\mathbf{Z}_i$ *for the* **z**-*parameter equivalent network.*

$$\mathbf{I}_2 = -\frac{\mathbf{z}_{21}\mathbf{I}_1}{\mathbf{z}_{22} + \mathbf{Z}_L}$$

and
$$\mathbf{I}_1 = \frac{\mathbf{E}_1 - \mathbf{z}_{12}\mathbf{I}_2}{\mathbf{z}_{11}}$$

or
$$\mathbf{E}_1 = \mathbf{z}_{11}\mathbf{I}_1 + \mathbf{z}_{12}\mathbf{I}_2 = \mathbf{z}_{11}\mathbf{I}_1 + \mathbf{z}_{12}\left(-\frac{\mathbf{z}_{21}\mathbf{I}_1}{\mathbf{z}_{22} + \mathbf{Z}_L}\right)$$

and
$$\mathbf{Z}_i = \frac{\mathbf{E}_1}{\mathbf{I}_1} = \mathbf{z}_{11} - \frac{\mathbf{z}_{12}\mathbf{z}_{21}}{\mathbf{z}_{22} + \mathbf{Z}_L} \tag{26.49}$$

For the output impedance, $\mathbf{E}_s = 0$, and

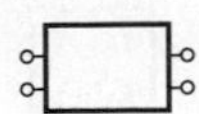

$$\mathbf{I}_1 = -\frac{\mathbf{z}_{12}\mathbf{I}_2}{R_s + \mathbf{z}_{11}} \quad \text{and} \quad \mathbf{I}_2 = \frac{\mathbf{E}_2 - \mathbf{z}_{21}\mathbf{I}_1}{\mathbf{z}_{22}}$$

or
$$\mathbf{E}_2 = \mathbf{z}_{22}\mathbf{I}_2 + \mathbf{z}_{21}\mathbf{I}_1 = \mathbf{z}_{22}\mathbf{I}_2 + \mathbf{z}_{21}\left(-\frac{\mathbf{z}_{12}\mathbf{I}_2}{R_s + \mathbf{z}_{11}}\right)$$

and
$$\mathbf{E}_2 = \mathbf{z}_{22}\mathbf{I}_2 - \frac{\mathbf{z}_{12}\mathbf{z}_{21}\mathbf{I}_2}{R_s + \mathbf{z}_{11}}$$

Thus,
$$\boxed{\mathbf{Z}_o = \frac{\mathbf{E}_2}{\mathbf{I}_2} = \mathbf{z}_{22} - \frac{\mathbf{z}_{12}\mathbf{z}_{21}}{R_s + \mathbf{z}_{11}}} \tag{26.50}$$

26.10 CONVERSION BETWEEN PARAMETERS

The equations relating the **z** and **y** parameters can be determined directly from Eqs. (26.22) and (26.31). For Eqs. (26.31a) and (26.31b),

$$\mathbf{I}_1 = \mathbf{y}_{11}\mathbf{E}_1 + \mathbf{y}_{12}\mathbf{E}_2$$
$$\mathbf{I}_2 = \mathbf{y}_{21}\mathbf{E}_1 + \mathbf{y}_{22}\mathbf{E}_2$$

The use of determinants will result in

$$\mathbf{E}_1 = \frac{\begin{vmatrix} \mathbf{I}_1 & \mathbf{y}_{12} \\ \mathbf{I}_2 & \mathbf{y}_{22} \end{vmatrix}}{\begin{vmatrix} \mathbf{y}_{11} & \mathbf{y}_{12} \\ \mathbf{y}_{21} & \mathbf{y}_{22} \end{vmatrix}} = \frac{\mathbf{y}_{22}\mathbf{I}_1 - \mathbf{y}_{12}\mathbf{I}_2}{\mathbf{y}_{11}\mathbf{y}_{22} - \mathbf{y}_{12}\mathbf{y}_{21}}$$

Substituting the notation

$$\Delta_\mathbf{y} = \mathbf{y}_{11}\mathbf{y}_{22} - \mathbf{y}_{12}\mathbf{y}_{21}$$

we have
$$\mathbf{E}_1 = \frac{\mathbf{y}_{22}}{\Delta_\mathbf{y}}\mathbf{I}_1 - \frac{\mathbf{y}_{12}}{\Delta_\mathbf{y}}\mathbf{I}_2$$

which, when related to Eq. (26.22a),

$$\mathbf{E}_1 = \mathbf{z}_{11}\mathbf{I}_1 + \mathbf{z}_{12}\mathbf{I}_2$$

indicates that

$$\mathbf{z}_{11} = \frac{\mathbf{y}_{22}}{\Delta_\mathbf{y}} \quad \text{and} \quad \mathbf{z}_{12} = \frac{\mathbf{y}_{12}}{\Delta_\mathbf{y}}$$

and, similarly,

$$\mathbf{z}_{21} = \frac{\mathbf{y}_{21}}{\Delta_\mathbf{y}} \quad \text{and} \quad \mathbf{z}_{22} = \frac{\mathbf{y}_{11}}{\Delta_\mathbf{y}}$$

For the conversion of **z** parameters to the admittance domain, determinants are applied to Eqs. (26.22a) and (26.22b). The impedance parameters can be found in terms of the hybrid parameters by first forming the determinant for $\mathbf{I}_1$ from the hybrid equations,

$$\mathbf{E}_1 = \mathbf{h}_{11}\mathbf{I}_1 + \mathbf{h}_{12}\mathbf{E}_2$$
$$\mathbf{I}_2 = \mathbf{h}_{21}\mathbf{I}_1 + \mathbf{h}_{22}\mathbf{E}_2$$

That is,

$$\mathbf{I}_1 = \frac{\begin{vmatrix} \mathbf{E}_1 & \mathbf{h}_{12} \\ \mathbf{I}_2 & \mathbf{h}_{22} \end{vmatrix}}{\begin{vmatrix} \mathbf{h}_{11} & \mathbf{h}_{12} \\ \mathbf{h}_{21} & \mathbf{h}_{22} \end{vmatrix}} = \frac{\mathbf{h}_{22}}{\Delta_\mathbf{h}}\mathbf{E}_1 - \frac{\mathbf{h}_{12}}{\Delta_\mathbf{h}}\mathbf{I}_2$$

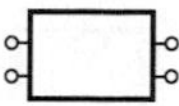

and $$\frac{\mathbf{h}_{22}}{\Delta_\mathbf{h}}\mathbf{E}_1 = \mathbf{I}_1 - \frac{\mathbf{h}_{12}}{\Delta_\mathbf{h}}\mathbf{I}_2$$

or $$\mathbf{E}_1 = \frac{\Delta_\mathbf{h}\mathbf{I}_1}{\mathbf{h}_{22}} - \frac{\mathbf{h}_{12}}{\mathbf{h}_{22}}\mathbf{I}_2$$

which, when related to the impedance parameter equation,

$$\mathbf{E}_1 = \mathbf{z}_{11}\mathbf{I}_1 + \mathbf{z}_{12}\mathbf{I}_2$$

indicates that

$$\mathbf{z}_{11} = \frac{\Delta_\mathbf{h}}{\mathbf{h}_{22}} \quad \text{and} \quad \mathbf{z}_{12} = -\frac{\mathbf{h}_{12}}{\mathbf{h}_{22}}$$

The remaining conversions are left as an exercise. A complete table of conversions appears in Table 26.1.

TABLE 26.1
Conversions between **z, y,** *and* **h** *parameters.*

To ↓ / From →	z		y		h	
z	$\mathbf{z}_{11}$	$\mathbf{z}_{12}$	$\frac{\mathbf{y}_{22}}{\Delta_\mathbf{y}}$	$\frac{-\mathbf{y}_{12}}{\Delta_\mathbf{y}}$	$\frac{\Delta_\mathbf{h}}{\mathbf{h}_{22}}$	$\frac{\mathbf{h}_{12}}{\mathbf{h}_{22}}$
	$\mathbf{z}_{21}$	$\mathbf{z}_{22}$	$\frac{-\mathbf{y}_{21}}{\Delta_\mathbf{y}}$	$\frac{\mathbf{y}_{11}}{\Delta_\mathbf{y}}$	$\frac{-\mathbf{h}_{21}}{\mathbf{h}_{22}}$	$\frac{1}{\mathbf{h}_{22}}$
y	$\frac{\mathbf{z}_{22}}{\Delta_\mathbf{z}}$	$\frac{-\mathbf{z}_{12}}{\Delta_\mathbf{z}}$	$\mathbf{y}_{11}$	$\mathbf{y}_{12}$	$\frac{1}{\mathbf{h}_{11}}$	$\frac{-\mathbf{h}_{12}}{\mathbf{h}_{11}}$
	$\frac{-\mathbf{z}_{21}}{\Delta_\mathbf{z}}$	$\frac{\mathbf{z}_{11}}{\Delta_\mathbf{z}}$	$\mathbf{y}_{21}$	$\mathbf{y}_{22}$	$\frac{\mathbf{h}_{21}}{\mathbf{h}_{11}}$	$\frac{\Delta_\mathbf{h}}{\mathbf{h}_{11}}$
h	$\frac{\Delta_\mathbf{z}}{\mathbf{z}_{22}}$	$\frac{\mathbf{z}_{12}}{\mathbf{z}_{22}}$	$\frac{1}{\mathbf{y}_{11}}$	$\frac{-\mathbf{y}_{12}}{\mathbf{y}_{11}}$	$\mathbf{h}_{11}$	$\mathbf{h}_{12}$
	$\frac{-\mathbf{z}_{21}}{\mathbf{z}_{22}}$	$\frac{1}{\mathbf{z}_{22}}$	$\frac{\mathbf{y}_{21}}{\mathbf{y}_{11}}$	$\frac{\Delta_\mathbf{y}}{\mathbf{y}_{11}}$	$\mathbf{h}_{21}$	$\mathbf{h}_{22}$

26.11 COMPUTER ANALYSIS

PSpice (DOS)

Determining A_v and Z_o for the Hybrid Equivalent Network
The computer analysis of this chapter will be limited to a practical session in entering controlled sources into the PSpice software package. The system to be analyzed is the loaded two-port hybrid equivalent model of Fig. 26.54. Note the insertion of a sensing voltage source to define the direction of $\mathbf{I}_1$ and the common reference point for both the input and output circuits. The desired output voltage $\mathbf{V}_L = \mathbf{E}_2$ will also be the controlling voltage for the VCVS of the input circuit.

Before examining the input and output files, let us first determine the voltage gain using Eq. (26.46).

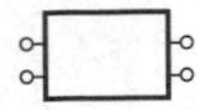

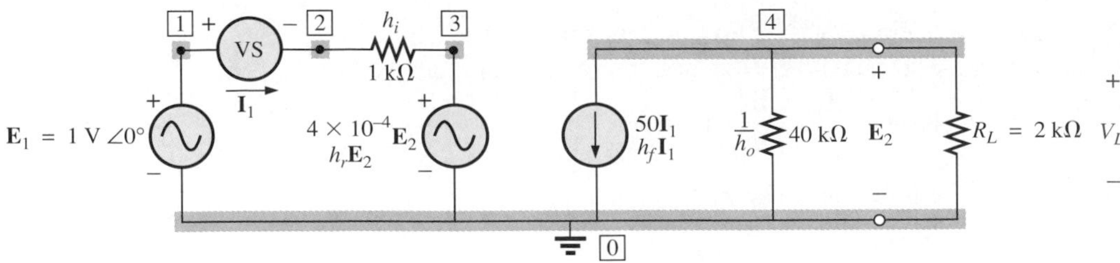

FIG. 26.54
Assigning the defining parameters for a PSpice (DOS) analysis of a hybrid equivalent network.

$$\mathbf{A}_v = \frac{-\mathbf{h}_f R_L}{\mathbf{h}_i(1 + \mathbf{h}_o R_L) - \mathbf{h}_r \mathbf{h}_f R_L}$$

$$= \frac{-(50)(2\text{ k}\Omega)}{(1\text{ k}\Omega)(1 + (25 \times 10^{-6}\text{ S})(2\text{ k}\Omega)) - (4 \times 10^{-4})(50)(2\text{ k}\Omega)}$$

$$= \frac{-100 \times 10^3}{(1\text{ k}\Omega)(1 + 50 \times 10^{-3}) - 40} = \frac{-100 \times 10^3}{1050 - 40} = -99.01$$

and $$\mathbf{A}_v = \frac{\mathbf{E}_2}{\mathbf{E}_1} = \frac{\mathbf{V}_L}{\mathbf{E}_1}$$

so that $$\mathbf{V}_L = \mathbf{A}_v\mathbf{E}_1 = (-99.01)(1\text{ V}\angle 0^\circ)$$
$$= \mathbf{99.01\text{ V}\angle 180^\circ}$$

The input and output files appear in Fig. 26.55 with the required entries and the request for the load voltage. The output file provides an exact match with the solution just obtained.

The last PSpice run of the section (and the text) will determine the output impedance for the network of Fig. 26.54 by setting the voltage

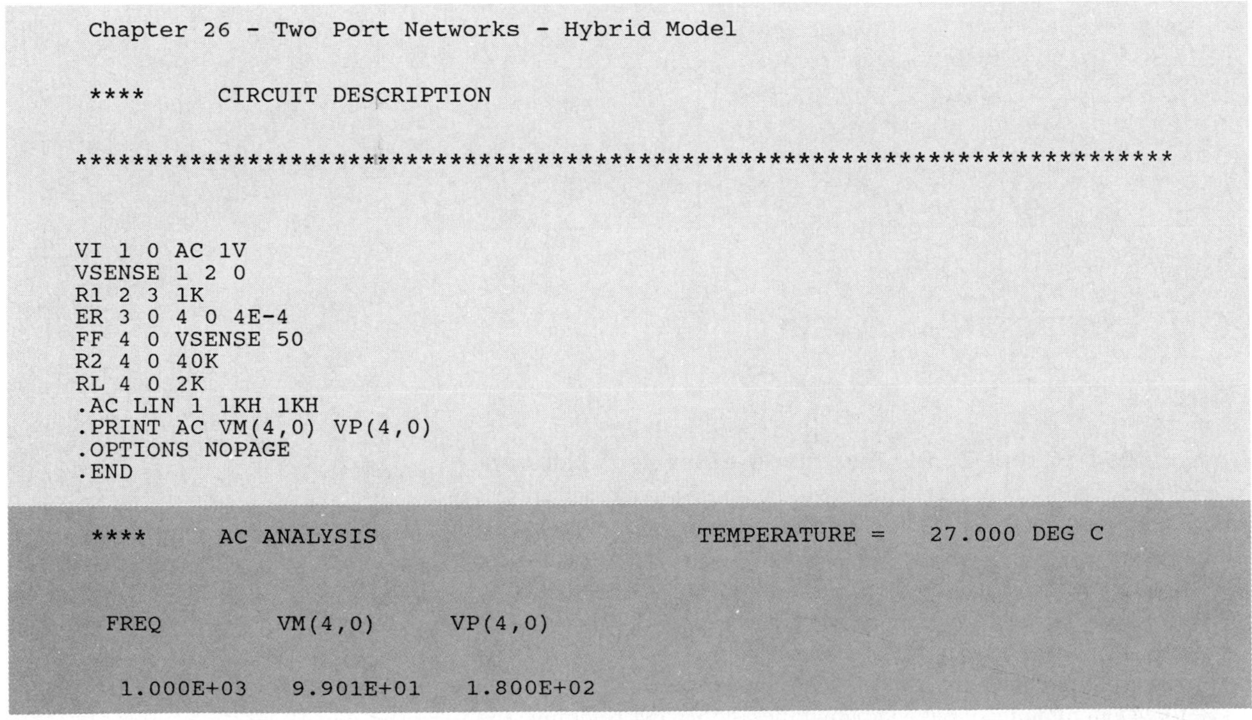

```
Chapter 26 - Two Port Networks - Hybrid Model

 ****     CIRCUIT DESCRIPTION

******************************************************************************

VI 1 0 AC 1V
VSENSE 1 2 0
R1 2 3 1K
ER 3 0 4 0 4E-4
FF 4 0 VSENSE 50
R2 4 0 40K
RL 4 0 2K
.AC LIN 1 1KH 1KH
.PRINT AC VM(4,0) VP(4,0)
.OPTIONS NOPAGE
.END

 ****     AC ANALYSIS                          TEMPERATURE =   27.000 DEG C

  FREQ        VM(4,0)      VP(4,0)

   1.000E+03   9.901E+01   1.800E+02
```

FIG. 26.55
Input and output files for the network of Fig. 26.54.

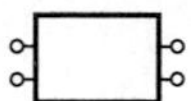

source to zero volts, as shown in Fig. 26.56. A current source of 1 A is applied, and the voltage $\mathbf{V}_s$ is determined; it has the same magnitude and angle as $\mathbf{Z}_o$ since

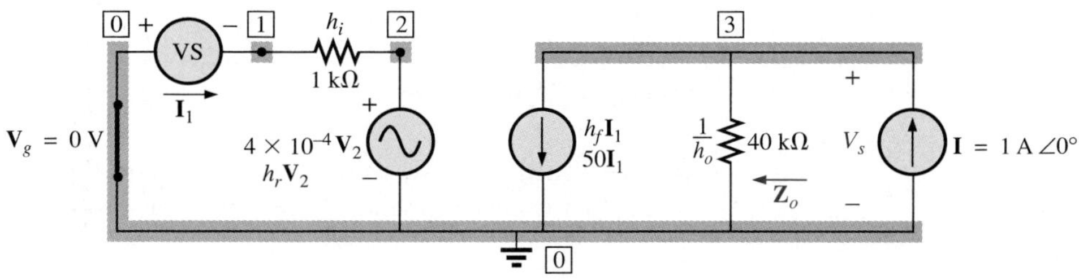

FIG. 26.56
Determining $\mathbf{Z}_o$ for the hybrid-equivalent network using PSpice (DOS).

$$\mathbf{Z}_o = \frac{\mathbf{V}_s}{\mathbf{I}} = \frac{\mathbf{V}_s}{1\text{ A}\angle 0°} = \frac{V_s}{1\text{ A}}\angle\theta_s - 0° = V_s\,\Omega\angle\theta_s$$

Before reviewing the input and output files, let us again determine the solution using the longhand approach to check the computer result.

Eq. (26.48):

$$\mathbf{Z}_o = \frac{1}{\mathbf{h}_o - \dfrac{\mathbf{h}_r\mathbf{h}_f}{\mathbf{h}_i + R_s}} = \frac{1}{25 \times 10^{-6}\text{ S} - \dfrac{(4 \times 10^{-4})(50)}{1\text{ k}\Omega + 0}}$$

$$= \frac{1}{25 \times 10^{-6}\text{ S} - 20 \times 10^{-6}\text{ S}} = \frac{1}{5 \times 10^{-6}\text{ S}} = \mathbf{200\ k\Omega}$$

The input file of Fig. 26.57 includes the applied current source of 1 A at zero degrees and uses a frequency of 1 kHz to establish the ac

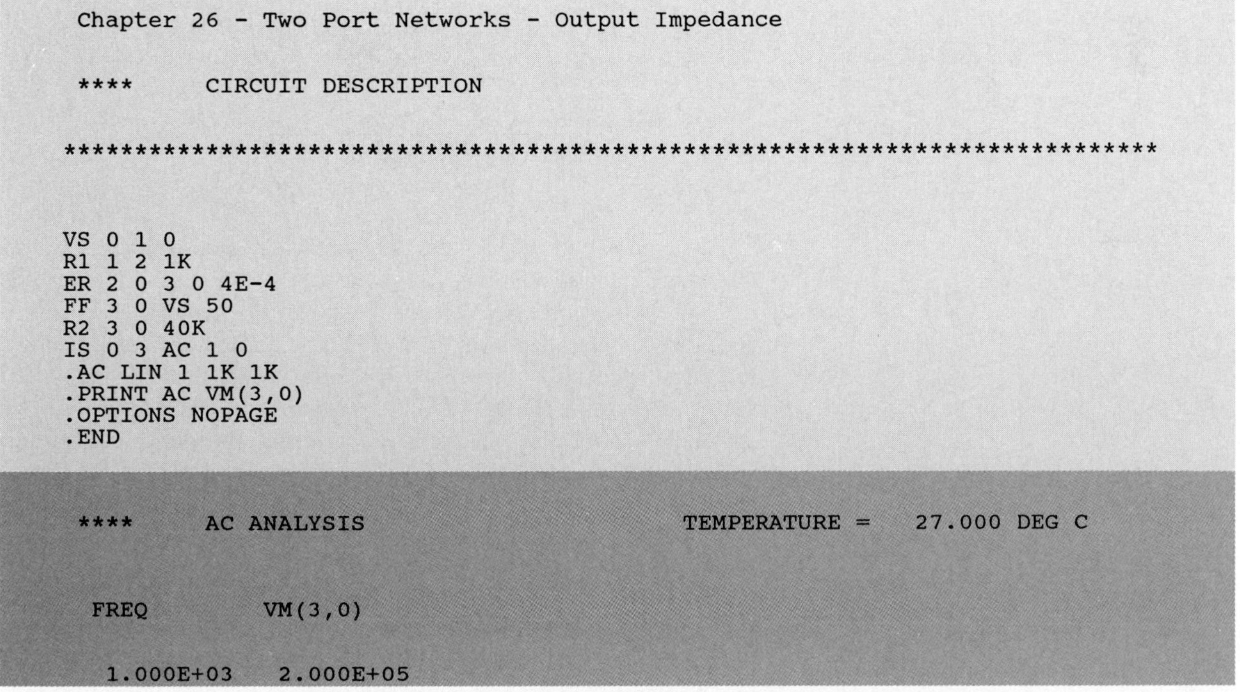

```
Chapter 26 - Two Port Networks - Output Impedance

****     CIRCUIT DESCRIPTION

****************************************************************************

VS 0 1 0
R1 1 2 1K
ER 2 0 3 0 4E-4
FF 3 0 VS 50
R2 3 0 40K
IS 0 3 AC 1 0
.AC LIN 1 1K 1K
.PRINT AC VM(3,0)
.OPTIONS NOPAGE
.END

****     AC ANALYSIS                          TEMPERATURE =   27.000 DEG C

  FREQ          VM(3,0)

  1.000E+03     2.000E+05
```

FIG. 26.57
Applying PSpice (DOS) to the network of Fig. 26.54.

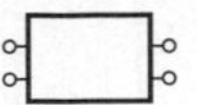

analysis. In particular, note the ability to carry the reference node through to the sensing voltage source and the need to redefine the nodes of Fig. 26.54. The magnitude of the resulting voltage $\mathbf{V}_s$ is again an exact match with the above solution.

PSpice (Windows)

Determining A_v The gain for the network of Fig. 26.54 will now be determined using schematics. Using a current-controlled current source defined by:

	Symbol	Library File	Schematic
CCCS	**F**	analog.slb	F1 [symbol] F

and a voltage-controlled voltage source defined by:

VCVS	**E**	analog.slb	E1 [symbol] E

the network of Fig. 26.58 can be developed.

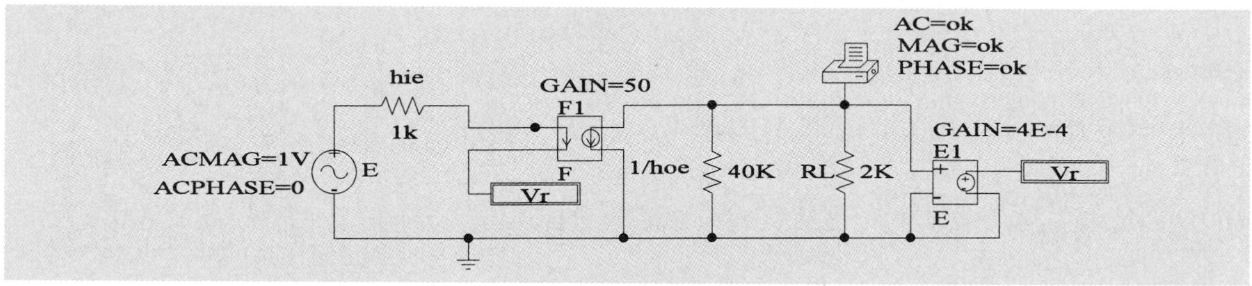

FIG. 26.58
Windows schematic for the network of Fig. 26.54.

As noted in earlier chapters, the current directions in the CCCS and the voltage polarities in the VCVS must be adhered to in the development of the network. Because of the size of the network and the presence of two controlled sources, it would be difficult if not confusing to make the necessary connections with the **WIRE** menu selection. For such situations the **GLOBAL** option under the **port.slb** library file will prove useful. Once chosen and in place, it must be assigned an attribute value (LABEL) such as **Vr** in Fig. 26.58. By choosing **GLOBAL** again and giving it the same **LABEL,** another point in the network can be assigned the same voltage level. In other words, the two points are connected, although the need for a direct wire is unnecessary. Any additional **GLOBAL** points with the same **LABEL** will also be assigned the same potential level.

Both controlled sources must be assigned a **GAIN** by double-clicking the element on the schematic and entering the desired level. If more than one VCVS is entered for the same schematic, be sure that the part

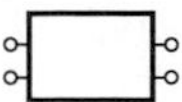

names are different, or an error message will surface upon simulation. In fact, for any network be sure that no two parts have the same label.

The output results appearing in Fig. 26.59 are the same as obtained earlier; that is, the magnitude of V_L is 99.01 V and the phase angle is 180°.

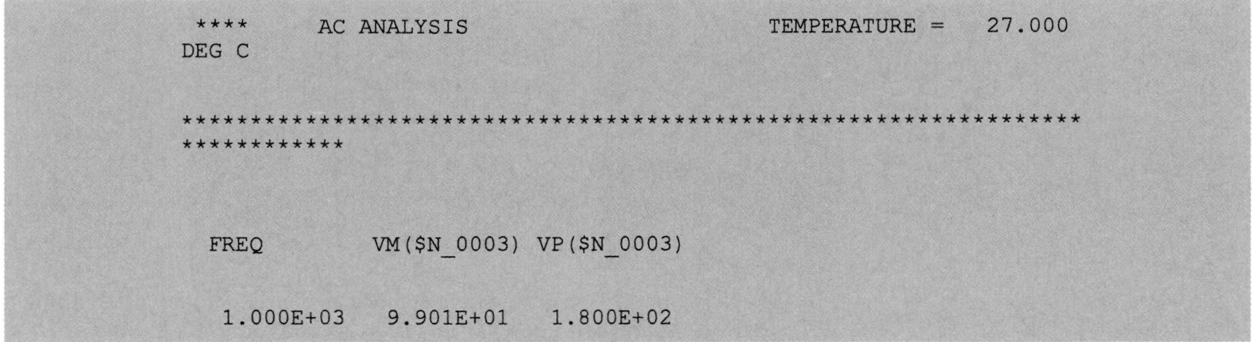

```
****        AC ANALYSIS                           TEMPERATURE =    27.000
DEG C

*************************************************************************
************

  FREQ         VM($N_0003) VP($N_0003)

   1.000E+03   9.901E+01    1.800E+02
```

FIG. 26.59
Output results for the Windows analysis of Fig. 26.54.

PROBLEMS

SECTION 26.2 The Impedance Parameters Z_i and Z_o

1. Given the indicated voltage levels of Fig. 26.60, determine the magnitude of the input impedance $\mathbf{Z}_i$.

2. For a system with $\mathbf{E}_i = 120\text{ V}\angle 0°$ and $\mathrm{I}_i =$ $6.2\text{ A}\angle -10.8°$, determine the input impedance in rectangular form. At a frequency of 60 Hz, determine the nameplate values of the parameters.

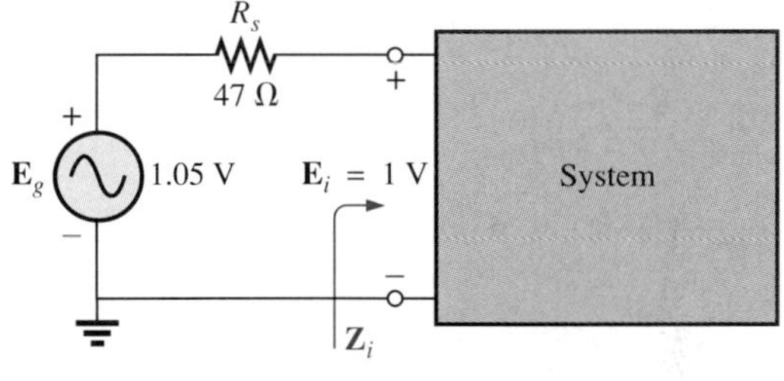

FIG. 26.60
Problem 1.

3. For the multiport system of Fig. 26.61:
   a. Determine the magnitude of $\mathbf{I}_{i_1}$ if $\mathbf{E}_{i_1} = 20$ mV.
   b. Find $\mathbf{Z}_{i_2}$ using the information provided.
   c. Calculate the magnitude of $\mathbf{E}_{i_3}$.

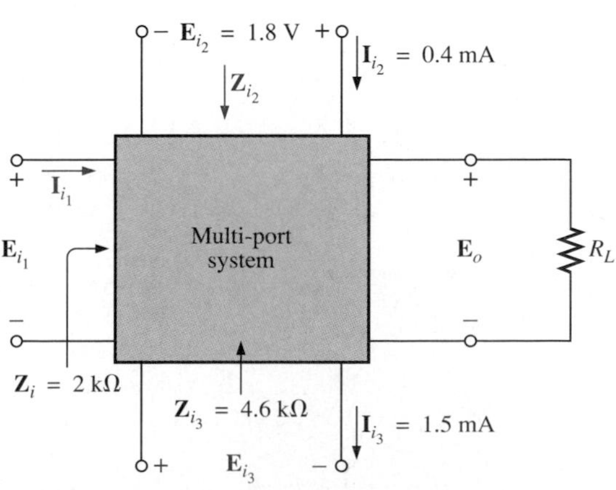

FIG. 26.61
Problem 3.

4. Given the indicated voltage levels of Fig. 26.62, determine $\mathbf{Z}_o$.

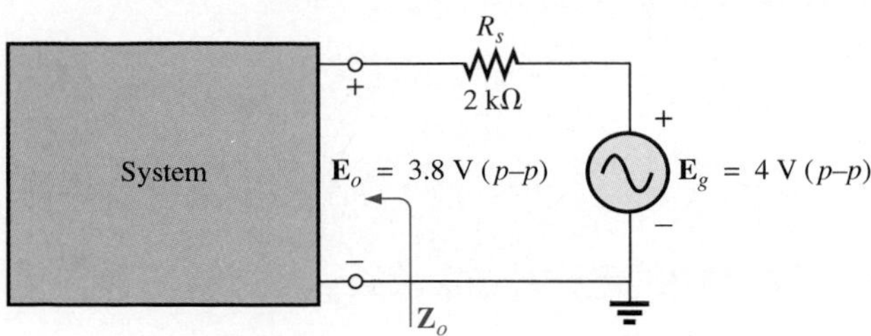

FIG. 26.62
Problems 4 to 6.

5. For the configuration of Fig. 26.62, determine $\mathbf{Z}_o$ if $e_g = 2 \sin 377t$ and $\upsilon_R = 40 \times 10^{-3} \sin 377t$, with $R_s = 0.91$ kΩ.
6. Determine $\mathbf{Z}_o$ for the system of Fig. 26.62 if E_g = 1.8 V (p-p) and E_o = 0.6 V rms.
7. Determine the output impedance for the system of Fig. 26.63 given the indicated scope measurements.

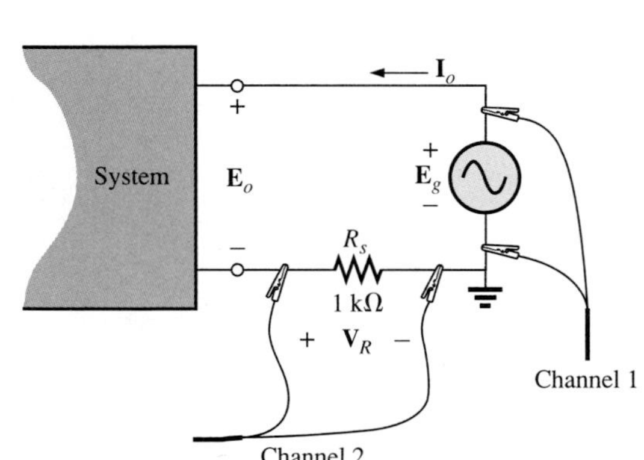

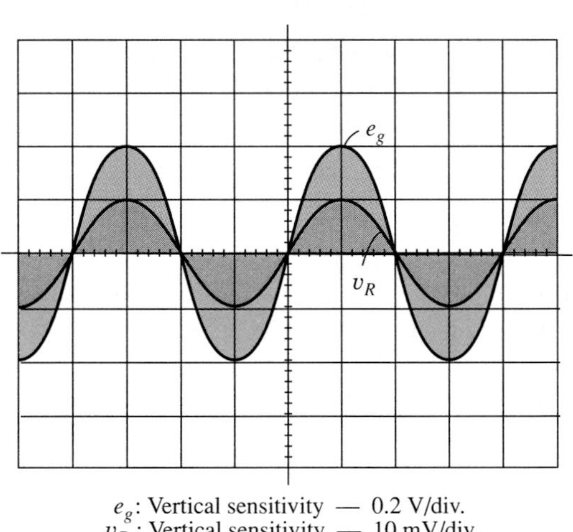

FIG. 26.63
Problem 7.

SECTION 26.3 The Voltage Gains $A_{\upsilon_{NL}}$, A_{υ}, and A_{υ_T}

8. Given the system of Fig. 26.64, determine the no-load voltage gain $\mathbf{A}_{\upsilon_{NL}}$.

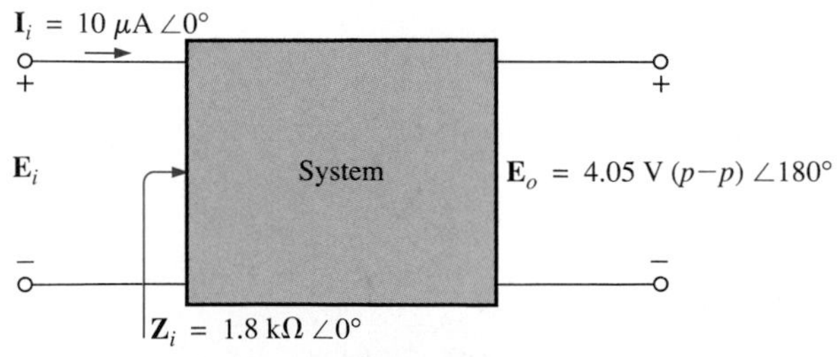

FIG. 26.64
Problem 8.

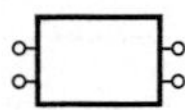

9. For the system of Fig. 26.65:

a. Determine $\mathbf{A}_v = \mathbf{E}_o/\mathbf{E}_i$.

b. Find $\mathbf{A}_{v_T} = \mathbf{E}_o/\mathbf{E}_g$.

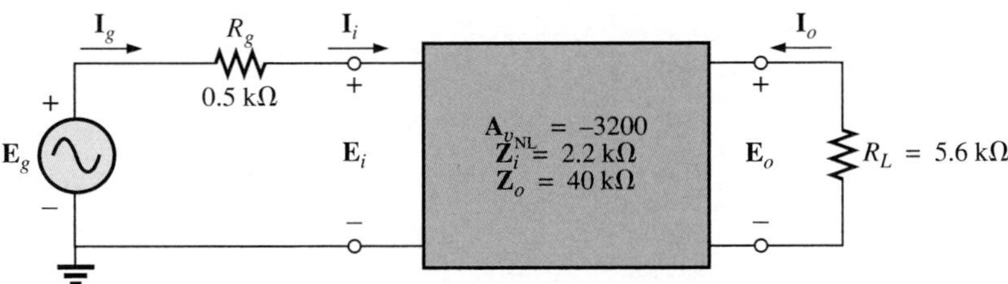

FIG. 26.65
Problems 9, 12, and 13.

10. For the system of Fig. 26.66(a), the no-load output voltage is −1440 mV, with 1.2 mV applied at the input terminals. In Fig. 26.66(b), a 4.7-kΩ load is applied to the same system, and the output voltage drops to −192 mV, with the same applied input signal. What is the output impedance of the system?

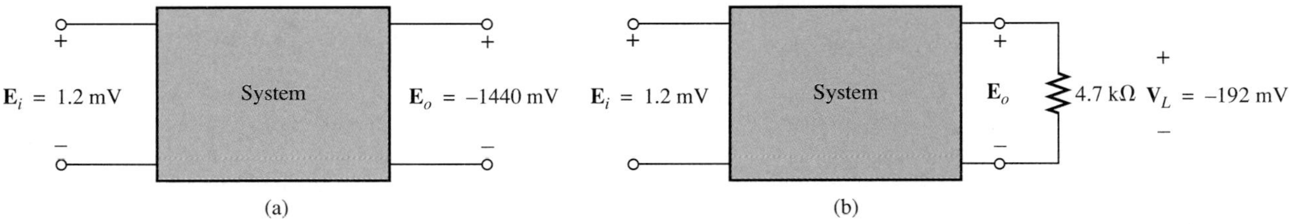

FIG. 26.66
Problem 10.

***11.** For the system of Fig. 26.67, if $\mathbf{A}_v = -160$, $\mathbf{I}_o = 4 \text{ mA} \angle 0°$, and $\mathbf{E}_g = 70 \text{ mV} \angle 0°$:

a. Determine the no-load voltage gain.

b. Find the magnitude of $\mathbf{E}_i$.

c. Determine $\mathbf{Z}_i$.

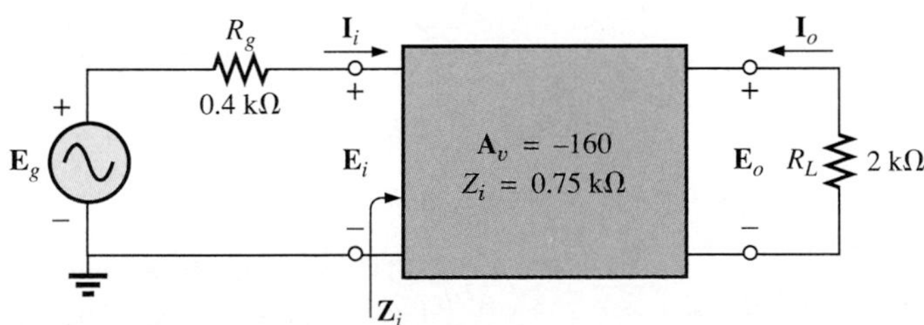

FIG. 26.67
Problems 11, and 14.

SECTION 26.4 The Current Gains A_i and A_{i_T}, and the Power Gain A_G

12. For the system of Fig. 26.65:

a. Determine $\mathbf{A}_i = \mathbf{I}_o/\mathbf{I}_i$.

b. Find $\mathbf{A}_{i_T} = \mathbf{I}_o/\mathbf{I}_g$.

c. Compare the results of parts (a) and (b) and explain why the results compare as they do.

13. For the system of Fig. 26.65:
 a. Determine A_G using Eq. (26.13) and compare with the result obtained using Eq. (26.14).
 b. Find A_{G_T} using Eq. (26.16) and compare to the result obtained using Eq. (26.17).

14. For the system of Fig. 26.67:
 a. Determine the magnitude of $\mathbf{A}_i = \mathbf{I}_o/\mathbf{I}_i$.
 b. Find the power gain $A_{G_T} = P_L/P_g$.

SECTION 26.5 Cascaded Systems

15. For the two-stage system of Fig. 26.68:
 a. Determine the total voltage gain $\mathbf{A}_{v_T} = \mathbf{V}_L/\mathbf{E}_i$.
 b. Find the total current gain $\mathbf{A}_{i_T} = \mathbf{I}_o/\mathbf{I}_i$.
 c. Find the current gain of each stage $\mathbf{A}_{i_1}$ and $\mathbf{A}_{i_2}$.
 d. Determine the total current gain using the results of part (c) and compare it to the result obtained in part (b).

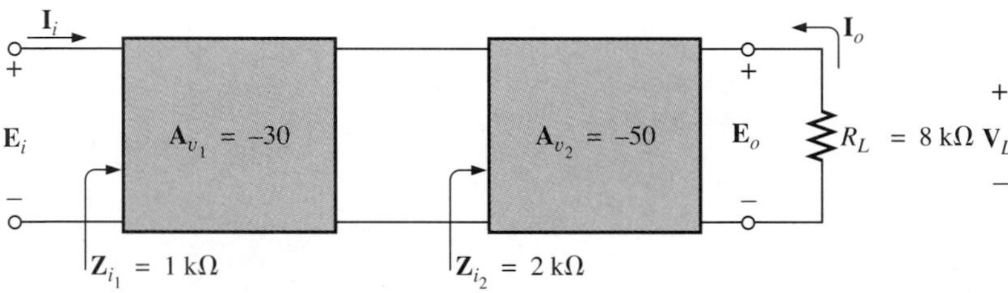

FIG. 26.68
Problem 15.

***16.** For the system of Fig. 26.69:
 a. Determine $\mathbf{A}_{v_2}$ if $\mathbf{A}_{v_T} = -6912$.
 b. Determine $\mathbf{Z}_{i_2}$ using the information provided.
 c. Find $\mathbf{A}_{i_3}$ and $\mathbf{A}_{i_T}$ using the information provided in Fig. 26.69.

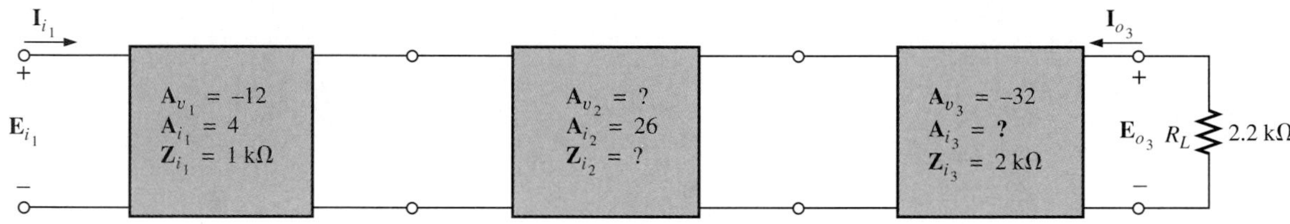

FIG. 26.69
Problem 16.

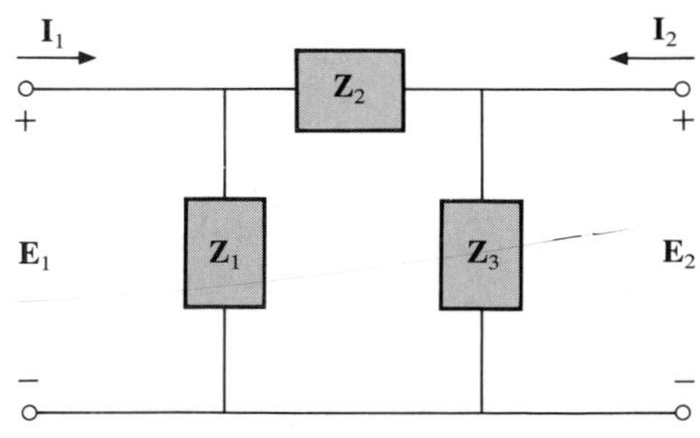

FIG. 26.70
Problems 17 and 21.

SECTION 26.6 Impedance (z) Parameters

17. a. Determine the impedance (**z**) parameters for the π network of Fig. 26.70.
 b. Sketch the **z** parameter equivalent circuit (using either form of Fig. 26.32).

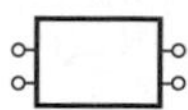

18. a. Determine the impedance (**z**) parameters for the network of Fig. 26.71.

b. Sketch the **z** parameter equivalent circuit (using either form of Fig. 26.32).

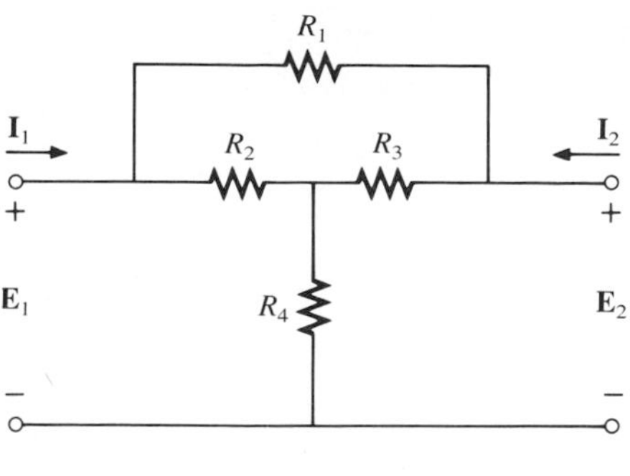

FIG. 26.71
Problems 18 and 22.

SECTION 26.7 Admittance (y) Parameters

19. a. Determine the admittance (**y**) parameters for the T network of Fig. 26.72.

b. Sketch the **y** parameter equivalent circuit (using either form of Fig. 26.44).

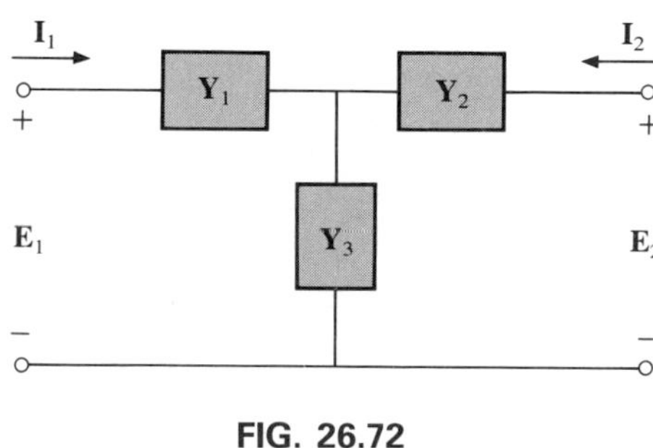

FIG. 26.72
Problems 19 and 23.

20. a. Determine the admittance (**y**) parameters for the network of Fig. 26.73.

b. Sketch the **y** parameter equivalent circuit (using either form of Fig. 26.44).

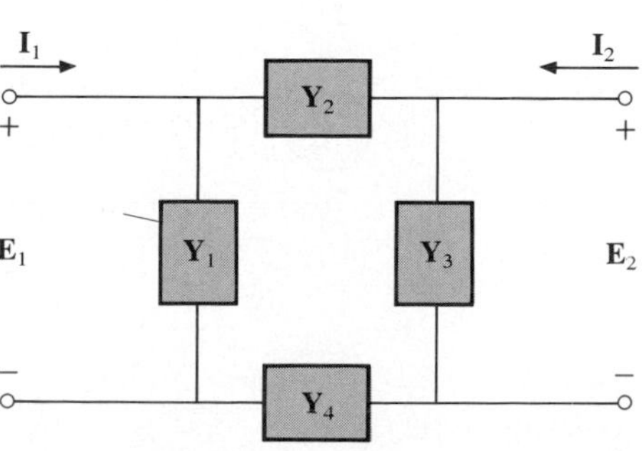

FIG. 26.73
Problems 20 and 24.

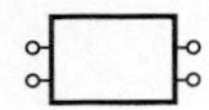

SECTION 26.8 Hybrid (h) Parameters

21. **a.** Determine the **h** parameters for the network of Fig. 26.70.
b. Sketch the hybrid equivalent circuit.

22. **a.** Determine the **h** parameters for the network of Fig. 26.71.
b. Sketch the hybrid equivalent circuit.

23. **a.** Determine the **h** parameters for the network of Fig. 26.72.
b. Sketch the hybrid equivalent circuit.

24. **a.** Determine the **h** parameters for the network of Fig. 26.73.
b. Sketch the hybrid equivalent circuit.

25. For the hybrid equivalent circuit of Fig. 26.74:
a. Determine the current gain $A_i = \mathbf{I}_2/\mathbf{I}_1$.
b. Determine the voltage gain $A_v = \mathbf{E}_2/\mathbf{E}_1$.

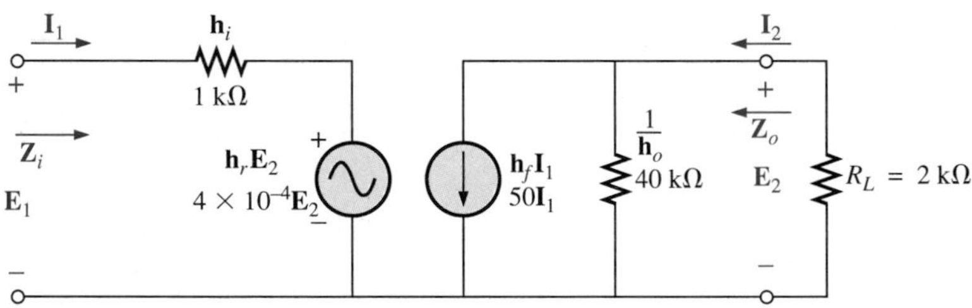

FIG. 26.74
Problems 25 and 26.

SECTION 26.9 Input and Output Impedances

26. For the hybrid equivalent circuit of Fig. 26.74:
a. Determine the input impedance.
b. Determine the output impedance.

27. Determine the input and output impedances for the **z** parameter equivalent circuit of Fig. 26.75.

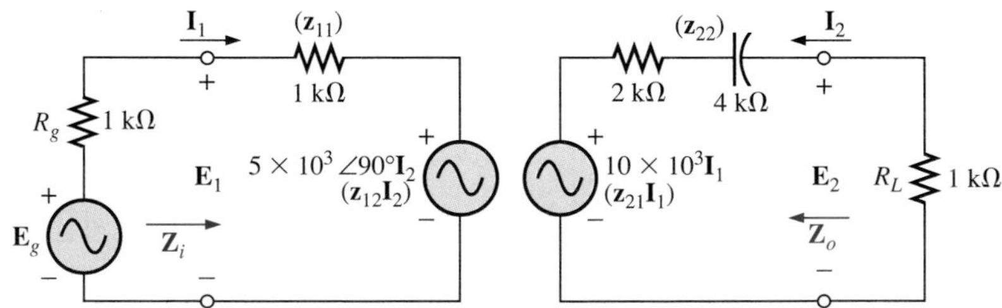

FIG. 26.75
Problems 27, 32, and 34.

28. Determine the expression for the input and output impedance of the **y** parameter equivalent circuit.

SECTION 26.10 Conversion Between Parameters

29. Determine the **h** parameters for the following **z** parameters:

$$\mathbf{z}_{11} = 4\text{ k}\Omega$$
$$\mathbf{z}_{12} = 2\text{ k}\Omega$$
$$\mathbf{z}_{21} = 3\text{ k}\Omega$$
$$\mathbf{z}_{22} = 4\text{ k}\Omega$$

30. a. Determine the **z** parameters for the following **h** parameters:

$$\mathbf{h}_{11} = 1\text{ k}\Omega$$
$$\mathbf{h}_{12} = 2 \times 10^{-4}$$
$$\mathbf{h}_{21} = 100$$
$$\mathbf{h}_{22} = 20 \times 10^{-6}\text{ S}$$

b. Determine the **y** parameters for the hybrid parameters indicated in part (a).

SECTION 26.11 Computer Analysis

PSpice (DOS or Windows)

31. For $\mathbf{E}_1 = 4\text{ V} \angle 30°$ determine $\mathbf{E}_2$ across a 2-kΩ resistive load between 2 and 2′ for the network of Fig. 26.34.

32. For $\mathbf{E}_g = 2\text{ V} \angle 0°$ determine $\mathbf{E}_2$ for the network of Fig. 26.75.

33. Determine $\mathbf{Z}_i$ for the network of Fig. 26.34 with a 2-kΩ resistive load from 2 to 2′.

34. Determine $\mathbf{Z}_i$ for the network of Fig. 26.75.

GLOSSARY

Admittance (y) parameters A set of parameters, having the units of siemens, that can be used to establish a two-port equivalent network for a system.

Hybrid (h) parameters A set of mixed parameters (ohms, siemens, some unitless) that can be used to establish a two-port equivalent network for a system.

Impedance (z) parameters A set of parameters, having the units of ohms, that can be used to establish a two-port equivalent network for a system.

Input impedance The impedance appearing at the input terminals of a system.

Output impedance The impedance appearing at the output terminals of a system with the energizing source set to zero.

Single-port network A network having a single set of access terminals.

Two-port network A network having two pairs of access terminals.

Appendixes

Appendix A

PSpice, C++, and MathCAD

PSpice

The PSpice software package employed throughout this text is derived from programs developed at the University of California at Berkeley during the early 1970s. SPICE is an acronym for Simulation Program with Integrated Circuit Emphasis. SPICE has undergone many changes since the early versions of circuit-analysis programs such as ECAP (Electrical Circuit Analysis Program) were developed. Although a number of companies have customized SPICE for their particular use, MicroSim has developed both a commercial package and a student version. The commercial or professional versions employed by engineering companies can be costly, so MicroSim encourages duplication of the student version in the educational community. The programs of this text were all run on a student version to ensure that they will work in a classroom environment. PSpice is one of a number of circuit-analysis and simulation programs that can be used to carry out either dc or ac analysis on essentially any type of circuit with practical elements. The commercial package is in general too extensive for most student applications and quite expensive, whereas the student package is designed to provide an exposure to this powerful software package at minimum cost. In fact, as mentioned above, MicroSim encourages duplication of its student version to ensure maximum distribution and ease of use.

The content of this text was designed to provide sufficient detail to take the user through the analysis without a supplementary text. However, MicroSim provides a library of manuals at minimum cost that could prove useful for details beyond the scope of the text. PSpice is currently provided in two formats: DOS and Windows. The former requires a line-by-line description of the network defined using the various nodes of the network. The latter requires an actual drawing of the network before the analysis can be performed. Using either format to enter the network, one should be aware that the results and network descriptions printed out by each system will be the same. In other words, the primary difference between the two formats is simply how the network is entered, not the manner in which the analysis is performed and the results provided.

Equipment Requirements

The analysis of this text was performed on the latest available version of PSpice, Version 6.2. It can be installed and run on the following platforms:

PC-compatible computers (IBM, COMPAC, TANDY, GATEWAY, etc., and compatible PC clones)

SUN SPARC stations running Sun OS 4.1.2 (Solaris 1.0.1)

SUN SPARC stations running Solaris 2.3 (Sun OS 5.3)

HP9000/700 workstations

Hardware Requirements

80386-based computer or better
At least 4 MB of RAM
Math coprocessor
At least 20 MB of hard-disk free space
A serial port (for nonnetworked versions) or a parallel port (for networked versions)
At least one high-density, 3.5″ disk drive
Microsoft MS-DOS 3.3 or later
Color or monochrome display and compatible adapter
For Windows applications, Windows Version 3.1 or later
CD-ROM drive for CD-ROM version

Documentation

The documentation is available on-line (CD-ROM) or in hard copy format (manuals). The hard copy format is available by calling the MicroSim sales office at (714) 837–3022 or (800) 245–3022.

Manuals available include the following:

Installation manuals (DOS or Windows)
Circuit Analysis User's Guide (DOS or Windows)
Circuit Analysis Reference Manual (DOS or Windows)
Tutorials, Application Notes, and Design Ideas Manual (for all platforms)
MicroSim Schematics (Windows)

Technical Support

Technical support is available at the following number from 8:30 A.M. to 5:00 P.M. Pacific Standard Time, Monday through Friday: (714) 837–0790.

MicroSim also provides a Bulletin Board Service (BBS). On the BBS you will find information regarding operation of the programs they offer and a message utility that you can use to let them know your comments and suggestions. They can be reached at (714) 830–1550 (1200–14.4K baud, N–8–1).

The Internet address is Tech.Support@MicroSim.com.

C++

The C++ version employed in this text is Borland C++ 4.0 available in both DOS and Windows on CD-ROM.

Hardware and Software Requirements

DOS Version 4.01 or higher
Windows 3.1 or higher, running in 386-enhanced mode
Hard disk with 2 MB of available disk space for a minimum CD-ROM installation (90 MB for a complete installation to your hard disk)
CD-ROM drive
At least 4 MB of extended memory (RAM)

Although the following items are not required, they can improve the system's response:

8 MB of RAM
Math coprocessor (if your writing programs use floating-point math); Borland C++ 4.0 emulates a math chip if you don't have one.

Documentation (CD-ROM Format Only)

User's Guide
Programmer's Guide
Library reference
DOS reference
Turbo Debugger User's Guide
Object Window Programmer's Guide
Object Windows Reference Guide
Prebuilt example programs
And more

MathCAD

This text employed the Student Edition of MathCAD Version 2.54 by Richard B. Anderson and published by Addison-Wesley Publishing Company, Inc., and The Benjamin/Cummings Publishing Company, Inc. The Student Edition was developed and programmed by MathSoft, Inc.

Information can be obtained by calling the Addison-Wesley Publishing Company at (617) 944–3700 at the Eastern Regional Office, Route 128, Reading, MA 01867.

Appendix B

CONVERSION FACTORS

To Convert from	To	Multiply by
Btus	Calorie-grams	251.996
	Ergs	1.054×10^{10}
	Foot-pounds	777.649
	Hp-hours	0.000393
	Joules	1054.35
	Kilowatthours	0.000293
	Wattseconds	1054.35
Centimeters	Angstrom units	1×10^{8}
	Feet	0.0328
	Inches	0.3937
	Meters	0.01
	Miles (statute)	6.214×10^{-6}
	Millimeters	10
Circular mils	Square centimeters	5.067×10^{-6}
	Square inches	7.854×10^{-7}
Cubic inches	Cubic centimeters	16.387
	Gallons (U.S. liquid)	0.00433
Cubic meters	Cubic feet	35.315
Days	Hours	24
	Minutes	1440
	Seconds	86,400
Dynes	Gallons (U.S. liquid)	264.172
	Newtons	0.00001
	Pounds	2.248×10^{-6}
Electronvolts	Ergs	1.60209×10^{-12}
Ergs	Dyne-centimeters	1.0
	Electronvolts	6.242×10^{11}
	Foot-pounds	7.376×10^{-8}
	Joules	1×10^{-7}
	Kilowatthours	2.777×10^{-14}
Feet	Centimeters	30.48
	Meters	0.3048
Foot-candles	Lumens/square foot	1.0
	Lumens/square meter	10.764
Foot-pounds	Dyne-centimeters	1.3558×10^{7}
	Ergs	1.3558×10^{7}
	Horsepower-hours	5.050×10^{-7}
	Joules	1.3558
	Newton-meters	1.3558

To Convert from	To	Multiply by
Gallons (U.S. liquid)	Cubic inches	231
	Liters	3.785
	Ounces	128
	Pints	8
Gauss	Maxwells/square centimeter	1.0
	Lines/square centimeter	1.0
	Lines/square inch	6.4516
Gilberts	Ampere-turns	0.7958
Grams	Dynes	980.665
	Ounces	0.0353
	Pounds	0.0022
Horsepower	Btus/hour	2547.16
	Ergs/second	7.46×10^9
	Foot-pounds/second	550.221
	Joules/second	746
	Watts	746
Hours	Seconds	3600
Inches	Angstrom units	2.54×10^8
	Centimeters	2.54
	Feet	0.0833
	Meters	0.0254
Joules	Btus	0.000948
	Ergs	1×10^7
	Foot-pounds	0.7376
	Horsepower-hours	3.725×10^{-7}
	Kilowatthours	2.777×10^{-7}
	Wattseconds	1.0
Kilograms	Dynes	980,665
	Ounces	35.2
	Pounds	2.2
Lines	Maxwells	1.0
Lines/square centimeter	Gauss	1.0
Lines/square inch	Gauss	0.1550
	Webers/square inch	1×10^{-8}
Liters	Cubic centimeters	1000.028
	Cubic inches	61.025
	Gallons (U.S. liquid)	0.2642
	Ounces (U.S. liquid)	33.815
	Quarts (U.S. liquid)	1.0567
Lumens	Candle power (spher.)	0.0796
Lumens/square centimeter	Lamberts	1.0
Lumens/square foot	Foot-candles	1.0

To Convert from	To	Multiply by
Maxwells	Lines	1.0
	Webers	1×10^{-8}
Meters	Angstrom units	1×10^{10}
	Centimeters	100
	Feet	3.2808
	Inches	39.370
	Miles (statute)	0.000621
Miles (stature)	Feet	5280
	Kilometers	1.609
	Meters	1609.344
Miles/hour	Kilometers/hour	1.609344
Newton-meters	Dyne-centimeters	1×10^{7}
	Kilogram-meters	0.10197
Oersteds	Ampere-turns/inch	2.0212
	Ampere-turns/meter	79.577
	Gilberts/centimeter	1.0
Quarts (U.S. liquid)	Cubic centimeters	946.353
	Cubic inches	57.75
	Gallons (U.S. liquid)	0.25
	Liters	0.9463
	Pints (U.S. liquid)	2
	Ounces (U.S. liquid)	32
Radians	Degrees	57.2958
Slugs	Kilograms	14.5939
	Pounds	32.1740
Watts	Btus/hour	3.4144
	Ergs/second	1×10^{7}
	Horsepower	0.00134
	Joules/second	1.0
Webers	Lines	1×10^{8}
	Maxwells	1×10^{8}
Years	Days	365
	Hours	8760
	Minutes	525,600
	Seconds	3.1536×10^{7}

Appendix C

DETERMINANTS

Determinants are employed to find the mathematical solutions for the variables in two or more simultaneous equations. Once the procedure is properly understood, solutions can be obtained with a minimum of time and effort and usually with fewer errors than when using other methods.

Consider the following equations, where x and y are the unknown variables and a_1, a_2, b_1, b_2, c_1, and c_2 are constants:

Col. 1		Col. 2		Col. 3	
a_1x	$+$	b_1y	$=$	c_1	**(C.1a)**
a_2x	$+$	b_2y	$=$	c_2	**(C.1b)**

It is certainly possible to solve for one variable in Eq. (C.1a) and substitute into Eq. (C.1b). That is, solving for x in Eq. (C.1a),

$$x = \frac{c_1 - b_1y}{a_1}$$

and substituting the result in Eq. (C.1b),

$$a_2\left(\frac{c_1 - b_1y}{a_1}\right) + b_2y = c_2$$

It is now possible to solve for y, since it is the only variable remaining, and then substitute into either equation for x. This is acceptable for two equations, but it becomes a very tedious and lengthy process for three or more simultaneous equations.

Using determinants to solve for x and y requires that the following formats be established for each variable:

Col. 1, Col. 2 / Col. 1, Col. 2

$$x = \frac{\begin{vmatrix} c_1 & b_1 \\ c_2 & b_2 \end{vmatrix}}{\begin{vmatrix} a_1 & b_1 \\ a_2 & b_2 \end{vmatrix}} \qquad y = \frac{\begin{vmatrix} a_1 & c_1 \\ a_2 & c_2 \end{vmatrix}}{\begin{vmatrix} a_1 & b_1 \\ a_2 & b_2 \end{vmatrix}} \tag{C.2}$$

First note that only constants appear within the vertical brackets and that the denominator of each is the same. In fact, the denominator is simply the coefficients of x and y in the same arrangement as in Eqs. (C.1a) and (C.1b). When solving for x, the coefficients of x in the numerator are replaced by the constants to the right of the equal sign in Eqs. (C.1a) and (C.1b), whereas the coefficients of the y variable are simply repeated. When solving for y, the y coefficients in the numerator are replaced by the constants to the right of the equal sign and the coefficients of x are repeated.

Each configuration in the numerator and denominator of Eqs. (C.2) is referred to as a *determinant* (D), which can be evaluated numerically in the following manner:

$$\text{Determinant} = D = \overset{\begin{matrix}\text{Col.} & \text{Col.}\\ 1 & 2\end{matrix}}{\begin{vmatrix} a_1 & b_1 \\ a_2 & b_2 \end{vmatrix}} = a_1b_2 - a_2b_1 \tag{C.3}$$

The expanded value is obtained by first multiplying the top left element by the bottom right and then subtracting the product of the lower left and upper right elements. This particular determinant is referred to as a *second-order* determinant, since it contains two rows and two columns.

It is important to remember when using determinants that the columns of the equations, as indicated in Eqs. (C.1a) and (C.1b), be placed in the same order within the determinant configuration. That is, since a_1 and a_2 are in column 1 of Eqs. (C.1a) and (C.1b), they must be in column 1 of the determinant. (The same is true for b_1 and b_2.)

Expanding the entire expression for x and y, we have the following:

$$x = \frac{\begin{vmatrix} c_1 & b_1 \\ c_2 & b_2 \end{vmatrix}}{\begin{vmatrix} a_1 & b_1 \\ a_2 & b_2 \end{vmatrix}} = \frac{c_1b_2 - c_2b_1}{a_1b_2 - a_2b_1} \tag{C.4a}$$

$$y = \frac{\begin{vmatrix} a_1 & c_1 \\ a_2 & c_2 \end{vmatrix}}{\begin{vmatrix} a_1 & b_1 \\ a_2 & b_2 \end{vmatrix}} = \frac{a_1c_2 - a_2c_1}{a_1b_2 - a_2b_1} \tag{C.4b}$$

EXAMPLE C.1 Evaluate the following determinants:

a. $\begin{vmatrix} 2 & 2 \\ 3 & 4 \end{vmatrix} = (2)(4) - (3)(2) = 8 - 6 = \mathbf{2}$

b. $\begin{vmatrix} 4 & -1 \\ 6 & 2 \end{vmatrix} = (4)(2) - (6)(-1) = 8 + 6 = \mathbf{14}$

c. $\begin{vmatrix} 0 & -2 \\ -2 & 4 \end{vmatrix} = (0)(4) - (-2)(-2) = 0 - 4 = \mathbf{-4}$

d. $\begin{vmatrix} 0 & 0 \\ 3 & 10 \end{vmatrix} = (0)(10) - (3)(0) = \mathbf{0}$

EXAMPLE C.2 Solve for x and y:

$$\begin{aligned} 2x + \ y &= 3 \\ 3x + 4y &= 2 \end{aligned}$$

Solution:

$$x = \frac{\begin{vmatrix} 3 & 1 \\ 2 & 4 \end{vmatrix}}{\begin{vmatrix} 2 & 1 \\ 3 & 4 \end{vmatrix}} = \frac{(3)(4) - (2)(1)}{(2)(4) - (3)(1)} = \frac{12 - 2}{8 - 3} = \frac{10}{5} = \mathbf{2}$$

$$y = \frac{\begin{vmatrix} 2 & 3 \\ 3 & 2 \end{vmatrix}}{5} = \frac{(2)(2) - (3)(3)}{5} = \frac{4-9}{5} = \frac{-5}{5} = \mathbf{-1}$$

Check:

$$\begin{aligned} 2x + y &= (2)(2) + (-1) \\ &= 4 - 1 = 3 \quad \text{(checks)} \\ 3x + 4y &= (3)(2) + (4)(-1) \\ &= 6 - 4 = 2 \quad \text{(checks)} \end{aligned}$$

EXAMPLE C.3 Solve for x and y:

$$\begin{aligned} -x + 2y &= 3 \\ 3\,x - 2y &= -2 \end{aligned}$$

Solution: In this example, note the effect of the minus sign and the use of parentheses to ensure the proper sign is obtained for each product:

$$x = \frac{\begin{vmatrix} 3 & 2 \\ -2 & -2 \end{vmatrix}}{\begin{vmatrix} -1 & 2 \\ 3 & -2 \end{vmatrix}} = \frac{(3)(-2) - (-2)(2)}{(-1)(-2) - (3)(2)}$$

$$= \frac{-6+4}{2-6} = \frac{-2}{-4} = \mathbf{\frac{1}{2}}$$

$$y = \frac{\begin{vmatrix} -1 & 3 \\ 3 & -2 \end{vmatrix}}{-4} = \frac{(-1)(-2) - (3)(3)}{-4}$$

$$= \frac{2-9}{-4} = \frac{-7}{-4} = \mathbf{\frac{7}{4}}$$

EXAMPLE C.4 Solve for x and y:

$$\begin{aligned} x &= 3 - 4y \\ 20y &= -1 + 3x \end{aligned}$$

Solution: In this case, the equations must first be placed in the format of Eqs. (C.1a) and (C.1b):

$$\begin{aligned} x + 4y &= 3 \\ -3x + 20y &= -1 \end{aligned}$$

$$x = \frac{\begin{vmatrix} 3 & 4 \\ -1 & 20 \end{vmatrix}}{\begin{vmatrix} 1 & 4 \\ -3 & 20 \end{vmatrix}} = \frac{(3)(20) - (-1)(4)}{(1)(20) - (-3)(4)}$$

$$= \frac{60+4}{20+12} = \frac{64}{32} = \mathbf{2}$$

$$y = \frac{\begin{vmatrix} 1 & 3 \\ -3 & -1 \end{vmatrix}}{32} = \frac{(1)(-1) - (-3)(3)}{32}$$

$$= \frac{-1+9}{32} = \frac{8}{32} = \mathbf{\frac{1}{4}}$$

The use of determinants is not limited to the solution of two simultaneous equations; determinants can be applied to any number of simultaneous linear equations. First we will examine a shorthand method that is applicable to third-order determinants only, since most of the problems in the text are limited to this level of difficulty. We will then investigate the general procedure for solving any number of simultaneous equations.

Consider the three following simultaneous equations:

$$\begin{array}{cccc}
\text{Col 1} & \text{Col. 2} & \text{Col. 3} & \text{Col. 4} \\
\hline
a_1x\ + & b_1y\ + & c_1z\ = & d_1 \\
a_2x\ + & b_2y\ + & c_2z\ = & d_2 \\
a_3x\ + & b_3y\ + & c_3z\ = & d_3
\end{array}$$

in which x, y, and z are the variables, and $a_{1,2,3}$, $b_{1,2,3}$, $c_{1,2,3}$ and $d_{1,2,3}$ are constants.

The determinant configuration for x, y, and z can be found in a manner similar to that for two simultaneous equations. That is, to solve for x, find the determinant in the numerator by replacing column 1 with the elements to the right of the equal sign. The denominator is the determinant of the coefficients of the variables (the same applies to y and z). Again, the denominator is the same for each variable.

$$x = \frac{\begin{vmatrix} d_1 & b_1 & c_1 \\ d_2 & b_2 & c_2 \\ d_3 & b_3 & c_3 \end{vmatrix}}{D}, \quad y = \frac{\begin{vmatrix} a_1 & d_1 & c_1 \\ a_2 & d_2 & c_2 \\ a_3 & d_3 & c_3 \end{vmatrix}}{D}, \quad z = \frac{\begin{vmatrix} a_1 & b_1 & d_1 \\ a_2 & b_2 & d_2 \\ a_3 & b_3 & d_3 \end{vmatrix}}{D}$$

where

$$D = \begin{vmatrix} a_1 & b_1 & c_1 \\ a_2 & b_2 & c_2 \\ a_3 & b_3 & c_3 \end{vmatrix}$$

A shorthand method for evaluating the third-order determinant consists simply of repeating the first two columns of the determinant to the right of the determinant and then summing the products along specific diagonals as shown below:

$$\begin{array}{c} \\ D = \\ \\ \end{array}
\begin{array}{ccc|cc}
 & & 4(-) & 5(-) & 6(-) \\
a_1 & b_1 & c_1 & a_1 & b_1 \\
a_2 & b_2 & c_2 & a_2 & b_2 \\
a_3 & b_3 & c_3 & a_3 & b_3 \\
 & & 1(+) & 2(+) & 3(+)
\end{array}$$

The products of the diagonals 1, 2, and 3 are positive and have the following magnitudes:

$$+a_1b_2c_3 + b_1c_2a_3 + c_1a_2b_3$$

The products of the diagonals 4, 5, and 6 are negative and have the following magnitudes:

$$-a_3b_2c_1 - b_3c_2a_1 - c_3a_2b_1$$

The total solution is the sum of the diagonals 1, 2, and 3 minus the sum of the diagonals 4, 5, and 6:

$$+(a_1b_2c_3 + b_1c_2a_3 + c_1a_2b_3) - (a_3b_2c_1 + b_3c_2a_1 + c_3a_2b_1) \qquad \textbf{(C.5)}$$

Warning: **This method of expansion is good only for third-order determinants!** It cannot be applied to fourth- and higher-order systems.

EXAMPLE C.5 Evaluate the following determinant:

$$\begin{vmatrix} 1 & 2 & 3 \\ -2 & 1 & 0 \\ 0 & 4 & 2 \end{vmatrix} \longrightarrow \begin{array}{c} \qquad\qquad\quad (-)\;\;(-)\;\;(-) \\ \left|\begin{array}{ccc|cc} 1 & 2 & 3 & 1 & 2 \\ -2 & 1 & 0 & -2 & 1 \\ 0 & 4 & 2 & 0 & 4 \end{array}\right. \\ \qquad\qquad\quad (+)\;\;(+)\;\;(+) \end{array}$$

Solution:

$$\begin{aligned} &[(1)(1)(2) + (2)(0)(0) + (3)(-2)(4)] \\ &\qquad\qquad -[(0)(1)(3) + (4)(0)(1) + (2)(-2)(2)] \\ &= (2 + 0 - 24) - (0 + 0 - 8) = (-22) - (-8) \\ &= -22 + 8 = \mathbf{-14} \end{aligned}$$

EXAMPLE C.6 Solve for x, y, and z:

$$\begin{aligned} 1x + 0y - 2z &= -1 \\ 0x + 3y + 1z &= +2 \\ 1x + 2y + 3z &= 0 \end{aligned}$$

Solution:

$$x = \frac{\left|\begin{array}{ccc|cc} -1 & 0 & -2 & -1 & 0 \\ 2 & 3 & 1 & 2 & 3 \\ 0 & 2 & 3 & 0 & 2 \end{array}\right.}{\left|\begin{array}{ccc|cc} 1 & 0 & -2 & 1 & 0 \\ 0 & 3 & 1 & 0 & 3 \\ 1 & 2 & 3 & 1 & 2 \end{array}\right.}$$

$$= \frac{[(-1)(3)(3) + (0)(1)(0) + (-2)(2)(2)] - [(0)(3)(-2) + (2)(1)(-1) + (3)(2)(0)]}{[(1)(3)(3) + (0)(1)(1) + (-2)(0)(2)] - [(1)(3)(-2) + (2)(1)(1) + (3)(0)(0)]}$$

$$= \frac{(-9 + 0 - 8) - (0 - 2 + 0)}{(9 + 0 + 0) - (-6 + 2 + 0)}$$

$$= \frac{-17 + 2}{9 + 4} = -\frac{\mathbf{15}}{\mathbf{13}}$$

$$y = \frac{\left|\begin{array}{ccc|cc} 1 & -1 & -2 & 1 & -1 \\ 0 & 2 & 1 & 0 & 2 \\ 1 & 0 & 3 & 1 & 0 \end{array}\right.}{13}$$

$$= \frac{[(1)(2)(3) + (-1)(1)(1) + (-2)(0)(0)] - [(1)(2)(-2) + (0)(1)(1) + (3)(0)(-1)]}{13}$$

$$= \frac{(6 - 1 + 0) - (-4 + 0 + 0)}{13}$$

$$= \frac{5 + 4}{13} = \frac{\mathbf{9}}{\mathbf{13}}$$

$$z = \frac{\begin{vmatrix} 1 & 0 & -1 \\ 0 & 3 & 2 \\ 1 & 2 & 0 \end{vmatrix}\begin{matrix} 1 & 0 \\ 0 & 3 \\ 1 & 2 \end{matrix}}{13}$$

$$= \frac{[(1)(3)(0) + (0)(2)(1) + (-1)(0)(2)] - [(1)(3)(-1) + (2)(2)(1) + (0)(0)(0)]}{13}$$

$$= \frac{(0 + 0 + 0) - (-3 + 4 + 0)}{13}$$

$$= \frac{0 - 1}{13} = -\frac{\mathbf{1}}{\mathbf{13}}$$

or from $0x + 3y + 1z = +2$,

$$z = 2 - 3y = 2 - 3\left(\frac{9}{13}\right) = \frac{26}{13} - \frac{27}{13} = -\frac{\mathbf{1}}{\mathbf{13}}$$

Check:

$$\begin{aligned} 1x + 0y - 2z &= -1 \\ 0x + 3y + 1z &= +2 \\ 1x + 2y + 3z &= 0 \end{aligned} \quad \begin{aligned} -\frac{15}{13} + 0 + \frac{2}{13} &= -1 \\ 0 + \frac{27}{13} + \frac{-1}{13} &= +2 \\ -\frac{15}{13} + \frac{18}{13} + \frac{-3}{13} &= 0 \end{aligned} \quad \begin{aligned} -\frac{13}{13} &= -1 \ \checkmark \\ \frac{26}{13} &= +2 \ \checkmark \\ -\frac{18}{13} + \frac{18}{13} &= 0 \ \checkmark \end{aligned}$$

The general approach to third-order or higher determinants requires that the determinant be expanded in the following form. There is more than one expansion that will generate the correct result, but this form is typically employed when the material is first introduced.

$$D = \begin{vmatrix} a_1 & b_1 & c_1 \\ a_2 & b_2 & c_2 \\ a_3 & b_3 & c_3 \end{vmatrix} = \underset{\text{Multiplying factor}}{a_1} \underbrace{\left(+\underbrace{\begin{vmatrix} b_2 & c_2 \\ b_3 & c_3 \end{vmatrix}}_{\text{Minor}}\right)}_{\text{Cofactor}} + \underset{\text{Multiplying factor}}{b_1} \underbrace{\left(-\underbrace{\begin{vmatrix} a_2 & c_2 \\ a_3 & c_3 \end{vmatrix}}_{\text{Minor}}\right)}_{\text{Cofactor}} + \underset{\text{Multiplying factor}}{c_1} \underbrace{\left(+\underbrace{\begin{vmatrix} a_2 & b_2 \\ a_3 & b_3 \end{vmatrix}}_{\text{Minor}}\right)}_{\text{Cofactor}}$$

This exapnsion was obtained by multiplying the elements of the first row of D by their corresponding cofactors. It is not a requirement that the first row be used as the multiplying factors. In fact, any *row* or *column* (not diagonals) may be used to expand a third-order determinant.

The sign of each cofactor is dictated by the position of the multiplying factors ($a_{,1}$, b_1, and c_1 in this case) as in the following standard format:

$$\begin{vmatrix} + \rightarrow & - & + \\ \downarrow & & \\ - & + & - \\ + & - & + \end{vmatrix}$$

Note that the proper sign for each element can be obtained by simply assigning the upper left element a positive sign and then changing sign as you move horizontally or vertically to the neighboring position.

For the determinant D, the elements would have the following signs:

$$\begin{vmatrix} a_1^{(+)} & b_1^{(-)} & c_1^{(+)} \\ a_2^{(-)} & b_2^{(+)} & c_2^{(-)} \\ a_3^{(+)} & b_3^{(-)} & c_3^{(+)} \end{vmatrix}$$

The minors associated with each multiplying factor are obtained by covering up the row and column in which the multiplying factor is located and writing a second-order determinant to include the remaining elements in the same relative positions that they have in the third-order determinant.

Consider the cofactors associated with a_1 and b_1 in the expansion of D. The sign is positive for a_1 and negative for b_1 as determined by the standard format. Following the procedure outlined above, we can find the minors of a_1 and b_1 as follows:

$$a_{1(\text{minor})} = \begin{vmatrix} \cancel{a_1} & \cancel{b_1} & \cancel{c_1} \\ \cancel{a_2} & b_2 & c_2 \\ \cancel{a_3} & b_3 & c_3 \end{vmatrix} = \begin{vmatrix} b_2 & c_2 \\ b_3 & c_3 \end{vmatrix}$$

$$b_{1(\text{minor})} = \begin{vmatrix} \cancel{a_1} & \cancel{b_1} & \cancel{c_1} \\ a_2 & \cancel{b_2} & c_2 \\ a_3 & \cancel{b_3} & c_3 \end{vmatrix} = \begin{vmatrix} a_2 & c_2 \\ a_3 & c_3 \end{vmatrix}$$

It was pointed out that any row or column may be used to expand the third-order determinant, and the same result will still be obtained. Using the first column of D, we obtain the expansion

$$D = \begin{vmatrix} a_1 & b_1 & c_1 \\ a_2 & b_2 & c_2 \\ a_3 & b_3 & c_3 \end{vmatrix} = a_1\left(+\begin{vmatrix} b_2 & c_2 \\ b_3 & c_3 \end{vmatrix}\right) + a_2\left(-\begin{vmatrix} b_1 & c_1 \\ b_3 & c_3 \end{vmatrix}\right) + a_3\left(+\begin{vmatrix} b_1 & c_1 \\ b_2 & c_2 \end{vmatrix}\right)$$

The proper choice of row or column can often effectively reduce the amount of work required to expand the third-order determinant. For example, in the following determinants, the first column and third row, respectively, would reduce the number of cofactors in the expansion:

$$D = \begin{vmatrix} 2 & 3 & -2 \\ 0 & 4 & 5 \\ 0 & 6 & 7 \end{vmatrix} = 2\left(+\begin{vmatrix} 4 & 5 \\ 6 & 7 \end{vmatrix}\right) + 0 + 0 = 2(28 - 30)$$

$$= \mathbf{-4}$$

$$D = \begin{vmatrix} 1 & 4 & 7 \\ 2 & 6 & 8 \\ 2 & 0 & 3 \end{vmatrix} = 2\left(+\begin{vmatrix} 4 & 7 \\ 6 & 8 \end{vmatrix}\right) + 0 + 3\left(+\begin{vmatrix} 1 & 4 \\ 2 & 6 \end{vmatrix}\right)$$

$$= 2(32 - 42) + 3(6 - 8) = 2(-10) + 3(-2)$$

$$= \mathbf{-26}$$

EXAMPLE C.7 Expand the following third-order determinants:

a. $D = \begin{vmatrix} 1 & 2 & 3 \\ 3 & 2 & 1 \\ 2 & 1 & 3 \end{vmatrix} = 1\left(+\begin{vmatrix} 2 & 1 \\ 1 & 3 \end{vmatrix}\right) + 3\left(-\begin{vmatrix} 2 & 3 \\ 1 & 3 \end{vmatrix}\right) + 2 + \left(\begin{vmatrix} 2 & 3 \\ 2 & 1 \end{vmatrix}\right)$

$$
\begin{aligned}
&= 1[6 - 1] + 3[-(6 - 3)] + 2[2 - 6] \\
&= 5 + 3(-3) + 2(-4) \\
&= 5 - 9 - 8 \\
&= \mathbf{-12}
\end{aligned}
$$

b. $D = \begin{vmatrix} 0 & 4 & 6 \\ 2 & 0 & 5 \\ 8 & 4 & 0 \end{vmatrix} = 0 + 2\left(-\begin{vmatrix} 4 & 6 \\ 4 & 0 \end{vmatrix}\right) + 8\left(+\begin{vmatrix} 4 & 6 \\ 0 & 5 \end{vmatrix}\right)$

$$
\begin{aligned}
&= 0 + 2[-(0 - 24)] + 8[(20 - 0)] \\
&= 0 + 2(24) + 8(20) \\
&= 48 + 160 \\
&= \mathbf{208}
\end{aligned}
$$

Appendix D

COLOR CODING OF MOLDED MICA CAPACITORS (PICOFARADS)

RETMA and standard MIL specifications

Color	Significant Figure	Decimal Multiplier	Tolerance ±%	Class	Temp. Coeff. PPM/°C Not More than	Cap. Drift Not More than
Black	0	1	20	A	±1000	±(5% + 1 pF)
Brown	1	10	—	B	±500	±(3% + 1 pF)
Red	2	100	2	C	±200	±(0.5% + 0.5 pF)
Orange	3	1000	3	D	±100	±(0.3% + 0.1 pF)
Yellow	4	10,000	—	E	+100 − 20	±(0.1% + 0.1 pF)
Green	5	—	5	—	—	—
Blue	6	—	—	—	—	—
Violet	7	—	—	—	—	—
Gray	8	—	—	I	+150 − 50	±(0.03% + 0.2 pF)
White	9	—	—	J	+100 − 50	±(0.2% + 0.2 pF)
Gold	—	0.1	—	—	—	—
Silver	—	0.01	10	—	—	—

Note: If both rows of dots are not on one face, rotate capacitor about axis of its leads to read second row on side or rear.

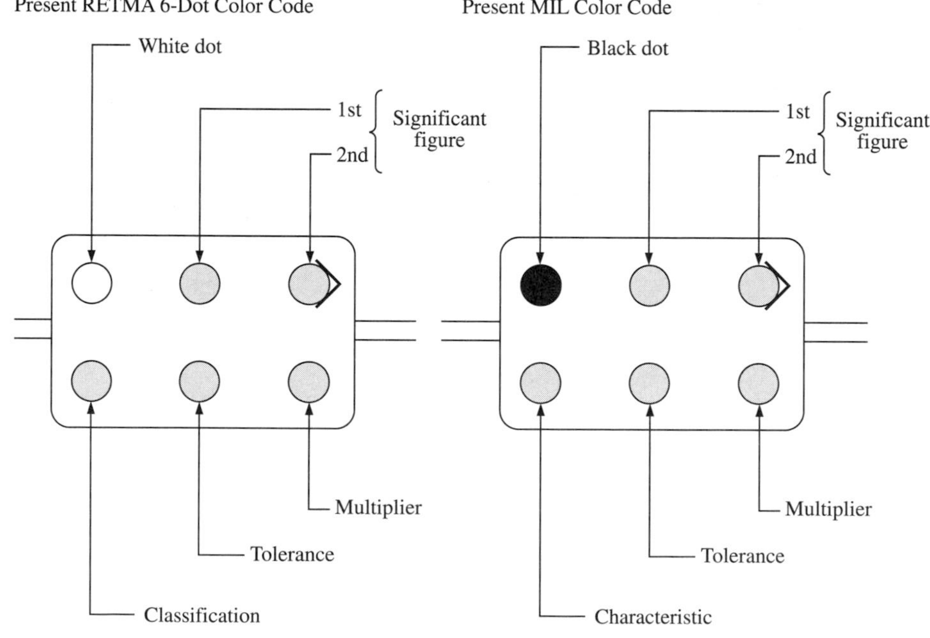

FIG. D.1

Appendix E

COLOR CODING OF MOLDED TUBULAR CAPACITORS (PICOFARADS)

Color	Significant Figure	Decimal Multiplier	Tolerance ±%
Black	0	1	20
Brown	1	10	—
Red	2	100	—
Orange	3	1000	30
Yellow	4	10,000	40
Green	5	10^5	5
Blue	6	10^6	—
Violet	7	—	—
Gray	8	—	—
White	9	—	10

Note: Voltage rating is identified by a single-digit number for ratings up to 900 V and a two-digit number above 900 V. Two zeros follow the voltage figure.

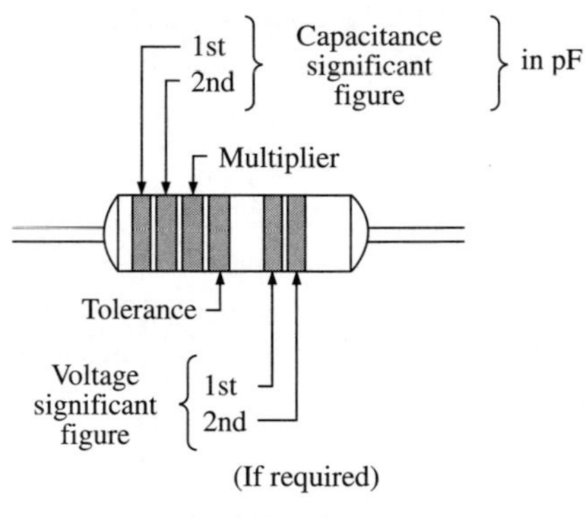

FIG. E.1

Appendix F

THE GREEK ALPHABET

Letter	Capital	Lowercase	Used to Designate
Alpha	A	α	Area, angles, coefficients
Beta	B	β	Angles, coefficients, flux density
Gamma	Γ	γ	Specific gravity, conductivity
Delta	Δ	δ	Density, variation
Epsilon	E	ϵ	Base of natural logarithms
Zeta	Z	ζ	Coefficients, coordinates, impedance
Eta	H	η	Efficiency, hysteresis coefficient
Theta	Θ	θ	Phase angle, temperature
Iota	I	ι	
Kappa	K	κ	Dielectric constant, susceptibility
Lambda	Λ	λ	Wavelength
Mu	M	μ	Amplification factor, micro, permeability
Nu	N	ν	Reluctivity
Xi	Ξ	ξ	
Omicron	O	o	
Pi	Π	π	3.1416
Rho	P	ρ	Resistivity
Sigma	Σ	σ	Summation
Tau	T	τ	Time constant
Upsilon	Υ	υ	
Phi	Φ	ϕ	Angles, magnetic flux
Chi	X	χ	
Psi	Ψ	ψ	Dielectric flux, phase difference
Omega	Ω	ω	Ohms, angular velocity

Appendix G

MAGNETIC PARAMETER CONVERSIONS

	SI (MKS)	CGS	English
Φ	webers (Wb) 1 Wb	maxwells $= 10^8$ maxwells	lines $= 10^8$ lines
B	Wb/m^2 1 Wb/m^2	gauss (maxwells/cm^2) $= 10^4$ gauss	lines/in.^{2} $= 6.452 \times 10^4$ lines/in.^{2}
A	1 m^2	$= 10^4$ cm^2	$= 1550$ in.^{2}
μ_o	$4\pi \times 10^{-7}$ Wb/Am	$= 1$ gauss/oersted	$= 3.20$ lines/Am
$\mathscr{F}$	NI (ampere-turns, At) 1 At	$0.4\pi NI$ (gilberts) $= 1.257$ gilberts	NI (At) 1 gilbert $= 0.7958$ At
H	NI/l (At/m) 1 At/m	$0.4\pi NI/l$ (oersteds) $= 1.26 \times 10^{-2}$ oersted	NI/l (At/in.) $= 2.54 \times 10^{-2}$ At/in.
H_g	$7.97 \times 10^5 B_g$ (At/m)	B_g (oersteds)	$0.313 B_g$ (At/in.)

Appendix H

MAXIMUM POWER TRANSFER CONDITIONS

Derivation of maximum power transfer conditions for the situation where the resistive component of the load is adjustable but the load reactance is set in magnitude.*

For the circuit of Fig. H.1, the power delivered to the load is determined by

$$P = \frac{V_{R_L}^2}{R_L}$$

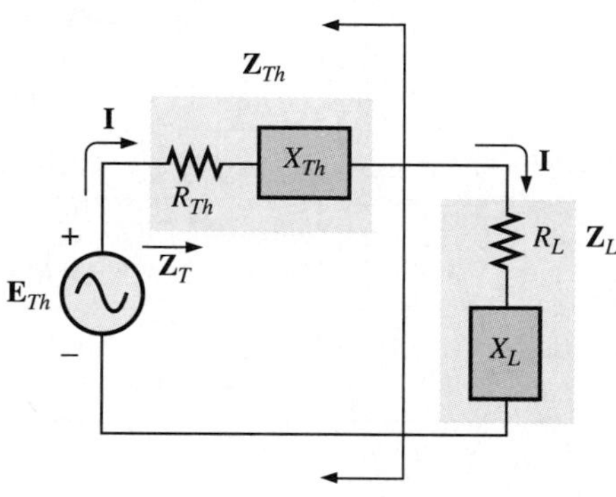

FIG. H.1

Applying the voltage divider rule:

$$\mathbf{V}_{R_L} = \frac{R_L \mathbf{E}_{Th}}{R_L + R_{Th} + X_{Th}\angle 90^\circ + X_L\angle 90^\circ}$$

The magnitude of $\mathbf{V}_{R_L}$ is determined by

$$V_{R_L} = \frac{R_L E_{Th}}{\sqrt{(R_L + R_{Th})^2 + (X_{Th} + X_L)^2}}$$

and

$$V_{R_L}^2 = \frac{R_L^2 E_{Th}^2}{(R_L + R_{Th})^2 + (X_{Th} + X_L)^2}$$

with

$$P = \frac{V_{R_L}^2}{R_L} = \frac{R_L E_{Th}^2}{(R_L + R_{Th})^2 + (X_{Th} + X_L)^2}$$

Using differentiation (calculus), maximum power will be transferred when $dP/dR_L = 0$. The result of the preceding operation is that

$$R_L = \sqrt{R_{Th}^2 + (X_{Th} + X_L)^2} \qquad \text{[Eq. (18.21)]}$$

*With sincerest thanks for the input of Professor Harry J. Franz of the Beaver Campus of Pennsylvania State University.

The magnitude of the total impedance of the circuit is

$$Z_T = \sqrt{(R_{Th} + R_L)^2 + (X_{Th} + X_L)^2}$$

Substituting this equation for R_L and applying a few algebraic maneuvers will result in

$$Z_T = 2R_L(R_L + R_{Th})$$

and the power to the load R_L will be

$$P = I^2R_L = \frac{E_{Th}^2}{Z_T^2}\,R_L = \frac{E_{Th}^2R_L}{2R_L(R_L + R_{Th})}$$

$$= \frac{E_{Th}^2}{4\left(\dfrac{R_L + R_{Th}}{2}\right)}$$

$$= \frac{E_{Th}^2}{4R_{av}}$$

with $$R_{av} = \frac{R_L + R_{Th}}{2}$$

Appendix I

ANSWERS TO SELECTED ODD-NUMBERED PROBLEMS

Chapter 1

5. 3 h
7. CGS
9. MKS = CGS = 20°C
K = SI = 293.15
11. 45.72 cm
13. **(a)** 15×10^3 **(b)** 30×10^{-3}
(c) 7.4×10^6 **(d)** 6.8×10^{-6}
(e) 402×10^{-6} **(f)** 200×10^{-12}
15. **(a)** 10^4 **(b)** 10
(c) 10^9 **(d)** 10^{-2}
(e) 10 **(f)** 10^{31}
17. **(a)** 10^{-1} **(b)** 10^{-4}
(c) 10^9 **(d)** 10^{-9}
(e) 10^{42} **(f)** 10^3
19. **(a)** 10^6 **(b)** 10^{-2}
(c) 10^{32} **(d)** 10^{-63}
21. **(a)** 10^{-6} **(b)** 10^{-3}
(c) 10^{-6} **(d)** 10^9
(e) 10^{-16} **(f)** 10^{-1}
23. **(a)** 0.006 **(b)** 400
(c) 5000, 5, 0.005
(d) 0.0003, 0.3, 300
25. **(a)** 90 s **(b)** 144 s
(c) 50×10^3 μs
(d) 160 mm **(e)** 120 ns
(f) 41.898 days **(g)** 1.02 m
27. **(a)** 2.54 m **(b)** 1.219 m
(c) 26.7 N **(d)** 0.1348 lb.
(e) 4921.26 ft
(f) 3.2187 m **(g)** 8530.17 yds
29. 670.62×10^6 mph
31. 2.045 s
33. 67.06 days
35. $900
37. 345.6 m
39. 47.29 min./mile
41. **(a)** 4.74×10^{-3} Btu
(b) 7.098×10^{-4} m^3
(c) 1.2096×10^5 s
(d) 2113.38 pints

Chapter 2

3. **(a)** 18 mN **(b)** 2 mN
(c) 180 μN
7. **(a)** 72 mN
(b) $Q_1 = 20\ \mu$C, $Q_2 = 40\ \mu$C
9. 3.1 A
11. 90 C
13. 0.5 A
15. 1.194 A > 1 A (yes)
17. **(a)** 1.248 million
(b) 0.936 million, sol. = (a)
19. 252 J
21. 4 C
23. 3.533 V
25. 5 A
27. 25 h
29. 0.773 h
31. 60 Ah:40 Ah = 1.5:1, 50% more with 60 Ah
33. 545.45 mA, 129.6 kJ
43. 600 C

Chapter 3

1. **(a)** 500 mils **(b)** 10 mils
(c) 4 mils **(d)** 1000 mils
(e) 240 mils **(f)** 3.937 mils
3. **(a)** 0.04 in. **(b)** 0.03 in.
(c) 0.2 in. **(d)** 0.025 in.
(e) 0.00278 in. **(f)** 0.009 in.
5. 73.33 Ω
7. 3.581 ft
9. **(a)** $R_{\text{silver}} > R_{\text{copper}} > R_{\text{aluminum}}$
(b) silver 9.9 Ω,
copper 1.037 Ω,
aluminum 0.34 Ω
11. **(a)** 21.71 $\mu\Omega$ **(b)** 35.59 $\mu\Omega$
(c) increases **(d)** decreases
13. 942.28 mΩ
15. **(a)** #8:1.1308 Ω, #18:11.493 Ω
(b) #18:#8 = 10.164:1 ≅ 10:1,
#18:#8 = 1:10.164 ≅ 1:10
17. **(a)** 1.087 mA/CM
(b) 1.384 kA/in.^{2}
(c) 3.6127 in.^{2}
19. **(a)** 21.71 $\mu\Omega$ (b) 35.59 $\mu\Omega$
21. 0.15 in.
23. 2.409 Ω
25. 3.67 Ω
27. 0.046 Ω
29. **(a)** 40.29°C **(b)** −195.61°C
31. **(a)** $\alpha_{20} \cong 0.00393$
(b) 83.61°C
33. 1.751 Ω
35. 142.86
41. −30°C:10.2 kΩ,
100°C:10.15 kΩ
43. 6.5 kΩ
47. **(a)** Red Red Brown Silver
(b) Yellow Violet Red Silver
(c) Blue Gray Orange Silver
(d) White Brown Green Silver
49. yes
51. **(a)** 0.1566 S **(b)** 0.0955 S
(c) 0.0219 S
57. **(a)** 10 fc:3 kΩ, 100 fc:0.4 kΩ
(b) neg. (c) no—log scales
(d) −321.43 Ω/fc

Chapter 4

1. 15 V
3. 4 kΩ
5. 72 mV
7. 54.55 Ω
9. 28.571 Ω
11. 1.2 kΩ
13. **(a)** 12.632 Ω **(b)** 4.1 MJ
17. 800 V
19. 1 W
21. **(a)** 57,600 J
(b) 16×10^{-3} kWh
23. 2 s
25. 196 μW
27. 4 A
29. 9.61 V
31. 0.833 A, 144.06 Ω
33. **(a)** 0.133 mA **(b)** 66.5 mAh
35. **(c)** ≅ 70.7 mA
37. **(a)** 12 kW
(b) 10,130 W < 12,000 W (yes)
39. 16.34 A
41. **(a)** 238 W **(b)** 17.36%
43. **(a)** 1657.78 W
(b) 15.07 A
(c) 19.38 A
45. 65.25%
47. 80%
49. **(a)** 17.9%
(b) 76.73%, 328.66% increase
51. **(a)** 1350 J
(b) *W* doubles, *P* the same
53. 6.67 h
55. **(a)** 50 kW **(b)** 240.38 A
(c) 90 kWh
57. $2.19

Chapter 5

1. **(a)** 20 Ω, 3 A
(b) 1.63 MΩ, 6.135 μA
(c) 110 Ω, 318.2 mA
(d) 10 kΩ, 12 mA

3. **(a)** 16 V **(b)** 4.2 V
5. **(a)** 0.388 A (CW)
(b) 2.087 A (CCW)
7. **(a)** 5 V **(b)** 70 V
9. 3.28 mA, 7.22 V
11. **(a)** 70.6 Ω, 85 mA (CCW), $V_1 = 2.8045$ V, $V_2 = 0.4760$ V, $V_3 = 0.850$ V, $V_4 = 1.870$ V
(b)–(c) $P_1 = 0.2384$ W, $P_2 = 0.0405$ W, $P_3 = 0.0723$ W, $P_4 = 0.1590$ W
(d) all $\frac{1}{2}$ W
13. **(a)** 225 Ω, 0.533 A
(b) 8 W
(c) 15 V
15. All V_{ab} (+ −)
(a) 66.67 V **(b)** −8 V
(c) 20 V **(d)** 0.18 V
17. **(a)** 12 V **(b)** 24 V
(c) 60 Ω **(d)** 0.4 A
(e) 60 Ω
19. **(a)** $R_s = 80$ Ω
(b) 0.2 W < $\frac{1}{4}$ W
21. $R_1 = 3$ kΩ, $R_2 = 15$ kΩ
23. **(a)** $R_1 = 0.4$ kΩ, $R_2 = 1.2$ kΩ, $R_3 = 4.8$ kΩ
(b) $R_1 = 0.4$ MΩ, $R_2 = 1.2$ MΩ, $R_3 = 4.8$ MΩ
25. **(a)** I (CW) = 6.667 A, $V = 20$ V
(b) I (CW) = 1 A, $V = 10$ V
27. **(a)** 20 V, 26 V, 35 V, −12 V, 0 V
(b) −6 V, −47 V, 9 V
(c) −15 V, −38 V
29. $V_0 = 0$ V, $V_4 = 10$ V, $V_7 = 4$ V, $V_{10} = 20$ V, $V_{23} = 6$ V, $V_{30} = -8$ V, $V_{67} = 0$ V, $V_{56} = -6$ V, I(up) = 1.5 A
31. 2 Ω
33. 100 Ω
35. 1.52%

Chapter 6

1. **(a)** 2, 3, 4 **(b)** 2, 3 **(c)** 1, 4
3. **(a)** 6 Ω, 0.1667 S
(b) 1 kΩ, 1 mS
(c) 2.076 kΩ, 0.4817 mS
(d) 1.333 Ω, 0.75 S
(e) 9.948 Ω, 100.525 mS
(f) 0.6889 Ω, 1.4516 S
5. **(a)** 18 Ω **(b)** $R_1 = R_2 = 24$ Ω
7. 120 Ω
9. **(a)** 0.8571 Ω, 1.1667 S
(b) $I_s = 1.05$ A, $I_1 = 0.3$ A, $I_2 = 0.15$ A, $I_3 = 0.6$ A
(d) $P_1 = 0.27$ W, $P_2 = 0.135$ W, $P_3 = 0.54$ W, $P_{del.} = 0.945$ W
(e) $R_1, R_2 = \frac{1}{2}$ W, $R_3 = 1$ W
11. **(a)** 66.67 mA **(b)** 225 Ω
(c) 8 W
13. **(a)** $I_s = 7.5$ A, $I_1 = 1.5$ A
(b) $I_s = 9.6$ mA, $I_1 = 0.8$ mA
15. 1260 W
17. **(a)** 4 mA **(b)** 24 V
(c) 18.4 mA
19. **(a)** $I_1 = 3$ mA, $I_2 = 1$ mA, $I_3 = 1.5$ mA
(b) $I_2 = 4$ μA, $I_3 = 1.5$ μA, $I_4 = 5.5$ μA, $I_1 = 6$ μA
21. **(a)** $R_1 = 5$ Ω, $R_2 = 10$ Ω
(b) $E = 12$ V, $I_2 = 1.333$ A, $I_3 = 1$ A, $R_3 = 12$ Ω, $I = 4.333$ A
(c) $I_1 = 64$ mA, $I_3 = 16$ mA, $I_2 = 20$ mA, $R = 3.2$ kΩ, $I = 36$ mA
(d) $E = 30$ V, $I_1 = 1$ A, $I_2 = I_3 = 0.5$ A, $R_2 = R_3 = 60$ Ω, $P_{R_2} = 15$ W
23. **(a)** $I_1 = 4$ A, $I_2 = 8$ A
(b) $I_1 = 2$ A, $I_2 = 4$ A, $I_3 = 1$ A, $I_4 = 1.333$ A
(c) $I_1 = 272.73$ mA, $I_2 = 227.27$ mA, $I_3 = 90.91$ mA, $I_4 = 500$ mA
(d) $I_2 = 4.5$ A, $I_3 = 8.5$ A, $I_4 = 8.5$ A
25. **(a)** $I = 4$ A, $I_2 = 4$ A, $I_1 = 3$ A
27. $R_1 = 6$ kΩ, $R_2 = 1.5$ kΩ, $R_3 = 0.5$ kΩ
29. $I = 3$ A, $R = 2$ Ω
31. **(a)** 6.13 V
(b) 9 V
(c) 9 V
33. **(a)** 4 V **(b)** 3.997 V
(c) 3.871 V **(d)** 3 V
(e) R_m large as possible
35. No! 4-V supply reversed

Chapter 7

1. **(a)** series: E, R_1, and R_4, parallel: R_2 and R_3
(b) series: E and R_1, parallel: R_2 and R_3
(c) series: E, R_1, and R_5; R_3 and R_4 parallel: none
(d) series: R_6 and R_7, parallel: E, R_1, and R_4; R_2 and R_5
3. **(a)** yes (KCL) **(b)** 3 A
(c) yes (KCL) **(d)** 4 V
(e) 2 Ω **(f)** 5 A
(g) $P_1 = 12$ W, $P_2 = 18$ W, $P_{del.} = 50$ W
5. **(a)** 4 Ω
(b) $I_s = 9$ A, $I_1 = 6$ A, $I_2 = 3$ A
(c) 6 V
7. $I_1 = 6$ A, $I_2 = 16$ A, $I_3 = 0.8$ A, $I = 22$ A
9. **(a)** 4 A
(b) $I_2 = 1.333$ A, $I_3 = 0.6665$ A
(c) $V_a = 8$ V, $V_b = 4$ V
11. **(a)** 5 Ω, 16 A
(b) $I_{R_2} = 8$ A, $I_3 = I_9 = 4$ A
(c) $I_8 = 1$ A **(d)** 14 V
13. **(a)** $V_G = 1.9$ V, $V_s = 3.65$ V
(b) $I_1 = I_2 = 7.05$ μA, $I_D = 2.433$ mA
(c) 6.268 V
(d) 8.02 V
15. **(a)** 0.6 A
(b) 28 V
17. **(a)** $I_2 = 1.667$ A, $I_6 = 1.111$ A, $I_8 = 0$ A
19. **(a)** 1.882 Ω
(b) $V_1 = V_4 = 32$ V
(c) 8 A ←
(d) 1.882 Ω
21. **(a)** 6.75 A
(b) 32 V
23. 8.333 Ω
25. **(a)** 24 A
(b) 8 A
(c) $V_3 = 48$ V, $V_5 = 24$ V, $V_7 = 16$ V
(d) $P(R_7) = 128$ W, $P(E) = 5760$ W
27. 4.44 W
29. **(a)** 64 V
(b) $R_{L_2} = 4$ kΩ, $R_{L_3} = 3$ kΩ
(c) $R_1 = 0.5$ kΩ, $R_2 = 1.2$ kΩ, $R_3 = 2$ kΩ
31. **(a)** yes **(b)** $R_1 = 750$ Ω, $R_2 = 250$ Ω
(c) $R_1 = 745$ Ω, $R_2 = 255$ Ω
33. **(a)** 1 mA **(b)** $R_{shunt} = 5$ mΩ
35. **(a)** $R_s = 300$ kΩ,
(b) 20,000
37. 0.05 μA

Chapter 8

1. 28 V
3. **(a)** $I_1 = 12$ A, $I_s = 11$ A
(b) $V_s = 24$ V, $V_3 = 6$ V
5. **(a)** 3 A, 6 Ω **(b)** 4.091 mA, 2.2 kΩ
7. **(a)** 8 A **(b)** 8 A
9. 9.6 V, 2.4 A
11. **(a)** 5.4545 mA, 2.2 kΩ
(b) 17.375 V **(c)** 5.375 V
(d) 2.443 mA
13. **(I)** CW: $I_{R_1} = 1.445$ mA;
down: $I_{R_3} = 9.958$ mA;
CCW: $I_{R_2} = 8.513$ mA
(II) CW: $I_{R_1} = 2.0316$ mA;
left: $I_{R_2} = 0.8$ mA;
CW: $I_{R_3} = I_{R_4} = 1.2316$ mA
15. **(d)** left: 63.694 mA
17. **(a)** CW: $I_{R_1} = -\frac{1}{7}$ A;
CW: $I_{R_2} = -\frac{5}{7}$ A
$I_{R_3} = \frac{4}{7}$ A(down)
(b) CW: $I_{R_1} = -3.0625$ A;
CW: $I_{R_3} = 0.1875$ A
$I_{R_2} = 3.25$ A(up)
19. **(I)** CW: $I_1 = 1.8701$ A;
CW: $I_2 = -8.5484$ A;
$V_{ab} = -22.74$ V
(II) CW: $I_2 = 1.274$ A;
CW: $I_3 = 0.26$ A;
$V_{ab} = -0.904$ V
21. **(a)** 72.16 mA, −4.433 V
(b) 1.953 A, −7.257 V
23. **(a)** All clockwise
$I_1 = 0.0321$ mA
$I_2 = -0.8838$ mA
$I_3 = -0.968$ mA
$I_4 = -0.639$ mA
(b) All clockwise
$I_1 = -3.8$ A
$I_2 = -4.2$ A
$I_3 = 0.2$ A
25. **(a)** Clockwise,
$I_1 = -\frac{1}{7}$ A, $I_2 = -\frac{5}{7}$ A
(b) Clockwise,
$I_1 = -3.0625$ A,
$I_2 = 0.1875$ A
27. **(I)** **(a)** Clockwise
(b) $I_1 = 1.871$ A,
$I_2 = -8.548$ A
(c) $I_{R_1} = 1.871$ A,
$I_{R_2} = -8.548$ A,
$I_{R_3} = 10.419$ A
29. $I_{5\Omega}$ (CW) = 1.9535 A,
$V_a = -7.26$ V
31. **(a)** All clockwise,
$I_1 = 0.0321$ mA,
$I_2 = -0.8838$ mA,
$I_3 = -0.968$ mA,
$I_4 = -0.639$ mA
(b) All clockwise,
$I_1 = 3.8$ A, $I_2 = -4.2$ A,
$I_3 = 0.2$ A
33. **(I)** **(b)** $V_1 = -14.86$ V,
$V_2 = -12.57$ V
(c) $V_{R_1} = V_{R_4} =$
$V_1 = -14.86$ V,
$V_{R_2} = V_2 = -12.57$ V,
$V_{R_3} = 9.71$ V (+ −)
(II) **(b)** $V_1 = -2.556$ V,
$V_2 = 4.03$ V
(c) $V_{R_1} = V_1 = -2.556$ V,
$V_{R_2} = V_{R_5} = V_2 = 4.03$ V,
$V_{R_4} = V_{R_3} = V_2 - V_1$
$= 6.586$ V
35. **(I)** $V_1 = 7.238$ V,
$V_2 = -2.453$ V,
$V_3 = 1.405$ V
(II) $V_1 = -6.64$ V,
$V_2 = 1.288$ V,
$V_3 = 10.676$ V
37. **(a)** $V_1 = 10.083$ V,
$V_2 = 6.944$ V,
$V_3 = -17.056$ V
(b) $V_1 = 48$ V, $V_2 = 64$ V
39. **(b)** **(I)** $V_1 = -14.86$ V,
$V_2 = -12.57$ V
(II) $V_1 = -2.556$ V,
$V_2 = 4.03$ V
(c) **(I)** $V_{R_1} = V_{R_4} = -14.86$ V,
$V_{R_2} = -12.57$ V
$V_{R_3} = V_1 + 12 - V_2$
$= 9.71$ V
(II) $V_{R_1} = -2.556$ V,
$V_{R_2} = V_{R_5} = 4.03$ V
$V_{R_3} = V_{R_4} = V_2 - V_1$
$= 6.586$ V
41. **(I)** $V_1 = -5.311$ V,
$V_2 = -0.6219$ V,
$V_3 = 3.751$ V
$V_{-5A} = -5.311$ V
(II) $V_1 = -6.917$ V,
$V_2 = 12$ V,
$V_3 = 2.3$ V
$V_{5A} = V_2 - V_1 = 18.917$ V,
$V_{2A} = V_3 - V_2 = -9.7$ V
43. **(b)** $V_{R_5} = 0.1967$ V
(c) no
(d) no
45. **(b)** $I_{R_s} = 0$ A
(c) no
(d) no
47. **(a)** 3.33 mA
(b) 1.177 A
49. **(a)** 133.33 mA
(b) 7 A
51. **(b)** 0.833 mA
53. 4.2 Ω

Chapter 9

1. **(a)** CW: $I_{R_1} = \frac{5}{6}$ A, $I_{R_2} = 0$ A,
CW: $I_{R_3} = \frac{5}{6}$ A
(b) E_1: 5.33 W, E_2: 0.333 W
(c) 8.333 W **(d)** no
3. **(a)** down: 4.4545 mA
(b) down: 3.11 A
5. **(a)** 6 Ω, 6 V
(b) 2 Ω: 0.75 A,
30 Ω: 0.1667 A,
100 Ω: 0.0566 A
7. (I) 2 Ω, 84 V (II) 1.579 kΩ,
−1.149 V
9. (I) 45 Ω, −5 V (II) 2.055 kΩ,
16.772 V
11. 4.041 kΩ, 9.733 V
13. **(I):** 14 Ω, 2.571 A,
(II): 7.5 Ω, 1.333 A
15. **(a)** 9.756 Ω, 0.95 A
(b) 2 Ω, 30 A
17. **(a)** 10 Ω, 0.2 A
(b) 4.033 kΩ, 2.9758 mA
19. **(I)** **(a)** 14 Ω
(b) 23.14 W
(II) **(a)** 7.5 Ω
(b) 3.33 W
21. **(a)** 9.756 Ω, 2.2 W
(b) 2 Ω, 450 W
23. 0 Ω
25. 500 Ω
27. 39.3 μA, 220 mV
29. 2.25 A, 6.075 V
35. **(a)** 0.357 mA **(b)** 0.357 mA
(c) yes

Chapter 10

1. 9×10^3 N/C
3. 70 μF
5. 50 V/m
7. 8×10^3 V/m
9. 937.5 pF
11. mica
13. **(a)** 10^6 V/m **(b)** 4.96 μC
(c) 0.0248 μF
15. 29,035 V
17. **(a)** 0.5 s **(b)** $20(1 - e^{-t/0.5})$
(c) 1τ: 12.64 V, 3τ: 19 V,
5τ: 19.87 V
(d) $i_C = 0.2 \times 10^{-3}e^{-t/0.5}$
$v_R = 20e^{-t/0.5}$
19. **(a)** 5.5 ms
(b) $100(1 - e^{-t/(5.5\times10^{-3})})$
(c) 1τ: 63.21 V, 3τ: 95.02 V,
5τ: 99.33 V
(d)
$i_C = 18.18 \times10^{-3}e^{-t/(5.5\times10^{-3})}$
$v_R = 60e^{-t/(5.5\times10^{-3})}$
21. **(a)** 10 ms

(b) $50(1 - e^{-t/(10\times10^{-3})})$
(c) $10 \times 10^{-3}e^{-t/(10\times10^{-3})}$
(d) $v_C \cong 50$ V, $i_C = 0$ A
(e) $v_C = 50e^{-t/(4\times10^{-3})}$
$i_C = -25 \times 10^{-3}e^{-t/(4\times10^{-3})}$

23. (a) $80(1 - e^{-t/(1\times10^{-6})})$
(b) $0.8 \times 10^{-3}e^{-t/(1\times10^{-6})}$
(c) $v_C = 80e^{-t/(4.9\times10^{-6})}$
$i_C = 0.163 \times 10^{-3}e^{-t/(4.9\times10^{-6})}$

25. (a) 10 μs (b) 3 kΩ (c) yes

27. (a) $v_C = 52\text{ V} - 40\text{ V } e^{-t/123.8\text{ms}}$
$i_C = 2.198\text{ mA } e^{-t/123.8\text{ms}}$

29. 1.386 μs

31. $R = 54.567$ kΩ

33. (a) $v_C = 60(1 - e^{-t/0.2\text{s}})$,
0.5 s: 55.07 V, 1 s: 59.596 V
$i_C = 60 \times 10^{-3}\, e^{-t/0.2\text{s}}$
0.5 s: 4.93 mA,
1 s: 0.404 mA
$v_{R_1} = 60\, e^{-t/0.2\text{s}}$
0.5 s: 4.93 V, 1 s: 0.404 V
(b) $t = 0.405$ s, 1.387 s longer

35. (a) 19.634 V
(b) 2.31 s
(c) 1.155 s

37. (a) $v_C = 3.275(1 - e^{-t/52.68\text{ms}})$
$i_C = 1.216 \times 10^{-3}\, e^{-t/52.68\text{ms}}$

39. (a) $v_C = 27.2 - 25.2\, e^{-t/18.26\text{ms}}$
$i_C = 3.04\text{ mA } e^{-t/18.26\text{ms}}$

41. 0–4 ms: 0.3 mA,
4–6 ms: 0.9 mA,
6–7 ms: 3 mA,
7–10 ms: 0 mA,
10–13 ms: −3.2 mA,
13–15 ms: 1.8 mA

43. 0–4 ms: 0 V,
4–6 ms: −8 V,
6–16 ms: 20 V,
16–18 ms: 0 V,
18–20 ms: −12 V,
20–25 ms: 0 V

45. $V_1 = 10$ V, $Q_1 = 60$ μC,
$V_2 = 6.67$ V, $Q_2 = 40$ μC,
$V_3 = 3.33$ V, $Q_3 = 40$ μC

47. (a) 56.54 V
(b) 42.405 V
(c) 14.135 V
(d) 43.46 V
(e) 433.44 ms

49. 8640 pJ

51. (a) 5 J
(b) 0.1 C
(c) 200 A
(d) 10 kW
(e) 10 s

Chapter 11

1. Φ: 5×10^4 maxwells,
5×10^4 lines, B: 8 gauss,
51.616 lines

3. (a) 0.04 T

5. 952.4×10^3 At/Wb

7. 2624.67 At/m

9. 2.133 A

11. (a) $N_1 = 60$ t
(b) 13.34×10^{-4} Wb/Am

13. 2.687 A

15. 1.35 N

17. (a) 2.028 A (b) ≅2 N

19. 6.1×10^{-3} Wb

21. (a) $B = 1.5(1 - e^{-H/700\text{ At/m}})$
(c) $H = -700 \log_e(1 - B/1.5\text{ T})$
(e) Eq: 40.1 mA

Chapter 12

1. 4.25 V

3. 14 turns

5. 15.65 μH

7. (a) 2.5 V (b) 0.3 V
(c) 200 V

9. 0−3 ms: 0 V, 3−8 ms: 1.6 V,
8−13 ms: −1.6 V,
13−14 ms: 0 V,
14−15 ms: 8 V,
15−16 ms: –8 V,
16−17 ms: 0 V

11. 0−5 μs: 4 mA, 10 μs: −8 mA,
12 μs: 4 mA, 12−16 μs: 4 mA,
24 μs: 0 mA

13. (a) 2.27 μs
(b) $5.45 \times 10^{-3}(1 - e^{-t/2.27\mu\text{s}})$
(c) $v_L = 12e^{-t/2.27\mu\text{s}}$
$v_R = 12(1 - e^{-t/2.27\mu\text{s}})$
(d) i_L: $1\tau = 3.45$ mA,
$3\tau = 5.179$ mA,
$5\tau = 5.413$ mA,
v_L: $1\tau = 4.415$ V,
$3\tau = 0.598$ V,
$5\tau = 0.081$ V

15. (a) $i_L = 0.882 \times 10^{-3}\,(1 - e^{-t/0.735\mu\text{s}})$,
$v_L = 6e^{-t/0.735\mu\text{s}}$
(b) $i_L = 0.882 \times 10^{-3}e^{-t/0.333\mu\text{s}}$
$v_L = -13.23e^{-t/0.333\mu\text{s}}$

17. (a) $i_L = 1.765\text{ mA} - 4.765\text{ mA } e^{-t/588.24\mu\text{s}}$
$v_L = 16.2\text{ V } e^{-t/588.24\mu\text{s}}$

19. (a) $i_L = -0.692\text{ mA} - 2.308\text{ mA } e^{-t/19.23\mu\text{s}}$
$v_L = 24\text{ V } e^{-t/19.23\mu\text{s}}$

21. 25.68 μs

23. (a) $i_L = 3.638 \times 10^{-3}\,(1 - e^{-t/6.676\mu\text{s}})$,
$v_L = 5.45\, e^{-t/6.676\mu\text{s}}$
(b) 2.825 mA, 1.2186 V
(c) $i_L = 2.825 \times 10^{-3}e^{-t/2.128\mu\text{s}}$
$v_L = -13.27\, e^{-t/2.128\mu\text{s}}$

25. (a) 0.243 V
(b) 29.47 V
(c) 18.96 V
(d) 2.025 ms

27. (a) 20 V
(b) 12 μA
(c) 5.376 μs
(d) 0.366 V

29. $i_L = -3.478\text{ mA} - 7.432\text{ mA } e^{-t/173.9\mu\text{s}}$
$v_L = 51.28\text{ V } e^{-t/173.9\mu\text{s}}$

31. (a) 8 H
(b) 4 H

33. L: 4 H, 2 H
R: 5.7 kΩ, 9.1 kΩ

35. $V_1 = 16$ V, $V_2 = 0$ V,
$I_1 = 4$ mA

37. $V_1 = 10$ V
$I_1 = 2$ A
$I_2 = 1.33$ A

39. $W_C = 360$ μJ
$W_L = 12$ J

Chapter 13

1. (a) 10 ms (b) 2 (c) 100 Hz
(d) amplitude = 5 V,
$V_{p\text{-}p} = 6.67$ V

3. 10 ms, 100 Hz

5. (a) 60 Hz (b) 100 Hz
(c) 29.41 Hz (d) 40 kHz

7. 0.25 s

9. $T = 50$ μs

11. (a) $\pi/4$ (b) $\pi/3$ (c) $\frac{2}{3}\pi$
(d) $\frac{3}{2}\pi$ (e) 0.989π (f) 1.228π

13. (a) 3.14 rad/s
(b) 20.94×10^3 rad/s
(c) 1.57×10^6 rad/s
(d) 157.1 rad/s

15. (a) 120 Hz, 8.33 ms
(b) 1.34 Hz, 746.27 ms
(c) 954.93 Hz, 1.05 ms
(d) 9.95×10^{-3} Hz, 100.5 s

17. 104.7 rad/s

23. 0.4755 A

25. 11.537°, 168.463°

29. (a) v leads i by 10°
(b) i leads v by 70°
(c) i leads v by 80°
(d) i leads v by 150°

31. (a) $v = 25 \sin(\omega t + 30°)$
(b)
$i = 3 \times 10^{-3} \sin(6.28 \times 10^3 t - 60°)$

33. $\frac{1}{3}$ ms

35. 0.388 ms

37. (a) 0.4 ms

(b) 2.5 kHz
(c) −25 mV

39. (a) 1.875 V (b) −4.778 mA

41. (a) 40 μs
(b) 25 kHz
(c) 17.13 mV

43. (a) 2 sin $377t$
(b) 100 sin $377t$
(c) 84.87×10^{-3} sin $377t$
(d) 33.95×10^{-6} sin $377t$

45. 2.16 V

47. 0 V

49. (a) $T = 40\ \mu$s, $f = 25$ kHz, $V_{av} = 20$ mV, $V_{eff} = 28.28$ mV
(b) $T = 100\ \mu$s, $f = 10$ kHz, $V_{av} = -0.3$ V, $V_{eff} = 0.212$ V

Chapter 14

3. (a) 3770 cos $377t$
(b) 452.4 cos $(754t + 20°)$
(c) 4440.63 cos $(157t - 20°)$
(d) 200 cos t

5. (a) 210 sin $754t$
(b) 14.8 sin$(400t - 120°)$
(c) 42×10^{-3} sin$(\omega t + 88°)$
(d) 28 sin$(\omega t + 180°)$

7. (a) 1.592 H (b) 2.654 H
(c) 0.8414 H

9. (a) 100 sin$(\omega t + 90°)$
(b) 8 sin$(\omega t + 150°)$
(c) 120 sin$(\omega t - 120°)$
(d) 60 sin$(\omega t + 190°)$

11. (a) 1 sin$(\omega t - 90°)$
(b) 0.6 sin$(\omega t - 70°)$
(c) 0.8 sin$(\omega t + 10°)$
(d) 1.6 sin$(377t + 130°)$

13. (a) ∞ Ω (b) 530.79 Ω
(c) 265.39 Ω (d) 17.693 Ω
(e) 1.327 Ω

15. (a) 9.31 Hz (b) 4.66 Hz
(c) 18.62 Hz (d) 1.59 Hz

17. (a) 6×10^{-3} sin$(200t + 90°)$
(b) 33.96×10^{-3} sin$(377t + 90°)$
(c) 44.94×10^{-3} sin$(374t + 300°)$
(d) 56×10^{-3} sin$(\omega t + 160°)$

19. (a) 1334 sin$(300t - 90°)$
(b) 37.17 sin$(377t - 90°)$
(c) 127.2 sin $754t$
(d) 100 sin$(1600t - 170°)$

21. (a) C (b) $L = 254.78$ mH
(c) $R = 5\ \Omega$

25. 318.47 mH

27. 5.067 nF

29. (a) 0 W (b) 0 W
(c) 122.5 W

31. 192 W

33. 40 sin$(\omega t - 50°)$

35. (a) 2 sin$(157t - 60°)$
(b) 318.47 mH (c) 0 W

37. (a) $i_1 = 2.828$ sin$(10^4t + 150°)$, $i_2 = 11.312$ sin$(10^4t + 150°)$
(b) $i_s = 14.14$ sin$(10^4t + 150°)$

39. (a) 5 ∠36.87°
(b) 2.83 ∠45°
(c) 16.38 ∠77.66°
(d) 806.23 ∠82.87°
(e) 1077.03 ∠21.80°
(f) 0.00658 ∠81.25°
(g) 11.78 ∠−49.82°
(h) 8.94 ∠153.43°
(i) 61.85 ∠−104.04°
(j) 101.53 ∠−39.81°
(k) 4326.66 ∠123.69°
(l) 25.495×10^{-3} ∠−78.69°

41. (a) 15.033 ∠86.19°
(b) 60.208 ∠4.76°
(c) 0.30 ∠88.09°
(d) 2002.5 ∠−87.14°
(e) 86.182 ∠93.73°
(f) 38.694 ∠−94°

43. (a) 11.8 + j 7
(b) 151.9 + j 49.9
(c) 4.72×10^{-6} + j 71
(d) 5.2 + j 1.6
(e) 209.3 + j 311
(f) −21.2 + j 12
(g) 7.03 + j 9.93
(h) 95.698 + j 22.768

45. (a) 6 ∠−50°
(b) 0.2×10^{-3} ∠140°
(c) 109 ∠−230°
(d) 76.471 ∠−80°
(e) 4 ∠0°
(f) 0.71 ∠−16.49°
(g) 4.21×10^{-3} ∠161.1°
(h) 18.191 ∠−50.91°

47. (a) $x = 4$, $y = 3$
(b) $x = 4$
(c) $x = 3$, $y = 6$ or $x = 6$, $y = 3$
(d) 30°

49. (a) 56.569 sin$(377t + 20°)$
(b) 169.68 sin $377t$
(c) 11.314×10^{-3} sin$(377t + 120°)$
(d) 7.07 sin$(377t + 90°)$
(e) 1696.8 sin$(377t - 120°)$
(f) 6000 sin$(377t - 180°)$

51. $i_1 = 2.537 \times 10^{-5}$ sin$(\omega t + 96.79°)$

53. $i_T = 18 \times 10^{-3}$ sin $377t$

Chapter 15

1. (a) 6.8 Ω ∠0°
(b) 754 Ω ∠90°
(c) 15.7 Ω ∠90°
(d) 265.25 Ω ∠−90°
(e) 318.47 Ω ∠−90°
(f) 200 Ω ∠0°

3. (a) 88×10^{-3} sin ωt
(b) 9.045 sin$(377t + 150°)$
(c) 2547.02 sin$(157t - 50°)$

5. (a) 4.24 Ω ∠−45°
(b) 3.04 kΩ ∠80.54°
(c) 1617.56 Ω ∠88.33°

7. (a) 10 Ω ∠36.87°
(c) **I** = 10 A ∠−36.87°, $\mathbf{V}_R$ = 80 V ∠−36.87°, $\mathbf{V}_L$ = 60 V ∠53.13°
(f) 800 W (g) 0.8 lagging

9. (a) 1660.27 Ω ∠−73.56°
(b) 8.517 mA ∠73.56°
(c) $\mathbf{V}_R$ = 4.003 V ∠73.56°, $\mathbf{V}_L$ = 13.562 V ∠−16.44°
(d) 34.09 mW, 0.283 leading

11. (a) 3.16 kΩ ∠18.43°
(c) 3.18 μF, 6.37 H
(d) **I** = 1.3424 mA ∠41.57°, $\mathbf{V}_R$ = 4.027 V ∠41.57°, $\mathbf{V}_L$ = 2.6848 V ∠131.57°, $\mathbf{V}_C$ = 1.3424 V ∠−48.43°
(g) 5.406 mW
(h) 0.9487 lagging

13. (a) 40 mH (b) 220 Ω

15. (a) $\mathbf{V}_1$ = 37.97 V ∠−51.57°, $\mathbf{V}_2$ = 113.92 V ∠38.43°
(b) $\mathbf{V}_1$ = 55.80 V ∠26.55°, $\mathbf{V}_2$ = 12.56 V ∠−63.45°

17. (a) **I** = 39 mA ∠126.65°, $\mathbf{V}_R$ = 1.17 V ∠126.65°, $\mathbf{V}_C$ = 25.86 V ∠36.65°
(b) 0.058 leading
(c) 45.63 mW
(g) $\mathbf{Z}_T$ = 30 Ω − j 512.2 Ω

19. $\mathbf{Z}_T$ = 3.2 Ω + j 2.4 Ω

25. (a) $\mathbf{Z}_T$ = 3 Ω + j 8 Ω, $\mathbf{Y}_T$ = 41.1 mS − j 109.5 mS
(b) $\mathbf{Z}_T$ = 60 Ω − j 70 Ω, $\mathbf{Y}_T$ = 7.1 mS + j 8.3 mS
(c) $\mathbf{Z}_T$ = 200 Ω − j 100 Ω, $\mathbf{Y}_T$ = 4 mS + j 2 mS

27. (a) $\mathbf{Y}_T$ = 538.52 mS ∠−21.8°
(c) **E** = 3.71 V ∠21.8°, $\mathbf{I}_R$ = 1.855 A ∠21.8°, $\mathbf{I}_L$ = 0.742 A ∠−68.2°
(f) 6.88 W
(g) 0.928 lagging
(h) e = 5.25 sin$(377t + 21.8°)$, i_R = 2.62 sin$(377t + 21.8°)$, i_L = 1.049 sin$(377t - 68.2°)$, i_s = 2.828 sin $377t$

29. (a) $\mathbf{Y}_T$ = 129.96 mS ∠−50.31°
(c) $\mathbf{I}_s$ = 7.8 A ∠−50.31°, $\mathbf{I}_R$ = 5 A ∠0°, $\mathbf{I}_L$ = 6 A ∠−90°

(f) 300 W
(g) 0.638 lagging
(h) $e = 84.84 \sin 377t$,
$i_R = 7.07 \sin 377t$,
$i_L = 8.484 \sin(377t - 90°)$,
$i_s = 11.03 \sin(377t - 50.31°)$

31. **(a)** $\mathbf{Y}_T = 0.416$ mS $\angle 36.897°$
(c) $L = 10.61$ H, $C = 1.326\ \mu$F
(d) $\mathbf{E} = 8.498$ V $\angle -56.897°$,
$\mathbf{I}_R = 2.833$ mA $\angle -56.897°$,
$\mathbf{I}_L = 2.125$ mA $\angle -146.897°$,
$\mathbf{I}_C = 4.249$ mA $\angle 33.103°$
(g) 24.078 mW
(h) 0.8 leading
(i)
$e = 12.016 \sin(377t - 56.897°)$,
$i_R = 4 \sin(377t - 56.897°)$,
$i_L = 3 \sin(377t - 146.897°)$,
$i_C = 6 \sin(377t + 33.103°)$

33. **(a)** $\mathbf{I}_1 = 18.09$ A $\angle 65.241°$,
$\mathbf{I}_2 = 8.528$ A $\angle -24.759°$
(b) $\mathbf{I}_1 = 11.161$ A $\angle 0.255°$,
$\mathbf{I}_2 = 6.656$ A $\angle 153.690°$

39. **(a)** $R_p = 94.73\ \Omega$,
$X_p = 52.1\ \Omega$ (C)
(b) $R_p = 4$ kΩ,
$X_p = 4$ kΩ (C)

41. **(a)** $\mathbf{E} = 176.68$ V $\angle 36.44°$,
$\mathbf{I}_R = 0.803$ A $\angle 36.44°$,
$\mathbf{I}_L = 2.813$ A $\angle -53.56°$
(b) 0.804 lagging
(c) 141.86 W
(f) $\mathbf{I}_C = 1.11$ A $\angle 126.43°$
(g) $\mathbf{Z}_T = 142.15\ \Omega + j\ 104.96\ \Omega$

43. $R = 4\ \Omega$, $X_L = 3.774\ \Omega$

Chapter 16

1. **(a)** 1.2 Ω $\angle 90°$
(b) 10 A $\angle -90°$
(c) 10 A $\angle -90°$
(d) $\mathbf{I}_2 = 6$ A $\angle -90°$,
$\mathbf{I}_3 = 4$ A $\angle -90°$
(e) 60 V $\angle 0°$

3. **(a)** $\mathbf{Z}_T = 3.87\ \Omega \angle -11.817°$,
$\mathbf{Y}_T = 0.258$ S $\angle 11.817°$
(b) 15.504 A $\angle 41.817°$
(c) 3.985 A $\angle 82.826°$
(d) 47.809 V $\angle -7.174°$
(e) 910.71 W

5. **(a)** 0.375 A $\angle 25.346°$
(b) 70.711 V $\angle -45°$
(c) 33.9 W

7. **(a)** 1.423 A $\angle 18.259°$
(b) 26.574 V $\angle 4.763°$
(c) 54.074 W

9. **(a)** $\mathbf{Y}_T = 0.099$ S $\angle -9.709°$
(b) $\mathbf{V}_1 = 20.4$ V $\angle 30°$,
$\mathbf{V}_2 = 10.887$ V $\angle 58.124°$
(c) 1.933 A $\angle 11.109°$

11. 33.201 A $\angle 38.89°$

13. 139.71 mW

Chapter 17

3. **(a)** $\mathbf{Z} = 21.93\ \Omega \angle -46.85°$,
$\mathbf{E} = 10.97$ V $\angle 13.15°$
(b) $\mathbf{Z} = 5.15\ \Omega \angle 59.04°$,
$\mathbf{E} = 10.3$ V $\angle 179.04°$

5. **(a)** 5.15 A $\angle -24.5°$
(b) 0.442 A $\angle 143.48°$

7. **(a)** 13.07 A $\angle -33.71°$
(b) 48.33 A $\angle -77.57°$

9. -3.165×10^{-3} **V** $\angle 137.29°$

11. $\mathbf{I}_{1k\Omega} = 10$ mA $\angle 0°$
$\mathbf{I}_{2k\Omega} = 1.667$ mA $\angle 0°$

13. $\mathbf{I}_L = 1.378$ mA $\angle -56.31°$

15. **(a)** $\mathbf{V}_1 = 19.86$ V $\angle 43.8°$,
$\mathbf{V}_2 = 8.94$ V $\angle 106.9°$
(b) $\mathbf{V}_1 = 19.78$ V $\angle 132.48°$,
$\mathbf{V}_2 = 13.37$ V $\angle 98.78°$

17. $\mathbf{V}_1 = 220$ V $\angle 0°$
$\mathbf{V}_2 = 96.664$ V $\angle -12.426°$
$\mathbf{V}_3 = 100$ V $\angle 90°$

19. (left) $\mathbf{V}_1 = 14.62$ V $\angle -5.86°$
(top) $\mathbf{V}_2 = 35.03$ V $\angle -37.69°$
(right) $\mathbf{V}_3 = 32.4$ V $\angle -73.34°$
(middle) $\mathbf{V}_4 = 5.677$ V $\angle 23.53°$

21. $\mathbf{V}_1 = 4.372$ V $\angle -128.66°$
$\mathbf{V}_2 = 2.253$ V $\angle 17.628°$

23. $\mathbf{V}_1 = -10.667$ V $\angle 0°$
$\mathbf{V}_2 = -6$ V $\angle 0°$

25. $-2451.92\mathbf{E}_i$

27. **(a)** No
(b) 1.76 mA $\angle -71.54°$
(c) 7.03 V $\angle -18.46°$

29. Balanced

31. $R_x = R_2R_3/R_1$
$L_x = R_2L_3/R_1$

33. **(a)** 11.57 A $\angle -67.13°$
(b) 36.9 A $\angle 23.87°$

Chapter 18

1. **(a)** 6.095 A $\angle -32.115°$
(b) 3.77 A $\angle -93.8°$

3. $i = 0.5\text{ A} + 1.581 \sin(\omega t - 26.565°)$

5. 6.261 mA $\angle -63.43°$

7. -22.09 V $\angle 6.34°$

9. 19.62 V $\angle 53°$

11. $\mathbf{V}_s = 10$ V $\angle 0°$

13. **(a)** $\mathbf{Z}_{Th} = 21.312\ \Omega \angle 32.196°$
$\mathbf{E}_{Th} = 2.131$ V $\angle 32.196°$
(b) $\mathbf{Z}_{Th} = 6.813\ \Omega \angle -54.228°$
$\mathbf{E}_{Th} = 57.954$ V $\angle 11.099°$

15. **(a)** $\mathbf{Z}_{Th} = 4\ \Omega \angle 90°$
$\mathbf{E}_{Th} = 4$ V + 10 V $\angle 0°$
(b) $\mathbf{I} = 0.5\text{A} + 1.118$ A$\angle -26.565°$

17. **(a)** $\mathbf{Z}_{Th} = 4.472$ kΩ $\angle -26.565°$
$\mathbf{E}_{Th} = 31.31$ V $\angle -26.565°$
(b) $\mathbf{I} = 6.26$ mA $\angle 63.435°$

19. $\mathbf{Z}_{Th} = 4.44$ kΩ $\angle -0.031°$
$\mathbf{E}_{Th} = -444.45 \times 10^3\mathbf{I} \angle 0.255°$

21. $\mathbf{Z}_{Th} = 5.099$ kΩ $\angle -11.31°$
$\mathbf{E}_{Th} = -50$ V $\angle 0°$

23. $\mathbf{Z}_{Th} = -39.215\ \Omega \angle 0°$
$\mathbf{E}_{Th} = 20$ V $\angle 53°$

25. $\mathbf{Z}_{Th} = 607.42\ \Omega \angle 0°$
$\mathbf{E}_{Th} = 1.62$ V $\angle 0°$

27. **(a)** $\mathbf{Z}_N = 21.312\ \Omega \angle 32.196°$,
$\mathbf{I}_N = 0.1$ A $\angle 0°$
(b) $\mathbf{Z}_N = 6.813\ \Omega \angle -54.228°$,
$\mathbf{I}_N = 8.506$ A $\angle 65.324°$

29. **(a)** $\mathbf{Z}_N = 9.66\ \Omega \angle 14.93°$,
$\mathbf{I}_N = 2.15$ A $\angle -42.87°$
(b) $\mathbf{Z}_N = 4.37\ \Omega \angle 55.67°$,
$\mathbf{I}_N = 22.83$ A $\angle -34.65°$

31. **(a)** $\mathbf{Z}_N = 9\ \Omega \angle 0°$,
$\mathbf{I}_N = 1.333$ A + 2.667 A $\angle 0°$
(b) 12 V + 2.65 V $\angle -83.66°$

33. $\mathbf{Z}_N = 5.1$ kΩ $\angle -11.31°$,
$\mathbf{I}_N = -1.961 \times 10^{-3}$ **V** $\angle 11.31°$

35. $\mathbf{Z}_N = 5.1$ kΩ $\angle -11.31°$,
$\mathbf{I}_N = 9.81$ mA $\angle 11.31°$

37. $\mathbf{Z}_N = 6.63$ kΩ $\angle 0°$
$\mathbf{I}_N = 0.792$ mA $\angle 0°$

39. **(a)** $\mathbf{Z}_L = 8.32\ \Omega \angle 3.18°$,
1198.2 W
(b) $\mathbf{Z}_L = 1.562\ \Omega \angle -14.47°$,
1.614 W

41. 40 kΩ, 25 W

43. **(a)** 9 Ω **(b)** 20 W

45. **(a)** 1.414 kΩ **(b)** 0.518 W

49. 25.77 mA $\angle 104.4°$

Chapter 19

1. **(a)** 120 W
(b) $Q_T = 0$ VAR, $S_T = 120$ VA
(c) 0.5 A
(d) $I_1 = \frac{1}{6}$ A, $I_2 = \frac{1}{3}$ A

3. **(a)** 400 W, −400 VAR(C),
565.69 VA, 0.7071 leading
(c) 5.66 A $\angle 135°$

5. **(a)** 500 W, −200 VAR(C),
538.52 VA
(b) 0.928 leading
(d) 10.776 A $\angle 21.875°$

7. **(a)** R: 200 W, L,C: 0 W
(b) R: 0 VAR, C: 80 VAR,
L: 100 VAR
(c) R: 200 VA, C: 80 VA,
L: 100 VA
(d) 200 W, 20 VAR(L),
200.998 VA, 0.995 (lagging)
(f) 10.05 A $\angle -5.73°$

9. **(a)** R: 38.99 W, L: 0 W, C: 0 W
(b) R: 0 VAR, L: 126.74 VAR, C: 46.92 VAR
(c) R: 38.99 VA, L: 126.74 VA, C: 46.92 VA
(d) 38.99 W, 79.82 VAR(L), 88.83 VA, 0.439 (lagging)
(f) 0.31 J
(g) $W_L = 0.32$ J, $W_C = 0.12$ J

11. **(a)** $\mathbf{Z} = 2.30\ \Omega + j\ 1.73\ \Omega$
(b) 4000 W

13. **(a)** 900 W, 0 VAR, 900 VA, 1
(b) $9\text{ A} \angle 0°$
(d) $\mathbf{Z}_1$: $R = 0\ \Omega$, $X_C = 20\ \Omega$
$\mathbf{Z}_2$: $R = 2.83\ \Omega$, $X = 0\ \Omega$
$\mathbf{Z}_3$: $R = 5.66\ \Omega$, $X_L = 4.717\ \Omega$

15. **(a)** 1100 W, 2366.26 VAR, 2609.44 VA, 0.4215 (leading)
(b) $521.89\text{ V} \angle -65.07°$
(c) $\mathbf{Z}_1$: $R = 1743.38\ \Omega$, $X_C = 1307.53\ \Omega$
$\mathbf{Z}_2$: $R = 43.59\ \Omega$, $X_C = 99.88\ \Omega$

17. **(a)** 7.81 kVA
(b) 0.640 (lagging)
(c) 65.08 A
(d) 1105 μF
(e) 41.67 A

19. **(a)** 128.14 W
(b) a–b: 42.69 W, b–c: 64.03 W, a–c: 106.72 W, a–d: 106.72 W, c–d: 0 W, d–e: 0 W, f–e: 21.34 W

21. **(a)** 5 Ω, 132.03 mH
(b) 10 Ω
(c) 15 Ω, 262.39 mH

Chapter 20

1. **(a)** $\omega_s = 250$ rad/s, $f_s = 39.79$ Hz
(b) $\omega_s = 3535.53$ rad/s, $f_s = 562.7$ Hz
(c) $\omega_s = 21{,}880$ rad/s, $f_s = 3482.31$ Hz

3. **(a)** $X_L = 40\ \Omega$
(b) $I = 10$ mA
(c) $V_R = 20$ mV, $V_L = 400$ mV, $V_C = 400$ mV
(d) $Q_s = 20$ (high)
(e) $L = 1.27$ mH, $C = 0.796\ \mu$F
(f) $BW = 250$ Hz
(g) $f_2 = 5.125$ kHz, $f_1 = 4.875$ kHz

5. **(a)** $BW = 400$ Hz
(b) $f_2 = 6200$ Hz, $f_1 = 5800$ Hz
(c) $X_L = X_C = 45\ \Omega$
(d) $P_{\text{HPF}} = 375$ mW

7. **(a)** $Q_s = 10$
(b) $X_L = 20\ \Omega$
(c) $L = 1.59$ mH, $C = 3.98\ \mu$F
(d) $f_2 = 2100$ Hz, $f_1 = 1900$ Hz

9. $L = 13.26$ mH, $C = 27.07$ nF
$f_2 = 8460$ Hz, $f_1 = 8340$ Hz

11. **(a)** $f_s = 1$ MHz
(b) $BW = 160$ kHz
(c) $R = 720\ \Omega$, $L = 0.7162$ mH, $C = 35.37$ pF
(d) $R_l = 56.25\ \Omega$

13. **(a)** $f_p = 159.155$ kHz
(b) $V_C = 4$ V
(c) $I_L = I_C = 40$ mA
(d) $Q_p = 20$

15. **(a)** $f_s = 11{,}253.95$ Hz
(b) $Q_l = 1.77$ (no)
(c) $f_p = 9{,}280.24$ Hz, $f_m = 10{,}794.41$ Hz
(d) $X_L = 5.83\ \Omega$, $X_C = 8.57\ \Omega$
(e) $Z_{T_p} = 12.5\ \Omega$
(f) $V_C = 25$ mV
(g) $Q_p = 1.46$, $BW = 6.356$ kHz
(h) $I_C = 2.92$ mA, $I_L = 3.54$ mA

17. **(a)** $X_C = 30\ \Omega$
(b) $Z_{T_p} = 225\ \Omega$
(c) $\mathbf{I}_C = 0.6\text{ A} \angle 90°$, $\mathbf{I}_L \cong 0.6\text{ A} \angle -86.19°$
(d) $L = 0.239$ mH, $C = 265.26$ nF
(e) $Q_p = 7.5$, $BW = 2.67$ kHz

19. **(a)** $f_s = 7.118$ kHz, $f_p = 6.647$ kHz, $f_m = 7$ kHz
(b) $X_L = 20.88\ \Omega$, $X_C = 23.94\ \Omega$
(c) $Z_{T_p} = 55.56\ \Omega$
(d) $Q_p = 2.32$, $BW = 2.865$ kHz
(e) $I_L = 99.28$ mA, $I_C = 92.73$ mA
(f) $V_C = 2.22$ V

21. **(a)** $f_p = 3558.81$ Hz
(b) $V_C = 138.2$ V
(c) $P = 691$ mW
(d) $BW = 575.86$ Hz

23. **(a)** $X_L = 98.54\ \Omega$
(b) $Q_l = 8.21$
(c) $f_p = 8.05$ kHz
(d) $V_C = 4.83$ V
(e) $f_2 = 8.55$ kHz, $f_1 = 7.55$ kHz

25. $R_s = 3.244$ kΩ, $C = 31.66$ nF

27. **(a)** $f_p = 251.65$ kHz
(b) $Z_{T_p} = 4.444$ kΩ
(c) $Q_p = 14.05$
(d) $BW = 17.91$ kHz
(e) 20 nF: $f_p = 194.93$ kHz, $Z_{T_p} = 49.94\ \Omega$, $Q_p = 2.04$, $BW = 95.55$ kHz
(f) 1 nf: $f_p = 251.65$ kHz, $Z_{T_p} = 13.33$ kΩ, $Q_p = 21.08$, $BW = 11.94$ kHz
(g) Network; $L/C = 100 \times 10^3$
part (e): $L/C = 1 \times 10^3$
part (f): $L/C = 400 \times 10^3$
(h) yes, $L/C \uparrow$, BW $\downarrow$

Chapter 21

1. **(a)** left: 1.54 kHz, right: 5.623 kHz
(b) bottom: 0.2153 V, top: 0.5248 V

3. **(a)** 1000 **(b)** 10^{12}
(c) 1585 **(d)** 1.096
(e) 10^{10} **(f)** 1513.56
(g) 10.023 **(h)** 1,258,925.41

5. 1.681

7. −0.301

9. **(a)** 1.845
(b) 18.45

11. 13.01

13. 38.49

15. 24.08 dB_s

19. **(a)** $0.1f_c$: 0.995, $0.5f_c$: 0.894, f_c: 0.707, $2f_c$: 0.447, $10f_c$: 0.0995
(b) $0.1f_c$: −5.71°, $0.5f_c$: −26.57°, f_c: −45°, $2f_c$: −63.43°, $10f_c$: −84.29°

21. $C = 0.265\ \mu$F,
250 Hz: $A_v = 0.895$, $\theta = -26.54°$,
1000 Hz: $A_v = 0.4475$, $\theta = -63.41°$

23. **(a)** $f_c = 3.617$ kHz,
f_c: $A_v = 0.707$, $\theta = 45°$,
$2f_c$: $A_v = 0.894$, $\theta = 26.57°$
$0.5f_c$: $A_v = 0.447$, $\theta = 63.43°$
$10f_c$: $A_v = 0.995$, $\theta = 5.71°$
$\frac{1}{10}f_c$: $A_v = 0.0995$, $\theta = 84.29°$

25. $R = 795.77\ \Omega \rightarrow 797\ \Omega$,
f_c: $A_v = 0.707$, $\theta = 45°$
1 kHz: $A_v = 0.458$, $\theta = 63.4°$
4 kHz: $A_v \cong 0.9$, $\theta = 26.53°$

27. **(a)** $f_{c_1} = 795.77$ Hz, $f_{c_2} = 1989.44$ Hz
f_{c_1}: $V_o = 0.656V_i$,
f_{c_2}: $V_o = 0.656V_i$
$f_{\text{center}} = 1392.60$ Hz: $V_o = 0.711V_i$
500 Hz: $V_o = 0.516V_i$,
4 kHz: $V_o = 0.437V_i$
(b) $BW \cong 2.9$ kHz, $f_{\text{center}} = 1.94$ kHz

29. **(a)** $f_s = 100.658$ kHz
(b) $Q_s = 18.39$, $BW = 5473.52$ Hz
(c) f_s: $A_v = 0.93$
$f_1 = 97{,}921.24$ Hz,
$f_2 = 103{,}394.76$ Hz,

$f = 95$ kHz: $A_v = 0.392$,
$f = 105$ kHz: $A_v = 0.5$
(d) $f = f_s$, $V_o = 0.93$ V,
$f = f_1 = f_2$, $V_o = 0.658$ V

31. **(a)** $Q_s = 12.195$
(b) $BW = 410$ Hz,
$f_2 = 5205$ Hz,
$f_1 = 4795$ Hz
(c) f_s: $V_o = 0.024V_i$
(d) f_s: V_o still $0.024V_i$

33. **(a)** $f_p = 726.44$ kHz (stop-band)
$f = 2.013$ MHz (pass-band)

35. **(a–b)** $f_c = 6772.55$ Hz
(c) f_c: -3 dB, $\frac{1}{2}f_c$: -6.7 dB,
$2f_c$: -0.969 dB,
$\frac{1}{10}f_c$: -20.04 dB,
$10f_c$: -0.043 dB
(d) f_c: 0.707, $\frac{1}{2}f_c$: 0.4472,
$2f_c$: 0.894
(e) f_c: 45°, $\frac{1}{2}f_c$: 63.43°, $2f_c$: 26.57°

37. **(a–b)** $f_c = 13.26$ kHz
(c) f_c: -3 dB, $\frac{1}{2}f_c$: -0.97 dB,
$2f_c$: -6.99 dB
$\frac{1}{10}f_c$: -0.043 dB,
$10f_c$: -20.04 dB
(d) f_c: 0.707, $\frac{1}{2}f_c$: 0.894,
$2f_c$: 0.447
(e) f_c: $-45°$, $\frac{1}{2}f_c$: $-26.57°$,
$2f_c$: $-63.43°$

39. **(a)** $f_1 = 663.15$ Hz, $f_c = 468.1$ Hz
$0 < f < f_c$: +6 dB/octave,
$f > f_c$: -3.03 dB
(b) f_1: 45°, f_c: 54.78°, $\frac{1}{2}f_1$: 63.43°,
$2f_1$: 84.29°

41. **(a)** $f_1 = 19{,}894.37$ Hz
$f_c = 1{,}989.44$ Hz
$0 < f < f_c$: 0 dB,
$f_c < f < f_1$: -6 dB/octave,
$f > f_1$: -20 dB
(b) f_c: $-39.29°$,
10 kHz: $-52.06°$,
f_1: $-39.29°$

43. **(a)** $f_1 = 964.58$ Hz,
$f_c = 7{,}334.33$ Hz
$0 < f < f_1$: -17.62 dB,
$f_1 < f < f_c$: +6 dB/octave,
$f > f_c$: 0 dB
(b) f_1: 39.35°, 1.3 kHz: 43.38°,
f_c: 39.35°

45. **(a)** $f = 180$ Hz $\cong -3$ dB,
$f = 18$ kHz: -3.105 dB
(b) 100 Hz: 97°,
1.8 kHz: 0.12° ≅ 0°,
18 kHz: $-61.8°$

47. $\mathbf{A}_v = -120/[(1 - j\,50/f)(1 - j\,200/f)(1 - jf/36\text{ kHz})]$

49. $f_c = 2$ kHz, $0 < f < f_c$: 0 dB,
$f > f_c$: -6 dB/octave

51. $f_1 = 1$ kHz, $f_2 = 2$ kHz,
$f_3 = 3$ kHz
$0 < f < f_1$: 0 dB,
$f_1 < f < f_2$: +6 dB/octave
$f_2 < f < f_3$: +12 dB/octave,
$f > f_3$: 13.06 dB

53. **(a)** woofer: 0.673, tweeter: 0.678
(b) woofer: 0.015, tweeter: 0.337
(c) mid-range: 0.998 ≅ 1

Chapter 22

1. **(a)** positive-going **(b)** 2 V
(c) 0.2 ms **(d)** 6 V **(e)** 6.5%

3. **(a)** positive-going
(b) 10 mV
(c) 3.2 ms **(d)** 20 mV
(e) 3.4%

5. V_2 of $(V_1 - V_2)/V = 0.1$ is
13.571 mV

7. **(a)** 120 μs **(b)** 8.333 kHz
(c) maximum = 440 mV,
minimum = 80 mV

9. prf = 125 kHz,
duty cycle = 62.5%

11. **(a)** 8 μs
(b) 2 μs
(c) 125 kHz
(d) 0 V
(e) 3.464 mV

13. 18.88 mV

15. 117 mV

17. $v_o = 4(1 + e^{-t/20\text{ms}})$

19. $i_C = -8 \times 10^{-3}e^{-t}$

21. $i_C = 4 \times 10^{-3}e^{-t/0.2\text{ms}}$
(a) $5\tau = T/2$ **(b)** $5\tau = \frac{1}{5}(T/2)$
(c) $5\tau = 10(T/2)$

23. $0 - T/2$: $v_C = 20$ V,
$T/2 - T$: $v_C = 20e^{-t/\tau}$,
$T - \frac{3}{2}T$: $v_C = 20(1 - e^{-t/\tau})$
$\frac{3}{2}T - T$: $v_C = 20e^{-t/\tau}$

25. $\mathbf{Z}_p = 4.573$ MΩ ∠−59.5°,
$\mathbf{Z}_s = 0.507$ MΩ ∠−59.5°

Chapter 23

1. **(a)** 120.1 V **(b)** 120.1 V
(c) 12.01 A **(d)** 12.01 A

3. **(a)** 120.1 V **(b)** 120.1 V
(c) 16.98 A **(d)** 16.98 A

5. **(a)** $\theta_2 = -120°$, $\theta_3 = 120°$
(b) $\mathbf{V}_{an} = 120$ V ∠0°,
$\mathbf{V}_{bn} = 120$ V ∠−120°,
$\mathbf{V}_{cn} = 120$ V ∠120°
(c) $\mathbf{I}_{an} = 8$ A ∠−53.13°,
$\mathbf{I}_{bn} = 8$ A ∠−173.13°,
$\mathbf{I}_{cn} = 8$ A ∠66.87°
(e) 8 A **(f)** 207.85 V

7. $V_\phi = 127$ V, $I_\phi = 8.98$ A,
$I_L = 8.98$ A

9. **(a)** $\mathbf{E}_{AN} = 12.7$ kV ∠−30°,
$\mathbf{E}_{BN} = 12.7$ kV ∠−150°,
$\mathbf{E}_{CN} = 12.7$ kV ∠90°
(b) $\mathbf{I}_{an} = 11.285$ A ∠−97.54°,
$\mathbf{I}_{bn} = 11.285$ A ∠−217.54°,
$\mathbf{I}_{cn} = 11.285$ A ∠22.46°
(c) $\mathbf{I}_L = \mathbf{I}_\phi$
(d) $\mathbf{V}_{an} = 12{,}154.28$ V ∠−29.34°,
$\mathbf{V}_{bn} = 12{,}154.28$ V ∠−149.34°,
$\mathbf{V}_{cn} = 12{,}154.28$ V ∠90.66°

11. **(a)** 120.1 V **(b)** 208 V
(c) 13.364 A **(d)** 23.15 A

13. **(a)** $\theta_2 = -120°$, $\theta_3 = +120°$
(b) $\mathbf{V}_{ab} = 208$ V ∠0°,
$\mathbf{V}_{bc} = 208$ V ∠−120°,
$\mathbf{V}_{ca} = 208$ V ∠120°
(d) $\mathbf{I}_{ab} = 9.455$ A ∠0°,
$\mathbf{I}_{bc} = 9.455$ A ∠−120°,
$\mathbf{I}_{ca} = 9.455$ A ∠120°
(e) 16.376 A **(f)** 120.1 V

15. **(a)** $\theta_2 = -120°$, $\theta_3 = 120°$
(b) $\mathbf{V}_{ab} = 208$ V ∠0°,
$\mathbf{V}_{bc} = 208$ V ∠−120°,
$\mathbf{V}_{ca} = 208$ V ∠120°
(d) $\mathbf{I}_{ab} = 86.67$ A ∠−36.87°,
$\mathbf{I}_{bc} = 86.67$ A ∠−156.87°,
$\mathbf{I}_{ca} = 86.67$ A ∠83.13°
(e) 150.11 A **(f)** 120.1 V

17. **(a)** $\mathbf{I}_{ab} = 15.325$ A ∠−73.30°,
$\mathbf{I}_{bc} = 15.325$ A ∠−193.30°,
$\mathbf{I}_{ca} = 15.325$ A ∠46.7°
(b) $\mathbf{I}_{Aa} = 26.54$ A ∠−103.31°,
$\mathbf{I}_{Bb} = 26.54$ A ∠136.68°,
$\mathbf{I}_{Cc} = 26.54$ A ∠16.69°
(c) $\mathbf{E}_{AB} = 17{,}013.6$ V ∠−0.59°,
$\mathbf{E}_{BC} = 17{,}013.77$ V ∠−120.59°,
$\mathbf{E}_{CA} = 17{,}013.87$ V ∠119.41°

19. **(a)** 208 V **(b)** 120.09 V
(c) 7.076 A **(d)** 7.076 A

21. $V_\phi = 69.28$ V, $I_\phi = 2.89$ A,
$I_L = 2.89$ A

23. $V_\phi = 69.28$ V, $I_\phi = 5.77$ A,
$I_L = 5.77$ A

25. **(a)** 440 V **(b)** 440 V
(c) 29.33 A **(d)** 50.8 A

27. **(a)** $\theta_2 = -120°$, $\theta_3 = +120°$
(b) $\mathbf{V}_{ab} = 100$ V ∠0°,
$\mathbf{V}_{bc} = 100$ V ∠−120°,
$\mathbf{V}_{ca} = 100$ V ∠120°
(d) $\mathbf{I}_{ab} = 5$ A ∠0°,
$\mathbf{I}_{bc} = 5$ A ∠−120°,
$\mathbf{I}_{ca} = 5$ A ∠120°
(e) 8.66 A

29. **(a)** $\theta_2 = -120°$, $\theta_3 = 120°$
(b) $\mathbf{V}_{ab} = 100$ V ∠0°,
$\mathbf{V}_{bc} = 100$ V ∠−120°,
$\mathbf{V}_{ca} = 100$ V ∠120°
(d) $\mathbf{I}_{ab} = 7.072$ A ∠45°,
$\mathbf{I}_{bc} = 7.072$ A ∠−75°,
$\mathbf{I}_{ca} = 7.072$ A ∠165°
(e) 12.25 A

31. 2160 W, 0 VAR, 2160 VA,
$F_p = 1$

33. 7210.67 W, 7210. 67 VAR(C), 10,197.42 VA, 0.707 leading
35. 7.263 kW, 7.263 kVAR, 10.272 kVA, 0.707 lagging
37. 287.93 W, 575.86 VAR(L), 643.83 VA, 0.4472 lagging
39. 900 W, 1200 VAR(L), 1500 VA, 0.6 lagging
41. $\mathbf{Z}_\phi = 12.98\ \Omega - j\ 17.31\ \Omega$
43. **(a)** 9237.6 V **(b)** 80 A
(c) 1276.8 kW
(d) 0.576 lagging
(e) $\mathbf{I}_{Aa} = 80\text{ A} \angle -54.83°$
(f) $\mathbf{V}_{an} = 7773.45\text{ V}\angle -4.87°$
(g) $\mathbf{Z}_\phi = 62.52\ \Omega + j\ 74.38\ \Omega$
(h) F_p (entire system) = 0.576, F_p (load) = 0.643 (both lagging)
(i) 93.98%
45. **(b)** $P_T = 5899.64$ W, $P_{\text{meter}} = 1966.55$ W
49. **(a)** 120.09 V
(b) $I_{an} = 8.492$ A, $I_{bn} = 7.076$ A, $I_{cn} = 42.465$ A
(c) 4928.5 W, 4928.53 VAR(L), 6969.99 VA, 0.7071 lagging
(d) $\mathbf{I}_{an} = 8.492\text{ A} \angle -75°$
$\mathbf{I}_{bn} = 7.076\text{ A} \angle -195°$
$\mathbf{I}_{cn} = 42.465\text{ A} \angle 45°$
(e) $\mathbf{I}_N = 34.712\text{ A} \angle -42.972°$

Chapter 24

1. (I) **a.** no **b.** no **c.** yes **d.** no **e.** yes
(II) **a.** yes **b.** yes **c.** yes **d.** yes **e.** no
(III) **a.** yes **b.** yes **c.** no **d.** yes **e.** yes
(IV) **a.** no **b.** no **c.** yes **d.** yes **e.** yes
7. **(a)** 19.04 V **(b)** 4.53 A
9. 71.872 W
11. **(a)** $i = 2 + 2.08 \sin(400t - 33.69) + 0.5 \sin(800t - 53.13°)$
(b) 2.508 A
(c) $v_R = 24 + 24.96 \sin(400t + 33.69°) + 6 \sin(800t - 53.13°)$
(d) 30.092 A
(e) $v_L = 16.64 \sin(400t + 56.31°) + 8 \sin(800t + 36.87°)$
(f) 13.055 V **(g)** 75.481 W
13. **(a)** $i = 1.2 \sin(400t + 53.13°)$
(b) 0.848 A
(c) $v_R = 18 \sin(400t + 53.13°)$
(d) 12.73 V
(e) $v_C = 18 + 23.98 \sin(400t - 36.87°)$
(f) 24.73 V **(g)** 10.79 W
15. $v_o = 2.257 \times 10^{-3} \sin(377t + 93.66°) + 1.923 \times 10^{-3} \sin(754t + 1.64°)$
17. $i_T = 30 + 30.27 \sin(20t + 7.59°) + 0.5 \sin(40t - 30°)$

Chapter 25

1. **(a)** 0.2 H
(b) $e_p = 1.6$ V, $e_s = 5.12$ V
(c) $e_p = 15$ V, $e_s = 24$ V
3. **(a)** 158.02 mH
(b) $e_p = 24$ V, $e_s = 1.8$ V
(c) $e_p = 15$ V, $e_s = 24$ V
5. 1.354 H
7. $\mathbf{I}_1 (R_1 + j X_{L_1}) + \mathbf{I}_2(j X_m) = \mathbf{E}_1$
$\mathbf{I}_1(j X_m) + \mathbf{I}_2(j X_{L_2} + R_L) = 0$
9. **(a)** 3.125 V **(b)** 391.02 μWb
11. 56.31 Hz
13. 400 Ω
15. $12{,}000t$
17. **(a)** 20 Ω
(b) 40 Ω
(d) $0.351\text{ A} \angle -6.71°$
(e) $28.1\text{ V} \angle -6.71°$
(g) 30 V
19. **(a)** $\mathbf{Z}_p = 280.71\ \Omega \angle -85.91°$
(b) $\mathbf{I}_p = 0.427\text{ A} \angle 85.91°$
(c) $\mathbf{V}_{R_e} = 8.54\text{ V} \angle 85.91°$
$\mathbf{V}_{X_e} = 17.08\text{ V} \angle 175.91°$
$\mathbf{V}_{X_C} = 136.64\text{ V} \angle -4.09°$
21. $\mathbf{Z}_i = 7980\ \Omega \angle 89.98°$
23. **(a)** 20 **(b)** 83.33 A **(c)** 4.167 A
(d) $a = \frac{1}{20}$, $I_s = 4.167$ A, $I_p = 83.33$ A
25. **(a)** $25\text{ V} \angle 0°$, $5\text{ A} \angle 0°$
(b) $80\ \Omega \angle 0°$ **(c)** $20\ \Omega \angle 0°$
27. **(a)** $\mathbf{E}_2 = 40\text{ V} \angle 60°$, $\mathbf{I}_2 = 3.33\text{ A} \angle 60°$, $\mathbf{E}_3 = 30\text{ V} \angle 60°$, $\mathbf{I}_3 = 3\text{ A} \angle 60°$
(b) $R_1 = 64.52\ \Omega$
29. $[\mathbf{Z}_1 + \mathbf{X}_{L_1}]\mathbf{I}_1 - \mathbf{Z}_{M_{12}}\mathbf{I}_2 + \mathbf{Z}_{M_{13}}\mathbf{I}_3 = \mathbf{E}_1$,
$\mathbf{Z}_{M_{12}}\mathbf{I}_1 - [\mathbf{Z}_2 + \mathbf{Z}_3 + \mathbf{X}_{L_2}]\mathbf{I}_2 + \mathbf{Z}_2\mathbf{I}_3 = 0$,
$\mathbf{Z}_{M_{13}}\mathbf{I}_1 - \mathbf{Z}_2\mathbf{I}_2 + [\mathbf{Z}_2 + \mathbf{Z}_4 + \mathbf{X}_{L_3}]\mathbf{I}_3 = 0$

Chapter 26

1. $Z_i = 986.84\ \Omega$
3. **(a)** $I_{i_1} = 10\ \mu$A
(b) $Z_{i_2} = 4.5$ kΩ
(c) $E_{i_3} = 6.9$ V
5. $Z_o = 44.59$ kΩ
7. $Z_o = 10$ kΩ
9. **(a)** $\mathbf{A}_v = -392.98$
(b) $\mathbf{A}_{v_T} = -320.21$
11. **(a)** $\mathbf{A}_{v_{NL}} = -2398.8$
(b) $\mathbf{E}_i = 50$ mV
(c) $\mathbf{Z}_i = 1$ kΩ
13. **(a)** $\mathbf{A}_G = 6.067 \times 10^4$
(b) $\mathbf{A}_{G_T} = 4.94 \times 10^4$
15. **(a)** $\mathbf{A}_{v_T} = 1500$
(b) $\mathbf{A}_{i_T} = 187.5$
(c) $\mathbf{A}_{i_1} = 15$, $\mathbf{A}_{i_2} = 12.5$
(d) $\mathbf{A}_{i_T} = 187.5$
17. **(a)** $\mathbf{z}_{11} = (\mathbf{Z}_1\mathbf{Z}_2 + \mathbf{Z}_1\mathbf{Z}_3)/(\mathbf{Z}_1 + \mathbf{Z}_2 + \mathbf{Z}_3)$,
$\mathbf{z}_{12} = \mathbf{Z}_1\mathbf{Z}_3/(\mathbf{Z}_1 + \mathbf{Z}_2 + \mathbf{Z}_3)$,
$\mathbf{z}_{21} = \mathbf{z}_{12}$,
$\mathbf{z}_{22} = (\mathbf{Z}_1\mathbf{Z}_3 + \mathbf{Z}_2\mathbf{Z}_3)/(\mathbf{Z}_1 + \mathbf{Z}_2 + \mathbf{Z}_3)$
19. **(a)** $\mathbf{y}_{11} = (\mathbf{Y}_1\mathbf{Y}_2 + \mathbf{Y}_1\mathbf{Y}_3)/(\mathbf{Y}_1 + \mathbf{Y}_2 + \mathbf{Y}_3)$,
$\mathbf{y}_{12} = -\mathbf{Y}_1\mathbf{Y}_2/(\mathbf{Y}_1 + \mathbf{Y}_2 + \mathbf{Y}_3)$,
$\mathbf{y}_{21} = \mathbf{y}_{12}$,
$\mathbf{y}_{22} = (\mathbf{Y}_1\mathbf{Y}_2 + \mathbf{Y}_2\mathbf{Y}_3)/(\mathbf{Y}_1 + \mathbf{Y}_2 + \mathbf{Y}_3)$
21. $\mathbf{h}_{11} = \mathbf{Z}_1\mathbf{Z}_2/(\mathbf{Z}_1 + \mathbf{Z}_2)$,
$\mathbf{h}_{21} = -\mathbf{Z}_1/(\mathbf{Z}_1 + \mathbf{Z}_2)$,
$\mathbf{h}_{12} = \mathbf{Z}_1/(\mathbf{Z}_1 + \mathbf{Z}_2)$,
$\mathbf{h}_{22} = (\mathbf{Z}_1 + \mathbf{Z}_2 + \mathbf{Z}_3)/(\mathbf{Z}_1\mathbf{Z}_3 + \mathbf{Z}_2\mathbf{Z}_3)$
23. $\mathbf{h}_{11} = (\mathbf{Y}_1 + \mathbf{Y}_2 + \mathbf{Y}_3)/(\mathbf{Y}_1\mathbf{Y}_2 + \mathbf{Y}_1\mathbf{Y}_3)$,
$\mathbf{h}_{21} = -\mathbf{Y}_2/(\mathbf{Y}_2 + \mathbf{Y}_3)$,
$\mathbf{h}_{12} = \mathbf{Y}_2/(\mathbf{Y}_2 + \mathbf{Y}_3)$,
$\mathbf{h}_{22} = \mathbf{Y}_2\mathbf{Y}_3/(\mathbf{Y}_2 + \mathbf{Y}_3)$
25. **(a)** 47.62 **(b)** −99
27. $\mathbf{Z}_i = 9{,}219.5\ \Omega \angle -139.4°$, $\mathbf{Z}_o = 29.07\text{ k}\Omega \angle -86.05°$
29. $\mathbf{h}_{11} = 2.5$ kΩ, $\mathbf{h}_{12} = 0.5$, $\mathbf{h}_{21} = -0.75$, $\mathbf{h}_{22} = 0.25$ mS

Index